P9-ECZ-731

MANUAL OF REMOTE SENSING

Second Edition
Volume II

MANUAL OF REMOTE SENSING

Second Edition

In two volumes

Volume II
Interpretation and Applications

Editor-in-Chief

Robert N. Colwell

Volume I Editor : David S. Simonett
Associate Editor : Fawwaz T. Ulaby

Volume II Editor : John E. Estes
Associate Editor : Gene A. Thorley

AMERICAN SOCIETY
OF
PHOTOGRAMMETRY

Library of Congress Cataloging in Publication Data
Main entry under title:

Manual of remote sensing.

Includes bibliographies and index.
1. Remote sensing. I. Colwell, Robert N.
II. American Society of Photogrammetry.
G70.4.M36 1983 621.36'78 83-6055
ISBN 0-937294-41-1 (v. 1)
ISBN 0-937294-42-X (v. 2)

PUBLISHED BY
AMERICAN SOCIETY OF PHOTOGRAMMETRY
210 Little Falls Street
Falls Church, Virginia 22046

Printed in the United States of America by
The Sheridan Press

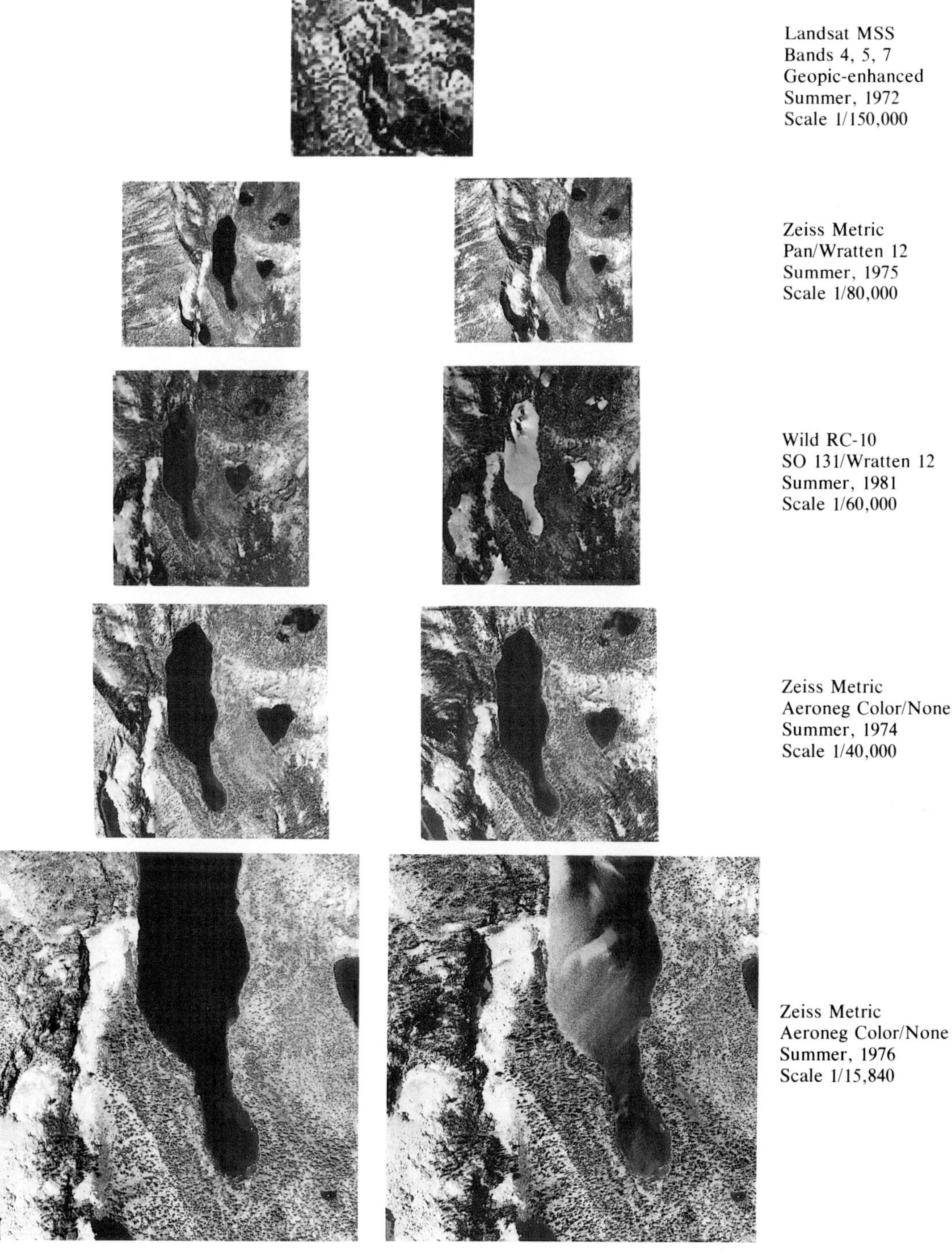

Shown here for the Enos Lake area on Idaho's Payette National Forest is a typical assortment of image types that recently have been evaluated in terms of their usefulness for the inventory and monitoring of various kinds of natural resources, as discussed on pages 1967–1969 of this volume. (Photos courtesy of Nationwide Forestry Applications Program of the U.S. Forest Service).

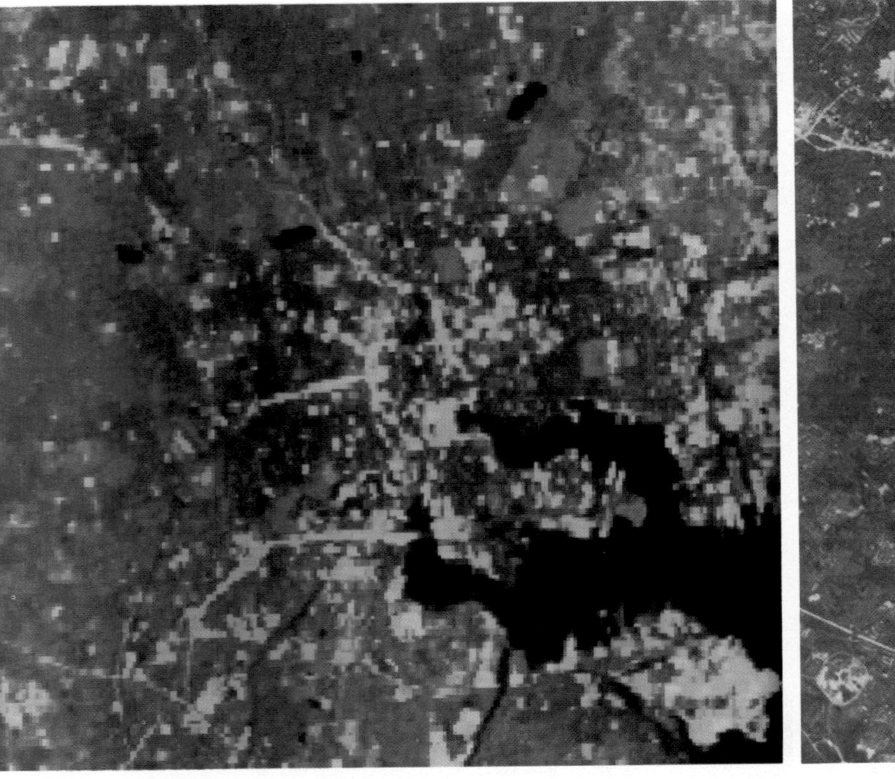

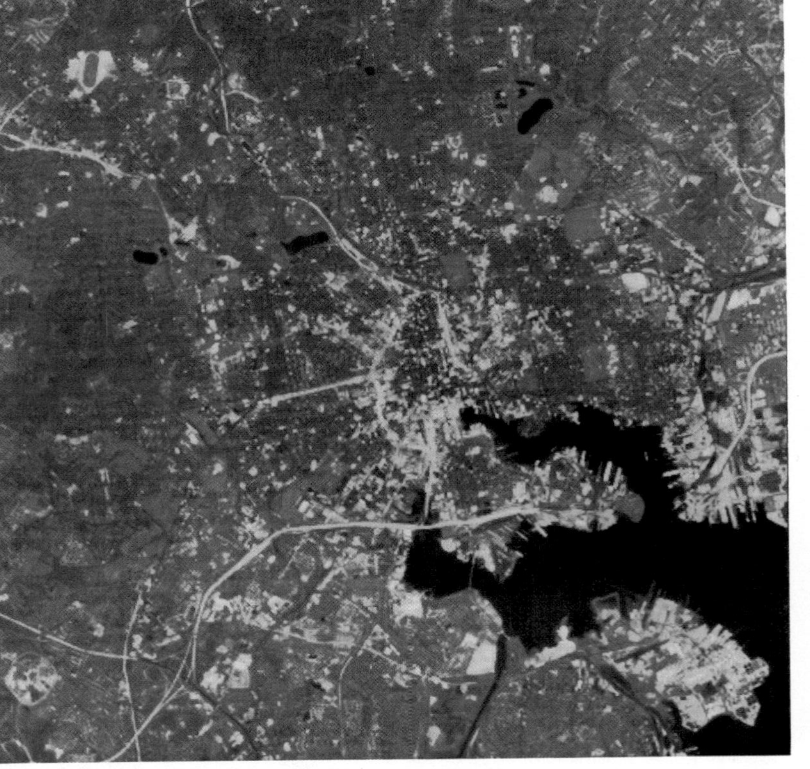

Left: Landsat 3 MSS imagery acquired on July 29, 1978 covering approximately 15 km (12 mi.) on a side, centered over Baltimore, Maryland. This false color composite image, having a spatial resolution of approximately 80 meters, was formed by using MSS bands 4 (green), 5 (red), and 7 (near infrared), printing them in blue, green, and red, respectively. *Right:* Landsat 4 Thematic Mapper imagery acquired on July 29, 1982, covering the same area. This false color composite image, having a spatial resolution of approximately 30 meters was formed by using TM bands 2 (green), 3 (red), and 4 (near infrared), printing them in blue, green, and red, respectively. In detecting and analyzing changes that occurred during the interim, one finds that the higher spatial resolution of the TM imagery can be very helpful. (Courtesy of NASA.)

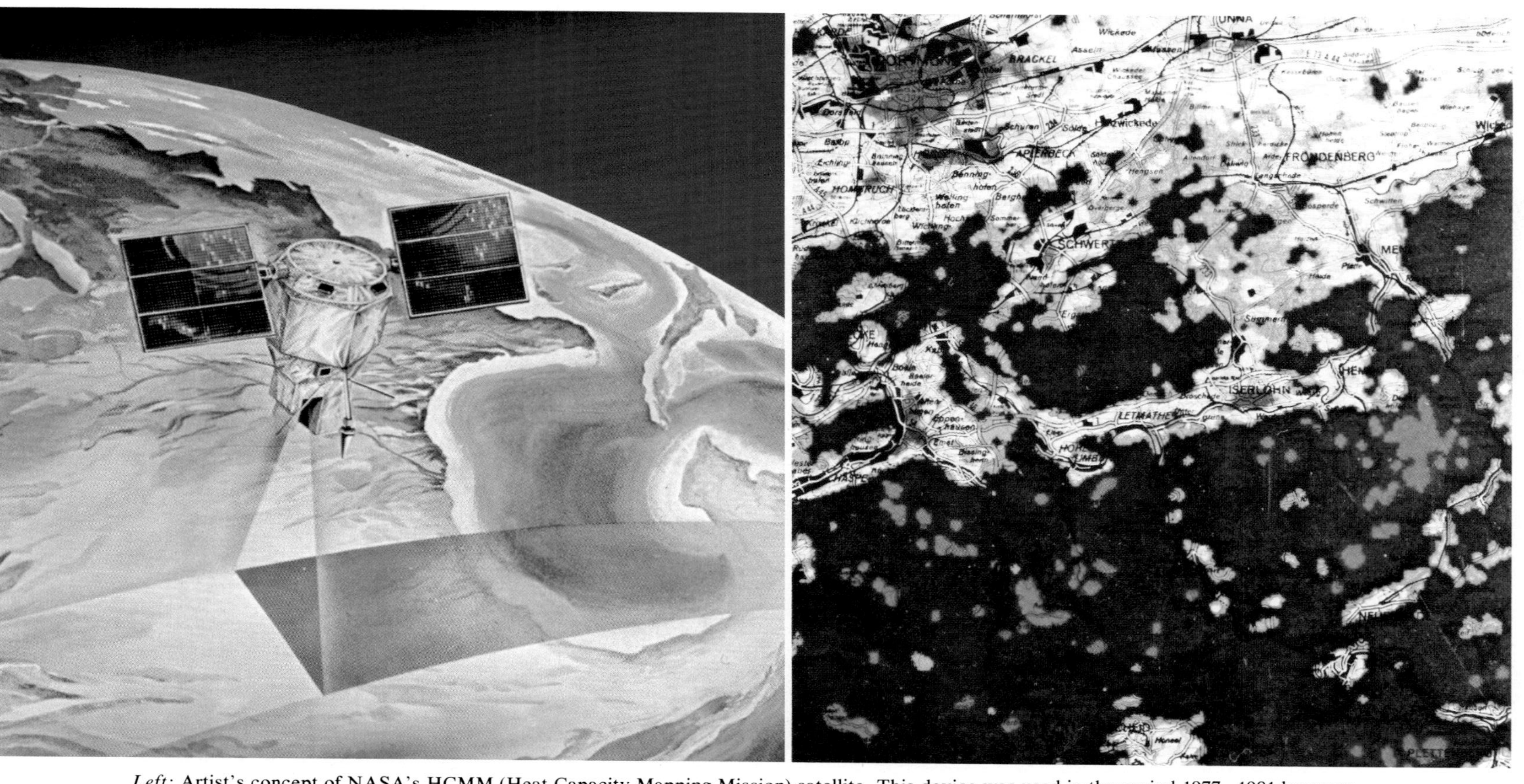

Left: Artist's concept of NASA's HCMM (Heat Capacity Mapping Mission) satellite. This device was used in the period 1977–1981 by seven countries of the European Community to carry out a joint investigation into the applicability of air-and-space remotely sensed thermal data to some typical European zones of agricultural, hydrological, and environmental interest. The HCMM measured the day/night temperature cycling of objects on the earth's surface. Such data were then related to evaporation and moisture in agricultural soils, to physiographic characters of the land and to anthropogenic heat release. *Right:* An overprinting of one of the HCMM scenes onto a thematic map for a typical European area that was included in this investigation. Red areas are warm; blue areas are cold; cyan and green areas are of intermediate temperatures. Further information about the HCMM and related remote sensing devices will be found in Chapters 1 and 13. (Courtesy of NASA Goddard Space Flight Center.)

Simulated SPOT satellite imagery of Toulouse, France and environs made by employing a false color infrared type of image enhancement. Approximate scale 1/100,000. Progressing from the top center of this figure in a clockwise direction, the annotated features, as determined in part by means of ''ground truth'' are: (1) a tree-lined canal; (2) and army complex including mess hall and barracks; (3) a bridge across the Garrone River leading to a major commercial district; (4) a municipal park and playing fields; (5) a track for horse racing events; (6) a large railroad station and associated switching yards; (7) concentrated small industry; (8) extensive hospital grounds and associated facilities; (9) a densely forested area; and (10), agricultural fields fringing the city limits. For details with respect to the SPOT satellite, see Chapters 1 and 31.

ALL OF THE REMAINING 4-COLOR FIGURES FOR THIS VOLUME APPEAR IN COLOR SIGNATURES THAT ARE LOCATED NEAR THE CENTER OF THE VOLUME. IN THE TEXT EACH SUCH FIGURE IS REFERRED TO AS A "COLOR FIGURE" AND IS DESIGNATED BY BOTH THE NUMBER OF THE CHAPTER TO WHICH IT PERTAINS AND THE FIGURE NUMBER WITHIN THE CHAPTER. THE COLOR PLATES BEGIN FOLLOWING PAGE 1754 AND APPEAR IN SEQUENCE BY CHAPTER AND FIG. NO.

FOREWORD

With the publication of the *Manual of Remote Sensing, Second Edition,* the American Society of Photogrammetry completes the ninth in a series of book projects aimed at advancing the science of photogrammetry and remote sensing. Since 1944, the Society has produced this series at an average rate of one major book every five years, each new book bringing forth a text that reflects the most modern aspects of the art. Considering that these landmark publications are produced entirely with volunteer authors and editors, the American Society of Photogrammetry can justifiably take pride in this remarkable series of achievements by its members. The list of major publications includes:

Manual of Photogrammetry, First Edition, 1944
Manual of Photogrammetry, Second Edition, 1952
Manual of Photographic Interpretation, 1960
Manual of Photogrammetry, Third Edition, 1966
Manual of Color Aerial Photography, 1968
Manual of Remote Sensing, First Edition, 1975
Handbook of Non-Topographic Photogrammetry, 1979
Manual of Photogrammetry, Fourth Edition, 1980
Manual of Remote Sensing, Second Edition, 1983

The publication of the second Edition of the Manual of Remote Sensing is especially gratifying because it represents a triumph over unfortunate circumstances that beset the project in its early stages. The formidable task of overseeing the preparation of the Second Edition, as Editor-in-Chief, was assigned originally to Professor Leonard Bowden of the University of California at Riverside. The talented Professor Bowden promptly proceeded to the appointment of editors and associate editors for the two volumes of the manual. The organization of the manual, chapter-by-chapter, was developed, a production time table was worked out and an author-editor was selected for each chapter. Then, in May 1979, came a sad and untimely event: Professor Bowden's sudden death of a heart attack. Although he died before the task had gained full momentum, Professor Bowden laid the groundwork for the project and his successor was able to build on that foundation. For the work that Professor Bowden accomplished in initiating the project, the American Society of Photogrammetry acknowledges a profound debt of gratitude and recognizes that the publication of this manual is, in considerable measure, a tribute to his initiative.

Following Professor Bowden's death, careful consideration was given to the selection of a successor with the competence and willingness to assume the formidable task of carrying on a major project initiated by someone else. Fortunately, a man with ideal qualifications and outlook was active in the society, and more fortunately still, he accepted the position when it was offered to him. Professor Robert N. Colwell of the University of California, Berkeley, brought to the position of Editor-in-Chief a distinguished technical background in photogrammetry and remote sensing plus a brilliant record as a writer and editor. As Editor-in-Chief of the society's *Manual of Photographic Interpretation,* published in 1960, he had acquired valuable experience in successfully managing the publication of that popular book. Above all, he had the talent and ability to adjust his outlook to prevailing conditions; this was a paramount consideration in maintaining the continuity and momentum that already existed within the team of some 300 people who were involved in preparing the Manual. Timely grants provided by the U.S. Geological Survey helped to defray certain editorial costs and also some of the costs entailed in preparing the chapter on Geological Applications. The

American Society of Photogrammetry hereby expresses its sincere thanks to the U.S.G.S. for this valuable assistance.

The extent of the trials and tribulations that Bob Colwell endured and conquered during the preparation of this Manual are completely known only to himself. Perhaps Executive Director Bill French and Publications Consultant Morris Thompson, who managed the business and production aspects of the project, have a substantial awareness of the problems that had to be solved. But that is all now history and the *Second Edition of the Manual of Remote Sensing* is now a reality. For this we, the presidents of the American Society of Photogrammetry during development of the Manual of Remote Sensing, Second Edition, congratulate and thank the Editor-in-Chief, the Volume Editors, the Associate Editors, the Author-Editors and all the contributors to this outstanding publication.

Francis H. Moffitt, 1979
Rex R. McHail, 1980
George J. M. Zarzycki, 1981
Allan C. Bock, 1982
William G. Hemple, 1983

PREFACE TO THE SECOND EDITION

As defined in Webster's dictionary, a "Manual" is "a book containing, in concise form, the principles, rules and directions needed for the mastery of an art, science or skill". That definition describes exactly what this 2-volume publication purports to be with respect to the art, science and skill known as "remote sensing". The term "concise" as used in the above definition connotes "the cutting out of all superficialities and the avoidance of elaboration". Nevertheless, Webster asserts that a Manual should be "only as summary as is compatible with an adequate statement of the available information". The authors of this Manual have attempted to be both concise and summary. The fact that, in so doing, they have produced a document of nearly 2700 pages indicates that the multifaceted field of remote sensing has truly come of age.

This maturing process has been due in no small measure to a perception by successive governing boards of the American Society of Photogrammetry over the years that progressively greater recognition should be given to various aspects of a fast-growing field that is now encompassed by the term "remote sensing—the progenitor of which was" photographic interpretation". Briefly stated: (1) in 1952 the Society added greatly to the prestige of this field by devoting an entire chapter in its Manual of Photogrammetry—Second Edition to the subject of "photographic interpretation; (2) in 1960 the Society gave this field another quantum jump in prestige by devoting an entire Manual to the subject of photographic interpretation; (3) by 1975 the Society acknowledged, through publication of its "Manual of Remote Sensing", that the advent of space age platforms, multispectral scanners, and computer-assisted analysis techniques was bringing about a far more encompassing science than had been dealt with in its comprehensive Manual of Photographic Interpretation; and (4) in 1978 the Society formally adopted the recommendation of several of its most visionary remote sensing scientists, including the late Professor Bowden, that work should begin immediately on the preparation of this, the definitive Second Edition of the Manual of Remote Sensing—an effort that was to require four more years to complete after Professor Bowden's untimely death. His initiative and the organizational ability that he demonstrated in getting this second edition underway are hereby gratefully acknowledged. In keeping with the previously mentioned tradition of high level support, two officials at Society headquarters who have greatly facilitated the preparation of this Manual are Morris M. Thompson and William D. French who have served, respectively, as Production Editor and Publications Director.

Throughout much of the time that was spent in producing this Manual the Editor-in-Chief had almost daily dealings with the highly competent Editors of Volumes I and II, Professors David Simonett and John Estes, respectively. In so doing, he gained great respect for their editorial abilities, their maturity of judgment and the great team spirit that they demonstrated in their dealings both with him and with the author-editors of the 36 chapters. Their jobs, in turn, were greatly facilitated by yet another pair of very competent and dedicated individuals, Professor Fawwaz Ulaby and Dr. Gene Thorley, who served as Associate Editors for Volumes I and II, respectively and whose valuable roles in producing this Manual are likewise gratefully acknowledged. Furthermore, our collective tasks were facilitated and the Manual's quality was significantly improved through the editorial reviews, chapter-by-chapter, that were made by members of the Society's Remote Sensing Applications Division. The uniformly high quality that will be found in the figures, tables and page layout throughout this Manual is due very substantially to the dedication and effectiveness of Jane Schott, the printer's representative and Accounts Director for The Sheridan Press.

But in the hierarchy that is required to produce a Manual such as this, the primary people in the trenches are the author-editors and contributing authors for the individual chapters. In a very major sense the high quality that will be found throughout this Manual is a tribute to the professional expertise and dedication of these individuals. We all are greatly indebted to them.

The potential usefulness of an index tends to increase with the size and comprehensiveness of the book to which it applies. It is especially gratifying to know, therefore, that this voluminous Manual contains an index of unusually high quality and thoroughness, thanks to the efforts of one dedicated individual, G. Carper Tewinkel,—a former president of the Society and an indefatigable indexer of its many publications.

There were many other friendly faces that I encountered in the trenches on a daily basis during the chapter-by-chapter editorial skirmishes. I refer to several coworkers of mine in the University's Space Sciences Laboratory, including James Hardin, Kevin Dummer and Betsy Ross. Their professional response to my many requests also is gratefully acknowledged.

When compared with the First Edition, this Second Edition will be found to have major parts that are entirely new, including the comprehensive chapter dealing with Geological Applications—a field in which very substantial advances have been made in recent years. Special thanks are hereby extended to Dr. Richie Williams and his team of more than fifty contributors for preparing the Geology Chapter under difficult circumstances. Most other parts of the Manual are very substantially new. There are some parts however, including those dealing with remote-sensing-related theory, instruments and techniques, which draw extensively on the text and illustrations that appeared in the First Edition of the Manual or even in its predecessor by 15 years—the Manual of Photographic Interpretation. In those limited cases in which the earlier material seemed to be the best, there obviously was little point in replacing it with something that was only second best, however novel it might be.

An overview of some of the most important recent advances in remote sensing technology is provided by the illustrations that appear on the dust jackets and in a preface of color photos for each of the two volumes. It has been possible to greatly increase the number of color photos throughout this Manual through cost savings achieved by placing them in "signatures". Within each chapter, the color figures have been numbered serially with all other figures, even though separated from them. Throughout the text, use of the term "Color Figure" is made where necessary to direct the reader to the nearest color signature.

Now that this tome finally has been completed, a glance at its table of contents will show even the uninitiated person that remote sensing, in its non-military aspects, is primarily useful in the discovery, evaluation, and development of the earth's natural resources and in the intelligent planning for human occupancy of the earth's surface. In the book of Genesis we learn that it was God's plan to give man "dominion over all the Earth . . . and to subdue it". This Manual presents an effort to describe one of the most valuable means to that end that man has yet devised. But quite probably more than a mere search of the Scriptures would be needed to convince the average contributor to this Manual, frustrated as he/she has been by tight deadlines and perplexed by rigid requirements imposed by an uncompromising editor, that preparation of The Second Edition of the Manual of Remote Sensing was in response to a Divine calling. Nevertheless, it is hoped that this, the published product, will convince both the contributors and the readers that the effort was eminently worthwhile.

Among the primary aims, slightly paraphrased, of the American Society of Photogrammetry are these: to advance knowledge concerning and stimulate interest in the science and art of photogrammetry and remote sensing; to provide means for the dissemination of new knowledge and information and thus to encourage the free exchange of ideas and intercourse among those contributing to the advancement of the art; to stimulate student interest in the fields of photogrammetry and remote sensing; and to foster a spirit of understanding and cooperation among the users of aerospace photography in the United States and throughout the world. It is hoped that those using this Manual will share the conviction of those who produced it, that it represents a furthering of all of these aims and that it is an excellent example of cooperative accomplishment that could only have been achieved under the sponsorship of the American Society of Photogrammetry.

Robert N. Colwell
Editor-in-Chief

Berkeley, California
January, 1983

PREFACE TO VOLUME II

Working as Editors of the Applications Volume of the Second Edition of the *Manual of Remote Sensing* has been a unique and thoroughly rewarding experience. Unique in the opportunity to work with so many authors of outstanding quality. Rewarding to be a part of and to see a volume of this scope and substance come together and be published.

When this work began we, as editors, made a number of decisions concerning the nature of this second edition of the Manual. We felt that the Applications volume should be both a revision and an update of previous editions. Because readers might not have access to the previous edition each chapter should stand alone as the definitive statement of the current utility of remote sensing in each application area. As such Chapter author-editors were given complete latitude to use all or none of the material from appropriate corresponding chapters in the first edition of the *Manual of Remote Sensing*. It was assumed that the reader would be an individual knowledgeable in a particular applications area who was interested in the information which could be gained from the use of remote sensing. Finally, we felt that authors should attempt, within the framework of their chapters, to provide information not only on the generic uses of remote sensing, not stressing any particular platform or sensor system, but also to provide the reader with a brief knowledgeable perspective on the future potential in each applications area.

It has taken nearly five years for this work to become a reality. For their efforts on this manual we would like to express our thanks to the many people who contributed to and helped in the production of this volume. We would particularly like to thank Ms. Susan Bertke and Charlene Sailer for their assistance with editorial and managerial tasks. Finally, we would like to remember Dr. Leonard W. Bowden, late, of the University of California, Riverside, Editor of Volume II of the First Edition of the Manual of Remote Sensing and Editor in Chief of this Edition until his passing. His counsel and his friendship have been and will be missed.

John E. Estes
Editor, Volume II

Gene A. Thorley
Associate Editor, Volume II

CONTENTS

VOLUME I (THEORY, INSTRUMENTS AND TECHNIQUES)

Chapter 1. The Development and Principles of Remote Sensing

Chapter 2. The Nature of Electromagnetic Radiation

Chapter 3. Matter-Energy Interaction in the Optical Region

Chapter 4. Matter-Energy Interaction in the Microwave Region

Chapter 5. Interaction Mechanisms Within the Atmosphere

Chapter 6. Photographic Systems for Remote Sensing

Chapter 7. Electro-Optical Non-Imaging Sensors

Chapter 8. Electro-Optical Imaging Sensors

Chapter 9. Radar Fundamentals and Scatterometers

Chapter 10. Imaging Radar Systems

Chapter 11. Passive Microwave Radiometry

Chapter 12. Landsat Satellites

Chapter 13. Microwave and Infrared Satellite Remote Sensors

Chapter 14. Meteorological Satellites

Chapter 15. Communication and Data Transmission Systems

Chapter 16. Orbital Mechanics for Remote Sensing

Chapter 17. Data Processing and Reprocessing

Chapter 18. Pattern Recognition and Classification

Chapter 19. Remote Sensing Software Systems

Chapter 20. Digital Hardware

Chapter 21. Image Geometry and Rectification

Chapter 22. Geographic Information Systems and Remote Sensing

Chapter 23. Ground Investigations in Support of Remote Sensing

Chapter 24. Fundamentals of Image Analysis: Analysis of Visible and Thermal Infrared Data

Chapter 25. Image Analysis-Microwave Region

VOLUME II (INTERPRETATION AND APPLICATIONS)

Chapter 26. Archaeology, Anthropology, and Cultural Resources Management

Chapter 27. Remote Sensing of Weather and Climate

Chapter 28. The Marine Environment

Chapter 29. Water Resources Assessment

Chapter 30. Urban/Suburban Land Use Analysis

Chapter 31. Geological Applications

Chapter 32. Engineering Applications

Chapter 33. Remote Sensing Applications In Agriculture

Chapter 34. Forest Resources Assessments

Chapter 35. Rangeland Applications

Chapter 36. Terrestrial Moons and Planets

CHAPTER 26

Archaeology, Anthropology, and Cultural Resources Management

Author-Editors: JAMES IAN EBERT and THOMAS R. LYONS

Contributing Authors: BRUCE W. BEVAN, EILEEN L. CAMILLI, SARAH DENNETT, DWIGHT L. DRAGER, ROSALIE FANALE, NICHOLAS HARTMANN, HANS MUESSIG, IRWIN SCOLLAR

GENERAL CONTENTS: Brief overview of remote sensing in archaeological methods; Historical overview: balloons and kites, photoarchaeology; Archaeological applications of remote sensing: aerial archaeological discovery, visual interpretation of cultural resources, archaeological survey and sampling, photogrammetry for recording and mapping cultural resources, in the laboratory: remote sensing and artifactual evidence; Anthropological remote sensing: anthropological remote sensing before 1972, The impact of Landsat; conclusion; references.

INTRODUCTION

The first edition of the *Manual of Remote Sensing* (Reeves 1975) did not have a chapter which discussed applications of remote sensing to archaeology and ethnology. These topics were discussed along with a potpourri of other subjects in Chapter 26. Most of the subjects covered in Chapter 26 of the first edition dealt with ways in which human activities are detected and measured. For example, modern settlement patterning in the Nile River Delta, the estimation of population in cities and towns using aerial photographs, the identification of cities from Landsat MSS imagery, and the mapping of "spoiled lands" in England were all discussed in that Chapter which is entitled: "People: Past and Present". As the reader can see these are not strictly archeological or anthropological applications and were not approached with the same perspective as archaeological or anthropological scientists would employ in their analysis. The present chapter, however, is written entirely from an anthropological/archaeological point of view, and the material presented herein is substantially different from that in Chapter 26 of the first edition of the *Manual.*

If this chapter is different from its 1975 counterpart, archaeology and anthropology have likewise changed in the period since the material for the first edition was assembled. Archaeology, in particular, has evolved to include an emphasis on cultural resource management. This facet of the discipline places importance on the use, study and conservation of cultural materials—sites, structures, and data—in the *present,* which is of course where they really exist. Cultural resource management is all that archaeology was before, but in addition it is pragmatic and demands accurate and justifiable measurement and documentation methods. Remote sensing can supply many such methods, and is increasingly accepted and employed by cultural resource managers. In addition to traditional, discovery-oriented remote sensing applications (which are still, it should be noted, quite valid in many situations), remote sensing methods and data have recently been used by archaeologists and anthropologists for several other important purposes. These include designing and collecting samples, precisely measuring and monitoring cultural structures and sites and the correlating of environmental zones and their contents with past and present human activities across large regional study areas. This chapter details some of these applications. In addition, case-study materials are also discussed which should help clarify the relevance of remote sensing to today's management of archaeological, anthropological and cultural resources.

One of the conceptual changes in archaeology and anthropology that is most relevant to understanding the increasing applicability of remote sensing to these studies involves the meaning of the term "site". Much of the historic view of an archaeological "site" deals with the "site specific" concept, that is it deals with some circumscribed activity center, such as the city of Pompeii, a battlefield, mud or masonry architectural remains, a stone quarry, or some other localized area that contains the physical remains of human presence. Yet there is a broader concept of "site" which is much more important in terms of the methodologies being employed today throughout the discipline, including in particular the methodology of remote sensing. Today an archaeological site is increasingly considered to be the physical environment of which man once was a part. Exampıes are not only the localized areas mentioned above but also large ecological segments such as forest and riverine areas, trade or communications routes and agricultural provinces.

It is difficult, of course, to find a single definition of an archaeological site that would be universally acceptable. However, one which is in harmony with the approaches of remote sensing and other aspects of archaeology as practiced currently is give by Knudson (1978): "Any area of the landscape that shows evidence of past human activity; a portion of the environment used by people". From another point of view, particularly that of the archaeological remote sensing specialist whose primary effort is the observation and interpretation of the environment, *a site is the area selected for archaeological investigation.*

If one accepts the definitions described above of a cultural site one can logically conclude that: (1) man is an integral part of the natural habitat, and the footprints of his past and passing qualify as areas of archaeological research; and (2) all his activities, products and impacts on the rest of the earth and its inhabitants are a part of the natural processes of evolutionary change.

For remote sensing purposes these are functional definitions. The approach to specifying what an archaeological site really is broadens, to a great extent, the traditional concept.

A BRIEF OVERVIEW OF REMOTE SENSING IN ARCHAEOLOGICAL METHODOLOGY

The Regional Perspective: The Environment

New definitions of the archaeological site, i.e. those regarding it as a part of the Earth's natural environment in which past (and present) people live and act, make a regional perspective particularly important as the beginning stage of cultural resource analysis. The megageographic relationships between cultural materials, activity loci and other features of the environment interpreted from small-scale (1:50,000 and smaller) space and aerial imagery enable the archaeologist to develop questions concerning prehistoric peoples which are potentially free from site-specific extrapolation.[1] Some will object to this view, saying that structures and archaeological materials cannot be seen in small-scale imagery. Yet studies discussed later in this chapter, such as Landsat-based cover-type stratification of a study area and the correlation of these strata and cultural resources, seem to suggest that site-specific studies may miss at least part of the point. People lived and live within our world system, and remote sensor data of relatively small scale, when employed in the early stages of cultural resources studies in any area, emphasize this.

There are many other cultural resource problems, however, that require greater resolutions and better definition than that provided by small-scale data. Relationships between specific characteristics of landforms, for instance, call for the use of medium-scale imagery; and, of course, the recording of specific cultural features that contain archaeological materials employs large-scale aerial or terrestrial photography, as will later be documented in connection with Figures 26-40 and 26-41. As the analysis of a study area proceeds from smaller to larger scale, complementary "layers" of the archaeological record are comprehended, analyzed, and documented.

Site Exploration

Exploration for sites with artifactual content has a long history dating back to and beyond the grave robbers of Egypt. Many discovery methods are today widely employed in a more disciplined approach to recovering the material and cultural treasures of the past. The standard method of discovery today is the on-the-ground survey inspection by a team of field archaeologists spread out at stated intervals and visually inspecting the ground for evidence of man's activities. This method also includes a description of the environmental setting and surface condition of the site, a collection of representative artifacts and a delineation of and the geographic location on some existing map of the survey area.

In recent years this technique has been aided and improved by the use of both oblique and vertical, medium- and small-scale aerial photographs. Site locations can be pinpointed on the aerial photography by the trained surveying crew and later transferred to existing maps with a considerably greater degree of locational accuracy than would be produced by guessing or interpolating positions on relatively small scale topographic maps. The imagery contains large amounts of environmental data as well which supplement the field notes and which, with later study, are used to develop predictive models for site type occurrences in unsurveyed areas.

Sampling

There has been much discussion in the literature on "one hundred percent" surveys of an area for archaeological remains or evidence. Inherent in some uses of this term is the assumption that, by walking over the landscape, a cadre of individuals can find 100% of the existing sites. It is apparent upon reflection that this is never truly the case since much of the archaeological record

[1] For the purposes of this chapter, the following definitions of various photographic and other remote sensor-data scale-ranges will be used:

Small-scale	1:50,000 and smaller
Medium-scale	1:15,000 to 1:50,000
Large-scale	larger than 1:15,000

It should be noted that these definitions may be found to be at variance with scale ranges defined by other authors. Yet they are appropriate to the applications discussed herein.

is buried within sediments, and inevitably some visible evidence is overlooked or ignored. Consequently a sampling procedure which is designed to identify representative numbers and types of recognizable sites within a project area is both practical and economical (Brown and Ebert, 1978). Remote sensing can add a valuable dimension to the formulation of archaeological sampling schemes as discussed at length later in this chapter.

Mitigation is the process of rescuing archaeological data prior to elimination or destruction due to causes other than planned research. Mitigation of archaeological sites is a requirement in those areas subject to surficial destruction or irreparable modification such as that associated with stripmines, dammed river valleys, highways, and pipelines.

A number of remote sensing techniques, in addition to the traditional techniques of collecting and excavating, are mitigating procedures. For instance, the sample mapping of a site can provide much important information. Tracing site elements and configurations on transparent overlays or aerial photographs of relatively large scale is one useful technique (Fig. 26-1). Maps derived from aerial photographs with the aid of horizontal and vertical control points established through ground survey then can provide planimetry and topography to virtually any desired scale and contour interval. Further, such maps can be taken to the field and supplemented rapidly with data derived from on-site inspection.

Terrestrial photogrammetry as applied to the mapping of standing walls is another form of mitigation. This method of site recording, along with conventional planimetric and topographic mapping, provides base line data for monitoring and measuring human or natural impact.

When mitigation takes the form of excavation, aerial photogrammetric mapping is a useful tool in planning, in developing data on volumes of material to be removed, and (during the process of excavation) in providing rapid and accurate maps of the exposed elements of the site itself.

Application of many of the techniques discussed in this chapter and in the listed references has led to the realization that a vast amount of archaeological data can be derived with the aid of such techniques without touching the physical remains of the archaeological record at all. Such an approach has been termed "non-destructive archaeology" (Lyons and Scovill, 1978) to differentiate it from the long-standing traditional and admittedly destructive process of excavating for the artifactual remains of human manufacture.

Over time, the methods of archaeological research have been greatly refined, as is to be expected in any viable discipline. In its early history, during the last century, archaeology was primarily an excavation process whatever its ultimate aims. This process was destructive of much information contained in the context from which man-made objects were exhumed, yet excavation became almost synonymous with the name of the discipline itself.

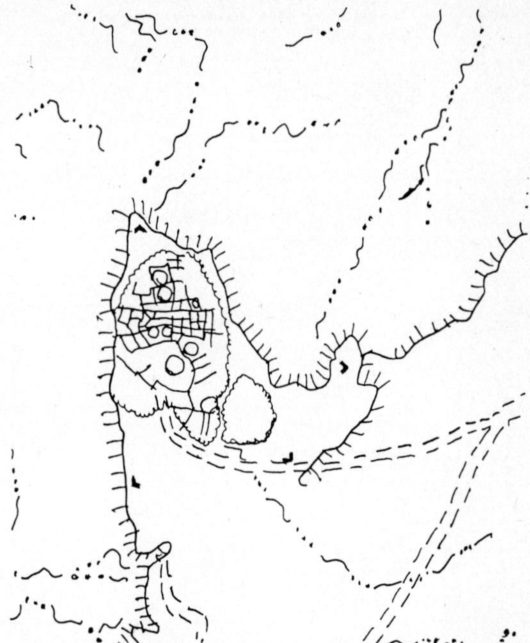

Fig. 26-1. Portion of a map of the El Rito site, a prehistoric pueblo village in New Mexico, traced monoscopically from 1:3000 scale black-and-white aerial photographs. The use of a stereoscope and magnifying eyepieces allowed the interpreter to discern 3-dimensional detail revealing the location of structural walls, drainages, scarps, and other aspects of the site; these were traced on an overlay sheet with one hand. Control panels placed prior to flying are marked with rough "L's." While the photographs contain the data to allow true photogrammetric mapping at a later date, this relatively crude method of mapping provided a fast, inexpensive way to supply archaeologists with maps of such ruins which were far more accurate than any previously in existence. Such maps are valuable for fieldwork or preservation planning purposes.

Four significant categories of archaeological procedures can be identified, *viz* destructive, moderately destructive, minimally destructive, and non-destructive. As with many related categories the lines of demarcation are not clear and the concept can best be illustrated as a continuum ranging from methods which inevitably destroy context and information inherent in the context, to methods which disturb no *in situ* data whatsoever (Fig. 26-2).

Given this continuum of methods developed over the history of the discipline of archaeology the following argument is made: A research design should be conceived, in terms of methodological approach, beginning at the non-destructive end of the spectrum and progressing, through the use of carefully chosen methods of increased data-destroying potential, to the ultimate and most destructive methods which serve as the final test of

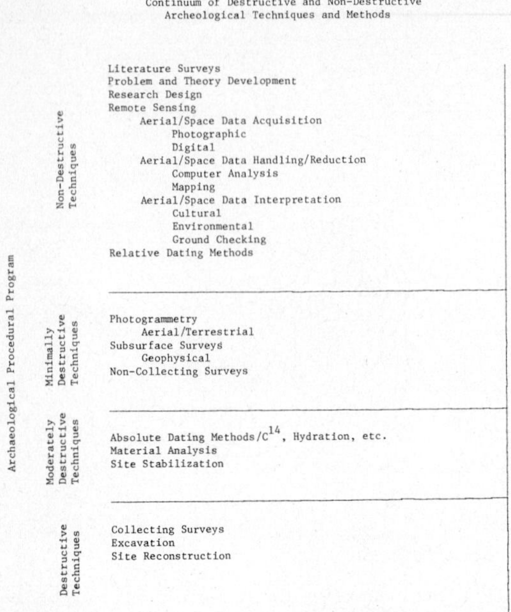

Fig. 26-2. Archaeological methods and techniques, the means by which the archaeologist studies and analyzes the record of the past, can be ranked along a continuum ranging from destructive to nondestructive. Remote sensing techniques are some of the least destructive available, and offer a range of resolution and data-collection capabilities virtually unparalleled by most other methods listed here.

ideas and concepts. Remote sensing data and methodology provide the archaeologist, anthropologist, architect and cultural resource manager with an available, cost-effective, and widely applicable means of pursuing such a design in the study and treatment of cultural resources.

HISTORICAL OVERVIEW[2]

The first archaeological site to be photographed from the air was the Bronze Age monument of Stonehenge in England; two aerial views were taken from a military balloon in 1906. At about the same period, box kites were successfully adapted by an archaeologist, Henry S. Wellcome, for photographing excavations in the Anglo-Egyptian Sudan. Technical improvements in cameras and aircraft gave impetus to archaeological study during World War I, and during the campaigns in the Near East, neither the British nor the Germans neglected the opportunity to take aerial photographs of archaeological sites.

The archaeological world is heavily indebted to the late O.G.S. Crawford for laying down the ground rules for the interpretation and publication of aerial photographs, and for inspiring students of archaeology to make use of aerial photographs. The turning point in the adoption of photointerpretation by archaeologists occurred in 1922, when Crawford and an associate were shown some Royal Air Force photographs containing the patterns of a Bronze Age field system. From this beginning, photographs that betray the presence of ancient sites were recognized as tools in archaeological exploration. In the interval between the two world wars, Crawford published many papers in *Antiquity,* an archaeological journal which he founded and edited until his death in 1957. His book *Wessex from the Air* (1928), written with Alexander Keiller, is a classic in photoarchaeology.

British archaeology was wedded to aerial photography by the work of Crawford in the 1920's, but photointerpretation has also been practiced by archaeologists in continental Europe, the Near and Far East, Africa, and the Americas. Unfortunately, not all parts of the world are covered by aerial photographs and only recently has spacecraft imagery become available.

Before World War II, most vertical and oblique photographs of archaeological sites were taken with hand-held cameras. Photogrammetric techniques were rarely used in interpreting these photographs. A change came during the war when a huge archive of pictorial documents was created by the almost total photographing of western Europe. Although these wartime photographs are not ideal for archaeological purposes, they contain much information of value to European archaeologists. Good details of many archaeological sites are visible in overlapping pairs of contact prints, or in stereomasics (Heath, 1957). Stereoscopic pairs of aerial photographs (when available) are now an indispensible part of the archaeological record.

In the Americas, aerial archaeological research did not maintain the same pace as in the Old World. The first known aerial photographs purposely taken of an archaeological site in the United States were those of Cahokia Mound, Madison County, Illinois, taken by Army fliers in 1921. Nine years later, Army fliers took photographs of an area in the Gila and Salt River valleys in central Arizona for use by the Smithsonian Institution in the study of prehistoric canals. Beginning in 1933, and up to World War II, Dache M. Reeves, then with the United States Army, took photographs of many earth-work sites in the Ohio Valley. This invaluable set of documents on American prehistory was deposited with the National Museum. During the same period, Lt. George R. Johnson, working with the Peruvian Naval Service, included many views of ar-

[2] In this section and in later portions of this chapter, a number of earlier publications by the American Society of Photogrammetry have been drawn on heavily. Special acknowledgement is made of the use of materials from Chapter 26 of the Manual of Remote Sensing, first edition.

chaeological monuments in his aerial photography of Peru. Subsequently, the Shippee-Johnson Peruvian Expedition discovered the "Great Wall of Peru", a pre-Inca construction near Chimbote, from the air. The imposing monuments of Middle America were also photographed from the air, and some sites were discovered by aerial expeditions of museums and educational institutions. In the first of the larger ventures, Col. Charles A. Lindbergh was the leader (Ricketson and Kidder, 1930). On a more modest scale, sites in the United States were photographed from light planes with hand-held cameras. Smithsonian Institution photographers took pictures of archaeological sites in the Missouri Basin for use in the emergency salvage program made necessary by the construction of dams, behind which huge areas of bottomland were to be flooded. Field archaeologists of the Smithsonian Institution's River Basin Surveys have discovered many sites by interpretation of aerial photographs taken for the Soil Conservation Service and other government agencies.

Following World War II, an extremely important paper titled: "The Technique of Air Archeology" was published in the Archaeological Journal by D. N. Riley (1946) which presented diagrams of such features as positive and negative crop marks, soil marks and shadow marks, and discussed how they allowed cultural features to be interpreted from aerial photographs. Riley's diagrams have been reproduced a number of times and now form the basis for much of the archaeological photointerpretation that is being done.

With the technological developments that occurred in the field of aerial photography as a result of the impetus induced by military reconnaissance during World War II, many countries, particularly the United States, began systemic aerial photography of vast stretches of land for purposes of topographic photogrammetric mapping. Much of this photography became available inexpensively to the general public, though often its scale was so small that little of archaeological significance was thought to be visible on it. Generally, archaeologists thought they had to obtain large-scale, usually oblique, hand-held photos to show their sites.

Photographs of individual archaeological sites are widely known in the literature and much has been learned from them as single entities. The sizes of residences and temples, approximations of the sizes of cities, and certain aspects of their economies are matters which an expert can determine from such a study without setting foot on the site. It has now become possible, with a number of such studies already made, to make comparisons among sites all over the world (Sjoberg, 1965). Technological, spatial, economic, social, and political patterns that are represented on photographs can be placed side by side for study. Not only the size and shape of a city are revealed, but also something is learned of the hinterland, which supported the city through its agriculture (Goguey, 1967). The sizes of these early cities may have been surprisingly large; for instance, it is estimated that Teotihuacan near Mexico City had about 100,000 inhabitants in the first millenium A.D.

The use of aerial photography in Western Europe has demonstrated the location of ancient "deserted villages", many but not all of them emptied by the Black Death of the 14th century. It has also been possible to trace the change of arable land to pasture land around the end of the 15th century (St. Joseph, 1966). Traces of man's activities in the Middle Ages, such as those found in England, consitute the commonest archaeological features visible on the surface of almost every country. These may be studied advantageously and with great precision by means of aerial photography (Bowen, 1962; Schellart, 1962). The English landscape, in particular, has furnished an excellent field laboratory for aerial archaeology practically from the time of the invention of the airplane (Figure 26-3). Sites include now-deserted villages that flourished in the 11th and 12th centuries, open field systems in ridges and furrows, boundary earthworks of medieval estates, and their accompanying moats and fishponds. Pockmarking the English countryside are the upturned and broken ground areas that mark the search for such raw materials as lime, clay, marl, and building stone—a search which was indelibly scarred the earth's surface. As St. Joseph states, the variety and number of applications of the aerial photography methods to remains of the Middle Ages, as well as to those of later times, have yet to be fully realized (St. Joseph, 1966). The same may be said for France (Chevallier, 1962) and other European countries.

Fig. 26-3. Roman villa at Ditchley, England. The foundations of the building and the boundary wall show as negative plant marks. A positive plant mark indicates the surrounding enclosure ditch. The walls were found to be about two feet thick, the ditch about ten feet wide. (Photograph by Major G. W. G. Allen; from Riley, 1946; courtesy Ashmolean Museum, Oxford.)

In site surveys using aerial photographs, the archaeologist has to check the suspected features on the ground, but at times this task is made very difficult because surface and subsurface features are obscured by vegetation. Through the use of various kinds of electrical sensitivity apparatus, such as resistivity circuits, electrical anomalies have been found that correspond with the images of features revealed on the photographic film. The value of combining aerial surveys with ground instrument surveys has been demonstrated in an experiment by Froelich Rainey (1974), of the University of Pennsylvania, at a Roman site near Kingscote, England.

New technology using aerial photography enables the investigator to examine the same terrain with different kinds of imagery. Edward Yost of Long Island University, has perfected a multiband camera using nine different wavelengths—six in visual, two in infrared, and one in the ultra-violet range—adapted in a Fairchild camera, which he used to find the location of protohistoric iron mines in western Africa. False-color prints enabled the investigator to pick out the objectives more clearly.

R. J. Rinker (1973a; 1973b) of the U.S. Army Topographic Engineers has perfected a technique for the location of cave openings, critical locales for the study of prehistory. Rinker's system, which consists of an infrared spectroscope and conventional camera apparatus, is flown over the site at a low altitude in a helicopter. It measures the difference between the ambient temperature and the temperature emitted from the cave opening; the difference is then recorded on an oscilloscope for permanent record. This method is useful in prehistoric-site surveys for the location of archaeological cave sites, and can be used in conjunction with the identification of talus slopes in front of once-occupied caves where differences in vegetation are observed. This latter confirming evidence is due to the fertilizing effect of human occupational debris on the front apron of palaeolithic and more recent cave sites.

BALLOONS AND KITES

Julian H. Whittlesey (1970), of the Whittlesey Foundation, has successfully adapted an 18 cubic meter hydrogen-filled balloon with special camera apparatus for archaeological photography in the Near East. The earliest cameras in his experiments were suspended about 45 m. above the surface, their shutters triggered by a pneumatic release from a hand bulb on the ground. The camera in this rig was fastened in a gimbal mount, which permitted true vertical-control photography and avoided the haphazard nature of other unmanned camera platforms. Whittlesey varied the camera altitudes from 10 to 600 m. according to the situation, and employed a variety of cameras and lenses in his experiments, including the Linhof, Hasselblad, and Graflex (Fig. 26-4). Whittlesey also adapted more sophisticated equipment to the mounts, always mindful of weight limitations. These items included an electrically advanced radio-controlled camera, hung in a lightweight gimbal specially constructed of magnesium. The camera was attached to the balloon above it by means of four converging lines. These four lines from the camera platform led obliquely below to the tethering line. As the balloon was, of necessity, tethered almost directly over the site, this line appears in all of the photographs. However, the tethering line is indistinct in the views, and does not detract from the archaeological information.

Fig. 26-4. Two views of the multiband aerial camera for archaeology. The camera is suspended in gimbals for tethered balloon flight. The package contains 3 Hasselblad EL500 radio-controlled cameras with Zeiss Distagon 50 mm lens and a strobe signal. (Courtesy Julian Whittlesey).

Views over water were obtained by Whittlesey by tethering the balloon to a boat. Ancient architectural remains, such as submerged Roman and ancient Grecian building walls and contructions, were clearly revealed. Such a site was found at Porte Chelli in the southern Argolid of Greece.

This on-site controlled photography enables the archaeologist to review the overall situation of his excavation on command and permits day-by-day observations. A stereoscope makes it possible for the archaeologist to interpret and map the site. Whittlesey also designed a multiband photography unit using synchronized cameras for archaeology which offered greatly increased capability by using a variety of equipment with the cameras, including different films and filters. These cameras were mounted in the gimbal mount system used with the single camera as described above and suspended from the balloon in the same manner.

Kites are still being used by some expeditions, including the University of Chicago excavations in Iraq. An adaptation of the box kite is the airfoil, or parafoil, which has been found to be good in steady winds up to 20 knots. Airfoils operate like box kites, but have no rigid structural members; rigidity is imparted when the device "inflates" with air upon launch. Such airfoils were originally invented to lift loads into the air in jungle areas, to be snagged by hooks trailing from aircraft. Whittlesey has flown a light-weight remotely controlled single-shot Graflex camera from this type of apparatus. A certain amount of care is required for the launching and recovery of the apparatus, especially in situations when the ground is uneven or the air currents are unsteady. The camera is drawn up to the parafoil by a system of cords that restrict the camera's movement and make it possible to target the camera precisely over the objective. This three-point tethering device allows the parafoil to shift with the winds, while the camera is held securely over the objective. Whittlesey also adapted this threepoint tethering system to a free balloon, allowing the latter to swing about independently of the camera. The camera system consisted of three EL 500 Hasselblad cameras equipped with 50 mm. lenses which, together with the holding frame and various operating controls and mechanisms, made for a large weight factor. However, the results obtained were compensatory. The great advantages of the multiband camera is that color as well as black-and-white or other films may be used at will according to the dictates and requirements of the archaeologist.

For lower elevation shots at less expense and more control in windy conditions, it was found that a camera bipod (Whittlesey, 1966) or two light-weight long collapsible rods with guy ropes holding them firmly in place forms an excellent camera platform. The camera is held in the true vertical position directly over the site, and the whole rig can be moved easily in any direction and to any unit of distance as the situation dictates. Any camera can be used by adapting it for the bipod. In the simpler kind of shutter release, the mechanism could be set on the ground with a timing device, and then hoisted up rapidly to its position in the bipod rig. Although only one picture at a time can be taken in this kind of set, its advantages are that it is a nearly foolproof method. For more sophisticated instrumentation, like that of Whittlesey, a radio-controlled Hasselblad was suspended about 11 to 12 m. above the ground. Members of the Whittlesey Foundation found that there was no photographic system that could equal this one in the recording of detailed architectural evidences such as mosaic floors. To record the placement of each bit of mosaic on a pen-and-ink drawing, for instance, would have been a long and expensive operation; but a photograph recorded the whole work in an instant, accurately and definitively. The advantage of total overview is nicely illustrated by the case of a somewhat demanding antiquities department abroad, which asked for a pictorial record of every item (including many thousands of potsherds) before an American expedition could return home at the end of the season. The director of the expedition cunningly obliged by laying out all of the specimens found that season in the courtyard, clambering up on the roof with his camera, and producing an instant record of every item. Because the requirement obviously was met by this variation of aerial photography, the expedition was given its leave.

PHOTOARCHAEOLOGY

Ever-increasing refinements in aerial observations of archaeological sites have appeared over the last 20–25 years, ranging from identifying Iron Age timber-framed houses in Scotland (Feacham, 1962) to detecting ancient archaeological traces of field systems in New Caledonia (Avias, 1967). But it seems that the most exciting advances in archaeological applications of remote sensing have been made recently in the United States. The newer and more sophisticated applications of aerial photography have been demonstrated by Strandberg (1967, 1974) in two different geographic parts of this country. In both instances, aerial photographic studies have revealed hitherto unknown phenomena, including prehistoric fishtraps and Indian sites. Strandberg coined the word, "photoarchaeology," for these studies.

Strandberg reported that the organization with which he has been associated (the Itek Data Analysis Center of Alexandria, Virginia, under contract to the U.S. National Park Service) undertook a field reconnaissance project in cooperation with the Smithsonian Institution in order to determine what type of aerial photographic film gives the best results for archaeology. The tests were run along the Missouri River south of Pierre, South Dakota. The locality is just south of the region where personnel of the River Basin Surveys, Smithsonian Institution, made extensive surveys and excavations prior to reservoir and dam construction by the government. Not only were all of the archaeologically-known sites spotted, but the analysis showed a wholly unsuspected, large, fortified, aboriginal Indian village site (Fig. 26-5). Its finding was surprising because both sides of the Missouri had been examined intensively by archaeologists for a least a quarter of a century.

Strandberg conjectured that Norsemen may have been responsible for the construction of this bastioned fort, sometime after the end of the 1st millenium A.D.; he cited the bastion construction,

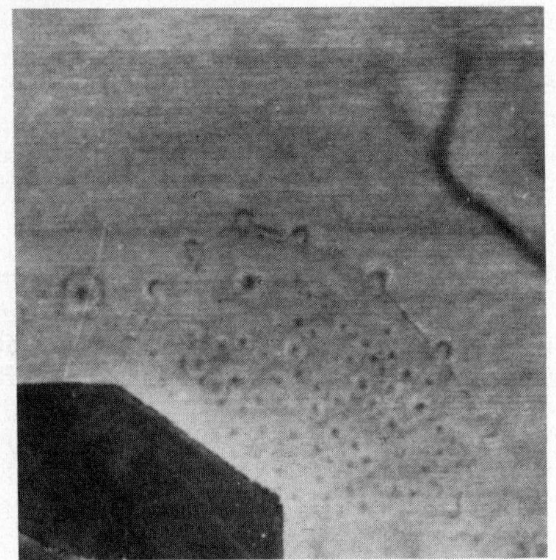

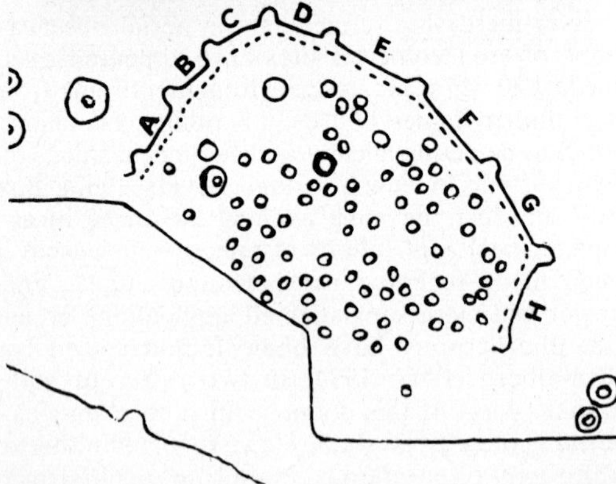

Fig. 26-5. This photo and accompanying sketch-map illustrate the present-day appearance of a fortified village site, classified as Pre-Arikara. The solid outer line on the sketch map marks the location of the moat. The dashed inner line marks the location of the palisade. The smaller circles indicate locations of older houses which were occupied at the time the moat and palisade were actively defended. The larger double circles are the locations of more recent Indian earth lodges. Residual traces of these features can be seen in the photo. The distances between the centers of the bastions are: A–124 feet, B–121 feet, C–129 feet, D–125 feet, E–209 feet, F–218 feet, G–195 feet, H–202 feet.

the "moat", and reputed historic evidence of blond and blue-eyed Mandan Indians in the area. Bastioned forts, however, are known to have existed in North America in aboriginal times with no apparent connection to white invaders. As to the physical anthropological evidence, it is known that early hunters and trappers ("squaw men"), took Indian mates. Thus, as with any archaeological data, the factual information made available by remote sensing is subject to the archaeologist's interpretation.

Strandberg's (1974) second project proved to be another excellent test case for aerial photography, and he demonstrated quite vividly the value of this method. In this instance, he undertook a contract with the Maryland Department of Natural Resources. The work involved the aerial photography of the mid-18th-century Fort Frederick site as part of a project to restore the fort (color Fig. 26-6A and B). Strandberg first obtained vertical infrared color photographs and had special color-transform black-and-white prints made from the original color infrared photos. An index map was first made, followed by the assembly of an index mosaic. From these a set of stereostrip mosaics was made.

In the preliminary assessment, a stereo-analysis was made of the entire study area from which a number of areas were chosen for more detailed investigations. Fortunately, the State of Maryland had already obtained black-and-white (pan-minus-blue) vertical aerial photography. A technique called "color slicing" was done to locate images as part of the color infrared analysis.

Strandberg's aerial photography investigations revealed potentially significant finds within Fort Frederick and in the surrounding region. He found some probable building foundations within the fort, and outside the fort three possible Indian village sites were located, one of which is evidently a major site. Extra dividends appeared in the form of two aboriginal Indian fishtraps of stone in the Potomac River. A metal detector was used to obtain additional data in on-site investigations made to support the findings from image analysis.

In 1969, the National Park Service of the United States Department of the Interior began another series of photoarchaeological experiments directed toward the ruins and roads of Chaco canyon under the direction of Thomas R. Lyons for the specific purpose of determining the application of remote sensing techniques to archaeology and cultural resource management. Included in the project were examinations of various aspects of remote sensing, including photointerpretation, photogrammetric mapping, analysis of emulsion differences, digital image analysis and terrestrial photogrammetric mapping and monitoring. Detailed descriptions of findings have been published elsewhere (Lyons, 1976; Lyons and Avery, 1977; Lyons and Hitchcock, 1977; Lyons and Ebert, 1978; Lyons and Mathien, 1980; Drager and Lyons, in press).

ARCHAEOLOGICAL APPLICATIONS OF REMOTE SENSING

AERIAL ARCHAEOLOGICAL DISCOVERY

In the United States, a number of early applications of remote sensing methods to archaeological exploration pointed up, in a spectacular way, the new perspective offered this field by aerial photographs, thus sparking interest within the science. One such experience was that of Jeffrey Parsons (Reeves, 1975) who, using aerial photography, discovered a vast network of raised fields used prehistorically in Colombia. The following account, much of which is by the author in his own words, serves as a fitting introduction to the sort of experimentation and research that excited the imagination of archaeologists in the 1960's about the potential of archaeological remote sensing.

In recent years, extensive tracts of pre-Colombian raised planting beds or ridged fields have been discovered in several parts of lowland tropical America (Parsons and Denevan, 1967). These ancient agricultural earthworks, suggesting a high degree of social organization and centralized authority, are chiefly associated with seasonally inundated flood plains that are exploited today only for extensive cattle raising. It is clear that in the past the flood plains were intensively cultivated, and that the man-made ridges provided improved drainages for crops. These agricultural earthworks, with associated raised causeways, were initially identified in the Llanos de Mojos of Bolivia's Oriente and in the San Jorge River flood plain of northern Colombia. More recently, similar features have been identified in the Guayas River valley, near Guayaquil, Ecuador; in the state of Apure on the Venezuelan Llanos; in coastal Surinam; and on the Yucatan peninsula. They are similar in some respects to the floating gardens or *chinampas* of the valley of Mexico and have striking counterparts, only recently discovered, in the poorly drained high Andean basins of Lake Titicaca and the Sabana de Bogota.

The ridged fields, which have not yet been dated, have been identified and delineated from aerial photography and from low-flying reconnaissance aircraft. From the ground, or from the dugout canoes that are the common means of transport in many of these areas, they are extraordinarily difficult to locate and it is not surprising that they have gone almost completely unnoticed until recently.

Their extent and nature become apparent only from the perspective of an aircraft. Although these discoveries have been of special interest to archaeologists and anthropologists, it is noteworthy that in virtually every case the ridged fields have been first identified and described by geographers familiar with the use of aerial photography and air reconnaissance. Parsons provides the following account of the discoveries:

> My own experience in Colombia may be worth recording. While doing field work on the Caribbean coast, we have heard numerous stories about curious parallel lines on the land in the lower Cauca and San Jorge valleys, corrugated landscapes of ridges of dikes, apparently, that the local inhabitants presumed to be the product of some unexplained natural process of sedimentation. Near Ayapel we had spent many hours in the saddle with vaqueros who professed to know just where to look for these lineaments, but they turned out to be a disappointing cluster of low mounds and ridges barely discernible under the tall Guinea grass and Para grass along the San Jorge River. They were interesting, but nothing to become excited about. On flights on commercial planes we had seen odd patterns, too, especially between Medllin and Cartagena, but informed Colombians assured us that they were caused by early gold dredging activities on the lower Cauca. William Denevan, than a Ph.D. candidate in geography at Berkeley, had been working in the seasonally inundated savannas of the Llanos de Mojos of eastern Bolivia where extensive aboriginal cultivation surfaces had been reported earlier in this century by the Swedish ethnologist Erland Nordenskiold. Some of Denevan's air photos, which came into our hands, looked pretty interesting. On a hunch, on the last day of our 1965 field season in Colombia, we visited the Instituto Geografico in Bogota to inspect the available air photo coverage of the north coast and especially the lower part of the San Jorge valley—well covered as a result of the petroleum exploration that had been carried out in the area. There they were, magnificent linear features, often delineated by sharp vegetation contrasts, almost as though someone had drawn a giant comb, first this way and then that way, across the low-lying lands adjacent to the sluggish river and its distributaries, behind the natural levees (Fig. 26-7 and 26-8). At first no one would believe what we described—not the archaeologists or the quaqueros (who had worked the area over somewhat for its gold-rich burial mounds), not the oil company geologists who had lived for months in field camps right in the midst of this strikingly patterned land, not the geographers at the university of the Instituto Geografico. So, after delineating some 170,000 acres, (68,000 hectares) of apparently ridged field landscapes from the photos, we went to the field for a look. There was no question about it. From the waterways or from horseback they were scarcely visible, but from a small aircraft chartered from the emergency strip at San Marcos a clearly delineated ridge-and-furrow landscape could be seen stretching out for miles across the floodplains. In June, in the first days of the rainy season, the humped backs of the ridges, typically some 20 feet across, were still accentuated by parched grass while the depressions between them, not yet inundated by the rising flood waters, remained

Fig. 26-7. Ridged fields in the San Jorge river valley, Colombia. The raised surfaces are in part delineated by trees. (Courtesy of James J. Parsons and William A. Bowen)

Fig. 26-8. Enlarged section of vertical airphoto showing geometric pattern of ridges along hyacinth-choked Cañon Los Angeles on San Jorge river flood plain, Colombia, near San Marcos. (Courtesy of Instituto Geografico Agustin Codazzi, Bogota)

green. From the air some looked very much like immense corrugated washboards. Others were in striking geometric patterns, or aligned at right angles to abandoned distributary channels of the San Jorge, like iron filings clinging to a magnet. There was no doubt but that we had discovered a major site of man's earlier occupancy of the area that had up to then gone unnoticed. But who built them and when? Mapped and described in a paper in the Geographical Review, (Parsons and Bowen, 1966) the story was widely disseminated by newspapers syndicated both in the U.S. and in Latin America. Volunteered suggestions as to their origin included—dredged tailings or mine dump refuse, fish ponds, irrigation ditches, immigrant trans-Pacific rice farmers, ancestors of the Lost Tribes of Israel. It was apparent that by the time the first Spaniards arrived in this area, in the 1530's, the ridges were no longer in use. There is no reference to them in the fairly extensive colonial chronicles of New Granada. Yet the enormity of the regimented human labor and the sophisticated landscape engineering that would have been required to build these cyclopian prehistoric features staggers the imagination. And what we see today may be, and probably is, but a small part of their original extent. If they were as old as we suspect—and the dense clay subsoils of the ridges suggests prolonged soils weathering—they might well have a crucial place in the history of the rise of American high cultures, the origins of which have been incresingly sought in the tropical lowlands and, following the University of Illinois anthropologist Donald W. Lathrap (1970, 1973), more specifically in the coastal plains of northern Colombia and perhaps of Ecuador (Parsons and Bowen, 1966).

The alluvial morphology of the San Jorge area suggests that these ridges may be of considerable age, for they appear especially to be associated with long abandoned meander scars, channels or canos now seasonally choked with water hyacinth. It is possible that a considerable extent of them could have been obliterated by sedimentation, perhaps assisted by the isostatic downwarping from sediment loading that must operate here very much as it does in the Mississippi Delta area and at the mouths of many other large rivers. Although ridges occur in geographical association with more recent archaeological sites (post 500 A.D.) this does not obviate the possibility that their construction might have been initiated by earlier peoples.

The discovery of the Ecuador oldfields was equally fortuitous (Parsons, 1969). We had been studying those of Colombia and were flying to Guayaquil on other business. Approaching the airport north of the city for landing, below the plane window, we unexpectedly observed a distinctively patterened landscape strikingly similar to that of the San Jorge Valley. Air reconnaissance with a cropduster who worked the nearby banana plantations and examination of air photography at the Instituto Geografico Militar in Quito unmistakably confirmed the existence of a minimum of 1,600 hectares (4,000 acres) of parallel ridges and swales mostly with a north-south orientation, lying north and east of the city and back from the river in patterns remarkably similar to those identified in Colombia and Boliva. Most of the ridges, which stood from 60 to 120 cm (2 to 4 ft). above the adjacent swales, supported a luxuriant second-growth forest while the lower land on either side of them was delineated by aquatic reeds and grasses.

Perhaps the most significant fact about the Magdalena and Guayas oldfields is that they lie immediately adjacent to what are, by a considerable margin, the two earliest pottery sites archaeologists have as yet identified in the New World (Puerto Hormiga, on the Canal de Dique near modern Cartagena, and Valdivia, on the coast of Ecuador) both of which have been dated by Carbon-14 at approximately 3000 B.C. These first pottery-making people were quite certainly efficient agriculturalists, dependent most probably on root crop cultivation, especially the high yielding yucca or manioc. Manioc leaves no archaeological traces in the humid tropics, but the widespread presence of grater teeth and large clay griddles with sharply up turned rims (*bundares*) is generally taken to indicate the presence of this crop. These earliest farmers also grew cotton, the bottle gourd, and cultivated fish poisons—crops related to manufacturing and to fishing—for there was clearly an early and very heavy dependence among such people on protein from costal and riverine sources (turtle, manatee, aquatic rodents, caiman, waterfowl). The flood plains and adjacent coasts of seasonally dry tropical lowland would seem to provide the ideal type of climate and topography that we have sought when looking for the sites where man first began his sedentary existence in Nuclear America. Lathrap, who, following Carl Sauer's cue, has imaginatively campaigned for the tropical lowlands as the hearth of New World high cultures, sees an expansive, aggressive root-crop complex generating increasing population on the flood plains of the Amazon, Magdalena and Guayas rivers until reaching the end of the available good alluvial land (Parsons, 1969). Only then would have come the massive move into the highlands where smaller but more diverse ecologic niches would have encouraged the further elaboration and differentiation of cultures.

Curiously, the only reference in the chronicles of the colonial period to aboriginal flood-plain cultivation on artifical ridges is from the Llanos of Venezuela and Colombia, but to date only a very limited extent of such features has been located in the vast area, chiefly in the state of Apure in Venezuela. Other types of pre-Columbian earthworks, especially causeways and earthen pyramids, are known from the Barinas area. Raised aboriginal planting beds have also been identified from air photography on the poorly drained coastal plains of Surinam, where they are threatened with destruction by the expansion of irrigated rice cultivation (Reeves, 1975, pg. 2004–2007).

Most recently, evidence of intensive ridged field cultivation of surprising extent has been reported from the Yucatan penninsula, in the Mexican states of Campeche and Quintana Roo as well as in Belize (Turner, 1974). These are still under study to determined their extent and possible relationship to the classic Maya culture. Again, reconnaissance from low-flying aircraft has been a crucial means of identifying these features.

In highland Mexico intensive cultivation of poorly drained inter-volcanic basins has been associated with the floating gardens of Xochimilco. Recently, air photography has revealed extensive evidence of similar features, long since abandoned, elsewhere in the valley of Mexico (Tenochititlan) as well as in the adjacent basin of Tlaxcala and Toluca. Similarly, ridged field landscapes of obvious pre-Columbian origin have been discovered on the poorly drained Sabana de Bogota in Colombia and along the shores of Lake Titicaca in highland Peru, again with the help of air photography (Figs. 26-9 and 10). The similarity of layout and design with those of the adjacent lowland floodplains is striking (Broadbent, 1968).

Fig. 26-9. Low oblique photo of ridged fields and causeway on seasonally inundated Llanos de Mojos in eastern Bolivia. (Courtesy of William Denevan)

VISUAL INTERPRETATION OF CULTURAL RESOURCES

There are essentially two ways of using aerial photography for visual interpretation of archaeological features. The first way, the way aerial photography has been used the most in archaeology, is to obtain aerial photographs, either vertical or oblique, of a known site and check to see if any features are visible on the photographs that are not visible on the ground. This procedure presupposes knowledge of the location of archaeological sites.

The second method requires no *a priori* knowledge of the location of an archaeological site and is often employed in a cultural resource area in which some development is pending. With the current emphasis on large-area cultural resources surveys, this method has been gaining in importance, because, through the acquisition of total aerial photographic coverage, later archaeological studies of the cultural resources can be conducted.

This type of investigation involves obtaining aerial photography of the area prior to conducting a large-area cultural-resources survey, either by purchasing existing aerial photography, or by contracting for new coverage of the area. In either case, due to the inherently destructive nature of archaeological excavations, the photography must be acquired before the field work begins. Since the area to be surveyed has probably had little or no previous archaeological investigation, photographs of the entire area should be obtained. For purposes of consistency, vertical photographs at the same scale should be acquired. Also, blanket coverage of the area should be obtained by flying consecutive flightlines with 60% forward lap and 15% sidelap.

In addition to providing the means for future study of the area, the aerial photographs assist in numerous aspects of the cultural resource manager's or scientist's efforts. The aerial photographs can be used in the planning stages to divide the survey area into zones for sampling; they can be taken into the field and used as maps to lay out sample areas or transects; they can be used in the field as aids in locating the position of sites which are encountered; or they can be interpreted prior to fieldwork for anomalous indications which can then be checked during fieldwork to determine their cultural affiliation, if any.

The general procedures of archaeological photointerpretation were first formalized as early as 1946 by D. N. Riley in a paper titled, "The Technique of Air-Archaeology." In this paper, he discusses the various types of marks visible on aerial photographs that allow interpreters to recognize cultural features. Riley lists shadow marks, soil marks, damp marks, frost and snow marks, and crop marks as the primary indicators of cul-

Fig. 26-10. Ridged fields along north shore of Lake Titicaca, Peru, exposed during time of exceptionally low water. (Courtesy of William Denevan and C. T. Smith)

tural features visible in large-scale photography (Riley, 1946). He also includes drawings and photographs showing how each type of mark is formed.

Subsequently, procedures for photointerpretation have undergone much refinement. These elements as well as the techniques and visual requirements of photointerpretation are discussed quite extensively in the archaeology chapter of the Manual of Photographic Interpretation (Amer. Society of Photogrammetry, 1960). They are discussed further in Chapter 24 of the present Manual. Spurr (1960) suggests that objects can be recognized by their shape, dimension, tone, texture, pattern, location, and associations. Avery (1977:24) agrees with this list, but adds shadows as also useful in identifying objects. In the field of archaeology, Lyons and Avery (1977) imply that the best photointerpreter will be one who is familiar with the cultural sequence and the environmental setting of a project area.

The interpretive process should be initiated by laying as much of the photography as possible onto a large table in the order in which it was acquired, that is, by flightlines. Sequential and overlapping photos and flightlines should be identified and the entire set of photos should be laid out as a large mosaic. If there are too many photos to work with at once, portions of the mosaic can be laid out. Mylar or clear acetate overlays should be made, identifying major cultural and environmental features such as roads, buildings, rivers, fences, forest boundaries, etc. This will allow the fieldworker to find the anomalous features which will be added to the map later. After known features have been identified a search of photography should be made to identify unknown features. In the field of archaeology, features that may have a cultural origin are of particular interest. People tend to build things in regular shapes, such as squares, rectangles, circles, or straight lines. Features of this type should be identified. Soil or vegetation patterns which appear out of place should also be identified. Features below the surface of the ground could be causing differences in soil or vegetation patterning and can only be discovered through field examination.

Once the mosaic has been examined to the satisfaction of the interpreter, consecutive photographs should be examined using a stereoscope (to create the illusion of depth) to determine the presence of topographic features which may be of interest but are invisible on the individual photographs. Because of the vertical exaggeration inherent in vertical stereoscopic pairs of photographs, small relief difference may be quite apparent when viewed through a stereoscope. Cultural features that are filled in or worn down may become visible when viewed stereoscopically. These features can be marked on the photographs and then transferred to the clear mosaic map after the photos have been replaced under the map.

If more than a single film emulsion type has been flown over a target area, or if photography from more than a single date is available, features may appear on one set of photos that are not visible on another. These features should also be added to the mosaic map. If photography of differing scales is used, it may be necessary to use mechanical aids, such as a Zoom Transfer Scope or a Map-O-Graph, to mark their locations on the mosaic map. (For a discussion of these and other types of imagery interpretation equipment, please refer to Chapter 24).

Following the creation of an interpretive map of the target area, both the map and the original photography can be used during the field phases of the survey. The aerial photographs are excellent field maps and often can be used to locate a site or a survey crew's location in the field much more accurately than can a topographic map of the same area. The interpretive map can be used to determine the location in the field of anomalous features identified on the photography. Samples can be taken in the field and their location can be identified on either the photos or the interpretive map, or both, for later laboratory analysis.

Low Budget Aerial Photography

Aerial photography can be a very economical way of getting an initial overall perspective of objects, phenomena and their spatial relationships; furthermore, as a reconnaissance tool, it can prove to be a highly cost-effective efficient way of exploring a large area for discovery purposes.

For the least expense, the first step is a search for existing photographs. Many government agencies and private companies have collections of aerial photographs. Copies of these can be purchased for much less than the cost of taking new photos. However, these photos may be at too small a scale and/or taken during the winter months and therefore may be unsuitable for many archaeological and anthropological applications.

If commercial photographs are not available, it is very easy to take one's own aerial photographs. The most expensive cost elements involved are the capital cost of the aircraft and its imaging system and their utilization costs. These costs will vary with the distance the aircraft must fly to the site, the number of flight lines needed for complete coverage of the study area and whether local weather conditions cause flight delay. However, this will insure that the quality of the aerial photography will meet the archaeological needs. For use as illustrations, oblique photos often can be better than vertical photos, and the user can select the desired viewpoint from the air. For reconnaissance applications, the time of photography can be critical. The time factor is best understood by a person doing field work onsite, observing from the ground the changes in plant growth, soil, and illumination with time of day, weather, and season. The quick reaction required for successful photography of this type cannot easily be done by someone distant from a site or study area.

All things considered, an airplane is the best aerial platform. For very low altitude photography requiring precise perspective control, a helicopter can be useful, even though it is several times more expensive. Balloons and kites, while excellent for carrying cameras, require special training for successful operation. For most applications, a high wing, propeller airplane is recommended. One should lease a plane with a window which can be completely opened in flight in order to eliminate light reflection from the window and image degradation caused by dirt and scratches. A researcher/user should also hire a pilot who has done this type of work before; the pilot can help by making tight turns around the area of interest, allowing oblique and near-vertical photos to be taken.

Conventional single lens reflex cameras having 35 mm or 2¼ inch film formats are excellent for archaeological aerial photography. A camera with a through-the-lens viewfinder will help the photographer keep the airplane struts and other hardware out of the photos. The focal length of the camera's lens and the flying height determine the scale of the photo; in many cases a flying altitude of several hundred meters and a normal to slight telephoto lens will be optimum. It is always advisable to take more than one camera; malfunctions are possible and this also allows different films to be easily used.

Normal color and black-and-white films are excellent for most applications; special cases are discussed in a later section. Often it will be best to emphasize the faster films. This will allow one to use a short exposure time; usually, the shorter the better, in order to minimize image blur caused by forward motion of the aircraft and by vibrations. If an automatic exposure camera is used, it is best to select the exposure time manually. Image blur is a major factor in aerial photography and exposure times shorter than 1/250 of a second can minimize degradation. Wind and turbulence can cause both the airplane and camera to bounce thus introducing noise into the imagery through instability of the aerial platform (Figure 16). Unfortunately, days having low haze and good visibility also often are accompanied by higher winds and turbulence. Exposure readings on the ground should be taken before the flight and compromises should be made between those readings and the ones measured from the aircraft, although they should be reasonably similar.

Temporal Effects in Aerial Photography for Archaeological Exploration

The success of aerial photographic reconnaissance is very dependent on season, weather, and illumination.

Buried architectural remnants, refilled earthworks, and cultural debris can alter the chemistry, moisture content, texture, and structure of the soil. In those instances in which the normal root zone of the area's vegetation intercepts archaeological features, the growth pattern of the vegetation can be affected. In particular, the maturation of vegetation, such as a field crop of grain, can change. Classic examples can be found in cultivated fields such as wheat or barley: buried wall remnants can advance the golden ripening stage while refilled earthworks can retard it. The color contrasts here can be quite large. However, in soils having little natural stratification or on irrigated land, these anomalous patterns will be minimal.

The time factor is also important in areas which currently are not cultivated. Buried features can be indicated by the clustering or pattern of different species of natural vegetation. Also, descendants of cultivated plants can linger in an area. These can often be most readily distinguished when the contrasting plants are in a flowering stage.

Trees can help or hinder such aerial surveys. In a forested area, certain tree species can indicate poor prior cultivation. However, trees more often obscure the view of surface details. In deciduous forests, this problem can be avoided somewhat through the taking of photography in the late fall or early winter after leaves have fallen but before snow covers the ground, or in early spring after snow melts but before trees leaf out.

The color of soil is quite time-dependent, due to its changing moisture content. Color contrasts, which can reveal buried features, are often most detectable after a rainstorm, at a time in which the soil is neither uniformly wet nor dry. Subtle topographic relief can be revealed after a light snow fall when the snow on slight ridges has melted.

These topographic patterns are also most visible under the low angle illumination of an early morning or late afternoon sun. However, the general terrain must be fairly level. At other times, shadows are undesirable, particularly for pictorial photography. For these applications, one can select times when the sky has a uniform, thin cloud cover; aerial photos taken before sunrise are also often useful.

Selecting Combinations of Films and Filters

Wise choice of camera films and lens filters often enhances the quality of vertical and oblique aerial photographs acquired for archaeological purposes. For pictorial photography, the goal is clarity of detail; reconnaissance photography usually requires the maximum contrast of subtle color differences between the object or phenomenan of interest and its scene background. Optimum selection of film/filter combinations can give minor, but worthwhile, improvements making images more valuable in the analysis process.

From the technical viewpoint, the problem is one of maximizing or minimizing the contrast of spectral radiance in a scene. In principle, narrow band filters are required. In practice, however, the natural variability of scenes and the requirement for adequate film exposure indicate that a wider spectrum should be selected. The following are some specific applications:

- The contrast of shadows can be maximized by matching a red filter with black-and-white panchromatic film. Even higher contrast is possible with black-and-white infrared film.
- If it is desirable to minimize shadow contrast, this can best be done with a blue filter and black-and-white film. The ultraviolet part of the spectrum will help also, provided that the film is sensitive to ultraviolet energy and the camera's lens is able to transmit this radiation.
- From time to time, film manufacturers have marketed films which are optimum for detecting objects underwater. If these are not available, one can use a blue or blue-green filter, whichever provides the closest match to the visible color of the water. No film or filter will help much in turbid water, although some help in this regard is offered by the work of Russian scientists, as reported in Chapter 2 of the Manual of Photographic Interpretation (Amer. Society of Photogrammetry, 1960).
- The choice of filter is usually not critical in the search for soil contrasts. If a specific contrast between soil types is desired, that contrast can be selected for maximization; otherwise, normal color film works very well. With black-and-white film, there can be some advantage in selecting the red part of the visible spectrum through the use of red filters.

• Vegetation contrasts are very important in archaeological exploration. While some contrasts are readily visible in any part of the optical spectrum, those contrasts resulting from differential moisture stress require special care. If a green filter is used, these contrasts usually disappear completely. In general, a red filter with black-and-white film makes them most apparent. Both color film and false color infrared film are also suitable for such photography.

• Water and green vegetation are easiest to distinguish from each other in the reflected infrared spectrum, using black-and-white infrared or color infrared film and filters.

Analytical Use of Oblique Aerial Photographs for Discovery and Mapping

In the United States, while some archaeologists and others choose to take oblique aerial photographs with hand-held cameras of the sites they study, vertical photographs are preferred for most cultural resource studies. One of the chief advantages of vertical photography is that such photographs offer a relatively faithful "map view" of the scene beheld, and of course stereoscopic coverage is easily obtained for photogrammetry. American archaeologists perhaps are fortunate that remote sensing was not employed in their field until after vertical aerial photography gained widespread use in engineering, and metric aerial cameras were readily available in almost all parts of the country.

This was not the case in Great Britain and Europe, where archaeologists prefer oblique photos for contrast, season and scale reasons. Thus techniques and methodologies employed in cultural-resources remote sensing in Europe are somewhat different than in this country. The earliest organized aerial archaeology was undertaken in England, followed shortly by aerial photography of ruins and their indications in France and Germany as much as fifty years ago. Such photography was, of course, exposed using hand-held cameras at an oblique angle. The resultant imagery represents a vast archive, estimated at over a million frames in Great Britain alone as well as several hundred thousand frames in other countries (Scollar 1977:347). This collection, and the work on which it is based, have provided impetus to continue taking and using oblique aerial photos for cultural resources purposes.

Oblique aerial photos have disadvantages when they are employed by the analyst/interpreter to solve some of the problems of the archaeologist or cultural resource manager. As previously indicated, however, they may have important advantages in certain cases as well, particularly in situations involving the sorts of archaeological remains and agriculturally-developed settings frequently found in the Old World. Europe has been the home of civilized peoples for millenia, and the structures they built have often had far more effect on the landscape than have sites in the New World. Forts, villages, and field-boundaries have etched their indications deeply into the environment in Europe and it is toward the discovery and mapping of these indications that most European aerial archaeology is directed. What is more, these indications are often easily discerned because of the contrast relative to present-day land use. Although the surface expression of many of Europe's ruins is obscured by agriculture, shallow foundations or buried traces of walls are revealed by crop marks, especially in well-drained soils. Crop marks are very obviously dependent on the type of crop being grown and the stresses under which the crop is developing, which makes agricultural treatment, weather history, and time of year extremely critical. The archaeologist, dependent as he is on crop mark discovery and identification of cultural resources, often must revisit an area again and again, taking photographs under all possible conditions and all angles of view. Some photographs may reveal one part or aspect of an ancient site, and others will show different aspects. At some sites, indications of ditches, earthworks, field boundaries, or walls too subtle to be seen in their entirety from the ground will be revealed by shadows, which appear only at certain sun angles and times of the year, again making it necessary to photograph these at exactly the right moment. The angle between the sun, the ruin, and the photographer also often is important, and when aerial obliques are exposed of a ruin by an experienced archaeological photographer these often have higher contrast and better visibility than verticals exposed at the same moment using the same film (Scollar, personal communication).

In any case, the European aerial archaeologist often must work with oblique aerial photos due to the great body of archival or "historic" photographs he is presented with. Because of the continuous and rapid pace of economic development in Europe, and concommitant land disturbance, some of these older photographs are actually superior to present-day views for defining the extent of archaeological sites. Unfortunately, many of the older photographs are of poor quality, having been taken with inferior lenses at slow shutter speeds, or having been scratched or having suffered image degradation over time. In addition, since they are oblique photos, it is difficult to use them in any simple way to produce planimetric maps of sites.

A pioneer effort to deal with these problems has been made by Irwin Scollar of the Rheinisches Landesmuseum in Bonn, West Germany. As director for the Laboratory of Field Archaeology there, Scollar was faced with the necessity of recording and mapping archaeological and historic sites in the Rhineland so that these could be avoided by land developers and construction projects; and also so that when and if excavation for mitigation purposes becomes necessary, the actual sites and features within them can be located and studied with a minimum of difficulty. Planning for cultural resource management purposes is particularly crucial in the Rhineland, perhaps the most heavily industrialized region in the world. A great many oblique aerial photographs are available, most of them having been taken during the last twenty years; these, of course, vary greatly in quality and scale. The Rheinisches Landesmuseum has been doing systematic aerial photography of archaeological sites since 1960 using cameras such as the Kodak K20, Williamson F117 and Fairchild F505 (Scollar 1979).

The data base, onto which archaeological site locations and outlines were to be transferred for planning purposes, was a series of some 4000 base-map sheets at a scale of 1:5,000 depicting the entire Rhineland. Early in Scollar's plotting attempts, hand-computational methods were used to "rectify" measurements taken from oblique aerial photos for subsequent plotting on maps; later, a telephone-line link with a computer was used to speed computations (Scollar 1975). Such calculations are, however, time-consuming and inaccurate or at least partially so, because only a few point locations can be measured from the oblique image, transformed by calculations, and transferred to the planimetric map. Such computations are doubly difficult when the terrain is hilly, as it is in much of the Rhineland.

In 1974, Scollar began to design specifications for an operator-interactive computer system to facilitate the rectification and mapping of measurements from oblique photographs, and to perform a number of additional image manipulation and enhancement tasks on these photographs as well. Acquisition of the computer and software was funded by the Volkswagen Foundation. Scollar's system was online by September 1976.

The image processing facility at the Rheinisches Landesmuseum is based on a PDP 11/10 computer for line buffering and control, coupled to a PDP 11/70 computer which performs calculations on image data. The PDP 11/70's capacity is also utilized for storage of collections data for the Museum in addiiton to image processing. Digital image input originates from an Optriconics P1000 drum scanner. Color and black-and-white video displays are viewed by the operator, who interactively chooses areas of interest and specifies operations to be performed on them. Although aerial photographs comprise the major data source used with this system, it has also been employed to enhance x-rays, specimen photos and multispectral scanner data. Operations performed by the system can be classed under two major headings: image enhancement and geometric correction/mapping.

Image enhancement is desirable because not all of the photographs available to the archaeologist are of consistent quality, contrast, sharpness, or freedom from "noise" (due to scratches, dirt, poor processing, etc.) and lens aberrations. In addition, digital enhancement can add information value to any photographic product by enhancing edges and filtering. The Rheinisches Landesmuseum system performs contrast enhancement, histogram modification or equalization, Fourier transform and non-linear spatial filtering, spherical lens aberration corection, homomorphic filtering, correction for motion blur, edge enhancement, and nonlinear filtering for removing dust spots and noise (for further discussion of such processes please refer to Chapter 24).

The geometric correction capabilities of the Rheinisches Landesmuseum system provide the solution to mapping archaeological indications and other visible features from oblique aerial photographs. Using ground control data from a limited number of points which appear in a set of photographs, the system creates a "pseudovertical" view of the site. The archaeological area of interest is indicated by the operator from his console; for instance, ground outlines of old fields or the walls of an ancient fort. This allows computations to be performed only on areas of interest, (rather than rectifying unnecessary modern details or environmental features which may not be of prime importance) and thereby speeds computations. Maps (1:1000 scale) are also scanned at 50-micron precision and stored digitally, separately from the digitized cultural resource locations. This allows the revision of cultural resource or map data independently; such data can, however, be conjoined and plotted together whenever necessary. In addition, photographic output of rectified views is provided by the system's Optronics P1500 film writer (Scollar 1977).

Four different views of one site in the Rheinland illustrate successive stages in image rectification, enhancement, and incorporation with the mapped data base. Figure 26-11 is an original oblique aerial photograph (Rheinisches Landesmuseum Archive Nr. HF24) showing crop marks which betray the location and show outlines of a temporary Roman marching camp, probably constructed in the late first or early second century AD, at Menzelen near Moers in the Rheinland. The original photo was taken in 1970 with Kodak Plus-X Aerecon film with an orange filter at approximately 500 meters altitude with a Fairchild F505 camera. In Figure 26-12, a pseudo-vertical representation of this original photograph has been computer-compiled. Through the use of four ground control points, the original was transformed into a matrix of 1500 × 1500 pixels (50 micrometer resolution) and reprinted by the drum printer.

Figure 26-13 shows the results of digital enhancement (Wallis Algorithm) performed on the geometrically-corrected matrix. Site details are far more obvious than in the original or the corrected view, allowing the recognition of details of interest by the operator. In Figure 26-14, the geometrically corrected view has been digitally superimposed over a portion of a 1:5000 (original scale) base map. Computing time for these steps

Fig. 26-11. Oblique aerial photograph of crop marks revealing outlines of a first- or second-century AD Roman walking fort near Menzelen near Moers in the Rhineland. This photograph was taken with Kodak Plus-X Aerocon film with an orange filter through the use of a Fairchild F505 camera equipped with a Schneider Xenar 180 mm lens. (Photography by Irwin Scollar, Rheinisches Landesmuseum Archive Nr. HF 24)

was approximately 30 minutes on the PDP 11/70 computer with an attached FT11-C floating point processor.

In addition to using the system for the improvement, enhancement and rectification of

Fig. 26-12. Using four well-spaced control points measured on the ground after the oblique photograph in Figure 26-11 was taken, the Rheinisches Landesmuseum system has produced this rectified, "pseudo-vertical" view of the Roman fort at a photographic resolution of 50 microns.

Fig. 26-13. Here, the geometrically-corrected matrix shown in Figure 26-12 has been further manipulated by the Rheinisches Landesmuseum system through the application of filtering algorithms. Contrast and edge-definition are greatly enhanced.

oblique aerial photography, Scollar has recently experimented, in conjunction with French remote sensing personnel (Centre de Recherche Geophysique, Garchy), with 10.5 micrometer infrared-scanner data. They have found that immediately following an abrupt temperature change early in the morning or at the onset of a cold front, thermal lag differences in the upper 30–50 cm of soil are strongly affected by micro-relief. Figure 26-15 shows a scan using the 10.5 micrometer scanner of about 4 km length at high resolution before enhancement (left) and after filtering for radio interference and enhancement (right). Temperature resolution is 0.1 degree C., and the temperature range is about 10°C. The flight altitude was roughly 400 m, and ground resolution at the nadir about 1 m. Figure 26-16 shows the results of rectification and filtering of a thermal scan taken in the Seine Valley.

Because of the special situations and problems encountered by European and British archaeologists and historians, the digital image enhancement, manipulation and rectification system at the Rheinisches Landesmuseum will undoubtedly serve as an example for the development of other, similar systems which should be in increasingly high demand in the future. When Scollar originally assembled his system, hardware and software costs were much higher than today; he estimates that a system with similar capabilities could be built now for 20–50% of what it cost at that time. Production models of image manipulation systems are also presently being designed by a number of firms, and this should reduce costs even further.

Again, although conditions in the United States and the rest of the Americas are somewhat different than in Europe (fewer large structures, less area

Fig. 26-14. By combining two digital data bases, one containing the unenhanced but geometrically-corrected version of original photograph HF 24 and the other a portion of a 1:5000 scale base map of the Rhineland, this figure provides locational information that can be used by land-use and other planners for the avoidance or study of the Menzelen fort.

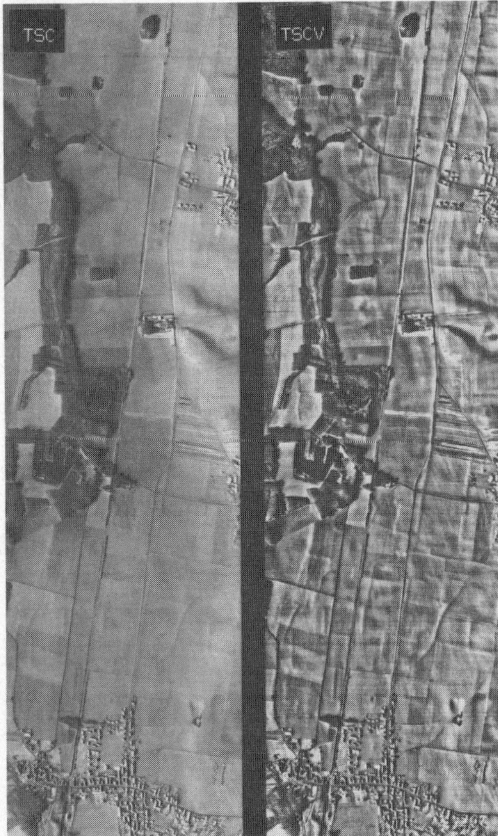

Fig. 26-15. Thermal scanner data can also be manipulated to the advantage of the archaeologist with the Rheinisches Landesmuseum system. This figure shows an original 10.5-micron thermal scan (left) of an area in the Seine Valley which has been used for farming for centuries; the original scan was made from an altitude of about 400 meters and shows some 4 km length at high resolution. In the adjacent scan (right), median filtering was applied to reduce radio interference, and enhancement and destriping were also undertaken. Old field boundaries are considerably more visible following such manipulation. (Scanning by Alain Tabbagh, Centre de Recherche en Geophysique, Garchy, France).

Fig. 26-16. An illustration of geometric correction of thermal scanner data. The right-hand image was made under very gusty wind conditions. In the left-hand strip, corrections for roll, pitch and yaw have been made using gyroscopic information; enhancement and geometric correction were also performed. The remaining slight wiggles visible in the road running through the picture are due to undersampling of the roll information which was available only once each scan line. (Thermal scanning courtesy of Allain Tabbaugh, Centre de Recherche en Geophysique, Garchy, France.)

planted in crops, and the extensive use of nitrogen fertilizers are a few such differences), there are probably a number of areas in the United States where methods similar to those outlined above would be profitable.

ARCHAEOLOGICAL SURVEY AND SAMPLING

The discovery of prehistoric and historic sites from the air is only a subset of the more standard and traditional means of discovery, viz. survey on the surface of the ground. Probably the great majority of archaeological data cannot be seen on aerial photographs or other remote sensor imagery. However, some evidence left by people in the past, such as the Anasazi roadway network discovered within the San Juan Basin in New Mexico, are too faint and extensive to be noticed easily on the ground, and can only be efficiently discovered from above. The same is true of other extensive features such as canal or raised field complexes and crop marks outlining ancient buried structures. It could be argued, of course, that surveying for archaeological materials while walking on the surface of the ground is simply "remote sensing" from about five feet altitude. Even when artifacts, features, or sites themselves cannot be seen on aerial photos or other remote sensor data, remote sensing methods can still help archaeologists in field survey and in the sampling and prediction of past materials.

Field Use of Remote Sensor Data

Cultural resource surveys are usually conducted in areas not well-traveled or easily accessible. Maps may be unavailable for such areas, and in any case, may contain incomplete and/or unreliable information. Aerial photographs are far more useful "maps" for the archaeologist in such circumstances, because minute details not recorded on maps such as trees, rocks and specific landforms can be noted and can aid in the exact determination of one's location. The use of an easily-made aerial photographic field kit carried in an air-photo map case allows prints to be fastened down and viewed stereoscopically, adding another dimension of information. One of the most useful applications of aerial photos, not only by archaeologists but anyone working out of doors, is simply as maps which enter into planning and field operations at every stage.

Pre-Field Logistics Planning

Most modern cultural resources surveys are competitively awarded, which requires that accurate estimates be made prior to any field activity of the area and ruggedness of terrain to be covered, the number of field crew members that will be required and their length of employment, and routes of travel and/or camps to be involved in the project. Some of these things depend upon knowledge of "unknown" resources in the area, and this will be discussed later under the section on sampling and prediction. Other field necessities, however, can be estimated and planned by careful study of terrain and distances using aerial photographs. Thus it is absolutely essential to search for and acquire aerial imagery in advance of the first stage of any project in which such data are to be seriously employed. This presents an interesting dilemma. Copying previously-acquired aerial photographs is relatively inexpensive, but existing imagery may be unsuitable and lag time between ordering and receiving imagery can be long, often as much as three months. It is more expensive, yet more expedient, to contract for aerial coverage of a target area, and it will guarantee high quality imagery which is in accordance with the project's specific requirements. Financial and temporal constraints must be evaluated in each case but one possible compromise may be to use a light aircraft and a hand-held camera to collect prefield aerial photos.

Marking Location of Sites and Features

Once the crew is in the field, it is necessary to determine the exact boundaries of the area which must be surveyed. Once located, the desired survey area is more or less intensively searched by the survey crew, and can be marked by the crew directly on the photographic prints as each portion of the area is completely surveyed. Aerial photographs carried by a hardworking crew in the field can become soiled or destroyed. Hence, it is wise to have a set of photos for each crew that will use them in a certain area as well as a back-up set and a clean set in the laboratory.

Traditionally, when a site is found, its location is determined either by noting landforms, by reading topographic contours from a map, or by triangulation on visible landmarks using a compass. Landforms can also be difficult to decipher if their extremities fall between grossly spaced contour lines. Triangulation with a compass is probably a better way to place sites on a map, but it depends not only on the accuracy and scale of the map, but the compass accuracy as well. When three points are used for triangulation with distant landmarks, the triangle in which the surveyed feature must fall can be very large on a map.

A far better solution is to use the same photographs, which served as a field planning aid and for finding one's way to the survey area, for marking the position of sites found. One precise and indelible way of marking site locations is to prick the photographic print with a needle mounted in a wooden handle and carried in the field kit (Loose and Lyons 1976). A site identification-number can be written on the back of the print to avoid obscuring photo details on the emulsion side. Once back at the laboratory, the pinprick location can be transferred to a map or to clean prints.

Photointerpretation of Sites in the Field

Although the discovery of cultural resources using remote sensing techniques was discussed previously in this chapter, it should be noted that there are advantages to doing some discovery-oriented interpretation in the field rather than in the laboratory.

The reason that photointerpretation in the field is advantageous is that it allows the archaeologist to employ a "feedback" process. Standing on an area positively identified as a cultural site on the ground, the fieldworker may be able to distinguish previously-ambiguous patterning in the photograph that would otherwise have gone unnoticed. The case of the patterning seen on the imagery can then be investigated firsthand—vegetation, soil coloration, or other factors not necessarily themselves cultural—and the recognition pattern refined. In some cases, the interpreter will extend the boundaries of the site by seeing patterning in the aerial photographs that is not apparent on the ground, then checking carefully to discover inconspicuous cultural debris beyond the previous site limits.

The chief problem in photointerpretation in the field is lack of proper instrumentation and a table to hold prints, combined with the difficulty of lighting prints to be viewed. A folding, metal plate and eight small magnets can be provided by a metal worker at minimum cost and can serve to anchor prints; these items should be painted flat black to minimize unwanted reflections. A pocket stereoscope, although only a semi-satisfactory substitute for tracking mirror models, can be used. Lighting is still most difficult, and it is the experience of one of the authors that the best solution to this is to wait for the proper time of day, when the sun is not too oblique but low enough that the interpreter's instruments and head cast less troublesome shadows. This, of course, limits time that can be spent interpreting photographs in the field. It is interesting to note in this regard that much of the result of archaeological surveying, carried out by crews on foot is also dictated by the angle of the sun (lithic materials being especially obvious at low sun angles, ceramics at higher sun angles).

Cultural Resources Sampling and Stratification

Sampling is the careful inspection (and re-inspection in multi-stage sampling) of a small amount of representative area for characteristics or parameters which are of interest, and which can then be projected to a larger area. The key here is the word "representative". This is where remote sensing comes to the aid of the cultural-resource manager and scientist—in insuring that sample areas are representative of the environmental variability within the total survey area as a whole, and to determine which sample plots can be compared and which cannot to arrive at sample statistics.

Although it is presently commonplace for cultural-resource managers and surveyors to speak of "100% surveys", and in fact "100% surveys" are required by law, in the United States and many other countries, prior to the destruction of areas by federally-funded land disturbance, all cultural resource surveys are in fact samples. It is best to realize this from the outset; for care must be taken to omit biases as consciously as possible from any sample. There are other advantages accruing from the realization that archaeological survey is actually sampling. Time and funding are always limited in any endeavor, and an explicit sample approach to cultural resources survey is more efficient than attempting a "100% survey". Sampling also allows the researcher to control explicitly for the many factors *other than the actual resources* that condition what is found during survey (e.g. vegetation cover, soil erosion, sun angle, weather, and the purpose of the survey) by accordingly adjusting sampling intensity and the parameters sampled. Finally, an explicit sampling approach to cultural resources assessment can serve as a remedy to the use of uniqueness and site-size as significance-criteria by allowing concentration on the range of resources within areas sampled, rather than on an extensive search for the most "significant" sites in a large tract.

Remote Sensing and Sample Stratification

After the choice of a study area has been made, a necessary second step in sampling is stratification. This involves dividing the study area into a number of subareas, from which a number of smaller samples will be drawn in turn.

Several methods by which a study area can be divided into strata have been suggested in the archaeological literature. Perhaps the simplest scheme is that of random sampling (Hill 1967). Other authors have suggested various types of stratified systematic samples (Binford 1964; Redman 1973; Rootenberg 1964). It must be pointed out that the term "stratified" is used by these authors in the sense of Berry and Baker (1968) as a method of dividing an area to be cluster-sampled into a number of arbitrary super-units, from which are drawn lesser units. The object of such an exercise is to insure the obtaining of an even dispersal of sample units across a study area, a distribution that does not necessarily result from random cluster sampling.

Such sampling strategies have been used with some success in cultural resources efforts, and are advantageous in "inventory surveys," since they are oriented toward the realistic goal of inspecting only a part of the total variability within a large area and projecting such data to a picture of the whole. There are problems with such stratification schemes, however. One problem that is frequently discussed is the determination of the proper sampling fraction—that is, the proportion of the total area that must be surveyed in order to accurately model the total population. In practice and in the literature, sampling fractions ranging from 10% to 50% or more have been invoked, in no instance with much logical justification. Unfortunately, neither for random samples nor for the sorts of stratified samples discussed above, can a "proper" sampling fraction be determined. This is because there is nothing against which the adequacy of a sample can be tested. Although several archaeologists have attempted to test a post-inventory sample against total-area site lists (Plog, 1968; Judge, Ebert and Hitchcock, 1975), purely inductive samples cannot be tested.

Stratification in which a parameter, or some variable, is held to be relevant to the significance of the resource base can also assist in developing an effective sampling scheme (Holmes, 1967). This approach uses *a priori* information to identify a group of cultural units that possess the same degree of diversity as the population from which they are drawn (Wood, 1955). An array based on the variable deemed important is constructed and then stratified into units. To achieve maximum sample precision, strata must be chosen so that their respective sampled averages are as different as possible, and their variances as small as possible. Clusters within strata, on the other hand, should be as internally homogeneous as possible, thereby minimizing within-cluster variances while maximizing between-cluster variances (Stuart, 1962).

Much of the material record left behind by past people results from adaptations to environmental necessities, and therefore differs from place to place. Remote sensing provides the techniques for dividing a study area into ecological zones, and in this way stratifying it in an informed way prior to the first sampling stage.

Environmental stratification for sampling is

often avoided by archaeologists, usually for one of two reasons. The first of these is that the archaeologist feels uncomfortable with the subject matter of the biologist. The bewildering array of Latin plant names, species, association, community distinctions, etc., need not concern the archaeologist, at least at the inductive levels of assessment. It is necessary only to divide the study area into areas that are different in some way to approach an informed preliminary stratification. Secondly, environmental zones distinguishable in the present may not be the same as they were in the past. Although it is likely that the specific composition of each zone varied under different climatic conditions in the archaeological past, the underlying determinants of zones and boundaries (i.e., drainage regimes, altitude and landforms) will have remained constant. Specific adaptations within each zone may be very different than they would be if hunter-gatherers lived there today, but zones and their boundaries in many cases were the same, and can be differentiated using contemporary aerial and space imagery (Ebert 1978).

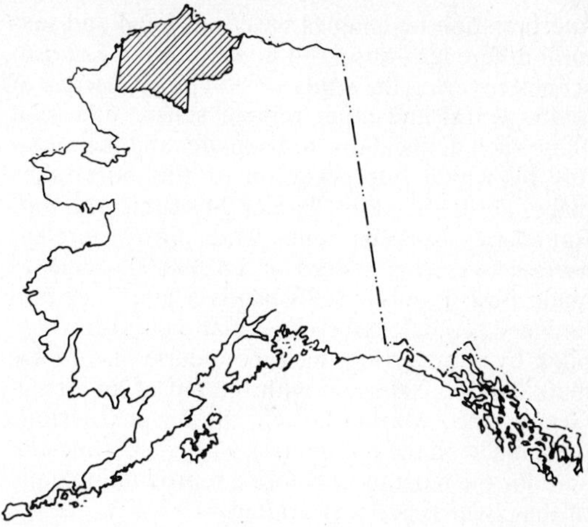

Fig. 26-17. Index map showing location of the National Petroleum Reserve in Alaska (NPRA) on Alaska's North Slope. Landsat imagery was used as the basis for ecologic/cover-type stratification which in turn served to orient sampling design there, as described in text.

Sample Stratification with Remote Sensor Data: An Arctic Example

One manner in which remote sensor data can be applied to the principles of sampling and stratification presented above is illustrated by a project carried out in the National Petroleum Reserve in Alaska (NPRA), by the Remote Sensing Division, National Park Service (Ebert 1978). In April and May of 1977, the National Park Service initiated the large-scale cultural resources assessment of the National Petroleum Reserve in Alaska (NPRA), an area of nearly 23×10^6 acres extending from the Arctic Ocean southward to the crest of the Brooks Range (Fig. 26-17). Since this area was soon to be opened by the Bureau of Land Management to petroleum exploration, time was a crucial factor, and the problems created by an enforced 2-year survey-time limit were further complicated by the extremely short 8–10 week summer field season dictated by the Arctic climate. Yet another constraint on fieldwork in NPRA was the extreme inaccessibility of much of the area, where wet tundra conditions allow travel only by air (largely only by helicopter) and foot. While area to be covered, personnel, and limitations with respect to time and funding are factors in the planning of any archaeological survey, these mild difficulties become severely-felt constraints in very large area surveys like that of NPRA. Accordingly, the Remote Sensing Division was requested to organize and supervise the application of remote sensing data and techniques to facilitate the already planned on-the-ground cultural resources survey.

Researchers involved in this effort envisioned a number of advantageous results accruing from the application of remote sensing within the framework of the NPRA cultural resources survey. First, in the manner discussed previously in this chapter, the use of aerial photographs and even Landsat visual products would provide a detailed "map" of the study area. Since the most detailed topographic maps in existence for most of the NPRA are still 1:250,000 scale topographic sheets with twenty-foot contours (not very useful on the essentially flat coastal plain, which comprises at least half of the NPRA), imagery of all scales would be invaluable for pre-field planning, determining the best transport routes, and in-field location of each site's spatial position once it was located.

Far more important, however, would be the use of remote sensor data and information derived from it in devising a sampling plan and in making projections, from the actual physical sample gathered during the 2-season survey, to arrive at an overall view of the cultural resources of the study area. It was expected that the crew could cover somewhat less than 5% of the 23,000,000 acres during the allotted time, and in fact the rigors of the field resulted in only about half of that percentage being analyzed. Clearly, this would be the case in any assessment project as ambitious as that planned for the NPRA. Nonetheless, the object of the project was to reach an understanding of the cultural resources there, and of their distribution and characteristics. Landsat color composite prints, it was decided, would serve as the primary basis for informed stratification of the area prior to sampling.

Ecologic/Cover-Type Stratification of the NPRA

The interpretation and delineation of sampling strata within the NPRA, which were termed "ecologic/cover-type zones" (Ebert 1978) proceeded according to traditional qualitative photointerpretation techniques based on tonal and textural differences observed in each of ten Landsat scenes covering the study area. The properties of multispectral and other remote sensor data that allow such delineations to be made, and the methods by which interpretation of this sort takes place, are detailed in Chapter 24 of this *Manual*. Boundaries between zones were drawn on clear acetate overlying scenes at 1:1,000,000-scale as made from Landsat MSS bands 5 and 7. A preliminary ecologic/cover-type map was then compiled by transferring these boundaries to a base map at 1:500,000 scale with the aid of an Art-O-Graph, Inc. Map-O-Graph. Some corrections were made on the preliminary worksheet (line closure for the most part) before a reproduction copy of the cover types was drafted.

While cover types were being interpreted from the imagery, it was noted that in many instances the boundaries of different types of strata were very sharp due to dramatic differences in topography, climate, soil or water economy over short distances. In contrast, there are also areas referred to as transitional zones, which are places of gradual change in cover-type (Kuchler 1967). In places where the interpreter believed transitional zones existed, hatched bars of equal width were used, suggesting the partial dominance of more than one type designation in an area. Prior to ground-truth checking, five ecologic/cover-type units were distinguished through the interpretation of 1:1,000,000 scale black-and-white bands 5 and 7 Landsat images:

A. Wet and/or Dry Sandy Surfaces
B. Moist Tundra
C. Very Moist/Wet Tundra
D. Wet Tundra
E. Brush

The distribution of these units is illustrated in Fig. 26-19.

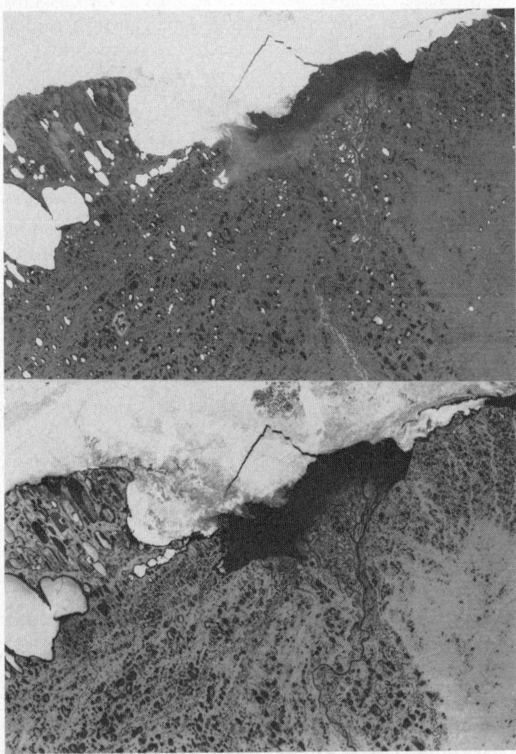

Fig. 26-18. Portion of black-and-white Band 5 (top) and Band 7 (bottom) Landsat 1:1,000,000-scale prints used for ecologic/cover-type mapping of the National Petroleum Reserve in Alaska (NPRA) for cultural resource sample stratification. The scene from which these images are derived is E-2539-21153, 14 July, 1976. Visual interpretation of tone and texture allowed the differentiation of zones different in drainage, soil composition, and vegetative cover; these properties, in turn, were related to the distribution and visibility of cultural resources found during ground survey, as discussed in the text. Much of the Beaufort Sea and many smaller inland lakes are ice-covered in this scene. The Band 5 image shows soil differences and soil saturation, while vegetation variation and the presence of open water areas are prominent in Band 7.

"Ground" Truth Checking

Any interpretations or conclusions reached through the use of remote sensor data, or for that matter any measurement procedure, should be checked for accuracy and consistency using independent methods. This sort of double checking has been referred to by those involved in remote sensing as "ground truth checking", a term which implies that data gathered from the air are tested by ground survey. With the advent of a wide range of different aerial and space platforms available to the scientist, ground truth has come to have a somewhat broader meaning, and includes the checking of data from one remote sensor platform using other remote sensor sources. This multistage approach to ground truth checking provides a wider range of information than simple ground survey. While some may object, it can be argued that actually walking on the ground is not necessary for ground truth checking if other data sources are available and sufficient.

Since the Landsat imagery, against which the revised small-scale classification was to be checked, was color-composite Landsat multispectral scanner data, color infrared (CIR) film was selected for field recording as it was compatible with the Landsat imagery. The value of CIR film lies in the fact that the contrasts between water, vegetation and other surface phenomena which are not obvious with normal color film are much enhanced in the infrared. Particularly important in the NPRA case is that water or wet areas tend to absorb infrared radiation, and thus appear black or dark blue; healthy vegetation reflects infrared energy more strongly than it does green energy and therefore it appears in various tones of pink and red, emphasizing the variation in textures and growth rates of the vegetation.

Fig. 26–19. Preliminary ecologic/cover-type map of the National Petroleum Reserve in Alaska compiled for cultural resources sampling purposes from Landsat 1:1,000,000 scale Band 5 and 7 visual images. This imagery allowed the identification of five major cover types which were thought to be important in controlling the distribution and function of habitation sites and activity areas in the past. These cover types served as sampling strata in the design of a sampling strategy for the study area. Legend notes are as folows: S = wet and/or dry sandy surface (beach); MT = moist tundra; VM = very moist to wet tundra; WT = wet tundra (standing water in places); B = Brush. Diagonal hatching denotes the dominance of two categories in an area; diagonally-hatched areas have been left unlabelled if their constitutents are easily determined from the map context; where the constituents of diagonally hatched areas are *not* easily determined from the map context, these constituents have been labelled in lower case (e.g. if the area is comprised of sand and moist tundra it would be labelled ''smt''; ''brush'' areas always occur in conjunction with one other category and where this is not easily determined from the map context, a lower case notation is used. (Interpreted and compiled by Galen W. Brown and James I. Ebert, Remote Sensing Division, Southwest Cultural Resources Center, Albuquerque, New Mexico.)

For this reason, it was determined that, in addition to inspecting parts of the NPRA area on the ground and from a helicopter, a series of handheld color and CIR photographs would be exposed from the helicopter at targets of interest. Since species of plants were not the target, boundaries between previously defined ecologic/cover-type zones were sought instead. Plant cover and drainage conditions were also inspected on the ground.

These data were supplemented, at the laboratory in Albuquerque, with interpretations derived from the high-altitude CIR transparency imagery acquired in the course of overflights in the summer of 1977 by the NASA/Johnson Space Center. This imagery, at scales of 1:60,000 and 1:120,000, was intermediate in scale between the ground and helicopter observation and imagery produced from the Landsat MSS data. This was invaluble in allowing the double checking of ''problem areas'' which could not be identified positively from Landsat imagery. An example of a problem area is one in which there is a lack of distinction, upon initial Landsat inspection, between areas of snow cover and limestone outcrops, both of which have essentially the same signature in the four Landsat bands. The greater resolution and stereo coverage provided by the NASA imagery allowed stereoscopic viewing of these areas and the determination of whether snow or limestone was being seen. A survey concentrating on a smaller area than NPRA could have included a secondary ecologic stratification directly from high-altitude imagery, but was deemed unfeasible in the regional project at hand.

Revised Small Scale Classification and Mapping

The second mapping stage followed immediately after field checking, and resulted in the compilation of a revised cover-type classification. Ten computer enhanced Landsat scenes were obtained in color composite form at scales of both

1:500,000 and 1:250,000 to be used in refining the interpretation.

With this imagery in hand, it became apparent that additional cover-types could be defined and that the location of boundaries between cover-types could be more accurately mapped. Since the practical on-the-ground visibility of archaeological data is often dependent upon whether vegetative cover is sparse enough to allow survey crews to see stone and other materials readily, one of the additional types was *Bare Rock with Alpine Tundra in Places*. Bridging the gap between this type and previously mapped units in Alpine Tundra, which occurs in large areas in the Brooks Range and immediate foothills. In band 7 imagery, rock outcrops can be identified from the color renditions of the rock as well as its internal structure (stratifications, metamorphic foliation, jointing, volcanic flow units), contacts with adjacent rock units, and its resistance to erosion (Sabins, 1978). Those rock outcrops which appear dark in the band 7 imagery could be distinguished from water bodies, which also appear black in band 7 because of the greater textural variation of bare rock as compared with flat water surfaces. "Ground truth checking", accomplished with the use of the Johnson Space Center (NASA) 1:30,000, 1:60,000, and 1:120,000 transparency imagery discussed in earlier parts of this section, revealed that many of the "bare rock" areas were not completely devoid of vegetation but hosted some sparse Alpine tundra.

The increase in delineated cover types to seven in the revised map (Figure 26-20), plus the inclu-

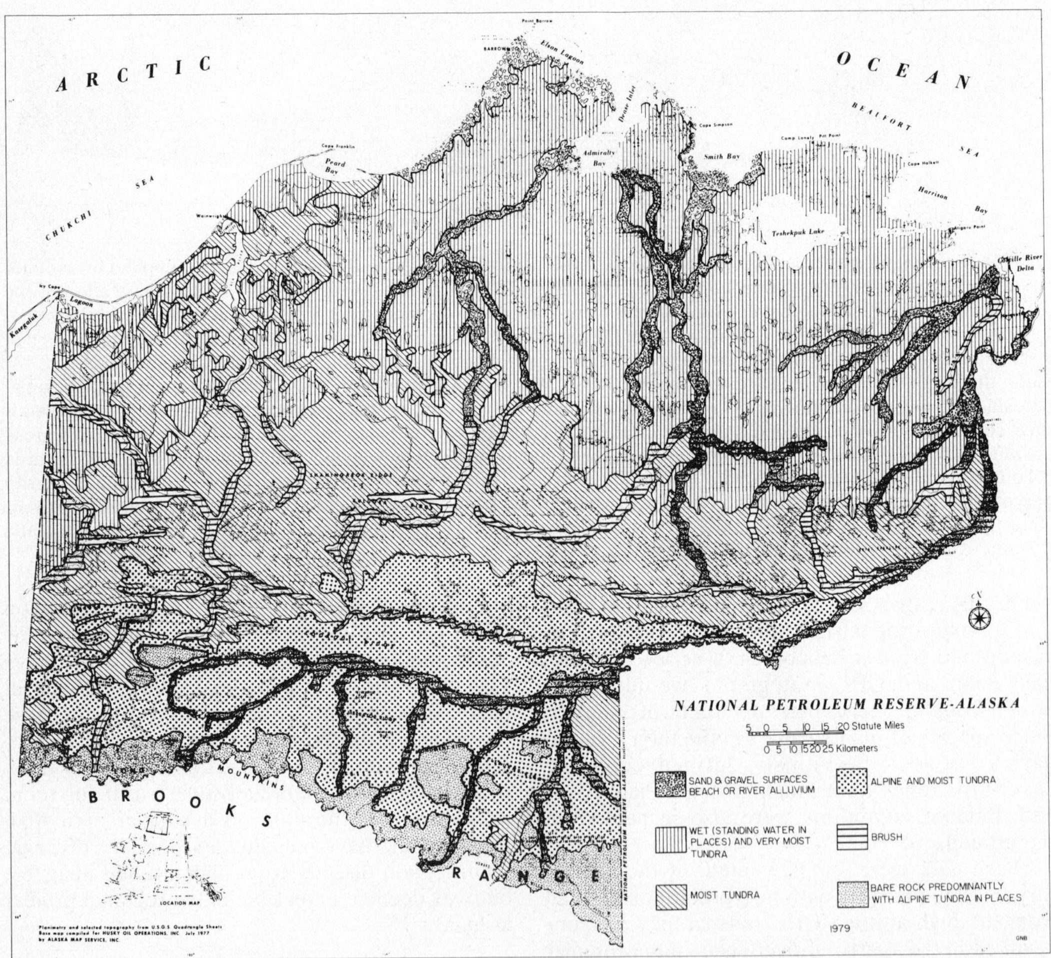

Fig. 26-20. Preliminary ecologic/cover-type map of the National Petroleum Reserve in Alaska, compiled from 1:250,000 scale color composite Landsat prints and spot-checked with 1:120,000, 1:60,000 and 1:30,000 scale color-infrared aerial photography as described in text. This delineation could serve as the basis for refined sampling stratification for cultural resources assessment in the study area. In addition, sampling fraction could be adjusted in the second stage on the basis of archaeological data collected on the ground during the first stage sample, as illustrated in Figs. 26-22 and 26-23. In this way, the results of remote sensor interpretation of natural areas (the strata) and of on-the-ground archaeological fieldwork could serve to reinforce and refine one another. (Planimetry and selected topography from U.S.G.S. quadrangle sheets. Base map compiled for Husky Oil Operations, Inc. by Alaska Map Service, Inc.)

sion of transitional categories, resulted in a total of 13 cover-type delineations or strata, enhancing the usefulness of the classification scheme for archaeological sampling purpose.

Large-scale Classification of Survey Areas

In order to relate actual cultural resource distributions observed in the field to the overall classification scheme, the areas covered by the NPRA field crews were reclassified using large-scale imagery, primarily Landsat EDIES computer-enhanced 1:250,000 scenes. It was necessary to define boundaries between ecologic/cover-types more accurately during this classification stage so that actual site locations could be precisely assigned to the proper class.

Upon the receipt of 1:250,000 USGS quad maps from the survey crews of the actual areas that they covered on the ground during 1977 and 1978, interpretation of these areas was begun. Although boundaries were to be more precise than previous mapping stages allowed, cover types were to remain consistent with the 1:500,000 map. Cover types were delineated from the 1:250,000 Landsat imagery directly onto acetate overlays. Because of the compatible scale of both the survey maps and the imagery, there was no need for enlargement or reduction; additionally Landsat imagery exhibits little relief displacement and is therefore relatively planimetric. The location of objects on the ground is reflected with tolerable map accuracy limits at 1:250,000 scale.

After boundaries delineating cover types were drawn on the overlays, individual site locations could be defined according to the land cover class in which the site was located. Figure 26-21 is a representation of one of the survey areas overlaid on a Landsat print for stratification. The actual area covered by each land cover class was measured with a planimeter, resulting in a total land area of 22,633,755.05 acres. The measurements were triple checked and averaged to achieve consistency. Any errors associated with these measurements could be the result of map scale errors (due to paper stretch, etc.), planimeter errors (which tend to be greater, percentage wise, for smaller measured areas), and planimeter operator errors.

Tabulation of Site Characteristics and Class Distributions

The correlations between ecologic/land cover classes and the nominal characteristics of sites, as well as the number of sites found in the course of survey of each cover type, were tabulated by computer. For maximum flexibility, each site was taken as a case, and site characteristics coded as "present" or "absent" for each case. The only non-binary variable used for each case was the code number of the ecologic/land cover class in which the case fell. Two types of site characteristics were recorded in the field and were available to the Remote Sensing Division during this tabulation: (1) landform/environmental variables, such as proximity to shores, hills, floodplains, etc. and (2) cultural classifications or descriptions, such as "isolate," "cairn," or the inclusion of bone, wood, or lithic material. A tabulation of each of these characteristics by ecologic/cover-type zone allowed the estimation of those zones in which different characteristics would occur with the greatest frequency, or the proportion of sites within each zone that could be expected to display specific characteristics.

Multistage Sampling Design

Stratifying the NPRA with reference to cover types, survey of the sampling strata, and tabulating site occurrence within each stratum constituted the first stage in what should ideally have been a multi-stage sampling design. Data collection in the first stage of a multi-stage sampling design serves two purposes: the gathering of data to answer questions about the nature of cultural resources in an area, and the gathering of data to determine whether the sample stratification used in the first data collection stage is suited to the purposes of the project. The adequacy of a sample stratification depends upon within-strata variances in measured parameters being as low as possible, and between-strata variances being as high as possible. That is, the strata must be adequate predictors of the phenomena in which one is interested.

First Stage Sample

It is assumed that prior to the first stage of sampling, the scientist knows little or nothing about the distribution of cultural resources in the area being studied. There is, therefore, equal probability that each subarea within the study area will contain sites of all types. This is of course improbable but is a "null" assumption upon which the first stage sampling fraction is based. In such a situation, it is necessary to distribute sample units among the strata in proportion to the stratum area (Fig. 26-22). This insures a representative coverage of each of the strata and provides information with which the stratification can be evaluated.

Second Stage Sample

The results of the first stage sample are used to determine the sampling fraction for each of the strata in the second stage sample. During the first stage sampling procedure, parameters of importance were sought and measured in the field. Cultural resource managers and scientists might, for instance, be interested in site size, age, number of lithic artifacts per site, or other scalar variables that can be subjected to statistical analysis. For each parameter, the standard deviation of its distribution in each stratum is calculated. The standard deviation of a sample distribution is a simple measure of the closeness of the extremities of that distribution to its mean, or

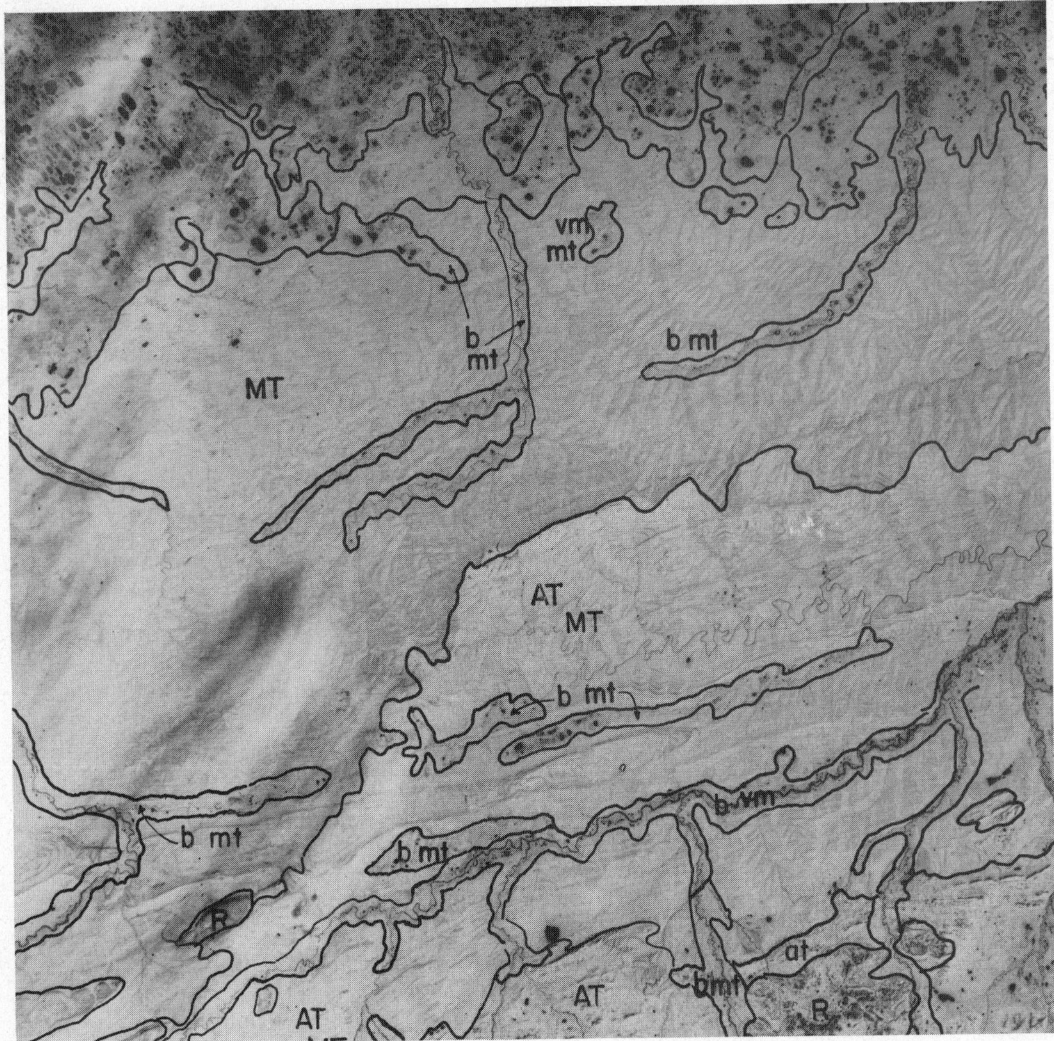

Fig. 26-21. A portion of a Landsat Band 7 print (Scene ID E-1744-21425, 6 August, 1974) used for preliminary ecologic/cover-type classification of the National Petroleum Reserve in Alaska. This classification was used as one data set for input into a sampling design for cultural resources in the NPRA, as explained in the text. Classes defined in this subscene are Bare Rock with Alpine Tundra (R), Alpine Tundra (AT), Moist Tundra (MT), Very Moist to Wet Tundra (VM), and Brush (B). The presence of two designators in lower case letters in one stratum indicates co-dominance or mixing of types. Differences in cultural resource occurrence and nature between strata indicated that in the NPRA survey, Landsat-derived classification was sufficient to explain much of the variation in cultural resources over space.

the range of variation of a parameter's values. A large standard deviation within a stratum means that the stratification does not inform or predict efficiently about values of the parameter in question; a small standard deviation means that the stratum is a good predictor of certain parameter values. Strata with small standard deviations need not be sampled heavily in the second sampling stage, for they are adequately defined and sampled. Those with large standard deviations, however, need to be sampled more heavily. In the second stage sample, sampling fractions are determined by the product of stratum area times standard deviation within each stratum (Fig. 26-23).

While this method of apportioning sample fractions is simple when applied to a single parameter, it becomes complex when multiple parameters are important. In some cases certain parameters correlate with one another, and in others a separate stratification may be needed to predict each parameter's distribution.

Remote Sensing Approaches to Cultural Resources Prediction

The prediction of the occurrence of specific sorts of cultural resources in certain places is closely tied to cultural resources sampling-procedures, but is undertaken not to arrive at a

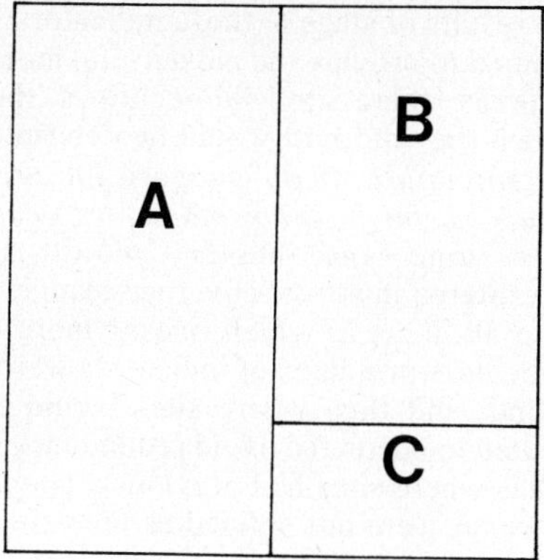

STRATUM	AREA (km^2)	PROPORTION OF TOTAL AREA	SAMPLING FRACTION
A	150	50%	50%
B	120	40%	40%
C	30	10%	10%

Fig. 26-22. Idealized first-stage sample allocation in which the sample fraction is distributed among sampling strata in proportion to the areas of those strata. In this example, if time and economic considerations allowed cultural resources survey of only 20 km^2, 10 km^2 would be surveyed within stratum A, 4 in stratum B, and one in stratum C. Strata are cost-effectively defined using remote sensor data, as described in the text.

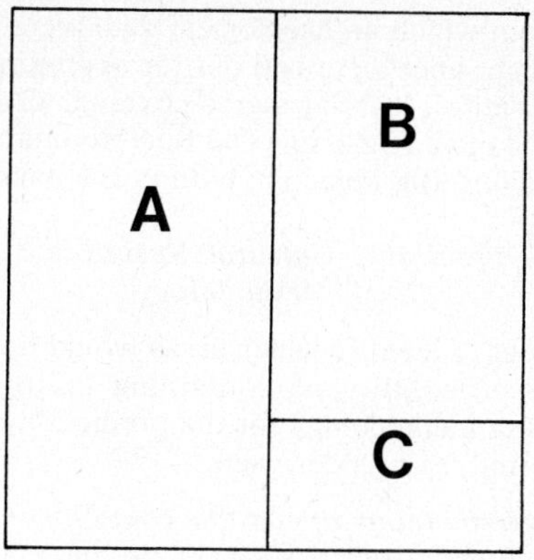

STRATUM	AREA (km^2)	STANDARD DEVIATION (sites/km^2)	Area x Std. Dev.
A	150	6.2	930 = 34%
B	120	12.9	1548 = 56%
C	30	9.5	285 = 10%
		Total	2763 = 100%

Fig. 26-23. Second-stage sample allocation in the same area as the idealized first-stage sample shown in Fig. 26-22. The second-stage sample fraction is distributed between strata on the basis of stratum area times the standard deviation of a parameter or characteristic of interest (site area, for instance). Where the standard deviation is high, the initial stratification has not adequately "explained" the distribution of parameter values, and more intensive sampling is required. Where it is low, the initial stratification was adequate and little additional sampling is needed. If, as in the first-stage sample, 20 km^2 could be surveyed in this stage, these would be distributed as follows: Stratum A, 6.8 km^2; Stratum B, 11.2 km^2; Stratum c, 2 km^2. This differs substantially from the first-stage sample fraction, and represents a refinement in sample design.

general view of the nature of resources within a study area, but to aid in the location of resources in which one is interested. Prediction of cultural resources is more suited to discoveries in non-threatened areas (for instance, National Parks) in response to specific problem orientations or needs, than in areas that are to be disturbed or destroyed. The first step in predicting the existence of cultural resources is to define the resource one is interested in; this may be "all cultural resources," or a finer categorization for specific purposes. Secondly, all of the known occurrences of the desired resource are examined for "indicators", i.e., independently measured phenomena (usually environmental) which correlate with the occurrence of known resources. The landscape is then searched for the occurrence of the indicator phenomena, and those places which are most like places where cultural resources have been discovered in the past become the most likely candidates for intensive survey.

An experiment in the prediction of cultural resources was carried out by the Remote Sensing Division of the National Park Service in 1979 in Shenandoah National Park. Aerial photography in the form of 1:12,000 scale color infrared transparencies was employed both in the measurement of environmental indicators correlating with known sites occurrence, and in the location of areas in which these indicators appeared but no cultural sites had been found.

Shenandoah National Park is a densely wooded tract covering part of the Blue Ridge Mountains in Virginia. The discovery of cultural resources is difficult there due to the almost impenetrable vegetation. In addition, there has been hesitation on the part of many archaeologists to even attempt to utilize aerial photography in this and many other Eastern areas due to the fact that the ground cannot be seen, for the most part, due to overstory and understory plants. The Remote Sensing Division project was designed to investigate the utility of aerial remote sensing as a cultural resource prediction tool in an Eastern woodland setting. An aerial overflight was planned for the spring of 1979 to take place at a time just before leafout of the forest canopy. This would allow the discrimination of conifers vs. deciduous trees; and since the understory vegetation typically begins to grow slightly prior to canopy leafout, it was thought that such imagery might allow classification in terms of subtle differences in understory species composition.

Rather than fly the entire park, an operation which would be expensive and unnecessary for the purpose of this experiment, it was decided to fly two "sample" areas at a scale of 1:12,000. These areas were chosen to cover survey transects from which archaeological data were available. Flight lines were laid out for approximately 36 linear miles, insuring stereo coverage of 10 mi^2 just north of and including Old Rag Mountain, and 4 mi^2 around Big Meadow, both in the park.

Predicting Cultural Resource Loci: Methodology

In order to identify which areas would have the highest probability of containing cultural resources, a methodology for the prediction of cultural resources was designed.

1) *Determination of possible environmental indicators.* Using baseline environmental data, archaeological information and the CIR transparencies, potential site location indicators were derived. This was accomplished by determining which environmental characteristics could be defined, located and measured using remote sensor data and/or other information in the form of geologic or vegetational maps and published reports.
2) *Formulation of taxonomies of different values or states of each indicator.* For each environmental indicator chosen for interpretation, a taxonomy of different values or states of that indicator was to be arrived at to allow objective measurements of the values of such possible variables. Wherever possible, previously existing taxonomies (e.g., slope types) or quantifiable measures were to be used.
3) *Devising recognition patterns for environmental indicators.* Explicit recognition patterns which allowed the consistent determination of which environmental indicators were present in a specific area seen on the image, as well as the state of each such indicator, were to be compiled through interpretation of the imagery.
4) *Pinpointing of site locations on aerial imagery.* The exact locations of the sites and materials previously found would be determined through the matching of topographic contours, and features shown on the maps of cultural resource occurrence, with photointerpreted details.
5) *Interpretation of environmental indicators from imagery.* Stereo-photointerpretation of the presence of the various environmental indicators chosen in stage 1, and states of each indicator according to the taxonomies devised in stage 2, would comprise this stage.
6) *Coding and tabulation of results of interpretation of known sites areas.* The presence of indicators, and their states, in known site location areas would be coded in computer-compatible form, punched on cards, and tabulated in this stage.
7) *Determination of useful cultural resource location indicators.* Through inspection of the results of stage 6, those indicators which seemed to provide the most useful indication of areas where sites did occur, vs. those in which they did not, would be determined.
8) *Interpretation of all imagery for areas in which useful location indicators occur.* In this stage, the imagery would be reinterpreted in a total coverage manner to locate all areas in which one or more of the indicators or values of indicators were to be found, and thus where sites would be expected to occur. To avoid redundancy, those areas where sites had previously been noted to occur were not defined as predicted location areas.
9) *Field-checking of interpreted predictive areas on the ground in Shenandoah National Park.*

Environmental Indicators and Their Measurement

A review was made of literature and available data pertaining to those environmental characteristics which might be defined, located and measured either on aerial photographs or through recourse to other information sources. Then an intuitive judgement was made as to which of these characteristics might have archaeological or anthropological significance—that is, which ones might reliably predict preferred cultural activity areas over those areas in which the activities might not have been pursued in the past. Those indicators chosen were:

1. Slope Type (Hack and Goodlett, 1960)
2. Slope Angle (Hack and Goodlett, 1960)
3. Aspect
4. Vegetation Type (Hack and Goodlett, 1960)
5. Vegetational Diversity
6. Soil Thickness (Hack, 1965)
7. Type of Surface Deposit (Hack, 1965)
8. Bedrock Type (Gaithright, 1976)
9. Proximity to geologic contacts, faults or shear zones (Gaithright, 1976)

Next, typologies or scales of measurement were devised or borrowed from previous studies for the description of the chosen environmental indicators. In most cases, recognition patterns for the different states or values of each indicator were inherent in taxonomic descriptions. A complete list of these taxonomies can be found in Ebert and Gutierrez (1980).

Image Interpretation Methodology

CIR imagery, in transparency form, was viewed with a Bausch and Lomb roll-film scanning-stereoscope. This instrument has the capability of enlarging the image up to the resolution limit of the film. This was extremely useful in detailed analyses of small areas as well as in making positive site identifications.

The first step in interpretation was familiarization with the type of terrain and topography in the study area. A form was developed for characterizing and recording important environmental characteristics of known site locations from the imagery. This generated the necessary data for performing computer analyses and predicting environmental settings characteristic of site locations in unsurveyed areas. Two interpreters performed the image interpretation simultaneously, one looking at the photos and making interpretations and vegetation diversity measurements, while the other was recording interpretations and verifying locational, geologic, and geomorphic data. All known historic and prehistoric sites within the study area were environmentally-characterized and data-recorded. Five-digit vegetation classification numbers were recorded for each site, representing the relative percentages of each of the five vegetation types within a 250 foot radius of the site. Each of the other environmental characterizations was also recorded on the data sheets for computer analysis.

Prediction of Cultural Resource Locations

Once data on the forms were tabulated and analyzed, several combinations of environmental factors appeared most useful in predicting preferred location for the known, previously surveyed sites. Certain environmental factors were consistently associated with either prehistoric or historic sites (site-type specific), and other factors seemed consistent for both types (not site-type specific).

Environmental Characteristics of All Sites:

- Slope—Slopes for both site types ranged from 0–5 degrees (80.4% of historic sites, 84.2% of prehistoric sites). Even areas on sideslopes were locally flat in the vicinity of the sites.
- Aspect—65.9% of prehistoric sites and 78.2% of historic sites faced from SE to SW, probably due to the milder temperatures experienced on these aspects during winter months.
- Proximity to Contact or Fault—55% of all prehistoric sites and 57% of all historic sites were located on or near a contact or a fault/shear zone. This is particularly significant in light of the fact that these linear features account for a very small proportion (about 5%) of the study area.

Environmental Characteristics Associated with Specific Site Types

- Slope Type—Historic and prehistoric sites are most often associated (46% and 50% respectively) with valley bottoms or footslopes; however, a relatively large proportion of historic sites (28%) were located on locally flat sideslopes.
- Surface Material—Historic sites were shown to occur primarily (87% of the time) on colluvial and alluvial deposits. Prehistoric sites did not appear to be preferentially located on any specific surface material.
- Bedrock—Historic sites were located on Old Rag Granite in almost 80% of cases. The Old Rag Granite has a well-developed soil mantle and a large number of springs or seeps. Prehistoric sites were evenly distributed on Catoctin (limestone) and Old Rag formations.

Soil mantle thickness was determined to be of little value in predicting sites due to difficulty in achieving an accurate determination of this variable from aerial photography; surficial deposits are a better measure of site location.

Once environments characteristic of potential site locations were determined, the imagery was re-examined in order to delineate the areas likely to contain sites but in which no sites were previously known to exist. Site prediction locations were then plotted on 7½-minute base maps.

Ground-Truth Checking in Shenandoah National Park

Ground-truth checking was carried out in Shenandoah National Park during the week of August 5–10, 1979. Selected predictive areas, determined as described above, were inspected intensively to determine if environmental site characteristics were accurately interpreted from the imagery and if cultural resources of material were contained within and around specific predictive areas.

Field reconnaissance was carried out by carefully locating predictive areas in the field, inspecting each area on foot for evidence of cultural resource occurrence, and noting the nature of environmental indicators on the ground. In addition to the noting of obvious surface occurrences, limited clearing of forest floor debris was undertaken. The geologic and geomorphic environment of each predictive area was examined and recorded to determine whether the environmental characteristics seen on the ground were interpreted correctly from the imagery; subjective judgements as to whether the areas were conducive to occupation were also made by the crew.

Cultural materials were recovered in 45% of the 20 predicted areas that were checked on the ground. These included several historic cabin structures, historic fields and associated artifacts, a previously unlocated 19th-century industrial site with a mill and mill pond, and one prehistoric site with a scatter of stone artifacts. Not all of the sites where cultural resources were predicted were visited, due to time constraints. While a success rate of 45% is less than totally successful, it is nonetheless remarkable in archaeological survey, especially in the Shenandoah area where the visibility of sites is low and blanket-coverage survey usually results in finding no sites at all. In addition, the best areas for finding sites—the most "ideal" places that sites would have been located—were excluded because sites had been found there in the past. Undoubtedly, if these areas had been resurveyed, additional cultural resources not found previously would have been located.

Remote Sensing as an Adjunct to Cultural Resources Mitigation

Remote sensing, as illustrated above, can be of much aid in the process of discovering, sampling and predicting the occurrence of cultural resources. There is another area of cultural resource management, however, where data recorded by remote sensing methods can be as helpful—mitigation. As discussed earlier in this chapter the concept of mitigation grows out of cultural resources management rather than problem-oriented archaeology; once cultural resources are discovered, a number of decisions about their exploitation or disposition must be made. The first decision is whether a cultural site is significant, i.e., whether it merits further treatment or can be ignored or destroyed without affecting archaeological knowledge or historic heritage values. Remote sensing provides a means by which cultural resources can be seen in a ''new perspective'', as is emphasized in the section of this chapter dealing with discovery of sites and materials. It can also be employed to illustrate relationships between one site and another, or a site and certain environmental variables. Such a perspective can often be employed productively in determining the significance of a resource.

If a resource is judged significant, then there are several alternative ways of treating it. A site or other material evidence of the past can be avoided and preserved; the argument often given for this alternative is that archaeologists and other scientists of the future will have more sophisticated questions to ask of cultural resources than today, and will do so with more sophisticated techniques and equipment. That there is some truth to this argument is illustrated by the increasing use of remote sensing in archaeology in recent years. If preservation is the course chosen for a cultural site, photogrammetric monitoring methods can be employed to insure that the site remains as it is, as described in the Photogrammetry section of this chapter.

Another alternative is the archaeological study of a site, usually through the agency of excavation. Archaeological excavation is inherently destructive, and a site, once excavated, is gone as surely as if it had been strip-mined. It is imperative that all possible classes of data be collected from an archaeological site before and during excavation. Such data include a wide range of environmental information, the precise position of features, artifacts and other site components, and the exact nature and methodology by which the excavation took place. Elsewhere in this chapter methods are suggested by which each of these tasks can be facilitated. There is, however, the very real possibility that we might today neglect to record some special sort of information which will be needed by archaeologists of the future to answer their questions. If the nature of such information cannot be forseen, how can the relevant data be recorded? Through aerial photography and other remote sensor data sources, information can be collected that is not even comprehended when it is gathered. Remote sensing gives the cultural resource manager and scientist access to a broad-spectrum, archival data source that can be referred to again and again in the future. Sites that are threatened with destruction by any means should be subjected to reasonably complete recording.

The Mitigation of "Ephemeral Cultural Features"

For some features remote sensing may be the only practical and necessary means of mitigation. While the term archaeology usually conjures up visions of excavated sites containing obvious structures and large amounts of artifactual materials, the advent of recently developed techniques, such as remote sensing as applied to cultural resource analysis has demonstrated that there are a number of other, less immediately obvious kinds of sites. Many of prehistoric (and historic) man's modifications of the natural landscape are ephemeral at best when viewed from the ground, and some of these features may have minimal artifactual associations. Nonetheless the information content of these features can be as high as or higher than that of excavated sites. Methods must be developed to extract the greatest possible information from such features. There are numerous examples of phenomena which are obscure or unseen from a proximate land-based perspective. The classic indicators of ephemeral cultural features are the soil marks, shadow marks and crop marks long identified and used for site discovery and location in Europe.

A particularly outstanding example of archaeological features not immediately or extensively visible on the ground is provided by the San Juan Basin prehistoric roadway system which connects Pueblo III Period sites in northwestern New Mexico, mentioned earlier in this chapter. Only a few miles of prehistoric roadways were known to exist at Chaco Canyon National Monument prior to the application of aerial photography. Beginning in the early 1970's intensive interpretation of aerial photos and other remote sensor data at the Remote Sensing Division of the Southwest Cultural Resources Center (then part of the Chaco Center) allowed the identification and mapping of more than 200 miles of these features extending throughout the San Juan Basin and beyond. Roadways to be interpreted and plotted were ground-truth checked in the field; in many cases, while short segments visible on the ground provided positive proof of the existence of the interpreted features, there was little or nothing to be seen along most of the roadway routes. Structural evidence associated with roadways include stairways, ramps, curbs along roadway margins, flagging or paving of their surfaces, and a number of diverse masonry structures, but these are for the most part restricted to very short segments or roadway, especially in the vicinity of large archaeological sites. It has been suggested that pottery and other archaeological evidence is distributed more densely along roadway routes than in similar but untravelled areas; and while this is almost certainly true, absolute sherd and artifact densities along roadways at any distance from prehistoric population centers are so low as to present serious problems in sampling and analysis.

Similar situations are presented elsewhere by agricultural fields and their boundaries, certain kinds of water-control features, organically-caused soil discoloration, and other cultural features which do not necessarily co-occur with artifactual evidence. While some of these may have aspects which are detectable on the ground, others may never be unambiguously detectable by digging, surface collecting, or even chemical or other physical analysis. This raises a number of very real problems which must be resolved if our responsibilities for the preservation, recording and analysis of cultural resources are to be conscientiously met.

One of the foremost of these problems is the determination of proper mitigation procedures. Perhaps the best way to approach any mitigation requirements is to question what kinds of information that are important to our understanding of the past would be lost if the particular site or feature were destroyed.

By far the most important information contained in such features as roadways is their location—where they are and where they go. This is precisely the information which is revealed to the maximum possible extent in remote sensor data. For this reason, identification and mitigation efforts should in all cases begin with remote sensor imagery; and in some cases the collection of remote sensor data may prove to be the only data collection necessary.

In most areas, previously flown U.S.G.S. or other mapping photography exists on which ephemeral features can be identified, if in fact they do exist in a project area. However, not all imagery is equally useful in revealing roadways or other features. As previously indicated, such factors as sun angle, scale, and season of year are extremely important. Remote Sensing Division studies indicate that black-and-white imagery at scales between approximately 1:20,000 and 1:30,000 is probably optimal for roadway identification if other factors such as sun angle and direction are correct; color infrared imagery can also aid greatly in the detection of road segments early in the vegetative growth season. In areas where roadways are suspected (and this must be taken to include the entire San Juan Basin and adjacent areas) all available imagery should be analyzed by a qualified interpreter to determine whether or not ephemeral prehistoric features occur.

If the determination is made that ephemeral locational features do exist in the study area, a second remote sensing stage is warranted. Just as in the use of any technique, the most useful applications can be made when the tool is suited to the purpose for which it is intended. Manipulation of scale, sun angle, photographic emulsion or sensor type, and other asepcts of the total remote sensing methodology must be carefully designed for maximum information gains. This can be assured only by carefully planning and by flying imagery specifically suited to the features that are to be detected. Such a stage is necessary even when portions of the features in question have been detected on preexisting imagery (Ebert and Lyons 1980).

PHOTOGRAMMETRY FOR RECORDING AND MAPPING CULTURAL RESOURCES

Contemporary archaeology, both prehistoric and historic, demands the total recording of all conceivable data pertaining to structures, sites and the placement of features and artifacts within the units of study. This presents something of a dilemma, for only when sites or structures are totally undisturbed can they contain "all" of the data inherent in them. Almost all data-recording techniques involve some changes in or disturbance of sites, however. In reality, all cultural resources are constantly subjected to changes of different sorts: those induced by natural forces, such as erosion, earthquakes, or other changes in the surface and subsurface of the earth; changes caused by human intervention such as those incurred when sites or materials in them are reutilized after being originally deposited; and other changes of a cultural origin caused by amateur or professional excavation of a site, or vandalism. It is, of course, only through professional excavation that attempts can be made at extracting as wide a range of information as possible from cultural resources.

Traditional archaeological, historical and architectural recording of structures and sites and their components began as, and for the most part continues to be, the drawing of maps and plans. Since the beginnings of cultural resource study, such recording has utilized proximal-methods of measurement, utilizing such measurement devices as tapes and range poles, and sighting equipment such as transits and alidades. These instruments, of course, are quite capable of accurately recording the relationships of chosen parts of a site or structure to each other, and their expert use has yielded much irreplaceable data. Unfortunately, when such methods are applied to the mapping of sites that are in the process of being lost forever (during archaeological excavation, for instance) they invariably result in the selection of only a specific range of data in which the researcher or excavator at hand is interested. Other data, which may be crucial to the explanation or the testing of hypotheses in the future, are lost, in many instances.

Remote sensing methods, primarily those employing photography, provide a more comprehensive means of recording data at a cultural resource site than drawing methods, because remote sensing records a range of information much of which is not even considered or understood at the time a site is recorded. Technically excellent and complete photography and/or other remote sensing recordings of cultural resources is an absolute necessity in modern archaeology, architecture, or other treatment and study of any cultural resources that may be imagined to be of significance now or in the future. It is the cultural resource manager's responsibility to insure that sites are so recorded and that remote sensing data are preserved.

Photogrammetry is the method by which this responsibility can be discharged. "Photogrammetry" has been defined and used in a number of different senses in recent years. In its broadest sense, it encompasses the "art, science and technology of obtaining reliable information about physical objects and the environment through the process of recording, measuring and interpreting photographic images and patterns of electromagnetic radiant imagery and other phenomena (American Society of Photogrammetry, 1980: 1249)", a definition fully compatible with the broadly defined term "remote sensing". In this section, however, "photogrammetry" will take on a somewhat narrower meaning, viz. that of recording and mapping the spatial relationships between objects and their indications (cultural, environmental, or otherwise), using photographic and other remote sensing methods.

Even under this narrower definition photogrammetry can be classified by its accuracy, usefulness, and point of view. Simple photographs can supply measurement information, for they produce an inclusive although somewhat distorted image of a scene. Even vertical aerial photographs, taken with metric cameras, contain distortions due to radial displacement and parallax; but for many cultural resources purposes such "errors" may be acceptable. In fact, these "problems" become assets when stereophotographs are employed, and a large part of this section will be devoted to the use of controlled stereophotography for the photogrammetric mapping of sites and structures. Stereophotogrammetry itself can be classified into two different sorts, aerial photogrammetry and terrestrial photogrammetry. These differ, of course, in the cameras' location and angle of view in that aerial photogrammetry employs aerospace photography, while terrestrial photogrammetry uses photos derived from cameras which are located on the ground. The principles involved in measurement from these two viewpoints are the same; the distinction is made primarily because terrestrial photogrammetry is a relatively new concept in cultural resources management and as such will be emphasized separately here.

Monoscopic Photography for Site Recording

The taking of site photography from terrestrial vantage points is a standard practice in archaeology. At different stages of excavation, when features or objects of interest are uncovered, most archaeologists pause to take a photograph that shows features or artifacts in the overall context of the site. Traditionally, 4″ × 5″ view cameras have been used by archaeologists in the field, although the advent of finer-grained fast films has led to the general acceptance of 35 mm cameras for site photography today.

The perspective from which archaeological site photos are taken is important, for differently-angled photographs offer a range of different viewpoints and utility. Although many photographs of in situ artifacts or features are taken at an oblique angle, some archaeologists have realized that a plan view is desirable; hence they attempt to take their photos with an axis perpendicular to the surface of interest (horizontally for profiles, vertically for floors or excavation surfaces). Vertical site photography has given rise to a large number of aids and devices for elevating a camera (and sometimes its operator as well) above a site. Many solutions to this problem have proven to be elaborate and cumbersome (cherry-pickers, 400-lb tripods, and tall ladders with supporting legs). One workable solution, in the form of a lightweight bipod device, has been invented and produced on a limited scale by Julian H. Whittlesey of the Whittlesey Foundation. The Whittlesey bipod (Fig. 26-24) is designed along the lines of a sailing ship's rigging, with two angled telescoping aluminum tubing legs guyed with two light ropes. The early models of Whittlesey's bipod extended to a height of 15 feet (Whittlesey 1966), and elevated essentially any 35 mm camera, holding it in a position roughly vertical to the floor of an area being studied. A later development was a much larger 30-foot bipod which could accomodate not only 35 mm cameras, but 2¼ inch × 2¼ inch cameras as well. Use of these bipods can produce high-quality "vertical" photos of a site; swinging the camera slightly from one side of the vertical axis to the other, easily accomplished by adjusting the guy wires, results in a slightly oblique stereo pair of the scene which is highly suitable for stereoviewing, if not for photogrammetry. Experiments have demonstrated that, by using the central parts of vertical bipod photography, the tracing of site features of interest often can be made to acceptable tolerances providing the area photographed is relatively flat. As in any measurement task, the minimum accuracy required for the task must be defined; if a high de-

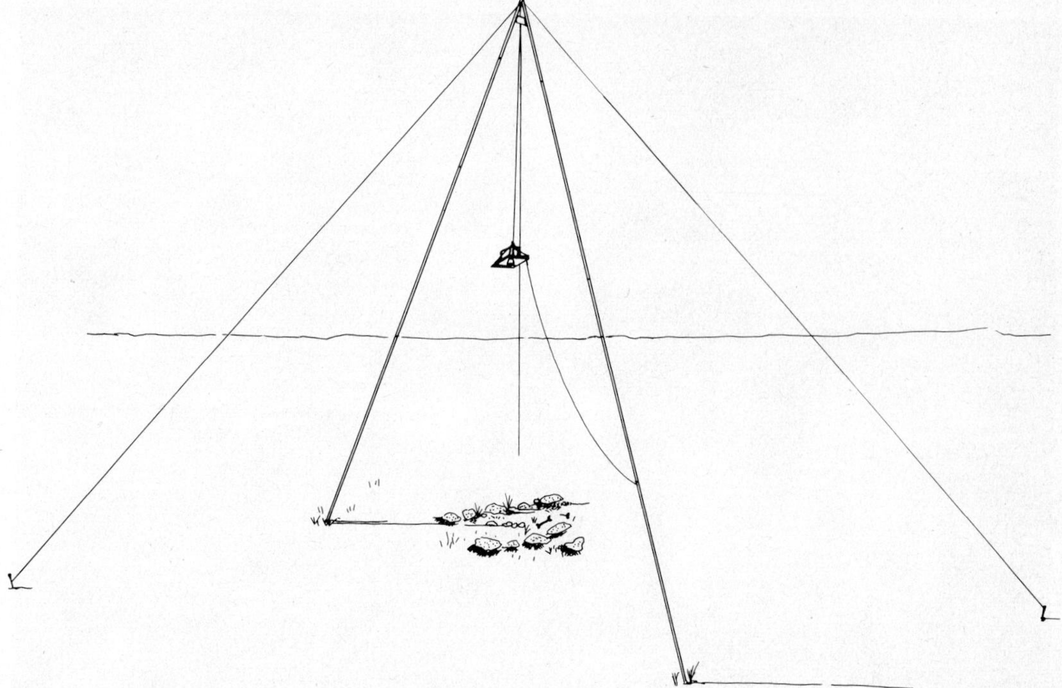

Fig. 26-24. Schematic drawing of the Whittlesey Bipod, designed to elevate a camera vertically over an archaeological site. Designed and manufactured by Julian Whittlesey, one of the pioneers of archaeological remote sensing, this device offers technical sophistication coupled with ''appropriate technology'' to the modern archaeologist.

gree of accuracy is not required, then stereo photogrammetric mapping may be unnecessary for plotting (Klausner, 1980; Boyer 1980).

Aerial Stereo Photogrammetry and Cultural Resources

At the close of the 1960's American archaeological remote sensing began rapidly to shift from being a totally interpretive, discovery process into something more complex and useful; this development was largely due to the introduction of aerial stereo photogrammetric mapping. Some of the first experiments in the use of such mapping to record archaeological sites took place in and around Chaco Canyon National Monument in northwestern New Mexico under the auspices of the Remote Sensing Division of the Southwest Cultural Resources Center, National Park Service. The responsibilities of the National Park Service include the preservation of many prehistoric and historic cultural resources, often structures, within its landholding. For this reason new methods for accurately mapping the ruins at Chaco Canyon National Monument were being sought. Within the Monument area lie the partially collapsed ruins of eighteen enormous pueblo structures, and many smaller ones, occupied, for the most part, between about 900 and 1200 A.D. Some of these structures were to be excavated for scientific study; others were being ''stabilized'', a process which entails at times partial reconstruction and modification of the original ruin. All pueblos at Chaco Canyon had been, and would continue to be, at the mercy of change from natural forces, visitors' impact, and other sources of change such as air pollution and vibration due to mining and the conversion of vast energy resources in the area. The Remote Sensing Division was required to find the most cost-effective means of recording these structures. Experiments began in 1972 using controlled, 1:3000 and 1:6000 black-and-white stereo aerial photographs. Figure 26-25 is an example of imagery used to map Penasco Blanco. The product of the mapping of this ruin is illustrated in Figure 26-26.

In addition, photogrammetric mapping principles can be applied to areas larger than single structures. Two large areas containing both prehistoric ruins and features of an extensive nature, were mapped microtopographically. The Kin Bineola area (Figs. 26-27 and 26-28) exhibits an extensive irrigation system now mapped and partially analyzed (Lyons, Pouls, and Hitchcock, 1972). Another large area, located at what appears to be the ''cultural center'' of Chaco Canyon, has been mapped in its entirety. This area contains four large and diverse ruins—Pueblo Alto, Pueblo Bonito, Chetro Ketl, and Pueblo del Arroyo—as well as dozens of roadways, trash deposits, walls, fields, smaller sites, and other structures.

At the other end of the spectrum of physical size, plots of spatial relationships within small sites can be arrived at quickly and economically by using photogrammetric methods and aerial or

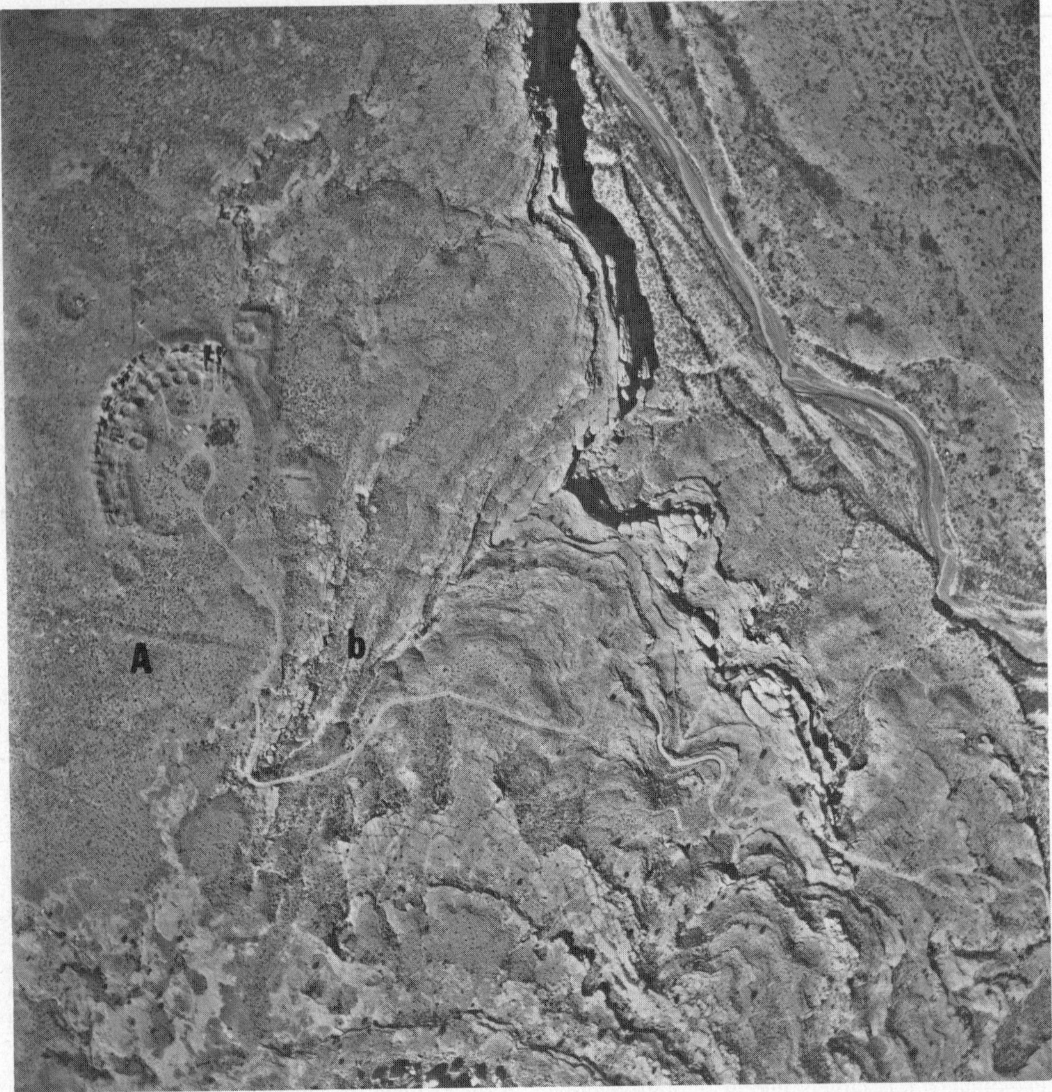

Fig. 26-25. Vertical aerial photograph of Penasco Blanco, one of the great pueblo ruins of Chaco culture in National Historical Park, which is located in northwestern New Mexico. The vestiges of a clearly-visible prehistoric roadway segment can be seen between points A and B.

ground-based photographic platforms. Both balloon- and bipod-platforms for vertical imagery have been experimented with by the Remote Sensing Division. The bipod platform seems most practical for recording small to medium-sized excavations, whereas a balloon platform appears to be more useful in the phased photography and mapping of ongoing excavations. Under such a procedure, a balloon-borne, radio-controlled camera might be permanently moored over a large site during excavation. As the excavation proceeded, photos would be taken sequentially; through the use of these images, the three-dimensional relationships between objects and features at a site could be determined and recorded with virtually no obligation of field time. The high resolution images provided by balloon-mounted cameras at low altitude might allow the identification of specific artifacts in some cases.

Accurate and easily referenced maps and dimensions can be an aid to the reconstruction and stabilization of ruins by indicating areas of natural destruction and structural weakness; they can also provide an estimate of the study's cost- and time-requirements by facilitating the estimation of volumes of material to be moved. Other uses of such data are less obvious. For instance knowledge of a building's room area (a quantity easily determined by calculations based upon photogrammetric data) may assist in estimating the number of inhabitants per structural unit. Some uses of photogrammetric map data are obvious; others will continue to appear as long as there are archaeologists with imaginations.

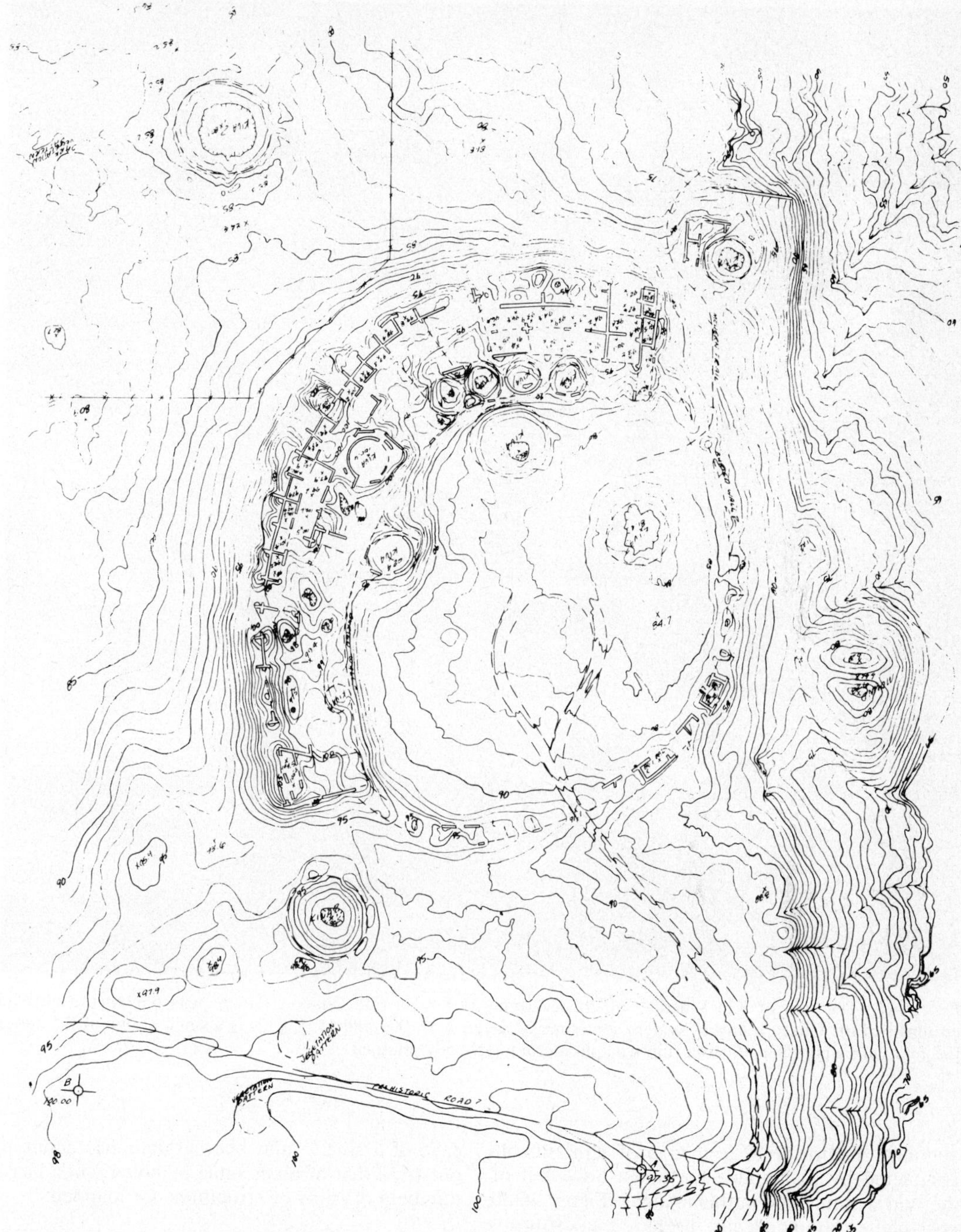

Fig. 26-26. A photogrammetric map of Penasco Blanco. Walls and depressions, as well as the ground surrounding the ruin, are drawn in one-foot contours.

Feasibility Studies of Archaeological Photogrammetry

The feasibility of remote sensing methods as contrasted with traditional methods in archaeology has been formally studied. Requisite imagery, limits of accuracy, types of data potentially available, archaeological applications of such data, and costs have been considered.

A cost analysis and comparison of a specific project approached through both traditional, on-the-ground methods and remote photogrammetric techniques is presented below. This example deals with the mapping of a large Pueblo III period ruin, Kin Bineola, near Chaco Canyon (Fig. 26-27). Kin Bineola is interesting in that its occupants apparently used an extensive irrigation network, and many of the walls in the ruin are still

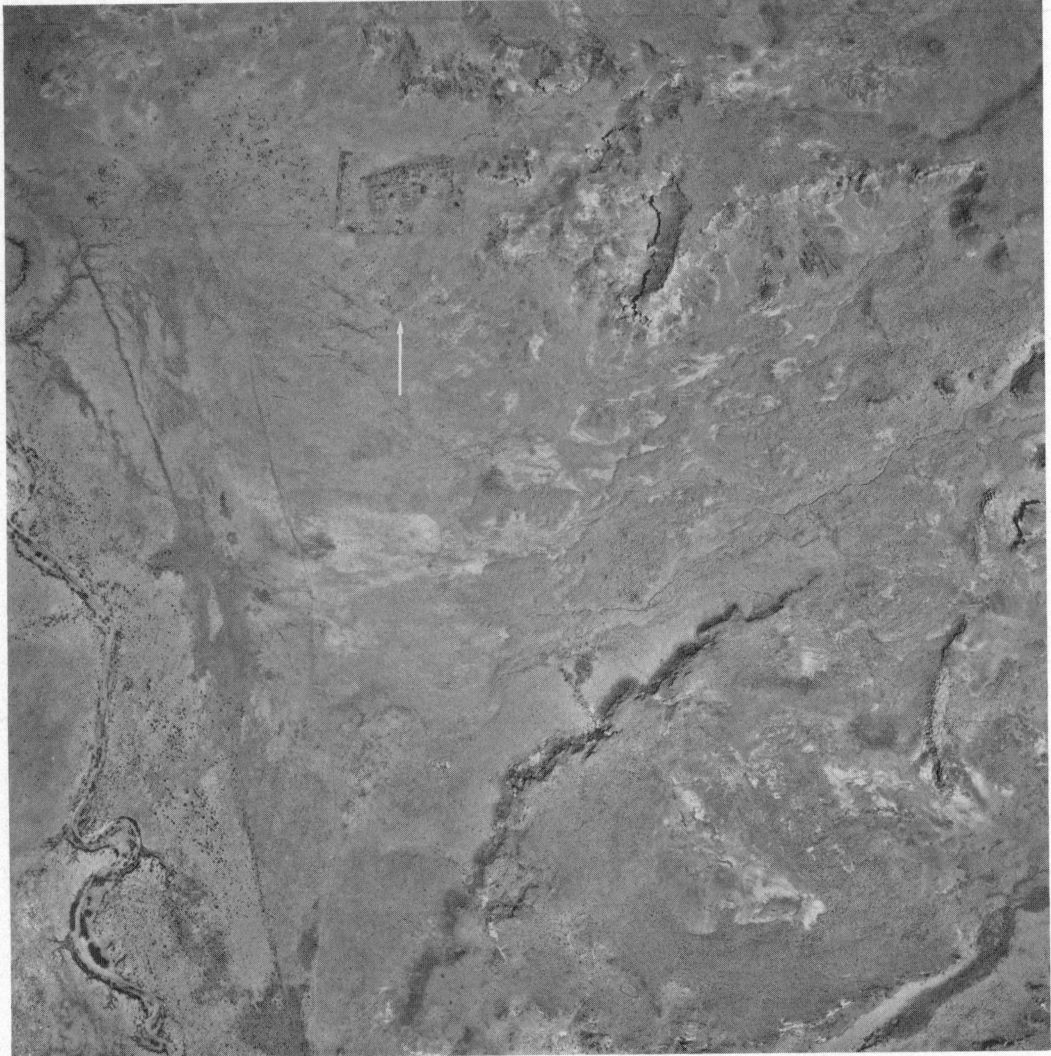

Fig. 26-27. The ruin of Kin Bineola, which lies some 21 km outside Chaco Canyon, was one of the outlying terminals of the Chaco roadway system sometime between A.D. 900 and 1175. This is a single frame of the stereo imagery with which Kin Bineola ruin was photogrammetrically mapped.

standing. Figure 26-29 is a map of Kin Bineola completed by plane-table methods; only half of the ruin was mapped in this manner. Figure 26-30 is a photogrammetrically prepared microtopographic map of the ruin. Comparison should be made between these maps in terms of accuracy, detail, and amount of extractable scalar information. In this instance, remote sensing has yielded a far superior product at a very comparable net cost. The full cost-to-product advantage inherent in the application of photogrammetric methods to the mapping of ruins, however, is not totally apparent until the mapping of a number of ruins is considered. While aerial-mapping costs may be comparable to those of plane-table mapping in the case of a single ruin, photogrammetric economy outstrips that of plane table mapping when larger numbers of ruins or structures are mapped.

Computer Manipulation of Photogrammetric Data

One useful outgrowth of photogrammetric mapping is based upon the "digitization" of points on a three-dimensional stereo model. To accomplish this, a controlled stereo pair is converted to diapositive glass plates and installed in a stereo plotter, so constructed that, when the interpreter has determined that he has correctly measured the location and height of an object, the

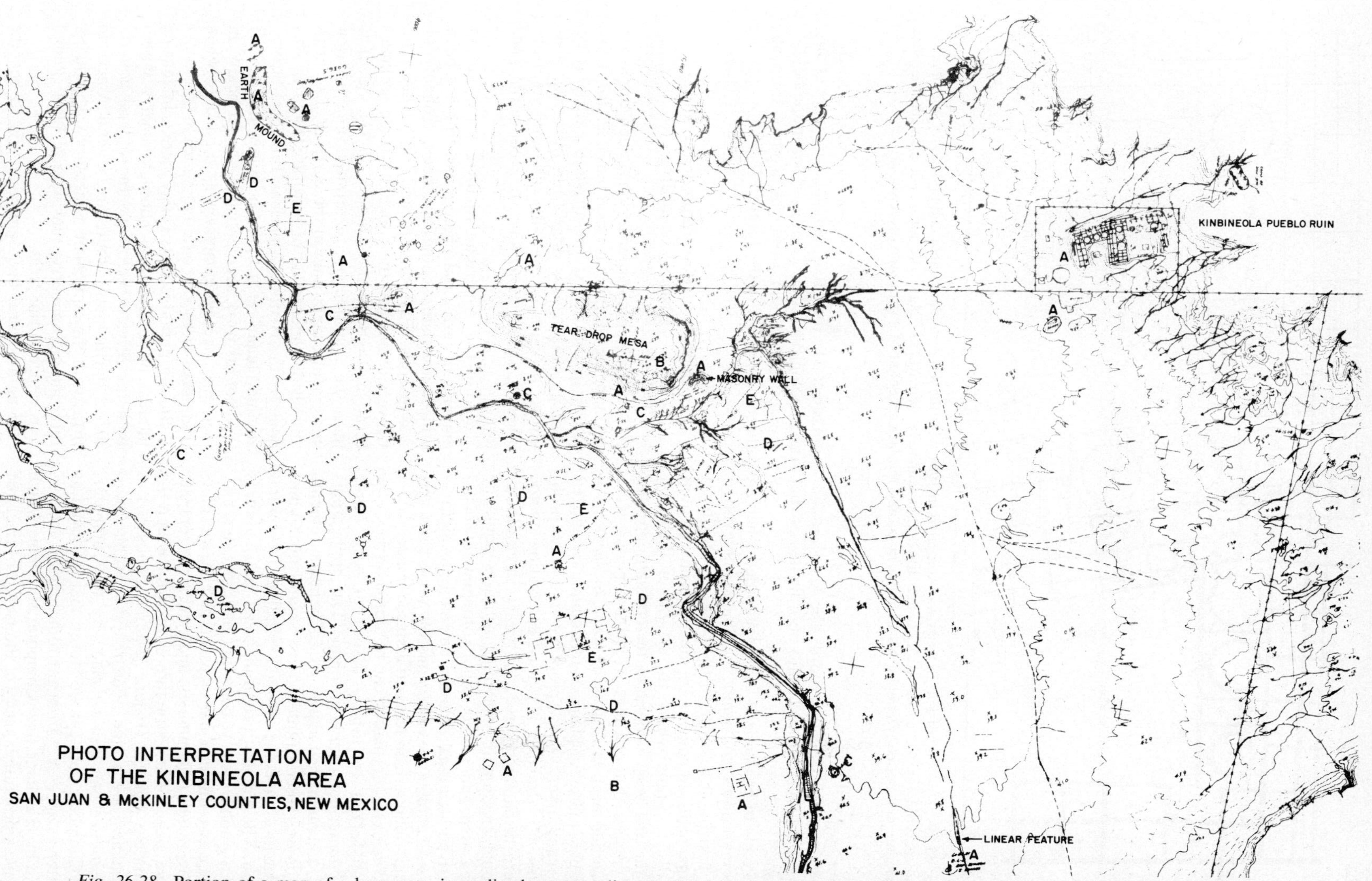

Fig. 26-28. Portion of a map of a large area immediately surrounding the Chaco ruin of Kin Bineola. The map was produced photogrammetrically. Accurate and detailed maps can serve as the basis of mapping and analyzing prehistoric irrigation systems or other large features.

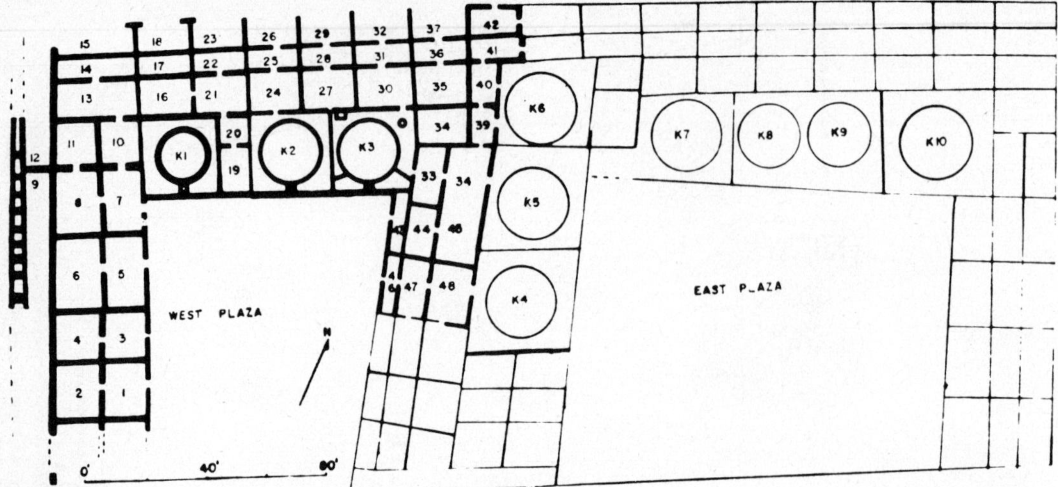

Fig. 26-29. This map of Kin Bineola was compiled in 1967 with the aid of a plane table and alidade. A 3-man crew was required for surveying work, which took approximately two weeks. Note the idealized, straight-line wall segments and perfectly circular kivas in the map, and the lack of detail of surrounding terrain.

machine automatically records the x,y (horizontal) and z (vertical, or height) coordinates of that point.

Points to be digitized can be recorded when an operator notes any significant change in elevation or horizontal bearing. In the case of the Chaco ruins, digitized points were located primarily atop walls, at spots where the height or direction of the wall being followed by the operator appeared to change dramatically. Additional points were recorded near the bases of walls to serve as a graphic reference plane. Points could be digitized in many other ways appropriate to diverse problems. For example, if phased excavation photos were being interpreted, the exact three-dimensional location of individual artifacts could be measured and stored. All photogrammetrically-derived locational data are amenable to digitization, including measurements obtained from vertical, oblique, and terrestrial stereo imagery.

Digitized information can be stored on cards, or transferred to tapes, discs, or any other computer-compatible medium. The applications of computer techniques to such measurements promises to constitute a most significant revolution in archaeological data handling interpretation. The simplest use of digitized information is as a graphic display; computer plotters can reproduce two-dimensional maps or planimetric plots (Figure 26-31) in minutes, and can create perspective views of any structure or area for which the data are digitized (Figure 26-32). Calculations of volumes, areas, and other relevant architectural quantities can also be made by the computer.

Computer data can be stored and retrieved as often as necessary from any remote point accessible by telephone line. Computer terminals which have a printout capacity of thousands of lines per minute can be used to obtain data and maps at any archaeological site or laboratory, or in the field—a practice that will surely be a much-used archaeological tool in the future.

Terrestrial Photogrammetry and Historic Recording

Terrestrial photogrammetry has been employed for mapping and architectural purposes for many years; in fact, most pre-airplane cartographic photogrammetry was done from terrestrial platforms, because of the difficulties in precisely orienting the balloon, kite, or other aerial platform.

Within the American historic preservation movement, terrestrial photogrammetry has been utilized less than it could have been, probably for two reasons. First, until recently only Ohio State University has had the capability to do architectural photogrammetry (Figure 26-33). Second, until the adaptive use/restoration/rehabilitation movement caught hold in the early 1970's, virtually all documentation of historic structures was done by the Historic American Buildings Survey summer teams at a cost far below the market value of the work accomplished—so far below, in fact, that photogrammetry was used only when traditional methods could not be used.

But this period is past. Today there is increasing interest in architectural photogrammetry within both the private sector and the federal government. Furthermore, there are several American firms with the equipment necessary for architectural photogrammetry, and with at least limited experience in this field.

If architectural photogrammetry is to become

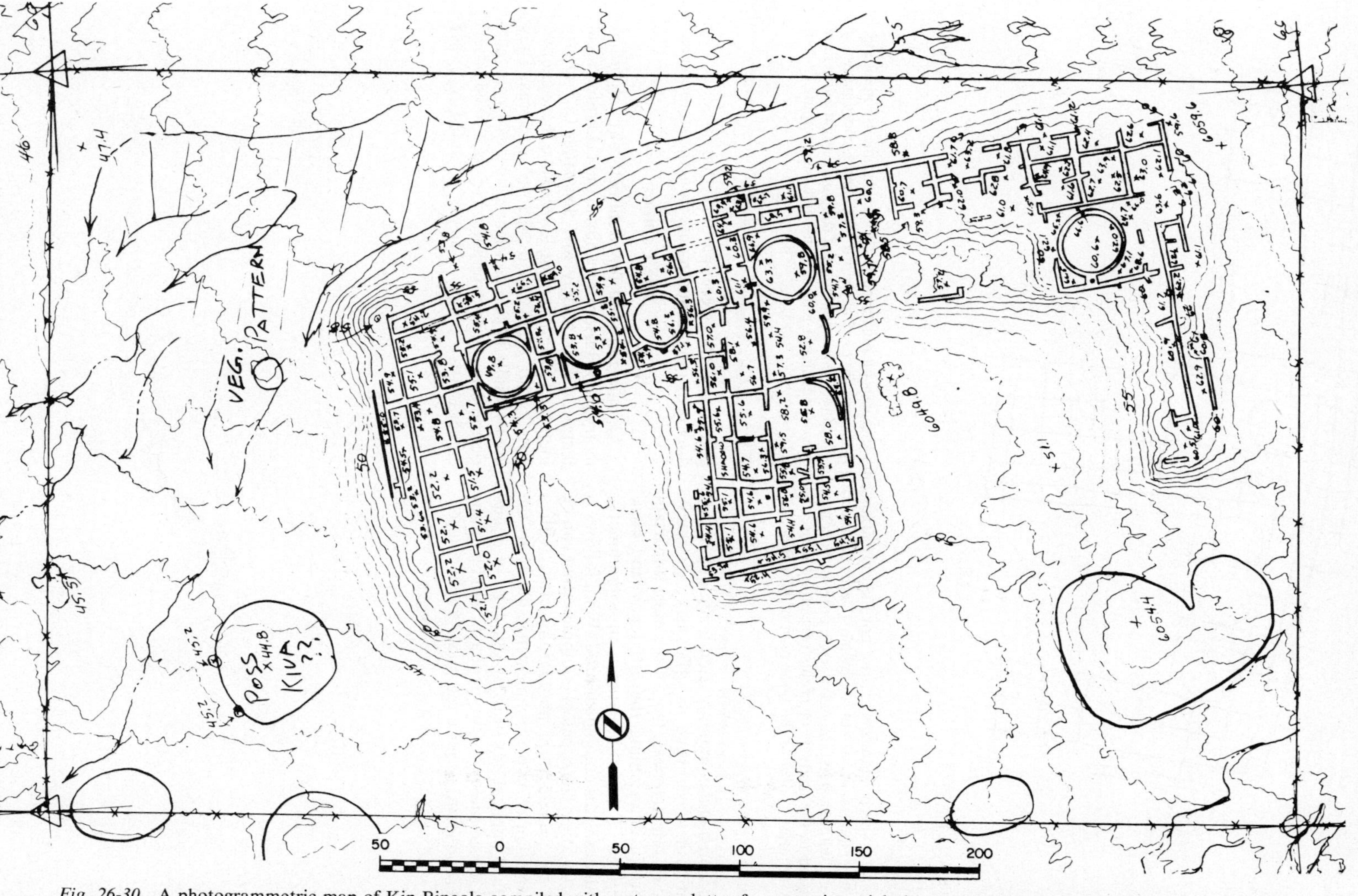

Fig. 26-30. A photogrammetric map of Kin Bineola compiled with a stereo plotter from metric aerial photographs. For approximately the same cost as was involved in the production of the map in Figure 29, this map provides the archaeologist with a far more accurate and complete representation of this Pueblo-III Period ruin at Chaco Culture National Historical Park in northwestern New Mexico.

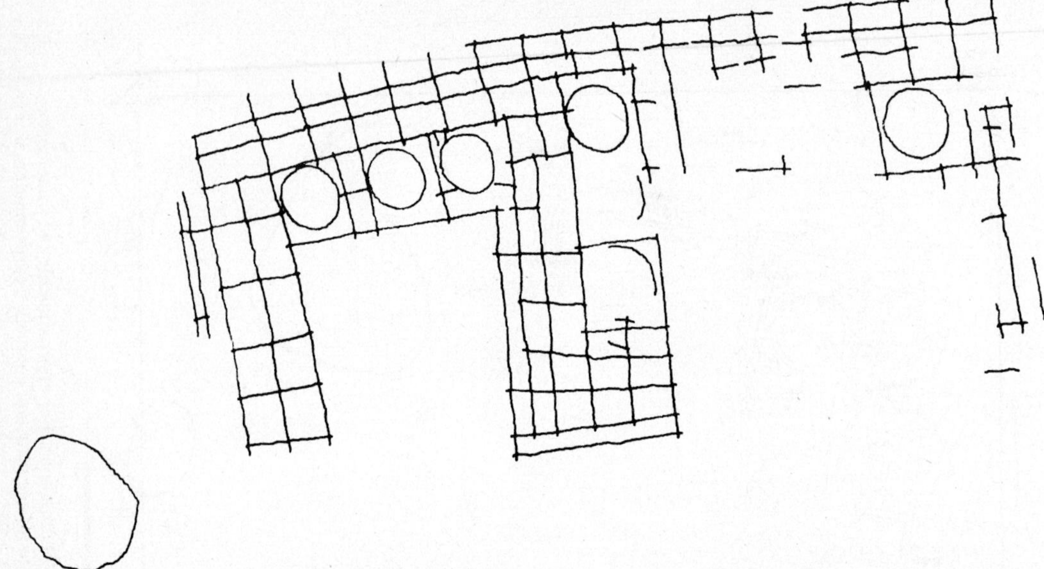

Fig. 26-31. Speed and accuracy are not the only arguments that can be mustered for photogrammetric mapping. Digitizing plotters, in current use by most aerial engineering firms, can be used to record x-, y- and z-coordinates for any point within a stereo model of a site or feature; computer plotting and manipulation of such data might add a new dimension to the analysis of prehistoric architecture. This two-dimensional plotter construct is a simplified—but highly accurate—view of the floor plan of Kin Bineola.

an accepted and commonly used technique in this country, it must prove its efficiency, accuracy, and cost-effectiveness. One of the most important factors in determining the efficiency and cost-effectiveness of architectural photogrammetry techniques is the equipment used. What follows is a discussion of the types of equipment that can be used for architectural photogrammetry, with a presentation of two examples that illustrate the practical implications of using one type of equipment versus another. Our emphasis will be on the cameras used to acquire the photographs (stereopairs) and the stereoplotters that produce the drawings and measurements from the stereopairs.

The Camera

All cameras used for architectural photogrammetry have lenses of a known and calibrated focal length and extremely low distortion (±5 micrometers). Cameras whose lenses have these characteristics are often called "metric" cameras. They are available in two configurations: as single metric cameras, or as "stereometric" or "fixed base" cameras (two single cameras mounted at either

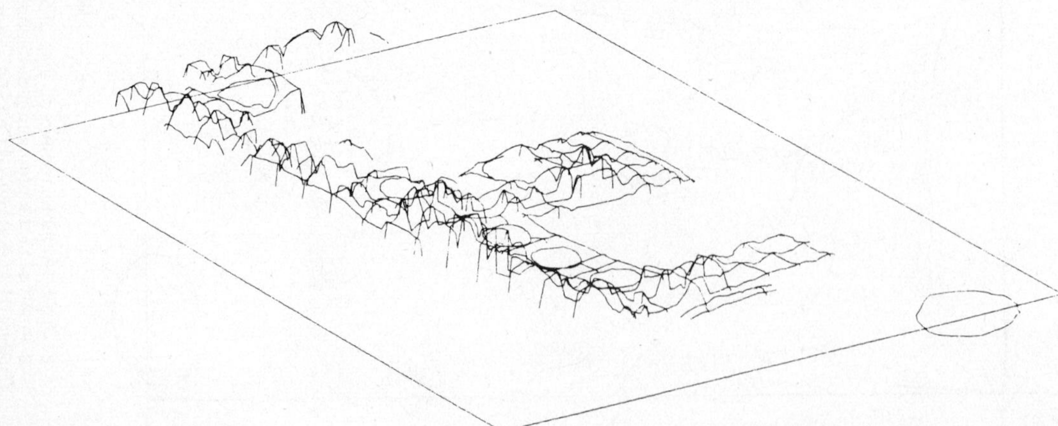

Fig. 26-32. The addition of a height (z) coordinate determined stereographically for each designated x- and y-intersect produces a three-dimensional view of the pueblo. Such a construct can be rotated to any angle with respect to any vantage point, and might be of help in reconstruction and stabilization of significant prehistoric resources.

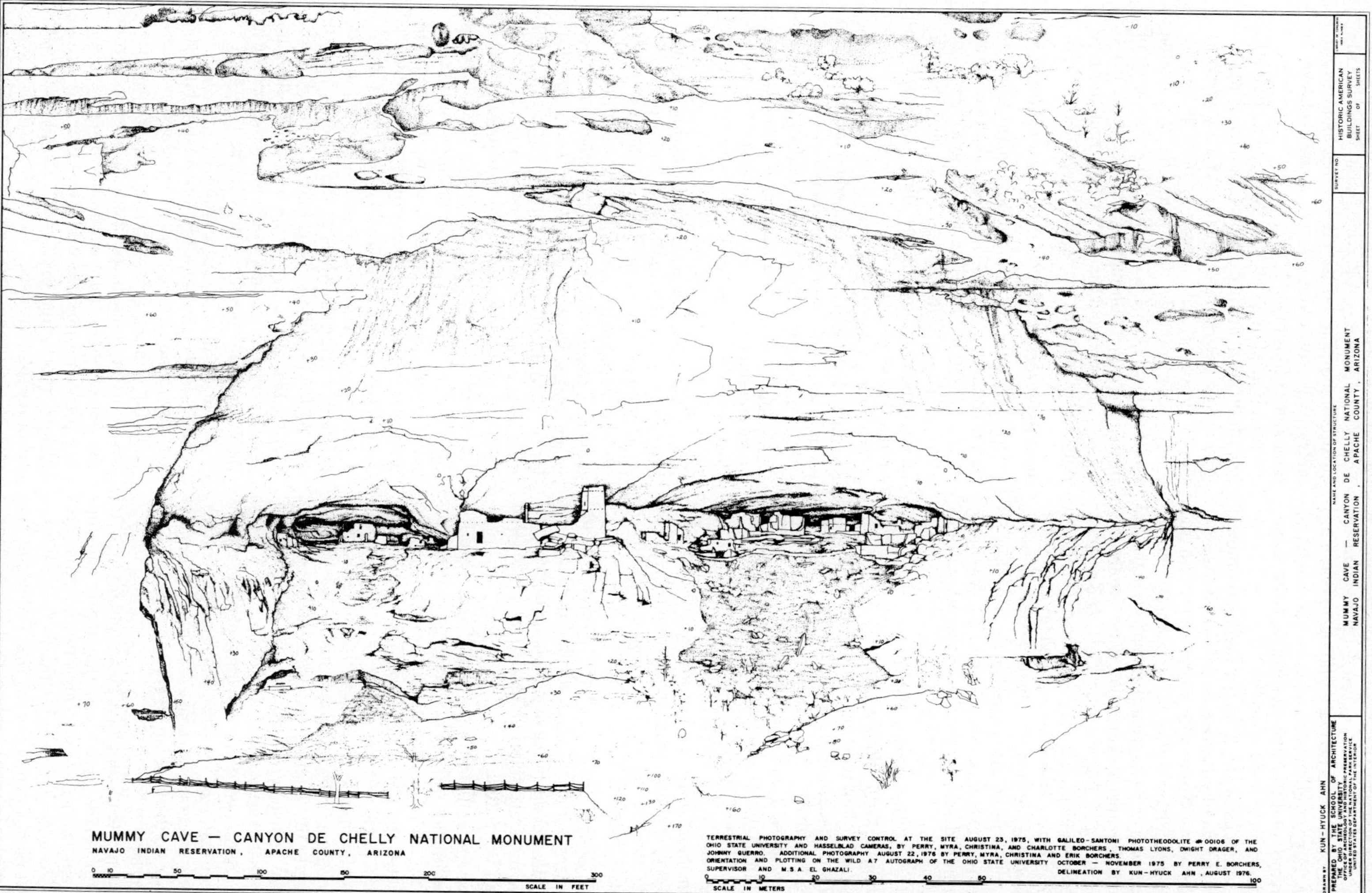

Fig. 26-33. Terrestrial photogrammetric map of Mummy Cave at Canyon de Chelly National Monument, Arizona. This elevation was produced from stereo terrestrial photos collected with a Galileo-Santoni phototheodolite and plotted by Perry E. Borchers, M.S.A. Ghazali and Kun-Hyuck Ahn on Wild A-7 Autograph at Ohio State University. (Courtesy Remote Sensing Division, National Park Service).

end of a basebar). Stereopairs are produced using a single metric camera by placing it at two successive camera stations—separated by a distance called the base—and taking a photograph at each position. For stereometric cameras, the length of the basebar of basetube determines the base, hence the name "fixed base".

Most varieties of single metric and stereometric cameras can be tilted above or below a horizontal reference plane. Some will accept sheet film and roll film in addition to the preferred glass plates. Wild Heerbrugg (Switzerland) cameras (single and stereometric) are perhaps the most commonly used worldwide, followed by Carl Zeiss (West Germany), Galileo Santoni (Italy), and Zeiss Jena (East Germany). Hasselblad (Sweden) and Kelsh (United States) also make single metric cameras that can be mounted in pairs to function as stereometric cameras. See Chapter 6, Volume I, for further discussion of terrestrial camera systems.

The Stereo Plotters

In architectural photogrammetry the stereo plotters most often used are of the "optical train" (also called "mechanical analog") type. This type takes two forms, one of which is specifically designed for close range and terrestrial photogrammetric applications; the other is designed principally or exclusively for aerial photogrammetric applications. In Europe, close range plotters are the form most often used in architectural photogrammetry, because they are considerably more versatile than aerial plotters. This versatility is due to two factors:

1) the shape and size of the plotter's model space, within which is formed the coherent optical model of the object photographed, and
2) the number and orientation of the planes that can be plotted.

The shape and size of the model space are the most significant differences between plotters designed for aerial work and plotters designed for close-range and terrestrial work. The model space on a typical close-range plotter is nearly cubic in shape while that of a typical aerial plotter is more like a rectangular box whose height is one-third or one-quarter the length of the other two sides. Furthermore, most aerial plotters are designed for base-to-distance ratios of approximately 1:3 and an overlap (between the individual photographs of the stereo-pairs) of 55–65%. Close-range plotters, with their larger model-space, can plot stereopairs with base-to-distance ratios between 1:3 and 1:20, and overlaps of 55–95%.

Aerial plotters can plot a plane roughly parallel to the film plane to produce planimetric and contour maps. Some are also able to plot cross-sections. Close-range and terrestrial plotters not only can plot planes parallel to the film planes but also can plot planes *perpendicular* to the film plane—for example, to produce vertical and horizontal sections of a building facade. Furthermore, most close-range plotters have tilt correctors that allow them to plot planes tilted relative to the film plane. The Wild Heerbrugg A40 Autograph, probably the most common plotter for close-range photogrammetry, can plot seven discrete planes relative to the camera film plane.

Projection plotters (Kelsh, Multiplex, Balplex) are often used in this country for aerial photogrammetry and have been used for close-range work. However, this type of plotter is considerably less versatile than a mechanical analog plotter. Because of its design (and the laws of physics) a projection plotter's model space is extremely limited, and the scale of the final drawing must be the same as the scale of the optical model.

Instrumentation: Applications in Architecture

How does the question of equipment translate into practical application? Let us take two examples of situations likely to be encountered in architectural applications and examine how the type of equipment influences the field work and the production of plots.

The first example is a two-story building, approximately 60 feet on a side and 50 feet high; the second, an elaborate cornice detail.

For the building, the desired product is a set of elevation drawings at a scale of 1″ = 4′. Using a photogrammetric system consisting of either a single metric camera or a stereometric camera, and a plotter designed for close-range work, the field work can be done in the following manner: Each of the four elevations of the structure is photographed, one stereopair to a side, for a total of four stereopairs for the entire structure. The field time required depends on whether one uses a single metric camera or a stereometric camera, but it should not exceed three hours total field time in any case. Ground control consists of two points in each stereopair. The four pairs are then processed and plotted. Mounting and aligning each pair in the close-range plotter will take approximately fifteen minutes each. Plotting will take about three hours per pair, to produce a manuscript of sufficient quality that, when inked, it will become a drawing equal in graphic quality to the measured drawings of the Historic American Buildings Survey. Total time for photography, plotting and inking will be approximately 20 hours per elevation. Note that by changing the gearing in the close-range plotter, a drawing at 1″ = 8′ or 1″ = 2′ can be as readily produced.

Thus, in the case of a small two story building, photogrammetric technicians producing stereopairs for plotting with a close-range plotter need not be overly concerned about the distance from the object. Also, they need only place two control points on the structure photographed. Technicians producing stereopairs for plotting with an aerial plotter must calculate the exact position of each camera station, must place three to six control points on the structure for each stereopair, and then must calculate the position in three dimensions of each of those control points. Use of an aerial plotter will necessitate four to eight times the number of stereopairs (depending on the desired scale of the final drawing) required when using a close-range plotter to cover the same area.

Now let us consider a decorative cornice, perhaps a full-blown Italianate cornice with brackets. The desired documentation includes an elevation drawing of a four-foot length of the cornice, as well as horizontal and vertical sections at intervals. These drawings would, of course, be very time-consuming to produce through the use of traditional hand-measuring techniques. Because of the depth (overhang) of the cornice, it will probably be impossible to plot an elevation on an aerial plotter at a meaningful scale (1″ = 1′, for instance) due to the limited size of the model space of such plotters.

Using a close-range plotter, one could produce from a single stereopair an elevation drawing at any scale of from 1″ = 4′ to 1″ = 2′, as well as vertical and horizontal sections in the same scales.

In less than ideal situations, the flexibility of plotters can determine whether or not the job may be possible. For example, ground control is not absolutely necessary with stereometric cameras if one uses a close-range plotter. This makes it possible to document fire-damaged buildings or other structures on which it would be otherwise unsafe to place ground control. The ability to plot from inclined photography makes it possible to record tall structures or buildings located on narrow streets, without resorting to cranes or boom trucks. In emergency situations, the ability to obtain proper documentation may depend totally upon the speed with which the field work can be completed.

The effective use of close-range equipment, while critical to success in architectural photogrammetry, will not be the only factor to determine the success or failure of the technique in the long run. A major factor is whether photogrammetrically produced drawings are clear, useful representations of architectural forms which accurately portray the different surfaces and edges of the structure, as well as the different materials and the boundaries between them. Not all architectural drawings need be done as elegantly as Historic American Buildings Survey drawings, but it is insufficient to render a Gothic cottage or a Greek Revival mansion using but a single line-width. It is important to note also that architects will usually perform a condition survey while completing the hand measuring of a structure; often this survey is required even with photogrammetrically-produced drawings. Thus, if the cost of the photogrammetrically-produced drawings and the condition survey together exceeds the cost of hand measuring, there will be little incentive to use photogrammetry.

Architectural photogrammetry holds a great deal of promise as a time- and cost-effective technique to aid architects in restoration and rehabilitation work, and in historic structures documentation—a promise that can only be fulfilled by using equipment specifically designed for such applications. While most users will be concerned particularly about cost (and somewhat less about time), they will be interested ultimately in the quality and usefulness of photogrammetrically-produced drawings. It is the quality of the final product that will determine whether architectural photogrammetry is tried only once or becomes routine practice.

Terrestrial Photogrammetry for Site and Rock Art Mapping

Terrestrial photogrammetric mapping can be applied in situations where it is neither efficient, nor perhaps possible, to produce and archaeological site map using aerial imagery or traditional mapping techniques. Close-range imagery permits the production of maps with smaller contour intervals, and a greater density of points, than maps produced from aerial photos. To date, close range photogrammetry has seldom been used to document archaeological sites. Scogings (1975) and Clouten (1974, 1976, 1977) report photogrammetric recording of South African petroglyphs and Australian pictographs, while in the United States, Perry Borchers has applied close range photogrammetry to documenting pueblo architecture.

More recently, the University of Texas at Austin, in conjunction with Dennett and Muessig, undertook a field test of the feasibility of using close range photogrammetric equipment to record two archaeological sites in Val Verde County, Texas (Turpin et al 1979). Figure 26-34 shows a small rock midden, delineated by fire cracked limestone blocks, which was chosen as the first site to be recorded. Scattered among the blocks are several dozen small artifacts, primarily flint chips and shell fragments. These artifacts were marked *in situ* by nails to which numbered tags had been attached. Control for the stereopairs was established by erecting three targets on a triangle of known dimensions. Three sets of horizontal, normal case stereopairs were taken with a tripod-mounted stereocamera placed in turn at three locations around the site. The planimetric and contour map in Figure 26-35 was prepared as a composite from the three stereopairs. Three cross sections of the site were also produced (Figure 26-36). Mapping this site by traditional means, using a transit and chain, alidade, or a grid, might take three to ten times longer than it took to produce the maps using close-range techniques. Moreover, traditional mapping requires that decisions be made in the field which determine what features will or will not be included in a final map. Close range imagery, which records complete dimensional information for the site, permits such

Fig. 26-34. Rock feature on Painted Canyon Flat, Val Verde County, Texas. The site has been prepared for photogrammetric recording with erection of ground control targets and tagging of artifacts. Reproduced from the right negative of one of three stereopairs of the feature.

decisions to be made in the laboratory, perhaps years after the original recording was completed.

At the second site, close range photogrammetry was applied to recording painted pictographs, which are located on a limestone canyon wall, also in Val Verde County, Texas (see Figure 26-37). Dennett and Muessig (1980) explained the shortcomings inherent in methods traditionally applied to rock art recording:

> Previous attempts to document rock art have depended on various graphic and photographic processes, which most archaeologists see to have several shortcomings. Many investigators have tried tracing, and while tracings fairly accurately capture the form and proportions of the pictographs, they fail to describe either the texture or the gross configuration of the surface on which the images have been executed. In addition, many pictograph panels are very large, or are located in positions unsuited for a clinging tracer, limiting the practicality of preparing a tracing. Moreover, the bulkiness of a full size tracing, usually done on thin cellophane, limits its manageability during analysis and for reproduction, not to mention during a windy day in the field.

The rock art site in Val Verde County was recorded with three stereopairs: the first depicts the entire site, the second shows the panel of paintings, and a third concentrates on a 5-meter segment of the panel. Color Figure 26-38 shows the view of this latter stereopair, and Figure 26-39 shows the planimetric and contour plot produced from the same stereopair. Two vertical cross sections were also plotted, showing clearly the curvature of the rock wall (see Figure 26-40).

Dennett and Meussig (1980) also note the advantages of photogrammetrically recording such sites:

> These plots provided very accurate scaled orthographic projections of the painted figures, as well as accurate information about the configuration of the supporting rock. The field work is rapid, and does not necessitate physical contact with the surface of the rock or the figures. The scaled plots, which we have produced in pencil on mylar and subsequently inked, are durable, easy to store, and inexpensively reproduced as diazo prints. When processed following procedures for archival permanence, the stereopairs form a lasting photographic record of the site, and can be viewed with a stereoviewer or reproduced as photographic prints.

Close-range photogrammetric equipment is generally designed with features which enable the user to record successfully and efficiently any of a variety of site configurations, including excavated sites, stratigraphic record trenches, surface scatters, standing or collapsed structures, and rock art sites. Features which contribute to the equipment's versatility include, in the cameras, the ability to take photographs inclined up or down

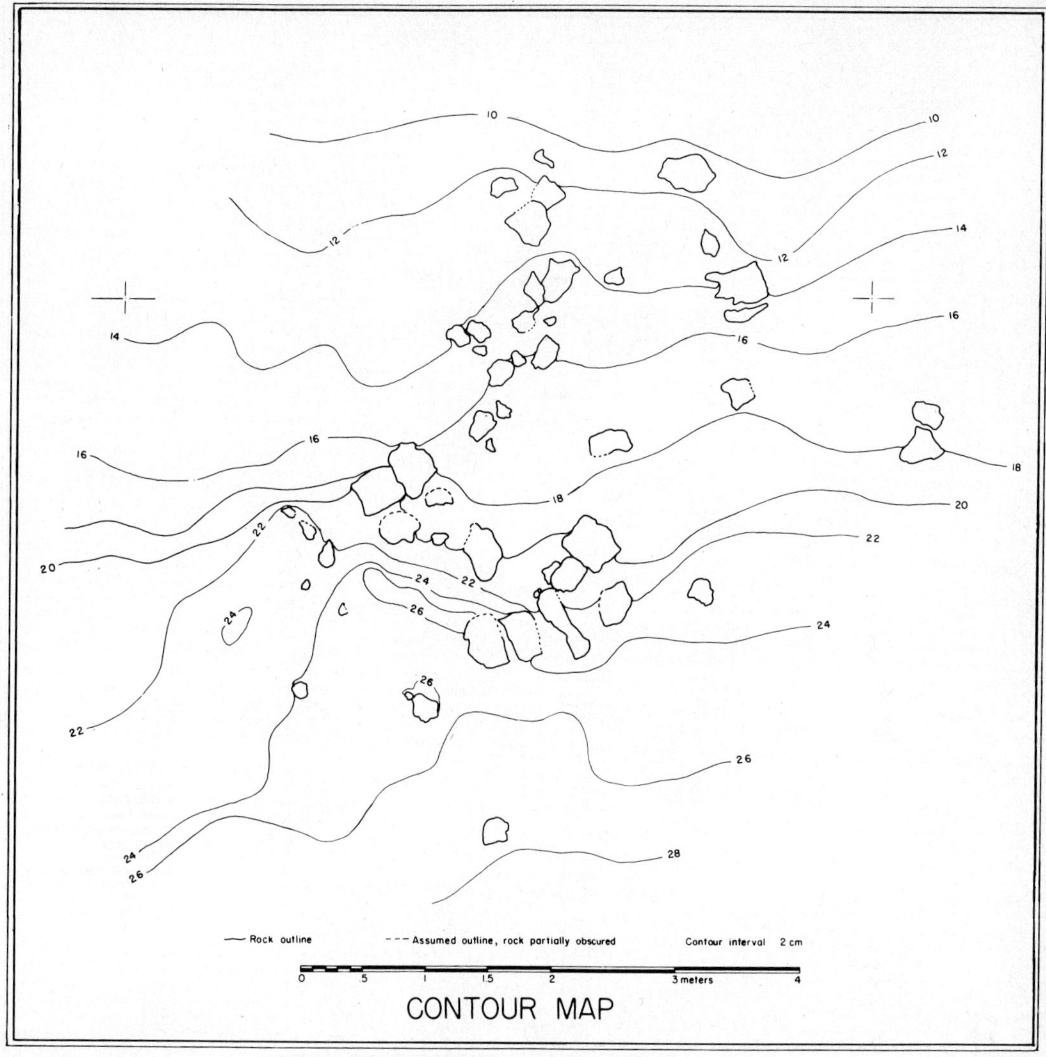

Fig. 26-35. This is a 2cm-contour map of the rock midden on Painted Canyon Flat plotted by Dennett, Muessig and Associated, Ltd., Iowa City. Reduced from a 1:20 plot.

from horizontal and the ability to vary the distance between the left and right camera positions. In close-range plotters, the shape of the model space is nearly cubic, unlike the flat model space of most aerial plotters. This important characteristic allows for the plotting of planes perpendicular or parallel to the film plane, making possible the ground-based recording of horizontal sites such as the rock scatter described previously.

Monitoring Cultural Resources Photogrammetrically

The concept of monitoring cultural resources stems from the acknowledgement that all sites and structures are in a constant process of change. Since that change cannot be prevented, and in fact may add to rather than detract from the value of cultural resources, it is desirable and necessary to be able to measure and characterize such change. Cultural resources preservation consists, then, not in stopping change but in insuring that the heritage value, information value, and overall integrity of the resources is preserved by monitoring these features, a process in which remote sensing methods and photogrammetry offer virtually the only solution.

Experiments are underway at the Remote Sensing Division of the National Park Service to develop methods for the incorporation of both aerial and terrestrial photogrammetry into monitoring of historic and prehistoric structures. The first requirement for monitoring using photogrammetry is the existence of controlled baseline data; that is, photogrammetric data from some point in the past. One of the Chaco Canyon National Monument pueblos, Kin Ya'a, which was mapped in the early 1970's, was chosen for the prehistoric subject of the Remote Sensing Division monitoring

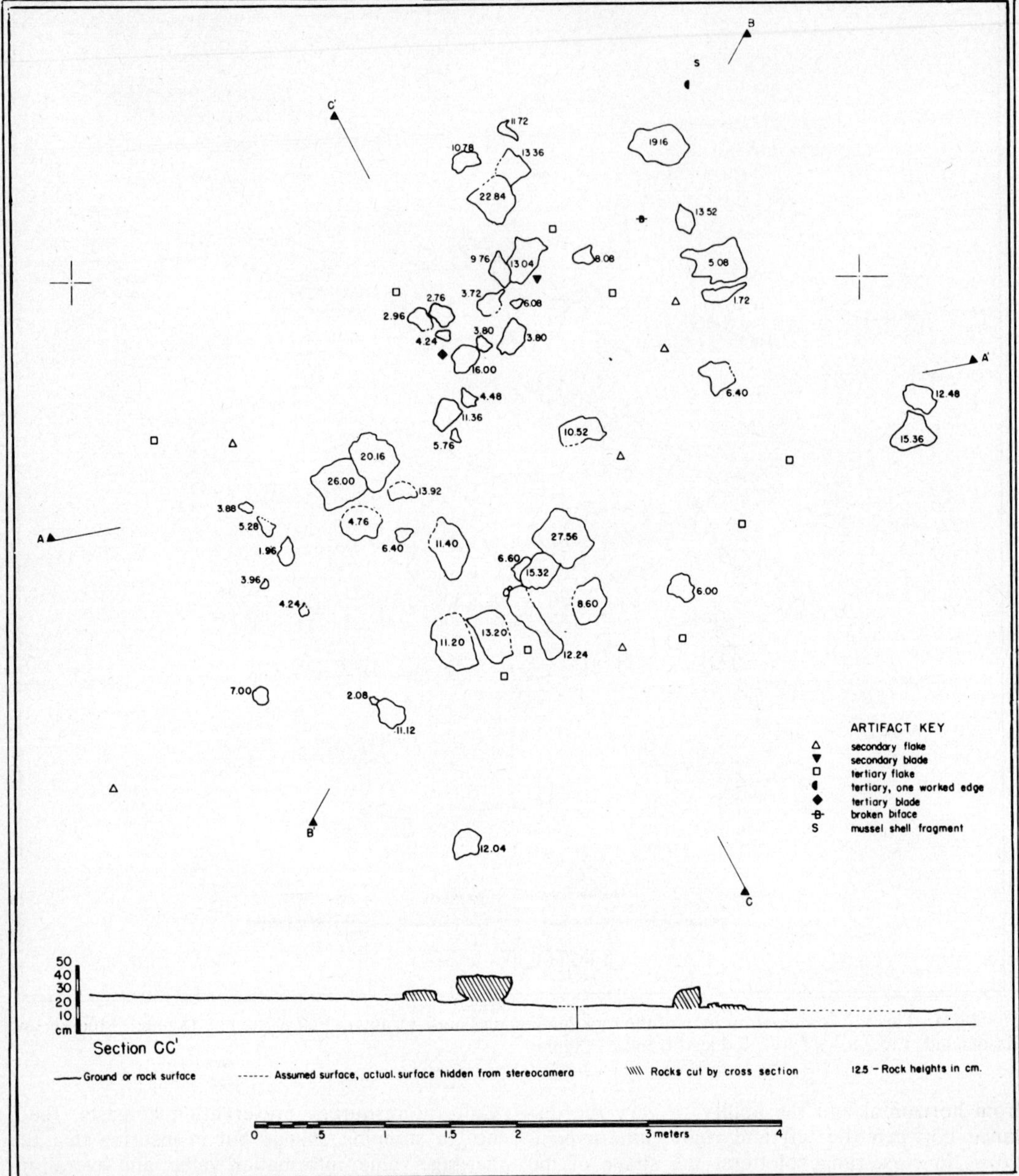

Fig. 26-36. Planimetric map and cross-section of the rock features showing artifact locations and the height (in cm) of each limestone block. Reduced from a 1:20 plot.

experiment for this reason (Figure 26-41). Another factor making Kin Ya'a an interesting subject for this experiment is the fact that change can be expected to have occurred since the original mapping. Several years after the original controlled stereo aerial photographs of Kin Ya'a were flown, a uranium ore body was determined to lie directly below the structure. Test holes were drilled all around the ruin to make the final determination of the extent of the ore, and present plans call for the use of a leaching process to extract the uranium. Although this will not require the excavation or destruction of the ruin, it will involve more boring and the use of heavy equipment and mining crews in the immediate site area.

At the outset of the monitoring experiment at Kin Ya'a, in the spring of 1980, the structure was again reflown subsequent to the setting and panelling of permanent control monuments. At the same time, stations for terrestrial photogrammetric cameras were permanently marked with monuments surrounding the structure. Terrestrial control points were marked by small threaded studs set into the mortar between stones (the

63,87

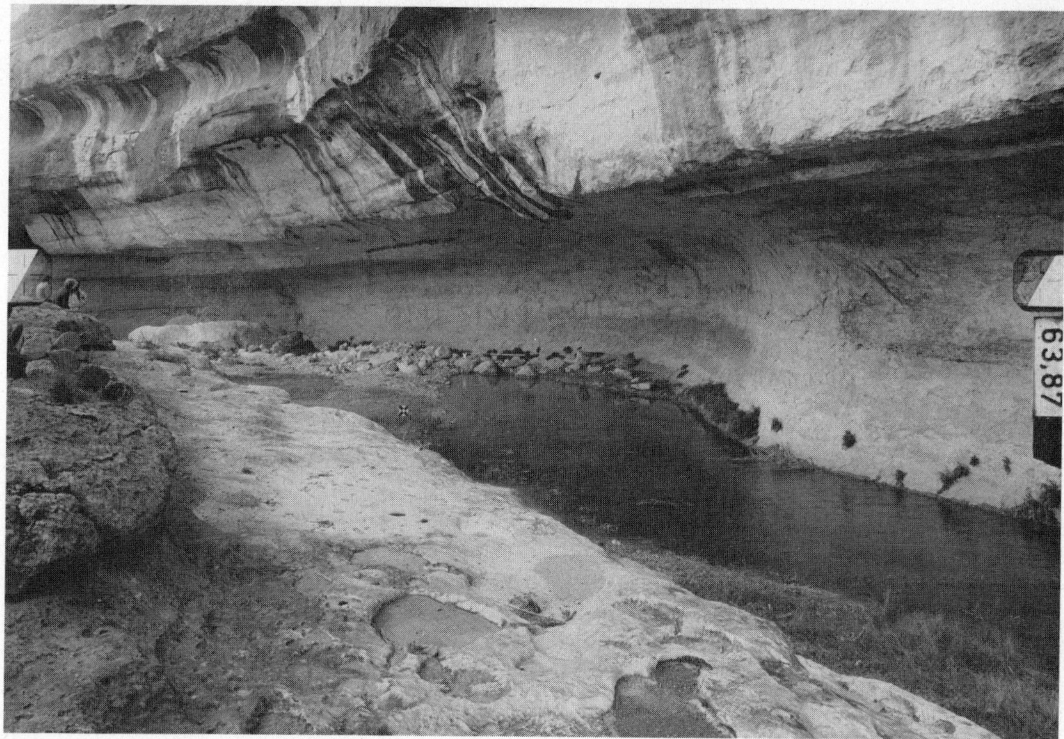

Fig. 26-37. General view, facing southwest, of Painted Rock Shelter in Val Verde County, Texas. Reproduced from the right negative of a stereopair.

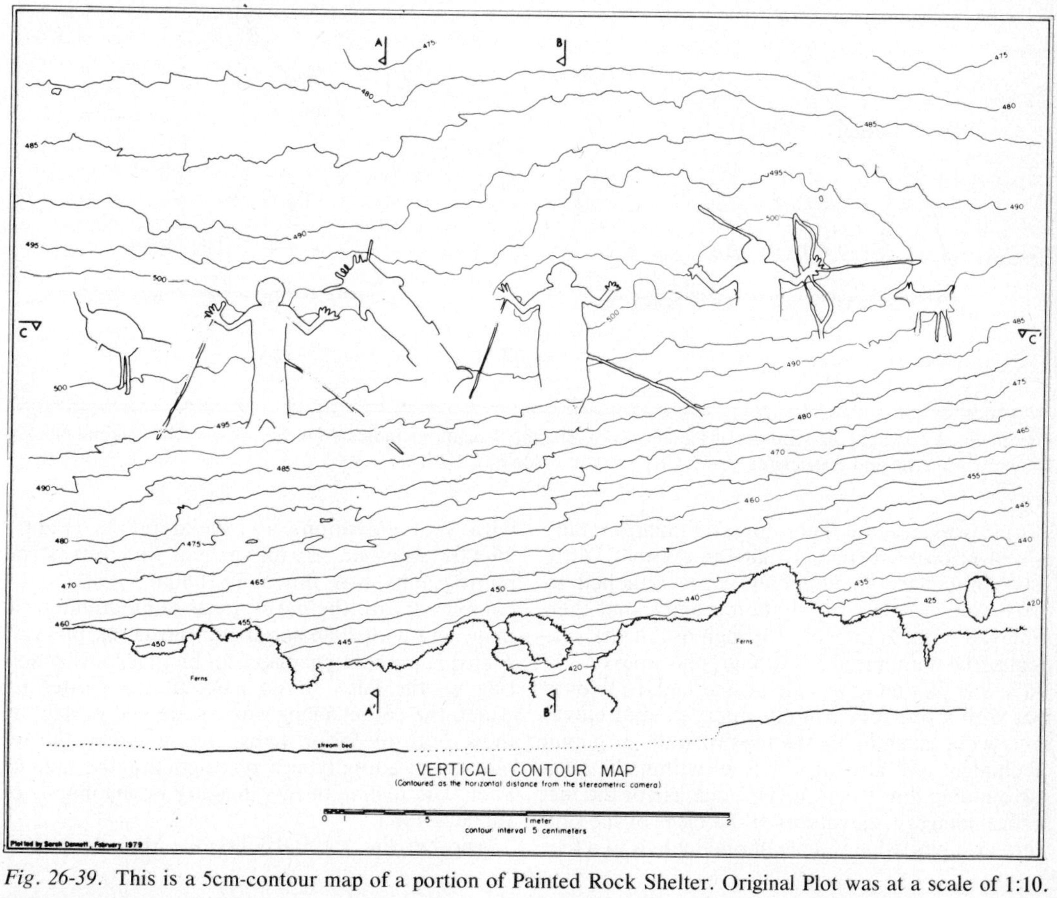

A
B
C
C'
A'
B'
Ferns
stream bed
VERTICAL CONTOUR MAP
(Contoured as the horizontal distance from the stereometric camera)
0 1 5 1 meter 2
contour interval 5 centimeters
Plotted by Sarah Dammett, February 1979

Fig. 26-39. This is a 5cm-contour map of a portion of Painted Rock Shelter. Original Plot was at a scale of 1:10.

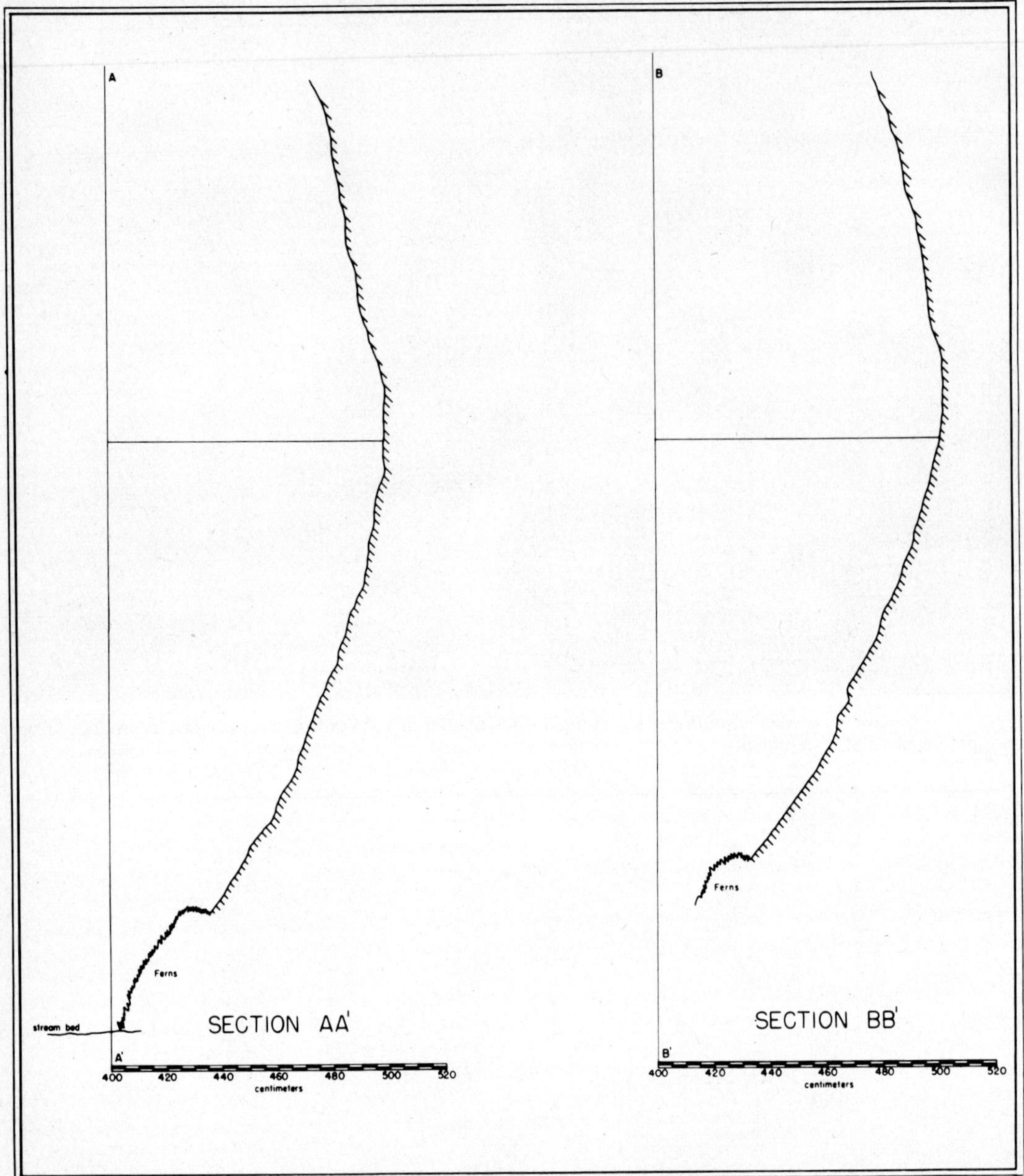

Fig. 26-40. Vertical cross-sections of Painted Rock Shelter. Locations indicated in Fig. 26-39. The original plot, by Dennett, Muessig and Associates, Iowa City, was at a scale of 1:10.

mortar was placed there for stabilization purposes) in order to insure that the original fabric was untouched; small targets were attached to these studs for terrestrial photography, and then removed without impact. Through use of the vertical aerial photographs, a topographic map of Kin Ya'a was plotted at a scale of one inch to twenty feet with a one foot contour interval; spot elevations were taken along the tops of walls at points of change, and also on surfaces within the area surrounding the structure. Through use of the terrestrial imagery, elevations of the faces of the ruin were also plotted at a scale of one inch to two feet with spot elevations at crucial points (Figure 26-42). Another set of vertical and terrestrial photographs was flown in January 1982 to be compared with the early aerial photogrammetric map, and with 1980 aerial and terrestrial plots. It is suspected that changes can be detected by setting up the later stereo pairs in the plotter on which the earlier maps were made and projecting these onto the prior maps, thus avoiding the replotting of a totally new photogrammetric map or plan. The testing of this imagery is ongoing as of January 1982.

Another site, the Barboursville Mansion in Vir-

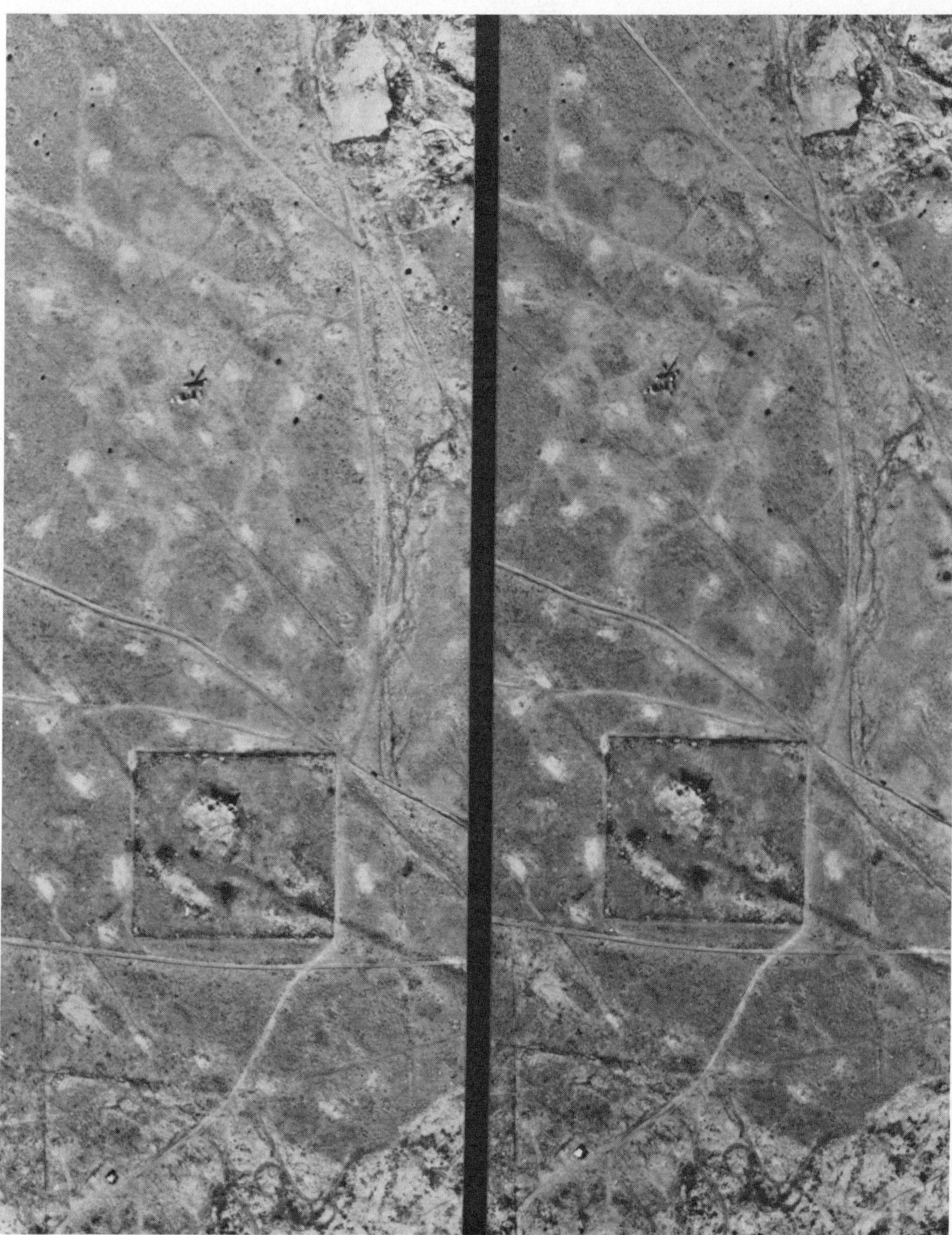

Fig. 26-41. Stereogram at a scale of 1:6000 of Kin Ya-a ruin in Chaco Canyon National Monument showing the fenced archaeological site, a trace of a prehistoric roadway feature running diagonally through the central plaza of the pueblo, and a network of drill pads and modern access roads in the immediate vicinity. This drilling activity, begun in early 1976, was directed toward defining the limits of a uranium ore body located beneath the site. Heavy equipment and the presence of laborers in the area of the site are potential sources of site disturbance. Current experiments by the Remote Sensing Division of the National Park Service in Albuquerque, described in the text, deal with the feasibility of monitoring changes in the site and structure using aerial and terrestrial photogrammetry (see Fig. 26-42). (Aerial photography by Koogle and Pouls Engineering, Inc., Albuquerque).

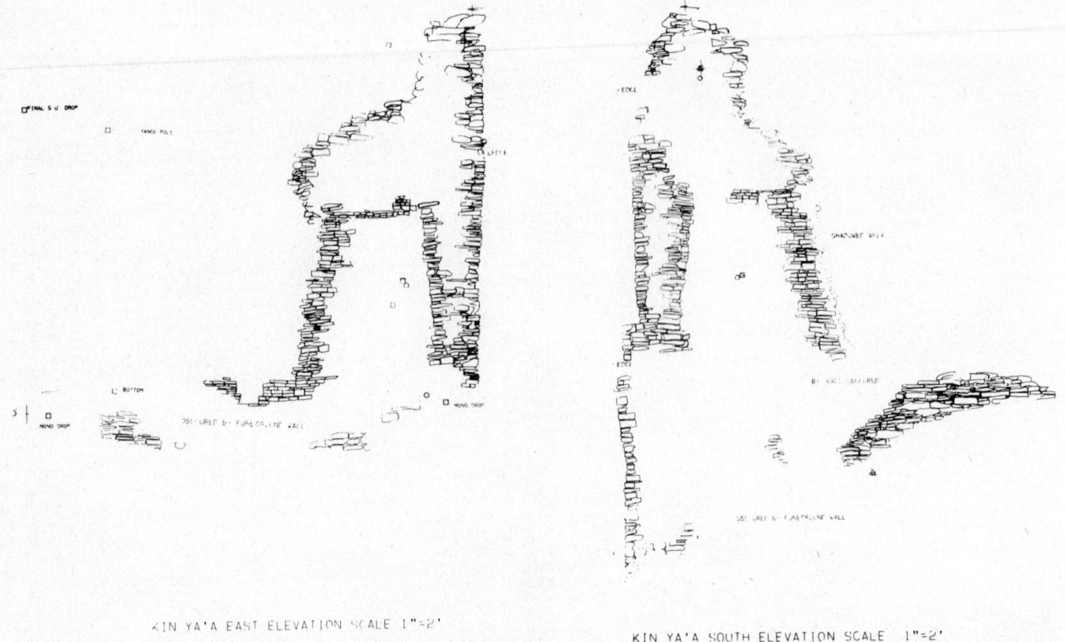

Fig. 26-42. East and south elevations of Kin Ya-a, a Pueblo-III Period ruin in Chaco Canyon National Monument, occupied for about 200 years centering about A.D. 1000. These elevation plots were produced from terrestrial stereo metric photos by Koogle and Pouls Engineering, Inc., Albuquerque, New Mexico. The data presented here are currently being used as baseline information which, compared with subsequent terrestrial and aerial photogrammetry of the ruin, will reveal structural changes caused by weathering or cultural activities in the area (see Fig. 26-41 for evidence of some cultural activities).

ginia, was chosen to illustrate the utility of aerial and photogrammetric data for monitoring historic structural changes. The Barboursville Mansion was designed and built by Thomas Jefferson, and burned on Christmas Day in 1844, leaving a roofless masonry shell and columns standing. This fragile structure is, of course, in a constant state of change primarily due to natural forces. Essentially the same procedures used to map Kin Ya'a were employed: control points, both vertical and horizontal, were permanently marked with unobtrusive monuments and then flagged for purposes of aerial and terrestrial photogrammetric photography. Both types of photography were obtained in the spring of 1980 and aerial maps (Figure 26-43) and terrestrial elevation-drawings (Figure 26-44) produced for comparison with future data. Metric stereo photography and photogrammetry were performed by Koogle & Pouls Engineering, based in Albuquerque, New Mexico.

Electronic Distance Measurement

One final means available to the archaeologist and architect for the recording of sites and structures should be noted here: the electronic distance measurement (EDM) device. This device, increasingly available today and already familiar to most photogrammetrists and mappers, bridges the gap between traditional surveying instruments and remote sensing, and should be recognized as useful (and in fact indispensible) to the archaeologist.

In Europe, EDM's are used extensively by both civil engineers and archaeologists in their measuring and mapping tasks. In the United States, surveyors and civil engineers rely heavily on EDM's, while archaeologists have not used the instruments except for experimental work. The success of surveyors and engineers in mapping large areas of land rapidly, accurately, and cheaply suggests that EDM mapping is an appropriate technique for archaeologists in the United States.

EDM's are, in effect, electronic measuring tapes. Some are more sophisticated than others, but they rely on timing the interval required for a modulated beam to travel from the instrument to a reflector, and back to the instrument. The most common design involves bouncing infrared energy off a reflector called a retroprism, although some systems are laser or microwave energy beams, or other sorts of reflecting units. Generally, EDM's are mounted on top of a theodolite, so that when the telescope of the theodolite is aimed at the retroprism, the EDM is also accurately aimed at the reflector. With this "single aim" system it is possible to determine the distance to a point (with the EDM), and the horizontal and vertical angles to the point (with the theodolite), very quickly. Obviously the point measured is the retroprism itself, but the retroprism is attached to a plumbing pole whose tip has been placed on the ground point to be measured. Several quick calculations with a pocket calculator will yield the elevation of the

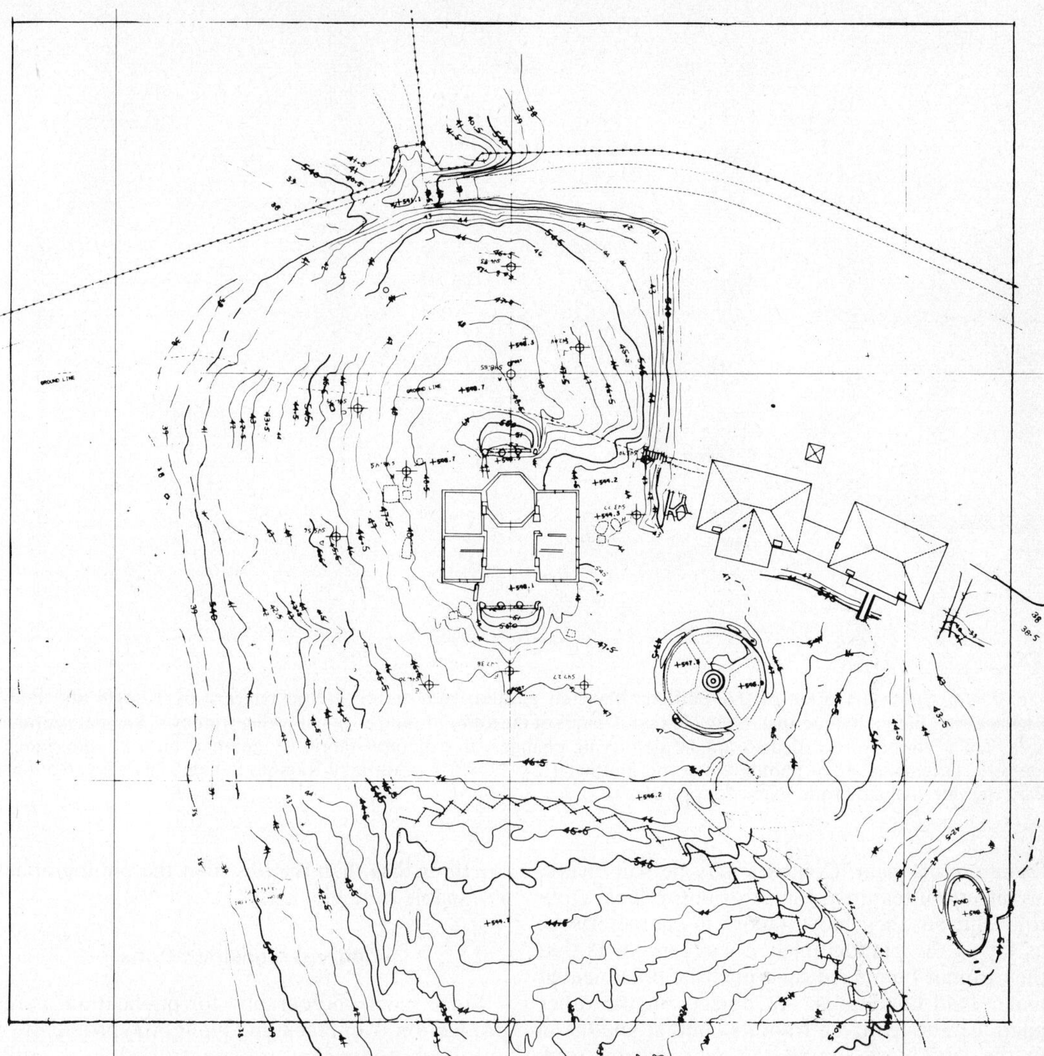

Fig. 26-43. Map produced by aerial photogrammetry of the Barbour Mansion in Barbourville, Virginia and its immediate setting. The barbour Mansion was designed by Thomas Jefferson and burned in 1844, leaving a roofless brick facade. This historic monument is fragile and susceptible to alterations induced by natural and recent cultural causes; the data in this figure and figure 26-44 are being used as baseline information for purposes of monitoring such changes. Metric photography and photogrammetry by Koogle and Pouls Engineering, Inc. (Courtesy Remote Sensing Division, National Park Service).

ground point and its position relative to other known points, such as a baseline or datum.

Compared to traditional methods of mapping archaeological sites, EDM mapping has some obvious advantages: the retroprism and plumbing pole are much easier to handle than a chain and rod, and the results are considerably more accurate (the accuracy of infrared EDM's is typically ±5 mm). While these advantages strongly suggest the use of EDM's for site mapping, the speed of the equipment and thus the rapidity with which the field portion of the mapping can be completed may be the primary advantage of the technique. Many "single aim" EDM's are capable of measuring three to four points every two minutes, as well as tracking on the retroprism as it is moved. Compared to mapping large areas with a transit, chain and rod, EDM mapping may be five to ten times faster. Given these advantages, an EDM may be effectively utilized for a number of archaeological mapping tasks: for example, mapping a large site (1 acre or more) in 0.5 meter contour intervals; laying out coordinate grid systems for excavation, surface collection, or reconnaissance; and piece-plotting sites with fairly low artifact density.

IN THE LABORATORY: REMOTE SENSING AND ARTIFACTUAL EVIDENCE

Archaeological evidence can be divided broadly into two categories: contexts, and the objects

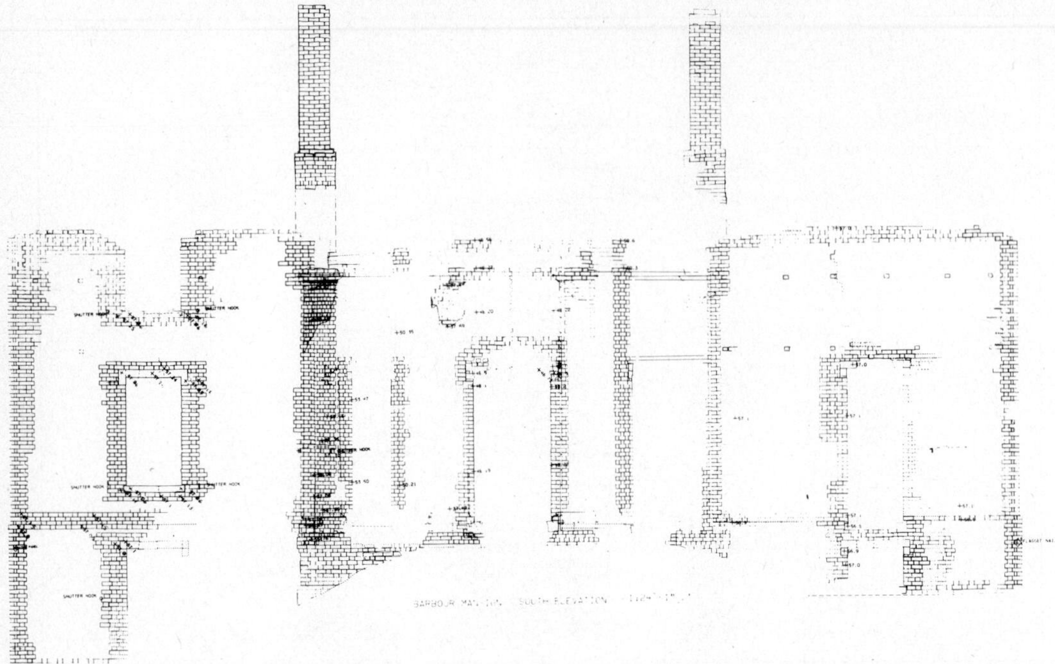

Fig. 26-44. South elevation of the Barbour Mansion plotted from stereo metric imagery by Koogle and Pouls Engineering, Inc., Albuquerque, New Mexico. Details of masonry around edges of walls, windows, and deteriorated areas are accurately recorded to aid in identifying changes in structural fabric by comparison with subsequent imagery. Original scale of photogrammetric plotting 1:24 (1″ = 2″). (Courtesy Remote Sensing Division, National Park Service, Albuquerque, New Mexico).

found within them. Contexts may be soil layers, architectural features, tombs or entire local environment; but their most important characteristic, as far as the photographer is concerned, is that they cannot be moved, and must be documented under field conditions. An object, on the other hand, can be removed from its context and taken to the laboratory for photography, where such variables as lighting, background and scale are almost entirely under the control of the photographer.

This very removal, however, deprives the object of some of its archaeological meaning. Its position and orientation within a context may have been recorded, so that its chronological or other significance is preserved; yet an object once removed from its context must, photographically speaking, stand alone. It is thus the photographer's responsibility to document the appearance of an object as accurately as possible, not only for the immediate purposes of illustration and publication, but also because in some cases the photograph will be the only permanent record for future researchers.

Finally, the fact that an object to be photographed can be arranged under the complete control of the photographer means that every aspect of the picture must be carefully considered. Objects do not fidget, do not sweat under the lights, and do not complain of unflattering likenesses; there can thus be no excuses if the results are anything less than the best that the photographer is capable of achieving.

General Considerations

Since most photography for publication or reproduction is in black-and-white, the problems of color photography are mentioned here only briefly. This section will concentrate mainly on the goals and techniques of producing black-and-white photographs.

As a rule, every portion of the object should be in sharp focus in the final photography, subject only to the limits set by the minimum aperture of the lens or by high degrees of magnification. Only when a particular portion of an object is to be emphasized, or when other specific effects are required, can less than the maximum attainable sharpness be justified. The background often is the most important element of an object photograph, but it should be the least noticed on the final print. Ideally, a background should be invisible, serving only to occupy the space on the print that is not taken up by the object itself. It should contrast in tone and brightness with the object, and the object should cast no shadow on it. Several techniques for achieving this result are outlined below. The way in which the object is lit determines the success or failure of the photograph. Primarily, the lighting should allow all portions of the object to be seen clearly. This is determined by the lighting ratio or contrast—the difference in brightness between the lightest and darkest areas of the image. For archaeological work, especially for publication, a lower contrast range is preferable, since detail will otherwise be lost in deep shadows or glaring highlights. While print contrast can be controlled somewhat in the darkroom, it is far easier to establish the correct contrast in the studio.

Hard lighting, which produces sharp transitions from light to dark, should be avoided except in a few specific cases. The hardness or softness of the lighting depends on the size of the light sources in relation to the size of the object: for instance, the sun is many times larger than the Earth, but its distance makes it seem almost a point-source of light, producing very hard, constrasty lighting.

Lighting looks most natural when it appears to come from a single source, such as the sun, or from a non-directional source such as an overcast sky (Fig. 26-45). What should be avoided is lighting that obviously comes from two or more distinct sources of approximately equal intensity; this produces multiple shadows and a confusing visual effect.

The goal in producing any black-and-white print, whether for publication or for display, is always to encompass the tonal range of the subject within the tonal range that the film/paper combination is capable of reproducing. Ansel Adams' Zone System is a highly sophisticated method of obtaining this, but it relies on individual development of each negative, thus adjusting the film's contrast range to fit the range of brightness present in the scene being photographed. In object photography, however, it is possible and much easier to rely on consistent processing of the film, and to adjust contrast by means of the lighting ratio, when the photograph is made. The final print should always show an even tonal scale, with no important areas of deep shadow or blank white. This ensures that significant detail throughout the object is preserved, and also allows for a slight increase in contrast usually associated with the reproduction process; areas of shadow or highlight that show barely perceptible detail on the final print may become obscured on a published plate.

Fig. 26-45. Minoan Bronze Age alabastron. A single lamp high on the right serves as the main light, with fill from a second diffused lamp. A green filter brings out the painted detail. Photographed for D. J. Hurst and P. P. Betancourt, Mediterranean Section (MS4478) by Nicholas Hartmann/MASCA. (Courtesy of the University Museum, University of Pennsylvania.)

Technical Considerations

A treatment as brief as this cannot pretend to be a complete manual of object photography; the authors of this chapter can offer only some general techniques from their experience.

Equipment

The preferred camera is the 35mm single-lens reflex; the specific brand used is relatively unimportant, provided certain criteria are met. The quality of the lenses, on which the quality of the image ultimately depends, must be the highest; the camera itself should be capable of non-automatic operation; and the range of accessories available should be large enough for both present and anticipated needs. Some half-dozen top quality camera systems meet these requirements, and the choice among them depends mostly on personal preference.

Although many studio photographers exclusively use large-format (4 × 5″ or larger) view cameras for object work, the 35mm SLR (single lens reflex) will fit the needs of the photographer/archaeologist or specialized archaeological photographer much better. The primary reason is size: an entire 35mm system, with lenses, filters, attachments and enough film for several days' work, can be carried or stored in the space occupied by a 4 × 5 inch view camera body alone. Along with the reduction in size goes a reduction in cost: the same 35 mm system will usually cost less than a good view camera *without* a lens. Nor is there necessarily any sacrifice in image quality, since modern lenses for 35mm cameras are routinely capable of resolving finer detail than most films can record. Finally, most 35 mm systems include a large range of close-up and other specialized accessories, and the smaller film size means that less space is needed for the studio, for processing, and for storage of negatives.

Other obvious equipment needs include a tripod and/or copy stand, a cable release, a selection of filters, lighting equipment and various backgrounds, reflectors and diffusers. Further equipment should be bought (or improvised) according to the individual photographers's needs.

The choice of lenses for object photography is fairly simple. A macro-type lens is ideal, since it will focus continuously from infinity to a reproduction ratio of 1:2 or 1:1 without attachments. Such lenses are also corrected for best results at ratios of 1:10 to 1:2, where much small-object photography is done. Similar results can be obtained using normal or telephoto lenses with extension tubes or *good* supplementary lenses, although different combinations will be needed for different magnifications, and the optical quality is slightly, but usually imperceptibly, less.

Film

Along with the recent improvements in lenses mentioned above, has come an improvement in film. Fast black-and-white films such as Kodak's Tri-X can produce results rivalling those of much slower and less tolerant emulsions. Again it is the practice of several of the authors of this chapter always to use the faster film unless there is a compelling reason to use something slower. Such reasons always have to do with the finer grain associated with slower films such as Panatomic-X, which should be used when prints larger than 11 × 14 inch will be needed. For more extreme enlargements, or when high magnification must be obtained in the darkroom rather than in the camera, such ultra-fine-grain films as H&W Control are useful despite their less satisfactory tonal scale. There also exist high-contrast copy films, which are useful when the subject's own contrast is extremely low.

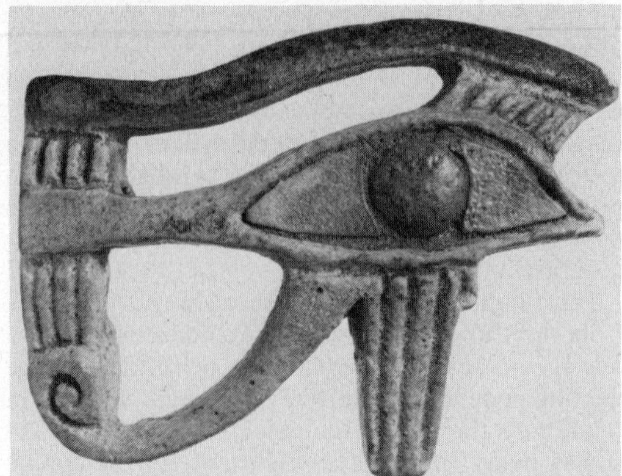

Fig. 26-46. Egyptian *wadjet* (Eye of Horus). Length 47mm. Lighting as in Fig. 26-47, with green filter. Photographed for S. J. Fleming by Nicholas Hartmann/ MASCA (Eq. 29-85-717). (Courtesy of University Museum, University of Pennsylvania.)

Backgrounds

For most publication photographs, a white or very light grey background is best, and there is a simple and effective way of obtaining this result. This system is useful primarily for small objects, where the camera can be mounted vertically on a copy stand, but it can be adapted for larger subjects and a horizontal set-up.

The object itself is placed on a sheet of clear plate glass that is in turn supported about 6 inches above the baseboard of the copy stand. The object is then lit from above, using the techniques explained below.

Below the glass is placed a sheet of plain white paper or card, which is evenly lit with a separate lamp, so that the background *receives* about 1 stop more light than the subject. The object lighting and the height of the glass should be adjusted so that any shadows cast by the object fall outside the field of view. This will produce an image in which the object appears to float on a background which is lighter than any portion of the object, and which shows no shadows (Fig. 26-46). The white background card should be only large enough to fill the field of view: too large a background can produce flare in the lens, and will spill light onto the edges of the subject, thus obscuring its outline (see Fig. 26-47 for a diagram of this set-up).

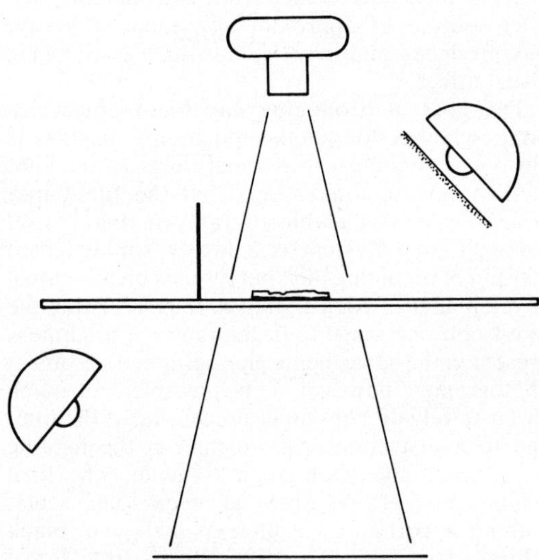

Fig. 26-47. Lighting diagram for Fig. 26-46: the object is placed on a sheet of plate glass above a white background. The main light is a single diffused lamp with reflector fill; the second lamp lights the background.

If a black background is needed, the best procedure is to place the object on or against a piece of black velvet (*not* velveteen). Any shadows cast by the object will be soaked up by the intense black of the material which, if the exposure is correct, will be so dark as to fall below the reproduction threshold of the film, permitting a dead-black background on the print (Fig. 26-48).

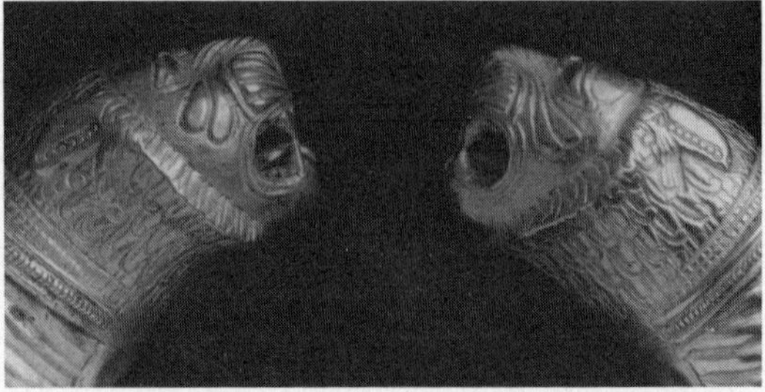

Fig. 26-48. Detail of gold arm-band, Northwest Iran. Area shown is 91 × 61mm. Diffused lighting with reflector cards, object placed directly on black velvet. Photographed for S. J. Fleming by Nicholas Hartmann/MASCA (NE 55-32-2). (Courtesy of the University Museum, University of Pennsylvania.)

Lighting

For most small-object work, one or two light sources should be sufficient: two movable-arm architect's lamps which have almost unlimited freedom of movement can be used. As mentioned above, the main concern is always to have either a single main light with subsidiary fill, or a totally diffused light, so that confusing multiple shadows are avoided. For smaller subjects, Alfred Blaker's diffuser-reflector method is best and simplest. In this system, the main light is a single lamp diffused with tracing paper, and a white card of appropriate size reflects light onto the shadowed side (Fig. 26-47). This produces a slightly directional lighting that nonetheless reveals detail all over the object (Fig. 26-46).

For larger objects, larger light sources will be needed. The expense and complexity of large studio lights can be avoided if a window is available. Indirect skylight (or sunlight diffused with a bed-sheet) can be used as the main light, with a large white or silvered reflector for fill: this is a larger diffuser-reflector system, producing the same even, slightly directional results as the smaller set-up described above (Fig. 26-49).

Several classes of objects pose special lighting problems; particularly troublesome are highly reflective objects such as metalware or enamel. Anything less than a completely diffused light will produce hot-spots or specular reflections and too high a contrast range. The solution is to either surround the object with a screen of diffusing material, such as cheesecloth or paper (the light tent), or to light each major subject plane with light reflected from a white card. In either case, direct lighting is avoided, and the effect of light from an open sky is obtained (Fig. 26-48).

For objects with extremely shallow relief, the light must graze the surface at almost 90 degrees, so as to bring up the detail. A white fill card or a second subsidiary light may also be needed (Fig. 26-50).

Transparent objects, such as glassware, present particular and often very complicated problems. In most cases, light should be transmitted from behind the object, by reflection from the background. However, a uniformly white background usually makes the object look blank and featureless, so black cards must be positioned outside the field of view, to give substance to the edges and planes of the object. Figures 26-51 and 26-52 show such a "brightfield" lighting, while Figures 26-53 and 26-54 show the inverse, "darkfield", rendering of the same subjects. Similar solutions can be used for many kinds of transparent objects.

Photomacrography

Depicting a very small subject might seem to be simply a matter of increasing the magnification to fill the frame, but other considerations soon intrude. Most importantly, as magnification increases, depth of field becomes more narrow, until at a certain point even the smallest available aperture will render only a portion of a three-dimensional object in focus. It is at this point that one must balance the desire to achieve as much magnification as possible on the film itself (keeping enlargement and grain at a minimum) with the ability to keep the entire subject in focus. Use of fine-grain film will enable one to use a lower magnification in the camera, preserving depth of field; the final magnification can then be obtained by

Fig. 26-49. Early Iron Age "beer strainer," Baq'ah Valley, Jordan. Window light with fill from silvered reflector sheet. The background is dark blue velvet, rendered as black with an orange filter. Photographed for P. McGovern by Nicholas Hartmann/MASCA (B80/I.32). (Courtesy of the University Museum, University of Pennsylvania.)

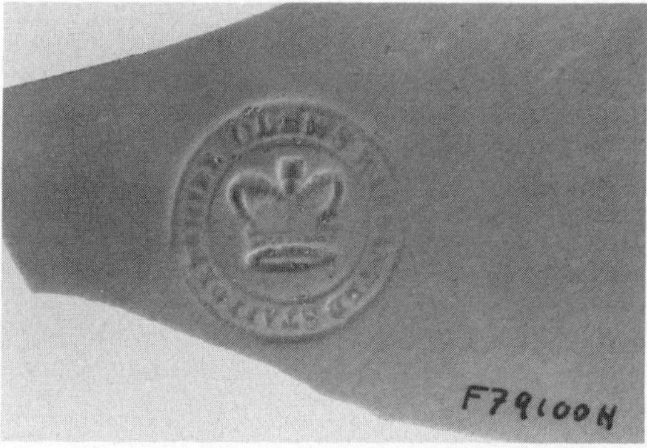

Fig. 26-50. Manufacturer's stamp, early 19th-Century plate, from Independence Hall. Diameter of stamp 23mm. Raking light from the left, with overall diffused light. Photographed for M. Parrington (MICA F79.100H) by Nicholas Hartmann. (Courtesy of the University Museum, University of Pennsylvania.)

Fig. 26-51. Obsidian blade from Santa Rita Corozal, Belize. Postclassic Maya Period. Length 46mm. Lit as in Fig. 26-52, photographed for A. and D. Chase, Department of Anthropology by Nicholas Hartmann/MASCA (P6B/6-4). (Courtesy of the University Museum, University of Pennsylvania.)

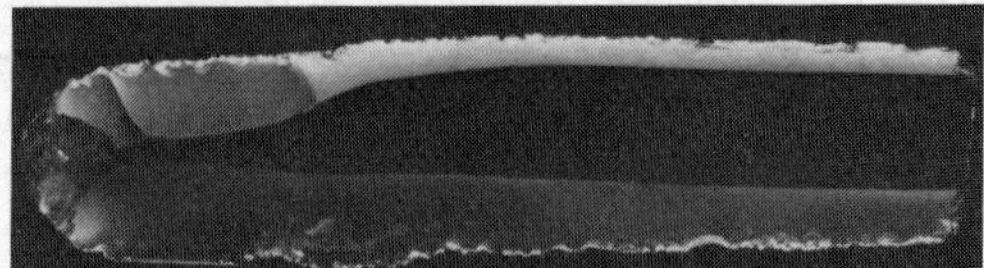

Fig. 26-53. Blade in Figure 26-51, with "darkfield" lighting.

enlarging only a portion of the negative, without an objectionable increase in grain.

Macro-type lenses, even with extension tubes, will give a magnification ratio on the print of only about 10× before grain begins to intrude, but it is possible to obtain up to 50× magnification on the print without resorting to extreme print enlargements, or to the use of a microscope. A wide-angle lens mounted on the camera in reverse, using adapters available for this purpose, will yield far more magnification than if it were mounted in the conventional way. Combined with a helical-mount extension tube or other means of making fine focus adjustments, this set-up will give magnifications comparable to those of a dissection microscope, with far better results (Fig. 26-55).

Color Object Photography

Most color pictures, for reproduction or for projection, are produced as transparencies. Since this involves a reversal rather than a negative process, several constraints apply that are not a concern in black-and-white photography. First of all, the slide cannot be enlarged or radically cropped; the composition therefore must be full-frame and exact. Secondly, color slide film has very little tolerance for other than correct exposure, so that color slides should always be bracketed at least half a stop above and below the "correct" exposure. Finally, color film will accentuate the contrast range of the subject: a well balanced lighting ratio—one that looks slightly flat in the viewfinder—will usually produce an open, easily readable image. The aim should be to strike a balance between highlights that will retain some density and shadows that are open enough to show detail.

In color pictures, there is no need for a tonal contrast between background and subject, since a difference in color will suffice to set them apart. However, a white or light-colored background produces too much glare around the object, and it is best to use dark, saturated colors. Reds or yellows tend to overpower the subject and are perceived as "advancing"; cooler colors such as blue or green or even black are preferable.

Dark-colored velvets are almost as effective as black in absorbing shadows cast by objects placed directly on them, provided the lighting is not too hard. A raised glass support with the background below can also be used, as for black-and-white,

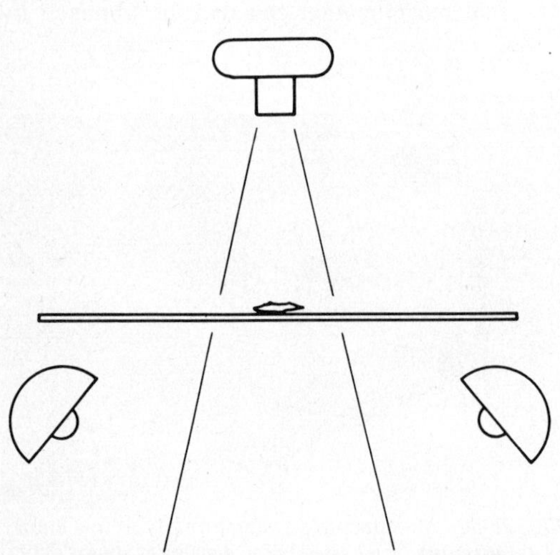

Fig. 26-52. Lighting diagram for Fig. 26-51. "Brightfield" back-lighting, with black cards placed to bring out the longitudinal facets.

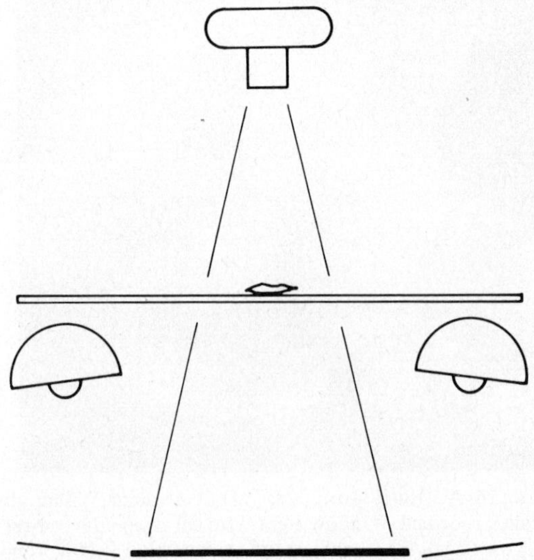

Fig. 26-54. Lighting diagram for "darkfield" back-lighting, as used in the photograph of a blade in Fig. 26-52.

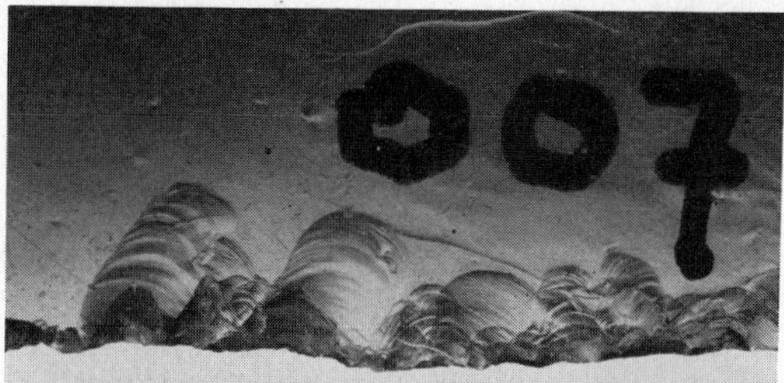

Fig. 26-55. Micro-chipping on obsidian blade from Santa Rita Corozal. Area shown is 6.7 × 10mm. Magnification was obtained with a reversed 35mm lens on a focusing extension tube, with "brightfield" lighting as in Fis. III-46a and III-46b. (P3B/5-2a). Photography by Nicholas Hartmann/MASCA. (Courtesy of the University Museum, University of Pennsylvania.)

but dark backgrounds will allow the glass to reflect anything directly above it, such as shiny surfaces of the camera or lens. A black light-shield should be fitted around the lens to block these reflections, which will otherwise be distractingly visible on the slide.

Selected Readings which expand upon the above discussions include:

—For object photography and related techniques:
  Alfred A. Blaker, *Handbook for Scientific Photography,* W. H. Freeman and Co. San Francisco (1977).
  Close-up Photography and Photomacrography, Eastman Kodak, Rochester (1977).

—For the Zone System and general black-and-white techniques:
  Ansel Adams, *The Negative, New York Graphic Society,* Boston (1948).
  Fred Picker, *Zone VI Workshop,* Amphoto New York (1974).

ANTHROPOLOGICAL REMOTE SENSING

Anthropology is one of the newest areas of study in which remote sensor data and methods are being applied. The use of remotely sensed data in this field has never been thoroughly covered in any publication directed toward the remote sensing specialist. For this reason, a comprehensive review of anthropological remote sensing and the directions it has taken in the last two decades is included here.

Use of remote sensing within anthropology in the United States can be divided into the period before and after 1972, the year the first Landsat satellite began to provide regular coverage of the earth's surface. Before 1972, aerial photography was used almost exclusively. Since 1972, anthropologists have explored the potential of earth resources satellite data extensively. Within the subfield of ethnography, satellite and especially Landsat data have sparked a revitalization of interest in remote sensing, perhaps because satellite data give a perspective on human societies which no other source has ever provided.

Current interest by ethnographers in the application of remote sensing techniques has precedents in both anthropology and related social science disciplines. Within anthropology, archaeologists have had early, continuing and intensive interest in remote sensing. Several recent publications, as well as this *Manual,* provide summaries and procedural guidelines for the archaeologist interested in remote sensing applications (Lyons and Avery 1977; Avery and Lyons 1978: Morain and Budge 1978). In contrast, ethnographers' use of remote sensing is much less developed, although a tradition of remote sensing applications within the discipline can be defined.

ANTHROPOLOGICAL REMOTE SENSING BEFORE 1972

According to Thomas Schorr (Schorr 1974b), who has traced the beginnings of aerial photography in anthropology, ethnographic use of aerial photographs began with a group of French scientists in the 1930's. If the title, "father of remote sensing in ethnography" is to be given to anyone, it should be to Marcel Griaule, who made "explicit ethnographic application" (Schorr 1974b: 166) of aerial photographs in research conducted in several former French colonies in Africa (Griaule 1937; 1946; 1949). Schorr relates the French ethnographers' interest to their closeness to French cultural geography and their resulting interest in spatial patterning of settlements, structures and land use. According to Schorr, work of geographers Brunhes (1925), Deffontaines (1948), Robequain (1929) and Gourou (1936) influenced French ethnographers. Judging by the number of entries in Schorr's bibliography, the most prolific user of aerial photography was Paul-Henry Chombart de Lauwe (1949; 1951; 1965; Chombart de Lauwe, *et al.* 1959; 1960) who studied patterns of structure and function existing between the "habitation complexes of different traditional cultures and their natural settings" (Schorr 1974b: 166). In addition to the French, Schorr mentions the work of German, British or British-trained and Soviet ethnographers, among them Dakeyne (1962), Gutkind (1952; 1956), and Porter (1956). In many of these studies ethnography was applied to practical problems, such as land use studies made in conjunction with development planning.

In constrast to the Europeans, American ethnographers were slow to integrate aerial photography into their research. Two reviews of technical tools in anthropology, one written in 1953 (Rowe 1953) and one in 1960 (Melbin 1960), mention only briefly the use of aerial photographs. In *Anthropology Today* Rowe (1953) noted the widespread use of aerial photographs by American archaeologists and French ethnographers, but did not describe any work by American ethnographers.

By the 1960's, things were beginning to change. Elmer Harp, an archaeologist familiar with remote sensing, commented on applications in ethnography as well as archaeology in a paper presented to one of several major annual remote sensing conferences in 1966. In that paper, he predicted that "new applications in ethnography would come as ethnographers developed ways to use remote sensing, with its regional, cartographic perspective, to study cultural systems and the ecological integration of those systems" (Harp 1966).

Most ethnographic use of aerial photography had been for mapping in conjunction with rural ecological studies. In a review of ethnographic studies of tropical ecology, Brookfield (1968) described the use of aerial photography to facilitate mapping of settlement and land-use distribution as well as environmental features. Many ethnographers probably used aerial photography in this way; but while publishing photographs and maps of their research communities, they made little reference to photointerpretation in their articles and monographs.

Using aerial photography for cartographic purposes provides baseline data necessary for ecological study much more rapidly and accurately than traditional ground surveys. However, the pioneering group of ethnographers of the 1960's went much farther than this. They began to delve into the potential of an aerial perspective. In doing this they found unanticipated benefits, e.g., new approaches to data gathering with native informants, new ways to take a census and to record population, land use and land tenure over large areas, and new ways to study social change and social organization. Examples of the benefits, accrued through the use of remote sensing to gain a regional perspective are discussed in the following summaries of some of these early studies.

Papago Indians, Arizona

Robert Hackenberg (1967) in a paper titled, "The Parameters of an Ethnic Group: A Method for Studying the Total Tribe", described using aerial photography to address an important methodological problem in anthropology. He wished to study the Papago Indians of Arizona, not from the viewpoint of a single community, but at the level of the entire society. This society was to be defined as a unit composed of individuals inhabiting a particular geographic space and interacting in a space-time continuum. Ideally, study of such a society would be based upon measurements made of every member, with patterns and variation described through frequency distributions linking individuals, attributes and events. Phenomena to be measured included aspects of demography, resource use and economics.

Aerial photographs were used in the construction of a data base which would permit quantitative description of demographic and economic realities of a defined unit. In this field research, Hackenberg gathered comprehensive genealogical and socioeconomic data for the Papago society, operationally defined as containing about 17,000 people in 1960 (Hackenberg 1967). In order to collect these data, maps of living areas and associated landholdings were made for the entire population by working with informants to interpret aerial photographs. Hackenberg also devised computer programs which would keep track of this population. This research would not have been possible, except with much time and expense, without the photographs as a tool to help locate, identify and measure the units of interest.

Water Supply Development and Agrarian Reform in Columbia

At about the same time, Thomas Schorr (1965; 1974a) was conducting research in the Cauca Valley of Colombia in order to describe social and environmental change accompanying water supply development and govenment-planned agrarian reform. A 25 km × 5 km area was studied using vertical photography from 1947, 1957 and 1959 at scales from 1:20,000 to 1:60,000, supplemented by oblique handheld aerial photography.

In what Schorr called aerial ethnography, he was able to: survey the area for ecological relations found to be contributing to the visible settlement patterns; classify settlement patterns; select communities representative of each settlement pattern for field investigation; and map settlements planimetrically and topographically. Interpretation of settlement patterns resulted in the division of the settlements into two zones: a zone of extensive agriculture along the spurs of ranges forming the valley walls; and a ribbon-shaped zone along the Cauca River where intensive agriculture was practiced on fertile, floodplain soils. Over time, the location of the river settlements had shifted, while the upland settlements were more permanent.

House counts and population estimates made as part of this study for the approximately 30 settlements in the area were a significant improvement over the aggregated figures available from the national census. Time series analysis of the several sets of photographs showed patterns of social and economic change. Proportions of land used in different kinds of agriculture and other productive uses were measured for differences over time. Physical improvements resulting from government programs were located. Indicators of change, such as powerlines or roofing material, were developed which could be used to analyze the modernization process in other areas of Colombia where photographic coverage was available.

In this study, photointerpretation enabled

Schorr to relate data on settlement, agriculture and environmental variables, such as soils, over a wide geographic range. Because photographs taken over a long period were available, changes accompanying modernization were charted. He pointed out that the photographs could not be used without detailed field work, but he used them to select the field work sites.

Schorr's comments on the relationship between the aerial photographs and social phenomena are instructive:

> The physical modifications and materials reconstructions that a population introduces in the landscape are tangible representations of those aggregate, human behavioral procedures that make up the epiphenomenon of organized social systems (Schorr 1974a)

These visible traces on the photographs create a "physical signature" representing what Schorr calls "total ecology". Because of the reduced scale and compressed time of the photographs and their ability to display interrelated patterns of information, they allow inference about processes of social change.

Ecology in the Philippines

Another important aerial ethnographic study was performed in the 1960's in Ifugao Province, Philippines, under the direction of Harold Conklin (1967; 1974). The region, an "unevenly chronicled but highly imageable arena of extraordinary hydraulic and agronomic performance" (1967), is characterized by a complex terraced agriculture with fields of irregular shape and size in an area of high relief. Conklin chose to concentrate on an area of about 100 square kilometers in which a population of 10,000 lived in about 300 dispersed settlements.

In order to describe land use and economy in a meaningful way, especially the details of agricultural production, Conklin conducted intensive field work and analyzed his data using componential analysis. As a result, Conklin had culturally valid terms and meanings for aspects of vegetation, topography and agriculture along with local definitions of changes in land tenure and status. By combining photointerpretation with an ethnoscientific approach, he described the structure of the Ifugao agricultural system in all its complexity. Using environmental and ethnographic studies, along with the techniques of photogrammetry and cartography, he published a series of maps at 1:5000 scale (Conklin 1972) showing the Ifugao landscape as it has been modified by its inhabitants.

Direction from Harvard

By 1969, a tradition of ethnographic remote sensing had existed for at least a decade. Another example of ethnographic remote sensing was the Harvard Chiapas Project directed by Evon Vogt (Vogt 1974b). In this project, a team of anthropologists worked with professional photointerpreters to explore the potential of aerial photography in aiding previously planned research, as well as for providing an avenue of new methodological approaches.

At Harvard University in May 1969, Vogt brought together a group of anthropologists experienced in the use of aerial photographs. This group included the members of the Harvard Chiapas Project, and others including Robert Hackenberg, Thomas Schorr and Priscilla and Conrad Reining. A followup symposium was held at the American Anthropological Association annual meeting that year. Many of the papers and discussions from the two meetings were published in 1974 in the volume *Aerial Photography in Anthropological Field Research* (Vogt 1974a), an excellent summary of developments during the 1960's.

One component of the Harvard Chiapas Project which demonstrates a range of benefits from using aerial photography is George Collier's (1974a) work. Results of this project included the development of techniques for photointerpretation with informants in the field and for mapping and censusing, as well as repeated demonstration of the impact of the aerial perspective in raising new methodological and theoretical questions.

In order to interpret Tzotzil Indian land use, Collier spent a preliminary period in the field learning the local microgeography. He described traversing the terrain along with informants, airphotos in hand, in order to inspect sites representing different land use. Along with this, systematic interviewing covered local concepts and labels for detailed aspects of the agrarian regime. Working with informants, he placed cultural labels on a photomosaic of the study area. Collier stated that only at this point was his ability to communicate with informants about the study area adequate for photointerpretation.

Following this preliminary phase, land use and tenure were mapped. "A year-by-year synopsis of land use and tenure of each of several hundred property parcels" (Collier 1974a) was accomplished in a short time, using only the airphotos and interviews in a laboratory setting. This survey was accurate, efficient and comprehensive and probably could not have been accomplished without Collier's using the photography in field and laboratory sessions with local residents. Through the use of this survey, a detailed house-by-house census covering genealogical, economic and social information was also completed rapidly. In several days, for a village of 670 population, Collier was able to:

> . . . Compile an exhaustive house-by-house census, to catalog ownership of several hundred garden plots, and to identify most of the community's sacred waterholes, mountains, caves and cross shrines (Collier 1974a)

Richard Price (1974) in another Chiapas Project study, also found that sharing the photointerpretation task with his informants facilitated rapid survey of land use.

Collier suggested that his data could be used to study social organization, for example the relationship between settlement location and either ecological variables, such as water location, or sociological ones, such as descent group solidarity. Intercommunity studies of such questions could be carried out using aerial photography as a comparable data base.

Methodological Comments

Most ethnographic photointerpretation before 1972 took place in the more-or-less traditional setting for the ethnographer i.e. rural, especially agricultural, communities. Aerial photographs were used successfully by scholars interested in questions relating to ecology or man-environment relationships, and correlate questions of settlement patterns, land tenure and the economics of land use.

Evon Vogt (1974b) in his summary of the Harvard Chiapas Project, refers to anticipated and unanticipated uses of aerial photography. His summary of the Chiapas project has broader relevance, since Hackenberg in the Southwest U.S. (1967), Schorr in Colombia (1974a), and Conklin in the Philippines (1967) pursued many of the same research interests and found many of the same benefits from aerial photography.

According to Vogt (1974b), some anticipated uses of aerial photography which bore fruit when applied in specific situations were: analysis of settlement patterns, census taking, mapping of land use and land ownership, sacred geography, and indices of cultural changes. Settlement-pattern analysis includes mapping and description as well as analysis of ecological and socioeconomic factors which are found to influence settlement location and its maintenance or change. Developments in census taking included refinement of rapid data-gathering techniques, involving preliminary work with local residents to establish a shared understanding for photointerpretation. Land use and land ownership maps were prepared with a speed and accuracy never before possible; again, photointerpretation had to be coordinated with ground checking and working with informants. The study of what Vogt called sacred geography showed for the first time the precise location of sacred places, their relationship to other geographic features and the relationship of social units to ceremonial areas. The development of indices of cultural change, through the interpretation of a succession of aerial photos taken over a period of time, went beyond the preliminary documentation of change (such as growth of irrigation or new roads) to the selection of specific features in the photo (such as roof type) which were useful as regionwide indicators of the process of social change.

As research projects continued, many anthropologists found unanticipated benefits of using aerial photography, including more precise ethnographic data, the savings of research time, and new ethnographic uses (Vogt 1974b). Advances in census data collection were largely anticipated. Such advances apply especially in cases where government statistics are nonexistent or too aggregated for ethnographic use. Also, a normal house-to-house census, unaccompanied by cross-checks on aerial photographs, will undoubtedly miss a proportion of the population. Precise data on location of any man-made or environmental features were relatively easy to collect; the photography allowed accurate, quantitative measures to be gathered more quickly than by previous techniques. Among the new ethnographic uses of aerial photography were studies of social organization such as location of descent groups or patterns of marriage or communication, and analysis over time of the impact of new technological developments (Vogt 1974b).

Perhaps the most interesting research area mentioned by Vogt was large-area ecological analysis. Aerial photography facilitates the analysis of a number of different kinds of phenomena over a large region, especially the study of systematic relationships between one kind of variable and another over space. A simple example of this would be the relationship of size or compactness of settlement to percent of land in a certain crop.

In summation, prior to the availability of data from the Landsat satellite and the accompanying excitement among a new group of ethnographers, there was a small but strong tradition of the use of aerial photography in ethnographic field research. this included a trend from using the photographs for mapping to letting the photographs become an integral part of research. At first, researchers used them to help answer questions previously posed, while later, photointerpretation spurred methodological and conceptual developments.

THE IMPACT OF LANDSAT

During the 1960's ethnographers found that the perspective offered by aerial photographs helped them to address a thorny theoretical and methodological problem: how to define and conduct research at the level of the entire society, tribe or cultural group. Small scale aerial photographs were visible proof that the community is only a small part of a wider network of communities occupying a specific geographic space. However, even with the depth of understanding gained by intensive single-community field work, ethnographics often could not definitively state how the specific community was related to other communities and/or to the region as a whole.

It was at a time when there was a growing interest among ethnographers in studying broader social, economic and ecological relationships that Landsat data first became available. Landsat's potential importance as a medium for studying entire societies was recognized early (Reining 1973). Three basic characteristics of Landsat data relate to its ethnographic potential: synoptic view, repetitive coverage and uniformity (National Academy of Sciences 1977). "Synoptic view" refers to the uniform view of a single large area (185 km by 185 km) on a Landsat scene. Each Landsat vehicle also repeatedly covers the same parts of the earth's surface every 18 days. As described in Chapter 12 of this *Manual,* Landsat scans each scene at approximately the same sun angle and at a uniformly vertical perspective, with the same spectral and spatial resolution for every image. To this list can be added the advantage of availability: imagery is relatively inexpensive, available to anyone, and coverage exists for most of the earth's land surface. (Any anthropologist who has tried to obtain aerial photography over a remote area will appreciate this advantage.) And, finally, the digital, statistical nature of the data holds potential for quantitative analysis and image-to-image comparison unparalleled in any other remote sensor data form (Hamlin 1977).

Priscilla Reining (1974a) has stated that the significance of Landsat data to ethnography lies in the ability to bridge the gap between the intensive field study and the universe of society which it represents. While field studies are necessary to understand social process, they have suffered most from the inability to extrapolate from them. When combined with aerial photography and traditional field work in selected sample sites, Landsat data make possible the quantitative study of such regional phenomena as entire agrarian systems and their changes, seasonally, annually and over longer time periods (Conant, Reining and Lowes 1975). The Landsat spacecraft, and the acquisition and processing of its data, are discussed further in Chapter 12.

An Historic Overview

The first Landsat satellite was launched in 1972. In that year a paper was presented by archaeologists, who were familiar with the use of aerial photographs, summarizing the potential of satellite data for both archaeology and ethnography (Lyons, Inglis and Hitchcock 1972). Since then, a revitalization of interest in the use of remote sensing has occurred among ethnographers, who are excited about the advantages of Landsat data. Since 1972, Landsat has been applied in a variety of ethnographic research. In this research there has been a progression from simple to more complex applications, from descriptive to more interpretive use of the images, from qualitative to quantitative analysis, and from papers exploring Landsat's potential to reports summarizing specific results.

Lyons, Inglis and Hitchcock (1972) forecasted that space imagery would provide two kinds of data for the ethnographers: environmental and cultural. Environmental information on physiography, drainage, erosion and so on, would provide a background for ecological analysis. Cultural information would include land use patterns, population and settlement patterns. The authors also referred to the use of imagery for mapping, for formulating hypotheses, for research design and for sampling. According to the authors, space-derived data held great potential for temporal and regional studies because of their quantitative, synoptic and repetitive nature.

Reining pioneered the combination of Landsat, aerial photography and ethnographic field work (Reining 1973; in press). Fieldwork was directed toward information on population and land use, for input into carrying capacity formulae which could synthesize relations between population and resources, and measure stress and potential in agrarian systems (Reining 1973; 1974a; 1974b; in press). For demographic estimation over wide areas, multipliers based on population density, village area and village type were established for sample villages (Reining 1974a; 1974b). Digital classification procedures with Landsat computer-compatible tapes were explored for extrapolation to the universe or entire society (Reining and Egbert 1975). Reining has had continuing interest in the development of area sampling procedures uniquely suited to the ethnographer, and in the use of Landsat to construct extensive maps of social groups.

In related work in the Sahel, Rosalie Fanale (1974a; 1974b) used repetitive Landsat imagery to analyze changing resource relationships for the Dogon of Mali during the Sahelian drought. The Dogon are farmers who, on much of the land they occupy, use a fallowing system for grain crops. Landsat was the primary data source for the definition of areas of similar agricultural practice, soils, vegetation and physiography, within the zone of Dogon occupation. An area was selected for further study in which villages and land use classes, such as crop-land fallow and wasteland (fallow land which did not regenerate over time) were mapped at 1:250,000 scale. Existing ratios of cropland to total arable land were compared with known ratios needed to maintain the agrarian system, given the production techniques of the time. This study was performed as part of a master's thesis (Fanale 1974b) and was limited by the inability to utilize digital Landsat data and to perform intensive field work. Results, however, suggested that with the inclusion of field work and digital analysis, quantitative description of the parameters of the Dogon or other similar agrarian systems would be a straightforward procedure.

More recently, investigators have been developing ways to identify and measure "small area events" (smaller than easily discernible on Landsat) and also have been studying region-wide human ecological phenomena (Francis Conant 1976; Conant and Cary 1977). Conant has also addressed how field data and Landsat analyses are best combined and has conducted research in which periods of field work are bracketed with analytical sessions so that results of each can influence the other.

Riddell (1978) is incorporating historic aerial photographs along with Landsata data, which he has digitally enhanced, in a retrospective study of changing land use patterns in Liberia. This study provides a time dimension which is rare, but important, in agrarian ecology. Riddell has measured the diminishing climax forest and changing percentages of land in fallow and crop; such measures, when combined with economic data, can be used to measure changing adaptive strategies of human populations.

Recently, ethnohistorians have become interested in the aerial perspective. At the 1978 meeting of the American Society of Ethnohistory, a symposium was held to discuss the potential of remote sensing in ethnohistoric research (Kruckman 1978). Most of the discussion at the symposium concerned the use of aerial photographs to locate historical sites (an application very similar to one in archaeology) and to reconstruct their environmental settings.

Landsat data constitute an important component of ongoing research on desertification, ecology and pastoral economy in many parts of the world. In the southwestern United States, Fanale (1978) has conducted research on the desertification process and how differences in environmental change relate to differences in land use and economy. In this study she applied area frame sampling procedures (Houseman 1975) and used remote sensing to select ground sites which are representative of regional variation.

Two conferences have been held in which ethnographers have met with members of other disciplines, including remote sensing specialists, to discuss the relevance and use of Landsat data within ethnography, especially for studies of human ecology in tropical and subtropical Africa, Asia and the Middle East (Conant, Reining and Lowes 1975; Conant 1978). A 1975 workshop organized by Conant and stimulated by Reining and Fanale's work, was an introduction to the Landsat system for a group of ethnographers, some of whom had had experience with aerial photography. Participants discussed the limitation of Landsat imagery that inhibit ethnographers from using it, especially its resolution, which is too coarse to capture the small area activities of interest to many ethnographers. Other topics discussed included: the need for field work to be part of a multistage design employing Landsat and other aerospace data; the need to study entire ecological or cultural areas; and the need to develop special signatures for shifting cultivation, settlement types and other imprints that are discernible on Landsat imagery.

The second conference in 1977, again organized by Conant, was titled, "The Use of Landsat Data in Studies of Human Ecology" and was held at the Wenner-Gren Foundation for Anthropological Research, the sponsoring organization. Conant has summarized the conference discussions in *Current Anthropology* (Conant 1978). Landsat's potential was described in terms of its capacity to "make repeated, standardized observations over a relatively large area, at times chosen to coincide with known or suspected seasonal and episeasonal changes in the environment" (Conant 1978) and, thus, to be a mechanism for spatial and temporal study of environment, subsistence activities and settlements. Discussions and working papers centered on topics such as: specific ecological questions needing research; problems with small-area phenomena; the need to work with scholars in other disciplines; ethical question; and the need to understand and apply machine processing.

Landsat and Human Agrarian Ecology

Work by Priscilla Reining (1973; 1974a; 1974b; in press) demonstrates applications of Landsat imagery to the study of agrarian systems. Reining used the concept of "carrying capacity" in a manner closely paralleling the use of the term by range managers (see chapter 35). "Carrying capacity" refers to the relationship between population and resources in a defined geographic area, a relationship which Reining expressed by a formula that describes the area required to support a given number of persons.

Critical to any such estimation are measures of area of cultivable land, population size and population density. Reining states:

> . . . The amount of land in a given soils/vegetation type, the amount of land within the soils type suitable to cultivation, the amount of land actually in cultivation (and the amount in reserve), the total area of a single culture or site-specific area, the boundaries between sites and, finally, the changes in such sites over time are the data required for carrying capacity estimates (Reining 1974a).

Some data are derived from field work in sample communities: cropping cycle, fallowing cycle, crops, population and economic statistics. Published statistical source and demographic census material may also provide data. But identification and mapping of villages and their areas, soil types, vegetation types and stages in the cropping and fallowing cycle can be identified on remote sensor data, after an initial field check to establish a correspondence between image characteristics and ground conditions. Because of its coverage, availability and machine processing potential, Landsat is far preferable to aerial photography when such estimates are to be made for wide areas.

Reining made carrying capacity estimates for select Hausa, Sonrai and Mossi villages in Niger and Upper Volta. Her procedures are detailed in the original report of her 1973 field research (Reining 1973), which is now being published by Ohio University Press (Reining in press). Various of Reining's papers also refer to this work and later developments from it (Reining 1974a; 1974b; Reining and Egbert 1975).

For example, for the Sonrai village, population data came from field work. Landsat and photointerpretation were used to estimate village area and to map soil types, fields and nonarable land for the village's domain. The area, and population of the village, the arable land needed to support the village population (given the parameters of the agrarian system) and the actual arable land were compared. For this village, the population of just under 500 occupied a land area of about 4 hectares with a density of approximately 150 persons per hectare, and cropped 300 hectares of land in 1973. Carrying capacity was estimated at a potential of 1000; the population, at less than half that, was well under carrying capacity.

In order to extend her estimates, Reining isolated the area of Sonrai occupation from Landsat imagery on the basis of characteristic settlement pattern and cropping (the Sonrai farm particular soil-types) and identified Sonrai villages. However, she realized that in order to estimate population and carrying capacity for the Sonrai as a whole, additional ground studies of a wider sample of communities were necessary. Since she was unable to perform this field work, she could only outline further procedures with an adequate sample. Population-density figures along with identification and measurement of all villages would be used to arrive at a society-wide population figure, while measurements of arable land and land in crop would be used to establish actual and potential carrying capacity.

Small Area Events

A problem with using Landsat imagery in many rural areas of the world is that most of the phenomena of interest to the ethnographer—settlements, field, roads, and so on—are too small to be discerned on such imagery. Reining (1974a) pointed out that villages as small as 250 meters in diameter and fields of 20 hectares were easy to identify with normal imagery; with enhanced imagery or digital data, actual resolution may be expanded to the 57 by 80 m. pixel size in which Landsat data are recorded. Even so, identification of features of relevance to the ethnographer may remain difficult.

Conant (Conant 1976; Conant and Cary 1977) has been studying the ecology and resource productivity of an area of Kenya in which small mountain settlements are dispersed among lands in slash and burn or swidden agriculture. Fields are small and irregular, with boundaries merging with natural features of the terrain (Conant and Cary 1977); a more difficult area to classify on Landsat could not be identified. Preliminary efforts to computer-classify the swidden fields met with only limited success; what was being classified was the mixture of actively cultivated land and fields in different stages of fallow (Conant and Cary 1977). In further work, Conant has been devising a way to use Landsat data to study this area despite the problems with classifying small-areas events.

In Landsat imagery, a pixel is a synopsis of all reflectance-producing events within a 57 m by 80 m area. Normal classification procedures cannot, of course, distinguish features smaller than this. However, a complex of ground events may be identified which corresponds to the smallest area measured on Landsat. Conversely, certain signatures on the imagery may indicate specific groupings of ground events. Conant calls the measurable signatures "leading indicators", or physical phenomena detectable from the satellite which indicate activities of interest on the ground (Conant 1976). Field work then must be performed to determine how "small area events" and physical phenomena combine to result in spectral characteristics on Landsat.

Machine Processing and the Ethnographer

As Hamlin (1977) has pointed out, Landsat data in visual form are nowhere near their full resolving power. Landsat is a statistical system, in which spectral reflectance for each of four bands is recorded, pixel by pixel, on Landsat computer-compatible tapes: only in its statistical, digital form are Landsat's spectral and spatial potential maximized.

There are other reasons for employing Landsat in digital form. Procedures for extrapolation from sample sites can be aided by machine classification of Landsat data. Problems in identifying small area events can be lessened when Landsat's maximum resolving power is used. Almost any quantitative measure that the ethnographer may want to make is relatively easy with machine classification and very difficult without it. In addition, time series analyses employing successive dates of imagery can be performed only with precise registration procedures, possible only with the digital data.

Some ethnographers have begun to use machine classification. Conant and Cary (1977) described an effort to classify swiddening. Reining and Egbert (1975) described a successful experiment in identifying Sonrai villages for a 30 by 30 km test site. Using the General Electric Image-100 processing system, they were able to correctly identify villages ranging in size from one to 100 pixels; classification was verified by comparison with low altitude aerial photography. Hamlin (1977) has written a review of automatic processing with Landsat for archaeologists and ethnographers, in which he outlines basic machine processing procedures (supervised and unsupervised classification, geometrical correction, coordinate transformation, image enhancement, and data base integration) and presents simple computer procedures and a series of programs as well. For more information the reader should refer to Chapter 24.

Although digital processing implies familiarity with aspects of computer programming, statistical anslysis, data management and image processing, and although machine processing is usually costly, the benefits of using Landsat data in digital form may outweigh the monetary costs or costs incurred in acquiring additional knowledge.

Applications of Remote Sensing to Navajo Ethnology

An interesting and recent example of the application of remote sensing data and methods to the ethnographic study of a present-day population is provided by a short summary of the work of Rosalie Fanale with the Navajo of the San Juan Basin, which contains the easternmost part of the Navajo Reservation and lies in northwestern New Mexico. In this study, remote sensing was employed to tie several aspects of the Navajo system—including their environment, the ways in which they perceive their surroundings, and their settlement pattern and mobility through time—into a coherent whole. A sampling design was engineered with the aid of remote sensor data, explicitly employing a frame for data collection in the field. Medium and small-scale aerial photographs were used to locate sample areas in the field. In addition, remote sensor data, particularly Landsat color composite images, were used as an aid to interviewing informants. Fanale's work stands as one of the best examples of the total integration of remote sensor data into an ethnographic research design.

The area of Fanale's study, in semi-arid northwestern New Mexico, has been occupied by the Navajo for several centuries (Brugge, 1980). The Navajo, who are pastoralists, have been blamed by many for degrading the rangeland in this area over the last 100 years. It is claimed that range degradation has resulted from raising animals in excess of the land's carrying capacity, and from improper land use techniques (Phelps-Stokes Fund 1939, Young 1961). U.S. government interest in Navajo land problems began in the late 19th century but it was not until the 1930's that the government became heavily committed to the Navajo livestock industry and range rehabilitation. Since the 1930's there have been a number of programs, most notably the Stock Reduction Program of the 1930's and 1940's; but the range remains overgrazed and degradation of land resources continues (Aberle, 1966; Bureau of Indian Affairs, 1971). The implication is that the Navajo have had little concern for their environment, little understanding of how their herds have affected the vegetation and little knowledge of how to manage the range.

Fanale's research was conducted, in part, to ascertain whether a way might be found to better explain Navajo land management and its history, by examining Navajo pastoralism and the changes in it within a regional framework. Preliminary to designing research, he compared the Navajo with other pastoralists groups, some of whom have developed complex systems of land management. Much of the existing knowledge of Navajo land use as published in studies of Navajo history, economy and social organization, was reviewed.

From this preliminary work, a group of postulates emerged: (1) that Navajo pastoralists, far from being unknowing or uncaring with regard to their land resources, may have had an indigenous land management system; (2) that such a system is most likely an adaptation to fluctuating, variable and unpredictable grazing resources; and (3) that a critical part of this land management system would have been herding mobility.

Fanale speculated that it might be possible, with the aid of remote sensing, to reconstruct this system of land management, no longer in use, whereby grazing mobility over a large and diverse area helped to maintain pasture quality and productivity. Such a system could be described both from the point of view of the knowledge and practices of individuals and families as well as in strict ecological terms. The history of Navajo pastoralism might then be told, even though land management has been limited as the result of exogenous factors, particularly govenment actions, over the last 100 years.

Consistent with the scope of the research questions being asked, the research design was regional in scale. Local level phenomena were related to region-wide processes, including historical processes; and changes at the local level were related to exogenous factors as a process over time. Both dimensions of time and space can be spanned effectively using remote sensing methods and techniques.

Preliminary stratification of the region to be studied proceeded, using Landsat 1:250,000 color composite data. It was determined that a scene imaged by Landsat-1 on August 16, 1973 would be best suited for the interpretation of informed ecological strata because that year had been exceptionally wet, and vegetative differentiation was more pronounced than on any other available scene of the area. Interpretation was visual, and followed the methods outlined elsewhere in this chapter under the subject of archaeological sampling design. Zones which appeared to be similar were segregated from those which appeared different, and zone boundaries were drawn on a 1:250,000 scale base map. Previously-collected environmental data (soils, vegetation, and landform information) were used in the lab to adjust and refine stratum boundaries, and intensive field-checking of these delineations followed.

Prior to ethnographic fieldwork, each stratum was divided into smaller sample units, and those units within each stratum to be sampled were chosen on the basis of several criteria.

At the local level, field work included interviewing and the collection of environmental data in selected sample units throughout the region. The procedures for dividing the region into smaller units, marking them on maps, photographing them from the air and selecting the final sample units followed the area sampling procedures developed by the U.S. Department of Agriculture for use with remote sensor data (Huddleston 1976, Houseman 1975). The sample units were small parcels of land of about 4 square miles. This size was based on a review of information from prior studies and was calculated to coincide with an area which one or two Navajo residential groups would be likely to hold as permanent range. Such a residence group is also the social unit through which a herd is usually managed. The final sample included 31 individual units or parcels of land.

A file of materials was prepared for each sample site to aid in location and orientation in the field, and in the recording of field data. Each file contained a topographic map, interpreted aerial photographs showing the sample unit and the location of settlement features within it, and other material, including overlays of land status and environmental classifications.

Although each site was bounded on the maps by roads or physiographic features, such as washes, it was difficult to decipher the precise boundaries of the unit on the ground. Without the interpreted aerial photography this task would have been impossible. Each sample unit was visited and, when possible, older residents within the units were interviewed.

The interviews were carefully planned but informally conducted. The open-ended approach resulted in rich social and cultural description despite the short time spent in each sample unit. Interviews concerned a resident's past land use practices and how these have changed. Other topics often were discussed, including priniciples of proper care of animals, definitions of types of grazing land and concepts of resource preservation. Interviews most often were taped and conducted through an interpreter. One especially useful technique was to ask informants to mark, on a 1:250,000 scale enlargement of a Landsat image, the places where their families used to go with their animals when they were young. Questions to informants were constructed to permit each person to tell an account giving his or her own emphasis and structure to local history.

CONCLUSIONS

The information presented in this chapter supports the conclusion that remote sensing methods and techniques are enjoying ever-widening adoption by archaeologists, anthropologists, and cultural resource managers. Since the beginnings of scientific concern with human behavior and evidence thereof—both in the past and the present—some more traditional forms of "remote sensing" have become important in these fields. Photography has always been recognized as indispensible for recording the archaeological record of the past, and has in fact been used by many ethnologists to record present behavior; surveying and other spatial measuring techniques have also always been part of the archaeologist's repertoire. Aerial photographs have long been employed for the detection and examination of sites, as well, and photogrammetry has been used to record and measure structures.

In the last decade, however, many new uses have been made of remote sensor data and techniques in the anthropological fields. Some of these have been based on "old" data sources such as aerial photography, while others have taken advantage of modern satellite platforms and electronic data-collection devices. These new archaeological and anthropological uses of remote sensing are primarily the result of new ways in which archaeologists and anthropologists are looking at their data. One of the best examples of this has been occasioned by the need to survey larger and larger areas, many in the western United States and due to energy resource exploitation. While it might be possible to totally survey small areas, it is obvious that sound, empirically-based sampling designs are necessary in very large areas, and remote sensing data can serve as the empirical basis for arriving at such designs. In the future, the need for assessment of cultural resources can only increase, and more refined methods will doubtless be needed; remote sensing will surely play a part in any such methods.

Possibly the most exciting frontiers of cultural-resources remote sensing are those areas in which the techniques and methods made available by remote sensing data join the theory and methods of the social sciences. Such areas include the use of remote sensing in archaeological and anthropological surveying and sampling, areas which have been emphasized in this chapter. As an aid to these archaeological and anthropological necessities, remote sensor data lend a unique perspective, never before available to the cultural scientist; in doing so they often point up some of the unstated fallacies under which we operate from time to time. Remote sensing is pointing the way by which the archaeologist and anthropologist can begin to employ true, controlled scientific sampling in their data collection, and this in turn may help revolutionize these fields. In addition, the wider overview given by remote sensor data, particularly those gathered from aerial or space platforms, along with the data needs of cultural resources management, is causing what appears to be a slow shift from the artifact or even physical "site" as a focus of analysis to a consideration of the environment in which past (and present) people exist as the proper object of study.

REFERENCES

Aberle, D., 1966, The Peyote Religion among the Navajo; Wenner-Gren Foundation for Anthropological Research, New York.

Agache, R., 1967, Recherche des moments favorable a la mise en evidence des vestiges archeologiques arases par l'agriculture dans le nord de la France in Secretariat de la Commission; V–9 to V–18, Vol. 14, Paris.

American Society of Photogrammetry, 1960, Manual of Photographic Interpretation, R. N. Colwell, ed; Falls Church, VA.

American Society of Photogrammetry, 1975, Manual of Remote Sensing; R. G. Reeves, ed.; Falls Church VA.

American Society of Photogrammetry, 1980, What photogrammetric engineering and remote sensing is; Photogrammetric Engineering and Remote Sensing, Vol. XLVI, No. 10, pp. 1249.

Avery, T. E., and T. R. Lyons, 1978, Remote sensing: practical exercises on remote sensing in archeology; Supplement No. 1 to Remote Sensing: A Handbook for Archeologists and Cultural Resource Managers., Washington, D.C., National Park Service. (U.S. Government Printing Office Stock Number, 024-000-00697-4).

Avery, T. E., 1977, Prehistoric and historic archaeology; Interpretation of Aerial Photographs, T. E. Avery, ed., Burgess Publishing Company, Third Edition pp. 179–204.

Avias, J., 1967, Apport de la photo-interpretation a la reconstitution des peuples disparus ou indigenes anciens en Nouvelle-Caledonie in Secretariat de la Commission, Paris; pp. V–19 to V–24 Vol. 16.

Bagrov, N. A., 1963, On the fluctuations of levels of closed lakes; Meteorology and Hydrology 6, pp. 41–46 (Translation in Soviet Hydrology 3, 1963, pp. 289–294).

Bates, Daniel G., and Susan H. Lees, 1977, The role of exchange in productive specialization; American Anthropologist 79(4), pp. 824–841.

Bennett, John, 1978, Social processes affecting desertification in developed economies; Paper presented at the Symposium, "Desertification: issues in measuring and monitoring the process with indicators" at the 1978 Annual Meeting of the American Association for the Advancement of Science.

Berry, Brian J. L. and A. M. Baker, 1968, Geographic sampling; Spatial analysis: a reader in statistical geography, Brian J. L. Berry and Duane F. Marble, eds.; Prentice Hall Englewood Cliffs, NJ.

Binford, Lewis R., 1964, A consideration of archaeological research design; American Antiquity 297 pp. 425–441.

Bond, G., 1963, Pleistocene environments in southern Africa; African Ecology and Human Evolution, F. C. Howell and F. Bourliere, (eds.); University of Chicago Press, Chicago, pp. 308–334.

Boocock, C. and J. J. Van Straten, 1962, Notes on the geology and hydrology of the Central Kalahari Region, Bechuanaland Protectorate; Transactions of the Geological Society of South Africa 65, pp. 130–132.

Borchers, Perry E., 1977, Photogrammetric recording of cultural resources; National Park Service Office of Archeology and Historic Preservation Publication No. 186, Washington, DC, Department of Interior.

Bowen, H. C., 1962, Air photographs and the study of ancient fields in England; Internat. Arch. of Photogramm., Vol. 14, pp. 411–418.

Boyer, W. Kent, 1980, Bipod Photogrammetry *in* Lyons, T. R. and F. J. Mathien, eds., Cultual resources remote sensing, Washington, DC: National Park Service, pp. 327–345.

Broadbent, Sylvia M., 1968, A prehistoric field system in the Chibcha Territory, Colombia; Nawa Pacha, 6, pp. 135–147.

Brookfield, H. C., 1968, New Directions in the study of agricultural systems in tropical areas; In Evolution and Environment, E. T. Drake, ed.; New Haven, Yale, pp. 41–419.

Brown, G. N. and J. I. Ebert, 1978, Ecological mapping for purposes of sample stratification in large-scale cultural resources assessment: The National Petroleum Reserve in Alaska; In Lyons, Thomas R. and James I. Ebert, eds.; Remote Sensing and Non-Destructive Archeology, pp. 53–63. Washington, DC, National Park Service, pp. 53–63.

Brugge, David M., 1964, Navajo land usage: a study in progressive diversification, In Clark S. Knowlton, ed., Indian and Spanish American adjustments to arid and semiarid environments. Contrib. No. 7 of the Committee on Desert and Arid Zone Research, Texas Technological College, Lubbock.

Brugge, D., 1980, A History of the Chaco Navajos; National Park Service. Washington, DC.

Brunhes, J., 1925, La geographie humaine; Volume III; la Vision Aerienne de la Terre, Paris, Alcan, vol. 3.

Büdel, J., 1957, The ice age in the tropics; Universitas 1, pp. 183–191.

BIA (Bureau of Indian Affairs), 1971, Navajo Area, Shiprock Agency, District 13., Soil and range inventory, Technical Report, Bureau of Indian Affairs, Branch of Land Operations, Shiprock, NM.

Bureau of Indian Affairs (BIA), 1978, Soil and range inventory for the San Juan Basin regional uranium study; By Rangeland Resources International, Inc., Mancos, CO., Report No. 9, San Juan Basin Regional Uranium Study, BIA, Albuquerque.

Butzer, Karl W., Glynn L. Isaac, Jonathan L. Richardson, and Celia Washburn-Kamau, 1972, Radiocarbon dating of East African lake levels; Science 175 (4027). pp. 1069–1076.

Chevallier, R., 1962, L'Archeologie Aerienne en France; Internat. Arch. of Photogramm., Vol. 14, pp. 401–407.

Chombart de Lauwe, P., (ed.), 1949, La Decouverte Aerienne du Monde; Paris: Horizons de France.

Chombart de Lauwe, P., 1951, Photographies Aeriennes; L'Etude de l'Homme sur la Terre, Paris, A. Colin.

Chombart de Lauwe, P., 1965, Des Hommes et des Villes; Paris, ed., Payot.

Chombart de Lauwe, P., et al., 1959, Famille et Habitation; Volume I. Science Humaines et Conceptions de l'Habitation, Paris: Centre National de la Recherche Scientifique.

Chombart de Lauwe, P., et al., 1960, Famille et Habitation. Volume II. Un Essai d'Observation Experimentale. Paris: Centre National de la Recherche Scientifique.

Clouten, Neville, 1974, The application of photogrammetry to recording rock art; Newsletter of the Australian Inst. of Aboriginal Studies, Canberra, Vol. 1, pp. 33–39.

Clouten, Neville, 1977, Further photogrammetric recordings of early man shelters, Cape York; Newsletter of the Australian Inst. of Aboriginal Studies, Canberra, Vol. 1, pp. 54–59.

Collier, G. A., 1974a, The impact of airphoto technology on the study of demography and ecology in Highland Chiapas: In Aerial Photography in Anthropological Field Research, E. Z. Vogt, ed.; Cambridge: Harvard University Press, pp. 78–93.

Collier, G. A., 1974b, The Kinprogram; On file at Department of Anthropology, Stanford University.

Conant, F. P., 1976, Satellite analysis of human ecosystems in the Sahel of East Africa; Proposal submitted to the National Science Foundation.

Conant, F. P., P. Reining and S. Lowes, (eds.), 1975, Satellite potentials for anthropological studies of subsistence activities and population change; Report and recommendations by the research workshop, May 27–30, 1975, Washington, D.C., American Association for the Advancement of Science.

Conant, F. P., 1978, The use of LANDSAT data in studies of human ecology; Current Anthropology 19(2): 382–384.

Conant, F. P., and T. K. Cary, 1977, A first interpretation of East African swiddening via computer-assisted analysis of 3 LANDSAT tapes; Proceedings, 1977, Machine Processing of Remotely Sensed Data Symposium, Purdue University, West Lafayette, Indiana, Purdue University, Laboratory for Applications of Remote Sensing, pp. 36–43.

Conklin, H. C., 1967. Some aspects of ethnographic research in Ifugao: Transactions, New York Academy of Science 30(1):99–121.

Conklin, H. C., 1972, Ethnographic atlas of Ifugao; A set of 8 maps of a central region of Ifugao, Scale 1:5000, New York, American Geographical Society.

Conklin, H. C., 1974, Ethnographic research in Ifugao; Aerial Photography in Anthropological Field Research, E. Z. Vogt, ed.; Cambridge, Harvard University Press, pp. 140–159.

Cooke, H. J., 1975, The palaeoclimatic significance of caves and adjacent landforms in western Ngamiland, Botswana; Geographical Journal No. 141, pp. 430–444.

Cooke, Ronald U. and Andrew Warren, 1973, Geomorphology in deserts; Batsford Press, London.

Crawford, O. G. S., and Alexander Keiller, 1928, Wessex from the air; Oxford, Clarendon Press.

Dakeyne, R. B., 1962, The pattern of settlement in central Nyanza, Kenya; The Australian Geographer Vol. 8, No. 4, pp. 183–191.

Deffontaines, P., 1948, Geographie et religions; Paris, Librairie Gallimard.

Dennett, Sarah and Hans Muessig, 1980, Archaeological applications for close-range photography; Technical Papers of the American Society of Photogrammetry, 46th Annual Meeting, March 9–14, 1980, St. Louis, Missouri, Falls Church, VA, American Society of Photogrammetry, pp. 335–341.

Downs, J. F., 1964, Animal husbandry in Navajo society and culture; Anthropology papers, University of California, Vol. 1, No. 1.

Drager, D. L., and T. R. Lyons, (eds.), in press, Remote sensing in cultural resource management: The San Juan Basin Project; Washington, Cultural Resources Management Division, National Park Service.

Dregne, Harold, 1976, Desertification: symptom of crisis; Desertification: Process, Problems, Perspectives, Paylore and Haney, eds.; Tucson, University of Arizona, Office of Arid Lands Studies.

Dury, G. H., 1973, Palaeohydrologic implications of some pluvial lakes in northwestern New South Wales, Australia; Geological Society of America, Bulletin 84, pp. 3663–3676.

Ebert, James I., 1977, Imaging remote sensor systems as sources of measured anthropological data; Report of the Remote Sensing Division, National Park Service, Albuquerque, NM.

Ebert, James I., 1978, Remote sensing and large-scale cultural resources management; Remote Sensing and Non-Destructive Archeology, Lyons and Ebert, eds.; Washington, DC, National Park Service, pp. 21–34.

Ebert, James I., M. C. Ebert and R. K. Hitchcock, 1976a, Anthropological, archaeological and ecological investigations in the Central District of Botswana, II: January–June 1975; Report on file at the Ministry of Local Government and Lands, Gaborone, Botswana and Department of Anthropology, University of New Mexico, Albuquerque.

Ebert, James I., M. C. Ebert and R. K. Hitchcock, 1976b, Atmospheric transmission of solar radiation: A note on one class of data being gathered by the University of New Mexico Kalahari Project; Botswana Notes and Records 8, Gaborone, Botswana, Fall 1976, pp. 299–300.

Ebert, James I. and Alberto A. Gutierrez, 1980, Cultural resources and remote sensing in the Eastern deciduous woodland: experiments at Shenandoah National Park; Paper delivered at the Second Conference on Scientific Research in the National Parks, Washington, DC, 1980. To appear in the Proceedings of that conference.

Ebert, James I. and Thomas R. Lyons, 1980, The detection, mitigation and analysis of remotely-sensed, "ephemeral" archeological evidence; Cultural Resources Remote Sensing, Lyons and Mathien eds.; Washington, DC, National Park Service, pp. 119–122.

Fairbridge, R. W., 1964, African ice-age aridity; Problems in Palaeoclimatology, Nairn, ed.; Interscience, London, pp. 356–363.

Fairbridge, R. W., 1968, Terraces, Lacustrine; Encyclopedia of Geomorphology, Van Nostrand Reinhold, New York City, pp. 1138–1140.

Fanale, R., 1974a, Analysis of settlement and land use patterns using remote sensor data; Paper presented at the 1974 Annual Meeting of the American Anthropological Association, Mexico City, (On file at Division of Remote Sensing, National Park Service, Albuquerque).

Fanale, R., 1974b, Utilization of ERTS-1 imagery in the analysis of settlement and land use of the Dogon of Mali; M. S. Thesis, Department of Anthropology, Catholic University.

Fanale, R., 1978, Ethnographic stratification of the San Juan Basin; Report No. 2, San Juan Basin Ethnographic Project, Division of Remote Sensing, National Park Service, Albuquerque.

Feacham, R., 1962, Timber structures revealed by aerial photography; International Archaeology of Photogrammetry, Vol. 14, pp. 419–424.

Gaithright, T. M., Geology of the Shenandoah National Park; Virginia Division of Mineral Resources, Charlottesville, VA, bull. 86.

Gardner, W. R., 1960, Soil water relations in arid and

semiarid conditions; in UNESCO Arid Zone Research XV, UNESCO, Paris, pp. 37–61.

Gourou, P., 1936, Les Paysans du Delta Tonquinois, Etude Humaine; Paris, Editions de l'Art et d'Histoire.

Goguey, R., 1967, Recherches aeriennes sur les structures archeologiques rurales et urbaines et leur influence sur le paysage actuel en Bourgogne; Secretariat de la Commission, eds., Vol. 16, pp. V–35 to V–40.

Griaule, M., 1937, L'emploi de la photographie aerienne dans la recherche scientifique; L'Anthropologie no. 40, pp. 469–476.

Griaule, M., 1946, Emploi de l'aviation dans la recherche ethnographique; Paris, National Congress of French Aviation.

Griaule, M., 1949, L'Ethnographie; decouverte aerienne du monde, Paris, Horizons de France, pp. 177–208.

Grove, A. T., 1969, Landorms and climatic change in the Kalahari and Ngamiland; Geographical Journal 135, pp. 191–212.

Gutking, E. A., 1952, Our world from the air: an international survey of man and his environment; London, Chatto and Windus. Gutking, E. A., 1956, Our world and the air: conflict and adaptation; Man's Role in Changing the Face of the Earth, W. L. Thomas, Jr., ed.; pp. 1–44.

Hack, J. T., 1965, Geomorphology of the Shenandoah Valley, Virginia and West Virginia, and origin of the residual ore deposits; US Geological Survey Professional Paper 484, Washington, DC.

Hack, J. T. and J. C. Goodlett, 1960, Geomorphology and forest ecology of a mountain region in the Central Appalachians, US Geological Survey Professional Paper 347, Washington, DC.

Hackenberg, R. A., 1967, The parameters of an ethnic group: a method for sutdying the total tribe; American Anthropologist Vol. 69, pp. 478–492.

Hamlin, C. L., 1977, Machine processing of LANDSAT data: an introduction for anthropologists and archeologists; Applied Science Center for Archeology, University Museum, The University of Pennsylvania, MASCA Newsletter No. 13(½) pp. 1–14.

Harding, S. T., 1935, Evaporation from large water surfaces based on records in California and Nevada; American Geophysical Union, Transactions Vol. 16, No. 2, pp. 507–511.

Harp, E., Jr., 1966, Anthropology and remote sensing; Proceedings of the fourth symposium on Remote Sensing of Environment, Ann Arbor, University of Michigan, Institute of Science and Technology, Willow Run Laboratories, pp. 729–739.

Harpending, H., and B. Davis, 1974, Environmental variation among hunter-gatherers; Paper presented at the Smithsonian Conference on the Application of Models in Theoretical Biology and Biogeography to Archeology and Anthropology, Washington, DC.

Heath, G. R., 1957, Improvements in the stereo-mosaic; Photogrammetric Engineering vol. 23, pp. 536–542.

Hill, James N., 1967, The problem of sampling; Chapters in The Prehistory of Arizona, III, Fieldiana Anthropology vol. 57, pp. 145–157.

Holmes, J., 1967, Problems in locational sampling; Annals of the Association of American Geographers vol. 57, pp. 757–780.

Houseman, E. E., 1975, Area frame sampling in agriculture; U.S. Department of Agriculture, Statistical Reporting Service, SRS no. 20. Washington, DC.

Huddleston, H. F., 1976, A training course in sampling concepts for agricultural surveys; U.S. Department of Agriculture, Statistical Reporting Service, SRS No. 21. Washington, DC.

Judge, W. James, James I. Ebert, and Robert K. Hitchcock, 1975, Sampling in regional archaeological survey; Sampling in archaeology, J. W. Mueller, ed.; Tucson, University of Arizona Press.

Kate, R. W., D. L. Johnson and K. J. Haring, 1977, Population, society and desertification; United Nations Conference on Desertification, Nairobi.

Klausner, Stephanie, 1980, Bipod photography: procedures for photographic mapping of archeological sites; Cultural Resources Remote Sensing, Lyons and Mathien, eds.; Washington, DC, National Park Service, pp. 293–326.

Knudson, S. J., 1978, Culture in retrospect: An introduction to archaeology; Chicago: Rand-McNally.

Kohler, M. A., T. J. Nordenson and D. R. Baker, 1959, Evaporation maps for the United States; United States Weather Bureau Technical Report 37.

Kruckman, L., 1978, The role of remote sensing in ethnohistorical research: Recent Research and New Directions; Paper presented at the 26th Annual Meeting of the American Society of Ethnohistory.

Kuchler, A. W., 1967, Vegetation mapping; London: Ronald Press.

Langbein, Walter B., 1961, Salinity and hydrology of closed lakes; US Geological Survey Professional Paper 412, U.S. Department of the Interior, Washington, D.C.

Lathrap, D. W., 1973, Gifts of the Cayman: some thoughts on the subsistence basis of Chavin; Variation in anthropology, Illinois Archaeological Survey.

Lathrap, D. W., 1970, The Upper Amazon; London, Thames and Hudson.

Lebedev, A. N., (ed.), 1968, Klimaticheskii spravochnik Afriki (Climate of Africa, Part I: Air Temperature, Precipitation); Gidrometeorologischeskoe Izdatelstvo, Leningrad.

Leopold, Luna B., 1951, Rainfall frequency: An aspect of climatic variation; American Geophysical Union, Transactions No. 32, pp. 347–355.

Loose, Richard W. and Thomas R. Lyons, 1976, Use of aerial photos in archeological survey along the lower Chaco River drainage; Remote Sensing Experiments in Cultural Resource Studies, Lyons, ed.; Cultural Resources Management Divison, National Park Service, Washington, DC, pp. 69–72.

Loucks, Orie L., 1970, Evolution of diversity, efficiency and community stability; American Zoologist Vol. 10, pp. 17–25.

Lyons, T. R., (ed.), 1976, Remote sensing experiments in cultural resource studies: Non-destructive methods of archeological exploration, survey, and analysis; Reports of the Chaco Center, No. 1, Albuquerque, National Park Service and University of New Mexico.

Lyons, T. R., and T. E. Avery, 1977, Remote sensing: A handbook for archeologists and cultural resource managers; Washington, D.C., National Park Service, (U.S. Government Printing Office, Stock Number 024-005-00688-5).

Lyons, T. R., and J. I. Ebert, (eds.), 1978, Remote sensing and non-destructive archeology; Washington, Cultural Resources Management Division, National Park Service.

Lyons, T. R., and R. K. Hitchcock, (eds.), 1977, Aerial remote sensing techniques in archeology; Reports of The Chaco Center, No. 2., Albuquerque, National Park Service and University of New Mexico.

Lyons, T. R., M. Inglis and R. K. Hitchcock, 1972, The application of space imagery to anthropology; Proceedings: Third Annual Conference on Remote Sensing in Arid Lands, Tucson, Office of Arid Lands Studies, The University of Arizona, pp. 244–265.

Lyons, T. R., and F. J. Mathien, (eds.), 1980, Cultural resources remote sensing; Washington, Cultural Resources Management Division, National Park Service.

Lyons, T. R., Pouls, B. G., and Hitchcock, R. K., 1972, The Kin Bineola irrigation study: an experiment in the use of aerial remote sensing techniques in archaeology; Third Annual Conference on Remote Sensing in Arid Lands Proceedings, Arizona University, Tucson, Chaco Center Control, No.1.

Lyons, Thomas R. and Douglas H. Scovill, 1978, Archeology and remote sensing: a conceptual and methodological stance; Remote Sensing and Non-Destructive Archeology, Lyons and Ebert, eds.; Washington, DC, National Park Service, pp. 3–20.

McGinnies, William G., 1976, Ecology of desertification; Desertification: Process, Problems, Perspectives, P. Paylore, ed.; Tucson, University of Arizona. Office of Arid Lands Studies.

Melbin, M., 1960, Mapping uses and methods; Human Organization Research, Adams and Preiss, eds.; pp. 255–266. pp. 255–266.

Miltrope, F. L., 1960, The income and loss of water in arid and semi-arid zones; UNESCO ARID ZONE RESEARCH XV, pp. 9–36.

Molineaux, C. E., E. E. Bliamptis, and J. T. Neal, 1971, A Remote-Sensing Investigation of Four Mojave Playas; Environmental Research Papers No. 352, Air Force Cambridge Research Laboratories, Bedford, Mass.

Morain, S. A., and T. K. Budge, 1978, Remote sensing: instrumentation for non-destructive exploration of cultural resources; Supplement No. 2 to Remote Sensing, a Handbook for Archeologists and Cultural Resource Managers, Washington, DC, National Park Service.

National Academy of Sciences, 1977, Resource sensing from space: prospects for developing countries; Report of the Ad Hoc Committee on Remote Sensing for Development, Board on Science and Technology for International Development Commission on International Relations. Washington, D.C., National Academy of Sciences/National Research Council.

Neal, James T., 1972, Playa surface features as indicators of environment; Playa Lake Symposium Proceeding, C. C. Reeves, Jr., ed.; ICASALS Publication No. 4, Texas Tech University, Lubbock, Texas, pp. 107–132.

Neal, James T., 1975, Playas and dried lakes: occurrence and development; Benchmark Papers in Geology 20, Halstead Press and Dowden, Hutchison and Ross, Inc., Stroudsburg, Pennsylvania.

Parsons, J., 1969, Ridged fields in the Rio Guayas Valley, Ecuador; American Antiquity, Vol. 34, pp. 76–80.

Parsons, J. J. and Bowen, William, 1966, Ancient ridged fields of the San Jorge River floodplain, Colombia; Geographical Review, No. 56, pp. 317–343.

Parsons, J. J. and Denevan, William, 1967, Pre-Columbian ridged fields, Scientific American No. 217, pp. 92–101.

Paylore, Patricia, 1976, Desertification: what, where, why, who; Desertification: Process, Problems, Perspective, Paylore and Haney, eds.; Tucson, University of Arizona, Office of Arid Lands Studies.

Pearson, K. C., 1965, Primary production in grazed and ungrazed desert communities of eastern Idaho; Ecology Vol. 46, No. 3, pp. 278–285.

Phelps-Stokes Fund, 1939, The Navajo Indian problem; New York.

Plog, Fred, 1968, Archaeological survey: A New Perspective; Unpublished M.A. Thesis, University of Chicago, Chicago, Ill.

Porter, P. W., 1956, Population distribution and land use in Liberia; Ph.D. dissertation, London School of Economics and Political Science.

Price, R., 1974, Aerial photography in the study of land use: a Maya Community; Aerial Photography in Ethnographic Field Research, E. Z. Vogt, ed., Cambridge, Harvard University Press, pp. 94–111.

Rainey, Froelich, 1974, Science and archaeology; Archeology, Vol. 27, No. 1, pp. 10–21.

Redman, Charles L., 1973, Multistage fieldwork and analytical techniques, American Antiquity No. 38, pp. 61–97.

Reeves, C. C., Jr., 1966, Pluvial lake basins in West Texas; Journal of Geology vol. 74, no. 3, pp. 269–291.

Reeves, Robert G., editor-in-chief, 1975, Manual of Remote Sensing; Falls Church, VA, American Society of Photogrammetry, vol. 2.

Reining, P., 1973, Utilization of ERTS-1 imagery in cultivation and settlement site identification and carrying capacity estimates in Upper Volta and Niger; Springfield, VA, National Technical Information Service.

Reining, P., 1974a, Human settlement patterns in relation to resources of less developed countries; Proceedings, COSPAR Meetings, Sao Paulo, Brazil, (On file at International Office, American Association for the Advancement of Science, Washington, D.C.).

Reining, P., 1974b, Use of ERTS-1 Data in carrying capacity estimates for sites in Upper Volta and Niger; Paper presented at the 1974 Annual Meeting of the American Anthropological Association, Mexico City, (On file at International Office, American Association for the Advancement of Science, Washington, DC.).

Reining, Priscilla, 1978, Handbook on desertification indicators; Based on the Science Association's Nairobi Seminar on Desertification, Washington, DC, American Association for the Advancement of Science.

Reining, P., in press, Carrying capacity with LANDSAT: cultivation and settlement identification in Upper Volta and Niger; Papers in International Studies, Africa Series, Athens: Ohio University Press.

Reining, P., and D. Egbert, 1975, Analysis with Image 100 of Sonrai, Niger villages for location and extent of more than one village; NASA Earth Resources Survey Symposium, Houston, Johnson Space Center, NASA.

Richardson, J. L., 1969, Former lake-level fluctuations—their recognition and interpretation; Mitteilungen Verein, Internat. Limnol. 17, pp. 78–93, Stuttgart.

Ricketson, Oliver Jr., and A. V. Kidder, 1930, An archaeological reconnaissance by air in Central America; Geographic Review Vol. 20, No. 2, pp. 177–206.

Riddell, J., 1978, Changing intensity in the use of tropical soils and possible modification in the moisture balance of the forest-savannah littoral: a West-African Case Study.

Riley, D. N., 1946, The technique of air-archaeology; Archaeological Journal Vol. 101, pp. 1–16.

Rinker, R. J., 1973a, Infrared thermal detection of caves; US Army Engr. Topo. Labs., Ft. Belvoir, Va., pp. 10.

Rinker, R. J., 1973b, An application of air photo analysis to a cave location study, US Army Topo. Engr. Labs., Ft. Belvoir, Va., pp. 18.

Robequain, C., 1929, Le Thanh Hoa: Etude geographique d'une province Annamite; Publications de l'Ecole Francasie d'Extreme-Orient, Paris, Van Oest, pp. 23–24.

Rogers, A. W., 1940, Pans (Review); Transaction of the South African Geological Society 22. pp. 55–60.

Roosen, R. G., R. J. Angoine, and C. H. Klemcke, 1973, Worldwide variations in atmospheric transmission 1; Baseline results from the Smithsonian observations, Bulletin of the American Meteorological Society 54, pp. 307.

Roosen, Robert G., Robert S. Harrington, James Giles, and Iben Browning, 1976, Earth tides, volcanoes and climate change; Nature Vol. 261, No. 5526, pp. 680–682.

Rootenberg,S., 1964, Archaeological field sampling; American Antiquity No. 38, pp. 61–97.

Rowe, J. H., 1953, Technical aids in anthropology: a historical survey; Anthropology Today, A. L. Krober, ed.; pp. 895–940.

Sabins, Floyd F., Jr., 1978, Remote sensing principles and interpretation; W. H. Frugman and Company, San Francisco, pp. 98.

St. Joseph, J. K. S., (eds.), 1966, The uses of air photography, nature and man in a new perspective; London: John Baker.

Schellart, A. I. J. M., 1962, Experiences with aerial photographs in the study of castles; International Archaeology of Photogrammetry, Vol. 15, pp. 425–431.

Scogings, D. A., 1975, Photogrammetric recording of petroglyphs; Wild Reporter, Heerbrugg, Switzerland: Wild-Heerbrugg, No. 9, pp. 15–16.

Scollar, Irwin, 1968, Computer image processing for archaeological air photographs; World Archaeology Vol. 10, No. 1, pp. 71–87.

Scollar, Irwin, 1975, Transformation of extreme oblique aerial photographs to maps or plans by conventional means or by computer; Aerial Reconnaissance for Archaeology, D. R. Wilson, ed.; Research Report No. 12, Council for British Archaeology.

Scollar, Irwin, 1977, Image processing via computer in aerial archaeology; Computers and the Humanities, Pergamon Press, Vol. 11, pp. 347–351.

Scollar, Irwin, 1979, Progress in aerial photography in Germany and computer methods; Aerial Archaeology, Edwards and Horne, eds; London: The Aerial Archaeology Foundation, No. 2, pp. 8–17.

Schorr, T. S., 1965, Cultural ecological aspects of settlement patterns and land use in the Cauca Valley, Colombia; Ann Arbor, University Microfilms.

Schorr, T. S., 1974b, A bibliography, with historical sketch; Aerial Photography in Anthropological Field Research, E. Z. Vogt, ed.; Cambridge, Harvard, pp. 163–188.

Schorr, T. S., 1974a, Aerial ethnography in regional studies: a reconnaissance of adaptive change in the Cauca Valley of Colombia, Aerial Photography in Anthropological Field Research, E. Z. Vogt, ed.; Cambridge: Harvard, pp. 40–53.

Sjoberg, Gideon, 1965, The origin and evolution of cities: Scientific American Vol. 213, No. 3, pp. 54–63.

Smith, George I., 1962, Subsurface stratigraphy of Late Quarternary deposits, Searles Lake, California; U.S. Geological Survey Professional Paper 450-C, pp. 65–69.

Spetzman, L. A., 1959, Vegetation of the Arctic Slope of Alaska; US Geological Survey Professional Paper 302-B. Washington, DC, US Government Printing Office.

Spooner, Brian, 1978, Problems in the development of social indicators of desertification; Paper presented at the Symposium, "Desertification: Issues in Measuring and Monitoring the Process with Indicator", at the 1978 Annual Meeting of the American Association for the Advancement of Science.

Spurr, Stephen H., 1960, Photogrammetry and Photo-Interpretation; Second Edition, New York, The Ronald Press Company.

Strandberg, C. H., 1967, Photoarchaeology, Photogrammetric Engineering, vol. 33, pp. 1152–1157.

Strandberg, C. H., 1974, Fort Frederick photoarchaeological study; Maryland State Department of Natural Resources, pp. 55.

Street, F. Alayne and A. T. Grove, 1976, Environmental and climatic implications of Late Quarternary lake-level fluctuations in Africa; Nature No. 261, pp. 385–390.

Stuart, Alan, 1962, Basic ideas of scientific sampling; Griffin's Statistical Monographs and Courses No. 4, Charles Griffin and Co., London.

Terhune, Elinor C., 1978, Plants as indicators of desertification; Paper presented at the Symposium, "Desertification: Issues in Measuring and Monitoring the Process with Indicators", at the 1978 Annual Meeting of the AAAS.

Turner, B. L., 1974, Prehistoric intensive agriculture in the Mayan lowlands; Science no. 185, pp. 118–124.

Turpin, Solveig A., Richard P. Watson, Sarah Dennett, and Hans Muessig, 1979, Stereophotogrammetric documentation of exposed archeological features; Journal of Field Archaeology Vol. 6, No. 3, pp. 329–337.

Twidale, C. R., 1972, Landform development in the Lake Eyre Region, Australia; Geographical Review 62, pp. 40–70.

United Nations, 1977, Report of the United Nations Conference on Desertification; Nairobi, 29 August–9 September 1977. A/CONF.-74/36.

Vogt, E. Z., 1974a, Aerial photography in anthropological field research; Harvard University Press, Cambridge.

Vogt, E. Z., 1974b, Aerial photography in Highland Chiapas ethnography; Aerial Photography in Anthropological Field Research, E. Z. Vogt, ed.; Harvard University Press, Cambridge, pp. 57–77.

Walter, H., 1971, Ecology of tropical and subtropical vegetation; translated by D. Mueller-Dombois, Oliver and Boyd, Edinburgh.

Whittlesey, Jultan H., 1970, Tethered balloon for archaeological photos; Photogrammetric engineering, Vol. 36, pp. 181–186.

Whittlesey, J. H., 1966, Bipod camera support, Photogrammetric Engineering, Vol. 32, pp. 1005–1010.

Wood, Walter F., 1955, Use of stratified random samples in a land-use study; Annals of the Association of American Geographers, Vol. 45, pp. 350–367.

Young, R. W., 1961, The Navajo Yearbook; The Navajo Agency, Window Rick, AZ.

Zar, Jerrold H., 1974, Biostatistical Analysis; Prentice-Hall, Englewood Cliffs.

CHAPTER 27

Remote Sensing of Weather and Climate

Author-Editors: DONALD R. WIESNET and MICHAEL MATSON

Contributing Authors: JIMMIE JOHNSON, ROBERT A. MC CLATCHEY, P. KRISHNA RAO, and ROBERT H. STEWART

GENERAL CONTENTS: Remote sensing of the earth's weather; remote sensing of the earth's climate; remote sensing of radiatively active atmospheric constituents; monitoring climate scale variability in the ocean from space; remote sensing of the cryosphere and hydrosphere; land surface processes; desertification; deforestation; urbanization; future trends; references.

INTRODUCTION

The remote sensing of weather and climate by polar-orbiting and geostationary satellites is regarded as the single most significant breakthrough for monitoring the Earth's weather and climate in the past quarter-century. It must be remembered, however, that each step in the collection of satellite weather data is accompanied by significant computer technology and software achievements that make the whole information-gathering system workable and manageable. Advances in telecommunications have been equally important in the development of weather-satellite information systems. This chapter will necessarily confine itself only to the state of the art of remote sensing of weather and climate.

Meteorological satellites are rapid and efficient methods of collecting global weather data. They provide accurate storm positions, storm sizes and intensities, cloud heights, upper air winds, precipitation estimates, numerical atmospheric sounding data, fog and stratus conditions, snow cover on the ground, haze and air pollution conditions, and a tropical storm (hurricane) tracking and monitoring capability.

Meteorological satellite images and data can be received anywhere in the world. Because of the international nature of both the science of meteorology and the technology of orbiting satellites in space, the World Meteorological Organization (WMO) in Geneva has become a focal point for coordinating the activities of the U.N. member countries. WMO has long sponsored a "World Weather Watch" Program which includes four geostationary satellites positioned strategically around the world over the equator to warn people of all nations of impending severe weather. These satellites are located at 104°E (operated by the Japanese Space Agency); at 70°W and 135°W (operated by the U.S. National Oceanic and Atmospheric Administration (NOAA)); and at 0° (operated by the European Space Agency). A fifth satellite at 75°E will be launched at a later date. As a result, about 90 percent of the population of the world will benefit from a more timely weather-warning system. All U.S. operational civilian meteorological satellite systems are operated by the National Earth Satellite Service (NESS), a component of NOAA. NESS will also operate the Landsat program starting in 1983.

Remote sensing of the earth's weather and climate comprise the first section of this chapter. This is followed by a discussion of remote sensing of atmospheric constituents. Monitoring climate scale variability in the ocean from space is described in detail next. Remote sensing of climatic aspects of the cryosphere (snow and ice), the hydrosphere, and the biosphere are subsequent sections.

Johnson and Vetlov (1977) list the hydrometeorological and geophysical parameters that can be determined from satellite observations (see Table 27-1).

REMOTE SENSING OF THE EARTH'S WEATHER

MACROSCALE

Today, our satellite systems provide more satellite data, at greater frequency, and covering greater areas, than ever before. A number of instruments designed to sense various parameters of meteorological and environmental interest have been tested experimentally. Currently, two sensors are being used operationally: The first senses in the visible spectrum and records the albedo or reflected light as imagery of the underlying clouds, ground or ocean surfaces. The second senses in the infrared spectrum and provides temperature data of the viewed cloud tops, terrain or ocean surfaces.

In the satellite data, each cloud type has a characteristic pattern and brightness or temperature return. Familiarity with these cloud characteristics will facilitate the analysis or interpretation of the satellite data.

TABLE 27-1

Hydrometeorological and Geophysical Parameters That Can Be Determined from Satellite/Observations

Atmosphere
- Mean temperatures of isobaric layers
- Total water-vapor content and its distribution by layers
- Total ozone content and its distribution by layers characteristics of aerosols
- Components of the radiation balance at the top of the atmosphere
- Wind speed and direction in the troposphere (at two or three levels)

Clouds
- Spatial distribution of cloud and its structure
- Height and temperature of cloud tops
- Phase composition (droplets, ice particles, or mixed) of the upper cloud layer
- Total water content of clouds
- Location of precipitation areas and their approximate intensity

Ocean Surface
- Temperature of the ocean surface
- Location of the major ocean-surface currents
- Degree of roughness of the ocean (e.g. waves) and the related surface wind fields
- Ice Conditions
- Location of polluted areas (of certain types) on the ocean surface

Land Surface
- Temperature of the land surface
- Degree of soil moisture
- Distribution of snow cover
- Characteristics of the soil and plant cover
- Location of areas of melting snow and ice.

By analyzing the type of clouds, the meteorologist can locate fronts, squall lines, jet streams, troughs, and ridges. Further, the meteorologist can access the stage and trend of development in mid-latitude and tropical storm systems. In many cases, it is possible to infer the presence of turbulence, the orientation of surface and upper-level winds, and the atmospheric stability.

To a trained satellite meteorologist, many clues to the current state of the atmosphere are apparent just by visual inspection. The broad-scale patterns stand out clearly. Not so apparent is the information implicit in the cloud patterns such as vertical velocity, vorticity, and divergence. Much progress has already been made in obtaining quantitative estimates of relative humidity, zero change in vorticity, and wind vectors, which can be used as observations for the computer forecast models. Some of these features will be illustrated.

Circulation Patterns

The various cloud elements can be used as targets to determine the circulation fields over both land and water. The geostationary satellites provide data at 30-minute intervals. Continuous satellite data provide an important tool for the monitoring of weather. The tracking of a storm's development intensification and demise can be operationally accomplished from the satellite images. Figures 27-1 and 27-2 show the tracking of a storm on November 26, 1975, for the period from 0600 GMT to 1700 GMT, using satellite data.

An elongated east/west area of cloudiness (Figure 27-1) is observed stretching from western Kansas eastward into Missouri. The cloudiness in western Kansas and Nebraska at (C) was associated with an upper level trough in the northern plains. In addition, the clouds to the east at (A) were associated with the positive vorticity advection (PVA) ahead of a vorticity center in the Oklahoma Panhandle. Passage of this easternmost system deposited snow over eastern Kansas with lighter amounts over Missouri. A second disturbance is located in the Four Corners area. The clouds associated with this disturbance are located at (B). Their configuration is influenced by the presence of mountainous terrain.

Continued development of the upper air distur-

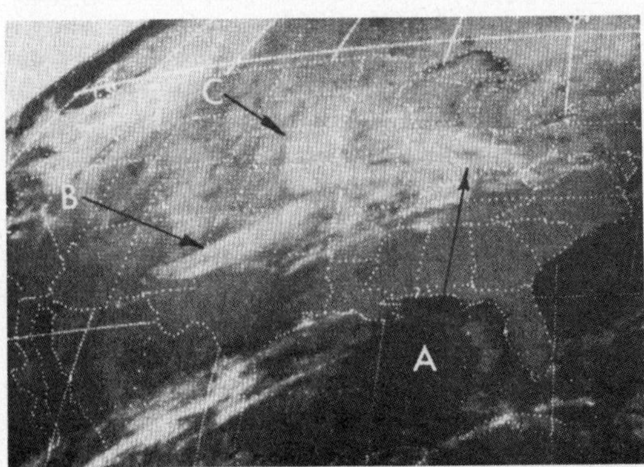

Fig. 27-1. SMS-1 IR 0600 GMT 26 Nov. 1975.

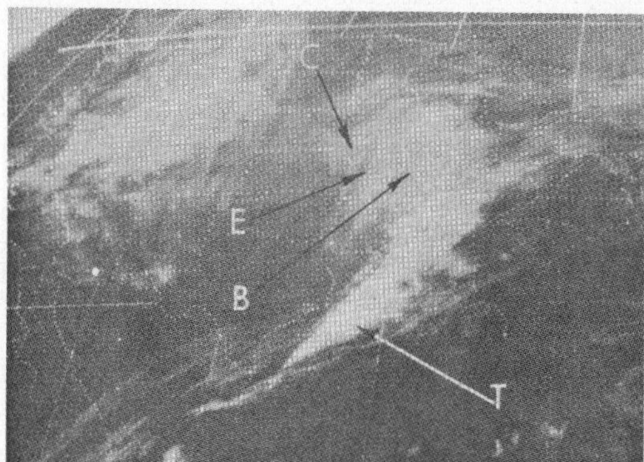

Fig. 27-2. SMS-1 IR 1700 GMT 26 Nov. 1975.

bance at (B) can be tracked in subsequent IR pictures. The clouds associated with the upper air disturbance (Figure 27-2) assume a more "comma-shape" configuration through the period. The sharp anticyclonic back edge of the comma at (E) separates the area of moderate and heavy snow to the east from lighter snow to the west (C). Thunderstorms at (T) develop as the surface low in southwest Louisiana intensifies and moves northeastward ahead of the trough axis.

Fronts, Troughs, and Ridges

Forecasting and analysis on a global scale is facilitated by satellite data that show the large synoptic-scale cloud patterns. Through these data, the meteorologist can assess current or impending changes such as the development of waves along a front in response to an approaching upper tropospheric vorticity maximum, the resulting new low center and frontal band and, finally, the features characteristic of dissipation.

Certain changes in the characteristics of frontal bands help to locate the positions of upper level ridges and troughs more precisely. These features can then be used concurrently with conventional data to locate changes in wind direction and vertical motion. A 500 mb trough can be positioned extending from the center of the vortex southward to the point where the frontal band—in visible imagery—becomes ragged or full of holes. If a comma-shaped cloud area (evidence of a vorticity maximum) is present, the trough line should be drawn to the west of the comma cloud.

Upper level ridge lines are determined from the distribution of middle and upper level clouds. A sharp ridge with small curvature will cause middle and high clouds produced upwind to end abruptly at the ridge line as the upward motion changes to downward. In this case, the ridge line would be at the forward edge of the clouds. Medium and broad ridges produce a less distinct forward edge to the clouds and cirrus usually spills out ahead of the ridgeline. Where this happens, the ridge is located westward of the cirrus edge, back where both middle and high clouds are present.

Figure 27-3, a 1-mile resolution visible geostationary satellite image of 1500 GMT, March 21, 1977, shows frontal cloud patterns extending from a low pressure center in the Northwest Atlantic Ocean to a developing frontal wave in Texas and Louisiana. Another band of frontal clouds and the associated low pressure center are developing along the northern Great Lakes and Plains States. An infrared geostationary-satellite image of this same area acquired nine hours later at 0000 GMT on March 22, 1977 (Figure 27-4) shows these same frontal cloud systems. This figure has the surface fronts and isobars superimposed on the imagery in white; streamlines of surface and weather symbols for precipitation are superimposed in black.

Jet Streams

Jet streams or wind maxima can be accurately located from satellite imagery. A large area of anticyclonically curved cirrus usually marks the equatorward side of the jet. On the poleward side, the cirrus cloud edge is quite sharp and can cast a shadow on the underlying cloud deck. If a shadow is not apparent on lower clouds, the edge of the cirrus shield can be located by noting the difference in the cloud texture; cirrus is the smoother cloud type in a frontal system. The jet-system core lies parallel to and within 100 km of the poleward edge of the cirrus shield. Once the jet stream has been located, the meteorologist can then infer the wind direction, wind shear, areas of potential clear air turbulence, and the horizontal temperature gradient.

Figure 27-5 is the same GOES infrared image shown in Figure 4, but this time the 300 mb isotach and isotherm analysis are superimposed. Note how the jet stream maximum lies just poleward of the cirrus clouds.

Extratropical Cyclone Development

Twice a day surveillance by the visible and infrared sensors on polar orbiting meteorologic satellites allowed meteorologists to monitor developing systems at 12-hour intervals. Now the geostationary satellites with their 24-hour surveillance allow hour-by-hour analysis of the stage of development of all storms.

The typical sequence of events for a developing cloud system are: The initial frontal wave, the deepening stage (cyclogenesis), the mature stage with or without secondary centers of secondary waves, and finally the decaying stage. Figure 27-6 is a sequenced set of SMS-1, 2-km visible images over the United States on August 19, 1977, from 1230 GMT through 2230 GMT showing the evolution of a system from the frontogenetic phase to the cyclogenetic stage. The first two images are 4 hours apart and the remaining images in the series are 2 hours apart. The first image at 1230 GMT shows what is sometimes referred to as the "baroclinic leaf" pattern over the Great Lakes (frontogenetic pattern). At this early stage of de-

Fig. 27-3. SMS-1 Visible 1 km 1500 GMT—March 21, 1977.

velopment, the northern side of the "leaf" cloud formation is rather smooth and has an "s" shape, while the southern side is ragged. Surface cyclogenesis is in progress by 1830 Z when the cloud pattern becomes "comma shaped". As cyclogenesis continues (2030 and 2230 Z) a dry slot digs into the south of the "comma head". The cloud edges become better defined as cyclogenesis is almost complete. However the low pressure center still remains rather intense at this stage.

Tropical Cyclones

Perhaps the earliest use of satellite data was for locating and tracking tropical systems. Today,

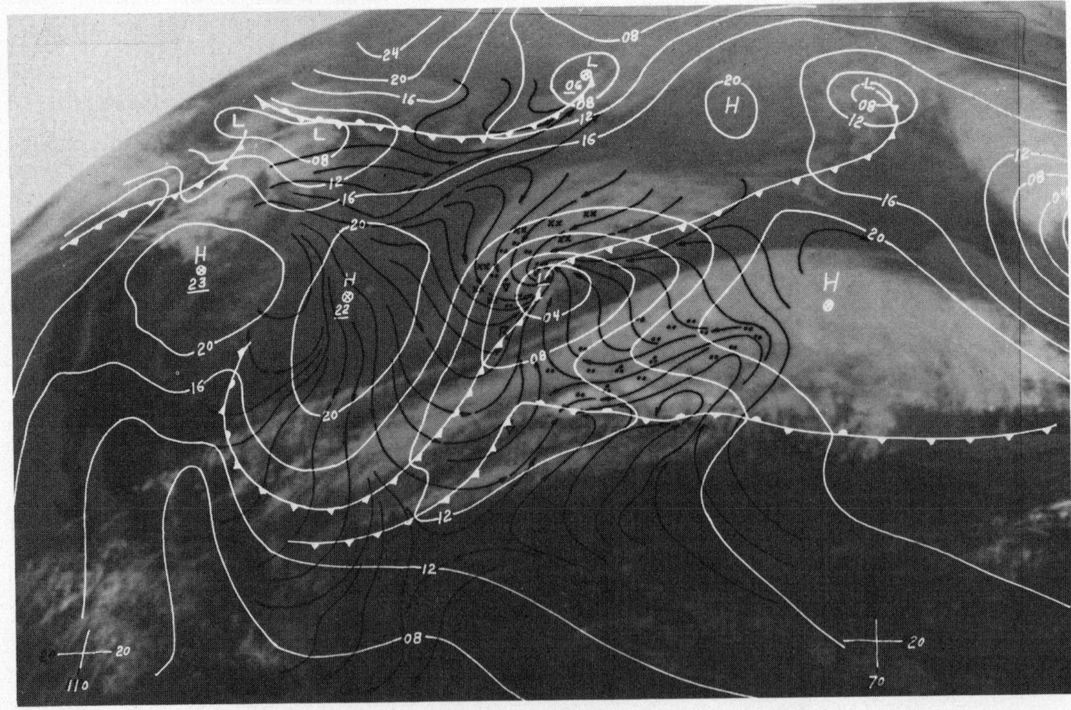

Fig. 27-4. SMS-1 Infrared 8 km 0000 GMT—March 22, 1977.

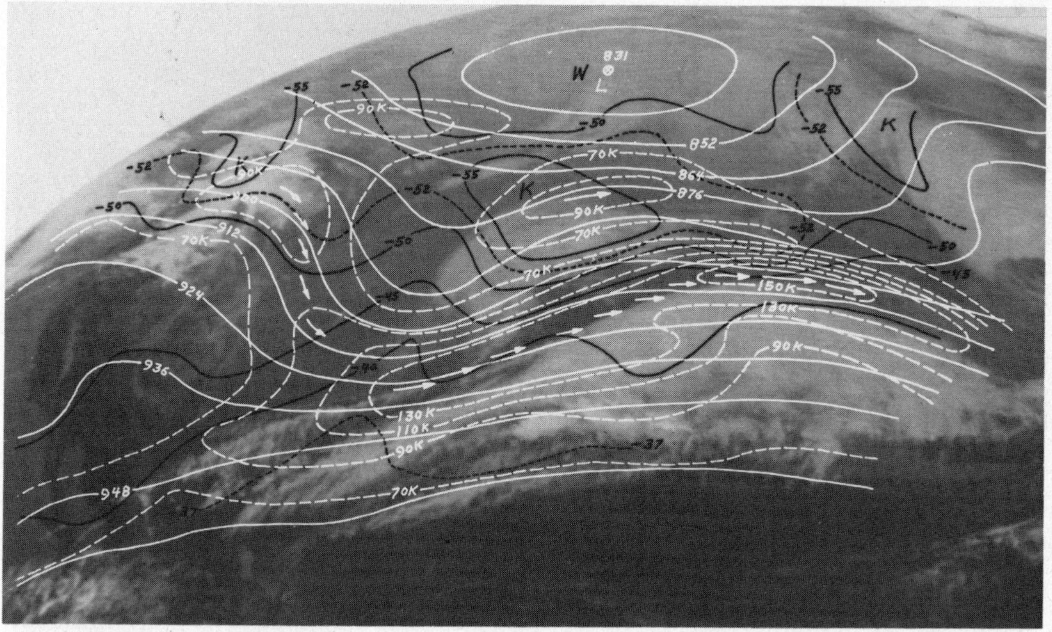

Fig. 27-5. SMS-1 Infrared 0000 GMT—March 22, 1977.

with the continuous coverage from the geostationary satellite, one can constantly monitor the progress of these dangerous storms.

Tropical cyclones undergo changes in their cloud patterns just as do the extratropical systems. From these variations, one can estimate the wind speed and central pressure of the system and make some judgment as to its continued development and movement. These estimates are based on the organization, size, and banding of the tropical system.

Tropical Storms are classified by the cloud features they exhibit. The classifications have been empirically determined by comparing the cloud patterns with measured storm data for a five-year period. A model of tropical cyclone development and weakening (Dvorak 1975) serves as guidance for the subsequent life expectancy of the storm.

The daily analysis of a cyclone entails the assigning to it of a T-number that describes the current intensity of the system in terms of its cloud features. The T-numbers range from T1 and T2 for developing, pre-storm patterns, to T5 through T8 for fully developed hurricanes or typhoons. The past history of cloud-pattern evolution and current intensity of the storm are then used with the model to predict the developmental trend over the next 24 hours.

Hurricanes and typhoons exhibit a great variety of cloud patterns, but most can be described as having a comma configuration. The comma tail is composed of convective clouds that appear to curve cyclonically into a center. As the storm develops, the clouds form bands that wrap around the storm center producing a circular cloud system that usually has a cloud-free, dark eye in its mature stage.

The use of enhanced infrared imagery in tropical cyclone analysis adds objectivity and simplicity to the job of determining tropical storm intensity from satellite observations. The infrared data not only afford continuous day/night storm surveillance but also provide quantitative information about cloud features that are related to storm intensity; thus, cloud-top temperature measurements and temperature gradients can be used in place of cloud features used qualitatively in visible data classification-techniques. To enable the analyst to make objective measurements of parameters related to cyclone intensity, the IR data are enhanced to display specified temperature ranges in discrete shades of gray. The temperature ranges assigned to these various shades of gray are shown in Figure 27-7.

Figure 27-8 shows three enhanced IR pictures per day of hurricane Anita (1977) along with the available central pressure measurements plotted against time. The pressure data were obtained from aircraft-reconnaissance observations.

The T-number, central pressure relationship is taken from Dvorak (1975). Ninety-two percent of the available pictures of Anita were quantifiable by either measuring the amount by which the cold clouds encircle the center or by using the surrounding temperature and eye criteria. Through a comparison of the daily pictures down the left side of Figure 27-8, Anita's cloud pattern evolution is evident from large changes in the cloud features related to intensity. There were two periods, however, on August 30 and 31 (Days "B" and "C", Figure 27-8) when little significant change in

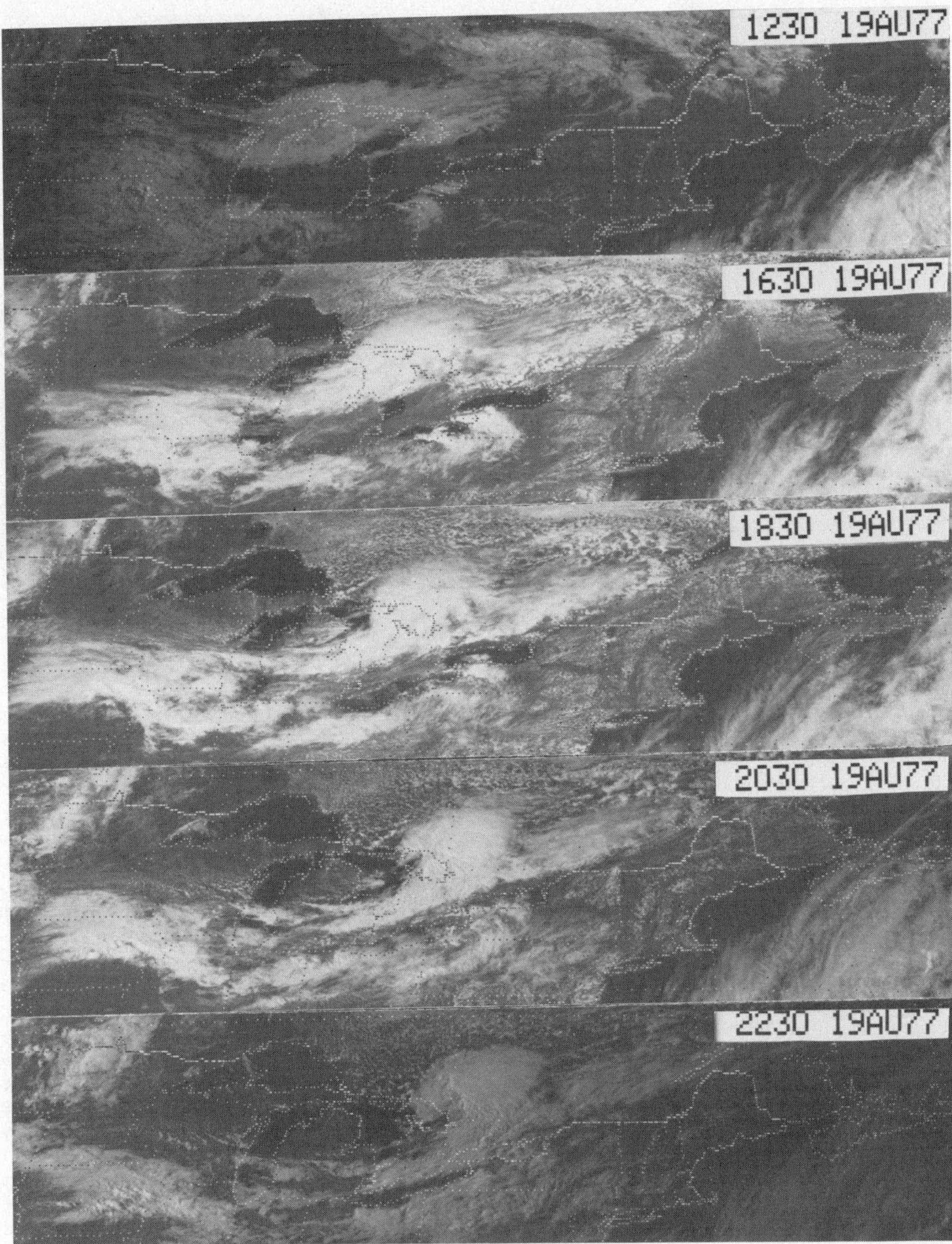

Fig. 27-6. SMS-1 2 km Visible Images 1230–2330 GMT—Aug. 19, 1977.

the cloud pattern was evident. On August 30 (Day B), the 0900Z and 1230Z pictures show that the cold high clouds are suppressed on one side of the storm pattern instead of the expected increased encirclement of the cloud-system center by cold clouds. During this period, the central pressure curve also shows a leveling out. On August 31 (Day C), just after the storm reached hurricane intensity, both the 0900Z and 1230Z pictures show signs of a "central cold cover" near the storm center accompanied by little change in cloud pattern. Again, no significant drop in the storm's central pressure occurred through the period. Note the contrast between these two periods of no pressure drop and the steady pattern evolution that takes place on August 29 and September 1 as

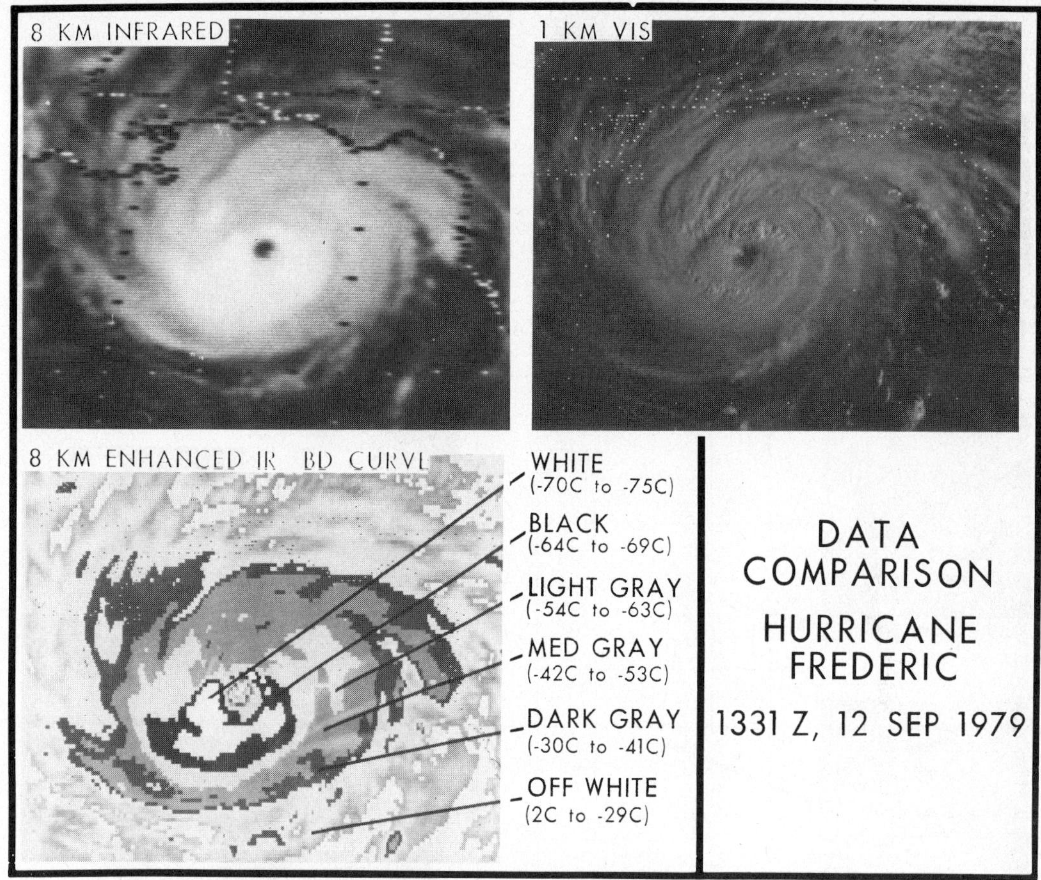

Fig. 27-7. SMS-2 Infrared 1331 GMT—September 12, 1979.

pressure falls steadily. On both of these days, the pattern shows significant evolution after 0900Z, which accompanies the increase of intensity.

Satellite Winds and Clouds

Satellite wind measurements, obtained from geostationary satellites, are based on cloud motions measured three times each day using satellite images. The measurements are made between 30°W and 180°W and 45°N and 45°S using the infrared images. Low-level satellite winds represent air motions near the 900 mb level. Upper-level satellite winds have an altitude assigned to them based on the equivalent black-body temperature of the cloud element being tracked. Figures 27-9 and 27-10 show satellite-derived winds for high and low levels on April 16, 1979.

Cloud heights can readily be obtained from geostationary-satellite infrared cloud-imagery. For accurate determination, the data must be calibrated and dealt with in digital form. Infrared data give the equivalent black-body temperature of cloud-top surfaces. This information, plus knowledge of the atmosphere's temperature structure where the clouds are observed, allows one to compute the actual cloud-top height. The temperature structure is obtained from plotted radiosondes, upper-air analyses, or forecasts. Since clouds composed of ice crystals are semitransparent to infrared radiation, when dealing with cirrus one must use estimates of cirrus-cloud transmissivity in order to obtain the proper cloud-top heights.

A cloud-top height display that uses upper air analysis-fields from the National Meteorological Center (NMC) and IR data from the geostationary satellites has been developed. Computer printer alphanumerics are used to produce a cloud-top height display by converting each VISSR IR 8-km pixel (IR sample) to temperature and then to height, based on the pressure/height/temperature profile at the given location from the NMC upper air analysis-field. The computer-printer alphanumerics produce a height display at about 8-km spatial resolution and 5,000-ft vertical resolution for heights at or above the 20,000-ft level.

An example of these analyzed cloud-top heights is shown in Figure 27-11 with the corresponding visible image shown in Figure 27-12. The contoured lines on Figure 27-11 represent radar intensity levels corresponding to light, strong, and intense precipitation. The computer program extracted IR data for the 2100 GMT and 2200 GMT

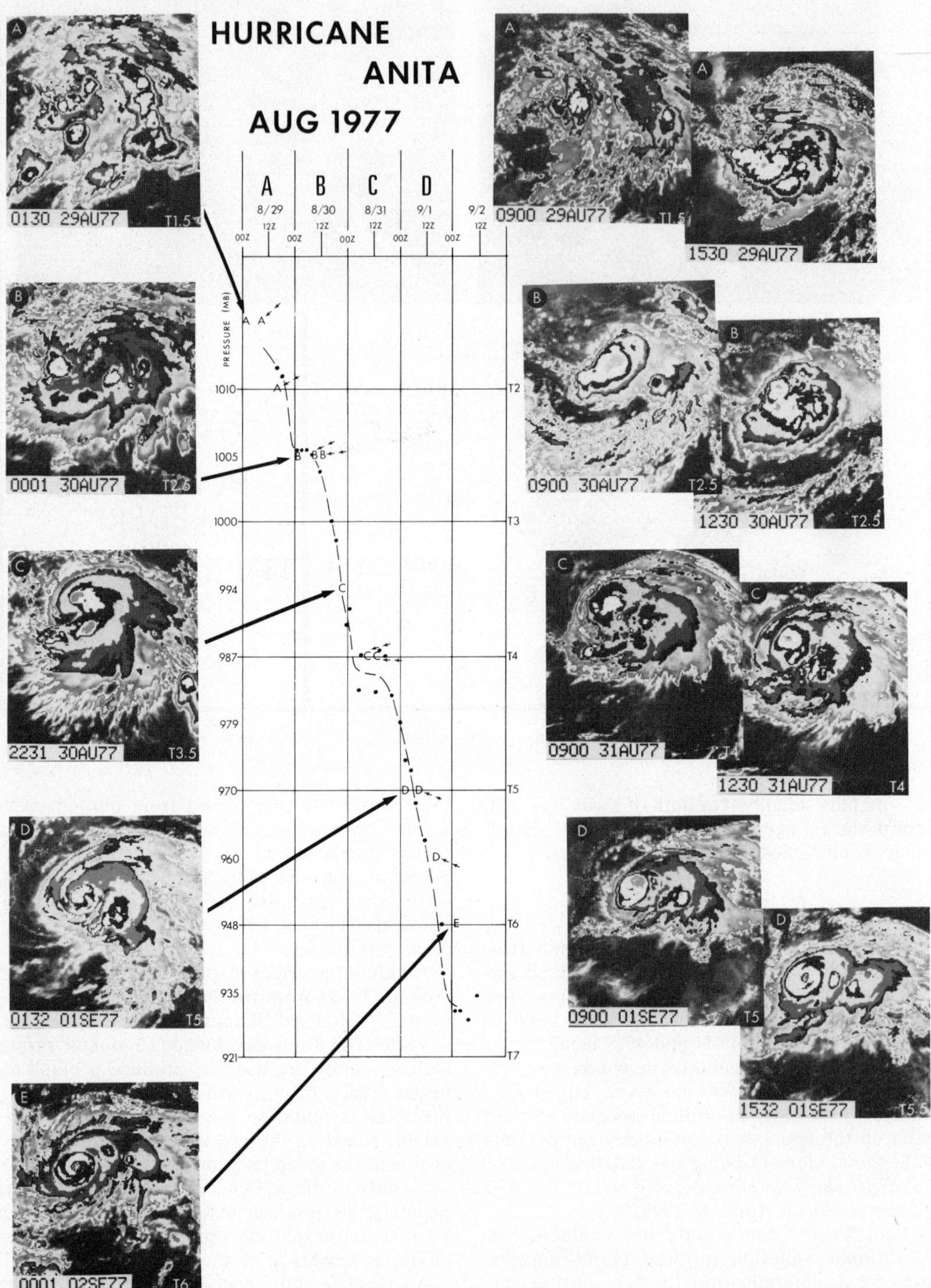

Fig. 27-8. Enhanced pictures of hurricane Anita (1977) are shown adjacent to the reconnaissance central pressure estimates. Pictures across the top marked "A" refer to the first day of development, those marked "B", the second day, etc. Times shown are GMT.

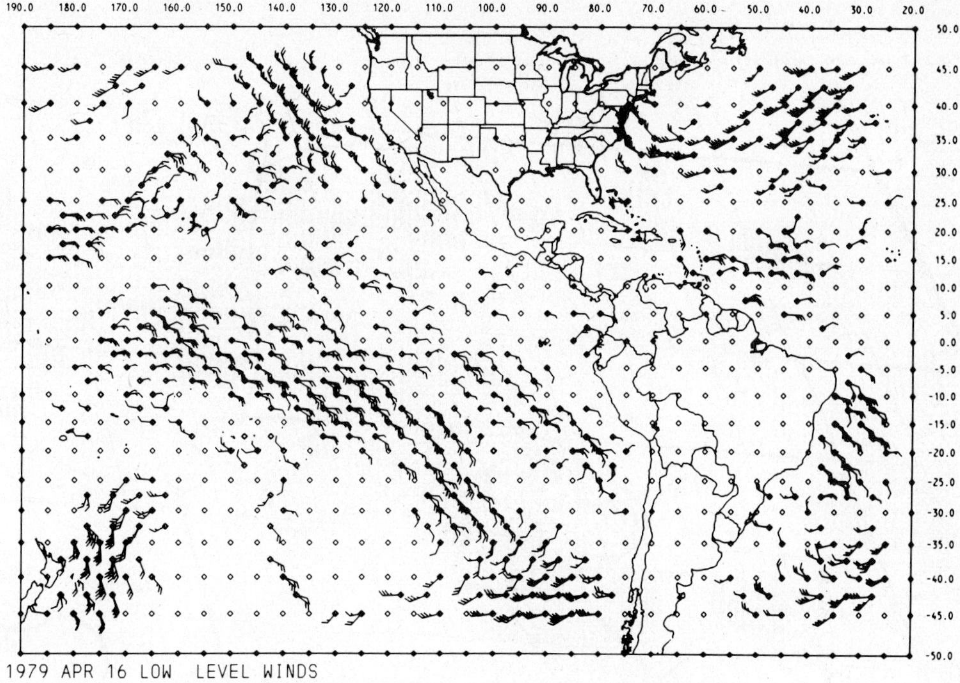

Fig. 27-9. GOES Derived Winds—Low Level, April 16, 1979.

picture times for two areas of approximately 1,000 km on a side. A height/temperature profile was then computed for each center latitude/longitude position using the latest NMC upper air data. Each IR-pixel temperature-value was converted to a height value, coded and stored in a printer-compatible file along with the radar observation codes of 2135 GMT and 2235 GMT.

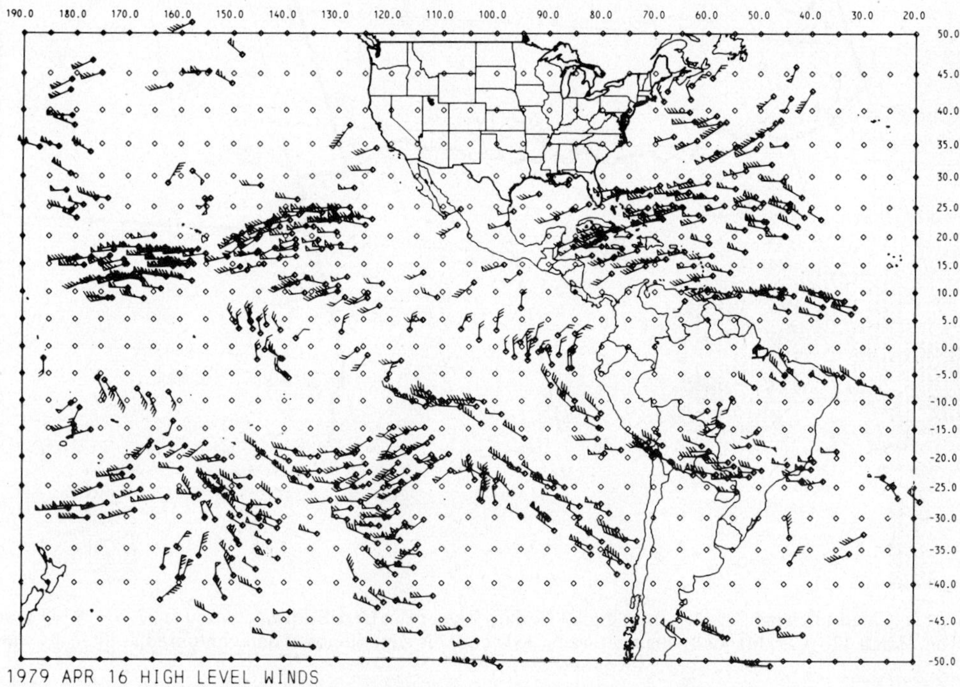

Fig. 27-10. GOES Derived Winds—High Level, April 16, 1979.

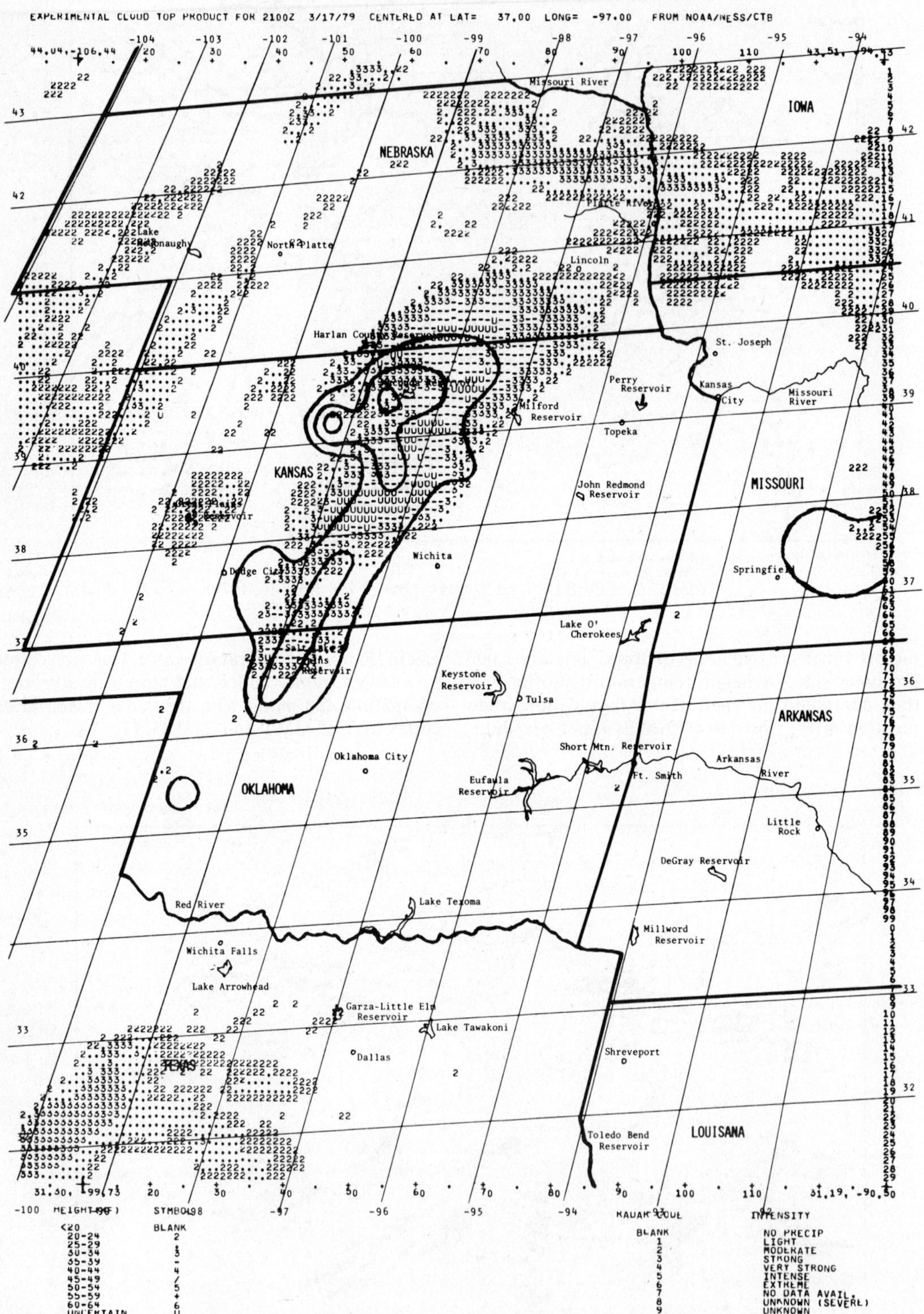

Fig. 27-11. Cloud-top height display from digital IR data taken from GOES-East centered at 37.ON, 97.OW at 2100 GMT on March 17, 1979. Manually digitized radar (SD code) is superimposed and contoured at intensity intervals 1–3–5.

Fig. 27-12. SMS-1 VIS 2100 GMT, March 17, 1979—One-kilometer resolution picture showing bounded test area.

Atmospheric Soundings

The TIROS N system satellites carry an instrument called the TIROS Operational Vertical Sounder (TOVS). Data from this instrument are operationally processed to provide global temperature and moisture profiles. The TIROS N system spacecraft, with their highly sophisticated sensors, provide significantly improved sounding products globally with the microwave channels permitting the derivation of soundings in cloudy areas.

The spacecraft provide global sounding data from their 14 orbits per day. The data, stored on the spacecraft, are transmitted to the Command and Data Acquisition ground stations. The data are then transmitted to the satellite control center in Suitland, Maryland, where they are entered into the computer for further processing. The soundings are produced at a 250-kilometer grid spacing with a provision for 500-kilometer grid spacing for reduced coverage.

Temperature soundings from the surface to 0.4 millibars are produced for various layers within this height range. Moisture soundings are produced for those atmospheric layers. A total output of the subsystem for a 24-hour period is about 8,400 soundings from one spacecraft.

Satellite temperature and moisture profiles are matched in time and space with radiosonde data acquired from the NMC. This is done twice daily using radiosondes at 0020 GMT and 1200 GMT. Every 12 hours, a statistical evaluation is performed comparing satellite derived soundings with radiosonde data to monitor the accuracy of the satellite soundings. Figure 27-13 shows a typical sounding obtained from the TOVS and compared with a radiosonde observation. The amount of data processed through this system is enormous. Figure 27-14 shows the amount of data obtained over just the Northern Hemisphere.

Precipitation Estimates

Techniques have been developed to estimate precipitation amounts by using geostationary-satellite images. Rainfall is inferred from satellite cloud-images using empirically derived relationships. Cloud area is measured within several temperature thresholds from a thermal infrared image. The amount of rain inferred is a function of the temperature in the cloud and the stage of development of the cloud.

Figure 27-15 shows an enhanced infrared geostationary-satellite image of Hurricane Anita on September 1, 1977. Four different temperature ranges are shown. The black indicates temperatures lower than −65°C; the white indicates temperatures higher than −20°C. The other temperatures lie within this range.

Figure 27-16 shows plots of the parameters derived for Hurricane Anita. Although the average rainfall varied little, the rain potential changed considerably because of the inverse rain potential on storm speed (more rain for slower storms.) The numbers on the rainfall potential graph indicate storm speed in knots.

Anita's rain computations were verified by ground truth data. The two-day average rainfall (September 2, 3) was 88 mm compared with the 102 mm average rain potential and 87 mm actual two-day accumulation.

MESOSCALE

Higher resolution data and geostationary satellites have increased our understanding (and to some extent our forecasting abilities) of mesoscale

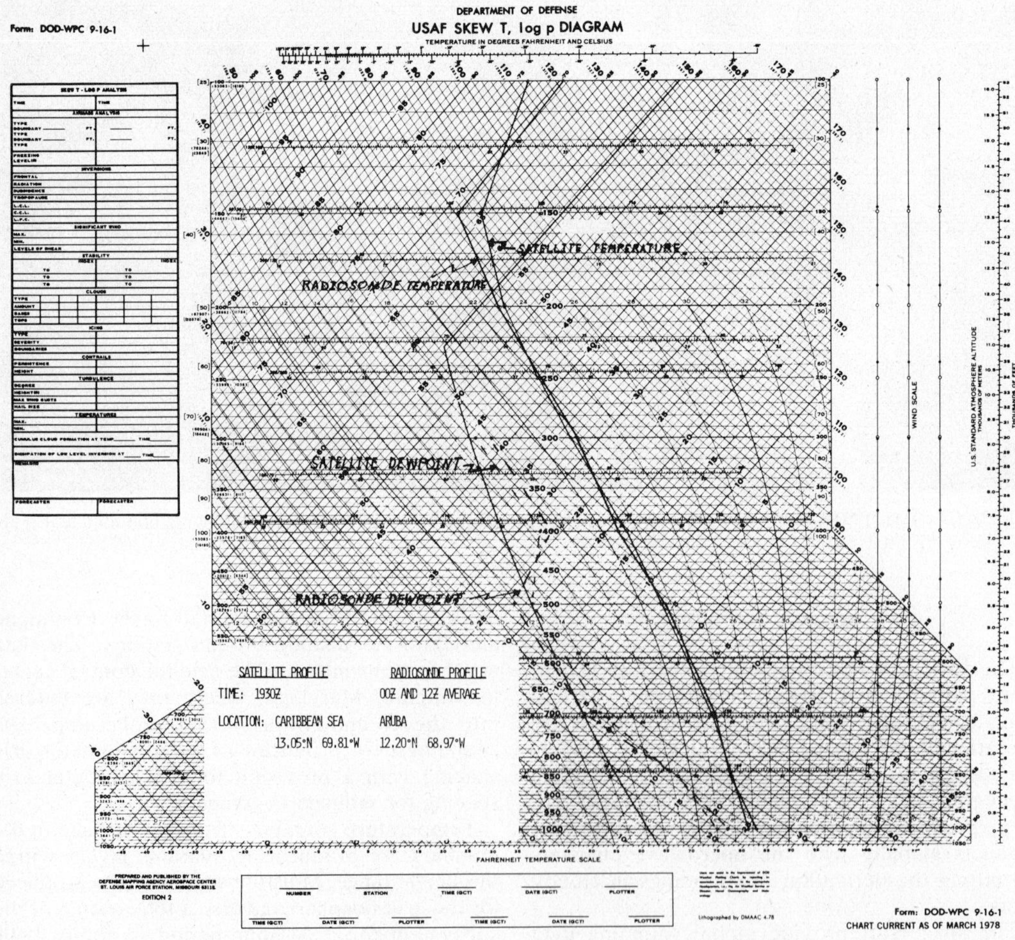

Fig. 27-13. Temperature and Humidity Profiles from TOVS.

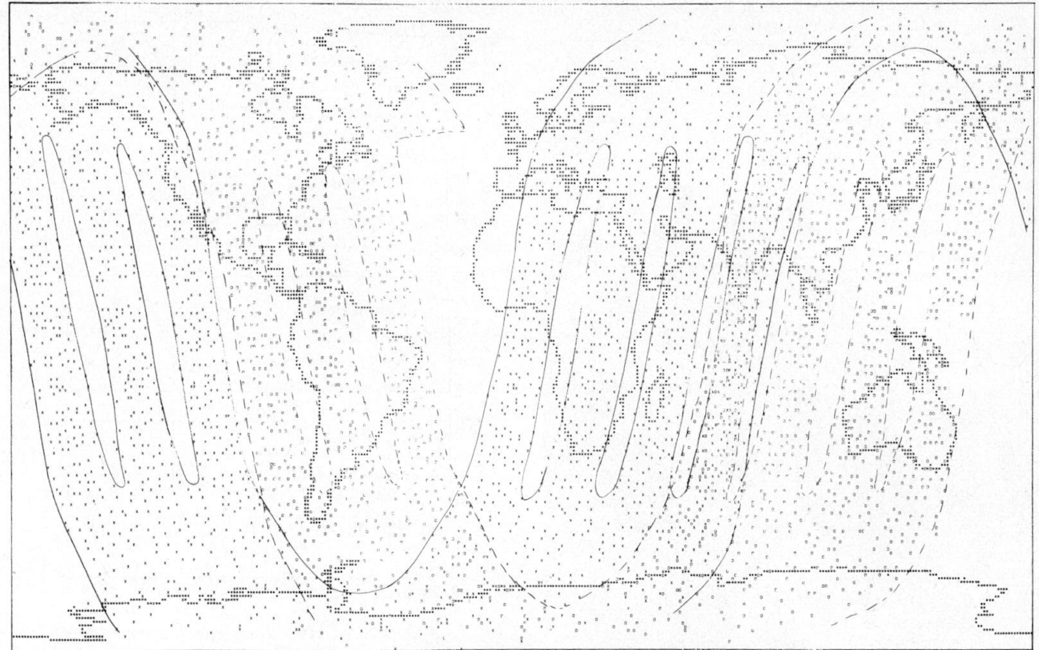

Fig. 27-14. Global sounding coverage from two polar-orbiting satellites.

weather systems. In particular, progress has been made in forecasting fog dissipation, afternoon thunderstorms, and lake effect storms. Some of these new, small-scale forecasting techniques are discussed in connection with the following figures.

Fog and Stratus Distribution and Dissipation

Fog and stratus areas often blanket large areas of the coastal and plains states. The actual areal extent and times of formation or dissipation are often difficult to assess from conventional reporting stations. Satellite data can quickly show the distribution of and changes in these low clouds.

Stratus clouds are uniformly textured clouds that vary from gray to white depending on their thickness and the angle of the sun. They commonly have irregular edges that conform to certain terrain features, such as river valleys, mountain foothills and coastal plateaus.

During the day there is a natural erosion or dissipation of the low clouds caused by daytime heating. Satellite pictures show that fog and low clouds dissipate first at their outer edges, gradually eroding into the center. Some variations of brightness, and thus of thickness, have been noted in these cloud areas. Gurka (1978) shows that the early morning bright areas rather than the darker parts near edges are usually the last to dissipate. He has devised a scheme by which one can predict the dissipation of radiation fog and stratus by measuring the brightness gradients present in the early morning pictures.

One example of this process is shown in Figures 27-17 and 27-18 with arrows pointing to the fog boundary. The fog at A on Figure 27-17 appears brighter than at B and is therefore either thicker or denser than that at B and should therefore dissipate at a slower rate. Valley fog will not necessarily dissipate in the manner described, particularly in deep, narrow valleys where the valley circulation will play a dominant role in dissipation (Spatola, 1972).

The process that leads to the erosion of fog from its outer edges inward can be explained in this way:

The sun heats the ground beyond the fog more than beneath the fog, and the air above the fog sinks to replace air within the fog that is spreading out to replace the rising air. This situation, which develops due to the differential heating of the early morning cloud-free and cloud-covered areas, is similar to the land/sea breeze effect (Purdom and Gurka, 1974). Furthermore, during the warmer months of the year, convective lines will commonly form along the early morning fog-boundary and at times move outward as the sinking air spreads out from the early morning fog-areas. This is similar to the behavior of the cloud lines that form at the edge of the sea-breeze front.

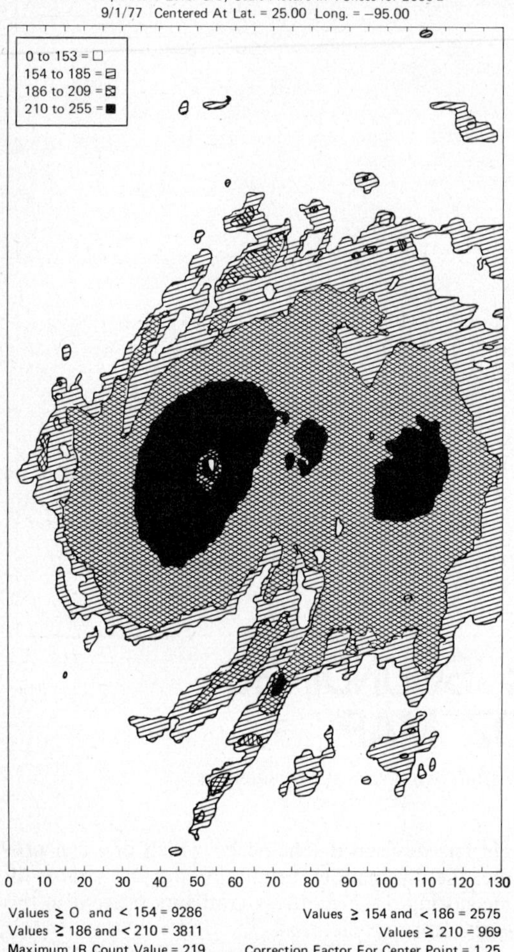

Fig. 27-15. GOES-2 IR 2000 GMT, September 1, 1977.

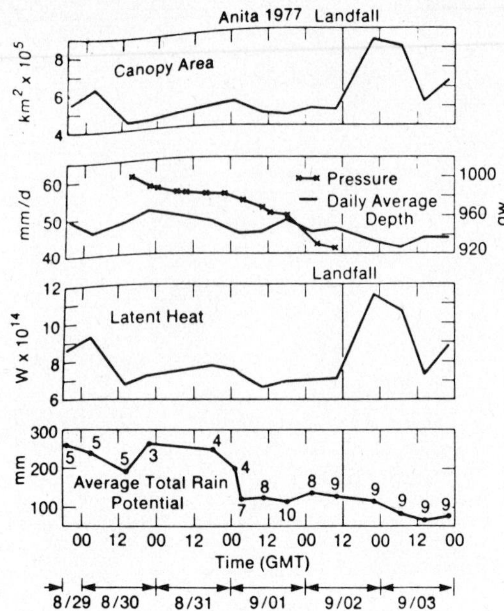

Fig. 27-16. Hurricane Anita Parameters.

The effect of an early morning fog-cover on afternoon convection can be seen clearly in an example presented by Gustafson and Wasserman (1976) (Figures 27-19, and 27-20). The fog and stratus, which are dissipating between 1530 and 1800 GMT, are at A on Figure 27-19.

As the fog begins to erode, a relatively bright line of cumulus forms along the edge of the early morning cloud-boundary where the ground has remained cool. By 1800 GMT, the effect of the fog in keeping the ground cooler is shown by the lack of cumulus at C (Figure 27-20).

Since fog erodes first at its outer edges, an important parameter that must be considered, in forecasting the time of fog dissipation at a point, is the location of the point relative to the fog boundary. The only tool available to the forecaster for determining this parameter is high-resolution visible satellite-imagery.

Lake Effect Storms

During the winter months, cold continental air passes over the warmer waters of the Great Lakes and produces a large field of fine cloud lines that extend over the adjacent land areas. These cloud lines are usually oriented parallel to the low-level winds. Once formed, they remain stationary and produce dreary skies and copious amounts of snow on adjacent land areas. Frequently these bands receive additional heating and pick up additional condensation nuclei as they pass over the large, industrialized centers such as Detroit; these cause an increase in the cloud-line size and could affect the cloud's downward excursion. Satellite data can quickly show details of the downwind extent and the area affected by these local storms.

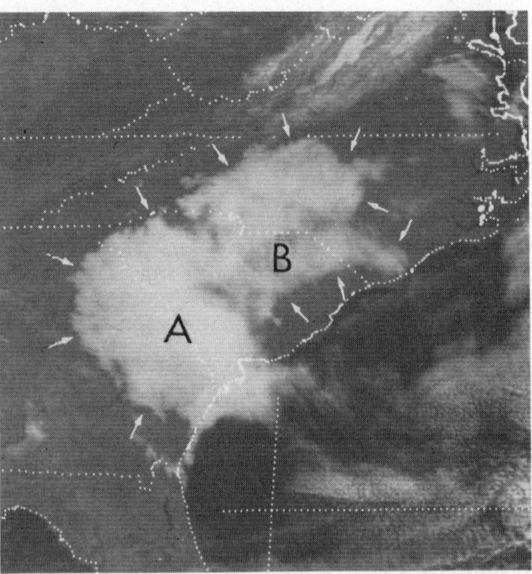

Fig. 27-17. SMS-1 Visible 2 km data, 1600 GMT 7 January 1975.

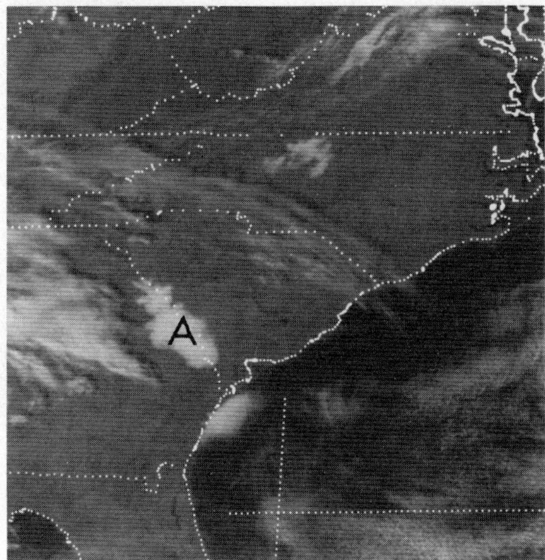

Fig. 27-18. SMS-1 Visible 2 km data, 2000 GMT 7 January 1975.

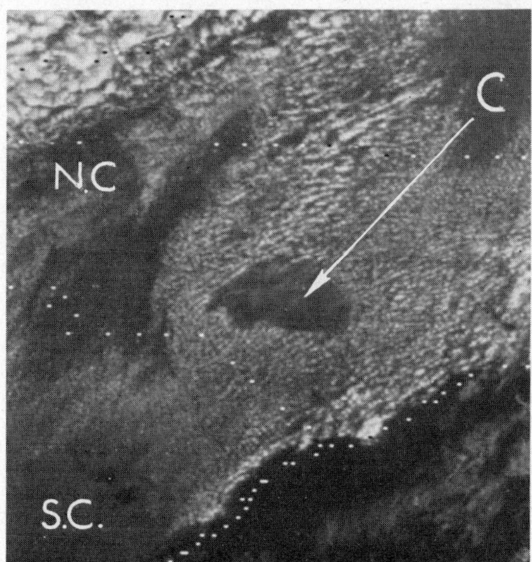

Fig. 27-20. SMS-1 Visible 1 km data, 1800 GMT 27 September 1975.

During the spring and early summer, the Great Lakes remain cooler than the surrounding shores. Increased diurnal solar radiation warms the land quickly and vertical mixing produces fair weather cumulus-fields about these water bodies. The effect of these cooler water bodies is apparent even under the influence of an organized system.

Cloud lines that form as a result of cold air advection over the Lakes often extend great distances, i.e. southward into Kentucky and Tennessee or eastward to the mid-Atlantic coastal states (Parmenter 1974). High-resolution imagery clearly delineates the areal distribution of these cloud lines and allows the user to assess the low-level wind shear and to monitor atmospheric changes through cloud development, dissipation, and orientation. Satellite data from December 17, 1975, illustrate this point.

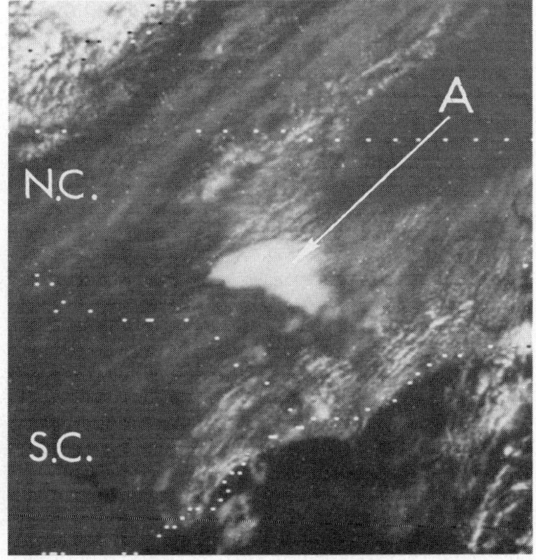

Fig. 27-19. SMS-1 Visible 1 km data, 1530 GMT 27 September 1975.

At 1500 GMT, December 17, a low was observed just north of Lake Huron with the attending frontal zone located just east of Lake Erie. Fog and stratus marked much of the front (F, Figure 27-21). Precipitation was principally located about the Great Lakes. Two small comma-cloud formations (G in Figure 27-21), composed of low and middle cloudiness, may represent two small vorticity centers propagating around the south side of the low. A long cloud line, H to H′, extends from the tail of the easternmost vortex westward across Lake Huron to Pellston, Michigan. Continuous light snow was reported along this line. Further south, a large area of lake-generated stratocumulus cloudiness (I, Figure 27-21) covers much of Michigan. Under light winds, the cloud lines are very close to-

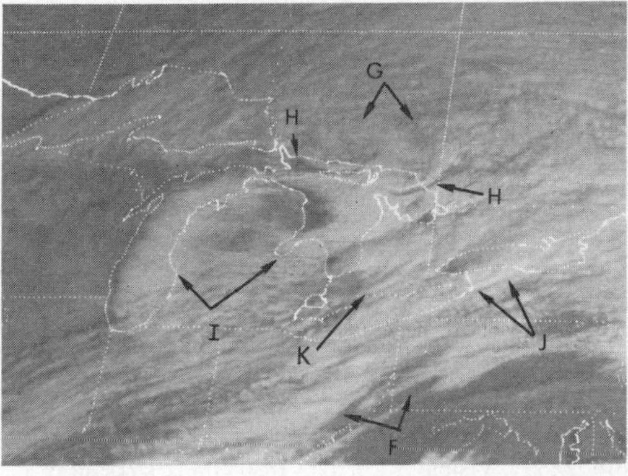

Fig. 27-21. SMS-1 Visible 2 km data 1500 GMT, December 17, 1975.

gether and appear as a general overcast. Snow showers were reported throughout the area. This low sun-angle view pinpoints two additional areas (J and K) where more vertically-developed cloud lines, similar to H, are located. Buffalo, New York, located along the northern edge of J (Figure 17-21), reported continuous light snow at this time.

Middle and high-level cloudiness that covered Lake Superior in Figure 27-21 moved northward during the following 3½ hours, revealing the distribution of cloudiness over this Lake (Figure 27-22). The orientation of cloud lines is governed by the low-level vertical wind shear and frictional effects. Here, an apparent northeasterly flow from Canada becomes northerly over the Lake and increasingly cyclonic on the southern shoreline (D to D′, Figure 27-22). Winds over Lake Michigan remain westerly and clouds cover the entire state of Michigan. These clouds dissipate along the west shore of Lake Huron reforming near the warmer lake center. The two cloud vortices (G), seen earlier, appear less organized by 1830 GMT; however, the long convective cloud line (H) is still apparent. Westerly flow throughout this area has allowed no meridional excursion of these clouds, concentrating continuous snow in a local area. Winds over eastern Lake Erie changed from south to southwesterly during this three-hour period, creating a condition that is increasingly favorable for heavy snows in Buffalo. Satellite imagery suggests that the convective cloud band seen at J, Figure 27-22, moved slowly northward and increased to the southwest due to favorable surface heating of the cold air over Lake Erie. Buffalo was reporting moderate snow at this time. A similar cloud formation can be seen over Lake Ontario (K).

During the spring and summer, the Great Lakes exhibit weather conditions that are influenced by warmer land and cooler water. Satellite data in Figure 27-23 show a deep low (986 mb) (P) located west of James Bay with a broad cloud band (Q−R) marking the frontal zone. Westerly

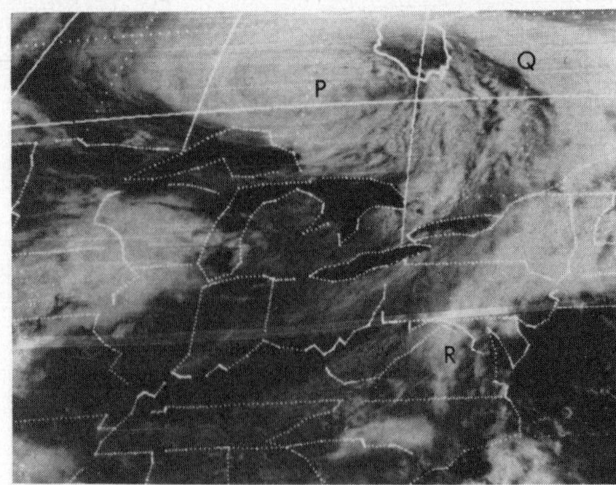

Fig. 27-23. SMS-1 Visible 4 km data 1500 GMT, 16 June 1975.

to southwesterly winds prevailed over the Lakes bringing warm air across the cooler Lakes. Although the air temperatures at 1500 GMT (Figure 27-23) were still in the 60s, the cooler air over the water effectively inhibited cumuliform cloud production and advection. This is most apparent over Lakes Huron and Erie at this time. In a later view, taken at 2000 GMT, Figure 27-24, the surface winds have become more westerly over Lakes Erie and Ontario. As a result, cooler, more stable, air is being advected along the axis of these Lakes, thus dissipating the cloudiness over Western New York and Pennsylvania (S). Further west, clouds ring the northern shores of Lakes Huron and Superior with a localized maximum (I) extending eastward from upper Michigan across southeastern Lake Superior.

Squall Lines and Thunderstorm Growth

Thunderstorms often form along squall lines along or ahead of a frontal zone. The development along these lines is affected by the stability of the

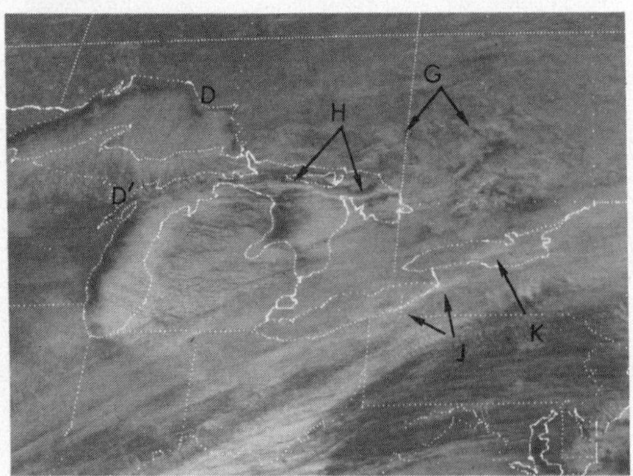

Fig. 27-22. Visible, 2 km data 1830 GMT, December 17, 1975.

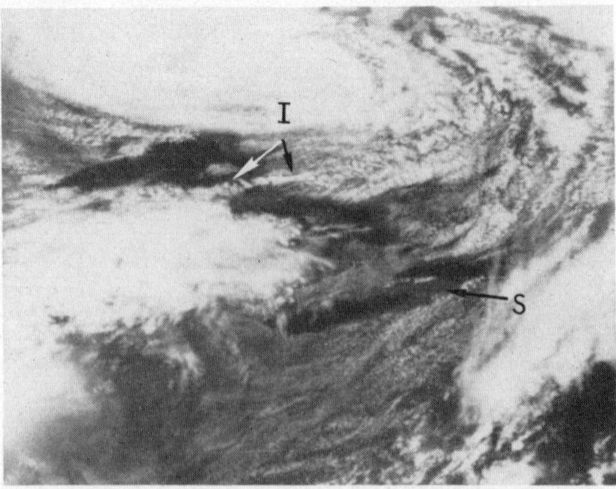

Fig. 27-24. SMS-1 Visible 4 km data 2000 GMT, 16 June 1975.

air through which the squall lines move. If a squall line passes over an early morning clear zone, the added instability will cause the thunderstorms to develop rapidly.

High resolution pictures have revealed the presence of long arc-shaped cumulus-congestus lines along the forward edge of thunderstorm clusters. These arcs are usually produced along the forward edge of out-rushing, rain-cooled air and mark the boundary of the mesohigh produced by the thunderstorm. These arcs accelerate away from the main showers. Arc lines often intersect old frontal boundaries or other arcs. Augmented lifting at such intersection points can result in intense, and often severe thunderstorms.

Geostationary satellite imagery gives meteorologists a unique view of convective development. It is the only meteorological observing tool that can simultaneously observe the evolution of clouds from the synoptic scale down to the cumulus scale. Satellite data can fill the spectral gap between large mesoscale events and the individual thunderstorm or thunderstorm system. The clouds and cloud patterns observed in a satellite image represent the integrated effect of ongoing dynamic and thermodynamic processes in the atmosphere. When that information is combined with more conventional data such as radar, surface and upper air observations, the convective scale interactions that are so vital in the formation and maintenance of deep convective activity may be analysed and better understood.

Satellite imagery presented a "birds eye" view of intense thunderstorm development on May 6, 1975. In Figure 27-25, low-level cumulus clouds in the moist air, can be seen as a long grey shadow (D) from central South Dakota southward through eastern Nebraska into north central Texas. The sharp western edge correlates to the dormant dry line. Note the line of thunderstorms (B) from southeast Missouri to western Iowa. Subsequent pictures through 1730 GMT show an increase in development and intensity of convective activity. In Figure 27-26, which was taken 33 minutes prior to the outbreak of a tornado at Omaha, Nebraska, very heavy thunderstorm activity can be seen from the Dakotas to northeast Oklahoma. The main squall line is near the western edge of the bright clouds, with cirrus blowoffs to the east and northeast of the line. Note also, several large thunderstorms (F) over eastern Texas. This activity is just east of the dry line near the intersection of the sub-tropical jetstream, as evidenced by the high clouds from the Texas Big Bend northeastward, and the low level moist tongue.

Differential Heating

The formation and distribution of convection, from the smallest fair weather cumulus to large cumulonimbus clusters, is often a function of surface heating alone. For instance, given uniform atmospheric conditions, fair weather cumulus will form over quickly heated land areas, leaving the cooler water surfaces such as lakes and rivers cloudfree. A similar pattern is created when low clouds persist over an area during the morning heating hours.

Where solar heating is the controlling mechanism for thunderstorm formation, an early morning cloud-cover will keep the underlying surface cooler. This cooler air, being more dense, sinks and spreads outward, lifting the warmer and more unstable air about its perimeter. This causes the thunderstorms to form along the perimeter of the early morning cloudiness. Even after the dissipation of the morning cloudiness, the surface heating lags behind and tends to keep this region free from convection for the remainder of the day. A squall line passing over an early morning clear area is often enhanced by this increased heating and instability.

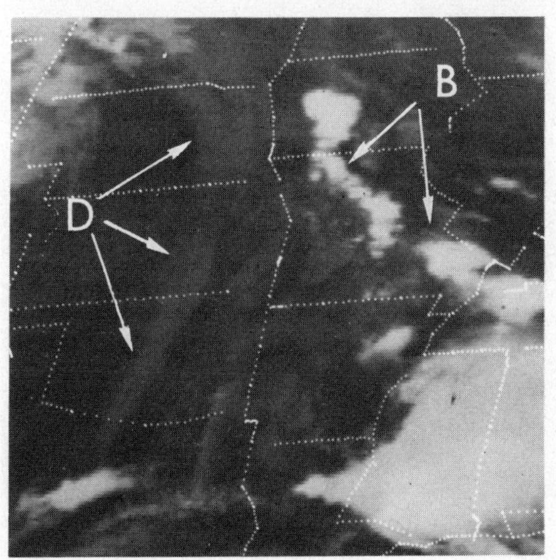

Fig. 27-25. SMS-1 IR 1501 GMT 6 May 1975.

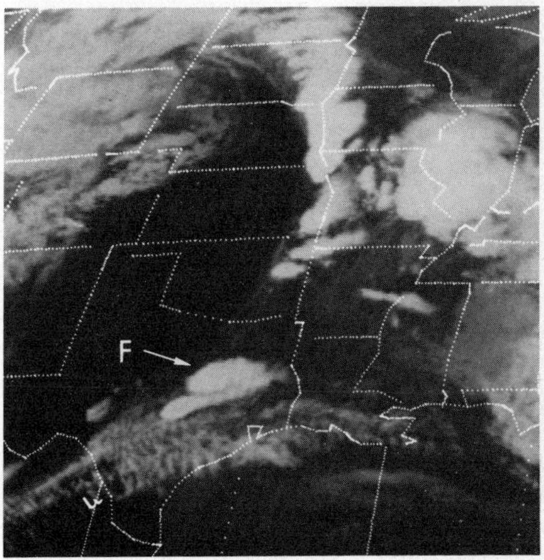

Fig. 27-26. SMS-1 IR 2100 GMT 6 May 1975.

The formation areas of afternoon convection are, in many cases, influenced by a variety of mesoscale features, such as differential heating, the sea breeze, and terrain. One such case occurred on July 17, 1977. The thunderstorm forecast that was available that morning showed a broad region of 60 to 70 percent thunderstorm probabilities over Virginia, the Carolinas, and Georgia. An important task for the forecaster is to refine synoptic scale forecasts by specifying the most likely areas of convection formation and movement. This task can be aided by using satellite imagery for analysis and for delineating regions and boundaries that might contribute to convection.

The National Weather Service surface analysis for 1200 GMT showed a broad area of low stratus and fog in the region of 60 to 70 percent thunderstorm probabilities. Differential heating owing to the fog can affect the formation of afternoon convection (Purdom and Gurka, 1974). The most likely formation area is on the clear side of the fog boundary, and the least likely area is near the center of the fog region. It is therefore important to be able to analyze the fog boundary correctly. The 1230 GMT GOES-1, 2-km, visible imagery (Figure 27-27) reveals the fog and stratus in central North Carolina. Fog and stratus on a scale this size have been shown to effect the evolution of afternoon convection over extensive areas.

By 1800 GMT, satellite imagery (Figure 27-28) indicates that all the convection is at or outside the earlier fog boundary, shown by the dashed line. Thunderstorms are developing at this time at C and D, Figure 27-28. The thunderstorms at C are forming in a warm tongue and surface convergence region. Terrain effects and differential heating contributed to the formation of thunderstorms at D which are located on the slopes of the Appalachian Mountains.

Differential heating induced by an area of morning fog had an effect on the formation areas of afternoon convection in this case. The sea breeze, terrain, and moisture maximum also influenced thunderstorm formation. Satellite imagery was used in this example as an aid for analysis and for locating boundaries that might contribute to convection.

It is important to note that solar heating was the main driving force behind convection in this case, with the synoptic scale situation being of less importance. Observations have shown that differential heating effects also contribute to the weather patterns and evolution in cases where the synoptic scale wind-field is stronger. However, in those cases, the interactions between the large and small scales are complex and difficult to forecast correctly. Therefore, differential heating as a tool in forecasting thunderstorms is used most reliably when the sun's heating is the major factor in convective development.

Mountain Effects

Under various conditions, mountains are barriers to low clouds or are the initiating points for middle- and high-level clouds. When the flow at the mountain top is perpendicular to the ridges and greater than 40 knots, wave clouds, suggesting turbulent conditions, can form downwind of the ridges.

Upslope motion of moist air along the eastern slopes of the Appalachians or the Rockies produces large areas of fog and stratus along the

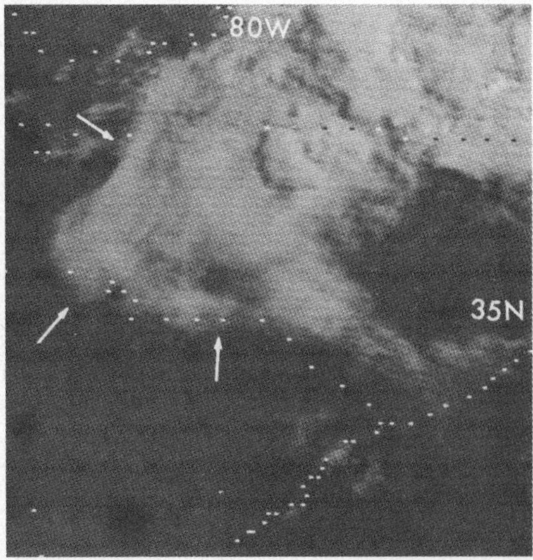

Fig. 27-27. GOES-1 2 km Visible Imagery 1230 GMT July 11, 1977.

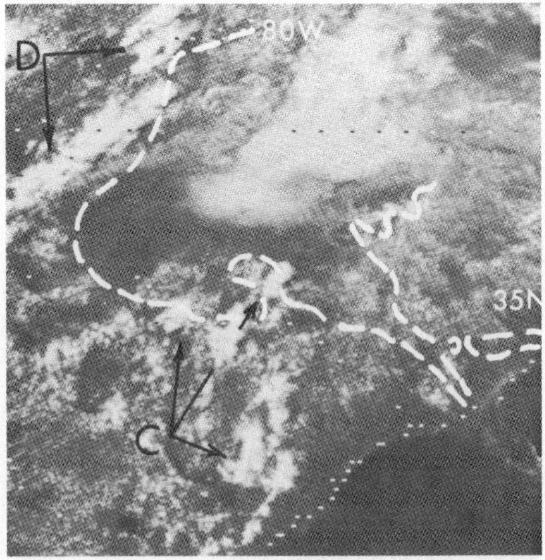

Fig. 27-28. GOES-1 1 km Visible Imagery 1800 GMT July 11, 1977.

windward sides of the lower slopes. In the fall and winter, strong surface highs, with low-level easterly flow, settle into the Plains States. This, together with long nights and strong nocturnal radiation, causes the fog to thicken and persist for days. The areal extent of this low cloudiness can easily be determined from the satellite data.

Satellite data have shown that wave clouds, short parallel stratocumulus or altostratus cloud bands, are quite frequently observed in the mountainous regions. A newer phenomenon is the role of the mountain peaks in generating and increasing middle and high cloudiness downwind from mountain ranges. This phenomenon is particularly noticeable when thin cirrus clouds enter the West Coast and become increasingly thicker and broader after crossing mountain ridges. The production of these clouds can affect the weather downwind by decreasing the solar heating or the amount of nocturnal radiation.

The 1-km visible imagery for 1330 GMT on April 27, 1976, (Figure 27-29) shows an extensive area of mountain wave cloudiness (A–A′) stretching from central Pennsylvania into southwestern Virginia. This wave cloud-pattern developed within a stratocumulus deck extending southward from the Great Lakes. The cloudiness was located in the region of northwesterly flow behind a large storm over the Canadian Maritime Provinces. The stratocumulus terminated abruptly where the low-level flow changed from cyclonic to anticyclonic curvature. A more cellular or convective pattern is present in the stratocumulus field north of Lake Erie and in portions of Ohio and Pennsylvania (B, Figure 27-29).

By late afternoon, the wavelength of the lee-wave clouds had increased markedly in Virginia and West Virginia. Part of this increase in wavelength can be attributed to the diurnal destabilization that occurred owing to daytime surface heating and stronger lower-level winds.

Air Pollution

During the summer months, large high-pressure areas with nearly calm and clear conditions become stationary over the East Coast. Under these conditions, natural atmospheric dispersion decreases and gaseous and particulate matter becomes concentrated in the lower layers of the atmosphere. The extent of the haze or pollution is best seen when the viewing angle of the satellite and the position of the sun are such that forward scattering in the atmosphere is maximized. If conditions are extremely hazy, normal land marks, such as the Great Lakes or East Coast, cannot be recognized.

Figure 27-30, a visible image for 1230 GMT, May 14, 1977, shows the contrast between clear and turbid air on either side of a front that extends across the mid-Atlantic states. This picture has

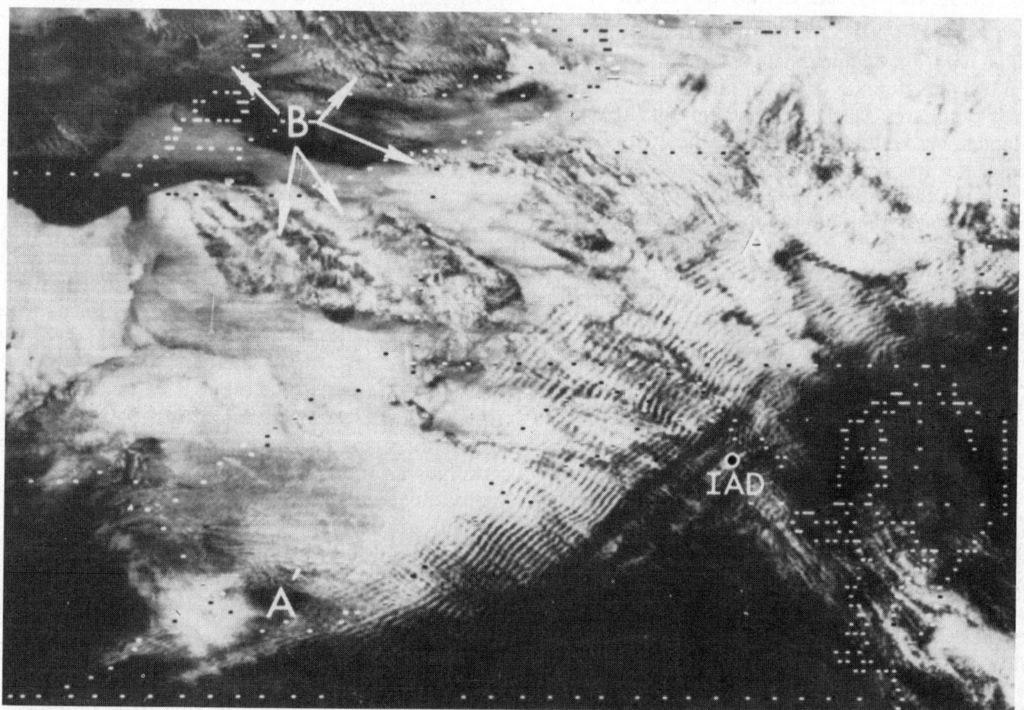

Fig. 27-29. GOES-1 1 km Visible Imagery 1330 GMT, 27 April 1976.

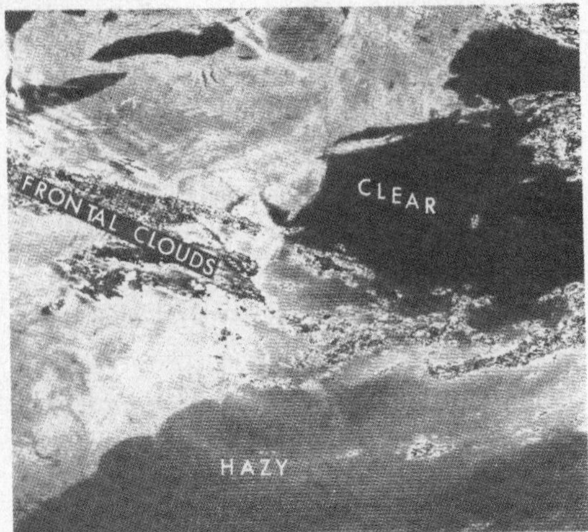

Fig. 27-30. GOES-1 Visible (2 km) 1230 GMT, May 14, 1977.

been artificially enhanced to accentuate the contrast. For this picture, the actual brightness difference between the clear and the hazy areas is estimated to be about 5 percent.

Satellite-observed areas of haze and pollution over the cental and eastern U.S. also are associated with the occurrence of fogs and reduced convection. Convection is often observed to be concentrated at the boundary between hazy and cleaner air.

REMOTE SENSING OF THE EARTH'S RADIATION BALANCE

The energy balance of the earth-atmosphere system plays a key role in understanding the earth's climate. All the components of the energy balance, viz. radiation, sensible heat, and latent heat, are important quantities and contribute greatly to the climate. Determination of the balance between these components and their temporal and spatial variations is essential because the non-uniform distribution of heating and cooling drives the motion of the atmosphere and the oceans. At present, only the radiation balance components have been successfully measured from space and they will be described in this chapter.

Solar Input

The climate of the earth is dependent on the radiation received from the sun, and some of the climate models have shown that a one percent change in the incoming solar energy can change the mean surface temperature of the earth by 2 degrees C (Bandeen et al., 1965). It is essential to know the daily variations in the "solar constant" in order to monitor and forecast weather and climate-related events.

In the past, solar constant measurements have been made primarily by ground-based instruments. However, with the introduction of rockets and satellites, instruments have been flown in space to measure the solar constant. Nimbus 6 and 7 satellites (Boldyrev and Vetlov, 1970; Gruber, 1977) carried special instruments to measure the incoming solar energy at the satellite altitude. Hickey et al. (1980) have described Nimbus 7 instrumentation and results, and only a brief summary will be provided here.

The instrument with which to measure the solar constant was part of the Earth Radiation Budget (ERB) instrument package on the NIMBUS 7 satellite. Each of the ten solar channels is independent and each sensor is a torridal-plated thermopile to which a cavity is mounted. The cavity is composed of an inverted cone within a cylinder, and the interior surfaces are coated with a black paint. The absorptivity of the unit is close to 0.999. An electrical heater attached to the cone is switched on once every two weeks during calibration sequences when the instrument views space. The radiometer has a 10-degree field of view, which allows the sun to fully irradiate the cavity for about three minutes during each 104-minute orbit. The zero-level irradiance is sensed while space is being viewed before and after each solar measurement and is subtracted to yield the net thermopile output. The solar irradiance is computed by applying the temperature-corrected sensitivity to the net thermopile output. The solar-constant value is this irradiance, corrected to mean earth-sun distance. The NIMBUS 7 measurements during a seven-month period (Nov. 1978–May 1979) provided a solar constant value of 1376.0 Wm^{-2} with an RMS deviation of ± 0.73 Wm^{-2}, and the maximum range was less than 3.3 Wm^{-2}. These results showed that it is possible to measure and monitor the solar constant from space with a precision of greater than 0.05 percent.

Radiation Budget

The dynamics of the earth-atmosphere system (weather and climate) are determined by the energy input to the system and the distribution, transformation, and storage in various forms. The earth-radiation budget measurements made at the top of the atmosphere involve most of the processes mentioned above. The reflectance and the upward flow of energy emitted by the earth-atmosphere system have large temporal and geographic variations and the net radiation also varies with time and space. The net radiation is the driving force for the atmospheric and oceanic circulation and the associated energetics.

Satellites can provide the required measurements of these radiation-budget parameters on a global scale. A number of satellites have been used in the past to measure these quantities; for example, Weinstein and Sumoi (1961) discussed the first measurements obtained from omnidirectional sensors from the Explorer 7 satellite in 1959. House (1965), Vonder Haar (1968), Vonder

Haar and Suomi (1971), and McDonald (1970) used the measurements from flat-plate radiometers to obtain radiation-budget parameters. These instruments provided only very low spatial-resolution data owing to their observing geometry. Rao (1964), Bandeen et al. (1965), Winston and Taylor (1967), Boldyrev and Vetlov (1970), Gruber (1977), Winston et al. (1979) have used data from narrow field-of-view sensors and higher spatial-resolution data to obtain earth-atmosphere radiation data. Because of the orbital parameters and narrow field of view of the instruments, most of the methods involved computing daily average values of radiant flux densities of reflected solar and emitted thermal radiation from respective single measurements of the radiances over each area.

The net radiation balance of the earth-atmosphere system can be represented by

$$N = (1 - A)\, I_o - E \qquad (27\text{-}1)$$

where

N = net radiation
A = albedo of the Earth-atmosphere system
I_o = the incoming solar radiation
E = outgoing longwave radiation

From the operational TIROS and NOAA series of satellites the albedo is determined from the visible channel by assuming that the reflectance in a narrow spectral interval (0.5 – 0.7 μm) is a good estimate of the full spectral reflectance (0.2 – 4.0 μm), that the observed reflectance is isotropic and independent of solar zenith angle, and that there is no diurnal variation of the reflecting surface. In the case of NIMBUS satellites, the albedo is determined from a wide spectral channel (0.3 – 4.0 μm) and corrections for the anisotropic reflection properties of the surfaces have been applied. However, no diurnal corrections were applied to the various reflecting surfaces. More details about derivation of albedo values from satellites have been described by Raschke (1968), Raschke and Bandeen (1970), and Gruber (1977). These time- and space-averaged albedo values, as derived from any of the methods mentioned above, seem to provide reasonable values. Figure 27-31 shows a global distribution of albedo derived from the NOAA operational satellites. It shows the average for a 45-month period (June 1974–Feb. 1978), and the shaded areas represent values above 30 percent, which generally correspond to clouds, snow, ice and bright desert regions. The top half of the figure shows the global distribution between 60 N and 60 S latitude. The lower left corresponds to the Arctic and the lower right to the Antarctic.

An estimate of the outgoing longwave radiative flux has been computed from the radiance measurements in the narrow spectral-window region (10–12 μm) by using regression models. This technique is primarily used to derive the outgoing longwave flux values from the TIROS and NOAA series of satellites. The regression technique considers different model atmospheres (about 100) covering a broad range of temperature and moisture as well as overcast and clear sky conditions. The regression equation is linear and explains about 98 percent of the variance. The daily mean outgoing radiation is estimated by averaging the daytime and nighttime observations. Similarly, from the NIMBUS satellites the outgoing longwave radiation has been computed from radiance measurements obtained in narrow spectral channels and by using a regression formula derived from model atmospheres. In all these studies corrections for the dependence on the zenith angle of measurements have been performed with a statistically derived "limb-darkening" function.

Figure 27-32 shows an example of the global distribution of outgoing longwave radiation derived from NOAA operational satellites. It shows the average for a 45-month period (June 1975–Feb. 1978). The shaded areas represent values less than 250 Wm^{-2}, and generally correspond to clouds, snow and ice. Figure 27-32 covers the same area and time as Figure 27-31. Over the oceans the shaded areas in both figures should be similar and should represent cloudy regions. As in the case of albedo, no diurnal variation of the outgoing longwave radiation has been considered in any of these studies.

From Equation (27-1) it is obvious that the net radiation flux N can be obtained if A, I_o, and E are known. Figures 27-31 and 27-32 show the distribution of A and E, and a similar global distribution can be shown for I_o. From these three quantities the distribution of the net radiative flux can be obtained for the same period as in Figures 27-31 and 27-32, as shown in Figure 27-33. The shaded areas represent radiative cooling and the clear areas denote warming. Most of the tropical latitudes, except Sahara and Arabia, between 20 N and 20 S latitude show radiative heating, indicating more radiant energy received than emitted. However, it is interesting to note radiative cooling over the Sahara desert and Arabia which arises from the highly reflecting and warm surface and relatively cloud-free conditions. Radiative cooling seems to be dominant at all latitudes beyond 30 N and 30 S which indicates more energy is emitted and reflected than is received.

There are certain limitations to these satellite-derived earth-atmosphere radiation-budget parameters, which have already been identified. The new generation Earth Radiation Budget instruments are being designed and mathematical models are being developed to consider the angular variation of the reflected solar radiation and to consider the anisotropic nature of the reflecting surfaces. These instruments are also being designed to measure the total outgoing longwave and reflected shortwave fluxes. These new types of measurements will become available in the late 1980's from the new generation of polar-orbiting satellites. It should also be emphasized that

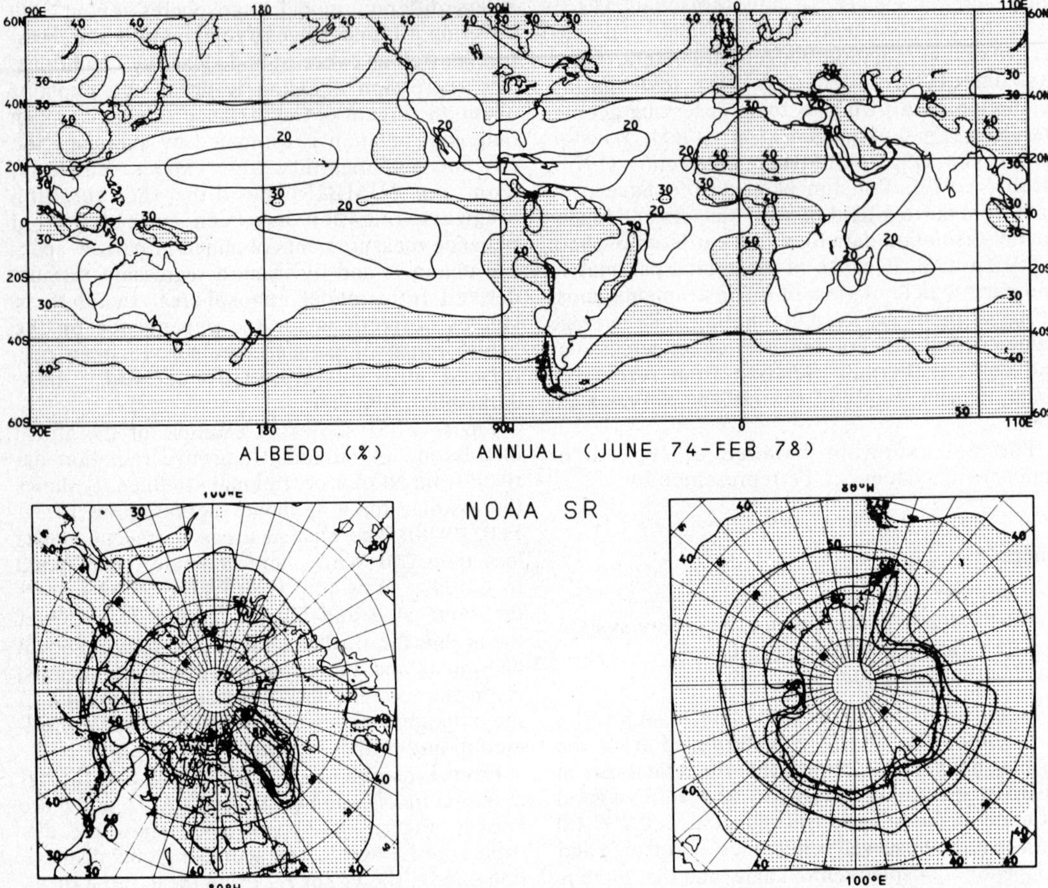

Fig. 27-31. The 45-month mean annual albedo (percent). The albedo values above 30 percent are shaded.

the satellite measurements provide only the radiation-budget quantities at the top of the atmosphere and do not provide adequate information about the radiation-budget parameters at the surface. However, techniques are being developed to estimate the surface quantities from the satellite measurements. In order to obtain total information on the earth-atmosphere radiation budget as a system, it is essential to integrate the satellite measurements with the surface-radiation measurements.

Some conclusions reached from the satellite-derived radiation-budget measurements can be summarized:

(a) The tropics between 30 N and 30 S latitude gain energy from space during all seasons. The only exceptions are Sahara, Arabia, oceanic regions between 10 S and 30 S latitude near South America and Western Africa.

(b) The equator-to-pole gradient of net radiation has its maximum change between summer and autumn in both hemispheres.

(c) The net annual radiation budgets for both the hemispheres seem to be in balance.

REMOTE SENSING OF RADIATIVELY ACTIVE ATMOSPHERIC CONSTITUENTS

The remote sensing of atmospheric constituents and other atmospheric physical parameters (such as temperature) requires a knowledge of the distributions and optical properties of all significant atmospheric constituents (See Chapter 5). From the point of view of remote sensing, the significance of any particular atmospheric constituent relates to its radiative properties, and not necessarily to its abundance. Table 27-2 provides a list of atmospheric constituents by abundance. This list is not exhaustive. There are new trace gases being observed in the atmosphere each year. However, the abundances and radiative characteristics of most of these species are small enough to ignore for current purposes. In Table 27-2, even though nitrogen and oxygen are the most abundant atmospheric molecules, their importance is small in terms of their interaction with the flow of radiation in the atmosphere because, in the normal isotope, both nitrogen and oxygen molecules are homonuclear and thus contain no net dipole moment. Any interaction between these species

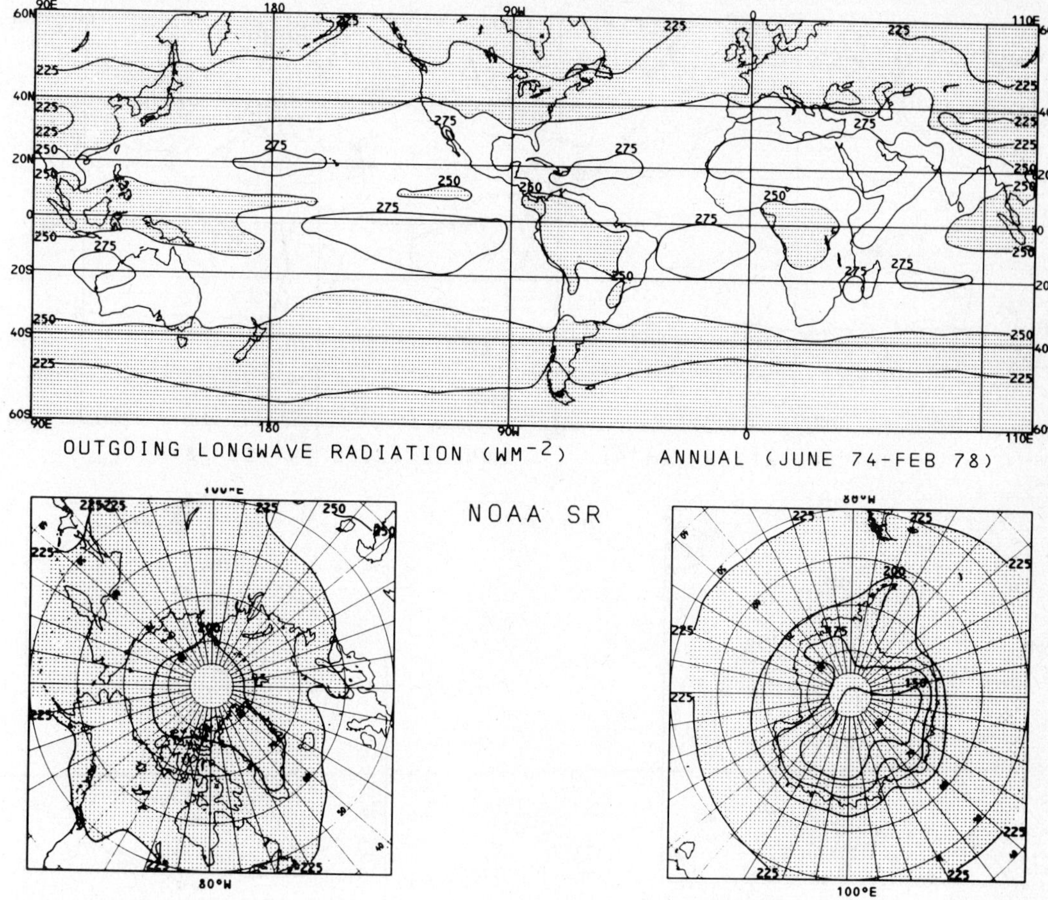

Fig. 27-32. The 45-month mean annual outgoing longwave radiation (Wm^{-2}). Values less than 250 Wm^{-2} are shaded.

and the radiation field must thus involve weaker interactions. Such interactions do occur and give rise to the atmospheric absorption by oxygen at 60 Ghz, near 7200 Å, and near 1.27 μm. Similarly, nitrogen has the broad (pressure-induced) feature centered near 4.3 μm in the infrared.

From the point of view of our discussion here, the most important constituents listed in Table 27-2 are water vapor, carbon dioxide, ozone and particulates (aerosols). The infrared and microwave portions of the spectrum are dominated by absorption due to water vapor. In addition, the thermal infrared is characterized by the absorption/emission of the important carbon dioxide bands centered near 4.3 μm and 15 μm and the ozone band at 9.6 μm. Methane, nitrous oxide, carbon monoxide and several other minor species complete the list of important naturally-occurring atmospheric molecules which we will consider here in relation to atmospheric absorption.

A discussion of radiatively-active atmospheric constituents would not be complete without considering aerosols. Several basic models will be described and optical properties will be discussed. The radiative properties of the molecular constituents of the atmosphere tend to be somewhat easier to describe than the corresponding properties of aerosols, for two reasons: 1) The vertical distributions of molecular constituents can either be assumed known (and modelled) from various measurements, or they can be measured for specific conditions. Aerosols, on the other hand, are highly variable and difficult to measure. 2) There are various kinds of aerosol material in the atmosphere and the chemical composition, in part, dictates the optical properties. Therefore, one not only needs to measure how many aerosol materials there are, but also to establish the chemical nature of each. The particle-size distribution is also important and must either be measured or estimated for a given situation.

It is convenient to consider separately the remote sensing of atmospheric temperature and moisture based on thermal emission measurements as a function of frequency and the remote sensing of atmospheric trace gases based on a measurement of either reflected sunlight or emission by the atmospheric constituent in question. As typically conceived, the remote sensing of at-

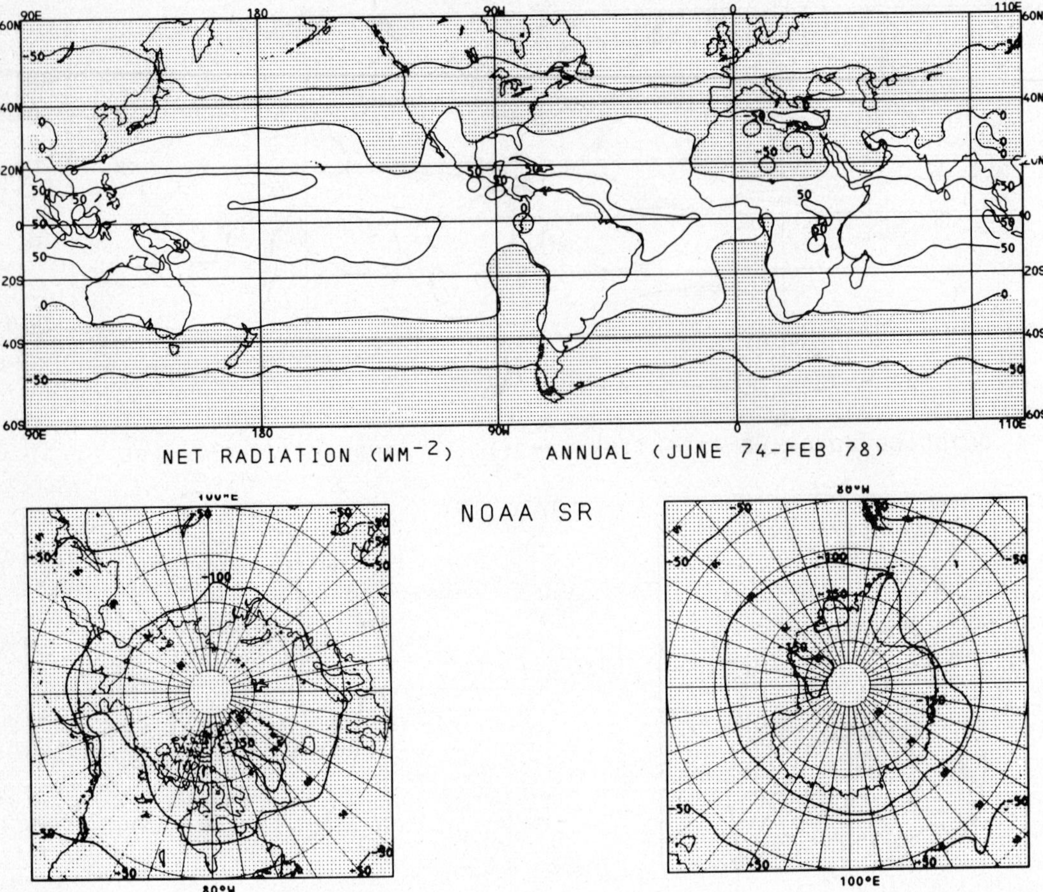

Fig. 27-33. The 45-month mean net radiation (Wm-[2]). Shaded areas represent negative values indicating net cooling.

mospheric temperature depends on the assumption that either CO_2 or O_2 is uniformly mixed in the atmosphere. The extent of validity of this assumption must be understood. Emission by CO_2 in either the 4.3 μm or the 15 μm band provides the data which can be interpreted in terms of the vertical temperature profile in the atmosphere. Emission in the microwave region by the 60 Ghz band of O_2 provides a similar capability.

The use of remote sensing to observe the existence and measure the abundance of trace gases in the atmosphere can be accomplished by either of two techniques. In absorption, a source (usually the sun) is observed with an instrument that has high spectral resolving power. The measurement is limited to paths from the point of observation toward the sun and is also limited to those wavelengths where sufficient solar energy is available. Alternatively, advantage can be taken of the process of thermal emission. The atmosphere itself is the source and again an instrument that has high spectral resolving power is required. Furthermore, the instrument needs to measure much lower levels of radiation. The observation of several absorption lines of the desired trace atmospheric gas must be separable from the many lines belonging to the more abundant (or more radiatively active) atmospheric molecules. This signature can, in principle, be used to determine the existence and the abundance of any trace gas whose spectral signature is known (or predictable).

TABLE 27-2

Atmospheric Constituents

Constituent	Fraction by Volume
Nitrogen (N_2)	0.78
Oxygen (O_2)	0.21
Argon (A)	0.0093
Carbon Dioxide (CO_2)	0.00033
Methane (CH_4)	1.6×10^{-6}
Nitrous Oxide (N_2O)	3.5×10^{-7}
Carbon Monoxide (CO)	7.5×10^{-8}
Ozone (O_3)	variable (10^{-8})
He, Ne, Kr	$<10^{-4}$
Water Vapor	variable ($<.03$)
Aerosols	
Dust	
Salt	variable
Liquid Water	

Molecular Constituents

As noted earlier, the species of primary interest in radiative transfer in the atmosphere are H_2O, CO_2 and O_3, followed by N_2O, CH_4, CO, O_2, and HNO_3. Additional trace gases of some radiative importance are also present, but they are not even observable except under extremely high spectral resolution and under favorable meteorological and geometric conditions. Using the information provided in the U.S. Standard Atmosphere, 1976, and the Handbook of Geophysics and Space Environments, AFCRL (see Valley, 1965), Table 27-3 has been constructed. This table represents three broad atmospheric models in terms of height, pressure, temperature, water vapor and ozone distributions. The water vapor in the troposphere has been derived from the relative humidity values provided in Valley, 1965. The stratospheric values (both dry and moist) have been taken from Table 21 of Goody, 1964. The dry stratosphere in these models is taken to have the constant mixing ratio of 3.2 ppmv and the moist stratosphere contains values that increase with altitude to over 24 ppmv in the vicinity of 25 km. A recent very thorough examination of the stratospheric water vapor problem by Penndorf (1978) suggests that the measurements giving rise to this moist stratospheric model may be in error. Penndorf's examination of a large body of measurements leads him to recommend the value of 4.15 ppmv as an appropriate average value for the stratosphere. This would require the substitution of stratospheric values in Tables 27-2 through 27-4 according to Eq. 27-2.

$$\int H_2O = \frac{c(ppmv)}{1.609} \int \text{air} = 2.579 \int \text{air} \quad (27\text{-}2)$$

Similarly, Table 27-3 provides ozone densities taken directly from the U.S. Standard Atmosphere, 1976. Ozone distributions contained in the Tropical and Subarctic Winter models have been inferred from data provided in Valley, 1965. The report of Penndorf (1978) critically examined available ozone as well as water-vapor data. An examination of the Penndorf report indicates a consistency between the ozone models presented here and those described in his report.

Both water vapor and ozone are highly variable in the atmosphere, so these models are only intended to represent mean conditions and might be used to estimate atmospheric radiative effects. If an accurate description of the extinction or emission of radiation by water vapor or ozone is required, it will be necessary to obtain measurements for any given set of conditions.

The vertical distributions of additional atmospheric gases of significance to atmospheric radiative transfer are listed in Table 27-4. Most of these results have been taken from the rather thorough discussion on this subject contained in the U.S. Standard Atmosphere, 1976. In some cases the results have been extended and interpolated in order to develop a more complete model. With the exception of CO_2, these species not only vary with altitude, but are also known to be variable at any given altitude. Thus, these values should only be taken as representative or mean profiles, which should be modified if actual measurements are available. Similarly, although the carbon dioxide concentration is listed as uniformly mixed at all altitudes to 100 km, it is known that the CO_2 concentration has seasonal variations and is increasing with time due to the combustion of fossil fuels. Furthermore, some measurements indicate larger values in the boundary layer in vegetated areas. Despite these variations, CO_2 can be taken to be uniformly mixed in the free atmosphere to good accuracy for most purposes of remote sensing.

Additional trace gases have been observed in the atmosphere. Some examples are NO, NO_2, SO_2, NH_2, HC_1. Although all of these species have, indeed, been observed, their abundances and vertical distributions in the atmosphere are not well known. Furthermore, their impact on the total radiation environment is small and in many cases narrow spectral features are blocked by stronger absorptions of more abundant atmospheric gases. Therefore, we will not discuss them further.

The calculation of atmospheric radiative properties requires both a description of the atmospheric path (constituent distributions, pressure, temperature) and a thorough knowledge of the molecular scattering and absorption properties of the radiatively active constituents. The absorption spectrum of the atmosphere is composed of many absorption bands as shown in Figure 27-34. These bands in turn are composed of many closely-spaced absorption lines. An effort to tabulate these detailed absorption-line parameters for seven atmospheric gases (H_2O, CO_2, O_3, N_2O, CO, CH_4, and O_2) is summarized by McClatchey, et al. (1973). This data compilation contains absorption-line frequencies, intensities, half-widths, and a number of additional parameters for each of some 150,000 molecular absorption lines covering the spectral region from 5000 Å to the microwave region. A similar (but much less extensive) data compilation described by Rothman, et al. (1978) is available containing data on a number of infrared bands of NO, SO_2, NO_2, and NH_3.

In general the use of remote sensing techniques for quantitative measurement of meteorological variables and trace gases requires the use of this detailed data base.[1] The calculation of atmospheric absorption (and emission) is limited by the accuracy of the absorption line parameters, the knowledge of the distributions of atmospheric gases along the path, and the shape of the absorption lines. Calculations are performed monochromatically over a closely-spaced frequency net and can be degraded to any spectral resolution

[1] This Data Compilation is available on magnetic tape by writing to the National Climatic Center, Federal Building, Asheville, N. Carolina.

TABLE 27-3

U.S. Standard Atmosphere, 1976

Ht (KM)	Pressure (MB)	Temp (K)	Density (G/CU-M)	Water Vapor Moist Stratos (G/CU-M)	Water Vapor Dry Stratos (G/CU-M)	Ozone (G/CU-M)
0	1.013E+03	288	1.225E+03	5.9E+00	5.9E+00	5.4E−05
1	8.988E+02	282	1.112E+03	4.2E+00	4.2E+00	5.4E−05
2	7.950E+02	275	1.607E+03	2.9E+00	2.9E+00	5.4E−05
3	7.012E+02	269	9.093E+02	1.8E+00	1.8E+00	5.0E−05
4	6.166E+02	262	8.194E+02	1.1E+00	1.1E+00	4.6E−05
5	5.405E+02	258	7.364E+02	6.4E−01	6.4E−01	4.6E−05
6	4.722E+02	249	6.601E+02	3.8E−01	3.8E−01	4.5E−05
7	4.111E+02	243	5.900E+02	2.1E−01	2.1E−01	4.9E−05
8	3.565E+02	236	5.258E+02	1.2E−01	1.2E−01	5.2E−05
9	3.080E+02	230	4.671E+02	4.6E−02	4.6E−02	7.1E−05
10	2.650E+02	223	4.135E+02	1.8E−02	1.8E−02	9.0E−05
11	2.270E+02	217	3.648E+02	8.2E−03	8.2E−03	1.3E−04
12	1.940E+02	217	3.119E+02	3.7E−03	3.7E−03	1.6E−04
13	1.658E+02	217	2.666E+02	1.8E−03	1.8E−03	1.7E−04
14	1.417E+02	217	2.279E+02	8.4E−04	8.4E−04	1.9E−04
15	1.211E+02	217	1.948E+02	7.2E−04	7.2E−04	2.1E−04
16	1.035E+02	217	1.665E+02	5.5E−04	3.3E−04	2.4E−04
17	8.850E+01	217	1.423E+02	4.7E−04	2.8E−04	2.8E−04
18	7.565E+01	217	1.217E+02	4.0E−04	2.4E−04	3.2E−04
19	6.467E+01	217	1.040E+02	4.1E−04	2.1E−04	3.5E−04
20	5.529E+01	217	8.891E+01	4.0E−04	1.8E−04	3.8E−04
21	4.729E+01	218	7.572E+01	4.4E−04	1.5E−04	3.8E−04
22	4.048E+01	219	6.450E+01	4.6E−04	1.3E−04	3.9E−04
23	3.467E+01	220	5.501E+01	5.2E−04	1.1E−04	3.8E−04
24	2.972E+01	221	4.694E+01	5.4E−04	9.4E−05	3.6E−04
25	2.549E+01	222	4.008E+01	6.1E−04	8.0E−05	3.4E−04
30	1.197E+01	227	1.841E+01	3.2E−04	3.7E−05	2.0E−04
35	5.746E+00	237	8.463E+00	1.3E−04	1.7E−05	1.1E−04
40	2.871E+00	250	3.996E+00	4.8E−05	7.9E−06	4.9E−05
45	1.491E+00	264	1.966E+00	2.2E−05	3.9E−06	1.7E−05
50	7.978E−01	271	1.827E+00	7.8E−05	2.1E−06	5.3E−05
70	5.220E−02	220	8.283E−02	1.2E−07	1.8E−07	4.3E−08
100	3.008E−04	210	4.990E−04	3.0E−10	1.0E−09	4.3E−11

consistent with a given sensor. An example of a high-resolution transmission spectrum is given in Figure 27-35. Clearly, the measurement of the spectral signature of a trace atmospheric gas will require this kind of spectral measurement and calculation capability. Not quite so obvious is the fact that even low spectral-resolution measurements, of the kind used to infer temperature and water vapor distributions from satellite platforms, require this kind of calculation technique in order to obtain the desired accuracy.

Atmospheric Aerosols

Atmospheric aerosol models are somewhat more complex to define than are the models of molecular constituents due to the greater variability of atmospheric aerosols in time and space. In general it is necessary to know (or measure) two different quantities in order to adequately describe a particular aerosol extinction in the atmosphere: We must know the complex index of refraction of the particles; and we must know the particle-size distribution. A knowledge of these two quantities is sufficient if we can assume spherical particles and apply Mie theory (see Van de Hulst, 1957). A number of aerosol models providing descriptions of these quantities can be found in the literature (c.f. Shettle and Fenn, 1975 and 1979; and Ivlev, 1967). These models are all similar as they are derived from a similar measurement base. Here, we have chosen the work of Shettle and Fenn, 1975 and 1979, as it most nearly provides a complete description of some representative models, including computed extinction coefficients. Both the refractive indices and the particle-size distributions depend strongly on relative humidity in the atmosphere. The higher liquid-water content of high relative-humidity conditions requires the use of a composite refractive index that is weighted more strongly by the refractive index of liquid water. Similarly, particles grow with increasing relative humidity and larger particles will impact extinction processes more strongly at longer wavelengths. We will not try to provide all that information here, but we will provide a limited set of models for a condition of 80 percent

TABLE 27-3—(*Continued*)

Tropical

Ht (KM)	Pressure (MB)	Temp (K)	Density (G/CU-M)	Water Vapor Moist Stratos (G/CU-M)	Water Vapor Dry Stratos (G/CU-M)	Ozone (G/CU-M)
0	1.013E+03	300	1.176E+03	1.9E+01	1.9E+01	5.6E−05
1	9.040E+02	294	1.071E+03	1.3E+01	1.3E+01	5.6E−05
2	8.050E+02	288	9.738E+02	9.3E+00	9.3E+00	5.4E−05
3	7.150E+02	284	8.771E+02	4.7E+00	4.7E+00	5.1E−05
4	6.330E+02	277	7.961E+02	2.2E+00	2.2E+00	4.7E−05
5	5.590E+02	270	7.213E+02	1.5E+00	1.5E+00	4.5E−05
6	4.920E+02	264	6.493E+02	8.5E−01	8.5E−01	4.3E−05
7	4.320E+02	257	5.856E+02	4.7E−01	4.7E−01	4.1E−05
8	3.780E+02	250	5.268E+02	2.5E−01	2.5E−01	3.9E−05
9	3.290E+02	244	4.697E+02	1.2E−01	1.2E−01	3.9E−05
10	2.860E+02	237	4.204E+02	5.0E−02	5.0E−02	3.9E−05
11	2.470E+02	230	3.741E+02	1.7E−02	1.7E−02	4.1E−05
12	2.128E+02	224	3.310E+02	6.0E−03	5.0E−03	4.3E−05
13	1.825E+02	217	2.930E+02	1.8E−03	1.8E−03	4.5E−05
14	1.556E+02	210	2.581E+02	1.0E−03	1.0E−03	4.5E−05
15	1.321E+02	204	2.256E+02	7.6E−04	7.6E−04	4.7E−05
16	1.116E+02	197	1.974E+02	6.5E−04	3.9E−04	4.7E−05
17	9.370E+01	195	1.674E+02	5.5E−04	3.3E−04	6.9E−05
18	7.890E+01	199	1.381E+02	4.6E−04	2.8E−04	9.0E−05
19	6.660E+01	203	1.143E+02	4.5E−04	2.3E−04	1.4E−04
20	5.650E+01	207	9.509E+01	4.3E−04	1.9E−04	1.9E−04
21	4.800E+01	211	7.925E+01	4.6E−04	1.6E−04	2.4E−04
22	4.090E+01	215	6.627E+01	4.8E−04	1.3E−04	2.8E−04
23	3.500E+01	217	5.619E+01	5.3E−04	1.1E−04	3.2E−04
24	3.000E+01	219	4.772E+01	5.5E−04	9.5E−05	3.4E−04
25	2.570E+01	221	4.051E+01	6.1E−04	8.1E−05	3.4E−04
30	1.220E+01	232	1.832E+01	3.2E−04	3.7E−05	2.4E−04
35	6.000E+00	243	8.602E+00	1.3E−04	1.7E−05	9.2E−05
40	3.050E+00	254	4.183E+00	5.1E−05	8.4E−06	4.1E−05
45	1.590E+00	265	2.090E+00	2.3E−05	4.2E−06	1.3E−05
50	8.540E−01	270	1.102E+00	8.4E−06	2.2E−06	4.3E−06
70	5.790E−02	219	9.211E−02	1.3E−07	1.8E−07	8.6E−08
100	3.000E−04	210	4.977E−04	3.0E−10	1.0E−09	4.3E−11

relative humidity consistent with the three atmospheric models provided in Table 27-3. (For additional information on the dependence of aerosol particles on relative humidity, see Shettle and Fenn 1979 and Hanel 1976). Table 27-5 describes seven different aerosol models for which extinction coefficients will be provided here. The first three models (Rural, Urban, and Maritime) are intended only for the boundary layer (lowest 2 km of the atmosphere) and the fourth model (Tropospheric) is intended for the remainder of the troposphere. The remaining models relate only to upper atmospheric regions. The terms, "dust-like", "water soluble", "oceanic", etc. under the heading, "complex index", refer to the complex index of refraction values given in Table 27-6. The terms "dust-like" and "water soluble" refer to the portions of collected aerosol that are insoluble and soluble in water, respectively. The term "oceanic" refers to a complex index of refraction of about 70 percent liquid water and 30 percent sea salt.

The complex indices of refraction corresponding to the fundamental aerosol types indicated in column 3 of Table 27-5 are provided as a function of wavelength in Table 27-6. The complex index of refraction for liquid water is also included, as it is necessary to use an appropriate weighted average of the dry aerosol material and liquid water to account for the actual aerosol particles at 80 percent relative humidity.

Table 27-7 provides the parameters used in Eq. 27-3 and 27-4 to represent the various particle-size distributions required in the aerosol models. Eq. 27-3 is the log-normal distribution and Eq. 27-4 is the Modified Gamma Function. Note that the parameters, as presented in Table 27-7, generate a particle-size distribution normalized to 1 particle per cubic centimeter. The actual number of particles in the lowest layer of the atmosphere will be determined by a measurement of the prevailing visibility (see below).

Through the use of appropriately weighted averages of the complex indices of refraction from Table 27-6 and the particle sizes described in Table 27-7, extinction coefficients have been computed according to the Mie theory. The results of these calculations are given in Tables 27-

TABLE 27-3—*(Continued)*

Subarctic Winter

Ht (KM)	Pressure (MB)	Temp (K)	Density (G/CU-M)	Water Vapor Moist Stratos (G/CU-M)	Water Vapor Dry Stratos (G/CU-M)	Ozone (G/CU-M)
0	1.013E+03	257	1.373E+03	1.2E+00	1.2E+00	4.1E−05
1	8.878E+02	259	1.194E+03	1.2E+00	1.2E+00	4.1E−05
2	7.775E+02	256	1.058E+03	9.4E−01	9.4E−01	4.1E−05
3	6.798E+02	253	9.372E+02	5.8E−01	6.8E−01	4.3E−05
4	5.932E+02	248	8.343E+02	4.1E−01	4.1E−01	4.5E−05
5	5.158E+02	241	7.459E+02	2.0E−01	2.0E−01	4.7E−05
6	4.467E+02	234	6.648E+02	9.8E−02	9.8E−02	4.9E−05
7	3.853E+02	227	5.906E+02	5.4E−02	5.4E−02	7.1E−05
8	3.308E+02	221	5.224E+02	1.1E−02	1.1E−02	9.0E−05
9	2.829E+02	217	4.538E+02	8.4E−03	8.4E−03	1.6E−04
10	2.418E+02	217	3.878E+02	5.5E−03	5.5E−03	2.4E−04
11	2.067E+02	217	3.315E+02	3.8E−03	3.8E−03	3.2E−04
12	1.766E+02	217	2.833E+02	2.6E−03	2.6E−03	4.3E−04
13	1.510E+02	217	2.422E+02	1.8E−03	1.8E−03	4.7E−04
14	1.291E+02	217	2.071E+02	1.0E−03	1.0E−03	4.9E−04
15	1.103E+02	217	1.769E+02	7.6E−04	7.6E−04	5.6E−04
16	9.431E+01	217	1.517E+02	5.0E−04	5.0E−04	6.2E−04
17	8.058E+01	216	1.300E+02	4.3E−04	2.6E−04	6.2E−04
18	6.882E+01	215	1.113E+02	3.7E−04	2.2E−04	6.2E−04
19	5.875E+01	215	9.529E+01	3.7E−04	1.9E−04	6.0E−04
20	5.014E+01	214	8.159E+01	3.7E−04	1.6E−04	5.6E−04
21	4.277E+01	214	6.976E+01	4.1E−04	1.4E−04	5.1E−04
22	3.647E+01	213	5.965E+01	4.3E−04	1.2E−04	4.7E−04
23	3.109E+01	212	5.099E+01	4.8E−04	1.0E−04	4.3E−04
24	2.649E+01	212	4.357E+01	5.1E−04	8.7E−05	3.6E−04
25	2.256E+01	211	3.721E+01	5.6E−04	7.4E−05	3.2E−04
30	1.020E+01	216	1.645E+01	2.9E−04	3.3E−05	1.5E−04
35	4.701E+00	222	7.371E+00	1.1E−04	1.5E−05	9.2E−05
40	2.243E+00	235	3.359E+00	4.1E−05	6.7E−06	4.1E−05
45	1.113E+00	247	1.570E+00	1.7E−05	3.1E−06	1.3E−05
50	5.719E−01	259	7.684E−01	5.8E−06	1.5E−06	4.3E−05
70	4.016E−02	246	5.694E−02	8.0E−08	1.1E−07	8.6E−08
100	3.000E−04	210	4.977E−04	3.0E−10	1.0E−09	4.3E−11

8a–g. Note that the extinction coefficients are normalized to 1 km^{-1} at 0.55 μm.

$$\frac{dN(r)}{dr} = \sum_{i=1}^{2} \frac{Ni}{ln(10) \cdot r \cdot \sigma_i \sqrt{2\pi}}$$

$$\exp - \left[\frac{(\log_{10} r - \log_{10} r_i)^2}{2\sigma_i^2} \right] \qquad (27\text{-}3)$$

$$\frac{dN}{dr} = n(r) = Ar \exp(-br^{\gamma}) \qquad (27\text{-}4)$$

Again, using the work of Shettle and Fenn (1975) Figure 27-36 has been constructed to represent two standard vertical distributions of the attenuation coefficient at 0.55 μm which, at the surface, correspond to visibilities (or Meteorological Ranges) of 5 km and 50 km respectively. Also contained in Figure 27-36 is the Rayleigh (or molecular) scattering profile at 0.55 μm for comparison. The Meteorological Range is defined as the distance over which the transmittance is reduced to 0.02 as indicated in Eq. 27-5.

$$\text{Transmittance} = 0.02 = \exp - [\sigma(\text{M.R.})]: \quad M.R. = 3.91/\sigma \qquad (27\text{-}5)$$

where σ = extinction coefficient at 0.55 μm
and $M.R.$ = Meteorological Range.

The background stratospheric curve is intended to represent a minimum atmospheric-aerosol loading based on all available measurements. The curve labelled "Volcanic" represents a mean atmospheric-aerosol model resulting from volcanic activity. Table 27-9 represents these final two vertical aerosol models in terms of "Scaling Factors" which, when multiplied by the extinction coefficients of Table 27-8, give the actual extinction coefficients per kilometer at any altitude in the model. The extinction coefficients from Table 27-8 that are intended for use with Table 27-9 are indicated in the right hand portion of the table. The maritime and urban extinction coefficients of Table 27-8 are intended for use only with the lowest 2 km of the models in Table 27-9.

TABLE 27-4

Vertical Distributions of Trace Gases

Ht(km)	CO_2(ppmv)	N_2O(ppbv)	CO(ppmv)	CH_4(ppmv)	HNO_3(ppbv)
0	322	270	0.19	1.50	0.0
1	322	270	0.18	1.50	0.0
2	322	270	0.17	1.50	0.0
3	322	270	0.16	1.50	0.0
4	322	270	0.15	1.50	0.0
5	322	270	0.13	1.50	0.0
6	322	270	0.11	1.50	0.0
7	322	270	0.09	1.50	0.0
8	322	270	0.07	1.50	0.0
9	322	270	0.06	1.50	0.1
10	322	270	0.05	1.50	0.3
11	322	264	0.04	1.48	0.8
12	322	257	0.04	1.46	1.2
13	322	250	0.04	1.44	1.4
14	322	220	0.04	1.42	1.6
15	322	190	0.04	1.40	1.8
16	322	160	0.04	1.38	1.9
17	322	130	0.04	1.36	2.0
18	322	100	0.04	1.34	2.1
19	322	100	0.04	1.32	2.3
20	322	100	0.04	1.30	3.0
21	322	100	0.04	1.26	3.7
22	322	100	0.04	1.22	4.2
23	322	100	0.04	1.18	5.2
24	322	100	0.04	1.14	6.0
25	322	100	0.04	1.10	3.8
30	322	100	0.04	0.90	2.6
35	322	100	0.04	0.75	0.2
40	322	100	0.04	0.60	0.0
45	322	100	0.04	0.43	0.0
50	322	100	0.04	0.25	0.0
70	322	100	0.04	0.25	0.0
100	322	100	0.04	0.25	0.0

The calculation of scattering-phase functions (the angular distribution of scattered radiation) can be directly calculated for these models by direct application of Mie theory to the particle-size distributions and complex indices of refraction. To tabulate such results would require large amounts of additional space; hence, such a tabulation is not provided here. However, Table 27-8 does contain an "asymmetry parameter", the mathematical form of which is provided in Eq. 27-6.

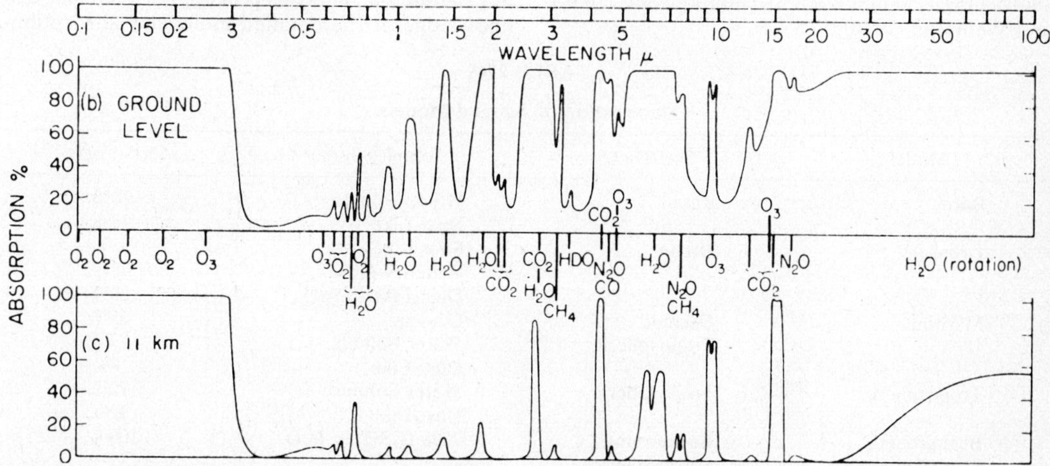

Fig. 27-34. Atmospheric gaseous absorption spectrum for a solar beam reaching the ground and 11 km altitude level. (Taken from Goody, 1964).

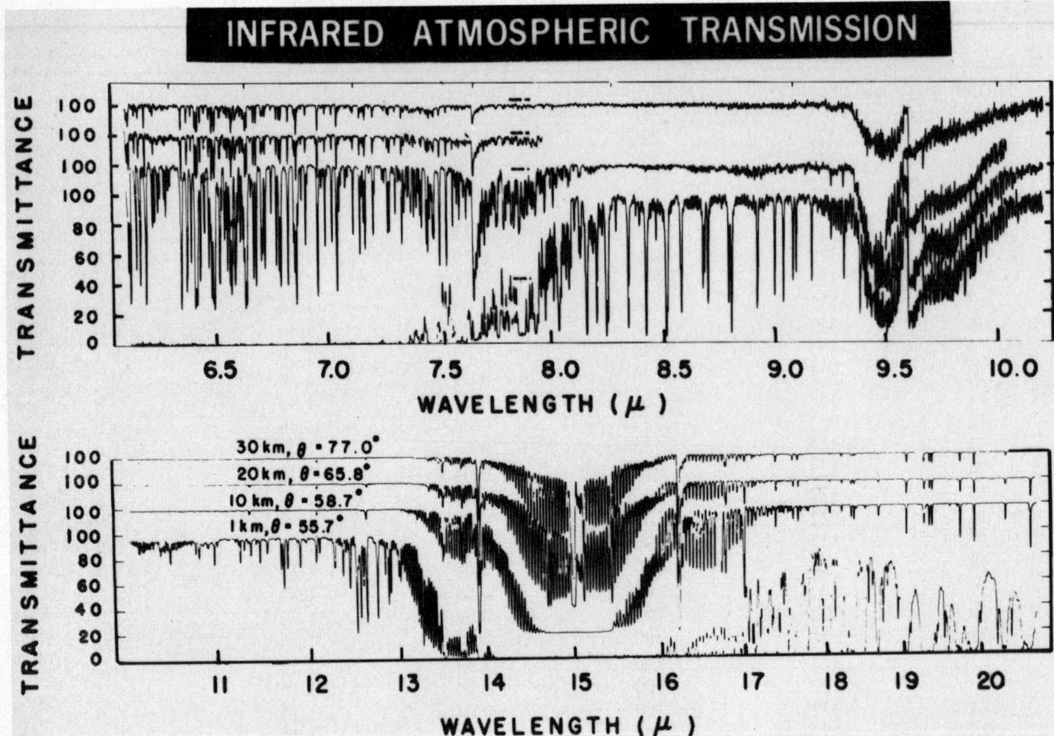

Fig. 27-35. Transmission measured from the indicated altitude observing the sun at the indicated solar zenith angle. Note that four spectra are displaced by 20% for ease of viewing. (From Murcray, D., 1970).

$$g = \frac{\int_{-1}^{+1} \cos\theta\, p(\theta)\, d\cos\theta}{\int_{-1}^{+1} p(\theta)\, d\cos\theta} \qquad (27\text{-}6)$$

where $p(\theta)$ is the phase function.

This quantity is of value when multiple scattering effects are important and there is interest in estimating the scattered-radiation field. This quantity can then be used in the application of various approximate radiative-transfer techniques such as the Eddington Approximation (see Shettle and Weinman, 1970).

We have attempted to provide here a sketch of the radiative properties of the atmosphere as they relate to the remote sensing of meteorological variables and trace atmospheric gases. In general the remote sensing of such quantities requires that we measure the atmospheric absorption or emission characteristics of one or more molecular components of the atmosphere. Whether the intended measurements are made at high spectral resolution or not, the accuracy requirements usually dictate that analysis procedures make use of a knowledge of the detailed molecular absorption-

TABLE 27-5

Description of Aerosol Models

Model	Size Dist.	Complex Index	Ext. Coef.
Rural	Rural	Water Soluble	70%
		Dust-Like	30%
Urban	Urban	Soot-Like	35%
		Water Soluble	45%
		Dust-Like	20%
Maritime	Oceanic	Oceanic	75%
	Tropospheric	Water Soluble	17%
		Dust-Like	8%
Tropospheric	Tropospheric	Water Soluble	70%
		Dust-Like	30%
Background Stratospheric	Background Stratospheric	75% H_2SO_4 in H_2O	100%
Volcanic	Volcanic	Volcanic	100%
Meteoric Dust	Meteoric Dust	Meteoric Dust	100%

TABLE 27-6a.

Refractive Index for the Different Aerosol Components

Wavelength (Microns)	Water Soluble	Dust-Like	Soot	Sea Salt	Water
.2000	1.530 −7.00E−02	1.530 −7.00E−02	1.500 −.350	1.510 −1.00E−04	1.396 −1.10E−07
.2500	1.530 −3.00E−02	1.530 −3.00E−02	1.620 −.450	1.510 −5.00E−06	1.362 −3.35E−08
.3000	1.530 −8.00E−03	1.530 −8.00E−03	1.740 −.470	1.510 −2.00E−06	1.349 −1.60E−08
.3371	1.530 −5.00E−03	1.530 −8.00E−03	1.750 −.470	1.510 −4.00E−07	1.345 −8.45E−09
.4000	1.530 −5.00E−03	1.530 −8.00E−03	1.750 −.460	1.500 −3.00E−08	1.339 −1.86E−09
.4880	1.530 −5.00E−03	1.530 −8.00E−03	1.750 −.450	1.500 −2.00E−08	1.335 −9.69E−10
.5145	1.530 −5.00E−03	1.530 −8.00E−03	1.750 −.450	1.500 −1.00E−08	1.334 −1.18E−09
.5500	1.530 −6.00E−03	1.530 −8.00E−03	1.750 −.440	1.500 −1.00E−08	1.333 −1.96E−09
.6328	1.530 −6.00E−03	1.530 −8.00E−03	1.750 −.430	1.490 −2.00E−08	1.332 −1.46E−08
.6943	1.530 −7.00E−03	1.530 −8.00E−03	1.750 −.430	1.490 −1.00E−07	1.331 −3.05E−08
.8600	1.520 −1.20E−02	1.520 −8.00E−03	1.750 −.430	1.480 −3.00E−06	1.329 −3.29E−07
1.0600	1.520 −1.70E−02	1.520 −8.00E−03	1.750 −.440	1.470 −2.00E−04	1.326 −4.18E−06
1.3000	1.510 −2.00E−02	1.460 −8.00E−03	1.760 −.450	1.470 −4.00E−04	1.323 −3.69E−05
1.5360	1.510 −2.30E−02	1.400 −8.00E−03	1.770 −.460	1.460 −6.00E−04	1.318 −9.97E−05
1.8000	1.460 −1.70E−02	1.330 −8.00E−03	1.790 −.480	1.450 −8.00E−04	1.312 −1.15E−04
2.0000	1.420 −8.00E−03	1.260 −8.00E−03	1.800 −.490	1.450 −1.00E−03	1.306 −1.10E−03
2.2500	1.420 −1.00E−02	1.220 −9.00E−03	1.810 −.500	1.440 −2.00E−03	1.292 −3.90E−04
2.5000	1.420 −1.20E−02	1.180 −9.00E−03	1.820 −.510	1.430 −4.00E−03	1.261 −1.74E−03
2.7000	1.400 −5.50E−02	1.180 −1.30E−02	1.830 −.520	1.400 −7.00E−03	1.888 −1.90E−02
3.0000	1.420 −2.20E−02	1.160 −1.20E−02	1.840 −.540	1.610 −1.00E−02	1.371 − .272
3.2000	1.430 −8.00F−03	1.220 −1.00E−02	1.860 −.540	1.490 −3.00E−03	1.478 −9.24E−02
3.3923	1.437 −7.00E−03	1.260 −1.30E−02	1.870 −.550	1.480 −2.00E−03	1.422 −2.04E−02
3.5000	1.450 −5.00E−03	1.280 −1.10E−02	1.880 −.560	1.480 −1.60E−03	1.400 −9.40E−03
3.7500	1.452 −4.00F−03	1.270 −1.10E−02	1.900 −.570	1.470 −1.40E−03	1.369 −3.50E−03
4.0000	1.455 −5.00E−03	1.260 −1.20E−02	1.902 −.580	1.480 −1.40E−03	1.351 −4.60E−03
4.5000	1.460 −1.30E−02	1.260 −1.40E−02	1.940 −.590	1.490 −1.40E−03	1.332 −1.34E−02
5.0000	1.450 −1.20E−02	1.250 −1.60E−02	1.970 −.600	1.470 −2.50E−03	1.325 −1.24E−02
5.5000	1.440 −1.80E−02	1.220 −2.10E−02	1.990 −.610	1.420 −3.60E−03	1.298 −1.16E−02
6.0000	1.410 −2.30E−02	1.150 −3.70E−02	2.020 − 620	1.410 −1.10E−02	1.265 −.107
6.2000	1.430 −2.70E−02	1.140 −3.90E−02	2.030 −.625	1.600 −2.20E−02	1.363 −8.80E−02
6.5000	1.460 −3.30E−02	1.130 −4.20E−02	2.040 −.630	1.460 −5.00E−03	1.339 −3.92E−02
7.2000	1.400 −7.00E−02	1.400 −5.50E−02	2.060 −.650	1.420 7.00E 03	1.312 3.21E 02
7.9000	1.200 −6.50E−02	1.150 −4.00E−02	2.120 −.670	1.400 −1.30E−02	1.294 −3.39E−02
8.2000	1.010 −.100	1.13 −7.40E−02	2.130 −.680	1.420 −2.00E−02	1.286 −3.51E−02
8.5000	1.300 −.215	1.300 −9.00E−02	2.150 −.690	1.480 −2.60E−02	1.278 −3.67E−02
8.7000	2.400 −.290	1.400 −.100	2.160 −.690	1.600 −3.00E−02	1.272 −3.79E−02
9.0000	2.560 −.370	1.700 −.140	2.170 −.700	1.650 −2.80E−02	1.262 −3.99E−02
9.2000	2.200 −.420	1.720 −.150	2.180 −.700	1.610 −2.60E−02	1.255 −4.15E−02
9.5000	1.950 −.160	1.730 −.162	2.190 −.710	1.580 −1.80E−02	1.243 −4.44E−02
9.8000	1.870 −9.50E−02	1.740 −.162	2.200 −.715	1.560 −1.60E−02	1.229 −4.79E−02
10.0000	1.820 −9.00E−02	1.750 −.162	2.210 −.720	1.540 −1.50E−02	1.218 −5.08E−02
10.5910	1.760 −7.00E−02	1.620 −.120	2.220 −.730	1.500 −1.40E−02	1.179 −6.74E−02
11.0000	1.720 −5.00E−02	1.620 −.105	2.230 −.730	1.480 −1.40E−02	1.153 −9.68E−02
11.5000	1.670 −4.70E−02	1.590 −.100	2.240 −.740	1.480 −1.40E−02	1.126 −.142
12.5000	1.620 −5.30E−02	1.510 −9.00E−02	2.270 −.750	1.420 −1.60E−02	1.123 −.259
13.0000	1.620 −5.50E−02	1.470 −.100	2.280 −.760	1.410 −1.80E−02	1.146 −.305
14.0000	1.560 −7.30E−02	1.520 −8.50E−02	2.310 −.775	1.410 −2.30E−02	1.210 −.370
14.8000	1.440 −.100	1.570 −.100	2.330 −.790	1.430 −3.00E−02	1.258 −.396
15.0000	1.420 −.200	1.570 −.100	2.330 −.790	1.450 −3.50E−02	1.270 −.402
16.4000	1.750 −.160	1.600 −.100	2.360 −.810	1.560 −9.00E−02	1.346 −.427
17.2000	2.080 −.240	1.630 −.100	2.380 −.820	1.740 −.120	1.386 −.429
18.0000	1.980 −.180	1.640 −.115	2.400 −.825	1.780 −.130	1.423 −.426
18.5000	1.853 −.170	1.640 −.120	2.410 −.830	1.770 −.135	1.443 −.421
20.0000	2.120 −.220	1.680 −.220	2.450 −.850	1.760 −.152	1.480 −.393
21.3000	2.060 −.230	1.770 −.280	2.460 −.860	1.760 −.165	1.491 −.379
22.5000	2.000 −.240	1.900 −.280	2.480 −.870	1.760 −.180	1.506 −.370
25.0000	1.880 −.280	1.970 −.240	2.510 −.890	1.760 −.205	1.531 −.356
27.9000	1.840 −.290	1.890 −.320	2.540 −.910	1.770 −.275	1.549 −.339
30.0000	1.820 −.300	1.800 −.420	2.570 −.930	1.770 −.300	1.551 −.328
35.0000	1.920 −.400	1.900 −.500	2.630 −.970	1.760 −.500	1.532 −336
40.0000	1.860 −.500	2.100 −.600	2.690 −1.000	1.740 −1.000	1.519 −.385

TABLE 27-6b.

Refractive Indices for the Different Types of Aerosols

Wavelength (Micron)	75% H2SO4	Volcanic	Meteoric
.2000	1.498 −1.00E−08	1.500 −7.00E−02	1.515 −1.23E−05
.2500	1.484 −1.00E−08	1.500 −3.00E−02	1.515 −2.41E−05
.3500	1.469 −1.00E−08	1.500 −1.00E−02	1.515 −4.18E−05
.3371	1.459 −1.00E−08	1.500 −8.00E−03	1.514 −5.94E−05
.4000	1.440 −1.00E−08	1.500 −8.00E−03	1.514 −9.95E−05
.4880	1.432 −1.00E−08	1.500 −8.00E−03	1.513 −1.81E−04
.5145	1.431 −1.00E−08	1.500 −8.00E−03	1.513 −2.13E−04
.5500	1.430 −1.00E−08	1.500 −8.00E−03	1.513 −2.61E−04
.6328	1.429 −1.47E−08	1.500 −8.00E−03	1.512 −3.99E−04
.6943	1.428 −1.99E−08	1.500 −8.00E−03	1.511 −5.30E−04
.8600	1.425 −1.79E−07	1.500 −8.00E−03	1.509 −1.02E−03
1.9600	1.420 −1.50E−06	1.500 −8.00E−03	1.506 −1.95E−03
1.3000	1.410 −1.00E−05	1.500 −8.00E−03	1.501 −3.72E−03
1.5360	1.403 −1.37E−04	1.490 −8.00E−03	1.495 −6.34E−03
1.8000	1.390 −5.50E−04	1.480 −8.00E−03	1.488 −1.06E−02
2.0000	1.384 −1.26E−03	1.460 −8.00E−03	1.482 −1.51E−02
2.2500	1.370 −1.80E−03	1.460 −8.00E−03	1.474 −2.24E−02
2.5000	1.344 −3.76E−03	1.460 −9.00E−03	1.467 −3.18E−02
2.7000	1.303 −5.70E−03	1.460 −1.00E−02	1.462 −4.10E−02
3.0000	1.293 −9.55E−02	1.480 −1.30E−02	1.456 −5.73E−02
3.2000	1.311 −.135	1.480 −1.40E−02	1.454 −6.94E−02
3.3923	1.352 −.159	1.490 −1.20E−02	1.454 −8.15E−02
3.5000	1.376 −.158	1.490 −1.10E−02	1.455 −8.82E−02
3.7500	1.396 −.131	1.500 −9.00E−03	1.459 −.103
4.0000	1.398 −.126	1.500 −7.00E−03	1.466 −.116
4.5000	1.385 −.120	1.520 −7.50E−03	1.485 −.131
5.0000	1.360 −.121	1.510 −9.00E−03	1.500 −.135
5.5000	1.337 −.183	1.510 −1.20E−02	1.508 −.132
6.0000	1.425 −.195	1.480 −1.50E−02	1.507 −.126
6.2000	1.424 −.165	1.460 −1.80E−02	1.504 −.124
6.5000	1.370 −.128	1.450 −2.40E−02	1.497 −.121
7.2000	1.210 −.176	1.440 −4.50E−02	1.469 −.119
7.9000	1.140 −.458	1.380 −7.20E−02	1.422 −.130
8.2000	1.200 −.645	1.340 −9.70E−02	1.395 −.142
8.5000	1.370 −.755	1.620 −.121	1.363 −.162
8.7000	1.530 −.772	1.950 −.170	1.339 −.182
9.0000	1.650 −.633	2.200 −.215	1.302 −.228
9.2000	1.600 −.586	2.230 −.240	1.281 −.273
9.5000	1.670 −.750	2.250 −.275	1.272 −.360
9.8000	1.910 −.680	2.800 −.304	1.310 −.450
10.0000	1.890 −.455	2.300 −.320	1.355 −.488
10.5910	1.720 −.340	2.200 −.305	1.419 −.547
11.0000	1.670 −.485	2.150 −.270	1.509 −.691
11.5000	1.890 −.374	2.050 −.240	1.847 −.634
12.5000	1.740 −.198	1.800 −.155	1.796 −.252
13.0000	1.690 −.195	1.760 −.148	1.711 −.219
14.0000	1.640 −.195	1.700 −.145	1.641 −.217
14.8000	1.610 −.205	1.650 −.157	1.541 −.198
15.0000	1.590 −.211	1.650 −.170	1.510 −.206
16.4000	1.520 −.414	1.750 −.200	1.478 −.467
17.2000	1.724 −.590	1.850 −.240	1.441 −.400
18.8000	1.950 −.410	2.000 −.305	1.354 −.557
18.5000	1.927 −.302	2.100 −.325	1.389 −.705
20.0000	1.823 −.235	2.250 −.318	1.803 −.765
21.3000	1.789 −.292	2.400 −.290	1.797 −.556
22.5000	1.870 −.315	2.500 −.350	1.661 −.592
25.0000	1.930 −.200	2.600 −.400	1.983 −.861
27.0000	1.920 −.180	2.500 −.430	2.023 −.666
30.0000	1.920 −.180	2.400 −.450	2.149 −.665
35.0000	1.900 −.190	2.300 −.520	2.146 −.380
40.0000	1.890 −.220	2.250 −.650	1.979 −.359

From: Shettle & Fenn—Optical properties of the atmospheric aerosols

TABLE 27-7

Parameters for Stratospheric Size Distributions (Normalized to 1 particle/cm³)

Type of Aerosol	N_1*	r_1	σ_1	N_2*	r_2	σ_2
	LOG NORMAL (Equation 27-3)					
Rural	0.999875	0.03μ	0.35	1.25×10^{-4}	0.5μ	0.4
Urban	0.999875	0.03μ	0.35	1.25×10^{-4}	0.5μ	0.4
Maritime—Oceanic Origin	1.	0.3μ	0.4	—	—	—
Tropospheric	1.	0.03μ	0.35	—	—	—
Meteoric Dust	1.	0.03μ	0.5	—	—	—
	MODIFIED GAMMA (Equation 27-4)					
	A	α	γ	b		
Background Stratospheric	324	1	1	18		
Volcanic	5,461.33	1	½	16		

*$N_1 + N_2 = 1$

line parameters of the more abundant atmospheric molecules. Trace gas identifications require the additional knowledge of the spectral parameters of the required trace gases. It is impossible to provide the detailed results here enabling the reader to calculate these effects. However, an estimate of atmospheric transmission and emission can be obtained at moderate spectral resolution by using the LOWTRAN transmission model described by McClatchey, et al., 1972 and Selby, et al., 1978.

The problem of atmospheric aerosols is primarily one of modifying the spectral signatures of the molecular absorption/emission processes. This may not always be the case as there are efforts being made to devise experiments to remotely sense aerosols in the atmosphere by measurement of scattered radiation. We have attempted to provide here a means of estimating the effects of aerosols on the radiation field being emitted or scattered by the earth or by the molecular constituents of the atmosphere. It is possible to do this owing to the slowly varying nature of the extinction relative to the molecular processes. An expanded version of the aerosol models provided here is available in the report by Shettle and Fenn, 1979 and will become part of the upcoming LOWTRAN 5 model (see Kniezys, et al., 1980).

MONITORING CLIMATE SCALE VARIABILITY IN THE OCEAN FROM SPACE

Investigations of the role of the ocean in determining climate have shown that some oceanic variables are correlated with short term (a few months to a few years) variability in weather patterns. Particularly notable are the studies by Namais (1978) and his colleagues that show that sea-surface temperature in the North Pacific influences the weather over North America, and studies by Bjerknes (1966) and Barnett (1977) demonstrating the connections between the strength of the Pacific trade winds and the "El Nino" phenomena along the Peruvian coast. These connections, while significant, are not well understood. That is, we cannot yet trace with confidence the causal connections linking Pacific temperature and Chicago rainfall; and except in a few geographic areas (Davis, 1976, 1978; Barnett, 1978), we cannot state with confidence what variable leads or lags another, or which oceanic variables can be readily used to predict climate. This leads us to suspect that we cannot yet specify with confidence those oceanic variables that must be measured to study climate. Presently, all we can say with certainty is that some variables should be important, but these are generally not measured, while other variables appear to be correlated with climate, but we know not why. (See also Chapter 28, dealing with Marine Resources).

The theoretically important variables are those describing the heat balance of earth. The oceanic contributions include the amount of heat stored in the upper layers of the ocean, the poleward transport of this heat, the fluxes of sensible and latent heat across the sea surface, including the radiant heating of the surface layers by the sun, the release of heat by rain and the wind stress at the sea surface (which both drives ocean currents and controls the fluxes across the surface).

Of these variables, all except wind stress and perhaps solar heating are difficult or impossible to measure directly from space. This forces us to observe more easily measured, but less direct variables such as sea-surface temperature, wind speed, atmospheric water content and cloud cover, all of which appear to be correlated with the primary variables.

This emphasis on secondary (proxy) variables

TABLE 27-8a

Attenuation Coefficients for RH80%, Rural Aerosol Model Normalized to an Extinction Coefficient = 1.00 KM-1, at a Wavelength = 0.55

Wavelength (Micron)	Extinction (KM-1)	Scattering (KM-1)	Absorption (KM-1)	Single Scat Alb.	Asymmetry Parameter
.200	2.071E+00	1.552E+00	5.190E−01	.7494	.7725
.300	1.715E+00	1.632E+00	8.282E−02	.9517	.7240
.337	1.580E+00	1.521E+00	5.816E−02	.9632	.7197
.550	1.000E+00	9.592E−01	4.081E−02	.9592	.6997
.694	7.609E−01	7.253E−01	3.569E−02	.9531	.6858
1.060	4.323E−01	3.907E−01	4.155E−02	.9039	.6650
1.536	2.535E−01	2.173E−01	3.622E−02	.8571	.6702
2.000	1.646E−01	1.494E−01	1.515E−02	.9079	.7181
2.250	1.468E−01	1.319E−01	1.484E−02	.8989	.7378
2.500	1.323E−01	1.160E−01	1.634E−02	.8765	.7653
2.700	1.341E−01	8.124E−02	5.281E−02	.6060	.8168
3.000	2.032E−01	6.626F−02	1.369E−01	.3261	.7661
3.392	1.287E−01	1.038E−01	2.495E−02	.8062	.7286
3.750	1.150E−01	1.062E−01	8.844E−03	.9231	.7336
4.500	1.048E−01	8.676E−02	1.806E−02	.8277	.7654
5.000	9.709E−02	8.129E−02	1.580E−02	.8372	.7735
5.500	8.916E−02	7.241E−02	1.675E−02	.8121	.7910
6.000	9.377E−02	4.558E−02	4.819E−02	.4861	.8303
6.200	9.712E−02	5.345E−02	4.367E−02	.5504	.8025
6.500	8.793E−02	5.780E−02	3.012E−02	.6574	.7957
7.200	8.604E−02	5.162E−02	3.442E−02	.6000	.7946
7.900	6.246E−02	3.318E−02	2.928E−02	.5312	.8468
8.200	5.601E−02	1.926E−02	3.675E−02	.3438	.8734
8.700	1.191E−01	5.694E−02	6.211E−02	.4783	.8831
9.000	1.259E−01	5.684E−02	6.911E−02	.4513	.6619
9.200	1.235E−01	4.869E−02	7.477E−02	.3944	.6994
10.000	8.741E−02	4.846E−02	3.895E−02	.5544	.7250
10.591	7.702E−02	4.210E−02	3.493E−02	.5465	.7449
11.000	7.269E−02	3.754E−02	3.515E−02	.5164	.7547
11.500	7.045E−02	3.175E−02	3.970E−02	.4365	.7665
12.500	7.444E−02	2.290E−02	5.154E−02	.3077	.7644
14.800	8.144E−02	1.903E−02	6.241E−02	.2337	.7265
15.000	8.809E−02	1.870E−02	6.940E−02	.2122	.7170
16.400	8.562E−02	2.361E−02	6.201E−02	.2757	.6769
17.200	8.961E−02	2.743E−02	6.213E−02	.3067	.6409
18.500	8.052E−02	2.440E−02	5.612E−02	.3030	.6442
21.300	7.675E−02	2.463E−02	5.212E−02	.3210	.6031
25.000	6.657E−02	2.051E−02	4.606E−02	.3081	.5854
30.000	5.748E−02	1.553E−02	4.194E−02	.2702	.5646
40.000	5.185E−02	1.092E−02	4.093E−02	.2106	.4977

From: Shettle & Fenn—Optical properties of the atmospheric aerosols

has important consequences for any program that seeks to monitor climate, and can lead to two distinctly different strategies for monitoring climatic variables. If we concentrate on measurement of the primary variable, we face a formidable program to understand the sources of error in the measurement and the way these errors are influenced both by other variables and by the sampling strategy. If, instead, we measure secondary variables, we face an equally difficult task of demonstrating the usefulness of the measurements to climatology and to understanding the conditions under which the measurements are useful. For example, is upwelling radiance, a measure of sea surface temperature, correlated with weather patterns in the same way or as significantly as the surface temperatures themselves? Can this be used as a substitute for the surface temperature?

Lastly, variables important to climate must be monitored in precisely comparable ways for many years. Typical periods of variability that are now of interest include seasonal and interannual variability, and their study will require decades of observations. This places stringent requirements on the calibration and intercomparisons of observations because no satellite operates for these periods. Thus, observations made by different instruments on different satellites must be comparable; otherwise, there is no way to distinguish the differences in response of the various instruments from changes in climate. This, in turn, requires that satellite instruments, their calibration and all

TABLE 27-8b

Attenuation Coefficients for 80% RH 80 Urban Aerosol Model: Normalized to an Extinction Coefficient = 1.00 KM-1, at a Wavelength = 0.55 μm

Wavelength (Micron)	Extinction (KM-1)	Scattering (KM-1)	Absorption (KM-1)	Single Scat. Alb.	Asymmetry Parameter
.200	1.964E+00	1.416E+00	5.485E−01	.7208	.7949
.300	1.640E+00	1.269E+00	3.710E−01	.7738	.7713
.337	1.524E+00	1.187E+00	3.373E−01	.7786	.7650
.550	1.000E+00	7.805E−01	2.195E−01	.7805	.7342
.694	7.771E−01	5.992E−01	1.779E−01	.7711	.7162
1.060	4.625E−01	3.329E−01	1.297E−01	.7197	.6873
1.536	2.869E−01	1.884E−01	9.857E−02	.6565	.6820
2.000	2.031E−01	1.251E−01	7.804E−02	.6158	.7131
2.250	1.798E−01	1.082E−01	7.166E−02	.6015	.7312
2.500	1.610E−01	9.308E−02	6.792E−02	.5782	.7583
2.700	1.561E−01	7.053E−02	8.560E−02	.4517	.8030
3.000	2.648E−01	6.838E−02	1.964E−01	.2583	.7171
3.392	1.545E−01	8.730E−02	6.723E−02	.5650	.7185
3.750	1.356E−01	8.243E−02	5.317E−02	.6081	.7400
4.500	1.222E−01	6.981E−02	5.317E−02	.5648	.7698
5.000	1.136E−01	6.475E−02	4.887E−02	.5699	.7778
5.500	1.050E−01	5.875E−02	4.623E−02	.5597	.7923
6.000	1.171E−01	4.142E−02	7.571E−02	.3536	.8142
6.200	1.175E−01	4.854E−02	6.897E−02	.4131	.7864
6.500	1.040E−01	5.102E−02	5.294E−02	.4908	.7867
7.200	9.765E−02	4.662E−02	5.103E−02	.4775	.7891
7.900	8.400E−02	3.708E−02	4.732E−02	.4394	.8147
8.200	8.055E−02	3.028E−02	5.026E−02	.3760	.8298
8.700	1.094E−01	4.773E−02	6.169E−02	.4362	.7276
9.000	1.134E−01	4.769E−02	6.572E−02	.4205	.7136
9.200	1.106E−01	4.211E−02	6.853E−02	.3806	.7361
10.000	8.702E−02	7.810E−02	4.892E−02	.4379	.7590
10.591	8.025E−02	3.229E−02	4.795E−02	.4024	.7729
11.000	7.884E−02	2.829E−02	5.054E−02	.3589	.7783
11.500	8.030E−02	2.365E−02	5.664E−02	.2946	.7808
12.500	9.101E−02	1.974E−02	7.127E−02	.2169	.7624
14.800	1.007E−01	1.975E−02	8.093E−02	.1962	.7094
15.000	1.039E−01	1.979E−02	8.409E−02	.1905	.7022
16.400	9.944E−02	2.217E−02	7.727E−02	.2229	.6714
17.200	9.888E−02	2.417E−02	7.472E−02	.2444	.6480
18.500	9.153E−02	2.269E−02	6.884E−02	.2479	.6417
21.300	8.246E−02	2.227E−02	6.019E−02	.2701	.6104
25.000	7.152E−02	1.930E−02	5.222E−02	.2699	5887
30.000	6.088E−02	1.553E−02	4.535E−02	.2551	.5651
40.000	5.252E−02	1.084E−02	4.168E−02	.2064	.5058

From: Shettle & Fenn—Optical properties of the atmospheric aerosols.

details of the processing of their data must be carefully documented. Such documentation has been sorely lacking in the past and leads to a need for reassessment of old data, none of which now extend back two decades, and a need for careful documentation and calibration of present and future data.

PRIMARY AND SECONDARY VARIABLES

The primary oceanic variables of interest to climate are those associated with the heat budget of the Earth. These are difficult to measure; hence our attention is forced towards less central but more practical measurements. To understand these measurement problems, it is useful to consider first the primary measurement, then the secondary measurements used as a substitute for the first, and finally the present measurements and the possibility of future improvements.

Heat Storage

During the Summer and Autumn at mid-latitudes, and all year round in the tropics, the sun warms the surface layers of the ocean. This stored heat can then be released to the atmosphere in Winter or Spring or transported by currents to other latitudes. Of primary importance to climate studies are estimates of heat stored and the amount of heat available for release to the atmosphere. This, in turn, requires a knowledge of temperature and stability as functions of depth in the ocean and a theoretical understanding of the

TABLE 27-8c

Attenuation Coefficients for RH80% Maritime Aerosol Model Normalized to an Extinction Coefficient = 1.00 KM-1, at a Wavelength = 0.55 μm

Wavelength (Micron)	Extinction (KM-1)	Scattering (KM-1)	Absorption (KM-1)	Single Scat Alb.	Asymmetry Parameter
.200	1.223E+00	1.092E+00	1.303E−01	.8935	.7954
.300	1.146E+00	1.131E+00	1.556E−02	.9864	.7782
.337	1.118E+00	1.108E+00	1.015E−02	.9909	.7752
.550	1.000E+00	9.936E−01	6.426E−03	.9936	.7717
.694	9.477E−01	9.423E−01	5.329E−03	.9944	.7721
1.060	8.754E−01	8.687E−01	6.616E−03	.9924	.7777
1.536	8.042E−01	7.970E−01	7.193E−03	.9911	.7872
2.000	7.293E−01	7.160E−01	1.334E−02	.9817	.8013
2.250	6.858E−01	6.785E−01	7.309E−03	.9893	.8089
2.500	6.217E−01	6.036E−01	1.810E−02	.9709	.8301
2.700	4.996E−01	4.013E−01	9.836E−02	.8031	.8844
3.000	6.795E−01	3.062E−01	3.733E−01	.4506	.8332
3.392	6.647E−01	5.677E−01	9.701E−02	.8541	.7557
3.750	5.926E−01	5.729E−01	1.966E−02	.9668	.7597
4.500	4.955E−01	4.443E−01	5.116E−02	.8968	.7823
5.000	4.467E−01	4.033E−01	4.344E−02	.9028	.7822
5.500	3.789E−01	3.418E−01	3.706E−02	.9022	.7944
6.000	3.593E−01	1.847E−01	1.745E−01	.5142	.8157
6.200	4.337E−01	2.690E−01	1.647E−01	.6203	.7712
6.500	3.702E−01	2.823E−01	8.784E−02	.7627	.7738
7.200	3.084E−01	2.396E−01	6.879E−02	.7770	.7784
7.900	2.643E−01	1.985E−01	6.586E−02	.7509	.7807
8.200	2.523E−01	1.844E−01	6.792E−02	.7308	.7800
8.700	2.490E−01	1.766E−01	7.250E−02	.7089	.7682
9.000	2.397E−01	1.664E−01	7.330E−02	.6943	.7659
9.200	2.277E−01	1.531E−01	7.452E−02	.6727	.7692
10.000	1.780E−01	1.073E−01	7.024E−02	.6054	.7780
10.591	1.532E−01	7.357E−02	7.961E−02	.4803	.7828
11.000	1.537E−01	5.471E−02	9.901E−02	.3559	.7776
11.500	1.679E−01	4.311E−02	1.248E−01	.2567	.7621
12.500	2.236E−01	4.490E−02	1.787E−01	.2008	.7115
14.800	2.835E−01	6.330E−02	2.202E−01	.2233	.6342
15.000	2.868E−01	6.453E−02	2.223E−01	.2250	.6294
16.400	2.908E−01	7 032E−02	2.205E−01	.2418	.5999
17.200	2.904E−01	7.448E−02	2.159E−01	.2565	.5854
18.500	2.781E−01	7.474E−02	2.034E−01	.2687	.5700
21.300	2.386E−01	6.585E−02	1.728E−01	.2759	.5512
25.000	2.021E−01	5.533E−02	1.468E−01	.2737	.5265
30.000	1.643E−01	4.259E−02	1.217E−01	.2592	.4996
40.000	1.494E−01	2.515E−02	1.243E−01	.1683	.4236

From: Shettle & Fenn—Optical properties of the atmospheric aerosols

depth to which winds will mix the ocean. However, the former is difficult to measure even using **in situ** measurements, and our understanding of the surface mixed layer is so poor that we cannot yet predict its behavior.

Sea Surface Temperature

Lacking an ability to measure heat storage from space, we resort to seeking significant correlations between climate and other more easily measured variables associated with heat storage. Namais (1978) noted that relatively warm and cold pools of surface water (defined as the deviation from the mean monthly temperature for the area) often extend over large portions of an ocean basin, that they may deviate by ±2°C from the normal temperature at mid-latitudes, and that the contrast in temperature between warm and cold pools of water can intensify storms. This changes the distribution of surface lows in winter and apparently influences weather patterns downwind of the oceanic area, particularly the position of waves in the westerlies.

In this example, sea-surface temperature serves as a crude indicator of heat storage, and Namais uses correlations to bridge the gap between an easily observed variable, surface temperature, and weather over North America, thus neatly avoiding the problems associated with measuring heat content.

Sea-surface temperature is measured from space by observing upwelling radiation from the surface in frequency bands or windows through

TABLE 27-8d

Attenuation Coefficients for 80%RH Tropospheric Aerosol Model Normalized to an Extinction Coefficient = 1.00 KM-1, at a Wavelength = 0.55 μm

Wavelength (Micron)	Extinction (KM-1)	Scattering (KM-1)	Absorption (KM-1)	Single Scat. Alb.	Asymmetry Parameter
.200	2.191E+00	1.663E+00	5.281E−01	.7590	.7667
.300	1.795E+00	1.731E+00	6.368E−02	.9645	.7176
.337	1.645E+00	1.603E+00	4.157E−02	.9747	.7128
.550	1.000E+00	9.737E−01	2.633E−02	.9737	.6879
.694	7.330E−01	7.112E−01	2.183E−02	.9782	.6690
1.060	3.644E−01	3.400E−01	2.443E−02	.9330	.6255
1.536	1.628E−01	1.434E−01	1.936E−02	.8811	.5769
2.000	6.469E−02	5.808E−02	6.608E−03	.8979	.5403
2.250	4.660E−02	4.014E−02	6.462E−03	.8613	.5167
2.500	3.401E−02	2.691E−02	7.099E−03	.7913	.4947
2.700	4.535E−02	1.586E−02	2.949E−02	.3498	.4703
3.000	1.189E−01	1.881E−02	1.001E−01	.1581	.4143
3.392	2.834E−02	1.865E−02	9.687E−03	.6582	.4190
3.750	1.647E−02	1.338E−02	3.096E−03	.8121	.3993
4.500	1.388E−02	7.103E−03	6.776E−03	.5118	.3563
5.000	1.073E−02	4.915E−03	5.820E−03	.4578	.3325
5.500	9.690E−03	3.234E−03	6.456E−03	.3338	.3095
6.000	2.550E−02	1.903E−03	2.360E−02	.0746	.2767
6.200	2.225E−02	2.287E−03	1.996E−02	.1028	.2751
6.500	1.466E−02	1.976E−03	1.268E−02	.1348	.2693
7.200	1.691E−02	1.486E−03	1.542E−02	.0879	.2464
7.900	1.434E−02	4.705E−04	1.387E−02	.0328	.2175
8.200	1.991E−02	1.989E−04	1.971E−02	.0100	.1992
8.700	3.129E−02	2.815E−03	2.848E−02	.0899	.2247
9.000	3.515E−02	3.126E−03	3.203E−02	.0889	.2215
9.200	4.074E−02	2.169E−03	3.857E−02	.0532	.2042
10.000	1.723E−02	1.041E−03	1.618E−02	.0605	.1952
10.591	1.515E−02	6.830E−04	1.447E−02	.0451	.1814
11.000	1.516E−02	5.379E−04	1.462E−02	.0355	.1726
11.500	1.789E−02	3.974E−04	1.749E−02	.0222	.1604
12.500	2.541E−02	2.682E−04	2.514E−02	.0106	.1398
14.800	3.270E−02	1.686E−04	3.253E−02	.0052	.1111
15.000	3.813E−02	1.745E−04	3.796E−02	.0046	.1065
16.400	3.038E−02	2.030E−04	3.017E−02	.0067	.1068
17.200	2.886E−02	2.479E−04	2.861E−02	.0086	.1086
18.500	2.552E−02	1.617E−04	2.536E−02	.0063	.0984
21.300	2.231E−02	1.257E−04	2.218E−02	.0056	.0888
25.000	1.937E−02	6.609E−05	1.930E−02	.0034	.0724
30.000	1.802E−02	3.100E−05	1.799E−02	.0017	.0549
40.000	1.791E−02	1.271E−05	1.790E−02	.0007	.0358

From: Shettle & Fenn—Optical properties of the atmospheric aerosols

which the atmosphere is relatively transparent. Three bands are important: 3.7 μm and 10.5 μm in the infrared, and 6.6 GHz in the super-high frequency radio band. Of course, none of these atmospheric windows is completely clear, and the signal from the surface is substantially degraded in passing through the atmosphere. In addition, other processes particularly at the surface further degrade the measurement.

Water, both as vapor and as clouds, is the primary source of error for the infrared observation. Water vapor partly absorbs the signal and reradiates energy at the colder temperatures typical of the atmospheric layers a few kilometers above the surface. Dense widespread clouds completely block radiation from the sea, and although they restrict the number of surface observations, they introduce no errors because their existence is obvious. Thin clouds and clouds smaller than the instrument's field of view are major concerns. Both lower the apparent sea temperature but not so much that it is obviously wrong. The problem is to detect their presence and then to correct or reject the erroneous measurement.

Several techniques have been proposed to handle the influence of clouds. The first notes that surface features tend to persist much longer than clouds, although persistent cloudiness occurs in some areas. All observations of a small surface area made over a period of several days are examined and, on the assumption that clouds can only reduce the observed temperature, the highest temperature in the set is assigned to the surface. A second technique uses observations of visible

TABLE 27-8e
Attenuation Coefficients for Background Stratospheric Aerosol Model (75% H2504) Normalized to an Extinction Coefficient = 1.00 KM-1, at a Wavelength = 0.55 μm

Wavelength (Micron)	Extinction (KM-1)	Scattering (KM-1)	Absorption (KM-1)	Single Scat. Alb.	Asymmetry Parameter
.200	1.487E+00	1.487E+00	2.046E−07	1.0000	.6804
.300	1.555E+00	1.555E+00	1.264E−07	1.0000	.7134
.337	1.515E+00	1.515E+00	1.095E−07	1.0000	.7253
.550	1.000E+00	1.000E+00	5.917E−08	1.0000	.7259
.894	7.063E−01	7.063E−01	8.780E−08	1.0000	.6943
1.060	2.886E−01	2.886E−01	3.700E−06	1.0000	.5918
1.536	9.992E−02	9.972E−02	1.990E−04	.9980	.4465
2.000	4.183E−02	4.055E−02	1.282E−03	.9694	.3223
2.250	2.728E−02	2.570E−02	1.574E−03	.9423	.2686
2.500	1.850E−02	1.560E−02	2.894E−03	.8435	.2233
2.700	1.334E−02	9.308E−03	4.030E−03	.6979	.1916
3.000	6.510E−02	6.321E−03	5.878E−02	.0971	.1580
3.392	8.927E−02	6.274E−03	8.300E−02	.0703	.1299
3.750	6.529E−02	5.1035−03	6.0195−02	.0782	.1108
4.500	4.764E−02	2.420E−03	4.522E−02	.0508	.0780
5.000	4.276E−02	1.449E−03	4.132E−02	.0339	.0629
5.500	5.807E−02	1.029E−03	5.704E−02	.0177	.0515
6.000	5.368E−02	1.052E−03	5.263E−02	.0196	.0454
6.200	4.391E−02	8.778E−04	4.304E−02	.0200	.0426
6.500	3.339E−02	5.526E−04	3.283E−02	.0166	.0379
7.200	4.457E−02	1.931E−04	4.437E−02	.0043	.0287
7.900	1.187E−01	5.007E−04	1.182E−01	.0042	.0222
8.200	1.471E−01	7.749E−04	1.463E−01	.0053	.0204
8.700	1.274E−01	9.640E−04	1.264E−01	.0076	.0206
9.000	9.289E−02	7.195E−04	9.217E−02	.0077	.0214
9.200	8.780E−02	5.804E−04	8.722E−02	.0066	.0202
10.000	5.020E−02	4.857E−04	4.971E−02	.0097	.0205
10.591	4.069E−02	2.720E−04	4.041E−02	.0067	.0169
11.000	5.736E−02	2.619E−04	5.710E−02	.0046	.0150
11.500	3.575E−02	2.576E−04	3.549E−02	.0072	.0157
12.500	1.974E−02	1.273E−04	1.962E−02	.0064	.0124
14.800	1.895E−02	4.946E−05	1.890E−02	.0026	.0983
15.000	1.953E−02	4.500E−05	1.948E−02	.0023	.0080
16.400	3.665E−02	3.751E−05	3.661E−02	.0010	.0063
17.200	4.152E−02	5.373E−05	4.147E−02	.0013	.0062
18.500	1.714E−02	3.782E−05	1.710E−02	.0022	.0062
21.300	1.619E−02	1.729E−05	1.617E−02	.0011	.0043
25.000	8.371E−03	1.073E−05	8.360E−03	.0013	.0034
30.000	6.321E−03	5.049E−06	6.316E−03	.0008	.0024
40.000	5.917E−03	1.557E−06	5.915E−03	.0003	.0013

From: Shettle & Fenn—Optical properties of the atmospheric aerosols

light and infrared radiation at 3.7 and 10.5 μm. Visible light is used to segregate cloud-free observations (pixels) from those contaminated by clouds. The infrared observation at 10.5 μm is used to estimate atmospheric water vapor, and this is then used to remove the influence of water vapor from the observations at 3.7 μm. A considerably more elaborate scheme based on the same techniques, but using many more frequencies between 3.7 and 15 μm to correct for clouds at several levels and for water vapor, has been proposed and tested by Chahine, Aumann and Taylor (1977). The latter two techniques both have promise, but neither has been extensively tested for its usefulness to climate.

Radio frequency observations of temperature are influenced by atmospheric water and water vapor, by foam and surface roughness and by reflected atmospheric radiation. Because of these multiple influences, radiometers typically make observations using two different polarizations at five frequencies for a total of ten independent observations. These are then used to determine all variables influencing the signal, including surface temperature and wind speed (the latter being a function of foam coverage).

In addition to the major variables given above, radiometers are sensitive to several less important influences. At 3.5 μm these include sun glint, aerosols and atmospheric gases other than water vapor; at 10 μm they include aerosols and atmospheric gases; and at radio frequencies solar and galactic radiation must be considered. In general, these influences can be either estimated or ignored without introducing significant error.

Finally, we note that even perfect observations

TABLE 27-8f

Attenuation Coefficients for Volcanic Aerosol Model Normalized to an Extinction Coefficient = 1.00 KM-1, at a Wavelength = 0.55 μm

Wavelength (Micron)	Extinction (KM-1)	Scattering (KM-1)	Absorption (KM-1)	Single Scat. Alb.	Asymmetry Parameter
.200	1.149E+00	7.006E−01	4.482E−01	.6099	.8272
.300	1.192E+00	1.079E+00	1.126E−01	.9055	.7148
.337	1.180E+00	1.095E+00	8.497E−02	.9280	.7076
.550	1.000E+00	9.473E−01	5.271E−02	.9473	.6978
.694	8.487E−01	8.079E−01	4.084E−02	.9519	.6886
1.060	5.302E−01	5.057E−01	2.452E−02	.9537	.6559
1.536	2.797E−01	2.648E−01	1.490E−02	.9467	.6062
2.000	1.455E−01	1.353E−01	1.018E−02	.9300	.5561
2.253	1.107E−01	1.020E−01	8.668E−03	.9217	.5255
2.500	8.635E−02	7.792E−02	8.429E−03	.9024	.4958
2.700	7.185E−02	6.343E−02	8.424E−03	.8828	.4729
3.000	6.075E−02	5.126E−02	9.493E−03	.8438	.4401
3.392	4.504E−02	3.761E−02	7.431E−03	.8350	.4015
3.750	3.401E−02	2.913E−02	4.879E−03	.8565	.3699
4.500	2.093E−02	1.775E−02	3.186E−03	.8478	.3125
5.000	1.539E−02	1.204E−02	3.347E−03	.7825	.2773
5.500	1.265E−02	8.689E−03	3.958E−03	.6970	.2472
6.000	1.021E−02	5.701E−03	4.507E−03	.5585	.2173
6.200	9.918E−03	4.671E−03	5.247E−03	.4710	.2054
6.500	1.042E−02	3.777E−03	6.646E−03	.3624	.1908
7.200	1.361E−02	2.497E−03	1.112E−02	.1834	.1623
7.900	1.789E−02	1.364E−03	1.652E−02	.0763	.1348
8.200	2.280E−02	9.941E−04	2.180E−02	.0436	.1233
8.700	2.919E−02	4.810E−03	2.438E−02	.1648	.1615
9.000	3.108E−02	6.025E−03	2.506E−02	.1938	.1757
9.200	3.231E−02	5.730E−03	2.658E−02	.1773	.1712
10.000	3.458E−02	4.490E−03	3.009E−02	.1298	.1521
10.591	3.181E−02	3.221E−03	2.859E−02	.1013	.1326
11.000	2.772E−02	2.603E−03	2.511E−02	.0939	.1230
11.500	2.473E−02	1.913E−03	2.282E−02	.0773	.1081
12.500	1.713E−02	8.970E−04	1.623E−02	.0524	.0801
14.800	1.566E−02	3.326E−04	1.532E−02	.0212	.0528
15.000	1.667E−02	3.181E−04	1.635E−02	.0191	.0514
16.400	1.649E−02	2.623E−04	1.620E−02	.0171	.0461
17.200	1.735E−02	2.855E−04	1.706E−02	.0165	.0446
18.500	1.772E−02	3.097E−04	1.741E−02	.0175	.0449
21.300	1.077E−02	2.270E−04	1.054E−02	.0211	.0415
25.000	1.052E−02	1.383E−04	1.038E−02	.0131	.0330
30.000	1.130E−02	5.899E−05	1.124E−02	.0052	.0198
40.000	1.330E−02	1.848E−05	1.328E−02	.0014	.0097

From: Shettle & Fenn—Optical properties of the atmospheric aerosols

do not necessarily measure temperature useful for climate studies, because the temperature sensed is that of the upper few millimeters or less of the surface layer. This may not be representative of deeper temperature (Saunders, 1967), being influenced by evaporation, back radiation from the atmosphere, and solar warming. Taken together, these can amount to 0.5 to 1.0°C, a significant fraction of the anomalous surface temperature of interest in relation to climate. More importantly, this signal has a large diurnal periodicity that can be aliased to long periods by the sun-synchronous satellites now being used to monitor surface temperature.

A test of the usefulness of infrared measurements of surface temperature made by satellites was performed by Barnett, Patzert, Webb and Bean (1979) for tropical Pacific areas. They compared sea-surface temperature along a line of longitude, measured by calibrated air-dropped expendable bathythermographs (AXBT's) with the Global Operational Sea Surface Temperature Computations (GOSSTCOMP) derived from NOAA satellite data. They found that (a) the two sets differed by 1 to 4°C, (b) the difference had a strong latitudinal and longitudinal structure, (c) the difference appears to be correlated with cloud cover and atmospheric water vapor and (d) the spatial and temporal structure of the difference is so similar to the expected climate signal that satellite observations, as presently analysed, cannot be used to observe the departures of sea-surface temperature from their historical mean (Figure 27-37).

TABLE 27-8g

Attenuation Coefficients for Meteroic Dust Aerosol Model Normalized to an Extinction Coefficient = 1.00 KM-1, at a Wavelength = 0.55 μm

Wavelength (Micron)	Extinction (KM-1)	Scattering (KM-1)	Absorption (KM-1)	Single Scat. Alb.	Asymmetry Parameter
.200	1.050E+00	1.050E+00	6.262E−04	.9994	.7173
.300	1.059E+00	1.057E+00	1.516E−03	.9986	.7039
.337	1.053E+00	1.051E+00	1.807E−03	.9983	.7020
.550	1.000E+00	9.949E−01	5.059E−03	.9949	.6908
.694	9.495E−01	9.416E−01	7.931E−03	.9916	.6872
1.060	8.146E−01	7.963E−01	1.827E−02	.9776	.6848
1.536	6.605E−01	6.232E−01	3.725E−02	.9436	.6891
2.000	5.438E−01	4.822E−01	6.161E−02	.8867	.6989
2.250	4.913E−01	4.160E−01	7.539E−02	.8466	.7046
2.500	4.468E−01	3.574E−01	8.943E−02	.7998	.7099
2.700	4.167E−01	3.162E−01	1.005E−01	.7588	.7133
3.000	3.806E−01	2.645E−01	1.161E−01	.6949	.7159
3.392	3.478E−01	2.147E−01	1.331E−01	.6172	.7134
3.750	3.280E−01	1.845E−01	1.435E−01	.5626	.7058
4.500	2.972E−01	1.509E−01	1.463E−01	.5077	.6827
5.000	2.751E−01	1.378E−01	1.373E−01	.5009	.6687
5.500	2.508E−01	1.262E−01	1.246E−01	.5031	.6583
6.000	2.262E−01	1.144E−01	1.118E−01	.5056	.6513
6.200	2.165E−01	1.094E−01	1.071E−01	.5053	.6494
6.500	2.025E−01	1.018E−01	1.007E−01	.5025	.6475
7.200	1.727E−01	8.264E−02	9.004E−02	.4786	.6467
7.900	1.491E−01	6.170E−02	8.737E−02	.4139	.6496
8.200	1.424E−01	5.237E−02	8.999E−02	.3678	.6506
8.700	1.408E−01	3.775E−02	1.031E−01	.2681	.6461
9.000	1.506E−01	3.151E−02	1.191E−01	.2093	.6334
9.200	1.640E−01	2.962E−02	1.344E−01	.1806	.6177
10.000	2.361E−01	4.059E−02	1.955E−01	.1720	.5327
10.591	2.448E−01	4.385E−02	2.010E−01	.1791	.5868
11.000	2.779E−01	5.296E−02	2.250E−01	.1906	.4632
11.500	2.507E−01	6.657E−02	1.842E−01	.2655	.4518
12.500	1.527E−01	5.990E−02	9.282E−02	.3922	.5121
14.800	9.603E−02	2.937E−02	6.666E−02	.3058	.5450
15.000	9.455E−02	2.633E−02	6.823E−02	.2784	.5467
16.400	1.458E−01	2.245E−02	1.233E−01	.1541	.4712
17.200	1.237E−01	1.822E−02	1.055E−01	.1473	.4853
18.500	1.835E−01	2.167E−02	1.618E−01	.1181	.3984
21.300	1.219E−01	2.356E−02	9.836E−02	.1932	.4070
25.000	1.292E−01	2.339E−02	1.058E−01	.1810	.3319
30.000	8.541E−02	1.783E−02	6.758E−02	.2088	.3427
40.000	4.108E−02	8.636E−03	3.245E−02	.2102	.3766

From: Shettle & Fenn—Optical properties of the atmospheric aerosols

In contrast to the accuracy of sea-surface temperature, the precision of many observations is very good, allowing oceanic features to be clearly seen. Numerous studies of ocean thermal features have defined the motion of eddies with scales of a hundred or so kilometers (the weather of the sea) and the instabilities of the strong ocean currents, such as the Gulf Stream and the Kuroshio (c.f. Bernstein, Breaker and Whritner, 1977; Legekis, 1977 and 1978). The latter studies may be important to climatology if the dynamics of the instability can be related to the mass transport of the major currents.

The accuracy of radio-frequency observations of temperature is less well known than that of infrared observations. At the present time, the accuracy of the observations made by SEASAT and NIMBUS-7 is on the order of 1 to 1.5°C, but improvements in the methods of analyzing these data are expected to reduce this error to less than 1°C.

All of this leads to the conclusion that, despite years of development, satellite observations of sea-surface temperature are not yet accurate enough to map temperature for climatological studies. However, they are precise enough to observe relative variations in temperature over distances of a few hundred kilometers. Thus spaceborne radiometers are used to map the positions of thermal features associated with strong currents.

Because of the difficulties of accurately measuring from space even such secondary variables as sea-surface temperature, we can consider

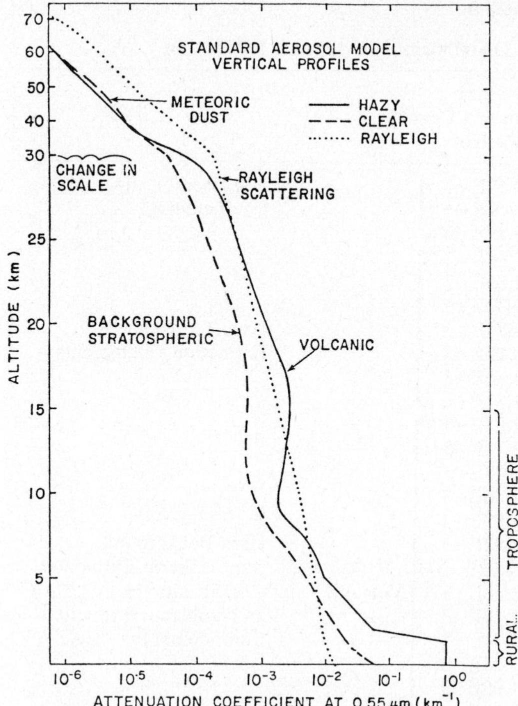

Fig. 27-36. Standard Aerosol Model Vertical Profiles.

using the related upwelling radiances at infrared or radio frequencies instead and correlating the radiances directly with climate. This is twice removed from heat content, but the measurement is accurate and unambiguous. This is an idea we will return to later.

Heat Transport

The **heat transported** by ocean currents is the product, averaged over depth, of temperature and the Lagrangian current. This integral is not easily observed even using **in situ** instruments, and cannot be measured from space. A somewhat cruder approximation to this quantity is given by the product of the surface geostrophic velocity times the surface temperature, on the assumption that both are representative of subsurface values down to the thermocline, that the average depth of the thermocline is known and that Eckman transport in the surface layers is small. A recent study of heat transport across 25°N in the doldrums of the Atlantic (Bryden and Hall, 1980) yielded consistent values using similar simplifications. Perhaps this scheme could be implemented using observations from space supplemented by some subsurface measurements, but the accuracy of the technique remains to be explored.

Currents

The geostrophic currents at the sea surface are observed by radar altimeters. In essence, the satellite altimeter measures the satellite's height above the surface at the same time its position in space (its orbit) is accurately measured. The difference in the two is the height of the sea surface relative to the earth's center. If the oceans are at rest, this height is the geoid, but if the oceans are moving, the currents produce slight variations in height about the geoid. These deviations, called surface topography are a measure of the strength of the surface currents.

The topography arises from the influence of earth's rotation on surface currents. The rotation produces a Coriolis force which must be balanced by a pressure gradient, on the assumption that forces due to friction and acceleration are relatively small. At the surface, the pressure gradient is seen as a slope, $\partial\zeta/\partial x$, in the height of the surface, ζ, relative to the geoid. More explicitly,

$$\partial\zeta/\partial x = 2\Omega \sin \phi \, v_0/g \qquad (27\text{-}7)$$

where x is a horizontal distance, v_0 is the geostrophic current at the surface, g is gravity, ϕ is latitude, and Ω is the rotational rate of earth in radians per second. Typically, $\partial\zeta/\partial x$ is quite small, varying from 10 cm/100 km to 150 cm/100 km; and its measurement requires centimeter accuracy in all quantities that enter the measurement of topography.

The accuracy of the technique depends on the accuracy of the geostrophic approximation, of the measurements of the satellite height and orbit, and of the geoid. The former is quite accurate for currents that persist for longer than a few days and that extend over distances greater than 20 to 30 km, provided the region is more than a few degrees from the equator where $\sin \phi = 0$.

The uncertainties in our knowledge of the satellite orbit and of the oceanic geoid are the major sources of inaccuracy in the observations of currents. Typically, the geoid varies by ±60 m over the world oceans, has undulations with wavelengths similar to those of ocean currents, is uncertain to ±1 m in most regions and is known to useful accuracy only in the northwest Atlantic. Orbits have similar uncertainty but vary only over long (10,000 km) wavelengths. In addition, a number of other phenomena influence the measurements of topography. These have typically one-tenth to one-hundredth the inaccuracy of orbits of the geoid, and their influence can be further reduced using readily available auxiliary information. Of these, the geophysical phenomena are most important because they can introduce systematic errors correlated with climate. These include (a) the electron content of the ionosphere and water vapor, both of which delay the altimeter signal, (b) atmospheric pressure, which depresses the sea surface, (c) ocean surface waves, which influence the altimeter's ability to detect the mean surface, and (d) non-geostrophic currents, which produce small changes in surface elevation.

The uncertainty in the geoid and in the satellite orbit has precluded measurements of mean currents, but fortunately the geoid does not change

TABLE 27-9

Scaling Factors Representing Vertical Distributions of Aerosol Extinction

Height	Clear (50 km Sea Level Visibility)		Hazy (5 km Sea Level Visibility)		
0.0	6.95E−2		7.57E−1		Uses Rural Extinction
1.0	2.58E−2		7.57E−1		Coefficients
2.0	9.70E−3		6.21E−2		
3.0	8.19E−3		3.46E−2		
4.0	6.43E−3		1.85E−2		
5.0	4.85E−3		9.30E−3		
6.0	3.54E−3		7.71E−3		Uses Tropospheric
7.0	2.30E−3		6.22E−3		Extinction Coefficients
8.0	1.41E−3		3.36E−3		
9.0	9.80E−4		1.81E−3		
10.0	7.87E−4		1.85E−3		
11.0	7.14E−4		2.11E−3		
12.0	6.63E−4		2.45E−3		
13.0	6.22E−4		2.80E−3		
14.0	6.45E−4		2.89E−3		
15.0	6.43E−4		2.92E−3		Uses Background
16.0	6.41E−4		2.74E−3		Stratospheric Extinction
17.0	6.01E−4	Background	2.46E−3	Volcanic	Coefficients for "CLEAR"
18.0	5.63E−4	Stratospheric	2.10E−3		and Volcanic Extinction
19.0	4.92E−4		1.71E−3		Coefficients for "HAZY"
20.0	4.23E−4		1.35E−3		
21.0	3.52E−4		1.09E−3		
22.0	2.96E−4		8.60E−4		
23.0	2.42E−4		6.60E−4		
24.0	1.90E−4		5.15E−4		
25.0	1.50E−4		4.10E−4		
30.0	3.32E−5		7.60E−5		
35.0	1.65E−5		2.45E−5		
40.0	8.00E−6		8.00E−6		
45.0	4.02E−6		4.02E−6		Uses Meteoric Dust
50.0	2.10E−6		2.10E−6		Extinction Coefficients
70.0	1.60E−7		1.60E−7		
100.0	9.30E−10		9.30E−10		

with time and orbits are smooth over long arcs; hence the altimeter can observe variations in surface currents over distances less than about 2,000 km by repeatedly traversing the same path on the sea surface.

Altimeter observations of ocean currents are only just beginning. The altimeter on Geos 3 operated for four years beginning in 1975 and was just able to observe the variability of strong currents, such as variation in the position of the Gulf Stream (Huang, Leitao, and Parra, 1978). The altimeter on Seasat was more sensitive and was clearly able to observe the Gulf Stream and associated eddies, but for only a short period of three months (Wunsch and Gaposhkin, 1980). Yet to be studied are the systematic errors in the measurement of currents, their correlation with variables of interest to climate, and their usefulness for monitoring ocean circulation.

The initial and still preliminary success of altimeters on Geos 3 and Seasat has stimulated planning for future altimetric satellites of improved accuracy that should be capable of monitoring ocean-surface currents. When these observations are coupled with programs that will substantially improve our knowledge of the geoid and orbits, and with subsurface observations of the ocean's internal density field made by in situ instruments, it will be possible, in principle, to monitor both the general circulation of the ocean and its transport of heat. Initially, these studies will concentrate on strong current systems such as the Gulf Stream and the Kuroshio or Somalia current, but later will extend to somewhat weaker but important flows such as the equatorial and circumpolar currents.

Fluxes of Sensible and Latent Heat and Momentum

The exchange of heat, water and momentum between the atmosphere and the ocean, especially the large-scale, long-term variability of the exchange, is fundamentally important to climatology. The exchanges, however, are extraordinarily difficult to measure using even the best possible instruments at the sea surface. The difficulty arises from the turbulent nature of the flow that

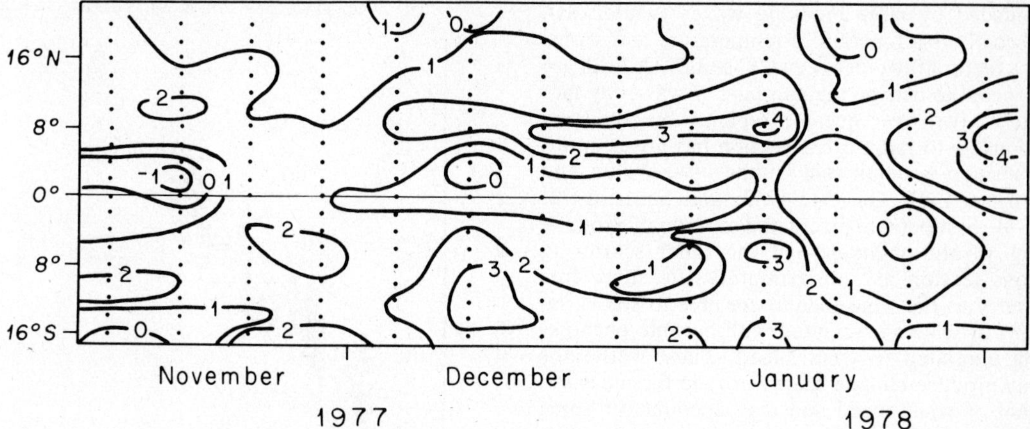

Fig. 27-37. Difference in degrees centigrade between sea surface temperature measured along 150°W by expendable bathythermographs (AXBT) and by infrared radiometers on NOAA satellites as analysed by the Global Operational Sea-Surface Temperature Computation (GOSSTCOMP). The difference is plotted as a function of time (1977–1978). From Barnett, et al., 1979.

carries these quantities and the difficulty of measuring the turbulent fluctuations, especially from not-quite-steady platforms at sea.

To avoid the problem of direct measurement of the fluxes, oceanographers and meteorologists tend to use equations relating fluxes to bulk properties of the flow that are easier to measure. These include mean wind velocity, air-sea temperature difference and relative humidity. These equations have been tested at a few places and times but generally for only light winds; little is known about fluxes in storms, even though storm fluxes may be one hundred to a thousand times larger than the exchanges during calmer conditions.

Despite the simplifications introduced by the bulk formulae, such simplications are of little use to remote sensing (except for the calculations of wind stress from wind velocity) because of the difficulty or even impossibility of observing air-sea temperature differences and relative humidity from space using existing techniques. There are possible solutions to the problem, but these are still very speculative, unproven, and may not work except in regions of very large fluxes. For example, the flux of sensible heat, which depends on air-sea temperature difference, might be estimated using sea temperatures measured by radiometers together with observations of the rate at which clouds form, augmented by the use of models of the atmospheric boundary layer. The technique may work when very cold air blows away from continents in winter, and would provide an index of the largest component of the sensible and latent heat budget of the northwest Atlantic and Pacific. Alternatively, fluxes of latent heat might be estimated from the bulk formula using surface temperature and wind speed and the specific water content of the lower kilometer of the atmosphere as observed by infrared radiometers.

Wind stress (the momentum flux) is the only flux that may be measurable from space, and because of the close relationship between stress and wind speed, these two parameters will be discussed together.

Wind Speed

The surface wind drives the ocean circulation, stirs up the upper layers of the ocean, and mediates the transfer of heat and water between the upper layers of the sea and the lower layers of the atmosphere, three processes important to climatology. In addition, wind speed is closely related to the flux of momentum to the sea, and it is the surface variable most accurately measured from space. Thus, it is one of the most important remotely-sensed oceanic variables.

The best measurements of wind velocity at the sea surface are made by radars which observe the scatter of centimeter-wavelength radio waves from the sea surface. The scatter is from small centimeter-wavelength wavelets on the surface, which are a function of the local wind velocity. Empirically, the amount of scatter per unit area of surface, the scattering cross section, is highly correlated with wind speed and direction at the surface.

In addition to the radar technique, cloud motion has been explored as a tool to observe wind velocity, but is less accurate than radar techniques. It can be used only in oceanic areas with suitable clouds and is basically an atmospheric problem. Therefore, it will not be discussed here.

The primary source of inaccuracy of the radar observations of wind occurs at the sea surface. The surface-wave energy or height is in equilibrium between the input from the wind (the quantity to be measured), the loss due to breaking, which is also a function of wind speed, and surface tension and viscosity. Surface films are surfactants, such as lipids, produced by marine microorganisms and carried to the surface by bubbles produced by large breaking waves (whitecaps). Viscosity is a function of temperature and varies by a factor of two between 0°C and 25°C. Both are expected to change the damping of waves by factors of two over oceanic regions, but their relationship to radar cross section has not been investigated in detail (see Huhnerfuss, Walter and Kruspe, 1977; Garrett, 1967; and Scott, 1972, 1975). Synthetic-aperture radars (which produce high resolution images of the radar scattering cross-section at approximately the same frequency as the wind radars) are able to see variations in scatter associated with oceanic phenomena unrelated to wind speed. These variations may produce climate scale errors in the measurement of wind speed and may account for some of the variability in the comparison between radar and surface observations of wind.

Radars to measure wind speed, called scatterometers, have flown on SKYLAB and Seasat, the latter being the more accurate and better documented instrument. Assessments of the performance of the Seasat radar indicate it can measure wind speed with an accuracy of ±1.5 m/s in speed and ±20° in direction over the range of 5 to 20 m/s (Jones, et al., 1979; Pierson, 1981), an accuracy consistent with that obtained earlier with sensitive aircraft radars (Jones and Schroeder, 1978) (Figure 27-38). These figures compare favorably with the accuracy of surface observations from ships. Satellites have the great advantage of being able to observe all the oceans, including the large remote areas in the southern hemisphere rarely visited by ships, particularly in wintertime. Yet to be investigated are the regional and long-term (climate scale) errors in the observations of wind, particularly those associated with sea-surface temperature and oceanic productivity; these are expected to be small. For example, the temperature dependence of the scatter (if it is important) could be accurately estimated from the known seasonal and regional variations in surface temperature.

Wind stress rather than velocity may be the quantity actually observed by wind radars. This hypothesis is based on the idea that wave energy is in balance between input from the wind and loss to breaking and viscosity, and that energy is directly related to momentum and only indirectly to wind speed. The hypothesis has not yet been tested, primarily because of the difficulty of measuring wind stress directly, but the distinction between stress and speed may not be of great practical importance. The formula relating stress τ to speed U is

$$\tau = \rho C_d U^2 \tag{27-8}$$

where ρ is air density, C_d is a drag coefficient, and U is wind speed measured at some fixed height, usually 10 meters. The drag coefficient is known to be nearly constant for most oceanic conditions, and varies slightly with wind speed, air-sea temperature difference and wave height. Thus τ and U should be related with sufficient accuracy that one may serve for the other except when the air-sea temperature difference exceeds a few degrees centigrade, a relatively rare occurence.

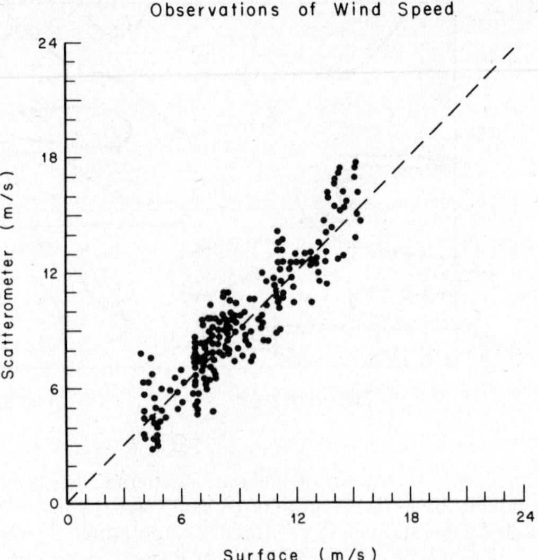

Fig. 27-38. Wind speed measured by the Seasat scatterometer compared with winds measured by carefully calibrated buoys (from Pierson, 1981).

Space-borne radar observations of wind are limited to those made during the three-month life of Seasat. However, the technique is of such value and usefulness that future space-borne radars are planned for operation in the mid-1980s.

Oceanic Rainfall

Total cumulative rainfall over the oceans is another primary variable of interest to climatologists. The latent heat released by rain in the intertropical convergence zone is a primary driving force for the tropical circulation (the Hadley circulation); rain drives storms such as hurricanes and modifies extra-tropical circulation, and the net flux of water (evaporation minus precipitation) plays an important role in the dynamics of some equatorial circulations through its modification of the ocean-density field. Because rainfall is most easily measured over the oceans, and because it plays a very important role in atmospheric processes over the ocean and in ocean dynamics, it is discussed here.

Rainfall is not directly measured by satellite instruments. Rather, radio-frequency radiometers, particularly those that observe radiation at frequencies near 19 GHz, measure total liquid water in the atmosphere over the ocean, and with a number of assumptions, this measurement is related to rainfall rate and ultimately to cumulative rainfall. The technique is still being developed, and considerable work is required before it will be useful for climatology.

The technique assumes (1) that the observation of radiation from water comes mostly from raindrops and not from cloud droplets; (2) that only liquid water, not ice, contributes to the signal; (3) that the height of the freezing level can be calculated from the surface temperature, knowing the lapse rate of a wet atmosphere; (4) that water exists as drops whose size distribution is a known function of rain rate; and (5) that each drop falls at its own terminal velocity (Wilheit, et al., 1977). These assumptions allow the observation of water content to be converted to rain rate averaged over the area seen by the radiometer, typically a circular area 20 to 30 km in diameter. If the rain cell-size and lifetime are known, the rain rate can be converted to rainfall.

The technique works only over the ocean. Over land the radiation from the surface dominates and atmospheric water is not measured. Over the ocean emission from the surface is low and the liquid water in the atmosphere is clearly sensed. Typically, the oceanic emissivity is less than 0.5, and the ocean surface has a very cold apparent temperature near 150 K.

The primary sources of uncertainty are (1) the unknown variability in the particle drop-size distribution; (2) the influence of vertical velocity on fall rates; and (3) radiation from other sources, including emission from the water surface, foam on the surface, and atmospheric oxygen and water vapor. Ulbrich and Atlas (1978) show that the drop-size distribution is best described by two parameters, and thus rainfall is not uniquely related to radio emissivity, the scatter in the correlation being approximately ±50 percent. If an additional parameter is used, the scatter is reduced to ±10 percent. Radiation from other sources is estimated from observations at other radio frequencies, observations produced by multifrequency radiometers such as those flown on Seasat and Nimbus 6 and 7. The relative importance of the other assumptions is being investigated, but in general they are not serious sources of concern.

The crude spatial resolution of radiometers is another major source of inaccuracy. The signal from rain is a very non-linear function of rainfall rate, and spatial averages of radiation from rainy and non-rainy areas do not yield average rainfall rates. To produce accurate estimates of rainfall rates, additional information about rain cell size is needed. This can come from higher radio frequencies, which have better spatial resolution, or from infrared or visible images of the convective cells producing rain.

There have been few tests of the accuracy of the radio-frequency techniques for measuring rainfall. Wilheit, et al. (1977) compared their measurements with rain estimated from a weather radar and estimated the accuracy of the technique to be ±50 percent. The comparison was crude and considerable work is necessary to evaluate the technique and to assess the importance of the major sources of error.

Convective rainfall may also be estimated from infrared measurements of the temperature of the tops of convective clouds; the higher and denser the cloud, the greater is the probability that it will produce heavy rain (Scofield and Oliver, 1977). The accuracy of the technique has been estimated over land, but not over the oceans.

Radio-frequency radiometers have flown on Seasat and on the Nimbus satellites beginning with Nimbus 5 in December 1972 and continuing through the latest on Nimbus 7 wich was still operating in 1981. These provide nearly continuous coverage, using at least two frequencies over an eight-year period, although with different types of instruments designed for different purposes. Thus, use of these observations requires combining data from different instruments, and for Nimbus 5 and 6, from two different instruments on the same satellite. Although this is possible in theory, it has not been done in practice. The atlas of oceanic rainfall for 1973–1974 produced by Rao, Abbot and Theon (1976) made use of observations only at a single frequency and suffers from computational errors, so it should be used only with considerable caution and only after consultation with the original authors.

MEASUREMENT STRATEGY

The correlations between oceanic and climatic variables that have been studied to date have used in situ and accurately measured data. In general, these correlations are weak, and oceanic variables explain only a small percentage, perhaps 10 to 30 percent of the variability in climate. This places stringent requirements on the accuracy of the observations used in the correlations. If the observations are in error by 10 to 30 percent, the correlation could be due entirely to systematic errors, a circumstance that not only leads to errors in interpretation but also to correlations that become weaker as the accuracy of the observations is increased. Thus, before observations made from space can be used in climate studies, it is necessary to show either that the observations are sufficiently accurate, with errors uncorrelated with other important indicators of climate, or that the observations plus errors correlate equally well with climate variability.

These two options lead to different strategies for collecting data from space. One concentrates on accuracy in order to tie the spacecraft observations to an understanding of the physical processes influencing climate. The other concentrates on consistent or repeatable observations in order to search for easily measured indices of climate variability, preferably indices which lead, if only by one season, variability in the climate elsewhere.

Regardless of the measurement scheme, programs to observe climate-scale variability from space must carefully consider the influence of orbits on the sampling of the oceanic variables in both time and space and must carefully document the accuracy and repeatability of the space-borne measurements.

Accurate Measurements

At first glance, satellite observations are assumed to be a replacement for in situ measurements of some variable of interest to climatology such as sea-surface temperature. If this view is taken, the measurement scheme must demonstrate that the satellite observation is an accurate indicator of the surface variable and that the observation is free of systematic errors produced by other geophysical variables, which themselves may be correlated with climate. The discussion in the last section demonstrates that this is a most difficult task. Almost all measurements observe a combination of several atmospheric and oceanic variables. These may be uncorrelated with the variable of interest but could be correlated with climate.

The idea can be explained using infrared observations of sea-surface temperature as an example. These infrared observations from space measure a combination of surface temperature and a weighted measure of the temperature of water vapor throughout the atmosphere. With suitable processing the infrared observations yield surface temperature with an accuracy of ±1 to 2°C, with the dominant source of uncertainty owing to atmospheric water. This accuracy must be compared with a climate signal of ±2°C in mid latitudes or ±4°C in some tropical areas. Clearly the error is on the order of 50 percent of the mid-latitude signal; more importantly, the source of error, water, is correlated with climate variability. For this method to be useful, the inaccuracy in the infrared measurements of sea-surface temperature must be reduced to perhaps one-fifth their present size, i.e. to about 0.8°C. To do this will require a major program using multifrequency radiometers in order to observe and correct for the influence of clouds and water vapor throughout the atmosphere, a program that is long, difficult and expensive.

Repeatable Measurements

The alternative to seeking ways to accurately measure some surface variable from space is to search for easily measured combinations of variables that can serve as indicators of climate. This shifts the difficulty away from increasing accuracy toward finding new and useful correlations to be substituted for the older but better understood correlations.

Again, infrared observations can serve as an example. The repeatable-measurement scheme accepts the observations of upwelling radiance as a combination of surface and atmospheric variables, seeks ways to make such observations accurately and repeatedly, and then searches for correlations between upwelling radiance and climate variability. That is, upwelling radiance instead of sea-surface temperature, is used as an indicator of climate. If this approach is taken, we then seek new correlations that are as useful as the old correlations involving surface temperature, knowing that this search for new correlations will not be easy, but that it will at least be easier than finding accurate ways of observing surface temperature from space. Certainly present infrared observations of upwelling radiance are both accurate and repeatable, with accuracy on the order of 0.1°C when made by suitably calibrated instruments.

The technique is limited mainly by lack of imagination. Can meanders in the Kuroshio, which are easily seen from space, be used to estimate its mass transport? Robinson and Taft (1972) and White (1975) show that the meander south of Tokyo is bistable and may depend on transport; Halliwell and Mooers (1979) show that wiggliness of the Gulf Stream is correlated also with transport, as is the loop current in the Gulf of Mexico (Molinari, et al., 1977). All these features are clearly seen from space, but can they be used? Similarly, is atmospheric liquid water content as useful as rainfall rate or total rainfall? Clearly, more investigations and imagination are needed to circumvent the inaccuracy of conventional spaceborne observations.

Calibration and Documentation

Regardless of whether a radiometer is used to observe either surface temperature or upwelling radiance, it must be well calibrated and its operation documented in order that radiometric observations from different satellites can be used over long periods of time (years to decades). The first step requires documentation of the instrument, its calibration before launch and in space, and all manipulations of its data used to derive its final output. If the output is radiance, the job is finished. If the output is surface temperature, the analysis of accuracy must include calibrations, using well documented and accurate surface observations of temperature obtained in studies such as that by Barnett et al. (1979). This holds true for wind speed, surface current and rainfall observations as well.

In the past, the documentation and assessment of accuracy have been lacking. This is unfortunate because good infrared measurements began with the Very High Resolution Radiometer (VHRR) on NOAA-2, launched in October, 1972, and many years of data now exist. How well does VHRR compare with its successor, the AVHRR, and with the scanning radiometer (SR) on the geostationary weather satellites? Do they measure precisely the same quantity? How are the data processed? Can data from various satellites be used interchangeably? These questions are mostly unanswered, but work has begun to find the answers. Similarly, can the data from various radio-frequency radiometers be combined to form time series of surface temperature, wind speed, ice cover, or atmospheric water and rain?

Sampling Schemes

The way in which variables are sampled can influence the interpretation of the signal. Two factors are important: The sampling can miss rare but important events, and periodic sampling can cause high frequency variability to appear to have low frequency aliasing.

The probability of missing a rare event depends both on (1) the lifetime of the event divided by the time interval between samples and (2) the probability that the event will occur at a sampling time, (unless the lifetime is long compared with the interval between samples). In the latter case, even rare events will always be observed. Because satellites observe oceanic variables at intervals of 12 to 36 hours, depending on the field of view of the particular instrument, and because oceanic features tend to persist for days, even rare events will usually be seen. However, oceanic variables forced strongly by the atmosphere, such as the diurnal variability of sea-surface temperature or currents in storms, and atmospheric variables such as surface winds and rainfall, will not be adequately sampled and some rare events will be missed. The probability of missing a particular event is greatest for intense local events such as thunderstorms, is less likely for regional events such as winds in the Gulf of Tuahantepec in winter, and even less likely for storms such as hurricanes and typhoons. Thus the variables most likely to be undersampled are rainfall and wind stress in storms, as well as surface temperature seen by infrared radiometers, a variable which is often obscured by clouds.

The importance of missing rare events depends on the strength of events in the tails of the probability density-distribution of the phenomena. For many phenomena the rare events are more important than the more frequent common events. For example, the transfer of energy to the sea by the winds depends on wind speed cubed; thus a hurricane with winds of 80 m/s is 4000 times more effective in stirring the ocean than are the 5 m/s trade winds found in the same latitudes. Thus one day of strong hurricane winds stirs the ocean more than weak trades acting for a year. Similarly, rainfall from a few large convective storms can dominate the rainfall distribution for a year. In both cases, rare events on the tails of the probability distribution must be observed in order to understand the role of the process for local climate.

Aliasing of diurnal variations into apparent seasonal variation is the second major sampling problem and is particularly important for measurements made by sun-synchronous satellites (Saunders and Hunt, 1980). These satellites observe at nearly the same local time each day, but the exact time can drift as the satellite orbit decays, and the time varies relative to local sunrise. Thus the daily heating cycle and daily variability in cloudiness are sampled at a variable number of hours after sunrise depending on season. This can, in principle, produce spurious, seasonal variability into the observations, but the subject has not been carefully explored.

Sun-synchronous satellites have yet another source of sampling error. Short duration events that tend to occur at the same local time can be missed. Tropical thunderstorms that begin in the afternoon and end in the evening are the most obvious example of this class of phenomena, and the diurnal migration of plankton and sea breezes are similar examples.

DATA SETS

Satellite observations of oceanic variables are relatively recent; many of the techniques are still being developed, and except for observations made by radio frequency and infrared radiometers, no long-duration sets of data exist.

Infrared Observations

Sea-surface temperature patterns were first seen from space by the High Resolution Infrared Radiometers (HRIR) on Nimbus 1 through 3 (1965–1970), but the images were noisy and hard to use. Later, clear images of surface temperature were produced by VHRR on the NOAA series of satellites (1972–1979) and by the VISSR on the geostationary GOES satellites (1974). The VHRR has now been replaced by the more sensitive AVHRR on the new NOAA series of satellites, and these continue to produce excellent images of sea temperature in cloud-free areas. Typically these instruments produce images with resolution of 0.2 to 0.5°C and 2 to 4 km every 12 hours, over the world oceans.

Radio Frequency Radiometers

Radio-frequency observations began with the Electrically Scanned Microwave Radiometer (ESMR) operating at 19.35 GHz and with the Nimbus E Microwave Spectrometer (NEMS) operating at five frequencies between 22 and 59 GHz, both carried on Nimbus-5, which was launched in 1974. Since then, all Nimbus satellites, plus Seasat, have carried various microwave radiometers. The Scanning Multifrequency Microwave Radiometers (SMMR) carried on Seasat (1978) and Nimbus 7 (1978–1981) are particularly important because they were specifically designed to observe sea temperature, rain and surface wind speed. For a description of the Nimbus program see Horan (1978). Typically these radiometers have a resolution of 50 km and view oceanic areas every 36 hours.

Wind Radars

Super-high frequency radars for measuring surface winds were operated for a few hours on SKYLAB and for three months in 1978 on Seasat.

The latter, the Seasat Scatterometer System (SASS), produced excellent maps of surface wind speed every three days with a resolution of 50 km and an accuracy of about ±1.5 m/s.

Altimeters

Radar altimeters for observing surface currents were flown on Geos 3 and Seasat. The former observed the oceans for four years (1975–1979), the latter for three months (July–October, 1978). These observed only stronger ocean currents and produced continuous observations along tracks that covered the ocean every 60 days with a grid having lines approximately 150 km apart, except that the grid was not completely filled by Geos.

CONCLUSION

The foregoing discussion raises many questions but provides few answers. Marine climatology is a new discipline, and its future direction is unclear. The same is true for measurements of climate scale variability from space. Both must develop together, the former to determine what must be measured, the latter to determine the best way for making the measurement. But despite this uncertainty, some trends are clear.

The primary oceanic variables of interest to climatology, with the exception of wind stress, cannot be measured from space. Instead, other related and more easily measured secondary variables are used; examples include using wind speed in place of wind stress, and sea surface temperature in place of heat storage.

The correlations between these oceanic variables that can be measured remotely and climate are weak, with surface observations explaining only 10 to 40 percent of the variability of climate. In addition, the range of variability of the oceanic variables is small. The dominant variation is seasonal, and deviations from this are smaller yet, on the order of ±2°C for surface temperature and ±2 m/s for wind speed.

The weak correlations and small signals place stringent demands on the accuracy of remote measurements, and indicate that even the remotely sensed secondary variables may not be useful. Not only must the inaccuracy of the measurement be small compared with the signal to be measured, but it must also be uncorrelated with other geophysical variables, which may themselves be correlated with climate.

Of the remotely sensed secondary variables, wind speed will be most useful, but routine observations of wind must await the next generation of ocean satellites, which are expected to carry wind radars. Next, surface currents measured by satellite altimeter will be useful, provided that our knowledge of the geoid and of satellite orbits continues to improve. Lastly, remote observations of sea-surface temperature will only be of limited use unless the accuracy of the observations is substantially improved. Surface temperature observations will be most useful for mapping the positions of those strong thermal fronts that are indicators of climate, for example, the position of the Kuroshio south of Tokyo.

The difficulty of making useful conventional observations from satellites demonstrates the need to find new, more easily measured indices of climate. These might include upwelling radiance instead of surface temperature, the positions of meanders in strong currents instead of mass transport, or the amount of liquid water in the atmosphere instead of rainfall.

REMOTE SENSING OF THE CRYOSPHERE AND HYDROSPHERE

THE CRYOSPHERE

The cryosphere—that portion of the earth's surface that is covered by snow and ice—gives rise to images of endless snow and ice-covered tracts of Polar terrain, Antarctic ice sheets, sea ice, icebergs, and mountain glaciers. These constituents of the cryosphere have long eluded the glaciologist's appetite for detailed data and knowledge of the physical properties and dynamic behavior of the permanently frozen areas of the planet. Remote sensing is clearly the optimum way to observe, study and measure the ice and snow of the Polar regions.

Indeed, when satellite cameras, radiometers, scanners, and return beam vidicons began to send back images of the transient and permanent snow cover in the 60s and 70s, geographers, climatologists and meteorologists soon realized how inexact their earlier attempts to depict and measure the transient snow cover, permanent snow, ice sheet extent, and sea-ice variations had been.

The cryosphere is dynamic, but it changes very slowly in response to climatic fluctuations and variations. On the other hand, areas of transient snow cover—i.e. areas in the snow transition zone (STZ), which expands and contracts seasonally but contrahemispherically—are much more dynamic. The annual extremes of snow cover may reach as far south as the Gulf States in severe winters as, for example, in 1979 (Wiesnet and Berg, 1981).

Currently, in the U.S., NOAA maps weekly snow cover of the Northern—but not the Southern—Hemisphere (Smigielski, 1981). Monthly charts are prepared from these weeklies, and the charts extend back to late 1966 (Wiesnet and Matson, 1976; Matson and Wiesnet, 1981). Microwave data from ESMR and some data from SMMR are used by NOAA for polar sea-ice and snow mapping, but these instruments have a reduced resolution compared with the current NOAA TIROS-N type of satellite (see Table 27-10). River-basin mapping of snow cover has become a popular and useful technique for operational hydrologists, especially in the western United States. The snow information is incorporated into flood forecast models for seasonal,

TABLE 27-10

Tiros-N Orbital Parameters and Sensor Characteristics

Orbital Parameters

Altitude	833 km
Period	102 minutes
Equatorial Crossing	1500 local time (northbound) 0730 local time (southbound)
Cycle	12 hr
Inclination	99°
Orbit	Sun-synchronous, near-polar
Coverage	Global with 2-satellite system at lower resolution

Sensor	Spectral Band	Nominal Resolution at nadir	Swath Width (km)	Coverage	Remarks
AVHRR[1]	0.55–0.9μm 0.72–1.1 3.55–3.93 10.5–11.5	1–4 km	2800	12 hr	Global coverage at 4 mi resolution; mostly direct read-out at higher resolution

For further information see the TIROS-N/NOAA A-G Satellite Series, NOAA Tech. Memo. NESS 95 from NTIS, 5285 Port Royal Rd., Springfield, Va. 22161.

[1] Advanced Very High Resolution Radiometer. In addition to the AVHRR, TIROS-N will have a TIROS operational Sounder (TOVS) system consisting of a High Resolution Infrared Radiation Sounder (HIRS), a Stratospheric Sounding Unit (SSU), and a Microwave Sounding Unit (MSU).

daily and annual forecasts (Hannerford and Brown, 1981; Rasmussen and Ffolliott, 1981; Shafer et al., 1981; and Ostrem, et al., 1981).

Sea Ice

The Navy/NOAA Joint Ice Center in Suitland, Md., produces global sea-ice analyses for civilian as well as Department of Defense users. Its Weekly Arctic Ice Analysis is put out as both a Western and an Eastern Arctic Chart at a scale of 1:11,110,000. These charts are a synthesis of shore-station ice reports, ship reports, aerial ice reconnaissance, and ice analyses from other sources, both foreign and domestic. In addition to the sources mentioned, the Joint Ice Center uses ice climatology, automated diagnostic aids and satellite imagery from NOAA, DMSP, and Nimbus 5 and 7 as analysis tools to improve and extend the surface-report data. Seven-day Ice Limit Forecasts are also produced.

For the Antarctic, owing to the extreme paucity of ground data, the Weekly Antarctic Ice Analysis chart is based primarily on satellite data. Despite a rather difficult logistics problem, a High Resolution Picture Transmission Satellite data-reception system has been installed at McMurdo Station, Antarctica, in order to gather NOAA high resolution (1.1 km) taped data from the TIROS-N series of satellites. Although its operations have been intermittent because of severe weather-induced breakdowns, this station is providing a great deal of heretofore unavailable data on Antarctica. The Japanese Antarctic Research Station at Syowa, on the northeastern Coast of Antarctica on the Consmonaut Sea, also installed a HRPT system in 1981. (Yamagouchi, personal communication, Oct. 1981.)

Landsat images have been converted into photomaps by the U.S. Geological Survey for much of Antarctica. However, Landsat's orbital configuration and field-of-view restrict it to only those areas north of 81°S. Fortunately the lower resolution NOAA satellites and the NASA Nimbus series can acquire imagery that includes the complete polar areas.

The bands most commonly used for sea-ice and polar studies are the visible, the near-IR, the thermal IR, and the passive microwave. However, active microwave radar is extremely useful for ice studies as has been demonstrated by aircraft. The day/night all-weather capability of radar makes it an optimum sensor for sea ice in Polar regions. Passive microwave sensors require far less power and are lighter in weight than active sensors, but are not capable of fine resolution. They also have an all-weather, day/night capability.

Visible-band sensors can be used to estimate ice concentrations, track large ice floes, and detect large polynas during daylight, as can thermal IR during the polar night. Passive microwave can penetrate cloud cover, detect new and multiyear ice, and estimate sea-ice concentrations (Zwally et al. 1981). The following are routinely detected; ice concentrations, age and stage, descriptive features (patches, fields, belts and strips), and ice limit and boundaries.

Lake Ice

Lake ice, though similar in appearance in many respects to sea ice, is quite different. As lakes freeze, the ice goes through stages of metamorphism from clear, transparent ice through a series of breakups, refreezings, snowcover melting, bottom thickening, and the ingesting of forest litter, dust, dirt, etc. from atmospheric sources.

In Canada many frozen lakes are used as aircraft landing strips, sites for drilling and, at times, roadways. The Great Lakes are closed to shipping for about three to four months each year because of ice. Accurate forecasts of ice conditions on these and other large navigable lakes are imperative so that ships and cargoes are not locked in for the winter or stopped enroute requiring ice-breaker rescue, which is costly and dangerous.

The NOAA TIROS-N satellites, with their multisensor package of visible, near-IR, and thermal-IR sensors, are well suited to provide considerable assistance to the lake-ice forecaster. Nevertheless, in the vicinity of these large lakes, surface obscuration by clouds is extremely common.

Lakes serve as indicators of winter severity; the number of days of ice cover varies from year to year in accordance with seasonal temperature variations. Some of the greatest ice concentrations in the Great Lakes occurred during the abnormally cold winters of 1976–77, 1977–78, and 1978–79. (Quinn, et al., 1978; Wartha-Clark, 1980). After monitoring Lake Erie ice for several years, Wiesnet (1979) was able to point out characteristic patterns of breakup as revealed by NOAA and Landsat images. Knowledge of these patterns can improve ice forecasting.

River Ice

Ice growth, decay, and movement on large rivers can be clearly discerned on Landsat imagery. These Landsat images and digital data from the computer compatible tapes are unequalled for studying ice dynamics on large rivers. However, because of the delay in data acquisition and the 18-day repeat cycle of Landsat, data from the NOAA and GOES satellites afford a better possibility of monitoring river-ice events in near-real time. McGinnis and Schneider (1978), in an extensive study of the April 1976 Ottawa River ice breakup in Canada, compare the data from all three satellites and point out each one's individual usefulness in monitoring river-ice breakup.

Operational hydrometeorologists find such information on river ice jams, ice covers or ice dams of great practical interest especially in remote regions, provided satellite data are received in near-real time; yet a serious effort to monitor and forecast river ice from satellite data has not been made in the U.S. to date, because of the lack of a quick-look capability for Landsat data.

Little has been written on remote sensing of river ice. Mashukov (1957) discussed aircraft reconnaissance techniques used to monitor ice bridges and dams on the Syr Dannya River in Russia for a seven-year period. Molchanov and Vlasov (1974) used aerial photography to measure the longitudinal profile of a river surface on an ice-jam sector. The Canadians have, in the seventies, made synthetic aperture radar flights over the St. Lawrence River to spot ice concentrations in an attempt to avoid ice jams at intakes for power plants along the river.

Dey et al. (1977) used NOAA 4 and 5 data to monitor the ice breakup on the Mackenzie River in Northern Canada during March–July in 1975–77, using photointerpretation techniques. The point of origin of the breakup was identified, and the breakup then moved simultaneously upstream and downstream from the point. Incidental to his snow-cover monitoring, Schneider (1977) monitored the ice breakup on the St. John River Basin using NOAA imagery.

Continental Ice Sheets

The ultimate in recorded climatic change is the initiation of an ice age. Large continental glaciers grow from accumulation centers and spread laterally over the landscape, burying it under thousands of feet of ice. Antarctica lies buried under the world's largest continental ice sheet; Greenland has the second largest ice sheet. The impact of continental ice sheets on climate is profound. Yet the severity of the climate and the human suffering, hardship and sacrifice that are required to collect **in situ** data have resulted in an inadequate data base on the climate of the Polar regions. Polar-orbiting meteorological and environmental satellites, such as the NOAA, Landsat, and Nimbus series, can provide and have provided much valuable data for climatological analysis.

For example, Gloersen and Zwally of NASA have published Nimbus-5 Electronically Scanning Microwave Radiometer (ESMR) images and data over Greenland, providing comprehensive brightness temperatures for the entire ice sheet, and revealing variations in surface brightness-temperature patterns of considerable interest to climatologists. ESMR brightness temperature mapping of Antarctica has also been provided by Zwally and Gloersen (1977). Wiesnet (1980) prepared a mosaic of Greenland from NOAA visible and IR data.

An even more pressing need for glaciologists, as well as climatologists, is the need for more accurate maps of the continental ice sheets, so that the extent and the mass balance of the ice sheet can be measured, and monitored for change detection. The current worldwide rise in sea levels is related by many to melting ice sheets, glaciers and the polar ice pack (Etkins and Epstein, 1982). Satellite-derived measurements will provide a sound benchmark for future estimates of change in these important climatic variables. Key programs are the National Science Foundation Office of Polar Programs, which supports U.S. Polar Research, and the U.S. Geological Survey Landsat photomaps of Antarctica. Another project of the

Geological Survey is the Satellite Image Atlas of the Earth's Glaciers (Williams and Ferrigno, 1981). This ambitious project will provide a benchmark of the world's glaciers during the 1970s. Landsat MSS and RBV images form the basis for the study. Climatologists will at last have an accurate compilation of source material on ice-sheet delineation and extent, ice shelves, sea ice, speed of outlet glacier movement, and areal changes in seaward margins of ice shelves and glaciers.

Studies in Greenland in the late 70s utilized Landsat and NOAA imagery to locate melting areas of the Greenland Ice Sheet in the Sounderstrom area relative to possible development of hydropower from this runoff. (Anker Wiedick, personal communication, 1979).

The third largest ice sheet in the world, located in Iceland, is called Vatnajokell. It has a surface area of about 8,300 km^2 as measured from Landsat images (Williams, 1976). Landsat images are a practical means of monitoring changes in ice sheet and glacier area, rate of movement and recession, as well as the study of ablation characteristics.

The importance of continental ice sheets—and indeed of snow and sea ice as well—to the earth climate system is great because of their high albedo and their thermal and radiative properties. Their interaction with other components of the system results in feedbacks that are possibly the most important interactions that determine the overall sensitivity of climate (Robock, 1982).

Mountain glaciers and permanent snow areas

Landsat imagery has proved to be very well suited to mountain glacier studies. The rate of glacier movement is commonly measured in centimeters per day; hence, only a few images widely spaced in time are required to measure rate of movement. The 80 m resolution of Landsat is adequate for most large mountain glaciers, but some small glaciers and permanent snow areas tend to be overlooked at this scale and are better monitored on foot or by aircraft. Meier (1974) demonstrated the detection of surging glaciers with Landsat imagery.

Although the advances and retreats of individual glaciers are of small or local climatic significance, the general advances of a large number of glaciers within a large region can be a more meaningful climatic indicator.

HYDROSPHERE

The hydrosphere is the realm of water that occurs on and in the earth, as distinguished from the solid part of the earth (the lithosphere). In a large sense it includes the water vapor of the atmosphere, the sea, and the ground or subsurface water (American Geological Institute, 1962). The inclusion of the water vapor in the atmosphere and its ultimate precipitation are items of great practical interest to the operational hydrologists, who deal in terms of daily runoff.

Precipitation measurements and estimates

Rainfall and snowfall are measured in terms of "depth." (Millimeters are the international units; the U.S. uses inches). Rain and snow gages catch and measure the precipitation. Since World War II, weather radars have scanned the skies for precipitation droplets out to a distance of about 125 miles from the station. Rainfall rate or intensity can thus be estimated. Today, satellite images also are used to estimate precipitation from visible, microwave and thermal-IR images.

Various techniques have been proposed (e.g., Scofield and Oliver, 1981; Woodley et al., 1981; Barret 1981; and Whitney and Herman, 1981) based on variations of clouds in space and time and their relation to precipitation intensity. These techniques apply chiefly to convective rainfall and require the frequent (30-minute) observations of the geostationary satellites.

Fowler et al. (1981) have demonstrated techniques using microwave radiometry to estimate precipitation. These microwave measurements have the advantage of actually detecting the droplets, but also the disadvantages of attenuation in heavy storms and background problems.

The Scofield-Oliver estimates for severe thunderstorms resulting from Hurricane Anita in September 1977 proved to be comparable to ground station observations. Thunderstorms over Mississippi's Pearl River basin in April 1979, were also estimated by this technique. The resulting isohyets were very close to those derived from ground observations.

In many areas of the world, long-term rainfall data are sparse to nonexistent. Faced with the need to build dams and reservoirs, engineers are hard pressed to design appropriate structures with little or no hydrometeorological data. The use of remote sensing for weather and climate perhaps reaches its greatest economic value in studies to plan capital improvements in developing countries where long-term climatologic and meteorological data are nonexistent.

Soil Moisture

The agriculturalists' need for soil-moisture measurements is easily recognized, but the need extends into the disciplines of hydrology and climatology as well. The agricultural needs pertain to crop-yield forecasting and irrigation scheduling; hydrological needs include rainfall/runoff computations for river-level forecasting, flood routing, and evaporation modeling; climatological needs pertain to desertification, continental heat balance, and global modeling of climate. Through a coupling of the satellite's ability for global coverage with the computer's ability for rapid processing of high volumes of data, the gathering of worldwide soil moisture information now seems not only manageable but highly desirable. Nevertheless, the task is non-trivial and has not yet been accomplished.

Schmugge and others (1981) have discussed **in situ** and remote-sensing methods for soil-moisture determination. Much of the following discussion is based on their excellent survey paper.

The dearth of point and point-profile data on soil moisture stems from the need until recently to rely almost entirely on difficult and time-consuming **in situ** methods for determining soil moisture in the field. Gravimetric (oven-drying), nuclear (neutron-scattering), gamma-ray attenuation, electromagnetic (electrical resistivity), tensiometric (soil-water tension), and hydrometric techniques are the field methods, each of which has its own advantages and disadvantages.

Table 27-11 (from Schmugge et al., 1981) lists and compares the remote-sensing approaches for soil-moisture sensing.

Because the relationship of the spectral reflectance of soil to the water content of the soil is complicated by soil texture, organic content, surface roughness, geometry of illumination, and spectral reflectance of the dry soil (Jackson et al., 1978) the reflected solar energy approach is not as promising as the thermal infrared and microwave techniques.

Measurement of the diurnal range of surface temperature has been demonstrated to be a function of soil moisture (Figure 27-39, Schmugge et al., 1981) in experiments at the U.S. Water Conservation Laboratory in Phoenix, Arizona (Idso et al., 1975). A wide range of temperature-amplitude/water-content relations was found for various soils and meteorological conditions. Schmugge, et al. (1978) were able to secure a single relationship for all soils investigated by expressing moisture values as a percent of field capacity (FC) where FC is the moisture content at the 1/3 bar pressure potential.

When a vegetative canopy covers the base soil, the diurnal range techniques cannot apply. However, using canopy temperature as an index, Jackson et al., (1977) established that a running sum of daily temperature values called "stress degree days" (SDD) has potential for irrigation scheduling. The approach has been confirmed for wheat (Millard et al., 1977) and SDD's have been correlated with yields of wheat and alfalfa (Idso et al., 1977; Reginato et al., 1978) and with other crops.

The emissive and reflective properties of a soil surface are a function of the soil moisture. These properties can be measured in the microwave portion of the electromagnetic spectrum. In addition, the ability of microwave radiation to penetrate cloud cover makes this portion of the spectrum attractive for soil-moisture remote sensing. Microwave remote sensing may be either passive (radiometry) or active (radar).

Passive microwave radiometry simply records the thermal emission from the surface. At microwave frequencies the intensity of the observed emission is proportional to the product of the temperature and emissivity, or the brightness temperature (T_B). The dielectric properties of a soil are strong functions of soil moisture. The emitted energy from a soil is a function of the emissivity, which is determined by the dielectric characteristics of the soil near the surface.

The effective sampling depth of microwave T_B values is only a few tenths of a wavelength, that is, about 2 to 5 cm for a 21 cm (L-Band) radiometer. (Wilheit, 1978). Surface roughness generally increases surface emissivity. The longer the wavelength, the greater the range of the dielectric constant for soils, the greater depth of soil contributing to the signal, and the greater the ability to penetrate vegetative canopy. This makes longer wavelength sensors better suited for passive mi-

TABLE 27-11

Comparison of Remote Sensing Approaches for Soil Moisture Sensing (from Schmugge et al., 1981)

Sensor	Advantages	Disadvantages	Noise Source
Thermal infrared (10–12 μm)	High resolution possible (400 m) Large swath Basic physics well understood	Cloud cover, limits frequency of coverage	Local meteorological condition Partial vegetative cover Surface topography
Passive microwave	Independence of atmosphere Moderate vegetation penetration	Poor spatial resolution (5–10 km at best) Interference from man-made radiation sources, limits operating wavelengths	Surface roughness Vegetative cover Soil temperature
Active microwave	Independence of atmosphere High resolution possible	Limited swath width Calibration of SAR	Surface roughness Surface slope Vegetative cover

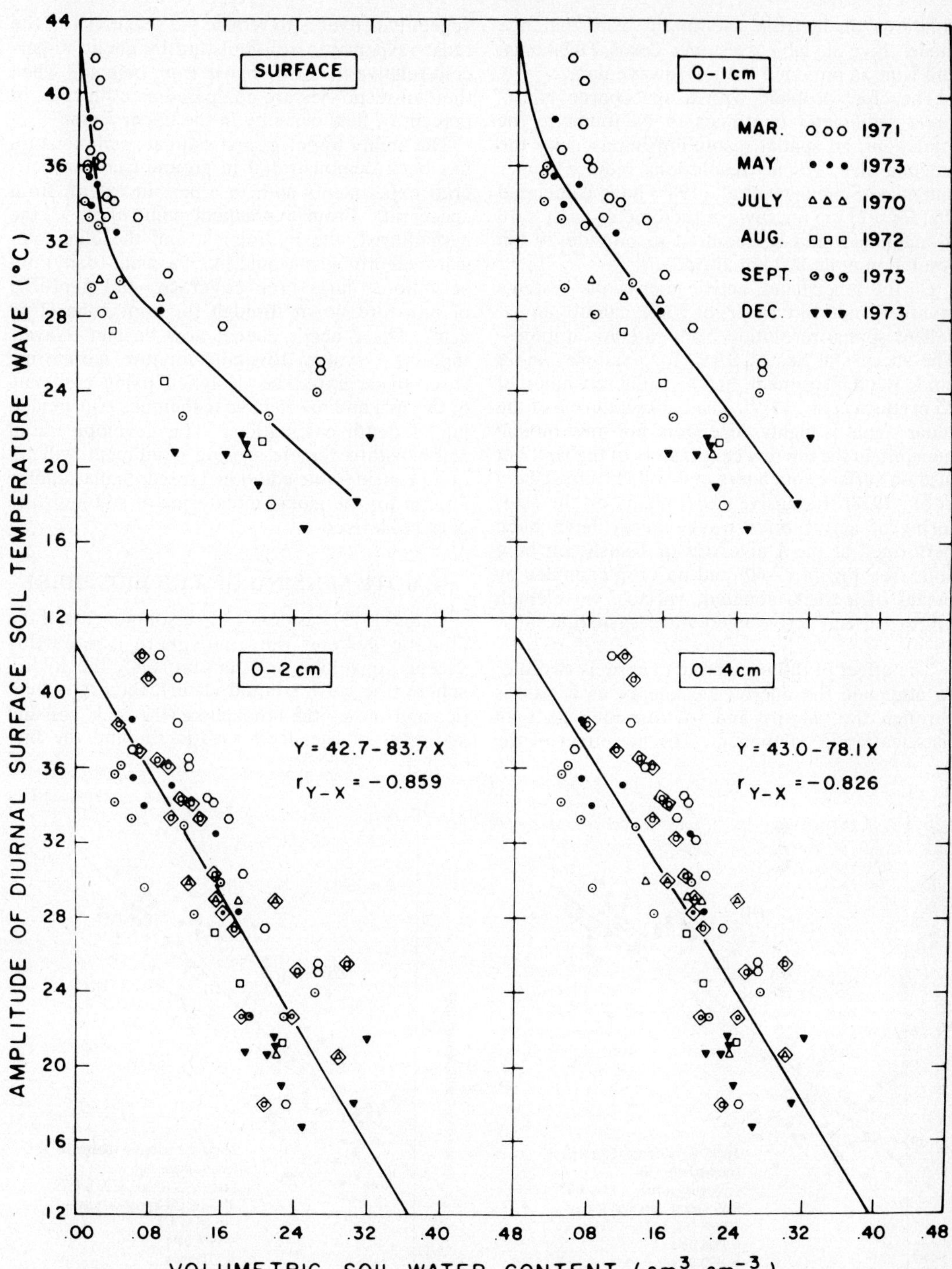

Fig. 27-39. Summary of results for the diurnal temperature variation versus soil moisture (Idso et al., 1975).

crowave soil-moisture remote sensing than sensors of short wavelength (e.g. 1.55 cm).

McFarland (1976) was able to show a good relationship between 21 cm brightness temperatures and the antecedent precipitation index (API) over portions of Texas and Oklahoma from the Skylab L-Band microwave radiometer. This instrument was a coarse resolution (115 km), non-scanning device, which was also used by Eagleman and Lin (1976), who compared T_B with soil moisture estimates over the 115-km "footprint." A correlation coefficient of 0.96 was obtained. The soil-moisture estimates were derived from a combination of ground-truth plus measurements and calculations made from a climatic water-balance model. Five Skylab passes over Texas, Oklahoma, and Kansas provided the microwave data.

The chief problem with a spaceborne microwave radiometer continues to be found in the limitations on spatial resolution imposed by the antenna size. Higher resolutions require longer antennas. Schmugge et al. (1981) have calculated that for a 21 cm microwave radiometer, a 10 × 10 m antenna would be required to provide 20 km resolution at an 800 km altitude.

On the other hand, active microwave systems (synthetic aperture radar, or SAR) can obtain excellent spatial resolution from satellite altitudes. The successful Seasat SAR, for example, an 18 cm L-Band instrument, had a spatial resolution of 25 m (Born et al., 1979). The backscattering of the radar signal is highly dependent not only on soil moisture in the top few centimeters of the soil, but also on surface roughness and soil texture (Ulaby et al., 1978). Extensive experiments on the scattering of active microwave energy have been performed at the University of Kansas on bare soils (see Figure 27-40) and on crop canopies by means of a truck-mounted, variable wavelength (30 to 1.6 cm) active microwave system (Ulaby, 1974).

The effect of the vegetative canopy is twofold; to attenuate the microwave energy as it passes through the canopy and to introduce its own backscattering component. The net effect of the vegetative cover is to reduce the sensitivity of the radar response to soil moisture by about 40 percent relative to the response from bare soil when the two responses are compared as a function of percent of field capacity in the upper 5 cm.

The ability to detect and estimate soil moisture has been demonstrated in ground-truth and aircraft experiments and, to a certain extent, from spacecraft. From a practical point of view, the agriculturist, the hydrologist and the climatologist-meteorologist would like frequent (daily) observations, large-area coverage, and a profile of moisture down through the root zone (1 to 2 m). These needs cannot now be met. Nevertheless, a system for soil moisture monitoring from space could be designed using elements of thermal and microwave techniques plus modeling of depth calculations. The development of soil-moisture remote-sensing equipment will require considerable additional research if a suitable system for the global monitoring of soil moisture is to be derived.

REMOTE SENSING OF THE BIOSPHERE

Gates (1979) has defined five components of the climatic system; the atmosphere (the earth's gaseous envelope and its aerosols), the hydrosphere (the earth's liquid water), the cryosphere (ice and snow), the lithosphere (the rock, soil and sediment of the earth's surface), and the bio-

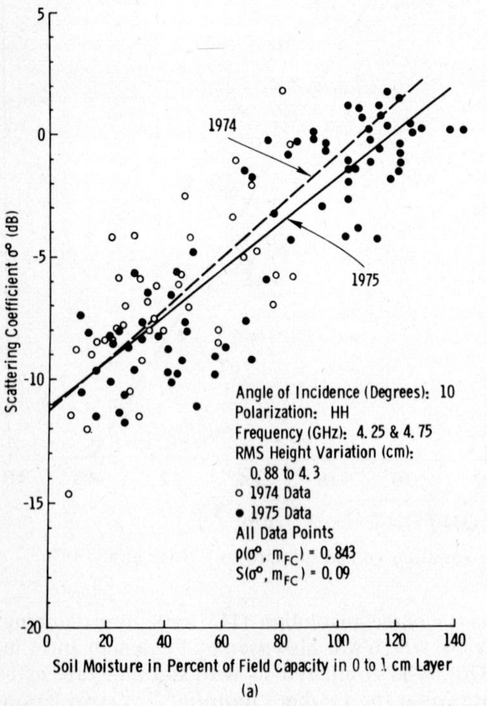

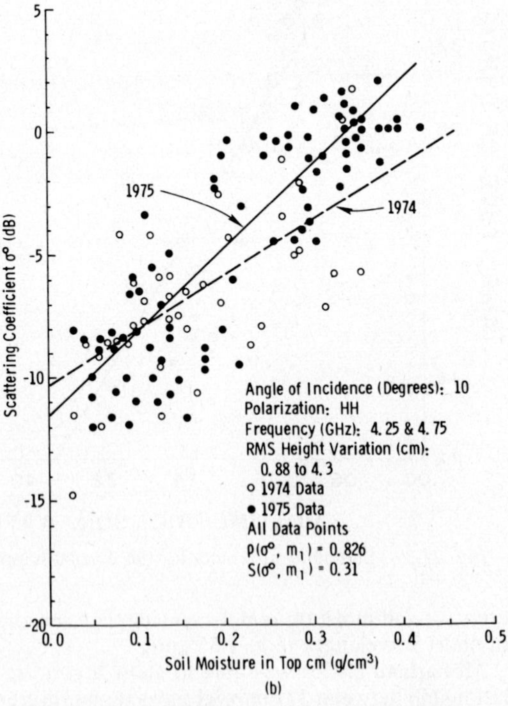

Fig. 27-40. Back scattering coefficient plotted as a function of soil moisture given (a) in percent of field capacity of top 1 cm and (b) volumetrically in top centimeter. The 1974 and 1975 bare-soil experiment data are combined (Batlivala and Ulaby, 1977).

sphere (the plant and animal life). The first three components are examined elsewhere in this chapter. The lithosphere regulates land temperature on diurnal and synoptic time scales and this variability may lead to changes in the surface radiation budget, already discussed. It is the purpose of this section to explore the climatic impact of biospheric alterations, specifically the consequences of desertification, deforestation and urbanization, and to examine remote sensing techniques amenable to monitoring these processes.

DESERTIFICATION

Warren and Maizels (1977) define desertification as "a sustained decline in the yield of useful crops from a dry area accompanying certain kinds of environmental change, both natural and induced." It is the earth's arid, semi-arid and subhumid lands which are vulnerable to desertification. Arid areas are characterized by insufficient rainfall for crop cultivation (below 200 mm) but enough to support sufficient vegetation for pastoral activity. Rainfall in semi-arid areas ranges from 200–600 mm and allows cultivation of drought-resistent crops. Subhumid areas may have intensive settlement, annual rainfall up to 800 mm, and a number of crop products, but are still susceptible to desertification arising from climatic or bio-geophysical degradation. The above classifications are seasonally, thermally, and latitudinally dependent.

What are the climatic consequences arising from the process of desertification? Most attempts to answer the question have been in the form of modelling experiments dealing with an albedo increase resulting from diminishing vegetative cover. Hare (1977) has provided an excellent summary of these modelling experiments, most of which agree with the results of Charney et al. (1975) and Charney (1975). Charney found the following bio-geophysical feedback mechanism: Lack of rainfall leads to lack of vegetation, and in turn, a higher albedo (bare soil vs. vegetation) which results in more reflected solar radiation (combined with emission of terrestrial radiation to space) producing a radiative sink of heat. The surrounding air mass loses heat radiatively, causing air to descend and compress adiabatically (to maintain thermal equilibrium). The relative humidity decreases and the dryer air mass results in a lack of rainfall. The entire process can also begin with the initial vegetative destruction, a process occurring through drought, cultivation, overgrazing, or soil desiccation. The alarming conclusion of Charney's hypothesis is that it represents a positive feedback, that is, one that reinforces itself. The climatic net result is a deficiency of rainfall over arid, semi-arid or sub-humid lands near the primary desertification zone, thereby spreading desertification to these areas.

Although there is not yet an operational remote sensing program to detect and monitor desertification, the United Nations (1977) has provided an excellent procedural outline of how such a system could be implemented:

"Global surveillance of the status of dryland ecosystems and of land use can be achieved most economically through the remote sensing power of specialized orbiting satellites. The Landsat system, already in operation, has this capacity. Landsat now provides imagery with a resolution of at least 50 meters with prospects that higher resolutions will soon be achieved.

The first step in the use of a satellite such as Landsat is to employ it for the identification and mapping of distinct units on the ground. This can be carried out in false color imagery (Landsat bands 4, 5, and 7) or in black-and-white. The information can be obtained in even more detail on a digital basis, using computer-compatible tapes. Much of the world's dry lands, perhaps 85 percent is already covered by Landsat. The mapping would define functional environment types as determined by their geology, landform and surface drainage, each type characterized by certain soils and vegetation cover. The characteristics of each unit would be established from imagery or tapes and supported by already existing information on geology, soils and vegetation. The findings would be validated on the ground, this so-called "ground truth" being determined by field sampling and traverses.

Initial demarcation of the topographical and soils units by a skilled photo-interpreter would be inexpensive, costing on the order of a few dollars per thousand hectares. The building up of ground truth would be a separate and continuing operation. Different combinations of boundaries would allow the information so obtained to be expressed in terms of a variety of references, such as pasture, land vegetation or salinity. The achievable resolution is adequate for general surveys of land status, for planning for extensive land use such as pastoralism, or as a first stage in the identification of likely areas for more intensive kinds of land use. The maps obtained could provide a framework for the interchange of experience among comparable environments.

To fix trends in dryland ecosystems, repeated monitoring on a uniform basis is required. This can be obtained from the Landsat system via remote sensing satellite-to-ground receiving stations. Each of these has an effective radius of about 2700 km but at the present time, only a part of the dry lands is properly covered. Access to a ground receiving station provides an opportunity for manipulating the data ouput to conform with local needs.

The storage, handling and reproduction of the data from the ground receiving station, and its integration with data from other sources, call for linked computer-based data systems which will generally form part of a national land-data

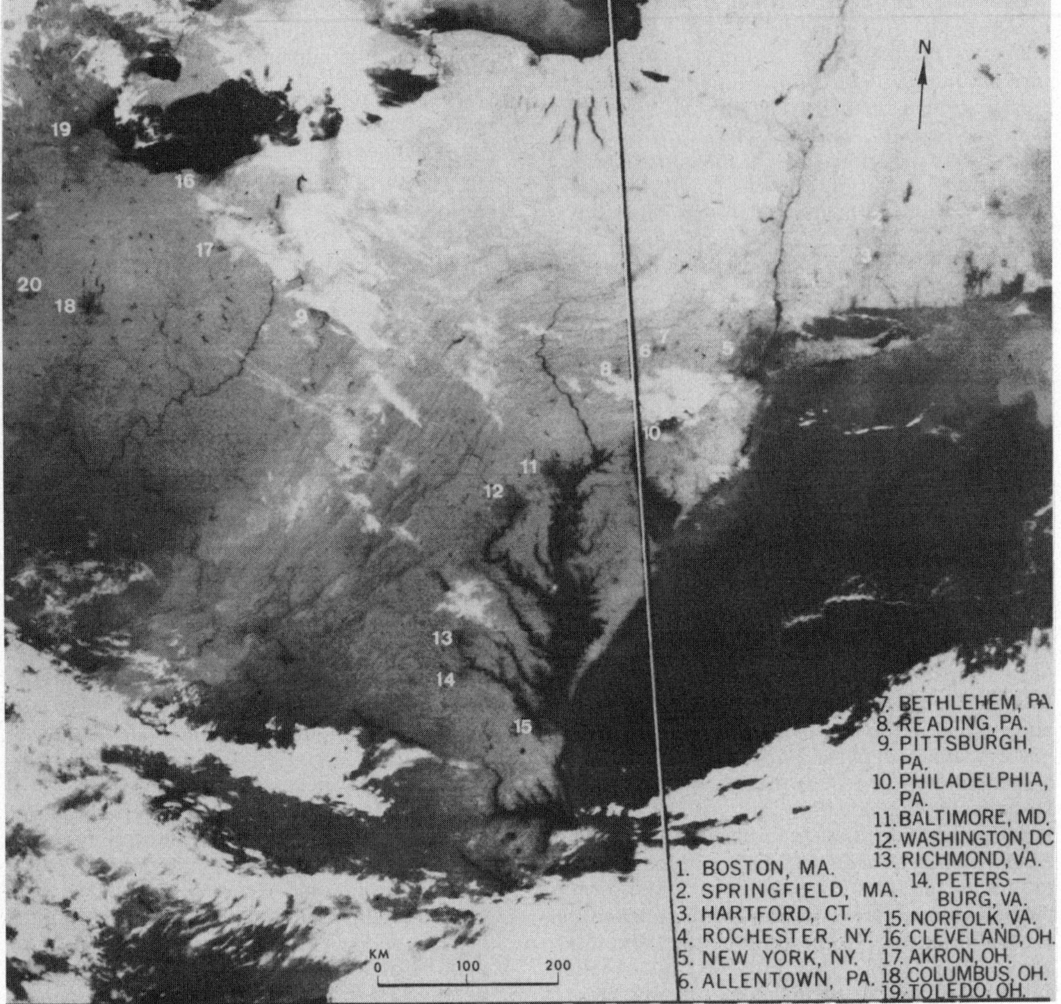

Fig. 27-41a. Urban heat islands identified from NOAA 5 VHRR thermal infrared data.

system. Information can be related to a given topographical unit or to another geographical subdivision by the use of a standard system, and experience suggests that a 1 to 2 km grid provides adequate definition for the general surveillance of dry lands. . . .

The evidence obtained from satellite data and other sources, linked through the data bank and supplemented with the information already available, should provide the basis for a number of important activities. The first of these should be the construction of a map showing types of desertification present and the relative vulnerability of the demarked ground units to further desertification. Then regional plans can be formulated for measures to combat desertification, linked with plans for improved land use, for the re-establishment of disturbed populations or for whatever else conditions call for. Following the regional plan, specific combative measures can be designed and sites selected for demonstration or pilot projects.''

DEFORESTATION

Approximately 11 percent of the globe and 38 percent of the continents are forest-covered. Depending on geographic location, removal of forest cover could have, at the least, local climatic effects. Potter et al. (1975) have modeled the removal of the tropical rain forest and arrived at the following sequence of events subsequent to deforestation; increased surface albedo, reduced surface absorption of solar energy, surface cooling, reduced evaporation and sensible heat flux from the surface, reduced convective activity and rainfall, reduced release of latent heat, weakened Hadley circulation, cooling in the middle and upper tropical troposphere, increased tropical lapse rates, increased precipitation in the latitude bands 5°N to 25°N and 5°S to 25°S, a decrease in the Equator-pole temperature gradient, reduced meridional transport of heat, and a decrease in precipitation between 45°N and 85°N and between 40°S and 60°S. This sequence supports an earlier

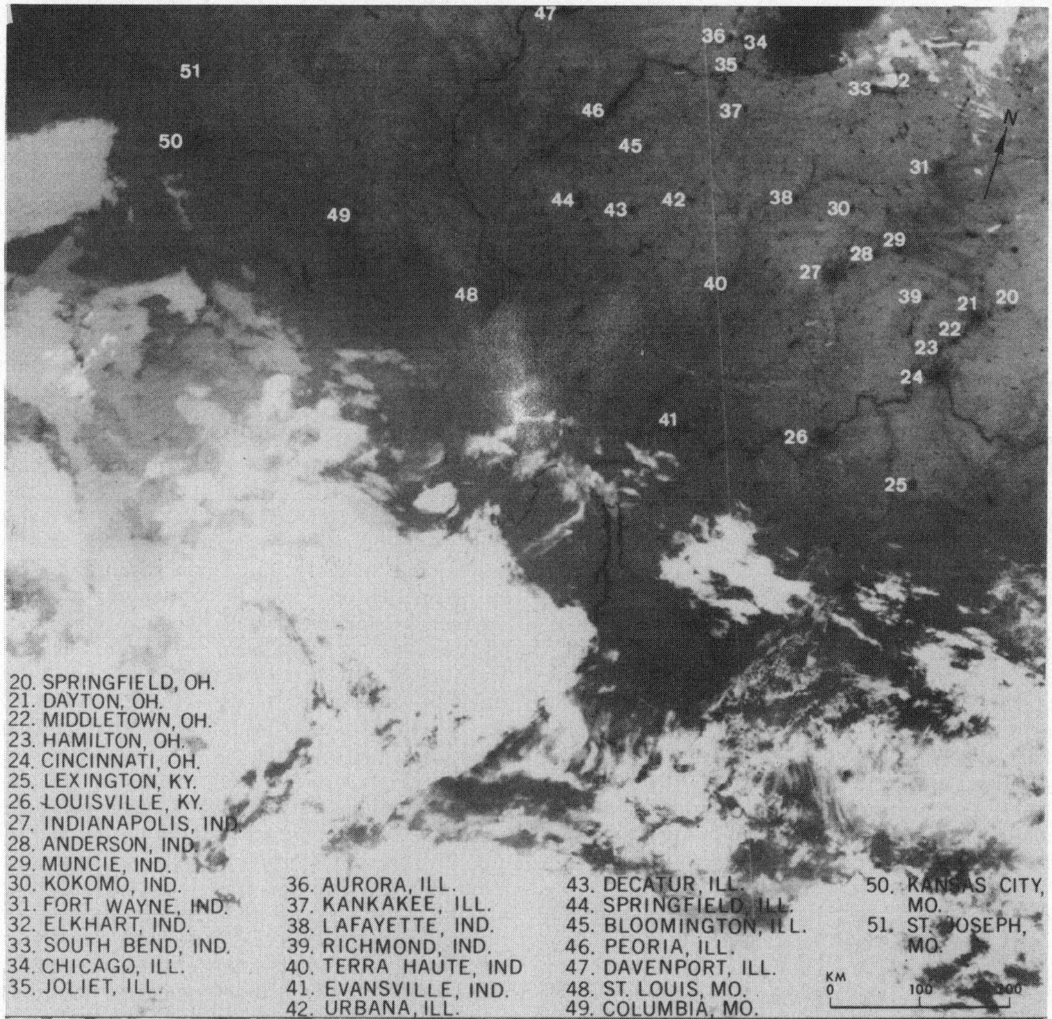

Fig. 27-41b. Urban heat islands identified from NOAA 5 VHRR thermal infrared data.

similar hypothesis proposed by Newell (1971). Baumgartner (1979) has examined the extreme case of a totally deforested earth. For such a case average annual global albedo would increase from 16.7 percent to 17.3 percent and the roughness parameter would decrease from a global average of 14.9 cm to 3.0 cm. The change in surface drag would shift the angle between the surface wind and the pressure field isobars, thereby influencing the general global atmospheric circulation. Schmitt et al. (1975) have studied the possibility of Northern Hemisphere forest removal as a causal mechanism for triggering an ice age. They found that the decrease of 1 percent in annual absorbed radiation associated with such deforestation was of the same order as Budyko's (1969) critical level of a 1.6 percent decrease necessary for triggering an ice advance to a mean latitude of 50°N. The study concluded, however, that such deforestation is insufficient in itself to trigger massive glacial advances.

Forests are also important in the carbon dioxide balance. Woodwell et al. (1978) have estimated that with a global deforestation rate of 1 to 2 percent per year the amount of carbon released annually to the atmosphere would be 5 to 10 $\times$ 10^{15} g, one to two times the amount currently released from fossil fuels. When considering only the tropical forests they found the amount of annual carbon release to be 3 to 6 $\times$ 10^{15} g. Pearman and Garratt (1972) have suggested that deforestation of tropical areas during this century could have increased the atmospheric CO_2 concentration by 0.3 ppm per year, about 40 percent of the observed increase. Farmer and Baxter (1974) concluded that between 1900 and 1920 there could have been a CO_2 increase of 14 ppm due to enlargement of world farmland at the expense of forests and grasslands.

In spite of the awareness of the climatic role that forests play, an integrated, operational system of assessing and classifying forests on a continental or hemispheric scale is still lacking. Regional documentation is available and is amply presented in Chapter 34. There is a climatic need, however, for such information on a larger geographic scale. A hemispheric baseline forest inventory should be established to reference future forest assessments and relate them to climatic changes. For the scale involved, satellite remote sensing holds the most promise for accomplishing the task and it could be carried out in a manner similar to that outlined for monitoring desertification.

URBANIZATION

Increased urbanization and industrialization in many cities since 1940 has increased the intensity and extent of the thermal anomalies that commonly are termed urban heat islands. The study of urban heat islands has become synonymous with urban climatology. Urban heat islands influence physiological comfort, cooling and heating requirements, air circulation and precipitation. With the trend towards urbanization increasing, the day is approaching when large megalopolises will exert regional influences on the surrounding environment.

Peterson (1969), Garstang et al. (1975) and Oke (1978) have provided a thorough review of the urban heat island phenomenon. Bornstein (1968) and Oke (1978) discuss the energy balances in rural and urban aras which lead to urban heat islands. The large heat capacity and high heat conductivity of urban building materials prevent rapid cooling of urban areas after sunset, contrasting with the situation in the rural environment. A variety of other factors, often equally important, enter into heat-island formation. Rapid runoff of precipitation and reduction of evapotranspiration in urban areas, plus the waste heat from residential and other buildings, including that from air conditioning in summer, are two such considerations. Landsberg (1970) also mentions the increased "roughness" of cities, with a resultant reduction of about 25 percent in wind speeds, and lower albedo. A tentative model of the effect of the trapping of insolation by tall buildings yielded an average daily decrease in albedo of 20 percent (Craig and Lowry, 1972). Terjung (1970) and Terjung and Louie (1973) have called attention to the increased absorption and reflection of solar radiation, and of emission and reradiation of terrestrial radiation by the vertical sides of tall buildings. Although Bornstein (1968) mentions the contribution of elevated layers of smoke, water vapor, carbon dioxide and sulfur dioxide to the development of nocturnal heat islands by absorption and reradiation of energy from urban surfaces, Atwater's (1972) model showed that this is a relatively minor factor when compared to the others. Rather, urban pollution reduces the incoming solar radiation by an average of about 15 percent according to Landsberg (1970). White et al. (1978) show a 4 percent net radiation loss in metropolitan St. Louis when compared to surrounding rural areas.

Only during the last decade have aircraft and satellite radiation data been used in studies of urban heat islands. Rao (1972) used 7.4 km resolution thermal measurements from the Scanning Radiometer (SR) on the Improved TIROS Operational Satellite (ITOS-1) to demonstrate that the New York City-Philadelphia-Baltimore-Washington, D.C., urban corridor can be roughly delineated with such data. Pease and Nichols (1976) used an airborne thermal radiation scanner to derive energy emission, surface albedo, energy absorption, and net radiation maps for Baltimore, Maryland. Carlson et al. (1977) performed a morning and evening surface temperature analysis for the Los Angeles area using 1 km resolution thermal infrared data from the NOAA 3 Very High Resolution Radiometer (VHRR). Their analysis showed that the highest morning temperatures were over the central business district and high-density residential areas. Scofield and Weiss (1977) used 30 minute sequences of 1 km resolution visible imagery and 8 km resolution digital thermal IR data from the Synchronous Meteorological Satellite (SMS) to illustrate that towering cumulus typically form over areas such as Washington, D.C., and Baltimore, Maryland, because of the low-level convergence created by the differential surface heating between urban and rural areas. Matson et al. (1978) showed the existence of over 50 urban heat islands in the midwestern and northeastern United States using the NOAA 5 VHRR thermal infrared data (see Figures 27-41a and 27-41b). Analysis of digital data from the satellite for selected cities yielded maximum urban-rural temperature differences ranging from 2.6 to 6.5°C. They also demonstrated that through computer enhancement and enlargement of the NOAA-5 satellite imagery, the urban heat islands of St. Louis, Washington, D.C., and Baltimore, can be depicted at a usable scale as large as 1:1,000,000 (see Figures 27-42 and 27-43). A numerical model by Dodd (1979) used the 500 m Heat Capacity Mapping Mission (HCMM) satellite surface temperature data to derive the moisture availability, thermal conductivity, and sensible and evaporative heat fluxes at the urban surfaces of Los Angeles and St. Louis. Price (1979) used the HCMM data to determine the excess radiated power from northeastern urban areas on June 6, 1978, and found that the New York metropolitan area was radiating 40,000 kW at the time of the satellite overpass.

Landsberg (1956), Scofield and Weiss (1977), Oke (1978) and also scientists of the St. Louis METROMEX project (Changnon, 1978; Huff and Vogel, 1978) have all documented the effects of major metropolitan areas on the local weather and climate, such as windfield distortion, recirculation of pollutants, snowmelt enhancement, and a catalyst for increased summer thunderstorms and summer precipitation. Murphy et al. (1976) have modeled the effect on atmospheric circulation of large so-called nuclear "energy parks" in ocean areas. Their results showed that for energy parks

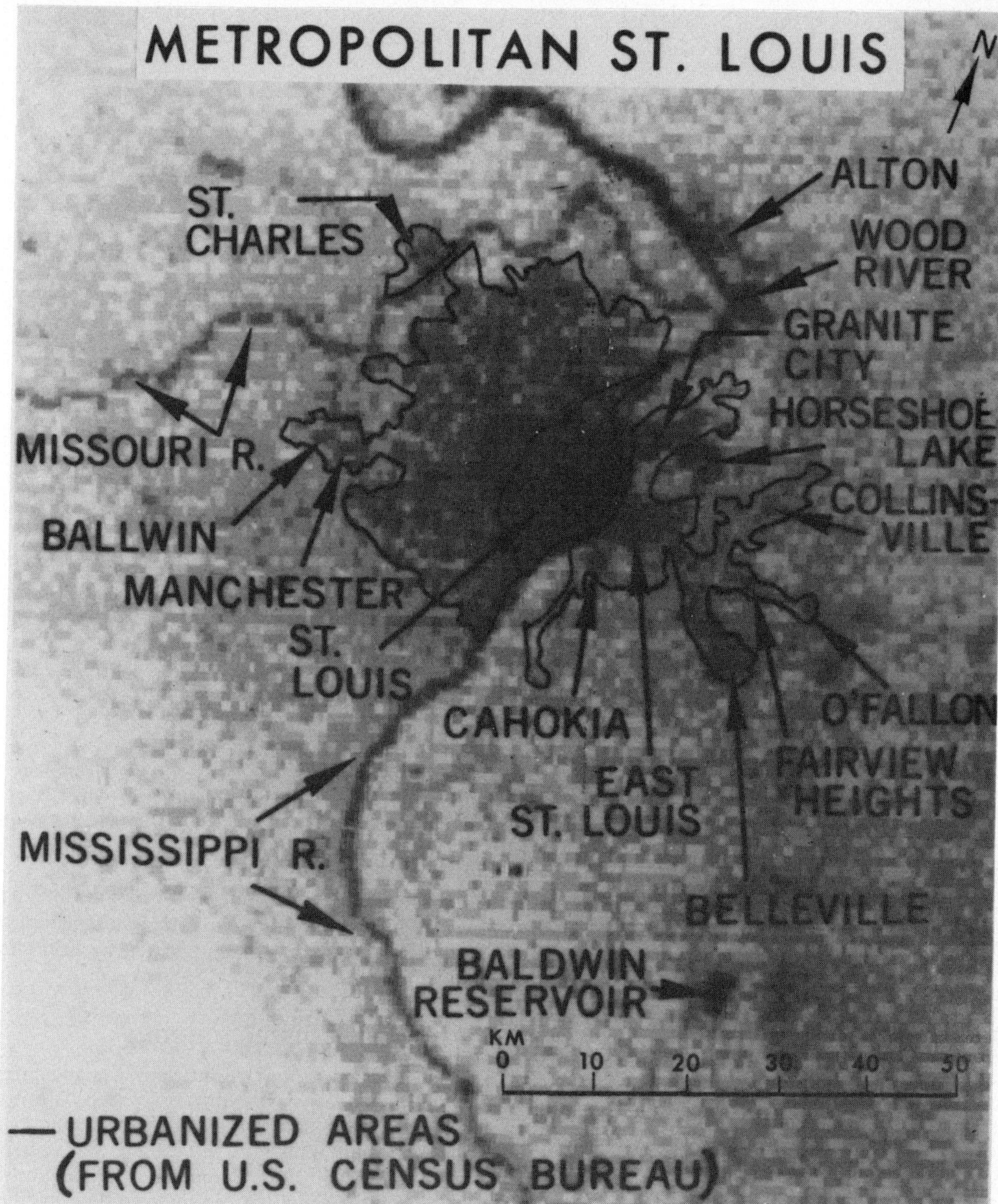

Fig. 27-42. Computer enhancement and enlargement of NOAA 5 VHRR image of St. Louis.

covering an area of 5° latitude and 10° longitude and with a total sensible heat output of 15×10^7 MW, pressure, temperature and rainfall were affected not only locally but over the hemisphere. Using the same model as Murphy et al. (1976) but varying the location, number, and sensible heat output of the energy parks, Williams et al. (1977a, b) have arrived at similar conclusions. They have recommended that case studies of urban heat islands be performed using general circulation models to study their effect on the global atmospheric system. If one examines an area of the Midwest similar in size to the energy parks of Murphy's study (e.g., Ohio, Indiana and Illinois), and assigns all 30 cities there (see Figure 27-41b) a heat output equal to St. Louis' estimated value of 1.6×10^4 MW (Maisel, 1975), a total heat output of about 5.0×10^5 MW is obtained. Since this heat output is smaller than that used in Murphy's models by a factor of 300, and the urban areas of Ohio, Indiana and Illinois are not concentrated in one area as are the energy parks of Murphy's study, it seems doubtful that the atmosphere could be significantly affected on a global scale. This is in agreement with similar conclusions by Washington and Chervin (1979) and Williams et

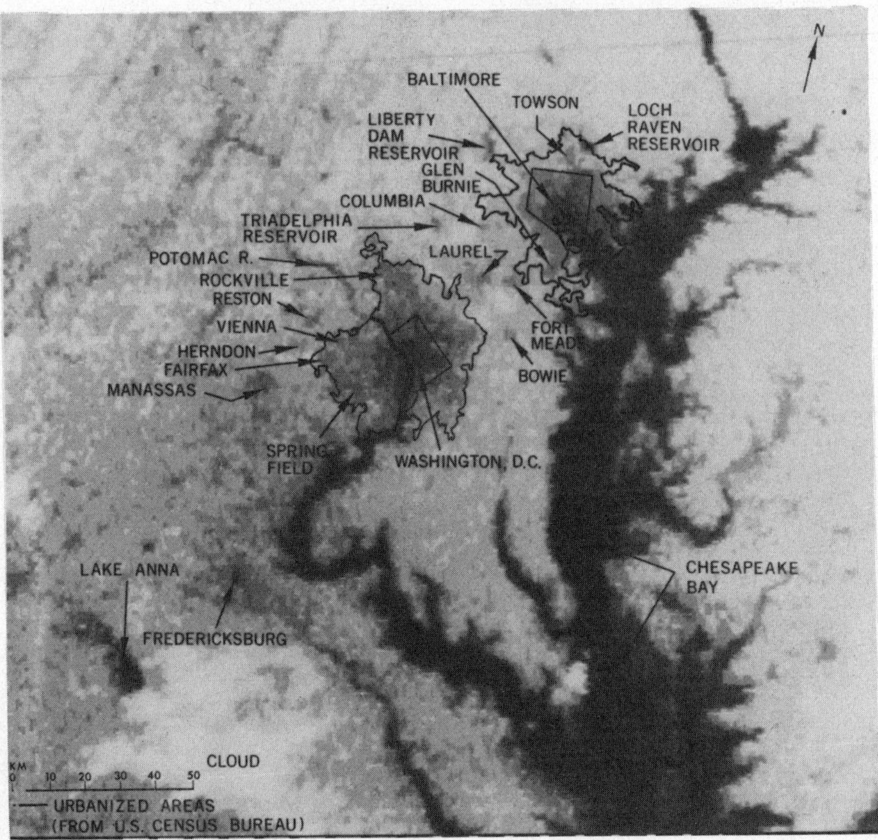

Fig. 27-43. Computer enhancement and enlargement of NOAA 5 VHRR image of Washington, D.C. and Baltimore, Maryland.

al. (1979). However, perhaps there should be an attempt to incorporate heat inputs of the magnitude and distribution associated with urban "corridors" into models of regional weather and climate. Figure 27-41a clearly shows the familiar urban "corridor" stretching from New York City to Washington, D.C. Smaller urban "corridors" also appear in the Midwest as shown in Figure 27-41b. Three of the most prominent ones are 1) Columbus-Springfield-Dayton-Middletown-Hamilton-Cincinnati; 2) Indianapolis-Anderson-Muncie-Fort Wayne; and 3) Urbana-Decatur-Springfield.

Aircraft and satellite thermal infrared sensors clearly provide a methodology whereby urbanization can be monitored and quantified. Variation of urban heat islands seasonally, and with varying meteorological conditions, could also be ascertained using the satellite data. Such data could be used as input for refining models dealing with urban micro-climatology and the regional climatology of large urban areas. Ground-based studies and satellite-based studies of urban heat islands will, however, continue to complement each other; the ground-based data providing atmospheric measurements, calibration for satellite sensors, and measurements for cloudy days; the satellite-based data, by providing instantaneous and comprehensive spatial information, alleviating the problem of the distribution and amount of equipment available for monitoring an urban island.

FUTURE TRENDS

As long as support continues for mesometeorology and for the study of severe storms, improvements in the monitoring of meteorological phenomena will be sought and can be anticipated. One new technique is called "nowcasting." "Nowcasting" is the real-time synthesis, analysis, and warning of significant (chiefly hazardous) local and regional weather based on a combination of observations from satellites, ground-based radar, event-gage sensors, and dense ground networks reporting through satellites. Flash-flood and tornado warnings should improve significantly in areas where nowcasting is introduced. Akin to this is the emerging technique of rapid repeat-cycle stereography from two geosynchronous satellites. Precipitation "estimates" from satellite visible and IR data will improve markedly and will be termed "measurements" in the eighties.

The GOES-VAS (Visible and IR Spin Scan Radiometer—VISSR/Atmospheric Sounder) launched in September 1980 on GOES 4 also offers an exciting new capability that promises to improve mesoscale prediction. It provides geosynchronous temperature soundings for the first time.

During 1979 an experiment originally called "the First GARP Global Experiment (FGGE)" and now called the "Global Weather Experiment (GWE)", utilized satellite data and surface observations to study synoptic-scale weather patterns in tropical areas where observations were virtually nonexistent. These GWE studies established that the variability in the tropics and southern hemisphere are far greater than earlier supposed. The information will be collated with other data and will aid in Global Climate Model (GCM) evaluation and improvement.

The future of intermediate and long-term global and regional climate prediction as a result of satellite remote sensing seems brighter. Recent work on hemispheric snow cover (Matson and Wiesnet, 1981) has provided an accurate new data base on northern hemisphere snow-cover from satellite records for the first time. Incorporation of these climatological data into GCM's and other models should reduce the assumptions implicit in dynamic climatological forecasting. They are also useful in evaluating model performance.

While progress in remote sensing from space has been "truly phenomenal" such things as subsurface ocean properties, snow depth and water equivalent, subsurface soil moisture, evapotranspiration, and river runoff continue to evade precise measurement.

The future of climatic forecasting is intimately bound up with satellite collection of global data. Snowcover, mentioned earlier, is now used in GCM's as a known input rather than as an assumed or climate-generated parameter. Sea ice is also known; sea surface temperatures, vegetation indices, and precipitation over the oceans can be determined from satellite remote sensors and used synergystically to study their individual and collective roles in climatic variation.

REFERENCES

American Geological Institute, 1962, Glossary of Geology and Related Sciences; 2nd ed., Amer. Geol. Inst., Washington, 325 pp., (72-p. supplement.).

Atwater, M. A., 1972, Thermal effects of urbanization and industrialization in the boundary layer; a numerical study; Bounder-Layer Meteorology, vol. 3, pp. 229–245.

Bandeen, W. R., M. Halev, and I. Strange, 1965, A radiation climatology in the visible and infrared from the TIROS meteorological satellites; NASA Tech Note D-2534, 30 pp.

Barnett, T. P., 1977, An attempt to verify some theories of El Nino; J. Phys. Oceanog., vol. 7, no. 5, pp. 633–647.

Barnett, T. P., 1978, The role of the oceans in the global climate system; in, Climate Change, John Gribbin, (ed.) Cambridge University Press, pp. 157–177.

Barnett, T. P., W. C. Patzert, S. C. Webb, and B. R. Bean, 1979, Climatological usefulness of satellite determined sea surface temperatures in the Tropical Pacific; Bull. Am. Meteorolog. Soc., vol. 60, no. 3, pp. 197–205.

Barrett, E. C., 1981, Satellite rainfall estimation by cloud indexing methods for desert locust survey and control; in Satellite Hydrology, M. Deutsch, D. R. Wiesnet, and A. Rango, eds., Amer. water Res. Assn., Minneapolis, MN, pp. 92–100.

Batlivala, P., and F. T. Ulaby, 1977, Estimation of soil moisture with radar remote sensing; Proceedings of the 11th International Symposium, Remote Sensing of the Environ. Environ. Res. Inst. of Mich., Ann Arbor, pp. 1557–1566.

Baumgartner, A., 1979, Climatic variability and forestry; World Climate Conference, World Meteorological Organization, Geneva, Switzerland, p. 273–279.

Bernstein, R. L., L. Breaker, and R. Whritner, 1977, California current eddy formation; ship, air, and satellite results; Science, vol. 195, pp. 353–359.

Bjerknes, J., 1966, The possible response of the atmospheric Hadley circulation to equatorial anomalies of ocean temperature; Tellus, vol. 4, pp. 820–829.

Boldyrev, V. G., and I. P. Vetlov, 1970, Space and time variability of outgoing radiation (from the METEOR. satellites data); Meteor. Gidrol., vol. 10, pp. 23–31.

Born, G. H., J. A. Dunne, and D. B. Lame, 1979, Seasat mission overview; Science, v. 204, no. 4400, pp. 1405–1406.

Bornstein, R. D., 1968, Observations of the urban heat island effect in New York City; Journal of Applied Meteorology, vol. 7, pp. 575–582.

Bryden, H. L., and M. M. Hall, 1980, Heat transport by currents across 25°N latitude in the Atlantic Ocean; Science, vol. 207, no. 22, pp. 884–885.

Budyko, M. I., 1969, The effect of solar radiation variations on the climate of the earth; Tellus, vol. 31, pp. 611–617.

Carlson, T. N., J. N. Augustine, and F. E. Boland, 1977, Potential application of satellite temperature measurements in the analysis of land use over urban areas; Bulletin of the American Meteorological Society, vol. 58, pp. 1301–1303.

Chahine, M. T., H. H. Aumann, and F. W. Taylor, 1977, Remote sounding of cloudy atmospheres: III Experimental verification; J. Atmos. Sci., vol. 35, no. 5, pp. 758–765.

Changnon, S. A., 1978, Urban effects on severe local storms at St. Louis; Journal of Applied Meteorology, vol. 17, pp. 578–586.

Charney, J., 1975, Dynamics of deserts and drought in the Sahel; Quarterly Journal of the Royal Meteorological Society, vol. 101, pp. 193–202.

Charney, J., P. H. Stone, and W. J. Quirk, 1975, Drought in the Sahara: a biogeophysical feedback mechanism; Science, vol. 187, pp. 434–435.

Craig, C. D., and W. P. Lowry, 1972, Reflection on the urban albedo; Conference on Urban Environment and Second Conference on Biometeorology, Philadelphia, American Meteorological Society, pp. 159–164.

Davis, R. E., 1976, Predictability of sea surface temperature and sea level pressure anomalies over the North Pacific Ocean; J. Phys. Oceanog., vol. 6, no. 3, pp. 249–266.

Davis, R. E., 1978. Predictability of sea level pressure anomalies over the North Pacific Ocean; J. Phys. Oceanog., vol. 8, no. 2, pp. 223–246.

Dey, B., H. Moore, and A. F. Gregory, 1977, The use of

satellite imagery for monitoring ice break-up along the MacKenzie River, N.W.T.; Arctic, vol. 30, no. 4, pp. 234–242.

Dodd, J. K., 1979, Determination of surface characteristics and energy budget over an urban-rural area using satellite data and a boundary layer model; M.S. Thesis, Department of Meteorology, Pennsylvania State University, 87 p.

Dvorak, Vernon F., 1975, Tropical analysis and forecasting from satellite imagery; Monthly Weather Review 103, pp. 420–430.

Dvorak, Vernon F., and Stanley Wright, 1977, Tropical cyclone intensity analysis using enhanced infrared satellite data; Reprinting 11th Technical Conference on Hurricanes and Tropical Meteorology, Miami American Meteorological Society, pp. 268–273.

Eagleman, J., and W. Lin, 1976, Remote sensing of soil moisture by a 21-cm passive radiometer, Jour. Geophys. Res., vol. 81, p. 3660.

Etkins, R., and E. S. Epstein, 1982, The rise of global mean sea level as an indication of climate change, Science, vol. 215, pp. 287–289.

Farmer, J. G., and M. S. Baxter, 1974, Atmospheric carbon dioxide levels as indicated by the stable isotope record in wood; Nature, vol. 247, pp. 273–275.

Fowler, M. G., H. Burke, K. R. Hardy, and N. K. Tripp, 1981, The estimation of rain rate over land from spaceborne passive microwave sensors, *in* Satellite Hydrology, M. Deutsch, D. R. Wiesnet, and A. Rango, eds., Amer. Water Res. Assn., Minneapolis, MN, pp. 101–108.

Garrett, W. D. 1967. Damping of capillary waves at the air-sea interface by organic surface-active material; J. Marine Res., vol. 25, no. 3, pp. 279–291.

Garstang, M., P. D. Tyson, and G. D. Emmitt, 1975, The structure of heat islands; Reviews of Geophysics and Space Physics, vol. 13, pp. 139–165.

Gates, W. L., 1979, The physical basis of climate; World Meteorological Organization, Geneva, Switzerland, Climate Conference, pp. 71–84.

Goody, R. M., 1964, Atmospheric Radiation I, Theoretical Basis, Oxford, Clarendon Press, 436 pp.

Gruber, A., 1977, Determination of the earth-atmosphere radiation budget from NOAA satellite data; NOAA Technical Report NESS 76, National Oceanic and Atmospheric Administration, U.S. Department of Commerce, Washington, D.C., 28 pp.

Gurka, James J., 1974, Using Satellite Data for Forecasting Fog and Stratus Dissipation; Preprints, 5th Conference on Weather Forecasting and Analysis, St. Louis, American Meteorological Society, pp. 54–57.

Gurka, James J., 1978, The role of inward mixing in the dissipation of fog and stratus; *Monthly Weather Review 106,* pp. 1633–1635.

Gustafson, A. V., and S. E. Wasserman, 1976, Use of satellite information in observing and forecasting dissipation of fog; Geophysics Publications vol. 12, no. 10, 22 p.

Halliwell, G. R., and C. N. K. Mooers, 1979, The space-time structure and variability of the shelf water-slope water and Gulf Stream surface temperature fronts and associated warm-core eddies; J. Geophys. Res., vol. 84, no. C12, pp. 7707–7725.

Hanel, G., 1976, The properties of atmospheric aerosol particles as functions of the relative humidity at thermodynamic equilibrium with the surrounding moist air, Advances in Geophysics, H. E. Landsberg, J. Van Miegham, eds., Academic Press, New York, vol. 19, pp. 73–188.

Hannaford, J. F., and A. J. Brown, 1981, Application of snow covered area to runoff forecasting in the Sierra Nevada, California; *in* Deutsch, M., Wiesnet, D. R., and Rango, A., eds., Satellite Hydrology, Amer. Water Res. Assn., Minneapolis, MN, pp. 165–172.

Hare, K. F., 1977, Climate and desertification; in, Desertification: Its Causes and Consequences, Pergamon Press, pp. 63–168.

Hickey, J. R., L. L. Stowe, H. Jacobowitz, P. Pellegrino, R. Maschoff, A. Arking, F. House, A. Ingersoll, and T. Vonder Haar, 1980, Solar constant determination from NIMBUS 7 ERB cavity radiometer observations; Submitted to Science.

Horan, J. J., 1978, NIMBUS: the vanguard of remote sensing; IEEE Spectrum, vol. 15, no. 9, pp. 36–43.

House, F. B., 1965, The radiation balance of the earth from a satellite; Ph.D. thesis, University of Wisconsin.

Huang, N. E., C. D. Leitao, and C. C. Parra, 1978. Large-scale Gulf Stream frontal study using GEOS-3 radar altimeter data; J. Geophys. Res., vol. 83, no. C9, pp. 4673–4682.

Huff, F. A., and J. L. Vogel, 1978, Urban, topographic and diurnal effects on rainfall in the St. Louis region; Journal of Applied Meteorology, vol. 17, pp. 565–577.

Huhnerfuss, J., W. Walter, and G. Kruspe, 1977, On the variability of surface tension with mean windspeed; J. Phys. Oceanog., vol. 7, no. 4, pp. 567–751.

Idso, S. B., 1975, The utility of surface temperature measurements for remote sensing of surface soil water status; Jour. Geophys. Research, vol. 80., no. 21, pp. 3044–3049.

Idso, S. B., R. D. Jackson, and R. J. Reginato, 1977, Remote sensing of crop yields; Science, vol. 196, pp. 19–25.

Ivlev, L. S., 1967, Aerosol model of the atmosphere; Prob. Fiz. Atmos., no. 7, Leningrad, pp. 125–160, translated by Foreign Science and Technology Center, Depart. of the Army, available from U.S. Nat. Tech. Information Service (AD 760-397).

Jackson, R. D., R. J. Reginato, and S. B. Idso, 1977. Wheat canopy temperature: a practical tool for evaluating water requirements. Water Resour. Res. vol. 13, pp 651–656.

Jackson, R. D., J. Cihlar, J. E. Estes, J. C. Hilmars, and C. Wiegand, 1978, Soil moisture estimation using reflected solar and emitted thermal radiation, NASA Conf. Publ. 2073, 219 p.

Johnson, D. S., and I. P. Vetlov, 1977, The Role of Satellites in WMO Programs in the 1980's; World Meterological Organization, World Weather Watch Planning Report No. 36, WMO No. 494, Secretariat of the World Meteorological Organization, Geneva, Switzerland, 69pp.

Jones, W. L., and L. C. Schroeder, 1978, Radar backscatter from the ocean: dependence on surface friction velocity; Boundary Layer Meteorol., vol. 13, pp 133–149.

Jones, W. L., P. G. Black, D. M. Boggs, E. M. Bracelente, R. A. Brown, G. Dome, J. A. Ernat, I. M. Halberstam, J. E. Overland, S. Peterkerych, F. J. Wenty, P. M. Woiceshyn, and M. C. Wurtele, 1979, SEASAT scatterometer: results of the Gulf of Alaska Workshop; Science, vol. 204, no. 29, pp 1413–1415.

Kneizys, F. X., E. P. Shettle, W. O. Gallery, J. H.

Chetwynd, L. W. Abreu, J. E. A. Selby, R. W. Fenn, and R. A. McClatchey, 1980, Atmospheric Transmittance/Radiance Computer Code Lowtran S. To be published.

Landsberg, H. E., 1970, Man made climatic changes, Science vol. 170, pp 1725–1734.

Landsberg, H. E., 1956, The climate in towns; *in* Man's Role in Changing the Face of the Earth, W. L. Thomas Jr., ed. Univ. of Chicago Press, pp 584–606.

Legeckis, R., 1977, Long waves in the eastern equatorial Pacific Ocean: a view from a geostationary satellite; Science, vol. 197, pp 1197–1181.

Legeckis, R., 1978, A survey of worldwide sea surface temperature fronts detected by environmental satellites; J. Geophys. Res., vol. 83, no. 9, pp. 4501–4522.

Maisel, M., 1975, Localized releases of heat and moister; Atmospheric Implications of Energy Alternatives, NCAR Forum, October 1974, National Center for Atmospheric Research, pp 11–13.

Mashukov, P. M., 1957, Ledovyi rezhim reki Syr-Dar ia po dannym aviarazyedok (The Ice Regime of the Syr Darya River according to Aerial Reconnaissance Data), Geofiz. Observatorii, vol. 15, no. 16, pp. 131–153 (text in Russian).

Matson, M., E. P. McClain, D. F. McGinnis, and J. A. Pritchard, 1978, Satellite detection of urban heat islands; Monthly Weather Review, vol. 106, pp. 1725–1734.

Matson, M., and Wiesnet, D. R., 1981, New data base for climate studies; Nature, vol. 289, no. 5797, pp. 451–456.

McClatchey, R. A., R. W. Fenn, J. E. A. Selby, F. E. Volz, and J. S. Garing, 1972, Optical Properties of the Atmosphere (Third Edition), ERP 411, AFCRL-72-0497. (Available from NTIS)

McClatchey, R. A., W. S. Benedict, S. A. Clough, D. E. Burch, R. F. Calfee, K. Fox, L. S. Rothman, and J. S. Garing, 1973, AFCRL atmospheric absorption line parameters compilation, ERP 434, AFCRL-TR-73-0096. (Available from NTIS, AD 762-904)

McDonald, T. H., 1970, Data reduction processes for spinning flat-plate satellite-borne radiometers; ESSA TECH. REPT., NESC 52.

McFarland, M. H. 1976, The correlation of Skylab L-band brightness temperatures with antecedent precipitation; Proc. Conf. on Hydrometeor. Amer. Meteor. Soc., Boston, pp. 60–65.

McGinnis, D. F., Jr., and S. R. Schneider, 1978, Monitoring river ice breakup from space; Photogramm. Eng. and Remote Sensing, vol. 44, no. 1, pp. 57–68.

Meier, M. F., 1973, Evaluation of ERTS imagery for mapping and detection of changes of snowcover on land and on glaciers; Symposium on Significant Results Obtained from the ERTS-1 Satellite, Goddard Space Flight Center, vol. 1, Sect. A., pp. 863–875.

Millard, J. P., R. D. Jackson, R. C. Goettleman, R. J. Reginato, S. B. Idso, and R. L. LaPado, 1977, Airborne monitoring of crop canopy temperatures for irrigation scheduling and yield predictions; Proc. 11th Internat. Sympos. on Remote Sensing, Environmental Res. Inst. Michigan, Ann Arbor, pp. 1453–1461.

Molchanov, A. K., and V. P. Vlasov, 1974, Studying ice jams from aerial photographs; Trans. of the State Hydrol. Inst., no. 219, Hydrometeorological Press, Leningrad, pp. 81–83, translation by CRREL, Rept. No. CRREL-TL-473, pp. 57–68.

Molinari, B. S., D. W. Behringer, G. A. Maul, and R. Legeckis, 1977, Winter intrusions of the loop current. Science, vol. 198, pp. 505–507.

Murcray, D., 1970, Unpublished data.

Murphy, A. H., A. Gilchrist, W. Hafele, G. Kromer and J. Williams, 1976, The impact of waste heat release on simulated global climate; Research Memorandum RM-76-79, International Institute of Applied Systems Analysis, Austria, 22 p.

Namais, J., 1978. Multiple causes of the North American abnormal winter 1976–1977. Monthly Weather Review, *106:* 279–295.

Newell, R. E., 1971, The Amazon forest and atmospheric general circulation; in Man's Impact on Climate, W. H. Matthews, W. W. Kellogg and G. D. Robinson, eds., MIT Press, Cambridge, Mass., pp. 457–460.

Oke, T. R., 1978, Boundary Layer Climates, Methuen and Co. Ltd., London, England, 372 p.

Ostrem, G., R. Anderson, and H. Odegaard, 1981, Operational use of satellite data for snow inventory and runoff forecast, in Deutsch, M., Wiesnet, D. R., and Rango, A., eds., Satellite Hydrology, Amer. Water Res. Assn., Minneapolis, MN, pp. 230–234.

Parmenter, Frances C., 1974, Observing and forecasting local effects from satellite data; Preprints, 5th Conference on Weather Forecasting and Analysis, St. Louis, American Meteorological Society, pp. 46–59.

Pearman, G. I., and J. R. Garratt, 1975, Global aspects of carbon dioxide; Search, vol. 3, pp. 67–73.

Pease, R. W., and D. A. Nichols, 1976, Energy balance maps from remotely sensed imagery; Photogrammetric Engineering and Remote Sensing, vol. 42, pp. 1367–1374.

Penndorf, R., 1978, Analysis of ozone and water vapor field measurement data, Report No. FAA-EE-78-29.

Peterson, J. T., 1969, The climate of cities: A survey of recent literature; U.S. National Air Pollution Administration, Publication no. AP-59, 48 p.

Pierson, W. J., 1981, Winds over the ocean as measured by the scatterometer on Seasat; in Oceanography From Space, J. F. R. Gower, ed., Plenum Publ. Corp.

Potter, G. L., H. W. Ellsaesser, M. C. MacCracken, and F. M. Luther, 1975, Possible climatic impact of tropical deforestation; Nature, vol. 258, pp. 697–698.

Price, J. E., 1979, Assessment of the urban heat island effect through the use of satellite data; Monthly Weather Review, vol. 107, pp. 1554–1557.

Purdom, James F. W., and James J. Gurka, 1974, The effect of early morning cloud cover on afternoon thunderstorm development; Preprints, 5th Conference on Weather Forecasting and Analysis. St. Louis, American Meteorological Society, pp. 58–60.

Quinn, F. H., R. A. Assel, D. E. Boyce, G. A. Leshkevich, C. R. Snider, and D. R. Wiesnet, 1978, Summary of Great Lakes weather and ice conditions, Winter 1976–77; NOAA Tech. Memo. ERL GLERL-20, 141 p.

Rao, P. K., 1964, Some seasonal variations of outgoing longwave radiation as observed by TIROS II and III satellites; Weather, vol. 29, no. 3, pp. 88–89.

Rao, P. K., 1972, Remote sensing of urban "heat islands'' from an environmental satellite; Bulletin of the American Meteorological Society, vol. 53, pp. 647–648.

Rao, M. S. V., W. B. Abbott, and J. S. Theon, 1976, Satellite-derived global oceanic rainfall atlas (1973 and 1974); NASA Special Publication SP-410, Washington: U.S. Government Printing Office.

Raschke, E., and M. Pasternak, 1968, The global radiation balance of the earth-atmosphere system obtained from radiation data of the meteorological satellite Nimbus 2; Space Research VIII, Amsterdam, N. Holland Publ. Co., pp. 1033–1043.

Raschke, E., and W. R. Bandeen, 1970, The radiation balance of the planet Earth from radiation measurements of the satellite Nimbus 2; J. Appl. Meteor., vol. 9, pp. 215–238.

Rasmussen, W. O., and P. F. Ffolliott, 1981, Prediction of water yield using satellite imagery and a snowmelt simulation model, *in,* Deutsch, M., Wiesnet, D. R., and Rango, A., eds., Satellite Hydrology, Amer. Water Res. Assn., Minneapolis, MN, pp. 193–196.

Reginato, R. J., S. B. Idso, and R. D. Jackson, 1978, Estimating forage crop production: A technique adaptable to remote sensing; Remote Sensing of the Environ., vol. 7, pp. 77–80.

Robinson, A. R., and B. A. Taft, 1972, A numerical experiment for the path of the Kuroshio; J. Marine Res., vol. 30, no. 1, pp. 65–101.

Robock, A., 1981, The use of snow and ice data in energy balance climate modeling; in Snow Watch, 1980, G. Kukla, A. Hecht, and D. Wiesnet, eds., Glaciology Data, Rept. GD-11, World Center A for Glaciology Snow & Ice, Boulder, Colo.

Rothman, L. S., S. A. Clough, R. A. McClatchey, L. G. Young, D. E. Snider, and A. Goldman, 1978, AFGL Trace Gas Compilation, Applied Optics, vol. 17, no. 4.

Saunders, P. M., 1967. The temperature of the ocean-air interface; J. Atmos. Sci., vol. 24, pp. 269–273.

Saunders, R. W. and G. E. Hunt, 1980, METEOSAT observations of diurnal variation of radiation budget parameters; Nature, vol. 283, pp. 645–647.

Schmitt, W. R., D. K. Stidd, and J. D. Isaacs, 1975, Ice ages and northern forests; Climate of the Arctic Proceedings, pp. 117–119.

Schmugge, T. J., B. Blanchard, A. Anderson, and J. Wang, 1978, Soil moisture sensing with aircraft observations of the diurnal range of surface temperature; Water Res. Bull., vol. 14, pp. 169–178.

Schmugge, T. J., T. J. Jackson, and H. L. McKim, 1981, Survey of in-situ and remote sensing methods for soil moisture determination, in Satellite Hydrology, M. Deutsch, D. R. Wiesnet, and A. Rango, eds., Amer. Water Res. Assn. Minneapolis, MN, pp. 333–352.

Schneider, S. R., 1977, Operational satellite assessment of snow cover and river ice in the Saint John Basin; World Meteor. Organiz., World Weather Watch, St. John Basin Pilot Proj. Task Force Rept. no. 6-2, 26 p.

Scofield, R. A., and C. E. Weiss, 1977, A Report on the Chesapeake Bay region NOWCASTING experiment; NOAA Technical Memorandum NESS 94, Washington, D.C., 55 p. (NTIS No. PB-277-102).

Scofield, R. A., and V. J. Oliver, 1981, A satellite derived technique for estimating rainfall from thunderstorms and hurricanes, in Satellite Hydrology, M. Deutsch, D. R., Wiesnet, and A. Rango, eds., Amer. Water Res. Assn., Minneapolis, MN, pp. 70–76.

Scott, J. C., 1972, The influence of surface-active contamination on the initiation of wind waves. J. Fluid Mech., vol. 56, no. 3, pp. 591–606.

Scott, J. C., 1975, The preparation of water for surface-clean fluid mechanics. J. Fluid Mech., vol. 69, no. 2, pp. 339–351.

Selby, J. E. A., F. X. Kneizys, J. H. Chetwynd, and R. A. McClatchey, 1978, Atmospheric Transmittance Radiance: Computer Code LOWTRAN 4; AFGL-TR-78-0053, ERP No. 626. (Available from NTIS, AD-A058-643).

Shafer, B. A., C. F. Leaf, and J. K. Marron, 1981, Landsat derived snow-cover as an input variable for snowmelt runoff forecasting in South Central Colorado, in Deutsch, M., Wiesnet, D. R., and Rango, A., eds., Satellite Hydrology, Amer. Water Res. Assn., Minneapolis, MN, pp. 218–224.

Shettle, E. P., and J. A. Weinman, 1970, The transfer of solar irradiance through inhomogeneous turbid atmospheres evaluated by Eddington's approximation, J. Atmos. Sci., vol. 27, pp. 1048–1055.

Shettle, E. P. and R. W. Fenn, 1975, Models of the atmospheric aerosols and their optical properties; AGARD Conference Proceedings No. 183, Optical Propagation in the Atmosphere, pp. 2.1–2.16, presented at the Electromagnetic Wave Propagation Panel Symposium, Lingby, Denmark, 27–31 Oct. 1975. (Available from NTIS, Acc. No. N76-29817)

Shettle, E. P. and R. W. Fenn, 1979, Models for the aerosols of the lower atmosphere and the effects of humidity variations on their optical properties; AFGL-TR-79-0214. (Available NTIS)

Smigielski, F., 1981, Northern hemisphere snow and ice charts of NOAA/NESS, in Kukla, G., Hecht, A., and Wiesnet, D. R., 1981, Snow Watch 1980, Glaciol. Data Report GD-11, World Data Center A for Glaciology Snow and Ice, 148 p.

Spatola, A. A., 1972, Climatology of Appalachian Valley fog at White Sulphur Springs, W. Va., and Nearby Stations During Months of Peak Fog Frequency; AFCRL-72-0054, Environmental Research Paper, No. 382.

Terjung, W. G., 1970, Urban energy balance climatology; a preliminary investigation of the city-man system in downtown Los Angeles; Geophysical Review, vol. 60, pp. 31–50.

Terjung, W. G., and S. Louie, 1973, Solar radiation and urban heat islands; Annals of the Association of American Geographers, vol. 63, pp. 181–207.

Ulaby, F. T., 1974, Radar measurement of soil moisture content; IEEE Trans. Antennas Propagat., vol. AP-22, pp. 257–265.

Ulaby, F. T., P. P. Batlivala, and M. C. Dobson, 1978, Microwave backscatter dependence on surface roughness, soil texture, I, bare soil; IEEE Trans. Geosci. Elect., vol. GS-16, pp. 286–295.

Ulbrich, C. W., and D. Atlas, 1978, The rain parameter diagrams: Methods and applications; J. Geophys. Res., vol. 83, no. C3, pp. 1319–1325.

United Nations, 1977, Desertification—an overview; Desertification: Its Causes and Consequences, Pergamon Press, pp. 1–62.

U.S. Government 1976, U.S. Standard Atmosphere; NOAA-S/T 76-1562, U.S. Gov't Printing Office, Wash., D.C., 20402.

Valley, S. L., ed., 1965, Handbook of Geophysics and Space Environments, AFCRL.

Van de Hulst, H. C., 1957, Light Scattering by Small Particles; John Wily & Sons, Inc., New York, 470 p.

Vonder Haar, T. H., 1968, Variations of the Earth's Radiation Budget: Ph.D. thesis, University of Wisconsin, 118 pp.

Vonder Haar, T. H., and V. E. Suomi, 1971, Measurement of the Earth's radiation budget from satellites during a five-year period. Part I. Extended time and space means; J. Atmos. Sci., 28, pp. 305–314.

Warren, A., and J. K. Maizels, 1977, Ecological change and desertification; Desertification: Its Causes and Consequences, Pergamon Press, pp. 169–260.

Wartha-Clark, J., 1980, Satellite observation of Great Lakes ice-winter 1978–79; NOAA Tech. Memo. NESS 112, 36 p.

Washington, W. M., and R. M. Chervin, 1979, Regional climatic effects of large-scale thermal pollution; Simulation studies with the NCAR general circulation model; Journal of Applied Meteorology, pp. 3–16.

Weinstein, M., and V. E. Suomi, 1961, Analysis of satellite infrared radiation measurements on a synoptic scale; Mon. Wea. Rev. vol. 89, pp. 419–428.

Winston, J. L., and V. R. Taylor, 1967, Atlas of World Maps of Long-Wave Radiation and Albedo; ESSA Tech. Dept. NESC 43, Washington, D.C., 32 pp.

Winston, J. L., A. Gruber, T. I. Gray, Jr., M. S. Varnadore, C. L. Earnest, and L. P. Manello, 1979, Earth-Atmosphere Radiation Budget Analyses Derived from NOAA Satellite Data June 1974. February 1978, Vol. I and II; NOAA/NESS, U.S. Dept. of Commerce, Washington, D.C.

White, J. M., F. D. Eaton, and A. H. Auer, 1978, The net radiation budget of the St. Louis metropolitan area; Journal of Applied Meteorology, vol. 17, pp. 593–599.

White, W. B., 1975, Secular variability in the large-scale baroclinic transport of the North Pacific from 1950–1970; J. Marine Res., vol. 33, no. 1, pp. 141–155.

Whitney, L. F., Jr., and L. D. Herman, 1981, A statistical approach to rainfall estimation using satellite data; in Satellite Hydrology, M. Deutsch, D. R. Wiesnet, and A. Rango, eds., Amer. Water Res. Assn., Minneapolis, MN, pp. 139–143.

Wiesnet, D. R., 1979, Satellite studies of fresh-water ice movement on Lake Erie; Jour. of Glaciology, vol. 24, no. 90, pp. 415–426.

Wiesnet, D. R., 1980, A satellite mosaic of the Greenland ice sheet, Proc. Riederalp Workshop, Sept. 1978, IASH-AISH Publ. no. 126, pp. 343–348.

Wiesnet, D. R., and C. P. Berg, 1981, The satellite record of the winter of 1978–79 in North America; in Deutsch, M., Wiesnet, D. R., and Rango, eds., Satellite Hydrology, Amer. Water Resources Assn., Minneapolis, MN, pp. 183–187.

Wiesnet, D. R., and M. Matson, 1976, A possible forecasting technique for winter snow cover in the Northern Hemisphere and Eurasia; Monthly Weather Rev. vol. 104, pp. 828–833.

Wilheit, T. T., A. T. Change, M. S. V. Rao, E. B. Rogers, and J. S. Theon, 1977, A satellite technique for quantitatively mapping rainfall rates over the ocean; J. App. Meteorol., vol. 16, no. 5, pp. 551–560.

Wilheit, T. T., Radiative transfer in a plane stratified dielectric, 1978, IEEE Trans., Geosci. Elect., vol. GE-16, pp. 138–143.

Williams, R. S., Jr., 1976, Vatnajokull Icecap, Iceland; in ERTS-1, A New Window on our Planet, R. S. Williams, Jr., and W. D. Carter, eds., U.S. Geological Survey Prof. Paper 929, pp. 188–193.

Williams, J., G. Kromer, and A. Gilchrist, 1977a, Further studies of the impact of waste heat release on simulated global climate: Part 1; Research Memorandum RM-77-15, International Institute for Applied Systems Analysis, Austria, 22 p.

Williams, J., G. Kromer, and A. Gilchrist, 1977b, Further studies of the impact of waste heat release on simulated global climate: Part 2; Research Memorandum RM-77-34, International Institute for Applied Systems Analysis, Austria, 32 p.

Williams, J., W. Hafele, and W. Sassin, 1979, Energy and climate: A review with emphasis on global interactions; World Climate Conference, World Meteorological Organization, Geneva, Switzerland, pp. 153–164.

Williams, R. S., Jr., and J. G. Ferrigno, 1981, Satellite image atlas of the earth's glaciers, in Deutsch, M., Wiesnet, D. R., and Rango, A., eds., Satellite Hydrology, Amer. Water Res. Assn., Minneapolis, MN, pp. 178–182.

Woodley, W. L., C. G. Griffith, and J. A. Augustine, 1981, Rain estimation over several areas of the globe using satellite imagery; in Satellite Hydrology, M. Deutsch, D. R. Wiesnet, and A. Rango eds., Amer. Water Res. Assn., Minneapolis, MN, pp. 84–91.

Woodwell, G. M., R. H. Whittaker, W. A. Reiners, G. E. Likens, C. C. Delwiche, and D. B. Botkin, 1978, The biota and the world carbon budget; Science, vol. 199, pp. 141–146.

Wunsch, C., and E. M. Gaposhkin, 1980, On using satellite altimetry to determine the general circulation of the oceans with application to geoid improvement; Reviews of Geophysics and Space Physics, vol. 18, pp. 725–745.

Zwally, H. J., and P. Gloersen, 1977, Passive microwave images of the polar regions and research applications; Polar Record, vol. 18, pp. 431–450.

Zwally, H. J., J. Comiso, C. Parkinson, W. Campbell, F. Carsey, and P. Gloersen, 1981, Antarctic sea ice cover from satellite passive microwave, in Snow Watch, 1980, G. Kukla, A. Hecht, and D. Wiesnet, eds., Glaciological Data, Rept. GD-11, World Data Center A for Glaciology Snow & Ice, Boulder, Colo., pp. 79–85.

CHAPTER 28

The Marine Environment

Author/Editors: ROBERT W. JOHNSON AND JOHN C. MUNDAY JR.
Contributing Authors: VIRGINIA CARTER, ANDREW J. KEMMERER, BRUCE M. KENDALL, RICHARD LEGECKIS, FABIAN C. POLCYN, JOHN R. PRONI, AND DONALD J. WALTER

GENERAL CONTENTS: Physical, chemical, biological and geological oceanography; pollution and water quality. Measurement systems: sonar; UV, visible and IR scanners and photography; lidar; microwave radiometers; radar. Parameters: surface temperature, salinity, currents, fronts and water masses, sea level, waves, sea ice, ocean color, chlorophyll, submerged vegetation, wetlands, bathymetry, coastal processes, suspended sediment, outfalls and dumping, oil pollution, water quality models. Locations: oceans, estuaries, coastal zones. References.

I. INTRODUCTION

Life and physical processes on the planet Earth are greatly affected by ocean and estuarine areas. Since water comprises about 70 percent of the Earth's surface, the oceans are important in the exchange of mass and energy within the terrestrial system, and between the Earth and our solar system, particularly the Sun. Operational and research satellite systems are now providing global synoptic measurements of climatic and environmental parameters that relate to the major processes of the oceans and atmosphere. Satellite systems can also provide measurements on local scales that are of interest to oceanography, and which expand the capacities provided by conventional shipboard techniques. For needs at higher spectral and spatial resolution, aircraft platforms may be used. These obtain remotely sensed measurements more amenable to coordination with surface measurements under the existing weather or other environmental conditions. Finally, conventional platforms, such as ships and buoys, may be used in combination with satellite data-relay systems to transmit data to other locations for analysis and interpretation.

Electromagnetic and acoustic measurements are usually made in discrete spectral ranges. Sensors may operate in the passive mode only, depending on emitted or reflected radiation from the sea as their signal source. Active sensors, in contrast, transmit and receive a reflected signal. The reader should consult Volume 1 (chapters 1–11) for details of these measurement systems. Table 28-1 lists classes of remote sensors and some of the oceanographic parameters that they measure. Acoustic sensing of the water column from the surface, a type of active remote sensing, is included because of its special use for the underwater medium.

Table 28-2 presents typical remote sensors, platforms, and spatial resolutions. Spatial resolution is, in general, a function of instrument spectral characteristics and distance (or altitude). Platforms range from satellites in geosynchronous orbits to surface ships that tow acoustic sounding devices. Specialized sensing systems such as submerged and/or buoy-mounted packages are discussed briefly.

Remote sensing by cameras and photo-visual techniques has historically been significant. Manual and photointerpretive techniques are discussed where they are important factors in the technology base; however, emphasis is placed on newer developments that use electronic sensors with analog or digital outputs, especially those that are amenable to radiofrequency data transmission and/or magnetic tape storage, and subsequent computer analysis.

Instrumentation and techniques for multifrequency and side-scanning sonar have been developed to provide information on material (i.e., suspended solids) and physical (i.e., thermal) gradients in the water column, as well as a new perspective on bottom-contour mapping. Integration of such water-column data with aerial and orbital remote sensing would appear to be in the developmental stage. Data-collection platforms and other means of bridging the air/water interface will be important factors in the development of combined radiance and sonar "remote sensing" systems.

Intervening atmospheric and water column effects have been limiting factors in the application of remote-sensing techniques to a number of topics of high interest. Investigation of these effects has been facilitated by expansion of near-surface spectral measurements from ships and low-altitude aircraft. The primary objective of these measurements is to isolate atmospheric and water-column factors, as well as to provide data in the same units (i.e., radiance) as high-altitude remote sensing data. These measurements have been a strong factor in the development of atmospheric and water-column radiative transfer models that have aided in the analysis and interpretation of remotely sensed data.

It should also be noted that an important role in

TABLE 28-1

Classes of Ocean Remote Sensors, Measurement Mechanism, and Parameters Measured

Remote sensor	Measurement mechanism	Typical parameters
Acoustic		
Active nadir and scanning	Reflectance and scattering	Particles in water column; bathymetry; thermal or salinity discontinuities; fish
UV, visible, near IR		
Passive	Reflectance, scattering, and fluorescence	Particles; chlorophyll *a*, fish, vegetation; water color, water masses, oil slicks; bathymetry, topography
Active (laser)	Fluorescence, reflectance, and scattering	Chlorophyll *a*; phytoplankton diversity; bathymetry; sea state; particles
IR		
Passive	Emission	Temperature
Microwave		
Passive (radiometer)	Emission	Temperature, salinity, wind speed
Active (radars and scatterometer)	Reflectance and scattering	Altitude; topography; wind velocity and direction; sea state; sea ice; oil slicks

oceanic remote sensing is played by transfer characteristics of the air-water interface at the marine surface. In many cases, high signal losses across this interface have historically limited application of particular types of sensors; however, improved *in situ* devices, combined with telemetry techniques, have expanded marine sensing capabilities (e.g., the use of sonar sounders and subsequent surface transmission to aircraft or satellites).

Developments in data reduction and analysis algorithms have recently displayed considerable maturity. These developments are important factors in the growth of systematic and widespread applications of remotely-sensed data to study the marine environment. In general, mature techniques measuring variables pixel by pixel are highlighted in the appropriate sections which follow. Statistical techniques, by contrast, are less developed and implemented. Nevertheless, their importance has been demonstrated particularly in microwave and radar systems applications.

Finally, as appropriate, we have used portions of the text and figures from the First Edition of the Manual. It is recommended that the First

TABLE 28-2

Typical Spectral Ranges, Platforms and Spatial Resolution*

	Platforms		
Spectral range	Ship	Aircraft (at 3 km)	Satellite
Acoustic (20–200 kHz)	1 m	—	—
UV, visible, near IR (300–1000 nm)			
Passive (spectrometer)	—	8 m	40 m
Active (laser)	—		
Altimeter			
Fluorometer		50 m	—
Infrared (1–20 μm)			
Passive (radiometer)	—	8 m	1 km
Microwave (1 mm–50 cm)			
Passive (radiometer)	—	10 m–1 km	10 km–150 km
Active (radar)	—		
PPI radar	—	—	—
Real aperture SLAR	—	~10 m	—
Synthetic-aperture SAR	—	~10 m	~25 m
Scatterometer	—		25 km
Altimeter	—	10 cm vertical	10 cm vertical

* Photography excepted.

Edition continue to be considered as a source of information, particularly for earlier studies which may be of interest to many users.

II. OCEANOGRAPHY

A. *PHYSICAL OCEANOGRAPHY*

The investigation of general ocean circulation has been one of the central themes of physical oceanography for over a century. Gravitational, Coriolis, viscous and inertial forces, when acting in a fluid medium that is subject to density variations, atmospheric pressure, and surface winds, produce a rich dynamic system that is continually yielding its secrets to intense study. With the development of remote sensing, synoptic repetitive measurement of key oceanographic variables has become available for large regions of the ocean surface, thus permitting new frontiers in the study of circulation to be advanced both locally and globally. In the sections following, aerial and orbital remote sensing are shown to have already contributed a significant new understanding of circulation. The potential for further gains is rich indeed.

1.0. Sea Surface Temperature

a. General Capabilities

Sea-surface temperature may be determined from emitted radiances in the thermal infrared or microwave spectral regions. The type of sensor is, of course, fixed by the spectral region, but the specific wavelength which is selected depends in many cases on multiple uses envisioned for the sensor. All remote measurements of sea-surface temperature are obtained from the top several cm due to strong water absorption in the infrared and microwave regions.

Infrared measurements may be made using radiometers and single-band or multispectral scanning sensors. These instruments were flown for many years on aircraft and were readily adaptable to meteorological and environmental satellites (e.g., NOAA, TIROS, and Landsat). Optimally, infrared sensors require clear and cloud-free atmospheric conditions but can operate effectively in either daylight or darkness. Without atmospheric correction, temperature differences of 1°C may be detected from orbit, and absolute values determined within 2° to 3°C. With atmospheric correction, absolute temperature values may be determined to within 1°C. Spatial resolutions of 8 m may be obtained from scanners aboard an airborne platform (at an altitude of 3 km), or of 1 km from a satellite platform.

Thermal measurements in the microwave spectral region are less affected by environmental conditions and may be used in cloudy weather as well as in light rain, fog, and other adverse conditions. Preflight and/or simultaneously obtained calibration data from nearby spectral bands may be used to obtain absolute temperatures within 1.0°C. Spatial resolutions range from 500 m for a sensor mounted on an aircraft to 150 km from satellite altitudes. Spatial resolutions of this magnitude (for microwave sensors) are usually acceptable for wide-area oceanographic studies, while infrared sensors with their better resolution are used for frontal features such as are found in the Gulf Stream or in coastal zones.

b. Aerial Methods

Traditionally, sea surface temperature is measured from bucket samples, engine intakes, or a ship-towed sensor. In addition, many investigators have used precision radiation-thermometer measurements obtained at the surface for comparison with remotely sensed data. Temporal data may also be collected by fixed buoys or platforms in which the data are stored internally until periodic readout or transmission to another location occurs, usually via a radio telemetry link.

Temperatures may be determined remotely from airborne platforms (Swift, 1980). Measurements are usually made along the line directly under the aircraft. Spatial distributions may be determined by flying several parallel lines, then developing a thermal distribution map by hand plotting or computer routine.

Also, using aircraft platforms, thermal infrared line scanners that provide a two-dimensional image of the scene may be used to determine temperature distributions over a swath along the aircraft flight line. In many cases, thermal information is obtained concurrently with photography or other visible spectral range data to aid in location and interpretation. For the marine environment this is especially desirable because of the lack of control points over open water. Thermal scanner calibration is discussed later in this chapter under Thermal Plumes.

c. Satellite Methods

Polar-orbiting and geosynchronous satellites have been used to measure earth-surface temperatures over regional, hemispheric, and global areas on a repetitive and timely basis using measurements in the infrared spectral range (McClain, 1980; Rao et al., 1971, 1972; Maul and Hansen, 1972). Major constraints to mapping surface temperatures by satellite infrared measurements include cloud cover, and absorption/emission by water vapor and carbon dioxide. In the case of water vapor, techniques are being developed to provide corrections by measurement of water vapor in the other spectral bands for in-flight calibration. Relative temperature differences between ship- and satellite-acquired sea-surface temperatures range from about 0.5° to 1.5°C. Absolute temperature differences vary from about 1.0° to 2.0°C. Satellites in use include U.S. operational spacecraft (e.g., NOAA, GOES, TIROS) and research spacecraft (e.g., Nimbus, Landsat) and, since the latter half of 1977, the Japan Meteorological Agency and European Space Agency geostationary satellites at 140°E and 0° longitude, respectively. These spacecraft are similar in operation and instrumentation to the GOES satellites, which are on station at 75°W and 135°W (WMO, 1975).

Microwave measurement of sea-surface temperature has been accomplished with the Seasat and Nimbus 7 satellites using the Scanning Multichannel Microwave Radiometer (SMMR). Hofer et al. (1981), Wilheit et al. (1980) and Lipes et al. (1979) provide further information. These and other Seasat-evaluation activities have demonstrated algorithms for sea-surface-temperature (SST) retrieval with accuracies of 1°C about negligible biases for both ascending and descending pass retrievals in open ocean when rain and sunlight are absent. These results come from about 125 spot comparisons with high-quality surface truth, accurate to a few tenths of a degree, in the north and west Pacific and in the Gulf of Alaska. About 20 comparisons in the tropical Pacific to lower quality surface truth show a similar scatter about a warm bias of a few tenths of a degree. These spot comparisons were made under favorable surface and atmospheric conditions over a wide range of SST (10–30°C).

An additional Seasat-evaluation activity involved separately mapping radiometer and ship SST observations data, using identical mapping procedures, for a one-month summertime period in the western North Pacific (Bernstein, 1982). Both data types are sufficient to construct monthly maps of SST and SST anomaly, which agree with each other to within 0.6–0.8°C rms. The "noise" level of the data used to construct these maps has also been computed, where noise is really a combination of instrument effects and unresolved but true spatial and temporal variability. For the selected area and time, such variability should be quite large compared with other oceanic areas. Still, the 1.3°C radiometer noise-level, thus defined, is well below that of the ship data, which is 1.9°C. In addition, the superior geographic coverage which is characteristic of satellite data makes the radiometer an excellent instrument for the large scale SST mapping purposes required for climate-related research.

d. Sensors and Experimental Results

(1) Radiometers—Infrared radiometers on aircraft are used on a routine basis to survey water-surface temperature distributions in limited areas. These line-measurement sensors are used individually or with other instruments for weather surveys, pollution-site location, calibration of research instruments, and comparison with other aircraft and satellite temperature-measurements. Sensitivities of 0.1°C and accuracies within 0.5°C are typically provided for ranges from −50° to +150°C. Fields of view range from 0.15° to 20°. The surface spatial resolution is a function of the particular configuration as well as the aircraft altitude.

Swift (1980) reviewed experiments conducted using an airborne nadir-viewing microwave radiometer for surface-temperature measurement. A precision 2.65 GHz radiometer was developed with the specific objective of measuring ocean temperature absolutely to within 1°K (Hidy et al., 1972). As a demonstration of instrument performance, the radiometer was flown down the length of the Chesapeake Bay in conjunction with extensive surface measurements (Blume et al., 1977). Surface measurements of salinity and temperature were used to calculate the brightness temperature of a flat ocean surface with corrections for sky radiation, antenna losses, and beam efficiency. The comparison between the calculated and measured surface-brightness temperature is summarized in the column labeled ΔT_B in Table 28-3. Note that, for the lowest windspeed of 1 kt (knots), the average difference between measured and calculated values is 0.3°K, thereby indicating that the surface temperature can be measured to within 1°K under calm conditions. The remaining data show that the difference between measurements and the calculations increases with windspeed and hence with surface roughness; consequently the data were used to generate an empirical windspeed correction for the nadir-viewing instrument. Further work involving satellite systems will have to closely examine the need for atmospheric corrections if accurate reliable repetitive measurements are to be achieved.

As a check on the accuracy of this empirical correction, a subsequent flight (Blume et al., 1978) was conducted at constant windspeed over the lower portion of the Chesapeake Bay, and compared with surface truth collected at 12 locations within the area surveyed. The absolute measurements are mapped in Figure 28-1 and agreed with the values of surface observations to within 1°K. Thus, the two data sets strongly suggest that a 1°K accuracy in the measurement of surface temperature can be obtained using microwave radiometry.

(2) Airborne Scanners—Thermal imaging line-scanners may be used to obtain spatial distributions of temperature differences in surface waters. Single overflights provide information on synoptic distributions that are not readily available from surface measurements, and sequential coverages may yield temporal distributions that provide an insight into discharge and receiving water dynamics. With appropriate calibration and atmospheric correction, accuracies better than 0.5°C may be obtained using airborne scanners. Further discussion and techniques for calibration are found later in this chapter in the section dealing with Thermal Plumes.

(3) Satellite Infrared Scanners

(a) Noise—Noise exists in all electronic systems, including satellite infrared-scanners. The noise in the electronic signal, referred to as the

TABLE 28-3

S-Band Radiometer and Sea Truth Data (Swift, 1980)

Date	Time	Target Area	Sea Truth: SS Temp. (°C)	Sea Truth: Sal. (‰)	Sea Truth: Wind speed (knots)	Calculated T_B (Kelvin)	Rad. T_B (Kelvin)	Bright Temp. Diff. ΔT_B (Kelvin)	Remarks
11-16-72	12:30	Sta. 1	13.1	25.2	3	101.0	102.0	1.0	h = 460 m
	12:15	2	13.4	24.2	2	101.2	101.8	0.6	h = 460 m
	12:00	3	12.7	23.2	3	101.1	101.7	0.6	h = 460 m
	13:55	4	13.2	20.9	1	101.5	101.7	0.2	h = 460 m
	15:10	5	12.7	19.8	1	101.4	101.9	0.5	h = 460 m
	17:00	6	12.4	17.0	1	101.5	101.7	0.2	h = 460 m
	18:25	7	11.9	16.0	1	101.4			restricted
	19:30	8	11.9	15.4	8	101.5			air traffic
11-17-72	07:35	9	11.0	13.6	16	101.1	103.4	2.3(1.8)	h = 230 m
	08:45	10	10.8	12.1	12	101.3	102.9	1.6	h = 230 m
	09:40	11	10.3	9.8	12	101.0	102.3	1.3	h = 230 m
	10:25	12	9.8	6.8	8	100.8	101.6	0.8	h = 230 m
	10:55	13	9.4	4.1	9	100.6	101.9	1.3	h = 230 m
	11:30	14	9.1	2.8	10	100.5	101.5	1.0	h = 230 m
	12:55	15	7.7	0.3	7	99.8	101.0	1.2	h = 230 m
	12:20	16	7.7	0.4	10	99.8	100.6	0.8	h = 230 m
		17					100.4		no ship data

noise-equivalent change in temperature (NEΔT), produces random or coherent two-dimensional noise patterns intermingled with thermal emission patterns from the earth. When the noise fluctuations are of comparable magnitude to the surface thermal gradient associated with fronts, it becomes difficult to distinguish them from each other.

(b) Cloud Cover—Couds often obscure the ocean surface. Since infrared radiation is ab-

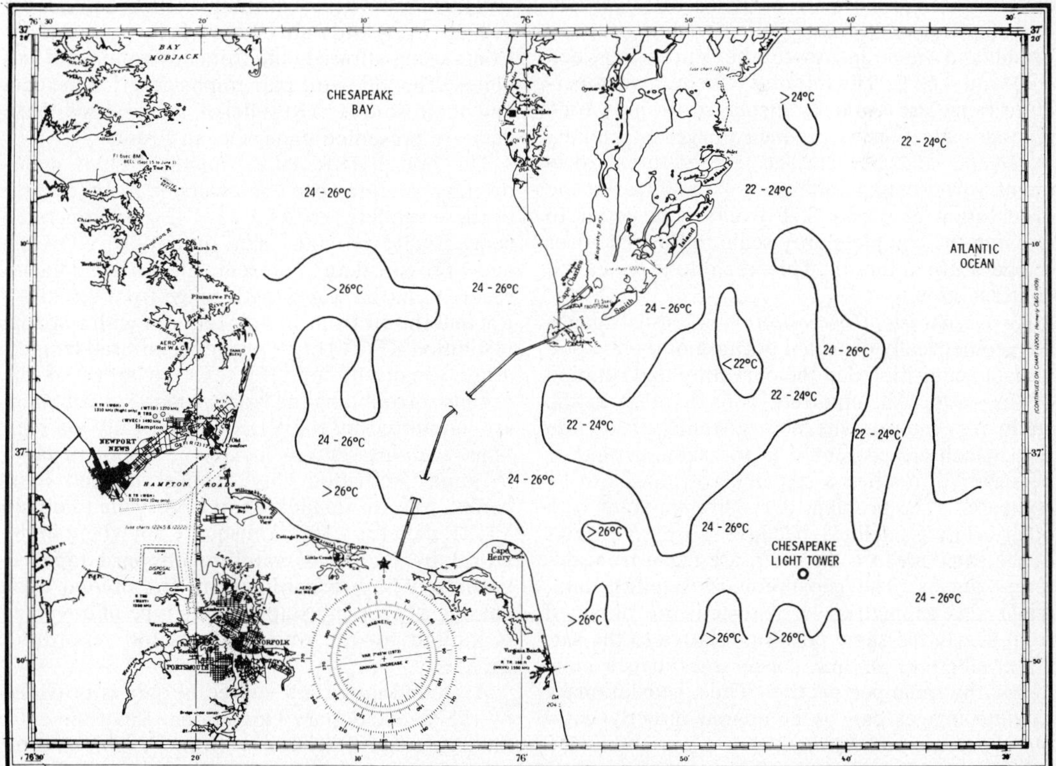

Fig. 28-1. Isotherms in lower Chesapeake Bay on August 24, 1976, in increments of 2°K (Blume *et al.*, 1978).

sorbed by water vapor, clouds are a severe limitation in infrared observations of sea surface temperature (SST) fronts. Thus, sequential views of SST fronts are usually intermittent in time and space. The clearest views of SST fronts are usually associated with the passage of atmospheric cold fronts that are followed by large masses of relatively cold, dry air. Cloud formations can sometimes be used to locate the SST fronts. For example, Legeckis (1977) has reported the suppression of clouds over the cooler water of the Equatorial Current in the Pacific.

(c) Atmospheric Attenuation—Thermal infrared radiation is absorbed by water vapor, aerosols, carbon dioxide, and ozone. Therefore, (depending on atmospheric conditions) the ocean temperature can appear to be 1°–10°C lower at the satellite than at the sea surface. The relative magnitude of temperature correction for each variable is 10°C for water vapor, 0.1°–1°C for aerosols, and 0.1°C for CO_2 and O_3. Although atmospheric absorption reduces the absolute values of sea-surface temperatures as recorded by the satellite, the relative distribution of ocean temperatures can be measured within the limitation of noise and atmospheric variations. To correct for the effects of atmospheric absorption, one must have access to the distribution of the moisture field in the atmosphere. Methods of resolving this problem for satellite data have been proposed by Smith et al. (1970), Braun (1971), Maul and Sidran (1973), and Cogan and Willand (1974).

(d) Image Enhancement—Meteorological satellite infrared-sensors are designed for night and day cloud observations. Their dynamic range is broad and responds to temperatures between −90° and +60°C. The infrared data in this temperature range are normally displayed on grey tone photographic film to produce images of clouds, water and land. Sea surface temperatures fall in the narrower dynamic range of 0° to 40°C. For the observation of ocean SST fronts it is better to reassign the complete grey scale to this narrower temperature interval. This technique is called contrast stretch.

(e) Geometric Corrections—Satellite images are geometrically distorted because of the method of data acquisition and the curvature and rotation of the earth. An important consideration in the study of ocean fronts is the availability of satellite data which are presented in the same geometric perspective to allow a sequence of images to be compared. The problem is readily apparent with polar-orbiting satellites, because successive views of the same area on the earth are made from different angles. The geostationary satellite data, while also geometrically distorted, are obtained from nearly the same position relative to the satellite subpoint, so that successive images have nearly the same perspective. Thus, geostationary satellite images can be compared directly without geometric corrections.

The degree to which polar-orbiting satellite data must be corrected geometrically depends on the accuracy required. For example, Landsat data used to improve maps have been geometrically corrected within subsample accuracy by Bernstein and Ferneyhough (1975). The Canada Centre for Remote Sensing produces geometrically corrected Landsat images which can be overlayed on a 1:50,000 scale topographic map. These images do require ground control points, so they can only be produced for coastal regions (Guertin, 1981). The earth curvature and rotation errors associated with Very-High Resolution Radiometer (VHRR) data can be removed approximately by applying a simple algorithm described by Legeckis and Pritchard (1976). Unfortunately, the VHRR data are not geometrically corrected on a regular basis. Therefore, it is necessary for users to acquire and reprocess digital satellite data whenever geometric corrections are appropriate.

Although standard geometric corrections can put the satellite data into a uniform perspective, the data are not yet accurately fixed geographically. Accurate fixing requires either precise satellite-navigation information or landmarks which can be identified on the image. Colvocoresses (1974) proposed the adoption of the Space Oblique Mercator (SOM) map projection for satellite imagery. Snyder (1978) derived the detailed transformation equations for the SOM, which will be implemented for Landsat D. The present projection used for Landsat data is the Hotine Oblique Mercator (HOM).

(f) Data Sources—Legeckis (1978) provided a survey of worldwide sea-surface temperature fronts using infrared data from environmental satellites. The following paragraphs describe the data and their sources. Results of analyses of these data are presented under Ocean Fronts.

The first VHRR data from the NOAA polar-orbiting satellites became available in 1972. Four of these satellites (NOAA 2, 3, 4, and 5, operated sequentially) provided data until 1978; NOAA 6 and 7 provide data at present. The visible data are recorded in the wavelength range from 0.6 to 0.7 μm and the IR from 10.5 to 12.5 μm with a spatial resolution (IFOV) of 1 km and an infrared temperature sensitivity of 0.5°C. Large areas of the oceans surrounding the North American continent are monitored by the VHRR twice daily via real-time receiving stations located at Wallops Island, Virginia; Redwood City, California; and Fairbanks, Alaska. In addition, it is possible to obtain VHRR data for a 3000-km square anywhere in the world by means of satellite onboard tape recorders, whose recordings are transmitted upon passing a receiving station. This type of coverage is limited by the onboard satellite recording-capability.

A second important source of data is provided by the Geostationary Operational Environmental Satellite (GOES) system with the Visible and Infrared Spin-Scan Radiometer (VISSR) as described by Bristor (1975), and more recently the VISSR Atmospheric Sounder (VAS). The data are available half-hourly in the visible band from 0.55 to 0.75 μm and in the infrared from 10.5 to 12.6 μm. The spatial resolution for the visible data is variable (IFOV, 1–8 km), whereas that for the infrared is fixed at 8 km. The infrared temperature-sensitivity is approximately 0.5°C. The GOES satellite revolves in the same direction as the earth so that it remains essentially stationary above the equator. The data, useful for ocean-front analysis (see Ocean Fronts), cover a scene 50° in latitude and longitude around the subpoint of the satellite. The United States presently operates GOES-5 and GOES-4 with subpoints located respectively at longitudes 75° and 135°W. Similar Japanese and European geostationary satellites are in operation (Fea, 1980; Fusco et al., 1980); the U.S.S.R. is expected to provide a fifth satellite somewhat later. This system of five geostationary satellites will allow continuous observations to be made of the entire earth from the equator to 60° latitude, both north and south.

In 1978 the NOAA satellite system was augmented by the next generation of NOAA polar orbiters, called TIROS-N (Ludwig, 1974). This system has the capability of global (4 km resolution) and regional (1 km resolution) observations of the oceans four times each day. Because of several improvements in the new AVHRR radiometer and in the data-transmission procedures, it is expected that the electronic noise will be reduced to an NEΔT of 0.2°C. This improvement should allow the detection of ocean fronts at both 1- and 4-km spatial resolution, and will greatly increase the quantity of useful data. Because the polar-orbiting satellite transmits data continuously, it is possible for anyone to receive the data, given the proper equipment. Research groups outside of the United States are taking advantage of this opportunity to record data in areas of their interest (Popham, 1977).

When the improved data became available from the VHRR and the VISSR instruments, the Environmental Data Service, an agency of NOAA, established a facility in Washington, D.C., to archive and distribute the satellite data. Presently, all visible and IR images are archived as photographic negatives. In addition, digital data are saved in limited quantities. For the study of ocean fronts, it is more useful to have digital data rather than the archived photographic images, because the archived photographs are usually not processed specifically for the optimum detection of ocean fronts. The digital data provide the flexibility to process the data for specific applications in both a qualitative and a quantitative format, as will be described in later sections. Some ocean-related data processing is done at the National Environmental Satellite Service (NESS). Data and products from data analyses are available from the Satellite Data Services Division of NOAA (NOAA, 1980).

2.0. Ocean Fronts, Currents, and Topography

a. Satellite Benefits and Constraints

Dynamic features of coastal and offshore waters are usually indicated by gradients in oceanographic parameters. These include temperature, salinity, suspended particulates, sea state, and surface topography. Several of these parameters may be observed by remote sensing. Although most techniques observe these parameters only at the surface, it is increasingly apparent from satellite data that subsurface dynamics frequently manifest at the surface. Also, more than one parameter is usually involved. Thus, measurements from different instruments on a given platform will indicate the same phenomena via different spectral bands or spectral ranges. An example is the Seasat detection of the Gulf Stream by microwave altimeter, imaging radar, and thermal infrared techniques.

Satellite data useful for the study of ocean gradients and fronts are intermittent both in time and space. This is partly due to natural constraints such as cloud cover and the seasonal variability of SST gradients. Another constraint is the quality of the electronic signal; the data from improved satellite systems, such as the TIROS-N, are of better quality. This should allow more detailed description of most SST fronts in cloud-free ocean areas.

Another important constraint that arises is in the selections that are made among the large amounts of satellite data that are available. The process of selecting and reprocessing data for specific applications in hydrology, meteorology, and oceanography can saturate the available facilities. For example, the GOES data have not been adequately utilized for ocean-front studies. The new TIROS-N satellite system has further increased the quantity of data available. The Nimbus 7 Coastal Zone Color Scanner (CZCS) is able to detect ocean-color differences that may reveal fronts, even when nearly isothermal surface temperatures prevent their recognition with infrared sensors. The short-lived Seasat, with 25 m resolution synthetic aperture radar (SAR), revealed sea state variations suggesting currents, eddies, and a mid-Atlantic coastal boundary layer, but even after 36 months only some Seasat SAR data have been thoroughly studied.

The extraction of useful information from the large quantity of available satellite data is a challenge to both scientists and resource managers. Landgrebe (1976) has pointed out that a systems-oriented approach is required in order to achieve the full potential of remotely sensed data. It may be useful for oceanographers to consider this approach in satellite studies of ocean fronts.

In the coastal zone, spatial resolution is as important as the choice of parameters. Historically, optical instruments mounted on aircraft platforms measuring temperature in the infrared and suspended solids in the visible have been used to study plumes and other coastal features. Results from studies along the Atlantic coast of the U.S. indicate that spatial resolutions of 80 m may be used to study plumes emanating from coastal estuaries and bays, while lower resolutions are sufficient for large oceanic features. In general, this indicates that aircraft and Landsat satellite platforms are effective for coastal zone investigations, while most other satellite systems (meteorological, oceanographic, etc.) provide large-area coverage that is applicable to oceanic studies (Klemas et al., 1974; Johnson and Harriss, 1980; Munday and Fedosh, 1981; Vukovich and Crissman, 1981).

b. Oceanic Fronts and Currents

(1) Seasonal Variability—From inspection of several years of satellite infrared-images, it is apparent that the detection of oceanic SST fronts is limited according to season. This appears to be related to the seasonal changes in solar insolation and air-sea interaction. During the winter months, ocean surface-layers tend to be well mixed, so that horizontal temperature gradients, such as those associated with currents or eddies, are easily recognizable at the sea surface. During the warmer season, however, nearly isothermal surface layers obscure these horizontal temperature gradients, which are known from hydrographic data to exist at depth. Furthermore, in tropical oceans these isothermal conditions persist during all seasons and, as a result, it has often been difficult to detect, by means of SST, front signatures in the tropics. From the point of view of VHRR monitoring, the ocean surface between the equator and 25° north or south latitude is nearly isothermal. Between latitudes 25° and 35° SST fronts can be observed from autumn to spring. For latitudes greater than 35° the SST fronts can be observed regardless of season, although ice cover is a limiting factor. There are exceptions to these generalizations that arise from locally induced upwelling or from advection by currents. Active microwave data, such as from Seasat, could significantly improve detection of fronts and currents in seasons when infrared contrasts are small.

(2) Satellite Detection—The VHRR scanners have been used to survey ocean fronts on a worldwide basis. Usually,visible images are used to find cloud-free areas of the ocean, and infrared images are used to detect SST fronts. When visible images are not available, it is usually possible to distinguish SST fronts from cloud patterns by pattern recognition and by the relative spatial immobility of SST fronts over a period of several days.

The areas of the world in which VHRR data have been obtained are shown in Figure 28-2. Areas 1-6 have been monitored continuously, twice per day, since 1972; all other areas have been monitored intermittently. Examples of VHRR imagery of SST fronts observed in each area are summarized in Table 28-4. Each image is uniquely identified by an orbit number and a sat-

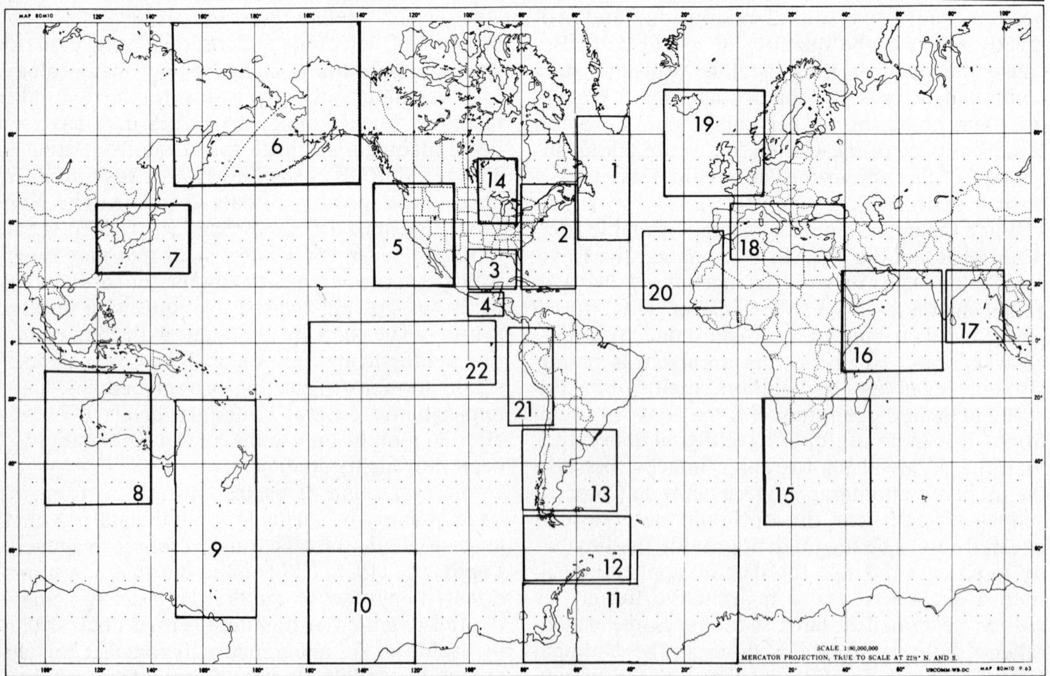

Fig. 28-2. The areas of the world (1-22) where sea surface temperature fronts have been observed in satellite thermal infrared images (Legeckis, 1978).

TABLE 28-4

Summary of Fronts Seen in VHRR Images From NOAA Polar-Orbiting Satellites (Legeckis, 1978)

Area	VHRR	NOAA	Date	Location
1	5086	5	Sept. 13, 1977	Greenland
	3542	4	Aug. 25, 1975	Labrador Sea
	5395	5	Oct. 8, 1977	Newfoundland
	3554	4	Aug. 26, 1975	North Atlantic Polar Front
2	4158*	5	June 30, 1977	Gulf of Maine
	3517*	4	Aug. 23, 1975	Cape Cod
	4531*	4	Nov. 12, 1975	Gulf Stream North of 35°N
	1282*	4	Feb. 26, 1975	Gulf Stream South of 35°N
	3235*	5	April 17, 1977	W. Atlantic Subtropical Front
3	4721*	3	Nov. 22, 1974	Loop Current
	406*	4	Dec. 18, 1974	Western Gulf of Mexico
4	5590*	4	Feb. 5, 1976	Gulf of Tehuantepec
5	1374	5	Nov. 17, 1976	E. Pacific Subtropical Front
	520	5	Sept. 9, 1976	West coast of U.S.
6	2120	4	March 25, 1975	Bering Sea
7	6841*	4	May 15, 1976	Kuroshio North of 35°N
	3817	5	June 3, 1977	Kuroshio South of 35°N
	1745*	4	April 4, 1975	Sea of Japan
8	6754	4	May 8, 1976	Western Australia
9	6552	5	Jan. 10, 1978	Eastern Australia
	5462*	5	Oct. 14, 1977	Eastern Australia
	660*	5	Sept. 21, 1976	Macquarie Ridge
10	1613	5	Dec. 6, 1976	Ross Sea
11	2917	5	March 22, 1977	Weddell Sea
12	859*	5	Oct. 7, 1976	Drake Passage
13	5339*	4	Jan. 16, 1976	Brazil—Falkland currents
	6929*	4	May 22, 1976	Brazil—Falkland currents
	6850	5	Feb. 3, 1978	Chile
14	3737*	5	May 27, 1977	Great Lakes of U.S.
15	2830*	5	March 15, 1977	West coast of S. Africa
	1530*	5	Dec. 30, 1976	Agulhas Current
16	589	5	Sept. 15, 1976	Somali
	2272	5	Jan. 29, 1977	Arabian Sea
17	2360	5	Feb. 5, 1977	Bay of Bengal
18	4607	5	Aug. 5, 1977	Mediterranean Sea
19	4248	5	July 7, 1977	England
	2346	5	May 22, 1975	Iceland
20	505	5	Sept. 8, 1977	Northwest Africa
21	3211	5	April 15, 1977	Peru

The areas are shown in Figure 28-2 and the location is used to identify the sea surface temperature fronts or associated currents. VHRR identifies the number of the image. All original VHRR images may be obtained from the Environmental Data Service in Washington, D.C.

* Images that have been computer reprocessed (geometrically corrected and enhanced for the temperature range at the SST front). Examples are shown.

ellite identification-number. The VHRR images shown here have been geometrically corrected and were enhanced for the temperature range in each specific frontal zone. The warmer water is represented by darker shades of gray. Images referenced in Table 28-4 but not shown here may be ordered from the NOAA Environmental Data Service. For each front the limiting factors which hindered the satellite observations and the magnitude of observed SST gradients are estimated whenever possible. In most cases there was no verification by simultaneous hydrographic measurements. However, the satellite-derived positions of the fronts appear to coincide with the general location of fronts reported by direct observations.

Several fronts occur in the Gulf of Maine (Legeckis et al., 1980). During spring and summer a distinct area of cooler water appears over the Georges Bank, as shown in Figure 28-3 for June 30, 1977. The SST gradient at this front is about 2°C in a distance of 2 km. This structure has been described from hydrographic surveys by Beardsley et al. (1977), and the position of the front appears to be oriented parallel to the bottom topography of the Grand Banks. By late summer the front becomes less distinct, as shown in Figure 28-4 for August 23, 1975. During the summer and fall it is also possible to detect cooler water along the coast of Maine and off the southern coast of Nova Scotia as shown in Figures 28-4 and 28-5. The frontal zone off Nova Scotia has been de-

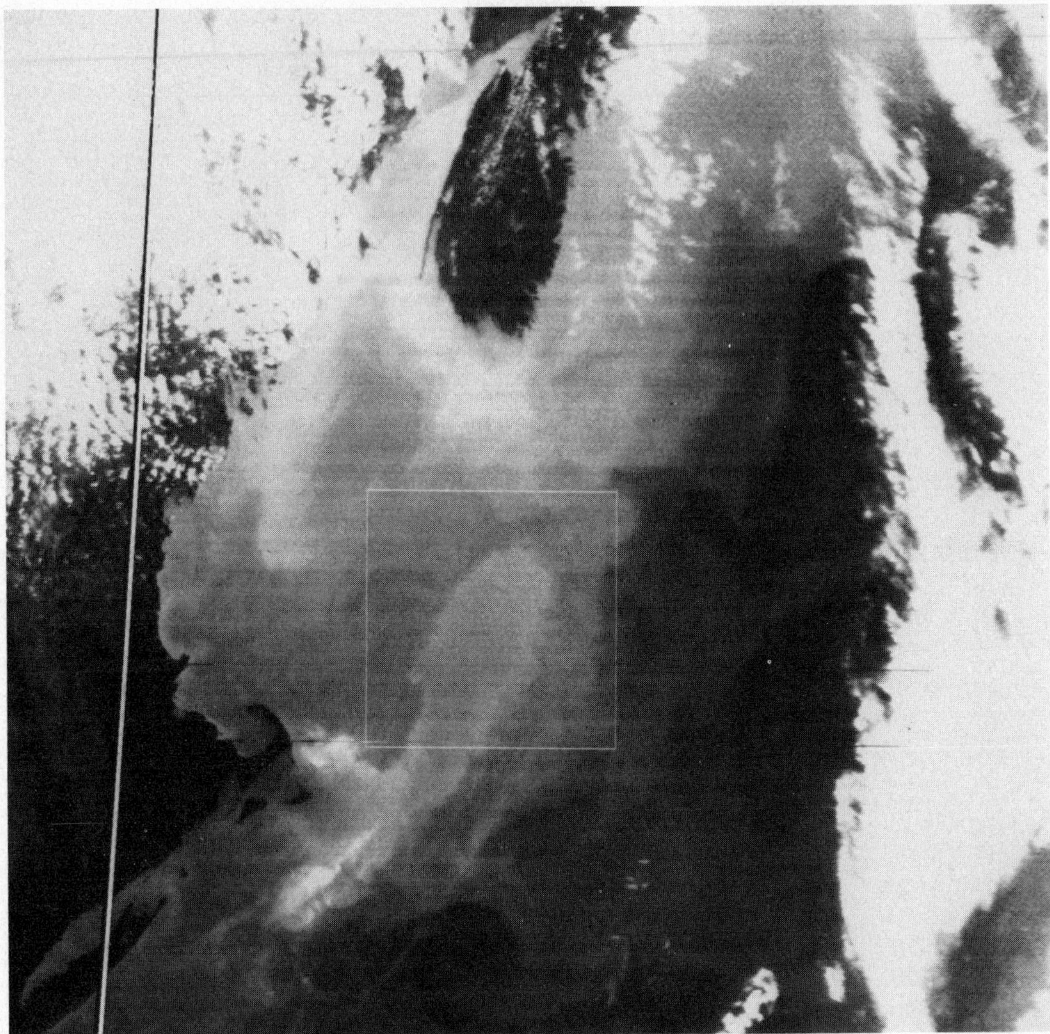

Fig. 28-3. VHRR IR image 4158 from NOAA 5 obtained on June 30, 1977. SST fronts appear off the coast of Nova Scotia and in the vicinity of the Georges Banks. Warmer water, associated with Gulf Stream eddies, appears as darker shades of gray south of the Gulf of Maine.

scribed by Fournier et al. (1977) and is due to coastal upwelling. During winter and spring, it is difficult to detect surface fronts in the Gulf of Maine because waters are isothermal and there is an increase in cloud cover.

North of Cape Hatteras, at latitude 35°N, the circulation off the east coast of North America is greatly influenced by the Gulf Stream, as shown in Figure 28-6. Both the northern and southern SST boundaries of this current can be identified in aircraft and satellite infrared data (Maul and Hansen, 1972; Stumpf, 1974). Also, large-scale (100–200 km) eddies form to the north (warm-core rings) and south (cold-core rings) of the current (Stumpf and Rao, 1975; Maul et al., 1978). These SST fronts can be observed throughout the year, limited only by cloud cover, but are most pronounced in the fall, winter, and spring and less well defined during the summer. Seasat Synthetic Aperture Radar (SAR) has been useful for tracking the warm-core rings (Lichy et al., 1981), but detection of cold-core rings has been less certain (Cheney, 1981).

The Gulf Stream south of Cape Hatteras at latitude 35°N is illustrated in Figure 28-7. Owing to seasonal variations of sea-surface temperature, this current can be observed in thermal data only between fall and spring. Air-sea temperature differences appear to determine the earliest date of detection. For example, the Gulf Stream was detected in 1974 during December, in 1975 during November, and in 1976 during September. In view of the extreme cold air temperatures during the winter of 1976 it appears that the early detection of the Gulf Stream in September 1976 was indicative of an atmospheric cooling trend.

The dynamics of the Gulf Stream are of particular interest. In Figure 28-7, downstream of

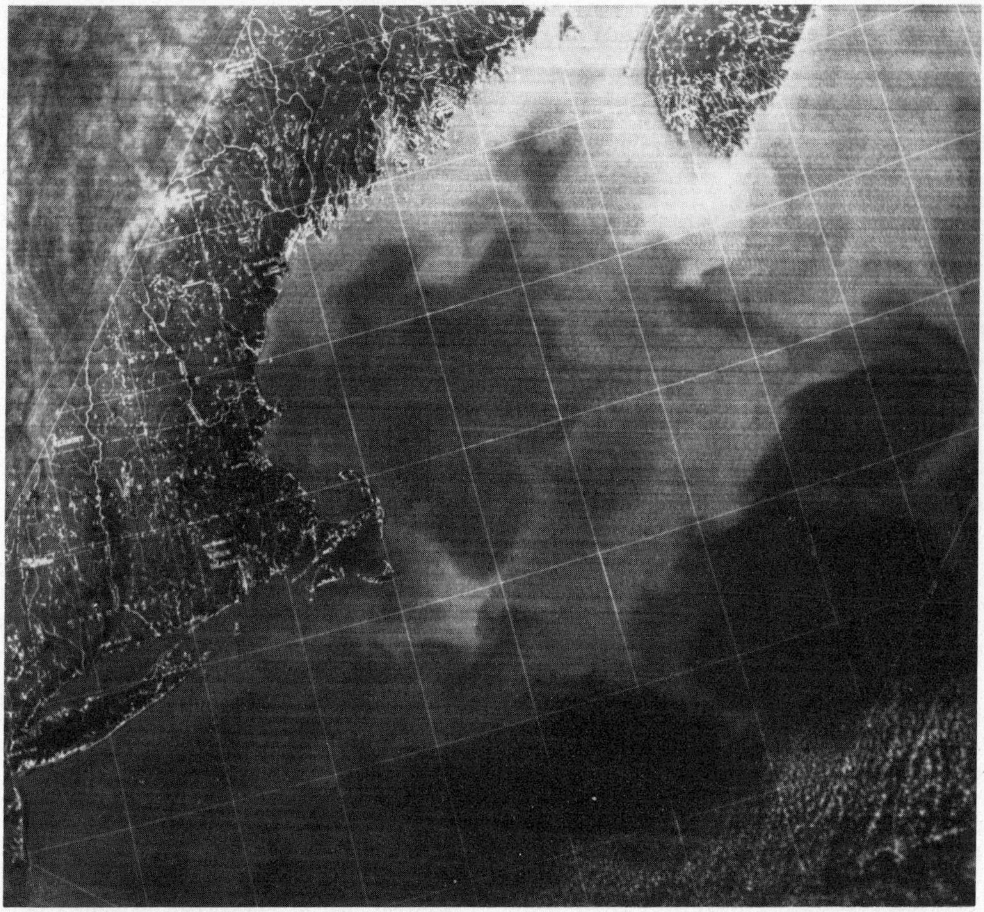

Fig. 28-4. VHRR IR image 3517 from NOAA 4 obtained on August 23, 1975. Cooler water (lighter shades of gray) appears south of Nova Scotia and south of Cape Cod. The warmest water, south of the Gulf of Maine, is associated with Gulf Stream eddies.

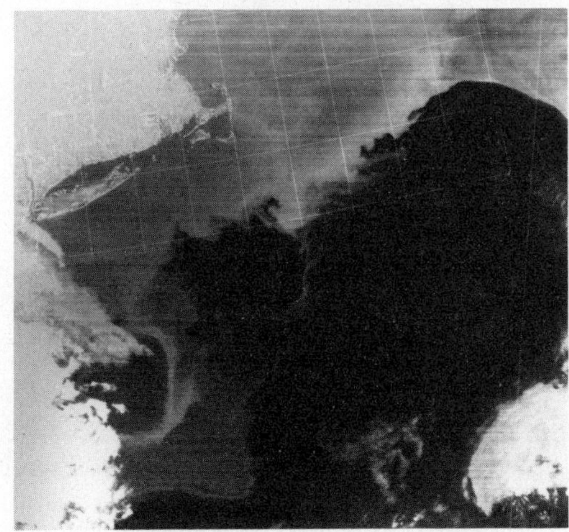

Fig. 28-5. VHRR IR image 4531 from NOAA 4 obtained on November 12, 1975. The Gulf Stream north of Cape Hatteras and south of Cape Cod appears as the warmest (darkest) water. Large-amplitude meanders as well as two warm core eddies are evident south of Cape Cod and Long Island.

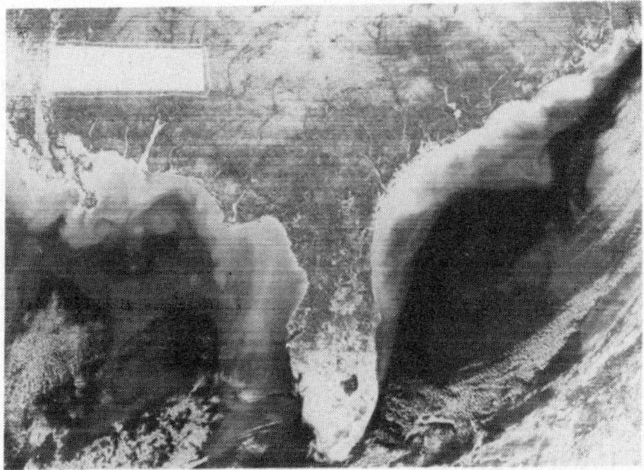

Fig. 28-6. VHRR IR image 1282 from NOAA 4 obtained on February 26, 1975. The seaward deflection of the Gulf Stream downstream of latitude 31°N is evident. This deflection extends about 100 km seaward of the 100-m isobath before turning landward.

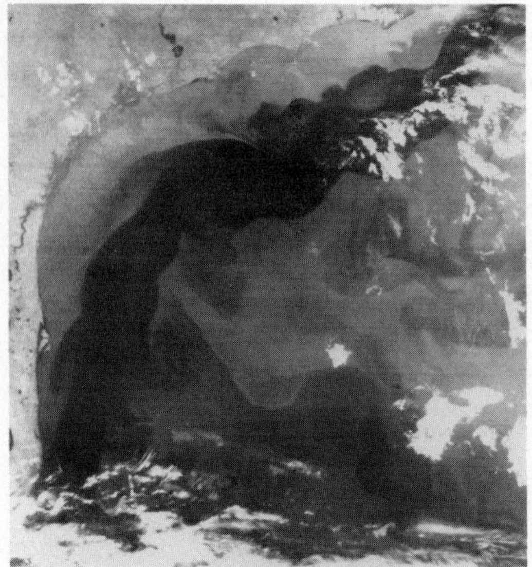

Fig. 28-7. VHRR IR image 3235 from NOAA 5 obtained on April 17, 1977. Note the Gulf Stream south of Cape Hatteras and the Atlantic Subtropical Front east of Florida.

latitude 31°N, there is a persistent seaward deflection of the current that is detectable in the satellite images (Legeckis, 1979). This deflection occurs downstream of a bulge in the continental slope off Charleston, South Carolina, suggesting a coupling between the current and bathymetry. Additional evidence of a coupling has been provided by the Seasat SAR (Hayes, 1981). On many occasions wavelike patterns resembling a leewave phenomenon are seen in both SAR and infrared data downstream of this change in bathymetry. Low-frequency wave motions were suggested by Brooks (1976) and by Orlanski and Cox (1973). Mollo-Christensen et al. (1980) interpreted similar waves as barotropic instability waves forced by tide-Gulf Stream interaction; they used satellite image measurements to estimate the Gulf Stream current speed.

The Subtropical Front in the western Atlantic, also illustrated in Figure 28-7 for April 1976, is usually only observed during the winter and spring months. The observations are limited by clouds and the persistence of excessive atmospheric water vapor. The front has been observed as far eastward as longitude 60°W between latitudes 24° and 30°N. The SST gradients usually do not exceed values of 2°C per km. Direct observations of this front have been reported by Voorhis (1969). An unexplained feature in the figure is the wavelike pattern; similar wavelike features were observed as far as longitude 60°W during February 1976.

Many fronts seen in satellite imagery are features of ocean circulations traditionally associated with boundary currents and upwelling. Others involve previously unresolved ocean circulation questions as, for example, the seaward deflection of the Gulf Stream in Figure 28-7 discussed earlier and now related to bathymetry. In Figures 28-8 and 28-9 the deep northward intrusion of the Loop Current in the Gulf of Mexico during the winter was detected for the first time in satellite imagery. According to Molinari et al. (1977), this event was previously thought to occur only during the summer. A comparison of the Brazil and the East Australia currents shows that large warm core eddies are shed at the southernmost extension of these currents. It is suggested that these currents terminate by shedding the eddies. The Agulhas Current, on the other hand, turns eastward over the Agulhas Plateau and maintains the appearance of a narrow meandering current, similar to the Gulf Stream and Kuroshio. Tseng et al. (1977) measured speeds and locations of the Brazil and Falkland currents with 152 Nimbus 5 images; they found good agreement with available surface information. Fish catches appeared to be higher when fronts were the most visible in the images.

Besides the use of satellite microwave and infrared techniques, ocean fronts may be detected in the visible spectral region through sun glint variability (Strong and DeRycke, 1973). Simultaneous thermal infrared and visible spectrum imagery from NOAA-2 proved that the Gulf Stream boundary from Florida to Cape Hatteras could be detected in visible spectrum imagery as a brightness contrast. This contrast presumably results from the effect of the Gulf Stream current on surface roughness and whitecaps, an effect long observed by mariners.

Further insight into ocean dynamics is being

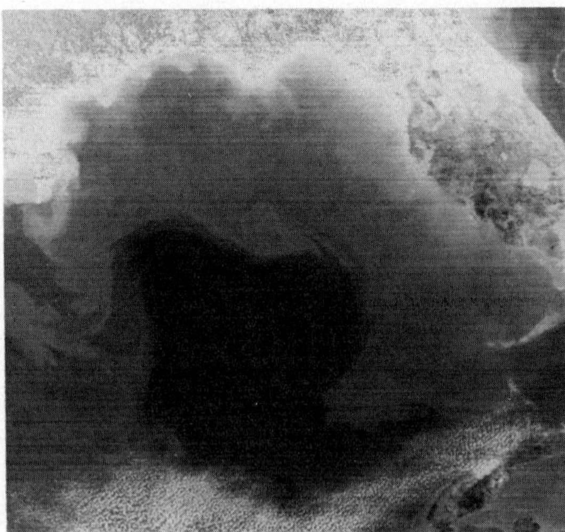

Fig. 28-8. VHRR IR image 4721 from NOAA 3 obtained on November 22, 1974. The Loop Current in the Gulf of Mexico is shown during its northward intrusion cycle. Three months later, a large (300 km) eddy pinched off from the Loop Current and moved westward in the Gulf. The sea surface temperatures in this image are shown in Fig. 28-9.

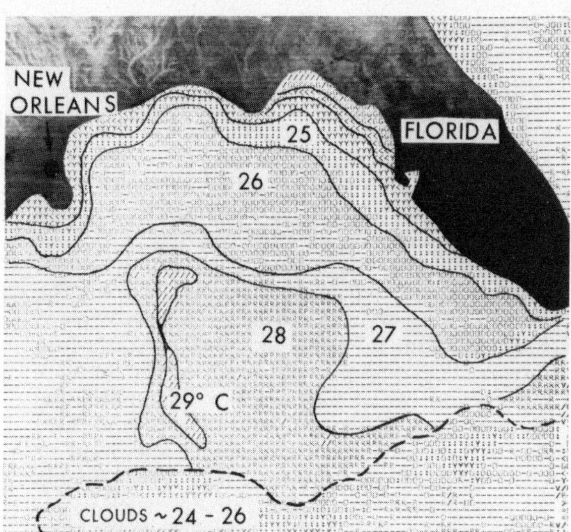

Fig. 28-9. Sea surface temperatures for the image in Figure 28-8 shown as a computer printout of characters. Each represents an average temperature value in a 10-km square. These temperatures are not corrected for atmospheric effects and should be several degrees lower than actual surface temperatures. The contours are used to separate areas of similar temperature. The warmest water (~29°C) appears at the western side of the Loop Current.

provided by satellite detection of internal waves (Apel et al., 1975, 1976). In addition to a direct measurement of wavelength from the image, the group velocity can be determined from the distance between packets generated during flood tides. Data on the group velocity and wavelength allow an inference that the product of residual gravity and depth of the water is changing; then, from the mapping of density, heat storage in the upper mixed layer may be inferred by attributing the density anomaly in the layer solely to temperature rather than to salinity (Mollo-Christensen and Mascarenhas, 1979). On this basis, Mascarenhas (1979) mapped heat storage on the northeastern continental shelf of North America from Landsat data.

c. Surface Topography

Remote measurement of ocean topography provides an independent method for analyzing ocean dynamic features such as currents, tides and other movements of water masses. This is accomplished with radar altimetry via statistical analysis of the time of return from a succession of radar pulses. Mather (1979) has noted that attempts to define low frequency or stationary sea-surface heights are flawed, due in part to errors in geoid determinations. Radar altimetry data from GEOS-3 were analyzed by Leitao et al. (1979) for comparison with other techniques (infrared and *in situ*) for the detection of the Gulf Stream surface boundaries. A sea surface profile from GEOS-3 pass 9216 on January 19, 1977, indicated elevation differences of about 0.5 m across an eddy and about 1.0 m across the Gulf Stream, with the higher elevation on the southern boundary, as shown in Figure 28-10. Results indicated that the combination of satellite infrared imagery and radar altimetry offers a potentially revolutionary tool for ocean dynamics studies. In particular, (1) the maximum surface temperature gradient shown in infrared imagery after image enhancement is a good indicator of the Gulf Stream northern surface boundary, (2) height measurements of the boundary obtained by radar altimetry agree with surface measurements of the boundary determined by depth-temperature profiles, and (3) the combination of infrared and altimeter data provides an excellent data base that indicates boundary geometry and permits computation of current dynamics. The potential limitations of the infrared imagery due, for example, to the presence of cloud cover and lack of temperature contrast, are not factors with radar altimetry, while the lack of area coverage by the altimeter is made up by the synoptic coverage of the infrared imagery.

Seasat, operating in the fall of 1978, also included a radar altimeter, similar to that used on GEOS-3. As a result of instrument refinements and experimental techniques, Seasat allowed a finer assessment of the capability to measure sea-surface elevations based on measurements from satellite platforms. Initial results indicate that the noise (precision) in the altimeter height measurement is on the order of 5 to 8 cm for the 4 m significant-wave-height (SWH) conditions that existed during the Bermuda calibration activity (Tapley et al., 1979). Comparison of Seasat and GEOS-3 mean-surface topography for the western North Atlantic is shown in Figure 28-11 (Leitao et al., 1980). Thus, experimental evidence indicates that satellite altimeters have adequate precision to contribute to quantitative oceanographic investigations and practical applications.

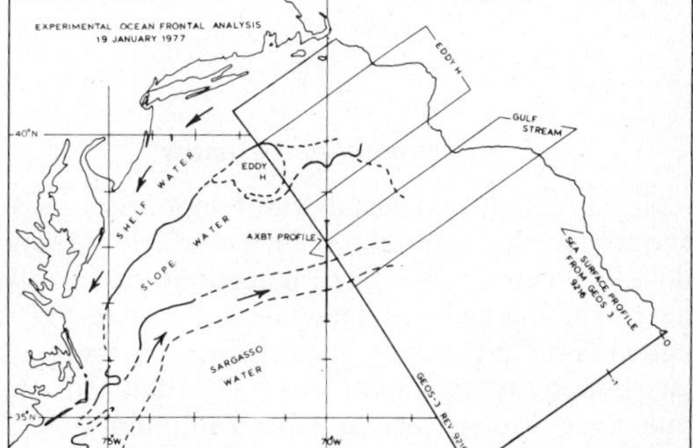

Fig. 28-10. Comparison between sea surface topographic profile from GEOS-3 pass 9216 and Experimental Ocean Frontal Analysis of January 19, 1977 (Leitao *et al.*, 1979).

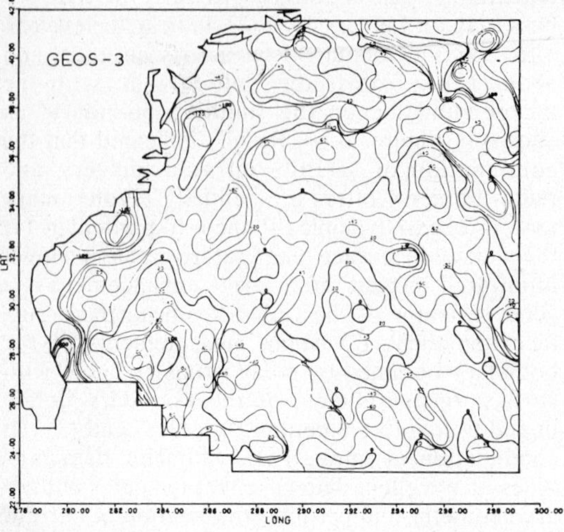

GEOS-3 Mean Sea Surface Topography for Jul-Sep 78

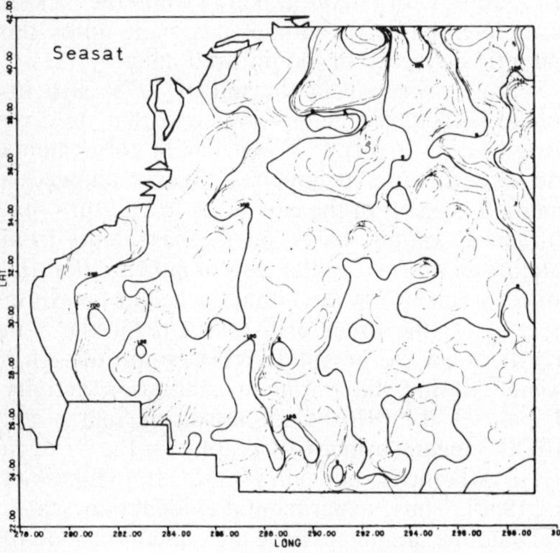

SEASAT Mean Sea Surface Topography for Jul-Sep 78

Fig. 28-11. Comparison of GEOS-3 and Seasat topographic data.

d. Instrumented Drogues

Measurement of coastal water motion is most frequently accomplished with Eulerian techniques, which involve fixed placement of current meters in the field of interest. Loss of current meters is quite common due to piracy, storms, or destruction by fishing trawls. The data so obtained are also subject to wave contamination in the upper 50 m of the water column, especially on the continental shelves. Circulation may therefore be studied by the Lagrangian approach, using current-following tracers. This approach is desired in some applications because it produces an actual trajectory of a water parcel, and therefore constitutes direct evidence of water displacement. For motions over large areas of the open sea, the Lagrangian approach is quite common, involving surface-drift bottles, bottom drifters, and instrumented drogues. Bottles and drifters unfortunately yield only start and end points. Instrumented drogues, in contrast, can be successfully tracked with data-collection systems (DCS) involving radio telemetry. Substantial ship time is involved if tracking is from the sea surface; a cheaper method is to use DCS on satellites. Ruzecki et al. (1976) put transponders for the French Eole satellite (launched at NASA Wallops Flight Station) on floats attached to subsurface drogue plates at depths of from 2 to 30 m, to track currents between Chesapeake Bay and Cape Hatteras over 30 day periods. Results confirmed the pattern inferred by Bumpus and Lauzier (1965) of a general southward flow terminating in an entrainment in the Gulf Stream, such as reported by Rao et al. (1971). Gulf Stream kinematics *per se* were studied by Kirwan et al. (1976) with a satellite-tracked drift buoy. The great potential of this technique for studying currents has recently been shown by McNally (1981), who obtained extensive data from satellite-tracked drifters in the eastern mid-latitude North Pacific (Figure 28-12).

3.0. Coastal Currents

a. Photogrammetry of Drifters

In coastal waters, aerial photography is an obvious choice for tracking a set of Lagrangian drifters, and photogrammetric analysis can yield accurate average current velocities. Such remote sensing has been a tool for circulation analysis in the coastal zone for over three decades (Cameron, 1952; Keller, 1963; Zdanovich, 1964; Ramey, 1968; Yeske et al., 1975). Such techniques were used by the National Ocean Survey (NOS) in a circulation survey of Boston Harbor and the Massachusetts Bay area (NOS, 1971). A strong advantage of remote sensing is that the spatial density of sampling points can be extremely high, and the temporal density is limited only by the number of data-collecting aerial overflights.

The types of targets that have been used include naturally occurring foam, white-painted plywood, polystyrene floats, cardboard sections, aluminum powder, and dye-emitting buoys. Ramey (1968) provided, in addition to wooden panels and aluminum powder, a trail of rhodamine dye to mark the entire length of a buoy-deployment pattern. Water vanes and sails can be suspended at desired depths beneath floats of rigid construction. In port regions which have heavy shipping traffic, consideration should be given to making targets a minimum hazard to navigation. Also, targets should, if possible, be biodegradable and inexpensive. A bibliography of drogues, drags, and sea anchors in historical use is provided by Monahan and Monahan (1973). Field operations have been well-described by Keller (1963) and further discussion is found under Waste Outfalls.

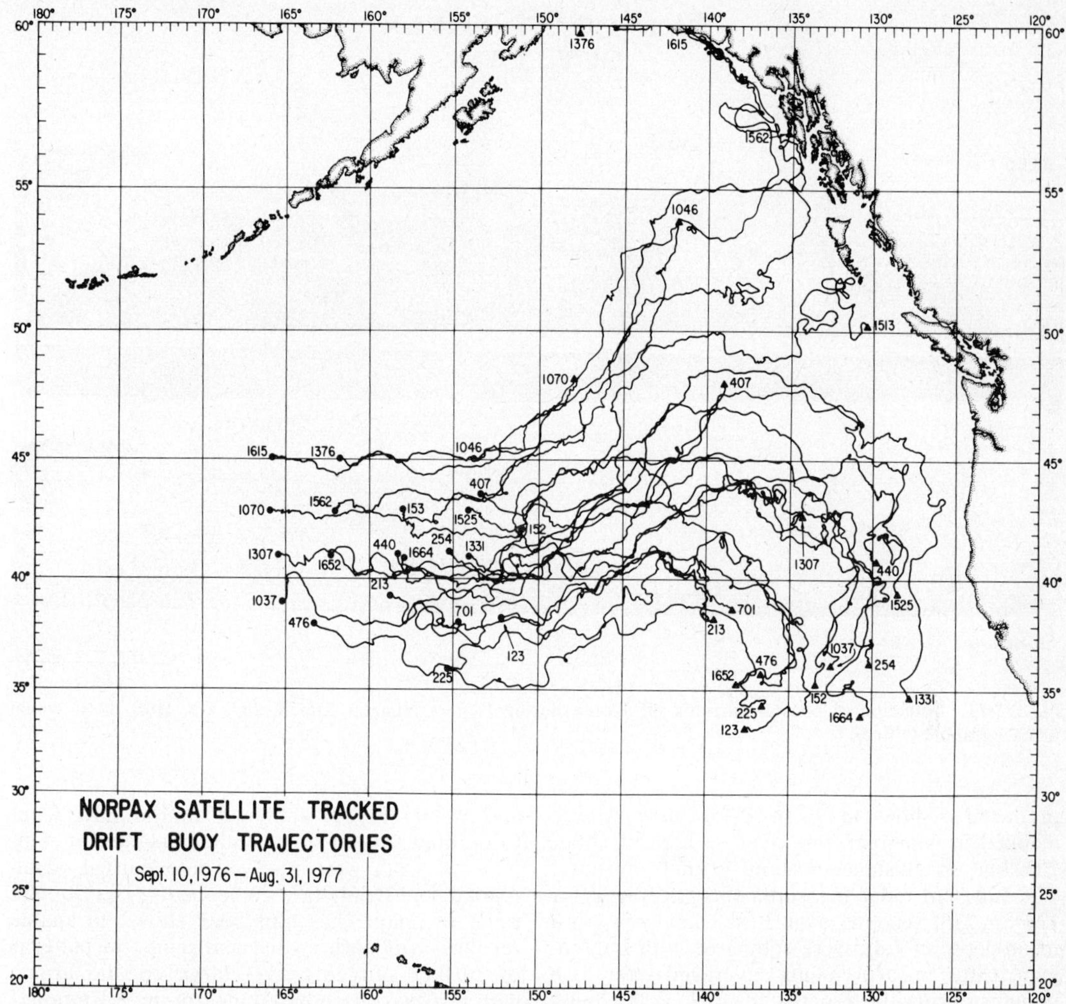

Fig. 28-12. Trajectories of drifters for the period October 1976 to August 1977. Solid dots indicate initial position; solid triangles indicate last position. Individual trajectories are identified by drifter ID numbers (McNally, 1981).

b. Infrared Scanner Imagery

Airborne infrared-scanner imagery has proven to be very useful in the detection and mapping of large coastal currents, many of which are not discernible on other types of imagery. During a study conducted in 1972 by the U.S. Naval Oceanographic Office in cooperation with the Republic of Korea, thermal infrared imagery was taken with an ARA-35 imager aboard an RB-57 aircraft. Imagery from flights at 10:00 a.m. and 3:00 p.m. on October 31 along the east coast of Korea is shown in Figure 28-13. The spiral cells represent cold water advecting seaward out of the embayed coast. Each turns landward at about 5 to 8 km off the coast toward the next downstream embayment, a pattern strikingly similar to meandering longshore currents driven by winds and waves (Sonu, 1972). The city of Pohang borders the bay near the cape, Changi-Kap.

c. Radar

Another means of mapping coastal currents, and hence water-mass movement, is a shore-based "ground wave" propagation-mode radar system (Barrick et al., 1977). The reflective principle of this system is shown in Figure 28-14. The reflected radar energy results from Bragg scattering by ocean waves in the illumination region. Wave trains with wavelengths $L = \lambda/2$, where λ is the radar wavelength, advancing and receding from the receiving antenna, produce Doppler frequency shifts of $2v/\lambda = 2(gL/2\pi)^{1/2}/\lambda = (g/\pi\lambda)^{1/2}$, where v is the wave train phase velocity, and g the gravitational constant. Any current superimposes an additional shift of $2v_c/\lambda$, where v_c is the mean effective current velocity radial to the radar. Using a frequency of 25 MHz ($\lambda = 12$ m), and two stations on the Florida coast, a computer map of the Gulf Stream current on October 20, 1976 was

Fig. 28-13. Detection of coastal currents off Korea during Project Nuggett Ranch, 1972 (O. Huh, U.S. Naval Oceanographic Office).

produced as shown in Figure 28-15. Currents were mapped in near real time over an area of about 2000 km^2, to a distance of about 70 km from shore. The Eulerian radar measurements, averaged for 128 (or 256) seconds over 3 × 3 km areas to a mean depth of 1 m, were compared with Lagrangian drifter measurements, averaged over 5–8 minutes (a curvilinear path 0.4–1.0 km) at a depth from 0 to 0.46 m. The agreement was better than 30 cm/sec. For 128-second integration times, the Doppler resolution is 5 cm/sec.

In further development of this technique, Frisch and Weber (1980) analyzed similar data from Cook Inlet, Alaska for periodic tidal-current components. Data were resolved into diurnal, semidiurnal, and steady-flow components. The Doppler radar technique has thus been shown to enable resolution of both spatial and temporal patterns of surface currents over large coastal areas, with just two shore-based measurement stations.

d. Nearshore Currents

In nearshore regions, circulation patterns associated with the breaker zone are difficult to dis-

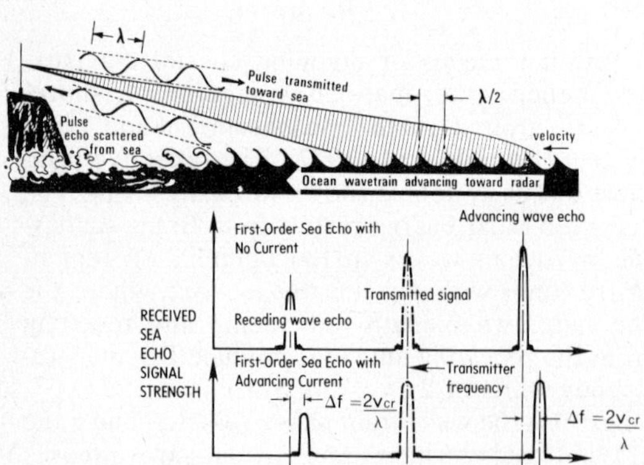

Fig. 28-14. Sketch showing the principles of first-order HF Bragg scatter from the sea, and resulting signal echo spectra without and with an underlying current (Barrick *et al.*, 1977).

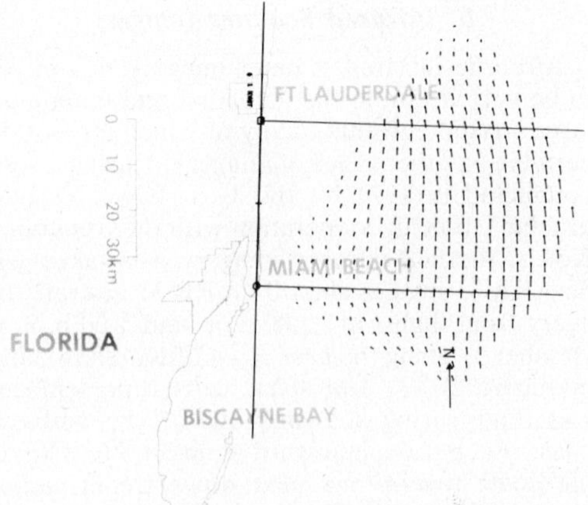

Fig. 28-15. Computer-generated map of 20 October 1976 as deduced by radar (Barrick *et al.*, 1977).

cern and measure using traditional techniques. Current meters are impractical here because of wave action and shallow depths. As an alternative, remote sensing can be an important tool. Dye or floats may be released at different times in the tidal cycle and followed as indicators of current direction and speed. Aerial photography used in conjunction with the releases will improve the quality of the derived current-velocity data.

Meandering currents, frequently produced by waves approaching at an angle to the beach, have been shown by Sonu (1972) to be associated with undulating or rhythmic bottom topography. To observe and monitor these circulation patterns and changes in bottom topography, Sonu employed airborne, radio-controlled cameras suspended from a tethered balloon at a constant altitude of 120 m (Sonu, 1969). Dye which had been injected into the water was tracked through the use of time-lapse photography.

Where information on tidal-current direction (apart from speed) is adequate for the purpose at hand, very useful characterization of the nearshore flow field (see Waste Outfalls) may be obtained with anchored buoys that release dye detectable from the air. Figure 28-16 shows a sequence of tidal-flow diagrams for a section of shoreline near Grandview, Virginia containing a notable prominence at the center. Twelve fluorescein dye-releasing buoys were anchored and photographed hourly from 2,000 ft altitude with 70 mm color film. Ebb and flood tide sequences revealed that the prominence is at the central position of a node in the flow field. The node migrates along the beach with every change in the tide (Munday et al., 1975).

e. Bay and River Effluents

Bay and freshwater effluents from bay and river mouths spread and diffuse into ambient marine water, controlling outflow patterns and determining the pattern of sediment deposition in shoals, bars, and various deltaic forms. The effluent and ambient water masses normally differ significantly in several parameters, especially salinity, temperature, and turbidity (Wright, 1970). As a result, various types of remote sensors can be utilized to monitor the temporal and spatial changes in outflow patterns. Color Figure 28-17 shows a color infrared photograph of an area near the Mississippi Delta illustrating the interface between river water and the Gulf of Mexico. Such images were analyzed by Huebner (1971) to plot the mixing and trajectory of the outflow of the Mississippi River. Photographs analyzed in com-

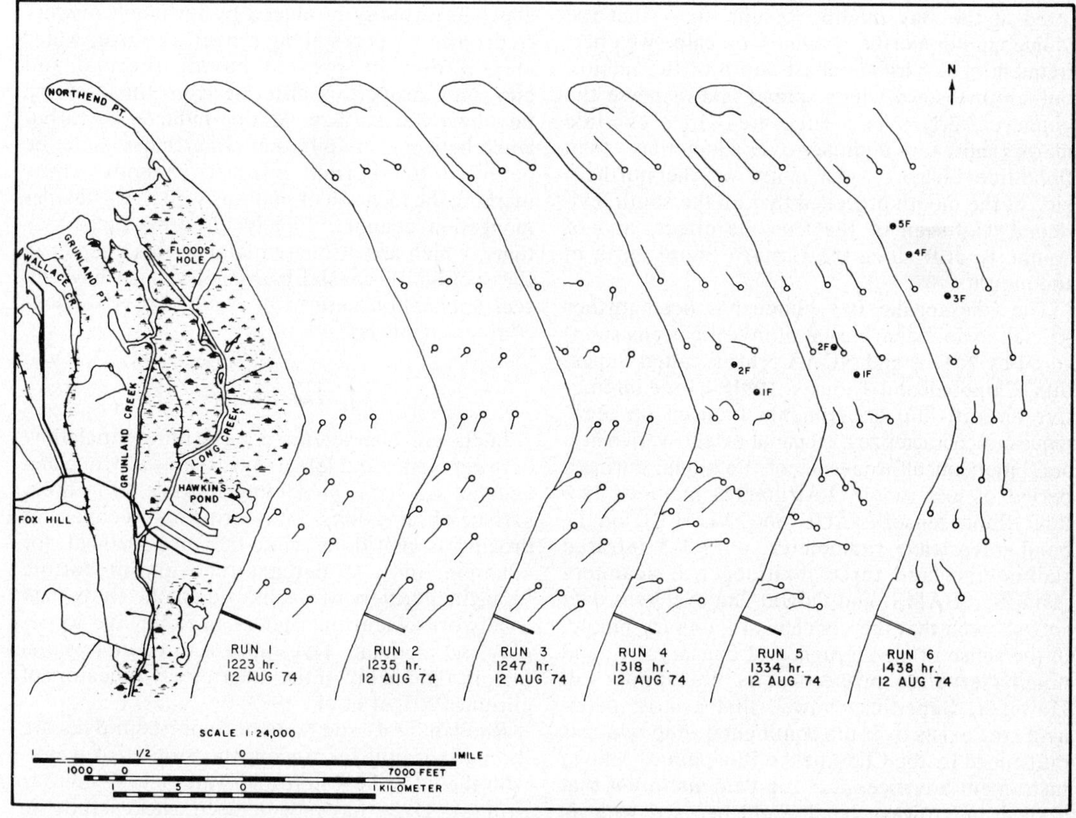

Fig. 28-16. Coastal currents mapped with dye tracks from fixed buoys at Grandview, Virginia. A nodal point progresses southward during ebb tide (Munday *et al.*, 1975).

bination with thermal-infrared scanner imagery are particularly fruitful in a study of mixing. Patterns in the thermal imagery are similar to those on the photographic imagery but, being dependent upon temperature and consequently upon buoyancy differences induced by mixing processes, they indicate not only horizontal but also vertical mixing. The utilization of remote-sensor imagery in studying the hydrodynamics of effluent plumes is discussed by Coleman et al. (1972); Wright et al. (1973); and Wright and Coleman (1971); see also the section on Waste Outfalls in this chapter.

Since these early studies, the use of multispectral scanners for quantitative definition of chlorophyll, suspended matter, and thermal distributions in bay effluents has been demonstrated in the New York Bight (Johnson et al., 1981). A coordinated aircraft-ship experiment was clearly necessary for full exploitation of the remote sensing data, and for accurate mapping of parameter distributions from limited point sampling on the surface.

The dynamics of the Chesapeake Bay plume have recently been studied by examination of 92 dates of Landsat MSS Band 5 images, with digital color-encoded density slicing using a television display-system (Munday and Fedosh, 1981). The plume was mapped from surface-turbidity discontinuities according to a polar coordinate grid centered at the Bay mouth. Results show that the plume, under northerly winds or calm weather, frequents the Virginia coast south of the mouth, but southwestern winds spread and disperse the plume easterly over a large area. Late ebb-tide images show turbid plumes over a larger area than flood-tide images. Tidal phase on the northern side of the mouth preceded that on the south; evidence was seen of the Coriolis effect, and of southerly drift along the Eastern Shore north of the mouth.

The Chesapeake Bay plume has been further studied in an extensive set of investigations sponsored by NASA and NOAA/NMFS called Superflux (Campbell and Thomas, 1981). Three interactive aircraft-ship experiments focused on techniques to characterize the spatial extent, variability, and biochemical properties of the plume during a period of low runoff. Instruments included two lidar fluorosensors (AOL and ALOPE), an L-band microwave radiometer, a PRT-5 infrared radiometer, and three multispectral scanners (MOCS, TBAMS, and OCS). The excellent data sets showed that the concept of a unitary plume, in the sense of a congruence of conservative and nonconservative properties, is not supported. However, Superflux showed that a large definable area exists over the continental shelf which is influenced by the Chesapeake Bay plume. Among instrument advances was the demonstration that a single microwave band could be used with an infrared radiometer to map salinity and temperature.

Koopmans (1971) has shown examples of the use of sequential multispectral photographs to measure quantitative changes in the morphology of river mouths. For such dynamic areas as beaches, inlets, and river mouths, measurements of change are essential to engineering plans for harbors, navigation improvements (Color Figure 28-18), flood control and land reclamation, and shoreline mapping. Wright et al. (1972) made use of various remote-sensing techniques in research on the stratification and bed-form characteristics in a tidal inlet (see also Shoreline Change).

Although large-river systems are used for transportation, irrigation, and recreation, many of the rivers are poorly charted. The relationship between channel-bottom topography and water-surface patterns can be qualitatively used to aid navigation on these rivers. Various remote-sensing techniques, such as low-angle oblique photography, nadir color-photography, and thermal-infrared scanner imagery, can provide information on surface-water patterns that are related to major features in channel topography. Studies on the Brahmaputra River in Bangladesh (Coleman, 1969) and the Mississippi River (Coleman and McIntire, 1971) have shown that water-surface roughness, turbidity contrasts, and turbulent boils can be correlated with channel topography. Vertical color and color infrared photography reveals patterns that result from differences in turbidity produced by turbulent mixing. In deep scour pools along a river's course, water masses may be present having thermal and buoyancy properties differing from those of the flowing water surface. Mixing induced by turbulence between the different water masses can be detected by thermal infrared scanners, thus marking the location of pools as well as the deeper navigation channel. Finally airborne Lidar systems, which are discussed later in this chapter in the section on coastal bathymetry, also have direct application here in providing cross-sections of river channels.

f. Tidal Inlets

Inlets are characterized by a differential flow between upper and lower layers, plus a strong and agitated sea state, presenting a problem for direct surface observations. Monitoring and recovery of drogues is best done when they are surficial; for example, only 18 percent recovery of bottom (dragging) drifters was achieved on one study, and a network of current-meters is expensive to deploy and maintain. Dyes appear to be more satisfactory than many of the other direct indicators of current (Wright et al., 1972).

Rhodamine-B and sodium fluorescein dyes are the most useful for tracing the motion of water, with the latter best in turbid water (see Estuarine Fronts). Dyes have been used successfully in ground-water transport studies, determination of stream flow as pollutant analogs, determination of estuarine circulation, and definition of longshore currents. For use in tidal inlets, concentrated dye released at the surface will be detectable visually from the air; however, detectability of dyes released at depth will be marginal. Dye leaving a source area becomes more diffuse and less detectable. Releases may be staggered according to various depths, or staggered according to time delays. A multiple-color release may be used to characterize different layers and surface areas.

A new method that shows some potential for measuring an inlet current has been demonstrated in limited experimentation with the Seasat Synthetic Aperture Radar (SAR) by Shemdin et al. (1980). Moving objects are displaced from their true positions due to the Doppler effect in SAR imagery. The magnitude of the Doppler displacement is $\Delta x = yv_y/(v_x - v_a)$ where v_a is the aircraft or satellite speed in the (flight) x-direction, v_x and v_y are the moving-object velocity components parallel and normal to the x-direction, respectively, and finally, y is the instantaneous distance from nadir to the object along the x-normal. By the sign convention, negative v_y produces a displacement opposite to the flight direction. A shear-current on the ocean surface produces a bright band in the SAR image on the side of positive displacement, and a corresponding dark band on the opposite side. By identification with Δx, the widths of these bands reveal the current velocity. For Grays Harbor, Washington, the ebb-tide current determined from Seasat SAR imagery was 1.0 ± 0.4 m/sec, compared to 1.4 m/sec predicted in tidal current tables.

g. *Estuarine Fronts and Circulation*

Remote sensing permits the mapping of the movement of localized estuarine fronts, as shown by Mairs and Clark (1973), Klemas et al. (1974), Klemas and Polis (1977), and Munday and Gordon (1977). Dynamics of fronts and convergence zones can be studied directly with photogrammetry by the proper choice and placement of tracers. Small dye-emitting buoys, for example, have several advantages. Drifting types were discussed in an early treatise by Zdanovich (1964), and fixed buoys by Welsh (1967).

If properly designed, drifting buoys will produce dye patches resolvable in aerial imagery at a scale up to 1:60,000 and have negligible wind drag (Gordon and Munday, 1977). Buoys may be deployed from vessels near fronts, in conjunction with aerial observations communicated to vessel crews by radio contact. Dye streams from anchored buoys reveal the short-term history of current directions near dynamic boundaries: the abrupt disappearance of dye streams pinpoints submergence zones and shows the link between submergence zones and water color-boundaries. Figure 28-19 shows fluorescein dye patches from anchored and drifting buoys in the vicinity of a dynamic convergence zone in Hampton Roads, Virginia.

In turbid water, uranine (sodium fluorescein) dye is the most visible, producing a highly visible green plume. The fluorescence peaks of uranine, rhodamine B, and rhodamine 6G are at 550, 590, and 600 nm, respectively. Wiedemann (1974) provides additional data on the spectral and physical properties of well-known dyes. Uranine can be fabricated into dye cakes by dissolving polyvinyl alcohol (PVA) in a minimum volume of hot water, then adding uranine powder to fix the concentration of PVA at 3–15 percent non-aqueous weight. The water volume controls viscosity for pouring and drying; the percentage of PVA controls the rate of dissolution and visibility: 3–6 percent cakes are easily imaged at 1:60,000 and last 3.5–5 hours. Thus, even on high-altitude photographs, patches of dye emitted by markers can be made visible when the markers and outfall plumes themselves are invisible due to small scale and/or atmospheric haze. Cakes are affixed to surface floats of anchored buoys, to surface-current followers, or to window-shade drogues (Terhune, 1968) consisting of 1 m^2 muslin sails suspended at the desired depth, the latter allowing remote sensing of currents at depth.

Remote sensing of dye-emitting markers to elucidate fronts and flow fields has been used for the siting of sewage outfalls (Figure 28-19) (Munday et al., 1978), study of nearshore dynamics (see Shoreline Change and Nearshore Currents), and measurement of vertical profiles of current in dredged shipping channels (Munday et al., 1980). In such applications, standard photogrammetric methods for determination of pathways of Lagrangian current followers are used, including digitizer measurement of buoy or dye coordinates on film, photogrammetric resection, and computer plotting of trajectories with or without interpolation.

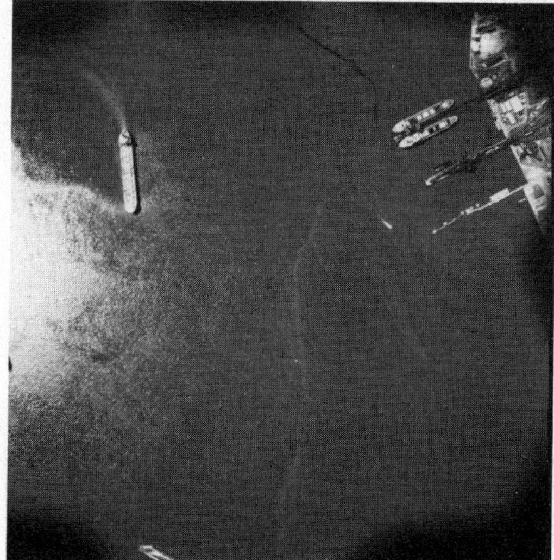

Fig. 28-19. Convergence zone and sewage outfall enhanced by dye buoys. Hi-speed Ektachrome, altitude 1500 m, September 9, 1974, Newport News, Virginia.

Over large areas, the main physical properties of interest in estuaries and bays are water movement, mixing processes, and the salinity and temperature distributions. The movement of sediments (in suspension or by saltation) is related to these conditions. In addition, basic physical parameters affecting flow and circulation include morphology, river inflow, and tidal dynamics. A typical circulation analysis will collate information on these parameters with the *in situ* collection and analysis of horizontal and vertical current, salinity, and temperature profiles, and possibly turbidity data. Inferences can be drawn from estuarine cross sections; however, more spatially comprehensive mapping provided from remote sensing would be helpful.

Satellite imagery and photography have been employed to study water mass circulation in the Delaware Bay and nearby coastal waters using turbidity (suspended solids) as a tracer (Klemas et al., 1974). Amos (1976) and Amos and Alföldi (1979) have also used suspended solids patterns derived from Landsat digital data to evaluate circulation patterns, sources of sediment, and sediment-transport budgets. The measurement of suspended-solids distributions and concentrations themselves are discussed under Suspended Inorganic Particulates, and estuarine circulation is discussed further under Water Quality and Circulation Models.

4.0. Sea State

a. Sea State and Wind

Remote sensing of the oceans, with passive and active microwave sensors, has been concentrated on the remote detection of surface wind-speed and wave conditions. The complex problems of sea state determination addressed here involve various electromagnetic interaction mechanisms dictated by the relative roughness and dielectric properties of the water surfaces. Theoretical models will be mentioned where helpful for describing the nature of the phenomena to be observed and the reported experimental measurements.

The outstanding feature of the sea's appearance is the presence of near-periodic waves caused by wind action. Wind waves often outlast or propagate into an area away from the wind system that generated them. Waves no longer driven by the wind, and in the process of decaying, are termed "swell." Swell is more regular and more even than sinusoidal wind waves and returns considerably less radar backscatter.

The restoring force of both wind waves and swell is gravity. The period of gravity waves may vary from about 1 to 30 seconds. The very small waves that are controlled instead by surface tension are termed "capillary waves"; they generally have periods less than about 0.1 second and wavelengths less than 2.5 cm. Waves of longer period and wavelength but for which surface tension cannot be neglected are termed "ultra-gravity" or "gravity-capillary" waves.

There is no simple correlation between large-scale surface roughness (gravity waves) and wind speed. Any given wind distribution is uniform over only a limited region of the sea. The distance over which a substantially constant wind blows is called fetch. For any given fetch, the wave amplitude increases with wind duration, and in time a steady state is reached at which the wave height remains constant and the sea is termed fully developed. Wave characteristics in the equilibrium state are thus a function of wind velocity and fetch. If the wind blows for a shorter length of time than the minimum duration, the wave height does not depend on the fetch but only on the wind velocity and duration. A complicating factor is instability at the sea/air interface, which is related to the sea/air temperature difference. Atmospheric instability results in a more turbulent air flow and a greater momentum transfer to the waves (i.e., wind stress). Some measurements indicate the mean wave height increases with sea/air temperature difference for a given wind speed (Brown, 1953). This rather involved air/sea interaction greatly complicates the problem of correlating radar backscatter with wind and large-scale surface roughness or sea state.

Oceanographers frequently characterize sea surface conditions with reference to standard scales such as the Douglas Scale or the Beaufort Number. Although several radar-measurement programs have characterized existing surface conditions with these scales, such simple scales do not provide adequate information for radar backscatter analysis. A preferred method of expressing the statistical properties of the sea surface is by the surface-energy density spectrum or surface power-spectrum. Figure 28-20 shows an example of a one-dimensional spectrum for fully developed seas that relates the square of the wave heights to the wave frequency. The maximum spectral energy tends toward lower frequencies with increasing wind speeds. The transition between gravity waves and capillary waves is normally assumed to be at 13.5 Hz, which corresponds to a wavelength of 1.7 cm. The phase velocity of capillary waves decreases with increasing wind speed, which is opposite to the behavior of gravity waves. Unfortunately, the difficulty in obtaining an accurate power spectrum for capillary waves causes most measured power spectra to be dominated by the larger gravity waves. However, radar backscatter is strongly dependent upon the small-scale surface roughness; consequently the available spectra are often only partially satisfactory for correlation with radar-backscatter measurements.

Perhaps because oceanographers are most often concerned with the high-energy gravity waves and less attention is given to the overriding ripples, the first models advanced to explain sea radar-return were erroneously related to these larger waves.

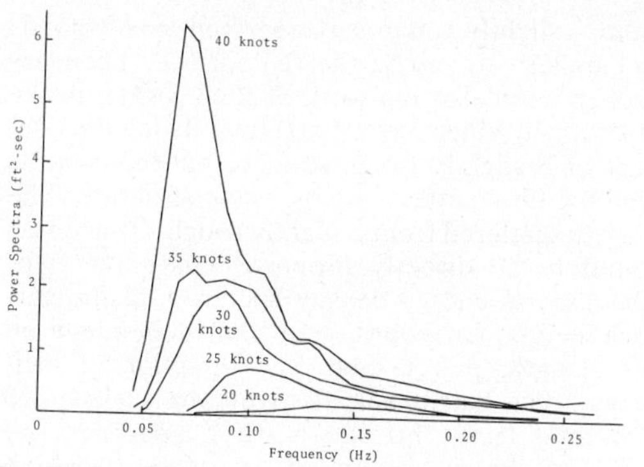

Fig. 28-20. Ocean wave power spectra (Moskowitz, 1964).

Now that the dominant influence of capillary waves at microwave frequencies has been recognized, several theoretical models have been advanced that offer plausible explanations for various aspects of the behavior of backscatter from sea surfaces. Nevertheless, the relationship between the sea surface and microwave backscatter is not yet well defined.

b. Radar Backscatter from the Sea Surface

The objective of active remote sensing of the ocean water-surface is to determine wave amplitudes and spectra, and surface-wind velocities and directions. As discussed, these are not mutually exclusive properties of the surface; this interdependence is itself a subject of extensive research. At the same time, many investigations of radar backscatter from a water surface have been made since World War II in an effort to determine the effects of surface roughness on radar return. Much of the earlier data are of only illustrative value, since the procedures varied from experimenter to experimenter and ground truth was sketchy or inferred. Goldstein (1946), who was perhaps the first to systematically study the phenomenon in a quantitative way, presents a thorough discussion of the results of these earlier investigations. Goldstein introduced the dimensionless quantity σ°, average radar cross section per unit area, as a scattering coefficient, which is the standard unit of measurement for radar backscatter from the sea surface. He found that: (1) there is pronounced polarization dependence of σ°, which depends on the roughness of the sea. In calm seas, horizontal polarization gives less return than vertical polarization; however, the difference between the two decreases in rougher seas. (2) There is a "critical incidence angle". Near vertical angles of incidence (0°), σ° drops slowly with angle. Beyond the critical angle, σ° decreases rapidly with increasing angle. Furthermore, the critical angle increases with decreasing radar wavelength. (3) The radar wavelength dependence of σ° varies between about λ^{-4} in calm seas to about λ° in rough seas, where λ is the free space wavelength of the incident radiation.

Measurements made by Macdonald (1956) and Wiltse et al. (1957) at various polarizations tended to agree with Goldstein's determinations. In addition, Macdonald showed that radar backscatter for vertical polarization decreases with wind direction in the following order: upwind, downwind, and crosswind. For horizontal polarization, this order can be interchanged depending upon frequency, angle of incidence, and sea state. Wiltse et al. compared return for various wind directions at a radar beam incidence of 60°, and found that, at this angle, σ° is almost independent of wave aspect for the three frequencies investigated.

Grant and Yaplee (1957), Ament et al. (1959), and Wiltse et al. (1957) studied σ° for a wide range of incidence angles, whereas most earlier measurements were only at grazing angles. Grant and Yaplee showed that, at wavelengths of 8.6 mm and 1.25 cm at angles above 10°, σ° increases with wind velocity. The return at 3.2 cm increased with wind velocity up to 10 to 15 knots, then decreased for higher wind velocities. However, σ° measured by Wiltse et al. (1957) at 3.13 cm, 1.25 cm, and 8.6 mm monotonically increased with wind velocity up to a maximum average velocity of 26 knots. The radar cross-section decreased with wind velocity at vertical incidence (0°) in both studies.

Extensive radar data with well documented ground observations have been reported by Guinard and Daley (1970), Krishen (1971), and Newton and Rouse (1972). Some of these data were recorded using multifrequency and dual polarized radar systems, and all were over a wide range of incidence angles and surface conditions. Data at P (75 cm), L (18.7 cm), C (6.3 cm), and X-band (3.3 cm) reported by Guinard and Daley appear to "saturate" for wind velocities in excess of 10 knots; however, data recorded at the higher frequencies varied more with wind speed than did the lower frequency data. Moore and Pierson (1971) showed that K-band (2.25 cm) returns do not saturate even for wind velocities as high as 40 to 50 knots, whereas L-band returns do not increase with wind velocity. The X-band measurements reported by Guinard and Daley showed the same trend as Moore and Pierson's K-band data, but with much less sensitivity to wind velocity.

Although the scattering coefficient depends, as seen above, on polarization, frequency, wind speed and wind direction, the most applicable function of σ° is its relation to the incidence angle. This relationship has generally three distinct regions: the quasi-specular region, the plateau or diffuse region and, beyond the critical angle, the interference region (Figure 28-21).

The quasi-specular region includes angles near normal incidence. In this region the radar cross-section is large, due primarily to specular reflection from facet-like surfaces oriented normally to the incident radiation. Since the maximum in-

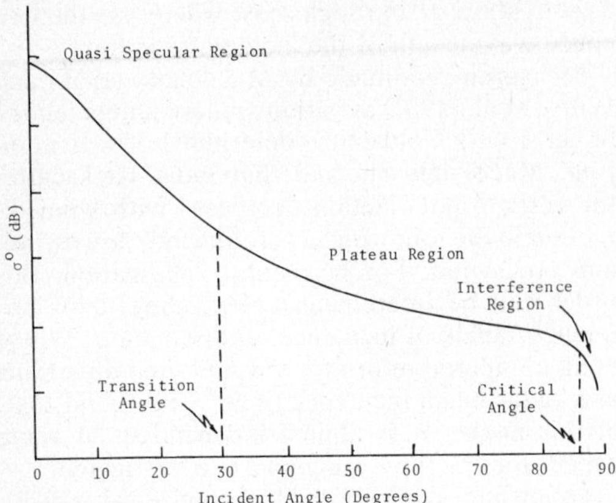

Fig. 28-21. Representative variation of radar $\sigma°$ with incident angle (Skolnik, 1969).

cluded angle at the crest of a wave can only reach approximately 120° before the wave breaks, there is a certain incidence angle above which specular return from near normal facets is very improbable. This angle is called the transition angle and is usually difficult to distinguish.

At larger incidence angles, where $\sigma°$ is relatively flat with angle, the curve displays a plateau or diffuse condition. Radar return measured at frequencies at X-band and higher, at incidence angles in the plateau region, have the greatest correlation with wind speed since return at the higher frequencies is strongly dependent on the small-scale surface structure of the sea. The scatterers in this region are mainly elements of the sea structure that have dimensions on the order of the incident-radiation wavelength and ride atop the larger wave structure. This region of the curve usually has a greater slope for horizontal polarization than for vertical polarization. The quasi-specular and plateau regions contain the range of incidence angles most useful for sea state and/or wind velocity studies from aircraft and satellites.

The plateau region and the interference region beyond it are separated by the critical angle as defined by Goldstein (1946). The critical angle is dependent on frequency, polarization, and sea state. Beyond this angle, $\sigma°$ decreases rapidly with increasing angle.

c. Scattering Models

Katzin (1957) visualized the sea surface as composed of randomly oriented facets of various sizes divided into two groups, large and small, relative to the incident wavelength. These groups were roughly analogous to gravity and capillary waves. The model employs a realistic physical picture of the surface but does not account for depolarization and the upwind-downwind ratio.

Wright (1968) and Bass et al. (1968) independently derived a theoretical relation for scattering from a slightly rough surface which corresponds to capillary waves on the sea surface. Their approach paralleled the work of Rice (1951), Peake (1959), and Valenzuela (1967) by utilizing the concept of a slightly rough surface, but included as well the tilting effect of large wave structure. The energy scattered from a slightly rough surface was found to be directly proportional to the two-dimensional energy density spectrum of the surface height variations, and also dependent on wavelength, polarization, and direction of incidence. The height variations on the surface are obtained from the ocean wave spectrum by utilizing the wave-number components which satisfy the Bragg scattering condition and lie in the equilibrium range of the spectrum as defined by Phillips (1966). The model predicts an upper bound to $\sigma°$ with increasing roughness. Guinard and Daley (1970) verified the predicted saturation using experimental data.

A detailed composite-surface model was developed by Fung and Chan (1969) utilizing the mathematical approach known as the Kirchhoff method to handle large waves vs. the small-perturbation method to handle small-sized scatterers. The advantage of this model is that the Kirchhoff method is applicable to the quasi-specular region, whereas the small-perturbation method best describes the diffuse region. Although its specific development has been questioned, the basic model advanced by Fung and Chan enables direct incorporation of the current work on definition of the frequency dependence and depolarization mechanism. Krishen (1971) also used a composite theory to describe experimental data. Krishen calculated the theoretical backscatter from large waves using the Kirchhoff method. Backscatter from the small-scale structure was calculated separately, such as had been done by Guinard and Daley. Krishen then added the average incoherent-scattering cross sections from the large-scale structure to that of the small structure to obtain composite-scattering cross sections. Numerous other models describing radar backscatter from the sea have been developed, but the above embody the basic approaches most often employed in addressing this problem.

d. Sea State from Airborne Radar

Many studies to determine characteristics of backscatter from the sea have been directed toward improved target detection on or near the sea surface. However, the goal of sea-state measurement necessitates a different experimental approach, as demonstrated in early measurements of surface winds conducted by the Naval Research Laboratory (NRL) and NASA.

NRL operates an airborne coherent-pulsed radar which is capable of transmitting a sequence of four frequencies, X-band (8.910 GHz), C-band (4.455 GHz), L-band (1.228 GHz), and P-band/UHF (0.428 GHz). At each frequency the system alternates the polarization of the transmitted signal between horizontal and vertical, receiving both the direct and cross-polarized backscatter. There are two pairs of antennas mounted back-to-back. One pair, the X- and C-band antennas, are circular parabolas with common boresite and equal azimuth beam-widths of 5°. The other pair, the L- and P-band antennas, are intermixed crossed dipole arrays with common boresite but unequal azimuth beamwidths, 12° for P-band and 5.5° for L-band. Because of the narrow-elevation beamwidth of the antennas, the angular data are recorded on multiple passes at three different altitudes.

Two experiments conducted by NRL are of particular interest. One was performed in July 1965 over low and moderate sea states in the vicinity of Puerto Rico. This study, reviewed by Guinard and Daley (1970), was initiated to verify the composite surface-model of Wright (1968) and Bass et al. (1968) over a wide variety of sea conditions. It indicated the need to measure sea return in high sea state conditions where nonlinear effects such as spray, foam, and shadowing might limit the model's effectiveness. Consequently, measurements were then made over an area of high seas in the North Atlantic Ocean in February 1969. In a study of both sets of measurements, Guinard et al. (1971) investigated the variation of radar cross section with wind velocity. Upwind X-band returns from the 60° incident angle for both polarizations showed the largest growth of radar cross-section with wind speed (Figures 28-22 and 28-23). Therefore, these particular parameters provide a conservative estimate of the location of the "saturation region" and, simultaneously, optimum conditions (within the limits of the NRL system) for wind-speed sensing. In the figures, two power laws were used to fit the data, a cube law for low wind speed and a square-root law for the higher wind speeds. The first domain extends up to a wind speed of about 10 knots; above that speed σ° rapidly approaches a saturation value.

Early measurements by NASA were made with a vertically-polarized K-band (13.3 GHz) radar

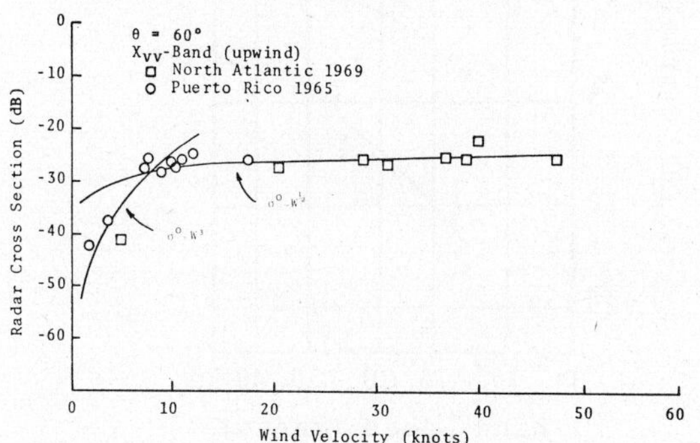

Fig. 28-22. Variation of radar cross section (X_{vv}) with wind speed (Guinard *et al.*, 1971).

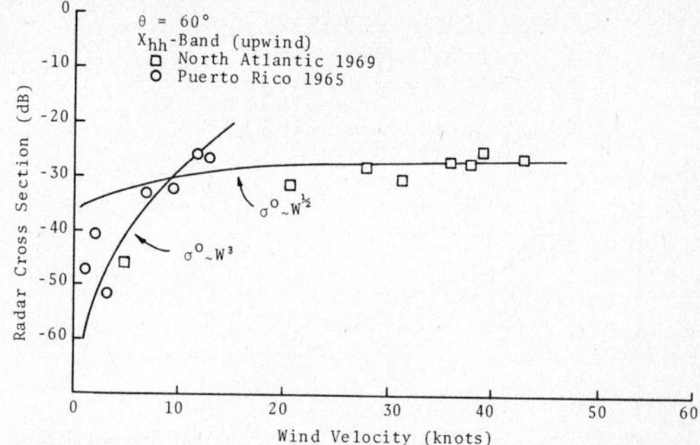

Fig. 28-23. Variation of radar cross section (X_{hh}) with wind speed (Guinard *et al.*, 1971).

system in March of 1968 and 1969 over the North Atlantic (Moore and Pierson, 1971). Measurements of σ° (normalized to 10° incidence) taken in the upwind direction as a function of wind speed (Figure 28-24) indicate that σ° increases with increasing wind speed for all angles above 10°. Figure 28-25 shows that crosswind values of σ° are less than upwind values and appear to vary differently with wind speed. Figure 28-26 shows that σ° continues to increase up to the maximum speed observed.

NASA also recorded sea-clutter measurements near Argus Isle, Bermuda, in Feburary 1979; over the North Atlantic in February 1971; and later in the radar scatterometer (RADSCAT) program (Jones et al., 1977). In addition, Texas A&M University directed a measurements program in the Gulf of Mexico in February 1971 using a 13.3 GHz radar similar to the NASA system. These 1971 data were analyzed by Newton and Rouse (1972) and compared with previously reported data (Figure 28-27). A significant conclusion was that, from the variance of σ° for any single wind condition,

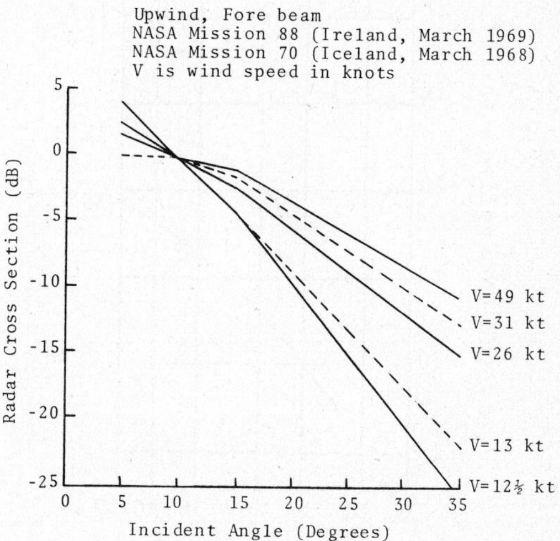

Fig. 28-24. Differential scattering coefficient of ocean at 2.24 cm normalized to 0 db at 10° (Moore and Pierson, 1971).

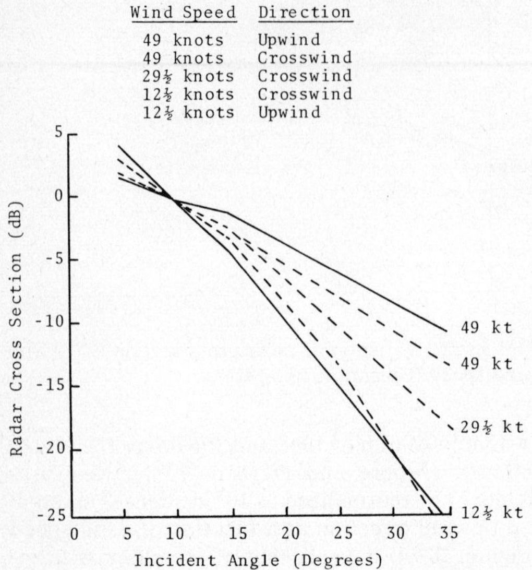

Fig. 28-25. Comparison of crosswind and upwind data, NASA Mission 88, Ireland, March 1969 (Moore and Pierson, 1971).

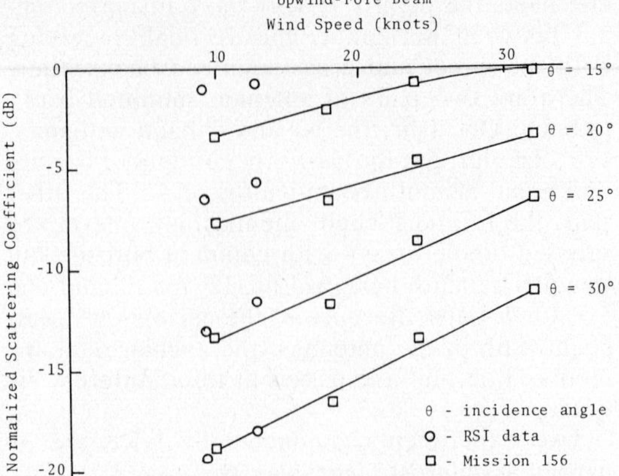

Fig. 28-27. Upwind normalized scattering coefficient measurements plotted against wind speed. NASA Mission 156 (Newton and Rouse, 1972).

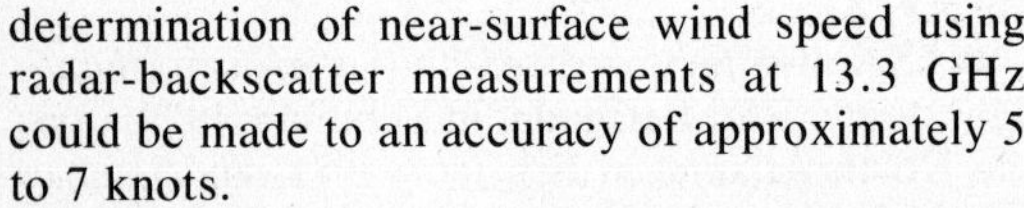

determination of near-surface wind speed using radar-backscatter measurements at 13.3 GHz could be made to an accuracy of approximately 5 to 7 knots.

Subsequently, a program was conducted at Marineland, Florida, in 1975 to derive wave spectra *per se,* instead of surface winds, from radar data. McLeish et al. (1980) used an X- and L-band airborne synthetic aperture radar to derive wave spectra, and compared results with wave spectra obtained from *in situ* devices. The latter included a pitch-roll buoy, orthogonally oriented current meters and pressure transducers on a sea sled, and an airborne laser wave profilometer. The mean wave direction from the radar data was in

WIND SPEED DEPENDENCE OF σ_0

NASA Mission 70 (Iceland 1968) and 88 (Ireland 1969)

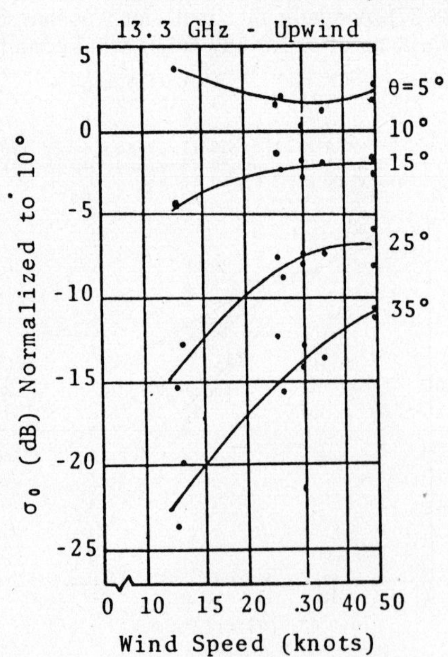

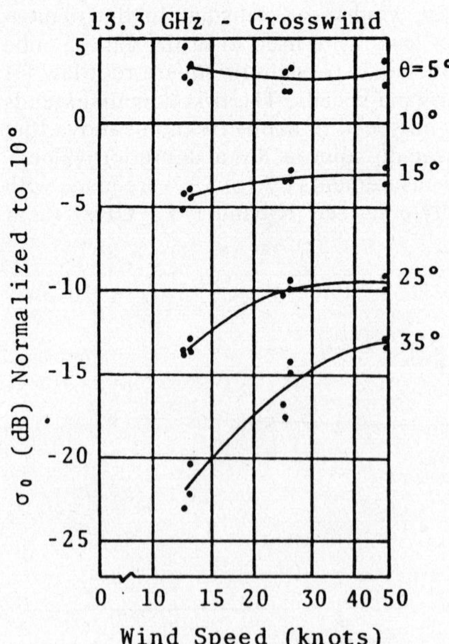

Fig. 28-26. Normalized radar scattering coefficient versus wind speed and direction at 13.3 GHz (Moore and Pierson, 1971).

good agreement with reference measurements, but radar-derived direction was more sharply peaked. Maxima of radar spectra agreed with maxima of reference wave-height spectra, but slope of the radar spectra at high frequency significantly differed from observed spectra of wave height, orbital velocity and surface slope.

e. Seasat Radar Measurements

Seasat-A, launched on June 28, 1978, was the first satellite dedicated to establishing the utility of microwave sensors for wide-area synoptic measurement of the Earth's oceans (Born et al., 1979). Objectives included measurement of sea state parameters such as sea-surface winds, wave heights, wavelengths and durations, and surface roughness. Other parameters such as sea-surface temperature measured from Seasat are discussed in other sections.

Primary sensors on Seasat for sea state determinations were the radar altimeter, Seasat-A Scatterometer System (SASS), the Scanning Multichannel Microwave Radiometer (SMMR), and the Synthetic Aperture Radar (SAR). Preliminary results from these sensors have been reported and will be summarized here. Some of these sensors have also been used on other satellites (e.g., the SMMR and its earlier version, the Electronically Scanned Microwave Radiometer on Nimbus 5 and 6). Their data have been used extensively in sea-ice experiments discussed under Sea Ice.

Jones et al. (1979) reviewed preliminary results from the SASS sensor, which was designed to provide synoptic, global, day-and-night measurements of ocean-surface wind vectors. For two operating modes in rain-free conditions, a limited number of comparisons with high-quality surface truth indicates that wind speed was determined to an accuracy of ±2 m/sec, and wind direction to ±20°, in two swaths 500 km wide on either side of the spacecraft. In addition, SASS correctly located high- and low-pressure centers via low winds that were within the swath. An operational capability for accurate location of such pressure centers would be a significant accomplishment, improving numerical weather prediction in data-poor areas.

Surface-wind conditions and sea-surface temperature may also be studied using SMMR data (Lipes et al., 1979). For sea-surface temperature, infrared sensors appear to provide better results (discussed under Sea Surface Temperature). In the case of surface winds, for the open ocean in rain-free cells of highest-quality surface truth, wind determinations gave standard deviations of 3 m/sec about a bias of 1.5 m/sec. The initial statistical evaluation is encouraging, particularly in view of the fact that the algorithms were not developed using Seasat data.

The Seasat SAR measurement of wave and ocean wind waves and swell was tested in a Gulf of Alaska Seasat verification experiment (Gonzalez et al., 1979). SAR ocean-imaging mechanisms have been examined by Raney (1980), Alpers et al. (1981), and Harger (1981). The Seasat SAR had a transmitted wavelength of 23.5 cm and 25 m resolution; gravity waves were imaged best when the waves traveled within 25° of the range direction. Results of the Alaska test were that SAR and surface data agreed to within ±15 percent in wavelength and ±25 percent in wave direction. Waves were 100 to 250 m in length, propagating in a direction predominantly across the satellite track, in sea states with significant wave height ($H_{1/3}$) in a range of 2 to 3.5 m. Image data were analyzed by optical Fourier transforms (OFTs) as in Figure 28-28.

In an investigation of SAR data for the east coast of the United States, Beal (1980) traced a well-organized, very low energy ocean-swell system from deep water across the continental shelf and into shallow water, and demonstrated the consistency of the SAR data with a comprehensive set of surface and aircraft measurements (Beal, 1981). Figure 28-29 shows this progression as revealed in a series of two-dimensional Fourier transforms, and in Figure 28-30 the composite data are compared to the location of the apparent wave generation source. Shuchman and Kasischke (1981) found reasonable agreement in further study between SAR-derived wavelength and direction data and wave-refraction model predictions off Cape Hatteras. A more detailed study is underway using the wave-climate model of Goldsmith et al. (1974). Early results indicate good agreement, even in the area known as Diamond Shoals where mixed seas are common. These results indicate that spaceborne imaging radar can measure wavelength and direction for waves of 100 to 200 m in deep water and coastal areas.

To extract the wave measurements, most SAR data are reduced to image format after processing with an optical correlator. Optically correlated images may be analyzed for wave direction and wavelength via enlargement and optical-visual methods, or analyzed via optical Fourier transforms (OFTs) using a laser apparatus. All-digital correlation, however, is now available for segments of SAR passes via computer processing of magnetic tape records. Digitally-correlated data may be further processed in the digital mode by digital fast Fourier transforms and maximum-entropy spectral analysis (Shuchman et al., 1979). The OFT and FFT results may be read directly for direction and wave number.

f. Optical Determination of Sea State

Badgley (1969) and Soules (1970) described in detail various surface and sub-surface features revealed in color pictures obtained by the astronauts in space. Such pictures obtained from space platforms can be used to infer sea state (Figure 28-31).

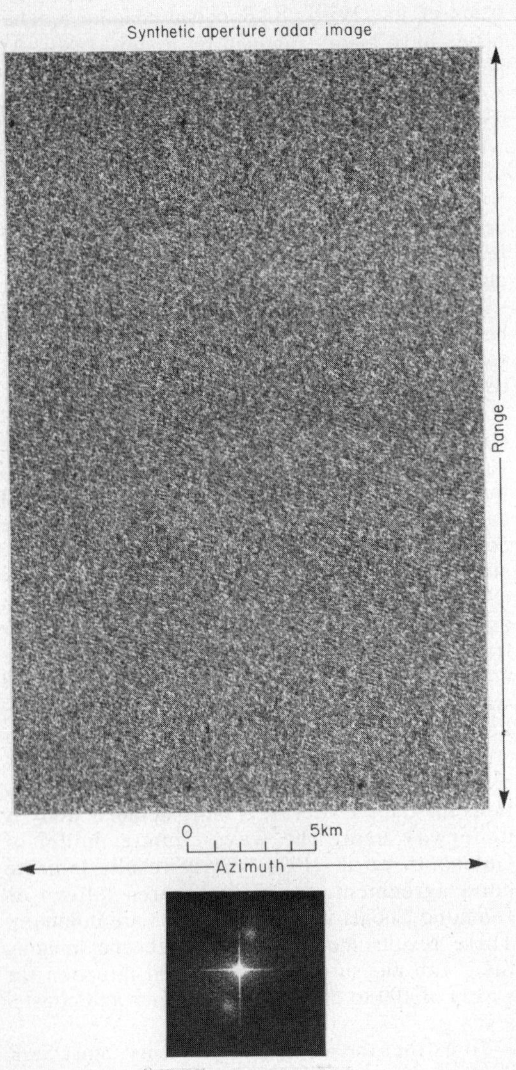

Fig. 28-28. Optical Fourier transform of Seasat SAR image of ocean waves (Gonzalez *et al.*, 1979).

McClain and Strong (1969) used sun-glitter patterns in the study of the near-surface wind field over oceanic areas. Pictures of sun glint from manned and unmanned satellites reveal remarkable variations in the brightness patterns as a function of sea state. Investigations by Cox and Munk (1954a,b), Rosenberg and Mullamma (1965), Duntley and Edgerton (1966), and Zdanovich and Semenchenko (1954 and 1963) showed clearly that (1) the reflectance at the specular point decreases with increasing sea state and (2) the width of the diffuse-reflection region around the specular point increases with increasing sea state.

Some sun-glitter pictures obtained from high altitude aircraft and spacecraft show irregularities (dark and bright patches) in the brightness field. In

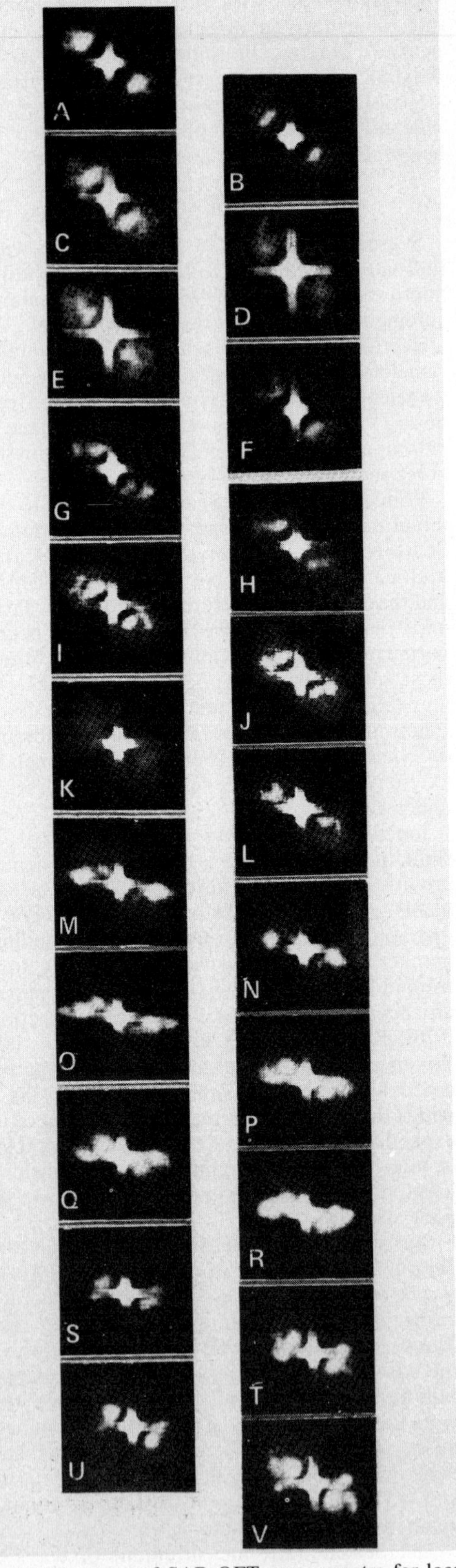

Fig. 28-29. A set of SAR OFT wave spectra for locations A through V on Figure 28-30.

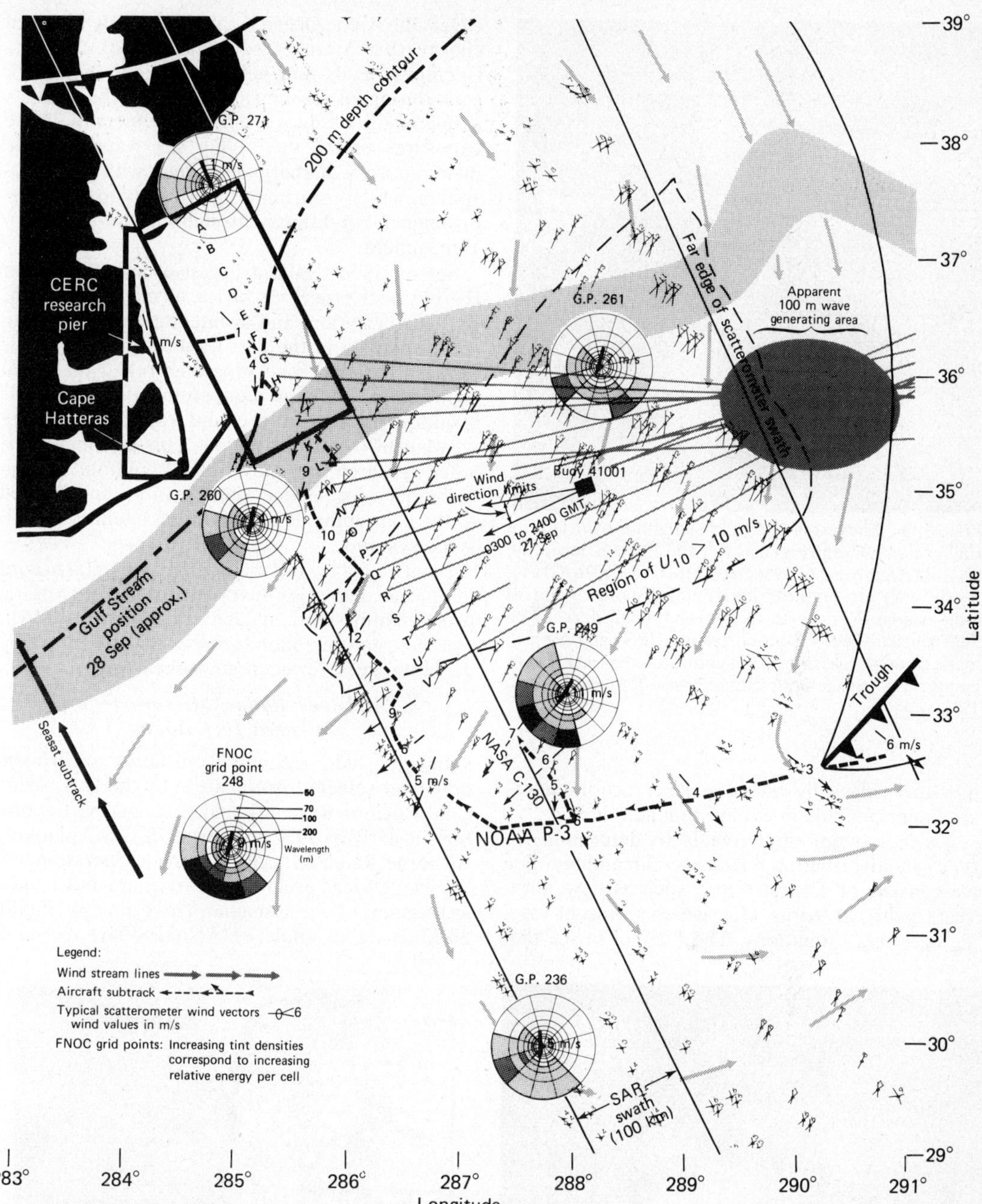

Fig. 28-30. Composite wind and wave conditions for Seasat pass 1339 (September 28, 1978), with spatially evolving wave vectors projected back to their apparent source (Beal, 1981).

Figure 28-32, obtained from an aircraft over the Gulf of Mexico, McClain and Strong (1969) attribute the bright and dark patches in the sun-glitter area to surface slicks. These slicks represent near-calm areas that extend several tens or hundreds of kilometers when the surface winds are very weak. The technique of deducing the surface-wind field from glitter data is not of great accuracy but is useful over very large oceanic areas where other sea-state data are sparse.

5.0. Sea Ice Distribution and Dynamics

a. Sea Ice Monitoring

Several objectives underlie efforts at monitoring sea ice. First, the operational demand for sea ice information is a result of the oil and natural gas exploration and developments in the Canadian Arctic. Remote detection of icebergs by radar and the measurement of the distribution and height of

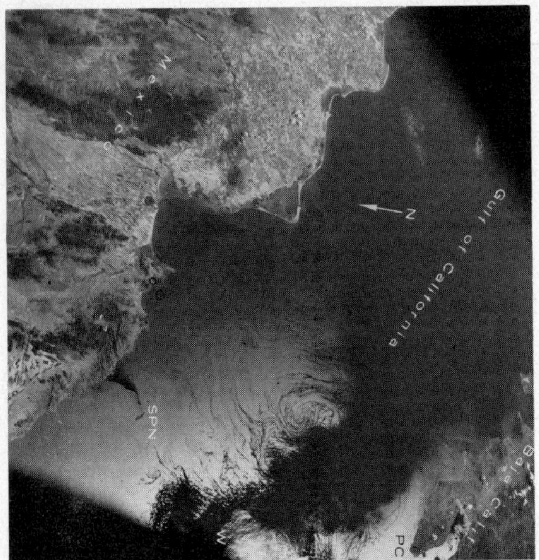

Fig. 28-31. Slicks and eddies in sun glitter over Gulf of California between Guaymas (G) and Punta Conception (PC) on October 13, 1968, at 1236 LST (1936 GMT). Swells (W) appear near horizontal specular point at right. Large dark swath in gulf and small dark patch between San Pedro Nolasco Island (SPN) and Mexico mark smooth sea zones, indicators of two sea-breeze circulations. Space-craft altitude was 226 km (Soules, 1970).

pressure ridges by airborne laser profilometers are important aids in exploration activities.

Another major objective is to determine ice types and distribution relative to climatology. Sea ice consists of several types, such as new, first-year, multiyear, bergs, etc., which represent varying ages and formations. The blanket of ice that covers the Arctic Ocean is a dynamic body of stored energy that varies greatly from year to year. Openings (leads and polynyas) occur in the ice pack during all seasons as a result of the stresses acting on the ice due to winds and currents. These exposures of the underlying sea water provide major sources of energy exchange with the atmosphere, which contribute significantly to the meteorological conditions throughout the Northern Hemisphere.

Sea-ice investigations are also concerned with the physical processes of ice formation, growth, drift, deformation, and disintegration. These data were obtained in the past by airborne reconnaissance programs; such programs relied on visual reconnaissance and consequently were constrained by the severe and rapidly changing weather of the Arctic region. This problem motivated research into more reliable remote sensing techniques for sea-ice identification, and has led to the continuing development of satellite surveillance systems.

Finally, sea ice is a major environmental hazard to marine activities involving offshore exploration, fishing, and marine transportation. All-weather real-time monitoring programs are a fundamental need for accurate forecasting.

b. *Airborne Radar, Microwave, and Infrared Techniques*

Imaging radar systems were flown experimentally over selected Arctic areas by the U.S. Army Cold Regions Research and Engineering Laboratory as early as 1957 and, by 1978, a Side-looking Airborne Radar (SLAR) was made operational in the ice-services program of Environment Canada, an Agency of the Canadian Government. Figure 28-33 is an example of AN/APS-94D (X-band)

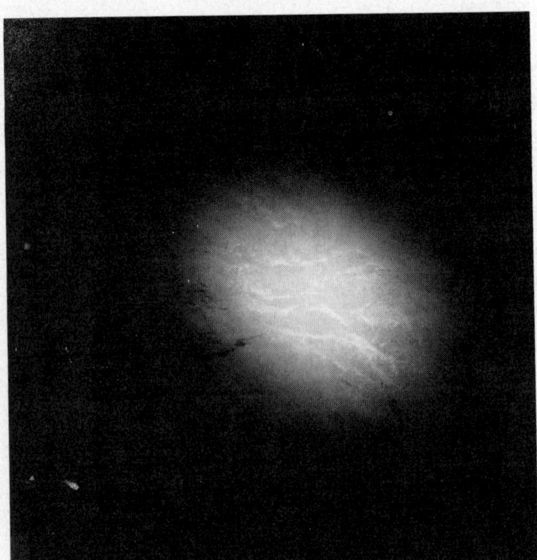

Fig. 28-32. NASA Convair 990 photograph (10 km altitude) of the Southern Gulf of Mexico at 1838 GMT on June 5, 1967 (McClain and Strong, 1969).

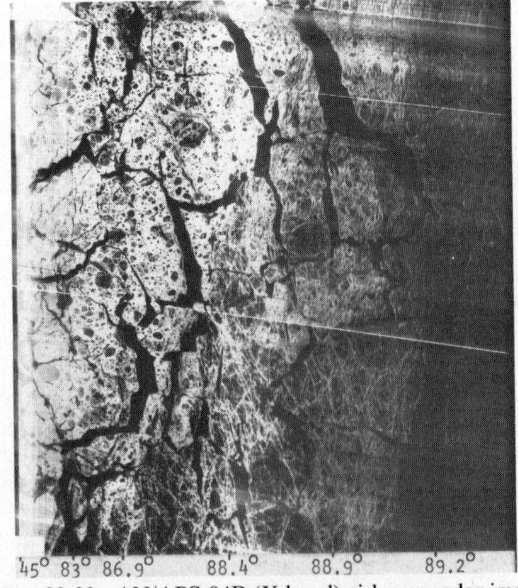

Fig. 28-33. AN/APS-94D (X-band) airborne radar imagery of the Beaufort Sea ice pack from 1100 m (Hengeveld, 1980).

radar imagery obtained over the Beaufort Sea ice pack at an altitude of 1100 m (Hengeveld, 1980). Regions of high radar return in the image are due to the rougher surfaces of multiyear ice. The darker regions are first-year ice containing rough and smooth regions, as well as ridges.

Much research has been directed towards sea-ice classification using passive and active microwave systems (radiometers, scatterometers, SLAR, and SAR). Rouse (1969) used 13.3 GHz (2.25 cm) VV-polarized airborne fanbeam-scatterometer data and established that the age and formation of sea ice strongly influences backscattered energy. The angular variation of the backscattering coefficient defined a unique signature for open water, smooth 1 m thick first-year (FY) ice, rough FY ice, and 4 m thick multi-year (MY) ice. First year ice contains salinities of 4–10 parts per thousand while brine in older ice dissolves out during summer melts: these changes affect the dielectric constant and the crystalline ice structure, leading to the variations in microwave response. The dependence of backscatter on ice type is illustrated in Figure 28-34 obtained from north of Barrow, Alaska, in the Beaufort Sea. Point (B) in the air photo denotes a boundary between ridged and smooth FY ice, (D) a boundary between smooth FY ice and MY ice at an open water crack, (I) a boundary between MY ice and smooth FY ice, and (K) a high-pressure ridge dividing two sections of smooth FY ice. From the profile shown at 25° incidence angle there is little differentiation between the area of pressure-ridged first-year ice labeled (K) and the old ice region (D through I). However, the old ice is clearly delineated utilizing data from the other angles. Analyzes of these data by Rouse and Schell (1970) showed that, through the use of various statistical classification methods, distinct clustering was achieved, indicative of highly accurate differentiation of ice type using a scatterometer. Parashar et al. (1977) tested a Bayesian classifier, based on the complete angular dependence of σ°_{VV}. Seven ice types were used as classes, corresponding to water, nilas, grey, grey-white, smooth FY, rough FY, and MY ice. The full set could be classified to 60% accuracy, and a coarse subset to 90%, consisting of water, thin ice, FY, and MY ice. Cross-polarized signatures were found by Gray et al. (1977) and Onstott et al. (1979) to yield superior ice signature separation. Further study of ice discrimination, using SLAR imagery, has been reported by Johnson and Farmer (1971), Loshchilov et al. (1978), and Hengeveld (1978).

Similar sea-ice identification potential has been noted with passive microwave sensors. Wilheit et al. (1972) and Gloerson et al. (1973) showed that new and old ice could be distinguished on the basis of emissivity between 10 and 60 GHz. Meeks et al. (1974) used dual-polarized microwave radiometers at 37 GHz to resolve open water, FY ice, second-year (SY) ice, MY ice, and refrozen melt ponds. Campbell et al. (1975) further defined limits of radiometric classification. Troy et al. (1980) have shown that multifrequency measurements can distinguish ice classes provided that the frequency spread is sufficiently great.

The joint USA/USSR Bering Sea Experiment (BESEX) conducted in February-March 1973, during the maximum ice extent, acquired the first detailed passive microwave data of the Bering Sea ice pack. Aircraft microwave imagery was interpreted using simultaneous surface-truth observations of ice type, concentration, metamorphosis, and motion. From March through April 1975, the Arctic Ice Dynamics Joint Experiment (AIDJEX) explored the utility of airborne microwave sensors for discrimination of the spatial and temporal characteristics of sea ice (Campbell et al., 1978). Results from microwave and infrared radiometers, imaging radars, a scatterometer, and mapping cameras were utilized as in Table 28-5. For the first time remote sensing was accomplished for a chosen large area (10^4 km^2) of Arctic Basin sea ice during all seasons of one year as the ice deformed, metamorphosed and drifted.

Characteristics of radar return from various ice parameters are very dependent on geographical location and icepack history. Therefore, interpretation of images and backscatter-coefficient records must be undertaken with detailed knowledge of the region in mind. The difficulties of establishing ice class signatures that may be extrapolated from one region to another have encouraged the development of classification methods involving a combination of scatterometer and radiometer measurements. During the Surveillance Satellite (SURSAT) Program of the Canadian government, the Convair 580 of the Canada Centre for Remote Sensing gathered nadir photography, 19.4 GHz (1.55 cm), H-polarized radiometer measurements at an incidence angle of 45°, and 13.3 GHz (2.25 cm) HV- and HH- polarized fanbeam scatterometer measurements at angles of 0–60° over the Beaufort Sea and the eastern Arctic area of northern Canada. The resulting data were analyzed to compile statistics on 11 World Meteorological Organization classes of sea ice, ranging from calm open water to MY ice and ice islands. Microwave signatures produced by this analysis are presented in Table 28-6. Figures 28-35 and 28-36 show the relations obtained between like- and cross-polarized backscattering coefficients and emissivity (Hawkins et al., 1980).

c. *Satellite Techniques*

Until recent years the primary source of ice data for forecasting was aerial reconnaissance. Satellite imagery was used only in climatology, due to its being unavailable in real time. Now, however, visible and infrared imagery is available daily from polar orbiting satellites, including NOAA, TIROS-N, and Landsat. McClain (1980)

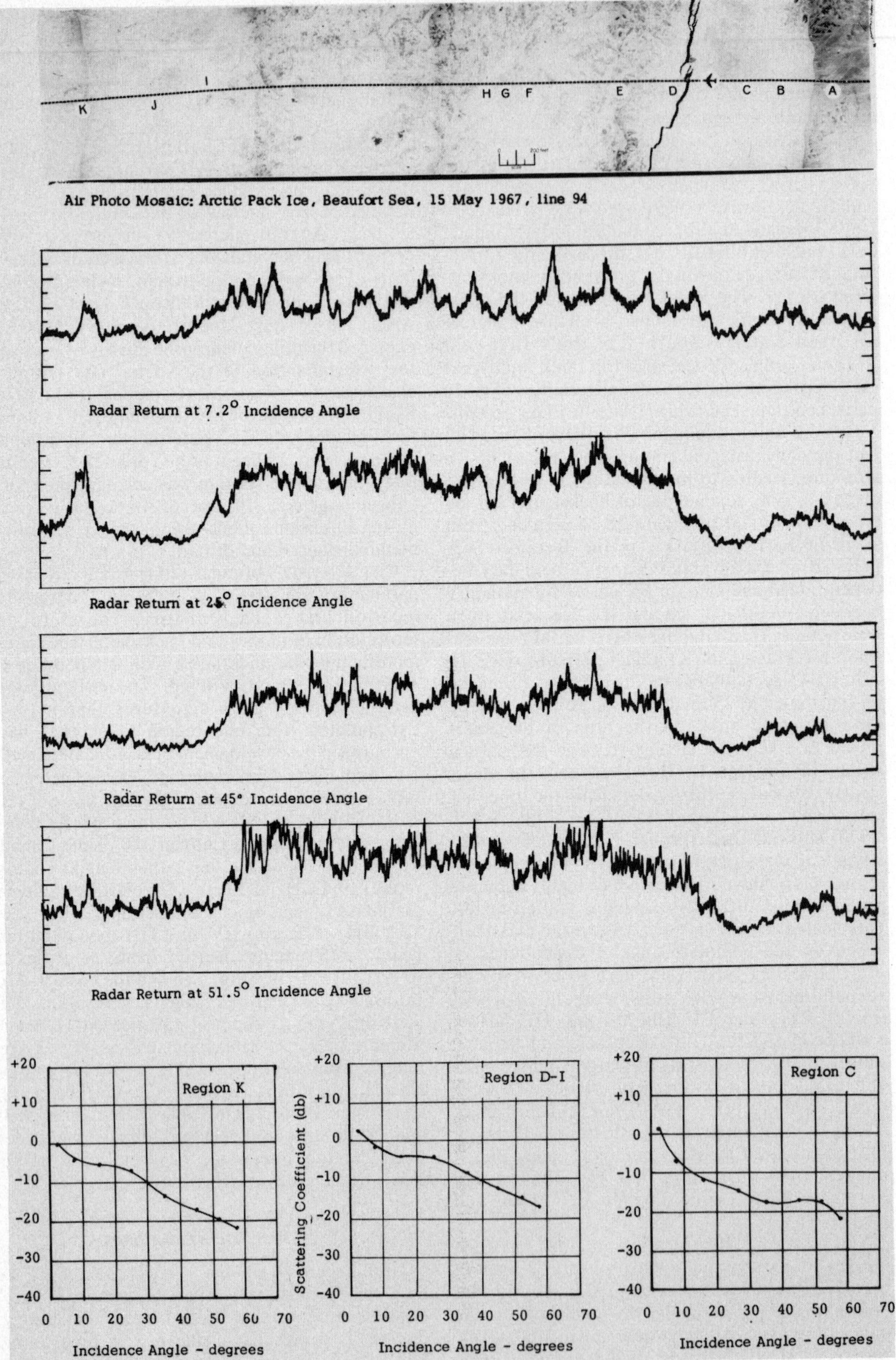

Fig. 28-34. 2.25-cm radar measurements of sea ice in Beaufort Sea, with photo mosaic for location reference (Rouse and Schell, 1970).

TABLE 28-5

Remote Sensing Platforms and Sensors in the AIDJEX Main Experiment (Campbell et al., 1978)

Sensor/sensor platform	CV 990, NASA	C-47, CCRS	Argus, DND	Flextrack
Radiometer	Aerojet	Aerojet		Aerojet
frequency	19 GHz	37 GHz		4.99, 13.4, 37 GHz
wavelength	1.8 cm	0.81 cm		6.0, 2.23, 0.81 cm
incident angle	nadir, ±55° starboard and port scanning	45° forward		H, V all three, nadir to 55° forward
resolution	500 × 500 m	15 × 15 m		1.5 × 1.5, 0.7 × 0.7, 0.4 × 0.4 m
Imaging Radar	JPL		Motorola AN/ADS94D	
type	synthetic aperture		real aperture	
frequency	1.215 GHz		9.2 GHz	
wavelength	24.5 cm		3.25 cm	
polarization	HH		HH	
incident angle	0–55° starboard		45–88° starboard and port	
resolution				
range	25 m		30 m	
swath	14 km		25 km	
Scatterometer		Ryan 720		
frequency		13.3 GHz		
wavelength		2.25 cm		
polarization		HH, HV, VV, VH		
incident angle		160° fore and aft		
resolution		15 × 15 m		
Infrared		Daedalus		PRT-5
wavelength		8–14 μm		8–14 μm
incident angle		155° starboard and port		nadir to 55° forward
Mapping Camera	RC-10	RC-10		Handheld Nikon
Altitude Data Collected	12 000 m	300 m	900 m	0 m (surface)

has reviewed the use of such imagery from operational and research satellites. Use of this imagery is transforming the ice-reconnaissance programs that are provided by the Atmospheric Environment Service in Canada, where ice forecasting is a major component of national environmental services. TIROS-N AVHRR imagery is enhanced to full resolution and scaled to match base charts for rapid analysis. Landsat imagery is of particular value in the Arctic summer when both northbound and southbound passes may be utilized; the large sidelaps permit ice features to be monitored for 4–5 days. An example of Landsat imagery used at Ice Forecasting Central in Ottawa is shown in Figure 28-37. Areas are classified into open water, very open pack ice (1–3 tenths), open (4–6), close to very close (7–9), and consolidated ice (10). An output product of NOAA 6 imagery is shown in Figure 28-38 with the corresponding analysis in Figure 28-39 (Mullane, 1980).

Thin new ice and small floes and icebergs are difficult to track, especially on AVHRR imagery. However, a limited age and thickness classification can be made by the trained analyst. The degree of roughness may be inferred for snow-free regions. These and other aspects of ice study with visible and infrared data are discussed by Mullane (1980).

To increase the extent of coverage now limited by cloud cover, a requirement exists for all-weather sensors. The microwave radiometer is most likely to help fill this need in the near future. For data from the Nimbus-5 Electronically Scanning Microwave Radiometer (ESMR-5), Campbell et al. (1980) described application of a time-lapse motion-picture technique for viewing short-term and long-term variations in the ice structure and motion. Figure 28-40 indicates the geographic area of the analysis. Due to the relatively complex interpretative procedures and results the interested reader should consult Campbell et al. for further information. The Nimbus-7 satellite was equipped with a scanning multichannel microwave radiometer (SMMR). The SMMR spatial resolution of 30 km is useful for ice surveillance, especially for a large area study.

Another possibility for all-weather capability is an orbital SAR. For an operational system involving orbital SARs, the number of satellites and the data-flow rates required would be formidable. However, the potential of SAR data is attractive. Thirty Seasat SAR data sets were obtained over

Microwave Signatures of Sea Ice from the CCRS Active-Passive Experiment, 1979.

Class (Thickness)	T (°C)	T_B (K)	ϵ	σ°_{hh} (dB)††† 8°	15°	25°	35°	45°	55°		σ°_{hv} (dB)†† 8°	15°	25°	35°	45°	55°
Water	0	107	0.35	7.7	−3.3	−21.3	−33.6	—	—	*	−7.8	−22.1	−35.4	−44.0	—	—
(0 cm)		±3	±0.01	±1.5	±1.6	±1.3	±0.9	—	—	†	±1.5	±1.5	±0.9	±1.2		
Grease	−2	135	0.46	−27.6	−36.3	−46.2	—	—	—		−39.2	−43.3	−45.3	—	—	—
		±4	±0.03	±1.8	±0.7	±0.6	—	—	—		±0.7	±0.5	±0.4			
Dark Nilas	−5	214	0.79	−11.4	−15.5	−20.6	−23.8	−25.8	−28.6		−24.9	−29.9	−31.4	−33.5	−34.7	−33.6
(<5 cm)		±7	±0.03	±3.4	±2.8	±2.5	±2.4	±2.3	±2.2		±2.9	±2.5	±2.4	±2.3	±2.2	±2.1
Light Nilas	−5	216	0.79	−2.8	−8.1	−14.1	−17.9	−20.3	−23.0		−17.1	−23.2	−25.3	−27.8	−29.5	−28.8
(5–10 cm)		±5	±0.02	±2.5	±2.0	±1.6	±1.6	±1.6	±1.8		±2.3	±1.7	±1.5	±1.5	±1.4	±1.5
Grey	−30	223	0.91	−5.9	−6.7	−10.7	−14.1	−15.1	−17.3	†	−19.0	−21.1	−21.8	−24.2	−24.7	−24.4
(10–15 cm)		±4	±0.02	±0.6	±0.9	±1.1	±1.2	±1.2	±0.8		±0.6	±1.6	±2.5	±2.2	±2.1	±1.7
	−5	213	0.78	1.4	−2.8	−8.2	−11.9	−13.8	−16.8		−12.5	−16.6	−18.0	−20.6	−21.9	−22.3
		±11	±0.05	±1.0	±0.7	±1.0	±1.2	±1.4	±1.3		±0.7	±1.1	±1.6	±1.6	±1.6	±0.9
	−2	214	0.78	0.4	−3.0	−8.4	−13.4	−16.4	−21.1		−13.2	−17.0	−17.2	−20.6	−22.1	−24.1
		±6	±0.02	±1.3	±1.0	±0.9	±0.9	±0.7	±1.0		±1.0	±0.6	±0.8	±0.9	±0.6	±0.9
Grey-white	−30	230	0.94	−7.9	−10.9	−14.1	−16.8	−17.3	−18.9		−19.4	−22.8	−22.5	−24.4	−24.8	−24.3
(5–30 cm)		±3	±0.01	±1.4	±1.3	±1.6	±1.7	±1.7	±1.6		±1.7	±2.8	±2.9	±3.0	±2.7	±2.4
	−5	253	0.94	−2.8	−5.5	−10.2	−14.8	−16.2	−18.9	†	−17.0	−19.1	−19.4	−22.4	−23.4	−23.5
		±2	±0.01	±0.6	±0.4	±0.5	±0.6	±0.5	±0.6		±0.9	±0.5	±0.6	±0.7	±1.1	±0.9
	−2	204	0.74	1.3	−4.6	−12.4	−18.0	−20.6	−22.0		−13.5	−19.9	−21.4	−25.1	−25.9	−25.4
		±6	±0.02	±1.4	±2.2	±2.5	±2.5	±2.5	±1.1		±1.4	±2.0	±2.1	±2.2	±2.0	±0.8
FY Smooth	−30	226	0.92	−9.0	−12.8	−16.8	−19.5	−20.2	−22.0		−21.3	−25.1	−25.3	−27.2	−27.6	−27.2
(30–200 cm)		±4	±0.02	±1.8	±1.6	±1.5	±1.5	±1.5	±1.5		±2.7	±3.3	±3.4	±3.6	±3.7	±3.7
	−5	225	0.95	−6.4	−8.7	−12.8	−16.4	−17.8	−20.2		−20.2	−22.5	−22.6	−25.1	−26.0	−25.8
		±5	±0.02	±1.8	±2.0	±2.0	±2.3	2.0	±2.1		±2.5	±2.2	±2.2	±2.3	±2.0	±2.1
	−2	240	0.88	−2.6	−8.0	−14.2	−18.8	−21.0	−24.3		−17.1	−22.6	−23.3	−26.6	−27.6	−27.7
		±9	±0.03	±2.1	±1.6	±1.9	±2.0	±2.1	±5.0		±1.6	±1.4	±1.8	±2.0	±2.0	±6.3
FY Rough	−30	227	0.93	−7.7	−11.5	−15.2	−17.6	−18.1	−20.6		−19.1	−22.5	−22.5	−24.4	−25.0	−25.5
(30–200 cm)		±3	±0.08	±1.5	±1.6	±2.0	±2.2	±2.3	±2.2		±2.2	±3.1	±3.3	±3.3	±3.3	±3.0
FY Wavebroken	−2	230	0.84	−2.2	−6.0	−10.7	−14.3	−15.7	−18.5		−15.9	−19.3	−19.5	−22.5	−23.2	−24.1
		±9	±0.04	±0.9	±1.1	±1.4	±1.6	±1.7	±2.0		±1.0	±1.2	±1.3	±1.4	±1.4	±1.7
SY	−30	208	0.85	−5.0	−6.9	−8.7	−10.2	−10.6	−12.4		−12.4	−12.9	−12.4	−14.2	−15.0	−14.9
(>200 cm)		±5	±0.02	±1.3	±1.7	±2.1	±2.3	±2.2	±2.2		±2.6	±2.9	±3.0	±3.0	±2.9	±2.8
MY Smooth	−30	192	0.77	−2.4	−4.1	−6.0	−7.8	−8.3	−10.3		−8.8	−9.0	−8.5	−10.5	−11.5	−11.8
(>200 cm)		±6	±0.03	±1.1	±0.9	±1.0	±1.1	±1.1	±1.2		±1.6	±1.7	±1.6	±1.6	±1.6	±1.5
Rough	−30	191	0.77	−2.3	−3.8	−5.5	−7.2	−7.6	−9.4		−8.2	−8.2	−7.7	−9.8	−10.7	−10.6
(>200 cm)		±8	±0.04	±1.3	±1.1	±1.2	±1.2	±1.2	±1.2		±1.8	±1.9	±1.8	±1.8	±1.7	±2.5
Ice island	−30	163	0.65	−0.7	−2.1	−4.0	−6.5	−6.4	−8.1		−6.3	−5.9	−5.6	−8.0	−8.6	−8.9
(>3000 cm)		±5	±0.2	±0.7	±1.0	±1.1	±1.1	±1.2	±1.0		±1.7	±1.9	±1.8	±1.8	±1.6	±1.5

Symbols

ϵ—Emissivity

T—Approximate ambient temperature

T_B—Measured brightness temperature at 19.4 GHz

Notes

* This water data is indicative of a very light wind situation

† Entries on this line represent limited statistics

†† Cross-polarized scattering coefficients may contain systematic biases which vary with incidence angle

††† Like polarized scattering coefficients contain biases no greater

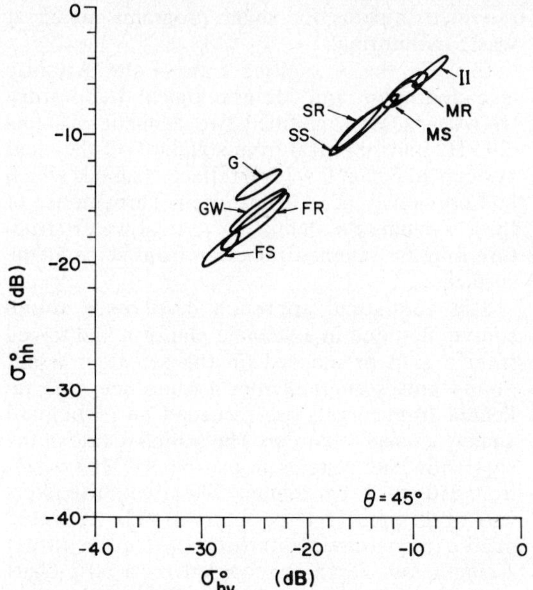

Fig. 28-35. Classification of sea-ice types by like- and cross-polarized Ku-band backscatter coefficients. Beaufort Sea, March 1979, incidence angle 45°. The ellipses are standard deviation contours for the following ice classes and subclasses: G—grey, GW—grey-white, FS—smooth first-year, FR—rough first-year, SS—smooth second-year, MS—smooth multi-year, MR—rough multi-year, and II—ice island, T3. Substantial overlap can be seen in the rough and smooth subclasses of old ice, and also GW and FR. Old ice is clearly separated from FY and younger subclasses using either polarization, but additional radar contrast is available with cross-polarization (Hawkins *et al.*, 1980).

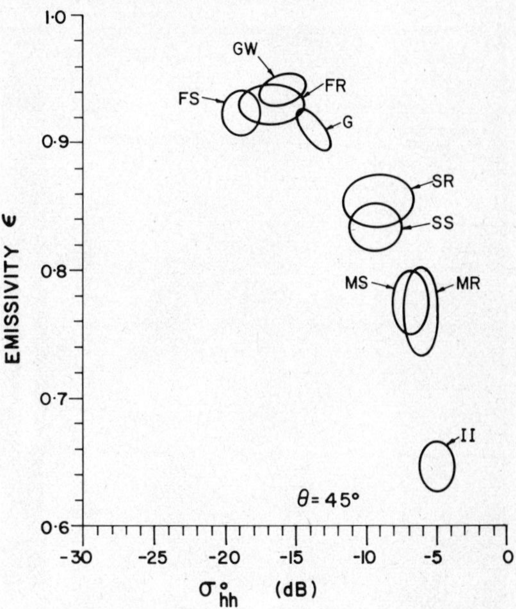

Fig. 28-36. Classification of sea-ice types by like-polarized Ku-band back-scatter coefficient and K-band microwave emissivity. Beaufort Sea, March 1979, incidence angle 45°. The advantage of a combined active/passive measurement for separating MY and SY classes and also in dividing rough and smooth subclasses is demonstrated by comparing Figures 28-35 and -36. Ambient temperature near −30°C (Hawkins *et al.*, 1980).

the western Arctic in 1978, coincident with airborne X-band SLAR missions. Analysis by H. Hengeveld reported by Vankoughnett et al. (1980) indicates that under summer-melt conditions, Seasat L-band SAR imagery of sea ice is relatively featureless, while airborne SLAR imagery reveals topography clearly. The difference is presumably due to the low SAR incidence angle. During freeze-up, the Seasat SAR differentiated old, FY, and young ice. Smooth SY floes showed a unique signature. Newly-formed ice reduced open-water sea clutter; open-water clutter was sometimes stronger than adjacent ice, the reverse of what is normally observed with SLAR. Because the dynamic range of FY and MY ice equals the range of new ice, ice classification based solely on L-band backscatter will be impossible, but shape and texture information will permit interpretation of a large number of ice types and features. An example of a digitally processed Seasat SAR image from Cambridge Bay, Northwest Territories, is shown in Figure 28-41.

d. Sea-Ice Information Products

In addition to the Canadian ice services program described above, a special-purpose receiving, interpretation, and distribution network has been developed in other areas, with visible and infrared images photointerpreted to develop sea-ice information products that are distributed by facsimile or mail (Dismachek, 1977). One is a composite weekly ice type and concentration chart covering the Bering, Chukchi, and the Beaufort Sea area seas bordering Alaska. NOAA also provides a twice-weekly, composite satellite analysis of ice conditions in the Great Lakes. Special VHRR stored-data coverage is regularly obtained over the Ross Sea area of Antarctica from November through January to assist U.S. Navy ice forecasters in their support of resupply missions. Similar coverage is obtained for other areas of the Antarctic and for the East Greenland Current area in support of U.S.A., Canadian, and British ice analysts. The U.S.S.R. has long made use of satellite images from U.S.A. and their own satellites as a navigation aid in the Arctic (Vasil'ev, 1968).

6.0. Water Column Measurements

a. Acoustical Systems and Applications

(1) Methods—Various phenomena in the ocean may be studied with sonar. The present discussion centers on systems emitting high-frequency sound, 20 kHz and above, and detecting backscattered acoustical energy for the study

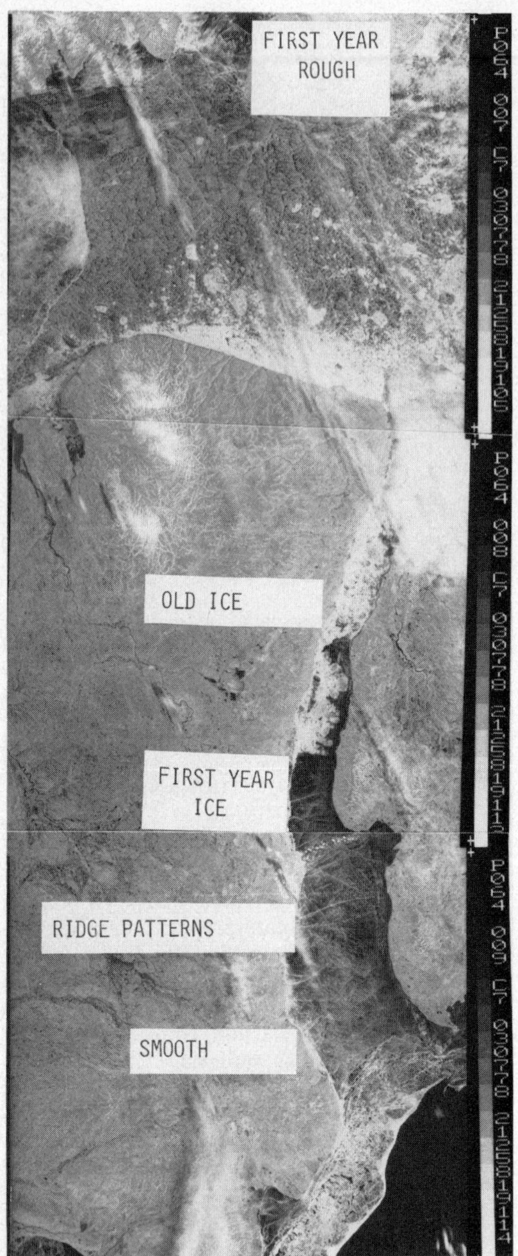

Fig. 28-37. Landsat MSS 6 image of ice features in the area of Prince of Wales Strait and McClure Strait, 3 July 1978 (Mullane, 1980).

of oceanic processes and for the study of oceanic pollution.

Most multifrequency acoustic systems used on ships were developed to look specifically at strong reflective targets on the bottom or within the water column, such as submarines and schools of fish. Consequently, these systems discounted as noise or reverberation the less reflective targets within the water column. With the advent of the 1970's, growing oceanic environmental concerns provided impetus for sonar programs aimed at waste monitoring.

In 1975 the Acoustics Lab of the Atlantic Oceanographic and Meteorological Laboratory (NOAA-AOML) modified two acoustic systems (20 kHz and 200 kHz) from standard off-the-shelf systems to detect low level reflected signals which had previously been disregarded. Through use of these systems the details of vertical water structure may be sampled remotely from ships at the surface.

The acoustical approach involves a sound source mounted in a suitable platform and towed from a ship or moored on the sea floor itself. Sound pulses emitted from a transducer and reflected from targets are recorded on a standard analog acoustic recorder. The sampling rate of the signal (on the systems in use by AOML) varies from 250 pulses per minute (240 msec pulse interval) to 25 ppm (2.4 sec interval). The reflected profile of acoustic intensity (reflective target strength) vs. depth is recorded on a strip-chart recorder. By inspection, the profile patterns may be interpreted for water-column stratification, the presence of particulates and biota, and even particulate motion. Applications to physical oceanography, waste dispersion, and biological distributions follow readily. Thus, sonar provides for rapid three-dimensional sampling which is very amenable to combination with other sampling techniques.

(2) Applications

(a) Internal Waves, Frontal Processes, and Bottom Resuspension—Field studies have been conducted using acoustics to study the intermixing of water masses, internal wave interaction on the continental slope, and resuspension of bottom sediment on the slope (Newman et al., 1977).

Internal waves have been observed on numerous occasions using acoustical remote sensing techniques. Figures 28-42 and 28-43 show internal wave oscillations occurring on the upper mixed-layer interface of the water column. Also shown in Figure 28-42 is an acoustically reflective interface which is an intrusion of colder slope water into warmer shelf water (Newman et al., 1977). The interaction that transpires when an internal wave group traverses a water-mass boundary such as shown is very poorly understood. It is possible that the flows associated with the internal waves may facilitate the exchange of water-mass parcels, nutrients, and particles across the intrusional boundary. Newman et al. have shown that internal wave packets on the New York continental shelf in 120 m of water cause resuspension of particulate matter. The turbulence created by passage of an internal wave is strong enough to lift bottom sediment into the water column, allowing sediment to be transported to new locations. The ability of internal waves to cause resuspension appears to be unaffected by the intrusion of colder water (see Figure 28-43).

Fig. 28-38. NOAA 6 image of the Labrador Coast, 20 April 1980.

(b) Particulate and Liquid Spilled Material—Proni et al. (1976a) found acoustics useful for detecting, tracking, and determining dispersion characteristics of the particulates associated with ocean dumping. A great advantage of acoustic systems is that they allow for immediate on-site analysis of subsurface phenomena, as well as detailed analysis later.

Sewage Sludge—New York City has been dumping sewage off the New Jersey and New York coasts for a great many years. A difficult problem associated with the studies of these dumping practices is to determine the physical fate of the material after it has been dispersed in the environment. One approach is to obtain chemical samples at depth, but the exact locations of samples in the plume are hard to determine. Acoustic methods permit determination of: 1) the actual depth at which chemical samples are obtained; and 2) the space-time evolution of the dumped material.

The density structure within the water column significantly affects the distribution of dumped sewage sludge. In Figure 28-44, obtained in three passes over a plume, it can be seen that acoustic data obtained from a well-mixed watermass in September show particulate material falling directly to the bottom. These data differ quite significantly from those of Figure 28-45 obtained (in two passes) during a summer experiment in highly stratified water. Here there is a horizontal flow outward from the plume axis (30 cm/sec for the

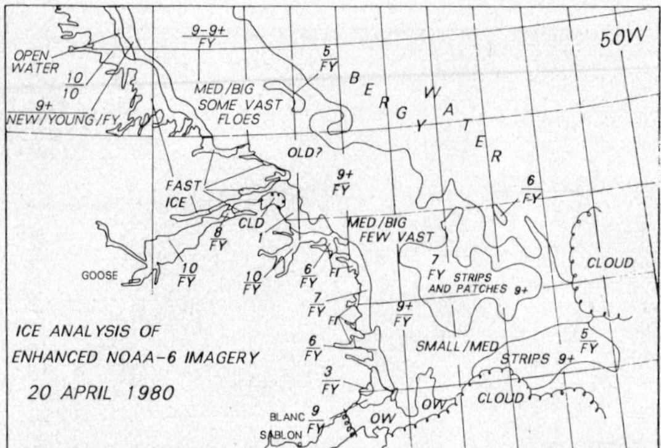

Fig. 28-39. Ice analysis of NOAA 6 image from Figure 28-38 (Mullane, 1980).

first half hour after dumping) on the bounds of the thermocline. This horizontal flow is likely a result of the momentum created by the dump itself. As the heavier particles are introduced they sink until the water-density interface is met. Their momentum then carries them radially outward producing the observed horizontal flow. Another result of the dump, as seen in Figure 28-45, is internal wave activity.

Dredge Spoil—Dredge spoil amounts to the largest volume of particulate matter disposed in the marine environment today. It also has the potential, depending on its origin, to be hazardous to the marine environment. The 20 kHz acoustic system described earlier was initially tested during a dredging operation in the Government Cut ship channel, Miami, Florida. This initial test was critical in fostering interest in the use of acoustics as a remote sensing tool in oceanography (see Newman et al., 1978).

Due to tidal flow a significant volume of suspended particulate matter was swept out of the ship channel and into the ocean after being suspended during a dredging operation. Figure 28-46 shows acoustic data obtained on a seaward passage out of the channel, with a temperature profile superimposed on the image. As observed with the aid of the temperature profile, much of the acoustically reflective material has accumulated at the thermocline. The acoustic patterns also reveal discrete reflecting layers within the thermocline region offshore of the major particulate plume (see Proni et al., 1976b, 1977).

Another dredging operation was studied with acoustical data in Lake Ontario. In October 1976, material from the Genesee River was dredged and moved to a dump site approximately 3 km northeast of Windsor Beach, Rochester, New York. At this time the lake was well mixed with a uniform temperature from surface to bottom (18–20 m) of 11°C. Figure 28-47 shows 200 kHz acoustical data obtained during three passes over the dumped material; the first pass occurred shortly

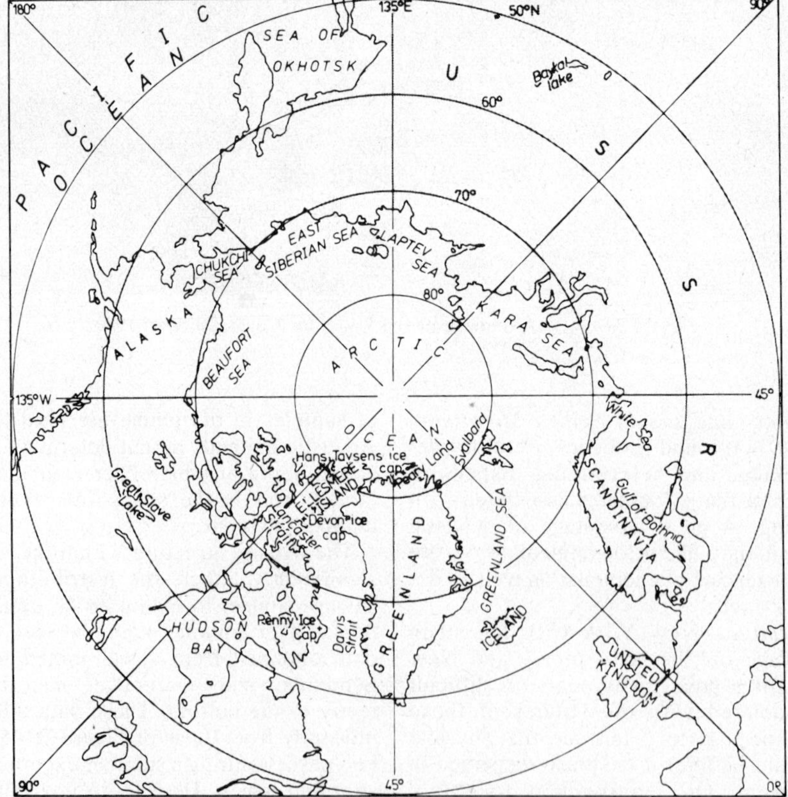

Fig. 28-40. A map of the arctic region included in time-lapse analysis of Nimbus-5 microwave radiometer data for ice dynamics (Campbell *et al.*, 1980).

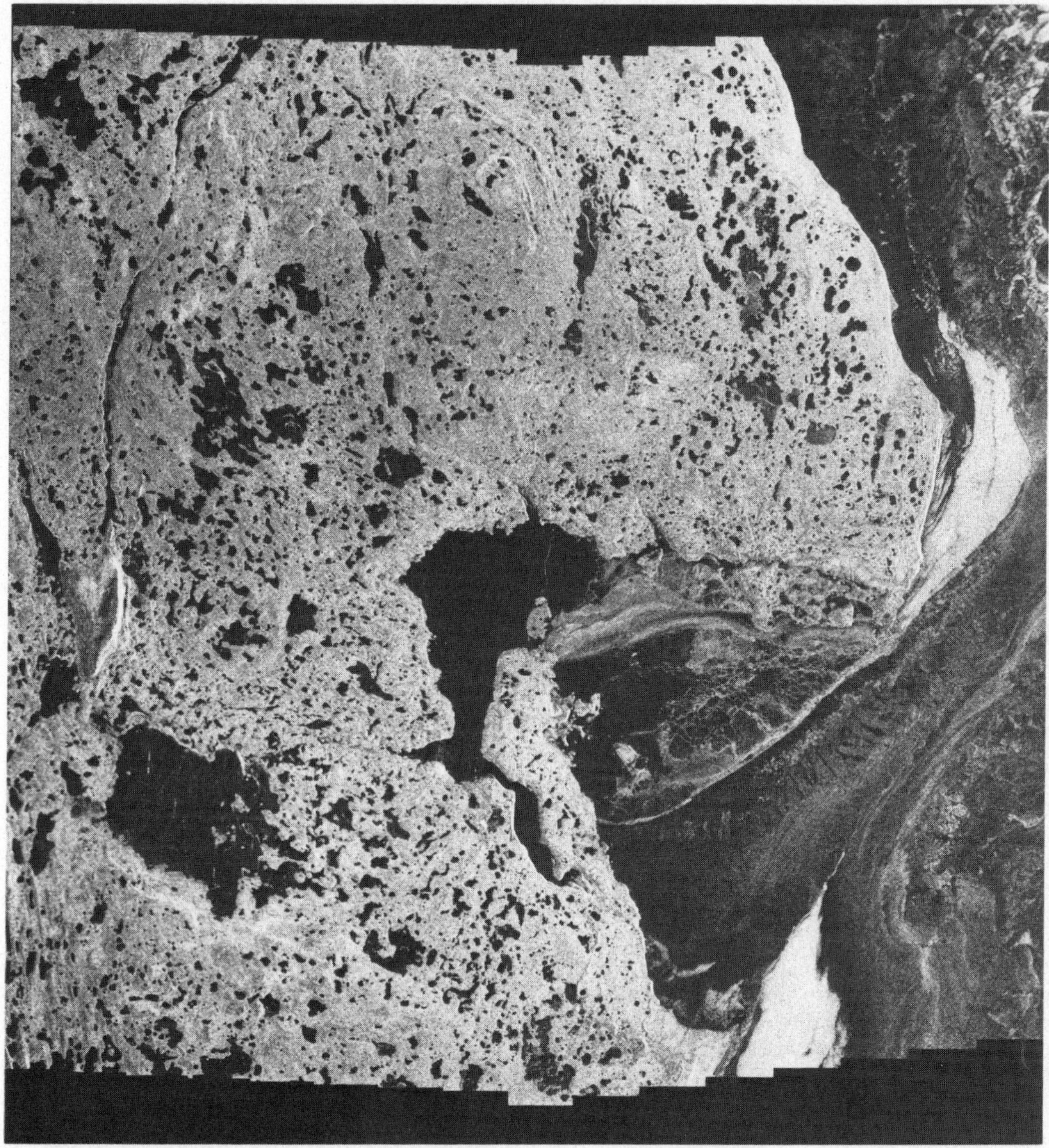

Fig. 28-41. Digitally-processed Seasat SAR image for ice analysis. Cambridge Bay, Northwest Territories, Canada (MacDonald, Dettwiler and Assoc., Ltd.).

after commencement of the dumping event, and two more passes occurred at the times shown. Note in each transect that the width of the plume increases with depth. Also note the existence of bottom surges in the 6 to 10 minute pass. These bottom surges are the result of turbulent mixing at the bottom of the plume. Evidently a transfer of momentum is occurring that creates a second plume moving upward. The third plume transect shows this surface-directed plume at a later time.

These phenomena had not been observed previously, but they could be an important mechanism for dispersion of dumped particulate material throughout the water column for shallow well-mixed waters. It would be interesting to analyze similar events occurring in well-stratified shallow water, such as that now used for disposal of dredge spoil off the northeast coast of New Jersey.

Pharmaceutical Waste—In 1978 a study was conducted off the northern coast of Puerto Rico designed to determine if pharmaceutical waste (less than 2 percent solid material) was detectable and trackable with an acoustical method.

In Figure 28-48, the first pass through a pharmaceutical waste dump shows that two distinct plumes are present. Later it was determined that the barge used two exit ports at its stern to expel the waste. As the majority of the plume is seen acoustically during a three minute pass, and the ship speed was 4 knots, the distance across the plume is calculated to be 350 m. During the study, passes were made over many regions of water

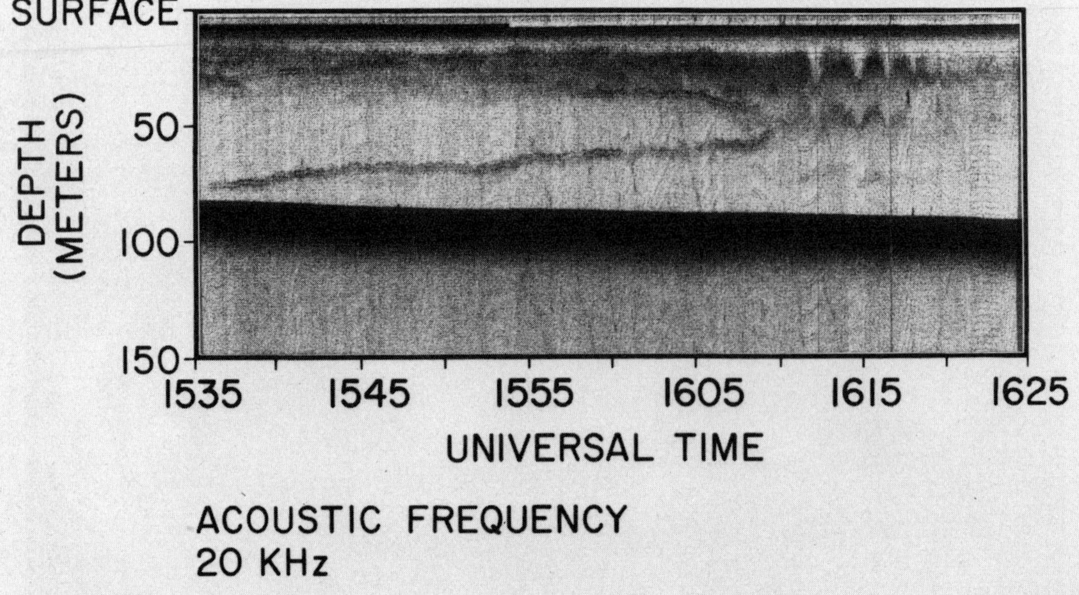

Fig. 28-42. Acoustic measurements at 20 kHz on 22 June 1976 off New York, indicating internal waves and cold water intrusion (Proni *et al.*, 1976a).

which had smaller concentrations than those shown in Figure 28-48. It was determined, after water column sampling and numerous acoustic transects of the dumping area, that much material still remained in the water column from dumps which had occurred up to 2 days prior to the study.

During this investigation numerous transects of various other waste plumes were made. In all these observations, the plumes dispersed in a northwesterly direction, probably due to southeast winds. However, the eastern boundaries of most of the plumes retained a larger concentration of particulates than elsewhere. The dumps occurred in an area of the deep ocean that may be fairly dynamic in winds and surface currents. Theories suggest that oceanic eddies may be present (Hansen, 1974) and other theories point to the existence of a current flowing northwest along the northern banks of the Antilles (Kort, 1972). Particulate dispersion may be enhanced and/or hindered depending on the presence or absence of these features.

Oil Spills—During the summer of 1979 the IXTOC I exploratory well blew out in the Bay of Campeche, Gulf of Mexico, spewing 10 k to 100 k bbl of oil per day into the surrounding water (see the section on Oil Pollution). NOAA subsequently

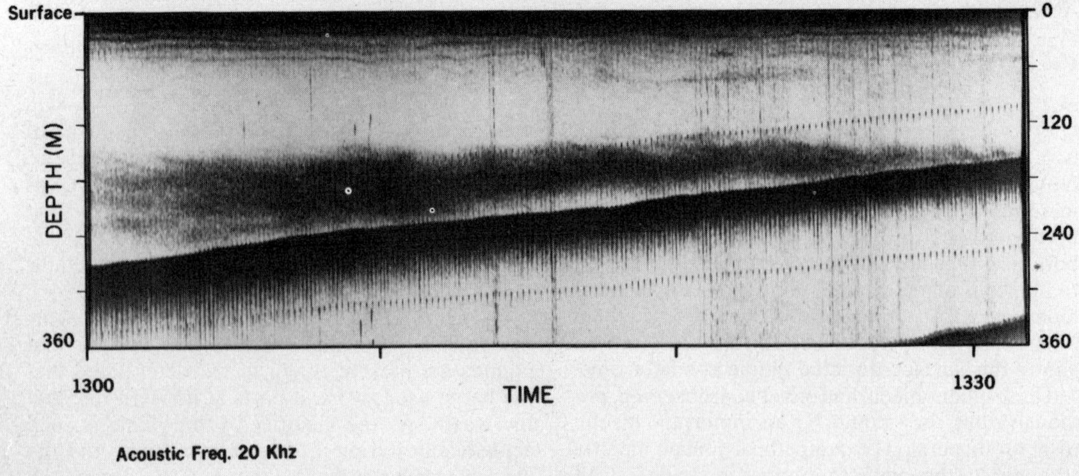

Fig. 28-43. Acoustic measurements at 20 kHz in May/June 1979 indicating internal waves in New York Continental Shelf water (Proni *et al.*, 1976a).

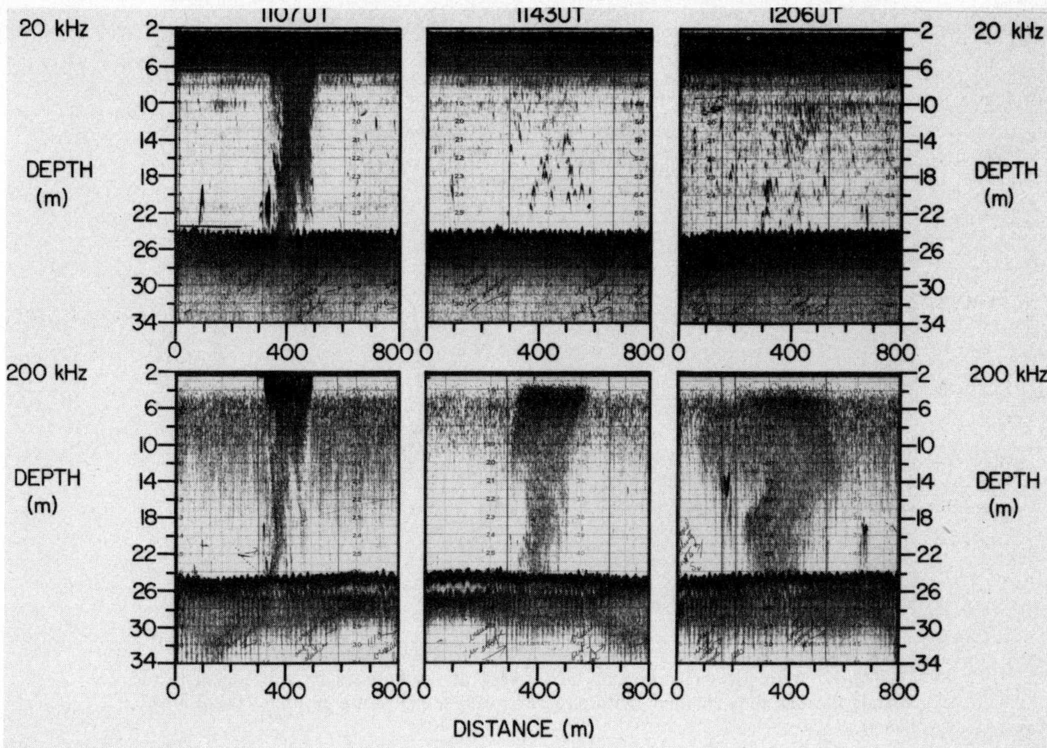

Fig. 28-44. Acoustic measurements at 20 and 200 kHz of particulates from a sewage sludge dump in the New York Bight on September 22, 1975. Well-mixed water (Proni *et al.,* 1976a).

organized a multidisciplinary field study to investigate the possible effects on the environment in the vicinity of the well and also in U.S.A. waters. Acoustical techniques were employed to detect and track subsurface particulate oil observed off the Texas Coast and captured in a net in the Port Mansfield Cut (Walter and Proni, 1980).

Initial observations using acoustics and chemical sampling within the plume showed some correlation between subsurface hydrocarbon concentrations and acoustic return. During the study, tropical storm Henry passed close by the area, causing a thorough mixing of the water column and making detection of horizontal distributions on density interfaces more difficult. However, after passage of the storm and the subsequent restratification of the water column, subsurface layered distributions of hydrocarbons appeared.

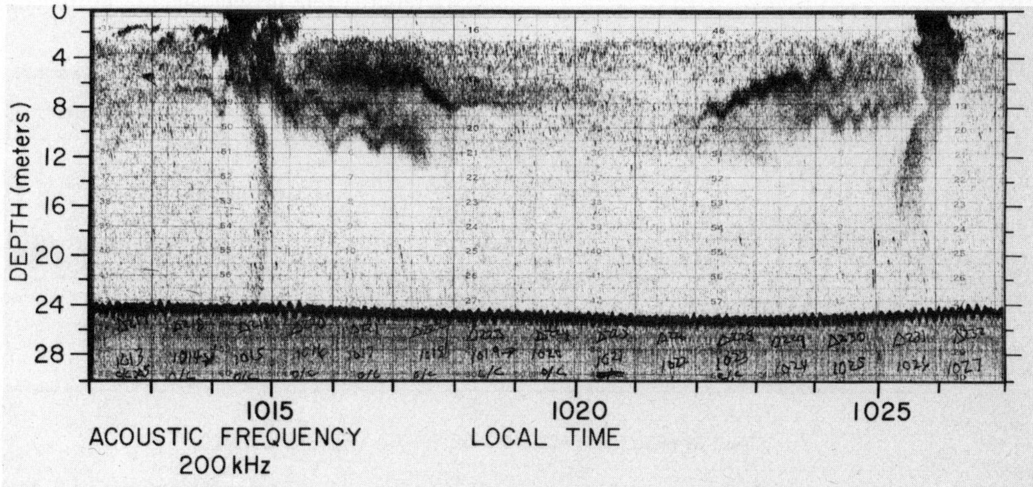

Fig. 28-45. Acoustic measurements at 200 kHz of particulates from a sewage sludge dump in the New York Bight on July 15, 1976. Horizontal flow of dump material in stratified water (Proni *et al.,* 1976a).

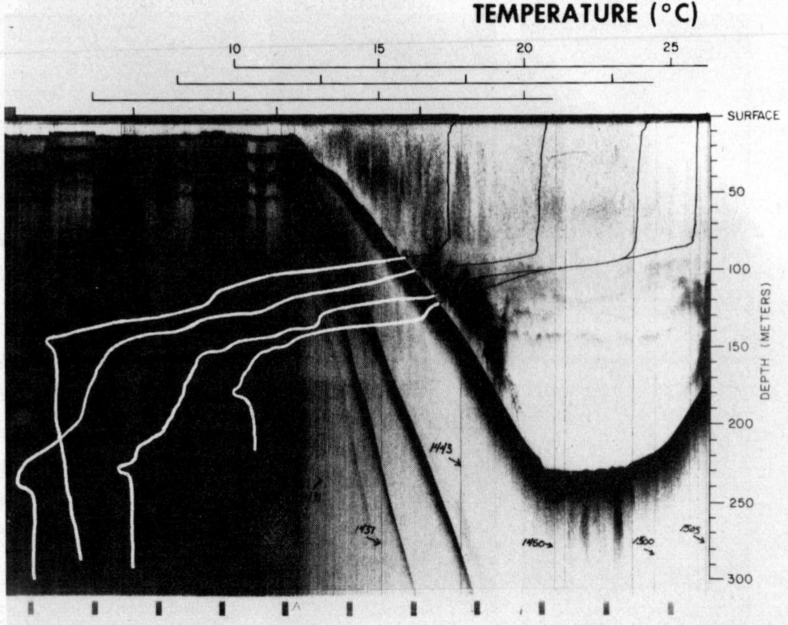

Fig. 28-46. Acoustic measurements at 20 kHz of particulates from dredging in the Miami, Florida ship channel, with superimposed temperature profiles (Proni *et al.*, 1976a).

Figure 28-49 shows a layer of increased acoustic return ranging from 9 to 20 m depth. Coincidental with this record, chemical samples showed elevated levels of hydrocarbons (Feist and Boehm, 1980). Unfortunately, this was only a short-term one-time study. Other tests are needed to prove the utility of acoustic remote sensing techniques for the detection of subsurface oil concentrations. Natural hydrocarbon seepage has previously been observed with acoustics (Geyer and Sweet, 1974); therefore, the available studies indicate that acoustic detection of spilled hydrocarbons is feasible.

b. Moored Data Buoys

Buoys have been used to make *in situ* measurements of profiles of oceanographic and weather parameters and to transmit these data to other locations for subsequent analysis and interpretation. The utility of satellites for data telemetry and dissemination has been demonstrated during the past several years. During 1978, a complement of 15 environmental moored data buoys acquired and disseminated over 37,500 synoptic weather messages. In addition, 25,000 wave spectra reports were disseminated from 10 buoys

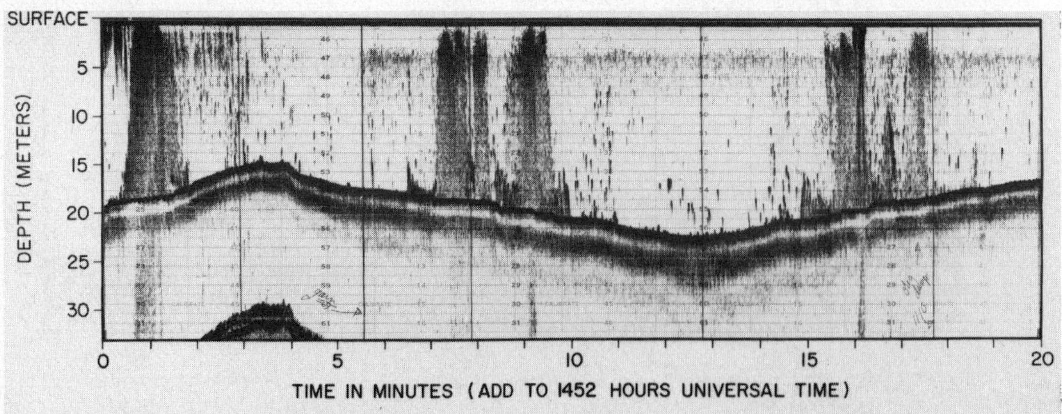

Fig. 28-47. Acoustic measurements of 200 kHz on 20 October 1976 showing dumped dredged material in Lake Ontario (Proni *et al.*, 1976a).

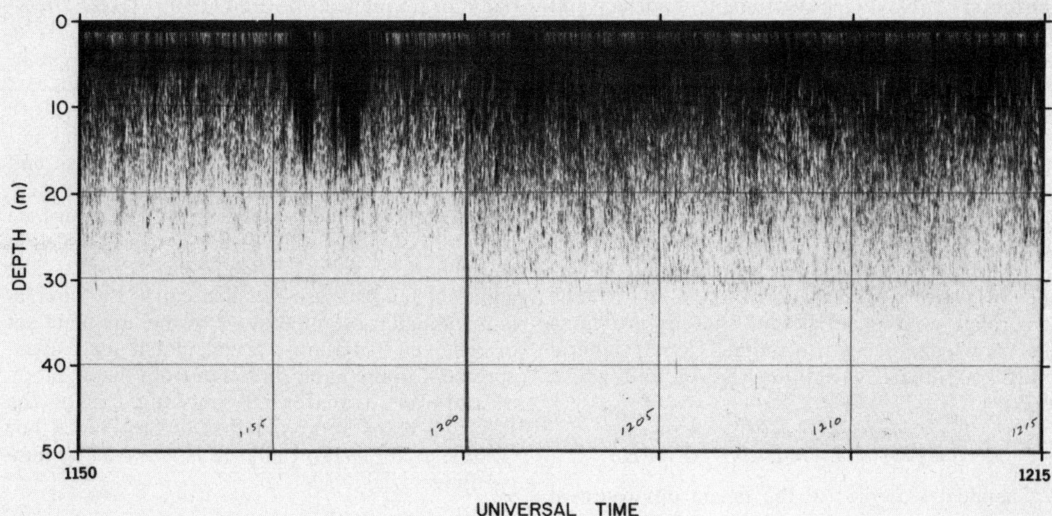

Fig. 28-48. Acoustic measurements of a pharmaceutical waste dump (less than 2 percent solid materials) off the northern coast of Puerto Rico on October 29, 1978 (Proni *et al.*, 1976a).

equipped with wave data systems. During 1979, up to 25 moored buoys were on station, all equipped with wave-data systems. Satellite-data telemetry has allowed the reliable delivery of high-quality data from various remote and often hostile ocean areas. These data buoys are having major impact by providing environmental data for:

- Improving weather forecasting
- Providing an initial data base for evolving weather-climate monitoring programs
- Augmenting ocean-climate monitoring studies
- Augmenting remote satellite measurement/monitoring systems
- Providing environmental base-line data for energy-resource assessment.

The launch of polar-orbiting satellites with data-collection systems (DCS) capable of data collection, data relay, and position determination from surface platforms has provided a global environmental measurement/monitoring capability at economic scales unachievable before. Lagrangian drifting buoys and moored buoys are exploiting DCS capabilities of various satellites (see Coastal

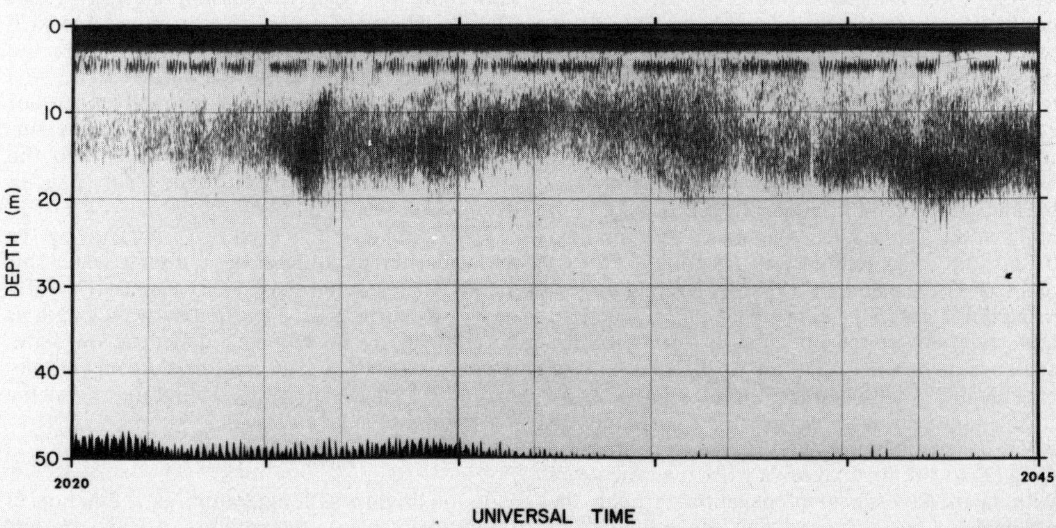

Fig. 28-49. Acoustic measurements on September 18, 1979, indicating crude oil from the IXTOC I (Bay of Campeche) blowout at depths from about 9 to 20 m (Proni *et al.*, 1976a).

Currents) by relaying measurement of surface and deep ocean currents over large spatial and temporal scales. During the Global Weather Experiment, drifting-buoy surface measurements were used as an ancillary data source for remote satellite measurements of vertical profiles of temperature and wind.

Improving the measurement/monitoring capabilities on a global basis will lead to a better understanding of oceanic and atmospheric climates. DCS-obtained data can contribute to design of a global observation system for routine long-range weather prediction, thereby providing the knowledge needed to reduce global vulnerability to climate variations (Kerut and Hass, 1979).

B. CHEMICAL OCEANOGRAPHY

Chemical variables of the ocean environment are nearly all beyond present remote sensing capabilities, except for salinity. This important exception is significant for study of density-driven features of circulation.

1.0. Salinity Measurements

The only direct technique for remote salinity measurement is the use of a microwave radiometer. However, preliminary study has indicated that a laser Raman detection of sulfate ion in sea water may indirectly yield salinity (Zimmerman, 1976).

a. Microwave Radiometer Technique

The results of several aircraft programs have demonstrated that water temperature, salinity, and oil spill thickness can be derived from passive microwave measurements with an accuracy that satisfies most user applications (Swift, 1980). In particular, a remote sensing technique to concurrently measure sea-surface temperature and salinity has been demonstrated with a dual-frequency microwave radiometer system (Blume et al., 1981). Accuracies in temperature of 1°C and in salinity of 1‰ for salinity greater than 5‰ were obtained after correcting for extraterrestrial background radiation, atmospheric emission and attenuation, sea-surface roughness, and antenna beamwidth. The radiometers, operating at 1.43 and 2.65 GHz, comprise a third-generation system using null balancing and feedback noise injection, and were developed specifically for obtaining sea-surface temperature and salinity maps of coastal and estuarine areas.

The atmosphere is essentially transparent to electromagnetic radiation at frequencies of 1 to 3 GHz (30 to 10 cm). Extensive work over the years with microwave-signal propagation through the atmosphere has indicated that the influence of clouds is small at these frequencies except under very severe storm conditions. Also, the background galactic-noise tends to decrease substantially for frequencies beyond about 1 GHz. Therefore, the frequency range from 1 to 3 GHz is well-suited for minimal extraterrestrial background radiation and atmospheric interference.

Nevertheless, accurate surface-temperature measurement by airborne radiometers in this microwave region requires detailed knowledge and correction of these effects. Their impact on measured brightness temperature of the ocean surface can still be of the order of a few degrees Kelvin and therefore must be taken into account. The apparent temperature T_R (also called the equivalent radiometric temperature of the complete set of received radiations) is calculated for a measurement in the nadir direction from the equation of radiative transfer. By making use of the Rayleigh-Jeans approximation to the Planck law (Blume et al., 1978), radiance is replaced by temperature:

$$\begin{aligned} T_R = T_B[1 - \tau(h)] + (1 - e)[1 - \tau(h)] \\ [(T_{\cos} + T_{\text{gal}})(1 - \tau_0) + T_{atm}] \qquad (28\text{-}1) \\ + \tau(h) < T_{atm} > + \Delta T_W + \Delta T_p \end{aligned}$$

The first term accounts for the attenuated emission T_B from the ocean surface. The attenuation factor $[1 - \tau(h)]$ with atmospheric opacity $\tau(h)$ at sensor altitude h is a small-opacity approximation to $\exp(-\tau(h))$. The second term in equation (1) accounts for reflected downwelling radiation, consisting of the temperature of the downward radiation of the extraterrestrial noise $(T_{\cos} + T_{\text{gal}})$ attenuated $(1 - \tau_0)$ by the entire atmosphere, and the downward radiation T_{atm} of the atmosphere itself, all reflected $(1 - e)$ by the ocean surface and attenuated $[1 - \tau(h)]$ by the intervening atmosphere up to the sensor. Kirchoff's law is assumed where water absorption a equals emissivity e. Upwelling atmospheric emission is given by $\tau(h)$ $\langle T_{atm} \rangle$, where $\langle T_{atm} \rangle$ is the averaged physical temperature of the intervening atmosphere between the sensor and the sea surface. The next term ΔT_w is the contribution due to the sea-surface roughness generated by shear forces of the surface winds. The last term ΔT_p is due to the antenna pattern deviating from the ideal "pencil" beam shape.

The brightness temperature T_B is given by the molecular temperature of the radiating water surface T_s and its emissivity e. The emissivity of a dielectric surface at a particular wavelength is determined by its complex dielectric constant, which for calm seawater is a function only of temperature T_s and salinity S. Therefore, T_B at the wavelength λ is given by:

$$T_B(\lambda) = e_\lambda\,(T_s, S)T_s \qquad (28\text{-}2)$$

Plots of brightness temperature as a function of salinity and surface temperature at 1.43 GHz and 2.65 GHz are given in Figures 28-50 and 28-51, respectively. The dual inversion of microwave brightness temperatures in two bands (L- and S-

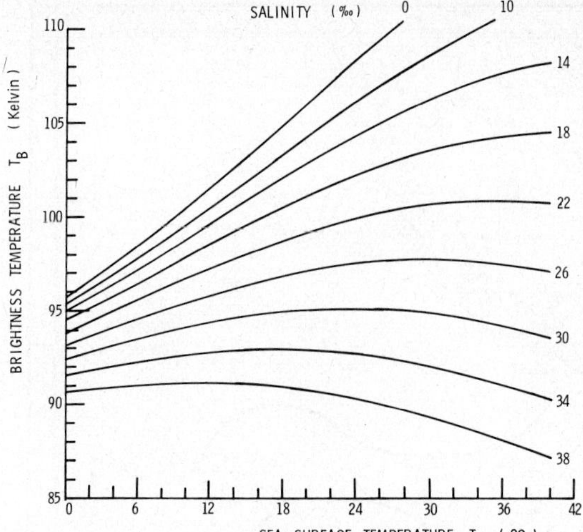

Fig. 28-50. Microwave brightness temperature at 1.43 GHz for normal incidence *vs.* molecular sea surface temperature for a smooth sea at different salinities (Kendall, 1980).

band) using derived regression equations produces values for both surface temperature and salinity. These measurements of "microwave brightness" radiation may be made over a wide range of environmental conditions (e.g., on cloudy and hazy days, and at night), thus maximizing the opportunities for coastal and estuarine monitoring.

Although the demonstrated absolute accuracy of the radiometer system is 1‰ for salinity above 5‰, and 1°C for temperature, the relative accuracy within a given data set is better than 0.5 ‰ and 0.5°C. The spatial resolution of these measurements is given by the antenna beam "footprint" for the given instrument, roughly one-third of the measurement altitude *h*.

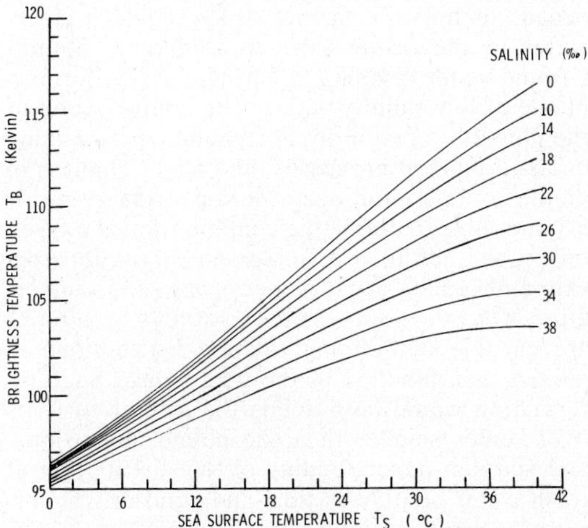

Fig. 28-51. Microwave brightness temperature at 2.65 GHz for normal incidence *vs.* molecular sea surface temperature for a smooth sea at different salinities (Kendall, 1980).

b. Coastal Applications

On August 24, 1976, the L- and S-band radiometer system was flown on an NASA Wallops Flight Center C-54 aircraft over the lower part of the Chesapeake Bay and adjacent Atlantic Ocean (Blume et al., 1978). High salinity gradients at the Bay-mouth mixing zone permitted verification of instrument performance over a wide range of salinity concentrations. Both radiometers were pre-flight calibrated with a liquid-nitrogen cryogenic antenna load placed under and closely coupled to the antenna. The measured radiometric data, together with latitude and longitude coordinates from the aircraft's inertial navigation system, were recorded on digital magnetic tape and computer processed. A comparison of the radiometrically measured values of surface temperature and salinity with those taken at sea-truth locations was performed. The resulting mean deviation error of −1.52‰ and 0.55°C, with standard deviations of 0.92‰ and 0.59°C for salinity and surface temperature, respectively, indicate that the desired accuracy of 1‰ for salinity and 1°C for temperature was achieved.

The calculated values of salinity and surface temperature were plotted as a function of geographic position. A contour map of the salinity distribution is shown by isohalines in 2‰ increments in Figure 28-52. It is seen that the higher continental shelf salinities are reduced considerably by mixing with fresh water outflow in the bay region. Higher salinity shelf-waters, because of tidal effects and the Coriolis force, favor the northern part of the Chesapeake Bay entrance and the inside of the Delmarva Peninsula. The lower salinity bay-waters exit through the southern part of the bay entrance. The contour map of the surface temperature in 2°C increments seen earlier in Figure 28-1 shows cooler shelf-temperatures giving way to warmer temperatures of the fresh water outflow in the region of mixing. The fine thermal structure of surface waters caused by solar radiation and river outflow has been averaged in order to show the general temperature distribution. Again, a southern outflow of the warmer Chesapeake Bay waters is indicated.

The Langley radiometer system was then applied to the detection of submarine fresh water springs (Blume et al., 1981). The first mission on February 4, 1978 consisted of coverage of three-fourths of the coastal areas around the island of Puerto Rico. During the second mission on February 6, 1978, special attention was directed to the northwest portion of Puerto Rico where several submarine springs had been reported. The surface area above these springs of fresh water exhibits temperature decreases of 4°C and salinities 5‰ lower than the surrounding seawater.

During data collection, white caps of breaking waves were present along reefs, and banking maneuvers of the aircraft were necessary to follow the jagged coastline. White caps have higher

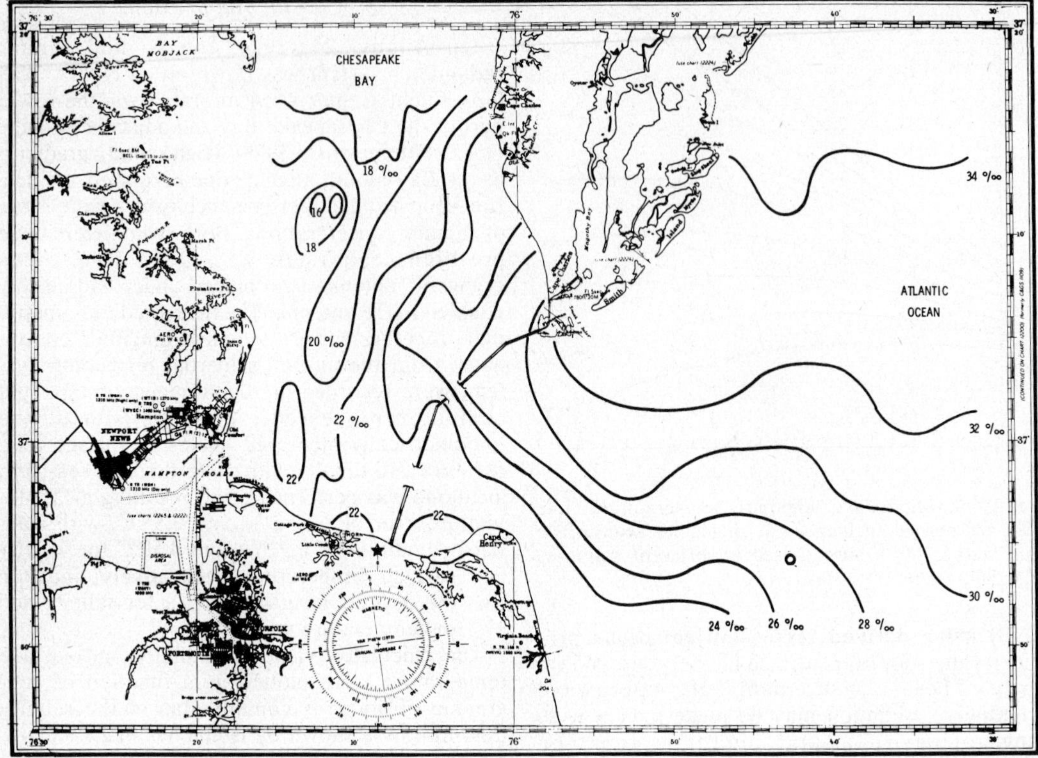

Fig. 28-52. Isohalines of the lower Chesapeake Bay in 2‰ increments on August 24, 1976 from microwave data (Blume *et al.*, 1978).

emissivity than calm water and thus increase the brightness temperature. Banking angles greater than 15° move close to the region of the Brewster angle and tend also to increase the brightness temperature. However, infrared measurements are only very slightly affected by white caps and high banking angles. Therefore, data were combined from L- and S-band microwave and infrared radiometers to resolve these influences and identify fresh water outflows (submerged and superficial). River and lagoon outflows were discriminated from the submarine outflows based on available maps.

An example of the salinity distribution is shown in Figure 28-53 along a flight line across the harbor of Arecibo. The main contributions of fresh water in this area are the Rio La Vega, the Rio Grande de Arecibo, and the drainage canals from the inland marshes. The mixing of fresh water and ocean water takes place in the bight. The transport of the mixed water masses appear to be in a westerly direction according to the salinity distribution as shown.

Spring locations reported previously correlated well with locations detected by the radiometers. Forty-four submarine fresh-water springs were identified, which indicates that springs are more numerous around Puerto Rico than had been thought earlier. The majority are located at the northwest and southeast portions of the coastline. Salinity and temperature patterns indicate the run-off direction. It may be possible to relate remotely measured salinity and temperature gradients to outflow rates.

The Georgia-South Carolina coast has large expanses of salt marsh with river mouths and tidal inlets at intervals of 10 to 20 km. Some of these openings carry large amounts of fresh water to the ocean, such as the Savannah River which ejects water to the ocean through Calibogue Sound. Ground water discharged during ebb tide forms a plume of low salinity water. The configuration of the plume is a result of poorly-understood mixing and entrainment processes, and major changes in plume configuration occur during a tidal cycle. It is impossible to define the configuration at a given time with traditional oceanographic tools aboard a ship and remote sensors measuring only temperature. Quasi-synoptic salinity surveys by aircraft are capable of defining salinity distributions of plumes as a function of the tidal phase. Such information would help define the role of currents over scales smaller than the plume dimensions, and sharpen understanding of the distribution of fresh water relative to tidal and wind-driven currents, as well as to inlet bathymetry.

On November 8, 1979, a series of 20 flight lines was flown by the NASA Wallops P-3 aircraft off the coast of Georgia to measure the salinity distribution with onboard microwave radiometers

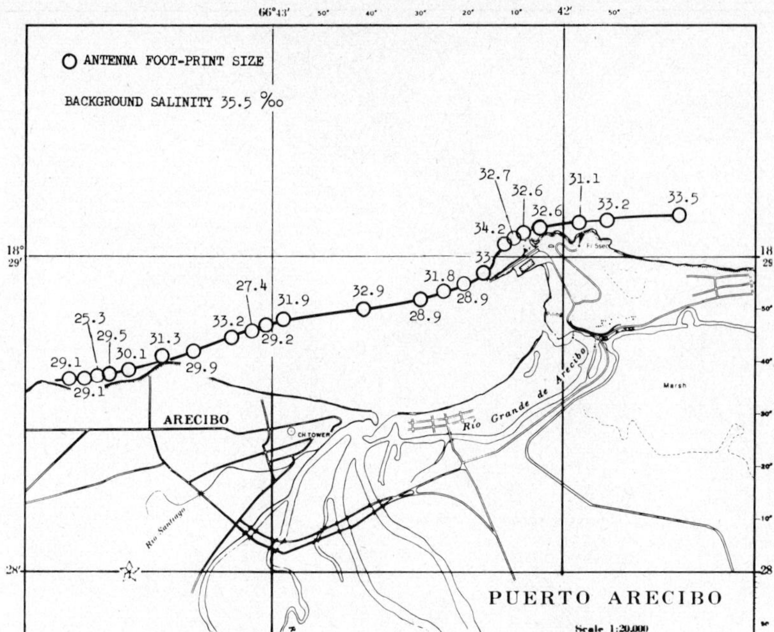

Fig. 28-53. Salinity distribution along a flight path across Arecibo Harbor, Puerto Rico, from microwave data (Blume *et al.,* 1981).

(Kendall and Blanton, 1981). The data were obtained from mid-ebb to one-half hour before slack. The flood phase was examined similarly 18 hours later on November 9. Figure 28-54 shows a salinity map obtained during the ebb phase from only the L-band radiometer, as the S-band radiometer data were biased by radar interference. Single-band salinity measurement was possible because the independently obtained sea-truth data for the measurement area showed only a small variation in surface temperature. Therefore, a constant value of surface temperature could be used in the data regression-equations without affecting the accuracy of salinity measurement. The accuracy was 1‰, verified by comparison with sea-truth data. The strong outflow of the Savannah River plume is clearly indicated by the isohalines in the figure.

C. BIOLOGICAL OCEANOGRAPHY

The various biological components of marine waters present a broad range of remote sensing problems. There are three major categories of interest. The first contains the widely distributed animal and plant cells suspended in the sea, called plankton. Much remote sensing is directed at the plant component, the chlorophyll-bearing phytoplankton, with their unique optical properties. Phytoplankton may be widely distributed horizontally and vertically in the water column. The second category contains plant material usually found in near-shore regions, including floating algae, rooted submerged aquatic vegetation, and emergent rooted plants located in tidally flooded wetlands. The third category of major interest contains commercial fish species.

Animals in the marine environment range in size from microscopic zooplankton (that may be remotely detected as particles in the water column) to large fishes and whales that may be detected individually. An interesting facet of the remote sensing of marine animals is that indirect measurements play a large role in many cases. Surface temperature measurements are used in locating tuna and anchovy in coastal upwelling areas, and oil slicks sometimes indicate schools of menhaden.

1.0. Chlorophyll Detection and Measurement

Chlorophyll *a* is an important aquatic parameter that has significance for water quality (see the section on Phytoplankton Blooms) and for marine productivity. Chlorophyll is addressed here as the primary photosynthetic pigment in phytoplankton particles at the base of the marine food chain. As such, the emphasis is on qualitative and quantitative assessments related to subsequent investigations of marine productivity (see pp. 324–362 in Odum, 1971, for a detailed discussion). The question of ocean color is treated first.

a. Ocean Color

(1) Fundamentals—The introduction of particulates to a water parcel modifies its optical properties via changes in scattering and absorp-

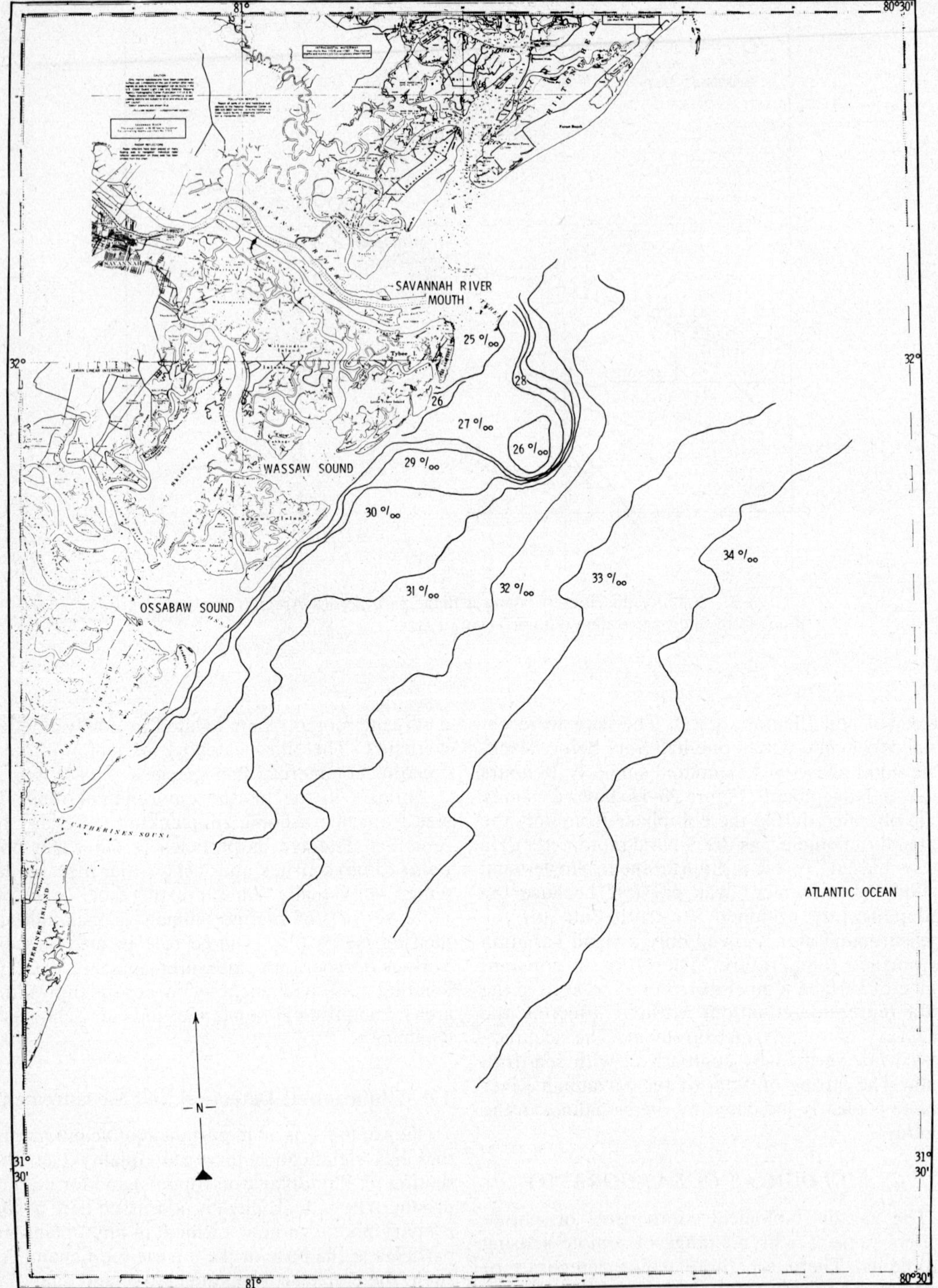

Fig. 28-54. Salinity map of Georgia coastal area during ebb tidal cycle on November 8, 1979 from microwave data (Kendall and Blanton, 1981).

tion. From early work by Rayleigh (1903), Mie (1908), and others (see Shifrin, 1951; Van de Hulst, 1957), it was established that the scattering coefficient has the form

$$b = N\pi\rho^2 f(2\pi\rho/\lambda) \qquad (28\text{-}3)$$

for N spheres of radius ρ at wavelength $\lambda = \lambda_0/n_0$ where n_0 is the refractive index of the medium, and λ the wavelength in vacuum. The function $f(x)$ for $x = 2\pi\rho/\lambda$ is an infinite series proportional to λ^{-4} for $x << 1$, and to λ^{-2} for $x \sim 1$; it approaches the limit 2 for $x >> 1$. In the limit of $x \rightarrow 0$ and of small n', where $n' = n/n_0$ is the index of refraction of the sphere relative to that of the medium, the function reduces to its first term, the familiar Rayleigh scattering proportional to λ^{-4}, applicable to molecular scattering. Thus, the spectral properties of oceanic scattering may be considered in terms of the exponent m in the power law λ^{-m} (similar to the Ångstrom exponent for atmospheric aerosols) for each type of water component. Measurements for filtered oceanic water suggest that molecular scattering in the sea (b_w) obeys $m = 4.3$ (Morel, 1974). For oceanic particles (b_p), values of m around 1 are representative (Morel, 1973). However, pigmented and absorbing particles have more complicated spectral scattering properties (Latimer, 1958; Duntley et al., 1974; Mueller, 1973; Kiefer et al., 1979) due to interaction between scattering and absorption.

The Rayleigh volume scatter is a function of angle θ where, for unpolarized light, $\beta(\theta) = 0.5(1 + \cos^2\theta)$, which is symmetrical about $\theta = \pi/2$. However, the volume scattering phase function $P(\theta)$ of typical oceanic particles (see Figure 28-55) and coastal sediments is strongly peaked in the forward direction; 60% of the scattered photons are found at $\theta < 10°$. This asymmetry has permitted development of simple expressions for diffuse volume reflectance from the sea (measured just above the sea surface). Gordon (1973) and Gordon

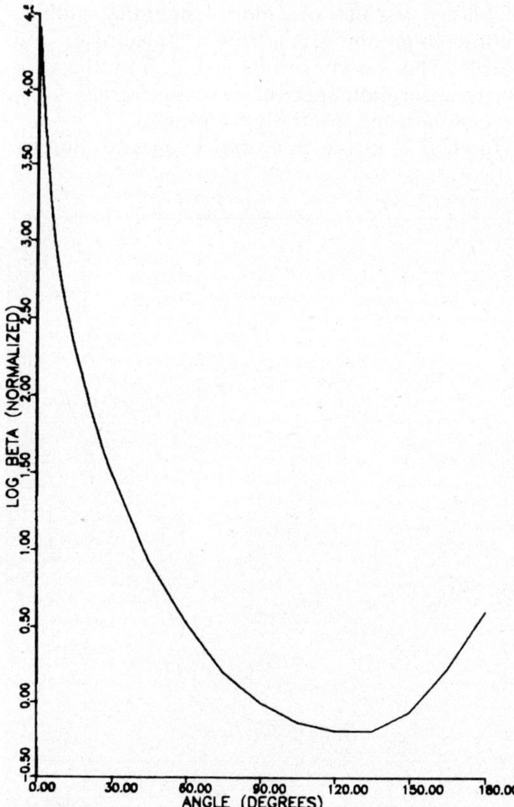

Fig. 28-55. Idealized volume scattering function for marine particulates normalized at 90° (Prieur, 1976).

et al. (1975) used a Monte Carlo technique to simulate radiative transfer; numerical results were fitted by a series, whose first term approximates rigorous multiple-scattering solutions to the equation of radiative transfer. The approximation, called the quasi-single-scattering approximation, is:

$$R = k_1 \frac{bB}{a + bB} = k_1 \frac{\omega_0 B}{1 - \omega_0(1 - B)} \quad (28\text{-}4)$$

where k_1 is a constant, b is the total scattering coefficient, B is the backscattering fraction $0 \leqslant B \leqslant 1$ obtained from:

$$B = \frac{2\pi}{b} \int_{\pi/2}^{\pi} \beta(\theta) \mathrm{Sin}\theta d\theta, \quad (28\text{-}5)$$

$\beta(\theta)$ ($sr^{-1} \cdot m^{-1}$) is the volume-scattering function from $\beta(\theta) = bP(\theta)$, a is the absorption coefficient, and ω_0 the single scattering albedo $\omega_0 = b/(a + b)$. The total attenuation coefficient $c = a + b$. The coefficients a and b are here defined for a collimated beam. The form of R, where R is a function of bB/a, was first suggested by Gamburtsev in 1924 (see Kozlyaninov, 1972), and then by Duntley (1942). A similar expression developed by Morel and Prieur (1977) with a successive order scattering method is

$$R = 0.33(bB/a)(1 + \Delta) \quad (28\text{-}6)$$

where Δ is a correction term smaller than 5%; Δ is only slightly dependent on wavelength in the visible part of the spectrum (Prieur, 1976). Gordon et al. (1975) found $k_1 = 0.3244$ for the sun at zenith, and 0.3687 for a uniform sky.

From these expressions it is seen that R can be calculated knowing k_1 and the inherent optical properties a, b, and B as a function of wavelength. The approximations given here are valid to better than 0.5% for $\omega_0 < 0.6$ (Gordon, 1973; Jain and Miller, 1977). For scattering by water molecules, $B = 0.5$; for marine particles a mean volume-scattering function proposed by Morel (1973) yields the value 0.012.

In water containing various inclusions, a and b are partitioned by $a = a_w + a_p + a_y$ and $b = b_w + b_p + b_y$, where w, p, and y stand for water, particles, and yellow substance, respectively. The particles may include inorganic particles, phytoplankton, zooplankton, and detrital debris. For different particulate components p_i, a_p and b_p may be obtained from $a_p = \sum_i m_i a'_{pi}$ and $b_p = \sum_i m_i b'_{pi}$, where m_i is concentration of component i, and a'_{pi} and b'_{pi} are the specific absorption and scattering coefficients. The complete spectral characterization of the absorption and scattering properties of all oceanic particulates would, by the above relations, permit a solution to the general problem of ocean color (see Jerlov, 1968; Jerlov and Nielsen, 1974; Williams, 1970; Mueller, 1973; Morel and Prieur, 1977; Gower, 1972, 1980; Hovis and Leung, 1977).

The absorption spectra of oceanic components obey more complicated laws than the scattering

TABLE 28-7

Spectral Coefficients for Pure Water Attenuation (c_w), Scattering (b_w), and Absorption (a_w) (Morel and Prieur, 1977)

λ (nm)	c_w	b_w (m^{-1})	a_w	λ (nm)	c_w	b_w (m^{-1})	a_w	λ (nm)	$c_w \simeq a_w$ (m^{-1})	b_w
380	0.030	0.0073	0.023	490	0.022	0.0024	0.020	600	0.245	0.00101
390	0.027	0.0066	0.020	500	0.028	0.0022	0.026	610	0.290	0.00094
400	0.024	0.0058	0.018	510	0.038	0.0020	0.036	620	0.310	0.00088
410	0.022	0.0052	0.017	520	0.050	0.0018	0.048	630	0.320	0.00082
420	0.021	0.0047	0.016	530	0.053	0.0017	0.051	640	0.330	0.00076
430	0.019	0.0043	0.015	540	0.058	0.0016	0.056	650	0.350	0.00071
440	0.019	0.0039	0.015	550	0.066	0.0015	0.064	660	0.410	0.00067
450	0.018	0.0035	0.015	560	0.072	0.0013	0.071	670	0.430	0.00063
460	0.019	0.0032	0.016	570	0.081	0.0013	0.080	680	0.450	0.00059
470	0.019	0.0029	0.016	580	0.109	0.0012	0.108	690	0.500	0.00055
480	0.021	0.0027	0.018	590	0.158	0.0011	0.157	700	0.650	0.00052

spectra because of the varying optical properties of the absorbing materials that are present. For water itself, spectral values of a_w are given in Table 28-7 (Morel and Prieur, 1977). A dissolved component which contributes absorption in coastal waters is the so-called yellow substance (or Gelbstoff). Højerslev (1979) has shown it is not correlated with plankton/detrital breakdown, nor with suspended organic matter or water temperature; instead, yellow substance and its fluorescent by-products are dependent on river run-off and salinity. Its absorption spectrum obeys $a(\lambda) = a(\lambda_0)exp(-0.014(\lambda - \lambda_0))$ for λ_0 between 280 and 450 nm. At 450 nm, $a = 0.212\ m^{-1}$ for a concentration of 1 mg/l. When its breakdown products are excited at 367 nm, fluorescence occurs at 490 nm.

For absorption due to particles, only average values can be obtained in field situations; representative spectra are shown in Figure 28-56. In the laboratory, spectral characteristics of chlorophyll-bearing phytoplankton have been reported by many investigators. For a_p of a single species of phytoplankton, Figure 28-57 shows a typical spectrum. Other species are represented in the literature. Reflectance measurements of various species, with or without the presence of other materials normally found with chlorophyll in the marine environment, have aided design of passive sensing techniques and field experiments. Figure 28-58 from Suits (1973) shows calculated phytoplankton reflectance in the presence of low concentrations of yellow substance and particulate material. Note the presence of: (1) absorption in the blue spectral region below 500 nm, peaking at 440 nm; (2) an absorption peak at 680 nm; (3) the so-called hinge point at about 500 nm, and (4) an increase of reflectance in the near-infrared >700 nm. Figure 28-59 indicates the relative spectral properties of inorganic particles and phytoplankton. The hinge point and other spectral characteristics of phytoplankton are manifested in laboratory studies by Yentsch (1960), Duntley et al. (1974), and Wilson and Kiefer (1979).

Field-reflectance data suggest a division of natural waters into several classes: 1) deep blue 'desert' waters with negligible phytoplankton; 2) blue-green waters with small amounts of phytoplankton; 3) turbid waters with inorganic particulates in dominance; and 4) turbid waters with phytoplankton in dominance (including algal blooms). Experimental reflectance spectra for some of these classes are shown in Figures 28-60 and 28-61. The last class, turbid algal waters, may display a variety of colors, including amber, yellowish-green, green-brown, turquoise, and purple. This variety results mainly from the variety in absorption spectra, as scattering *per se* is, by comparison, spectrally featureless.

In class 2, as the chlorophyll concentration in-

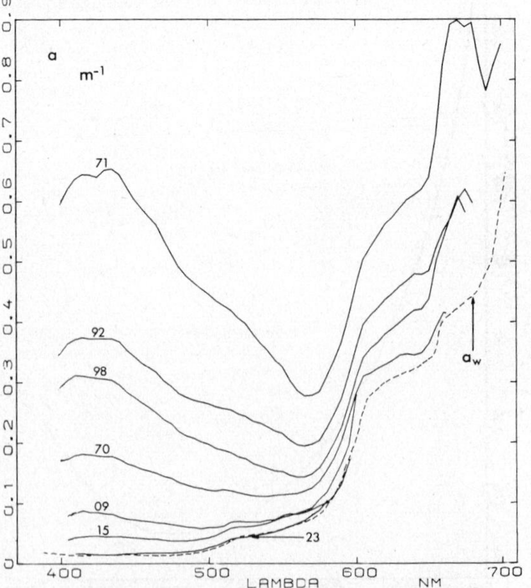

Fig. 28-56. Spectral values of absorption coefficient (m^{-1}) for stations in chlorophyll-rich green oceanic waters (Morel and Prieur, 1977).

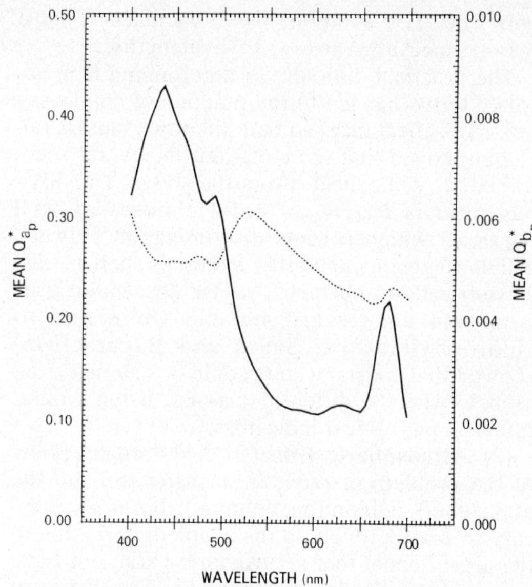

Fig. 28-57. Mean values of diffuse absorption and scattering coefficients for laboratory grown *Monochrysis lutheri* (Kiefer *et al.*, 1979).

creases, blue reflectance decreases, finally showing a minimum near 440 nm, the Soret absorption-band for chlorophyll *a in vivo*. A maximum is formed in the green band near 570 nm. Between the blue and the green a cross-over tends to develop, called by Duntley (1963) the hinge point, at which reflectance remains nearly constant. Many examples that confirm the hinge point in field situations are now in the literature. Figure 28-62 shows uncorrected reflectance spectra obtained at 100 m altitude (4 × 100 m FOV) by an aerial scanner off the coast of British Columbia (Borstad et al., 1980; see also Bressett, 1974). The

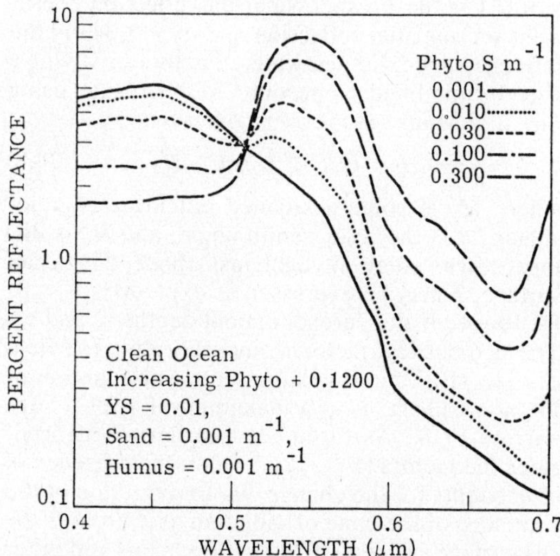

Fig. 28-58. Calculated reflectance spectra with increasing phytoplankton concentration (Suits, 1973).

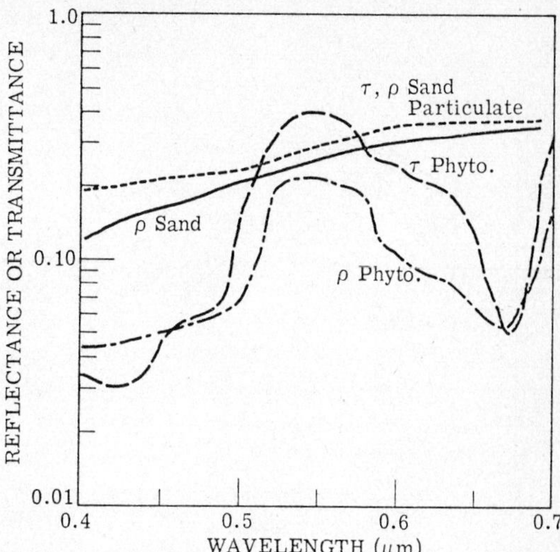

Fig. 28-59. Relative spectral properties of sand and phytoplankton (Suits, 1973).

reflectance peak at 685 nm is due to chlorophyll *a* fluorescence (Neville and Gower, 1977).

Reflectance spectra for single species of phytoplankton may be computed after laboratory measurement of absorption and scattering properties (Duntley et al., 1974; Kiefer et al., 1979). Computed spectra may be compared with field spectra, but promising results should be expected only when a single species dominates in the field situation.

(2) Inversion Methods—For remote sensing the problem is to derive concentrations of oceanic components from measured reflectance or upwelling radiance spectra. From Morel (1980), if there are L absorbing components (phytoplankton) and

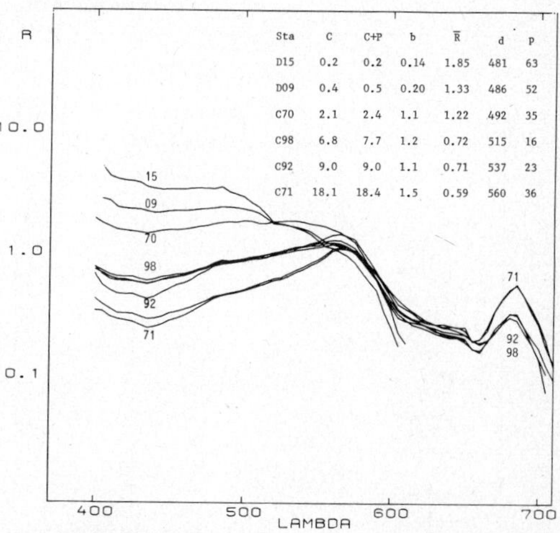

Fig. 28-60. Field reflectance spectra for chlorophyll-rich green water stations of Figure 28-56. C—chlorophyll *a* mg/m³; C + P—chlorophyll *a* plus phaeophytin; b—scattering coefficient at 550 nm; R̄—average reflectance ratio in percent; λ_d and p—CIE dominant wavelength and purity (Morel and Prieur, 1977).

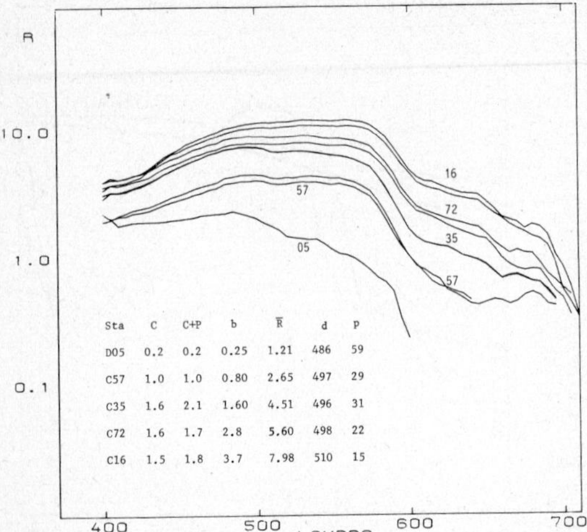

Sta	C	C+P	b	$\bar{R}$	d	P
D05	0.2	0.2	0.25	1.21	486	59
C57	1.0	1.0	0.80	2.65	497	29
C35	1.6	2.1	1.60	4.51	496	31
C72	1.6	1.7	2.8	5.60	498	22
C16	1.5	1.8	3.7	7.98	510	15

Fig. 28-61. Field reflectance spectra for waters dominated by inorganic particles. See legend of Figure 28-60 (Morel and Prieur, 1977).

as well a non-absorbing particulate fraction, there are $2L + 1$ unknowns: L phytoplankton concentrations and $L + 1$ backscattering coefficients. The phytoplankton absorption spectra are presumed known as they can be measured in the laboratory, using an opal-glass, wet-filter, or integrating-sphere technique to eliminate the effects of cellular scattering. Relative absorption spectra are obtained quite easily, but absolute spectra are obtained only with care. The scattering spectra are even more difficult to obtain. At sea, perhaps only the total scattering b is obtained as sea truth. If a spectral law is assumed for the phytoplankton scattering via a choice of m, and B is presumed, then the $L + 1$ scattering unknowns are reduced to

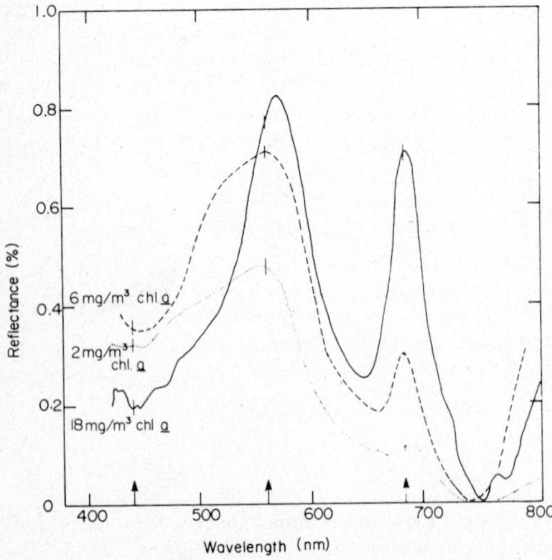

Fig. 28-62. Reflectance spectra (uncorrected) for three locations off British Columbia with varying chlorophyll concentrations (average concentration in 0-5 m layer) (Borstad *et al.*, 1980).

only one. For solution, the reflectance R must then be measured at $L + 1$ wavelengths.

The practical difficulty of determining R in absolute terms has led to the practice of employing ratios of reflectance, so that unknown factors (including noise) that are covariant in several spectral bands will cancel. From the above, one obvious ratio is R_{440}/R_{560}. Morel chooses as well R_{440}/R_{400}, which permits discrimination between cellular pigments and other materials such as (dissolved) yellow substance, whose absorption rises from 440 nm toward 400 nm while that of chlorophyll drops. Smith and Baker (1978) showed that coastal waters could be characterized as belonging to different classes, using similar ratios as bio-optical indicators.

(3) Atmospheric Effects—A thorough review of the problem of radiative transfer through the atmosphere is found in Volume I, but it is necessary to briefly touch on the problem here, due to the great impact that atmospheric variations have on remote sensing of ocean color and its constituent causes. The impact is large because ocean diffuse-volume reflectance is very small compared to the atmospheric contributions to radiance reaching the sensor at altitude. Also, the concentrations of chlorophyll a and other components which are of interest to oceanographers are often small: for chlorophyll a, concentrations over the open ocean less than 0.5 mg/m³ are frequent.

For a nadir view, the radiance N above the ocean at the sensor is

$$N = N_p + T\,(N_w + N_g + N_s), \qquad (28\text{-}7)$$

where N_p is path radiance, T is atmospheric transmission, N_w is upwelling water volume radiance above the surface, N_g is sun glint radiance, and N_s is reflected skylight radiance. N_w and R, diffuse volume-reflectance below the surface, are related by $N_w = H(1 - r_u)\,R/\pi n_0^2$, where H is downwelling irradiance above the surface, $r_u = 0.021$ is the Fresnel water-air reflection coefficient for specular reflection and $n_0 = 1.343$ is the refractive index of sea water. Factors involving r may be modified to account for sea state using Cox and Munk (1955). H is given by:

$$H = H_0 T^{\sec\theta} \operatorname{Cos}\theta + H_d \qquad (28\text{-}8)$$

where H_o is solar irradiance outside the atmosphere, θ is the solar zenith angle, and H_d is diffuse downwelling skylight just above the ocean surface. T may be expressed as $\exp(-\alpha(\tau_R + \tau_A))$, for Rayleigh and aerosol optical depths τ_R and τ_A. The geometrical factor α, incorporating the field of view (FOV) and volume scattering function of the atmosphere, is at a maximum of 1 for a very narrow FOV. As formulated earlier, R incorporates the factors $(1 - r_u)/n_o^2 \simeq 0.5$; the presence of n_o^2 accounts for the change due to refraction of the f-number of the cone of radiation as it crosses the sea surface. Austin (1974) discusses this and other water surface effects on the remote sensing of spectral radiance from below the sea surface.

N_g and N_s are complicated functions of atmospheric aerosols, sea state (wind velocity), and viewing geometry. N_p may be resolved into N_R and N_A, the Rayleigh and aerosol radiance contributions, respectively. Through atmospheric models these may be related to τ_R and τ_A (Jain and Gluckstein, 1981).

The factor in r_u is minimized by avoidance of high solar angles and high sea states. It may be estimated, via appropriate spectral assumptions, from N as measured in the reflective infrared bands where R is essentially zero.

Detailed correction algorithms to remove the atmospheric influences from N have been proposed by Gordon and Clark (1980), Morel (1980), Viollier et al. (1980), and Sturm (1981). Unfortunately, each proposed algorithm requires presumptions about aerosol spectral-properties, despite the fact that aerosol concentration and spectral properties are highly variable. Simpler approaches that also require presumptions may, therefore, be effective, such as the chromaticity method used by Munday et al. (1979) for Landsat. Gordon and Clark (1980) assessed the accuracy that is possible in measurement of ocean chlorophyll-concentrations using a detailed atmospheric correction algorithm, with a chlorophyll-retrieval algorithm of the form $\log C = a + b \log (R_1/R_2)$; they concluded that errors less than $\pm$ 1/4 $\log_{10}C$ are achievable over a wide range of pigment concentrations for the waters studied thus far. Their algorithm is being used on data from the Coastal Zone Color Scanner of Nimbus 7. Tassan (1979) used sensitivity analysis to compare retrieval algorithms. Errors of 40–80% of C between 0.2 and 10 mg/m^3 for the ratio algorithm dropped to 15% with a difference expression, $\log C = a + b (R_1 - R_2)$. Sturm (1979) found differences up to 50% when comparing the Gordon and Clark (1980) and Morel (1980) algorithms.

b. Passive Aerial Techniques

Early airborne investigations of chlorophyll spectral characteristics in oceanic environments were conducted by Clarke et al. (1969). Typical reflectance curves for two different chlorophyll concentrations are shown in Figure 28-63. The hinge point and other features of these curves led to successful use of ratios of spectral radiances for chlorophyll detection and measurement. Airborne multi-spectral scanners with numerous narrow spectral bands allow optimal selection of bands for such ratios. Arvesen et al. (1971), McNeil et al. (1976), and Miller et al. (1976) have used ratios to locate and measure chlorophyll concentrations. Other investigators have used multiple regression techniques for the quantitative mapping of chlorophyll *a* concentrations, thereby correlating aircraft multispectral scanner data with point-sampled data (Johnson, 1978). Figure 28-64 shows a mapping of chlorophyll concentrations in the New York Bight on April 13, 1975. The instrument was the NASA Ocean Color Scanner (OCS), the aircraft prototype of the Nimbus 7 Coastal Zone Color Scanner (CZCS) discussed below. In subsequent experiments, continuous surface measurements were obtained by fluorometry with a moving vessel in radio contact with the craft (Johnson et al., 1981). A further step would be telemetry of chlorophyll data to the aircraft to aid in the selection of flight lines.

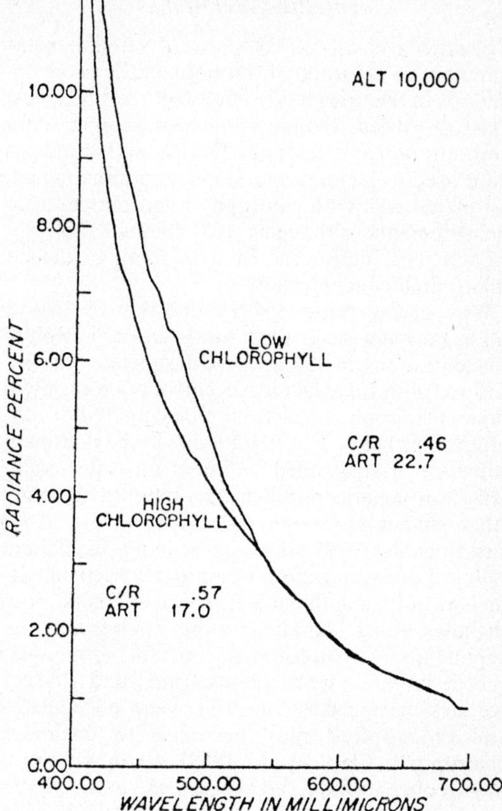

Fig. 28-63. Upwelling radiance spectra obtained over Georges Bank, Nova Scotia (Clarke *et al.*, 1969).

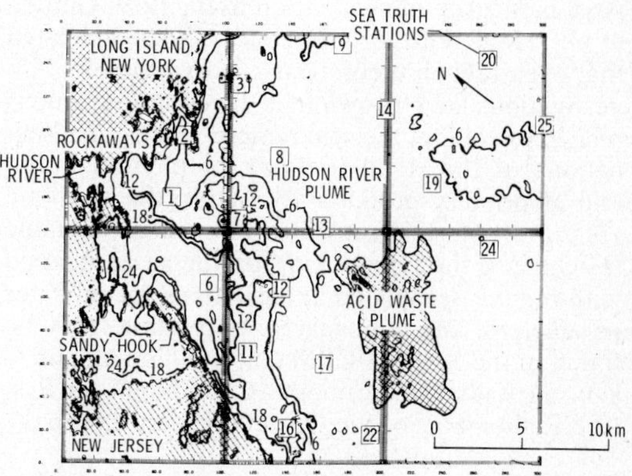

Fig. 28-64. Quantitative distribution of chlorophyll *a* in the New York Bight apex on April 13, 1975 (Johnson, 1978).

c. Satellite Techniques

Szekielda et al. (1975) reported satellite measurement of chlorophyll (phytoplankton) concentrations in the upwelling regions off the northwest coast of Africa. Photographic and scanner data from aircraft and satellite (Skylab and Landsat) were used to locate ocean color-variations which correlated well with chlorophyll and temperature measurements, although a two-channel ratio approach was ineffective in determining surface chlorophyll concentrations.

Bowker and Witte (1977) found that the correlation between suspended solids and chlorophyll concentrations in the lower Chesapeake Bay interfered with Landsat measurement of chlorophyll alone. Chlorophyll *a* could not be detected in concentrations below 5 to 10 $\mu g/\ell$ due to the dominant influence of suspended sediment on reflectance. This is a general problem encountered in many other studies (Johnson, 1978; Kim et al., 1979; Tassan et al., 1979) and is severe in marine waters with large concentrations of inorganic particulates, such as in the confluence of strong currents, or in shallows along shorelines experiencing current-dependent resuspension of bottom sediments. Where inorganic particulates from such sources are lacking, the correlation between particulates and chlorophyll may be used to estimate chlorophyll (Clark et al., 1980). Even if chlorophyll concentrations are very high, as in phytoplankton blooms with values up to 1000 $\mu g/\ell$, simultaneous variations in chlorophyll and inorganic particulates will still impede accurate chlorophyll measurements (Munday and Zubkoff, 1981).

With an atmospheric correction algorithm devised by Gordon and Clark (1980), and prelaunch calibration factors, data from the CZCS on the Nimbus 7 spacecraft have been processed for chlorophyll concentrations in Florida waters, as seen in Color Figure 28-65 (Hovis et al., 1980). However, unless concurrent shipboard measurements are used to confirm these distributions, the results should be interpreted in the light of the limitations discussed above with regard to chlorophyll-particulate mixtures (Sorenson, 1979; Gordon et al., 1980). Further, Hojerslev (1980) indicated that even though theoretical and experimental investigations have shown that the euphotic zone is related to color of the sea, remote sensing determinations of the standing stock of phytoplankton and of primary productivity within the euphotic zone are not satisfactory at this time. Valuable insight into the interlocking problems associated with remote sensing data, algorithms for in-water parameters, and atmospheric correction, may be found in the Sixth IUCRM (Inter Union Commission on Radio Meteorology) Colloquium on Passive Radiometry of the Ocean (Gower, 1980).

d. Laser Fluorosensors

Considerable effort has been directed during the last decade toward laser fluorosening of water quality parameters. Airborne laser fluorosensors emit downwelling radiation pulses that enter the water column and induce fluorescence in plankton and other pigmented particles, and in some dissolved organic materials. Upwelling fluorescence emission that survives attenuation in the water column and the atmospheric path is collected by a large-aperture telescope and converted to a detectable signal by a photodetector. The laser fluorosensor equation as derived by Browell (1977) yields the concentration of a vertically distributed fluorophore n_f in terms of measured, system, and environmental factors:

$$n_f = \left(\frac{P_F H^2}{P_L}\right)\left(\frac{4\pi\Delta_F}{T\eta_{tr}\eta_{\text{rec}}}\right) \times \left(\frac{\mu_w^2 \exp[H(\beta_L + \beta_F)]}{(1 - R_w)^2}\right)\left(\frac{k_L + k_F}{\sigma_f}\right), \text{mg/m}^3 \tag{28-9}$$

where

P_F = peak detected fluorescence emission power at fluorescence wavelength *F, W;*

P_L = peak detected laser output power at laser wavelength *L, W;*

H = distance from sensor to fluorescent target, m;

T = effective collection area of telescope receiver, m^2,

Δ_F = ratio of fluorescence emission bandwidth to effective filter bandwidth, with a minimum value of 1;

η_{tr} = laser (transmitter) efficiency;

η_{rec} = telescope (receiver) efficiency;

R_w = specular reflectance for the air-water interface at normal incidence for visible spectrum;

μ_W = refractive index of water for visible spectrum;

β_L = atmospheric beam attenuation coefficient at laser wavelength *L*, m^{-1};

β_F = atmospheric beam attenuation coefficient at fluorescence wavelength *F*, m^{-1};

k_L = effective attenuation coefficient for water at laser wavelength *L*, m^{-1};

k_F = effective attenuation coefficient for water at fluorescence wavelength *F*, m^{-1}; and

σ_f = effective *in vivo* fluorescence cross section for the fluorophore (chlorophyll *a*) for complete fluorescence band centered at wavelength *F* when excited at laser wavelength *L*, m^2/mg.

Variable water-attenuation coefficients have until recently limited the accuracy obtainable in measurement of n_f. Bristow et al. (1981) discuss the derivation of the above equation, giving particular attention to the water attenuation coefficients and the fluorescence cross section σ_f. A method for correction of variations in the water attenua-

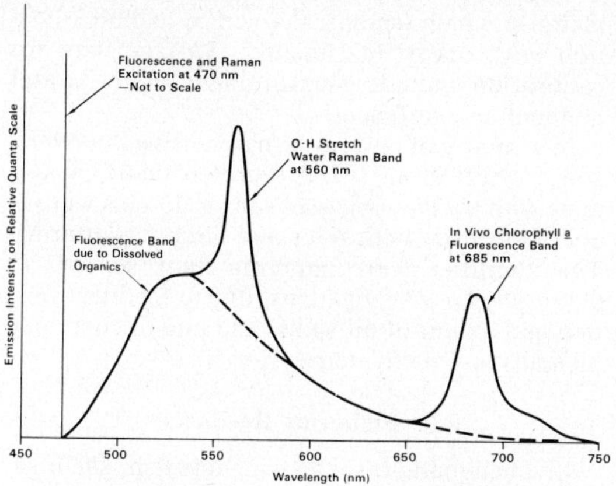

Fig. 28-66. Schematic showing spectral emission typical of many fresh and marine waters produced by excitation at 470 nm (Bristow *et al.*, 1981).

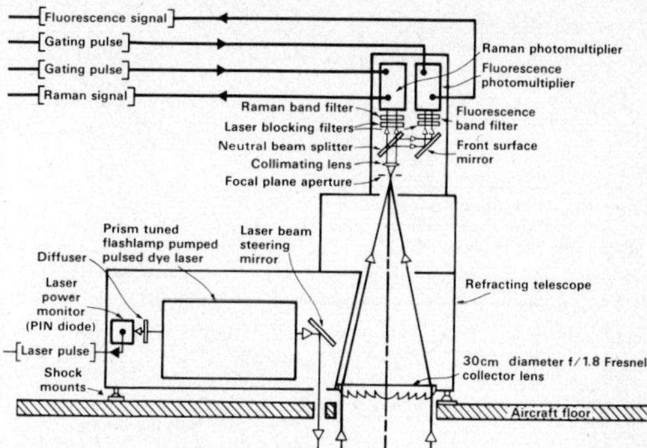

Fig. 28-67. Optical diagram of airborne laser fluorosensor for monitoring chlorophyll fluorescence and water Raman signals (Bristow *et al.*, 1981).

tion coefficients, now used by Bristow et al., Hoge and Swift (1980a), and others, is the normalization of results against laser-induced water Raman emission, from the 3418 cm^{-1} O-H vibrational stretching mode of liquid water. When excited at 470 nm, water Raman emission occurs at 560 nm as illustrated in Figure 28-66, in the same region as Gelbstoff emission. Careful selection of the excitation wavelength is required to position the Raman emission away from fluorescence peaks of substances under study, but near enough that spectral variations in attenuation coefficients may be neglected.

A diagram of the instrument used by Bristow et al. is shown in Figure 28-67, with digitized output in Figure 28-68 (the Raman pulse is delayed 3 μsec to separate it from the fluorescence pulse). Results in Figure 28-69 show success in measurement of chlorophyll *a* to an accuracy of roughly 20 percent in fresh waters.

The variability in fluorescence cross-section, σ_f, according to species, light history, cell vigor, and excitation wavelength, makes surface calibration necessary. The variation in σ_f with excitation wavelength is shown in Figure 28-70 for four species of phytoplankton. The species dependence, due to species variations in pigment composition and inter-pigment energy transfer within the photosynthetic apparatus, has been exploited in a multispectral laser technique for remote discrimination of algal color groups. This technique is implemented in an Airborne Lidar Oceanographic Probing Experiment (ALOPE) system developed and laboratory tested at the NASA Langely Research Center (Brown et al., 1978). The instrument is a unique four-color dye-laser system pumped by a single linear xenon lamp. Fluorescent dyes, which lase at 454, 539, 598, and

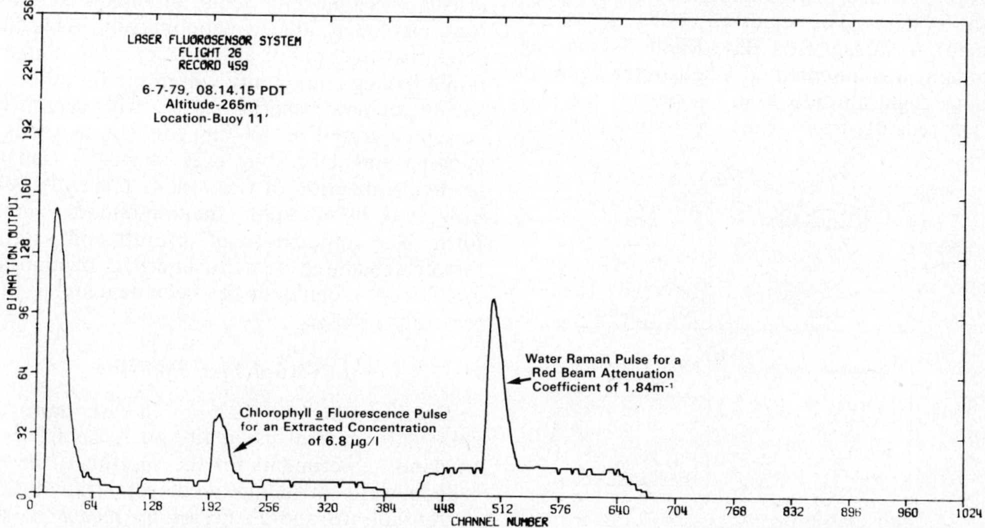

Fig. 28-68. Digital laser fluorosensor record, showing time sequence of airborne laser fluorosensor signals obtained over Lake Mead, Nevada on 7 June 1979 for a measured chlorophyll concentration of 6.8 μg/l. Raman pulse delayed 3 μ sec (Bristow *et al.*, 1981).

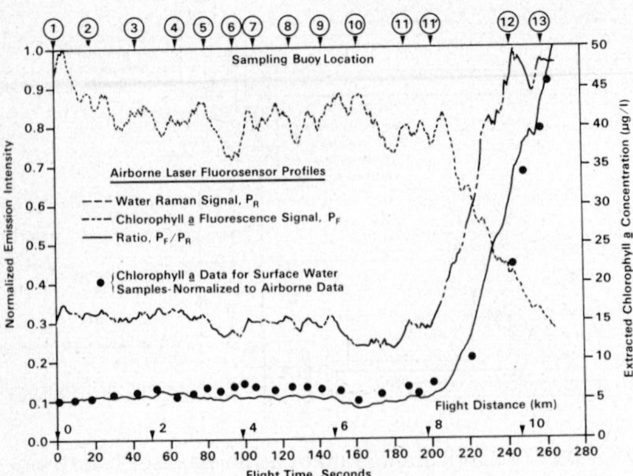

Fig. 28-69. Airborne laser fluorosensor profiles of P_F, P_R, and P_F/P_R for flight of 7 June 1979 over Las Vegas Bay. Profiles smoothed using a moving 5-point average scheme. Also shown are the chlorophyll *a* values for the 28 ground truth samples, which have been normalized to the laser fluorosensor profile for P_F/P_R (Bristow *et al.*, 1981).

617 nm, are the active media in four individual dye lasers, which are separately activated using a rotating intercavity shutter. Induced fluorescence from chlorophyll *a* in the water is collected by a telescope, passed through a narrow band-pass filter (centered at 685 nm), digitized, and recorded. Results of laboratory tests using mixtures of species from each of four algal color groups (blue-green, golden brown, green, and red) have produced correlation coefficients (between samples and ALOPE measurements) of 0.994 for total chlorophyll *a* and from 0.81 to 0.88 for the individual color groups.

Results from a NASA/University of Rhode Island/EPA experiment suggest the applicability of the ALOPE system for monitoring total chlorophyll *a* and phytoplankton diversity in estuarine systems (Farmer et al., 1980). In an experiment in Narragansett Bay, Rhode Island, the instrument was mounted on a helicopter platform (nominal flight altitude 30 m); however, the system has recently been evaluated from an aircraft

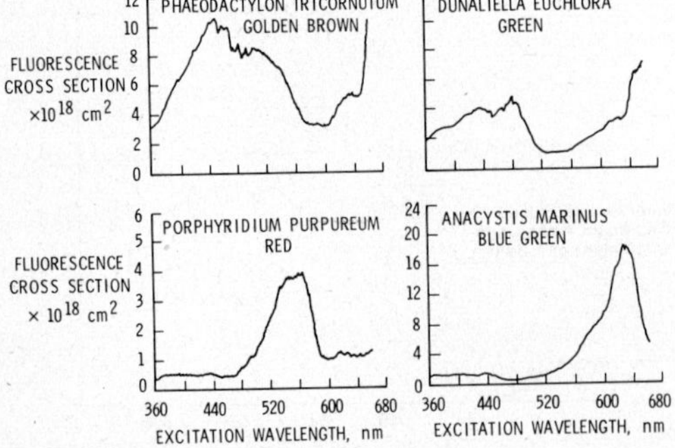

Fig. 28-70. Variation of fluorescence cross section, σ_f, with excitation wave length for four species of phytoplankton (Brown *et al.*, 1978).

platform which increases coverage and coordination with other instruments. Surface data for calibration include measurement of the water attenuation-coefficient.

In a survey of current research using fluorosensors, Jarrett et al. (1979) reported that four systems with a wide variety of sensor designs were in operation from helicopter and aircraft platforms. Two facilities were applying fluorosensors to chlorophyll *a* investigations, one to the identification and extent of oil spills, and one to combined oil spills and bathymetry.

2.0. Fisheries Resources

Living aquatic resources include fish, shellfish, aquatic plants and algae, aquatic mammals, and turtles (Table 28-8), and can be categorized into three environmental regimes—marine, coastal and estuarine, and fresh water regions. Fisheries resources are those living aquatic resources being utilized by a fishery for either recreational or commercial purposes.

Remote sensing of aquatic animals can be either direct or indirect (Maughan, 1969; Drennan, 1970; Laurs, 1972). The latter involves acquisition of ancillary environmental data that can be related to the distribution and abundance of the resource. Direct remote sensing, the observation of the animals themselves, is limited to the few upper meters of the water column when the individual or groups of animals (schools or pods) are near the surface. An important application of direct remote sensing is the finding of fish, both commercial and recreational (see Tomczak, 1977).

In the early 1970's, approximately 70 million tons of conventional types of fish were harvested annually; from a conference on fisheries management sponsored by the Food and Agriculture Organization of the United Nations in 1973, the consensus was that only about 30 million additional tons can be produced economically. This indicates that the goal of all elements of the international fishing community will in the future be the search for new fisheries stocks. This search will be concentrated on looking for new methods of locating and harvesting present stocks, and upgrading utilization of fish stocks presently being harvested, by marketing them in a more valuable form. The application of aircraft and satellite remote-sensing systems to improve location and assessment techniques has been developing since the early 1960's.

a. Visual Fish Spotting

Wood and McGee (1925) in an early paper discussed the idea of using aircraft to spot fish. In England, experiments for the spotting of herring schools were carried out on behalf of the Ministry of Agriculture and Fisheries by means of seaplanes operating from Felixstowe in 1921 and from Plymouth in 1923. The results were negative, but operations were limited and conditions too unfa-

TABLE 28-8

Representative Fish Types Observable From Low Level Aircraft (after Stevenson and Pastula, 1971)

ATLANTIC OCEAN AND MEDITERRANEAN SEA		*PACIFIC OCEAN AND INDIAN OCEAN*	
Eastern	*Northern (Continued)*	*Eastern*	*Eastern (Continued)*
Fish:	Fish:	Fish:	Fish:
Spanish sardine (*Sardinella aurita*)	ladyfish (*Elops saurus*)	basking shark (*Cetorhinus maximus*)	ocean sunfish (*Mola mola*)
herring (*Sardinella eba*)	blue runner (*Caranx crysos*)	white shark (*Carcharodon carcharias*)	striped bass (*Morone saxatilis*)
Spanish mackerel (*Scomberomorus maculatus*)	tarpon (*Megalops atlantica*)	northern anchovy (*Engraulis mordax*)	Pacific saury (*Cololabis saira*)
yellowfin tuna (*Thunnus albacares*)	herring (*Clupea harengus*)	Pacific sardine (*Sardinops sagax*)	swordfish (*Xiphias gladius*)
skipjack tuna (*Katsuwonus pelanis*)	Atlantic mackerel (*Scomber scombrus*)	Pacific bonito (*Sarda chiliensis*)	striped marlin (*Tetrapturus audax*)
pilchard (*Sardinops trachurus*)	butterfish (*Poronotus triacanthus*)	jack mackerel (*Trachurus symmetricus*)	Mammals:
Northern	Atlantic menhaden (*Brevoortia tyrannus*)	Pacific mackerel (*Scomber japonicus*)	gray whale pilot whale Blackfish (killer whale)
Fish:	*Mediterranean Sea*	Pacific barracuda (*Sphyraena argentea*)	Porpoise and dolphin Seals and sea lions
thread herring (*Opisthonema oglinum*)	Fish:	yellowtail (*Seriola dorsalis*)	
Spanish mackerel (*Scomberomorus maculatus*)	Spanish sardine (*Sardinella aurita*)	white seabass (*Cynoscion nobilis*)	Invertebrates: Squid Jellyfish
bluefish (*Pomatomus saltatrix*)	Atlantic mackerel (*Scomber scombrus*)	bluefin tuna (*Thunnus thynnus*)	*Western and Indian Oceans*
gulf menhaden (*Brevoortia patronus*)		albacore tuna (*Thunnus alalunga*)	Fish:
		yellowfin tuna (*Thunnus albacares*)	pilchard (*Sardinops pilchardus*)
		skipjack tuna (*Katsuwonus pelamis*)	sardine (*Sardinella fimbriata*)
		jacksmelt (*Atherinopsis californiensis*)	mackerel (*Rastrelliger kanagurta*)

vorable to permit any definite conclusion of general application to be drawn.

Bullis (1967) stated "The use of aircraft by the fishing industry has become widespread over the past twenty years, and a technology of sorts has been developed by highly specialized commercial fish spotters. Some of our major fisheries have reached a point of near dependence on this technology—the menhaden fishery, for example, and to an extent the tuna fishery."

Roithmayr (1971) reported 34 types of fish that can be detected at night. These types of fish are located throughout the oceans of the world, but most species are coastal inhabitants. Leatherwood et al. (1972) include several aerial photographs in their guide to identification of whales, porpoises and dolphins, and report locating pods of whales and schools of porpoises and dolphins from low-flying aircraft. From these references it can be seen that a broad range of surface-appearing fish and mammals can be detected from low-altitude platforms (Color Figure 28-71).

Visual detection of fish schools from aircraft is limited to about 3 km above the surface, with the great bulk of fish-spotting done from 150 to 950 m. Several types of utility aircraft are used, with the general trend towards the use of twin engine aircraft, preferred because of greater cruising range and improved safety factors in operations over water.

Commercial fish-spotters, working in direct support of the fishermen in coastal waters (up to 15 km from land), still use single engine aircraft because of the greater maneuverability and slower air speeds. This allows the fish-spotter to hover over the target fish school until the harvesting operation is completed. Most aircraft used for fish-spotting on a regular basis are modified to maximize the amount of time in the air by the addition of extra fuel tanks, pilot relief facilities, and additional surface-to-plane communications.

Squire (1972) summarized the elements in a visual fish-spotting system: "Specific observation of a fish school has three phases: (1) distinguishing a school, (2) identifying the species, and (3) estimating the weight of the school. The detection of near-surface schools during the day is dependent upon the pilot's ability to distinguish subtle color and light-intensity differences in the water. Detection of schools at night is possible only during the dark period of the moon and depends on the pilot's ability to discern gradation of light intensity. Bioluminescence of planktonic organisms agitated by schooling fish indicates by a dull glow the location and size of the school. Species are identified during the day on the basis of a combination of two or more of the following characteristics: color of school or individual fish, shape of school, and behavior and size of individuals within the school. At night, species identification is based on shape of the luminous area and behavior of the schooling fish under undisturbed conditions, or by the behavior of the school after being subjected to a stimulus from an external source such as a flash from the aircraft's landing light."

The efficiency of using aircraft for detecting living marine resources has been thoroughly documented (Cushing et al., 1952). In the United States the services of the fish-spotter are vital to the success of menhaden, tuna, and anchovy fisheries.

Remote observation has also been used by fishery biologists to gain information on distribution and abundance of pelagic, near-surface schooling fish (Squire, 1972; 1973). Sette (1949) investigated aerial surveys for sardine abundance off southern California. Jones and Sund (1967) found that aerial survey for tuna was about 2½ times more efficient than a vessel at locating fish schools, while the U.S. Navy found that aircraft were 20 times better than ships for biological observations (Levenson, 1968). Despite several studies demonstrating the applicability of aerial surveys for evaluating fish stocks (Squire, 1972; Roithmayr and Wittman, 1972), aerial techniques have not been implemented operationally for biological study.

Marine mammals, especially dolphins, have been subjects of major visual remote sensing surveys for about the last five years. Surveys are prompted by public concern about the status of marine-mammal populations throughout all coastal regions of the United States and in the eastern Tropical Pacific. Various visual techniques have been evaluated to identify acceptable methods for computing reliable estimates of stock size (Leatherwood et al., 1978).

b. Photography of Fish Schools

Pelagic fish stocks constitute a sizeable portion of the U.S. fish production and several underutilized species have been identified as having tremendous potential for increased production (Bullis and Carpenter, 1968). Historically, investigations of their distribution and abundance have been from slow-moving ships covering only small portions of the ocean surface.

Studies have shown that pelagic fish lend themselves well to aerial photographic sensing (Bullis, 1967; 1970; Benigno, 1970) for distribution and abundance, but it must be emphasized that photography will provide only relative-abundance estimates. The principal advantage of photography over other survey techniques is large area coverage for nominal cost.

Photographic data may also be used to calibrate satellite data. In studies initiated in July 1972, photographically detected menhaden (*Brevoortia patronus*) distribution patterns were correlated with Landsat MSS imagery. Based on selected oceanographic parameters and commercial catch information, menhaden-distribution prediction models were developed (Kemmerer et al., 1973; Ewing, 1973). The results achieved would not have been possible without the timely and synoptic distribution information acquired through photographic sensing.

(1) Photographic Parameters—As in terrestrial photography, aerial photography of the marine environment requires clear, cloudless skies for optimum results. "Noise", from diffusely scattered light that is reflected from the surface of the water, blurs imagery and reduces contrast. Unlike terrestrial photography, high solar elevation angles are detrimental to marine photography. Once the solar angle approaches the acceptance angle of the lens, approximately 50° for a 6-inch lens and 9-inch format, the mirrored image of the sun or its halo will occur on the photograph.

At low solar angles, available light restricts photography. The amount of light entering the water through the surface at angles less than about 20° from the horizontal decreases rapidly, as does the incident light per unit square area. Also, marine photography demands more film speed because of the relatively high absorption of light by water. An increase in exposure as great as 2.5 f-stops over ordinary exposure values for terrestrial photography has been found necessary for maximum penetration and school detection.

Tests of different cameras, films, and filter combinations have shown that the most effective system for detecting fish schools in usual coastal waters is a 9-inch format aerial mapping camera, equipped with 6-inch lens that has been color- and near-infrared-corrected, and supplied with color infrared film and a minus-blue antivignetting filter. No one film is best suited for all water conditions and species of fish that may be encountered. Benigno and Kemmerer (1973) concluded that color infrared film (Color Figure 28-72) produces the best results under most conditions.

Multispectral photography is useful for the selection of acceptable film and filter combinations for photography in specific circumstances. A flexible and useful system consists of four 70 mm (film width) cameras equipped with 80 mm (focal length) lenses clustered in a mount, boresighted, and connected to take four photographs simultaneously. Table 28-9 lists film and filter combinations for many applications.

In 1975, a multispectral photography experiment was conducted off the Bahamas to select an acceptable film and filter combination for surveys of giant bluefin tuna (*Thunnus thynnus*). Four 70 mm cameras were clustered and mounted in a twin-engine aircraft flown at an altitude of 304 m and at an airspeed of about 130 km/hr over schools of bluefin tuna. Most of the fish encountered were at depths of 10 to 15 m in waters optically similar to offshore oceanic waters. The blue and green portions of the light spectrum produced good quality photography; green was preferred due to better contrast. The red and near infrared portions failed to image the tuna due to a rapid attenuation of longer wavelengths in water (NFEL, 1975).

A second multispectral photography test was conducted in May 1976 over waters 40 km off the coast of California to identify an acceptable combination of film and filters for imaging bottlenosed (*Tursiops truncatus*) and whitebelly (*Delphinus delphis*) dolphin (NFEL, 1976). Contrast measurements made on photographic transparencies with an image analyzer showed that the red band was optimum, the green band was acceptable, and the blue and near infrared bands were unacceptable.

(2) Photographic Interpretation—Where aerial photography is employed to gain information about the distribution and abundance of surface schooling fishes, an assumption is commonly made that the surface area of fish schools is related to biomass. This assumption, however, has not been adequately tested. Limited investigations suggest that the assumption may be valid for some species. At present, fish biomass or census estimates from photography should be used with reserve.

Significant correlation has been found to exist between surface area and fish-school biomass of thread herring (*Opisthonema oglinum*), although the statistics are based on a limited number of data points (Stevenson and Pastula, 1971). Aerial photographs of herring schools were acquired just prior to their capture by commercial purse seining vessels (Figure 28-73). School surface-area was measured by planimetry on the photographic imagery, and school weight was taken from the commercial vessel's catch logs.

Data from repetitive photographic missions in the Mississippi Sound show that within-day, day-to-day, and seasonal variations in size distribution (surface area) and mean school-size of menhaden remain fairly constant with time. The results indicate that surface area of menhaden schools may correlate with biomass.

Under most circumstances several species of pelagic fish will occur in the same geographical area. Resource managers and industry are concerned with these stocks on a species-by-species basis. Thus, there is a need to identify fish species remotely. Unless quite large, individual fish are not resolved at an efficient photographic scale, preventing detection of identifying characteristics. Investigations to determine how to identify species through characteristics of the school have had to rely on visual techniques used by professional fish spotters. These observers utilize a

TABLE 28-9

Summary of Film and Filter Combinations for Multispectral Photography of Marine Animals

Kodak* Panchromatic Film	Kodak* Wratten Filter	Approximate Spectral Sensitivity (nm)
Tri-X (2403)	98	400–500 (blue)
Tri-X (2403)	61	500–600 (green)
Tri-X (2403)	29	600–700 (red)
Infrared (2424)	89B	700–900 (near IR)

* Reference to trade name does not imply endorsement.

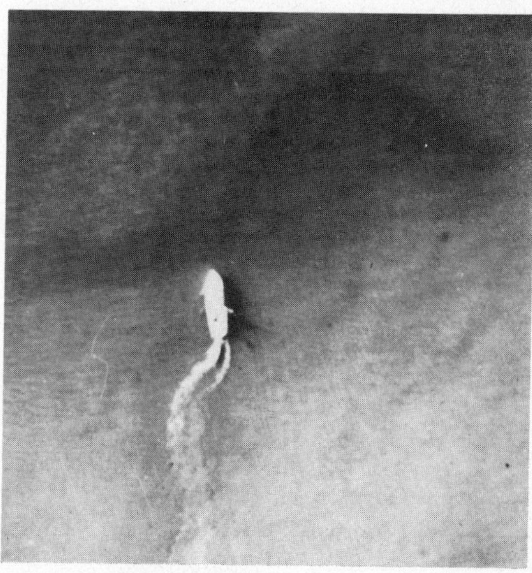

Fig. 28-73. School of thread herring (*Opisthonema oglinum*) photographed just prior to capture by commercial purse seining vessel (Stevenson and Pastula, 1971).

wealth of subjective information to identify species, and the synthesis of this information appears too complex to reconstruct for use in remote sensing. Moreover, much of the success of professional fish spotters appears to stem from their experience concerning the occurrence and distribution of fished species. Those criteria reportedly used most often for species identification are school behavior, texture, and color. Although configuration and texture are professed to be used in identification, how these parameters are used has not been defined. The more objective criterion for remotely identifying pelagic fish schools appears to be color, or more correctly, spectral reflectivity. Loya and Graves (1968) found differences in the spectral reflectivity of 15 species tested. Figure 28-74 depicts spectral signatures of three species of pelagic schooling fish. The usefulness of color film as a spectral radiometer for exploiting the spectral differences is minimized by its narrow exposure latitude, the selective spectral absorption of water, and the difficulty in predicting color-reproduction characteristics. Multispectral photography may be more successful.

c. *Low Light Level Image Intensification*

Bioluminescence is produced by thousands of species of marine animals and plants, including plankton. The lanternfishes and euphausid shrimp that predominate in the deep scattering layer possess luminous organs. Direct observations and the use of sensitive underwater photometers reveal the universal occurrence of bioluminescence in the oceans, the phenomenon occurring in all temperate seas, particularly during warm seasons. It occurs more often and with greater intensity throughout the year in tropical and subtropical seas. Studies in the Atlantic and Indian oceans, and in the Mediterranean Sea, have shown luminescent organisms always present where measurements were made with underwater photometers (Clarke and Wertheim, 1956; Clarke and Breslau, 1959; Clarke and Kelly, 1964).

The major concentrations of luminescing organisms are in the upper 100 m, usually within the lighted zone, and in waters where most pelagic fishes abound. Population densities of luminescent organisms vary considerably. The maximum concentration reported for Bahia Fosforescencia, Puerto Rico, was 7,600 cells per liter (Clarke and Breslau, 1960); cell densities as high as 220,000 per liter were found in Oyster Bay, Jamaica, in the West Indies (Seliger et al., 1962). The luminous dinoflagellate, *Pyrodinium bahamense,* is the most abundant organism in both bays.

Most bioluminescence in the sea is caused by dinoflagellates which emit light when stimulated. The light is produced by an enzyme-catalyzed oxidation of luciferin to an excited chemiluminescent form. When stimulated, each molecule of luciferin releases one photon. If sea water is stirred, the luminous discharges of individual cells look like sparkling crystals. If the water is agitated rapidly, the points of light emitted by dinoflagellates fuse into a bright glow. Turbulence resulting from the swimming motion of fishes provides the mechanical stimulation that outlines their bodies with light and produces a luminous trail.

The development of tactical night-vision devices for military use stimulated interest toward night detection of fish schools. Research by the U.S. Army Night Vision Laboratory at Ft. Belvoir, Virginia, produced the 2.7 kg Starlight Scope illustrated in Figure 28-75. The scope is sensitive to a wide spectrum of light from ultraviolet to infrared; it produces a green image that is magnified and free of scanning lines; the image is produced on a small screen that is viewed through the eyepiece.

Feasibility tests were conducted starting in 1968 from commercial fishing vessels, fixed wing aircraft, and a helicopter. The starlight scope was coupled to a closed circuit television camera as

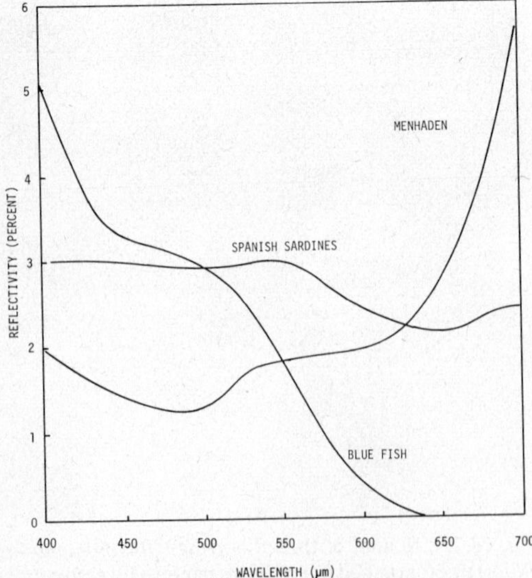

Fig. 28-74. Spectral signatures of three species of pelagic fishes (Stevenson and Pastula, 1971).

Fig. 28-75. The Starlight Scope uses an image intensifier tube which amplifies reflected starlight more than 40,000 times (Roithmayr and Wittmann, 1972).

illustrated in Figure 28-76 and located in the pilot house of a commercial mackerel seiner. After the fishermen had surrounded a school of Spanish mackerel with the net, all lights were extinguished and the images of the luminescing fish were detected with the scope and recorded on videotape (see Figure 28-77). The school is approximately 16 m from the scope and is 2 to 3 m below the surface. Each moving fish is outlined by a luminous trail produced by light from stimulated organisms in the fish wake. About 3,200 kg of Spanish mackerel were landed after the tests were completed.

Such feasibility tests using several types of low-light sensors demonstrate that: (1) bioluminescence associated with fish schools makes them conspicuous, (2) the school perimeter is usually well defined, (3) schools can be detected with the sensor and television camera from aircraft at 1500 m altitude, and (4) the school image can be recorded on videotape. A prototype low-light sensor camera and mount are shown in Figure 28-78, with the schematic in Figure 28-79.

The system in Figure 28-79 detected and recorded numerous schools of saury, anchovy and euphausid shrimp in surveys off the U.S. Pacific Coast in 1971. Thirty-seven night flights were made at altitudes ranging from 300 to 1200 m along predetermined track lines. Small commercial catches of saury were made by ten Japanese fishing vessels during the surveys. The videotape data showed that anchovy schools were brilliant, small in size, and circular in outline, while saury were dim, large sized, and irregularly shaped.

Further tests in 1971 in the Mississippi Sound resulted in the detection of more than 1000 schools of menhaden during one night. Luminescing schools as seen on the television screen are shown in Figure 28-80. A commercial spotter pilot estimated the large schools contained 30 to 35 tons each, while smaller schools contained 3 to 5 tons each. The results indicate that the low-light sensor detects schools that are invisible

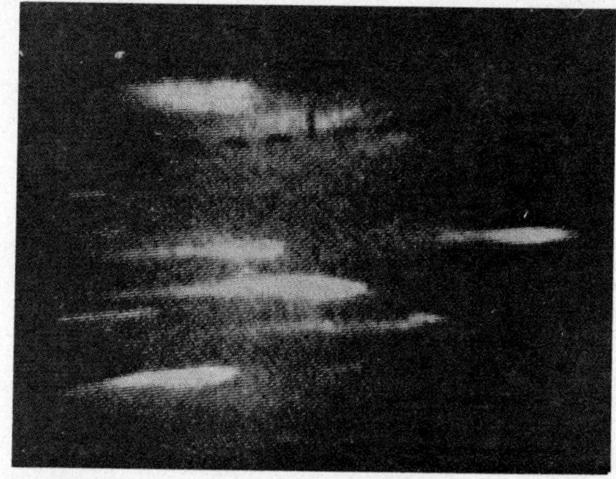

Fig. 28-77. A school of Spanish mackerel showing associated bioluminescence as detected with the Starlight Scope and seen on the television screen. The school is 2 to 3 m below the surface; its distance from the sensor is approximately 16 m (Roithmayr and Wittmann, 1972).

to the naked eye and records schools too numerous to be counted by an observer. Analysis of videotape data showed that schools covered as much as 0.005 km^2 in a total search area of 0.64 km^2.

A more recent system consists of a television camera fitted with an intensifier tube that amplifies light about 120,000 times, a 12.7 cm television monitor, a video tape recorder, and a power source (Vanselous, 1977). This system is used from aircraft at altitudes of 1000 m and at speeds of 185 km/hr during dark-moon periods. An improved version with better optics and an added scanning capability is in use for locating anchovy schools off Baja, California (Stevenson, 1975).

d. Laser Fish Finders

Recent advances in pulsed, high-powered laser systems operating in the blue-green portion of the visible spectrum suggest the use of airborne lasers

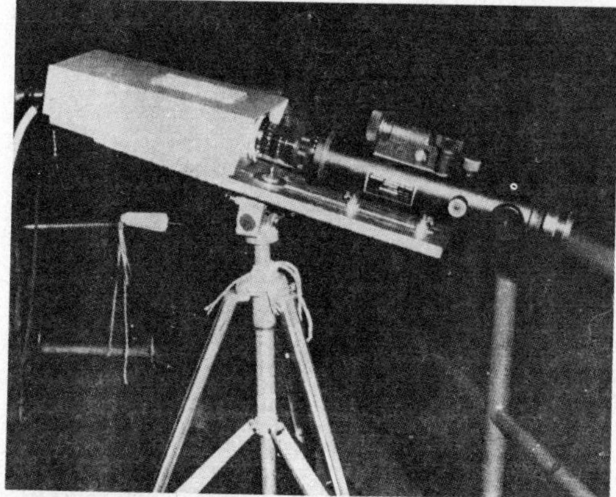

Fig. 28-76. The Starlight Scope coupled to a Plumbicon television camera for detecting liminescing Spanish mackerel schools (Roithmayr and Wittmann, 1972).

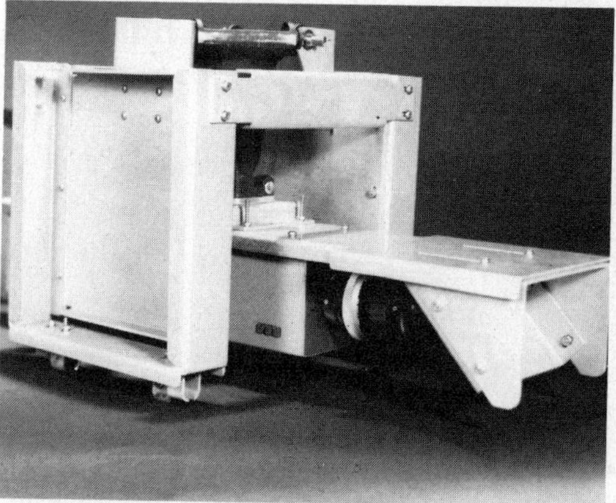

Fig. 28-78. Low light level image camera and aircraft mount (Roithmayr and Wittmann, 1972).

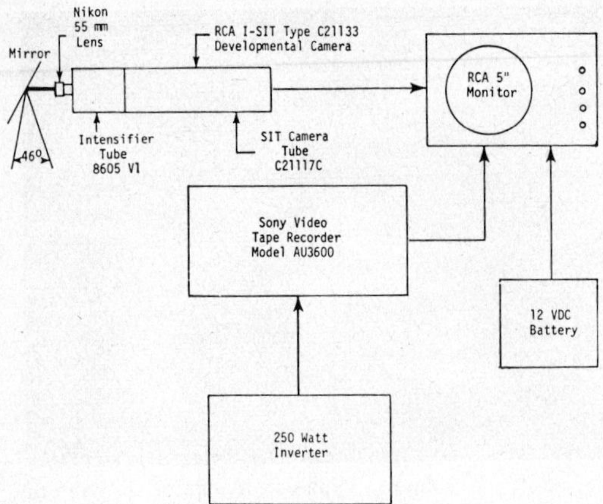

Fig. 28-79. Schematic showing the low light level image of intensification system (Roithmayr and Wittmann, 1972).

for evaluation of living marine resources. Computer simulations, combined with limited laboratory tests, indicate that a 2000 kw laser system scanning a 75 m swath width can detect fish at depths of 16 m from altitudes of approximately 1700 m. Reflected laser power received at an airborne detector varies as a function of laser power, beam radius, incidence angle, wind speed and direction, fish depth, water attenuation, and aircraft altitude (Murphree et al., 1974, 1976).

e. Radio Tracking

A form of direct remote sensing being used more frequently in studies of movement and distribution patterns of marine mammals and sea turtles is radio tracking. This requires the attachment of a small transmitter to an animal for transmissions to a receiver located in a surface vessel, aircraft, or spacecraft. In salt water, the transmitter antenna must be above the water surface.

Radio tracking of surfacing animals such as dolphins and sea turtles has depended extensively on the tracking technology developed for terrestrial animals. Most of these systems have a range of only about 18 km and a life of about 40 days. Sea water switches are generally used to activate the transmitters when the animals surface. Examples of tracking experiments for marine mammals are given by Evans (1971, 1974) and Kaufmann and Irvine (1975), and for sea turtles by DeBlanc (1979).

The limited range of most radio transmitters for animal tracking studies prompted investigators to develop transmitters which could be located by satellites (Grandy et al., 1977). The satellite systems which are utilized include the Random Access Measurement System (RAMS) on Nimbus-6 and more recently the ARGOS system aboard TIROS. Both are similar in that they enable suitable transmitters to be located several times a day with accuracies approaching a few kilometers.

A satellite transmitter weighing approximately 900 gm has been developed for dolphin studies in the Pacific (Jennings et al., 1979). Designed especially to operate for at least one year, the transmitter is equipped with a sea-water switch and a clock to optimize periods of transmission. Tests of the transmitter harnessed to the dorsal surface of a captive bottlenosed dolphin have been conducted and field tests on wild animals also have been scheduled.

f. Fishing Vessel Surveillance

Adoption of the Fishery Conservation and Management Act of 1976 by the United States, coupled with the launch of Seasat in 1978, prompted an investigation to determine if fishing vessel surveillance could be achieved from space as an aid to fisheries management and enforcement. The Seasat Synthetic Aperture Radar (SAR) was the sensor of primary interest.

Two pre-launch field tests were conducted in 1976 over concentrations of foreign fishing vessels operating off the east coast of the United States (Georges Bank) and in the Bering Sea (Woods and Ivey, 1977). These tests provided aerial SAR data of fishing vessels simultaneous with aerial photography and monitoring by other aircraft and surface vessels.

Analytical emphasis has remained on aerial SAR data due to the loss of Seasat several months after launch. Data processed digitally have demonstrated that vessels can be reliably detected and located by measurements of speed and direction (NFEL, 1978). Current emphasis is on vessel-classification techniques.

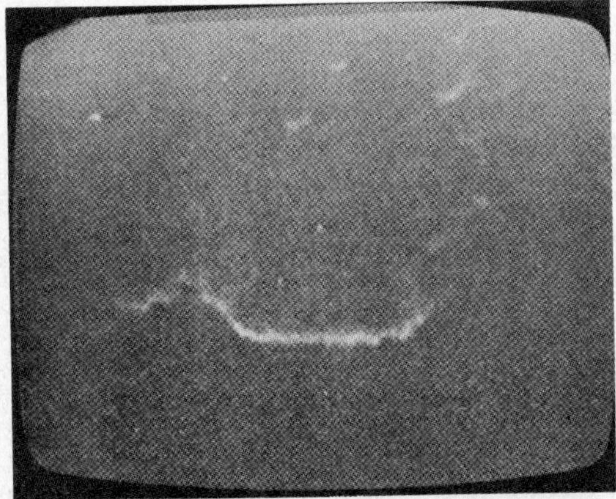

Fig. 28-80. Low light sensor video data showing schools of Gulf menhaden from 915 m. The horizontal field of view is 530 m (Roithmayr and Wittmann, 1972).

g. Indirect Methods of Fishery Assessment

Fishery assessment may be assisted by measurement of aqueous parameters which affect distribution and abundance. There exists a wealth of literature dealing with environmental effects on fish. Laevastu and Hela (1970) reviewed the influence of various environmental factors, noting that fish respond to the total environment and that it is difficult to attribute their spatial distribution to any single factor. However, most remote-sensing studies have been conducted in a manner that did not permit the luxury of complete suites of remote, conventional oceanographic, and fishing data. The common approach has been to investigate the relationship between a certain fish stock and a single remotely measured parameter. The options include temperature, salinity, chlorophyll, turbidity, and sea state, all of which may be measured remotely. Water surface temperature is the easiest to measure, if it is tolerable to restrict the observation to a few microns of depth. The physics of the measurement technique is discussed under Sea Surface Temperature.

(1) Temperature—Probably the best documented use of remote temperature measurements in fisheries is in connection with the Pacific Coast tuna fishery. Pearcy (1973) discussed the relationship between the albacore-tuna fishery and surface temperature patterns near the Columbia River. In summer during the fishing season, surface temperatures are controlled by upwelling of cold water along the coast and the influx of relatively warm Columbia River water. The first catches of the season were associated with the Columbia River plume. As the season progressed there was a northward migration of albacore, with highest catches along the oceanic edge of the plume in 15.5°C water. Surface temperatures were measured weekly by airborne thermal radiometers. Temperature contour-maps developed from these measurements provided the position of the river plume and upwelling. Location of the fish was determined by the interaction of these water masses as well as by the temperature *per se*.

(2) Salinity—Techniques for remote measurement of salinity are not as operational (see Salinity). Accuracies using microwaves are of the order of 1‰, useful for study of estuarine and nearshore fisheries. But the expense and complexity of instrumentation, and the relatively recent development of this technology, have constrained its application to fisheries studies.

(3) Chlorophyll—Methods to remotely measure chlorophyll concentrations received much attention during the 1970's (see Chlorophyll). Chlorophyll concentration is considered an index of basic biological productivity and in this sense is related to fishery production. Gower (1972), referring to a National Academy of Sciences summer study in 1969, noted that a chlorophyll concentration above 0.2 mg/m^3 indicates presence of sufficient plankton to support a commercially viable fish concentration; therefore a map of chlorophyll concentrations would be of great use in the fishing industry. Aside from this there is a paucity of studies explicitly defining, either in time or space, the relationship between chlorophyll concentration and fishery resources. Nevertheless, the implicit relationship between chlorophyll and fish makes this area of study one of continuing interest.

(4) Water Clarity—Water clarity has an effect on the distribution of certain fish stocks. Kemmerer and Benigno (1973) found a correlation between sightings of menhaden in the Mississippi Sound and turbid water patterns imaged by Landsat and verified by Secchi extinction depths. Remote sensing of turbidity could prove to be a valuable tool in fishery studies (see the section on Turbidity).

(5) Sea State—A possible use of remote-sensing technology lies in assessment of sea states in proposed fishing areas. Laevastu and Hela (1970) cited several examples of the effect of waves on fish. Both passive (Cox and Munk, 1954a,b) and active (Moore, 1966) methods are available for assessment of wave conditions (see Sea State). Quasi-real-time surveys of sea state in fishing areas from satellites or aircraft might be effectively employed in fishing operations.

(6) Summary—The utility of indirect remote-sensing methods applied to fisheries may be summarized as follows:

1) Remotely measured sea-surface temperature is the most used and effective technique;
2) Remote measurement of chlorophyll concentration as an index of basic biological productivity must ultimately be related to fishery production; thus, it must be considered for more research;
3) Remote salinity measurements as applied to nearshore and estuary fisheries offer interesting possibilities;
4) Application of indirect methods is a relatively new and rapidly developing art and much may be expected from it within the next decade;
5) Inadequate spatial resolution from orbit necessitates the use of indirect methods for applying remotely sensed data.

h. Satellite Methods

Several satellite-related fishery investigations have been completed that relied on indirect forms of remote sensing. They are the ERTS-1 Menhaden Experiment, the Skylab-3 Gamefish Experiment, the Landsat Menhaden and Thread Herring Resources Investigation, and the Seasat-A Surface Layer Transport Investigation.

(1) ERTS-1 Menhaden Experiment—This 1972 experiment concentrated on the Mississippi Sound (8700 km^2 in the northern Gulf of Mexico). Menhaden distribution and abundance data were acquired concurrently with three Landsat passes during the fishing season, augmented with temperature and salinity data from aerial sensors. An extensive vessel-sampling operation involving up to 144 stations also was conducted. Because menhaden are surface-schooling pelagics, aerial photography was used to obtain the distribution and abundance data.

Surface data were compared with menhaden distribution, producing correlation coefficients (r) of −0.22 for salinity ($n = 195$; $p > 0.99$), −0.15 for Forel-Ule color ($n = 13$; $p > 0.90$), and −0.15 for Secchi disc turbidity ($n = 195$; $p > 0.95$). No significant correlation was found with temperature. The only field operation where Landsat MSS data were sufficiently free from cloud contamination for analysis was on August 7, 1972. Menhaden school locations were superimposed on images from the four MSS bands and the images were density sliced. Density levels in Band 5 (600−700 nm) were moderately correlated with menhaden distribution (Kemmerer et al., 1974); no useful information was apparent in any of the other three bands.

(2) Skylab-3 Gamefish Experiment—This experiment was conducted over a 2-day period in 1973 to determine the value of remotely sensed data for inferring distribution patterns of oceanic game fish. The study area (18,000 km^2) was located between Pensacola and Panama City, Florida. Aircraft and Skylab-3 remotely sensed data were obtained simultaneously with game-fish distribution and abundance data collected during a white-marlin fishing tournament. Oceanographic parameters correlating with white-marlin distribution included temperature ($r = 0.41$; $p > 0.99$; $n = 44$), and chlorophyll *a* and *c* ($r = 0.21$; $p > 0.90$; $n = 44$). Correlation coefficients for salinity, turbidity, and color were −0.15, 0.13, and −0.18, respectively; none were significant at 90 percent.

Radiance data from the Skylab-3 ocean color sensor (S-192) were compared with white-marlin distribution. These data were limited to relatively small areas due to cloud cover. Empirical multiple regression models were developed with white-marlin distribution as the dependent variable against oceanographic parameters and S-192 band radiances. Marlin distribution against temperature and radiance values from four bands (440−520, 490−560, 640−760, and 750−900 nm) yielded a correlation coefficient of 0.89 ($n = 11$). Distribution against only oceanographic parameters (temperature, salinity, and turbidity) was less precise ($r = 0.50$; $n = 22$). Results suggest that the combination of satellite-measured ocean color and temperature might be useful for inferring white-marlin distribution (Savastano, 1975).

(3) Landsat Menhaden and Thread Herring Investigation—A Landsat Menhaden and Thread Herring Investigation from 1975 to 1977 was designed to verify the earlier Menhaden experiment, and in particular to emphasize ocean color for inferred fish-distribution patterns. Three study areas were located in the Mississippi Sound and off Louisiana, comprising most menhaden habitats in the northern Gulf of Mexico. Menhaden were emphasized because adequate distribution data for thread herring could not be acquired during periods of Landsat coverage.

A significant departure from the earlier experiment was almost total reliance on the menhaden industry (approximately 80 vessels and 40 spotter aircraft) for fish distribution and abundance data. Scientific observers were aboard selected fishing vessels during overpasses to sample sites of menhaden capture for salinity, temperature, turbidity, color, and chlorophyll *a*. Two and three research vessels separately acquired surface information during overpasses for development of chlorophyll and turbidity algorithms.

Oceanographic preferences of menhaden were estimated through comparisons of data obtained at the capture locations. Unfortunately, very few samples were collected from areas without menhaden, preventing comparisons of oceanographic conditions in areas with and without menhaden, because the fishing vessels remained near the fish. Therefore, the data were examined for spatial and temporal consistency. Secchi disc depth and Forel-Ule color measurements were relatively consistent at locations of menhaden capture, and somewhat different where sampled by the research vessels. Little or no consistency was found with any of the other parameters.

Because of the apparent relationship between menhaden distribution and water color and Secchi disc depth, an attempt was made to use Landsat radiances for inferring the menhaden distribution. Direct attempts to infer color and Secchi depth from the radiances were relatively unsuccessful. Correlation coefficients for menhaden distribution against radiance values in each spectral band generally were significant at levels exceeding 90 percent (Table 28-10).

Several approaches were evaluated for classification of Landsat MSS data into maps of inferred menhaden distribution. Essentially, they consisted of supervised classification into high and low probability menhaden areas. Table 28-11 summarizes results from six classification attempts. The July 19, 1976 classification was unique in being designed to demonstrate the value of a satellite-aided fishery harvest and assessment

TABLE 28-10

Correlation Coefficients for the Relationship Between Menhaden Distribution and Landsat MSS Spectral Bands (Kemmerer, 1979b)

Spectral Range (nm)	Mississippi Sound		Louisiana
	May 20	June 25	July 24
500−600	0.65**	0.46*	0.42**
600−700	0.74**	0.82**	0.36*
700−800	0.67**	0.69**	0.28*
800−1100	0.61**	0.30*	0.20
Sample Size	36	18	33

* Significant at the 90 percent confidence level.
** Significant at the 99 percent confidence level.

TABLE 28-11

Summary of Landsat Classification Results for Menhaden Distribution (Kemmerer, 1979b)

Date	Day of Landsat Coverage		Day After Landsat Coverage	
	Number of Menhaden Schools	Classif. Accuracy	Number of Menhaden Schools	Classif. Accuracy
May 20, 1975	53	91	19	68
June 23, 1975	18	83	23	74
July 24, 1975	30	87	30	84
July 19, 1976	14	86	20	70
July 27, 1976	11	91	3	0
July 28, 1976	3	100	–	–
Totals	129	88	95	74

system, with near-real-time processing and analysis of Landsat MSS data into high probability fishing areas for distribution to the fishing fleet.

The reason for a menhaden-water color relationship is uncertain, but postulates have been advanced (Kemmerer, 1979b). It is suggested that the menhaden respond to total light intensity, and possibly to spectral intensity (the literature contains many examples of fish responses to varying light levels). In the morning, when light intensity was low, most fish were caught near shore in relatively clear water. As light intensity increased, with the rising sun, the fish moved offshore into deeper and more turbid waters (the Secchi depth average was 1.1 m at the time of overpasses). Catch rates declined during this period and then increased in the late afternoon (after 1500 hours), when catches were further offshore and in clearer waters. At night, the schools tended to fractionate and move inshore. Average daily capture locations were significantly further offshore when the skies were overcast than when they were clear.

(4) Seasat Surface Layer Transport Investigation—Over 90 percent of the coastal finfish and shellfish catch in the northern Gulf of Mexico is composed of shrimp, menhaden, and croakers, which spawn offshore and depend on water movement to transport their eggs and larvae to estuarine nursery grounds. Unfavorable offshore surface currents during critical spawning periods can significantly affect recruitment of these stocks. Nelson et al. (1977) were successful in demonstrating how surface-transport estimates derived from wind-field computations could be used to forecast menhaden recruitment along the Atlantic coast. Estimates of surface transport, therefore, could serve as a basis for yield models designed to provide harvest forecasts in advance of fishing seasons.

This investigation was designed to use Seasat microwave scatterometer (SASS) data. Because SASS measures backscatter coefficients that relate directly to wind stress, these coefficients can be used in Ekman and Sverdrup type transport-models for estimates of surface-layer transport (Brucks and Leming, 1977).

To evaluate the accuracy of aircraft acquired SASS data and to aid in refinement of the geophysical algorithms for the Seasat SASS, sonic anemometer measurements of wind stress were compared to airborne scatterometer measurements of ocean roughness. These comparisons indicated good agreement between the two data sets, although the scatterometer tended to overestimate wind stress during periods of strong surface winds (Brucks et al., 1980). The Seasat SASS data are now being evaluated for estimates of surface transport in the northern Gulf of Mexico.

3.0. Aquatic Plants

a. Submerged Aquatic Vegetation

Systems of submerged aquatic vegetation (SAV) serve multiple roles in coastal ecosystems, providing a habitat for many sessile and slow-moving invertebrate species as well as the more motile shrimp and crabs. SAV beds provide refuge for these organisms from predators. In addition, the productivity in plant and associated algal components is near that of many cultivated terrestrial crops; SAV detritus contributes greatly to the coastal food chain. Leaves and roots of SAV also bind sediments and baffle currents, thereby stabilizing the bottom and hindering erosion and sediment loss (Orth et al., 1979).

Aerial photography has been used worldwide for mapping SAV and investigating associated problems (Edwards and Brown, 1960; Kelly and Conrod, 1969; Kolipinski and Higer, 1970; Ogrosky, 1978; Macomber and Fenwick, 1979). Orth et al. (1979), in a study of long-term change, have monitored and mapped the lower Chesapeake Bay SAV on almost a yearly basis since 1975. It is noteworthy that in this program remote sensing has largely replaced an earlier reliance solely on field sampling.

The method consists of acquiring photography,

transferring delineations of SAV beds to maps, measuring bed areas and plant density, and compiling data in a computer data base. Species identifications are made via field sampling. For the aerial photography, careful planning is needed to ensure complete bed coverage and to include land-based control points. A preliminary visual aerial survey will permit elimination of flight lines over bare sand or mud bottoms. The scheduling of flights must provide for low tide (±0.5 m) as predicted in tide tables, with adjustment for any wind set-up. Flight direction aligned with the tide propagation ensures the lowest tidal stage. If windy conditions cannot be avoided, then offshore winds are preferred, in order to minimize sea state and wind-generated water turbidity. Simultaneously, atmospheric haze must be minimal, with no more than thin overcast or scattered broken cloud cover. These conditions maximize the SAV-bottom contrast (see Color Figure 28-81). The solar elevation angle should be between 20° and 40° to minimize sun glitter. Finally, the season of coverage should be when growth stage ensures maximum bed delineation.

The choice of flight altitude is a trade-off between frame coverage and spatial resolution. Orth and Gordon (1975) found that a scale of 1:30,000 permitted discrimination of dense 1 meter patches of grass. Later SAV mapping was at an altitude of 3660 m (12,000 ft) with a standard 6-inch focal length lens to produce a scale which matched that of the 1:24,000 topographic maps. Color film is preferred because, although SAV discrimination is based primarily on tone and texture, there are frequently extra clues in the color dimension which give color film a slight advantage over black-and-white film (see the discussion by Austin and Adams, 1978). Furthermore, color film offers a great advantage in shoreline and other coastal applications as part of an archival film data-base. To ensure optimal photographic densities for bed discrimination, test exposures should be put on each roll and preprocessed for selection of the best development time and temperature.

During the data reduction, an estimate of percent cover within each bed may be made visually by comparison with an enlarged crown-density scale (see Figure 28-82) such as used for estimates of forest tree crown cover from aerial photography. Bed boundaries are transferred to base maps with a zoom transfer scope and then to translucent mylar copies of the base maps. The latter step permits a variety of overlay options during production of final maps. Bed areas may be measured with a manual or electronic planimeter/digitizer. For development of a computer data-base, a useful approach is to digitize bed boundaries according to a sorting by Universal Transverse Mercator (UTM) grid squares. Orth et al. (1979) used 1 km squares; for each square a 7-field data block was put on tape containing, by field, UTM coordinates divided by 100 km (2 × 2 digits), UTM coordinates for one corner of the 1 km square (2 × 2 digits), topographic map index, waterway code, resource code, bed area within the square, and (x,y) coordinates of the bed perimeter within the square.

Although aerial photography is the usual choice for SAV mapping, there are some instances of the use of multispectral scanners. For example, aerial scanners were used in conjunction with cameras in several experiments in Lake Ontario to study distribution and biomass of the submerged aquatic *Cladophora* (Wezernak et al., 1974; Wezernak and Lyzenga, 1975).

b. Kelp Bed Surveys

Jensen et al. (1980) reviewed the historical use of aerial photography (black-and-white, color, and color infrared) for kelp bed surveys, and some of the implications of Landsat imagery, and airborne and satellite radar. Landsat is of interest due to low cost (to the user) of its repetitive data collec-

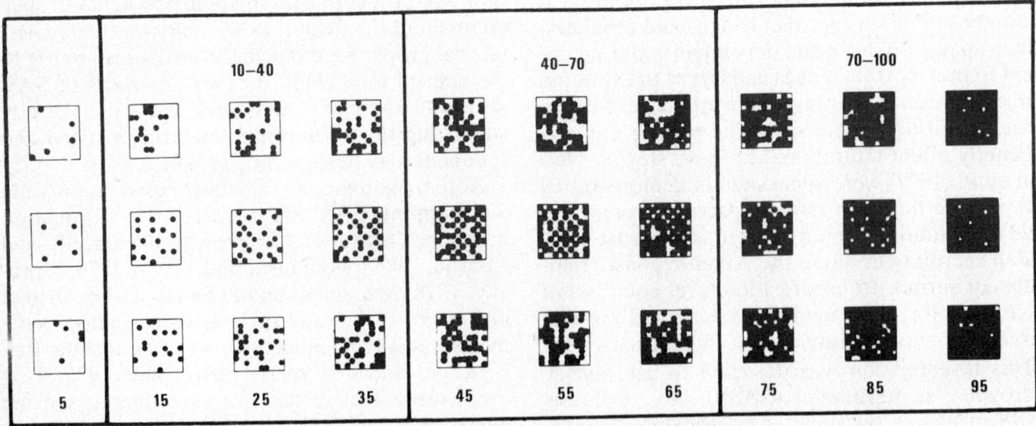

Fig. 28-82. Forestry crown-density scale used to estimate percent cover of SAV in Figure 28-81.

tion, while radar provides the potential for all-weather surveys. Under most conditions Landsat data appeared to provide results that correlated well with aerial photography, even though Landsat consistently underestimated the areas of kelp beds for a variety of reasons that were data-set specific. In general, underestimates from Landsat were due to a photointerpreter's ability on high resolution photography to identify less dense kelp areas, which Landsat confused with the ocean background, and to non-ideal turbidity and water depth during Landsat passes. High altitude color-infrared photography and X-band radar imagery were shown to provide aerial data on *Macrocystics pyrifera* at approximately the same level of accuracy as conventional large scale inventories. L-band satellite radar imagery exhibited potential for kelp monitoring. However, additional L-band research must determine if kelp can be discriminated from the ocean background under calm sea state. Landsat data may provide accurate statistics if consistent underestimation is offset using a simple linear equation. Given these results, multispectral sensors offer potential for operational monitoring of renewable kelp resources.

4.0. Marine Wetlands

a. Inventory and Mapping Needs and Problems

The U.S. Fish and Wildlife Service (FWS) 1956 inventory of wetlands (Shaw and Fredine, 1956) estimated that there were 29.9 million hectares of wetland in the conterminous United States; 1.62 million hectares were fresh coastal and 2.14 million hectares were saline. Most wetlands less than 16.2 hectares in size were excluded from the survey. Since 1956, many hectares have been lost as a result of increased pressure for housing, industrial development and agricultural development. This loss, especially in highly industrialized states such as New Jersey, New York, and Delaware, has increased the impetus for development of mapping, inventory, and monitoring techniques. As a result, all of the 23 coastal states have ongoing or completed wetland inventories.

Wetland definition and classification are prerequisites to mapping and inventory. Because wetlands are dynamic ecosystems with a complex interrelationship of hydrology, soils, and vegetation, legal definitions of wetland tend to lack clarity. There is no uniformity from state-to-state. The definition of "coastal wetlands" may include all wetlands, all "tidal" wetlands, or only vegetated (marsh, swamp) wetlands lying within the "coastal zone".

Individual states have modified existing wetland definition- and classification-systems or developed their own because no existing system has proved adequate. The use of different systems means that the resulting inventory products are not mutually compatible. They have different scales, formats, minimum mapping units (MMU), and classes, and do not allow for valid comparisons of acreage, wetland type, or vegetative composition on a regional or national basis.

When wetland maps are utilized for regulatory-permitting procedures, accurate boundaries are critical. The wetland/upland boundary and the mean high water line (MHWL; see Tidal Boundaries) boundary are most often used for legal purposes. MHW generally marks the boundary between state-owned and privately-owned, state-regulated property. McEwen et al. (1976) discuss the difficulties of using MHW as a legal boundary.

A severe problem involved in wetland management is that of overlapping federal, state and local responsibility. The National Oceanic and Atmospheric Administration (NOAA), the U.S. Army Corps of Engineers, the Environmental Protection Agency (EPA), and the FWS all have varying degrees of authority and involvement with coastal wetlands, and their responsibilities, in turn, overlap those of state and local agencies.

b. State and Federal Photographic Mapping Techniques

A broad review of wetland remote sensing techniques and programs was provided by Carter (1977). Coastal wetlands can be mapped with black-and-white (B/W) panchromatic, B/W multispectral, natural color or color infrared (CIR) photographs, and digital multispectral scanner (MSS) data. Generally, where species composition is important, investigators have recommended CIR photography (Color Figure 28-83).

Most state inventories are based upon aerial photographs and supplemented by ground surveys and previously published maps, reports, and soil surveys. The New England coastal states use B/W photographs as a basic data source. The third Rhode Island inventory was accomplished using 1:12,000 leaves-off B/W aerial photographs without field checking (MacConnell, 1974). Maps were prepared by stereoscopic transfer to 1:24,000 scale USGS base maps.

In New York and New Jersey, color IR photographs were the principal data source (Brown, 1977). In New Jersey, dominant wetland species, the wetland/upland boundary, and the biological MHWL were delineated on 1:24,000 scale photo base-maps using manual transfer techniques. In New York, a similar procedure was used with field checking to map six wetland categories: (1) Coastal Fresh Marsh, (2) High Marsh or Salt Meadow, (3) Intertidal Marsh, (4) Coastal Shoals, Bars, and Mudflats, (5) Formerly Connected Tidal Wetlands, and (6) Littoral Zone.

In Virginia, coastal wetland mapping was done on a county-by-county basis using field survey techniques that were supplemented with photographs where available. The Virginia law, unlike the law of most other states, defines tidal wetlands as "the area between mean low water and the elevation above mean low water equal to the factor 1.5 times the mean tidal range." At least one of 40 designated "wetland plant species" must also be present. Twelve marsh community types were defined in a classification system developed specifically for this inventory (Silberhorn et al., 1974).

Georgia, Mississippi, and Louisiana are using high-altitude photography, Alabama and Florida are using both high- and low-altitude photography, and Texas is using Landsat data.

On a national basis, the U.S. Geological Survey (USGS) maps coastal wetlands as a base category on the standard topographic map series. The U.S. Fish and Wildlife Service (USFWS) is presently inventorying coastal wetlands as part of a new National Wetland Inventory (NWI) using a new classification system (Cowardin et al., 1976) that will provide for comparison of map and inventory products at the national, regional, and state levels. The Cowardin system is a hierarchical system based primarily on a vegetation, soils and water regime. The highest level is the Ecological System—all coastal (tidal) wetlands fall into the Marine or Estuarine systems. Subtidal and Intertidal subsystems are further divided into classes; there are nine classes in the Estuarine System (see Table 28-12): (1) Neretic water/benthos, (2) Aquatic Bed, (3) Reef, (4) Rocky shore, (5) Beach/Bar, (6)

TABLE 28-12
Wetland Classification System for Marine and Estuarine Ecological Systems (from Cowardin et al., 1976)

Ecological Subsystem	Class	Subclass	Dominant Type (Examples)
Intertidal	Forested Wetland	Dead	
		Needle-leaved Evergreen	*Chamaecyparis thyoides*
		Broad-leaved Evergreen	*Rhizophora mangle*
		Needle-leaved Deciduous	*Taxodium distichum*
		Broad-leaved Deciduous	*Nyssa sylvatica*
	Scrub/Shrub Wetland	Dead	
		Needle-leaved Evergreen	young *Chamaecyparis thyoides*
		Broad-leaved Evergreen	*Myrica cerifera*
		Needle-leaved Deciduous	young *Taxodium distichum*
		Broad-leaved Deciduous	*Iva frutescens*
	Emergent Wetland	Succulent	*Salicornia*
		Broad-leaved	*Peltandra virginica*
		Narrow-leaved	*Spartina alterniflora*
	Rocky Shore		Littorine/Lichen Limpet Balanoid *Mytilus/Mitella* Fucoid Laminaroid Coralline
	Beach/Bar		*Uca, Ocypode*
	Flat	Organic	
		Vegetated	*Zostera marina*
		Fine	
		Coarse	
	Reef	Worm	*Sabellariid*
		Mollusc	*Crassostrea*
Subtidal	Reef	Worm	*Sabellariid*
		Mollusc	*Crassostrea*
	Aquatic Bed	Floating	*Eichornia crassipes*
		Floating-leaved	*Nuphar advena*
		Submergent vascular	*Zostera marina*
		Submergent algal	*Halimeda*
	Neritic Water/ Benthos	Worm	*Spirochaetopterus*
		Sponge	*Hippospongia*
		Mollusc	*Pecten, Rangia*
		Crustacean	
		Ascidian	*Cnemidocarpa*
		Alcyonarian	*Muricea*
		Barren	

Flat, (7) Emergent Wetland, (8) Scrub-shrub Wetland, and (9) Forested Wetland. The Marine System has the first six of the above classes. The following is a brief summary of recommended film types for the nine classes:

—*Forested Wetland; Scrub-shrub Wetland; Emergent Wetland:* These classes are best identified and mapped using CIR photographs. B/W photographs are a good second choice. Separation into subclasses and lower levels is good with CIR; species may be identified in many cases, even with high-altitude photographs. Brown (1977) discusses the operational use of CIR.

—*Aquatic Bed:* Aquatic beds include both surface and submerged vascular and macroalgal vegetation. High turbidity in some coastal areas masks the water penetration effect of both color, CIR, and B/W film. Color film is the first choice for detection and mapping of SAV (see the section on Submerged Aquatic Vegetation). Color IR is poor for subsurface detection, but excellent for surface vegetation or submerged vegetation with surface leaves (kelp). A combination of color and CIR films, at 1:10,000 scale or greater, is recommended for the inventory of the maximum number of macroalgal classes.

—*Rocky Shores:* Color IR is the best choice for discriminating vegetation zones (mostly macroalgae) on rocky shores. Where vegetation is sparse, B/W film appears satisfactory.

—*Beach/Bar; Flat:* These classes can be mapped with almost any film; however, color infrared and more specifically black-and-white infrared films clearly show the land/water interface or tidal boundary. This boundary may be difficult to interpret from color film when turbidity is high in adjacent waters.

—*Neretic Water/Benthos:* This class appears as open water on most film. In order to separate aquatic beds from the unvegetated substrates, color film is generally required. Bottom sediments such as sand may be identified with low-altitude photographs.

—*Reef:* Reefs are mostly subtidal and, for this reason, color film is better than CIR or B/W in most cases (see the section on Bathymetry).

The NWI is using available aerial photography to prepare wetland maps at 1:100,000 for the entire United States, including Puerto Rico and the Virgin Islands. The minimum mapping unit (MMU) for this task is approximately 0.5 ha; the update interval is 8–10 years.

Coastal wetlands are also being inventoried and mapped as part of the USGS Land Use and Data Analysis (LUDA) program. The LUDA system Level II classes are Forested and Nonforested Wetlands (Anderson et al., 1976)—only vegetated classes are considered wetlands. High-altitude CIR photographs are the basic data source for the 1:250,000- and 1:100,000-scale map products. The MMU for this inventory is 40 acres; the recommended update interval is 5 years.

Aerial photography is obviously the main mapping method at present. Film types and scales vary; flight scheduling for low tide and correct season is important. Aerial photographs will remain a major data source for the foreseeable future because of their high resolution, and the low cost of manual photointerpretation versus automated digital mapping from scanner data.

c. Use of Satellite Data

There are no documented operational wetland programs that use satellite data for mapping. Some coastal wetland classification and mapping has been done experimentally with Skylab photographs. The use of Landsat data is extensive but generally still in the experimental phase, with coastal marshes the most studied. The program closest to being operational at present is in Texas, where natural vegetative associations have been mapped at 1:250,000 scale using Landsat digital data (FWS, 1976; Finley et al., 1979).

The potential of both Landsat images and digital data for gross mapping of coastal marshes has been demonstrated often. Anderson (1975) and Carter et al. (1973) used Landsat images to map wetland vegetation in the Chesapeake Bay area and along the Georgia-South Carolina coast, and to classify marshes by salinity type in South Carolina. Tidal stage affected wetland reflectance characteristics. Also, seasonal data were needed to identify and delineate fresh water tidal marshes. Extensive wetland alterations (dredge disposal, canal construction, road building) were easily monitored using images. Digital data were employed by Reeves (1973) in unsupervised classification of marsh and water bodies of Galveston Island, Texas; Carter and Schubert (1974) used digital data, field spectral reflectance data, and a look-up table approach to map wetland vegetation classes, mud flats and spoil at 1:20,000 scale in Chincoteague Bay, Virginia. Carter (1976) correlated area measurements from Chincoteague with local primary productivity figures to estimate annual bio-production. Klemas et al. (1974, 1975) used digital data to classify coastal vegetation in Delaware, achieving accuracies as high as 80 percent. Bartlett et al. (1977) incorporated both field spectral measurements and atmospheric corrections in classification procedures. Butera (1976) used aircraft and satellite MSS data to determine Louisiana marsh salinity zones from indicator plant species.

Many problems remain to be solved before Landsat will be considered an operational tool for use by wetland managers. Constraints include the necessity for high geometric accuracy and the need for better spatial, spectral, and radiometric resolution. Map scales must be 1:24,000 or larger. Landsat image data presently meet National Map Accuracy Standards (NMAS) at scales of 1:500,000 or smaller (McEwen and Schoonmaker,

1975); Landsat digital data must be geometrically corrected to meet NMAS at larger scales. The methodology for geometric correction and sampling is now well established (Rifman and McKinnon, 1974; Caron and Simon, 1975; Rowan et al., 1974; Bernstein and Stierhoff, 1976).

The principal limitation of Landsat is the coarse spatial resolution. The nominal resolution is 0.45 ha but the effective resolution is at least 1.4 times as large, about 0.70 ha, due principally to the inevitable misregistration of small wetland areas from instantaneous fields-of-view during satellite overpasses. Penney and Gordon (1975) found in preliminary studies that Landsat failed to reliably discriminate small wetlands that comprise over half of Virginia's wetland area. Ground inventories show that three out of four non-oceanside Virginia wetlands are smaller than 0.40 ha—that is, smaller than even the nominal Landsat resolution.

Also, many wetlands (especially inland wetlands) have large species diversity, and/or sharp spatial zonation. These two characteristics, in conjunction with limited spatial resolution from orbit, may produce a non-specific spectral signature for classification by the present Landsat series. Even if spatial resolution is sufficient to allow detection of the wetland area as a whole, the blurring of spatial distributions of species within the area may degrade the outcome to a confusion between different wetland types or to confusion with other types of land cover. Therefore, satellite spatial resolution must be improved to deal with the spatial texture of wetland spectral signatures.

Clearly, Landsat data are useful and reliable for spectral discrimination of large, vegetated coastal marshes, consisting of single species, but detailed classification of more complex wetlands will require further research and perhaps new spectral bandwidths. The desirable nominal spatial resolution for direct interface with regional, national, and state inventories is approximately 0.25 ha or less (<40 m). Landsat digital data may be useful for updating present inventories, and for inventory in Alaska where available map coverage is 1:63,360 scale and smaller and the area is vast. Limitations may be solved by the Thematic Mapper (TM) of Landsat D, which will have six spectral bands (0.45–0.52, 0.52–0.60, 0.63–0.69, 0.76–0.90, 1.55–1.75, and 10.40–12.50 μm), a spatial resolution (bands 1 to 5) of 30 m, and a radiometric resolution of 8 bits (256 levels). The TM should provide maps meeting present inventory requirements; however, the maps will not have boundaries accurate enough for legal or regulatory use. Improvement should be seen in separation of different wetland types. For detailed ecological studies, however, aerial data will continue to be necessary (see Kolipinski et al., 1969; Weisblatt, 1972; Egan and Hair, 1973; Reimold et al., 1973).

d. Wetlands Regulation

The question of whether remote sensing is useful operationally in wetlands *regulation* has been addressed in Virginia. The 1972 Virginia Wetlands Act mandated not only a wetlands classification and inventory, but also a permit system requiring approval for all wetland alterations. Established classification systems were found too broad in scope to be suitable; therefore, a new system based on common plant communities was devised. Maps at 1:24,000 scale showing 12 types of wetlands as small as 0.25 acre form the inventory, and are used by local wetland boards, and state and federal agencies (Silberhorn, 1978). Despite early literature lauding the use of color infrared aerial photography in the inventory process, Virginia studies showed that for its classification system, plant community discrimination based primarily on photographs would be unsatisfactory because of film density variations due to leaf position, plant distributions, and the large number of species per acre (Whitman and Marcellus, 1973). Consequently, photographs were used only for preliminary classification on topographic maps using a zoom transfer scope; such maps were used in the field for extensive on-site plant community identifications. The regulatory program that is based on the resulting inventory has succeeded in reducing wetland losses in the state from 450 to 25 acres per year (Silberhorn, 1978). Low altitude color photography is used regularly in evaluation of permit applications. The use of photography in place of site visits has reduced overlapping effort by several state and federal agencies.

e. Species Discrimination

Some wetland species discrimination is possible with color, color infrared, and multispectral photography. Early analyses based on spectral reflectance data were reported by Anderson (1969, 1970), Anderson and Carter (1972), and Carter and Anderson (1972).

The spectral properties of canopies of different marsh plants are produced by a combination of the optical properties of individual vegetative components (leaves, stems, etc.), and the effects of plant growth form, density and height, tidal stage, and soil type. Reimold et al. (1972) and Pfeiffer et al. (1973) measured the reflectance of individual leaves of *Spartina alterniflora* and *Juncus roemerianus* and found significant differences, particularly in the infrared (0.7–1.4 μm), where *Juncus* leaves exhibit much higher reflectance. Canopy reflectance varied with canopy height. Data on *S. alterniflora, Spartina patens,* and *Distichlis spicata,* suggest the importance of canopy structure (Bartlett, 1979). Leaf reflectances are comparable for these three species, yet stands of *S. alterniflora* are spectrally distinct from the other two.

In temperate regions, there is a seasonal change in reflectance signatures as both canopy structure and leaf optical characteristics change. Carter and Schubert (1974) measured canopy reflectance over several marsh cover types (*S. alterniflora, S. patens, Iva frutescens,* and organic mud flat) between May and October. All four classes were spectrally distinct in October, although contrast was sometimes greater during other months. Optimal spectral discrimination between *S. alterniflora* and *S. patens* with *D. spicata* occurred in December, when more rapid senescence of *S. patens* and *D. spicata* in response to approaching winter caused reflectance signatures to diverge. Lowest contrast occurred in May and June. Thus, traditional reliance on the growing season for data collection and analysis of wetlands vegetation at temperate latitudes is subject to re-evaluation.

f. Biomass and Productivity

Stroud and Cooper (1968) used CIR aerial photography to map acreage of major community types in a North Carolina marsh. Combined with data from harvesting, acreage data led to productivity values for several species. Net primary productivity for the whole marsh was estimated to be 1534 kcal/m^2/yr. Reimold et al. (1972) found that general biomass classes could be established for *S. alterniflora* using tonal variations in CIR aerial photography.

Field spectral measurements in the Landsat MSS bands using a radiometer have confirmed that biomass is related to canopy reflectance (Bartlett, 1979). Figure 28-84 (Bartlett and Klemas, 1981) shows a linear regression (r^2 = 0.81) between emergent green biomass (measured by harvest) and the ratio of canopy reflectance in MSS bands 7 (0.8–1.1 μm) and 5 (0.6–0.7 μm). MSS 5 (red) reflectance is inversely correlated with the percentage of biomass in the canopy due to chlorophyll absorption. MSS 7 (infrared) reflectance, where plant leaves have low absorption and high reflectance, was positively correlated with canopy height. Near-infrared reflectance is sensitive to the number of stacked leaf layers within a canopy (Gausman et al., 1976).

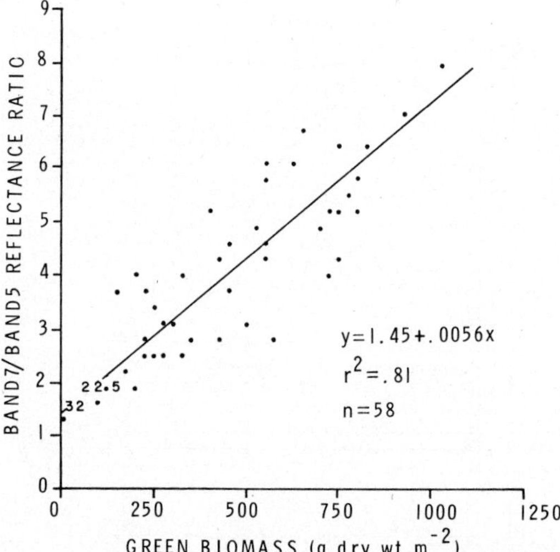

Fig. 28-84. Landsat MSS band 7/band 5 reflectance ratio as a function of green biomass of *Spartina alterniflora* (Bartlett and Klemas, 1981).

Thus, either aerial photography or field or orbital radiometry may provide useful estimates of emergent green biomass. The speed of sampling that is inherent in remote sensing will allow more cost-effective and extensive sampling than is now possible by harvesting measured quadrats. Reduced accuracy from orbit will be offset somewhat by the more effective characterization of large areas from orbit than can be achieved from harvesting.

In those situations where satellite data yield plant-community classification maps of satisfactory accuracy, results may be converted to productivity maps by assigning a productivity value P_i to each plant community i. The average productivity P of a marsh is then obtained from $P = \sum_i A_i P_i / \sum_i A_i$ where A_i is the area of the ith community. Butera (1979) has proceeded one step further, toward nutrient transport analysis, by analyzing Landsat MSS data for distance from each marsh pixel to the nearest water. A measure of relative nutrient availability is then expressed as the ratio of productivity to distance.

D. GEOLOGICAL OCEANOGRAPHY

Oceanographers intrigued with the three-dimensional form of the coastal and near-shore submarine landscape use traditional photography or sonar for mapping and change detection. To supplement these traditional techniques, new remote sensing instruments have recently become operational, including multispectral scanners and lidar. Together they provide for a powerful attack on topics of interest in geological oceanography that can be addressed from altitude, including sediment budget and transport, bathymetry, and coastal processes. Meanwhile, the importance of aerial photography for the historical record of shoreline change is not diminished by the growing use of the newer remote data-collection techniques. However, in the abyssal sea, the great depths put the ocean bottom beyond all aerial and orbital remote sensing. There, the only technique of interest is side-scan sonar, which reveals bottom topography in exciting detail.

1.0. Sediment Budget and Transport

Suspended sediment spatial distributions, detailed according to size, shape, and composition, are important factors in budget estimates related to exchange of sedimentary materials between coastlands, near-shore waters, and deep-water areas. The use of remote sensing to measure surface suspended-sediment concentrations, with application to sediment-budget studies, as well as circulation and water-quality models, is discussed in detail later (see Suspended Inorganic Particulates, and Water Quality and Circulation Models). Here, it may be summarized that methods have been developed for measuring surface suspended-sediment concentrations using Landsat digital data (when size, shape, and composition are more or less uniform). The results are contributing to sediment transport and circulation studies based on suspended materials as a tracer. Amos (1976) and Amos and Alföldi (1979) have used Landsat digital data calibrated by field sampling to study the sediment budget of the Bay of Fundy, Canada.

2.0. Bathymetry

Traditional techniques of remote bathymetry include acoustic sounders (sonar) from vessels, and visible spectral range photographic and scanner data from aircraft and satellite platforms. These are being complemented by the emerging use of lasers and side-scan sonar techniques. Laser systems using time-gating avoid the requirement for direct surface contact that is inherent in acoustic methods, but side-scan sonar systems provide three-dimensional bottom mapping analogous to side-looking radar systems for terrestrial surfaces. This section will omit discussion of the commercially available depth sounders that are widely used for navigation.

a. Photogrammetric Bathymetry

Exploitation of submarine resources has altered traditional mapping priorities and generated the development of new and improved mapping techniques. Color aerial photography, with its remarkable clear-water penetration characteristics and dramatic presentation of submerged detail, is a basic tool of the National Ocean Survey for mapping the seabed in shoal areas and waters of moderate depths. The vertical accuracy required for nautical charting can be obtained through the use of precision photogrammetric techniques. Although single stereoscopic models will occasionally suffice for the compilation of bathymetric relief around small islands or inlets, extensive offshore shoal areas require blocks of overlapping strips of aerial photographs to bridge the zones between vertical control points.

(1) General Approach—Aerial photogrammetry for bathymetry requires several minor departures from the method normally used when the bundle of light rays passes solely through the atmosphere. The effect of refraction at the water-air interface must be taken into account in aerotriangulation for imaged points underwater. The solution of this problem requires a mathematical model for two-media refraction.

To be theoretically correct and entirely acceptable for photogrammetric triangulation, the mathematical model must be based upon the actual underwater position of a point rather than its refracted, or apparent, position. This is because there is no single apparent object position that will satisfy the condition of collinearity on the several photographs that image each underwater point. The use of actual positions provides that both land-based and underwater geodetic control points can be treated indiscriminantly during aerotriangulation. The direct method for treating refraction is straightforward for the vertical control points—only the index of refraction, nominal flying height, and water depth at the time of photography need be known in order to accurately determine the coordinate corrections for measured image coordinates on near-vertical aerial photography. Sea state effects are automatically treated as image-coordinate residuals in the analytic aerotriangulation solution. The influence of random wave slopes, periods, and heights on the accuracy of depth measurements has been statistically investigated for the geometric conditions of block aerotriangulation, and found to be tolerable for moderately calm seas and aerial photography with approximately 70 percent overlap.

(2) The Mathematical Model—The basic mathematical model for correcting image coordinates for refraction of underwater points is illustrated in Figure 28-85, and for nadir photographs has the form

$$\Delta d = dh(1\text{-}1/a)/(H\text{-}h) \qquad (28\text{-}10)$$

where

- Δd is the required correction in meters for the image point. Its sign is always negative (correction toward the photo center).
- d is the radius of the image point, in meters, from $(x^2 + y^2)^{1/2}$.
- h is the depth in meters (negative in sign) of the underwater point at the time of photography.
- H is the flying height in meters.
- a is the ratio of tangents of the angles of refraction (r) and incidence (i) (Rinner, 1969) obtained from

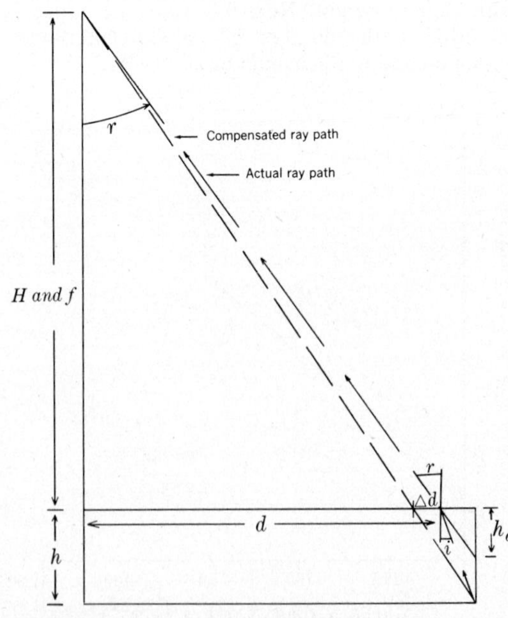

Fig. 28-85. Radial image displacement due to water surface refraction (Umbach and Harris, 1972).

$$a = (\tan r)/(\tan i) = [n^2 + (n^2 - 1)(\tan^2 r]^{1/2} \quad (28\text{-}11)$$

where:

n is the index of refraction for rays passing from water into air; $n = 1.340$ for seawater.

f is the camera focal length in meters; $\tan r = d/f$ for a vertical photograph.

The apparent depth of an underwater point (h_a) is given by Rinner (1969) as $h_a = h/a$. Eq. (28-11) for $n = 1.340$ becomes:

$$a = [1.7956 + (0.7956) \tan^2 r]^{1/2} \quad (28\text{-}12)$$

To obtain the refraction-corrected values (x',y') of photo image coordinates (x,y),

$$x' = x(1 - \Delta d/d) \text{ and } y' = y(1 + \Delta d/d) \quad (28\text{-}13)$$

(3) Underwater Vertical Control Points— The change in the values of Eq. 28-11, and its effect on (Δd) caused by 1° or 2° tilt in the aerial photograph, are insignificant when $H > 100\ h$. Therefore, the refraction-compensated coordinates of all vertical control points can be determined prior to analytic aerotriangulation using Eqs. 28-10, 28-12 and 28-13. This should be done following coordinate refinement for film and camera distortion and for comparator error.

(4) Underwater Photogrammetric Points— To be theoretically correct, the unknown depths (h) of test points should be solved as additional unknowns in the aerotriangulation process using Eq. 28-10. Points of unknown depth should, through an iterative process, be treated for the effects of refraction in the same way that underwater control points are treated. This is the only way that the principle of collinearity used in aerotriangulation can be strictly applied without treating the systematic effect of differential refraction as accidental errors during minimization of image-coordinate residuals. The interactive approach can be incorporated into present block-solution computer programs as a special purpose case.

Present block-aerotriangulation programs, together with compilation instruments and side computations for converting the mean apparent depths of points to mean true depths, will yield sufficient accuracy for the application of photogrammetry to inshore regions. This approximate approach (without strict correction for refraction) generates error residuals on the plate coordinates of underwater control points that appear unrealistically large. However, the resulting error of water-depth determination is less than 2 percent of the depth when aerotriangulation is performed for side- and forward-lap of 65 to 70 percent.

(5) Stereoscopic Compilation—Tewinkel (1963) and Meijer and Groenveld (1964) demonstrated that, when depth curves and spot depths are to be mapped with a stereoplotter, the apparent depths of model control points, which are the product of the aerotriangulation, should be used for leveling the model just as if there were no water refraction. The true depth of these and all other points within the leveled stereomodel can be obtained by applying variable depth-correction factors. Figure 28-86 shows a typical set constructed by applying the Meijer method to 1:15,000 scale photography at a 2× model scale with a photographic endlap of 70 percent. Note that, in the absence of sunspots within the neat model, the entire area can be measured twice

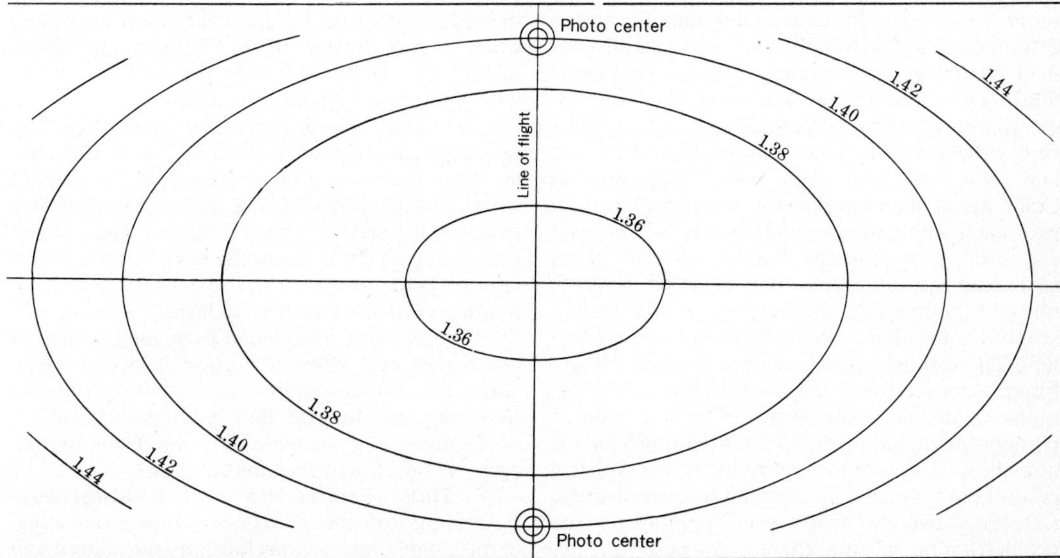

Fig. 28-86. Model of depth correction factors for RC-8 photographs with 70% overlap. a. Multiply measured depth by factors to get true depth. b. Divide depth of contour desired by these factors to get depth setting required (Umbach and Harris, 1972).

using alternate flight lines. These two independent sets of depth determinations may be averaged for the final manuscript compilation.

If small-scale high-altitude bridging photography of certain areas of interest does not show sufficient detail to provide control points for at least alternate models of large-scale photography, a secondary analytic aerotriangulation would be required using the low-altitude photography. Should extensive areas fall into this category, the work may be expedited by omitting the setting of each small-scale stereomodel for the determination of the mean true depths of pass points and supplementary vertical control points. The mean true depths of all imaged points can be computed from the output of the small-scale aerotriangulation. The mean true depth (h_m) valid for both sets of photographs is then determined by the following formula, which involves the determination of the mean value (a) for each point in the aerotriangulation:

$$h_m = h_a(\Sigma a_i)/n \qquad (28\text{-}14)$$

where

- h_a is the mean apparent depth in meters of the point, as determined in the block aerotriangulation.
- a_i are the individual values of (a), as defined in Eq. 28-12, for each photo image of the underwater point.
- n is the number of photos on which the point is imaged.

The resulting value (h_m) is then used as (h) for the treatment of these points as underwater control points in an independent block aerotriangulation using the larger scale photography.

(6) Evaluation Field Test—To evaluate the accuracy of block aerotriangulation and stereoscopic compilation for underwater mapping and to determine if the metric concessions critically detract from the final product, a full-scale operational test was performed at a small shoal area off the southern coast of Puerto Rico. Figure 28-87 shows the available geodetic control and the actual flight lines for 7 strips and 40 photographs. Color aerial photography for water penetration was taken with tandem mounted Wild RC-8 aerial cameras. One camera had a 420-nm glass antivignetting filter and color-reversal positive film to accommodate the wide range of expected spectral transmission found in turbid coastal waters. The second camera, equipped with a 740 nm filter, was used for synchronous infrared photography to delineate the waterline at the time of photography. Although infrared photography is not absolutely essential, it reduces the vertical control requirements by enabling the stereoscopic transfer of discrete "water level" points along the beach from the infrared to the color photography, even in heavy surf.

Experience indicates that a scale of 1:15,000 from a flight altitude of 2,300 m is optimum for the accurate stereomapping of the ocean bed in this particular region. The principal factors involved in the selection of the optimum scale are the spacing of photogrammetric points required to span areas devoid of bottom imagery and the vertical accuracy required in the stereoscopic compilation process.

Photographic endlap and sidelap of approximately 70 percent were specified to optimize the balance of error residuals caused by sea-surface roughness and by the effect of a smaller than normal base-to-height ratio on the geometry of the block aerotriangulation and stereoscopic compilation. Large overlaps were also essential for stereocoverage free of sun glint; photography was avoided during hours around the local noon. Preflight computation of the size and shape of the solar reflection on the aerial photographs is a necessary part of flight planning over water bodies.

(7) Control and Tidal Datum—Only four corner horizontal control stations are normally used for the National Ocean Survey's Block Solution; however, additional control can provide the basis for a more meaningful accuracy evaluation. Abundant vertical control may be obtained along the entire surf zone via transfer from infrared to color photography. In order to reference the results to a meaningful and common datum, tidal observations are required during the aerial photography.

Every effort should be made to acquire photography during, or very close to, mean low water. Among the advantages are greater clarity of bottom detail. However, this requires that the tidal datum must be determined prior to photography, and tide observers are required at intervals along the shoreline during photography so that the flight mission can be informed of tidestages. A further constraint is imposed by confining photography to those few times during the year when mean low water occurs during suitable illumination. Other factors are discussed under the section dealing with Submerged Aquatic Vegetation.

Tidal observations during aerial photography are required to correlate depths of subsurface objects and underwater target panels, if the block of aerial photography does not include the shoreline as a usable vertical control. The five square symbols in Figure 28-85 mark the sites of underwater control points used on the test project. Surface floating vertical control is undesirable because of the false parallax introduced by a combination of horizontal and vertical motion between exposures. In addition, moored panels have a tendency to "porpoise" or dive into moving surf and become physically unstable (see the Tidal Boundaries section for further information).

(8) Data Processing and Evaluation—Following routine laboratory film-processing, aerotriangulation was performed and underwater contour maps were prepared. The National Ocean Survey "Block Analytical Aerotriangulation Solution" was applied to the aerial photography fol-

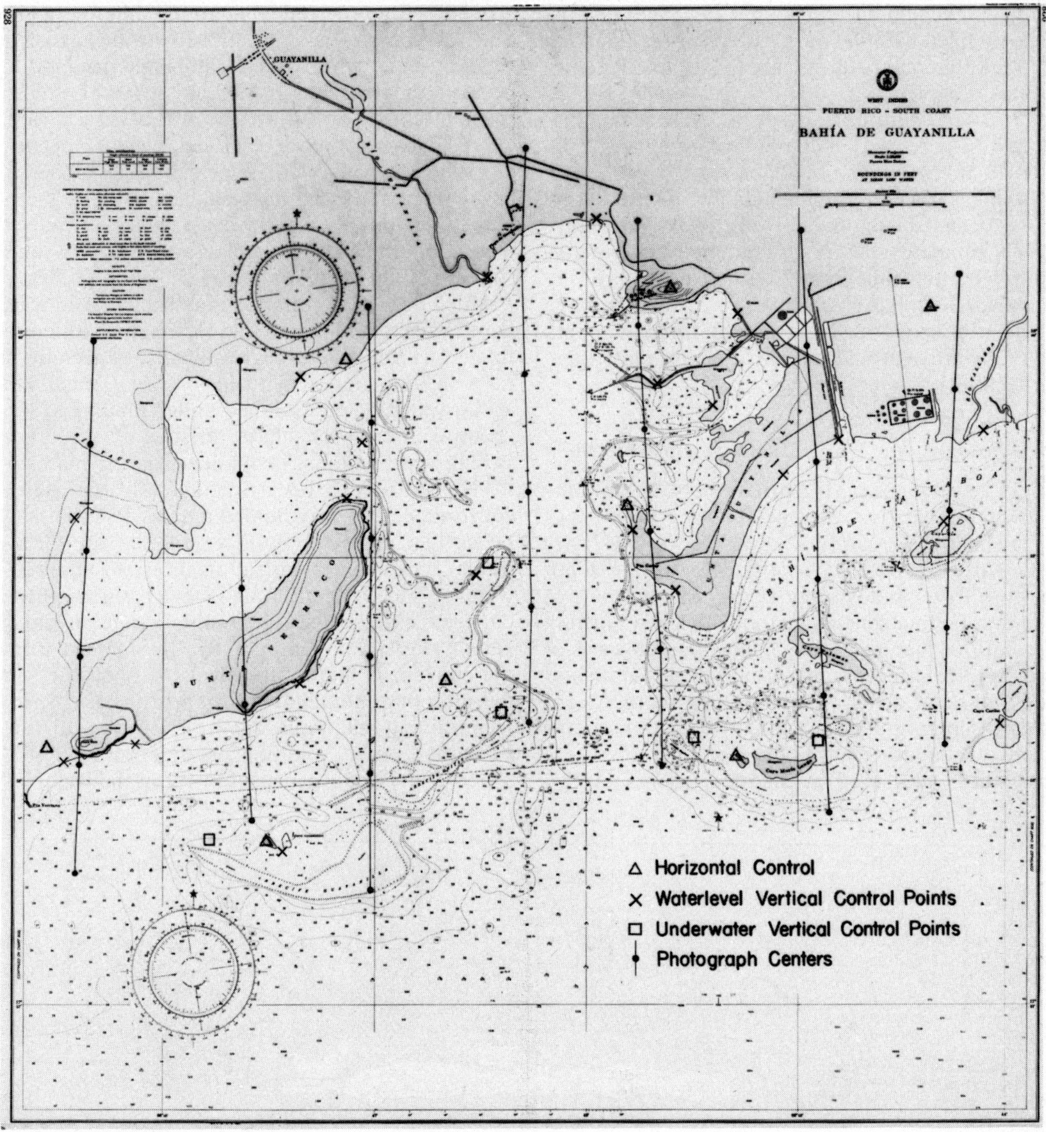

Fig. 28-87. Flight plan and geodetic control layout for verification of method of photogrammetric bathymetry (Umbach and Harris, 1972).

lowing the treatment of plate coordinates for displacement due to sea-surface refraction as previously described. For technical aspects of block aerotriangulation, see Keller and Tewinkel (1969).

To evaluate the accuracy and validity of this approach, several solutions were made under varying control constraint conditions. Initially, a "free" photogrammetric block-aerotriangulation solution was performed. This solution is a unique adjustment to a minimum of horizontal and vertical control that permits no deformation of the photogrammetric geometry. Other control conditions were carried through the solution and solved for as unknowns. All available control was finally appropriately weighted and introduced as a geometric constraint during the second solution. The root mean square errors (RMSE) of the plate residuals for the "free" and constrained control configurations were 8.6 and 10.4 μm, respectfully.

Routine block solutions over land masses normally yield approximately 5 μm for the RMSE of plate residuals. The larger RMSE for bathymetric plate residuals is attributable to a combination of the enforcement of a single apparent depth for each point, a larger standard error of pointing on underwater objects, and the refractive effects of sea-surface roughness. The RMSE of the differences between photogrammetrically determined depths and the true depths of withheld control in the "free" solution was 0.55 m, with a maximum difference of 1.4 m. The RMSE in the constrained solution, using all available control, was 0.18 m, with a maximum difference of 0.4 m. These statistics, coupled with detailed studies of residual and depth difference patterns, show the practicability of this approach.

(9) Map Compilation—A properly scaled diagram of the refraction-compensation curves (Figure 28-86) was positioned beneath a Wild B-8 working surface to provide data for vertical datum corrections during model compilation. Figure 28-88 compares the photogrammetrically determined bottom topography with that acquired by standard hydrographic survey methods. A large number of spot soundings and intermediate depth curves were omitted from the figure for clarity and to facilitate comparison.

b. Photographic Water Transparency Methods

(1) Multispectral Approach—As an outgrowth of hazards experienced during World War II in shallow uncharted waters, Moore (1947) developed a transparency method for water depth plotting from photography alone. This was one of the first remote-sensing methods that did not rely on bottom features (save in the limited case of stereo in extremely transparent waters with good textural relief detail). Basically, the method consists of determining the optical extinction coefficient of the water column from the apparent brightness measured at identical points in two specially filtered (red and green) photographs. The spot depths are determined by means of a ratioing calculator, which compares the brightness in the two bands to known water color-transmission. Results with ±10% accuracy to a depth of about 6 m are achieved over homogeneous sandy bottoms of the type similar to subaerial beach used for brightness calibration. The technique is limited to calm seas, relatively clear water, bright skies without cloud patterns, and sun angles between 30° and 55°.

Much deeper penetration may be achieved with a multispectral photographic technique with more numerous and finer spectral bands, especially if used in conjunction with an *in situ* (immersed) spectroradiometer. Detailed understanding of the spectral attenuation characteristics of the intervening water column is obtained via the increase in the number of spectral bands, and the use of subsurface spectroradiometer data. Results may be further improved by increasing the aircraft altitude, and using polarization filters to suppress water surface reflections and wave patterns; these changes extend the acceptable range of solar zenith angles for photography. In simplest form the concept is merely to select the band (i.e., in the blue through yellow portion of the EM spectrum) giving maximum depth penetration for the prevailing water conditions as determined by the spectral attenuation measurement in the water. A

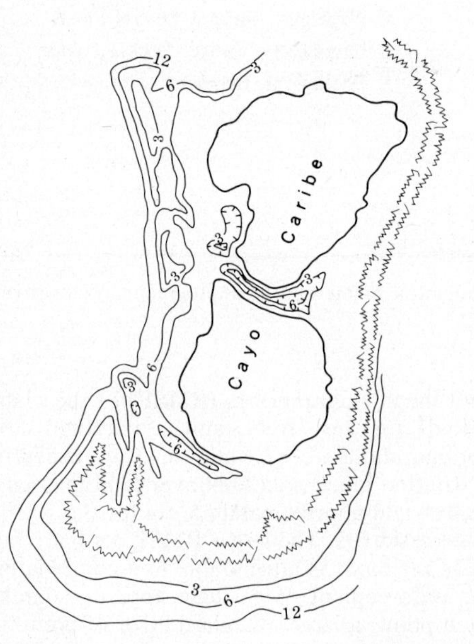

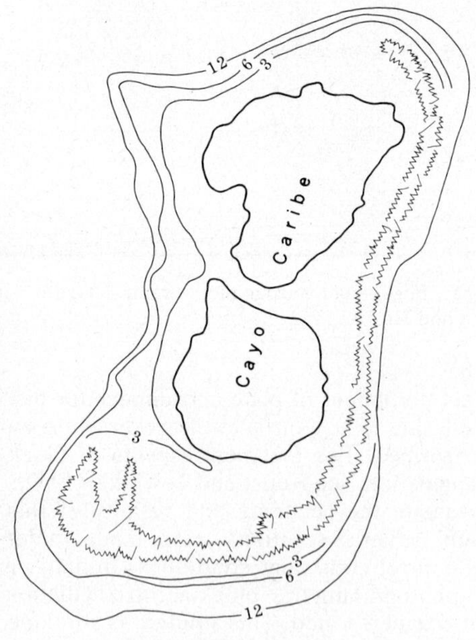

Fig. 28-88. Comparison of depth curves by photogrammetric and conventional methods (Umbach and Harris, 1972).

substitution of the selected band for the green band in Moore's transparency method permits mapping to a greater depth in any given body of water. This approach is essentially an optimization of the Moore technique. Ross (1969) obtained penetration to 25 m using a pair of bands at 460–510 nm and 510–560 nm. Yost and Wenderoth (1970) determined that the best band was 493–543 nm.

A further refinement is to treat each band as a monochromatic "sounding" with a characteristic extinction range beyond which no light is reflected. The overlay of a series of dyed, sharpened-contrast transparencies, one derived from each color band, produces a false color map where each hue represents a discrete depth interval. Relatively narrow-band filters with no spectral overlap must be used to avoid spectral redundancy between adjacent bands. Also, a precision photographic laboratory is necessary for repeatable results, with control of the effective gammas of the different photographic bands employed.

A less quantitative alternative is the use of color film, with a yellow filter to remove the effect of its blue sensitivity. Since attenuation by scattering in relatively clear waters is largely due to Rayleigh scattering in the water column with a $\lambda^{-4.3}$ dependence (see Ocean Color), blue light is rarely image-forming in terms of bottom topography. Hence, its subtraction provides improved contrast in directly-imaged bottom features, permitting recognition of morphological patterns and more sophisticated stereo interpretation for a complex bottom morphology. Around 1970, both GAF and Kodak marketed special-purpose color films with altered spectral sensitivities. The GAF emulsion (Current, 1969) omitted the blue-sensitive layer and had an ASA of 1000. Tests by Vary (1969) showed a depth penetration in the Bahamas to 150 feet. The Kodak film was termed water-penetration film and had two narrow spectral bands in the blue-green. The high speed of these color films and the removal of blue sensitivity, which eliminated the need for a filter, permitted image formation under lighting conditions impossible with normal color film. In turbid coastal waters, however, with only a diffuse bottom reflectance, such color films offer no inherent advantage over comparable speed panchromatic films. Experimentation with water-penetration film in Chesapeake Bay waters showed that it produced low chromatic contrast and was by comparison less useful for mapping shallow bottom features than ordinary color film. For bottom photography, chromatic washout due to the spectral attenuation in a double traverse of the air and water columns restricts the image-forming light to a band of 200 nm (Hodder, 1973). In turbid waters, the presence of significant quantities of scatterers, in size ranges comparable to the wavelength, further restricts the usable bandwidth, producing a need in multispectral techniques for even narrower bandpass filters.

(2) Temporal Image Merging—In addition to multispectral methods, a multitemporal approach termed "statistical temporal image merging" may be used (Hodder, 1971, 1972). Multiple images of the same site obtained under random conditions on different days are combined so that the time-varying portion (sediment, clouds, etc.) of each image averages out and the stable portion (bottom features) reinforces the information. With removal of the normal limitation to water clarity on any single day, the resolution should improve roughly as the square root of the number of independent images merged. This approach may be combined with the multispectral approach, using narrow band interference filters (20 nm) across the possible range of water-penetration wavelengths. The bands need to be chosen according to the selective transmission of turbid coastal waters; the narrow bandwidths eliminate spectral redundancy allowed by wide-band filters such as the Wratten series. Even without a priori knowledge of the water column's properties, an optimum band can be chosen from each day's record to support the statistical/temporal image merging. Only a relative correlation of averaged film density to water depth will be obtained unless, of course, some limited soundings are available for calibration. In this case, Moore's two-band spectral ratioing may be generalized to multiple bands, with each available band tested as a base for ratioing.

The mechanization of this technique was described in detail by Hodder et al. (1971). Here the basic steps will be outlined. Depth calibration must be employed on each temporal record. Initially each given frame of film exposed over the water should exhibit a relatively simple brightness (film density) to depth relation. In turbid coastal waters, the image forming light for bottom photography at the maximum water depths varies in center wavelength from blue to yellow as a function of sediment load as well as size and shape of the constituent scatterers. By comparison of the patterns in the imagery to the "true" bathymetry obtained by soundings, as judged by the formation of coherent patterns that match the known bathymetry, it is evident that the depth penetration is a function of wavelength. Testing revealed differences even between the 580 nm and the adjacent 550 nm band, which demonstrates the necessity of employing narrow-band filters for enhancing photographic brightness/depth contrasts. The 500 nm band showed discrete patterns closely correlated to differences in the bottom material. With such details in consideration, a "synthetic" image may be prepared as a merged set of selectively thresholded bands, taken simultaneously to yield an optimum image for the later temporal image merging. Features can be selectively emphasized or subtracted from each single-date multispectral image set. Synthetic images from the various dates are then combined and depth-calibrated. These manipulations are performed on digitized image data.

(a) Implementation and Results—A computer program statistically merges the digitized photographic data from successive flights over the same target site. This program computes local surfaces by linear regression techniques and then joins them into the complete contour map by smoothing and blending. Results are illustrated in Figure 28-89, which depicts four bands of imagery from a single day, as well as the bathymetric map resulting from merging of the optimal synthetic image from four different days. Here the computer-derived contours agree with the available bathymetric charts and soundings out to the 18 m contour.

(b) Calibration Methods—A critical hypothesis to test is whether suspended sediment contributes dominantly to the backscattered light. The problem is that higher bottom reflectances in shallower water may be confounded with increased backscatter due to an increase in suspended sediment. Even in the simplest of inlets a bidirectional flow periodically occurs, producing a similar complexity in the transport and spatial distribution of the suspended sediment (particularly the fines, which produce the greatest chromatic effect in scattering attenuation). Further, even at the same measurement station for a given site, a marked change in spectral transmissity of the intervening water column may occur at different times. This is, of course, due to the temporal change of size, type, and concentration of suspended sediment. Thus, the very spatial randomness of water transparency with time is exploited in the statistical/temporal image merging technique, but it provides a complex calibration problem. The classical stereoscopic approach, in contrast, depends on imaging of bottom textural detail (not simply changes in returned energy, as is the case for temporal image merging), which is generally satisfied in shallow water.

An approach to remote depth calibration for multispectral film-density levels is to build a catalog of characteristic reflectance spectra, or modulation transfer function (MTF) data, for the sites of interest, and then to use these to relate depth to film density. A simpler technique to implement in shallow zones where the bottom is visible is stereoscopic measurement of depth as discussed earlier. Spot depths may be extracted for calibration of spectral densities of the imagery, in order to determine the spectral attenuation coefficient with respect to a "true" depth. Results may then be extrapolated over the remainder of the target area assuming a homogeneous water body.

Even provided that a remote calibration technique as described above can be implemented, the degree of homogeneity of the waters over the area being mapped must be defined. In the simplest case one might employ thermal infrared mappers to infer current or water-mass boundaries. If inhomogeneity is suspected, separate calibration of each homogeneous region should be considered.

c. Wave Refraction

Surface wave-train morphology may be used to infer water depth. Wave refraction techniques re-

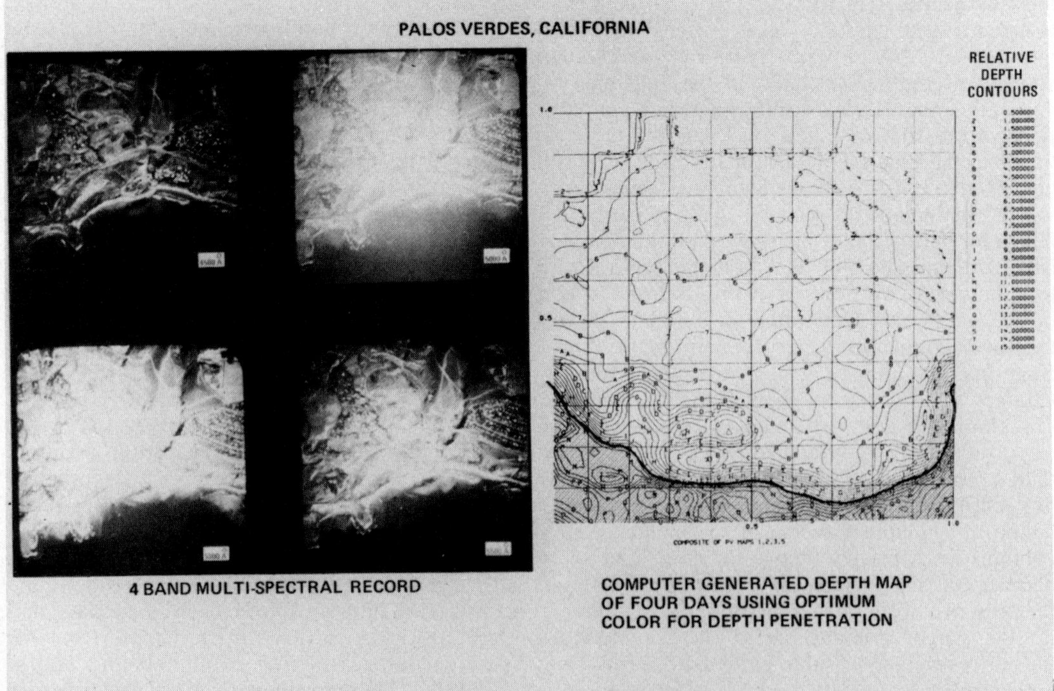

Fig. 28-89. Multiband imagery and resulting bathymetry (Hodder *et al.*, 1971).

quire only the imaging of the water surface and are hence unrelated to water transparency, relying only on the laws of hydrodynamics. While they may be applied to regions where water depth changes slowly, they are unfortunately invalid for those coastal waters where depth changes rapidly with respect to 2 or 3 wavelengths (by Shannon's theorem), since the assumption of stationarity of the wave trains is lost. In such waters the confidence level for the inferred depth (if, indeed, a measure of wave characteristics can be made at all) is reduced drastically. This method therefore deos not offer a strong alternative to optical measurements in shallow waters, unless the waters are turbid and circumstances otherwise prevent direct soundings from vessels. Around Cape Hatteras, where shoals interfere with vessel operations, where sedimentology is dynamic, and where optical-spectrum methods are less viable due to mixed seas, wave-refraction methods can provide a check on other bathymetric measurement techniques. Shuchman et al. (1979) used Seasat SAR data analyzed by OFT methods to infer Cape Hatteras bathymetry, obtaining fair agreement with sounding data.

d. Multispectral Scanner Bathymetry

In addition to the above methods, a multispectral scanner (MSS) may be used for bathymetric determinations. In the MSS method (Polcyn et al., 1970), ratios of signal returns from several bands of the spectrum are used. In Figure 28-90, five spectral bands are displayed from the area near Caesar Creek and extending out to Pacific Reef in the Florida Keys. In the blue band (400–440 nm) the light penetration is not optimum and a low contrast image is formed. In the green bands (500–520 and 550–580 nm) the underwater detail is clearer, and in the red band (620–680 nm) only the shallow features are displayed because of the higher absorption of the water at these wavelengths. Finally, in the near-infrared band (800–1000 nm) only land-water boundaries are mapped.

The sensor output voltage (V) in each band is, from spectral properties of light penetration, (Polcyn et al., 1970; Brown et al., 1971):

$$V = K\rho H e^{-\alpha Z(\cos^{-1}\theta + \cos^{-1}\phi)} \tag{28-15}$$

where atmospheric effects are neglected.

From a ratio of two bands:

$$Z = \frac{1}{(\alpha_2 - \alpha_1) f(\theta,\phi)} \ln \frac{V_1 K_2 \rho_2 H_2}{V_2 K_1 \rho_1 H_1} \tag{28-16}$$

Here the depth of the water is equated to variables that either are easily measured by the scanner system itself or can be estimated. The α_i are water extinction-coefficients weight-averaged over the two bands, the ρ_i are reflectances of the bottom

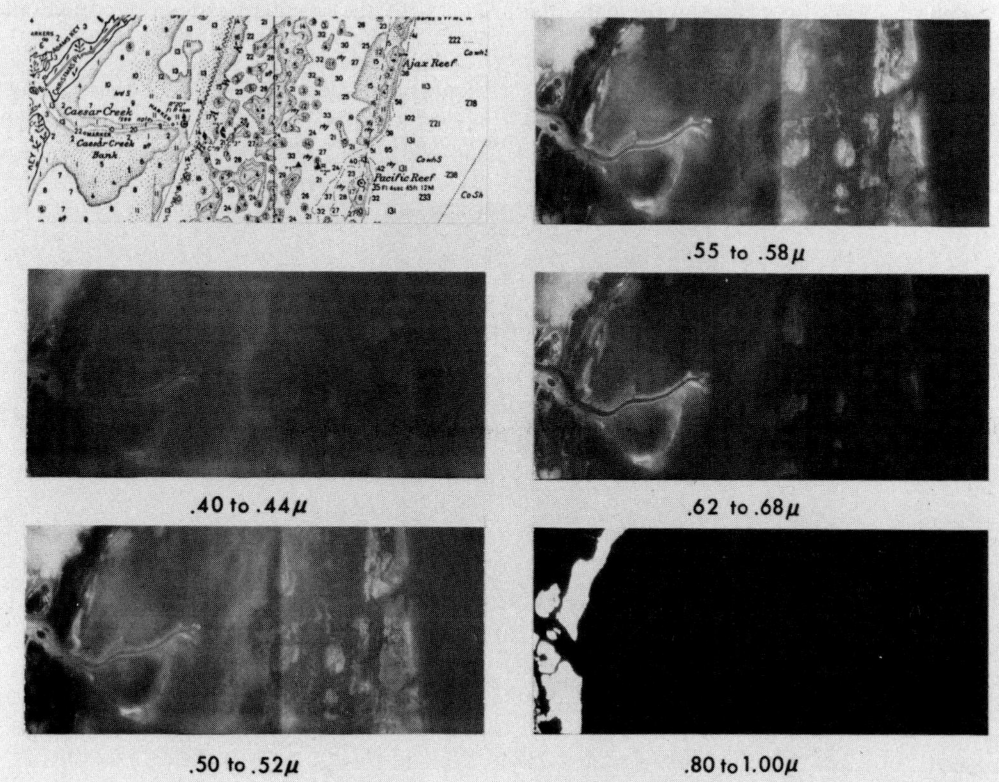

Fig. 28-90. Airborne multispectral scanner records from the Florida Keys illustrating penetration depth as a function of wave length (Polcyn *et al.*, 1970).

material, and H_i is solar irradiance just beneath the air-water interface. K_i represents scanner constants that are known from tests of the system, and θ and ϕ are angles of incidence and reflection. Accurate results require normalization against system noise and illumination change; for each scanline, the sensors should be referenced to both an internal lamp and the solar radiation arriving at the top of the aircraft.

Computations may be performed on a digital computer. Scattering and surface reflection effects are removed by making use of deep-water points lacking a bottom reflection. Typical results are shown in Figure 28-91 where each symbol signifies a particular depth range. The outline of Caesar Creek can be seen in the data, and for comparison, a black-and-white strip map of the area is shown.

e. Satellite Bathymetry of Shallow Seas

In order to meet the need for updating navigational charts around the world, the Landsat MSS is now used for bathymetric surveys (Polcyn and Lyzenga, 1979). Landsat MSS in polar orbit detects the sunlight reflected from the ocean surface and from shallow sea bottoms and, through computer analysis of MSS digital tape data, water depths may be estimated for each pixel, 80 m on a side. The Landsat series has been shown to be useful in updating charts in two ways (Hammack, 1977; Polcyn, 1976): first, through the detection and location of new reefs and shoals hazardous to shipping, and second, for preparation of new charts showing the water depth at 7,500,000 pixels for each scene covering 10,000 square miles. Each Landsat scene is formed in 25 seconds. The near-polar orbit repeats a given ground track every 18 days, and each day a track overlaps a track from the previous day by amounts that reach 60% at high latitudes, giving nearly complete coverage of the Earth. Of the four MSS bands, the green and red bands, MSS 4 and 5, are used for estimating depth. The two infrared bands, MSS 6 and 7, are generally used for delineating land/water boundaries.

(1) Data Tape Processing—While the standard method of processing Landsat data may be acceptable for most terrestrial applications, specialized processing is needed for oceanographic applications. These applications require that a maximum amount of radiometric information be preserved during processing, even at the cost of greater complexity in data processing on the part of the user. Oceanic volume reflectance is much smaller than terrestrial reflectance in the Landsat MSS bands, so that information about the ocean is contained in the lowest digital levels of MSS sensor output. To preserve the radiometric information, the user should avoid data tapes containing data already subjected to enhancement procedures such as contrast enhancement (contrast stretch with scene and band-dependent thresholds), haze removal via dark pixel subtraction, and edge enhancement. Such procedures are suitable for image enhancement but not for radiometric analysis of digital data. Therefore, raw data or data with the standard radiometric calibration alone should be obtained.

Also, some types of geometric corrections should be avoided. Corrections, if desired, may be applied by the user to computer compatible tapes

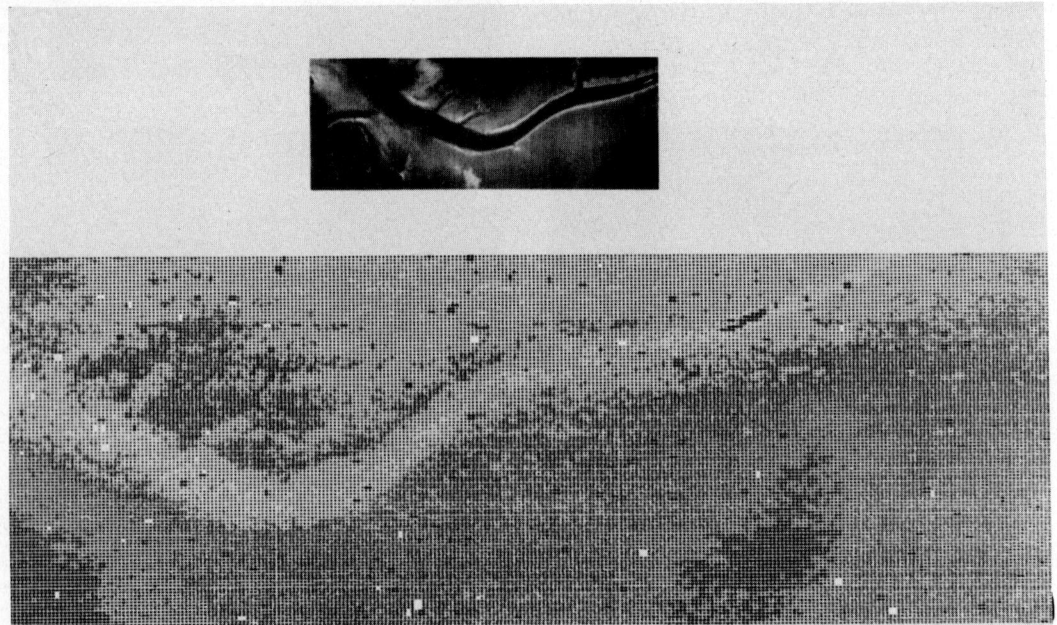

Fig. 28-91. Processed airborne MSS data for the region of Figure 28-89 in print-out and image form (Polcyn *et al.*, 1970).

(CCTs) of raw data as received, using any of several software packages already in the public domain. The first step in the correction process is usually the insertion of artificial pixels to equalize scan line lengths. If adequate ground control is available to the user, it will be better to avoid data tapes already adjusted for line length anomalies, and carefully apply ground control for this adjustment on a scene-by-scene basis.

Resampling for the purpose of making geometric corrections is accomplished by several methods, including nearest-neighbor and cubic convolution methods. The nearest-neighbor method, although it cannot make corrections to less than one-half pixel accuracy, preserves all of the original data values and has essentially no effect on spatial frequencies. The nearest-neighbor method is also the only method that can be safely applied to data that have been classified or processed in such a manner that output data values are not linearly related to radiance, such as in signal compression. Cubic convolution, in contrast, replaces each digital value by means of a $(\sin x)/x$ filtering of the neighboring 4×4 pixel array of values, which obviously alters the original values. This is a serious compromise of radiometric accuracy for ocean data which, to begin with, span only a few digital levels. For best results, a user should obtain raw (uncorrected) data tapes and carry out geometric correction only once and only when necessary. This is especially true for scenes where ground control points have been established and are available to the user.

(2) Depth Extraction—The most useful measurement for observing the first 30 m of depth is the solar irradiance attenuation by water. MSS bands 4 and 5 have been used to measure depths through a physical model relating the observed signal to a number of variables either known a priori or found by measurement of water transparency and bottom reflectance.

Algorithms for calculating the depth ranges and accuracies have been developed under sponsorship of the Defense Mapping Agency (DMA), NASA, and NOAA. In a milestone experiment conducted in 1975, the maximum depth detectable in the Bahama test range was 40 m (130 ft) using the high gain mode of Landsat for an October solar elevation. The high gain setting provides for an increase in sensitivity of about 3×. The highest accuracy for the depth measurement was 2.5% when the key variables of water attenuation and bottom reflectivity were known. Typical accuracies of 10% were achieved at depths of 22 m (70 ft) based on the assumption of stable values away from the control points (Polcyn, 1976).

The data in Table 28-13 give the digitization-limited maximum depths for both low- and high-gain Landsat 2 data for three different solar elevations. Assumed constants were 30 percent bottom reflectance, and attenuation coefficients of $K = 0.075\ m^{-1}$ (MSS 4) and $0.325\ m^{-1}$ (MSS 5). The form of the algorithm is

TABLE 28-13

Maximum Depths from Landsat 2 (Polcyn, 1976)

	MSS 4			MSS 5		
	30°	45°	60°	30°	45°	60°
Low Gain						
$V_0 - V_\infty$	24	33	40	30	43	52
Z_{max} (m)	21.0	23.2	24.6	5.2	5.8	6.1
High Gain						
$V_0 - V_\infty$	79	111	136	100	141	172
Z_{max} (m)	29.1	31.4	32.8	7.1	7.6	7.9

$$Z_{max} = \frac{1}{2k} \ln (V_0 - V_\infty) \qquad (28\text{-}17)$$

where V_∞ is the signal from deep water, and V_0 is a constant calculated from known parameters.

Inspection of the table readily shows the improvement possible with Landsat sensors operating in the high gain mode (on the order of 33 percent better). Definition of Z_{max} was based on one digital count above the deep water signal. Statistically, an extended shoal area at the maximum depth will be detectable because of the number of observations made spatially. For very small shoal areas, it may not be possible to measure areas 30 m deep with a consistently high accuracy. However, signals 4 and 5 digital counts higher than deep water levels will consistently give good detectability to 22 m. But improvement in these values is possible. In a 1977 study sponsored by NOAA, the optimum band location and band widths were investigated for bathymetric applications (Lyzenga and Polcyn, 1978; Lyzenga et al., 1976). By operating in a slightly bluer region than MSS 4, and using two bands instead of one, one can make improvements through the use of two band correlation and/or ratio techniques. Landsat D will carry the current type of MSS, as well as a six channel sensor with higher spatial resolution called the Thematic Mapper (TM) with wavelengths suitable for making improved water depth measurements. The higher spatial resolution of the TM on Landsat D (35 m) will be useful for locating smaller rock pinnacles, coral heads, or sea mounts. An indication of the impact that this higher resolution will have may be obtained from inspection of Landsat 3 return beam vidicon (RBV) data that have roughly 40 m resolution. RBV data permit an assessment of the utility of space data to locate potentially hazardous shoals.

One of the principal advantages of Landsat as a source of data for bathymetric mapping is its repetitive coverage of a given area at 18-day intervals. This repetition not only permits the identification and separation of permanent features from transient effects, such as water quality and atmospheric variations, but it also may be used to reduce dependence on surface measurements by allowing the extraction of water-attenuation parameters from an analysis of scenes at different tidal stages. In multitemporal processing (see the section on Temporal Image Merging) a composite data set is prepared in which corresponding pixels from the separate data sets are brought into registration with each other. Direct registration is generally preferred because it is computationally more efficient and easier to apply (since only relative control information is needed), and this information can be obtained in a semiautomatic process directly from the data. After two or more Landsat data sets have been spatially registered, several types of multitemporal processing techniques can be applied. These include temporal averaging, change detection and removal of transient effects such as clouds (Lyzenga and Polcyn, 1979).

Multidate data analysis for the Bahama test range has been successfully carried out using low- and high-gain data sets. Figure 28-92 shows MSS 4 data from two dates; the low-gain data set was scaled upward to overlay the high-gain data. Figure 28-93 shows the results of transient cloud suppression for the combined frames. Taking maximum penetration as the depth for a signal that is one standard deviation below the band mean, one finds that the penetration depths for MSS 4 were 21.9 m for the low-gain data set, 24.8 m for the high-gain data, and 25.4 m for the multitemporal data set. The corresponding figures for MSS 5 were 5.6, 6.1, and 6.2 m, respectively. Deeper penetration can be obtained by spatial filtering, but at the cost of decreasing the spatial resolution. The increased accuracy and penetration depth obtained by multitemporal processing is accompanied by the benefit of obtaining a cloud-free image.

(3) Mapping Results—A recent test of the methodology was made for selected portions of the Chagos Archipelago. The Defense Mapping Agency issued a corrected chart No. 61610 of this area, based on interpretation of Landsat data acquired on April 16, 1976. A new reef was detected approximately 9 miles east of Speaker's Bend and was named Colvocoresses Reef. With techniques tested in the Bahamian Photobathymetric Calibration Range, computer depth charts were prepared using basic physical parameters without recourse to knowledge of water depth in the area (Doak et al., 1979). Figure 28-94 shows a grey map presentation of the depths for Colvocoresses Reef. Color-coded depth charts were prepared for this area as well as for other reefs in the Archipelago. Verification of results is currently underway through the effects of the Joint Services Chagos Research Expedition.

f. Coastal Bathymetry by Airborne Lidar

Development of airborne laser technology during the 1970s has produced airborne systems capable of depth resolution of better than 20 cm (Kim et al., 1975). Hoge et al. (1980) demonstrated that an airborne pulsed neon laser-system provides accurate depth measurements in clear water up to 10 m deep and in turbid water up to 4.6 m deep. The helicopter-mounted Airborne Oceanographic Lidar (AOL) measurements in Figure 28-95 compared favorably to bathymetric data obtained simultaneously by a NOAA/NOS vessel. Both profiles indicate high relief on the bay floor of the

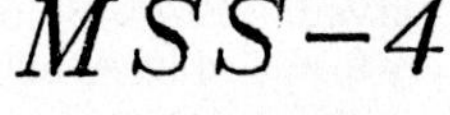

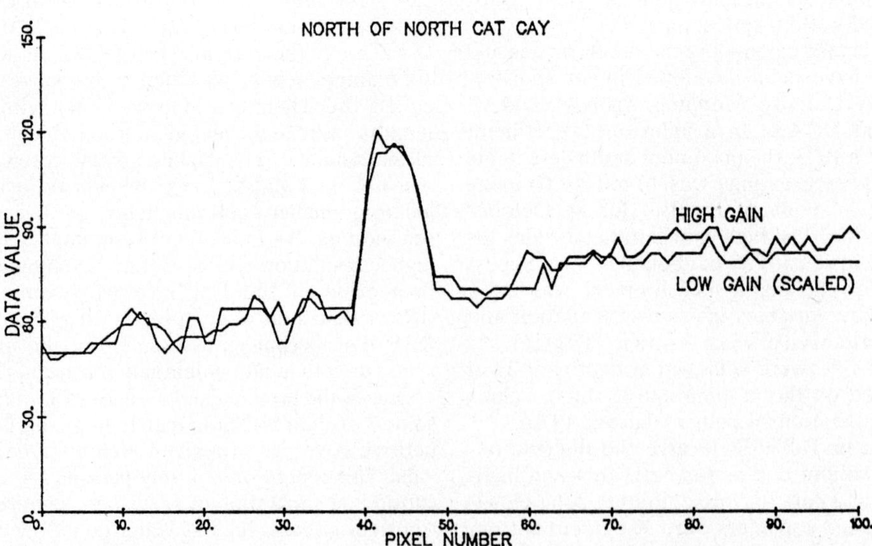

Fig. 28-92. Scaled low gain and high gain Landsat MSS 4 data north of North Cat Cay, Great Bahama Bank (Polcyn and Lyzenga, 1979).

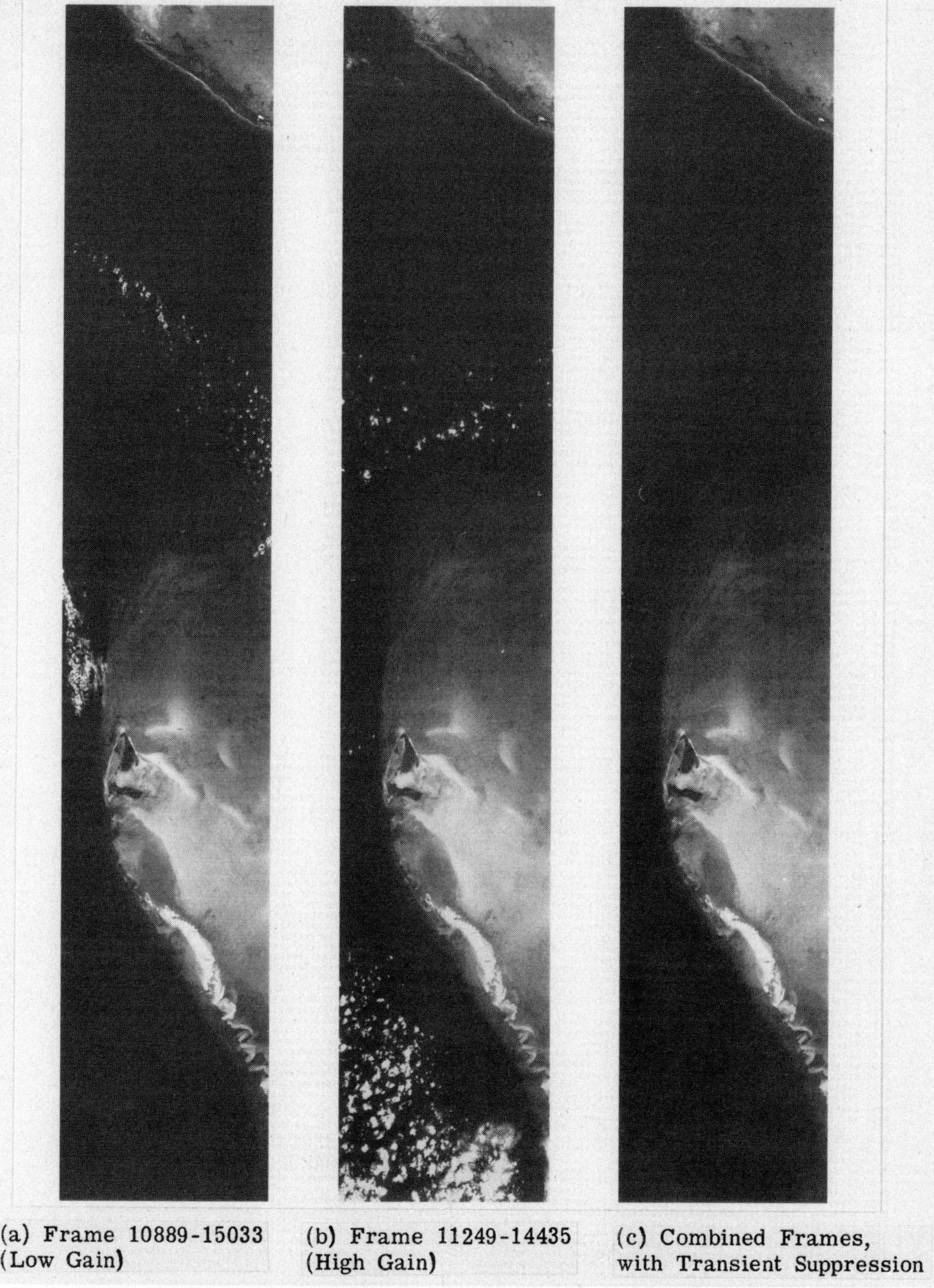

Fig. 28-93. Raw and processed Landsat MSS 4 data for frames 10889-15033 (strip 3) and 11249-14435 (strip 3) (Polycyn and Lyzenga, 1979).

test area. A sequence of pulses, horizontally separated by about 0.3 m, can be identified on the steep faces of the submarine features. The AOL technique showed good precision, particularly at depths less than 3 m, with variations from 7 cm to 20 cm rms as determined by least-squares fitting a second-order polynomial throughout the data. This method of depth measurement provides a means of more rapid coverage than is presently available from conventional vessel techniques.

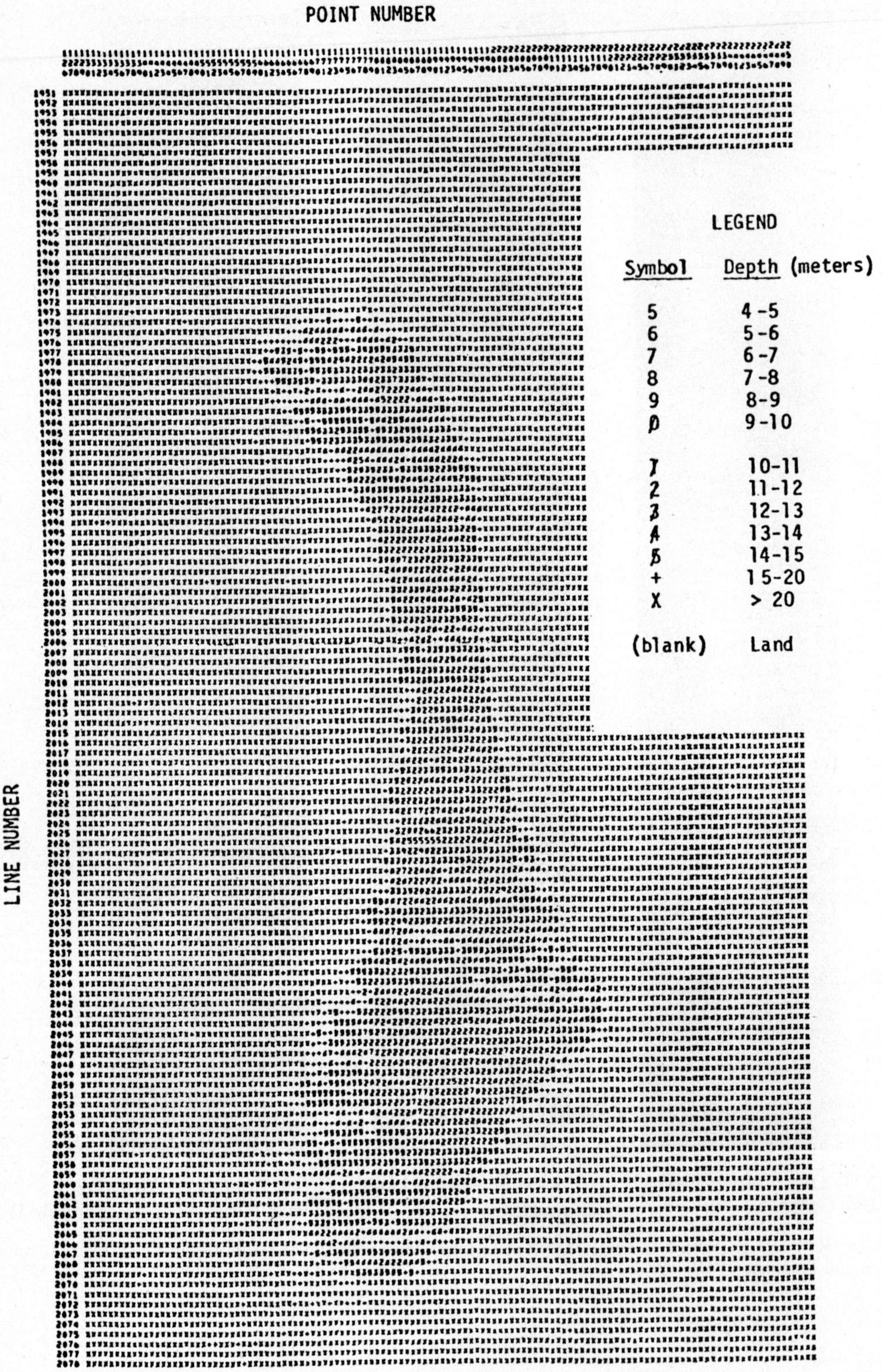

POINT NUMBER

LINE NUMBER

LEGEND

Symbol	Depth (meters)
5	4-5
6	5-6
7	6-7
8	7-8
9	8-9
Ø	9-10
1̸	10-11
2̸	11-12
3̸	12-13
4̸	13-14
5̸	14-15
+	15-20
X	> 20
(blank)	Land

Fig. 28-94. Calculated depth values from Landsat for Colvocoresses Reef (Polcyn and Lyzenga, 1979).

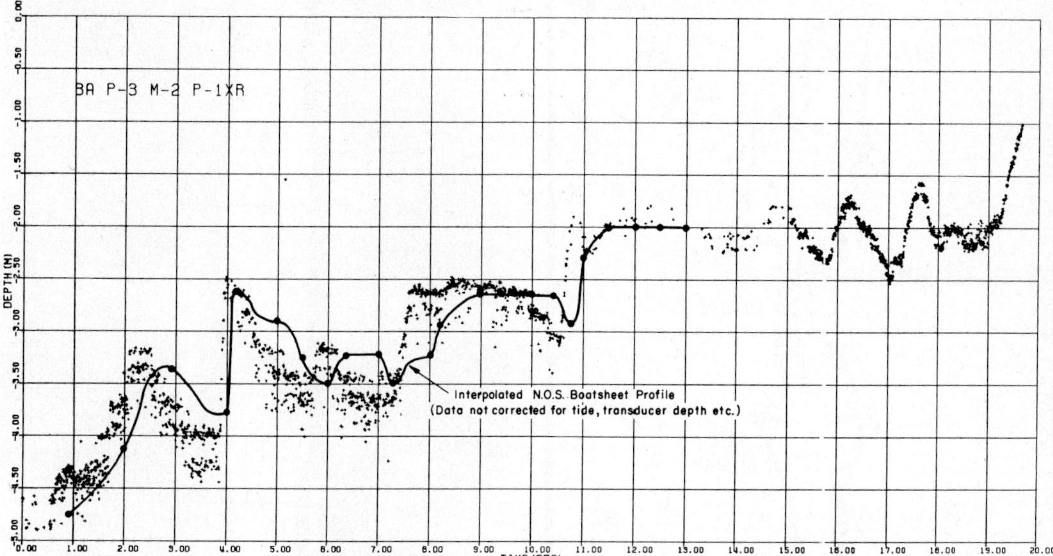

Fig. 28-95. Cross-section comparison of NOAA launch data with helicopter-mounted lidar (AOL) data in the Chesapeake Bay (Hoge *et al.*, 1980).

g. Bathymetric Mapping by Side-Scan Sonar

Prior and Coleman (1980) applied side-scan sonar surveys, together with seismic and fathometer data, to study submarine landslides and other features in the Mississippi River Delta region. Slopes were 0.2° to 1.5° in water depths of 5 to 100 m. Side-scan sonar is an acoustic approach in which a transducer "fish" is towed behind an exploration vessel on a steel cable that acts both as a tow line and as a transmission cable to a dual-channel recorder on the vessel (Belderson et al., 1972; Flemming, 1976; see also Acoustical Systems). The reflected sound waves provide a two-dimensional image similar to SAR data, or to photographs when the sun is low and behind the camera. Accurate depth and Loran (position) measurements allow parallel ship transects to be developed into a continuous bathymetric mapping. Figure 28-96 shows a line drawing of a mud flow gulley, and Figure 28-97 shows a side-scan sonar image of one position of a gulley. This two-dimensional fathometer approach to bottom bathymetry is a valuable tool for detailed subsurface definition, as well as a potential technique for verification of optical bathymetry.

3.0. Coastal Processes and Mapping

a. Utility of Aerial and Orbital Data

The availability of aerial photography beginning in the 1930s transformed the study of coastal processes. The aerial photograph provided a hard-copy data base showing relationships and patterns in coastal features that are difficult to discern from ground level except by long and tedious groundwork. The particular value of aerial photography lies in the rapid acquisition of volumes of metric data from an environment that is very dynamic.

Orbital scanner imagery, having become available only in the last decade, is not yet as useful because of its limited spatial resolution; however, photography from orbit, where available, should have resolution sufficient for many applications. The continued acquisition of orbital scanner imagery will, in time, provide a data base suitable at least for analysis of large-area dynamics of sediment transport and long-term change in entire coastlines. Other new sensors, such as radar, have found use in near-shore wave dynamics (see the Sea State section) relating to refraction and wave climate at the shoreline. But aerial photography continues to be the most popular technique for coastal data acquisition. It is so prevalent and familiar, as one of the most widely used data sources for coastal engineering, planning, and research, that it tends to be taken for granted. Basic articles in this field were collected by El-Ashry (1977).

b. Shoreline Change

The state of beach systems, and their rates of change, are the focus of coastal studies. Aerial photographs and satellite imagery are major sources for this information, along with ground surveys and existing maps and charts.

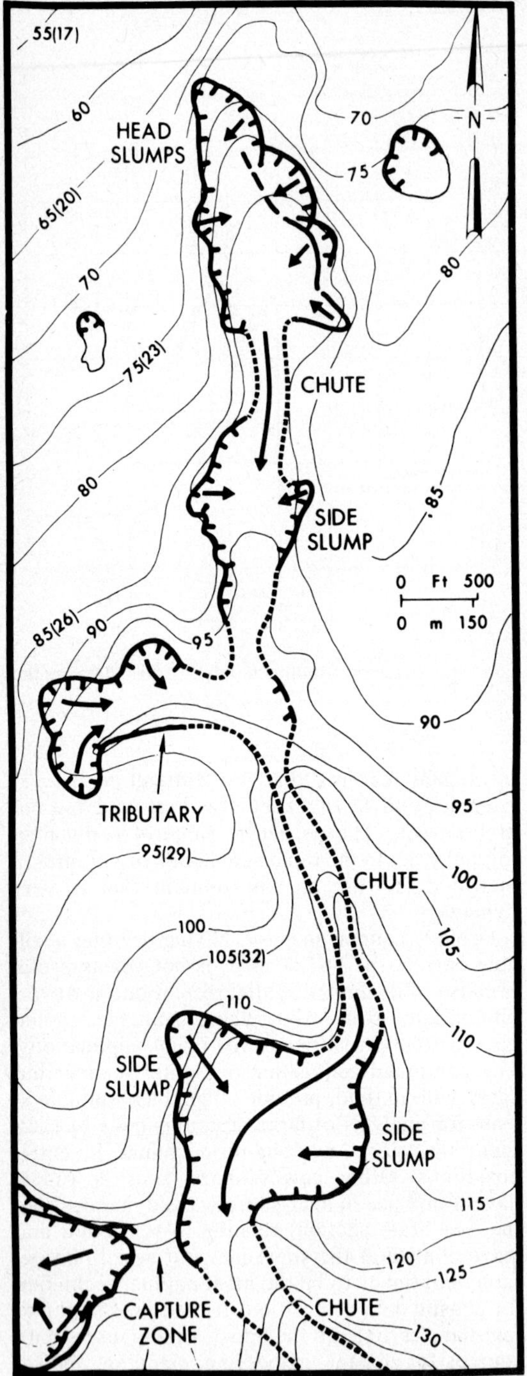

Fig. 28-96. Line drawing from side-scan sonar of a submarine mudflow gully with adjacent slumps (Prior and Coleman, 1980).

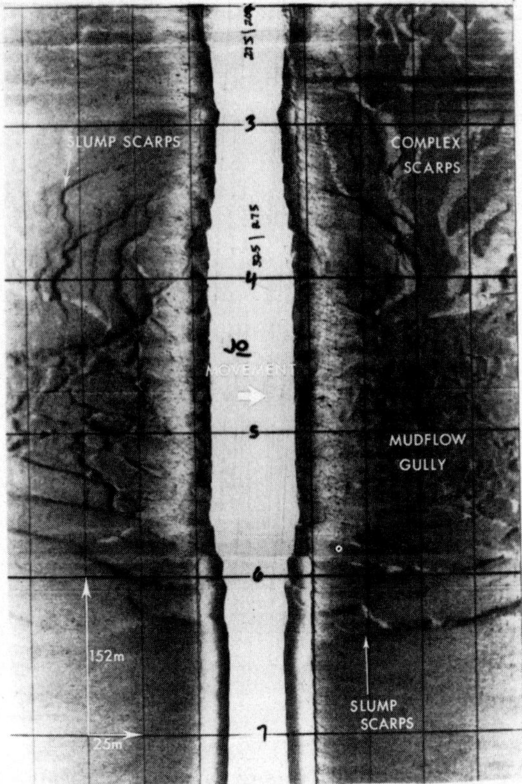

Fig. 28-97. Side-scan sonar image of mudflow gully emerging from within a major area of slumps and escarpments (Prior and Coleman, 1980).

Coastal processes have long been studied using aerial photography. Examples of earlier work may be found in McCurdy (1947), Shepard (1950), and Sonu (1964). McCurdy characterized the various types of coasts and the coastal features that contributed to each type, showing how bars, beaches, tombolos, and spits could be recognized on aerial photographs. Much of such early work was qualitative due to the lack of field measurements concomitant with the aerial photography. In more recent studies, with better field calibration, shoreline erosion and accretion have been measured using aerial photography (El Ashry and Wanless, 1968; Langfelder et al., 1970). Tanner (1978) has recently compiled standards for measuring erosion and accretion. Sequential aerial photography has been used to quantify changes in shoreline erosion, accretion, and orientation, as well as overwash distribution (Zeigler and Ronne, 1957; Boc and Langfelder, 1977; Hosier and Cleary, 1977). Relict beach ridges are revealed by topographic relief and patterns of coastal vegetation.

Ground survey, of course, provides the highest quality data in terms of accuracy (Dolan et al., 1978). However, only some areas along any stretch of coast will have been surveyed recently enough to bypass the need for new data. Historical data are usually sparse, lacking in control, and hard to find; thus, even when new survey data are available, the measurement of shoreline changes may be limited to comparisons of very short intervals of time, which produces results of limited accuracy where rates of change are low to average. Maps and nautical charts are often available and these may extend back to the mid-1800s. In the United States these may be sought from the U.S. Geological Survey, NOAA's National Ocean Survey and occasionally from other federal and state agencies. It is well known, however, that earlier charts are of questionable accuracy. Also, their spatial and temporal coverage is sporadic.

Aerial photographs are a better source of data in many cases. Precision photogrammetry is hardly possible from most historical imagery but, compared to early charts, the accuracy from use of early photography is welcome. However, photography is available in general only back to the 1930s. In using such photography, one finds that the usual factors and distortions must be considered, including camera interior-orientation (related to radial scale variations and resolution), exterior orientation (including tilt), and scale variation due to relief. A frustrating problem along undeveloped shorelines is the lack of well-defined, even recognizable, control points. For measurement of shoreline change, these must be traceable through time on a series of photographs.

Where control points are available, shoreline change may be determined once the position of the shoreline is fixed. A simple method for fixing the shoreline in a succession of aerial photographs was presented by Moffitt (1969). Four control points, whose state plane-coordinates (X, Y) are known, are located in the first photograph. Their elevation differences must be minor. The image coordinates (x,y) and state plane-coordinates are related by the projectivity equations for a plane:

$$X = \frac{a_1x + b_1y + d_1}{a_4x + b_4y + 1}$$

$$Y = \frac{a_2x + b_2y + d_2}{a_4x + b_4y + 1} \qquad \text{(28-18a)}$$

The eight unknown coefficients $a_1, a_2, \ldots, d_2$ can be solved for the first photograph by four pairs of equations, one pair for each control point. Then, points of interest along the shoreline whose state plane-coordinates are initially unknown can be found using the above equations. If the photograph is not badly tilted, the equations may be simplified to

$$X = a_1 + a_3x - a_4y$$

$$Y = a_2 + a_3y + a_4x \qquad \text{(28-18b)}$$

After solution for the first photograph, common points on the second may be oriented to the state coordinate-system, and so on through the photographic series. If the absolute position of the shoreline is not required, an even simpler approach may be employed, that of direct scaling from correlated features to the shoreline.

Due to tides and waves, the diurnal position of the shoreline varies, and for gently sloping beaches, the horizontal variation of the interface of beach and water can reach 60 m. Therefore, tidal stage and beach profile must be known to adjust the shoreline to a standard condition. Dolan et al. (1980) suggest that the high water line (not the swash line) is relatively insensitive to tidal phase over a few tidal cycles. With corrections for seasonal tidal variations, the high water line may be used during photointerpretation to standardize tidal stage, permitting accurate measurement of long-term rates of change. Lueder (1954) discusses photographic methods of locating and measuring the high water line, as well as other beach zones and beach characteristics.

Finally, the spatial resolution of the photography itself must be considered, which depends on photographic scale and the MTF of the film employed. All factors combined, the accuracy of shoreline-change measurements depends on the time interval between photographs, the rate of shoreline change and its variance, the adequacy of corrections for tidal stage and beach profile, and the film scale and MTF.

The film to use depends mainly on the potential supplementary uses for the photography. For shoreline change only, black-and-white infrared film exposed through a deep red filter provides excellent delineation of the land-water interface. When, however, the photography may also be employed for analysis of coastal vegetation, color infrared film (with the usual yellow filter) is much more desirable. Natural color film should be avoided because the land-water interface is not well-distinguished in the visible region of the spectrum.

The choice of altitude depends on many factors, especially cost. Detailed mapping will require scales as large as 1:5,000, which translates to an altitude of 2,500 ft (762 m) for a 6-inch lens mapping-camera.

Satellite sensors such as the Landsat MSS or RBV, or the Seasat SAR, have been suggested for shoreline mapping. The RBV with 30 m resolution may be useful, but only in regions lacking recent photography and suffering severe rates of change. Discussion of the uses of orbital remote sensing for the coastal environment may be found in Gierloff-Emden (1977). Coastal engineering applications of satellite data from civilian satellites will be out of the question until much better spatial resolutions are available.

The mere mapping of a shoreline and its rate of change will not be sufficient information for understanding coastal processes and specifying appropriate methods for shoreline stabilization. For these purposes information on near-shore water flow is required. Methods of measurement are discussed in the Nearshore Currents section.

c. Coastal Landforms and Topography

In highly dynamic coastal regimes, the vertical dimension in mapping will be quite important to both coastal planners and coastal geologists.

Topographic maps are periodically updated, with changes in cultural features shown as revisions. Topography is not revised as often; thus, topography shown on many maps of coastal dune areas may be inaccurate. In a study of Currituck Spit, Virginia/North Carolina, Hennigar (1980) found that available maps revised in 1973 contained topography compiled in 1952. Using historical photography, Hennigar determined that every dune that existed in 1952 has since metamorphosed into a parabolic dune and migrated many tens of meters. In the absence of stabilizing vegetation, some dune crests have been migrating at more than 10 m per year. Temporal changes in one large dune are shown in Figure 28-98. Useful data were obtained at low cost with stereo photographs and a parallax bar; photogrammetric quality mapping using plotters would of course have been desirable, but was impossible in this case because of the lack of historical ground-control data.

In the absence of stereo aerial photography, considerable information on the dynamics of topography may still be inferred from a historical photographic sequence. Boulé (1976) analyzed available imagery of Fisherman Island, Virginia, at the northern terminus of the mouth of the Chesapeake Bay. He was able to deduce the age and history of relict shorelines from changes in the vegetative patterns and island morphology. Field visits showed that these relict shorelines are recognizable at ground level as distinct ridges with characteristic vegetational communities.

From experience it is clear that depositional features (bars, spits, keys, washovers, beach ridges, and dunes) and erosional features (wave cut terraces, sluices, sea cliffs, and stacks) show more clearly and are easier to study from aerial photographs than from the ground. In addition to the photographic applications already mentioned, estimates of the amount of sediment deposited and eroded can be determined photogrammetrically (El-Ashry and Wanless, 1967). Applications to coastal engineering are numerous, and include assessment of the impact of shoreline-control measures such as the construction of breakwaters, groins, and other structures used to trap littoral drift materials.

Photographic sequences confirm the well-known fact that beach widths vary with the seasons, and that storms are the cause of the greatest changes, through high winds, waves, and tides. Most storm-induced changes are gradually smoothed and obliterated; thus, in time a coast may regain its former outline and width. However, a long-lasting change may be produced by the opening and closing of beach inlets. Hennigar (1979) noted that on Currituck Spit, Old Currituck Inlet opened before 1507 and closed in 1728–1730. New Currituck Inlet opened 8 km southward in 1713 and closed in 1828. These changes had a long-term impact on the salinity and biota in Back Bay behind the spit. Inlets and washovers are, therefore, of special interest in beach photointerpretation.

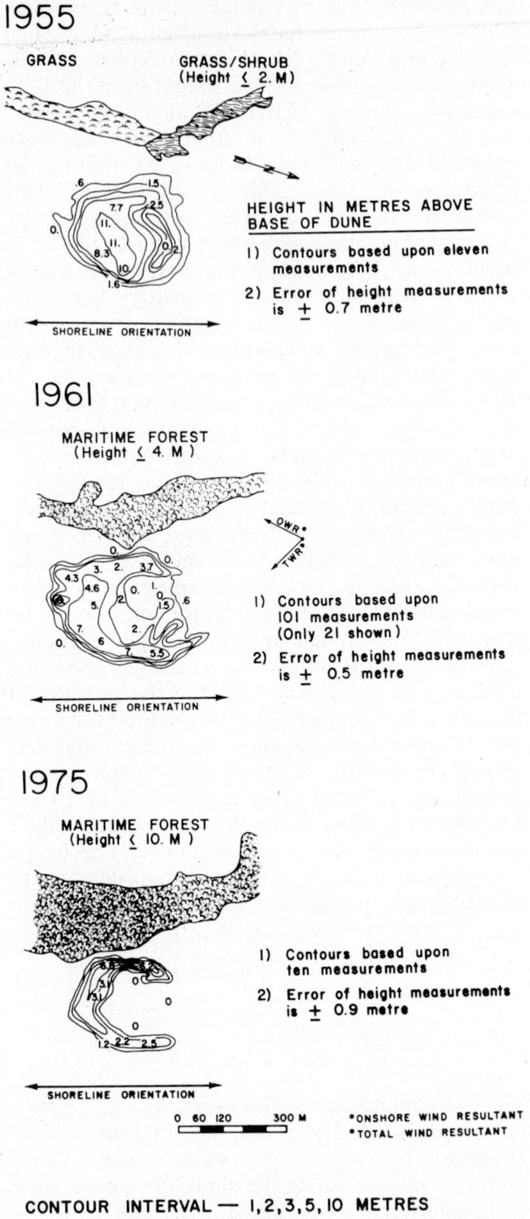

Fig. 28-98. Temporal changes in the morphology of Penny's Hill, Currituck Spit, Virginia/North Carolina, from parallax bar measurements on a historical photo sequence (Hennigar, 1980).

Although aerial photographs are the most important source of remote sensing data for coastal geomorphology, radar has distinct advantage along cloudy coastlands. Most of the features seen in aerial photographs can be interpreted from radar imagery. Contrasted with the dark tones of coastal water, the tones of mudflats, backshore swamps, mangrove canopies, beach faces, surf zones, and vegetated beach ridges are progressively lighter. Comparison of radar and photographic images of the same area reveals a distinct and unique pattern for some features resulting from cultural activity. Therefore, careful texture analysis of such patterns on radar imagery partially compensates for its inferior spatial resolution.

d. Tidal Boundaries

Demarcating and mapping boundaries related to tidal datum planes present difficult problems, particularly in areas where the boundaries are obscured by some type of vegetation. Both the mean high-water and mean low-water lines play important roles in coastal boundary determinations. Efforts in the past were focused on the mean high-water line due to its importance in the control of wetlands and in coastal zone management. However, today a higher priority is given by the National Ocean Survey to mapping the mean low-water line, for use in defining the U.S. Territorial Sea (3-mile), Jurisdictional Zone (12-mile) and Contінguous Zone (200-mile) boundaries (Collins, 1978). Only a relatively small part of the total length of the mean low-water line is obscured by vegetation, making it a minor problem in comparison to the mean high-water line.

The mean high-water line in dense mangrove and in coastal marsh is almost invariably obscured by vegetation, especially in data acquired with remote sensors. In such environments, terrestrial operations are extremely difficult, if not totally impractical. Together, these factors create problems for the cartographer, engineer, and surveyor (Jones and Shofnos, 1961; Brewer and Heywood, 1972; Brewer and Crabat, 1972; Fitzgerald, 1972; Guth, 1972; and Jones, 1972).

Regardless of the type of area or vegetation involved, there are three primary elements to be considered in the presurvey planning, and subsequent execution of the field-survey and data-acquisition: (1) establishment of tidal datums, (2) ground-data acquisition, and (3) remote-sensor data acquisition.

(1) Tidal Datums and Ground Control— Important tidal datums and tidal datum lines are defined as follows:

Mean high water (MHW) and mean low water (MLW): The average height of the high waters, and mean low waters, respectively, over a 19-year period. For shorter periods of observation, corrections are applied to eliminate known variations and reduce the results to the equivalent of the 19-year values.

Mean high-water line (MHWL) and mean low-water line (MLWL): The lines drawn on a map or chart to represent the intersection of the mean high-water datum with the shore, and the mean low-water datum with the shore, respectively; on the ground both are imaginary lines.

The availability of a datum in the area of the investigation is a primary prerequisite. However, the existence of this datum does not preclude the necessity to establish and/or expand the tidal datum network to provide local datums in many sections of the investigation area. This will depend upon the complexity and size of the area under investigation and the proximity of the known datum(s) to the area.

An analysis of datum requirements must be made well in advance, and any necessary tidal observations must be part of the initial phase of the field work. An essential procedure in establishing tidal datums is to reference them in some manner to facilitate their use and future recovery. Proper referencing is achieved by using a tide staff, in conjunction with observations, and differential leveling to connect the staff to three or more permanent or temporary bench marks.

After the tidal datums have been established and referenced, differential levels, based on either the tide staff or the bench marks, are run to points on the tidal datum line of interest. Each point is marked with a stake to preserve, at least temporarily, its identity for placement of targets, positioning later on a suitable map or photograph, or fixing by traverse for subsequent plotting. The position of the line as staked can then be plotted by plane table methods either directly on a rectified color photograph or on a stable-base copy of an appropriate map.

Identification of the mean high water line on the ground may be accomplished, after tidal datums have been established, by relatively inexperienced personnel. Various features indicate the location of the MHWL. Those features were, in a New Jersey project (Hull et al., 1973), a change in the elevation of the marsh at the MHWL as small as 0.06 m, sometimes as much as 0.24 m, and usually abrupt. Another indicator was the feel of the soil under foot: the soil below the MHWL was soft, while that above was much more firm. The third indicator was the inshore edge of a vegetative community, recognizable by the vigor with which individual plants developed during the growing season. Vigorous growth does not necessarily mean that plants attained the same or even nearly the same size everywhere. Also, the types of plants in these communities are not significant for boundary purposes. The important factor is the existence of this vegetation and its generally consistent relationship to the MHWL. Placing sole dependence upon this relationship to demarcate and map the MHWL may be misleading at times, because indicator vegetative communities may, under favorable conditions, grow below the MHW datum. Sand and shell deposits along the edge of marsh exposed to wave action may smother vegetation and eliminate the indicators.

(2) Remote Sensing Data—The inshore edge of the vegetative community described above is generally identifiable quickly and easily on color infrared (CIR) and black-and-white infrared photography, and occasionally on thermal infrared scanner imagery. The immediate objective is to find a tone or hue in the remote sensor data that will denote the MHWL in relation to objects on the ground, by correlation with known temperature differences or spectral reflectance characteristics at the land-water interface. Targets of a nature compatible with the characteristics of the sensor involved may be placed on the tidal datum line, as established on the ground, in order to provide for its positive identification in acquired data. Such targets may be of any material that gives high spectral or tonal contrast with the background.

Remote sensor data-acquisition requires a considerable amount of coordination between air and ground personnel, particularly for the collection of those types of ground data that must be acquired simultaneously with flight operations. The flight, in turn, must be coordinated such that aerial data are acquired when the tide reaches a predetermined level that is equivalent to mean high or low water.

CIR is preferred for aerial photography intended for use in mapping tidal boundaries. It should be obtained when the stage of tide is at or just below MHW. In dense mangrove areas, the mean high-water line is neither detectable nor identifiable on vertical color or CIR photographs. Black-and-white infrared may be used when boundaries are free of vegetation. Field verification is therefore essential to the production of tidal boundary maps to acceptable accuracy standards.

e. Offshore Bars and Underwater Features

One of the striking aspects of coastal photography is the display of underwater features in shoal water. The depth to which submerged features may be photographed depends on the spectral intensity of illumination as a function of angle, the reflection characteristics of the water surface and bottom, and the optical properties of the water column. Under ideal conditions, features at a considerable depth may be photographed: Sharkov and Kudritsk (1956) obtained orthochromatic photographs of the bottom as deep as 20 m in the Caspian Sea; Dietz (1947) reported visibility to 55 m at Bikini Atoll. The photographic films and exposures to use for maximum penetration of coastal waters were investigated by Yost and Wenderoth (1970). They developed an isoluminous technique for removing brightness differences from aerial photographic images of coastal waters. The technique involves the color additive projection of sandwiches of negatives and positives that have been obtained in different spectral bands. This color-enhancement technique allows the colors of surface and submerged features to be visually discriminated and appreciated in the absence of the usual brightness contrasts that impair color discrimination (Color Figure 28-99). The mathematics of this and other isoluminous techniques for photographic and digital data were analyzed by Munday and Alföldi (1975). Yost and Wenderoth (1970) found that the best photographic spectral band for water penetration in northeastern United States and Gulf of Mexico waters was the green band (493–543 nm). Good results were obtained independent of solar angle and aircraft altitude, as long as sun glint and excess wind-waves were avoided. The photographic exposure to use for optimal penetration of coastal waters should be chosen to produce an image optical density of unity for the water mass of interest. Multispectral color-additive photography permits greater chromatic separation of underwater color targets than normal color photography.

For murky waters, information about submerged features can be obtained by indirect methods. Breaking waves may indicate shallow reefs and banks. Deeper features cause wave refraction or wave foreshortening. Along some coasts the presence of the large kelp, *Macrocystis,* indicates rock bottoms with depths typically ranging from 8 to 30 m.

Offshore bars delineated on an aerial photograph indicate by their orientation and spacing the predominant vector of long-period swells, as bars are caused by the interference of approaching swell and reflected swell (Clos-Arceduc, 1964) (see the Bathymetry section).

III. COASTAL POLLUTION AND WATER QUALITY MEASUREMENT

The coastal zone contains that unique environmental triple point, where the water, land, and atmospheric components of the terrestrial surface converge and interact. Special hydrodynamic regimes are produced by the mixing of fresh with salt water, resulting in features of the coastal zone that nourish population growth in all levels of the biological kingdom, including man. Man, however, introduces pollution. Because of the variety in pollutants and in the manner and location of release, the natural complexity of the coastal zone is compounded to a level that makes tracing of pollutant pathways and pollution control extremely difficult.

Remote sensing is one avenue for meeting this challenge. It provides a means for locating, identifying, and mapping certain coastal features and pollutants. Such activity can be of assistance in efforts to study the natural system and the impact of pollutants, with a view toward pollutant monitoring and ultimately toward control. However, the mixing waters of the coastal zone create a remote sensing problem with multivariate dimensions of the highest degree. Horizontal and vertical gradients in currents, temperature, salinity, turbidity, and biological activity are compounded by point and non-point releases of pollutants, both conservative and non-conservative, of great biochemical variety and complexity.

Nevertheless, investigators have succeeded in defining topics of broad scope for which remote sensing can be useful. One is point-source pollution, encompassing ocean dumping, waste outfalls, and oil pollution. Another is non-point pollution, which produces higher turbidity, algal blooms from nutrient enrichment, and increased chemical concentrations. A third topic is the growing application of remote sensing to circulation and water-quality models.

A. POINT SOURCE POLLUTION

1.0. Ocean Dumping

Pollution in the coastal zone includes plumes from ocean dumping of waste materials, such as sewage sludge, acid, and other chemicals. These materials often have unique spectral characteristics that may be used and exploited for remote sensing. In some circumstances there is also a unique spatial signature.

Early literature describes interpretation of aerial photography for plume location, and tracing of plume movement and dispersion (see Waste Outfalls). Since the launch of Landsat-1 in 1972, multi-spectral scanner imagery from orbit has proved to be useful. Fontanel et al. (1973) detected industrial-waste plumes between France and Corsica containing iron, titanium oxide, and sulphuric acid, using Landsat MSS 4 images; Wezernak and Roller (1973) detected ocean dumping on the east United States coast using MSS 5 data; comparison with MSS 6 data indicated plume depth via the shorter penetration depth of 750 nm radiation in water. Klemas and Philpot (1981) summarize a series of studies in which drift and dispersion of acid iron-waste plumes 64 km off the Delaware coast were investigated using Landsat imagery (Figure 28-100), current drogues, and ship data. Magnitudes of plume drift velocities were compatible with velocities of current drogues; the direction of drift showed seasonal variation, and was influenced by storms when dispersion was accelerated. These results were obtained mainly through interpretation of image data. In cases of low signal to noise ratios, it was necessary to resort to digital enhancement of multispectral digital data on Landsat computer-compatible tapes.

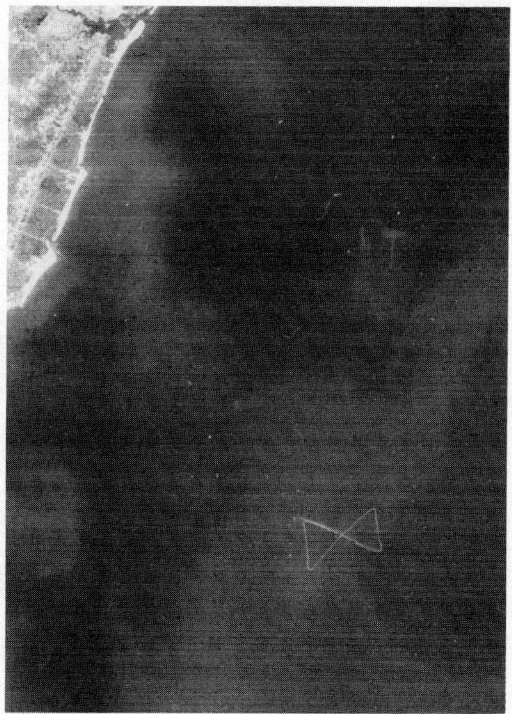

Fig. 28-100. Acid-iron waste plume from a vessel dump off Delaware visible in Landsat MSS 5 image of 28 August 1975 (Klemas and Philpot, 1981).

The discrimination between various components contributing to upwelling radiance is made possible by the spectral separability of pollutants and their brightness differences from clear water. Plumes in the coastal zone, due to scattering by suspended particulates, usually produce higher upwelling radiance than emanates from clear water.

Recent studies summarized by Johnson and Ohlhorst (1981) emphasize laboratory and field spectral experiments to calibrate the remote measurement of single-pollutant concentrations. Again, ocean dumping on the east United States coast has provided a target of opportunity (Figure 28-101). NASA Modular Multispectral Scanner (M2S) data in eleven visible and near infrared channels analyzed by Johnson et al. (1977) in Figure 28-102 show the spectral brightness of acid and sewage plumes, relative to background, from 400 to 1000 nm. Acid shows a clear backscattering peak at 600 nm. Similar plume-ratio spectra from a laboratory spectrometer are shown in Figure 28-103. Other laboratory studies (Lewis, 1977) show that the acid-waste signature is produced by soluble ferric iron being precipitated as ferric hydroxide upon dilution with sea water. The reaction is initiated at a pH of about 2.8. Due to color changes as the pH threshold is crossed, the acid plume will show a variety of spectra: Figure 28-104 shows rapid-scan spectrometer spectra obtained over an acid dump; laboratory transmission spectra for dilutions of the same acid are shown in Figure 28-105. For dilute concentrations away from the pH threshold, the acid concentration can be measured remotely at one wavelength (or by a ratio of band radiances, which normalizes against noise that is simultaneously affecting the given channels); Figure 28-106 shows iron concentration as a function of a two-color radiance ratio. For high iron concentrations near and above the pH threshold, the measurement of concentration would require a multispectral approach because of the pH-controlled color changes. For dumps of sewage sludge, laboratory data in Figure 28-107 (Witte et al., 1977) indicate that single wavelengths would suffice for measurement of suspended-solids concentration. A multispectral multiple regression equation may improve accuracies of measurement; such an approach led to the map in Figure 28-108 (Johnson, 1980). The above spectral considerations, of course, must be

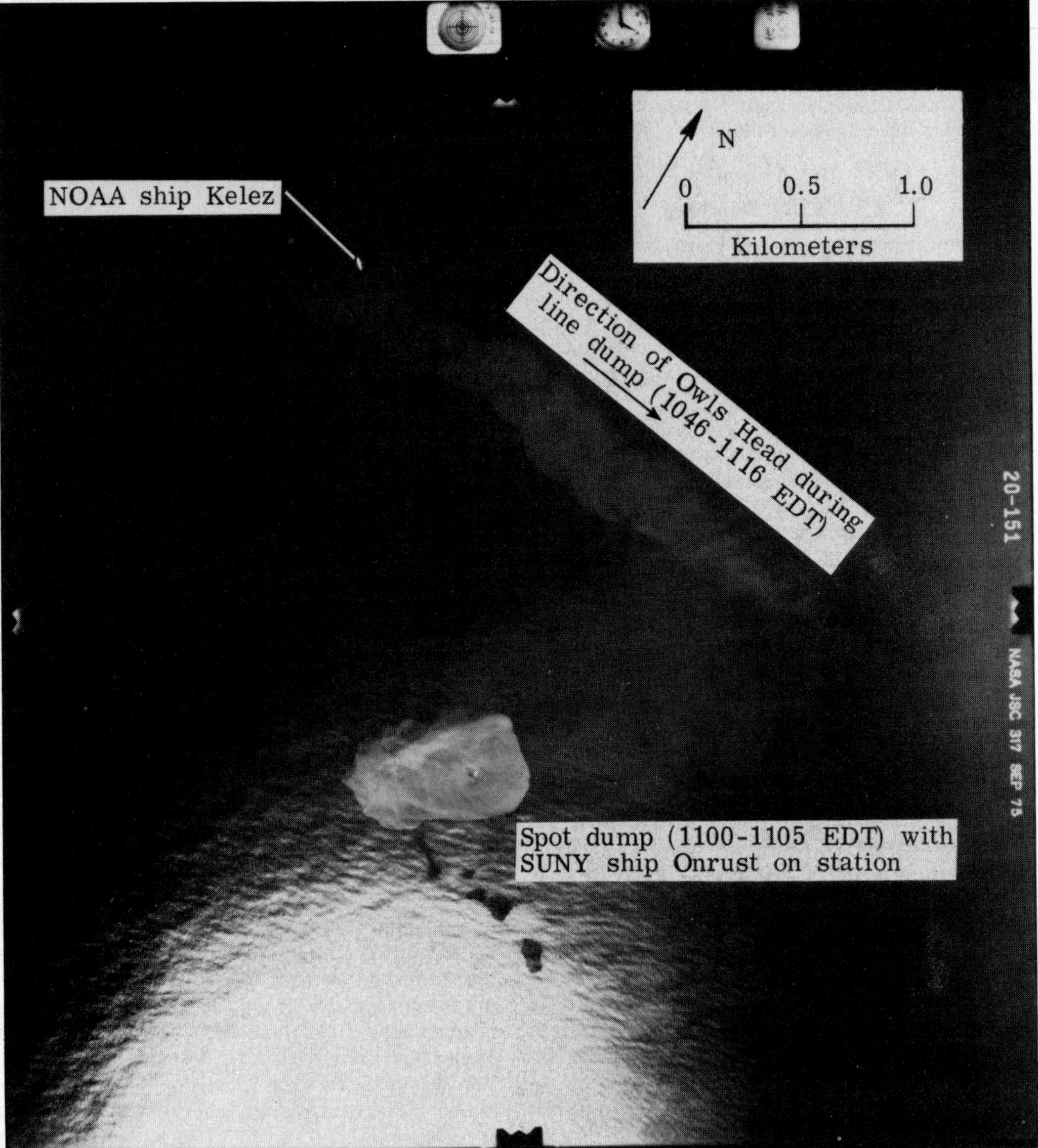

Fig. 28-101. Plumes from line and spot dumping of sewage sludge in the New York Bight on 22 September 1975 1 hour after dumping (Johnson *et al.*, 1977).

supplemented when high altitude remote sensing is involved, because of the variable spectral effect of changes in atmospheric path radiance.

In the case where multiple pollutants influence ocean spectra, eigenvector analysis may be used to extract the various spectral signatures. Simonds (1963) describes the operations by which eigenvectors are obtained from a covariance matrix generated from a set of spectra of n components. From n eigenvectors V_i, the magnitude of spectral band j of any spectrum contributing to the data set, can be reconstructed by the equation:

$$x_j = \overline{x}_j + a_1V_1 + a_2V_2 + \ldots + a_nV_n \qquad (28\text{-}19)$$

where $\overline{x}_j$ is the average magnitude of band j for the data set, and a_i is the scalar multiple of V_i. Usually most of the total variance is accounted for by p vectors where $p << n$. The first three eigenvectors (normalized) from spectra over an acid-waste dump obtained by the NASA Multichannel Ocean Color Sensor (MOCS) are shown in Figure 28-109 (Grew, 1977). The first vector accounts for 96.5 percent of the variance in the data set. A scatter plot of pairs of coefficients a_1 and a_2, measured in different parts of the plume, shown in Figure 28-110, reveals two linear regions with a transition point. The transition point corresponds to the pH-controlled reaction-threshold discussed earlier. Eigenvector analysis has also

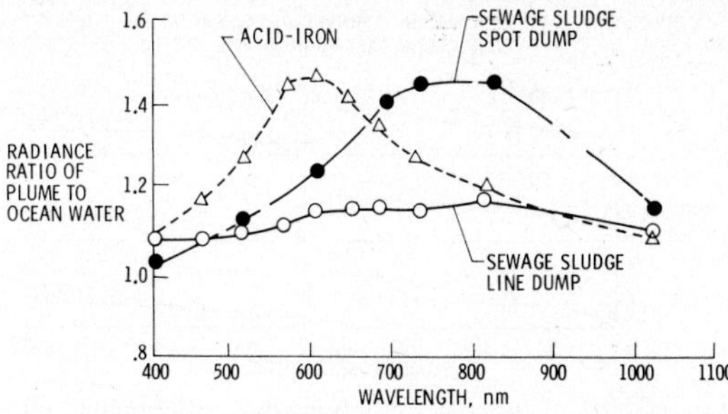

Fig. 28-102. Spectral brightness of sewage and acid-iron dumps relative to background from aerial multispectral scanner data (Johnson *et al.*, 1977).

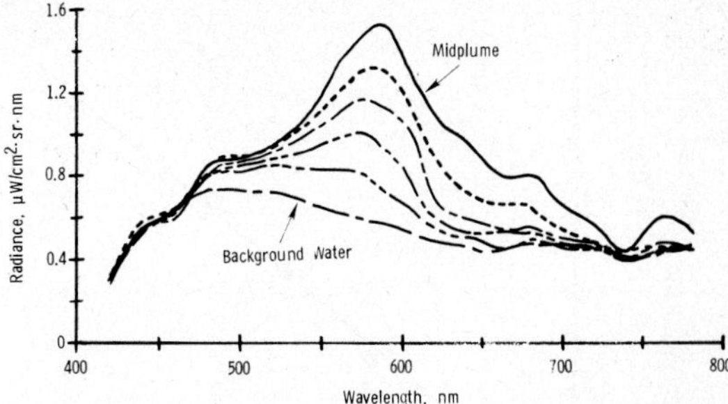

Fig. 28-104. Rapid-scan spectrometer spectra across an acid-iron waste dump at sea (Lewis, 1977).

been used by Klemas and Philpot (1981), where acid plumes were partly masked by clouds or sediment, to separate the variables and reject cloud and sediment pixels from the plume, resulting in an enhancement of the plume (Figure 28-111).

2.0. Waste Outfalls

a. Plumes and Pollutants

For outfalls discharging waste into waters nearer shore, the background (receiving) waters are more turbid than the clear waters subject to ocean dumping. Greater difficulties in remote sensing of pollutants should be expected as background turbidities increase. There will, however, be some outfalls in clear-enough water to allow plume discrimination from orbit by simple grey-level analysis of single Landsat MSS bands (Figure 28-112).

Generally, however, a multispectral technique from aircraft will be required. In this way the spectral properties that characterize different parcels of water can be fully exploited for discrimination of the various pollutants and natural water constituents that may be present. Although it will rarely be possible to positively identify individual pollutants, at least points of discharge may be detected, plus direction of flow and rate of dispersion.

Industrial discharges mapped in aerial multispectral studies by Polcyn and Wezernak (1970) are shown in Figure 28-113. The scanner collected data in 17 channels; four are shown. The 0.44–0.46 μm band detects oil films. The 0.58–0.62 μm band emphasizes a steel-plant discharge from a strip mill as a dark tone, and a pickling liquor discharge as a bright tone. The 0.72–0.80 μm band reveals the waste-treatment outfall for the city of Detroit. Finally, the 8–14 μm thermal band discloses the warm water discharges from a power plant, the steel plant and the Rouge River inflow. Spectral reflectances computed from such data for various pollutants are displayed in Figure 28-114. The results show that multispectral studies can be used to discriminate broad classes of colored pollutants.

Many pollutants that discolor water can be detected using color photography. Even when receiving waters are highly polluted, discharges at or near the surface are usually visible for at least a short distance from the release point. The use of color infrared instead of color photography is recommended when the photographic altitude is above roughly 6000 m, in order to minimize the effects of atmospheric haze, or when it is desired to discriminate floating or very shallow vegetation, algal blooms, and shoreline vegetation. If possible, simultaneous bore-sighted color and color infrared photography is preferred. The scale of photography should be adequate for detecting and identifying features associated with pollutant

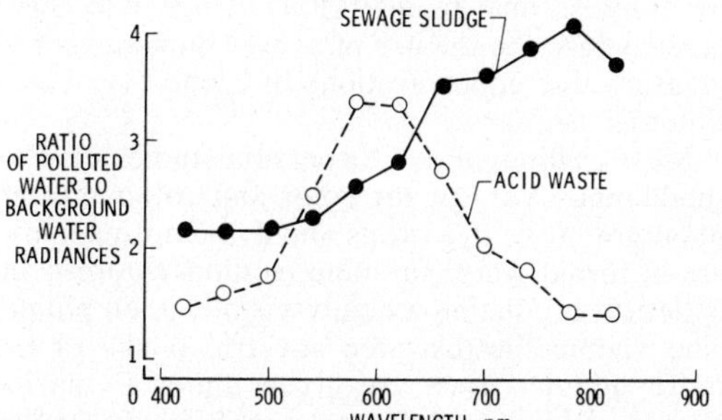

Fig. 28-103. Laboratory spectral brightness of sewage and acid-iron wastes (Johnson *et al.*, 1977).

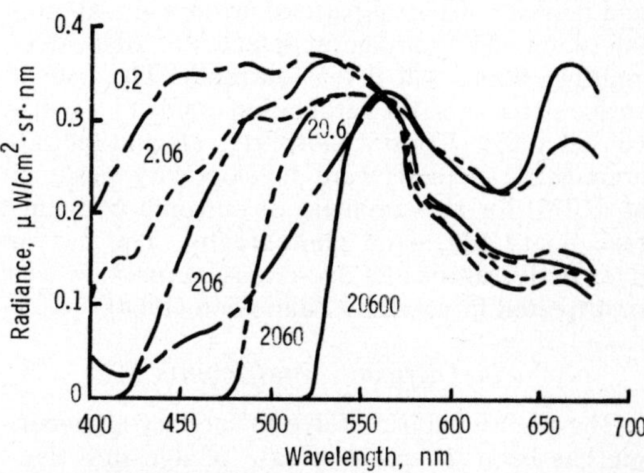

Fig. 28-105. Normalized spectral radiance calculated from laboratory transmission data for different acid-iron waste concentrations (mg/l) (Lewis, 1977).

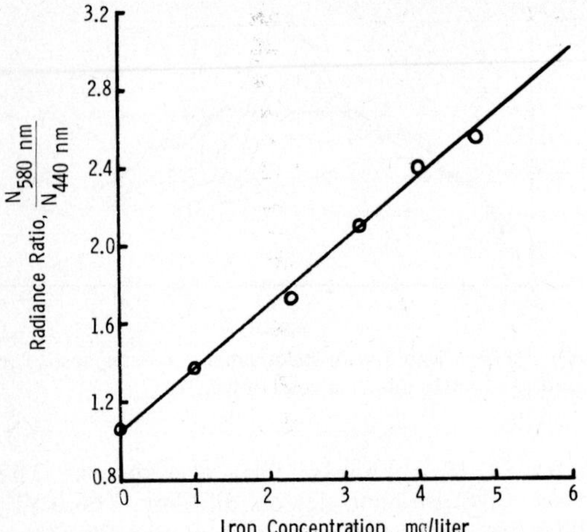

Fig. 28-106. Radiance ratio (580 nm/440 nm) for different acid-iron concentrations (Lewis, 1977).

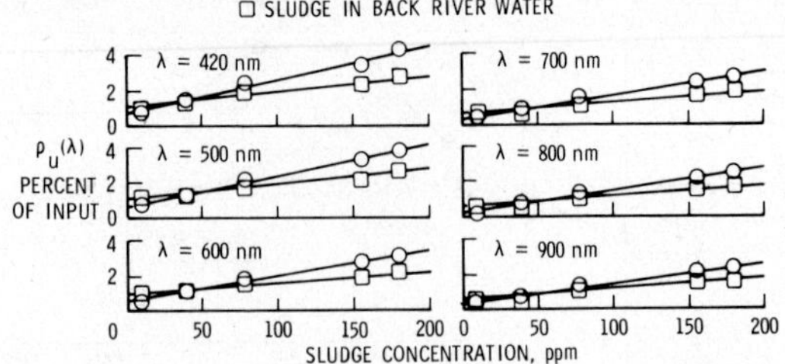

Fig. 28-107. Linearity of laboratory reflectance with sewage sludge concentration at various wavelengths (Witte *et al.*, 1977).

outfalls and spills, such as storage tanks, pipelines, sewage treatment plants, raw material and waste holding facilities, and shipping docks. The scale range most suitable for such interpretations is usually 1:5,000 to 1:10,000. Such detailed photography may conveniently follow reconnaissance mapping at smaller scales such as 1:50,000 (Vass and vanGenderen, 1978).

The timing and frequency of photographic and scanner coverage depend on the pollution source and the objective of monitoring. Some outfalls discharge only at certain times of day. In other cases, tidal action of receiving waters alters pollutant dispersion and hence visibility, requiring selection of appropriate tidal-current phases. In this regard, low tides reduce the depth over submerged outfall nozzles.

Remote sensing of outfalls can be enhanced by the use of dyes in the effluent, free releases of dye, and/or dye-releasing markers, which enhance the flow field in the outfall region. The dye-marker techniques are particularly suited for outfall analysis: dye streams from anchored buoys placed at an outfall site show the direction of tidal flow and indicate dispersal patterns even when an outfall plume itself may be undetectable. Also, dye-emitting floats will reveal currents. The use of surface markers for circulation analysis is discussed under Coastal Currents; elegant photogrammetric methods were developed by Yeske et al. (1975) for measurement of currents with surface floats and aerial photography. The use of dyes in dye drops and dye-releasing markers was first treated in detail by Zdanovich (1964).

b. Diffusion Coefficients

The traditional use of dye in sanitary engineering has been the point release of dye in a dye-dispersion study for measurement of diffusion coefficients, with dye concentrations measured by boat-borne fluorometers (Okubo, 1971). Dye concentrations can be measured instead by photography and reduced to diffusion coefficients, but concentrations of dye two orders of magnitude higher are required, generally making dye expenses prohibitive. Continuous dye releases have been detected down to a few parts per billion on aerial photography.

To obtain diffusion coefficients, a simple diffusion equation is employed, *viz.*,

$$\frac{\delta W}{\delta t} = \sum_{i=x,y,z} \left[\frac{\delta}{\delta i}\left(D \frac{\delta W}{\delta i} \right) - \frac{\delta}{\delta i}(V_i\, W) \right] + S \qquad (28\text{-}20)$$

where W is concentration, V velocity, D diffusion coefficient, and S the sources and sinks. More detailed approaches are indicated in Okubo (1965). Neglecting S and the z-dependence, the two-dimensional solution for a point-source release is:

$$W(x,y,t) = W_{max} \exp - [x^2/2\sigma_x^2 + y^2/2\sigma_y^2] \qquad (28\text{-}21)$$

where σ_i^2 is the variance in direction i, and the coordinate system is presumed to move with the centroid of the dye patch. The expression represents an ellipse $x^2/a^2 + y^2/b^2 = 1$ with major and minor axes $a^2 = 2\sigma_x^2 \ln (W_{max}/W)$ and $b^2 = 2\sigma_y^2 \ln (W_{max}/W)$. Measurement of a and b as a function of time (the boundary of the patch being taken as an appropriate ratio of W_{max}/W) yields the diffusion coefficients via:

$$D_i = \Delta\sigma_i^2/2\Delta t \qquad (28\text{-}22)$$

The application of aerial photography to acquire the patch measurements is thoroughly discussed by Burgess and James (1971). Hoge and Swift (1980a) describe the use of a laser fluorosensor to measure dye concentrations of 2 ppb (see Laser Fluorosensors).

Most traditional dye-dispersion studies employ rhodamine dye, but for color and color infrared photography of dye drops and dye-emitting markers in turbid water, uranine (sodium fluorescein) is better, producing a highly visible green plume. The visible fluorescence spectral peaks of the more popular dyes are given under Estuarine Fronts. Additional references to diffusion studies include Betz (1968), Costin et al. (1963), Costin (1965), Foxworthy (1965), Gunnerson (1965),

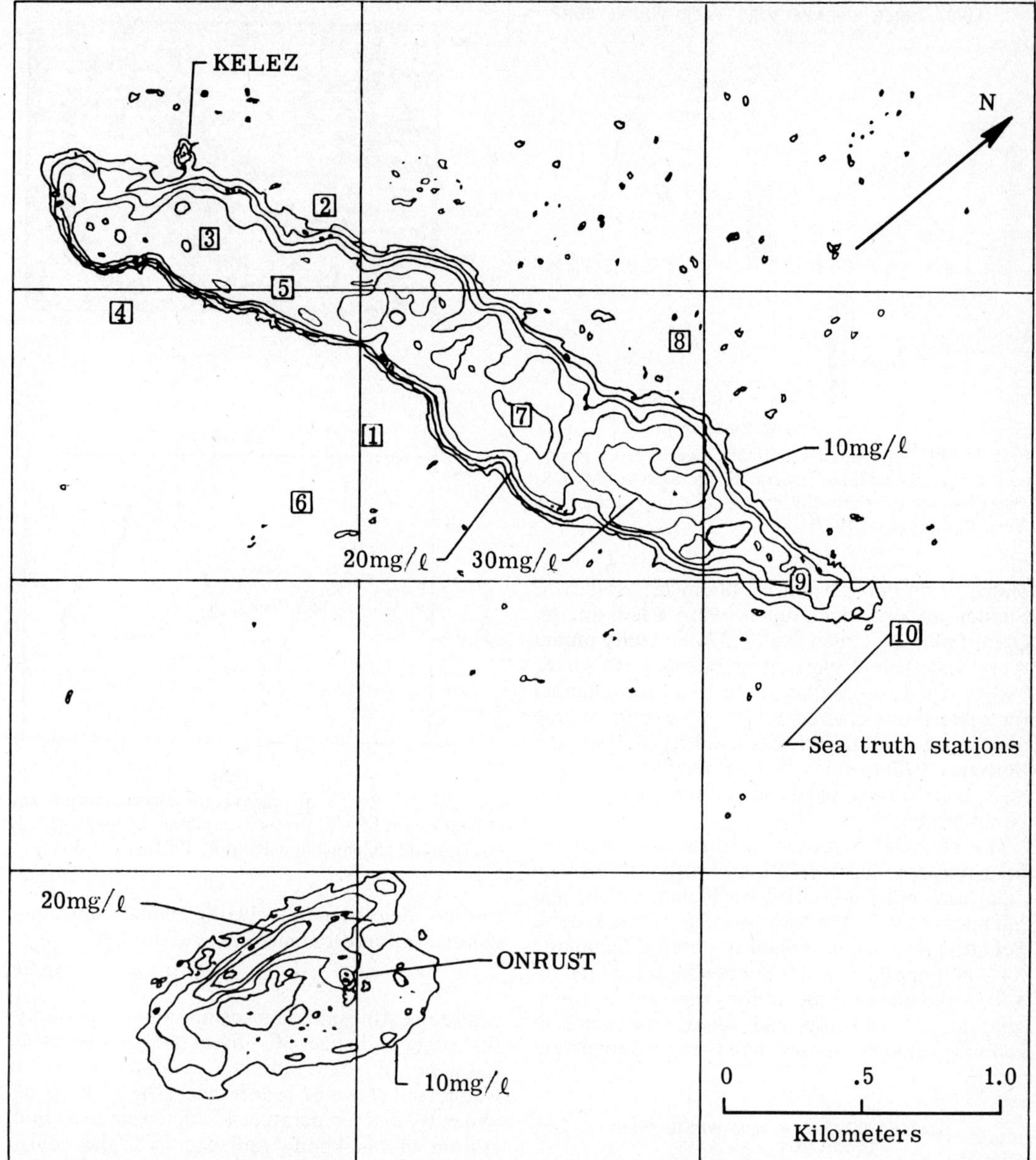

Fig. 28-108. Quantitative distribution of suspended solids in the dumps of Figure 28-101 from a multispectral multiple regression analysis (Johnson, 1980).

Huebner and Bouma (1973), Ichiye (1962, 1963), Ichiye et al. (1964), Katz et al. (1965), Kisiel (1965), Okubo (1965), Rienert (1965), and Wright and Collings (1964). Some of these are collected in Ichiye (1964).

c. *Thermal Plumes*

Some effluents are hotter than receiving waters, making thermal infrared scanner-imagery particularly useful for analysis. Thus, thermal imagery is widely used for detailed studies of plumes from low altitude. Overflights provide synoptic information on spatial distributions that are not readily available from surface measurements, and subtidal coverage yields temporal distributions as a function of tidal phase. For regional analysis, orbital imagery from thermal scanners on Nimbus and NOAA satellites may reveal areas subjected to industrial and urban thermal loads.

Fang and Parker (1976) and Kuo and Talay (1979) studied heated water discharges from the Surry nuclear power station on the tidal James River, Virginia, using combinations of data from boat surveys and the 8–14 μm thermal band from a multispectral scanner. Results of a survey conducted three hours after high tide on May 17, 1977 are shown in Figure 28-115. The effect on the

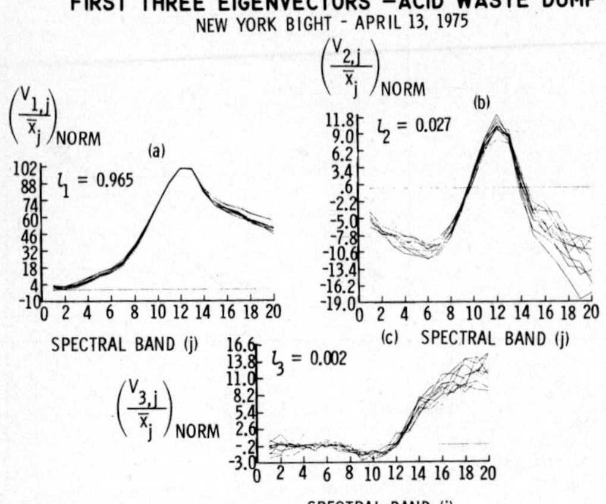

Fig. 28-109. First three significant eigenvectors generated from Multichannel Ocean Color Sensor (MOCS) data obtained over an acid-iron waste dump in the New York Bight (Grew, 1977).

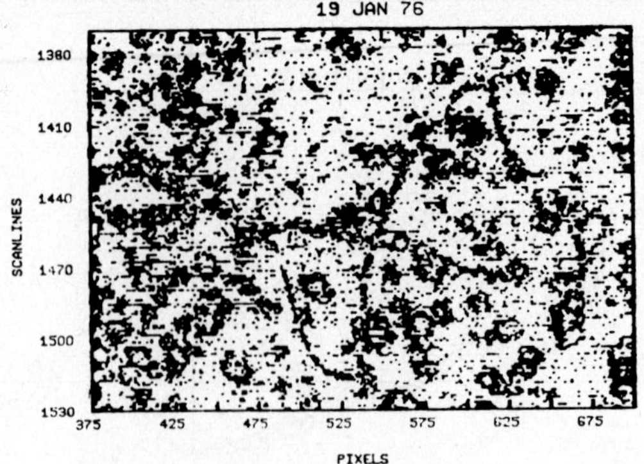

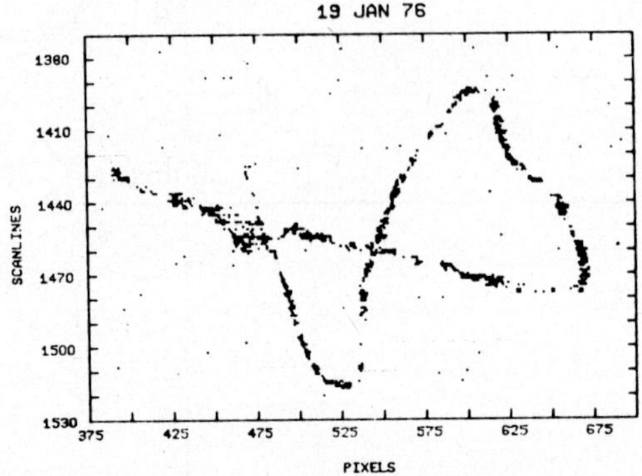

Fig. 28-111. Result of eigenvector discrimination and enhancement of acid-iron waste plume against a cloudy background (Klemas and Philpot, 1981).

plume of an ebb current of 40 cm/sec is shown. Similar airborne techniques were used on the Great Lakes by Green et al. (1977) to study plume types and their dispersion in inland fresh water bodies. On Lake Michigan the 8–14 μm channel (in some cases filtered to 10–12 μm to reduce atmospheric influence) was used by Polcyn and Stewart (1970) to study thermal mixing. In the images, the effects of winds on circulation may often be discerned.

(1) Thermal Scanner Calibration—Methods for environmental correction of airborne thermal data have been presented by Pickett (1966) and Saunders (1967). For airborne thermal scanners, Schott (1979) has described in detail a technique for wholly airborne calibration. The technique involves the use of a model for atmospheric transmission, sky radiation and water reflectance to calibrate remotely sensed data without concurrent

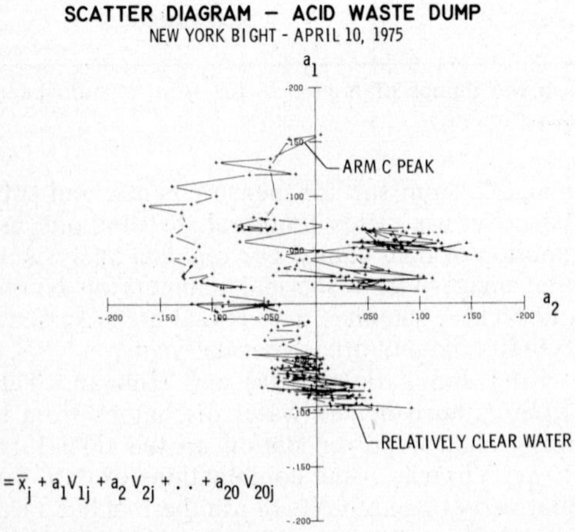

Fig. 28-110. Scatter plot of the first two eigenvector coefficients in different parts of an acid-iron waste plume (Grew, 1977).

surface measurements. In the model, upwelling thermal infrared radiance W is given by:

$$W = \tau \epsilon W_T + W_a + W_s R \qquad (28\text{-}23)$$

where τ is atmospheric transmission, ϵ emissivity, W_T target radiance, W_a direct and scattered air radiance, W_s sky radiance reflected from the water, and R water reflectance. The relation between W_T and temperature T is determined by integration of blackbody radiance over the sensor bandpass, as in

$$W_T = \int_{\lambda_1}^{\lambda_2} 2hc^2\lambda^{-5}(e^{h\nu/kt} - 1)^{-1}d\lambda. \qquad (28\text{-}24)$$

Appropriate values can be obtained from standard blackbody tables. R and ϵ, as a function of angle, are known and tabulated in the literature. Exploiting the dependence of terms in W on height and view angle, aerial measurements at several altitudes and look angles permit a complete solution for T.

Thermal images obtained for this purpose during a height profile are shown in Figure 28-116. Tests of this technique showed substantial improvements over measurements using internal instrument-calibration techniques. Without model corrections, the mean error was 3.23°F with a standard deviation of 1.25°F, while, with atmo-

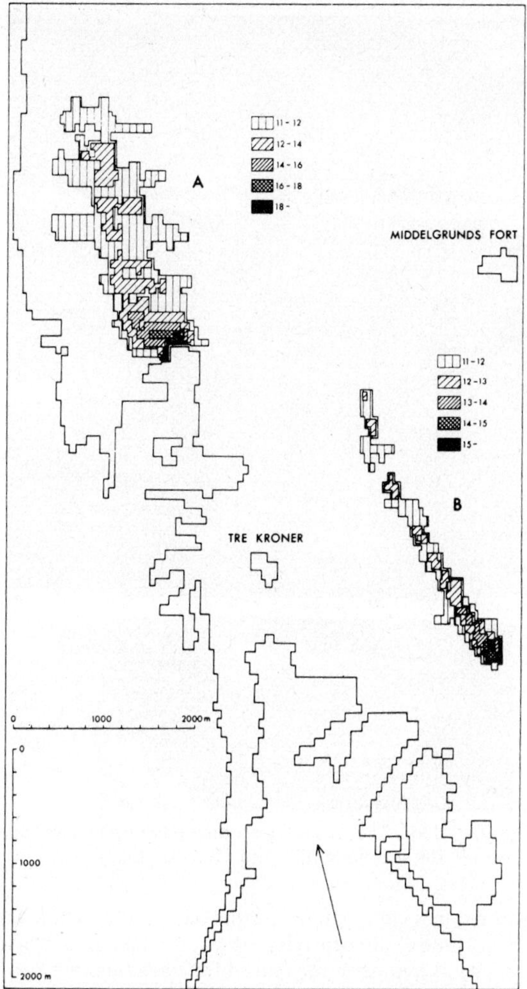

Fig. 28-112. Copenhagen waste plumes from Landsat MSS 5, 27 August 1973 (Helldén and Åkersten, 1977).

spheric and other corrections, the mean error was reduced to 0.70°F with a standard deviation of 0.59°F. Figure 28-117 shows the effect of the technique on the aerial temperature measurement, which may be compared to the 0.5°K accuracy of temperature measurement from orbit (illustrated earlier in Figure 28-9).

Some effluents contain materials more buoyant than receiving waters, such as oils present in sewage that form oil slicks downstream of outfalls. The presence of oil slicks may be detected (as discussed below) via thermal imagery, microwave radiometry, and radar even when they are too thin for detection in the visible spectrum.

3.0. Oil Pollution

Of all the chemical materials spilled into the marine environment, oil is among the most notorious and ubiquitous. Other materials have been locally more important in some cases, such as mercury in Minimata Bay, Japan, and kepone in the James River, Virginia. The U.S. Coast Guard lists thousands of chemicals in the Chemical Hazards Response Information System (CHRIS) handbook, and some amount of these and hundreds of new chemicals produced each year may ultimately contribute to marine pollution. However, concerning remote sensing, oil is the most studied, and only preliminary studies have begun on the great majority of other pollutants (Washburn and Sandness, 1977). From results of numerous investigations of the potential for remote sensing of spilled materials, a general conclusion is that unattended remote monitoring is not yet possible for any more than a handful of pollutants, and these only in cases where the given pollutant is the only substance whose concentration is varying. On the other hand, sensors on spacecraft used in the planetary exploration

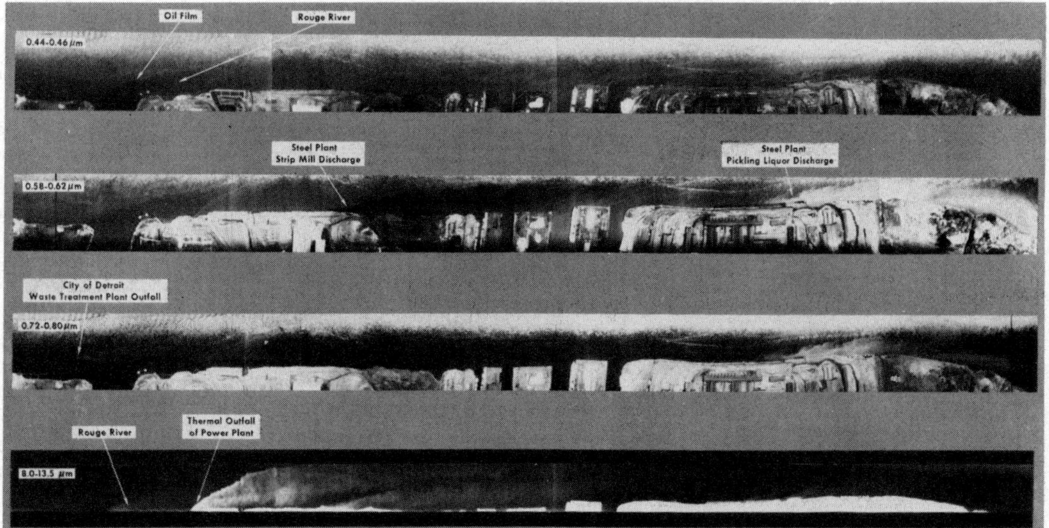

Fig. 28-113. Coastal discharges detected with airborne multispectral scanner data (Polcyn and Wezernak, 1970).

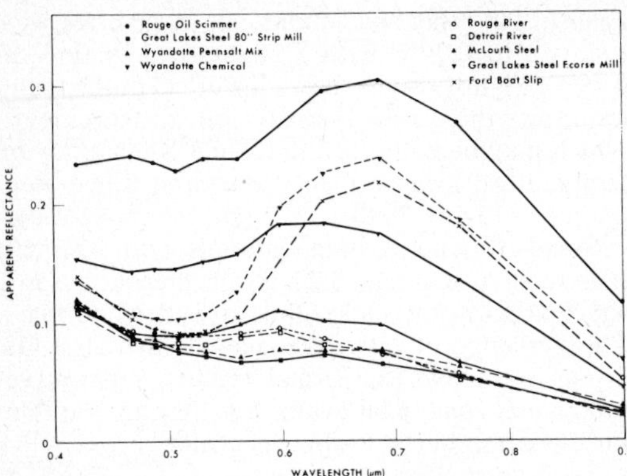

Fig. 28-114. Spectral features of major pollutants detected in Figure 28-113 (Polcyn and Wezernak, 1970).

program suggest that sustained effort might yield techniques involving narrow spectral bandwidths and sophisticated instruments for monitoring many specific pollutants. The potential cost has thus far precluded such effort for pollutants other than oil.

Emphasis on remote sensing of oil pollution is justified because of the wide use of oil, the dependence on oceanic tankers for its distribution, and its frequent release into the marine environment. The fact that oil typically rises to or spreads across the surface of a water body has contributed to development of successful techniques applied to surveillance activities, and to assessment and cleanup activities. These goals require different capabilities: surveillance aims at small to modest spills where detection and subsequent legal proceedings are most important, while cleanup typically deals with massive spills where detection is easiest and volume measurement and transport are most important.

a. Visible Region Spectral Properties

Theoretical and laboratory tests have established the electromagnetic properties of oil films on water. Based on such data, single band and multispectral detection techniques have been designed for the ultraviolet to microwave portions of the spectrum (Estes and Senger, 1972). Dielectric

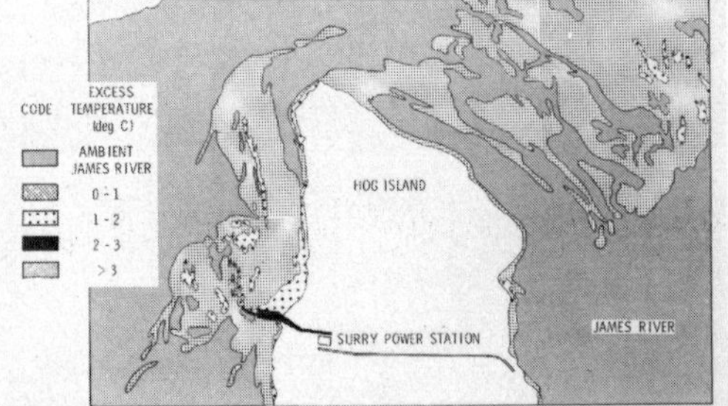

Fig. 28-115. Thermal plume from the Surry nuclear power station on the tidal James River, Virginia, 17 May 1977 (Kuo and Talay, 1979).

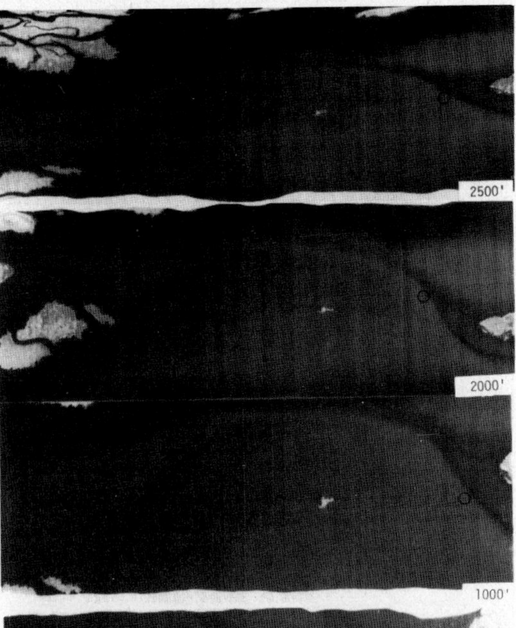

Fig. 28-116. Thermal images obtained during an altitude profile for scanner calibration (Schott, 1979).

constants and optical properties, such as refractive index, absorptivity, fluorescence, and emissivity, have been measured for varieties of petroleum and its products as well as naturally occurring biological oils. Field tests have considered environmental factors such as ambient air and water temperatures, salinity, sea state, and turbidity.

In the visible and ultraviolet region, different oils have unique spectral signatures, yielding sometimes brighter and sometimes darker than background radiances, depending on the wavelengths of illumination and observation, the type of oil, and the background water clarity. Thus, color and brightness contrasts, which are determined for chosen oils overlying turbid mid-latitude waters, may have to be revised for clear tropical waters. An oil-slick spectral radiance model of Stewart et al. (1970), which predicts, with increasing wavelength, a change from positive to negative oil/water contrast in the blue region, was confirmed by multispectral scanner imagery shown in Figure 28-118, taken over an uncapped well in the Santa Barbara Channel, California. The white patch at the left at longer wavelengths is due to dispersant. Similar results were obtained in aerial photography of field releases in the York River, Virginia (Munday et al., 1971) shown in Figure 28-119. The near ultraviolet

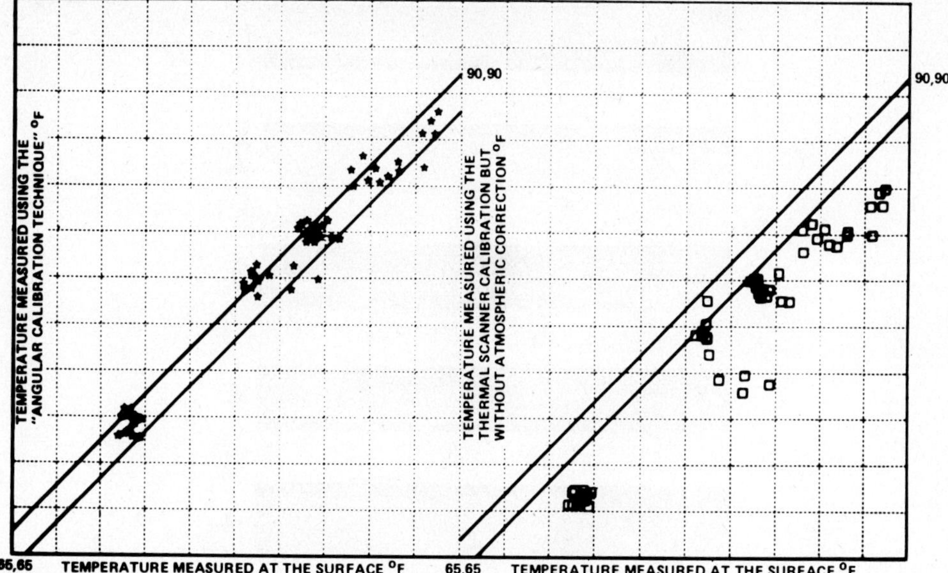

Fig. 28-117. Comparison of temperature measurements from surface sampling and a calibrated thermal scanner. The lines represent a 1°F error envelope (Schott, 1979).

is seen to be excellent for imaging slick edges and thin regions of various fuel oil slicks, while the blue band is next best, and satisfactory for measurement of slick areas. For delineating thick regions, the green band is best. No band was found to distinguish thick and thin regions for No. 2 fuel oil, but such oil is distinguishable from Nos. 4 and 6 by lacking negative contrast in the blue and green bands. Fuel oils usually show interference rings at slick edges, up to a depth of a few wavelengths, permitting an estimate of minimum spill volume. Where thick regions are distinguishable by color contrast, the minimum thickness is on the order of 1 mm. Menhaden oil is distinguishable from fuel oils because it remains in small thick lenses surrounded by a colorless microlayer. Based upon such visible region data, color film is recommended as an all-purpose record for most situations; quick-look records can be obtained from television cameras with high sensitivity in the blue and green regions, while low light-level television extends the capability into periods of darkness (Worsfold et al., 1975). Narrow-band multispectral scanners offer all the available spectral capability, except in dim light. All visible-region passive sensors are, of course, limited to available light and clear weather.

b. Thermal

Scanners with a thermal channel offer, as well, the capability for a temperature map of an oil slick, day or night. Such a map, however, must be interpreted carefully. First, over water, the 8–14 μm thermal band includes atmospheric- and surface-reflection contributions to the signal received, according to the expression (also see Schott, 1979, discussed under Waste Outfalls)

$$L = \epsilon_w L_{w,T} + (1 - \epsilon_w) L_s \qquad (28\text{-}25)$$

where total radiance L is a function of water emissivity ϵ_w, water black-body radiance at surface temperature T, and downwelling sky radiance L_s. Other contributions to L such as path radiance may be neglected for low altitude surveys. The range of quoted values for ϵ_w is 0.970–0.993, which contributes uncertainty of about 1°K to the term in $L_{w,T}$. The term in L_s adds a variable contribution that depends on atmospheric water vapor and carbon dioxide content; the size of the contribution is 0–2°K. Pickett (1966) offers a method for determining the size in specific cases.

Proceeding to the thermal signal over an oil slick, an emissivity value for the oil in question is required. Values for specific oils are sparse; one determination for a thin film of petroleum oil on water yielded 0.972 compared to 0.993 for clear water (Buettner et al., 1964; Buettner and Kern, 1965). Theoretical data for such an oil slick, isothermal with water, yield an apparent radiometric temperature that is cooler than the water temperature by 1.0–1.4°K. However, field data (Munday et al., 1971) yielded differences of as much as 2.7–4.0°K, with the apparent radiometric oil temperature decreasing as slick thickness increased up to roughly 50 μm. Evidently, the oil value of 0.972 is too large, and the value to use depends on oil-film thickness (for very thin slicks, the remote instrument partially "sees through" the slick to the underlying water). It has been suggested that thermal infrared data might be correlated with oil-slick thickness. Such a correlation appears possible via the indicated emissivity/thickness dependence, but the usual surface-temperature variations in estuarine waters will introduce ambiguity. Also, emissivity values

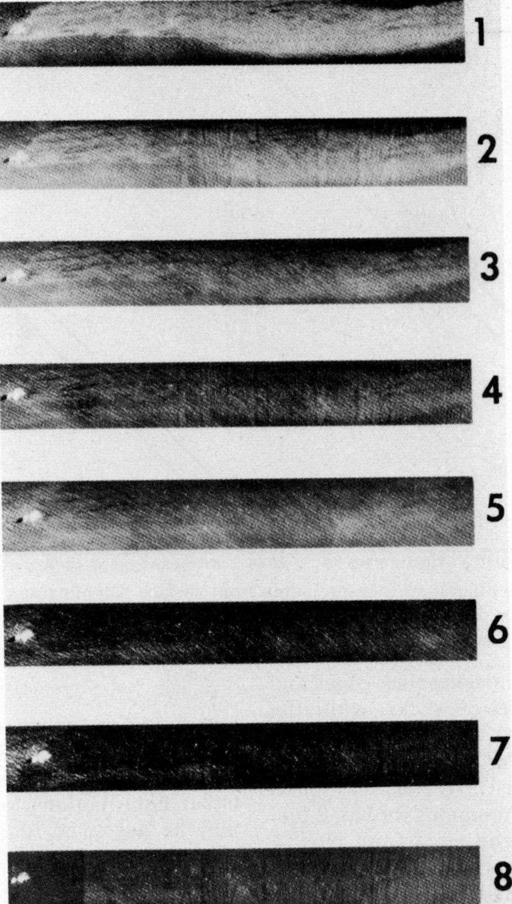

Fig. 28-118. Multispectral scanner radiance records of the Santa Barbara oil slick, 7 March 1969. Altitude 600 m at 0815 hours. Bands 1-8 in μm: 0.32–0.38, 0.41–0.43, 0.43–0.455, 0.455–0.47, 0.50–0.52, 0.545–0.58, 0.63–0.68, 0.75–0.855 (Stewart *et al.,* 1970).

and their thickness dependencies may be oil specific.

c. Microwave

Greater success has been reported for determination of oil-slick thickness using a microwave radiometer (Hollinger and Mennella, 1973; Hollinger, 1974), which offers the advantage of being operable in more adverse weather conditions. In laboratory experiments, radiometer data at two wavelengths showed the expected cycle of interference effects according to $\lambda/4$ as the slick thickness was increased (Figure 28-120). Studies conducted in the Chesapeake Bay then indicated that thickness measurement was possible under field conditions. In Figure 28-121, the isotherms for 31.0 and 19.3 GHz were used to infer the thickness contours. These were in fair agreement with surface measurements of thickness, which are difficult to obtain and not always representative of average values over large areas due to patchiness.

d. Radar

In the realm of active remote sensing systems, radar is quite effective in detecting oil slicks (Guinard, 1971). Oil smooths surface capillary waves and reduces radar backscatter, especially for viewing angles beyond 20°. However, slicks caused by spilled oil cannot be discriminated from those caused by wind patterns, thin ice, or natural oils. Because the mechanism of detection is indirect via surface roughness, thickness determination is necessarily crude and dependent on sea state; for detection *per se,* seas must be neither very calm nor very rough. However, radar offers many advantages. For example, use of a side-looking radar permits slick detection at long range over wide areas. Also, radar provides all-weather day/night operations unaffected by clouds or darkness. Finally, radar can be arranged to produce quick-look records on dry-silver film.

e. Fluorosensors

In the visible and ultraviolet regions of the spectrum, active sensors can detect oils by exploiting oil fluorescence. Fluorescence-emission spectra of several oils spilled at sea measured by a laser fluorosensor are seen in Figure 28-122. Such spectra show that broad classification of oils is feasible with fluorosensors, via

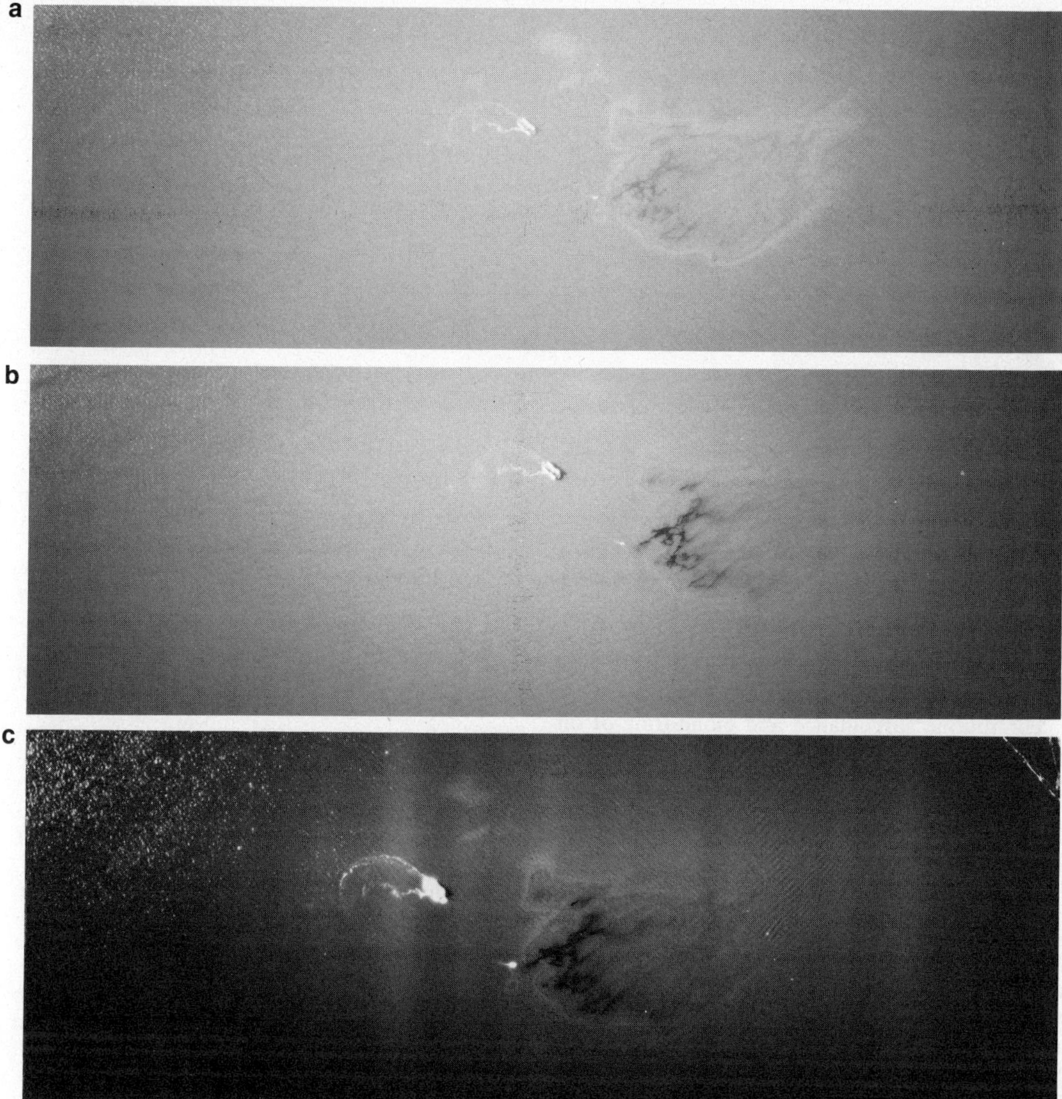

Fig. 28-119. Multispectral photography of No. 4 oil slick, 6 November 1970, York River, Virginia. Altitude 600 m, 150 gal. a. Kodak 2403, Wratten 47; b. Kodak 2403, Wratten 57; c. Kodak 8443, Wratten 12 (Munday *et al.*, 1971).

type-specific emission-peak wavelengths that vary from 400 to 600 nm (Fantasia and Ingrao, 1974). Most results from field work have been more modest at the level of oil detection (see Kim and Ryan, 1974).

In the field situation, there may be confusion between oil and other targets. Marine waters contain natural materials that also fluoresce, including Gelbstoff and chlorophyll *a,* but the former fluoresces in a larger bandwidth, and chlorophyll *a* fluorescence has a peak at a longer wavelength (685 nm for chlorophyll *a in situ*); furthermore, spatial gradients and texture patterns of natural materials vary from those of oil.

A specialized fluorosensor, the Fraunhofer Line Discriminator, operating in an airborne imaging mode, has also detected oil slicks and provided data for slick morphology and volume (Watson et al., 1977). Of the wavelengths sensed (486.1, 589.0, and 656.3 nm) the blue wavelength (486.1 nm) had the greatest sensitivity for 29 different crude oils, due to the broad blue peak in the fluorescence emission spectrum. The quantum yield of oil fluorescence varies with the thickness for oils with specific gravities less than approximately 0.875. It should, therefore, be possible to measure slick thickness in the field using fluorosensors, given knowledge of the specific gravity.

A different approach is a thickness measurement via depression of water Raman backscatter. Hoge and Swift (1980b) report potential for success with this technique for thicknesses between 0.05 and 5 μm using an airborne 337 nm nitrogen laser.

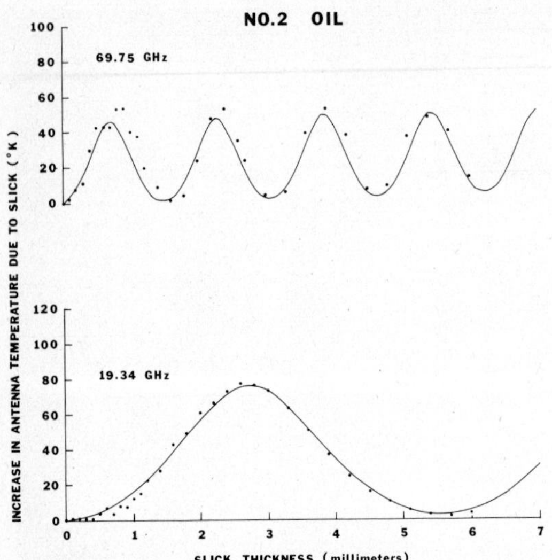

Fig. 28-120. Comparison of theoretical and measured microwave brightness temperatures of a No. 2 oil slick in a tank test (Hollinger, 1974).

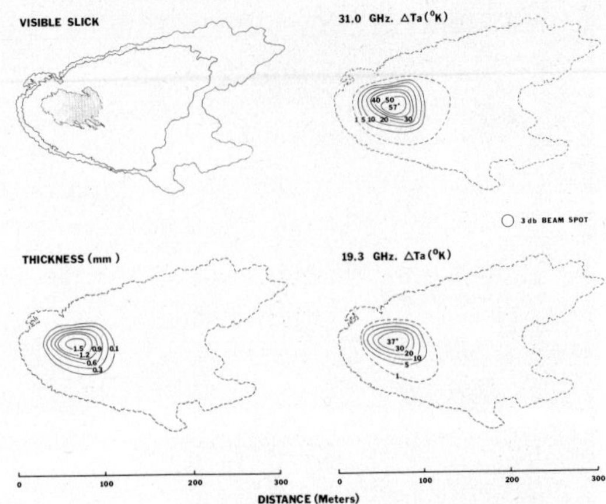

Fig. 28-121. Field verification of microwave measurement of No. 2 oil slick thickness. Volume 630 gallons spilled near the Chesapeake Light Tower off Virginia on 11 July 1972. Upper left drawing from photograph 1 hour after release. Simultaneous radiometer data at two frequencies (right) used to derive thickness contours at lower left. Surface-measured thickness at centroid: 2.4 mm (Hollinger, 1974).

f. Surveillance Systems

For routine surveillance and monitoring of oil pollution, a multi-sensor system is required to achieve all-weather and illumination capability. Several countries have developed airborne surveillance systems. The U.S. Coast Guard initiated a program in 1968 and is currently developing a fourth generation system called AIREYE, the Airborne Remote Instrumentation System (Manning et al., 1980). It will be carried on six Falcon 20G jet aircraft routinely monitoring the estuarine and coastal environment. AIREYE will include a side-looking airborne radar, viz. a new generation AN/APS-94 SLAR designated as AN/APS-131 with 200 kw peak power. A detection swath-width of 80 km is possible. Targets detected by the SLAR can be overflown and imaged by a 3-channel infrared/ultraviolet scanner. Two 7.6–13.8 μm infrared bands provide for noise equivalent temperature-sensitivity of 0.2 and 0.02°K; the UV channel operates at 320–400 nm. An aerial-reconnaissance camera, KS-87B, with 11.4 cm square film format, is configured for nadir or 30° depression angle operation. To provide nighttime identification of polluting vessels, an active gated television (AGTV) is being developed, capable of detecting 320 candela at 21 km. It will allow zoom magnification to a 2.3° field of view. In the active mode, a one-watt pulsed lead vapor laser will provide illumination at 722.9 nm. The AGTV will permit stop action and either manual or computer-controlled gimbal tracking.

g. Satellite Detection of Oil

Many instances of satellite detection of oil slicks have been investigated. The implications are significant, of course, for the future when pointable sensors with high spatial resolution will be on board operational satellites with global coverage; at present, spatial resolution from geosynchronous satellites is too coarse for surveillance, while polar-orbiting satellites with better spatial resolution have inadequate repeat cycles and data turnaround-times for routine use. Other problems include the paucity of spectral channels, their wide bandwidths, variable oil/water radiance contrasts, especially in and out of sunglint regions, and small signal to noise ratios. The noise sources that are particularly troublesome for orbital oil detection include electronic noise, path radiance, and variable particulate and plankton loading of the underlying water (Manning et al., 1980).

The Ixtoc 1 oil well blowout in Campeche Bay near the Yucatan Penninsula of Mexico on June 3, 1979, provided an opportunity for testing the utility of satellite data in operational detection and tracking of oil slicks. Given the high rate of release, estimated variously as 10,000 to 100,000 bbls per day, and the duration until capping on March 24, 1980, this spill is the largest in history. Five satellites detected the spill: GOES, Nimbus 7, TIROS-N, and Landsats 2 and 3. The satellites all presented similar data-analysis problems, namely coarse resolution, sun-angle variations, and weather limitations. The Coastal Zone Color Scanner on Nimbus 7 was unreliable in slick detection: oil was briefly visible for just a few minutes of each cloud-free day in sunglint regions. Only a small portion of the dispersed oil was detected. For such large releases, remote detection is hardly important, but volume and tracking data are badly needed. The CZCS, and similar coarse spatial-resolution sensors, do not meet these needs. In the TIROS-N and GOES visual imag-

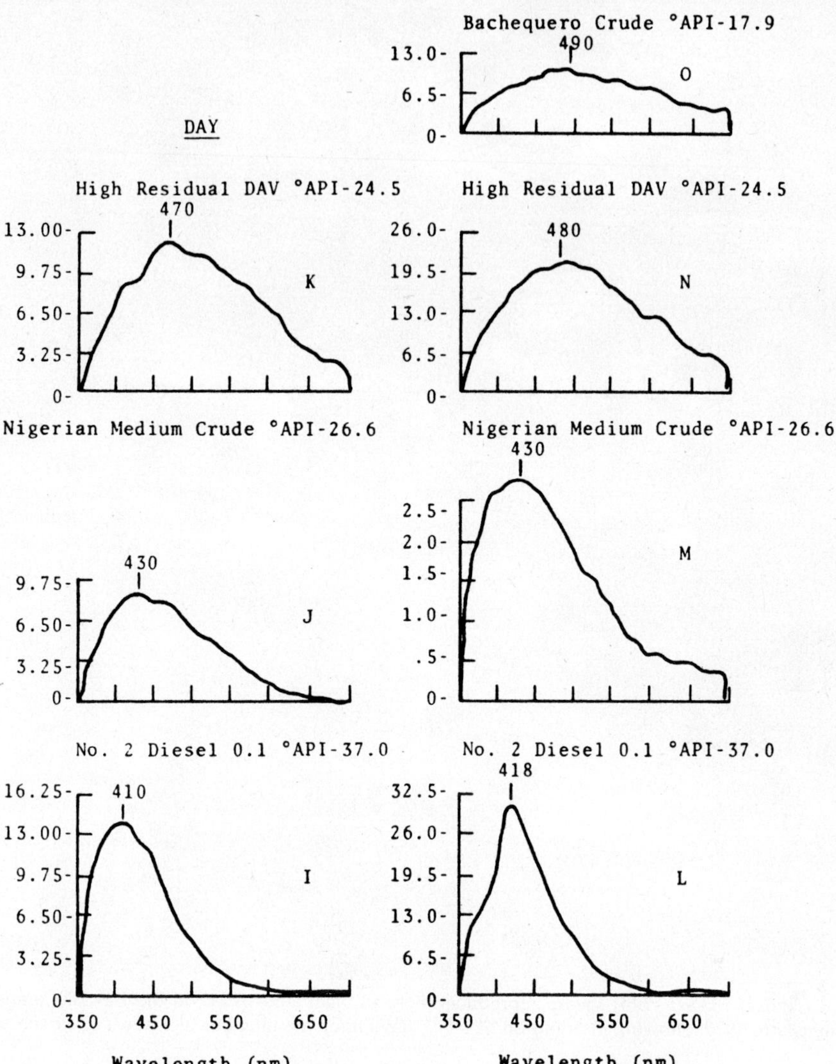

Fig. 28-122. Fluorescence emission spectra of sea water and oil spilled at sea measured by a system consisting of a laser, telescope, spectrometer, and image dissector (Fantasia and Ingrao, 1974).

ery, the oil formed a light grey-toned crescent originating at the well-site and corresponding with locations determined by aerial reconnaissance. A time series of GOES data discriminated oil slick from clouds. In Landsat MSS imagery, the slicks were visible as dark streaks, both in and out of sunglint areas, and as light-toned areas of mousse (a thick, frothy light-brown sea water emulsion of oil) (Hayes, 1980).

Ixtoc oil slicks in a mosaic of color enhancements of Landsat MSS scenes are shown in Figure 28-123. The MSS data were superior to single-band high-resolution (30 m) RBV data. MSS data produced higher contrast, and allowed enhancement in color addition, while the higher spatial resolution of RBV data necessitated the handling of too much data for the operational objective at hand. The MSS data from six months of coverage were useful in monitoring oil discharge and transport over thousands of square kilometers of the Gulf of Mexico, leaving no doubt that the oil washing ashore in Texas during August, 1979, did, in fact, originate from Ixtoc 1 (Deutsch et al., 1980). This finding helped eliminate any ambiguities rising from less-synoptic aerial surveys. Also, Landsat data confirmed flow directions of nearshore currents used by the Coast Guard in oil-drift prediction models (Hayes, 1980).

Although Seasat was inoperative by the time of Ixtoc 1, analysis of Seasat SAR data in the Santa Barbara Channel indicates oil slick detection. A pattern of low signal radar return offshore of Coal Oil Point corresponds fairly closely to the known

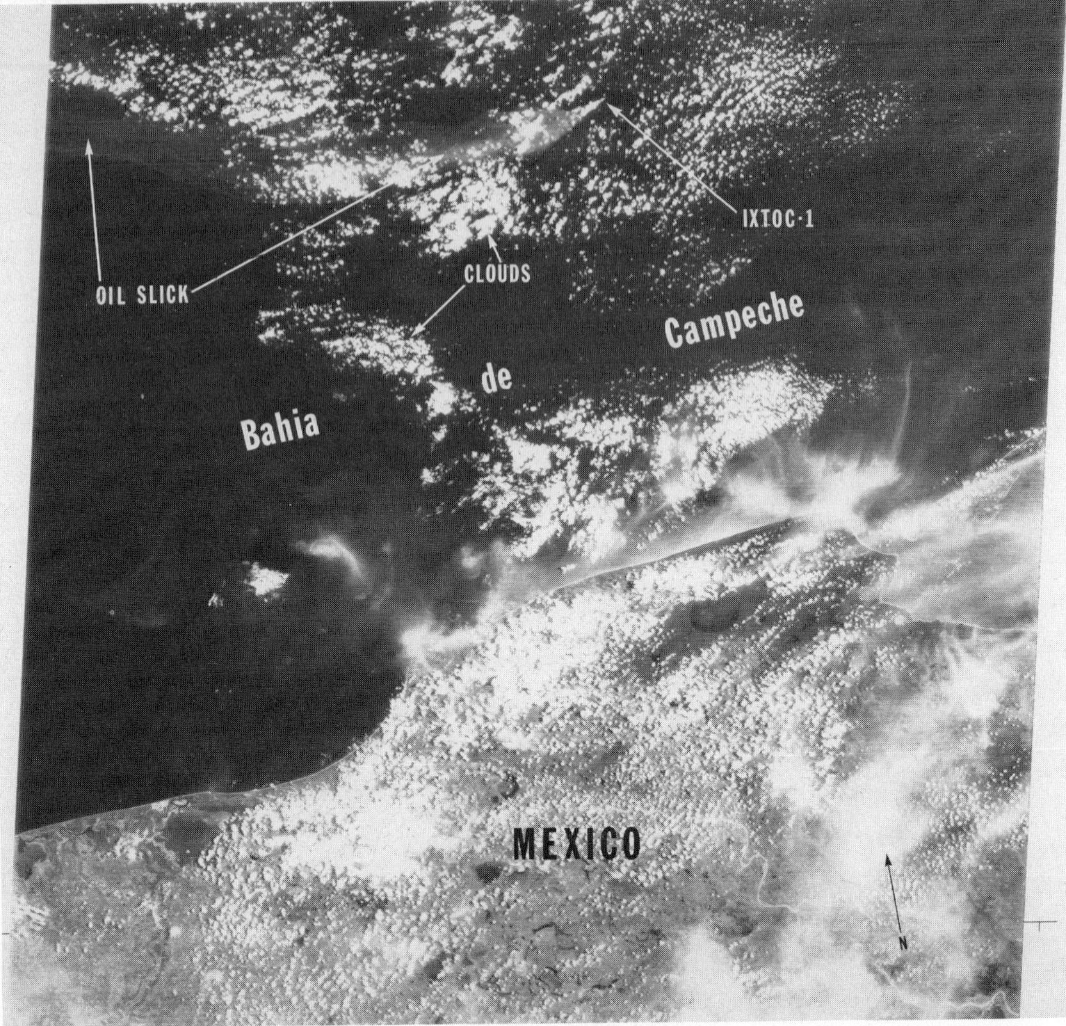

Fig. 28-123. Landsat MSS 2 color scene, reproduced here in black-and-white, in the Bay of Campeche, Gulf of Mexico, August 16, 1979, showing distribution of oil 52 days after the blowout of Ixtoc 1 (Deutsch et al., 1980).

location of linear natural oil seeps (Wilson et al., 1979) detectable by Landsat (Deutsch and Estes, 1980).

B. NON-POINT WATER QUALITY VARIABLES

Substances found in marine and coastal waters have both natural and artificial origins, and may be transported large distances, undergoing mixing, dilution, and dispersion. Great effort is involved in discovering points of origin for introduced materials. Those pollutants with dispersed points of origin, termed non-point pollution, figure in many environmental issues affecting resources and use of the coastal zone. Their remote sensing is often of interest; however, the problems (unrelated to remote sensing) in discriminating transport pathways and origins have had the consequence of limiting somewhat the intensity of effort toward development of suitable remote sensing techniques.

Toxic chemicals, nutrient chemicals (as in urban waste and agricultural soil runoff), and inorganic sediments are the principal classes of input pollutants. These affect and give rise to unnatural fluctuations of natural marine substances, producing, for example, red tides. All of the above materials are under consideration here.

Because chemical solutes are generally colorless and distributed through the water column, they are beyond the application of most optical techniques allowing sensing beneath the water surface. There may be developments in the future of Raman scattering techniques suitable for some non-point variables. For the present, however, some remote sensing is possible because colorless chemicals as well as bacteria often are adsorbed onto surfaces of suspended sediment particles. Chemical concentrations can then be tied to the sediment concentrations measured by remote sensing.

1.0. Turbidity

Suspended sediment and phytoplankton particles, in contrast to most dissolved chemicals, are easily detected by optical techniques. Through scattering and absorption, they affect the general clarity of water and hence the optical depth, as well as the color. Thus, the simplest methods of mere detection of suspended particulate material are optical measurements of turbidity, via Secchi disk depth or a device calibrated in Jackson Turbidity Units (JTUs), but these approaches make no distinction among possible causes of reduced clarity. Secchi depths are, as well, subjective. Consequently, they are held in less esteem for exacting analysis than measurements for specific types of particles. Nevertheless, such measures of turbidity are often obtained in field work in the marine environment.

Many water-quality monitoring agencies still find it convenient to characterize water bodies by Secchi disc depth and various other measures of turbidity or water clarity, instead of by filterable suspended particulate-matter concentration in mg/l. The U.S. Geological Survey has recommended that the use of turbidity measures be discontinued altogether (Pickering, 1976). However, it is likely that marine use of Secchi depth and other measures of turbidity will continue indefinitely. As Kozlyaninov (1980) has shown, measurements with a standard white disk, if made in overcast weather with a slight sea, can be used to calculate the approximate value of the total attenuation coefficient (α) for the band 520–540 nm. The simple relationship $\alpha = 3.0/z_b$, where z_b is the limiting depth of visibility of the disk, will yield α to within 20%.

Therefore, it must be noted that little study has been done in the marine environment on remote measurements transformed into Secchi depth or JTUs. Most work is on suspended sediment as discussed below. However, fresh water studies indicate that Landsat data and Secchi depth have correlation coefficients of about 0.9 (Rogers et al., 1975). Also, a correlation will usually be found between suspended solids and JTUs. Hence, there is every indication that remote measurement of Secchi depth or JTUs can be made operational in the marine environment. Due to the frequent success of measuring suspended sediment in both fresh and marine waters using Landsat MSS data, it is widely held that broad intervals of turbidity can be discriminated with Landsat even without surface calibration.

In view of the fact that remote quantitative measurement of turbidity is considered operational by some investigators (the same cannot yet be said for types of sediment and/or phytoplankton species), it is surprising that remote sensing of turbidity is not in wider use.

A field study in Lake Mead, Nevada, directed toward remote sensing of turbidity, was conducted by Holyer (1978). Surface upwelling radiance was measured in ten spectral bands between 400 and 1000 nm from a vessel. Data were obtained simultaneously with water sampling at the surface and at the Secchi extinction depth. Radiance data were corrected for skylight reflection and wind-generated waves to yield volume reflectance R. Particle size, not type, was found to be the dominant factor influencing correlations between R and turbidity. Both nonfilterable residue (mg/l) and nephelometric turbidity (NTU) were modeled as a function of R via second order polynomials in one or more spectral bands. Results approached the U.S.A. EPA accuracy-requirement of $\sigma^2 = 0.05$ (roughly 23 mg/l error at 100 mg/l). Using combined clay, silt, and fine-sand data, a universal NTU algorithm was obtained in two wavelengths, 652 and 782 nm, which may be successful in spatial extrapolation.

Application of Holyer's algorithms to atmospherically-corrected Landsat data over the Fraser River estuary of British Columbia, Canada by Aranuvachapun and LeBlond (1981) gave greater errors than expected. A logarithmic algorithm for suspended sediment concentration gave better results in this instance.

2.0. Suspended Inorganic Particulates

Typical inorganic sediments have scattering/absorption coefficient (b_p/a_p) ratios much larger than unity throughout the visible spectrum; hence the input of sediments to a water parcel will increase the volume reflectance R, and the water will appear brighter in contrast to clear water (see section on Ocean Color). In addition, optical depth τ will decrease, where τ is defined as:

$$\tau = \int_0^{z_0} c(z)dz, \qquad (28\text{-}26)$$

for the total attenuation coefficient c as a function of depth from the surface to depth z, related to optical transmission T for a narrow collimated beam by $T = e^{-\tau}$.

The above suggests two approaches to remote sensing of suspended inorganic particulates. One is laser backscatter-measurement of return-signal attenuation with depth; the other is a measurement of total upwelling radiance either under passive solar or active laser illumination. Both approaches yield data that can be transformed into either a turbidity value or a concentration of particulates.

Various types of inorganic particulates, with different size distributions and particulate shapes, yield different volume scattering functions and magnitudes of b_p and a_p, as in Figure 28-124 (Ghovanlou et al., 1977; Gupta and Ghovanlou, 1978). Therefore, each suspended sediment situation requires individual calibration, and where numerous types of particulates are simultaneously

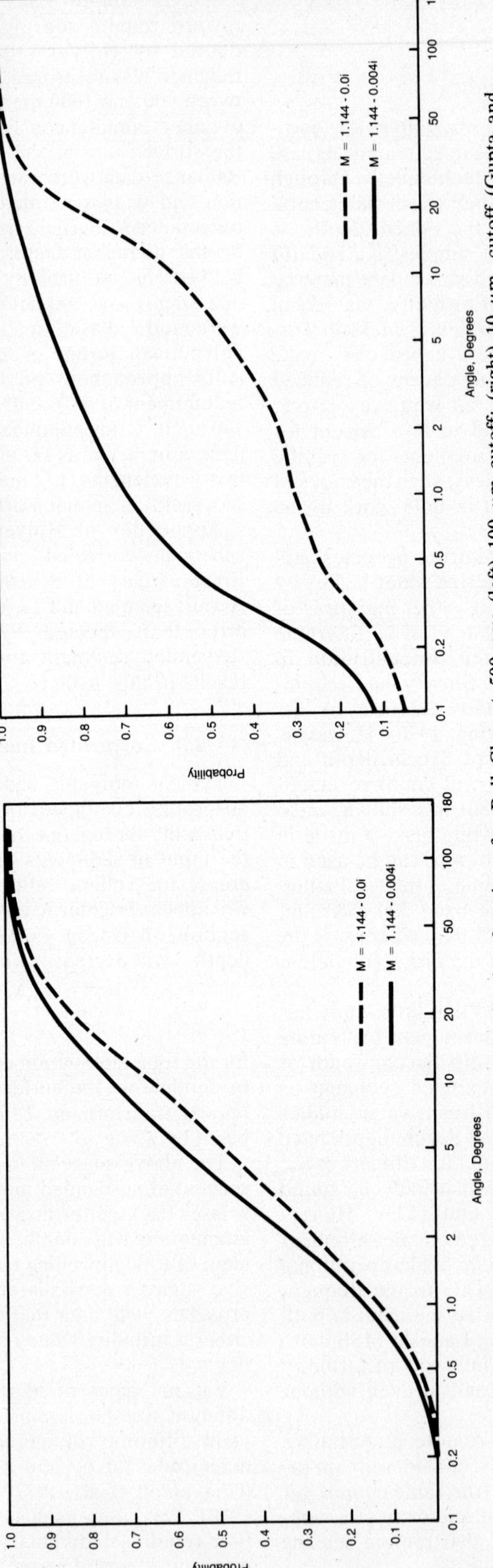

Fig. 28-124. Volume scattering probability distribution functions for Ball Clay at 500 nm: (left) 100 μm cutoff; (right) 10 μm, cutoff (Gupta and Ghovanlou, 1978).

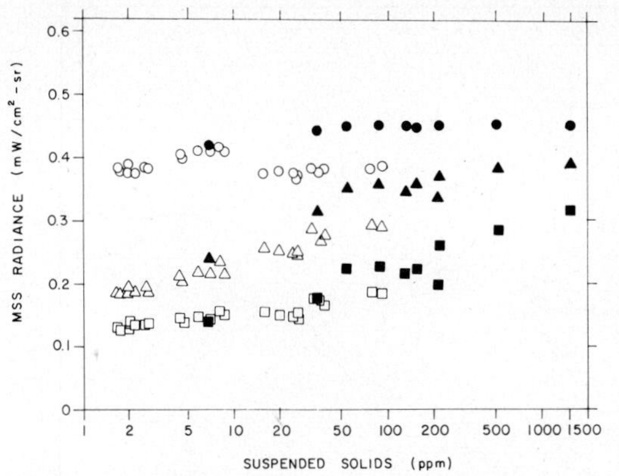

Fig. 28-125. Landsat MSS 4, 5, and 6 radiance versus suspended solids concentration. Minas Basin, Bay of Fundy, Nova Scotia. Three scenes; solar angle correction (Munday and Alföldi, 1979).

varying, a single-band measurement cannot be reduced to particulate concentration of any one type. Multispectral methods involving numerous narrow bands are required to unravel even a small number of components. A measurement of total particulate concentration in such a case may be noisy, and even a turbidity value should be viewed with caution.

It is usual to take b_p and a_p to be proportional to the concentration of the component in question; the detailed relation can be obtained from Shifrin (1951). Then, for $b_p >> b_w$, the volume reflectance R (see section on Ocean Color) obeys $R = S/(m_1 + m_2S)$, for concentration of sediment S, where the m_i are constants. The expression for R asymptotically approaches $1/m_2$, and as seen is non-linear with S. Ideally, the relation for low values of sediment concentration is linear. Figure 28-125 shows Landsat MSS 5 radiance versus S, and Figure 28-126 the same data after transformation to $1/R = m_2 + m_1/S$, which yields a straight line (Munday and Alföldi, 1979).

The best wavelengths for passive solar measurement of suspended sediment are between 550 and 650 nm; shorter wavelengths introduce greater atmospheric noise, while longer wavelengths suffer high absorption by water, restricting data collection to the upper few cm of the water column.

The measurement of suspended sediment can be accomplished with aerial scanner data as shown earlier in Figure 28-108 (Johnson, 1975, 1980), or from Landsat data (Bowker et al.,1975; Alföldi and Munday, 1978) using MSS5 or band ratios. The coastal environment necessitates that particular attention be paid to atmospheric variations, if data from different dates are to be compared. With a chromaticity technique used to standardize variable atmospheric path radiance over the scene (Munday et al., 1979), Amos and Alföldi (1979) measured suspended sediment levels in the macrotidal Bay of Fundy, Nova Scotia, producing thematic maps as shown in Figure 28-127.

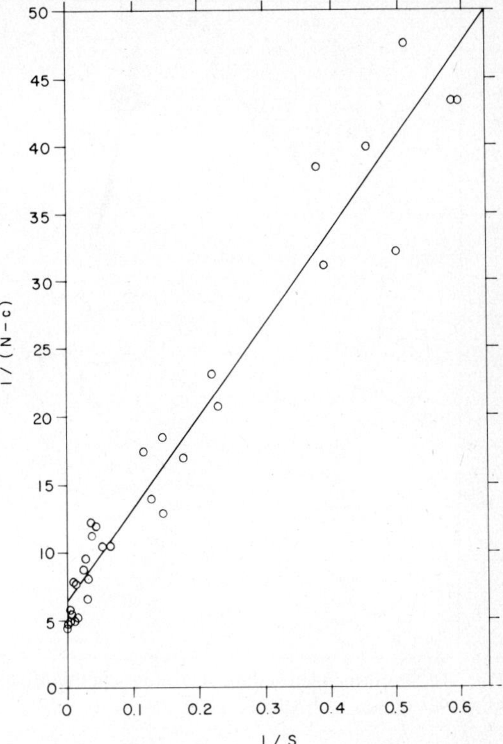

Fig. 28-126. MSS 5 data from Figure 28-125 transformed for test of $R = S/(m_1 + m_2S)$. Pearson correlation coefficient $r = 0.98$ (Munday and Alföldi, 1979).

The measurement of S can also be accomplished with photographs, which have the advantage of being easily obtained. However, numerous steps are required in the photographic data-reduction process in order to account for variables that affect the measurement accuracy. Intrinsic to the photographic operation itself is the non-linear DlogE relation, plus vignetting, azimuthal irregularities of lens transmission in many cameras, and the complexity of measuring emulsion densities in panchromatic film and color film to high accuracy (Scarpace, 1978). Extrinsic factors include the influence of solar elevation angle on sunglint patterns and subsurface illumination geometry, and the azimuthal variation of sunglint radiance (Bressette, 1977).

The accuracies in measurement of S obtainable by remote sensing have rarely been investigated in detail. Most investigators have calculated a regression between S or $\log_e S$ and upwelling radiance, with or without multiband ratios and normalization, and used the regression to construct an S contour map. The accuracy for the predicted S data has usually been expressed only as a correlation coefficient for the data used in obtaining the regression. The Pearson product-moment correlation coefficient is used for convenience, but its applicability may be questioned. There are insufficient radiance values at each S value to permit a test of normality for the radiance

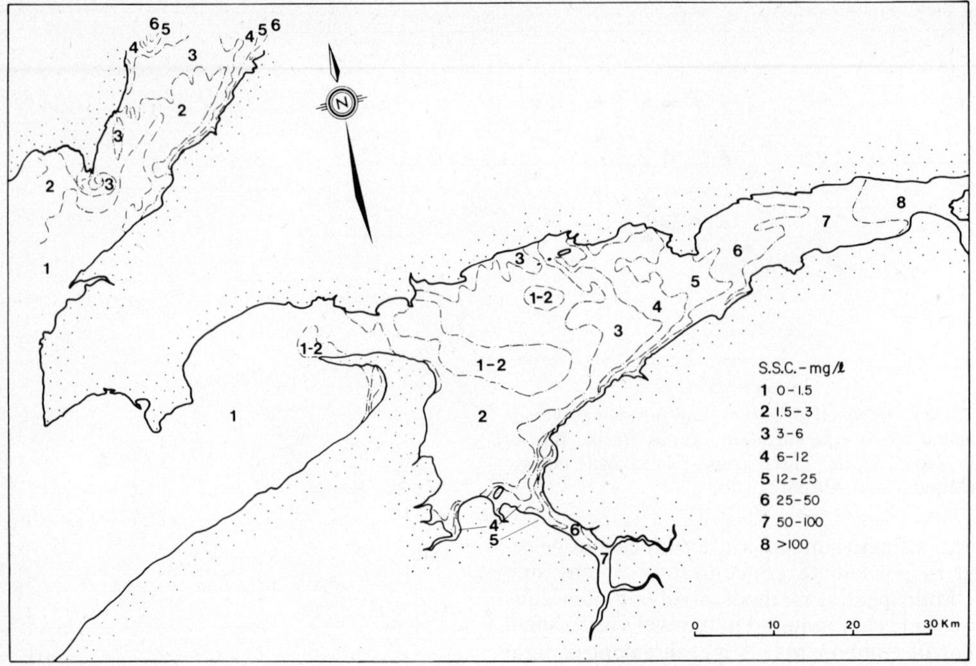

Fig. 28-127. Suspended sediment contours in the Minas Basin, Bay of Fundy, Nova Scotia, from Landsat on 3 May 1974 (Amos and Alföldi, 1979).

values; also, some relationships involved are non-linear, such as the radiance-versus-S relation, and atmospheric transfer; therefore, for a particular S value, the associated radiance values could be skewed. Proceeding nevertheless with Model I regression statistics (Sokal and Rohlf, 1969), the statistical standard error of prediction associated with Landsat measurement of S can be determined from:

$$Error = \pm\, t_{(0.05,\ n-2)}\, s_{y.x} \sqrt{\left[(1/m) + (1/n) + (X - \bar{X}_i)^2/\Sigma(X_i - \bar{X}_i)^2\right]} \quad (28\text{-}27)$$

where X is the point estimate of $\log_e S$, t is the t-statistic at the 95 percent confidence level, $s_{y \cdot x}$ is the root mean square deviation from regression, m is the number of pixels averaged for each point measurement, n is the number of samples, and X_i and $\bar{X}_i$ are the point and mean values of $\log_e S$. For 95 points in the Bay of Fundy obtained over nine Landsat passes, the belt of estimated error at 1 mg/l was −24 and +31%, at 20 mg/l, −19 and +24%, and at 1000 mg/l, −24 and +32% (Munday et al., 1979).

One source of error is sensor-system noise. For Landsat, a sensor noise of ±1 digital count in each band would result in errors of 100 percent at low S values. This potential error is reduced in usual practice, as most investigators employ pixel-averaging to suppress scanner-detector striping. For sensors with better radiometric resolution and better signal-to-noise ratios, such as the Coastal Zone Color Scanner, more accurate measurement of S should be obtainable.

Another source of error is associated with inhomogeneous distribution of suspended materials in water. This error is inherent in all types of data collection, whether for single or multiple dates, where surface information and remote sensing data are being compared. The greater the within-pixel inhomogeneity, the greater the error. Inhomogeneity has two effects: First, the collected sample may not be representative of the "average" particulate concentration in the field of view. Second, due to nonlinearity between radiance and S, an "average" radiance measurement inverted to S produces an underestimate of the true value. Increasing inhomogeneity in the distribution of a given amount of particulates results in a worsening underestimate of its average concentration in the field of view. The magnitude of these effects cannot be evaluated without knowledge of the particular spatial distribution of suspended material under consideration. For the Nimbus 7 Coastal Zone Color Scanner, Doerffer (1979) found this error for North Sea data was 30 percent.

A significant consideration in measuring S from total upwelling radiance is the effective depth of measurement. The penetration depth is defined as the depth above which 90 percent of the diffusely reflected (total upwelling) irradiance originates. This depth is denoted z_{90}. For elastic scattering processes (no energy loss, and hence no wavelength change), the penetration depth is approximately the inverse of the attenuation coefficient of downwelling irradiance E_d, measured at the mean of the concentration over the penetration depth (Gordon and McCluney, 1975). Esti-

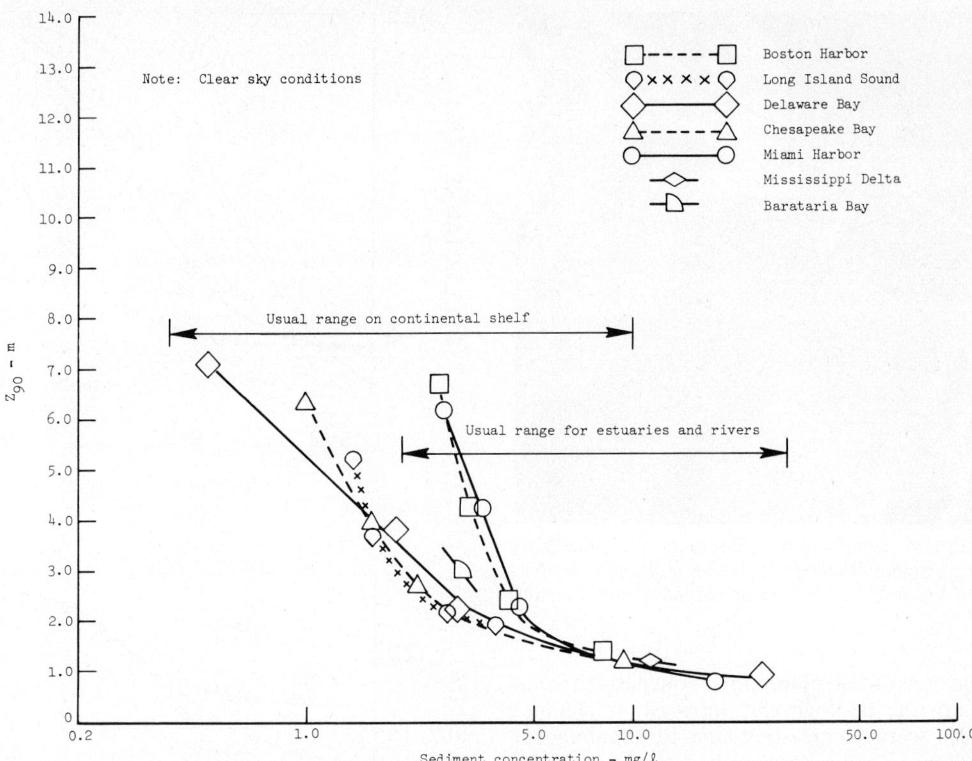

Fig. 28-128. Penetration depth Z_{90} at 540 nm derived from laboratory measurements with suspended sediment concentrations in various natural waters (Whitlock, 1976).

mates derived from laboratory data of the penetration depths for various sediment levels are shown in Figure 28-128 (Whitlock, 1976).

It is desirable to understand in detail the impact of highly absorbing particulates and dissolved substances on the measurement of scattering inorganic particulates. If the absorption and scattering properties of each of the components present is known then, by the equations given under the section on Ocean Color, the volume reflectance R can be calculated. In turbid coastal waters, the necessary spectral data are usually lacking. These data must be obtained before even the applicability of the simple expressions for R can be examined, namely that the single-scattering albedo $\omega_o < 0.6$.

The use of lasers has opened the attractive possibility of measurement of the depth profile of S. It was mentioned earlier that laser determination of the attenuation depth could be used for measuring S; such an approach requires gating of the return signal to allow assignment of reflected laser radiance to a sequence of layers in the water column, but does not require measurement of absolute radiance from each depth. Success in laser studies has led, however, to near-operational measurement of absolute radiance levels versus depth (Hickman et al., 1975) which permits, via the same considerations as above, the quantitative determination of the depth profile. The obvious advantages of lasers include their day/night capability, and the monochromatic nature of the radiation involved. The latter simplifies the laboratory work required for appropriate calibration of the technique. Multispectral dye lasers should provide power in resolving multiple-particulate components.

3.0. Phytoplankton Blooms and Nutrient Input

One of the more dramatic events in the marine environment is the red tide. This phenomenon originates as a natural bloom of reddish dinoflagellates, which is exacerbated by nutrient enrichment from non-point pollution. Toxins produced in some red tides are poisonous to marine animals, and also to unwary swimmers and consumers of contaminated seafood. Remote sensing is of interest for direct investigation of bloom phenomena, and for indirect indication of nutrient pollution.

In clear waters the remote sensing of blooms is accomplished as discussed earlier for chlorophyll. In turbid waters the remote sensing problem is complex, as the presence of multiple components, some of which are spectrally diverse, makes precise measurement of either chlorophyll or cell concentrations somewhat difficult using existing techniques. Although dinoflagellate blooms in sediment-laden Virginia waters consistently produced reflectance decreases in visible bands (see Figure 28-129), correlation between photographic

Fig. 28-129. Dinoflagellate bloom in the tidal York River, Virginia. Patches of *Gymnodinium splendens* darken the bright turbid water (Munday and Zubkoff, 1981).

optical density and chlorophyll concentration was poor. In the photographic infrared, reflectance changes were inconsistent due to simultaneous variations in chlorophyll and inorganic suspended-sediment concentrations. The impact of such variations on reflectance is illustrated by the reflectance-model simulations of Figure 28-130 (Munday and Zubkoff, 1981). Constant reflectance will be observed, even with changes in sediment S and chlorophyll C, if they obey

$$\frac{dC}{dS} = \frac{a_s'b_w - b_s'(a_w + a_y) + C(a_s'b_c' - b_s'a_c')}{b_c'(a_w + a_y) - a_c'b_w + S(a_s'b_c' - b_s'a_c')} \quad (28\text{-}28)$$

where a and b are specific absorption and scattering coefficients for water w, sediments s, chlorophyll particles c, and yellow substance y.

Areas of low reflectance on satellite and aerial imagery and photography of coastal waters have traditionally been interpreted as representing low sediment loads. Nearshore scouring due to tidal dynamics generally produces higher sediment loads and higher reflectance values. However, because blooms generally cause reflectance decreases in turbid waters, blooms and low sediment loads will be indistinguishable in imagery of poor spatial and spectral discrimination. In imagery of high spatial resolution, blooms might be distinguished by their texture, which is more complex than sediment distributions due to cell motility. Also, blooms may have sharper concentration gradients.

Despite the above limitations, some useful data on blooms can be obtained with simple aerial survey. Low altitude reconnaissance in Virginia's Chesapeake Bay has shown that blooms occur in all seasons, and throughout the bay's tributaries and nearshore zones (Zubkoff et al., 1979). Some satellite observations of blooms have been made

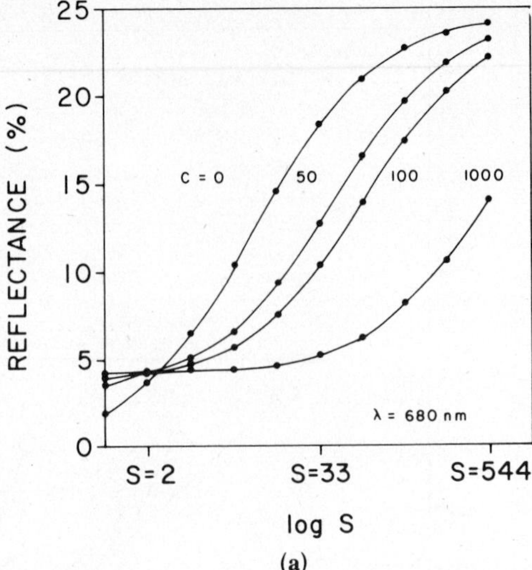

(a)

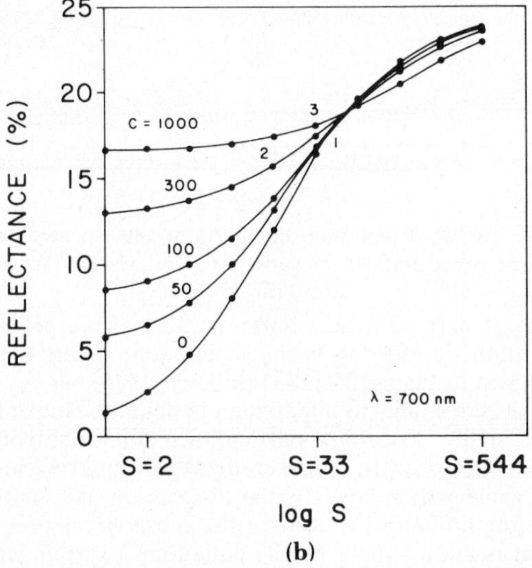

(b)

Fig. 28-130. Reflectance model simulations of the impact of simultaneously varying chlorophyll and sediment concentrations. C = chlorophyll *a* in μg/l; S = sediment in mg/l. a. 680 nm. b. 700 nm (Munday and Zubkoff, 1981).

as well. Striking examples of blooms of the blue-green *Nodularia spumigena* and *Aphanizomenon flos-aquae* have been detected in the Baltic Sea by Landsat (see Figure 28-131; Ulbricht et al., 1978; Horstmann et al., 1978). Ostrom (1976) has used such data in a preliminary analysis of biomass and the Baltic Sea nitrogen budget.

4.0. Chemical Solutes and Yellow Substance

Colorless chemicals that are present in the water column cannot be detected directly by remote sensors. Situations may occur, however, when concentrations of such chemicals are correlated with suspended particulate concentrations. This may result from chemical adsorption

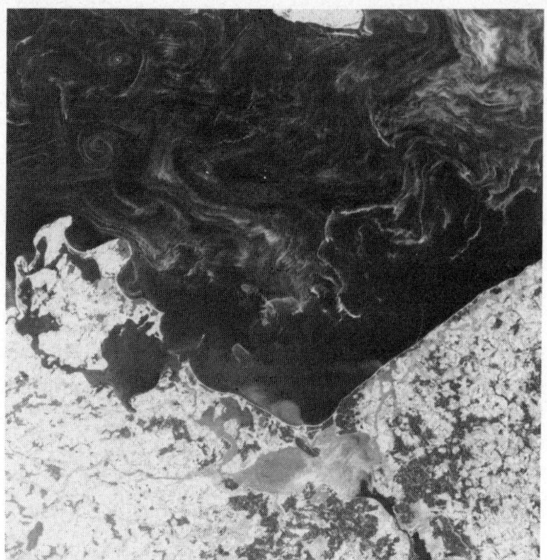

Fig. 28-131. Blue-green algal blooms in the Baltic Sea from Landsat 7 August 1975. Contrast-enhanced MSS 4 image (Horstmann *et al.*, 1978).

on particulates, dissolution of absorbed or adsorbed chemicals from particulates, or input of both to the water column from a common origin. Remote measurement of turbidity may then be calibrated for chemical concentration, as successfully accomplished in Saginaw Bay, Michigan (Rogers et al., 1976).

The well-known colored solute of nearshore marine waters called yellow substance has been characterized (see section on Ocean Color), and will interfere with accurate remote measurement of chlorophyll and turbidity unless corrections are made. Yellow substance is in large measure the result of erosion and runoff from the land. Particles are carried in streams and are deposited. Yellow stuff and its fluorescent by-products are, therefore, dependent upon runoff and salinity and may be used as an indicator of fresh water input.

C. WATER QUALITY AND CIRCULATION MODELS

A well-established practice in the marine environment is the use of hydrodynamic and water quality models for engineering decisions and water-pollution control. To acquire the necessary hydrodynamic understanding, basin-wide circulation may be studied using physical hydraulic models and numerical models. Both types are constructed using hydrographic data obtained in field surveys, using vertical strings of current meters, and sampling for vertical profiles of salinity and temperature. Some engineering decisions are based directly on the hydrodynamic models, such as design and placement of new facilities (see section on Waste Outfalls). The models are also used to acquire understanding of hydrodynamic characteristics of coastal waters as a prerequisite for developing appropriate water-quality models; the dynamics of tidal circulation require that water-quality models, whether steady-state or time-dependent, be properly formulated, calibrated and verified. Hydrodynamic models are also used *in toto* in the construction of water quality models by adding equations for pollutant sources and sinks, and pollutant-dispersion factors (Chen, 1978b). In the case of oil pollution, oil-slick spreading laws are incorporated (see Fay, 1969; Hoult, 1972).

1.0. Water Quality Models

Remote sensing has contributed to model construction and verification, and its role is increasing. For models predicting concentrations and flushing times of pollutants in small estuaries and coastal embayments, the model by Ketchum (1951) may be used. It describes the exchange between various parts of an estuary as a result of tidal oscillations, and permits the calculation of the average distribution of fresh and salt water. The Ketchum model has been modified to include both conservative and nonconservative pollutants, and then applied to various embayments in Chesapeake Bay, Virginia.

Necessary inputs to the model include the low tide and high tide volume of water in the estuary, and the local intertidal volume (the fraction of the total tidal prism in each part of the estuary at high tide). It is important to have high accuracy in these inputs since the estuary is divided into segments that represent the excursion of a water particle during a tidal cycle. This is done by requiring that the low-tide volume of each segment be equal to the high-tide volume of the next landward segment.

Low-tide and high-tide volumes can be calculated grossly from mean tidal range and topography, and consequently an estimate of the tidal prism can be made. However, low accuracy results from the fact that shoreline topography is not well-known. However, if an accurate measurement of water-surface area is available as a function of tidal phase measured with tide gauges, the total tidal prism can be calculated, and intertidal volumes much more accurately determined.

Black-and-white infrared film with a very deep red filter (Wratten 25a) records only near-infrared energy. Since water reflects very little energy in the infrared, whereas land and especially sandy shorelines have high reflectance, the land-water contrast is good. The film can be used with conventional mapping equipment to make water-surface maps at various tidal stages (Jones, 1957). It has been used successfully for resolving the time dependent storage function for an inlet-marsh-bay system in Virginia (Gordon et al., 1973). The advancing water can be easily discerned as it covers the lower land features, and the total area-time relationship can be determined, yielding the tidal prism. For this purpose, low altitude 70 mm coverage is convenient and inexpensive. Data on water areas can be extracted from processed film by directly digitizing a projected image.

2.0. Hydrodynamic Numerical Models

Further application of remote sensing to calibration and verification of models has recently been emerging. As discussed by Hill and Graham (1979), Landsat may be used for: 1) finite-element model grid-spacing, 2) boundary delineation, 3) calibration, and 4) verification.

a. Grid Spacing

In development of numerical models, the well-known Courant-Friedrichs-Levy (CFL) Condition $\Delta t \leqslant \Delta s/\sqrt{2gh}$ specifies a constraint on the finite element grid spacing Δs for a chosen time-step interval Δt. But Δs should also be chosen such that $D_H \; \Delta t/(\Delta s)^2 <$ constant (where D_H is eddy-mass diffusivity in the horizontal direction), in order to suppress numerical dispersion. Landsat maps of suspended sediment concentrations will reveal the quantitative magnitude of concentration gradients in different regions of water bodies of interest. The maximum gradient will be one indicator of the magnitude of D_H, in that sharp gradients suggest low values of D_H and vice versa. It is important to note that density fronts inhibit the mass transfer that normally accompanies turbulent diffusion. Hence, interpretation of sediment concentration gradients from Landsat maps, or from any other data source, should discount the influence of density fronts when they are present. The same bank of Landsat data used for estimation of D_H can be used to find probable estuarine density-fronts, which appear in Landsat imagery as isolated sharp turbidity-boundaries. Such tend to recur at certain places during similar tidal phases (Klemas and Polis, 1977).

b. Boundary Delineation

Inspection of Landsat images and Landsat sediment-concentration maps helps in setting appropriate locations for boundaries between large water bodies of interest, and places of influx and efflux. Boundary locations would be chosen if possible at places where many Landsat scenes indicate a small concentration gradient along the boundary. Ideally, the cross-sectional mass transport would be uniform.

c. Calibration

Landsat 3 can provide values of suspended-sediment concentration over a water body at a spatial resolution of 80 m (or 480 m if 6 × 6 pixel averaging is used to reduce detector striping). Furthermore, the Landsat data are collected for large water bodies in a few minutes. For a finite-element Chesapeake Bay storm-surge model, Chen (1978a) used element lengths of 5 to 10 km, and a time step of 4 minutes. Thus, both the spatial and temporal resolutions inherent in Landsat data are finer than those employed in large-area numerical models. Note that other sources of calibration data are far inferior in terms of spatial and temporal resolution.

Because celerity of a disturbance $\sqrt{(gh)}$ is generally greater than water transport q, numerical computation is more efficient when the hydrodynamic and water-quality components of estuarine modeling are treated separately (Chen, 1978b). For the water-quality component, calibration proceeds with setting of initial values of suspended-sediment concentration at the nodes of the model, and subsequent adjustment of model coefficients. Landsat sediment-concentration maps can be used for initial values.

d. Verification

During model verification, computed solutions are compared with independent field data to determine how well the model approximates the prototype. One Landsat pass can be used to initiate the model, and another pass, chosen from an appropriate set of tidal phase, fresh-water inflow, season, and wind conditions, can then be used to assess whether reasonable agreement has been achieved after running the model to a new condition after several tidal cycles, or to steady-state.

Assessing the degree of agreement can proceed quantitatively by direct comparison of model output values with Landsat maps. The assessment can also proceed qualitatively, by comparison of gradients and concentration lines between the water-quality model with the Landsat maps, and as well by comparison of tidal-current vectors from the source hydrodynamic model with streamflow lines in the Landsat maps and scene images.

These concepts are not yet in widespread use, but two instances may be cited. A finite-element time-dependent hydrodynamic model for Apalachicola Bay, Florida, was used to provide input to a dispersion (mass balance) model for an idealized pollutant representing turbidity. The model was run to steady state (35 tidal cycles) and produced isocontours of concentration. These were in fair agreement with digitally-contoured Landsat turbidity-patterns for similar inflow conditions (Figure 28-132 and Color Figure 28-133). In the course of different model runs, a range of values for the dispersion coefficient (3–40 m²/sec) was tried; the value providing the best match between model output and Landsat data was 5 m²/sec, illustrating that Landsat data may be used in sensitivity analysis of model input constants (Hill and Graham, 1979; Graham et al., 1978).

A second example involves a finite-difference time-dependent model (Greenberg, 1977) for the macrotidal Bay of Fundy, Nova Scotia, which experiences an average tide of 12 m. This model was used to simulate the most intense conditions of sedimentation that might result from construc-

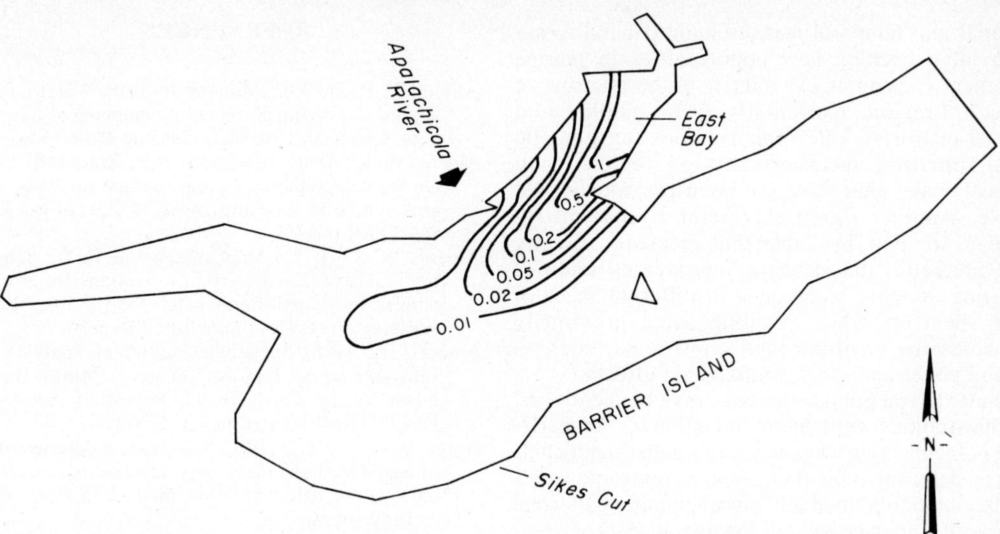

Fig. 28-132. DISPER-1 Model simulation of idealized suspended sediment concentrations after 35 tidal cycles in Apalachicola Bay, Gulf of Mexico. Tidal phase angle 0°. Compare with Figure 28-133 (Hill and Graham, 1979).

tion of a tidal power barrage. Landsat maps of quantitative suspended-sediment concentrations from spring conditions of maximum turbidity were used to initiate the model. The model was then adjusted until, after 7 cycles, it approached summer conditions of sediment dynamics. The pattern of contours in Figure 28-134 agreed well with sediment contours derived from a summertime Landsat image in Figure 28-127 (Amos and Greenberg, 1980; Munday et al., 1979). The results showed that only minor sedimentation within the headpond region will occur; such a conclusion has major engineering consequences for the feasibility of the barrage, and was based in part on Landsat MSS data.

IV. CURRENT STATUS AND OUTLOOK

Remote sensing will play an ever increasing role in investigations of the marine environment, as evidenced by steady increase in the use of combinations of traditional techniques with remote sensing systems. The continuing development of new remote measurement techniques, complementary and parallel development of microcomputer technology, and sensors with improved

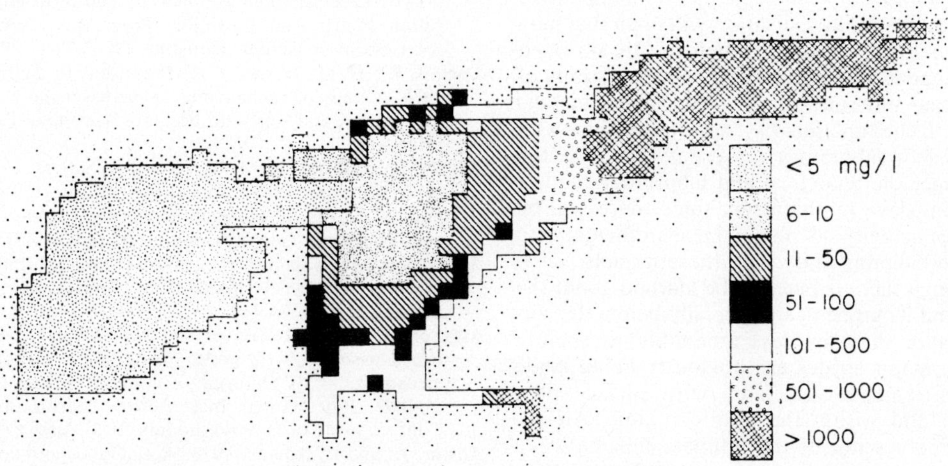

Fig. 28-134. Greenberg (1977) model simulation of suspended sediment concentrations after 7 tidal cycles in the Minas Basin, Bay of Fundy, Nova Scotia. High tide condition. Initiated with Landsat-measured concentrations. Compare with Figure 28-127 (Amos and Greenberg, 1980).

spatial and temporal measurement-capability are providing exciting new opportunities in marine science. Expanding capabilities in the microwave spectral-region, particularly from satellite and aerial platforms, offer opportunities for synoptic and repetitive measurements on regional and global scales that have not been previously possible. Another significant factor is that instruments are now available that measure, directly or indirectly, the same or interlocking features by one or more techniques in different parts of the spectrum, thus avoiding some previously troublesome environmental constraints, such as cloud cover and other atmospheric effects.

Several emerging techniques may be mentioned as illustrations, without any intention of summarizing or highlighting what may be the most significant: First, determination of oceanic currents and water masses by combined and complementary thermal infrared and passive microwave measurements, in conjunction with precision altimetry. Second, measurement of sea surface temperatures by infrared radiometers (a generally accepted technique) and the complementary development of microwave thermal-measurements with calibration using data from nearby spectral bands. Finally, significant development of multifrequency sonar-sensors and analysis techniques which, in combination with systems such as time-gated laser systems, will provide information on coastal zone and oceanic underwater features more readily than previously available, and in nearly pictorial detail. The resulting data should stimulate significant model development.

These measurement systems may not be readily available to all investigators or users of the marine environment, due to hardware and data-reduction constraints. However, it is clear that, as the aircraft and satellite experiments continue, there will be opportunities available throughout the marine community to participate and/or subsequently to analyze data from such systems. With repetitive coverage, lines of investigation will open that may not have been considered before. In short, it would appear that analyses and investigations of the marine environment in the future can deal more with the functional system rather than with analytical or laboratory representations. At the same time, the analytical and laboratory models should improve in character, since there will be more opportunity to obtain measurements with which to calibrate and verify these models.

Although there is much to be learned about the ocean and its processes, especially below the surface where they are less amenable to remote sensing, major strides are obviously being made in measurements associated with surface phenomena and within the photic region. An enhanced perception of the intimate link between subsurface oceanic phenomena and their surface manifestations is recently apparent, due entirely to the magnificent view provided by orbiting satellites.

REFERENCES

Alföldi, T. T. and J. C. Munday Jr. 1978. Water quality analysis by digital chromaticity mapping of Landsat data, Canadian J. Remote Sensing 4:108–126.

Alpers, W. R., D. B. Ross and C. L. Rufenach. 1981. On the detectability of ocean surface waves by real and synthetic aperture radar. J. Geophys. Res. 86(C7):6481–6498.

Ament, W. S., R. C. MacDonald and R. D. Schewbridge. 1959. Radar terrain reflections for several polarizations and frequencies. Naval Res. Lab., Washington, D.C. (Unpublished Report).

Amos, C. L. 1976. Suspended sediment analysis of seawater using Landsat imagery, Minas Basin, Nova Scotia. Paper 76-1C, Report of Activities, Part C, Geol. Surv. Can., p. 55–60.

Amos, C. L. and T. T. Alföldi. 1979. The determination of suspended sediment concentration in a macrotidal system using Landsat data. J. Sedim. Petr. 49:0159–0174.

Amos, C. L. and D. A. Greenberg. 1980. The simulation of suspended particulate matter in the Minas Basin, Bay of Fundy—a region of potential tidal power development. Proc. Canadian Coastal Conf., Natl. Res. Council, Ottawa, p. 2–20.

Anderson, J. R., E. E. Hardy, J. T. Roach and R. E. Witmer. 1976. A land-use and land-cover classification system for use with remote-sensor data. U.S. Geol. Survey Prof. Paper 964, 28 p.

Anderson, R. R. 1969. The use of color infrared photography and thermal imagery in marshland and estuarine studies. Second Ann. Earth Res. Progr. Rev., Vol. 3, NASA Johnson Space Center, Houston, TX.

Anderson, R. R. 1970. Spectral reflectance characteristics and automated data reduction techniques which identify wetland and water quality conditions in the Chesapeake Bay. Third Ann. Earth Res. Progr. Rev., Vol. 3, NASA Johnson Space Center, Houston, TX.

Anderson, R. R. 1975. ERTS-1 investigation of wetland ecology. Final Rept. NASA-21752. The American Univ., Washington, D.C., 105 p.

Anderson, R. R. and V. Carter. 1972. Wetland delineation by spectral signature analysis and legal implication. Fourth Ann. Earth Res. Progr. Rev., NASA Johnson Space Center, Houston, TX, 3:78-1–78-9.

Apel, J. R., H. M. Byrne, J. R. Proni and R. Sellers. 1976. A study of oceanic internal waves using satellite imagery and ship data. Remote Sensing of Env. 5:125–135.

Apel, J. R., J. Proni, H. M. Byrne and R. Sellers. 1975. Near-simultaneous observations of intermittent internal waves on the continental shelf from ship and spacecraft. Geophys. Res. Lett. 2:128–131.

Aranuvachapun, S. and P. H. LeBlond. 1981. Turbidity of coastal water determined from Landsat. Remote Sensing of Env. 84:113–132.

Arvesen, J. C., E. C. Weaver and J. P. Millard. 1971. Rapid assessment of water pollution by airborne measurement of chlorophyll content. Paper No. 71-1097. Proc. Amer. Inst. Aeron. Astron. Joint Conf. Sensing of Env. Pollutants, Palo Alto, CA.

Austin, A. and R. Adams. 1978. Aerial color and color infrared survey of marine plant resources. Photogramm. Eng. Remote Sensing 44:469–480.

Austin, R. W. 1974. The remote sensing of spectral radiance from below the ocean surface. In N. G. Jerlov and E. S. Nielsen (eds.), Optical Aspects of Oceanography, Academic Press, NY, p. 316–344.

Badgley, P. C. 1969. Oceans from space. Gulf Publ. Co., Houston, TX.

Barrick, D. E., M. W. Evans and B. L. Weber. 1977. Ocean surface currents mapped by radar. Science 198:138–144.

Bartlett, D. S. 1979. Spectral reflectance of tidal wetland plant canopies and implications for remote sensing. Ph.D. Diss., Univ. Delaware, Newark.

Bartlett, D. S. and V. Klemas. 1981. In situ spectral reflectance studies of tidal wetland grasses. Photogramm. Engl and Remote Sens. 47(12):1695–1703.

Bartlett, D. S., V. Klemas, R. H. Rogers and N. J. Shah. 1977. Variability of wetland reflectance and its effect on automatic characterization of satellite imagery. Proc. 43rd Ann. Mtg., Amer. Soc. Photogramm., Falls Church, Va., p. 70–89.

Bass, F. G., I. M. Fuks, A. I. Kalmykov, I. E. Ostrobsky and A. D. Rosenberg. 1968. Very high frequency radio wave scattering by a disturbed sea surface. IEEE Trans. on Antennas and Propagation, AP-16(5):554–568.

Beal, R. C. 1980. Spaceborne imaging radar: monitoring of ocean waves. Science 208:1373–1375.

Beal, R. C. 1981. Spatial evolution of ocean wave spectra. *In* Spaceborne Synthetic Aperture Radar for Oceanography, Johns Hopkins Univ. Press, Baltimore, MD, p. 110–127.

Beardsley, R. C., B. Butman and M. A. Noble. 1977. The water structure, mean currents, and shelf/slope water on the New England Continental Shelf and Georges Bank region (abstract). Eos Trans. AGU 58(9):887.

Belderson, R. H., N. H. Kenyon, A. H. Stride and A. R. Stubbs. 1972. Sonographs of the sea floor. Elsevier, Amsterdam, 185 p.

Benigno, J. A. 1970. Fish detection through aerial surveillance. Techn. Conf. Fish Finding, Purse, Seining, and Aimed Trawling, Reykjavik, FAO, FKK:FF/70/78, p. 13.

Benigno, J. A. and A. J. Kemmerer. 1973. Aerial photographic sensing of pelagic fish schools: a comparison of two films. Proc. Fall Conv. Symp. on Remote Sensing in Oceanography, Amer. Soc. Photogramm., Falls Church, VA, p. 1032–1040.

Bernstein, R. 1982. Sea surface temperature mapping with the Seasat microwave radiometer, J. Geophys. Res. (in press).

Bernstein, R. and D. Ferneyhough, Jr. 1975. Digital image processing. Photogramm. Eng. Remote Sensing 41(12):1465–1476.

Bernstein, R. And G. C. Stierhoff. 1976. Precision processing of earth image data. Amer. Sci. 64(5): 500–508.

Betz, H. T. 1968. The remote measurement of Rhodamine B concentration when used as fluorescent tracer in hydrologic studies. Publ. No. NR 05-00-000-0101, NASA Johnson Space Center, Houston, TX.

Blume, H-J. C., A. W. Love, M. J. Van Melle and W. W. Ho. 1977. Radiometric observations of sea temperature at 2.65 GHz over the Chesapeake Bay. IEEE Trans. Antennas Propag. AP-25:121–128.

Blume, H-J. C., B. M. Kendall and J. C. Fedors, 1978. Measurement of ocean temperature and salinity via microwave radiometry. Boundary-Layer Meteorol. 13:295–308.

Blume, H-J. B., B. M. Kendall and J. C. Fedors. 1981. Multifrequency radiometer detection of submarine freshwater sources along the Puerto Rican coastline. J. Geophys. Res. 86(C6):5283–5291.

Boc, S. J. and L. J. Landgelder. 1977. An analysis of beach overwash along North Carolina's coast. Dept. Civil Eng. Rept. No. 77-9, North Carolina St. Univ. at Raleigh, 17 p.

Born, G. H., J. A. Dunne and D. B. Lane. 1979. Seasat mission overview. Science 204:1405–1406.

Borstad, G. A., R. M. Brown and J. F. R. Gower. 1980. Airborne remote sensing of sea surface chlorophyll and temperature along the outer British Columbia coast. Proc. 6th Canadian Symp. Remote Sensing, Halifax, Can. Aeron. Space Inst., Ottawa, p. 541–549.

Boulé, M. 1976. Geomorphic interpretation of vegetation on Fisherman Island, Virginia. M.S. Thesis, Virginia Inst. Mar. Sci., Gloucester Point, VA, 125 p.

Bowker, D. E., W. G. Witte, P. Fleischer, T. A. Gosink, W. J. Hanna and J. C. Ludwick. 1975. An investigation of the waters in the lower Chesapeake Bay. Proc. 10th Intl. Symp. Remote Sensing of Env., Ann Arbor, MI, p. 411–420.

Bowker, D. E. and W. G. Witte, 1977. The use of Landsat for monitoring water parameters in the coastal zone. Proc. AIAA Joint Conf. Satellite Applications to Marine Technology. Amer. Inst. Aeron. Astron., New Orleans, LA, p. 193–198.

Braun, C. 1971. Limits on the accuracy of infrared radiation measurement of sea-surface temperature from a satellite. NESS Techn. Memo. 30, NOAA, Washington, D.C.

Bressette, W. E. 1974. An optical filtering system for remote sensing of phytoplankton and suspended sediment. Proc. 1974 Earth Env. Res. Conf., Philadelphia, PA.

Bressette, W. E. 1977. Effect of detector threshold, location of the sun and flight altitude upon spectral variations in remote sensing over water. Remote Sensing of Earth Res., Univ. Tennessee Space Inst., Tullahoma, 6:67–88.

Brewer, R. K. and H. R. Cravat. 1972. Baseline establishment for positioning Federal-State boundaries. Proc. 4th Ann. Offshore Techn. Conf.

Brewer, R. K. and A. K. Heywood. 1972. Coastal boundary mapping. Proc. 38th Ann. Meeting, Amer. Soc. of Photogramm., Falls Church, VA.

Bristor, C. L. 1975. Central processing and analysis of geostationary satellite data. Tech. Memo. 64, Nat. Env. Satell. Serv., Nat. Oceanic and Atmos. Admin., Washington, D.C.

Bristow, M., D. Nielsen, D. Bundy and R. Furtek. 1981. Use of water Raman emission to correct airborne laser fluorosensor data for effects of water optical attenuation. Appl. Opt. 20(17):2889–2906.

Brooks, D. A. 1976. Long waves trapped by the Cape Fear continental shelf topography: A model study of their propagation characteristics and circulation patterns. Rep. 76-2. Center Mar. and Coastal Stud., N.C. State Univ., Raleigh.

Browell, E. V. 1977. Analysis of laser fluorosensor systems for remote algae detection and quantification. NASA Tech. Note D-8447, Washington, D.C.

Brown, P. R. 1953. Wave data for the eastern North Atlantic. Mar. Obs. 23(160):94.

Brown, C. A., Jr., F. H. Farmer, O. Jarrett, Jr. and W. L. Staton. 1978. Laboratory studies of *in vivo* fluorescence of phytoplankton. Proc. 4th Joint Conf. Sensing of Env. Pollutants, Amer. Chem. Soc., Washington, D.C., p. 782–788.

Brown, C. L., F. C. Polcyn, A. N. Sellman and S. R. Steward. 1971. Water-depth measurement by wave refraction and multispectral techniques. Willow Run Lab. Rept. No. 31650-31-T, Univ. Michigan, Ann Arbor.

Brown, W. W. 1977. Wetlands mapping in New Jersey and New York. Proc. Amer. Soc. Photogramm. 43rd Ann. Mtg., Falls Church, VA, p. 381–395.

Brucks, J. T. and T. D. Leming. 1977. Seasat-A wind stress measurements as an aid to fisheries assessment and management. Proc. AIAA Joint Conf. Satellite Applications to Marine Technology, Amer. Inst. Aeron. Astron., New Orleans, LA.

Brucks, J. T., W. L. Jones and T. D. Leming. 1980. Comparison of surface wind stress measurements: airborne radar scatterometer versus sonic anemometer. J. Geophys Res. 85(9):4967.

Buettner, K. J. K. and C. D. Kern. 1965. The determination of infrared emissivities of terrestrial surfaces. J. Geophys. Res. 70(6):1329–1337.

Buettner, K. J. K., C. D. Kern and J. F. Cronin. 1964. The consequences of terrestrial surface infrared emissivity. Proc. 3rd Symp. Remote Sensing of Env., Ann Arbor, MI, p. 549–561.

Bullis, H. R., Jr. 1967. A program to develop aerial photo technology for assessment of surface fish schools. Proc. 20th Ann. Session, Gulf and Caribbean Fisheries Inst., p. 40–43.

Bullis, H. R., Jr. 1970. World fishery technology. California Mar. Res. Comm. CALCOFI Rep. No. 14, p. 57–66.

Bullis, H. R., Jr. and J. S. Carpenter. 1968. Latent fishery resources of the Central West Atlantic region. The future of the fishing industry in the United States, Univ. Washington Pub. in Fisheries, Seattle, New Series 16:61–64.

Bumpus, D. and L. M. Lauzier. 1965. Surface circulation on the Continental Shelf off eastern North America between Newfoundland and Florida. Serial Atlas of the Marine Env. Folio (7). Amer. Geog. Soc., NY.

Burgess, F. J. and W. P. James. 1971. Airphoto analysis of ocean outfall dispersion. Env. Prot. Agency 16070ENS06/71, Washington, D.C., 287 p.

Butera, M. K. 1976. A technique development for the determination of Louisiana marsh salinity zones from vegetation mapped by multispectral scanner data: a comparison of results from satellite and aircraft data. Earth Res. Lab. Rept. No. 161, NASA Johnson Space Center, Houston, TX, 54 p.

Butera, M. K. 1979. Computer-implemented remote sensing techniques for measuring coastal productivity and nutrient transport systems. *In* Satellite Hydrology, Amer. Water. Res. Assoc., Minneapolis, MN., p. 522–532.

Cameron, H. L. 1952. The measurement of water current velocities by parallax methods. Photogramm. Eng. 18:99–104.

Campbell, J. W. and J. P. Thomas (eds.). 1981. Chesapeake Bay Plume Study: Superflux 1980. NASA Conf. Publ. 2188, 515 p.

Campbell, W. J., R. O. Ramseier, W. F. Weeks, and P. Gloersen. 1975. An integrated approach to the remote sensing of floating ice. Proc. 3rd Can. Symp. on Remote Sensing, Edmonton, Alberta, p. 39–72.

Campbell, W. J., J. Wayenberg, J. B. Ranseyer, R. O. Ramseier, M. R. Vant, R. Weaver, A. Redmond, L. Arsenault, P. Gloersen, H. J. Zwally, T. T. Wilheit, A. T. C. Chang, D. Hall, L. Gray, D. C. Meeks, M. L. Bryan, F. T. Barath, C. Elacki, F. Leberl and T. Farr. 1978. Microwave remote sensing of sea ice in the AIDJEX Main Experiment. Boundary-Layer Meteorology 13:309–337.

Campbell, W. J., R. O. Ramseier, H. J. Zwally and P. Gloersen. 1980. Arctic sea-ice variations from time-lapse passive microwave imagery. Boundary-Layer Meteorology 18:99–106.

Caron, R. H. and K. W. Simon. 1975. Attitude time-series estimator for rectification of spaceborne imagery. J. Spacecraft and Rockets 12(1):27–32.

Carter, V. 1976. Application of remotely-sensed data to wetland studies. Proc. 19th COSPAR Mtg., VI-4, 9 p.

Carter, V. 1977. Coastal wetlands: the present and future role of remote sensing. Proc. 11th Intl. Symp. Remote Sensing of Env., Ann Arbor, MI, p. 301–323.

Carter, V. and R. R. Anderson. 1972. Interpretation of wetlands imagery based on spectral reflectance characteristics of selected plant species. Proc. 38th Ann. Mtg., Amer. Soc. Photogramm., Falls Church, VA, p. 580–587.

Carter, V. and J. Schubert. 1974. Coastal wetlands analysis from ERTS MSS digital data and field spectral measurements. Proc. 9th Intl. Symp. Remote Sensing of Env., Ann Arbor, MI, p. 1241–1260.

Carter, V., J. W. McGinness and R. R. Anderson. 1973. Mapping northern Atlantic Coastal marshlands, Maryland-Virginia, using ERTS-1 imagery. Remote Sensing of Earth Res., Univ. Tennessee Space Inst., Tullahoma, 11:1101–1120.

Chen, H. S. 1978. A storm surge model. Vol. II. A finite element storm surge analysis and its application to a bay-ocean system. Spec. Rept. 189, Virginia Inst. Marine Sci., Gloucester Point, 149 p.

Chen, H. S. 1978. A mathematical model for water quality analysis. Proc. Conf. Verification of Mathematical and Physical Models in Hydraulic Engineering. Amer. Soc. Civil Eng., NY, p. 350–357.

Cheney, R. E. 1981. A search for cold water rings. *In* Spaceborne Synthetic Aperture Radar for Oceanography, Johns Hopkins Univ. Press, Baltimore, MD, p. 161–170.

Clark, D. K., E. T. Baker and A. E. Strong. 1980. Upwelled spectral radiance distribution in relation to particulate matter in sea water. Boundary-Layer Meteorology 18(3):287–298.

Clarke, G. L. and L. R. Breslau. 1959. Measurements of bioluminescence off Monaco and Northern Corsica. Bull. de L'Institut Oceanographique, Monaco, 56(1147):31.

Clarke, G. L. and L. R. Breslau. 1960. Studies of luminescent flashing in Phosphorescent Bay, Puerto Rico and in the Gulf of Naples using a portable bathyphotometer. Bull. de L'Institute Oceanographique, Monaco, 57(1171):32.

Clarke, G. L. and M. G. Kelly. 1964. Variations in transparency and in bioluminescence on longitudinal transects in the Western Indian Ocean. Bull. de L'Institute Oceanographique, Monaco, 64(1319):20.

Clarke, G. L. and G. K. Wertheim. 1956. Measurements of illumination of great depths and at night in the Atlantic Ocean by means of a new bathyphotometer. Deepsea Res. 3(3):189–205.

Clarke, G. L., G. C. Ewing and C. J. Lorenzen. 1969. Remote measurement of ocean color as an index of biological productivity. Proc. 6th Int. Symp. Remote Sensing Env., Ann Arbor, MI, p. 991–1002.

Clos-Arceduc. 1964. Aerial photography and the study of coastal deposits. Transl. from La photographie aerienne et l'etude des depots prelittoraux, Etud. Photo-Interpretation 1:1–53.

Cogan, J. L. and J. H. Willand. 1974. Mapping of sea

surface temperature by the NOAA-2 satellite. Environ. Res. Techn. Inc., Monterey, CA, 72 p.

Coleman, J. M. 1969. Brahmaputra River: channel processes and sedimentation. Sed. Geol. 3(2/3):131–239.

Coleman, J. M. and W. G. McIntire. 1971. Transiting coastal river channels. Int. Hydrog. Rev. 48(1): 13–43.

Coleman, J. M., L. D. Wright and J. N. Suhayda. 1972. Density gradients at river mouths. Naval Research Rev. 25(10):9–20.

Collins, J. 1978. Cost benefits of photobathymetry. Proc. 44th Ann. Mtg., Amer. Soc. Photogramm., Falls Church, Va., p. 166–172.

Colvocoresses, A. P. 1974. Space oblique mercator. Photogram. Eng. 40(8):921–926.

Costin, J. M. 1965. Dye tracer studies on the Bahama Banks. Symp. Diffusion in Oceans and Fresh Waters, Lamont Geol. Obs., Columbia Univ., NY, p. 68–70.

Costin, J. M., P. Davis, R. Gerard and B. Katz. 1963. Dye diffusion experiments in the New York Bight. Tech. Rep. No. CU-2-63, Lamont Geol. Obs., Columbia Univ., NY.

Cowardin, L. M., V. Carter, F. C. Golet and E. T. LaRoe. 1976. Interim classification of wetland and aquatic habitats of the United States. U.S. Fish and Wildlife Serv., Washington, D.C., 109 p.

Cox, C. and W. Munk. 1954a. Measurement of the roughness of the sea surface from photographs of sun's glitter. J. Optical Soc. of America 44:838–850. 44:838–850.

Cox, C. and W. Munk. 1954b. Statistics of the sea surface derived from sun glitter. J. Marine Res. 13:198–227.

Cox, C. and W. Munk. 1955. Some problems in optical oceanography. J. Mar. Res. 14:63–78.

Current, I. B. 1969. A blue-insensitive Anscochrome aerial film. Proc. 35th Ann. Mtg. Amer. Soc. Photogramm., Falls Church, VA, p. 43–54.

Cushing, D. H., F. Devold, J. Marrs and F. Kristjonsson. 1952. Some modern methods of fish detection. FAO Fish. Bull. 5(95):119.

DeBlanc, D. P. 1979. Evaluation of radio tracking for headstart marine turtles. Special Report, NOAA/NMFS/Nat. Fish. Eng. Lab., NSTL Sta., MS.

Deutsch, M. and J. E. Estes. 1980. Landsat detection of oil from natural seeps. Photogramm. Eng. Remote Sensing 46(10):1313–1322.

Deutsch, M., R. R. Vollmers and J. P. Deutsch. 1980. Landsat tracking of oil slicks from the 1979 Gulf of Mexico oil well blowout. Proc. 14th Intl. Symp. Remote Sensing of Env., Ann Arbor, MI, p. 1197–1211.

Dietz, R. S. 1947. Aerial photographs in the geological study of shore features and processes. Photogramm. Eng. 13(4):537–545.

Dismachek, D. C. (ed.). 1977. National Environmental Satellite Service Catalog of Products. NOAA Techn. Memo NESS 88, Washington, D.C., 102 p.

Doak, E., D. Lyzenga and F. Polcyn. 1979. Remote bathymetry by Landsat in the Chagos Archipelago, ERIM Report 135900-1-T, Ann Arbor, MI.

Doerffer, R. 1979. The effect of patchiness within single picture elements of an ocean color scanner. Proc. Workshop EURASEP Ocean Color Scanner Expts. 1977, Joint Res. Centre, Ispra, Italy, p. 83–88.

Dolan, R., B. Hayden and J. Heywood. 1978. A new photogrammetric method for determining shoreline erosion. Coastal Eng. ASCE 2:21–39.

Dolan, R., B. P. Hayden, P. May and S. May. 1980. The reliability of shoreline change measurements from aerial photographs. Shore and Beach 48(4):22–29.

Drennan, K. L. 1970. Some potential applications of remote sensing in fisheries. Proc. Symp. Remote Sensing in Marine Biology and Fishery Resources, Remote Sensing Ctr., Texas A&M Univ., Austin, p. 25–65.

Duntley, S. Q. 1942. Optical properties of diffusing materials. J. Opt. Soc. Am. 32:61–70.

Duntley, S. Q. 1963. Light in the sea. J. Opt. Soc. Am. 53:214–233.

Duntley, S. Q. and C. F. Edgerton. 1966. The use of meteorological satellite photographs for the measurement of sea state. Final Report No. II-2, Contract No. NObs, 86012, U.S. Navy Bureau of Ships Project FAMOS, LaJolla, CA, p. 129.

Duntley, S. Q., R. W. Austin, W. H. Wilson, C. F. Edgerton and S. E. Moran. 1974. Ocean color analysis. Ref. 74-10, Scripps Inst. Oceanogr., San Diego, CA, 67 p.

Edwards, R. W. and M. W. Brown. 1960. Aerial photographic method for studying distribution of aquatic macrophytes in shallow waters. Ecol. J. 48:161–163.

Egan, W. G. and M. E. Hair. 1973. Automated delineation of wetlands in photographic remote sensing. Proc. 7th Intl. Symp. Remote Sensing of Env., Ann Arbor, MI, p. 2231–2251.

El-Ashry, M. T. (ed.). 1977. Air photography and coastal problems. Dowden, Hutchinson & Ross, Stroudsburg, PA, 425 p.

El-Ashry, M. T. and H. R. Wanless. 1967. Shoreline features and their changes. Photogramm. Eng. 33(2):184–189.

El-Ashry, M. T. and H. R. Wanless. 1968. Photointerpretation of shoreline changes between Capes Hatteras and Fear (North Carolina). Mar. Geol. 6:347–379.

Estes, J. E. and L. W. Senger. 1972. The multispectral concept as applied to marine oil spills. Remote Sensing of Env. 2:141–163.

Evans, W. E. 1971. Orientation behavior of Delphinids: radio telemetric studies. An. of N.Y. Academy of Science 188:142–160.

Evans, W. E. 1974. Radio telemetric studies of two species of small Cetaceans. *In* The Whale Problem, Harvard University Press, Cambridge, MA.

Ewing, G. C. 1973. A preliminary assessment of ERTS imagery for marine resources. Symp. Significant Results from ERTS-1, Vol. II Summary of Results, NASA Goddard Space Flight Center, Greenbelt, MD, p. 146–150.

Fang, C. S. and G. S. Parker. 1976. Thermal effects of the Surry nuclear power plant on the James River, Virginia. Part VI. Results of monitoring physical parameters. Special Report 109, Virginia Inst. Marine Science, Gloucester Point.

Fantasia, J. F. and H. C. Ingrao. 1974. Development of an experimental airborne laser remote sensing system for the detection and classification of oil spills. Proc. 9th Intl. Symp. Remote Sensing of Env., Ann Arbor, MI, p. 1711–1745.

Farmer, F. H., C. A. Brown, Jr., O. Jarrett, Jr., J. W. Campbell and W. L. Staton. 1980. Remote sensing of phytoplankton density and diversity using an airborne fluorosensor. *In* Advanced Concepts in Ocean Measurements for Marine Biology, F. P. Diemer, F. J. Vernberg, and D. Z. Mirkes (eds.), Univ. of South Carolina Press, Columbia.

Fay, J. A. 1969. The spread of oil slicks on a calm sea.

In Oil on the Sea, D. Hoult (ed.), Plenum Press, NY, p. 53–64.

Fea, M. 1980. Some research uses of Meteosat and early results. ESA Meteosat Data Management Dept., Darmstadt, West Germany, ESA Bull. 21:32–37.

Feist, D. L. and P. D. Boehm. 1980. Subsurface distributions of petroleum from an offshore well blowout—the IXTOC I Blowout, Bay of Campeche. IXTOC I Symposium.

Finley, R. J., S. McCulloh and P. Harwood. 1979. Landsat classification of coastal wetlands in Texas. *In* Satellite Hydrology, Amer. Water Res. Assoc., Minneapolis, MN, p. 453–462.

Fitzgerald, I. Y. 1972. Remote sensing in a circulatory survey of Boston Harbor. Seminar Operational Remote Sensing, Amer. Soc. of Photogramm., Falls Church, VA.

Flemming, B. W. 1976. Side-scan sonar: a practical guide. Intl. Hydrogr. Rev. 53(1).

Fontanel, A., J. Guillemot and M. Guy. 1973. First ERTS-1 results in southeastern France: geology, sedimentology, pollution at sea. Symp. Significant Results from ERTS-1, NASA Goddard Space Flight Center, Greenbelt, MD, p. 1483–1511.

Fournier, R. O., J. Marra, R. Bohrer and M. VanDet. 1977. Plankton dynamics and nutrient enrichment of the Scotian Shelf. J. Fish. Res. Bd. Can. 34(7):1004–1018.

Foxworthy, J. E. 1965. Multi-dimensional aspects of eddy diffusion determined by dye diffusion experiments in coastal waters. Symp. Diffusion in Oceans and Fresh Waters, Lamont Geol. Obs. of Columbia Univ., NY, p. 71–73.

Frisch, A. S. and B. L. Weber. 1980. A new technique for measuring tidal currents by using a two-site HF Doppler radar system. J. Geophys. Res. 85(3): 485–493.

Fung, A. K. and H. L. Chan. 1969. Backscattering of waves by composite rough surfaces. IEEE Trans. on Ant. and Prop., AP-17(5):590–597.

Fusco, L., R. Lunnon, B. Mason and C. Tomassini. 1980. Operational production of sea-surface temperatures from Meteosat image data. ESA, Meteosat Data Management Dept., Darmstadt, West Germany. ESA Bull. 21:38–43.

Gandy, W. F., T. M. Vanselous, and J. G. Jennings. 1977. Tracking marine animals by satellite. Proc. ARGOS Mtg., Paris, France.

Gausman, H. W., R. R. Rodriques and A. J. Richardson. 1976. Reflectance of cotton leaves and their structure. Agron. J. 68:295–296.

Geyer, R. A. and W. E. Sweet. 1974. Naturally occurring hydrocarbons in the Gulf of Mexico and Caribbean Sea. Ocean 74 Proc., IEEE Publishing Co., New York, 1:289–300.

Ghovanlou, A. H., J. N. Gupta, R. G. Henderson and L. Poole, 1977. Radiative transfer model for remote sensing of suspended sediments in water. Proc. 4th Joint Conf. Sensing of Env. Pollutants, Amer. Chem. Soc, Washington, D.C., p. 789–794.

Gierloff-Emden, H. G. 1977. Orbital remote sensing of coastal and offshore environments. Walter de Gruyler, Berlin, 176 p.

Gloersen, P., W. Nordberg, T. J. Schmugge and T. T. Wilheit. 1973. Microwave signatures of first-year and multi-year ice. J. Geophys. Res. 78:3564–3572.

Goldsmith, V., W. D. Morris, R. J. Byrne and C. H. Whitlock. 1974. Wave climate model of the Mid-Atlantic shelf and shoreline (Virginian Sea). NASA SP-358, Washington, D.C., 146 p.

Goldstein, H. 1946. The frequency dependence of radar echoes from the surface of the sea. Phys. Rev. 69:695.

Gonzales, F. I., R. C. Beal, W. E. Brown, J. F. R. Gower, D. Lichy, D. B. Ross, C. L. Rufenach and R. A. Shuchman. 1979. Seasat synthetic aperture radar: Ocean wave detection capabilities. Science 204:1418–1421.

Gordon, H. H. and J. C. Munday, Jr. 1977. Lagrangian drifter design for the determination of surface currents by remote sensing. Remote Sensing of Earth Resources, Univ. Tennessee Space Inst., Tullahoma, 6:89–107.

Gordon, H. H., M. E. Penney and R. J. Byrne. 1973. Remote sensing applications in marine science programs at VIMS. Virginia Inst. Marine Sci., Gloucester Point, 79 p.

Gordon, H. R. 1973. Simple calculation of the diffuse reflectance of the ocean. Appl. Opt. 12:2804–2805.

Gordon, H. R. and D. K. Clark. 1980. Atmospheric effects in the remote sensing of phytoplankton pigments. Boundary-Layer Meteorology 18:299–313.

Gordon, H. R. and W. R. McCluney. 1975. Estimation of the depth of sunlight penetration in the sea for remote sensing. Appl. Opt. 14(2):413–416.

Gordon, H. R., O. B. Brown and M. M. Jacobs. 1975. Computed relationships between the inherent and apparent optical properties of a flat homogeneous ocean. Appl. Opt. 14:417–427.

Gordon, H. R., J. L. Mueller and W. A. Hovis. 1980. Phytoplankton pigments from the Nimbus-7 Coastal Zone Color Scanner: comparison with surface measurements. Science 210:63–66.

Gower, J. F. R. 1972. A survey of the uses of remote sensing from aircraft and satellites in oceanography and hydrography. Pacific Marine Science Rept. 72-3, Dept. of Env., Ottawa, Canada.

Gower, J. F. R. (ed.). 1980. Passive radiometry of the ocean. Proc. 6th I.U.C.R.M. Coll., Reidel, Dordrecht, Holland, 355 p.

Graham, D. S., J. M. Hill and B. A. Christensen. 1978. Verification of an estuarine model for Apalachicola Bay, Florida. Proc. Conf. Verification of Mathematical and Physical Models in Hydraulic Engineering, Amer. Soc. Civil Eng., NY, p. 237–245.

Grant, C. R. and B. S. Yaplee. 1957. Backscattering from water and land at centimeter and millimeter wavelengths. Proc. IRE, 45:976–982.

Gray, A. L., J. Cihlar, S. Parashar and R. Worsfold. 1977. Scatterometer results from shorefast and floating sea ice. Proc. 11th Intl. Symp. Remote Sensing Environ., Ann Arbor, MI, p. 645–657.

Green, T., R. Madding and F. Scarpace. 1977. Types of thermal plumes in coastal waters. Water Res. 11:123–125.

Greenberg, D. A. 1977. Effects of tidal power development on the physical oceanography of the Bay of Fundy and Gulf of Maine. *In* Fundy Tidal Power and the Environment, Daborn (ed.), Acadia Inst., Canada, p. 200–232.

Grew, G. W. 1977. Signature extraction of ocean pollutants by eigenvector transformation of remote spectra. Proc. 4th Joint Conf. Sensing of Env. Pollutants, Amer. Chem. Soc., Washington, D.C., p. 659–666.

Guertin, F. E. and E. Shaw. 1981. Definition and potential of geocoded satellite imagery products. Proc. 7th Canadian Symp. Remote Sensing, Winnepeg, Manitoba, 11 p.

Guinard, N. W. 1971. The remote sensing of oil slicks.

Proc. 7th Intl. Symp. Remote Sensing of Env., Ann Arbor, MI, p. 1005–1026.

Guinard, N. W. and J. C. Daley. 1970. An experimental study of a sea clutter model. Proc. IEEE 58:543–560.

Guinard, N. W., J. T. Ransone and J. C. Daley, 1971. Variations of the NRCS of the sea with increasing roughness. J. Geophys. Res. 76(6):1525–1538.

Gunnerson, C. G. 1965. Limitations of Rhodamine-B and Pontacyl Brilliant Pink-B as tracers in estuarine waters. Symp. Diffusion in Oceans and Fresh Waters, Lamont Geol. Obs. of Columbia Univ., NY, p. 53–68.

Gupta, J. N. and A. H. Ghovanlou. 1978. Radiative transfer in turbid water. Soc. Photo-Opt. Instr. Eng., Redondo Beach, CA, 160(Ocean Optics V):132–147.

Guth, J. E. 1972. Coastal boundary mapping. Proc. Symp. Coastal Mapping, Amer. Soc. of Photogramm., Falls Church, VA.

Hammack, J. C. 1977. Landsat goes to sea. Photogramm. Eng. Remote Sensing 43:683.

Hansen, D. V. 1974. A Lagrangian Buoy experiment in the Sargasso Sea. *In* Proc. Free Drifting Buoys, AIAA Drift Buoy Symp. NASA Langley Res. Center CP-2003:175–192 (Available NTIS).

Harger, R. O. 1981. SAR ocean imaging mechanisms. *In* Spaceborne Synthetic Aperture Radar for Oceanography, Johns Hopkins Univ. Press, Baltimore, MD, p. 41–52.

Hawkins, R. E., C. E. Livingstone, A. L. Gray, K. Okamoto, L. D. Arsenault, D. Pearson and T. L. Wilkinson. 1980. Single and multiple parameter microwave signatures of sea ice. Proc. 6th Canadian Symp. Remote Sensing, Halifax, Can. Aeron. Space Inst., Ottawa, p. 217–229.

Hayes, R. M. 1980. Operational use of remote sensing during the Campeche Bay oil well blowout. Proc. 14th Intl. Symp. Remote Sensing of Env., Ann Arbor, MI, p. 1187–1191.

Hayes, R. M. 1981. Detection of the Gulf Stream. *In* Spaceborne Synthetic Aperture Radar for Oceanography, Johns Hopkins Univ. Press, Baltimore, MD, p. 146–160.

Hellden, U. and I. Akersten. 1977. Landsat digital data for water pollution and water quality studies in southern Scandinavia. Proc. 11th Int. Symp. Remote Sensing of Env., Ann Arbor, MI, p. 875–884.

Hengeveld, H. G. 1978. Operational applications of SLAR over Canadian ice covered waters. W.M.O. Workshop, Remote Sensing of Sea Ice, Washington, D.C.

Hengeveld, H. G. 1980. Utilization and benefits of SLAR in operational ice data acquisition. Proc. 6th Canadian Symp. Remote Sensing, Halifax, Canad. Aero. Space Inst., Ottawa, Ontario, p. 81–88.

Hennigar, H. F. 1979. Historical evolution of coastal sand dunes on Currituck Spit, Virginia-North Carolina. M.A. Thesis, Virginia Inst. Mar. Sci., Gloucester Point, 121 p.

Hennigar, H. F. 1980. Quantification of changes in coastal topography using simple parallax measurements. Photogramm. Eng. Remote Sensing 46(1): 71–76.

Hickman, G. D., A. H. Ghovanlou, E. J. Friedman, C. S. Gault and J. E. Hogg. 1975. A feasibility study for a remote laser water turbidity meter. NASA Langley Res. Center CR-132376. SPARCOM Inc., Alexandria, VA, 35 p.

Hidy, G. M., W. F. Hall, W. N. Hardy, W. W. Ho, A. C. Jones, A. W. Love, M. J. Van Melle, H. H. Wang and A. E. Wheeler, 1972. Development of a satellite microwave radiometer to sense the surface temperature of the world oceans. NASA CR-1960.

Hill, J. M. and D. S. Graham. 1979. Using enhanced Landsat images for calibrating real time estuarine water quality models. *In* Satellite Hydrology, Amer. Water Res. Assoc., Minneapolis, MN, p. 603–614.

Hodder, D. T. 1971. Improved techniques for bathymetric plotting in turbid coastal waters. Second Coastal and Shallow Water Res. Conf., Office of Naval Res., Washington, D.C.

Hodder, D. T. 1972. A monograph on application of ulticolor dye tracers and multispectral photography to coastal water mapping. Techn. Rept. ONR Contract N00014-72-C-0149, SD 72-SA-0071, North American Rockwell.

Hodder, D. T. 1973. Evaluation of water penetration for bottom sediment mapping in Little Harbor, Catalina Island. 1973 ASP/ACSM Fall Conv. Symp. Remote Sensing in Oceanography, Amer. Soc. Photogramm, Falls Church, VA.

Hodder, D. T., P. B. Chandler and G. A. McCue. 1971. Coastal bathymeric plotting. Final Rept. ONR Contract N00014-71-C-0370, SD 71-763, North American Rockwell.

Hofer, R., E. G. Nkoju and J. Waters. 1981. Microwave radiometric measurements of sea surface temperature from the Seasat satellite. First Results. Science 212:1385–1387.

Hoge, F. E. and R. N. Swift. 1980. Absolute tracer dye concentration using airborne laser-induced water Raman backscatter. Appl. Opt. 20(7):1191–1202.

Hoge, F. E. and R. N. Swift. 1980. Oil film thickness measurement using airborne laser-induced water Raman backscatter. Appl. Opt. 19(19):3269–3281.

Hoge, F. E., R. N. Swift and E. B. Frederick. 1980. Water depth measurement using an airborne pulsed neon laser system. Appl. Opt. 19:871.

Højerslev, N. K. 1979. On the origin of yellow substance in the marine environment. Proc. Workshop EURASEP Ocean Color Scanner Expts. 1977, Joint Res. Centre, Ispra, Italy, p. 13–28.

Højerslev, N. K. 1980. Water color and its relation to primary productivity. Boundary-Layer Meteorology 18:203–220.

Hollinger, J. P. 1974. The determination of oil slick thickness by means of multifrequency passive microwave technique. Naval Res. Lab. CG-D-31-75, Washington, D.C.

Hollinger, J. P. and R. A. Mennella. 1973. Oil spills: measurements of their distribution and volumes by multifrequency microwave radiometry. Science 181:54–56.

Holyer, R. J. 1978. Toward universal multispectral suspended sediment algorithms. Remote Sensing of Env. 7:323–338.

Horstmann, U., K. A. Ulbricht and D. Schmidt. 1978. Detection of eutrophication processes from air and space. Proc. 12th Intl. Symp. Remote Sensing of Env., Ann Arbor, MI, p. 1379–1386.

Hosier, P. E. and W. J. Cleary. 1977. Cyclic geomorphic patterns of washover on a barrier island in southeastern North Carolina. Env. Geol. 2:23–31.

Hoult, D. P. 1972. Oil spreading on the sea. Ann. Rev. Fluid Mechanics, v. 4.

Hovis, W. A. and K. C. Leung. 1977. Remote sensing of ocean color. Optical Eng. 16(2):158–166.

Hovis, W. A., D. K. Clark, F. Anderson, R. W. Austin,

W. H. Wilson, E. T. Baker, D. Ball, H. R. Gordon, J. L. Mueller, S. Z. El-Sayed, B. Sturm, R. C. Wrigley and C. S. Yentsch. 1980. Nimbus-7 Coastal Zone Color Scanner: System description and initial imagery. Science 210:60–63.

Huebner, G. L. 1971. Planning and implementation of remote sensing experiments. Final Report, RSC-756-1, Remote Sensing Center, Texas A&M University, Austin.

Huebner, G. L. and A. H. Bouma. 1973. Environmental impact assessment of shell dredging in San Antonio Bay, Texas. Final Report, Proj. 870, Texas A&M Res. Found., Austin.

Hull, W. V., I. Y. Fitzgerald, R. K. Brewer and C. I. Thurlow. 1973. Remote sensing for demarcating and mapping obscured tidal boundaries. NOAA, National Ocean Survey, Rockville, MD.

Ichiye, T. 1962. Studies of turbulent diffusion of dye patches in the ocean. J. Geophys. Res. 67:3212–3216.

Ichiye, T. 1963. Some comments and prospectives on dye diffusion experiments. Tech. Note (AEC and ONR), Lamont Geol. Obs., Columbia Univ., NY, 11 p.

Ichiye, T. (ed.). 1964. Symposium on diffusion in oceans and fresh waters. Lamont Geol. Obs., Columbia Univ., NY, 159 p.

Ichiye, T., H. Iida and N. B. Plutchak. 1964. Analysis of diffusion of dye patches in the ocean. Tech. Repts. CU-8-64 (ONR) and CU-9-64 (AEC), Lamont Geol. Obs., Columbia Univ., NY, 16 p.

Jain, S. C. and S. A. Gluckstein. 1981. Water quality remote sensing. Contract Rept., Moniteq Ltd. MTR 81–20, Concord, Ontario, Canada, 97 p.

Jain, S. C. and J. R. Miller, 1977. Algebraic expression for the diffuse irradiance reflectivity of water from the two-flow model. Appl. Opt. 16:202–204.

Jarrett, O., Jr., C. A. Brown, J. W. Campbell, W. M. Houghton and L. R. Poole. 1979. Measurement of chlorophyll *a* fluorescence with an airborne fluorosensor. Proc. 13th Symp. Remote Sensing of Env., Ann Arbor, MI, p. 703–710.

Jennings, J. C., W. F. Gandy and T. M. Vanselous. 1979. Development of a satellite linked marine mammal transmitter system. Proc. ARGOS Meeting, France.

Jensen, J. R., J. E. Estes and L. Tinny. 1980. Remote sensing techniques for kelp surveys. Photogramm. Eng. Remote Sensing 46:743–755.

Jerlov, N. G. 1968. Optical Oceanography. Elsevier, Amsterdam, Holland.

Jerlov, N. G. and E. S. Nielsen (eds.). 1974. Optical aspects of oceanography. Academic Press, NY, 494 p.

Johnson, J. D. and L. D. Farmer. 1971. Use of side-looking air-borne radar for sea ice identification. J. Geophys. Res. 76(9):2138–2155.

Johnson, R. W. 1975. Quantitative suspended sediment mapping using aircraft remotely sensed multispectral data. Proc. NASA Earth Res. Survey Symp., NASA Johnson Space Center, Houston, TX, p. 2087–2098.

Johnson, R. W. 1978. Mapping of chlorophyll *a* distributions in coastal zones. Photogramm. Eng. Remote Sensing 44(5):617–624.

Johnson, R. W. 1980. Remote sensing and spectral analysis of plumes from ocean dumping in the New York bight apex. Remote Sensing of Env. 9:197–209.

Johnson, R. W. and R. C. Harriss. 1980. Remote sensing for water quality and biological measurements in coastal waters. Photogramm. Eng. Remote Sensing 46(1):77–85.

Johnson, R. W. and C. W. Ohlhorst. 1981. Applications of remote sensing to monitoring and studying dispersion in ocean dumping. *In* Ocean Dumping of Industrial Wastes, B. H. Ketchum, D. R. Kester and P. K. Park (eds.), Plenum Press, NY, p. 175–191.

Johnson, R. W., G. S. Bahn and J. P. Thomas. 1981. Synoptic thermal and oceanographic parameter distributions in the New York Bight Apex. Photogramm. Eng. Remote Sensing 47(11):1593–1598.

Johnson, R. W., C. W. Ohlhorst and J. W. Usry. 1977. Location, identification, and mapping of sewage sludge and acid waste plumes in the Atlantic coastal zones. Proc. 4th Joint Conf. Sensing of Env. Pollutants, Amer. Chem. Soc., Washington, D.C., p. 644–647.

Jones, A. C. and P. N. Sund. 1967. An aircraft and vessel survey of surface tuna schools in the Lesser Antilles. Commercial Fish. Rev. 29(3):41–45.

Jones, B. G. 1957. Low-water photography in Cobscook Bay, Maine. Photogramm. Eng. 23(2):338–342.

Jones, B. G. and W. Shofnos. 1961. Mapping the low-water line of the Mississippi Delta. Internat. Hydrog. Rev.

Jones, D. A. 1972. Role of Federal government in coastal mapping. Proc. Symp. Coastal Mapping, Amer. Soc. Photogramm., Falls Church, VA.

Jones, W. L., L. C. Schroeder and J. L. Mitchell. 1977. Aircraft measurements of the microwave scattering signature of the ocean. IEEE Trans. Antennas and Propagation, AP-25(1):52–61.

Jones, W. L., P. G. Black, D. M. Boggs, E. M. Bracalente, R. A. Brown, G. Dome, J. A. Ernst, I. M. Halberstam, J. E. Overland, S. Peteherych, W. J. Pierson, F. J. Wantz, P. M. Woiceshyn and M. G. Wurtele. 1979. Seasat scatterometer: results of the Gulf of Alaska workshop. Science 204: 1413–1415.

Katz, B., R. Gerard and M. Costin. 1965. Response of dye tracers to sea surface conditions. Tech. Rep. CU-22-65 (AEC), Lamont Geol. Obs., Columbia Univ., NY.

Katzin, M. 1957. On the mechanism of sea clutter. Proc. IRE, 45:44–54.

Kaufmann, J. H. and A. B. Irvine. 1975. Movement and activities of Atlantic bottle nosed dolphins, *Tursiops truncatus*. Cont. Rept, U.S. Marine Mammal Commission Contract No. MM4AC-004.

Keller, M. 1963. Tidal current surveys by photogrammetric methods. Tech. Bull. No. 22, U.S. Coast and Geodetic Survey, Washington, D.C., 20 p.

Keller, M. and G. C. Tewinkel. 1969. Block analytic aerotriangulation. ESSA Tech. Rept. 35, National Ocean Survey, Rockville, MD.

Kelly, M. G. and A. Conrod. 1969. Aerial photographic studies of shallow water benthic ecology. *In* Remote Sensing of Ecology, Univ. Georgia Press, Athens, p. 173–184.

Kemmerer, A. J. 1979a. Behavior patterns of Gulf menhaden (*Brevortia patronus*) inferred from fishing and remotely sensed data. *In* Fish Behavior and Fisheries Management: Capture and Culture (in press).

Kemmerer, A. J. 1979b. Remote sensing of living marine resources. Proc. 13th Int. Remote Sensing Environment, Ann Arbor, MI, 2:729–738.

Kemmerer, A. J. and J. A. Benigno. 1973. Relationships between remotely sensed fisheries distribution information and selected oceanographic parameters in the Mississippi Sound. Symp. Signif. Results from ERTS-1, NASA Goddard Space Flight Center, Greenbelt, MD, IB:1685–1695.

Kemmerer, A. J., J. A. Benigno, G. B. Reese and F. C. Minkler. 1973. A summary of selected early results from the ERTS-1 menhaden experiment. Contr. No. 246, NMFS Southeast Fisheries Center, Pascagoula, MS, p. 31.

Kemmerer, A. J., J. A. Benigno, G. B. Reese and F. C. Minkler. 1974. Summary of selected early results from the ERTS-1 menhaden experiment. Fishery Bull. 72(2):375–389.

Kendall, B. M. 1980. AIAA Paper No. 80-1953. AIAA Conf. Sensor Systems for the 80's. Colorado Springs, CO.

Kendall, B. M. and J. O. Blanton. 1981. Microwave radiometer measurement of tidally induced salinity changes off the Georgia coast. J. Geophys. Res. 86(C7):6435–6441.

Kerut, E. G. and G. Haas. 1979. Geostationary and orbiting satellites applied to remote ocean buoy data acquisition. Proc. 13th Intl. Symp. Remote Sensing of Env., Ann Arbor, MI, p. 519–533.

Ketchum, B. H. 1951. The exchange of fresh and salt water in tidal estuaries. J. Mar. Res. 10.

Kiefer, D. A., R. J. Olson and W. H. Wilson. 1979. Reflectance spectroscopy of marine phytoplankton. Part 1. Optical properties as related to age and growth rate. Limnol. Oceanogr. 24(4):664–672.

Kim, H. H. and P. T. Ryan (eds.). 1974. The use of lasers for hydrographic studies. Proc. Symp., NASA Wallops Station, Wallops Island, VA, 279 p.

Kim, H. H., C. R. McClain, L. R. Blaine, W. D. Hart, L. P. Atkinson and J. A. Yodes. 1979. Ocean color studies from a U-2 aircraft platform. NASA TM 80574, Washington, D.C.

Kim, H. H., P. O. Cervenka and C. B. Lankford. 1975. Development of an airborne laser bathymeter. NASA TN D-8079, Washington, D.C., 39 p.

Kirwan, A. D., Jr., G. McNally and J. Coehla. 1976. Gulf Stream kinematics inferred from a satellite-tracked drifter. J. Phys. Oceanogr. 6:750–755.

Kisiel, C. C. 1965. Dye studies in the Ohio River. Symp. Diffusion in Oceans and Fresh Waters, Lamont Geol. Obs., Columbia Univ., NY, p. 28–52.

Klemas, V. and W. D. Philpot. 1981. Remote sensing of ocean-dumped waste drift and dispersion. *In* Ocean Dumping of Industrial Wastes, Plenum Press, NY, p. 193–211.

Klemas, V. and D. F. Polis. 1977. Remote sensing of estuarine fronts and their effects on pollutants. Photogramm. Eng. Remote Sensing 43(5):599–612.

Klemas, V., D. S. Bartlett, W. Philpot, R. H. Rogers and L. Reed. 1974. Coastal and estuarine studies with ERTS-1 and Skylab. Remote Sensing of Env. 3(3):153–174.

Klemas, V., M. Otley, M. Philpott, C. Wethe and R. Rogers. 1974. Correlation of coastal water turbidity and circulation with ERTS-1 and Skylab imagery. Proc. 9th Intl. Symp. Remote Sensing of Env., Ann Arbor, MI, p. 1289–1318.

Klemas, V., D. S. Bartlett and R. H. Rogers. 1975. Coastal zone classification from satellite imagery. Photogramm. Eng. Remote Sensing 51(4):499–514.

Kolipinski, M. C. and A. L. Higer. 1970. Detection and identification of benthic communities and shoreline features in Biscayne Bay using multiband imagery. Third Ann. Earth Res. Prog. Rev., NASA Johnson Space Center, Houston, TX, 3:41-1–41-16.

Kolipinski, M. C., A. L. Higer, N. S. Thomson and R. F. J. Thomson. 1969. Inventory of hydrobiological features using automatically processed multispectral data. Proc. 6th Intl. Symp. Remote Sensing of Env., Ann Arbor, MI, p. 79–96.

Koopmans, B. N. 1971. Interpretacion de fotografias aereas en mofrologia costera. Centre Interamericano de Fotointerpretacion, Ministerio de Obras Publicas, Bogota, Columbia.

Kort, V. G. 1972. New data on the dynamic structure of the western boundary currents in the tropical Atlantic. Rd. D-V Ren. Cons. Perm. Int. Explor. Mer. 162:276–279.

Kozlyaninov, M. V. 1972. The basic relationships between the hydro-optical parameters. *In* K. S. Shifrin (ed.), Optics of the Ocean and the Atmosphere, Nauka, p. 5–24.

Kozlyaninov, M. V. 1980. Calculation of the visibility of a standard white disk. Transl. from Russian. Oceanology 20(2):215–217.

Krishen, K. 1971. Correlation of radar backscattering cross sections with ocean wave height and wind velocity. J. Geophys. Res. 76(27):6528–6539.

Kuo, C. Y. and T. A. Talay. 1979. Remote monitoring of a thermal plume. NASA TM 80125, Washington, D.C.

Laevastu, T. and I. Hela. 1970. Fisheries oceanography. Fishing News (books), LTD, London.

Landgrebe, D. 1976. Computer-based remote sensing technology: A look to the future. Remote Sensing of Env. 5:229–246.

Langfelder, L. J., D. B. Stafford and M. Amein. 1970. Coastal erosion in North Carolina. Proc. Amer. Soc. Civil. Eng., J. Waterways and Harbors Div. 96 (WW2)-7306:531–545.

Latimer, P. 1958. Apparent shifts of absorption bands of cell suspensions and selective light scattering. Science 127(3288):29–30.

Laurs, R. M. 1972. The needs of fishing fleet operators in terms of marine ecology, fish detection, communications, meteorology navigations aids. 5th U.S.-European Conf. "Eurospace", San Francisco, CA, p. 20.

Leatherwood, S., W. E. Evans and D. W. Rice. 1972. The whales, dolphins, and porpoises of the Eastern North Pacific: A guide to their identification in the water. NUC TP 282, Naval Undersea Res. and Devel. Center, San Diego, CA, p. 175.

Leatherwood, S., J. R. Gilbert and D. G. Chapman. 1978. An evaluation of some techniques for aerial censuses of bottle nosed dolphins. J. Wildlife Mgmt. 42:239–250.

Legeckis, R. 1977. Long waves in the Eastern Equatorial Pacific Ocean: A view from a geostationary satellite. Science 197(4309):1179–1181.

Legeckis, R. 1978. A survey of worldwide sea surface temperature fronts detected by environmental satellites. J. Geophys. Res. 83(C9):4501–4522.

Legeckis, R. 1979. Satellite observations of the influence of bottom topography on the seaward deflection of the Gulf Stream off Charleston, South Carolina. J. Phys. Oceanogr. 9(3):483–497.

Legeckis, R. and J. Pritchard. 1976. Algorithm for correcting the VHRR imagery for geometric distortions due to the earth's curvature and rotation and roll attitude errors. Tech. Memo. 77, Nat. Env. Satell. Serv., Nat. Oceanic and Atmos. Admin., Washington, D.C.

Legeckis, R., E. Legg and R. Limeburner. 1980. Comparison of polar and geostationary satellite infrared observations of sea surface temperatures in the Gulf of Maine. Remote Sensing of Env. 9:339–350.

Leitao, C. D., N. E. Huang and C. G. Parra. 1979. A note on the comparison of radar altimetry with infrared and *in situ* data for the detection of the Gulf Stream surface boundaries. J. Geophys. Res. 84(B8):3969–3973.

Leitao, C. D., N. E. Huang and C. G. Parra. 1980. Sea surface topography from Seasat and Geos-3. Mar. Techn. 451–457.

Leuder, D. R. 1954. A method for estimating beach trafficability from aerial photographs. School of Civil Eng. Techn. Rept. 6(5), Cornell Univ., Ithaca, NY.

Levenson, C. 1968. Factors affecting biological observations from the ASWEPS aircraft. Inf. Rept. No. 68-102, U.S. Navy Oceanogr. Off., Washington, D.C., p. 6.

Lewis, B. W. 1977. Relation of laboratory and remotely sensed spectral signatures of ocean-dumped acid waste. Proc. 4th Joint Conf. Sensing of Env. Pollutants, Amer. Chem. Soc., Washington, D.C., p. 654–658.

Lichy, D. E., M. G. Mattie and L. J. Mancini. 1981. Tracking of a warm water ring. *In* Spaceborne Synthetic Aperture Radar for Oceanography, Johns Hopkins Univ. Press, Baltimore, MD, p. 171–182.

Lipes, R. G., R. L. Bernstein, V. J. Cardone, K. B. Katsaros, E. G. Njoku, A. L. Riley, D. B. Ross, C. T. Swift and F. J. Wentz. 1979. Seasat scanning multichannel microwave radiometer: Results of the Gulf of Alaska Workshop. Science 204:1415–1417.

Loshchilov, V. S., A. D. Masanov and I. G. Serebrennikov. 1978. The use of SLAR for the mapping of sea ice and the study of its dynamics. W.M.O. Workshop Remote Sensing of Sea Ice, Washington, D.C.

Loya, B. R. and C. D. Graves. 1968. Fish identification by remote sensing. Report 11435-6001-RO-00, TRW Systems Group, Redondo Beach, CA.

Ludwig, G. H. 1974. The future polar orbiting environmental satellite system. Proc. IEEE Electron. and Aerosp. Syst. Conf., Inst. Electr. and Electr. Eng., NY, p. 498–502.

Lyzenga, D. and F. Polcyn. 1978. Analysis of optimum spectral resolution and band location for satellite bathymetry. Rept. No. 128200-1-F, Env. Res. Inst. Michigan, Ann Arbor, MI.

Lyzenga, D. and F. Polcyn. 1979. Techniques for the extraction of water depth information from Landsat digital data. Rept. 129900-1-F, Env. Res. Inst. Michigan, Ann Arbor, MI.

Lyzenga, D., C. Wezernak and F. Polcyn. 1976. Spectral band positioning for purposes of bathymetry and mapping bottom features from satellite altitudes. Rep. 115300-5-T, Env. Res. Inst. Michigan, Ann Arbor, MI.

MacConnell, W. P. 1974. Remote sensing land-use and vegetative cover in Rhode Island. Coop. Ext. Serv. Bull. No. 200, Univ. Rhode Island, Kingston, 93 p.

MacDonald, F. D. 1956. The correlation of radar sea clutter on vertical and horizontal polarization with wave height and slope. Conv. Rec. IRE, 4:29–32.

Macomber, R. T. and G. H. Fenwick. 1979. Aerial photography and seaplane reconnaissance to produce the first total distribution inventory of submerged aquatic vegetation in Chesapeake Bay, MD. Proc. 45th Ann. Mtg., Amer. Soc. Photogramm., Falls Church, VA, p. 498–503.

Mairs, R. L. and D. K. Clark. 1973. Remote sensing of estuarine circulation dynamics. Photogramm. Eng. 39(9):927–938.

Manning, A. P., Jr., J. R. White and R. R. Vollmers. 1980. Current status of remote sensing for oil pollution control in U.S. coastal waters. Proc. 14th Intl. Symp. Remote Sensing of Env., Ann Arbor, MI, p. 249–268.

Mascarenhas, A. S., Jr. 1979. Characteristic of upper heated oceanic layer from satellite observations. M.S. Thesis, Mass. Inst. Techn., Cambridge, MA.

Mather, R. S. 1979. The analysis of Geos 3 altimeter data in the Tasman and Coral seas. J. Geophy. Res. 84(B8):3853–3860.

Maughan, P. M. 1969. Remote sensor applications in fishery research. Marine Techn. Soc. J. 3(2):11–20.

Maul, G. A. and D. V. Hansen. 1972. An observation of the Gulf Stream surface front structure by ship, aircraft, and satellite. Remote Sensing of Env. 2:109–116.

Maul, G. A. and M. Sidran. 1973. Atmospheric effects on ocean surface temperature sensing from the NOAA satellite scanning radiometer. J. Geophys. Res. 78:1909–1916.

Maul, G. A., P. W. deWitt and S. R. Baig. 1978. Geostationary satellite observations of Gulf Stream meanders: infrared measurements and time series analysis. J. Geophys. Res. 83(C12):6123–6135.

McClain, E. P. 1980. Passive radiometry of the ocean from space—An overview. Boundary-Layer Meteorol. 18:7–24.

McClain, E. P. and A. E. Strong. 1969. On anomalous dark patches in satellite-viewed sunglint areas. Monthly Weather Rev. 97:875–884.

McCurdy, P. G. 1947. Manual of Coastal Delineation from Aerial Photographs. Pub. No. 592, U.S. Navy Hydrogr. Off., Washington, D.C., 140 p.

McEwen, R. B. and J. W. Schoonmaker. 1975. ERTS color image maps. Photogramm. Eng. Remote Sensing 41(4):479–487.

McEwen, R. B., W. J. Kosco and V. Carter. 1976. Coastal wetland mapping. Photogramm. Eng. Remote Sensing 42(2):221–232.

McLeish, W., D. Ross, R. A. Shuchman, P. G. Teleki, S. V. Hsiao, O. H. Shemdin and W. E. Brown, Jr. 1980. Synthetic aperture radar imaging of ocean waves: comparison with wave measurements. J. Geophys. Res. 85(C9):5003–5011.

McNally, G. J. 1981. Satellite-tracked drift buoy observations of the near-surface flow in the eastern mid-latitude North Pacific. J. Geophys. Res. 86(C9):8022–8030.

McNeil, W. R., K. P. B. Thomson and J. Jerome. 1976. The application of remote spectral measurements to water quality monitoring. Can. J. Remote Sensing 2(1):48–58.

Meeks, D. C., G. A. Poe and R. O. Ramseier. 1974. A study of microwave emission properties of sea ice AIDJEX 1972. Final Rept. 1786FR-1, Aerojet Electrosystems Co., Azusa, CA, 140 p.

Meijer, W. and O. J. Groenveld. 1964. Formula for conversion of stereoscopically observed apparent depth of water to true depth. Photogramm. Eng. 30:1037.

Mie, G. 1980. Beitrage zur Optik truber Medien, speziell kolloidalen Metall-losungen. Ann. Physik. 25:377.

Miller, J. R., K. S. Gordon and D. Kamykowski. 1976. Air-borne water-colour measurements off the Nova Scotia coast. Can. J. Remote Sensing 2(1):42–47.

Moffitt, E. H. 1969. History of shore growth from aerial photographs. Shore and Beach 37(1):23–27.

Molinari, R. L., S. Baig, D. W. Behringer, G. A. Maul and R. Legeckis. 1977. Winter intrusions of the Loop Current. Science 198:505–507.

Mollo-Christensen, E. and A. S. Mascarenhas, Jr. 1979. Heat storage in the oceanic upper mixed layer inferred from Landsat data. Science 203:653–654.

Mollo-Christensen, E., P. Cornillon and A. S. Mascarenhas, Jr. 1980. Method for estimation of ocean current velocity from satellite images. Science 212(4495):661–662.

Monahan, E. C. and E. A. Monahan. 1973. Drogues, drags and sea anchors. Sea Grant Tech. Rep. No. 36, Univ. Michigan, Ann Arbor, 21 p.

Moore, J. G. 1947. The determination of the depths and extinction coefficients of shallow water by air photography using colour filters. Royal Soc. (London), Philos. Trans. 240(816).

Moore, R. K. 1966. Radar scatterometry—an active remote sensing tool. Proc. 4th Int. Symp. Remote Sensing of Env., Ann Arbor, MI.

Moore, R. K. and W. J. Pierson, Jr. 1971. Worldwide oceanic wind and wave predictions using a satellite radar-radiometer. J. Hydronautics. 5:52–60.

Morel, A. 1973. Diffusion de la lumière par les eaux de mer. Résultats expérimentaux et approched théorique. *In* Optics of the Sea. AGARD Lect. Ser. 61, p. 31.1–31.76.

Morel, A. 1974. Optical properties of pure water and pure sea water. *In* N. G. Jerlov and E. S. Nielsen (eds.), Optical Aspects of Oceanography, Academic Press, New York, p. 1–24.

Morel, A. 1980. In-water and remote measurements of ocean color. Boundary-Layer Meterology 18:177–201.

Morel, A. Y. and H. R. Gordon. 1980. Report of the working group on ocean color. Boundary-Layer Meteorology 18:343–355.

Morel, A. and L. Prieur. 1977. Analysis of variations in ocean color. Limnol. Oceanogr. 22:709–722.

Moskowitz, L. 1964. Estimates of the power spectra for fully developed seas for wind speeds of 20 to 40 knots. J. Geophys. Res. 69(24):5161–5174.

Mueller, J. L. 1973. The influence of phytoplankton on ocean color spectra. Ph.D. Thesis, Oregon State Univ., Corvallis.

Mullane, T. F. 1980. Operational use of satellite imagery in the Canadian ice program. Proc. 6th Canadian Symp. Remote Sensing, Halifax, Can. Aeron. Space Inst., Ottawa, p. 17–32.

Munday, J. C., Jr. and T. T. Alföldi. 1975. Chromaticity changes from isoluminous techniques used to enhance multispectral remote sensing data. Remote Sensing of Env. 4:221–236.

Munday, J. C., Jr. and T. T. Alföldi. 1979. Landsat test of diffuse reflectance models for aquatic suspended solids measurement. Remote Sensing of Env. 8:169–183.

Munday, J. C., Jr. and M. S. Fedosh. 1981. Chesapeake Bay plume dynamics from Landsat. *In* Chesapeake Bay Plume Study: Superflux 1980, NASA Conf. Publ. 2188:79–92.

Munday, J. C., Jr. and H. H. Gordon. 1977. Progress toward a Circulation Atlas for application to coastal water siting problems. *In* Application of Remote Sensing to the Chesapeake Bay Region, NASA Conf. Publ. 6:345–358.

Munday, J. C., Jr. and P. L. Zubkoff. 1981. Remote sensing of dinoflagellates blooms in a turbid estuary. Photogramm. Eng. Remote Sensing 47(4): 523–531.

Munday, J. C., Jr., W. G. MacIntyre and M. E. Penney. 1971. Oil slick studies using photographic and multispectral scanner data. Proc. 7th Intl. Symp. Remote Sensing of Env., Ann Arbor, MI, p. 1027–1043.

Munday, J. C., Jr., R. J. Byrne, C. S. Welch, H. H. Gordon, J. D. Boon III, D. K. Stauble, D. Baker and E. P. Ruzecki. 1975. Applications of remote sensing to estuarine problems. Virginia Inst. Mar. Sci., Gloucester Point, 168 p.

Munday, J. C., Jr., C. S. Welch and H. H. Gordon. 1978. Outfall siting with dye-buoy remote sensing of coastal circulation. Photogramm. Eng. Remote Sensing 44(1):87–96.

Munday, J. C., Jr., T. T. Alföldi and C. L. Amos. 1979. Bay of Fundy verification of a system for multidate Landsat measurement of suspended sediment. *In* Satellite Hydrology, Amer. Water Res. Assoc., Minneapolis, MN, p. 622–640.

Munday, J. C., Jr., C. S. Welch and H. H. Gordon. 1980. Estuarine circulation from dye-buoy photogrammetry. Ports '80, Amer. Soc. Civil Eng., NY, p. 417–428.

Murphree, D. L., C. C. Taylor and R. W. McClendon. 1974. Mathematical modeling for the detection of fish by an airborne laser. AIAA J. 12(12):1686–1692.

Murphree, D. L., R. W. McClendon, B. W. Ward and F. Glaum. 1976. Mathematical modeling of fish detection with a scanning airborne laser. Miss. State Univ.

National Ocean Survey. 1971. Photographic and thermal remote sensing survey of Boston Harbor surface currents. NOAA, Washington, D.C., 16 p.

National Fisheries Engineering Laboratory. 1975. A study of remote sensing techniques for detection and enumeration of giant bluefin tuna. Spec. Report, NOAA, NMFS, NSTL Sta., MS.

National Fisheries Engineering Laboratory. 1976. Test report: porpoise film test. Spec. Report, NOAA, NMFS, NSTL Sta., MS.

National Fisheries Engineering Laboratory. 1978. Synthetic aperture radar data analysis of fishing vessels. Spec. Report. NOAA, NMFS, NSTL Sta., MS.

National Oceanic and Atmospheric Administration. 1980. Satellite data users bulletin 2(1). Science Inf. Center, Rockville, MD.

Nelson, W. R., M. C. Ingham and W. E. Schaaf. 1977. Larval transport and year-class strength of Atlantic menhaden: *Brevoortia tyrannus*. Fishery Bull. 75(1):23–41.

Neville, R. A. and J. F. R. Gower. 1977. Passive remote sensing of phytoplankton via chlorophyll *a* fluorescence. J. Geophys. Res. 82:3487–3493.

Newman, F. C., J. R. Proni, J. J. Tsai, C. A. Lauter, R. L. Sellers, D. J. Walter and W. P. Damman. 1978. Preliminary acoustic studies of dumped dredge material. Env. Effects Lab., U.S. Army Corp of Eng., Waterways Exp. Sta., MS.

Newman, F. C., J. R. Proni and D. J. Walter. 1977. Acoustic imaging of the New England shelf-slope water mass interfaces. Nature 269(5631):790–791.

Newton, R. W. and J. W. Rouse, Jr. 1972. Experimental measurements of 2.25 cm backscatter from sea surfaces. IEEE Trans. Geosc. Electron. GE-10(1).

Odum, E. P. 1971. Fundamentals of Ecology. Third Edition. W. B. Saunders Co., Philadelphia, PA.

Ogrosky, C. E. 1978. Aerial photographic methods for the detection of submerged vegetation in coastal New Jersey. Proc. Amer. Soc. of Photogramm. Fall Mtg., Falls Church, VA, p. 423–430.

Okubo, A. 1965. A theoretical model of diffusion of dye patches. Symp. Diffusion in Oceans and Fresh Waters, Lamont Geol. Obs., Columbia Univ., NY, p. 74–85.

Okubo, A. 1971. Oceanic diffusion diagrams. Deep Sea Res. 18:789–802.

Onstott, R. G., R. K. Moore and W. F. Weeks. 1979. Surface-based scatterometer results of arctic sea ice. IEEE Trans. GE-17, p. 78–85.

Orlanski, I. and M. D. Cox. 1973. Baroclinic instability in ocean currents. Geophys. Fluid Dyn. 4:297–332.

Orr, D. G. 1968. Multiband-color photography. *In* Manual of Color Aerial Photography, Amer. Soc. of Photogramm., Falls Church, VA, p. 441–450.

Orth, R. J. and H. H. Gordon. 1975. Remote sensing of submerged aquatic vegetation in the lower Chesapeake Bay, Virginia. Final Rept. NASA-10720. Virginia Inst. Mar. Sci., Gloucester Point, 62 p.

Orth, R. J., K. A. Moore and H. H. Gordon. 1979. Distribution and abundance of submerged aquatic vegetation in the lower Chesapeake Bay, Virginia. Virginia Inst. Mar. Sci., Gloucester Point, VA. Report No. 600/8-79-029/SAVI. Office of Res. and Dev., EPA, Washington, D.C.

Öström, B. 1976. Fertilization of the Baltic by nitrogen fixation in the blue-green algae *Nodularia spumigena*. Remote Sensing of Env. 4:305–310.

Parashar, S. K., R. M. Haralick, R. K. Moore and A. W. Biggs. 1977. Radar scatterometer discrimination of sea ice types. IEEE Trans. GE-15(2): 83–87.

Peake, W. H. 1959. Interaction of electromagnetic waves with some natural surfaces. IRE Trans. on Antennas and Propagation (Special Supplement), AP-7:5324–5329.

Pearcy, W. G. 1973. Albacore oceanography off Oregon—1970. Fishery Bull. 71(2):489–504.

Penney, M. E. and H. H. Gordon. 1975. Remote sensing of wetlands in Virginia. Proc. 10th Int. Symp. Remote Sensing of Env., Ann Arbor, MI, p. 495–504.

Pfeiffer, W. J., R. A. Linthurst and J. L. Gallagher. 1973. Photographic imagery and spectral properties of salt marsh vegetation as indicators of canopy characteristics. Proc. Symp. Remote Sensing in Oceanography, Amer. Soc. Photogramm., Falls Church, VA, p. 1004–1016.

Phillips, O. M. 1966. The dynamics of the upper ocean. Cambridge Univ. Press.

Pickering, R. J. 1976. Measurement of "turbidity" and related characteristics of natural waters. U.S. Geol. Survey Open File Report 76-153, Reston, VA, 7 p.

Pickett, R. L. 1966. Environmental corrections for an airborne radiation thermometer. Proc. 4th Symp. Remote Sensing of Env., Ann Arbor, MI, p. 259–262.

Polcyn, F. C. 1976. NASA/Cousteau ocean bathymetry experiment. ERIM Report No. 118500-1-F, Env. Res. Inst. Michigan, Ann Arbor, MI.

Polcyn, F. C. and D. R. Lyzenga. 1979. Landsat bathymetric mapping by multitemporal processing. Proc. 13th Int. Symp. on Remote Sensing of Env., Ann Arbor, MI, p. 1269–1276.

Polcyn, F. C. and S. R. Stewart. 1970. Results of remote sensing survey of thermal discharges. Four State Enforcement Workshop., Env. Res. Inst. Michigan, Ann Arbor.

Polcyn, F. C. and C. T. Wezernak. 1970. Pollution surveillance and data acquisition using multispectral remote sensing. Water Res. Bull., J. Amer. Water Res. Assoc. 6(6):920–934.

Polcyn, F. C., W. L. Brown and I. J. Sattinger. 1970. The measurement of water depth by remote sensing techniques. Report No. 8973-26-F, Infrared and Optics Lab., Willow Run Lab., Univ. Michigan. Ann Arbor.

Popham, R. W. 1977. Workhorse in space. NOAA Mag 7(3):34–37.

Prieur, L. 1976. Transfert radiatif dans les eaux de mer. Application a la determination de parametres optiques caracterisant leur teneur en substances dissoutes et leur contenu en particules. D. Sci. thesis, Univ. Pierre et Marie Curie, 243 p.

Prior, D. B. and J. M. Coleman. 1980. Submarine landslides on the Mississippi River Delta-Front slope. Geoscience and Man XIX:41–53.

Proni, J. R. and H. B. Stewart, Jr. 1978. Acoustic techniques for ocean pollution studies and their easy transferability to developing nations. Interscience 3(1):24–27.

Proni, J. R., F. C. Newman, R. L. Sellers and C. Parker. 1976a. Acoustic tracking of ocean dumped sewage sludge. Science 193:1005–1007.

Proni, J. R., F. C. Newman, D. C. Rona, D. E. Drake, G. A. Berberian, C. A. Lauter, Jr. and R. L. Sellers. 1976b. On the use of acoustics for studying suspended oceanic sediment and for determining the onset of the shallow thermocline. Deep-Sea Res. 23:831–837.

Proni, J. R., F. C. Newman, E. R. Meyer, H. B. Stewart, Jr., D. J. Walter, R. Sellers and C. A. Lauter, Jr. 1977. On the use of acoustics in applied oceanographic and coastal engineering problems with emphasis on the ocenaic transport of particulate material. Thalassia Jugoslavica 13(3/4): 289–393.

Ramey, E. H. 1968. Measurement of ocean currents by photogrammetric methods. Tech. Memor. 5, U.S. Coast and Geodetic Survey, Washington, D.C., 18 p.

Raney, R. K. 1980. SAR processing of partially coherent and sinusoidally dynamic ocean waves. Proc. 6th Canadian Symp. Remote Sensing, Halifax, Canada Aero. Space Inst., Ottawa, Ontario, p. 243–248.

Rao, P. K., W. L. Smith and R. Koffler. 1972. Global sea-surface temperature distribution determined from an environmental satellite. Monthly Weather Rev. 100:10–14.

Rao, P. K., A. E. Strong and R. Koffler. 1971. Gulf Stream and Middle Atlantic Bight: complex thermal structure as seen from an environmental satellite. Science 173:529–530.

Rayleigh, L. 1903. Scientific papers. Cambridge Univ. Press, New York.

Reeves, C. A. 1973. Comparison of two unsupervised classification algorithms for delineation of coastal features of Galveston Island, Texas. NASA Johnson Space Center LEC-0460, Houston, 6 p.

Reimold, R. J., J. L. Gallagher and D. E. Thompson. 1972. Coastal mapping with remote sensors. Proc. Coastal Mapping Symp., Washington, D.C., p. 99–112.

Reimold, R. J., J. L. Gallagher and D. E. Thompson.

1973. Remote sensing of tidal marsh. Photogramm. Eng. 39(5):477–488.

Reinert, R. L. 1965. Near-surface oceanic diffusion from a continuous point source. Symp. Diffusion in Oceans and Fresh Waters, Lamont Geol. Obs., Columbia Univ., NY, p. 19–27.

Rice, S. O. 1951. Reflection of electromagnetic waves from slightly rough surfaces: Com. Pure Appl. Math. 4:351–378.

Rifman, S. S. and D. M. McKinnon. 1974. Evaluation of digital correction techniques for ERTS images. TRW Systems Group Final Rept. 20534-6003-TU-02, NASA Goddard SFC, Greenbelt, MD, p. 1-1 to 4-14.

Rinner, K. 1969. Two media photogrammetry. Photogramm. Eng. 35:275.

Rogers, R. H., N. J. Shah, J. B. McKeon and V. E. Smith. 1976. Computer mapping of water quality in Saginaw Bay with Landsat digital data. Proc. 42nd Ann. Mtg., Amer. Soc. Photogramm., Falls Church, VA, 13 p.

Rogers, R. H., N. J. Shah, J. B. McKeon, C. Wilson and L. Reed. 1975. Application of Landsat to the surveillance and control of eutrophication in Saginaw Bay. Proc. 10th Int. Symp. Remote Sensing of Env., Ann Arbor, MI, p. 437–446.

Roithmayr, C. M. 1971. Airborne low-light sensor detects luminescing fish schools at night. Comm. Fish. Rev. 32(12):42–51.

Roithmayr, C. M. and F. P. Wittman. 1972. Low light level sensor development for marine resource assessment. 8th Ann. Conf. and Exp., Marine Tech. Soc., Washington, D.C., p. 277–288.

Rosenberg, G. V. and Y. A. R. Mullamma. 1965. Some possibilities of determining wind speed over an ocean surface using observations from artificial earth satellites. Atmospheric and Oceanic Physics Series, Izvestiya of the Academy of Sciences, U.S.S.R. 1(3):282–290.

Ross, D. S. 1969. Experiments in oceanographic aerospace photography; I—Ben Franklin spectral filter tests. Philco-Ford Corp. Space and Re-entry Syst. Div. TR-DA2108 (U.S. Naval Oceanogr. Off.), Palo Alto, CA, 170 p.

Rouse, J. W., Jr. 1969. Arctic ice type identification by radar. Proc. IEEE, 57(4):605–611.

Rouse, J. W., Jr. and J. A. Schell. 1970. Radar studies of Arctic ice. Tech. Rept. TSC-20, Remote Sensing Center, Texas A&M Univ., College Station.

Rowan, L. C., P. H. Wetlaufer, A. F. H. Goetz, F. C. Billingsley and J. H. Stewart. 1974. Discrimination of rock types and detection of hydrothermally altered areas in south-central Nevada by the use of computer-enhanced ERTS images. Prof. Paper 883, U.S. Geol. Survey, Reston, VA, 35 p.

Ruzecki, E. P., C. Welch, J. Usry and J. Wallace. 1976. The use of the EOLE satellite system to observe continental shelf circulation. Offshore Techn. Conf., No. 2592:697–708.

Saunders, P. M. 1967. Aerial measurement of sea surface temperature in the infrared. J. Geophys. Res. 72(16):4109–4117.

Savastano, K. J. 1975. Application of remote sensing for fishery resources assessment and monitoring. Inv. No. 240, Contr. T-8217B, NASA Johnson Space Center, Houston, TX.

Scarpace, F. L. 1978. Densitometry on multi-emulsion imagery. Photogramm. Eng. Remote Sensing 44(10):1279–1292.

Schott, J. R. 1979. Temperature measurement of cooling water discharged from power plants. Photogramm. Eng. Remote Sensing 45:73–761.

Seliger, H. H., W. G. Fastie, W. R. Taylor and W. D. McElroy. 1962. Bioluminescence of marine dinoflagellates. I. An underwater photometer for day and night measurements. J. Gen. Physiol. 45(5):1003–1017.

Sette, O. E. 1949. Methods of biological research on pelagic fisheries resources. Indo-Pac Fish Counc., Proc. 1st Mtg., p. 132–138.

Sharkov and Kurditsk. 1956. Application of aerial method for geological investigation of sea bottom. Moscow Acad. of Sci. Aeromethod Lab., U.S.S.R.

Shaw, S. P. and C. G. Fredine. 1956. Wetlands of the United States. Circ. 39, U.S. Fish and Wildlife Serv., Washington, D.C., 67 p.

Shemdin, O. H., A. Jain, S. V. Hsiao and L. W. Gatto. 1980. Inlet current measured with Seasat-1 Synthetic Aperture Radar. Shore and Beach, October, p. 35–37.

Shepard, F. P. 1950. Photography related to investigation of shore processes. Photogramm. Eng. 16(5):756–769.

Shifrin, K. S. 1951. Scattering of light in a turbid medium. NASA Techn. Transl. F-477, Washington, D.C., 1968, 212 p.

Shuchman, R. A. and E. S. Kasischke. 1981. Refraction of coastal ocean waves. *In* Spaceborne Synthetic Aperture Radar for Oceanography, Johns Hopkins Univ. Press, Baltimore, MD, p. 128–135.

Shuchman, R. A., E. S. Kasischke, A. Klooster and P. L. Jackson. 1979. Seasat SAR coastal ocean wave analysis: a wave refraction and diffraction study. Final Rep. 1386002F, Env. Res. Inst. Michigan, Ann Arbor, MI.

Silberhorn, G. M. 1978. Virginia's wetland inventory: an essential tool for local coastal zone management. *In* Wetlands Functions and Values: The State of Our Understanding, Amer. Water Res. Assoc., Minneapolis, MN, p. 93–100.

Silberhorn, G. M., G. M. Dawes and T. A. Barnard, Jr. 1974. Coastal wetlands of Virginia interim report No. 3: guidelines for activities affecting Virginia wetlands. Virginia Inst. Mar. Sci. SRAMSOE 46, Gloucester Pt., 52 p.

Simonds, J. L. 1963. Applications of characteristic vector analysis to photographic and optical response data. J. Opt. Soc. Amer. 53:968.

Skolnik, M. I. 1969. A review of radar sea echo. NRL 2025, Naval Res. Lab., Washington, D.C., p. 12.

Smith, R. C. and K. S. Baker. 1978. Optical classification of natural waters. Limnol. Oceanogr. 23: 260–267.

Smith, W. L., P. K. Rao, R. Koffler and W. R. Curtis. 1970. The determination of sea-surface temperatures from satellite high resolution infrared window radiation measurements. Mon. Weather Rev. 98:604–611.

Snyder, J. P. 1978. The space oblique mercator projection. Photogramm. Eng. Remote Sensing 44(5): 585–596.

Sokal, R. R. and F. J. Rohlf. 1969. Biometry. W. H. Freeman, San Francisco, CA, 776 p.

Sonu, C. J. 1964. Study of shore processes with aid of aerial photography. Photogramm. Eng. 30(6): 932–941.

Sonu, C. J. 1969. Tethered balloon for study of coastal dynamics. Proc. Symp. on Earth Obs. from Balloons, Amer. Soc. of Photogramm., Fall Church, VA, p. 91–103.

Sonu, C. J. 1972. Field observation of nearshore circulation and meandering currents. J. Geophys. Res. 77(18):3232–3247.

Sorensen, B. M. 1979. The North Sea Ocean Color Scanner Experiment. 1977 Final Report. Joint Research Centre, Ispra, Italy.

Soules, S. D. 1970. Sun glitter viewed from space. Deep Sea Res. 17:191–195.

Squire, J. L. 1972. Apparent abundance of some pelagic marine fishes off the southern and central California coast as surveyed by an airborne monitoring program. Dept. of Commerce Fisheries Bull. 70(3): 1005–1019.

Squire, J. L. 1973. NOAA workshop on the application of aerospace remote sensing to fisheries problems. NOAA/NMFS, Mississippi Test Facility, Bay St. Louis, MS, p. 58.

Stevenson, M. R. 1975. A review of some uses of remote sensing in fishery oceanography and management. IEEE Ocean 1975: 467–471.

Stevenson, W. H. and E. J. Pastula. 1971. Observations on remote sensing in fisheries. Commercial Fish Rev. 33(9).

Stewart, S., R. Spellicy and F. Polcyn. 1970. Analysis of multispectral data of the Santa Barbara oil slick. Publ. 3340-4-F, Willow Run Lab., Univ. Michigan, Ann Arbor, 57 p.

Strong, A. E. and R. J. DeRycke. 1973. Ocean current monitoring employing a new satellite sensing technique. Science 182:482–484.

Stroud, L. M. and A. W. Cooper. 1968. Color-infrared aerial photographic interpretation and net primary productivity of a regularly-flooded North Carolina salt marsh. Dept. Botany, North Carolina St. Univ. at Raleigh, 86 p.

Stumpf, H. G. 1974. A satellite-derived experimental Gulf Stream analysis. Mariners Weather Log 18:149–152.

Stumpf, H. G.and P. K. Rao. 1975. Evolution of Gulf Stream eddies as seen in satellite infrared imagery. J. Phys. Oceanogr. 5(2):388–393.

Sturm, B. 1979. First results of CZCS data analysis on EURASEP test sites. Proc. Workshop EURASEP Ocean Color Scanner Expts. 1977, Joint Res. Centre, Ispra, Italy, p. 141–150.

Sturm, B. 1981. Atmospheric correction algorithm. Nimbus Experiment Team, 15th CZCS Mtg., Monterey, NOAA/NESS, Washington, D.C., App. D.

Suits, G. 1973. Preliminary results of water reflectance calculations using AQUACAN. Unpubl. memor., Env. Res. Inst. Mich., Ann Arbor.

Swift, C. T. 1980. Passive microwave remote sensing of the ocean—A review. Boundary-Layer Meteorol. 18:25–54.

Szekielda, K., D. J. Suszkowski and P. S. Tabor. 1975. Skylab investigation of the upwelling off the northwest coast of Africa. Proc. Earth Res. Survey Symp., NASA Johnson Space Center, Houston, TX, p. 2005–2022.

Tanner, W. F. (ed.). 1978. Standards for measuring shoreline changes. Dept. Geol., Florida St. Univ., Tallahassee, 87 p.

Tapley, B. D., G. H. Born, H. H. Hagar, J. Lorell, M. E. Parke, J. M. Diamante, B. C. Douglas, C. C. Good, R. Kolenkiewicz, J. G. Marsh, C. F. Martin, S. L. Smith III, W. F. Townsend, J. A. Whitehead, H. M. Byrne, L. S. Fedor, D. C. Hammond and N. M. Mogard. 1979. Seasat altimeter calibration: initial results. Science 204:1410–1412.

Tassan, S. 1979. The usefulness of model sensitivity analysis in remote sensing of marine water quality. Proc. Workshop EURASEP Ocean Color Scanner Expts. 1977, Joint Res. Centre, Ispra, Italy, p. 155–170.

Tassan, S., B. Sturm and E. Diana. 1979. A sensitivity analysis for the retrieval of chlorophyll contents in the sea from remotely sensed radiances. Proc. 13th Int. Symp. Remote Sensing of Env., Ann Arbor, p. 713–727.

Terhune, L. D. B. 1968. Free-floating current followers. Fish. Res. Bd. Canada Tech. Rep. 85, 30 p.

Tewinkel, G. C. 1963. Water depths from aerial photographs. Photogramm. Eng. 29:1037.

Tomczak, A. H. 1977. Environmental analyses in marine fisheries research. FAO Fisheries Techn. Paper No. 170, 141 p.

Troy, B. E., J. P. Hollinger, R. M. Lerner and M. M. Wisler. 1980. Measurement of the microwave properties of sea ice at 90 GHz and lower frequencies. J. Geophys. Res. 86(C5):4283–4289.

Tseng, Y. C., H. M. Inostroza and R. Kumar. 1977. Study of the Brazil and Falkland currents using THIR images of Nimbus V and oceanographic data in 1972 to 1973. Proc. 11th Int. Symp. Remote Sensing of Env., Ann Arbor, MI, p. 859–871.

Ulbricht, K. A., D. Schmidt and U. Horstmann. 1978. Mass appearance of blue green algae in the Baltic Sea: evaluation of multispectral Landsat scenes by image processing. In Proc. Int. Conf. Earth Observation from Space and Management of Planetary Resources, Toulouse, France, p. 77–79.

Umbach, M. J. and W. D. Harris. 1972. Photogrammetric bathymetry. NOAA, Natl. Ocean Survey, Rockville, MD.

U.S. Fish and Wildlife Service. 1976. Existing state and local wetland surveys. Vol. 2. Washington, D.C., 453 p.

Valenzuela, G. R. 1967. Depolarization of EM waves by slightly rough surfaces. IEEE Trans. Ant. Prop. 5(4):A0–15.

Van de Hulst, H. C. 1957. Light Scattering by Small Particles. Wiley, New York.

Vankoughnett, A. L., R. K. Raney and E. J. Langham. 1980. The Surveillance Satellite Program and the future of microwave remote sensing. Proc. 6th Canadian Symp. Remote Sensing, Halifax, Can. Aeron. Space Inst., Ottawa, p. 9–16.

Vanselous, T. M. 1977. Fishery engineering advancements—a 5-year SEFC progress report. Mar. Fish. Rev. 39(4):12–24.

Vary, W. E. 1969. Remote sensing by aerial color photography for water depth penetration and ocean bottom detail. Proc. 6th Int. Symp. Remote Sensing of Env., Ann Arbor, MI, p. 1045–1059.

Vasil'ev, K. P. 1968. Use of meteorological satellite data as a navigation aid. Problems of Satellite Meteorology, Trudy No. 36, Hydrometeorological Research Center of the USSR, 46-57 (Israel Program for Scientific Translations, Cat. No. 5669, 1970).

Vass, P. A. and J. L. vanGenderen. 1978. Monitoring environmental pollution by remote sensing. Proc. 12th Int. Symp. Remote Sensing of Env., Ann Arbor, MI, p. 219–234.

Viollier, M., D. Tanre and P. Y. Deschamps. 1980. An

algorithm for remote sensing of water color from space. Boundary-Layer Meteorology 18:247–268.

Voorhis, A. D. 1969. The horizontal extent and persistence of thermal fronts in the Sargasso Sea. Deep Sea Res. Suppl. 16:331–337.

Vukovich, F. M. and B. W. Crissman. 1981. Monitoring the Chesapeake Bay using satellite data for Superflux III. *In* Chesapeake Bay Plume Study: Superflux 1980, NASA Conf. Publ. 2188:93–110.

Walter, D. J. and J. R. Proni. 1980. Acoustic observations of subsurface scattering during a cruise at the IXTOC I blowout in the Bay of Campeche, Gulf of Mexico. IXTOC I symposium.

Washburn, J. F. and G. A. Sandness. 1977. Detection, identification, and quantification techniques for spills of hazardous chemicals. Proc. 11th Int. Symp. Remote Sensing of Env., Ann Arbor, MI, p. 1629–1635.

Watson, R. D., M. E. Henry, A. F. Theisen, T. J. Donovan and W. R. Hemphill. 1977. Marine monitoring of natural oil slicks and manmade wastes utilizing an airborne imaging Fraunhofer line discriminator. Proc. 4th Joint Conf. Sensing of Env. Pollutants, Amer. Chem. Soc., Washington, D.C., p. 667–671.

Weisblatt, E. A. 1972. Multispectral discrimination of deltaic environments. Dept. of Geography, Louisiana State Univ., Baton Rouge, 148 p.

Welsh, J. G. 1967. A new method of measuring coastal surface currents with markers and dye dropped from an aircraft. J. Mar. Res. 25(2):190–197.

Wezernak, C. T. and D. R. Lyzenga. 1975. Analysis of *Cladophora* distribution in Lake Ontario using remote sensing. Remote Sensing of Env. 4:37–48.

Wezernak, C. T. and N. Roller. 1973. Monitoring ocean dumping with ERTS-1 data. Symp. Significant Results from the ERTS-1, NASA, Washington, D.C., p. 635–642.

Wezernak, C. T., D. R. Lyzenga and F. C. Polcyn. 1974. *Cladophora* distribution on Lake Ontario. Env. Prot. Agency 660/3-74-028, Washington, D.C.

Whitlock, C. H. 1976. An estimate of the influence of sediment concentration and type on remote sensing penetration depth for various coastal waters. Tech. Memor. X-73906, NASA Langley Res. Center, Hampton, VA, 17 p.

Whitman, R. I. and K. Marcellus. 1973. Textural signatures for wetland vegetation. Proc. Fall Conv. Symp. on Remote Sensing in Oceanography, Amer. Soc. Photogramm., Falls Church, VA, p. 979–991.

Weidemann, H. 1974. The use of fluorescent dyes for turbulence studies in the sea. In N. G. Jerlov and E. S. Nielsen (eds.), Optical Aspects of Oceanography, Academic Press, NY, p. 257–288.

Wilheit, T., A. T. C. Chang and A. S. Milman. 1980. Atmospheric corrections of passive microwave observations of the ocean. Boundary-Layer Meteorol. 18:65–77.

Wilheit, T., W. Nordberg, J. Blinn, W. Campbell and A. Edgerton. 1972. Aircraft measurements of microwave emission from arctic sea ice. Remote Sensing of Env. 2:129–139.

Williams, J. 1970. Optical Properties of the Sea. U.S. Naval Inst. Ser. Oceanogr., Annapolis, MD, 123 p.

Wilson, M. J., P. E. O'Neill and J. E. Estes. 1979. Satellite detection of oil on the marine surface. *In* Satellite Hydrology, Amer. Water Res. Assoc., Minneapolis, MN, p. 593–602.

Wilson, W. H. and D. A. Kiefer. 1979. Reflectance spectroscopy of marine phytoplankton. Part 2. A simple model of ocean color. Limnol. Oceanogr. 24(4):673–682.

Wiltse, I. C., S. P. Schlesinger and C. M. Johnson. 1957. Back-scattering characteristics of the sea in the region from 10 to 50 KHz. Proc. IEEE, 45:220–228 and 244–246.

Witte, W. G., J. W. Usry, C. H. Whitlock and E. A. Gurganus. 1977. Laboratory upwelled spectral signature measurements of secondary treated sewage sludge for remote sensing of ocean dumping. Proc. 4th Joint Conf. Sensing of Env. Pollutants, Amer. Chem. Soc., Washington, D.C., p. 648–653.

Wood, H. and G. McGee. 1925. Aircraft experiments for the locating of herring shoals in Scottish waters. Fishery Board of Scotland, Scientific Investigators.

Woods, E. G. and J. H. Ivey. 1977. Fisheries imaging radar surveillance test (FIRST) Bering Sea. Proc. Amer. Inst. of Astron. and Aeron., New Orleans, LA. LA.

World Meteorological Organization. 1975. Information on Meteorological Satellite Programmes Operated by Members and Organizations WMO 411, Geneva, Switzerland, 52 p.

Worsfold, R. D., J. A. Allen and B. E. Fretts. 1975. The use of television for remote sensing. Remote Sensing of Env. 4:5–35.

Wright, J. W. 1968. A new model for sea clutter. IEEE Trans. on Antennas and Propagation, AP-16(2): 217–223.

Wright, L. D. 1970. Circulation, effluent diffusion and sediment transport, mouth of South Pass, Mississippi River Delta. Tech. Rept. 84, Coastal Studies Inst., Louisiana State Univ., Baton Rouge, 56 p.

Wright, L. D. and J. M. Coleman. 1971. Effluent expansion and interfacial mixing in the presence of a salt-wedge, Mississippi River Delta. J. Biophys. Res. 76(36):8649–8661.

Wright, L. D., J. M. Coleman and J. N. Suhayda. 1973. Periodicities in interfacial mixing. Tech. Rept. 133, Coastal Studies Inst., Louisiana State Univ., Baton Rouge, p. 127–135.

Wright, L. D., C. J. Sonu and W. V. Kielhorn. 1972. Water-mass stratification and bed form characteristics in East Pass, Dostin, Florida. Marine Geol. 12:43–58.

Wright, R. R. and M. R. Collings. 1964. Fluorescent tracing techniques and applications to hydrologic studies. Am. Water Works Assoc. J. 56(6).

Yentsch, C. S. 1960. The influence of phytoplankton pigments on the color of seawater. Deep-Sea Res. 7:1–9.

Yeske, L., F. Scarpace and T. Green. 1975. Measurement of lake currents. Photogramm. Eng. Remote Sensing 41(5):637–646.

Yost, E. and S. Wenderoth. 1970. Remote sensing of coastal waters using multispectral photographic techniques. Sci. Eng. Res. Group TR-10, Long Island Univ., Greenvale, NY, 210 p.

Yost, E. and S. Wenderoth. 1970. Remote sensing of coastal environments using multispectral photographic techniques. Proc. Symp. Hydrobiology, Miami Beach, FL, Amer. Water Res. Assoc., p. 274–302.

Zdanovich, V. G. 1964. Methods for studying ocean currents by aerial survey. Akademiya Nauk SSSR. Translated from Russian by the Israel Program for Scientific Translations (Jerusalem, 1967). TT66-51148. Available CFSTI. 212 p.

Zdanovich, V. G. and I. V. Semenchenko. 1954. Statistics of the sea surface derived from sun glitter. J. Mar. Res. 13:198–227.

Zdanovich, V. G. and I. V. Semenchenko. 1963. Applications of aeromethods for investigation of the ocean. U.S. Air Force Foreign Techn. Div. Transl. AD662–577.

Zeigler, J. M. and F. C. Ronne. 1957. Time lapse photography—an aid to studies of the shorelines Naval Res. Rev. No. 4, Off. Naval Res., Washington, D.C.

Zimmerman, A. V. 1976. Research and investigation of the radiation induced by a laser beam incident on seawater. NASA Langley Res. Center CR-145149. Chesapeake College, Wye Mills, MD. 93 p.

Zubkoff, P. L., J. C. Munday, Jr., R. G. Rhodes and J. E. Warinner. 1979. Mesoscale features of summer (1975–1977) dinoflagellate blooms in the York River, Virginia (Chesapeake Bay). *In* Toxic Dinoflagellate Blooms, Proc. 2nd Int. Conf., Elsevier/North-Holland, NY, p. 279–286.

CHAPTER 29

Water Resources Assessment

Author-Editor: VINCENT V. SALOMONSON

Contributing Authors: THOMAS J. JACKSON, JAMES R. LUCAS, GERALD K. MOORE, ALBERT RANGO, THOMAS SCHMUGGE, and DONNA SCHOLZ.

GENERAL CONTENTS: Subsurface water, groundwater, shallow sand and gravel aquifers, other aquifers, soil moisture, evapotranspiration, crop-water stress, antecedent precipitation index, snow and ice mapping and monitoring, snowmelt-runoff modeling, surface-water studies, surface-water area, turbidity, water quality, relative trophic status, land cover observations, flood plain delineation, hydrogeomorphic analysis, wetlands mapping, irrigated lands, urban hydrology, future trends, references.

INTRODUCTION

Water, a fundamental substance for sustaining life itself, is a key factor in maintaining agricultural production, energy production, and other activities at optimum levels. The assessment of the water resources of an area and the change of these resources with time means, in its simplest sense, the assessment of the individual components of the hydrologic cycle. The quantification of these components involves estimating the amount of water stored in the various environments in which the hydrologic cycle occurs (atmosphere, the surface of the Earth, and the layers beneath the surface of the Earth) and the fluxes of water in and out of these environments. Assessing the quantity and quality of water is another aspect of water resources assessment that is of great and growing importance as the population of the Earth increases, thereby placing greater stress on existing water supplies.

Hydrologic processes are phenomena that vary rapidly in space and time. In the past, measurements of these hydrologic processes have been accomplished primarily by *in-situ* or point measurements. This has required either a highly dense network of *in-situ* observations or, more frequently, an assumption of uniformity or adherence to a priori knowledge of the variability in space and time. The latter assumption creates uncertainties in the water resources assessment that become less tolerable, as their value increases, because of greater use and stress on water resources.

Remotely sensed data have the inherent properties of being able to provide synoptic observations with high observational density over relatively large areas. However, what is normally measured by *in-situ* devices used in hydrology, and what is observed by remote-sensing devices, may not be exactly the same. One effort in evaluating and using remotely sensed data has been that of determining the correlation between *in-situ* measurements and remotely sensed measurements and quantifying the added information value of remotely sensed data. This assumes that remotely sensed measurements may be less accurate at a given point, but, because of the greater number of observations, may produce greater accuracy over large areas. Remote-sensing measurements, as opposed to conventional *in-situ* measurements, generally require careful and knowledgeable interpretation and involve much larger data processing capabilities. The need for such capabilities stems from the pressure, developed by increased information requirements, associated with increased population and progress in enhancing the quality of life. That need has caused hydrologists and water resource managers to evaluate remote sensing as, at least and in general, an ancillary tool with which to increase their data bases in a timely and cost-effective manner.

At present, it may be assumed that there are three broad categories for using remote sensing in hydrologic studies. In the first category, simple qualitative observations are made. A visual observation or interpretation from a photo that water from factory effluent into a stream has a different color than stream water would be an example. Such an observation naturally suggests a site for the collection of a water sample. As a rule, such observations can be used for guiding the placement of *in-situ* observations, assisting in interpolation between point measurements, or assisting in the extrapolation of estimates beyond the bounds of *in-situ* observation networks.

In the second category of remote-sensing utilization, geometric form dimensions, patterns, geographic location, and distribution are the types of information derived. Area, shape, length, and identification of features such as land cover categories, based on multispectral or texture classification, are provided. The quantitative analyses of a drainage basin and channel network and the geographic location of fractures, faults, lineations, etc., fall into this category. The identification and quantification of the acreage covered by land cover categories that influence watershed runoff, evapotranspiration (ET), and soil-moisture storage would also be involved. Remotely sensed data can be used to detect and map such features more quickly and effectively over larger areas than ground-based methods often permit.

A third category is the use of remote sensing for direct estimation of a hydrologic parameter through the development of correlation between the remotely sensed observation and a corresponding *in-situ* measurement technique. Examples include the estimation of soil moisture, snow depth or water equivalent, sediment load in a water body, and precipitation.

The data provided by remote sensing still represent new measurement techniques for much of hydrology and water resources assessment. In the 1960's or even in the early 1970's, most of the published data fit into the first category of remote-sensing utilization. However, in the 1970's, many applications of remotely sensed data falling into the second category were developed. The frequency of results from the Landsat, airborne multispectral systems, and improved camera systems reflect this progress. In addition, the literature also shows progress in the third category of remote-sensing utilization. Tables 29-1a and 1b list several data sources in the United States, which indicate the kinds of remote-sensing data available, or potentially available, for use in water resources assessment.

The discussion provided herein either draws heavily from the water resources chapter provided in the first edition of the "Manual of Remote Sensing" (Meyer and Welch, 1975) or should be used in combination with the discussion in the first edition to provide a more complete understanding of the use of remote sensing for water resources applications. The discussion contained in the first edition relative to remote sensing of the chemical composition of water, water quality, gaining and losing streams, ground-water discharge, and identification of potential land collapse from sink-hole areas is not repeated or amplified here.

A principal contribution of this second-edition discussion on water resources will be to extend the discussion of the application of remote sensing to water resources studies to include data available from satellite sources. Particular emphasis has been placed on the use of Landsat data because of its pervasiveness and demonstrated utility for a wide variety of applications in water resources assessment.

Another key reference source for an overview of the progress in using satellite observations is the *Proceedings of the Fifth Pecora Symposium on Remote Sensing*, (Deutsch et al, 1981), which features satellite hydrology. For example, pages 38 to 40 in that reference provide a table summarizing applications, data sources, and precautions or limitations to be aware of in using satellite data in water resources assessment. This chapter will complement and extend sources such as those just noted by discussing in greater detail the principles and techniques involved in applying remotely sensed data. This discussion will also describe the availability of data and will use examples of observations from recent remote sensors existing on spacecraft and aircraft to illustrate the precisions, accuracies, and other measures of application progress for water resources assessment and hydrological studies.

SUBSURFACE WATER

Hydrologic remote sensing is defined as the study of the Earth's water resources by the use of electromagnetic radiation (EMR) that is either reflected or emitted from its surface in wavelengths ranging from 0.3 micrometers to 3 meters. Within this wavelength interval, most remote-sensing instruments record spatial variations in EMR coming from the Earth's surface in different spectral bands. Hydrologists must understand the characteristics of landscape cover and the effects of the elements in the energy path on electromagnetic energy to properly evaluate remote-sensor data.

The delineation of radiometric patterns displayed on imagery is referred to as image analysis. (For more details on image processing, see Chapter 17 on Data Preprocessing and Processing, Chapter 18 on Classification, and Chapter 24 on Photo-interpretation.) Image interpretation involves the identification of radiometric patterns on imagery that correspond to landforms, drainage, and cover type. An image analyst with training in ground-water hydrology must analyze landscape patterns on imagery to interpret geomorphology. Geomorphologic relationships are analyzed to develop structural and stratigraphic interpretations. Ground-water interpretations are developed through the analysis of structural and stratigraphic relationships.

Geohydrological analysis from imagery is one of the most complex uses for remotely sensed data. Its complexity results from the fact that the object of study, groundwater, is not portrayed directly on the data. Most applications of remotely sensed data, however, deal with surface phenomena, because it is the surface that is being photographed or imaged by satellite or aircraft. The geohydrologist commonly infers subsurface hydrological conditions from surface indicators such as areal geological features and structures, vegetation, streamflow characteristics, soils and soil-moisture anomalies, vegetative types and distribution, discontinuous ice cover on streams, differential snowmelt, springs, and many others.

In dealing with ground-water exploration, the term "aquifer" must be noted. The definition of an aquifer in a hydrologic sense, however, requires descriptions of the parameters of transmissibility and storage coefficient as well as information on depth of the water table or the altitude of the artesian head throughout the aquifer system. Remote-sensor data cannot be used to directly obtain these types of aquifer properties. Only in-

TABLE 29-1a

Selected Remote-Sensing Systems Available in the United States and Applicable for Water Resources Monitoring and Hydrological Studies

Vehicle/Sensor	Spectral Bands	Nominal Spatial Resolution	Appropriate Image/Scene Areal Coverage	Frequency of Coverage	Period of Data Availability	Data Center
Operational						
Ground-Based Radar	cm Wavelengths	Variable or grids 5-10 km square	10^3-10^4 km^2	10^1-10^4 sec	Many years	Silver Spring, MD (NOAA/NWS)
Aircraft Gamma Radiation Flights	Emission from ^{238}U, ^{232}Th, ^{40}K	≈600 m from 300 m altitude	≈5 km Flight lines	Variable	1972 to present	Silver Spring, MD (NOAA/NWS)
NOAA/VHRR	0.6- 0.7 μm 10.5-12.5 μm	0.9 km	Subcontinent	1/day-visible 2/day-IR	1972 to 1978	Suitland, MD (NOAA/NESS)
ESSA-NOAA/ AVCS-SR	Visible (AVCS) 0.6- 0.7 μm (SR) 10.5-12.5 μm	4 km	Subcontinent	1/day-visible 2/day-IR	1966 to 1978	Suitland, MD (NOAA/NESS)
SMS-GOES/ VISSR	0.55- 0.7 μm 10.5 -12.5 μm	1 km	1/3rd of globe (Western Hemisphere)	Several times per day	1974 to present	Suitland, MD (NOAA/NESS)
Tiros-N/AVHRR	4.5 bands: visible, near infrared, thermal infrared	1.1-4.0 km	Subcontinent	12-24 hours	1978 to present	Suitland, MD (NOAA/NESS)
Research and Development						
Landsat 1, 2, and 3 MSS	0.5-0.6 μm 0.6-07 μm 0.7-0.8 μm 0.8-1.1 μm	80 m	34,000 km^2	Once every 18 days	1972 to present	EROS Data Center Sioux Falls, SD
NASA Medium Altitude Aircraft	Multispectral scanners and microwave instrumentation	Widely varying characteristics, contact NASA/Ames Research Center				
High-Altitude NASA Aircraft	Visible and infrared photography	10 meters (approx.)	400-900 km^2	Variable	Occasional coverage in selected areas for 10 years or more	EROS Data Center Sioux Falls, SD
Skylab-EREP/ Multispectral Cameras, Spectometers	Visible and near-infrared, thermal infrared	10-70 m	10,000-30,000 km^2		1973, 1974 (three flights)	EROS Data Center
Skylab-EREP Microwave Scatterometer-Radiometer	2.2 cm	11 km	0-48° incidence angles	Variable	1973, 1974	EROS Data Center or NASA/Johnson Space Center
L-band Radiometer	1.4 GHz	124 km		Variable	1973, 1974	EROS Data Center or NASA/Johnson
Nimbus 1-7	Multispectral radiometers	4-55 km	Subcontinent	Daily in selected periods	Discontinuous coverage since 1964	Greenbelt, MD (NASA/Goddard)
Nimbus 5, 6 Microwave Radiometers	1.55 cm 0.86 cm	30 km	Subcontinent	Daily	Discontinuous coverage since 1972	Greenbelt, MD (NASA/goddard)
Heat Capacity Mapping Mission (HCMM)	0.5- 1.1 μm 10.5-12.5 μm	500 m	700-km swath	3 days	1978-1980	Greenbelt, MD (NASA/Goddard)
Nimbus 7/SMMR	6.6 GHz 10.69 GHz 18.0 GHz 22.2 GHz 37.0 GHz	92 X 144 km 57 X 88 km 34 X 53 km 28 X 43 km 17 X 26 km	1000-km swath	3 days	1978 to present	Greenbelt, MD NASA/Goddard
Seasat A/SMMR	(See Nimbus G)					Jet Propulstion Lab,
Seasat A/SAR	1.35 GHz	25 m	100-km swath	(See remarks)	June-October 1978	Pasadena, CA, and Suitland, MD (NOAA/NESS)

TABLE 29-1b

Planned or Considered Spacecraft Missions That Would Have Features Relevant to Hydrology and Water Resources

Vehicle/Sensor	Spectral Bands	Spatial Resolution	Image/Scene Areal Coverage	Revisit Interval	Remarks
Planned					
Landsat D/ Thematic Mapper (1982)	0.45– 0.52 μm 0.52– 0.60 μm 0.63– 0.69 μm 0.76– 0.90 μm 1.55– 1.75 μm 2.08– 2.35 μm 10.40–12.55 μm	30 m 30 m 30 m 30 m 30 m 30 m 120 m	185 × 185 km	16 days	Second Generation Earth Resources Satellite System (Landsat-D′ to be launched 1–3 Years after Landsat-D launch)
Considered					
Imaging Radar (1980's)	—X, C, or L band SAR's —To be flown on early shuttle flights	25–100 m	100–200 km swath	TBD	—Flight of an L-band SAR occurred on early shuttle flight in 1981 —Later advanced SAR might happen in mid-1980's time frame
Multichannel Passive Microwave (1980's)	K to P-band	20–50 km	≈1000 km	≈3 days	Large antenna technology involved
High Resolution & Pointable Observing Systems	Landsat bands	10–30 m	Very selected areas	Variable	Solid-state technology will be involved
Geosynchronous Satellite	Landsat bands	100–500 m	Local to subcontinent areas	Fraction of a day	Not likely until at least 1990's

ferences can be made regarding the subsurface through surficial expression of the aquifer. General principles of photographic interpretation, sometimes assisted by digital processing of image data, may be applied to images to recognize image patterns favorable for ground-water occurrence. Some detected patterns directly imply the presence of shallow sands and gravels. Other patterns indicate rock types or structures favorable for ground-water occurrence. When selecting imagery for a hydrologic analysis, factors such as the scale of analysis and temporal effects on imagery must be addressed. These are discussed in Chapter 18.

In using remote sensing for operational ground-water studies, one should realize that costs and time are required in using remote-sensing techniques and data. These items should be included in a study plan. However, experience has shown that the total cost and time usually will be lower through the use of remote sensing.

The benefits that accrue in the use of remote sensing are usually greatest when this information source is applied at the beginning of a study. Remotely sensed data inherently and relative to other data sources provide a large-area synoptic view with high observational density. This allows data obtained at a point on the ground to be extrapolated or extended over a much larger area presuming the association of the point measurement with a remotely sensed feature can be established. To permit interpretation of remotely sensed data and association with ground-based information or other knowledge, access to or purchase of some special equipment may be necessary and therefore should be included in any exploration or study plan.

Ground-water exploration methods may be divided into two categories: regional and local. Within these categories, several methods listed in Tables 29-2a and 2b can be used. High-altitude aircraft and satellite-based observations are generally most applicable to regional exploration, whereas low-altitude aerial photography and ground-based remote sensing apply to local studies. Regional studies would typically cover areas of 10^2 km and larger.

Of the various sources of data listed in Tables 29-1a and 1b, those sources providing photography or imagery in the visible and near infrared are almost exclusively used for ground-water exploration with the possible exception being the thermal infrared as discussed by Meyer and Welch (1975). Therefore, key sources of data are high- and low-altitude aircraft photography, Skylab photography, and photography available from

TABLE 29-2a

Regional Exploration Methods

Method	Procedures	Personnel Requirements	Area Coverage Rate	Results
Interpret Satellite Images or Mosaic	Analyze tones, textures, shapes, patterns, location, and association.	One scientist, in office	7,000 to 34,000 km^2/day	Interpretation of lithology, structure, and ground-water occurrence based on landforms, drainage patterns, land use, soil tones, and vegetation types and patterns
Geologic Reconnaissance	Examine lithologies, orientations, and fracture patterns.	One scientist, in field	250 km^2/day	Generalized geologic map and sections showing lithology, stratigraphy, and structure
Hydrologic reconnaissance	Inventory largest wells and springs. Examine rock types, orientations, and fracture patterns.	One scientist, in field	100 km^2/day	Generalized hydrologic map showing aquifers, aquitards, and areas of recharge and discharge
Magnetic Prospecting	Measure differences in the Earth's magnetic intensity	One scientist and pilot for aerial studies	250 to 2,500 km^2/day	Calculation of sedimentary rock thickness and interpretation of basement structure
		One scientist and helper for ground studies	2.5 to 250 km^2/day	
Gravity Prospecting	Measure differences in the Earth's gravitational force.	One scientist, one surveyor, and two helpers	2.5 to 250 km^2/day	Interpretation of shallow structures and lateral changes in rock density

other manned spacecraft missions such as the Apollo and Mercury programs and Landsat imagery. In the following discussion, while a primary emphasis is placed on Landsat imagery applications, the discussion applies to imagery from other sources as well.

SHALLOW SAND AND GRAVEL AQUIFERS

Most well-sorted sands and gravels are fluvial deposits, either in the form of stream-channel deposits and valley fills or as alluvial fans and bajadas. The remainder are cheniers, beach ridges, beaches, and some wind-deposited dunes. The shapes, patterns, tones, and textures that infer shallow aquifers are those that indicate coarse materials and a near-surface water table. Keys to detection of shallow sand and gravel aquifers which may be seen on Landsat images include:

a. Shape or form
 —Stream valleys, particularly broad valleys with low-stream gradients.
 —Underfit valleys represented by topographically low elongate areas.
 —Natural levees (levees themselves may be fine-grained materials).
 —Meander loops defining location and relative thickness of point bars.
 —Meander scars in lowlands, oxbow lakes, arcuate dissection of upland areas.
 —Arc deltas (coarsest materials) and other deltas.
 —Drainage-line offsets, changes in drainage patterns; or changes in size or frequency of meanders (may be caused by faults or cuestas or by other deposits of coarser-grained materials).
 —Braided drainage channels and scars.
 —Alluvial fans, coalescing fans, and bajadas.
 —Cheniers, beach ridges, and dunes.
 —Aligned oblong areas of different natural vegetation representing old offshore bars or dissected beaches.

b. Pattern
 —Drainage patterns infer lithology and degree of structural control; drainage density (humid regions) and drainage texture (arid regions) infer grain size, compaction, and permeability.
 —Snowmelt: If everything else is equal, anomalous early melting of snow and greening of vegetation show areas of ground-water discharge.
 —Distinctive types of native vegetation commonly show upstream extensions of drainage patterns, areas of high soil moisture, and landform outlines (humid

TABLE 29-2b

Local Exploration Methods

Method	Procedures	Personnel Requirements	Area Coverage Rate	Results
Interpret Aerial Photographs	Analyze tones, textures, shapes, patterns, location, and association.	One scientist, in office	100 to 500 km^2/day	Interpretation of lithology, structure, and ground-water occurrence based on landforms, drainage patterns, land use, soil tones, and vegetation types and patterns.
Detailed Hydrologic Mapping	Inventory all wells and springs. Determine nature of ground-water occurrence. Determine aquifer characteristics.	One scientist, in field	10 km^2/day	Inforamtion on ground-water occurrence, well yields, and water quality. Description or model of targets for test drilling.
Detailed Geologic Mapping	Examine lithologies, orientation, and fracture patterns. Trace contacts.	One scientist, in field	2.5 km^2/day	Detailed geologic map and sections, showing lithology and structure; based on rock outcrops, soils, and vegetation.
Electrical Prospecting	Measure Earth's electrical potential and resistivity.	Two scientists and two to four helpers	2 to 12 km^2/day	Interpretation of vertical stratigraphy, lateral changes in lithology, and structure.
Seismic Prospecting	Measure travel time of elastic energy.	Two scientists and one to four helpers	0.5 to 5 km^2/day	Interpretation of vertical stratigraphy, lateral changes in lithology, and structure (including locations of rock fractures).
Shallow Test Drilling	Soil auger or jetting rig.	One scientist, one driller, and two helpers	0.5 to 5 km^2/day	Lithology, porosity, and permeability of unconsolidated materials.
Deep Test Drilling	Cable-tool or rotary rig and geophysical logs.	One scientist, one driller, and two helpers	0.1 to 1 km^2/day	Lithology, porosity, and permeability of subsurface materials. Well yield and aquifer characteristics.

regions); abrupt changes in land cover type or land use infer landforms that may be hydrologically significant but do not have a characteristic shape.

—Elongated lakes, sinuous lakes, and aligned lakes and ponds represent remnants of a former stream valley.

—Topographically low, elongated, aligned areas represent abandoned stream valleys.

c. Tone

—Soil type: Fine-grained soils are commonly darker than coarse-grained soils.

—Soil moisture: Wet soils are darker than dry soils.

—Type and species of native vegetation: Vegetation is generally well adapted to type and thickness of soil, drainage characteristics, and seasonal period of saturation of root zone.

—Land use: For example, in areas of periodic flooding, native vegetation tends to occur in low lands and agriculture on uplands.

—Anomalous early or late seasonal growth of vegetation usually occurs in areas of high soil moisture, such as where the water table is close to the land surface.

d. Texture

—Uniform or mixed types and species of native vegetation.

—Contrast between sparse vegetation on topographic highs and denser vegetation in low (wetter) areas.

—Mixed land use, representing local differences in topography, soil type, or drainage characteristics, for example.

All the image keys discussed above have been detected on Landsat images. Locally, however, other keys may be important, and it should be realized that many hydrologically significant features are too small to be visible on Landsat images. yet, for many types of images acquired at other scales, the recognition features discussed above also hold.

Landscape features that have a characteristic shape or form can be used to detect shallow aquifers with reasonable confidence. Best results are generally obtained by manual interpretation of either a band-7 (near-infrared, 0.8 to 1.1 μm) or a color-composite image. A low sun-elevation angle (e.g., November 1 through February 15 in the norther hemisphere) is desirable to necessary. In areas with dark-colored soils, a thin snow cover enhances the topographic shadowing produced by a low sun-elevation; on light-colored soils, snow may obscure fine topographic detail.

Imaged patterns as described above have a lower relative reliability than landforms as indicators of shallow aquifers on Landsat images. Nevertheless, large areas can be examined for distinctive patterns in short periods. Landsat images also provide a good format for the detection of patterns covering large areas and for delineation of major boundaries. Yet, fewer drainage channels are typically visible on Landsat images (visible drainage lines vary with season and sun-elevation angle) than on aerial photographs; however, gross drainage patterns can be delineated quickly. Band-7 images are generally best for detecting streams, lakes, and ponds, but lakes with emergent vegetation may be recognized more easily on the band-5 (red light, 0.6 to 0.7 μm) image. Color-composite images are best for the study of snow-melt and vegetation patterns. Digital processing is commonly helpful to enhance small local differences in vegetation and snow-melt. Many local vegetation patterns that indicate ground-water springs, however, are below the resolution limits of the scanner.

Unless ground-truth information is available, image tones and textures have the lowest relative confidence as indicators of shallow aquifers. In some cases, however, combinations of tones and textures form patterns that, in turn, become distinctive indicators. Digital processing of Landsat spectral data using an interactive man-and-machine system has proved useful for enhancement and classification. An alternative is manual interpretation of either a color-composite image or black-and-white images. Textures must generally be detected and delineated manually, although the complex spectral signatures of some areas are distinctive and may be classified by machine processing. The main problem with the use of image tone and textures for groundwater analysis is that an area underlain by a shallow aquifer may have a variable appearance; no single indicator occupies an area large enough to form a diagnostic pattern. An underfit valley in early spring, for example, may have some areas that show the dark tones of high soil moisture abutting other areas that have an anomalous early growth of native vegetation. Such cases can be enhanced by machine processing, but the significant pattern (the valley) must generally be recognized visually. Early spring and fall images are generally best for detection of differences in soil type, soil moisture, and type and species of vegetation.

Boundaries that are gradational on the ground are generally more clearly defined on Landsat images. Hydrologically significant landforms, patterns, tones, and textures can be delineated most easily on Landsat images. On the other hand, more detail is visible on aerial photographs. In some areas, Landsat images can be used to select locations for test wells. In other areas, Landsat images are best used for selecting locales that can be examined in more detail on aerial photographs or on the ground. Color Figure 29-1, when compared with Figure 29-2, illustrates some results from a study that used tones and patterns on Skylab photography to assess depth to highly saline groundwater (Kruck and Kantor, 1975). Regions in which the groundwater is within 5 to 7 meters of the surface appear light blue in the simulated color-infrared Landsat image (Fig. 29-1), whereas areas that are more reddish in tone have groundwater that is 10 meters or more from the surface.

In the previous discussion, the predominant emphasis has been on the use of visible and near infrared or reflected solar radiation for locating shallow aquifers. Several studies have been conducted using thermal infrared remote sensing techniques for determining sub-surface soil moisture and water table conditions. Early work includes a study by Myers and Heilman (1969) linking surface-soil temperatures to subsurface-soil conditions, such as water-table distribution. Moore and Myers (1972) showed that a correlation between radiometric temperature and depth to shallow (1.5–4.5 m) aquifers exists; they indicated that detection of shallow aquifers using pre-dawn thermal infrared imagery is possible. Abdel-Hady et al. (1971) conducted an experiment using sand and silty-clay soil columns, where the water table was controlled to a depth of four feet to determine the effects of shallow water tables on surface emittance. Results from this study indicated that ground water table differences that occur within 0 to 4 feet of the surface were detectable. Myer et al. (1975) reported on results of two models which related remotely-measured surface-temperature differences to subsurface temperatures that were found to be correlated to the presence of shallow water tables. This study indicated that if a subsurface thermal-anomaly exists, (e.g. a shallow water table) a surface temperature difference will be manifested.

Huntley (1978) and Heilman and Moore (1981) describe more recent research involving the use of thermal infrared data for finding shallow aquifers. It would appear at this time that thermal infrared imagery obtained just before sunrise under low wind-flow conditions and near the autumnal equinox will provide the most information in this regard. The results would apply to aquifers occurring at depths where the diurnal temperature variation of the overlaying soil is significant. Experience in the South Dakota region indicates application for depths of 3 meters or less. To obtain a reasonable correlation (e.g., 0.8), a model of the

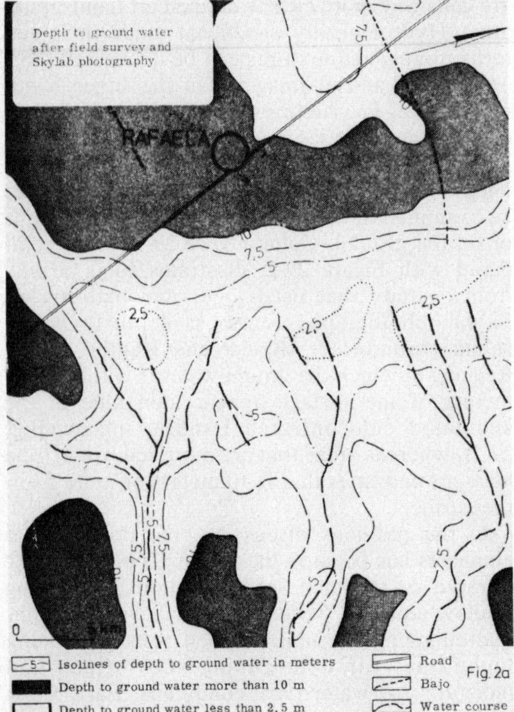

Fig. 29-2. Depth to ground-water analysis based on field survey and Skylab photography. Compare with Color Figure 29-1. (Kruck and Kantor, 1975).

conditions involved must be used that corrects the thermal infrared observations for variations in land cover (Moore et al, 1981).

Most recently, the thermal inertia concept has been integrated into remote sensing research. Thermal inertia represents the resistance of a material to a change in its temperature. Variations in a soil's moisture content are reflected in changes in its thermal inertia. It is expected that the soil moisture profile characteristics of a well-drained area will differ from that of an area having a shallow water table. Some of the first work done using thermal-inertia remote sensing was that of Watson (1971, 1979). His work deals with the differentiation of geologic materials in arid zones using algorithms developed for calculating thermal inertia. Kahle and Gillespie also used the thermal inertia concept for distinguishing between various geologic materials in arid zones (Kahle et al, 1976; Gillespie and Kahle, 1977). Kahle et al. worked on an improved theoretical model and algorithm for calculating thermal inertia values, which incorporates the transfer of latent and sensible heat between the ground and atmosphere in addition to radiative transfer. Results of this work showed that the thermal inertia concept allowed the distinction of material with similar reflective characteristics, but different thermal properties. Pratt and Ellyett (1978) modelled thermal inertia to distinguish between various soil types. Their work has shown that the mineralogical content of a soil has little influence on the thermal inertia; however, variations in soil moisture do influence the thermal inertia. The residual moisture contents of a clay soil will be higher after a drying period than those of a sandy soil.

A study conducted by Ezra et al. (1982) successfully used remotely-sensed thermal inertia mapping-techniques to distinguish regional patterns of shallow water tables in a semi-arid irrigated agricultural environment.

Apparent thermal inertia can be estimated, in part, by measurements of the diurnal temperature amplitude and albedo (Price, 1977). Diurnal temperature differences can be measured from aerial thermal infrared data acquired during the period of minimum and maximum temperature (approximately 0400 and 1400 respectively). Albedo can be approximated from broadband visible and near-infrared aerial photography ($0.5-1.1\ \mu m$) acquired coincidently with the daytime thermal data. From these data it is possible to determine apparent thermal inertia, which gives a relative measure of differences in soil thermal-inertia characteristics (Pratt and Ellyett, 1979). Results of this research indicate that, due to the complex environmental system, it is not feasible to reliably delineate regional patterns of very shallow water tables through the use of remote sensing techniques presently available for measuring apparent thermal inertia. There are some features identified within specific fields, however, which may be directly related to the local presence of shallow water tables. It is also shown that, by using the apparent thermal inertia mapping technique, differences in evaporative cooling, due to soil-moisture effects between and within fields, can be clearly seen.

The 1977 launching of the experimental Heat Capacity Mapping Mission (HCMM) has allowed very large scale regional studies to be made using thermal inertia mapping techniques for soil moisture and drainage-related problems.

OTHER AQUIFERS

Virtually, all consolidated rocks contain groundwater at shallow depths, but ground-water abundance depends not only on the rock type but also on the amount and intensity of fracturing. When attempting to identify rock types on Landsat images, one should have experience or prior knowledge of an area to ensure acceptable results. The following keys can be used, however, to detect certain aquifers and differences in rock types on Landsat scenes

a. Outcropping rock type
   —Landforms; topographic relief.
   —Outcrop patterns; banded patterns for sedimentary rocks (outlined by vegetation in humid regions), plateau or homocline for basalt flows, or curving patterns for folded beds.

—Shape of drainage basins.
—Drainage patterns, density, and texture.
—Fracture types and symmetry (as inferred by lineaments).
—Relative abundance, shape, and distribution of lakes.
—Tones and textures: difficult to describe; best determined by study of known examples.
—Types of native land cover.

b. Lineaments
—Continuous and linear stream channels, valleys, and ridges; discontinuous but straight and aligned valleys, draws, swags, and gaps.
—Elongate or aligned lakes, large sinkholes, and volcanoes.
—Identical or opposite deflections (such as doglegs) in adjacent stream channels, valleys, or ridges; alignment of tributary junctions.
—Elongate or aligned patterns of native vegetation, thin strips of relatively open (may be rights-of-way), or dense vegetation.
—Alignment of dark or light soil tones.

All the foregoing keys have been seen on Landsat images, but the list is not intended to be all-inclusive. Other factors could be important locally.

Lineaments are all types of natural straight-line features on images. It is well to remember that lineaments are not necessarily rock fractures and that they do not necessarily localize ground-water occurrence. A few lineaments can be correlated with faults; the physical nature of most other lineaments, not correlated with faults, must be investigated by indirect means (generally by comparing lineament trends with joint trends) because of the small scale of Landsat images (few landmarks are visible) and residual materials covering many rock outcrops.

Further discussion involving the association between lineaments and ground-water occurrences is needed to more fully understand the basis and probability of success in applying remote-sensing image interpretation. An association of lineaments and ground-water occurrence is dependent on the presence of dense heterogeneous aquifers, rock fractures, and the surface expression of these fractures. Dense heterogeneous aquifers (Rassmussen, 1963) do not yield water from the rock fabric under gravity drainage. The only space for storage and movement of groundwater is in fractures enlarged by brecciation, weathering, solution, or corrosion. Corrosion in dense rocks occurs by solution of the rock cement (generally 5 to 10 percent of the fabric) and physical removal of loose grains by turbulent flow.

It is logical that many fractures that localize the occurrence of ground-water also have an expression at land surface. The same processes of weathering, solution, and corrosion operate on land surfaces, as do additional processes such as frost wedging and mass wasting. Therefore, a fracture that is a plane of weakness for enlargement by groundwater may be represented by a topographic depression, a different soil tone, or a vegetation anomaly on the land surface.

Many fractures are vertical; in this case lineaments may represent favorable locations for water wells. Other fractures are oblique; well locations must be offset to intersect the fractures below the water table but at a shallow depth (nearly all fractures are progressively smaller at increasing depths). In some limestone terrains, and probably in other areas of bedded sedimentary rocks, sheetlike solution cavities form along bedding planes. If the rocks are flat-lying, the hydrologically significant openings are nearly parallel to the land surface. In these cases, however, vertical joints commonly control the locations and trends of the horizontal sheetlike cavities; the same vertical joints may have a topographic, tonal, or vegetation expression at the surface.

A number of reports (Lattman and Nickelson, 1958; Hough, 1960; Boyer and McQueen, 1964; Leuder and Simons, 1962; Lattman and Parizek, 1964; Trainer, 1967; Trainer and Ellison, 1967; Moore et al, 1969, Powell et al, 1970; Sonderegger, 1970; Siddiqui and Parizek, 1971; Tomes, 1975; and Moore, 1976) have shown a useful correlation between lineaments detected on aerial photographs and the occurrence of groundwater in dense, fractured limestones. Siddiqui and Parizek (1971), for example, found that wells on fracture traces in Pennsylvania had a median productivity 55 times greater than wells between fracture traces and nearly 10 times greater than randomly located wells. In contrast, however, Moore (1976) using Skylab photographs covering central Tennessee, found that a majority of the favorable locations for high yield wells in central Tennessee cannot be determined by lineament detection and mapping. Therefore, it would appear that, in areas underlain by heterogeneous aquifers, the best results are obtained by combining lineament detection and mapping with data and information collected by conventional means.

The best image formats for detecting lineaments are the same as for landforms; many lineaments are enhanced by a low sun-elevation angle. Manual interpretation is necessary, and unaided manual interpretation commonly produces adequate results. Recent work (e.g., Goetz and Billingsley, 1973; Pincus, 1969; Rowan et al, 1974; Schowengerdt et al, 1981; and Zall and Russell, 1981) has shown, however, that machine processing such as contrast stretching, Fourier transforms, high spatial-frequency filtering, and fine-scale detail-enhancement are helpful. Such enhancements may be necessary in otherwise featureless areas.

The inference of deeper deposits of sand and

gravel is controversial and has not been confirmed by test drilling. However, the hypothesis is interesting and provides a potentially important interpretation key for Landsat images: coarse-grained aquifers should occur at a depth beneath lineaments that is visible on Landsat images. A number of previous workers (e.g., Gilbert, 1882; Russell, 1885; Crosby, 1885; Fisk, 1947; Mollard, 1957; Plafker, 1964; Martin et al, 1973; Drake and Vincent, 1975; O'Leary and Simpson, 1975; and Withington, 1976) have detected and mapped lineaments in thick unconsolidated materials; almost all these workers note that many lineaments are formed by drainage lines and patterns. Evidence relating the origin and orientation of lineament sets and systems to fractures in underlying consolidated rocks is impressive, although not conclusive. It is thought that lineaments represent fractures that propagated upward from basement rocks during depositional periods (instead of at some undetermined later date). If lineaments controlled stream-channel positions during depositional periods as well as at present, then coarse stream-channel materials may be found at depth in alluvium-filled valleys, glaciated terrains, and coastal-plain deposits—beneath lineaments on the present land surface.

It is important that some of the most important image characteristics that have been learned in the use of imagery for groundwater be summarized. For example, one of the most important things that has been learned since Landsat imagery became available is that time of year is critical for obtaining the maximum geologic and hydrologic information from the images. The exact best time depends on local conditions and weather patterns as well as interpretation objectives. Suggested best periods for use of Landsat images (actual best time may vary because of local conditions or weather patterns) in ground-water explorations are:

a. Low sun-elevation angle
 —Best: November 20–January 20 (<25° at 35°N latitude)
 —Good: November 1–February 15 (<30° at 35°N latitude)
 —Fair: October 5–March 1 (<35° at 35°N latitude)
b. Maximum area of bare soil
 —March–May (spring crops)
 —October–December (winter crops)
c. Drainage patterns
 —Best: November 20–January 20 (low sun angle)
 —Good: April–May (high stream stages)
d. Soil-moisture patterns
 —Best: April
 —Good: March–May
e. Snowmelt patterns
 —Best: March
 —Good: November–April
f. Native vegetation types and differences
 —Best: October, April
 —Good: September–November; April–May
g. Small lakes and ponds
 —Best: After heavy, intense showers
 —Good: March–May; after seasonally heavy rains
h. Native-vegetation density: June–July (April–May for annual western grasses)
i. Areas beneath deciduous overstory: December–March
j. Irrigated crops versus dry-land farming: Dry period; crop grown enough to cover bare soil; crops of one type or several types with distinctive signatures (spectral reflectances)
k. Lithologies in glaciated terrain: May
l. Desert vegetation: April

The black-and-white Landsat images that generally contain the most ground-water information are bands 7 and 5. However, band-6 (near-infrared energy, 0.7 to 0.8 μm) images are useful for detection of soil-moisture patterns, and some information on vegetation may be extracted from band-4 (green light, 0.5 to 0.6 μm) images. Because of these factors, a color-composite image is generally the single most useful Landsat product for ground-water interpretations.

The selection of transparencies or prints, as well as the plan for reproducing and viewing the images, is entirely a matter of personal preference. Transparencies must be projected (overhead projector for nominally-sized 9-inch film and lantern slide projector for 70-mm film) or viewed on a light table. Many prints may also be viewed by transmitted light to avoid glare from overhead lights. Some scientists believe that transparencies have more detail than prints, but there is no inherent reason why this should be so.

A vertical view of the image is best for detection of most features. Viewing at a low angle and slowly rotating the image may enhance lineaments. Viewing at a distance decreases resolution and decreases the distraction of cultural patterns; this may help to detect large features or regional trends. Similarly, a projected image may be viewed close to the screen or at some distance, as well as in focus or out of focus.

In general, various forms of products, processing, and enhancement are possible when employing either photographic or digital methods. Manual analysis and interpretation of photographic or image products are recommended before proceeding with digital processing. The advantages of digital processing must be weighed against the higher costs involved. At present landforms, patterns, and lineaments usually can only be satisfactorily detected and recognized by the human eye through use of the principles of photointerpretation (see Chapter 24). A final manual interpretation for the features related to groundwater exploration will be necessary regardless of the processing and enhancement methods.

SOIL MOISTURE AND EVAPOTRANSPIRATION

APPROACH AND DATA ASSESSMENT

Moisture in the upper layers of the soil profile is an important portion of the total water balance of the earth-atmosphere system and is an important parameter in many disciplines related to hydrology such as weather, climate, and agriculture. In hydrology, the moisture content of the soil is important for partitioning rainfall into its runoff and infiltration components.

The soil layer that is usually considered in hydrology and related disciplines is that which can interact with the atmosphere through evapotranspiration. This layer constitutes the root zone of plants and therefore depends on the type and stage of maturity in plant cover, but is typically 1 to 2 meters. The moisture content of this soil layer fluctuates in response to precipitation (input) and the evapotranspiration (output).

The use and application of remotely sensed data for soil-moisture determination is still very much under development. However, much research has been accomplished, which shows clearly that observations are sensitive to variations in soil moisture. In addition, the fundamental physical principles influencing remotely sensed observations are well understood. What remains is to couple research with the applications experience already gained to improve the precision and accuracies attainable by the scientific and applications communities. It is clear at this time that remotely sensed data can be acquired and interpreted in such a way as to provide an ancillary data source complementing conventional hydrological observations. There have been many reviews of the status of remote sensing for soil moisture over the past few years. Some helpful and representative examples are those by Schmugge (1978), Schmugge et al (1980), Idso et al (1975), and Ulaby et al (1974). This section will provide a description of the basic principles, conditions, and methods that should be considered in using remotely sensed data for soil-moisture determination and some representative examples to illustrate the points made.

It is well to have in mind at least a conceptual model of the factors, dimensions, and processes taking place when interpreting or applying remotely sensed data. Some of these are illustrated in Figure 29-3 taken from Saxton et al (1974). Figure 29-3 suggests that the reader understand that he is dealing with a soils-vegetation system that is bounded on one side by the atmosphere and on the other side by the deep layer of the soil and earth mantle. A conservation-of-mass principle is involved where the fluxes in and out of the system

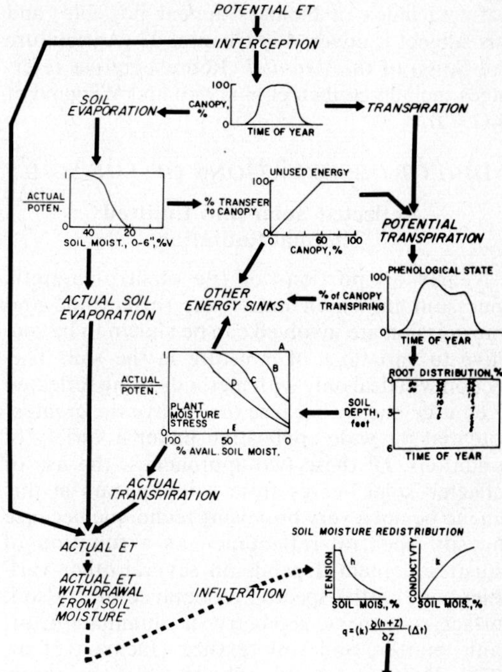

Fig 29-3. Flow chart of SPAW model (Saxton et al, 1974).

involve not only the precipitation (*P*), evaporation (*E*), and transpiration (*T*) as previously noted but also surface runoff (*R*), lateral subsurface flow (*L*), percolation (*Q*), and capillary rise from lower levels of the soil (*C*). These are all expressed in the following equation for the change in soil moisture (*SM*):

$$\frac{d(SM)}{dt} = P - R + L + E - T + C - Q \qquad (29\text{-}1)$$

In general, most remotely sensed measurements of moisture in the soil either directly respond to soil-moisture content in the upper few centimeters (10^0 to 10^1 cm) or use the response of plants as related to their ability to transpire, thereby indirectly providing an indication of the soil-moisture status in the root zone (10^0 to 10^2 cm). Remotely sensed observations sensitive to soil moisture, and any other series of soil-moisture observations for that matter, must be sensitive to the relationships indicated in Figure 29-3 and Eq. 29-1 in order to be properly applied.

Thus, either direct observations of the soil by remote sensors or indirect estimates using plant condition appear to be useful in soil-moisture observations. The precise amount of vegetation, limiting a direct approach or required to use an indirect approach, is not yet determined. In any case, it is clear that some estimate of the amount and type of vegetation present is necessary for accurately and precisely applying remote-sensing observations. Remotely sensed observations of leaf-area index or biomass appear possible, and this subject is covered in Chapter 33 (Agriculture and Soils) of this *Manual*. Representative references include Holben et al (1980) and Wiegand et al (1979).

DIRECT OBSERVATIONS OF THE SOIL

Reflected Solar and Emitted Thermal Radiation

Nearly all portions of the electromagnetic spectrum in which emission, reflection, and transmission are involved can be shown to be sensitive to variations of moisture in the soil. This section will deal only with methods using reflected or emitted energy because these have the greatest potential for wide application under a variety of conditions. Of these two approaches, the use of reflected solar energy from soils appears at this time to be not a very promising technique because the soil spectral reflectance as a function of moisture content depends on several other variables such as the spectral reflectance of dry soil, surface roughness, geometry of illumination, organic matter, and soil texture (Jackson et al, 1978). These factors, plus the fact that the observations represent only a very thin surface layer (>1 mm), limit the utility of solar reflectance measurements for soil-moisture determinations.

In other portions of the electromagnetic spectrum, such as in the emitted thermal radiation regions between 3 to 15 μm and the microwave regions between 1 to 52 cm, water has strong influence on the thermal and dielectric properties of soils that make the remote sensing of soil moisture possible. Based on these properties, a considerable number of results are available that fall into the third category of remotely sensed data utilization mentioned in the introduction to this chapter.

For thermal infrared measurements, the observations are usually limited to the 8- to 12-μm regions of the spectrum where the effects of the atmosphere are minimized, but not negligible, and the emitted energy is near the maximum for terrestrial surfaces. The time rate of change of soil temperature is a function of internal and external factors. The internal factors are thermal conductivity (K) and heat capacity (C), which can be combined to obtain the thermal inertia $P = (KC)^{1/2}$. Since K and C are strong functions of soil moisture, the thermal inertia and resultant soil temperatures will be dependent on soil moisture. A detailed discussion of thermal inertia is given in Chapter 33. The external factors are those meteorological ones that affect the energy balance at the surface of the Earth, such as incoming solar radiation, ambient air temperature, relative humidity, and wind. If the variability in these external factors can be accounted for or held reasonably constant, the time rate of change of remote estimates of soil temperature or differences in surface temperature are related to soil moisture (Idso et al (1975; Jackson et al 1978).

Empirical results illustrating the potential for such measurements are shown in Figure 29-4 and also in Figure 33-56 in the agriculture and soils chapter. Figure 29-4 shows how the temperature of a field or soil varies as a function of time and moisture content. Figure 33-56 illustrates the correlation observed between the diurnal temperature at the surface of the soil and soil moisture over various layers of the soil profile up to 4 cm thick. In general, results such as these indicate that the basic attributes of remotely sensed observations (ability to rapidly survey a relatively large area with high repetition) can be used for obtaining indications of soil-moisture status, but they should be carefully interpreted to allow for differences in meterological conditions and soil type.

Microwave

As suggested previously, the dielectric properties of water strongly influence the dielectric properties of soils. The use of microwave measurements for soil-moisture estimations is possible because of these properties. The dielectric properties of soil, as influenced by the presence of water, indicate that both microwave reflection and emission may be related to soil-moisture variations. Figure 29-5 shows the dependence of a soil's dielectric properties on moisture content at two different microwave wavelengths (1.55 and 21 cm). The breakpoint in these curves depends on soil texture being at lower moisture contents for lighter soils and at higher moisture contents for heavier soils (Wang and Schmugge, 1980). In addition to dielectric dependence, microwave measurements are also sensitive to the polarization of the emitted or reflected energy, the roughness of the surface, the observation or incidence angle of the emitted or reflected energy, and the amount of vegetation present that acts as an obscuring layer or a "screen" of the radiation. Some effects of the atmosphere are involved, but, at the longer wavelength (75 cm) used for soil-moisture sensing, these effects are usually small except where heavy precipitation occurs. The negligible effects of the atmosphere make microwave mea-

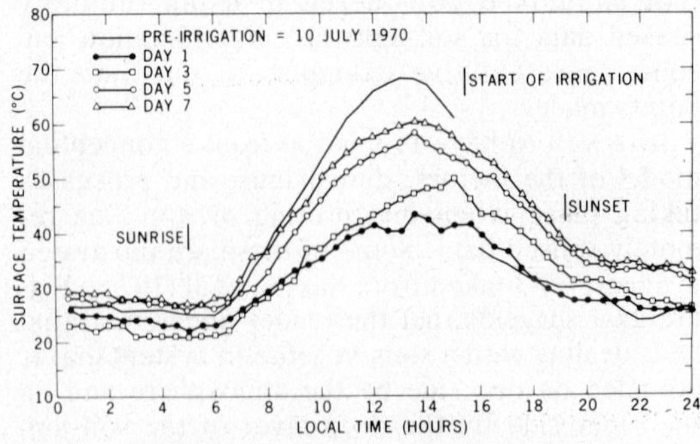

Fig. 29-4. Diurnal surface temperature variation. Data from U.S. Water Conservation Laboratory, Phoenix, Arizona. (NASA/GSFC).

surements attractive for soil-moisture estimation as well as other applications. The following paragraphs describe some of the basic factors necessary to consider in using microwave data for soil-moisture determination and give some illustrations of the effects produced by variations in these factors. Again, additional material on this topic is provided in Chapter 33.

Passive Microwave

The range of the dielectric constant presented in Figure 29-5 produces a change in emissivity from greater than 0.9 for a dry soil to less than 0.6 for a wet soil, assuming an isotropic soil with a smooth surface. This change in emissivity for a soil has been observed by truck-mounted radiometers in field experiments (Poe et al, 1971; Blinn et al, 1972; Newton, 1977; Njoku and Kong, 1977; Wang et al, 1980a; Wang and Choudhury, 1981; Newton and Rouse, 1980; Njoku and O'Neill, 1982; and Schanda et al, 1978) and by radiometers in aircraft (Schmugge et al, 1974; Estes et al, 1978; Burke et al, 1979; Barton, 1978; Choudhury et al, 1979; and Basharinov and Shutko, 1978) and satellites (Schmugge et al, 1977; McFarland, 1976; and Eagleman and Lin, 1976). In no case were emissivities as low as 0.6 observed for real surfaces. It is believed that this is primarily due to the effects of vegetation and surface roughness, both of which generally increase surface emissivity.

Significant improvements in the understanding of the effects of individual scene parameters on the relationship of brightness temperature to soil moisture have been achieved using ground-based measurements acquired during controlled experiments (refer to references in previous paragraph). The results of these field measurements are described in detail in Chapter 33. However, demonstration of the potential of passive microwave sensors for estimating soil moisture on an operational basis must be performed with aircraft and spacecraft sensors that integrate large areas of natural terrain. A series of aircraft experiments performed over the last several years by a number of investigators has demonstrated the sensitivity of microwave radiometers to soil moisture in agricultural terrain. Skylab and Nimbus satellites have also provided significant results for very large areas. Examples of results from these experiments are presented later in this chapter.

An example of aircraft data is presented in Figure 29-6. Results from nine flights during the 3-year period, 1976 through 1978, over a test site in Hand County, South Dakota, are compared with regression results for data obtained over the Phoenix and Imperial Valley areas in 1973 and 1975 (Schmugge, private communication). In each case, the correlation between soil moisture in the top 2.5 cm and observed T_B was >0.85. These data were for a range of surface conditions including fallow fields, wheat, alfalfa, and pasture. The scatter in the aircraft data presented in Figure 29-6 could be attributed to surface roughness variations and uncertainty of ground measurements. The standard deviation of the ground measurements is represented by the error bars in Figure 29-6. The number of samples ranged from 6 to 29, depending on the length of the fields. This difficulty in making accurate ground measurements has hampered the determination of the accuracy for remote-sensing techniques.

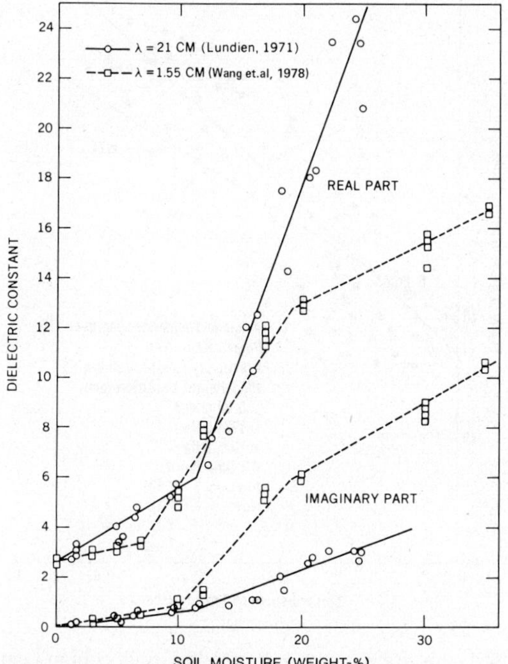

Fig. 29-5. Dependence of the soil's dielectric constant on its moisture content (Schmugge, 1978).

Active Microwave Response to Soil Moisture

Analogous to the optical reflectivity of terrain, the backscattering coefficient (σ^0) describes the scattering properties of terrain in the direction of the illumination source. The scattering behavior of terrain is governed by the geometrical and

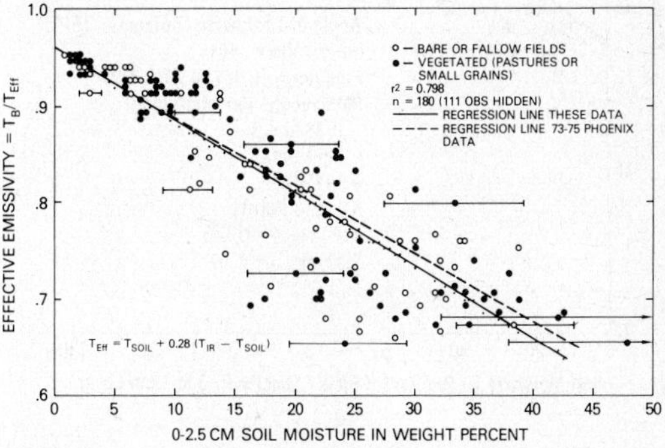

Fig. 29-6. Airborne 21-cm microwave radiometer emissivity data versus soil-moisture content for an area in Hand County, South Dakota.

dielectric properties of the surface (or volume) relative to the wave properties (wavelength, polarization, and angle of incidence) of the incident radiation. Recall from Figure 29-5 that the dielectric constant of a soil-water mixture is strongly dependent on its water content. Thus, in general, σ^0 of terrain is dependent on the soil-moisture content of an effective surface layer whose thickness is governed by the penetration properties of the terrain at the wavelength used. This thickness will be approximately the same for active and passive microwave systems. However, in addition to its dependence on soil moisture content, σ^0 is generally a function of the surface (or volume) roughness and vegetation or snow cover (if not bare). The variations of σ^0 with soil moisture, surface roughness, incidence angle, and observation frequency have been studied extensively in ground-based experiments conducted at the University of Kansas, using a truck-mounted 1- to 18-GHz active microwave system (Batlivala and Ulaby, 1977).

In addition to surface roughness, another soil variable that has exhibited an influence on the σ^0 response to moisture is soil texture. Figure 29-7 presents plots of two linear regression lines based on experimental measurements acquired in 1974 at a test site near College Station, Texas, and in 1975 at a site near Lawrence, Kansas (Ulaby et al, 1978). The 1974 soil was Miller clay with 49-percent clay content, whereas the 1975 soil was Eudora silt loam with only 17.2-percent clay content. The two regression lines show a substantial difference in sensitivity (slope). A similar difference in sensitivity due to soil texture was observed by Schmugge (1980) in a study of the effects of texture on passive microwave response to soil moisture. Airborne data acquired over test sites located near Phoenix, Arizona, and in Imperial Valley, California, showed a weaker sensitivity to moisture content of heavy soils (high clay content) than for light soils. To incorporate soil texture in the microwave response to soil moisture, the latter parameter was expressed in terms of percent of field capacity m_{fc}. The same conversion to percent of field capacity used by Schmugge (1980) was applied to the radar data for 1974 and 1975, and the resulting regression lines, shown in Figure 29-7a, are in better agreement than those shown in Figure 29-7b. Although these results suggest that the dependence of σ^0 on soil texture can be removed by expressing moisture content in percent of field capacity, it is apparent that additional experiments that cover a wide range of soil texture are needed.

Progress is being made in the application of active microwave measurements to soil moisture. Results from ground-based experiments such as those described by Ulaby and Batlivala (1976), Ulaby et al (1977), and Ulaby et al (1978), which indicate that C-band observations are superior to other wave-lengths, have been corroborated by

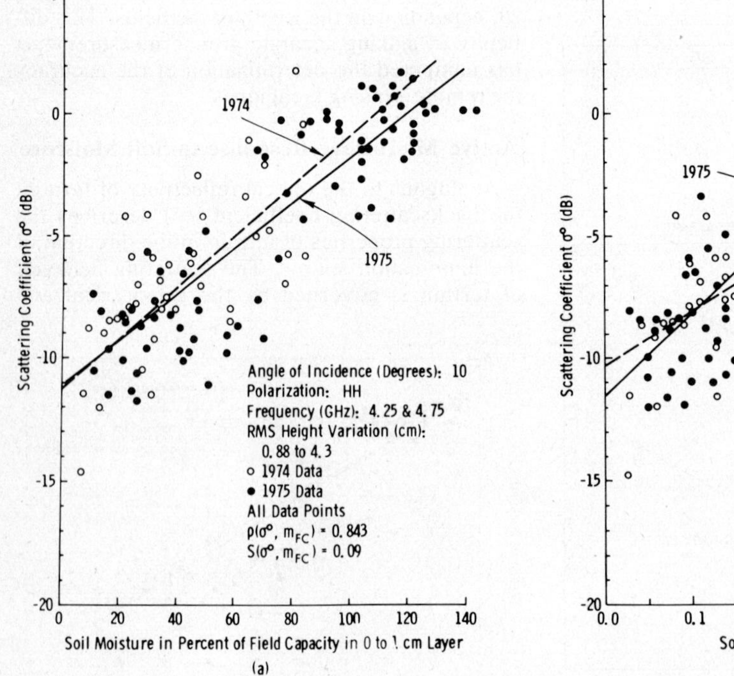

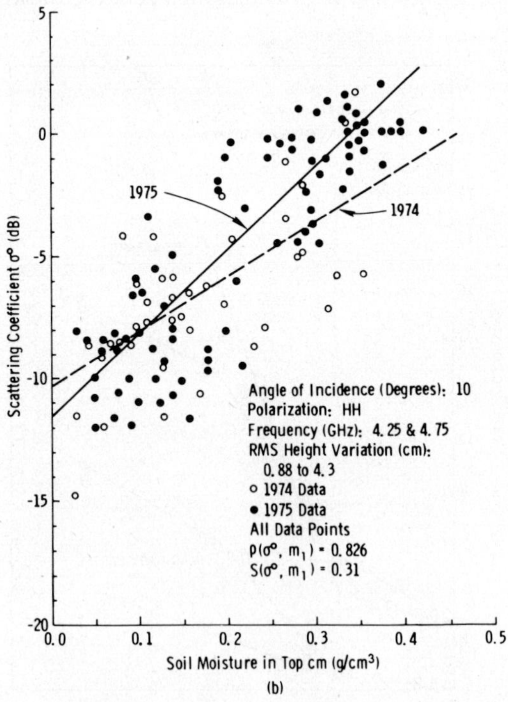

Fig. 29-7. Backscattering coefficient plotted as a function of soil moisture given (a) in % of field capacity of top 1 cm and (b) volumetrically in top cm. The 1974 and 1975 bare soil experiment data are combined (Batlivala and Ulaby, 1977).

airborne scatterometer measurements (e.g., Jackson et al, 1980). Some preliminary results from the Seasat Synthetic Aperture Radar have also been encouraging and are described in a later section.

Gamma Radiation Attenuation Technique

Gamma-radiation attenuation-techniques for the determination of soil moisture are based on the difference between the natural terrestrial gamma-radiation flux measured for comparatively wet and dry soils. The presence of moisture in the soil causes an effective increase in soil density resulting in an increased attenuation of the gamma flux for relatively wet soil and a correspondingly lower flux at the ground surface. The gamma flux near the ground originates primarily from the ^{40}K, ^{238}U, and ^{208}Tl radioisotopes in the soil. In a typical soil, 91 percent of the gamma radiation is emitted from the top 10 cm of the soil, 96 percent from the top 20 cm, and 99 percent from the top 30 cm (Zotimov, 1968). Other sources of radiation that contribute to the measured gamma flux include the daughter products of radon gas in the atmosphere (^{214}Bi), high-energy cosmic particles (i.e., greater than 3.0 MeV), and trace sources of radioactivity within the aircraft and the detection system itself (Bissell and Peck, 1974).

With this method, reliable, real-time mean areal soil-moisture measurements can be made for the upper layer of soil if both background and current uncollided terrestrial gamma-count rates and background soil-moisture data are available.

To test the approach, an aircraft carrying a gamma-ray detector and counter was flown over test sites in North Dakota and Minnesota (Carroll, 1981). The aircraft was at an altitude of 150 m and measured the natural terrestrial gamma radiation from a swath of 300 m. Therefore, the radiation data collected over flight lines of 15 to 20 km are mean areal measurements over approximately 4.5 to 6 km^2. This ability to measure mean areal soil moisture is attractive in the light of the extensive ground sampling required to estimate mean areal values with accuracy.

The aircraft data were compared with 0- to 20-cm gravimetric measurements made at 25 to 30 locations on each flight line. These ground data were used to determine background soil-moisture conditions and to serve as an independent check on the airborne soil-moisture computations. Results of the comparison of airborne estimates of soil moisture and gravimetric measurements for 47 flight lines yielded an r-squared value of 0.87.

Because of radiation sources present in the atmosphere, it is necessary to fly at low altitudes to minimize the counts from extraneous sources. This makes the technique impractical for regions with significant relief. Also, the spatial resolution is twice the aircraft altitude or about 300 m, and at least 5 seconds of data are required to obtain sufficient counts. Therefore, the spatial resolution is limited to about 300 by 300 m even for low-altitude aircraft. With these types of limitations in mind, the gamma-ray technique can provide useful areal averages of surface soil moisture.

Estimation of Water Content in the Root Zone

Remote-sensing techniques can provide estimates of the surface-soil water content. Through the use of microwave techniques, soil-moisture estimates can be obtained for soil depths of 5 to 10 cm. However, this depth is still shallow since the root zone for many plants is a 1- to 2-m zone. Jackson (1980) proposed a method of estimating profile soil-moisture from surface measurements. He assumed that, at some time during the day, the hydraulic potential would be near enough to equilibrium so that the soil water content profile could be calculated from a knowledge of the water characteristic and the surface moisture content. Jackson, used a simulation model to calculate profiles for several soil textures. The model produced error-bounds for various surface layer thicknesses, time of day, and conditions of rainfall and evaporation. He concluded that, if the water content of the surface 10 cm is known, the average water content in the 0.1- to 1-m layer could be estimated with a standard error of 0.04 cm^3. This was for a wide range of surface-moisture conditions in which the average moisture in this deeper layer was 0.25 cm^3. These optimum results were obtained using the surface water contents measured before dawn when the moisture profile most closely approximated the hydraulic equilibrium assumption. These results were for bare soil situations that produce the most extreme soil-moisture profiles. It is expected that the soil-moisture profiles under vegetated conditions would be less extreme and therefore yield better results.

This is one example of efforts to use the surface water-content measurement to obtain information concerning the water status of the root zone. Efforts are continuing to be directed toward adding a function to simulate water absorption by roots within the profile. Other work includes the use of remote-sensing data in water budget models (e.g., Bernard et al, 1981). Time series measurements that detect stress (e.g., thermal infrared techniques) can be used to infer root zone water contents (Idso and Ehrler, 1976).

INDIRECT ESTIMATE OF SOIL MOISTURE

All the approaches previously discussed were primarily dealing with conditions when little or no vegetation was present. If a sufficient amount of vegetation (not yet precisely defined) is present, the condition of the vegetation may be observed and may possibly serve as an indicator of soil-moisture status in the root zone.

The temperature of a crop canopy compared

with the ambient air temperature has been shown to be related to crop stress (Jackson et al, 1978). For example, if the canopy temperature is several degrees lower than the ambient air temperature, it is transpiring freely and being cooled, indicating that moisture is readily available in the root zone. However, if the canopy temperature is equal to or greater than the ambient air temperature, it is quite likely that the plant is under considerable stress, which can be associated with soil moisture. Jackson et al (1981) have developed a crop water stress index (CWSI) that illustrates this in the form of results such as those shown in Figure 29-8.

The crop water stress index is derived from energy balance considerations. The associated equation, where $T_c - T_A$ is the canopy (c) − air temperature (A) difference, is:

$$T_c - T_A = \frac{\gamma(1 + r_c/r_a)}{\Delta + \gamma(1 + r_c/r_a)} \cdot \frac{r_a R_n}{\rho c_p} - \frac{(e_s - e)}{\Delta + \gamma(1 + r_c/r_a)} \qquad (29\text{-}2)$$

where

γ = psychrometric constant (Pascals C^{-1})
Δ = slope of the saturated vapor pressure-temperature relation (Pascals C^{-1}) evaluated at the mean to T_c and T_A
r_c = crop resistance (s m^{-1})
r_a = aerodynamic resistance (s m^{-1})
R_n = net radiation (W m^{-2})
ρ = density of air (kg m^{-3})
c_p = specific heat of air (J kg^{-1} C^{-1})
e_s = saturated vapor pressure of the air (Pascals) at temperature T_A
e = actual vapor pressure of the air (Pascals)

Equation 29-2 shows that $T_c - T_A$ is dependent on the vapor pressure deficit and the net radiation that can be estimated from incoming solar radiation data. The remaining term to be evaluated is r_a. One way of accomplishing this for a crop is to measure $T_c - T_A$ when the plants are no longer transpiring. Under these conditions, $r_c \to \infty$ and Eq. 28-2 becomes

$$T_c - T_A = r_a R_n/\rho c_p \qquad (29\text{-}3)$$

from which r_a can be calculated. Equation 29-3 is then solved for the ratio r_c/r_a that can be used in the relation:

$$TR/ET_p = (\Delta + \gamma)/[\Delta + \gamma(1 + r_c/r_a)] \qquad (29\text{-}4)$$

to obtain the ratio of transpiration (TR) to potential evapotranspiration (ET_p). The ratio TR/ET_p, or its complement 1-TR/ET_p, is defined as the crop water stress index. The index is calculated from a one-time measurement of surface temperature, air temperature (wet and dry bulb), with an estimate of net radiation. Under uniform environmental conditions, it may be possible to relate the ratio to the average daily evapotranspiration.

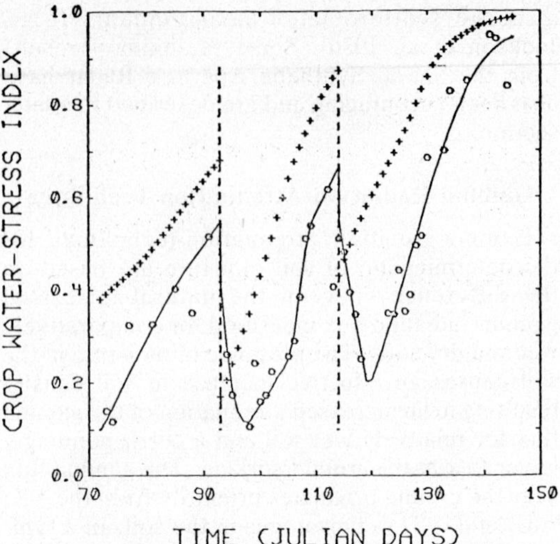

Fig. 29-8. The crop water stress index (CWSI) as a function of Julian days for a wheat plot that received two post-emergence irrigations. Circles represent the data points, and the solid line was drawn by eye to show the trend of the data points. The plus (+) symbols represent the extractable water used from the 0- to 1.1-m depth (Jackson et al, 1981).

An airborne or spaceborne infrared scanner may effectively "map" ground temperatures. Thermal imagery of agricultural fields has been presented by Blad and Rosenberg (1976), Heilman et al (1976), Bartholic et al (1972), and Millard et al (1978). A recent example is provided in Color Figure 29-9 (Hatfield et al, 1980). This image was acquired on May 3, 1979, from an aircraft overflight of a 50-km transect in the Sacramento Valley of California using a Texas Instruments RS-25 Scanner (10.5 to 12.5 μm). As can be illustrated by the color scale accompanying the image, the range of temperatures was over 20°C (22° to 44°C). To be useful as a soil-moisture status estimation technique, imagery at frequent intervals is needed, perhaps several times per week during periods of high evapotranspiration. The cost of operating an appropriate airborne system and its auxiliary computer requirements may be quite high, but may be economical for large irrigation districts or other institutions that have responsibility for water distribution over thousands of hectares.

With repeated acquisition of imagery, temperatures of individual fields can be compared with other fields of known water status, with bare soil areas or with ponds or streams to correlate the field's temperature with its soil-moisture status. In this case, air temperatures and vapor pressures above the crop would probably not be available for references, but field-to-field comparisons on a particular day would compensate for day-to-day environmental changes that cause temperature differences unrelated to stress conditions.

Satellite systems are capable of measuring surface temperatures from space (Price, 1980). Equation 29-2 serves as a guide as to when temperature differences due to irrigation or soil-moisture differences might be detectable. In arid and semiarid areas where water vapor in the air is low, the vapor pressure deficit will be the dominant term in Eq. 29-1, and the canopy-air temperature difference will be negative. In humid areas, the radiation term will dominate, and surface temperatures will be equal to or warmer than the air. On this basis, large temperature differences between fields in the Eastern United States are not expected, whereas large differences may be evident for the irrigated valleys of the Western United States.

More recently, Cihlar (1980) has studied the interrelationships of soil water on remotely measured surface temperatures, including plant canopy temperatures. The results of this study seem to show that plant canopy temperatures relate well to soil moisture in the upper layers of the soil when there is a full canopy cover obscuring the ground. Furthermore, any such relationships appear to be specific to the crop involved. Therefore, in attempting to make absolute estimates of soil moisture, as opposed to relative and qualitative observations in space or time, such factors should be carefully considered. Further discussion on the use of surface-temperature measurements versus soil moisture or crop observations versus soil-moisture status are provided by Soer (1980), Byrne et al (1979), and Heilman and Moore (1981b).

When plants are subjected to high levels of moisture stress, their reflectance in the near infrared (0.7 to 1.2 μm) tends to decrease, because of wilting and deterioration of the plant leaf structure. If stress conditions exceed the plant's tolerance, the plant may die, thereby exposing more of the soil surface and lowering the reflectance in the near infrared for that particular portion of the scene. Therefore, crop status over large regions can be monitored, and a general assessment of moisture conditions can be ascertained. Thompson and Wehmann (1979, 1980) have discussed the use of Landsat data for this purpose. For more information on moisture stress in crops the reader should consult Chapter 33.

In the middle infrared, the reflectance of plant canopies is related to leaf water absorption. Therefore, data taken in these spectral regions may be related to soil-moisture status in the root zone. Tucker (1980) discusses this possibility and relates it to the potential usefulness of such observations from Landsat-D.

Evapotranspiration is very closely related to soil moisture and therefore is included in this section. Certainly, it is a major component of the water balance of the soil-moisture profile (Eq. 29-1). The actual evapotranspiration is often estimated by calculating the potential evapotranspiration and multiplying by a crop coefficient. The identification of crops may be done by remote sensing (e.g., Landsat or airborne imagery) and used to obtain this crop coefficient. Although not much has been accomplished in these general areas, Jackson et al (1981) have suggested that crop coefficients could be estimated using a spectral index such as a ratio of bands in the near infrared and visible. When applying this approach to a wheat crop in Arizona, they found that the results closely paralleled the crop coefficient for small grains developed by Jensen (1974).

DATA AVAILABILITY

There are very little data available that can be used by the hydrologist directly for soil-moisture estimation using remote sensing. The previous sections have shown the status of remote sensing and soil moisture or evapotranspiration as being very much in the development stages. An extensive data base of soil-moisture values derived by remote sensing is not currently available. Hand-held, ground-based radiometers operating in the reflected solar and thermal infrared have been applied in research studies to measure parameters related to soil moisture and evapotranspiration. In addition, there are instruments that obtain thermal infrared, reflected solar, and microwave measurements when operated from various aircraft, including the NASA aircraft listed in Table-1a. A basic tenet of soil-moisture remote-sensing research is that, if these instruments were flown and concurrent ground truth obtained, it may be possible to use remote sensing to extend ground-truth benchmark measurements to fill in the detail between the point measurements.

Only a very few spacecraft measurements related to soil moisture have been available. These will be described in the next section through the use of illustrative results. In the future, the use of a thermal band on the operational environment satellites, including the geosynchronous satellites, may permit soil-moisture-related observations to be derived over large areas. Microwave instruments flown for meterological or oceanographic purposes may also provide observations related to soil moisture, but the future availability of remote-sensing observations primarily configured or tailored for obtaining soil-moisture observations is not likely for some time. The future of remote sensing of soil moisture, at least from space, will heavily depend on the development of information-extraction methods applied to data acquired either by airborne instrumentation or by instruments installed for another primary purpose on spacecraft.

APPLICATION EXAMPLES AND DISCUSSION

The preceding discussion of the applicability of various parts of the electromagnetic spectrum for determining soil moisture and evapotranspiration, as well as associated remarks on data analyses methods and data availability, have illustrated many application examples. However, to illustrate some of the ways in which data from spacecraft have been examined to assess their ability for soil moisture estimation, the following examples are provided.

21-cm Results from Skylab

In experiments conducted from tower- and aircraft-platforms, it has become apparent that, for the passive microwave approach, longer wavelength systems are more effective for soil-moisture sensing. The data acquired from the 21- and 2.2-cm radiometers on Skylab and the 15.5-cm Electrically Scanning Microwave Radiometer (ESMR) on Nimbus 5 provide an opportunity for analyzing the effectiveness of such systems for soil-moisture sensing. (Refer to Chapter 13, Microwave and Infrared Satellite Remote Sensors, for specifics on the Skylab radiometers.)

An indication of the greater sensitivity of a 21-cm radiometer for sensing soil-moisture variations is given in Figure 29-10, in which data at the 1.55-, 2.2-, and 21-cm wavelengths are presented for a swath across north central Texas on June 5, 1973 (Eagleman and Lin, 1976). The 21-cm radiometer shows a 45°K decrease in T_B, which Eagleman and his coworkers had related to increased soil moisture, whereas the 2.2-cm radiometer shows only a 15°K decrease and the 1.55-cm radiometer shows essentially none. Primarily, the implication is that the shorter wavelength systems are prevented from observing any soil-moisture change by the vegetation cover, as was noted earlier, whereas the 21-cm system is able to observe soil-moisture changes through the naturally occurring vegetation of the area.

With a footprint on the ground as large as that of the 21-cm Skylab radiometer (115 km), the acquisition of realistic surface measurements for soil moisture is impossible. Therefore, a surrogate index or parameter for soil moisture is used. Based on the obvious but complex relationships between precipitation, evapotranspiration, surface and subsurface runoff, and soil moisture, the precipitation history over an area is commonly used to infer the soil moisture. Many models for characterizing the precipitation history have been devised. One of the simplest in concept and computation is (Linsley et al, 1958):

$$API = \left(\sum_{i=1}^{n} \right) K^i P_i \qquad (29\text{-}5)$$

where

API = antecedent precipitation index
p_i = daily precipitation for each day from n days previous to the current day

The parameter K, which is less than unity, characterizes the loss of moisture from the soil caused by evapotranspiration and subsurface runoff, and is a function of soil type, slope, season, and vegetative cover. The values normally are empirically assigned in the range of 0.85 to

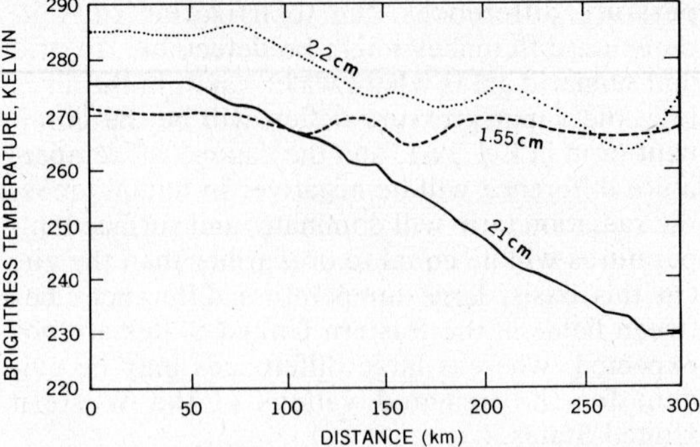

Fig. 29-10. Variations of the Skylab 2.2-cm and 21-cm (Eagleman and Lin, 1976) and the Nimbus-5 1.55-cm brightness temperatures for a 300-km swath across north central Texas on June 5, 1973.

0.95. The K value may either be constant or vary as a function of time.

McFarland (1976) made a comparison of *API* with the S-194 T_B values for a 930-km pass on June 11, 1973, starting over the middle of the Oklahoma panhandle and ending at the Texas-Louisiana border at 32°N. The S-194 T_B values over the study area ranged from 229.8° to 275.2°K. These values for an assumed emitting skin depth temperature of 298°K (the approximate air temperature along the ground track at the time of the data pass) would produce an emissivity range from 0.77, for very moist terrain, to 0.92 for very dry terrain, both vegetated. Beyond the study area, the brightness temperatures decreased to 95°K over the Gulf of Mexico for an emissivity of 0.31 (water temperatures were assumed to be near 300°K for airborne PRT-5 thermal infrared readings).

The study area includes conditions ranging from the loose sandy soils and sparse vegetative cover of the high plains of the Texas and Oklahoma panhandles to the tight clay soils and heavily vegetated terrain of eastern Texas where the lowest T_B values were observed. Weather conditions at the time of the pass at 1518 to 1520 Greenwich Mean Time (GMT) (1018 to 1020 Central Daylight Time (CDT)) varied from thin broken cirrus over the Texas and Oklahoma panhandles to multilayered overcast conditions from just south of the Red River to the Louisiana border. Precipitating moderate thunderstorms, with an areal coverage of approximately 30 percent, were occurring from the Fort Worth area to near 100 miles south-east along the ground track. Their rainfall amounts were generally light since the cells, as determined from weather radar film, were 3 to 5 miles in diameter and were moving toward the north at 20 knots. The air temperatures along the track ranged from 294° to 299°K.

The precipitation totals for the first 11 days of June 1973 ranged from zero in the Texas and Oklahoma panhandles to nearly 25 cm (10 inches) in the Dallas area. To eliminate the influence of very high daily point values of precipitation in the calculation of the API, the maximum daily rainfall for the API calculation was arbitrarily set at 5.08 cm (2.0 inches). The physical rationale for this assumption is that amounts more than 5.08 cm contribute to immediate runoff, but probably do not contribute to increased soil moisture. The API was calculated for each of the 180 precipitation-reporting stations of the National Oceanic and Atmospheric Administration (NOAA) climatological network along the ground track, using the rainfall data of the preceding 11 days. Comparisons of T_B with an *API* calculated by using the preceding 5-day precipitation yielded poor agreement. This is because most of the precipitation that contributed to the *API* values occurred more than 5 days before the Skylab pass, which indicates that the radiometer is responding to rain effects even before this 5-day interval.

The arithmetic average of the *API* for the 115-km diameter footprint, coincident with the position of the spacecraft for every third data point, was then calculated for correlation with the S-194 T_B. After trials within the range of 0.85 to 0.95, the value of the parameter K was set to 0.9, which resulted in the correlation of *API* to S-194 brightness temperatures. The S-194 T_B at every third data point and the footprint *API* are displayed in Figure 29-11. The spatial overlap between adjacent points presented in this figure is about 87 percent. For this 930-km pass, there are only eight independent sensor footprints. The trend which develops when the plot of these points is analyzed indicates that there is a strong relationship between the S-194 t_B and *API;* the correlation coefficient between the two for those points with $T_B < 270°K$ is -0.97.

In an analysis of other Skylab data, Wang and

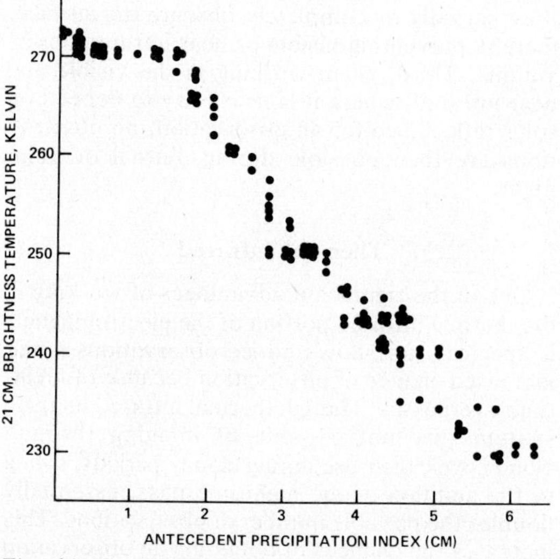

Fig. 29-11. Antecedent precipitations index (cm) (McFarland, 1976).

Schmugge (unpublished results, 1980) studied two pairs of parallel passes that were separated by 5 days. One pair was in June (4 and 9) 1973 and the other pair was in September (13 and 18) 1973. The June passes went from the northwest corner of Montana down to the southeast corner of Missouri. There were heavy rains along the eastern portion of the pass before the June 4 pass. T_B values as low as 220°K were observed over eastern Nebraska. Five days later, the T_B had increased to the 260° and 270°K range. The September passes started over western Texas at the Mexican border and ended south of Lake Michigan. T_B values on both passes ranged from 260° to 270°K, at the southwestern end, to 220°K, again over Oklahoma and Kansas where the *API* values were greater than 1.5 inches. A comparison of T_B and *API* values was made for the data from these four passes, where the *API* values for each separate footprint were calculated using the approach of McFarland described earlier. For these four passes, there were a total of 73 non-overlapping footprints, plus another eight from the pass analyzed by McFarland. The correlation coefficient for these data was 0.84. This agreement is quite acceptable considering the wide range of terrain and vegetation conditions included in the data and the large variability of rainfall that occurs over a 115-km footprint.

In an attempt to provide a more quantitative relationship between S-194 T_B and soil moisture, Eagleman and Lin (1976) incorporated the meterological data into a moisture budget model to obtain estimates of soil moisture. These calculations were calibrated using several hundred ground measurements of soil moisture that were made within 1 day of the Skylab pass. The soil-moisture information from these two sources was used to prepare soil moisture contour maps, from which average values of soil moisture over a footprint could be calculated. They obtained data for 11 footprints from five Skylab passes over Texas and Kansas. The scatter plot of T_B and their 0- to 2.5-cm average soil-moisture values are presented in Figure 29-12. These data yield a correlation coefficient of -0.96, which again is very good considering the large footprint of the sensor.

The S-194 results indicate that a 21-cm radiometer definitely responds to surface moisture variations resulting from recent rainfall (i.e., either standing water or moisture in the surface layer of the soil). On the basis of ground-based observations, it is presumed that the radiometer is primarily responding to surface soil-moisture variation. Therefore, it is clear that these spaceborne microwave radiometers have been able to detect moisture effects of recent rainfall, which provides an indication as to the geographical extent and amount of precipitation. In this mode, radiometers could be used in conjunction with the visible-IR imagers to better estimate rainfall over land. To develop the applications of spaceborne radiometers for more quantitative es-

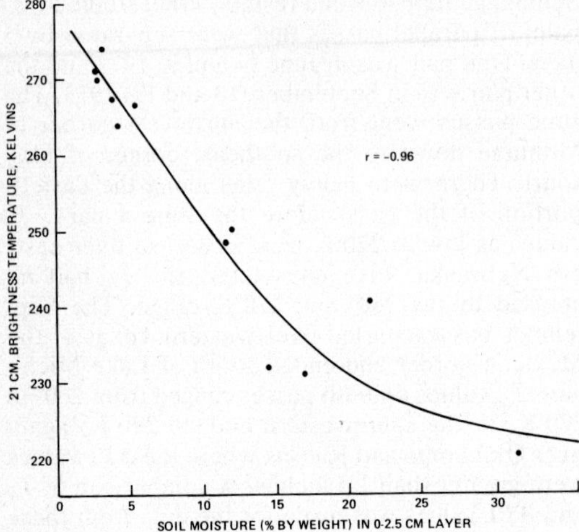

Fig. 29-12. Scatter plot of T_B and their 0- to 2.5-cm average soil-moisture values for five Skylab passes over Texas and Kansas.

timates of soil moisture for applications to crop yield, water-shed runoff- and evapotranspiration-models will require considerably more data than those available from the Skylab mission. In retrospect, if the S-194 sensor had been operated during the entire 9-month period that Skylab was active, a much larger data set could have been obtained for the types of analyses discussed here.

Seasat Synthetic Aperture Radar

The L-band Synthetic Aperture Radar (SAR) on Seasat has provided a number of high resolution (25 m) images over land areas. These are presently being analyzed for their hydrological content. An example is given in Figure 29-13, which is a SAR image over central Iowa obtained at 13:30 GMT on August 16, 1978. The area covered is 100 by 100 km at 80-m resolution. Ames, Iowa, is in the lower left corner of the scene. The field patterns are clearly evident and result from different tillage practices or conditions producing different surface roughnesses. The brighter tone on the left side (west) is due to rainfall, and bright streaks heading off to the northeast are the result of rain cells that moved out from the main rain system. The increased wetness is apparent in this scene.

SNOW AND ICE MAPPING AND MONITORING

APPROACH/DATA ASSESSMENT

Snow and ice on the Earth's surface are very distinctive features that can be detected readily with a variety of remote-sensing techniques. Various wavelength bands are sensitive to differing snow or ice properties; therefore, careful consideration should be made of the observational capabilities of each region of the electromagnetic spectrum before deciding on the techniques to be used in a particular situation.

Visible and Near Infrared

In the visible and near infrared region of the spectrum, several advantages are associated with observing snow and ice. There is a strong reflectivity contrast between snow and non-snow areas, which facilitates the mapping of areas covered by snow. This striking contrast exists for most forms of ice; however, the ease of mapping ice cover is not at quite the same level as that of snow because of the low reflectivity of certain ice types.

The Landsat bands have certain advantages associated with this inherent high snow/no snow contrast. The large contrast appears very distinctively in bands 4 and 5, with the latter band being preferable because of less atmospheric disturbance. The near-infrared bands (bands 6 and 7) do not possess such strong snow or ice reflectivity contrasts, but differences in reflectivity noted in these bands have been traced to the occurrence of surface melting, which is important for runoff or breakup forecasting. Beyond the range of Landsat 1, 2, and 3 capabilities in the middle near infrared (e.g., 1.55 to 1.75 μm), the reflectance of snow falls drastically until it is less than that of surrounding non-snow surfaces and, more importantly, much less than cloud reflectance. The advantage here is the potential for automatic discrimination of snow and clouds, which can significantly improve the determination of snow-covered area.

Certain disadvantages in the visible and near infrared should also be considered. First, only surface conditions are observed, which are just a part of the total information content needed for the management of snow and ice. Subsurface snow and ice properties are at least equal in importance to surface features. Second, clouds are frequently present in snow and ice regions, and they partially or completely obscure the surface, thereby preventing visible or near infrared observations. Third, when working in the visible and near infrared, where it is necessary to depend on solar reflectance for an observation, no observations are then possible during diurnal or polar night.

Thermal Infrared

One of the significant advantages of working in the thermal infrared portion of the electromagnetic spectrum for snow and ice observations is the increased chance of observation because of nighttime overpasses. Though thermal infrared imaging systems are not capable of imaging through cloud-cover, their use during cloudy periods, owing to the addition of the nighttime pass, essentially doubles the possible number of observations. This increases the chances of obtaining an observation during a cloud-free period. In the polar regions, this advantage further extends to making thermal

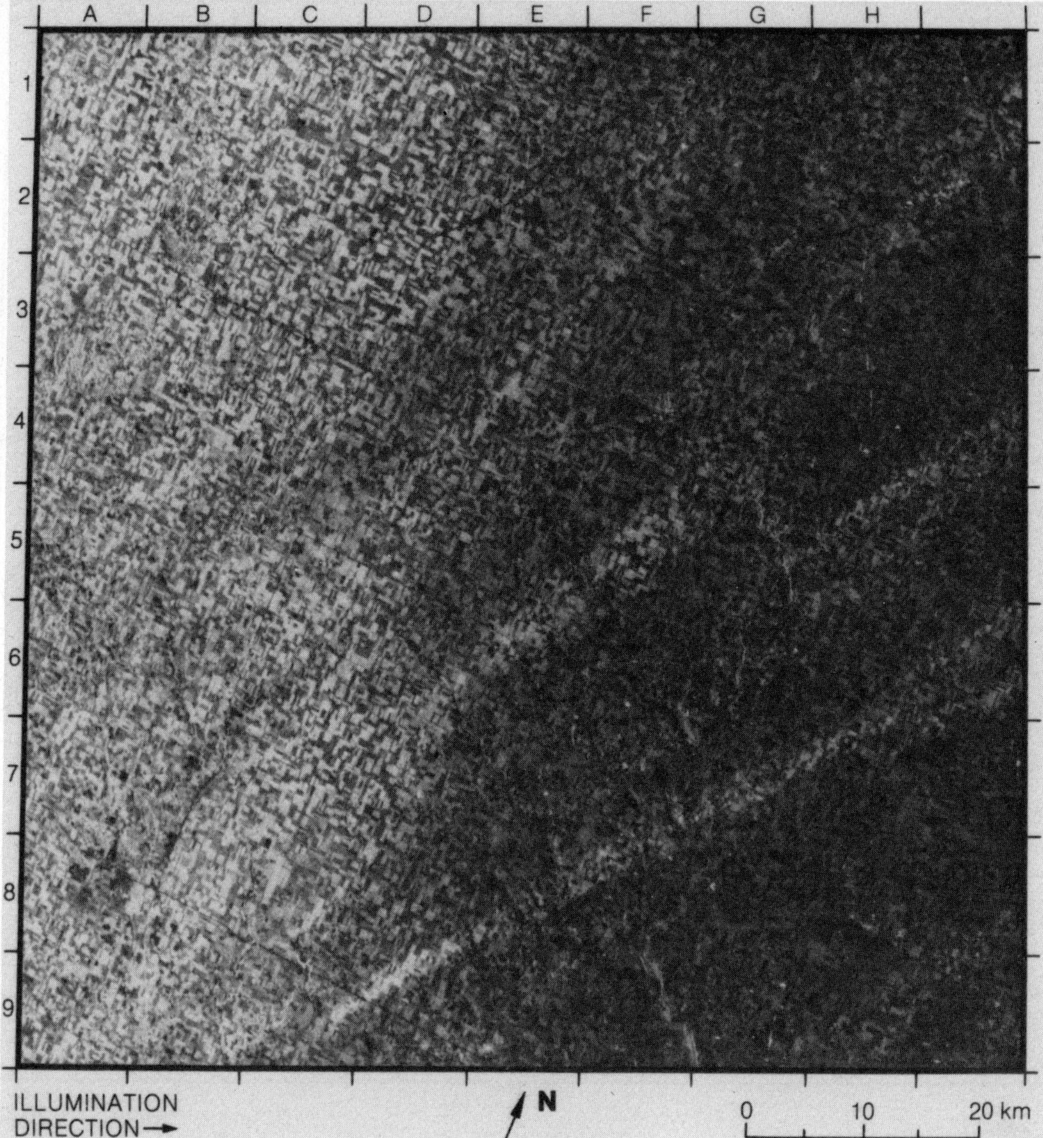

Fig. 29-13a. Plains of central Iowa SAR image. On August 16, 1978, the Seasat satellite passed over central Iowa near Ames (A7) and obtained this image. During the late evening of the previous day, a large frontal storm had moved into the vicinity from the west and northwest. Typical mid-latitude frontal convective storms occurred for several hours approximately 10 to 12 hours before the acquisition of the image. Prior to the passage of the satellite, the storm broke up and several isolated cells moved to the east and northeast, distributing rain along the respective ground tracks. The result is seen in this image where the light areas to the west reflect the general location of the major frontal precipitation activity and the several bright paths across the Iowa farmland denote the tracks of the isolated convective rain cells. The several dark areas are those where precipitation was very light or absent. Other items of interest are the patchwork pattern on the rolling topography, which is intensely farmed, and the rough texture of the more steeply inclined and unfarmed river courses (e.g., at A3 and G5 in Fig. 29-13a). Although many railroads crisscross the imaged area, the few that are readily perceptible are those whose orientation parallels the ground track of the satellite, (e.g., at H4 in Fig. 29-13a). (Seasat image from Rev. 723.) (Ford et al, 1980).

observations throughout the winter when cloud conditions permit. Observations of the snow-pack in the thermal band can be used to delineate the area covered by snow because of the temperature contrast with the snow-free areas, especially in the daytime during spring snowmelt. In addition, the temperature of the snowpack can never rise above 0°C so that observation of the surface temperature can be used, first, to identify when the snowpack surface reaches 0°C during the day and could possibly be melting and, second, to observe when the 0°C temperature is maintained diurnally indicating a possible ripe snowpack.

When consideration is given to the use of satel-

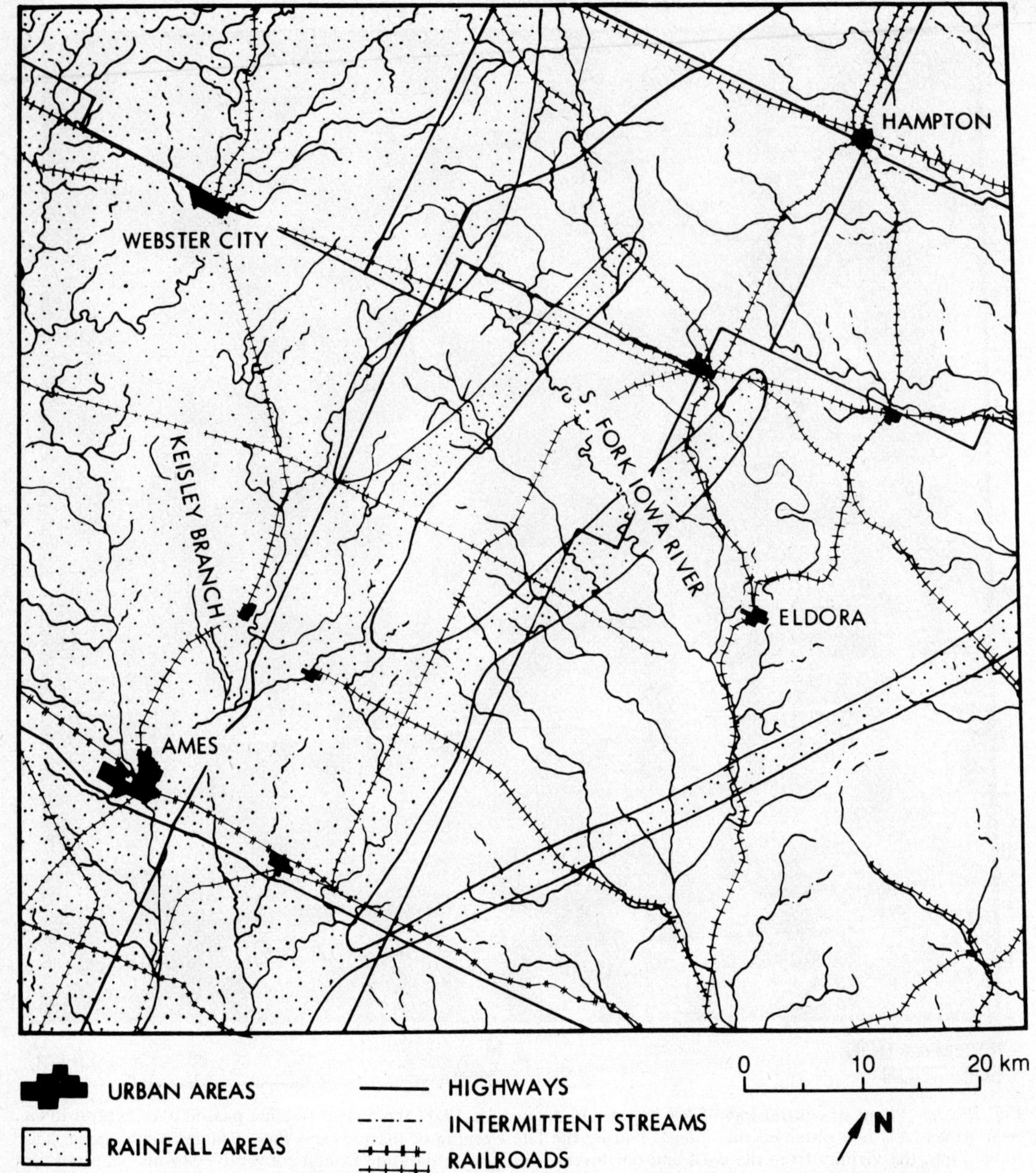

Fig. 29-13b. Plains of central Iowa SAR image (Continued).

lite thermal infrared data for snowpack monitoring, it is found that the spatial resolution is generally poorer than that of the visible and near infrared, thus resulting in a less detailed monitoring capability. The best resolution and most reliable thermal infrared data have been those from the HCMM satellite with 500-m resolution (Price 1980). Visible data have had as much as an order of magnitude better resolution (40 m on Landsat 3 RBV). Except during spring snowmelt or ice-breakup conditions, the contrast between snow and ice and surrounding surfaces is usually much less in the thermal than in the visible and near infrared. This directly results in a greater error in mapping of snow- or ice-cover conditions. As is true of the visible and near infrared data, only a surface attribute of the feature (in this case, temperature) is recorded, and little or no information content with depth is available. Finally, unless care is taken, significant problems can arise in mountainous terrain where variations in temperature with elevation can be confused with snow cover temperature differences.

Microwave

When working in the microwave wavelength region, several very significant and important advantages are evident for the glaciologist. Emission and reflection of microwave radiation from snow and ice surfaces is strongly affected by subsurface properties, thereby permitting the possibility of inferring information with depth. Another important advantage of the microwave region is that, depending on wavelength, microwave radiation will penetrate clouds and most precipitation, thus providing an all-weather observational capability. This is very significant in snow and ice regions where clouds frequently obscure the surface. Through the use of active microwave techniques, resolution from space can be as good as 10 m, which is sufficient for detailed analysis of snow and ice properties.

Passive microwave resolution from space is inherently poor because of the large antenna sizes required, but improvements in the next few years are foreseen. Microwave interaction with the snow and ice is extremely complex, especially in the active microwave case, and the result is that data interpretation is extremely difficult. This basic complexity is further confused by the rapid changes in snow and ice characteristics, such as crystal size and liquid-water content, that are possible under varying climatic conditions. The dielectric constants of water and ice are so drastically different that even a little melting will cause a strong microwave response. Because of the uncertainties in the microwave interactions, significantly more ground information is needed for microwave snow and ice studies than for comparable visible, near infrared, and thermal infrared studies.

Integrated Approach

Because of the significant changes in remotely sensed properties that can result from the complex physical processes prevalent in natural snow and ice fields, reliance on only one type of sensor is not recommended. Rather, a multisensor approach should provide an improved monitoring capability. It is foreseen that the following combination may eventually provide a nearly complete characterization of the snowpack.

Visible sensor(s) would provide regular coverage of snow- and ice-covered areas and permit one to determine the date of disappearance of snow or ice from a particular area. In addition, the visible spectrum would be used to obtain snow and ice reflectivity information to be used in determining the distribution or radiation input in energy balance studies. Near-infrared bands would be used to delineate areas of surface melting and for automatic discrimination between snow and clouds. Thermal infrared would be used to delineate the areas of the pack ready to melt and perhaps to infer the ripeness of the snowpack. Nighttime thermal observations would also be used to improve the chance of cloud-free observations. Passive microwave radiometers would be used to give regional and even continental observations of snow-cover extent during cloudy conditions to supplement the operational visible observations and, as resolutions improve, the observations could be used on basins larger than 10,000 km^2. In addition, because of the surface penetration capability, the passive microwave would be used on the high plains and in large basins to obtain measurements of snowpack water-equivalent and liquid-water content (at the surface and with depth). For ice, this technique would be used to estimate ice thickness on large water bodies. The active microwave sensors would be used in a manner similar to the passive sensors with the major differences being their use in smaller basins, in mountainous terrain, and over smaller water bodies.

Utilization of this integrated approach would then enable the scientist to almost completely characterize the snowpack and greatly improve his understanding of ice conditions as well. Snow properties available for hydrological predictions include: snow-covered area; areal snow water equivalent; status of melting; and presence and/or quantity of liquid water. If appropriate microwave frequencies are used, snow properties available would also include information on the hydrological condition of the underlying soil for areas as small as several km^2 up to areas as large as the central high plains. For ice, it will be possible to determine: ice type and condition; ice thickness; ice-covered area; and areas where the ice is actively melting.

ACQUISITION METHODS AND ALTERNATIVES

Utilization of remote-sensing data for assessing snow and ice conditions can be achieved using multistage observations, especially in the research phase. The eventual goal, when possible, is to operationally observe the various parameters from space. This goal is necessitated as these applications require the large area, synoptic view on a highly repetitive basis. These are capabilities inherently provided by sensors on automated "free-flyer" spacecraft. To realize such an observational capability, development of fundamental understanding through a hierarchical process (i.e., from the ground upward) is advocated. First, conventional snow and ice observations must be made so as to understand as completely as possible the characteristics of these materials. In addition, these observations will serve as ground truth for the multistage remote-sensing measurements. To provide the most detailed and controlled remote-sensing observations, ground-based remote-sensing platforms should be used. The next stage would be an extension of the ground-based results over larger areas using aircraft sensors, thus permitting observation of a greater variety of snow and ice conditions. An aircraft platform also serves to provide a transition from detailed local understanding to space observations. The first phase of space observation entails the research testing of the suitability of an optimum sensor configuration. Evaluation of the results can lead to the development of an operational spaceborne system, supplemented by limited conventional, ground-based, and aircraft observations. The observed data can then be used in conjunction with a wide variety of models, permitting calculation of the critical snow and ice parameters.

Conventional Measurements

Currently, conventional snow and ice measurements can be separated into manual and automatic observations. In manual snow surveys, the snow depth, water equivalent, and density of the snowpack are periodically (usually once a month during the snowmelt season) measured at preestablished snow courses. The snowline can also be observed from the ground, providing an approximation of the snow-covered area. More quantitative ground observations are obtained by digging a snow observation pit and by characterizing the temperature, density, water equivalent, crystal type and size, hardness, and layering of the snow profile. Measurements of the liquid water content of the snowpack at various levels can also be accomplished using conventional methods. Centrifuge, calorimeter, and capacitance techniques are all possible. The most field-reliable technique is now the freezing calorimeter (Jones et al, 1980). Capacitance techniques developed for research operations are recommended for use when available. For ice observations, various ice coring and ice-thickness kits have been developed by the Corps of Engineers (Ueda et al, 1975). Observations of ice type and cover from the ground are extremely difficult because of lack of viewing perspective.

In recent years, automated snow gages have been used at remote sites to provide high-frequency observations. Most commonly used gages for measurement of snow water equivalent are the pressure pillow (or tanks) and the radioactive profiler. Depending on the type of data relay system used, the snow water-equivalent measurements can be obtained, usually upon command. Regular daily measurements are commonly available, supplemented by much shorter period readings during emergency situations. Microwave telemetry of the remotely collected data using landlines has been common in the past, but more recently, relay of the data has been accomplished using ionized meteor layers and satellite data-collection systems. The U.S. Soil Conservation Service SNOTEL (SNOw TELemetry) system in the Western United States operates using the former system (Barton and Burke, 1977), and NOAA provides data relay capabilities using their Geostationary Operational Environmental Satellite (GOES) series of satellites (see Chapter 14, Meteorological Satellites). The use of automated sensors and various data relay methods has encouraged the increased use of snowmelt models that require highly temporal and real-time data.

Ground-Based Platforms

Probably the most reliable ground-based remote-sensing platform available is man himself. He can remotely observe in the visible spectrum, but there are outside factors that subjectively influence his interpretation of these data. To assist him in being more objective, hand-held instruments have been developed to record spectral characteristics in the visible, near infrared, thermal infrared, and even the microwave intervals. These instruments are most generally used for vegetation studies, but can also be used for measurement over snow. These techniques are best suited for trips into remote regions where access is only possible by skis or over-snow vehicles. Radar systems that sit on top of the snow (or slightly above it), as well as transmitters buried in the snowpack by Boyne and Ellerbruch (1979), have been used. Such techniques are especially useful in particular problem areas (e.g., avalanche hazard areas).

The use of these instruments in the hand-held mode is limited because of a lack of viewing perspective (sometimes the snow can be higher than the observer) and little control on angle of observation. As a result, it has been common to place instrumentation on a fixed platform or on a mobile truck facility. This permits the capability of looking down on the snowpack at carefully controlled distances above the snow and at repeatable viewing angles. Truck facilities provide the additional capability for moving to sites with varying snow conditions. Most truck systems used in snow experiments have used various configurations of active and passive microwave instruments such as described by Stiles and Ulaby (1980) and Chang et al (1980). Visible and infrared capabilities have been added to provide an ancillary observing capability. In conducting a remote-sensing experiment using truck-mounted instruments, the objective is to obtain a detailed understanding of the remote-sensing snow-properties interaction over a very small area. To accomplish this, an intensive program of conventional snow measurements must be undertaken to characterize the snowpack properties.

Although the same radiometers and scatterometers used from truck systems for snow are applicable for ice studies, the logistics of such ice studies are more complicated. In addition, various other instruments have been used. A short-pulse radar system has been tested on the Great Lakes by Vickers et al (1974) and has proved to be successful. The phasing of the return waves reflected from different surfaces (e.g., air-ice versus ice-water) was carefully measured to infer the ice thickness. Similar techniques using radio-echo devices have been used for glacier soundings (Watts and England, 1976).

There are limitations to ground-based observations. First, movement from one area to another, even with trucks, is particularly difficult during snow conditions. As a result, only limited observations of varying snow conditions and localities are possible. Second, selection of a representative location for truck measurements can be difficult; therefore, extension of the truck measurements to larger areas may be questionable. Finally, the resolution from a truck system is generally too high for operational snow hydrology applications and is more suited to detailed snow-physics studies.

Airborne Observations

To overcome some of the limitations of the ground-based observations, airborne remote sensing has been used as the next logical step in the hierarchical approach. Use of an aircraft platform allows the more effective collection of data for a variety of snow conditions and geographical areas. In the advocated hierarchical approach, aircraft observations are used to extend the higher resolution, specific-area ground-based truck measurements to larger and more representative areas. Aircraft-borne sensor resolution is generally lower than that of the truck measurements and therefore is more indicative of the kind of resolution that may eventually be possible in operational applications. Therefore, the aircraft level of sensing provides an intermediate step to eventual satellite observations.

Low Altitude Observations

Light aircraft for observations from less than a few thousand feet have been used for two primary purposes: visual snow-cover observations and gamma-ray estimates of the snow water-equivalent. In the first case, observations of the snowline or snow-cover extent are made at 1000 to 2000 feet. The snowline observed can be recorded on a previously prepared base map that can be later converted into snow-covered area. Supplemental information, such as melting condition of the snowpack and apparent snow depth inferred from the appearance of natural landscape features, can also be recorded. It is also possible to obtain a more precise estimate of the snow depth by aerial-oblique observations of specially designed snow stakes (Peterson, 1977). In addition, there are methods that have been used over small areas to measure the snow depth using photogrammetric techniques (Cooper, 1965). These low-altitude observations can be used to supplement or calibrate satellite observations, or they can be used to obtain the same data on basins too small for effective satellite application.

The detection of natural emissions of gamma radiation from the Earth has also been used to measure the water-equivalent of snow. Gamma-ray detection must be carried out at low altitudes (about 150 m) because of the significant atmospheric attenuation of gamma-ray radiation. Background gamma-radiation of the soil is obtained before snow falls and then periodically throughout the snow season; subsequent flights are flown to measure the gamma radiation through the attenuating snow cover. The degree of attenuation is related to the snow water-equivalent through various calibration curves. NOAA's National Weather Service has an operational program to obtain such data on the shallow snowpacks of the high plains of the upper Midwest United States (Carroll and Vadnais, 1980). Although results have been encouraging, there are significant drawbacks to this method. It cannot be used in mountainous terrain because of safety factors, and deep mountainous snowpacks also drastically attenuate the gamma radiation, thus further limiting its usefulness. In fact, it has been estimated that the method cannot be used to measure snow with a water equivalence of more than 40 cm (Bissell and Peck, 1974). Interpretation of the data becomes further confused when the soil-moisture level between the fall calibration-flight and the snow-season flights changes significantly. Because of the low altitude and narrow data swath-width, the method is confined to measurements of limited index lines. However, this technique seems to have significant importance for calibration of areal water-equivalent measurements using microwave sensors mounted on aircraft or space platforms.

Medium Altitude Observations

A variety of sensors can be flown on heavy-duty medium-altitude aircraft such as the NASA C-130 and Convair-990. Because of their load-carrying capability, they may carry both passive and active microwave instruments, plus sensors such as multispectral scanners, mapping cameras, and thermal infrared radiometers. For snow and ice studies, this ability to carry a multisensor complement is very useful because, although the microwave sensors provide data of primary importance, the other sensors can provide useful ancillary information. For snow- and ice-experiments, these aircraft usually operate at between 2000 and 25,000 feet in coordination with ground-data acquisition.

Because of the all-weather capability of microwave observations, aircraft-mounted systems have been used for ice measurements over large navigable water bodies such as the Great Lakes. For example, a cooperative program between the U.S. Coast Guard, National Weather Service, and NASA has been conducted during the winter on the Great Lakes to provide ice charts that show ice type and thickness (Schertler et al, 1975). Ice charts derived using microwave data are transmitted to ships in near real-time so that the information can be used for navigation to avoid ice areas that will impede vessel transit. Two microwave instruments are used. A side-looking airborne radar (SLAR) at 3-cm wavelength is mounted aboard a C-130 aircraft, and a swath width of 54 miles (100 km) is used for interpretation of ice type. A support aircraft carries a second instrument, an S-band (1.8 GHz) short-pulse radar, for measuring ice thickness. Combining the area-wide ice-type measurements with discrete flight lines of ice-thickness data permits the preparation of summary ice-charts that allow vessels navigating the lakes in winter to pick the path of least resistance.

High Altitude Observations

Aircraft operating above 40,000 feet provide the logical transition to spacecraft operations. Because of the high altitude, the fields of view of the sensors are large and somewhat analogous to those seen from space. Owing to power and weight constraints on efficient high-altitude operations, however, aircraft operating at these altitudes are limited to carrying fewer instruments than the larger platforms operating at lower altitudes. Probably the best known high-altitude research airplanes are the NASA U-2 and RB-57. Even from 60,000 feet, where these planes typically operate, photographic images over a snow-study basin of 228 km^2 may need to be mosaicked to get meaningful areal snow-cover data. For operational purposes, such platforms could not be used because of excessive costs and a slow data turnaround-time. However, for research purposes, the high-altitude approach can be of significant value. Aircraft imagery from 8.5 km over an experimental basin (40 km^2) in Switzerland was shown to be useful for application of a snowmelt model for runoff simulations (Martinec, 1972). In addition, 3-cm radar is now available from the RB-57, which should help in the testing of microwave capabilities for snow water-equivalent determinations.

Although significant contributions will result from the high-altitude observations in snow and ice research, it is expected that the major aircraft observing-platform will remain at medium altitudes. Use of airplanes at this level provides the greatest versatility in type and amount of sensors carried as well as flexibility of flight operations for coordination with ground-truth efforts. However, in snow and ice research, it is likely that ancillary data-acquisition will be required and obtained in specific situations using multilevel aircraft flights.

Spaceborne Observations

Although ground-based and aircraft remote-sensing observations of snow and ice are necessary steps in developing a working observational capability, the satellite perspective is the most advantageous for a number of reasons. Large-area viewing is maximized from space as is the capability for regular repetitive observation. Because of the vastness of the Earth's snow and ice resources, monitoring from space is the only efficient and effective alternative.

Currently, there are many satellites providing visible, near infrared, and thermal-infrared observations useful in snow and ice research. Landsat has provided a large amount of useful data on the areal extent of snow cover on basins as small as 10 km^2, and these data have been successfully used for snowmelt-runoff estimation (Rango and Peterson, 1980). For ice research, Landsat has been found to be useful in tracking ice-flow motions (Rango et al, 1973) and in identyfying ice cover, type, and distribution (McMillan and Forsyth, 1976; Sydor, 1976). At 80-m resolution, the most useful Landsat multispectral scanner (MSS) bands are 5 and 7. Both thermal and visible types of imagery have recently become available at 500-m resolution from the HCMM satellite. Although not yet fully investigated, it is believed that the HCMM data can be used successfully for snow mapping on basins of several hundred km^2 and larger. Nighttime thermal observations of ice cover will also be possible.

Although at present the Landsat and HCMM satellites provide only research data, there are several operational satellites with snow and ice applications. NOAA satellites possess both the Very High Resolution Radiometer (VHRR) and the Advanced VHRR (AVHRR) with about 1-km resolution in the visible and thermal infrared (VHRR) and the near infrared (AVHRR). For snow-cover mapping, it has been found that these sensors are applicable on basins greater than approximately 1000 km^2. They are also valuable in the same way for observing larger scale ice features. The primary advantage of using the NOAA satellites for snow and ice applications is that repetitive coverage is available twice every 24 hours. Further expanding on this capability is the GOES series of satellites, which provide geosynchronous coverage with observational frequencies of up to once every 5 minutes. This temporal frequency has been employed in the study of severe storms. Such observational frequencies are especially suited to environmental hazard monitoring (e.g., for snowmelt floods or real-time ice navigation). The Defense Mapping Satellite Program (DMSP) run by the U.S. Air Force provides coverage similar to the NOAA satellites but with observations available every 6 hours through the use of multiple satellites. However, these data are not widely available outside the military community.

Microwave data from space have been available on a number of experimental platforms. Those with particular significance to snow and ice research have been the Nimbus series of satellites and the Geodetic Earth Orbiting Satellite (GEOS). Nimbus 5 ESMR at 1.55-cm wavelength, Nimbus 6 ESMR at 0.81-cm wavelength, and Nimbus 7 SMMR at 0.81-, 1.36-, 1.66-, 2.8-, and 4.54-cm wavelengths have been used in experimental snow and ice studies. In snow studies, it has been shown that, because of the poor resolution of these radiometers (25 to 100 km), applications are currently limited to flat, open regions such as the high plains of the United States. Even in these conditions, it is apparent that the microwave data are highly correlated to snow depth (and water equivalent) and are very sensitive to liquid water in the snowpack (Rango et al, 1979). In addition, the Nimbus microwave data have been used to map the polar-ice distributions and their changes throughout various seasons. Several significant discoveries involving open-water areas in pack ice, sea-ice persistence in particular areas, ice age, and ice-edge dynamics have resulted from the Nimbus data (Gloersen et al, 1978). The radar altimeter on GEOS has also been used effectively to map the elevation of the Greenland ice sheet (Brooks et al, 1978). Such knowledge, obtained periodically, could be used to improve the knowledge of the mass balance of glaciers and ice sheets.

ANALYSIS METHODS FOR HYDROLOGY

Snow Cover Analysis

There are several ways that remote-sensing data of snow fields can be handled to facilitate integration of snow-cover areal extent. Because of the sharp contrasts between snow and non-snow areas, photointerpretation has found widespread application. Two major approaches will be discussed here, viz., image interpretation using a zoom transfer scope and use of an image interpretation grid. In addition, because of the multispectral nature of large amounts of remote-sensing data, interpretation of snow cover using digital processing will be discussed.

Zoom Transfer Scope Example

Figure 29-14 is an example of a portion of a Landsat scene taken during spring snowmelt-conditions in the Wind River Mountains of Wyoming on June 28, 1976. The outlined area references the Wind River Mountain Range. Several streams fed by snowmelt flow from these mountains. Figure 29-15 presents a schematic to show the location of the region with respect to state borders and national parks. The Dinwoody Creek

Fig. 29-14. Landsat band-5 image of snow in the Wind River Mountains on June 28, 1976, with study basin boundaries. Figure covers approximately 80 × 110 kilometers.

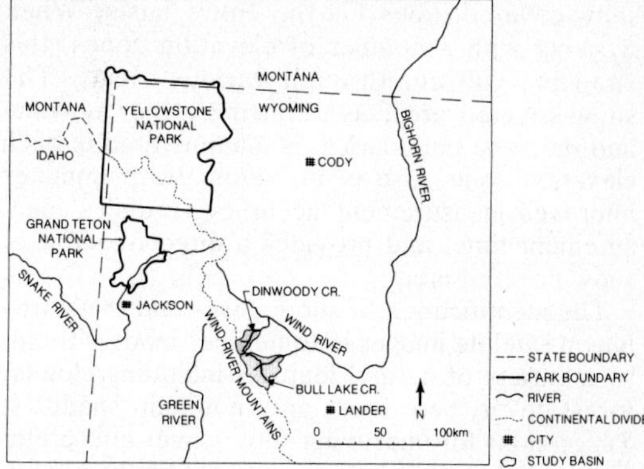

Fig. 29-15. Location map of the Dinwoody and Bull Lake Creek study basins in northwestern Wyoming, USA.

basin will be used for this example. Details are shown in Figure 29-16. Drainage area for Dinwoody is 228 km^2, elevation ranges from 1981 to 4202 m and, for this study, four elevation zones are delineated. A 1:250,000 scale base-map overlay similar to that shown in Figure 29-16 is compiled for the snow-cover mapping. This base map will include certain points of reference plus the basin boundary, stream network and gauge, and elevation zones. Usually, a transparency form of Figure 29-14 at 1:1,000,000 scale is mounted on the zoom transfer scope for analysis. The zoom transfer scope is used to precisely register the satellite imagery and the watershed map boundaries. This instrument superimposes the imagery on base-map overlays at various scales through a system of mirrors, lenses, and scale adjustments with an additional capability for removing image distortion. (See Chapter 17 for more information on image distortion and Chapter 24 for use of the zoom transfer scope. When registration has been achieved, the snow cover on the Landsat scene at 1:1,000,000 scale is mapped on the watershed overlay at 1:250,000 scale. This is generally done by tracing the line separating snow from non-

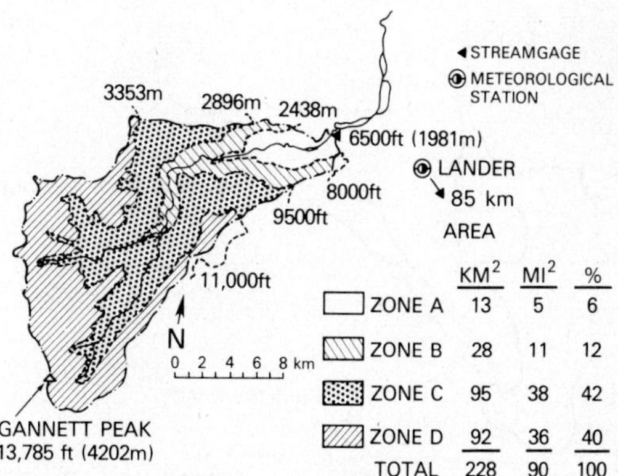

Fig. 29-16. Elevation zones and areas of the Dinwoody Creek basin.

snow-covered areas for the entire basin. When working with a number of elevation zones, this snowline will cut through various zones. The snow-covered area, as defined by the snowline and the zone boundaries, is planimetered in each elevation zone. Use of an automatic planimeter improves measurement accuracy, reduces measurement time, and provides a direct readout of snow-covered area.

The identification of snow cover and the snowline in satellite images is sometimes made difficult by a variety of natural features including clouds, forest cover, bare rock, and mountain shadows. Suggestions for improving snow-cover interpretation in such a situation are presented in handbook form by Bowley et al (1980). In some areas of the world, such as Scandinavia, the snow cover does not retreat up mountain slopes in a gradual and contiguous fashion. Rather, the disappearance of the snow is irregular without a well-delineated snowline (Anderson and Odegaard, 1980). In these instances, other techniques such as grid-unit analysis or digital interpretations may be required.

Figure 29-17 shows the snowline traced on the 1:250,000 scale Dinwoody Creek base map for June 28, 1976. As determined by planimetering, the snow cover percentage by zone is 0 percent (zone A), 0 percent (zone B), 47 percent (zone C), and 99 percent (zone D). These data are plotted in Figure 29-17 along with the data from Landsat for the remaining available dates during snowmelt in 1976. A snow-cover depletion-curve for each zone can then be obtained as shown in Figure 29-18. These curves can be used for comparing snow-cover retreat in other years, for obtaining daily zonal snow-cover values for use in modeling (Rango and Martince, 1979), and for graphically predicting snowmelt runoff (Moravec and Danielson, 1980).

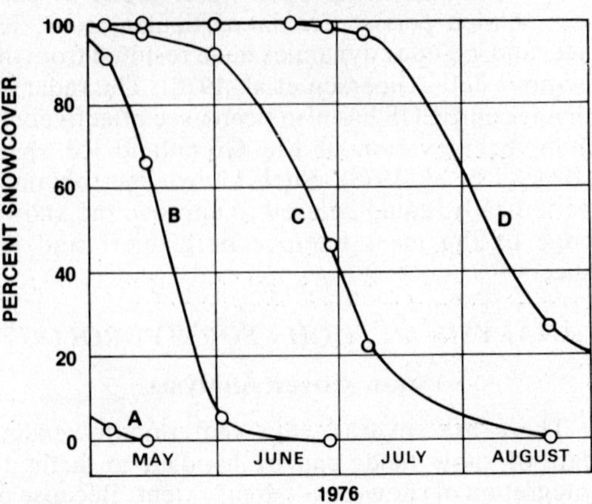

Fig. 29-18. Depletion curves of snow-covered area in zones A, B, C, and D of the Dinwoody Creek watershed (May through August 1976).

Grid Analysis Method

A more detailed snow-cover analysis, using photointerpretation, results when a snow-interpretation grid system is used. The use of grid units for the manual interpretation of snow cover was reported by Katibah (1975). The grid system permits an analyst to make decisions for discrete parts of the basin as to the areal extent of snow based on visual appearance as well as on other factors such as density and type of vegetation cover, elevation, aspect, bare rock, and shadows. The snow cover in each of the grid units can then be classified according to a variety of snow-cover measures. Katibah (1975), for example, used the following five-level system:

Code	Snow Cover Class	Midpoints
1	No snow present	0.00
2	0 to 20% of grid snow covered	0.10
3	20 to 50% of grid snow covered	0.35
4	50 to 98% of grid snow covered	0.74
5	98 to 100% of grid snow covered	0.99

The percent of snow cover in the entire basin can then be determined by summing the snow cover in each of the grid units. Such techniques are considerably more time-consuming than the delineation of the snowline previously described, but may provide a more effective means of measurement in areas of discontinuous snow such as in Scandinavia. These grid-unit techniques are also well suited for special snow interpretations such as inference of snow water-equivalent. Figure 29-19 shows the superpositioning of a 1-km^2 grid over the Dinwoody Creek basin in Wyoming. Interpretation of the date of snow disappearance from each grid unit was required in this study. Figure 29-20 presents the average grid-unit elevation and the 1976 date of snow disappearance in the grid, as determined by using Landsat data (Martinec and Rango, 1981). The grid-unit approach can easily be adapted to the particular ap-

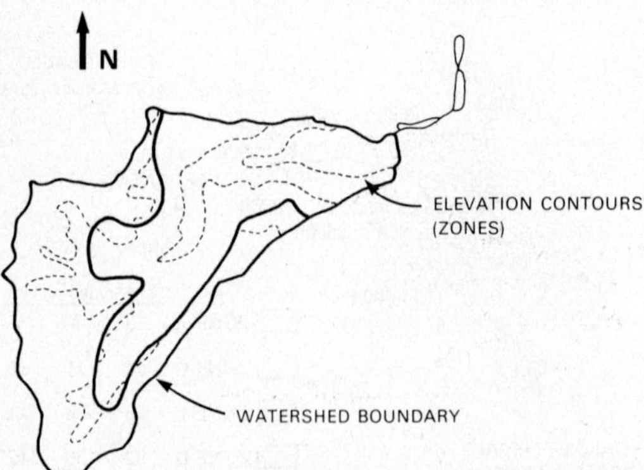

Fig. 29-17. Snowline traced on 1:250,000 scale Dinwoody Creek base map using June 28, 1976 Landsat data.

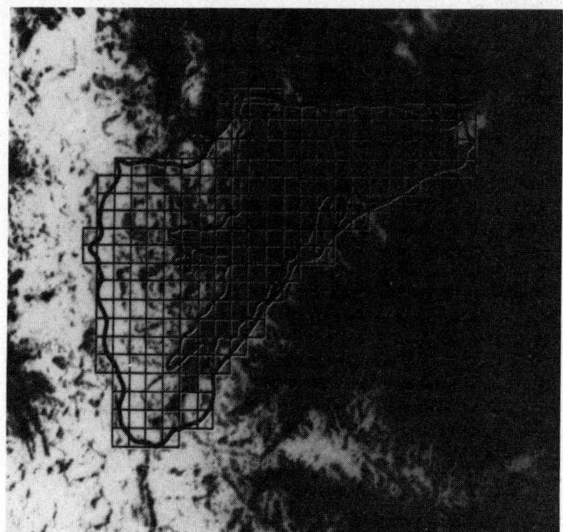

Fig. 29-19. Landsat band-5 view of the Dinwoody Creek, Wyoming, watershed on June 21, 1974, showing basin boundary, snowline, and snow interpretation grid. Each grid square is 1 kilometer by 1 kilometer.

plication under investigation. In addition, such accounting and analysis methods can be readily adapted or converted to digital-analysis techniques.

Digital Methods

In general, because snow area is such an easily identifiable parameter using satellite data, the reliance on digital methods is not necessary. This is especially true when only a few basins are being analyzed. In such instances the previously outlined photointerpretation techniques are entirely adequate. When analysis of numerous basins is required, however, digital techniques may then be preferred. Supervised classifications of snow are easily accomplished as has been shown by Rango and Itten (1976). The man-interaction capability of various computer systems is vital in snow-classification work to assist in delineating and separating snow in trees, in shadows, from bare rock, and from the clouds. For determining snow area, the two major subclasses of hydrologic importance are dry and wet snows. Landsat classifications tend to be too detailed, sometimes providing 20 or more categories of dry snow alone. Therefore, these detailed categorizations must be lumped into one or two classes of dry snow by the image analyst.

One advantage of classifying snow using digital techniques involves the capability of merging the snow data with conventional topographic data. This permits the categorization of snow not only by area but also with elevation. Certain snowmelt models require the input of snow cover by elevation zone that could be automated in this case. Bartolucci et al (1975) successfully merged and registered digital elevation data with Skylab and Landsat data and produced area estimates of snow cover by elevation zone in the San Juan Mountains of Colorado.

Another capability associated with working in the digital mode is the possibility of automatic discrimination between snow and clouds if a band at 1.55 to 1.75 μm is available. Because of the strikingly low reflectance of snow versus clouds in this band, a multispectral classification of snow separate from clouds is possible. Such capabilities have been demonstrated by Bartolucci et al (1975) and Rango and Itten (1976).

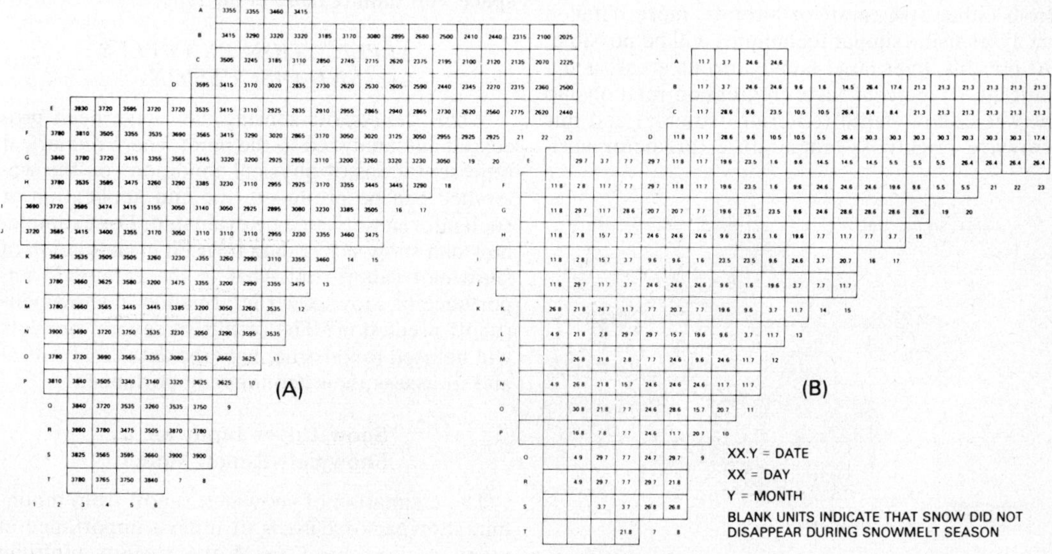

Fig. 29-20. Dinwoody Creek, Wyoming, snow interpretation grid showing (A) average grid unit elevation in meters and (B) estimated date of snow disappearance from each grid unit in 1976 using Landsat (Martinec and Rango, 1981).

Microwave Snow Depth Analysis

Digital data from microwave scanners on certain satellites, such as Nimbus 5 through 7, can be computer-processed to yield microwave-brightness temperature data averaged over various size-subdivided study areas. At this point, photointerpretation can be used to superimpose the brightness temperatures, conventional snow-ground measurements, and base maps for analysis. Figure 29-21 shows a study area in southern Alberta and Saskatchewan, Canada of about 5.5° latitude by 11° longitude that was used in a microwave snow depth analysis (Rango et al, 1979). Four general vegetation types are prevalent over this relatively flat area. To make the cover conditions as uniform as possible for the snow depth and brightness-temperature data analysis, the study was limited to the areas of short- and high-grass prairies.

Snow-depth values in inches for each of the numbered stations in Figure 29-21 were used to draw isonivals that were then averaged over each 1- by 1-deg grid block in the short- and high-grass prairie areas. The ESMR data of Nimbus 6 were computer-averaged over the same 1- by 1-deg grid elements. Both vertically and horizontally polarized brightness-temperature data were used in simple linear regression analyses with the snow-depth data. Figure 29-22 illustrates the snow depth versus the vertically polarized brightness-temperature data for Nimbus 6 and the resulting significant (at the 0.002 level) regression line and statistics (Rango et al, 1979). The Nimbus 6 data are from the daytime pass on March 15, 1976, and the snow-depth data are from readings on the same day. Because of the poor resolution of the ESMR data (about 25 to 30 km), homogeneous flat areas should be used to simplify data analysis. As improved-resolution sensors become available, from either spacecraft or aircraft, more detailed analyses using similar techniques will be possible. At present, over large homogeneous areas, a significant (95 percent level) regression relationship between snow depth (criterion variable) and microwave brightness-temperature (predictor variable) can be developed. The estimation of snow depth under dry-snow conditions is a definite possibility in such study areas. If further data sets are acquired, these estimates could be expanded over a greater range of snow depth. Similar specific regression relationships would have to be derived for use in other snow studies. A more profitable approach in estimating snow water-equivalent or depth in specific areas is to use a microwave-scattering model in combination with the brightness temperatures to predict snowpack properties. Such experiments have recently been successfully carried out with truck, aircraft, and spacecraft data (Chang et al, 1981).

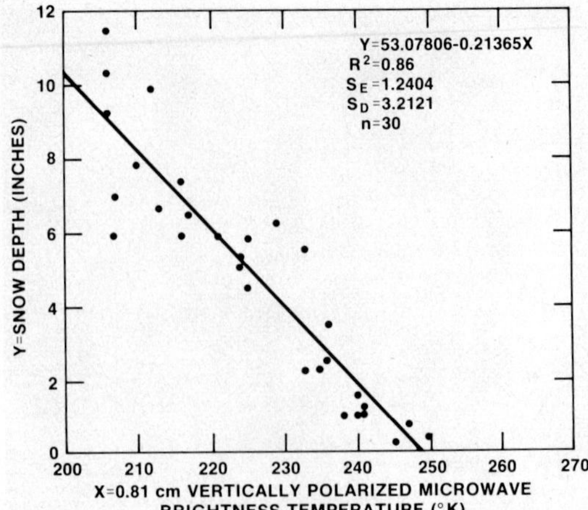

Fig. 29-22. Nimbus 6 vertically-polarized microwave brightness-temperature versus snow depth on the Canadian high plains. Nimbus 6 data from daytime pass March 15, 1976, summarized by 1-degree latitude-longitude grid; snow depth data from May 15, 1976, summarized over the same grid; data included from short- and high-grass prairie areas only.

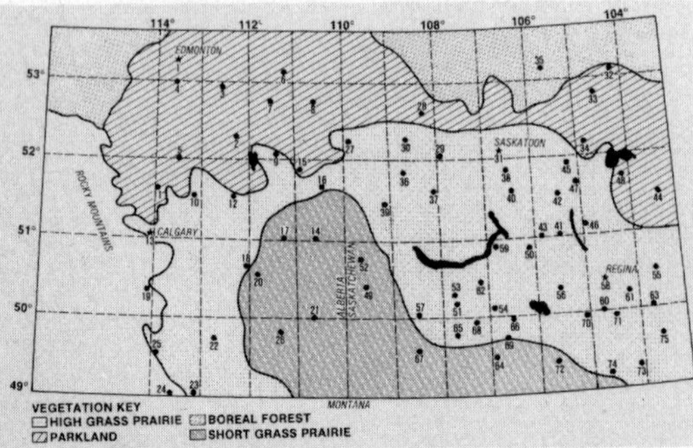

Fig. 29-21. Canadian high plains snow/microwave study area (Rango et al, 1979). Figures for snow depths are in inches.

APPLICATION EXAMPLES AND DISCUSSION

When the remote-sensing data have been processed and analyzed to the point where numerical representations of physical conditions on the watershed can be produced, the end application of such information must be evaluated. For example, how can snow-cover data be used in simulation of snowmelt runoff and what is the economic importance of snow-cover information in snowmelt-runoff prediction? The results of several projects will be used to illustrate how the data can be used and to assess their economic importance.

Snow Cover Input for a Snowmelt-Runoff Model

The estimation of snowmelt runoff from mountain snowpack-basins is of utmost importance in many regions because of the various multiple water uses (e.g., irrigation, hydroenergy, water supply, and flood control) that have been developed. The importance has been magnified in recent years because the demand for water in many areas now exceeds the average water supply. The use of hydrological models to improve the management of the available water in snowmelt areas has increased in the last 15 years. During this same period, significant advances in the development of remote-sensing technology have been made with particular application to the mapping of snow cover.

A snowmelt-runoff model was designed by Martinec (1970) to operate under conditions in which snowmelt is the major contributor to runoff and in mountain basins with substantial elevation range. Originally, the model was developed and tested on small experimental basins in Europe, ranging in size from 2.65 to 43.3 km^2. The small size of these basins and the frequency of cloudiness in the European Alps led to the use of aircraft photography for determining a critical model-input variable (i.e., the percentage of the drainage basin covered by snow).

To be operationally useful, a snowmelt model must function on basins considerably larger than those used by Martinec (1970) in development of the snowmelt model. In the United States, operational-type basins in snowmelt areas are typically about several hundred km^2 and larger as defined by the streamgages and diversion structures. The use of aircraft coverage for these large basins is not economically feasible, but since 1972, the availability of high-resolution satellite data has made snow-cover mapping possible on a repetitive basis. Thus, the use of satellite-derived snow-cover data and the foregoing snowmelt model could potentially permit estimating snowmelt runoff on operational-size basins.

The snowmelt-runoff model used and shown in Figure 29-23 is simple in concept and requires only the following daily inputs for approximately 500-m elevation zones: the extent of snow cover, temperature (degree-days), and precipitation amount. The deterministic approach used in the original development of the model facilities its application in new conditions and areas.

The areal extent of snow-cover constitutes the basic information for the day-to-day computation of snowmelt runoff by the model. For forecasting, it would be best to have areal snow-water volume of the basin as an input, but this is extremely difficult to obtain, especially on large basins. This input may eventually become possible using remote-sensing techniques; however, snow cover can be now used as an index of the snow reserves for runoff simulations. Satellite data can be analyzed to produce snow cover, and the data can be plotted against time during the snowmelt season. In Figure 29-15, Landsat snow-cover data during snowmelt in 1976 on the Dinwoody Creek basin in Wyoming was plotted by elevation zone, and snow-cover depletion curves were drawn by eye for each zone. The shape of these curves tends to be relatively constant, but their location

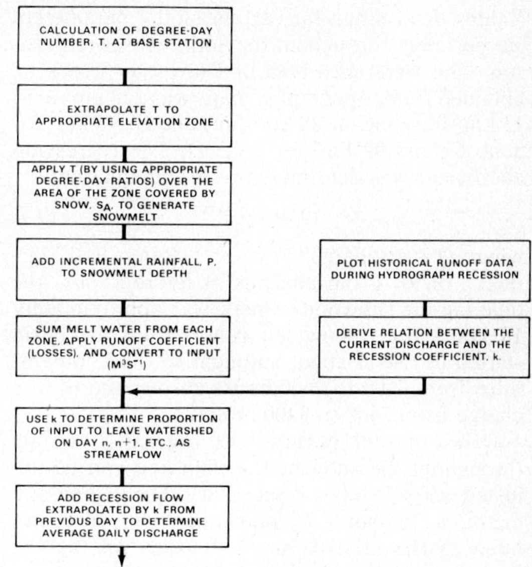

Fig. 29-23. Flow diagram for a snowmelt-runoff model.

will be displaced from year to year depending on snow and climatic conditions. The snow cover data for each day can then be estimated from the depletion curves and used as input to the snowmelt model.

Based on the use of four elevation zones for Dinwoody Creek, the model equation takes the following form:

$$Q_{n+1} = c_n \Big\{ \left[a_{An}(T_n + \Delta T_{An})\, S_{An} + P_{An}\right] \frac{A_A \cdot 10^{-2}}{86400} + \left[a_{Bn}(T_n + \Delta T_{Bn})\, S_{Bn} + P_{Bn}\right] \frac{A_B \cdot 10^{-2}}{86400} + \left[a_{Cn}(T_n + \Delta T_{Cn})\, S_{Cn} + P_{Cn}\right] \frac{A_C \cdot 10^{-2}}{86400} + \left[a_{Dn}(T_n + \Delta T_{Dn})\, S_{Dn} + P_{Dn}\right] \frac{A_D \cdot 10^{-2}}{86400} \Big\} (1 - k_n) + Q_n\, k_n \qquad (29\text{-}6)$$

where

Q = average daily discharge (m^{3}s^{-1})
c_n = runoff coefficient
a_n = degree-day factor (cm·°C^{-1}·d^{-1})
T_n = measured number of degree-days
T_n = correction by the temperature lapse rate (°C·d)
S_n = snow coverage (100% = 1.0)
P_n = precipitation contributing to runoff (cm)
A = area (m^2)
k_n = recession coefficient
n = index referring to the sequence of days
A,B,C,D as subscripts, refer to the four elevation zones
$\frac{10^{-2}}{86400}$ converts cm·m^2 per day to m^{3}s^{-1}

Values determined for certain model parameters are pertinent throughout the entire snowmelt season. The watershed area of Dinwoody Creek as obtained from topographic maps was 228 km² with 13 km² for zone A, 28 km² for zone B, 95 km² for zone C, and 92 km² for zone D. The regression coefficient was determined from the equation:

$$k = 0.884Q^{-0.0677} \qquad (29\text{-}7)$$

where Q is discharge in m³ sec⁻¹ (Rango and Martinec, 1979). From analysis of hydrographs, the time lag for Dinwoody Creek was approximately 18 hours (i.e., snowmelt runoff from the basin started to rise at about midnight so that temperatures from 0600 to 0600 hours correspond to discharge from 2400 to 2400 hours).

Other model parameters logically change throughout the snowmelt season and can be adjusted every 15 days, if necessary. The degree-day factor, a, is shown by Martinec (1980) to be related to the relative snow density, ρr, by the equation:

$$a = 1.1\ \rho r \qquad (29\text{-}8)$$

where ρr is equal to $\rho s/\rho w$ or the density of snow divided by the density of water. Thus, the general seasonal increase in ρr could be used as an index for the increase in a. For the Dinwoody Creek basin, a was gradually increased from 0.35 on April 1 to 0.60 on September 30 in zone D. In view of the length of the snowmelt season, seasonal variations of other parameters must be evaluated. Based on sparse information on climate (Barry and Chorley, 1970), the temperature lapse-rate was estimated to be higher in the Wind River Mountains than in the Alps and to vary from 0.85°C per 100 m in April to 0.95°C per 100 m in July to 0.80°C per 100 m in September. A lack of direct measurements prohibits the actual determination of the lapse rate. Regional differences between the meteorological station at Lander airport and the mountainous basin were accounted for by subtracting up to 2°C from the Lander data. The runoff coefficient was also assumed to vary during the season; it was estimated to be in the range from 0.85 in April to 0.75 in July to 0.90 in September.

Air temperature expressed in degree-days was used in this simple model as an index of snowmelt. For the Dinwoody Creek basin, temperature data from the airport were used. The number of degree-days for each 24-hour period was determined by summing the hourly temperatures and dividing by 24 and using 0°C as the base temperature. Temperatures below the freezing point were regarded as 0°C. Maximum and minimum temperatures could also have been used in this degree-day determination. The degree-day figures refer to the 24-hour periods starting at 0600 hours. These temperature data were extrapolated to the hypsometric mean elevation of the respective elevation zones by the previously discussed temperature lapse rate. The resulting degree-days were used for calculating snowmelt. Extrapolation errors could be minimized if the temperature were measured in the basin near the mean elevation.

Daily precipitation amounts at Lander were used to satisfy the model input requirements. Lacking an acceptable method for extrapolating the precipitation data both horizontally and vertically, the Lander data were used as zonal inputs as recorded. Again, measurement of precipitation in the basin would greatly aid in the application of the model for snowmelt-runoff simulation. For the final snow-covered area determinations, the sequence of precipitation events at Lander was used for identification of late season transient snowfall, temporarily causing an increase in snow-covered area but not contributing to the snowmelt hydrograph.

When all the necessary input data were prepared on a daily basis, Eq. 29-1 was used to calculate snowmelt depths by zones and to transform these values to runoff by the previously mentioned recession techniques. In the Dinwoody Creek basin, snowmelt starts about April 1 and continues well into September. The model was used to simulate daily streamflow from April 1 to September 30.

The following example calculations are for July 2 and 3, 1976, for Dinwoody Creek using Eq. 29-6:

Example 1: July 2, 1976, calculation using July 1 input data:

$$\begin{aligned} Q_{\text{July 2}} = 0.75\ \{&[0.55\ (23.91 - 7.21)\ 0 + 0]\ 1.505 \\ &+[0.50\ (23.91 - 11.41\ 0 + 0]\ 3.241 \\ &+[0.45\ (23.91 - 15.98)\ 0.385 \\ &\quad + 0]\ 10.995 \\ &+[0.40\ (23.91 - 19.69]\ 0.98 \\ &\quad + 0]\ 10.648\} \\ &(1 - k_n) + Q_n k_n \end{aligned}$$

where

$$\begin{aligned} k_n &= 0.884\ Q_n{}^{-0.0677} \\ Q_n k_n &= Q_n \cdot 0.884\ Q_n{}^{-0.0677} \\ &= 0.884\ Q_n{}^{0.9323} \\ Q_{\text{July 1}} &= 12.662\ \text{m}^3\text{sec}^{-1} \end{aligned}$$

$$Q_{\text{July 2}} = 0.75\{0 + 0 + 15.106 + 17.614\}\ (1 - 0.884 \cdot 12.662^{-0.0677}) + 0.884 \cdot 12.662^{0.9323}$$

$$Q_{\text{July 2}} = 15.698\ \text{m}^3\text{sec}^{-1}$$

Example 2: July 3, 1976, calculation using July 2 input data:

$$\begin{aligned} Q_{\text{July 3}} = 0.75\ \{&[0.55\ (16.39 - 7.21)\ 0 \\ &\quad + 0.127]\ 1.505 \\ &+[0.50\ (16.39 - 11.41)\ 0 \\ &\quad + 0.127]\ 3.241 \\ &+[0.45\ (16.39 - 15.98)\ 0.355 \\ &\quad + 0.127]\ 10.995 \\ &+[0.40\ (16.39 - 19.69)\ 0.98 \\ &\quad + 0]\ 10.648\} \end{aligned}$$

$$(1 - (0.884 \cdot 15.698^{-0.0677})) + 0.884 \cdot 15.698^{0.9323}$$

$$Q_{\text{July }3} = 0.75 \{0.191 + 0.412 + 2.117 + 0\} (0.266) + 11.517$$

$$Q_{\text{July }3} = 12.060 \text{ m}^3\text{sec}^{-1}$$

Based on this snowmelt-runoff calculation method, daily snowmelt runoff was calculated with the model given only the actual April 1 discharge with no updating with actual discharge data thereafter. Snowmelt runoff was simulated for the April 1 to September 30 snowmelt season for 1976 and 1974 and compared with actual discharge. The computed and measured snowmelt-runoff hydrographs are shown in Figure 29-24. Comparison on a seasonal basis for 1976 indicates a 5-percent difference between 6-month volumes for the computed and measured flows. For 1974, this volumetric difference is only about 1 percent. Such differences are quite reasonable when compared with conventionala (nonremote-sensing) simulation procedures. Daily differences in simulated and actual runoff were evaluated using the Nash-Sutcliffe (1970) R^2 value. For 1976, R^2 was 0.86; for 1974, it was 0.83. This indicates that about 85 percent of the variation in the actual daily runoff values is explained by this modeling approach (Rango and Martinec, 1979).

Though the use of remotely sensed snow-cover data and conventional temperature and precipitation data, the model developed by Martinec (1975) can be used to simulate runoff accurately on mountainous basins as large as 500 m^2 (Rango, 1980a). The location of an automatic temperature-precipitation observation station in the basin would further improve the simulation accuracy. Landsat provides an effective means for obtaining the critical snow-cover input parameter required by the snowmelt model.

The determination of model parameters and variables on a rational basis, such as in this model with little or no optimization, facilitates its application in new basins. Although only streamflow simulation was attempted in this study, the use of this model for discharge forecasts has great potential. A means of extrapolating the snow-cover depletion curves based on more commonly observed parameters, such as temperature and snow water-equivalent, would be required. Coupling this with either forecasts of temperature or statistical temperature data would permit use of the model as a water management tool. The fact that the model can provide reasonable flow simulations for a 6-month period without any updating by actual discharge measurements further indicates the possible application to ungaged watersheds.

Economics of Snow Cover Data for Snowmelt-Runoff Prediction

To test the operational usefulness of a particular type of new data, such as satellite snow-cover information, a quasi-operational demonstration test of the satellite snow-cover data by operational water-management agencies is required. The objectives of this testing are to determine whether the satellite snow-cover data can effectively be incorporated into water-management agency procedures for snowmelt-runoff prediction and whether runoff prediction improvements can result. In addition, the timelines and suitability of the snow-cover data should be evaluated.

A snow-mapping demonstration project was begun in 1975 and completed in 1979. Four different study areas were chosen for the project to ensure variability of snow conditions, vegetation characteristics, cloud cover, and snowmelt climatology and to more easily extrapolate project results to other snowmelt regions. The study areas were designated as Arizona, California, Colorado, and the Pacific Northwest; and the specific watershed studies are shown in Figure 29-25. Agencies participating in the experiment included the U.S. Geological Survey (USGS) and the Salt River Project in Phoenix; the California Department of Water Resources (DWR) in Sacramento; the Soil Conservation Service (SCS), Bureau of Reclamation, and the Colorado Division of Water Resources in Denver; and the Bonneville Power Administration, Corps of Engineers, and the National Weather Service in Portland. In addition,

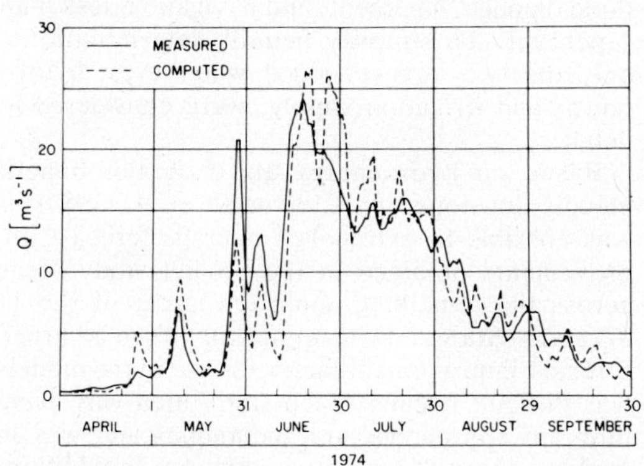

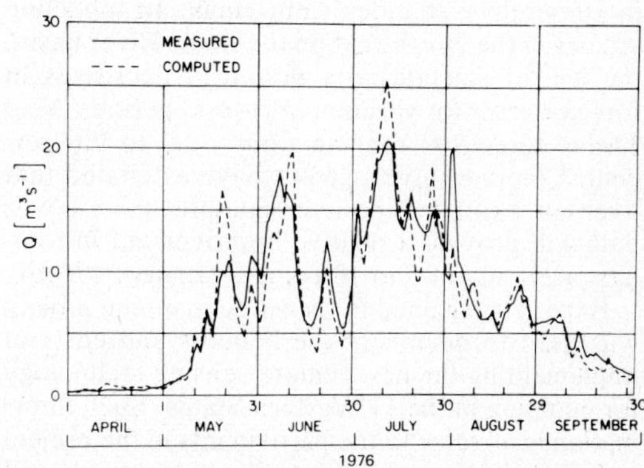

Fig. 29-24. Computed and measured runoff in the Dinwoody Creek basin during the snowmelt season.

Fig. 29-25. Study watersheds for the snow mapping ASVT Project (Rango, 1980b).

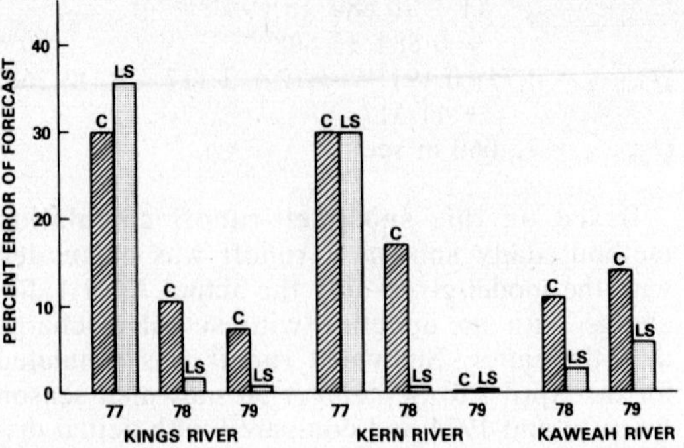

Fig. 29-26. Seasonal snowmelt-runoff forecast error using conventional (C) and Landsat snowcover (LS) procedures in Southern California on May 1 (Rango, 1980b).

the National Environmental Satellite Service (NESS) provided NOAA satellite data in support of the study-center investigations.

The results of the project indicate that snow-cover information is a valuable ancillary data source for improving snowmelt-runoff predictions. When satellite snow-cover data were tested in both empirical seasonal runoff estimation and short-term modeling approaches, definite potential for reducing forecast error was evident. For example, over a 3-year period, errors in seasonal streamflow estimates for three basins in California were reduced from 15 to 10 percent by using satellite snow cover data. A comparison of forecast error resulting from both conventional and Landsat snow-cover seasonal regression procedures used on May 1 is shown in Figure 29-26 for the three test basins from 1977 to 1979. During the severe drought year of 1977, both methods had large percentage errors, but the actual differences in streamflow volumes were small. In modeling studies in the Northwest on the Boise River basin, the use of satellite data produced decreases in forecast error for various short-term periods (3- to 14-day forecasts) ranging from −2.0 to 9.6 percent. Colorado investigators have estimated that over an extended period, satellite snow-cover data will provide a relative improvement in forecast accuracy of 6 to 10 percent (Rango, 1980b).

Experience gained in the snow-mapping project was used to document the benefits and costs of implementing the new remote-sensing technology for mapping in the 11 Western States. Such information is of value to the participants in the project to justify continuation in an operational mode and to other interested personnel in water-management agencies so that they can make decisions on whether or not to embark on a program to implement this new technology. In order to focus on the most important water uses in the evaluation, a sizing of the market for the 11 Western States was performed. This involved estimation of the upper-bound dollar values for water for various uses and also the importance of snowmelt to total streamflow in each of the USGS hydrologic sub-regions in each of the 11 states. It was found that these states use an average of 2,235 million acre feet (MAF) per year for hydropower generation. At an average alternate energy cost of \$3.20/AF, the total value of the hydropower generated is \$7.15 billion for the 11 Western States (Castruccio et al, 1980). The average amount of streamflow resulting from snowmelt was determined to be 68 percent, which, when multiplied by the total value, yields \$4.86 billion for the contribution of snow to hydropower. Similar upper-bound value calculations indicate that the contribution of snowmelt to irrigation water is worth \$1.74 billion a year for the 11 Western States. Similar calculations for other water uses resulted in the following percentages of the total value resulting from snowmelt water supply: hydropower, 65 percent; irrigation supply, 22 percent; municipal and industrial supplies, 9 percent; flood damage, 4 percent; and navigation, less than 1 percent. To simplify benefit determinations, only the two largest valued water uses, hydropower and irrigation supply, were considered in detail.

Based on hydroenergy and irrigation-benefit models developed by Castruccio et al. (1980), it was possible to extrapolate average forecast improvements obtained in the project study areas across the remaining applicable areas of the 11 Western States. Determination of an average forecast-improvement factor for use in the models was difficult because each study area was using different approaches and techniques and was in various stages of project completion. In addition, the Landsat snow-cover data are not truly operationally available at this time. Therefore, existing results and experience of the project investigators had to be relied on to arrive at an estimate. For the benefit-cost study, the conservative estimate of the Colorado investigators was used (i.e., a 6- to 10-percent relative improvement in forecasting using satellite snow-cover data could be expected). Based on an existing forecast error of 18 percent, this would be an absolute improvement of 1 to 2 percent. In order to be on the conservative side, the hydroenergy and irrigation benefit-models were run with a 6-percent relative improvement in forecasting accuracy. The later results of actual tests performed in California, as shown in Figure 29-26, and in the Northwest indicate that the 6-percent relative improvement value is definitely on the conservative side and valid for the experience so far in the demonstration project. Use of this value of 6 percent as an input parameter to the models resulted in a computed total average annual satellite snow-cover benefit of $38 million in the 11 Western States; $10 million a year for hydroenergy, and $28 million a year for irrigation (Castruccio et al, 1980).

In order to estimate the cost for operational satellite snow-cover mapping and runoff prediction consideration must be given to four major components: satellite data products, image interpretation, data implementation, and equipment. Satellite development, launch, and research were not considered part of the project costs for this determination. The average winter-season costs in a study based on the snow-mapping project experience were $400 for image data products, $800 for image interpretation, and $600 for implementation of the data into the forecast scheme. The major piece of equipment necessary would be a zoom transfer-scope for image interpretation with an average annual equipment cost of $250. This total of $2050 would work out to be approximately $0.23/km^2 a year. For those portions of the entire 11 Western States impacted by snowmelt runoff (2,195,250 km^2), the average annual benefit-cost ratio of using satellite-derived snow-cover information would be 38,000,000:505,000 or 75:1 (Castruccio et al, 1980). This figure would result without any improvements in existing satellites, interpretation systems, and forecasting models.

The snow-mapping project has proved that satellite snow-cover data can be used for reducing snowmelt-runoff forecast error in a cost-effective manner. However, the use of the satellite data is not fully functional because there is an excessive time lag between Landsat data acquisition and receipt by the user. A truly operational application will only be possible in the 11 Western States when the turnaround time is reduced to 72 hours for all data, and the water management agencies can be assured of a continuing supply of operational snow cover data from space. In addition, the frequency of coverage from Landsat is not optimum. NOAA satellite data are available with the required frequency and quick turnaround, but, since the data resolution is much poorer than that of Landsat, these data are more difficult to manipulate for snow mapping. However, by using both Landsat and NOAA in combination, it would be possible to obtain sufficient snow-cover data for use in operational snowmelt-runoff forecasting.

SURFACE WATER

DATA ASSESSMENT

Remotely sensed data can be used to effectively monitor various hydrologic aspects of lakes, rivers, and wetlands. Differences in the appearance of surface water are readily apparent on aerial photographs and multispectral scanner imagery. The feasibility of using remotely sensed data for information extraction must be assessed by considering the principles of light and water interaction. Spectral characteristics of water vary with wavelength and are the result of not only the molecular nature of water but also impurities within the water body. Delineation of the location and spatial extent of surface water is most successfully carried out by using data acquired in the near-infrared or microwave wavelengths. Visible wavelength data (0.4 to 0.7μm) may provide information on certain physical conditions within lakes, rivers, and wetlands because there is a significant amount of radiation in this wavelength region (see also chapter 3). Thermal observations may depict the surface temperature (upper few microns) of a water body. The physics governing the emission of energy from a water body in the thermal and microwave portions of the spectrum is relatively straightforward and is described in chapters 3 and 4. Water bodies tend to reflect microwave energy in a way that is derivable from the Fresnel equations with backscattered energy generally being dependent on the roughness of the surface. Discussion of the physics involved is also provided in chapter 4. The effects of water and its constituents in reflecting, scattering, and absorbing solar radiation are somewhat more complicated and need to be discussed in more detail because of application in surface-water studies.

Sunlight (solar radiation and diffuse sky radiation) entering a surface-water body is subjected to depletion as a result of: (1) absorption and scattering by pure water, and (2) scattering, diffraction, and reflection by suspended particles in impure water (McCluney, 1976; Gordon, 1974). A summary of the variables that can affect remote sensing of physical water-quality characteristics (Moore, 1978) is as follows:

a. Time of year—The Earth receives 7 percent more energy from the sun on January 1 than on July 1 because of its orbit.
b. Sun-elevation angle—More solar energy is specularly reflected from water surfaces at low sun-elevation angles than at high angles. Also, the path length of solar energy through the atmosphere is longer at low

sun-elevation angles, and more solar energy is absorbed and scattered.

c. Aerosol and molecular content of atmosphere—These constituents determine the amount of solar energy absorbed and scattered by the atmosphere. Some energy, received by a satellite, is backscattered before reaching the water surface.

d. Water-vapor content of the atmosphere—Water vapor affects energy absorption at near infrared and thermal infrared wavelengths.

e. Specular reflection of skylight from water surface—Specularly reflected skylight is received by a satellite. The intensity and wave-length distribution of this energy depends on atmospheric scattering, which produces skylight.

f. Roughness of water surface—A rough surface may produce more or less specular reflection than a smooth surface. At high sun-elevation angles, the area of sun glint may be within the satellite field of view.

g. Film, foam, debris, or floating plants on water surface—Although these features may not be resolved on a satellite image, they contribute to the spectral characteristics of the measured signal.

h. Water color—Dissolved colored materials increase absorption of solar energy in water.

i. Water turbidity—The concentration, size, shape, and refractive index of suspended particles determine turbidity and increase the amount of energy backscattered in water bodies.

j. Reflectance and absorbance characteristics of suspended particles—Particles may be inorganic sediments, phytoplankton, zooplankton, or a combination of these. When present in high concentrations, particles affect the spectral distribution of backscattered energy.

k. Multiple reflections and scattering of solar energy in water—The spectral results of these processes are difficult to predict, but they may be important.

l. Depth of water and reflectance of bottom sediments—Water clarity determines the importance of bottom reflectance. Solar energy may not reach bottom in turbid water.

m. Submerged or emergent vegetation—Vegetation may change bottom reflectance, obscure the water surface, or contribute to the spectral characteristics of the measured signal.

The basic principle of light and water interaction is expressed by the equation:

$$I_0 = I_{SR} + I_A + I_B \qquad (29\text{-}9)$$

where

I_0 = solar energy that reaches the water surface

I_{SR} = flux that is specularly reflected at the water surface

I_A = flux absorbed by the water

I_B = flux backscattered to the water surface and thereby available for remote detection

Specular reflection is equal at all wavelengths and changes the absolute level of a remotely measured flux. Relative spectral signatures are only slightly affected by spectral reflection of incident solar energy and diffuse skylight, but corrections are necessary for quantitative calculations under hazy conditions and for sun-elevation angles less than 30 degrees. Solar energy that is not specularly reflected is refracted downward at the water surface and is affected by absorption and scattering. Absorption and backscatter are highly influenced by inorganic and organic substances within the water body and produce distinctive spectral signatures. Unlike specular reflection, the amounts of absorption and backscatter from a water body are highly dependent on the wavelength interval being sensed by the recording instrument.

The recorded remotely-sensed signal is that part of the backscattered energy that returns to the water surface. If a water body is relatively clear and shallow, solar energy is reflected from the bottom and can be detected. The water depth that permits detection of the bottom depends on water color, turbidity, bottom reflectance characteristics, and the intensity of incident light. Intensity of the measured signal can be correlated with water depth if the reflectance characteristics of the bottom materials and the color and turbidity of the water are uniform within the area of interest.

Penetration of radiation into pure water is described by means of an extinction coefficient, k, which takes into account the effects of both scattering and absorption. A parallel (unscattered) beam of radiation of wavelength (λ) passing through water a distance, dx, is reduced in intensity by an amount of dI, which is proportional to the intensity and to the distance, dx, or

$$dI = -kIdx \qquad (29\text{-}10)$$

where k has dimensions of L^{-1} (L is length and is dependent on wavelength (λ)). If the intensity of the radiation at $x = 0$ is I_0, then for some distance, x, $I = I_0e^{-kx}$. Figure 29-27 shows the values of k ranging from 0.186 to 2.65 mm. From the values shown in this figure, it can be seen that the reduction in intensity of sunlight after passing through very thin layers of water will be quite high.

Theories incorporating the concept of the extinction coefficient suggest that the basic color of water of an infinitely deep pure-water body would be blue. However, values of the extinction coefficient in Figure 29-27 are valid only for pure water. The introduction of suspended organic and inorganic materials introduces further scattering, absorption, and reflection by these particles. Scattering by small (in comparison with the wavelength of light) particles is wavelength-selective

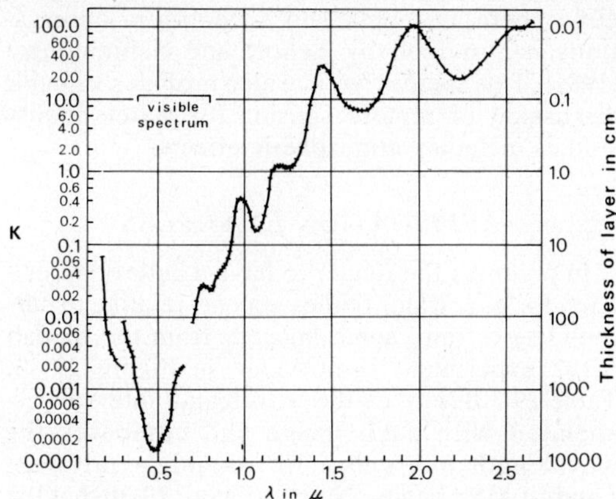

Fig. 29-27. Absorption coefficient for pure water (pure sea water) for parallel radiation, wavelength range of 0.186 to 2.65 μm (after Defant, 1961).

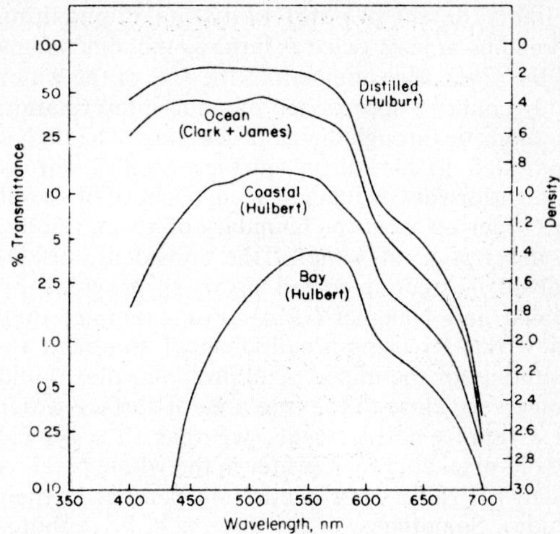

Fig. 29-28. Spectral transmittance for 10 m of various water types (after Specht et al, 1973).

and Rayleigh's law applies. Small wavelengths are scattered most, and the basic blue color of the water remains relatively unchanged. As particle size increases, scattering becomes independent of wavelength so that a color shift of water toward the green begins to occur. With increasing turbidity, water color shifts even more toward yellowish green until the color of the water approaches or becomes that of the natural color of the particles creating the turbidity.

Discoloration of the water can also occur as a result of the presence of life forms, such as algae. In these instances, the spectral response of the predominant life form will basically determine the water color. In shallow waters, the natural color of the bottom can be detected, and water color will be altered toward bottom color.

Theoretical considerations involving the concept of the extinction coefficient suggest that the best depth penetration of water in the visible and near infrared range could be made by data obtained near 0.5 μm. Values of the transmittance (defined as the ratio, I/I_0) of various water types at 10-meter depths are illustrated in Figure 29-28 (Hulbert, 1945; Clarke and James, 1939; Specht et al, 1973). Figure 29-28 shows that the maximum transmittance of light in distilled and clear ocean water occurs in the spectral range between 0.44 and 0.54 μm. Peak light-transmittance in distilled or clear water lies near 0.48 μm (Duntley, 1963). In more turbid waters or in waters containing substances such as dissolved organic materials and plankton, the peak wavelength of light transmittance shifts toward the longer wavelengths (Pestrong, 1968; Polcyn, 1970). For more information on marine phenomena the reader should consult Chapter 28.

Pestrong studied the relative ability of light at various wavelength spans ranging from 0.40 to 0.44 μm, 0.46 to 0.50 μm, 0.52 to 0.55 μm, 0.55 to 0.59 μm, 0.60 to 0.63 μm, 0.65 to 0.71 μm, and 0.7 to 0.83 μm to penetrate marine water in a marsh environment. He found the greatest degree of penetration to be in the 0.55- to 0.36-μm wavelength span. Polcyn found that, for data taken in the Florida Keys, maximum penetration occurred in the wavelength span of 0.5 to 0.58 μm.

In contrast to the relatively high transmittance characteristics of water bodies in the visible wavelengths, nearly all incident energy is absorbed in the near-infrared portion of the spectrum. Very little near-infrared energy is backscattered from water; therefore, water bodies have a significantly lower reflectance than terrestrial features throughout this portion of the spectrum, as illustrated in Figures 29-28 and 29-29. Because of this phenomenon, surface water appears black on reflective infrared images except for cases of extremely high near-surface turbidity. The areal extent of water bodies is relatively easily and accurately delineated on remotely sensed data acquired in the near infrared.

When an attempt is made to measure the area covered by a water body, accuracy is found to be a function of the field of view or pixel size. To

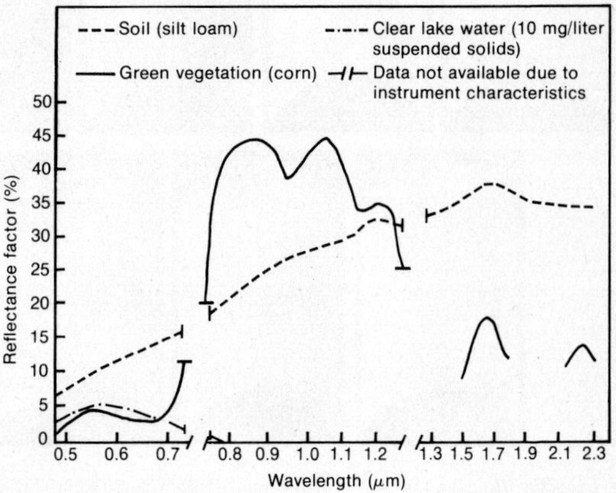

Fig. 29-29. Spectral reflectance characteristics of green vegetation, soil, and clear water.

reliably detect a water body, its dimensions should be at least twice as large as the dimensions of the pixel. Measurement of the size of the water body could be approached by photointerpretation methods or through digital processing. The digital approach to measuring surface area is not as straightforward as it might seem because of mixed or border pixels at the boundary of a water body. A description of some of the considerations involved in treating mixed pixels is provided by Malila and Nalepka (1974). For example, they show that by using a multichannel approach accounting for the mixed pixel problem, they could come very close to the true area of surface water in a representative scene, whereas, if a typical whole-pixel approach (water in the whole pixel) is used, a sizable error occurs (14 percent in their study). Sometimes, in any approach, be it photointerpretation or digital processing, some small water bodies are not detected, and the area of water in a scene is underestimated. In this case, some investigators have been successful in empirically adjusting low-resolution estimates for an apparent systematic bias in a region with many small water bodies mixed in with larger water bodies (Gilmer et al, 1980; Work and Gilmer, 1976).

Another significant consideration when one attempts to assess the condition or constituent loading in water bodies is the effect on the observations of the intervening atmosphere. An analytical approach requires adjusting for water-vapor absorption, haze and Rayleigh-scattering in the atmosphere using the approach equations for radiative transfer in the atmosphere (Chandresekhar, 1960; Maul and Gordon, 1975; Dozier and Frew, 1981). Practical approaches to these problems are discussed in other chapters. Another example in which corrections for the effects of atmosphere are applied to 8- to 14-μm observations is provided by Schott and Schimminger (1981). This same reference also provides valuable discussion of remote sensing for water quality studies including atmospheric effects.

APPLICATION EXAMPLES

In order for the reader to have a better appreciation for how water bodies appear, relative to surrounding terrain, some imagery from the Skylab S-192 experiment is provided in Figure 29-30. Table 29-3 describes the wavelength intervals associated with each image and compares the wavelength intervals with roughly equivalent Landsat MSS bands. Note in Figure 29-30 that the greatest contrast exists in the 0.98- to 1.80-μm image corresponding to band 7 of the Landsat MSS. For delineating surface-water area, Landsat band 7 is clearly the optimum choice and is only limited for this purpose by the attendant spatial resolution. Figure 29-31 is a Landsat MSS band taken over the southern Wisconsin and Illinois region. It simply illustrates, for a larger area, how spaceborne imagery can be used to determine the number and distribution of lakes. In attempting to make estimates of the total surface water area, one must keep in mind the considerations previously described. Because of the repetitive nature of satellite coverage, spaceborne observations are particularly suited for monitoring dynamic changes in surface parameters in remote areas or areas that are difficult to access. This capability allows areas that may need water for development to be surveyed and the availability of surface water to be assessed. An example of such a survey was performed for an area in Southwest Iran by Krinsley (1976). The area studied is shown in two Landsat false-color composites (Color Fig-

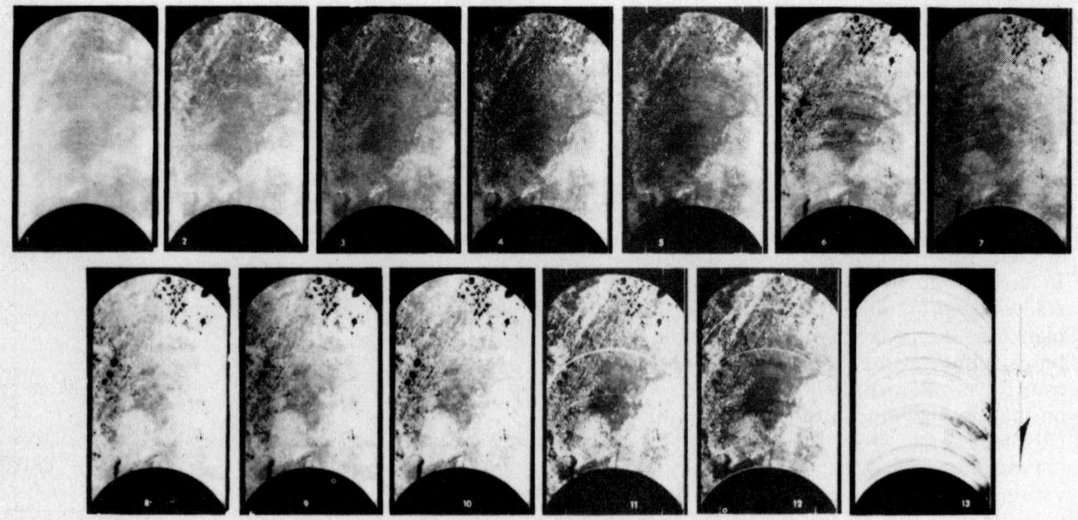

Fig. 29-30. Skylab S-192 Imagery of the Florida Green Swamp; SL/2 T-6 Pass 10, June 13, 1973 (Coker et al, 1975).

TABLE 29-3

Skylab S-192 and Landsat-1 MSS Bands (Coker et al, 1975)

S-192		Landsat-1	
Band No.	Band (microns)	Band No.	Band (microns)
1	0.41– 0.46		
2	0.46– 0.56		
3	0.52– 0.56		
4	0.56– 0.61	4	0.5–0.6
5	0.62– 0.67	5	0.6–0.7
6	0.68– 0.76	6	0.7–0.8
7	0.78– 0.88		
8	0.98– 1.08	7	0.8–1.1
9	1.09– 1.19		
10	1.20– 1.30		
11	1.55– 1.75		
12	2.10– 2.35		
13	10.20–12.50		

ures 29-32a and 32b). Contained within this image are two rather large playas that vary greatly in the amount of water stored during the course of a year. Color Figure 29-32a shows the Shiraz Playa (upper left of image) and the Neriz Playa (upper center to upper right) in March 1973 at or near its peak annual extent and volume; Color Figure 29-32b shows these two playas at their minimum storage condition. Having some knowledge of the average depth of these playas and combining it with measurements of surface area from Landsat, it was possible to estimate the approximate volume of water stored from February to May 1973. The surface area changes between September 1972 and August 1973 are shown in Figure 29-33. The Landsat images also show the extent of agriculture (red tones associated with irrigated vegetation) that might be augmented if this water supply were further exploited.

Another illustrative example of using spaceborne imagery to monitor water-storage changes in an area that is difficult to access is provided by Higer et al (1976). Their study concerned the amount of water retained in four impoundment areas in the Everglades water basin. This basin falls generally in the center of the portion of the Florida panhandle shown in Color Figure 29-34 and extends to the upper right where impoundment area number 1 (the apparent enclosed feature near the upper right of the image) exists. In a study that combined depth measurements obtained by the data collection system provided by Landsat 1 and measurements of surface-water area, a water management model based on an ability to estimate volumetric changes has been developed to assist in water budget studies for this Everglades area. The use of density changes in the imagery, plus depth measurements, is illustrated in Figure 29-35. This effort is instructive from the point of view of combining depth- and area-measurements to get water volume. It also illustrates that caution must be applied in wetlands areas to adjust appropriately for vegetation obscuring or being mixed in the surface-water area. Some photo-interpretation and judgment by a person with regional and hydrologic knowledge must be employed to achieve meaningful results. If simple density slicing or pure pixel classification were applied to this area, the vegetation signature distributed throughout would tend to make the estimates of surface-water area and volume too low.

The next level of sophistication in applying spaceborne imagery or remote-sensing imagery, in general, beyond estimating the area covered by water would be to assess conditions within the water body as expressed in turbidity or color changes. Pollutants (both inorganic sediments and organic constituents) may affect the reflective or emissive properties of water bodies and thus become detectable by multispectral scanning instruments. Observations in the visible and thermal infrared portions of the spectrum are optical measures of physical water quality parameters. Both qualitative and quantitative estimates of water quality may be derived from multispectral data.

When qualitative or relative estimates of water quality are being made, it is adequate to simply detect and delineate the area of pollution. Relative turbidity, expressed as tonal changes, is easily detected on Landsat multi-spectral images. Table 29-4 can be used for interpreting tones and hues in terms of sediment concentration in the upper layers of water bodies. Such keys are useful for examining turbidity plumes in lakes and estuaries to determine which of several water bodies is most turbid, where sediment is being deposited, or what sorts of circulations are occurring in the upper layers.

Figure 29-36 shows a lake in Utah as it appears in the four Landsat MSS bands (Strong, 1974; Strong, 1976). This figure shows not only regions of high turbidity in bands 4, 5, and 6, but also the location and extent of an algal bloom occurring in the lake at the time of the overpass. Color Figure 29-37 shows a portion of the Mississippi River Delta region along the southern Louisiana coast. In this figure, the very turbid water discharging from the Mississippi River has light blue hues, whereas slightly turbid to clear waters result in dark blue to black hues. Other research efforts that have used images to assess circulation patterns and relative sediment or constituent concentrations include Strong (1978), Klemas et al (1973), Finley and Baumgardner (1980), and Klemas et al (1974). All of these studies clearly indicate that remote-sensing imagery can be used for turbidity and color studies related to pollutants as an aid to conventional techniques for interpolating between spots where surface truth, point-specific measurements have been made, or

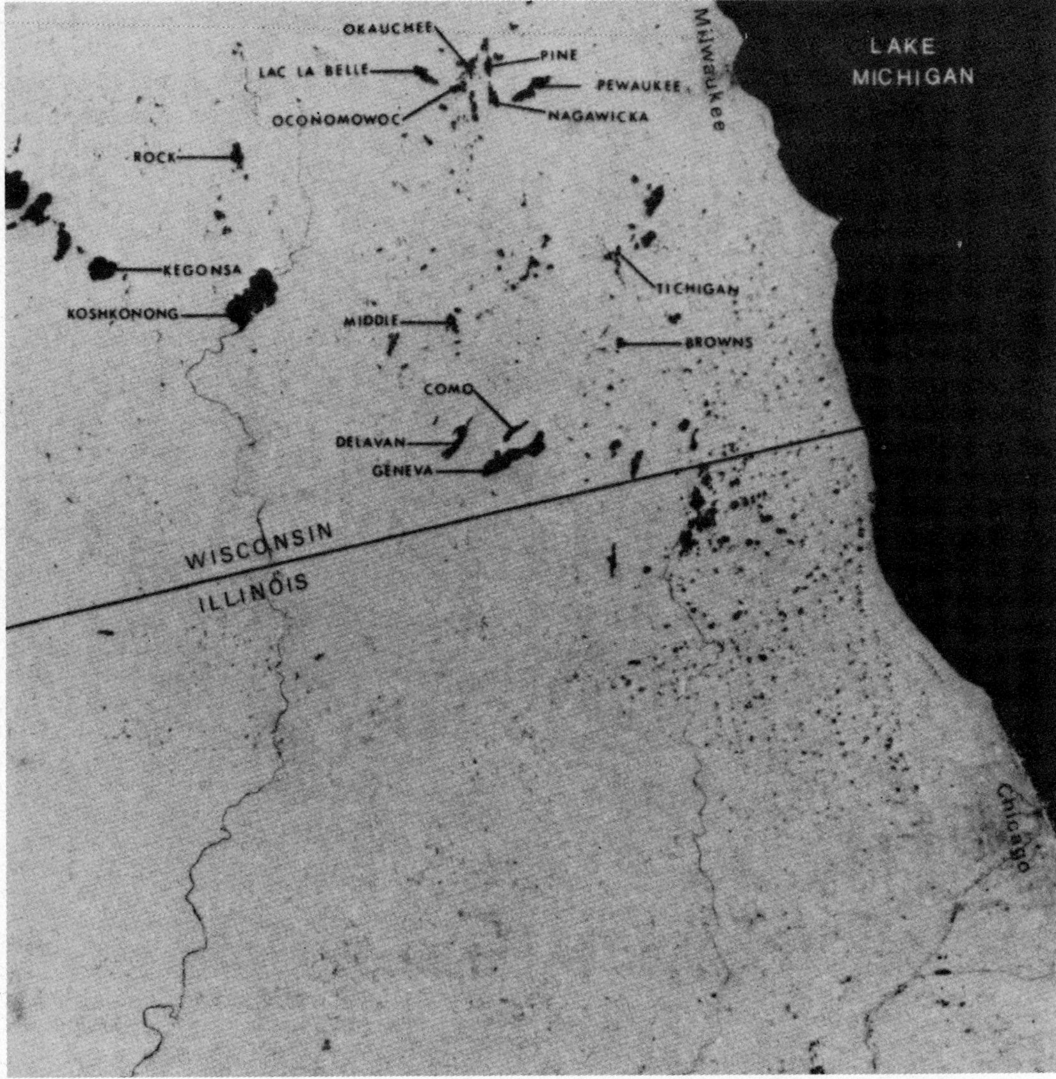

Fig. 29-31. Reproduction of an EROS Data Center band 7 print of Scene 1017-16093 (August 9, 1972). Water bodies stand out in stark contrast to the lighter-toned land features. The labeled lakes, excluding Lake Michigan, were sampled by the National Eutrophication Survey during 1972. The photograph covers a distance of approximately 185 kilometers along each edge (Boland and Blackwell, 1975).

as a means of extrapolating limited surface surveys to broader areas not normally covered by conventional techniques.

As a specific example of the point just made in the preceding paragraph, Moore et al (1974) suggested a procedure for using Landsat photographic images to classify relative lake turbidity and color and thereby reduce the necessity for field-sampling and analysis of water. In this study, the tones of ten prairie lakes in South Dakota were matched with the gray scale at the bottom of a Landsat scene; this was done for each of the four black-and-white images that constituted the Landsat scene. Lakes with different tones on any image were assigned to a different water class. The result was that the ten lakes fell into five classes. The authors concluded that, if a ground-based sampling program were developed, this procedure would be useful.

A report by Scarpace et al (1979) suggests a procedure for using Landsat digital data to assess the trophic status of inland lakes. This study resulted in a classification of approximately 250 lakes in south-central and southeastern Wisconsin. It was determined that a minimum of three dates of Landsat data are needed to adequately classify a lake. Results of regressions between Landsat data and trophic class data supplied by the Wisconsin Department of Natural Resources were excellent. The results of the investigation was a semiautomatic data-acquisition and data-handling system which, when used in conjunction

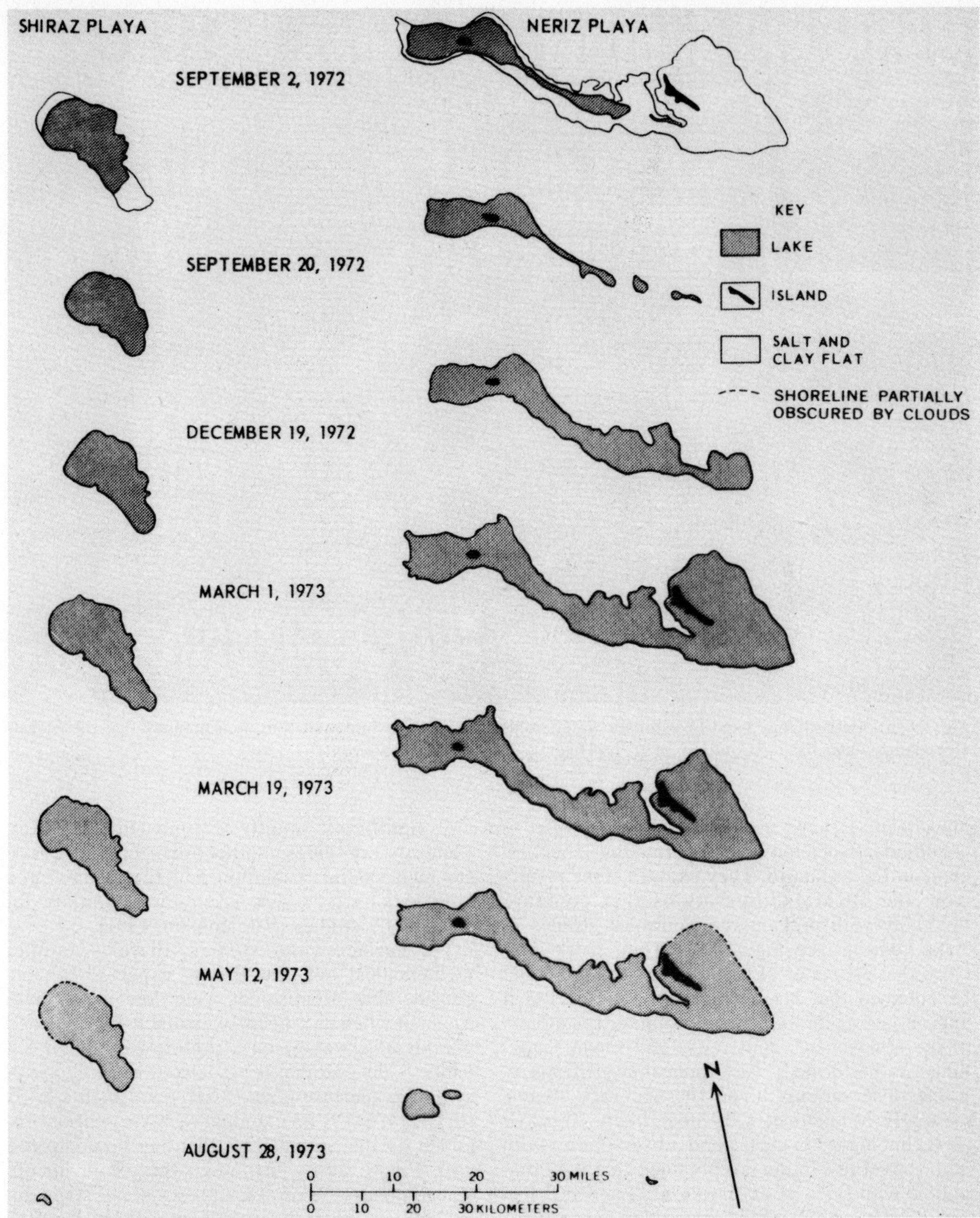

Fig. 29-33. Diagram showing lake fluctuations at Shiraz and Neriz Playas (Krinsley, 1976). Compare with Landsat color composites of Color Figs. 29-32a and 29-32b.

with an analytical categorization scene, could classify the relative trophic levels of all of the significant lakes in the State of Wisconsin.

The previous edition of the "Manual of Remote Sensing" discussed the use of thermal remotely-sensed observations (Meyer and Welch, 1975). Several examples were given showing that relative and absolute temperature estimates could be made that depicted plumes and circulation patterns in water bodies. Since that time, observations from spaceborne instruments have been studied. These spaceborne observations generally apply to relatively large (10 km^2 and larger) lakes because of the low spatial resolution of the sensors. Some of the most interesting observations have been obtained over the Great Lakes. The highest spatial-resolution observations come from the HCMM (see Price, 1980). Schott and Schim-

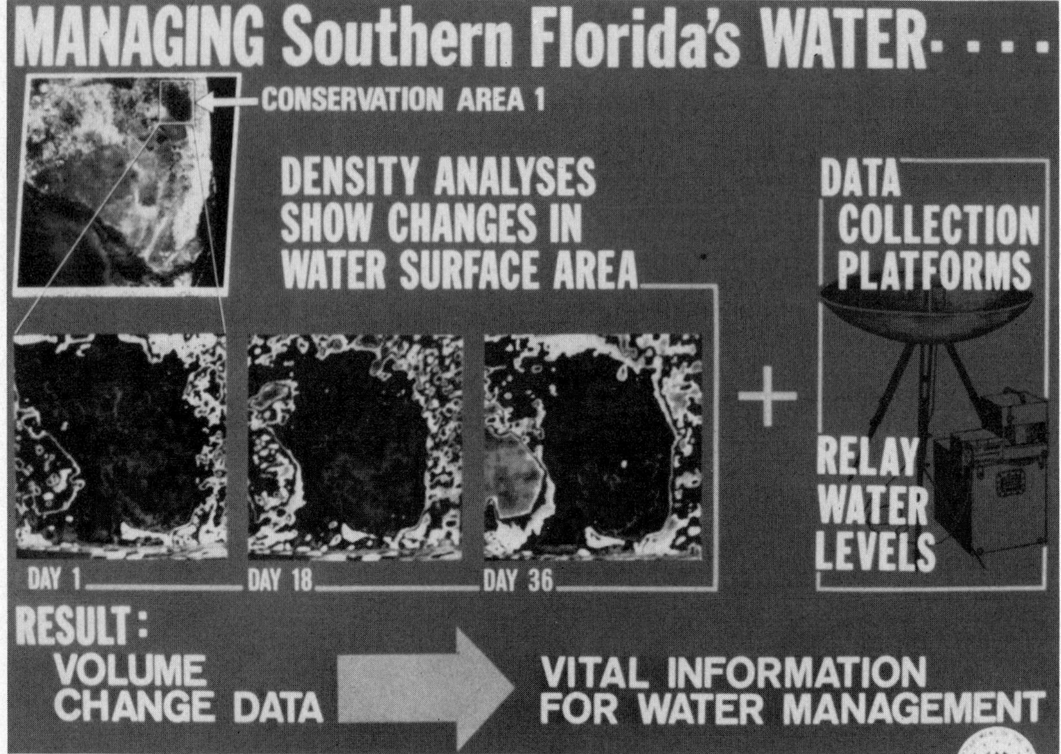

Fig. 29-35. Methodology used to estimate water volume changes in impoundment area number 1 of the Florida Everglades water basin. (Courtesy of A. L. Higer, U.S. Geological Survey.)

minger (1981) have used this sensor to observe conditions associated with a thermal-bar phenomenon on Lake Ontario. They showed some results from the HCMM observations that compare favorably with previous conventional studies of Lake Ontario (see Fig. 29-38). Their work indicates that estimates of surface temperature could be obtained that are accurate to about 1.1°C if appropriate calibration, accounting for the effects of the atmosphere, emissivity, and sensor capability, is performed. The applicability of spaceborne observations, given the accuracy that is presently obtainable, will most likely improve as technology advances, and observations with high spatial resolution are obtained that are compatible with Landsat observations. Landsat 3 had a thermal band, but it did not function properly for any significant length of time (Price, 1981). Landsat-D provides another opportunity to evaluate high spatial resolution and thermal infrared observations to assess accuracy and utility for water body studies (Salomonson, 1981).

For surface-water studies, there is another technological advance that is expected to hold considerable promise for water resources management when data become available for extended periods of coverage on a routine basis. This advance is the one that is associated with the use of synthetic aperture radar (SAR) data. Figure 29-39 (Bryan, 1981) is an example of SAR imagery obtained by the Seasat SAR that can be compared with Figure 29-34. Although this is an uncontrolled mosaic, several features seem to stand out more clearly than those in Figure 29-34. A sketch

TABLE 29-4

Qualitative Estimate of Relative Turbidity of Water from Landsat Images (Moore, 1978)

Relative Turbidity	Tone of Image				Hue of Color Composite
	Band 4	Band 5	Band 6	Band 7	
None	Dark	Dark	Black	Black	Black
Slight	Medium	Dark	Black	Black	Dark blue
Moderate	Light	Medium	Dark	Black	Medium blue
Heavy	Light	Light	Medium	Dark	Light blue
Very heavy	Light	Light	Light	Medium	White

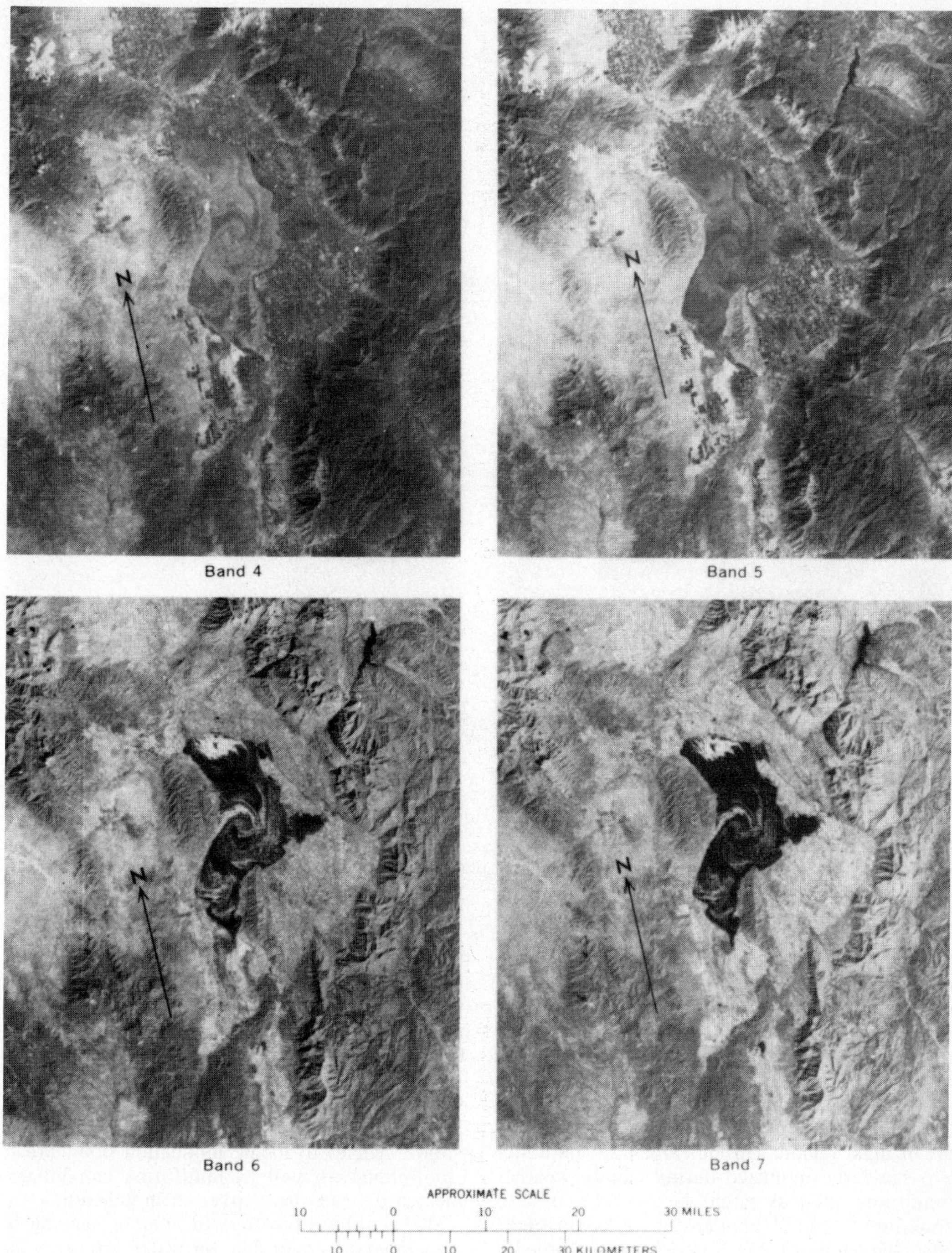

Fig. 29-36. Comparison of a part of Landsat-1 image 1051-17420, bands 4 to 7, showing algal blooms in Utah Lake. Band 4, penetrates the water to show the greatest amount of turbidity; band 5, high turbidity with more distinct shoreline; band 6, turbidity pattern less distinct, but sediment and algae visible; and band 7, shoreline clearly delineated, slight water penetration, and distinctive pattern of algal bloom. (Strong, 1976).

of some of the drainage features that seem to stand out is provided by M. L. Bryan of the Jet Propulsion Laboratory, Pasadena, California (private communication) and is shown in Figure 29-40.

Some further comparisons of Seasat SAR and airborne SAR data with Landsat data over the Wind River Mountain region of Wyoming have been accomplished by Foster and Hall (1981). They point out that when lakes are calm and smooth, they act as specular reflectors and appear dark in the imagery. When there are surface

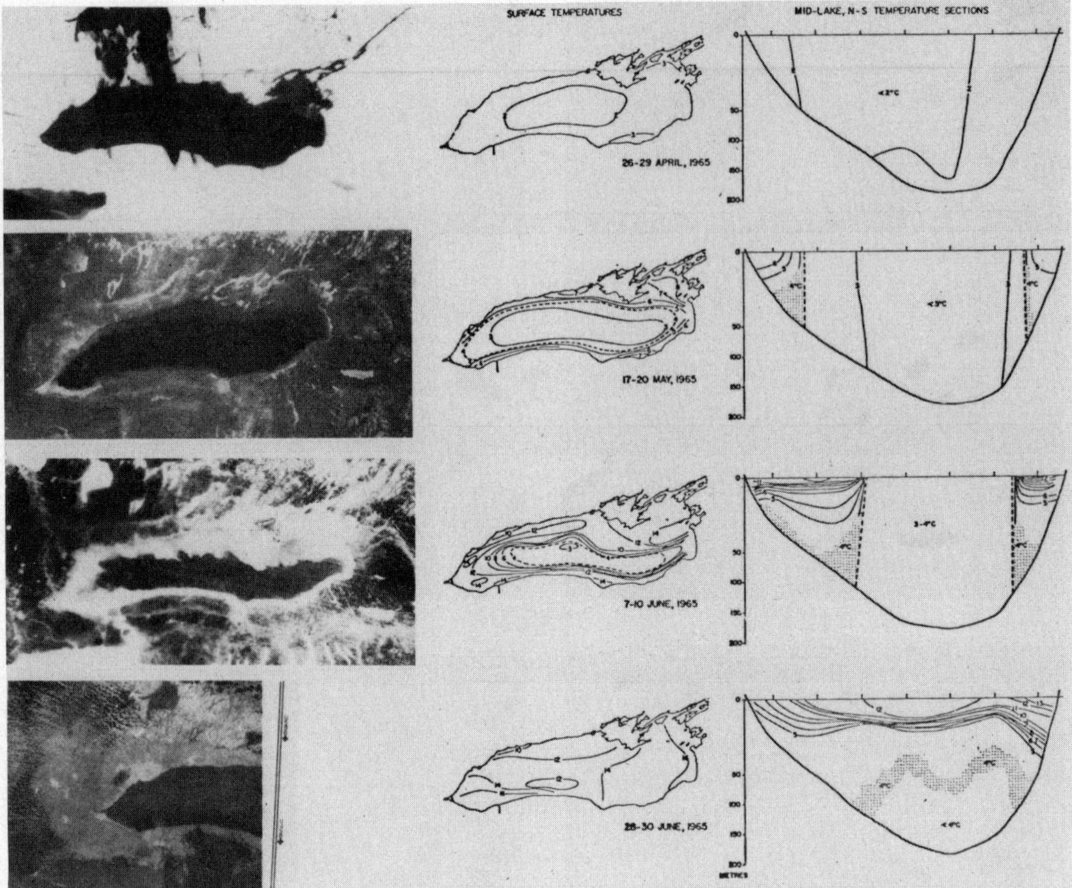

Fig. 29-38. Observations of surface temperature on Lake Ontario as seen in Heat Capacity Mapping Mission Imagery (on left) as compared with more conventional observations obtained in 1975 (Schott and Schimminger, 1981).

waves caused by wind action, they will appear bright in contrast to Landsat imagery in band 7 where they will continue to appear dark. Vegetation on the surface will also appear bright, as can be seen in the Everglades areas of Figures 29-39, because it is rough and backscatters the microwave energy. In these studies radar has some advantage because of its nearly all-weather and day or night remote-sensing capability. Its utility is particularly optimized during cloudy weather conditions such as might be associated with flooding. It would appear that even in clear weather conditions, the SAR imagery afforded a complementary data source in unfrozen surface-water studies.

LAND/SURFACE COVER[1]

Information with respect to land cover has two different uses in hydrologic and water resources studies. First, the characteristics of the land surface, both soil and vegetation, affect or control hydrologic processes such as infiltration, runoff rates, and evapotranspiration. Various hydrologic parameters are estimated from the land cover. The second use for land-cover information is as an indicator of water-related activities. Water-demand studies estimate irrigation-water use from land cover. Wetlands-mapping programs use vegetation species to locate inundation boundaries. Floodplains as well as landforms can be delineated by using land cover as an indicator.

The general advantage of remote sensing in providing land-cover data for water resources investigations is that, when the proper sensor-platform-frequency combination is used, the system can provide parameter estimates for individual hydrologic-response units over geographically large areas in a cost-effective manner.

Generally, most studies follow a two-step procedure when using remotely sensed data. First, the data are classified into a land cover, or land use, category. Then, a representative value of the hydrologic or water-related parameter for that category is applied. Some studies have developed

[1] For more information on remote sensing of land use, landcover parameters the reader should consult Chapter 30.

Fig. 29-39. Seasat SAR imagery over Florida (Bryan, 1981).

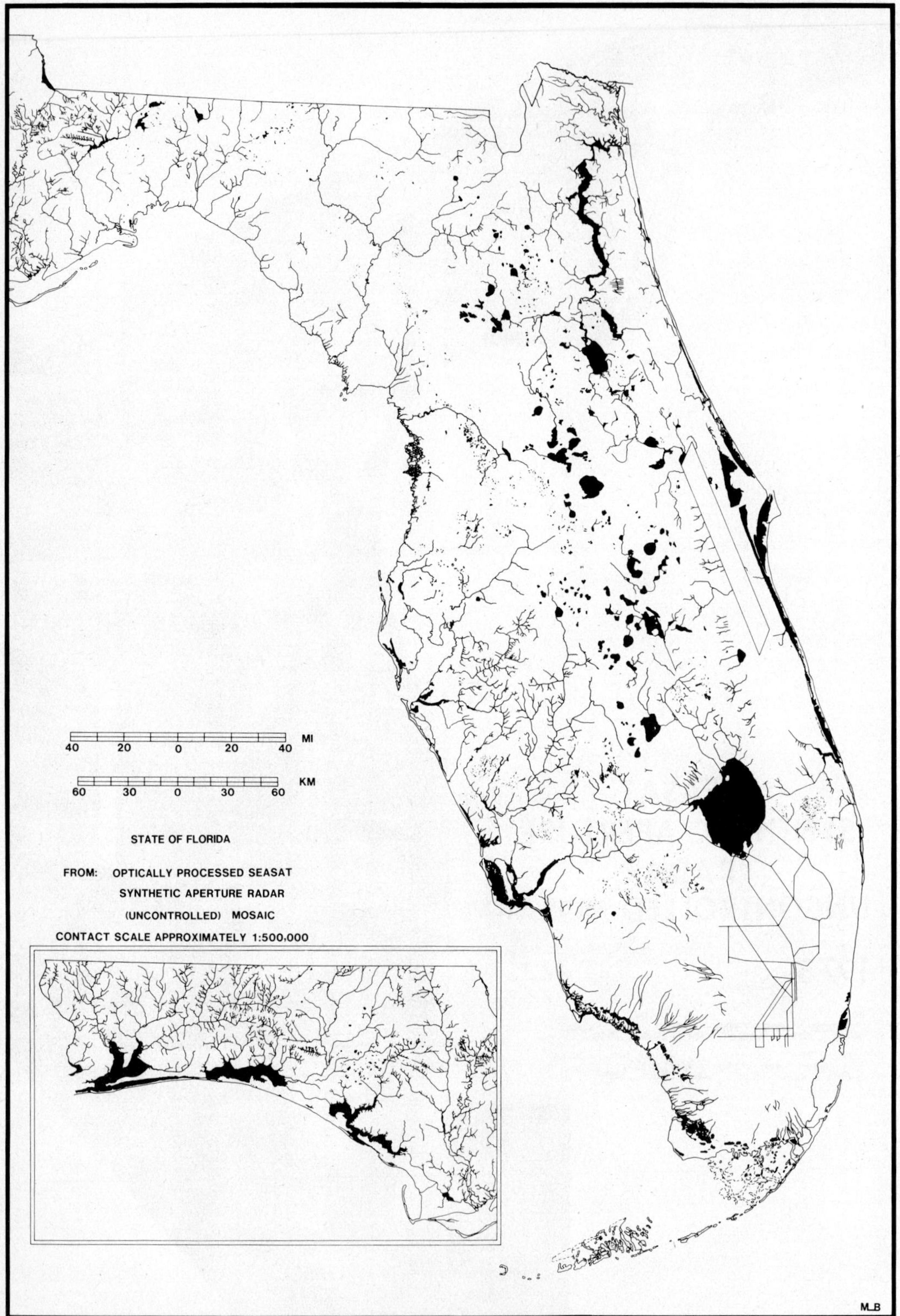

Fig. 29-40. A sketch of water bodies and drainage features obtained from Seasat SAR imagery in Figure 29-39. (Courtesy of M. L. Bryan, Jet Propulsion Laboratory.)

direct relationships between the remotely sensed data and the parameters.

DATA COLLECTION AND ACQUISITION CONSIDERATIONS

Selection of the acquisition method or data source should entail consideration of the following problem characteristics:

a. Variability of the land cover
b. Size of the decision- or modeling-unit that the data will be used to represent
c. Required accuracy
d. Spatial extent of the study area

These should be matched to a system that satisfies the requirements in the most cost-effective manner.

A sensor system's performance in estimating land cover depends on the spectral bands that it measures and the ground-resolution element over which it averages reflectance. More spectral bands and higher resolution usually require greater expenditures.

Perhaps the most widely used type of remotely sensed data for land cover in water-resources studies is that obtained within the visible and near infrared region of the spectrum. One reason for choosing this type of data is that it is relatively easy to interpret since the near-infrared energy that is sensed, like visible light, is reflected energy. Another reason for the frequent use of near infrared energy in remote sensing is the relative ease with which it may be acquired by users. Many commercial services exist that will collect photographic data for specific projects. Also, large amounts of data have been archived. Black-and-white small-scale photography has been collected by the USGS and USDA for many years. In recent years, NASA has collected many large-scale color-infrared images throughout the United States. Landsat MSS data, which include two near-infrared bands, have been regularly collected since 1972.

Data collected within other spectral regions have been used only in limited or experimental studies involving land cover identification for water resources. Thermal infrared and microwave data are more difficult to obtain and interpret. Electromagnetic energy-matter interactions within these spectral regions are dependent on the thermal and electrical properties of the target, whereas those in the visible and near infrared are most dependent on the surface reflection characteristics. The thermal and electrical properties are influenced by many factors, such as soil moisture, which can confuse interpretations of land cover. However, for other types of hydrologic studies, these measurements can be valuable.

As previously indicated, for most studies, visible and near infrared data will usually be the most cost-effective. Another data-acquisition question is the choice of the platform. There are three general categories of platforms available: low-altitude aircraft, high-altitude aircraft, and spacecraft. For very specific problems requiring detailed current information, the most readily available source is to contract for the flying of low-altitude photography to the user's specifications. Sometimes, however, existing photography that is available from the U.S. Department of Agriculture will suffice, especially if the area is not very dynamic. If large changes are occurring, however, it is likely that the black-and-white photography from aircraft surveys will quickly become out of date.

A drawback to low-altitude platforms in many water resources-related land-cover studies covering several ten or hundreds of square kilometers is that a great number of individual images or data points must be collected, processed, and interpreted when such large areas are under investigation. Sometimes it is necessary to sacrifice detail to reduce the data volume or to produce more synoptic data. In these cases, high altitude and spacecraft platforms can be used. In most of these situations, it will be necessary to use data or platforms that are already available.

One of the most readily available data-collection alternatives for visible and near infrared data is the Landsat satellite. Landsat data have only limited spectral coverage, and spectral signatures are integrated over a fairly large ground resolution element, 0.45 ha. The advantage of this type of data is that it is available on a relatively frequent basis since 1972 and can be obtained in image or digital form. Landsat data can be used to define most types of cover through a careful selection of the time of year. Optimum usefulness of such data can be expected in areas of land cover with low spatial variability. When the size of ground-cover elements is close to that of the sensor resolution element, however, there are significant problems because of the mixing of individual cover responses.

Depending on the scale of study, Landsat data may have a limited information content for a given pixel because of the factors previously discussed, and therefore some error is to be expected. It is interesting to note, however, that substantial errors in estimating parameters for individual ground resolution elements can become insignificant if these estimates are used to determine an average parameter for a larger decision- or modeling-unit. Jackson and McCuen (1979) examined the problem of estimating the percent of impervious area from Landsat data. They compared estimates based on Landsat data and high resolution photography for model units ranging from one to thirty-six ground resolution elements. They found that the expected error of the parameter estimate decreased as the number of ground resolution elements used increased, in a manner similar to sampling from a normal distribution. Jackson et al (1977) examined the differences between estimates of percent of impervious area as obtained using conventional- vs Landsat-data for three levels of aggregation in a hydrologic model. These results, summarized in Table 29-5,

TABLE 29-5

Fourmile Run Error Analysis of Impervious Area Percentage (Jackson et al, 1977)

Level of Analysis (1)	Number of Samples (2)	Average Size in Acres (3)	Standard Deviation of Size, in Acres (4)	Size Range, in Acres (5)	Standard Error, as a Percentage (6)
Subwatersheds	179	80	60	5.6–264.3	14.9
Inlets	70	167	120	6.7–539.8	11.0
Junctions	20	603	617	29.1–2,891.8	6.5
Note: 1 acre = 4,046 m²					

show that, as the average model unit size increases, the error decreases. Their conclusion was that the level of error was unacceptable below the junction level, or model units of 1 mi².

DATA ANALYSIS

Because of the prevalent use of visible and near infrared data for identifying water-resources land cover, these data are emphasized in the following discussion.

That portion of the radiant energy of the sun that reaches the surface of the earth is partially reflected. The proportion of reflected energy is dependent on certain physical properties of the land cover and the wavelength being analyzed. Cover differentiation based on spectral data requires that the categories of interest have unique patterns of spectral response. Within other spectral regions, spectral measurements are dependent on more complex interactions that are difficult to isolate. Sometimes, however, these types of measurements can be used to improve the types of land cover identified using visible and near infrared data. An example of this is the use of thermal measurements to locate fields associated with irrigation in conjunction with reflected data.

Distinguishing between general land cover categories such as grass, trees, bare soils, and water is possible to some degree with almost any type of sensor. However, it is easier to do if the spectral bands to maximize separation are carefully selected. Color-infrared film and multispectral measurements within the visible and near infrared regions are most useful.

More refined land-cover classification schemes require more spectral and ancillary data such as antecedent conditions and crop calendars. Carter and Schubert (1974) developed a scheme for classifying wetland-vegetation species that considered the growth cycle of the species involved. They obtained ground-level spectral-reflectance measurements for each species over several months. These measurements were at the same wavelength intervals as used by the Landsat 2 sensor systems. The results of the measurement program are illustrated in Figure 29-41. Measurements in the two visible bands (Figs. 29-41a and 29-41b) indicate that there is little difference in the spectral responses of most species of wetland vegetation at any time; however, all the species have significantly different responses from those of soils. In band 7 (Fig. 29-41d), the changes in spectral response by month for wetland vegetation species and also for soil are seen to be significant, and many of the species are separable at several times of the year. Carter and Schubert (1974) concluded from these data that, by using October spectral measurements of bands 5 and 7, it would be possible to reliably separate all of the species.

Other examples of the use of ancillary data to obtain the best land-cover identification scheme includes the work conducted by Draegar (1977) and Heller and Johnson (1979). In these studies, Landsat data were used to estimate irrigated acreage for water-resources allocation and planning programs. Crop calendars were necessary in these studies because irrigated and dryland farms were not spectrally separable with Landsat data, except in the late portion of the growing season.

Identifying soil types on bare soil areas is difficult. Under ideal conditions, however, individual soils often have unique spectral responses. A typical set of spectral reflectance data for various soil types is shown in Figure 29-42a. How-

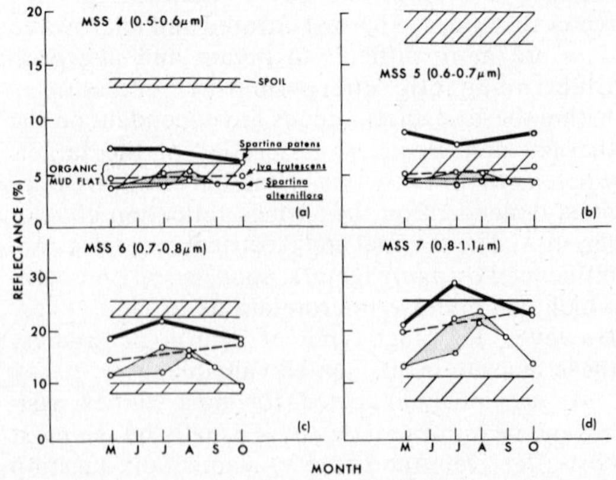

Fig. 29-41. Measured seasonal reflection of wetland land covers (Carter and Schubert, 1974).

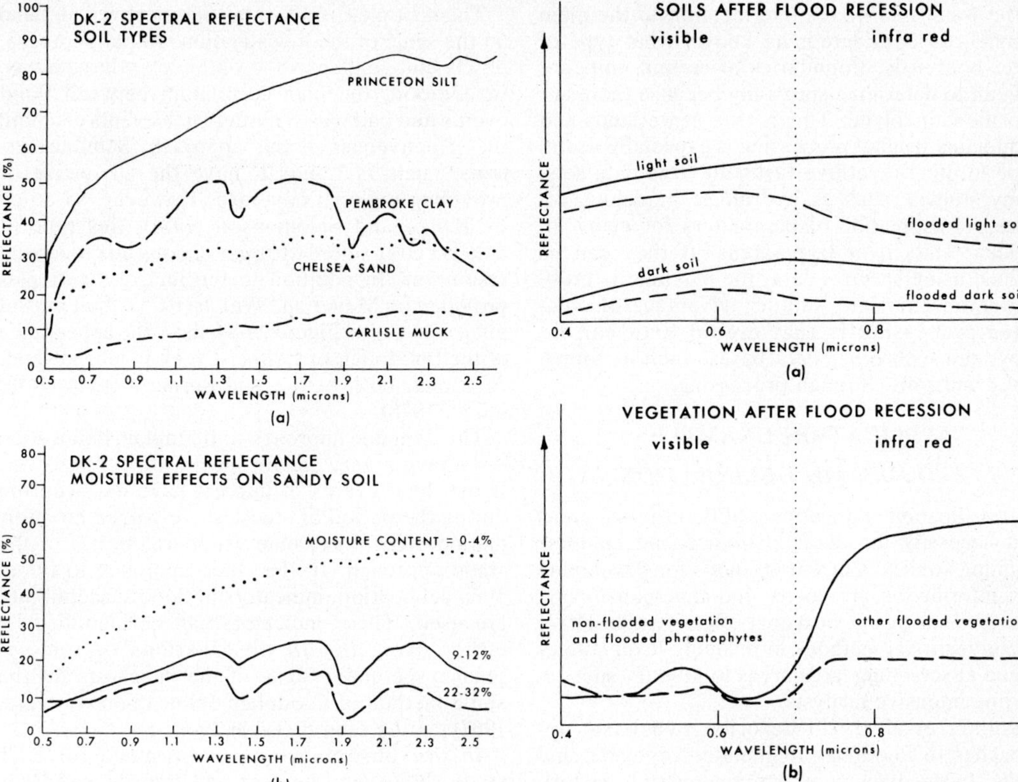

Fig. 29-42. Measured spectral reflectance of soils (Hoffer, 1971).

Fig. 29-43. Effects of flooding on the relative reflectance of soils and vegetation (Hoyer et al, 1974).

ever, plowing and moisture greatly influence the spectral reflectance and must be quantified before soils can be identified. Figure 29-42b illustrates the influence of soil moisture on reflectance.

The influence of moisture on soil reflectance is very useful in other types of studies, primarily flood inundation mapping and floodplain delineation. Studies conducted by Hoyer et al (1974) and others have shown that flooding causes an overall reduction in soil reflectance, adds pools of low-reflectivity water, and creates a moisture stress on the vegetation that reduces reflectance, thereby helping the image analyst to locate flooded areas days after the actual event. Figures 29-43a and 29-43b illustrate the general effects of excess water on reflectance.

Many hydrologic investigations involve urban land-covers and uses. Although urban land-covers differ in their absolute levels of spectral response, they are similar in their patterns within the visible and near-infrared regions. Jackson (1975) used this fact to develop a procedure for estimating the percent of impervious area directly from Landsat data. Using data from a period when most pervious covers had a high ratio of near infrared response to visible response, he developed the following equation:

$$\text{Percent of impervious area} = \text{EXP}(1.45 - 2.16(\text{band 7/band 4}) \qquad (29\text{-}11)$$

Basically, impervious covers, excluding bare soils, have low ratios and pervious covers have high ratios. Mixtures fall in between these low and high ratios. McKeon et al (1978) tested this procedure in another study and obtained good results.

Direct parameter estimation from multispectral data has also been performed by Blanchard (1975). He developed a procedure for estimating the Soil Conservation Service (SCS) runoff curve number (CN) directly from Landsat data within one river basin under certain conditions.

Numerous approaches to data analysis have been used to define land cover for water resources and hydrologic investigations. In certain types of studies, photointerpretation is required. Sometimes digital processing is performed and then analyzed visually. Problems involving the identification of boundaries usually require this approach (i.e., floodplain delineation, interpretation of geomorphology, and wetlands mapping). Defining this type of data generally requires the identification of characteristic patterns of the land cover. For instance, a good indicator of a floodplain is an oxbow lake. When such a feature is analyzed spectrally, all the investigator knows is that it is water. However, when the spatial pattern of the water and its relative location to the main channel are considered, he knows what type of water body it is. Boundaries of certain units are difficult to determine spectrally because there are transitions involved. This is true in wetlands and floodplains. Digital processing is especially useful in performing repetitive tasks on large data sets. Many studies, such as hydrologic modeling, require the estimation of parameters for many individual units over large areas. If they can be defined using spectral data, the machine is probably more efficient. Another advantage of computer processing is that digital data can be integrated with other data bases, such as topography and soils, through processing.

APPLICATION EXAMPLES

FLOODPLAIN DELINEATION

Identification and mapping of flood-prone areas is a necessity for flood insurance and land-use planning studies. Current methods for developing this information are costly and time-consuming. Remote sensing can be a cost-effective alternative to conventional methods in planning-level studies aimed at screening large areas to identify sites requiring intensive analysis.

Sollers et al (1978) describe two basic approaches to floodplain delineation: dynamic and static. In the dynamic or actual approach, historical evidence of flooding is used to delineate the extent of inundation. By observing the evidence of events of differing frequency, an inundated area-frequency relationship can be developed. Ground observations of such evidence are scarce and unreliable since one large flood can wipe out the evidence of all smaller events (Sollers et al, 1978). In addition, ground observations represent only a few points on a continuous boundary. If these data must be obtained over a large area, they present a very formidable task.

Remote sensing is a particularly attractive alternative when the dynamic approach is used because, through the proper selection of a sensor and platform, synoptic coverage of the extent of flooding can be obtained very quickly. Flood-inundation mapping using remotely sensed imagery has been successful on many rivers and in different regions, using color-infrared photography, Landsat data, and, to a limited degree, thermal infrared data. Examples include studies reported by Rango and Anderson (1974), Hoyer et al (1974), Moore and North (1974), Deutsch and Ruggles (1974), Rango and Salomonson (1974), Williamson (1974), Morrison and Cooley (1973), Wiesnet et al (1974), Robinove (1978), and Phillipson and Hafker (1981). The successful application of this approach derives from the distinctive changes induced by additional water in an area. These include increased soil moisture, moisture-stressed vegetation, and standing water, all of which result in reduced reflectivity that lasts up to 2 weeks after an event.

The best platform for this approach will depend on the scale of the investigation. Aerial photography is quite expensive to obtain over large areas. In addition, the high correlation between flood events and bad weather after such events can limit the effectiveness of this approach. Satellite systems, such as Landsat, have the advantage of providing regional coverage. However, as noted by Rango and Salomonson (1974), this type of data is better suited to larger regions due to spatial resolution. In addition to the imagery examples provided by Meyer and Welch (1975), the Landsat images in Color Figure 29-44 show the appearance of the floodplain of Cooper Creek in Queensland, Australia, both before and during a flood (Robinove, 1978).

The dynamic approach to floodplain delineation does have a very significant drawback, however. If the flood events of interest have not occurred during the period of study, there will be no information. Therefore, many support the use of the static approach. In the static approach to floodplain delineation, indicators of flood susceptibility are used. These indicators can be identified by either using *in-situ* observations or remote-sensing methods. Some of the indicators for the static method of floodplain delineatiion (Burgess, 1967) can be seen in Table 29-6.

In-situ observations can be made for each study; however, the cost and time required can become excessive. Floodplain delineation can also be based on previously compiled ancillary data such as soil maps. In some cases, soil maps are highly effective; however, they are not available for all areas.

Delineation of floodplains using the static approach and remotely sensed data is based on the identification of indicators through their multispectral response and the spatial pattern of these responses. Not all of the foregoing indicators can be identified. What can be defined will depend on the spectral measurements and the spatial resolution. For instance, the indicators that can be defined using Landsat data (Landsat floodplain indicators by Rango and Anderson, 1974) are seen in Table 29-7.

Harker and Rouse (1977) digitized color-infrared photographs obtained at an altitude of 10,675 m over a watershed in Texas and used computer analyses to classify the data into one of four categories: low vegetation not in the floodplain; low vegetation in the floodplain; trees not in the floodplain; and trees in the floodplain. Floodplain boundaries were then drawn using the classified output. These boundaries were compared with those developed using four other data sources: the U.S. Army Corps of Engineers 100-year flood records, soils data, topography, and aerial photographs. Good correlations generally exceeding 0.8 were found with all data except the boundaries developed from the soils. The best results were obtained in the low-vegetation areas.

Rango and Anderson (1974) examined the po-

TABLE 29-6

Environmental Indicators For Static Floodplain Mapping

a. *Class I—Meteorologic*
Climate: seasonality, temperature, and precipitation (distribution and intensity)
Storm type
Storm direction
Orographic effect

b. *Class II—Physiographic*
Bedrock composition
Bedrock structure
Upland physiography
Watershed shape
Valley orientation
Floodplain shape
Proportion of watershed in floodplain
Stream drainage pattern and density
Discharge to next-order stream, much larger stream, lake, tidewater
Underfit streams

c. *Class III—Geomorphic*
Floodplain type: meander plain, cover plain, or bar plain
Alluvial fans
Terraces
Levees: active or abandoned
Point bars and swales
River bars
Deltas
Channel configuration
Channels: abandoned, oxbows, or filled
Backswamp areas
Marshes
Former lake beds
Dunes
Landslides
Springs and seeps
Depth to water table
Swamping
Valley trenching

d. *Class IV—Topographic*
Relative elevation
Slope
Slope changes
Slope complexity
Microrelief

e. *Class V—Pedologic*
Structure
Texture
Organic content
Salinity
Depth
Vertical uniformity
Permeability
Drainage
Erosion: type, extent, and severity
Deposition: type, distribution, and freshness

f. *Class VI—Socioeconomic*
land use (rural and urban): character, intensity, changes over time, boundaries, alternatives, and limitations
Building: type, location, and condition
Transportation: type, location, and condition
Flood-alleviation measures
Channel constrictions
Repair work
Farm characteristics: size, type, and location
Management practices
Crop condition
Vegetation removal: logging and land clearing
Revegetation

TABLE 29-7

Landsat Floodplain Indicators

a. Upland physiograph
b. Watershed characteristics such as shape, drainage density, etc.
c. Degree of abandonment of natural levees
d. Occurrence of stabilized sand dunes on river terraces
e. Channel configuration and fluvial geomorphic characteristics
f. Backswamp areas
g. Soil-moisture availability (could also be a short-term indicator of flood susceptibility)
h. Soil differences
i. Vegetation differences
j. Land-use boundaries
k. Agricultural development
l. Flood alleviation measures on the floodplain

tential of Landsat imagery for floodplain delineation by using images of an area of the Mississippi River obtained before flooding that occurred in 1973. They used 1:100,000 scale images and successfully delineated the boundaries of the area's artificial levee systems, soil differences, agricultural and vegetation patterns, upland boundaries, backwater areas, and special flood-alleviation measures in urban areas.

In a related study, Sollers et al (1978) used both aircraft and Landsat multispectral data to delineate the floodplain for a watershed in Pennsylvania. They found that it was easier to use the Landsat data because of the problems of data reduction involved with the aircraft data. The best indicator was the floodplain soils. They observed good agreement in agricultural and developed areas but, in forested areas, agreement was poor because of the obscuring vegetation.

Another approach to the identification of flood-susceptible areas has been applied by Struve and Hudson (1979) and Wilcox and Struve (1979). Their studies were concerned with developing stage-area relationships for the evaluation of flood protection and land clearing programs. They classified digital Landsat data and registered it with digital elevation data on a 100- by 100-m grid. This was used as input to a computer model that yielded an area inundated-versus-streamflow relationship by land use. They found that the cost of this approach was comparable to conventional methods. The advantage of the Landsat approach was that it greatly reduced the time required for production of the final product. In addition, digitizing of the elevation data would not be necessary in the future, which would reduce the cost of updating the data base.

HYDROGEOMORPHIC ANALYSIS

The approach to hydrologic analysis emphasizes the physical structure and geometry of the watershed in describing or predicting the water and sediment yield. Land-cover parameters are valuable in these studies as direct inputs and also as indirect inputs by inferring the basin characteristics that include drainage density and type; stream length, sinuosity, and slope; and basin area, shape, width, length, slope, and aspect. A summary of some relationships between basin physiography and basin-runoff response is presented in Table 29-8. A further discussion of this table can be found in Gregory and Walling (1973).

Drainage network information is generally obtained from topographic maps, which are usually dated. Remote sensing, due to its more current nature, can often provide better and more reliable information than these maps. Rango et al (1975) reported on a study in which they compared conventional methods with several types of remotely sensed data on a number of watersheds in the United States. They found that watershed area, shape, and channel sinuosity estimates from Landsat images at a 1:100,000 scale were comparable to data obtained from 1:62,500 scale topographic maps, particularly in well-dissected terrain with moderate vegetation. In flat terrain or heavily forested terrain, results were comparable to 1:250,000 scale maps.

Limited tests have been conducted to evaluate the use of radar and thermal infrared imagery for estimating hydrogeomorphic parameters. Cannon (1973) obtained real aperture, cross-polarized, K-band radar imagery at two altitudes, 3050 and 6100 m, over the Mill Creek test area in Oklahoma. Data obtained from topographic maps and low-level photography were compared with those taken from imagery with a 15-m resolution. These comparisons showed that the accuracy and reliability of radar estimates of area, perimeter, order, bifurcation, ratios, and stream length are highly dependent on the size of the watershed, its relief, and the distribution of vegetation. For the scale of imagery used, Cannon concluded that a watershed would have to be at least 78 km^2 to use this type of data. He also obtained thermal infrared imagery over the same area at three different times of day: predawn, midmorning, and midafternoon. These were obtained with an aircraft from three altitudes: 1800, 3555, and 6400 m. Drainage information from topographic maps was compared with information developed from ther-

TABLE 29-8

Examples of Relationships Between Basic Characteristics and Basin Response (Gregory and Walling, 1973)

Index of Response	Related to	Area
Streamflow characteristic (Y) 71 indixes of streamflow used for station including 2 for low flow, 3 for duration of daily flows, 6 for flood peaks, 8 for flood volumes, and 13 for annual and monthly means.	Drainage area (A) Average slope of mean channel between points 10 and 85% of distance upstream from gauging site (S) Main channel length (L) Percentage of total area in lakes, ponds, swamps (St) Elevation (E) Percentage of area under forest (F) Values of potential maximum infiltration in inches during annual flood under average soil conditions (S_i) Snow index (S_n) Maximum 24-hour precipitation expected to be exceeded an average once every 2 years ($1_{24.2}$)	U.S.A. Four areas: Eastern (Potomac) Central (Kansas) Southern (Louisiana) Western (California)
Runoff per unit area (M)	Drainage density (Dd) Percentage of swamped area (P)	U.S.S.R.
Runoff/rainfall ratio (R)	Area (A) Basin length (LB) Relief (H) Perimeter (P) $R = -95.604+29.238A -15.36LB+0.184H -9.377P$	New Zealand
Hydrograph time (Base)	Mainstream length Distance from outlet to Centre Channel roughness Mainstream slope	Australia New South Wales
Lag time (L_t)	Length from outlet to center of gravity of source area L_{sa} Average width of source area W_{sa} Average slope of source area S_{sa} Drainage density (Dd) Where source area = half of watershed with highest land slope	U.S.A.
Low flow per unit area	Percent of catchment cleared of trees and brush	U.S.A. (Virginia)
Annual sediment yield	Precipitation intensity Average annual runoff Erosion factor Annual precipitation Area Average slope	U.S.A.
Reservoir sedimentation	Slope factor Age Gross erosion Capacity – inflow ratio Nonincised channel density Watershed shape	U.S.A. (Illinois)

mal images. It was found that the predawn images were the most useful and estimates were accurate and reliable, even for the smallest watersheds evaluated, 7 km^2.

Baker et al (1974) and Sharma and Jain (1978) have also tested remotely sensed hydrogeomorphic parameters against conventional estimates. Baker et al (1974) used Skylab S-190B imagery with low- and high-altitude photography to define channel systems and land cover for a watershed in Texas. They found that the most useful data were those derived from high-altitude aircraft color-infrared photography.

Sharma and Jain (1978) evaluated the use of 1:60,000 aerial photographs in estimating the runoff factors of several watersheds in India. The runoff factor is used in conjunction with the watershed area to predict the water yield from a rainfall event; it is estimated using relief, soil permeability, and vegetation. The relief parameter is determined from the local relief, drainage density, and ruggedness of the watershed. Sharma and Jain used aerial photographs and very large scale topographic maps to estimate all these data and developed a runoff factor of 0.54 for the basin. Then, for a period while streamgage records were available, they predicted the discharge in the stream using the available rainfall data. When these predictions were compared with the observed streamflows, it was found that, during periods of low flow, the discharge predictions were too high. This was attributed to the fact that the aerial photo-based runoff factor was determined for one time during the year, when, in reality, it varies temporally. Runoff factors estimated from the observed rainfall and runoff varied between 0.4 and 0.75.

WETLANDS MAPPING

Wetlands are an important component of the hydrologic system. Increased public concern over the environment in recent years has led to the enaction of national, state, and local legislation for the protection of both inland and coastal wetlands. Carter (1978) points out that a common first step in these programs is to identify and inventory wetland areas. Some of the important kinds of information that must be obtained from these surveys, as listed by Reimold and Linthrust (1975), are the boundaries, vegetation type, production patterns, and water movements.

Because of the inadequacy of and inaccuracies in existing map coverages, the dynamic nature and areal extent of the wetlands, and time constraints, many agencies have used remote sensing as a tool in wetlands mapping. The optimal platform for obtaining the data will depend on the ultimate use of the data. For an inventory, high-altitude aircraft or satellite imagery may be suitable. However, when regulation and the delineation of boundaries commensurate with 1:24,000 scale maps are the ultimate objectives, it will be necessary to use larger scale aerial photography.

Most large-scale applications involve delineation of the wetlands in terms of the upland boundary and mean high-water mark, as well as the major species of vegetation. Generally, the vegetation is used to identify the boundaries since the species of vegetation is usually indicative of the depth and duration of inundation (Carter and Schubert, 1974). Because of the importance of vegetation identification, most applications require the use of seasonal multispectral data that include visible and near infrared measurements.

Carter (1978) has prepared an excellent summary of coastal wetlands studies using aerial photographs, which is presented in Table 29-9 including an evaluation of the particular media used in the study. Landsat images and digital data have also been applied to coastal wetland studies. Carter et al (1973) accomplished the mapping of the Chesapeake Bay's upper wetlands boundary and gross vegetation types using 1:250,000 images. They found that much useful information could be extracted from 1:125,000 images; however, 1:24,000 images were of little value. Anderson et al (1973) performed a similar investigation with comparable results on the coastal areas of South Carolina and Georgia.

Carter and Schubert (1974) used Landsat digital data supported by seasonal ground-level spectrometer measurement to map the wetlands of the Chincoteague Bay in Virginia at a 1:20,000 scale. In an extension of this work, Carter (1976) used the data base to estimate the annual local primary productivity. Klemas et al (1975) used both Landsat digital data and Skylab photographs to study the coastal zone of Delaware. Computer analysis of the Landsat digital data produced a classification that had an accuracy of 80 percent. Landsat digital data were used by Morrow and Carter (1978) to classify three wetland areas in Alaska into the classification system used by the U.S. Fish and Wildlife Service.

Most of the applications of remote sensing to inland wetlands have used aerial photography, rather than digital image analysis techniques on MSS data. Because of the heterogeneity of wetland areas, due to diverse plant communities as well as to their location within an environment of varying moisture content, spectral signatures are often overlapping, which results in confusion for a classifier. Examples include the studies conducted by Shima et al (1976), Gammon and Carter (1979), and Carter et al (1979).

One of the most extensive studies of inland wetlands was conducted by Carter et al (1977) on the Great Dismal Swamp located in Virginia and North Carolina. In this study, seasonal high- and low-altitude color-infrared photographs and Landsat digital data were evaluated. The purposes of the investigations in this area included the management and manipulation of the surface and ground water and man-made influences. The investigators believed that the use of conventional data-collection methods would be too expensive and time-consuming and would be too difficult to

TABLE 29-9

Selected Research Projects in Remote Sensing of Coastal Wetlands (Carter, 1978)

Authors	Type of Study	Data Source	Conclusions
Cameron (1950)	Seaweed survey	Low-altitude filtered black-and-white panchromatic, color	Aerial photography is a rapid method to survey seaweed to 4 fathoms depth. Color is superior to black-and-white. Two species can be differenciated.
Olson (1964)	Tidal fresh marsh vegetation studies	Low-altitude black-and-white panchromatic, color, and color negative	General types of marsh vegetation, broad ecological zones, and many plant species and associations can be recognized. There are little differences in accuracy between the three types of photographs.
Stroud and Cooper (1968)	Tidal salt marsh vegetation and primary productivity	Low-altitude color and color infrared	Acreages of community types may be determined with color infrared. Estimates of primary productivity can be made.
Pestrong (1969)	Tidal salt marsh vegetation and geomorphology	Low-altitude multiband photographs, color, color infrared, and black-and-white panchromatic	Color infrared is best for vegetation discrimination.
Anderson (1969)	Tidal fresh and brackish marsh vegetation	Low-altitude color infrared from June and September, thermal infrared	Color infrared is superior to color for marshland plant delineation, especially with increase in altitude. Thermal infrared can be used to identify certain species. Seasonal changes in vegetation affect interpretation.
Kelly and Conrad (1969)	Shallow water benthic communities	Low-altitude color, black-and-white photographs, satellite photographs	Location, depth, and quantitative characteristics of bottom biotoa, sediments, and morphology were identified. Density of algae and sea grass can be estimated.
Carter and Anderson (1972)	Tidal marsh spectral reflectance	Low-altitude color infrared, high-altitude color infrared, spectral reflectance measurements	Spectral reflectance differences between marsh plants depend on density, leaf orientation, background water or soil, and sediment deposited on leaves.
Cibula (1972)	Fresh to brackish tital marsh vegetation	Aircraft multispectral scanner	Classification of plant communities is possible with automatic pattern recognition techniques and limited ground calibration.
Guss (1972)	Tidal salt marsh vegetation and tidal boundaries	Low-altitude color, color infrared, and multispectral photographs	Color-combined multispectral photography is recommended to differenciate species and to locate near high water line.
Reimold et al (1973)	Tidal salt marsh vegetation and primary productivity	Low-altitude color-infrared thermal imagery	Color infrared increases accuracy of primary productivity estimates for large marsh areas.
Anderson and Wobber (1973)	Tidal marsh vegetation and boundaries	Low-altitude color infrared and color	Pilot study shows that biological discrimination techniques can be applied to operational wetland species and boundary mapping.
Klemas et al (1974b)	Tidal marsh vegetative communities	High-altitude color infrared	Delaware's coastal wetlands were inventoried using manual interpretation and multispectral analysis.

TABLE 29-9—Continued

Authors	Type of Study	Data Source	Conclusions
McEwen et al (1976)	Tidal marsh boundaries and species	High-altitude color infrared	Vegetation interpretation is practical, economic, and accurate for establishing upper wetland boundary. Interpretations were made from color infrared to orthophoto base.
Shima et al (1976)	Tidal fresh marsh	Low-altitude color infrared	Low-altitude color infrared is good for mapping fresh-water marshes if season is considered and systematic field checking is done. Extrapolation from one area to another is difficult.
Martin (1975)	Tidal marsh map products	High-altitude multispectral black-and-white	Color orthophotoquads may permit direct interpretation of coastal features without transfer of interpreted data from photographs to orthophotobase.

interpret for such a large, diverse, and inaccessible area. Seasonal low- and high-altitude color-infrared photography was obtained and used to develop wetlands maps at a 1:24,000 scale. The analysts found that imagery obtained during vegetative dormancy conditions (i.e., without leaves) allowed the identification of the wetland boundary, the areas covered by water, the drainage pattern, the location of coniferous vegetation, and the classification of the understory vegetation. Photographs obtained during the growing season were then used to define deciduous vegetation. The classifications developed were used to interpret Landsat data. In the future, the data base will be updated using Landsat data.

IRRIGATED LANDS

Management of water supplies for irrigation is one of the most critical water-related problems in the Western United States. Scarce supplies and water-rights laws require that agencies know both what the total demand will be and what the distribution of demand for irrigation will be. Because of the vast areas involved, time constraints, and yearly changes, many agencies have begun using remote sensing as an aid in planning and decision-making. Two of the most important types of information required in these studies are crop types and whether or not fields are irrigated. From this information, statistical estimates can be made of the water demand.

Inventories of irrigated cropland have been conducted in Oregon (Draeger, 1977), Nebraska (Eucker et al, 1975 and Hoffman et al, 1975), California (Estes et al, 1975), Idaho (Heller and Johnson, 1979), Kansas (Lidster et al, 1979); Colorado (Heimes and Thelin, 1979) and in India (Thirvengadachari, 1981). Landsat data were used in all of these investigations.

In the study conducted by Draeger (1977), regional water-use agreements required a periodic inventory of all irrigated lands. These inventories are then compared with limitations, and, when they become close, a detailed survey and regulations are initiated. Draeger used Landsat images at a 1:250,000 scale for two dates during the growing season and performed a manual interpretation and sampling to estimate the acreage. It was necessary to use a midsummer image to identify all farmlands. A second image from late summer was then used to separate dryland and irrigated crops. Although precise estimates of the accuracy of this approach were not available, sampling indicated good performance. The cost of this approach was $1422, which compares with an estimated $20,093 if conventional methods were used. The user agency concluded that for the proposed application, which was a general estimate of total water-demand, the Landsat approach was highly cost-effective.

Some investigators have attempted to use more refined crop classifications and to estimate the spatial distribution of water demand. In California, the Kern County Water Authority uses a water-accounting model that simulates water storage and movement to assist in forecasting water supply and demand, and, ultimately, allocation and pricing (Estes et al, 1975; Estes et al, 1978). Table 29-10 lists the model inputs that could be estimated using remote-sensing techniques. Estimates of cropland must be developed for each of 251 spatial modeling units with an average size of 15 km^2. Several conventional methods for estimating cropland and irrigated acreage were compared with estimates derived from high-altitude color-infrared photography and different Landsat images. All the remote-sensing approaches were found to be highly cost-effective. A summary of the costs and time requirements for each procedure is shown in Table 29-11. Irrigated cropland estimates derived from the remotely

TABLE 29-10

Kern County Water Agency: Critical Water Accounting Model Inputs Amenable to Remote Sensing Techniques (Estes et al, 1975)

External Quantities	Definition	Source(s)	Remote Sensing Capabilities (Identify-Measure)
Agriculture Usage			
Gross irrigated acres	Total amount of irrigated acreage	Periodic air surveys, modified in districts	Irrigated croplands
Unit agricultural consumptive use	Acre-feet per acre irrigation requirement by individual crops	Department of Water Resources experimentation with individual crops	Crop identification
Surface and Ground-water Movement			
Volume of moisture deficient soil	Volume of unsaturated soil	Calculated from field work (soil surveys)	Soil moisture
% to perched water table	% of node overlying perched water table x nodal deep percolation	Field investigations	Perched water table areas
External Quantity Not Yet Incorporated into Model			
Soil salinity	Salinity of soil as measured by electro-conductivity (Ece)	Field investigations	Salinity damage assessment, soil salinity prediction

sensed data were combined with crop water-use information to estimate annual usage in water districts. When compared with agency estimates, these had greater than 90-percent accuracy.

Heller and Johnson (1979) conducted a study to evaluate the use of Landsat images in a program of inventorying irrigated croplands. A major objective of the investigation was to develop quantitative land cover information for a model of the Snake River Basin in a computer compatible format. Specific objectives of the study, in order of priority (Idaho Department of Water Resources, 1978), were to:

a. Identify irrigated land.
b. Measure acreages of irrigated lands.
c. Identify crop types, measure acreage by crop type, and estimate irrigation water demand.
d. Identify and measure annual changes in acreages of irrigated lands.
e. Identify sources of irrigation water.
f. Identify methods of irrigation.
g. Identify and measure acreages of range and dryland (non-irrigated) farming, and identify urban and other non-agricultural lands.

Although the authors were able to accomplish the first two tasks using Landsat data, they were unsuccessful in identifying the other items. However, they believed that the Landsat data were very valuable in developing a regional data base.

Finally, a study is being conducted in the high plains area of Colorado that will use the spatial distribution of irrigated acreage determined from Landsat data to estimate the annual volume and areal distribution of ground-water withdrawal for irrigation. Preliminary tests reported by Heimes and Thelin (1979) indicate that estimates will be highly accurate.

HYDROLOGIC LAND COVER

Most of the models used to support hydrologic investigations require input parameters that are defined in terms of land cover. Developing land cover information on a detailed, computer-compatible basis over large areas is a difficult and time-consuming task if conventional methods are used. The data available are often out of date or expensive to obtain if ground survey or manual interpretation of low-altitude aerial photographs is used. The material which follows in this section, presents several applications of remote sensing in estimating land cover-related hydrologic parameters.

Urban Hydrology

Colwell (1970) and Root and Miller (1971) used multispectral remote sensing to determine the spectral signatures of the land cover categories used in urban hydrology. Root and Miller (1971)

TABLE 29-11

The Cost-Effectiveness of Terrestrial Versus Remote-Sensing Techniques for Cropland Inventories (Estes et al, 1975)

Cropland Mapping Agency (Technique)	Cost for Inventoring all 3 Districts	Total Hectares (Acres) in all 3 Districts	Total Cropland Hectares (Acres) Inventoried	Time Required to Inventory all 3 Districts
		Cost for each 4047 ha (10,000 acres)		
Lost Hills, Semitropic Wheeler Ridge Water Districts (Terrestrial)	\$ 3,000	184,533 (455,991) \$66.00	116,833 (288,701) \$104.00	240 hr at \$12.50 per hr
Department of Water Resources (DWR) (Lowflight and Terrestrial	\$ 1,424[a]	184,533 (455,991) \$31.25	116,833 (288,701) \$49.32	230 hr at \$6.20 per hr
GRSU - 3 Districts U-2 1:125,000 (Highflight Inventory)	\$40.00[a]	184,533 (455,991) \$.87	118.067 (291,750) \$1.37	8 hr at \$5.00 per hr to compile basemap
GRSU - 3 Districts Landsat 1:1,000,000 (Multidate, Band 5 or 7)	\$82.50[a]	184,533 (455,991) \$1.81	116,833 (288,701) \$2.85	16.5 hr at \$5.00[b] per hr
GRSU - 3 Districts Color-Composite Landsat 1:1,000,000 (Multidate, Multiband Color-Composite Inventory)	\$67.50[a]	184,533 (455,991) \$1.48	116,883 (288,701) \$2.35	13.5 hr at \$5.00[c] per hr

[a] Does not include aircraft mobilization or the cost of photography.

[b] Each district was inventoried by making 5 single date, single band (5 or 7) overlays and by compositing these to form the cropland map. The dates examined required a mean interpretation time of 1 hr. Therefore, the three districts required a total of 15 separate Landsat maps at 1 hr each totaling 16.5 hr.

[c] Each district was inventoried by making 3 cropland overlays from 3 different Landsat color composites. The color composites were produced by registering and photographically combining multidate and multiband images into a 2- to 3-color composite. The district map represented a composite of these 3 interpreted color composites. Therefore, the 3 districts required a total of 9 separate Landsat maps at 1.5 hr each totaling 13.5 hr.

considered the relationship of the data to the percent of impervious area. This is one of the most important parameters used in urban hydrologic models.

One of the first investigations into the utility of Landsat data in urban hydrologic studies was a land-cover inventory conducted on the Anacostia River Basin located in Maryland. Ragan and Jackson (1975) compared the land-cover determinations and the estimates of percent of impervious areas as obtained using conventional and Landsat-based techniques in this 342-dm^2 watershed.

For the conventional analysis, low-level aerial photographs obtained between 1971 and 1972 (scale 1:4800) were overlain with a transparent grid. Points in each grid cell were photointerpreted into one of ten land cover categories. The percent of impervious area was determined for each individual cell by sampling. Approximately 94 man-days were required to complete the interpretation and compilation of the 92,400 samples in the 929 cells. This time estimate also includes transformation of the data into computer-based maps of land cover and imperviousness.

Later, Landsat data obtained in mid-1973 were analyzed on the GE-Image 100 system to classify the watershed land cover for each 0.45-ha pixel into one of the first seven categories listed in Table 29-12. Color slides of the cathode-ray tube display, a summary of the number of pixels in each category, and an alphanumeric printout of the results were obtained. Table 29-12 summarizes the land cover classifications obtained using Landsat and conventional approaches. Excellent agreement is evident from these results. As was pointed out by Ragan and Jackson (1975), since the Landsat data were obtained over a year after the photography, the Landsat estimates should show more of the area in the urban categories.

In the Landsat approach, the percent of impervious area was estimated by first assigning a representative land cover average value to each category. These average values were then weighted by

TABLE 29-12

Percent of Watershed Devoted to Specified Land Use (Anacostia River Basin) (Ragan and Jackson, 1975)

Lands Use	Large-Scale Aerial Photo Percent	Landsat Percent
Forested Areas	30.7	27.0
Highly Impervious	4.9*	6.5
Grassed Areas	8.5	10.4
Residential	44.9	43.5
Streets and Highways	9.9	5.5
Bare Land	N.C.**	0.4
Stream	1.0	N.C.**
Pond or Pool	0.1	N.C.**
Unclassified Pixels		6.7

* Industrial-commercial-parking lot
** N.C.—Not classified

the percent of the area assigned to the particular land cover class and summed to obtain the watershed impervious value. Using the Landsat-based approach, a value of 25.1 percent was obtained. This compares with a value of 23.5 percent from the conventional analysis. Because of the increased urbanization noted earlier, the agreement was considered excellent.

Perhaps the most significant aspect of the Anacostia investigation was the cost-difference between the two methods. As mentioned earlier, the aerial photo method required 94 man-days. The Landsat-based approach required less than 4 man-days. It is apparent from these results that, at this scale, significant cost savings in data collection and analysis are possible with the sacrifice of very little accuracy when Landsat data are used to define land cover.

This basic work in relating Landsat data to urban land cover was extended to comprehensive model comparisons using both planning and design models. The planning model used was the Surface Treatment Overflow and Runoff Model (STORM) developed by the U.S. Army Corps of Engineers (1976). It uses simplified representations of processes to facilitate the screening of many alternatives. STORM is a continuous simulation model that can be used for runoff quantity and quality analyses. Only the quantity aspects will be considered here.

Two alternative methods can be used for translating rainfall into runoff in a specific watershed. One method uses the SCS equation and curve numbers described previously. The other method uses the following relationship:

$$R = CA(P-D) \qquad (29\text{-}12)$$

where

R = runoff volume for the time increment (1 hour)
A = area
P = rainfall volume
D = available depression storage
C = runoff coefficient

C represents the fraction of excess rainfall that becomes runoff. The percent of impervious area (IMP) has a strong influence on C. An empirical relationship of the form:

$$C - C_P\,(100-IMP) + C_I\,(IMP) \qquad (29\text{-}13)$$

is used to express the effect.

Depression storage is recovered through evaporation. An upper limit of D_{max} is set for this parameter, and it is related to the watershed characteristics.

Conventional applications of STORM usually calibrate C_P, C_I, and D_{max} using recent streamflow records. When these records are unavailable, empirical values are used. Jackson et al (1976) developed a procedure for estimating the parameter values in ungaged watersheds on a regional basis that was compatible with Landsat data. In that study, C_P and C_I were determined for the region and the following equation was developed for C:

$$C = 0.0032(100-IMP) + 0.0086(IMP) \qquad (29\text{-}14)$$

Following a similar procedure for D_{max}, they obtained:

$$D_{max} = 0.0025(100-IMP) - 0.0006(IMP) \qquad (29\text{-}15)$$

Equations are considered valid only for events on the scale of the annual peak series in the Baltimore-Washington, D.C. region. In these equations, the percent of impervious area is estimated using the land-cover averaging technique described previously.

The conventional and Landsat-based approaches to estimating the parameters of STORM were compared in a detailed case study of the Fourmile Run Watershed, Jackson et al (1977). This 50.5-km^2 watershed in Northern Virginia was an excellent choice for testing because of the extensive precipitation, streamflow, and physiographic data base available. The comparison study was coordinated with the efforts of the local planning organization that was developing a stormwater management plan. As a part of their effort, workers in that organization had developed detailed land-cover and imperviousness data using low-level aerial photography and ground surveys. They had also contracted with a consulting firm to use hydrologic models to develop flood frequency information for planning and stormwater allocations for future design studies. The design-model investigation is described in the following section. The planning-model study used STORM and focused on the upper 37-km^2 portion of the watershed, which had been gaged between 1951 and 1972. In the conventional approach, the consultant used the recent streamflow record and the percent of impervious area determined by the planning agency to estimate the model parameters, C = 0.57 and D_{max} = 0.25 inch. The parameter values were used with precipitation data from several of the large events used in calibration to simulate the peak discharges shown in Figure 29-45. A standard-error estimate of 15.6 percent was obtained when the results were compared with the observed peak discharges. Since these events were also used in calibration, the error estimate is optimistic.

A flood-frequency relationship was generated for the watershed by using the calibrated STORM model to translate 52 years of hourly precipitation records into an equivalent discharge record. The annual peak discharge series was defined and used to determine the flood frequency. The curve is shown in Figure 29-46.

Landsat data processed with the aid of a GE-Image 100 were used to define the land cover of both the gaged and the entire watershed. In addition, an alphanumeric printout of the land cover of each pixel was obtained. From these data, the percent of impervious area of the gaged watershed was estimated as 39 percent. This compares with the value of 35 percent obtained by the planning organization. Next, Eqs. 29-13 and 29-14 were used to estimate the parameters of STORM, $C = 0.53$ and $D_{\max} = 0.18$ inch.

Peak events used to evaluate the conventional version were also simulated using the Landsat-based parameters. As shown in Figure 29-45, the results agree very well with those obtained using the conventional approach. The standard error of estimate is 17.2 percent, which is an increase of only 1.6 percent. In addition, since none of the events were used in the estimation of the Landsat-based parameters, the error estimate is a more valid verification of model reliability than the optimistically biased conventional value. Finally, the parameters were used along with the 52-year precipitation record to generate the flood-frequency curve shown in Figure 29-46.

A comparison of the model outputs and parameter values obtained using the two procedures in-

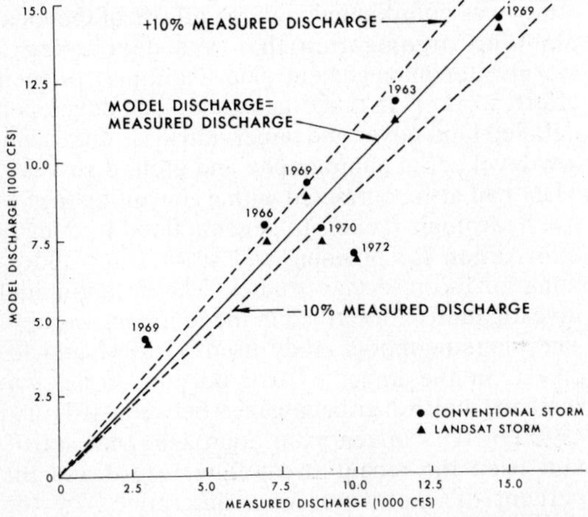

Fig. 29-45. Fourmile Run observed and predicted stream discharges (Jackson, et al, 1977).

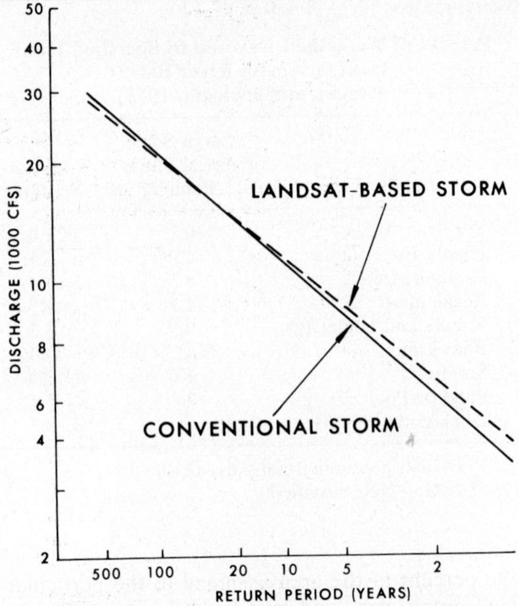

Fig. 29-46. Fourmile Run flood-frequency relationships (Jackson et al, 1977).

dicates that the Landsat-based approach is as reliable as the conventional procedure. Collection and processing of the land-cover and imperviousness data used in the conventional approach required 110 man-days at a total cost of $14,000. The Landsat-based approach required approximately 7 man-days at a total cost of $2350. The conclusion drawn from this investigation was that, for planning-level hydrologic investigations, Landsat data and computer analysis are a highly cost-effective approach for developing land-cover information.

The studies conducted by the planning agency and the consultant on the Fourmile Run watershed included the use of a detailed hydrologic design model. As described in the paper by Fitch et al (1976), for this level of analysis, a more detailed description of the hydrologic response was required to establish a stormwater-management program. The design-model study used basic units or subwatersheds with an average size of 31.5 ha.

The hydrologic design model used was the Water Resources Engineers Model (WREM), a variant of the EPA Stormwater Management Model. WREM consists of three component programs: Land Use Management, Runoff, and Transport. In the Land Use Management Program, the land-use data are converted to impervious ground-cover for each subwatershed. The Runoff Program translates a design storm event into a hydrograph and routes this hydrograph to a collection point called an inlet. Several subwatersheds are collected at an inlet. In the Transport Program, these inlet hydrographs are collected and routed through the major drainage system.

A study was conducted by Jackson et al (1977)

to evaluate the impact of using Landsat data on the parameter estimates and outputs of WREM. The basis of comparison was the land-cover and imperviousness data collected by the planning agency using conventional methods (this procedure was described in the previous section on STORM) and the design hydrographs obtained when these data were used in WREM.

In the Landsat-based approach, the conventional method was essentially replaced by the following procedure. Landsat data were classified into six categories: Single Unit Residential, Moderately Impervious, Highly Impervious, Bare Soil, Forest and Bare Soil, and an alphanumeric-scaled computer printout of these results. Next, a subwatershed boundary map was overlain on the computer printout. For a particular subwatershed, the number of pixels in each land-cover category was counted to obtain a percentage for each class. The representative average percent-of-impervious-area value for each land cover class was then used to obtain the imperviousness of the subwatershed. This process was repeated for each of the 179 subwatersheds.

All the parameters used in the conventional analysis were retained except for the impervious-area values. The consultant operated WREM with the modified values to obtain the Landsat-based design hydrographs for each subwatershed and at various locations within the watershed.

Percent-of-impervious-area estimates were compared at three levels. First, the estimates for each of the 179 subwatersheds were compared, assuming that the conventional approach yielded a perfect value. The average size of the subwatershed was 31.5 ha, and the standard error or estimate for the Landsat approach was 14.9 percent. Next, the level of comparison was for the inlet areas. These areas consisted of one or more subwatersheds. For these 70 areas, the average size was 75 ha. The standard error of estimate for the Landsat-based percent-of-impervious-area was 11 percent. Finally, the estimates were compared for junction areas. Junctions are stream reaches. Fourmile Run had 20 junction areas with an average size of 278 ha. The standard error of estimates for these areas was 6.5 percent.

A very strong trend was noted between the size of the area and the expected error. These results indicate that, for small units, the error encountered when using Landsat data to estimate the percent-of-impervious-area may be unacceptable. If a minimum acceptable standard error of estimate were specified, it could be determined, based on the size of the units to be used in a study, whether or not Landsat data would be acceptable.

Peak discharges of the hydrographs for the subwatersheds were compared. A relationship between error and unit size was observed. The error was considered too large in all cases. However, the errors for the individual units cancel out as the flows are routed downstream. The error in estimating peak discharge of the watershed design hydrograph was 7.4 percent of the conventional analysis result (26,368 cfs).

From this design model study, it was concluded that, when subwatersheds less of than several hundred hectares are required in analysis, the Landsat-based approach produces an unacceptable error rate. Analysis indicated that the major cause of this problem is the limited resolution of the Landsat sensor and the spectral bands it measures. However, future satellite systems with improved spatial resolution and more appropriate spectral band selections should increase the utility of this type of data in design studies.

Finally, other studies on the application of remote sensing of hydrologic parameters in urban areas have been conducted by Merry and McKim (1977), McKeon et al (1978), Algazi and Suk (1977), and Cermak et al (1979).

Section 208 Studies

Under Section 208 of the Federal Water Pollution Control Act, Amendments of 1972, grants were provided to regional water quality planning agencies to undertake and develop area-wide water quality plans. These studies, which must be completed within 2 years of initiation, require the identification of all point and nonpoint sources of pollution and the consideration of alternative structural and nonstructural solutions to current and future problems.

Nonpoint-source pollution is closely related to land cover and use. The planning agencies usually use models to simulate pollutant loads that require the land use or cover as an input. Some of the organizations are using Landsat data to define the inputs.

There are two major attractions in using digital Landsat data in these model studies. First, the data can be collected, analyzed, and implemented within a short period. Considering the limited time allotted to the 208 studies, it seems imperative that the information base be developed as quickly as possible to allow more time for analysis and plan formulation.

The second advantage of using Landsat data is the ability to work from a large-area data base. In this approach, an entire area is classified, and then information on individual watersheds is extracted as needed from the data base. Schecter (1976) describes the use of the Landsat data to develop ten land cover categories for the Triangle J Council of Governments region (1750 mi^2) in North Carolina. The Council stores the Landsat-defined land cover data on its own computers and then interfaces these data with digital files containing soil, slope, and rainfall to estimate the areal distribution of the quantity and quality of runoff using the EPA Storm Water Management Model (SWMM). It was estimated by Schecter (1976) that the cost of developing the land-cover data file from the Landsat base was $4.00 per square mile.

Another Section 208 investigation that used

Landsat data was conducted by the Ohio-Kentucky-Indiana Regional Council of Governments (OKI) (EPA, 1977). OKI developed estimates of the quantity and quality of runoff within its 3000-mi² area of jurisdiction using STORM for urban areas and a Rural Runoff Model (OKI, 1975) for all other areas. Both of these models required land cover as an input.

OKI chose to use Landsat data to develop an inventory of land use that would be directly input to the Rural Model and that would supplement data from other sources in urban areas. A private firm was contracted (Rogers et al, 1975) to perform the Landsat analysis and to provide OKI with a land cover map set at a 1:60,000 scale. The consultant classified the data into eleven categories, which were later reduced to three general classes by OKI. A digital overlay system was used to enter the boundaries of each of the 233 watersheds in the region into the computer system to define the cover for each automatically. A similar procedure was performed for each of the nine counties in the region. OKI received their products within 90 days at a total cost of $20,000, approximately $7.40 per square mile.

McKeon et al (1978) and Ragan et al (1978) reported a study conducted for the Metropolitan Washington Council of Governments (COG) that involved developing land-cover information from Landsat data for a 6100-km² region. The data base was to be used for nonpoint-source pollution studies and other planning activities. The investigators found that, although there was a large amount of land-use information available, it was in many formats, since they were compiled by the individual jurisdictions. Because it would have been a major undertaking to assemble these data into a common format, the planning agency decided to use remotely sensed data to develop their data base. The contractors supplied COG with land-cover maps and summary statistics for 109 watersheds and 95 policy-analysis districts. All of this work was performed within a total of 30 days for about $3.28 per square kilometer.

Soil Conservation Service Models

The Soil Conservation Service (SCS) has developed hydrologic models based on the material described in the "National Engineering Handbook," No. 4 (SCS, 1972). One computer-based model is called the TR-20 (SCS, 1969). Although the procedure was originally developed for use in rural and agricultural watersheds, it has been adapted for use in urban areas (SCS, 1975).

All the SCS models compute direct runoff through an empirical equation that requires the rainfall and a watershed coefficient as inputs. The watershed coefficient is called the curve number (CN) and represents the runoff potential of the land-cover soil complex. As described in SCS (1972), the relationships between land cover, hydrologic soil class, and CN are based on statistical analyses of observed situations and empiricism. The relationships are usually presented in a table format.

Two approaches have been evaluated for estimating CN's using remotely sensed data. One involves the use of remotely sensed data to determine land cover alone; the other attempts to estimate the CN itself without ancillary soils data.

The first approach was evaluated by Ragan and Jackson (1980) on a portion of the Anacostia River Basin. A local planning agency had used 1:4800 aerial photographs, detailed soil maps, and field surveys to define the CN of the watershed and to generate synthetic flood-frequency information. Both high-altitude color-infrared photographs and Landsat digital data were used to estimate land cover of the watershed. These data were combined with general soils data to estimate the CN. Both estimates were very close to the conventionally determined value. Cost savings were significant when using the remotely sensed data. The local government agency reported that a total of 160 man-days were involved in developing the same information that was estimated using the remotely sensed data in 3 man-days.

An important point of this study was that the SCS CN estimation procedure had to be modified to make the land cover categories compatible with those that can be estimated using Landsat data. This revised CN table is shown in Table 29-13. Slack and Welch (1979) and Cermak et al (1979) conducted similar investigations using Landsat data. As reported in Bondelid et al (1980), modifications of the basic SCS CN table can also be made to adapt it for use with the U.S. Geological Survey Land Use Map Series.

Blanchard (1975) has attempted to use remotely sensed data to directly estimate the CN without using ancillary soils data. In one phase of this study, he used Landsat data collected over several watersheds in Oklahoma during a fall dry period. He related the measured CN's on these watersheds to various Landsat-band combinations and observed fairly good correlation as shown in Figure 29-47. Blanchard concluded that this procedure would work well only in areas under dry

TABLE 29-13

Runoff Curve Numbers for Land Cover Delineations Definable from Landsat (Ragan and Jackson, 1980)

	Hydrologic Soil Group			
Land Use Description	A	B	C	D
Forest Land	25	55	70	77
Grassed Open Space	36	60	73	78
Highly Impervious (Commercial, Industrial, and Large Parking Lot)	90	93	94	95
Residential	60	74	83	87
Bare Ground	72	82	88	90

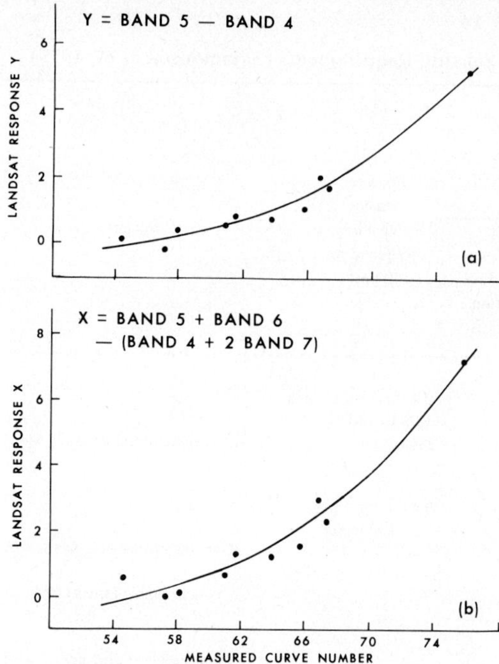

Fig. 29-47. The relations of MSS data from Scene 1058 to measured watershed-runoff curve numbers (Blanchard, 1975).

dormant conditions and minimal atmospheric interference. In addition, the procedure allows the estimation of the CN at only one time of the year. It will vary in agricultural areas with the crop calendar. These factors will limit the application of this approach.

Spanner et al (1982) employed a data base approach to demonstrate the potential of using various data sources, including Landsat Multispectral Scanner data, digitized USDA Soil Conservation Service soil maps, digitized NOAA precipitation isopluvial maps, and U.S. Geological Survey Digital Elevation Model (DEM) topographic data, to map potential soil loss using the USDA SCS Universal Soil Loss Equation (USLE). Data within this geographic information system were processed to derive coefficients with respect to crop management, soil erodability, rainfall, slope gradient and the length of slope coefficients of the USLE (Equation 29-12) for a mixed agricultural and rangeland region in Ventura County, California. Accuracies for the intermediate data sets ranged from 85 percent to 100 percent. The Pearson product-moment correlation coefficient, R, of the geobased soil loss model, as compared with the manual USLE model, was 0.90. This figure is significant to the .0001 level for a stratified random sample within the 7.5-minute-quadrangle study area. The registered intermediate data sets and predicted soil-loss images graphically indicated locations of high predicted soil loss, thereby indicating where further investigations should be conducted by the Soil Conservation Service.

Others

A variety of other types of investigations have been conducted that analyzed the utility of remotely sensed data for estimating hydrologic land cover parameters. Salomonson et al (1975) report a study conducted to evaluate the sensitivity of model simulation to parameter estimate error in order to determine the suitability of remote-sensing requirements for future systems. A version of the Stanford Watershed Model was used on three diverse watersheds. A summary of their results is presented in Table 29-14. All the values are referenced to an allowable ±5 percent variation in runoff.

Holtan et al (1977) used computer-aided Landsat data from two dates to estimate the land cover for input to the U.S. Department of Agriculture Hydrography Laboratory Model.

Remote-sensing techniques have been used to identify hydrologic source areas. These areas are used in partial-area hydrology models that are based on the principle that both surface and subsurface runoff are spatially concentrated at certain hydrologically active portions of a drainage area. As noted by Ishaq and Huff (1974), one problem with this approach to hydrologic modeling has been the identification and mapping of the source areas. Ishaq and Huff (1974) attempted to delineate source areas using low-altitude color-infrared photography on one watershed in Wisconsin. They found that such areas could be identified using densitometric measurements at 0.55 and 0.675 micrometers. Only a limited verification of this procedure was performed. Soil-moisture samples collected at one area identified as a source area were compared with those collected at another identified as a nonsource area. Values observed at the source-area site were significantly and consistently higher than those of the nonsource area.

Soils are more important than land cover in determining whether a site is a source area. When there is little soil exposed, the foregoing remote-sensing approach will not be successful. This was recognized by Ishaq and Huff (1974) and was later explored by Engman and Annett (1977). Engman and Annett (1977) found that the approach might work in areas where the land use is indicative of the soils, which is common in some regions.

OUTLOOK FOR THE FUTURE AND CONCLUSIONS

This chapter has reviewed and described the techniques developed and the progress made up to the present time in applying remote sensing to water resources management. As indicated in the introduction, much of the progress that has been made since the early 1970's has shown clearly that remote sensing, particularly of the type that produces multispectral images, can provide large-region, repetitive observations from which one can obtain highly useful quantitative measures of

TABLE 29-14

Summary of Sensitivity Analysis Results and Remote-Sensing Requirements (Salomonson et al, 1975)

(1)	(2) Town Creek, Alabama Watershed ($365\ km^2$)		(3) Alamosa Creek, Colorado Watershed ($277\ km^2$)		(4)	(5)
Watershed Model Input Parameter	Reference Value	Input Variation Giving ±5% Variation in Runoff	Reference Value	Input Variation Giving ±5% Variation in Runoff	Remote-Sensing Capability Required for $40\ km^2$	Remarks
Fraction of Impervious Area (FIMP)	$36.5\ km^2$	±20%	$27.7\ km^2$	±10%	0.4- to 0.8-km^2 area 130- to 180-m resolution	$e = \frac{rK}{\sqrt{A}}$ r = spatial resolution
Fraction of Water Surface Area (FWTR)	$36.5\ km^2$	±20%	$27.7\ km^2$	±10%	0.4-0.8 km^2 area 1.30-180 meter resolution	A = area K = data processing factor = 0.5 e = acceptable error, ±10%
Forested Fraction of Watershed (FFOR)			0.40	±10%	±0.04	Value applies during period with snow on ground (winter-spring)
Maximum Volume of Interception (VINTMR)	0.4 cm	Very insensitive ±100%	0.4 cm	Very insensitive 80% = +5% R.O.	Ability to distinguish grassland, moderate forest cover, and heavy forest cover	Appears possible with Landsat
Overland Flow Roughness Coefficient (OFMN)	0.05	±40%	0.35	No sensitivity to storm runoff	Level II Land use	Some categories are possible with Landsat. High altitude aircraft data are applicable
Overland Flow Surface Length (OFSL)	473 m	±40%	314 m	No sensitivity to storm runoff	± 10 to 20 m	High-altitude aircraft or Skylab Earth Terrain Camera
Overland Flow Surface Slope (OFSS)	0.062	±50%	0.34	No sensitivity to storm runoff	Percent slope, ±3%	Stereo photography or radar altimetry
Fraction of Incident Radiation Reflected by Snow (FIRR)			0.60 (snow)	±3.5%	±2 to 3% albedo	Requires anisotropy model and reflectance versus albedo relationships
Lower Zone Capacity – Soil Moisture (LZC)	10 cm	±7 to 8%	15 cm	±6 to 8%	± cm or ±7%	Microwave offers a possibility or thermal infrared
Precipitation			3 to 8 cm	±10%	±0.3 to 0.8 cm	Cloud-type identification or radar along with ground-based benchmark observations
Evaporation			50 cm/yr	±20%	±10 cm/yr	A heat balance approach or an empirical evaporation to land use and vegetation type, density, and stress approach may be successful

the area, shape, length, and related attributes of hydrologic features. Specific examples include the determination of snow cover, the making of land cover observations useful in storm water management planning, and the modeling of effects due to changes in land cover, drainage basin characteristics, and hydrogeological surface features related to ground-water exploration.

Water resource managers and hydrologists can expect further improvements will be made in the next decade that will enhance the utility of remote sensing. A key advancement will be provided through the use of Landsat-D data. Landsat-D, now called Landsat-4 since its launch in July of 1982, has several new capabilities of obvious utility in water resources management and hydrological studies (See Table 29-1, Salomonson 1981, and Chapter 12). Early TM products from the Thematic Mapper, such as those shown in Chapter 12, show that the nominal 30 meter spatial resolution of that system offers clear advantages over the previously available Multispectral Scanner (MSS) for mapping land cover features of interest to hydrologists. Through the use of digital processing and enhancement the TM data should provide results including imagery that can be accurately used to provide and update thematic information at 1:62,500 and 1:24,000 map scales. Water resources planning activities associated with urban and suburban environments and flood plain management, are among the types of activities that should greatly benefit from the Thematic Mapper data. When this improved capability is combined with the inherent capabilities of highly repetitive, large area, uniform data provided by space-borne imaging systems, resource management agencies should find remote sensing to be a highly valuable tool. Further improvements in spatial resolution expected with the use of solid state technology in the late 1980's and 1990's (See Table 29-1b) should further improve the imaging and mapping capability provided by space-borne and high altitude airborne systems.

Improvements in the spectral and radiometric capabilities of systems observing in the 0.4–15 μm regions of the spectrum should allow improved capability in doing thematic mapping and in making quantitative observations of change in spectral properties that could be related to relative or absolute measures of land cover, surface water, or snow cover conditions. The Landsat-D TM 1.55 to 1.75-μm band, will allow clouds to be more easily separated from snow-reflectance of snow in that band. The same band, together with the 2.08- to 2.35-μm band, may assist in identifying clay soils because of the sensitivity to hydroxyl ion absorption that exists in the 2.08- to 2.35-μm band. These same bands are also expected to respond to changes in leaf-water content of the vegetative cover and therefore should be useful, for example, in inferring soil-moisture status. If effects of the atmosphere on image quality can be appropriately removed, the blue band (0.45 to 0.52 μm) should assist greatly in water quality and bathymetric studies. The thermal band on the TM will be interesting in terms of its utility in land-use mapping and crop water-status studies, but much remains to be learned about how to apply information in the thermal infrared (Price, 1981). The improved spatial and spectral capabilities of the Landsat-D TM are expected to be very useful for geological studies, in general, and therefore can be expected to make a significant impact on hydrogeological studies related to ground-water exploration and mapping.

The use of remote sensing for water resources management and hydrological studies will be considerably enhanced if it can be demonstrated that, through remote sensing, one can accurately observe fundamental hydrologic variables such as soil moisture, precipitation, snow water, water equivalent, and free-water content. As has been shown in this chapter, very significant advances in applying the microwave part of the spectrum for these purposes have been made. It is believed that, when appropriate microwave systems are developed for operation on aircraft and spacecraft, observations made by such systems will be very useful in streamflow-runoff estimation and water-balance studies, particularly for large, remote, or inaccessible areas.

Because of the inherently high observational density and spatially specific characteristics of remotely sensed data, such data can complement the typically more accurate, but point-specific, measurements of hydrologic parameters (e.g., soil moisture, precipitation, snow-moisture equivalent, etc.) provided by conventional observation systems. In addition to our learning to process and interpret remotely sensed data, we must further develop techniques to merge these data with conventional data through the efficient use of database management systems and geo-referenced information systems. Agencies such as the U.S. Army Corps of Engineers (Corps of Engineers, 1979), are making progress in this area, but much remains to be done to effectively use remote sensing in water resources planning, management, and modeling.

Much research and development work also needs to be done in developing watershed management, runoff, or hydrogeological exploration models that will make better use of remote sensing. Most runoff models, for example, are now being developed around the use of point observations that must be assumed to be representative, in that they are averaged, or "lumped", over large parts of the watershed area being studied. New remote sensing-based models and methods need to be developed that can take into account and quantify the spatial variability of features within a watershed. Research is beginning on this type of "distributed parameter" modeling and significant results should be obtained in the 1980's. If the ability to acquire, process, and interpret relevant remote-sensing observations continues to improve at near the pace it has over the last two decades, some of the magnitudes of the storages and fluxes of water and the process involved in the hydrological cycle over regional, continental, and global scales will be far better understood before the end of the century. Studies such as those suggested by the publications of Eagleson (1978), L'vovich (1979), and the United Nations Educational, Scientific, and Cultural Organization (UNESCO, 1978) may be greatly facilitated. Such studies involve linking hydrological, meterological, oceanographic, and climate studies and are of great potential importance because the presence of water and the associated mass, energy budgets, dynamics, and physical processes all must be considered in deriving information of maximum use to the managers of water resources. Furthermore the possibility of our better understanding how water, which is so fundamental to the existence of life itself, is distributed and how it is changing in quanity and quality, plus the practical ramifications of increasing our knowledge in this area, makes the pursuit of such a scientific goal very exciting.

Finally, it must be recognized that the manner in which institutional arrangements evolve in the future will greatly affect the use of remotely sensed data. There must be continual dialogue and transfer of knowledge both ways between the personnel and agencies, on the one hand, having mandated responsibilities for managing water resources and, on the other hand, those scientists, engineers, and technologists who are developing systems and techniques applicable to acquiring, processing, interpreting, and applying remote sensing. A key example of an action that would very beneficially involve all of the parties just noted would be that of establishing, in all countries concerned, an operational remote-sensing organization that would be committed to the regular and sustained providing of remotely sensed data to water resources management agencies and hydrologists. Only when such an organization exists, can those agencies and their personnel make suitably strong educational, funding, and procedural commitments. Yet, such commitments are necessary for effectively applying modern remote sensing technology to the improved management of the world's finite water resources.

REFERENCES

Abdel-Hady, M., 1971, Depth to groundwater table by remote sensing; J. Irrigation Proc. Am. Soc. Civil Eng. v. 97, (No. IR3), pp. 355–367.

Algazi, V. R., and Suk, M., 1977, Satellite land use acquisition and applications to hydrologic planning models: Proceedings of the 11th International Symposium on Remote Sensing of Environment, Environmental Research Institute of Michigan, Ann Arbor, Michigan, p. 1171–1181.

American Society of Photogrammetry, 1975, Manual of Remote Sensing, Chapt. 19, Water Resources Assessment, pp. 1479–1552.

American Society of Photogrammetry, 1960, Manual of Photographic Interpretation, Chapt. 10, Photo Interpretation in Hydrology and Watershed Management, pp. 539–560.

Andersen, T., and Odegaard, H., 1980, Application of satellite data for snow mapping: Report No. 3, Norwegian National Committee for Hydrology, Oslo, 55 p.

Anderson, R. R., 1969, The use of color infrared photography and thermal imagery in marshland and estuarine studies: Second Annual Earth Resources Aircraft Program Status Review, NASA Manned Spacecraft Center, Houston, Texas, Sept. 16 to 18, 1969, p. 40-3 to 40-29.

Anderson, R. R., Carter, V., and McGinnis, J. W., Jr., 1973, Mapping Southern Atlantic coastal marshlands, South Carolina-Georgia, using ERTS imagery: Remote Sensing of Earth Resources, v. II, Tullahoma, Tennessee.

Anderson, R. R., and Wobber, F. J., 1973, Wetlands mapping in New Jersey: Photogrammetric Engineering, v. 39, no. 4, p. 353–358.

Baker, V. R., Holz, R. K., and Hulke, S. D., 1974, A hydrogeomorphic approach to evaluating flood potential in central Texas from orbital and suborbital remote sensing imagery: Ninth International Symposium on Remote Sensing of Environment, Environmental Research Institute of Michigan, Ann Arbor, Michigan, p. 629–645.

Barry, R. G., and Chorley, R. J., 1970, Atmosphere, Weather, and Climate: Holt, Rinehart, and Winston, Inc., New York, 320 p.

Bartholic, J. F., Namken, L. N., and Wiegand, C. L., 1972, Aerial thermal scanner to determine temperatures of soils and crop canopies differing in water stress: Agronomy Journal, v. 64, p. 603–608.

Bartolucci, L. A., Hoffer, R. M., and Luther, S. G, 1975, Snowcover mapping by machine processing of Skylab and Landsat MSS, *in* Operational applications of satellite snowcover observations: NASA Special Publication 391, South Lake Tahoe, California, p. 295–311.

Barton, I. J., 1978, A case study comparison of microwave radiometer measurements over bare and vegetated surfaces: Journal of Geophysical Research, 83, p. 3515–3517.

Barton, M., and Burke, M., 1977, SNOTEL: An operational data acquisition system using meteor burst technology: Proceedings of the 45th Annual Western Snow Conference, Albuquerque, New Mexico, p. 82–89.

Basharinov, A. Ye., and Shutko, H. M., 1978, Determination of the moisture content of the earth's cover by superhigh frequency (microwave) radiometric method: A Review, Radioteknika in Electronika, NASA Technical Translation, v. 23, p. 1778–1791.

Batlivala, P., and Ulaby, F. T., 1977, Estimation of soil moisture with radar remote sensing: Proceedings of the 11th International Symposium on Remote Sensing of Environment, p. 1557–1566.

Bernard, R., Vauclin, M., and Vidal-Madjas, D., 1981, Possible use of active microwave remote sensing data for prediction of regional evapotranspiration by numerical simulation of soil water movement in the unsaturated zone: submitted for publication in Water Resources Research.

Bissell, V. C., and Peck, E. L., 1974, Measurement of snow at a remote site: Natural radioactivity technique, *in* Advanced concepts and techniques in the study of snow and ice resources: National Academy of Sciences, Washington, D.C., p. 604–613.

Blad, B. J., and Rosenberg, N. J., 1976, Measurement of crop temperature of leaf thermocouple, infrared thermometry, and remote sensed imagery: Agronomy Journal, v. 68, p. 635–641.

Blanchard, B. J., 1975, Remote sensing for prediction of

watershed runoff: NASA Earth Resources Survey Symposium, Houston, Texas, p. 279–307.

Blinn, J. C., and Quade, J. C., 1972, Microwave properties of geological materials: Studies of penetration depth and moisture effects: Fourth Annual Earth Resources Program Review, NASA, Johnson Space Center, Houston, Texas Jan. 17–21, 1972.

Boland, D. H. P., and Blackwell, R. J., 1975, The Landsat-1 multispectral scanner as a tool in the classification of inland lakes: Proceedings of the NASA Earth Resources Survey Symposium, Houston, Texas, NASA TM-58168, p. 419–442.

Bondelid, T. R., Jackson, T. J., and McCuen, R. H., 1980, Comparison of conventional and remotely sensed estimates of runoff curve numbers in southeastern Pennsylvania: Proceedings of the Annual Meeting of the ASP, St. Louis, Missouri.

Bowley, C. J., Barnes, J. C., and Rango, A., 1980, Satellite snow mapping and runoff prediction handbook: NASA Technical Paper 1829, National Aeronautics and Space Administration, Washington, D.C.

Boyer, R. E., and McQueen, J. E., 1964, Comparison of mapped rock fractures and airphoto linear fractures: Photogrammetric Engineering and Remote Sensing, v. 30, no. 4, p. 630–635.

Boyne, H. S., and Ellerbruch, D. A., 1979, Microwave measurement of snow stratigraphy and water equivalence: Proceedings of the 47th Annual Western Snow Conference, Sparks, Nevada, p. 20–26.

Brooks, R. L., Campbell, W. J., Ramseier, R. O., Stanley, H. R., and Zwally, H. J., 1978, Ice sheet topography by satellite altimetry: Nature, 274 (5671), p. 539–543.

Bryan, M. L., 1981, Optically processed Seasat radar mosaic of Florida: Photogrammetric Engineering and Remote Sensing, v. 47, no. 9, p. 1335–1337.

Burke, W. J., Schmugge, T., and Paris, J. J., 1979, Comparison of 2.8 and 21 cm microwave radiometer observations over soils with emission model calculations: Journal of Geophysical Research, v. 84, p. 287–294.

Burgess, L. O. N., 1967, Airphoto interpretation as an aid in flood susceptibility determination: International Conference on Water for Peace, v. 1, no. 4, U.S. Government Printing Office, p. 867–881.

Byrne, G. F, Begg, J. E., Flemming, P. M., and Dunin, F. X., 1979, Remotely sensed land cover temperature and soil water status—a brief review: Remote Sensing of Environment, v. 8, p. 291–305.

Cameron, H. L., 1950, The use of aerial photography in seaweed surveys: Photogrammetric Engineering, v. 16, no. 4, p. 493–501.

Cannon, P. L., 1973, The application of radar and infrared imagery to quantitative geomorphic investigations: Proceedings of the Remote Sensing of Earth Resources Second Annual Conference, Tullahoma, Tennessee, p. 503–527.

Carroll, T. R., and Vadnais, K. G., 1980, Operational airborne measurement of snow water equivalent using natural terrestrial gamma radiation: Proceedings of the 48th Annual Western Snow Conference, Laramie, Wyoming, 10 p.

Carroll, T. R., 1981, Airborne soil moisture measurement using natural terrestrial gamma radiation: Soil Science, v. 132, no. 5, p. 358–366.

Carter, V., 1978, Coastal wetlands: Role of remote sensing: Proceedings of the Symposium on Technical, Environmental, Socioeconomic and Regulatory Aspects of Coastal Zone Management, San Francisco, California.

Carter, V., and Anderson, R. R., 1972, Interpretation of wetlands imagery based on spectral reflectance characteristics of selected plant species: Proceedings of the American Society of Photogrammetry, 38th Annual Meeting, Washington, D.C., March 12–17, 1972, p. 580–595.

Carter, V., Malone, D., and Burbank, T. H., 1979, Wetland classification and mapping in western Tennessee: Photogrammetric Engineering and Remote Sensing, v. 45, no. 3, p. 273–284.

Carter, V., McGinness, J. W., and Anderson, R. R., 1973, Mapping northern Atlantic coastal marshlands, Maryland-Virginia, using ERTS-1 imagery: Remote Sensing of Earth Resources, Univ. of Tennessee, Space Institute Tullahoma, Tennessee, edited by F. Shahrokhi, v. 11, p. 1011–1020.

Carter, V., and Schubert, J., 1974, Coastal wetlands analysis from ERTS MSS digital data and field spectral measurements: Ninth International Symposium on Remote Sensing of Environment, Ann Arbor, Michigan, April 15–19, 1974, Proceedings, Environmental Research Institute of Michigan, p. 1241–1260.

Carter, V., Garrett, M. K., Shima, L., and Gannon, P., 1977, The Great Dismal Swamp: Management of a hydrologic resource with the aid of remote sensing: Bulletin of Water Resources, v. 13, no. 1, p. 1–12.

Carter, W. D., 1976, Structural geology and mineral-resources inventory of the Andes Mountains, South America: *in* William, R. S., Jr., and Carter, W. D., 1976, ERTS-1, A new window on our planet: USGS Professional Paper 929, 362 p.

Castruccio, P. A., Loats, H. L., Lloyd, D., and Newman, P. A. B., 1980, Cost benefit analysis for the operational applications of satellite snowcover observations (OASSO), *in* Operational applications of satellite snowcover observations: NASA-Conference, Publication 2116, Sparks, Nevada, p. 239–254.

Cermak, R. J., Feldman, A. D., and Webb, R. P., 1979, Hydrologic land use classification using Landsat: Technical Paper No. 67, USAGE, Hydrologic Engineering Center.

Chandresekhar, S., 1960, Radiative transfer: Dover Publications, New York, New York, 393 p.

Chang, A. T. C., Foster, J. L., Hall, D. K., Rango, A., and Hartline B. K., 1981, Snow water equivalent determination by microwave radiometry: NASA Technical Memorandum 82074, Goddard Space Flight Center, Greenbelt, Maryland, 18 p.

Chang, A. T. C., Rango, A., and Shine, J. C., 1980, Remote sensing of snow properties by passive microwave radiometry: GSFC truck experiment in microwave remote sensing of snowpack properties: NASA Conference Publication 2153, Fort Collins, Colorado p. 169–185.

Choudbury, B. T., Schmugge, T., Newton, R. W., and Chang, A., 1979, Effects of surface roughness on microwave emissions from soils: Journal of Geophysical Research, v. 8A, p. 5699–5706.

Cibula, W. G., 1972, Application of remotely sensed multispectral data to automated analysis of marshland vegetation: Inference to the location of breeding habitats of the salt marsh mosquito (Aedes sollicitans): NASA Earth Resources Laboratory Report No. 020.

Cihlar, J., 1980, Soil water and plant canopy effects on remotely measured surface temperatures: International Journal of Remote Sensing, v. 1, p. 167–173.

Clark, G. L., and James, H. R., 1939, Laboratory analysis of the selective absorption of light by sea water: Journal of the Optical Society of America, v. 29, no. 2, p. 43–53.

Coker, A. E., Higer, A. L., Rogers, R. H., Shaw, N. J., Reed, L., and Walker, S., 1975, Automatic categorization of land-water cover types of the Green Swamp, Florida, using Skylab multispectral scanner, S-192, data: Proceedings of the NASA Earth Resources Survey Symposium, Houston, Texas, NASA TM X-58168, p. 479–506.

Colwell, J. E., 1970, Multispectral remote sensing of urban features: Final Report, Contract USGS 14-08-0001-11968, Infrared and Optics Lab, Willow Run Labs, Univ. of Michigan, Ann Arbor, Michigan.

Cooper, C. F., 1965, Snow cover measurement: Photogrammetric Engineering and Remote Sensing v. 31, no. 7, p. 66–619.

Corps of Engineers, 1979, Determination of land use from Landsat imagery: Applications to hydrologic modeling: Research Note No. 7, The Hydrologic Engineering Center, U.S. Army Corps of Engineers, Davis, California, 71 p.

Crosby, W. O., 1885, Notes on joint structure: Boston Society of National History Proceedings, v. 25, p. 243–248.

Defant, A., 1961, Physical Oceanography: Pergamon Press, New York, v. 1., 52 p.

Deutsch, M., and Ruggles, F., 1974, Optical data processing and projection applications of the ERTS-1 imagery covering the 1973 Mississippi River Valley Floods: Bulletin of Water Resources, v. 10, no. 5, p. 1023–1039.

Deutsch, M., Wiesnet, D. R., and Rango, A., eds., 1981, Satellite hydrology: American Water Resources Association, Minneapolis, Minnesota, 727 p.

Dozier, J., and Frew, J., 1981, Atmospheric corrections to satellite radiometric data over rugged terrain: Remote Sensing of Environment, v. 11, p. 191–205.

Draeger, W. C., 1977, Monitoring irrigated land acreage using Landsat imagery; an application example: Proceedings of the International Symposium on Remote Sensing of Environment, 11th, v. 1, p. 515–524.

Drake, B., and Vincent, R. K., 1975, Geological interpretation of the greater part of the Michigan basin, *in* Cook, J. J., ed., 10th International Symposium on Remote Sensing of Environment, Ann Arbor, Michigan, Oct. 6–10, 1975, Proc.: Environmental Research Institute of Michigan, v. 2, p. 933–948.

Duntley, S. Q., 1963, Light in the sea: Journal of the Optical Society of America, v. 53, no. 2, p. 214–233.

Eagleman, J., and Lin, W., 1976, Remote sensing of soil moisture by a 21-cm passive radiometer: Journal of Geophysical Research, v. 81, 3660 p.

Eagleson, P. S., 1978, Climate, soil, and vegetation: Water Resources Research, v. 14, no. 5, p. 705–776.

Engman, E. T., and Annett, J. R., 1977, Remote sensing applications to a partial area model: Final Report for NASA, Goddard Space Flight Center, Greenbelt, Maryland.

Environmental Protection Agency, 1977, 208 area-wise quality management planning case histories: EPA-625/8-77-001, Environmental Research Information Center, Cincinnati, Ohio.

Estes, J. E., Jensen, J. R., Tinney, L. R., 1978, Remote sensing of agricultural water demand information: A California study. Water Resources Research 14, p. 170–176.

Estes, J. E., Jensen, J. R., Tinney, L. R., and Rector, M., 1975, Remote sensing inputs to water demand modeling: Proceedings of the NASA Earth Resources Survey Symposium, p. 2585–2620.

Estes, J. E., Simonett, D. S., Tinney, L. R., and Ezra, C. E., 1978, Remote sensing detection of perched water tables: a pilot study: California Water Resources Center, University of California, Cont. No. 175, ISSN-0575-4941.

Eucker, C. C., Hoffman, R. O., Edwards, D. M., and Walling, G. P., 1975, Application of remote sensing to crop identification and water use: Proceedings of the International Seminar and Exposition on Water Resources Instrumentation, v. 1, p. 303–311.

Ezra, C. E., Bonn, F., and Estes, J. E., 1982, The feasibility of thermal inertia mapping for detection of perched water tables in semi-arid irrigated lands: Presented at the First Thematic Conference: Remote Sensing of Arid and Semi-Arid Lands, Cairo, Egypt, January 19–25, 1982, (in press).

Finley, R. J., and Baumgardner, R. W., Jr., 1980, Interpretation of surface water circulation, Arkansas Pass, Texas, using Landsat imagery: Remote Sensing of Environment, v. 10, p. 3–22.

Fisk, H. N., 1947, Geology of the Mississippi Valley region: Tulsa Geological Survey Digest, v. 15, p. 50–55.

Fitch, W. N., Hartigan, J. P., and Iwanski, M. L., 1976, Urban flooding response to land use change: Proceedings of the Symposium on Urban Hydrology, Hydraulics and Sediment Control, University of Kentucky, Lexington, Kentucky.

Ford, J. P., Blom, R. G., Bryan, M. L., Daily, M. I., Dixon, T. H., Elachi, C., and Xenos, E. C., 1980, Seasat views North America, the Carribean, and Western Europe with imaging radar: JPL Publication 80-67, Jet Propulsion Laboratory, Pasadena, California, p. 118–119.

Foster, J. R., and Hall, D. K., 1981, Multisensor analyses of hydrologic features with emphasis on the Seasat SAR: Photogrammetric Engineering and Remote Sensing, v. 47 no. 5, p. 655–664.

Gammon, P. T., and Carter, V., 1979, Vegetation mapping with seasonal color infrared photographs: Photogrammetric Engineering and Remote Sensing, v. 45, no. 1, p. 87–97.

Gilbert, G. K., 1882, Post-glacial joints: Journal of American Society, 3d ser., v. 23, p. 25–27.

Gillespie, A. R., and Kahle, A. B., 1977, Construction and interpretation of a digital thermal inertia image: Photogrammetric Engineering v. 43, no 8, pp. 983–1000.

Gilmer, D. S., Work, E. A., Jr., Colwell, J. E., and Rebel, D. L., 1980, Enumeration of prairie wetlands with Landsat and aircraft data: Photogrammetric Engineering and Remote Sensing, v. 46, no. 5, p. 631–634.

Gloersen, P., Zwally, H. J., Chang, A. T. C., Hall, D. K., Campbell, W. J., and Ramseier, R. O., 1978, Time-dependence of sea-ice concentration and multiyear ice fraction in the Arctic Basin: Boundary Layer Meterology, v. 13, p. 339–359.

Goetz, A. F. H., and Billingsley, F. C., 1973, Digital image enhancement techniques used in some ERTS (Landsat) application problems, *in* Freden, S. C., Mercanti, E. P., and Becker, M. A., eds.: 3d Earth Resources Technology Satellite-1 (Landsat-1) Symposium, Washington, D.C., Dec. 10–14, 1973,

Proc.: U.S. National Aeronautics and Space Administration Report, SP-351, sec. B, p. 1971–1992.

Gordon, H. R., 1974, Spectral variations in the volume scattering function at large angles in natural waters: Journal of the Optical Society of America, v. 64, p. 733–775.

Gregory, K. J., and Walling, D. E., 1973, Drainage basin form and process—a geomorphological approach: John Wiley and Sons, New York.

Guss, 1972, Tidelands management mapping for the coastal plains region—Proceedings of Symposium on Coastal Mapping, June 5–8, 1972: American Society of Photogrammetry, Falls Church, Virginia, p. 243–262.

Harker, G. R., and Rouse, J. W., Jr., 1977, Flood-plain delineation using multispectral data analysis: Photogrammetric Engineering and Remote Sensing, v. 43, no. 1, p. 81–87.

Hatfield, J. L., Millard, L. J., Reginato,S R. J., Jackson, R. D., Idso, S. B., Pinter, P. J., Jr., and Goettelman, R. C., 1980, Spatial variability of surface temperature as related to cropping practice with implications for irrigation management: Proceedings of the 14th International Symposium on Remote Sensing of Environment, San Jose, Costa Rica, Environmental Research Institute of Michigan, Ann Arbor, Michigan.

Heilman, J. L., Kanemasu, E. T., Rosenberg, N. J., and Blad, B. L., 1976, Thermal scanner measurement of canopy temperature to estimate evapotranspiration: Remote Sensing of Environment, v. 5, p. 137–145.

Heilman, J. L., and Moore, D. G., 1981, Ground-water applications of the heat capacity mapping mission: Satellite Hydrology, eds., Deutsch, M., Wiesnet, D. R., and Rango, A.: American Water Resources Association, Minneapolis, Minnesota, p. 446–452.

Heilman, J. L., and Moore, D. G., 1981, HCMM detection of high soil moisture areas: Remote Sensing of Environment, v. 11, p. 73–78.

Heimes, F. J., and Thelin, G. P., 1979, Development and application of Landsat-derived irrigated cropland maps for water-use determination in the high plains: Symposium on Remote Sensing for Irrigated Lands, Sioux Falls, South Dakota.

Heller, R. C., and Johnson, K. A., 1979, Estimating irrigated land acreage from Landsat imagery: Photogrammetric Engineering and Remote Sensing, v. 45, no. 10, p. 1379–1386.

Higer, A. L., Cordes, E. H., and Coker, A. E., 1976, Water management model of the Florida Everglades: ERTS-1, A new window on our planet, U.S. Geological Survey Paper 929, Williams, R. S., and Carter, W. D., eds.: U.S. Government Printing Office, Washington, D.C., p. 159–161.

Hoffman, R. O., Edwards, D. E., Wallin, G., and Burton, T., 1975, Remote sensing instrumentation and methods used for identifying center pivot irrigation systems and estimating crop water use: Proceedings of the International Seminar and Exposition on Water Resources instrumentation, v. 1, p. 312–317.

Hoffer, R. M., 1971, Remote sensing potentials for resource management: Biological Effects in the Hydrological Cycle: Purdue University, West Lafayette, Indiana, p. 211–227.

Holben, B. C., Tucker, C. J., and Fun, C. J., 1980, Assessing leaf area and leaf biomass with spectral data: Photogrammetric Engineering and Remote Sensing, v. 45, no. 5, p. 651–656.

Holtan, H. N., Ormsby, J. P., and Fisher, F. T., 1977, Application of a Maryland version of USDAHL-74 to a watershed in Prince George's County: Maryland-Proceedings of the Workshop on Watershed Research in Eastern North America, Edgewater, Maryland.

Hough, V. N. D., 1960, Photographic techniques applied to the mapping of rock joints: West Virginia Geological and Economic Survey Report Inv., no. 19, 21 p.

Hoyer, B. E., Hallberg, G. R., and Taranik, J. V., 1974, Summary of multispectral flood inundation mapping in Iowa, Public Information Circular 7: Iowa Geological Survey, Iowa City, Iowa.

Hulbert, E. O., 1945, Optics of distilled and natural water: Journal of the Optical Society of America, v. 35, no. 11, p. 698–705.

Huntley, D., 1978, On the detection of shallow aquifers using thermal infrared imagery: Water Resources Research, v. 14, p. 1075–1083.

Idaho Department of Water Resources, 1978, Idaho irrigated cropland inventory: Final Report, Boise, Idaho.

Idso, S. B., Schmugge, T. J., Jackson, R. D., and Reginato, R. J., 1975, The utility of surface temperature measurements for the remote sensing of soil water status: Journal of Geophysical Research, v. 80, p. 3044–3049.

Idso, S. B., and Ehrler, W. L., 1976, Estimating soil moisture in the root zone of crops: A technique adaptable to remote sensing: Geophysical Research Letters 3, p. 23–25.

Ishaq, A. M., and Huff, D. D., 1974, Application of remote sensing to the location of hydrologically active (source) areas: Proceedings of the 9th International Symposium on Remote Sensing of Environment, Environmental Research Institute of Michigan, Ann Arbor, Michigan, p. 653–666.

Jackson, R. D., Cihlar, J., Estes, J. E., Heilman, J. L., Ralke, A., Kanemasu, E. T., Millard, J., Price, J. C., and Wiegand, C., 1978, Soil moisture estimation using reflected solar and emitted thermal radiation: Chapter 4 of Soil Moisture Workshop: NASA Conference Publication 2073, 219 p.

Jackson, R. D., Idso, S. B., Reginato, R. J., and Pinter, P. J., Jr., 1981, Canopy temperature as a crop water stress indicator: Water Resources Research, v. 17, p. 1133–1138.

Jackson, R. D., Salomonson, V. V., and Schmugge, T. F., 1980, Irrigation management-future techniques: Proceedings of the Second National Irrigation Symposium, Oct. 20–23, 1980, Lincoln, Nebraska, p. 197–212.

Jackson, T. J., 1975, Computer aided techniques for estimating the percent of impervious area from Landsat data: Proceedings of the Workshop on Environmental Applications of Multispectral Imagery, ASP, Ft. Belvoir, Virginia.

Jackson, T. J., 1980, Profile soil moisture from surface measurements: Journal of Irrigation and Drainage Division, IR-2, American Society of Civil Engineers, p. 81–92.

Jackson, T. J., and McCuen, R. H., 1979, Accuracy of impervious area values estimated using remotely sensed data: Bulletin of Water Resources, v. 15, no. 2, p. 436–446.

Jackson, T. J., Ragan, R. M., and Fitch, W. N., 1977, Test of Landsat-based urban hydrologic modeling: Journal of the Water Resources Planning and Management Division, American Society Civil Engineers, v. 103, no. WRI, p. 141–158.

Jackson, T. J., Ragan, R. M., and Shubinski, R. P., 1976, Flood frequency studies on ungaged urban watersheds using remotely sensed data: Proceedings of the National symposium on Urban Hydrology, Hydraulics and Sediment Control, Lexington, Kentucky, p. 31–39.

Jackson, T. J., Chang, A., and Schmugge, 1981, Aircraft active microwave measurements for estimating soil moisture: Photogrammetric Engineering and Remote Sensing, v. 47, no. 6, p. 801–805.

Jensen, M. E., ed., 1974, Consumptive use of water and irrigation waste requirements: Irrigation and Drainage Division, American Society of Civil Engineers, 215 p.

Jones, E. B., Rango, A., and Howell, S., 1980, Measurement of liquid water content in a melting snowpack using cold calorimeter techniques, *in* Microwave remote sensory of snowpack properties: NASA Conference Publication 2153, Fort Collins, Colorado, p. 41–67.

Kahle, A. B., Gillespie, A. R., and Goetz, A. F. H., 1976, Thermal inertia imaging: A new geologic Mapping tool: Geophysical Research Letters, 3, pp. 26–28.

Katibah, E. F., 1975, Operational use of Landsat imagery for the estimation of snow areal extent, *in* Operational applications of satellite snowcover observations: NASA Special Publication 391, South Lake Tahoe, California, p. 150–160.

Kelly, M. G., and Conrad, A., 1969, Aerial photographic studies of shallow-water benthic ecology, Johnson, P. L., ed., Remote sensing in ecology: University of Georgia Press, Athens, Georgia, p. 173–184.

Klemas, V., Bartlett, D., Philpot, W., Rogers, R., and Reed, L., 1974, Coastal and estuarine studies with ERTS-1 and Skylab: Remote Sensing of Environment, v. 3, p. 153–174.

Klemas, V., Bartlet, D. S., and Rogers, R. H., 1975, Coastal zone classification from satellite imagery: Photogrammetric Engineering and Remote Sensing, v. 51, no. 4, p. 499–514.

Klemas, V., Borchardt, J. F., and Treasure, W. M., 1973, Suspended sediment observations from ERTS-1: Remote Sensing of Environment, v. 2, p. 205–221.

Klemas, V., Daiber, F. C., Bartlett, D., Crichton, O. W., and Fornes, A. O., 1974, Inventory of Delaware's wetlands: Photogrammetric Engineering, v. 40, no. 4, p. 433–440.

Krinsley, D. B., 1976, Lake fluctuations in the Shiraz and Neriz Playas of Iran, ERTS-1, A new window on our planet, U.S. Geological Survey Professional Paper 929, Williams, R. S., and Carter, W. D., eds.: U.S. Government Printing Office, Washington, D.C., p. 143–149.

Kruck, W., and Kantor, W., 1975, Hydrogeologic investigations in the Pampa of Argentina: Proceedings of the NASA Earth Resources Survey Symposium, Houston, Texas, National Aeronautics and Space Administration, Johnson Space Center, p. 2183–2197.

Lattman, L. H., and Nickelson, R. P., 1958, Photogeologic fracture-trace mapping in Appalachian Plateau: Bulletin of American Association of Petroleum Geologists, v. 42, no. 9, p. 2238–2245.

Lattman, L. H., and Parizek, R. R., 1964, Relationship between fracture traces and the occurrence of ground water in carbonate rocks: Journal of Hydrology, v. 2, no. 2, p. 73–91.

Lidster, W. A., Schmer, F. A., Ryland, D. W., and More, D. G., 1979, Remote-sensing technique for determining water table depths in irrigated agriculture: Proceedings of the Fall Technical Meeting, ASP Phoenix, Arizona, p. 821–838.

Linsley, R. K., Jr., Kohler, M. A., and Paulhus, 1958, Hydrology for Engineers: McGraw-Hill Book Company, New York, New York, 340 p.

Lueder, D. R., and Simons, J. H., 1962, Crustal fracture patterns and ground water movement: White Plains, N.Y., Geotechnics and Resources Inc., Report to Bureau of State Services, U.S. Department of Health, Education, and Welfare Final Report of grant WP-53, 148 p.

L'vovich, M. I., 1979, World water resources and their future (translation by American Geophysical Union): Lithocrafters, Inc., Chelsea, Michigan, 415 p.

Malila, W. A., and Nalepka, R. F., 1974, Advanced processing and information extraction techniques applied to ERTS-1: Proceedings of the Third Earth Resources Technology Satellite Symposium, Freden, A., Mercanti, E., Becker, M., eds., NASA SP-351, National Aeronautics and Space Administration, D.C., p. 1743–1772.

Martin, J. A., Rath, D. L., and Allen, W. H., 1973, Geologic ground and drainage patterns from ERTS-1 (Landsat-1) imagery in northern Missouri, *in* Anson, Abraham, ed., Symposium on Management and Utilization of Remote-Sensing Data, Sioux Falls, S. Dakota, Oct. 29–Nov. 1, 1973, Proc.: American Society of Photogrammetry, p. 333–341.

Martin, S. E., 1975, Color orthophotoquads produced from black-and-white film, *in* Workshop for Environmental Applications of Multispectral Imagery: American Society of Photogrammetry, Falls Church, Virginia, p. 299–303.

Martinec, J., 1970, Study of snowmelt-runoff process in two representative watersheds with different elevation range: IAHS Publication No. 96, Symposium of Wellington (N.Z.), p. 29–39.

Martinec, J., 1972, Evaluation of air photos for snowmelt-runoff forecasts, Proceedings of the Banff Symposium: The role of snow and ice in hydrology: IAHS-AISH Publication No. 107, p. 91–92.

Martinec, J., 1975, Snowmelt-runoff model for streamflow forecasts: Nordic Hydrology, v. 6, no. 3, p. 145–154.

Martinec, J., 1980, Hydrologic basin models: Remote sensing application in agriculture and hydrology, Fraysse, G., ed.: A. A. Balkema, Rotterdam, p. 447–459.

Martinec, J., and Rango, A., 1981, Aerial distribution of snow water equivalent evaluated by snow cover monitoring: Water Resources Research, v. 17, no. 5, p. 1480–1488.

Maul, G. A., and Gordon, H. R., 1975, On the use of the Earth Resources Technology Satellite (Landsat-1) in optical oceanograph: Remote Sensing of Environment, v. 4, p. 95–128.

McCluney, W. R., 1976, Remote measurement of water color: Remote Sensing of Environment, v. 5, p. 3–33.

McEwen, R. B., Kosco, W. J., and Carter, V., 1976, Coastal wetland mapping: Photogrammetric Engineering and Remote Sensing, v. 42, no. 2, p. 221–232.

McFarland, M. H., 1976, The correlation of Skylab L-band brightness temperatures with antecedent precipitation: Proceedings of the Conference of Hydrometerology, American Meteorological Society, Boston, Massachusetts, p. 60–65.

McKeon, J. B., Reed, L. E., Rogers, R. H., Ragan, R. M., and Wiegand, O. C., 1978, Landsat-derived cover and imperviousness categories for metropolitan Washington: An Urban/Non-urban Computer Approach, ASP Annual Meeting, Washington, D.C.

McMillan, M. C., and Forsyth, D., 1976, Satellite images of Lake Erie ice: January-March 1975, NOAA Tech Memo. NESS-80, 1976.

Merry, C. F., and McKim, H. L., 1977, Applications of remote sensing in the Boston urban studies program: CRREL Report 77-13, USACE Cold Regions Research and Engineering Laboratory, Hanover, New Hampshire.

Meyer, W., and Welch, R. I., 1975, Water resources assessment: Manual of remote sensing, Reeves, R. G., Anson, A., and Landen, D., eds.: American Society of Photogrammetry, Falls Church, Virginia, p. 1479–1551.

Millard, J. P., Jackson, R. D., Goettelman, Reginato, R. J., and Idso, S. B., 1978, Crop water stress assessment using an airborne thermal scanner: Photogrammetric Engineering and Remote Sensing, v. 44, no. 1, p. 77–85.

Mollard, J. D., 1957, Aerial mosaics reveal fracture patterns on surface materials in southern Saskatchewan and Manitoba: Oil in Canada, v. 26. p. 1840–1864.

Moore, D. G., Heilman, J. L., Tunheine, J. A., Western, F. C., Heilman, W. E., Beutler, G. A., and Ness, S. D., 1981, Evaluation of HCMM data for assessing soil moisture and water table depth: Final Report, NASA Grant NAS5-24206, South Dakota State University, Brookings, South Dakota, 199 p.

Moore, G. K., 1967, Lineaments on Skylab photographs—detection, mapping, and hydrologic significance in central Tennessee: U.S. Geological Survey Open-File Report 78-196, 81 p.

Moore, G. K., 1978, Satellite surveillance of physical water quality characteristics, *in* International Symposium on Remote Sensing of Environment, Michigan, 12th, 1978, Proc., Ann Arbor, Michigan: Environmental Research Institute of Michigan, p. 445–462.

Moore, G. K., Burchett, C. R., and Bingham, R. H., 1969, Limestone hydrology in the upper Stones River basin, central Tennessee: Tennessee Div. Water Resources, Research Ser. 1, 58 p.

Moore, G. K., and North, G. W., 1974, Flood inundation in the Southeastern United States from aircraft and satellite imagery: Bulletin of Water Resources, v. 10, no. 5, p. 1082–1096.

Moore, D. G., and Myers, V. I., 1972, Environmental factors affecting thermal ground water mapping: Remote Sensing Inst., SDSU, Brookings, S.D.

Moore, D. G., Wehde, M. E., and Myers, V. I., 1974, A guide for optical processing and use of ERTS-1 (Landsat-1) MSS data for analysis of surface water: A practical approach, Brookings, South Dakota, South Dakota State University, Remote Sensing Instru., Report SDSU-RSI-73-12, 21 p.

Moravec, G. F., and Danielson, J. A., 1980, A graphical method of stream runoff prediction from Landsat derived snowcover data for watersheds in the upper Rio Grande basin of Colorado, *in* Operational applications of satellite snowcover observations: NASA Conference Publication 2116, Sparks, Nevada, p. 171–183.

Morrison, R. B., and Cooley, M. E., 1973, Assessment of flood damage in Arizona by means of ERTS-1 imagery: Proceedings of the Symposium of Significant Results obtained from ERTS-1, v. 1, New Carrollton, Maryland, p. 755–760.

Morrow, J. W., and Carter, V., 1978, Wetland classification on the Alaskan north slope: 5th Canadian Symposium on Remote Sensing, Victoria, British Columbia, Canada.

Myer, C. R., Tunheim, J. A., and Moore, D. G., 1975, Heat flow temperature model for remotely mapping near surface water tables by thermography: Proc. S.D. Acad. Sci., v. 54, pp. 23–32.

Myers, V. I., and Heilman, M. D., 1969, Thermal IR for soil temperature studies: Photogrammetric Engineering, pp. 1024–1032.

Nash, J. E., and Suttcliffe, J. V., 1970, River flow forecasting through conceptual models; Part I—A discussion of principles: Journal of Hydrology, v. 10, no. 3, p. 282–290.

Newton, R. W., 1977, Microwave remote-sensing and its application to soil moisture detection: Technical Report RSC-81, Remote Sensing Center, Texas A&M University, College Station, Texas, University Microfilms no. 77–20, 398 p.

Newton, R. W., and Rouse, J. W., 1980, Microwave radiometer measurements of soil moisture content: IEEE Transactions on Antennas and Propagation, AP-28, no. 5, p. 680–686.

Njoku, E. G., and Kong, J. A., 1977, Theory for passive microwave-remote sensing of near-surface soil moisture: Journal of Geophysical Research, v. 82, p. 3103–3118.

Njoku, E. G., and O'Neill, P. E., 1982, Microwave radiometeric measurements of soil moisture in Kern County, California: accepted for publication in IEEE Transactions on Geoscience and Remote Sensing.

Ohio-Kentucky-Indiana Regional Council of Governments, 1975, A method for assessing rural non-point sources and its application in water quality management: WH-554, Water Resources Planning Division, EPA, Washington, D.C.

O'Leary, D., and Simpson, S., 1975, Lineaments and tectonism in the northern part of the Mississippi embayment, *in* Cook, J. J., ed., 10th International Symposium on Remote Sensing of Environment, Ann Arbor, Michigan, Oct. 6–10, 1975, Proc.: Environmental Research Inst. Michigan, v. 2, p. 965–973.

Olson, D. P., 1964, The use of aerial photographs in studies of marsh vegetation: Maine Agricultural Experiment Station, Bulletin 13, Technical Series, 62 p.

Peterson, N., 1977, Aerial snow depth markers: Snow Survey Training School, Soil Conservation Service, Stateline, Nevada, 9 p.

Pestrong, R., 1968, The evaluation of multispectral imagery for a tidal marsh environment; Office Naval Research Contract NONR-4430 (00), Technical Report.

Pestrong, R., 1969, Multiband photos for a tidal marsh: Photogrammetric Engineering, v. 35, no. 5, p. 453–467.

Phillipson, W. R., and Hafker, W. R., 1981, Manual versus digital Landsat analysis for delineating river flooding: Photogrammetric Engineering and Remote Sensing, v. 47, no. 9, p. 1351–1356.

Pincus, H. J., 1969, Analysis of remote-sensing displays by optical diffraction, *in* Cook, J. J., ed., 6th International Symposium on Remote Sensing of Environment, Ann Arbor, Michigan, Oct. 13–16, 1969, Proc.: Environmental Research Institute of Michigan, v. 1, p. 261–274.

Plafker, G., 1964, Oriented lakes and lineaments of northeastern Bolivia: Bulletin of the American Geological Society, v. 75, p. 503–522.

Poe, G. A., Stogryn, A., and Edgerton, A. T., 1971, Determination of soil moisture content using microwave radiometry: Final Report No. 1684 FR-1, DOC Contract 0-35239. Aerojet—General Corporation, El Monte, California, 169 p.

Polcyn, F. C., 1970, Measurement of water depth by multispectral ratio techniques: NASA Ann. Earth Resources Program Rev., 3rd, v. 3.

Powell, W. J., Copeland, C. W., and Drahovzal, J. A., 1970, Delineation of linear features and application to reservoir engineering using Apollo 9 multispectral photography: Alabama Geol. Survey Inf. Ser. 41, 37 p.

Pratt, D. A. and Ellyett, C. D., 1978, Image registration for thermal inertia mapping and its potential use for mapping of soil moisture and geology in Australia: Proceedings of the 12th International Symposium on Remote Sensing of Environment, Environmental Research Institute of Michigan, Ann Arbor, Michigan, pp. 1207–1217.

Pratt, D. A., Ellyett, C. D., Mcglaughlin, E. C., and McNabb, P., 1978, Recent advances in the application of thermal infrared scanning to geological and hydrological studies: Remote Sensing of Environment, v. 7, no. 2, pp. 177–184.

Price, J. C., 1980, NASA to launch heat capacity mapping mission: Journal of the British Interplanetary Society, v. 31, p. 131–136.

Price, J. C., 1981, The contribution of thermal data in Landsat multispectral classification: Photogrammetric Engineering and Remote Sensing, v. 47, n. 2, p. 229–236.

Ragan, R. M., and Jackson, T. J., 1975, Use of satellite data in urban hydrologic models: Journal of the Hydraulics Division, American Society of Civil Engineers, v. 101, no. HY12, p. 1469–1475.

Ragan, R. M., and Jackson, T. J., 1980, Runoff synthesis using Landsat and the SCS model: Journal of the hydraulics Division, American Society of Civil Engineers, v. 106, no. HY5.

Ragan, R. M., and Rogers, R. H., 1978, Use of Landsat satellite remote sensing for regional environmental planning and management: XV Convention, Pan American Federation of Engineering Societies, Santiago, Chile.

Rango, A., 1980a, Remote sensing of snow-covered area for runoff modeling, IAHS-AISH Publication No. 129: Proceedings of the Oxford Symposium on Hydrological Forecasting, p. 291–297.

Rango, A., 1980b, Operational applications of satellite snow cover observations: Bulletin of Water Resources, v. 16, no. 7, p. 1066–1073.

Rango, A., and Anderson, A. T., 1974, Flood hazard studies in the Mississippi River basin using remote sensing: Bulletin of Water Resources, v. 10, no. 5, p. 1060–1081.

Rango, A., Chang, A. T. C., and Foster, J. L., 1979, The utilization of spaceborne microwave radiometers for monitoring snowpack properties: Nordic Hydrology, v. 10, p. 25–40.

Rango, A., Foster, J., and Salomonson, V. V., 1975, Extraction and utilization of space acquired physiographic data for water resources development: Bulletin of Water Resources, v. 11, no. 6, p. 1245–1255.

Rango, A., Greaves, J. R., and De Rycke, R. J., 1973, Observations of arctic sea ice dynamics using the Earth Resources Technology Satellite (ERTS-1): Arctic v. 16, no. 4, p. 337–339.

Rango, A., and Itten, K. I., 1976, Satellite potentials in snow cover monitoring and runoff prediction; Nordic Hydrology, v. 7, no. 4, p. 209–230.

Rango, A., and Martinec, J., 1979, Application of a snowmelt-runoff model using Landsat data: Nordic Hydrology, v. 10, p. 225–238.

Rango, A., and Peterson, R., eds., 1980, Operational applications of satellite snowcover observation: NASA Conference Publication 2116, National Aeronautics and Space Administration, Sparks, Nevada, 302 p.

Rango, A., and Salomonson, V. V., 1974, Regional flood mapping from space: Water Resources Research, v. 10, no. 3, p. 473–484.

Rassmussen, W. C., 1963, Permeability and storage of aquifers in the United States: International Association of Scientific Hydrology, Committee of Subterranean Waters, Publication No. 64. p. 148.

Reimold, R. J., Gallagher, J. L., and Thompson, D. E., 1973, Remote sensing of tidal marsh: Photogrammetric Engineering, v. 39, no. 5, p. 477–488.

Reimold, R. J., and Linthurst, R. A., 1975, Use of remote sensing for mapping wetlands: American Society of Civil Engineers Transportation Engineering Journal, v. 101, no. TE2, p. 189–198.

Robinove, C. J., 1978, Interpretation of a Landsat image of an unusual flood phenomenon in Australia: Remote Sensing of Environment, v. 7, p. 219–225.

Rogers, R. H., Wilson, C. L., Reed, L. E., Shah, N. J., Akeley, R., Mara, T. G., and Smith, V. E., 1975, Environmental monitoring from spacecraft data: Proceedings of the Symposium on Machine Processing of Remotely Sensed Data, Laboratory for Applications of Remote Sensing, West Lafayette, Indiana, p. 3B–11, 3B–20.

Root, R. R., and Miller, L. D. 1971, Identification of urban watershed units using remote multispectral sensing: Completion Report Series, No. 29, Environmental Resources Center, Colorado State University, Ft. Collins, Colorado.

Rowan, L. C., Wetlaufer, P. H., Goetz, A. F. H., and others, 1974, Discrimination of rock types and detection of hydrothermally altered areas in south-central Nevada by the use of computer-enhanced ERTS (Landsat) images: U.S. Geological Survey Professional Paper 833, 35 p.

Russell, I. C., 1885, Geologic history of Lake Lahontan: U.S. Geological Survey Mon. v. 11, 288 p.

Salomonson, V. V., Ambaruch, R., Rango, A., and Ormsby, J. P., 1975, Remote sensing requirements as suggested by watershed model sensitivity analyses: Proceedings of the 10th International Symposium on Remote Sensing of Environment, Environmental Research Institute of Michigan, Ann Arbor, Michigan, p. 1273–1284.

Salomonson, V. V., 1981, The early 1981 view of Landsat-D progress: Proceedings of SPIE—The International Society of Optical Engineering, Bellingham, Washington, v. 278, p. 50–59.

Saxton, R. E., Johnson, H. P., and Shaw, R. H., 1974, Modeling evapotranspiration and soil moisture: Transaction of the American Society of Agricultural Engineers, v. 17, no. 4, p. 673–677.

Scarpace, F. L. Holmquist, K. W., and Fisher, L. T., 1979, Landsat analysis of lake quality: Photogrammetric Engineering and Remote Sensing, v. 45, no. 5, p. 623–633.

Schanda, E., Hafer, R., Wyssen, D., Musy, A., Meylan, D., Morzier, C., and Good, W., 1978, Soil moisture determination and snow classification with microwave radiometry: Proceedings of the 12th International Symposium on Remote Sensing of Environment, Manila, Philippines.

Schecter, R. N., 1976, Resource inventory using Landsat data for area-wide water quality planning: Proceedings of the Symposium on Machine Processing of Remotely Sensed Data, Laboratory for Application of Remote Sensing, West Lafayette, Indiana.

Schertler, R. J., Mueller, R. A., Jirberg, R. J., Cooper, D. W., Chase, T., Heighway, J. E., Holmes, A. D., Gedney, R. T., and Mark, H., 1975, Great Lakes all-weather ice information system: Proceedings of the 10th International Symposium on Remote Sensing of Environment, Ann Arbor, Michigan, p. 1377–1404.

Schmugge, T. J., 1978, Remote sensing of surface soil moisture: Journal of Applied Meteorology, v. 17, p. 1549–1557.

Schmugge, T. J., 1980, Effect of texture on microwave emision from soils: IEEE Transactions on Geoscience and Remote Sensing, GE-18, no. 4, p. 353–361.

Schmugge, T. J., Gloersen, P., Wilheit, T., and Geiger, F., 1974, Remote sensing of soil moisture with microwave radiometers: Journal of Geophysical Research, v. 79, p. 317–323.

Schmugge, T. J., Meneely, J. M., Rango, A., and Neff, R., 1977, Satellite microwave observations of soil moisture variations: Bulletin of Water Resources, v. 13, 265 p.

Schmugge, T. J., Jackson, T. J., and McKim, H. L., 1980, Survey of methods for soil moisture determination: Water Resources Research, v. 16, no. 6, p. 961–979.

Schott, J. R., and Schimminger, E. W., 1981, Data use investigations for Applications Explorer Mission A (Heat Capacity Mapping Mission)—HCMM's role in studies of the urban heat island, Great Lakes thermal phenomena and radiometric calibration of satellite data: Final Report, NASA Contract NAS5-24263. 128 p.

Schowengerdt, R., Babcock, E. M., Ethridge, L., and Glass, C. E., 1981, Correlation of geologic structure inferred from computer-enhanced Landsat imagery with underground water supplies in Arizona: Satellite Hydrology, Deutsch, M., Wiesnet, D. R., and Rango, A., eds.: American Water Resources Association, Minneapolis, Minnesota, p. 387–397.

Sharma, K. P., and Jain, S. E., 1978, Hydrological investigations of Tons Basin using aerial photographs: Irrigation and Power, v. 35, no. 2, p. 269–278.

Shima, L. J., Anderson, R. R., and Carter, V. P., 1976, The use of aerial color infrared photography in mapping the vegetation of a fresh-water marsh: Chesapeake Science, v. 17, no. 2, p. 74–85.

Siddiqui, S. H., and Parizek, R. R., 1971, Hydrogeologic factors influencing well yields in folded and faulted carbonate rocks in central Pennsylvania: Water Resources Research, v. 7, no. 5, p. 1295–1312.

Slack, R. B., and Welch, R., 1980, Soil Conservation Service runoff curve number estimates from Landsat data: Bulletin of Water Resources, v. 16, no. 5, p. 887–893.

Soer, G. J. R., 1980, Estimation of regional evapotranspiration and soil moisture conditions using remotely sensed crop surface temperatures: Remote Sensing of Environment, v. 9, p. 27–45.

Soil Conservation Service, 1969, Computer program for project formulation-hydrology: Technical Release No. 20.

Soil Conservation Service, 1972, National engineering handbook: Section 4, Hydrology.

Soil Conservation Service, 1975, Urban hydrology for small watersheds: Technical Release No. 55.

Sollers, S. C., Rango, A., and Henninger, D. L., 1978, Selecting reconnaissance strategies for floodplain surveys: Bulletin of Water Resources, v. 14, no. 2, p. 359–373.

Sonderegger, J. L., 1970, Hydrology of limestone terrains, photogeologic investigations: Geological Survey of Alabama Bull, 94C, 27 p.

Spanner, M. A., Strahler, A. H., and Estes, J. E., 1982, Soil loss prediction in a geographic information system format: Proceedings, 16th International Symposium on Remote Sensing of Environment, (in press).

Specht, M. R., and others, 1973, New color film for water-photography penetration: Photogrammetric Engineering, v. 39, p. 359–369.

Stiles, W. H., and Ulaby, F. T., 1980, Microwave remote sensing of snowpacks: NASA Contractor Report 3263, Goddard Space Flight Center, Greenbelt, Maryland, 404 p.

Strong, A. E., 1974, Remote sensing of algal blooms by aircraft and satellite in Lake Erie and Utah Lake: Remote Sensing of Environment, v. 3, p. 99–107.

Strong, A. E., 1976, Algal blooms in Utah Lake: ERTS-1, A new window on our planet, U.S. Geological Survey Professional Paper 929, Williams, R. S., and Carter, W. D., eds.: U.S. Government Printing Office, Washington, D.C. p. 270–272.

Strong, A. E., 1978, Chemical whitings and chlorophyll distributions in the Great Lakes as viewed by Landsat: Remote Sensing of Environment, v. 7, p. 61–72.

Stroud, L. W., and Cooper, A. W., 1968, Color infrared aerial photographic interpretation and net primary productivity of a regularly flooded North Carolina salt marsh: Water Resources Research Institute of the University of North Carolina, Report No. 14, 81 p.

Struve, H., and Judson, F. E., 1979, Development of stage area tables for the Yazoo backwater area using Landsat data: USACE Remote Sensing Symposium, Reston, Virginia.

Sydor, M., 1976, Western Lake Superior ice: U.S. Geological Survey Professional Paper 929, Washington, D.C. p. 169–172.

Thiruvengadackari, S., 1981, Satellite sensing of irrigation patterns in semiarid areas: An Indian study: Photogrammetric Engineering and Remote Sensing, v. 48, no. 10, p. 1493–1499.

Thompson, D. R., and Wehmann, O. A., 1979, Using Landsat digital data to detect moisture stress: Photogrammetric Engineering and Remote Sensing, v. 45, no. 2, p. 201–207.

Thompson, D. R., and Wehmann, O. A., 1980, Using Landsat digital data to detect moisture stress in corn-soybean growing region: Photogrammetric Engineering and Remote Sensing, v. 46, no. 8, p. 1087–1093.

Tomes, B. J., 1975, Use of Skylab and Landsat in a geohydrological study of the Paleozoic section, west-central Bighorn Mountains, Wyoming, *in* Proceedings of the NASA Earth Resources Survey Symposium, Houston, Texas: National Aeronautics and Space Administration Publication, NASA TM X-58168, v. I-D, p. 2167–2182.

Trainer, F. W., 1967, Measurement of the abundance of fracture traces on aerial photographs: U.S. Geological Survey Professional Paper 575-C, p. C184–188.

Trainer, F. W., and Ellison, R. L., 1967, Fracture traces in the Shenandoah Valley, Virginia: Photogrammetric Engineering, v. 33, no. 2, p. 190–199.

Tucker, C. J., 1980, Remote sensing of leaf water content in the near infrared: Remote Sensing of Environment, v. 10, no. 1, p. 23–32.

Ueda, H., Sellmann, P., and Abele, G., 1975, USA CRREL snow and ice test equipment Special Report 146: U.S. Army Cold Regions, Research and Engineering Laboratory, Hanover, New Hampshire, 17 p.

Ulaby, F. T., 1974, Radar measurements of soil moisture content: IEEE Transactions on Antennas and Propagation, AP-22, p. 257–265.

Ulaby, F. T., and Batlivala, P. P., 1976, Optimum radar parameters for mapping soil moisture: IEEE Transactions on Geoscience Electronics, GS-14, p. 81–93.

Ulaby, F. T., Batlivala, P. P., and Dobson, M. C., 1978, Microwave backscatter dependence on surface roughness, soil moisture and soil texture: Part 1—bare soil: IEEE Transactions on Geoscience Electronics, GS-16, p. 286–295.

Ulaby, F. T., Bradley, G., Dobson, M. C. and Bare, J. E., 1977, Analysis of the active microwave response to soil moisture, Part II: Vegetation-covered ground: RSL Technical Report 264-19, University of Kansas Center of Research, Inc., Lawrence, Kansas.

Ulaby, F. T., Cihlar, J., and Moore, R. K., 1974, Active microwave measurements of soil water content: Remote Sensing of Environment, no. 3, p. 185–203.

UNESCO, 1978, World water balance and the water resources of the earth: Studies and reports in hydrology, No. 25: United Nations Educational, Scientific, and Cultural Organization, New York, New York, 663 p.

U.S. Army Corps of Engineers, 1976, Urban storm water runoff "STORM," computer program 723-58-L2520: Hydrologic Engineering Center, Davis, California.

Vickers, R. S., Heighway, J. E., and Gedney, R. T., 1974, Airborne profiling of ice thickness using a short pulse radar *in* Advanced concepts and techniques in the study of snow and ice resources: National Academy of Sciences, Washington, D.C., p. 422–431.

Wang, J. R., and Schmugge, T. J., 1980, An empirical model for the complex dielectric permittivity of soils as a function of water content: IEEE Transactions Geoscience on Remote Sensing, GE-18, p. 288–295.

Wang, J. R., Shiue, J. C., and McMurtrey, J. E., III, 1980a, Microwave remote sensing of soil moisture content over bare and vegetated fields: Geophysical Research Letters, 7, p. 801–804.

Wang, J. R., and Choudhury, B. J., 1981, Remote sensing of soil moisture content over bare field at 1.46 Hz frequency: Journal of Geophysical Research 86, No. cb, p. 5277–5282.

Watson, K., 1971, Application of thermal modeling in the geological interpretation of IR images: Proceeding of the 7th International Symposium on Remote Sensing of Environment, Environmental Research Institute of Michigan, Ann Arbor, Michigan, pp. 2017–2041.

Watson, K., 1979, Regional thermal inertia mapping to discriminant geologic materials: Proceedings of the 13th International Symposium on Remote Sensing of Environment, Environmental Research Institute of Michigan, Ann Arbor, Michigan, pp. 79–80.

Watts, R. D., and England, A. W., 1976, Radio-echo sounding of temperate glaciers, ice properties and sounder design criteria: Journal of Glaciology, v. 17, no. 75, p. 39–48.

Wiegand, C. L., Richardson, A. J., and Kanemasu, E. T., 1979, Leaf area index estimates of wheat from Landsat and their implications for evapotranspiration and crop modeling: Agronomy Journal, v. 71, p. 336–342.

Wiesnet, D. R., McGinnis, D. F., and Pritchard, J. A., 1974, Mapping of the 1973 Mississippi River Floods by the NOAA-2 satellite: Bulletin of Water Resources, v. 10, no. 5, p. 1040–1049.

Wilcox, R. G., and Struve, H., 1979, Project-induced land clearing Tensas River project: USACE Remote Sensing Symposium, Reston Virginia.

Williamson, A. N., 1974, Mississippi River flood maps for ERTS-1 digital data: Bulletin of Water Resources, v. 10, no. 5, 1050–1059.

Withington, C. W., 1976, Basement tectonics of the Atlantic coastal plain as seen from ERTS-1 (Landsat-1) imagery (abs.), *in* Hodgson, R. A., Gay, S. P., Jr., and Benjamins, J. Y., eds.: Proceedings of the First International Conference on the New Basement Tectonics, 1974: Utah Geological Association Publication 5, 262 p.

Work, E. A., and Gilmer, D. S., 1976, Utilization of satellite data for inventorying prairie ponds and lakes: Photogrammetric Engineering and Remote Sensing, v. 42, p. 685–694.

Zall, L., and Russell, O., 1981, Ground-water exploration programs in Africa: Satellite hydrology, Deutsch, M., Wiesnet, D. R., and Rango, A., eds.: American Water Resources Association, Minneapolis, Minnesota, p. 416–425.

Zotimov, N. V., 1968, Investigation of a method of measuring snow storage by using the gamma radiation of the earth: Soviet Hydrology Selected Paper, English Translation, no. 3, p. 254–266.

CHAPTER 30

Urban/Suburban Land Use Analysis

Author-Editor: JOHN R. JENSEN
Contributing Authors: M. LEONARD BRYAN, STEVEN Z. FRIEDMAN, FLOYD M. HENDERSON, ROBERT K. HOLZ, DAVID LINDGREN, DAVID L. TOLL, ROY A. WELCH, JAMES R. WRAY

GENERAL CONTENTS: The spectral nature of urban land use; the spatial nature of urban land use; the temporal nature of urban land use; land use and land cover classification systems and accuracy assessment; selected urban land use applications; population estimation; housing quality data; monitoring energy conservation, utilization and production in urban areas; urban and suburban information for emergency situations; references.

INTRODUCTION

The goal of this chapter is to identify the fundamental spectral, spatial, and temporal nature of urban and suburban land use/land cover[1] as recorded by remote sensing systems and how imagery of these phenomena might be analyzed to solve urban problems. However, no attempt is made to exhaust the universe of multi-disciplinary urban applications to which remote sensing might be applied. Such an activity is best addressed by extensive bibliographies already developed (Todd, 1978).

Discussion commences by addressing the nature of urban land uses in relation to remote sensing. Three sections identify the spectral, spatial, and temporal characteristics of urban phenomena which should be understood to obtain useful urban/suburban information from imagery. Theoretical and practical relationships are identified based on findings from basic and applied urban remote sensing research.

This is followed by a section on urban and suburban remote sensing land use and land cover classification schemes and accuracy assessment. Classification schemes and methods of assessing classification accuracy are important because terms must be defined and acceptable criteria stipulated if professionals are to use successfully the remotely sensed urban information.

The sections referred to above provide fundamental information necessary to understand the practical use of remotely sensed urban and suburban information. This chapter concludes with selected urban applications at the international, national, regional and local levels which demonstrate the state-of-the-art of applied remote sensing in important topical areas.

[1] The term 'land use' refers to the cultural use of the land, while the term 'land cover' describes the actual materials present on the surface of the earth. For example, the land cover of a forested area may be various hardwood tree species while the primary land use may be recreation.

THE SPECTRAL NATURE OF URBAN LAND USE

Remote sensing of urban environments is basically concerned with recording and interpreting an image produced by radiant flux, Φ, which exits from a source area on the ground toward the sensor. To measure this phenomena, radiance, L, is computed as

$$L = \frac{d\,\Phi}{dA \cos\theta\, dw} \qquad (30\text{-}1)$$

where radiant flux, $d\,\Phi$, leaves a small surface area, dA, in a specific solid cone angle, dw, and direction, θ, toward the sensor (Suits, 1975). This is normally measured in watts per steradian per square meter, i.e., $W\, sr^{-1}m^{-2}$ (refer to Chapter 2 for detailed discussion). Such a quantity is important yet does not describe the spectral distribution of the flux incident to or reflected by the urban/suburban scene. Therefore, consider first the simple hemispherical reflectance formula

$$\rho = \frac{\Phi_r}{\Phi_e} \qquad (30\text{-}2)$$

where Φ_r is the radiant flux reflected and Φ_e is the radiant flux incident to the surface measured in $W\, m^{-2}$. This relationship can be extended to represent the diffuse wavelength-dependent reflectance of a surface in the form

$$\rho_\lambda = \pi L_\lambda / E_\lambda . \qquad (30\text{-}3)$$

In this equation, the spectral distribution of the reflected radiant flux, ρ_λ, is dependent on the reflected radiance exiting the target in a specified direction toward the sensor $L_\lambda(W\, sr^{-1}m^{-2}\lambda^{-1})$ and the spectral distribution of the incident irradiation $E_\lambda(W\, m^{-2}\lambda^{-1})$. This ratio is often used to characterize the spectral signature of a feature.

The word "signature" refers to a distinguishing characteristic. In remote sensing, the spectral signature of a feature comprises a set of values for the reflectance of the feature measured at specific

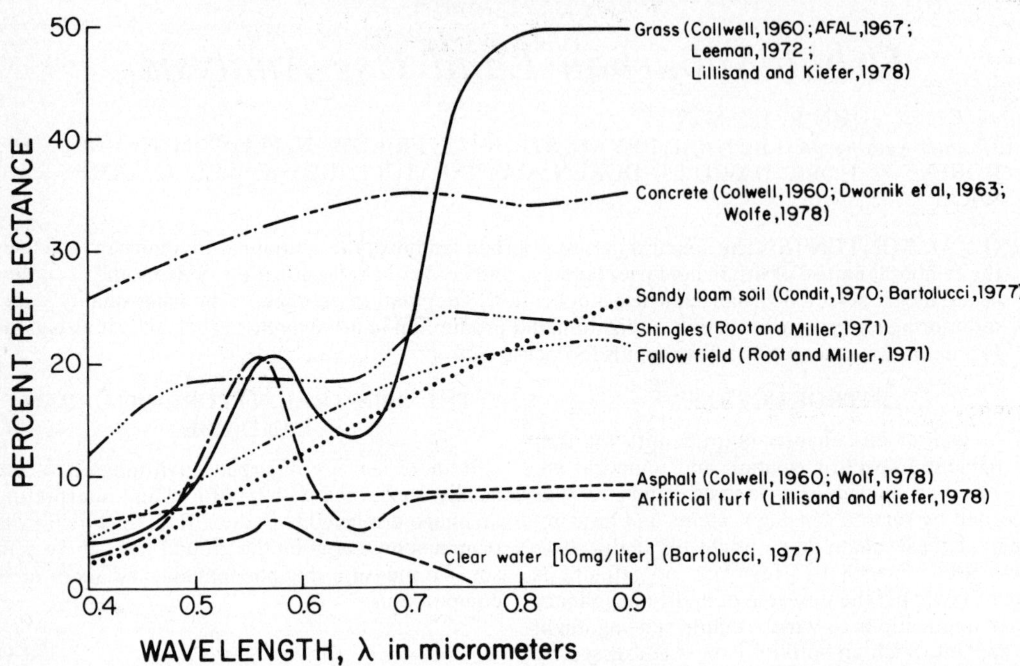

Fig. 30-1 Spectral reflectance curves for selected urban phenomena. The spectra are averages of results discussed in numerous studies. In all cases, data were obtained using a spectroradiometer. [Source: Jensen]

wavelength intervals. These spectral reflectance statistics are often used in remote sensing to aid in characterizing and understanding why urban features appear as they do on remotely sensed imagery. It should be emphasized, however, that the spectral signature is not constant for a given urban/suburban feature. It is dependent on the spectral distribution of the incident radiant flux onto the feature, on geometric relationships between the incident energy and sensor angle-of-view, on atmospheric effects, and on the physical properties of the feature (Slater, 1975).

To obtain an appreciation for the spectral nature of urban phenomena, spectral reflectance curves for selected phenomena present in urban scenes are shown in Figure 30-1. Spectra such as these were obtained from measurements using a spectroradiometer. The measurements give the distribution of reflectance of a material (e.g. grass) as a function of wavelength. Unfortunately, numerous man-made materials have roughly the same spectral reflectance properties. Thus, the spectral signature of urban phenomena obtained by remote sensors is often not unique, and accurate identification might require using other elements of image interpretation in addition to simply spectral response.

To gain an appreciation of the spectral signature of urban phenomena it is useful to evaluate the images of urban phenomena recorded by a variety of sensor systems including aerial photography, multispectral scanner imagery, and microwave imagery. The discussion does not include analysis of ultraviolet photography or passive microwave data as imagery from such regions is difficult to obtain and there exist few practical urban applications to date.

URBAN LAND COVER ANALYSIS WITH VISIBLE AND NEAR INFRARED IMAGERY

Vertical aerial photography is the principal remote sensing medium used for urban applications. The community of users is comprised of national, regional and local public agencies (public works, planning agencies, etc.) and private enterprise (consulting firms, photogrammetric firms, utility companies, etc.). Although this community is almost exclusively dependent on the use of aerial photography, it has not organized itself into a regionally or nationally visible entity to demand systematic aerial photography from the public sector. Basically, applications addressed by this user community require very high spatial resolution imagery. In the 1980's this community's photographic needs will continue to be met primarily by private enterprise. However, it may be supplemented by: the National High-Altitude Aerial Photography Data Base (USGS, 1980); or

Fig. 30-2 Panchromatic (Plus-x Aerographic) stereopair of the Jefferson National Expansion Memorial Historic Site in St. Louis, Missouri. The original scale was 1:3000 (1″ = 250′) flown at an altitude of 1500 feet using a 6″ focal length lens. [Source: National Park Service, Denver Service Center]

the Large Format Space Camera to be flown on the Space Shuttle (Doyle, 1978).

It would be futile to attempt to describe how all urban and suburban land use and land cover appear when recorded on aerial photography. Therefore, the approach will be to evaluate some of the fundamental physical phenomena shown in Figure 30-1 in terms of how they are recorded on panchromatic, multiband, color and color-infrared photography. This will provide general insight as to why these phenomena and others which are a combination of such materials appear as they do.

Panchromatic Aerial Photography

First consider the aerial photography stereopair of the Jefferson National Expansion Memorial in St. Louis, Missouri (Figure 30-2). Such photography is sensitive to wavelengths from $0.4-0.7\ \mu m$ and records objects in their true tonal density minus any color. Note that healthy green vegetation such as the grass surrounding the Memorial produces dark shades of grey (30-3). This is because approximately 80 to 90 percent of the incident blue, and red radiant flux, Φ, is absorbed by chloroplast pigments in green plants (Hoffer, 1978). This phenomena is commonly characterized on the graph of spectral reflectance as the chlorophyll absorption bands or regions (Figure 30-1).

Concrete structures are recorded in light tones because approximately 20−35 percent of the blue, green, and red radiant flux is reflected from such surfaces (Figure 30-1). Good examples are freeways and multiple story car garages (Figure 30-4). Heavily used concrete roads with frequent stops, however, accumulate oil and other sediment which eventually absorb approximately 90 percent of the visible incident flux resulting in a shift from light to dark tones. Similarly, asphalt paving, such as that used for tennis courts or tarred rooftop surfaces, absorbs 90 percent of the flux in the blue, green, and red resulting in darker tones (Figure 30-1).

Water may or may not be present in any substantial amount in urban environments. When bodies of water such as rivers are present (Figure 30-2), the spectral response will usually be in medium to dark grey tones due to almost 95 percent of the blue, green, and red radiant flux being absorbed (Figure 30-1). However, the relative turbidity, depth to bottom, and composition and color of the bottom will affect the eventual tone on film. The tone of water in reflecting ponds and swimming pools is often unpredictable due to these variables.

Evaluation of the spectral characteristics in Figure 30-1 reveals that these fundamental materials found in urban environments may be distinguishable from one another in the visible and near-infrared region. However, is this list of materials the kind of information required to make coherent decisions in urban areas? Of course the answer is no. The primary difficulty is that materials such as concrete, asphalt, roofing materials, plaster, paint, etc., are configured by man to produce a diverse array of land uses which unfortunately may produce similar spectral responses when imaged with the relatively coarse spectral resolution of certain sensor systems.

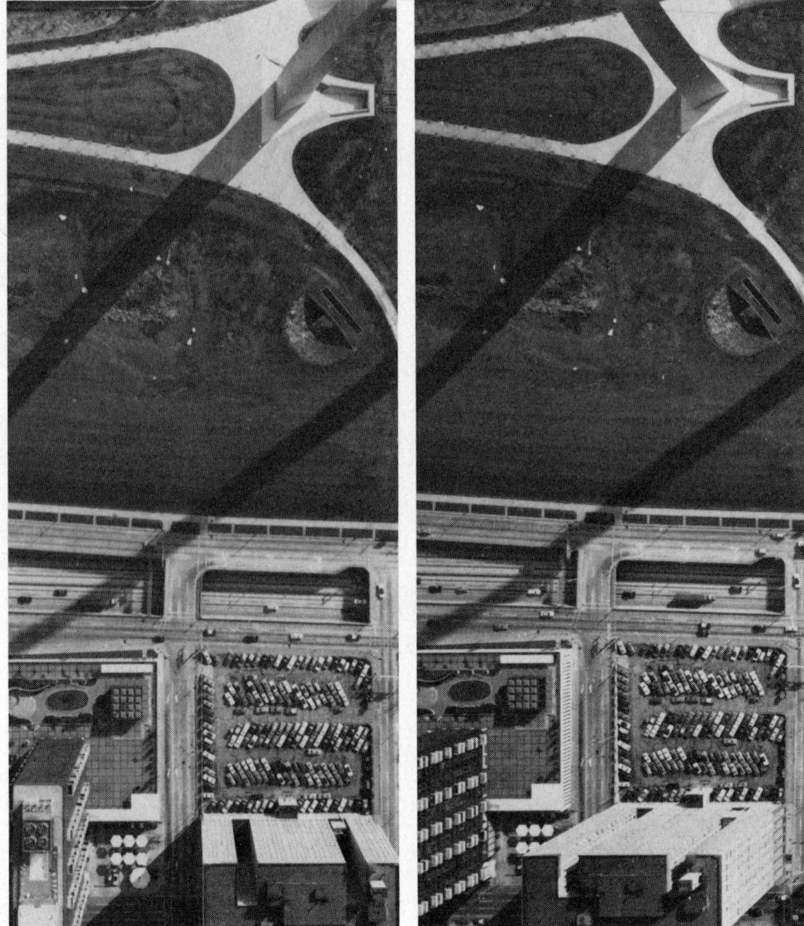

Fig. 30-3 A stereopair highlighting the vegetation surrounding the Jefferson National Expansion Memorial.

Thus, in such instances, it may be difficult to discriminate one specific urban class of information from another based solely on spectral characteristics (Estes and Simonett, 1975). The remainder of this section on panchromatic photography will briefly touch upon other elements of image interpretation which must be used to extract the useful urban information available in all types of aerial photography obtained over urban areas.

The importance of size, shape (geometry), image displacement, and texture/pattern information contained in aerial photography cannot be overemphasized. These elements of image interpretation are essential to accurate identification and classification of urban features for any application beyond simple regional reconnaissance. The profusion of heterogeneous and often confusing tonal (spectral) information may be sorted out by careful evaluation of these additional characteristics in the image. In particular, an interpreter's ability to measure the size, perimeter, area, volume, and absolute elevation of objects on aerial photography greatly improves his or her ability to discriminate one set of features from another and judge their significance. For example, the height of any feature above or below the local datum (e.g. the Jefferson Memorial or the railroad and freeway underpasses) can be easily computed using fundamental stereoscopic relationships (See Chapter *24*). In addition, the accurate plan position, perimeter, and area measurements can be readily computed.

Shadows produce distinctive spectral signatures. They are caused by an interposed opaque body which selectively absorbs or reflects the incident radiant energy. Shadows provide useful information concerning the silhouette of urban features. There is usually information in shadow areas recorded on panchromatic (Figure 30-4) or color film. This is generally not the case when using infrared film.

Texture is defined as the characteristic placement and arrangement of spectral tones, and can be a very valuable aid in identifying urban features. If this element is coupled with an analyst's ability to identify a pattern to this texture, then

Fig. 30-4 A stereopair highlighting concrete and asphalt surfaces near the Jefferson National Expansion Memorial. Note the tone of concrete roads, buildings, and roof top parking lots. The tennis court has an asphalt surface.

interpretation may be further enhanced. For example, the window lattices of hotels in Figures 30-3 and 30-4 enable the analyst to differentiate the side of the structure from the roof and also provide an impression of building height. Similarly, Busch Stadium has a unique size, shape, texture and pattern (spiral motif) which make it easily identifiable to persons who have seen such a structure before (Figure 30-4).

Multiband Aerial Photography

The use of such photography for urban area analysis is based on the assumption that useful information can be obtained by imaging a scene in discrete regions of the electromagnetic spectrum (Colwell, 1965). Multiband cameras are usually configured with the following broad bands: blue (0.4–0.5 μm), green (0.5–0.6 μm), red (0.6–0.7 μm), and infrared (0.7–0.9 μm). However, this can be modified using various film and filter combinations.

Multiband photography of an area near Century City, Los Angeles, is shown in Figure 30-5. Blue, green, red and white color panels located in the parking lot across from the hotel provide an appreciation of how urban features are recorded in discrete wavelength regions on multiband photography. Figure 30-5 A is the ''blue'' image. Note the washed out appearance of the landscape due to atmospheric scattering. Each of the bands (blue, green, and red) provides a unique record of the spectral reflectance in the corresponding region. However, since very few features in the urban scene are pure blue, green, or red, different proportions of these colors are reflected from most features. Consequently, such multiband black-and-white aerial photography in the visible region is interesting to look at yet difficult to interpret. Most interpreters forget that each image is produced by a finite sample of energy within a specific wavelength region and that the result should not be interpreted as if it were a panchromatic image.

The information provided by the black-and-white infrared image is quite different from that

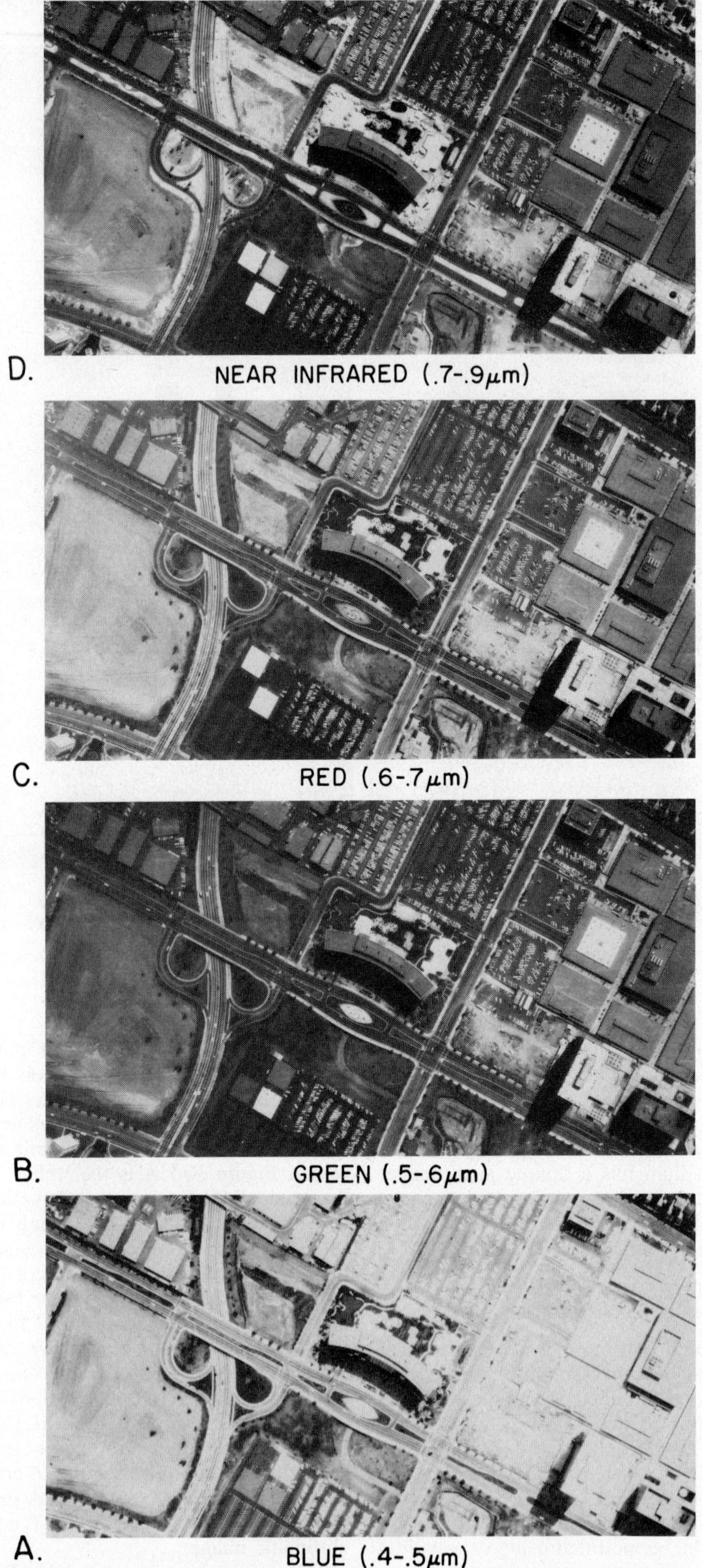

Fig. 30-5 Aerial photography of an area near Century City, Los Angeles, California, obtained on June 15, 1976. Note the targets on the ground across the street from the hotel.

provided by the visible images (Figure 30-5 D). Healthy green vegetation is recorded in bright tones because vegetation reflects much more of the incident radiant flux in the infrared region than in the blue, green and red regions (Figure 30-1). In the infrared region, the contrast between natural features such as bare soil, grass, water, etc. and urban structures such as asphalt, concrete, etc. are generally enhanced (Swain and Davis, 1978). Unfortunately, shadow areas in black-and-white infrared imagery are often devoid of useful urban information. For example, note the loss of information in the hotel's shadow area. This constraint is important in central business districts (CBD's) where tall buildings result in extreme shadow conditions. The problem is compounded by photography acquired at times other than when the sun is at its zenith. This produces excessively long shadows which can obscure parts of the scene.

Color Aerial Photography

Color aerial photography is used frequently in urban applications. In the case of color aerial photography obtained in the region from 0.4–0.7 μm, the scene is imaged in tones and hues which approximate the color balance with which the real-world is viewed. Unfortunately, in many downtown urban areas much of the scene is composed of concrete or asphalt. It is not surprising, therefore, that much color aerial photography of CBDs contains a substantial amount of grey tones even though it is a color image. About the only truly colorful features in such CBDs are the brightly colored building facades (visible due to building relief displacement relative to the photograph's nadir), occasional green vegetation, swimming pools, and multi-colored automobiles.

The use of color aerial photography in suburban applications is more effective. For example, consider photography of Goleta, California, where color provides important clues to an interpretation of specific suburban landscape features (Color Figure 30-6). The green avocado orchard, 9-hole golf course, and carefully manicured lawns are readily identifiable. The brown bare soil contrasts with the grey concrete highways and parking lots. Numerous swimming pools are easily identified. Interpretation of such imagery for suburban applications is straightforward because the color balance parallels the interpreter's real-world experience.

Color-Infrared Aerial Photography

This photography is usually filtered to record the radiant flux reflected from the scene in the region from 0.5–0.9 μm. As a result of the dyes used to develop the film (refer to Chapter 6), the resulting aerial photography is a false-color rendition of the original scene and should be interpreted carefully. It would be ideal if the color balance shift was such that all green features would record as blue on color infrared photography, all red features would image as green, and all features reflecting relatively large amounts of near-infrared radiant flux (0.7–0.9 μm) in the urban environment would image in bright red or magenta hues. Unfortunately, the human ability to predict the resultant signature of many urban features is poor because the real-world experiences gained as interpreters has not been sufficient to understand how much infrared radiant flux is reflected from urban landscapes. Human beings simply have never seen reflected infrared energy, so only gradually can an awareness of the relationships be developed.

Fortunately, there are certain general urban phenomena which can be expected to image relatively consistently on color infrared photography. For example, homogeneous vegetated surfaces such as lawns, parks, etc., usually produce bright magenta (red) hues (color Figure 30-7 and 30-12A) due to the high reflectance of the infrared radiant energy by these surfaces and the high absorptance in the green and red regions (Figure 30-1). Such vegetated surfaces contrast sharply with other surrounding man-made materials in urban environments such as asphalt and concrete which reflect relatively equal proportions of green, red, and infrared energy. This results in what is generally termed a "steel-grey" spectral signature for man-made structures on color infrared imagery and is especially evident when such imagery is obtained a small scales (color Figure 30-7).

Non-turbid water almost completely absorbs the longer wavelength infrared energy resulting in dark blue to black signatures for homogeneous, deep water bodies. Thus, the increased contrast between vegetation, water, and man-made structures is an important incentive for using color-infrared aerial photography for urban and suburban applications.

There are drawbacks, however. First, due to the lack of understanding of how much near-infrared energy is reflected from certain urban features (often these are painted surfaces), odd colors may represent what appear to be familiar features. Interpretation of such phenomena simply takes practice and clear thinking in order to identify these features consistently. The second drawback is the loss of information in shadow areas already discussed.

Overriding these negative elements is the fact that both black-and-white and color infrared photographs usually record only the green, red, and near-infrared energy and not the highly scattered blue light which can degrade panchromatic and normal color photography of urban areas. The ideal time of day to obtain photography of urban environments is near noon when excessive shadows can be avoided. Unfortunately, this is precisely the time when the smog and other forms of atmospheric particulate matter may reach their highest concentrations and severely degrade urban image quality. Thus, the haze penetration capability of infrared photography is important.

Aerial photography applied to urban problems is not constrained to low altitude data collection. High altitude and satellite platforms can provide aerial photography suitable for many urban applications. In addition, a regional synoptic view is obtained. For example, Lins (1976) compared the level III land cover classification of Fairfax, Virginia, obtained from high altitude color infrared photography (1:130,000) and SKYLAB S-190B natural color photography (1:970,000). Through the use of high altitude color infrared imagery as a substitute for ground truth it was determined that the SKYLAB color photography interpretation was 83 percent correct. Thus, even orbital aerial photography such as that obtained by the SKYLAB S-190B sensor system can provide sufficient spectral (and spatial) information to discriminate numerous Level III categories of urban land use.

As discussed in Chapters 24 and 25 aerial photography can be analyzed digitally as well as manually. Several researchers have digitized aerial photography of urban and suburban environments and then performed digital image processing to classify features (Hsu, 1978; Piech et al, 1977). Although these studies are promising, there are drawbacks. The main problem is that most of the classifiers used rely on the spectral (tonal) variable to provide the necessary information for accurate interpretation. What is usually lacking is the diverse background of the interpreter who provides the context for the study (i.e. a purpose) and can simultaneously evaluate site, association, tone, geometry (shape), texture, pattern, size, shadow etc., during the interpretation process. It is unlikely that all these diverse talents can be programmed into a computer to perform the type of detailed analyses required for most urban applications. Consequently, intensive urban inventories will probably continue to be done visually using aerial photography.

Multispectral Scanner Systems

Multispectral scanner systems (MSS), have been used to obtain a significant amount of urban information. Such sensors operate from aircraft (Clark, 1979) or from satellite platforms such as the Landsat satellites (Chapter *12*). This discussion focuses on the Landsat MSS system because it has provided a worldwide image data base for numerous urban applications (Carter and Gardner, 1977). There are limitations, unfortunately, to the applicability of such imagery to urban analysis and these will be discussed.

Multispectral scanning systems, like multiband aerial photography, record the radiant flux exiting from earth surfaces in specific spectral regions. In the case of the Landsat series (1, 2, and 3) the spectral resolution of the four bands was band 4 green (0.5–0.6 μm), band 5 red (0.6–0.7 μm), band 6 infrared (0.7–0.8 μm), and band 7 infrared (0.8–1.1 μm). These relatively broad spectral bands were not designed to be optimum for urban applications. Nevertheless, the following statements describe the general urban applicability of these bands (Colvocoresses, 1977):

- MSS band 5 is the fundamental band for indicating boundaries between natural and cultural features
- MSS bands 6 and 7 have demonstrated operational value in the near-infrared for vegetation, open water (especially shorelines), and cultural delineation. However, the two bands are highly redundant in information content.

As noted in the section on multiband photography, care must be taken when interpreting individual band images because tonal responses cannot be interpreted as if one were analyzing panchromatic photography. Consequently, much manual interpretation of urban environments using Landsat data has been concerned with analysis of color composites. Such composites are normally produced by filtering band 4 with yellow light, band 5 with magenta light and either band 6 or band 7 with cyan light. This produces a color-infrared false-color image with the urban areas exhibiting the characteristic 'steel-blue' tone previously discussed. Color composites of Seattle, Washington and Austin, Texas, demonstrate the contrast of the urban/suburban steel-blue signature versus the magenta hue of the natural vegetation (color Figure 30-8). It is obvious that the urbanized area can be differentiated from other general level I land cover classes. But what about more specific land cover classes nestled within urban areas (i.e., level II and III classes)? Basically, there is a case for lowered expectations when using Landsat imagery for detailed urban applications.

To understand the limitations inherent in both the manual and machine-assisted analysis of Landsat imagery for urban applications it is necessary to appreciate that there exist interrelated system and environmental constraints. The system constraints include an instantaneous-field-of-view (IFOV) of approximately 79 m^2 per picture element (pixel) and the relatively coarse spectral resolution previously discussed (Taranik, 1978). These two system constraints effectively control the level of spatial and spectral information made available for image analysis. The environmental constraint is equally important. Urban environments are notoriously heterogeneous in their assemblage of surface material. This may result in a high spatial frequency of change in land cover over short distances. The reflectance from land cover materials such as grass, rooftops, concrete driveways, asphalt, etc., within the IFOV are integrated to produce a single discrete pixel reflectance value (Smedes et al., 1975).

Ideally, the reflectance value produced by the diverse surface materials is related to the dominant land use of the area. Unfortunately, this is often not the case. On the other hand, there are certain urban phenomena which are remarkably homogeneous for extended areas such as the roof tops of large shopping complexes, industrial parks, parking lots, or recreation parks. Over these surfaces the IFOV will obtain fairly uniform reflectance values which again will hopefully represent a dominant land use class. If the surface materials integrated within the dimension of the pixel produce a unique reflectance value for a specific land cover (e.g. single-family residential) then the ability to discriminate between this class and all others based solely on spectral properties is improved. However, if different land covers within the IFOV are integrated and produce reflectance values similar to a totally different land cover then discrimination will be difficult.

To appreciate the nature of conducting machine-assisted urban inventories from Landsat, an analysis of land cover in Goleta, California, is presented. Goleta is situated at the northern limit of the Los Angeles-Ventura-Santa Barbara urban conurbation (color Figure 30-12). The area is comprised of a diversity of suburban and rural land uses including: 1) natural vegetation, primarily schlerophyll chaparral; 2) agriculture, dominated by mature citrus and avocado orchards; 3) single-family residences; 4) multiple-family residences; 5) commercial complex and barren regions; and 6) water surfaces. Although experiencing a growth slowdown due to a water moratorium, the area is representative of many urban-sprawl environments where agriculture and natural vegetation are displaced by residential housing and commercial establishments. A May 4, 1974, Landsat image of a portion of Goleta was used for the analysis (Figure 30-9).

The result of a supervised collection of training-class statistics for each of the land cover classes is displayed in Figure 30-10 for bands 4, 5, and 6 (Jensen, 1979a). These same statistics are plotted in three-dimensional feature space in Figure 30-11 to facilitate an appreciation of the spectral characteristics of these land uses (Jensen, 1979b). Evaluation of the parallelepipeds shown in Figure 30-11 provides insight as to why there may be difficulty when trying to classify certain land use classes at the urban fringe. First, it can be seen that the water is clearly separable from all other classes because of its high absorptance in the green, red, and near-infrared regions. Next, note that there is good separation between the natural vegetation and orchards primarily in band 5 (Figure 30-11A). Evidently, both reflect approximately equal amounts of green wavelength energy. This causes band 4 to contribute little to discriminating between the two classes. Likewise, both land covers reflect approximately equal amounts of near-infrared radiant energy (Figure 30-11E). Consequently, the natural vegetation and orchard classes in Goleta are separable from one another primarily due to the difference in chlorophyll absorption in band 5 and they are also separable from all other classes. This is not always the case. Other studies have reported confusion between agricultural land uses versus residential land use (Christenson and Lachowski, 1976; Williams and Borden, 1977).

Now the heart of the urban land use classification problem using Landsat multispectral scanner data must be addressed. First, note that single-family residential, multiple-family residential, and commercial complex/barren land uses all reflect approximately equal amounts of incident infrared radiant energy (Figure 30-11E). Information from this region alone would not allow adequate discrimination among these three classes. Also, there is substantial signature overlap in the near-infrared region between the urban and the vegetated surfaces (Figure 30-11E). However, reflected radiant energy in the green and red bands (4 and 5 respectively) allows discrimination between these three urban classes to a certain degree. Note in Figure 30-11A that the single-family residential category is almost totally separable from the multiple-family residential category using bands 4 and 5. Multiple-family residential's signature is very similar to that of commercial complex/barren in band 4, yet there is a substantial amount of separation in band 5. This stair-stepped increase in reflectance as one progresses from single-family to multiple-family residential to commercial complex/barren is assumed to be due to the percentage of vegetation present in each class. Basically, single-family residential land cover in Goleta, California, is relatively heavily vegetated and absorbs much of the green and red radiant energy while still reflecting much of the infrared energy. Multiple-family residential land-uses are usually apartment complexes with greater roof and parking lot areas, yet with some green vegetation. Thus, the integrated pixel value of the multiple-family residential class represents less absorption in the green and red (bands 4 and 5 respectively). Commercial complex/barren landscapes reflect much of the green, red and infrared energy resulting in higher values in all three bands. A maximum-likelihood classification of the scene (color Figure 30-12B) is in relatively good agreement with the land cover classes visible on an enlarged color infrared photograph of the study area (Figure 30-12A).

Additional level II and III urban categories would tend to cluster in the relatively tight feature space area just described. Consequently, the use of digital image-processing techniques to discriminate between additional urban classes based solely on spectral properties should be discouraged. This has caused numerous researchers to lump certain level II urban classes when working with Landsat data. For example, consider the classification of Seattle, Washington and Austin, Texas (color Figure 30-8). Here, only three classes of information are analyzed, i.e. residential, industrial/commercial, and water (General Electric, 1978). However, the adequate separation of residential, commercial/industrial, improved open

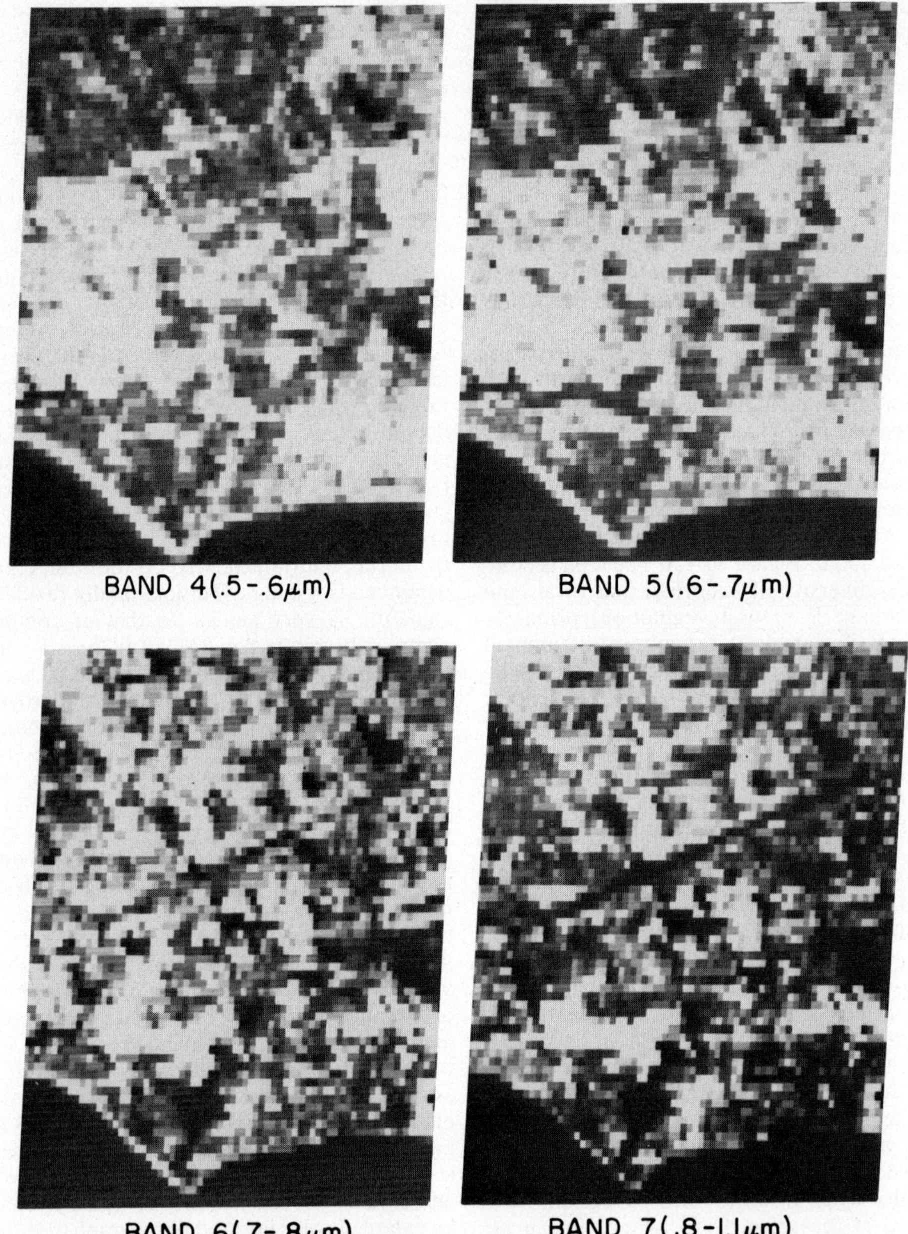

Fig. 30-9 Landsat images of Goleta, California, obtained on May 4, 1974. [Source: Jensen, 1978].

space, water, forested, and agricultural land has also been reported (Todd et al, 1978). Such analyses are useful for regional applications but often do not provide the specific information required for local urban decision-making.

Of interest is whether the signatures of urban phenomena remain relatively consistent for similar classes of land cover from one region to another. Very little research has been conducted on this topic. Also, additional research is required to identify how the following variables affect urban/suburban signatures: lot size; type of construction materials; spacing and orientation of structures; latitude (sun angle effects); and atmospherics. Such research will provide a better theoretical understanding of how urban land cover spectral signatures vary geographically.

URBAN LAND COVER ANALYSIS WITH THERMAL INFRARED IMAGERY

One of the early workers in remote sensing of the environment, Dana C. Parker (1933–1969), frequently began his lectures with the admonition

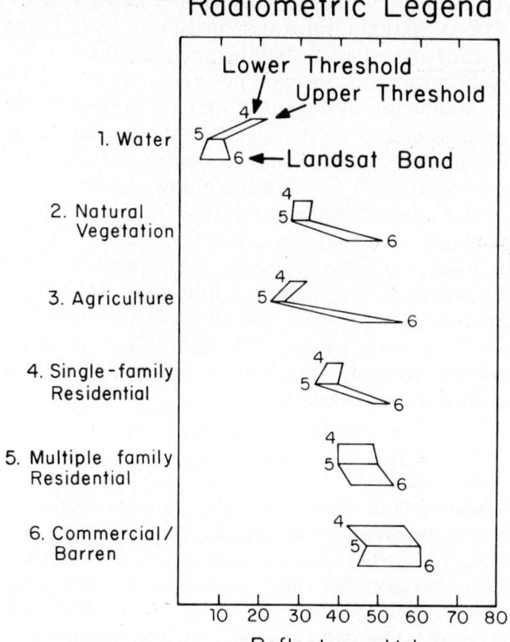

Fig. 30-10 Radiometric interpretive legend developed using training statistics obtained from the May 4, 1974, Landsat image of Goleta, California. Six classes of land use are investigated. [Source: Jensen, 1979]

to "think thermal." This advice is valuable when dealing with remote sensing of urban/suburban landscapes because much of modern day human activity is concerned with the production (space heating), elimination (refrigeration), control (cooking), and movement (steam pipes) of thermal energy. In each of these activities thermal energy is lost to the environment causing cities to become islands of concentrated heat.

The term "heat island" is frequently used by climatologists and meteorologists to refer to the concentration of urban heat in relation to the cooler surrounding rural landscape (Pease and Nichols, 1976). This heat island effect is illustrated by the patterns in Figure 30-13, a thermal infrared image of Ann Arbor, Michigan. The temperature gradient from the central business district (CBD) to the suburbs and rural areas is clearly evident from the brighter tones representing warmer temperatures to the darker toned cooler temperatures. The thermal diversity in the urban area is the result of man intermingling organic materials such as grass, trees, etc. with inorganic building materials. Thus, there exists great thermal variation in time and space especially in urban/suburban environments. These strong and varied thermal responses provide opportunities to study the urban scene by the use of heat-sensitive sensors. To appreciate how remote sensing is beneficial to urban problems it is necessary to consider fundamental environmental and system parameters which effect the spectral response of the thermal infrared detector. A detailed urban application of thermal infrared imagery is provided in the section on energy applications in this chapter. For more information on infrared radiation and the characteristics of sensor systems operating at infrared wavelengths the reader should consult Chapters 3 and 13 in Volume I of this *Manual*.

Spectral Considerations

The infrared portion of the electromagnetic spectrum may be divided into three parts: (1) the *near* or reflected infrared 0.7–1.1 μm, (2) the

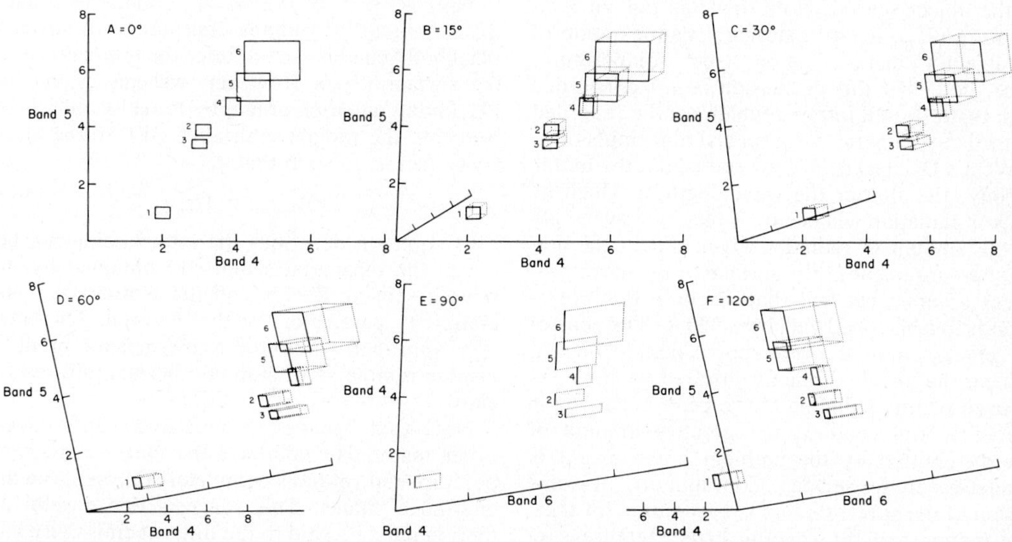

Fig. 30-11 Parallelepipeds of the training statistics shown in figure 30-10. Note the degree of overlap among the urban categories (4, 5, and 6) and the complete separation of vegetation (2, 3) and water (1). [Source: Jensen, 1979]

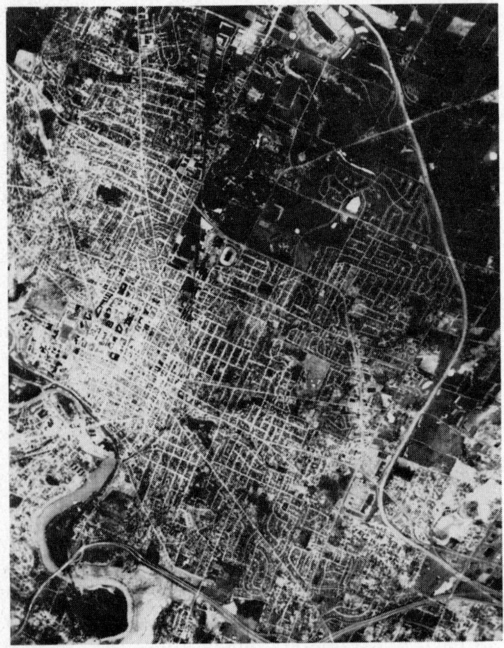

Fig. 30-13 Thermal infrared image of Ann Arbor, Michigan, obtained at night (8-14 μm). Note the thermal gradient from downtown to rural areas. [Courtesy, Daedelus Inc.]

middle infrared, in which both *reflected* and *emitted* infrared radiation moves through an atmospheric window at 3.5–5.5 μm, and (3) the *far* infrared where emitted radiation produced by the molecular activity of materials moves through a second atmospheric window at 8–14 μm. The emitted radiation in the middle and far infrared regions of the electromagnetic spectrum is imaged by sensors as "heat" or the *radiant* temperature of the object sensed. Note that this radiant temperature (T_{rad}) is an "external" manifestation of an object's kinetic (T_{kin}) or "true" temperature. Also, note that this discussion is *not* concerned with the reflected infrared which can be recorded by ordinary cameras using special film emulsions.

Wein's Displacement Law states that the hotter a body, the shorter the wavelength at which its peak of radiation will occur. A plot by wavelength of the amount of radiation given off by the ambient environment at the surface of the earth produces a somewhat bell shaped curve that corresponds to a blackbody at about 300 K. The peak of this energy curve occurs at approximately 9.6 μm, or near the middle of the far-infrared window. As a result of this radiation curve peak occurring in the 8–14 μm window, the greatest amount of energy emitted by the ambient environment is available here for sensing. Consequently, it is this region of the spectrum that is best suited for thermal mapping of the general terrain features, including urban and suburban landscapes, if the goal is a general view of the thermal environment.

If, however, the purpose of the thermal mapping is to gain information about hot, point sources of urban radiation such as steel mills, trash burning, or thermal power plants etc., then the middle infrared window may be more valuable because the power peak of the radiation for these hotter sources would shift to shorter wavelengths, or toward the 3.5–5.5 μm window (Lowe, 1978). In this middle-infrared window more energy is available to sense hot point sources and it can be used to aid in separating them thermally from the ambient background radiation. Therefore, depending on the kind of urban/suburban information required, an analyst might want to select thermal imagery obtained in either the middle- or far-infrared-window.

Emissivity Considerations

Materials in urban and suburban environments do not behave as blackbodies (i.e., they are not perfect absorbers and emitters of energy). Rather, the materials emit only a fraction of the energy emitted from a blackbody at the equivalent temperature. The ratio of the emitting radiant flux from a real material (F_r) to the radiant flux of blackbody (F_b) is the material's emissivity, ξ,

$$\xi = F_r/F_b \qquad (30\text{-}4)$$

Typical emissivity values for some common materials found in urban environments are shown in Table 30-1.

The emissivity of an object is important when measuring radiant temperatures in urban and suburban environments. In particular, recall from Chapter 2 that according to the Stephan-Boltzman Law ($W = \sigma T_{kin}^4$), the total radiant exitance from the surface of a blackbody (W) varies as the fourth power of absolute temperature (T_{kin}) when multiplied by the Stephan-Boltzmann constant σ ($5.6697 \times 10^{-8}\ W\ m^{-2}\ K^{-4}$). Thus, the remote measurement of radiant emittance, *W*, from a blackbody can be used to infer the temperature of the surface (T_{kin}). However, we can extend the blackbody radiation principles to real materials by reducing the radiant emittance (*W*) by the emissivity factor, ξ, such that

$$\mathrm{W} = \xi\ \sigma\ \mathrm{T}_{\mathrm{kin}}^4 \qquad (30\text{-}5)$$

This equation describes the interrelationship between the measured signal (*W*) obtained by the remote sensing device and the temperature and emissivity parameters of the material. This indirect approach to temperature measurement is used in remote sensing in the thermal infrared region.

Note that because of emissivity differences, urban materials could have the same kinetic temperature and yet have completely different radiant emittance values. This suggests that careful attention must be paid to the differnt emissivity values for phenomena in urban and suburban environments if valid conclusions are to be drawn

TABLE 30-1

Typical Emissivity Values for Selected Materials

Material	Buettner and Kern (1965)	Lillisand and Kiefer (1979)	Underwood, Houston and Hazard (1980)
Soil (wet)		.95	
Soil (dry)		.92	
Ice		.96	
Snow		.85	
Window glass		.94	.94
Paint (average 16 colors)		.94	
Wood and brick			.95
Stucco			.93
Buffed stainless steel		.16	
Quartz sand	.91		
Granite, rough	.90		
Granite	.82		
Durite	.86		
Obsidian	.86		
Feldspar	.87		
Cilica sandstone	.91		
Dolomite, polished	.93		
Dolomite, rough	.96		
Basalt, rough	.93		
Concrete	.97		
Asphalt	.96		
Water	.99		
Water with thin petroleum film	.97		

from the interpreted remote sensor data. For example, the emissivity of roof surfaces must be known to determine their kinetic temperature. Unpainted sheet-metal roofs have a very low emissivity (.16) while other roof materials generally have an emissivity from .88 to 0.94. Kinetic temperatures are usually underestimated if emissivity effects are not incorporated into the analysis.

The radiant temperature output (T_{rad}) from a remote sensor system may be related to the kinetic or true temperature of an object (T_{kin}) by the lationship:

$$T_{rad} = \xi^{1/4} T_{kin} \tag{30-6}$$

Thus, for any given object the radiant temperature recorded by a remote sensing device will be less than the kinetic temperature of the object.

Thermal Conductivity, Capacity, and Inertia Considerations

Temperature extremes and rates of temperature variation for materials are influenced by the thermal conductivity, capacity, and inertia of materials. *Thermal conductivity, K,* is the measure of the rate at which heat passes through a material (expressed as cal cm^{-1} sec^{-1} $°C^{-1}$). *Thermal capacity, c,* is the ability of a material to store heat (expressed in cal g^{-1} $°C^{-1}$). *Thermal inertia, P,* is a measure of the thermal response of a material to temperature changes (expressed as cal cm^{-2} $sec^{-1/2}$ $°C^{-1}$). It is calculated from the relationship

$$P = (K\ p\ c)^{1/2}. \tag{30-7}$$

Thermal inertia increases with an increase in conductivity, capacity and density, *p,* with the density of the material having the greatest impact. (Sabins, 1978)

So what is the significance of these parameters for urban and suburban applications of thermal infrared imagery? These parameters, and especially thermal inertia, provide insight into why certain cultural and natural features appear as they do on thermal infrared imagery. Basically, materials with higher thermal inertia have more uniform surface temperatures, day and night, than materials with lower thermal inertia. For example, the thermal inertia of water is somewhat greater than that of soils and rocks. Consequently during the day, water bodies have a cooler surface temperature than soils and rocks. At night the surface temperatures are reversed (i.e. a thermal crossover takes place) with soils and rocks becoming cooler than water. The reason is that convection currents maintain a relatively uniform temperature at the surface of a water body. Convection does not operate to transfer heat in soils and rocks. Therefore, heat from solar flux is concentrated near the surface of these solids during the day causing a higher surface temperature. At night the heat is radiated to the atmosphere and is not replenished by convection currents in these solid materials, causing surface temperatures to be lower than in adjacent water bodies.

Trees are usually present in residential areas and generally appear cooler (darker) than their surroundings on daytime images and warmer (lighter) on nighttime imagery (Figure 30-14). This

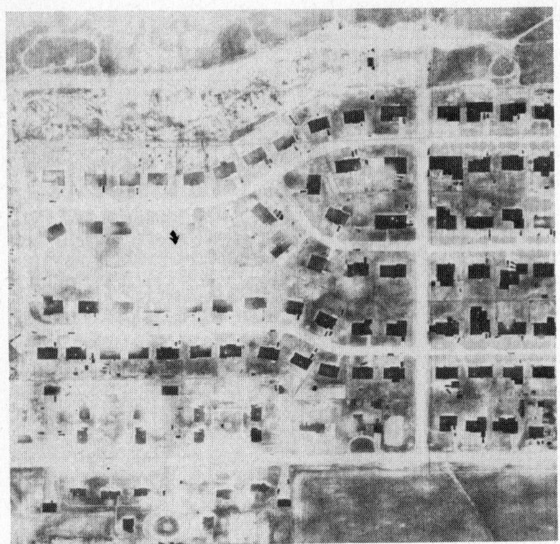

Fig. 30-14 The arrow identifies a concentration of trees within the residential area which are warmer than surrounding areas in this nighttime thermal infrared image (8-14 μm). [Courtesy, Texas Instruments]

Fig. 30-15 Roads and driveways are recorded in relatively bright tones on this nighttime thermal infrared image (8-14 μm). Also, note the range of temperatures associated with different rooftops. [Courtesy, Texas Instruments]

is because green deciduous vegetation transpires water vapor during the day lowering the leaf temperature. At night there is a cross-over as the trees have a relatively high thermal inertia compared to surrounding dry soil which rapidly radiates its heat to the atmosphere. Paved roads and parking lots appear relatively warm (light) both day and night (Figures 30-14 and 30-15) since the pavement surfaces heat up to temperatures higher than the surroundings during the daytime and tend to lose heat relatively slowly at night. Thus, they have a relatively high thermal inertia which may result in brighter returns on both day and night imagery. Damp ground, such as watered grass, may appear cooler than dry ground, both day and night, owing in part to the cooling effect as water is evaporated from the surface of the vegetation. As rooftop temperatures and resultant signatures (Figure 30-15) are one of the most important urban applications of thermal infrared imagery this topic will be discussed separately in the energy-application section.

From this brief discussion it is obvious that accurate interpretation and application of thermal infrared imagery to urban problems is dependent on a working knowledge of a material's thermal characteristics including its general diurnal temperature variation.

Geometric Considerations

When thermal infrared imagery is being used for urban applications it is important to remember that scanner distortions exist which are more complex than those inherent in aerial photography. As a consequence thermal infrared imagery is rarely used as a precision mapping tool unless systematic rectification is performed. Even then it may not be possible to eliminate one-dimensional relief displacement.

Basically, unless it is geometrically rectified, thermal infrared imagery may exhibit severe scale distortion in the direction perpendicular to the flight line. A good example is shown in Figure 30-16 and Figure 30-17 where a portion of New York City is recorded on panchromatic photography and thermal infrared imagery.

Evident in Figure 30-17 is one-dimensional relief displacement. Instead of displacement being radial from the nadir point as in a vertical photograph (Figure 30-16), it is perpendicular to the flight line. In some instances this displacement is useful in urban environments as it affords a side view of buildings and other tall features. In other cases, tall structures may obscure the view of important phenomena, just as on aerial photographs.

In recent years it has become possible to rectify some thermal infrared imagery to an orthogonal projection based on principles of differential rectification (Otepka, 1978). For example, consider the thermal infrared images of Vienna, Austria, rectified to be congruent with existing large scale maps (Figure 30-18). In this case the imagery may be compared with other collateral information in map format, thus increasing its utility. Additional insight into the use of thermal infrared imagery for urban problems will be provided in the energy-application section of this chapter.

URBAN LAND COVER ANALYSIS WITH RADAR IMAGERY

The all-weather capability of radar sensors initially prompted much interest in their possible application for urban analysis. As research has continued, however, the relationship between sensor system and environmental parameters has proven to be a formidable research area especially in

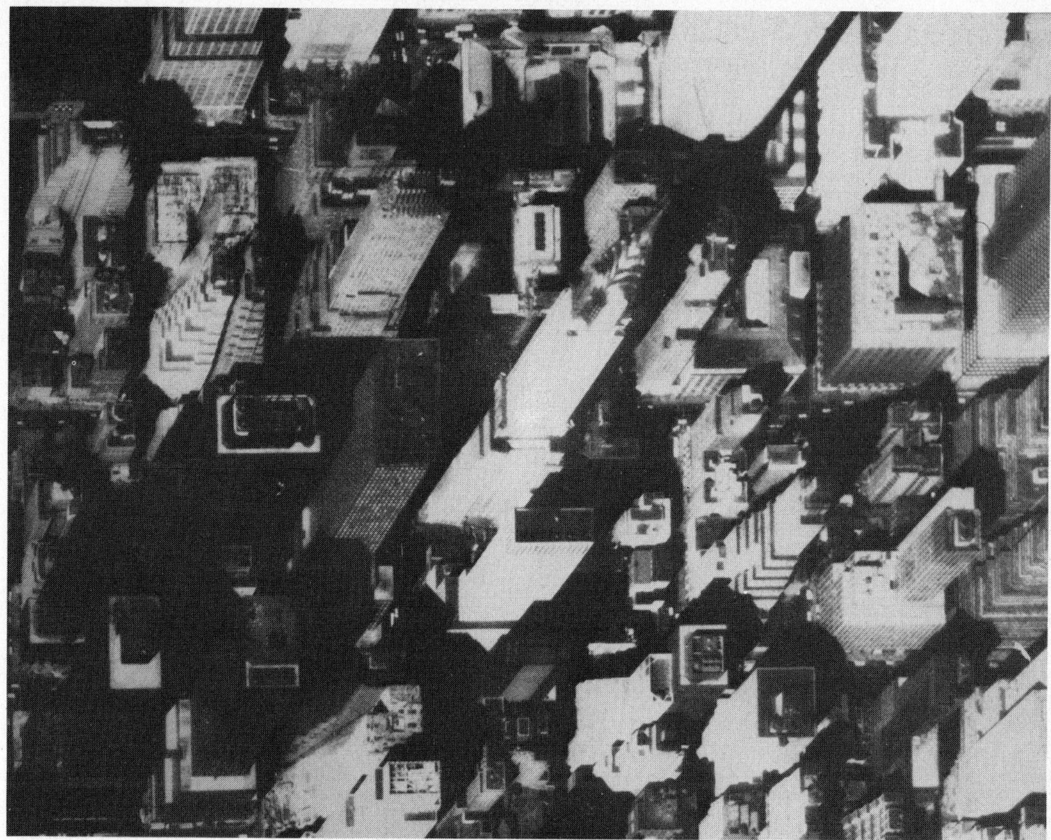

Fig. 30-16 Panchromatic aerial photography of downtown New York City. Note the radial displacement and existence of shadows. [Source: Texas Instruments]

urban and suburban environments. A major stumbling block has been the difficulty of acquiring radar imagery to evaluate its potential for urban applications. For many years K, Ka, and Ku band (0.75 to 2.4 cm), dual-polarized, real-aperture airborne radar-devices were the only radar systems available to the civilian scientific community. Later, X-band (2.4 to 3.75 cm) and L-band (15 to 30 cm) synthetic aperture airborne systems were developed and used in examining various parameters of the urban environment. In 1978, the launch of a single polarized (HH), L-band (25 cm) synthetic aperture radar (SAR) aboard the Seasat satellite provided imagery which was digitally and optically processed to investigate the potential contribution of space imaging radar systems for urban applications. Additional SAR missions, dubbed by NASA as Space Imaging Radar (SIR) systems, are planned for Space Shuttle sorties in the 1980's (Chapters 13 and 25).

Fundamental System and Environmental Parameters

Aerial photography and Landsat imagery are useless when urban and suburban areas are shrouded in cloud cover for extended periods of time. Microwave sensors, however, if shown to be appropriate to the given application, may be used to image through such cover, providing the potential for nearly all-weather monitoring. This can be especially important in emergency conditions, such as during floods or hurricanes, when it is necessary to obtain an accurate picture of the urban devastation in order to make thoughtful decisions (Matthews, 1975). Unlike other sensor systems, radar does not record in any simple way the spectral nature of urban and suburban features. Rather, signatures eventually recorded on film are the result of complex interactions between terrain features and actively produced microwave energy. Specifically, the image signatures are dependent on the interrelationships among wavelength, polarization, and depression angle of the radar beam and by the dielectric constant and surface roughness of the terrain features. For a complete discussion of these parameters see Chapter 9.

In order to discriminate one urban or suburban class or feature from any other there must be a significant difference between them in either surface roughness or dielectric constant. Generally, rougher surfaces relative to the size of the incident

Fig. 30-17 Nighttime thermal infrared image of downtown New York City (8-14 μm) in early 1980. Note the one-dimensional relief displacement perpendicular to the flight direction. Compare with figure 19 to identify thermal anomalies on building facades and rooftops. [Source: Texas Instruments]

wavelength energy will produce stronger radar returns (backscattering) and be recorded as brighter tones on the imagery. In addition to the surface roughness parameter, which is usually measured in centimeters, bright radar returns are also produced by metallic targets and by *corner reflectors* that are formed by three planar surfaces intersecting at right angles. Regardless of the incident angle at which a radar wave enters the cavity of a corner reflector, it is reflected directly back toward the antenna. Man-made structures such as buildings and bridges often function like corner reflectors.

It is generally hypothesized that a relationship exists among the following:

1) the amount of vertical relief (a surrogate for surface roughness) of materials,
2) the wavelength of the incident microwave energy, and
3) the depression angle which may govern the feasibility of discriminating between certain phenomena.

While this relationship has been documented for certain geologic rock units (Schaber et al, 1976), little research has been conducted to document the radar return produced by urban features that may be meters or hundreds of meters in height. Nevertheless, some very general comments can be made concerning the probable signature of urban and suburban features if we keep in mind the fact that the geologic work was based on the average surface roughness of small-scale vegetation, cobble, and sand particles etc., whereas the urban land cover may incorporate *both* a small-scale surface roughness and corner reflector component. Table 30-2 lists some general vertical dimensions for urban and suburban features.

Through use of the modified *Rayleigh criterion*, established by Peake and Oliver (1971) and employed by Schaber et al (1976), it is possible to predict the tonal response of the radar return of certain phenomena if the vertical relief of the object (h), the antenna depression angle (γ) and wavelength of incident microwave energy (λ) are known.

The basic equations are:

$$h_{smooth} < \frac{\lambda}{25 \sin \gamma} \tag{30-8}$$

$$h_{rough} > \frac{\lambda}{4.4 \sin \gamma} \tag{30-9}$$

Fig. 30-18 Rectified thermal infrared imagery of Vienna, Austria (8-14 μm), with planimetric detail superimposed. Note the bright return from roads and parks in this nighttime imagery. [Source: Otepka, 1978]

where h_{smooth} and h_{rough} establish the limiting boundaries in centimeters for smooth and rough surfaces.

Thus, if X-band (λ = 3 cm) radar imagery is obtained, as for an urban area west of Ottawa, Ontario, Canada (Figure 30-19) at a 25 degree depression angle in the center of the range (γ = 25), then the expected limiting values are as shown in Table 30-3. Evaluation of predicted and observed signatures for the materials discussed in Table 30-2 are presented in Table 30-3B. Basically, the predicted signatures are in reasonable agreement with the observed signatures present in the Ottawa scene. Water and roads image in dark tones, as predicted, because they reflect very little energy back toward the sensor. Grass and other vegetation were recorded in intermediate to bright tones. Since vegetation with a high moisture content (i.e. a higher dielectric constant) tends to reflect rather than transmit incident energy, the return may be brighter for such surfaces as is the case for numerous agricultural fields shown in Figure 30-19. The areas of residential and commercial land use produced bright returns as they exhibited substantial local relief and often acted as corner reflectors.

The aforementioned procedure was based on techniques used for assessing the surface roughness of small-scale features with heights measured in centimeters. Therefore, water, paved roads and grass are portrayed correctly as they are small-scale surface roughness features. However, the

TABLE 30-2

Average height of typical features found in urban and suburban environments

Feature	Height	
Water	.25 cm	(.1″)
Pavement or level bare soil	.50 cm	(.2″)
Grass	10 cm	(4″)
Single story buildings and trees (residential)	3.65 m	(12′)
Multiple story buildings (commercial)	9.14 m+	(30′+)

extrapolation to residential and commercial environments with heights measured in meters is not fully established especially in regard to the degree to which buildings act as corner reflectors. Precise models have yet to be developed for these macro-scale phenomena in urban environments.

The previous discussion was primarily concerned with the interrelationship of wavelength, depression angle, and surface roughness characteristics. It is important to evaluate another parameter which influences the radar return in urban environments, i.e. azimuth. Several investigators have noted the effects of radar azimuth (or look direction) and object orientation on signal response (Moore, 1969; Lewis et al, 1969; Waite et al, 1978; Bryan, 1979; and Henderson, 1980a). Specifically, when large flat, rectangular surfaces such as building facades are oriented perpendicular (orthogonal) to the impinging signal, the structure acts like a corner reflector enhancing the like-polarized return and generating a bright, specular response. Thus, it would be expected that a radar image of a rectilinear street grid obtained with a look direction at either points 1 or 2 in Figure 30-20 would produce a bright return. Conversely, relatively lower returns would be expected from an antenna located at position 3 or 4. This cardinal effect is especially pronounced in the CBD, along linear commercial concentrations and arterials, and some orthogonally oriented residential areas. Consequently, the same type of land cover may look different in various parts of the radar imagery solely as a function of its orientation to the flightline.

A good example of this phenomenon is provided in the work by Bryan (1979). In this study, L-band SAR data of the Los Angeles metropolitan area was obtained at 20 m resolution and HH polarization (Figure 30-21). Analysis of the flight-line mosaic revealed dissimilar radar returns from similar land use classes. Figure 30-22 depicts four such study areas with associated street maps keyed to Figure 30-21. Table 30-4 defines the radar azimuth for each study site. A detailed discussion of each study area is available in Bryan (1979). The lack of correlation between the actual land use (commercial and residential units) and the radar image tone is apparent. These urban data suggest that the angles at which the radar energy strikes a reflecting surface is a major factor governing the amount of the backscatter (brightness) in the radar image. What is interesting is the apparent consistency of the relationship between the radar azimuth angle, the street pattern, and the change in overall grey tone. Figure 30-23 is a graph of the grey level and θ (defined as the angle between the radar azimuth and street orientation as referenced to true north). For areas where θ was less than 10° the radar image is bright (116 to 193; average = 152). In areas where θ was greater than 10° the image brightness was considerably

Fig. 30-19 "Kanata" urban area west of Ottawa, Canada, imaged using an X-band (HH) radar with a swath width of 6 km and depression angles of 30° (near range) and 20° (far range). Resolution was approximately 3 m in original. [Source: INTERTECH of Ottawa, Ontario, Canada]

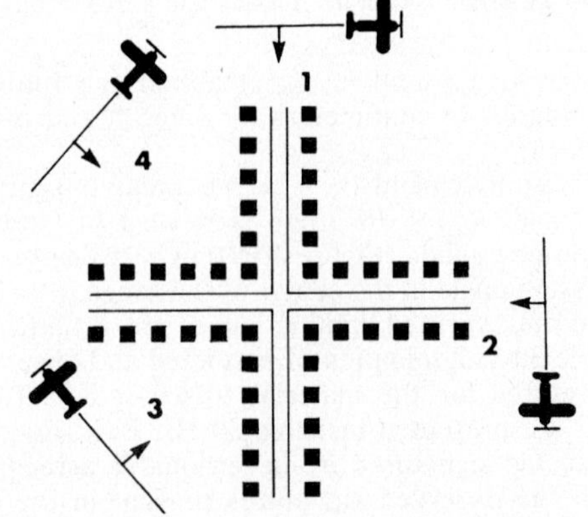

Fig. 30-20 Position of buildings in a rectilinear street grid system and potential radar azimuth direction. [Source: Bryan]

TABLE 30-3

A) Limiting Values of Vertical Relief (h) for Surface Roughness Categories with a Depression Angle (γ) of 25 Degrees

Roughness Category	X-band (3 cm)	Predicted Signature
Smooth	h < .28 cm	Dark
Intermediate	h = 0.28 to 1.62 cm	Intermediate
Rough	h > 1.62 cm	Bright

B) X-Band Response of Different Values of Vertical Relief with a Depression Angle (γ) of 25 Degrees

Material	h (cm)	Observed Signature	Predicted Signature
Water	.25	Dark	Dark
Pavement or bare soil	.50	Dark	Dark
Grass	10	Intermediate to bright	Bright
Single story and trees	3.65 m	Bright	Bright
Multiple story	9.14 m	Bright	Bright

Fig. 30-21 Mosaic of the Los Angeles, California, metropolitan area obtained using the NASA L-band (1,215 Ghz) radar operated by the Jet Propulsion Laboratory. [Source: Bryan]

lower (93 to 128; average = 115). The critical value for θ, at which brightness changed, was approximately 10° to 15°. This critical value of θ was independent of the land use (with respect to the commercial and residential categories) examined in this study.

Thus, for the Los Angeles area, the angle at which the change in backscatter was most pronounced was in the neighborhood of 10° to 15°. Radar range (as governed in part by the depression angle or incidence angle) was apparently of minor importance in controlling the backscatter from flat urban areas having numerous walls. Therefore, an awareness of both the linearities and the orientation of the cultural features is critical for an accurate interpretation of radar urban data from either aircraft or spacecraft.

With respect to spacecraft data, this need may be more critical than for aircraft data because in the latter case it is possible to change the vehicle flight track to compensate for the orientation of cultural features which are of particular importance. For example, consider the Seasat L-band image (HH) of the same study area (Figure 30-24). It exhibits almost opposite tonal responses from the Burbank and San Gabriel study areas. This is a function of the Seasat antenna azimuth angle being fixed. It therefore selectively enhanced those cultural features oriented orthogonal to its look direction.

In addition to considering radar azimuth (or look direction), depression angle, wavelength, and polarization to be system parameters, we should not overlook environmental parameters which influence the return of urban and suburban environments. The interpretability of settlements as influenced by environmental factors was ad-

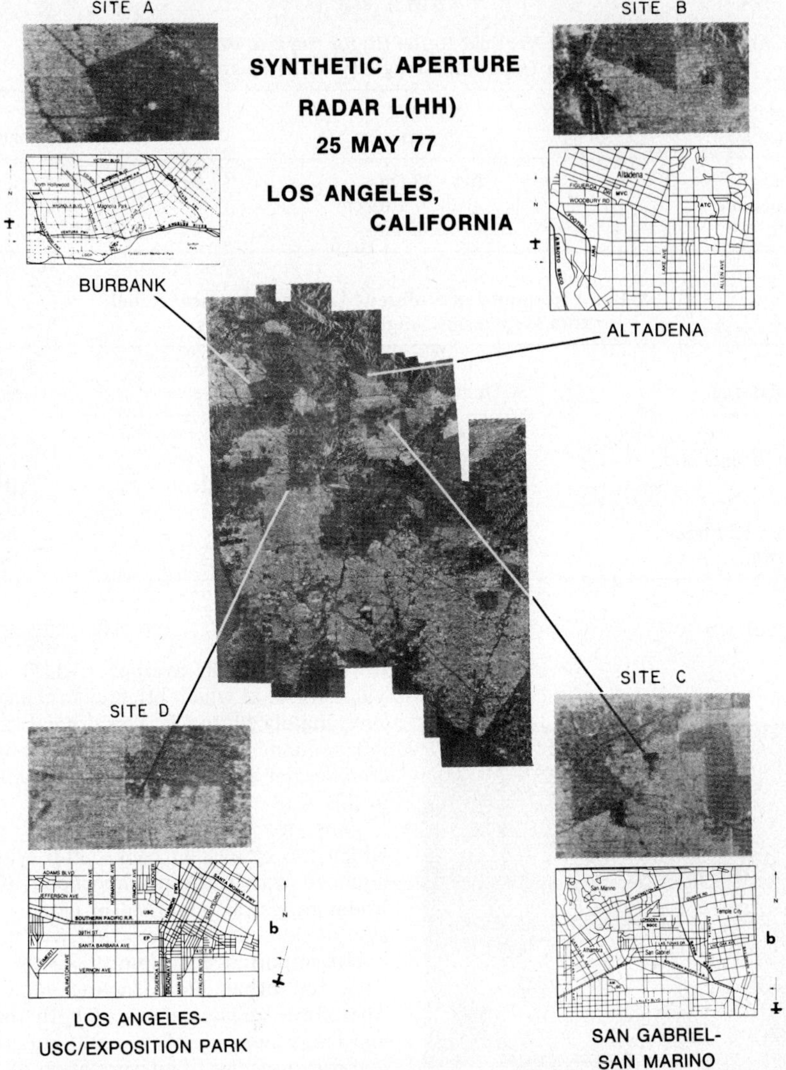

Fig. 30-22 These four study areas in the Los Angeles L-band data set demonstrate the effect of radar azimuth angle on cultural data. [Source: Bryan]

dressed by Henderson (1979) and Henderson and Anuta (1979) using X- and K-band imagery of the Midwest, Western, and the Northeastern United States (Figures 30-25 to 30-27). Henderson found that the settlement pattern and vegetation canopy of the Northeast were significant factors in *reducing* urban area visibility on radar imagery. Although local conditions certainly influenced a small urban area's detectability, Henderson and Anuta (1979) reported that, on a regional basis, land cover and landform conditions did not significantly affect visibility but that population size, vegetation, and settlement location in the range direction did.

As scientists continue to analyze urban areas with various radar and other remote sensing systems, the unique but complementary attributes of each sensor type become more apparent. The environment and system-related parameters that affect signal response are being viewed as something more than simply a radar anomaly. With the advent of digitally processed Landsat and SAR imagery, the texture component available from radar is receiving increased attention as an added feature element in merged data sets (Wu, 1979). In addition, the influence of environmental changes on these parameters is also receiving attention along with the effects of image scale and time of year (NASA, 1980).

Settlement Detection and Infrastructure Delineation

With the system and environmental parameters affecting the radar return briefly addressed, it is useful to review the development of settlement

TABLE 30-4

Locations of Study Sites, Keyed to Figure 30-22. All Measurements are Referenced to True North (Source: Bryan 1979)

Site	Run	Aircraft Track	Radar Azimuth Angle	Location
A	12	000°	090°	Burbank
B	8	000°	090°	Altadena
C	7	179°	269°	San Marino & San Gabriel
D	11	188°	278°	Los Angeles

detection and infrastructure delineation using radar. Among the first urban radar-studies was that of Simpson (1969) who examined residential location, size, and shape. In this study built-up areas in New England were delineated by several interpreters. One hundred percent of all cities with over 7,000 people, 80 percent of all settlements over 800 population, and 40 percent of settlements between 150–800 population could be identified. Sabol (1969), working with nineteen urban areas across the United States, concluded that there was an obvious relationship between the areal extent of urban population centers and their population. Henderson and Anuta (1979) examined K- and X-band imagery of various scales over fifteen study areas in the United States to explore system and environmental factors influencing settlement detectability. Results indicated that settlements having a population of greater than 1,000 could be detected with greater than 90 percent accuracy at scales of 1:400,000 or larger but that smaller settlements could not be consistently identified even at scales of 1:200,000.

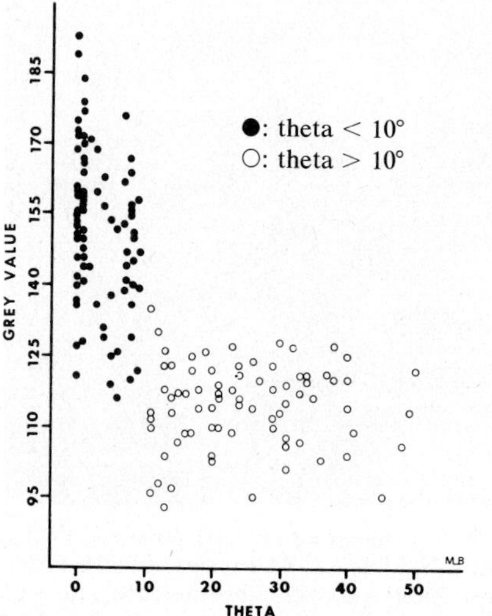

Fig. 30-23 Plot of grey value of SAR image versus theta, the angle between the radar azimuth and the street orientation.

Fig. 30-24 HH polarized digitally processed L-band Seasat SAR image of the Los Angeles metropolitan area obtained on 21 July 1978. Port facilities, airports, major roadways, commercial activity, and extent of urban build-up are quite apparent. However, the bright rectangle in the upper right center of the image (Burbank) is a function of house orientation to the look direction and not a distinct residential land cover. At L-band wavelengths wave patterns in the ocean appear as grey tone variations. The effect of radar foreshortening is also apparent in the appearance of the mountains. [Source: Bryan]

Although a settlement's location in the range direction did affect detectability, the effect was not predictable. Of the environmental characteristics examined, only vegetation was discovered to have a significant effect on detectability.

Other workers have directed their attention to the visibility and appearance of specific components of the urban infrastructure on radar imagery. Moore (1969) inspected HH-HV polarized K-band imagery of the Chicago area for transportation network and major land use analysis. He found that major linear transportation elements could be defined and that the HV mode proved most useful. While gross patterns of industrial open space and residential land use could be mapped, Moore was unable to identify commercial land use. However, he stressed the necessity of further work and continued system design. Lewis et al (1969) employed sixty-eight interpreters and multiple-polarized K-band imagery of several geographic areas in evaluating the detectability of linear cultural features (e.g. railroads, powerlines). Among their conclusions was that the high return from such features was dependent upon the polarization configuration of the radar system and the orientation of the feature relative to the flight path. HH polarization was found to be

Fig. 30-25 Westinghouse K_a-band, HH polarized, real aperture radar image of Syracuse, New York, obtained in July, 1966. High signal return is evident from the central business district, commercial arterials, and industrial concentrations around the lake. The airport runways are black in contrast to the surrounding grass and bright return from the terminal area. Vegetation masks much of the residential area but is still detectable by the geometric street pattern. [Source: Henderson]

more effective when the linear features were parallel to the flight path but otherwise the HV polarization was preferable.

Accuracies with which urban land cover could be detected were calculated by Bryan (1974) for study areas in Detroit using multi-frequency, multi-polarized (X and L band, HH and HV polarization) SAR imagery. Two-hundred random points were selected and the associated land cover noted. Residential areas were correctly identified 72 percent of the time with other features averaging above 55 percent. Although institutions and parks were frequently confused with open space and adjacent residential areas, Bryan concluded that, overall, radar did demonstrate potential for providing morphological and spatial data of urban areas. In a subsequent effort Bryan (1975) asked 685 students to identify but not delimit urban land cover types in southern Florida using multichannel radar imagery. The purpose was to define

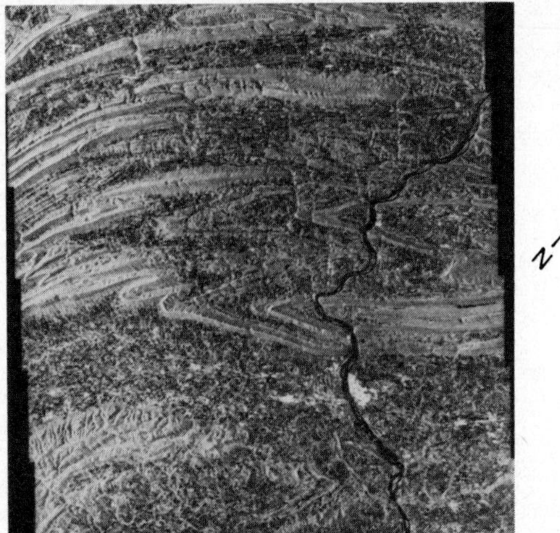

Fig. 30-26 HH polarized digitally processed L-band Seasat SAR image of Harrisburg, Pennsylvania. Although much of the transportation net is not visible at this small scale the commercial-industrial sections of the urban area are evident by their bright return. Vegetation again masks much of the residential area as it did in Figure 30-25. On the original image, bridges, oil tank farms and several small settlements could be identified. [Source: Henderson]

the degree of interpreter training requisite for urban radar-studies. Residential areas, linear transportation features, water, and some commercial activities were easily identified and all ground scenes could be differentiated at the 5% level of significance. Bryan concluded that for some, but not all, urban scenes little image in-

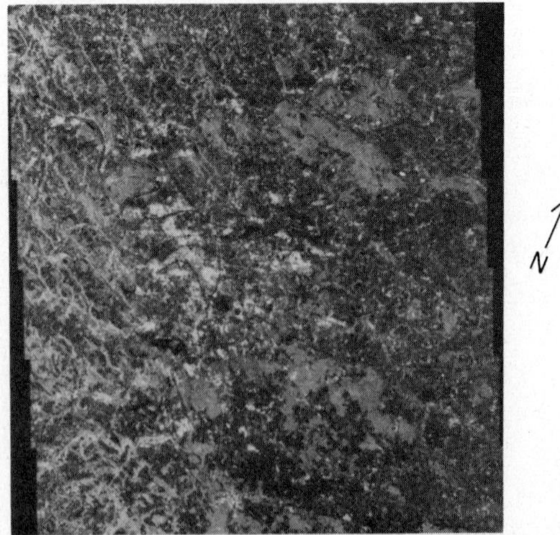

Fig. 30-27 Comparing this HH polarized digitally processed L-band Seasat SAR image of Paris, France, with Figures 30-21 and 30-22 illustrates the effect changing environment and settlement pattern have on land cover identifiability. That familiarity with the area would definitely be an asset in extracting urban data is readily apparent. [Source: Henderson]

terpretation experience was necessary, but that further analysis of other regions and classification schemes was still required.

Henderson et al (1980) employed digitally processed Seasat SAR L-band imagery of Denver, Colorado to evaluate the role of space platform radars for mapping urban land-cover. They observed that in this environment such data could be used to delimit the rural-urban fringe as well as single-family residential areas, open space, major commercial-services, and commercial-industrial concentrations but that only segments of the transportation network could be discerned.

To evaluate the types of urban data interpretable at different scales, Henderson (1980b) employed digital enlargements of a Seasat SAR scene of Denver, Colorado (Figure 30-28). At a scale of 1:41,000 urban land cover classification ranged from 77% accuracy in the inner city commercial zones to almost 94% in new residential areas and urban fringe growth areas. The most identifiable land-cover categories were residential, industrial-commercial, open space, and water. Through use of the same categories an 87.9% accuracy was attained for the imagery generated at a scale of 1:131,000 (Figure 30-29A-D).

Land Cover Category Overview

The potential contribution of radar imagery for urban area analysis has been examined by researchers from diverse perspectives. To place these results in a more uniform light and to provide an indication of present capabilities it is useful to summarize this work using Level II categories.

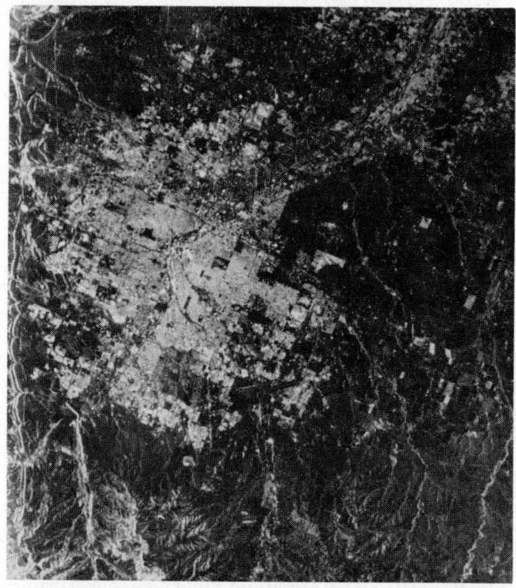

Fig. 30-28 Digitally processed Seasat L-band SAR images of Denver, Colorado. The influence of environment on urban land cover visibility is apparent on this Seasat SAR image compared with those of Figures 30-25 and 30-26. At this scale the limits of Denver's urban growth can be delimited rather easily. [Source: Henderson]

Residential

Single-family residential housing can be delimited on radar imagery although older residential areas in the inner city may be less distinct owing to vegetation cover than newer housing in the urban fringe and suburban areas. Similar residential areas may also appear dissimilar as a function of orientation to the flightline (Figure 30-24). Multi-family housing is identifiable but less precise due to the inherent diversity of shapes, sizes, arrangements, configurations, and spatial locations.

Commercial and Services

The Central Business District and commercial core of an urban area can be located, but the visibility of smaller business districts and shopping centers in suburban areas is a function of their orientation to flightline, size, and surrounding land cover. Commercial strip developments, in general, can be detected by the bright, linear signal response. Institutional, religious, health, and educational components of this category cannot be precisely identified with current radar resolutions (on the order of 10–15 meters). In such instances, the buildings and grounds comprising such activity generally appear on the imagery as dark grey, low return areas. Consequently, these land cover types can be grouped into Commercial and Services or Open Space categories.

Industrial

Heavy industrial concentrations are discernable as bright, high return areas adjacent to major transportation arteries, but light industry and manufacturing activities are often confused with commercial and service functions in the inner city area.

Transportation, Communications, and Utilities

Major roadways, although not continuously visible, can be inferred on radar imagery as can many smaller residential streets, but the visibility of the latter is particularly a function of look direction. Generally, railroads also have a higher return than roads. Railroad assembly yards and roundhouses can be observed as can airports, docks, and harbor facilities. However, the brightness and visibility of bridges and railroads is a function of polarization and orientation to the flightpath. Communications and utilities are not consistently visible, but since transmission towers and buoys often act as corner reflectors such land cover frequently can be inferred. Cross-country transmission and utility routes can also be detected if their right-of-way provides sufficient

a. SOUTHWEST QUADRANT OF DENVER, COLORADO

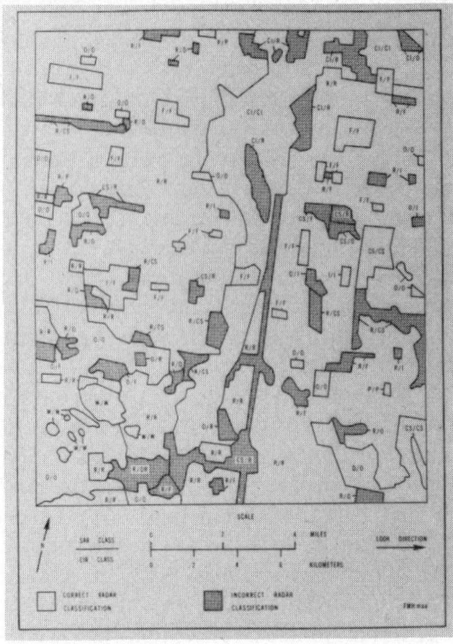

b. CLASSIFICATION EVALUATION

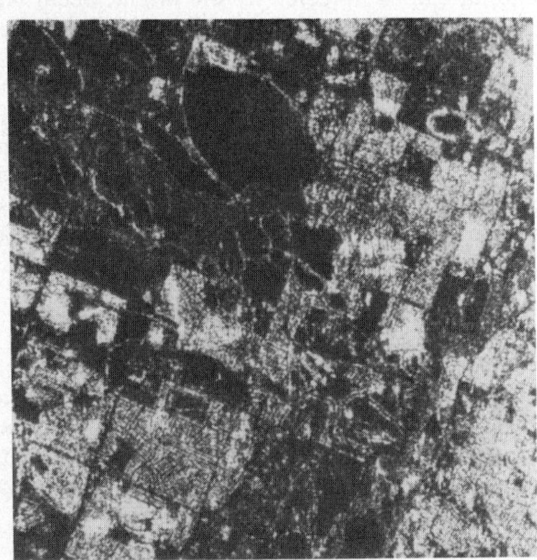

c. ENLARGEMENT OF SOUTHWEST PORTION OF "a" ABOVE

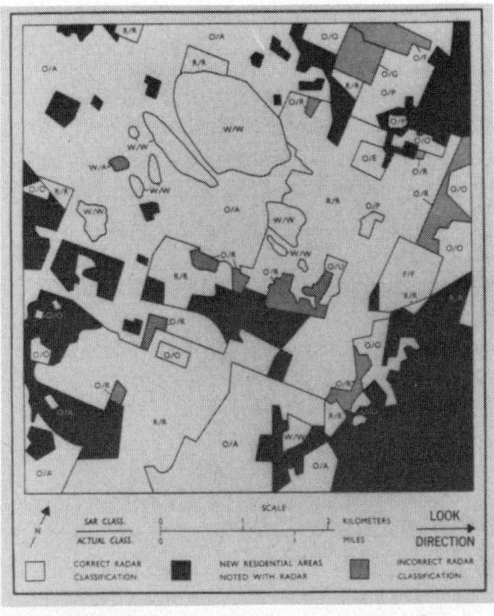

d. CLASSIFICATION EVALUATION

Fig. 30-29 A) Southwest quadrangle of Denver. The central business district; commercial industrial core; older, interior residential areas, and newer suburban growth on the fringe are visible. B) A land cover map derived from A. At this scale almost 88% accuracy was achieved in a complex mix of land cover types. C) The lower left portion of A was enlarged to produce the area observed in C and mapped in D. In addition to locating new residential development, the interpretation accuracy was over 93%. (A) = agricultural; (F) = recreational; (G) = cemetery; (O) = open; (C) = commercial; (E) = extractive; (I/CI) = commercial-industrial; (P) = public; (R) = residential; (T) = transportation; (U) = utilities; (W) = water; CC5 = commercial-services. [Source: Henderson]

contrast to surrounding land cover (e.g. open, linear areas through forested terrain).

Industrial and Commercial Complexes

Industrial parks and warehousing activities on the urban fringe can be detected but may be confused with commercial and services and multi-family housing.

Mixed Urban or Built-up Land

By definition this category is one within which individual land uses cannot be resolved into transportation, commercial, industrial, or residential activities. Although this category can be delimited at times on radar imagery its tone/texture response may be similar to that of other land cover categories in other parts of the city.

Other Urban or Built-up Land

The golf courses, zoos, urban parks, cemeteries, and undeveloped land comprising this category are not separable on radar imagery but generally appear as dark open-space areas.

In summary, the urban land cover types most identifiable with current radar systems are: residential (including older and newer housing); major transportation networks; commercial core and large strip developments; and commercial-industrial concentrations.

Finally, it should be remembered that the appearance and identification of urban land cover types is a function of environment and radar system interaction. Radar imagery does possess potential for urban land cover analysis, but its precise tonal and textural attributes are not yet known. The vegetation canopy may mask underlying activity; similar activity may evidence different tonal and textural response due solely to its orientation relative to the flight path, look angle, polarization and/or wavelength. Land cover that is readily observable in one environment or region may or may not be observable in another environment containing a different settlement pattern, economic activity mix, and morphology. These factors must be defined if the potential of radar imagery for urban applications, as a single sensor and as a component to merged remote sensing data sets, is to be realized.

THE SPATIAL NATURE OF URBAN LAND USE

Urban land use, consisting of buildings, transportation networks, businesses, parks and a variety of mixed uses, represents high frequency detail. Thus, remotely sensed data with high spatial resolution may be required for urban studies. In order to determine the spatial resolution necessary for specific types of urban studies, it is appropriate to examine the sizes, densities, and contrasts of features typically encountered in urban environments around the world, (Welch, 1982).

The relative size of urban built-up areas for cities of 50,000 to 500,000 population varies according to geographic region (Table 30-5; Tobler, 1969; Lo and Welch, 1977). In comparison to the United States, the urban structures of other countries are smaller and/or more closely spaced to accommodate equivalent populations in the reduced land areas (Table 30-6). Thus, detailed studies of the urban land use of foreign areas are likely to require image data of much higher spatial resolution than is necessary for comparable areas in the United States or Canada.

TABLE 30-5

Relative Sizes of Urban Built-Up Areas 50,000 to 500,000 Population

Country	Relative Size
United States	1.0
Canada	.60
Sweden	.15
Japan	.11
China	.03

Contrast between urban land use/cover classes is also a factor which determines the spatial resolution requirements for detailed urban area analysis. Typically, urban areas in the United States include parks; transportation routes bordered by trees or grass; ponds or streams; and buildings constructed from wood, metal or brick materials which contrast significantly with lawns or main thoroughfares. However, in less developed countries, exposed areas of vegetation and water may be of reduced size and buildings are likely to be constructed from brick or wood, which blends with the natural landscape. Thus, the contrast is reduced and, again, higher spatial resolutions are required.

In summary, the spatial frequencies (size and density) and contrasts of urban structures for the extreme examples of the United States and China may be approximated as shown in Figure 30-30. This illustration graphically depicts why resolution requirements for urban studies will vary with region.

SPATIAL RESOLUTION REQUIREMENTS FOR URBAN STUDIES

The standard method for the preparation of urban land use/cover maps in most countries is based on the visual interpretation of image data—with large- to medium-scale aerial photographs (of 1:10,000 to 1:60,000 scale) obtained with photogrammetric camera systems as the primary data source. Such aerial photographs usually have image resolutions of about 20–30 line pairs per

TABLE 30-6

Approximate Relative Building Density of Urban Built-Up Areas

Country	Approximate Relative Building Density/Unit Area
United States	1.0
Canada	1.6
Sweden	6.6
Japan	9.1
China	33.3

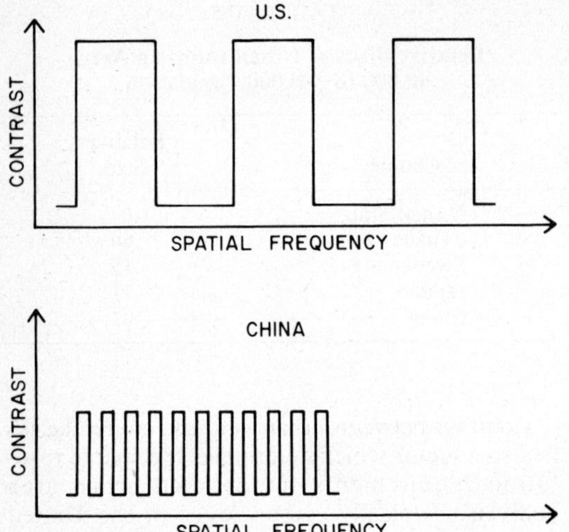

Fig. 30-30 Relative spatial frequency and contrast for urban built-up areas in the United States and China.

millimetre (Lpr/mm) for low-contrast objects which equate to ground resolutions of 0.3 to 3.0 m. High altitude photographs typical of those recorded on color infrared film with photogrammetric cameras mounted in U-2, RB-57 or Lear Jet aircraft, on the other hand, produce photographic resolutions of 15–20 Lpr/mm for low-contrast objects which, for the normal photoscales of 1:80,000 to 1:130,000, are equivalent to ground resolutions of between 4 and 9 m.

In the United States, the U.S. Geological Survey has developed a land use/cover classification system, for use with remotely sensed data (Table 30-7), which is comprised of successively finer levels of detail (Anderson, et al., 1976). Level I classes can be mapped from Landsat image data, whereas information typical of Levels II, III, and IV requires the use of high-, medium-, and low-altitude photographs, respectively. One of the criteria for the mapped classes is that a minimum accuracy of interpretation of 85 percent be maintained. Thus, although indistinct, there is a systematic relationship between the spatial resolution of remotely sensed data and its suitability for urban land use mapping tasks (Figure 30-31).

It is possible to approximate the relationships between representative urban-land parcel sizes and the measures of resolution for satellite and aircraft sensor systems as shown in Figure 30-32. In order to facilitate comparisons, all resolution values are given in terms of instantaneous-field-of-view (IFOV) or the approximate equivalent dimensions for photographic systems.[2] Experience has shown that for land parcels to be accurately identified and classified by visual interpretation, they must be several times larger than the sensor IFOV or data pixel. In digital classifications based on the spectral properties of objects, large IFOV's are likely to integrate the reflectance properties of various objects in the urban scene (e.g. grass, concrete buildings). Again, reliable classifications generally require land areas of sufficient dimensions to incorporate several pixels.

In order to illustrate this type of classification-related problem, it is appropriate to consider the relationships between the size of pixels and average urban land parcels (Figure 30-33). For example, if it is assumed that a minimum of four pixels are required for the reliable identification or classification of the basic land parcels, a sensor with an IFOV of 5 to 10 m is required for the Asian environment, whereas 30 m may be adequate for analyses of some urban areas in the United States (Welch, 1981; Clark, 1980). Empirical studies of Landsat MSS and RBV data of China and Sudan confirmed that resolutions of about 5 m IFOV or better are desirable for mapping the urban features in these areas (Welch and Pannell, 1975; 1982; Abdulla, 1980). Studies of urban areas in the United States from Skylab photography revealed that an IFOV of better than 10 m is required for detailed studies of Level II and III classes (Lins, 1976). Neither the MSS nor RBV Landsat-3 data are of adequate resolution for detailed studies of urban areas (IFOV's of about 79 and 40 m respectively). Attempts to create an image of improved resolution and interpretability, by merging digital data for a black-and-white RBV image with that of a MSS color composite of the same area, have met with some success (Figure 30-34), but at increased cost and complexity of processing (Lauer and Todd, 1981).

MINIMUM MAPPING UNIT

The discussion to this point has focussed on the ability to detect, identify and classify urban land use patterns recorded on remotely sensed data. While this is perhaps a most critical aspect of conducting studies of urban areas, it is also important to consider the minimum size of parcels that can be represented on a map product of a given scale. The minimum size threshold is governed by the following factors:

a. the ability to consistently recognize, classify and delineate parcels;
b. the time available for map compilation and the complexity of the map product desired by the given user; and
c. the cartographic inaccuracies involved in the portrayal and measurement of small parcels.

[2] The photographic ground resolution for a low-contrast target may be converted to an approximate IFOV by dividing the ground resolution value by 2.4. For example, a high-altitude aircraft photograph of 1:100,000 scale with an image resolution of 20 Lpr/mm for low-contrast targets and a corresponding ground resolution of 5 m would have an equivalent IFOV of $5/2.4 \simeq 2$ m.

TABLE 30-7

Land Use and Land Cover Classification System for Use with Remote Sensor Data [Anderson et al., 1976].

1 Urban or Built-up Land	11 Residential.
	12 Commercial and Services.
	13 Industrial.
	14 Transportation, Communications, and Utilities.
	15 Industrial and Commercial Complexes.
	16 Mixed Urban or Built-up Land.
	17 Other Urban or Built-up Land.
2 Agricultural Land	21 Cropland and Pasture.
	22 Orchards, Groves, Vineyards, Nurseries, and Ornamental Horticultural Areas.
	23 Confined Feeding Operations.
	24 Other Agricultural Land.
3 Rangeland	31 Herbaceous Rangeland.
	32 Shrub and Brush Rangeland.
	33 Mixed Rangeland.
4 Forest Land	41 Deciduous Forest Land.
	42 Evergreen Forest Land.
	43 Mixed Forest Land.
5 Water	51 Streams and Canals.
	52 Lakes.
	53 Reservoirs.
	54 Bays and Estuaries.
6 Wetland	61 Forested Wetland.
	62 Nonforested Wetland.
7 Barren Land	71 Dry Salt Flats.
	72 Beaches.
	73 Sandy Areas other than Beaches.
	74 Bare Exposed Rock.
	75 Strip Mines, Quarries, and Gravel Pits.
	76 Transitional Areas.
	77 Mixed Barren Land.
8 Tundra	81 Shrub and Brush Tundra.
	82 Herbaceous Tundra.
	83 Bare Ground Tundra.
	84 Wet Tundra.
	85 Mixed Tundra.
9 Perennial Snow or Ice	91 Perennial Snowfields.
	92 Glaciers.

Experience by U.S.G.S. personnel involved in the development of a land use and land cover classification-system for use with remotely sensed data has indicated that parcels with dimensions of less than 2 × 2 mm at the map scale should not be included on the final product (Anderson et al., 1976). Smith (1978) in a discussion of the preparation of urban land use maps of 1:50,000 scale for cities in Britain from aerial photographs of 1:60,000 scale, mentions that although parcels as small as 0.25 ha (50 × 50 m) could be identified in some instances, consistent recognition of five basic urban land use classes could only be undertaken for 5 ha parcels (225 × 225 m). A five hectare parcel was found to adequately represent the major urban land use categories and provided a 4.5 × 4.5 mm mapping unit. In another application, changes in urban land use at the fringes of the large cities in Britain were monitored with Landsat digital data. Minimum parcels of change, consisting of at least 2.5 ha, were required for consistent identification. No attempt was made in this application to classify the types of land use change due to the inadequate spatial resolution of the Landsat data (Carter and Smith, 1980).

In the United States, the U.S. Geological Survey is preparing 1:100,000 and 1:250,000 scale land use maps from high altitude aerial photography and requires a minimum size parcel of 4 ha for urban land use categories (Mitchell et al., 1977). Consequently, at the final map scales of 1:100,000 and 1:250,000 the minimum mapping unit measures 2 × 2 mm and 0.8 × 0.8 mm, respectively. Relationships between minimum parcel size and map scale may be further examined in Figure 30-35.

SUMMARY

The analysis of urban land use requires a sensor system capable of recording high frequency, low-

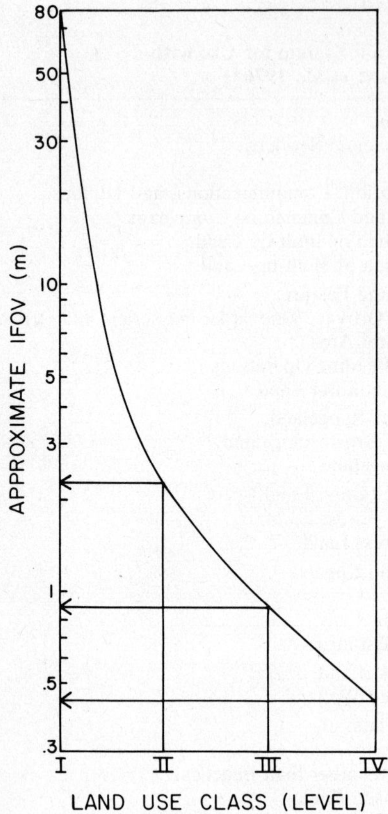

Fig. 30-31 Resolution (IFOV) as a function of USGS land use classification mapping requirements.

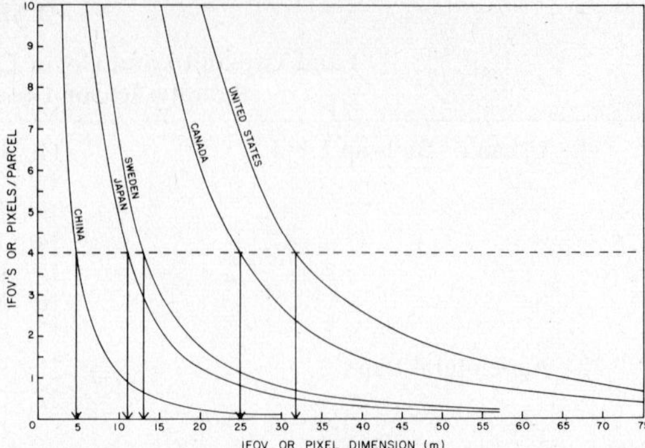

Fig. 30-33 Number of IFOV's contained within the typical land parcel sizes of different countries plotted as a function of IFOV (or pixel) dimension.

contrast detail with resolutions from about 0.5 to 10 m (IFOV). The most useful data for visual interpretation will be produced by sensor systems with IFOV's of better than 5 m. Comparable thresholds for digital data remain to be determined; but classifications based on the spectral reflectance properties of urban areas may not improve significantly as the IFOV is reduced. The minimum parcel size to be mapped should have dimensions equivalent to about 2 × 2 mm at the map scale. Under normal circumstances, the minimum size parcel should be several times larger than the sensor-system resolution element.

THE TEMPORAL NATURE OF URBAN LAND USE

ASSUMPTIONS CONCERNING TEMPORAL RESOLUTION

Three aspects of temporal resolution should be considered in most remote sensing-related scientific endeavors. First, phenomena may exhibit a developmental cycle which is predictable, e.g. a residential housing tract goes through numerous stages of predictable development. Second, there exist optimum times during the cycle when

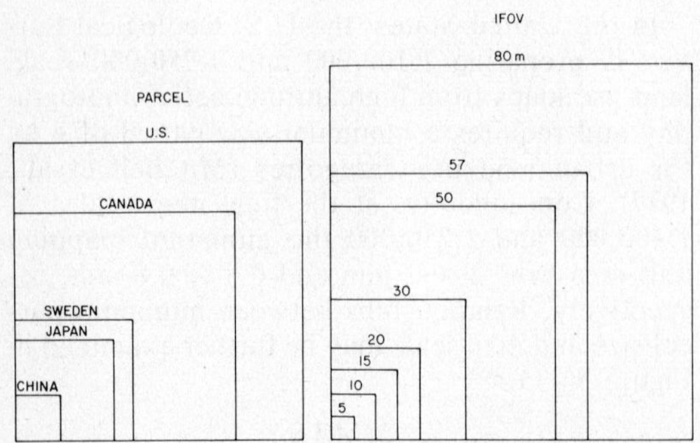

Fig. 30-32 Comparison of representative urban land parcel sizes for different countries with the IFOV's of current or proposed satellite sensor systems.

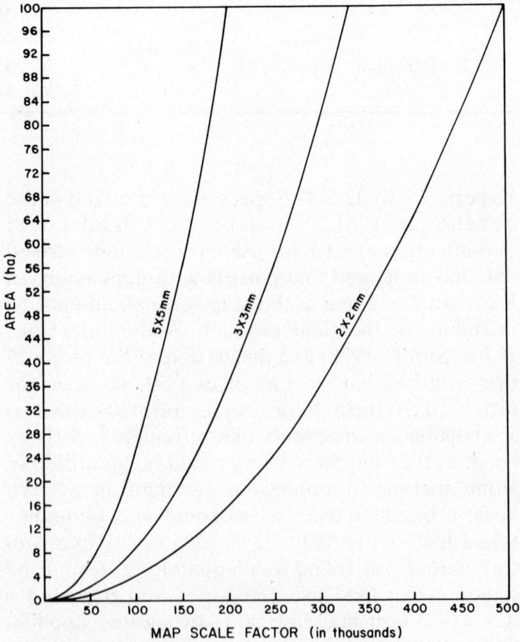

Fig. 30-35 Relationships between area in hectares for square parcels of 2 × 2, 3 × 3 and 5 × 5 mm (minimum mapping units) and map scale.

phenomena should be sampled to obtain useful information. Remote sensing often plays an important role in this data collection capacity. Finally, there may exist a lag in time between data collection and data availability for analysis which may preclude real-time monitoring of the phenomena. Improper consideration of any of these temporal resolution characteristics can result in inaccurate or untimely analysis of the data.

This section identifies numerous urban applications and their temporal data requirements as related to remote sensing data-collection. A temporal window for each of the applications (Table 30-8) is identified, based on empirical findings. There are urban applications which require only a single date inventory, e.g. certain types of commercial or industrial site selection. This discussion, however, emphasizes remote monitoring of urban phenomena and processes as they develop through time.

Land Use Change Detection

Numerous urban applications simply require binary (change/no change) land use change information. A good example is the Bureau of the Census which must update the boundaries of all urbanized areas within every SMSA every 5–10 years (Toll et al., 1980; Royal, 1980; Jensen, 1981). Such binary information is obtainable from the Landsat system and is reported in detail in the Bureau of the Census application section of this chapter.

Mapping the change in land use from various rural to Level II and III urban categories is often done every 5–7 years for areas within SMSAs (Table 4 in Anderson, 1977; Table 2 in Milazzo, 1980). However, when local and regional administrators require detailed Level IV urban land-use change information, the update cycle appears to be 3–5 years (Anderson, 1976; Miller and Miller, 1976; General Electric, 1974).

Critical Environmental Area Assessment

For land use change mapping in sensitive environmental areas, such as water bodies, wetlands or beaches, a 7- to 10-year update cycle is suggested (Milazzo, 1980). However, to detect environmental impact due to rapid development, numerous agencies require information on these environmental areas at 2–5 year intervals (Miller and Miller, 1976).

The demand for energy escalates man's impact on environment whenever energy resources are available. Typical land uses in this category include strip mining, and quarrying. These phenomena are usually inventoried every 2–5 years (Mausel, 1979).

City Infrastructure Characteristics

Planning agencies require housing stock and public works information, such as building height, number of floors, volume, perimeter, area, rooftop types etc. Large scale aerial photography continues to provide a substantial amount of this information (Lo, 1971; 1973a; 1973b; Moore, 1970; Henderson, 1979; Turner and Steiner, 1979; Ryerson, 1980). This information is usually collected every 5–10 years.

There exist two major areas where remote sensing has made substantial contributions to transportation research (Lindgren, 1979). First, a highway condition- and impact-analysis usually takes place every half year or yearly. Second, traffic and parking surveys (Wellar, 1973) require data on an hourly basis. In terms of both logistics and costs involved, such high temporal resolution is difficult to achieve for extended periods of time using aircraft platforms.

Socioeconomic Characteristics

Remote sensing may aid population estimation, especially for developing countries. Such national inventories take place generally every 10–20 years (Lo and Welch, 1977). At the regional and local levels, however, the time between inventories is often reduced to 5–10 years (Holz et al, 1969; Hsu, 1971; Lindgren, 1971; Deuker and Horton, 1971; Anderson and Anderson, 1973; Clayton and Estes, 1980). An extended discussion of this topic is found in the application section of this chapter.

Quantitative information obtained from high resolution imagery provides surrogate information for quality-of-life studies (Marble and Horton, 1969; Moore, 1970; Davies et al, 1973; Henderson and Utano, 1975; Higgs, 1975; McCoy and Metivier, 1973). Information is typically collected every 5–10 years.

Energy Utilization and Conservation

Several studies have monitored national and regional energy demand using remote sensing (Clayton and Estes, 1979; Welch and Zupko, 1980; Welch, 1980). Ideally this is done every 2–5 years. At the local level, remote sensing surveys have been conducted every 1–3 years for selected cities to determine solar photovoltaic potential (Angelici et al, 1980), or heat loss from residential and commercial property (Brown, 1978; Colcord, 1981). In addition, the estimation of solid-waste generation by urban areas is aided by multiple date analysis of aerial photography (Garofalo and Wobber, 1974). The energy-application section of this chapter elaborates on these topics.

Photogrammetric Mapping Requirements

Large scale topographic base mapping in urban areas is normally updated every 10 to 20 years (El-Baz, 1980). Topographic maps for civil engineering public works purposes are normally updated every 1–5 years (Wellar, 1973). Cadastral (property and tax) maps must be updated every 1–2 years (Wellar, 1973; Veign and Reeves, 1973; El-Baz, 1980).

TABLE 30-8

Selected Urban Applications and Their Temporal Data Requirements (Source: Jensen).

Application	Temporal Requirement															
				Years												
	Hourly	Daily	Weekly	.5	1	2	3	4	5	6	7	8	9	10	15	20
Land Use Change Detection																
• rural versus urban									—	—	—	—	—	—		
• Level II and III									—	—	—					
• Level IV							—	—	—							
Critical Environmental Areas Assessment																
• stable sensitive natural environments											—	—	—	—		
• dynamic, sensitive environments						—	—	—	—							
• dynamic energy resource areas						—	—	—	—							
City Infrastructure Characteristics																
• building characteristics									—	—	—	—	—	—		
• transportation route studies				—	—											
• traffic and parking facilities	—	—														
Socioeconomic Characteristic																
• national population estimation														—	—	—
• regional and local population estimates									—	—	—	—	—	—		
• quality of life indications									—	—	—	—	—	—		
Energy Utilization and Conservation																
• energy demand						—	—	—	—							
• energy production potential					—	—	—									
• energy waste					—	—	—									
Photogrammetric Mapping Requests																
• topographic base mapping														—	—	—
• topographic public works mapping					—	—	—	—	—							
• cadastral					—	—										
Meteorological Characteristics																
• daily weather	—	—														

Meteorological Characteristics

Hourly measurements of air pollution, cloud cover, temperature, and precipitation are generally required for large metropolitan regions. Again, although they may be difficult to obtain and costly to analyze for extended periods such data are available from NOAA geosynchronous satellites and the remainder from *in situ* devices.

Summary

From an examination of the preliminary information presented in Table 30-8 it is obvious that the majority of urban applications require data collection in cycles ranging from 1 to 10 years. Obvious anomalies are transportation and meteorological applications which require extremely high temporal resolution. This suggests that, for urban applications in general, a sensor system which systematically monitors the environment once every year on a near-anniversary date for change detection purposes, would be sufficient for most urban applications of remotely sensed data.

URBAN/SUBURBAN LAND USE AND LAND COVER CLASSIFICATION SYSTEMS AND ACCURACY ASSESSMENT

CLASSIFICATION SYSTEMS

The function of a classification system is to provide a framework for organizing and categorizing information. Assuming that classification systems are a necessary component of effective urban and suburban land use/land cover management and planning, it is important to point out that there is no ideal land use/land cover classification-scheme and it is unlikely one will be developed which is universally acceptable. Nevertheless, there exist numerous classification schemes which are functional at the national level (U.S. Executive Office of the President, 1957; Urban Renewal Administration et al., 1965; Ryerson and Gierman, 1975; Anderson et al., 1976) and at regional, and local levels (Michigan Land Use Classification and Referencing Committee, 1975; Gautam, 1976; Baker, et al., 1979).

Major points of difference between various urban/suburban classification schemes are their 1) emphasis, and 2) ability to incorporate information obtained using remote sensing. For example, the U.S. Geological Survey classification system (Anderson et al., 1976), is "resource" oriented in contrast with various "people or activity" oriented systems such as the *Standard Land Use Coding* (SLUC) *Manual* (Urban Renewal Administration et al., 1965). The U.S.G.S. rationale is that "although there is an obvious need for an urban-oriented land-use classification system, there is also a need for a resource-oriented classification system whose primary emphasis would be the remaining 95 percent of the United States land area." The U.S.G.S. classification system addresses this need with eight of the nine Level I categories (Table 30-7) treating land area that is not in urban or built-up categories (Anderson et al., 1976). The system is designed to be driven primarily by the interpretation of land use and land cover data obtained at various scales and resolutions (Table 30-9). Conversely, the SLUC system is land use "activity" oriented and primarily dependent on *in situ* observation and enumeration to obtain remarkably specific land use information even insofar as to the contents of individual buildings. Obviously, there exists the need to merge the two approaches to produce a hybrid urban/suburban classification system which would incorporate both land use interpreted from remotely sensed data, when applicable, and the very precise (and expensive) *in situ* land use information when necessary.

A hybrid classification scheme such as this was prepared and adopted for the State of Michigan (Michigan Land Use Classification and Referencing Committee, 1975). Although optimized for Michigan land use, it can serve as a general guide to the logic and preparation of detailed urban/suburban classification schemes developed elsewhere. Like the U.S.G.S. system, the Michigan system is an integrated and consistent system, permitting aggregation from the lowest level upward to successively higher levels, and in certain instances the disaggregation of any higher level into lower levels. This reduces cost and permits increased benefits as multidisciplinary users access the same database and avoid duplication of

TABLE 30-9

Classification Level	Typical Data Characteristics
I.	LANDSAT (formerly ERTS) type of data.
II.	High-altitude data at 40,000 ft. (12,400 m) or above (less than 1:80,000 scale).
III.	Medium-altitude data taken between 10,000 and 40,000 ft. (3,100 and 12,400 m) (1:20,000 to 1:80,000 scale).
IV.	Low-altitude data taken below 10,000 ft. (3,100 m) (larger than 1:20,000 scale).

effort. Basically, the nine Level-I categories of the U.S.G.S. system were adopted without change. However, a complete system through Levels II, III, and IV was produced consistent with Michigan environmental and cultural conditions.

The Michigan system is land-cover oriented up to and including Level III, making it amenable to remote sensing data analysis. It is both cover and activity oriented at Level IV so as to better meet user needs at the regional and local levels. In developing Level-IV categories for urban and built-up categories, the SLUC and Standard Industrial Code (SIC) were used as much as possible because local users of urban information already used, or were familiar with, these systems. SLUC and SIC system numbers were included for cross-referencing purposes in the classification.

The major characteristics of the Michigan classification system, including the minimum mapping units, are shown in Table 30-10. The urban and built-up portion of the classification system is found in Appendix A.

LAND USE CLASSIFICATION ACCURACY ASSESSMENT

To make remote sensing-derived land-use maps and associated statistics useful to the user community there must be a method for quantitatively assessing classification accuracy. This usually requires the collection of some *in situ* data to determine how well the remote sensing inventory has been performed. Then, it is necessary to select a method which will determine if the classification accuracy is acceptable.

Various methods for assessing land-use map accuracy have been proposed (Hord and Brooner, 1976; Van Genderen and Lock, 1977; Fitzpatrick-Lins, 1978; Hay, 1979; and Ginevan, 1979). Unfortunately, considerable disagreement exists among the authors of these studies as to the optimum methodology. Consequently, this section elaborates on a single procedure developed by Fitzpatrick-Lins (1978, 1980) which seems adequate for providing realistic assessments of urban/suburban land-use map accuracy. It is one of the methodologies adopted for evaluating the accuracy of land cover maps and associated maps prepared by the U.S. Geological Survey (Fitzpatrick-Lins and Chambers, 1977). For more recent information on procedures being employed by U.S.G.S. personnel, the reader should consult Rosenfield et al., 1982.

Land-use maps of the Greater Atlanta Region (Figure 30-36) compiled at a scale of 1:100,000 from high altitude aerial photography were chosen to test quantitatively the accuracy of a Level II classification. Most assessments of classification accuracy require the user to first identify a level of desired classification accuracy. This is used to determine the number of points to be sampled. In this instance, the criteria established by Anderson et al. (1976) were used which stated

TABLE 30-10
Characteristics of the Michigan Land Use Classification System

Criteria	Level I	Level II	Level III	Level IV
Primary Use:	National	National	State & Regional	Local
Mapping Range:	1:250,000 to 1:1,000,000	1:125,000 to 1:250,000	1:50,000 to 1:125,000	1:24,000 to 1:50,000
Minimum Mapping Area:	235-3700 acres	60-235 acres	10-60 acres	2-10 acres
Classification Basis:	cover	cover	cover/activity	cover/activity
System Origin:	USGS	USGS	Michigan	Michigan
Data Source:	Chiefly dependent upon image interpretation	Chiefly dependent upon image interpretation	Combination of image interpretation and ground truth collection	Combination of image interpretation and ground truth collection

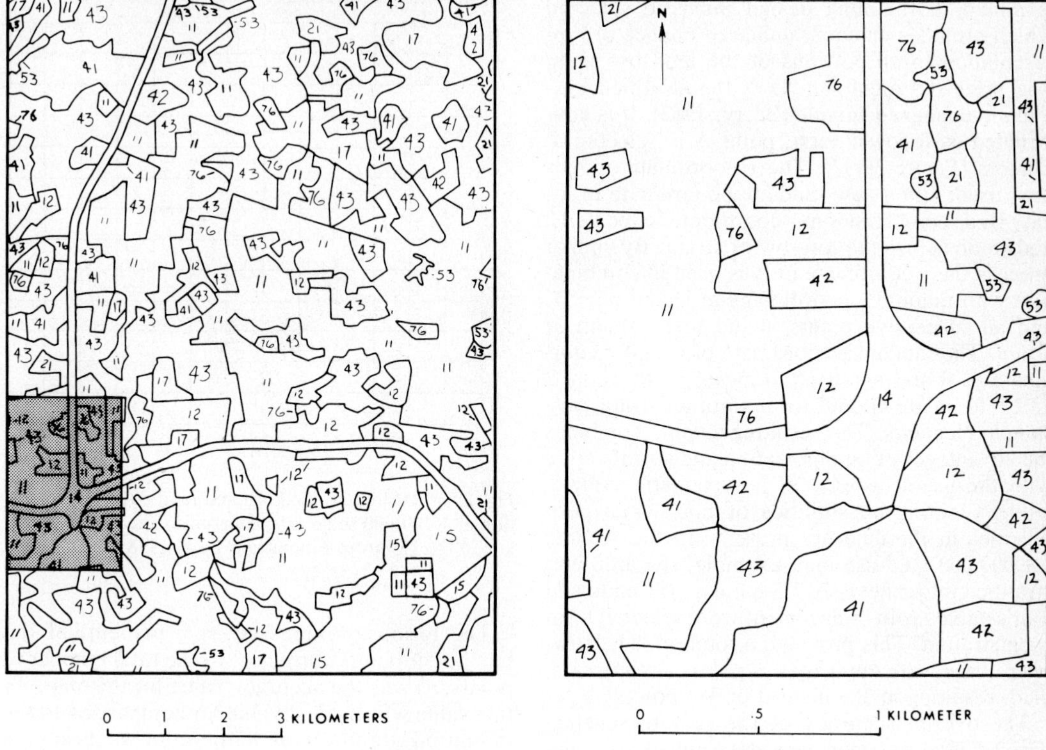

Explanation

11 Residential	21 Cropland and Pasture
12 Commercial and Services	41 Deciduous Forest
14 Transportation, Communications, and Utilities	42 Evergreen Forest
15 Industrial and Commercial Complexes	43 Mixed Forest
16 Mixed Urban or Built-Up Land	53 Reservoirs
17 Other Urban or Built-Up Land	76 Transitional Areas

Fig. 30-36 Land use and land cover maps of a portion of the Greater Atlanta Region. On the left is a 1:100,000 scale land use map fitted to the U.S. Geological Survey 7.5-min. Chamblee topographic quadrangle, Georgia. On the right is a 1:24,000 scale land use map which depicts in detail the shaded portion of the map on the left. [Source: Fitzpatrick-Lins, 1978, 1980]

that: 1) the land use- and land cover-maps must be at least 85 percent accurate, and 2) the accuracy of the interpretation must be approximately equal for most categories. The appropriate number of points to be sampled must then be determined.

The ideal number of points to be tested in the land-use map was determined from the formulas for the binomial probability theory. The formula for the number of points selected was

$$N = \frac{4pq}{E^2} \qquad (30\text{-}10)$$

where p is the expected percent accuracy, q is the difference between 100 and p, E is the allowable error, and N is the number of points to be sampled (Snedecor and Cochran, 1967).

For a sample where the expected accuracy was 85 percent at an allowable error of 4 percent (2 standard deviations of 2 percent), the number of points necessary for reliable results was

$$N = \frac{4(85 \times 15)}{4^2}$$

or 319 points. Fewer points would be sampled if the accuracy were assumed to be greater than 85 percent or the standard deviation acceptable were larger. The narrow limits of a 2-percent standard deviation were selected because the methods of sampling involved very little field work and, therefore, should be as precise as possible to offset any procedural errors. (Fitzpatrick-Lins, 1980).

With expected map accuracies of 85 percent and an acceptable error of 10 percent, the sample size for a map would be at least 50. According to Van Genderen and Lock (1977), the smallest sample size for meaningful results is 20 points, even if the sample is error free.

Once the number of points to be sampled is determined, a sampling design must be selected which provides an equal unbiased chance of representation to all portions of the land use map. The preferred areal sample is the stratified systematic unaligned sample (Berry, 1962). It is constructed as follows: First, point A is selected at random (Figure 30-37). The x coordinate of A is then used with a new random y coordinate to locate B, a second random y coordinate to locate E, and so on across the top row of strata. By similar process the y coordinate of A is used in combination with random x coordinates to locate point C and all successive points in the first column of strata. The random x coordinate of C and y coordinate of B are then used to locate D, those of E and F to locate G, and so on until all strata have sample elements. The resulting sample combines the advantages of randomization and stratification with the useful aspects of a systematic sample, while avoiding possibilities of bias when periodicities in the data are present. In the Atlanta 1:1,000,000 land-use map example, the area was stratified into ninety-six 7.5-minute grid units and four samples with replacement were selected from each stratum. This provided a total of 384. However, three were duplicates of points already sampled, resulting in 381 instead of 384 points.

The overall accuracy of the remote sensing-derived land-use map was determined by comparing it with the known land use at each sample location. The initial method of determining the accuracy was empirical in that the percentage of accuracy was a simple ratio of correct interpretation to the total number of points sampled. The information is normally displayed as a contingency table (Figure 30-38). For the overall classification, 343 (90 percent) of 381 points were classified correctly.

Accuracy Determination for the Entire Map

The 90 percent accuracy statistic was useful but did not report how confident one should be that the map actually exceeded the 85 percent criterion. To determine this, the following methodology was used.

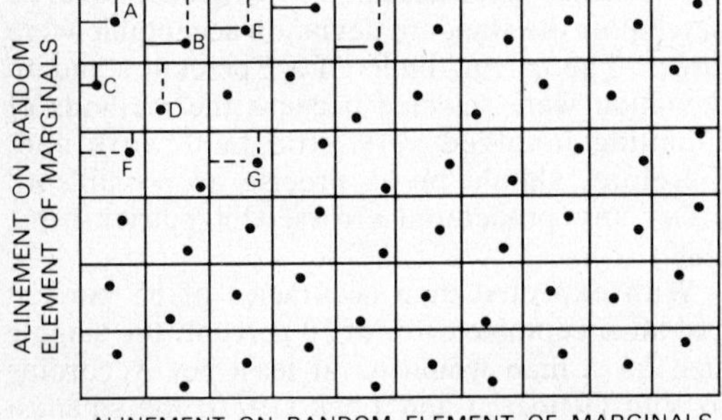

Fig. 30-37 A stratified systematic unaligned sample. [Source: Berry, 1962]

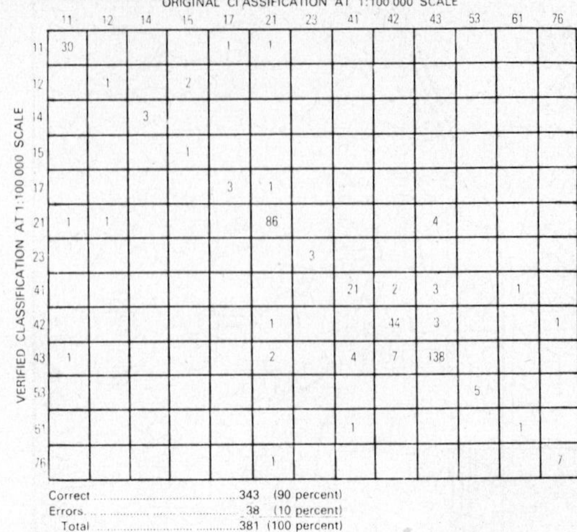

ORIGINAL CLASSIFICATION AT 1:100 000 SCALE (columns); VERIFIED CLASSIFICATION AT 1:100 000 SCALE (rows)

	11	12	14	15	17	21	23	41	42	43	53	61	76
11	30				1	1							
12		1		2									
14			3										
15				1									
17					3	1							
21	1	1				86				4			
23							3						
41								21	2	3		1	
42						1			44	3			1
43	1					2		4	7	138			
53											5		
61								1				1	
76						1							7

Correct .... 343 (90 percent)
Errors .... 38 (10 percent)
Total .... 381 (100 percent)

Fig. 30-38 Matrix of land use and land cover classification at 1:100,000 scale as compared to the verified classification. [Source: Fitzpatrick-Lins, 1978; 1980]

The ratio, p (expressed as a percent) of the number of points correct, r, to the total number of points, n) was the accuracy value for the map. As this value was the test value for comparison to the minimum standard of 85-percent accuracy, a one-tailed test was appropriate. The 95-percent one-tailed lower confidence limit for a binomial distribution was obtained from the equation (derived from Snedecor and Cochran, 1967, p. 211):

$$p = \hat{p} - \{1.645\ \sqrt{\hat{p}\hat{q}/n} + 50/n\}, \quad (30\text{-}11)$$

where p = the accuracy of the map expressed as a percent;
$\hat{p}$ = the sample value of p or r/n expressed as a percent;
$\hat{q}$ = $100 - \hat{p}$; and
n = the sample size.

If the p value exceeded the 85-percent criterion at the lower confidence limit, it was possible to accept (with 95-percent confidence) that the map met or exceeded the accuracy standards. When the 90 percent ($\hat{p}$) overall accuracy statistic for 381 points (n) was analyzed, the 95 percent one-tailed lower confidence limit was 87 percent. Thus, it exceeded the 85 percent accuracy criterion.

Accuracy Determination for the Individual Categories

The contingency table showed how each category sample was interpreted and where misinterpretation occurred (Figure 30-38). Whereas a single accuracy statement with a given lower confidence limit was sufficient to describe the entire map, two ways of expressing the percent accuracy of individual categories were required, namely, the analysis of commission and omission errors. The reader should analyze the error distribution in

Figure 30-38 to see how "points correct" and "points total" were computed for the five major land use and land cover categories summarized in Table 30-11.

To determine the true 95 percent confidence limits for each category, the following two-tailed test was applied:

$$p = \hat{p} \pm \{1.96 \sqrt{\hat{p}\,\hat{q}/n} + 50/n\} \quad (30\text{-}12)$$

where $\hat{p}$ was again the value of the true portion correct, expressed as a percentage in Table 30-11AB, and n was the number of points sampled.

The results of the five categories examined for errors of commission are shown in Table 30-11A. Only Mixed Forest Land (43) exceeded the specified accuracy of 85 percent at the lower confidence limit. Residential (11) and Cropland and Pasture (21) at 83 percent for the lower confidence limit approached the criterion of 85 percent accuracy. A larger sample for all categories except for Mixed Forest Land would give more reliable results and would narrow the confidence interval.

The results of examining errors of omission are shown in Table 30-11B. The accuracies of four of these categories approached or exceeded the criterion of 85-percent accuracy. Category 41 was less accurate, yet the range in accuracy was so great that it would be necessary to test several more points for more precise results. The overall accuracy of the land use and land cover maps exceeded the criterion of 85-percent accuracy at the scale of 1:100,000.

Such procedures allow land use maps derived from remote sensing to be quantitatively evaluated to determine overall and individual category-classification accuracy. Improved methods for assessing classification accuracy continue to be developed (Rosenfield, 1981). Their proper use will enhance the credibility of using remote sensing information when making decisions.

SELECTED URBAN LAND USE APPLICATIONS USING REMOTELY SENSED DATA

Urban and suburban land use information obtained from remotely sensed data may be used for diverse planning activities at the local, regional, national and international levels. The following sections first identify the major urban land-use data-collection activities taking place in the United States including the U.S. Geological Survey Land Use and Land Cover Mapping program and the U.S. Bureau of the Census mapping of urban expansion for Standard Metropolitan Statistical Areas (SMSA). Following these topics, the collection of timely socioeconomic statistics from remotely sensed data is addressed including 1) population estimation at the regional and national level, 2) assessment of housing characteristics and environmental quality at the regional and local level, and 3) energy conservation and planning activities at the regional and local level for urban areas. A final case study demonstrates the utility of accurate, up-to-date urban and suburban land use information in the Three Mile Island emergency.

TABLE 30-11A

Major Land Use and Land Cover Categories on the Atlanta Map Analyzed for Errors of Commission (Fitzpatrick-Lins, 1980).

Category	Points Correct	Points Total	Percent Correct	95-Percent Confidence Limits (Percentage)
11 Residential -----	30	32	94	83–100
21 Cropland and Pasture.	86	92	93	83–100
41 Deciduous Forest Land.	21	26	81	64–98
42 Evergreen Forest Land.	44	53	83	72–94
43 Mixed Forest Land.	138	148	93	89–98

TABLE 30-11B

Major Land Use and Land Cover Categories on the Atlanta Map Analyzed for Errors of Omission

Category	Points Correct	Points Total	Percent Correct	95-Percent Confidence Limits (Percentage)
11 Residential -----	30	32	94	83–100
21 Cropland and Pasture.	86	92	93	87–99
41 Deciduous Forest Land.	21	27	78	61–95
42 Evergreen Forest Land.	44	49	90	81–99
43 Mixed Forest Land.	138	152	91	86–96

THE LAND USE AND LAND COVER MAPPING PROGRAM OF THE U.S. GEOLOGICAL SURVEY

INTRODUCTION

Since 1974, the U.S. Geological Survey (USGS) has engaged in nationwide base-line mapping of land use and land cover using remotely sensed data. Maps have been printed primarily at a scale of 1:250,000. However, as 1:100,000 scale base maps have become available, they have been used for special applications and for mapping certain areas. These scales are appropriate for mapping land use and land-cover data on a nationwide basis within a practical time frame, and with an acceptable degree of standardization, accuracy, and level of detail.

Land use and land-cover mapping and data compilation along with related research and development being carried out by the Geological Survey can be divided into the following major activities (Place, 1977):

1. Preparation, and release to open file, of land use and land cover maps for use with the 1:250,000 scale base maps and the new 1:100,000 scale base maps as these become available (Figure 30-39).
2. Preparation, and release to open file, of associated maps showing political units (counties and states), hydrologic units (drainage areas), census county subdivisions including census tracts, and areas of Federal land ownership.
3. Preparation and sale of magnetic tapes containing data obtained by digitizing in polygon format the land use and land cover and associated maps. Documented software, necessary for the effective use of the digital data, is also provided.
4. Compilation and publication of land use and land cover statistics by political units, hydrologic units, census county-subdivisions, and areas of Federal land ownership. Such statistics are made available mainly by states. However, statistics may also be made available for such area units as Standard Metropolitan Statistical Areas and major river basins.
5. Experimentation with and demonstration of land use and land cover mapping at scales larger than 1:100,000 for specific applications.
6. Experimentation with LANDSAT data to obtain consistent mapping results and measurement of spatial and temporal changes in land use and land cover.
7. Research on and development of a Geographic Information Retrieval and Analysis System (GIRAS) for handling land use and land cover data in conjunction with environmental, socio-economic, demographic, and other data (Mitchell et al., 1977).

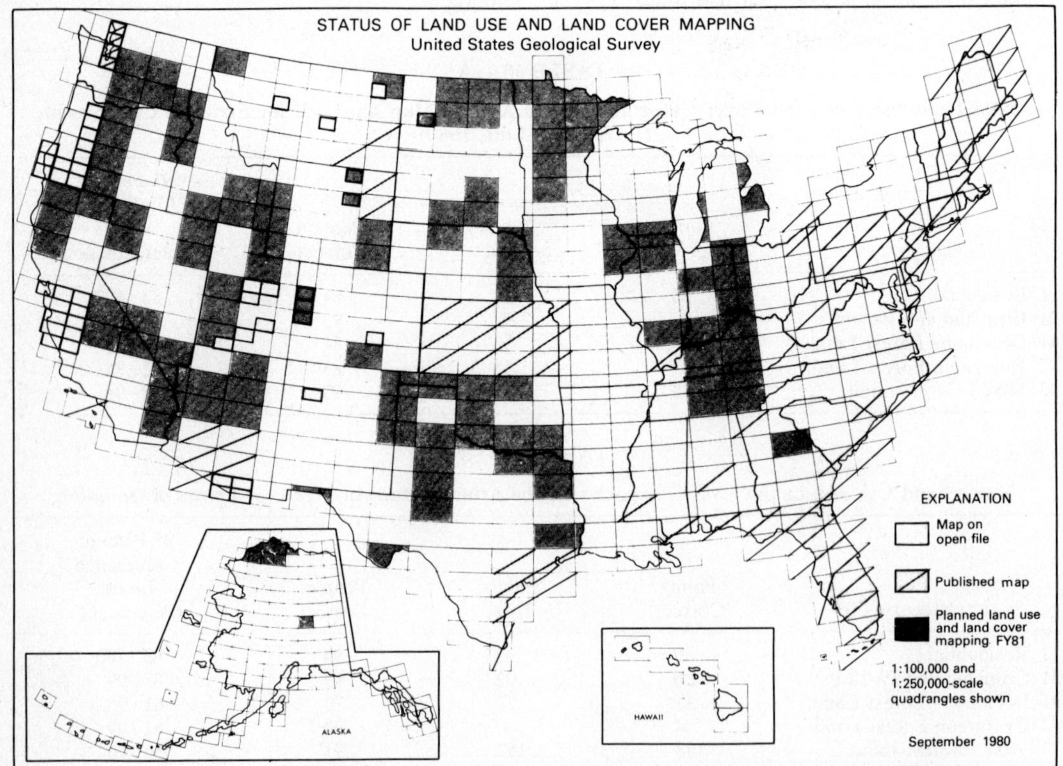

Fig. 30-39 Status of land use and land cover mapping in the United States. [Source: U.S. Geological Survey]

8. Analytical and interpretative studies on land use patterns, problems, and trends, and the application of land use and land cover maps and data to the study and solution of "real world" problems.

By October 1981, land use and land cover mapping was completed for more than two million square miles or nearly 60 percent of the Nation. Priority was given to the mapping of urban regions, coastal areas, energy production areas, and other areas undergoing rapid urbanization. Availability of suitable source materials played a role in the selection of areas to be mapped first. Completion of mapping for the entire nation is anticipated in the 1986–1987 time period.

Land use and land cover data are categorized according to the classification system presented in USGS Professional Paper 964 (Anderson et al., 1976), together with other established specifications. The minimum mapping unit is 10 acres (4 ha) for the urban or built-up uses, water areas, transitional areas in an urban situation, confined feeding operations, certain other types of agricultural land, and strip mines, quarries, and gravel pits. All other categories are delineated with a minimum unit of 40 acres (16 ha).

Compilation Procedures

The Geologic Survey's land use, land cover and associated maps are compiled and produced at the USGS Regional Mapping Centers using the following procedures (Loelkes et al., 1977; Milazzo, 1980).

Search for Source Materials

Before a map is compiled, a search is made to determine if source materials are available which meet the requirements for land use and land cover mapping. These requirements are: a) aerial photographs must not be more than 3 years old; b) cloud or snow cover must be 10 percent or less; c) photographs must be 1:80,000 scale or smaller; and d) photographs must be capable of being tilt-corrected. If source material is available, authorization is given for compilation of the land use and land cover and associated maps.

The source materials most commonly used for compilation are:

1. Color-infrared photographs taken by RB-57 or U-2 aircraft at an altitude of 60,000 to 70,000 feet, scale 1:120,000 to 1:130,000.
2. Black-and-white panchromatic photographs taken from an aircraft at an altitude of about 40,000 feet. This type of photography is obtained under contract by the Geological Survey and is quad-centered[3] at a scale of approximately 1:80,000.
3. Special materials available to the Geological Survey.

To fill in small areas not having photographs meeting requirements outlined previously, other sources such as aerial photographs available from the Agricultural Stabilization and Conservation Service of the U.S. Department of Agriculture are used.

In addition to aerial photography, much research has been conducted on the use of Landsat imagery for land use and land cover mapping. The frequency of coverage, availability, and low cost make Landsat MSS imagery a uniquely appealing source for land use and land cover map-update. However, it does possess certain formidable limitations for operational use. Studies undertaken in southern Arizona (Place, 1974) and elsewhere (Fitzpatrick-Lins, 1978b) using optical analysis of Landsat MSS imagery have shown that it is not wholly appropriate for Level II land use and land cover mapping (Milazzo, 1980). Many USGS Level II land use and land cover categories cannot be identified and mapped from the Landsat imagery alone.

A somewhat more promising application of Landsat data for land use and land cover mapping is offered by computer-aided analysis of Landsat MSS digital data. Computer processing makes possible the examination and classification of large amounts of data quickly. Since the data are already in a digital format, they form the basis of a geographic information system capable of providing simultaneous area measurement and tabulation. Thus, the need to convert a graphic land use and land cover map into digital format to derive area statistics is eliminated.

One problem, however, lies in converting spectral signatures into valid land use and land cover categories. It is unrealistic to expect a land use and land cover classification developed from computer-classified Landsat data to match category for category or to merge with the USGS Level II classification system which is designed for human compilers using other remotely sensed sources.

Finally, the Landsat RBV imagery appears to be a promising source for land use and land cover mapping roughly equivalent to the small-scale, high-altitude aerial photographs. Although no thorough evaluation has been conducted, the imagery viewed thus far suggests promise for use as a photographic source for direct mapping in cases where the resolution permits reliable discrimination of discrete Level II land use and land cover categories.

Precompilation Field Activities

This task, prior to compilation, consists of an investigation of a region to determine if land use and land cover patterns differ from others previously mapped. This investigation is carried out by a team of two geographers working together. The

[3] In "quad-centered" photography the middle photograph of a stereo triplet of 1:80,000 scale photographs is so positioned that its photographic nadir falls at or near the center of a given 7.5 minute quadrangle sheet.

precompilation task begins with the planning of aircraft traverses.

Generally, a traverse is flown at an altitude of 500 to 1,000 feet. The combination of low-altitude and low-airspeed allows the team to acquire oblique photographs, and to plot on 1:250,000 scale maps the land use and land cover sites they have selected as part of the background information to be used in briefing the compilers. The map compilers are then briefed thoroughly on the characteristics of the land use and land cover and on any anomalies known to exist in the region.

Interpretation and Compilation

The land use and land cover map is compiled to portray the Level II categories of the land use and land cover classification. Compilation specifications for the land use and land cover maps are documented in Loelkes (1977). The task of interpreting the vertical aerial photographs and transferring the interpretation to the correct geographic location on the base manuscript map is accomplished either by photogrammetric techniques or by direct transfer methods using optical enlargement. When the photogrammetric method is used, compilation is at the scale of the planimetric base. This is the standard stereographic method; the stereo model is placed in the plotter, scaled and leveled, and a pantograph is used to scribe the map.

Quality Control

Quality control consists of an in-depth review of the map manuscript to determine if category definitions and specifications for interpretation and compilation have been consistently adhered to. A complete cartographic review of the manuscript map includes reviewing the accuracy of interpretation, proper placement of identification numbers, adherence to specified minimum sizes for the land use and land cover categories, use of specified line weights, and any other necessary cartographic review. If the compiler and the reviewers cannot resolve differences in interpretation, then the areas in question are marked for possible checking in the field.

After a final review and approval, final reproductions are made and the maps are released to open file in the appropriate Mapping Center, along with a copy of the planimetric base. Copies are sent to the various state cooperators and USGS outlets as required.[4]

Documentation Procedures

Documentation is the organization of remotely sensed data used by geographic regions in order to establish a set of photointerpretation keys that can be used to assist in the identification of specific land use and land cover categories within each region. Documentation of land use and land cover compilation from remotely sensed data is needed in order to obtain interpretation consistency, permit accurate compilation updating, and provide a photointerpretation guide for those who use land use and land cover maps.

The regional documentation framework presently being employed is a combination of that employed by Küchler (1964) in his map of "Potential Natural Vegetation of the Conterminous United States" and Marschner's (1950) "Major Land Uses in the United States." Küchler's classification system is used for areas where the natural cover is dominant and Marschner's for areas where the cover has changed and where agricultural and related land use activities prevail. Within each subregion, source photography is selected and registered to specific sample sites that depict both the unique and the more generally characteristic land use and land cover signatures typifying the subregion. The photography ranges from ground 35 mm photographs to high-altitude aircraft photographs and in some cases satellite images. When each land use and land cover category within the various subregions has been documented at least one time, documentation of compilation is complete.

LAND USE AND LAND COVER MAP UPDATE PROCEDURES

An essential requisite to better use of the Nation's land is current information on land use and land cover and on the rates and trends of *changes*—hence, the need for a program of map update (Anderson, 1977). Major considerations in planning a nationwide program for updating USGS land use and land cover maps are:

- How often should maps be updated?
- What remotely sensed source materials should be used for detecting and compiling changes in land use and land cover?
- What base maps should be used for presenting data on land use and land cover changes?
- What methods should be used for identifying and mapping changes?
- What procedures and what formats should be used for updating maps?

Recommendations for updating land use and land cover maps in accordance with the goals of the USGS land use and land cover mapping program have been offered (Milazzo, 1980). In summary, the recommendations are as follows:

- Initiate periodic overview "photoinspections" of regions using available remotely sensed source materials to identify changes, with inspection scheduled at a 5-year mini-

[4] In addition to these products, for several years the USGS land use/land cover have been published in the "L-series," in two colors, base information in one color and land use polygons in the other.

mum interval. The actual frequency of land use map update is, however, dependent on the amount and type of land use and land-cover change identified in the photoinspection.

- It is expected that small-scale, high-altitude photographs will continue to be the mainstay of future land use and land cover map compilations, and it has been recommended that they continue to be one of, if not, the primary photographic sources for updating land use and land cover maps. Large-scale photographs are to be used in a limited and complementary way when more suitable smaller scale sources do not exist or when a more detailed land use and land cover classification is required for special applications. The use of Landsat digital data in updating land use and land cover maps is dependent on the development of improved digital data analysis techniques. At present, Landsat digital data are appropriate for monitoring gross land use and land cover changes and for assisting in the update selection process for manually compiled land use and land cover maps. The data do not constitute a primary source for such an update, but function in a complementary role.

- The 1:250,000-scale base maps will be used to update land use and land cover maps of those areas where either 1:100,000-scale bases are not available or where there are regions in which it is desirable to continue the presentation of map data at 1:250,000-scale. The 1:100,000-scale base maps will be used for updating those maps already compiled at that scale or for updating existing 1:250,000-scale land use and land cover maps for areas in which 1:100,000-scale bases are available and the regional location or the nature of the land use and land cover data make conversion to the larger scale base desirable.
- The "original map"-to-"new source material" method will be the principal land use and land cover change-detection technique to avoid land-cover change mapping-error (Figure 30-40). It will be supplemented by using the original compilation of remotely sensed source material where needed and when available.
- Maps will be updated using the "updated original" approach in which only the outdated portions of the original map are replaced with new data. This will result in a complete land use and land cover map format. Updated statistics on land use and land cover will be provided. A separate overlay showing only the polygons of land use and land cover change may be provided as an optional associated map supplement (Figure 30-41). Recompilations will be conducted only when necessary.

These recommendations represent a foundation for the development of a strategy for updating maps in the nationwide land use and land cover mapping program. As will be demonstrated in the next section, there exist other users who do not require systematic nationwide coverage of all Level II categories and who have different change detection requirements and approaches to the problem.

U.S. BUREAU OF THE CENSUS URBANIZED AREA PROGRAM

INTRODUCTION

An important aspect of the decennial census conducted by the Bureau of Census is the location

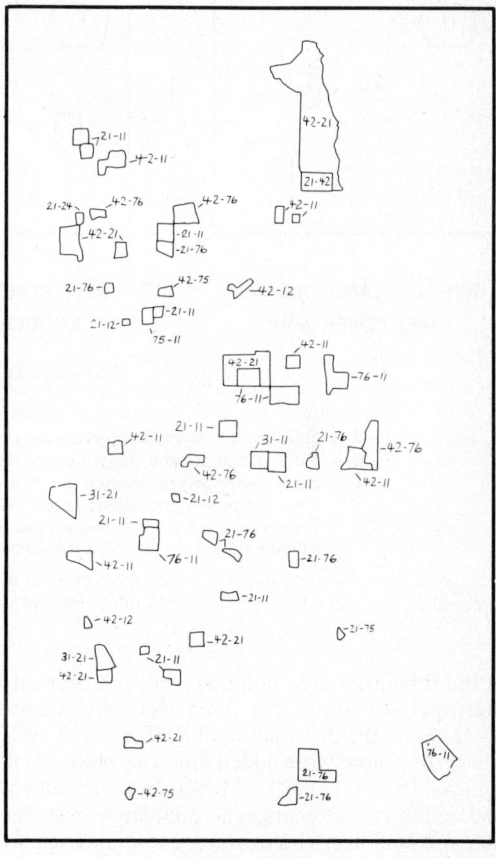

Fig. 30-41 Example of a land use and land cover change map showing only the polygons of land use and land cover change occurring between 1972 and 1978 for a portion of the Ocala, Florida, 1:100,000 scale topographic map. Changes were derived from a comparison of the 1972 land use and land cover map with 1978 aerial photographs. Polygons are identified by two Level II land use and land cover codes indicating the former and the new categories, respectively. [Source: Milazzo, 1980]

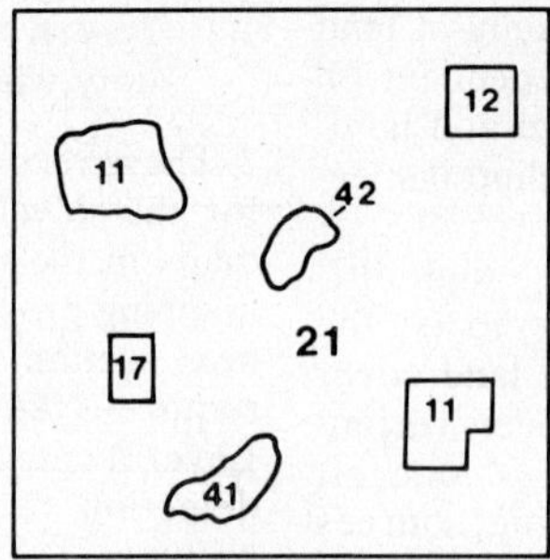

ORIGINAL REMOTELY SENSED SOURCE MATERIAL

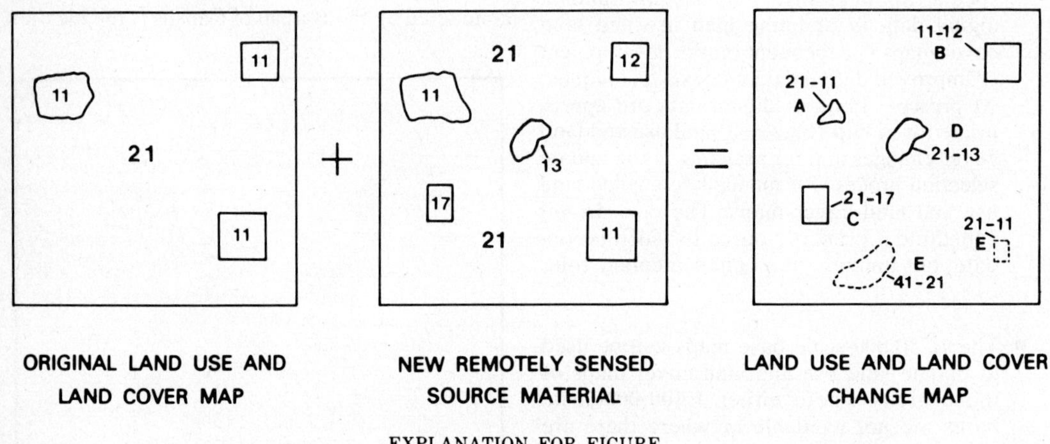

ORIGINAL LAND USE AND LAND COVER MAP

NEW REMOTELY SENSED SOURCE MATERIAL

LAND USE AND LAND COVER CHANGE MAP

EXPLANATION FOR FIGURE

A – False change polygon mapped due to boundary delineation error in original map compilation
B – False change polygon mapped due to interpretation error in original map compilation
C – False change polygon mapped due to polygon omission in original map compilation
D – True change polygon mapped falsely due to polygon omission in original map compilation
E – True change polygon not mapped due to error in original map compilation (dashed lines indicate valid change polygon not delineated on change map)

Fig. 30-40 Possible land use and land cover change mapping errors resulting from an original map-to-new source material change detection approach. [Source: Milazzo, 1980]

of the urbanized area boundary in each Standard Metropolitan Statistical Area (SMSA). Figure 30-42 shows the distribution of SMSAs as of 1976. Thirty-five more were added after the 1980 census bringing the total to 323. SMSAs are reviewed and updated to reflect changes in population distribution and land use. Urbanized area boundaries are used as funding criteria for a number of federal programs, including General Revenue Sharing, Urban Mass Transit, and the Federal Highway Act.

Delineation of an urbanized area boundary requires an initial step of identifying an urban fringe zone (Figure 30-43). To establish a fringe zone an "outer line" is delineated that includes all possible areas of urban expansion within the SMSA. The area between the outer line and the previously determined urban boundary from the most recent census is called the "urban fringe zone." Census Bureau statistics are then compiled within the urban fringe, and are analyzed by small geographic areas called "enumeration districts". Each district is analyzed in a 'yes' or 'no' mode to determine whether or not it should be included within the final urban boundary. The most important criterion for inclusion within the final urban boundary is whether or not the population density for an enumeration district is greater than 1,000 persons per square mile.

In times past, Census Bureau geographers typically used aerial photography to perform this task. However, when timely photography was not available for each of the more than 300 SMSAs, other data sources were used such as local land use maps. These unsystematic data sources often

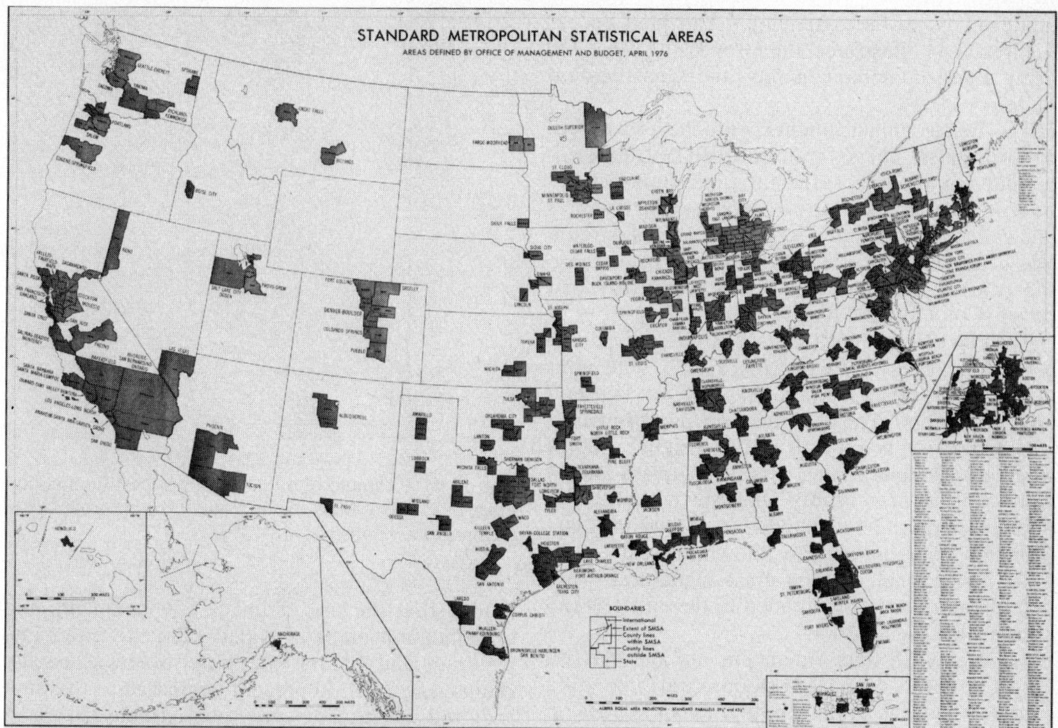

Fig. 30-42 The distribution of Standard Metropolitan Statistical Areas as of 1976. [Source: Office of Management and Budget]

resulted in a non-standard data base. Bureau of the Census personnel were interested in the use of Landsat data for this task because it provides synoptic coverage for most urban areas, enabling the geographers to work with fewer products than when using other data sources. Landsat imagery represents a standardized data base with repeatable spatial (geometric), spectral, and temporal characteristics. The repetitive data collection allows analysts to select "current" imagery prior to each census. Finally, the Landsat digital data may be processed using a variety of manual and computer-assisted analysis techniques.

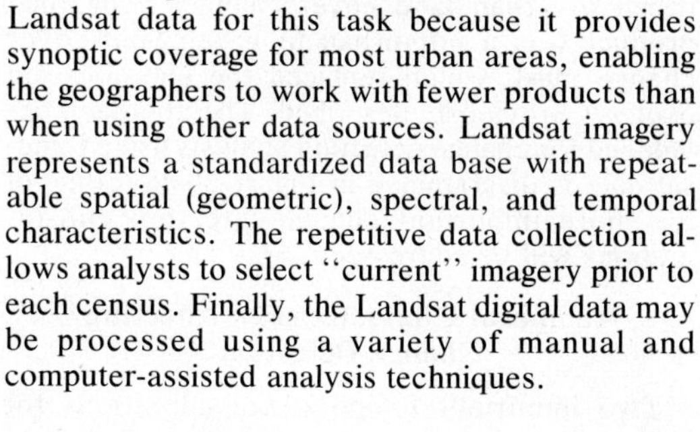

URBAN FRINGE ZONE
OUTER LINE
URBAN EXPANSION SINCE 1970
INNER LINE
1970 URBANIZED AREA
ENUMERATION DISTRICTS

Fig. 30-43 Establishing a fringe zone around an urbanized area. [Source: Toll]

To evaluate the usefulness of satellite imagery, a joint venture between NASA/Goodard Space Flight Center and the Bureau of the Census was begun. Initially, Landsat image enhancement and classification maps were produced. The project then began analyzing the change between dates using change-detection algorithms. Additional research evaluated the use of geographic information system technology as an aid to identifying urban change and identifying the urban boundary. Finally, various types of imagery other than Landsat were evaluated. Results of these studies are discussed in subsequent sections.

URBAN AREA DELINEATION USING LANDSAT IMAGE ENHANCEMENTS AND CLASSIFICATION MAPS

The first two years of the study focused on the development of procedures to delineate urban area expansion from Landsat data (Christenson and Lachowski, 1976; Lachowski et al., 1979). Emphasis was placed on producing Landsat image enhancements and classification overlays for five selected SMSAs (Richmond, Va.; Seattle, Wa.; Austin, Tx.; Boston, Mass.; and Orlando, Fla.). The image enhancements provided Census Bureau geographers with a graphic display of urban areas from which urban boundaries were visually identified. Furthermore, computer-aided classification of Landsat digital data, by training on known areas of urban growth, produced thematic maps on the extent of urban land cover. Figure 30-44 illustrates the processing flow for using Landsat enhancements and classification maps.

The image enhancements were necessary because Landsat images from the EROS Data Center did not provide sufficient tonal contrast between urban and non-urban land cover for outer line delineation. A stepwise linear contrast stretch was generally applied to the raw data to increase the contrast between urban and non-urban land cover (Color Figure 30-45).

The classification of Landsat multispectral data provided Census Bureau geographers with non-urban versus urban land use maps for use in urban area delineation. The classification algorithm used was feature-space partitioning (Figure 30-46) where MSS band values (usually bands 5 and 7) were plotted in a scatter diagram and then interactively partitioned with a cursor until the classification corresponded to known ground conditions. Once final training class signatures were determined, they were applied to the entire urban area (Color Figure 30-47).

Landsat MSS data enhancements and classifications were compared to conventional data sources, such as aerial photography, for use in outer line delineation (Lachowski et al., 1979 and Davis, 1978). Census Bureau geographers, working with Landsat enhancements and classifications for five selected urban areas, were able to define the geographic location of the outer line to include almost all possible expanses of urban land cover, with few errors of having non-urban land cover within the outer line. Comparisons of results obtained using large scale aerial photography versus Landsat data showed that similar outer line locations were delineated. Individual structures such as shopping centers, commercial buildings, or residential houses could not be discerned on the Landsat image enhancements; however, general patterns of urban sprawl could usually be delineated. In areas where rural and urban land cover were mixed, the delineation was difficult. For these areas geographers preferred working with the classification overlays. The Bureau concluded that the use of image enhancements and classification maps provided adequate outer line accuracy and that the interpreter time was usually reduced by over 50 percent compared to the time required when using aerial photography.

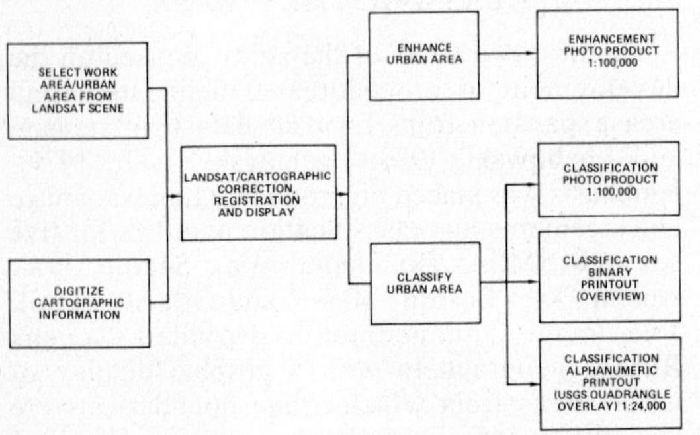

Fig. 30-44 Flow diagram for the processing of Landsat data for use in urban area delineation. [Source: Lachowski et al., 1979]

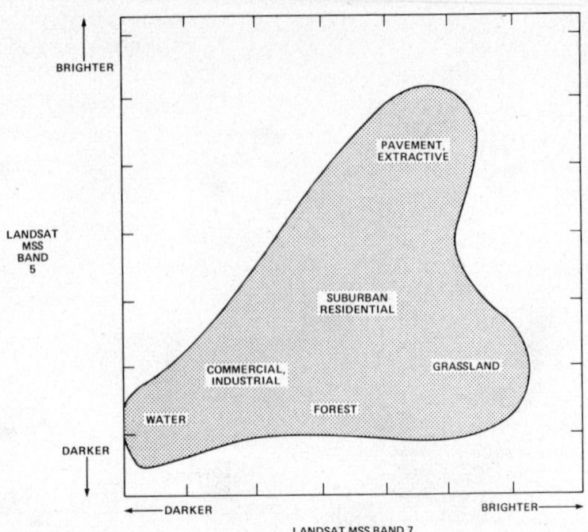

Fig. 30-46 Example of a feature space partitioning diagram of Landsat MSS bands 5 and 7 for an urban area. [Source: Lachowski et al., 1979]

CHANGE DETECTION OVERVIEW

Digital change detection procedures have been investigated to delineate urbanizing areas. Multitemporal Landsat data were used to detect non-urban to urban land cover changes. The final product was a non-urban to urban land cover change mask which replaced the classification overlays previously described. The approach developed for change detection analysis using Landsat data is diagrammed in Figure 30-48. Some of the more important change-detection considerations will be addressed.

Temporal Characteristics Affecting Change Detection

Two important temporal considerations for change detection analysis are the times of the year and the number of Landsat dates to use in the analysis. In one study, five Landsat scenes were evaluated throughout the vegetative season for a test site north of Denver, Colorado, to determine the optimum times of the year to conduct change detection (Riordan, 1980). There was adequate contrast between non-urban and urban land cover using any of the dates; however, the August date provided slightly better results. It is generally believed that the optimum times of the year to conduct change detection are environment dependent. Some of the factors which influence this decision are discussed in the environmental characteristics section.

The number of dates used in the change detection analysis is dependent on the types of change one is attempting to identify. For example, con-

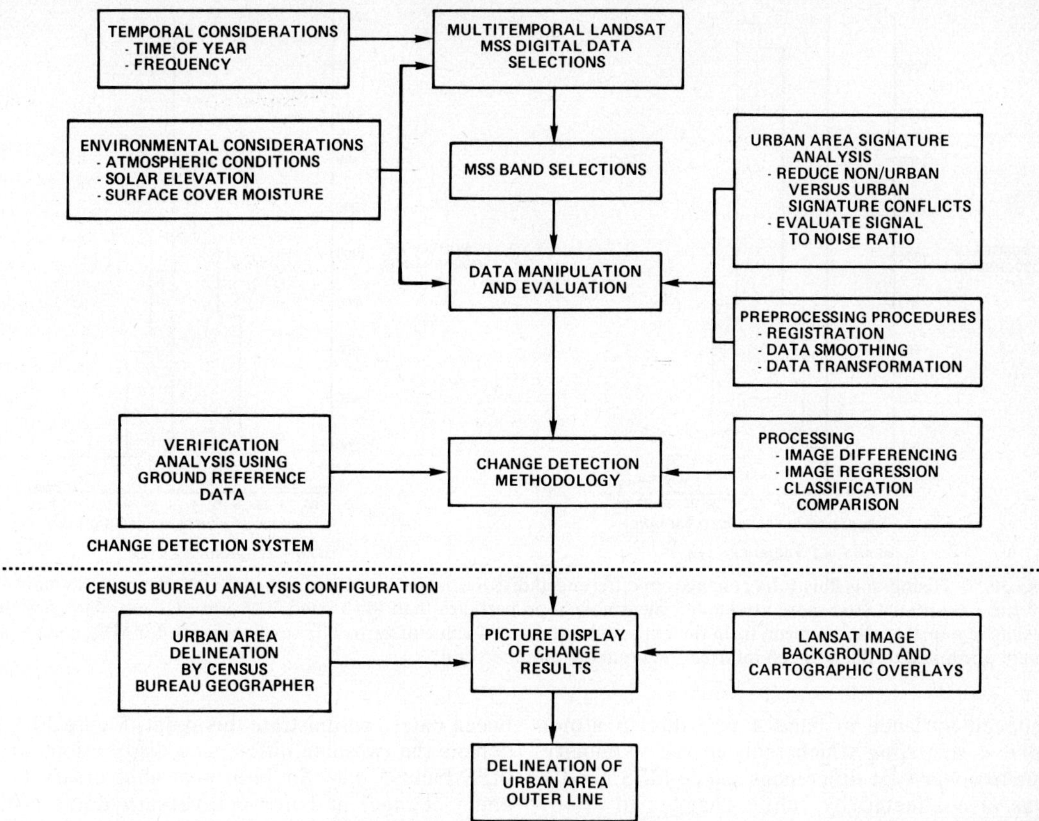

Fig. 30-48 Flow diagram of the urban change detection system for use in urban area delineation. [Source: Toll]

sider the radiance changes which occur during non-urban to urban land-cover development as shown in Figure 30-49. During the early stages of residential development (up to and including building construction) the spectral contrast between agriculture and urban development are pronounced. However, as landscaping within the subdivision matures, the spectral contrast is reduced (Toll et al., 1980). To adequately monitor all of these types of change several Landsat images would have to be analyzed. However, for Census Bureau purposes generally only two scenes were required because it was only necessary to detect non-urban to urban land cover change and not the specific stages of development.

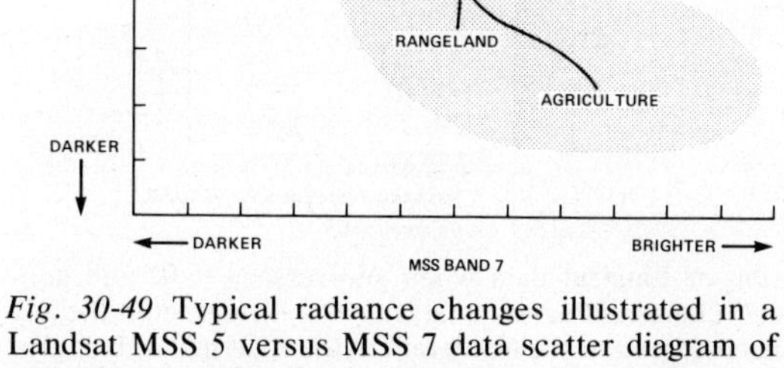

Fig. 30-49 Typical radiance changes illustrated in a Landsat MSS 5 versus MSS 7 data scatter diagram of non-urban to urban land cover conversion in Denver, Colorado. [Source: Toll]

Environmental Characteristics Affecting Change Detection

Several factors affecting the change in radiance between dates are not associated with land-cover change. This, of course, frequently results in change-detection error. Of special importance are the adverse effects caused by differences in atmospheric conditions, sun angle, and soil moisture.

Atmospheric changes between dates, even when not pronounced, may cause changes in the mean and variance of Landsat digital data. This is especially true for MSS band 4 data where the shorter wavelength radiant flux is attenuated far more than the longer wavelength radiant flux recorded in MSS band 7. Atmospheric effects for MSS band 4 and MSS band 7 data are illustrated in Figure 30-50. Note that image-differenced MSS band 4 data had a smaller variance and a larger deviation from zero than MSS band 7 data. The

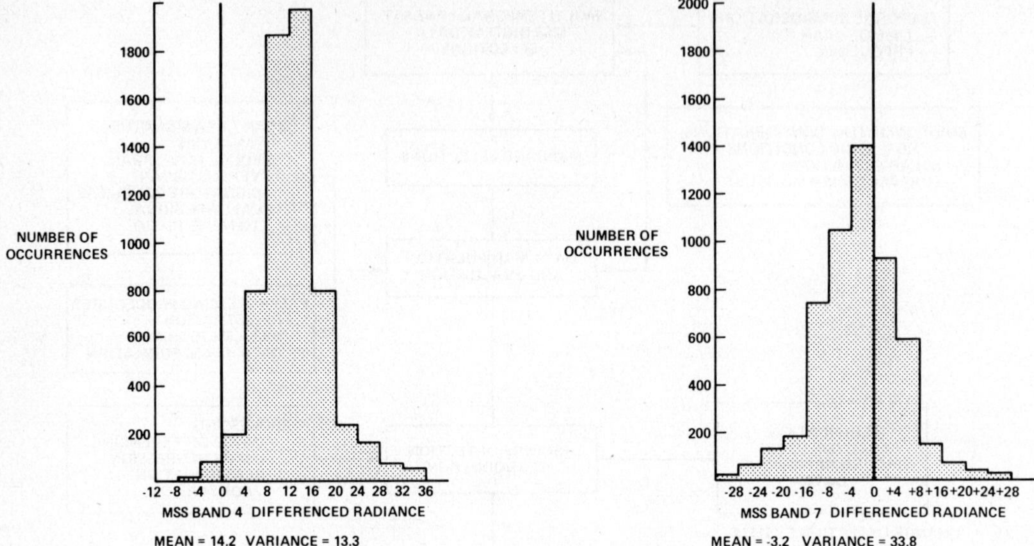

Fig. 30-50 Histograms illustrating changes in differenced distribution as a result of atmospheric effects. MSS band 4 (.5 μm–.6 μm) data are more attenuated by atmospheric particles than MSS band 7 (.8 μm–1.1 μm) data, which results in a further shift in mean from the expected differenced value of zero. The small variance for MSS band 4 is likely attributed to the reduced information content. [Source: Toll]

reduced variance in band 4 was due to atmospheric scattering which reduced the variability due to cover type differences. Since MSS band 4 was also affected by subtle changes in atmospheric conditions, there was a larger difference in means between dates than observed for the other MSS bands. These results were typical for all of the scenes studied.

Analysis of Landsat histograms indicated that there were changes in the mean and variance when non-anniversary Landsat scenes were used. One of the primary agents responsible for this condition was the change in solar elevation between dates. To illustrate this point, Figure 30-51 depicts the two-date differenced distributions for MSS band 5 data for both near-anniversary (1° angle change) and non-anniversary dates (10° angle change). There was a shift in mean from the expected value of zero and also an increase in variance for the non-anniversary dataset. The shift in mean from zero was attributed to changes in irradiance between dates due to differences in solar-elevation angles. The increased variance was believed to be due to the increased effects of shadows in those scenes having a large difference in sun angle. Of course, some of the increase in

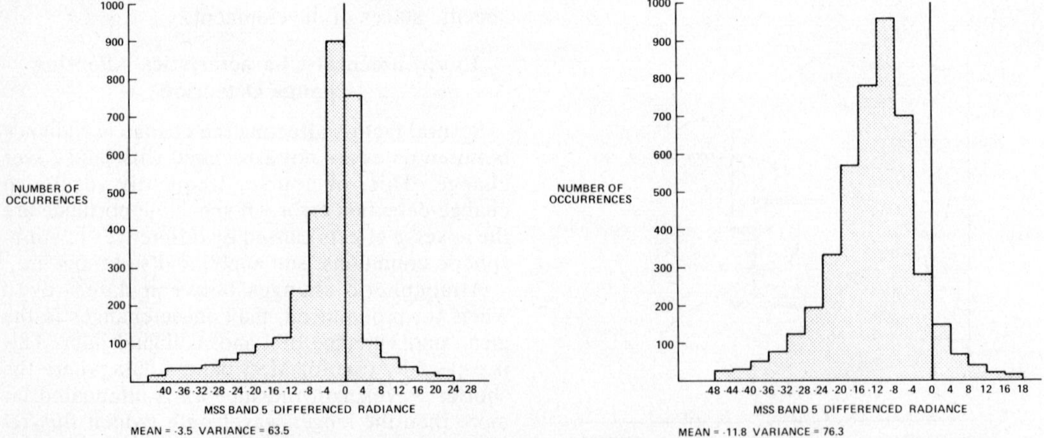

Fig. 30-51 Histograms illustrating changes in the distribution of Landsat data when anniversary (left) and non-anniversary dates (right) were used in an "image differencing" change detection analysis. The histogram to the left had a 1 degree difference in solar elevations between differenced dates and the histogram to the right had a 10 degree difference. A shift in mean from zero was observed as a result of differences in solar elevations (i.e., irradiance intensities). In addition, increasing solar elevation differences resulted in spreading the histogram. [Source: Toll]

variance in non-anniversary dates must also be explained by changes in vegetative cover between dates.

Soil moisture affects the radiance means and variances of Landsat data. Analysis of results for Austin, Texas, attributed change-detection error to soil-moisture differences between the two dates of Landsat data (Stauffer and McKinney, 1978). Figure 30-52 illustrates change in the mean and variance for Landsat MSS band 7 data when precipitation occurred prior to the Landsat overpass. During the four days prior to the May 1, 1976, Landsat overpass, the study area received 2.0 cm of precipitation in comparison to the May 10, 1976 Landsat scene that received only 0.1 cm of precipitation prior to the overpass. Increases in precipitation prior to the overpass resulted in an increased mean of 19 digital counts and a variance of 123 for MSS band 7 data. The overall lower radiance values for the May 1 date were attributed to the increased soil water content. These changes were also typical for MSS bands 5 and 6.

Three environmental factors affecting change-detection results were identified including: 1) differences in atmospheric conditions; 2) differences in sun angle; and 3) differences in soil moisture. The impact of these factors may be partially reduced by carefully selecting the MSS data. Anniversary dates should be used whenever possible to reduce problems from sun-angle differences and vegetation-phenology changes. In addition, climatological data for the week preceding the overpasses should be examined to ensure that soil-moisture differences are not a factor.

MSS Band Options

For urban change-detection analysis with Landsat data, any or all of the four MSS bands may be used. However, because Landsat MSS data are highly correlated, not all bands are usually required for analysis. For example, results for urban change detection in Denver, Colorado, indicated that each individual channel performed equally well when subjected to the same change-detection algorithm (Toll et al., 1980). However, results for change detection in Richmond, Virginia, concluded that the use of MSS band 5 data provided the highest urban change-detection accuracy (Toll et al., 1980). The optimum band or bands used in change detection are environment dependent and should be selected after evaluating the environmental factors previously discussed.

CHANGE DETECTION PREPROCESSING PROCEDURES

Registration

Multiple date images of the same geographic area must be precisely registered to one another to perform accurate urban change-detection. A slight spatial misregistration between scenes will result in substantial change-detection error. All of the Landsat MSS data used in these studies were rectified using a digital image rectification system

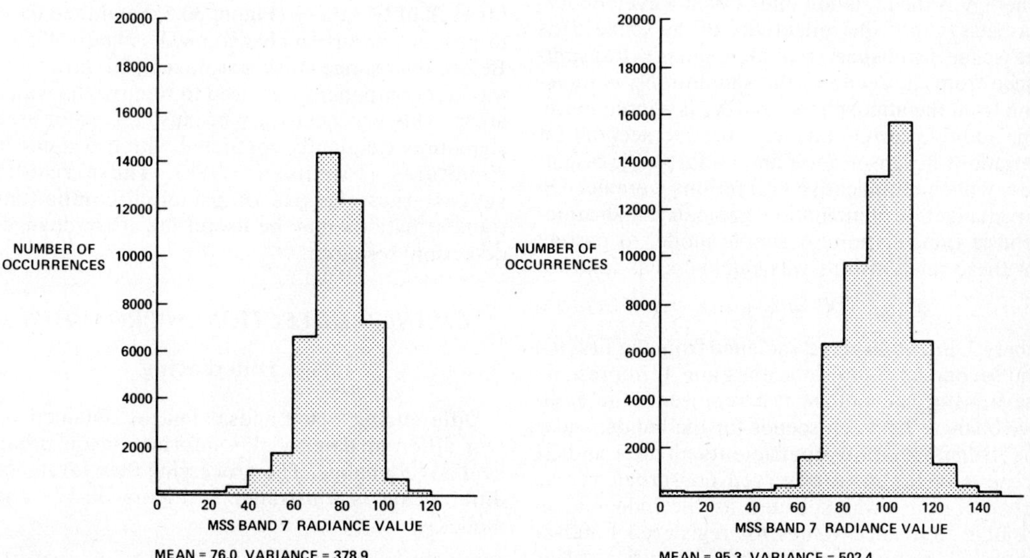

Fig. 30-52 Histograms illustrating the effect of surface soil moisture on the Landsat data mean and variance. For the four previous days of the Landsat overpass, the May 1, 1976, Landsat scene (the histogram to the left) received 2.0 cm of precipitation in comparison to the May 10, 1976, Landsat scene (the histogram to the right) that received 0.1 cm of precipitation. In fact, as a result of this difference there was a change in radiance mean of 19 and variance of 123 between dates for MSS Band 7. The overall reduced radiance values for the May 1 data were primarily attributed to the irradiance absorptance due to the increased water content within the surface cover. In addition, as seen by the smaller variance for the May 1 data, the amount of information was reduced. [Source: Toll]

(DIRS) which provided registration to within sub-pixel accuracy (Van Wie and Stein, 1976). Analysis of misregistration between scenes for 16 evenly distributed control points showed typical scene-to-scene misregistration errors of less than 0.20 pixels. In areas with few edges for comparison, however, registration errors greater than 0.75 were apparent (McKinney, 1979).

Data Transformations

Adverse effects from differences in atmospheric conditions, sun angles, and surface-cover moisture were previously identified as factors affecting the radiance changes between dates. To adjust for these differences, several transformations were applied to the Landsat data including the following model:

$$X^T_{ij}\,(t_2) = X^U_{ij}\,(t_2) - (\bar{x}\,(t_2) - \bar{x}\,(t_1)) \quad (30\text{-}13)$$

where X^T is the second-date (t_2) Landsat scene transformed for each pixel at line (i) and column (j); X^U is the second-date (t_2) Landsat scene untransformed for pixel (i), (j); and $\bar{x}$ are the mean values for the first (t_1) and second (t_2) dates. The Landsat date with the most land-cover contrast (an indication of image quality) was used as time 1. This transformation was performed for each pixel in each band.

A second data transformation used was based on the general transfer equation of the radiance to the satellite sensor:

$$N_{\lambda_i}\,c_i = 1/\pi\,(\rho H \tau) + N_a + N_n \quad (30\text{-}14)$$

where N is the radiation intensity at wavelength λ_i for class c_i; ρ is the reflectivity of the scene; H is the scene irradiance; τ is atmospheric transmittance from the scene to the satellite; N_a is radiation from the atmosphere; and N_n is source noise. The additive corrections were used to account for variations in sensor noise and atmospheric brightness while multiplicative corrections were needed for changes in illumination geometry and atmospheric transmission. A simple model to correct for these relationships was calculated as follows:

$$X\,(t_2) \cong X^T = a + b\,X\,(t_1) \quad (30\text{-}15)$$

where X is the observed radiance from the first (t_1) and second (t_2) Landsat scenes and X^T represents the transformed radiance; a represents the additive changes between scenes for the bands; and b the transformation coefficients. The Landsat scene with the most observed non-urban versus urban contrast was selected as the independent variable. Radiance values for registered Landsat features which remained constant, such as clear deep water, parking lots, and large commercial buildings were used for training. These data-transformation techniques increased change-detection accuracy for some areas and decreased it for others (Royal, 1980). The error was the result of reducing spectral contrast between non-urban and urban cover while adjusting for sensor noise, illumination geometry, and atmospheric effects.

To effectively monitor urban land-cover changes one must discriminate between non-urban and urban land cover. Unfortunately, results have shown that urban-area spectral signatures are often similar to non-urban signatures. To reduce the signature conflict a principal-component transformation was applied in certain instances. For a test site in the fringe area of Washington, D.C., the usefulness of a principal-component transformation was assessed for a two date (i.e., 8 band) Landsat MSS data set in order to reduce a residential versus agriculture conflict (Williams and Borden, 1977). Both principal-component and raw Landsat data were used to define training areas in a supervised classification. Results showed a reduction in the amount of residential land cover misclassified as agriculture in agriculture test sites (43 to 17 pixels) when using the principal components. The number of occurrences in which residential land-cover was misclassified remained the same for both approaches.

In an analysis of principal component results for Orlando, Florida, the first principal component provided discrimination between water and land cover and the second principal component provided discrimination between non-urban and urban land cover (Friedman, 1978). Image enhancements of both the Landsat MSS band 5 and second principal component for Orlando are shown in Figures 30-53 and 30-54. The second principal component in a threshold analysis was used to discriminate non-urban from urban cover for each of two dates (Figure 30-55) and then used to produce an urban change-mask (Figure 30-56). Before the change mask was developed, however, the first component was used to remove the water areas. This was necessary because the water area signatures frequently conflicted with urban-cover signatures (Friedman, 1978). These results suggest that the use of principal-component transformations may be useful for urban change-detection research.

CHANGE DETECTION APPROACHES

Image Differencing

Differencing two Landsat images obtained on two different dates yields information on urban land cover change. The processing flow for image differencing is illustrated in Figure 30-57. The model is:

$$D_{ij} = X_{ij}\,(t_1) - X_{ij}\,(t_2) \quad (30\text{-}16)$$

where X represents the radiance value of the desired Landsat band which is either raw or transformed at time 1 (t_1) and time 2 (t_2); D is the resultant (i.e., difference) image-radiance value for the pixel at line (i) and column (j). The expected value

Fig. 30-53 Enhanced Landsat imagery of MSS Band 5 data for the Orlando, Florida, SMSA, April 18, 1975. [Source: Friedman, 1978]

of *D,* given no change of land cover, is zero. Any radiance changes between time 1 and time 2 result, of course, in differenced radiance values departing from zero. Almost all urban land-cover changes showed radiance changes departing from zero. A significant radiance change is indicated when the differenced values exceed a specified threshold (*T*). A weighting of one-half the dynamic range (e.g., 128) is normally added to the differenced data to remove negative numbers.

Image-differencing techniques represent a computationally efficient change-detection processing approach; however, there often are inadequacies. The three factors which commonly cause problems are: 1) slight spatial misregistration between scenes (possibly greater than one pixel) which can result in change detection errors (e.g. through the erroneous presumption that a change in scene brightness represents true change rather than misregistration); 2) locating the threshold between radiance change and no radiance change pixels is often arbitrary; and 3) radiance changes are often the result of phenomena other than urban land cover changes (e.g., agriculture changes, reservoir level fluctuations, etc.). Change-detection errors resulting from spatial misregistration can be reduced by accurately registering Landsat data by procedures described previously.

To establish a threshold (*T*) between radiance change pixels and no-radiance change pixels, either of two threshold techniques have been found to be adequate. For the first technique an empirical approximation of *T* is used. This consists of interactively adjusting the threshold until the desired balance of change from no-change radiance is obtained (Figure 30-58). The threshold is interactively adjusted by training on large areas of observed or predetermined urban land cover change until an appropriate estimate of *T* is obtained (Royal, 1980; Jensen, 1981). This *T* value is then used for the entire urban area.

A statistical approach developed by Ingram et al (1981) also may be used to specify *T*. Through the use of an estimate of the within-scene variability as input into a univariate equation, an estimate of *T* can be determined using

$$T = \bar{y} \pm t_{\alpha,\, n-1} \cdot s_w \sqrt{\frac{n+1}{n}} \qquad (30\text{-}17)$$

where *T* is defined by the upper and lower thresholds, s_w is the within-scene standard deviation, $\bar{y}$ is the mean difference of the within-year averages, *n* is the sample size from which the standard de-

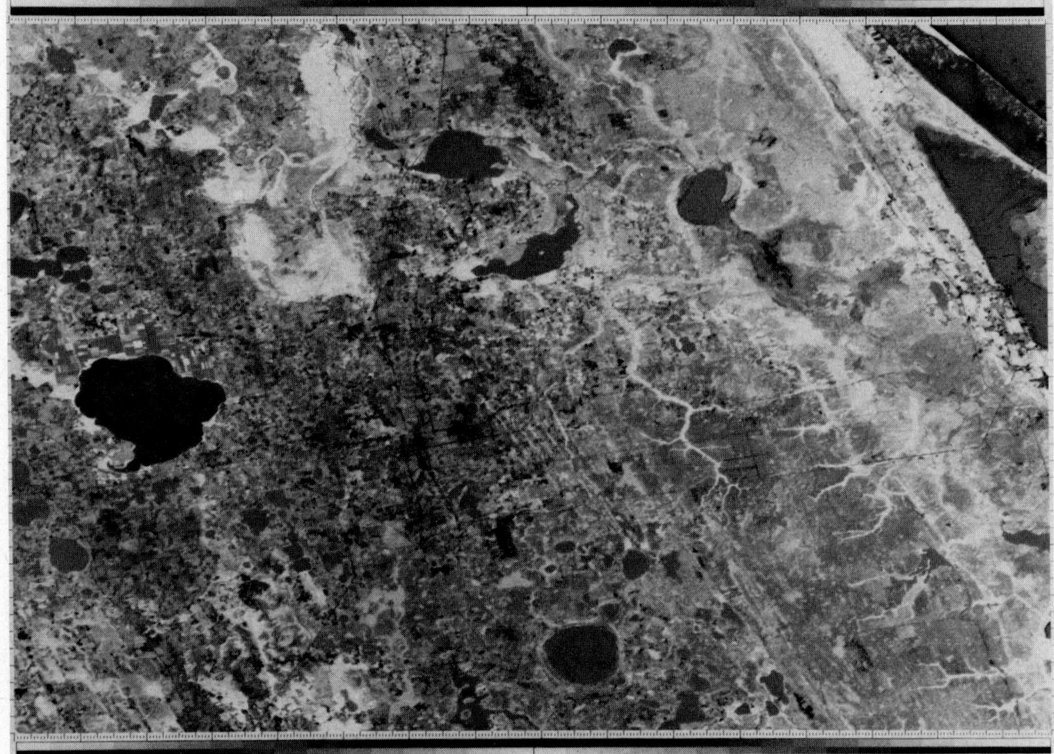

Fig. 30-54 Enhanced Principal Component Transformation of the Orlando, Florida SMSA. [Source: Friedman, 1978]

viation is selected, and t is selected from a t-distribution with n-1 degrees of freedom. The within-scene variation is estimated for both time 1 and time 2 by calculating the variance from differenced data sets that occur on consecutive Landsat overpasses. The larger of the two standard deviations is used. However, if the standard deviations have been determined as homogeneous they are combined to provide a better estimate. If assumptions of statistical independence and normality are met, a pixel with a differenced value at the tail side of the T, has a specified confidence interval.

Changes detected by image differencing are not restricted to the urban versus non-urban categories. For example, agricultural land-use changes may cause large spectral differences which may be mistaken for non-urban to urban land use change.

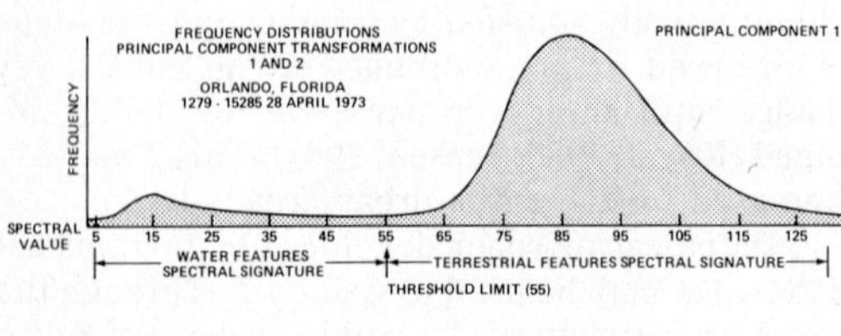

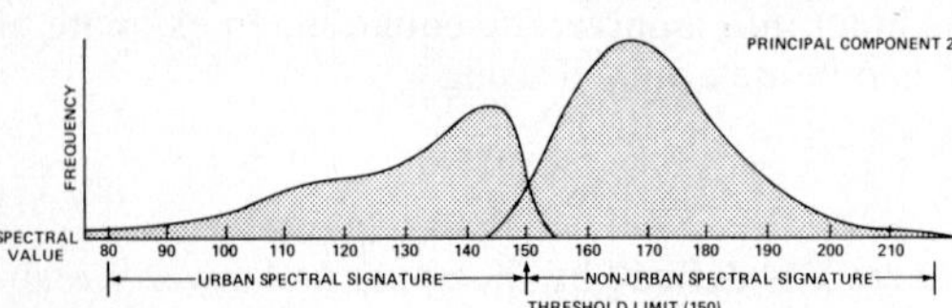

Fig. 30-55 Histograms of the first two principal components of Landsat MSS data, illustrating differences in cover features. [Source: Friedman, 1978]

Image Regression

A variation of the image differencing technique is to use a least squares transformation (i.e., image regression) between dates to reduce adverse effects from differences in atmospheric conditions or sun angles. The model is

$$e_{ij} = Y_{ij}\,(t_2) - (a + b\,X_{ij}\,(t_1)) \qquad (30\text{-}18)$$

where Y is the radiance value of the pixel (i-line, j-column) of the Landsat band(s) at time 2 (t_2); X is the radiance value of the pixel at time (t_1); e_{ij} is the residual of the regression; a defines the intercept; and b adjusts for differences in variance. When $a = 0$ and $b = 1$, then the image regression method is similar to image differencing. The regression technique accounts for differences in the mean and variance between radiance values for different dates. The expected value of e_{ij} is zero with values departing from zero indicating radiance changes. To establish a threshold for the radiance

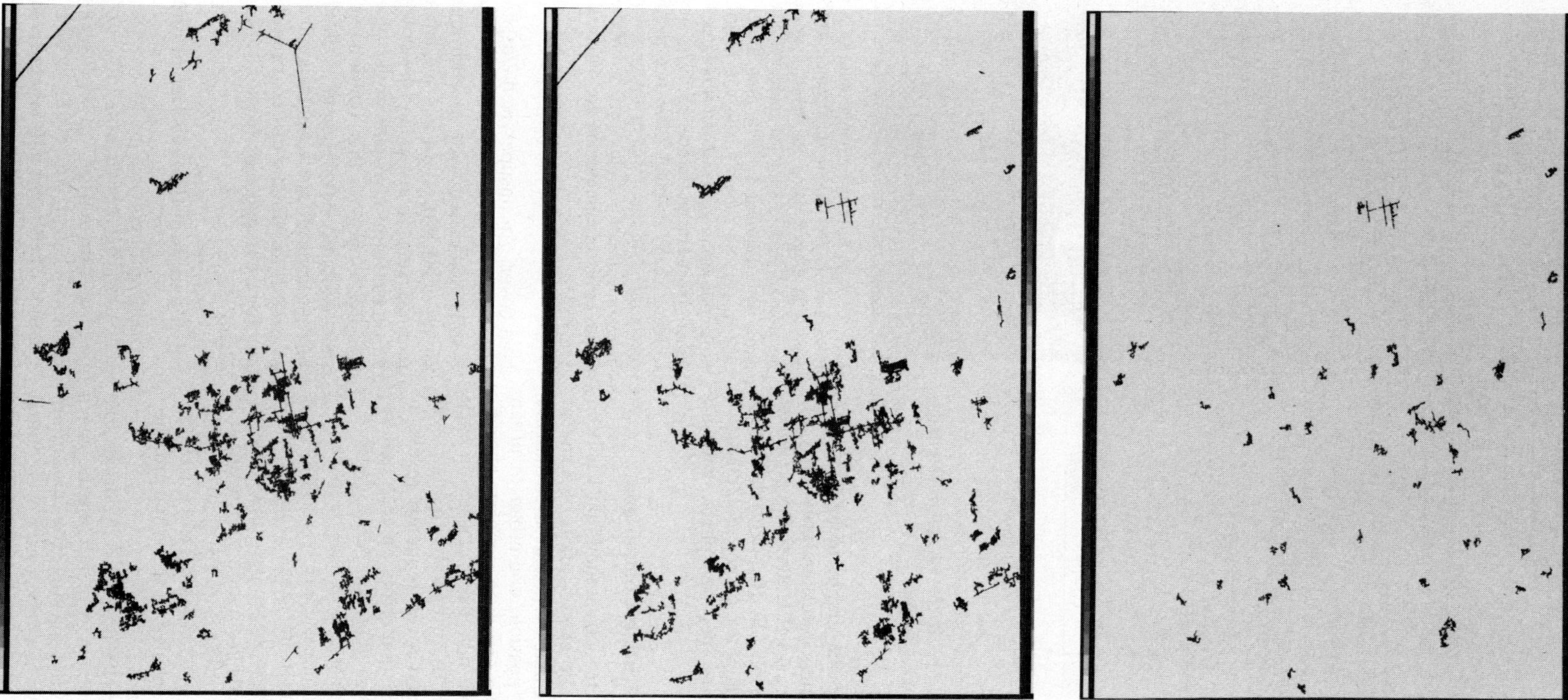

Fig. 30-56 The outcome of the threshold of urban from non-urban cover for the second principal component for both 1973 (left) and 1975 (middle) after water features were removed through analysis of the first principal component. Areas of urban change are illustrated in the picture to the right. [Source: Friedman, 1978]

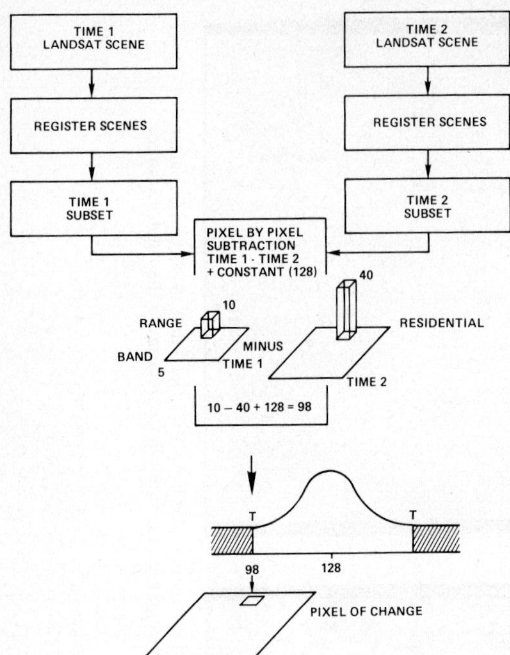

Fig. 30-57 Flow diagram of the image differencing change detection scheme. [Source: Toll]

Fig. 30-58 A land use change map of a portion of the Fitzsimmons 7½ minute quadrangle near Denver, Colorado, produced using the band 5 image differencing method applied to Landsat imagery. The two Landsat dates were October 1, 1976, and Septebmer 30, 1978. The imagery for the two dates was registered to a UTM projection using ground control points. The results are superimposed on an October 15, 1978, aerial photograph (original scale 1:52,000). The two symbols represent pixels found in the lower or upper tails of the change histogram. The thresholds were determined empirically. Some errors of commission are outlined. [Source: Jensen, 1981]

changes in *e,* either the empirical or statistical thresholding techniques described previously for image differencing are used. Unfortunately, image regression often results in the degradation of non-urban to urban land use change-detection accuracy while reducing the atmospheric and sun angle effects.

Post-classification Comparison

A third change-detection method that has been evaluated requires the classification of Landsat data for two or more dates. The classified data are then compared for possible non-urban to urban cover changes. In this approach, accurate classification of urban from non-urban classes on each data set is essential. Unfortunately, the heterogeneity of the urban land cover frequently results in overall high classification errors on each date resulting in poor change detection performance.

For example, in one study a feature-space partitioning classification procedure was used to discriminate non-urban from urban land cover on two anniversary dates. A comparison of classification results yielded information on non-urban to urban land-cover changes (Toll et al., 1980). Overall, high change detection errors were reported and were attributed to the urban area heterogeneity problem.

Work by Riordan (1980) reduced the urban area misclassification problem by identifying the stages of urban conversion having the maximum spectral contrast. All new urban areas were found to have a period of construction activity when the excavated areas contrasted well with surrounding land cover. An unsupervised classification of the study area showed that the excavated areas were readily associated with spectral clusters (Figure 30-59). This procedure has the advantage of detecting a stage in urban land cover which is distinguishable from other land cover types. It may be possible to monitor that location over a period of time to identify development. A disadvantage may be that several Landsat scenes must be analyzed to ensure the finding of excavated areas prior to any regrowth and then to monitor those areas for changes.

Evaluation of Change-Detection Techniques

Image differencing and image-regression procedures require the indentification of thresholds to obtain information on the areas exhibiting radiance change. An advantage of using image regression over image differencing is that regression procedures can adjust for between-data environmental differences in the variance. Unfortunately,

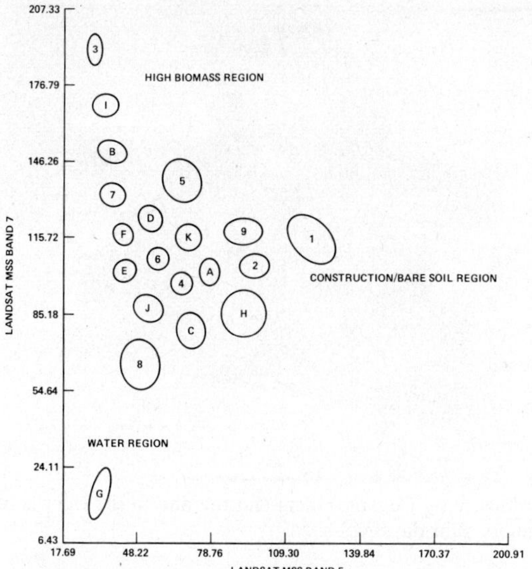

Fig. 30-59 Plot of spectral clusters, band 5 versus band 7, Fort Collins, Colorado, region showing the construction/bare soil, water, and high biomass clusters. [Source: Riordan, 1980]

regression procedures are more costly to implement than image differencing procedures. In addition, image regression frequently results in the degradation of non-urban to urban land cover change information while reducing the atmospheric and sun angle effects. Finally, both image differencing and image regression techniques often identify changes other than non-urban to urban.

Results from post-classification comparison, on the other hand, provide the non-urban to urban land cover change of interest. In addition, multitemporal problems from sun angle and atmospheric condition differences usually are not a severe a problem. This is because the classification comparison requires the analysis of only a single date at a time. A disadvantage of post-classification comparison is that the categorization of urban land cover on each individual date is often inaccurate due to the urban area heterogeneity problem discussed previously. This often results in substantial change detection error.

Determining urban change by detecting and then monitoring construction activity also can be useful. Unfortunately, this post-classification method requires significant additional work, since several Landsat scenes must be processed to detect and monitor construction activity.

THE INTERFACE OF REMOTE SENSING AND GEOGRAPHIC INFORMATION SYSTEMS FOR IDENTIFYING URBANIZING AREAS

The determination of urban boundary locations can be made more efficient by interfacing remote sensing technology with geographic information systems. (Mills and Dwyer, 1979; Lowe, 1978; Knapp et al., 1979; Shelton and Estes, 1979; Estes, 1981.) The use of a geographic information system provides the ability to analyze Landsat results and census information in conjunction with one another. Using this configuration, Census Bureau geographers have interactively determined the location of final urban boundaries. For example, consider the following study of Orlando, Florida.

The Orlando, Florida, SMSA underwent rapid population growth in the 1970's. In order to map or delineate the urban expansion, Landsat data and census information, such as population density statistics and census unit-area boundaries, were integrated and analyzed (Friedman, 1978; Davis and Friedman, 1979). Figure 30-60 illustrates the integration of census-tract boundaries with Landsat imagery. However, Enumeration District boundaries or other census unit-area boundaries could also be used.

It was found that an effective way to analyze results by census unit-area was to utilize choropleth maps showing population densities or urban extent (Figure 30-61). The choropleth maps were compiled from population statistics obtained from previous census records. To update these maps, digital urban land-cover information obtained from analysis of current Landsat data was integrated with the choropleth maps (Figure 30-62). Tabular statistics such as those given in Table 30-12 were used to assist in deciding whether or not the census unit-area was urban.

The analysis of Census Bureau information and Landsat data in an automated geographic information system proved to be superior to manual approaches for delineating urbanized area boundaries. The success of this urban update procedure depended upon the accuracy and reliability of input data.

Fig. 30-60 Census Tract boundaries are combined with a Landsat MSS band 5 enhancement for the Orlando, Florida, SMSA. [Source: Friedman]

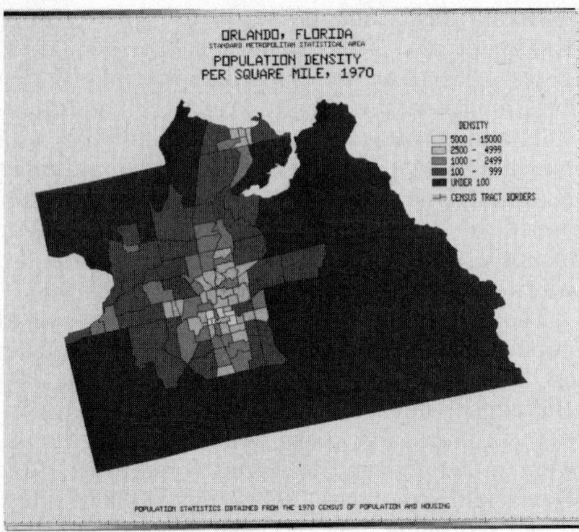

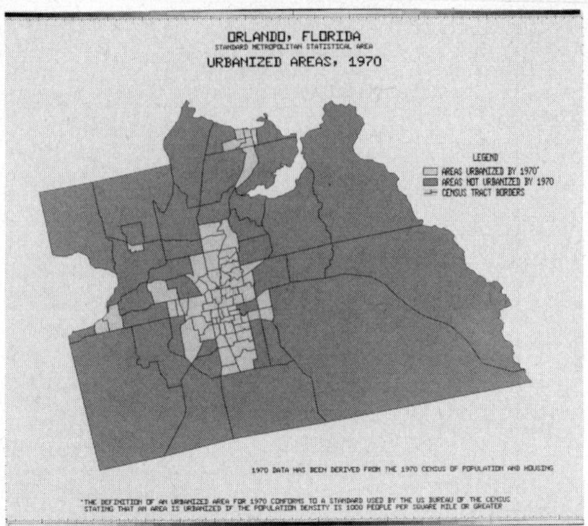

Fig. 30-61 The choropleth map to the left is of population density by Census Tracts and the one to the right is of non-urban versus urban also by Census Tracts for the Orlando, Florida, SMSA.

THE USE OF SENSORS IN ADDITION TO LANDSAT MSS FOR URBAN AREA DELINEATION

Census Bureau geographers currently utilize aerial photography to delineate the outer line of an urban area. Because of data acquisition problems and the non-digital nature of aerial photography, the NASA/Census Bureau project evaluated the utility of Landsat MSS data. Landsat MSS data were the only digital remote sensing data routinely available for evaluation. However, because of the potential of other remote sensing systems to meet Census Bureau geographer goals, additional systems were also examined resulting in: 1) an assessment of the Landsat-D program; 2) an evaluation of the Space Shuttle program; and 3) an evaluation of Seasat Synthetic Aperture Radar (SAR) data.

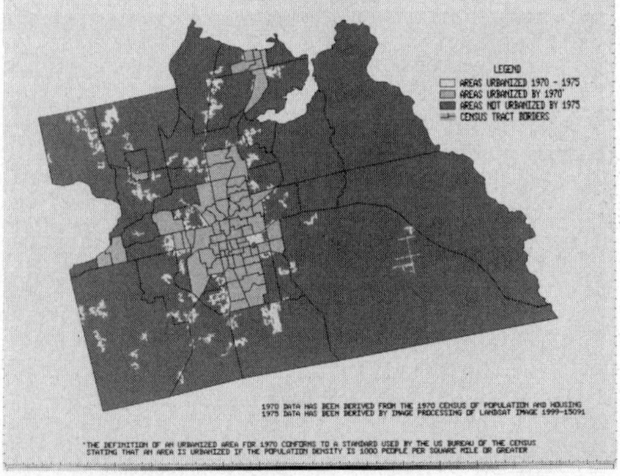

Fig. 30-62 The integration of change detection results from Landsat imagery with non-urban versus urban choropleth maps for Orlando, Florida. [Source: Friedman]

New remote sensing capabilities in the Landsat-D program will be provided by the Thematic Mapper (TM). This includes an increased spatial resolution (30 m for TM versus 79 m for MSS), additional channels (7 for TM versus 4 for MSS), and improved radiometric accuracy that will enhance the signal-to-noise ratio. These changes may improve the quality of data for use in urban area delineation. Figure 30-63 illustrates a simulation of TM data (TM channel 7 not shown because of sensor noise problems) compared with MSS data for a study area in the fringe zone of Denver. Because of the higher spatial resolution, the TM data provided a more detailed view of the urban fringe zone. Royal (1980) discusses the Landsat-D program and related Census Bureau applications.

Padron et al (1980) reviewed the proposed Space Shuttle system payload for potential applications by Census Bureau geographers. Of special importance were the Shuttle Imaging Radar (SIR) and the Large Format Camera (LFC). Two prototypes of SIR are planned to be flown: SIR-A and SIR-B. SIR-A, flown in the fall of 1981, carried an L-band (25cm) SAR system for surface-observation experiments utilizing residual Seasat SAR modules. For more information on SIR and SAR systems the reader should consult Chapter 25. SIR-B is a multifrequency synthetic aperture radar derived from military and geologic investigations with possible applications for urban mapping.

The Large Format Camera (LFC) is a high performance metric camera that may provide uniform, timely, and synoptic coverage of certain SMSAs. From a nominal altitude of 300 km, each 23 cm × 46 cm frame will cover 225 × 450 km on the surface with a ground resolution of between 14 and 25 m depending on the choice of the film.

SUMMARY

Various areas of remote sensing applications were evaluated for use in the urban boundary program at the U.S. Bureau of the Census. During the initial stages of the study, results of Landsat image enhancements and classifications were demonstrated to be as accurate for delineating urban areas as results obtained using aerial photography. Similarly, urban change-detection results demonstrated that urban change masks were also useful in delineating urban areas. Emphasis was placed on the development of various change-detection processing approaches such as image differencing, image regression, and classification comparison. In addition, an attempt was made to improve overall accuracy using preprocessing techniques such as principal-component analysis. Finally, a geographic information system was used to integrate census statistics and change-detection boundaries and to provide the flexibility necessary to input additional data bases. Other sensor systems such as the Landsat Thematic Mapper and the Shuttle System payload will provide alternatives to Landsat MSS data during the 1980's for monitoring the expansion of urban areas.

POPULATION ESTIMATION

INTRODUCTION

The need by governments at all levels for timely, accurate demographic data is well recognized. The U.S. decennial Census of Population, for example, is mandated by the Constitution to provide the basis for allocating seats in the U.S. House of Representatives as well as in scores of state legislatures and city councils. The results are also used as a basis for distributing federal aid to state and local governments. With so much at stake, many government officials have found a decennial census to be inadequate; intercensal estimates are needed. Congress has discussed the idea of a less-detailed census five years after each national census but as yet no legislative action has been taken. Cost is the major constraint; the cost of the 1980 census to the American tax payers was over $1 billion.

In the Third World, population data are no less needed than in the U.S. Unfortunately, few national censuses have been undertaken and, of those that have, many have been grossly inaccurate. Inadequate maps, poorly trained enumerators, a highly mobile (often nomadic) population, vast distances, and limited financial resources have made it difficult to conduct the type of census common to the more industrialized nations. Still, as inaccurate as the data may be, they are often the only data available, so year after year they continue to be used as the basis for government policies.

With such a demand for population data, it would seem that a variety of remote sensing techniques would have been developed to help meet this need. However, if the existing remote sensing literature can be viewed as a valid indicator, such techniques have not been developed, or employed, to any great extent. Their application has been limited largely to Third World countries where census population data are virtually nonexistent.

As a source of cost data, the literature has been even less revealing. Relative costs can be inferred from the few published studies but specifics are lacking. Still, in terms of both timeliness and costs, remote sensing would seem to offer great utility for acquiring certain types of population and housing data.

POPULATION ESTIMATION TECHNIQUES

On the basis of published research, three techniques have been developed for estimating population from remote sensing imagery. As Henderson (1979) has pointed out, population estimates are not obtained directly but rather are inferred from various housing and land-use characteristics which can be identified on the imagery. The three techniques, therefore, are referred to by the surrogate measures which they employ: the dwelling-unit, the land use or area-density, and the built-up area. The three techniques differ in terms of the accuracy of results, the usefulness of the results, and the extent to which they have been successfully applied.

Dwelling Unit Technique

The use of dwelling units as a surrogate for estimating population has been shown to be valid (Starsinic and Zitter, 1968). In its simplest form this technique can be expressed as follows:

$$\text{Estimated Population} = (\text{Number of Dwelling Units}) \times (\text{Average Family Size}) \quad (30\text{-}20)$$

A number of studies (Green, 1956; Hadfield, 1963; Binsell, 1967; and Lindgren, 1971) have demonstrated how dwelling-unit estimates can be derived from aerial photographs. Using various scales of imagery ranging from 1:4,500 (Hadfield) to 1:20,000 (Lindgren), dwelling-unit estimates have been derived on the basis of such criteria as form and structure of roof, number of chimneys, number of stories, relative size of structure, presence of sidewalks and footpaths, presence of garages, carports, driveways and parking areas, and amount and quality of vegetation. Although results have differed slightly in terms of the degree of accuracy, the trends have been consistent. For example, numbers of residential structures have consistently been identified with great accuracy (99%+), although the tendency has been to slightly underestimate. Single-family detached dwellings have tended to be overestimated, while estimates of all dwelling units have tended to be underestimated by varying degrees (Lindgren 3%; Green 7%; Hadfield 10%; and Binsell 15%).

TABLE 30-12

Urbanized Land Cover Statistics for the Orlando SMSA
Orlando, Florida
1970 and 1975

1970 Statistics Based on the 1970 Census of Population and Housing
1975 Statistics Based on Urban Change Detection from Landsat Imagery

	Census Tract Number	Total Acres Per Tract	1970 Population Statistics		1970 Urbanized Land Cover Statistics		1975 Urbanized Land Cover Statistics		Urbanized Land Cover Change between 1970 and 1975		
			Numbers of People	Density Per Mile Squared	Acres	Pct.	Acres	Pct.	Acres	Pct.	Major
102	No Tract	1,161,155	0	0.0	0	0.0	0	0.0	0	0.0	
66	166.00	89,141	2,224	16.1	0	0.0	6	0.0	6	100.0	
91	212.00	64,934	2,131	21.0	0	0.0	0	0.0	0	0.0	
67	167.00	157,858	7,379	29.9	0	0.0	1,483	0.9	1,483	100.0	
71	177.00	76,679	3,817	31.9	0	0.0	2,406	3.1	2,406	100.0	
86	207.00	21,558	1,687	48.6	0	0.0	671	3.1	671	100.0	
89	210.00	11,803	943	51.1	0	0.0	684	5.8	684	100.0	
79	179.00	30,373	3,533	74.4	0	0.0	2,026	6.7	2,026	100.0	
70	170.00	24,780	3,345	86.4	0	0.0	1,351	5.5	1,351	100.0	
92	213.00	26,304	3,834	87.8	0	0.0	0	0.0	0	0.0	
78	178.00	30,054	4,600	98.0	0	0.0	1,694	5.6	1,694	100.0	
68	168.00	41,116	5,369	99.1	0	0.0	2,864	7.0	2,864	100.0	
48	148.00	15,705	3,373	137.5	0	0.0	1,278	8.1	1,278	100.0	
93	214.00	6,035	1,650	175.0	0	0.0	636	10.5	636	100.0	Yes
95	216.00	11,830	3,519	190.4	0	0.0	1,253	10.6	1,253	100.0	Yes
87	208.00	12,496	3,889	199.2	0	0.0	1,388	11.1	1,383	100.0	Yes
35	135.00	5,125	1,717	214.4	0	0.0	442	8.6	442	100.0	
95	216.00	4,349	1,461	215.0	0	0.0	294	6.8	254	100.0	
75	175.00	14,725	5,823	253.1	0	0.0	1,327	9.0	1,327	100.0	
65	165.00	9,319	4,067	279.3	0	0.0	0	0.0	0	0.0	
64	164.00	7,420	3,959	341.5	0	0.0	687	9.3	687	100.0	
49	149.00	5,100	3,908	364.9	0	0.0	739	14.5	739	100.0	Yes
72	172.00	1,749	1,007	368.5	0	0.0	0	0.0	0	0.0	
24	124.00	3,575	2,610	467.3	0	0.0	1,260	35.3	1,260	100.0	Yes

90	211.00	3,545	2,345	513.6	0	0.0	0	0.0	0	0.0	
31	131.00	1,503	1,259	535.9	0	0.0	1,048	69.7	1,048	100.0	Yes
105	221.00	4,526	3,865	546.5	0	0.0	726	16.0	726	100.0	Yes
47	147.00	3,437	3,031	564.3	0	0.0	662	19.2	662	100.0	Yes
50	150.00	5,014	4,558	581.8	0	0.0	146	2.9	146	100.0	
94	215.00	5,223	4,840	604.3	0	0.0	285	5.4	285	100.0	
101	222.00	5,517	5,693	660.4	0	0.0	480	8.7	480	100.0	
77	177.00	2,795	3,197	732.1	0	0.0	72	2.6	72	100.0	
96	217.00	4,795	5,620	764.5	0	0.0	489	10.4	489	100.0	Yes
51	151.00	5,075	6,381	798.4	0	0.0	130	2.6	130	100.0	
45	145.00	1,792	2,127	800.0	0	0.0	403	23.7	403	100.0	Yes
69	169.00	3,943	5,371	871.0	0	0.0	1,374	34.9	1,374	100.0	Yes
36	136.00	4,247	5,901	874.1	0	0.0	801	7.1	801	100.0	
41	141.00	2,484	4,045	1,042.2	2,484	100.0	2,484	100.0	0	0.0	
23	123.00	3,628	6,458	1,142.5	3,629	100.0	3,629	100.0	0	0.0	

The authors also found that while errors varied considerably from block to block, overestimations over large areas acted to compensate for underestimations; final totals, therefore, tended to be quite accurate.

In general, the dwelling-unit technique works more effectively in rural areas than in urban. It would accordingly appear to have greatest applicability to Third World countries where populations are still predominantly agrarian and rural. As a typical example, Allan and Alemayehu (1975) applied this technique to a rural farming area of Ethiopia where population data had never been acquired. With 1:20,000-scale black-and-white photos as their data source, the authors employed a plot-sampling method commonly used in forest inventorying. For the seventy-three stereo photos covering the study area, the principal point of each was selected as the center point for a sample plot. Four concentric circular plots, each double the area of the preceding one, were drawn about the 73 principal points. Dwelling units were counted within each sample plot and multiplied by an average household size of five. The results (Table 30-13) revealed that, as the sample plot size increased, population estimates decreased, probably as a result of fewer sampling errors. Unfortunately there was an absence of ground truth so results could not be confirmed. Even in the absence of such data, however, the technique appeared capable of producing useful population estimates.

Critical to the dwelling unit method are the figures for average family (or household) size. While Allan and Alemayehu used an approximate figure of 5 throughout their entire study area, a more effective technique was employed by Lo and Chan (1980) in their study of the Sheung Shui—Fan Ling area in the New Territories of Hong Kong. Lo and Chan developed a typology of dwellings on the basis of aerial photographic and field analyses. Four distinct patterns of rural dwellings were identified and average household sizes determined for each (Figure 30-64). Whereas average household sizes are frequently derived from field sampling, Lo and Chan concluded that, whenever possible, data from previous censuses should be used since they provide more precise estimates of average household size than field surveys.

Not only can the dwelling-unit technique be employed as a substitute for census-taking–in some cases it can also be employed to test the reliability of certain census data. Eyre *et al.* (1970) found two types of errors when they compared census results in Jamaica with aerial photos. First, they discovered areas for which population data either had been lost or were never gathered and so were omitted from the census. Second, they found areas for which population figures were given but in which no one had been living at the time of the census. Similarly, Schulze (1969) found that con-

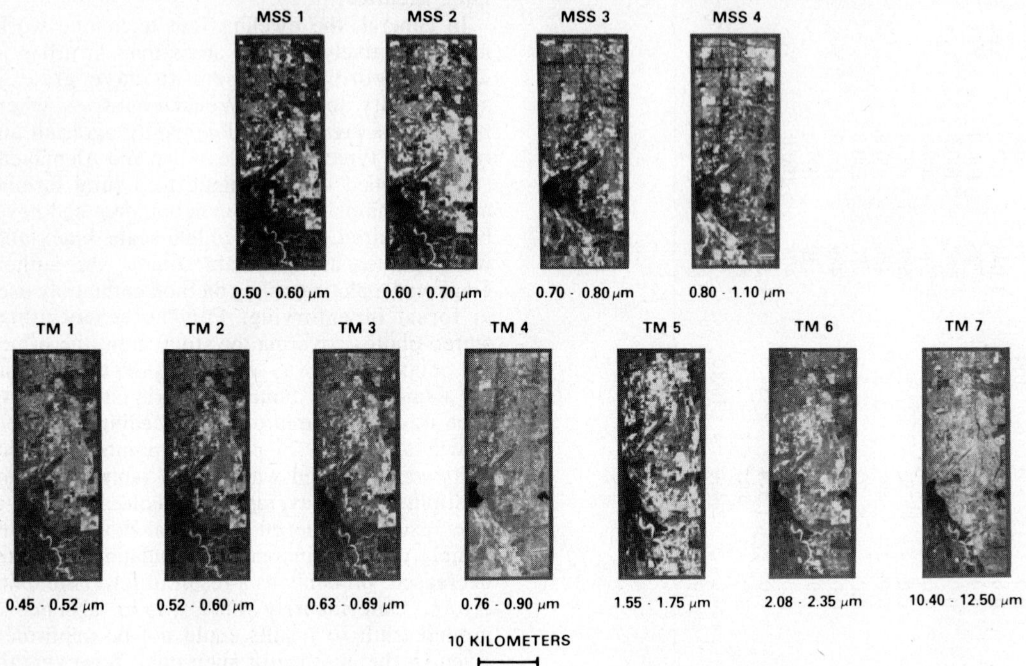

Fig. 30-63 A visual comparison of Landsat MSS bands 1-4 and simulated Landsat TM bands 1-6 for an urban study area in Denver, Colorado. [Source: Toll]

ventional population-estimation techniques were producing inaccurate data for a nearly inaccessible region of South Africa. The dwelling-unit technique provided results that suggested conventional techniques were underestimating the population of the region by as much as 113%. Whenever possible, then, it would seem wise to acquire aerial photos simultaneously while conducting a census; this would be especially true for the Third World.

While the dwelling-unit method seems particularly appropriate in Third World settings, it is not without usefulness in more urbanized societies. An example involves the preparation of intercensal population estimates. Such estimates are needed as often as every two or three years in unusually fast-growing areas. But, with the high cost of field surveys, earlier census data often continue to be used even when known to be outdated. Under these circumstances the dwelling-unit approach would provide state and local agencies with a cost-effective alternative.

In a specific application Hsu (1971) used the dwelling-unit technique to produce choropleth maps of population changes in Atlanta between 1952 and 1968. The 1952 map of population density was derived primarily by counting dwellings on appropriate topographic maps and multiplying the totals by an average household size obtained from census data. Since his data were acquired on the basis of one-fourth square mile grid-cells, values were calculated as:

$$\text{Population Density per square mile} = (\text{Persons Per Household}) \times (\text{Housing Counts}) \times 4 \qquad (30\text{-}21)$$

A 1968 population density map was produced from dwelling counts made from aerial photos (1:5,000) and multiplied by average household sizes derived from local census data. On the basis of these data, a third chloropleth map was produced representing the magnitude of change occurring per square-mile cell between 1952 and 1968. The author felt that this type of information would be of considerable utility to local planners.

For all the apparent advantages to the dwelling-unit approach, it does contain several important drawbacks. For one thing, it is a tedious method. Estimating dwelling units over large areas is a laborious task, although fortunately in Western societies it is becoming possible to integrate housing data from other sources (building permits, utility connections) into the calculations. In the Third World, however, such data are sel-

TABLE 30-13

Estimated Population by Sample Plot Size

Relative Plot Area	Plot Diameter on Photos (cm)	Total Area of All Sample Plots On the Ground (km²)		No. of Plots		Assumed Population Per Household	Sampling Error of Mean Households Per Plot	Mean Enumerated Households Per Plot	Estimated Population	
		(km²)	% Total Study Area	In Sampled Area	In Total Study Area				Total for Study Area	Average Density for Study Area Persons/km²
1	1.000	2.2934	1.1%	73	6356	5	± 26.1%	2.726	86,630	434
2	1.414	4.5867	2.3%	73	3178	5	± 21.7%	5.055	80,325	402
4	2.000	9.1735	4.6%	73	1589	5	± 16.5%	8.932	70,965	355
8	2.828	18.3469	9.2%	73	794	5	± 12.6%	16.000	63,520	318

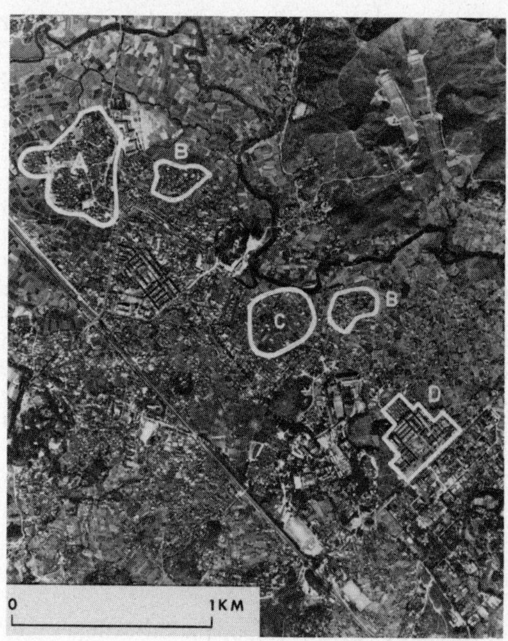

Fig. 30-64 Panchromatic aerial photograph of Sheung Shui-Fan Ling Area of the New Territories, Hong Kong taken on 24 December 1975. Four patterns of rural dwellings are identified: (A) the regular and compact pattern of traditional village houses; (B) the irregular but dense cluster of temporary shacks; (C) the isolated pattern of the new village houses; (D) the regular pattern of mixed multistoried buildings and traditional village houses of market towns. Average household sizes are 3.1 for A, 3.3 for B and 4.4 for C and D. Original scale was 1:25,000. [Source: Lo, 1980]

dom available. Eyre et al (1970) have pointed out the problem of enumerating dwellings in tropical areas where such dwellings are commonly constructed in the shade of large trees. This can also be a problem in the humid middle latitudes if care is not taken to acquire aerial photography during the period when deciduous trees are leafless. In the Northern Hemisphere this period extends from about mid-October to mid-May.

There are further problems of a basically photointerpretive nature—for example, the confusion of non-residential structures with residential structures. Eyre et al. (1970) found in Jamaica that seasonal and temporary shelters were often mistakenly identified as permanent dwellings. Lo and Chan (1980) mention the confusion of farm buildings with residential structures in the New Territories of Hong Kong.

Multi-family dwellings pose another problem entirely, although a number of previously mentioned studies including those of Binsell (1967) and Lindgren (1974) have demonstrated that relatively accurate dwelling-unit estimates can be made for selected areas. However, to attain consistently accurate dwelling-unit estimates in areas of predominantly multi-family structures requires both a high degree of photointerpretation skill and a familiarity with the area(s) under investigation.

A final comment may be in order concerning types of film to be employed. Most of the studies based on the dwelling-unit approach have used black-and-white photography. Clearly black-and-white photos are the least expensive and, over vast areas where the numbers of individual photos may be numbered in the hundreds, cost is an important consideration. However, under certain circumstances, color infrared imagery may be more useful. In urban areas where haze and pollution may be a problem the haze-penetration capability of color infrared imagery can justify the added expense. Also, in certain areas of heavy vegetation such as the northeastern U.S. the color contrast between natural vegetation and residential structures on such imagery may dramatically improve the accuracy of dwelling counts.

Land Use/Area Density Technique

This method derives population estimates by measuring various types of residential land uses and multiplying the results by average population densities for each type. Collins and El-Beik (1971) were among the first to employ this technique when they undertook an experiment to estimate population in the city of Leeds, England. Limiting themselves to three "common" residential types (semi-detached, terraced, and back-to-back) which they could identify with greater than 90% accuracy, the authors selected a series of enumeration districts comprised exclusively of one or the other of these house types. For the districts consisting exclusively of semi-detached houses, population estimates were derived by the dwelling-unit method. For those composed exclusively of either terraced or back-to-back housing, however, the square footage of such housing was estimated by measuring it in strip lengths on the photos. Totals were divided by what the authors called the photographic factor (PF) to obtain population estimates. The PF was an average population-density per square-foot figure, derived on the basis of census data and Ordnance Survey maps.

To determine the accuracy of their estimates, Collins and El-Beik compared their data to census statistics which, unfortunately, had been acquired two years prior to the date of the aerial photos; the accuracy figures were, therefore, somewhat skewed. Overall, the population estimates for districts containing exclusively terraced houses exceeded the census figures by 0.8%, while estimates for districts of back-to-back houses exceeded census figures by only 0.3%. The most varied results were obtained for the districts containing semi-detached houses and in which the dwelling-unit method was employed. On the whole these latter estimates were 6.4% less than comparable census figures.

Kraus et al (1974) applied this technique to four California cities in order to see what effect size of city would have on the results. Employing 70 mm black-and-white imagery acquired at a contact scale of 1:600,000 and which was enlarged optically to about 1:40,000, the authors mapped land use according to the following system.

R_1 — Single-family residence
R_m — Multi-family residence
T_P — Trailer park residence
C — Commercial/industrial

Population densities for the three residential land use types were derived from census block data. Population estimates for each city were computed in the following manner:

$$P = (A_{R_1} \cdot D_{R_1}) + (A_{Rm} \cdot D_{Rm}) + (A_{R_{tp}} \cdot D_{R_{tp}}) \quad (30\text{-}22)$$

Where P is the total estimated population: A_{R_1}, A_{Rm}, $A_{R_{tp}}$ are the areas devoted to each land use type; and D_{R_1}, D_{Rm}, $D_{R_{tp}}$ are the population densities for each land use type.

The results when compared with census data (Table 30-14) proved very favorable. The authors concluded that if larger-scale mapping were conducted (1:20,000), if a larger number of land use categories were employed, and if some population density figures were derived for commercial land use, resulting overall population estimates would be even more accurate.

Thompson (1975) has suggested that with all the land use data being gathered by the U.S. Geological Survey's Census City Project, it may even be possible to produce population estimates for small areas within cities. To test this theory Thompson selected areas in Washington, D.C. and compared changes in the amount of residential land use between 1970 and 1972; all residential categories were collapsed into a single category. Differences in the amount of residential land use

TABLE 30-14

Population Estimation Data

Information Category	Fresno	Bakersfield	Santa Barbara	Salinas
Estimated Population	235,270	170,226	142,790	54,866
Actual Population	259,028	180,263	133,437	58,896
Percent Error	9.17% (−)	5.57% (−)	7.00% (+)	6.84% (−)

were multiplied by an average population density (generated from 1970 census data) to determine net population gain or loss during the two year period. The results displayed an underestimation of about 5%.

The land use/area density technique, like the dwelling-unit technique, is not without its difficulties. To provide accurate estimates, the land use/area density technique requires detailed land-use information. If cities continue to improve the quality and accessibility of their land-use information, the technique will become more dependable. However, if acreages of various land-use types are to be measured directly from aerial photos, population estimates provided by this technique probably will prove unreliable, although Collins and El-Beik (1971) found their measures of various housing types from aerial photos differed by only negligible amounts from comparable ones obtained from Ordnance Survey maps.

Built-Up Area Technique

This method of population estimation is based upon the relationship between the built-up area of a settlement and its population—a relationship examined by Nordbeck in The Law of Allometric Growth (1965). An early test of this relationship was undertaken by Holz et al. (1969) in the Tennessee River Valley area where a file of sequential aerial photos was available for use. The authors, employing step-wise linear regression, tested the relationship of four variables to population size, including built-up area, transportation links, population of nearest large settlement, and distance of nearest large settlement. Their results revealed marked variations between population and area for different sized settlements. They concluded that, as the population of a settlement increases, variables other than the size of the built-up area become increasingly necessary for accurately estimating a settlement's population.

Ogrosky (1975) later applied this same test to cities in the Puget Sound area but used high altitude (1:135,000-scale) color infrared transparencies for acquiring the necessary data. Again a high degree of linear association was found between the region's population and the four variables. Contrary to the findings of Holz et al., in this study the size of an urban area was found to be by far the best single estimator of urban population.

With the relationship between built-up area and size of population strengthened by Ogrosky's research, Lo and Welch (1977) applied this approach to cities of Mainland China using measurements of built-up areas from 70 mm Landsat transparencies. In spite of the fact that the transparencies were dated 1972–1974 and the population statistics with which the results were compared were dated 1970, the approach appeared to work well. Because of the small-scale of the LANDSAT imagery, however, the technique may be limited to cities of at least 500,000 population.

In the three examples mentioned, measurements of built-up areas were made manually. Anderson and Anderson (1973) have compared such measurements made by human interpreters with those derived by an analog-digital imaging processing system. They found that the population estimates obtained from both techniques were similar to those appearing in published documents.

CONCLUSION

All three population-estimation techniques have applicability in certain situations. The dwelling unit is the most useful in rural areas and particularly in Third World settings when census data are in short supply. The land use/area density technique may be most applicable in the U.S., where increasingly detailed land-use information is being acquired by cities and towns. And finally, the built-up area technique is applicable only to large cities but could conceivably complement the dwelling-unit technique by providing population estimates for Third World cities. What these techniques need is a broader appreciation of their capabilities for acquiring population data. If planners were better aware of these techniques, they would almost certainly apply them with greater frequency and with increasingly more favorable results.

HOUSING QUALITY DATA

INTRODUCTION

Within the United States there are scores of federal, state, and local agencies responsible for gathering and analyzing housing-quality data. While methods of analyzing housing data have become increasingly more sophisticated, primarily through the use of high-speed computers, methods of acquiring such data have changed little; they remain laborious and extremely expensive. Although few housing agencies have even considered employing remote sensing techniques for acquiring housing data, research results in this area have been encouraging.

Basically, the data acquired by housing agencies are of two types. One is concerned with the structural condition and quality of individual residential units, while the other is related to the residential or neighborhood environment. Intuitively, remote sensing techniques would appear to have much greater potential regarding the residential environment than the condition of individual residences.

Characteristics of Individual Units

National data on individual residential units are collected by the U.S. Bureau of the Census through its decennial *Census of Housing* and by the American Public Health Association on a more or less continuous basis. Since 1973 the

Census Bureau has also been publishing an *Annual Housing Survey* based upon a sample of approximately 75,000 housing units drawn from 15 SMSAs. The data acquired in this survey include such items as tenure and race of occupant, number and type of rooms, condition of kitchen and bathroom facilities, type of sewage disposal, and condition of structure. In general, remote sensing techniques can be of little help in providing information of this type. However, the previous discussion on population-estimation techniques revealed that information could be provided on the estimated number of dwelling units, the number of floors, architectural style, number of chimneys, presence of porches, fire escapes, carports and garages, and the amount of yardspace and parking space. While these data cannot necessarily be thought of as a substitute for the survey data, they can certainly supplement them.

Neighborhood/Environmental Quality

Remote sensing can be an effective means for providing data on residential quality, particularly at the neighborhood and community levels. Research in this area has generally followed one of two approaches. The first approach has been to compare data on environmental quality acquired by remote sensing with data acquired by more conventional ground-survey methods. The second has been to test the relationship between housing density and a variety of socioeconomic factors. Both approaches have demonstrated some success.

In one example of the comparative approach, Marble and Horton (1969) examined housing-quality data gathered by the Los Angeles County Health Department and concluded that at least 21 of the variables utilized by the Department were potentially measureable by remote sensing. Of these, seven were shown to be sufficient for discriminating between various housing quality classes:

1. On-street parking
2. Loading and parking hazards
3. Street width
4. Hazards from traffic
5. Refuse
6. Street grade
7. Access to buildings

When these variables were analyzed on aerial photos, it was found that the seven could be reduced to four (street width, on-street parking, street grade, and hazards from traffic) with only a slight reduction (82% to 78%) in the accuracy of assigning blocks to housing quality classes.

Howard et al (1974) employed a similar approach comparing residential-quality data extracted from color-infrared imagery (1:6,000) with those gathered by the American Public Health Association (Table 30-15). Of the thirty-seven environmental criteria used in determining residential quality, thirty-one could be acquired by remote sensing. Several of these criteria were applied to a sample of Denver neighborhoods to determine residential quality. While the authors found the approach "promising", they failed by their own admission to provide any cost comparisons between the remote sensing and APHA systems.

Concerning the relationship between housing density and socioeconomic conditions, Metivier and McCoy (1971) observed that density of single-family housing was an effective surrogate for poverty. In a follow-up study (1973) the same authors employed simple correlation and regression analyses to examine the relationships between density of single-family housing and five socioeconomic variables: percent owner-occupied, percent renter-occupied, house value, average rent, and median family-income. Tests of significance showed relationships between density and income, density and house value, and density and average rent to be significant at the 0.01 level. A similar study undertaken by Henderson and Utano (1975) corroborated these results. Density of single-family housing displayed a strong correlation with average rent, average house value, median family-income, and average number of rooms.

The ability to evaluate housing quality by remote sensing has important implications for the field of public health where recent studies have demonstrated an association between poverty and disease. Accordingly, Rush and Vernon (1975) employed low-level color photography (1:6,000 and 1:12,000) to acquire data on the density and quality of selected residential areas. These data, as well as census data on average rent, housing value, percent renter-occupied, and property value, were compared to a variety of health statistics to determine the degree of relationship. Although neither set of data was found to be superior, data acquired by remote sensing were found to be as useful as census data in determining health outcomes.

CONCLUSION

To this point most research on housing quality analysis has involved single-family housing. Certainly further research is required in terms of multi-family housing. Another weakness of the research in this area has been the failure to provide estimates of comparative costs of remote sensing methods versus the more conventional ground-survey methods. Nevertheless, the results of available research have been sufficiently encouraging to warrant their being brought to the attention of housing authorities.

MONITORING ENERGY CONSERVATION, UTILIZATION, AND PRODUCTION IN URBAN AREAS

INTRODUCTION

The depletion of fossil-fuel energy reserves has prompted interest in the development of energy

TABLE 30-15

Environmental Criteria Used in City Condition Determination

	Criteria Used by APHA	Criteria Extractable from Imagery
A. LAND CROWDING		
1. Coverage by Structures	X	X
2. Residential Building Density	X	X
3. Population Density	X	
4. Residential Yard Areas	X	X
5. Building Frontages		X
6. Multiple versus Single Unit Structures		X
B. CONDITION OF PRIVATE FREE SPACE		
7. Landscaping		X
8. Condition of Grassed Areas		X
9. Presence of Litter or Garbage		X
C. NON-RESIDENTIAL LAND USES		
10. Areal Incidence of Non-Residential Uses	X	X
11. Linear Incidence of Non-Residential Uses	X	X
12. Specific Non-Residential Hazards and Nuisances	X	X
13. Smoke Incidence	X	X
14. Hazards to Morals and Public Peace	X	
15. Non-Structure-Supporting Land (Utilized)		X
16. Non-Structure-Supporting Land (Unutilized)		X
17. On-Street Parking		X
D. HAZARDS AND NUISANCES FROM TRANSPORTATION SYSTEM		
18. Street Traffic	X	X
19. Railroads and Switchyards	X	X
20. Airports	X	X
21. Alleyways		X
E. HAZARDS AND NUISANCES FROM NATURAL PHENOMENA		
22. Surface Flooding	X	X
23. Swamps and Marshes	X	X
24. Uneven Ground	X	X
F. INADEQUATE UTILITIES AND SANITATION		
25. Sanitary Sewage System	X	
26. Public Water Supply	X	
27. Streets and Sidewalks	X	X
28. Condition of Parkways		X
G. INADEQUATE BASIC COMMUNITY FACILITIES		
29. Elementary Public Schools	X	X
30. Public Playgrounds	X	X
31. Public Playfields	X	X
32. Other Public Parks	X	X
33. Public Transportation	X	
34. Food Stores	X	
H. OTHER		
35. Architectural Style of Dwellings		X
36. Unpaved versus Paved Parking Lots		X
37. Lack of Curbing along Parkways		X

management and conservation plans. To be effective, these plans must be broadly based, covering all aspects of the energy cycle from the availability of energy resources to energy-consumption practices. Understanding the energy budget of urbanized areas is of primary importance, since large quantities of energy are required to support the operation of complex urban systems.

Due to complexity of internal structure, in addition to the vast areas covered by modern urban regions, difficulty is frequently encountered in collecting data for urban energy-planning. The data collection problem is further compounded since many urban regions are served by a multitude of public agencies which obtain energy from independent suppliers. Still, the data must be obtained in a uniform and consistent format. One possible source of such data is remote sensing.

Remote sensing is being used in certain instances as an integral part of urban energy-conservation programs. For example, remote sensing in the thermal infrared region may be used to identify buildings with damaged or insufficient insulation. Damaged pipelines used as conduits for transferring heat from generating sources to buildings have been identified as well. These studies have not been limited to the analysis of heat loss of just a few structures. Entire towns have been analyzed to determine their overall energy efficiency and to identify structures which require modifications to reduce needless heat loss.

Effective conservation measures alone cannot solve the energy problem. Fossil fuels are limited and efficient use of the resource will only forestall their eventual depletion. There is a need to investigate and develop new alternative sources of energy. Several suitable energy replacements have been proposed including nuclear, solar, geo-thermal, biomass, and wind-driven power. Research into the utility of these new power sources has been highly speculative, sometimes promising great quantities of power. However, in many of these studies the spatial perspective of the problem was not addressed. Because alternative power sources may be very land-intensive, geographical constraints must be considered when determining their true effectiveness.

Remote sensing technology may provide the regional perspective needed for these energy-replacement studies. Furthermore, when combined with collateral data, remotely sensed imagery may be used to indicate the potential effectiveness of new energy resources. Two applications which will be discussed are indicative of current procedures that employ remotely sensed data for energy planning applications. One application covers the potential for generating electrical energy from solid waste. Not only was it found that significant amounts of electrical energy could be produced from solid waste, but it has been indicated that the problem of refuse disposal could be solved (Garofalo and Martin, 1977). Another research project dealt with the potential utility of solar photovoltaic energy as a source of electricity. It was determined that solar power might be a very useful resource, potentially supplying nearly 50 percent of the electrical-energy demand for a portion of Los Angeles, California.

New schemes must be devised for the management of these data collected for energy management and planning. Inevitably, data will be obtained from many sources and a variety of formats and techniques will need to be devised to integrate remote sensing and collateral data types. Geographic information systems will likely become the basis for this data assimilation. Eventually complex models will be developed from which future energy demand in reference to changes in urban land use can be analyzed and understood (Clayton and Estes, 1979). Thus, the role of remote sensing in urban energy-applications should continue to grow.

THERMAL INFRARED DETECTION OF HEAT LOSS

Thermal infrared imagery obtained over urban and suburban environments has been used to identify sources of heat loss. These studies have included locating structural leaks in buildings, pipelines, and large-building environmental control systems where heat energy is dissipated needlessly. Several studies have reported positive results enabling homeowners, businesses, and governmental agencies to identify and remedy sources of unwanted heat loss.

Qualitative Analysis of Thermal Infrared Imagery

Thermal infrared imagery may be obtained to monitor relative rooftop temperatures for the purpose of identifying structures with excessive radiative heat loss (Figure 30-65). Frequently such structures, depicted as lighter toned features on positive thermal imagery, have insufficient attic insulation, allowing heat to escape through the ceiling to the attic. Since adequately insulated ceilings will not radiate excessive amounts of heat into the attic, corresponding rooftops (as depicted by house number 2, Figure 30-65) are relatively cool and appear dark on the imagery. Warmer roofs appear in lighter tones. Small light-toned patches may be related to structural failures or insulation voids while the overall insulation factor for the rooftop is good (illustrated by house number 1). On a typical image, many houses are found to be relatively well insulated (houses 3, 4,

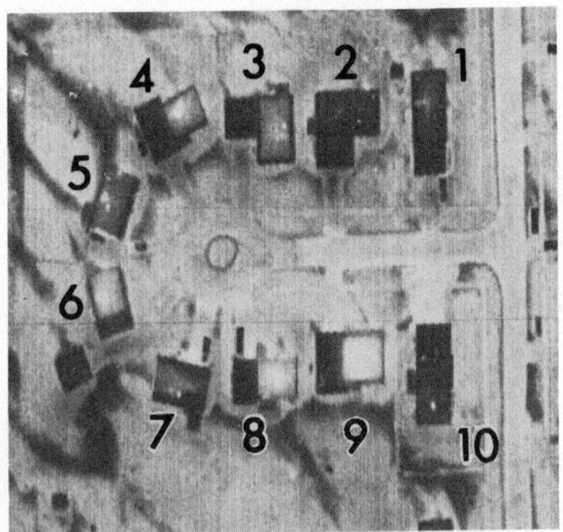

Fig. 30-65 A thermal infrared image (8–14 μm) of a residential area in central Iowa. Various examples of heat loss may be identified through analysis of rooftop signatures. Houses 1, 2 and 10 appear to be well insulated, while houses 6, 8, and 9 appear to be poorly insulated. [Source: Texas Instruments Company and Iowa Utility Association]

5, and 7). However, some structures are found to have severe insulation deficiencies (houses 6, 8, and 9). Heat loss not associated directly with rooftop radiation can be identified from the imagery as well. For example, when garage doors are left open (as in house number 10), excessive heat being radiated laterally into the environment can be detected while rooftop temperatures appear to be relatively cool.

Close examination of thermal infrared imagery (Figure 30-18) can yield additional information. Many houses exhibit dark rims at roofline perimeters. These featues are roof eaves which lie outside the heated exterior of the house. Being exposed to the ambient environment on both top and bottom surfaces, eaves cool quickly at night. Carports and garages are generally unheated and appear darker (cooler) than heated living spaces. Frequently, the warmest features depicted on the images are hot spots located on roofs where heating systems vent energy to the surrounding environment.

It is possible to display the black-and-white thermal infrared images in color through the use of color enhancement techniques (color Figure 30-66). When color-enhanced images are being interpreted it is important to realize that the hottest apparent temperatures are usually depicted as the brightest or lightest toned features. The coolest objects are depicted as darker toned features, such as the house located in the right-center of Figure 30-66 (see arrow). A ground investigation revealed that his house was unoccupied for several weeks before the overflight.

Ground-based thermal-infrared images can also be obtained (Figure 30-67 and Color Figure 30-68). Hot spots attributed to a variety of phenomena including poor sidewall insulation, uninsulated doors and windows, and structural leaks can be identified. Although extreme detail can be obtained from a ground-based system, extensive use of such a system is unlikely due to the time and costs involved in doing extensive surveys of residential areas in this manner.

Quantitative Analysis of Thermal Infrared Imagery

Most thermal infrared studies have been based on qualitative analysis. These are considered both successful and valuable for energy-conservation programs. However, quantifiable techniques for measuring heat loss are needed as well (Hazard, 1979). To date, elaborate models have been devised but not extensively tested (Brown, 1978; Schott, 1978; Bowman and Jack, 1979; Madding, 1979). The models are quite complex, requiring detailed collateral information pertaining to building structure, thermal conductivity of building materials, and local meteorological conditions occurring at the time of thermal sensing. With the proper ancillary data collected, it might be possible to obtain a quantitative estimate of a building's heat loss and of the amount of insulation present in a structure (Brown, 1978). So much collateral information is needed, however, that without the assistance of building residents or automatic collection devices, implementation of these models would be difficult (Schott, 1978).

Fig. 30-67 A ground based thermal infrared image (8–14 μm) used to analyze building heat loss which is not associated with rooftop dissipation. The warmest spots appear as light toned features. [Source: Daedalus Enterprises, Inc.]

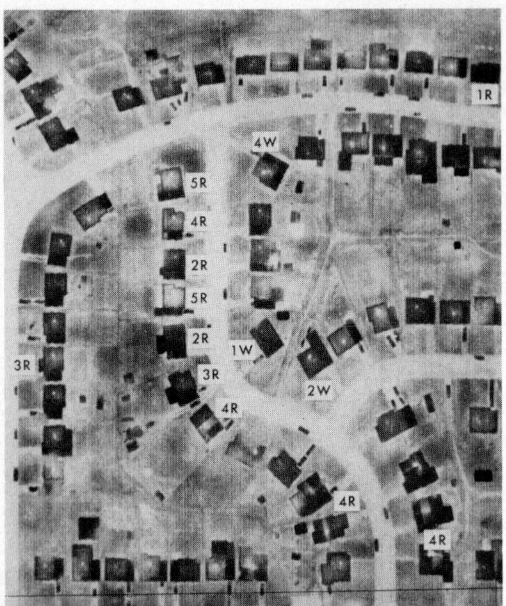

Fig. 30-69 R-value densities of insulation determined from the thermal infrared imagery. [Source: Energy Measures Corporation]

In one study it was possible to obtain generalized heat-loss estimates by correlating rooftop-brightness values on thermal infrared imagery (Figure 30-69) with a calibrated step-wedge (Underwood et al., 1980). The step-wedge was prepared by optical comparison of diffuse film density levels to known surface temperatures. The measured surface temperature and average R-values of amount of insulation were computed for five step-wedge levels (Table 30-16). It was estimated that the BTU loss for each grey level was within ±10 percent of the true heat loss 90 percent of the time (Underwood et al., 1980). The Fort Worth imagery depicts some subtle features of interest (Figure 30-69) because, to a lesser extent, driveways and sidewalks are still warm as a result of the thermal inertia of objects heated during the daytime. Cooler lawns and yard terraces contrast sharply with streets and driveways. There is also considerable variation among the thermal signatures of individual yards due primarily to differences in the moisture content of soils, the amount of vegetation cover, soil type and compaction, and radiation from the sidewalls of houses and other structures in the area. Close analysis of house 5R (located in the center of the image) revealed that a great amount of radiant energy was lost to the surrounding environment from the walls of the house (probably more than from the roof alone). These bright responses along the side walls of houses have been referred to as *blooming, shining,* or *flaring*.

A significant amount of research has been conducted to determine the emissivity of roof surfaces in order to more adequately determine their kinetic temperature and thermal resistance. Emissivity pertains to the amount of radiation emitted by a material in relation to emissions of a blackbody at the same temperature (refer to equation 30-4).

Most roofing materials do not vary greatly in emissivity. Generally, the emissivity values range between 0.88 and 0.94 (Lillesand and Kiefer, 1979), yet this difference can be detected by carefully adjusting the thermal scanner. One major exception to the emissivity range noted is the response of bare or unpainted metalic roofing-materials which characteristically exhibit very low emissivity values (0.22 or lower) and generally appear quite cool (dark) on thermal images. In reference to the thermal image of Fort Worth (Figure 30-69) a number of small, very black rectangles are visible in some backyards. These are small metal-roofed storage sheds. Automobiles parked out of doors also appear quite cold on night time imagery because their metal surfaces have low emissivities. Careful analysis of Figure 30-69 reveals several cars that appear "grey" and are recorded as relatively warm features. These signatures result from the residual built-up heat within the auto after it has recently been driven.

Because of the similarity in emissivity of all but metal roofing-materials (Table 30-1), some researchers feel that emissivity differences are not as significant a variable in heat loss studies as, (1) the cold night sky, (2) roof angle, (3) wind speed, (4) attic ventilation, and (5) moisture conditions on the roof surface e.g., snow, frost, ponded water (Hazard, 1980).

TABLE 30-16

Fort Worth Roof Calibration Code 1978–1979

Code	Diffuse Density (IR Neg)	Surface Temp.	Average R-Value
1-R	.65	−2.92°C	R23
2-R	1.18	−1.55°C	R19
3-R	1.42	0°C	R11
4-R	1.56	+3.2°C	R08
5-R	1.79	+6.0°C	R04

Source: Energy Measures Corporation, Austin, Texas

Optimal Conditions for Thermal Infrared Surveys

Thermal infrared imagery used for urban and suburban heat loss studies is usually acquired during rather specific environmental conditions (Table 30-17). Ideally, the survey is conducted very late at night to minimize effects of solar insolation. During early morning hours just before dawn, maximum contrast between the cooling ambient background and rooftops can be obtained. Ambient air temperatures should be cold (not to exceed 2°C) and sky conditions must be clear because an overcast sky traps radiation and returns some of it to the ambient environment, and especially to flat roof surfaces oriented toward the sky.

Thermal-resistance factors of the roof-sky and roof-attic interface must be considered. The thermal resistance of a radiation barrier, in this case the roof, consists of the barrier resistance plus the thermal resistance of the air film on both sides of that barrier (Hazard and Underwood, 1979). Inside air movements are generally not extensive, and thermal resistance of the inside air film can be discounted. However, outside wind-speeds will seriously affect the apparent temperature sensed by a thermal detector. As wind speed increases, the difference in apparent thermal temperature of objects in the environment decreases, and their spectral signatures become similar (Figure 30-70). There are differing opinions on what maximum wind-speed will prohibit residential thermal sensing. A general consensus suggests that wind

TABLE 30-17

General Considerations for Heat Loss Surveys

	A. ENVIRONMENTAL
1. TIME:	The nighttime thermographic overflight should not begin until four hours after sunset (Cihlar et al, 1977) or one hour after the expected nighttime low (Lawrence, 1977).
2. CLOUDS:	There should be little or no cloud cover (Lawrence, 1977; Cihlar et al, 1977) in order that night cooling will provide maximum radiant temperature contrasts. (This criterion has been disputed by Bowman and Jack (1979) who believe that exact quantitative analysis of rooftop temperature can only be determined under total overcast).
3. TEMPERATURE:	Low ambient nighttime temperatures are optimal, and should not exceed 3°C (Lawrence, 1977). Optimum temperature is near 0°C (Lawrence, 1977; Cihlar, et al, 1977). At temperatures below 0°C ice may form, causing an insulating effect. Daytime maximums should be below 5°C (Lawrence, 1977). If temperature is higher, the nighttime flight must occur later after sunset.
4. DEWPOINT:	The temperature-dewpoint spread should be at least 3°C (Lawrence, 1977) to reduce the probability of the formation of frost or dew.
5. WINDS:	A consensus holds that winds should not exceed 7.5 m.p.h. Calm air is most favorable (Lawrence, 1977).
6. INVERSION:	Temperature inversions create problems in interpretation and should be avoided (Lawrence, 1977).
	B. TECHNICAL
1. PHOTOGRAPHY:	Accompanying aerial photography is essential for referencing when analyzing thermograms (Lawrence, 1977). The photography should be collected on or near the date of thermal sensing (Cihlar et al, 1977).
2. CALIBRATION:	Blackbody settings are adjusted at the operator's discretion, and are best coordinated with some known ground temperatures. No constant temperature range is required.
3. ALTITUDE:	Thermal sensing should be conducted at altitudes which range between 360 and 490 meters (Cihlar et al, 1977).
4. ROOF CONDITIONS:	Roofs should be relatively free of standing water, snow, or ice. These features will greatly alter the thermal emission of a rooftop. If the rooftop is covered by ice, snow, or water, the mission should be deferred.
5. METEOROLOGICAL DATA:	Accompanying meteorological data collected for the day of the flight can be most useful in later interpretation and analysis of the thermographic imagery (Cihlar et al, 1977).
6. PUBLIC AWARENESS:	When extensive surveys are being conducted, public awareness of the project is beneficial, providing much needed information about the nature of individual residences and structures (Budge and Morain, 1978).

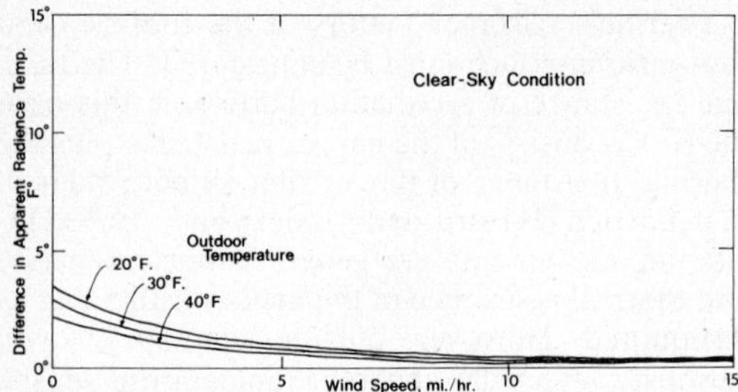

Fig. 30-70 Air movement will affect the sensed apparent temperature of an object. As wind speeds increase, differences in apparent temperatures decrease.

speeds should be less than 7.5 miles per hour, although satisfactory results can be obtained with winds up to ten miles per hour. At wind speeds greater than ten miles per hour, radiance temperature curves for almost all levels of insulation merge so closely that it is almost impossible to make valid estimates of heat loss, or make valid measurements of signature variations between roof surfaces (Brown, 1978; Burch, 1979).

An ideal situation for a heat-loss survey is immediately after the passage of a winter cold-front. Under clearing skys, the cold, dry air behind the cold front is very stable, exhibiting little turbulence in the atmospheric column between the airborne detector and the earth. This condition presents the largest thermal contrast between roof surfaces for thermal detection studies. Even under these ideal conditions, however, such surveys may be hampered by large amounts of precipitation deposited during the passage of the cold front. Moisture accumulations on roof tops in the form of snow, frost, ice, or ponded water may create signature anomalies that may be confused with hot air leaks, insulation voids or other signature variations.

Problems Affecting the Thermal Infrared Surveys

Thermal measurements are often made with the assumption that structures have flat or only slightly sloping rooftops. Although the condition occurs quite frequently in west coast housing, structures located in other parts of the United States have steeply pitched or gabled roofs for the purpose of minimizing the snow load. These structural features cause additional interpretation and measurement problems. Steep roofs frequently have heated air trapped along ridge lines which may cause *hot* linear signatures along ridges which may not be related to structural heat loss. However, these ridge flares may also indicate structural leaks where insulation has failed.

Another problem affecting the quantification of thermal infrared surveys is the orientation of the thermal scanner or radiometer in relation to the flat or pitched plane surfaces of roofs. Frequently one side of a rooftop, the side most directly facing the scanner platform, will appear to be relatively warmer than the side facing away from the scanner. Since in most cases it can be assumed that both sides of the rooftop are radiating the same amount of heat, the differential must be attributed to factors related to scanner geometry and look angle. These variations in the thermal signatures across a rooftop can be partially explained through Lamberts' Law of Cosines. The law states that, for any radiating plane surface that is a perfectly diffusing black body, the intensity of the emitted radiation at a receiving point varies only as a function of the angle between the normal to the surface and the line of sight to that surface (Hackforth, 1960; Tuyahov, 1972). Although roofs are not black bodies, the condition seems to apply to analysis of steeply pitched roofs (Figure 30-71). However, this situation is partially offset since as the angle of line of sight increases, the instantaneous-field-of-view increases, resulting in increased areal coverage for pixels sensed and a greater aggregate temperature. At a look angle of 60°, fifty percent less energy is sensed per unit area. However, since twice the unit area is sensed in comparison to normal, the total energy reaching the sensor platform will be the same for both the normal and 60° look angles (Figure 30-72).

Scanner geometry is not the only factor affect-

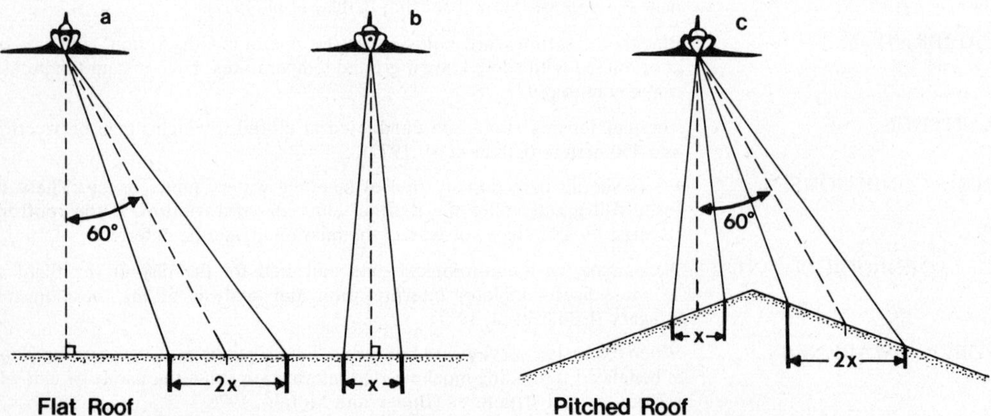

Fig. 30-71 Roof pitch in relation to viewing angle will affect sensed thermal values.

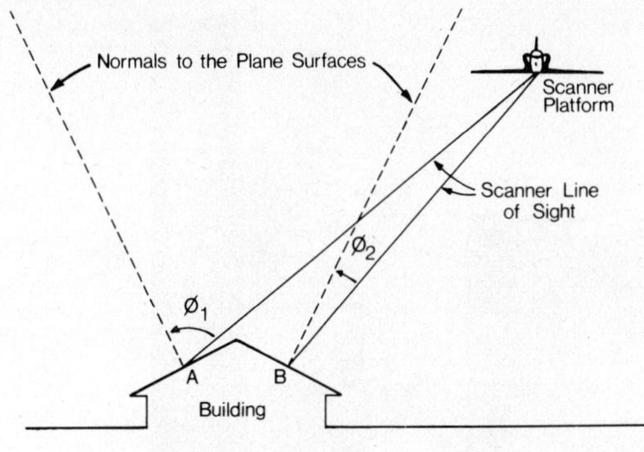

Fig. 30-72 Actual surface area viewed by a detector varies with scan angle. At a scan angle of about 60°, the detector would receive approximately fifty percent of the radiation received from the same surface at a normal angle. However, since the area viewed would be twice as large, the image signature would remain constant. [Source: Texas Instruments]

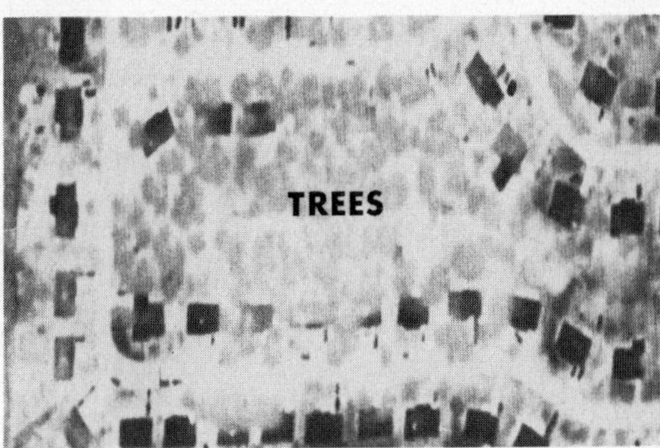

Fig. 30-73 The presence of trees will mask and obscure the spectral signatures from rooftops. [Source: Texas Instruments]

ing sensed temperature readings. Flat roofs lose heat more rapidly to the nighttime sky compared to the sloping roofs. Sloping roofs tend to gain additional heat as a result of sidewall flaring. A variation in roof slope from 0.33 to 0.83 may result in a change in apparent roof temperature of 0.6°C (Brown, 1978). Consequently, the measurement of heat loss from a house with a sloping roof is considerably more difficult and less accurate than for a flat roof (Brown and Cihlar, 1978).

Another problem affecting the interpretation of nighttime thermal imagery is the masking or obscuring effect resulting from the relatively warm signature of trees (Figure 30-73). This interpretation problem is important in forest regions where large trees frequently cover and obscure rooftops.

Some Typical Heat Loss Studies

A number of thermal infrared studies have been conducted for the purpose of conserving heat energy through minimization of structural heat loss. NASA is surveying all Federal buildings to determine structural deficiencies or inadequacies in rooftop insulation (Bowman and Jack, 1979). The Canadian government has undertaken an extensive program to locate structural deficiencies and problems associated with heat loss from large buildings (Figures 30-74 and 30-75), heat distribution lines, and single family dwellings (Cihlar et al., 1977). The Canadian Center for Remote Sensing and the Ontario Center for Remote Sensing have been active in demonstrating the use of thermal infrared imagery for heat-loss studies. They have examined the varying signatures of different insulation materials and the probable causes of structural leaks which contribute to heat loss.

Some of the more intriguing studies are the community heat-loss projects which in most cases have been sponsored by local utility companies as part of their conservation programs. They require a significant amount of public involvement and demonstrate how the remote sensing community can help improve local living standards. In one characteristic application, the entire town of Farmington, New Mexico, was surveyed (Budge and Morain, 1978; Budge and Inglis, 1978). Prior to the survey, an extensive public awareness campaign was conducted, designating the date of the overflight and stressing the importance of such a study. Questionnaires were distributed to the public in order to learn about different housing conditions such as the type and amount of attic insulation, the average monthly energy bill for heating, and other vital information. After the overflight, the thermal data were processed to obtain both black-and-white (Figure 30-76) and pseudo-color (Figure 30-77) images for analysis. Trained analysts employed by the local utility company met with interested residents for the purpose of identifying anomalies associated with damaged or inadequate insulation.

Conclusions

The interpretation of patterns on thermal infrared imagery of urban areas is not a simple task and cannot always be related easily or accurately to the actual kinetic (true) temperature of the object being sensed. Images of urban areas are best used with collateral information such as other forms of remote sensing imagery and maps. Ground-based measurements are particularly valuable, especially if these can be made at the time of overflight.

MODELING THE REGIONAL USE OF ELECTRICITY

Urbanized areas are usually well illuminated at night. From multiple sources such as indoor, outdoor, and street lighting, sufficient light energy is radiated into space to be recorded by satellites equipped with sensitive imaging systems (Figure

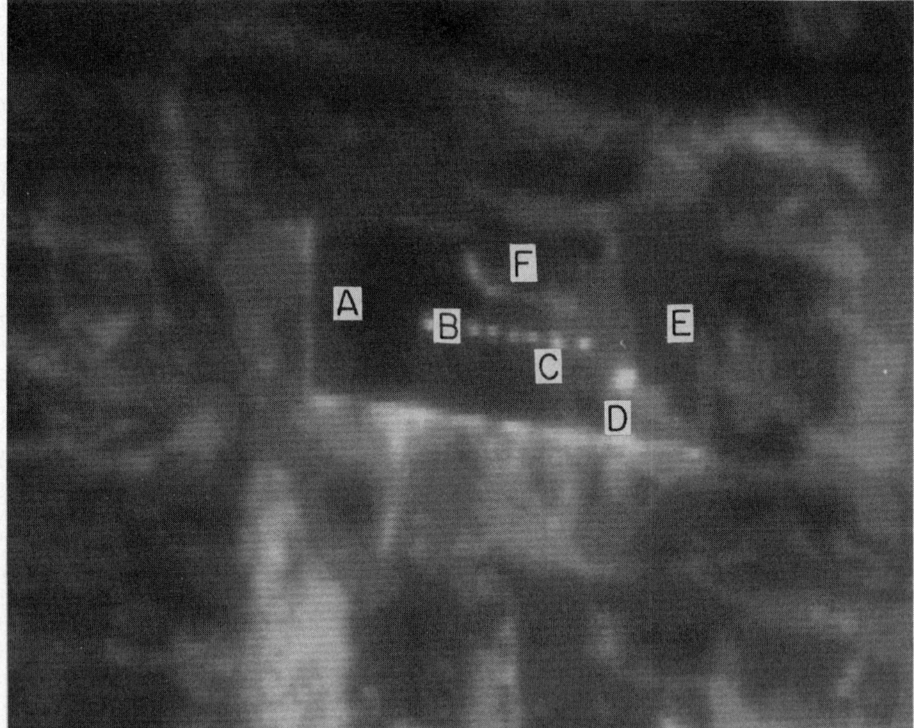

Fig. 30-74 A thermal infrared image and accompanying aerial photograph of a heavy machinery depot. Several structural features such as skylights (c), smokestacks (d), and a penthouse (e), can be identified. A structural anomaly (f) was attributed to water collection at a low spot on the roof. If left unattended, this problem could result in damage to roof structure or insulation. [Source: Canada Center for Remote Sensing]

30-78). Since 1972, the U.S. Air Force has maintained surveillance of the nighttime sky under the Defense Meteorological Satellite Program (DMSP). The light-sensitive sensors on board DMSP satellites have recorded gas flares associated with "burn off" at oil fields, flood lights used by the Japanese fishing fleet for nighttime squid-fishing, as well as illumination patterns of urbanized areas (Croft, 1977).

DMSP satellites provide synoptic coverage of

Fig. 30-75 A thermal infrared image and accompanying aerial photograph of a hospital. The upper portion of the structure (a) has no anomalies. The very bright patch (b) has been attributed to an uninsulated portion of the roof, and a bright spot at the bottom of the structure (c) was attributed to failure around a drain. [Source: Canada Center for Remote Sensing]

Fig. 30-76 A thermal infrared image of a residential area in Farmington, New Mexico. [Source: Technology Application Center, University of New Mexico]

the earth, facilitating regional remote-sensing applications that require nocturnal imagery (Figure 30-79). Of the four sensors that DMSP satellites are equipped with, two channels have been most useful. An infrared band (8–13 μm) has been applied to meteorological applications such as cloud-movement studies. The other channel, imaging the visual and near-infrared wavelengths (0.4–1.1 μm), has been useful for monitoring radiant light sources, most of which are associated with human activities (Croft, 1979). The visual and near-infrared wavelength radiometer is capable of recording data at two resolution levels, 3.6 km and 0.6 km. The nominal resolution of 0.6 km was found to be more sensitive to weaker light sources and best suited to urban monitoring (Welch and Zupko, 1980).

Estimation of Regional Energy Consumption

A method using DMSP imagery for monitoring regional consumption of electrical energy was devised (Welch, 1980). The procedure involved making densitometric measurements from film transparencies of DMSP data to compute the volume of radiant energy emitted by urbanized areas. Each city, or *illuminated urbanized area* (IUA), was defined in three dimensional space (Figure 30-80). The perimeter of the IUA was defined by the (x,y) plane, while the brightness of each IUA was expressed as a z value. Thus, IUA illumination was depicted as a three-dimensional block diagram (Figure 30-81).

Volumetric measurements of aggregate brightness were made to derive a relative ranking of IUAs based on the amount of radiant light energy emitted. However, quantitative estimation of total electrical energy consumption of any IUA from analysis of DMSP data required the use of ancillary data. A 1970 population distribution map (U.S. Bureau of the Census, 1973) was used to verify whether the depiction of IUAs on DMSP imagery conformed to actual population-distribution patterns. Population counts from the 1970 census and power-consumption figures (1975 data) for thirty-five selected urban areas were used to determine if IUA irradiance correlated with urban population counts and electrical energy consumption figures.

The perimeters of several IUAs were compared

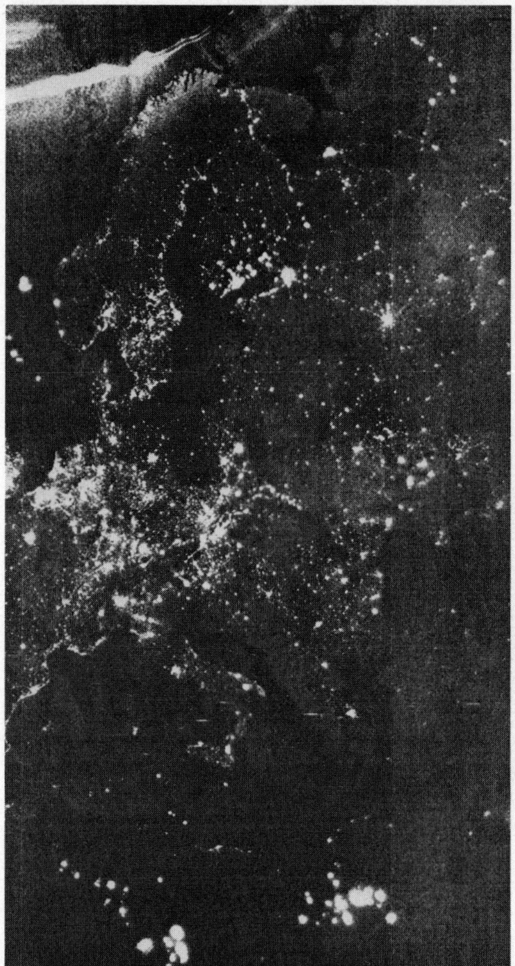

Fig. 30-78 A nighttime DMSP image covering most of Europe and northern Africa obtained on November 28, 1978. The major cities of Europe are clearly visible. With a connect-the-dot approach, several important land transportation routes as well as the coast lines of Mediterranean, North, and Baltic Seas can easily be identified. The brightest spots, appearing at the bottom of the image and again in the North Sea, are natural gas flares associated with burn off at major oil fields. Another interesting feature, the aurora borealis, illuminates the upper-left portion of the image. [Source: United States Air Force Defense Meteorological Satellite Program (DMSP). Film transparencies archived for NOAA at the University of Wisconsin Space Science and Engineering Center]

to the mapped boundaries of urban areas depicted on the 1970 population-distribution map (Figure 30-82). Visual analysis indicated that the imagery and the map correlated well. Exact planimetric measurements of several urban areas were made, and it was observed that IUAs were slightly larger than corresponding urban areas as defined by the Bureau of the Census. The variation was attributed to a combination of the following factors: 1) halation caused by the DMSP sensor system, 2) urban expansion since 1970, and 3) non-conformance of IUAs to exact urban boundaries (Welch and Zupko, 1980). Further quantitative analysis of the size difference was determined to be useless because of the small scales of both kinds of imagery used for the comparison.

With verification that IUA perimeters conform to urban area boundaries, the relationship between IUA emittance and population counts or power consumption data were investigated. Scatter diagrams were constructed from analysis of thirty-five sample cities (Figure 30-83). There was a strong linear relationship exhibited for population versus kWh (power consumption), population versus IUA dome volume, and kWh versus IUA dome volume (Welch, 1980). Regression analysis also suggested that there was a strong correlation between urban population, electrical power consumption, and IUA dome volume.

The researchers concluded that small scale DMSP imagery could be an effective tool for regional analysis of electrical energy consumption (Welch and Zupko, 1980). Estimates of power consumption for all IUAs within a region and power consumption of an entire region could be based on the sampling procedure used for analysis of the thirty-five test sites. However, digitally formatted DMSP data would be needed as a base for deriving precise quantitative figures for IUA volume.

Digital Analysis Problems

Analysis of digital DMSP data indicated that most pixels composing an IUA are saturated and that detail above the saturation level is lost (Croft, 1979). Inspection of a digital image covering the northeastern United States bears our that fact (Figure 30-84). All pixels depicted as white are saturated. The presence of so many saturated pixels prevents the analyst from integrating the light output from cities (Croft, 1979). The severity of this problem can also be seen through inspection of the three-dimensional block diagrams representing select IUAs (see Figure 30-81). All urban areas exhibit the same maximum apparent brightness (expressed as elevation). Additionally, owing to the saturation of the DMSP sensor system the major portion of each area was depicted as a plateau.

The pixel-saturation problem limits the utility of DMSP imagery but possible modification of sensor output could provide more useful data. Then, with the addition of collateral data, DMSP data (or other nighttime imagery) might be useful for monitoring regional energy-utilization and facilitate conservation of limited energy resources.

MODELING THE CONVERSION OF ELECTRICITY FROM SOLID WASTE

The disposal of massive quantities of solid waste generated within urban regions has become an issue of public concern. Available space for

Fig. 30-79 DMSP imagery is well suited to analysis of urbanized areas in the United States. Major urban centers stand out strikingly against the black background on this February 17, 1979 image. [Source: United States Air Force Defense Meteorological Satellite Program (DMSP)]

sanitary land fill is limited, and current air-quality standards prohibit incineration programs in most urban areas. Consequently, suitable remedies for the disposal or treatment of solid waste are being considered. One proposed solution has been to convert solid waste either directly into electrical energy or into a fuel-oil substance which could be utilized for generating electrical energy. The feasibility of undertaking such a waste-conversion program can be assessed by estimating the potential replacement value of energy derived from solid waste in relation to local energy demand.

As of 1980, more than one billion tons of solid waste were produced annually in the United States, potentially yielding nearly 1.3 billion barrels of oil. However, much of this resource material cannot be collected economically because

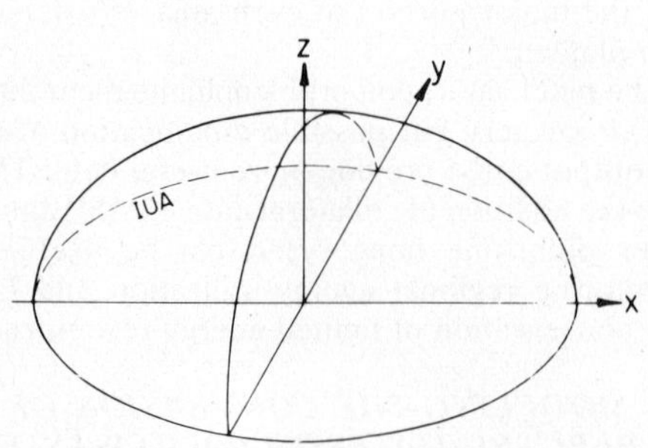

Fig. 30-80 An Illuminated Urbanized Area (IUA) can be defined in three dimensional space. The perimeter of the IUA is defined by the (x, y) plane, while brightness is expressed as a z value. [Source: Welch, 1980]

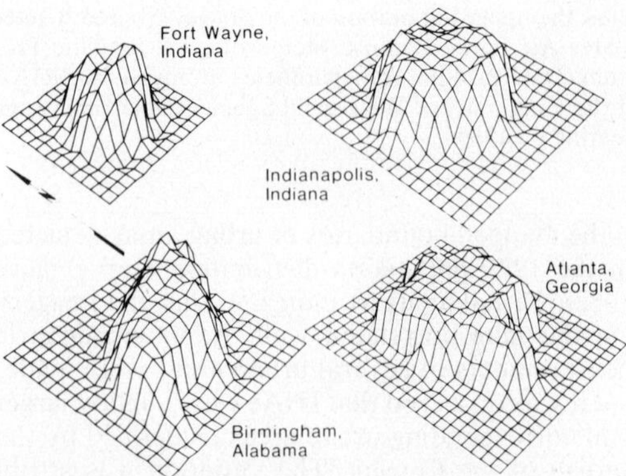

Fig. 30-81 Four IUAs represented as three dimensional block diagrams. [Source: Welch, 1980]

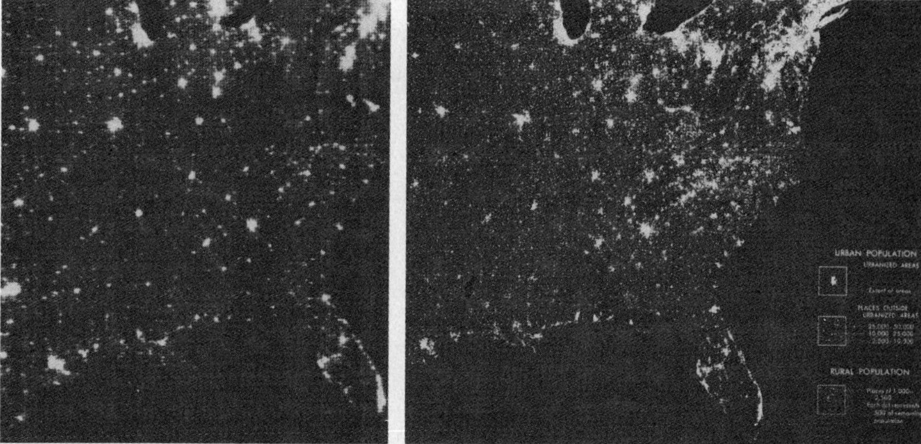

Fig. 30-82 There is a striking similarity in the depiction of urbanized areas between DMSP imagery and a population distribution map produced by the U.S. Bureau of the Census. [Source: Welch and Zupko, 1980]

most solid waste is sparsely distributed across the country. In order to determine where solid-waste conversion programs would be feasible, areas of solid-waste concentration must be located. Remote sensing technology has proven to be useful for this purpose by locating waste producers and tabulating the volume of waste produced within defined regions. The results of the research have indicated that enough solid waste is available in urban places to make conversion programs feasible.

One study covering a portion of Tampa, Florida, demonstrated that sources of solid waste could be identified and that reliable estimates of the volume of retrievable waste could be made (Garofalo and Wobber, 1974). Small scale color-infrared photography was used to identify and map distinct categories of waste producers. The categories were selected to differentiate between waste producers based upon the average amount of solid waste produced. The number of waste producers in each category were summed, and the total volume of available solid waste was estimated by multiplying the summations by a factor referred to as a *waste multiplier*. Waste multipliers are the average quantity of waste generated by a waste source, resident, or employee per unit of time.

A more extensive study was undertaken to determine if solid waste could replace current resources utilized for the generation of electrical power within urban communities (Garofalo and Martin, 1977). For this study, the waste potential of Salem, New Jersey, was estimated. Salem was selected because it was a small, self-contained urban community with a population approaching 8,000 people. Site analysis was facilitated by the availability of color-infrared aerial photography at 1:12,000 (Figure 30-85) and 1:80,000 scales.

Calculation of Current Energy Demand

Accurate figures describing Salem's annual electrical energy demand were not readily avail-

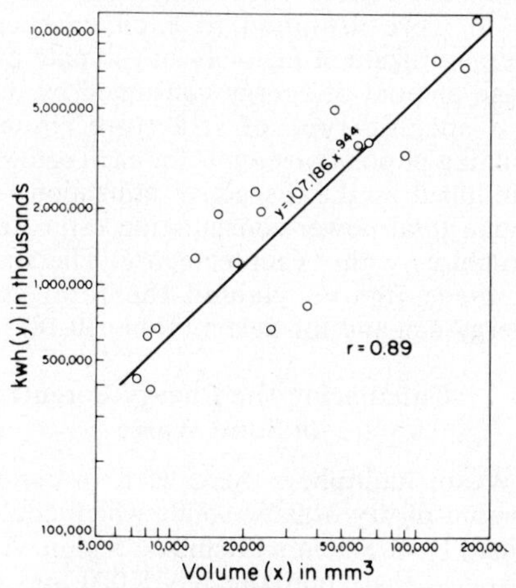

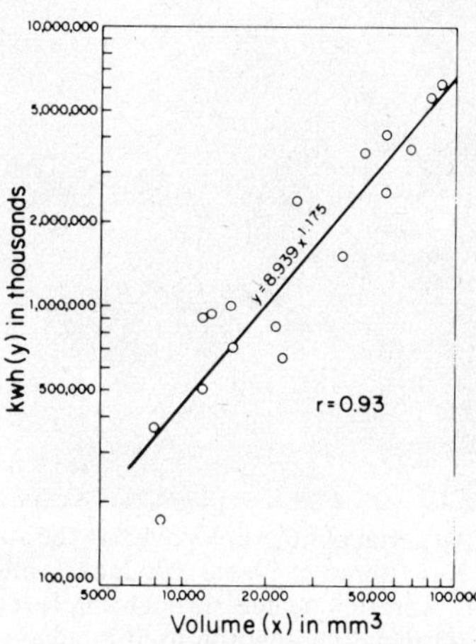

Fig. 30-83 A linear correlation exists between electrical energy consumption and IUA volume.

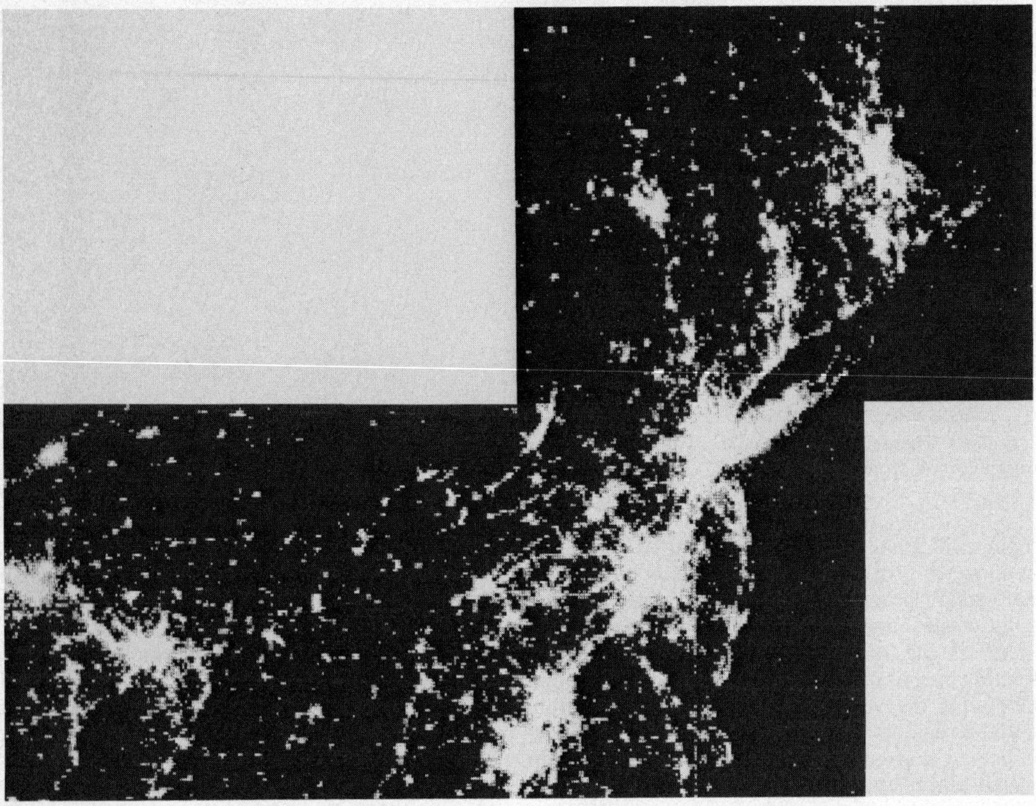

Fig. 30-84 Most pixels comprising IUAs are saturated, precluding precise quantitative analysis. When IUAs are closely spaced, as in the northeast United States, analysis of volume becomes impossible. [Source: SRI International]

able from the local utility company. Instead, the values were estimated through a process which involved the analysis of aerial photography and average power use factors. The process began with analysis of the 1:12,000 scale imagery to produce a map depicting the distribution of buildings. Four categories of structures were identified: industrial, commercial, single family dwellings, and multi-family dwellings (Figure 30-86). Through detailed analysis, the number of structures were inventoried and the total areal coverages of each category were computed. Then, unique energy-utilization factors (Federal Power Commission, 1970) were assigned to each category. The energy-utilization factor is an estimate of the average amount of energy consumed by occupants of a specific type of structure. Finally, the building-population counts for each category were multiplied by the respective utilization factors to obtain total power-consumption estimates for all structures within each category. The summation of these figures yielded the total aggregate energy-demand for Salem (Table 30-18).

Fig. 30-85 An aerial photograph covering the study area of Salem, New Jersey. Waste producers and waste source areas were delineated through interpretation of this image. (Original photograph used in analysis was a color-infrared image at the scale of 1:12,000.) [Source: Garofalo and Martin, 1977]

Calculating the Energy Potential of Solid Waste

Waste multipliers were used to estimate the amount of dry organic solids which could be obtained from Salem's urbanized region. It was determined that approximately 8,000 tons could be made available annually (Table 30-19). Given that each ton of solid waste could yield 1.25 barrels of

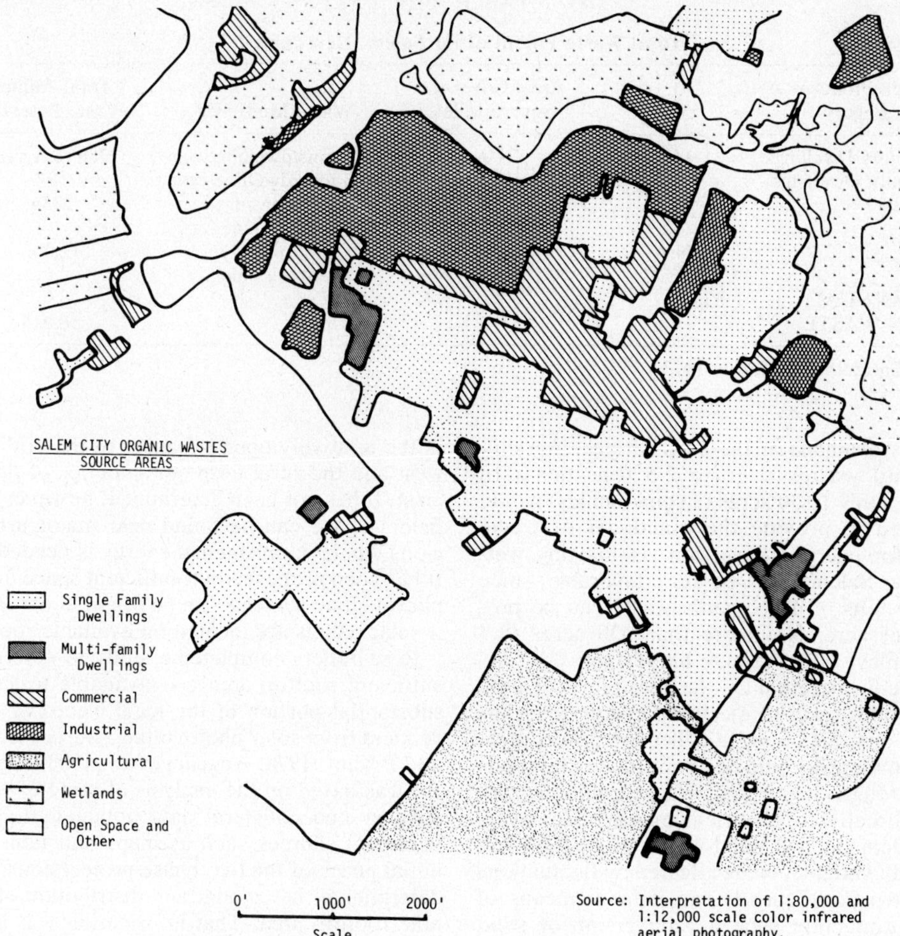

Fig. 30-86 A land use map depicting waste producers and waste source areas was made through analysis of 1:12,000 and 1:80,000 scale color-infrared imagery. [Source: Garfalo and Martin, 1977]

fuel oil after processing (Anderson, 1972), 10,000 barrels of oil could be obtained from Salem's solid waste annually. That volume of oil is equivalent to 45 billion Btu; and since one kilowatt hour (kWh) is equivalent to 3,412 Btu, approximately 13 billion kWh could be derived from solid waste each year. If the conversion process were actually implemented, eighteen percent of Salem's annual electrical energy demand could be met by this renewable resource.

The researchers believed that a minimum of twenty-five percent of Salem's energy demand would have to be met by solid waste conversion before the process could become economically feasible. It was determined that the seven percent shortcoming could be filled from a variety of rural

TABLE 30-18

Total Annual Aggregate Energy Demand
Salem, New Jersey

Waste Producer Category	Number of Units	kWh
Single Family Dwellings	1600	38,560,000
Multi-Family Dwellings	900	12,510,000
Commercial Facilities	200	6,537,800
Industrial Facilities	14	17,612,000
TOTAL DEMAND (kWh)		75,219,800

(After Garofalo and Martin, 1977.)

TABLE 30-19

Total Waste Potential for Salem, New Jersey

Waste Producer Category	No. of Units	Areal Coverage Square Miles	Waste Multiplier	Total Annual Waste Potential
Single Family Dwellings	1600	N/A	1.43 tons/dwelling/year	2288 tons/year
Multi-Family Dwellings	900	N/A	0.66 tons/dwelling/year	594
Commercial	N/A	0.20	38.83 tons/day/ square mile	2834
Industrial	N/A	0.30	116 tons/day/ square mile	12,702
TOTAL ANNUAL URBAN WASTE				18,418

(After Garofalo and Martin, 1977.)

resources available at the urban fringe. Agricultural solid wastes average 2.5 tons per acre annually. Only 1,830 acres (3 square miles) would be needed to provide the remainder of waste needed for energy conversion. Harvesting wetlands near Salem might be more favorable, since up to 10 tons of dry organic waste can be produced per acre. Not more than 430 acres (0.7 square miles) would require harvesting.

Final consideration of the feasibility of converting solid waste to electrical energy must include an analysis of waste-collection procedures. If waste materials could be obtained at continuous levels throughout the year, refuse could be converted directly to electrical energy. However, energy demand and solid-waste production are seasonally variable, being affected by fluctuations in climate. Consequently, an indirect means of energy production such as conversion of solid waste to crude oil would be more beneficial. The oil could be burned when needed and could be easily transported away from the primary source area when warranted. Thus, energy conversion plants could be located at major waste production or collection centers. While most numerical values for waste potential and electrical energy demand were estimated indirectly, the research suggests that interpretation of aerial photography can be used to assess regional energy demand and the potential for deriving electrical energy from solid waste.

MODELING THE USE OF PHOTOVOLTAIC ENERGY SOURCES

Generation of electrical power from photovoltaic power sources (solar cells) can exploit another renewable resource of the future which might help meet urban energy requirements. Currently, major problems preclude the use of photovoltaic technology on a large scale. They include the high costs for manufacturing and the large land areas required for energy collection. However, photovoltaic technology is advancing rapidly, and is becoming less expensive and more efficient (Wrigley and Storti, 1978). Spatial limitations would not prevent the use of solar photovoltaics in the relatively open expanses of the arid Southwest and the rural farm communities of the Midwest. It has not been determined, however, if sufficient space can be found near major urban regions where supplemental energy is needed most. It has been proposed that sufficient space for solar photovoltaic systems can be found in urban areas if solar panels are located on available rooftops.

Researchers completed a study to determine if sufficient rooftop area was available to supply a substantial portion of the local electrical energy demand from solar photovoltaic sources (Angelici and Bryant, 1978; Angelici et al., 1980). The project was based on the analysis of remotely sensed imagery and collateral data obtained from conventional sources such as maps and tables. The initial phase of the two phase project consisted of determining the amount and distribution of available rooftop area. That information was eventually used to estimate the photovoltaic potential of the study area. In the second phase, solar-potential estimates were compared to actual power-consumption figures to determine if the solar photovoltaic potential of the urban area was significant.

A study area covering sixty-five square miles of the San Fernando Valley, a suburban residential region of Los Angeles, California, was selected for analysis. The principal land-cover features consist of single-family dwellings, multi-family dwellings, strip and concentrated commercial districts, and some light industry and manufacturing. The study area was selected because minimal periods of cloud cover are experienced annually, making it well suited to photovoltaic technology.

Data Collection and Processing Problems

The study could only be completed by synthesizing data obtained in various scales and formats. Land-use maps at 1:24,000 scale were provided by an agency of the State of California. False-color infrared and other types of aerial imagery were obtained for rooftop-area estimation. Current electrical power demand was determined from statistics and maps supplied by the local public utility. Finally, data pertaining to the efficiency of solar technology were obtained from solar-power researchers. Not only was data collection a major problem, but, techniques for storage and integration of these data had to be considered as well. The researchers utilized the Image Based Information System (IBIS) for data processing (Bryant and Zobrist, 1977). IBIS is a subset of the VICAR image processing system (Seidman and Smith, 1979).

Determination of Solar Photovoltaic Potential

Since photovoltaic panels require maximum solar exposure for efficient operation, rooftop area assessments were limited to measuring south-facing and flat rooftop areas only. Direct measurement of these rooftop areas would have been impractical given the size of the study area. Consequently, image processing technology was used to determine land cover and to estimate the relative proportion of land covered by flat and south-facing rooftops.

Land Use Mapping

Land use maps (1:24,000) of the study area were obtained by photointerpretation of medium scale imagery that was supplemented by ground survey and verification (Figure 30-87). The land use maps were encoded using a coordinate digitizer and transformed into digital image format. To facilitate data handling, the five land use images were combined to form a single digital image (Color Figure 30-88). In digital format, the areal coverage of each land use type in relation to the total area of the study site could easily be determined through histogramming procedures.

Estimation of Available Rooftop Area

The relative proportion of south facing and flat rooftops to total area for each major land use category was determined through multistage sampling of low altitude aerial photography for sixty-two randomly selected sample sites within the study area (Figure 30-89). First, all flat and south-facing rooftops were delineated on imagery covering each sample site (Figure 30-90). Then, the boundaries of the major land use categories were superimposed on the same imagery (Figure 30-91), and a tally of available rooftop to total area was compiled for each land use category. The total areal coverage and usable rooftop areas were summed for all sixty-five sample sites, and the ratio of available rooftop area to total area was determined for each individual land use class (Figure 30-92).

Assessment of Solor Photovoltaic Potential

The modeling of solar photovoltaic potential involved three steps. First, incoming solar radiation was determined based on data from field tests and models from a Solar Photovoltaic Project at Jet Propulsion Laboratory. Cloudiness and haze conditions were considered in deriving incoming radiation figures on a monthly basis. The available solar energy values were scaled by the efficiency factor of solar photovoltaic cells. Research indicated that current technology could yield up to 14.2 percent efficiency (Wrigley and Storti, 1978), however, it was assumed for this study that ten percent efficiency was commercially available. The results of these calculations were expressed in kilowatt hours of photovoltaic potential per square foot. Finally, the photovoltaic potential factor was multiplied by available rooftop area for each land use category to determine their respective photovoltaic potentials. The solar potential of each category was aggregated to yield an estimate of the photovoltaic potential for the entire San Fernando Valley. It was determined that, on an annual basis, 2.5 million megawatt hours of electrical energy could be derived from photovoltaic sources.

Determination of Actual Power Consumption

To determine the potential energy replacement value available from photovoltaic sources required that comparisons be made to actual energy consumption values. A major portion of the study area's energy needs must be satisfied by solar power to justify the expenditures involved with conversion to that energy source. Consequently, the photovoltaic energy projections were compared with energy consumption figures supplied by the local power utility, the Los Angeles Department of Water and Power (DWP). The DWP supplies electrical power in the San Fernando Valley test site through a network consisting of 533 feeder areas (Figure 30-93).

The DWP, as with all utility companies, has been primarily concerned with meeting peak power demand for a mainline feeder, determined by monitoring amperes, as opposed to aggregate energy produced to meet requirements over a period of time. Thus, for the purposes of this study, it was necessary to compute the monthly and bihourly aggregate demand expressed in kilowatt hours (kWh) from peak amperage demand readings. The computation of monthly kWh for each mainline feeder service area was accomplished by averaging weekly peak amperage figures for each feeder and using these figures as an apportioning coefficient to the power substation total annual kWh demand, the only actual power demand which existed.

Assessment of Replacement Value

The replacement value of electricity derived from solar photovoltaic sources was modeled through a comparative study between energy demand and solar energy potential for each individual feeder area. The comparison was facilitated

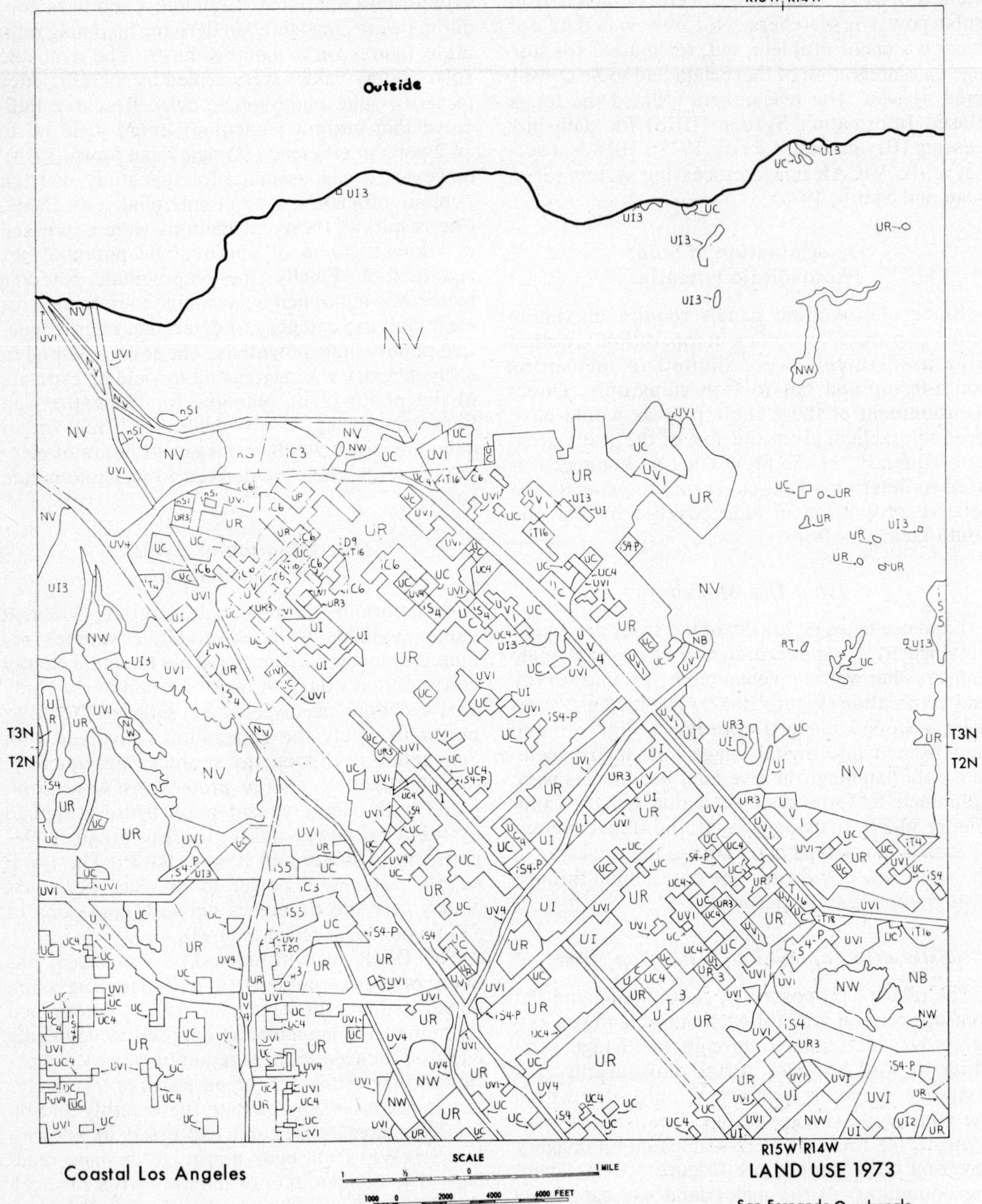

Fig. 30-87 One of five land use maps prepared by the California Department of Water Resources. [Source: The Jet Propulsion Laboratory.]

by IBIS software which enabled the expression of solar potential for each individual feeder area. Available rooftop area (Figure 30-94) and potential photovoltaic energy (Figure 30-95) could be modeled in the same format as actual power demand (Figure 30-96). Through merging of these data, the potential replacement value of solar photovoltaic energy was determined (Figure 30-97). The thematic maps used as illustrations depict conditions which occurred during September, 1978. Additional maps were produced for each month of that year. Feeder areas exhibited both deficit and surplus conditions. When the entire study area was considered 47.9% of the total annual energy demand could be obtained from solar photovoltaic sources (Table 30-20). The

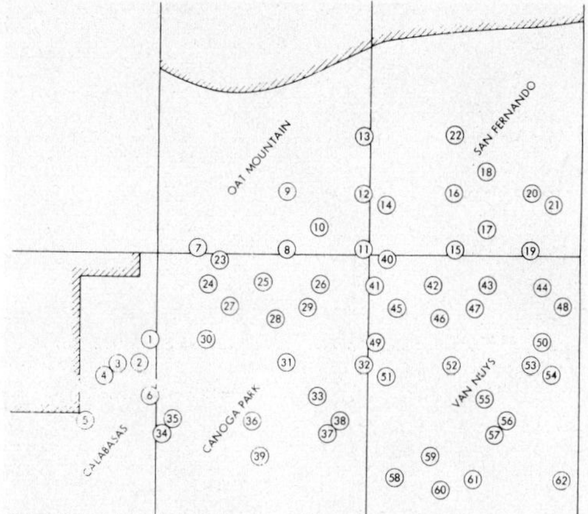

Fig. 30-89 Sixty-two sample sites were selected to determine the percentage of available rooftops. [Source: JPL]

study clearly indicated that photovoltaic power could supply a significant amount of the electrical demand for the San Fernando Valley region.

Conclusions

Through the use of digital image processing and information systems technology, many data sets were combined for analysis. Although the original formats of the source material were incompatible, they were reformatted and interfaced in the digital framework. These data were then compared and modeled to yield new information pertaining to the replacement value of photovoltaic technology. In future applications, new parameters pertaining to solar power research, such as solar power storage systems, the costs of energy conversion ver-

Fig. 30-90 The usable rooftop areas were delineated on low altitude imagery for all sixty-two sample sites. [Source: JPL]

Fig. 30-91 Land use information was overlayed on the same imagery to determine the amount of available rooftop space per each individual land use category. [Source: JPL]

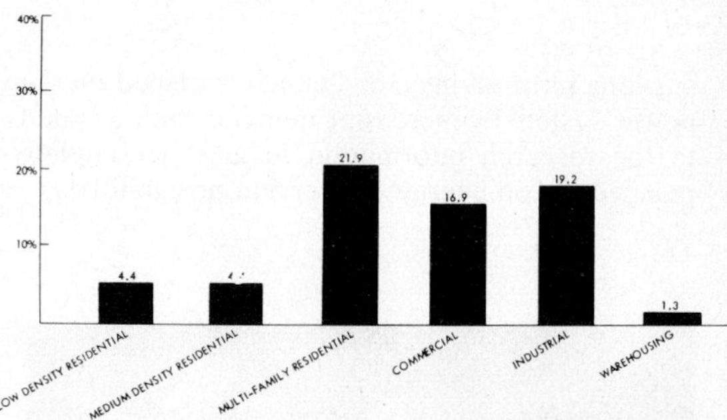

Fig. 30-92 After rooftop areas for all sixty-two sample sites were tabulated, the percentage of available rooftop area per land use category was determined.

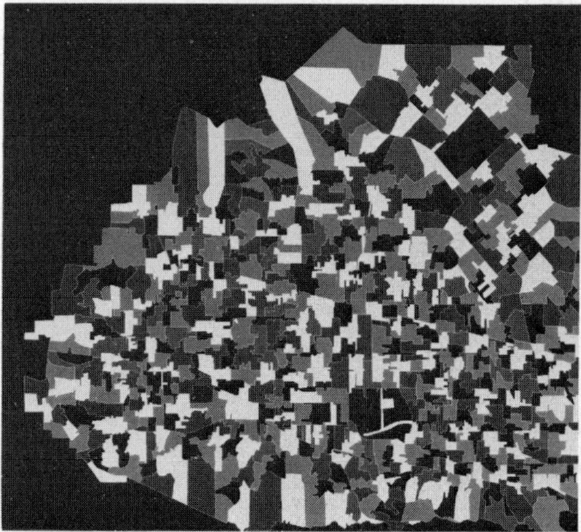

Fig. 30-93 Electric feeder areas serving the study area. [Source: JPL]

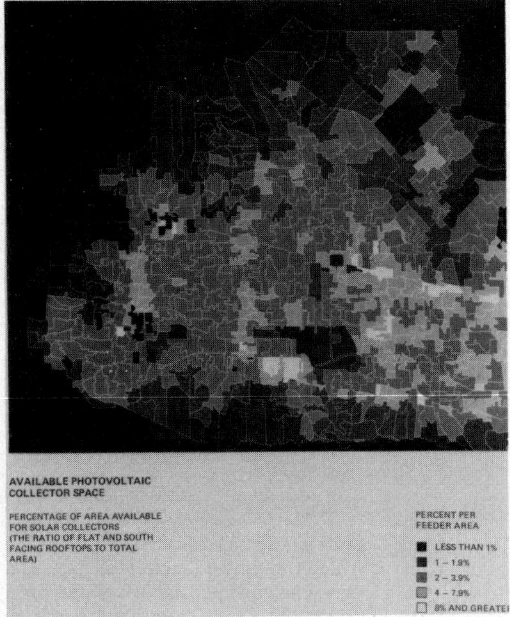

Fig. 30-94 Percentage of available rooftop area in each Feeder Area. [Source: JPL]

sus long term savings, or the stress placed on the power system by increasing demand can be added to the research information to gain a complete perspective on energy conversion possibilities.

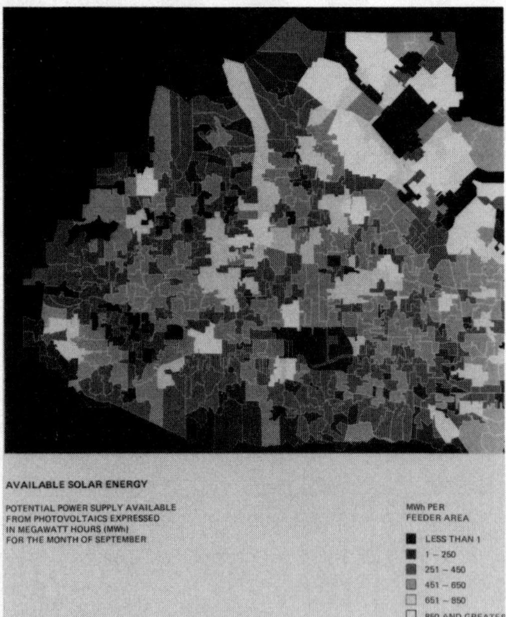

Fig. 30-95 Potential photovoltaic power (kWh) within each mainline feeder service area of the San Fernando Valley, near Los Angeles, during the month of September. [Source: JPL]

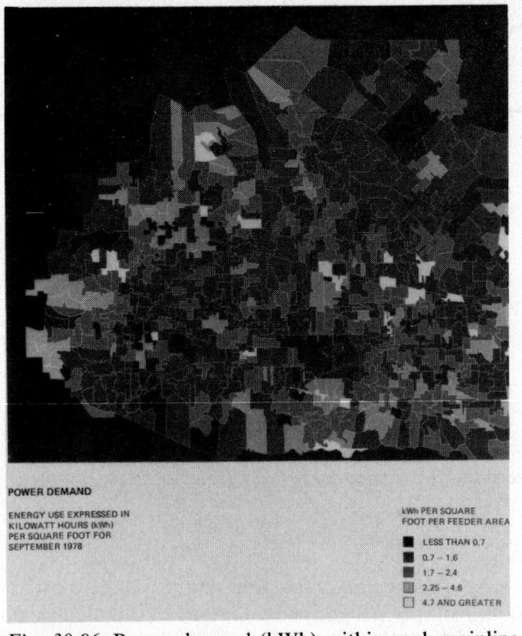

Fig. 30-96 Power demand (kWh) within each mainline feeder service area of the San Fernando Valley, near Los Angeles, during September 1978. [Source: JPL]

URBAN AND SUBURBAN LAND USE INFORMATION FOR EMERGENCY SITUATIONS: THE THREE MILE ISLAND EXAMPLE

A nuclear accident occurred in March, 1979, at the Three Mile Island powerplant on the Sus-

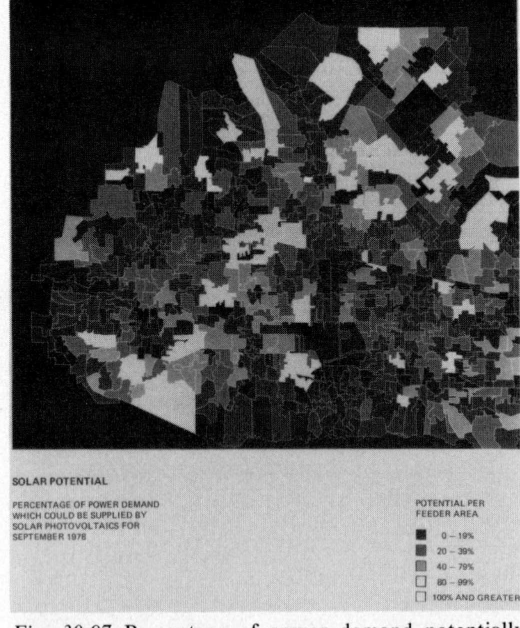

Fig. 30-97 Percentage of power demand potentially supplied by solar photovoltaics, for each mainline feeder service area of the San Fernando Valley, near Los Angeles, during September. [Source: JPL].

TABLE 30-20

Monthly Summary of Modelled Photovoltaic Potential for the San Fernando Valley, Los Angeles, 1978.

Month	Solar Potential (kWh)	Actual 1978 Demand (kWh)	Percent Supplyable From Photovoltaics
January	165,320,256	403,431,168	41.0%
February	179,381,904	358,191,360	50.1%
March	229,350,416	382,626,816	59.9%
April	215,650,528	368,013,312	58.6%
May	219,944,848	413,149,696	53.2%
June	217,750,992	467,941,632	46.5%
July	240,564,704	532,511,744	45.2%
August	239,841,216	502,435,328	47.7%
September	227,553,376	471,772,160	48.2%
October	196,069,200	435,819,520	45.0%
November	173,290,576	402,258,432	43.1%
December	163,511,568	411,375,360	39.7%
Annual Total	2,268,229,584	5,149,526,528	47.9%

quehanna River, southeast of Pennsylvania's capitol at Harrisburg. The accident raised questions about where such plants are located and what geographic area an accident might affect.

Using river water to cool the turbines and the nuclear reactor, the Three Mile island thermal nuclear powerplant occupies an insular location in an area of relatively low population density. Even so, it is at the center of a triangle formed by populous electricity-consuming metroplitan centers at Harrisburg, York, and Lancaster. Small residential areas do lie nearby, but extensive open water, agricultural land, and forest land dominate the immediate vicinity. Surburban Middletown, however, lies less than 5 miles north. Off to the southeast—and usually downwind—lies industrial Lancaster in the heart of a rich agricultural area.

An accurate land use and land cover map of the plant site and vicinity that is keyed to political units and census statistical areas is one tool which a layman can understand and which the politician, planner, and utility company can use in real-time decision making. Color Figure 30-98 is a reduced reproduction of such a map. An unannotated version was produced in a single evening by a computer-driven plotter. At one scale, it overlayed a USGS topographic base map which provided other essential information, such as roads, mountains, and drainage lines. The legend, also prepared by computer, identified by color and shading pattern 19 categories of Level II land use or land cover. In general, the smaller land use polygons in darker tones are urban and built-up areas. The larger polygons in lighter tones are agricultural and forested areas, or water bodies. The rings spaced at an interval of 5 miles were used in analysis, planning, and *assessment of potential impact*. The grid lines spaced at an interval of 10 km are in the Universal Transverse Mercator (UTM) rectangular coordinate system. They were used for location control in mapping and analysis by computer, but locations can also be expressed in geographic coordinates.

The land use and census area map was constructed from a statewide digital data base completed in 1978 as a cooperative effort between the State of Pennsylvania's Department of Environmental Resources and the U.S. Department of Interior's Geological Survey. The land use data were compiled primarily from color infrared aerial photography. The data base also included information on topography, census county subdivisions, political units, hygrologic units, and Federal Land ownership. The U.S.G.S. Geographic Information Retrieval and Analysis System (GIRAS) made possible the 1) mapping by computer, 2) correlation of land use and area measurement with census and other data, 3) updating of all data sets, and 4) retrieval and analysis of data. The system was used as a tool to assess the potential impact of a serious man-made hazard, and to provide graphic and statistical data to prepare contingency plans. The necessity to have accurate urban and suburban land use information in a readily accessible data base will increase as man continues to impact the environment (Wray, 1980).

APPENDIX A

A CLASSIFICATION SYSTEM FOR URBAN AND BUILT-UP LAND USE[5]

1 URBAN & BUILT UP

Urban and built-up land comprises areas of intensive use with much of the land covered by

[5] Level I and II definitions were taken from Anderson, James R., Ernest E. Hardy, and John T. Roach, *A Land-Use Classification System for Use with Remote-Sensor* Data, U.S. Geological Survey, Circular 671, pp. 7–15. Level III and IV definitions are either from existing functional definitions or were developed by Committee. Standard Land Use Coding Manual (SLUC) categories and, when appropriate, Standard Industrial Code (SIC) categories are included with definitions for reference (Michigan Land Use Classification and Referencing Committee, 1975).

structures. Included in this category are cities, towns, villages, strip developments along highways, transportation, power, and communications facilities, and such isolated units as mills, mines, and quarries, shopping centers, and institutions.

As development progresses, small blocks of land of less intensive or non-conforming use may be isolated in the midst of built-up areas and will generally be included in the (1) category. Agricultural, forest, or water areas on the fringe of urban and built-up areas will not be included except where they are part of low-density urban development. The urban and built-up land category takes precedence over others when the criteria for more than one category are met. Thus, residential areas that have sufficient tree cover to meet forest land criteria will be placed in the residential category.

11 *Residential*

Residential land uses range from high density, represented by the multiple-unit structures of urban cores, to low density, where houses are on lots of more than an acre, on the periphery of urban expansion. Linear residential developments along transportation routes extending outward from urban areas should be included as residential appendages to urban centers, but care must be taken to distinguish them from commercial strips in the same locality. Residential development along shorelines is also linear and sometimes extends back only one residential parcel from the shoreline to the first road.

Areas of sparse residential land use will be included under another category. In some places, the boundary will be clear where new housing developments abut against intensively used agricultural areas, but the boundary may be vague and difficult to discern when residential development is sporadic, or occurs in small isolated units over an extended period of time in areas of mixed or less intensive uses. A careful evaluation of density and the overall relation of the area to the total urban complex must be made.

Residential sections may also be included in other use categories where they are integral parts of the other use. Housing on military bases, at colleges and universities, living quarters for laborers near a work base, or lodging for employees of agricultural field operations or resorts are often difficult to identify and may be placed within the institutional, industrial, agricultural, or commercial categories.

111 Multi-family, medium- to high-rise

Includes all multi-family and apartment structures of 4 or more stories, Included are apartments, condominiums, and the like, whether in complexes or as single structures. It is inclusive of lawns, parking areas, any small-area recreational facilities. These areas should also be included under categories 112, 113, 114 and 115.

1111 *High density*

Development containing an average gross density of 35 or more dwelling units per acre.

1112 *Medium dwelling*

Development containing an average gross density of more than 20 dwelling units per acre, but less than 35 dwelling units per acre.

1113 *Low density*

Development containing an average gross density of 20 or less dwelling units per acre.

112 Multi-family, low-rise

Similar to 111 except that it is for structures of 3 or less stories. Duplexes are not included, but townhouses are.

1121 *High density, apartment*

Apartment development containing an average gross density of 18 or more dwelling units per acre.

1122 *Medium density, apartment*

Apartment development containing an average gross density of more than 12 dwelling units per acre but less than 18 dwelling units per acre.

1123 *Low density, apartment*

Apartment development containing an average gross density of 12 or less dwelling units per acre.

1124 *High density, townhouse*

Townhouse development containing an average gross density of 12 or more dwelling units per acre.

1125 *Medium density, townhouse*

Townhouse development containing an average gross density of more than 8 dwelling units per acre, but less than 12 dwelling units per acre.

1126 *Low density, townhouse*

Townhouse development containing an average gross density of 8 or less dwelling units per acre.

113 Single-family/duplex

This category includes areas having detached single and two-family structures used as a permanent dwelling more than two rows wide except for those strip developments connected to larger residential areas. Associated structures may include tool sheds, garages, garden sheds, etc.

1131 *High density*
Development or grouping containing an average gross density of 6 or more dwelling units per acre.

1132 *Medium density*
A development or grouping containing an average gross density of more than 3 dwelling units per acre, but less than 6 dwelling units per acre.

1133 *Low density*
A development or grouping containing an average gross density of 3 or less dwelling units per acre.

[1134] *Non-farm residence* (see category 291 for farmsteads)*
A dwelling located in a rural or urban-rural fringe area and occupied by a non-farming family.

1135 *Mobile home*
Single or several mobile homes not located in a mobile home park.

1136 *Seasonal dwelling*
A dwelling occupied only during a specific season of the year.

1139 *Other*
Any single family or duplex structure not covered above.

114 Strip residential
Predominantly residential development located in a linear pattern along a shoreline or road containing a minimum of 5 or more structures and not more than two rows wide. Land use directly behind the strip itself should be unrelated to residential land use.

1141 *High density, shoreline*
Continuous strip residential grouping wherein are located 8 or more structures per 1000 linear feet.

1142 *Medium density, shoreline*
Continuous strip residential grouping wherein are located more than 2 structures per 1000 linear feet, but less than 8 structures per 1000 linear feet.

1143 *Low density, shoreline*
Continuous strip residential grouping wherein are located 2 or less structures per 1000 linear feet.

1144 *High density, roadside*
Continuous strip residential grouping wherein are located 8 or more structures per 1000 linear feet.

1145 *Medium density, roadside*
Continuous strip residential grouping wherein are located more than 2 structures per 1000 linear feet, but less than 8 structures per 1000 linear feet.

1146 *Low density, roadside*
A continuous strip residential grouping wherein are located 2 or less structures per 1000 linear feet.

115 Mobile home parks (SLUC 140)[6]
Area of land used for a grouping of mobile homes shall be classed into this category. Usually these areas will include mobile homes in numbers over three. Related services and recreational spaces are to be included. Single mobile homes will be classed as of 113.

1151 *High density*
A development or grouping containing an average gross density of 12 or more dwelling units per acre.

1152 *Medium density*
A development or grouping containing an average gross density of more than 6 dwelling units per acre, but less than 12 dwelling units per acre.

1153 *Low density*
A development or grouping containing an average gross density of 6 or less dwelling units per acre.

116 Group and transient quarters
A structure which is used for housing, whether permanent or temporary, of a number of *unrelated* individuals. This differs from apartments in that residents tend to interact with each other for meals, care, etc. All of these quarters could also logically go under other uses (i.e., hotels—commercial, dormitory—educational), but they are classed here because all have predominantly residential characteristics in common.

1161 *Rooming and boarding houses*
Generally operated on a commercial basis, renting rooms to five or more persons not related to the proprietor, with or without board. (SLUC 121)

[6] Reference is to the Standard Land Use Classification Manual Code (SLUC) or Standard Industrial Code (SIC) category. Where a cross-reference exists, it is indicated.

1162 *Membership lodging*
Includes organizational private hotels, organizational lodging houses and membership residence dormitories. Does not refer to religious organizational facilities. (SLUC 122)

1163 *Residence halls and dormitories*
Buildings containing nurses and students residential facilities, including adjacent lawn and parking. (SLUC 123)

1164 *Retirement homes and orphanages*
Includes residential facilities for the aged, orphaned, or indigent. (SLUC 124)

1165 *Religious quarters*
Includes convents, monasteries, abbeys, rectories, parish houses and parsonages. (SLUC 125)

1166 *Residential hotels*
Non-organizational residential or apartment hotels operated as a facility wherein guests reside on a semi-permanent basis. (SLUC 130)

1167 *Hotels, tourist courts, motels*
Transient lodging facilities operated on a commercial basis, by day or week. (SLUC 151)

1168 *Migrant quarters*
Season dwellings used for housing seasonal workers.

1169 *Other*
Group and transient uses not covered above. (SLUC 190)

119 Other residential
All other obviously residential land uses not covered above.

12 *Commercial, Services & Industrial*
Commercial areas are those used predominantly for the sale of products and services. They are often abutted by residential, agricultural, or other contrasting uses which help differentiate them. The principal components of the commercial-use category are urban central business districts; shopping centers, usually in suburban and outlying areas; commercial strip/developments along major highways and access routes to cities; and resorts. The main buildings, secondary structures, and areas supporting the basic use are all included—office buildings, warehouses, driveways, sheds, parking lots, landscaped areas, and waste-disposal areas.

Commercial areas may include some noncommercial uses too small to be separated out. Central business districts often include some institutions, such as churches and schools, and commercial strip developments may include some residential units. These are not separated out unless they exceed one-third the total commercial area or are large enough to stand alone. Recreational areas are not segregated as such at Level II but may cause some problems in identification. Recreational facilities that form an integral part of an institution should be included in the institutional category. A self-contained sports area, on the other hand, such as a stadium for professional events, is commercial.

Education, religious, health, correctional, and military facilities are also found in this category. All buildings, grounds, and parking lots that compose the facility are included within the institutional unit, but areas not specifically related to the purpose of the institution should be placed in the appropriate category. Auxiliary land use, particularly residential, commercial and services, and other supporting land uses on a military base would be included in the institutional subcategory, but agricultural areas not specifically associated with correctional, educational, or religious institutions are placed in the appropriate agricultural category. Small institutional units, for example, many churches and some secondary and elementary schools, will not meet the minimum area requirements and will be included within another category, usually residential or commercial.

121 Primary/central business district (CBD)
The main commercial service center in a city. Each city usually has only one CBD, and it can be identified as being the most densely constructed urban portion of a city. It can be further identified as being normally located at the confluence of the major transportation network.

1211 *Commercial*
Commercial retail establishments. (SLUC 5) *1/p. 57*

1212 *Services*
Includes business, financial, personal, professional and repair service establishment. (SLUC 6) *1/p. 57*

1213 *Health*
Includes hospitals and medically related clinics (SLUC 651) *2/p. 57*

1214 *Education*
Includes all types of public and private institutions where educa-

tion is the primary use. (SLUC 68) *3/p. 57*

1215 *Religious*
Includes all forms of churches, synagogues and retreats. (SLUC 691) *4/p. 57*

1216 *Correctional*
Includes all types of detention and correction facilities. (SLUC 674) *5/p. 58*

1217 *Military*
Self-explanatory. (SLUC 675) *6/p. 58*

1218 *Government administration and services*
Includes city halls, police stations, post offices, and the like. *7/p. 58*

1219 *Other commercial, services, and institutional*
Self-explanatory

122 Shopping center
Usually a single structure, or a group of structures, containing a large amount of floor space and a variety of commercial and service establishments. They are identified by the large common parking lot, usually larger in area than the structure group itself. These are often referred to as neighborhood, community or regional shopping centers.

1221 thru 1229 Use level IV categories 1221 through 1229 as listed under 121 (e.g. 1221 commercial, 1222 services, etc.).

123 Strip development
Linearly patterned commercial service area only one building deep occurring with a minimum of five establishments. This land use should be backed by a non-related land use.

1231 thru 1239 See 121 above.

124 Secondary/neighborhood business district
Relatively compact groups of stores, institutional structures, and/or services outside of the CBD. These uses should be located on major streets and surrounded by non-commercial uses. Parking is either in several scattered small store lots or municipal lots.

1241 thru 1249 See 121 above.

125 Other commercial and services
Single or joint establishments which do not fit the above definitions, but can be recognized and mapped at level III (e.g., large single stores such as K-Mart).

1251 *Grain, feed and seed*
1252 *Livestock sales*
1253 *Other commercial*
1254 *Other services*

126 Other institutional
Large single institutional land uses, normally in a non-urban setting, such as a prison or military base.

1263 thru 1269 See 121 above.

127 Indoor cultural, public assembly, and recreation
Indoor facilities for cultural activity, recreation, and public assembly, such as planetariums, auditoriums, and tennis courts. Outdoor cultural, public assembly, and recreational lands (and their accompanying incidental buildings) are under Category 19, ''open and other''.

1271 *Indoor cultural*
Facilities such as libraries, museums, art galleries, planetariums, aquariums, and historic buildings. *8/*

1272 *Indoor public assembly*
Facilities such as movie theaters, other theaters, auditoriums, exhibition halls, arenas, and fieldhouses, multiple purpose civic centers. *9/p. 58*

1273 *Indoor recreation*
Facilities such as meeting, social, performance, class, craft rooms, courts (tennis, etc.), ice skating rinks, bowling alleys, dance halls, gymnasiums, swimming pools, multiple purpose recreation centers. *10/p. 59*

13 *Industrial*
Industrial areas include a wide array of uses from light manufacturing and industrial parks to heavy manufacturing plants. Identification of light industries—those focused on design, assembly, finishing, and packaging of products—can often be based on the type of building, parking, and shipping arrangements. Light industrial areas may be, but are not necessarily, directly in contact with urban areas; many are now found at airports or in relatively open country. Heavy industries are raw materials such as iron ore, lumber, or coal. Included are steel mills, pulp or lumber mills, electric power generating stations, oil refineries and tank farms, chemical plants and brickmaking plants. Stockpiles of raw materials, large power sources, and waste product disposal areas are usually visible, along with transportation facilities capable of handling heavy materials.

131 Primary metal production (SLUC 33) (milling, smelting, forging)
This category covers the production of metals from natural mineral ores. The production excludes mining and transportation, but does include storage areas, furnaces, reducers, crushers, etc. Also included are power plants used either exclusively or primarily for metal production. These areas are usually quite large with rail or boat facilities and generally produce large amounts of heat.

1311 *Blast furnaces and steel mills* (SLUC 3311) (SIC 3312)
1312 *Steel wire products* (SLUC 3313) (SIC 3315)
1313 *Cold rolled sheet, strip and bars* (SLUC 3314) (SIC 3316)
1314 *Steel pipe and tube* (SLUC 3315) (SIC 3317)
1315 *Iron and steel foundries* (SLUC 332) (SIC 332)
1316 *Smelting and refining* (SLUC 333, 334) (SIC 333, 334)
1317 *Rolling, drawing and extruding* (SLUC 335) (SIC 335)
1318 *Nonferrous foundries* (SLUC 336) (SIC 336)
1319 *Other primary metal industries* (SLUC 339) (SIC 339)

132 Petrochemicals (SLUC 28, 29, 31) (storage, refining, etc.)
This category includes refineries, plastic and rubber production, asphalt plants, and similar establishments. They are usually identified by the large number of tanks, pipelines, and associated rail or boat transportation. Small asphalt plants may be associated with sand and gravel operations.

1321 *Industrial inorganic and organic chemicals* (SLUC 281) (SIC 281)
1322 *Plastics and manmade fibers* (SLUC 282) (SIC 282)
1323 Biological, medicinal and pharmaceutical drugs (SLUC 283) (SIC 283)
1324 Soap and detergents (SLUC 2841, 2842, 2843) (SIC 2841, 2842, 2843)
1325 *Perfumes, cosmetics and toilet preparations* (SLUC 2844) (SIC 2844)
1326 *Paints, varnish, enamels, agricultural chemicals, explosives, and allied products* (SLUC 285, 286, 287, 289) (SIC 285, 286, 287, 289)
1327 *Petroleum refining and related industries* (SLUC 291, 292, 299) (SIC 291, 292, 299)
1328 *Tire and rubber manufacturing* (SLUC 211, 312, 313, 314, 319) (SIC 301, 302, 303)
1329 *Other petrochemical industries* (SIC 306, 307)

133 Primary wood processing (SLUC 24, 26) (lumber, pulp, paper)
Here, only processing which starts with wood (logs) are considered not the final product. An establishment which, in one operation, starts with logs and ends with paper *is* included, but one that starts with pulp will not.

1331 *Logging camps and logging contractors* (SLUC 24) (SIC 241)
1332 *Sawmills and planing mills* (SLUC 242) (SIC 242)
1333 *Millwork, plywood and related products* (SLUC 243) (SIC 243)
1334 *Wooden container manufacturing* (SLUC 244) (SIC 244)
1335 *Pulp mill* (SLUC 261) (SIC 261)
1336 *Paperboard mill* (SLUC 263, 264) (SIC 263, 264)
1337 *Paperboard containers and boxes* (SLUC 265) (SIC 265)
1338 *Building board and building paper* (SLUC 266) (SIC 266)
1339 *Other primary wood processing*

134 Stone, clay, glass (SLUC 32) (cement, brick etc.)
This category is concerned with the manufacture of cement (dry, not ready mix), bricks, clay tiles, glass, and other types of products. These establishments are usually located near their source of raw materials, such as an open pit extractive area, and near large uses of water and stockpiles of sand or stone. A sand and gravel operation is not included with this land use.

1341 *Flat glass manufacturing* (SLUC 321)
1342 *Glass and glassware* (SLUC 322)
1343 *Cement* (hydraulic (SLUC 323) (SIC 324)
1344 *Structural clay products, brick* (SLUC 324) (SIC 325)
1345 *Pottery and related products* (SLUC 325) (SIC 326)
1346 *Concrete, gypsum and plastic products* (SLUC 326) (SIC 327)
1347 *Cut stone and stone products* (SLUC 327) (SIC 328)
1348 *Abrasive, asbestos and miscellaneous nonmetallic mineral products* (SLUC 328) (SIC 329)
1349 *Other stone, clay, glass products*

135 Metal fabrication (SLUC 34) (secondary manufacturing)
These firms use primary metals and re-

shape them for production of a final marketable item. No forging occurs, but rolling, casting and machining will occur. An example of this industry is automobile manufacturing, where frames, bodies, etc., are made from steel. This use is characterized by single story flat roofed buildings, with no major heat production occurring—except for energy production.

1351 *Ordnance and accessories* (SLUC 341) (SIC 191, 192, 193, 194, 195, 196, 199)
1352 *Machinery (except electrical)* (SLUC 342) (SIC 351, 352, 353, 354, 355, 356, 357, 358, 359)
1353 *Electrical machinery* (SLUC 343) (SIC 361, 362, 363, 364, 365, 366, 367, 369)
1354 *Motor vehicle equipment* (SLUC 3441) (SIC 371)
1355 *Aircraft and parts* (SLUC 3442) (SIC 372)
1356 *Ship and boat building* (SLUC 3443) (SIC 373)
1357 *Railroad equipment* (SLUC 3444) (SIC 374)
1358 *Other transportation equipment* (SLUC 3445, 3449) (SIC 375, 379)
1359 *Other metal fabrication*

136 Non-metal fabrication (SLUC 21, 22, 23, 25, 27, 35)
Plants which created a product using secondary processed materials (non-metal) such as furniture production which uses lumber, not logs, or printing which uses paper, not logs, for pulp.

1361 *Food and kindred products* (SLUC 21) (SIC 21)
1362 *Textile mill products* (SLUC 22) (SIC 22)
1363 *Apparel, fabrics, leather* (SLUC 23) (SIC 23)
1364 *Furniture and fixtures* (SLUC 25) (SIC 25)
1365 *Newspaper, books, commercial publishing and printing* (SLUC 271) (SIC 271)
1366 *Bookbinding and printing trade services* (SLUC 277) (SIC 278)
1367 *Engineering and scientific instruments* (SLUC 351, 352) (SIC 381)
1368 *Optical, surgical, and dental instruments* (SLUC 353, 354, 355, 356, 357) (SIC 383, 384, 385, 386, 387)
1369 *Other non-metal manufacturing*

139 Other
Industrial uses not covered in the above categories.

14 *Transportation, Communication & Utilities*
Transportation routes greatly influence other land uses, and many land-use boundaries are outlined by them. The types and extent of transportation facilities in a locality determine the degree of access and affect both the present and potential use of the area.

Highways and railways are characterized by areas of activity connected in linear patterns. The highways include areas used for interchanges, limited access right-of-way, and service and terminal facilities. Rail facilities include stations, parking lots, roundhouses, repair and switching yards, and related areas, but overland track is not included unless six or more tracks are joined to give sufficient width for delineation at a scale of 1:250,000. Spur connections from an active line are included in the appropriate industrial or extractive category.

Airports, seaports, and major lakeports are isolated areas of high utilization, usually with no well-defined intervening connections, although some water ports are connected by canals. Airport facilities include the runways, intervening land, terminals, service buildings, navigation aids, fuel storage, parking lots, and a limited buffer zone. Small airports, such as those on rotatable farmland, heliports, and land associated with seaplane bases are not included. Port areas include the docks, shipyards, drydocks, locks, and watercourse-control structures.

Communications and utilities areas involved in transport of water, gas, oil, electricity and areas used for airwave communications are also included in this category. Pumping stations, electric substations, and areas used for radio, radar, or television antennas are the major types.

Small facilities, or those associated with an industrial, commercial, or extractive land use, are included within the larger category with which they are associated. Long-distance gas, oil, electric, telephone, water, or other transmission facilities rarely constitute the dominant use of land over which they pass. If these uses are dominant and meet the minimum width criteria, they may be identified as transportation uses.

141 Air transportation (SLUC 431, 439)
Includes all facilities directly connected with air transport, whether it be commercial, municipal, or private. These high utilization areas include the runways, intervening land, terminals, ser-

vice buildings, plane hangars, navigation aids, fuel storage areas, parking lots and a limited buffer zone. Most of the airports will be surrounded by a high perimeter fence, visible on high altitude imagery, which will clearly define the areas to be included.

1411 *Commercial aviation*
1412 *General aviation*
1413 *Military*
1419 *Other*

142 Rail transportation (SLUC 411, 412)
This category includes all facilities which would be connected with rail transportation; overland track (which has a width which can be delineated at a scale of 1:60,000), roundhouses, repair and switching yards, and related areas. Also included in this category are the accompanying and necessary rights-of-way.

1421 *Active yard*
1422 *Station or terminal*
1423 *Abandoned right-of-way*
1424 *Active track*
1429 *Other*

143 Water transportation (SLUC 441, 449)
This category includes those areas related to water transportation, excluding the water. The major components of this category are the port areas, docks, shipyards, drydocks, and locks.

1431 *Port facilities*
1432 *Lock and dam*
1439 *Other*

144 Road transportation (SLUC 451, 452)
This category includes all roads and road rights-of-way associated with the roads (including the median), bridges, rest areas, and weighing stations. Also included are truck and bus terminals. This does not include parking areas which are connected with a particular use (i.e., CBD, residential, factory).

1441 *Limited and controlled access expressway*
1442 *5 or more lanes*
1443 *4 lanes*
1444 *3 lanes*
1445 *2 lanes*
1446 *Truck terminal*
1447 *Bus terminal*
1449 *Other*

145 Communications
Those areas associated with radio, radar, television, telegraph, telephone, etc., are included into this category. Small facilities or those associated with an industrial, commercial, or extractive land use are included within the category which they are associated with. Long distance transmission facilities rarely constitute the dominant use of land over which they pass. If these uses are dominant, and can be recognized from the imagery alone, they may be identified as a communication land use.

1451 *Telephone* (SLUC 471)
1452 *Telegraph* (SLUC 472)
1453 *Television* (SLUC 474)
1454 *Radio* (SLUC 473)
1459 *Other* (SLUC 479)

146 Utilities
Areas associated with the transport of gas, oil, water, or electricity are included into this category. Pumping stations, electric substations, etc., will constitute the major components of this category. Small facilities and those associated with an industrial, commercial, or extractive land use are included into the appropriate category. Long distance pipelines, etc., rarely constitute the dominant use of land over or under which they pass. If these uses are dominant and can be recognized for the imagery alone, they may be identified as a utility use.

1461 *Electrical production and transmission* (SLUC 481)
1462 *Gas storage and transmission* (SLUC 482)
1463 *Petroleum storage and transmission* (SLUC 4911, 4912)
1464 *Solid waste disposal and transfer* (SLUC 485)
1465 *Sewage treatment and transmission* (SLUC 484)
1466 *Water treatment and transmission* (SLUC 483)
1469 *Other*

[15] *Map Industrial Parks under appropriate category in Commercial Services & Institutional (12) or Industrial (13)*

16 *Mixed*
This category is used for a mixture of second-level urban uses in larger cities (generally more than 50,000 inhabitants) where no one use predominates. In any category, as much as one-third intermixture of another use is allowed without changing the basic classification, but where the intermixture is greater, where several uses, though each is less than one-third, are included, or where individual second-level units may be too small to be separated, although the aggregate of such uses may be large, the mixed category is used.

This category typically includes developments along transportation routes and the smaller cities, towns, and built-up areas where separate land uses may not be distinguishable. Residential, commercial, industrial, institutional, and occasionally other land uses may be included. Farmsteads intermixed with strip or cluster settlements will be included within the built-up land, but other agricultural land uses should be excluded.

17 *Extractive*

Extractive land encompasses both surface and subsurface mining operations, such as sand and gravel pits, stone quarries, oil and gas wells, and metallic and nonmetallic mines. In size, these activities range from the unmistakable giant strip or pit mines covering vast areas to the often unidentifiable gas wells less than a foot square. Surface structures and equipment may range from a minimum of a loading device and trucks to extended areas with access roads, processing facilities, stockpiles, equipment sheds, and numerous vehicles. Spoil material and slag heaps are usually found within a short trucking distance of the major mine areas and may be the key indicator of underground mining operations. Uniform identification of all these diverse extractive uses is extremely difficult from remote sensor data alone.

Industrial complexes where the extracted material is refined, packaged, or further processes are included in the industrial category even if the plant is adjacent to the mine. Areas of future reserves are included in the appropriate present-use category, agricultural or forest land, regardless of the expected future use. Unused pits or quarries that have been flooded are placed in the water category if the water body is larger than 40 acres. Areas of tailings, abandoned pits and quarries, and stripmined areas may remain barren for decades unless steps are taken to hasten the establishment of vegetation. Until vegetative cover is established, such parcels remain in the extractive category.

171 Open pit

Extractive activities which are primarily carried out upon the surface of the earth through the creation of a large pit.

1711 *Sand and gravel*
1712 *Rock quarry*
1713 *Coal* (strip)
1714 *Iron*
1715 *Other*

172 Shaft

Extractive activities primarily carried out underground; portions of this activity covered under the barren land category include disturbed land and development waste rock.

1721 *Salt*
1722 *Iron*
1723 *Gypsum*
1724 *Copper and associated metals*
1729 *Other*

173 Wells

This category includes the areas used for the extraction of oil and natural gas and other minerals from the sub-strata. In the case of one individual well, the area immediately surrounding the well is all that is to be placed into this category. Care must be taken not to confuse these wells with water wells.

1731 *Oil*
1732 *Gas*
1733 *Salt*
1739 *Other*

179 Other extractive

Extractive uses not covered in the above categories.

19 *Open & Other*

Open land consists of land used for outdoor cultural, public assembly and recreational purposes. Examples would be zoos, botanical gardens, fairgrounds, some parkland, ski areas and cemeteries. Open land may be in intensive or extensive use. It may include structures incidental to the outdoor land uses.

191 Outdoor cultural

Outdoor cultural includes such facilities as botanical gardens, arboretums, and zoos.

1911 *Botanical gardens and arboretums* (SLUC 7123)
1912 *Zoos* (SLUC 7124)
1919 *Other*

192 Outdoor public assembly

This category includes such facilities as movie theaters, auditoriums, exhibition halls, arenas and fieldhouses, and multiple purpose civic centers.

1921 *Amphitheatres* (SLUC 7211)
1922 *Drive-in movies* (SLUC 7213)
1923 *Stadiums* (SLUC 7221)
1924 *Racetracks* (SLUC 7223)
1925 *Fairgrounds* (SLUC 7311)
1926 *Amusement parks* (SLUC 7312)
1929 *Other*

193 Outdoor recreation

All recreation facilities and areas which are basically on open land. They may, however, include incidental buildings such as shelters, toilets, beach change areas, etc. Does not include rangeland, forest, water, wetland, and barren lands within parks or recreation areas. These would be classified in categories 3, 4, 5, 6, and 9, respectively, at level I, at the third, fourth, and fifth levels. Ownership and institutional characteristics such as park boundaries would be indicated by a separate mapping legend.

1931 *Landscaped and aesthetic areas 11/p. 59*
Includes land maintained as with lawn, plantings, paving, beaches, ornamental objects, and walks. They occur in parks or as institutional or estate grounds, malls, etc. They provide open space relief or environmental embellishment. They may also provide casual walking, relaxing, open air and passer-by enjoyment, social encounter or assembly.

1932 *Play, games, and athletics 12/p. 59*
Includes playlots, playgrounds, and athletic fields. They have been developed for active play and recreation, for different age groups. The category as a whole has a wide variety of possible installations for free or organized play, ranging from swings and sandboxes to softball and baseball diamonds. Bleachers or grandstands may be included, except for those of such scale as to justify inclusion under category 127 as places of public assembly.

1933 *Sports areas 13/p. 59*
Includes a variety of uses such as miniature and regular golf courses, golf driving ranges, off-road recreational vehicle trails, tracks and areas, shooting ranges, etc. Athletics is covered under 1932. Conceivably, some hunting and fishing land uses could be in categories 2, 3, 4, 5, 6 and 9, with institutional or functional identification at a lower level.

1934 *Other land-based activities 14/p. 59*
All recreation facilities and areas which are basically on open land. They may, however, include incidential buildings such as shelters, toilets, beach change buildings, etc. This does not include rangeland, forest, water, wetland or barren land within parks or recreation areas. These would be classified in categories, respectively, at Level I, at the third, fourth, and fifth levels. Ownership and institutional characteristics such as park boundaries would be indicated by a separate mapping legend.

1935 *Water-dependent recreation 15/p. 60*
This category includes such activity areas as swimming beaches and pools, marina and yacht basins and boating sites.

194 Cemeteries

199 Other

REFERENCES

Abdulla, A. E., 1980, The Potential of Remotely Sensed Satellite Data for the Topographic Mapping of Sudan at Small Scales: Masters Thesis, Department of Geography, University of Glasgow, 116 p.

Adeniyi, Peter O., 1980, Land Use Change Analysis Using Sequential Aerial Photography and Computer Techniques: *Photogrammetric Engineering and Remote Sensing,* V. 46, p. 1147–1164.

AFAL, 1967, *Data Compilation of Target and Background Characteristics:* compiled at the University of Michigan for the Air Force Avionics Lab at Wright Patterson Air Force Base, Dayton, Ohio, 300 p.

Allan, J. A., and T. Alemayehu, 1975, Rural Population Estimates from Air Photographs: an example from Wolamo, Ethiopia: *ITC Journal,* V. 1, p. 85–100.

Anderson, D. E., and P. N. Anderson, 1973, Population Estimates by Humans and Machines: *Photogrammetric Engineering,* V. 39, p. 147–154.

Anderson, L. L., 1972, *Energy Potential From Organic Wastes: A Review of the Quantities and Sources,* Bureau of Mines, Information Circular No. 8549, 16 p.

Anderson, J. R., E. E. Hardy, J. T. Roach, and R. E. Witmer, 1976, A Land Use and Land Cover Classification System for Use with Remote Sensor Data: *Geol. Survey Prof. Paper 964,* 28 p.

Anderson, J. R., 1976, Land Use and Land Cover Map and Data Compilation in the U.S. Geological Survey: *Proceedings,* Second Annual W. T. Pecora Symposium, Falls Church: American Society of Photogrammetry, p. 2–12.

———, 1977, Land Use and Land Cover Changes—A Framework for Monitoring: *Journal of Research,* U.S. Geological Survey, V. 5, p. 143–153.

Angelici, G. L., and N. A. Bryant, 1978, Solar Potential Inventory and Modelling: *Proceedings,* Sixteenth Annual Conference of the Urban and Regional Information System Association, p. 442–453.

Angelici, G. L., N. A. Bryant, R. K. Fretz, and S. Z. Friedman, 1980, *Urban Solar Photovoltaics Potential: An Inventory and Modelling Study Applied to the San Fernando Valley Region of Los Angeles:* Jet Propulsion Laboratory, California Institute of Technology, JPL Technical Report 80–43, 36 p.

Baker, R. D., J. DeSteiguer, D. Grand, and M. Newton, 1979, Land-Use/Land-Cover Mapping from Aerial Photographs: *Photogrammetric Engineering and Remote Sensing,* V. 45, p. 661–668.

Bartolucci, L., B. F. Robinson, and L. F. Silva, 1977, Field Measurements of the Spectral Response of Natural Waters: *Photogrammetric Engineering and Remote Sensing,* V. 43, p. 595–598.

Berry, B., 1962, *Sampling, Coding, and Storing Flood Plain Data:* U.S. Department of Agriculture Handbook #237, 27 p.

Binsell, Ronald, 1967, Dwelling Unit Estimation from

Aerial Photography: Northwestern University, Department of Geography, 37 p.

Bowman, R. L., and J. R. Jack, 1979, *Feasibility of Determining Flat Roof Heat Losses Using Aerial Thermography:* NASA, Technical Memorandum No. 79152, 17 p.

Brown, R. J., 1978, Infrared Scanner Techology Applied to Building Heat Loss Determination: *Canadian Journal of Remote Sensing,* V. 4, p. 1–9.

Bryan, M. L., 1974, Extraction of Urban Land Cover Data from Multiplexed Synthetic Aperture Radar Imagery: *Proceedings,* Nineth International Symposium on Remote Sensing of Environment, Ann Arbor, Michigan, p. 271–288.

———, 1975, Interpretation of an Urban Scene Using Multi-Channel Radar Imagery. *Remote Sensing of Environment,* V. 4, p. 49–66.

———, 1979, The Effect of Radar Azimuth Angle on Cultural Data: *Photogrammetric Engineering and Remote Sensing,* V. 45, p. 1097–1107.

Bryant, N. A., and A. L. Zobrist, 1977, IBIS: A Geographic Information System Based on Digital Image Processing and Image Raster Data Type: *IEEE Transactions on Geoscience Electronics,* V. GE-15, p. 152–159.

Budge, T. K., and M. H. Inglis, 1978, Residential Heat Loss Mapping of Farmington, New Mexico Using Airborne Thermal Scanning: *Proceedings,* American Society of Photogrammetry Fall Technical Meeting p. 82–91.

Budge, T. K., and S. A. Morain, 1978, *Neighborhood Heat Loss Mapping by Airborne Thermal Scanner,* Technology Application Center, Institute for Applied Research Services, University of New Mexico, Technical Report TAC TR 78-010, 48 p.

Buettner, K. J. K. and C. D. Kern, 1965, *Journal of Geophysical Research,* V. 70, p. 1333.

Burch, D. M., 1979, The Use of Aerial Infrared Thermography to Compare the Thermal Resistance of Roofs: *NBS Technical Note 1107,* U.S. Department of Commerce, 1 p.

Carter, P., and W. E. Gardner, 1977, An Urban Management Information Service Using Landsat Imagery: *Photogrammetric Record,* V. 9, p. 157–171.

Carter, P., and T. F. Smith, 1980, Monitoring Urban Growth in the U.K. from Landsat Satellite Data: *International Archives Photogrammetry,* Hamburg, V. B10, p. 476–485.

Christenson, J. W., and H. M. Lachowski, 1976, Urban Area Delineation and Detection of Change Along the Urban-Rural Boundary as Derived from Landsat Digital Data: *Proceedings,* American Society of Photogrammetry, Seattle, Washington, p. 28–33.

Christenson, J. W., J. B. Davis, V. J. Gregg, A. M. Lachowski, and R. L. McKinney, 1977, Landsat Urban Area Delineation: Intralab Projects 75–3, NASA/Goddard Space Flight Center, Greenbelt, Maryland, 50 p.

Christenson, J. W., 1977, *Preliminary Design Requirements for a Census/Urbanized Area Application System Vertification and Transfer:* Final Report NASA contract NAS5-23412, Goddard Space Flight Center, Greenbelt, Maryland, 96 p.

Christenson, J. W., D. L. Dietrich, H. M. Lachowski, M. L. Stauffer, and R. L. McKinney, 1979, Urbanized Area Analysis Using Landsat Data: *Harvard Library of Computer Graphics/1979 Mapping Collection,* Harvard University, Cambridge, Massachusetts, V. 4, p. 5–11.

Cihlar, J. R. J. Brown, G. Lawrence, J. H. Barry, and R. B. James, 1977, Use of Aerial Thermography in Canadian Energy Conservation Programs: *Proceedings,* Eleventh International Symposium on Remote Sensing of Environment, p. 1179–1206.

Clark, J., 1979, Landsat-D Thematic Mapper Simulation in an Urban Area Using Aircraft Multispectral Scanner Data: *Remote Sensing Quarterly,* V. 1, p. 17–32.

———, 1980, The Effect of Resolution in Simulated Satellite Imagery on Spectral Characteristics and Computer-Assisted Land Use Classification: Report 715–22, Jet Propulsion Laboratory, Pasadena, California, January 30, 86 p.

Clayton, C., and J. E. Estes, 1979, Distributed Parameter Modelling of Urban Residential Energy Demand: *Remote Sensing Quarterly,* V. 1, p. 106–115.

———, 1980, Image Analysis as a Check on Census Enumeration Accuracy: *Photogrammetric Engineering and Remote Sensing,* V. 46, p. 757–764.

Colcord, J. E., 1981, Thermal Imagery Energy Surveys: *Photogrammetric Engineering and Remote Sensing,* V. 47, p. 237–240.

Collins, W. G., and A. H. A. El-Beik, 1971, Population Census With the Aid of Aerial Photographs: An Experiment in the City of Leeds: *Photogrammetric Record,* V. 7, p. 16–26.

Colvocoresses, A. P., 1977, Proposed Parameters for an Operational Landsat: *Photogrammetric Engineering and Remote Sensing,* 43, p. 1139–1145.

Colwell, R. N., 1960, The Photo Interpretation Picture in 1960: *Photogrammetric Engineering and Remote Sensing,* V. 16, p. 292–314.

———, 1965, The Extraction of Data from Aerial Photographs by Human and Mechanical Means: *Photogrammetria,* V. 20, p. 211–228.

Condit, H. R., 1970, The Spectral Reflectance of American Soils: *Photogrammetric Engineering,* V. 36, p. 955–966.

Cooley, W. W., and P. R. Lohnes, 1971, *Multivariate Data Analysis:* John Wiley and Sons, New York, New York.

Croft, T. A., 1977, *Nocturnal Images of the Earth from Space:* Stanford Research Institute, SRI Project No. 5593, Order No. 68197, 109 p.

———, 1979, *The Brightness of Lights on Earth at Night, Digitally Recorded by DMSP Satellite:* SRI International 57 p.

Davies, S., A. Tuyahov, and R. K. Holz, 1973, Use of Remote Sensing to Determine Urban Poverty Neighborhoods: *In* Holz, R. K., ed., *The Surveillance Science:* Boston, Houghton Mifflin Co., p. 386–390.

Davis, J. B., 1978, Test Phase Evaluation Report on the Census-NASA ASVT: Draft Report, Geography Division, Bureau of the Census, Washington, D.C., 50 p.

Davis, J. B., and S. Z. Friedman, 1979, Assessing Urbanized Area Expansion Through the Integration of Landsat and Conventional Data: *Proceedings,* of the American Society of Photogrammetry, 45th Annual Meeting, Washington, D.C., p. 776–791.

Dayal, H. H., and B. A. Khairzada, 1976, The First National Demographic Survey of Afghanistan: The Role Played by Air Photos and Photo-Counting Techniques: *The ITC Journal,* V. 1, p. 84–97.

Deuker, K., and F. Horton, 1971, Urban Change Detection Systems: Status and Prospects: *Proceedings,* Seventh International Symposium on Remote Sensing of Environment, p. 1523–1536.

Doyle, F. J., 1978, The Next Decade of Satellite Remote Sensing: *Photogrammetric Engineering and Remote Sensing,* V. 44, p. 155–164.

Dwornik, S. E., D. G. Orr, and L. M. Young, 1963, *Reflectance Curves of Soil, Rocks, Vegetation and Pavement:* Report No. 1746R, USAERDL, Ft. Belvoir, Virginia.

El-Baz, F., 1980, A System of Advanced Cartographic Technology for Developing Nations: *Proceedings,* Fourteenth International Symposium on Remote Sensing of Environment, p. 701–712.

Estes, J. E., and D. S. Simonett, 1975, Fundamentals of Image Interpretation: in the *Manual of Remote Sensing,* L. W. Bowden, and E. L. Pruitt, eds., Falls Church: American Society of Photogrammetry, p. 869–1076.

Estes, J. E., 1981, Remote Sensing and Geographic Information Systems Coming of Age in the Eighties: Proceedings of the Seventh Annual Pecora Symposium on Remote Sensing, Sioux Falls, South Dakota (in press).

Eyre, L. A., B. Adolphus, and M. Amiel, 1970, Census Analysis and Population Studies: *Photogrammetric Engineering,* V. 36, p. 460–466.

Federal Power Commission, 1970, *The 1970 National Power Survey:* Part IV, Technical Advisory Committee Reports, p. 404.

Fitzpatrick-Lins, K. and M. J. Chambers, 1977, Determination of Accuracy and Information Content of Land Use and Land Cover Maps at Different Scales: *Remote Sensing of the Electro Magnetic Spectrum,* V. 4, p. 41–48.

Fitzpatrick-Lins, K., 1978A, Accuracy and Consistency Comparisons of Land Use and Land Cover Maps made from High-Altitude Photographs and Landsat Multispectral Imagery: *U.S. Geological Survey Journal of Research,* V. 6, p. 23–40.

———, 1978B, Accuracy of Selected Land Use and Land Cover Maps in the Greater Atlanta Region, Georgia: *U.S. Geological Survey Journal of Research,* V. 6, p. 169–173.

———, 1980, The Accuracy of Selected Land Use and Land Cover Maps at Scales of 1:250,000 and 1:100,000: Geological Survey Circular 829, 24 p.

Ford, K., 1979, *Remote Sensing for Planners:* Rutgers: Center for Urban Polich Research, 219 p.

Friedman, S. Z., 1978, Report for Phase I of the Census Urbanized Area Application System Verification and Transfer: Jet Propulation Laboratory, Pasadena, California, 91 p.

———, 1980, *Mapping Urbanized Area Expansion Through Digital Image Processing of Landsat and Conventional Data:* Jet Propulsion Laboratory, Pasadena, California, Publication 79–113, (March), 90 p.

Garofalo, D., and F. J. Wobber, 1974, Solid Waste and Remote Sensing: *Photogrammetric Engineering and Remote Sensing,* V. 40, p. 45–59.

Garofalo, D., and K. R. Martin, 1977, Regional Energy Availability from Conversion of Solid Waste: *Photogrammetric Engineering and Remote Sensing,* V. 43, p. 727–738.

Gautam, N. C., 1976, Aerial Photo-Interpretation Techniques for Classifying Urban Land Use: *Photogrammetric Engineering and Remote Sensing,* V. 42, p. 815–822.

General Electric, 1974, *Definition of the Total Earth Resources System for the Shuttle Era:* Philadelphia: General Electric Co. Space Division, chapter 8.

———, 1978, Synopsis of Census Urban Area ASVT Baseline Processing: Interium Report to National Aeronautics and Space Administration, Greenbelt: General Electric Co. (October), 39 p.

Ginevan, M. E., 1979, Testing Land-Use Map Accuracy: Another Look: *Photogrammetric Engineering and Remote Sensing,* V. 45 (10), p. 1371–1377.

Green, N. E., 1956, Aerial Photographic Analysis of Residential Neighborhoods: An Evaluation of Data Accuracy: *Social Forces,* V. 35, 142–147.

Gregory, S., 1963, *Statistical Methods and the Geographer:* (2nd ed.), New Jersey: Humanities Press, p. 82–100.

Hackforth, H. L., 1960, *Infrared Radiation:* McGraw-Hill, 303 p.

Hadfield, S. A., 1963, Evaluation of Land Use and Dwelling Unit Data Derived from Aerial Photography: Urban Research Section, Chicago Area Transportation Study, Chicago, Illinois.

Haralick, R. M., 1979, Statistical and Structural Approaches to Texture: *Proceedings,* IEEE, V. 67, p. 786–804.

Hay, A. M., 1979, Sampling Designs to Test Land-Use Map Accuracy: *Photogrammetric Engineering and Remote Sensing,* V. 45, p. 529–533.

Hazard, W. R., 1979, *Aerial Analysis of Neighborhood Boundaries:* Energy Measures Corporation, 11 p.

———, and S. A. Underwood, 1979, *Modeling Heat Loss with Aerial Thermography.* Final Report, D.O.E., Contract No. EM-78-C-01-4148, 123 p.

Henderson, F. M., and J. Utano, 1975, Assessing Urban Socio-economic Conditions with Conventional Air Photography: *Photogrammetria,* V. 31, p. 81–89.

Henderson, F. M., 1979, Housing and Population Analyses: *In* Ford, K. (ed.), *Remote Sensing for Planners:* New Brunswick, N.J., Rutgers University, Center for Urban Policy Research, p. 135–154.

———, 1979, Land Use Analysis of Radar Imagery: *Photogrammetric Engineering and Remote Sensing,* V. 45 (3), p. 295–307.

——— and M. A. Anuta, 1979, Settlement Detection with Radar Imagery: *Joint Proceedings of the ASP-ACSM 1979 Fall Technical Meeting,* p. 89–104.

Henderson, F. M., 1980a, Influence of Radar Azimuth Angle on Small Settlement Detection: International Society Archives of Photogrammetry, V.B7, p. 404–411.

———, 1980b, Extracting Urban Data from Seasat SAR Imagery: The Merit of Image Enlargements and Density Slicing: International Archives of Photogrammetry, V.B7, p. 411–421.

———, S. W. Wharton, and D. L. Toll, 1980, Preliminary Results of Mapping Urban Land Cover with Seasat SAR Imagery: *Technical Papers 1980 ACSM-ASP* Convention, ASP 46th Annual Meeting, p. 310–318.

Higgs, G. K., 1975, Multispectral Approach to Urban Neighborhood Analysis and Delineation: *Proceedings,* American Society of Photogrammetry, p. 444–465.

Hoffer, R. M., 1978, Biological and Physically Considerations in Applying Computer-Aided Analysis Techniques to Remote Sensor Data: Chapter 5 in P. Swain and S. Davis (eds.), *Remote Sensing—The Quantitative Approach:* New York: McGraw-Hill, p. 277–289.

Holz, R. K., D. L. Huff, and R. C. Mayfield, 1969,

Urban Spatial Structure Based on Remote Sensing Imagery: *Proceedings,* Sixth International Symposium Remote Sensing of Environment, V. 2, p. 819–830.

Holz, R. K., 1977, Cultural Features Imaged and Observed from SKYLAB 4: Chapter 8 in *SKYLAB Explores the Earth,* Washington: NASA, 225–242.

Hord, R. M. and W. Brooner, 1976, Land-Use Map Accuracy Criteria: *Photogrammetric Engineering and Remote Sensing,* V. 42, p. 671–677.

Howard, W. A., L. C. Harold, L. B. Driscoll and L. R. LaRerriere, 1974, Residential Environmental Quality in Denver Utilizing Remote Sensing Techniques: University of Denver Publications in Geography, #74–1, 175 p.

Hsu, S, 1971, Population Estimation: *Photogrammetric Engineering and Remote Sensing,* V. 37, p. 449–454.

———, 1978, Texture-Tone Analysis for Automated Land-Use Mapping; *Photogrammetric Engineering and Remote Sensing,* V. 44, p. 1393–1404.

Ingram, K. J., E. M. Knapp and J. W. Robinson, 1981, Change Detection Technique Development for improved Urbanized Area Delineation, Technical Memorandum 81/6087, Computer Sciences Corporation, Silver Springs, M.D. 75 pages

Jackson, M. J., P. Carter, T. F. Smith, and W. G. Gardner, 1980, Urban Land Mapping from Remotely Sensed Data: *Photogrammetric Engineering and Remote Sensing,* V. 46, p. 1041–1050.

Jenkins, G. M., and D. G. Watts, 1968, *Spectral Analysis and its Applications:* Holden-Day, Inc., San Francisco, 525 p.

Jensen, J. R., 1978, Digital Land Cover Mapping Using Layered Classification Logic and Physical Composition Attributes: *The American Cartographer,* V. 5, p. 121–132.

Jensen, J. R., F. A. Ennerson, and E. J. Hajic, 1979, An Interactive Image Processing System for Remote Sensing Education: *Photogrammetric Engineering and Remote Sensing,* V. 45, p. 1519–1527.

Jensen, J. R., 1979a, Computer Graphic Feature Analysis and Selection: *Photogrammetric Engineering and Remote Sensing,* V. 45, p. 1507–1512.

———, 1979b, Spectral and Textural Features to Classify Elusive Land Cover at the Urban Fringe: *The Professional Geographer,* V. 31, p. 400–409.

———, 1980, *Urban Area Change Detection Procedures with Remote Sensing Data:* Report contract NAS5-26129, Goddard Space Flight Center, Greenbelt, Maryland (December), 50 p.

———, 1981, Urban Change Detection Mapping Using Landsat Digital Data: *The American Cartographer,* V. 8, p. 127–147.

Knapp, E. M., C. O. Justice, and K. J. Ingram, 1979, The Integration of Landsat Data with Ancillary Data for Improved Environmental Analysis: Current Status and Future Directions: Handout for the 17th Annual Conference of the Urban and Regional Information Systems Association, San Diego, California, 10 p.

Kraus, S. P., L. W. Senger, and J. M. Ryerson, 1974, Estimating Population from Photographically Determined Residential Land Use Types: *Remote Sensing of Environment,* V. 3, p. 35–42.

Kuchler, A. W., 1964, Potential Natural Vegetation of the Conterminous United States: (Map scale 1:3, 168,000), New York: American Geographical Society Special Research Publication No. 36.

Lachowski, H., C. Croteau, D. Dietrich, R. Fries, R. Hicks, C. Peterson, A. Royal, and J. Davis, 1979, Test Phase Processing Census Urban Area Application System Verification and Transfer: Final Report NASA 5-96354, Mod. No. 256, General Electric Co., Beltsville, Maryland, 50 p.

Lauer, D. T. and W. J. Todd, 1981, Land Cover Mapping with Merged Landsat RBV and MSS Images: ASP Fall Technical Meeting, September 1981, San Francisco, CA, p. 68–89.

Lawrence, G. R., 1977, Detection of Heat Loss From Buildings Through Aerial Thermography Applications and Methodology: *Proceedings,* Fourth Canadian Symposium on Remote Sensing, p. 220–226.

Lewis, A. J., H. C. MacDonald, and D. S. Simonett, 1969, Detection of Linear Cultural Features with Multipolarized Radar Imagery: *Proceedings,* 6th International Symposium on Remote Sensing of Environment, Ann Arbor, Michigan, p. 879–893.

Lillesand, T. M., and R. W. Kiefer, 1979, *Remote Sensing and Image Interpretation:* John Wiley and Sons, 396 p.

Lillesand, R. L., 1972, Techniques for Change Detection: *IEEE Transactions on Computers,* C-21, p. 654.

Lindgren, D. T., 1971, Dwelling Unit Estimation with Color-IR Photos: *Photogrammetric Engineering,* V. 37, p. 373–377.

———, 1979, Chapter 9: Transportation Planning: *In* Ford, Kristina (ed.), *Remote Sensing for Planners,* Rutgers: Center for Urban Policy Research, p. 155–168.

Lins, H. F., 1976, Land-Use Mapping from Skylab S-190B Photography: *Photogrammetric Engineering and Remote Sensing,* V. 42, p. 301–307.

Loelkes, G. L., Jr., 1977, Specifications for Land Use and Land Cover and Associated Maps: *U.S. Geological Survey Open-File Report 77-555,* 82 p.

Loelkes, G., L. Hardin, E. Jessen, E. Napur, R. Johnson, and W. Good, 1977, The Compilation Process for Land Use and Land Cover and Associated Maps: *Remote Sensing of the Electromagnetic Spectrum,* V. 4, p. 20–40.

Lo, C. P., 1971, A Typological Classification of Buildings in the City Centre of Glasgow from Aerial Photographs: *Photogrammetria,* V. 27, p. 135–157.

———, 1973a, Cartographic Presentation of Three-Dimensional Urban Information: *The Cartographic Journal,* (December), p. 77–84.

———, 1973b, The Use of Orthophotographs in Urban Planning: *Town Planning Review,* V. 44, p. 71–87.

———, and R. Welch, 1977, Chinese Urban Population Estimates: *Annals, Association of American Geographers,* V. 47, p. 246–253.

———, 1979, Surveys of Squatter Settlements with Sequential Aerial Photography—A Case Study in Hong Kong: *Photogrammetria,* V. 35, p. 45–63.

———, and F. F. Chan, 1980, Rural Population Estimation from Aerial Photographs: *Photogrammetric Engineering and Remote Sensing,* V. 46, p. 337–345.

Lowe, D. S., 1978, Use of Landsat in Computer Data Bases: *Proceedings,* Conference on Computer Mapping Software and Data Bases: Application and Dissemination, Harvard University, Cambridge, Massachusetts, 10 p.

Madding, R. P., 1979, *Thermographic Instruments and*

Systems: Board of Regents, University of Wisconsin, 132 p.

Mallon, H. J., and J. Y. Howard, 1971, An Assessment of Remote Sensor Imagery in the Determination of Housing Quality Data: Washington, D.C., U.S. Department of Commerce, NTIS Report PB-211-380.

Marble, D. F., and F. E. Horton, 1969, Extraction of Urban Data from High and Low Resolution Images: *Proceedings,* Sixth International Symposium on Remote Sensing of Environment, p. 807–818.

Marschner, F. J., 1950, Major Land Uses in the United States: (Map scale 1:5,000,000), U.S. Department of Agriculture Research Series.

Matthews, M. L. and L. M. Miller, 1979, *A Bibliography: Change Detection with Remote Sensing Imagery:* Remote Sensing Center, Texas A & M University, College Station, Texas, 5 pages

Matthews, R. E., 1975, ed., *Active Microwave Workshop Report:* NASA: Washington, D.C., p. 126–147.

Mausel, P. W., L. Alger and R. Herner, 1979, Identification of Surface-Disturbed Features through ISURSL Non-Parametric Analysis of Landsat MSS Data: Proceedings, Machine Processing of Remotely Sensed Data, Purdue Univ., West Lafayette, Ind., p. 172–182.

Mausel, P. M., W. J. Todd, and M. F. Baumgardner, 1974, An Analysis of Metropolitan Land Use by Machine Processing of Earth Resources Techology Satellite Data: *Proceedings,* Association of American Geographers, Washington: Association of American Geographers, p. 54–57.

McCoy, R. M., and E. D. Metivier, 1973, House Density Versus Socioeconomic Conditions: *Photogrammetric Engineering,* V. 39, p. 43–49.

McKinney, R. L., 1979, Cartographic Considerations for the Integration of Landsat Digital Imagery with Existing Spatial Data: *Proceedings,* American Society of Photogrammetry.

Metivier, E. D. and R. M. McCoy, 1971, Mapping Urban Poverty Housing from Aerial Photographs: Proceedings of the Seventh International Symposium on Remote Sensing of Environment, Ann Arbor, 1563–1569.

Michigan Land Use Classification and Referencing Committee, 1975, *Michigan Land Cover/Use Classification System:* Lansing: Office of Land Use, Department of Natural Resources, State of Michigan (July), 60 p.

Milazzo, V., R. Ellefson, and D. Schwarz, 1977, Updating Land Use and Land Cover Maps: *Remote Sensing of the Electromagnetic Spectrum,* V. 4, p. 103–116.

Milazzo, V., 1980, A Review and Evaluation of Alternatives for Updating U.S. Geological Survey Land Use and Land Cover Maps: *U.S. Geological Survey Circular 826,* 19 p.

Miller, C. A., and L. D. Miller, 1976, *County Resource Land Use Planning Activities in the United States:* Remote Sensing Center, Texas A & M University, College Station, Texas, 264 p.

Mills, R. F., and J. L. Dwyer, 1979, Operational Analysis of a Geographic Information System in New Jersey: Presented at the *Second Annual International User's Conference on Computer Mapping Software, Hardware, and Data Bases,* Harvard University, Cambridge, Massachusetts, 10 p.

Mitchell, W. B., S. C. Guptill, K. E. Anderson, R. G. Fegeas, and C. A. Hallam, 1977, GIRAS: A Geographic Information Retrieval and Analysis System for Handling Land Use and Land Cover Data: *Geol. Survey Prof. Paper 1059,* 16 p.

Moore, E. G., 1969, Side-Looking Radar in Urban Research: A Case Study: *U.S. Geological Survey Interagency Report—NASA-138,* U.S.G.S. Open File Report, 24 p. (NTIS No. N69 16108).

———, 1970, Application of Remote Sensors to the Classification of Aerial Data at Different Scales: A Case Study in Housing Quality: *Remote Sensing of Environment,* V. 1, p. 109–121.

Mullens, R. H., Jr., 1969, Analysis of Urban Residential Environments Using Color Infrared Aerial Photography: An Examination of Socioeconomic Variables and Physical Characteristics of Selected Areas in Los Angeles Basin: Studies in Remote Sensing of Southern California and Related Environments, U.S. Department of Interior Contract 14-08-0001-10674, Technical Report IV., 30 p.

Mumbower, L. E., and J. Donoghue, 1967, Urban Poverty Study: *Photogrammetric Engineering,* V. 33, p. 610–618.

Naago, M., and T. Matsuyama, 1979, Edge Preserving and Smoothing: *Computer Graphics and Image Processing,* V. 9, p. 394–407.

NASA, 1980 ERSAR—Earth Resources Synthetic Aperture Radar—Program Definition Workshop Report: Pasadena: National Aeronautics and Space Administration, 55 p.

National Academy of Sciences, 1976, *Resources and Environmental Surveys from Space with the Thematic Mapper in the 1980's,* National Academy of Sciences, Washington, 122 p.

Nordbeck, S., 1965, The law of Allometric growth: Ann Arbor, University of Michigan, Discussion Paper 7, Michigan Inter-University Community of Mathematical Geographers.

Norwood, V. T., 1974, Balance Between Resolution and Signal to Noise Ratio in Scanner Design for Earth Resources Systems, *Proceedings* of the Society of Photo-optical Instrumentation Engineers, V. 51, Scanners and Imagery for Earth Observation, August 19–20, San Diego, California, p. 37–42.

Ogrosky, C. E., 1975, Population estimates from satellite imagery: *Photogrammetric Engineering and Remote Sensing,* V. 41, p. 707–712.

Otepka, G., 1978, Practical Experience in the Rectification of MSS Images: *Photogrammetric Engineering and Remote Sensing,* V. 44, p. 459–467.

Padron, R. I., J. A. Royal, W. C. Dallam, and W. K. Stow, 1980, A Review of the Capabilities of the Proposed Shuttle Payloads for Monitoring Urban Expansion: NASA 5-25707, General Electric Co., Beltsville, Maryland.

Peake, W. H. and T. L. Oliver, 1971, The Response of Terrestrial Surfaces at Microwave Frequencies: Ohio State University Electro-science Lab. 2240-7, Technical Report AFAL-TR-70-301, Columbus, Ohio.

Pease, R. W. and D. A. Nichols, 1976, Energy Balance Maps from Remotely Sensed Imagery: *Photogrammetric Engineering and Remote Sensing,* V. 42, p. 1367–1374.

Piech, K. R., D. W. Gaucher, J. R. Schott, and P. G. Smith, 1977, Terrain Classification Using Color Imagery: *Photogrammetric Engineering and Remote Sensing,* V. 43, p. 507–513.

Place, J. L., 1974, Change in land use in the Phoenix (1:250,000) quadrangle, Arizona, between 1970 and 1973—ERTS as an aid in a nationwide program for mapping general land use: Proceedings *Earth Resources Technology Satellite-1 Symposium,* Washington, D.C., V. 1, p. 393–423.

———, 1977, The Land Use and Land Cover Map and Data Program of the U.S. Geological Survey: An Overview: *Remote Sensing of the Electromagnetic Spectrum,* p. 1–9.

Rabchevsky, G. A., 1977, Temporal and Dynamic Observations from Satellites: *Photogrammetric Engineering and Remote Sensing,* V. 43, p. 1515–1518.

Riordan, C. J., 1980, *Non-Urban to Urban Land Cover Change Detection Using Landsat Data,* Report #1 NASA Contract NAS5-25696, Earth Resources Department, Colorado State University, Fort Collins, 40 p.

Robinson, J. W., 1979, *A Critical Review of the Change Detection and Urban Classification Literature,* Technical Memorandum 79/6235, Computer Sciences Corp., Silver Springs, Md., 50 p.

———, 1980, Change Detection Methodology for Urbanized Area Delineation: in Appendix A, Technical Memorandum in Preparation, Computer Sciences Corporation, Silver Springs, Maryland.

Root, R. R. and L. D. Miller, 1971, *Identification of Urban Watershed Units Using Remote Multispectral Sensing,* Environmental Resources Center, Completion Report Series No. 29, Colorado State University, Fort Collins, Colorado, 50 p.

Rosenfield, G. H., 1981, Analysis of Variance of Thematic Mapping Experiment Data: *Photogrammetric Engineering and Remote Sensing,* V. 47, in press.

Rosenfield, G. H., K. Fitzpatrick-Lins, and H. S. Ling, 1982, Sampling for Thematic Map Accuracy Testing: *Photogrammetric Engineering and Remote Sensing,* V. 48, No. 1, p. 131–137.

Royal, J. A., 1980, *Change Detection Method Development Census Urban Area Application Pilot Test,* Final Report contract NAS5-25707, General Electric Company, Beltsville, Md., (May), 74 p.

Rubingh, J. and D. Carlson, 1979, *Agricultural Land Conversion in Colorado,* Colorado Department of Agriculture, Denver, 95 p.

Rush, M., and S. Vernon, 1975, Remote Sensing and Urban Public Health: *Photogrammetric Enginneering and Remote Sensing,* V. 41, p. 1149–1155.

Ryerson, R. and D. Gierman, 1975, A Remote Sensing Compatible Land Use Activity Classification: *Technical Note 75-1,* Ottawa: Canada Centre for Remote Sensing, (May), 18 p.

Ryerson, R. A., 1980, Land Use Information from Remotely Sensed Data: Users Manual 80-1, Ottawa: Canada Centre for Remote Sensing, (February), 30 p.

Sabins, F. S., 1978, Remote Sensing Principles and Interpretation. San Francisco, W. H. Freeman and Company, 426 p.

Sabol, J., 1969, The Relationship Between Population and Radar-Derived Area of Urban Places: in the *Utility of Radar and Other Remote Sensors in Thematic Land Use Mapping from Spacecraft,* D. S. Simonett (ed.), Annual Report U.S. Geological Survey Interagency Report—NASA-140, January, p 44–75.

Schaber, G. G., G. L. Berlin, and W. E. Brown, 1976, Variations in Surface Roughness within Death Valley, California: Geologic evaluation of 25-cm wavelength radar images: *Geological Society America Bulletin,* V. 87, p. 29–41.

Schott, J. R., 1978, Principles of Thermal Infrared Remote Sensing for Heat Cost Determination: *Proceedings,* American Society of Photogrammetry Fall Technical Meeting, p. 457–467.

Schulze, R. E., 1969, A Comparison Between Official Population Data and an Aerial Photograph Population Survey in the Tugela Location: *South African Geographical Journal,* V. 51, p. 123–132.

Seal, H. L., 1964, *Multivariate Statistical Analysis for Biologists,* Methven and Co., Ltd., London, England.

Seidman, J. B. and A. Y. Smith, 1979, *VICAR Image Processing System: Guide to System Use,* Jet Propulsion Laboratory, JPL Technical Report 77-37 Revision 1, 69 p.

Shelton, R. L. and J. E. Estes, 1979, Integration of Remote Sensing and Geographic Information Systems: *Proceedings,* 13th International Symposium on Remote Sensing of Environment, Ann Arbor, Michigan.

Shepard, J. R., 1964, A Concept of Change Detection: *Photogrammetric Engineering and Remote Sensing,* 30, p. 649.

Simpson, R. B., 1969, APQ-97 Imagery of New England: A Geographic Evaluation: *Proceedings of 6th International Symposium on Remote Sensing of Environment,* Ann Arbor, Michigan, p. 909–925.

Slater, P. H., 1975, Photographic Systems for Remote Sensing: Chapter 6 in R. G. Reeves (Ed.) Manual of Remote Sensing, V. 1, p. 237.

Smedes, H. W., R. L. Hulstrom, and K. J. Ranson, 1975, The Mixture Problem in Computer Mapping of Terrain: Improved Techniques for Establishing Spectral Signatures Atmospheric Path Radiance, and Transmittance: *Proceedings,* NASA Earth Resources Survey Symposium, Houston: NASA, p. 1099–1156.

Smith, T. F., 1978, National Survey of 'Developed Land' Using Air Photography and Satellite Data: *Proceedings,* International Symposium on Earth Resources and the Endangered Environment, Freiburg, FRG, p. 917–927.

Snedecor, G. W. and W. F. Cochran, 1967, *Statistical Methods,* Ames: Iowa State University Press, p. 202–211, and 516–517.

Southern District Office, 1975, *Coastal Los Angeles Land Use Study, 1973,* District Report, State of California, Department of Water Resources, 10 p.

Starsinic, D. E., and Zitter, M., 1968, Accuracy of the Housing Unit Method in Preparing Population Estimates for cities: *Demography,* V. 5, 475–484.

Stauffer, M. L. and R. L. McKinney, 1978, *Landsat Image Differencing as an Automated Land Cover Change Detection Technique,* Report contract NAS5-24350, General Electric Co., Beltsville, Md., (Aug), 30 p.

Suits, G. H., 1975, The Nature of Electromagnetic Radiation; Chapter 3 in R. G. Reeves (Ed.) Manual of Remote Sensing, V. 1, p. 59.

Swain, P. H. and S. M. Davis, 1978, *Remote Sensing: The Quantitative Approach,* McGraw Hill, New York, 396 p.

Taranik, J. V., 1978, Principles of Computer Processing of Landsat Data for Geologic Applications: *U.S. Geological Survey Open File Report,* NO. 78-117, 50 p.

Texas Instruments, 1978, *Aerial Infrared Thermograms and Residential Heat Loss,* Ecological Services, Texas Instruments Corporation, 34 p.

Thompson, D., 1975, Current Population Estimation Using Land Use Data Derived from High Altitude

Photography: *Proceedings,* Association of American Geographers, V. 7, p. 237–243.

Tobler, W. R., 1969, Satellite Confirmation of Settlement Size Coefficients: *Area,* V. 1, (3), p. 31–34.

Todd, W. J., 1977, Urban and Regional Land Use Change Detected by Using Landsat Data: *Journal of Research,* U.S. Geological Survey, 5, p. 529–534.

———, 1978, *A Selective Bibliography: Remote Sensing Applications in Land Cover Inventory Tasks,* Sioux Falls: Technicolor Graphics Services (April), 37 p.

Todd, W. J., R. N. Hall, C. C. Henry, B. L. Lake, 1978, Metropolitan Land Cover Inventory Using Multiseasonal Landsat Data: *USGS Open File Report 78-378,* Sioux Falls: EROS Data Center, 26 p.

Toll, D. L., J. A. Royal, and J. B. Davis, 1980, Urban Area Update Procedures Using Landsat Data: *Proceedings,* American Society of Photogrammetry, (Oct), 12 p.

Tucker, C. J., 1978, A Comparison of Satellite Sensor Bands for Vegetation Monitoring: *Photogrammetric Engineering and Remote Sensing,* V. 44, p. 1369–1380.

Turner, H. and D. Steiner, 1979, Digital Image Processing Techniques to Extract Metric Data on Buildings from Shadows on Simulated Air Photos: *Photogrammetria,* 35, p. 141–160.

Tuyahov, A. J., 1972, *Remote Sensing of a Barrier Island: Padre Island, Texas* and *Remote Sensing as Information System for the Urban Environment,* M. A. Reports, Department of Geography, The University of Texas at Austin, 163 p.

Underwood, S. A., A. G. Houston, W. R. Hazard, 1980, *Performance Evaluation: Estimates of Structural Heat Loss From Interpretation of Aerial Thermography,* 1 p.

Urban Renewal Administration, Housing and Home Finance Agency, and Bureau of Public Roads, 1965, *Standard Land Use Coding Manual, A Standard System for Identifying and Coding Land Use Activities,* Washington; D.C., 111 p.

U.S. Bureau of the Census, *Population Distribution, Urban and Rural, in the United States: 1970,* United States Maps GE-70, No. 1, U.S. Government Printing Office, Washington, D.C.

U.S. Executive Office of the President, Bureau of the Budget, 1957, *Standard Industrial Classification Code,* Washington, D.C.

U.S. Geological Survey, 1980, High Altitude Photography Program: *Landsat Data Users Notes,* V. 11 (March), 2 p.

U.S. Geological Survey, 1980, Transition to Operational Landsat System Planned: *Landsat Data Users Notes,* V. 14 (Sept), p. 8.

Van Genderen, J. L. and B. F. Lock, 1977, Testing Land-Use Map Accuracy: *Photogrammetric Engineering and Remote Sensing,* Vol. 43, p. 1135–1137.

Van Wie, P. and M. Stein, 1976, A Landsat Digital Image Rectification System: Proceedings, Machine Processing of Remotely Sensed Data, Purdue Univ., West Lafayette, Ind., p. 18–26.

Veign, J. L. and F. B. Reeves, 1973, A Case for Orthophoto Mapping: *Photogrammetric Engineering,* V. 34, p. 1059–1064.

Waite, W. P., H. C. MacDonald, D. N. Tolman, C. A. Barlow, and M. Borengasser, 1978, Dual Polarized Long Wavelength Radar for Discrimination of Agricultural Land Use: *Proceedings of American Society of Photogrammetry Technical Meeting,* p. 595–607.

Weismiller, R. A., S. J. Kristof, D. K. Scholz, P. E. Anuta, and S. A. Momin, 1977, Change Detection in Coastal Zone Environments: *Photogrammetric Engineering and Remote Sensing,* 43, p. 1533–1539.

Welch, R., 1980, Monitoring Urban Population and Energy Utilization Patterns from Satellite Data: *Remote Sensing of Environment,* V. 9, p. 1–9.

Welch, R., 1981, Spatial Resolution and Geometric Potential of Planned Earth Satellite Missions: *Proceedings of the Fifteenth International Symposium on Remote Sensing of Environment,* Ann Arbor, Mich., p. 1275-1283.

Welch, R., 1982, Spatial resolution requirements for urban studies: *International Journal of Remote Sensing,* V. 3, in press.

Welch, R., and C. W. Pannell, 1975, Landsat Investigations of Recent Urban Land Use Changes in Northeast China: *Proceedings of the Tenth International Symposium on Remote Sensing of Environment,* Ann Arbor, Mich., p. 373–382.

Welch, R. and C. W. Ponnell, 1982, Comparative resolution of Landsat-3 MSS and RBV image data of China: *The Photogrammetric Record,* v. 10, p. 575–586.

Welch, R., and S. Zupko, 1980, Urbanized Area Energy Utilization Patterns from DMSP Data: *Photogrammetric Engineering and Remote Sensing,* V. 46, p. 201–207.

Wellar, B. S., 1973, Remote Sensing and Urban Information Systems, *Photogrammetric Engineering,* V. 34, p. 1041–1050.

Williams, D. L. and F. Borden, 1977, A Reduction in AG./Residential Signature Conflict Using Principal Components Analysis of Landsat Temporal Data: *Proceedings,* American Society of Photogrammetry, p. 230–39.

Wolfe, W. L. and I. Zissis, 1978, *The Infrared Handbook,* Washington: U.S. Government Printing Office, p. 3–97 to 3–101.

Wray, J. R., 1980, Land Use and Land Cover in the Greater Pittsburgh Region, PA., 1973: (Map scale 1:125,000), Reston: U.S. Geological Survey.

———, 1980, Computer-Plotted Map of Land Use and Land Cover, Three Mile Island and Vicinity, with Census Tracts: (Map scale 1:350,000), Reston: U.S. Geological Survey.

Wrigley, C., and G. Storti, 1978, *Development of an Improved High Efficiency Thin Silicon Solar Cell,* Solarex Corporation, Report No. SX/115/3Q, Rockville, Maryland.

Wu, S. T., 1979, An Improvement in Land Cover Classification Achieved by Merging Microwave Data with Landsat Multispectral Scanner Data: *Proceedings American Society of Photogrammetry,* St. Louis: p. 293–309.

Chapter 31

Geological Applications

Author-Editor: RICHARD S. WILLIAMS, JR.

Contributing Authors: JOHN M. AARON, MICHAEL J. ABRAMS, RICHARD W. BIRNIE, HERBERT W. BLODGET, DONOVAN R. BOWLEY, THOMAS BREWER, RONALD L. BROOKS, WM. DOUGLAS CARTER, PAT S. CHAVEZ, JR., MALCOLM M. CLARK, DAVID H. COUPLAND, WILLIAM E. DAVIES, MORRIS DEUTSCH, JAMES I. EBERT, CHRISTOPHER D. ELVIDGE, JOHN P. FORD, JOSEPH R. FRANCICA, JULES D. FRIEDMAN, ALEXANDER F. H. GOETZ, DAVID P. GOLD, WILLIAM R. HEMPHILL, YNGVAR W. ISACHSEN, ANNE B. KAHLE, MARGUERITE J. KINGSTON, WILLIAM S. KOWALIK, STEPHEN P. LEATHERMAN, BAERBEL K. LUCCHITTA, RONALD J. P. LYON, THOMAS R. LYONS, JOHN W. M'GONIGLE, NED MAMULA, JR., STUART E. MARSH, CAROL L. MOLNIA, ELLIOT C. MORRIS, JAMES T. NEAL, DENNIS W. O'LEARY, STEVE ONYSKO, DONALD G. ORR, MELVIN H. PODWYSOCKI, LAWRENCE C. ROWAN, DONALD B. SEGAL, SHIRLEY L. SIMPSON, C. SCOTT SOUTHWORTH, DAVID D. STELLER, RONALD W. STINGELIN, HAROLD SVENSSON, ARNOLD F. THIESEN, CHARLES M. TRAUTWEIN, ROBERT K. VINCENT, BARRY VOIGHT, and ROBERT D. WATSON[1]

GENERAL CONTENTS: Definition of geological remote sensing; historical considerations; interrelationship of geologic remote sensing with other disciplines; types and availability of remotely sensed data for geological investigations; analysis techniques; applications to geologic subdisciplines: geologic mapping, economic geology, engineering geology, geologic hazards, glaciology, geomorphology, marine geology, geobotany, archaeological geology, astrogeology; future trends; references.

INTRODUCTION

Material contained in this chapter assumes that the reader has a solid background in the basic principles of geology and especially in those subdisciplines of geology which can make effective use of remotely sensed data. Although photogeologic techniques and methods are every bit as important now as they were to geologists working with the first stereopairs of aerial photographs, many textbooks are available on photogeology that cover the subject in far greater depth and breadth than can be done in this chapter (see also the Photogeology section of the Analysis Techniques section of this chapter).

This chapter represents an evolutionary step forward from Chapter 4, Photo Interpretation in Geology, in the American Society of Photogrammetry's *Manual of Photographic Interpretation* (Colwell, 1960) and Chapter 16, Terrain and Minerals: Assessment and Evaluation, in the first edition of the *Manual of Remote Sensing* (Reeves et al., 1975a). Chapter 4, Photo Interpretation in Geology (Tator et al., 1960), encompassed 174 pages written by 15 geologists. The chapter contained many excellent stereopairs of aerial photographs (selected to illustrate various types of landforms and geological phenomena), presented a geologic analysis of many of the aerial photographs, and included a good bibliography.

Chapter 16, Terrain and Minerals: Assessment and Evaluation (Reeves et al., 1975a) encompassed 245 pages written by 52 geologists. Some of the text and illustrations from Chapter 4 of the *Manual of Photographic Interpretation* were also included, so the 15 authors of the earlier work were noted as part of the chapter authorship. The introductory part of the chapter emphasized geomorphology and could serve as an introductory text on various aspects of that subject. The remainder of the chapter was devoted to a variety of remote sensing studies built around a series of aerial photographs, a single Landsat image, an aerial thermograph, a radar image, or a particular sensing system. Both of the geologically oriented chapters in the *Manual of Photographic Interpretation* and the first edition of the *Manual of Remote Sensing* were primarily based on the analysis of aerial and/or satellite photographs or images by established photogeological techniques.

Geologic remote sensing has grown so rapidly over the past decade that it is difficult to do justice to the technology in a single chapter or even an entire book. Excellent textbooks now exist on the subject of remote sensing in geology and related subjects. Three textbooks, in particular, provide solid reference material for geologists involved in remote sensing: *Remote Sensing: Principles and Interpretation* by Floyd Sabins (1978a), *Remote*

[1] For key to authorship of each heading and subheading of this chapter, see p. 1952–1953.

Sensing and Image Interpretation by Thomas Lillesand and Ralph Kiefer (1979), and *Remote Sensing in Geology* by a number of contributors, including the editors, Barry Siegal and Alan Gillespie (1980).

The present chapter is divided into six parts to assist the reader to understand geologic remote sensing technology in all its ramifications. The introductory material, in the first part, provides an historical account of how the technology evolved to where it is today and how geologic remote sensing relates to other disciplines. Even though a geologist may be working in only one aspect of geologic remote sensing technology, it is important that he or she be familiar with the other aspects of the technology: theory, laboratory studies, field investigations, surveys by aircraft, and satellite sensing systems and surveys. Knowledge of the limitations of sensing systems and image processing techniques is also imperative to avoid making errors in the geological analysis of remotely sensed data.

The second part of the chapter is devoted to a review of those spacecraft and aircraft sensors and data availability that are of specific value to geological investigations. Non-imaging and high-resolution systems are also discussed by various authors. Examples of specific data coverage, using Cape Cod, Massachusetts and environs as the primary area, will give geologists an excellent review of the types of remotely sensed data and related information now available for many areas.

The third section of the chapter reviews optical and digital techniques for image processing that are currently being used by geologists. Photogeological analysis remains the key technique for most geologists. Digital image processing may be used to increase the interpretability of a particular image or for specialized geological applications. Few geologists have access to the sophisticated digital image analysis and related instrumentation available to Charles Trautwein for his research on the Nabesna A-4 Quadrangle of Alaska. His work, however, provides an excellent example of some of the more advanced techniques presently being applied to remotely sensed data and is a measure of how far geological remote sensing technology has progressed during the past 20 years. (Compare Trautwein's research on Combined Data Sets as presented here with Chapter 4 in the *Manual of Photographic Interpretation,* for example.)

The fourth part of the chapter is the largest and most important element in the entire chapter. The approach used is the "case history" one, in which concise text and illustration(s) are presented for each of the principal subdisciplines of geology (and related fields), and where the use of remotely sensed data is relevant to the type of scientific investigation. Most of the 52 authors involved in the preparation of the chapter have contributed, directly or indirectly, to the fourth section. Well documented case histories are provided by geologists who are active research scientists in the particular subdiscipline being discussed. Because this chapter is in a *Manual,* each contributing author has stressed the practical rather than the theoretical aspects of geological remote sensing as applied to each topic. Major topics covered include the application of remote sensing technology to geologic mapping, economic geology, engineering geology, geologic hazards, glaciology, geomorphology, marine geology, and astrogeology. Key references for each case history are provided, so that the reader can delve more deeply into a particular topic than is possible in this chapter. Each of these major topics is worthy of a separate book, and some already have textbooks available.

Neither astrogeology nor glaciology was covered in either the *Manual of Photographic Interpretation* (Colwell, 1960) or the *Manual of Remote Sensing* (Reeves et al., 1975a). A small chapter on archaeology appeared in both the above; while this edition of the *Manual of Remote Sensing* includes full chapters on archaeology (and anthropology) and extraterrestrial applications (including astrogeology). Glaciology is discussed in the present chapter of the *Manual,* in Chapter 27, Weather and Climate; Remote Sensing of the Earth's Weather, and in Chapter 29, Water Resources Applications. This expanded coverage is another measure of the explosive growth in remote sensing technology as applied to geology and related disciplines. A new technological development first appears as a few papers, then as a chapter in a book, and then as one or more textbooks.

The fifth section of this chapter provides insight on future trends in geologic remote sensing. Frederick J. Doyle, President of the International Society of Photogrammetry and Remote Sensing, has provided a comprehensive up to date review of future spacecraft and sensors in two recent works (Section 17.10, Future Satellite Remote Sensing Systems, *in* Light et al., 1980; and Doyle 1982), and much of the fifth section is drawn from his review papers and the Office of Technology Assessment report (1982). Additional glimpses on geologic remote sensing in the future are provided by Goetz and Rowan (1981) and the National Academy of Sciences (1982).

The sixth section of the chapter provides 1,057 selected references to the key literature in geological remote sensing. Because of the rapid growth of the geologic remote sensing literature, a separate bibliographic volume would be needed to encompass the full scope of the available literature. For example, the bibliography compiled by Wm. Douglas Carter and Lawrence C. Rowan for their International Geological Correlation Programme (IGCP) Project 143, "Remote Sensing and Mineral Exploration," now (mid-1982) has over 1,500 entries. Chapter 4, Photo Interpretation in Geology (Tator et al., 1960), in the *Manual of Photographic Interpretation* (Colwell, editor, 1960), had 1,309 references, although some redundancy occurred because of topical grouping.

Chapter 16, Terrain and Minerals: Assessment and Evaluation (Reeves et al., 1975a) in the first edition of the *Manual of Remote Sensing* had 475 references (Williams, 1978a).

The application of remote sensing technology to the geological sciences is a continuing challenge because of the scope and diversity of geological phenomena to be understood and/or monitored on Earth and other planets and moons. During the past 20 years geological remote sensing technology has evolved from the photogeological stage to the digital image processing stage, from the local to the regional and global perspective, from the black-and-white aerial photograph to a wide variety and choice of single or sequential aerial and/or satellite images and photographs of a given area. Before the advent of Landsat who could have predicted that a global study of glaciers and other types of phenomena would become possible in such a short period of time (Williams and Ferrigno, 1981)? High-flying aircraft, Landsat, and other types of Earth observation satellites now make such studies a reality. Similar satellites and sensors have flown by or orbited many of the planets and moons of the Solar System, thereby providing geologists with other worlds to explore, study, and monitor (see the section of this chapter on Astrogeology, and also see Chapter 36, Extraterrestrial Applications). Some day, perhaps even by the end of this century, the first spacecraft to explore planetary systems of other stars will be launched with various types of remote sensors. Sophisticated remote sensing devices will eventually bring to the inhabitants of the planet Earth their first direct knowledge of other planetary systems in the Universe. There is, of course, much more to explore, study, understand, and monitor about our own planet, Earth, and aerial and satellite photographs and images will provide some of the data needed to better understand its geological nature (National Academy of Sciences, 1982). In this chapter, the author-editor and the other 51 authors have endeavored to capture some of the scientific enthusiasm and creativity that has been applied to the solution of various geologic problems in the attempt to gain a better understanding of geologic phenomena through the use of remotely sensed data. This chapter, then, will provide a guide for geologists to many of the types of geological studies, local, regional, and global, that can be accomplished with remotely sensed data.

DEFINITION OF GEOLOGICAL REMOTE SENSING

Geological remote sensing can be defined as the analysis of data acquired by sensors which record energy: The energy may be recorded (1) in one or more wavelengths of the electromagnetic spectrum, and (2) from acoustic (sonar) devices, to measure or monitor various physical and/or chemical characteristics, static or dynamic, of the geological environment of our planet and other planets and moons of our Solar System. Remote sensing technology, as applied to the geological sciences, is a multifaceted tool that, when properly applied to certain types of geological studies, enables geologists to more effectively and accurately carry out such investigations.

To a few geologists, geologic remote sensing is an overly sophisticated term that has replaced or superseded the older term photogeology. Geologic remote sensing, however, encompasses far more than simply using stereopairs of black-and-white aerial photographs as a substitute for a line map or as a means of studying and then mapping a particular area. For many areas in the United States a suite of image data, in various formats, now exists in various archives, as noted in the section of this chapter on "Types and Availability of Remotely Sensed Data for Geological Investigations". Such data may have been acquired by satellite or aircraft sensors. Studies of aerial and satellite data also may involve field studies and related data acquisition, laboratory investigations, and theoretical studies. Perhaps the biggest change that has taken place in geologic remote sensing during the past decade or two is the proliferation of sources and availability of remotely sensed data. As Goetz and Rowan (1981) noted, the ready availability of Landsat data to geologists is a quantum leap forward from the preceding black-and-white aerial photograph era.

Electromagnetic Spectrum Regions Relevant to Geology

Figure 31-1 shows the various regions of the electromagnetic spectrum of interest to geologists, as well as the type of remote sensors or remote sensing systems used to record energy, either passively or actively, from each region. For a more comprehensive discussion of electromagnetic radiation please refer to Chapter 2 of Volume I of this *Manual*. Photographic sensors are covered in Chapter 6; Electro/Optical Non-Imaging Sensors, in Chapter 7; Electro/Optical Imaging Sensors, in Chapter 8; Radar Imaging Systems, in Chapter 10; and Passive Microwave Radiometry in Chapter 11, Volume I of this *Manual*.

Conventional black-and-white aerial photographs, used in pairs stereoscopically, are the type of remotely sensed data most used by field geologists. Most geologists do not realize, however, that such aerial photographs only encompass part of the visible region (about 0.51 μm to 0.68 μm) because of the spectral insensitivity of panchromatic aerial film at the longer wavelength (greater than about 0.70 μm) and the use of a Wratten 12 ("minus-blue") haze filter to eliminate the shorter wavelengths (better haze penetration).

Color and color-infrared aerial photographs are available to geologists for many areas of the United States. From 1972 to 1982, the Landsat 1,2, and 3 spacecraft produced over 1,000,000 scenes with two bands in the visible (0.5–0.6 μm and 0.6–0.7 μm), and two bands in the near-infrared (0.7–0.8 μm and 0.8–1.1 μm) part of the electromagnetic spectrum.

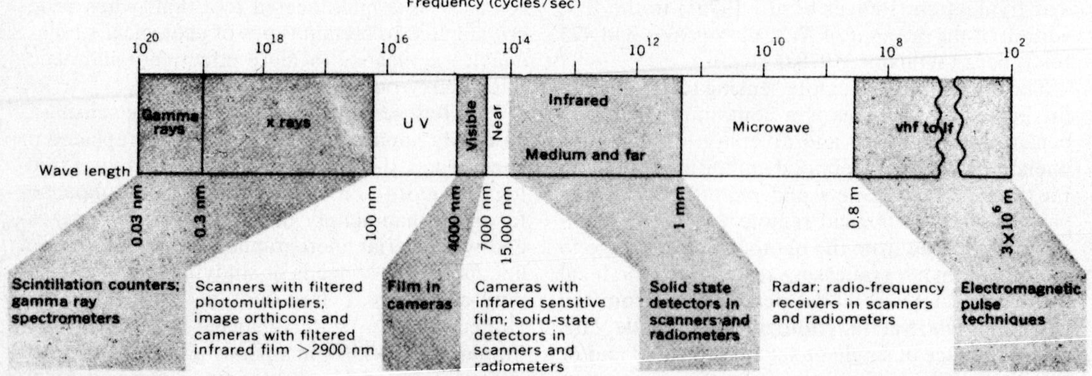

Fig. 31-1. The electromagnetic spectrum and remote sensing systems and sensors used to record energy at various wavelengths. The regions of particular interest to geologists are concentrated in the visible, near-infrared, thermal infrared, and microwave (see also Goetz and Rowan, 1981). (Adaptation from Figure 6 in Williams and Carter (1976), which was modified from Parker, D. C., and Wolff, M. F., 1965.)

Black-and-white aerial photographs and color and color-infrared aerial photographs of the United States are readily available from the Earth Resources Observation Systems (EROS) Data Center; the Aerial Photographic Field Service, Agricultural Stabilization and Conservation Service; and the National Oceanic and Atmospheric Administration's (NOAA's) National Ocean Survey. Landsat images are also readily available from the EROS Data Center. Consequently, the type of aircraft or satellite photography or imagery most readily available to geologists is generally limited to the visible and near-infrared regions of the electromagnetic spectrum.

Theoretical Studies

Our understanding of the potential of remote sensing for geologic applications has continually improved as our comprehension of the interaction of matter and electromagnetic energy has increased (see also Chapters 2, Electromagnetic Radiation; 3, Energy Interactions, Visible to Thermal Region; and 4, Matter/Energy Interactions, Microwave Region). Detailed studies of the spectral signatures of minerals, rocks, soils, and vegetation in the visible and infrared have guided the development of remote sensing from the panchromatic film cameras of four decades ago to current narrow band-width spectral scanners. Theoretical studies of the spectral properties of materials (Hunt, 1980; Aronson and Emslie, 1975; Colwell et al., 1963) have demonstrated the potential and limitations of spectra acquired in the visible, near-infrared, and mid-infrared regions of the electromagnetic spectrum.

These studies have demonstrated that many minerals have characteristic reflection spectra and display unique absorption features that can be used to identify the material. Studies of the reflection properties of Earth materials (Brennan and Bandeen, 1970; Coulson, 1966) have shown that, under varying illumination conditions, many natural surfaces do not reflect light isotropically. However, this anisotropic behavior can be avoided if viewing and illumination geometries in the field or from aircraft or satellite sensors are regulated (see also Chapter 5, Matter/Energy Interactions within the Atmosphere). More detailed studies in the laboratory and in the field of the reflection properties of Earth surfaces and materials have increased our understanding of what can be successfully discriminated or identified.

Laboratory Studies

Laboratory studies of the visible and near-infrared reflectance of minerals, rocks, and soils were initiated in the late 1960s by Graham R. Hunt and John W. Salisbury (with the U.S. Geological Survey since 1981). They pioneered in this research at U.S. Air Force Cambridge Research Laboratories, and Hunt continued the work at the U.S. Geological Survey from 1975 until his death in 1980. During the past decade Hunt and a number of co-workers measured and published the reflectance spectra of hundreds of varieties of pure minerals and rock types (Hunt, 1977, 1978, 1979, and 1980; Hunt and Salisbury, 1970 and 1971, 1974, 1975, 1976a, b, and c; Hunt and Wynn, 1979; Hunt and Ashley, 1979; Hunt and Evarts, 1981; Hunt et al., 1971a and b, 1972, 1973a and b, 1974a and b). This work gave geologists the perspective to view remote sensing as an extension of classical laboratory spectroscopy, and to recognize that surface materials could be discriminated and often identified on remote sensing imagery by knowledge of known reflectance properties of the constituents of rocks and soils. A number of recent studies have employed the insights afforded by this work in their analysis of Landsat and airborne scanner data (Ashley and Abrams, 1978; Goetz and others, 1975; Rowan et al., 1974). These studies have demonstrated that lithologies can be discriminated by knowledge of variations in the spectral properties of the mineral constituents, and iron-oxide and clay minerals can be identified as a result of their characteristic spectral absorption bands.

Similar laboratory studies of the mid-infrared reflectance properties of Earth materials (Lyon, 1964, and 1965; Vincent et al., 1975) have demonstrated that this region can also be used to discriminate silicate and carbonate mineralogy. Application of these laboratory studies to remote sensing have been limited by the lack of multichannel optical-mechanical scanners which operate in the mid-infrared. However, a number of studies (Vincent and Thompson, 1972; Kahle et al., 1980) have demonstrated discrimination of lithologies employing the known spectral behaviour of rocks and minerals in the mid-infrared.

Field Studies and Data Acquisition

Smith (1943) and Tator et al. (1960), in their early photogeologic studies, recognized the importance of coincident field investigations to help identify surface features. With the development of multispectral photography (Cronin, 1967; Brown et al., 1967) and imagery, and with the growing recognition by geologists of the value of laboratory spectra, various kinds of field radiometers (Friedman et al., 1969a) and spectrometers (Cronin, 1967; Cronin et al., 1968; Cronin et al., 1971) were brought into use. These instruments allowed geologists to measure the reflectance characteristics *in situ,* of rocks, soils, and vegetation and the thermal emission from geothermal areas and volcanoes (Friedman et al., 1969a) and use these measurements to better analyze remotely sensed data.

The proper use, calibration, and correlation of these instruments to remote sensing data has been investigated by Duggin (1980), Marsh (1978), Marsh and Lyon (1980), and Robinson and Biehl (1979). Cronin (1967) was one of the pioneers in developing spectroradiometers for the measurement of spectra from geologic materials in the field *in situ*. Cronin (1967) did his initial work with a Pritchard photometer. In the late 1960s, using a triradiometer designed and built by R. H. Noble at the University of Arizona, Cronin made field spectral measurements in nine different bands, of the Sun, sky, and sample in succession, from the near-ultraviolet, through the visible, and into the near-infrared. The nine bands were selected to conform to the film-filter combinations of the Itek nine-lens aerial camera, as discussed later in the section on Marine Geology.

Field equipment has now progressed from broad-band (100 nm), such as these early field radiometers and the Landsat-band radiometers, to narrowband (4 nm) spectroradiometers that measure from the ultraviolet through the near-infrared. Figure 31-2 is a photograph of a high spectral resolution (4 nm) field spectroradiometer. The instrument employs a grating-spectrometer, a Si-PbS detector, and microprocessor electronics to measure and digitally record the spectral reflectance of the surface from 0.3 μm to 3.0 μm. The Jet Propulsion Laboratory has also developed an advanced portable spectroradiometer which is undergoing testing.

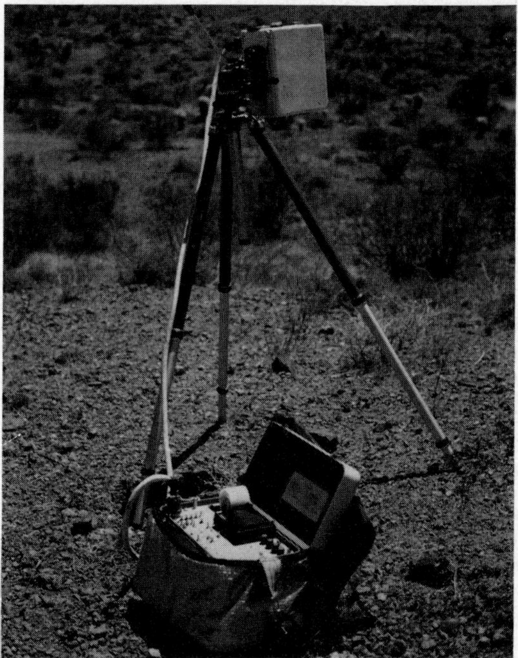

Fig. 31-2. Field portable spectroradiometer developed by Wm. Collins (Geophysical Environmental Research, Inc.) employs a grating-spectrometer and microprocessor electronics to measure and digitally record the surface reflectance from 0.3 to 3.0 μm. The instrument has a very high spectral resolution, 0.004 μm. (Photograph by Stuart E. Marsh).

Remote sensing studies which have employed field spectra began in the late 1960s (Cronin, 1967; Cronin et al., 1968), continued during the 1970s (Longshaw, 1974; Raines and Lee, 1974), and have become more numerous as their value to improved discrimination of rock types and soils has been recognized. Lyon (1975) demonstrated the utility of field spectral measurements used in conjunction with Landsat data for improved reconnaissance mineral exploration. Abrams et al. (1977) employed field reflectance measurements to better discriminate hydrothermal alteration from aircraft scanner data. Conel et al. (1980) and Blom et al. (1980) employed field spectra to improve discrimination of uranium alteration phenomena in sedimentary units and discriminate plutonic rocks. Marsh (1978) used an Hyperion balloon system as a platform for aerial radiometric measurements in conjunction with ground radiometric and near-synchronous Landsat multispectral scanner (MSS) measurements, at several geologic field sites.

More advanced aircraft and satellite sensors are now in use which measure spectral regions beyond the visible wavelengths. Field or laboratory studies of the spectral properties of minerals, rocks, and soils are vitally important for a complete understanding of the imagery acquired from these new aircraft and satellite scanners. For a discussion of general aspects of "Ground Investigations in Support of Remote Sensing," please refer to Chapter 23 of this *Manual*.

Aircraft Data Acquisition

Most geologists use aerial photographs which have been acquired for purposes other than geological mapping. Most black-and-white aerial photographs archived by federal or state agencies or by private companies were acquired for topographic mapping or monitoring of land use. Since the late 1960s the National Aeronautics and Space Administration (NASA) has acquired a large number of color and color-infrared aerial photographs of various "test sites" or regions of the United States. NASA, in particular, has acquired an enormous variety of other types of remotely sensed data in support of various research projects or as part of the development of satellite remote sensing systems. Nearly all of the aerial photography acquired by federal or state agencies can be retrieved from various archives, as discussed later in this chapter in the section on Types and Availability of Remotely Sensed Data for Geological Investigations.

Specialized types of aerial imagery, such as multiband aerial photography, aerial thermography (thermal infrared imagery or TIR), side-looking airborne radar (SLAR or SLR), and microwave imagery were usually acquired for specific research projects and are not generally archived. Therefore, for most geologists, aerial photography, usually black-and-white but now more commonly including color or color infrared (Smith and Anson, 1968), remains the most likely type of remotely sensed data available for a given area.

Outside the United States, only a few countries permit essentially unrestricted access to the aerial photographic data base of the country. Canada, the United Kingdom, and the Nordic countries are among the few countries which, like the United States, maintain aerial photographic archives available to the public. Most countries consider aerial photographs to be classified from a national security standpoint.

For geologists interested in or involved in the actual acquisition of aerial data, an excellent review of airborne platforms (balloons and aircraft) is provided by Colvocoresses (1974 and 1975). The routine use of balloons as a platform for an aerial camera was pioneered by Whittlesey (1970) in his archaeological research. Marsh (1978) used an Hyperion balloon and gimbaled camera mount designed and built by Julian H. Whittlesey. The helicopter also is occasionally used as a platform for remote sensing experiments. The airborne work with the Fraunhofer line discriminator (FLD) instrument, for example, made extensive use of helicopters (see section on Luminescence of this chapter).

Direct acquisition of aerial photographs by geologists from an aircraft is accomplished by using either a high-wing aircraft to acquire oblique aerial photographs or by using a variety of aircraft outfitted with an aerial mapping camera in a gyrostabilized mount to acquire overlapping frames of vertical aerial photographs. Oblique aerial photographs of a particular field site or geological phenomenon can be most easily obtained with a standard 35-mm single lens reflex (SLR) camera loaded with black-and-white panchromatic, black-and-white infrared, color, or color-infrared film and used with the recommended filter(s) (Fleming and Dixon, 1981) (see color Figures 31-42, 31-142, and 31-217).

The acquisition of vertical aerial photographs (or other types of aerial imagery) of a particular site requires that the geologist be familiar with the operation of the aircraft and the remote sensing system to be used and also to be involved in mission planning and mission execution. Chapter 6 of this *Manual* provides a summary of Photographic Sensors, the sensors most apt to be used by geologists. Chapters 8 and 10 provide information of Electro/Optical Imaging Sensors and Radar Imaging Systems, respectively. Chapter 2, Procurement of Aerial Photography (Colwell et al., 1960) in the *Manual of Photographic Interpretation* (Colwell, editor, 1960) and also Chapter 4, Aerial Cameras (Livingston et al., 1980) and Chapter 5, Aerial Photography (Brew and Neyland, 1980) in the *Manual of Photogrammetry* (Slama et al., 1980) provide additional information on the acquisition of remotely sensed data from aircraft.

When referring to aerial photographs, one frequently uses the terms low-, medium-, and high-altitude aerial photography (after Colvocoresses, 1974). For the purposes of this chapter, however, low-altitude aerial photographs are those acquired at less than 9 km (large scale: >1:50,000); medium-altitude aerial photographs are those obtained from 9 to 15 km (medium scale: 1:50,000 to 1:100,000); and high-altitude aerial photographs are those acquired from aircraft platforms above 15 km to about 25 km (small scale: <1:100,000). The term ultrahigh-altitude photography is sometimes used to refer to satellite photography. Sabins (1978a) offered new definitions for large-scale (>1:50,000), medium-scale (1:50,000 to 1:500,000), and small-scale (<1:500,000) imagery, taking into account satellite imagery.

Satellite Data Acquisition

For most geologists the direct acquisition of satellite data is not possible, and remotely sensed data of a specific area are obtained from one of several archives described later in this chapter. Geologists should be knowledgeable of satellite orbits, field of view, spatial resolution, and spectral sensitivity of various sensors, image processing procedures, and other aspects of satellite remote sensing in order to be able to properly use remotely sensed data from satellite platforms. Colvocoresses (1974 and 1975) provides a good review of satellite platforms. Additional information on future platforms is provided by Doyle (1982). The Landsat Data User's Handbooks (NASA, 1971; and USGS, 1979) give a good review of the Landsats 1, 2, and 3 series of spacecraft and sensors. Similar handbooks are available for all of the spacecraft and sensors which provide data of interest to geologists. The section of this chapter on Types and Availability of Remotely Sensed Data for Geological Investigations provides a concise review of such spacecraft and sensors.

Concerned geologists should, however, endeavor to participate in the design of remote sensing devices that may be carried on future spacecraft. They can do this by providing recommendations to NASA's Office of Space Science and Applications (OSSA, formerly called Office of Space and Terrestrial Applications or OSTA), sitting on appropriate committees, or serving on proposal-review groups. In some cases geologists may become principal investigators on a specially designed remote sensing system. Two geologists, Alexander Goetz of the Jet Propulsion Laboratory (California Institute of Technology) and Lawrence Rowan (U.S. Geological Survey), were coprincipal investigators representing a team of scientists participating in the Shuttle multispectral infrared radiometer (SMIRR) experiment carried on the second Space Shuttle mission (see Section on Radiometry of this chapter; Taranik and Settle, 1981; and Williams, 1982). They recommended the design for SMIRR, were involved in the aircraft test surveys, participated in the Space Shuttle mission, and are presently conducting field work to support the analysis of the SMIRR data.

With the availability of the Space Shuttle, geologists may have a greater opportunity to participate in the design of specialized sensors. Opportunities may also exist for participating in actual Space Shuttle missions to operate various sensors in a laboratory environment or as a mission payload specialist. For example, two geologists, Kathryn D. Sullivan, the first woman to be selected as a member of the Explorer's Club, and Anthony W. England, are scientist-astronauts assigned to the L. B. Johnson Space Center in Houston, Texas. Eventually there may be a resident geologist as a member of the permanent staff of an orbiting space station.

HISTORICAL CONSIDERATIONS

This section is designed to give geologists a brief historical perspective on the development of remote sensing, especially the key turning points as the technology progressed from early field photography, through aerial photography and imagery, to the imagery and photography obtained by sensors carried onboard satellites. A more comprehensive discussion of the historical development of remote sensing is given in Chapter 1, Development and Principles of Remote Sensing, in this *Manual;* Chapter 1, The Development of Photo Interpretation (Quackenbush, 1960) in the *Manual of Photographic Interpretation* (Colwell, 1960); and Chapter 2, History of Remote Sensing (Fischer et al., 1975), in the *Manual of Remote Sensing* (Reeves et al., 1975a).

Terrestrial Photography

Many geologists use a 35-mm camera to document geological features during a field mapping project or for laboratory research. The choice of films includes different varieties of black-and-white (panchromatic), black-and-white infrared, color, and color-infrared films. A choice also can be made as to which of various filters will be used. Numerous books or booklets have been published on field photography. The Kodak Company, however, has the widest variety of books, booklets, and circulars about photography and the types of films and filters to be used in the field, in the laboratory, and on aircraft.

The U.S. Geological Survey has a 100-year history of involvement in terrestrial (ground) photography and was one of the pioneering organizations to employ photographers to document early expeditions to the western United States. The early wet-plate (collodion) photographic equipment was so cumbersome that considerable effort went into the acquisition of early field photographs. This difficulty contrasts sharply with the ease in which geologists acquire field photographs today.

The National Archives and Records Service (NARS) of the General Services Administration (GSA) archives all wet-plate type field photography of the U.S. Geological Survey from the late 1800s. Approximately 23,000 dry-plate (glass) negatives were acquired until the 1930s by the U.S. Geological Survey during geologic, topographic, and hydrologic field surveys. These photographs, along with approximately 100,000 nitrate-base, cut-film negatives, which span the time period from the early 1900s to the beginning of World War II, are under the jurisdiction of the U.S. Geological Survey Photo Library in Denver, Colorado. Some modern field photographs of the U.S. Geological Survey are also stored in the Survey's Photo Library. The collection of nitrate base film negatives covers an enormous variety of field sites and subject matter in U.S. and foreign areas. Both NARS and the U.S. Geological Survey Photo Library have numerous photograph albums which provide access to the wet-plate, dry-plate, and nitrate-base film negatives collection by scientists and the general public. For additional information on this important historical collection of terrestrial photographs, the reader is referred to the report by Taylor (1979).

Aerial Photography

Although aerial photographs taken from balloons were available for a few areas (especially of cities) in the late 1800s and early 1900s, systematic aerial photographic mapping of the United States did not begin until the mid-1930s (USGS, 1981g). A comprehensive inventory was compiled by Taylor and Spurr (1973) of historic aerial photographs, acquired in the 1930s and early 1940s by five federal agencies: Agricultural Stabilization and Conservation Service, Soil Conservation Service, U.S. Forest Service, U.S. Geological Survey (mostly of the New England region only), and U.S. Bureau of Reclamation. The authors stated that, "These photographs, of about 85 per cent of the contiguous land in the United States, provide a unique record of the physical and cultural landscape of the country during the period just before World War II. They constitute an important reference aid for studying changes in the natural environment and in rural and urban development that have occurred since then." These historical aerial photographs have special value to geologists (and geographers) interested in studying areas which have been subjected to change, either natural or cultural. Older aerial photographs of coastal areas, for example, provide a permanent record for comparison with natural changes in the coast or coastal engineering projects (Anders and Leatherman, 1982). See also three later sections in this chapter, entitled respectively, "Mini-Atlas of Image and Related Data Coverage of Cape Cod, Massachusetts, and Environs;" "Coastal Engineering;" and "Coastal Studies." Studies of river valleys, areas of slope failure, and estuarine areas, *prior* to extensive development, are possible with these historic aerial photographs.

Since World War II most areas of the United States have been photographed many times, some with a wide variety of aerial film and filter combinations. The black-and-white aerial photograph has been the mainstay of aerial photographic surveys, however, because of its use in map making (aerial photogrammetry). This type of aerial photograph, used in stereopairs, is the one most used by geologists in the field.

Beginning in the 1960s the National Aeronautics and Space Administration (NASA) embarked on a major aircraft survey program (NASA Earth Resources Survey Program) to survey test sites for various principal investigators working on pre-Landsat geological remote sensing projects. Different remote sensing systems were carried aboard the U-2C, RB-57F, NC-130B, and NP-3A aircraft. For documentation of U.S. Geological Survey-NASA cooperative geological remote sensing during this period see Carter (1967) and Table 31-24. Additional aerial survey missions have been carried out by NASA during the Landsat era in support of Landsat investigations or to test new sensors for future spacecraft. All of the NASA aerial photographic data, whether black-and-white, color, or color infrared is archived at the EROS Data Center of the U.S. Geological Survey. Most of the NASA aerial surveys were site specific, although large areas (e.g., the entire State of Arizona) were photographically surveyed in some cases. In 1978, NASA began high altitude (20,000 m) aerial surveys of Alaska, with its U-2C and RB-57F aircraft, using black-and-white (1:125,000-scale) and color infrared film (1:62,500-scale). NASA has also acquired extensive panoramic photography from the U-2 aircraft with the Itek optical bar (KA-80) camera with a focal length of 610 mm. In 1980, the U.S. Geological survey and other federal agencies began a long-term "high-altitude" (13,000 m) repetitive aerial survey program (National High-Altitude Photography or NHAP) of the 48 conterminous states using black-and-white (1:80,000-scale) and color infrared (1:58,000) film (Williams and Taranik, 1981). In time, then, geologists will have ready access to black-and-white or color infrared aerial photographs of the entire United States (see section on Types and Availability of Remotely Sensed Data for Geological Investigations in this chapter for more specific information on current availability of aerial photography of the U.S.).

In addition to the high-altitude aerial photography acquired by NASA with its RB-57F and U-2 aircraft, the Lockheed SR-71 (YF-12) is currently operated by the U.S. Air Force as a high-altitude (25 km to 36 km), long range strategic reconnaissance aircraft. According to Colvocoresses (1974 and 1975), the SR-71 "is capable of worldwide reconnaissance operations including aerial photographic missions and multiple forms of remote sensing." High performance, specialized imaging systems, including panoramic cameras, are among the remote sensing instrumentation probably carried by the SR-71, but these photographic records are, for the most part, classified from the standpoint of military security. Classified aerial photography and related types of remotely sensed data represent an important historical and scientific record of natural and cultural conditions of the Earth's surface. Many scientists believe that these data should be properly archived and released for scientific use at the earliest opportunity (Hamlin and Goodenough, 1978; Ross, 1978; Mathisen, 1978).

Aerial Imagery

In the early and mid-1960s, various types of remote sensors, that had been developed for tactical and strategic reconnaissance were applied by a number of organizations for the airborne acquisition of environmental data. Firms such as HRB-Singer, Inc. and Texas Instruments, both developers of airborne infrared scanners for the Department of Defense, conducted thermal infrared surveys under contract to various government agencies and commercial companies. The University of Michigan, drawing on its extensive experience in airborne infrared technology gained from Department of Defense contracts, also provided services to different government agencies. The Terrestrial Sciences Laboratory of the Air Force Cambridge Research Laboratories initiated a program in airborne geological remote sensing (Neal, 1962; Bliamptis, 1967), to evaluate the usefulness of acquiring geological and geophysical data (Figure 31-3).

In February 1962, The University of Michigan, with the support of the U.S. Navy's Office of Naval Research (whose personnel coined the term "remote sensing"; see Williams, 1972b), U.S. Army Research Office, and U.S. Air Force Cambridge Laboratories, held the first in a series of Symposia on "Remote Sensing of Environment." The first symposium was attended by 71 people, including many geologists and other scientists who, recognizing the value of this new technology to scientific investigations, devoted much of their subsequent professional careers to fostering the growth of the technology.

The Proceedings volume of the first Symposium on Remote Sensing of Environment (University of Michigan, 1962), gave scientists, not associated with reconnaissance technology and its applications, the first clear indication of the types of data that could be acquired with the "new" sensors (Parker and Wolf, 1965). The University of Michigan Symposia (see Table 31-25 in the Applications of Remote Sensing Technology to Geologic Subdisciplines section of this chapter), coupled with the relaxation in the security classification of imagery and sensors, have provided the main impetus for applications in many fields (Stringham and Williams, 1970).

Perhaps the first aerial thermograph to be declassified was one of Nittany Mountain, Pennsylvania. Lattman (1963) used this image in his pioneering paper on the "Geologic Interpretation of Airborne Infrared Imagery." Fischer et al. (1964) published the first thermal infrared study of an active volcano. The papers by Lattman and by Fischer et al. were followed within a few years by a proliferation of papers, the result of Dr. Harold Brown's (then Secretary of the United States Air Force) signature on and the U.S. Air Forces' release of two key documents in 1967, one providing a mechanism ("Order of Merit") for the declassification of some types of optical-mechanical scanners (United States Air Force, 1967), the other providing an easy means for the declassification of images from such scanners. Between the time of the first Michigan Symposium in 1962 and the U.S. Air Force action in 1967, much discussion took place within the military-intelligence community and between the supporters of a greater role for scientific remote sensing. Many of the scientists who attended the first Michigan Symposium had ties to both groups. Dr. Brown, as Director, Defense Research and Engineering, and later as Secretary of the U.S. Air Force, was the first high-level government official to support a move toward declassification. A U.S. Air Force Scientific Advisory Board Report (U.S. Air Force, 1966) also played a role in this significant and far-reaching policy change.

The first Michigan Symposium, the publication of the paper by Lattman, and the declassification procedures approved by Dr. Brown provided the foundation and opened the door for the explosive growth of a technology that is generally referred to as geologic remote sensing. The military services, several government agencies, and many

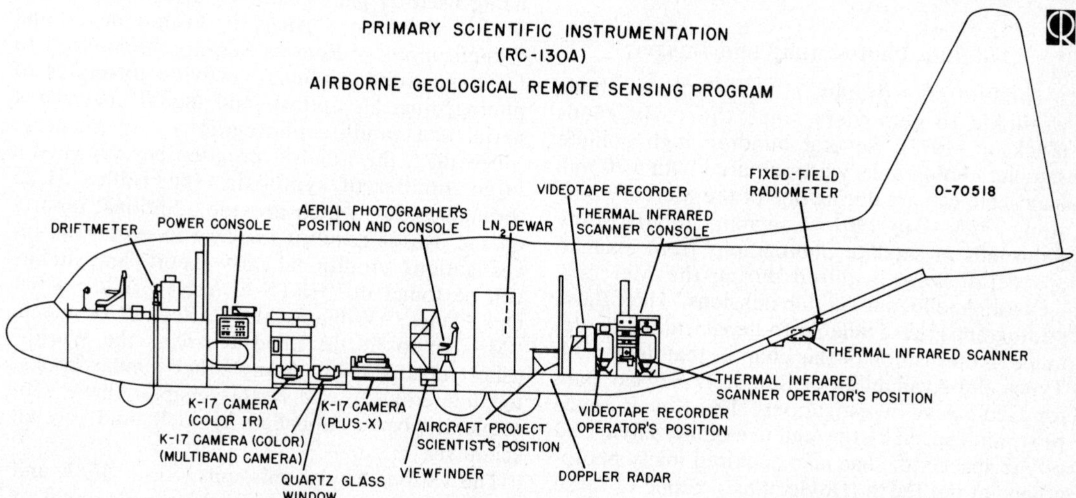

Fig. 31-3. Schematic diagram of the interior configuration and scientific instrumentation onboard the specially equipped Air Force Cambridge Research Laboratories RC-130A Hercules aircraft. This type of aircraft was used in the airborne geological remote sensing program of the Terrestrial Sciences Laboratory between 1965 and 1971. The actual survey aircraft used was the prototype ski-equipped C-130 aircraft and was designated an RC130A/D (Serial No. 50021).

universities now have extensive research programs which are based on remote sensor technology. Nearly all organizations, both U.S. and non-U.S., involved in the environmental sciences, that is, the sciences in which field measurements are an integral part of the research (e.g., geology, forestry, agriculture, geography, oceanography, hydrology, etc.), have active research programs in remote sensing.

Private survey companies now routinely acquire non-conventional types of imagery for specific applications. Optical-mechanical (multispectral and infrared) scanners and radar imaging systems are the most likely types of instruments used in non-conventional aerial surveys. An optical-mechanical scanner can acquire line-scan images in the ultraviolet, visible, near-infrared, and thermal infrared parts of the electromagnetic spectrum. Side-looking radar imaging systems, either synthetic aperture radar (SAR) or real aperture radar (SLAR), have been used in aerial surveys of the United States and foreign areas. The largest such survey carried out to date was Projecto RADAM, which was initially designed for a resource inventory of the Amazon Basin but was later extended to cover all of Brazil.

For geologists interested in aerial imagery, it is often impossible to locate imagery flown for a specialized purpose. This is true, for example, of most of the aerial thermography acquired. A little more radar imagery, however, is available. Since 1980, the U.S. Geological Survey has been acquiring limited amounts of radar imagery of the United States which is being archived at the EROS Data Center. In addition, radar imagery, that has been acquired for military and proprietary interests, is also available, although often at considerable expense.

Satellite Photography and Imagery

Satellite photography of the Earth has been available to geologists since the early 1960s (Dickson, 1977). Several hundred high oblique satellite photographs were acquired with a 70-mm hand-held camera during one of the Mercury missions (MA-4) in 1961 (Lowman, 1980). Many thousands of satellite photographs from manned spacecraft were acquired during the Mercury, Gemini, Apollo, and Skylab missions. All of these photographs are available to geoscientists as detailed in the section of this chapter that deals with Types and Availability of Remotely Sensed Data for Geological Investigations. The Soviet space program, especially through use of the Salyut and Soyuz spacecraft, has also acquired many photographs of the Earth (Doyle, 1982). Some of these photographs have been published in the Soviet scientific literature (e.g., *Earth Research from Space,* a journal published by the USSR Academy of Sciences as reported by Carter, 1981).

Most of the hand-held photographs acquired by the manned spacecraft program have limited value to geological studies because of the oblique views. Documentation of dynamic geological phenomena, such as volcanic eruptions (Friedman and Heiken, 1977; Simkin and Krueger, 1977) is an exception to this general rule. Two manned missions, Apollo 9 and Skylab, carried sensors specially designed for geologic remote sensing investigations and produced high quality and useful data. The multispectral terrain photography experiment (SO65) on Apollo 9, which used four Hasselblad cameras and four different film-filter combinations, acquired 90 sets of photographs on 70-mm film between 3 and 13 March 1969. The success of the Apollo 9 experiment set the stage for both the Landsat program and the Skylab project.

Landsats 1, 2, and 3, by mid-1981, had acquired over 1,000,000 scenes of the Earth's surface in four different bands (NASA, 1971; USGS, 1979). On July 23, 1982, the program completed a decade of successful acquisition of multispectral scanner (MSS) images and higher resolution Landsat 3 return beam vidicon (RBV) images of the Earth. Probably the most important aspect of the Landsat program has been that all images are readily available, from U.S. and foreign archives, to all types of users. The motto of the Landsat Program, "For the Benefit of All Mankind," has been fulfilled and stands as a landmark achievement among the various national space programs (Hood, n.d.). Many geologists, because of the data accessibility, routinely use Landsat images in their investigations, either in film format or as computer compatible tapes (CCT's). The section of this chapter entitled: "Applications of Remote Sensing Technology to Geologic Subdisciplines," provides the reader with many good examples of how Landsat images, in different formats, are being used by geologists (see also Short, 1982). The sections on, "Analysis Techniques," and "Applications of Remote Sensing Technology to Geologic Subdisciplines" provide examples of photogeological, optical, and digital analysis of aerial and satellite photography and imagery. Since 1972, the Landsat program has spawned a large number of symposia (see Tables 31-25 through 31-30), which provide additional reports on the use of Landsat imagery in geological investigations. Additional early reports and studies can be found in "ERTS-1, A New Window on Our Planet" (Williams and Carter, 1976) and in "Mission to Earth: Landsat Views the World" (Short et al., 1976). Many other similar books, too numerous to cite in this chapter, have subsequently been published by U.S. and foreign scientists.

The various Skylab missions (SL-2, SL-3, and SL-4) were also valuable in the development of geological remote sensing technology. The Earth Resources Experiment Package (EREP) (NASA, 1974) included three important sensing systems: the S-190A multispectral photographic camera, the S-190B Earth terrain camera, and the S-192 multispectral scanner. Additional details on each of these systems is provided in NASA's "Skylab Earth Resources Data Catalog" (1974) and in the Section on Types and Availability of Remotely Sensed Data for Geological Investigations of this chapter. The Skylab mission also acquired a number of hand-held camera (70-mm) frames (mostly oblique views) of the Earth. NASA's Special Publication, SP-380, "Skylab Explores the Earth" (1977) includes several reports on results of the hand-held camera experiments.

Other U.S. civilian satellites which provide image data of the Earth include the heat capacity mapping mission (HCMM); Seasat synthetic aperture radar (SAR), the Nimbus-7, which carries the coastal zone color scanner (CZCS), and the Space Shuttle (OSTA-1), which carried the shuttle imaging radar system (SIR-A) and the shuttle multispectral infrared radiometer (SMIRR). The meteorological satellites have spatial resolution too coarse (0.5 km to 1 km or more) to offer much value to geologists, except for structural analysis of large areas or when monitoring large scale dynamic geological phenomena (e.g., volcanic eruptions, changes in the margins of the ice shelves and floating glacier tongues in Antarctica, etc. For further details see the sections of this chapter on "Volcanic Hazards," "Glaciologic Hazards," and "Glaciology," respectively, and Chapter 14, Meterological Satellites, in this *Manual*). Information on the HCMM, CZCS, SIR-A, SMIRR, and Seasat SAR data characteristics and availability is provided in the section on "Types and Availability of Remotely Sensed Data for Geological Investigations" of this chapter.

Another potential source of high resolution satellite photography and imagery is to be found in the remote sensing systems on past, current, and future military-reconnaissance satellites. Colvocoresses (1974) briefly discussed references to the "so-called reconnaissance/surveillance satellites of the film-return type" (Driscoll, 1971; Klass, 1971) and mentions the SAMOS satellite series of the U.S. and the Cosmos satellite series of the U.S.S.R. A good early review of the relationship between military-reconnaissance and civilian satellites is given by Orhaug and Dyring (1973) in the Swedish journal, *Forskning och Framsteg*. Figure 3 in their article is reproduced here in modified form as Figure 31-4, in which "Big Bird" and SAMOS are compared to other types of manned and unmanned civilian spacecraft. The first SAMOS was launched in ca. 1967 (Orhaug and Dyring, 1973), and the early military-reconnaissance satellites were probably vidicon cameras rapidly followed by development of the film-return variety to achieve maximum spatial resolution of cultural and natural features. According to Light et al. (1980), the U.S.S.R. has depended primarily on film-return satellites for surveillance of the Earth.

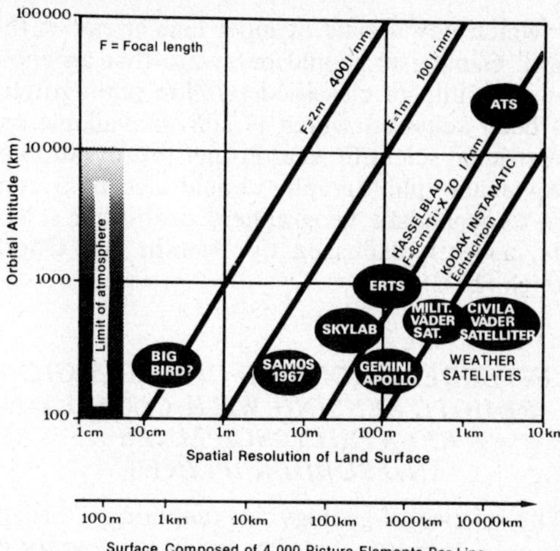

Fig. 31-4. Comparison of known or probable characteristics and capabilities of military-reconnaissance satellites and manned and unmanned civilian spacecraft (modified from Orhaug and Dyring, 1973) (Note: Earth Resources Technology Satellite or ERTS was the early name of Landsat).

Until 1978 the U.S. government had not officially announced the existence of a military-reconnaissance satellite program. On 1 October 1978, however, President Jimmy Carter, in a speech at Kennedy Space Center, stated that, "Photo reconnaissance satellites have become an important stabilizing factor in world affairs in the monitoring of arms agreements" (*Aviation Week and Space Technology,* 1978). The best review of the present "Relationships Between the Civilian and National Security Space Programs" is presented in Chapter 6 of the Office of Technology Assessment's report on "Civilian Space Policy and Applications" (1982). Other relevant articles on the subject include those by Dyring (1975), Ross (1978), and Mathisen (1978). It should be recognized that a satellite provides an exceptionally stable platform and that, with an appropriate image-motion compensation system, long enough focal length, superior optics, and high resolution film, the potential resolving power of such a system is very high indeed. Generally speaking, however, high resolution of a particular area is gained at a loss in wide area coverage.

Since ca. 1967 the United States has acquired an enormous amount of satellite photography from various series of military reconnaissance satellites. This photography, which has high value to some types of specialized geologic and geographic investigations, is only available to scientists willing to seek a Top Secret (Special Access: National Technical Means) security clearance. The main deterrent to its use is the fact that the photography cannot be published or discussed with scientific colleagues because of its classified condition. Such photography would be especially valuable for certain types of global environmental and geological studies, especially as a complementary data set to the existing Landsat imagery which now exists of most land areas of the world. Geologists should be aware that an enormous quantity of classified satellite photography has been acquired which is still unavailable for unrestricted scientific use. Proper preservation of this valuable photography should also be a concern to geologists, geographers, and other scientists, as was discussed by Hamlin and Goodenough (1978).

INTERRELATIONSHIP OF GEOLOGIC REMOTE SENSING WITH CLOSELY RELATED DISCIPLINES AND SUBDISCIPLINES

The science of geology is exceptionally broad. Not only does it include an enormous variety of subdisciplines but those subdisciplines also provide the bridge or interface between completely independent sciences. The American Geological Institute (1980) recognizes ten major categories of geology and its closely related disciplines, each of which can presently be subdivided into 4 to 18 subdisciplines. The major categories, with the number of subdisciplines indicated in parentheses, are as follows: geology (18), economic geology (5), geochemistry (10), geophysics (9), paleontology (10), hydrology (4), engineering geology (6), oceanography (6), planetology (5), and other (9). The "other" category includes remote sensing.

All of the major categories of geology and its closely related disciplines use remote sensing technology in one or more of their subdisciplines. Case histories relating to many of these subdisciplines are presented in a later section, Applications of Remote Sensing Technology to Geologic Subdisciplines, of this chapter. For example, under the section on Economic Geology, case histories are provided in which remote sensing technology is used in the search for coal, metals, non-metals, and oil and gas. Hydrology is well covered in a separate chapter in this *Manual,* Applications in Archaeology and Anthropology (Chapter 26), Marine Resources Applications (Chapter 28), Engineering Applications (Chapter 32), and Extraterrestrial Applications (Chapter 36).

In the section of this chapter entitled: "Applications of Remote Sensing Technology to Geologic Subdisciplines," special emphasis is given to those subdisciplines of geology and closely related disciplines which have made effective use of geological remote sensing technology: geologic mapping, economic geology, engineering geology, geologic hazards, glaciology, geomorphology, marine geology, geobotany, archaeological geology, and astrogeology. Geological remote sensing studies of the Earth and astrogeological studies of the moons and other planets of the Solar System, using data acquired from spacecraft, either in transit or in planetary orbit, share a special kinship to the technology of remote sensing. Remote determination of geological characteristics on the basis of geometric form, spectral response, and spatial distribution from images are common to both. Comparative planetology studies give rise to new geological concepts, thereby enriching both terrestrial and extraterrestrial geology (Williams, 1978b). Digital image enhancement techniques, so commonly used with Landsat images (see the section on Analysis Techniques of this chapter) have their genesis in techniques applied by the Jet Propulsion Laboratory (California Institute of Technology) to images of the Moon, Mars, and, more recently to the Galilean moons of Jupiter and the moons of Saturn (see Chapter 36, Extraterrestrial Applications, of this *Manual* and a later section of this chapter on Astrogeology). The U.S. Geological Survey's Branch of Astrogeologic Studies has also contributed to the development of digital image processing techniques for geological and astrogeological studies.

Because of the use of geological remote sensing technology in so many subdisciplines of geology and related disciplines, there is considerable cross-fertilization and enrichment among these disciplines and subdisciplines. Digital image processing and enhancement techniques (Moik, 1980), first applied to images of extraterrestrial bodies, have been used in the enhancement of Landsat images of the Earth. Now this same technology is being applied to the enhancement of side-scan sonar images of the sea bottom, substantially adding to our knowledge of the morphology of the sea floor, as discussed in the Marine Geology section of this chapter; see also Figures 31-64 through 31-66. Remote sensing technology will eventually provide geologists with an accurate view of the morphology of the Earth's surface, in both its terrestrial and submarine aspects. It is the 75% of the Earth's surface that lies beneath the sea that offers an area of great challenge to geologists as remote sensing technology matures.

TYPES AND AVAILABILITY OF REMOTELY SENSED DATA FOR GEOLOGICAL INVESTIGATIONS

The introduction to this chapter was designed to give geologists an adequate review of the various facets of geological remote sensing, its historical development, and its interrelationship to other disciplines. The objective of this section is to review briefly the various imaging sensor systems used on aircraft and spacecraft and provide specific information on the availability, to geologists, of the great variety of remotely sensed data now available from these diverse systems. Non-imaging and/or high-resolution systems will also be discussed by various scientists active in their use. The last part of this section is a mini-atlas of the types of imagery and related data potentially available for any given area of the United States, using Cape Cod, Massachusetts, and environs as a representative area.

REMOTELY SENSED IMAGE DATA FROM AIRBORNE PLATFORMS

Types of Photographic Images

Aerial photographs used by geologists may be either oblique (high or low) or vertical. In the years following World War II, trimetrogon aerial photography (one vertical, two oblique-right and left) was acquired of large, previously unmapped areas,—especially polar areas. Geologists mapping in Antarctica (Smith, 1967) still use high oblique aerial photographs because, except for Landsat images, they are the only image data available of many areas (Figure 31-5; see also Figure 31-190).

Vertical aerial photographs may be black-and-white panchromatic (Figure 31-41), black-and-white infrared (Figure 31-43), color (Color Figure 31-45), color infrared (Color Figure 31-46), or multiband (Figure 31-238). Various film-filter combinations can be used for special applications, such as in archaeological geology (Williams, 1975) or in water penetration studies (Vary, 1967 and 1969); (See also the Section on Marine Geology of this chapter). Aerial photographs may be acquired at low altitude (less than 9 km) (Figure 31-41), medium altitude (9 km to 15 km) (Figure 31-9, Color Figure 31-10, and Figure 31-48), or high altitude (15 km to 25 km) (Color Figures 31-49 and 31-50). Most black-and-white aerial photographs are acquired by an aerial mapping camera which carries film 25.4 cm (10 in) wide, giving a 23-cm (9-in.) image frame format, and which has a 152.4-mm (6-in.) focal length. The denominator of the photographic scale which results from the use of such a camera is always twice the altitude in feet of the survey aircraft above the terrain. For example, at 3050 m (10,000 ft.) above the terrain, the photograph scale is 1:20,000, and at 12,000 m (40,000 ft.), the photograph scale is 1:80,000 (See also Weeden and Bolling, 1980).

Fig. 31-5. High oblique, trimetrogon aerial photograph (right oblique) looking NNE at Ross Island, Antarctica. The Erebus Ice Tongue and Mount Erebus Volcano (smoking) are in the background. The Ross Ice Shelf is on the right. This U.S. Navy aerial photo No. 217 was acquired on 14 November 1959 from an altitude of 4,900 m (16,000 ft) above sea level. (From Scientific Committee on Antarctic Research (SCAR) aerial photographic archives, U.S. Geological Survey).

Aerial photographs of the United States acquired by governmental and private organizations now number in the tens of millions. Aerial photographs of the United States, acquired at various dates, altitudes, scales, and camera orientations, using various film emulsions and filters, are readily accessible to the geologic community through archives maintained by the U.S. Geological Survey (USGS) and the Agricultural Stabilization and Conservation Service (see Table 31-2). Other aerial photographs for use in geologic investigations include the USGS and National Ocean Survey (NOS) low altitude photography (Figures 31-41 and 31-43), USGS medium altitude mapping photography (Figure 31-9, Color Figure 31-10, and Figure 31-48), and high altitude photography acquired by the National Aeronautics and Space Administration (NASA) (Color Figures 31-49 and 31-50). The most significant new aerial photography program for geologic and many other applications is the National High Altitude Photography (NHAP) Program. The NHAP Program is acquiring 1:80,000-scale black-and-white and 1:58,000-scale color infrared photographs, simultaneously, of parts of the conterminous United States each year, so that full coverage of the United States will be available by the mid- to late-1980s. The NHAP Program then calls for the conterminous United States to be rephotographed at 6-year intervals. Figure 31-6 depicts completed and contracted NHAP coverage as of 1 June 1982. The NHAP photography will provide geologists with a data base comparative to satellite-acquired data for spectral, spatial, and temporal analysis, including stereoscopic coverage not yet available from most satellite images. Figure 31-9 and Color Figure 31-10 presents examples of the black-and-white and color-infrared NHAP photography. Information about aerial photography acquired by the Bureaus of Reclamation, Indian Affairs, and Land Management, NASA's Ames Research Center and the L. B. Johnson Space Center, the National Space Technology Laboratories, the

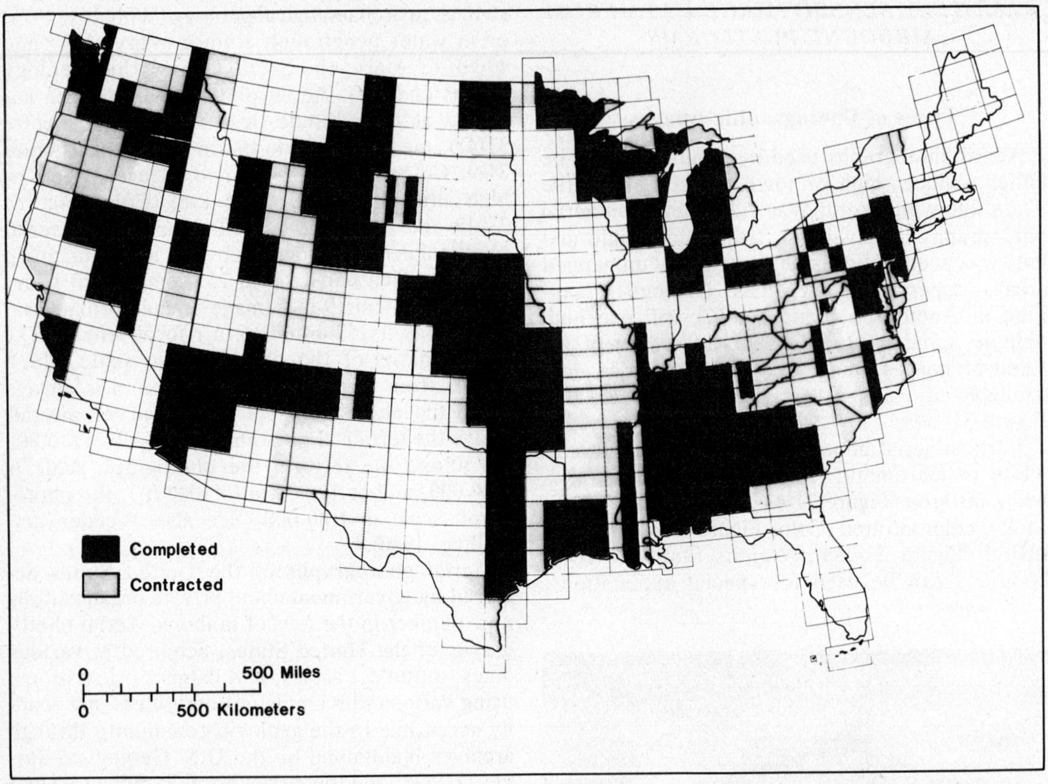

Completed
Contracted
0 500 Miles
0 500 Kilometers

Fig. 31-6. Generalized index map, as of June 1, 1982, showing the status of completed and contracted coverage of black-and-white and color-infrared National High Altitude Photography.

Environmental Protection Agency, U.S. Army Map Service, U.S. Navy, and U.S. Air Force, as well as information about aerial photography held by the Department of Agriculture, Tennessee Valley Authority, State and municipal agencies, and by some commercial organizations is also included in the USGS data bases (Kroeck, 1976). Figure 31-7 is an index map depicting NASA high-altitude aerial photography acquired over the State of Alaska in an aerial mapping program carried out from 1978-1980. (Color Figure 31-8 is a NASA high-altitude color-infrared aerial photograph of

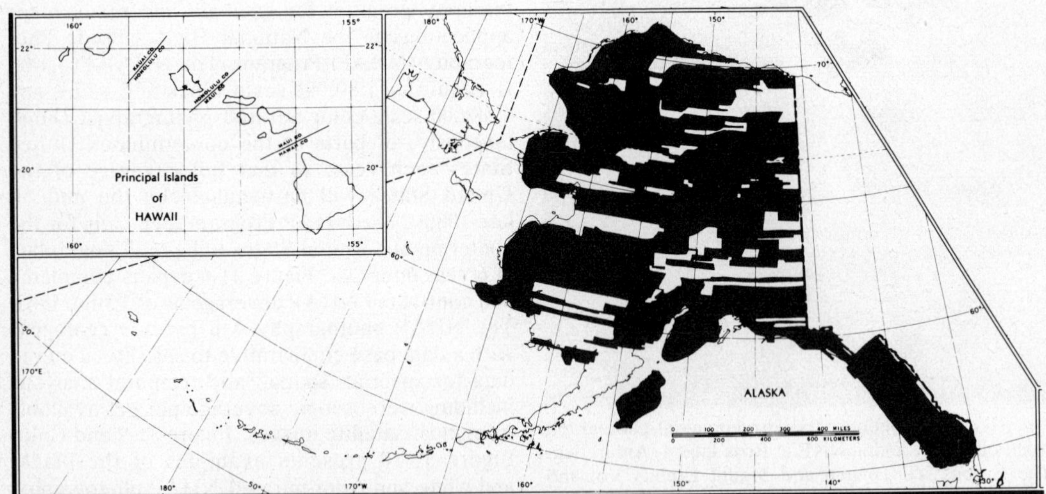

Fig. 31-7. Index map showing National Aeronautics and Space Administration high altitude aerial photography acquired over the State of Alaska from 1978–1980. The photography was acquired simultaneously in two forms: black-and-white, scale 1:120,000, and color infrared, scale 1:60,000. From information provided by Bureau of Land Management, Alaska State Office, Anchorage, Alaska.

the terminus of the Alsek Glacier, Alaska, acquired on 21 June 1978, as part of this program.

Availability of Photographic Images

The National Ocean Survey is the source for aerial photographs acquired by agencies of the U.S. Department of Commerce. The Aerial Photography Field Office of the Agricultural Stabilization and Conservation Service is the source for aerial photographs acquired by agencies of the U.S. Department of Agriculture (e.g., U.S. Forest Service, Soil Conservation Service).

The U.S. Geological Survey, Earth Resources Observation Systems (EROS) Data Center (EDC), Sioux Falls, South Dakota, is the source for aerial photographs acquired by bureaus of the Department of the Interior and the National Aeronautics and Space Administration. The EROS Data Center's User Services Section, the USGS's National Cartographic Information Center (NCIC) offices, and State NCIC affiliates provide information and assistance on requests for aerial photographs. "A Guide to Obtaining Information from the USGS 1982," by Clarke et al. (1982) is an invaluable reference for geologists and others; this guide is updated and published at frequent intervals and distributed free upon request. Another useful reference booklet is the Map Data Catalog (U.S. Geological Survey, n.d.). The types of formats for standard aerial photography and indexes provided by the EROS Data Center are listed in Table 31-1.

Three reference data bases have been established by the USGS to facilitate user requests for aerial photographs. The INORAC (*In*quiry, *Or*der and *Ac*counting) system is a computerized data-processing system designed to process and control the aircraft and satellite data archived at EDC. The Federal agencies listed previously, with the exception of the Department of Agriculture, Department of Commerce, and the Tennessee Valley Authority are included in the INORAC Main Image File (MIF) and may be accessed by agency, year, scale, film type, and recording technique (vertical, panoramic, low and high oblique). The Aerial Photography Summary Record System (APSRS) is a computerized system developed by the USGS to help users determine photographic coverage. APSRS is administered by NCIC and is constantly updated with information provided by contributing agencies and organizations. APSRS data are provided as graphics (maps), microfiche, and written answers to customer queries. A low cost system called Micrographic Indexes has been developed by the USGS for further assistance. Using a microfiche viewer, the user can obtain information about aerial photographs, such as date acquired, film emulsion, and scale. NOS photoindexes, ASCS photoindexes, USGS mapping photoindexes, NASA high-altitude photography coverage diagrams, and the current NHAP data are contained on Micrographic Indexes, and are retrievable by geographic region.

TABLE 31-1

Film formats and product materials for standard black-and-white, color, and color infrared aerial photographs and also black-and-white aircraft data indexes archived at the U.S. Geological Survey's EROS Data Center

Image Size	Product Format
AERIAL MAPPING (black-and-white, color, and color infrared)	
22.9 cm (9.0 in.)	Paper
22.9 cm (9.0 in.)	Film Positive
22.9 cm (9.0 in.)	Film Negative*
45.7 cm (18.0 in.)	Paper
68.6 cm (27.0 in.)	Paper
91.4 cm (36.0 in.)	Paper
PHOTO INDEXES (black-and-white only)	
A—15.4 × 30.5 cm (10 × 12 in.)	Paper
B—Other	Paper
NASA RESEARCH (black-and-white, color, and color infrared)	
55.8 mm (2.2 in.)	Film Positive
55.8 mm (2.2 in.)	Film Negative*
11.4 cm (4.5 in.)	Paper
11.4 cm (4.5 in.)	Film Positive
11.4 cm (4.5 in.)	Film Negative*
22.9 cm (9.0 in.)	Paper
22.9 cm (9.0 in.)	Film Positive
22.9 cm (9.0 in.)	Film Negative*
22.9 × 45.7 cm (9 × 18 in.)	Paper
22.9 × 45.7 cm (9 × 18 in.)	Film Positive
22.9 × 45.7 cm (9 × 18 in.)	Film Negative*
45.7 cm (18.0 in.)	Paper
68.6 cm (27.0 in.)	Paper
91.4 cm (36.0 in.)	Paper

* Film negatives do not exist for color products.

Orthophotoquads are products derived from 7.5-minute quad-centered airphotos and are available from NCIC offices. Each is available as a diazo print, photographic print, or opaque film (See Figure 31-48). The 1:24,000- or 1:25,000-scale orthophotoquads provide an inexpensive planimetric base for geologic field mapping, and can cost 90 percent less than comparable non-stereoscopic coverage with conventional aerial photographs.

The following U.S. Geological Survey leaflets provide additional information on aerial photographs and related products:

How to Order Aerial Photographs (U.S. Geological Survey, 1981a)

Understanding Color-Infrared Photographs and False Color Composites (U.S. Geological Survey, 1981b)

The Aerial Photograph Summary Record System (U.S. Geological Survey, 1981c)
Using APSRS Microfiche (U.S. Geological Survey, 1981d)
Micrographic Indexes (U.S. Geological Survey, 1981e)
Advance Materials (U.S. Geological Survey, 1981f)
Looking for an Old Aerial Photograph (U.S. Geological Survey, 1981g)
Alaska from Space (U.S. Geological Survey, 1981h)
Intermediate-Scale Base Maps (U.S. Geological Survey, 1981i)
A Selected Bibliography on Maps, Mapping, and Remote Sensing (U.S. Geological Survey, 1981j)
Slope Mapping (U.S. Geological Survey, 1981k)
Looking for An Old Map (U.S. Geological Survey, 1981l)
How to Order Maps on Microfilm (U.S. Geological Survey, 1981m)
How to Order Landsat Images (U.S. Geological Survey, 1981n)
Manned Spacecraft Photographs and Major Metropolitan Area Photographs and Images (U.S. Geological Survey, 1981o)
Map Scales (U.S. Geological Survey, 1981p)
Map Accuracy (U.S. Geological Survey, 1981q)
National Cartographic Information Center (U.S. Geological Survey, 1981r)
MiniCatalog of Map Data (U.S. Geological Survey, n.d.)
Map Data Catalog (U.S. Geological Survey, n.d.)
Understanding Maps and Scale (U.S. Geological Survey, n.d.)

Table 31-2 provides addresses for the seven principal federal archives which contain aerial photographs of interest to geologists: National Archives and Records Service (General Services Administration), Library of Congress, EROS Data Center or NCIC (USGS), USGS Photographic Library, Agricultural Stabilization and Conservation Service (USDA), NOS, and Defense Intelligence Agency. Another excellent source book for access to various archives is *Everyone's Space Handbook, A Photo/Imagery Source Manual* (Kroeck, 1976). An excellent collection of aerial photographs provided by agencies of the Federal Government is included in an appendix to the chapter on Analysis Techniques of the U.S. Army's *Remote Sensing Applications Guide* (Department of the Army, 1979).

TABLE 31-2

Principal federal archives of aerial photographic data

Historical Photographs (Aerial and Ground)

National Archives and Record Service (NARS)
General Services Administration
8th and Pennsylvania Avenue, N.W.
Washington, D.C. 20408

Library of Congress
10 First Street, S.E.
Washington, D.C. 20540

Photographic Library
U.S. Geological Survey
Box 25046, Federal Center
Denver, Colorado 80225

*Aerial Photographs of the United States and a Few Foreign Areas**

Earth Resources Observation Systems (EROS) Data Center
U.S. Geological Survey
User Services Section
Sioux Falls, South Dakota 57198

or

National Cartographic Information Center (NCIC)
U.S. Geological Survey (507)
Reston, Virginia 22092

United States Aerial Photographs

Aerial Photography Field Office
Agricultural Stabilization and Conservation Service
U.S. Department of Agriculture
2222 West 2300 South
P.O. Box 30010
Salt Lake City, Utah 84125

U.S. Coastal Aerial Photographs

National Ocean Survey, C-3415
U.S. Department of Commerce
Rockville, Maryland 20852

Aerial Photographs of Foreign Areas

Defense Intelligence Agency
Department of Defense
ATTN: RTS-3
Washington, D.C. 20301

* Foreign area coverage is extremely limited and is mainly restricted to NASA Earth resources survey aircraft missions.

Types of Non-Photographic Images

Airborne non-photographic imaging sensors for geologic applications include optical-mechanical line scanners and active microwave systems (Lowe, 1980). Multispectral and infrared optical-mechanical line scanners can operate in the ultraviolet, visible, or infrared region of the electromagnetic spectrum and employ reflective and/or refractive optical elements for imaging. Active microwave imaging systems are called radars (acronym of Radio Detection and Ranging) and can be real aperture or synthetic aperture radar (SAR) side-looking airborne or satellite radar (SLAR or SLR).

Optical-mechanical scanners were developed to acquire imagery at wave-lengths longer than the spectral sensitivity of photographic film. Once used primarily for thermal infrared imagery, line scanners are now used in multispectral sensing in the ultraviolet, visible, near-infrared, and thermal infrared regions of the electromagnetic spectrum from airborne and space platforms. The operating wavelength region of a multispectral scanner

(MSS) is determined by the spectral response of its detectors and optical filters employed. Multispectral scanners use an oscillating mirror to scan the scene radiance perpendicular to the direction of flight. The scanner rate is adjusted to provide adjacent or overlapping scan lines as the sensor moves forward. The output video signal from the detector is used to generate multiband imagery. Tape-recorded MSS acquired data can be digitally processed by computer for identification and classification of features using spectral and spatial pattern-recognition techniques. Further information on multispectral scanners can be obtained from NASA (1973), Drummond (1972), Lowe et al. (1975), Lowe (1968), Slater (1980), and Chapter 8 in this *Manual*.

Airborne optical-mechanical infrared line scanners can acquire data in the infrared region of the electromagnetic spectrum (0.7–300 μm). Reflective and emitted (thermal) infrared radiation data provide spectral information about the physical properties of the Earth's surficial materials not duplicated in any other region of the electromagnetic spectrum. The 8- to 14-μm wavelength region provides the maximum intensity of the Earth's radiant energy flux and is preferred to the 3- to 5-μm atmospheric window (Sabins, 1978a) except for studies of geothermal areas and volcanoes as discussed in the sections of this chapter entitled "Geothermal Exploration and Related Surveys" and "Remote Sensing of Volcanic Hazards" and illustrated in Figure 31-13. The principle and operation of optical-mechanical thermal infrared line scanners are described by Lowe et al. (1975), Sabins (1978a), Siegal and Gillespie (1980), and Chapter 8 in this *Manual*. Velocity vector and attitude (roll, pitch, and yaw) of the airborne sensor create geometric distortions in optical-mechanical line scan imagery and should be rectified to achieve geometric accuracy (Williams and Ory, 1967).

Radar imaging systems operate in the microwave region of the electromagnetic spectrum (0.83 cm to about 133 cm). Radar is an active system; i.e., it uses its own electrical energy to illuminate the terrain. Radar can operate independent of cloud cover and sun illumination, and thus is considered an all-weather day-and-night system. High-resolution radars direct a narrow beam of energy at right angles to the flight path of the reconnaissance aircraft; hence they are termed side looking. The illumination (depression) angle and illumination direction (look) can be controlled and may be oriented to enhance subtle topographical features through shadowing effects.

Radar return is dependent on the physical properties of the terrain, specifically surface roughness, composition, and moisture content. Variations in grey tone and the high spatial resolution of radar imagery provide geologists a valuable tool for structural and lithologic (based on drainage) mapping. Of most significance, radar provides a means of rapid geologic reconnaissance over large regions which have predominant cloud cover (e.g., the Amazon Basin and the Caribbean). Radar systems can be either real aperture or synthetic aperture radar (SAR) and differ primarily in the method for achieving resolution in the azimuth direction. Real aperture or "brute force" systems use an antenna of maximum length to produce a narrow angular beam width in the azimuth direction. The SAR systems employ a small antenna (synthetic length) that transmits a broad beam which, however, produces the effect of a much narrower beam. Table 2 in Moore and Sheehan (1981) provides an excellent comparative review of real and synthetic aperture radar systems and products. Further information on radar systems and their use in geologic investigations may be obtained in Feder (1960 and 1962), Skolnik (1962), Hackman (1967), Dellwig et al. (1968), Stafford (1969), Rydstrom (1970), Moore et al. (1975), Sabins (1978a), MacDonald (1980), and Chapters 9, Radar Fundamentals and Scatterometers, and 10, Radar-Imaging Systems in this *Manual*.

Another part of the microwave band of the electromagnetic spectrum is also used by geologists with sensors which record microwave energy emitted by the Earth's surface at wavelengths of 1 mm to about 30 cm. Such wavelengths are longer than the thermal infrared but shorter than what is generally used in active radar systems (Schmugge, 1980). Early research was done on volcanic areas in the western United States (Oberste-Lehn, 1970a) and on various geologic materials (Oberste-Lehn, 1970b). During the 1970s considerable research was accomplished with airborne and satellite passive microwave sensors in the study of sea ice and glaciers. Sea ice in polar areas has been the subject of a long term, intensive study with passive microwave sensors operating on the Nimbus series of spacecraft (Gloersen and et al., 1973; Zwally and Gloersen, 1977; Zwally et al., 1983) especially with the Nimbus 5 electronically scanned microwave radiometer (ESMR) system which operates at a wavelength of 1.55 cm. Higher resolution passive microwave satellites are under consideration for the future (Staelin and Rosenkranz, 1978). Another type of newly developed imaging system is the Fraunhofer Line Discriminator (FLD). Although such a system was originally developed as a fixed-field sensor, recent developments have created a wide area imaging system. Geological applications of the FLD system will be discussed later in this chapter in a subsection entitled "Luminescence."

Availability of Non-photographic Images

The acquisition of airborne multispectral scanner, infrared line scanner, passive microwave radiometry, and radar data is so specialized that such data are usually available only from a limited number of organizations who either own or lease systems for this purpose. Because initial procurement of airborne nonphotographic data is also costly, commercial aerial survey films or specialized archives should be consulted to determine whether suitable data already exist.

Airborne multispectral scanner data are available commercially from the Environmental Research Institute of Michigan (ERIM), Bendix Aerospace Systems, Daedalus Enterprises, Inc., Actron Industries, Texas Instruments, Inc., and a few other companies. Specific characteristics of the various systems are tabulated for reference in Lowe et al. (1975), various chapters in Siegal and Gillespie (1980), and Chapter 8 of this *Manual*. Several airborne multispectral scanners have been used at NASA centers to acquire simulated thematic mapper sensor data to aid in the evaluation of the Thematic Mapper (TM), the principal sensor on Landsat 4. The L. B. Johnson Space Center (JSC), Houston, Texas; the Earth Resources Laboratory (ERL) of the National Space Technology Laboratories (NSTL), Bay St. Louis, Mississippi; and the Ames Research Center, Moffett Field, California, are the NASA centers active in the acquisition of airborne multispectral scanner data. The NASA Ames Research Center received a modified version of the Daedalus AADS 1260 airborne thematic mapper (ATM) in early 1982. The Daedalus AADS 1268 ATM has the capability of recording 11 different spectral bands from the ultraviolet to the thermal infrared. NASA's Johnson Space Center and the National Space Technology Laboratories have been using the 8-channel NS001 multispectral scanner in their research program. Table 31-3 provides NS001 and the new Daedalus AADS 1268 ATM characteristics and the TM equivalents. Since 1978, over 17,400 km (9,400 nautical miles) or 38 hours of Thematic Mapper Simulator (TMS) data have been acquired. About 42 percent of these data have been processed to a computer compatible tape (CCT) format and are available from JSC. Status reports on NS001 flight history and test sites are available upon request from JSC or NSTL. Color Figure 31-11 presents a TMS band 6 image and a generated false color composite. Characteristics of the NS001 TMS are described by Richard et al. (1978).

The EROS Data Center has received Daedalus DS-1260 multispectral scanner and Thematic Mapper Simulator aircraft data acquired and generated by GeoSpectra Company (See Color Figure 31-12) in connection with a project to simulate various spatial resolutions of multispectral data. TMS data with resolutions of 10, 20, 30, 40 and 80 meters were simulated over the Table Rock, Wyoming, geology test site (Color Figure 31-12). Uncorrected and resampled digital tapes (9-track, 1,600-bpi) and photographic imagery are available through the ERØS Data Center.

Commercially available airborne infrared line scanners and their operating and performance characteristics are tabulated by Lowe et al. (1975), and Lowe (1980). Thermal infrared imagery is lacking for most areas; therefore data acquisition by contractor or investigator must be procured at a cost approximately three times that of conventional aerial photography (Sabins, 1978a; Pálmason et al., 1970). Thermal infrared surveys can be either single flight lines or multiple flight lines for mosaic coverage (Williams and Ory, 1967). Fundamentals and techniques of infrared technology are provided by Holter et al. (1962), Wolfe (1965), and Wolfe and Zissis (1978). Information on conducting airborne infrared surveys for geologic applications is provided by Sabins (1978a), Lillesand and Kiefer (1979), and Settle (1981). An airborne infrared image acquired over Mount Hekla volcano, Iceland, in 1968, is presented in Figure 31-13.

Currently two radar systems are available for commercial surveys: the Aeroservice-Goodyear electronic mapping system, GEMS, a synthetic aperture radar (SAR) system, and the Motorola Aerial Remote Sensing, Inc, MARS, a real aperture radar system. Nonmilitary SAR systems for research use are currently operated by NASA's Jet Propulsion Laboratory (three), NASA's L.B. Johnson Space Center, and the Environmental Research Institute of Michigan (two). Airborne

TABLE 31-3

Comparison of NASA's NS001 Airborne Multispectral Scanner System and the Daedalus AADS 1268 ATM with the Landsat 4 Thematic Mapper Channels and Spectral Bands

NS001 Channel Characteristics

NS001 Channel	Landsat 4 Thematic Mapper (TM) Bands	Spectral Region (μm)
1	1	0.45–0.52
2	2	0.52–0.60
3	3	0.63–0.69
4	4	0.76–0.90
5		1.00–1.30
6	5	1.55–1.75
7	7	2.08–2.35
8	6	10.40–12.50

Daedalus AADS 1268 ATM Characteristics

AADS 1268 ATM Channel	Landsat 4 Thematic Mapper Bands	Spectral Region (μm)
1		0.42–0.45
2	1	0.45–0.52
3	2	0.52–0.60
4		0.605–0.625
5	3	0.63–0.69
6		0.695–0.75
7	4	0.76–0.90
8		0.91–1.05
9	5	1.55–1.75
10	7	2.08–2.35
11	6	8.5–13

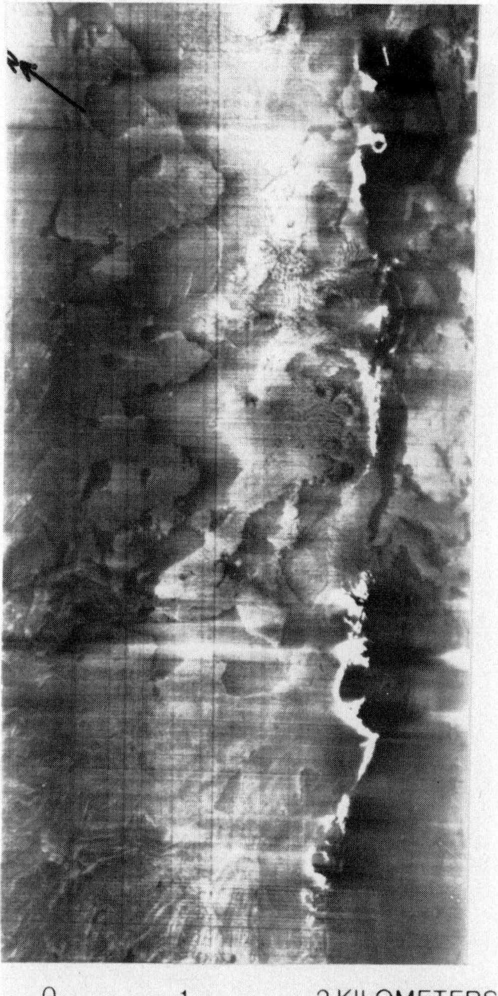

Fig. 31-13. An airborne thermal infrared line-scanner image acquired on 23 August 1968 and 2236 hr GMT at an altitude of 1500 m over Mount Hekla, Iceland. The image was obtained with the University of Michigan's Institute of Science and Technology M1A1 optical-mechanical scanner, using an unfiltered InSb detector. Light areas represent thermal infrared emission from the northeast-trending eruption fissure (Friedman et al., 1969a and c). Curvilinear light-toned "hot" spots indicate craters; dark (cool) areas southeast of the fissure row are perennial firn fields and small glaciers.

real aperture radar systems for research are currently operated by NASA's Langley Research Center and the University of Kansas. Characteristics of these systems are tabulated in Moore et al. (1975), Lowe (1980), and Sabins (1978a). Radar images of most of the United States which have been acquired by or for the Strategic Air Command are archived and distributed to the public through the Goodyear Aerospace Corporation, Department 408A, Litchfield Park, Arizona 85340. Inquiries should specify longitude and latitude of the area of interest. The scale of the original image film ranges from 1:100,000 to 1:600,000 and can be photographically enlarged. Resolution of the airborne SLAR images ranges from 3 to 15 meters. The quality and usefulness of these radar images for geological investigations are extremely variable, however.

Radar surveys of specific geographic regions have been conducted by commercial firms and are available for public purchase. For example, high resolution SAR coverage of the western overthrust belt, acquired by Aeroservice-Goodyear, is available as image strips, contact prints, and cronapaque mosaics ranging in scale from 1:250,000 to 1:1,000,000. Color Figure 31-14 is a map of the conterminous United States depicting commercially available radar data.

All radar images of Alaska acquired by either the GEMS or MARS radar systems, under House Bill H-10464, Amendment No. 37, are available to the public from the EROS Data Center (EDC) at the cost of reproduction. Further information on airborne real aperture and synthetic aperture SLAR, for geologic applications, can be obtained in Moore and Sheehan (1981), and Pascucci et al. (1981). Figure 31-15 presents airborne real aperture and synthetic aperture SLAR image examples.

IMAGING SENSOR SYSTEMS ON SPACECRAFT

Spacecraft of Interest to Geologists

In the past 20 years, the United States has launched many spacecraft and satellite systems that have acquired imagery of the Earth of significant value to geologists. Remote sensing technology has evolved so rapidly, however, that a comprehensive listing of the various data bases does not exist. In this section, an effort is made to provide to the geologist a thumbnail sketch of the characteristics of the various spacecraft sensor systems and, more importantly, the availability of data from such systems. To facilitate the use of this *Manual,* these data have been tabulated in a form that will provide quick access and reference.

Imaging sensors on manned spacecraft, Earth observation satellites, and the Space Shuttle are classified as photographic and nonphotographic. The most important spacecraft systems will be emphasized because of their significant use in geologic remote sensing, although other systems (non-imaging) will be discussed briefly. Spacecraft systems that acquire useful Earth resources data are presented in Table 31-4.

Geologic applications of meteorological satellites include monitoring effusive and explosive volcanic eruptions, such as, thermal emission from lava flows (Williams and Friedman, 1970) and volcanic plume studies. They also include preparation of small-scale cartographic products of polar regions (Berg et al., 1982; Wiesnet and

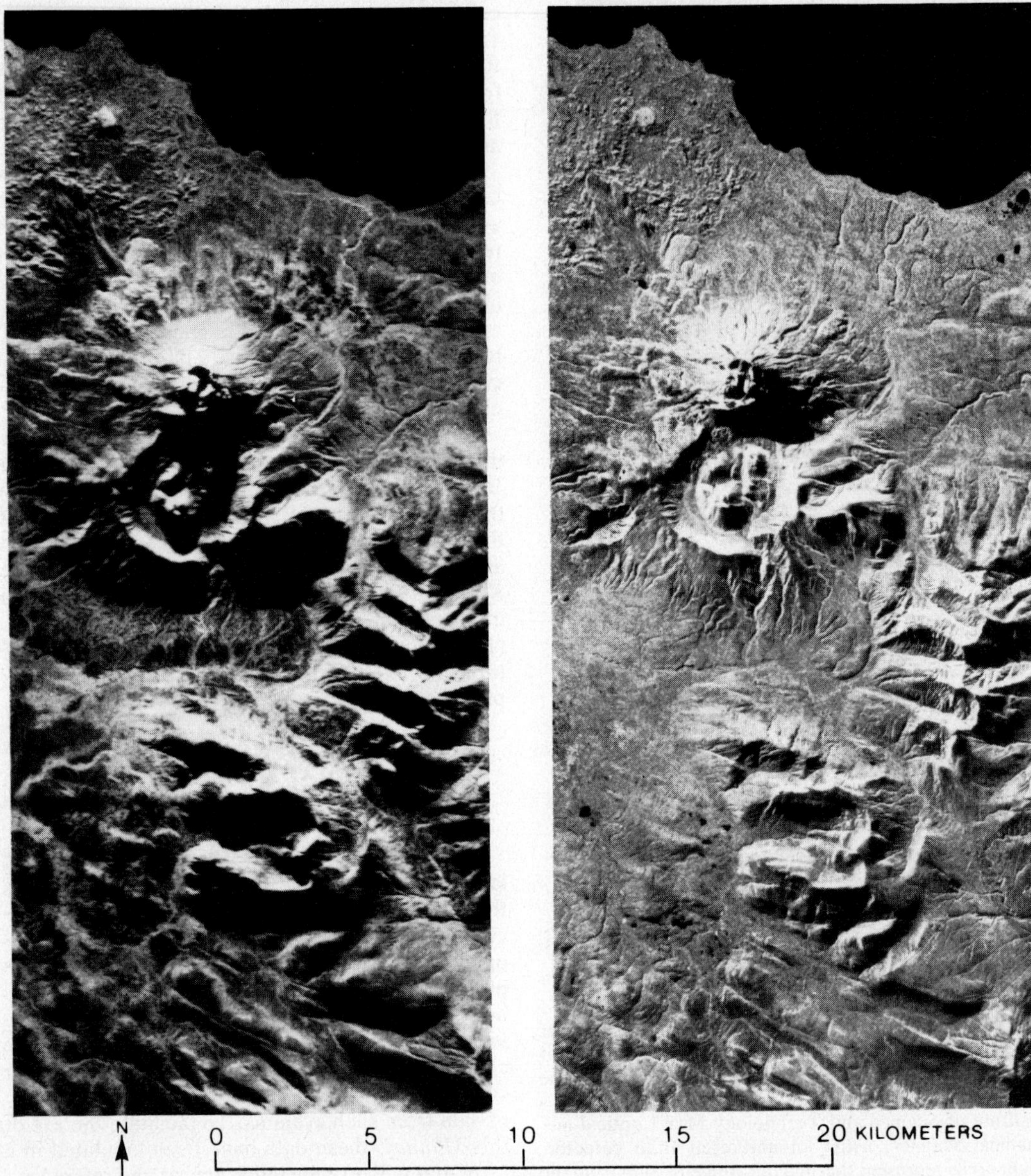

Fig. 31-15. Portions of MARS real aperture and Goodyear-Aeroservice synthetic aperture SLAR imagery of the Ugashik Quadrangle, Alaskan Peninsula. Both systems acquired X-band (2.4–3.8 cm) "horizontal-transmit" and "horizontal-receive" (HH) polarized radar with shadows depicting the respective south and southeast look directions. The imagery is centered on Mount Peulik (N57°45′, W156°25′, elevation 1,475 m (4,835 feet)) and the surrounding terrain which consists of the Chinitna and Naknek (sandstone, siltstone, shale, and conglomerate) Formations of Upper Jurassic age (Albert, 1982). Becharof Lake, to the north, acts as a specular reflector. (Images courtesy of John Jones, U.S. Geological Survey.)

TABLE 31-4
Spacecraft which acquire data of value to geologists

IMAGING SENSORS	
PHOTOGRAPHIC	NON-PHOTOGRAPHIC
Manned Spacecraft	*Earth Resources Observation Satellites*
Mercury	Landsat
Gemini	Skylab
Apollo	HCMM
Skylab	Seasat
	Nimbus 7
Non-U.S. Spacecraft	*Space Shuttle (OSTA-1)*
Cosmos	SIR-A
Meteor	*Moderate-Resolution (≤1 km) Meteorological Satellites*
Salyut	
Soyuz	NOAA 2-5
	ITOS/TIROS N
	NOAA 6
	DMSP
NON-IMAGING SENSORS	
Geodynamic Satellites	
LAGEOS	
Magsat	
Space Shuttle (OSTA-1)	
SMIRR	

Scott, 1982); however, the disadvantage is the currently limited spatial resolution of the picture elements (pixels) (1 km). Details of the synoptic and daily coverage aspects of meteorological satellites are discussed by McClain (1980) and in Chapters 14 and 27 of this *Manual*.

Non-U.S. spacecraft and satellite activities in geologic remote sensing are primarily limited at present to space programs of the Union of Soviet Socialist Republics (USSR), although many Earth resources satellite systems are under development by the European Space Agency (ESA), and are discussed in the "Future Trends" section of this chapter. Very little information is readily available about USSR remote sensing activities, and the data are not generally available outside the Soviet Union. Multispectral sensors and television camera systems have been employed on the unmanned Cosmos and Meteor satellites. Salyut and Soyuz photographic sensors (especially the MFK-6 multispectral camera system) have been used on the manned satellites. Chapter 17 (Light et al., 1980) in the *Manual of Photogrammetry* (Slama et al., 1980) and the Landsat Data User's Notes (USGS, 1982) discuss these systems in detail.

Non-imaging geodynamic satellites include LAGEOS (laser geodynamic satellite) and Magsat (magnetic field satellite). Theoretically, LAGEOS was designed to provide precise measurements (between 20 mm and 0.1 m) of rates and direction of movement of continental plates. Information on LAGEOS can be obtained through NASA's Goddard Space Flight Center, ATTN: Code 942, Greenbelt, Maryland 20771. Magsat is an outgrowth of earlier experimental measurements from the Polar Orbiting Geophysical Observatory (POGO) and Orbiting Geophysical Observatory (OGO) satellites. Through the deployment of a scalar magnetometer and a vector magnetometer, mission objectives are to (1) obtain an accurate quantitative description of the Earth's main magnetic field, (2) provide data and models to update and refine magnetic charts and maps of the world, and (3) compile global scalar and vector crustal magnetic anomaly maps. Settle and Taranik (1982a) and Mayhew (1982) discuss the characteristics and applications of Magsat.

The following sections will focus on photographic and non-photographic imaging sensors of interest to geologists. A brief introduction to each satellite will be followed by information sources and examples of data. Characteristics of each mission, and sensors utilized have been tabulated and listed. Data availability, products and distribution centers have also been tabulated, to assist geologists in the acquisition of data from various archives.

Types, Availability, and Examples of Photographic Images from Satellites

Geologic remote sensing from a space based platform began with the launch of Mercury, Gemini, Apollo and Skylab manned spacecraft which employed automatic and hand-held photographic cameras (Lowman, 1980). Specific flights, dates, duration, camera systems, film, quantity of photographs, and general regions covered have been tabulated for reference in Tables 31-5 through 31-9. Table 31-10 lists the respective data

TABLE 31-5
Satellite photographs from the Mercury Missions (see Figure 31-16)

Flight	Date	Duration	Camera	Film	Quantity of Photographs	Region Covered	Comments
MA-4	1961		Automatic	70-mm	Several hundred high obliques	Many of North Africa	Unmanned
MA-8	1962		Hand-held with 80-mm lens	70-mm	—	—	First manned
MA-9	1963	34 hours	Hand-held with 80-mm lens	70-mm	29 good to excellent	Southeast Asia and Tibet	Second manned

TABLE 31-6

Satellite photographs from the Gemini Missions (see Figure 31-17)

Flight	Date	Duration	Camera	Film	Quantity of Photographs	Region Covered
GT-3	Mar 23, 1965	4 hr 52 min 3 revolutions	Hasselblad 500 C 80-mm lens	70-mm Kodak Ektachrome	25	Southwestern United States
GT-4	Jun 3, 1965 to Jun 6, 1965	97 hr 56 min 62 revolutions	Hasselblad 500 C 80-mm lens	70-mm Kodak Ektachrome	100 of planned 219 of additional sites	Worldwide
GT-5	Aug 21, 1965 to Aug 28, 1965	190 hr 55 min 120 revolutions	Hasselblad 500 C 80-mm lens	3 rolls Ektachrome, 1 roll Anscochrome	250 of Earth and clouds, 175 excellent quality	Worldwide
GT-6-A	Dec 15, 1965 to Dec 16, 1965	25 hr 51 min 16 revolutions	Hasselblad 500 C 80-mm lens	Kodak Ektachrome	193 total photographs	Worldwide
GT-7	Dec 4, 1965 to Dec 18, 1965	330 hr 35 min 206 revolutions	Hasselblad 500 C and 250-mm Zeiss sonnar lens	Kodak Ektachrome	250 usable photographs out of 429 total	Worldwide
GT-8	Mar 16, 1966 to Mar 17, 1966	10 hr 42 min 7 revolutions	Hasselblad 500 C with 80-mm lens	Kodak Ektachrome	19 total	Pacific region
GT-9	Jun 3, 1966 to Jun 6, 1966	72 hr 21 min 45 revolutions	Hasselblad super-wide angle-C with 38-mm lens	Kodak Ektachrome	360 total	South America, Africa
GT-10	Jul 18, 1966 to Jul 21, 1966	70 hr 46 min 44 revolutions	Hasselblad super-wide angle-C with 38-mm lens	Kodak Ektachrome	526 total	Worldwide
GT-11	Sep 12, 1966 to Sep 15, 1966	71 hr 17 min 44 revolutions	Hasselblad super-wide angle-C with 38-mm lens	Kodak Ektachrome (improved)	238 total	Worldwide
GT-12	Nov 11, 1966 to Nov 15, 1966	94 hr 34 min 59 revolutions	Hasselblad super-wide angle-C with 38-mm lens	Kodak Ektachrome (improved)	415 total	Worldwide

and information centers for each spacecraft as well as formats and types of Apollo, Gemini, and Skylab photographic products available.

Photographic coverage from the manned spacecraft missions is sporadic but can be accessed geographically through the various data centers or through the following publications. Skylab photographic data coverage is contained in the Skylab Earth Resources Catalog (NASA, 1974). Lowman's chapter (1980) in *Remote Sensing in Geology* (Siegal and Gillespie, 1980), and Sabins (1978a), provide a comprehensive description of these missions and their applications. Geologic applications of manned spacecraft photography are described by Lowman (1964, 1965, 1967, 1969a, 1969b, and 1973; Lowman et al.,

TABLE 31-7

Satellite photographs from the Apollo Missions (see Figure 31-18)

Flight	Date	Duration	Camera	Film	Quantity of Photographs	Region Covered
Apollo 6	1967	—	Automatic	70-mm color film	One complete revolution of high quality overlapping vertical photographs	Southern United States to central Africa
Apollo 7	Oct 11, 1968	163 revolutions	S005 70-mm hand-held Hasselblad 500 EL 80-mm lens	70-mm color film	200 of geologic importance	Worldwide
Apollo 9	Mar 3, 1969 to Mar 13, 1969	10-day mission	S005 70-mm hand-held Hasselblad 500 EL 80-mm lens	70-mm color film	390 outstanding quality/912 total	Worldwide
			and S065 (multispectral terrain photography) Hasselblad 500 EL 80-mm Zeiss planar lens	70-mm color film	90-4 picture sets acquired	Southern United States and northern Mexico
Apollo 11	July 1969	—	S005 Hasselblad 500 EL	70-mm	—	—

TABLE 31-8

Satellite photographs from the Skylab Missions (see Color Figure 31-19, Color Figure 31-20, Figure 31-21, Color Figure 31-51, and Figure 31-52)

Flight	Date	Duration	Camera	Film	Quantity of Photographs	Region Covered
Mission 1	May 14, 1973	Launch				
SL-2	May 25, 1973 to Jun 22, 1973	28 Days	S-190A Experiment Multispectral Camera S-190B Experiment Earth Terrain Camera	70-mm 11.4-cm	5275 frames from 11 Orbits	50°N. to 50°S. Latitudes
			S-192 Experiment Multispectral Scanner		—	50°N. to 50°S. Latitudes
SL-3	Jul 28, 1973 to Sep 25, 1973	59 Days	S-190A S-190B S-192	70-mm 11.4-cm	13,429 frames from 44 orbits	50°N. to 50°S. Latitudes
SL-4	Nov 16, 1973 to Feb 8, 1974	84 Days	S-190A S-190B S-192	70-mm 11.4-cm	17,000 frames from 1,214 orbits	50°N. to 50°S. Latitudes

1967 and 1973) and by Lee and Weimer (1975). Individual collections of Gemini, Apollo, and Skylab photographs of the world have been published by NASA (1967, 1968a, 1970, 1977).

Types, Availability, and Examples of Non-Photographic Images from Satellites

Landsat, Heat Capacity Mapping Mission (HCMM), Seasat, and the various Nimbus satellite systems have all acquired non-photographic images of the Earth which have provided significant data for geologic applications. In addition to these Earth observation satellites, other vehicles including Skylab (a manned space station) and the Space Shuttle *Columbia's* Office of Space and Terrestrial Applications (OSTA-1) payload (the Shuttle Imaging Radar system (SIR-A) also acquired non-photographic images of the Earth. Characteristics of the Skylab S-192 multispectral scanner and data availability are provided in Table 31-11. See also NASA (1974) for additional information. Because of image processing problems, very little of the Skylab S-192 data is available, however.

The Landsat system consists of two independent sensors, the multispectral scanner (MSS) (Taranik, 1978a) and the return beam vidicon (RBV) (See Table 31-12). By mid-1981, as previously indicated, over 1 million Landsat scenes of

TABLE 31-9

Characteristics of the S-190A and S-190B cameras used on the Skylab Missions

S-190A
SKYLAB MULTISPECTRAL CAMERA

Camera No.	Spectral Band (μm)	Film Type	Ground Resolution (m)
1	0.7 to 0.8	Black and white infrared	73–79
2	0.8 to 0.9	Black and white infrared	73–79
3	0.5 to 0.88	Color infrared	73–79
4	0.4 to 0.7	High-resolution color	40–46
5	0.6 to 0.7	Black and white	30–38
6	0.5 to 0.6	Black and white	40–46

S-190B
EARTH TERRAIN CAMERA

Filter Band Pass (μm)	Film Type	Estimated Ground Resolution (m)
0.4–0.7	Color	21
0.5–0.7	High-resolution black and white	17
0.5–0.88	Color infrared	30
0.5–0.88	High-resolution color infrared	23

TABLE 31-10

Availability of photographic data from manned spacecraft

For Mercury:

National Aeronautics and Space Administration
Office of Public Affairs
Audio Visual Services
Washington, D.C. 20546
Tel. No.: 202-755-8366

For Apollo, Gemini, or Skylab:

U.S. Geological Survey		NASA
National Cartographic Information Center (NCIC) 507 National Center Reston, VA 22090 Tel. No.: 703-860-6045 FTS: 928-6045	EROS Data Center Sioux Falls, SD 57198 Tel. No.: 605-594-6511 FTS: 784-7511	Technology Application Center University of New Mexico Albuquerque, NM 87131 Tel. No.: 505-277-3622

Standard Film Formats and Types of Black and White or Color Photographic Products Available from the EROS Data Center

Image Size	APOLLO/GEMINI Nominal Scale	Product Format*
55.8 mm (2.2 in.)	Variable	Film Positive
55.8 mm (2.2 in.)	Variable	Film Negative**

Image Size	SKYLAB S190A Nominal Scale	Product Format*
55.8 mm (2.2 in.)	1:2,850,000	Film Positive
55.8 mm (2.2 in.)	1:2,850,000	Film Negative**

Image Size	SKYLAB S190B Nominal Scale	Product Format*
11.4 cm (4.5 in.)	1:950,000	Film Positive
11.4 cm (4.5 in.)	1:950,000	Film Negative**

* Prints at various scales are available only as custom order products and are priced accordingly.
** Film negatives do not exist for these color products.

the world had been acquired. Four Landsat satellites have been launched since July 1972; of the first three, however, only the Landsat 3 and Landsat 4 spacecraft remain in operation. The digital MSS data (Color Figure 31-22) complimented by the Landsat 3 RBV high-resolution (30-m) panchromatic image data (Figure 31-23) are used for energy and mineral resource exploration, monitoring of dynamic phenomena, and shallow seas mapping. The Landsat Data User's Handbooks (NASA, 1971; and U.S. Geological Survey, 1979) and the Landsat Data User's Notes, published bi-monthly by the EROS Data Center (See Table 31-17) provide information on recent developments of the Landsat program. The U.S. Geological Survey has established many National Cartographic and Information Center (NCIC) offices and NCIC State affiliate offices to support users' requests (Clarke et al., 1982). Further details of the Landsat system and its geological applications can be obtained in Short et al. (1976), Williams and Carter (1976), and in Chapter 12 of this *Manual*.

The Heat Capacity Mapping Mission (HCMM), launched on 26 April 1978, was the first civilian spacecraft specifically designed to test the feasibility of measuring thermal variations in the Earth's surface temperature in order to infer its surface characteristics. The HCMM spacecraft carried a special thermal sensor, the heat capacity mapping radiometer (HCMR), which operated with two channels: one in the visible and near-infrared, the other in the thermal infrared region. Preliminary findings from HCMM investigators, reporting on various analyses and applications of the HCMR data to geologic research, indicate that the mapping of geologic structure, lithologic discrimination, and possible detection of surface seepage of hydrocarbon traps at depth are possible. Over 37,600 standard image products were obtained in the 28 months of flight operation and can be geographically accessed on microfiche in the HCMM Data User's Handbook (NASA, 1980). Maps depicting HCMM computer compatible tape data of day and night passes and the location of day/night registered pairs are also available. Current information is available through the HCMM Data User's Bulletin which is issued periodically by NASA's Goddard Space Flight Center (See Table 31-17). See also the reports by Kahle et

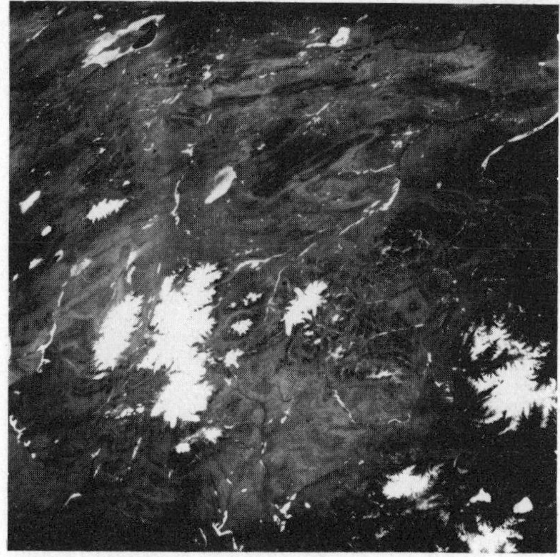

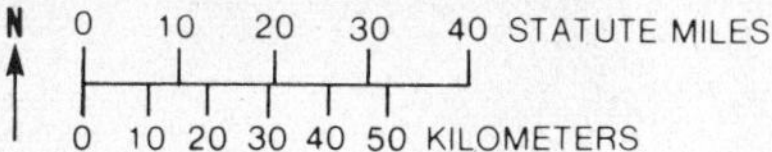

Fig. 31-16. Mercury 9 (MA-9) color photograph (reproduced here in black-and-white) taken in 1963 over the Tibetan Plateau, China (People's Republic of China), by L. Gordon Cooper with a 70 mm Hasselblad camera having an 80-mm lens from an altitude of 162 km (100 miles). Orbital photographs such as this, stimulated interest in geologic mapping from space. (Photograph ID No. MA9-22 courtesy of NASA.)

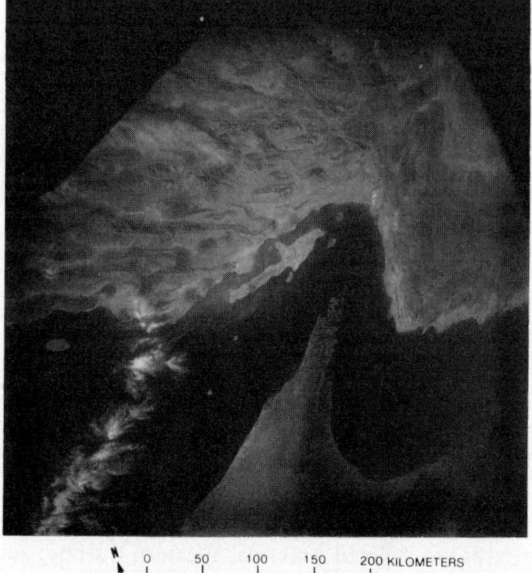

Fig. 31-17. Gemini 12 color photograph (reproduced here in black-and-white) acquired on 11 November 1966 over the Strait of Hormus, separating the Persian Gulf and the Gulf of Oman. The photograph was taken with a Hasselblad camera from an altitude of 250 km. Sand dunes and volcanic mountains are seen on the Arabian Peninsula. The intensely folded Zagros Mountains of Iran are observed north of the Strait. (Photograph ID No. S-66-63082 courtesy of NASA.)

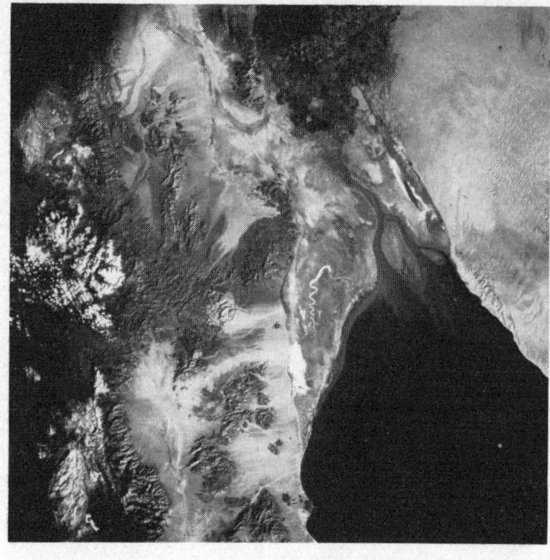

Fig. 31-18. Apollo 9 color photograph (reproduced here in black-and-white) acquired 3 March 1969 using four Hasselblad cameras mounted on a window frame from an altitude of 240 km, over the Colorado River Delta at the Gulf of California. To the west of the Gulf are granitic and rhyolitic terrain with gravel alluvial fans. Sand dunes are located east of the Gulf. (Photograph ID No. AS9-26A-3781A courtesy of NASA.)

TABLE 31-11

Characteristics of the Skylab S-192 Multispectral Scanner (MSS) and data availability

Band Number	Wavelength μm	Spectral Band
1	0.41 to 0.45	Blue
2	0.44 to 0.52	Blue-green
3	0.49 to 0.56	Green
4	0.53 to 0.61	Green
5	0.59 to 0.67	Red
6	0.64 to 0.76	Red-near-IR
7	0.75 to 0.90	Near-IR
8	0.90 to 1.08	Near-IR
9	1.00 to 1.24	Near-IR
10	1.10 to 1.35	Near-IR
11	1.48 to 1.85	Near-IR
12	2.00 to 2.43	Near-IR
13	10.20 to 12.50	Thermal IR

Data Archive: EROS Data Center
U.S. Geological Survey
Sioux Falls, South Dakota 57198

Tel. No. 605-594-6511
FTS: 784-7511

Standard Film or CCT Formats of Skylab S-192 MSS Image Products Available:
12.7-cm × 12.7-cm black-and-white film prints or transparencies (positive or negative), color composite prints or positive transparencies from up to three bands, and computer compatible tapes.

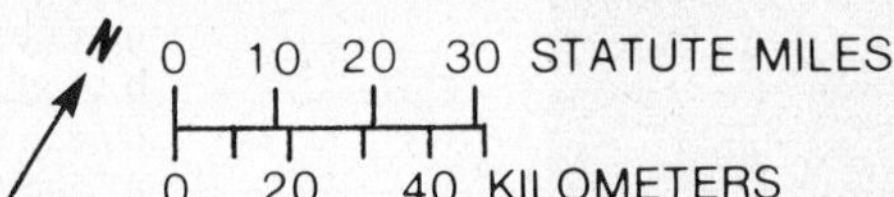

Fig. 31-21. Hand-held color photograph (reproduced here in black-and-white) acquired January 1974, during the Skylab 4 visual observations project, of the Wyoming overthrust belt. The area covered includes Yellowstone National Park, the Wind River Range, and the Absaroka Range. The photograph was acquired at an altitude of 435 km during the winter to provide low Sun angle and light snow cover enhancement of the surface structure and topography. (Photograph ID No. SL4-138-3846 courtesy of NASA.)

al. (1981b) and Watson et al. (1981) for discussion of geologic applications of thermal inertia mapping using HCMM data. Table 31-13 provides information of the characteristics of the HCMM satellite and data availability. Figure 31-24 is an example of HCMM images.

Seasat, launched 29 June 1978, was the first civilian space platform to acquire synthetic aperture radar (SAR) imagery of the Earth. Although the mission was primarily designed for oceanographic monitoring, there were many data-acquisition passes over North America, Western Europe, and the Caribbean Region (Ford et al., 1980) which provided significant new data for geologists. The Seasat SAR acquired L-band (23.5-cm wavelength) imagery of approximately 25-m resolution, with look directions of northeast (ascending node) and northwest (descending node). Optically processed Seasat SAR data have been used to produce uncontrolled mosaics of California, Pennsylvania, Florida, Jamaica, and the United Kingdom (Bryan, 1981) and Iceland (Hunting

TABLE 31-12

Characteristics of the Landsat 1, 2, and 3 spacecraft and sensors and data availability

LAUNCH DATES:

*Landsat 1—23 July 1972. Operation ended on 6 January 1978.
Landsat 2—22 January 1975. Operation ended on 8 February 1982.
Landsat 3—5 March 1978. Still in limited service (as of July 1982). Operation is expected to be terminated on 31 March 1983.

ORBITAL ELEMENTS:

Orbit: Circular, near polar.
Inclination: 99.09°.
Altitude: 919 km.
Coverage: 82°N to 82°S.
Period: 103 minutes with crossing of Equator at 0930 hours, local time.
Cycle: 18 days. Note: Landsat 3 followed Landsat 2 by 9 days.

SENSORS: *Return Beam Vidicon Cameras (RBV)* (see Figure 31-23)

	Wavelength (μm)	Pixel Spatial Resolution (m)	Image Format/Comments
Landsats 1 and 2, three RBV's:			
Band 1	0.475–0.575 (blue-green)	80	Simultaneous view from 3 cameras of a scene 185 km × 185 km with 14-percent sidelap at Equator and 10-percent forward lap along orbital track. Coverage similar to the single MSS image. Frame format.
Band 2	0.580–0.680 (yellow-red)		
Band 3	0.690–0.830 (red-IR)		
Landsat 3, two RBV's	0.505–0.750 (panchromatic into near-IR)	30	2 side-by-side, slightly overlapping images 98 km × 98 km (4 RBV images coincide with a single MSS frame). Frame format.

Multispectral Scanner (MSS) (See Figure 31-22)

	Wavelength (μm)	Pixel Spatial Resolution (m)	Image Format/Comments
Landsats 1, 2, & 3			
Band 4	0.50–0.60 (green)	80	185-km strip images have 10 percent forward lap and 14 percent sidelap at Equator, and these increase toward the poles.
Band 5	0.60–0.70 (red)		
Band 6	0.70–0.80 (near-infrared)		
Band 7	0.80–1.1 (near-infrared)		
Landsat 3 only:			
Band 8	10.4–12.5 (thermal IR)	Range of thermal sensitivity: 260°–340°K	Thermal sensor never operated properly. Only a few scenes available of a limited number of areas.

Data Archive: EROS Data Center
Sioux Falls, South Dakota 57198
Tel. No.: 605-594-6511
FTS: 784-7511

Standard Film or CCT Formats of Landsat MSS and RBV Image Products Available:

Image Size	Standard Landsat Images Nominal Scale	Product Format
55.8 mm (2.2 in.)	1:3,369,000	Film Positive
55.8 mm (2.2 in.)	1:3,369,000	Film Negative
18.5 cm (7.3 in.)	1:1,000,000	Paper
18.5 cm (7.3 in.)	1:1,000,000	Film Positive
18.5 cm (7.3 in.)	1:1,000,000	Film Negative
37.1 cm (14.6 in.)	1:500,000	Paper
74.2 cm (29.2 in.)	1:250,000	Paper

(*Until the launch of Landsat 2, Landsat 1 was called ERTS 1 (Earth Resources Technology Satellite 1).

TABLE 31-12 (CON'D)

Computer Compatible Tapes (CCT)

Tracks	b.p.i.	Format
9	1,600	Tape Set
9	6,250	Tape Set

NOTE: Film products are available in black-and-white or color, except no color is available in the film negative format. Landsat 3 RBV images are available only as 1:500,000-, 1:250,000-, and 1:125,000-scale, black-and-white, standard products at square sizes of 18.5 cm, 37.1 cm, and 74.2 cm, respectively. Also, not all Landsat 3 RBV scenes are available in the CCT format.

Fig. 31-23. Part of a Landsat 3 return beam vidicon (RBV) image (Landsat ID No. 30556-17300-Subscene B) acquired on 12 September 1979 at approximately 1030 hr local time of the Salt Lake City region. Although the broad band (0.505−0.750 μm) image lacks the four-band, multispectral capability of the MSS system, the higher resolution of 30 m, as compared to 79 m of the MSS image, provides increased spatial discrimination. Tailings ponds, waste dumps, and excavation benches of the Bingham Canyon mine are readily observed. The calibrated reseau marks (+) permit high fidelity reconstruction of the internal sensor geometry, thereby providing a superior planimetric base map.

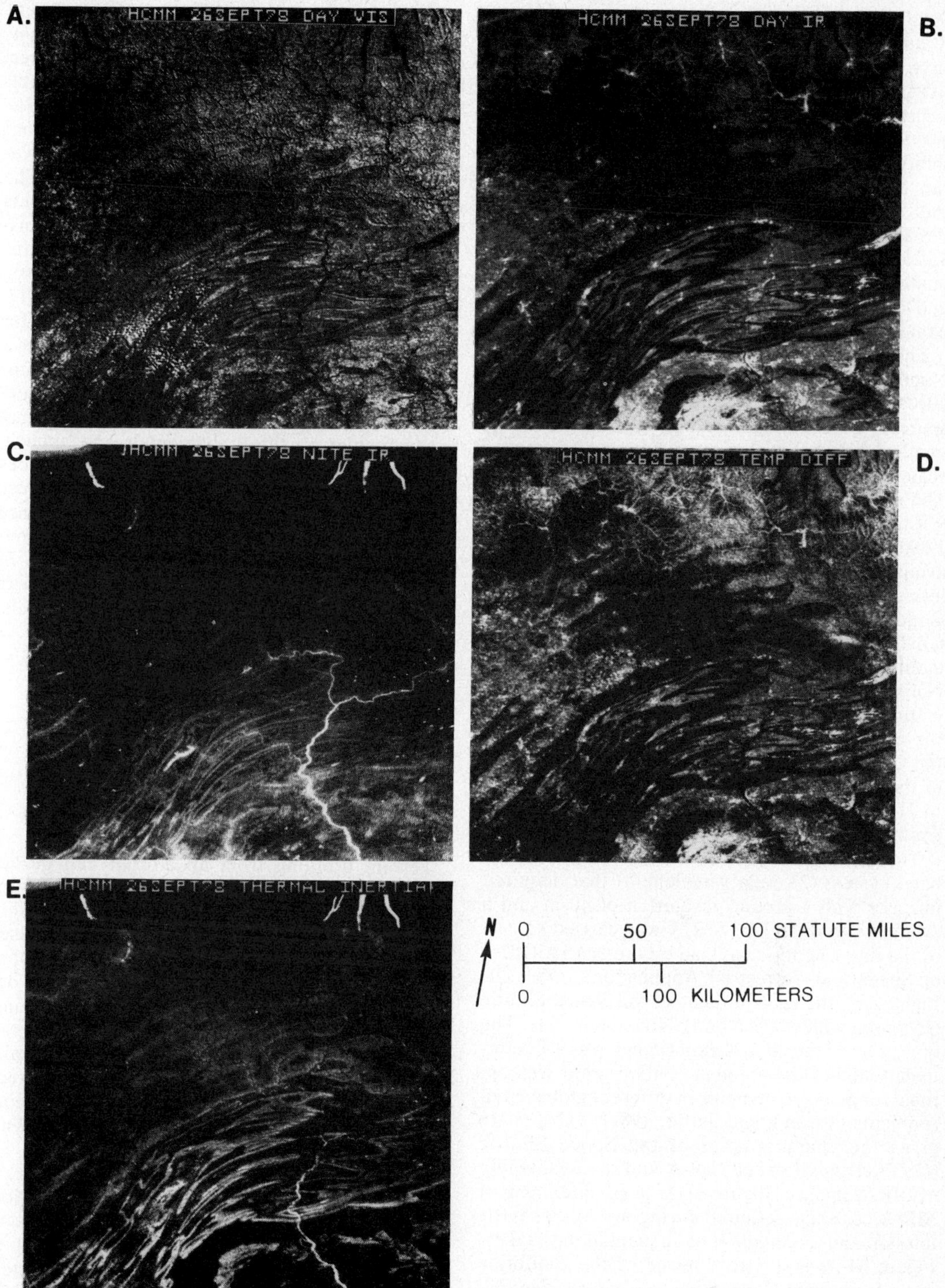

Fig. 31-24. Subscenes of HCMM data acquired on 26 September 1978 over the Appalachian Mountains of central Pennsylvania. The day visible (A) and the day infrared (B) scenes were acquired at approximately 1330 hr EST; the night infrared (C) scene was acquired at approximately 0230 hr EST on the same day. The temperature difference (ΔT) (D) was determined by subtracting the radiometric temperatures observed by the HCMM during the day and night passes. Thermal inertia (E) is a derivative product of the day and night thermal images (ΔT values). The marked difference in the composition of the surface material of the folded Appalachian strata and the glacial deposits around the Finger Lakes region can be observed. Urban heat-emitting centers are observed in both the day and night thermal infrared scenes. The ridges of the Appalachian Mountains, composed of quartzite and dolomite (covered with forest; Lattman, 1963), and regional groundwater patterns exhibit high thermal inertia at night.

Geology and Geophysics, Ltd., n.d.). Over 280 90-km X 90-km digitally correlated subscenes have been produced and are available from NOAA's Environmental Satellite, Data, and Information Service (see Table 31-14). Table 31-14 provides information on characteristics and availability of Seasat SAR data. Additional information can be obtained through the Satellite (Seasat) Data User's Bulletin (see Table 31-17), Ford et al. (1980), Elachi (1980), and Wu et al. (1981). Figures 31-25 and 31-26 are index maps of the approximate Seasat SAR coverage of North America and of Alaska, respectively. Figure 31-27 is an example of Seasat SAR images.

The Coastal Zone Color Scanner (CZCS) on Nimbus 7, is the only sensor in orbit that is specifically designed to study living marine resources. Potential applications include the study of algal blooms, water mass boundaries, mesoscale circulation patterns, and regional structure. The six-band multispectral scanner acquires data with 800-m resolution and a swath width of 1,800 km. Information on the satellite and data is available through the Nimbus 7 Data User's Bulletin which is issued periodically by NASA's Goddard Space Flight Center (see Table 31-17). CZCS data catalogs covering 1979 and 1980 monthly orbits are available through NOAA's National Environmental Satellite, Data, and Information Service (NASA, 1978a). Information on the characteristics and availability of data from the CZCS are given in Table 31-15. A discussion of the applications of CZCS images can be found in Hovis et al. (1980). Figure 31-28 provides an example of two CZCS images.

The Shuttle Imaging Radar (SIR-A) is an L-band system (23.5-cm wavelength) that acquired imagery with a ground resolution of 40 m and a swath width of 50 km. SIR-A was carried as part of the first scientific payload on the NASA Office of Space and Terrestrial Applications (OSTA-1) flight; i.e., the second flight of the Space Shuttle *Columbia* which began on 12 November 1981. The objective of the SIR-A experiment was to determine whether the antenna configuration was optimal for geologic mapping in different global environments (Taranik and Settle, 1981). Table 31-16 gives the characteristics of the Space Shuttle (OSTA-1) mission and SIR-A and the availability of SIR-A images. Figure 31-29 is an index map of SIR-A coverage acquired during the 2-day shuttle mission and a tabulation of sequential data takes. Figure 31-30 is a SIR-A image of the California coast from Point Conception to Ventura. See Figure 31-166 in the Engineering Geology section of this chapter for a Seasat SAR image of part of the same coastal area.

Characteristics of spacecraft and satellite systems and their non-photographic imaging sensors are tabulated for reference. Launch dates, orbital elements, sensor wavelength, resolution, and data format are included.

There are three major Federal Government facilities for the archiving and distribution of satellite data. These are: The EROS Data Center[1] U.S. Geological Survey; National Environmental Satellite, Data, and Information Service, National Oceanic and Atmospheric Administration; and National Space Science Data Center, National Aeronautics and Space Administration. The data centers and respective satellite data products have been provided for reference after the discussion of each satellite remote sensing system in this section. Specific satellite data user's handbooks, notes, and data catalogs are listed in Table 31-17. Additional information on various archives for satellite image data can be found in *Everyone's Space Handbook, A Photo/Imagery Source Manual* (Kroeck, 1976). The U.S. Army's *Remote Sensing Applications Guide* is another informative pamphlet on the fundamentals and applications of remote sensing, including appendices which identify available remote sensing services (Department of the Army, 1979). A new reference manual entitled, *A Guide to Environmental Satellite Data,* is a comprehensive description of U.S. environmental satellites and their sensors, with appendices on respective data products and distribution centers (Cornillon, 1982).

NON-IMAGING AND/OR HIGH-RESOLUTION SYSTEMS

Airborne Systems

Profilometry

Terrain profiling from an aircraft can be accomplished by measuring the round-trip time from pulse transmission to pulse return of a nadir-oriented pulsed radar or pulsed laser. The laser sensor's measurement precision (±3 cm) is superior to the radar sensor's precision (±20 cm) because of the laser's significantly smaller beamwidth.

In order to determine terrain elevations, the aircraft's motion must be continuously monitored, so that any deviations from level flight can be corrected. The airborne laser in use at NASA's Wallops Flight Center is coupled with a vertical accelerometer, the use of which provides data for removing the effect of short-term aircraft vertical motion from the measurements (Krabill et al. 1980). The accelerometer, although subject to bias and drift over a period of several minutes, is extremely sensitive to short-term vertical

[1] Note: On 1 October 1982, the National Environmental Satellite, Data, and Information Service of the National Oceanic and Atmospheric Administration assumed responsibility for the acquisition, archiving, processing, and distribution of all Landsat data at the EROS Data Center.

TABLE 31-13

Characteristics of the Heat Capacity Mapping Mission (HCMM) spacecraft and the Heat Capacity Mapping Radiometer (HCMR) sensor and data availability

LAUNCH DATE: 26 April 1978

ORBITAL ELEMENTS:

Orbit:	Circular, sun-synchronous.
Altitude:	620 km (540 km from 23 February 1980, until data termination on 31 August 1980)
Inclination:	97.6°
Coverage:	Day/night passes over given area within 12 hours at 35° latitude and poleward; within 36 hours other latitudes. Real time only of the United States including Alaska, southern Canada, northern Mexico, Europe, and eastern Australia.
Cycle:	16 days

SENSOR: Heat Capacity Mapping Radiometer (HCMR) (Previously flown on the Nimbus 5 spacecraft)

	Wavelength	Instantaneous field of view	Swath	Range	NER/NEΔT (Noise equivalent radiance/ Noise equivalent temperature difference)
Visible-near infrared channel	0.55–1.1 μm	500 m	716 km	0–100% Albedo	0.2 milliwatt/centimeter² (NER)
Thermal channel	10.5–12.5 μm	600 m	716 km	260°–340°K	0.3°K at 280°K (system NEΔT; 0.4°K at 280°K)

Data Archive: National Space Science Data Center
World Data Center-A
Goddard Space Flight Center
National Aeronautics and Space Administration
ATTN: Code 601
Greenbelt, Maryland 20771
Tel. No.: 301-344-6695

Standard Film or CCT Formats of HCMM Image Products Available:

1:4,000,000-scale black-and-white images on 241-mm print paper or positive or negative transparencies:

- Day visible
- Day thermal infrared
- Night thermal infrared
- Temperature difference (Night vs. Day)
- Thermal inertia

Computer compatible tapes (CCT's): 9-track, 800 or 1,600 bpi

motions of the aircraft. The bias and drift effects are removed by fitting the altitude measurement to three known ground points near the beginning, middle, and end of each flight line. If specific flight lines are to be flown, ground crews may provide flight alignment targets such as large tethered balloons.

The profiling results of a particular flight line in the Wolf River Basin near Memphis, Tennessee, are depicted in Figure 31-31. The laser-derived profile is compared with a photogrammetrically-derived profile; the agreement between the two profiles was ±12-cm-rms (root mean square) error over open ground, and ±50-cm rms discrepancy in the forested portion, where a degraded photogrammetric accuracy would be anticipated. Figure 31-31 also depicts another feature of airborne profilers; tree canopy heights can be measured, because the canopy reflects a portion of the energy back to the aircraft, while the remainder of the energy continues to the forest floor to be subsequently reflected back to the aircraft.

The U.S. Geological Survey and the Charles Stark Draper Laboratory are cooperating in the development of another laser and inertial guidance based airborne system called the Aerial Profiling of Terrain System (APTS) (Henriksen et al, 1980; Chapman, 1979 and 1982). The research project began in 1974 and is directed toward combining an inertial navigation system, three accelerometers and three gyros, with a laser profiler. The inertial unit provides continuous three-dimensional position data, so that when combined with the profiler measurement, the position and elevation of each profile point can be calculated. The accuracy goals are ±15 cm vertically and ±60 cm horizontally. To meet these strict specifications, a laser tracker for periodically obtaining update information was added to the system as shown in Figure 31-32. The pulsed laser tracker will automatically lock-on and track retroreflectors previously positioned on the ground. The APTS is to be installed in a twin-engine aircraft early in 1983, and flight tests are to begin during that spring.

Luminescence

Luminescence is the property of some materials to emit light when excited by external stimuli such

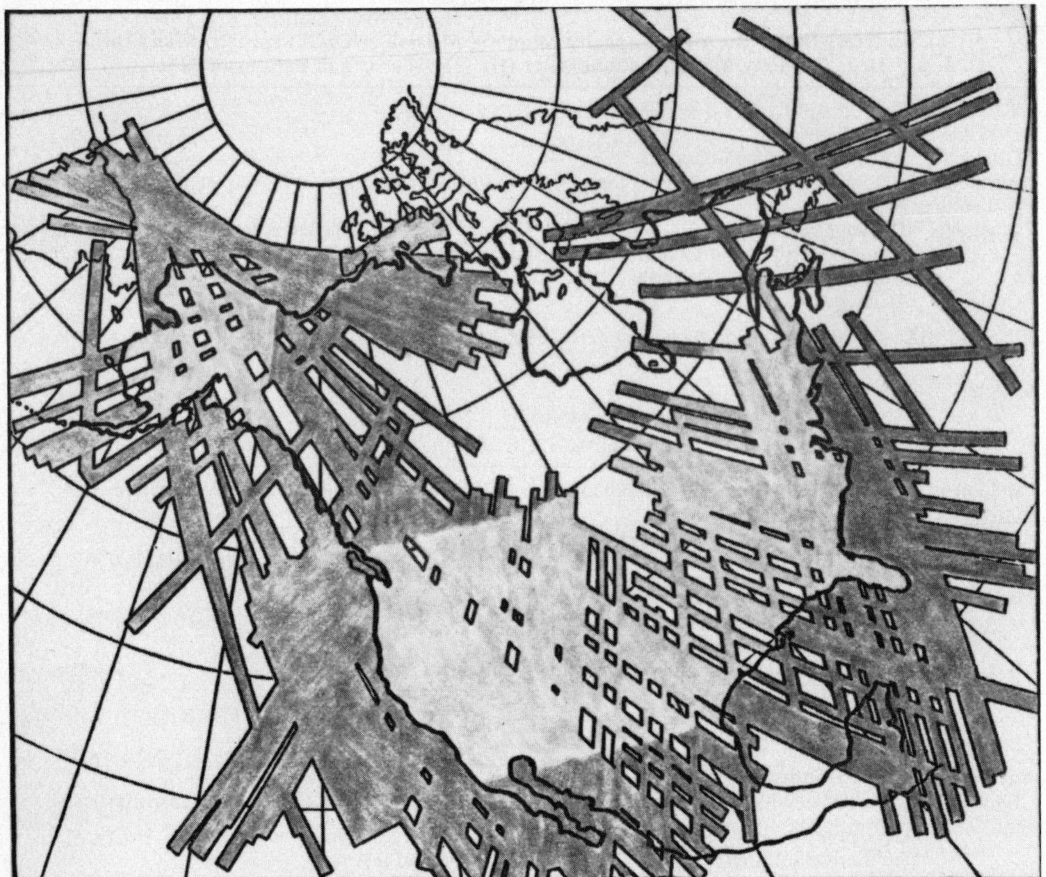

Fig. 31-25. General coverage of Seasat Synthetic Aperture Radar (SAR) over the North American continent from the 29 June 1978 launch until the 10 October 1978 termination of the mission. United States coverage portrays ascending (southeast to northwest) and descending (northeast to southwest) satellite tracks. (Index map courtesy of John Jones, U.S. Geological Survey.)

as ultraviolet or visible radiation. Portable ultraviolet lamps were used in the 1940s and 1950s to prospect for luminescing minerals, notably scheelite ($CaWO_4$) and minerals of uranium. The method is awkward, however, because the lamps are of low power, thereby limiting their effective range to a few centimeters or tens of centimeters, and the work must be performed at night to avoid sunlight which would obscure any lamp-induced luminescence.

The Fraunhofer line[2] discriminator (FLD) is an airborne electro-optical instrument which permits detection during the day of materials stimulated to luminesce by solar ultraviolet and visible radiation. An airborne FLD has been tested by the U.S. Geological Survey in collaboration with the National Aeronautics and Space Administration, other Federal agencies, and industrial groups. This instrument uses glass spacer Fabry-Perot filters of narrow bandwidth (<0.1 nm) to isolate Fraunhofer lines of interest (Hemphill, 1981).

To measure the luminescence of a mineral, the FLD computes, from reflected light, the ratio of the central intensity of a Fraunhofer line to the solar continuum a few tenths nanometers distant; it then compares this ratio with either a conjugate measurement in skylight and direct sunlight, or with a previously determined reference voltage or artificial light source. Luminescence is indicated where the ratio measured from the target material exceeds the ratio measured from skylight and direct sunlight, or simulated by a reference voltage or artificial light source (Plascyk, 1975).

Procedures employing a laboratory fluorescence spectrometer permit prediction of the detectivity of a target material with an FLD prior to mounting an airborne survey. Luminescence spectra of the material are corrected for

[2] Fraunhofer lines are dark lines in the solar spectrum caused by selective absorption of light by gases in the relatively cool upper part of the solar atmosphere.

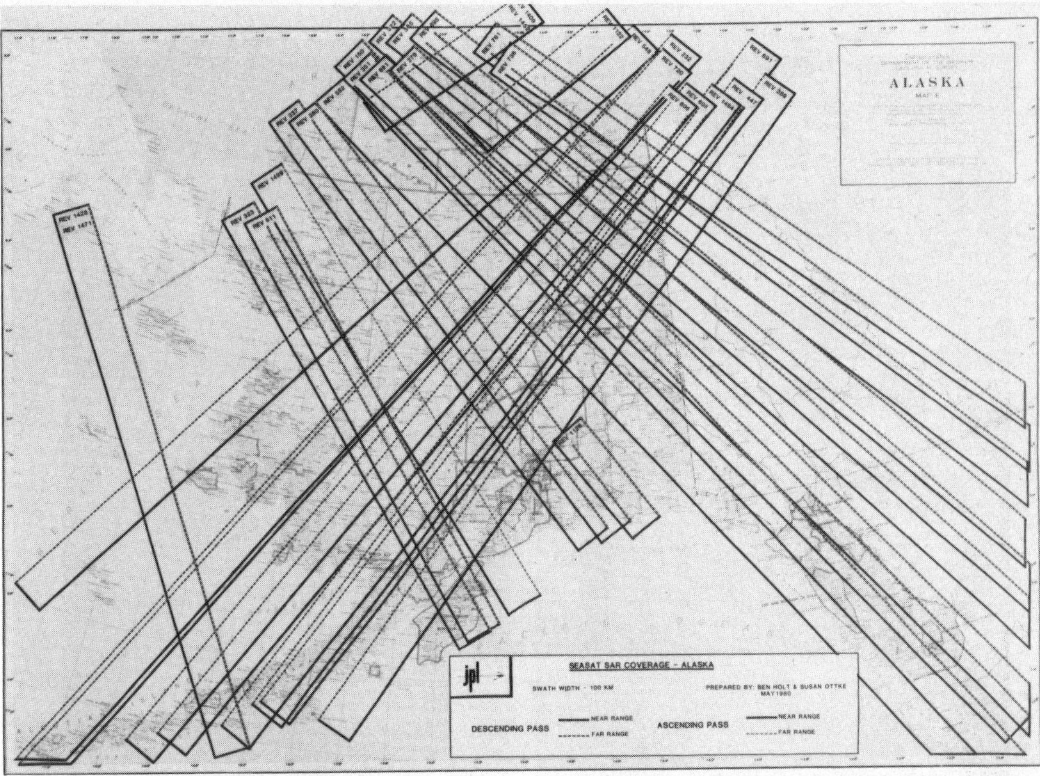

Fig. 31-26. Seasat SAR coverage map of Alaska depicting ascending (southeast to northwest) and descending (northeast to southwest) passes with respective look directions of northeast and northwest. (Index map courtesy of Ben Holt, Jet Propulsion Laboratory.)

wavelength variation in the spectrometer source, optics, and detector; a solar correction permits the luminescence spectra to be expressed in terms of their intensity under natural (daylight) illumination. By a comparison of these measurements with the spectra of a luminescence standard, such as rhodamine WT dye, the luminescence of the material at several Fraunhofer lines may be expressed in terms of rhodamine dye equivalency. The FLD detectivity may be assessed at each Fraunhofer line, and the optimum line for field observation of the material may be selected. Chapter 4 (Janza et al., 1975) in the *Manual of Remote Sensing* (Reeves et al., 1975b), and Watson (1981) describe the apparatus and procedures for performing these measurements in the laboratory, and the instrumentation and solar corrections that must be applied. The U.S. Geological Survey's FLD is sensitive to rhodamine WT concentrations as low as 0.1 part per billion (ppb) (Plascyk and Gabriel, 1975), a sensitivity that approaches levels heretofore achievable only with laboratory fluorometers and spectrometers.

Although the FLD was designed for helicopter operation in a non-imaging radiometer mode, the instrument is now routinely used aboard fixed-wing aircraft where integration with a line scan imaging system permits acquisition of two-dimensional image swaths about 1.5 km wide from aircraft altitudes of 2.5 km. Because of the coarse spatial resolution (45 m instantaneous field of view), absence of gyro-stabilization of the FLD-scanner mount, and other factors, the resulting images are of limited operational use. However, FLD images, such as shown in Color Figure 31-33 do provide insight in suggesting that airborne surveillance of luminescence with an operational version of the FLD could provide valuable information in exploring for certain types of mineral commodities.

Radiometry

Radiometry refers to the measurement of the intensity of electromagnetic energy reflected or emitted from objects within the field of view. Lowe et al. (1975) described radiometers, including spectrometers and polarimeters, that were invented during the early stages of remote sensing research. In the following brief section, only those radiometric instruments that are most useful for

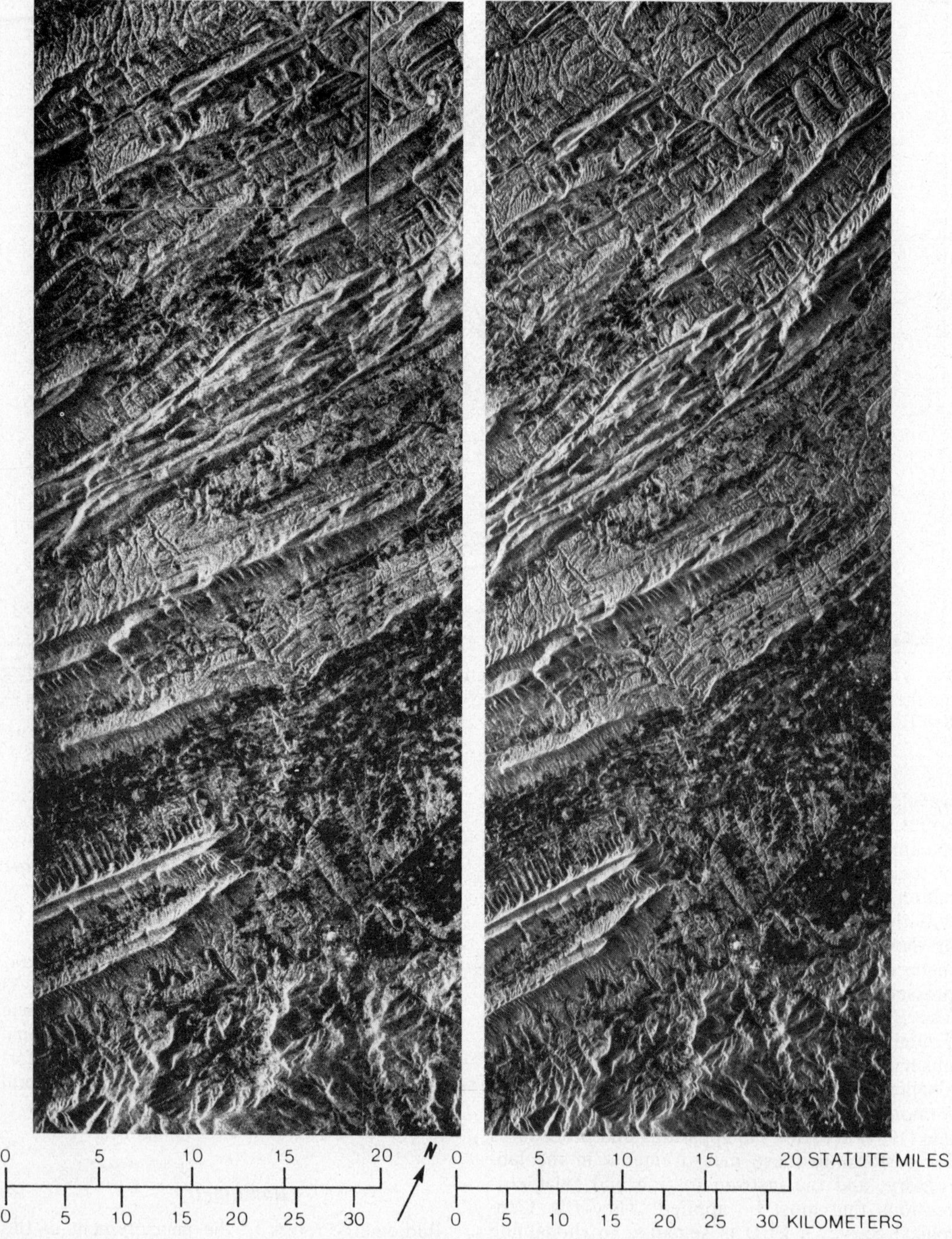

Fig. 31-27. A Seasat synthetic aperture radar (SAR) optically (left) and digitally (right) correlated images of the Valley and Ridge Province of Virginia. These images were acquired 23 July 1978, as the satellite ascended from Cape Hatteras, North Carolina, to Lake Erie. The Seasat SAR data were acquired on revolution 0378 with a look direction of northeast and a look angle of 20.5° (off the vertical). The images are centered on the Shenandoah Valley of Virginia, which is composed of sedimentary rocks of Ordovician age and is bounded to the east by the Blue Ridge Anticlinorium, composed of igneous and metamorphic rocks of Precambrian to Cambrian age. To the west are the folded Appalachian Mountains, composed of sedimentary units of Cambrian through Devonian age. The Seasat SAR image ID numbers are 051A (left) and 03780247 (right).

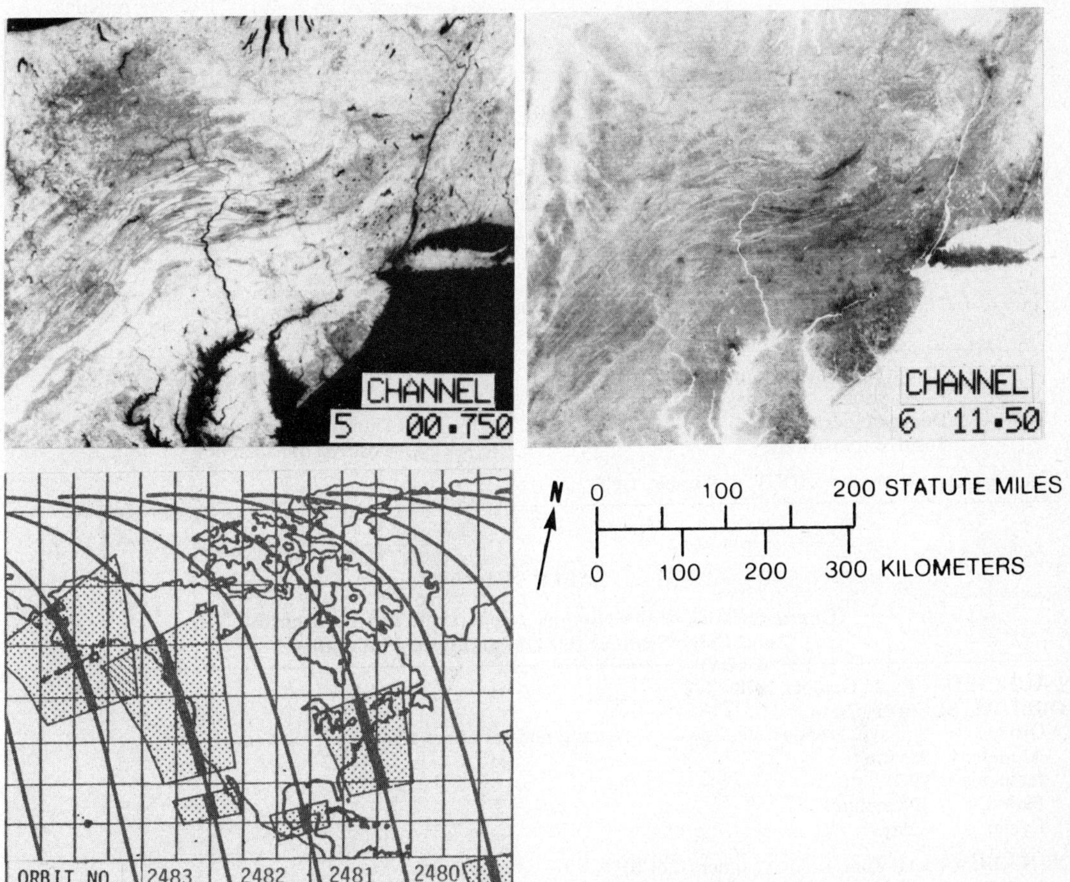

Fig. 31-28. Two Nimbus 7 Coastal Zone Color Scanner (CZCS) images, channels 5 and 6, simultaneously acquired on 21 April 1979 (orbit 2480) at approximately 1110 hrs local time, over the eastern seaboard region of the United States. This enlargement of channels 5 and 6, two of the six CZCS bands (See Table 31-15 and Figure 31-63) are centered on 0.75 μm (0.7–0.8 μm) and 11.5 μm (10.5–12.5 μm), respectively. A data catalog index map for 21 April 1979 is also included.

TABLE 31-14

Characteristics of the Seasat spacecraft and Synthetic Aperture Radar (SAR) sensor and data availability

LAUNCH DATE: 29 June 1978
OPERATION ENDED: 10 October 1978
ORBITAL ELEMENTS:

Orbit:	Nearly circular
Altitude:	790.17 km ± 50 m
Inclination:	108° nominal, 104°–108° range
Period:	100.75 minutes
Orbits per day:	14.3
Cycle:	152 days

SENSOR: Synthetic aperture radar (SAR) (data limited to 60 min/day; direct readout only)

Frequency	Wavelength	Polarization	Spatial Resolution	Swath Width/ Field of View	Antenna Depression Angle
1274.8 gigahertz	L-band, 23.5 cm	Horizontal, Horizontal (HH)	25 m at 4-look directions.	100-km swath on one side of spacecraft.	70° from the horizontal.

Data Archive: National Environmental Satellite, Data, and Information Service
National Oceanic and Atmospheric Administration
Washington, D.C. 20233
Tel. No.: 301-763-8111
FTS: 763-8111

Standard Film or CCT Formats of Seasat SAR Image Products Available:

Optically Processed:	Digitally Processed:
70-mm format (black and white):	(90-km × 90-km coverage)
Paper print	Paper print
Duplicate negative	Duplicate negative
Positive transparency	Positive transparency

Computer compatible tapes (CCT's): 9 track, 1,600 bpi

TABLE 31-15

Characteristics of the Nimbus 7 spacecraft and the Coastal Zone Color Scanner (CZCS) and data availability

LAUNCH DATE: 24 October 1978
ORBITAL ELEMENTS:

Orbit:	Sun-synchronous, near polar; ascending node at about 1200 hours local time.
Altitude:	955 km
Inclination:	99.3°
Period:	104 minutes
Cycle:	6 days

SENSOR: Coastal Zone Color Scanner (CZCS)

Wavelength of Bands (μm)	Pixel Spatial Resolution	Swath	Measurements
0.43–0.45	800 m	1800 km	Chlorophyll absorption
0.51–0.53			Chlorophyll distribution
0.54–0.56			Gelbstoffe (yellow substance)
0.66–0.68			Chlorophyll concentration
0.70–0.80			Surface vegetation
10.5–12.5			Surface temperature
			Diffuse attenuation coefficient

Data Archive: National Environmental Satellite, Data, and Information Service
National Oceanic and Atmospheric Administration
Washington, D.C. 20233
Tel. No.: 301-763-8111

Standard Film or CCT Formats of CZCS Image Products Available:
Image format is 25.4 × 25.4 cm (10 × 10 in) for black-and-white prints or positive or negative film transparencies which includes all 6 spectral bands. Each band measures 3.8 × 6.4 cm (1½ × 2½ in) of an area 700 × 1,636 km acquired during 2 min of sensor operation.

Computer compatible tapes (CCT's): 9 track, 1,600 bpi, containing three 2-min scenes.

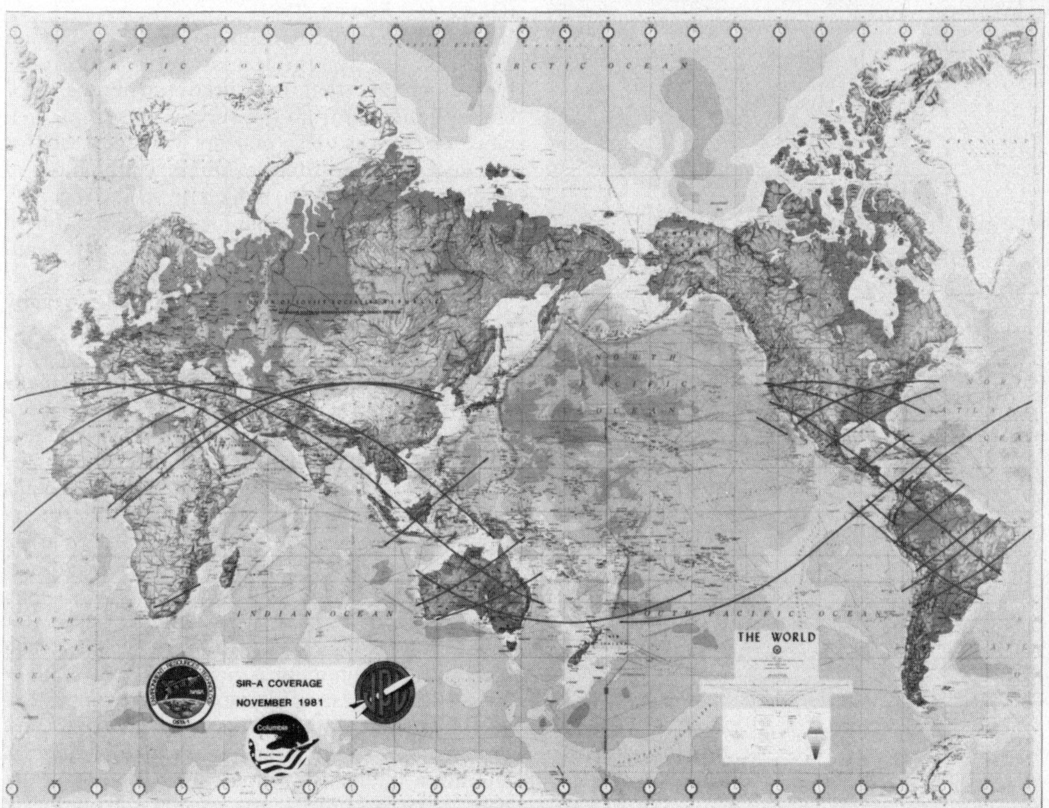

Fig. 31-29. Index map to Shuttle Imaging Radar (SIR-A) image acquisition showing the 26 data takes acquired during the 12–14 November 1981 flight of the Space Shuttle *Columbia*'s OSTA-1 mission. (Index map courtesy of J. P. Ford, Jet Propulsion Laboratory.)

making measurements in the field or from aircraft and satellite platforms are discussed.

Laboratory studies of the spectral reflectance of minerals and rocks have demonstrated that some important minerals can be identified by analyzing the wavelength positions of electronic and vibrational absorption bands in the 0.4 to 2.5 μm region (Hunt, 1977 and 1979; Hunt and Ashley, 1979). This capability was extended to the field by the invention of digitally recording field-portable spectroradiometers (Goetz et al., 1975; Goetz and Rowan, 1981), although some important limitations are imposed by the atmosphere, solar flux levels, and weathering of the mineral constituents (Goetz and Rowan, 1981).

Identification of the most diagnostic absorption features led to the construction of spectroradiometers that measured the intensity of energy reflected from the surface in several specific bandpasses. This concept was the basis for the design of the Shuttle Multispectral Infrared Radiometer (SMIRR), which is discussed in a later section of this chapter, and a hand-held ratioing radiometer (HHRR) (Goetz and Rowan, 1981). Because the filters in the HHRR are interchangeable, this instrument is applicable to many geologic, as well as other, remote sensing problems.

High-spectral resolution measurements made using an airborne spectroradiometer have been used to map mineralogical differences remotely, because contrasts in the spectra are due to the presence of different absorption features (Collins et al., 1981). Figure 31-34 shows two airborne reflectance spectra recorded for altered rocks in the Marysvale, Utah, mining district. Kaolinite is characterized by an absorption feature that consists of a doublet having the minimum near 2.20 μm (Hunt et al., 1973a). Alunite also has a doublet, but the minimum is near 2.17 μm (Hunt et al., 1971b). These mineralogical differences are useful in studying hydrothermally altered zones, because they are commonly indicative of different alteration intensities. Other distinctions that appear to be possible at the spectral resolution of this type of system (8 nm) include montmorillonite, muscovite, jarosite, and carbonate minerals. Measurements of high-resolution airborne spectra have also proven to be very useful for detecting spectral differences in vegetation that is stressed by anomalously high metal contents (see also the section of this chapter on Geobotany).

A digitally recording portable field emittance spectrometer has been devised by the Jet Propulsion Laboratory, California Institute of Technology of Pasadena, California, and is being tested and calibrated (A.F.H. Goetz, oral communication, 1982).

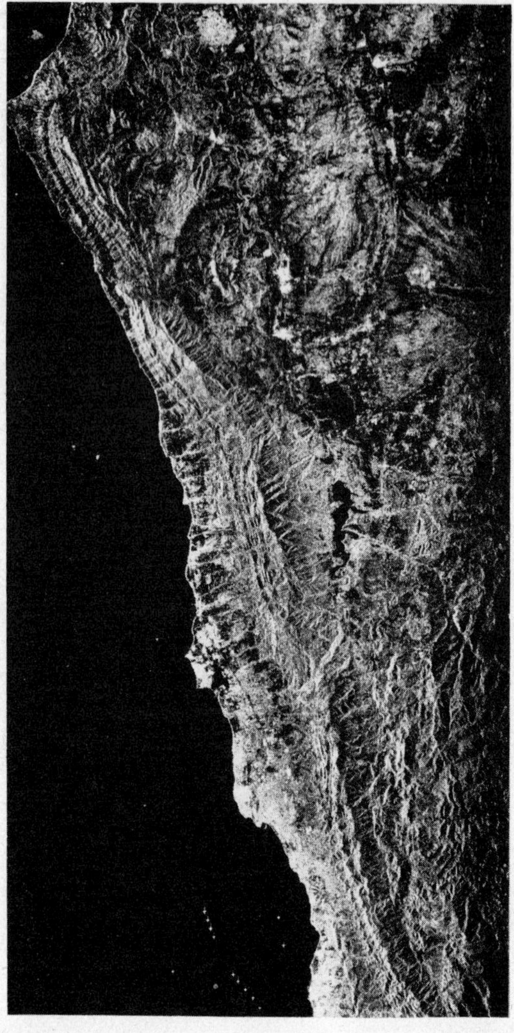

Fig. 31-30. This Shuttle Imaging Radar (SIR-A) image was acquired on 13 November 1981, at approximately 1448 hr local time. It shows the California coast from Point Conception across Santa Barbara to Ventura. The optically correlated L-band (23.5-cm wavelength) SIR-A image, centered on 34°30′ north latitude, 120°00′ west longitude, was acquired from an altitude of 243 km on data take 24C. Stratification of the folded sedimentary rock layers in the east-west trending Santa Ynez Range is prominently displayed on the image. The north margin of the range is abruptly truncated by the Santa Ynez Fault. The two aligned rows of four and five bright spots, respectively, in the Santa Barbara Channel represent oil drilling platforms, as discussed in the Engineering Geology section of this chapter. Other bright spots in the ocean are produced by radar returns from vessels in the area. (Image Courtesy of J. P. Ford, Jet Propulsion Laboratory.) (See also Figure 31-166 for a Seasat SAR image of part of the same coastal area.)

Satellite Systems

Profilometry

The nadir-oriented radar altimeters onboard the Seasat (Townsend, 1980) and GEOS-3 (Leitao et

TABLE 31-16

Characteristics of the Space Shuttle (OSTA-1) mission and the Shuttle Imaging Radar (SIR-A) and data availability

LAUNCH DATE: 12 November 1981, OSTA-1 (30-hr mission)
ORBITAL ELEMENTS:
Orbit: Circular
Altitude: 245 km
Inclination: 38–40°
Coverage: 40°N to 30°S Latitude

SENSOR: Synthetic aperture radar (SAR)

Frequency	Wavelength	Polarization	Spatial Resolution	Swath Width/ Field of View	Antenna Depression Angle
1.278 gigahertz	L-band, 23.5 cm	Horizontal, Horizontal (HH)	40 m at 7-look directions	50 km	43° off horizontal

Data Archive: National Space Science Data Center
World Data Center-A
Goddard Space Flight Center
National Aeronautics and Space Administration
ATTN: Code 601
Greenbelt, Maryland 20771
Tel. No.: 301-344-6695

Standard Film or CCT Formats of SIR-A Image Products Available:
Optically Processed:
125-mm wide film
(black and white)
1:500,000-scale

TABLE 31-17

Handbooks and periodic newsletters related to remote sensing satellites of interest to geologists

Satellite	Handbooks	Newsletters
Landsat	*Landsat Data Users Handbook* (NASA, 1971; and U.S. Geological Survey, 1979)	*Landsat Data Users Notes* U.S. Geological Survey EROS Data Center Sioux Falls, South Dakota 57198
Skylab	*Skylab Earth Resources Data Catalog* (NASA, 1974)	
HCMM	*HCMM Data Users Handbook* (NASA, 1980)	*HCMM Data Users Bulletin* NASA Goddard Space Flight Center HCMM Data Applications ATTN: Code 902 Greenbelt, Maryland 20771
Seasat		*Satellite Data User's Bulletin (Seasat)* NOAA National Environmental Satellite, Data, and Information Service Washington, D.C. 20233
Nimbus 7	*The Nimbus 7 Users' Guide* (NASA, 1978b)	*Nimbus 7 Data Users Bulletin* NASA Goddard Space Flight Center Nimbus Data Applications ATTN: Code 902 Greenbelt, Maryland 20771 *Nimbus 7 CZCS Data Catalog* (Issued monthly by NASA. Same address as listed under Nimbus 7 Data Users Bulletin) (NASA, 1978a)

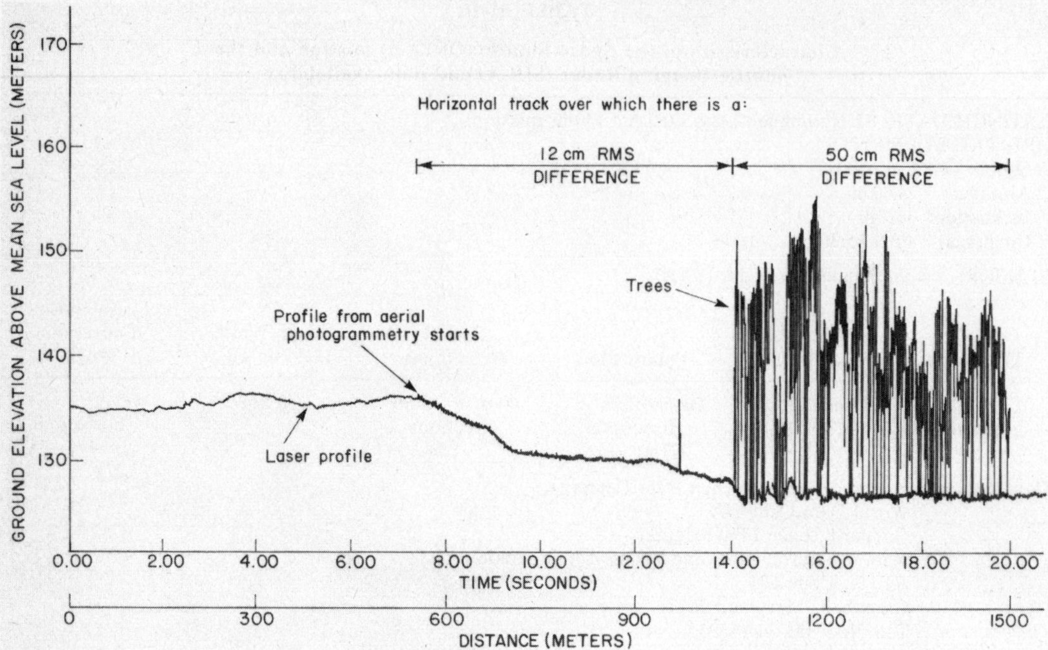

Fig. 31-31. Comparison of airborne laser with photogrammetrically-derived profiles in the Wolf River Basin, Tennessee.

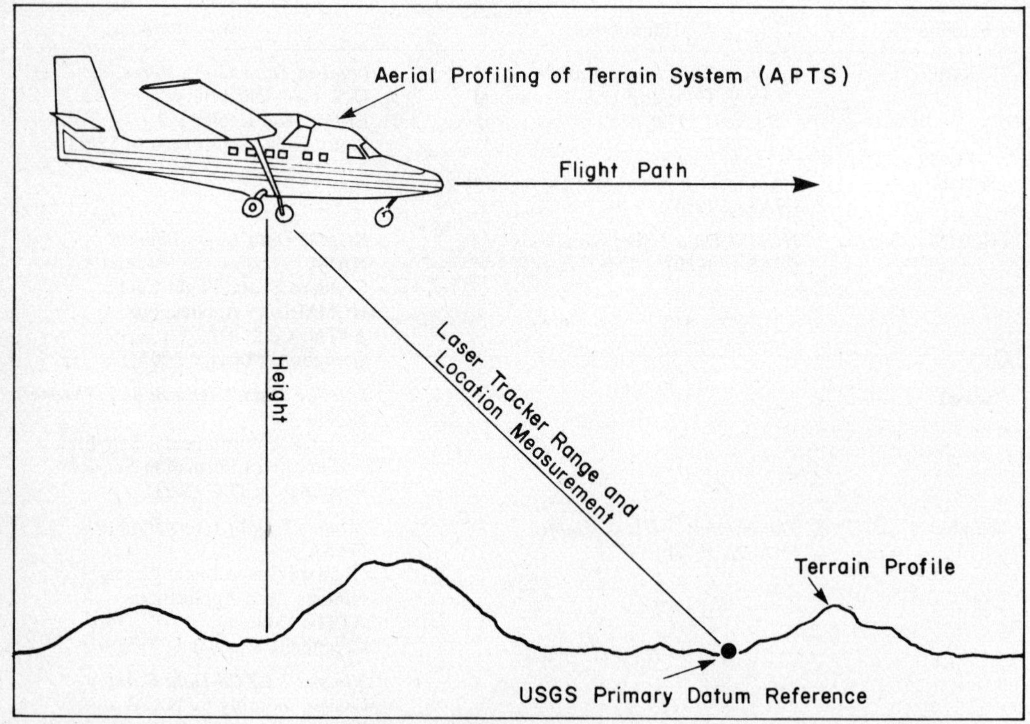

Fig. 31-32. Aerial profiling of terrain system (APTS) operating concept. (Diagram courtesy of Lowell E. Starr, U.S. Geological Survey).

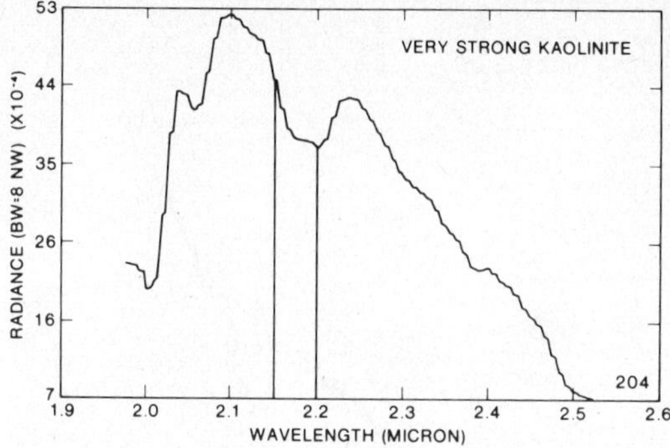

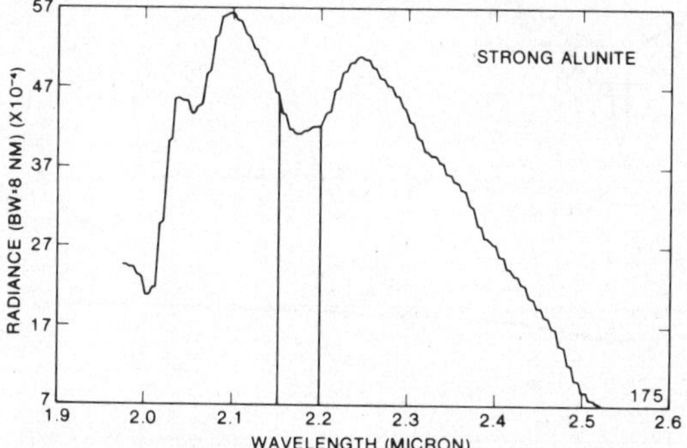

Fig. 31-34. High-resolution spectra obtained by an airborne spectroradiometer for intensely altered rocks in the Marysvale, Utah, mining district. Vertical lines at 2.17 μm and 2.20 μm for references. BW = bandwidth in nm. (Unpublished data from W. Collins and S. Chang, written communication, 1982.)

al., 1975) satellites were specifically designed to study the surface of the Earth's oceans. Before launch, the applicability of either altimeter to profiling over land was not known; the primary question was whether the altimeters' servo tracking systems would respond quickly enough to the rapid slope changes of many types of terrain. Furthermore, it was believed, before launch, that the backscatter from the terrain would not be sufficiently strong or coherent to permit the tracker to sense the return pulse. Analyses have shown that, for geographic areas of low relief, the satellite-altimeter-derived surface elevations have an accuracy of ± 1 m with respect to the reference ellipsoid and are applicable to: terrain profiling (Brooks, 1981), land subsidence studies (Krabill and Brooks, 1979), ice sheet mapping (Brooks, 1982; Brooks et al., 1978, and 1982).

The data rate for each altimeter was 10 per sec. As the velocity for each satellite was approximately 7 km per sec, the resultant terrain elevations are spaced about 700 m apart along each groundtrack. A surface elevation corresponding to each altimeter height measurement may be computed by algebraically subtracting the measured vertical distance between satellite and terrain A from the computed satellite height above terrain based on orbital computation H. This surface height is measured with respect to the orbit reference ellipsoid and, in order to achieve surface heights referenced to mean sea level E, the geoid-ellipsoid separations G must be subtracted as:

$$E = H - A - G.$$

The orbital computations generally provide H accurate to within a few meters. The error in H has a long wavelength and is considered constant for a particular geographic area. To compensate for the orbit-to-orbit differences in H, each satellite altimeter pass is zero-set on level terrain or on a water surface of a known elevation.

While the GEOS-3 radar altimeter onboard tracking system responded reasonably well to terrain undulations, the Seasat radar altimeter was very sluggish. Brooks (1982) has developed a Seasat retracking algorithm which rectifies the terrain elevations in the Seasat altimeter data archive.

Land subsidence studies utilizing the satellite altimeter measurements have been performed by comparing the altimeter-derived profiles with profiles from presubsidence large-scale maps. Such a comparison in the Houston, Texas, area is shown in Figure 31-35.

An important contribution of satellite altimetry has been the determination of surface elevations for remote portions of the Earth such as deserts and ice sheets. The altimeter measurements have resulted in contour maps of unprecedented accuracy in such remote areas. An example is the topographic map of a part of Greenland shown in Figure 31-36. This map was produced by combining all available GEOS-3 and Seasat measurements. The surface "terraces" on the figure were discernable from the altimeter data. The existence of such terraces had previously been observed by glaciologists (Carl Benson, oral communication), but their extent over such long distances was not previously realized.

Radiometry

The Shuttle Multispectral Infrared Radiometer (SMIRR) is a spectroradiometer covering the region 0.5–2.5 μm in 10 channels (Figure 31-37) that acquired data from 100-m-diameter spots along the subspacecraft ground track. SMIRR was carried as part of the science payload aboard the second flight of the Space Shuttle *Columbia* on 12–14 November 1981, (Taranik and Settle, 1981). The objectives of the SMIRR experiment were to: (1) obtain spectroradiometric measurements from orbit of land surfaces in 10 wavelength channels known to be useful for rock and mineral identification; (2) determine the spectral response of known rock types under different climatic conditions worldwide; (3) establish the utility of orbital narrow-band radiometry in the 2.0 to 2.5 μm region for direct identification of

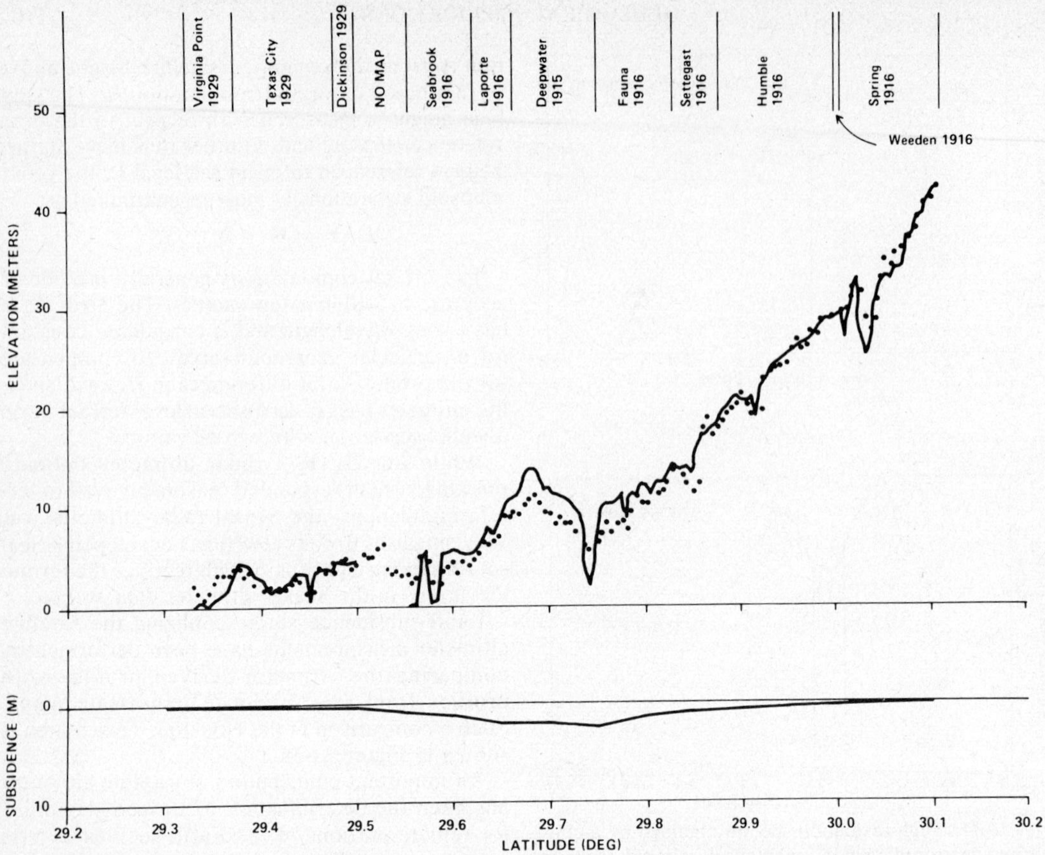

Fig. 31-35. Comparison of altimeter (dots) and pre-subsidence maps (solid line) of the area around Houston, Texas. Altimeter measurements were acquired in November 1976. The bottom profile shows subsidence between 1943 and 1973 as measured by the U.S. Geological Survey.

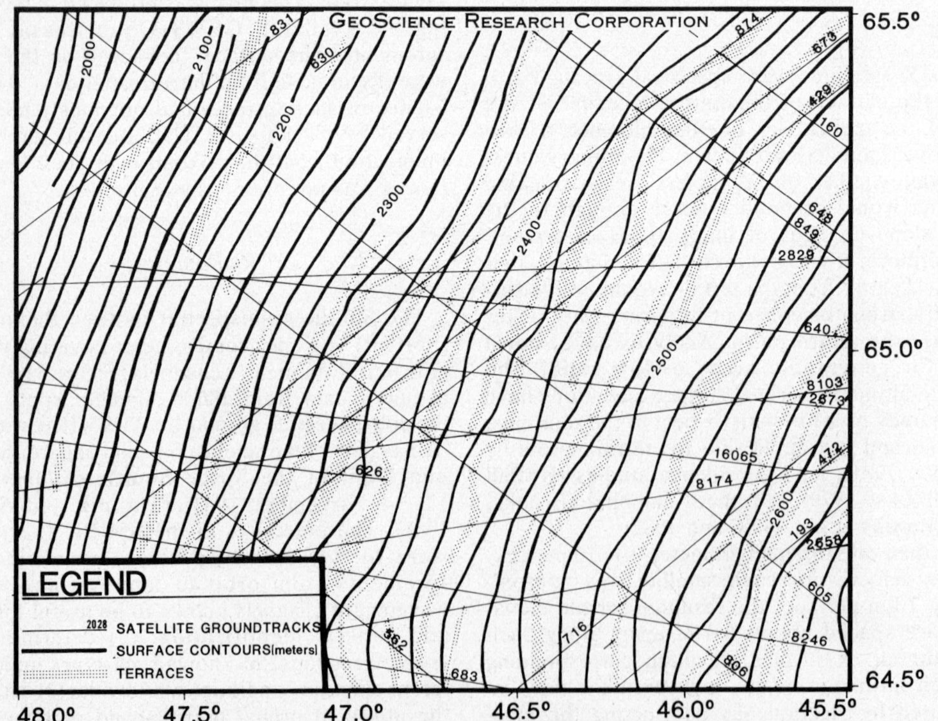

Fig. 31-36. Satellite altimeter-derived surface elevation contours and discerned terraces for a portion of the Greenland ice sheet. Elevations are referenced to the ellipsoid (a = 6,378.137km, f = 1/298.257).

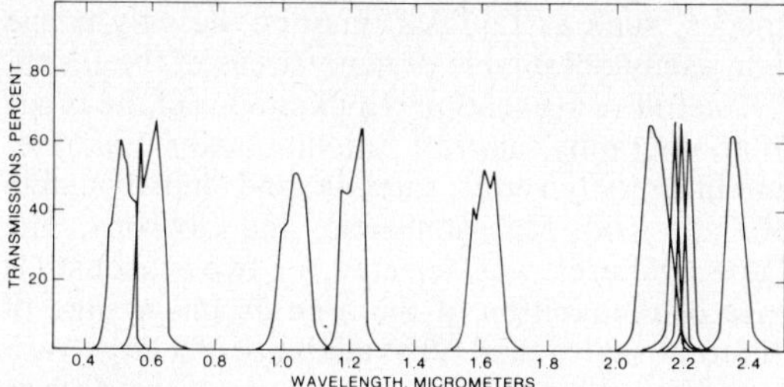

Fig. 31-37. Bandpasses for the 10 SMIRR spectral filters. (Goetz et al., 1982.)

rocks and minerals and establish the requirements for direct identification of rocks and minerals; and (4) establish the requirements for future narrow-band imaging systems (Goetz and Rowan, 1981).

The main elements of SMIRR consist of two bore-sighted 16-mm cameras, coaligned radiometer optics, a spinning filter wheel, two thermoelectrically cooled detectors, and associated timing and signal-conditioning electronic systems. Data were recorded on the Shuttle payload tape recorder. Preflight and inflight calibrations were achieved using calibrated light sources (Goetz et al., 1982).

The shortened mission of *Columbia* yielded 186 minutes of SMIRR data out of an originally planned 305 minutes. Cloud-free coverage was obtained in the eastern United States, Mexico, southern Europe, North Africa, the Middle East, and China. Planned coverage of Australia, South America, and South Africa was not possible because of the unfavorable lighting conditions resulting from the 2-hour launch delay. Approximately 70 minutes of data, equivalent to 400,000 spectra, were obtained under totally cloud-free conditions.

Analysis of the SMIRR data requires normalization using a ground target for which the reflectance can be established in the 10 SMIRR spectra from SMIRR spectra recorded in the laboratory (Figure 31-38) for known mineral and rock standards (Goetz et al., 1982). In the examples described below, the ground target used for normalization is a sample of dune sand (KS-11) collected in western Egypt along orbit 16 (Goetz et al., 1982).

The area traversed in western Egypt consists of moderately rugged upland areas of Cretaceous, Paleocene, and Eocene sedimentary rocks that are surrounded by extensive areas of generally northwest-oriented dunes, gravel deposits, and playas (Egyptian Geological Survey, 1981). Lithologic units of particular interest in this study include carbonate rocks and the Nubia Sandstone, which is commonly kaolinitic. These rocks contain minerals that may be characterized by absorption bands in the 2.0 to 2.5 μm region (Goetz and Rowan, 1981).

The SMIRR spectra shown in Figure 31-39 (A and B) illustrate the spectral contrast observed for exposures of carbonate rock (A), and kaolinitic sandstone (B), respectively, in western Egypt (Rowan et al., 1983). The carbonate rocks are distinctive in the mean SMIRR spectrum (Figure 31-39A), as in the SMIRR laboratory spectrum for the carbonate standard (Figure 31-38), because a CO_3-absorption band causes low reflectance in the 2.35-μm channel. The kaolinitic sandstone is marked by low reflectance in the 2.20-μm channel because of the presence of a hydroxyl overtone absorption band that is centered near 2.20 μm (Figures 31-38 and 31-39A). SMIRR spectra obtained for a playa deposit (not shown in Figure 31-39 (A and B)) indicate the presence of kaolinite, which was substantiated by mineralogical analysis. The presence of montmorillonite was also indicated in some of the Quaternary gravel deposits (Goetz et al., 1982) but has not been verified.

The results of this preliminary analysis of SMIRR data obtained along orbit 16 in Egypt substantiate the initial conclusion (Goetz et al., 1982) concerning the identification of carbonate rocks and kaolinite. In addition, many of the exposed lithologic units appear to be distinguishable in the SMIRR data for considerable distances, because they are composed of minerals that give rise to characteristic absorption features that can be detected through the analysis of such data. However, detailed field sampling and mineralogical analysis in the laboratory are needed for verification of these conclusions.

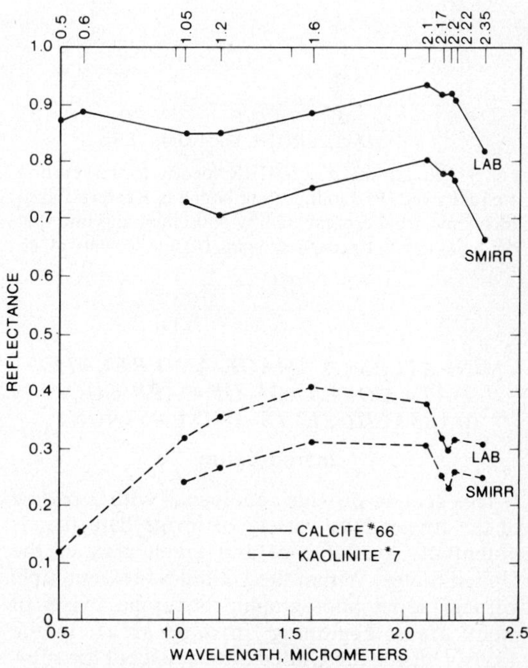

Fig. 31-38. Laboratory and SMIRR ground tests of the same samples. (Goetz et al., 1982.)

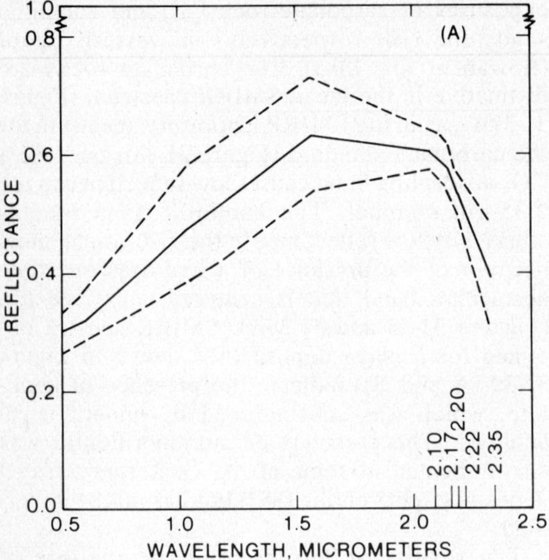

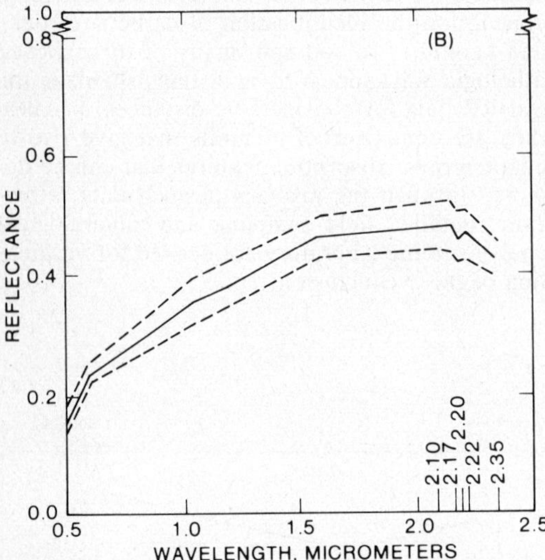

Fig. 31-39. Normalized SMIRR spectra for (A) carbonate rocks and (B) kaolinitic sandstone in western Egypt. Mean spectrum represented by solid line, and one standard deviation by dashed lines (From Rowan et al., 1983).

MINI-ATLAS OF IMAGE AND RELATED DATA COVERAGE OF CAPE COD, MASSACHUSETTS AND ENVIRONS

Introduction

This section provides geologists with a review of the tremendous variety of image data that is potentially available of any given area of the United States. Within the United States, multiple vertical aerial photographic coverage exists of many areas, beginning, in most areas, in the 1930s. Outside the United States, except for a few progressive countries, vertical aerial photographs are difficult, if not impossible to obtain. Satellite images, such as Landsat, may be the only image data easily obtainable of many areas of the Earth.

The following section is a mini-atlas of the types of photography (aircraft, satellite, and submarine) and imagery (aircraft, satellite, and ship) available of Cape Cod, Massachusetts, and environs. The Cape Cod area was selected for two reasons: (1) ease of recognition of the area on image data of different scales, and (2) availability of a large variety of image data because of the use of this area as a test site by NASA and other organizations. Figure 31-40 is an index map to prominent geographic features in the Cape Cod area (from Oldale, 1976).

Photographic Data of Cape Cod and Environs

Figure 31-41 is a low-altitude, large-scale black-and-white, vertical aerial photograph acquired by the U.S. Geological Survey of part of Monomoy Island on 21 February 1974. The narrow neck of Monomoy Island has been breached several times in the last two decades, including a 1500-m-wide breach in the early February 1978 storm (Figure 31-43; see also Color Figure 31-45). Color Figure 31-42 is a high oblique color aerial photograph of the breach taken by Richard W. Kelsey on 12 July 1978, 5 months after the severe storm of 6-7 February 1978. This photograph is typical of oblique aerial photographs taken by geologists from a light aircraft when documenting geologic phenomena (static or dynamic) in the field (See also the Economic Geology section of this chapter.)

Figure 31-43 is a low-altitude, large-scale, black-and-white infrared, vertical aerial photo-

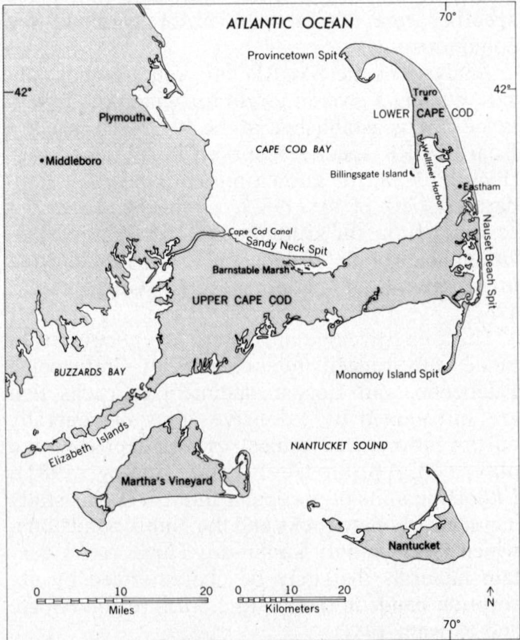

Fig. 31-40. Index map of Cape Cod, Massachusetts and environs (from Oldale, 1976).

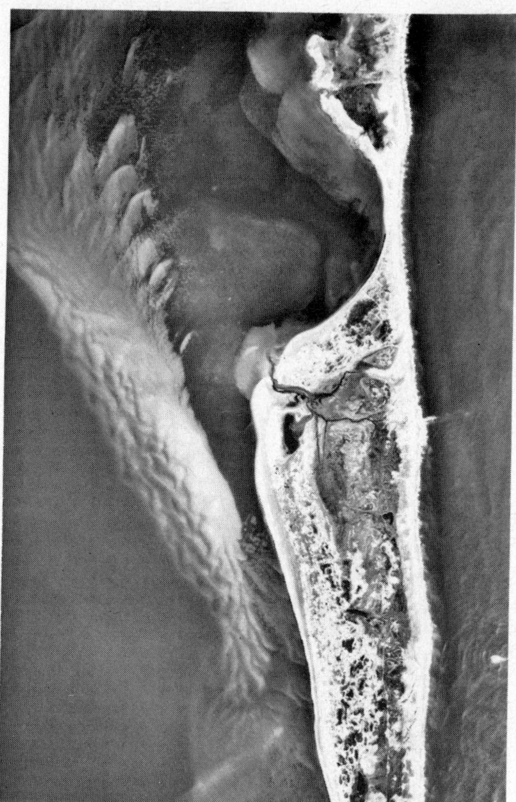

Fig. 31-41. A low altitude, large scale (original contact scale of 1:24,000), panchromatic (black-and-white), vertical aerial photograph of part of Monomoy Island, Cape Cod, Massachusetts. This aerial photograph (No. GS-VDJT, roll 1, frame 55), was acquired by the U.S. Geological Survey on 21 February 1974 from an altitude of 3,660 m with a 152.7-mm (6-in) focal length camera. Kodak Plus-X Aerographic film was used with a Wratten No. 12 haze ("minus-blue") filter. This combination of film and filter limits the spectral sensitivity of conventional black-and-white aerial photography to a wavelength range extending from 0.51 μm to about 0.68 μm (See Figure 31-44). Compare with Figures 31-43, Color Figure 31-45, and Color Figure 31-54, before and after the storm breach occurred at the narrow neck of Monomoy, as predicted by Oldale et al. (1971).

graph acquired by the National Ocean Survey of the breach of Monomoy Island on 12 July 1978. Black-and-white infrared aerial film provides very little water penetration (Figures 31-44, 31-235, and 31-238). It is, therefore, an excellent film to use (with appropriate filters) to document subaerial landforms in shallow water areas. Figure 31-41, the standard panchromatic aerial photograph, shows considerably more detail of the shallow bottom, and there is considerable confusion for the geologist in deciding what is subaerial (dry land) and what is submarine. This ambiguity is not present in the black-and-white infrared aerial photograph. (See also Figure 31-238, a set of aerial photographs of the Florida Keys taken with the Itek nine-lens aerial camera.)

Color Figure 31-45 is a low-altitude, large scale color aerial photograph acquired by the National Ocean Survey of the breach area on Monomoy Island on 19 August 1978. (See also Figure 31-57.) Good discrimination of types of vegetation cover,

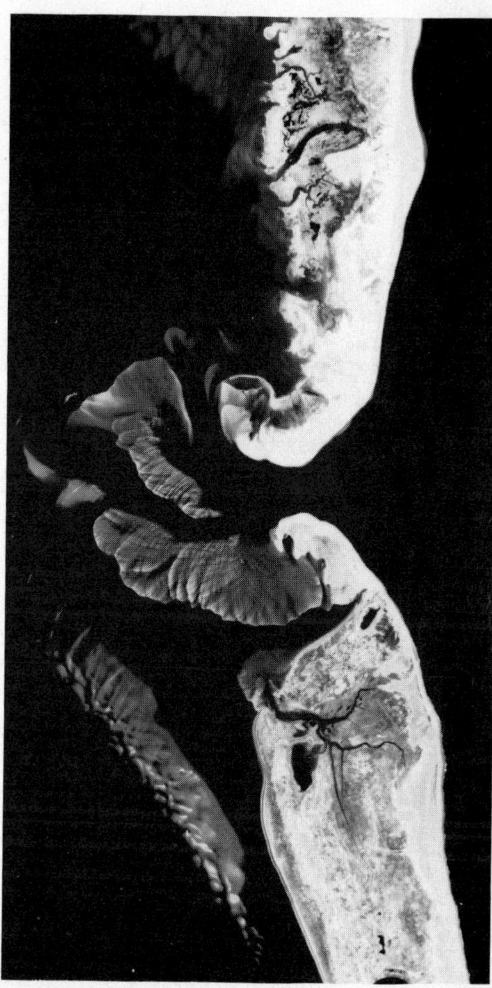

Fig. 31-43. A low altitude, large scale (original contact scale of 1:20,000), black-and-white infrared aerial photograph of the breached area of Monomoy Island, Cape Cod, Massachusetts. This aerial photograph (No. NOS12JUL78ER, Frame 1221) was acquired by the National Ocean Survey on 12 July 1978, 5 months after the breach occurred during a severe coastal storm on 6–7 February 1978 (Williams, 1979a). The photograph was acquired from an altitude of 3,050 m with a 152.7 mm (6 in) focal length camera with Kodak Infrared Aerographic film with a Wratten 89B filter. The combination of film and filter limits the spectral sensitivity to about 0.71 μm to 0.88 μm, severely limiting any water penetration (Figure 31-44). This type of film-filter combination is excellent for delineating the water-land boundary by eliminating confusion in attempting to discriminate between dryland and shallow water submarine features. (See also Figures 31-236 and 31-238). (See Color Figure 31-42 for an oblique aerial photograph of the breached area and environs acquired on the same day.)

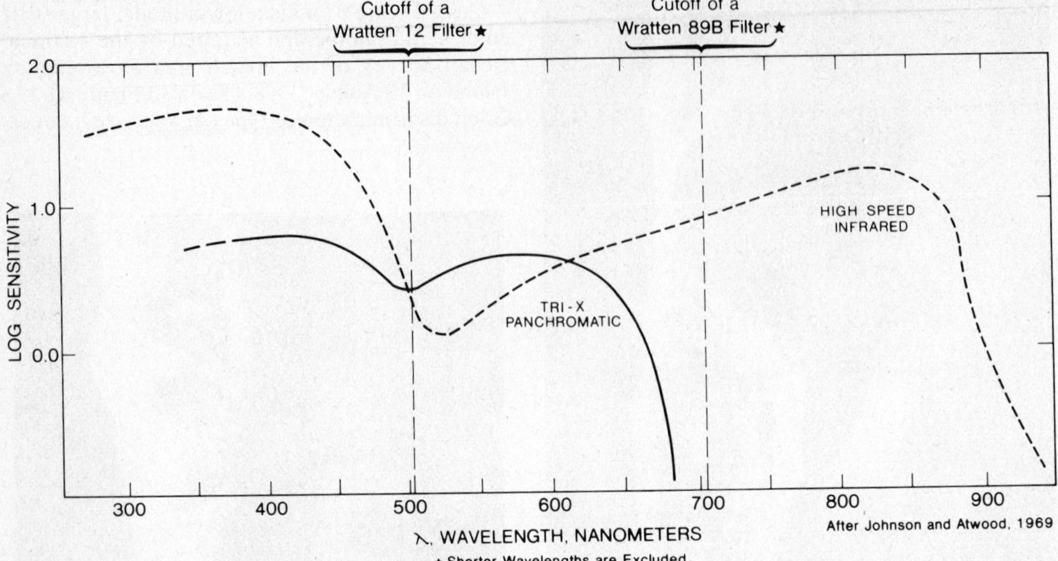

Fig. 31-44. Graph showing the spectral sensitivity, expressed as the reciprocal of exposure in ergs/cm², of panchromatic (black-and-white) and black-and-white infrared aerial films when used with a Wratten 12 filter and Wratten 89B filter, respectively (from Johnson and Atwood, 1969).

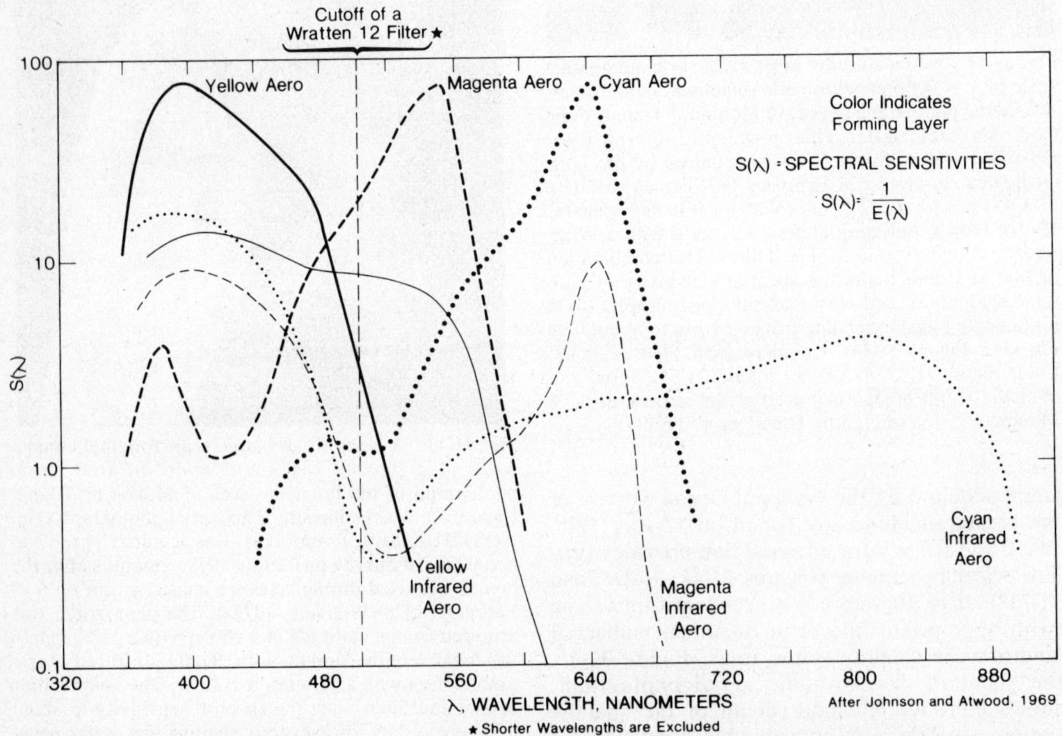

Fig. 31-47. Graph showing the spectral sensitivity of color Ektachrome and color Ektachrome infrared aerial films (from Johnson and Atwood, 1969). Color aerial photography is usually flown unfiltered (no HF filters), thereby achieving (unlike on black-and-white aerial photography which almost always uses a Wratten 12 "minus-blue" filter) a spectral sensitivity equivalent to that of the human eye. To get proper color rendition, color infrared aerial photography must use a Wratten 12 filter. Note that, unlike filtered black-and-white infrared aerial photography (see Figure 31-44), the color infrared aerial photograph is sensitive to the green and red parts of the visible spectrum.

unvegetated (dunes and beach) areas, and submarine features is characteristic of this type of aerial photography.

Color Figure 31-46 is part of a low-altitude, large-scale color-infrared aerial photograph of the northern tip of Monomoy Island and the southern tip of Nauset Spit. It was acquired on 7 May 1969 by Air Force Cambridge Research Laboratories. Excellent discrimination of vegetation types, particularly in wetlands areas is possible with this type of aerial photography. Although color infrared aerial photography is sensitive to the near-infrared, it is also sensitive to the visible spectrum, to about 0.51 μm, because the film is used with a Kodak Wratten 12 filter, the same filter (called a haze-penetrating or "minus-blue" filter) as is used with panchromatic aerial photography (Figure 31-47). This is the reason why submarine features are visible on this "infrared" aerial photograph.

Figure 31-48 is a part of the 1:25,000-scale orthophotoquad (map) of Monomoy Point, Massachusetts 7.5-minute quadrangle. The orthophotograph was prepared from a medium-altitude (12,200 m), small-scale (1:80,000), quad-centered black-and-white aerial photograph (0.51–0.69 μm) acquired with a Wratten 12 "minus-blue" filter on 1 April 1977.

Color Figures 31-49 and 31-50 are high-altitude, small-scale color and color infrared aerial photo-

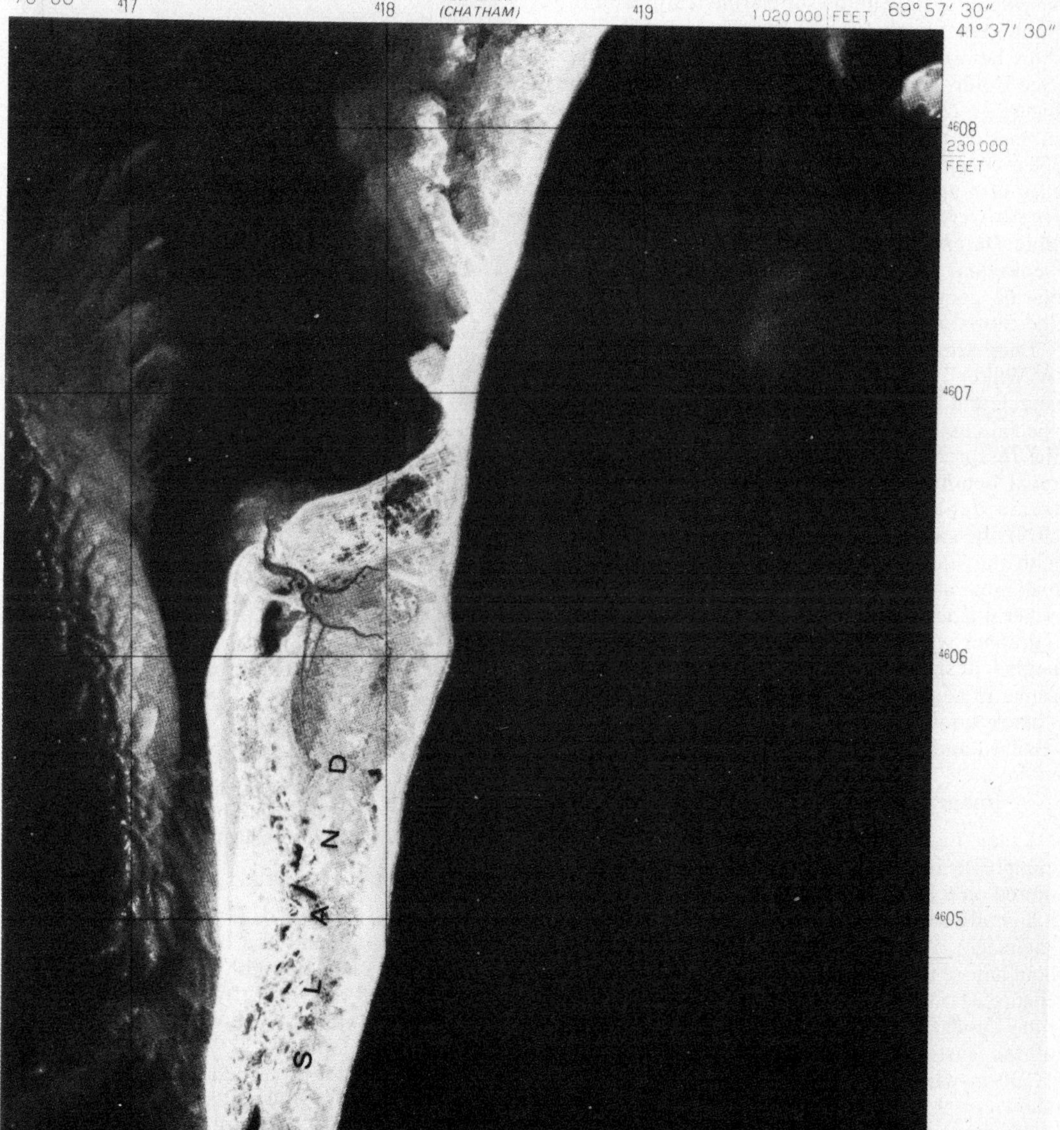

Fig. 31-48. Part of the northeastern section of the 1:25,000-scale, orthophotoquad (map) of the Monomoy Point, Massachusetts, 7.5-minute quadrangle (U.S. Geological Survey, 1977b). The orthophotograph was prepared from a medium-altitude, small-scale, quad-centered, black-and-white aerial photograph acquired on 1 April 1977. The photoimagery was rectified by optical scanning techniques.

graphs, respectively, of the southeastern part of Cape Cod, including part of Monomoy Island and Nauset Spit. Note how much better the thin cirrus clouds are "penetrated" with the color infrared film. These two aerial photographs were obtained simultaneously by NASA's RB-57F aircraft on 13 September 1969.

Color Figure 31-51 is a high-resolution color satellite photograph of the southeastern part of Cape Cod which was acquired with the S-190B camera on the Skylab 3 mission. Figure 31-52 was taken with the S-190A multispectral camera and is a panchromatic (black-and-white) photograph, one of the six film-filter combinations available from the camera. Both satellite photographs were acquired on 12 September 1973. Note the extraordinary detail visible in the submarine features of the Nantucket Shoals area, a famous graveyard for ships before completion of the Cape Cod Canal. (See Color Figure 31-54, Figure 31-59, and Color Figure 31-228.) Color Figure 31-53 is a group of six submarine photographs of the seafloor in 200 to 300 m of water at the edge of the Continental Shelf east of Cape Cod, Massachusetts. The photographs were acquired from the research submersible *Diaphus* by John M. Aaron of the U.S. Geological Survey, during studies of the submarine geomorphology of the Continental Shelf and Slope offshore from Cape Cod.

Deep sea photography has also been perfected by Walter H. John of the U.S. Naval Ocean Research and Development Activity (NORDA) with special cameras. A stereoscopic camera onboard the research submersible *Alvin* is used to document benthic fauna and geologic features at extreme depths (National Geographic Society, 1979). In studies of geologic features associated with the East Pacific Rise, including hydrothermal, mineral-encrusted chimneys which spew forth mineral-laden water, Emory Kristof, a staff photographer with the National Geographic Society, helped design a special motion picture camera frame to acquire 16-mm color photographs of the "black smokers" and other dynamic phenomena (Ballard and Grassle, 1979).

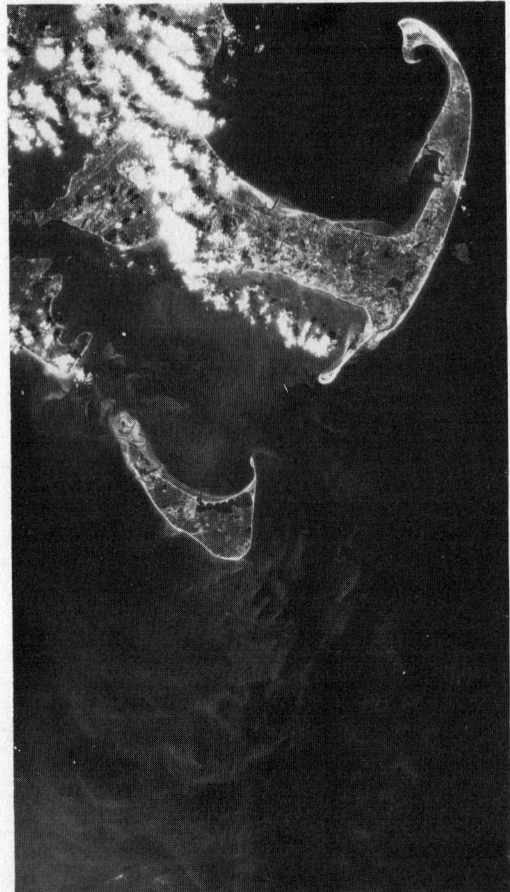

Fig. 31-52. A camera No. 6 panchromatic (black-and-white) satellite photograph of the southeastern part of Cape Cod. The photograph, taken with a camera having a focal length of 152.7 mm (6 inches), was acquired during the Skylab 3 mission with the S-190A multispectral camera system (six film-filter combinations) on 12 September 1973 at 1411 hr EST (EROS Data Center Accession No. G30A04213300). The film is Kodak SO-022 (panatomic black-and-white), exposed through a BB filter, which limits the spectral band to the wavelength range from 0.5 μm to 0.6 μm. The estimated ground resolution is from 40 m to 46 m with this film-filter combination. (See Table 31-9 for the film-filter combinations used on the other five cameras). Compare with Color Figure 31-51, a satellite photograph from the S-190B Earth Terrain Camera acquired at the same time.

Imagery of Cape Cod and Environs

Color Figure 31-54 is a standard MSS color composite image of Cape Cod and environs acquired on 6 July 1976 (compare with Figure 31-228, a digitally enhanced version of the identical Landsat image). This Landsat image shows Monomoy Island before the breach occurred in February 1978. Figure 31-55 is a part (subscene B) of the wintertime Landsat 3 return beam vidicon (RBV) image of the eastern part of Cape Cod in which the 1,500-m-wide breach of Monomoy is clearly shown on 9 March 1978 (Williams, 1979a). Figure 31-56 is part of subscene D, the next subscene south of Figure 31-55. Note the improvement in spatial resolution (30-m pixels) as compared with the Landsat 1 MSS image having 79-m pixels in Color Figures 31-54 and 31-228. Figure 31-57 is part of an enlargement of a Landsat 3 RBV image taken on the same day (19 August 1978) as the National Ocean Survey vertical color aerial photograph shown in Color Figure 31-45. The four Landsat 3 RBV subscenes (A, B, C, and D) of image 30167-14444 have been mosaicked and are being used for a 1:100,000-scale, experimental image map of the New Bedford quadrangle and as the image base for a 1:100,000-scale experimental "Geologic Map of Cape Cod" by the U.S. Geological Survey (Williams, 1979b).

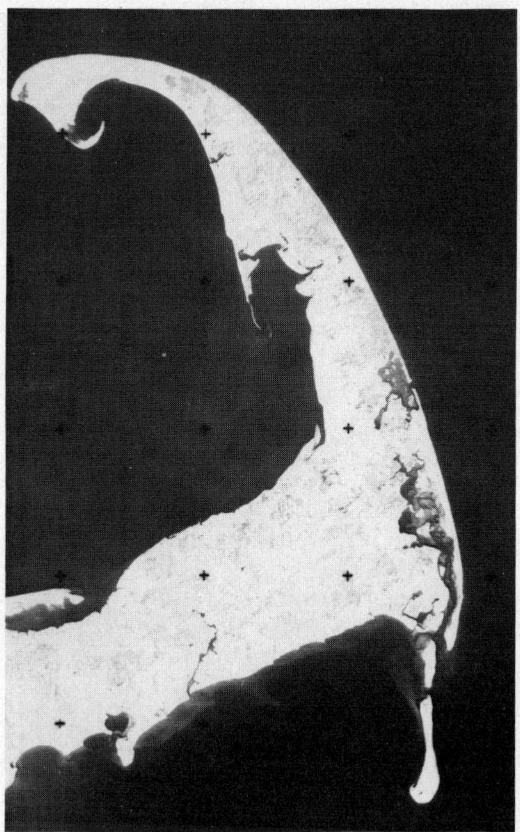

Fig. 31-55. Part of a wintertime Landsat 3 RBV image (ID No. 30004-14435-B) acquired on 9 March 1978 which clearly shows the 1,500-m-wide breach which occurred during a severe winter storm on 6–7 February 1978 (Williams, 1979a). The Landsat 3 RBV, with a pixel size of 30 m (compared to the 79-m pixel of the Landsat MSS image), records energy in a broad spectral band (0.505 to 0.750 μm) similar to panchromatic aerial photographs (Williams, 1979b).

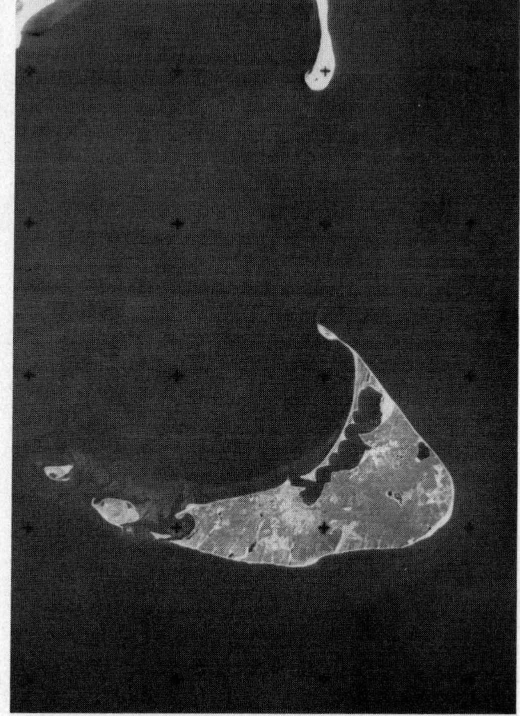

Fig. 31-56. Part of a Landsat 3 RBV image (30004-14435-D) of Nantucket Island, Cape Cod, Massachusetts, acquired on 9 March 1978. Note the improvement in spatial resolution of features on Nantucket Island (30-m pixel) when compared with the Landsat MSS images (79-m pixel) in Color Figures 31-54 and 31-228.

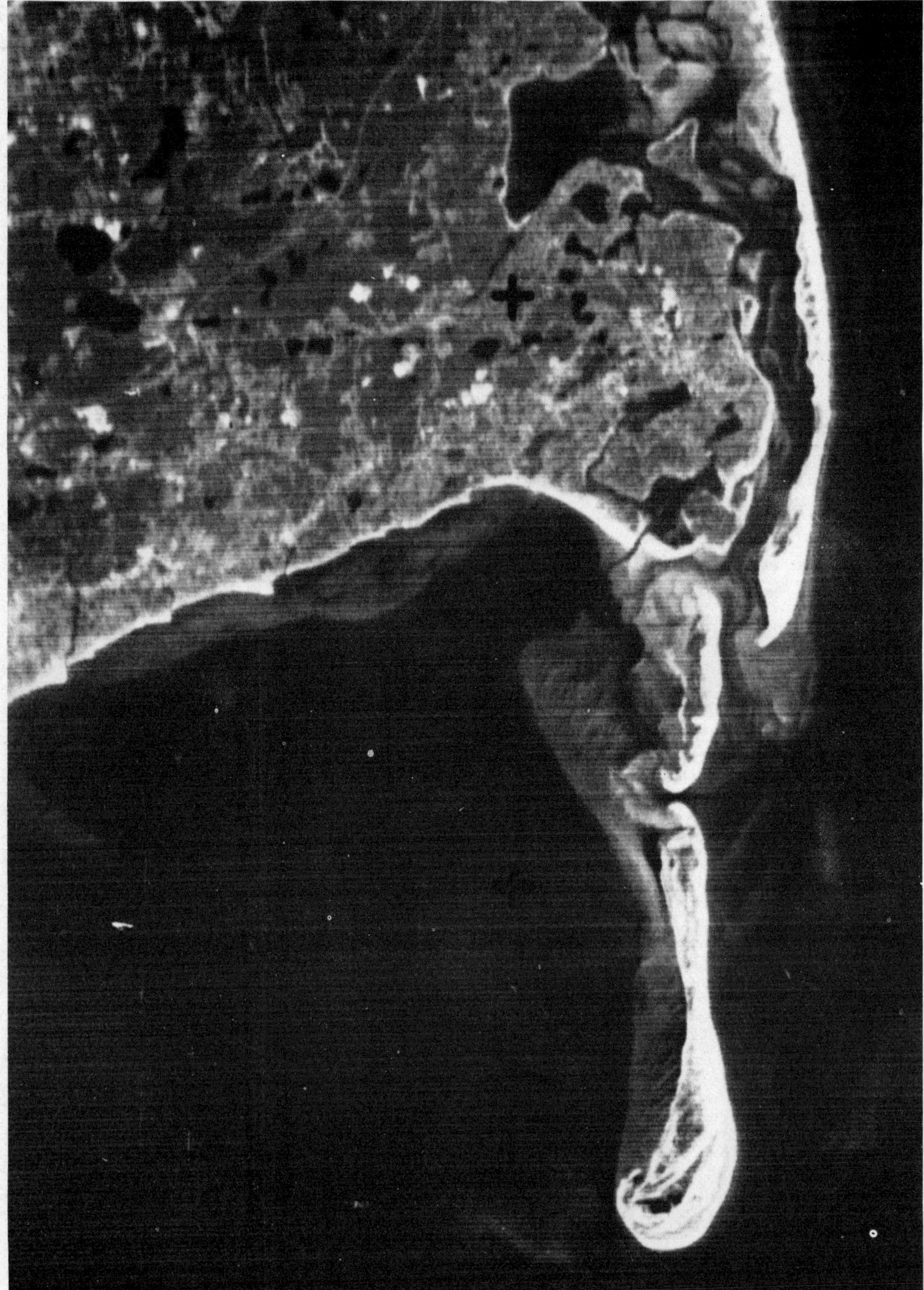

Fig. 31-57. Part of an enlargement (to 1:125,000) of a Landsat 3 RBV image (30167-14444-D) of southeastern part of Cape Cod, Massachusetts, including Monomoy Island and Nauset Spit. Compare with Color Figure 31-45, a vertical color aerial photograph (1:20,000) acquired by the National Ocean Survey on the identical date (19 August 1978). The four Landsat 3 RBV subscenes (A, B, C, and D) of this image have been mosaicked by the U.S. Geological Survey for the 1:100,000-scale experimental image map base of the New Bedford Quadrangle and for use as the base for an experimental "Geologic Map of Cape Cod" (Williams, 1979b).

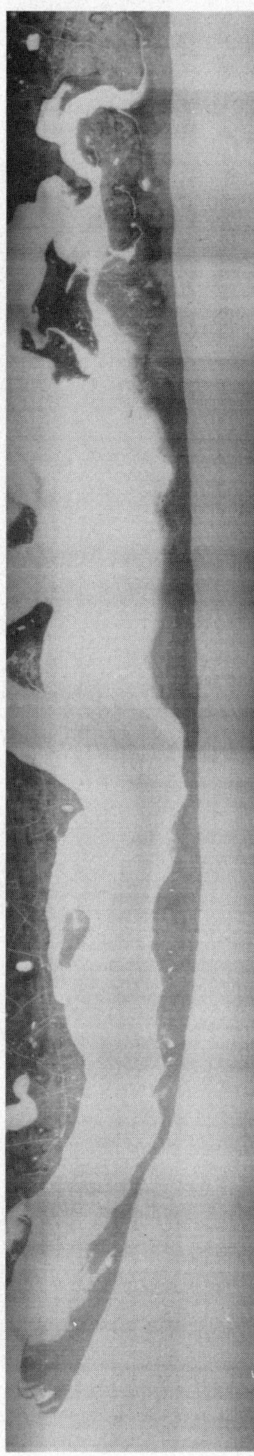

Fig. 31-67. Aerial thermograph (thermal infrared image) of Nauset Spit, Cape Cod and environs. It was acquired on 2 May 1969 at 0005 hr EDT from an altitude of 3,660 m by the U.S. Air Force Cambridge Research Laboratories. The InSb detector was used unfiltered, recording energy emitted by the Earth's surface from about 3 to 5.5 μm. The original scale of the aerial thermograph was about 1:80,000.

Figure 31-58 is a Seasat synthetic aperture radar (SAR) image of part of Nantucket Island and environs, Cape Cod, Massachusetts, which was acquired on 27 August 1978. The Seasat SAR system operated at a wavelength of 23.5 cm and did not penetrate water (Reed, 1981); however, the presence of submarine features is manifested at the surface by a physical process not entirely understood. Figure 31-59 is part of a National Ocean Survey hydrographic chart (Georges Bank and Nantucket Shoals) showing the Nantucket Shoals area (A) and a comparison between part of the 27 August 1978 Seasat SAR image (B) shown in Figure 31-58 and a Landsat 1 MSS band 5 image (1724–14472) (C) acquired on 17 July 1974. Figure 31-60 is a side-looking airborne radar (SLAR) image mosaic of the State of Massachusetts and environs, including Cape Cod.

Figures 31-61 and 31-62 are day-visible and day-infrared (thermal) images, respectively, from the heat capacity mapping radiometer (HCMR) onboard the heat capacity mapping mission (HCMM) spacecraft of parts of New England, New York, and New Jersey, including coastal and offshore areas on 21 April 1979.

Figure 31-63 is a set of six images of the northeastern United States and coastal areas acquired on 21 April 1979 by the 6-band (channel) coastal zone color scanner (CZCS).

Figure 31-64 is an index map to part of Cape Cod for the two side-scan sonographs (sonar images), Figures 31-65 (SS-1) and 31-66 (SS-4), which show morphologic details of the sea bottom between Martha's Vineyard and the Elizabeth Islands (SS-4) and southwest of Martha's Vineyard offshore from Cape Cod (SS-1) (O'Hara and Oldale, 1980). (See also the Marine Geology section of this chapter.) Side-scan sonar equipment produces images of the seafloor from a ship in a manner analogous to the acquisition of side-looking radar (SLAR) images of the Earth's surface from either an aircraft or a satellite.

Remote sensing technology will increasingly be used to image the 75 percent of our planet which is covered by ocean, either by side-scan sonar images of the deeper parts or for greater detail in shallower areas, or by aerial photographs or Landsat images of the shallowest areas. Another instrument, successfully used to image the so-called mineral-spewing "black smokers" and unique marine benthic fauna along the East Pacific Rise and the Galápagos Rift, is a color video system, based on charge-coupled devices (CCD's), developed by RCA and Benthos, Inc. The images generated by this imaging system are viewed and tape recorded within the research submersible *Alvin* (National Geographic Society, 1979; Ballard and Grassle, 1979).

Figure 31-67 is an aerial thermograph (thermal infrared image) acquired by U.S. Air Force Cambridge Research Laboratories on 2 May 1969, showing the southeastern part of Cape Cod, including most of Nauset Spit.

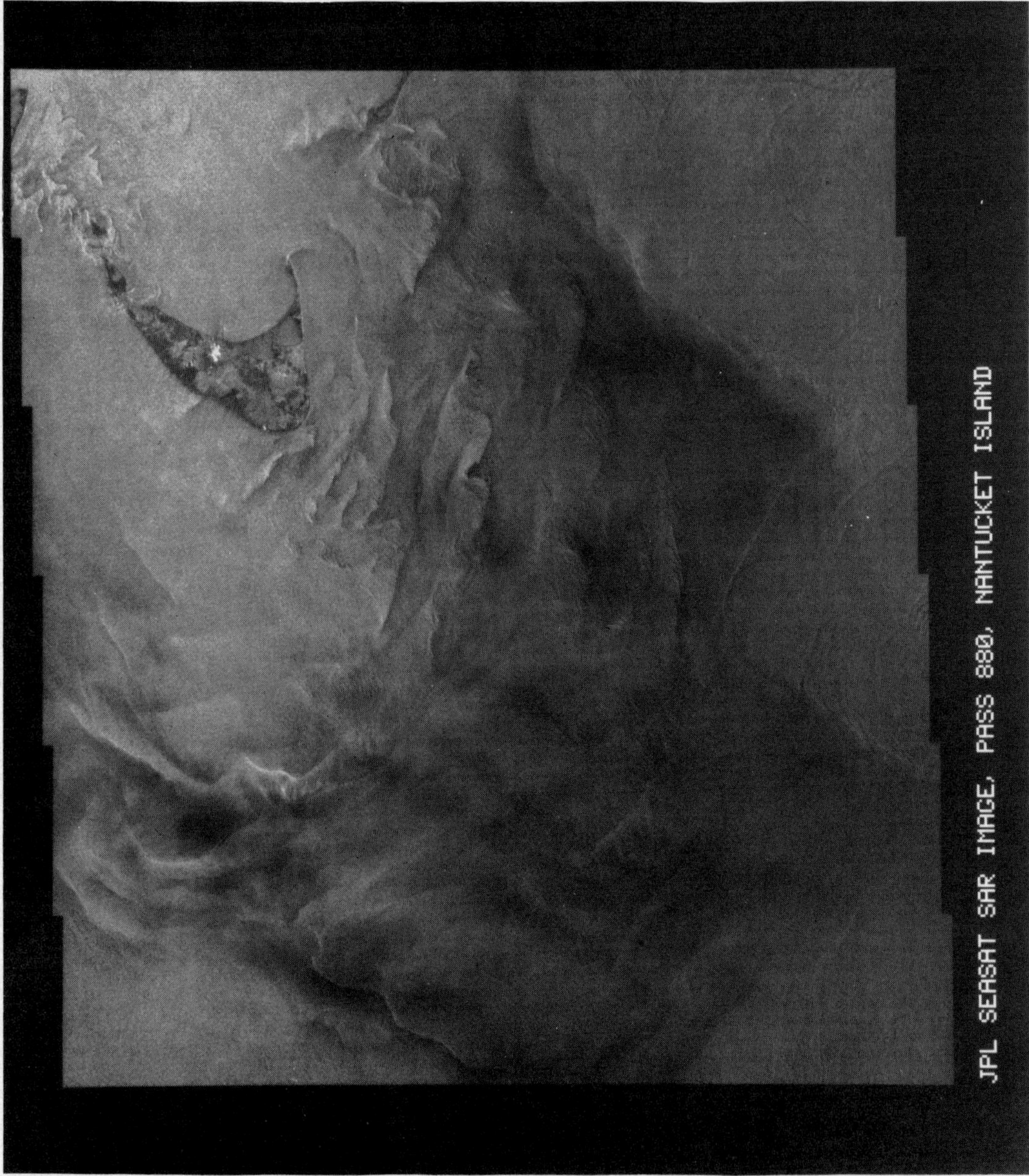

Fig. 31-58. Seasat synthetic aperture radar (SAR) image of Nantucket Island, part of Monomoy Island, and the Nantucket Shoals area on 27 August 1978 (Seasat Revolution 880). The Seasat SAR system operated at a wavelength of 23.5 cm and did not penetrate water. The surface winds at the time of image acquisition were 10 knots from the west-southwest with a sea running about 1.7 m (Reed, 1981). (Seasat SAR image courtesy of Charles L. Reed, Defense Mapping Agency, from a Jet Propulsion Laboratory processed image.)

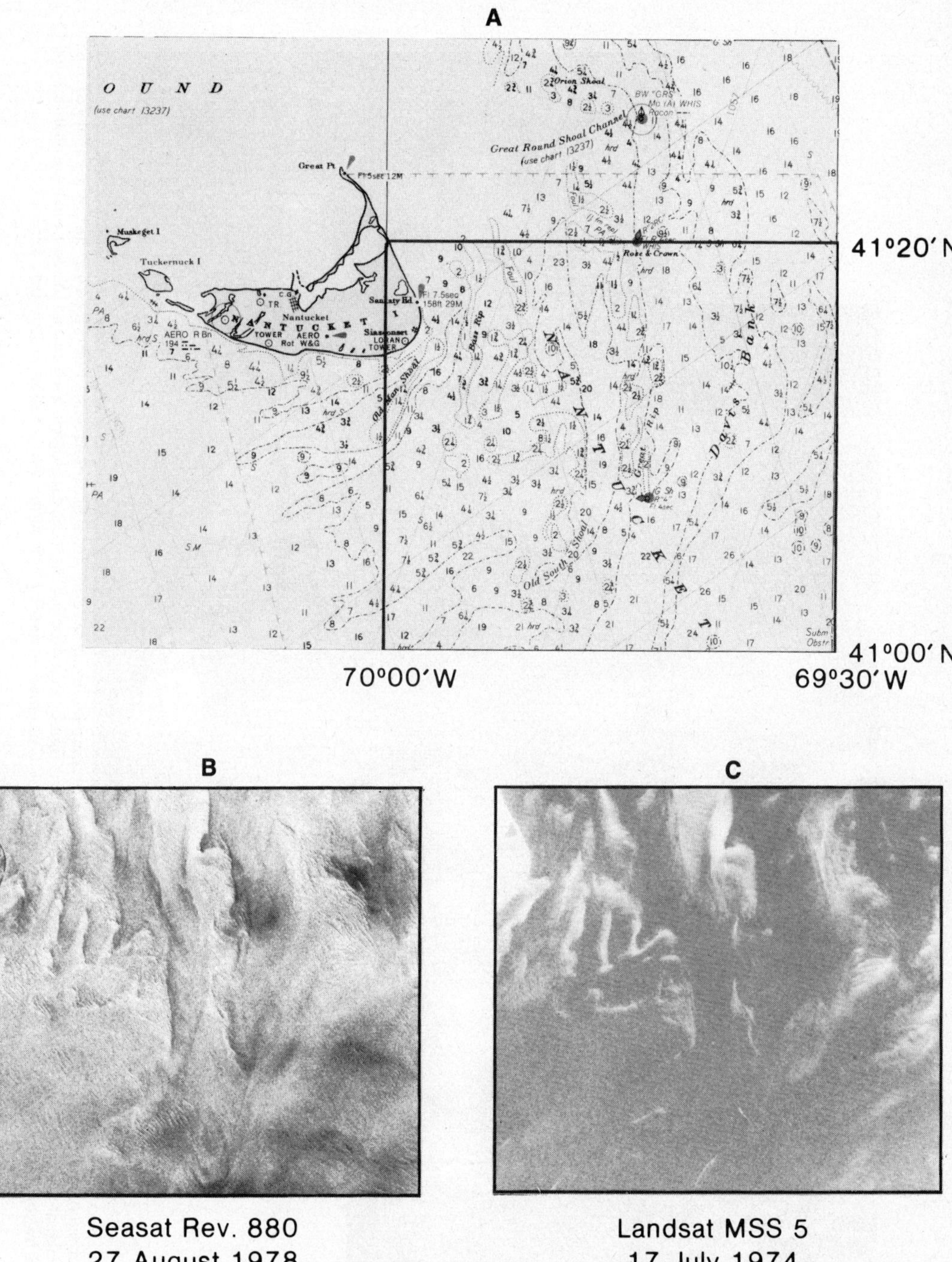

Fig. 31-59. Part of a 1:400,000-scale National Ocean Survey hydrographic chart (NOS 13200; 24th edition, 28 November 1981) of Georges Bank and Nantucket Shoals area (A) and a comparison between the Seasat SAR image (B), which was acquired on 27 August 1978 (Seasat Revolution 880) (See also Figure 31-58), and a Landsat 1 MSS band 5 image (1724-14472) (C) acquired on 17 July 1974. Each image covers an area 43 km by 38 km. The NOS chart, the Seasat SAR image, and the Landsat image were provided by Robert A. Schuchman and David P. Lyzenga, Environmental Research Institute of Michigan, from research carried out by Charles Reed (1981).

Fig. 31-60. An uncontrolled mosaic of Massachusetts and environs compiled from several strips of side-looking (west-looking) airborne radar (SLAR) images. The original mosaic scale of 1:500,000 has been reduced in this figure. Prepared for the U.S. Geological Survey and the National Aeronautics and Space Administration with the Cooperation of the U.S. Army by the Grumman Aircraft Engineering Corporation in October, 1968.

Fig. 31-61. A day-visible heat capacity mapping radiometer (HCMR) image of parts of New England, New York, and New Jersey acquired on 21 April 1979 from the heat capacity mapping mission (HCMM) spacecraft. Compare with Figure 31-62, a day-infrared (thermal) image of the same scene. The day-visible image records energy in the 0.55- to 1.1-μm spectral region with an instantaneous field of view of 500 m.

Fig. 31-62. A day-infrared (thermal) HCMR image from the HCMM spacecraft of part of the New England area and environs, the same area as shown in Figure 31-61 (a day-visible HCMR image) both of which were acquired simultaneously on 21 April 1979. The day- (and night-) infrared (thermal) image records energy in the 10.5 to 12.5 μm spectral region with an instantaneous field of view of 600 m. Note the different spectral responses in the two images of the Adirondack Mountains and offshore marine currents, especially the complex boundary between the warmer Gulf Stream and the colder Labrador Current.

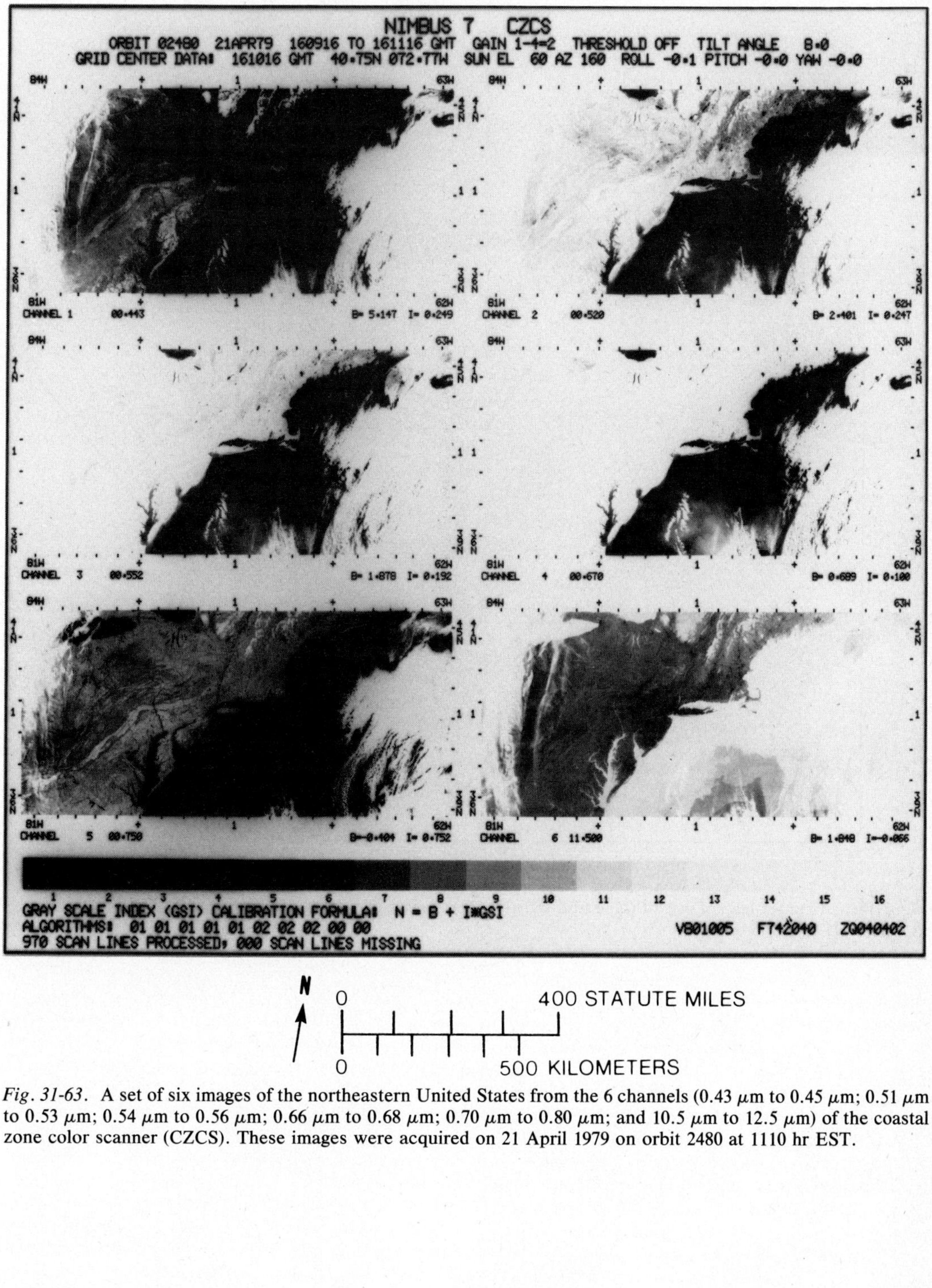

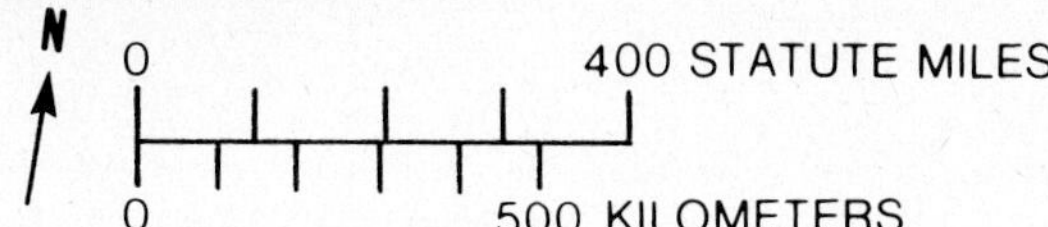

Fig. 31-63. A set of six images of the northeastern United States from the 6 channels (0.43 μm to 0.45 μm; 0.51 μm to 0.53 μm; 0.54 μm to 0.56 μm; 0.66 μm to 0.68 μm; 0.70 μm to 0.80 μm; and 10.5 μm to 12.5 μm) of the coastal zone color scanner (CZCS). These images were acquired on 21 April 1979 on orbit 2480 at 1110 hr EST.

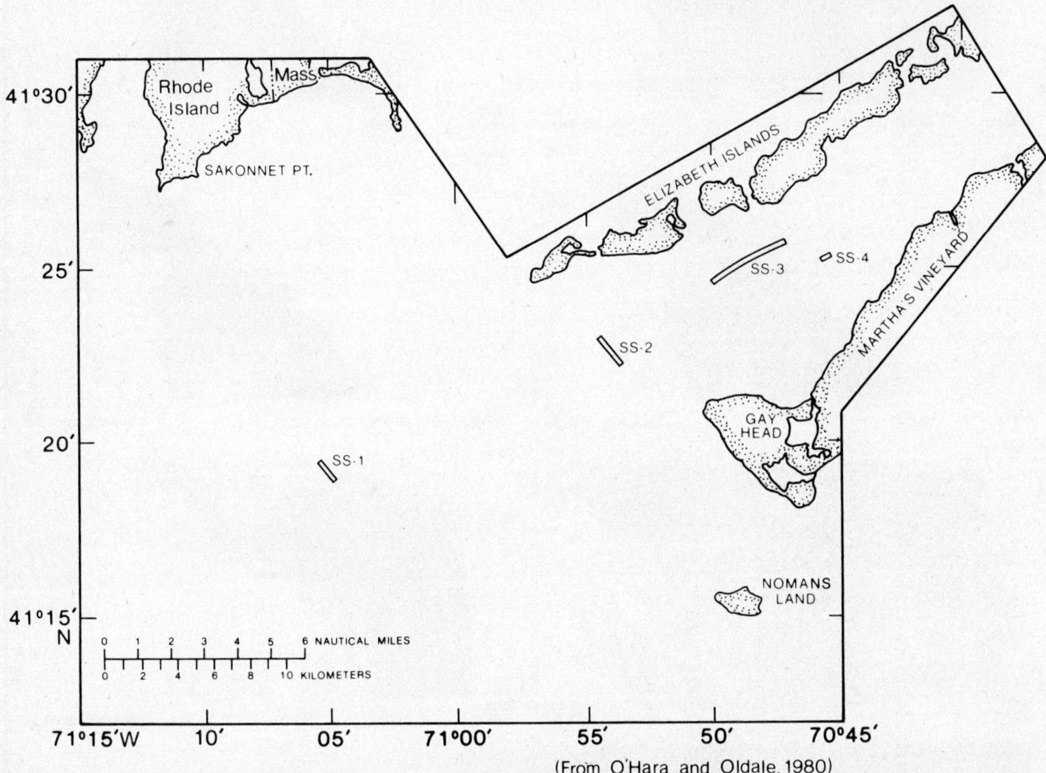

Fig. 31-64. An index map of part of Cape Cod showing locations of the side-scan sonographs (sonar images) shown in Figures 31-65 and 31-66.

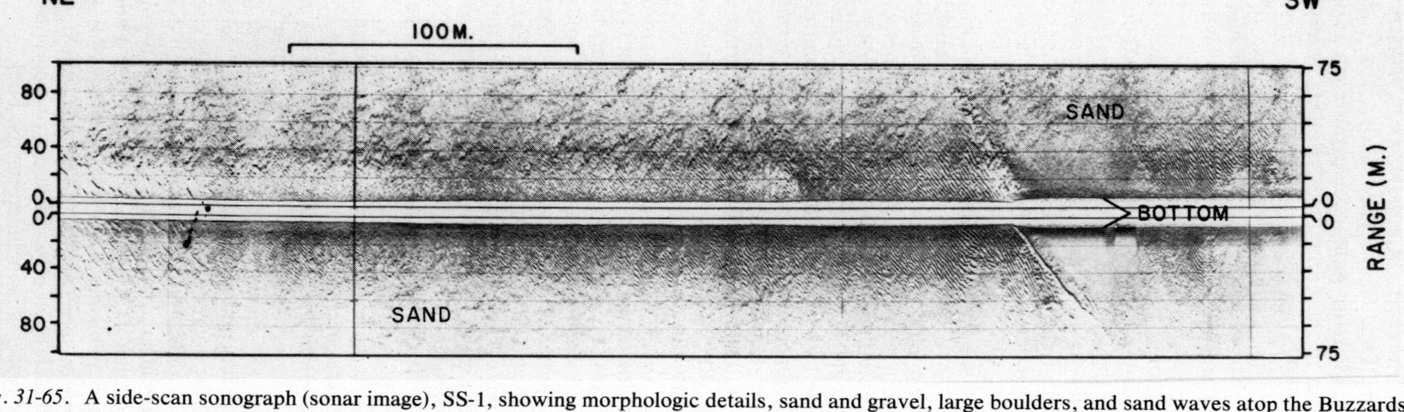

Fig. 31-65. A side-scan sonograph (sonar image), SS-1, showing morphologic details, sand and gravel, large boulders, and sand waves atop the Buzzards Bay morainal ridge in eastern Rhode Island Sound, southwest of Martha's Vineyard, Cape Cod, Massachusetts (from O'Hara and Oldale, 1980). (See Figure 31-64 for location.) (Side-scan sonographs courtesy of D. W. O'Leary, U.S. Geological Survey.)

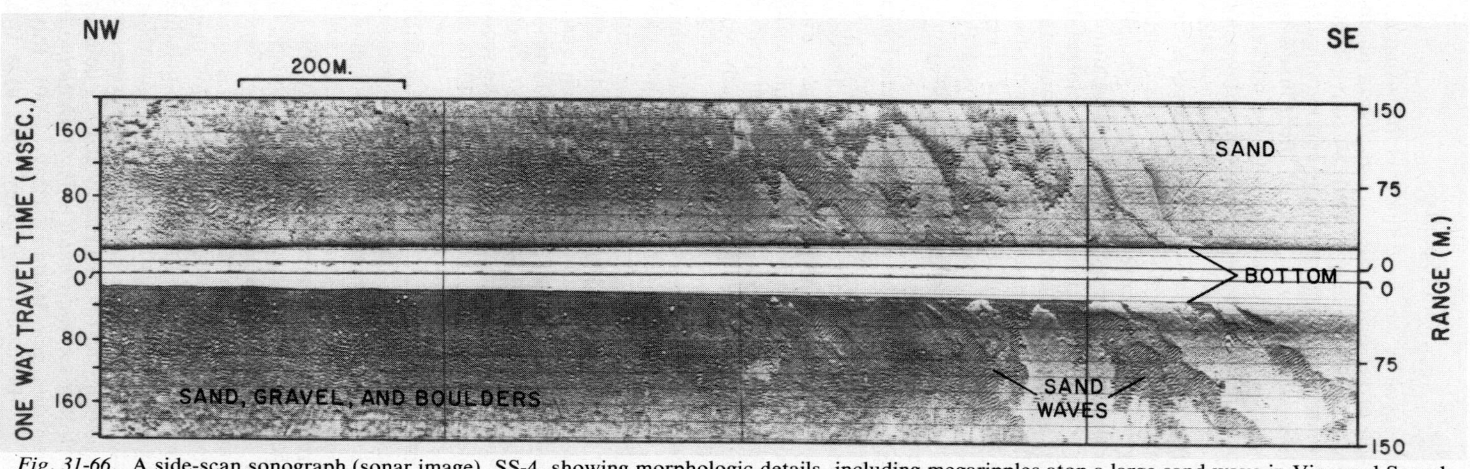

Fig. 31-66. A side-scan sonograph (sonar image), SS-4, showing morphologic details, including megaripples atop a large sand wave in Vineyard Sound, between Martha's Vineyard and the Elizabeth Islands, Cape Cod, Massachusetts (see Figure 31-64 for location).

ANALYSIS TECHNIQUES

PHOTOGEOLOGY AND GEOLOGICAL REMOTE SENSING

Photogeology

Photogeology can be defined as the visual extraction of geological information from a photograph or image (Robinove, 1963) by conventional photo-interpretation instruments and techniques. Traditionally the instrument most used by geologists is the stereoscope, either the pocket or mirror variety (or a more sophisticated version such as the mirror stereoscope with binoculars; the Bausch and Lomb SIS stereoscope is one example). The technique used is direct visual analysis of stereopaired photographs (usually vertical black-and-white aerial) of the area under study. Geologic information is usually plotted directly on the photograph (or on a transparent overlay) and then transferred later to a printed line map or orthophotoquad map by any of several optical devices; e.g., Saltzmann Rectifying Projector, Bausch and Lomb Zoom Transfer Scope, sketchmaster, etc.

Many pioneering photogeologists obtained their photo-interpretation training in military service during the World War II era. Prof. A. J. Eardley, a geologist at the University of Michigan, published *Interpretation of Geologic Maps and Aerial Photographs* (Eardley, 1941a). Later that year he published an expanded revision of his early work called *Aerial Photographs: Their Use and Interpretation* (Eardley, 1941b). At about the same time, another geologist, H. T. U. Smith, published a similar book, *Aerial Photographs and Their Applications* (Smith, 1943). All three of these books laid the groundwork for what was to become an important tool in many types of geological investigations: photogeology. Photogeology became a very popular tool in the search for oil and gas, especially during the late 1940s and 1950s. This occurred because it was viewed as an economical way of mapping possible oil- and gas-bearing geologic structures visible on stereopaired aerial photographs or aerial photographic mosaics (Smith and Renfro, 1955; Lattman, 1959; Pressman, 1960; and Kelly, 1961).

Chapter 4, Photo Interpretation in Geology, by Ben Tator et al. (1960) in the *Manual of Photographic Interpretation* (Colwell, 1960) is a superb basic text in photogeology. It also contains an excellent bibliography with 1309 citations, including many duplicate entries due to subject grouping. Another excellent book, which also serves as a photogeology text book in many geology courses, is Dick Ray's *Aerial Photographs in Geologic Interpretation and Mapping* (Ray, 1960). Ray's book has been reprinted many times as a U.S. Geological Survey Professional Paper. In 1961, Miller and Miller published a fine textbook, *Photogeology,* which is also recommended for the serious student of photogeology. Lattman and Ray's (1965) excellent field handbook, *Aerial Photographs in Field Geology,* one of the field manuals in Holt, Rinehart and Winston's "Geologic Field Techniques Series," is highly recommended for photogeological work in the field. Also in the early 1960s, Eugene Avery, a forester, published his first edition of *Interpretation of Aerial Photographs* (Avery, 1962). This book is also recommended to students of photogeology, because it can be used for self instruction.

In addition to the foregoing, we refer the reader to the many existing handbooks and textbooks, including those just discussed, for some excellent sets of stereopaired aerial photographs, numerous examples of the use of photogeology in various types of geological investigations, and a comprehensive discussion of instrumentation used in photogeology, as well as the principles and use of various types of stereoscopes and mosaicking techniques. With the wealth of published material available in photogeology, it would be not only redundant but presumptuous to attempt to do justice to the subject of photogeology in this chapter. Before we turn our attention to the metamorphosis of photogeology into geologic remote sensing, the reader should be aware of the large variety of United States and non-United States catalogs, atlases, or other compendia of aerial photographs which show different types of landforms either as single or stereopaired aerial photographs. Ray's U.S. Geological Survey Professional Paper (1960) has already been mentioned; two other U.S. Geological Survey Professional Papers contain selected aerial photographs of the United States (Denny et al., 1968) and outside the U.S. (Warren et al., 1969). The U.S. Department of Agriculture released a set of five airphoto atlases of the rural United States (Baker and Dill, 1970a, 1970b, 1971a, 1971b, and 1971c), which contains a large number of aerial photographic mosaics and stereopairs. Jack Mollard's book, *Landforms and Surface Materials of Canada: A Stereoscopic Airphoto Atlas and Glossary* (Mollard, 1974), is a sterling photographic overview of the variety of landforms in Canada. His book contains over 600 stereo triplets and stereopairs which cover terrain analysis for engineering, geological, and environmental studies. By the late 1960s, enough color and color infrared aerial photographs had become available that the American Society of Photogrammetry published its first *Manual of Color Aerial Photography* (Smith and Anson, 1968). An excellent selection of single aerial photographs of geologic features is included in that book.

Geological Remote Sensing

Beginning in the early 1960s with the First Symposium on Remote Sensing of Environment

(University of Michigan, 1962) and gathering steam with the general declassification of most airborne thermal infrared images (aerial thermography) in 1967 (see the section on "Historical Considerations" of this chapter), a gradual evolution began to what is now known as geological remote sensing. This evolution was accelerated by the greater availability of optical-mechanical scanners and the development of new types of aerial cameras in the 1960s. Also of great importance was the success of the early space missions, especially the Apollo 9 SO65 multiband camera experiment (Lowman, 1980). At the national level (National Academy of Sciences, 1969) and also at the international level (Working Group on Geology, Canada, 1971), there was a drive for the launch of the first Earth Resources Technology Satellite (ERTS 1) beginning in the late 1960's (Dickson, 1977). All of these developments likewise signaled a change in photogeology and the development of a larger activity—geological remote sensing.

Geological remote sensing, as we know it today, encompasses the principles and techniques of photogeology and much more. Geological remote sensing embraces the optical (manual) and digital analysis of images from a vast array of remote sensing devices, mostly passive but some active, from the ultraviolet through the visible, near-infrared, thermal infrared, and into the microwave wavelengths (see the Introduction section and Figure 31-1 of this chapter). Although much of the current emphasis in geological remote sensing is on spectral analysis of digital images, this emphasis is related to both the spectral character of different rock types *and* to the fact that most digital image data are available only in two-dimensional form. In the not too distant future, multispectral remote sensing satellites will be launched that are capable of imaging the Earth's surface in stereo. The current national high altitude photography (NHAP) program, designed to repetitively photograph the United States, will make high quality stereoscopic aerial photographs, both panchromatic and color infrared, available to geologists for regional studies (see the section of this chapter on "Future Trends"). Until more photographs and images capable of being analyzed in stereo become more generally available, many geologists will find that the conventional vertical, black-and-white stereopaired aerial photographs are the most useful and cost-effective tool to support field studies.

The growth of geological remote sensing technology has seen the periodic publication of textbooks on the subject. Beginning with the Second Edition of Avery's book, *Interpretation of Aerial Photographs* (Avery, 1968), there has been a steady inclusion of remote sensing techniques in addition to or in place of conventional photogeological techniques. Each of the books noted in this section contains one or more sections devoted to optical (manual) and/or digital image analysis techniques.

In addition to Chapter 16, Terrain and Minerals: Assessment and Evaluation (Reeves et al., 1975), the first edition of the *Manual of Remote Sensing* contains an excellent chapter on the "Fundamentals of Image Interpretation" (Estes et al., 1975). The book by Barrett and Curtis (1976), *Introduction to Environmental Remote Sensing,* also includes several sections on image analysis. Bill Smith's book, *Remote Sensing Applications for Mineral Exploration* (Smith 1977), contains some preliminary work on digital image analysis in addition to the traditional approach of using direct visual analysis.

In the United States, there were three excellent textbooks published in 1978, 1979, and 1980 which address the image analysis subject. Floyd Sabins (1978), in his book, *Remote Sensing: Principles and Interpretation,* devoted an entire chapter to digital image processing. Tom Lillesand and Ralph Kiefer (1979) in their book, *Remote Sensing and Image Interpretation,* interweave the subject of image interpretation (optical and digital techniques) throughout the 10 chapters of the volume.

For geologists, perhaps the best book available is the one edited by Barry Siegal and Alan Gillespie (1980), *Remote Sensing in Geology*. Two of the four sections of the book are devoted to image processing and enhancement (two papers) and interpretive techniques (five papers). In the image processing section, the paper by Skaley (1980) on "Photooptical Techniques of Image Enhancement," and especially the paper by Gillespie (1980), "Digital Techniques of Image Enhancement," are recommended to geologists. The paper by Soha et al. (1976), "Computer Techniques for Geological Applications," provides an excellent earlier summary of techniques directed at geologists. Two U.S Geological Survey reports on digital processing of images for geological studies are by Taranik (1978) and by Condit and Chavez (1979).

The book by V. R. Slaney (1982), *Landsat Images of Canada - A Geological Appraisal,* includes a brief section on analytical methods and another section on MSS image interpretation. Also in 1981, Floyd Sabins published a *Remote Sensing Laboratory Manual* in three parts as a companion volume to his book *Remote Sensing: Principles and Interpretation* (1978). The lab manual includes interpretation exercises on examples of different types of remotely sensed data and a section on digital image processing. Nick Short's book, *The Landsat Tutorial Workbook: Basics of Satellite Remote Sensing* (1982), includes a section on "Photointerpretation of Landsat Images" and an appendix on "Principles of Computer Processing of Landsat Data."

In the remainder of this section, which deals with "Analysis Techniques," emphasis will be placed on analysis of multispectral data through either optical (manual) or digital image processing techniques for geological investigations. For a more comprehensive discussion of the geological applications of such techniques, the reader is referred to the many textbooks just discussed. Two other books, which are devoted to the general principles of digital image processing, are also highly recommended: the IEEE book, *Digital Image Processing for Remote Sensing,* which was edited by Ralph Bernstein et al. (1978) and the superb book, *Digital Processing of Remotely Sensed Images,* by Johannes Moik (1980). Chapters 17, "Data Preprocessing and Processing"; 21, "Image Geometry and Correction"; and 24, "Manual and Digital Image Analysis in the Visible and Infrared Regions" of this *Manual* should also be consulted.

In addition to the books just discussed of special interest to geologists, there are a number of general books on remote sensing technology that should also be cited: Robert Rudd's *Remote Sensing: A Better View* (Rudd, 1974), Dorothy Harper's, *Eye in the Sky: An Introduction to Remote Sensing* (Harper, 1976), Joseph Lintz and David Simonett's, *Remote Sensing of Environment* (Lintz and Simonett, 1976), Philip Swain and Shirley Davis', *Remote Sensing: The Quantitative Approach* (Swain and Davis, 1978), and Benjamin Richason's, *Introduction to Remote Sensing of the Environment* (Richason, 1978).

In the following section is a brief discussion on atmospheric and radiometric corrections of image data acquired from Earth-orbiting satellites. Such corrections are necessary, when imaging the Earth through a turbid atmosphere, for precise studies of the spectral reflectance of rock units.

MULTISPECTRAL DATA

Optical Image Processing

Introduction

The effective use of Landsat or other types of multispectral scanner (MSS) data depends to a large degree not only on the skill of the geologist in selecting, analyzing, and interpreting the data for his or her specific needs, but in selecting the best possible processing technique or combination of techniques, for the specific application and budget. Within a few years after the launch of Landsat 1 on 23 July 1972, the geological community ably demonstrated the importance of a common, global data base for a wide variety of geoscience applications, such as regional geologic mapping, minerals exploration, geothermal exploration, volcanology, hydrogeology, engineering geology, marine and coastal geology, glacial geology, glaciology, and identification and assessment of geologic hazards.

It seemed, during the early years of the Earth Resources Survey Program, that Landsat and other satellite data would become a near-universal tool for geoscientists around the world. In fact, using the MSS data provided by Landsat 1 geologists, working on a wide variety of applications, made remarkable contributions to the geologic knowledge and literature worldwide. A listing of even a representative set of references is far beyond the scope of this section; but the literature is already replete with this type of information. Probably the best single source of information readily available to the geological community concerning accomplishments of the early years of the Landsat program is provided by Williams and Carter, (1976) in "ERTS-1, A New Window on Our Planet." Chapter 2 of this valuable reference presents a cross-section of preliminary results of geologic studies based upon Landsat 1 data. In his introduction to that chapter, Fischer (1976, p. 48–49) lists four geologic hypotheses supporting the design criteria for Landsat 1. Essentially, these hypotheses are as follows:

1. Certain large features (including glacial features, faults and other lineaments, and volcanic features) are present on the surface of the Earth which, because of their size and subtle expression, have gone unrecognized in conventional ground and aerial surveys, but can be seen on images having sufficient areal coverage and spectral uniformity.
2. Color or multispectral images, properly recorded and processed to preserve color uniformity, are useful in mapping distributions of rock types and alteration products.
3. Some environmental features are only intermittently visible, depending on angle of illumination and on snow, water, or vegetation distributions.
4. Some geologic processes, such as sedimentation and geologic motion, can be better understood if viewed in a "time-lapse" mode.

Fischer then went on to cite the various applications made of the Landsat data and "the tremendous amount of geologic information that is being derived from the relatively straightforward use of ERTS data . . ." He also stated in the same introduction, "Full use of the multispectral qualities of ERTS, however, *requires* (emphasis added) that the interpretations be made using the digital magnetic tapes (CCT's) of the data instead of the analog images." Without so stating specifically, therefore, he points out the digital and optical options in processing Landsat multispectral and multitemporal data. However, as has proved to be common thereafter in the remote sensing community, no alternative to digital processing is suggested if enhancements of detail, features, boundaries, or colors are needed, because they do not appear or are not distinct on the raw or standard data, black-and-white, or color products distributed by the EROS Data Center. Enhancement of some features can be obtained at times when the sun angle, growing conditions, land use, or weather will produce a radiometric response favorable for interpretation of that particular feature. The repetitive data coverage provided by Landsat, therefore, can commonly be used to advantage as an aid to interpretation in lieu of special optical or digital processing for feature enhancement. The purpose of this section is to demonstrate to the geoscientist a variety of optical enhancement techniques and data selection procedures for enhancement of geological features, that, in some cases, could eliminate the necessity for very costly, and commonly unavailable, digital processing of Landsat MSS data.

A Rationale for Data Processing

The trend in Landsat data production during the late 1970s and early 1980s by the remote-sensing and data user communities has been almost exclusively toward digital image processing, which requires, of course, the interim production of computer-compatible tapes. Although there is no doubt that digital analysis is required for some very important geologic applications, such as detection of mineral alteration zones and for automatic, quantitative areal classification of various surface features, the use of film analog imagery should not be ruled out for many equally valuable uses. This is particularly important, because many geoscientists simply do not have the technical capability or hardware to perform digital processing. For many applications the geoscientist requires only improved image interpretability and does not need quantitative measurements of radiance or area, or improved cartographic accuracy. Furthermore, the limited funding available for many low budget projects automatically precludes the possibility for digital processing, and hence tends to eliminate the use of Landsat data and modern remote sensing technology.

Landsat MSS data have been shown to be tremendously versatile and to be an especially powerful tool for multiple types of applications. When faced with a situation where the standard Landsat MSS film data do not depict adequate information content, and digital processing is not considered feasible, the photogeologist has learned from experience that optional photo-optical techniques can quite often be employed or devised that may provide the desired data products. Custom photographic processing, photo-optical contrast enhancing, negative-positive masking, additive color viewing of spectral and temporal data, generation of special purpose spectral composites, and color coding are techniques that may be used to good advantage by geoscientists. Optical enhancements can depict features not easily discernable on the raw data; they can sharpen boundaries between various surface features, depict surface detail with greater clarity, and simplify and improve time-change analysis. Illustrated examples are provided in this section to demonstrate some of the photo-optical processing options that have been applied to geological problems and to encourage greater use of Landsat data by geologists throughout the world, especially where high-technology facilities are unavailable.

Regional Landsat Mosaics

The ability to "see" and integrate extremely large tracts of land in a single image or mosaic of images is perhaps the most useful attribute of images acquired from space. Such mosaics are useful to the geologist for general purpose geologic mapping, and in the preparation of tectonic, metallogenic, and hydrogeologic maps. They are especially useful to the exploration geologist in developing strategies that will aid in the discovery of new mineral or energy resources.

Optional Techniques for Preparation of Regional Mosaics

In the report on the results of a meeting of an *ad hoc* committee, established by the U.S. Geological Survey for the Geological Society of America, to determine the most appropriate technique for producing a Landsat mosaic of the entire North American Plate (Carter, 1983), three options for producing the mosaic were considered by the committee. The options and their advantages and disadvantages follow:

1) Construct a conventional photo-mosaic using existing Landsat imagery: The advantages of this option are minimum cost and least time needed for preparation. The disadvantages are that the product would be of the lowest quality and of marginal utility outside of satisfying the commitment. There is no advance of technology nor is existing technology used to full capability.
2) Construct a conventional photo-mosaic employing *digitally enhanced* (emphasis added) existing scenes. This would produce an improved product employing well understood technology at little risk. The disadvantages stated by the Committee are that the mosaic would "have no spin-off benefits" and the maximum quality of the product would not be achieved, nor would the highest state of the art of present technology be employed.
3) Prepare an all-digital mosaic in which image enhancement and physical joining of images and tone balancing between images are done entirely by computer. This would produce the best product with the greatest number of potential immediate products to serve additional needs. The latest computer technology of greatest flexibility would be employed and the project could be used as a basis to stimulate new advances in the art of digital image processing, digital cartography, and spatial data applications. The disadvantages are the high costs involved (about 5 times the

cost of the photographic product), the high degree of risk involved, and the extended time needed to produce the mosaic.

Custom Photographic Mosaic of the Central Rift Valley of Kenya

A fourth option, or a variation of the second option listed above, is possible in the preparation of Landsat regional mosaics. A conventional photo-mosaic can also be prepared employing imagery that has been optically enhanced by custom photographic processing. At the request of the U.S. Agency for International Development and the Survey of Kenya, the U.S. Geological Survey was asked to prepare a Landsat mosaic of the Central Rift Valley to be used as an aid in regional geologic, geothermal, and ground-water investigations. The desire was to produce a mosaic of maximum utility at the lowest possible cost, employing 70 mm film data, the same data as are already on hand at the Regional Remote Sensing Facility in Nairobi. The Facility maintains an archive of 70-mm Landsat 1, 2 and 3 MSS data covering East Africa. The Survey of Kenya and the Water Resources Ministry were particularly interested in this project, not only to aid their investigations, but to determine the possibility for their effectively using the existing archive of 70-mm Landsat imagery. To compile the mosaic, it was decided to prepare photographically enhanced, rather than digitally enhanced, images.

An extremely important—but sometimes overlooked—consideration in the application of Landsat data for any purpose is the careful selection of the best available data. Because the Equator roughly bisects the north-south trending valley, the sun angle of the Landsat imagery is always relatively high. Whereas discrimination of types of vegetative cover is generally enhanced by a high Sun angle, any terrain that suggests geologic structure is naturally enhanced by shadowing if the Sun is low in the horizon. To achieve the maximum terrain enhancement and minimum saturation caused by high desert albedo, data collected at times of minimum Sun angle should be used, if available. For equatorial areas such data would have been acquired in the December–January or June–July time frames, depending on whether the specific area was north or south, respectively, of the Equator. Fortunately, an excellent set of 13 scenes that included the entire Rift Valley was collected by Landsat 2 on 24–26 January 1976.

Cloud coverage over the Valley itself was negligible. However, there was extensive cloud cover to the east of the Central Rift Valley along a line between Mount Kenya and Mount Kilimanjaro. This demonstrates the need for a microfiche or quick-look capability in order to determine the geographic distribution of the cloud cover, and not merely the percentage of the scene covered by the clouds.

The Sun angle ranged from 43° to 46° above the horizon. Because all of the data were collected over a 3-day period, and because no significant atmospheric changes occurred, solar illumination of the scene was essentially uniform, although no attempt was made to measure or compare radiance differences between any of the scenes.

Through use of the individual 70-mm positive transparencies, 1:1,000,000-scale false-color composite transparencies were produced by sequential direct exposure on color Ektachrome film of bands 4, 5, and 7 through blue, green, and red filters, respectively (the so-called standard false color composite). False-color composites for some of the same scenes produced as standard data products by the EROS Data Center, as well as digital enhancements, were examined. It was observed that in printing the scenes, there was a tendency to saturate the red tones. This in effect obscured the lineaments indicative of the rift structure and some of the drainage patterns as well. For the intended geologic applications of the scene, the red saturation severely reduced the relevant information content of the imagery.

A custom photographic specialist produced original, false-color composites of each scene on Ektachrome positive film transparencies at a scale of 1:1,000,000 for all 13 scenes (Table 31-18). He then printed the first of 13 direct positive-to-positive Cibachrome prints at scales of 1:1,000,000—carefully limiting the band 7 or red density. The print was mounted on a base prepared from the 1:1,000,000-scale Operational Navigation Charts published by the Defense Mapping Agency. To begin, he selected a scene from near the center of the mosaic. One by one he printed the remaining 12 Cibachrome prints, matching each as closely as possible in color and tone to the adjoining scenes.

The individual images were "butt-edged" with the result that the boundaries of each scene can be seen on the resulting mosaic (Color Figure 31-68).

TABLE 31-18

Landsat 2 MSS Images Used to Prepare a 13-Scene Mosaic of the Central Rift Valley of Kenya, East Africa (Color Figure 31-68)

Path, Row	Date	Landsat I.D. No.
180,59	24 January 76	2367-06571
180,60	24 January 76	2367-06573
180,61	24 January 76	2367-06580
180,62	24 January 76	2367-06582
181,57	25 January 76	2368-07020
181,58	25 January 76	2368-07022
181,59	25 January 76	2368-07025
181,60	25 January 76	2368-07031
181,61	25 January 76	2368-07034
181,62	25 January 76	2368-07040
182,57	26 January 76	2369-07074
182,58	26 January 76	2369-07081
182,59	26 January 76	2369-07083

No attempt was made to use curvilinear join lines, such as is commonly done in the case of mosaics prepared from aerial photographs, because vignetting is not a problem with Landsat data due to the very narrow look angle of the sensors and the essentially equal Sun illumination over the entire scene. The use of aerial mosaicking techniques by many geologists and others indicates a failure to take advantage of the design characteristics of the Landsat sensor system.

Whereas the join lines can be obscured digitally (at considerable cost), this may also be accomplished by using special film mosaicking techniques successfully employed by the U.S. Geological Survey's National Mapping Division. Join lines, however, do not impair the utility of mosaicked image data. In fact, they have a positive effect of delineating the data coverage from each image and are readily correlated with the Landsat scene index map. The mosaic provides a basis for selection of relatively small subscenes where more sophisticated digital processing is needed for such purposes as pixel by pixel analysis, spectral separation of alteration zones, or quantitative spatial classifications. It is most unlikely that such analyses would be needed for the entire region covered by any mosaic.

Cibachrome prints were employed because they are prepared directly by a positive process, thus eliminating the necessity for internegatives, and improving the interpretability of the scene for extracting maximum relevant detail. The Cibachrome paper employs a silver-dye bleach principle for maximum color distinction and brightness, preserves a high level of spatial resolution, and is relatively scale stable.

Following the preparation of the Central Rift Valley mosaic, the U.S. Geological Survey, on an experimental basis, prepared color separations of the mosaic and lithographs for distribution. Comparisons of digitally enhanced and photo-optically enhanced imagery used to make mosaics reveal the high degree of quality that is achievable employing custom photographic processing.

A Comparison of Various Types of Image Enhancement by Digital and Optical Processing

The EROS Data Center, as part of its continuing effort of Landsat image quality improvement, developed a program, originally referred to as the EROS Digital Image Enhancement System (EDIES), to provide computer-enhanced Landsat images. EDIES was the test-bed system for the present EROS Digital Image Processing System (EDIPS). To illustrate the vast improvement possible by digital-image enhancement techniques, the EROS Data Center prepared a display employing Landsat 1 image 1700-17422 collected over the Needles, California, area on 23 June 1974. The display consisted of two 1:500,000-scale photographically produced images: one, a standard false-color composite print, the other a digitally enhanced print. Color Figures 31-69A and B are parts of the northeastern quadrant of the standard Landsat MSS false-color composite image product and the digitally enhanced image product, respectively, photographed directly from the display board. In addition to the corrections for mirror velocity, Earth rotation, and aspect ratio made on the digitally enhanced and corrected image, the digital-image enhancement also consisted of radiometric restoration (destriping), sampling geometric restoration, contrast enhancement, synthetic line generation, detector misregistration correction, and high frequency edge enhancement (U.S. Geological survey, 1977c). The digitally enhanced image product was clearly superior to the standard image product. Surface texture, especially in the mountainous rock outcrop areas and on the alluvial fans, was sharply defined. Roads and drainage areas were more distinct, colors were brighter, and better color separation was achieved.

As an additional comparison, however, the EROS Program Office contracted with John Milne of Technical Photo, Inc., to prepare a custom Cibachrome print at the same scale for the same scene employing a set of the standard 1:3,369,000-scale positive transparencies of Landsat MSS bands 4, 5, and 7 purchased from the EROS Data Center. The resulting custom print was also clearly superior to the standard product, and in some ways compared favorably with the digitally enhanced print: color separation was equal or superior to the digitally enhanced image. Vegetative patterns, especially in agricultural areas, were more distinct, probably because of the absence of pixel clustering for edge enhancement. The sharpness of roads and drainage patterns approached that of the digital image. Surface texture, while not as pronounced as on the digitally enhanced image, was greatly superior to that depicted by the standard product, and quite adequate for field interpretation by the geologist. Color Figure 31-69C is a photograph of the same portion of the optically enhanced image.

Despite the fact that none of the geometric corrections or destriping can be done photographically, nor the fine texture or edge enhancement duplicated, the custom, or photographically enhanced image can provide a data base that is clearly superior to the standard product for geologic and land use mapping and also can provide a quick and relatively inexpensive alternative to digital image enhancement. Also, although the cartographic accuracy of the photographically enhanced image is not equal to that of the geometrically corrected images produced digitally, the user needs only to acquire already published topographic maps as needed for his or her application.

Multispectral Composite Images

Probably the most useful and extensively used product produced from the Landsat program has been the Landsat MSS false-color composite image. It is most commonly produced by combining MSS bands 4, 5, and 7 on to color film with exposures made through blue, green and red filters, respectively. A similar, but not completely identical, false-color composite image can also be made by substituting the MSS band 6 data for MSS band 7.

Relatively few investigators, however, deviate from the standard false-color composite image to enhance selected features or to depict features in different colors. A notable example of a non-standard color composite is given by the National Geographic Society (1976), which published a Landsat mosaic of the United States depicting vegetation in green in an attempt to achieve a "natural color." The mosaic is not in natural color, however, because it was prepared merely by projecting the MSS band 7 or reflected infrared data in green rather than red. MSS band 5, which best depicts vegetation and actually represents red reflected radiation, was projected in red; band 4, or reflected green radiation, was projected in blue. Therefore, any color composite image containing reflected solar infrared radiation from band 6 or 7 is by definition a false-color composite image. Non-standard false-color composite images can be prepared either optically or digitally. False-color composite images that have had custom photographic processing applied to them can be displayed by using color additive projection onto a rear-view, ground-glass screen. This is but one of several methods used to optically produce false-color composite images.

The "additive color viewer" is a particularly useful and potentially valuable piece of equipment. It is a relatively simple and inexpensive (when compared to computer hardware) multi-image projector and viewer. Various types of lithology, terrain, water, moisture, land use, suspended sediment, and shallow bottoms of water bodies have varying ranges of reflectance of solar energy (depicted by film density) and, therefore, are not always definitively interpretable from a single band, black-and-white image. Compositing of data from two or more of the MSS bands provides the geologist with a capability to separate certain selected features with "color-coded" representations.

The additive color viewer enables the investigator to examine any 2-, 3- or 4-band combination of negative or positive, original or reprocessed transparencies—each projected through a blue, green, or red filter, or by unfiltered light—under controlled illumination intensities for each. A wide variety of band, filter, and illumination intensities is therefore possible. There is no necessity to prepare film or prints in order to examine the results as is the case with photographic or diazo processing. Any useful data product generated on the screen can readily be photographed by the geoscientist and the results recorded on 35-mm slides or other format. Slater (1975) ably describes the production of color by the additive process. Those readers concerned with the concepts and fundamentals of remote sensing are urged to refer to his paper and to Chapter 6, Photographic Sensors, of the present *Manual* (also written by Slater) as a technical basis for an explanation of the enhancement techniques described herein.

A Spectral Data Corporation additive—color viewer was used for the preparation of various spectral and temporal composites of interest to the marine, environmental, or ground-water geologist. Wenderoth and Yost (1974) describe the design, operation, and application of the viewer for earth resources applications. Space limitations prevent an adequate description herein of the equipment, which is a valuable tool for the geologist employing remote sensing technology, but which, in the author's experience, has not been used to its full potential by many facilities. The reader is referred to Chapter 24 of this *Manual* for a description of additive color viewers and related instrumentation.

A common deficiency in film reproductions of 70-mm, electron beam recorder (EBR)-generated Landsat MSS standard products for additive-color viewing is the variation in density and density range of the four bands of data. This deficiency also may impair the quality or information content of false color composite images prepared by a photo lab technician who employs equal times of color film exposure for each band. In many cases the density range of MSS bands 4 and 6 differs significantly from those of MSS bands 5 and 7 on the standard products distributed by the EROS Data Center. It should be understood that the variation in density between bands that is discussed here is a function of the exposure times used in the reproduction of the imagery from the original film transparency and not a density difference caused by the spectral data acquired by the sensor. When the annotation grey scale associated with each image varies in density from band to band, the variation is caused by exposure time differences during reproduction. When, for example, MSS band 4 is of high film density and MSS bands 5 and 7 are of low film density, the resulting "standard" false-color composite image may be the same as a 2-band composite, because inadequate MSS band 4 information has been transmitted through the blue filter to depict the surface characteristics recorded by the sensor. In some cases cited below, the 70-mm transparencies were photo-optically reprocessed in order to contrast stretch the density range for selected land or water areas (Falconer et al., 1975). Additionally, the magnitudes of the contrast stretches were controlled to make all bands compatible in density range as required for additive color projection. It is often possible, therefore, to reprocess the standard Landsat film to enhance desired detail without the necessity of resorting to the purchase and computer processing of computer-compatible tapes (Falconer et al., 1981a).

Special Purpose Spectral Composite Images

Flood Plain and Wetlands Enhancement for a Río Paraná, Argentina, Scene

An attempt was made to show maximum flood plain and water-distribution detail along the Río Paraná in the vicinity of Rosario, Argentina, again employing the additive color viewer and unreprocessed 70-mm standard positive transparencies of Landsat image 2340-13082 acquired on 28 December 1975. A highly detailed, enhanced image was composed on the screen of the viewer, and a small portion of the area was photographed with a 35-mm camera equipped with a closeup lens. Color Figure 31-70A is the resulting enlargement of the photographed portion of the image. To maximize discrimination between land and water, vegetation was shown in yellow by a variation of the composite described above. Blue was used to depict water surfaces and greatly intensified by projecting both MSS bands 4 and 5 in blue light. MSS band 6 was shown in green light, and MSS band 7 in red light. The intricate water detail is clearly shown in bright blue. The rectangular fields are readily delineated by a closeup lens. The limits of the flood plain are also clearly depicted. Use of this composite in conjunction with topographic and geologic maps along with well log data made possible the rapid preparation of accurate piezometric profiles by Argentine hydrogeologists.

Despite the fact that the source data consisted of 70-mm transparencies and that the composite was recorded on 35-mm film, the resulting data product is of exceptionally high spatial resolution. This demonstrates that high resolution data products can be produced by techniques other than digital image processing.

Depiction of Oil from Blowout in the Gulf of Campeche, Mexico

With the growth of offshore oil production the possibility of oil blowouts poses potential geologic hazards. The detection of oil then becomes of interest to the environmental geologist concerned with cleaning up the affected waters or geological engineers attempting to plug the well. On 3 June 1979 one of the most serious and widespread environmental pollution incidents in offshore oil drilling history began with the explosion and subsequent oil blowout from the Ixtoc 1 well in the Bay of Campeche in the southern Gulf of Mexico. Unofficial estimates of the discharge ranged from 20,000 to 100,000 barrels per day. Numerous oil slicks on many different Landsat images were depicted on digitally enhanced products (Deutsch et al., 1980).

Color Figure 31-70B is a previously unpublished print made from a 35-mm slide copy of the screen of the additive color viewer containing the 70-mm positive transparencies of Landsat scene 30502-15532, acquired on 20 July 1979. After considerable experimentation, an especially vivid depiction of the numerous slicks of oil floating on the water between the well and the coastline was obtained. Two high contrast positive transparencies of MSS band 5 and one of MSS band 6 were prepared. The composite was made by projecting the information content of an MSS band 4 negative transparency in white light, one of the MSS band 5 positives in blue light and one in green light, and the MSS band 6 positive in red light. The sea appears white to light gray, and the oil slicks appear as dark gray. The sediment plumes discharging from the land mass appear blue, and imagery of the land surface is severely degraded. The same degree of contrast enhancement and the same negative-positive composite probably would not be optimum for all situations, but the additive color viewer provides the ease and flexibility for the experienced investigator to look at literally scores of possible combinations of negatives, positives, densities, and illuminations in a matter of minutes and to select the composite containing the most relevant information.

Color Separation of Reefs and Oil Slicks in the Gulf of Suez

During the month of October 1972, a break occurred in underwater pipe of an oil drilling platform in the Gulf of Suez, and several studies were made to see if the resulting oil spill could be detected by the Landsat 1 multispectral scanner (Otterman et al., 1974; and Deutsch et al., 1977). Employing a set of negative and positive 70-mm transparencies of Landsat image 1109-07493, acquired on 9 November 1972, the investigators made an attempt to depict the extent of the floating oil using the additive-color viewer. The 70 mm transparencies were photo-optically reprocessed in order to increase the contrast within the density range of the water area. In the unreprocessed positives, the water areas were too dense to show water detail when projected on the additive color viewer.

The reprocessed images were placed in the viewer, and a variety of false color composites, utilizing different band, color, and negative-positive combinations, were prepared and examined. The best combination in terms of clearly depicting oil slicks from water was found to be the reprocessed MSS band 4 positive transparency projected through a green filter, composited with the unreprocessed MSS band 7 negative transparency projected in white light (Color Figure 31-

70C). In this negative-positive mask, water appeared in tones of light green and the oil slicks as white. The composite provided an unanticipated dividend in that turbid waters along the shoreline and submerged reefs offshore appeared in green tones. The exposed reef on the lower right appeared as a dark green feature—the same as the land area. Detail over the land area, which was of no particular interest for this study, was grossly degraded.

Interpretability of the data is clearly improved by employing a multiband rather than a single band analysis. Multispectral analysis makes it possible to depict water features in separate colors from other features that have similar densities in one, two, or three—but not all four—of the bands of the multispectral scanner.

Temporal Composite Images for Analysis of Dynamic Features

Depiction of New Lava Flow at Mount Nyiragongo, Zaire

In January 1977, according to reports published by the French Press, Mount Nyiragongo in eastern Zaire violently erupted. The various press accounts differed as to the number of casualties and the amount of damage caused by a fast moving lava flow heading southward in the direction of the city of Goma on the north shore of Lake Kivu. The city with its 65,000 inhabitants is located about 20 km south of the caldera of Mount Nyiragongo.

Examination of the Landsat MSS image, 2049-07351, acquired by Landsat 2 on 12 March 1975, revealed the presence of lava flows in parts of Virunda National Park north and west of Mount Nyiragongo. The image also showed quite clearly a smoke plume originating in the crater of the volcano and blowing in a southwesterly direction for about 35 km over Lake Kivu. A history of the volcanic activity at Mount Nyiragongo is provided by Tazieff (1979). A detailed description of the 1977 eruption was also provided by the same author (Tazieff, 1976–77).

At the request of the U.S. Information Agency, NASA attempted to collect post-eruptive imagery in this normally cloud-covered region. Usable imagery was eventually collected by Landsat 2 on 6 February 1978. Examination of the raw data for MSS image 21111-07053 clearly showed two new lava flows, one pointing southward directly at Goma, and the other eastward, and then turning south.

A complete four-band set of the 70-mm (scale 1:3,369,000) black-and-white Landsat MSS positive transparencies was obtained from the EROS Data Center. They were the least expensive data available. Through use of the additive color viewer of the Spectral Data Corporation, two false-color composites were generated for both the pre-eruption and post-eruption scenes on the viewer screen. The first was the "standard" false-color composite, in which the MSS band 4 image is projected onto the screen through a blue filter, MSS band 5 through a green filter, and MSS band 7 through a red filter. The lava flows on both scenes could be readily seen, although they are somewhat obscured because of the vigor and density of the covering vegetation, which resulted in a deep magenta background for the entire scene. To minimize the effects of the dense vegetation along the margins of the southern half of the valley, a second false-color composite was generated for the two scenes. MSS bands 5, 6, and 7 were exposed sequentially through blue, green, and red filters, respectively. The vegetation then appeared in tones of yellow to brown depending on the film density, because MSS band 5, which best depicts vegetation, became the complement of the MSS band 6 (printed in green) and MSS band 7 (printed in red) combination. The lava flows showed much more clearly through the yellow of the special composite than through the red of the standard composite. The vegetative response then appears as yellow, which offers a much more contrasting background to the dark gray to black lava flows. Color Figures 31-71A and B were prepared from 50- × 50-mm (2- × 2-in) slides of 35-mm photographs of the viewer screen. A rationale for application of this composite for geologic studies is given by Falconer et al. (1981b), who pointed out the potential advantages it provided for hydrogeologic exploration and mapping in the Lake Ontario Basin.

The temporal composite (Color Figure 31-71C) was generated on the viewer screen by projecting the MSS band 6 and 7 pre-eruption images in red and MSS bands 6 and 7 of the post-eruption images in green. The composite, which normally would be predominantly yellow because of the mix of red and green light, was photographed through a yellow filter resulting in a white to light gray background. The pre-existing flows are shown in black. The long reddish-black lava flow to the west of the volcano also turned out to be a pre-existing flow. It appears red, because it was obscured by haze on 12 March 1975, when Landsat 2 acquired the pre-eruption image. The new lava flows are shown brightly in red, both originating from flanks of the volcano below the caldera.

Monitoring of Snowline Dynamics in the Cascade Mountains, United States and Canada

Landsat has provided the glacial geologist with a unique opportunity to delineate and measure snow and ice cover for a variety of applications in engineering, sedimentation, exploration, and even trafficability or navigation. The repetitive coverage of the Landsat series has added the all-important capability of observing and measuring dynamics. Meier and Evans (1975) in their study of the snowline and glacial dynamics in a portion of Cascade Mountains along the United

States–Canadian border used Landsat 1 data to determine the extent of ice and snow at the end of the winter of 1972–73, and a second set collected at the end of the summer of 1973 to determine the areal extent and distribution of glacial ablation and residual snowpacks. In the same paper they described a variety of processing and analytical techniques to acquire the desired information.

A number of attempts to prepare a single time-change composite that would clearly and definitively depict the change in areal extent of snow and ice were ineffective. The compositing of single band data through a technique such as was used to map flooded areas (Deutsch and Ruggles, 1978) failed to similarly show the ice-distribution and snowline changes, primarily because of the wide range of film densities at both times of collection and because of the areal—rather than linear—distribution of the observed phenomena.

After considerable trial and error a temporal composite was devised that fully met the study objectives: For an end-of-the-winter Landsat MSS image (1258-18322; 7 April 1973), a spectral composite was printed on a Cibachrome transparency, employing the film positives of MSS band 5 projected through a blue filter, MSS band 6 projected through a green filter, and MSS band 7 projected through a red filter (Color Figure 31-72A). This is the same type of spectral composite as described previously in the discussion of the new Mount Nyiragongo lava flow (Color Figure 31-71), except that the Cascade composite was prepared photographically, employing standard color laboratory techniques and equipment rather than an additive color viewer. The composite was designed to show vegetation in tones ranging from yellow to brown, depending on vegetation density, thereby eliminating the presence of reds or magenta to depict any surface features.

For the end-of-the-summer Landsat MSS image scene (1420-18303; 16 September 1973), only MSS band 5 was used. The negative of MSS band 5 was projected through a red filter onto a Cibachrome transparency. Snow, being the brightest feature on the positive film, now became the most dense or the reddest feature on the negative. The exposure was gradually increased to the point where all of the features depicted in lower film densities on the negative became saturated or white (Color Figure 31-72B). The only features remaining in red on the negative were snow and ice plus some very small areas of sand bars along the river channels, which are readily classified by simple pattern recognition—a difficult feat for the computer, but an almost automatic function for the human brain.

The areal change in snow and ice cover was vividly depicted by simply compositing the transparencies of the end-of-the-winter positive spectral composite, (on which no features appear as red or magenta) with the end-of-the-summer single band negative, which shows the ice and snow in bright red over the white winter ice and snowfields (Color Figure 31-72C). Accurate registration of the two images was easily accomplished merely by merging common features shown on the original film negative for the end-of-the-summer scene. Although a description of the compositing technique has not previously been published, the American Society of Photogrammetry printed the composite on the cover of the February 1979 issue of *Photogrammetric Engineering and Remote Sensing,* accompanied with a cover photo caption (American Society of Photogrammetry, 1979).

Conclusions

The manner in which Landsat multispectral scanner data are processed may pose an important limitation to the application of space technology by the geologic community. Emphasis in recent years has been on high technology digital processing of computer compatible tapes. For the geoscientist whose information requirements exceed those contained on the raw or standard data products, but who lacks access or funds for use of digital processing systems, a number of optical enhancement alternatives exist that could fulfill some of his or her information needs.

The versatility of Landsat multispectral data opens the possibility for the geoscientist to employ custom photographic processing and custom mosaicking in the preparation of a wide variety of special purpose spectral, temporal, negative-positive, or multiband-single band composites as alternatives to digital image processing. An important element in the pursuance of any project in which Landsat data can be effectively used is the determination of processing techniques likely to provide an information content adequate for the needs of the geoscientist. Complete reliance on standard data products or high technology digital processing as the only two Landsat processing options will effectively rule out the possibility for many geoscientists to employ Landsat data and remote sensing technology despite many potentially beneficial applications, such as described in this section.

Atmospheric and Radiometric Corrections

Radiometric and atmospheric corrections are applied to remove the influences of extrinsic sensor and atmospheric variables. The goal is to obtain a corrected data set whose attributes are more directly related to the intrinsic surface properties of interest.

Atmospheric and sensor simulation models have provided a wealth of information about the radiometric and atmospheric effects expected in remotely-sensed data (Turner et al., 1971; Turner, 1975 and 1979; Malila et al., 1976; Dave, 1980; and Kahle et al., 1981). However, it may not be possible to calculate the desired corrections from such models for specific applications, because the requisite system and atmospheric parameters are often unknown. Methods are available to estimate several of the unknown parameters either directly from the digital or from judicious field measurements of surface standards.

Visible and Near-Infrared Wavelengths

Absolute radiometric calibration of multispectral sensors requires that the spectral and intensity response of the instrument be known precisely. Because the stability of a sensor system may vary with time and imaging conditions it is necessary to repetitively calibrate the instrumentation to detect both short- and long-term variations. To perform an absolute radiometric calibration, the sensor must view sources of known spectral radiance, which are generally standards established by the National Bureau of Standards (NBS). The Landsat detectors were designed to view a calibration light source and dark level once per scan line and the solar intensity once per orbit. The solar irradiance is used to monitor the consistency of the on-board light source. This continual calibration of the Landsat system insures the stability of the sensor response. However, the absolute calibration factors (USGS, 1979, p. AE16) which can be used to convert digital values to radiances for Landsat 1 and 2 were determined at pre-launch. They have not been updated; therefore, their accuracies are in doubt (Kowalik, 1981).

Most radiometric calibration techniques correct scene radiance without regard to atmospheric path radiance and scattering conditions during data acquisition. Compensation must also be made for these atmospheric parameters, if the airborne or satellite data are to provide an accurate quantitative representation of the surface spectral character.

Geologic studies with remotely sensed data often include hilly or mountainous terrain where illumination varies appreciably from pixel to pixel because of the topography. Ratios of different wavelength bands are commonly used to subdue that variability and to enhance subtle spectral reflectance differences between surface-cover types (Vincent, 1973; Rowan, et al., 1974; Abrams, 1980; and Goetz and Rowan, 1981). Additive terms inherent in remotely-sensed data introduce a complication when ratio variables are created. This is explained to first order by the following equation which describes a measurement by a vertically-viewing sensor:

$$L_C = \frac{S_c T_c \rho_c \,(Hd_c \text{ COS } \alpha + Hsky_c)}{\pi} + S_c\,(Lp_c - Of_c) \quad (31\text{-}1)$$

(These parameters are described in Table 31-19).

In a ratio of two different wavelengths, the COS α (surface topographic orientation) term will cancel only if the two additive terms in (31-1) are negligible or are removed by a correction before ratioing. The two additive terms are referred to hereafter as term A—the skylight, and term B—the path-radiance minus the system offset. Modeling of Equation 31-1 for Landsat data of a semi-arid terrain under a high Sun elevation angle (56°, for example), shows that term *B* can cause significant mis-interpretation problems in ratio images (Kowalik, 1981). At lower sun angles the problem becomes more severe. For example, Figure 31-73 shows how the Landsat 5/4 ratio (red/green light) varies with COS α for four different rock types according to Equation 31-1. For the model conditions, most topographic surfaces lie at COS α greater than 0.35. If no corrections are applied, Figure 31-73 shows that larger ratio values will occur on well illuminated slopes (high COS α). This causes confusion between these rock types, because well illuminated surfaces of the tuff, granodiorite, and albitic quartz monzonite (COS $\alpha \geq 0.85$) have ratio values similar to poorly illuminated areas of the limonitic quartz monzonite (COS $\alpha \leq 0.50$). After a correction for term B, the ratio's variability with COS α is significantly reduced, these rock types are better separated, and the ratio values are more directly proportional to

TABLE 31-19

Explanation of Parameters Given in Equation (31-1)

L_c	= the radiance, measured by the sensor, in integer units of Digital Number
S_c	= The multiplicative gain factor of the sensor system (DN/radiance units = DN/mw/cm^2 sr)
T_c	= the vertical transmittance of the atmosphere from ground to sensor (dimensionless)
Hd_c	= the direct illumination upon a surface at the bottom of the atmosphere and perpendicular to the sun vector (mw/cm^2)
$Hsky_c$	= the illumination due to skylight upon the surface (mw/cm^2)
COS α	= the surface orientation term (dimensionless); α is the angle between the sun vector and the perpendicular to the surface
ρ_c	= the surface reflectance, the ratio of the reflected energy to the incident energy, assuming Lambertian reflectance properties; i.e., ρ_c/π *sr* units of the irradiance are assumed to be reflected (dimensionless)
Lp_c	= the radiance scattered by the atmosphere into the sensor (mw/cm^2 sr), path radiance
Of_c	= the offset value, above which the radiance measured by a sensor is calibrated to read positive digital number units (mw/cm^2 sr)

Note: The subscript *c* refers to a single wavelength band.

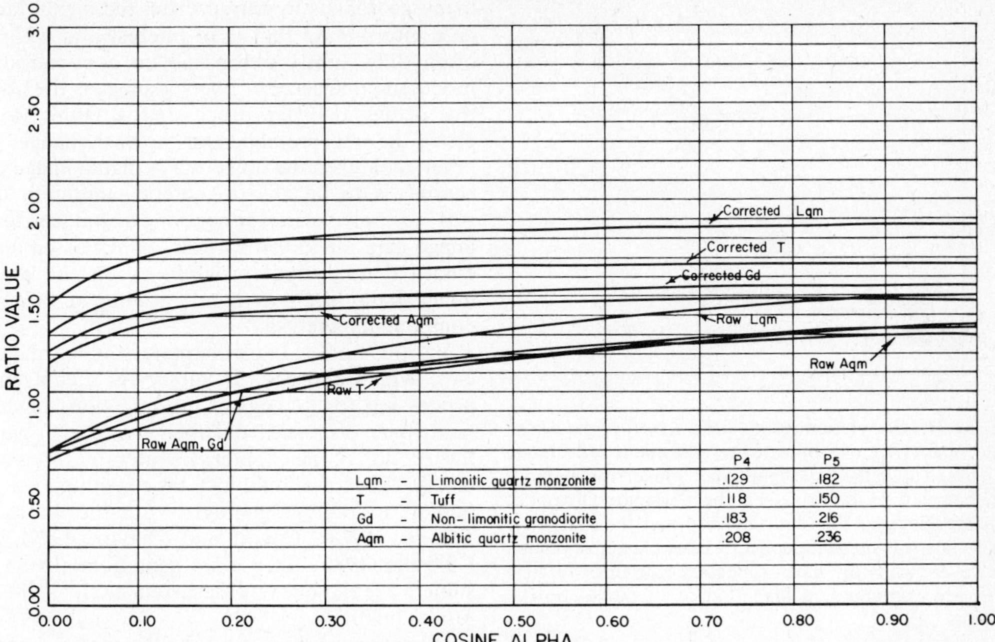

Fig. 31-73. Modeled Landsat 5/4 ratio values as a function of surface orientation (COS α) for the Yerington District of western Nevada on 1 June 77 (Sun elevation angle = 56°) during a Landsat 2 overpass. Curves for four different rock types are shown before and after correction for term B in equation 31-1. Well illuminated topographic slopes occur at high COS α. A steep slope of 35° facing directly away from the incoming sunlight has a COS α value of 0.35.

the surface reflectance characteristics than before. Failure to use a correction for term B has often caused well illuminated, non-limonitic sand dunes and well illuminated, non-limonitic, massive, silicic, plutonic rock outcrops to be erroneously interpreted as limonitic areas in Landsat ratio images (Kowalik, 1981). Additive term A in Equation 31-1 causes the remaining curvature in Figure 31-73, after a correction for term B. Under these modeled conditions, term A has a small adverse effect, because most topographic surfaces lie at larger COS α than 0.35.

Methods for estimating term B from remotely sensed data include the dark object subtraction (Vincent, 1973) or histogram minimum technique (Chavez, 1975), the covariance matrix technique (Switzer et al., 1981), and the dark/light target method (Lyon et al., 1975). Kowalik (1981) and Switzer et al. (1981) have compared estimates from these techniques for various Landsat data sets. Ahern et al. (1979) have extended the dark object subtraction method to map path-radiance variations across a Landsat image having a series of clear water lakes distributed around the image.

Studies into the absolute quantitative relationship or statistical correlation between surface and satellite spectral data have been limited (Abrams, 1978; Marsh and Lyon, 1980). These studies have employed the dark/light target method to correct satellite digital numbers to a satellite equivalent reflectance which empirically compensates for atmospheric effects. Marsh and Lyon (1980) showed that the satellite equivalent reflectance had a root-mean-square-error of ± 4 percent when compared with surface measurements. Though the absolute equivalency of the surface and satellite data could not be shown to be better than $+4$ percent reflectance, analysis of the statistical correlation of the data indicated that the surface and satellite data were highly correlated (Figure 31-74). Correlation coefficients (r) of at least 0.95 were determined for the four Landsat bands. These correlation coefficients for the corrected satellite-equivalent reflectance (light/dark target method) were 25% higher than when surface reflectance was correlated with raw satellite digital numbers.

Thermal Infrared Wavelengths

To achieve an accurate measure of the temperature of the surface of the Earth from the radiance measurements obtained with an airborne or satellite thermal sensor, the system should be continually calibrated against an on-board black-body calibration source. This calibration will allow the user to relate received radiance to an apparent brightness temperature. This apparent brightness temperature must then be corrected for the emissivity of the surface, atmospheric attenuation of the upward radiation, and atmospheric emission of radiation upward into the sensor and reflected upward to the sensor. To achieve a radiometri-

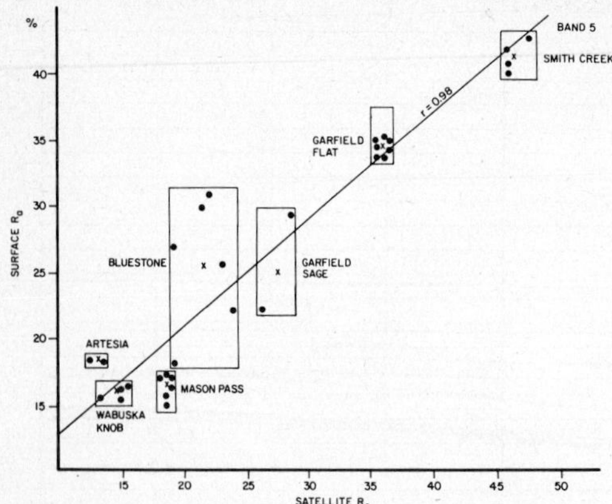

Fig. 31-74. Correlation results for the surface measured reflectance (R_a) and the satellite equivalent reflectance (R_s) based upon the dark/light target conversion method. Site names refer to test sites in the Yerington District of western Nevada. The data indicate that the individual pixels at a test site are a noncorrelated cluster of points. However, when the mean reflectance of all of the test sites are compared, a high degree of correlation is present.

cally and atmospherically calibrated temperature, it is thus necessary to measure or model a variety of system and meterologic parameters.

A number of studies have been conducted to investigate the effects of atmospheric radiative transfer on satellite measurements of surface temperature (Diermendjian, 1960; Wark et al., 1962; Kunde, 1965; Smith et al., 1970; Anding et al., 1971; Maul and Sidran, 1973; and Kahle et al., 1981). These studies make use of standard atmosphere models such as the LOW-TRAN-5 model developed at the Air Force Geophysical Research Laboratory (Kneizys et al., 1980). Use of these techniques has been shown to markedly improve the accuracy of apparent brightness temperatures recorded by satellite thermal infrared scanners.

Even after thermal data are corrected for system and atmospheric parameters, it is often difficult to directly relate changes in surface temperature to changes in surface type because of the pronounced effect of variations in surface albedo, slope, azimuth, soil moisture, and micrometerology on the temperature of the target. The recently developed technique of thermal inertia modeling (Watson et al., 1971; Watson, 1975; Kahle, 1977; Kahle et al., 1981; Gillespie and Kahle, 1977; and Price, 1977), has allowed construction of thermal inertia images which can be directly related to a single body property of surface materials.

Digital Image Processing and Enhancement for Geology

Introduction

Digital processing and enhancement of multispectral images are used directly or indirectly by many geologists to extract from the digital image data information that is of interest and which is often only subtly expressed on conventionally processed images. For many geologists the objective of digital image processing is simply to improve the interpretability of the basic image. For other geologists the objective of digital image processing is to be able to combine different data sets, to apply various processing techniques to the image data, or to use various sophisticated mathematical analytical procedures on the digital image data, such as band ratioing or principal components analysis.

Various books, book chapters, and journal articles, which address the subject of digital image processing for geological investigations have already been discussed and cited in an earlier part of this section of the chapter. Chapters 17, 21, and 24 of this *Manual* should also be consulted. Of particular interest to geologists are the works by Taranik (1978), Condit and Chavez (1979), and Gillespie (1980). Two other general books on the subject are the ones by Bernstein et al. (1978) and Moik (1980).

Digital processing and enhancement of Landsat computer compatible tape (CCT) data can markedly improve the interpretability of a standard Landsat MSS image, either a single band image or a false-color composite image. The routine use of the EROS Digital Image Processing System (EDIPS) by the U.S. Geological Survey's EROS Data Center began about a year after the launch of Landsat 3 (5 March 1978) or on 1 February 1979. Prior to this time all standard Landsat photographic images distributed by the EROS Data Center were processed by the electron-beam image recorder (EBR) at NASA's Goddard Space Flight Center as 70-mm positive or negative film transparencies and then shipped to the EROS Data Center for archiving. Any CCT's needed for such pre-1 February 1979 Landsat images need to be ordered from NASA through the EROS Data Center.

The EBR-processed Landsat image often does not show all the spectral information recorded by the multispectral scanner (MSS) in one or more of the four bands. This is particularly true of Landsat scenes of areas of high reflectivity, such as desert areas, snow- and ice-covered polar and temperate regions and glaciers (See Color Figures 31-176 and 31-201). As a result, many perfectly good Landsat images are only marginally useful because of the poor information content in the EBR-generated film product. Many geologists are not aware that a variety of digital processing and enhancement techniques can be applied to CCT's of such images resulting in an extraordinary improvement in the interpretability of many of these Landsat images (see Color Figures 31-75 and 31-76).

In late 1976, the EROS Data Center began to offer computer-enhanced Landsat images, in which the enhanced image was "produced from digitally preprocessed Landsat computer compatible tapes (CCT's), on a laser beam film recording system (U.S. Geological Survey, 1977c)." As previously indicated, the image processing system used to create computer enhanced images was called the EROS Digital Image Enhancement System (EDIES), the test-bed or precursor design to the operational EROS Digital Image Processing System (EDIPS) (Ragland and Chavez, 1977). Table 31-20 is a list of the image processing algorithms used in the EDIES process (U.S. Geological Survey, 1977c).

The EDIPS uses the same EDIES algorithms, modifications of the EDIES algorithms, and new algorithms, especially those which improve image geometry. It should be noted that many of the specialized algorithms used in digital image processing had their genesis in the Image Processing Laboratory of the California Institute of Technology's Jet Propulsion Laboratory (JPL), a result of JPL's pioneering research on digital processing and enhancement of images of other planets and moons of the Solar System, and from digital image processing research by the U.S. Geological Survey's Branch of Astrogeologic Studies.

Color Figure 31-75 is a standard EDIPS Landsat MSS false color composite image (10102-10361) of the Western Sahara in the border area between Algeria, Mauritania, and Morocco, northwestern Africa. The image was acquired on 2 November 1972 and was originally processed as a 70-mm film transparency on NASA's EBR. Color Figure 31-75, however, was processed as a standard EDIPS product. The resulting image is "washed out" and is similar to the Landsat MSS false color composite generated from the EBR-generated 70-mm film transparencies. Color Figure 31-76 is a custom-processed EDIPS Landsat MSS false color composite image (10102-10361; 2 November 1972) of the identical scene shown in color Figure 31-75. Color Figure 31-76 is a striking example of how properly processed CCT data can yield Landsat images with markedly improved interpretability. Another spectacular custom-processed EDIES image of the Vatnajökull area of Iceland is shown in Color Figure 31-201. Spectral information contained in the CCT data is often not visible on images resulting from either "standard" EBR- or EDIPS-processed images, and the geologist should always be aware of this fact.

Another way in which digital image processing can be used to improve the interpretability of Landsat MSS images is to combine different types of image data or other types of data (see section in this chapter on *Combined Data Sets*). Color Figure 31-77 is a Landsat MSS false color composite image (5137-17505, 3 September 1975) of the San Francisco area, California, combined with a Landsat 3 RBV image (30166-18065; 18 August 1978). By registering the 79-m-pixel MSS bands 4, 5, and 7 data with the 30-m-pixel RBV data, the information content of the MSS image was augmented by the 2.6-fold improvement in spatial resolution provided by the RBV data. Color Figure 31-78 is a set of three Landsat images: MSS, RBV, and merged MSS and RBV of part of the city of San Francisco, the enlarged southwestern portion (one-eighth) of Color Figure 31-77.

TABLE 31-20

EDIES Processing Algorithms

1. Radiometric restoration (destriping)
2. Sampling geometric restoration
3. Earth rotation correction
4. Detector misregistration correction
5. Contrast enhancement
6. Edge enhancement
7. Synthetic line generation

The following two sections describe the use of digital image processing techniques to enhance lineaments on MSS images, by Pat S. Chavez, Jr., and to process MSS data for mapping geologic units, by Melvin H. Podwysocki and Donald B. Segal. These are two applications to geology which are well documented by many investigators. More specific uses of these digital image processing techniques can be found in the various case histories described in the section of this chapter dealing with *Applications of Remote Sensing Technology to Geologic Subdisciplines*. The final part of this section of the chapter, by Charles M. Trautwein, addresses the subject of *Combined Data Sets*.

Enhancements for Lineament Identifications

The identification of lineaments has important ramifications in geology, because such features can represent faults and fracture zones where offset has not occurred (Sabins, 1978). These features, in turn, can signify potentially hazardous or economically important environments. For example, they can serve as conduits for ground water at the present time, or faults and fracture systems may have acted as favorable settings for ore deposition in the geologic past. Faults can also identify zones of seismic risk, areas of subsidence caused by the rapid withdrawal of ground water, or structures associated with hydrocarbon reservoirs. This section describes several image processing methods that can be used to enhance linear features on digital MSS images for visual interpretation. Additional information regarding the mapping of lineaments using Landsat MSS data can be found in papers by Gold and Parizek (1976) and Babcock and Sheldon (1976). Also, Saunders and Hicks (1976) discuss the application of lineament mapping in different areas of geologic research, and a paper by Bechtold et al. (1976) describes the use of images with different resolutions to map lineaments in a coal mining area.

For details on the statistical treatment of lineament information the reader is referred to Wheeler and Stubbs (1976) and Haman (1976). Examples of mapping lineaments using aeromagnetic data are given by Schwab and Frohlich (1976) and Barosh (1976).

The digital processing techniques discussed in this section are designed to generate high quality photographic products for the geologist to use for visual interpretation. This section does not describe any of the work that has been done in the area of automatic edge detection, however, most of which is not applicable to geology. For a discussion on the various methods used to automatically detect edges, see Abdou (1978). An excellent comparison between linear, semi-linear, and non-linear edge detectors is given by VanderBrug (1976). A paper by Gurney (1980) discusses the problems of threshold selection in automatic edge detection, especially in the mapping of edges that are at or below the resolution of the imaging system.

Definition of Terms

There has been confusion in the literature regarding the meaning of the terms used to describe the linear features that occur on remotely sensed images. The definitions presented by O'Leary et al. (1976) are followed in this section. *Lineament* refers to "a mappable, single or composite linear feature of a surface, whose parts are aligned in a rectilinear or slightly curvilinear relationship and which differs distinctly from the pattern of adjacent features and presumably reflects a subsurface phenomenon." The term linear "is an adjective that describes the linelike character of some object or objects" (O'Leary et al. 1976).

Enhancement Methods

Various techniques have been used to enhance lineaments found on photographs and remote sensor images. Besides the use of digital imaging systems, optical processing methods, such as the Fourier transformation, have been used to enhance linear features on photographs (Wildey, 1967; Andrews, 1970; and Lepley, 1976). A simple non-digital device that can also be used to enhance lineaments on photographs and other images is the Ronchi ruling, as pointed out by Pohn (1970). The Ronchi ruling is an inexpensive coarse diffraction grating which has closely spaced parallel black lines on glass or mylar. By a slow rotation of it in front of a photograph or image, those linear features that are perpendicular to the grating lines become enhanced and those that are parallel to the grating become suppressed (i.e., directional filtering). A good example of the use of the Ronchi ruling is given by Offield (1975). Offield used the Ronchi ruling to identify several sets of linear features which "provided a basis for the definition of major structural zones and, potentially, for the subdivision of the Colorado Front Range into domains of uniform fracture character."

In the late 1960s and early 1970s digital image processing techniques began to be widely used to help in the analysis of remotely sensed data acquired from various planetary (extraterrestrial and terrestrial) digital imaging systems. Various methods were developed to specifically extract geologic information from digital images (Rowan et al., 1974; Goetz et al., 1975; Berlin et al., 1976). Some of the information-extraction techniques were developed to enhance high frequency digital information contained in the digital image for structural analyses.

The spatial information in an image can be considered as being composed of low and high frequencies. The low frequency component is usually represented by large areas with constant brightness, which in the case of Landsat multispectral scanner (MSS) data, usually represents the albedo or color information. High frequency information consists of brightness changes over a short spatial dimension that occur because of contrast in slope attitude or topographic features or contrast in brightness at boundaries between different geologic units.

Various digital image processing techniques, such as the Fast Fourier Transformation (FFT) (Andrews, 1970), have been developed to suppress the low frequency component and enhance the high frequency component. This particular technique can be very time consuming, especially on some of the miniprocessing systems, unless a special purpose array processor or hardwired system is used. The need for a special purpose system to do FFT's becomes even more important as the size of the digital image array increases due to increased sensor resolution. Examples include Seasat synthetic aperture radar (SAR) images, Landsat 3 return beam vidicon (RBV) images, and the Landsat 4 thematic mapper (TM) images.

A digital spatial filtering technique that is much more efficient than that of taking the transform of a digital image was developed by the Jet Propulsion Laboratory (JPL) to process planetary images (Seidman, 1972). The filter works with data in image space and gives equal weight to every picture element (pixel) within the filter window or kernel used. Use of equal pixel weights allows the running average used in the spatial filter to be updated from pixel to pixel with very simple and fast addition and subtraction instructions. This particular filtering technique is known as a subtractive boxcar filter, because it works by subtracting the average value of the kernel being used from the digital value of the pixel at the center of the kernel (Seidman, 1972; Gillespie, 1976; Chavez et al., 1976). The size of the rectangular kernel used is often determined by the size of the linear features that the geologist is interested in mapping. The high frequencies or "details" that are enhanced can vary from very fine information that is close to the resolution limit of the imaging system to structural information hundreds of pixels long.

In the remainder of this section, examples of some digital processing techniques that can be used for linear feature enhancement and mapping will be shown. The two test areas used for demonstration purposes (southwest Jordan and Montrose, Colorado) are each covered by one-fourth of a Landsat MSS image. The data presented in this section are derived from two previous studies. In the Jordan test site, high-pass spatial filters were used with different size kernels to enhance medium- and large-sized linear features and topographic patterns. The first difference of the image, which approximates the first derivative, was also used to enhance very small linear features and patterns.

For the test site in Colorado, an edge enhancement technique is used to show how very small linear features can be enhanced while still preserving the albedo information contained in the image. This area is also used to compare the Landsat MSS image with a shaded relief image derived from Defense Mapping Agency (DMA) digitized topographic data. Finally, the DMA topographic data are combined with the MSS digital image to generate a Landsat stereo pair.

The subtractive spatial filter technique was used by Chavez et al. (1976) to analyze a portion of a Landsat MSS image of an area in southwest Jordan. The region studied was selected because of its arid climate and the wide range of rock types encountered within a very small area, as shown in the geologic map (Color Figure 31-79); and, because of the complex structure containing numerous fault systems (Figure 31-80A). Geologic units range in age from Precambrian to Quaternary, and include metamorphic, plutonic, volcanic, sedimentary, and alluvial materials (Bender, 1967, 1968a, and 1975a). The most prominent geologic feature is the Wadi Araba, an extension of the East African-South Turkey rift system (Bender, 1974). The Wadi Araba is a depression approximately 1,200 m below the level of surrounding terrain.

The MSS digital data for the area were subjected to two types of image processing: image correction and image enhancement. Image corrections eliminated undesired artifacts and distortions, rendering "clean" data bases; such data bases incorporate corrections for atmospheric scattering, Sun elevation, destriping, missing-lines, and geometric errors. Details about these corrections are given by Chavez et al. (1977). Additional information can also be found elsewhere in this Manual in Chapters 17, Data Preprocessing and Processing; 21, Image Geometry and Correction; and 24, Manual and Digital Image Analysis in the Visible and Infrared Regions. The image correction stage can be an important step in linear feature enhancement because some of the techniques applied (e.g., destriping and missing-line removal) improve the results for visual analysis. The image enhancements used were aimed at extracting the

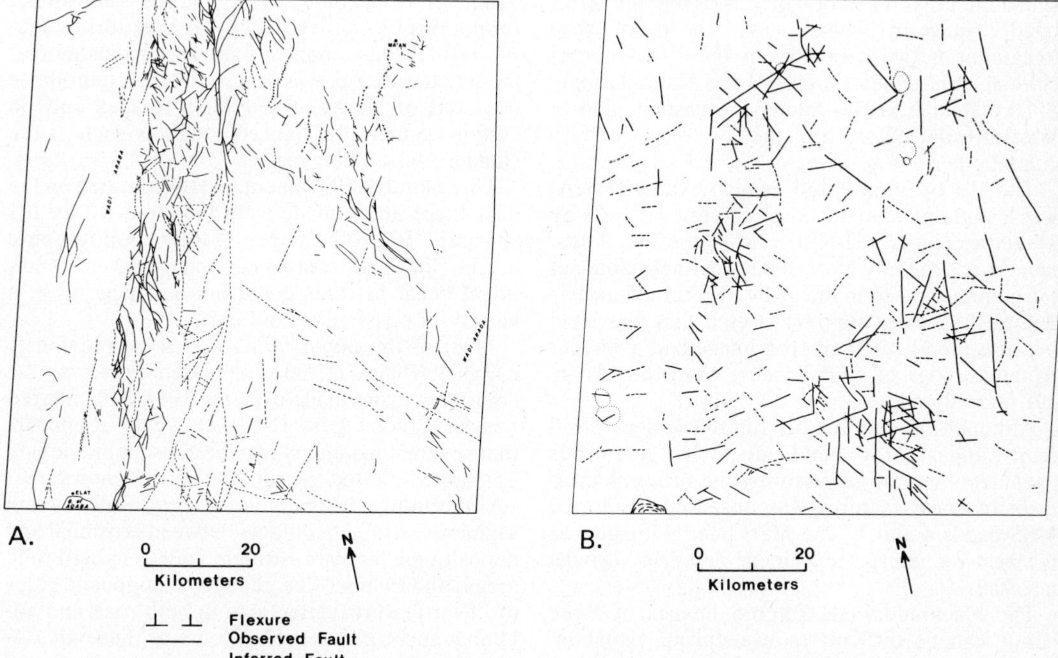

Fig. 31-80. Comparison of faults and flexures mapped on a geologic map of southwest Jordan and lineaments mapped from an enhanced Landsat image of the same area. *A*. Mapped faults and flexures of southwest Jordan and adjacent areas on the *Geologische Karte von Jordanien,* Aqaba-Ma'an, 1:250,000 (Bender, 1968b). *B*. Lineaments (not seen in A.) mapped from Landsat image (1342-07430; 30 June 1973) enhanced by digital spatial filtering. Solid lines = strong lineament expression, dotted lines = subtle lineament expression.

structural information in the image at several different frequencies. The techniques used were the spatial filtering technique discussed above and first differencing. The first difference is computed simply by taking the difference of two adjacent pixels in either the horizontal, vertical, or diagonal direction. The first difference of the image, which approximates the first derivative for first and second order functions, was employed to highlight the "structural fabric" or the maze of very small linear features that are at the resolution limit of the imaging system.

A spatial filter will generally enhance features that are less than half the size of the kernel being used and suppress features that are more than half the kernel size. As a preliminary step, two-dimensional spatial filtering was used to generate images that enhanced both regional and intermediate-level linear features. The kernel sizes used for linear feature enhancement were 101 by 101 pixels (approximately 8,000 meters/side) and 31 by 31 pixels (approximately 2,500 meters/side). Kernel sizes were kept equidimensional to ensure that all directions would be weighted equally. This type of a kernel also helps to reduce the introduction of linear artifacts that can be seen in directional digital filtered images as described by Gillespie (1976).

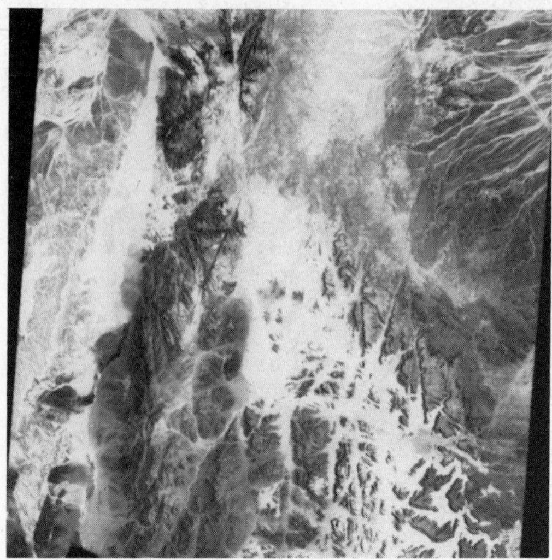

Fig. 31-81. MSS band 7 of one-fourth of a Landsat image (1342-07430; 30 June 1973) of southwestern Jordan and adjacent areas. A linear contrast stretch has been applied to the image after image correction techniques were used to remove geometric distortions and photometric artifacts.

Examination of the enhanced images revealed numerous linear features in all rock types, on alluvial fans along the eastern fringe of Wadi Araba, on the floor of Wadi Araba, and in unconsolidated sediment areas east of Quiveira (see the MSS band 7 image in Figure 31-81). The most dense lineament network occurs in the Precambrian crystalline basement, a part of the Nubo-Arabian Shield (Bender, 1974), along the eastern side of Wadi Araba. Short lineaments are especially common here.

The 101 by 101 filtered image (Figure 31-82A) was less grainy than the image filtered with a 31 by 31 kernel (Figure 31-82B) and, therefore, lineaments are generally more clearly defined. Some of the graininess seen in the 31 by 31 filtered image is due to the harder contrast stretch that was used because the 31 by 31 filtered image had a smaller dynamic range of digital values than the 101 by 101 filtered image.

Although the 101 by 101 filtered images of all four Landsat MSS bands displayed numerous lineaments in various terrains, the best enhancement of linear features was observed in filtered MSS bands 4 and 7. The MSS band 4 image was the best for interpreting linear detail in surficial material.

The black-and-white filtered images of three bands can be difficult to use during visual interpretation because often there can be linear features that will show-up only in one or two bands (e.g., linear features due to vegetation differences might show-up only in bands 6 and 7). To minimize this problem, three filtered images can be used to make a color composite. The 101 by 101 filtered images of MSS bands 4, 5, and 7 were incorporated into a single false-color composite image (Color Figure 31-83). This was done by using the spatially filtered images of MSS bands 4, 5, and 7 through blue, green, and red color filters, respectively, and printed as a color image. Analysis of this composite product indicated that the linear patterns were more detailed than those depicted on unprocessed MSS images and on certain computer enhanced products such as the standard false color image.

Three kinds of lineaments were identified on the false-color image made from the three 101 by 101 spatially filtered images: faults, topographic crests, lithologic contacts, and flexures. Additional linear features could not be categorized; a variety of possibilities did exist.

Most of the major faults that were previously mapped (Figure 31-80A) are identifiable on the color composite made from the 101 by 101 filtered images (Color Figure 31-83). However, it appears that several lineaments could represent previously unmapped faults. Some of the indicators that seem to imply this include: straight wadi channel segments, straight contacts between erosional and depositional features, straight valleys in hard rock areas, and distinct hue changes on opposite sides of linear features detectable in both rock and adjoining surficial materials. Some of these are depicted in Figure 31-80B.

The image generated by the horizontal first difference technique (Figure 31-82C) proved to be an excellent data source for obtaining a perspective of the area's "structural fabric" (see also Figure 31-110, a derivative Landsat MSS image of

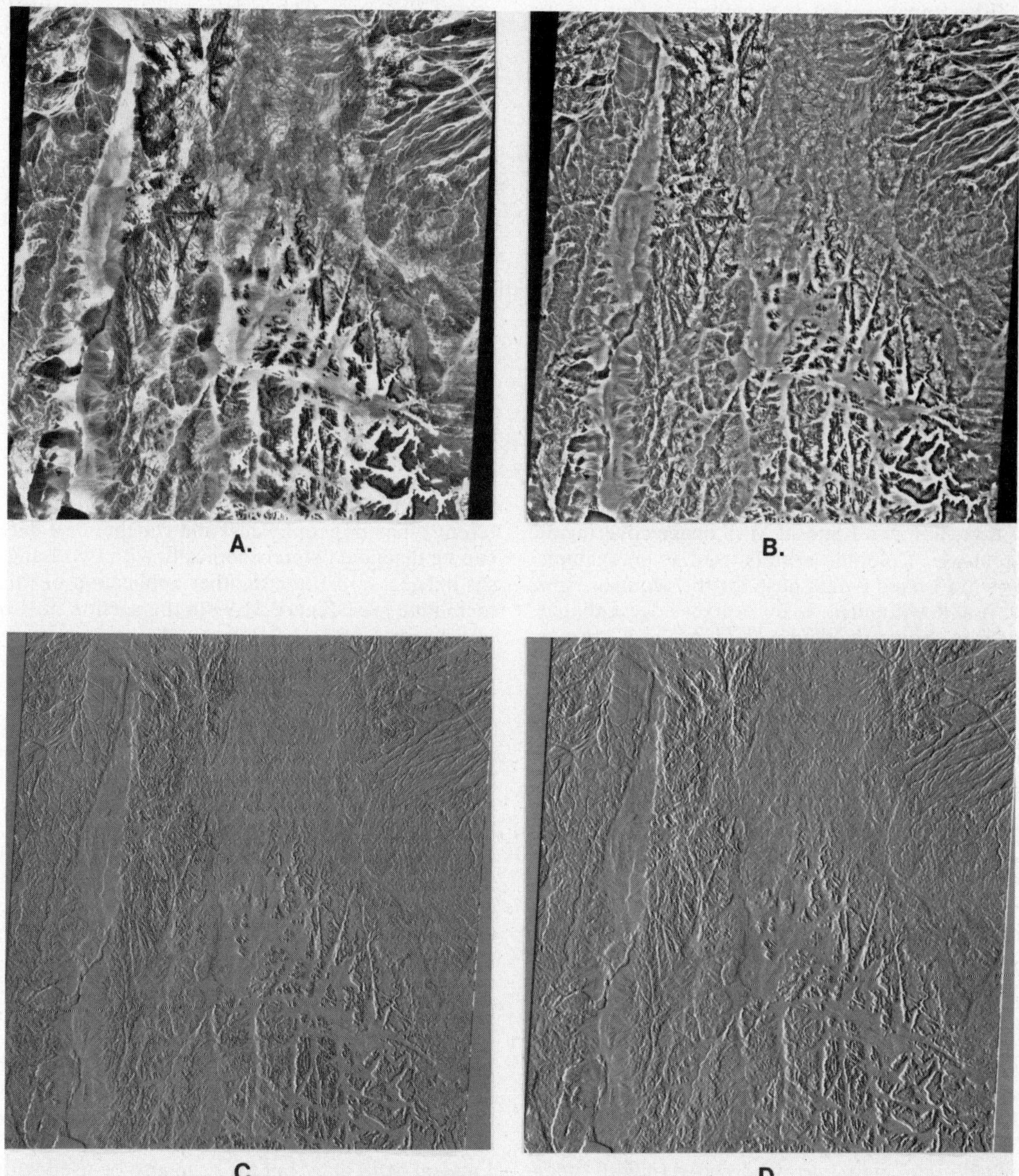

Fig. 31-82. Group of four Landsat MSS band 7 images of southwestern Jordan (1342-07430; 30 June 1973) to which various digital spatial filtering techniques have been applied. *A.* MSS band 7 filtered image using a 101 by 101 kernel with a linear contrast stretch applied to the filtered product. *B.* MSS band 7 filtered image using a 31 by 31 kernel with a linear contrast stretch applied to the filtered product. *C.* Horizontal first difference image of MSS band 7 with a linear contrast stretch. *D.* Same image product as the one shown in Figure 31-82C, but with a 3- by 3-pixel smoothing filter applied to the horizontal first difference after a linear contrast stretch.

northeastern Iceland). This type of linear pattern is not detectable on standard MSS images or on those images that have been computer processed to enhance large lineaments or gross lithologies. Using a very small (3 by 3 or 5 by 5) smoothing filter on the first difference can help to suppress random noise and to better connect various trends together for better discrimination (Figure 31-82D).

Because a regular high-pass spatial filter removes most of the albedo information from an image, its use is not always advisable. For example, for a soil scientist or geologist interested in mapping drainage patterns, where both lithology and very fine structural information could be important, the output of a regular high-pass spatial filter might not be an optimum product to use. A standard false-color composite with edge enhancement might be better.

Edge enhancement is a technique that can be considered a first order correction to the modulation transfer function (MTF) of an image (Chavez, 1975). It improves the fine detail/definition in the standard false-color composite image which does contain albedo information that is needed for mapping lithology. Edge enhancement is often achieved through application of the spatial filtering technique discussed above. A spatially filtered product is generated using a very small kernel, usually ranging from 3- by 3- to 9- by 9-pixels; the original image is then added to the filtered image (i.e., a spatial filter is used with 100 percent addback). The kernel size that should be used for edge enhancement must be selected based on the number of edges or how "busy" the particular image is. A method was developed by Chavez and Bauer (1982), which uses the standard deviation of the horizontal first difference to select the kernel size that should be used.

A portion of a Landsat MSS image covering the Montrose, Colorado area is used as an example. The MSS band 7 data base for the Montrose area with and without a 5- by 5-pixel edge enhancement is shown in Figures 31-84A and C, respectively. The image processed through the 5-by 5-pixel spatial filter with zero percent addback (i.e., the spatial filter output before being added to the original image) is shown in Figure 31-84B.

Shaded Relief

If the primary objective of the photointerpreter is to interpret structure only from topographic information, it might be better, in certain cases, to consider using digitized topographic data rather than Landsat MSS products. Even spatially filtered images will generally show linear features that are caused by non-topographic information (i.e., cultural features), and can make it difficult to map only topographically caused lineaments. The advantage of using the digitized topographic data, such as the data available from the National Cartographic Information Center (NCIC) of the U.S. Geological Survey, is that the information contained in the digital array is related only to topography. The digital topographic data can be used to generate shaded relief images (Batson and others, 1975). Such images can be generated very efficiently on most computers, and several images representing a shaded relief image from three different Sun azimuths with two or three Sun elevation angles for each azimuth (e.g., 45, 30, and 20 degrees) can be produced. By the making of shaded relief images from three different Sun azimuths, lineaments that have a particular directional trend will be close to optimally enhanced in one of the resultant images. An example of a shaded relief image made from the standard Defense Mapping Agency (DMA) tapes, having an approximate 60-m spatial resolution, is shown in Figure 31-85. Currently, digital elevation model (DEM) data, also available from NCIC, are being synthesized with an approximate spatial resolution of 30 meters. These data, however, are available for only limited areas.

Digital topographic data can also be used to generate a Landsat stereoscopic pair (Batson et al., 1976). By identifying control points in both the topographic data and Landsat images and using digital geometric correction procedures, such as those discussed in Chapter 21 (*Image Geometry and Correction*) of this *Manual,* one can register the digital topographic data to the Landsat data. Once the registration has been completed, parallax can be introduced to the Landsat data in the horizontal direction. Parallax is either to the left or right based on the height difference represented in adjacent pixels. Figures 31-86A and B are a Landsat stereoscopic pair generated from an image of the Montrose, Colorado, area; a 51- by 51-pixel high-pass spatial filter was applied to the images.

Combined stereoscopic data sets, in which different types of geophysical and geochemical data can be displayed stereoscopically with two Landsat images, constitute another application of this technique (See Figure 31-99 in the section of this chapter on *Combined Data Sets*).

Summary

There are many different digital image-processing techniques that can be used to enhance or map linear features. This section has covered only a few of these techniques which are mostly designed to generate a high-quality image product for use in direct visual analysis. Spatially filtered and first-differenced products have been successfully used to enhance linear features. They are both produced by simple techniques which can be implemented efficiently on a digital computer. Digitized DMA and DEM topographic data are becoming more widely used in various digital processing techniques. For linear feature enhancement, these data sets can be used to either generate shaded relief images or to create a synthetic Landsat MSS stereo-pair. The techniques discussed in this section were applied only to Landsat MSS data, but they can also be used with other data sets, such as Landsat RBV data, digitized aerial photographs, radar images, and sidescan-sonar images.

Selected Methods for Digital Processing of Multispectral Scanner Data for Mapping Lithologic Units

Introduction

This section will describe several digital image enhancement techniques for distinguishing and identifying geologic units based on the spectral properties of geologic materials. Spectral properties, which can be derived from either airborne or satellite multispectral scanner (MSS) data, serve as a basis for lithologic mapping, because they allow the geologist to discriminate and perhaps even identify geologic units that appear

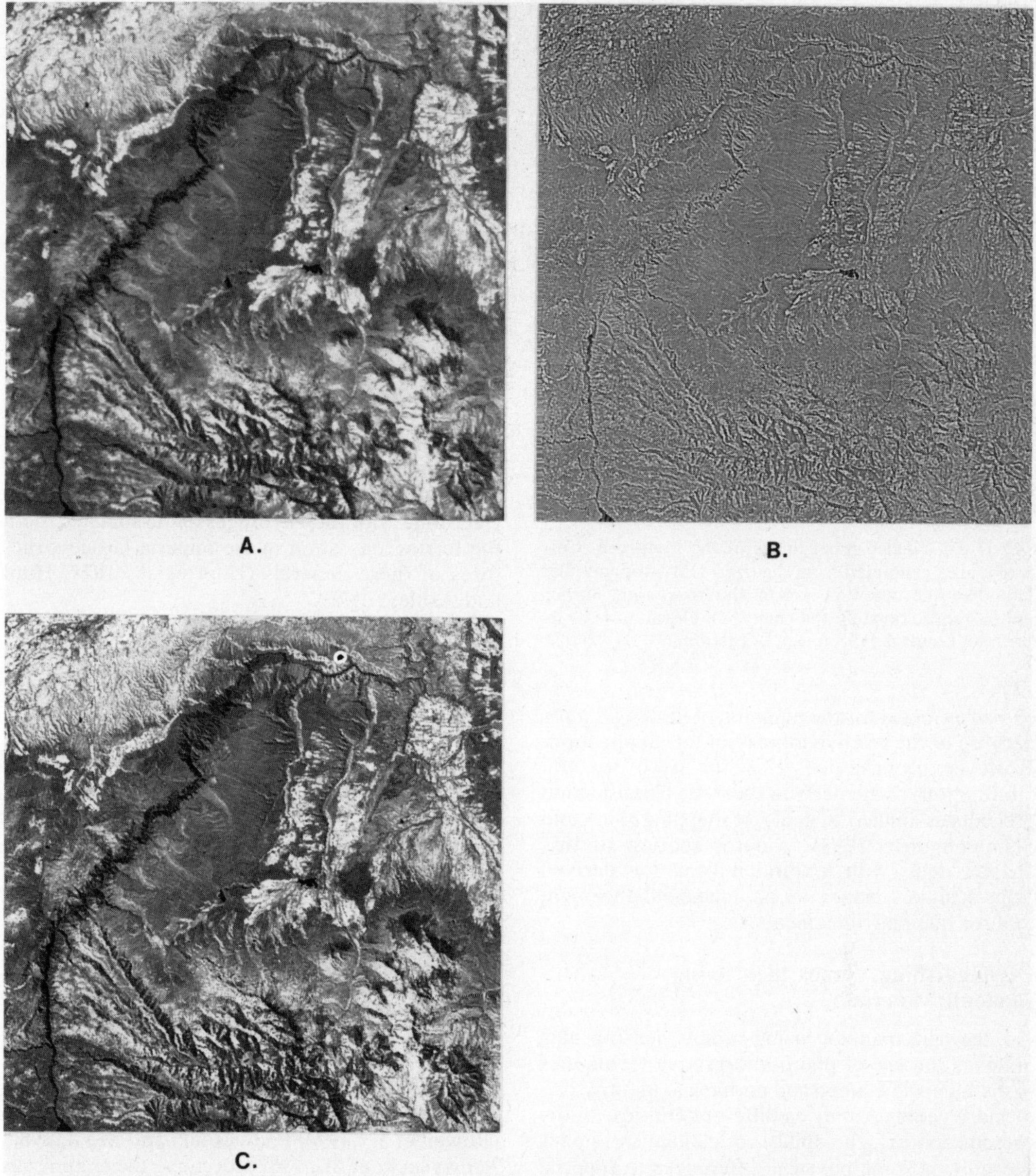

Fig. 31-84. One-fourth of a Landsat MSS image (1407-17190; 3 September 1973) of the Montrose, Colorado, area to which different digital image enhancement techniques have been applied. *A*. Landsat MSS band 7 image to which a linear contrast stretch has been applied without an edge enhancement. *B*. The same image processed through a spatial filter. The kernal size of the spatial filter was 5- by 5-pixels. *C*. Edge-enhanced version of A, generated by adding the filtered and original images (filter with 100 percent addback).

on an image. Several procedures commonly applied to MSS data will be discussed in detail, and examples from arid to semiarid regions will be presented. Procedures to be discussed include contrast enhancement, band ratioing, and principal component and inverted principal component analyses. References will be given to additional examples in the literature. Because of space constraints in this *Manual,* it is not possible to cover all possible enhancement procedures. However, several other procedures which have documented geologic applications will be referenced with a brief discussion of their utility. Digital thematic classification procedures, which also can be used for lithologic mapping, will not be discussed here, because this section is devoted primarily to digital image enhancements of scanner data to which photointerpretive techniques commonly are applied. Textural information, which also can be derived from the scanner data, will not be dis-

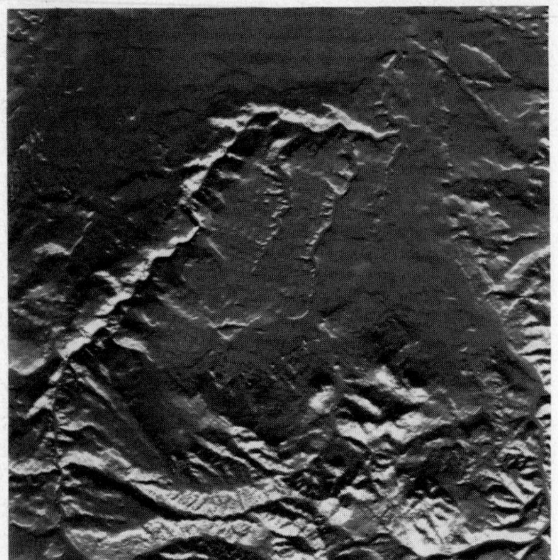

Fig. 31-85. Shaded relief image of the Montrose, Colorado, area generated from digitized DMA topographic data. The Sun direction is from the lower right corner and has approximately the same Sun elevation (35°) as does the Landsat MSS image of the area.

cussed as an aid for the mapping of lithologic units because of the relative infancy of this application. However, promise does exist for use of textural information, particularly in thematic classification procedures applied to geologic mapping (Shih and Schowengerdt, 1983). A later section of this chapter deals with textural information derived from scanner images as an image-enhancement tool for mapping lineaments.

Distinguishing Versus Identifying Geologic Materials

Lithologic mapping using remote-sensing data involves the use of photo-interpretive techniques to distinguish between and perhaps to identify two or more geologic units as different entities. In the present context, the ability to distinguish implies recognizing some physical differences in spectral or spatial information on the image. Geologists are able to recognize and map units on an image based on a number of criteria, such as differences in overall brightness, spectral contrasts between two or more compared scanner bands, and spatial information, such as drainage density or pattern and spatial patterns of individual units. With few exceptions, however, only intuitive guesses may be made as to the cause of those contrasts without consulting previously published data or visiting the area on the ground.

A goal of geologic mapping is to actually identify geologic materials in terms of their age, lithology, and structural relationships. Remote sensing can do nothing to determine absolute age, but relative ages can be inferred from stratigraphic position and other spatial relationships can often be interpreted from an image. Within limits, some lithologies can also be identified. To identify a geologic material implies the recognition of specific physical properties which allow the analyst to predict the composition of the geologic unit.

Special Properties of Geologic Materials

Several physical properties of geologic materials can be derived from remote-sensing data. Ferric iron-bearing oxides and oxyhydrides, such as goethite, hematite, lepidocrocite, and other ferric iron-bearing minerals, such as jarosite, have distinctive broad absorption bands centered in the ultraviolet, visible, and near-infrared portions of the electromagnetic spectrum. These minerals often are referred to by the field term limonite (Blanchard, 1968). The absorption bands associated with limonitic minerals are caused by electronic vibrational processes associated with the ferric-iron cation in the mineral lattice structures of these minerals (Hunt et al., 1971; Hunt and Ashley, 1979).

In situ field spectra for some representative types of rocks and vegetation are presented in Figure 31-87. The overall depression in the spectral response curves of limonitic materials (spectra 2, 3, and 4, Figure 31-87) below 0.5 μm is caused by an intense ferric iron absorption band centered in the ultraviolet region. A non-limonitic rock (spectra 1, Figure 31-87) lacks the absorption. Other spectral features attributable to the presence of ferric iron occur as an inflection in the curve at 0.65 μm (spectra 4, Figure 31-87), and broad, shallow absorption bands in the 0.85- to 0.95-μm region (spectra 2 and 4, Figure 31-87). The depth and position of the absorption band in the last region will depend upon the species of ferric iron-bearing limonitic mineral present.

The identification of limonitic rocks utilizing Landsat and aircraft MSS data is being done routinely as an exploration method for delineating potential areas of hydrothermally altered rocks (Rowan et al., 1974; Podwysocki and Segal, 1980; Podwysocki et al., 1982), because these minerals often are found either as the primary minerals or secondary weathering products in hydrothermally altered rocks. Hydrothermally altered rocks commonly contain pyrite formed as part of the alteration process which then weathers to form limonitic minerals. However, not all limonitic rocks are hydrothermally altered; limonite also may result from diagnetic processes or secondary weathering of iron-bearing minerals in unaltered rocks. Additionally, some hydrothermally altered rocks are bright and non-limonitic (bleached). Thus, errors of omission and commission occur in mapping hydrothermally altered rocks based solely on limonite absorption.

Minerals containing Al-O-H, Mg-O-H (Hunt et al., 1973) or the carbonate radical (Hunt and Salisbury, 1973) in their mineral lattice structures

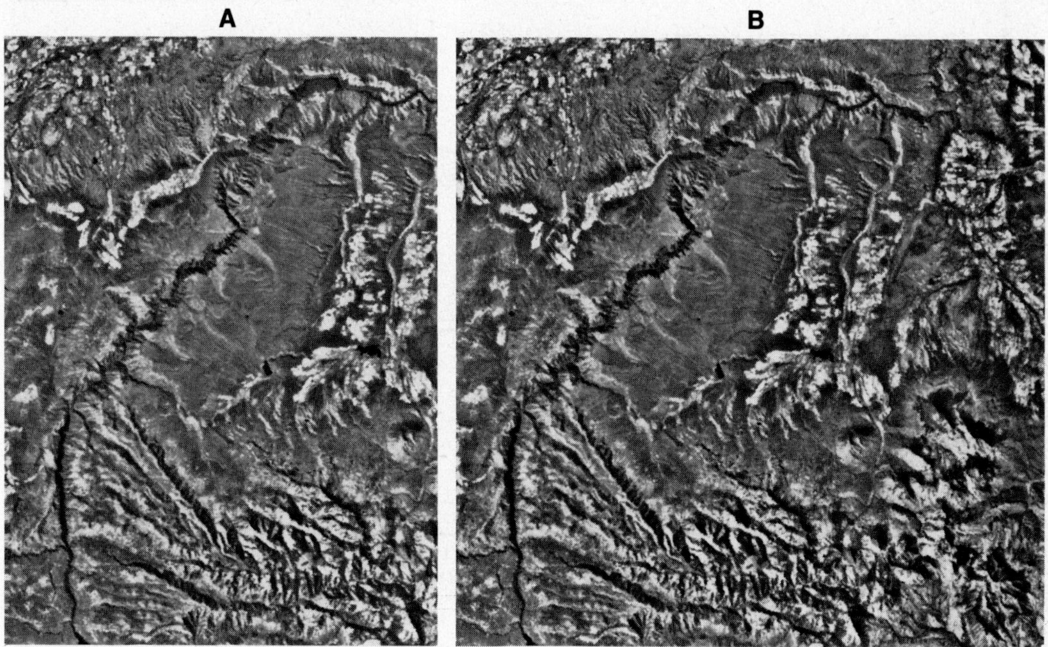

Fig. 31-86. One-fourth of a Landsat MSS image of the Montrose, Colorado, area (1407-17190; 3 September 1973) in which synthetic stereoscopy has been created from digitized topographic data (1:250,000-scale) of the Defense Mapping Agency (DMA). *A.* Left-hand image of the Landsat MSS stereoscopic pair made by using the digitized DMA topographic data. The digital topographic data, geometrically registered to the Landsat MSS image, was used to introduce the parallax into the Landsat MSS image. *B.* Right-hand image of the Landsat MSS stereoscopic pair. The stereo pair images were both processed with a 51- by 51-kernel spatial filter to enhance terrain contrasts.

have narrow and oftentimes distinctive absorption bands in the 2.1-μm to 2.5-μm region. These bands are caused by overtones and bending moments of absorption bands which are centered in the mid-infrared (Hunt and Salisbury, 1970a, 1970b). Carbonates, clay minerals such as kaolinite, montmorillonite, talc, and other minerals, such as jarosite, alunite, pyrophyllite, and sericite, have such diagnostic bands. The exact positions and intensities of the absorption bands often can be diagnostic of specific minerals, but can only be determined with spectroradiometers capable of 0.01 μm spectral resolution. Present imaging multispectral scanners are incapable of this spectral resolution. Spectrum 4 (Figure 31-87) shows such an absorption band located in the 2.2-μm region associated with a hydrothermally altered volcanic rock containing alunite and kaolinite. Note that unaltered rocks (spectra 2 and 3, Figure 31-87) show little or no absorption in this region. Clays and other minerals diagnostic of hydrothermally altered rocks have been mapped using aircraft MSS data (Ashley and Abrams, 1980; Rowan and Kahle, 1982; Podwysocki et al., 1982a). Furthermore, these authors have been able to discriminate between several types of altered rocks. The identifications are not unique, because clay minerals also are found in sedimentary rocks. However, the ability to sense these absorption bands from aircraft and from the Landsat 4 Thematic Mapper (TM) does give the geologist another tool for identifying lithologies or at least restricting the number of lithologies which must be considered.

Two mid-infrared phenomena bear mentioning, although less work has been carried out in this portion of the electromagnetic spectrum compared to the visible and near-infrared. Minerals containing Si-O, Al-O-H, Mg-O-H, and C-O bonds in their mineral lattice structures show distinctive absorption bands in the 6.0-μm to 12.0-μm region (Hunt and Salisbury, 1974, 1975, and 1976). Kahle and Rowan (1980) have shown the usefulness of absorption bands in the mid-infrared in discriminating rocks containing varying amounts of silica.

Thermal inertia, which is a measure of a material's response to a cyclical period of heating and cooling, is affected by the density and thermal conductivity of the sensed material. Thermal inertia, as sensed by optimal consecutive day and night passes utilizing scanner bands in the visible and mid-infrared, has been used to map geologic units. Ultramafic rocks have characteristic thermal inertias (Pohn et al., 1974) and dolomites have been distinguished from linestones and granites (Rowan et al., 1970) using derivative images of thermal inertia.

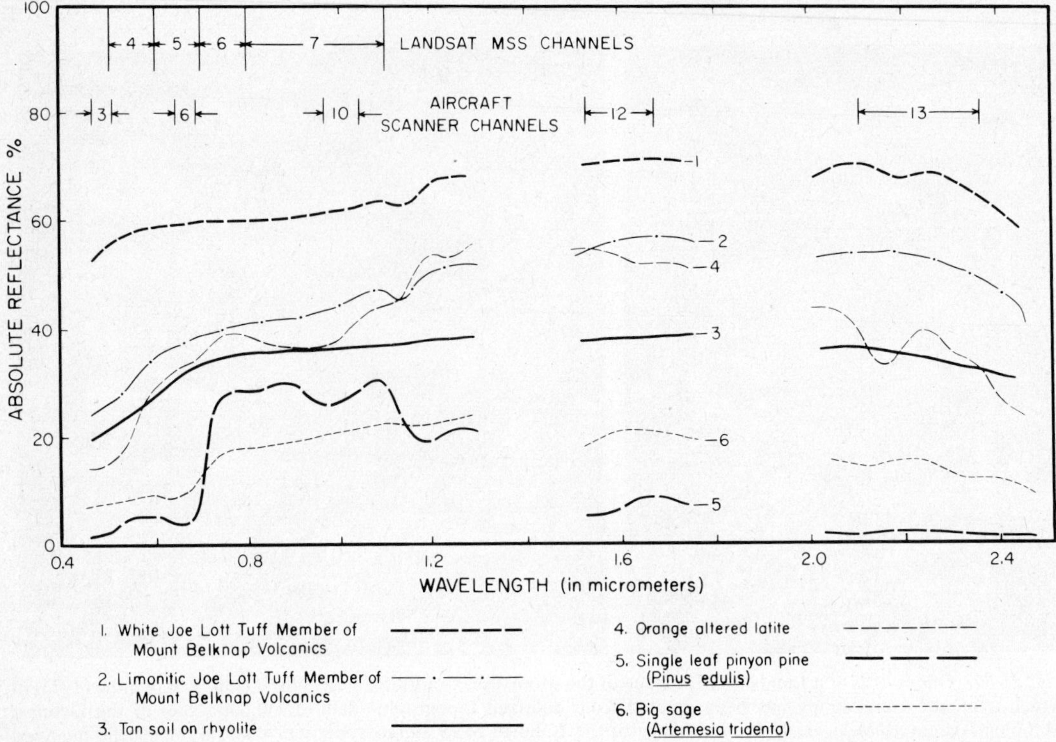

Fig. 31-87. Representative *in situ* field spectra of materials discussed in this section. Spectra were collected using a U.S. National Aeronautics and Space Administration-Jet Propulsion Laboratory portable field reflectance spectrometer (Goetz et al., 1975).

All of the above discussed phenomena are distinctive for an individual or group of individual materials, but because of instrument limitation, no attempt has been made to utilize all these physical phenomena at any one time. With present-day scanner systems, we are only able to uniquely identify limonite. Other identifications are limited to recognition of groups such as the carbonate- or hydroxyl-bearing lithologies. High spectral resolution, non-imaging spectro-radiometers have made identifications for several individual minerals in these groups from airborne (Podwysocki et al., 1983) and spaceborne (Rowan et al., 1982) platforms.

The following sections describe several selected digital image-enhancement techniques useful for both distinguishing between and identifying specific rock-forming material. Examples will be discussed which use either Landsat data of whole scenes or airborne MSS data for more areally limited areas in south-central Utah.

Examples of Lithologic Mapping

Color Figure 31-88A is a digitally processed standard Landsat false-color-composite image of an area centered on the Richfield, Utah, 1 degree × 2 degree quadrangle. The image, which has been contrast stretched using a three-point linear stretch for each MSS channel, is a digitally merged composite of two scenes; the upper 75 percent of this composite scene is derived from the southern part of scene 1699-17353, whereas the lower 25 percent of the scene is derived from the northern part of scene 1699-17355. The images were collected consecutively on 22 June 1974. The digital data for this composite scene were preprocessed to convert the seven-bit values of MSS bands 4, 5, and 6 to eight-bit data by multiplying their values by a factor of two; MSS band 7 data were converted from six-bit to eight-bit by multiplying by a factor of four. The data were then destriped to eliminate the modulo-six line striping present in the Landsat 1 data using a histogram equalization technique. The images were contrast stretched and geometrically corrected prior to recording the images on film. A sketch map of the prominent physiographic features on the image is shown in Figure 31-88B.

The eastern quarter of the scene, including the Sevier and Markungut and associated plateaus, and the Tushar Range lie within the High Plateaus

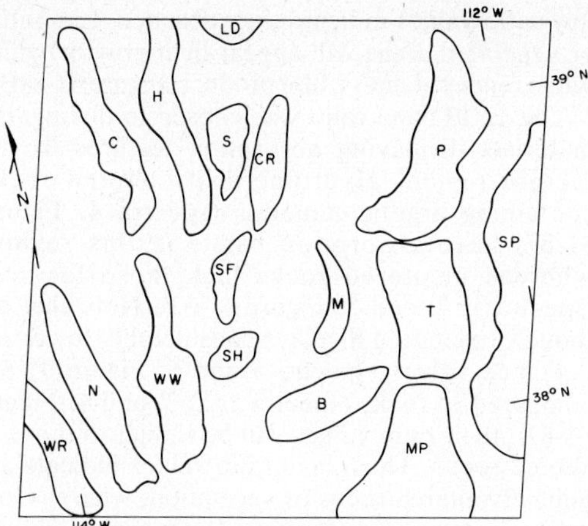

Fig. 31-88B. Sketchmap of physiographic features shown in Color Figure 31-88A. Key to symbols: *B* - Black Mountains; *C* - Confusion Range; *CR* - Cricket Range; *H* - House Range; *LD* - Little Drum Mountains; *M* - Mineral Mountains; *MP* - Markungut Plateau and other plateaus; *N* - Needles Range; *P* - Pavant Range; *S* - Sevier Lake; *SF* - San Francisco Mountains; *SH* - Shauntie Hills; *SP* - Sevier Plateau; *T* - Tushar Range; *WR* - White Rock Mountains; *WW* - Wah Wah Mountains.

subprovince of the Colorado Plateaus province. These physiographic entities form a transition zone between the flatlying rocks to the east, and the Basin and Range province, which occupies the western three-quarters of the scene.

The geology of the area encompassed by the image is quite complex and would be difficult to portray on a map at the scale of the image. The northern half of the scene is underlain principally by sedimentary rocks of Paleozoic age in the west (Confusion, House, and Cricket Ranges) and by Paleozoic and Mesozoic rocks in the east (Pavant Range). The few igneous rocks that occur in the northern area are upper Tertiary to Quaternary volcanics in the sedimentary basin fill (*A* and *B*, respectively, Figure 31-88A) and isolated Mesozoic intrusives in the Ranges (*C*, Figure 31-88A). South of an east-west line defined by the junctions of the Pavant and Tushar Ranges on the east and the Confusion and Wah Wah Ranges on the west, exposed rocks are predominantly igneous, except for the northern portions of the Needles and Wah Wah Ranges, which are primarily of sedimentary origin. The igneous rocks are predominantly volcanic, consisting of Tertiary volcanic vent complexes and extensive regional ashfall units. The only exceptions are the central portion of the Mineral Mountains, which contains a Tertiary granitic batholith, and several small Tertiary stocks scattered throughout the ranges. Most of the southern ranges are extensively mineralized and contain many exposures of rocks that have been hydrothermally altered to argillic, advanced argillic and sericitic stages. Examples of altered areas occur in the Shauntie Hills, Wah Wah, Needles and Tushar Ranges (*A1, A2, A3,* and *BRC,* respectively, Color Figure 31-88A). The Sevier and Markungut and associated plateaus contain mostly Mesozoic sedimentary rocks with some Tertiary volcanic vent and flow facies.

The sedimentary rocks for most of the region have been affected by large-scale eastward thrusting during the Mesozoic. Late Tertiary Basin and Range normal faulting, oriented mainly north-south, has affected both the igneous and sedimentary rocks.

As has been discussed in detail earlier, the Landsat MSS is only able to identify limonitic materials, that is, those materials exhibiting red, pink, orange, yellow and brown colors associated with some ferric iron-bearing minerals. Those materials often will show on standard false-color-composite Landsat images as shades of light yellow, yellow-brown, and occasionally green. Examples of limonitic materials are shown at *A1, E,* and *F* (Color Figure 31-88A).

Band Ratioing

Spectral band ratioing is a proven technique which allows identification of geologic materials based on the recognition of diagnostic absorption bands. Band ratioing is useful because it minimizes the effects of topographic slope, aspect, and albedo differences between rocks and enhances the sometimes subtle differences in reflectivity between bands which are diagnostic of various surficial materials. This technique is most efficiently applied when a rationale exists for the choice of specific bands. Application of this method in a "shotgun" fashion should be discouraged, because it leads to an overwhelming number of individual ratio images and an even greater number of combinations of color-ratio composites. Band ratioing has been used successfully with Landsat data for identifying rocks containing limonite (Rowan et al., 1974; Goetz et al., 1975; Blodget et al., 1978; see also the section on "Lithology" of this chapter). The method also has been applied to aircraft MSS data for detecting sericitized and argillized rocks (Rowan and Kahle, 1982; Podwysocki et al., 1982). Hydrothermally altered rocks, particularly volcanic rocks, are characterized by minerals such as sericite, jarosite, alunite, and clay minerals, which have relatively narrow absorption bands in the 2.2-μm region (Hunt, 1981).

Multispectral scanner data collected by a U.S. National Aeronautics and Space Administration 24-channel airborne scanner were analyzed for an area near Marysvale, Utah. The study area, which includes the Big Rock Candy Mountain (*BRC,* Color Figure 31-88A) in the Tushar Range, is underlain primarily by middle to upper Tertiary intrusive and extrusive rocks derived from several local volcanic centers. Many of these volcanics have been hydrothermally altered by sulfuric acid-rich waters associated with multiple periods of volcanism. The scanner channels used in this analysis are shown in Figure 31-87. Band ratios were specifically chosen to accentuate differences in the spectra of the rock and vegetation curves shown in Figure 31-87. The choices made allowed discrimination of bleached white argillic rocks, argillic rocks containing limonite, limonite rocks containing no argillic component, rocks containing neither an argillic nor limonitic component, and vegetation.

A 6/10 band ratio was chosen to distinguish the differences between rocks, which have relatively gently sloping spectral curves between the two bands (spectra 1, 2, 3, and 4, Figure 31-87), and vegetation, which has a marked increase in reflectance between channels 6 and 10 (spectra 5 and 6, Figure 31-87). Low ratio values indicate vegetation, whereas high ratio values indicate absence of vegetation. Contrast stretching and recording the ratio values onto film (Figure 31-89A) yield black to dark-gray shades for low-ratio values where vegetation is present, with lighter gray to white shades for increasingly higher ratio values where vegetation is absent.

A 12/3 ratio was selected to define limonitic rocks, which have a marked falloff in reflectance in channel 3 compared to channel 12, because of ferric-iron absorption. Nonlimonitic rocks lack this characteristic ferric-iron absorption band (compare spectra 2, 3, and 4, Figure 31-87, with spectra 1, Figure 31-87). Note that vegetation also displays a relatively high value in channel 12 and a low value in channel 3. High-ratio values indicate the presence of limonite or vegetation, whereas low-ratio values indicate their absence. Limonitic or vegetated areas will appear light gray or white when rendered onto a film product (Figure 31-89B).

The 12/13 band ratio was chosen to distinguish materials displaying absorption features in the 2.2-μm region. Hydrothermally altered rocks containing argillic minerals (spectra 4, Figure 31-87) have absorption bands in this region, whereas unaltered rocks lack these features (spectra 1, 2, and 3, Figure 31-87). Note that although vegetation displays considerably lower reflectance values (spectra 5 and 6, Figure 31-87) compared to rocks (spectra 1, 2, 3, and 4, Figure 31-87), their ratio values will be similar to those of altered rocks. Thus, high-ratio values indicate argillically altered rocks or vegetation, whereas low ratios imply their absence. Both argillically altered rocks and vegetation will appear in light gray to white shades when processed to film (Figure 31-89C).

The 12/3 and 12/13 band ratios individually cannot be used to map limonite or argillic rocks because of the ambiguity related to vegetation. However, use of the 6/10 band ratio in a color-ratio-composite (CRC) image will alleviate the confusion.

The CRC image (Color Figure 31-90) was produced by color-additive projection of the 6/10 ratio in green, the 12/3 ratio in blue and the 12/13 ratio in red. The image reveals: 1) areas of bleached, hydrothermally altered rocks as yellow, 2) limonitic argillic rocks as white, 3) limonitic, unaltered rocks as cyan (light blue-green), and 4) spectrally flat rocks, those rocks containing neither limonite nor Al-O-H absorption bands, as green. Vegetation, depending upon the type, is depicted in hues of red, orange, and magenta. The

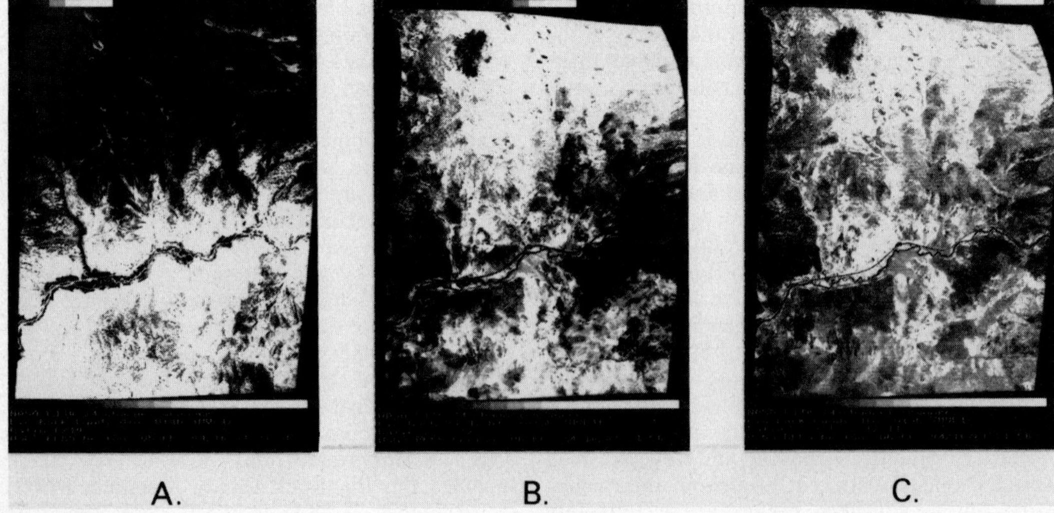

Fig. 31-89. Black-and-white ratio images derived from aircraft scanner data for a region centered on the Big Rock Candy Mountain north of Marysvale, Utah. West is at the top of the image. Resolution at nadir is about 10 m. Length of the image (west to east) is approximately 7 km. *A.* 6/10 band ratio; *B.* 12/3 band ratio; and *C.* 12/13 band ratio.

nearly black areas are caused by mixed pixel signatures of spectrally flat rocks and some vegetation. Greater interpretive detail is given by Podwysocki et al. (1982). Thus, in the CRC image, limonite was uniquely identified because of its characteristic spectral response. Argillically altered rocks were distinguished from other rocks on the image, but field checking was necessary to ascertain that they were related to hydrothermally altered rocks and not to some other type of clay-rich rocks, such as shales. By careful selection of spectral bands for analysis, this processing technique has allowed the geologist to identify limonitic rocks, argillized rocks, argillized and limonitic rocks, and rocks containing neither limonitic nor argillic components.

Principal Component Transformation

The principal component (PC) transformation, also known as the Karhunen-Loéve transformation, has been utilized in image processing for nearly a decade (Taylor, 1974; Fontanel et al., 1975; Merembeck et al., 1977; Blodget et al., 1978). Unlike band ratioing, which requires a careful choice of band combinations to emphasize specific spectral characteristics, the PC transformation requires no *a priori* information and thus can be applied in a less directed fashion. It is commonly applied to all available bands of scanner data at one time. The resultant output images emphasize distinctions between surface materials, but because each new output image is a linear additive combination of all the input channels, the individual images or their color composites often are not interpretable in terms of a sensed phenomenon, as is possible with ratio images. Therefore, the method is best used as a tool for distinguishing between rather than identifying lithologies. However, as will be demonstrated, an evaluation of the transformation values sometimes will allow identification of a physical factor that manifests itself in the resultant image.

A PC transformation is a unique mathematical transformation based on either the scene covariance or correlation matrix, which produces new variables known as components or axes, that are linear combinations of the original variables; each component contains data uncorrelated with the other components. The mathematics of this transformation are discussed in Chapter 17, Data Preprocessing and Processing, of this *Manual.* The components are ordered in the transformation so that the first component contains the majority of the total scene variance, with latter components containing less and less of the total scene variance. Figure 31-91A, a scatter plot of two hypothetical scanner bands (X1 and X2), shows the typical high correlations found in Landsat data. The PC transformation creates new uncorrelated axes, such that the first component (PC1) accounts for the majority of the scene variance; the second axis (PC2) is chosen orthogonal to the first. Each new axis or component can be referred to as some additive combination of the original channels X1 and X2. Only two channels are shown in this example for the sake of simplicity. The graphic analogy would hold for any number of input channels, but would be difficult to portray beyond three variables. Note that the second component (PC2) contains less variance (less scatter along the PC2 axis). For image analysis of these components, the variances are equalized by contrast stretching to produce the spherical distribution shown in Figure 31-91B. This step is

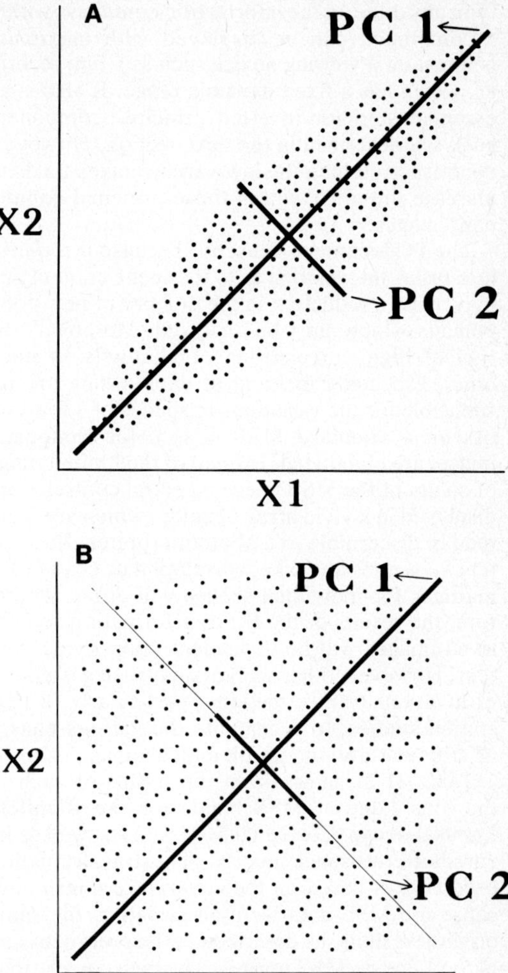

Fig. 31-91. A graphic representation of the principal component transformation and the inverted principal component transformation, on two hypothetical scanner bands X1 and X2, showing the decrease in contributed variance by the second component PC2 (A) and the effect of equalizing the variance by contrast stretching both components (B).

typically done so the information contained within a component can be displayed with maximum contrast on a viewing device such as a film recorder, which has a fixed dynamic range. It also is an essential step for inverted principal component analysis, described in the next section. However, contrast stretching the low variance axes tends to increase the noise within those principal component images.

The PC technique is useful, because it requires no *a priori* information as to the scene content and may allow a reduction in the number of new components which must be analyzed. Moreover, the typical high correlations of channels to each other, especially in Landsat data, which are responsible for the variations in shades of a few colors on a standard MSS false-color-composite image, are eliminated. Instead of the limited range of colors of the MSS image, spectral contrasts are displayed in a vivid array of colors which are more readily discernible to a photointerpreter. Because no two scenes are alike, correlation or covariance matrices for individual scenes will differ. Therefore, the results of the PC transformation and derived images will be dependent upon scene content. However, other workers have suggested use of invariant transformations, particularly in vegetation studies, to accentuate the various phases of crop maturation (Kauth and Thomas, 1976).

Table 31-21 summarizes the means of each of the four Landsat MSS bands for the Richfield Landsat image (Color Figure 31-88A) as well as its variance-covariance matrix. The transformation used here is based on the covariance matrix, because the MSS data were all scaled to the same brightness units, as described in the preprocessing of the Landsat MSS images. The high correlations between MSS bands, which are evident in the correlation matrix (Table 31-22), also are evident in

TABLE 31-21

Means and variance-covariance matrix for the four Richfield, Utah, Landsat MSS image bands shown in Color Figure 31-88A. Individual channel variances are located along the main diagonal (the trace) of the matrix; all values off the trace are covariances.

MSS	4	5	6	7
4	1618.47			
5	1952.90	2446.25		
6	1442.82	1780.64	1450.70	
7	996.58	1224.42	1080.20	880.11
CHANNEL MEAN	103.50	114.22	118.26	105.50
STANDARD DEVIATION	40.23	49.46	38.09	29.67

TABLE 31-22

Correlation matrix for the four Richfield, Utah, Landsat MSS image bands shown in Color Figure 31-88A.

MSS	4	5	6	7
4	1.00			
5	0.98	1.00		
6	0.94	0.95	1.00	
7	0.84	0.83	0.96	1.00

the black-and-white images of the four MSS bands (Figures 31-92A-D). The relatively low correlation of MSS bands 4 and 5 to MSS band 7 is most likely caused by marked contrasts between the spectra of vegetation (spectra 5 and 6, Figure 31-87) and geologic materials (spectra 1, 2, 3, and 4, Figure 31-87). this is evident by comparing the contrasts of the cultivated fields and the vegetation in the high mountainous terrain in MSS bands, 4, 5, and 7 (Figures 31-92A, 31-92B, and 31-92D, respectively). Table 31-23 shows the transformation (eigenvector) matrix based on the scene variance-covariance matrix and the resultant principal component images (Figure 31-93A-D) generated from the eigenvector matrix. The high degree of redundancy shown in the four MSS bands (Figure 31-92A-D) is in sharp contrast to the four principal components (Figure 31-93A-D) derived from the MSS data.

Because of the decrease in scene variance accounted for in higher order components, these components tend to look more noisy and bear less information in the eyes of the photointerpreter. This inherent attribute of the PC transformation may lead the interpreter to discard the higher order components. Before doing so, however, the interpreter should make a visual evaluation, because statistical "out-liers", which may be responsible for only a small portion of the total scene variance, may be found in these axes and may be of geologic significance. Thus, although the first three components contain the largest proportion of the total variance and therefore ought to convey the most information, for reasons to be stated later, a color-additive color-composite image was created from components 2, 3, and 4, using blue, green, and red, respectively (Color Figure 31-94). The choice of colors was done subjectively, based on which combination produced the best in the interpreter's eye.

The geologist can then use photointerpretive techniques, including comparison with other types of images, in order to map lithologic units. By association with the false-color-composite image (Color Figure 31-88A), vegetation appears red-brown. Valley-fill materials south of the Little Drum Mountains are magenta, whereas most valley fill to the south is yellow to green. On the standard false-color-composite image, the Ter-

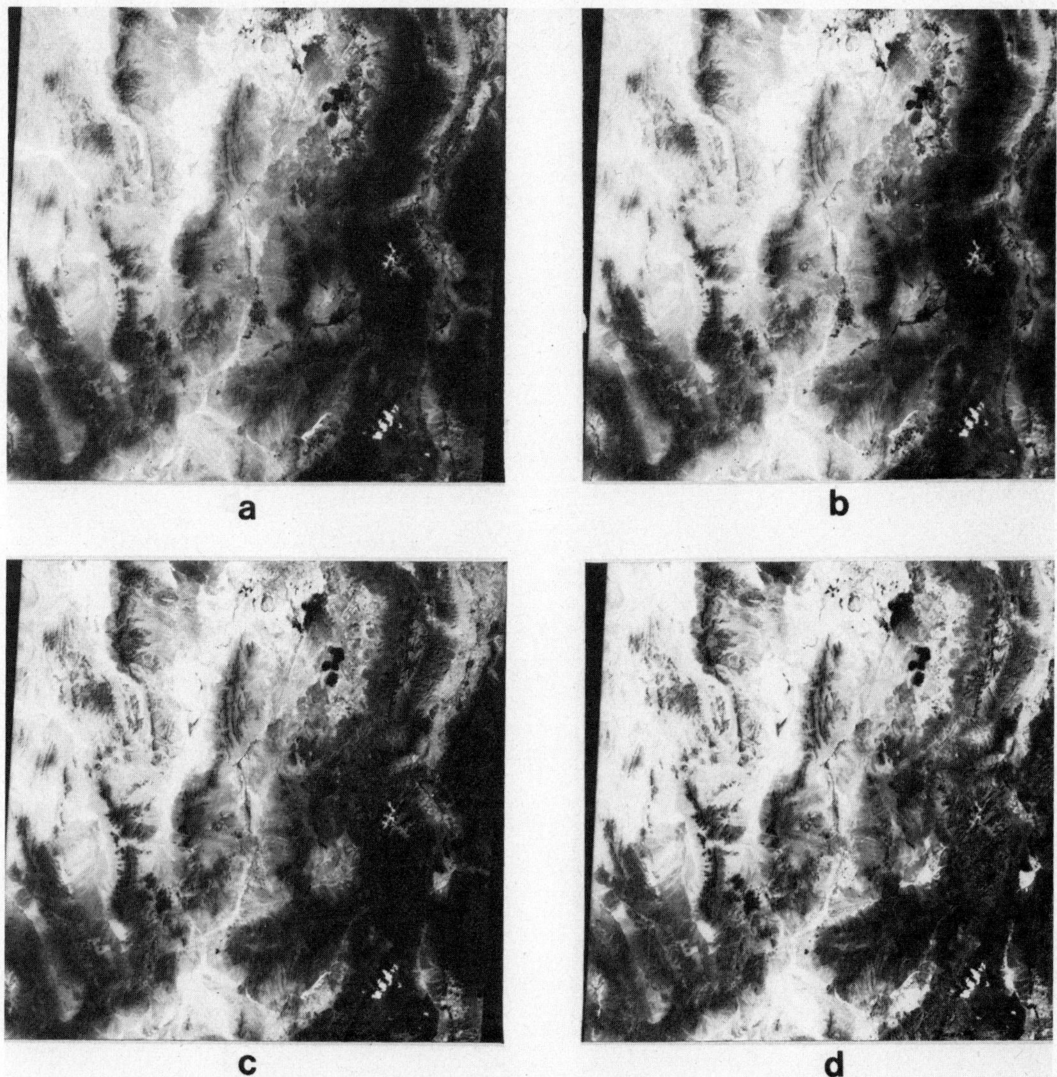

Fig. 31-92. Images of the individual MSS bands for the Richfield, Utah, Landsat scene (1699-17353/5; 22 June 1974). *A*. MSS band 4; *B*. MSS band 5; *C*. MSS band 6; and *D*. MSS band 7.

TABLE 31-23

Eigenvector matrix, eigenvalues, and percent variance accounted for in each principal component of the four Richfield, Utah, Landsat MSS image bands shown in Color Figure 31-88A.

MSS CHANNEL	PC1	PC2	PC3	PC4
4	0.5077	0.3362	−0.7896	0.0759
5	0.6262	0.4320	0.6083	0.2263
6	0.4806	−0.3838	0.0700	−0.7854
7	0.3452	−0.7437	−0.0397	0.5712
EIGENVALUE	6070.91	275.91	36.23	12.48
PERCENTAGE	94.92	4.31	0.57	0.20

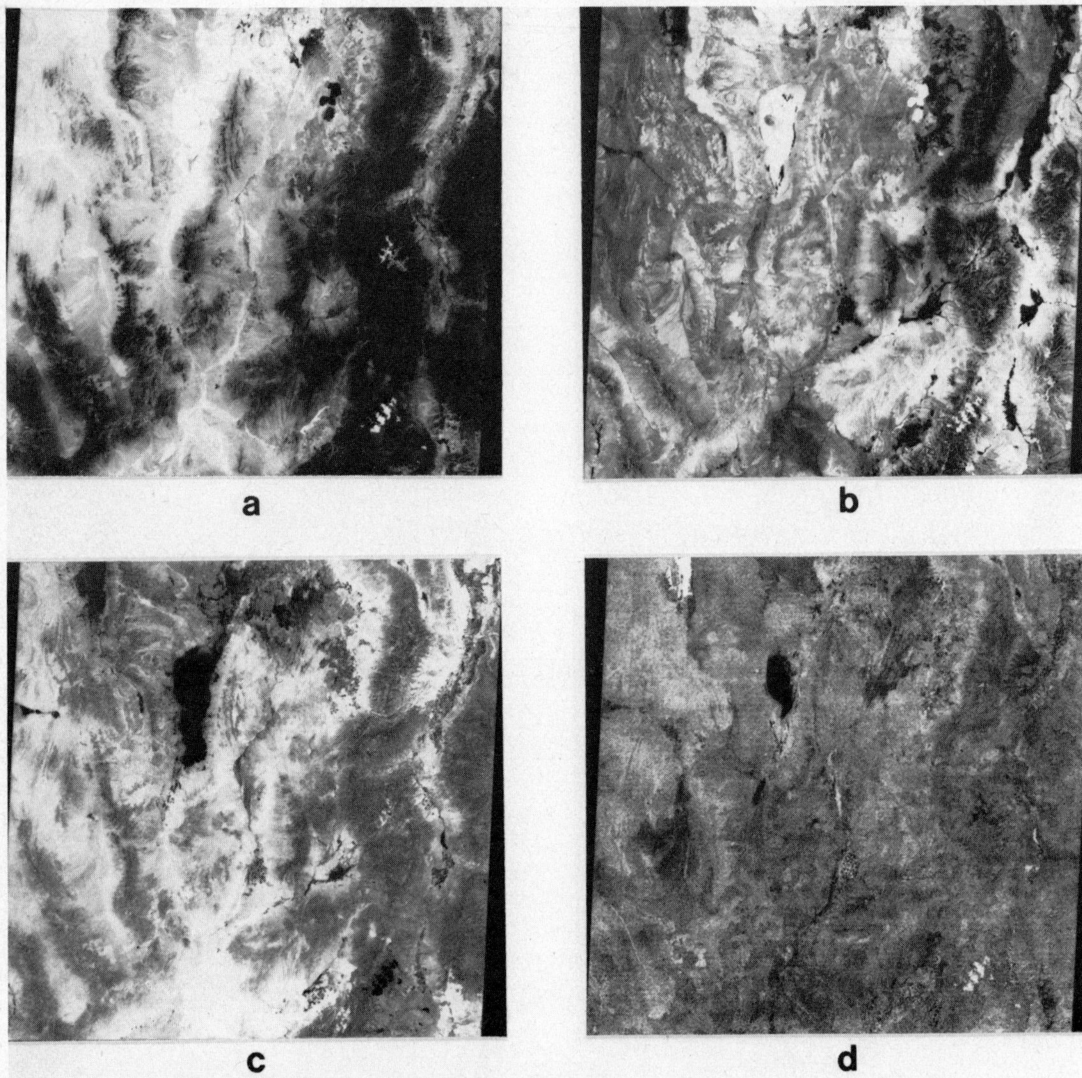

Fig. 31-93. Principal component (PC) images of the individual axes for the Richfield, Utah, Landsat scene (1699-17353/5; 22 June 1974). *A.* PC 1; *B.* PC 2; *C.* PC 3; and *D.* PC 4.

tiary Flagstaff Limestone (*F,* Color Figure 31-88A), a sequence of tan carbonates and medium red siltstones, appears brown, and bright hydrothermally altered volcanic rocks in the Shauntie Hills (*A1,* Color Figure 31-88A) appear white to light yellow, yet both these areas appear light green in the PC color composite (*A1* and *F,* Color Figure 31-92). These two areas, although disparate in their brightnesses, appear to have a common factor in their reflectance values as deduced from their common color in the PC color-composite image. Many other color variations in the mountain ranges abound in this image but are too small to point out on this full Landsat scene. Others would be apparent if one of the other combinations of components were chosen.

Although the PC images provide useful media on which the photointerpreter can distinguish different units, the ascribing of a physical attribute to each of the distinguished units on the images often is difficult because the new variables are combinations of the original four MSS bands. However, some materials can be identified utilizing the eigenvector matrix, the images, and knowledge of the spectral responses of sensed materials. The following discussion is an attempt to show that individual component images and the resultant color-composite image based on them often can be related to specific attributes of the sensed materials.

The eigenvector of the first principal component (PC1, Table 31-23) contains nearly equal

Fig. 26-6a. Fort Frederick, Maryland. These infrared color stereo multiband photos were taken of Fort Frederick, at 600 ft. altitude. Note the differences in the appearance of possible foundation patterns in the parade ground. (Courtesy of Carl H. Strandberg.)

Fig. 26-6b. Natural color convergent oblique stereogram taken from north to south of Fort Frederick, Maryland at an altitude of 1000 ft. (Courtesy of Carl H. Strandberg).

Fig. 26-38. A portion of Painted Rock Shelter, showing the figures executed in the Red Monochrome style.

Fig. 27-39. Landsat-1 computer-enhanced, color-composite image (1:2,000,000) of Vatnajokull, Iceland. Its areal extent on 22 September 1973 was 8,300 km² according to measurements made for a 1:250,000 enlargement of this image (1435–12070). (Williams, 1976).

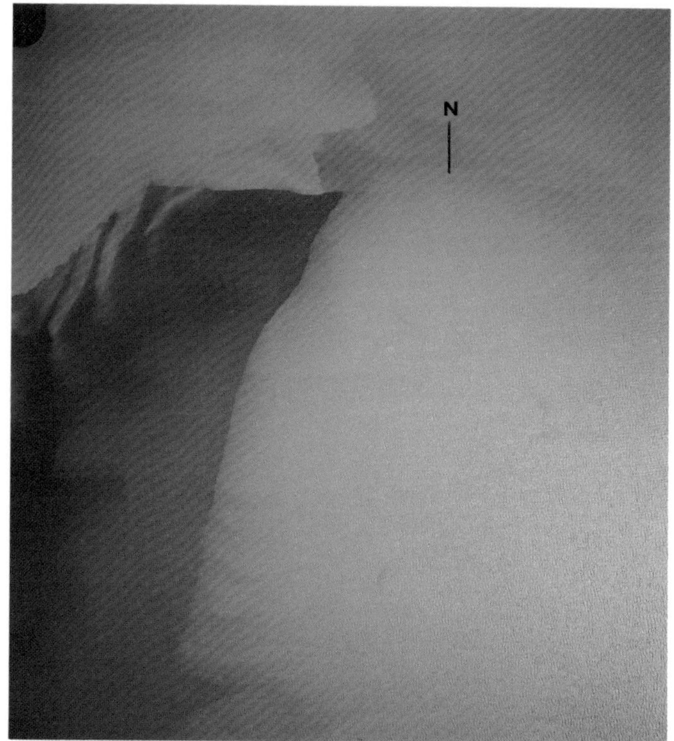

Fig. 28-17. Ektachrome Infrared photograph of an area near the Mississippi Delta. Altitude 450 m. Camera Wild RC-8, 1/250 at f/5.6. NASA Mission 159. Note the mixing of three water masses—ocean water, water from West Pass, and partially mixed water from the South Pass (Heubner, 1971).

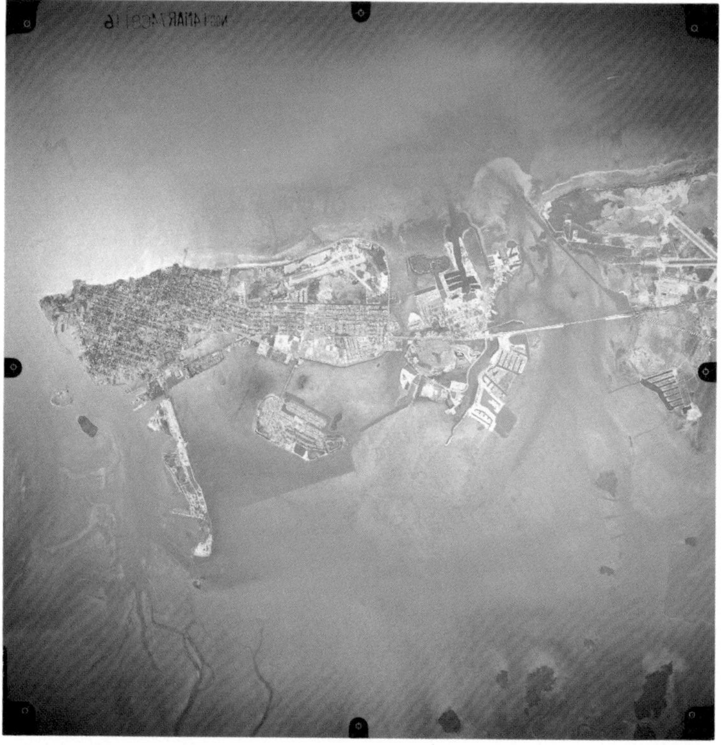

Fig. 28-18. Example of aerial photography for harbor planning at Key West, Florida. USC&GS Aerial Photograph No. 60-S-413 taken on March 11, 1960 at 1:40,000 scale on Anscochrome FPC-132 film (Orr, 1968).

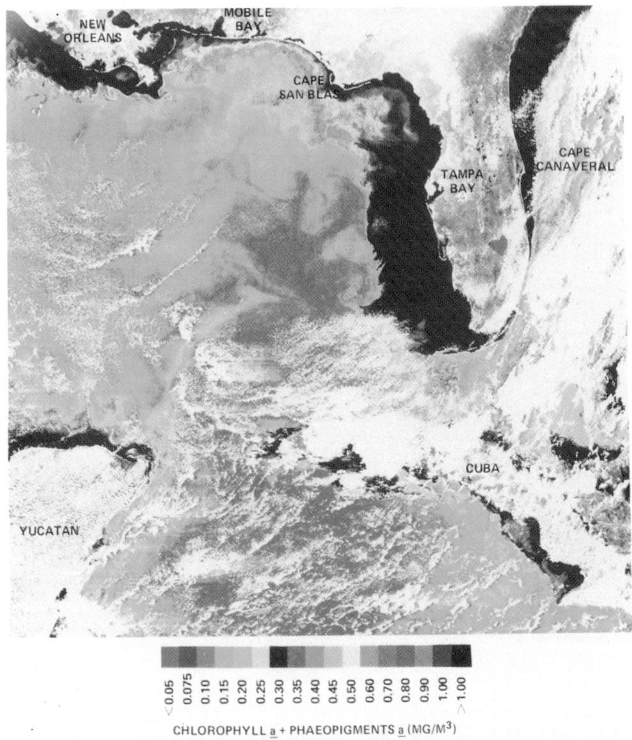

Fig. 28-65. A map of phytoplankton pigments in the Gulf of Mexico on 2 November 1978 derived from the Nimbus-7 CZCS data of orbit 130. Open water picture elements (pixels) are coded according to the color scale to show estimated chlorophyll *a* + phaeopigments *a*. Cloud and land pixels are represented in quasi-natural colors (Hovis *et al.*, 1980).

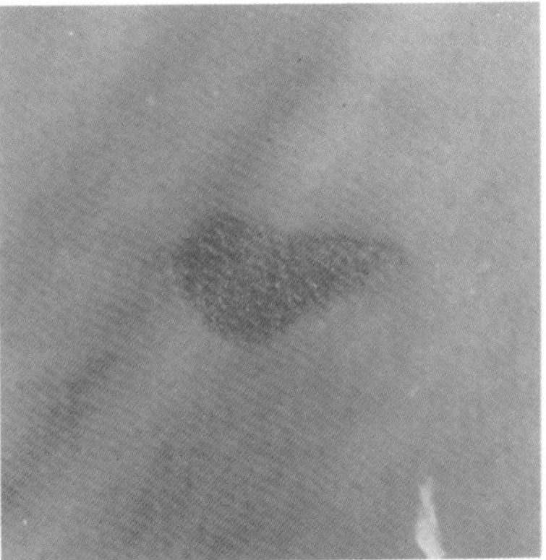

Fig. 28-72. Aerial photograph of schooling ladyfish (*Elops saurus*) taken on Ektachrome Infrared film (Stevenson and Pastula, 1971).

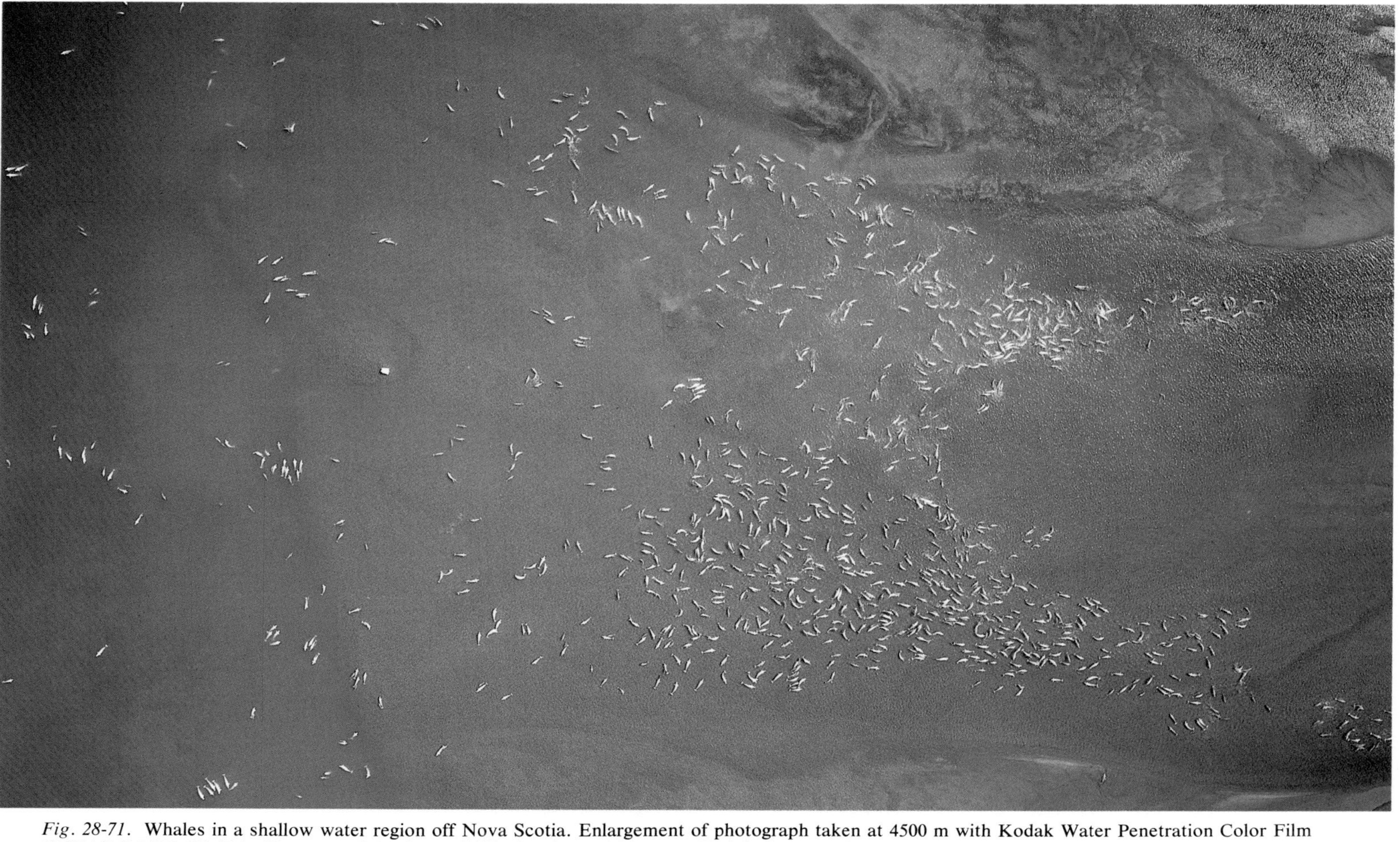

Fig. 28-71. Whales in a shallow water region off Nova Scotia. Enlargement of photograph taken at 4500 m with Kodak Water Penetration Color Film (ESTAR BASE) SO-224. (Courtesy of Quebec Wildlife Service and Eastman Kodak Company.)

Fig. 28-81. Submerged aquatic vegetation as imaged from 150 m altitude in the Chesapeake Bay. A stand of *Zostera marina* with Kodak VPS 2107 (Orth *et al.,* 1979).

Fig. 28-83. Coastal wetlands in New Jersey imaged from an altitude of 1500 m with color infrared film (Carter, 1977).

Fig. 28-133. Classified Landsat image, 26 February 1975, of Apalachicola Bay under low wind and near ebb-tide slack (similar to condition of Figure 28-132) (Hill and Graham, 1979).

A

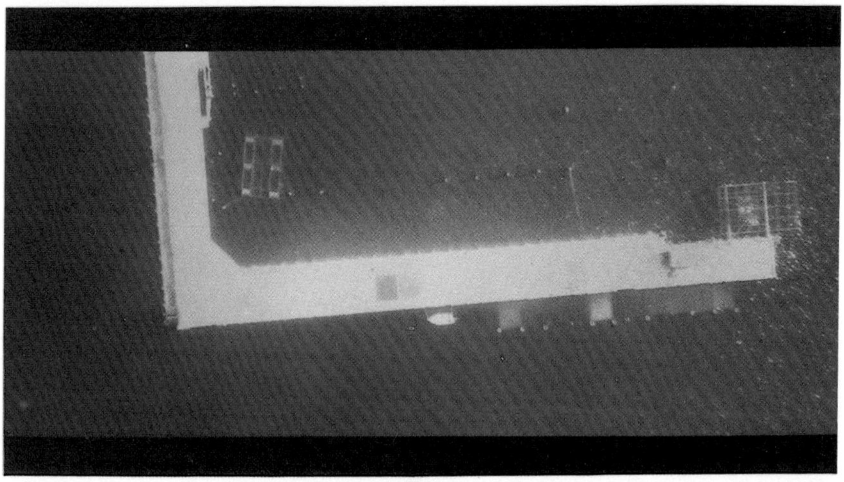

B

C

Fig. 28-99. Use of isoluminous multispectral reproductions to allow simultaneous discrimination of surface and submerged targets. a. Isoluminous color-additive rendition; b. Ektachrome film exposed for maximum underwater detail; c. normal color-additive rendition (Yost and Wenderoth, 1970).

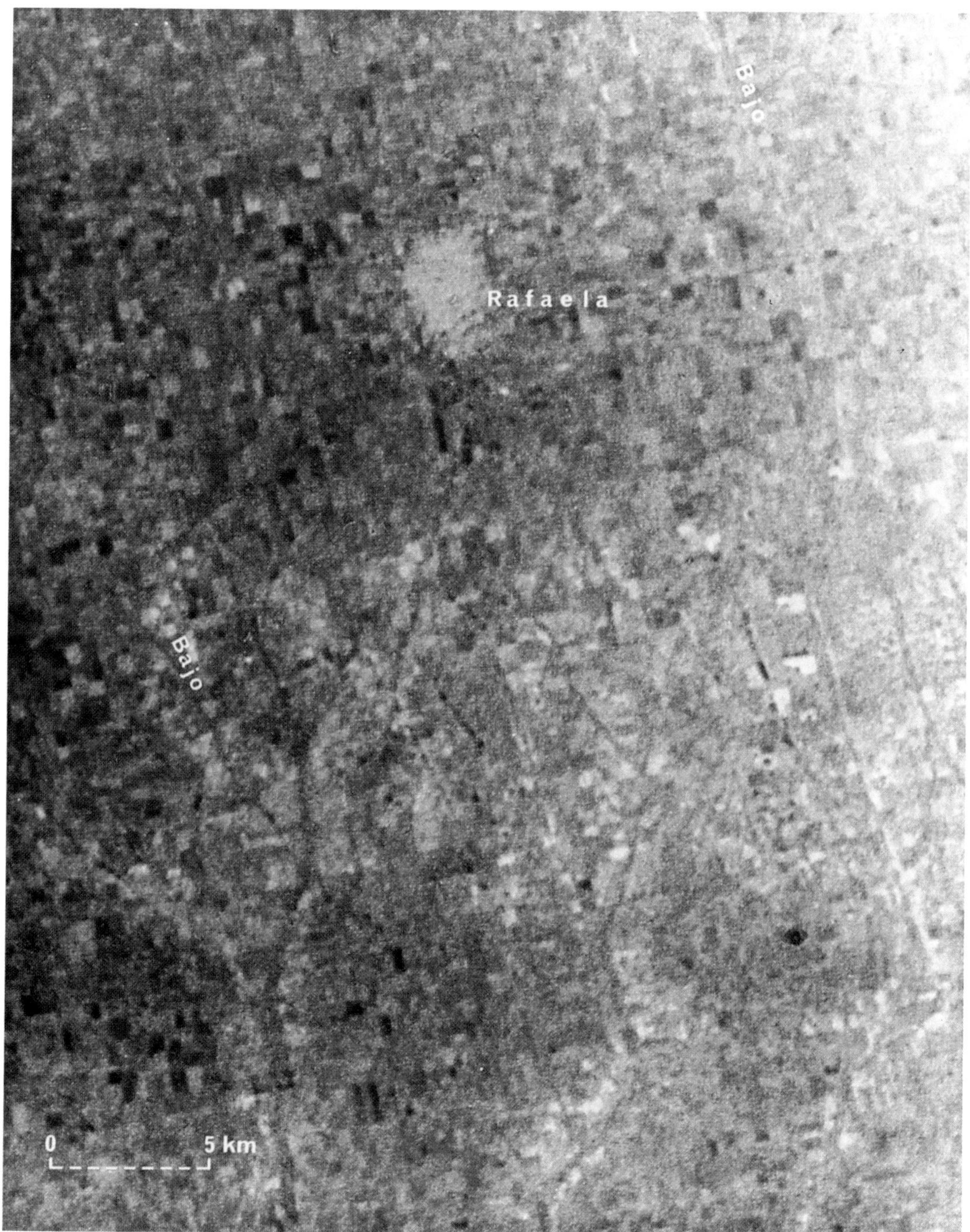

Fig. 29-1. Skylab infrared image reflecting vegetation differences that are associated with varying ground-water depth. Compare with black-and-white Figure 29-2. (Kruck and Kantor, 1975).

Fig. 29-9. Color-enhanced airborne thermal infrared image of a 50-km transect in the Sacramento Valley taken at an altitude of 3300 m by a Texas Instruments RS-25 Scanner observing in the 10.5 to 12.5-μm band. Temperature range from 20°C to 44°C is seen at bottom.

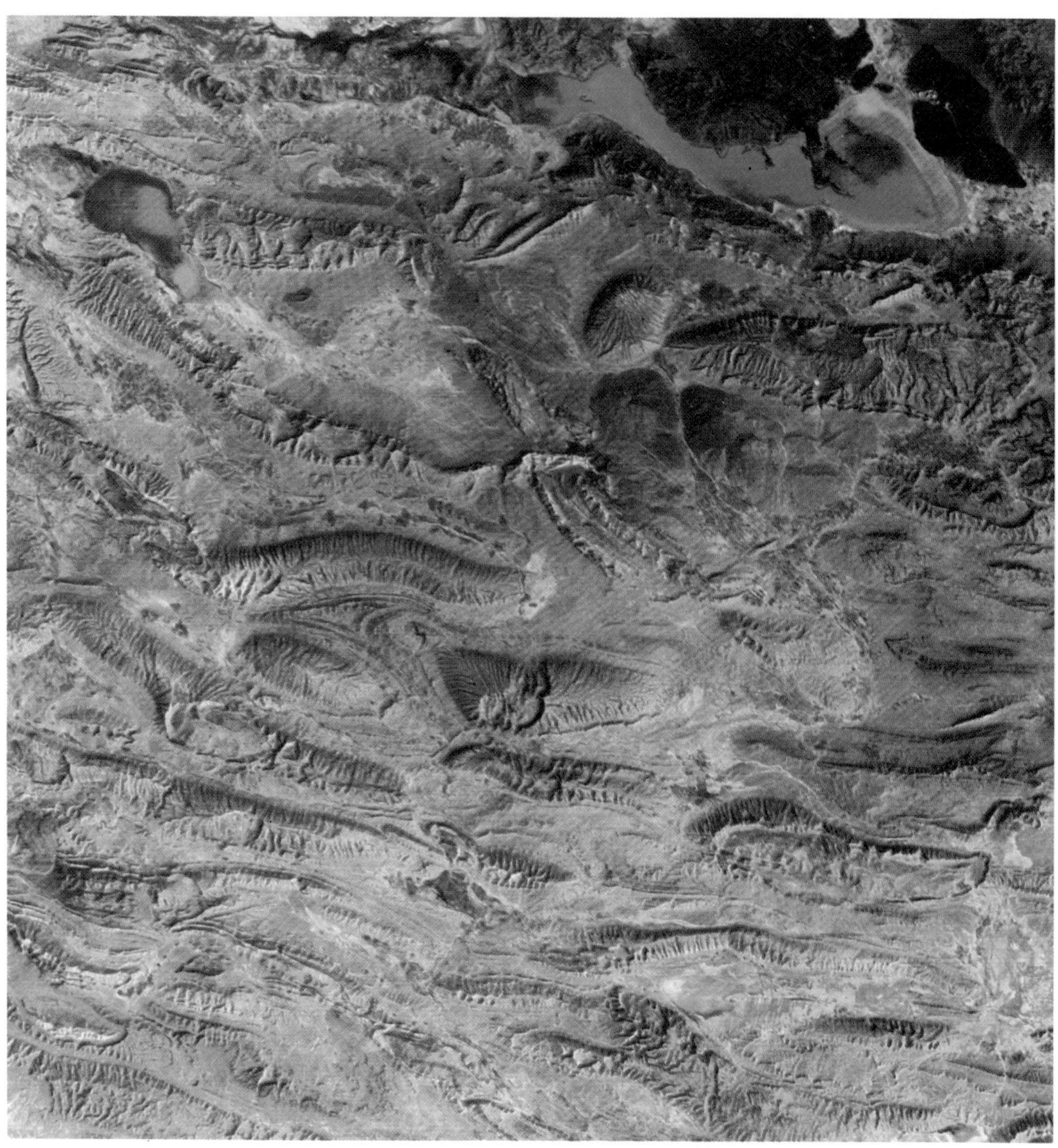

Fig. 29-32a. Color-composite Landsat-1 image of the Shiraz and Neriz Playas of Iran on March 1, 1973 (1221-06293). The Shiraz Playa is in the upper left of the image, and the Neriz Playa is in the upper center to the right (Krinsley, 1976).

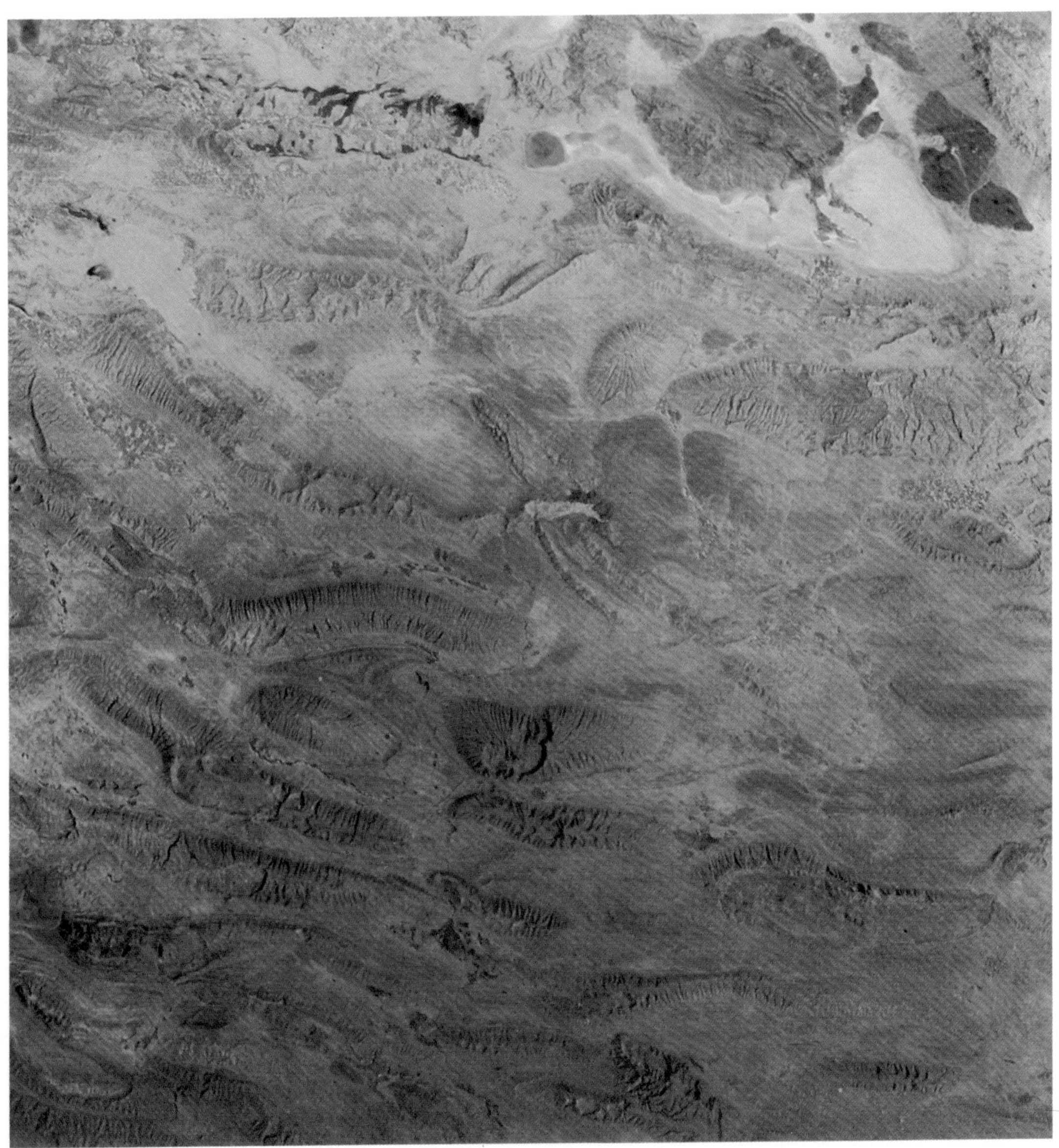

Fig. 29-32b. Color-composite Landsat-1 image of the Shiraz and Neriz Playas of Iran on August 28, 1973 (1401-06280) (Krinsley, 1976).

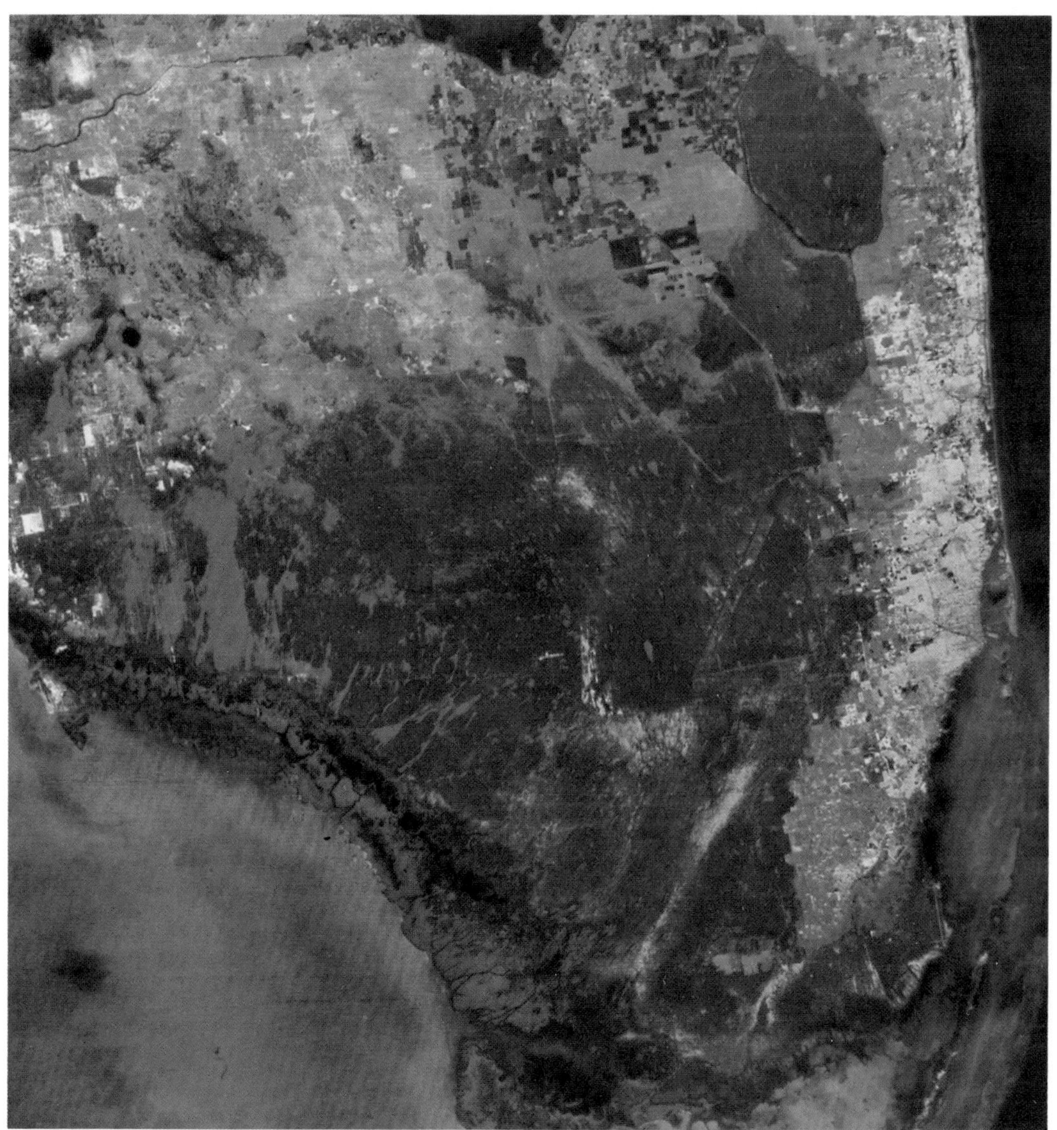

Fig. 29-34. Color-composite Landsat-1 image of the Everglades National Park area of Florida (1242-15240) (Higer et al, 1976).

Fig. 29-37. False-color composite Landsat image of southeastern Louisiana (30052-1500).

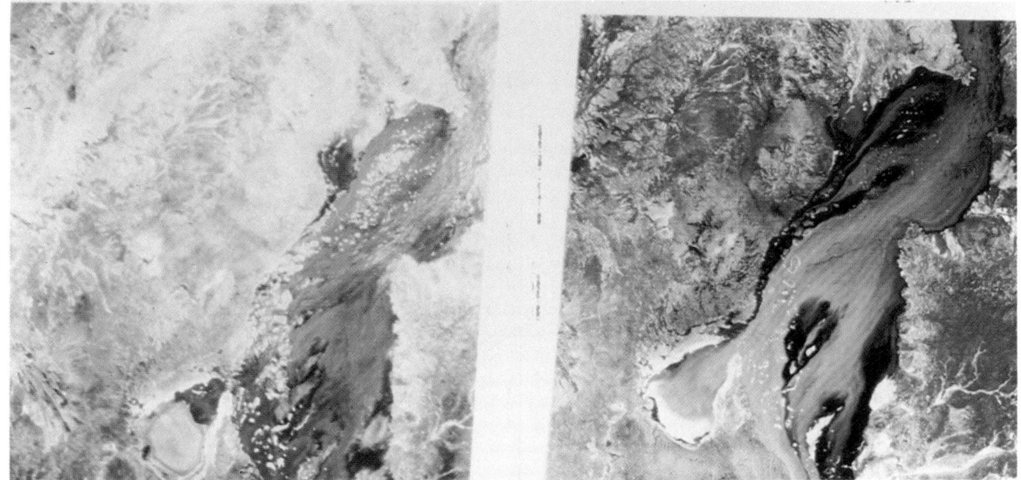

Fig. 29-44. Appearance of Cooper Creek floodplain in Queensland, Australia, in July 1973 and February 1974 in Landsat imagery. (Courtesy of C. Robinove, U.S. Geological Survey.)

Fig. 30-6. Color aerial photography of an urban area in Goleta, California. [Source: Pacific Western Aerial Surveys]

Fig. 30-7. High altitude color infrared photography of Santa Barbara County, California. The original scale was 1:130,000. The image was obtained by a NASA Ames U-2 aircraft. Color Figure 30-8 see next page.

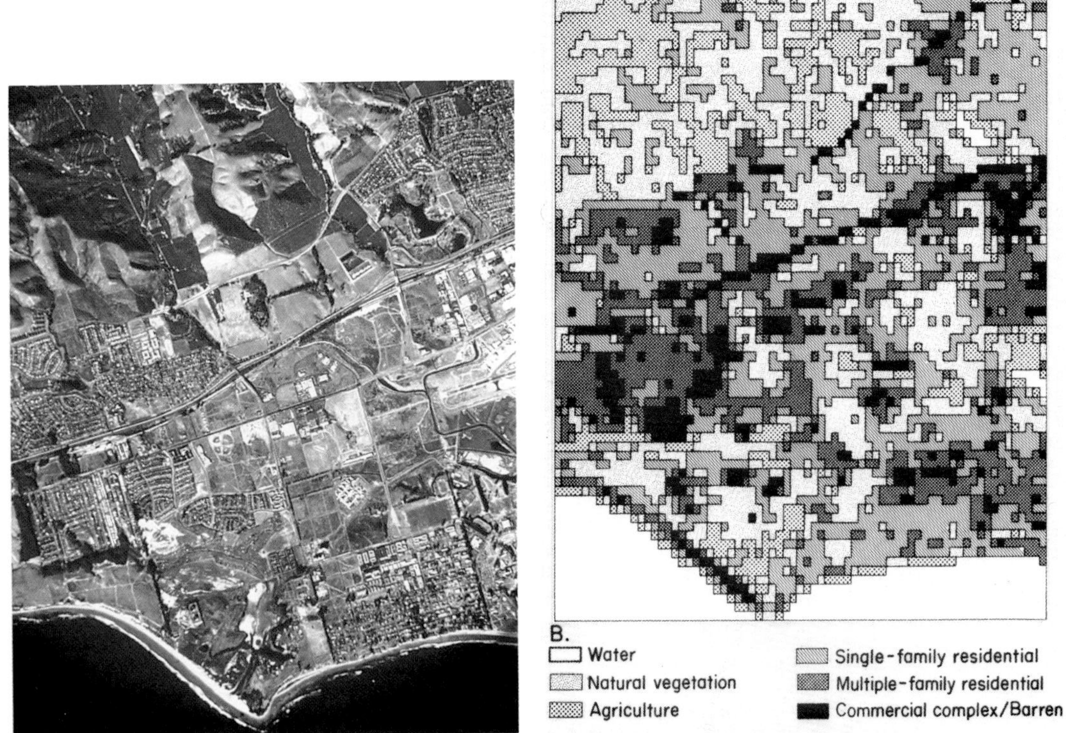

Fig. 30-12 High altitude color infrared photography and Landsat land cover classification of a part of Goleta, California.

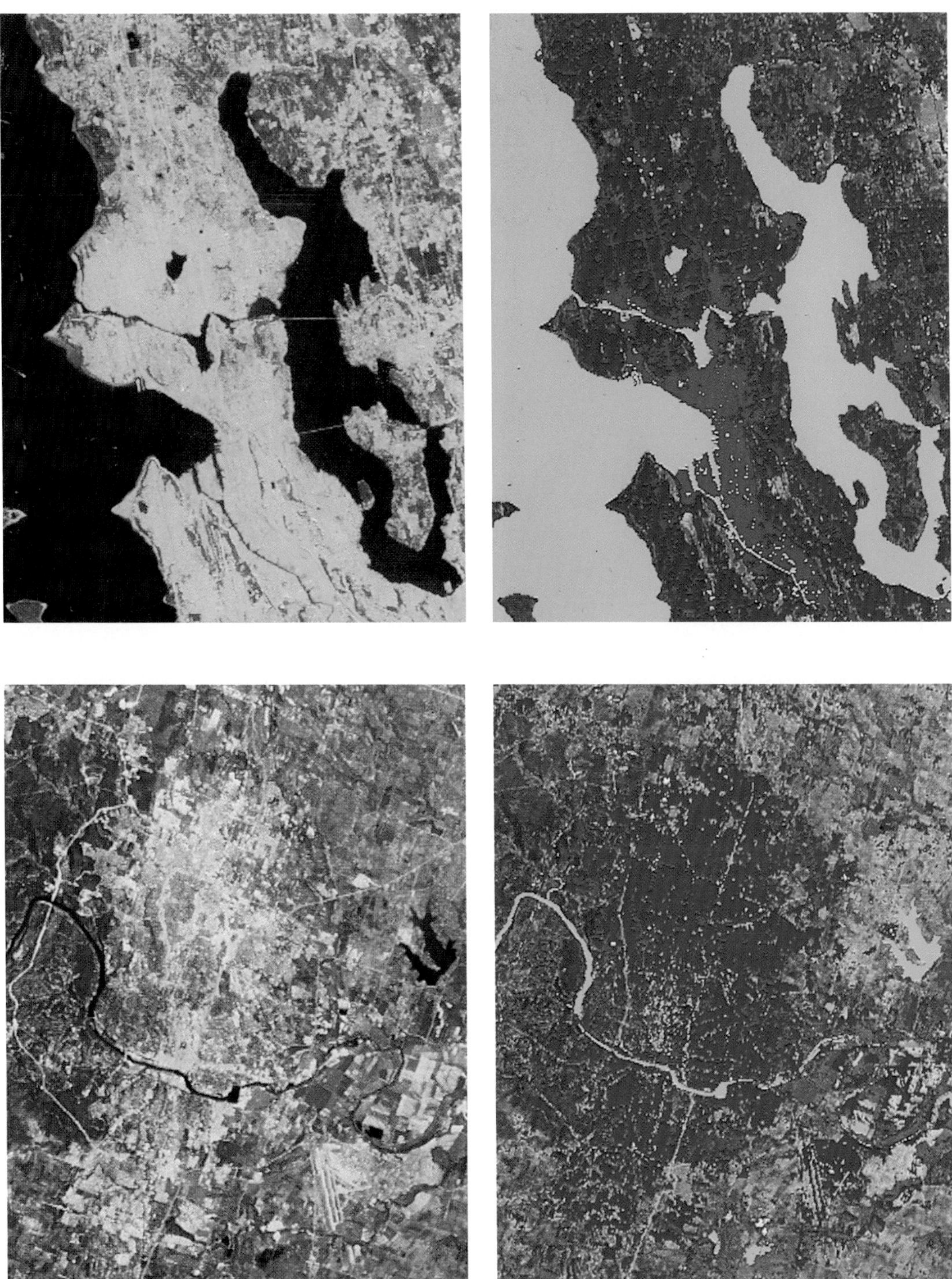

Fig. 30-8 Landsat color composites and classification maps of Seattle, Washington, and Austin, Texas. [Source: Christenson, 1977].

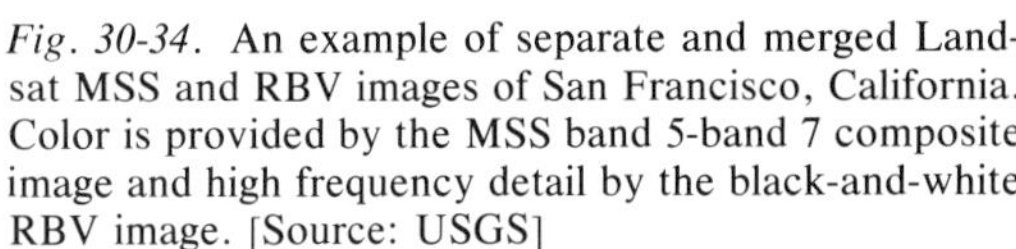

Fig. 30-34. An example of separate and merged Landsat MSS and RBV images of San Francisco, California. Color is provided by the MSS band 5-band 7 composite image and high frequency detail by the black-and-white RBV image. [Source: USGS]

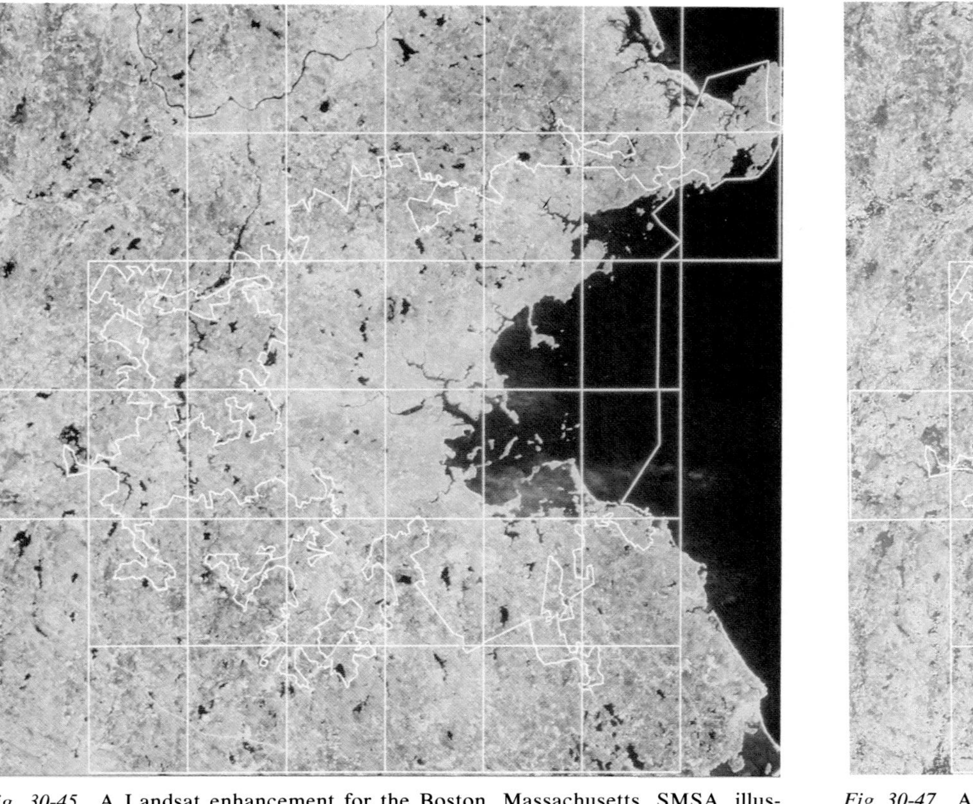

Fig. 30-45. A Landsat enhancement for the Boston, Massachusetts, SMSA, illustrating sprawl. Included are 7½′ quadrangle map boundaries and the 1970 urban boundary.

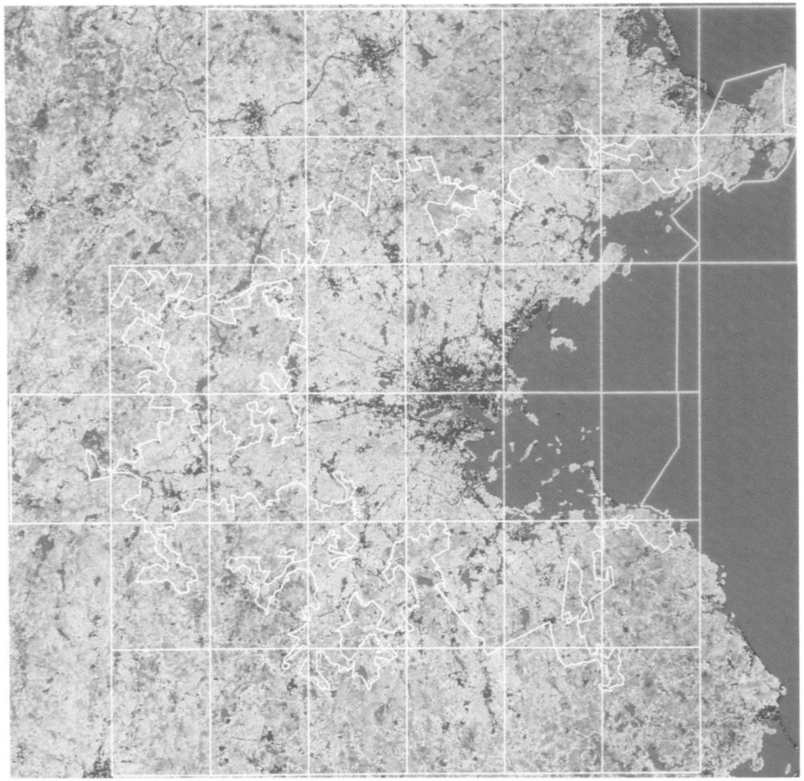

Fig. 30-47. A classification of Boston, Massachusetts, urban land cover with a Landsat enhancement for background. Included are 7½′ Quadrangle map boundaries and the 1970 final urban boundary.

Fig. 30-66 A *pseudo-color* thermal infrared image (8–14 μm) of a residential district in Wichita, Kansas. Black depicts the coldest features, while increasingly warmer areas are depicted by the color sequence blue, magenta, red, orange, and yellow. The arrows denote vacant lots. [Source: Texas Instruments]

Fig. 30-68 A ground based pseudo-color thermal infrared image of the same structures shown in Figure 30-67. The progression from coolest to warmest is black, magenta, blue, aqua, green, orange, red and white. [Source: Daedalus Enterprises, Inc.]

Fig. 30-77 A pseudo-color image of a residential area in Farmington, New Mexico. The color scale for depiction of temperature is provided on the image. The warmer rooftops are tan in color, while cooler rooftops are blue or magenta. [Source: Technology Application Center, University of New Mexico]

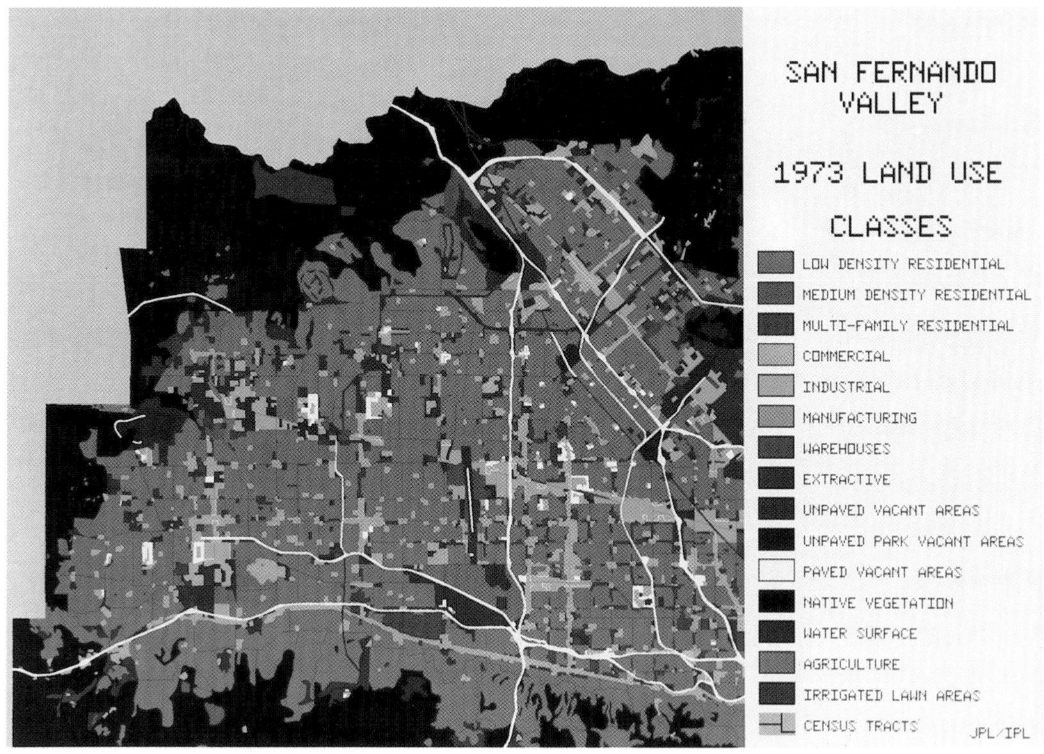

Fig. 30-88 This composite digital land use map of the San Fernando Valley test area was prepared from five 1:24,000 land use maps. [Source: Angelici, 1978]

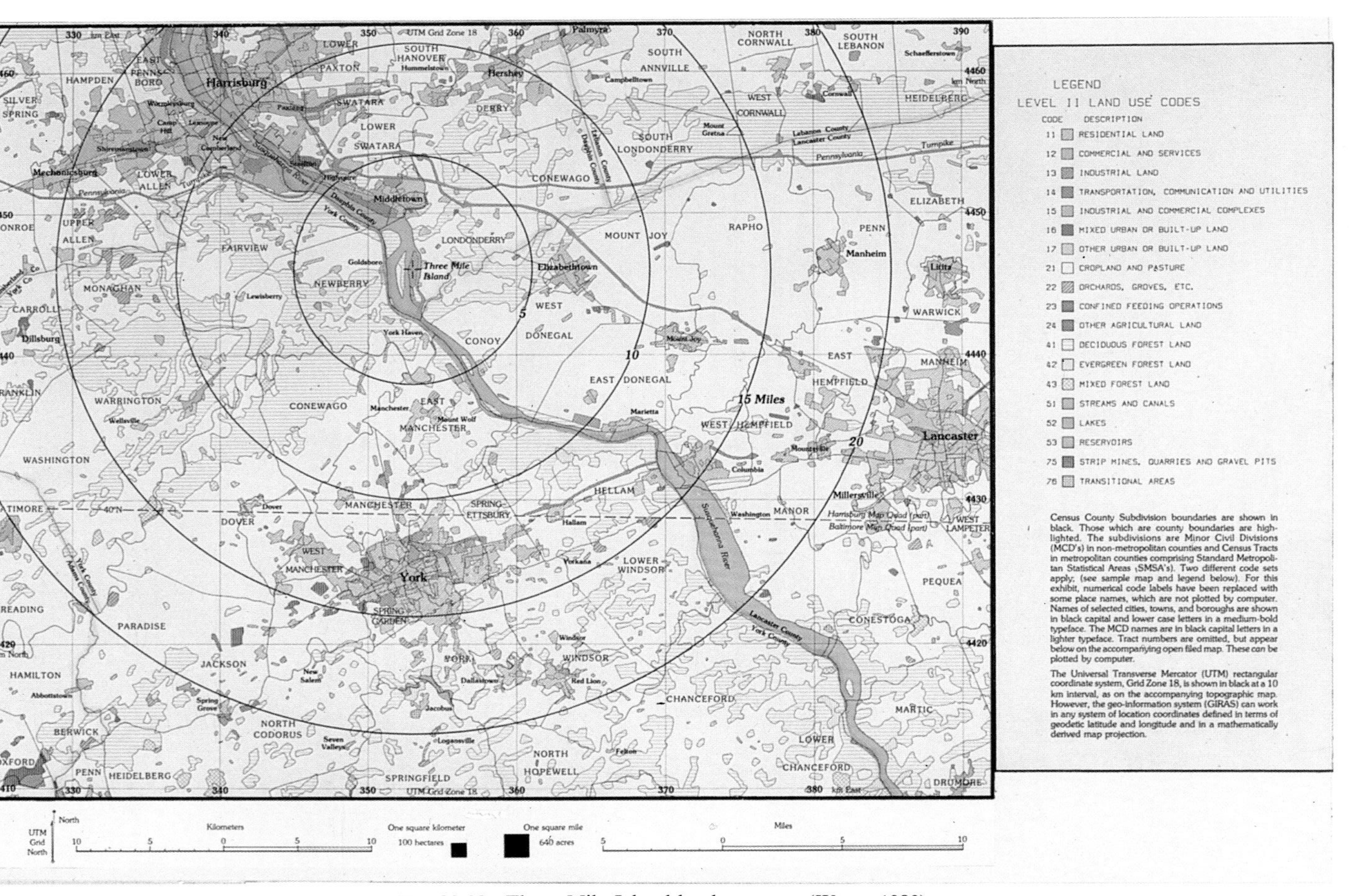
LEGEND
LEVEL II LAND USE CODES
CODE DESCRIPTION
11 RESIDENTIAL LAND
12 COMMERCIAL AND SERVICES
13 INDUSTRIAL LAND
14 TRANSPORTATION, COMMUNICATION AND UTILITIES
15 INDUSTRIAL AND COMMERCIAL COMPLEXES
16 MIXED URBAN OR BUILT-UP LAND
17 OTHER URBAN OR BUILT-UP LAND
21 CROPLAND AND PASTURE
22 ORCHARDS, GROVES, ETC.
23 CONFINED FEEDING OPERATIONS
24 OTHER AGRICULTURAL LAND
41 DECIDUOUS FOREST LAND
42 EVERGREEN FOREST LAND
43 MIXED FOREST LAND
51 STREAMS AND CANALS
52 LAKES
53 RESERVOIRS
75 STRIP MINES, QUARRIES AND GRAVEL PITS
76 TRANSITIONAL AREAS
Census County Subdivision boundaries are shown in black. Those which are county boundaries are highlighted. The subdivisions are Minor Civil Divisions (MCD's) in non-metropolitan counties and Census Tracts in metropolitan counties comprising Standard Metropolitan Statistical Areas (SMSA's). Two different code sets apply; (see sample map and legend below). For this exhibit, numerical code labels have been replaced with some place names, which are not plotted by computer. Names of selected cities, towns, and boroughs are shown in black capital and lower case letters in a medium-bold typeface. The MCD names are in black capital letters in a lighter typeface. Tract numbers are omitted, but appear below on the accompanying open filed map. These can be plotted by computer.
The Universal Transverse Mercator (UTM) rectangular coordinate system, Grid Zone 18, is shown in black at a 10 km interval, as on the accompanying topographic map. However, the geo-information system (GIRAS) can work in any system of location coordinates defined in terms of geodetic latitude and longitude and in a mathematically derived map projection.
UTM Grid Zone 18
Three Mile Island
Harrisburg
Middletown
Hershey
Elizabethtown
Lancaster
York
Susquehanna River
5
10
15 Miles
20
UTM Grid North
North
Kilometers
One square kilometer
100 hectares
One square mile
640 acres
Miles

Fig. 30-98. Three Mile Island land use map (Wray, 1980).

Fig. 31-8. NASA high altitude color infrared aerial photograph (Kodak SO-193 film with a Wratten 12 filter) of part of the terminus of the Alsek Glacier and the Alsek River, southeast of Yakutat, Alaska. This photograph (1:65,000-scale) was acquired by a National Aeronautics and Space Administration Ames U-2C aircraft on 21 June 1978 from an altitude of 19,800 m, with an RC-10 camera having a 305 mm (12 in.) focal length, in support of the aerial mapping program in Alaska. (EROS Data Center, Accession No. 578002618 ROLL, frame 4898). The spectral range of the film-filter combination is 0.51 μm$-$0.9 μm.

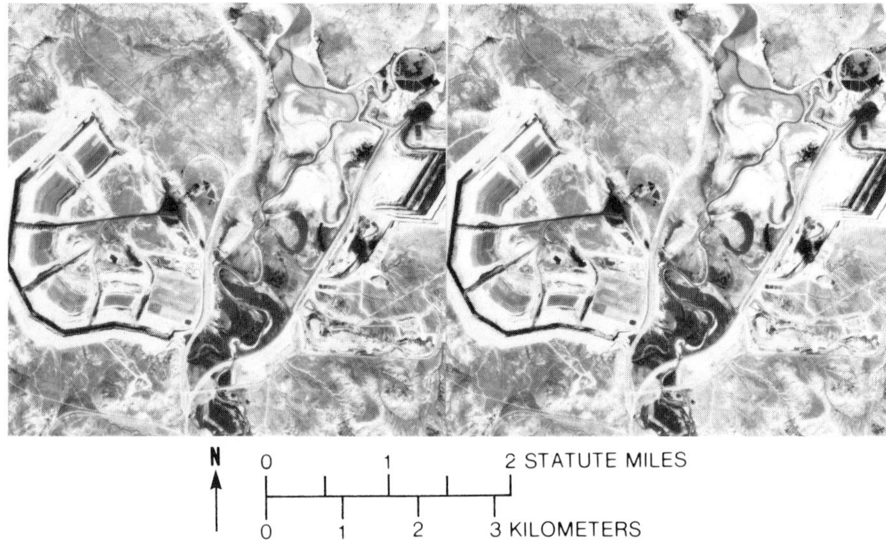

Fig. 31-9. This National High-Altitude Photography (NHAP) vertical black-and-white stereopair of the Decker-Birney coal strip mine, southeast Montana, was acquired on 4 September 1980 at a scale of 1:80,000 from an altitude of 12,200 m (40,000 feet). The photography was flown in a north-south flight direction with the exposure centered at N. 45°00′; W. 106°50′ over the southern part of the USGS 7.5′ quadrangle designated as Decker, Montana. Such photographs have a base to height ratio of 0.57. It also has 60 percent forward overlap and 30 percent sidelap in order to provide complete lateral stereoscopic coverage (Photograph I.D. No. HAP 80, 164-112).

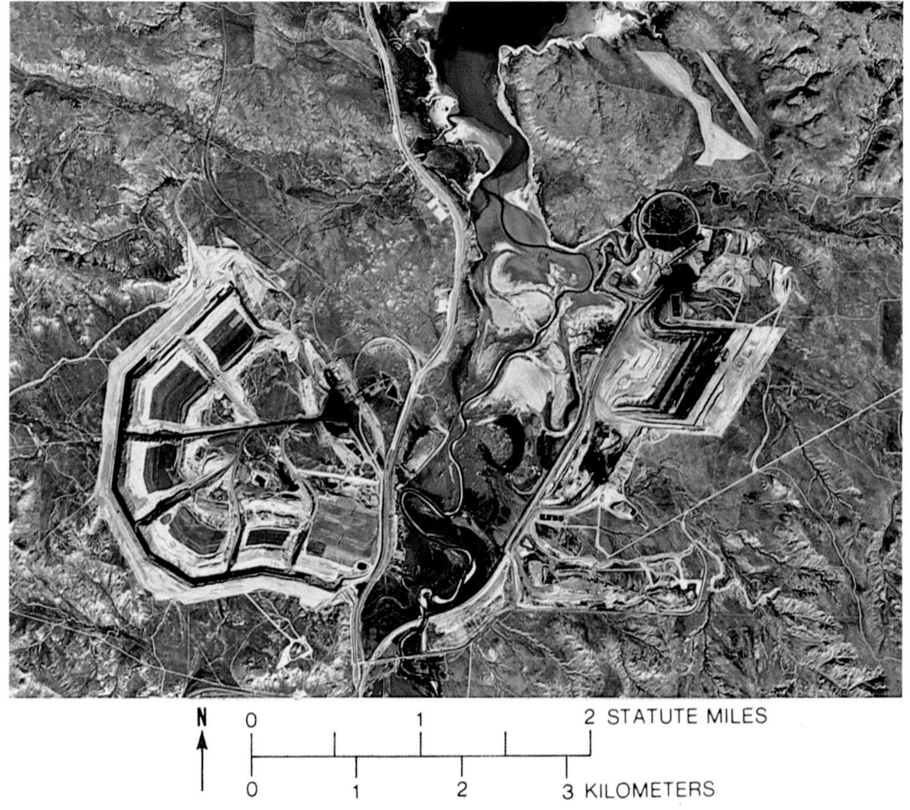

Fig. 31-10. This quadrangle-centered National High-Altitude Photography vertical color-infrared aerial photograph, scale 1:58,000, of the Decker-Birney coal strip mine, Montana, was acquired on 4 September 1980, simultaneously with the black-and-white photograph shown in Figure 31-9. (Photograph ID No. HAP 80, 223-167).

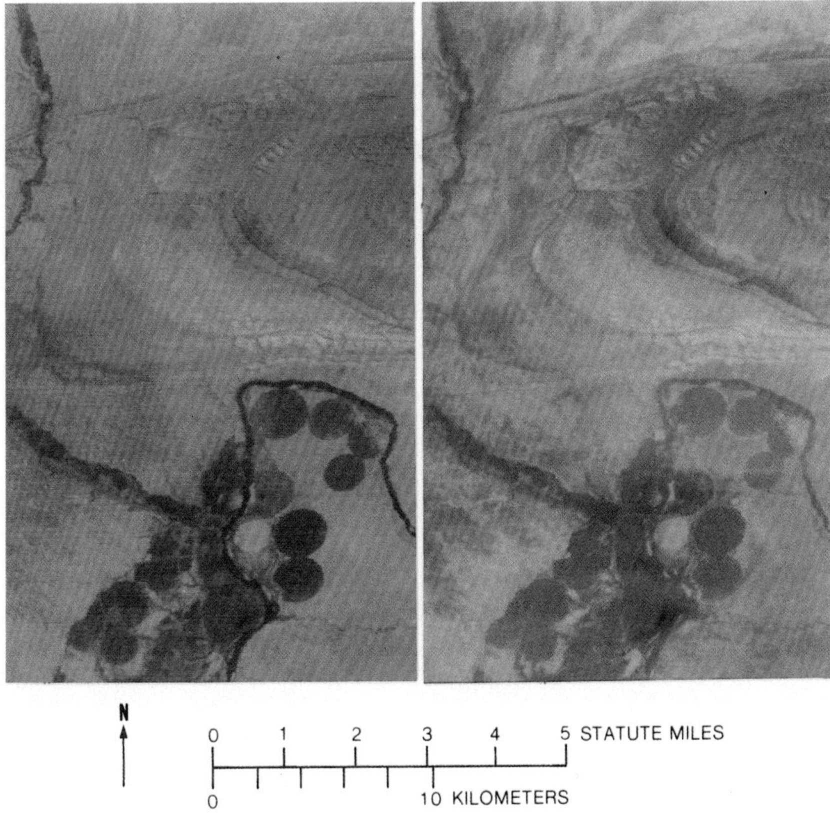

Fig. 31-11. A channel 6 (TM band 5; 1.55 μm–1.75 μm) Thematic Mapper Simulator (TMS) image and a false color composite image (bands 5, 6, and 7 in blue, green, and red, respectively) acquired by the NS001 of the Split Mountain area, Uinta Basin, northeast Utah. (Image courtesy of John Dwyer, EROS Data Center, U.S. Geological Survey.)

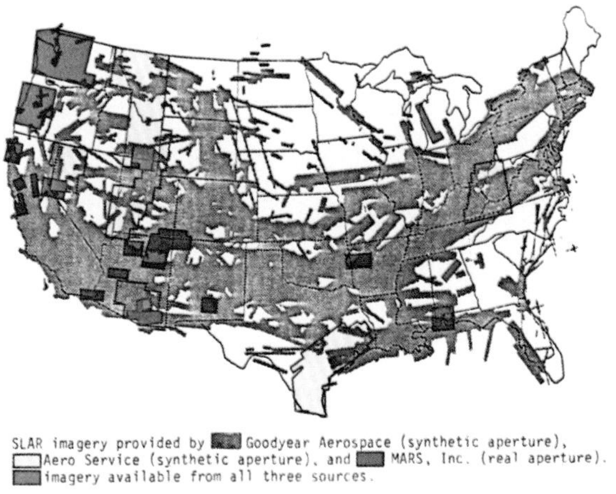

SLAR imagery provided by Goodyear Aerospace (synthetic aperture),
Aero Service (synthetic aperture), and MARS, Inc. (real aperture).
imagery available from all three sources.

Fig. 31-14. Generalized map depicting partial coverage of SLAR imagery of the conterminous United States provided by commercial organizations. (Courtesy of John Jones, U.S. Geological Survey.)

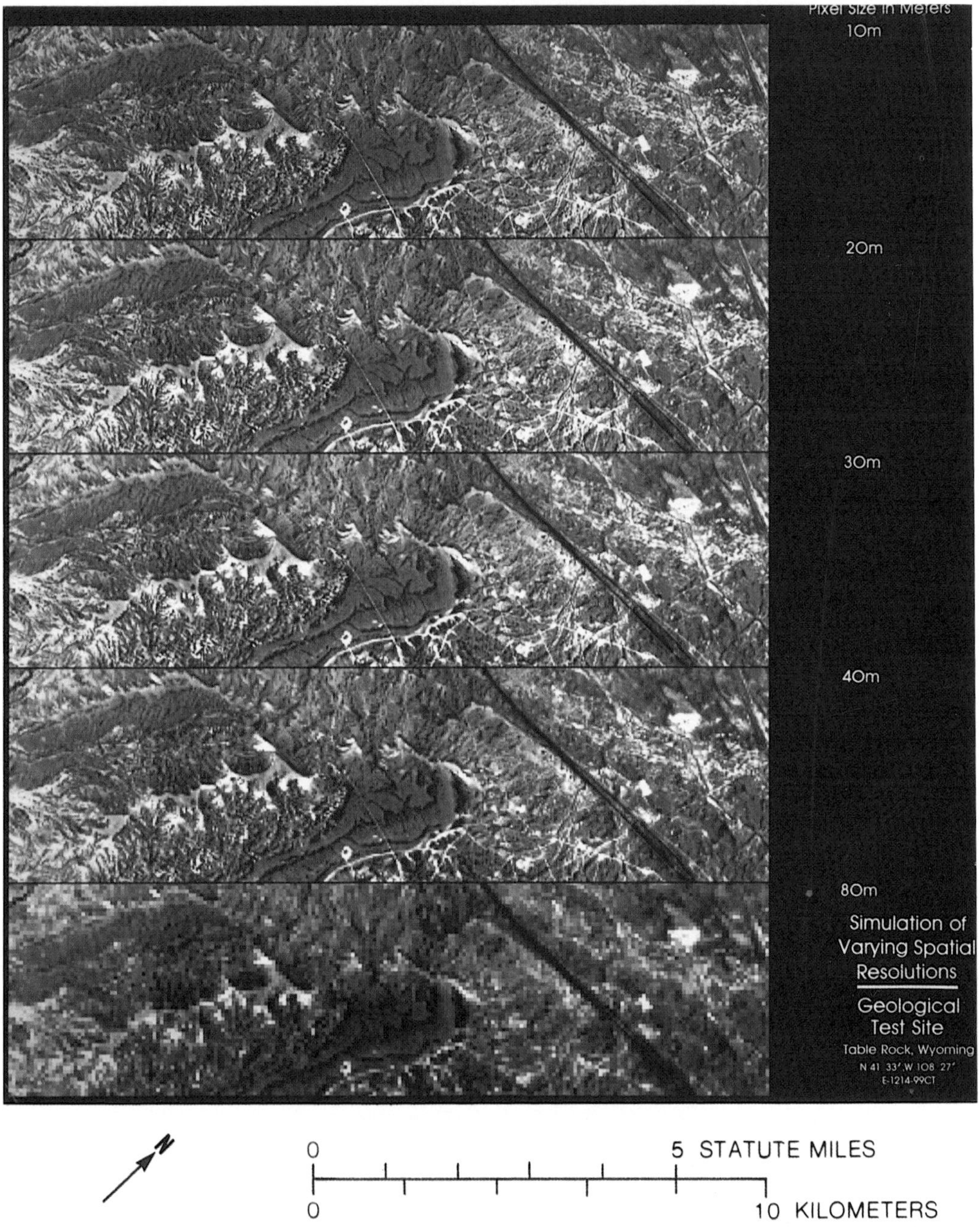

Pixel Size in Meters

10m

20m

30m

40m

80m

Simulation of Varying Spatial Resolutions

Geological Test Site

Table Rock, Wyoming

N 41 33' W 108 27'

E-1214-99CT

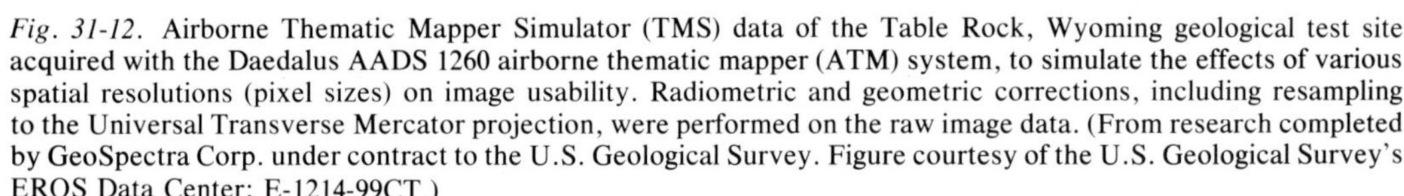

Fig. 31-12. Airborne Thematic Mapper Simulator (TMS) data of the Table Rock, Wyoming geological test site acquired with the Daedalus AADS 1260 airborne thematic mapper (ATM) system, to simulate the effects of various spatial resolutions (pixel sizes) on image usability. Radiometric and geometric corrections, including resampling to the Universal Transverse Mercator projection, were performed on the raw image data. (From research completed by GeoSpectra Corp. under contract to the U.S. Geological Survey. Figure courtesy of the U.S. Geological Survey's EROS Data Center; E-1214-99CT.)

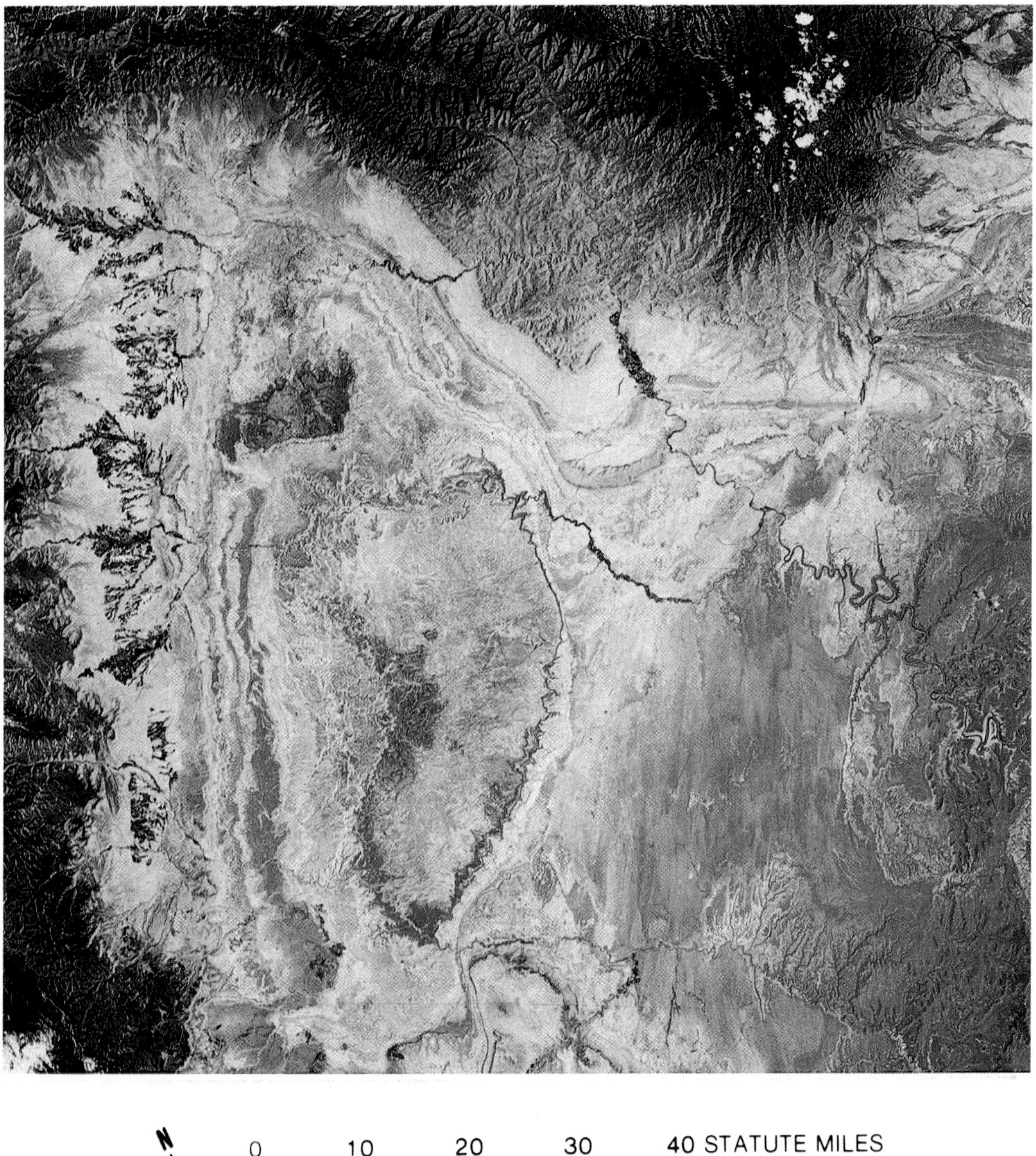

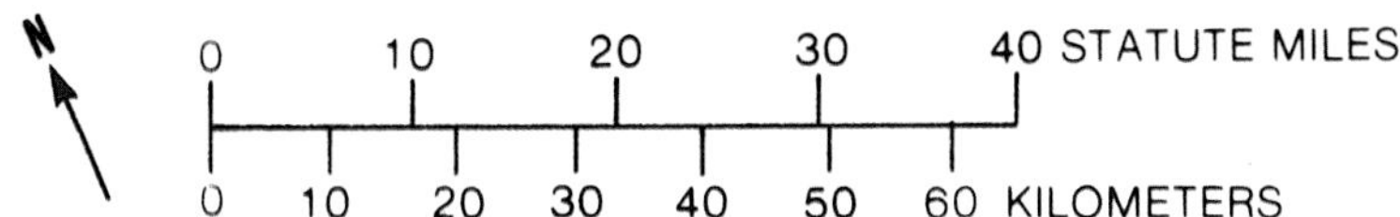

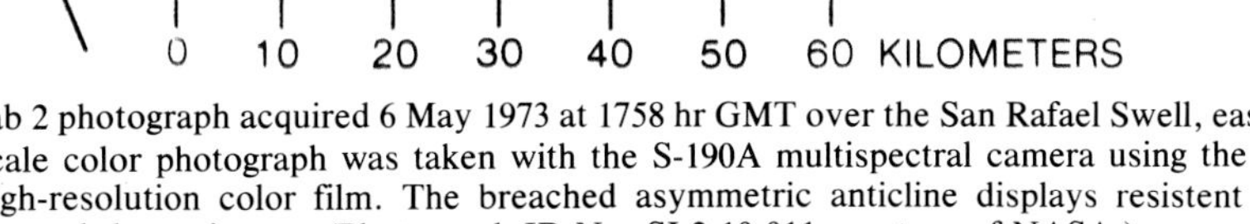

Fig. 31-19. Skylab 2 photograph acquired 6 May 1973 at 1758 hr GMT over the San Rafael Swell, east-central Utah. This 1:425,000-scale color photograph was taken with the S-190A multispectral camera using the 0.4- to 0.7-μm bandpass and high-resolution color film. The breached asymmetric anticline displays resistent strata forming hogback ridges around the perimeter. (Photograph ID No. SL2-10-011 courtesy of NASA.)

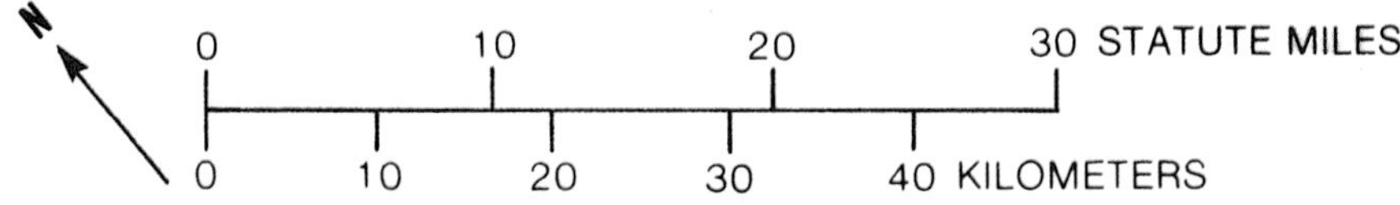

Fig. 31-20. Skylab 3 photograph acquired 8 August 1973 at 16 hr 01 min 34 sec GMT over the Bingham Canyon, Utah, open-pit copper mine. This 1:475,000-scale (original scale) color photograph was taken with the S-190B Earth terrain camera using a 0.4- to 0.7-μm bandpass filter and high-resolution color film. The 21-meter ground resolution can be compared with the 79- and 30-meter resolution image data acquired by the Landsat 2 MSS and Landsat 3 RBV systems, respectively, in Color Figure 31-22 and Figure 31-23. (Photograph ID No. SL3-83-300 courtesy of NASA.)

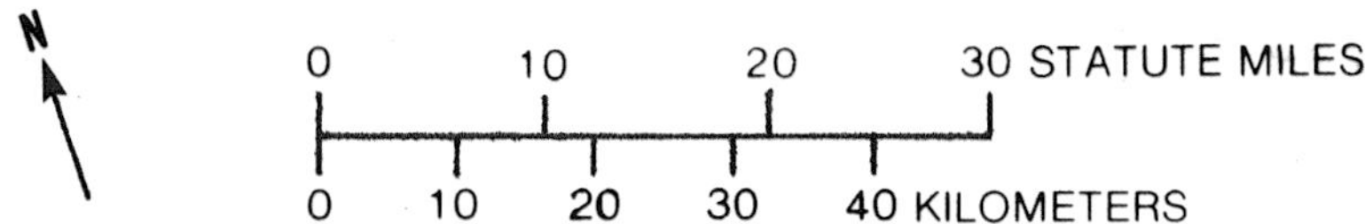

Fig. 31-22. Part of a Landsat 2, multispectral scanner (MSS) false-color composite image acquired on 28 August 1980 (Landsat ID No. 22045-17301) at approximately 1030 hr local time of the Salt Lake City, Utah area. The false-color composite image was generated from MSS bands 4, 5, and 6 (rather than the usual band 7), using a blue, green and red filter, respectively. The Wasatch and Uinta Mountains, to the east, mark the transition into the Basin and Range Province. The Bingham Canyon open-pit copper mine is identified as disturbed (bright) terrain in the center of the Oquirrh Mountains. The Great Salt Lake, with salt evaporation facilities, and Utah Lake, with dispersed algal blooms (Strong, 1976), are also visible in this scene.

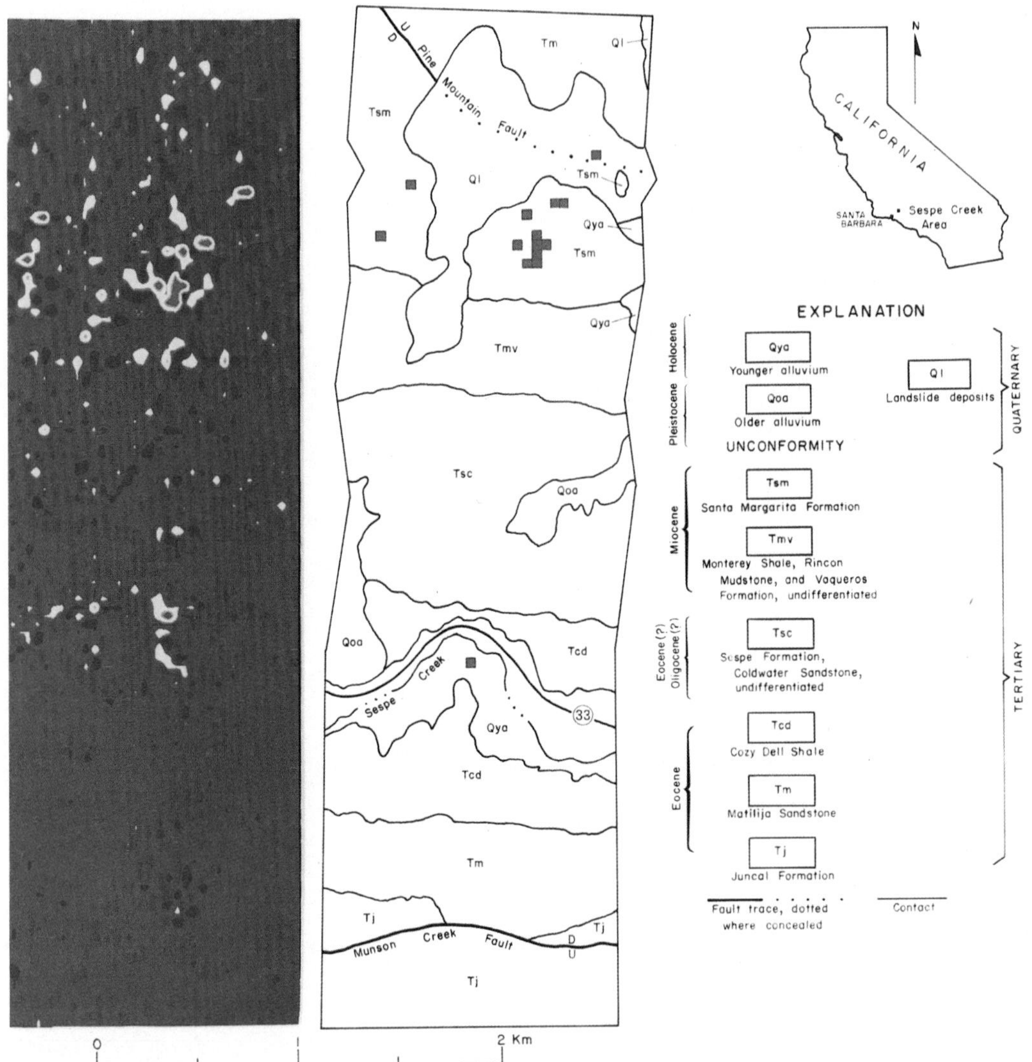

Fig. 31-33. Luminescence image and geologic map of the Sespe Creek area, Ventura County, California. The image was acquired at 1114 hr PST on 6 November 1979, with the FLD imaging system operating at 486.1 nm. Red areas on both the image and the geologic map represent luminescence that exceeds 3.9 ppb rhodamine dye equivalency. Their position correlates well with the occurrence of the phosphatic-gypsiferous Santa Margarita Formation (Tsm), and corroborates laboratory measurement of significant luminescence of samples of phosphate rock and gypsum collected from that Formation. Lower luminescence levels, shown in yellow (2.2–3.8 ppb rhodamine dye equivalency) are believed to be luminescent detritus transported downslope from the Santa Margarita Formation, southward toward Sespe Creek. Luminescence values greater than 2.2 ppb rhodamine dye equivalency are rare south of Sespe Creek, where the Santa Margarita Formation is absent. Outlined shape of the geologic map is irregular in order to accommodate distortion in the luminescence image caused by aircraft roll and drift. (Geologic map from Vedder et al., 1973).

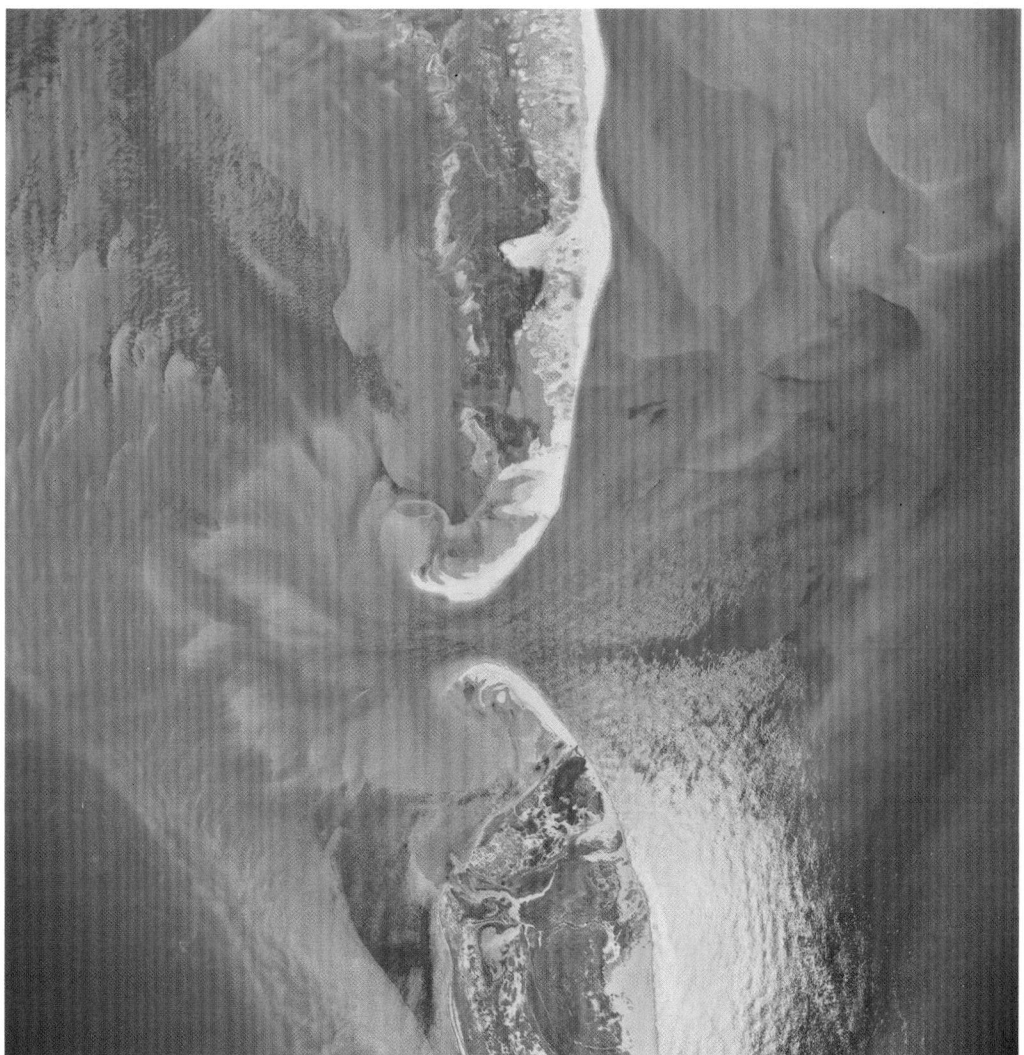

Fig. 31-45. Low-altitude, large scale (original contact scale of 1:20,000), color aerial photograph of the breach area on Monomoy Island, Cape Cod, Massachusetts. This photograph (No. NOS19AUG78BC, frame 6805) was acquired by the National Ocean Survey on 19 August 1978 from an altitude of 3,050 m with a 152.7 mm focal length (6 in) camera. Natural color photography such as this is generally flown unfiltered, and employs the only aerial film which has the same spectral sensitivity (Figure 31-47) as the human eye, thereby faithfully recording the scene as a human would view it. Compare with Figure 31-57, a Landsat 3 RBV image acquired on the identical date. (For Color Figure 31-42, see next page.)

Fig. 31-42. A high oblique color aerial photograph looking northeast across the breach of Monomoy Island towards the southern tip of Nauset Spit, Cape Cod, Massachusetts. Color aerial photograph No. 78-712-25C by Richard W. Kelsey, Chatham, Massachusetts, was taken on 12 July 1978 from an altitude of about 700 m (See Figure 31-43, a vertical aerial photograph taken on the same day). This type of oblique aerial photograph is typical of those used by geologists in documenting static or dynamic geological phenomena with a hand-held 35 mm camera in a light aircraft.

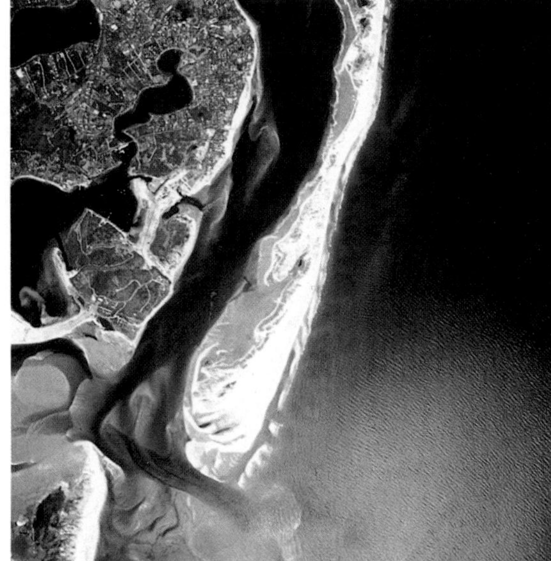

Fig. 31-46. Part of a low altitude, large scale (original contact scale of 1:24,000) false color infrared aerial photograph of the northern tip of Monomoy Island and the southern tip of Nauset Spit. This photograph (No. AFCRL, Roll 48, Frame 18) was acquired by U.S. Air Force Cambridge Research Laboratories on 7 May 1969, at 1205 EDT, from an altitude of 7,320 m with a K-17 aerial camera (focal length of 305 mm (12 in)). The water level inside Nauset Spit was imaged at 24 min after low tide (mean tidal range of 1 m); the water level outside Nauset Spit was imaged at 2 hr, 24 min after low tide (mean tidal range of 2 m).

Fig. 31-50. High altitude, small scale color infrared aerial photograph of the southeastern part of Cape Cod. Aerial photograph No. NASA/MSC103SEP69, roll 8, frame 6303. Compare with Color Figure 31-49, a high altitude, small scale color aerial photograph acquired at the same time and note the better penetration of cirrus clouds by this color infrared aerial film. This photograph was acquired on 13 September 1969 at 1020 hr EST from an altitude of 18,300 m by NASAs RB-57 F, with a 152.7 mm (6 in) focal length RC8 camera, using Kodax SO-117 film, on Mission 103, Site 176.

Fig. 31-49. High altitude, small scale color aerial photograph of the southeastern part of Cape Cod. Aerial photograph No. NASA/MSC103SEP69, roll 5, frame 5975. Compare with Color Figure 31-50, a high altitude, small scale color infrared aerial photograph acquired at the same time. This color aerial photograph (Kodak 2448 film) was acquired on 13 September 1969 at 1020 hr EST from an altitude of 18,300 m by NASA's RB-57F, with a 152.7 mm (6 in) focal length RC8 camera, on Mission 103, Site 176.

Fig. 31-51. A high resolution color satellite photograph of the southeastern part of Cape Cod. The photograph was acquired during the Skylab 3 mission with the S-190B Earth Terrain Camera having a focal length of 457 mm (18-in) on 12 September 1973 at 1411 hr EST (EROS Data Center Accession No. G30B08631300). The film is Kodak SO-242 and was exposed unfiltered to record energy between about 0.4 μm and 0.7 μm (see Figure 31-47). The estimated ground resolution is 21 m. Note the submarine morphologic detail in the submarine shoals around Nantucket Island. Compare with Figure 31-52, a satellite photograph from the S-190A multispectral camera acquired at the same time.

Fig. 31-54. Standard Landsat MSS color composite image (ID No. 5444-14084) of Cape Cod and environs acquired on 6 July 1976. Compare with Figure 31-228, a digitally enhanced version of the identical Landsat image. Landsat MSS images are acquired in four different spectral bands from a platform orbiting 919 km above the Earth. This particular Landsat image shows Monomoy Island before the 1,500 m breach occurred on 6–7 February 1978. (See Figure 31-55.)

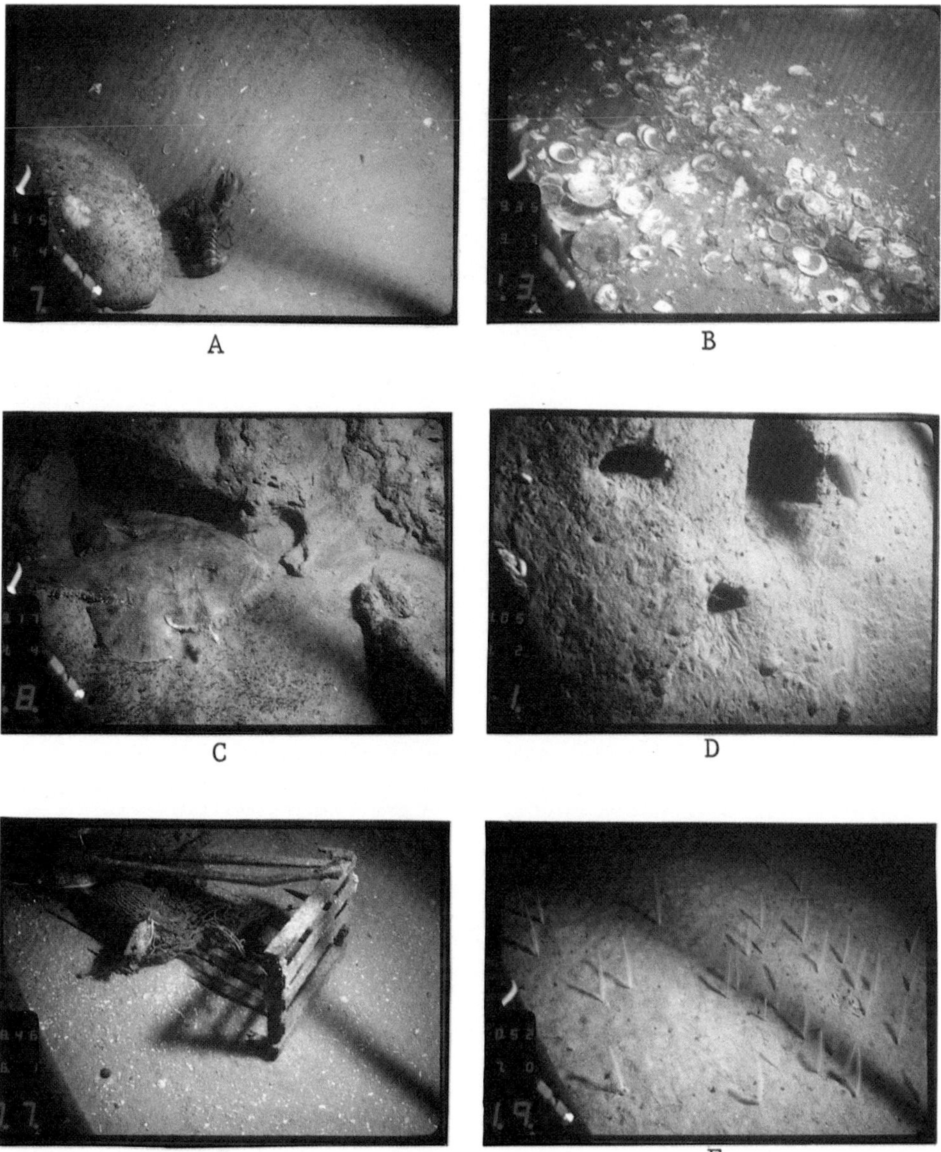

Fig. 31-53. A group of submarine color photographs of various geomorphic features at the edge of the Continental Shelf at the head of Lydonia Canyon (See Figures 31-239, 31-240, and 31-241), east of Cape Cod, Massachusetts. A. View of a boulder, transported to its present position by ice (glacial erratic) or by meltwater streams discharging into the ocean during lower sea level in late Wisconsinan time. Submarine currents tend to scour sediments around boulders creating a hospitable place for lobsters (*Homarus americanus*). B. Concentration of pelecypod and other shells. C. Clay slump blocks related to possible mass movement down the Continental Slope. A goosefish (*Lophius piscatorius*) is in the left center of the photograph. D. The exposed clay substrate on the Continental Slope is a preferred location for creation of burrows by the tilefish (*Lopholatilus chamaeleonticeps*) and benthic animals. E. Human artifacts such as this partially disintegrated lobster trap are common on the Continental Shelf. F. Benthic marine organisms, such as these sea pens (*Pennatula*), also disturb sediments of the seafloor. All color photographs (Kodak Ektachrome film using a Benthos camera with strobe) courtesy of John M. Aaron, U.S. Geological Survey, taken during the period 23 July to 9 August 1978 in 200–300 m of water from the research submersible *Diaphus* during studies of the submarine geomorphology of the Continental Shelf and Slope offshore from Cape Cod, Massachusetts. Such studies are important in assessing the stability of the seafloor in connection with the planned development of potential offshore hydrocarbon resources.

Fig. 31-68. A Landsat false-color image mosaic of the Central Rift Valley of Kenya, east Africa, prepared from 13 custom-processed Landsat MSS images for the U.S. Agency for International Development and the Survey of Kenya by the U.S. Geological Survey to be used as an image base map for regional geologic, geothermal, and ground-water investigations. The thirteen Landsat 2 MSS images were acquired during the period 24-26 January 1976 (see Table 31-18 for list of Landsat images). Cibachrome prints were mosaicked by John Milne, Technical Photo, Inc., at a scale of 1:1,000,000 and fitted to 1:1,000,000-scale U.S. Air Force Operational Navigation Charts published by the Defense Mapping Agency (U.S.).

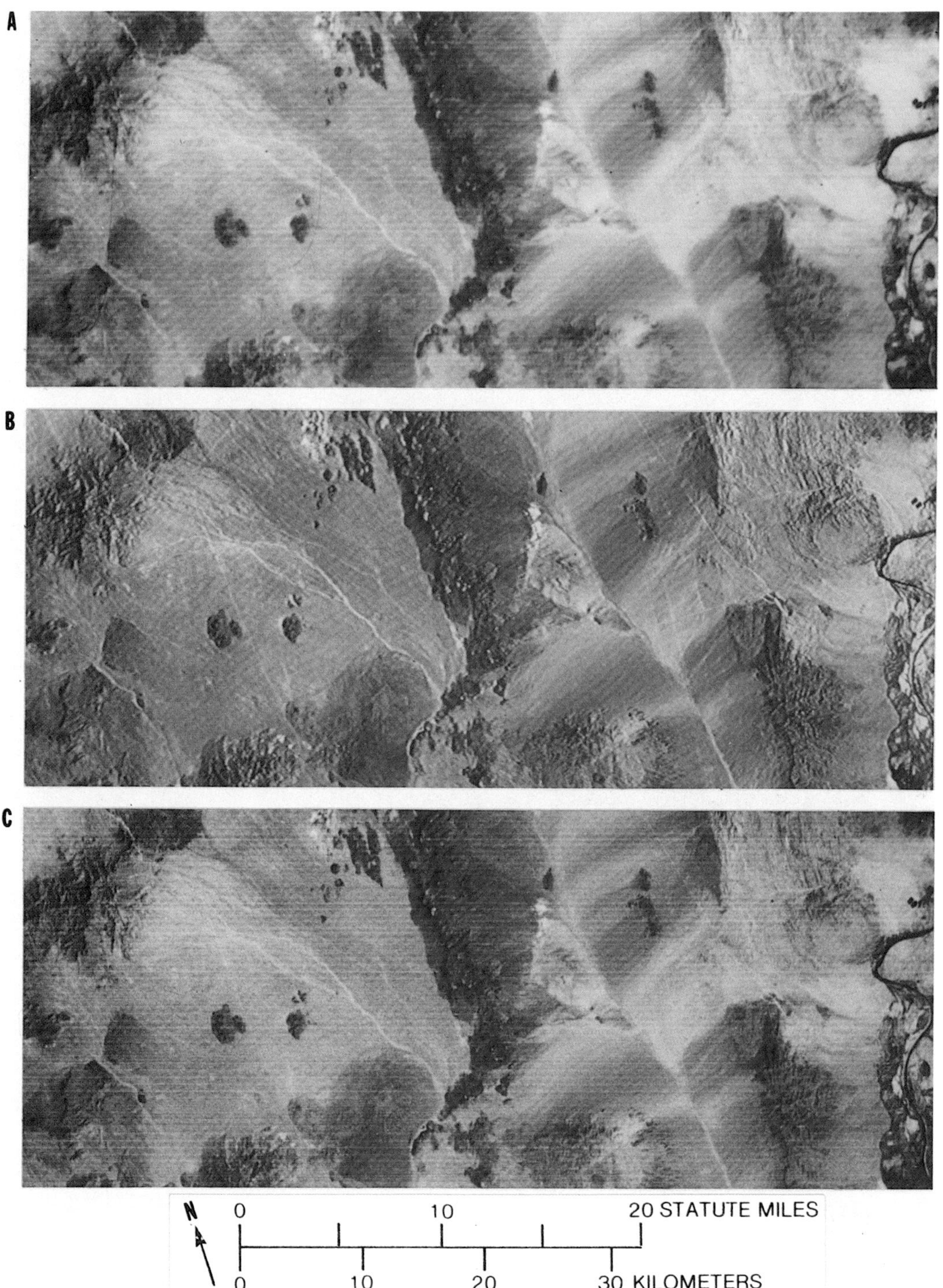

Fig. 31-69. Part of a Landsat 1 MSS false color composite image (1700-17422) of the Needles, California, area and environs acquired on 23 June 1974, in which three different methods of preparing Landsat MSS false-color composite images are compared: *A*. Standard MSS false-color composite from the EROS Data Center; *B*. Digitally enhanced MSS false-color composite, also from the EROS Data Center; and *C*. Custom-processed photographically enhanced MSS false-color composite by Technical Photo, Inc. The original scale of the image was 1:500,000.

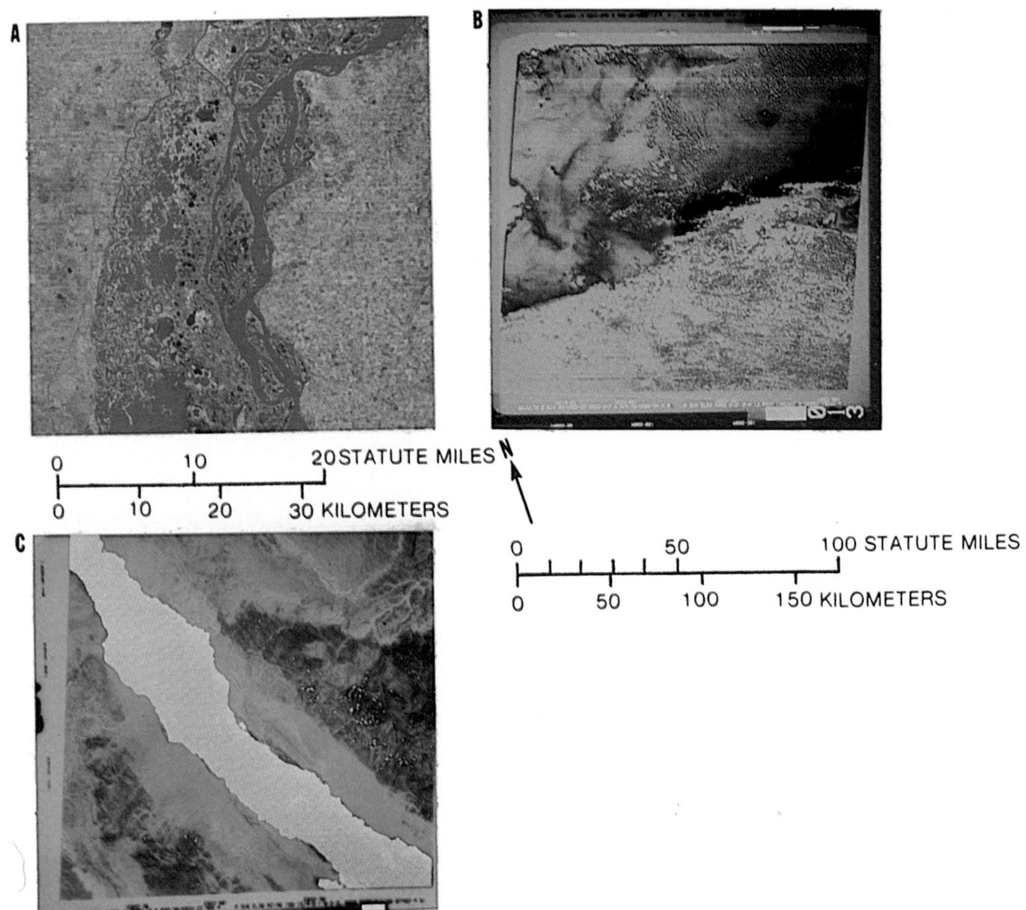

Fig. 31-70. Group of three special purpose spectral composite images prepared from reprocessed 70-mm Landsat MSS transparencies, on an additive-color viewer to demonstrate practical geoscience applications. *A*. Landsat 2 MSS "false" color composite image (2340-13082; 28 December 1975) of the Río Paraná, Argentina, area, with MSS bands 4 and 5 positives in blue, MSS band 6 positive in green, and MSS band 7 positive in red to maximize the distinction between land and water. *B*. Landsat 3 MSS "false" color composite image (30502-15532; 20 July 1979) of the Gulf of Campeche, Mexico area with MSS band 4 negative in white, MSS band 5 positive in blue; another MSS band 5 positive in green, and an MSS band 6 positive in red, to maximize the distinction between oil slicks and water. *C*. Landsat 1 MSS "false" color composite image (1109-07493; 9 November 1972) of the Gulf of Suez area with reprocessed MSS 4 band positive in green and an unreprocessed band 7 negative in white, to maximize the distinction between oil slicks, water, and reefs.

A

B

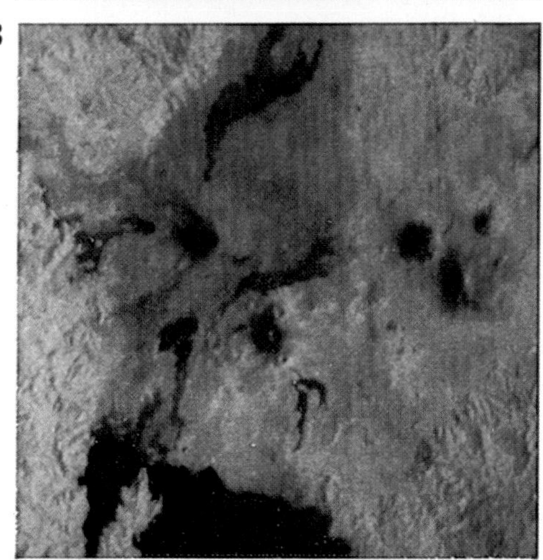

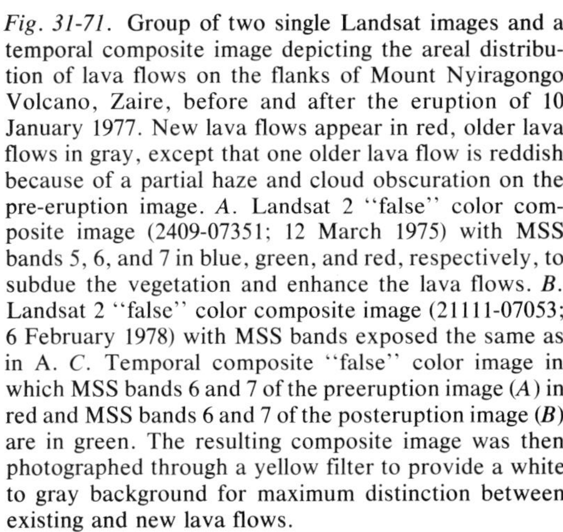

Fig. 31-71. Group of two single Landsat images and a temporal composite image depicting the areal distribution of lava flows on the flanks of Mount Nyiragongo Volcano, Zaire, before and after the eruption of 10 January 1977. New lava flows appear in red, older lava flows in gray, except that one older lava flow is reddish because of a partial haze and cloud obscuration on the pre-eruption image. *A*. Landsat 2 "false" color composite image (2409-07351; 12 March 1975) with MSS bands 5, 6, and 7 in blue, green, and red, respectively, to subdue the vegetation and enhance the lava flows. *B*. Landsat 2 "false" color composite image (21111-07053; 6 February 1978) with MSS bands exposed the same as in A. *C*. Temporal composite "false" color image in which MSS bands 6 and 7 of the preeruption image (*A*) in red and MSS bands 6 and 7 of the posteruption image (***B***) are in green. The resulting composite image was then photographed through a yellow filter to provide a white to gray background for maximum distinction between existing and new lava flows.

C

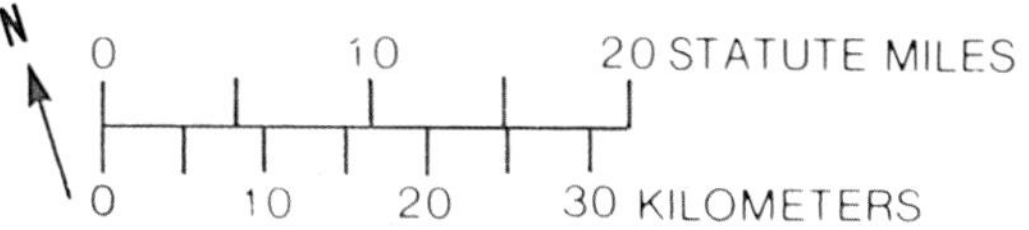

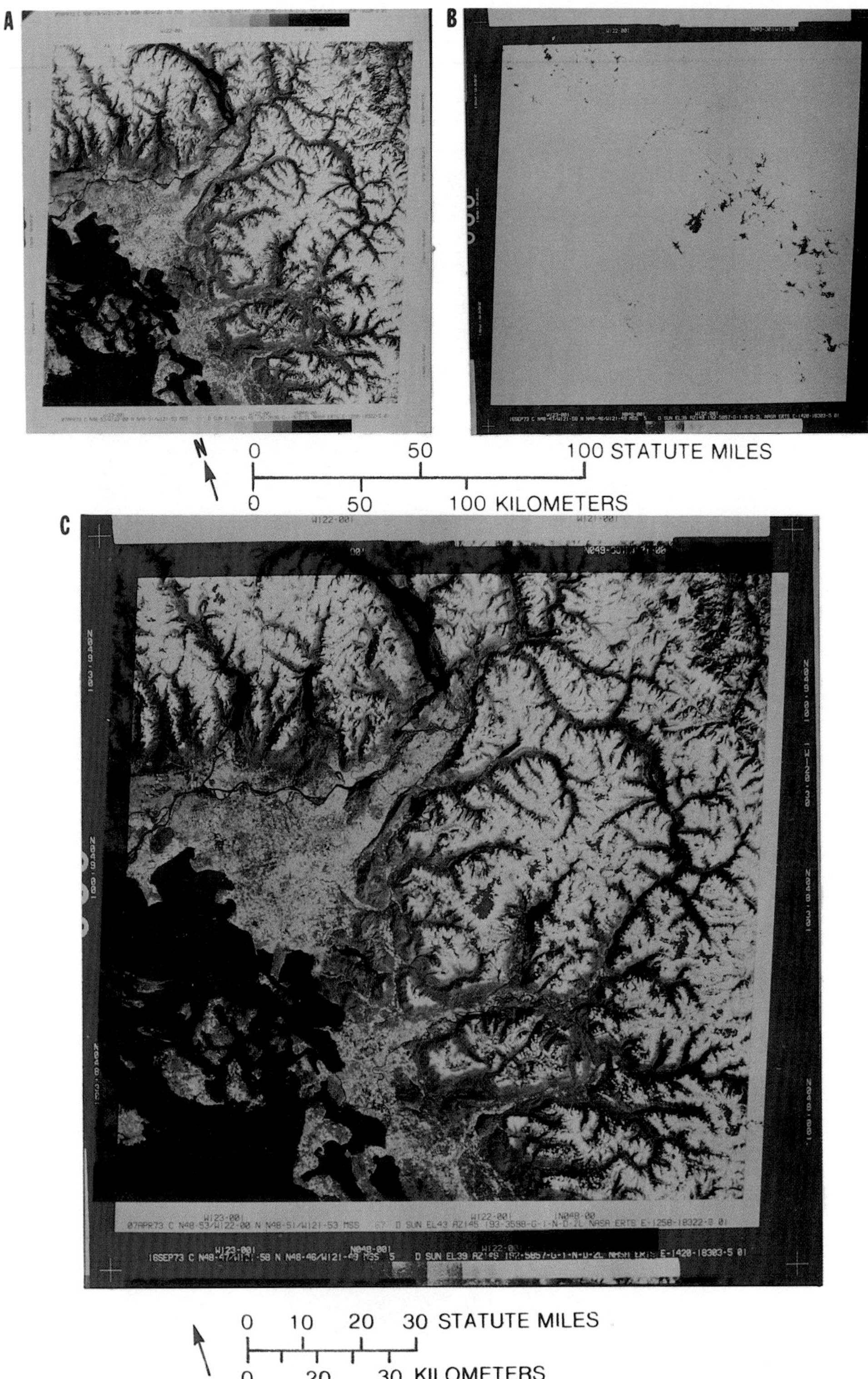

Fig. 31-72. Group of two single Landsat images and a temporal composite image showing the seasonal fluctuation during 1973 in snow cover in the Cascade Mountains of the United States and Canada. *A*. Landsat 1 "false" color composite image (1258-18322; 7 April 1973) with MSS band 5 positive through a blue filter, MSS band 6 positive through a green filter, and MSS band 7 positive through a red filter, showing snow cover at the end of winter. *B*. Landsat 1 "false" color composite image (1420-18303; 16 September 1973) with MSS band 5 negative in red, showing snow cover, snow-covered glaciers, and glaciers at the end of the summer. *C*. Temporal composite "false" color image in which the end-of-winter image (A) and the end-of-summer image are combined to depict the maximum and minimum seasonal mountain snow-pack on a single image.

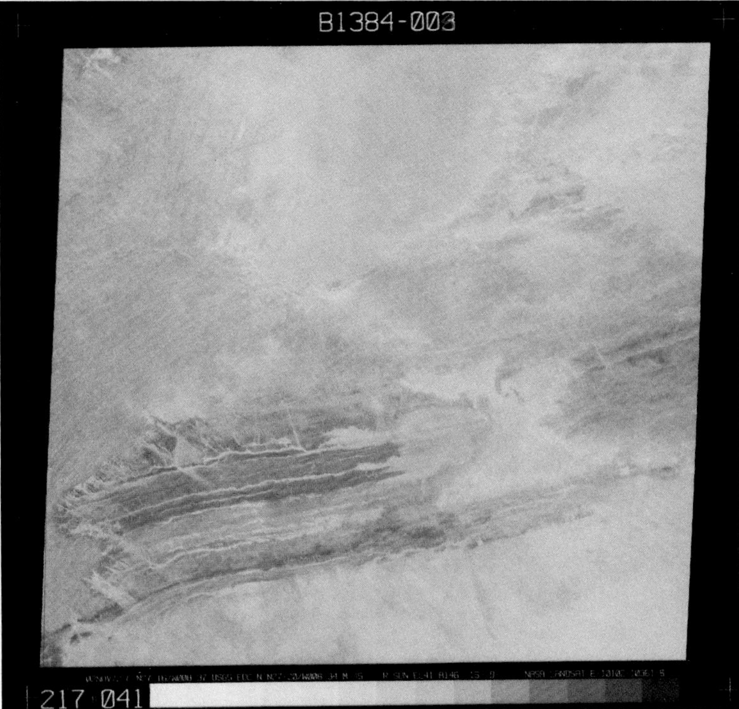

Fig. 31-75. Standard EROS Digital Image Processing System (EDIPS) Landsat MSS false color composite image (10102-10361; 2 November 1972) of the western Sahara, in the border area between Algeria, Mauritania, and Morocco, northwestern Africa. This Landsat MSS false-color composite image is similar to the one produced from 70-mm film transparencies. Compare this "washed out" image with Color Figure 31-76, a custom-processed EDIPS image.

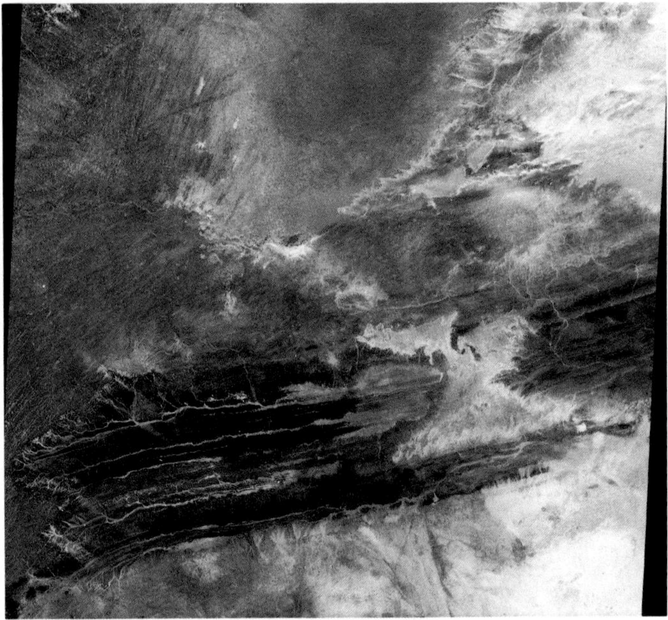

Fig. 31-76. Custom-processed EDIPS Landsat MSS false-color composite image (10102-10361; 2 November 1972) of the western Sahara, in the border area between Algeria, Mauritania, and Morocco, northwestern Africa. Special contrast stretches have been applied to each of the three MSS bands (4, 5, and 7) used to make the false-color composite to markedly increase the interpretability of the Landsat image. Compare this image with Color Figure 31-75, a standard EDIPS image. Spectral information contained on the CCT's is often not visible on images resulting from either "standard" EBR or EDIPS processing.

Fig. 31-77. A special Landsat MSS false color composite image (5137-17505; 3 September 1975), of the San Francisco area, combined with a Landsat 3 RBV image (30166-18065; 18 August 1978). The 79-m pixels of MSS bands 4, 5, and 7 were registered to the 30-m pixels of the single RBV band, increasing the spatial resolution by 2.6 times but retaining the spectral sensitivity of the MSS false-color composite image. (Specially merged Landsat MSS/RBV image courtesy of the U.S. Geological Survey's EROS Data Center; PAO No. E-1012-99CT).

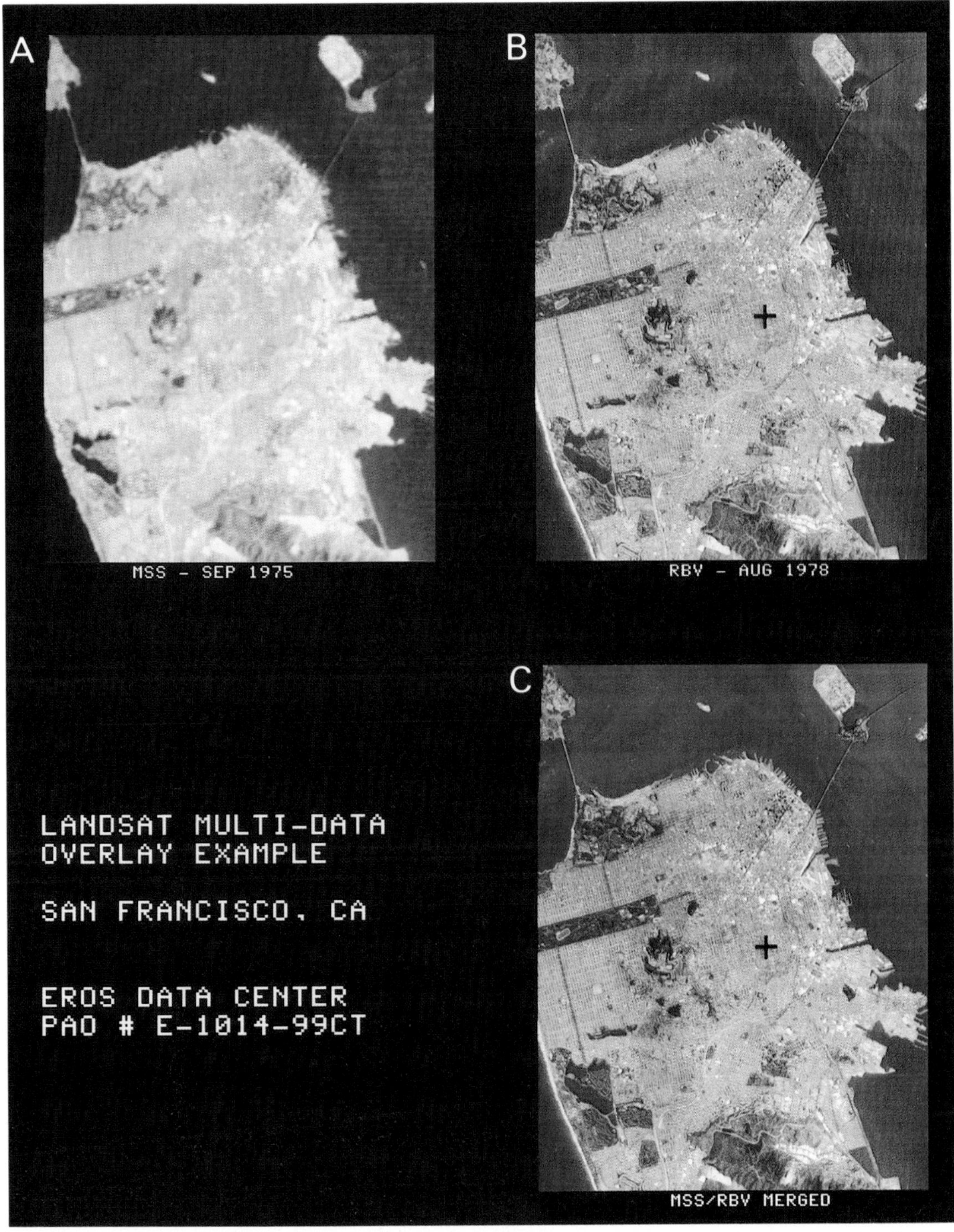

Fig. 31-78. Group of 3 Landsat images which cover an enlarged portion of the southwestern corner of Color Figure 31-77. *A*. Landsat MSS false-color composite image (5137-17505; 3 September 1975) in which 79-m pixels of MSS bands 4, 5, and 7 are combined. *B*. Landsat 3 RBV image composed of 30-m pixels (30166-18065; 18 August 1978). *C*. Combination of images in *A* and *B* in which the MSS false-color composite image (*A*) is merged with the RBV image (*B*) to create a higher resolution Landsat false color composite image of the city of San Francisco. (Special group of Landsat images courtesy of the U.S. Geological Survey's EROS Data Center; PAO No. E-1014-99CT).

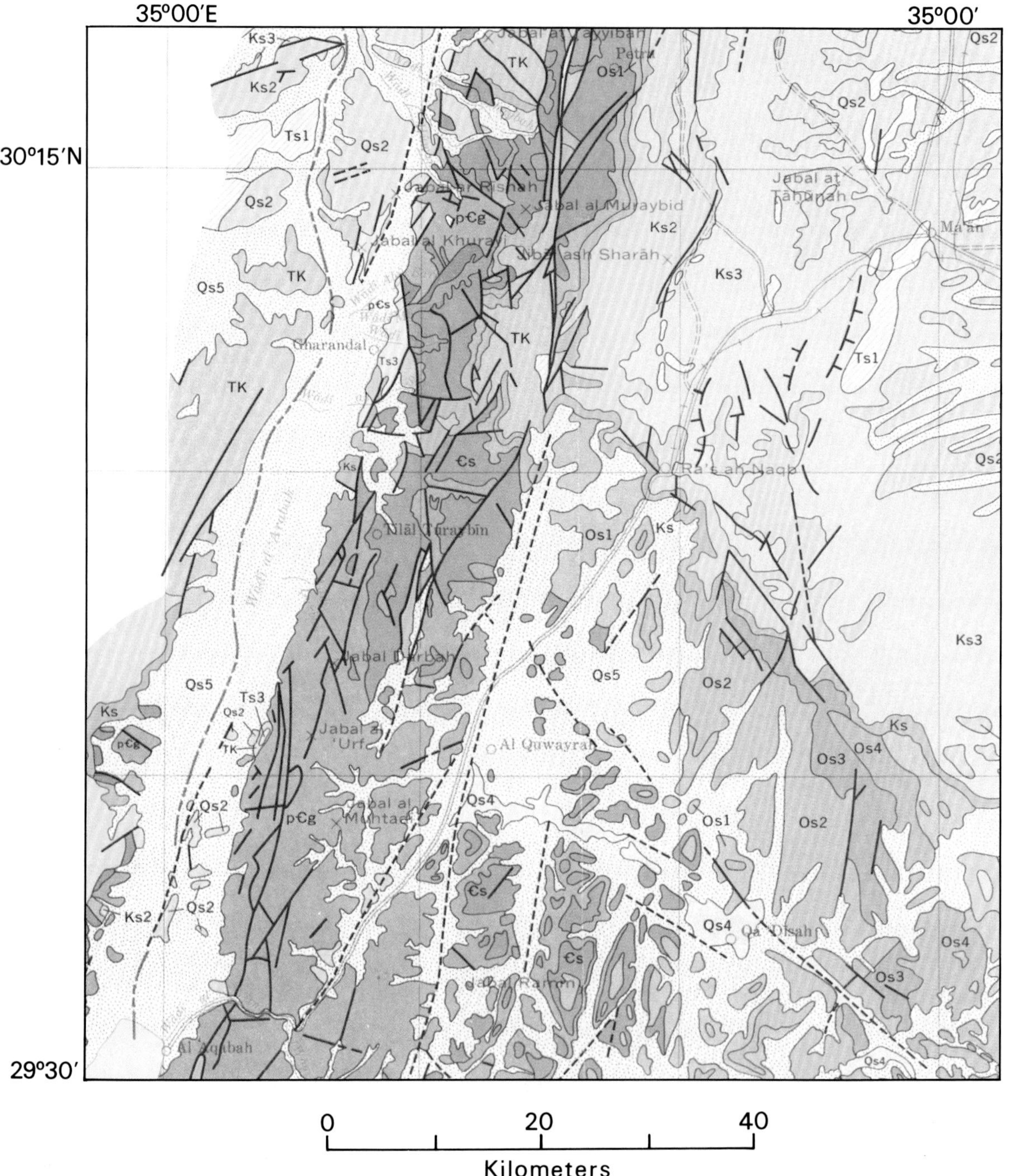

Fig. 31-79. Geologic map of southwest Jordan and adjacent areas compiled at an original scale of 1:500,000 and explanation of map symbols (from Bender, 1975b).

EXPLANATION

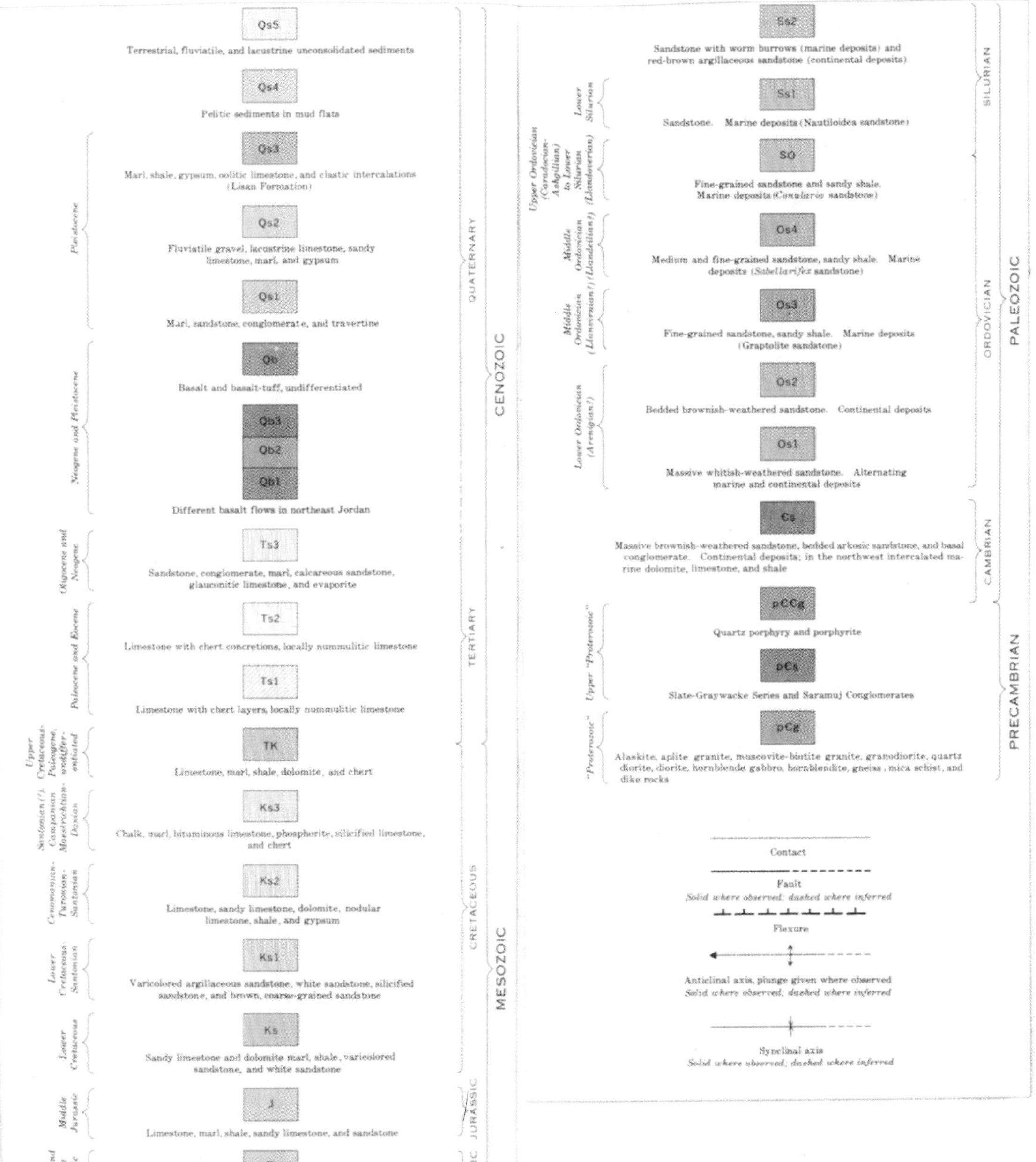
Qs5
Terrestrial, fluviatile, and lacustrine unconsolidated sediments
Qs4
Pelitic sediments in mud flats
Qs3
Marl, shale, gypsum, oolitic limestone, and clastic intercalations (Lisan Formation)
Qs2
Fluviatile gravel, lacustrine limestone, sandy limestone, marl, and gypsum
Qs1
Marl, sandstone, conglomerate, and travertine
Pleistocene
Qb
Basalt and basalt-tuff, undifferentiated
Qb3
Qb2
Qb1
Different basalt flows in northeast Jordan
Neogene and Pleistocene
QUATERNARY
Ts3
Sandstone, conglomerate, marl, calcareous sandstone, glauconitic limestone, and evaporite
Oligocene and Neogene
Ts2
Limestone with chert concretions, locally nummulitic limestone
Ts1
Limestone with chert layers, locally nummulitic limestone
Paleocene and Eocene
TERTIARY
CENOZOIC
TK
Limestone, marl, shale, dolomite, and chert
Upper Cretaceous-Paleogene, undifferentiated
Ks3
Chalk, marl, bituminous limestone, phosphorite, silicified limestone, and chert
Santonian(?), Campanian Maestrichtian-Danian
Ks2
Limestone, sandy limestone, dolomite, nodular limestone, shale, and gypsum
Cenomanian-Turonian-Santonian
Ks1
Varicolored argillaceous sandstone, white sandstone, silicified sandstone, and brown, coarse-grained sandstone
Lower Cretaceous-Santonian
Ks
Sandy limestone and dolomite marl, shale, varicolored sandstone, and white sandstone
Lower Cretaceous
CRETACEOUS
J
Limestone, marl, shale, sandy limestone, and sandstone
Middle Jurassic
JURASSIC
Ꞧ
Limestone, sandy limestone, sandstone, shale, and locally gypsum
Lower and Middle Triassic
TRIASSIC
MESOZOIC
Ss2
Sandstone with worm burrows (marine deposits) and red-brown argillaceous sandstone (continental deposits)
Ss1
Sandstone. Marine deposits (Nautiloidea sandstone)
Lower Silurian
SILURIAN
SO
Fine-grained sandstone and sandy shale. Marine deposits (Conularia sandstone)
Upper Ordovician (Caradocian-Ashgillian) to Lower Silurian (Llandoverian)
Os4
Medium and fine-grained sandstone, sandy shale. Marine deposits (Sabellarifex sandstone)
Middle Ordovician (Llandeilian?)
Os3
Fine-grained sandstone, sandy shale. Marine deposits (Graptolite sandstone)
Middle Ordovician (Llanvernian?)
Os2
Bedded brownish-weathered sandstone. Continental deposits
Os1
Massive whitish-weathered sandstone. Alternating marine and continental deposits
Lower Ordovician (Arenigian?)
ORDOVICIAN
Ꞓs
Massive brownish-weathered sandstone, bedded arkosic sandstone, and basal conglomerate. Continental deposits; in the northwest intercalated marine dolomite, limestone, and shale
CAMBRIAN
PALEOZOIC
pꞒg
Quartz porphyry and porphyrite
pꞒs
Slate-Graywacke Series and Saramuj Conglomerates
Upper "Proterozoic"
pꞒg
Alaskite, aplite granite, muscovite-biotite granite, granodiorite, quartz diorite, diorite, hornblende gabbro, hornblendite, gneiss, mica schist, and dike rocks
"Proterozoic"
PRECAMBRIAN
Contact
Fault
Solid where observed; dashed where inferred
Flexure
Anticlinal axis, plunge given where observed
Solid where observed; dashed where inferred
Synclinal axis
Solid where observed; dashed where inferred

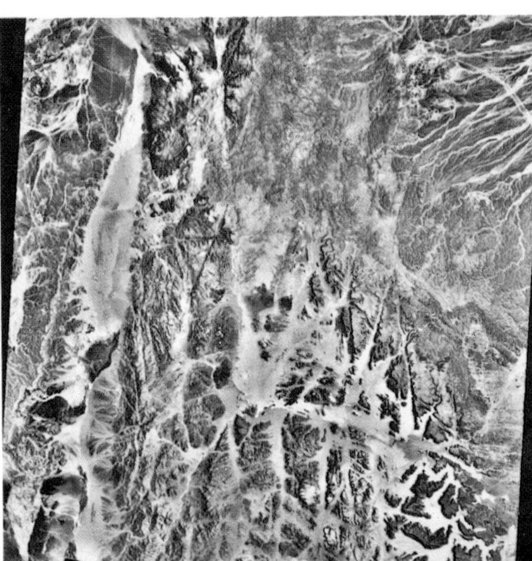

Fig. 31-83. False-color composite of the 101 by 101 kernel filtered Landsat MSS bands 4, 5, and 7 images which have been processed through blue, green, and red color filters, respectively. The three spatial filtered images had a linear contrast stretch applied before compositing.

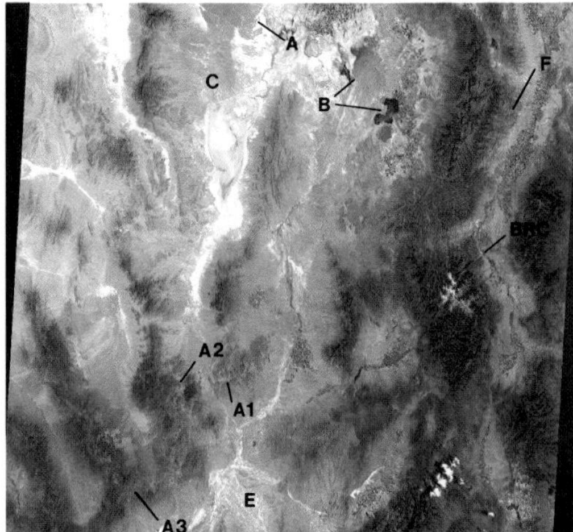

Fig. 31-88A. A standard color-additive, false-color-composite Landsat image (1699-17353/5; 22 June 1974) of the Richfield, Utah, scene, showing MSS bands 4, 5, and 7 in blue, green, and red, respectively. Limonitic rocks of medium brightness (*F*) appear yellow to yellow-brown on this image. Bright limonitic materials (*E, A1*) have a pale yellow to white cast. Intermediate (*A*) and basaltic (*B*) volcanic rocks show as shades of dark gray to black. Key to symbols: *A1, A2, A3,* and *BRC*—areas of hydrothermally altered rocks, *C*—Granitic intrusive of Jurassic age, *E*—Unconsolidated basin fill of the Escalante desert, *F*—Tertiary Flagstaff Limestone.

Fig. 31-90. Color-additive color-ratio-composite derived from the ratio images of Figure 31-89, showing the 6/10, 12/3, and 12/13 ratios in green, blue, and red, respectively. Big Rock Candy Mountain (*BRC*) is a bleached argillically altered area. Key to colors: Yellow - bleached argillically altered; White - limonitic and argillically altered; Cyan - limonitic, non-argillized rocks; Grenn - Materials with no significant absorption bands; Red, magenta and orange - vegetation.

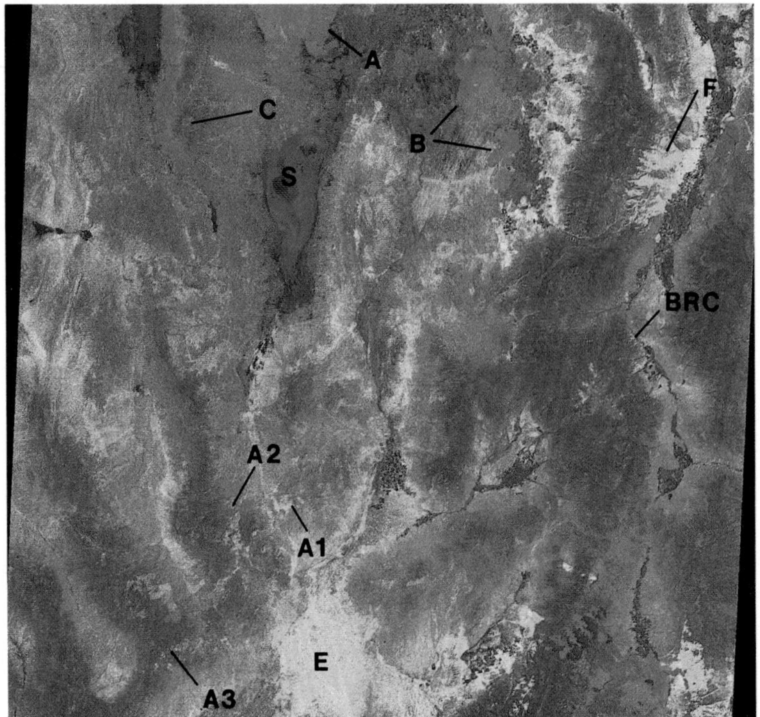

Fig. 31-94. A color-additive, color-composite Landsat MSS image (1699-17353/5; 22 June 1974) of principal components 2, 3, and 4 shown in blue, green, and red, respectively, for the Richfield, Utah scene. Note that limonitic rocks appear green irrespective of their brightness (compare *A1* and *F* of Color Figures 31-88A and 31-94). Intermediate to basaltic volcanic rocks appear purple to magenta (*A* and *B*, respectively, Color Figures 31-88A and 31-94). The bright Sevier Lake playa (*S*) also appears magenta.

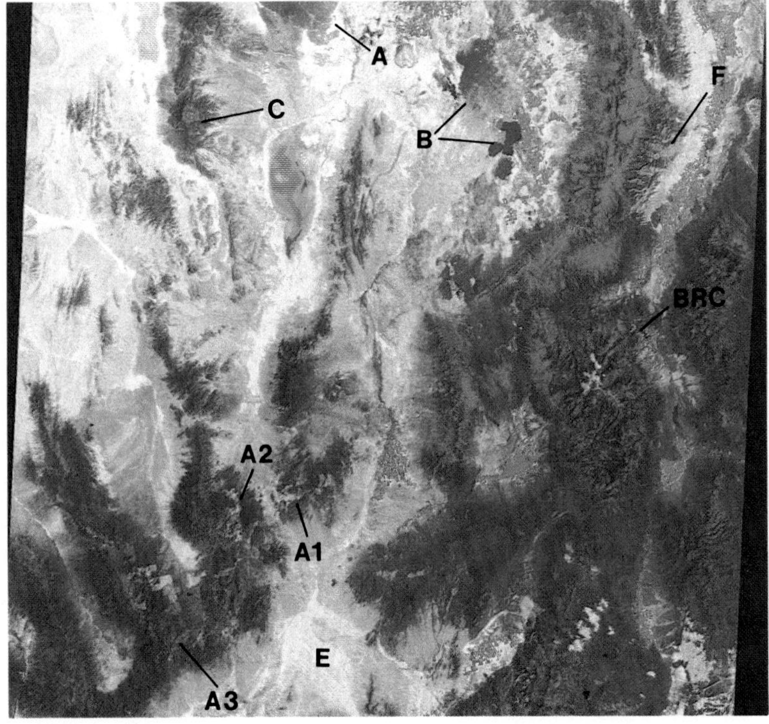

Fig. 31-95. Color-additive color-composite Landsat MSS image (1699-17353/5; 22 June 1974) of inverted principal components 1, 2, and 4, shown in blue, green, and red, respectively, for the Richfield, Utah scene. Annotation is the same as used in Color Figure 31-88A.

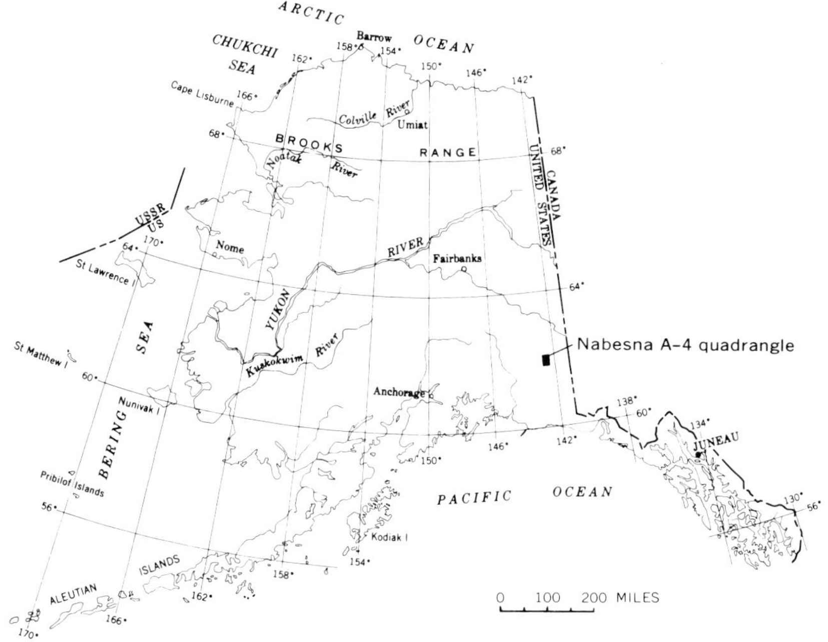

INDEX MAP OF ALASKA
SHOWING LOCATION OF AREA

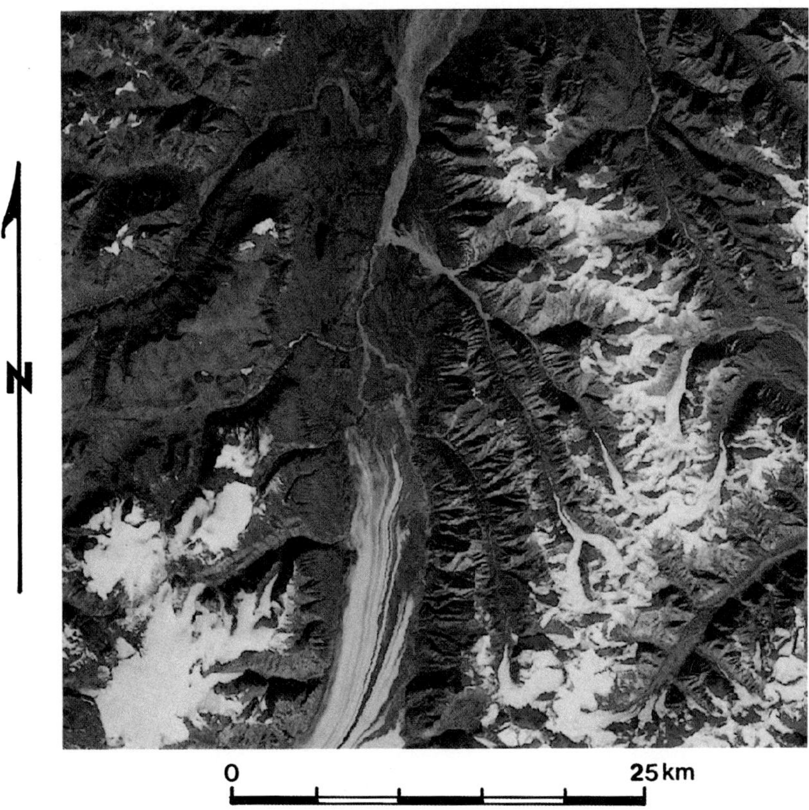

Fig. 31-96. Index map of Alaska showing the location of the Nabesna A-4 quadrangle and Landsat MSS false-color composite image (1422-20210, 18 September 1973) of the area showing characteristics of the terrain, including the terminus of the Nabesna Glacier.

MAP TITLE NABESNA	NUMERICAL IMPORTANCE OF DATA VARIABLE AND SUB—CLASSIFICATION											
DATA VARIABLE	**10**	**9**	**8**	**7**	**6**	**5**	**4**	**3**	**2**	**1**	**0**	**OFF**
10 Gravity (milligals)	-130	-128	-126	-124	-122	-120	-118	-116	-114	-112	-110	>-105
9 Aeromag (gammas)	4800	4850	4900	4950	5000	5050	5100	5150	5200	5250	5300	>5400
8 Copper/Chrome (PPM/PPM)	–	–	–	–	–	–	–	–	–	–	–	<1.0
7 Copper (PPM)	2000	1585	1259	1000	794	631	501	398	316	251	200	<185
6 Slope (%)	–	–	–	–	–	–	–	–	–	–	–	<10%
5 Topo (BETA<)	–	–	–	–	–	–	–	–	–	–	–	Shadow
4 MSS 5/MSS 4	2.0	1.9	1.8	1.7	1.6	1.5	1.4	1.3	1.2	1.1	1.0	<1.0
3 Geology (Within Intrusives)	T_p	K_{np}	K_{mt}	K_{ng}	T_{hd}	K_{dd}	–	–	–	–	–	–
2 Geology (Distance to any Intrusives)	<1Km	–	1Km	–	2Km	–	3Km	–	4Km	–	5Km	>6Km
1 Mineral Occurrence (Distance to any Occurrence)	<1Km	–	1Km	–	2Km	–	3Km	–	4Km	–	5Km	>6Km

Fig. 31-100. Application of the quantitative model developed for assessing porphyry-type copper potential in the Nabesna quadrangle. Description shows ranking of model parameters, their range limits, and weighting factors. Imaged results show model-defined areas superposed on Landsat MSS data (warmer colors denote areas having higher potential). Porphyry copper systems identified include Orange Hill (A), Bond Creek (B), Nikonda Creek (C), and Cross Creek (D).

Fig. 31-101. Color-composite image of the East Tintic Mountains, Utah, based on principal component transformation of MIR multispectral data. *A,B* = quartzite; *C,D,E* = interbedded sandstone, limestone, quartzite, shale, dolomite, and chert; *F* = silicified rocks; *G* = mine areas; *H* = Dragon mine; *I* = argillized rocks; *J* = quartz latite and quartz monzonite; *K* = latite and monzonite; *L* = vegetated areas; *M* = calcitic quartz latite; *N,O* = carbonate rocks; *P* = hydrothermal dolomite; *Q* = mine tailings and ponds (From Figure 3 in Kahle and Rowan, 1980).

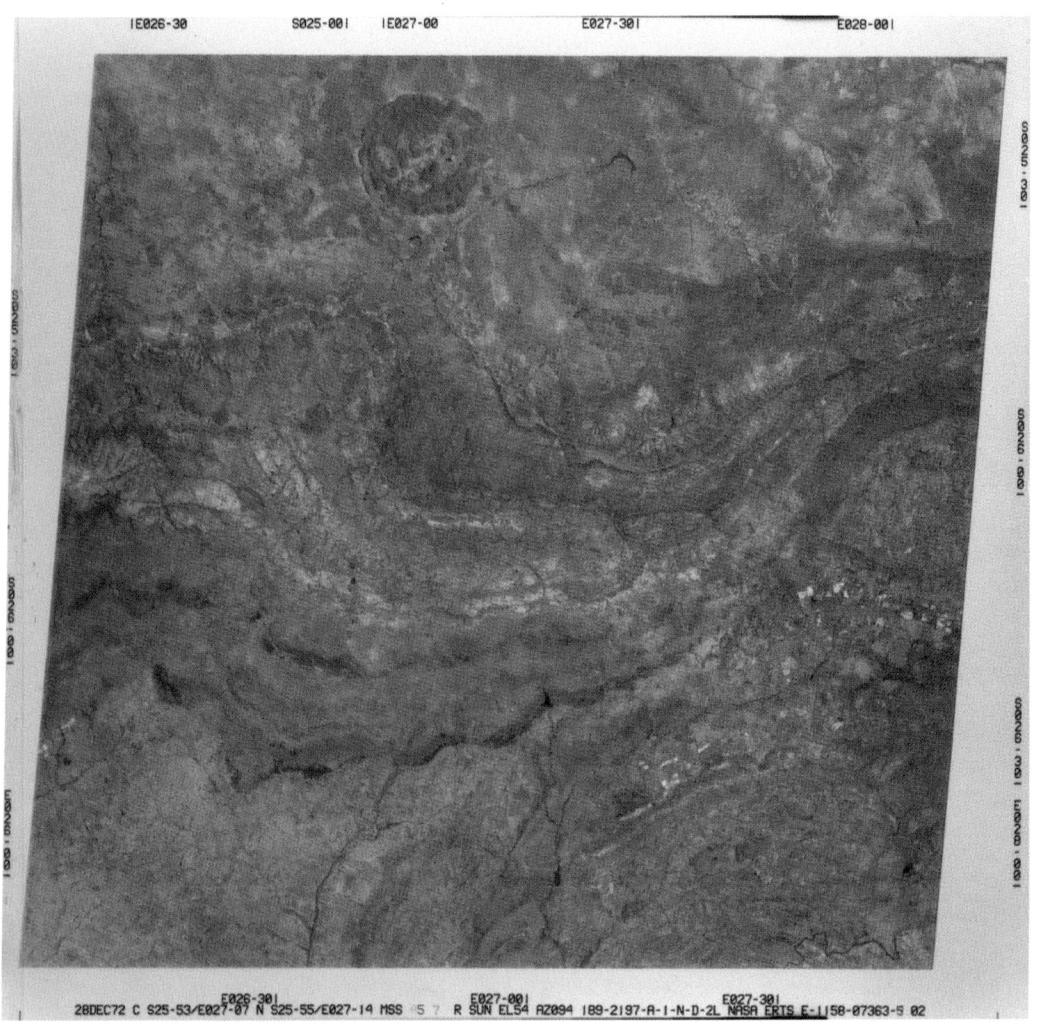

A

Fig. 31-116. Landsat images of the Ventersdorp area, Witwatersrand region, South Africa. A. Image (1158-07363; 28 December 1972) acquired after the end of the rainy season. B. Image (1050-07355; 11 September 1972) acquired during an extended dry period. Note how the differences in rock types are more evident on A than B because of the different geobotanical response. (Fig. 31-102 is two pages hence).

1S025-00 E027-00I E027-30I E028-00I

E026-30I 027-30G

11SEP72 C S25-57/E027-08 N S25-59/E027-14 MSS R SUN EL42 AZ055 189-0691-A-1-N-D-2L NASA ERTS E-1050-07355-4 01

B

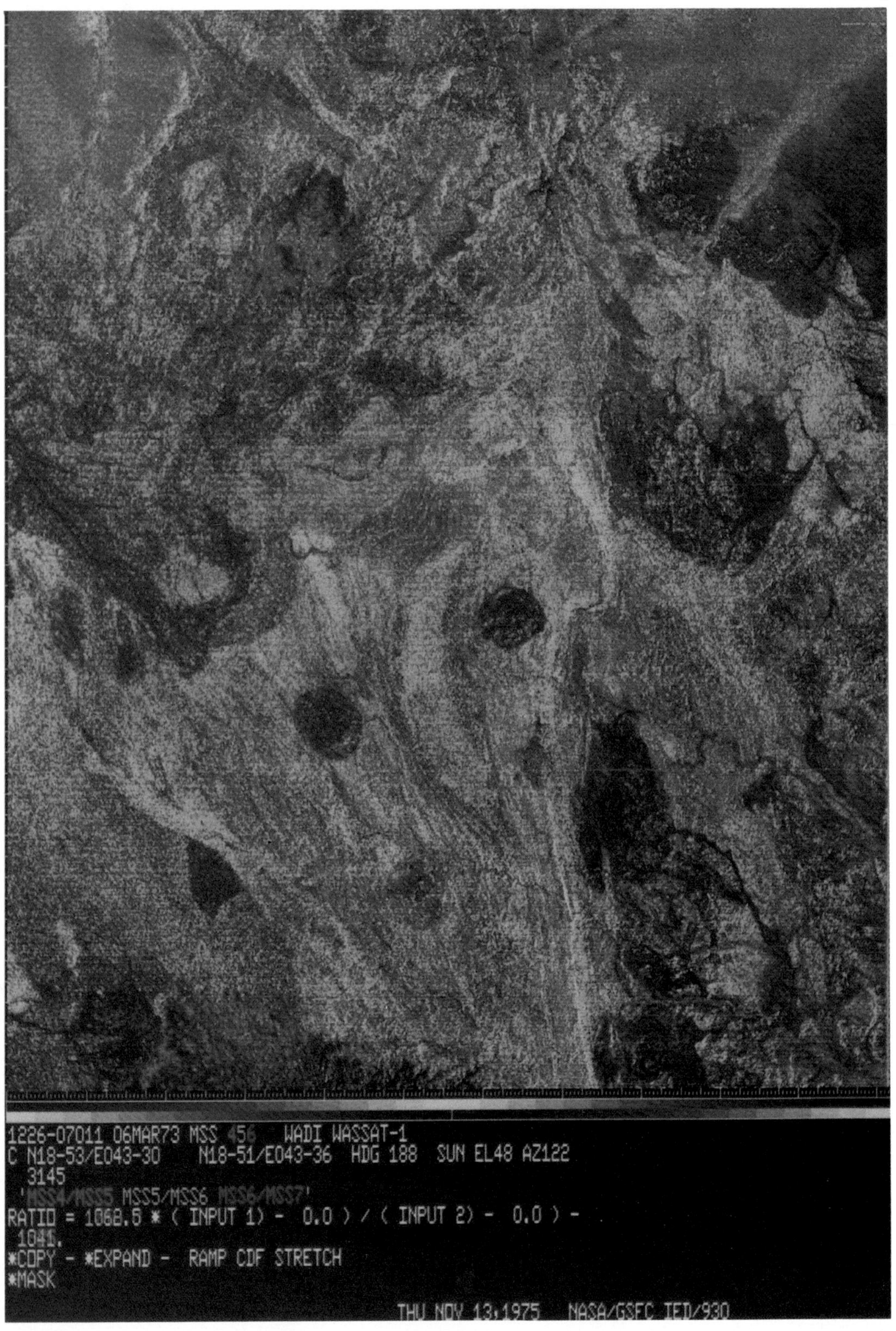

Fig. 31-102. Contrast-stretched, band-ratio Landsat MSS color-composite image of the Asir area, Saudi Arabia (digital data from Landsat scene 1226-07011; 6 March 1973).

Fig. 31-118. Landsat MSS color composite image (2107-01121; 9 May 1975) that shows the Yeelirrie uraniferous calcrete deposit in western Australia.

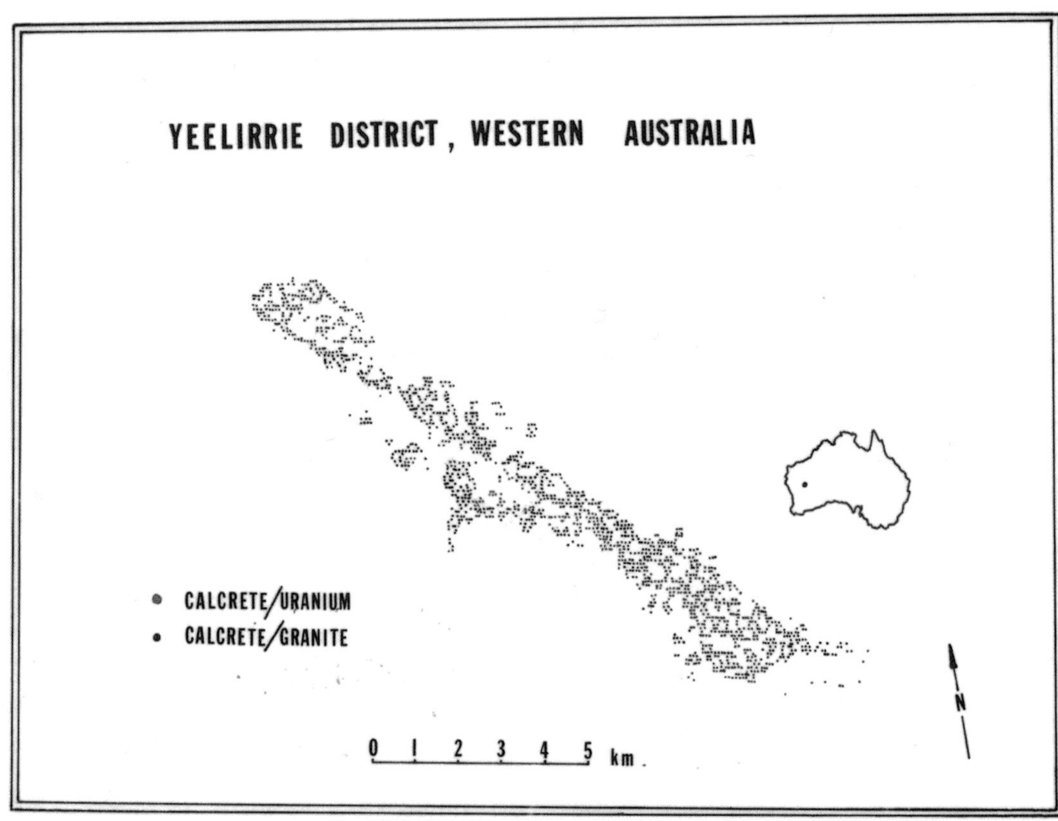

Fig. 31-119A. Digital map derived from spectral analysis of a Landsat image (2107-01121; 9 May 1975) of the Yeelirrie Mine (Color Figure 31-118), western Australia, showing the distribution of mineralized (red) and unmineralized (black) calcrete. Field spectra of the Yeelirrie uraniferous calcrete deposit were used in developing a supervised classification scheme for Landsat MSS digital data. Modification of map furnished by R. Cary and M. Longman, Australian Minerals Company, Ltd., Perth, Australia. (Modified map drafted by C. S. Southworth, U.S. Geological Survey).

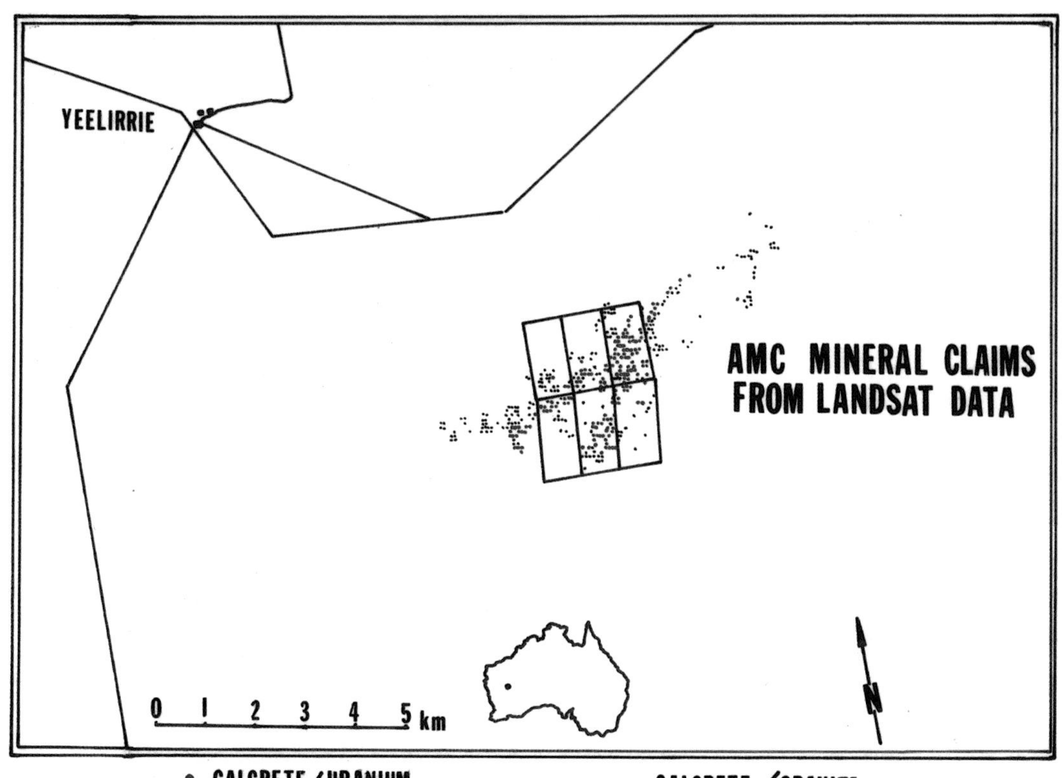

Fig. 31-119B. Digial map, derived from spectral analysis of the Landsat image (2107-01121; 9 May 1975) of the Yeelirrie Mine (See Color Figure 31-118), of the Australian Minerals Company, Ltd. mineral claims, showing mineralized areas in red, unmineralized areas in black, Yeelirrie Mining District, western Australia. Modification of map furnished by R. Cary and M. Longman, Australian Minerals Company, Ltd. (AMC, Ltd.), Perth, Australia. (Modified map drafted by C. S. Southworth, U.S. Geological Survey).

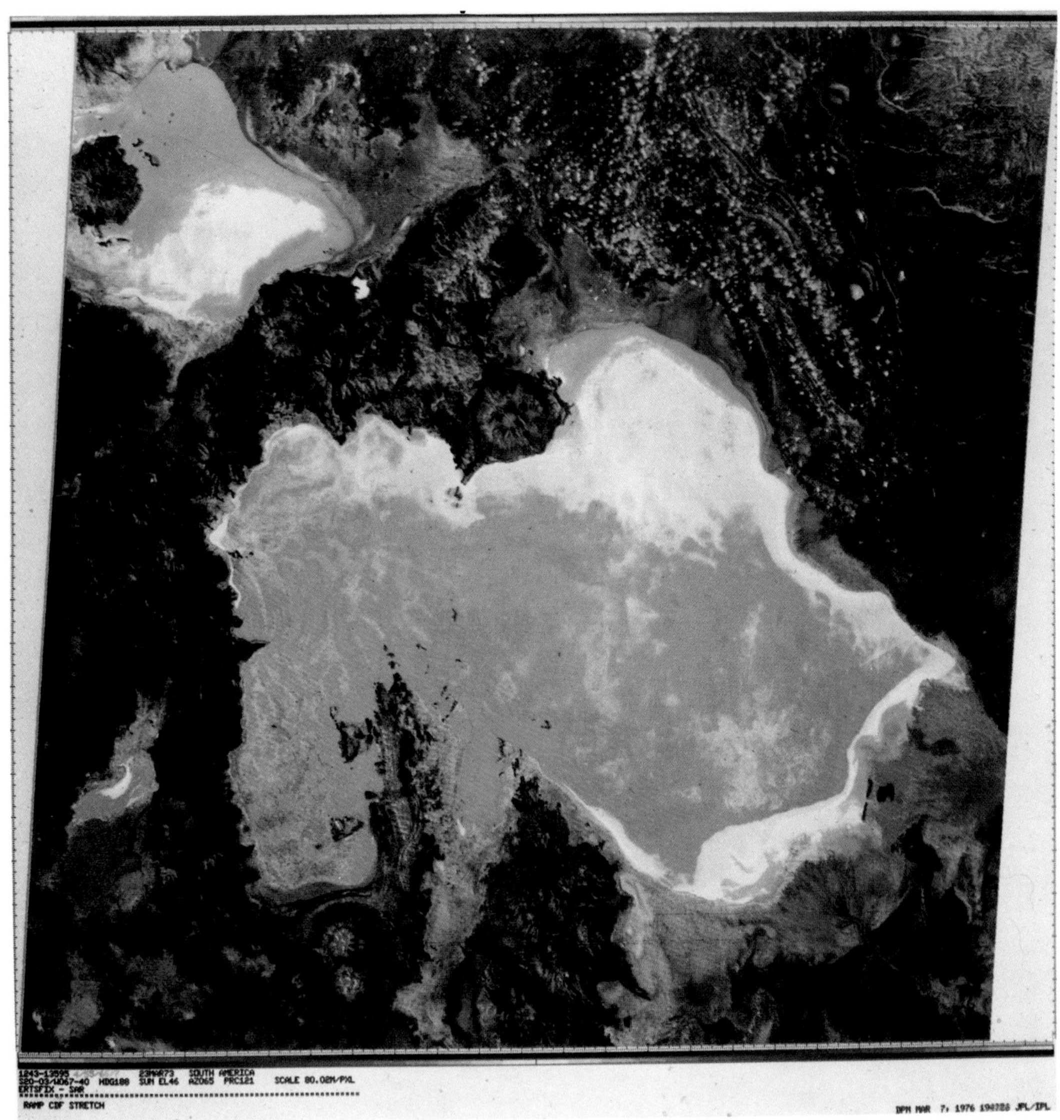

Fig. 31-120. Color band-ratio Landsat image (1243-13595; 23 March 1973) of the Salar de Uyuni area, southern Bolivia, showing limonite alteration in orange-brown; water in yellow; sodium chloride in light blue, pink, and white; gypsum (sodium sulfate) and gypsiferous soils in green. MSS band ratios 4/5, 5/6, and 6/7 were composited with yellow, red, and blue filters, respectively, to achieve this "false" color composite image. Image courtesy of Alexander F. H. Goetz, Jet Propulsion Laboratory.

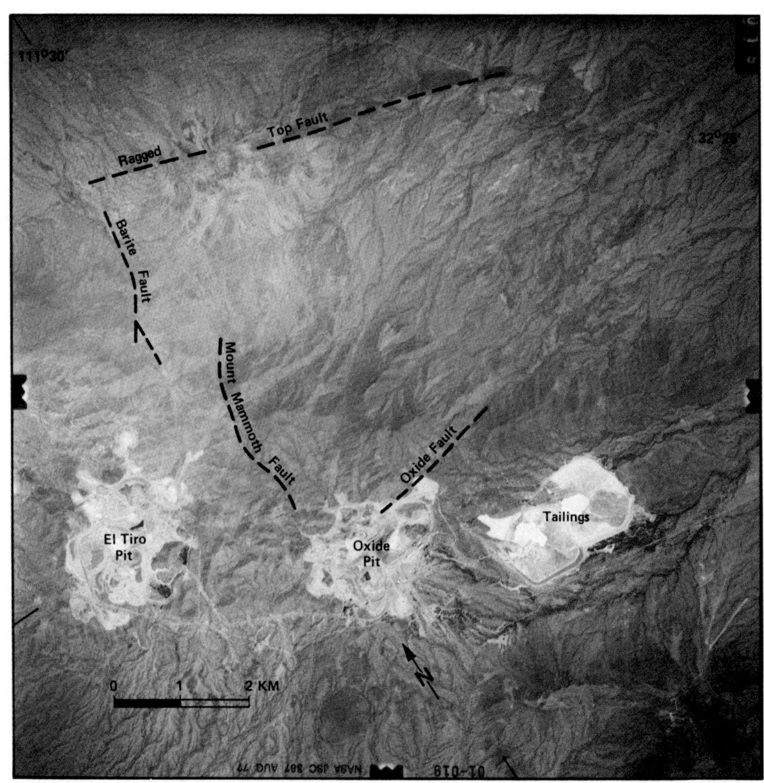

Fig. 31-122. Vertical color aerial photograph of the Silver Bell porphyry copper Mining District, southern Arizona. Oxide and El Tiro open pits, tailings dumps and ponds, and major faults are prominent features identified on the photograph. NASA RB-57 photograph acquired on 18 August 1977 during JSC Mission 367; Roll 1, Frame 19. (EROS Data Center No. 6367000100019). Original scale of photograph, 1:51,000.

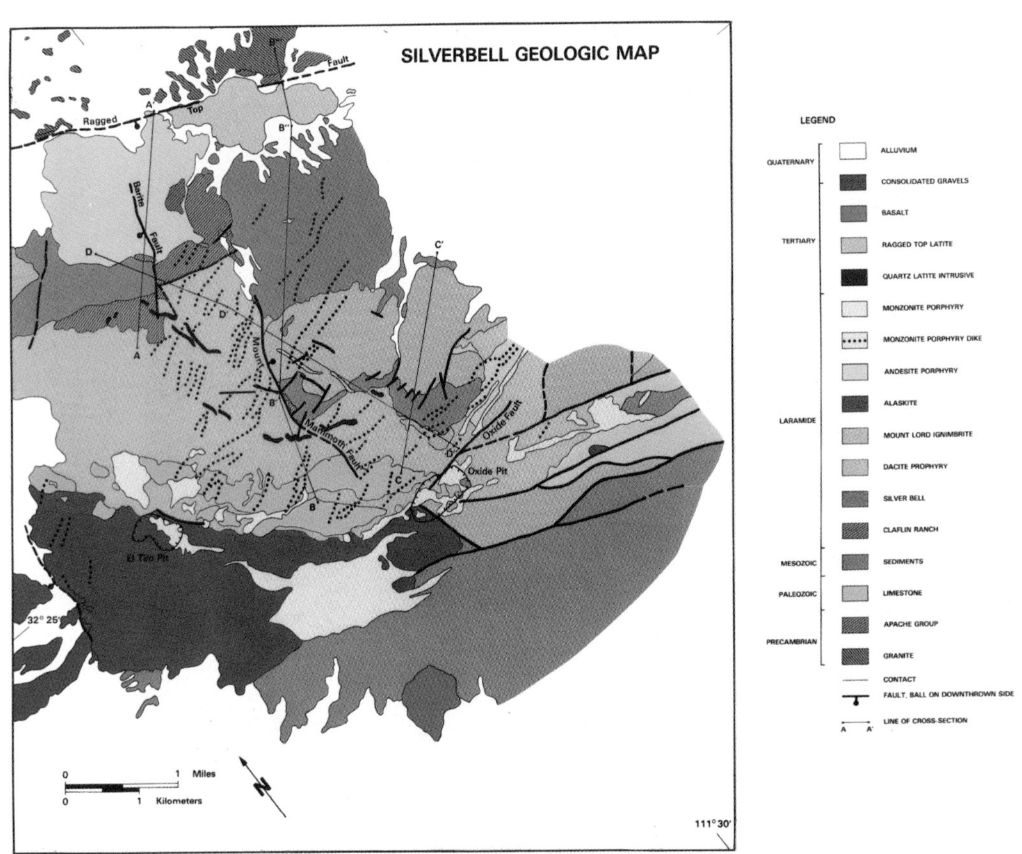

Fig. 31-123. Geologic map of the Silver Bell, Arizona area. The map was compiled by D. Brown (Texasgulf) from an unpublished ASARCO map, Texasgulf mapping, and published sources.

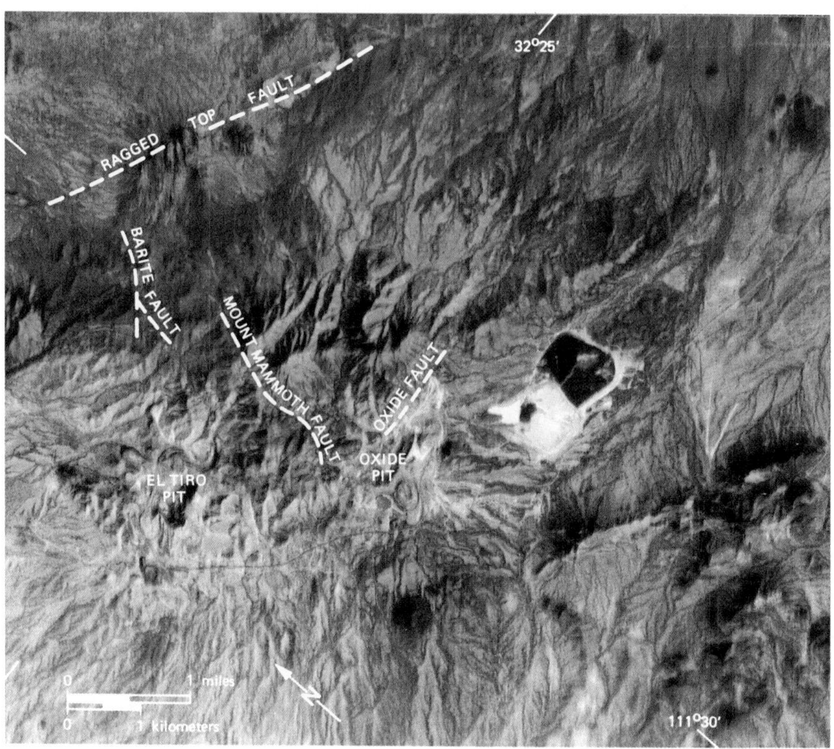

Fig. 31-124. Color-ratio composite image of the Silver Bell porphyry copper Mining District; data acquired October, 1978, by NASA's NS-001 Thematic Mapper Simulator aircraft scanner. Band ratios 0.66 μm/0.56 μm, 0.83 μm/1.15 μm, and 1.65 μm/2.2 μm are displayed in green, blue and red, respectively. This combination highlights the phyllic alteration zone in yellow-orange colors. Image courtesy of the Jet Propulsion Laboratory.

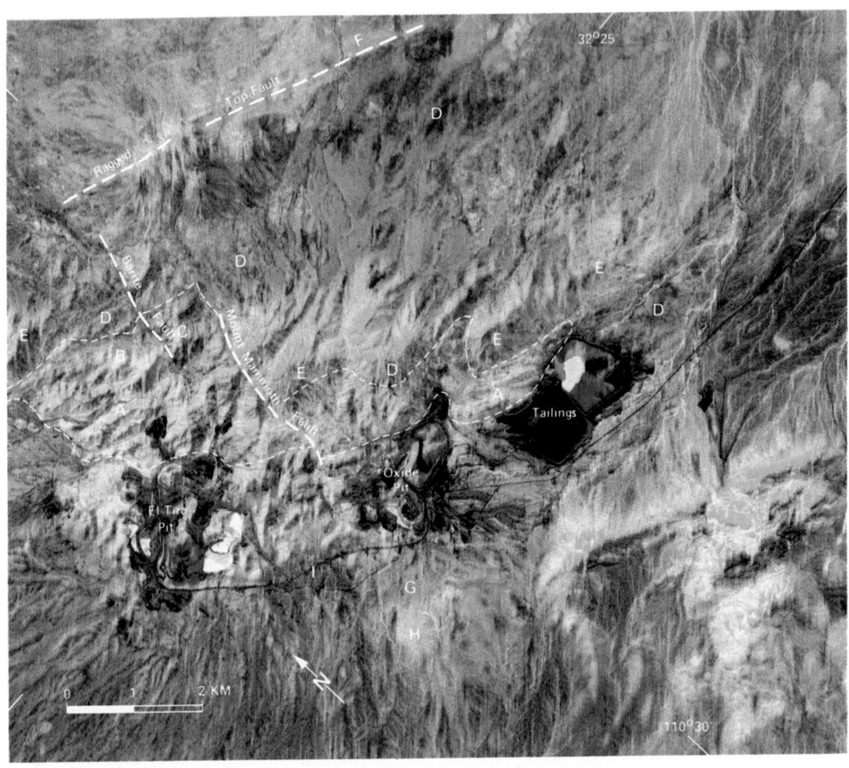

Fig. 31-125. Computer enhanced NS-001 Thematic Mapper Simulator aircraft scanner data for the Silver Bell area. The data were processed to maximize the separation of the geologic units exposed in the area. The red, blue and green components represent different linear combinations of the original seven wavelength bands. Alteration zones and mapped geologic units are separable on the image. Image courtesy of the Jet Propulsion Laboratory.

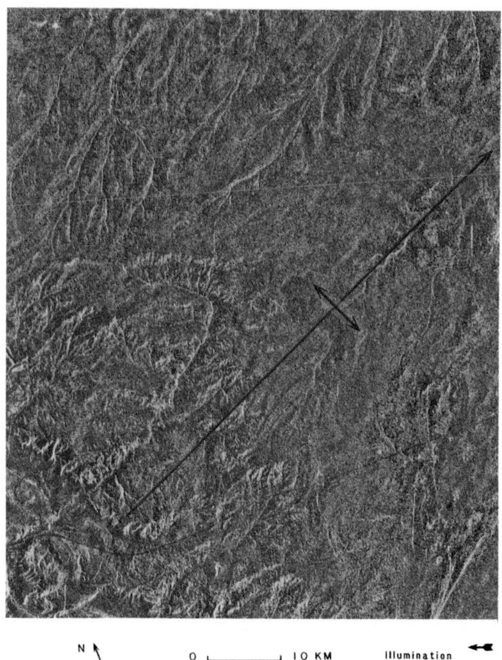

Fig. 31-132. Seasat radar image in Figure 31-136 enhanced by filtering and color encoding to reveal fold structure (Wamsutter Arch). Fold axis shown by black line. Arrowheads denote directions of dip and plunge of fold. Image courtesy of the Jet Propulsion Laboratory.

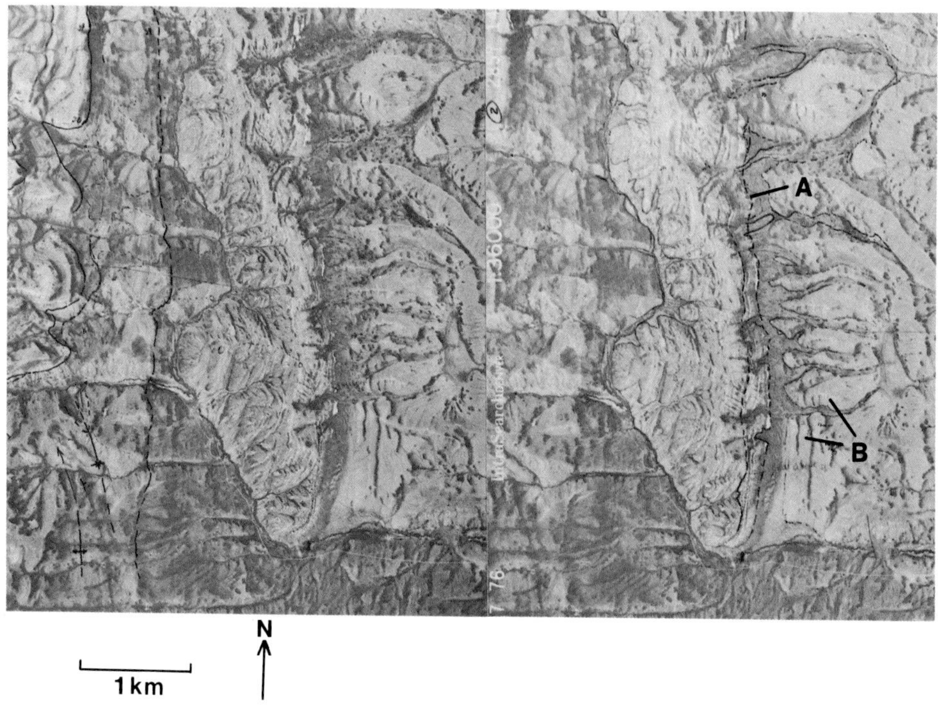

Fig. 31-140. Stereoscopic pair of vertical, color aerial photographs showing part of the area included in Figure 31-139A, parts of the Elkol and Warfield Creek 7½-minute quadrangle maps, southwestern Wyoming. Color aerial photographs no. 2857-149 (41°38′45″ North latitude; 110°36′18″ West longitude) and 2857-150 (41°38′56″ North latitude; 110°38′41″ West longitude) courtesy of IntraSearch, Denver, Colorado. Aerial photographs acquired on 27 May 1976 at 12 noon m.s.t. from a mean altitude of 5,486 m above the terrain with a 152 mm-focal length camera. The original scale of this photography (obtained by using Kodak 2445 aeronegative film) was 1:36,000.

Fig. 31-143. Stereoscopic pair of oblique, 35 mm-, color-infrared aerial photographs of the same area as shown in Color Figure 31-140. The sandstone and marly beds in the Cretaceous Hilliard Shale (area B) show up more plainly than they do on standard color aerial photographs (compare with area B shown on Color Figure 31-140). Aerial photographs by John W. M'Gonigle, U.S. Geological Survey.

Fig. 31-145. Terrestrial color photograph (looking north) of the hills along the Powder River, about 15 kilometers southwest of the color stereopair shown in Color Figure 31-146A. Note the banded appearance of the hillside, the bright orange clinker, and the coal outcrop (greyish) in the center of the picture. Photograph by Carol L. Molnia, U.S. Geological Survey.

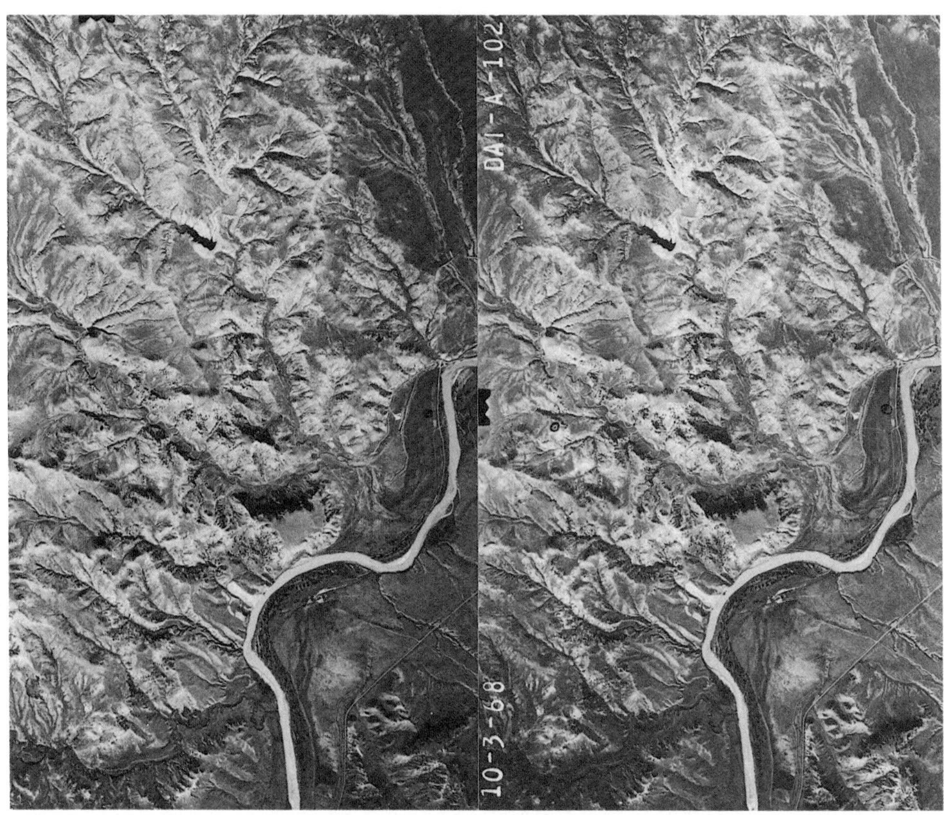

Fig. 31-146A. Stereoscopic pair of vertical color aerial photographs of an area along the Powder River, near the Wyoming-Montana border. Note the extensive areas of red clinker and the light-colored underlying rocks. Color aerial photographs no. DAI-A-1028 and DAI-A-1029 (44°55′ North latitude; 106°02′ West longitude) courtesy of IntraSearch, Denver, Colorado. Aerial photographs acquired on 3 October 1968 at 1255 p.m. m.s.t. from a survey altitude of 4,815 m above mean sea level with a Zeiss aerial camera (152.45-mm focal length). The original scale of the color aerial photographs was 1:24,000.

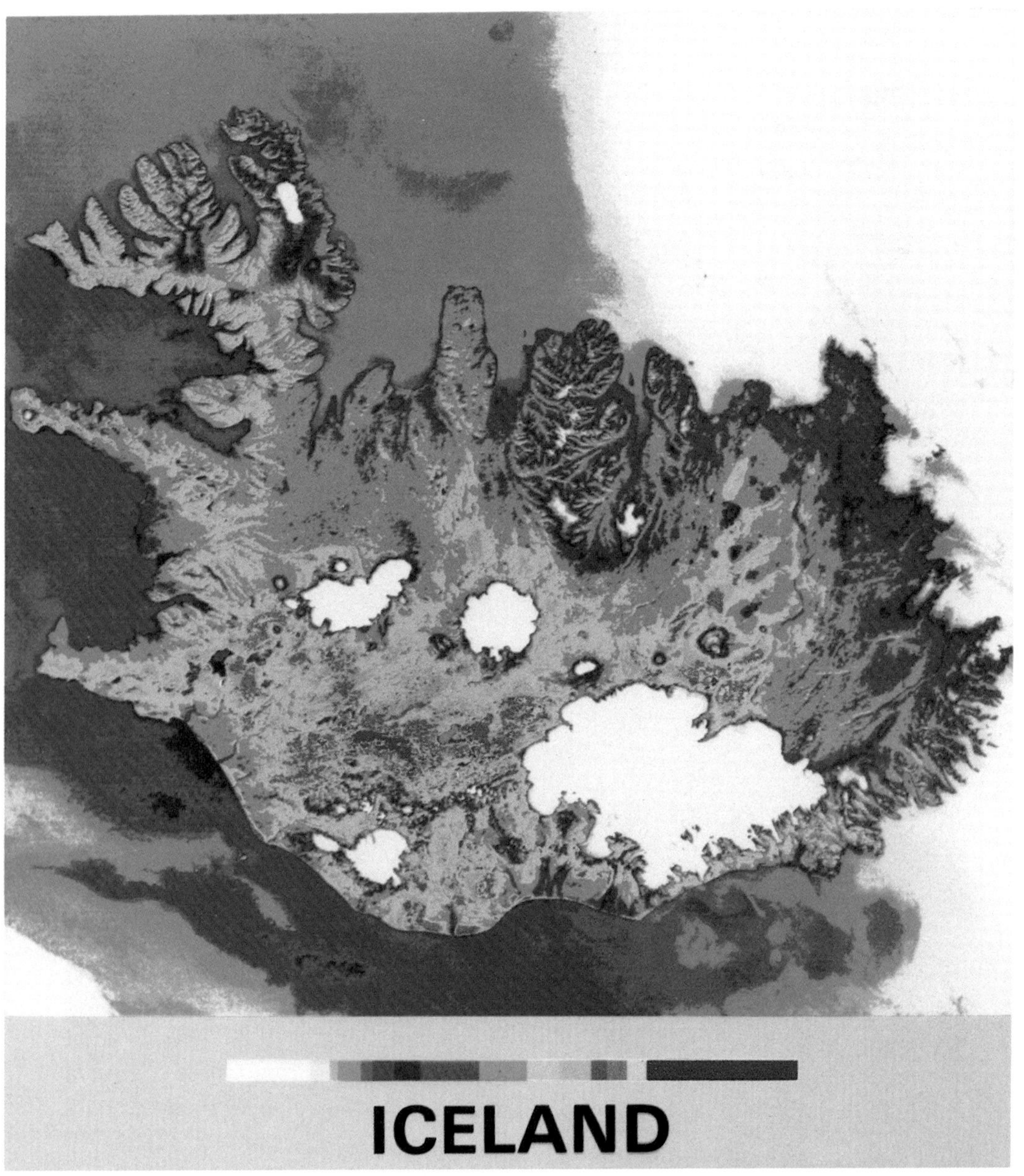

Fig. 31-149. Color-coded, daytime, Heat Capacity Mapping Mission (HCMM) thermal infrared image (10.5–12.5 μm) of Iceland acquired by the Lannion, France, satellite receiving station in July 1980 (Aviation Week and Space Technology, 1980). Clouds (white) are visible south, east, and northeast of Iceland, but most of the island is clear. Icecaps and snowpack are depicted in white, the largest white area being the Vatnajökull icecap (8,300 km^2) in southeast Iceland (See Color Figures 31-197, 31-201, 31-202 and Figures 31-198, 31-204, and 31-205). The green, orange and red colors indicate the apparent radiometric temperature of vegetation (mostly in coastal lowlands and interior valleys), bare soil; tephra, aeolian and glacial deposits; and rock outcrops, including extensive areas of lava flows. The apparent radiometric temperature is the result of a complex interrelationship of emissivity and temperature of each of these different materials. Solar heating of the ground in this satellite thermograph effectively masks any thermal emission from geothermal areas. If an effusive volcanic eruption had been occurring during the time of image acquisition it would have been recorded (Williams and Friedman, 1970a). Preliminary research by R. S. Williams, Jr. (USGS) and M. Matson (NOAA) with nighttime satellite thermograms of Iceland from NOAA 6 advanced very high resolution radiometer (AVHRR) has recorded thermal emission from effusive eruptions of Hekla Volcano (18 April 1981) (Figure 31-13) and in the Krafla area (10 February 1981), but no unambiguous determination of geothermal activity could be achieved with the 1 km pixel resolution of the NOAA 6 AVHRR sensor. For additional information about the HCMM sensor and spacecraft see Table 31-13 in this chapter, Taranik (1981), and Price (1981).

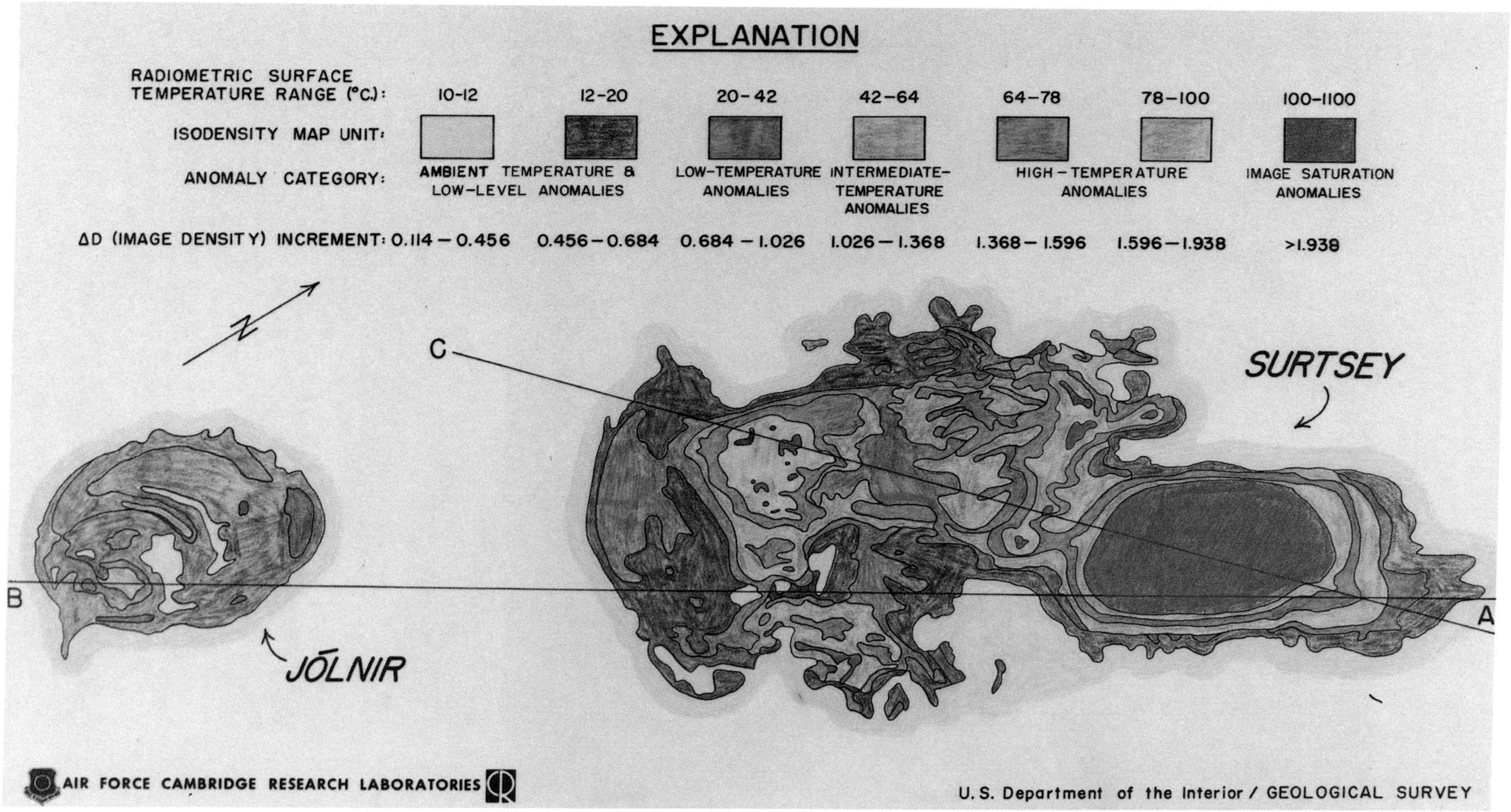

Fig. 31-150. Isodensitometric map of Surtsey Volcano and its satellite volcano, Jólnir, Vestmannaeyjar, Iceland, showing a correlation between image density and radiometric surface temperature. M1A1 aerial thermograph (See Figure 31-151) acquired on 19 August 1966 at 1745 UT (See Icelandic local time) from an altitude of 1525m, with a filtered InSb detector (4.5–5.5 μm). Image courtesy of U.S. Air Force Cambridge Research Laboratories. From Friedman and Williams (1968).

Fig. 31-154. Panoramic color aerial photograph of industrial lagoon. This impoundment, which is used for neutralization of acidic waters, has a high albedo caused by the lime used for the neutralization. The rectangular, well defined rim around the lagoon is typical of lined impoundments. Sludge has been removed from the lagoon with a dragline, which results in a pattern of pools within the lagoon and sludge piles at the edge. Fallen and standing branchless dead trees border the site. Questions which arise when evaluating a site for potential hazards such as this are: What is responsible for the dead vegetation? Where has the sludge been taken after removal from the lagoon? Has the impoundment been breached by dragline operations? What are the directions of surface runoff and groundwater movement from the site relative to surface and ground water supplies? A less obvious question concerns the history of the site. Various industrial activities have occurred in this vicinity for about 150 years. Old impoundments and landfills now obscured by development can be seen on older photographic coverage of the area. (Photographic information: EPA/EPIC no. 0097, 28 November 1979, 12 noon e.s.t., survey altitude about 150 m; center coordinates: 42°31′39″ N. latitude, 71°09′15″ W. longitude).

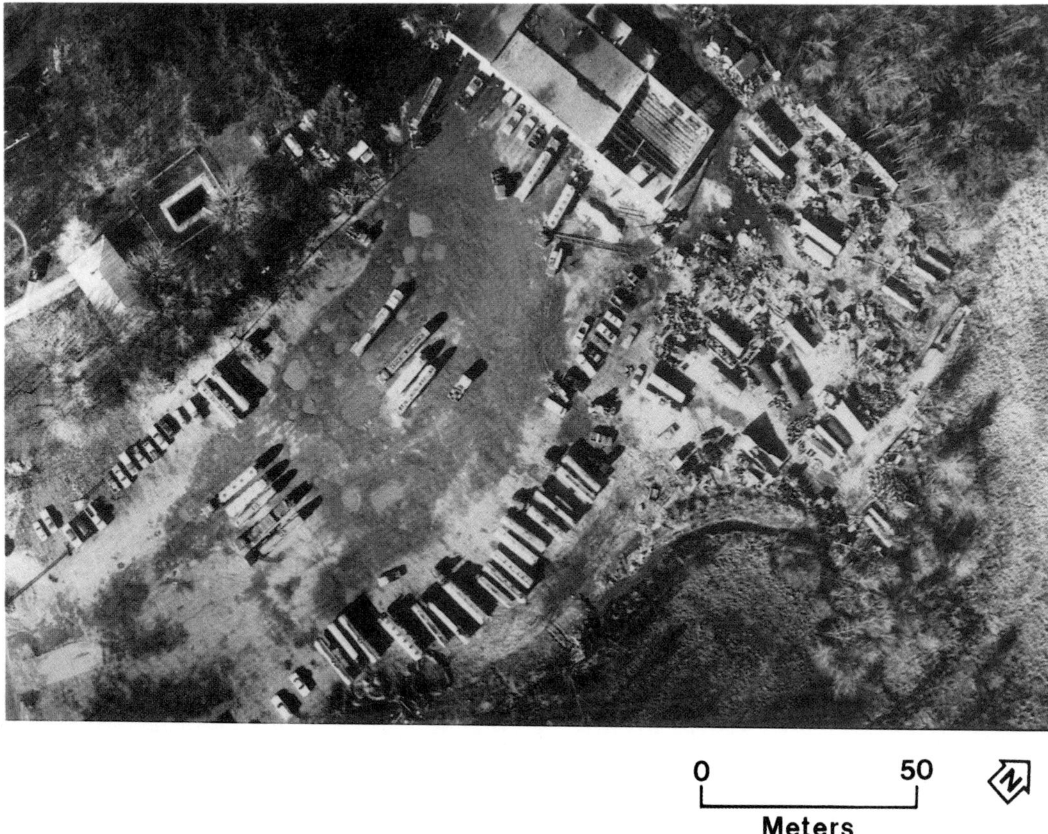

0 50

Meters

Fig. 31-155. Panoramic color aerial photograph of tank-truck washing facility. Tank trailers are indicated by filler-ports on top of the tanks and by semicircular shadows. An unlined, shadowed lagoon in the woods immediately adjacent to the plant contrasts with the swimming pool at the nearby residence. Note the small irregular lagoon behind the trailer at the left margin of the photograph and questionable drainage into a swamp to the left of the littered area. The unpaved truck park shows poor drainage. Fifty-five gallon drums are visible in the lower left of the littered area; these show the honeycomb pattern typical of drums standing on end. (Photographic information: EPA/EPIC no. 0192, 28 November 1979, 12 noon e.s.t., survey altitude about 150 m; center coordinates: 43°32′12″ N. latitude, 71°14′14″ W, longitude.)

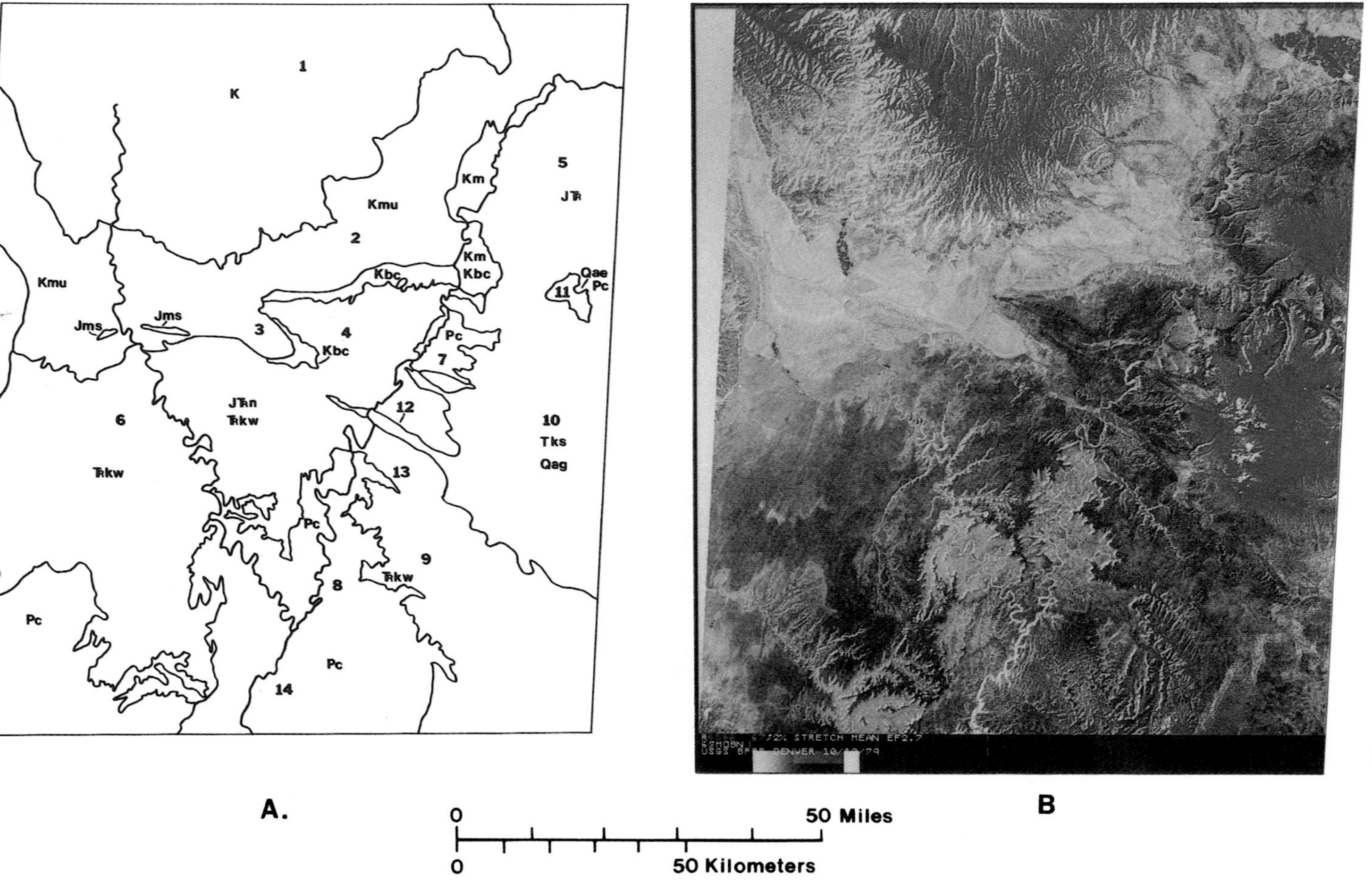
1
K
Kmu
2
Km
5
JTR
Km
Kbc
Kbc
Qae
Pc
11
Kmu
Jms
Jms
3
4
Kbc
Pc
7
JTRn
TRkw
6
12
10
Tks
Qag
TRkw
13
Pc
9
TRkw
8
Pc
Pc
14
A.
B
0
50 Miles
0
50 Kilometers

Fig. 31-157A. Geologic sketch map prepared from a 1:800,000-scale (original scale), ratioed, Landsat MSS color composite image (5165-17030; 1 October 1975) showing geologic map units visible in the scene: *Qag*, Quaternary alluvial gravels in the La Sal Mountains; *Qae*, alluvium; *Tks*, Soda syenite porphyry of Tertiary age; *K*, Cretaceous clastic rocks of Book Cliffs and the Kaiparowits Plateau, including the upper part of the Mesaverde Group and the Castlegate Sandstone; *Kmu*, upper Shale Member of the Mancos Shale of Cretaceous age; *Km*, Mancos Shale, undifferentiated; *Kbc*, Burro Canyon Formation (uraniferous); *Jms*, Salt Wash Member of Morrison Formation of Jurassic Age (uraniferous); JT_RN, Navajo Sandstone; T_RKW, Kayenta Formation and Wingate Sandstone group of Triassic age (redbeds); *Pc*, Cutler Formation of Permian age (redbeds). Features of special geologic interest are as follows: (*1*) Book Cliffs and Kaiparowits Plateau, partly tree covered. Rock units include the Green River Formation of Tertiary age, the Mesaverde Group of gray sandstone and the Castlegate Formation of the Book Cliffs. (*2*) Grand Valley. Main rock unit is the Upper Member of the Mancos Shale of Cretaceous Age, north of Yellowcat structural dome. (*3*) Courthouse syncline. Rock exposures include those of (*2*) above. (*4*) Region of Salt Valley and Yellowcat structural dome. (*5*) Uncompahgre Plateau. Redbeds of Jurassic and Triassic age. (*6*) Redbeds of Kayenta Formation and the Wingate Sandstone group of Jurassic and Triassic age; includes cliff-forming Wingate Sandstone and Entrada Sandstone. (*7*) Richardson amphitheater. Rock exposures are redbeds of the Cutler Formation of Permian age. (*8*) Redbeds of Cutler Formation near Hatch Point and Gibson salt dome which is not exposed at the surface. (*9*) Plateau formed on Wingate Sandstone Group and bounded by cliff-type escarpment. (*10*) Exposed intrusive syenites and alluvial gravels of the La Sal laccolithic complex. Heavily tree covered. (*11*) Fischer Valley. Cutler Formation of Permian age and alluvium (*Qae*). Includes exposed highly contorted residual caprock of the Paradox Member salt beds of the Hermosa Formation; constitutes the top of a deformed diapir. (*12*) Spanish Valley with the Moab fault system extending northwest of the Colorado River. Red indicates agricultural land. These valleys, underlain by an elongated salt diapir, are bounded on their southwest side by the north-west-striking, high-angle Moab fault. (*13*) Cane Springs anticline and underlying salt diapir exposing Cutler Formation (*Pc*) redbeds, and cut by northeast-striking high-angle faults. (*14*) Concentric horst-and-graben terrain of the Needles fault zone, which possibly forms part of a large circular subsidence structure in this region. (*15*) B. Landsat MSS color composite image (5165-17030; 1 October 1975) of the Paradox basin area, Utah. This specially enhanced image was created by computer processing of CCT's to increase the contrast in MSS bands, 4, 5, 6, and 7, preparation of three band-ratioed images (4/5, 4/6, and 6/7), and projection of the latter through red, blue, and green filters, respectively, onto color film (the color image is reduced from its original scale of 1:800,000).

Fig. 31-156. Panoramic color aerial photograph of a drum-washing and storage facility and landfill dumps. The pattern of drums stored on their sides is typical of this method of storage. The storage area is unpaved; site drainage is poor and uncontrolled. The top of the photo illustrates two contrasting types of landfill operations. The large dump on the left is a municipal landfill. Typical characteristics are piles of cover material left by dump trucks along the working face, striations left by bulldozers during covering operations (further enhanced by erosion on the working face), and different degrees of vegetation on the working face, which shows the pattern of use. Not obvious here, but present on many municipal landfill sites, are a dump-keeper's shed and a separated metal pile. The smaller landfill on the right is for industrial sludge. Note especially the lack of relief above the surrounding terrain, the subdued shape of the dumped sludge piles, and the limited cover material. (Photographic information: EPA/EPIC no. 0093, 28 November 1979, 12 noon e.s.t., survey altitude about 150 m; center coordinates: 42°31′25″ N. latitude, 71°08′55″ W. longitude.)

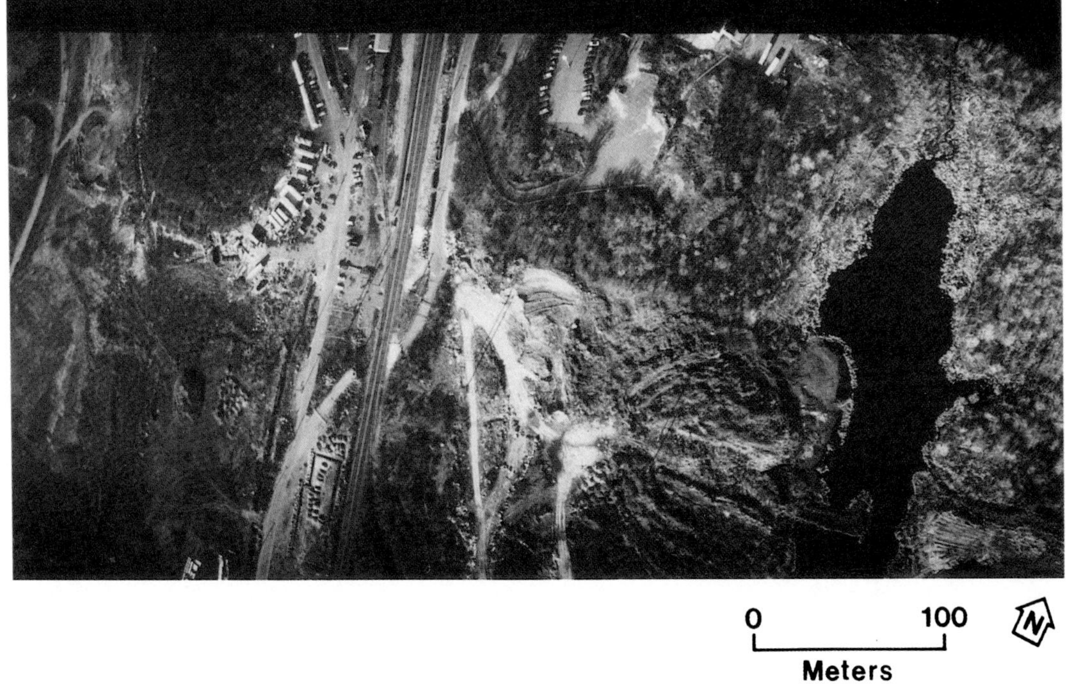

Fig. 31-158. Panoramic color aerial photograph which shows an area in which considerable dumping has occurred over the past 100 years. Recent activity is evident along the power line which crosses the lake on a man-made spit. The striated surface is characteristic of covering operations on recent landfills. The area above the spit contains an older landfill with a more diffuse striation pattern and considerable vegetative cover. The brown deltaic form to the right of the spit is of unknown character. Several casual, recent filling operations are evident at the top of the photo. (Photographic information: EPA/EPIC no. 0083; 28 November 79, noon e.s.t.; survey altitude about 150 m; center coordinates: 42°31′15″ N. latitude, 71°08′45″ W. longitude.)

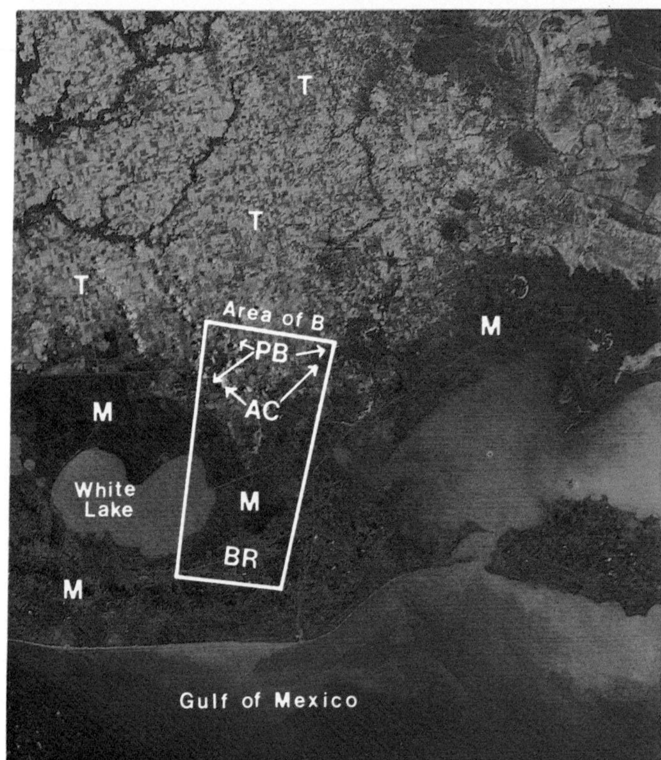

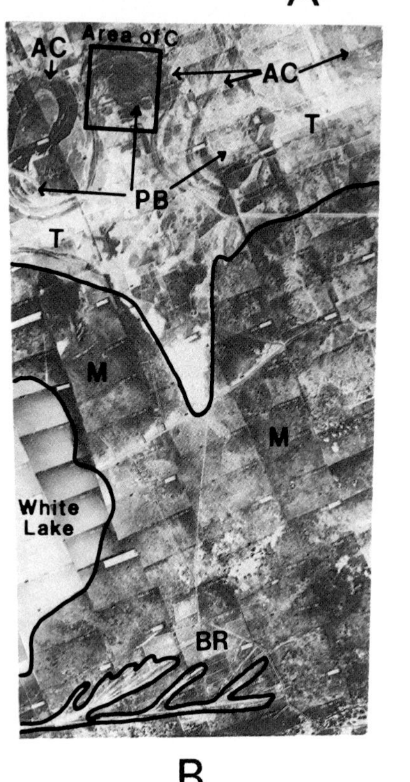

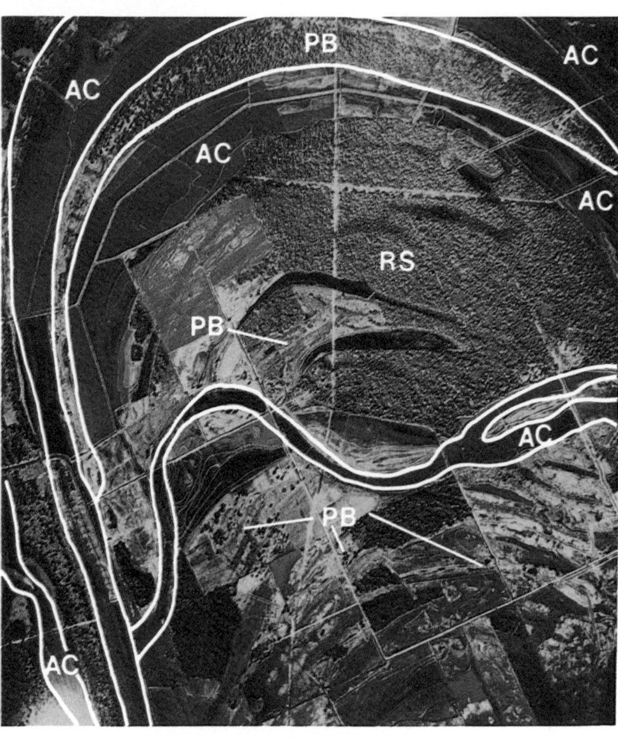

Fig. 31-161. Three examples of images of the same area at different scales showing landforms, drainage features, and cover-type patterns that relate to potential occurrence of construction materials on the coastal plain of southern Louisiana. *A*. A portion of a Landsat MSS false-color composite image (30054-16013; 28 April 1978), showing abandoned channels, point bars, a Pleistocene terrace, beach ridges (cheniers), and marsh areas. *B*. A Department of Agriculture aerial photomosaic which displays the same landforms at a larger scale. *C*. A portion of a large-scale color-infrared aerial photograph which was acquired over the area on 19 March 1969. The relationship between area coverage and image detail is evident by comparing the three images.

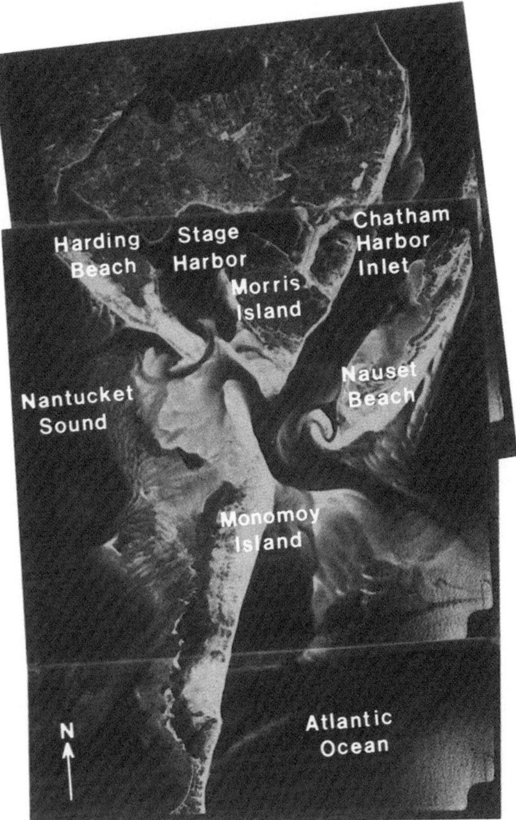

Fig. 31-165. Uncontrolled color aerial photographic mosaic (three photographs) of the Stage Harbor—Harding Beach—Monomoy Island—Nantucket Sound area of southeastern Cape Cod, Massachusetts. The aerial photographs were acquired by the U.S. Coast and Geodetic Survey on 6 October 1963 to show details of the submarine morphology in the area bounded by Morris Island, Stage Harbor, Harding Beach, Nantucket Sound, and Monomoy Island. The 1:20,000-scale aerial photographs were obtained during the 1964–65 project to determine where to locate a new channel from Nantucket Sound into Stage Harbor across Harding Beach. (See Figure 31-164 A and B. Figures 31-162 and 31-163 are on the following two pages.)

Fig. 31-162. A high-altitude (18,900m) color aerial photograph acquired on 16 October 1975 by a NASA RB-57 aircraft of the Western Energy Company's Rosebud Mine in southeastern Montana. (NASA JSC Mission no. 325, Roll 6, Frame 64.) (Approximate scale, 1:51,200.) Compare with Color Figure 31-163, an NHAP, color-infrared aerial photograph acquired about 5 years later.

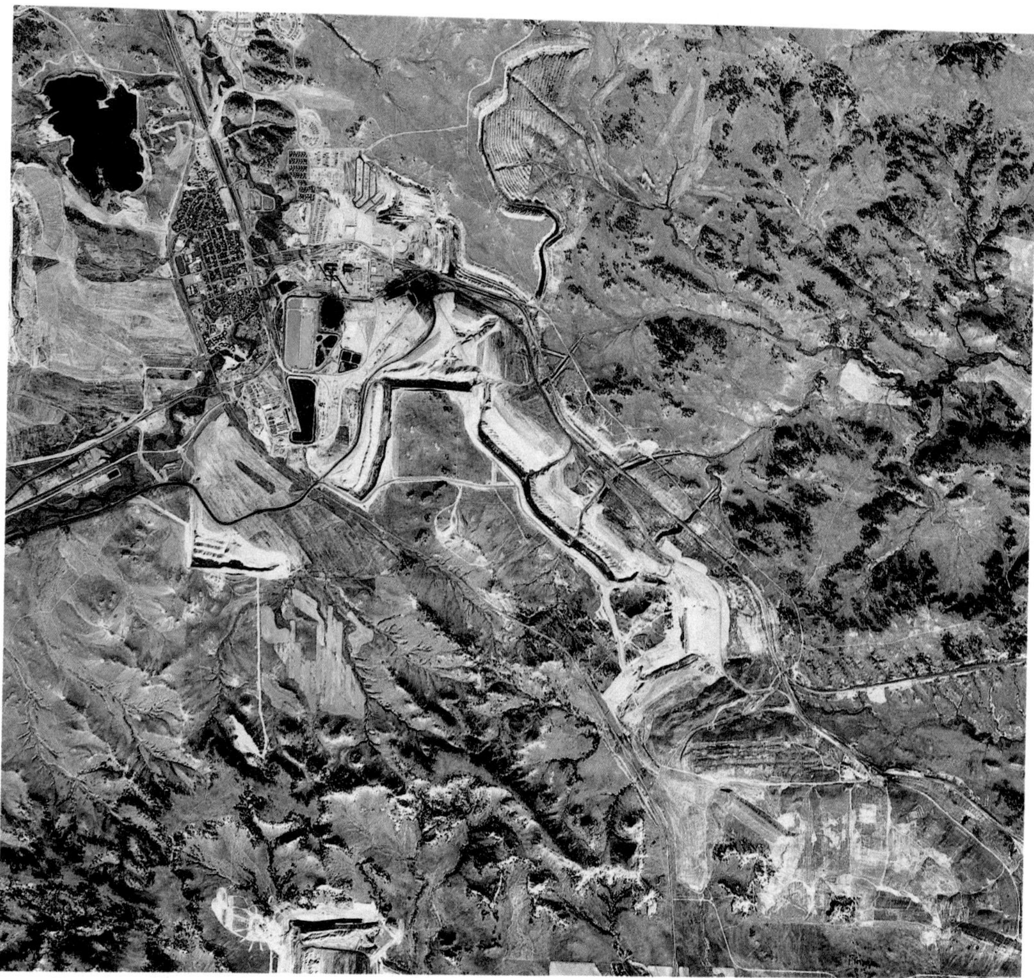

Fig. 31-163. A medium-altitude (11,710m) color-infrared aerial photograph of the Western Energy Company's Rosebud Mine in southeastern Montana, which was acquired on 14 September 1980 by the National High Altitude Photography (NHAP) program. (Photograph No. 247–49; 450612 HAP 80 F) (Approximate scale, 1:64,800). Compare changes with Color Figure 31-162, a NASA color-infrared aerial photograph acquired about 5 years earlier.

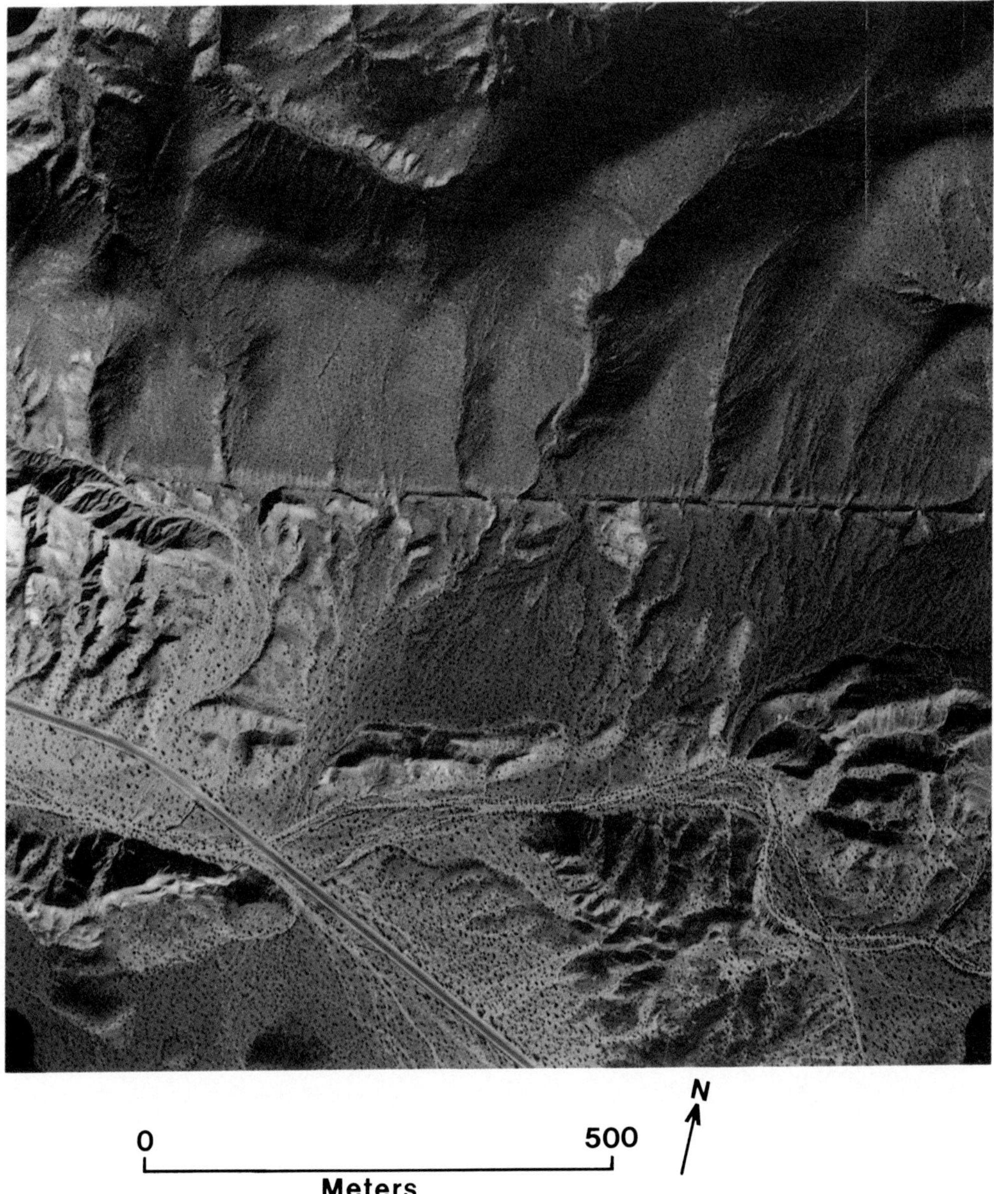

Fig. 31-168. Color aerial photograph of the Garlock fault south of Searles Lake (location shown on Figure 31-167). The scale of 9-by 9-in. color negative is about 1:5,000; this scale is useful for detecting branching and secondary faults (when present) and for measuring small offsets of channels and ridges. Color may help correlate offset gravels with distinctive sources (not present on this photograph) and emphasize lithologic contrasts across the fault. The fault has created a conspicuous hillside valley that here is locally several meters deep; the valley is bounded on the downslope side by a nearly continuous low shutter ridge that dams or diverts channels from upslope across the fault. Photographs of this scale are also used for detailed geomorphic and tectonic analysis; low Sun angle enhances definition of surface relief without major loss of color contrast. (U.S. Geological Survey color aerial photograph No. GSGF 1–139, 17 January 1976: Sun elevation, 18°; Sun azimuth, 133°).

Fig. 31-174. Color infrared aerial photograph of the northeastern part of Heimaey, Vestmannaeyjar, Iceland, showing the new land to the northeast and east (+2.5 km² addition to the island of Heimaey) and the encroachment of lava flows on the eastern one-third of the town of Vestmannaeyjar. Tephra (volcanic ash) blankets other parts of the town, except where cleanup operations have already removed the tephra. In color infrared photography, healthy vegetation is recorded in red rather than green. Photo acquired by the NASA NP-3A aircraft, with a Wild-Heerbrugg RC-8 camera, 152 mm focal length, from an altitude of 1,370 m on 17 August 1973; NASA Mission 253, Roll 37, Frame 130.

Fig. 31-176. Specially processed Landsat false-color composite image (1392-12191; 19 August 1973) of southwestern Iceland produced by the Image Processing Laboratory of the Jet Propulsion Laboratory (See Figure 13.21 in Abrams and Siegal, 1980). It is also similar to Color Figure 31-202 (and to Figure 13 in Soha et al, 1976). The original histograms of this image are trimodal, permitting snow- and ice-covered, water, and other land areas to be processed individually. Zonal contrast stretching was applied to each of the three component MSS bands (4, 5, and 7) to encompass the full brightness range. MSS band 4 is projected through a yellow filter, MSS band 5 through a red filter, and MSS band 7 through a blue filter. The effects of three recent volcanic eruptions can be seen in the image: Surtsey in 1963–67, Hekla in 1970, and Eldfell on Heimaey in 1973. *1*. Reykjavík, capital of Iceland. *2*. Rich farming area of the lower Thjórsa valley. *3*. Surtsey, a small (2.5 km^2) volcanic island off the south coast. *4*. Heimaey, an island to the northeast in the same archipelago, which was the scene of an extremely damaging volcanic eruption in early 1973. Nearly half the fishing port of Vestmannaeyjar was destroyed or damaged (Williams and Moore, 1973). The new land area on the east side of Heimaey can be seen. The tephra-fall pattern at *5* and new lava flows at *6* are from the 1970 volcanic eruption from the famous Mount Hekla. The 1970 tephra contained high concentrations of fluorides and contaminated grasslands that were grazed by sheep. At least 7,000 head were lost in the area because of fluoride poisoning. Several glaciological features are also prominant (See also the Section on "Glaciology" of this chapter). The glacier-margin lake, Hvítárvatn *7*, is distinctive with its powder-blue color (here shown with a pattern of parallel lines) characteristic of turbidity caused by glacial sediment. The snowline is delineated on the southern edge of Hofsjökull *8*, Langjökull *9*, Mýrdalsjökull *10*, and Eyjafjallajökull *11* icecaps. Note the absence of vegetation at *12* from an area where one of the outlet glaciers of Hofsjökull has receded. A prominent nunatak at *13* can be seen in the center of Langjökull. The snow-capped shield volcano, Skjaldbreidur, can be seen at *14* and tectonic fissures that extend to the southwest. Sediment plumes from the mouths of rivers laden with glacial rock flour are prominent, particularly the one at the mouth of Thjórsa *15*. The circular caldera form, light-colored rhyolitic rocks, and lighter colored altered-ground areas of Iceland's largest geothermal area, Torfajökull at *16*, contrast sharply with the surrounding darker basaltic rocks or glacial deposits derived from such rocks. The altered ground of the Reykjanes geothermal area is just barely discernible at *17* (Williams, 1976b). (Landsat image courtesy of Michael J. Abrams, Jet Propulsion Laboratory.)

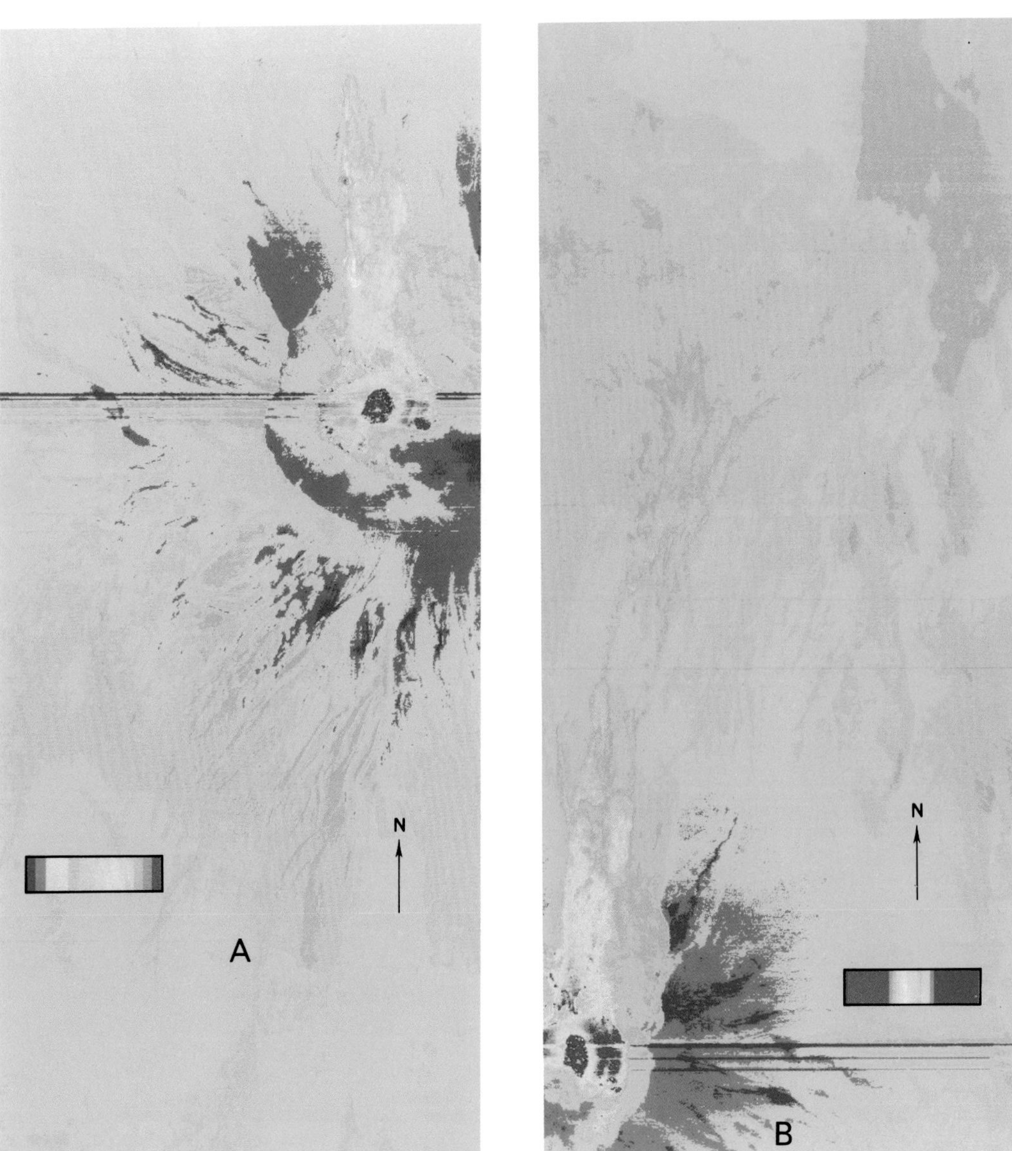

Fig. 31-179. Vertical themal infrared images of Mount St. Helens, RS-14A scanner; 8–14 μm region; 4,572 m flight altitude. U.S. Geological Survey calibrated color-coded images processed by computer; 3 mrad (milliradian) spatial resolution. Vertical stripe on side of image indicates color-coded temperature of upper blackbody of RS-14A scanner. *A.* 13 August 1980, 1941–1943 PDT, showing crater, amphitheater, and pyroclastic-flow deposits. *B.* 13 August 1980, 1948–1951 PDT, showing crater, amphitheater, and pyroclastic-flow deposits. (From Figure 173 C and D in Friedman et al., 1982b).

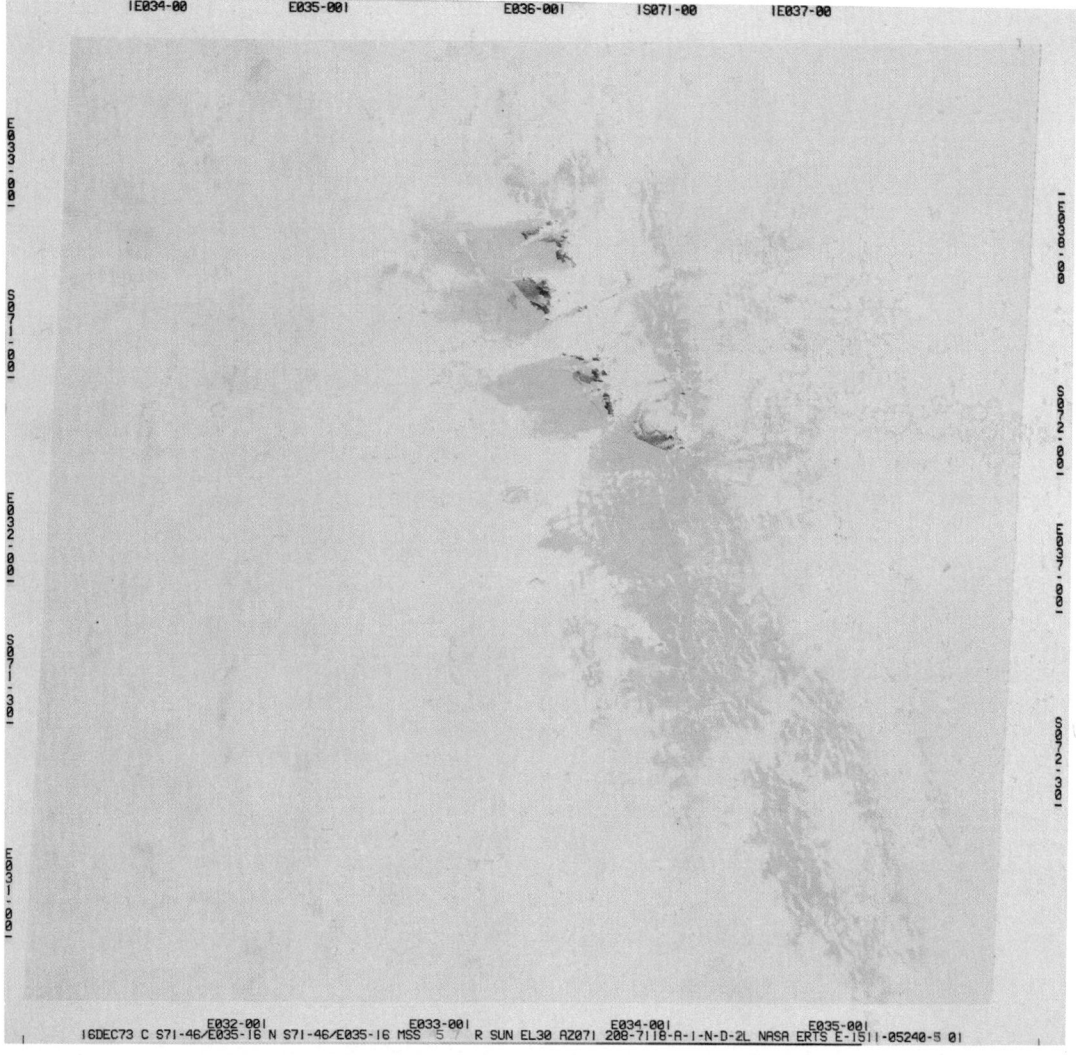

Fig. 31-196. Landsat MSS color composite image (1511-05240; 16 December 1973) of the Queen Fabiola (Yamato) Mountains area, East Antarctica, showing extensive areas of ''blue ice'' (bare glacier ice) around nunataks and associated morainic debris. Landsat (and NOAA) images can be used to delineate areas of blue ice in Antarctica, some of which have proven to contain extraordinary accumulations of meteorites. Since 1969, Japanese scientists have collected 4,813 meteorites within the blue-ice areas shown on this image or about 25 percent of the extant world-wide collection of meteorites.

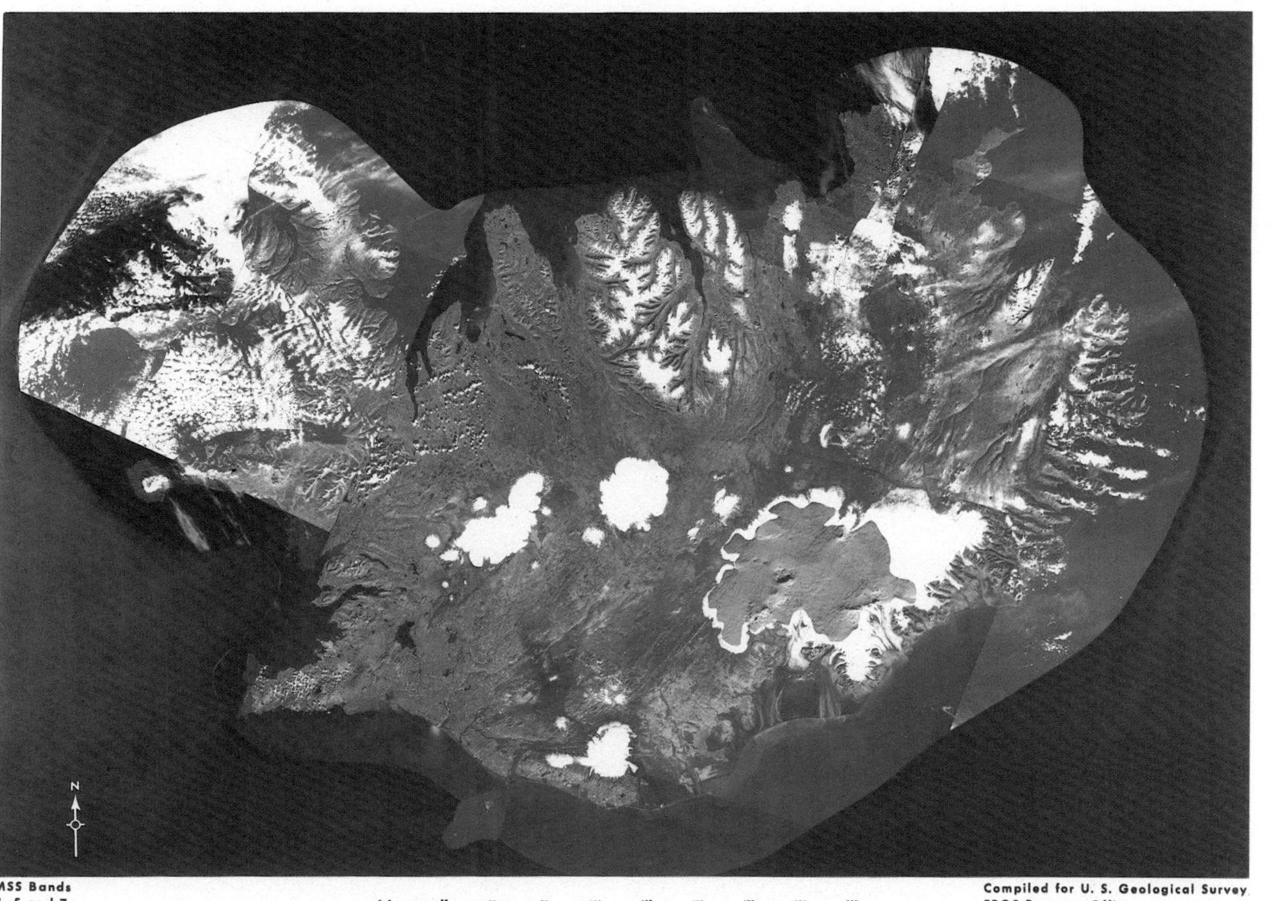

Fig. 31-197. Landsat 1 false color mosaic (MSS bands 4, 5, and 7) of Iceland. The ice caps of Vatnajökull (See also Color Figures 31-201, 31-202, and Figure 31-204), Hofsjökull (also partly shown on Color Figure 31-201), Langjökull (Color Figure 31-176), and Mýrdalsjökull-Eyjafjallajökull (Color Figure 31-176) are well depicted. (See Figure 31-198 for an index map to the principal glaciers and geological features of Iceland.) This uncontrolled mosaic was prepared by General Electric's Space Division (Photographic Engineering Laboratory) in 1974 for the U.S. Geological Survey from the best Landsat images available of Iceland through 1973.

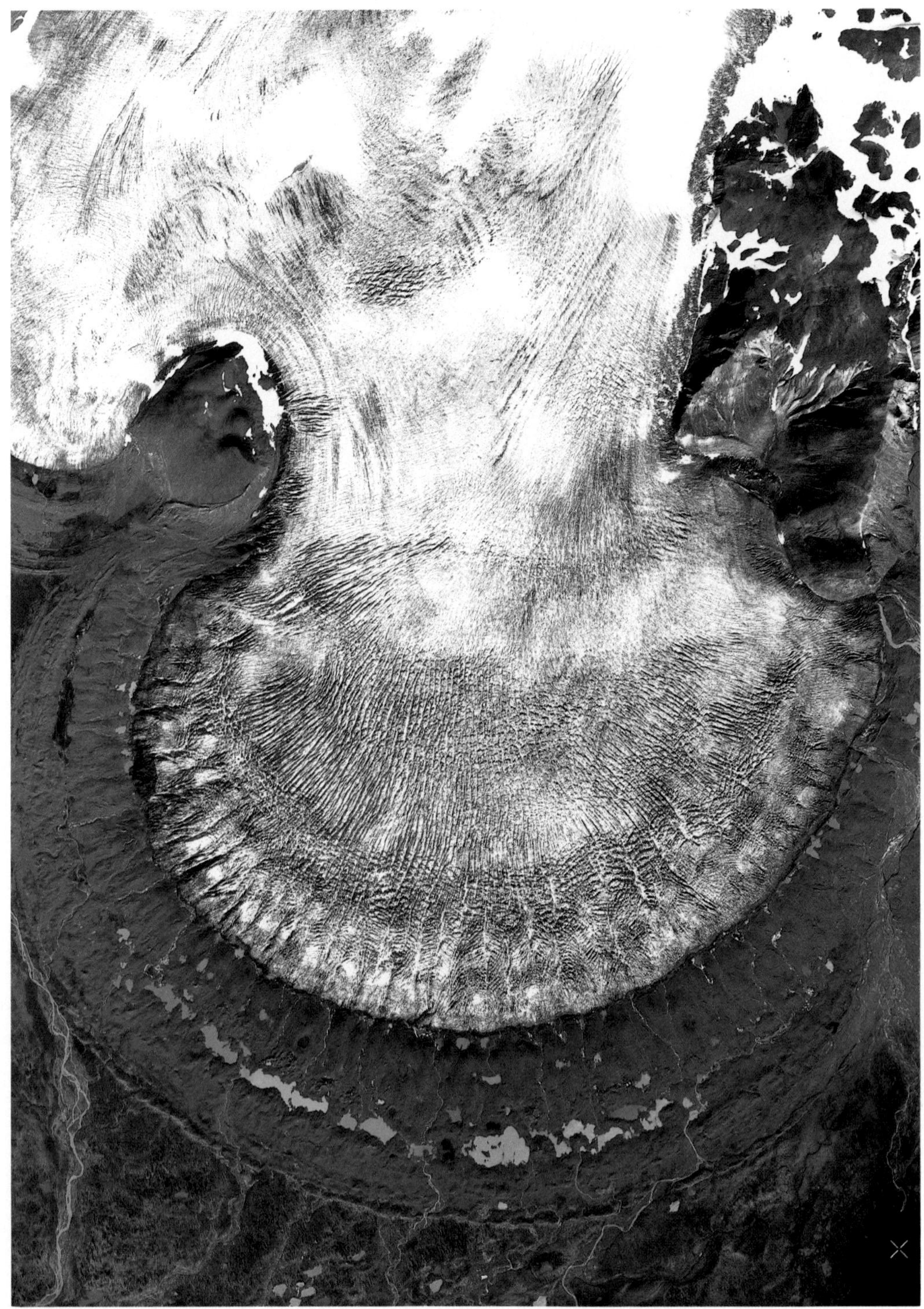

Fig. 31-199. Vertical color-infrared aerial photograph of Múlajökull, an outlet glacier from the Hofsjökull ice cap in central Iceland. The aerial photograph was acquired on the same day as Landsat images 1492-12185 and 1492-12191 (Color Figure 31-176), 19 August 1973, during NASA Mission 253, Test Site 714 (Roll 42, Frame 146). The complex nature of the crevasse pattern suggests that Múlajökull is a surging glacier and is recovering from a surge (Meier, written communication, 1982). Vegetation has crossed the outer terminal moraine of several concentric moraines and is colonizing the deglaciated terrain. (See Color Figure 31-201 and Figure 31-204 for a comparison with other Landsat images of Múlajökull). The photograph was acquired with an RC-8 camera with a focal length of 152 mm at altitude above the terrain of 3,300 m. Approximate scale of 1:20,000.

Fig. 31-201. Landsat MSS false color composite image (1426-12070; 22 September 1973) of Vatnajökull ice cap and environs (including part of Hofsjökull, all of Múlajökull, and part of Mýrdalsjökull), Iceland. MSS bands 4, 5, and 7 were enhanced by special computer processing (See Table 31-20) and projected through yellow, red, and blue filters, respectively. Called a triple, piecewise linear-stretch process, the three zones (bright snow, light vegetation, and dark water or bare soil/rock), identified from computer-generated histograms (See caption for Color Figure 31-176), were independently stretched linearly to capture their full brightness range. Each zone in each of the three bands was recombined to create three new digitally enhanced bands. The digital data were then converted to analog form (film transparency) by a laserbeam image recorder and composited to produce the image shown here. Considerable detail in the bright (snowcovered) areas on Vatnajökull is achieved, although not as pronounced as the low-Sun-elevation angle image of Vatnajökull shown in Figure 31-204. The standard MSS false color composite of this image was published as a 1:500,000-scale Landsat image map (U.S. Geological Survey, 1976). From a 1:250,000-scale enlargement of this image, the areal extent of Vatnajökull on 22 September 1973 was measured as 8,300 km^2 (Williams and Ferrigno, 1983; Björnsson, 1980b; Williams, 1983a, 1983b). (Image courtesy of Lincoln Perry, EROS Data Center, U.S. Geological Survey).

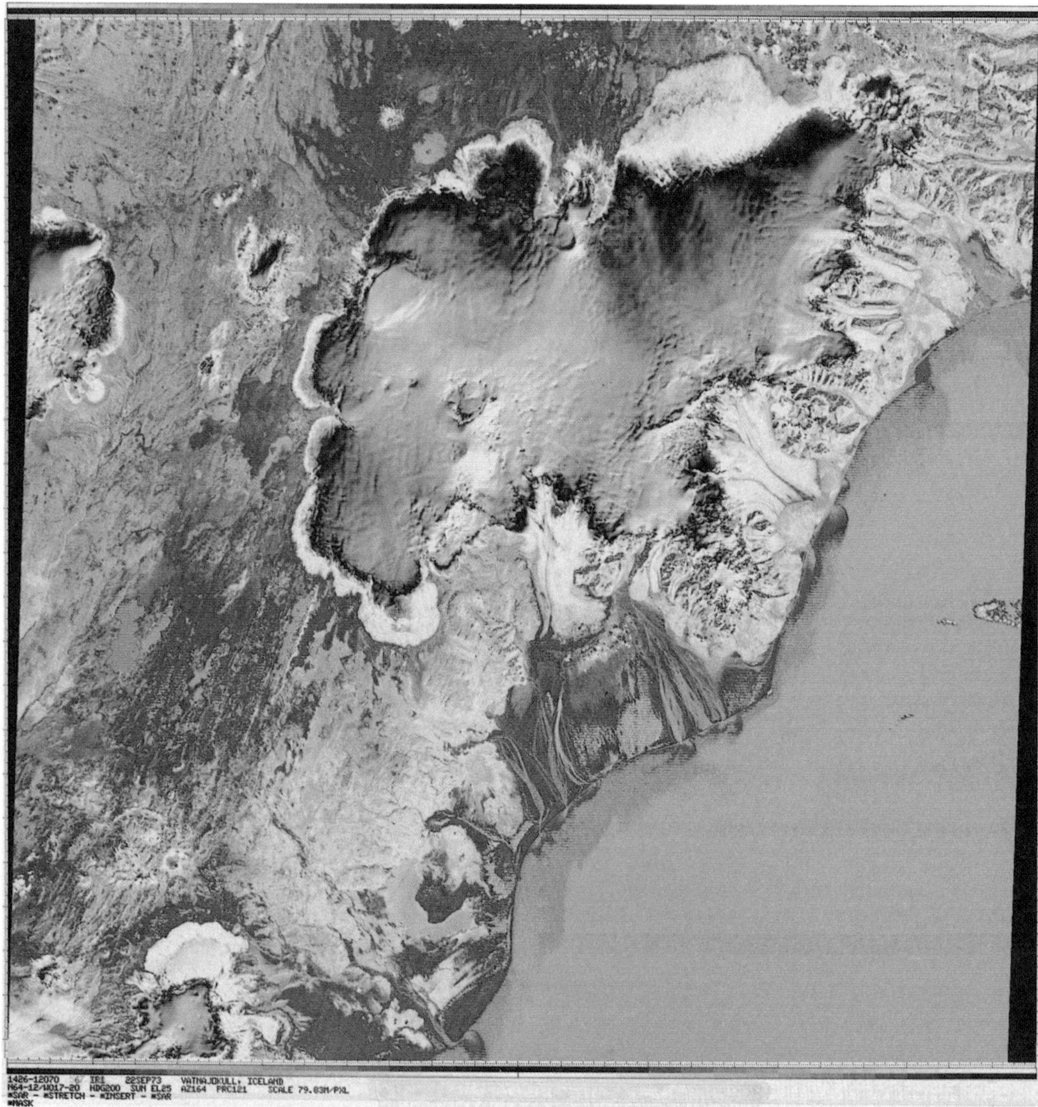

Fig. 31-202. Landsat MSS false color composite image (1426-12070; 22 September 1973) of Vatnajökull ice cap and environs, Iceland. MSS bands 5, 6, and 7 were enhanced by special computer processing and projected through yellow, red, and blue filters, respectively. The original computer-generated histograms of the spectral characteristics of the image fell into three groupings (trimodal), thereby permitting snow-covered areas, vegetation, and bare rock or soil to be processed independently. Each of these three zones was stretched in each of the three MSS bands to encompass the full range of brightness and then composited by the Image Processing Laboratory of Jet Propulsion Laboratory (Soha et al, 1976). This type of computer-enhanced image of Vatnajökull (and Mýrdalsjökull, Langjökull, and Hofsjökull, as well; See Color Figure 31-176) appears to provide a good correlation of glacier color and the zones in the accumulation and ablation areas of a glacier portrayed schematically in Figure 31-203. On Vatnajökull the colors apparently correlate as follows: light blue, bare glacial ice; orange, superposed ice; black, soaked zone; dark gray, percolation zone; light gray, dry-snow zone. Note how much bare glacier ice is exposed on Mýrdalsjökull compared with Vatnajökull, suggesting why the former has been more reduced in area over the past several decades (See Table 31-39). Landsat-type images may ultimately prove to have important scientific value in a global system of glacier monitoring by delineating the approximate position of the firn line at the end of the summer melt season (Østrem, 1975; Gunnar Østrem, 1982, verbal communication). The use of Landsat images to carry out global monitoring of glaciers is an extension of the research by Meier and Post (1962), who first described the use of equilibrium line altitudes (ELA's) and accumulation area ratios (AAR's), derived from aerial photographs to infer regional distribution of glacier mass balances. (Landsat image courtesy of Michael J. Abrams, Jet Propulsion Laboratory).

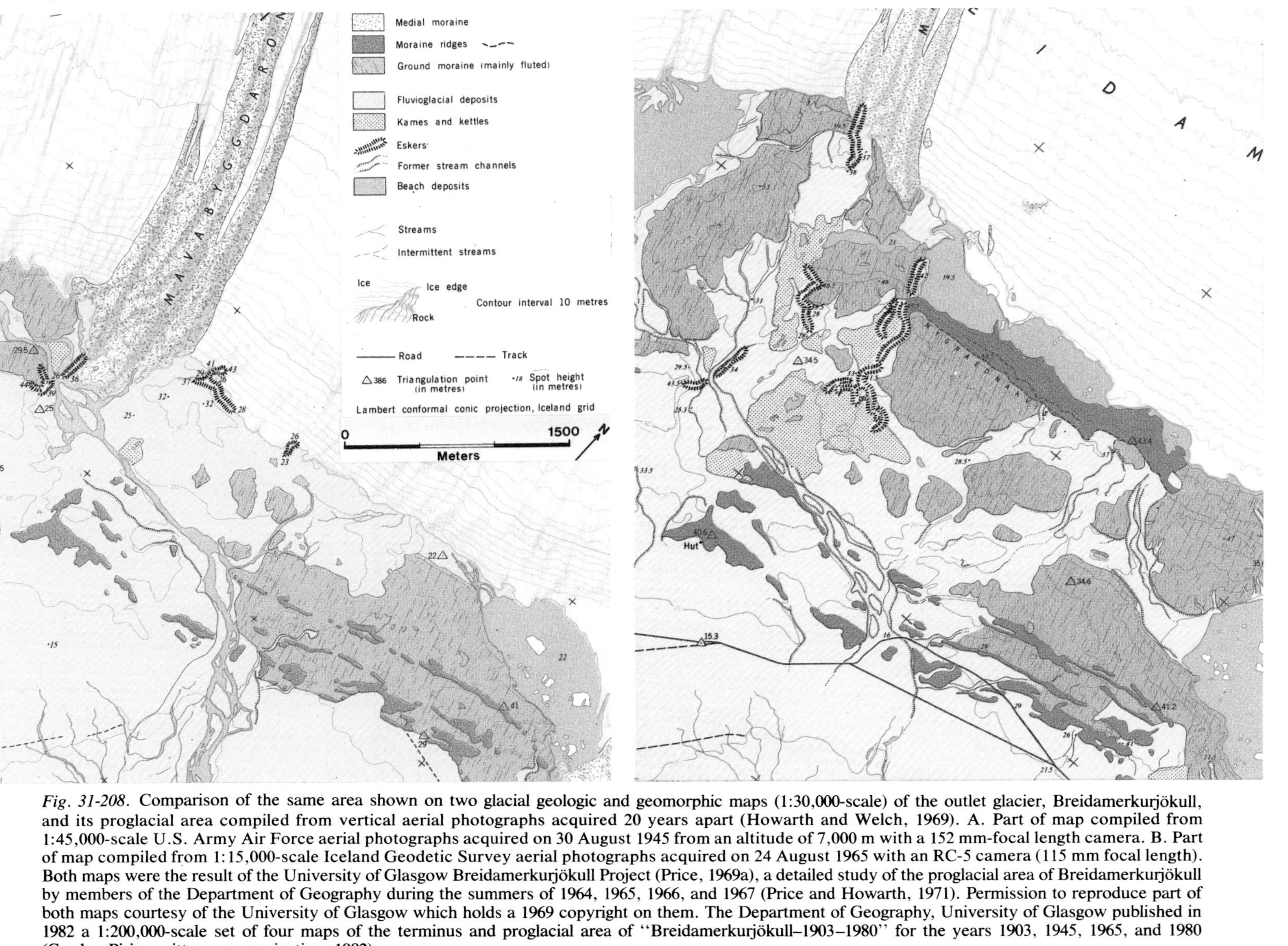

Fig. 31-208. Comparison of the same area shown on two glacial geologic and geomorphic maps (1:30,000-scale) of the outlet glacier, Breidamerkurjökull, and its proglacial area compiled from vertical aerial photographs acquired 20 years apart (Howarth and Welch, 1969). A. Part of map compiled from 1:45,000-scale U.S. Army Air Force aerial photographs acquired on 30 August 1945 from an altitude of 7,000 m with a 152 mm-focal length camera. B. Part of map compiled from 1:15,000-scale Iceland Geodetic Survey aerial photographs acquired on 24 August 1965 with an RC-5 camera (115 mm focal length). Both maps were the result of the University of Glasgow Breidamerkurjökull Project (Price, 1969a), a detailed study of the proglacial area of Breidamerkurjökull by members of the Department of Geography during the summers of 1964, 1965, 1966, and 1967 (Price and Howarth, 1971). Permission to reproduce part of both maps courtesy of the University of Glasgow which holds a 1969 copyright on them. The Department of Geography, University of Glasgow published in 1982 a 1:200,000-scale set of four maps of the terminus and proglacial area of "Breidamerkurjökull–1903–1980" for the years 1903, 1945, 1965, and 1980 (Gordon Pirie, written communication, 1982).

Fig. 31-210. Color terrestrial photograph, taken in August 1964, of a palsa, a peat-covered frost mound (5m high) in the Varangerfjord area of northern Norway. Photograph by Harald Svensson, Institute of Geography, University of Copenhagen.

Fig. 31-212. Color terrestrial photograph, taken in 1968, of part of an active ice wedge below a polygon furrow in Adventdalen, Spitsbergen (Svalbard), Norway. Photograph by Harald Svensson, Institute of Geography, University of Copenhagen.

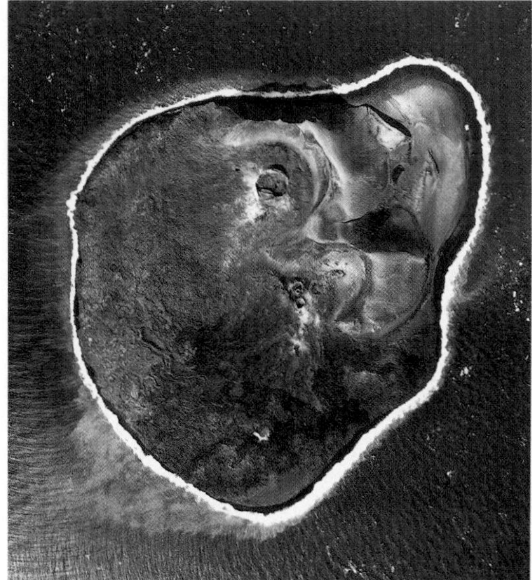

Fig. 31-216-H. Vertical color aerial photograph of Surtsey, Vestmannaeyjar, Iceland, showing its geomorphic development on 3 April 1968. The photograph was acquired during NASA Mission MX-70, Roll 1, Frame 6972, from an altitude of 3,050 m with an RC-8 camera (152 mm focal length).

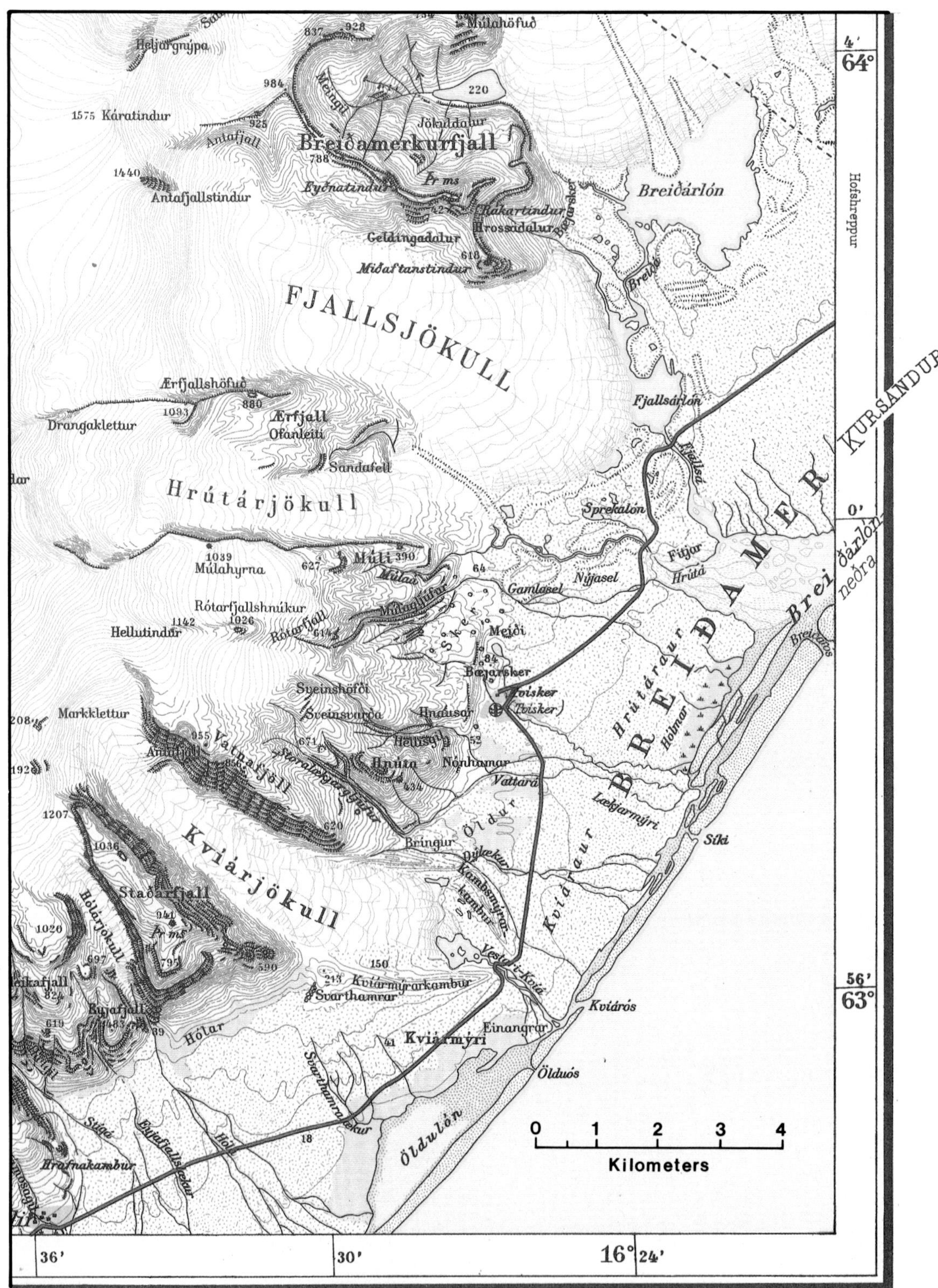

Fig. 31-219. Topographic map of the Kvíárjökull outlet glacier, Iceland, and its terminal moraine published by the Iceland Geodetic Survey in 1972. This 1:100,000-scale map (Blad 87, Öraefajökull) is based on the 1904 Danish Geodetic Institute map, which was compiled by plane-table survey methods. This 1972 version of the map was revised photogrammetrically from vertical aerial photographs acquired during the period 1961–1969. Note that the terminus of Kvíárjökull has retreated still further from its position on 30 August 1945 (See Figure 31-218).

Fig. 31-227. A Landsat 3 MSS false-color composite image (30347-14450; 15 February 1979; 0945 e.s.t.) of Cape Cod, Massachusetts and environs, showing the enormous areal extent of sea ice in the harbors and bays of Cape Cod (See Figure 31-40 for an index map of the area). The extreme sea ice conditions were due to northwesterly winds and subfreezing temperatures from a week-long Arctic air mass. These severe meteorological conditions produced especially heavy concentrations of pack ice in the southeastern part of Cape Cod Bay, including an area between Barnstable and Wellfleet Harbors that was 10 km wide by 30 km long, with a 2-m high pressure ridge along the shoreline. Nauset Harbor, Pleasant Bay, Chatham Harbor, Stage Harbor, and a 5-km wide area west of Monomoy Island were also frozen solid; Buzzards Bay had a heavy ice cover except for a shore lead on the NW side of the bay. Nantucket Island was surrounded by drifting ice floes and frazil ice, with the heaviest, solid concentrations extending along the entire northwesterly shore from Muskeget Island to Great Point, and into Nantucket Sound for a distance of about 10 km. A discontinuous sea ice cover also extended about 40 km N. (mostly floe ice) and S. (mostly frazil ice) of Nantucket Island. During a 13-hour ferry and U.S. Coast Guard cutter trip (normally a 3-hour transit time) on February 15, 1979, between Woods Hole and Nantucket, ice ridges 3 m high were encountered, and ice was 4 m thick at Brant Point just inside the riprap jetties at the entrance to Nantucket Harbor. Although only field observations can verify the actual sea ice thickness, the Landsat image was indispensible in providing: (1) the true areal extent of the sea ice in the harbors and coastal waters of Cape Cod; (2) a pictorial view of what past, bad sea ice years (such as in 1857, 1918, and 1937) must have been like; and (3) an historical record depicting those areas on Cape Cod which could benefit from better design of future coastal engineering structures because of heavy sea ice accumulation. (See Color Figure 31-228 for a summertime (6 July 1976) Landsat image of Cape Cod, Massachusetts, and environs.)

Fig. 31-228. Landsat MSS false color digitally enhanced composite image (5444-14084; 6 July 1976) of Cape Cod, Massachusetts and environs. The outer Cape Cod coastline is characterized by relatively high eroding glacial cliffs (Leatherman et al., 1981) with well defined spit systems on each end. To the north, the prominent peninsula, a complex spit known as the Province Lands, is distinguished by the large amount of open sand. To the south, the Nauset Spit system is shown to consist of two embayments, three spit segments, and two islands (See also Figure 31-34). Nauset Spit-Eastham and -Orleans, enclosed Nauset Harbor, and Nauset Beach serves as the outer barrier for Pleasant Bay. The recent (1978 vintage) inlet at Inward Point divided Monomoy into two islands (See also Figures 31-35, -36, -37, -39, -43, -44, and -48; and Color Figures 31-42, -45, -46, -49, -51, -53, and -227). This Landsat image is also available in poster format from the Eastern National Monument and Park Association (1980).

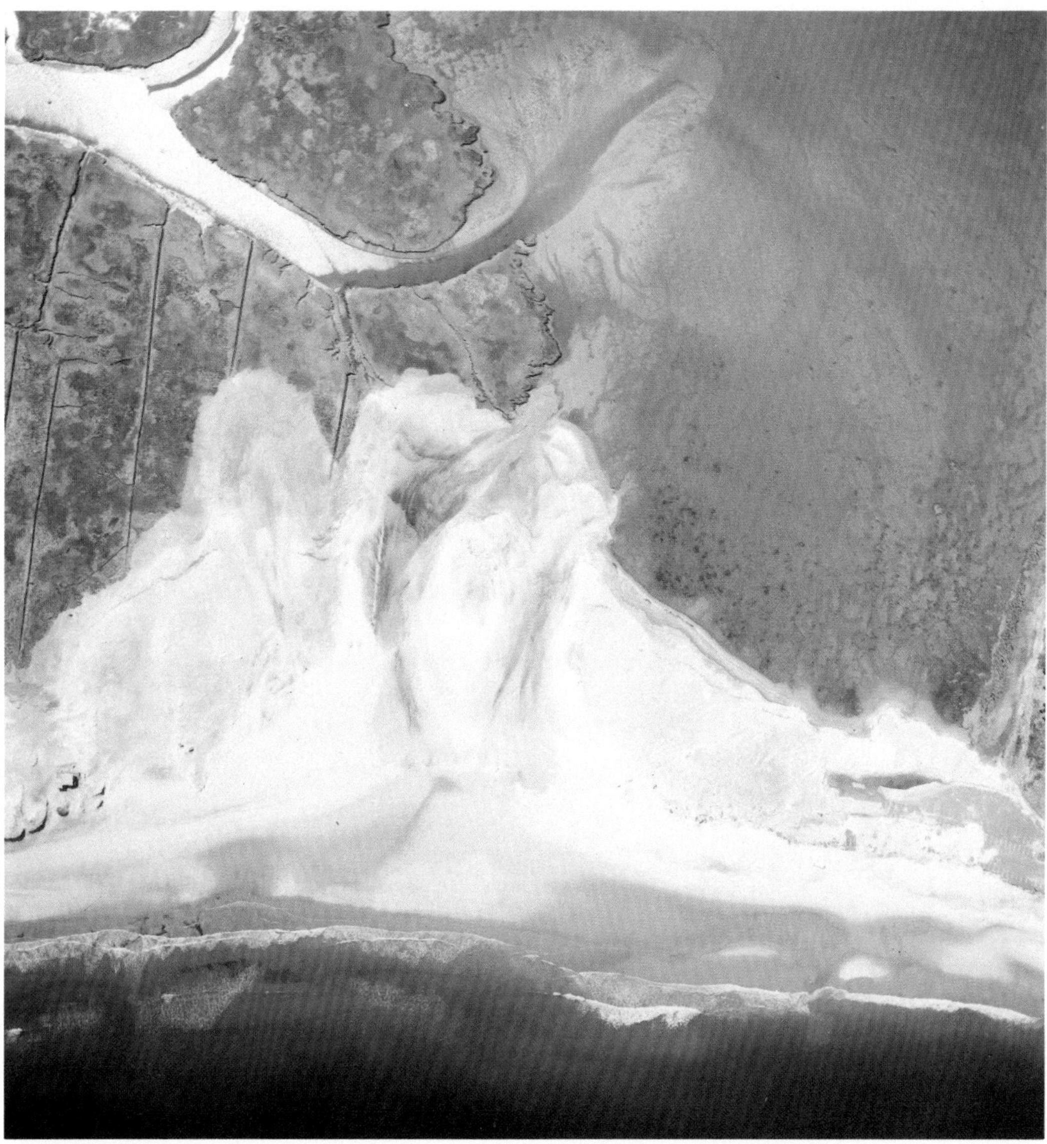

Fig. 31-232. A color vertical aerial photograph acquired by the National Ocean Survey on 31 March 1978, in the month following the February Blizzard. This 1:3,000-scale photograph clearly illustrates the extent of massive washover deposits emplaced on the salt marsh surface following total dune breaching. The once highly utilized Coast Guard Beach bathhouse and parking lot facility were completely destroyed, representing a costly loss to the National Park Service. Higher energy overwash surges crossed at the shoreline embayment during the storm (near photo center). Beach recovery has nearly healed the scar, but the pattern of wet and dry sand still denotes the presence of the former shoreline reentrant. The cohesive nature of the underlying salt marsh deposit prevented possible opening of a new inlet into Nauset Harbor. Natural color aerial photography may have been used instead of color infrared aerial photography because of better water penetration capability and delineation of the nearshore bar system. (NOS aerial photograph No. CCNB-LA-1-01C.)

Fig. 31-235. Vertical color aerial photograph of Cayo Icacos, Puerto Rico, showing water penetration to a depth of 20 m with unfiltered (0.36–0.68 μm) color aerial film (See also Figure 31-39). Original scale of the aerial photograph, 1:10,300, U.S. Coast and Geodetic Survey photograph no. S-9733 acquired on 6 February 1962. Camera focal length, 152.29 mm. Survey altitude, 1,570 m.

Fig. 31-242. Color photograph of sparse to barren area on mineralized quartz monzonite at the Kalamazoo porphyry copper deposit, near San Manuel Pinal County, Arizona. The altered quartz monzonite is barren and red stained. In the background is a hill bearing the normal vegetative cover of the area on Tertiary rhyodacite. Soil analyses on the barren altered area revealed 216 ppm Zn (±73) and 664 ppm Cu (±307). The vegetation anomaly is caused by one or more of the following factors: high Cu concentrations, high pH due to sulphide weathering, or extremely adverse conditions induced by argillic alteration. (Soil analyses courtesy of Newmont Exploration, Ltd.)

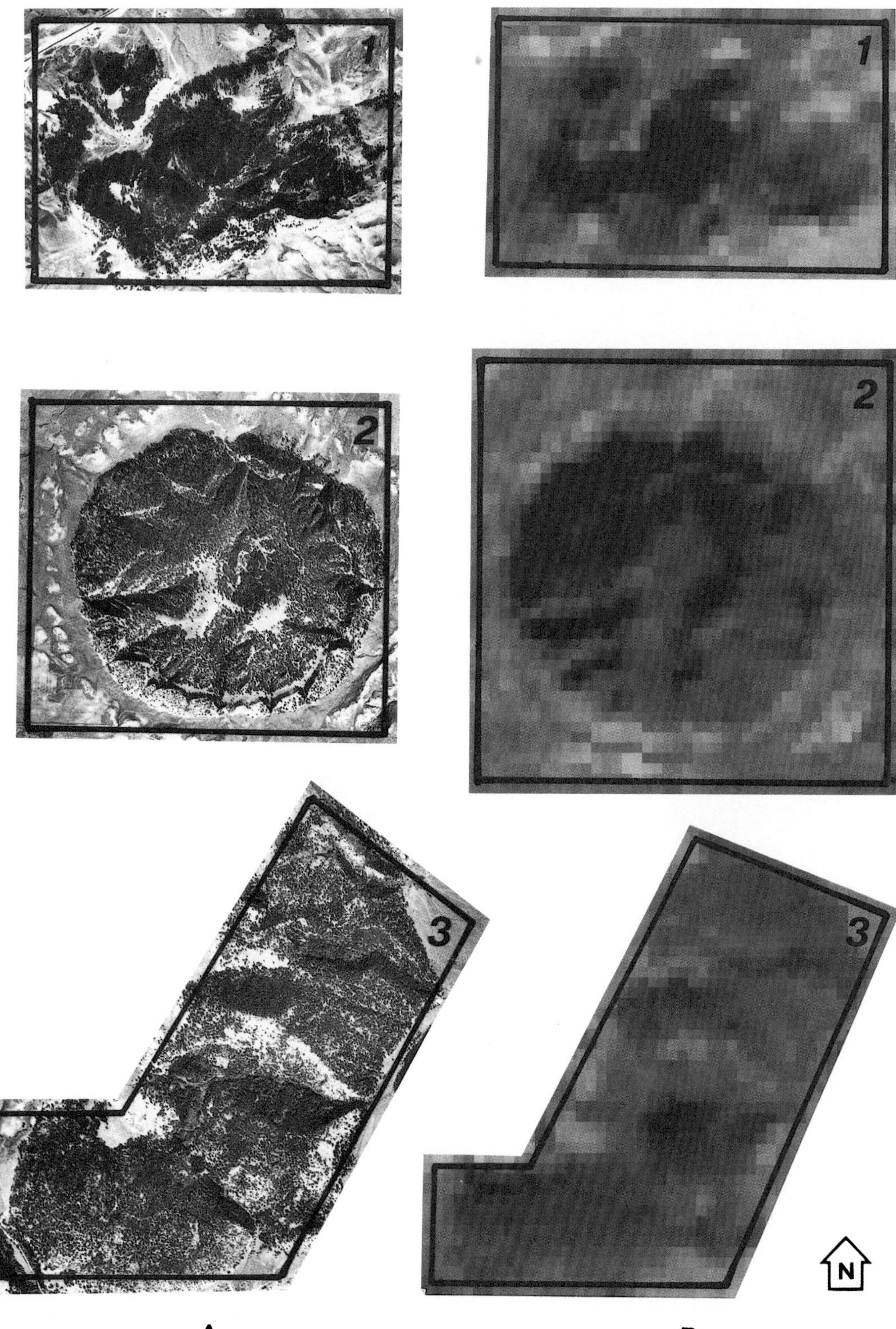

Fig. 31-244. Enlargements of three areas on a Landsat 2 MSS false color composite image (2169-81655; 16 September 1979) of the Sundance, Wyoming area, taken from the cathode ray tube (CRT) of a GE Image 100. A comparison of areas 1, 2, and 3 (shadows excluded) shows that areas 2 and 3 have higher reflected radiance (brighter tones) caused by the spectral contribution from soil and sedimentary rocks as compared to the densely forested igneous rock in area 1. The three areas on the Landsat image (right) are compared with the same three areas on a vertical, black-and-white aerial photograph of the same area (Figure 31-243). The Landsat image has been enlarged to a scale of about 1:35,000, revealing the individual pixels (79 m) which make up a standard Landsat MSS scene (about 7.5×10^6 pixels per scene).

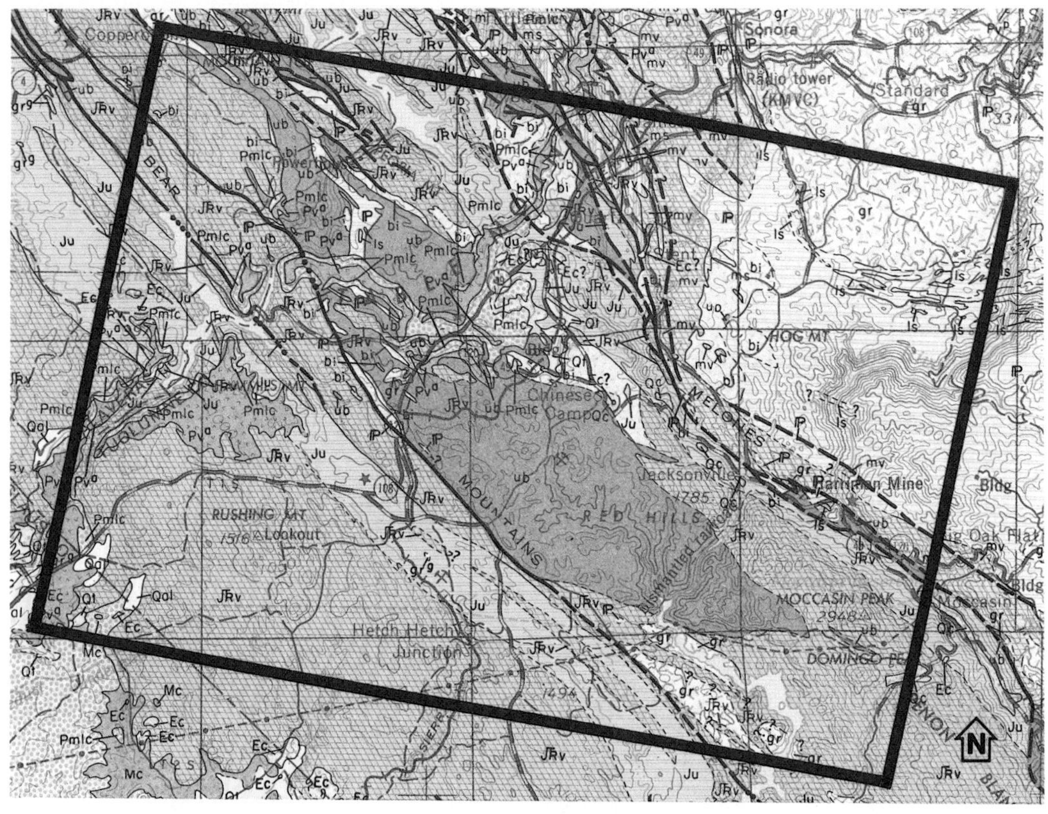

EXPLANATION

Sedimentary and Metasedimentary Rocks

Qal	Alluvium
Qsc	Stream channel deposits
Qt	Quaternary nonmarine terrace deposits
Qc	Pleistocene nonmarine
Pmlc	Middle and/or lower Pliocene nonmarine
Mc	Undivided Miocene nonmarine
Ec	Eocene marine
Tc	Tertiary nonmarine
Ju	Upper Jurassic marine
ms	Pre-Cretaceous metasedimentary rocks
IP	Paleozoic marine

Igneous and Meta-Igneous Rocks

Pv^a	Pliocene volcanic: andesite
bi	Mesozoic basic intrusive rocks
ub	Mesozoic ultrabasic intrusive rocks
JRv	Jura-Trias metavolcanic rocks
mv	Pre-Cretaceous metavolcanic rocks

Fig. 31-246. Geologic Map of the Red Hills area, California. California Division of Mines. Scale, 1:250,000.

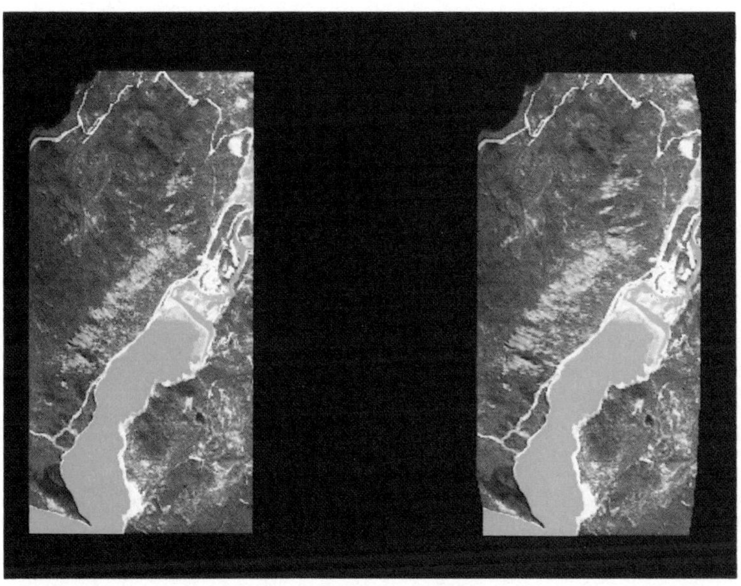

Fig. 31-260. Stereoscopic pair of simulated SPOT images of the Gun Lake area of the Canadian Rockies (in British Columbia). The area covered is 18 km^2, with a simulated 20-m pixel resolution (bands 1, 2, and 3 of SPOT combined as a false-color composite (Simard, 1981)).

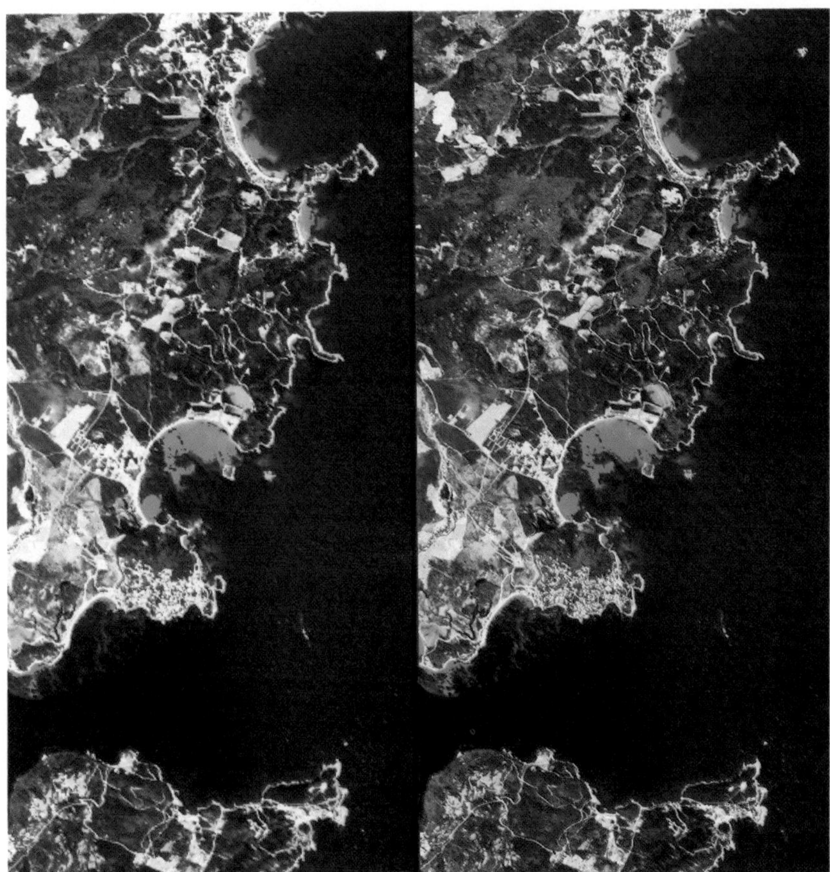

Fig. 31-261. The SPOT satellite will produce digital imagery in two spectral bands in the visible and one in the near-infrared with a pixel resolution of 20 m. In the panchromatic mode a 10-m pixel resolution will be achieved. Both modes of imagery will cover an area with dimensions of 60 × 60 km. These two simulated subscenes, acquired with the 9-channel Daedalus multispectral airborne scanner, cover an area 5.35 × 12 km in the Gulf of Porto Vecchio area, Corsica. The left-hand scene simulates a standard multispectral SPOT image, in which four pixels for each of the three spectral bands have been aggregated into one 20-m pixel to achieve a false color composite image. The right-hand image is a higher resolution panchromatic image digitally combined with the multispectral data to achieve greater resolution (Figure caption modified from CNES, n.d.; Image courtesy of CNES. The original scale of the figure was 1:62,500). (See also Color Figures 31-77 and 31-78 for digitally combined Landsat MSS and Landsat 3 RBV images of the San Francisco area, California.)

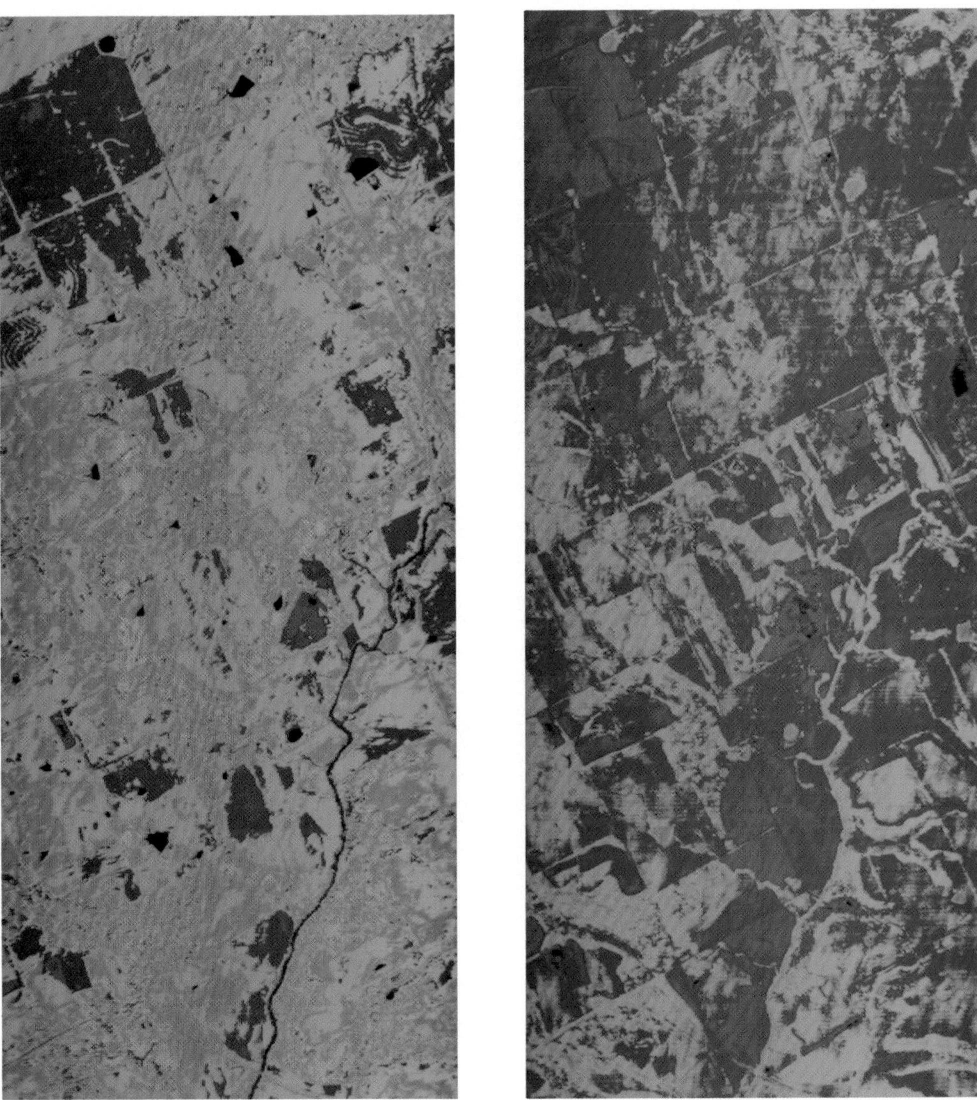

Fig. 32-8. Color enhanced thermal infrared imagery and a photomap showing generalized depths to groundwater or bedrock from a test site in Jefferson County, Kansas.

Fig. 32-26. Color infrared photograph showing a potash plant and surrounding terrain, along with clues to a 1200 gallon a minute groundwater supply for potash mine and plant development. (Color IR photo A37465-49, taken in June 1977.)

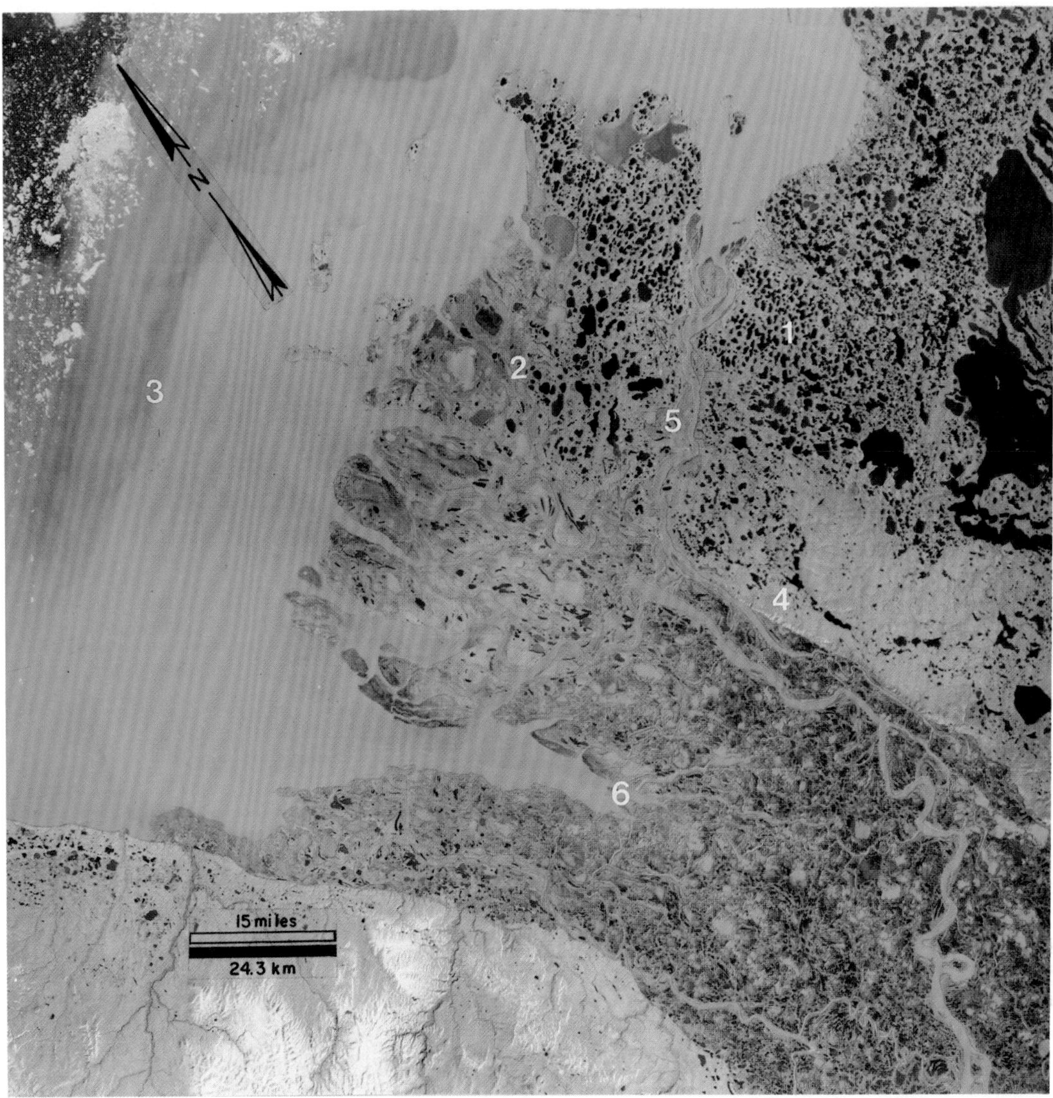

Fig. 32-32. Landsat image used in regional planning studies, Mackenzie Delta, Northwest Territories. Remote sensing studies involved tracking the movement of sea ice, mapping linears to assess evidence of recent faulting, assessing offshore sedimentation, locating sand and gravel sources, and locating river and channel crossing sites, and pipeline and road routes. (Landsat image E-1764-20103-8, acquired August 26, 1974.)

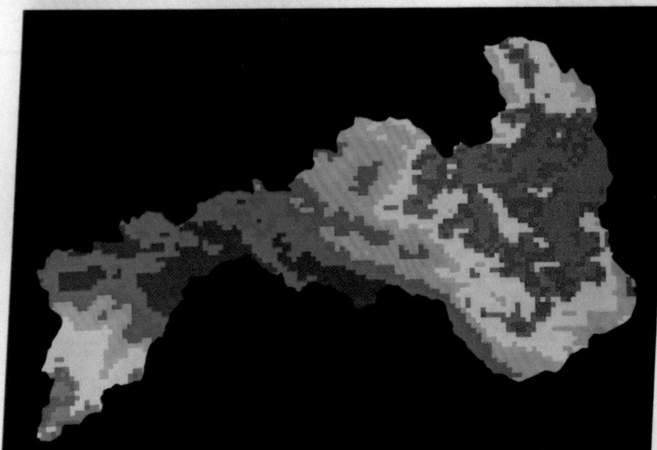

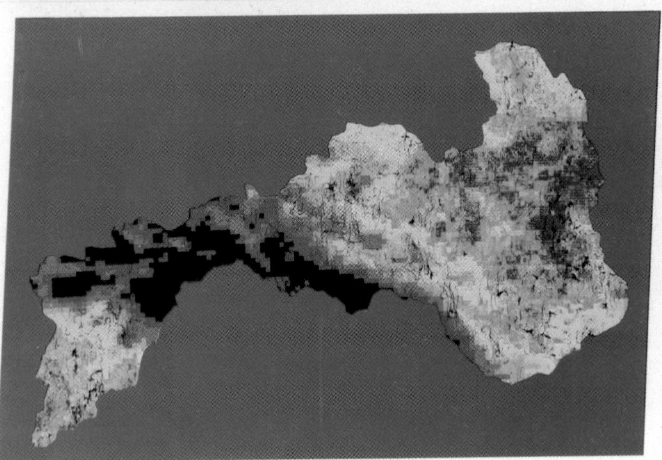

Color Fig. 32-38. These three color-coded maps show the area distribution of certain input parameters to the evapotranspiration model for the Middle Fork of the Feather River in California's Sierra Nevada. In each instance NOAA-4 satellite infrared data were used as input, as discussed in the text. The *temperature* map at the *top* is color coded as follows:

Color	*Temperature in °F*
Dark Blue	0–31
Blue	32–45
Purple	46–61
Blue Green	62–68
Green	70–78
Yellow Green	79–84
Yellow	85–88
Yellow Red	90–92
Red	93–96
Lavender	97–108

The *net radiation* map in the *middle* is color coded as follows:

Color	*Net radiation in ly/day*
Dark Blue	0–250
Blue-Green	260–360
Yellow-Green	370–410
Green	420–430
Yellow-Brown	440
Yellow	450
Red-Brown	460
Yellow-Red	470
Lavender-Red	480
Red	490–600

The *potential evapotranspiration* map at the *bottom* is color coded as follows:

Color	*ETp in inches of water*
Red	0.00–0.05
Orange	0.06–0.11
Yellow-Orange	0.12–0.15
Yellow	0.16–0.19
Yellow-Green	0.20–0.22
Dark Green	0.23–0.24
Green	0.25–0.26
Blue-Green	0.27–0.28
Dark Blue	0.29–0.33
Blue	0.34–0.41

Another case study that pertains primarily to the northeast part of this area appears in Chapter 34. In that study however, the objectives were different. Hence the output products were derived from somewhat different kinds of remotely sensed data, acquired on different dates, and employed a different color coding.

Color Fig. 32-37. Landsat color composite made from bands 4, 5, and 7, showing the entire Feather River Watershed, within which studies were performed relative to water yield forecasting.

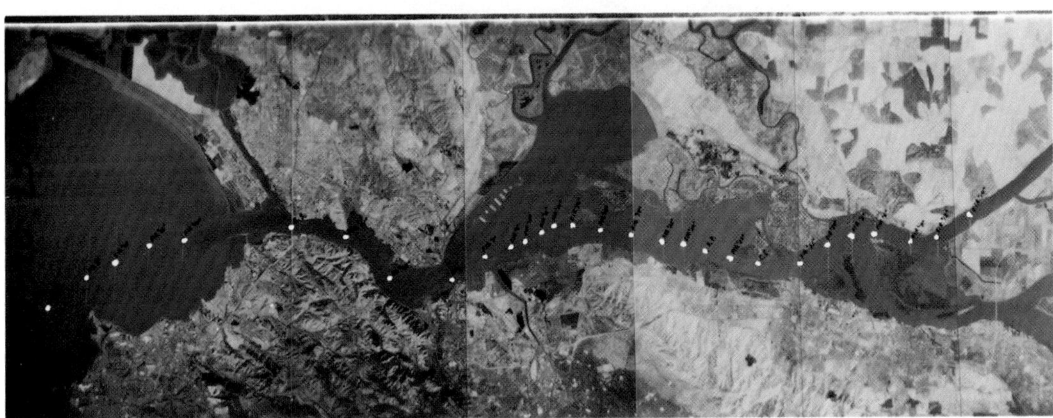

Color Fig. 32-54. Location of water quality sample sites within the study area. The sites were numbered from 1 to 29 from left to right. Even on the original, high-quality color infrared photographs from which this mosaic was made, it was not possible to detect water quality differences of the type portrayed on Color Figures 32-56 and 32-57 for this same area, and as of essentially the same time. (After Khorram, 1981a).

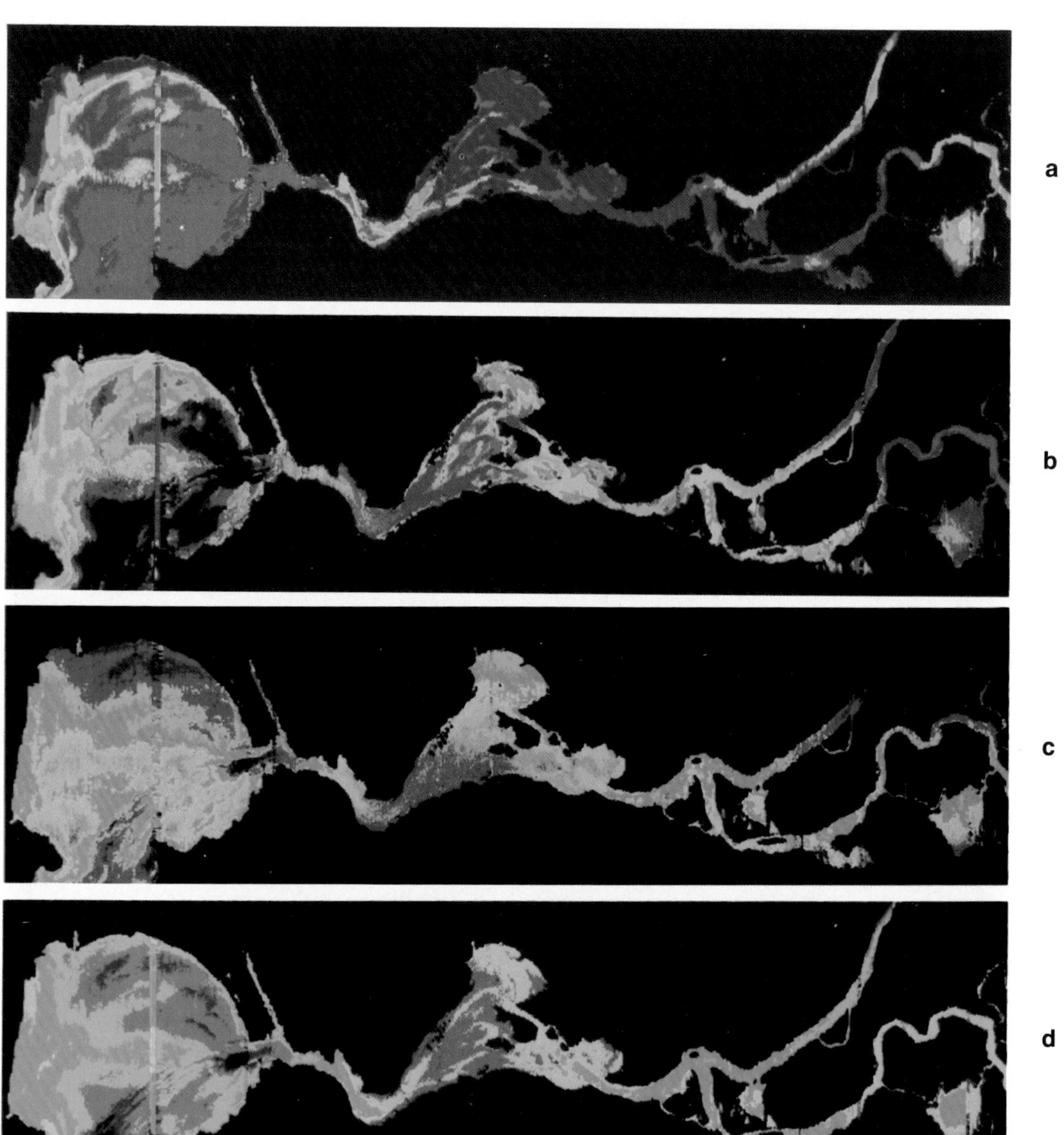

Fig. 32-56. See caption for this figure on next page.

Color Fig. 32-56a. Salinity distribution as derived from OCS data for the Delta region; San Pablo Bay is on left. After Khorram 1981a

Legend

Color	*Salinity, Percent*
Black	0.0
Dark Blue	0.1–0.9
Blue	1.0–1.9
Light Blue	2.0–3.9
Green	4.0–5.9
Yellow	6.0–7.9
Yellow-orange	8.0–9.9
Orange	10.0–11.9
Light Red	12.0–13.9
Red	14.0–15.9
Brown-red	16.0–22.9

Color Fig. 32-56b. Chlorophyll distribution as derived from OCS data for the Delta region; San Pablo Bay is on left. After Khorram 1981a

Legend

Color	*Chlorophyll, mg/l*
Black	.000
Dark Blue	.001–.004
Blue	.005–.009
Light Blue	.010–.014
Green	.015–.019
Yellow	.020–.024
Yellow-orange	.025–.029
Orange	.030–.034
Light Red	.035–.039
Red	.040–.044
Brown-red	.045–.084

Color Fig. 32-56c. Suspended solids distribution as derived from OCS data for Delta region; San Pablo Bay is on left. After Khorram 1981a

Legend

Color	*Suspended Solids, mg/l*
Black	0.0
Dark Blue	0.1–9.9
Blue	10.0–19.9
Light Blue	20.0–29.9
Green	30.0–39.9
Yellow	40.0–49.9
Yellow-orange	50.0–59.9
Orange	60.0–69.9
Light Red	70.0–79.9
Red	80.0–89.9
Brown-red	90.0–99.9

Color Fig. 32-56d. Turbidity distribution as derived from OCS data for the Delta region; San Pablo Bay is on left. After Khorram 1981a

Legend

Color	*Turbidity, mg/l SiO_2*
Black	0.0
Dark Blue	0.1–9.9
Blue	10.0–19.9
Light Blue	20.0–29.9
Green	30.0–39.9
Yellow	40.0–49.9
Orange	50.0–59.9
Red	60.0–69.9

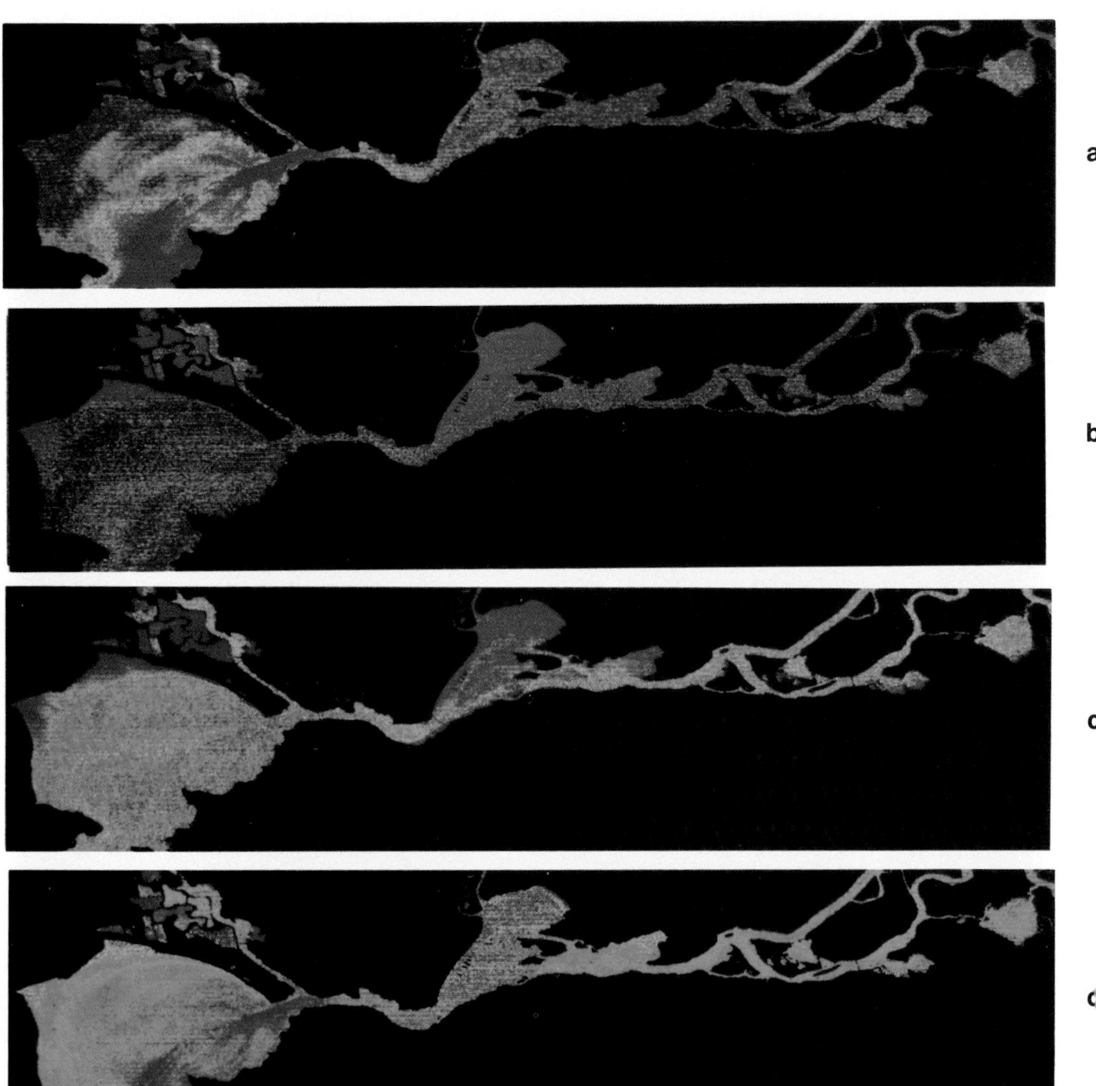

Fig. 32-57. See caption for this figure on next page.

Color Fig. 32-57a. Salinity distribution as derived from the enhanced Landsat MSS data for the Delta region; San Pablo Bay is on left. (After Khorram 1981b)

Legend

Color	*Salinity, %*
Black	0.0
Dark Blue	0.1–0.9
Blue	1.0–1.9
Light Blue	2.0–3.9
Green	4.0–5.9
Light Green	6.0–7.9
Yellow	8.0–9.9
Yellow-orange	10.0–11.9
Orange	12.0–13.9
Light Red	14.0–15.9
Red	16.0–17.9
Brown	18.0–25

Color Fig. 32-57b. Chlorophyll distribution as derived from the enhanced Landsat MSS data for the Delta region; San Pablo Bay is on left. (After Khorram 1981b)

Legend

Color	*Chlorophyll, mg/l*
Black	0.000
Dark Blue	0.000–.004
Blue	.005–.009
Light Blue	.010–.014
Green	.015–.019
Light Green	.020–.024
Yellow	.025–.029
Yellow-orange	.030–.034
Orange	.035–.039
Light Red	.040–.044
Red	.045–.049
Brown	.050–.100

Color Fig. 32-57c. Suspended solids distribution as derived from the enhanced Landsat MSS data for the Delta region; San Pablo Bay is on left. After Khorram 1981b

Legend

Color	*Suspended Solids, mg/l*
Black	0.0
Dark Blue	0.1–9.9
Blue	10.0–19.9
Light Blue	20.0–29.9
Green	30.0–39.9
Light Green	40.0–49.9
Yellow	50.0–59.9
Yellow-orange	60.0–69.9
Orange	70.0–79.9
Light Red	80.0–89.9
Red	90.0–100

Color Fig. 32-57d. Turbidity distribution as derived from the enhanced Landsat MSS data for the Delta region; San Pablo Bay is on left. After Khorram 1981b

Legend

Color	*Turbidity, mg/l SiO_2*
Black	0.0
Dark Blue	0.1–9.9
Blue	10.0–19.9
Light Blue	20.0–29.9
Green	30.0–29.9
Light Green	40.0–49.9
Yellow	50.0–59.9
Yellow-orange	60.0–69.9
Orange	70.0–79.9
Light Red	80.0–89.9
Red	90.0–100

Color Fig. 32-58. Salinity map of the study area, derived from Landsat digital data (After Khorram et. al. 1982.) Note the San Francisco Peninsula and bridges spanning San Francisco Bay.

Legend

Color	*Salinity in parts per thousand*
Blue	0–5.0
Light Blue	5.1–9.8
Light Green	9.9–14.8
Green	14.9–20.0
Yellow	20.1–24.9
Red	25.0–29.9
Brown	30.0–35.0

Color Fig. 32-59. Turbidity map of the study area, derived from Landsat digital data (After Khorram et. al. 1982.)

Legend

Color	*Turbidity in NTU*
Dark Blue	0–1.4
Blue	1.5–3.5
Light Blue	3.6–5.6
Light Green	5.7–7.6
Green	7.7–10.4
Yellow	10.5–15.9
Orange	16.0–20.7
Red	20.8–25.4
Brown	25.5–73.5

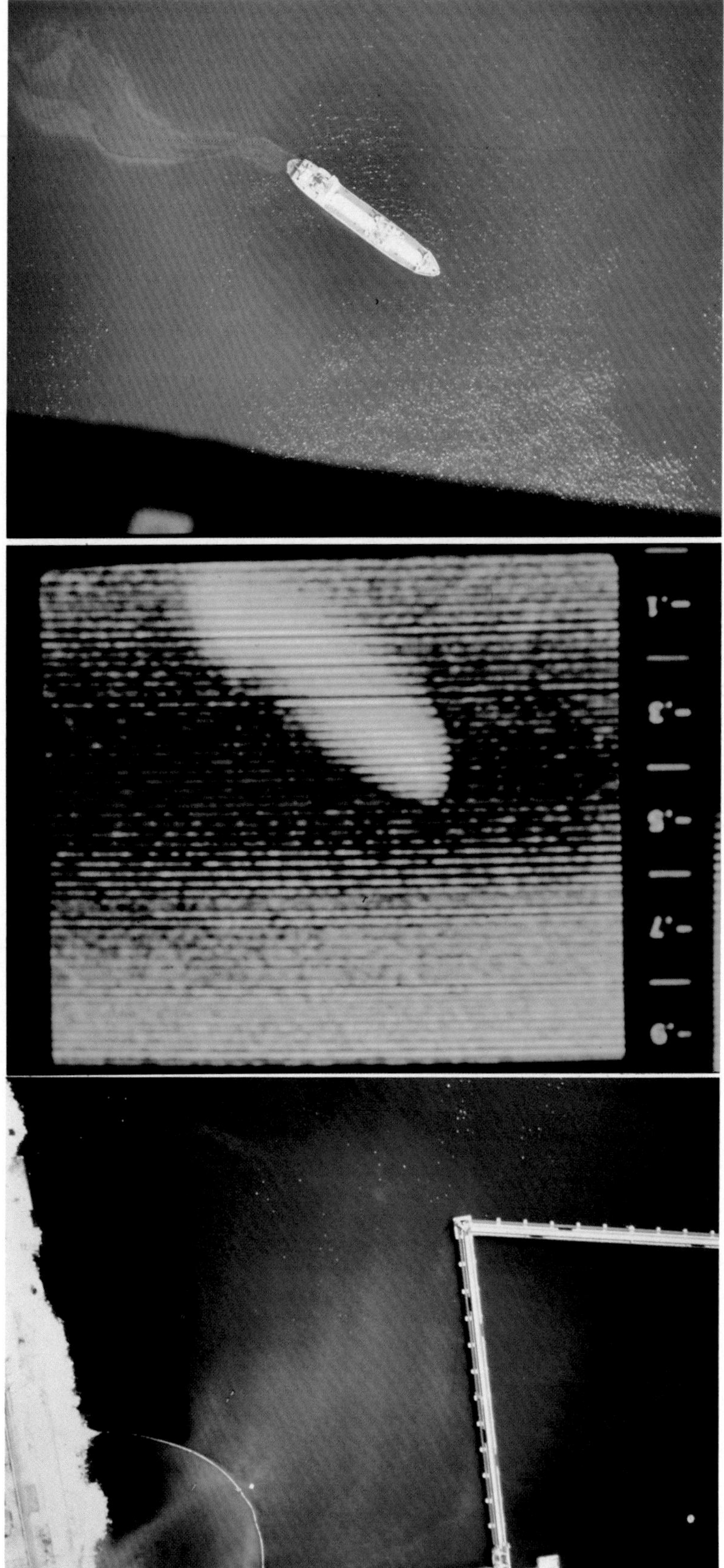

Fig. 32-60. Three examples of water quality mapping in Italy are shown here: *Top:* This false color photo, taken from an altitude of 450 m (1500 ft.) shows an illegally discharging vessel in Augusta Bay, Sicily. *Middle:* This thermogram, taken from an altitude of 225 m (750 ft.) shows an illegally discharging vessel, also in Augusta Bay, Sicily. *Bottom:* This false color photo, taken from an altitude of 450 m (1500 ft.) shows an industrial discharge into eastern Sicily's coastal waters. (After Geraci, 1981).

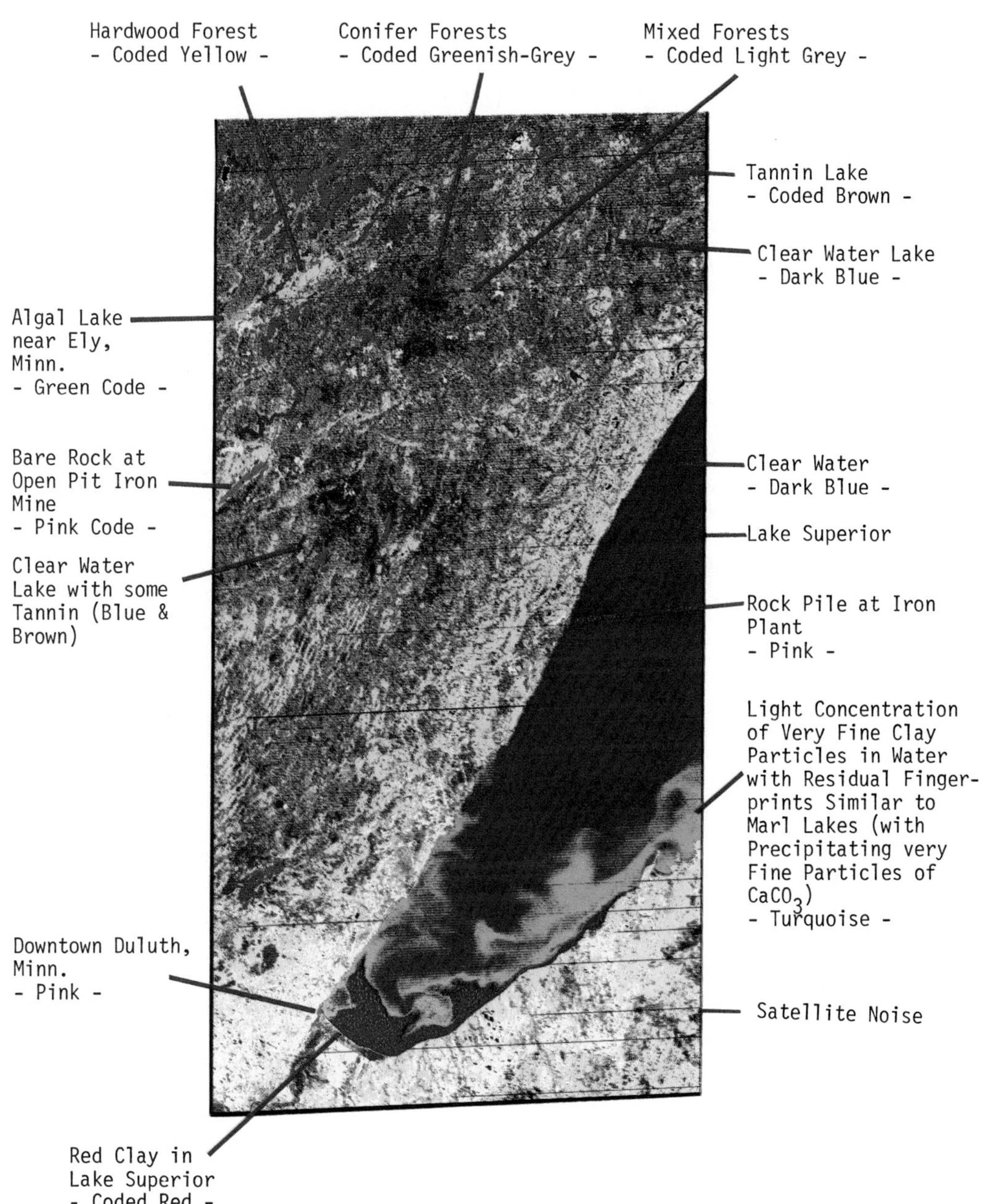

Color Fig. 32-62a. Part of one of the three coded maps showing one of the color schemes used. 12 August 1972.

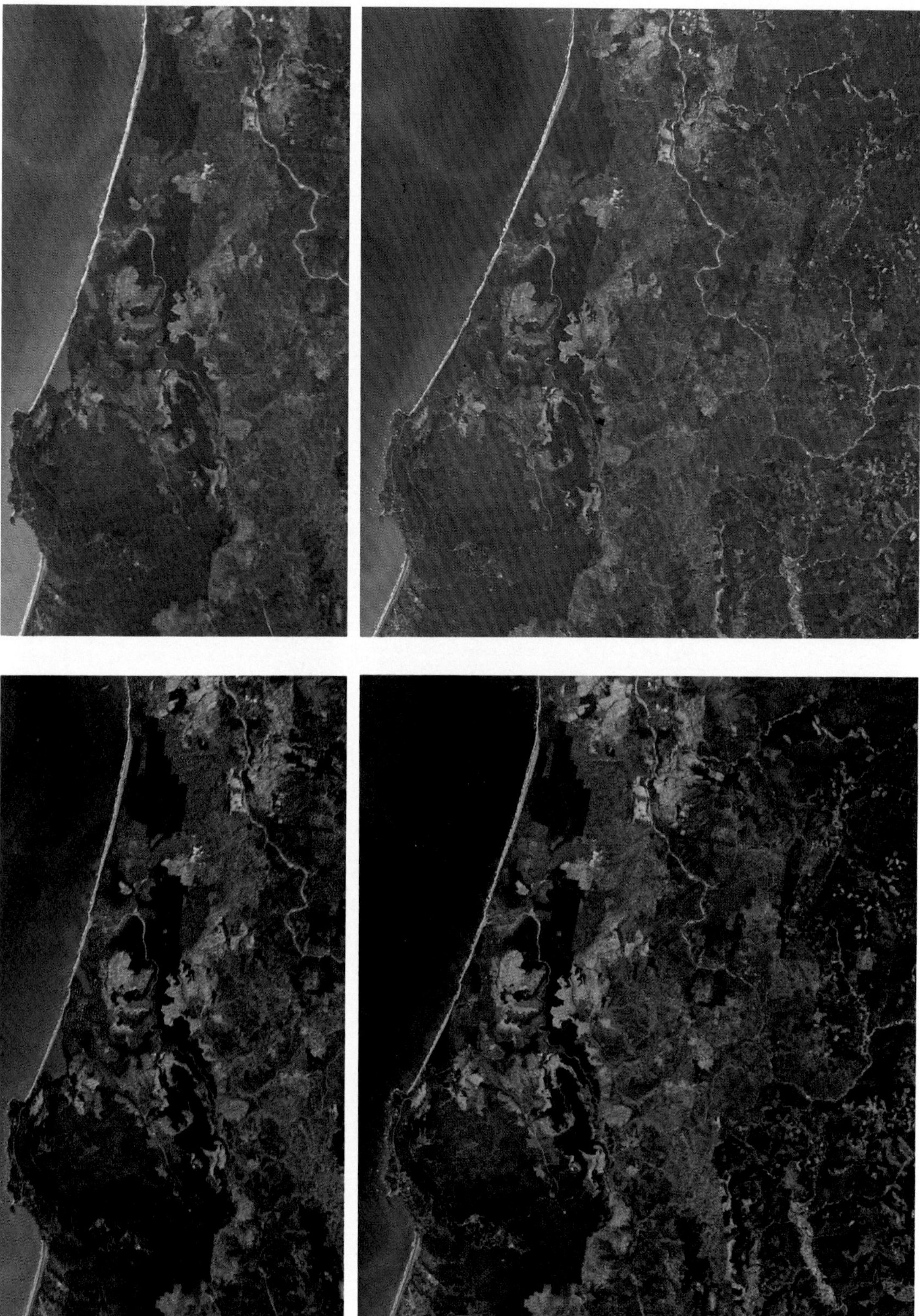

Fig. 32-77. These two matching space-photo stereograms of Redwood National Park and its environs were taken by Skylab astronauts in June, 1973 from an altitude of 270 miles and are reproduced here at a scale of approximately 1/500,000. All of the virgin stands of Coast Redwood trees (*Sequoia sempervirens*) within this scene appear darker than other vegetation on the *top* (ektachrome) stereogram and even more distinctly so on the *bottom* (color infrared) stereogram. Figures 32-77 through 32-88 all deal with portions of the area shown here. Collectively, these photos emphasize the extent to which various kinds of imagery can highlight specific engineering problems and also how they can suggest means by which engineers might best solve those problems.

Fig. 32-79. This color infrared photograph shows a portion of Redwood National Park and its environs as imaged on October 6, 1972 from an altitude of 65,000 feet through the use of a U-2 aircraft, and a precision aerial camera having a 6″ focal length and a frame size of 9″ × 9″. From the engineer's standpoint this photography provides a highly favorable combination of (a) wide area coverage, (b) good spatial resolution and (c) adequate stereoscopic parallax for use in perceiving the steepness of slopes—a factor of great importance in assessing each area's susceptibility to slipouts.

A
B
C
D
A
F
G
H
A
I
B

Fig. 32-85. These nine photos serve to document both the valuable nature of the trees in and around Redwood National Park and their susceptibility to damage from excessive erosion.

A. View into an undisturbed *old-growth* forest area which is representative of much of the Redwood Creek Unit of the Park.

B. View into a *second growth* forest area near the Redwood Creek Unit of the Park. This indicates the ability of some areas, such as certain parts of the proposed buffer strip, to recover following careful logging.

C. Undercutting of the stream bank has caused the fall of several valuable old growth trees, including the one shown here. An extensive crack in the butt log of this tree (where man is standing) has resulted in much of the wood being made unusable for lumber production. If the tree had been felled by man rather than by nature, the direction of fall probably could have been controlled so that little or no damage to the butt log would have resulted.

D. Hummocks of unstable soil are clearly seen here on the grassy slope. Undercutting of the stream bank by Redwood Creek also is apparent. Note the arcuate scarps at the head of landslides, probably triggered by bank erosion.

E. This photo clearly shows the magnitude of gully erosion and debris deposition in a representative tributary to Redwood Creek.

F. Portion of Redwood Creek showing the contrast between old growth stand conditions on the left and previously cutover lands on the right and below "A". Also note large amounts of sand and gravel in stream bed and undercutting of the stream bank at many points.

G. Stand of old growth trees in the Park along Redwood Creek. Note how gravel and silt deposits raise and broaden the stream bed channel. This in turn causes an undercutting of the toe of the slope which borders the stream, thereby threatening old growth forests along the creek. See also Figure H.

H. Same bend of Redwood Creek as shown in Figure G. It is apparent from a study of the banks of the stream that, during peak runoff, the waters of Redwood Creek swirl around the roots of some of the world's largest trees and also deposit large amounts of sediment at the base of these trees. It is important to indicate that mostly fine sediment is deposited within the grove and that this has been going on for thousands of years. This is one of the most dramatic examples available of the threat posed to the big trees by stream channel aggradation.

I. Oblique photo showing mouth of Tom McDonald Creek at A and old tractor logging area at B. Big Tree Grove and Redwood Creek are in left of photo.

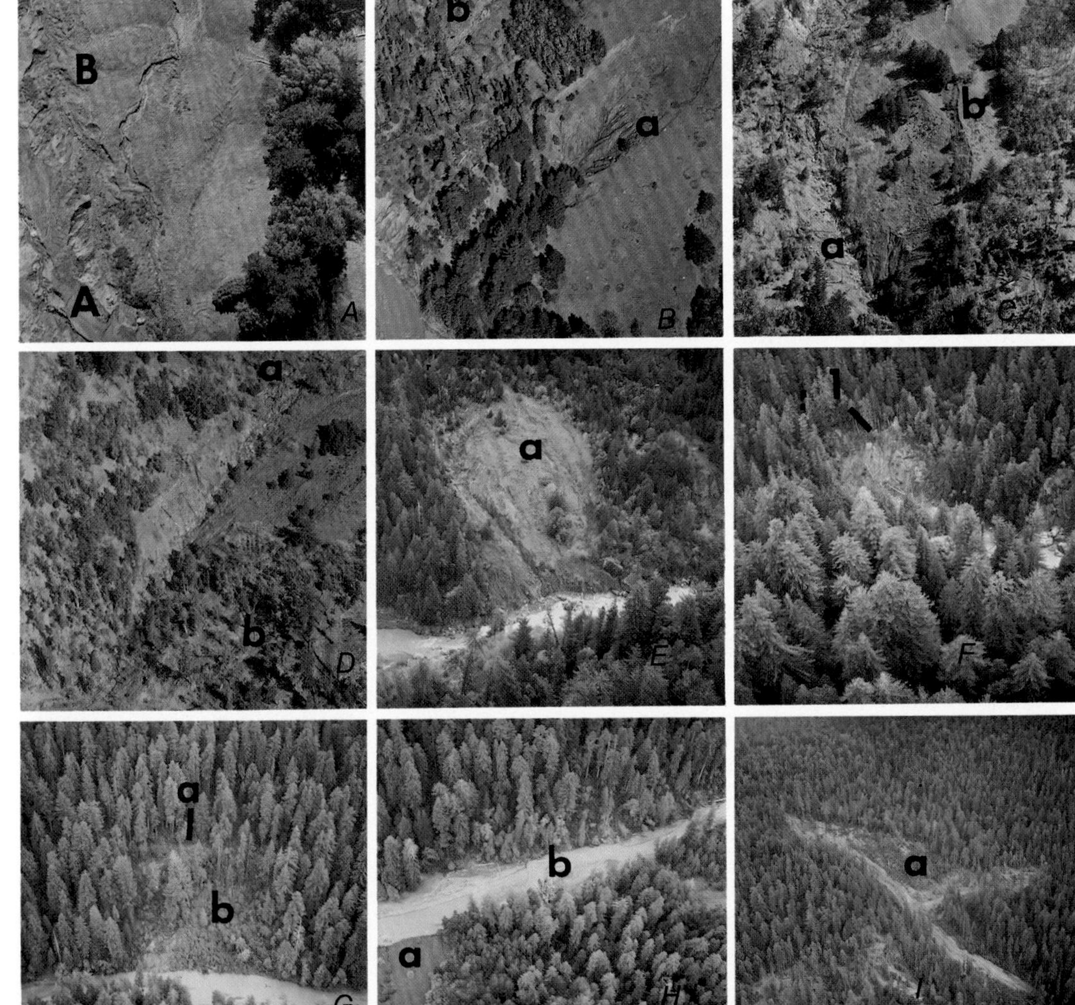

B
A
A
b
a
B
b
a
C
a
b
D
a
E
F
a
b
G
b
a
H
a
I

Fig. 32-86. These low altitude aerial oblique photos document the fact that areas in and around Redwood National Park are highly susceptible to slipout and erosion problems, even in the absence of any significant disturbance by man.

A. This photo shows massive soil movements and gully erosion patterns along slope A-B.

B. Note at "a" the headwall of a recent land slump and subsequent surface erosion of exposed soil. Note also that the slope from "a" to "b" is the upper limit of an unstable slope near the stream channel of Redwood Creek (lower left corner of the photo). There is accelerated erosion on the surface and many tree stems are seen to be leaning at various angles indicating an unstable slope condition. Also note cracks showing lateral margins and arcuate crown of a larger active slump upslope from "a".

C. Another example of soil erosion and massive soil movement on an unstable slope. Note the very deep gully at "a" and the headwall of soil movement from right to left and downslope at "b".

D. Another view of the same area. Note at "a" the top of the ravine which shows extreme cases of gully erosion. At "b" the presence of many leaning trees indicates existing soil movements below the ground surface.

E. A streamside prairie on Redwood Creek just north of the south boundary of the Park. Note here how the bank is being undercut by Redwood Creek. On the slope at "a" are many deep gullies and evidence of surface erosion. Across the creek in the immediate foreground notice many leaning trees, a phenomenon indicative of subsurface soil movements which may contribute to a major slide in this area of the Park, regardless of whether the adjacent area is being logged.

F. At "1" is the headwall of a land slump which could have been caused by wearing away of the toe of the slope adjacent to the stream bed. As with most other examples in this series, serious erosion problems are seen to exist here independent of whether logging is occurring in the area.

G. At "a" note the exposed line of soil which marks the headwall of a small land slide which is occurring even in the absence of logging. At "b" the land mass has moved downslope towards the stream. Note the jumbled position of tree stems on the slope which indicates movement of the soil.

H. At "a" is an unstable prairie slope adjacent to Redwood Creek just south of the Park boundary. At "b" there is a log jam in the channel of Redwood Creek, which diverts the water course, during periods of high runoff, into the opposite bank. Such diversion causes bank erosion which can be clearly seen on this photo.

I. Although a portion of the area shown here recently was subjected to selective cutting, virtually all of the slipout and associated erosion that is centered around "a" occurred before men had disturbed the area significantly.

b
a
A
a
B
a
C
D
b
a
E
b
a
F
a
G
a
H
I

Fig. 32-87. Evidence that man's engineering activities are likely to aggravate erosion problems in and around Redwood National Park can be dramatically seen in this set of low altitude oblique aerial photos.

A. Road failure at "a" has resulted in gravel deposits being formed in the stream channel at "b". Increases in stream velocity caused by water which must go around this gravel deposit may undercut the bank across from "b", especially at times of peak runoff.

B. Area at "a" shows massive slope movement and subsequent surface erosion. Note deep gullies, jackstrawed trees and a road which has been so completely washed out as to be virtually irreparable.

C. Debris accumulating in stream channels will collect and cause log jams such as this road-crossing at "a". One can also see an impressive amount of stream bank erosion and gullying of roadcuts and fills, below "a" on this photo.

D. Ground photo of tractor logging. Note skid trails up and down the slope and islands of relatively undisturbed soil where clusters of stumps are located.

E. This is an area several miles upstream from the Park. Note that a bridge has washed out at "a". At "b" a large mass of unstable soil is seen to be moving into Redwood Creek and is being carried downstream. Also note how clearcut logging has gone on adjacent to the channel of Redwood Creek providing the opportunity for logs and slash to be introduced directly into the channel of Redwood Creek.

F. The unstable slope of a prairie along Redwood Creek can be seen to be moving downhill betwen "a" and "b". Soil movement threatens the road at "a". It also threatens to contribute much soil and rock material and other debris which can be carried into the Park itself by Redwood Creek during periods of high water. For example, note the failure of the fill along the lower road.

G. The lower right half of the area shown in this low altitude oblique aerial photo has been clearcut, greatly increasing the prospects of soil erosion as compared with the remaining area of uncut timber.

H. Major highway upstream from the Park boundary. The cut slope at "a" is nearing failure causing a potentially hazardous situation to traffic on the highway.

I. This photo was taken looking across a large fill on the downslope of a logging road where extreme surface erosion of material is taking place.

a
A
B
1
a
C
a
D
b
E
F
2
1
G
H
I

Fig. 32-88. Possible engineering-related measures that can be taken in and around Redwood National Park, and the value of low-altitude aerial oblique photos for documenting such measures are illustrated in this figure.

A. Selective logging area at "a" and uncut buffer at "b" adjacent to Redwood Creek. Both measures are designed to minimize slipout and erosion problems. This area is immediately upstream from the Park.

B. A look into a timber stand following selective logging of the area. Some silviculturists believe that the leaving of this level of stand density after logging usually is preferable to clear-cutting and may be more aesthetically palatable to critics of timber harvesting in the Redwood region.

C. Ground photo of a heavily cut area at edge of Redwood National Park. Although there is a large debris cone that was deposited on the road at "a", where yellow truck is parked, the combination of gentle slopes and the leaving of some trees to form a residual stand may minimize further erosion and slipout problems in this area.

D. As can be seen from this photo, several acres of nearly flat land are being cleared for a log deck and landing. If this engineering activity were undertaken in areas of steeper terrain there would be much more serious erosion problems. In the background, tractor logging took place a few years ago. To the right of "a" is an older logged area which is regenerating mostly to brush and some mixed conifer species.

E. Large scale color photo which indicates the more highly erodible nature of the soils with which engineers must contend in this area. Note many gullys, gravel deposits and cuts in the road bed surface at b and between a and b, even though the slopes here are only moderately steep.

F. Large scale color infrared photo of same area as in Figure E. A careful comparative study of these photos will show that color infrared photography does help to identify significant features including sources of underground water (springs) which contribute to the lack of soil stability in this region. In this photo example the downslope direction is upward on the page.

G. The area shown in the foreground was harvested and tractor-logged in 1956 and now is covered, for the most part, with a dense stand of brush and small hardwood trees. On close inspection some young conifers can be seen starting to rise above the alders. At area 2 in the background, is evidence of more recent tractor logging. Large old growth redwood trees along Redwood Creek and inside the present Park boundary can be seen in the bottom quarter of this photo.

H. Plantation of seedlings along banks of a cable logging area. Even though this area is in a region of unstable soils, this practice tends to reduce large amounts of soil movement over the surface.

I. Same area as Figure H. Color infrared photographs such as this one can be useful in determining seedling survival rates in plantations. Dead or dying trees would appear green in color on these photos against the red color of the healthy vegetation. The early identification of such trees can greatly facilitate their timely replacement.

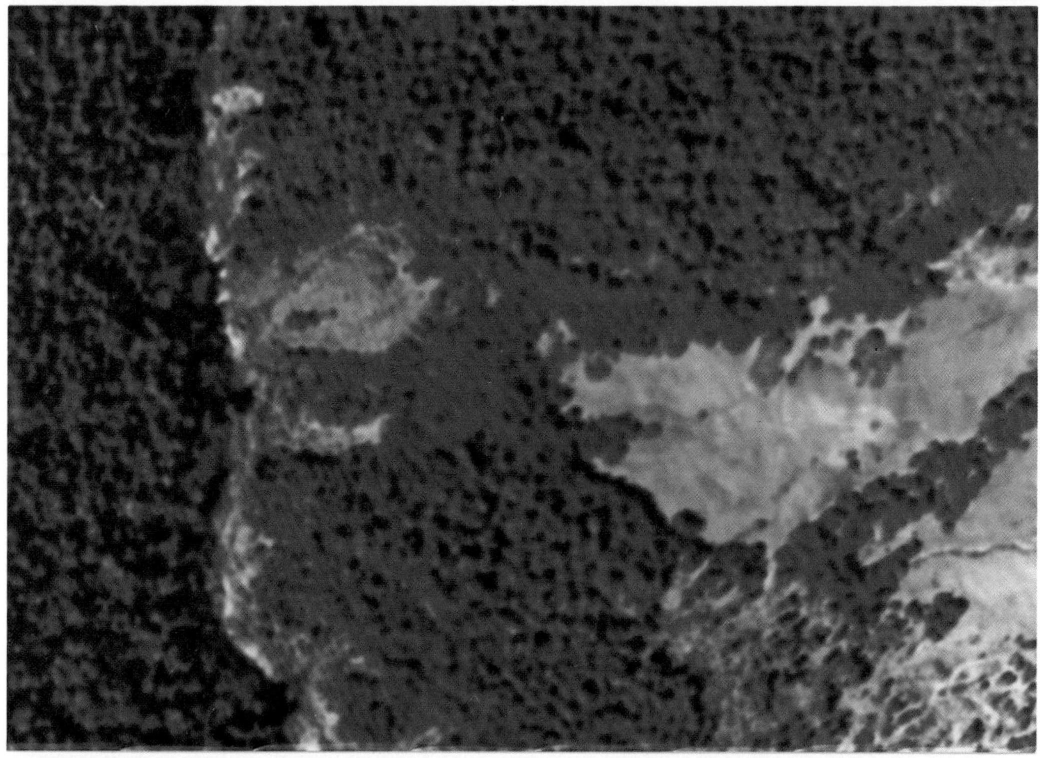

Fig. 32-80. These greatly enlarged reproductions show part of a color infrared photograph as imaged on April 4, 1973 from an altitude of 65,000 feet through the use of a U-2 aircraft and a large format camera having a 24-inch focal length and a frame size of 9″ × 18″. The areas shown correspond to portions of the lower right quadrant of Figure 32-79. Prairie and forest areas presently undergoing soil mass movement are clearly seen as are forested areas which have experienced sliding in the recent past, one of which also appears in Figure 32-82. Note how the increased spatial resolution of this image type allows improved detection and identification of recently-stabilized (hardwood-dominated) slide areas. Also among the features which are more readily identifiable are: active slumping in prairies, slides adjacent to road cuts, stand density, water resources and characteristic of the road network. The original transparency of this photography, enlarged to a scale of 1:15,840, is a very useful map base on which to display management problems and resource features.

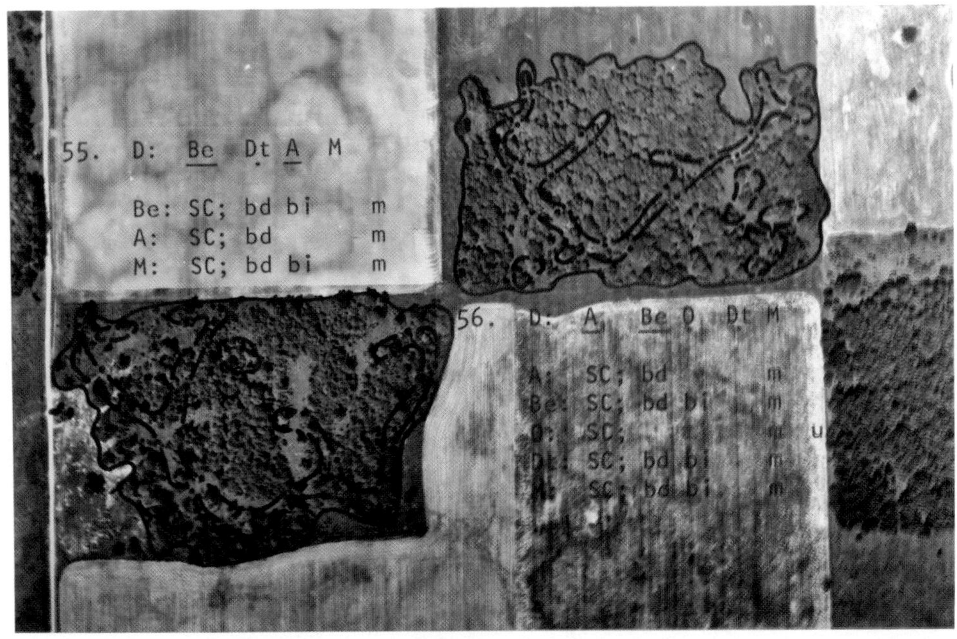

LEGEND

forest type
stand no.
species:
dominant —
co-dominant •
>symbol

55 D : Be Dt A M

stressed species
Be : SC ; bd bi m
A : SC ; bd m
M : SC ; bd bi m

probable cause of stress:

bd biotic, disease
bi biotic, insect
a abiotic
m management
u unknown

stress distribution:

I isolated •
S scattered ·—·
C concentrated ~~
>symbol

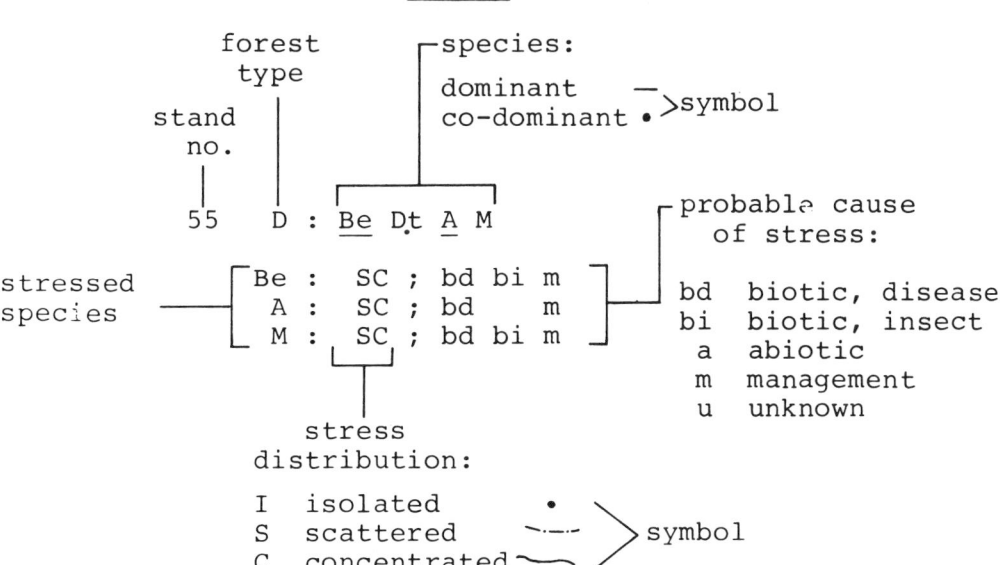

Color Fig. 32-100. Natural vegetation—stress condition parameters, showing the amount of information that is obtained by remote sensing techniques. (Courtesy Ontario Ministry of Transportation and Communication.)

Color Fig. 32-101. Artist enhanced photomontage of proposed I-66 section in Rosslyn, Virginia, adjacent to Washington, D.C. (Courtesy Eastern District Federal Office, Federal Highway Administration.)

Fig. 33-4. Ektachrome infrared photograph of an agricultural area in the Lower Rio Grande Valley, Texas.

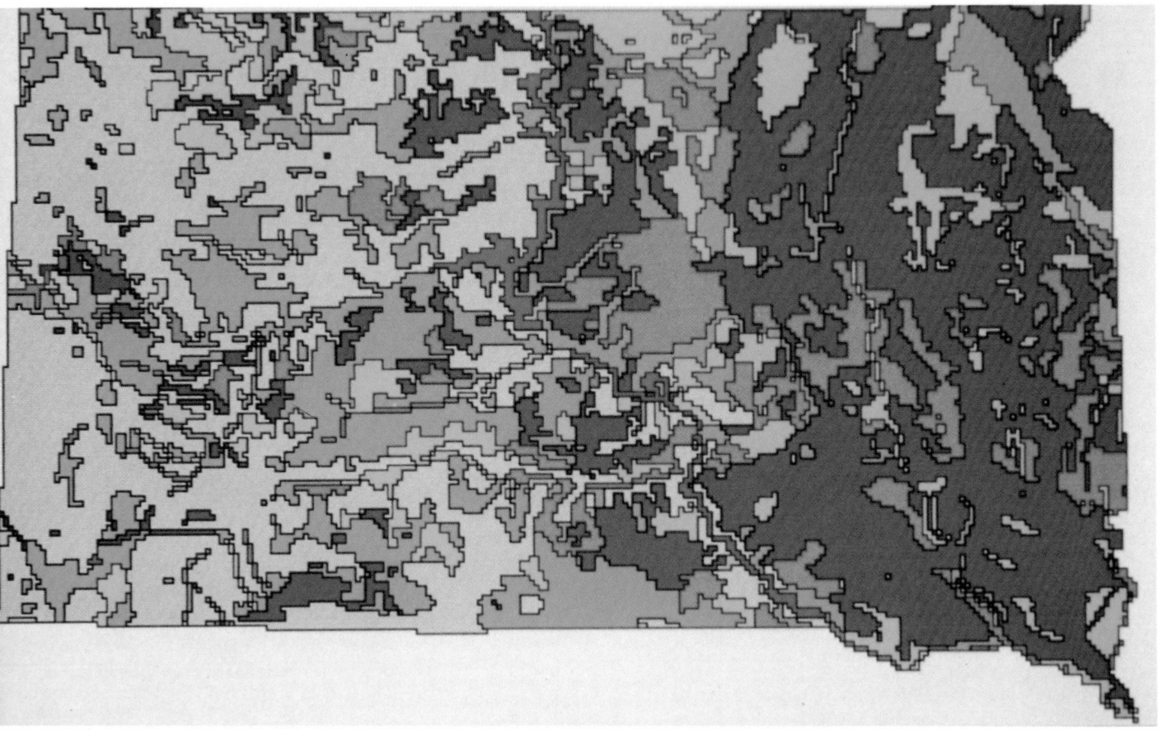

Fig. 33-20. Cropland use intensity from Landsat. Red color is 70–100% cultivated. Pink 50–70%, Light Blue 30–50%, Green 5–30%, and Yellow <5%. Water is Purple. This figure represents approximately 20 million hectares in South Dakota.

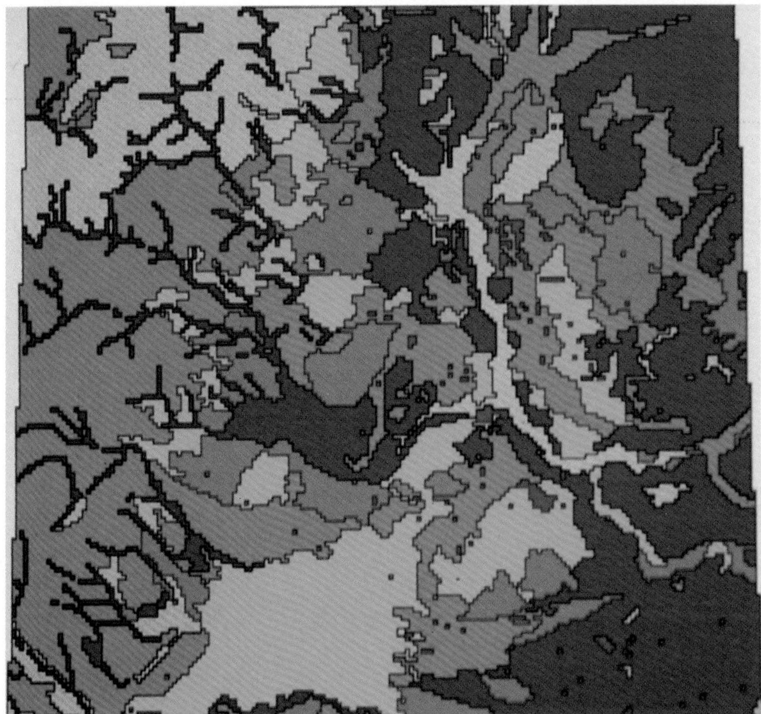

Fig. 33-21. Cropland use intensity from Landsat. Red color is 70–100% cultivated. Pink 50–70%, Light Blue 30–50%, Green 5–30%, and Yellow <5% cultivated. Water is Purple. This figure represents approximately 25 million hectares in the Manchurian Plain of China.

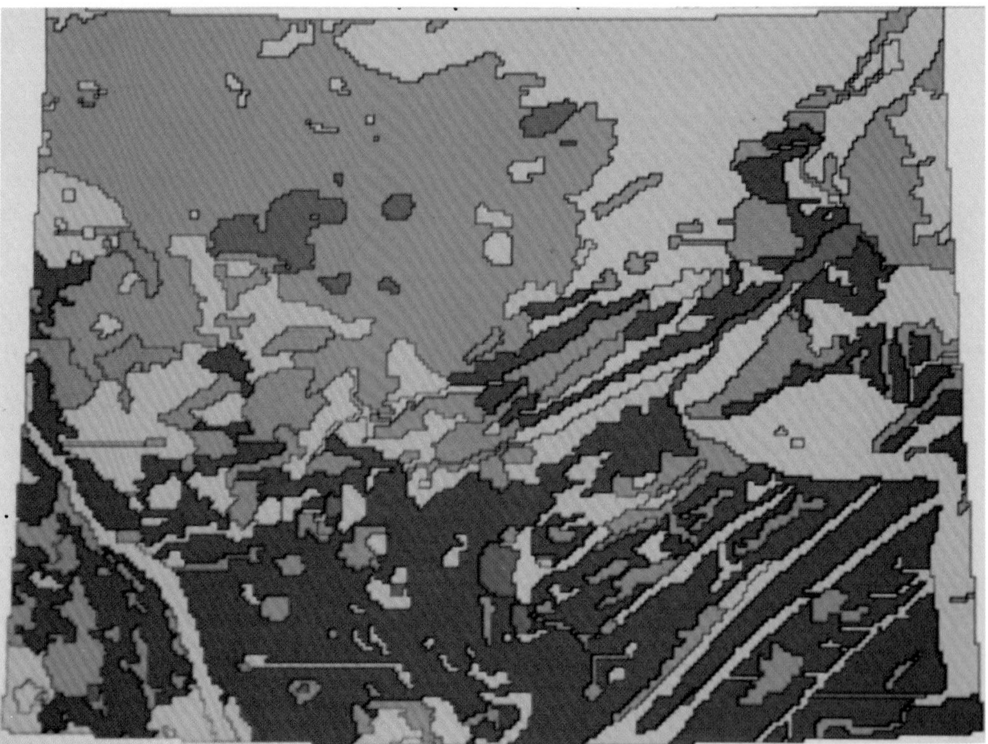

Fig. 33-22. Cropland use intensity from Landsat. Red color is 70–100% cultivated. Pink 50–70%, Light Blue 30–50%, Green 5–30%, and Yellow <5% cultivated. Water is Purple. This figure represents approximately 25 million hectares in the West Siberian Plain of the U.S.S.R.

Fig. 33-33. Color infrared positive print of an area at Nixon, Texas. Arrow points to the characteristic ''pinkish'' image of silverleaf sunflower plants as compared with the adjacent darker magenta image of another plant sphere.

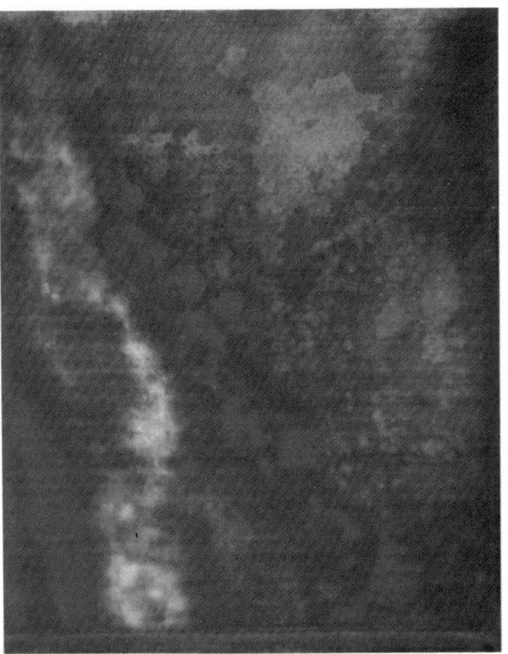

Fig. 33-39. Positive print of a cotton field showing magenta-appearing healthy cotton, a saline area on the left, and cotton root rot areas on the right. Photo by Ronald L. Bowen.

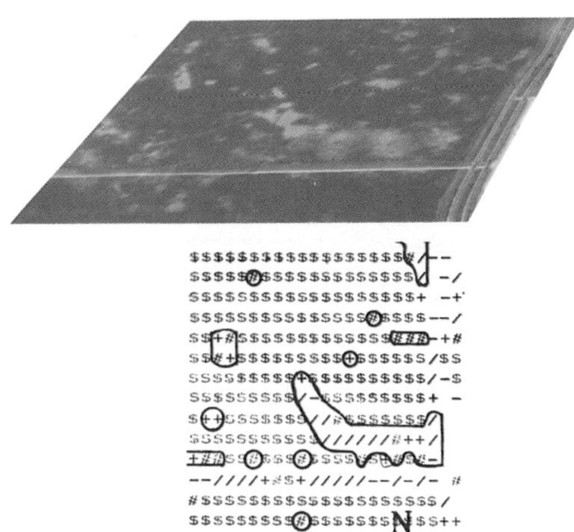

Fig. 33-41. Upper photo is a positive print of an infrared transparency showing areas of white-appearing chlorotic sorghum and magenta appearing normal (N) sorghum. Lower photo is a printout of Landsat-1 band 5 data; chlorotic areas corresponding to those in the upper photo are encircled. Digital counts corresponding to the printout symbols are: \$ = 30 to 33, # = 33 to 36, + = 36 to 39, / = 39 to 42, − = 42 to 45, and blank = ≥45.

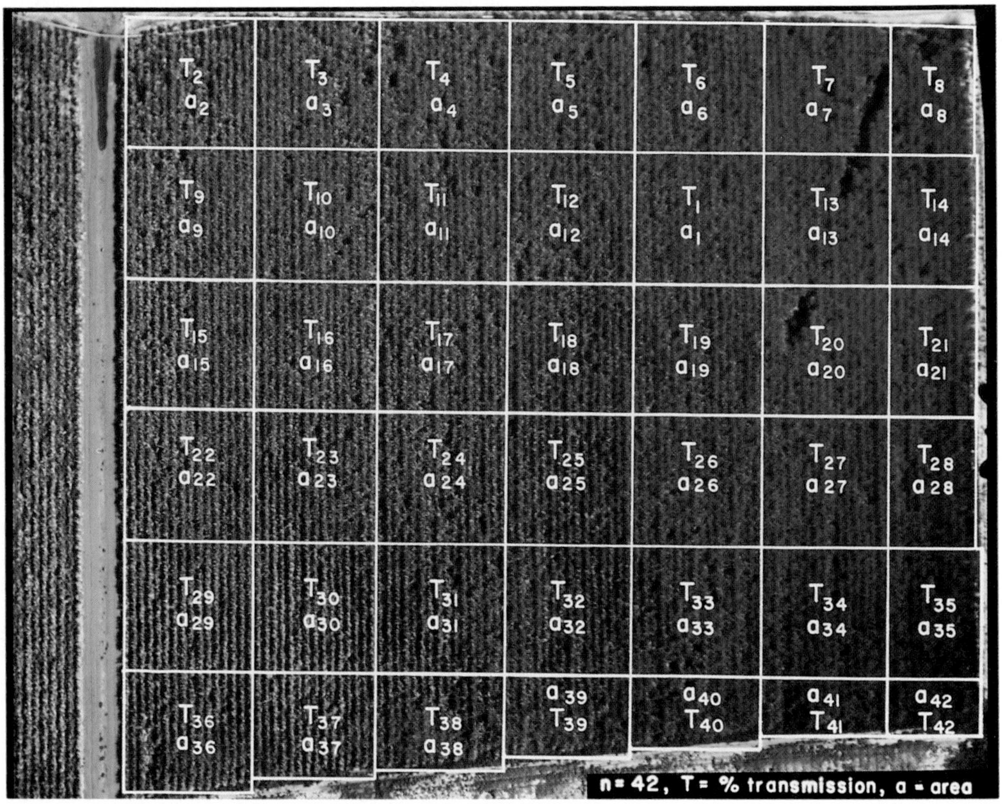

Fig. 33-40. A positive print of an infrared transparency showing a sugarcane field which has been subdivided into areas differing in percent vegetative cover.

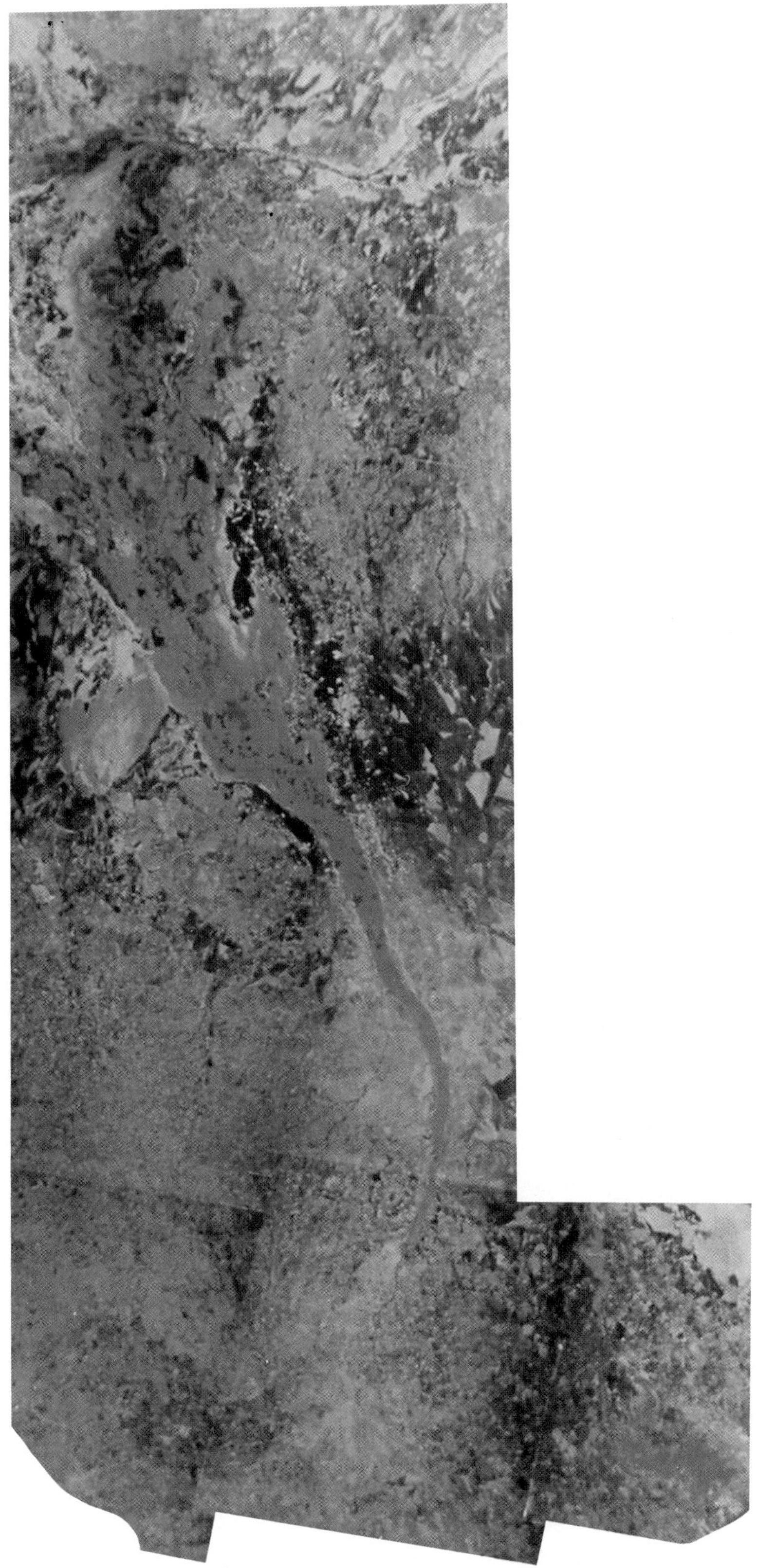

Fig. 33-49. CIR mosaic of the Sudd region of the Upper Nile River in Sudan.

Fig. 33-68. Pseudo-colored thermal imagery of the Phoenix test site acquired at 5:53 A.M., April 1, 1976. The six differentially-irrigated wheat plots appear in the center and are identified as 1 through 6 going from bottom to top, which is west to east. Bare soil adjoins these plots to the east and alfalfa to the west. South of the bare soil is another field of alfalfa; and west of it is another field of wheat. Red = 6−7°C, Orange = 5−6°C, Violet = 4−5°C, Yellow = 3−4°C, Green = 2−3°C, Lt. Blue = 1−2°C, Md. Blue = 0−1°C, Dk. Blue = (−1)−0°C.

Fig. 33-69. Pseudo-colored thermal imagery of the Phoenix test site acquired at 2:06 P.M., April 1, 1976. Red = 40–42°C, Orange = 38–40°C, Violet = 36–38°C, Grey = 34–36°C, Yellow = 32–34°C, Lt. Green = 30–32°C, Aqua = 28–30°C, Dk. Green = 26–28°C, Md. Blue = 24–26°C, Dk. Blue = 22–24°C, White = 20–22°C.

Fig. 33-70. Pseudo-colored imagery of the difference between P.M. and A.M. surface temperature measurements, April 1, 1976, Phoenix. White = 41–43°C, Red = 38–40°C, Orange = 35–37°C, Yellow = 32–34°C, Lt. Green = 29–31°C, Dk. Green = 26–28°C, Lt. Blue = 23–25°C, Dk. Blue = 20–22°C, Violet = 17–19°C, Black = 14–16°C.

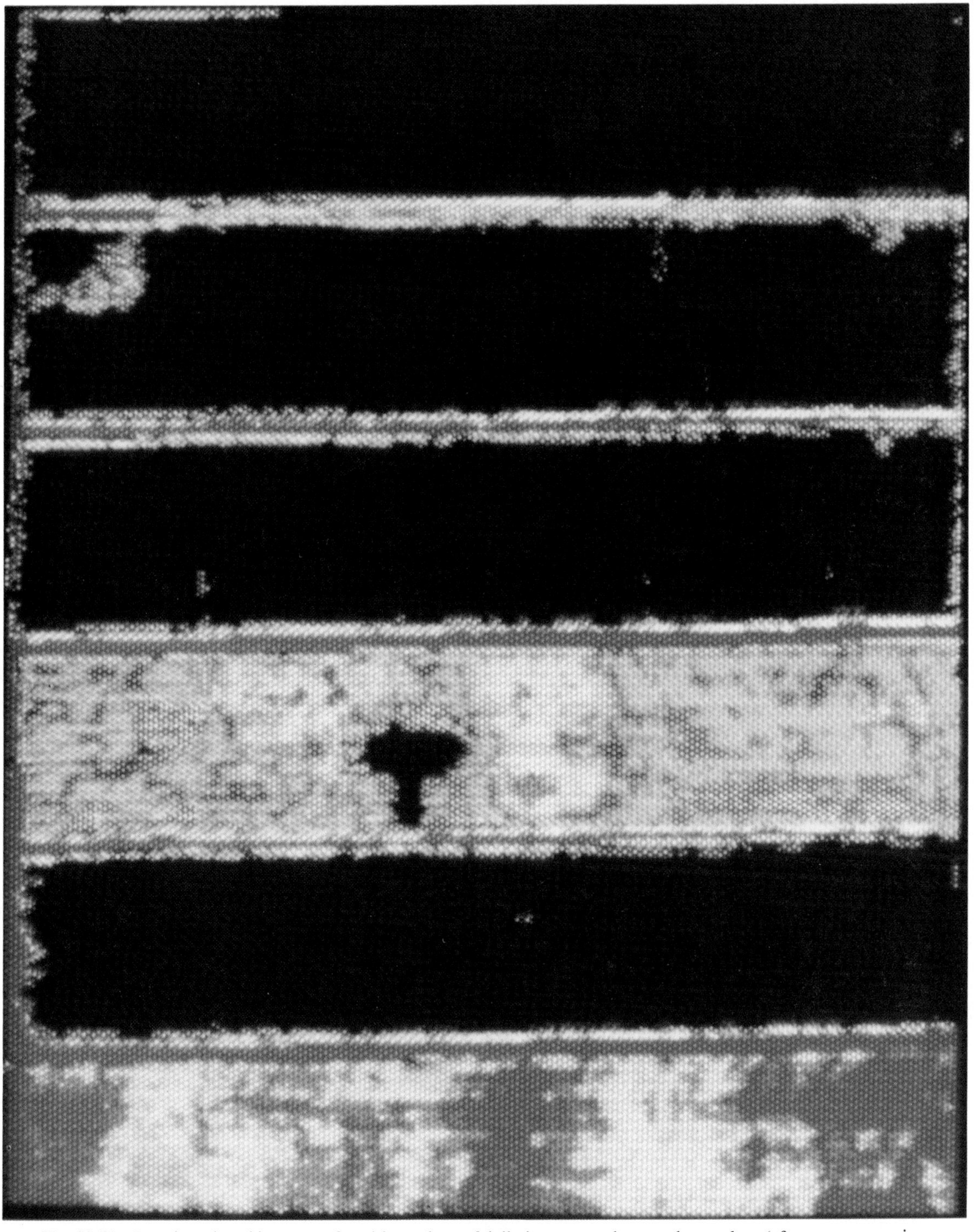

Fig. 33-71. Pseudo-colored imagery of positive values of daily incremental stress degree days (afternoon crop minus air temperatures), April 1, 1976, Phoenix. Red = 8°C, Orange = 7°C, Yellow = 6°C, Lt. Green = 5°C, Dk. Green = 4°C, Lt. Blue = 3°C, Dk. Blue = 2°C, Violet = 1°C, Black = 0°C.

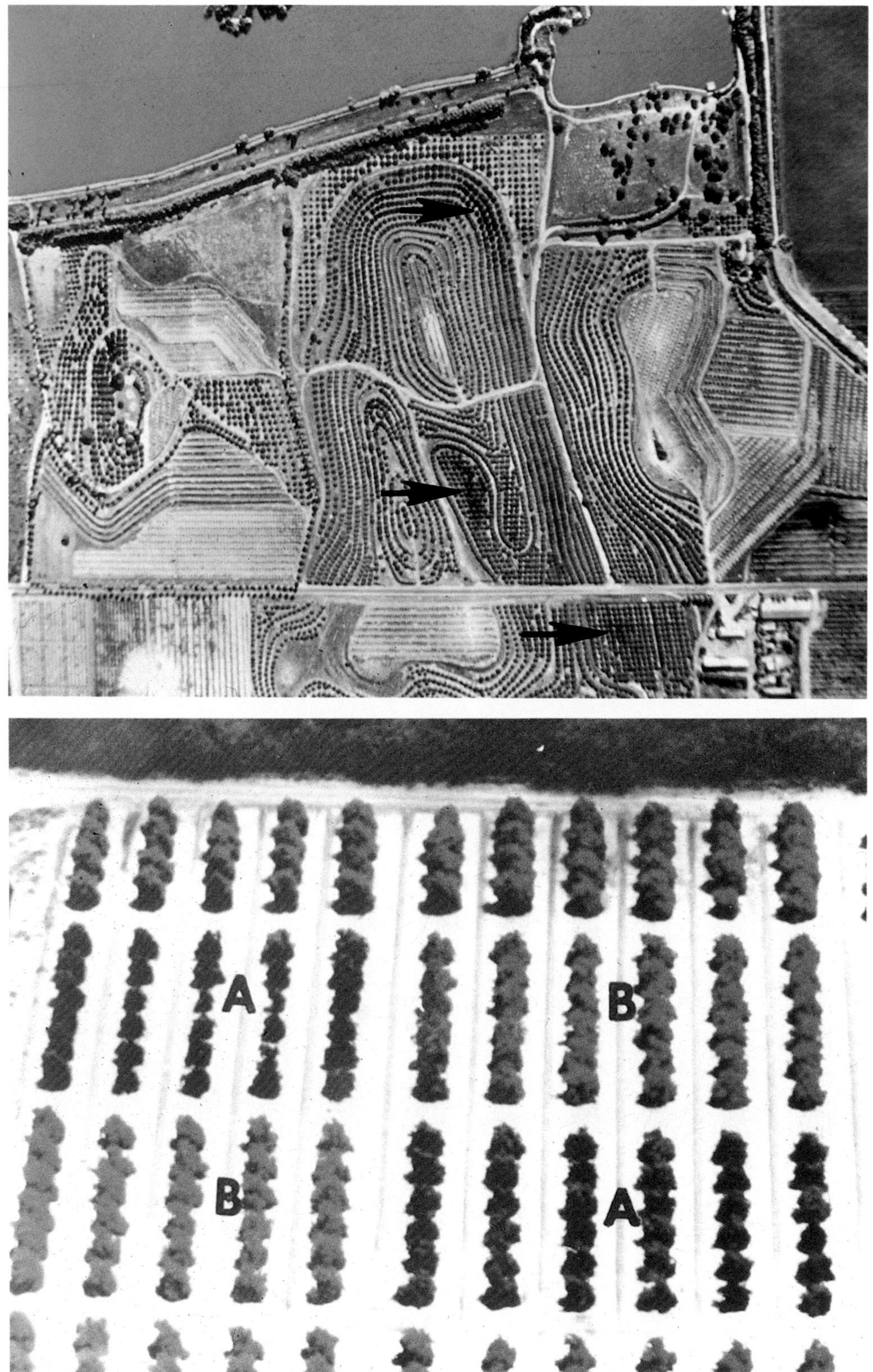

Fig. 33-74. Upper photo (a) shows brown soft scale infested citrus plots at "A" and uninfested plots at "B". (Scale: 1:750). Lower photo (b) shows heavily infested portions of citrus grove. (Scale 1:10,000). Both are CIR photos.

Fig. 33-76. Enclosed area in upper photo (a) shows brown soft scale detectable on citrus trees with less than one percent of their leaves infested. (Scale 1:2,000). Lower photo (b) shows citrus trees in corner of grove that are heavily infested (more than 15 percent of the leaves) with brown soft scale. (Scale; 1:5,000). CIR photo.

Fig. 33-77. Upper picture (a) shows a heavy infestation of citrus blackfly at arrow. (Scale: 1:10,000). Lower picture (b) is of a grove infested with citrus mealy bug showing mottled appearance of sooty-mold deposits. (Scale: 1:800). Both are CIR photos.

Fig. 33-78. Upper photo (a) shows typical flattened circular mound of the harvester ant. Lower photo (b) shows a pasture heavily infested with harvester ants. (Scale 1:5,000).

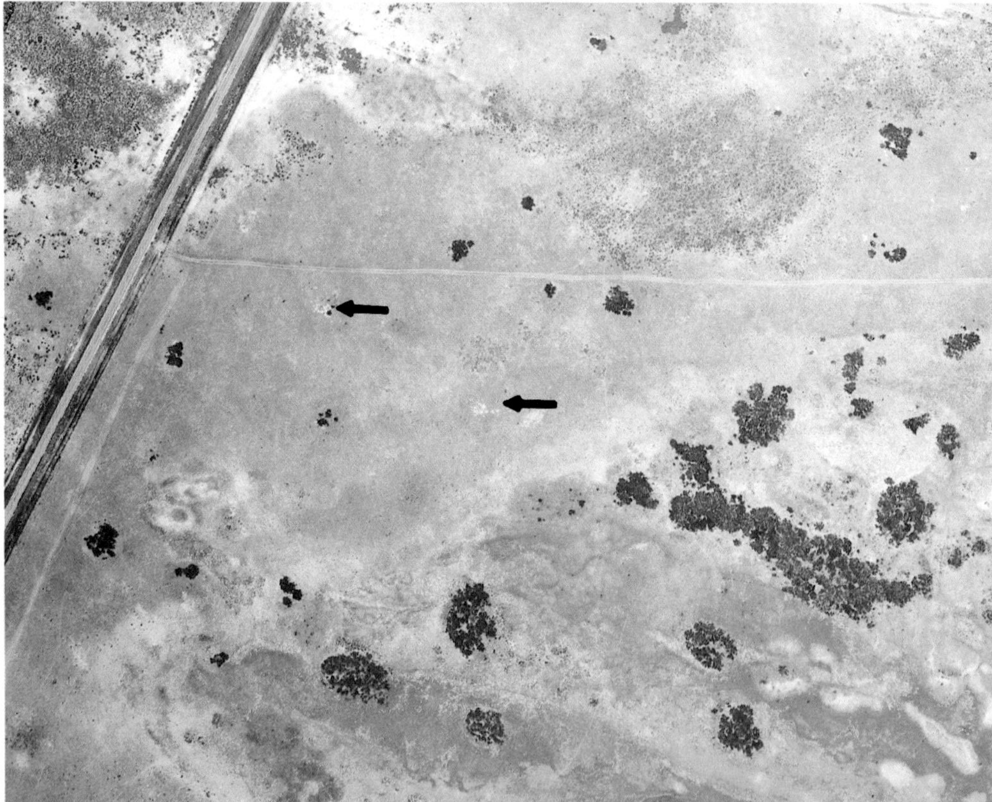

Fig. 33-79. Upper photo (a) is a close-up of a Texas leaf cutting ant mound showing its irregular shape. Lower photo (b) shows irregular shape and light color of Texas leafcutting ant mounds. (Scale: 1:5,000).

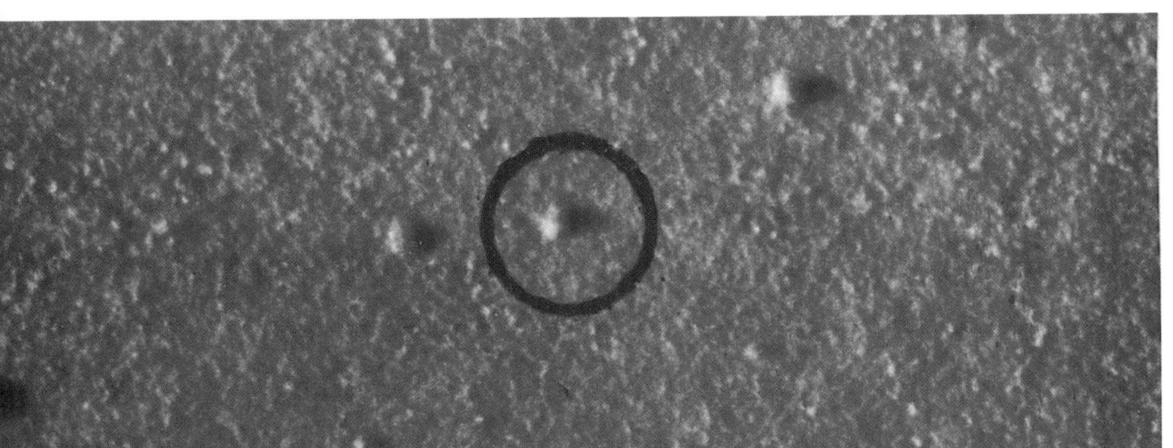

Fig. 33-80. Photo shows an imported fire ant mound photographed during low sun angle to the mound. (Scale 1:3,000).

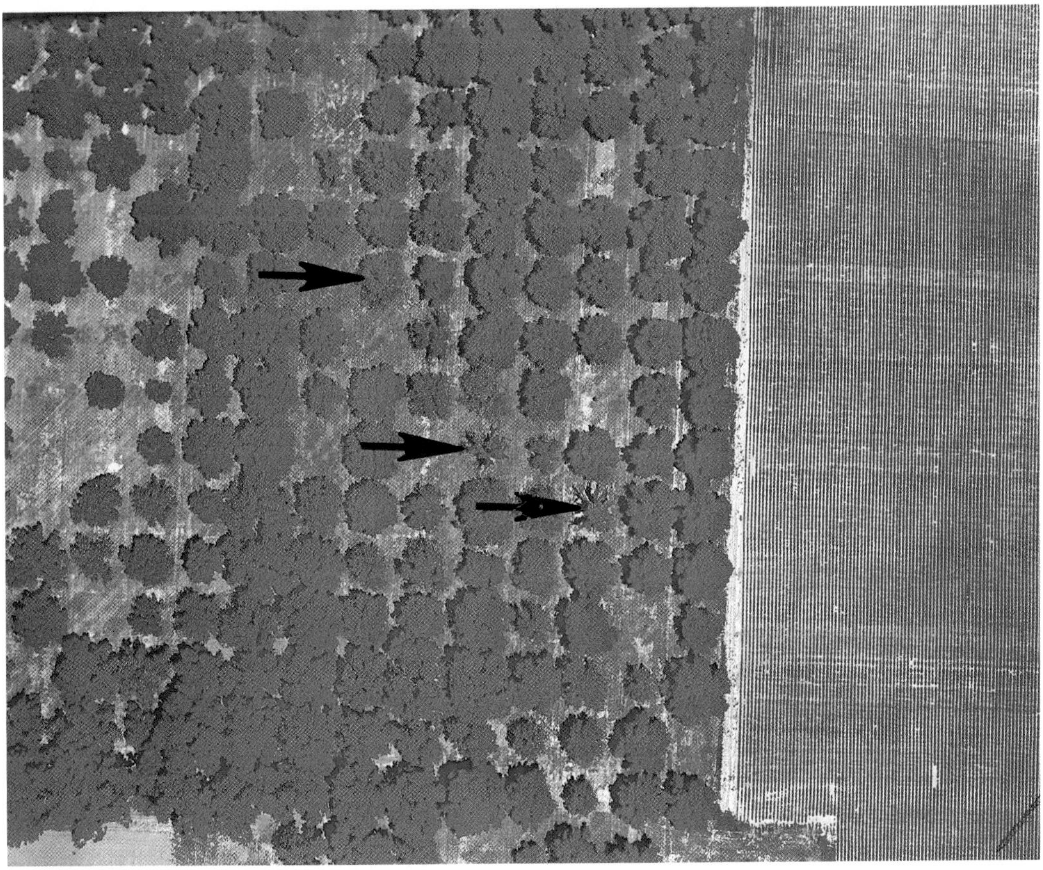

Fig. 33-81. Infrared aerial photograph of pecan orchard containing trees infested with Prionus root borers (at arrow). (Scale: 1:2,500).

Fig. 33-83. Photo shows pecan trees prior to budbreak while all other trees are in full foliage. (Scale: 1:10,000).

Fig. 33-84. Photo shows coffee trees growing beneath larger shade trees. (Scale: 1:10,000).

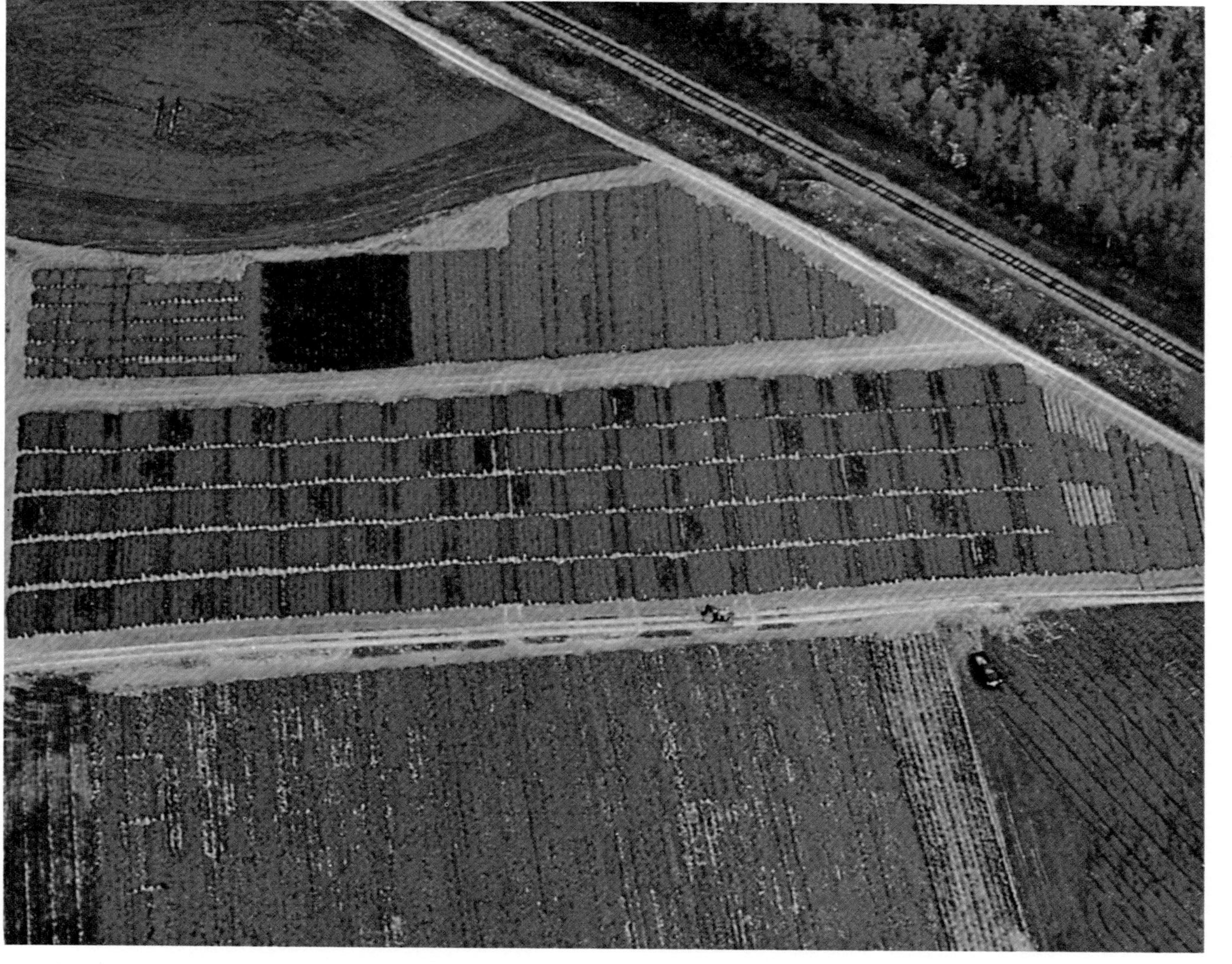

Fig. 33-86. Ektachrome Aero Infrared, type 8443 shows healthy potato plant foliage as red and the diseased leaves as a dark green to black color. This is the same area as that shown in Figure 33-85 and was taken at the same time. (Photo furnished by Prof. G. Cooper and F. Manzer, Depart. of Bot. and Pl. Path., Univ. of Maine.)

Fig. 33-87. Pecan grove showing trees infected with Clitocybe root rot (at arrows). (Scale: 1:2,500).

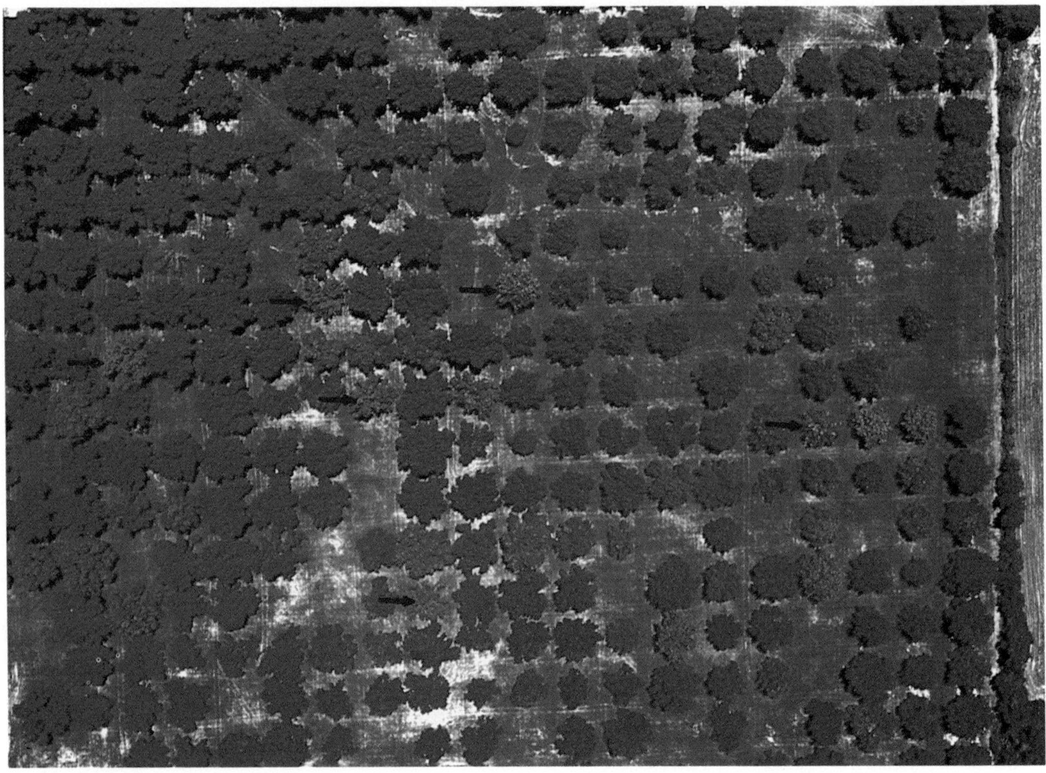

Fig. 33-88. Pecan grove showing trees infected with bunch disease (at arrows). (Scale 1:2,500).

Fig. 33-89. Citrus grove showing trees infested with the burrowing nematode (at arrows).

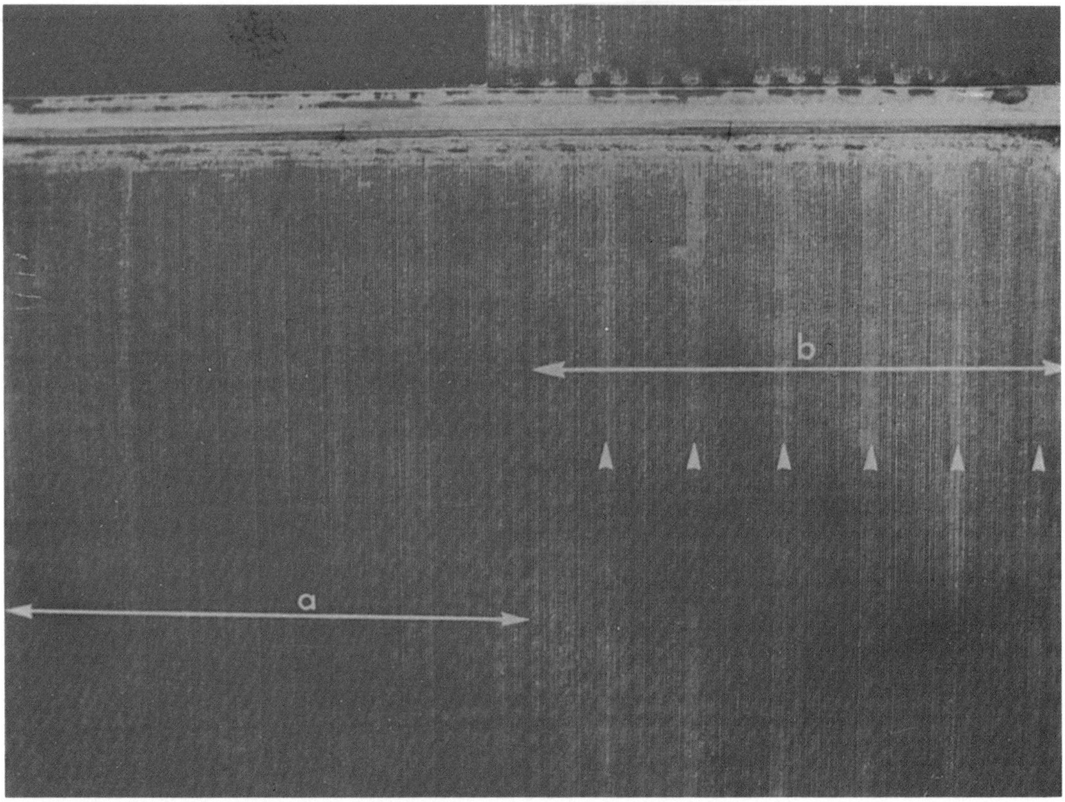

Fig. 33-90. Cotton field showing area rotated to grain sorghum the previous year at "a", and area planted to cotton the previous year at "b" with 24 rows treated with nematocide and 6 rows (at arrows) untreated.

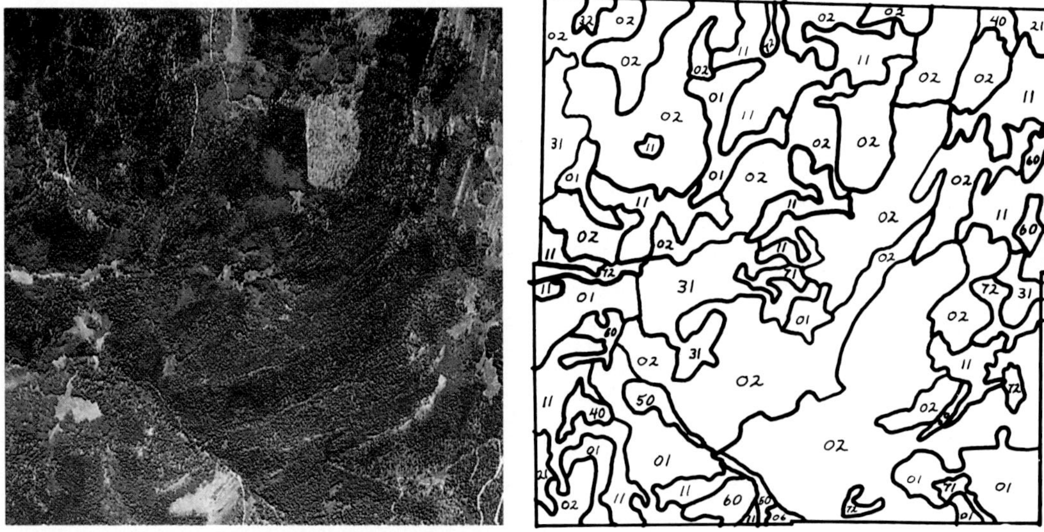

Fig. 34-2. A portion of a false color infrared aerial photograph, original scale 1:24,000, is illustrated on the left. The photograph was taken over Grand County, Colorado on July 7, 1977. A trained image analyst mapped the forest land as seen on this photograph using the classification system shown in Table 34-2. Map results are illustrated on the right. (Photo and map courtesy of the U.S. Forest Service, USDA.)

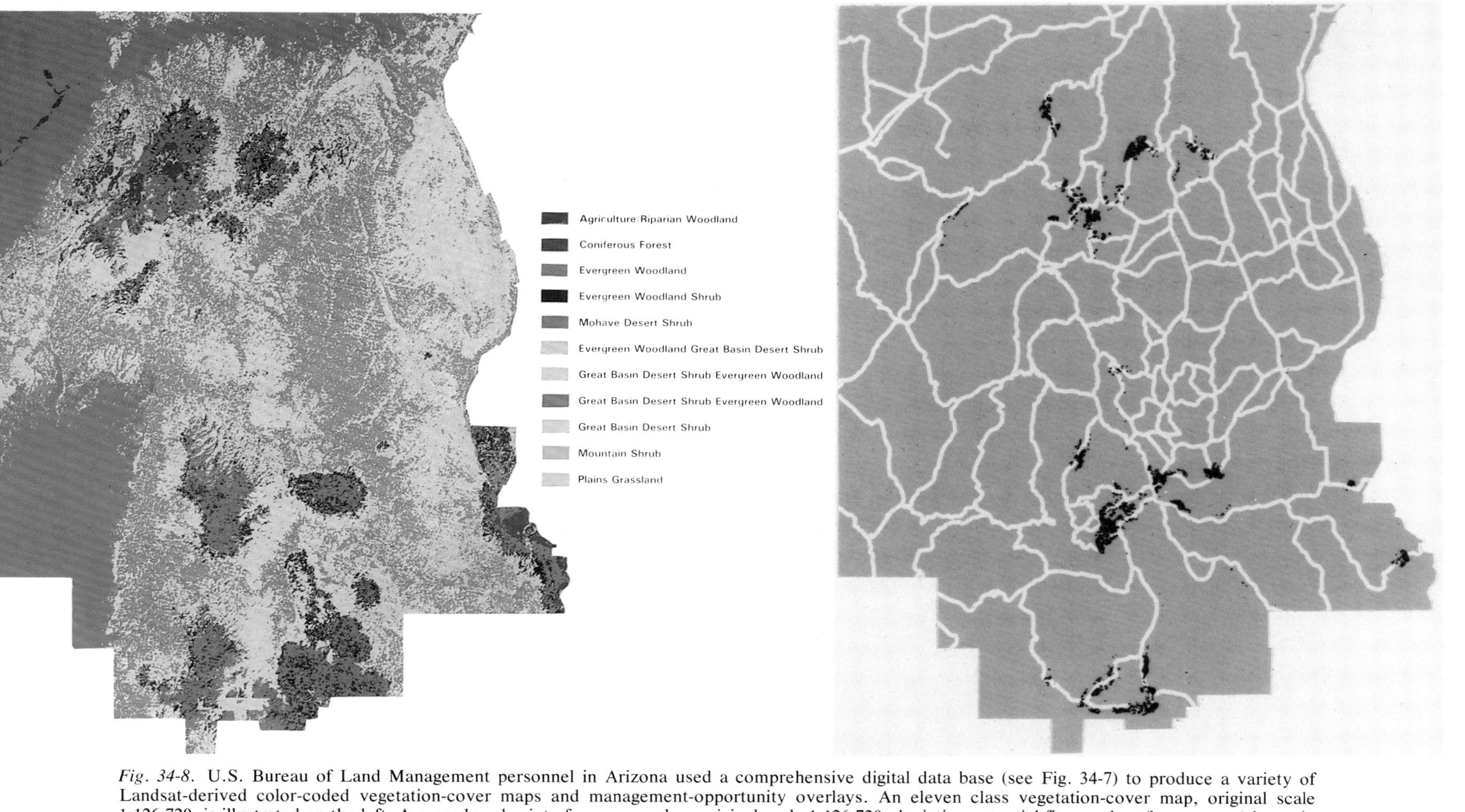

Fig. 34-8. U.S. Bureau of Land Management personnel in Arizona used a comprehensive digital data base (see Fig. 34-7) to produce a variety of Landsat-derived color-coded vegetation-cover maps and management-opportunity overlays. An eleven class vegetation-cover map, original scale 1:126,720, is illustrated on the left. A reproduced print of a map overlay, original scale 1:126,720, depicting potential firewood cut/burn areas (shown in black) near roads (shown in white) is illustrated on the right. The overlay was prepared from a digital data base in which model parameters were Landsat-derived vegetation cover-type (evergreen woodland), elevation (1500 to 1860 meters), slope (0–15%), aspect (variable), and road net (within 400 meters of road). (Map products courtesy of U.S. Bureau of Land Management and U.S. Geological Survey, USDI.)

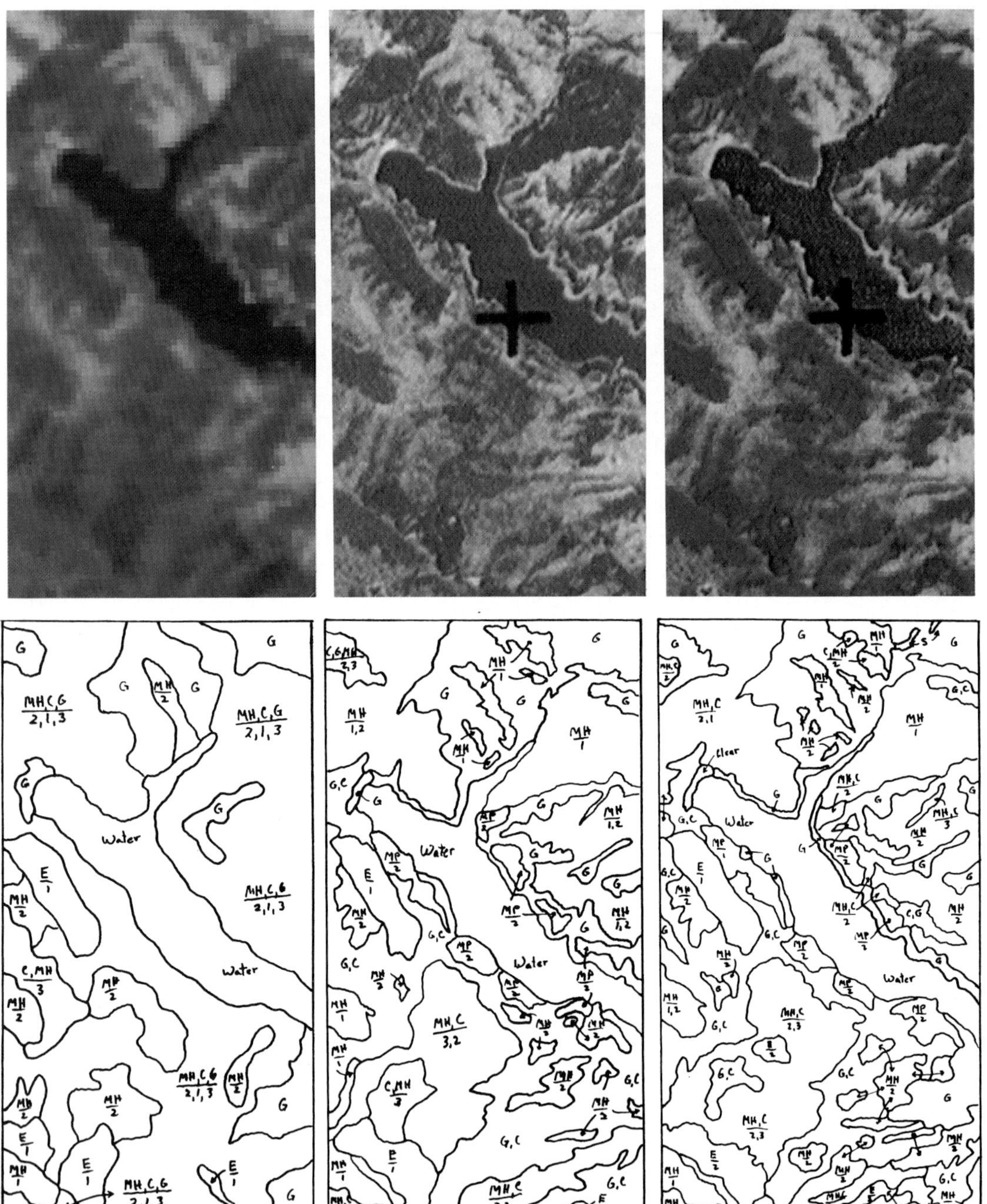

Fig. 34-11. The wildland area adjacent to San Pablo Reservoir, California is shown on Landsat imagery. An MSS color-composite image is at the top left, an RBV black-and-white image is at the top center and an RBV/MSS merged color composite image is at the top right. All three images were photographically enlarged to a scale of approximately 1:47,000 for purposes of analysis. Each map overlay was derived from the image above it and depicts vegetation type (MP = Monterey pine, E = Blue gum eucalyptus, S = Willow, MH = Mixed hardwoods, C = Chaparral, G = Grassland) and vegetation density (1 = Dense (80+%), 2 = Semidense (50–80%), 3 = Open (20–50%), 4 = Very open (5–20%), 5 = Unstocked (<5%). When more than one type class (numerator) or density class (denominator) occurred in a single stand, each was noted by decreasing order of dominance. See text for further explanation. (Photos and maps courtesy of U.S. Geological Survey, USDI.)

Fig. 34-22. Colored equidensity image of a forest surrounded by agricultural land. (Courtesy of the Institut für Forsteinrichtung und Forstliche Bertriebswirtschaft, Freiburg, Germany.)

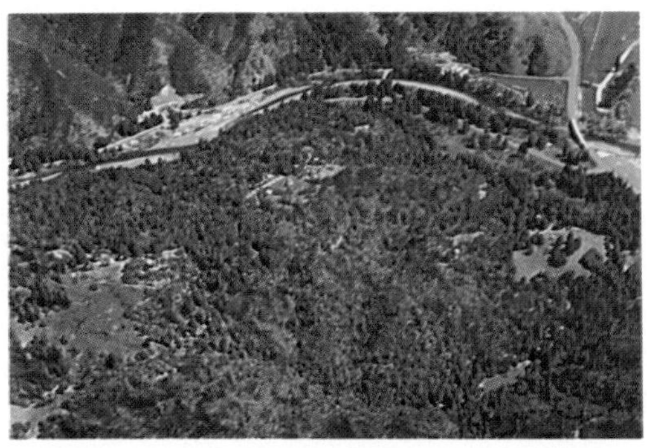

Fig. 34-31. Ektachrome infrared photography illustrating the ease with which fire perimeters (blue-black) and islands of unburned vegetation (red) can be delineated. (Courtesy of the Forestry Remote Sensing Laboratory, University of California.)

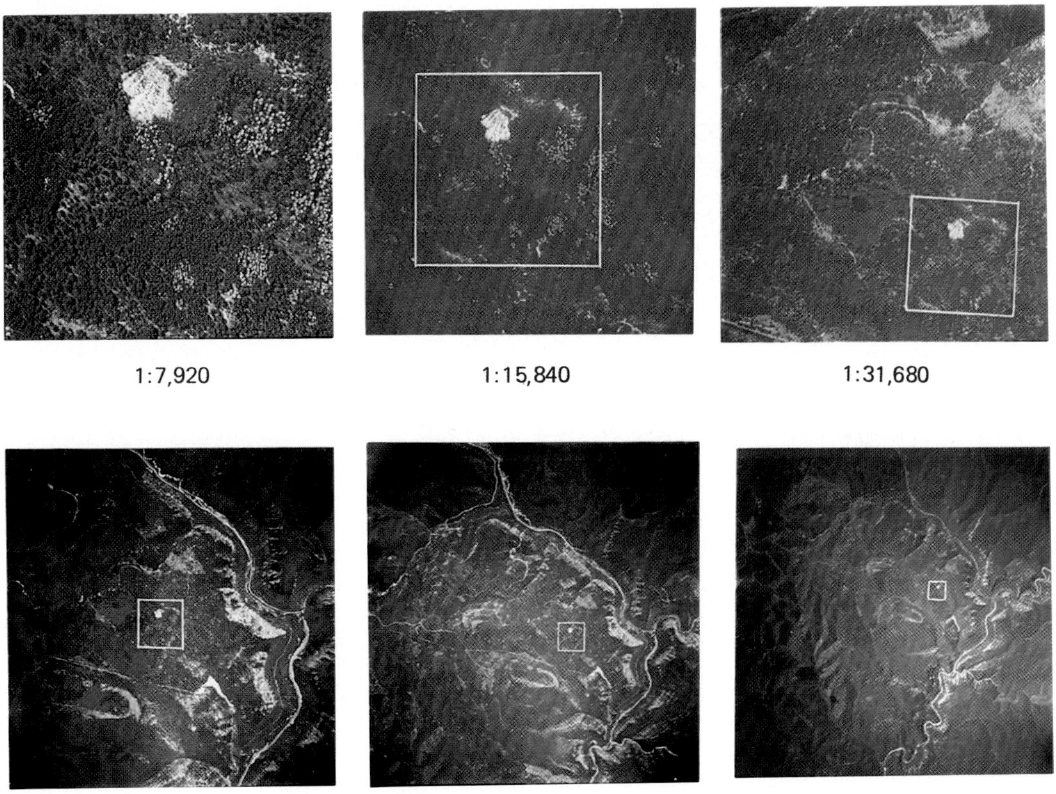

1:7,920 1:15,840 1:31,680

1:63,360 1:120,000 1:174,000

Fig. 34-32. The effect of scale on the detection and mapping of mountain pine beetle infestations in Ponderosa pine near Lead, South Dakota. CIR aerial transparencies taken with 70-mm J. A. Maurer KB-8A cameras with 150mm Schneider Xenotar f 2.8 lens for larger scales and 38 mm Zeiss Biogen lens for smaller scales. The white squares etched on the small-scale photos refer to the coverage of the 1:7,920 scale. (Photos courtesy of Forest Service, USDA.) (For Color Figure 34-27, see next page.)

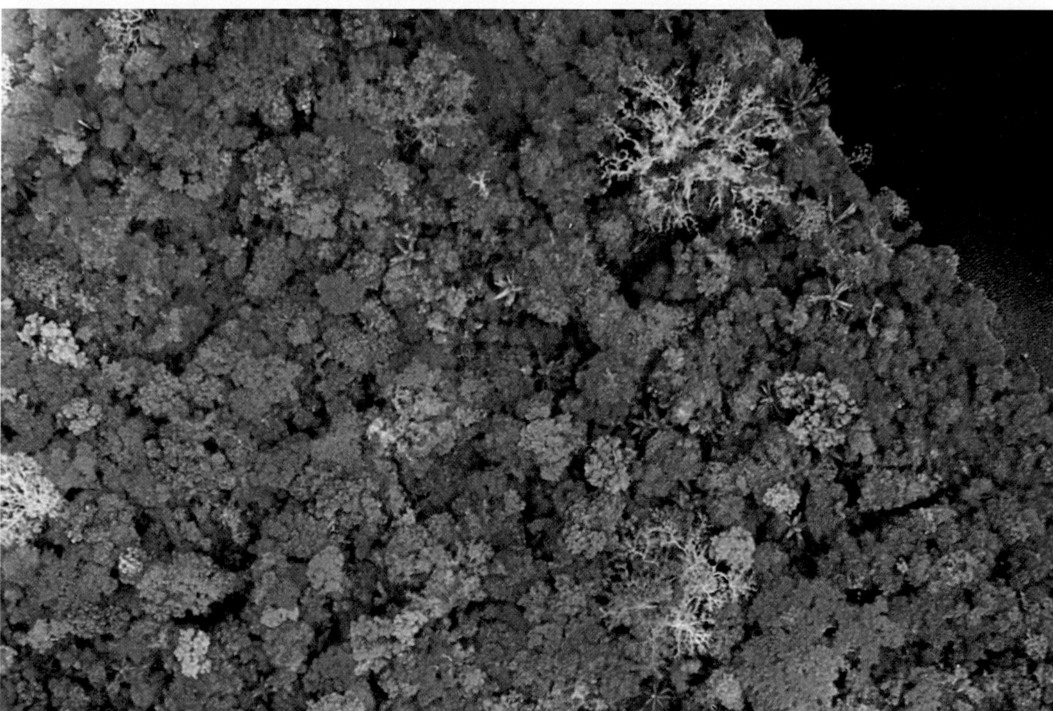

Fig. 34-27. Aerial photography of tropical-forest land in Surinam. Note heterogeneous composition, leafless emergents in their dormant stage, different palms with their characteristic crown structures (some of which indicate wetter soil conditions), and the wide varieties of crown structure, tone, and color. Stereo observation would reveal open upper canopy and denser lower canopy. *Top.* Agfa color CN-17 with an HF-3 filter. *Bottom.* Kodak Ektachrome infrared aero 8443 with a Wratten-12 filter. (Photos courtesy of the International Institute for Aerial Survey and Earth Science, Enschede, the Netherlands.)

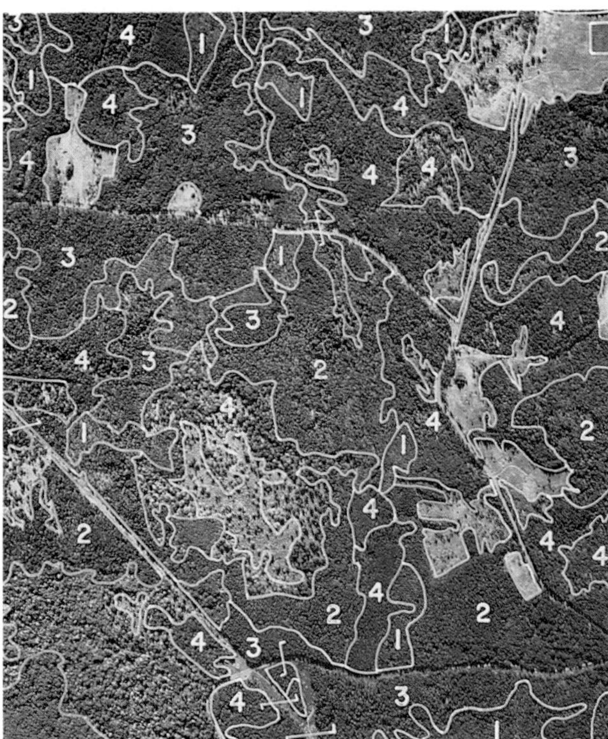

Fig. 34-38. Color-infrared photography (Nova Scotia, September 11, 1971) shows mortality of conifers following attacks by spruce budworm. Levels of damage are: 1—mortality over 50 percent; 2—25 to 50 percent mortality; 3—10 to 25 percent mortality; 4—1 to 10 percent mortality. The main species are red spruce and balsam fir. (Photo courtesy of the Canadian Forestry Service.)

Fig. 34-37. Three levels of spruce-budworm defoliation on balsam fir near Ely, Minnesota. Ansochrome D200, original scale 1:1,600, Maurer KB-8 camera f 2.8 lens: A. No visible defoliation; all conical-shaped crowns are green; B. Moderate to heavy defoliation; most crowns are yellow-red to brown; C. Severe defoliation; in addition to heavy browning of foliage, many gray, dead crowns are present. (Photos courtesy of Forest Service, USDA.)

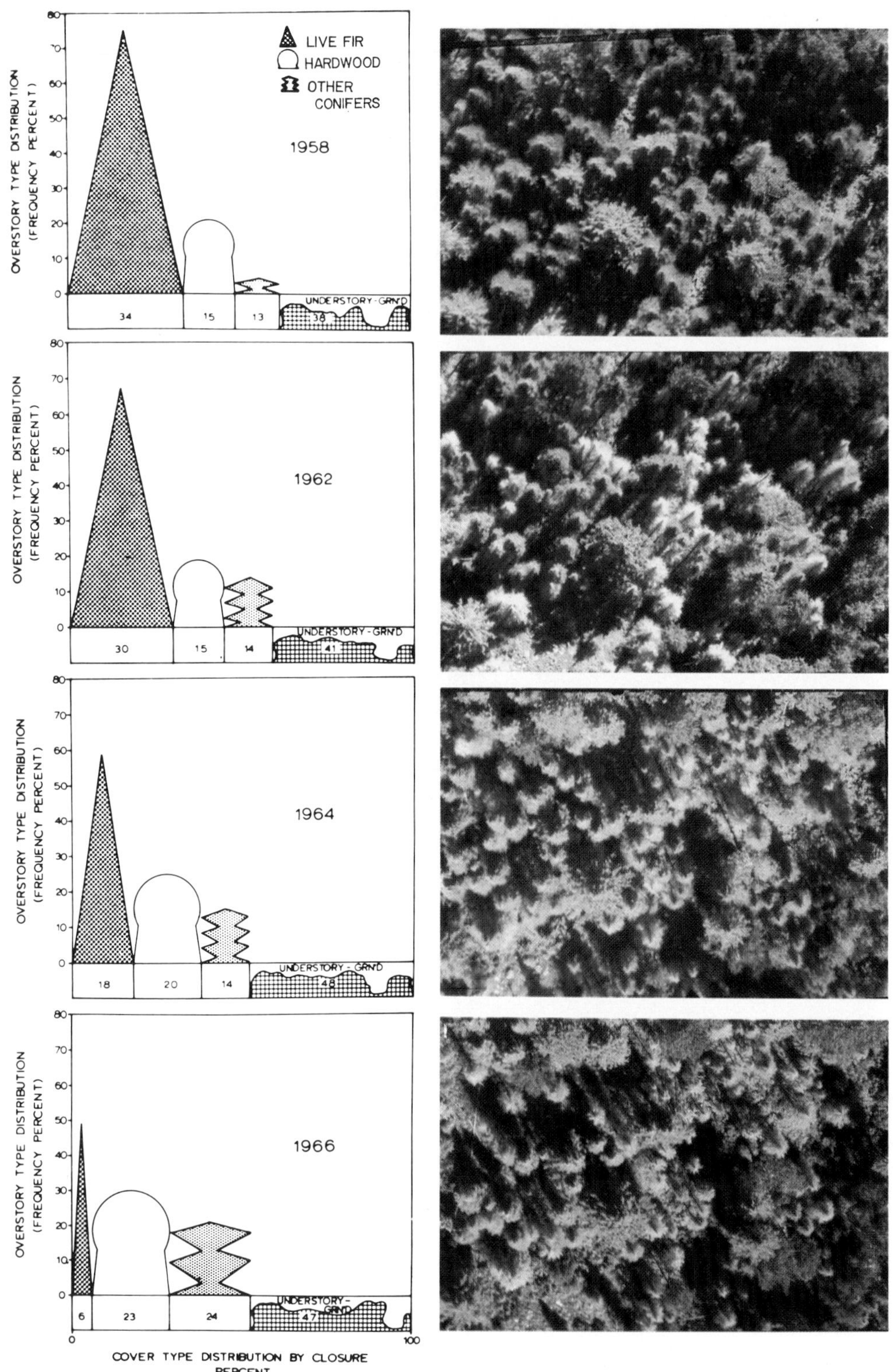

Fig. 34-39. Photography for a 0.4 hectare plot showing cover type distribution by frequency percent, and cover closure percent. Color photographs show the changes that have occurred. Color photographs made from Anscochrome D/200 reproduced in black-and-white. (Photos and diagrams courtesy of Forest Service, USDA.)

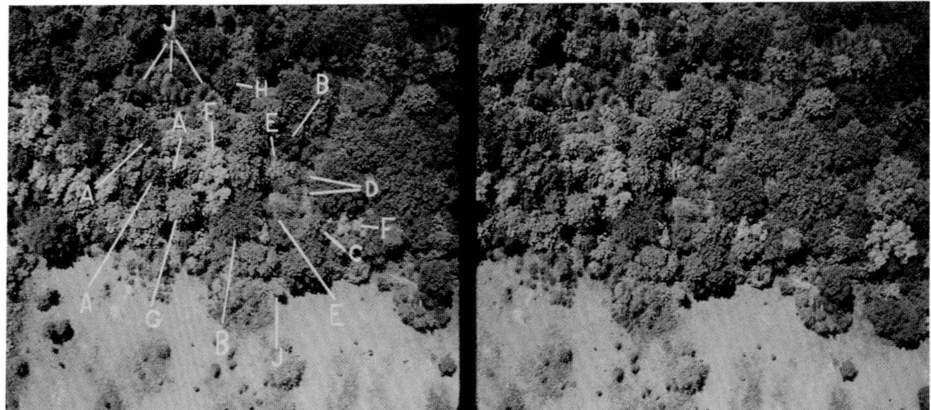

Fig. 34-41. Stereogram of hardwood stand near Poughkeepsie, New York. Species indicated are: (A) healthy ash (Fraxinus americana), (B) healthy American elm (Ulmus americana) (C) moderately affected ash, (D) and (E) dead American elm, (F) healthy black willow (Salix nigra), (G) healthy sugar maple (Acer saccharum), (H) healthy red maple (Acer rubrum), (J) paper birch (Betula papyrifera) with leaves skeletonized and brown from a leaf miner. (Photos courtesy of Forest Service, USDA.)

Fig. 34-44. SO_2 damage to vegetation can sometimes be clearly outlined on photographs at very small scales. Detection was made on a CIR 4× enlargement from a contact scale of 1:160,000. Three damage zones can be identified: (1) total-kill zone, denuded of vegetation; (2) heavy-kill zone, where all trees are gone, but some lower vegetation remains; (3) medium-injury zone, forested, but with many openings in the crown canopy which are caused by mortality. (Photo courtesy of the Canadian Forest Service.)

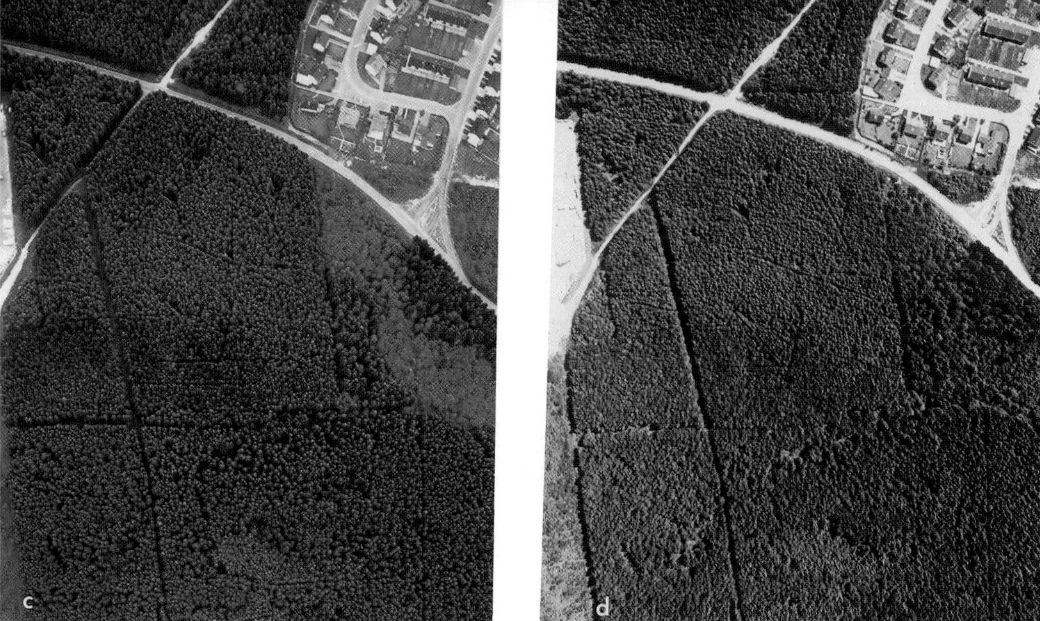

Fig. 34-45. (a) Black-and-white IR, (b) panchromatic, (c) color IR, (d) color, (original scale 1:5,000) photos of a West German spruce stand affected by fluoride gases emanating from a brick plant to right of photos. The darkened area in the middle of the photos is trees with needle losses in crown interior from the previous year's fumigation. (Courtesy of Abteilung Luftbildmessung und Interpretation, Institut für Forsteinrichtung und Fortsliche Betriebswirtschaft, Freiburg, Germany.)

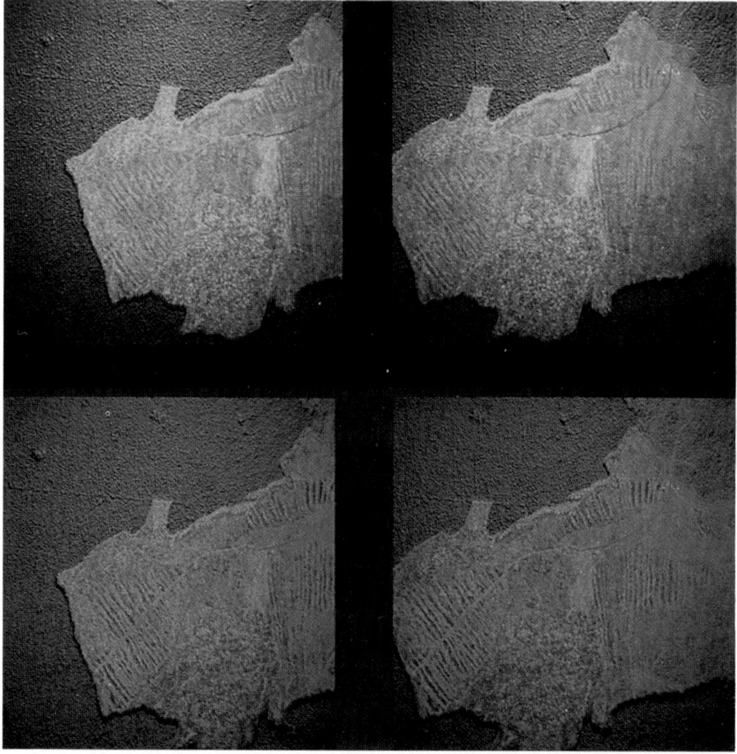

Fig. 34-48. Black-spruce logging depletion and slash-disposal survey in Koochiching County, Minnesota (70-mm, color and color infrared, original scale 1:16,000, Aug. 3, 1971). (Photos courtesy of School of Forestry, University of Minnesota.)

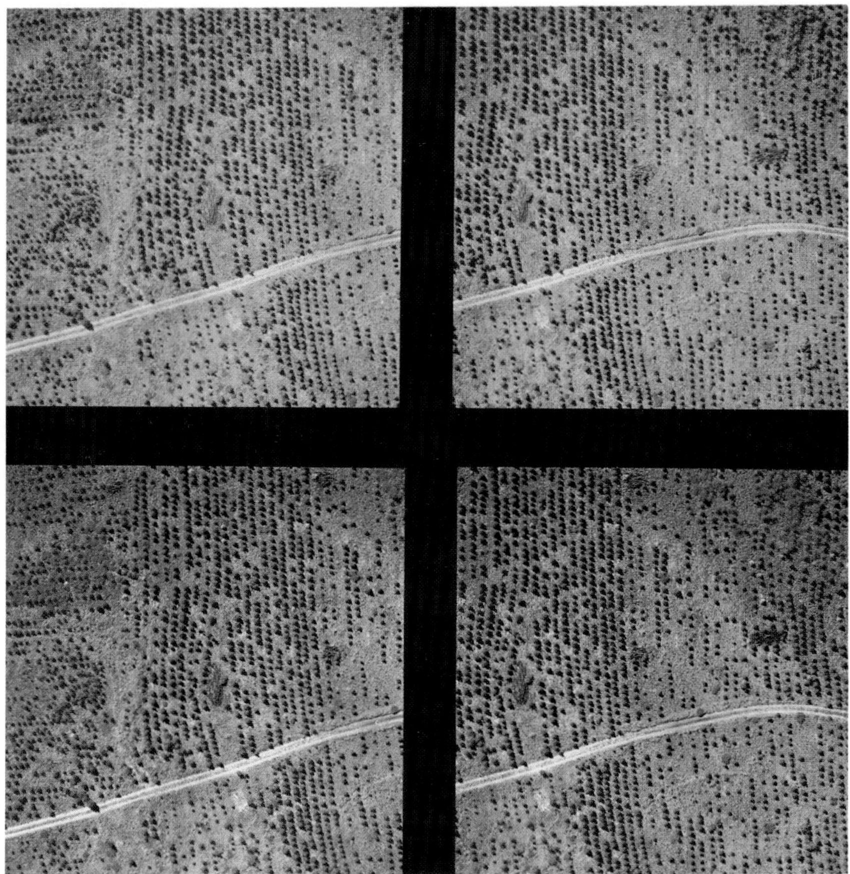

Fig. 34-49. Pine-plantation survey at Willow River, Minnesota (70-mm, color and color infrared, 1:2,000 original scale, Aug. 28, 1971). Average tree height = 2 to 3 meters. Survival pattern is obvious at this scale, and also easily discernible at 1:16,000 scale. (Trees with yellowish crowns were killed by Armillaria root rot). (Photos courtesy of School of Forestry, University of Minnesota.)

Fig. 34-52. Example of Landsat imagery showing several spectrally heterogeneous areas used for developing training statistics using the "multi-cluster blocks" techniques (from Hoffer, et al., 1975b).

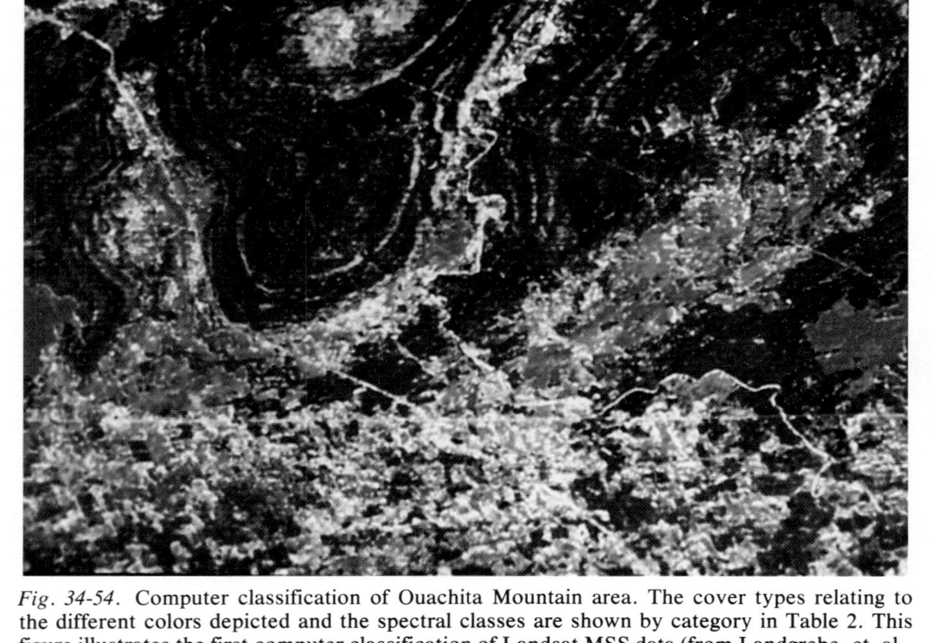

Fig. 34-54. Computer classification of Ouachita Mountain area. The cover types relating to the different colors depicted and the spectral classes are shown by category in Table 2. This figure illustrates the first computer classification of Landsat MSS data (from Landgrebe, et. al., 1972).

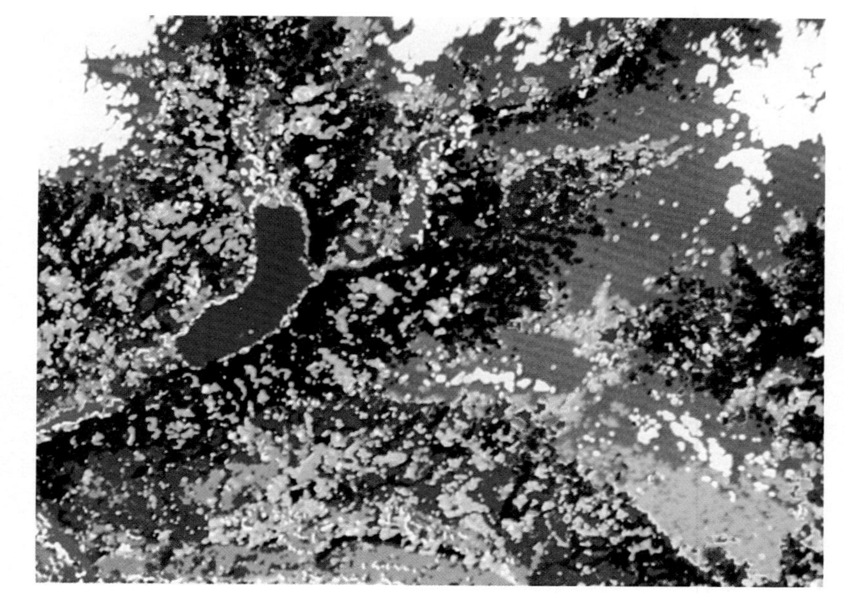

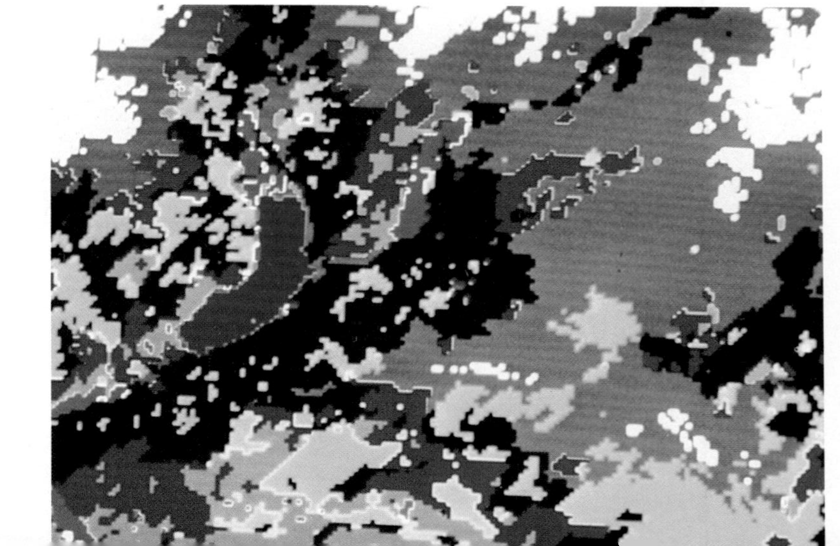

Fig. 34-59. Comparison of ''Per Point'' (*Top*) and ''ECHO'' (*Bottom*) classification results for forest cover types. The ECHO classifier utilizes both the spectral and spatial information content in the data, resulting in a classification map that does not have the ''salt-and-pepper'' effect of per-point classifiers

Fig. 34-61. Areal distribution of Daily Potential Evapotranspiration over the study area for Spring 1978.

Color	*Daily Potential Evapotranspiration in inches of water*
Black	0–.10
Blue	.11–.15
Light Blue	.16–.20
Green	.21–.25
Dark Green	.26–.30
Yellow	.31–.40
Orange	.41–.50
Pink	.51–.54

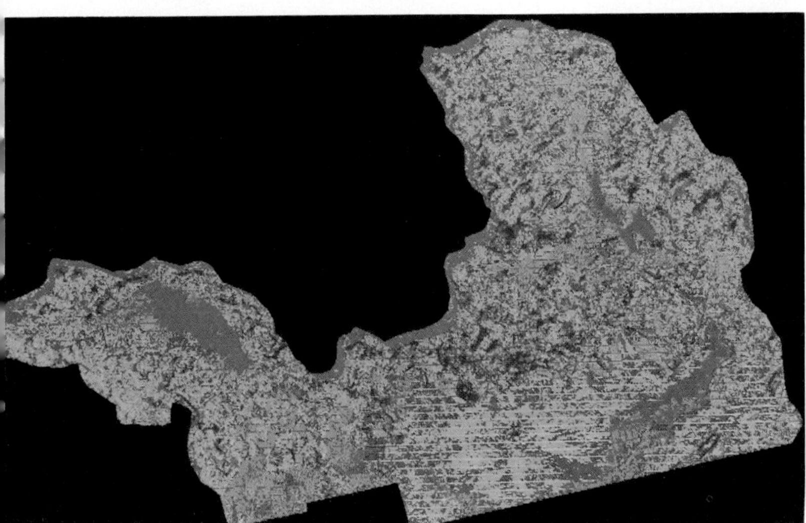

Fig. 34-62. Average Daily Net Radiation Map of the Upper Middle Fork of the Feather River Watershed for Spring, 1978. Identification key:

Color	*Average Daily Net Radiation in ly/day*
Blue	251–303
Light Blue	304–350
Green	351–359
Yellow	360–400
Orange	401–452
Red	453–503
Pink	504–582

Fig. 34-63. Vegetation/terrain map for the Middle Fork of the Feather River Watershed. Identification key:

Color	*Vegetation/ Terrain Type*
Lavender	Sagebrush
Purple	Sagebrush/ Bare Soil
Off White	Bare Soil
White	Rock
Orange	Agricultural
Yellow	Marsh/Mesic Grassland
Red	Mixed Conifer
Orange/Red	Red Fir
Light Blue	Riparian Hardwoods
Blue/green	Mesic Brush
Dark Green	Brush
Light Green	Oakwood/ Xeric Brush
Dark Blue	Water

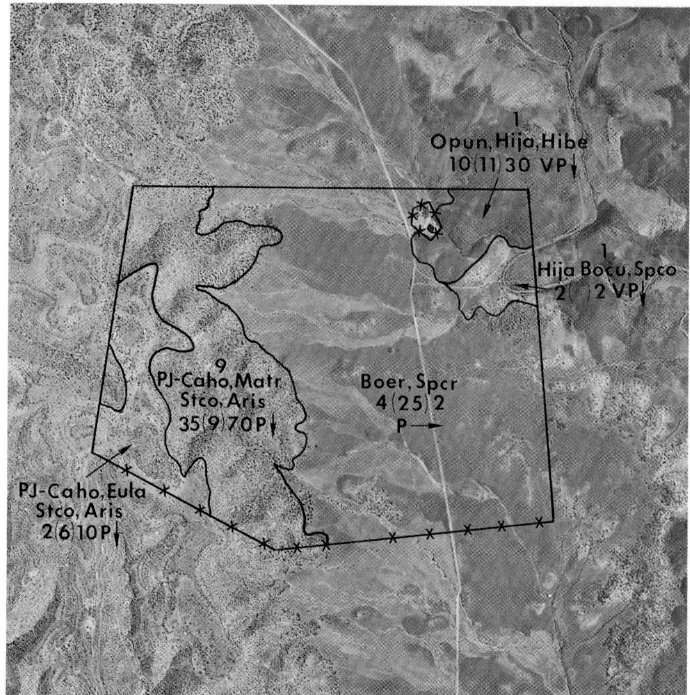

Fig. 35-6. An example of mapped plant community classes for range inventories using color aerial photographs (scale 1:24,000). The first numeric symbol indicates the broad vegetation type, the alpha codes identify specific plant community sub-types within a much broader vegetation class represented by the initial numeric code, and the lower alpha/numeric codes indicate ground conditions and apparent range condition and trend. (Photo courtesy U.S. Forest Service)

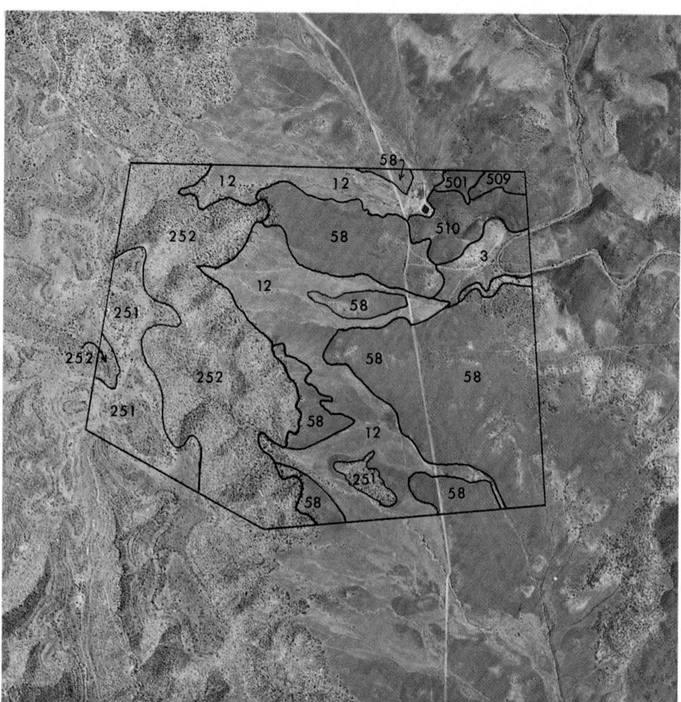

Fig. 35-7. An example of mapped soil units associated with the vegetation of Figure 35-6. Note how the soil types correspond to the colors in the aerial photograph (photoscale 1:24,000). The numeric legend for this map corresponds to soil types as follows: 3-Tobler fine sandy loam; 12-Anthony sandy loam; 58-Connville fine sandy loam; 251-Mescal loam; 252-Retriever very sandy loam and rock, outcrop; 501-Rimrock stony clay; 509-Graham & House Mtn. Soils (0–20% slope); and 510-Graham & House Mtn. Soils (20–60% slope)

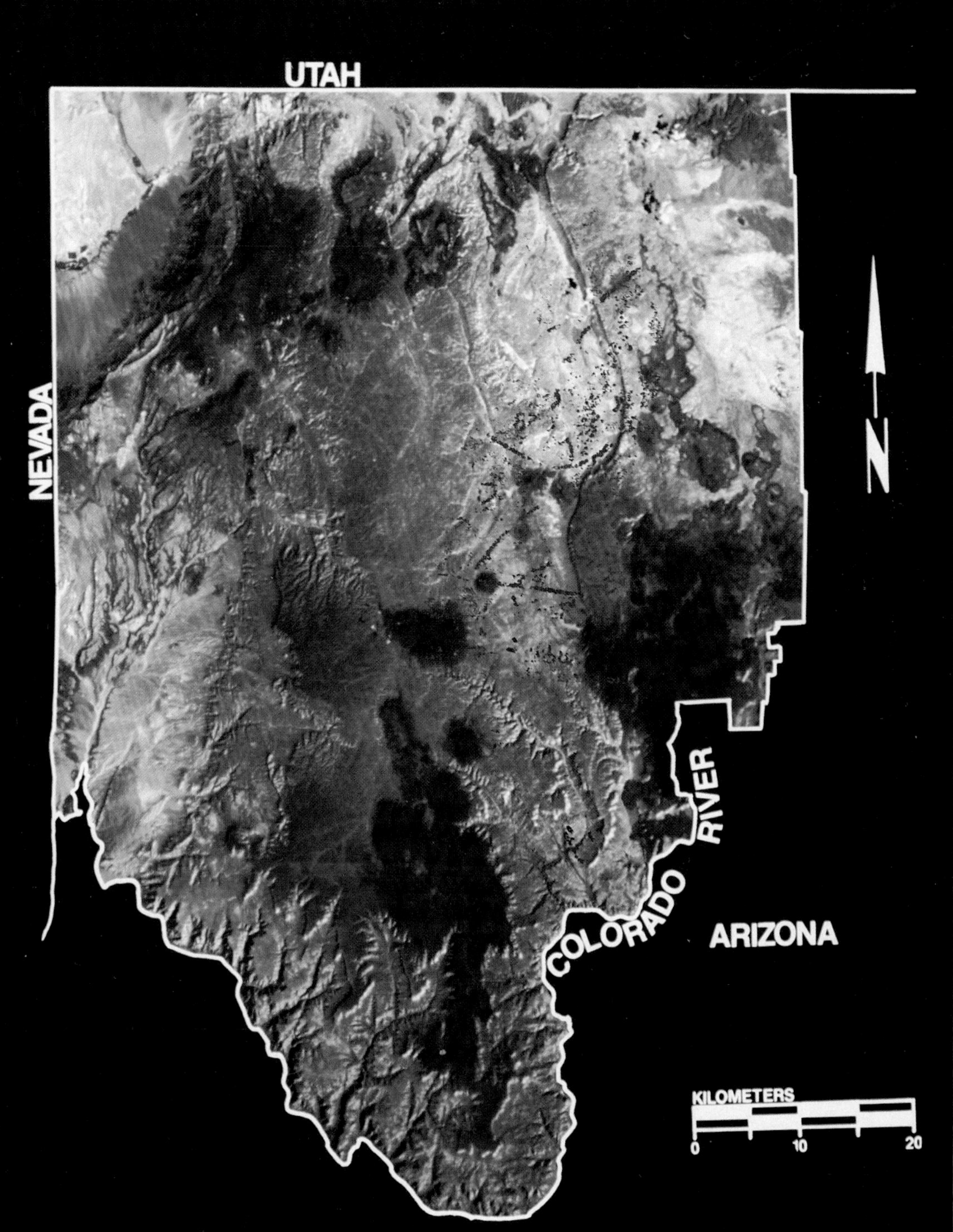

Fig. 35-8. Part of a Landsat false color composite (scene ID-2947-17074) acquired on August 26, 1977, showing an area of northwestern Arizona. The vegetation of the area includes Mohave and Great Basin desert shrub, grassland, mountain shrub, deciduous woodland, evergreen woodland, and coniferous forest (Rohde and Miller, 1981). Vegetation maps produced from a standard and an enhanced false color composite image of this scene appear in Figures 35-9 and 35-10 respectively. (Photo courtesy of EROS Data Center, Sioux Falls, S. Dakota and BLM, Denver, Colorado)

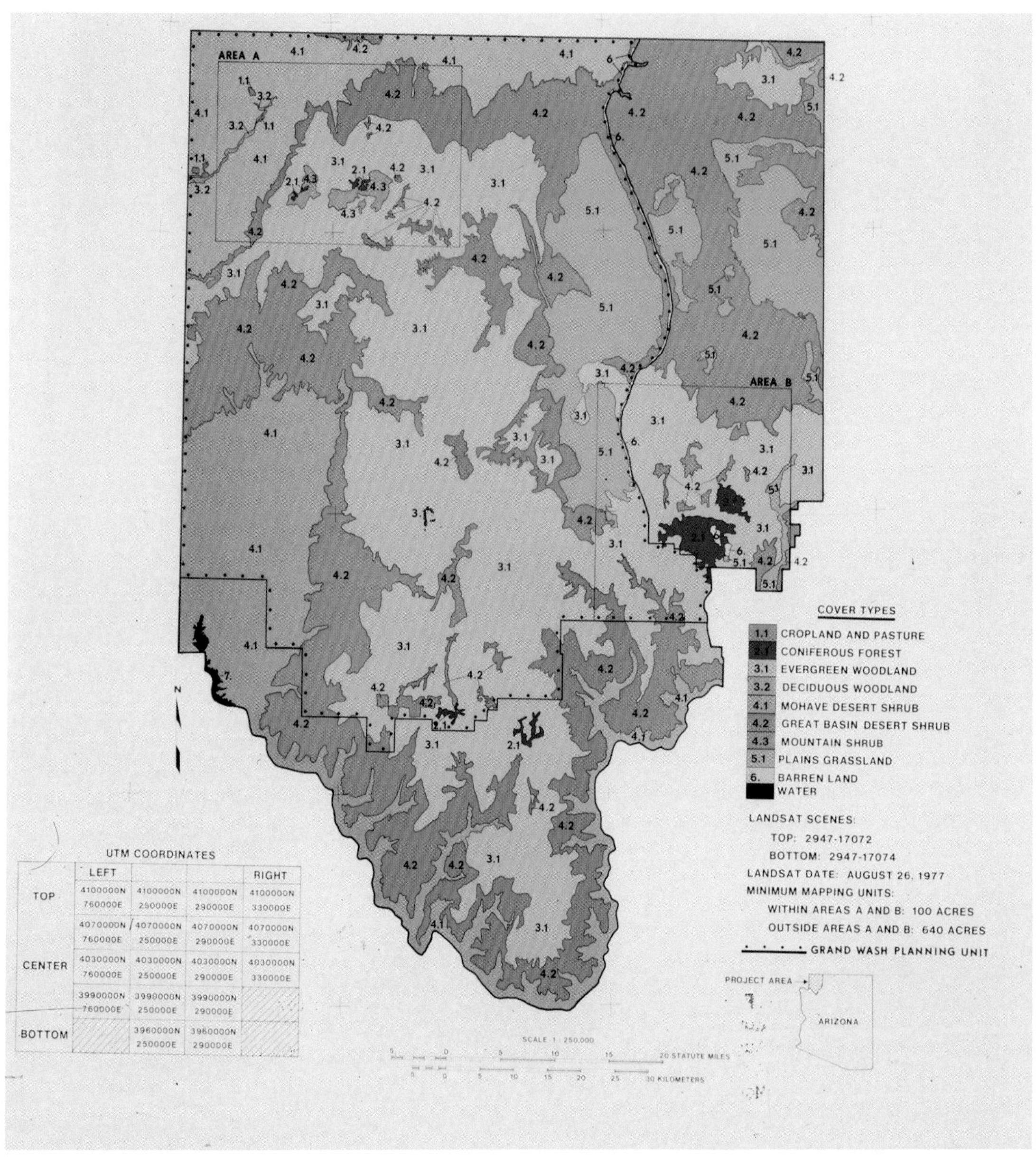

Fig. 35-9. Vegetation cover type map interpreted manually from a standard Landsat false color composite image at a scale of 1:250,000 (from Bonner, 1979). Compare with Figure 35-8.

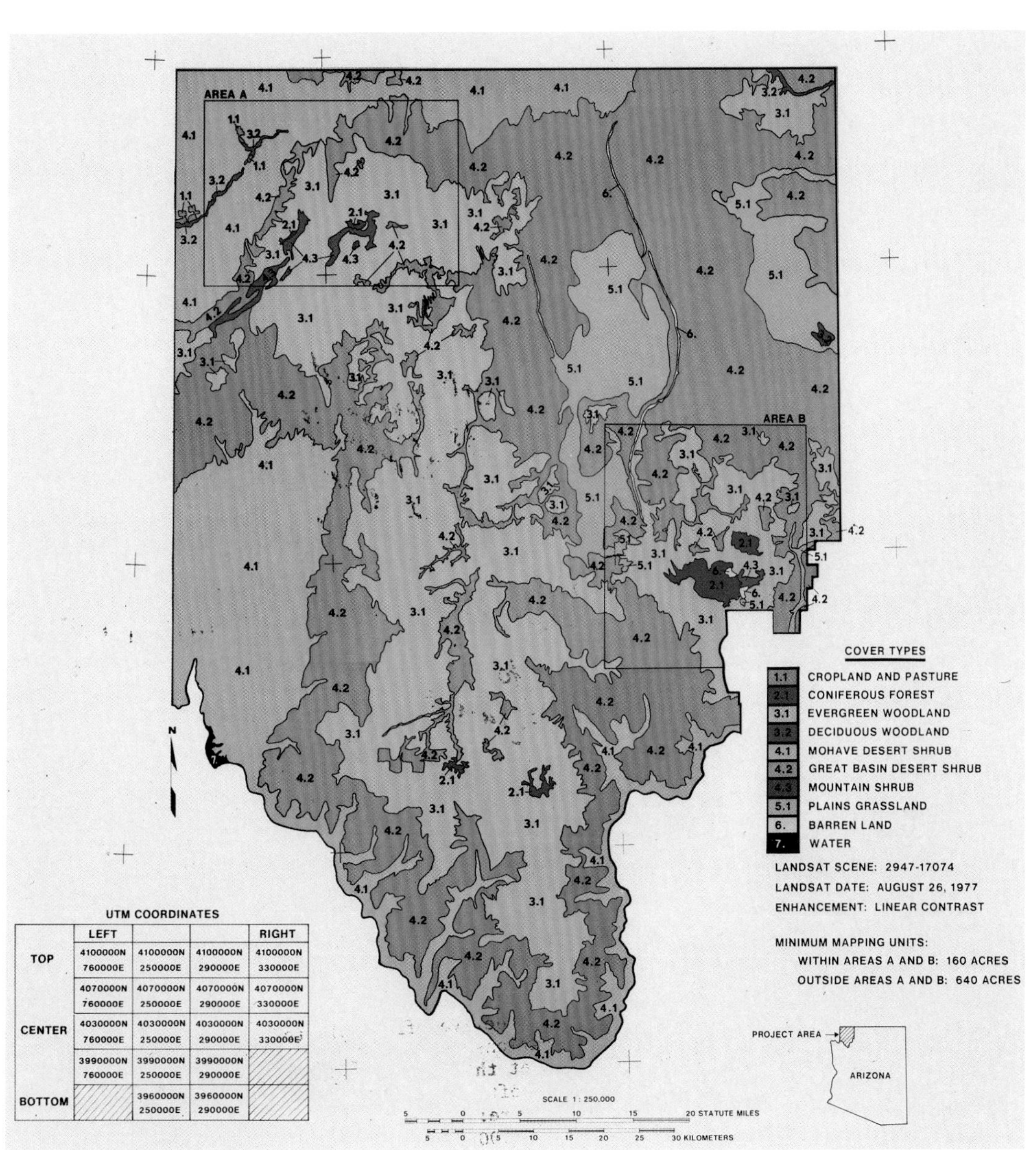

Fig. 35-10. Vegetation cover type map interpreted manually from an enhanced Landsat false color composite image at a scale of 1:250,000 (Bonner, 1979).

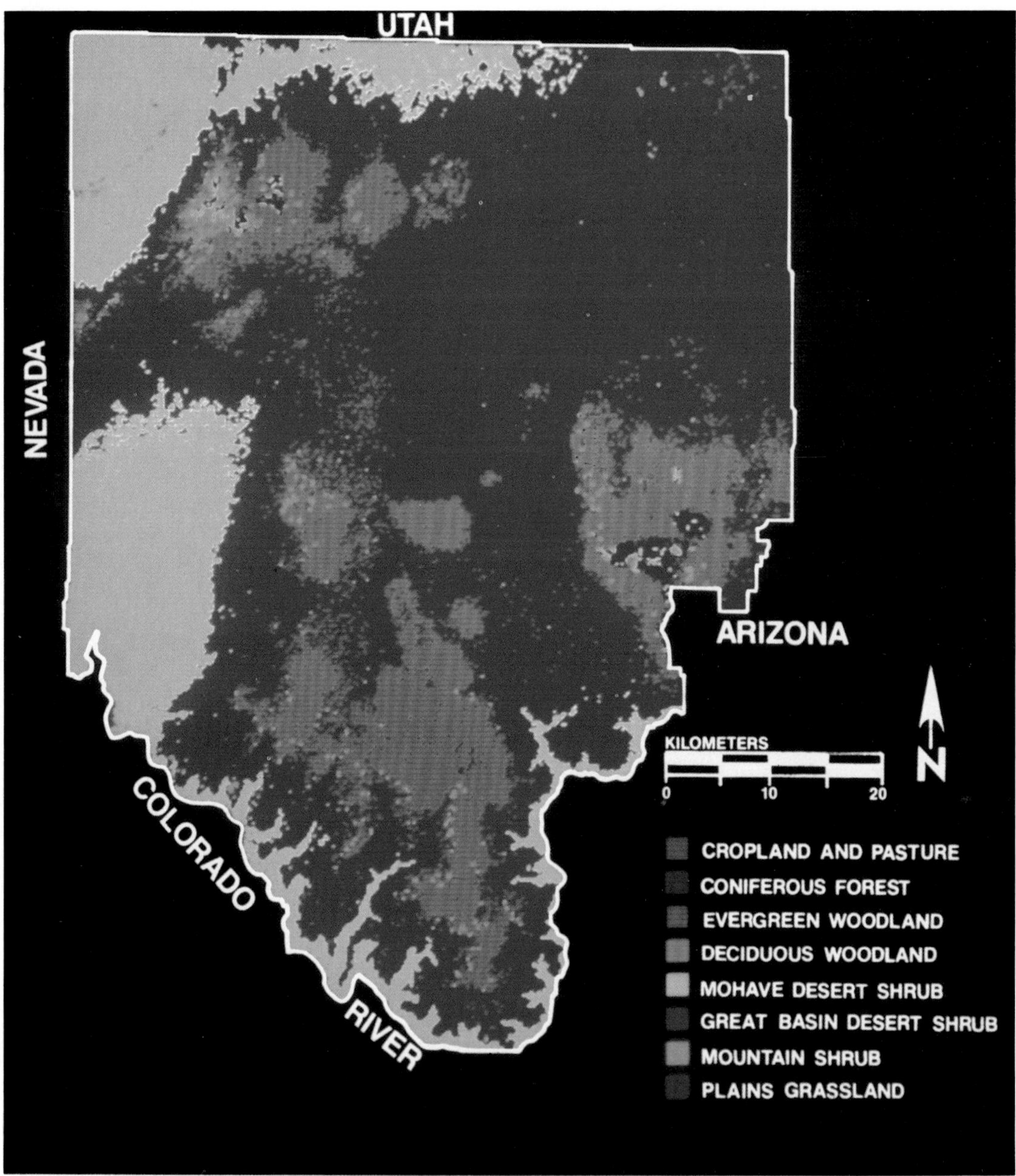

Fig. 35-13. Land cover map for northwestern Arizona produced from digital analysis of Landsat data (see Figure 35-8) in which 83 Landsat derived spectral classes represent eight land cover classes. This map is a final product in which digital terrain data have been used in post classification sorting. The application of this technique has improved overall classification accuracy from 54% to 73%. (Photo courtesy of USGS EROS Data Center, Sioux Falls, S.D. and U.S. Bureau of Land Management, Denver).

Fig. 35-14. Color aerial photograph (original photo scale of 1:6000) of a sample unit used in a land cover mapping project in northwestern Arizona to estimate acreages of land cover types and to determine classification accuracy of the cover map in Figure 35-13. The sample unit is partitioned into 64 photo plots which correspond in area to a Landsat picture element. Information about the cover type for each pixel was determined from photo interpretation. The predominant vegetation type is Evergreen Woodland. (Photo courtesy of USGS EROS Data Center, Sioux Falls, S. Dakota and U.S. Bureau of Land Management, Denver)

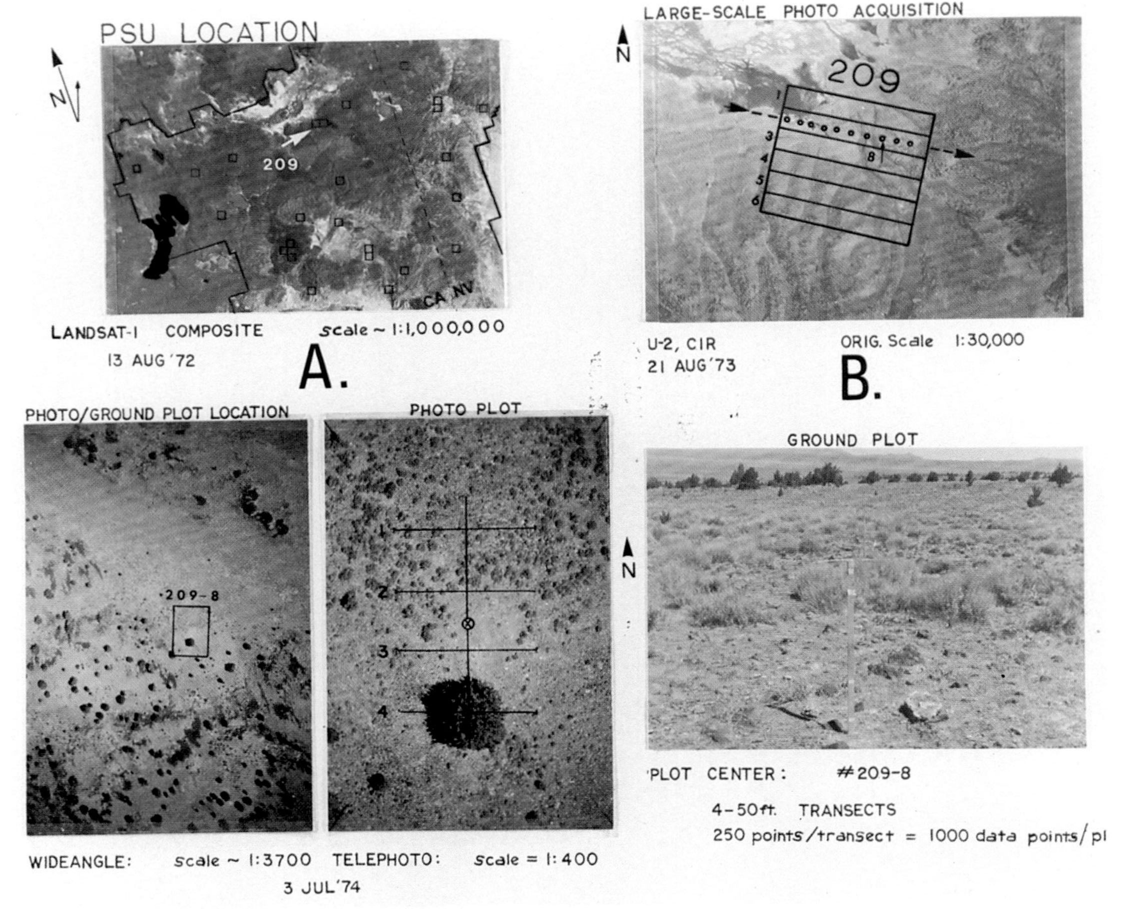

Fig. 35-16. Examples of Landsat imagery and aerial photographs on which forage production estimates were made in Northeastern California. Estimates were also made on the ground. Each level of data, to estimate forage production, corresponds to a sample stage in the multistage sampling frame shown in Figure 35-15. (Photo courtesy of Remote Sensing Research Program, Space Sciences Laboratory, Univ. of Calif., Berkeley).

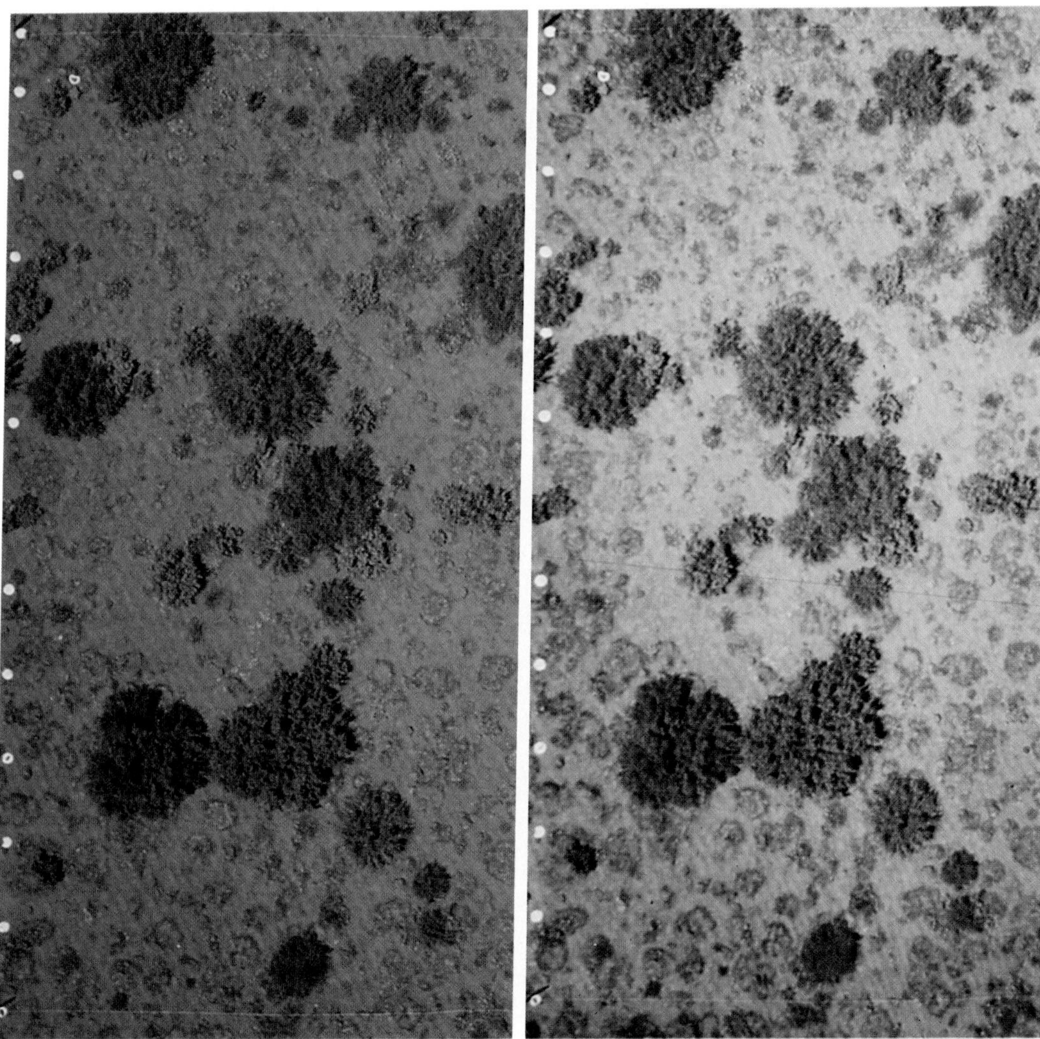

Fig. 35-20. Large-scale 70mm color and color infrared aerial photographs taken July 14, 1978, within Lassen National Forest, northeastern California. The dominant woody plants can be identified equally well on both film types at this time of season. Note, however, that the color infrared photo is better for distinguishing healthy from unhealthy plants and live from dead plant material. In this plant community color photographs taken later in the fall were better for species identification. North is towards the bottom of the photographs. The paper plates mark the ends of line transects used to estimate canopy cover. The plates are spaced 5 feet apart.

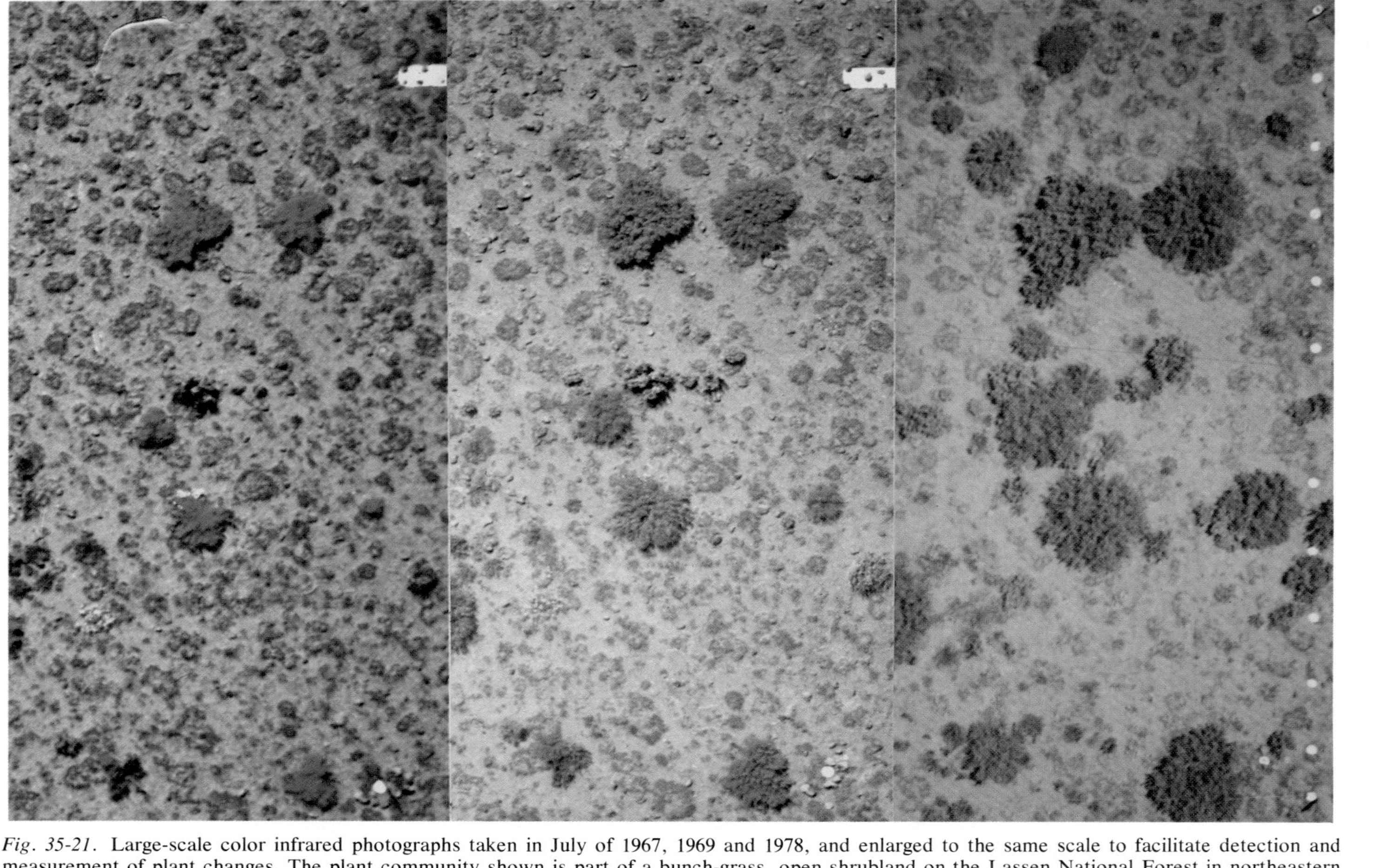

Fig. 35-21. Large-scale color infrared photographs taken in July of 1967, 1969 and 1978, and enlarged to the same scale to facilitate detection and measurement of plant changes. The plant community shown is part of a bunch-grass, open shrubland on the Lassen National Forest in northeastern California. The dominant woody plants include: A-Big sagebrush, (*Artemisia tridentata*); B-Bitterbrush, (*Purshia tridentata*); C-Rabbitbrush, (*Chrysothamnus sp*); E-Buckwheat, (*Eriogonum umbellatum*); and L-Wherry, (*Leptodactylon pungens*). See Fig. 35-22 for locations of woody species. The dominant herbaceous graminoids comprising the understory, but which are not labelled include. Idaho fescue, (*Festuca idahoensis*); Carex, (*Carex rossii;* Squirreltail (*Sitanion hystrix*); and Needle and Thread grass, (*Stipa occidentallis*). These photographs show only a portion of a larger sample site measuring approximately 52 × 75 feet. The 7 inch paper plates shown in the 1978 photo mark the ends of line transects used to estimate canopy cover of the plants in the study area. The paper plates are spaced 5 feet apart. North is towards the bottom of the photo.

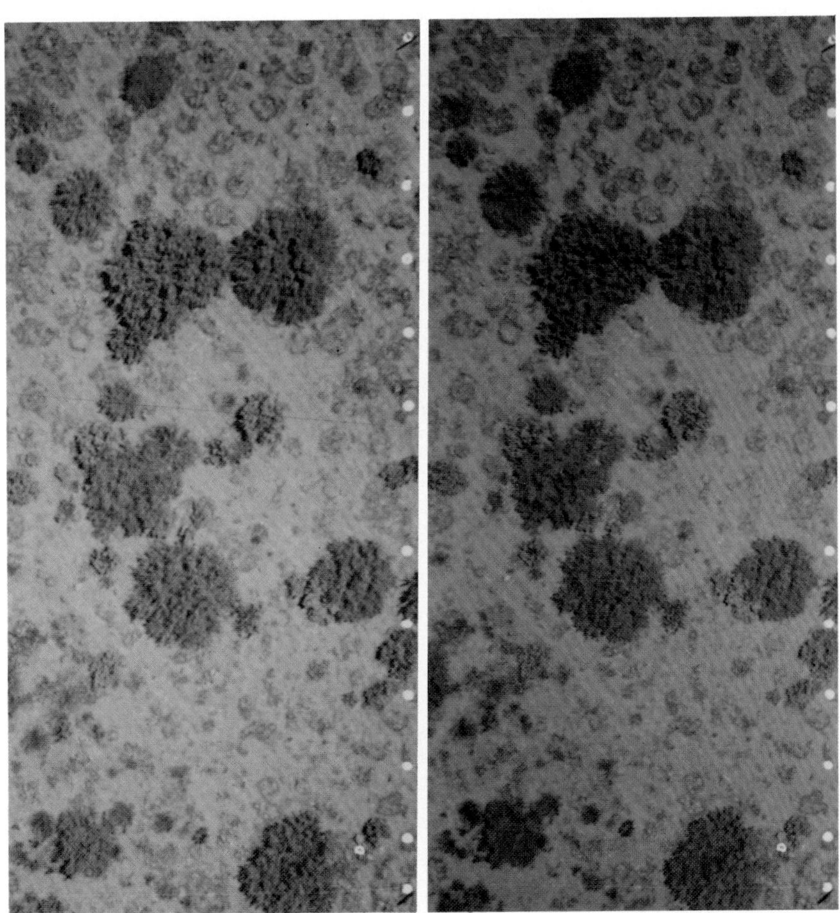

Fig. 35-23. Stereo photographs from large-scale, 70mm color infrared aerial photographs taken July 14, 1978. The area shown corresponds to that seen in Figure 35-20 and 35-21. Stereoscopic viewing improves plant detection and plant species identification. The tallest shrubs are only 3 feet high.

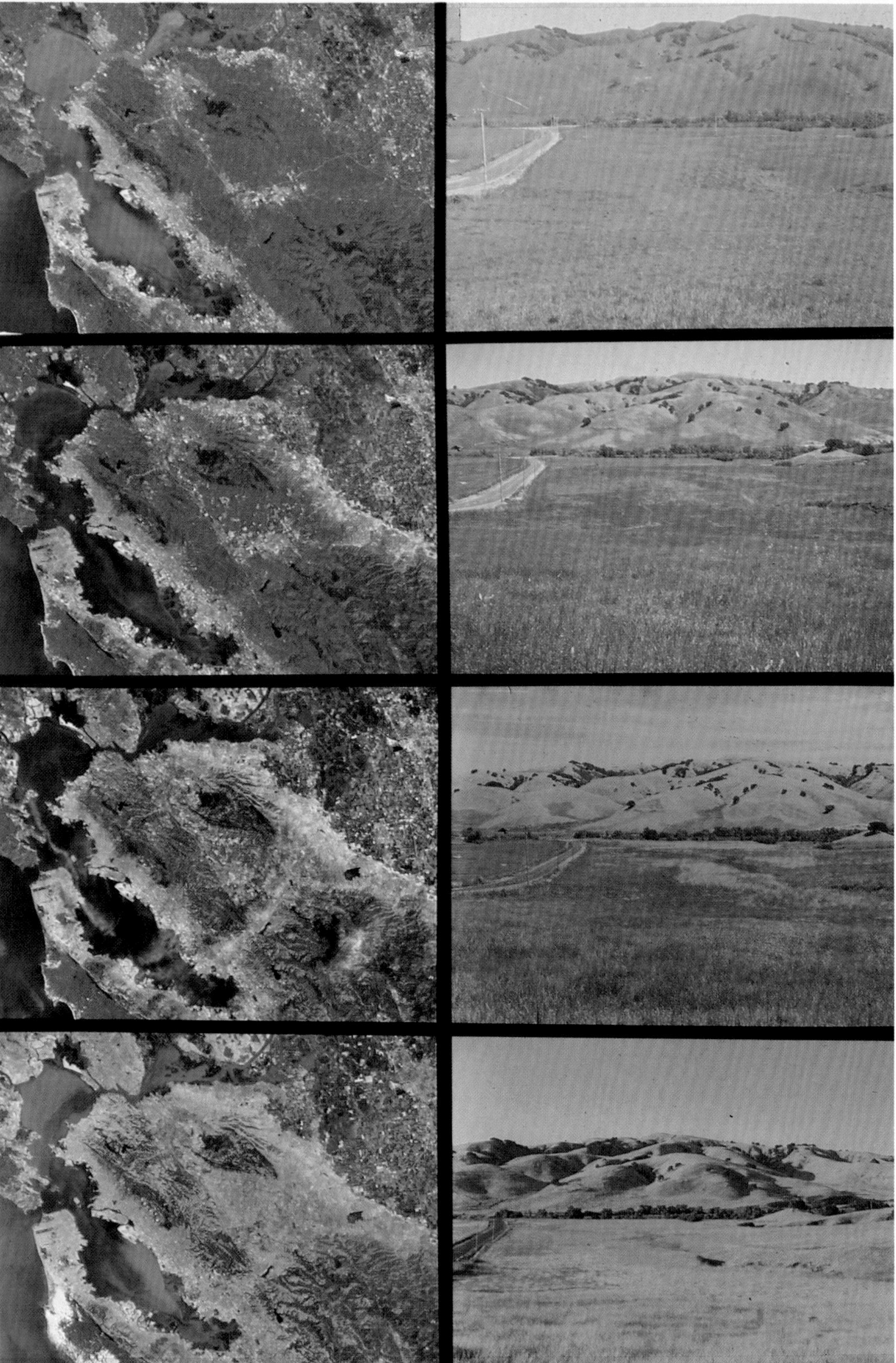

Fig. 35-25. Portions of four Landsat false color composites showing the area around San Francisco Bay, California. Note the changes in appearance of the annual grassland, the dominant vegetation type in the undeveloped areas, as it progresses through its normal annual cycle. Photo A was taken April 4, 1973, Photo B was taken May 10, 1973, Photo C was taken May 28, 1973, and Photo D was taken June 15, 1973. The grasslands reach maximum greenness and foliage production near the early part of April. By May, grasslands on shallow soil and in regions of less annual rainfall have dried and by late May to early June most of the grasslands have dried. Changes in the ratios of MSS Bands 7 and 5 data are indicators of changes in green biomass during the season for a given site and changes in relative differences in green biomass between regions for a given date.

Fig. 35-28. Typical elk habitat in the Blue Mountains, northeast Oregon. The twelve or more light-toned objects are elk. The forest in the background provides thermal protection for the animals, while the clear cut area in the foreground provides forage (from study by Isaacson et al, 1982).

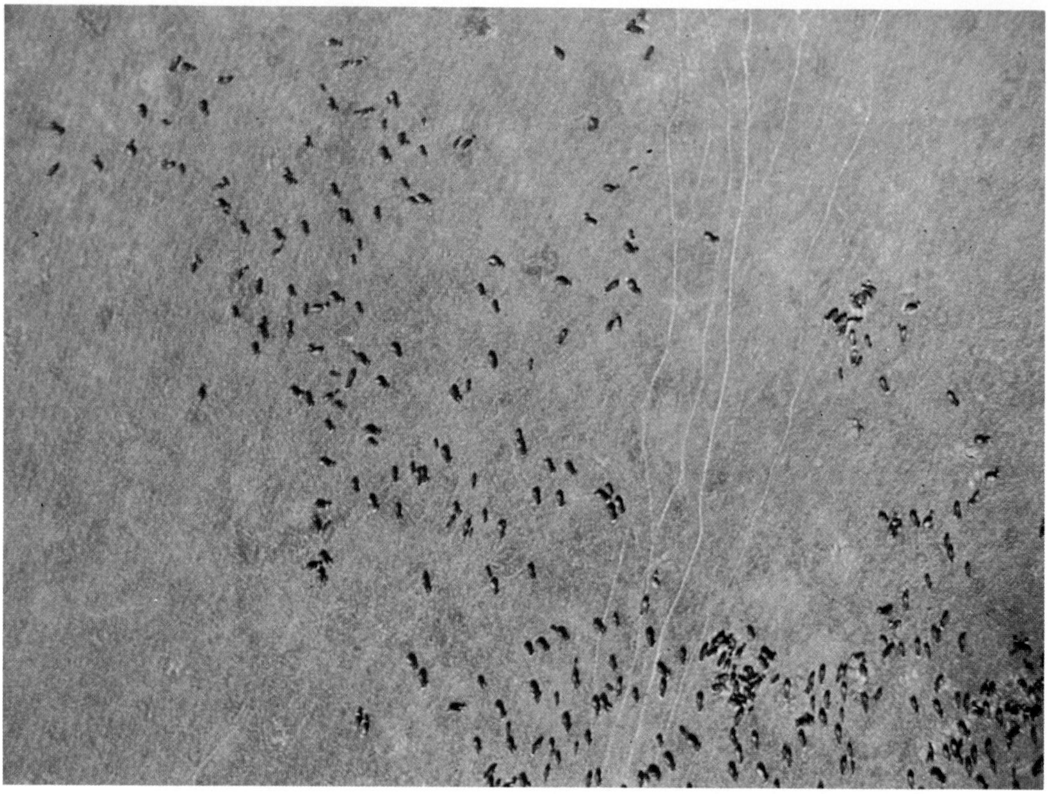

Fig. 35-30. Photo A (top) is an oblique color aerial photograph taken with a 35mm camera showing wild ungulates near Masa Mara in the northern Serenghetti Plains of southern Kenya. The majority of the animals are wildebeast. Photo B shows a group of shoats (mixture of sheep and goats) which compete with wild animals for forage. (Photos courtesy of Kenya Regional Ecological Monitoring Unit, Nairobi).

Fig. 35-31. The 9 × 9 inch panchromatic photograph, at the approximate scale of 1:5000, and the 70mm color, aerial photograph at a scale of 1:2500, illustrate the two levels from which estimates of livestock numbers are made by photo interpreters. Estimates of livestock numbers made from the panchromatic photo are adjusted based upon the ratio of animals counted on the color photograph and the corresponding area of the panchromatic photo. (Photo from Roberts and Colwell, 1968).

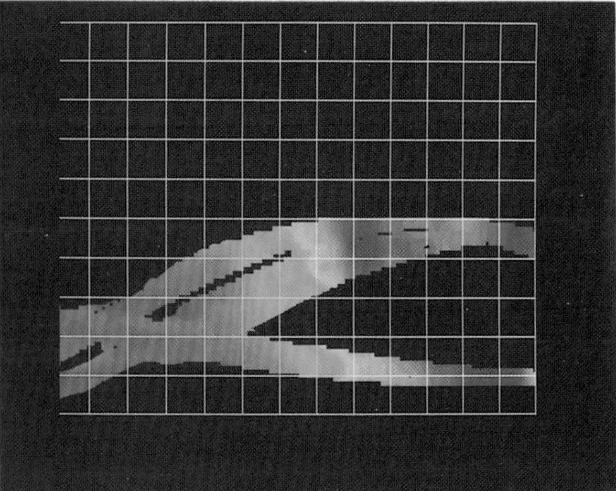

Fig. 36-16. Color-coded map depicting the concentration of Th on the lunar surface, based on Apollo 15 and 16 gamma ray spectrometry data. Blue areas correspond to low Th abundance, while red areas correspond to higher amounts. The projection is in simple cylindrical form and the grid spacing is 10° in both latitude and longitude. The region displayed extends from about 125° W. long. to the prime meridian, and from 80° N. to 20° S. lat. The boundary between terrae and maria roughly corresponds to the change from blue to green. The maria have a higher Th content than the terrae.

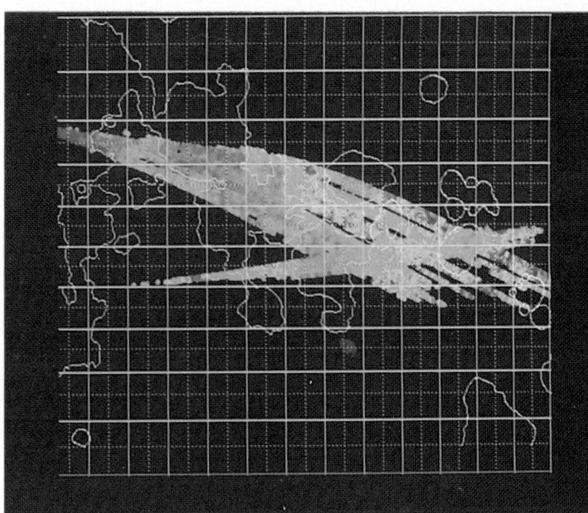

Fig. 36-17. Color-coded map of the Al/Si ratio of the lunar frontside, based on data from the Apollo 15 and 16 x-ray fluorescence experiments. Grid spacing is at 5° in both latitude and longitude. The boundary between maria and terrae is included as a white line. Blue areas correspond to low values, while red areas correspond to higher values. The maria are depleted in Al relative to Si as compared to the terrae. The frame extends from about 15° W. long. to 110° E. long., ±50° in lat. about the equator, and is in a Mercator projection.

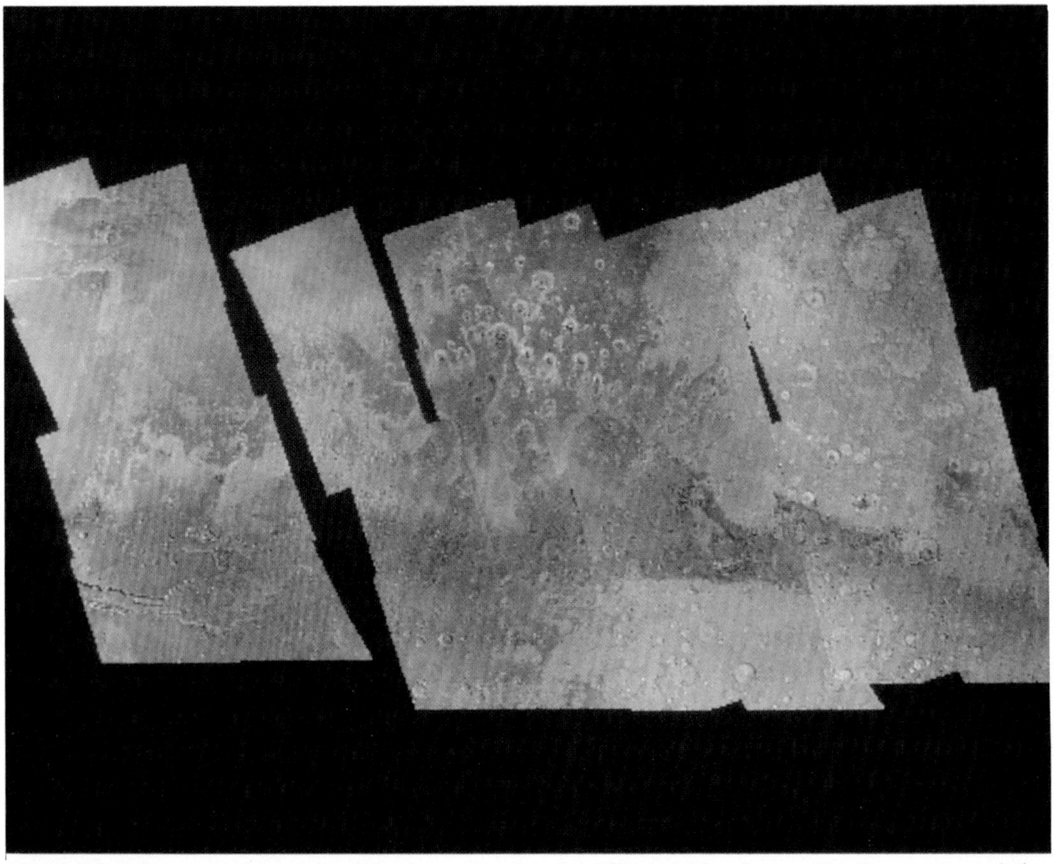

Fig. 36-27. Color composite mosaic of Martian surface based on Viking Orbiter frames. Significant color variations can be seen, especially within darker areas. The mosaic transects about 80 degrees in longitude, centered at the equator. Courtesy, E. Strickland, Washington University.

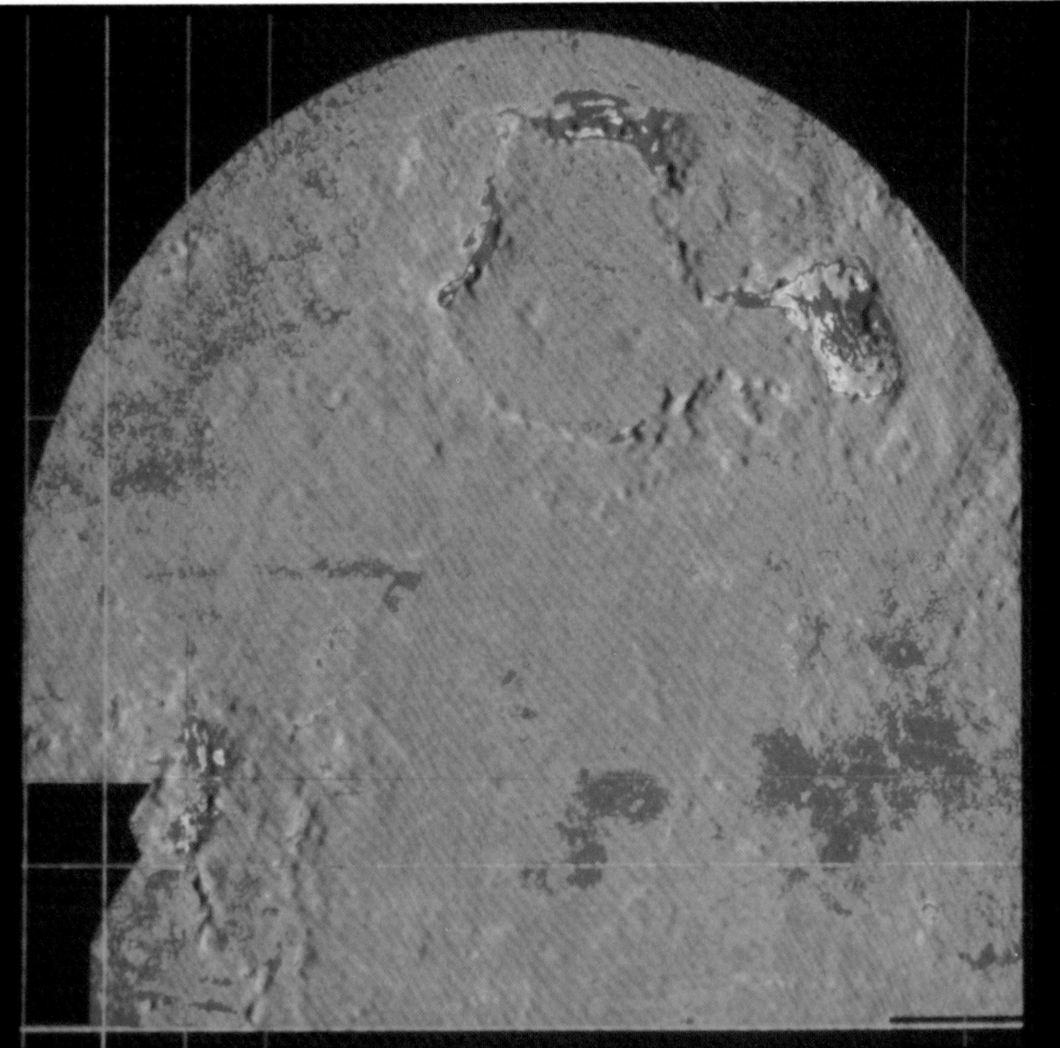

Fig. 36-30. A color-coded version of the 12.6 cm radar reflectivity of the Venusian surface has been digitally overlayed onto a shaded relief map depicting topography. The simulated sun is from the west at 20° above the horizon and the elevation data are from the Pioneer-Venus radar mapper experiment (Pettengill et al., 1980). Reflectivity data are from the Arecibo observatory (Campbell and Burns, 1980). Blue areas correspond to low values of reflectivity, yellow and red areas to intermediate values, and white areas to high values. Reflectivity values are proportional to the degree of surface roughness of a few tens of centimeters, except at low latitudes, where meter-scale roughness is also important. The lateral resolution of the Arecibo data varies from 10 to 20 km, while the radar mapper data have a footprint width that averages 100 km. The image extends from about 260° W. long. to 30° E. long., and from about 80° N. to 0° in latitude. Vertical and horizontal lines on the left and bottom are spaced 10° apart. Maxwell is a radar and topographically rough area located in the upper right. Lakshmi Planum is the radar and topographically smooth area in the upper middle part of the frame. Beta Regio is the linear trough that is rough in the radar data. There is clearly a correlation between relief and radar roughness on Venus.

Fig. 36-32. Voyager 1 global color views of the four Galilean satellites. Proceeding clockwise from upper left, the satellites are Io, Europa, Ganymede, and Callisto. After Smith et al. (1979).

Fig. 36-35. Voyager 1 color enhanced frame showing a volcanic plume extending about 160 km above the surface of Io. JPL public information frame P-21305C.

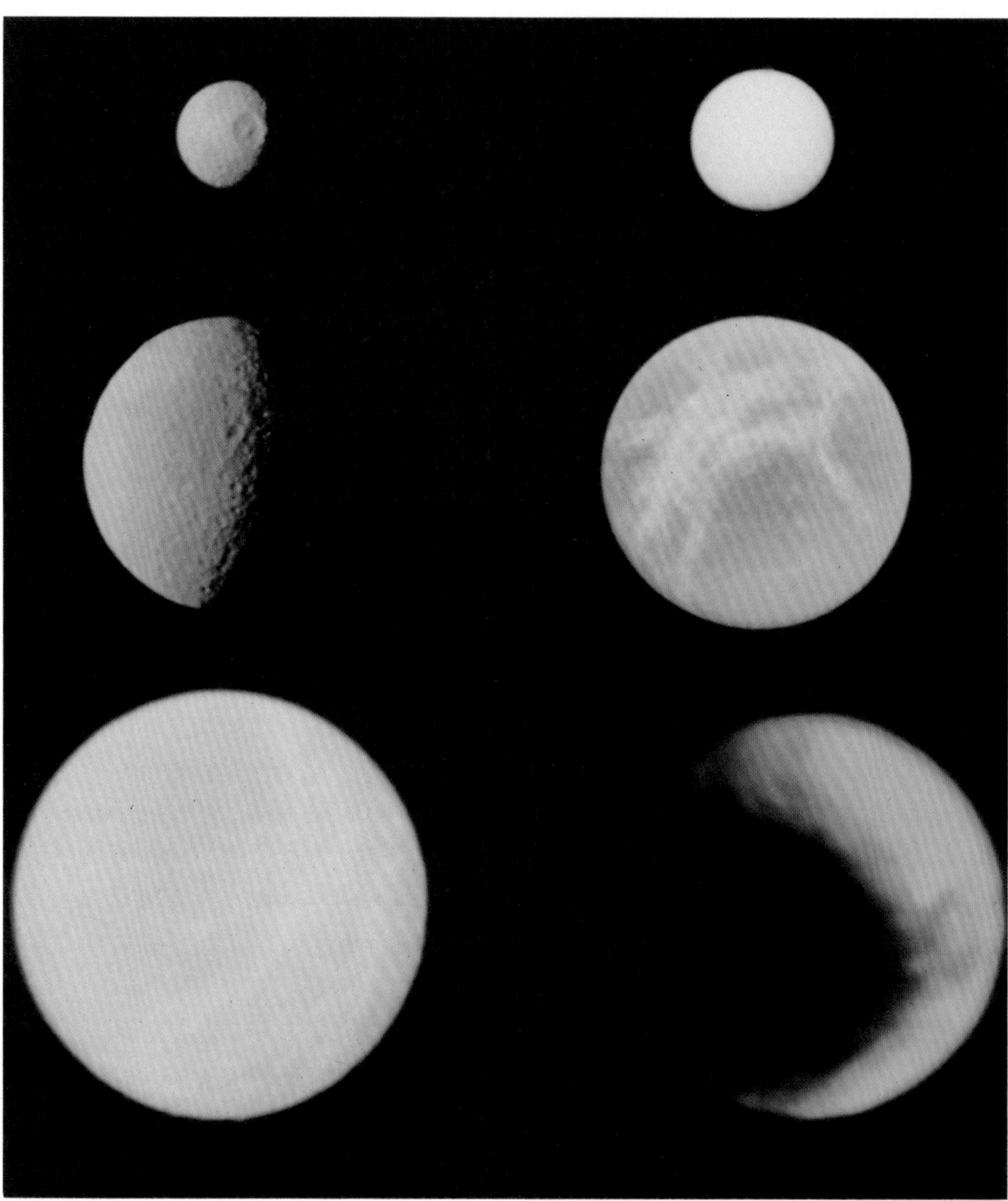

Fig. 36-39. Composite view from Voyager 1 data of the larger icy satellites of Saturn. The satellites are, proceeding clockwise from the upper right: Enceldaus, Tethys, Dione, Rhea, Iapetus, and Mimas. After Smith et al. (1981).

loadings of each of the four MSS bands and accounts for nearly 95 percent of the total scene variance. The term "loading" is used here to refer to the proportion of each Landsat MSS band pixel summed to create the new pixel value for a given principal component. The equal positive loadings have been interpreted as a brightness component in the transformation of Landsat data (C. J. Robinove, U.S. Geological Survey, 1982, personal communication), a fact which is qualitatively ascertained by examination of the first component image (Figure 31-93A). At first glance, the image appears similar to any of the four Landsat MSS bands (Figure 31-92A-D).

Principal component two (PC2, Table 31-23), accounting for 4.3 percent of the total scene variance, contains nearly equal positive loadings of Landsat MSS bands 4 and 5 contrasted against (read the negative signs as contrasts) a nearly equal negative loading of MSS band 6 and a more negative loading of MSS band 7. Basically, this component contrasts MSS bands 4 and 5 against MSS bands 6 and 7, and suggests that materials showing strong contrasts between these two pairs of bands, will be emphasized in the resultant image. The materials that exhibit the strongest contrasts will show the greatest separation in gray shades on the image. Examination of the spectral response curves of vegetation (spectra 5 and 6, Figure 31-87) shows that vegetation contains such spectral contrasts, whereas rocks lack them (spectra 1, 2, 3, and 4, Figure 31-87). The image of the second component (Figure 31-93B) shows that the upland surfaces, which contain dense stands of coniferous and deciduous vegetation, and the lowland cultivated areas, stand out in dark shades. Valley bottoms, which contain sparse grasses and sagebrush, are shown in shades of light gray to white.

Principal component three (PC3, Table 31-23) contains a large positive loading of MSS band 5 contrasted against a negative loading of MSS band 4 and accounts for nearly 0.6 percent of the total scene variance. The contribution of MSS bands 6 and 7 are insignificant, because their loadings are nearly zero. The third component is interpreted as defining materials that have a strong contrast between MSS bands 4 and 5 as compared to materials that exhibit little contrast between these two bands. Examination of spectra 2, 3, and 4 (Figure 31-87) shows that limonitic materials exhibit such contrasts between the two bands, whereas non-limonitic rocks (spectra 1, Figure 31-87) lack this contrast. Examination of the third principal component image (Figure 31-93C) shows known areas of limonitic rocks as shades of light gray to white (compare with *F* and *A1* in Color Figure 31-88A).

Principal component four (PC4, Table 31-23) contains a relatively small positive loading of MSS band 5 and a larger positive loading of MSS band 7 contrasted against a large negative loading of MSS band 6; it accounts for 0.2 percent of the total scene variance. The contribution of MSS band 4 to this component is insignificant. The interpretation of this component is enigmatic and may be attributable to two physical phenomena. The strong contrast of MSS band 6 to MSS band 7 may relate to the broad and shallow absorption band associated with limonitic rocks in the MSS band 7 portion of the spectrum (spectra 2 and 4, Figure 31-87). In addition, a stronger contrast exists between MSS bands 6 and 7 for vegetation (spectra 5 and 6, Figure 31-87). The relatively small but significant contribution of MSS band 5 renders this interpretation questionable. The obvious contrast in the fourth principal component image (Figure 31-93D) between the white and black playas in the northwest and north-central portions of this image, respectively, suggests that they most likely affected this axis. This component tentatively is interpreted as attributable to vegetation with perhaps some contribution from limonite. Detailed statistical training and field checking would be necessary to determine the exact attributes of this component.

The color-composite image of principal components 2, 3, and 4 can now be examined with the above interpretations in mind (Color Figure 31-94). This combination was chosen to subdue effects caused primarily by differences in brightness (component one was excluded for this reason) and to enhance the spectral contrasts that were interpreted from the eigenvector matrix for the lower order components. Component one could have been included, but it would have required exclusion of one of the other components from the color composite. Naturally, the interpretation of an image containing component one also would be different.

In Color Figure 31-94, limonitic areas are shown in bright green, and areas interpreted tentatively as vegetated and limonitic (the contribution of the fourth component) are shown in yellow to yellow-green. Dark rocks, such as basalts, which are barren of vegetation and which contain no limonitic coatings, and would thus be considered spectrally flat (lacking absorption bands), appear purple to magenta. Bright playa surfaces (and in one instance, clouds) are bright red. Note, however, that some playa surfaces are shown in magenta (*S*, Color Figure 31-94), similar to basalts. Use of the first principal component (brightness) would allow the photointerpreter to differentiate basalts from playas. This suggests that the purple to magenta colors of the composite image represent geologic materials containing no limonite, irrespective of their brightness. Cultivated areas in the valleys as well as natural stands of coniferous and deciduous vegetation in the uplands appear red-brown. Other colors may be interpreted from this image with careful evaluation of the eigenvectors, although *a priori* information concerning some geologically known areas is most useful. Without a doubt, the various units discussed are vividly portrayed in this rendition.

The principal component transformation need

not necessarily be restricted to full Landsat scenes. The process is easily applied to smaller portions of image data. Naturally, eigenvectors will differ; hence their interpretations also will differ. Other potential variations of this technique might include exclusion of certain types of material from the gathered statistics, such as vegetation. Then the resultant process might be better able to portray minor differences in the spectral responses of materials in more apparent color differences.

Inverted Principal Component Analysis

The inverted principal component transformation is an additional step beyond the initial principal component transformation. Mathematically, it is a transformation of the principal component data back into the domain of the original variables utilizing the inverse of the principal component matrix (Soha and Schwartz, 1978). This amounts to reapplying the principal component transformation on the normalized (contrast stretched) principal component variables using a row for column transpose of the eigenvector matrix. In graphic terms, the inverted principal component transformation involves re-rotating from the principal component axes back to the original channel axes (Figure 31-91B). The advantages of this process include: 1) maximizing the spectral separability of the sensed materials as in the principal component transformation because of the normalization applied to the principal component data, 2) subduing the "noise" inherent in the later components of the PC transformation, and 3) making the resultant images more interpretable in a phenomenological context.

Color Figure 31-95 illustrates the image generated from the inverted principal component transformation for the area corresponding to Color Figure 31-88A. The color-additive image consists of inverted principal components 1, 2, and 4 (analogous to Landsat MSS bands 4, 5, and 7) displayed in blue, green, and red, respectively. Note that the colors of Color Figures 31-88A and 31-95 are similar, but those in Color Figure 31-95 are exaggerated. Most significantly, areas containing limonitic rocks are enhanced due to the large degree of spectral contrast between limonitic and non-limonitic rocks. Areas containing bright limonitic rocks are displayed as yellow, and darker limonitic rocks are displayed in shades of green. Typically these colors will show in Landsat MSS false-color-composite, contrast-enhanced images only for those limonitic rocks of medium brightness. Bright limonitic rocks are seldom recognized, because they appear white on the standard enhanced image. Areas of non-limonitic rocks are shown in shades of blue. Note that some of these interpretations could have been made from the original MSS false-color composite (Color Figure 31-88A), but the colors were more subdued. Kahle and Rowan (1980) applied the same technique to mid-infrared thermal data to distinguish between rocks of varying silica content (see also the section on "Lithology" of this chapter).

Other Enhancement Techniques for Lithologic Mapping

Because of limited space, not all possible techniques can be discussed in depth; however, others bear mentioning. Density slicing involves the assignment or arbitrarily chosen colors to individual gray levels, or ranges of gray levels, in a digital data set. The process is applied to a single channel of data (i.e., a single band, band ratio, principal component, etc.), and is best used where a particular data set displays a specific phenomenon. As an example, a band ratio of Landsat MSS bands 5/6 can be used as a sensitive indicator of vegetation in semiarid to arid terrains, where low ratios indicate increased vegetation density. Raines et al. (1978) showed an excellent correlation of the MSS bands 5/6 ratio values to vegetation density. The changes in vegetation density, in turn, were related to relatively small differences in clastic sedimentary rock facies. A density slice of a band ratio of 1.6μm/2.2μm proved useful for distinguishing between several types of hydrothermally altered rocks (Ashley and Abrams, 1980; Podwysocki et al., 1982).

Another technique is masking, which may be used to eliminate a specific unwanted phenomenon from consideration. Examples applied to band ratioing techniques are discussed by Segal and Podwysocki (1980), Chavez et al. (1982), and Podwysocki et al. (1982). Merged data sets (see the section on "Combined Data Sets" in this chapter of the *Manual*) provide an excellent method for superposing the effects of disparate data sets. Ancillary information such as aeromagnetic, topographic, gravity, or geochemical data, among others, can be merged with multispectral image data. The ancillary data then can be used to create a digital stereo model, which can provide the photointerpreter with another useful tool to analyze the spatial associations of these disparate data. Compound band ratios, which require a judicious use of ratioed band ratios in order to emphasize a specific phenomenon, have proven successful for locating limonitic rocks in highly vegetated terrains, (Segal, 1982).

Summary

Adaptations of the discussed and other enhancement procedures are limited only by the ingenuity of the user, but it should be emphasized that there should be a sound basis for the application of a particular technique. Most of these techniques are based on the premise that lithologic units can be distinguished and sometimes identified because of several known spectral properties of geologic units. Geologists have made use of physical properties which include the presence of: 1) absorption bands in the ultraviolet, visible and near-infrared portion of the electromagnetic spectrum to detect limonite, 2) absorption bands in the 2.1-μm to 2.5-μm region to detect minerals containing A1-O-H or Mg-O-H associations, or the carbonate radical, 3) absorption bands in the mid-infrared region to detect minerals containing C-O, Si-O, A1-O-H, and Mg-O-H associations in their lattice structures, and 4) characteristic thermal inertias associated with some geologic materials.

Several techniques have been demonstrated and others referenced in this section using spectral properties for mapping geologic materials. Band ratioing can be used for identifying some geologic materials based on their characteristic spectral contrasts. Principal component transformations can be utilized to distinguish materials by maximizing their separability on images, but identification of the specific phenomenon associated with the individual components oftentimes is difficult because of the "combining nature" of the new variables or components. The inverted principal component transformation can be used to distinguish and identify materials, because it maximizes spectral contrasts as do principal components, but also because it transforms the data back into the domain of the original MSS channels. Density slicing, image masking, merged data sets, and compound band ratios can be used to enhance lithologic mapping capabilities.

COMBINED DATA SETS

In many geological applications, remotely sensed data are used with other types of geoscience data. Comparisons of both manually and digitally analyzed image data with geological, geophysical, and geochemical map data are an important component of the interpretive process. Combinations of image- and map-formatted data sets can be accomplished manually by superposing one data set on another using transparent overlays or optical image processing equipment; digital processing techniques offer the most latitude, however, in combining and analyzing diverse types of geoscience data (Fischer et al., 1978).

A digital data base incorporating spatially encoded geologic, geophysical, geochemical, topographic, and remotely sensed data was developed at the EROS Data Center of the U.S. Geological

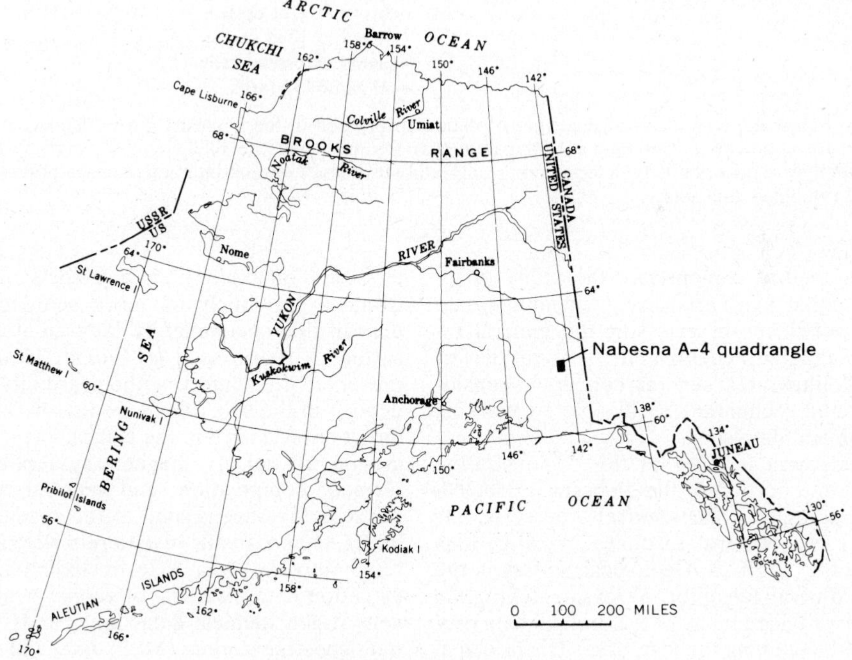

INDEX MAP OF ALASKA
SHOWING LOCATION OF AREA

Fig. 31-96. Index map of Alaska showing the location of the Nabesna A-4 quadrangle and Landsat MSS false-color composite image (1422-20210, 18 September 1973) of the area showing characteristics of the terrain, including the terminus of the Nabesna Glacier.

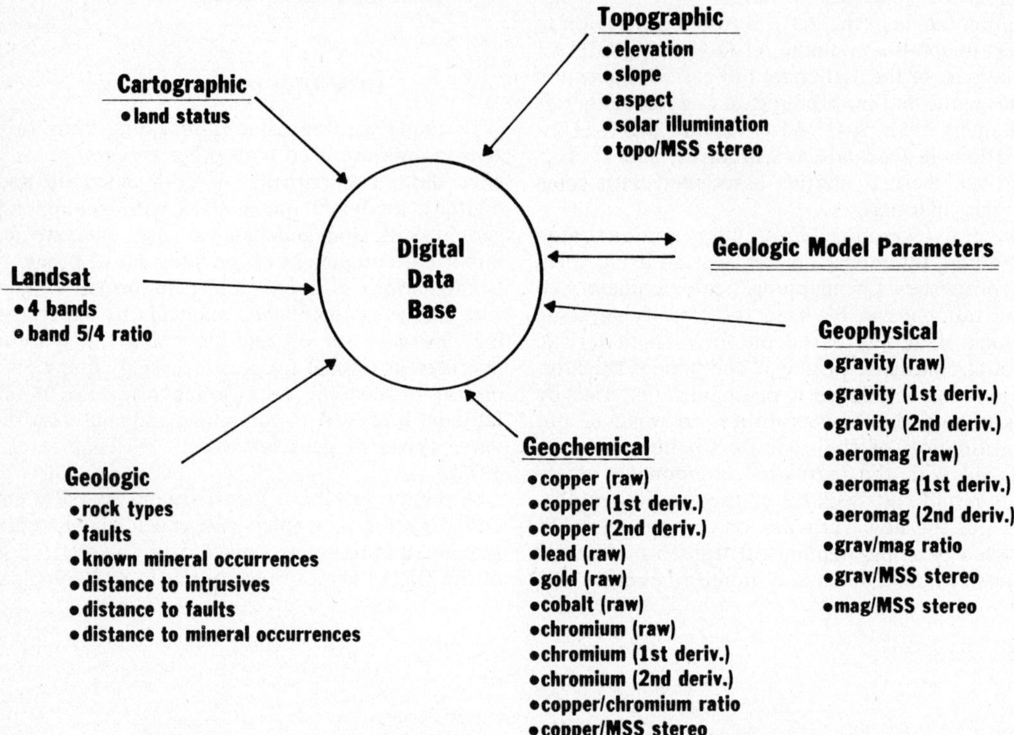

Fig. 31-97. Schematic diagram showing types of data incorporated in the Nabesna digital data base to establish mineral resource potential. Geologic model parameters for assessing the likelihood of porphyry-type copper mineralization were quantitatively determined through statistical analyses and interactive image processing of individual and combined data sets.

Survey in 1980 to demonstrate the utility of interactive digital image processing and multivariate analysis techniques in assessing the mineral resource potential of Federal lands. An overview of this work illustrates several concepts, considerations, and techniques applicable to working with combined data sets.

The study area selected for this demonstration is located in a portion of the Nabesna 1:250,000-scale quadrangle in east-central Alaska (Color Figure 31-96). Mineral resource investigations carried out by the U.S. Geological Survey in the Alaskan Mineral Resource Assessment Program (AMRAP) provided many of the original data necessary for developing the data base. The principal objectives of the study were, first, to develop a quantitative model that numerically described the physical and chemical attributes of porphyry-type copper-molybdenum mineralization in the area and, second, to apply the model to assess the spatial favorability for this type of mineral occurrence within the area.

In mineral resource assessment, much the same as in mineral exploration, a geologist seldom confronts a previously unknown economic mineral deposit that is completely exposed at the Earth's surface. Criteria used for initially identifying the presence of, or inferring the possibility of, such a deposit in the subsurface are usually indirect. Indirect criteria used in the Nabesna study included gravimetric density, magnetic susceptibility, trace element composition, and spectral reflectance, because they are related to the specific types of data selected. In all, 16 data sets were initially entered into the data base, including topographic elevation data, two sets of geophysical data, five sets of geochemical data, four bands of Landsat multispectral scanner (MSS) data, and three types of geologic data. From these data sets, 21 additional data sets were derived through digital processing (Figure 31-97).

The topographic data, geochemical data, and Landsat MSS data were acquired in a digital format. All other data were digitized in geographic information system (GIS) format as sets of points, lines, and polygons from available 1:250,000-scale

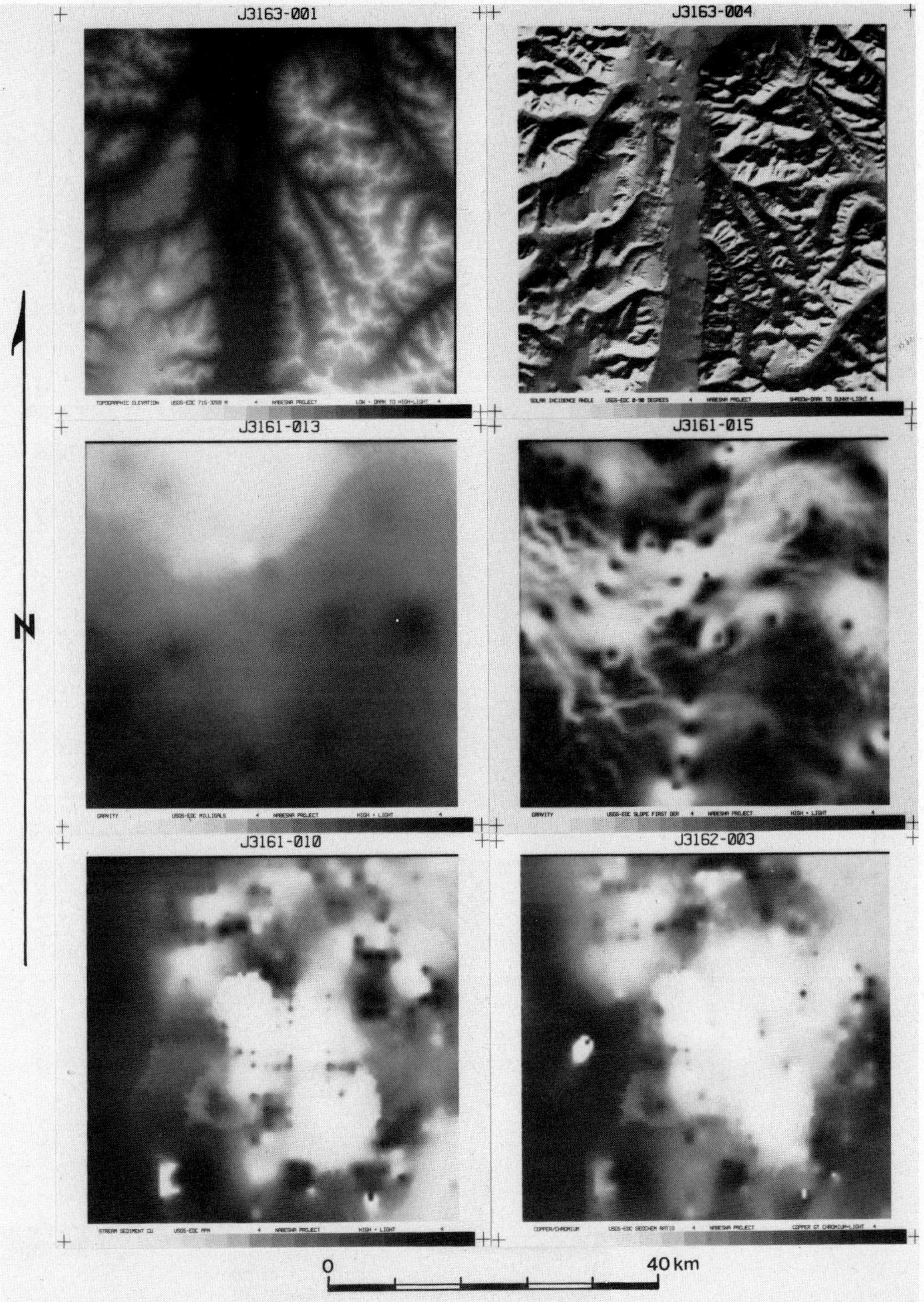

Fig. 31-98. Examples of original and digitally processed image data used to develop model parameters: topographic elevation data (top left) and processed shaded relief image (top right), Bouguer gravity data (center left) and processed first derivative image (center right), and geochemical copper data (bottom left) and processed copper/chromium ratio image (bottom right).

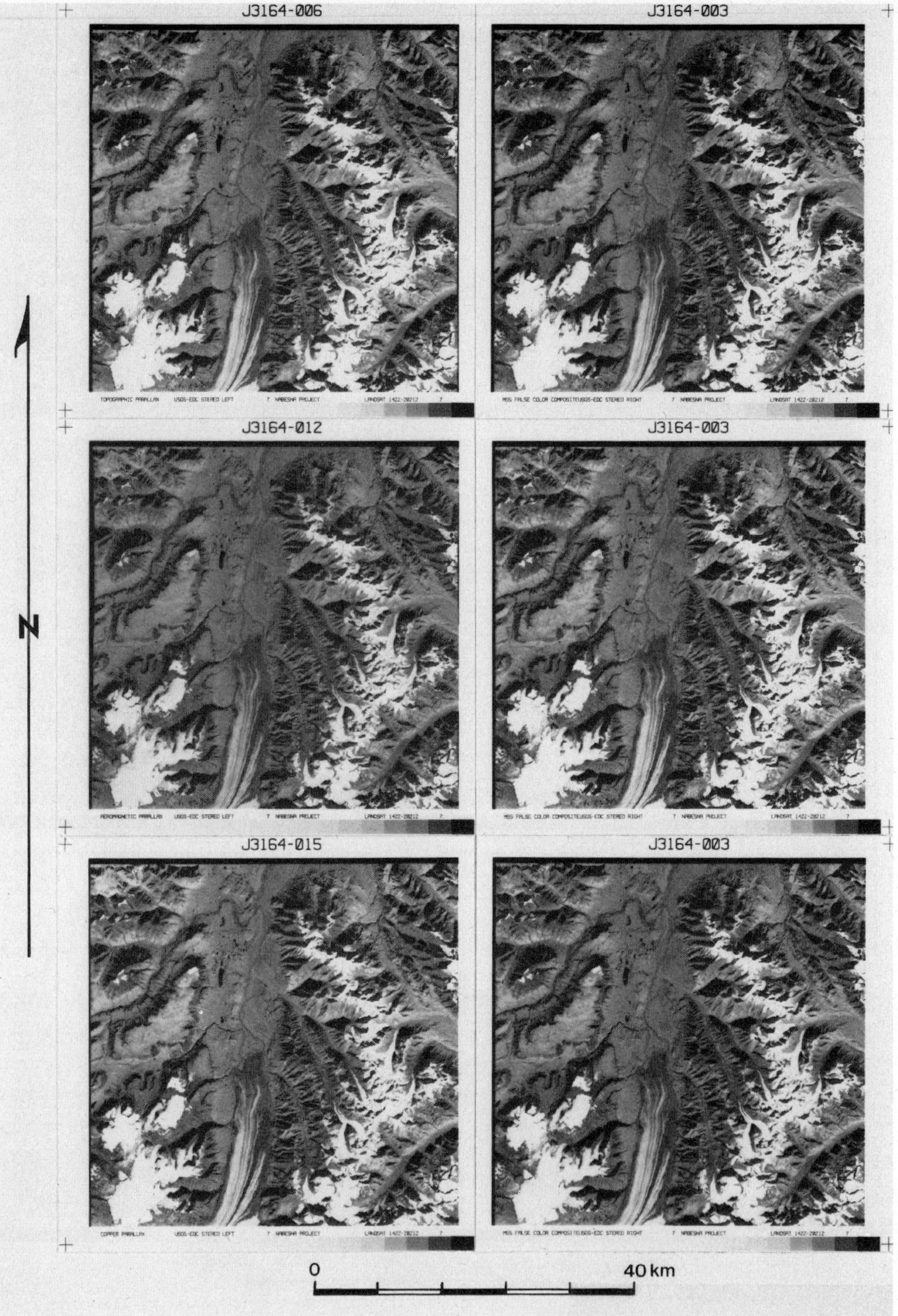

Fig. 31-99. Examples of stereoscopically combined data sets. Parallax has been digitally introduced into the Landsat MSS band 7 image data as a function of topographic elevation data (top stereo pair), residual aeromagnetic data (center stereo pair), and geochemical copper data (bottom stereo pair).

MAP TITLE NABESNA		NUMERICAL IMPORTANCE OF DATA VARIABLE AND SUB–CLASSIFICATION											
DATA VARIABLE		10	9	8	7	6	5	4	3	2	1	0	OFF
10	Gravity (milligals)	-130	-128	-126	-124	-122	-120	-118	-116	-114	-112	-110	>-105
9	Aeromag (gammas)	4800	4850	4900	4950	5000	5050	5100	5150	5200	5250	5300	>5400
8	Copper/Chrome (PPM/PPM)	–	–	–	–	–	–	–	–	–	–	–	<1.0
7	Copper (PPM)	2000	1585	1259	1000	794	631	501	398	316	251	200	<185
6	Slope (%)	–	–	–	–	–	–	–	–	–	–	–	<10%
5	Topo (BETA<)	–	–	–	–	–	–	–	–	–	–	–	Shadow
4	MSS 5/MSS 4	2.0	1.9	1.8	1.7	1.6	1.5	1.4	1.3	1.2	1.1	1.0	<1.0
3	Geology (Within Intrusives)	T_p	K_{np}	K_{nt}	K_{ng}	T_{hd}	K_{dd}	–	–	–	–	–	–
2	Geology (Distance to any Intrusives)	<1Km	–	1Km	–	2Km	–	3Km	–	4Km	–	5Km	>6Km
1	Mineral Occurrence (Distance to any Occurrence)	<1Km	–	1Km	–	2Km	–	3Km	–	4Km	–	5Km	>6Km

Fig. 31-100. Application of the quantitative model developed for assessing porphyry-type copper potential in the Nabesna quadrangle. Description shows ranking of model parameters, their range limits, and weighting factors. Imaged results show model-defined areas superposed on Landsat MSS data (warmer colors denote areas having higher potential). Porphyry copper systems identified include Orange Hill (A), Bond Creek (B), Nikonda Creek (C), and Cross Creek (D).

maps. All data sets were registered to a Universal Transverse Mercator (UTM) map base, resampled to a 50-meter ground-equivalent grid cell size, and, where deemed physically valid, interpolated to create continuous data arrays. In this configuration, the data base could be used for both statistical and image analyses of incorporated data, development and application of a quantitative geologic model, and generation of final output maps that maintain the locational control necessary for establishing mineral potential.

Interactive digital processing techniques applied to the spatially encoded data within the data base included both image and statistical analyses. The processing sequence began with the univariate analysis of each initial data set. In this stage of analysis, features or attributes of each set that are most likely to be associated with porphyry copper occurrences were identified and quantified. Included in this stage were histogram analysis, spatial filtering, and derivative analysis of continuous data.

After all data sets were individually analyzed to select threshold values and to extract anomalies using techniques described by Sinclair (1976) and Miesch (1981), they were displayed in a merged or overlay mode to determine the degree of spatial correlation of anomalies. In this stage of analysis, model parameters defined by individual and combined data sets were established. Finally, data integration techniques, including ratioing and multivariate analyses, as described by Davis (1973) and Koch and Link (1970), were applied to quantify the numeric range of each parameter to be used in the quantitative model.

Representative image products generated during the progressive stages of analysis are shown in Figures 31-98 and 31-99. Topographic elevation data were used to create solar illumination data shown as a shaded relief image in Figure 31-98 (top). These data, in turn, were used to mask out shadowed terrain when evaluating Landsat MSS band ratios. Bouguer gravity data, used to define variations in the density of rock bodies, were processed to create gradient data shown as a first derivative image in Figure 31-98 (center). Gradient data, in turn, were used to evaluate the geometry of geologic contacts and structural trends based on density contrast and source depth. Stream-sediment copper data were statistically analyzed to establish a 185-ppm threshold value separating anomalous and background populations in the data set. In conjunction with chromium data, copper data were used to create copper/chromium ratio data shown in Figure 31-98 (bottom). The ratioed data were used to differentiate copper values associated with mafic-igneous sources from those associated with felsic-igneous sources.

Image products generated to establish spatial correlations between data sets are exemplified in Figure 31-99. Through the use of a technique described by Batson et al. (1976), the combination of Landsat MSS data with topographic, aeromagnetic, and copper data was achieved by digitally distorting the Landsat data geometry as a function of a calculated "relief displacement" from the data set to be merged. The distorted Landsat data, when imaged and stereoscopically viewed with an undistorted Landsat image, present a three-dimensional perspective of the data used to distort the imagery (i.e., topographic, aeromagnetic, copper, or any other continuous data set) that is geographically registered to the base image.

The sequence of univariate analysis, data-set merging, and integration techniques resulted in a model consisting of ten quantified parameters that, together, physically and chemically described probable porphyry-type copper-molybdenum centers. The resultant model included four regional parameters and six local parameters. These parameters were categorically weighted with respect to their relative importance in describing a favorable geologic environment, and the model range of each parameter was scored on the basis of its goodness of fit with its statistically defined optimum. In its application, the model was used as the basis for establishing a composite score for each grid cell within the data base; each grid cell, in this instance, had a set of parameter values that were determined by prior processing of the original, spatially registered data sets.

Regionally, Bouguer gravity, residual aeromagnetic, and geochemical copper and chromium parameters were successfully employed in describing four areas of high porphyry copper-molybdenum potential, two of which contain known porphyry-type deposits. Locally, parameters developed from topographic slope and aspect, Landsat, and ground-based geologic data were used to identify specific porphyry copper targets within the areas of high regional potential. Model format and imaged results are shown in Color Figure 31-100.

In this case study, the combination of data sets was an integral part of the interpretive process—from the standpoint of data base design, through analytic procedures, to the resultant geologic model. Although there is nothing conceptually new in the approach to working with diverse sets of data (Missallati et al., 1979), digital processing technology currently offers an efficient means of handling them. Specific techniques mentioned in this brief summary are only a sampling of the possibilities (Agterberg, 1981; Singer and Mosier, 1981); their utility can only be realized when sound geologic reasoning is applied in combining, analyzing, and interpreting remotely sensed data with any other type of geoscience data.

APPLICATIONS OF REMOTE SENSING TECHNOLOGY TO GEOLOGIC SUBDISCIPLINES

INTRODUCTION

This part of Chapter 31 is devoted to the presentation of different case histories relevant to various subdisciplines of geology, in which remotely sensed data have been effectively used to better understand a geologic problem, map an area, document or monitor geologic phenomena, and so forth. The major subdisciplines canvassed are: geologic mapping, economic geology, engineering geology, geologic hazards, glaciology, geomorphology, marine geology, geobotany, archaeological geology, and astrogeology. Each of the many authors has endeavored to combine text, illustrations, and key references to the relevant literature, so that the reader will have a well illustrated, yet concise introduction to each topic addressed. As with the rest of the chapter, the authorship of each subsection, section, or part of the chapter is listed at the end of the chapter. Readers are encouraged to correspond with the author(s) for additional information about specific topics.

The scientific literature on various aspects of geologic remote sensing has increased severalfold since the publication of the first edition of the *Manual of Remote Sensing* (Reeves et al., 1975 a and b), and the referenced literature in this first *Manual* predates 1974. As noted previously in this chapter, an international bibliography on "Remote Sensing and Mineral Exploration" (International Geological Correlation Programme Project 143), compiled by W. D. Carter and L. C. Rowan of the U.S. Geological Survey, had exceeded 1,500 citations by mid-1982, more references than given in the entire geology chapter (Tator et al., 1960) of the *Manual of Photographic Interpretation* (Colwell, 1960) and more than three times the number given in the geology chapter (Reeves et al., 1975a) of the first edition of the *Manual of Remote Sensing* (Reeves et al., 1975b).

The 1,057 references at the end of this chapter have all been cited in the text. They provide selective coverage of the geological remote sensing literature. The reader is also referred to available bibliographies, such as Carter (1967) and Krumpe (1976), and to the proceedings volumes of various regular or special symposia, to regular scientific journals and newsletters, or to proceedings or abstracts volumes of annual meetings of professional societies.

Also of special interest to geologists are the numerous Technical Letters and Interagency Reports to the National Aeronautics and Space Administration (NASA) which were prepared by the U.S. Geological Survey (USGS) or its contractors during the period 1964–1970. They were prepared in cooperation with the Earth Resources Program Office of NASA. Those reports of interest to geologists are listed in Table 31-24. An annotated list of the technical letters was prepared by the U.S. Geological Survey (Carter, 1967). Several of the reports listed in Table 31-24 were subsequently published in scientific journals or in USGS publications. Copies of the original technical letters are available either from the U.S. Geological Survey Library, Reston, Virginia, or from the Donald Lee Kulow Memorial Library at the U.S. Geological Survey's EROS Data Center, Sioux Falls, South Dakota. The technical letters represent an important body of geological remote sensing research, the results of which pointed the way to the Landsat program. Llaverias (1970) and Llaverias and Lowe (1970) also prepared two comprehensive remote sensing bibliographies for Earth resources for 1966–1967 and 1968, respectively, in preparation for the Landsat effort.

Table 31-25 is a chronological list of the International Symposia on Remote Sensing of Environment, a series of remote sensing symposia first begun at the University of Michigan in February 1962, and, since 1978, also held periodically in various foreign cities. Table 31-26 is a chronological list of all of the principal special symposia on remote sensing which have been sponsored or co-sponsored by the American Society of Photogrammetry between 1970 and 1982.

Table 31-27 is a chronological list of the William T. Pecora Memorial Symposia held in Sioux Falls, South Dakota, between 1975 and 1982. Table 31-28 is a selected list of miscellaneous remote sensing symposia, the first six of which were specifically directed at geological remote sensing. The four International Conferences on the New Basement Tectonics are of particular interest to structural geologists and economic geologists. The Canadians held seven symposia on remote sensing between 1972 and 1981 and have scheduled an eighth in 1983. The Remote Sensing Society of England, in addition to publishing a periodic newsletter and the *International Journal of Remote Sensing* (see Table 31-31), has sponsored a number of topical symposia in remote sensing, including the proceedings volume for the Eighth Annual Conference of the Remote Sensing Society, "Geological and Terrain Analysis Studies" (Allan and Bradshaw, 1981). The latter proceedings volume, as well as the other eight volumes available from the Remote Sensing Society, are given in Table 31-28.

Table 31-29 is a chronological list of Landsat symposia sponsored by the National Aeronautics and Space Administration. Table 31-30 is a chronological list of Committee on Space Research (COSPAR) plenary meetings on studies of Earth resources from orbiting space platforms from 1973 to 1982. Each COSPAR symposium has produced a proceedings volume, and the last three meetings have yielded special proceedings volumes on remote sensing and mineral exploration. These publications are also noted in Table 31-30. Table 31-31 is a selected list of professional societies involved in geological remote sensing and associated journals. Also noted in Table 31-31 are other periodical remote sensing publications or newsletters relevant to geologists.

The geological remote sensing literature is now so diverse that there is a considerable problem in keeping abreast of new developments and findings. A careful and comprehensive literature search is necessary before beginning a new geological study in which remotely sensed data will play a part in the investigation. Tables 31-24 through 31-31 provide information on the source of much of the available literature, but traditional scientific journals should also be canvassed. Specific case histories in the rest of this part of the chapter will provide a good window on the key relevant literature for each topic.

GEOLOGIC MAPPING

Introduction

Geologic mapping is the principal method used by geologists to understand the geologic nature of the planet. Geologic mapping in a given area involves an understanding of the character and nature of the landforms (geomorphology), the lithologic characteristics and stratigraphic age of the rocks, and the structural and tectonic setting of the rocks. The use of geologic remote sensing in geomorphology is discussed in a later section of this chapter and by Lowman and Lattman (1980); remote sensing of lithology is discussed in this section, in a previous part ("Analysis Techniques") of this chapter, and by Abrams and Siegal (1980); the use of remote sensing techniques for structural and tectonic studies is discussed later in this section and by Gold (1980).

Aerial photographs have been used by most field geologists ever since they became available during the 1930s and especially after World War II (Lattman and Ray, 1965; Mekel, 1970). For a more extensive discussion on photogeology the reader is referred to the previous section of this chapter on "Photogeology and Geological Remote Sensing." Many excellent books and articles are available on the use of aerial photographs for geologic mapping projects (Ray, 1960; Miller and Miller, 1961; Lattman and Ray, 1965; Mekel, 1970; and Dueholm, 1979, to cite just a few).

The availability of Landsat MSS images of most land areas of the Earth has provided new opportunities for photogeologists. Many areas can now be studied and mapped geologically for which neither adequate maps nor vertical aerial photographs were available. Landsat images are also being used as image map bases on which the geology is overprinted, such as maps of the Yemen Arab Republic (Grolier and Overstreet, 1978), Antarctica (Wolmarans and Krynauw, 1981), Cape Cod, Massachusetts (Williams et al., 1983a), and other areas. On these maps the Landsat image is being used as a substitute for conventional line maps, which may or may not be available at the needed scale for a given area. Table 31-32 shows the Landsat image maps currently available from the U.S. Geological Survey. Williams et al. (1982a) report on the availability of Landsat image maps of Antarctica, geologic mapping in Antarctica in which Landsat images are being used as the map base, and other uses of Landsat images by the Antarctic Treaty nations.

Landsat images have been used for broad regional geomorphic studies, such as McKee's classic study of sand seas of the world (McKee, 1979),

TABLE 31-24

List of Selected U.S. Geological Survey Technical Letters and/or Interagency Reports to NASA Issued During the Period 1964–1970 Under a USGS-NASA Cooperative Research Project in Geological Remote Sensing

Technical Letter No.	Title	Author(s)
2 (1964)	Geologic Reconnaissance Report of the Pisgah Crater, California Area (12 p.)	S. J. Gawarecki
3 (1964)	Interim Report of Ultraviolet Absorption and Stimulated Luminescence Investigations Being Undertaken in Cooperation with the National Aeronautics and Space Administration	
	Part I—Ultraviolet Video Imaging System	W. R. Hemphill S. J. Gawarecki
	Part II—Spectral Distribution of Ultraviolet Stimulated Luminescence	W. A. Fischer D. L. Daniels
	Part III—Measurement of Ultraviolet Reflectance	W. A. Fischer R. Gerharz
4 (1965)	Preliminary Geologic Map—Pisgah Crater and Vicinity, California	T. W. Dibblee, Jr.
5 (1965)	Preliminary Results of Aerial Infrared Surveys at Pisgah Crater, California	W. A. Fischer J. D. Friedman T. M. Sousa
6 (1965)	Ultraviolet Absorption and Luminescence Investigations Progress Report	W. R. Hemphill S. U. Carnahan
7 (1965)	Topographic Studies of Pisgah Crater, California	R. E. Altenhofen J. K. Oman T. M. Sousa
8 (1965)	Reflectance Measurements in the 0.6 to 2.5 Micron Part of the Spectrum (11 p.)	R. Gerharz W. A. Fischer
9 (1965)	Preliminary Geologic Map of the Mono Craters Quadrangle, California	R. W. Kistler
10 (1965)	Laboratory Tests of the (HRB-Singer, Inc.) Reconofax IV Infrared Imaging System (Report is Classified CONFIDENTIAL)	W. R. Hemphill G. R. Boynton P. W. Philbin M. E. O'Neal, III
11 (1966)	Geologic Map of the Pisgah and Sunshine Cone Lava Fields (5 p.)	W. S. Wise
12 (1966)	Geologic Map of the Mono Craters Area, California (11 p.)	J. D. Friedman
13 (1966)	Infrared Spectral Emittance of Rocks from the Pisgah Crater and Mono Craters Areas, California (20 p.)	D. L. Daniels
14 (1966)	Summary of Significant Results of Remote Sensing Studies in 1965 (15 p.)	W. D. Carter
15 (1966)	A Millimeter Wavelength Interferometer Spectrometer	G. Wyntjes N. O. Young
16 (1966)	Geological Evaluation of Radar Imagery, Oregon Coast (13 p.)	P. D. Snavely, Jr. H. C. Wagner
17 (1966)	Evaluation of Ektachrome and Multiband Photography in Caliente Range, California (6 p.)	J. G. Vedder E. W. Wolfe
18 (1966)	Evaluation of Enhanced Multiband Photography, San Andreas Fault Zone, Carrizo Plain, California	R. E. Wallace
19 (1966)	Geologic Evaluation of Radar Imagery of the Central Part of the Oregon High Cascade Range (11 p.)	D. A. Swanson
20 (1966)	Composition of Basalt Flows at Pisgah Crater, California: Preliminary Data (29 p.)	J. D. Friedman
21 (1966)	Lake Surveying Techniques in the Geological Survey—Progress Report (7 p.)	H. E. Skibitzke C. J. Robinove
22 (1966)	Time, Shadows, Terrain, and Photointerpretation (8 p.)	R. J. Hackman
23 (1966)	Geological Appraisal of Radar Imagery of Southwestern Oregon (3 p.)	W. P. Irwin
24 (1966)	Photogeologic Interpretation of Gemini IV Color Photograph: Baja, California (3 p.)	R. Tabor
25 (1966)	Evaluation of Radar Imagery of Highly Faulted Volcanic Terrane in Southeast Oregon (7 p.)	G. W. Walker
26 (1966)	Application of Radar Imagery to a Geologic Problem at Glacier Peak Volcano, Washington (2 p.)	R. Tabor
27 (1966)	Geologic Evaluation of Radar Imagery of Flights 100-B and 100-C Across the Central Sierra Nevada, California (4 p.)	P. C. Bateman
28 (1966)	Radar Imagery of Twin Buttes Area, Arizona Test Site 15 (4 p.)	J. R. Cooper
29 (1966)	Radar Imagery: Salton Sea Area, California (7 p.)	E. W. Wolfe

30 (1966)	Preliminary Evaluation of Radar Imagery of Yellowstone Park, Wyoming (16 p.)	R. L. Christiansen
		K. L. Pierce
		H. J. Prostka
		E. T. Ruppel
31 (1966)	Ultraviolet Instrumentation for Orbital Remote Sensors	H. T. Betz
32 (1966)	Laboratory Measurement of Ultraviolet Reflection (2200–7000 A) and Stimulated Emission of Rocks and Rock-Forming Minerals (57 p.)	H. Watts
33 (1966)	Geological Studies of the Earth and Planetary Surfaces by Ultraviolet Absorption and Stimulated Luminescence (9 p.)	W. R. Hemphill
		R. Vickers
34 (1966)	Gemini V Color Photography of Salton Sea Area, California (4 p.)	E. W. Wolfe
35 (1966)	Evaluation of High Resolution Infrared Radiometry (HRIR) Imagery (15 p.)	M. A. Conti
36 (1966)	The Effect of Prolonged Ultraviolet Radiation on the Intensity of Luminescence (7 p.)	D. L. Daniels
37 (1966)	Preliminary Ultraviolet Reflectance of Some Rocks and Minerals from 2000 A to 3000 A (45 p.)	A. N. Thorpe
		C. M. Alexander
		F. E. Senftle
38 (1966)	Geological Evaluation of Radar Imagery, Southwestern and Central Utah (10 p.)	L. S. Hilpert
39 (1966)	Interpretation of Ultraviolet Imagery of the Meteor Crater, Salton Sea and Arizona Sedimentary Test Sites (28 p.)	W. R. Hemphill
40 (1966)	Geologic Interpretation of the Gemini V Photograph of the Salt Range—Potwar Plateau Region, West Pakistan (13 p.)	W. R. Hemphill
41 (1966)	Possible Application of Remote Sensing Techniques and Satellite Communications for Earthquake Studies (3 p.)	R. E. Wallace
		D. B. Slemmons
42 (1966)	Use of Infrared Imagery in Study of the San Andreas Fault System, California (14 p.)	R. E. Wallace
		R. M. Moxham
43 (1966)	Geologic Utilization of Gemini Color Photography of Duba Area, Saudi Arabia (3 p.)	R. F. Johnson
		J. A. MacKallor
44 (1966)	Preliminary Report on Radar Imagery of Cedar City–Iron Springs Area, Utah (8 p.)	P. L. Williams
45 (1966)	Geologic Evaluation of Radar Imagery: San Andreas Fault Zone from Stevens Creek, Santa Clara County to Mussel Rock, San Mateo County, California (8 p.)	R. D. Brown, Jr.
46 (1966)	An Evaluation of the Gemini IV Color Photos of the Gulf of California–Central Texas Area (7 p.)	H. Drewes
47 (1966)	Geologic Evaluation of Radar Imagery of the Near Spanish Peaks Region, Colorado (5 p.)	R. B. Johnson
48 (1966)	Geological Evaluation of Radar Imagery, Appalachian Piedmont, Harford and York Counties, Maryland and Pennsylvania (6 p.)	D. L. Southwick
49 (1966)	Geological Evaluation of Radar Imagery, North-Central Nevada (8 p.)	R. J. Roberts
50 (1966)	A Preliminary Evaluation of Airborne and Spaceborne Remote Sensing Data for Hydrologic Uses (5 p.)	C. J. Robinove
51 (1966)	Applications of Remote Sensor Data to Cartographic Programs (7 p.)	W. Sibert
52 (1966)	Geologic Investigations of Remote Sensing Techniques, Final Report to NASA FY 1966 (20 p.)	W. D. Carter
53 (1966)	Evaluation of Nimbus Vidicon Photography, Southwest France and Northeast Spain (7 p.)	E. G. Hasser
54 (1966)	Potential Time-Cost Benefits from Use of Orbital-Height Photographic Data in Cartographic Program (16 p.)	L. E. Starr
		W. Sibert
55 (1966)	Evaluation of Nimbus Imagery, Sonora-Baja California	E. G. Hasser
56 (1966)	Geological Evaluation of Nimbus Vidicon Imagery, Northwest Greenland (6 p.)	W. E. Davies
57 (1966)	Liquid Nitrogen Blackbody for Spectral Emittance Studies (7 p.)	D. L. Daniels
		A. E. Stoddard
58 (1966)	Geological Evaluation of Radar Imagery in Southern Utah	R. J. Hackman
59 (1966)	Analysis of Earth Orbiter Test Site Program in Relation to U.S. Mineral Needs (11 p.)	W. D. Carter
60 (1966)	Extent of Relict Soils Revealed by Gemini IV Photographs (10 p.)	H. Drewes
		R. Morrison
61 (1966)	Hydrologic Interpretation of Nimbus Vidicon Image, Great Salt Lake, Utah (3 p.)	D. C. Hahl
		A. H. Handy
62 (1966)	Radar Imagery—Meteor Crater, Arizona (14 p.)	G. G. Schaber
63 (1966)	Preliminary Studies of Soil Patterns Observed in Radar Images—Bishop Area, California (5 p.)	M. F. Sheridan
64 (1966)	Geologic Evaluation of Nimbus Vidicon Photography, Chesapeake Bay–Blue Ridge (6 p.)	C. R. Lewis
		W. E. Davies
65 (1966)	Dispersive Multispectral Scanning: A Feasibility Study (53 p.)	J. Braithwaite
66 (1966)	Status Report of Infrared Investigations July 1, 1966–September 30, 1966 (9 p.)	R. M. Moxham
67 (1966)	Integrated Landscape Analysis from Radar Imagery	N. R. Nunnally
68 (1966)	Hydrologic Evaluation of Gemini Photographs of Fringes of the Sahara Africa (13 p.)	J. R. Jones
69 (1966)	Gemini Photography Evaluation (12 p.)	R. H. Nugent
		L. E. Starr
70 (1966)	Measurement of Luminescence by the Fraunhofer Line-Depth Method (20 p.)	H. T. Betz
71 (1967)	Infrared Imagery of Lordsburg–Silver City Area, New Mexico (13 p.)	W. F. Pratt
72 (1967)	Geologic Interpretation of Infrared Images of the Pond Oreille Area, Idaho (10 p.)	J. E. Harrison

TABLE 31-24—con't

Technical Letter No.	Title	Author(s)
73 (1967)	Ultra-High-Altitude Photography Compilation Evaluation (6 p.)	R. H. Nugent
74 (1967)	Aerial Infrared Surveys in Water Resources Studies (20 p.)	R. M. Moxham
75 (1967)	Use of Infrared Imagery and Color Photography in the Study of Missile Impact Craters (7 p.)	H. J. Moore D. Cummings D. E. Gault
76 (1967)	Infrared Imagery of Part of the High Cascade Range and McKenzie River Valley, Oregon (19 p.)	D. A. Swanson
77 (1967)	Geologic Evaluation of Infrared Imagery of Highly Faulted Terrain in Southeast Oregon (17 p.)	G. A. Walker
78 (1967)	Use of Infrared Imagery in Circulation Studies of the Merrimack River Estuary, Massachusetts (11 p.)	D. R. Wiesnet J. E. Cotton
79 (1967)	Resolution Study (8 p.)	R. H. Nugent H. B. Loving
80 (1967)	Thermal Surveys of Lake Erie in the Cleveland and Toledo Areas, Ohio–Michigan	R. M. Moxham H. Skibitzke
81 (1967)	Radar Imagery: Parmachenee Lake Area, West-Central Maine (14 p.)	D. S. Harwood
82 (1967)	Thermal Anomalies and Geologic Features of the Mono Lake Area, California, As Revealed by Infrared Imagery (84 p.)	J. D. Friedman
83 (1968)	Geological Evaluation of Infrared Imagery Eastern Part of Yellowstone National Park, Wyoming and Montana (46 p.)	H. S. Smedes
84 (1967)	Radar Images—San Francisco Volcanic Field, Arizona, a Preliminary Evaluation (17 p.)	G. G. Schaber
85 (1967)	Geologic Evaluation of Infrared Imagery of the Twin Buttes Area, Arizona	J. R. Cooper
86 (1967)	Annotated Bibliography of USGS Technical Letters—NASA on Remote Sensing Investigations Through June 1967 (46 p.)	W. D. Carter
87 (1967)	A Photo-Mosaic of Western Peru from Gemini Photography (10 p.)	J. A. MacKallor
88 (1967)	An Engineering Feasibility Study of an Orbiting Scanning Radiometer—Final Report (83 p.)	J. Braithwaite A. W. Krause W. L. Brown
89 (1967)	Ektachrome and Ektachrome Infrared Photography of the Twin Buttes Area, Arizona (9 p.)	J. R. Cooper
90 (1967)	Additional Infrared Spectral Emittance Measurement of Rocks from the Mono Crater Region, California (11 p.)	D. L. Daniels
91 (1967)	Far Infrared Luminescence (21 p.)	A. E. Stoddard
92 (1967)	Visible and Ultraviolet Reflectance and Luminescence from Various Saudi Arabian and Indiana Limestone Rocks (67 p.)	H. V. Watts
93 (1968)	Evaluation of Infrared Imagery Applications to Studies of Surficial Geology—Yellowstone Park (35 p.)	K. L. Pierce
94 (1968)	Film Density Analyzers for Infrared Investigations (19 p.)	R. M. Turner
95 (1967)	Survey of Lunar Geology (69 p.)	J. J. Cook
96 (1968)	Vacuum Ultraviolet Measurements of Reflectance and Luminescence from Various Rock Samples (44 p.)	H. V. Watts
97 (1967)	Comparison of a UV Scanner/Photomultiplier with an Image Orthicon (14 p.)	H. Goldman
98 (1968)	Hydrologic Evaluation of Infrared Imagery of Great South Bay, Hempstead Bay Region of Long Island, New York	E. J. Pluhowski
99 (1968)	Infrared Survey of the Pisgah Crater Area, San Bernardino County, California (52 p.)	S. J. Gawarecki
100 (1968)	Ultraviolet Absorption and Luminescence Studies Progress Report for the Period April to December 1967 (54 p.)	W. R. Hemphill
101 (1968)	The Remote Measurement of Rhodamine B concentration When Used as Fluorescent Tracer in Hydrologic Studies (48 p.)	H. T. Betz
102 (1968)	Structural Geologic Interpretation from Radar Imagery (24 p.)	R. G. Reeves
103 (1968)	Imagery of Craters Produced by Missile Impacts (38 p.)	H. J. Moore II R. Kachadoorian J. F. McCauley
104 (1968)	Distinction Between Bedrock and Unconsolidated Deposits on 3-5 Infrared Imagery of the Yellowstone Rhyolite Plateau (7 p.)	R. L. Christiansen
105 (1967)	Infrared Imagery and Radiometry Summary Report (66 p.)	R. M. Moxham
106 (1968)	Evaluation of Radar and Infrared Imagery of Sedimentary Rock Terrane, South-Central Yellowstone National Park, Wyoming (12 p.)	W. R. Keefer
107 (1968)	Synoptic Temperature Measurements of a Glacier Lake and its Environment (14 p.)	W. J. Campbell
108	Never completed.	
109 (1968)	Geothermal Studies of Yellowstone National Park (Test Site II), Wyoming (11 p.)	D. E. White R. O. Fournier L. J. P. Muffler L. H. Truesdale

110 (1968)	Aerial Infrared Images of the Geysers Geothermal Steam Field and Vicinity, Sonoma County, California (13 p.)	R. M. Moxham
111 (1968)	Preliminary Soil Classification Map of Southwestern U.S. and Mexico from Space Photography (14 p.)	R. B. Morrison
112	Never completed.	
113 (1968)	Geologic Evaluation of Thermal Infrared Imagery, Caliente and Temblor Ranges, Southern California (24 p.)	E. W. Wolfe
114 (1968)	Gemini Mosaic Along the Thirty-Second Degree of Latitude from Baja California to Central Texas (11 p.)	J. A. MacKallor
115	Never completed.	
116	Never completed.	
117 (1968)	Making Color Infrared Film a More Effective High Altitude Sensor (31 p.)	R. W. Pease
121 (1968)	Application of Ultraviolet Reflectance and Stimulated Luminescence to the Remote Detection of Natural Materials (36 p.)	W. R. Hemphill
122 (1968)	Gridding of Near Vertical Unrectified Space Photographs (46 p.)	R. H. Rapp
123 (1968)	Aerial Infrared Surveys at the Geysers Geothermal Steam Field, California (48 p.)	R. M. Moxham
124 (1968)	Preliminary Evaluation of Infrared and Radar Imagery, Washington and Oregon Coasts (13 p.)	P. D. Snavely
125 (1968)	Potential of Radar Remote Sensors As Tools in Reconnaissance Geomorphic, Vegetation and Soil Mapping (14 p.)	D. S. Simonett
126 (1968)	Land Evaluation Studies with Remote Sensors in the Infrared and Radar Regions (31 p.)	D. S. Simonett
127 (1968)	Remote Detection of Solar Stimulated Luminescence (24 p.)	W. R. Hemphill
128 (1968)	Preliminary Remote Sensing of the Delaware Estuary (24 p.)	R. W. Paulson
129 (1968)	Selected Bibliography of Remote Sensing (34 p.)	R. B. Hones
133 (1969)	Geologic Evaluation of Radar Imagery, Caliente and Temblor Ranges, Southern California (17 p.)	E. W. Wolfe
134 (1968)	Bibliography of Remote Sensing of Earth Resources for Hydrologic Applications (75 p.)	R. K. Llaverias
135 (1968)	Thermal Infrared Imagery in Urban Studies	B. S. Wellar
149 (1969)	Microwave Radiometric Studies and Ground Truth Measurements of the NASA/USGS Southern California Test Site (95 p.)	A. T. Edgerton
150 (1969)	Multispectral Techniques for General Geological Surveys Evaluation of a Four-Band Photographic System	D. F. Crowder
165 (1970)	Geologic Evaluation of Anomalies Between Like-Polarized and Cross-Polarized K-Band Side-Looking Radar Imagery of Yellowstone National Park (24 p.)	G. M. Richmond
166 (1969)	Detection of Thick Surficial Deposits on 8–14 μm Infrared Imagery of the Madison Plateau, Yellowstone National Park (8 p.)	H. A. Waldrop
175 (1969)	Test of Data Compilation Procedures for the Fraunhofer Line Discriminator (35 p.)	G. E. Stoertz
177 (1969)	Rapid Heat-Flow Surveying of Geothermal Areas, Utilizing Individual Snowfalls As Calorimeters (21 p.)	D. E. White
179 (1970)	Geologic Interpretation of a Radar Mosaic of Yellowstone National Park (16 p.)	H. J. Prostka
181 (1970)	Geologic Studies of Yellowstone National Park Imagery Using An Electronic Image Enhancement System (24 p.)	H. W. Smedes
189 (1970)	Remote Sensing Bibliography for Earth Resources, 1966–67 (127 p.)	R. K. Llaverias
200 (1970)	Mapping Terrestrial Radiation Emission with the RS-14 Scanner	R. W. Pease R. H. Alexander S. R. Pease
203 (1970)	Remote Sensing Bibliography for Earth Resources (246 p.)	R. K. Llaverias D. G. Lowe

TABLE 31-25

Chronological List of the International Symposia on Remote Sensing of Environment

Symposium Number	Symposium Theme	Dates	Site	Proceedings Publ. Date
* First		13–15 Feb 1962	Ann Arbor, Michigan	1962*
Second		15–17 Oct 1962	Ann Arbor, Michigan	1963
Third		14–16 Oct 1964	Ann Arbor, Michigan	1965
Fourth		12–14 Apr 1966	Ann Arbor, Michigan	1966
Fifth		16–18 Apr 1968	Ann Arbor, Michigan	1968
Sixth		13–16 Oct 1969	Ann Arbor, Michigan	1969
Seventh		17–21 May 1971	Ann Arbor, Michigan	1971
** Eighth		2–6 Oct 1972	Ann Arbor, Michigan	1972**
Ninth		15–19 Apr 1974	Ann Arbor, Michigan	1974
Tenth		6–10 Oct 1975	Ann Arbor, Michigan	1975
Eleventh		25–29 Apr 1977	Ann Arbor, Michigan	1977
Twelfth		20–26 Apr 1978	Manila, Philippines	1978
Thirteenth		23–27 Apr 1979	Ann Arbor, Michigan	1979
Fourteenth		23–30 Apr 1980	San Jose, Costa Rica	1980
Fifteenth		11–15 May 1981	Ann Arbor, Michigan	1981
	First Thematic Conference: Remote Sensing of Arid and Semi-Arid Lands	19–27 Jan 1982	Cairo, Egypt	1982 (in press)
*** Sixteenth		2–9 Jun 1982	Buenos Aires, Argentina	1983
	Second Thematic Conference: Remote Sensing for Exploration Geology	6–10 Dec 1982	Ft. Worth, Texas	1983
Seventeenth		9–13 May 1983	Ann Arbor, Michigan	1983
Eighteenth		1–5 Oct 1984	Paris, France	1985
	Third Thematic Conference: Remote Sensing for Exploration Geology	16–19 Apr 1984	Colorado Springs, Colorado	1984

Also see "Proceedings of the International Symposia on Remote Sensing of Environment, Indexes and Abstracts," ERIM, Volume 1 and 2, 1981.
* Symposia 1-7 cited under University of Michigan
** Symposia 8- cited under Environmental Research Institute of Michigan (ERIM)
*** Symposium cancelled. Papers selected for presentation published in the special proceedings volumes.

a recent study of the deserts of China (Walker, 1982), and a continuing study of the Earth's glaciers (Williams and Ferrigno, 1981). Geologic remote sensing, whether by aerial photographs or by satellite images (Swanson, 1978) and photographs, has been especially useful to geologic mapping in lesser developed countries, and a number of books, symposia, or journal volumes have presented this important application of the technology (Hood, n.d.; National Academy of Sciences, 1977; Collins and Van Genderen, 1978; Allan and Bradshaw, 1981; Lynn and Allan, 1981; and Fontanel and Voûte, 1982).

Manned-spacecraft photography, such as from the Skylab project, has sometimes been used by geologists for geologic mapping (NASA, 1977), but the usual oblique view and limited areal coverage severely limit their usefulness for geologic studies. Muehlberger et al. (1977) effectively used Skylab photographs in a structural and tectonic analysis of areas in Africa, the Middle East, New Zealand, Chile, Mexico, and Guatemala. Silver et

TABLE 31-26

Chronological List of American Society of Photogrammetry-Sponsored Remote Sensing Symposia

Symposium Title	Sponsor	Date	Place	Proceedings Publ. Date
International Symposium on Photography and Navigation	ASP	December 1970	Columbus, Ohio	1970
Operational Remote Sensing	ASP	February 1972	Houston, Texas	1972
Symposium on Coastal Mapping	ASP	June 1972	Washington, D.C.	1972
Remote Sensing in Oceanography	ASP	October 1973	Orlando, Florida	1973
Management and Utilization of Remote Sensing Data	ASP	November 1973	Sioux Falls, S. Dakota	1973
1975 ASP Fall Technical Meeting and the International Symposium on Land Use Applications of Remote Sensing	ASP	October 1975	Phoenix, Arizona	1975
Second Annual William T. Pecora Memorial Symposium	ASP	October 1976	Sioux Falls, S. Dakota	1976
Digital Terrain Models Symposium	ASP	May 1978	St. Louis, Missouri	1978
Thermosense I	ASP	October 1978	Chattanooga, Tenn.	1978
7th Color Workshop Proceedings	ASP	May 1979	Davis, California	1979
Machine Processing of Remotely Sensed Data Symposium	ASP	June 1979	West Lafayette, Indiana	1979
Thermosense II	ASP	November 1979	Albuquerque, New Mexico	1979
Auto-Carto IV	ASP	November 1979	Reston, Virginia	1979
8th Color Workshop Proceedings	ASP	April 1981	Luray, Virginia	1981
Pecora VII Symposium	ASP	October 1981	Sioux Falls, S. Dakota	1982
Auto-Carto V	ASP	August 1982	Washington, D.C.	1982

TABLE 31-27

Chronological List of William T. Pecora Memorial Symposia

Symposium Title	Sponsor	Date	Place	Proceedings Publ. Date
First Annual W. T. Pecora Memorial Symposium	American Mining Congress, ASP, USGS, AAPG, AAG, GSA, SEG	October 1975	Sioux Falls, S. Dakota	1977 USGS Professional Paper 1015
2nd Annual William T. Pecora Memorial Symposium: Mapping with Remote Sensing Data	ASP & USGS	October 1976	Sioux Falls, S. Dakota	1977
3rd Annual William T. Pecora Memorial Symposium	ASP & AAPG	October 1977	Sioux Falls, S. Dakota	None
Pecora IV—Application of Remote Sensing Data to Wildlife Management	National Wildlife Federation	October 1978	Sioux Falls, S. Dakota	1978
Fifth Annual William T. Pecora Symposium: Satellite Hydrology	American Water Resources Association	June 1979	Sioux Falls, S. Dakota	1981
Pecora VI: Integration of Remote Sensing into the Exploration Process	NASA, AAPG, USGS, SEG	April 1980	Sioux Falls, S. Dakota	None
Pecora VII Symposium: Remote Sensing: An Input to Geographic Information Systems in the 1980's	ASP, AAG, NCGE	October 1981	Sioux Falls, S. Dakota	1982

al. (1977) used Skylab photographs in an analysis of regional geological features of southwestern North America. McKee et al. (1977) used Skylab photographs in their continuing research on the geomorphic nature of the sand seas of the world (McKee and Breed, 1976). Lowman (1980), one of the pioneer users of satellite photographs for geologic studies, provides an excellent review of the history of the use of such data. The use of photographs acquired during the Apollo-Soyuz Test Project for geological mapping has been documented by various authors in NASA Special Publication 412 (El-Baz and Warner, 1979).

Lithology

Lithologic Mapping

Laboratory data have been used to establish both a physical and an empirical basis for remote sensing, allowing theory to be related to observable spectral features of pure and mixed substances, including minerals, rocks, and soils. Field-acquired spectral data bridge the gap between laboratory data and remote sensing aircraft and satellite scanner images. This understanding of the spectral information in remotely sensed image data has led to image processing techniques designed to select and display that information required for the solution of a particular problem. These techniques have application to a wide variety of problems in geologic mapping and resource exploration.

The visible and near-infrared (VNIR) spectra of rocks and minerals have been discussed by numerous authors. A good review of the use of laboratory and field spectra of rocks and minerals will be found in the introductory part of this chapter. One of the largest collections of spectra is given in a series of papers by Hunt and Salisbury (1970, 1971, 1976a, 1976b) and by Hunt et al. (1971a, 1971b; 1972; 1973a, 1973b; 1974a, 1974b) which contain the spectra of over 200 minerals and 150 rock samples. The most common components of minerals and rocks, namely silicon, aluminum, and oxygen, do not possess energy levels such that transitions between them can yield spectral features in the visible and near-infrared range. Consequently, no direct information concerning the bulk composition of geological materials is available in this range. However, considerable indirect information is available, because the crystal structure imposes its effect upon the energy levels, and therefore upon the spectra of specific ions present in the structure (Burns, 1970; Hunt, 1977; Hunt, 1980). In naturally occurring geological materials most spectral information in the visible and near-infrared reflectance regions is dominated by the very common presence of iron.

Landsat multispectral data, at wavelengths between 0.5 and 1.1 μm, enable us to determine uniquely only the presence or absence of "limonitic" rocks (iron oxides) and the presence or absence of vegetation. The presence of limonite is inferred from measurements of the Fe^{+3} charge transfer band, whose long wavelength wing lies between 0.4 and 0.6 μm, and the $Fe^{+2}-Fe^{+3}$ electronic transition whose absorption band lies between 0.8 and 1.0 μm in reflectance spectra. The presence of vegetation is inferred from the chlorophyll structure at 0.5 and 0.65 μm. These determinations are most easily accomplished using band ratioing techniques (see also the "Analysis Techniques" part of this chapter).

TABLE 31-28

Selected List of Miscellaneous Remote Sensing Symposia

Symposium Title	Sponsor	Date	Place	Proceedings Publ. Date
1st International Conference on the New Basement Tectonics	Utah Geological Society	June, 1974	Salt Lake City, Utah	1976
2nd International Conference on Basement Tectonics	Basement Tectonics Committee, Inc., USGS, NASA	July 1976	Newark, Delaware	1979
3rd International Conference on Basement Tectonics	Basement Tectonics Committee, Inc., USGS, NASA	July 1978	Durango, Colorado	1981
4th International Conference on Basement Tectonics	Basement Tectonics Committee, Inc., NASA, USGS, Geological Survey of Norway, Norway Petroleum Society	July 1981	Oslo, Norway	(in press)
1981 International Geoscience and Remote Sensing Symposium (IGARSS '81)	IEEE	June 1981	Washington, D.C.	1981
1982 International Geoscience and Remote Sensing Symposium	IEEE	June 1982	Munich, Germany	1982
Remote Sensing of Earth Resources Conferences 1-8	University of Tennessee	Mar 1972	Tullahoma, Tennessee	1972
		Mar 1973		1973
		Mar 1974		1974
		Mar 1975		1975
		Mar 1976		1977
		Mar 1977		1977
		Mar 1978		1980
		Mar 1979		1980
Symposium on Machine Processing of Remotely Sensed Data	Purdue University	Oct 1973	West Lafayette, Indiana	1973
		Jun 1975		1975
		Jun-Jul 1976		1976
		Jun 1977		1977
		Jun 1979		1979
		Jun 1980		1980
		Jun 1981		1981
		Jul 1982		1982

U.S. Army Corps of Engineers Remote Sensing Symposium		Nov. 1981	Nashville, Tennessee	1982
5th International Conference on Basement Tectonics	U.S. Army Basement Tectonics Committee, Inc.	16–20 Oct. 1983	Cairo, Egypt	—
Canadian Symposia on Remote Sensing				
1st		Feb 1972	Ottawa, Ontario	1972
2nd		Apr 1974	Guelph, Ontario	1974
3rd		Sep 1975	Edmonton, Alberta	1975
4th		May 1977	Québec City, Québec	1977
5th		Aug 1978	Victoria, B.C.	1978
6th		May 1980	Halifax, N.S.	1980
7th		Sep 1981	Winnipeg, Manitoba	1981
8th		May 1983	Montréal, Québec	1983
Proceedings of the Remote Sensing Society (England):				
Fundamentals of Remote Sensing (1964)—microfiche				
Edited by W. G. Collins and J. L. Van Genderen				
Remote Sensing Data Processing (1975)—microfiche				
Edited by J. L. Van Genderen and W. G. Collins				
Land Use Studies by Remote Sensing (1976)				
Edited by W. G. Collins and J. L. Van Genderen				
Monitoring Environmental Change by Remote Sensing (1977)				
Edited by J. L. Van Genderen and W. G. Collins				
Remote Sensing Applications in Developing Countries (1978)				
Edited by W. G. Collins and J. L. Van Genderen				
Remote Sensing and National Mapping (1979)		Remote Sensing of Earth Resources Conferences University of Tennessee Tullahoma, Tennessee		
Edited by J. A. Allan and R. Harris				
Coastal and Marine Applications of Remote Sensing (1980)				
Edited by A. P. Cracknell				
Geological and Terrain Analysis Studies (1981)				
Edited by J. A. Allan and M. Bradshaw				
Matching Remote Sensing Technologies and Their Applications (1981)				
Edited by D. Lynn and J. A. Allan				

Table 31-29

Chronological List of NASA-Sponsored Landsat Symposia

Title	Dates	Place	Publication Date
Symposium on Significant Results Obtained from Earth Resources Technology Satellite-1	March, 1973	Greenbelt, Maryland	1973
Earth Resources Technology Satellite-1 Symposium Proceedings	September 1973	Greenbelt, Maryland	1973
3rd Earth Resources Technology Satellite-1 Symposium	December 1973	Washington, D.C.	1974
NASA Earth Resources Survey Symposium	June 1975	Houston, Texas	1975

While the *unique* identification of other rock materials is not possible, considerable success can be achieved in discriminating among many rock units and delineating boundaries between them. This allows large areas to be "mapped," with only minimal field checking to identify the mapped units. These units do not necessarily correspond to traditional geologic units, being based on spectral (and hence compositional) differences rather than differences in composition, age, and origin. To achieve maximum separation of units, it is often desirable to apply one of several different image processing techniques, such as principal component transformation or construction of color ratio composites (Goetz and Rowan, 1981; see also the "Analysis Techniques" part of this chapter).

By extending the wavelength range of scanners further into the infrared than the range measured by the present Landsat multispectral scanners (<1.1 μm), one finds it possible to identify additional surface materials. Through use of the Thematic Mapper Simulator aircraft scanners, with additional bands centered near 1.6 and 2.2 μm, it is possible to recognize the presence of hydrous minerals from the Al-O-H vibrational overtones of these minerals, located at wavelengths greater than 2.0 μm (Abrams et al., 1977; Rowan et al., 1977; Hunt and Ashley, 1979; Abrams et al., 1983). Recognition of such hydrous phases allows detailed mapping of hydrothermal alteration which is usually characterized by the presence of hydrous minerals, such as kaolinite, sericite, and alunite.

Use of very narrow wavelength bands in the region between 2.0 and 2.5 μm will allow specific identification of many more minerals. The most common bands in this wavelength region involve the OH stretching mode and water. In spectra of minerals and rocks, bands will appear depending upon the local cationic coordination of the OH in the mineral. The fundamental OH stretching mode near 2.77 μm may form combination tones with other fundamentals, including lattice and vibrational modes. In particular, such combinations with fundamental Al-OH or Mg-OH bending modes produce features near 2.2 or 2.3 μm, respectively (Hunt, 1980). Precise determination of the shape and location of these features by the use of very narrow wavelength bands allows identification of minerals that contain hydroxyl groups (muscovite, kaolinite, etc.) or water (montmorillonite, gypsum, and quartz). In addition to the water and OH features, the carbonate minerals also display absorption bands between 1.6 and 2.5 μm, which are due to combinations and overtones of the four fundamental internal vibrations of the planar CO_3^{-2} ion. Narrow band spectroscopy again will allow identification of the carbonate materials. An experimental narrow band profiling instrument, the Shuttle Multispectral Infrared Radiometer (SMIRR), was flown on the second Space Shuttle flight (see the section on "Satellite Radiometry" of this chapter. The SMIRR instrument has 10 medium and narrow band filters designed to test the feasibility of using narrow band spectrometery for mineral identification from space. Preliminary analysis of these data indicate that kaolinite, montmorillonite, and carbonates have been identified (Goetz et al., 1982).

In the middle infrared (MIR) (5–40 μm) region of the spectrum, the emission spectrum of the surface material can be measured. Laboratory measurements of middle infrared spectra of rocks and minerals show many diagnostic features. The region between 8 and 14 μm holds the most promise for remote sensing, because this is an excellent spectral window in the Earth's atmosphere and is also the region of maximum thermal emission at terrestrial surface temperatures (Vincent, 1975). Within this spectral range, the most prominent spectral features are due to silicon-oxygen (Si-O) stretching vibrations. These features change wavelength and intensity with varying composition and structure (Lyon, 1965; Hunt and Salisbury, 1974, 1975, 1976; Vincent et al., 1975). Lyon and Patterson (1966) indicated that these features (known as reststrahlen bands in reflec-

Table 31-30

Chronological Records of Committee on Space Research (COSPAR) Plenary Meetings Dealing with Studies of Earth Resources from Space Platforms

Meeting	Place	Date	Pertinent Publications
XVI	Konstanz, Federal Republic of Germany	23 May–5 June 1973	Approaches to Earth Survey Problems Through Use of Space Techniques, COSPAR, Akademie Verlag Berlin-1975.
XVII	Sao Jose dos Campos and Sao Paulo, Brazil	June–July 1974	COSPAR Advances in Space Research
XVIII	Varna, Bulgaria	1975	COSPAR Advances in Space Research
XIX	Philadelphia, Pennsylvania	1976	COSPAR Advances in Space Research
XX	Tel Aviv, Israel	1977	COSPAR Advances in Space Research
XXI	Innsbruck, Austria	1978	COSPAR Advances in Space Research
XXII	Bangalore, India	1979	*Remote Sensing and Mineral Exploration* by W. D. Carter, L. C. Rowan and J. F. Huntington, eds. COSPAR Advances in Space Exploration, vol. 10, 173 p. (1980).
XXIII	Budapest, Hungary	1980	*Remote Sensing-1980* by A. B. Kahle, G. Weill and W. D. Carter—COSPAR Advances in Space Research, vol. 1, no. 10, 314 p. (Pergamon Press, Oxford, and New York)
XXIV	Ottawa, Canada	May 1982	*Remote Sensing and Mineral Exploration-1982* by W. D. Carter and L. C. Rowan, eds.—COSPAR Advances in Space Research (in press) (Pergamon Press, Oxford, and New York)

Table 31-31

Selected List of Remote Sensing Societies and Published Newsletters or Journals

Society/Organization	Country	Newsletter/Journal	Frequency
International Society of Photogrammetry and Remote Sensing	International	*Photogrammetria*	Quarterly
American Society of Photogrammetry	U.S.A.	*Photogrammetric Engineering and Remote Sensing*	Monthly
Remote Sensing Society	England	Newsletter/*International Journal of Remote Sensing*	Periodic/ Quarterly
Elsevier Scientific Publishing Co.		*Remote Sensing of Environment*	Quarterly
Canadian Remote Sensing Society	Canada	*Canadian Journal of Remote Sensing*	Periodic
Academy of Sciences of the U.S.S.R.	U.S.S.R.	*Soviet Journal of Remote Sensing*	Bimonthly
Murray Felsher	U.S.A.	Washington-Remote Sensing Newsletter	Monthly
Canada Centre for Remote Sensing	Canada	Remote Sensing in Canada (a newsletter published in English and French)	Monthly
U.S. Geological Survey/NOAA	U.S.A.	Landsat Data Users Notes	Bi-monthly

tance spectra of polished samples) shift to shorter wavelengths as silica content increases. The most complete set of laboratory spectra of rocks available for the MIR are the transmission and reflection spectra of Hunt and Salisbury (1974, 1975, 1976).

The use of spectral emittance data in the middle infrared for mineral identification was demonstrated with NASA's 24-channel scanner (Kahle and Rowan, 1980; Kahle et al., 1980) following earlier demonstrations with 2-channel data (Vincent and Thomson, 1972a; Vincent et al., 1972). The 24-channel scanner has 6 channels between 8 and 13 μm. A principal component color composite image (Color Figure 31-101), using three of the bands, was created from data acquired over Tintic, Utah, a semi-arid, geologically complex area of high relief and moderate vegetation (Kahle and Rowan, 1980). In general, the red colors in the image represent rocks in which quartz was a major constituent, while green indicates nonsilicate rocks (carbonates) and vegetation. The more intense the red, the higher the quartz content of a unit, illustrating the dominant effect of the Si-O spectral features in this wavelength region. It is significant that quartz monzonite and quartz latite appear pinkish in the image, while monzonite and latite appear blue. This image helped demonstrate the promise this spectral region holds for geologic applications.

Computer-Enhanced Landsat Imagery for Lithologic Discrimination in Southwestern Saudi Arabia

Other sections of this chapter (for example, "Analysis Techniques" and the previous section) describe techniques useful for directly or indirectly extracting lithologic information from various types of aircraft or satellite image data. Identification of rock classes, however, can frequently be expanded or refined by using computer-enhanced digital multispectral data. Although such multispectral data classes are often unfamiliar to scientists working only with aerial photographs, they are commonly used by geologists who apply satellite data for geologic mapping. Multispectral scanner (MSS) spectral reflectance data provided by the NASA Landsat satellites are generally the most used, because they are widely accessible. A complete description of the Landsat series is provided in Chapter 12 of this *Manual* and in NASA (1971) and U.S. Geological Survey (1979).

Landsat data are available in both analog (image) and digital format. The imagery has been widely applied in regional geological mapping, for which the synoptic view is ideal for showing regional perspective. Applications are documented in proceedings of various remote sensing conferences (Miller, 1975; Blodget, 1981; etc.) in remote sensing texts (Sabins, 1978a; Siegal and Gillespie, 1980; etc.) and the technical geologic literature (Halbouty, 1980; Beall and Squyres, 1980; Gathright, 1982; etc.). The most frequent application of satellite data has been the extension of conventional aerial photographic interpretation techniques to Landsat imagery to exploit the advantage of the regional field of view.

More recently, however, digital Landsat data have been computer-processed in a variety of ways to enhance specific visual characteristics (see Color Figures 31-75 and 31-76). Processing techniques include sharpening feature edges and lineaments, increasing the contrast and range of tone/hue, and modification of the MSS data to make specific surface materials more readily identifiable. The major classes of algorithms that have been developed are described in the "Analysis Techniques" part of this chapter.

Early research (Rowan et al., 1974; Blodget et al., 1975; and Blodget, 1977) convincingly demonstrated that appropriately processed Landsat MSS data can greatly increase the ability to identify geologic materials; in particular, a wide variety of rock types and hydrothermally altered materials can be discriminated in desert terrains where ex-

Table 31-32

Landsat image maps available from the U.S. Geological Survey as of early 1983

Image Map Title	Format	Scale	Publication Date
Chesapeake Bay and Vicinity Winter 1976–77	Color mosaic	1:500,000	1978
Upper Chesapeake Bay	Color image, enhanced & precision-processed	1:250,000	1977
Upper Chesapeake Bay	Color image, precision-processed	1:500,000	1976
Upper Chesapeake Bay	Color image, analog processed	1:500,000	1973
Florida	Color mosaic	1:500,000	1975
Pensacola Bay	Color image	1:500,000	1977
Lake Seminole	Color image	1:500,000	1977
Apalachee Bay	Color image	1:500,000	1977
Okefenokee Swamp	Color image	1:500,000	1977
Gulf Hammock	Color image	1:500,000	1977
Lake George	Color image	1:500,000	1977
Charlotte Harbor	Color image	1:500,000	1977
Lake Okeechobee	Color image	1:500,000	1977
Sanibel Island	Color image	1:500,000	1977
The Everglades	Color image	1:500,000	1977
Florida Keys	Color image, precision-processed for water enhancement	1:500,000	1977
Georgia	Color mosaic	1:500,000	1976
Arizona	Sepia mosaic, with cultural features, duotone	1:500,000	1975
Arizona	Black and white mosaic	1:500,000	1975
Phoenix	Sepia mosaic, with cultural features	1:250,000	1975
New Jersey	Color mosaic	1:500,000	1973
Medicine Bow River, Wyoming/Colorado	Color image	1:500,000	1979
Dry Fork Cheyenne River, Wyoming	Color image	1:500,000	1979
Pumpkin Creek, Montana	Color image	1:500,000	1979
United States, Conterminous	Black and white, band 5	1:5,000,000	1974
	Black and white, band 7	1:5,000,000	1974
Wenatchee, Washington	Landsat image mosaic combined with planimetric base	1:250,000	1979
Antarctica			
Ellsworth Mountains	Blue-tone mosaic	1:500,000	1976
Victoria Land Coast	Blue-tone mosaic	1:1,000,000	1976
McMurdo Sound Region	Black and white image	1:250,000	1975
McMurdo Sound Region	Black and white image	1:500,000	1975
McMurdo Sound Region	Blue-tone image	1:1,000,000	1975
Vatnajökull, Iceland	Color image (fall)	1:500,000	1976
Vatnajökull, Iceland	Black and white image (winter)	1:500,000	1977
Yemen	Color mosaic I-1143-A Geographic I-1143-B Geologic	1:500,000	1978
Berry Islands, Bahamas	Color image (Hydrographic-Information)	1:500,000	1980
Dyersburg, Tennessee/Missouri/ Kentucky/Arkansas	Color image (Landsat 4 TM)	1:100,000	1983
Las Vegas/Nevada/Arizona/California	Color mosaic	1:250,000	1983
New Bedford, Massachusetts	Black and white mosaic (Landsat 3 REV)	1:100,000	1983

tensive areas of bedrock are exposed. Optimum enhancements were created where Landsat MSS band ratio data (4/5, 5/6, and 6/7) from a high Sun angle summer scene were contrast enhanced using a ramp, cumulative distribution function (CDF) contrast stretch. These data were then combined into a color composite image. A computer-enhanced Landsat image constructed in this manner was used to demonstrate the capabilities of this technology in Saudi Arabia (Color Figure 31-102).

The area included in this image is located in the Asir Mountains, about 100 km north of Najran on the Yemen Arab Republic border. The highly dissected terrain is moderately rugged, and local relief exceeds 500 m. Vegetation in this typical low-latitude desert environment is extremely sparse and generally restricted to poorly developed soils.

Geologically, the area is on the southeastern margin of the Arabian Shield (U.S. Geological Survey, 1970). The basement rocks consist of a complex sequence of generally metamorphosed, interlayered volcanic and sedimentary Precambrian assemblages. These are locally intruded by igneous rocks ranging in composition from gabbro to syenite and in age from Precambrian to Cambrian. The volcanic rocks in the area range in composition from andesite to rhyolite and in texture from agglomerate to thick massive flows of lithic tuff. The volcanic rocks are commonly interlayered with sedimentary strata that variously include sandstone, conglomerate, graywacke, shale, and limestone. The basement rocks are, in part, unconformably overlain by recent unconsolidated alluvial and aeolian sands and Cambro-Ordovician sandstone. The sandstone has been eroded from much of the western parts of the test area and is now commonly observed only in isolated buttes, where remnants cap the basement rocks.

Figure 31-103 is a geologic map modified from that portion of the Asir quadrangle (Brown and Jackson, 1959) covering the area of the ratio color-composite image (Color Figure 31-102). The three dark-blue sub-circular spectral units just south of the center of the image correspond closely with the peralkalic granite intrusions (*gp*) identified on the map. In addition, all other (*gp*) intrusions that are mapped in the lower two-thirds of the test area can be identified equally well. In the northern part of the scene, however, similar granites are complexly associated with basic and ultrabasic intrusive rocks (*bu*), and both lithologies are in part overlain by the Wajid sandstone (*€Ow*); thus, the margins of most of the individual rock units are not easily defined spectrally. The calc-alkalic granites (*gr*) are equally well-defined spatially, but they cannot be visually distinguished from the peralkalic intrusions on the enhanced imagery. Overstreet and Rossman (1970), who mapped the area at a scale of

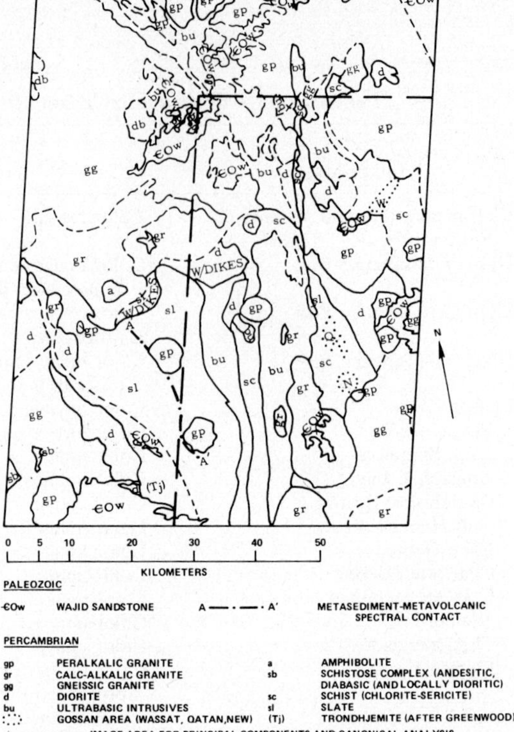

Fig. 31-103. Geologic map of test area (modified from Brown and Jackson, 1959) shown in Color Figure 31-102.

1:100,000, described both units as biotite granites, but noted that those corresponding to the (*gr*) tended to be porphyritic; this characteristic is also described by Greenwood (1980). The spectral signatures would therefore be expected to be similar and, in fact, they are. The subtriangular intrusion which is bluish-black, near the center of the southwest quadrant of the enhanced image, was mapped by Greenwood (1980) and described as a granodiorite/quartz monzonite. This outcrop area was not shown on the Asir quadrangle (Brown and Jackson, 1959), even though intrusions of similar size had been mapped. In addition to the excellent spectral correlation that exists with the granite intrusions, many other equally good correlations can be made with other rock types. For example, schistose diabase (*sb*) is clearly defined as light greenish-blue areas at the southwestern image margin. The amphibolite (*a*) that intrudes the granite just west of the image center is pinkish buff on the enhanced imagery, whereas diorite (*d*) shows up as bluish green. The purple spectral unit at the center of the image corresponds to diorite intruded by felsic dikes. Gneissic granite (*gg*) situated in the southwest quadrant appears dark green. All of the rock units defined by Brown and Jackson (1959) correspond exceptionally well spatially, and spectrally, over the area.

Although the rock units mapped as schists (*sc*) also have good spatial correspondence to the ratio-composite image, they are made up of three locally distinctive spectral units. In addition, the slate (*sl*), comprising the host rock for the two western circular intrusions on the same map, is also depicted as three discrete spectral units. In the case of both the schist and slate units, the three chromatically well-defined spectral units are pale green, cream, and finely mottled purple-blue and green. The most clearly defined spectral contact within the slate corresponds to the A-A′ line on Figure 31-103. Note that the pale-green units on the image also cannot be separated by visual inspection from the mapped ultrabasics on the basis of hue.

When these spectral units were compared with Greenwood's geologic map (1980), however, the lithologic relationships were immediately evident. All of the pale-green hues, including the ultrabasics of Brown and Jackson (1959), coincide with mapped metamorphosed volcanic rocks. On the other hand, the off-white hues correspond to metamorphosed sedimentary rocks, and the mottled units are described as metavolcanics intruded by diabase sills. Therefore, the A-A′ line marks the contact between metasediments and metavolcanics. The well-defined purple unit, which was noted earlier to correspond with a diorite intruded by felsic dikes, was remapped by Greenwood (1980) as gabbro; the spatial distribution of this gabbro remained the same. The granites just west of the north arrow (N) designation, (*gg* and *gp*), are shown on the image as dark green. Greenwood mapped these intrusions as quartz monzonites.

A similar dark green color dominates the unit (*gg*) mapped in the southwest corner of the image and is slightly lighter than the trondhjemite (Tj) unit. Greenwood shows this area (*gg*) to consist largely of trondhjemite and quartz diorite. As noted earlier, Precambrian terrain is locally overlain by the Wajid sandstone (*€Ow*). These isolated units are spectrally defined by several shades of distinctive salmon-pink.

It is readily apparent that, although some spectral ambiguities can be found when the enhanced imagery is compared with the earlier published maps, these discrepancies were largely resolved when the geology was more completely understood. If vegetation exceeds 10 percent of the surface area, however, the vegetational reflectance masks that of the rocks, and thus this technique is limited to desert environments.

Similar processing was subsequently accomplished for extensive areas to the north and east of the test area, and the resultant images have served to resolve stratigraphic mapping problems (Blodget et al., 1978; and Blodget and Brown, 1982) and to complement traditional field mapping of the Abha quadrangle (Greenwood, 1981) at 1:250,000-scale.

Structure and Tectonics

Application of Remote Sensing to Structural Geology and Tectonics

Modern remote sensing systems have the ability to provide instantaneous views of the surface of the Earth at different scales for selected wavelengths of reflected and emitted radiation[1]. With the right combination of filters or processing of the imaged data (band ratioing, contrast stretching, density slicing, etc.), selected types of information can be extracted from or enhanced in an image. Gathering data on a scale consistent with the size of the feature being analyzed is an important consideration in the application of remote sensing to structural and tectonic studies (see Table 31-33). The size relationships of structures, structural domains and the limits of structural homogeneity, and analogy with the structural interpretation of known terrains and models of deformation, should provide the correct framework for extrapolation of structures observed on small-scale satellite images, such as Landsat MSS images. Direct mapping of large structural features, in contrast to mapping primary and secondary discontinuities on a mesoscopic scale (man-size), requires an overview approach involving landforms, drainage patterns, land-use, and any variations in lithology that are manifested as changes in tone and texture on photographs and images. Each scale contains information that is unique and pertinent to that scale. Such information may not always be preserved or transferred faithfully into mosaics or, during the compilation of more detailed quadrangle maps, into regional maps. The aerial photograph is the most appropriate base for the direct mapping of macroscopic-scale structures (faults, folds, craters, calderas, volcanoes, dikes, etc.) and other features of uncertain origin (fracture traces, lineaments, etc.), while images from satellite platforms and airborne radar systems are used extensively as a base for interpretation and analysis of the megascopic scale "tectonic" features, such as fold mountain belts, basins, island arcs, rift valleys, large plutons, lineaments, and physiographic and tectonic provinces (see summary in Table 31-33).

Morphotectonics (Hills, 1961) is the study of landforms of regional or tectonic significance, using topography as the primary criterion, and it is the basis for interpreting geologic structures from small-scale remotely sensed images. The primary discontinuities of most interest to the structural geomorphologist are: the unconformity,

[1] Some systems are designed to measure the intensity of body forces (gravity or magnetic) as well as elevation (relief profiling); see sections of this chapter on airborne (Figure 31-31) and satellite profilometry (Figures 31-35 and 31-36).

Table 31-33

Optimum scales and corresponding platforms for mapping deformation and geotectonic phenomena*

Structural Scale and Order	Natural Scale and Size of Feature	Types of Phenomena	Types of Maps	Platform and Medium of Observation
Gigascopic 1	1:50 million >1000 km	Continental or oceanic plates, seismic plates	Globes; world-wide and continental geologic, tectonic and seismic maps.	Space probes; vidicon, scanner, radiometer (visible, IR & gamma ray) Various satellites (visible & IR) Seismic networks (long waves)
Megascopic 2	1:1 million 10 km to 1000 km	Mountain belts, basins, island arcs, rift valleys structural provinces, plutons, megalineaments, craters	Large globes; continental, national, state geologic, tectonic and seismic maps.	Landsat vidicon photography & scan imagery (visible & IR) Skylab & intelligence satellites; photography and scan imagery (visible & IR)
Macroscopic 3	1:1000 10 m to 10 km	Folds, faults, lineaments, craters volcanoes, dikes, fracture traces, etc.	Regional, geological and structural maps; fabric diagrams.	Aerial photographs and scan imagery (visible, IR, radar)
Mesoscopic 4	1:1 1 cm to 10 m	Folds, faults, cleavage, joints bedding, geologic contacts.	Detailed geologic and structural maps, fabric diagrams.	Life-size field work (visible IR, radar, gamma rays) seismic, acoustic, gravity, and magnetic stations
Microscopic 5	1000:1 10 microns to 1 cm	Micro-fractures, deformation lamellae, grain size.	Micro-fabric orientation diagrams.	Microscope (visible, IR, electron, etc.)
Submicroscopic 6	100 million: 1 1 angstrom to 10 microns	Lattice defects.	Crystal structure charts	X-ray, electron microscope

* Larger-scale maps are produced from the higher orders of observation and smaller-scale maps from the lower orders. The arrows on the right indicate the range in scale: (a) up into the unshaded region by synthesis (integration and mosaicking) of the primary observations, and (b) downward into the shaded region by analysis (enlarging and enhancing to the limit of resolution). (From Gold et al., 1974; Gold, 1980. Compiled initially from various sources: Brock, 1960; Carey, 1962; Turner and Weiss, 1963).

bedding surface or formational boundary that separates rocks of differential degradation rates, and joint orientation and density, upon which are developed drainage systems of contrasting patterns. Through the use of optical and microwave remote sensing systems, discontinuities between bedrock and/or their soils may be detected by differences in color or tone, texture, or in the type and abundance of the vegetation. The secondary structural discontinuities of interest involve local changes in the state of the host rocks imposed during deformation. The difference in chemical and physical properties may be induced by comminution, differential strains in fault or shear zones, or by the creation of voids (joints, tension cracks, etc.) or interconnecting fractures to yield zones of high porosity and permeability. These discontinuities generally weather more readily than the country rocks and are characterized by an alignment of negative topographic features, such as straight valley segments, swales, and wind and water gaps; cemented fracture zones may often be expressed as resistant ridges or dike-like features.

The relationship of landforms to specific bedrock structural style is illustrated in the map and block-diagram cross section (Figure 31-104) of the Zagros Mountains of western Iran (Oberlander, 1965; Twidale, 1971; Gold, 1980). Note the common development of breached anticlines, and of topographic valleys in the crests of those anticlines in which the "resistant" lithologic "strut" unit is underlain by a less resistant rock type. These conditions are promoted by the preferential development of tension fractures in the hinge zone of the anticlines. In layered rocks with differential erosive properties, the symmetry/asymmetry aspects of ridges (hogbacks and cuestas) and of the tributary streams in the trellis drainage regimes yields information on the attitude of the beds (see Figure 31-105). In a more deeply eroded orogenic belt such as the Appalachian fold mountains, there is an even more remarkable match between the major physiographic and tectonic provinces (Figure 31-106A). Here the mega- and macroscopic scale folds are manifest as sinous ridges, underlain by Tuscarora quartzite (Silurian), and as subsidiary ridges underlain by the less resistant Oswego sandstone (Ordovician). As in the Zagros Mountains, inversion of structural relief and topography is common, so that synclinal ridges alternate with anticlinal valleys. These are the normal geomorphic expressions of lithology and structure in orogenic belts, and it is only these structural features that commonly are represented on tectonic maps. Not all the "cross-strike" topographic features[2] apparent in Figure 31-106B can be explained by conventional "lithology and primary structure" control. Many geologists (Gold, 1980; Hills, 1961; Hobbs, 1911; O'Leary et al., 1976a; Sonder, 1938; Wheeler et al., 1974) believe that such cross-strike structural discontinuities are associated with steeply dipping, transgressive fracture zones. There is no apparent displacement of strata across the lineaments, and some have been

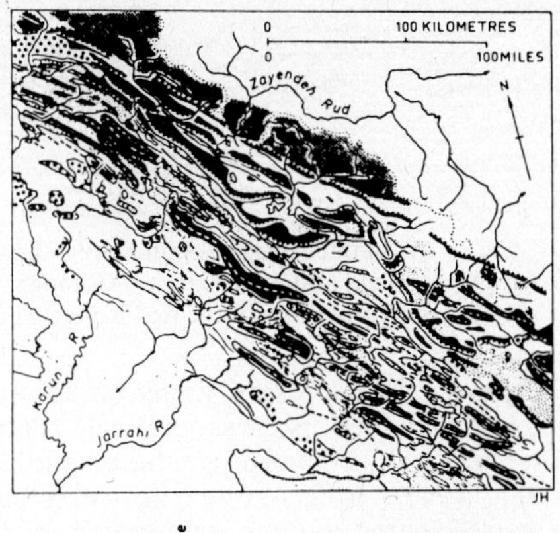

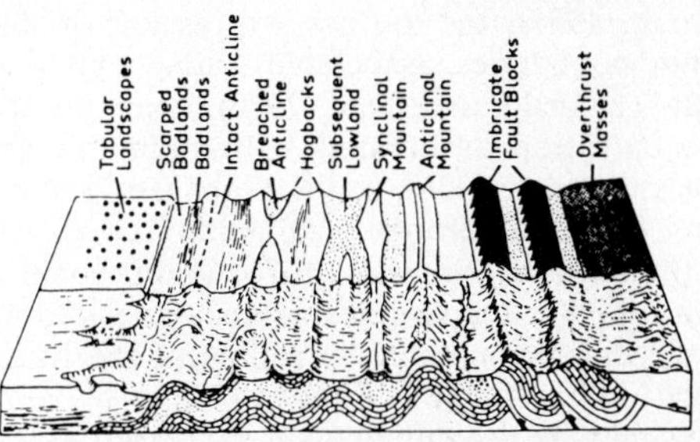

Fig. 31-104. The Tertiary and Mesozoic rocks exposed in the Zagros Mountains of western Iran offer a good example of the relationship of specific landforms to structure. An arid climate, relatively recent deformation (Pliocene), a simple fold style, and the presence of several massive and resistant formations, have preserved table lands, hogbacks, intact and breached anticlines, anticlinal valleys and synclinal mountains, elongate domes and basins, and fault block mountains. To the southeast, these rocks have been pierced by extrusive salt domes. Diagrams after Twidale (1971) as modified from Oberlander (1965).

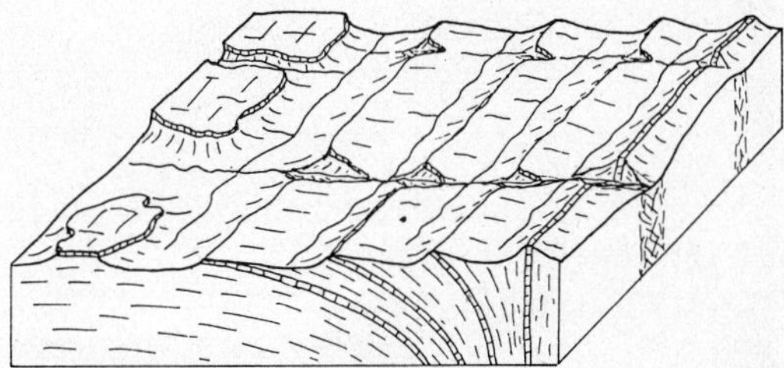

Fig. 31-105. Varied expressions of primary and secondary discontinuities in a region of stratified rocks. Note the effect of the change in dip of the resistant beds on the landforms. The cross-strike lineaments are shown conceptually to be controlled by zones of increased fracture density. (After Gold, 1980).

[2] Because of their relatively linear surface expression, they are referred to as *lineaments* (Hobbs, 1911) or *fracture traces* (Lattman and Parizek, 1964) depending on their length (greater or less than 1.6 km respectively).

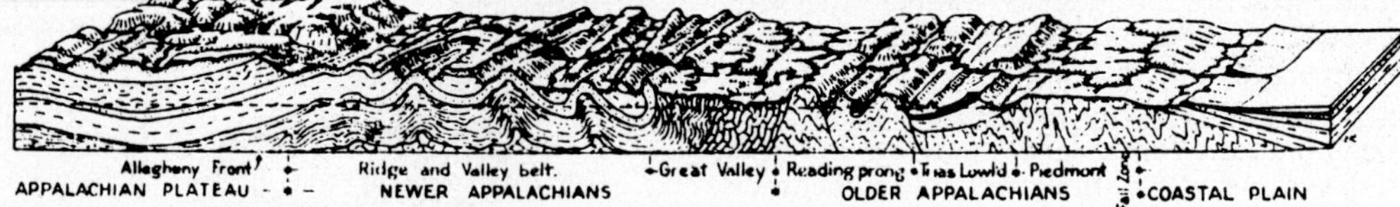

Fig. 31-106A. Morphotectonic section across the Appalachian orogenic belt showing the relationship between the physiographic and tectonic provinces. Although the section is not to scale, it would represent a section from approximately northwest to southeast through Pennsylvania. (After Strahler, 1963).

demonstrated to represent zones of increased fracture density on a macroscopic scale (Parizek, 1975) or regionally between rigid basement blocks (Sonder, 1947), hinge zones between basins (Brock, 1957), the surface expression of dikes, strain boundaries associated with stepped tear faults (Kowalik and Gold, 1976), zones of stress concentration associated with regmatic shear zones (Sonder, 1956), or the propagation of basement fractures through sedimentary cover rocks to the surface (Norman, 1976; Podwysocki and Gold, 1979). Those lineaments associated with displacement undoubtedly represent the trace of faults or fault zones. Although these linear features may be discontinuously expressed as alignment of wind and water gaps or straight stream or river segments, and tend to be masked by the dominant lithostructural elements in folded terranes, they are the main morphotectonic element of the shield and platform areas. The reader is referred to the paper by O'Leary et al. (1976a) for a detailed discussion on definitions and nomenclature of linear features, and to Lattman (1958), Lattman and Parizek (1964), Parizek (1975, 1976a, 1976b), and Gold (1980) for the subsurface character of lineaments and fracture traces and their significance in ground water prospecting. A suggested nomenclature involving scale of linear traces without obvious displacements is given in Table 31-34.

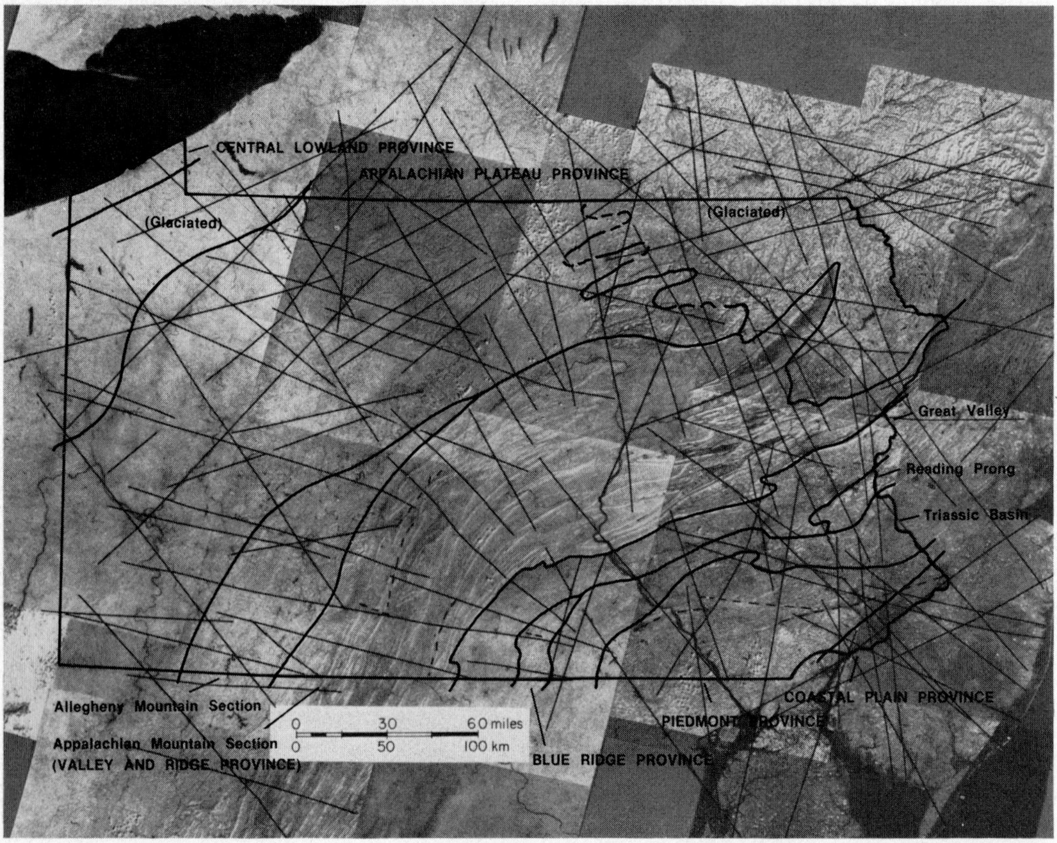

Fig. 31-106B. Megalineament map of the State of Pennsylvania and environs plotted on a Landsat MSS band 7 image mosaic. The dashed lines represent known faults that exhibit displacements on this scale. The physiographic provinces—the Allegheny Plateau to the north and west, the curved Appalachian folded belt through the central section, and the Piedmont to the southeast—show up well. The dark crescent-like area in the northeast part of the state is the "anthracite coal basin" around Scranton and Wilkes-Barre, also known as the Lackawanna syncline.

Table 31-34

Scale of some linear features on maps, photographs, and images which exhibit no obvious displacement in the field* (After Gold, 1980)

Surface Feature	Size	Mapping Base	Possible Structures
Joint traces	Centimeters to tens of meters	Outcrop maps, orientation diagrams, large scale aerial photographs	Bedding
Fracture traces	Approx. 100 m to 1.6 km	Aerial photographs, large scale topographic maps	Narrow, steeply dipping zones of joint concentrations up to 33 m wide
Lineaments	(a) Short: 1.6 to 10 km	Topographic maps, small scale aerial photographs	Broad zones of up to a few km wide of disrupted rocks, including concentrations of narrow fracture zones
	(b) Intermediate: 10 to 100 km	Topographic maps and relief models (1:250,000), high altitude imagery	
	(c) Long: 100 to 500 km	Satellite imagery, small scale relief models	Petrographic provinces and aligned volcanic centers, rift valleys, aulacogens, and continental sutures as much as 100 km wide
	(d) Megalineaments: 500 km	Satellite imagery, and mosaics of Landsat imagery	

* Some ot these features may be dikes or faults with little or no strike separation.

In an analysis of lineaments it should be realized that the orientations and density measured on one scale are not necessarily the same as those mapped on another scale. Long lineaments that were mapped from the Landsat mosaic (Figure 31-106B) of Pennsylvania (Figure 31-106C) have a consistent west-northwest trend, and a radial "fan-like" attitude about the bend in the Appalachian orocline. The orientation/frequency relationships of short lineaments, mapped on the individual Landsat images of Pennsylvania, are summarized into cells in Figure 31-107, and they reveal a variation in orientation coincident with the major structural and physiographic provinces (Kowalik and Gold, 1976). An integration of the small cell data into a synoptic diagram generally reveals an oblique relationship of fracture data with scale; this relationship has been rationalized (Gold et al., 1973) in terms of a second order shear model (Moody and Hill, 1956). In a detailed study of linear features on three scales along the Tyrone-Mount Union lineament in central Pennsylvania, an orthogonal pattern, coincident in directions, is apparent in the mesoscopic (joint traces) and megascopic (lineaments) scale data, but not in the macroscopic (fracture traces and short lineaments) scale features (Canich, 1976). The latter exhibit preferred orientations conjugate to the dominant lineament direction of 135°; these geometric relationships are portrayed in hierarchial scale form in Figure 31-108.

TN
West
East

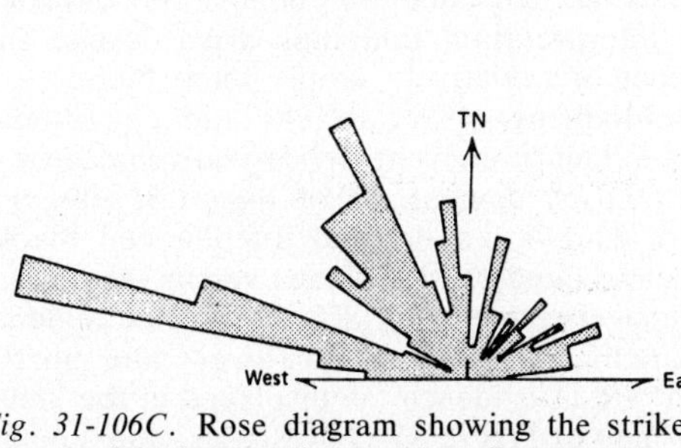

Fig. 31-106C. Rose diagram showing the strike frequency of the intermediate and long lineaments mapped on the Landsat 1 mosaic of Pennsylvania and adjacent areas. The longer cross-strike lineaments account for the dominant west-northwest and north-northwesterly trends; some of these strike across physiographic province boundaries and are still discernible in recent glacial and coastal plain sediments. The shorter lineaments tend to "fan" with the Pennsylvania orocline.

Direct observations from remotely sensed images on a megascopic scale have led to revisions of some morphotectonic concepts, viz., (1) the cratons are less stable than originally supposed; (2) lineaments and semi-rigid crustal blocks are common in and to all cratons; (3) many continental basins are bounded by lineaments; (4) lineaments are not limited by tectonic province boundaries or geologic age, and appear to be rejuvenated fractures along ancient zones of weakness; (5) long linear fractures and semi-rigid crustal blocks probably are manifestations of intraplate deformations, and they should be considered along with fold mountain belts, plateaus, basins, crustal arches, major faults and rift systems, suture zones, and tectonic province boundaries in any tectonic analysis.

The main forum for papers on lineaments (*senso lato*) are in the proceedings volumes of the International Conference on Basement Tectonics, and in the monthly journals such as *Photogrammetric Engineering and Remote Sensing, American Association of Petroleum Geologists Bulletin,* and the *Geological Society of America Bulletin.* The Proceedings of the First International Conference on the New Basement Tectonics was edited by R. A. Hodgson, S. Parker Gay, Jr., and J. Y. Benjamins, and published by the Utah Geological Association, Pub. #5, in 1976. The Proceedings of the Second International Conference on Basement Tectonics, was edited by M. H. Podwysocki and J. L. Earle, and published by the Basement Tectonics Committee, P.O. Box 5868, Denver, Co. 80201, in 1979. The Proceedings of the Third International Conference on Basement Tectonics was published in 1981. The Proceedings of the Fourth Conference is in press (see Table 31-28).

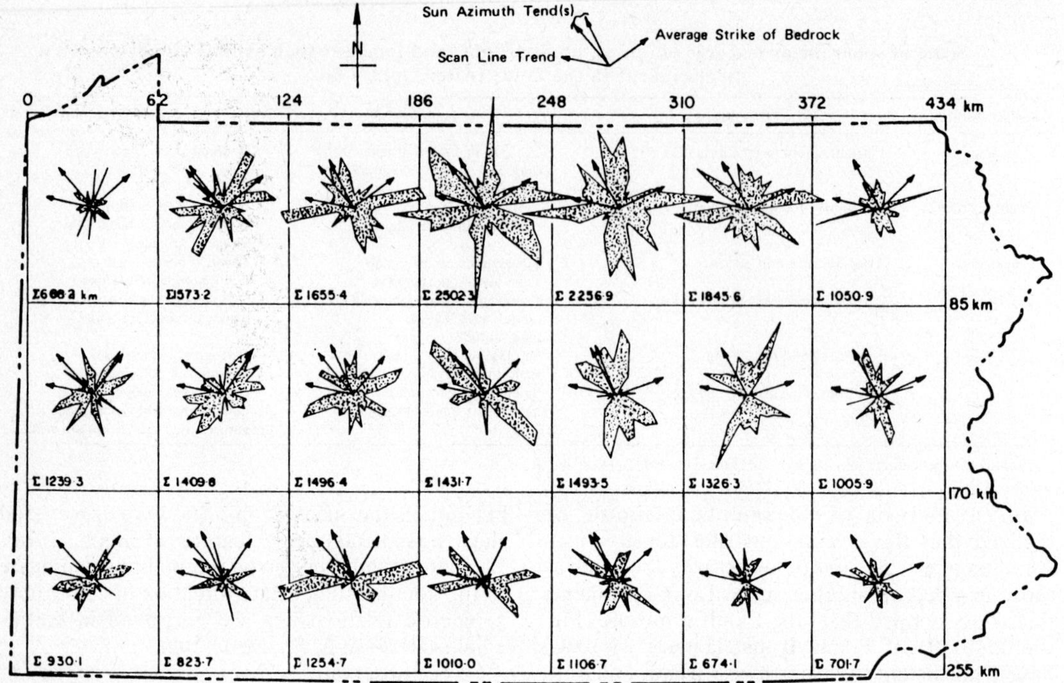

Fig. 31-107. Intermediate length lineament orientations summarized in rose diagrams for cells (62 km by 85 km) on a grid across Pennsylvania. The scan line direction (west-northwest arrow), sun azimuth (2 arrows for cells on overlapping images with different sun angle), and general strike of bedrock (northeast trending arrow) are superimposed on each rose diagram. Note the general paucity of lineaments near these arrows, due to inherent biases (scan line effect and sun azimuth-shadow effects) in the system and the artificial filtering of lineaments coinciding with bedding strike during the mapping process. The density of lineaments in each cell can be judged by the sum of their lengths, recorded in kilometers in the lower corner. (From Kowalik and Gold, 1976).

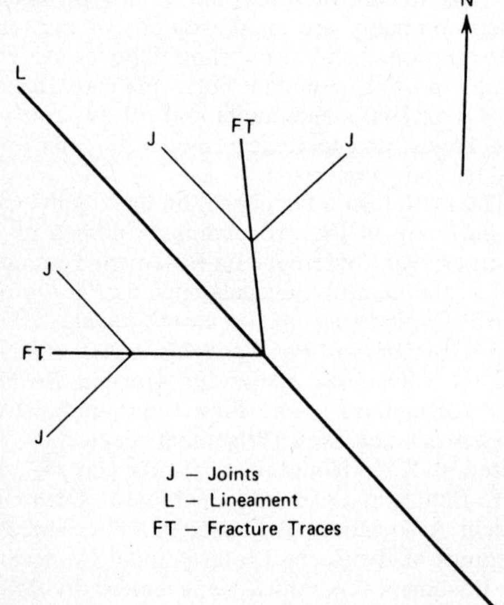

Fig. 31-108. Geometric relationship between "fractures" mapped on meso-, macro-, and megascopic scales in the vicinity of the Tyrone—Mount Union lineament in central Pennsylvania. The strike lineaments and fold axes trend northeast, and have been omitted from the drawing. (After Canich, 1976).

Structure and Tectonics of Northeastern Iceland

Studies of Landsat imagery of northeastern Iceland provide examples of some of the advantages and problems associated with derivative image enhancement (Chavez et al., 1977a; see also the "Analysis Techniques" section of this chapter), demonstrate the necessity for adequate field verification, and illustrate complexities of structural interpretation that may arise despite the presence of a relatively simple image fabric.

Highlighting of rectilinear and curvilinear structural fabric elements by derivative imagery is illustrated by comparison of Figure 31-109 with Figure 31-110, an original unenhanced black-and-white Landsat MSS image versus the derivative image, respectively. Two pervasive regional lineament trends, west-northwest and north-south, are in evidence, as displayed in the lineament map of Figure 31-111 (Mamula and Voight, in press). A third northeasterly trend is locally present.

Field investigations have proven essential to interpret adequately the derivative image. For example, boundaries of dark patches in the neovolcanic zone of axial rifting in northeast Ice-

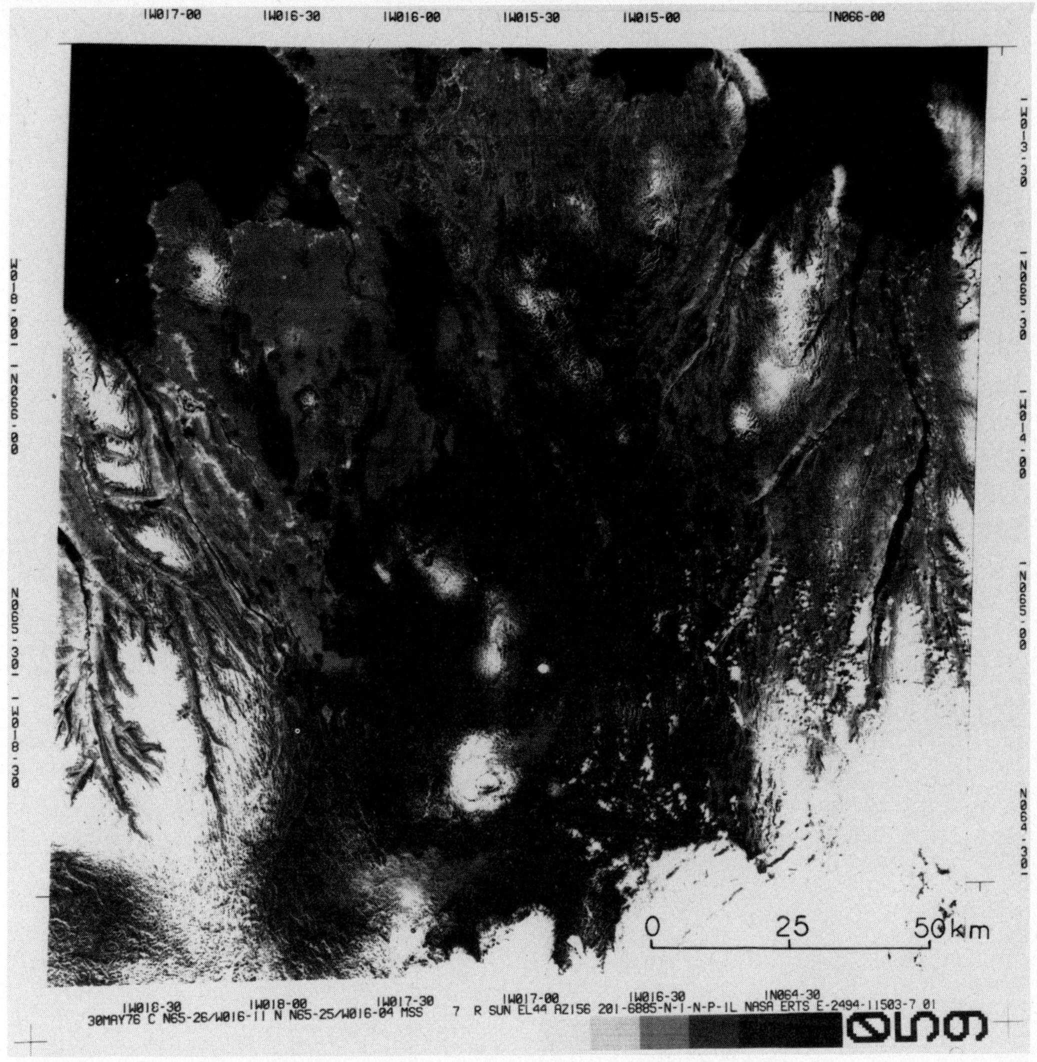

Fig. 31-109. Standard band 7 MSS image of northeastern Iceland acquired on 30 May 1976. The area covered by this image (2494-11503) represents the approximate eastern half of the total area studied.

land show up as pronounced linear features on the derivative image (Figure 31-110) and were, therefore, mapped as lineaments with a north to northeast trend. These dark patches, originally interpreted as young lava with flow boundaries controlled by fault scarps, are shown by field study to be localized patches of aeolian deposits, which blanket lava flows. Very little vegetation is present (Kristján Saemundsson, written communication, 1982). These features, emphasized by the derivative enhancement technique, are thus nontectonic in origin. Their trends merely reflect the predominance of strong winds from the south and southwest.

The strongly expressed structurally significant lineaments, with west-northwest and north-south trends, form intersection angles of about 50°–60°, suggesting the possibility of simple interpretation as conjugate shears. According to one conjugate shear model, the west-northwest set can be considered as right-lateral Riedel R shears, and the north set considered as left-lateral R′ shears, with a wide region involved in (approximately) simple shear parallel to the direction of plate tectonic spreading (Figure 31-112). Field study nevertheless demands a more complex view of the matter, for within the neovolcanic zone, north-south trending lineaments clearly represent extensional features, fissures, extensional faults, and crater rows, directly related to active spreading. If, over broad domains of Plio-Pleistocene crust, the north-south trending lineaments are assumed to represent the conjugate (R′) shear set, then this strike-slip deformational mode must be superimposed on preexisting structural surfaces of primary extensional formation.

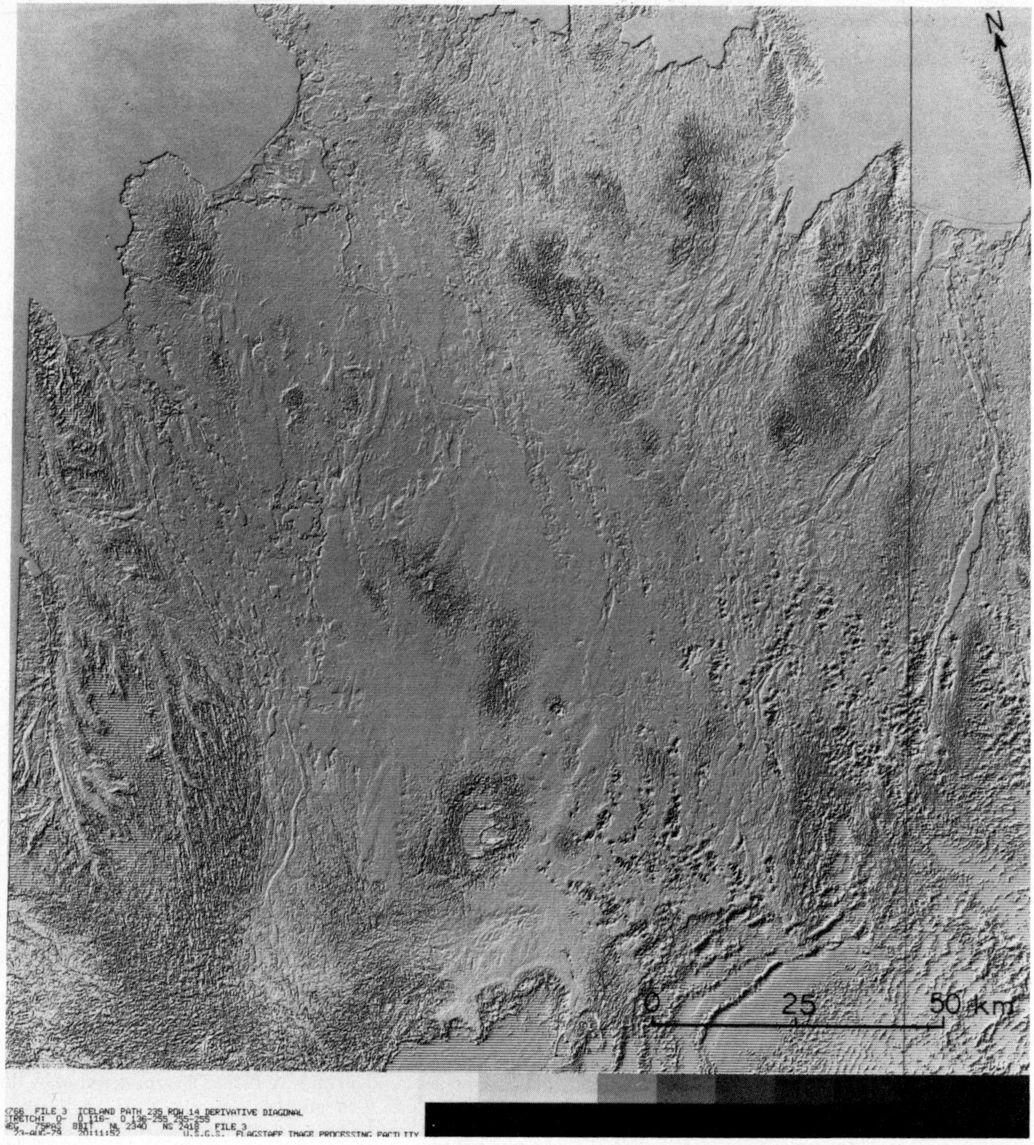

Fig. 31-110. Derivative image enhancement of a Landsat band 7 MSS image (2494-11503) acquired on 30 May 1976. Image courtesy of the U.S. Geological Survey's Flagstaff Image Processing Facility.

An alternative view holds that *both* lineament sets are of primary extensional origin, with the north-south trending set due to axial rifting, and the west-northwest set due to thermoelastic contraction of the spreading, cooling lithospheric plate (Turcotte, 1974; Jefferis and Voight, 1981). Thermoelastic contraction causes extension fracturing highly oblique to the rift zone axis, which in the diverging plate occurs near the edge of, and beyond, the active neovolcanic boundary. A problem with this interpretation is that thermal contraction-induced fissuring might be expected to be perpendicular to the axial rift zone, which is not the case for the west-northwest lineaments; the mathematical physics of the problem is, however, sufficiently obtuse so as to render this point uncertain, and it seems possible that contraction fissuring could follow approximately the direction of plate spreading in oblique-spreading ridges. Once formed, extensional-fissure zones could certainly accommodate strike-slip motion with reorientation of stress. Fissure zones originating by extension could thus strongly influence the subsequent development of transforms.

Field evidence is not yet decisive on the matter of interpretation of lineament fabric. Evidence at present favors pronounced (on the order of 10 km) right-lateral shearing on relatively narrow bands of west-northwest-trending structures near Husavík, at the Tjörnes Fracture Zone boundary. and at Axarfjördur, where the spreading along the Kráfla and Askja fissure swarms is transformed into right-lateral shears (Figure 31-111, at locations *H,A;* Saemundsson, 1974; Björnsson et al.,

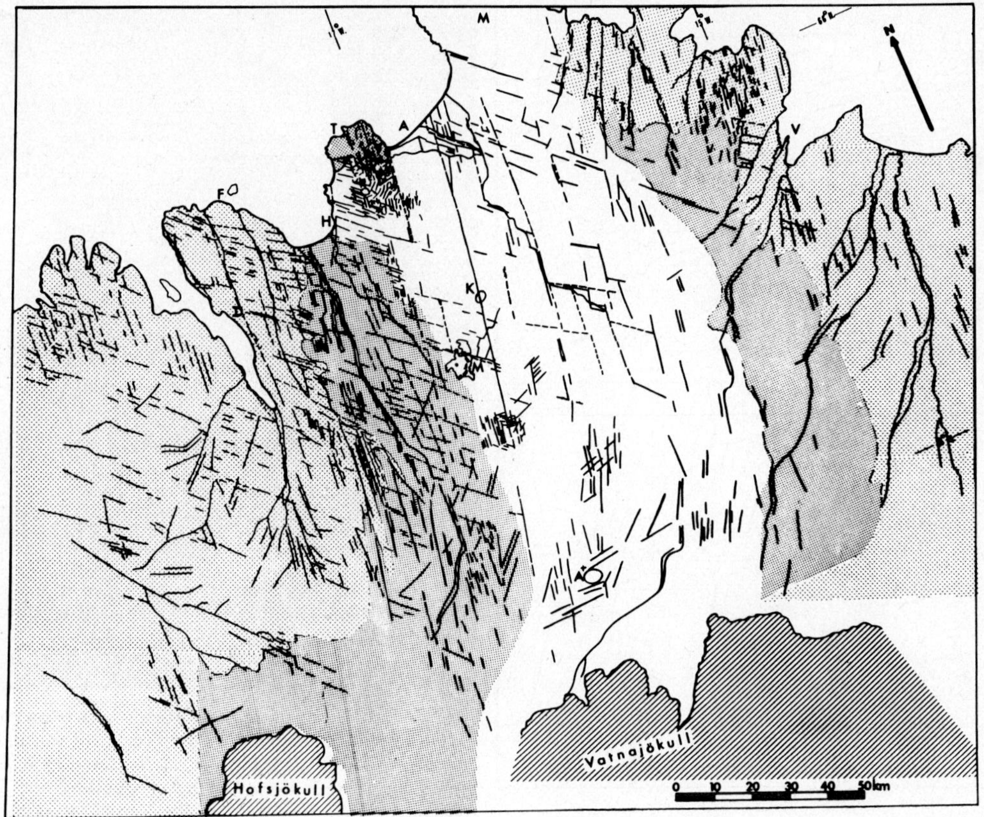

EXPLANATION

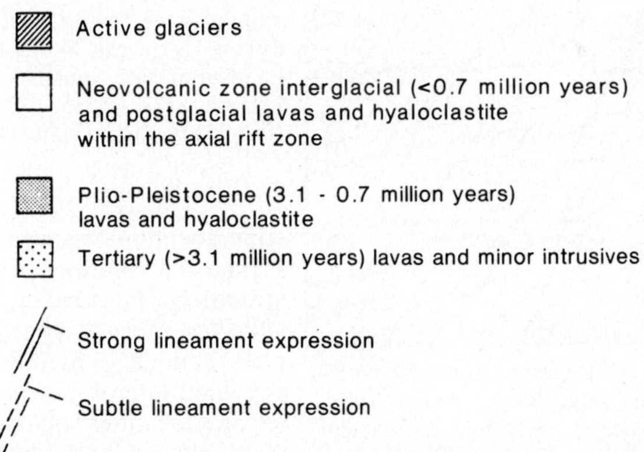

Fig. 31-111. Derivative image map of lineaments, northeastern Iceland. Southern boundary of Tjörnes Fracture Zone, connecting the axial rift zone of northeast Iceland with the offshore Kolbeinsey Ridge, is commonly regarded as the WNW-trending Husavík fault system (*H*). Geographic features are identified as follows: *D*, Dalsmynni; *F*, Flatey; *H*, Husavík; *T*, Tjörnes; *A*, Axarfjördur; *K*, Krafla; *M̀*, Mývatn; *A*$^{+}$. Askja; *M*. Melrakkaslétta; *V*. Vopnafjórdur.

1977; Mamula and Voight, in press). Furthermore, a 90° bend of dikes related to right-lateral west-northwest transform shearing with 15 km or so displacement has been mapped as a 10 km-wide band in the peninsula south of Flatey (Figure 31-111, location *F*). Elsewhere, however, there is lack of evidence for west-northwest shearing; for instance, although right-lateral shearing was proposed for a strongly expressed lineament through Dalsmynni (Figure 31-111, location *D;* Einarsson, 1976), subsequent geologic and paleomagnetic mapping rules out decipherable strike-slip displacement.

In summary, the north-south-trending set reflects structures originating by extensional tectonics; subsequent shearing may have occurred at

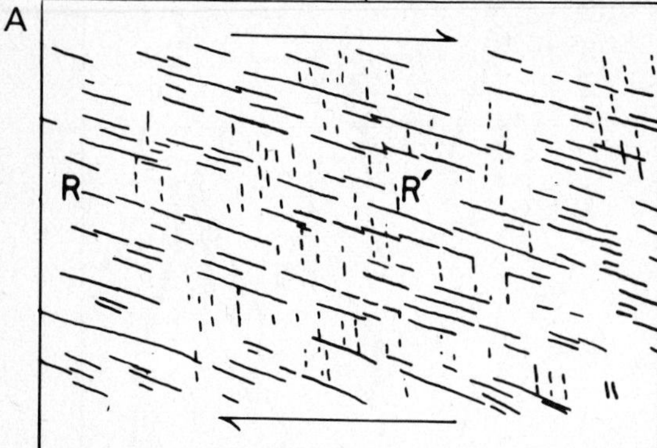

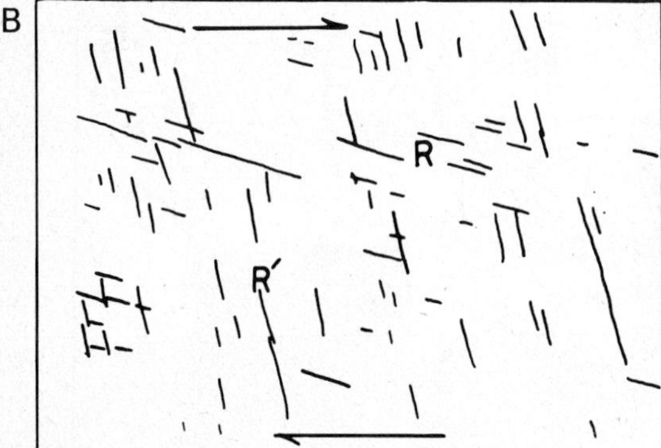

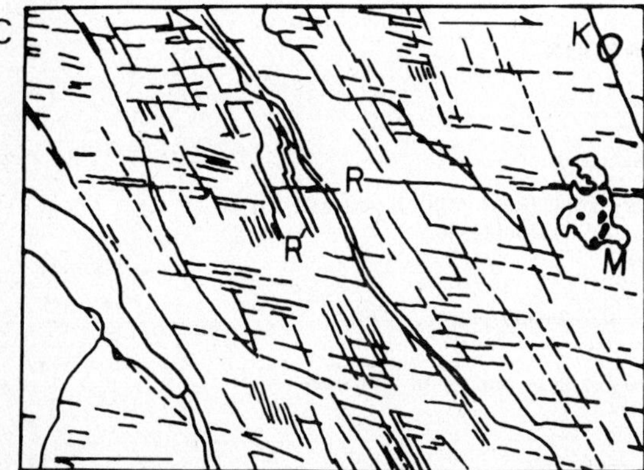

Fig. 31-112. Comparison of lineament trends of northeastern Iceland to simple shear Riedel models (after Mamula and Voight, in press). A. Shear experiment (modified from Hoeppener et al., 1969; see also Freund, 1974) in which the R and R′ shears form at an angle of approximately 15° and 75° from the direction of applied simple shear as indicated along top and bottom boundaries of diagram. B. The results of a similar shear experiment (modified from Freund, 1974) in which the conjugate Riedel shears formed at an angle of approximately 12–15° and 80° to the direction of applied simple shear as indicated on the diagram. C. Lineament map of a part of northeast Iceland taken from Figure 31-111. Top and bottom boundaries are oriented at an azimuth of 100°, the inferred sea-floor spreading direction. The orientation and relative magnitude of the R and R′ shears in the examples given suggest a similarity between Riedel shears of many orders of magnitude difference. *M* = Mývatn, *K* = Krafla.

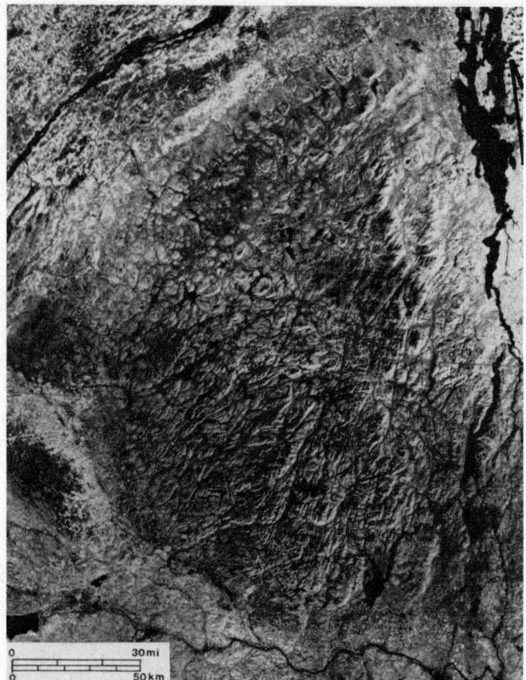

Fig. 31-113. Landsat mosaic of northern New York State, a region dominated by the breached Adirondack Mountains dome, recognizable as the large dark, oval area of forested mountains. Surrounding the Proterozoic metamorphic rocks of the core of this dome are gently dipping Paleozoic strata expressed in lighter tones by agricultural meadows. The waffle-iron pattern in the southern half of the dome is produced by arcuate east-west ridges and valleys which are crossed by more prominent north-northeast- to northeast-trending linear valleys. Note the North arrow in upper right corner. Landsat mosaic compiled from the following MSS band 7 images: 1079-15115, 1079-15122, 10 October 1972; and 1080-15174, 1080-15180, 11 October 1972.

some locations because of reorientation of stress. Primary formation of west-northwest trending structures is uncertain, but some role seems plausible for thermal contraction near the divergent plate boundary. Significant strike-slip shearing in the right-lateral sense has been documented at several localities, but can be discounted at others. Thus, despite the relatively simple geometric arrangement of lineament trends in northeast Iceland, the interpretive question has proven to be complicated and is not yet completely resolved. Lineament maps nevertheless continue to provide a useful guide to field studies.

Mapping Brittle Structures

Landsat multispectral scanner (MSS) images provide the most useful single tool for initiating a regional analysis of brittle deformation, because they provide synoptic views of large areas at a constant, low-azimuth Sun angle, thus creating an apparent relief map. Such a "map" accentuates linear geomorphic features, such as straight valleys, shorelines, and escarpments, most of which result from differential erosion along faults and other fracture systems.

A striking example of such a Landsat "relief map" is that of northern New York (Figure 31-113). The dark oval area defines the breached Adirondack dome, a mountainous core of highly fractured, high-grade metamorphic rocks of Proterozoic age. This heavily forested core contrasts with the lighter toned farmlands developed on gently dipping Paleozoic strata that surround the uplift. The dome is bordered by Lake Champlain on the east, the St. Lawrence River on the northeast, and the Mohawk River on the south.

In the central and southern part of the uplift can be seen an east-west, arcuate pattern of ridges and valleys that reflect differential erosion of a tightly folded sequence of granitic gneisses and metasedimentary rocks. More pronounced than this pattern, however, is that produced by the many north-northeast- to northeast-trending linear valleys that clearly cross these areas of earlier ductile deformation, and must be largely fracture-related.

A four-stage investigative procedure for extracting and identifying geological information from Landsat image products has been detailed by Isachsen (1973) in connection with studies of brittle deformation in the Adirondack Mountains.

Stage I: Identification of all spectral signatures (i.e. points, lines, and areas which may be geologically linked) and tracing them onto clear acetate or mylar overlays.

Stage II: Office identification of spectral signatures using existing information (e.g., geologic and topographic maps, individual aerial photographs, index mosaics of aerial photographs, and other remotely sensed data) in order to identify them as either: a) clearly nongeological (i.e., man-made features); b) clearly geological (from existing geological maps); c) new signatures that are probably geologically linked (e.g., linear drainage features, aligned volcanic cones, etc.); such features should thereafter be referred to by appropriate geomorphic names rather than by the vague, catch-all term "lineament."

Stage III: Evaluation of remaining anomalies by observation and handheld oblique aerial photography from a low-altitude aircraft for initial ground location and identification, followed by conventional ground study of representative examples.

Stage IV: Compilation of all available geological and remotely sensed data to produce an enhanced general geological map or a brittle-structures map.

Applying this procedure to the Adirondack region, a map was made of all linear features clearly visible on Landsat MSS band 7 images, at a scale of 1:1,000,000 (Isachsen, 1973, 1974) and, later, at 1:250,000 (Isachsen and McKendree, 1977). The latter analysis produced the majority of lineaments seen in Figure 31-114. Subsequent stereoscopic examination of high-altitude color-infrared aerial photographs (positive transparencies) at a scale of 1:130,000 showed additional linear features which were not apparent on Landsat MSS images. These shorter lineaments have northwesterly trends (Figure 31-114) and, because they are parallel to the southeasterly direction of solar illumination, are not highlighted on the satellite images.

After compilation of this map from Landsat images and aerial photographs, all identified lineaments were evaluated using topographic and geologic maps to determine their character, and to eliminate the following: 1) man-made linear elements, such as railroads, transmission lines, canals, and forest property lines that had been enhanced by logging practices; 2) lithological lineaments produced by either normal lithological contacts or foliation trends, rather than by brittle deformation; and 3) features which could not be validated on either topographic or geologic maps. This screening procedure left lineaments that are clearly recognizeable as natural geomorphic features, such as straight stream valleys or stream segments, meandering streams with overall linear trends, long narrow lakes, straight lacustrine shorelines, and linear escarpments. Subtle linear vegetation strips in areas of low relief generally were found to define straight stream courses. It is important to emphasize that although some kind of fracture control is a likely explanation for these linear features, objectivity dictates against the application of terms such as faults, joint zones, shear zones, and fracture zones without ground identification. Any extrapolations from *valid* geomorphic descriptions to *hypothetical* structural features should be clearly identified as tentative speculations.

To produce a 1:250,000-scale map of New York State that would show the maximum amount of information describing brittle deformation, the map of linear geomorphic features was embellished by the addition of all previously mapped faults, indicating type, attitude, relative movement, and amount of displacement, and locations of breccias (Isachsen and McKendree, 1977). A drainage map was used as a base, because drainage patterns show a sensitive response to brittle deformation. A reproduction of the Adirondack part of this map, much reduced and using only selected drainage features, is shown as Figure 31-114. It continues to serve as a guide to field studies aimed at determining the nature, distribution, and origin of brittle structures in the basement rocks exposed in the Adirondacks, their relationship to the doming, their movement history, and the significance of their mapped extensions into overlying Phanerozoic strata to the south and west (Isachsen and McKendree, 1977).

Although the great majority of brittle structures are linear, important exceptions exist. A notable example is a circular drainage pattern, 10 km in

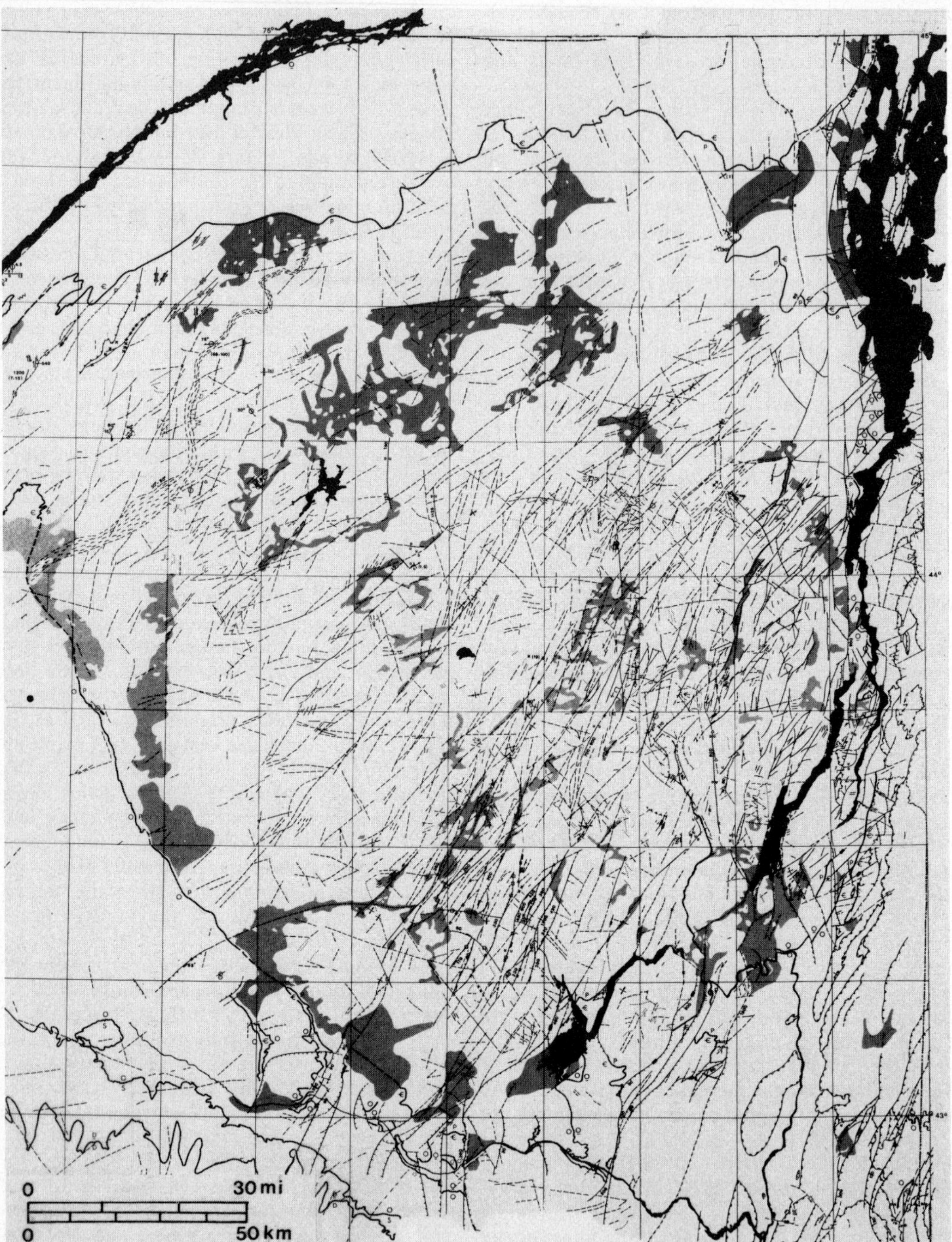

Fig. 31-114. Brittle-structure map of area shown in Figure 31-113, with geological period boundaries shown by a heavy line. Map combines lineaments from 1:250,000-scale Landsat MSS band 7 images and 1:130,000-scale high-altitude color-infrared aerial photographs with mapped faults and fracture systems described in the geological literature. Selected drainage is shown in black for reference with Figure 31-113, and areas of extensive cover in gray. Reduced from maps by Isachsen and McKendree (1977).

diameter, visible 50 km south of the Adirondacks on Landsat MSS images of the Catskill Mountains. Geological studies and gravity modeling suggest that this circular structure is a buried meteorite crater (Isachsen et al., 1977).

ECONOMIC GEOLOGY

Introduction

In the 10 years that have passed since the launch on 23 July 1972 of the first Earth Resources

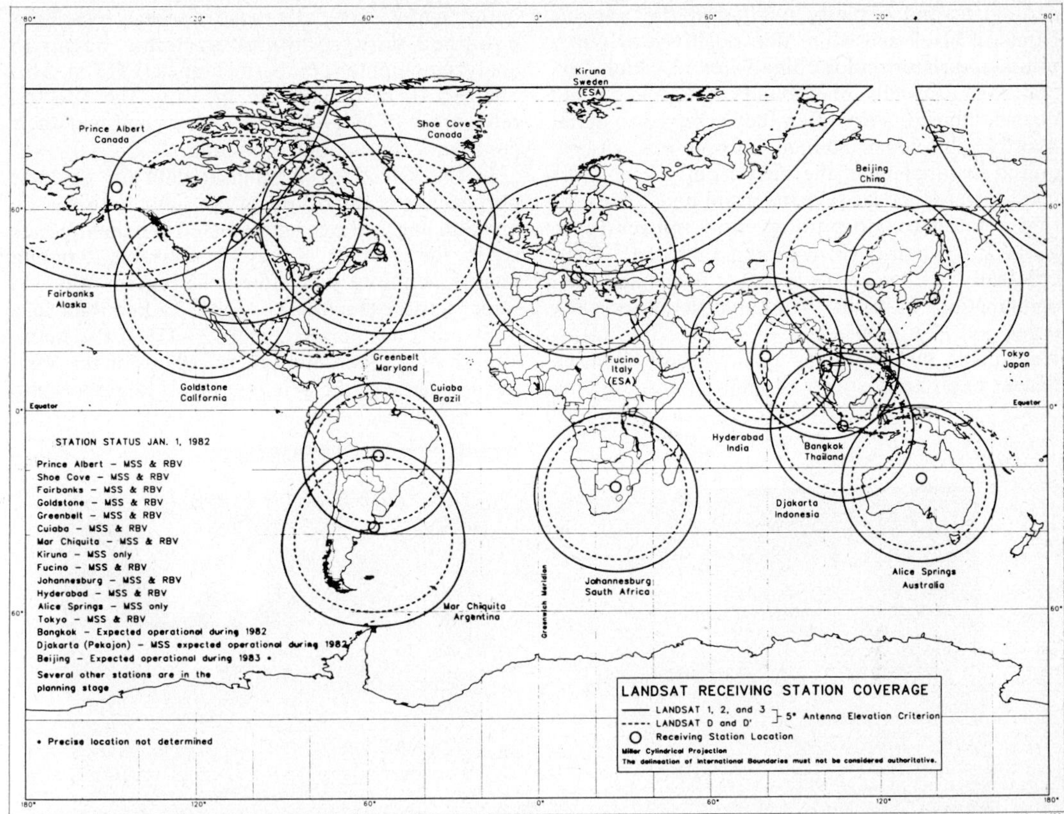

Fig. 31-115. U.S. Geological survey index map showing reception radius for Landsats 1, 2 and 3, and the smaller one for Landsat 4, of existing or planned Landsat receiving stations. The reception range for the French SPOT satellite will fall about midway between the Landsat 1, 2, and 3 ellipse (solid line) and the Landsat 4 ellipse (dashed line).

Technology Satellite (ERTS-1), now referred to as the Landsat series, geologists throughout the world have been learning how to use space data in conjunction with conventional aerial photography. This has been especially true of those geologists who have the responsibility for finding and developing mineral and energy resources: economic geologists. It has taken a considerable amount of time, an entire decade in fact, for Landsat to acquire cloud-free images of most of the Earth's land surface. In fact, there are still a few areas of northern South America, southern South America (especially in Patagonia), Central America, parts of the Caribbean region, and tropical regions of Central Africa and Asia that are outside the direct receiving range of existing Landsat receiving stations (Figure 31-115) still not covered or for which existing data are extremely poor because of cloud cover. Nevertheless, economic geologists have shown considerable interest in utilizing these data in the search for mineral and energy resources. The reasons for this interest are many, but perhaps the foremost are that Landsat provides synoptic views of large areas under uniform lighting conditions, and that these views are presented in formats that are reliable, understandable, and flexible.

Each Landsat image covers approximately 185 × 185 km (34,225 km^2) at uniform scales and can be obtained in the form of film transparencies or paper prints as black-and-white or color infrared images in a variety of formats (see Table 31-12). For economic geologists familiar with photogeological analysis, the Landsat data relate closely to their previous experiences with interpretation of aerial photographs (Ray, 1960) and, therefore, does not represent a major change in analysis procedure. The scale of Landsat data, however, is significantly different, and some geologists have found it difficult to make a transition to extremely small scales. Scales can be selected at 1:3,369,000, 1:1,000,000, 1:500,000, 1:250,000, and 1:125,000. In addition, Landsat images can be obtained as computer compatible tapes (CCT's) enabling one to conduct image enhancement, band ratioing, and spectral measurement of rock types, alteration zones, and other geologic features at scales up to 1:25,000, if desired (see the "Analysis Techniques" section of this chapter).

Most geologists began their evaluation of Landsat data by studying individual images of areas where they were familiar with the terrain and geology. They used the approach of most photogeologists and visually evaluated the various bands in black-and-white film positives or paper prints and, later, color composites in which several bands (usually MSS bands 4, 5, and 7) had been combined. Very often they referred to aerial photographs when detailed information was required to supplement the view from space. Most studies were focused on patterns of linear features that related to morphology and indirectly to geologic structures (Rowan and Lathram, 1980). Gradually, they used the separate bands and then combinations of bands to distinguish certain rock types and alteration zones in which crude ratios were made by overlaying positive and negative images of different spectral bands. It was not long before many, especially those in research facilities equipped with computer systems, began to analyze computer compatible tapes (CCT's). This enabled them to actually measure the spectral reflectance of individual rock types and map their continuity throughout the scene.

In the selection of Landsat data for geologic purposes it is important to consider both season and Sun angle. If one is interested in mapping rock types, for example, it may be advisable to obtain images acquired just after a December wet and rainy season (Figure 31-116A), rather than in a September dry period (Figure 31-116B) as demonstrated by Viljoen and others (1975) for the Ventersdorp area of the Witwatersrand Region, South

Fig. 31-117. Landsat images of northern California. A. Image (1275-18290; 24 April 1973) acquired during a high solar elevation angle (54°). B. Image (1167-18283; 6 January 1973) acquired during a low solar elevation angle (22°). The latter image greatly accentuates topographic relief. At higher latitudes, where large seasonal variations in solar elevation angle occur, the effect is even more pronounced, and makes studies of subtle geomorphic features possible. Astrogeologists and astronomers have long used this technique in studies of the Moon and other bodies in the Solar System (see Figures 31-253 and 31-257). The next time you view the Moon through a telescope or high-power binoculars, note how accentuated are the landforms on the Moon near the terminator as compared with those near the center of the disc.

Africa. In the post-wet season imagery, rock types are much more clearly displayed by the spectral response of the overlying vegetation than during the dry season.

If one wishes to map structural geologic features, black-and-white, low Sun angle images are adequate and preferable to those of high Sun angle. This scene of northern California (Figure 31-117A) during a springtime high Sun angle (54°) period is useful in showing drainage patterns and other surface features. Its counterpart (Figure 31-117B), obtained during a wintertime low Sun angle (22°) period, accentuates topographic relief and is more useful in displaying structural patterns observed as linear features which generally relate to fault and fracture zones. Such features generally control drainage patterns. With greater availability of radar (SLAR) imagery, geologists will be using radar images for structural and geomorphic studies (Albert, 1982) (see also Figure 31-15).

Geologic mapping, using visual and digital methods for analyzing Landsat data has generally taken advantage of the benefits provided by low Sun angle illumination. Francica et al. (1980), however, point out that the accuracy of digital mapping of lithology is improved in areas of high terrain relief such as the Ladakh Himalayas by using summer time (July) high-angle Solar illumination (58°). The amount of shadow produced by steep slopes is considerably reduced at this time, and the ability to measure the surface reflectance of rock and soil types is considerably improved. Six major rock classes were identified, measured, and mapped in this project: 1) ophiolite; 2) the Ladakh intrusives, which include biotite-hornblende granite and associated granodiorities, norites, and gabbros; 3) The Ladakh molasse, an interbedded complex of conglomerates, sandstones, and mudstones; 4) the Dras Volcanics, consisting primarily of andesites and volcano-clastics; 5) the Indus flysch, a thick sequence of contorted silty shales; and 6) a wide variety of unconsolidated Quaternary deposits.

Satellite images and image mosaics have, for

the first time, enabled geologists to undertake revisions of geologic maps in areas such as those in Bolivia, for which no base maps and aerial photographs exist (Brockmann, 1978). In addition, new maps, such as lineament maps, have been constructed for many nations, such as the United States (Fischer et al., 1976; Rowan and Lathram, 1980); Bolivia (Brockmann, 1980), Finland (Mikkola, 1982), and the USSR (Shcheglov, 1979). For Bolivia, special maps were published to show the relationships of lineaments to intrusive rocks and mineral belts of the Andes (Brockmann, 1980) and these maps should be of great use in undertaking national mineral exploration programs.

The Landsat program began with three U.S. receiving stations and a satellite carrying two tape recorders having a capacity of 500 hr each. Canada and then Brazil were the first nations to construct their own stations at Prince Albert, Saskatchewan, and Cuiaba, Brazil, respectively. These ventures were found to be so useful that stations have subsequently been or are being built at several other locations including Shoe Cove, Newfoundland (later deactivated in the fall of 1982); Mar Chiquita, Argentina; Alice Springs, Australia; Fucino, Italy; Kiruna, Sweden; Johannesburg, South Africa; Beijing, China; Hyderabad, India; Bangkok, Thailand; Djakarta, Indonesia; and Tokyo, Japan (see Chapter 12, Landsat Satellites, and Figure 31-115). The importance of these stations to each of these nations is based heavily on their needs to inventory and monitor their natural resources and especially to improve their mineral and energy resource base.

It has taken 10 years to reach a reasonable level of sophistication in the use of satellite imagery by a few economic geologists throughout the world. Several books (Williams and Carter, 1976; Smith, 1977; Woll and Fischer, 1977) or chapters of books (Rowan and Lathram, 1980) have been published that document the progress of remote sensing applications to mineral and energy exploration. Most compilations, however, are disappointing in that they either were compiled too early in the history of satellite remote sensing, when there were only very few case histories available, or took so long in preparation that they are somewhat out of date with respect to present remote sensing technology.

Most of the petroleum and mineral industry, because of the intensely competitive nature of the business, is reticent about documenting the means by which a successful exploration effort was achieved, especially the application of a new technology such as remote sensing. The two William T. Pecora Memorial Symposia dealing with the petroleum and mineral industries (3rd and 6th) are the only Pecora Symposia (Table 31-27) not to have proceedings volumes. For whatever reason, therefore, there are only a few well documented examples of mineral or energy resource discoveries in which satellite data served as a principal source of information leading to the discovery. Some of the best that could be found are included in the "Mineral Exploration" section of this chapter. The reader is also referred to the proceedings volumes of the Second Thematic Conference: "Remote Sensing for Exploration Geology," of the 16th International Symposium on Remote Sensing of Environment, held in Fort Worth, Texas, on 6–10 December 1982, too late for inclusion in this chapter (see Table 31-24).

The effective dissemination of information about the use of remotely sensed data for mineral exploration to the international community has improved markedly. This relatively rapid development has been assisted in part by increased communication by mail, but also through regional workshops and symposia developed under the International Geological Correlation Programme (IGCP) of the International Union of Geological Sciences (see Table 31-29). Its IGCP Project 143 "Remote Sensing and Mineral Exploration" (Carter et al., 1980; Kahle et al., 1981a) involves about 500 geologists from 85 countries, and from this nucleus of scientists it was possible to stimulate research and transfer the technology to others. It is these few who must continue to press forward further development of these and other new exploration techniques, if they are to meet the global mineral and energy resource demands of the future. This is not to imply that remote sensing is "the panacea" for economic development, for it is but one of several new tools which are now available to the world that will assist in the quest for an adequate mineral and energy resource supply and perhaps make life somewhat better than it has been in the past for a greater part of the world population.

Experimental satellite and aircraft data have been acquired by NASA and other organizations. Medium-altitude aerial photographs from the National High Altitude Photography (NHAP) program are available for parts of the contiguous United States (Figure 31-6) and Alaska (Figure 31-7). Radar imagery is also becoming more available (Figure 31-15). All have proven to be a useful supplement to Landsat. The color and color-infrared photography from U-2 and RB-57 high-altitude aircraft; a variety of data from the Skylab and Shuttle experiments (Settle and Taranik, 1982b); imaging radar and radar altimetry from Seasat; the coastal zone color scanner (CZCS) from Nimbus-7; and the visible and thermal infrared data from the Heat Capacity Mapping Mission (HCMM) have literally inundated the geologic community with new, experimental data (see the frontispiece illustrations of this *Manual;* also see the "Types and Availability of Remotely Sensed Data for Geological Investigations" section of this chapter). Because such satellites are experimental, however, there was no provision by NASA to make the data readily available to the general public as is the case with Landsat data. NASA used its usual approach by selecting a limited number of principal investigators to analyze the data and report on the results in the scientific literature. Magsat, which orbited between November 1979 and June 1980, provided global magnetic data that are still being analyzed. The resulting maps have recently been described by Settle and Taranik (1983). They confirm, and show an improvement in, our knowledge of deep crustal magnetic anomalies in comparison to the first experimental maps derived from the POGO (Polar Orbiting Geophysical Observatory) of the 1960s.

The POGO results reported by Regan et al., (1975) and Regan (1977) put into relative perspective some of the major known magnetic anomalies of the world. They show, for example, that the Kursk Magnetic Anomaly (KMA) of west central Russia, famous for its banded-iron ore production, is the only anomaly which exceeded 8 gammas from space altitudes (orbits 400 to 1500 km high). The Bangui anomaly, in the Central African Republic, on the other hand, is a copper-rich area having a negative anomaly of ⁻6 gammas. The demonstration by Green et al., (1978) of the feasibility of combining digital Landsat data with digital aeromagnetic data to form stereoscopic models of magnetic anomalies in a Landsat pictorial display (see Figure 31-99 for a similar example in Alaska) suggests that it should be possible to model larger areas of several scenes in mosaic form and use satellite magnetic data to develop three-dimensional models. Such modeling should improve our understanding of the relationships between extensive surface geologic features and deep crustal anomalies.

The Shuttle Multispectral Infrared Radiometer (SMIRR) data and Shuttle Imaging Radar (SIR-A) data collected in 1981 are just being analyzed as this chapter is being written (see earlier section of this chapter on SMIRR and a later discussion in this section on SIR-A). Several rock types have been identified using SMIRR data in desert regions of Egypt, Mexico, and Spain by Goetz et al., (1982). The preliminary results from this experiment appear to be excellent and show considerable advancement over previous systems. Landsat 4, launched on 16 July 1982, will improve exploration capabilities by adding higher resolution data (30-m pixel) and additional spectral bands (1.55–1.75 μm and 2.08–2.35 μm) designed to aid in the discrimination of hydroxyl clay minerals associated with surface alteration often due to the presence of ore deposits. As this text is being written, the first data from both the Landsat 4 multispectral scanner (MSS) and thematic mapper (TM) sensors have been processed and appear to exceed design specifications. (see frontispiece figures of this *Manual*).

While these experimental data still need to be evaluated in the context of their applications to mineral and energy exploration, the future is bright. Exploration geologists have, through the Geosat Committee, Inc., had the opportunity to make their studies and recommendations recognized by NASA and other agencies. Space systems have been designed to meet some of their requirements. It is now their responsibility to work with the new data and test it in practical exploration environments, such as with the nearly completed Joint NASA/Geosat test case project (Abrams et al., 1983). The Silver Bell porphyry copper test site is discussed later in this section.

Although much of the emphasis and many of the results of investigations reported in this section describe the use of Landsat, Seasat, and other satellite data and computer interpretation, we remind the reader that methods used in conventional visual analysis of aerial photography, including the use of stereoscopes, such as the Kern PG-2 photogrammetric plotter, remain important tools for the field geologist. In fact, they account for a major part of the use of remote sensing data and equipment in operational mapping programs of the U.S. Geological Survey and most other mapping agencies, organizations, and companies. For this reason, two articles are included in this section on the use of aerial photographs for geologic and coal-bed mapping in Wyoming.

The Economic Geology section of the chapter begins with this "Introduction" and is followed by specific case histories in the following four sub-sections: Mineral Exploration, Petroleum Exploration, Coal Exploration, and Geothermal Exploration and Related Surveys. The last topic reviews of the use of aerial thermography in geothermal exploration and to other types of surveys, including active volcanoes.

As one evaluates the various case histories and discussion of topics that follow, one should keep clearly in mind that the technology of remote sensing from aircraft and space platforms, while still in its infancy, is developing rapidly and its contribution to the role of the economic geologist is significant. First and foremost are the Landsat data that provide the geologist with the opportunity to study almost any land area of the Earth. The Landsat spectral data provide the geologist with information from which can be drawn inferences as to the type of materials that lie within the scene. As new bands are added, as they were to Landsat 4 and SMIRR, the ability to discriminate materials remotely will increase. As our ability to use these data sources grows, especially when combined with other geophysical parameters (Chavez et al., 1979; Podwysocki et al., 1982b, 1983a), such as aeromagnetics, gravity, and geochemistry, our understanding of the Earth's crust, its structure, chemical composition, and anomalous conditions that relate to the location of ore deposits (Carter, 1982) or petroleum reservoirs, will also improve.

The economic geologist, therefore, must continue to be involved in developing the use and applications of remotely sensed data from aircraft and spacecraft. Ideas and data requirements must be provided to the scientists and engineers who build the instruments and the vehicles on which they are carried. Such an interchange will lead to better systems that can improve our ability to measure and map the surface of the Earth and thereby help discover and develop new mineral and energy resources.

Mineral Exploration

Examples of Significant Mineral Resource Exploration Activities and Discoveries Based on the Use of Landsat Data

Nickeliferous laterite, West Irian Jaya, Indonesia

The use of Landsat data by commercial mineral exploration companies is rarely publicized in the scientific literature because of the proprietary nature of the information. Several government/industry endeavors, however, have provided useful demonstration projects in which both visual and digital methods of analyzing Landsat data have been used. Taranik et al. (1978) tested the use of Landsat data for investigating nickeliferous laterite deposits of Gag Island in West Irian Jaya, Indonesia, as a result of discussions between C. D. Reynolds of U.S. Steel International and William D. Carter of the U.S. Geological Survey. At the time, Carter was interested in a Landsat tropical forest test area, in which metal deposits were reasonably well known but relatively undisturbed by mining activity. Reynolds noted that U.S. Steel had recently explored Gag Island by geological mapping, geophysical surveys, and drilling and had shipped about 10,000 metric tons of ore-grade material for mining evaluation and metallurgical testing. Reynolds provided color aerial photographs and terrestrial color photographs (35-mm slides) of local scenes to James V. Taranik at the EROS Data Center to aid in the analysis. Contrast-stretch enhancement and supervised parallelepiped analysis methods successfully showed the distribution of nickeliferous laterite on Gag Island. Subsequently, similar studies conducted of Kawe Island, in the same Landsat image but in apparently unexplored areas, showed indications of the presence of laterite soils covering a significantly smaller area.

Porphyry Copper Deposits, Saindak Area, Western Pakistan

Using an analysis technique in which the spectral characteristics of a known mineralized area are used to analyze other areas of the Landsat image, Schmidt (1976) analyzed a Landsat image (1109-08054; 9 November 1972) of the Saindak area of the Chagai district of western Pakistan, where a known porphyry copper deposit with propylitic alteration had been mapped by Ahmed et al. (1972) and Schmidt (1973). Using spectral characteristics of the known site as a guide, Schmidt was able to detect 19 areas in adjacent regions having similar characteristics. Field observations indicated that 5 of the 19 sites contained surface manifestations of mineralized rock in sufficient abundance to warrant further exploration. Dykstra and Birnie (1979) later repeated, refined, and added to the results of this work.

Copper and Molybdenum Deposits, Sonora, Mexico

Lineament studies of a Landsat mosaic of Mexico showed a high degree of correlation between significant ore deposits and intersections of major lineaments (Salas, 1975). His study was followed by a joint U.S. Geological Survey-Government of Mexico geological team that used Landsat data, geochemical sampling and analysis, and aeromagnetics to identify and subsequently to drill and prove a major polymetallic copper-molybdenum ore deposit (Alcaparroso), in Sonora, Mexico (Raines et al, 1980b). Additional research in the same region was discussed by Turner et al., 1982.

Alluvial Tin Deposits, Rondonia Area, Western Brazil

Visual analysis of single band and color composite images remains a major effort by many geologists. Keighley et al. (1980), consulting geologists, reported that MSS band 7 and Landsat MSS color-composite images were used to distinguish tin-bearing from other granite intrusive masses in the Rondonia area of western Brazil. The tin-bearing granites occur as domes and ring complexes ranging in diameter from 1 to 18 km along tectonic zones of weakness that strike N.45°–65°E. and that extend a distance of 2000 km, from eastern Bolivia to east of the Río Xingu. Paleochannels of tributaries to the Amazon and Madiera River systems crossing this trend have been mapped from Landsat images, and likely sites have been identified for exploratory trenching and drilling. At least four ages of channels have been identified, and the oldest, at depths of 50 meters, have been found to contain alluvial and eluvial placer tin. Current production from these deposits is 6000 metric tons of tin per year, currently valued at U.S. $100 million. Estimates of reserves indicate that production can be doubled in the near future.

Uranium-Thorium-Zinc-Copper-Nickel Occurrence, Kentucky

Geochemical anomalies that may be further evidence of the existence of the east-west trending 38th parallel lineament have been discovered in Wolfe, Powell, and Menifee Counties, Kentucky. A major east-west linear trend is discernable on Landsat images of this area. Stream-water, stream-sediment, and outcrop samples collected along the northeast-southwest-trending Corbin sandstone outcrop belt show anomalous concentrations of U, Th, Zn, Cu, and Ni only in parts of the belt that skirt the 38th parallel lineament.

Landsat studies also show that the anomalies are closely associated with the intersections of the four major linear trends present in eastern Kentucky. This association, as well as the occurrence of a uraniferous kimberlite pipe, suggests that the anomalies resulted from ascending fluids that utilize these lineaments as conduits (Richers, 1981).

Uranium Mineralization, Western Australia

R. Cary and M. Longman (written communication, 1978), geologists on the staff of the Australian Minerals Company, Limited, of Perth, Australia, collected field spectra of the Yeelirrie uraniferous calcrete deposit, first discovered in 1972 (Premoli, 1976). Cary and Longman used their field spectra to guide their supervised classification analysis of a Landsat image (2107-01121; 9 May 1975; Color Figure 31-118) until they were satisfied that they could produce a reasonably accurate digital map of the known mine area (Color Figure 31-119A). They then used the reflectance data as a basis for searching the remainder of the Landsat scene for areas having similar surface reflectance (similar materials). A small area was located about 7 km southeast of the Yeelirrie homestead on which they staked six mining claims (Color Figure 31-119B). A drilling program was initiated in which 85 percent of the drill holes cut calcrete, but which found only low grade uranium mineralization.

The Use of Multispectral Image Band Ratios in Mineral Exploration

Early studies by Robert K. Vincent, Geospectra Corporation; Lawrence C. Rowan, U.S. Geological Survey; Alexander F. H. Goetz, Jet Propulsion Laboratory (California Institute of Technology); and others have demonstrated that ratioing Landsat MSS spectral bands can be useful in displaying areas of hydrothermal alteration in which limonite occurs in association with and is, in many places, the surface expression of mineralized ground. Vincent (1973) found that a band 5/6 ratio enhanced the black-and-white display of the iron-rich Phosphoria Formation in The Atlantic City district of Wyoming. Rowan et al. (1974 and 1976), using Landsat MSS band ratios of 4/5, 5/6, 6/7, were able to define several hydrothermally altered areas in the vicinity of the Goldfield Mining District, Nevada, where limonitic alteration was exposed (see also Rowan and Lathram, 1980). These results led to an operational project under the Conterminous United States Mineral Resource Assessment Program (CUSMAP) in which ratio-derived alteration maps covering 1° by 2° quadrangles at a scale of 1:250,000 were produced by computer analysis (Podwysocki et al., 1983a, 1983b). The hydrothermally altered localities shown on the maps were then visited in the field to determine the composition of the hydrothermal alteration. Those areas having sericite, montmorillonite, kaolinite, and other hydroxyl clay minerals in association were identified as possible targets for exploration and further detailed geochemical and geophysical studies.

Similar band ratios of a Landsat image covering the Salar de Uyuni, the largest evaporite basin in southern Bolivia, were made by A. F. H. Goetz in 1978 (Color Figure 31-120). Alteration zones appeared orange-brown in the ratio image and marked the vents of volcanoes within the area. Sodium chloride deposits of the Salar de Uyuni and neighboring Salar de Empexa appeared light blue, pink, and white. Gypsiferous soils and gypsum (sodium sulphate)-rich evaporite basins were green and were easily distinguishable from sodium chloride deposits (Carter et al., 1983).

Airborne Remote Sensing Applications to Porphyry Copper Exploration

Introduction

As part of the Joint NASA/Geosat test case project (Abrams et al., 1983), Michael J. Abrams (Jet Propulsion Laboratory) and David M. Brown (Texasgulf Western, Inc.) examined the applications of airborne multispectral remotely sensed data, using new spectral bands designed for the Landsat 4 Thematic Mapper (TM), specifically for porphyry copper exploration and for general geologic mapping. One of the areas studied was the Silver Bell Mining District in southern Arizona (Figure 31-121). This district is located in the southwestern desert and is characterized by moderate relief, sparse vegetation, and excellent rock exposures. Mining operations date back to the 1860s, when high-grade copper deposits were discovered in tactites. In the 1940s drilling activities revealed the presence of large low-grade disseminated copper mineralization. In 1954, American Smelting and Refining Company (ASARCO) began mining operations from the El Tiro and Oxide open pits, producing copper, molybdenum, and silver from enriched chalcocite ore.

General Geology

Formations ranging in age from Precambrian to Recent are exposed near Silver Bell. The more resistant of these, Paleozoic limestone and Tertiary volcanics, predominate in the scattered peaks and ridges comprising the Silver Bell Mountains (See the color aerial photograph, Color Figure 31-122, and geologic map, Color Figure 31-123, of the area). Porphyry copper mineralization occurs along the southwest flank of these mountains in hydrothermally altered igneous rocks. These are principally intrusives considered to be components of the Late Cretaceous-Paleocene Laramide Revolution.

For most of its length the zone of alteration strikes west-northwest; indirect evidence suggests that a fault representing a line of profound struc-

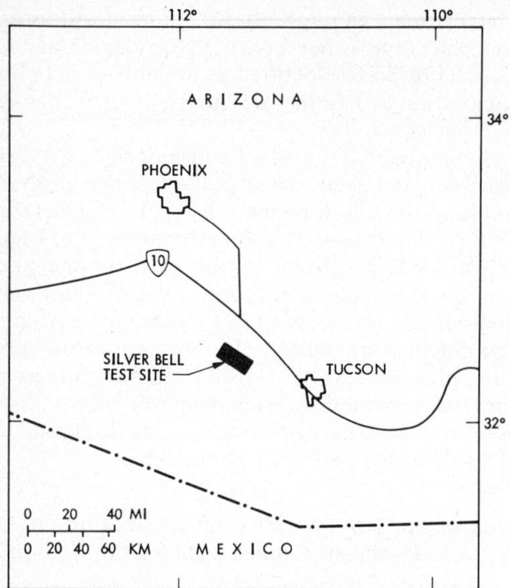

Fig. 31-121. Index map showing the location of the Silver Bell porphyry-copper deposit in southern Arizona.

tural weakness existed in this position before Laramide time (Richard and Courtwright, 1966). A parallel fault, the Ragged Top fault, lies to the north and shows considerable vertical displacement, juxtaposing Precambrian rocks to the north against Tertiary rocks to the south.

The beginning of Laramide activity was marked by the intrusion of alaskite into the Paleozoic and Cretaceous sedimentary rocks. An interval of erosion followed, resulting in deposition of the Claflin Ranch and Silver Bell units. The next events were intrusion of an elongate stock or sill of dacite porphyry and parallel faulting along the major structural line. Monzonite stocks and east-northeast-striking dikes were then emplaced near this line. Alteration and sulfide mineralization took place next, again controlled by the main structural line. Mid-Tertiary emplacement of latite intrusives, dikes, and basaltic flows followed the ore-forming events.

Remote Sensing Data

Aircraft multispectral scanner data were obtained from NASA's NS-001 Thematic Mapper Simulator in October, 1978, at an altitude of 5,000 m, resulting in an instantaneous field of view (IFOV) of 12 m. The seven wavelength bands of this instrument are given in Table 31-3. This scanner has the same seven spectral bands as the Landsat 4 Thematic Mapper except for an additional band at 1.00–1.30 μm.

A color-ratio composite image (Color Figure 31-124) was produced by displaying band ratios 3/2, 4/5, and 6/7 as green, blue, and red, respectively. These ratios were selected to display spectral characteristics of iron oxide and clay minerals that are associated with hydrothermal alteration processes. The 3/2 ratio highlights the presence of iron oxide minerals that have a strong fall-off in reflectance from 1.0 μm towards shorter wavelengths because of an intense ferric iron absorption band in the ultraviolet. The 6/7 ratio highlights the presence of hydrous minerals because of their strong absorption band near 2.2 μm and a steep fall-off in reflectance from 1.6 μm towards longer wavelengths due to a major OH^- absorption band near 2.7 μm. In Color Figure 31-124, surface materials with iron oxides will have a large green color component; those having hydrous minerals will have a large red component; where both occur together, a yellow or orange color will result. The most striking feature in the ratio picture is the yellow-orange zone which trends west-northwest for 6 km from the tailings pond, through the two pits, then veers to the north. This feature corresponds to the phyllic alteration zone, mapped by ASARCO from detailed and extensive field work and laboratory analyses of rock samples. The color is due to spectral features of surface materials caused by the presence of limonite after pyrite and of sericite and kaolinite. The green areas along the eastern part of the Ragged Top fault are limonitic Precambrian rocks; the green patches southwest of the Oxide pit are hematitic Cretaceous arkoses; both of these areas are unaltered and are separable from altered rocks by their lack of hydrous minerals.

Additional computer processing was performed to evaluate the separability of all the geologic units exposed in the area. Field measurements of spectral reflectance were obtained for 18 altered and unaltered rock types in the 0.45 to 2.45 μm wavelength region. The NS-001 wavelength bands were convolved with the spectra to produce equivalent scanner band values. These values were then input into a stepwise linear discriminant function algorithm to determine the optimal linear combinations of the original wavelength variables to separate the sampled rock types (Jennrich, 1977). The first three output-transformed variables were combined to produce a color image (Color Figure 31-125). The area outlined with short dashes is the mapped extent of the dacite porphyry. The different colors in the outlined area indicate variable mineralogical compositions: the light blue areas (*A*) correspond to phyllically altered dacite; the orange area (*B*) corresponds to propylitically altered dacite; the purple area (*C*) and green-blue areas represent weakly altered to unaltered dacite, and dacite interbedded with other rock types, respectively. The dark blue areas (*D*) are outcrops of the Silver Bell unit; green areas (*E*) correspond to Claflin Ranch unit outcrops; other, lighter green areas (*F* and *G*) are limonitic Precambrian rocks and hematitic Cretaceous arkose; the light blue area (*H*) is an outcrop of Tertiary basalt. These are only some of the units which are separable in this image. A detailed interpretation reproduces virtually all the mapped geologic contacts; in addition, several zones of alteration are discriminable as separate units.

This example illustrates the feasibility of using remotely sensed data to recognize hydrothermal alteration associated with a porphyry copper deposit, and the superb mapping information that can be extracted from the data. The Thematic Mapper on Landsat 4 will provide the exploration geologist with a new and improved geologic tool.

Petroleum Exploration

Introduction to the Use of Remote Sensing Technology in the Search for Oil and Gas

Since 1945, petroleum geologists have frequently used aerial photographs as an important tool in exploring for surface manifestations of structural features that might lead to the discovery of new oil and gas reservoirs (see also the "Photogeology and Geological Remote Sensing" section of this chapter). Stereoscopic pairs of aerial photographs have been especially helpful in estimating the dip of strata to an accuracy of 1° to 2°. Trollinger (1968) was one of the first to describe the advantage of the synoptic view provided from space altitudes, when he studied color satellite photographs of the Delaware Basin of west Texas acquired by the Gemini astronauts (see Table 31-6). He found that for the first time he could see surface manifestations of deep crustal structures that, until then, had been recognized only in geophysical and borehole data of the region.

Saunders et al. (1973) conducted systematic visual studies of ERTS 1 (Landsat 1) image mosaics covering several oil and gas provinces in west Texas, New Mexico, Colorado, and Montana-Wyoming. They demonstrated that, in many cases, there was a correlation between surface lineaments expressed in the images and the locations of subsurface oil and gas fields. Similar results were obtained by Collins et al. (1974) in western Oklahoma.

Floyd Sabins, in his textbook, *Remote Sensing, Principles and Interpretation,* devotes several pages to oil exploration (Sabins, 1978b). Bentz and Gutman (1977) discussed the importance of Landsat to petroleum exploration in foreign areas. Peterson (1979) addressed the significance of lineaments in oil and gas exploration. Venkataramanan (1979) discussed the application of Landsat imagery to petroleum exploration in India. Gathright (1982) and Blodget (1981) addressed the use of Landsat images in the search for oil and gas fields in the Appalachian Mountains.

Miller (1977) demonstrated the value of visual analysis of Landsat images to petroleum exploration in foreign, less well-mapped regions, such as in the Sudan and Kenya. Lineaments in Sudan and Kenya that he interpreted from Landsat images, he defined as part of a regional fracture system at the north end of the Lamu Embayment. He associated these features with the East African rift system and interpreted them to be a failed, immature triple junction with one arm opening into the Rudolf Trough and a second arm extending into the Ogaden Basin. Although much of the area is covered by Quaternary deposits of merging alluvial fans, a swampy area along the Ewaso N'Giro drainage system suggested subsidence along what may have been a structural trough or downwarp. The presence of a local trough along the Ewaso N'Giro was confirmed by magnetometer and gravity surveys as well as reflection seismic surveys. Unconfirmed information made available by J. Vandenakker (oral communication, 1982) indicates that the Chevron Oil Company has drilled and successfully developed oil resources in the region; reportedly it is also planning a pipeline to Port Sudan.

Maurin and Riguidel (1978), of the French Petroleum Company, TOTAL, of Paris, France, prepared a comprehensive treatise (in French) on the use of various digital enhancement and analysis methods to assist geomorphic and structural mapping of petroleum exploration targets in Tunisia. A Landsat 1 image (1199-09305; 7 February 1973) of the Kef Si AEK oil field area, Tunisia, was contrast-stretched and filtered by a 5 by 5 pixel array "boxcar" filter. This technique enhanced the Turonian fold-and-fault system which trends northeast across the more northerly trending Djibel Mrhila anticline. Oil and gas deposits were found to occur at the intersection of these two structural features. The treatise by Maurin and Riguidel (1978) provides much of the mathematical background that geologists need in order to work with Landsat CCT's. Similarly, Taranik (1978b) prepared a paper entitled "Principles of Computer Processing of Landsat Data for Geologic Applications." A detailed publication on digital image analysis methods was published by Johannes Moik (1982) of the NASA Goddard Space Flight Center (See also the "Analysis Techniques" section of this chapter).

Halbouty (1976), in a comprehensive article in the *Bulletin of the American Association of Petroleum Geologists,* provided many examples of the visual analysis of Landsat data in petroleum exploration. A few examples of the applications of experimental digital image analysis were also cited. Four years later, Halbouty (1980) published a second study of 15 giant oil and gas fields of the world, and pointed out that if Landsat data had been available earlier, the data could have been of significant help in discovering and developing at least 13 of them. He indicated that the systematic study of Landsat images, especially those of remote and poorly mapped areas of the world, could help cut the costs of exploration significantly. These articles contributed significantly to the exploration process by encouraging major oil companies, and some independents, to invest considerable time and effort in developing their own expertise in the use of Landsat data and in establishing special in-house research groups and digital image processing laboratories. Furthermore, it created interest that led to the development of the Geosat Committee, Inc., a consortium of more than 100 mineral and petroleum exploration firms and consultants who participate in joint government (NASA and USGS) and industry development of remote sensing techniques. The objective of the Geosat Committee, Inc., is to foster the development of remote sensing technology and to prepare its constituency for the effective use of new data that have higher resolution and more or different spectral bands, such as the Landsat 4 Thematic Mapper data, first available in July 1982. The Geosat Committee, Inc., in association with NASA, contracted with the Jet Propulsion Laboratory, for a major remote sensing study of three petroleum fields, three copper deposits, and two uranium deposits, as part of the "Joint NASA/Geosat Test Case Project" (Abrams et al., 1983). This report of more than 2000 pages, illustrated with over 200 color plates and maps, will become available by mid-1983. Synopses of two of the joint NASA/Geosat test case projects are presented in the "Economic Geology" section of this chapter. These are (1) an analysis of airborne remote sensing of a porphyry copper test site in Arizona (see the preceding "Mineral Exploration" section of this chapter), and (2) the following description from Harold Lang's work (Lang, 1982) on analysis of airborne remote sensing of the Lost River gas field, West Virginia:

> "The Lost River gas field is typical of many fields in the Ridge and Valley Province of the Appalachian Mountains. The Devonian sandstone and shale reservoir has fracture porosity and forms an anticlinal trap. Analysis of a 1:500,000-scale Landsat image (2815-14560; 16 April 1977) by lineament interpretations prepared by two independent interpreters demonstrates that subjectivity in recognizing individual lineaments may be obviated by lineament density (number of lineaments per unit area) mapping. Such maps show, for example, that the Lost River gas field is located beneath an area of high lineament density. Lineament density mapping at 1:48,000 scale, using aircraft-acquired Landsat 4 Thematic Mapper Simulator (TMS) data, yields lineament density isopleths that mimic subsurface structural contours of the Lost River gas field. These results suggest an approach for using lineament density mapping as an exploration tool and demonstrate that lineament density mapping could have been used to help find the Lost River gas field, if such information had been available prior to field development" (Lang, 1982).

The following two sections describe the use of radar and Landsat images in the quest for petroleum reservoirs. The first section presents examples of the use of Seasat SAR and SIR-A images in the Appalachian Mountains and in Wyoming for petroleum exploration. The second section discusses the use of a specially processed Landsat image, in conjunction with seismic surveys, to discover a concealed fault in the Bay County area of Michigan.

The Use of Radar in the Search for Hydrocarbons

The primary advantage of using radar images in exploring for hydrocarbons lies in the ability of the imaging systems to: (1) enhance subtle topographic features in heavily vegetated areas; and, (2) to penetrate cloud cover. Radar images, therefore, have been used for geological exploration mainly in areas of tropical rainforest. For example, the regional distribution of faults and linear patterns in western Irian Jaya, Indonesia, was first seen on airborne radar images acquired in 1974 with a radar operating in the X-band (2.8-cm wavelength). Analysis of the regional structural pattern suggested that plate tectonism was responsible for the fragmentation that has separated the oil-producing Salawati Island area from the mainland (Froidevaux, 1978 and 1980). For a general reference on the geological interpretation of radar images, the work by Mekel (1972), constituting one of the chapters in the ITC Textbook of Photo-Interpretation, is recommended.

The synthetic-aperture radar (SAR) experiment on the Seasat satellite provided in 1978 the first spaceborne radar images of North America available to the scientific community. The imaging radar was operated at the 23.5-cm wavelength (L-band) with a look angle (the supplementary angle to the depression angle) of about 20° and a maximum resolution of 25 m (when images have been digitally processed). Images of the heavily forested terrain of the southern Appalachians (Figures 31-126 and 31-127) reveal a pronounced enhancement of linear topographic features in the area. The potential for hydrocarbon traps is favorable in zones of fracture porosity, particularly at the intersection of lineaments. The enhancement on the images of extensive linear topography, and of short linear features less than 10 km long, results from the high sensitivity of the Seasat SAR to change of surface slope. This high sensitivity is also responsible for the strong geometric distortion on the images. Foreslopes having a magnitude equal to or greater than the radar look angle are obliterated by layover. Linear features that have substantial topographic relief from end to end are geometrically rotated on the images. Corresponding images from the Landsat multispectral scanner (MSS) and return beam vidicon (RBV) camera show no geometric distortion of this type. Linear topography on images acquired with the Landsat sensor systems is enhanced by

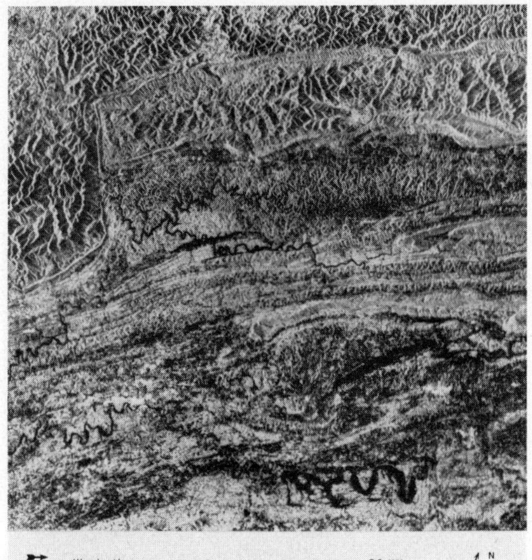

Fig. 31-126. Seasat radar image of the folded Appalachians and Pine Mountain overthrust sheet, Tennessee-Kentucky-Virginia. The image has been optically correlated from digitally recorded data and is a digitally mosaicked composite from two Seasat radar images: Seasat Rev 407, acquired 25 July 1978, and Seasat Rev 163, acquired 12 June 1978. The spatial resolution is approximately 30 m; the scene center is at 36°15′ N. latitude, 83°47′ W. longitude. The illumination direction is N. 67°30′ E., and the orbital altitude is 795 km. The Seasat radar operated at a 23.5-cm wavelength; polarization was parallel or horizontal-horizontal (HH). The two separate images are archived by NOAA's National Environmental Satellite Data and Information Service (Table 31-14). Image mosaic courtesy of the Jet Propulsion Laboratory.

Fig. 31-127. Seasat radar image of the folded Appalachians and Pine Mountain overthrust sheet, Tennessee-Kentucky-Virginia (Seasat Rev 874, acquired on 27 August 1978). Optically correlated from digitally recorded data, the digitally mosaicked image corresponds to Seasat coverage in part of Figure 31-126. The spatial resolution is 30 m; the illumination direction is N. 67.5° W. Image courtesy of the Jet Propulsion Laboratory.

solar shadowing, when the Sun elevation is below about 30° (see Figures 31-117B, 31-204, and 31-225). This condition occurs only for images acquired between November and February in temperate latitudes. Mapping from a suitable Landsat MSS image (Figure 31-128) and from the area common to the Seasat SAR images (Figures 31-126 and 31-127) has shown that the small-scale linear topographic features are preferentially enhanced on the radar images (Ford, 1980). The extent of the enhancement is shown on the histograms of lineament frequency versus strike in Figure 31-129. In this instance, the preferred orientations of the lineaments are more readily interpreted from the SAR data. The histograms show in each case that lineament perception is comparatively reduced in the direction of scene illumination. In the case of Seasat this deficiency is offset by the dual directions of scene illumination that were obtained with the imaging radar system, as a result of passes having been made over an area during the descending and ascending orbits of the spacecraft. The regional distribution and

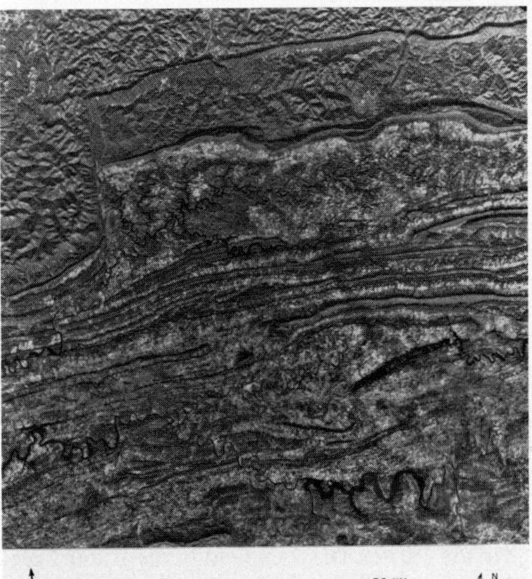

Fig. 31-128. Landsat MSS band 6 image (1858-15300) of the folded Appalachians and Pine Mountain overthrust sheet, acquired on 28 November 1974, corresponding to Seasat radar coverage in Figure 31-126. Image is digitally processed and contrast-stretched to enhance linear features. Pixel resolution is 79 m, Sun elevation is 25°, and the azimuth is 151° (N. 29° W.). Specially processed Landsat image courtesy of the Jet Propulsion Laboratory.

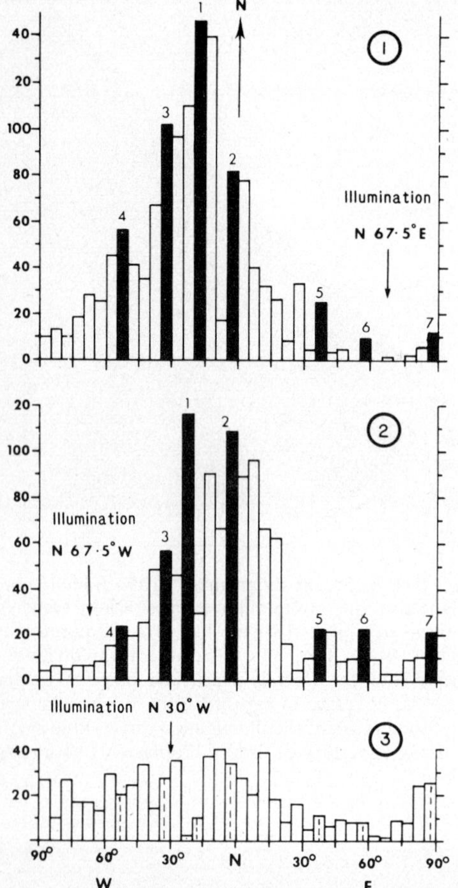

Fig. 31-129. Histograms of lineament frequency versus strike for short lineaments mapped from corresponding areas on Seasat SAR and Landsat MSS images: (1) from the Seasat SAR image in Figure 31-126, (2) from the Seasat SAR image in figure 31-127, and (3) from Landsat MSS image in Figure 31-128. Frequency on vertical axis, strike on 5° increments east and west of north. Solid bars are frequency maxima interpreted from the two Seasat SAR images. These maxima are not apparent from the Landsat MSS image (Figure 31-128).

Fig. 31-130. Seasat radar image subscene of Bitter Creek area, Patrick Draw oilfield, southwest Wyoming. Image is digitally correlated from digitally recorded data (Seasat Rev 789; acquired on 21 August 1978). Resolution is 25 m; subscene center is at 41°45′ N., 108°30′ W.; the illumination direction is N.66° W. Image courtesy of the Jet Propulsion Laboratory.

alignment of the short lineaments mapped from the SAR images, and the relationship of the lineaments to known structures and geophysical trends, provide a basis for further locating faults and fracture traces.

In the Patrick Draw Oilfield in the Bitter Creek area of southwest Wyoming the Seasat SAR backscatter is dominated by slope effects from the numerous small drainages of low relief near the Continental Divide and by surface scattering from the vegetation and rocks in the interchannel areas. The slopes produce bright returns that have a high spatial frequency on the radar image (Figure 31-130). Variations in the density of the predominantly sagebrush vegetation cover tend to produce medium to dark returns that have a low spatial frequency on the image. Systematic changes in the low-frequency distribution of the medium- to dark-gray levels are obscured on the image by the high-frequency distribution of the bright returns.

SAR images consist of three basic spectral components (Figure 31-131). Daily (1983) has shown a method of enhancing both the high- and the low-frequency components that contain useful information by filtering and color encoding. Low-frequency spatial variations on the image that represent changes in vegetation density are displayed by hue. High-frequency spatial variations associated with the small drainages are retained as intensity (Color Figure 31-132). The changes in vegetation represented by the hues on the image are influenced by soil moisture, soil composition, slope, and subtle surface characteristics of the Tertiary bedrock section. Thus the hues outline the plunging structure of the Wamsutter Arch; this feature is not evident on the unenhanced Seasat SAR image (Figure 31-130). The structure is not readily apparent at the surface in the field, though it is known from subsurface records. Retention of the high-frequency spatial

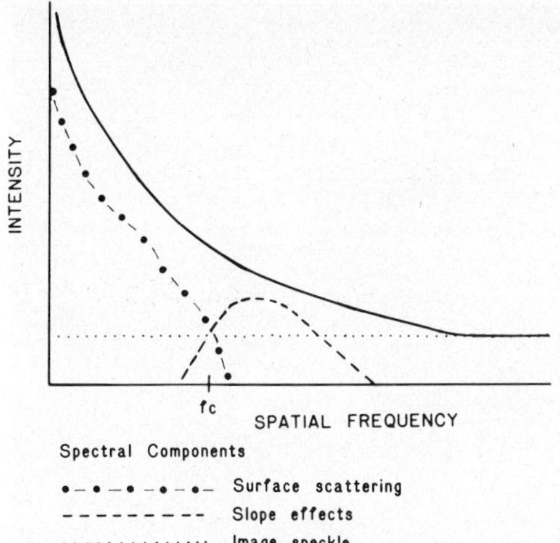

Fig. 31-131. Typical power spectrum of a SAR image, showing spatial distribution of spectral components. Cutoff frequency, fc, that provides the basis for filtering surface scattering components from slope components, is scene dependent.

variations, as intensity, enhances linear topographic features on the image.

The Shuttle Imaging Radar (SIR-A) experiment provided a second generation of spaceborne scientific radar images of the Earth in 1981 (Elachi and others, 1982; McCauley and others, 1982). The images were acquired at the same wavelength that was used to operate the Seasat SAR, but with a look angle of about 47° and resolution of 40 m. SIR-A coverage of the oil and gas producing Appalachian Plateau area in southeast Kentucky, and the adjacent Ridge and Valley area of southwest Virginia (Figure 31-133) enhances extensive linear topography and short linear features on the image. The strike and distribution of the latter relative to known faults and magnetic trends suggest that many of them are structural in origin (Ford, 1982). On corresponding Seasat SAR images the extensive features are perceptible but the short linear features are strongly distorted or impossible to locate. This contrast results primarily from the difference in the SIR-A and the Seasat SAR look angle. The look angle and the slope of the terrain govern the local incidence angle, which determines the factors that dominate the radar backscatter (Figure 31-134). At the steeply sloping surfaces in southeast Kentucky the SIR-A backscatter is dominated by slope effect. This effect enhances the small-scale topography of the area. In contrast, the Seasat SAR backscatter is dominated by layover. This layover obliterates the small-scale topography. Lineament mapping from a corresponding Landsat 3 RBV image (Figure 31-135) shows equivalent perception of the major and minor features seen on the SIR-A image (Figure 31-133), but the small-scale topography that strikes near-parallel to the scene illumination of the RBV image is suppressed.

Analysis of SIR-A images of foreign areas has already yielded some important scientific results. Charles Elachi of Jet Propulsion Laboratory and his colleagues have discovered that Seasat SIR-A images can penetrate from 1 to 5 m of dry sand in Egypt and the Sudan, thereby revealing geomorphic and structural characteristics of the concealed bedrock (Elachi et al., 1982). This discovery has important implications for the geological exploration of the sand seas of the Earth and those of Mars. In another paper, Elachi (1982) summarized the use of radar images of the Earth from space and provided several excellent examples of such images, including color-enhanced radar images similar to Color Figure 31-132.

Application of Landsat Imagery to Petroleum Exploration in Bay County, Michigan

Introduction

The following case history of the use of Landsat in petroleum exploration is excerpted from a previously published paper by Vincent and Coupland (1980). It is a good example of the use of computer enhancement of a Landsat image to emphasize a subtle lineament in an area of the Michigan Basin where 200 m of glacial drift blanket the underlying bedrock. Seismic data confirmed the existence of a fault and a favorable structure for oil and gas accumulations along this fault on the southeastern margin of a graben structure.

Geologic Setting in Relation to Oil and Gas Potential of the Michigan Basin

The Michigan Basin, an intracratonic basin within the North American lithospheric plate, underlies most of the Southern Peninsula of the State of Michigan. Most of the consolidated sedimentary rocks in the basin are of Paleozoic age (600-270 million years old), though some are of the late Jurassic Period (180-160 million years old), and lie unconformably on Pennsylvania sediments (Permian and Triassic rocks are missing). Overlying these consolidated sedimentary rocks are unconsolidated glacial deposits of Pleistocene age (≤1 million years old). The depth of sedimentary rocks overlying crystalline basement rocks of Precambrian age (≥600 million years) in Gratiot County, near the central part of the basin, was found to be approximately 5300 m by a McClure Oil Company deep test hole in 1976. Although oil or gas has been found in every formation in the Michigan Basin (including small pockets of oil and gas in the glacial drift), the most prolific formations have been the Devonian-aged Traverse, Dundee, and Detroit River Formations, the Silurian-aged Niagran Formation (reefs), and

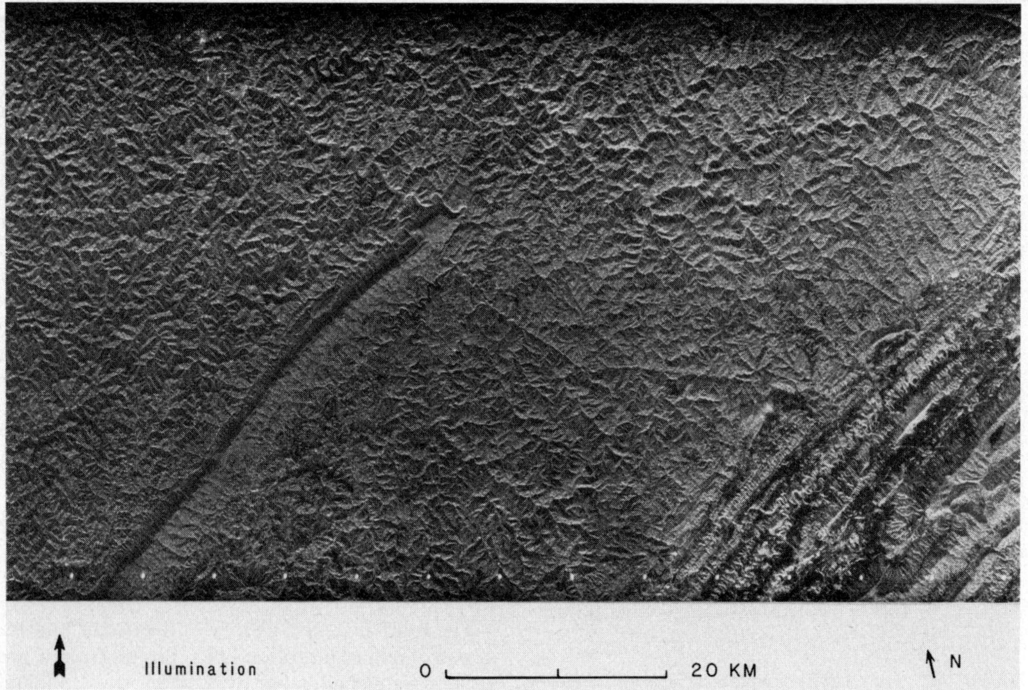

Fig. 31-133. Shuttle radar image of the Appalachian Plateau in southeast Kentucky and southwest Virginia. Image is optically correlated from optically recorded data, acquired on SIR-A Data Take 24A on 13 November 1981. The spatial resolution is 40 m; the scene center is at 37°10′ N., 82°16′ W.; the illumination direction is N.18°E. Image courtesy of the Jet Propulsion Laboratory.

the Ordovician-aged Trenton and Black River Formations. All these are almost exclusively carbonate rocks. The Trenton-Black River Group are the producing horizons of the Albion-Scipio Trend

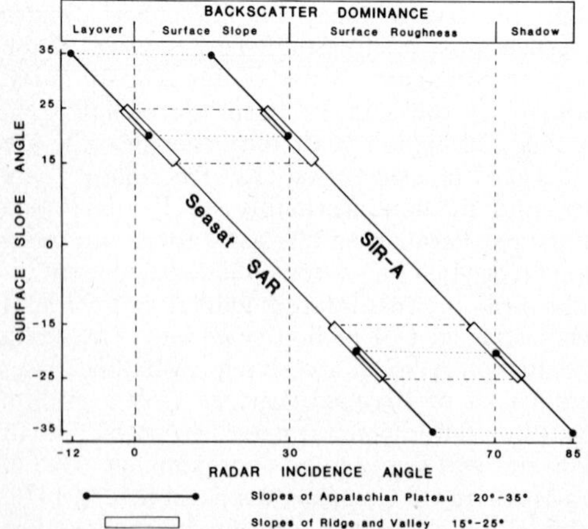

Fig. 31-134. SIR-A and Seasat SAR incidence angles plotted against range of surface slope angle common in the Appalachian Plateau and Ridge and Valley area, Kentucky-Virginia, showing range of factors that dominate backscatter.

in southern Michigan, which has produced over 116 million barrels of oil and over 180 billion cubic feet of gas since its discovery in 1957.

It has long been assumed that Pleistocene glacial deposits, which blanket virtually all the bedrock in the Southern Peninsula of Michigan to varying thicknesses (up to approximately 300 m), would render faults and other structural features in the bedrock invisible. This assumption has previously discouraged the use of satellite imagery or aerial photography in reconnaissance exploration for underlying structures favorable for the accumulation of hydrocarbons.

Geologic Analysis of Landsat Imagery

Analysis of Landsat images of the Michigan Basin was begun in 1975. First, a 1973 color mosaic of Michigan, produced photographically by General Electric from standard Landsat images (unenhanced), was used to map glacial features, quarries, sinkhole lakes, and shoreline features (Drake and Vincent, 1975). Some linear features up to several tens of km long were noted on this mosaic, some of which did not appear to be of glacial origin. Linear features were mapped from a contrast-stretched color composite image (Landsat MSS bands 4, 5, and 7) of the Saginaw Bay

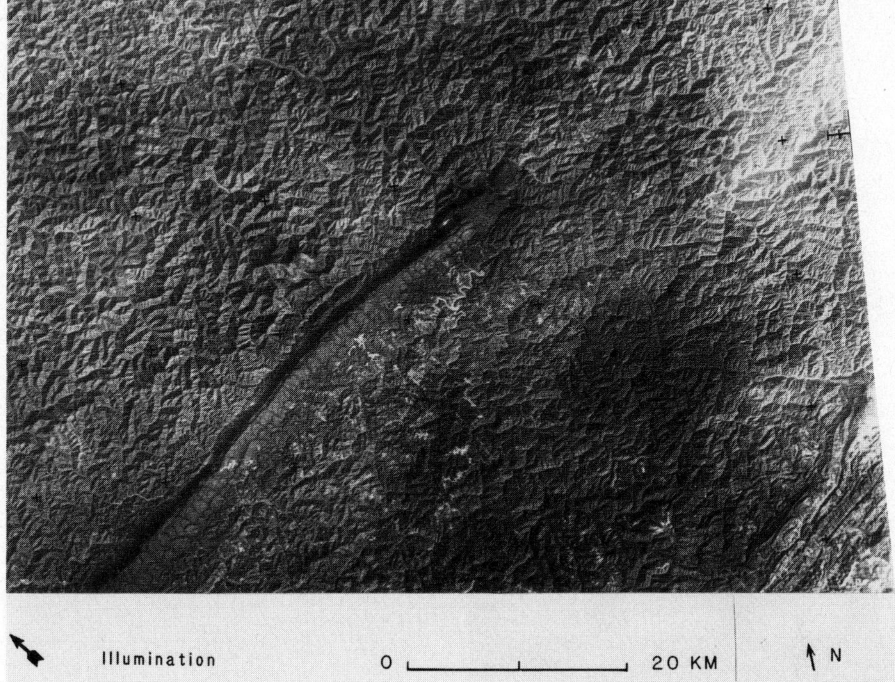

Fig. 31-135. Landsat 3 RBV subscene (30319-15321-D) of the Appalachian Plateau, Kentucky-Virginia, acquired on 18 January 1979, corresponding to SIR-A coverage in Figure 31-133. The pixel resolution is 30 m; the solar elevation angle is 24°. The image is illuminated from the southeast (azimuth of 146° measured clockwise from the north).

area in east-central Michigan produced from a computer compatible tape. Figure 31-136 shows a contrast-stretched MSS band 7 image of that frame. On the basis of some of those linear features plus publicly available well log data, geophysical anomalies, topographic maps, and bathymetric contour maps of Saginaw Bay, it was hypothesized that a graben trends northeast under Saginaw Bay from the central part of the Southern Peninsula of Michigan (Drake, 1976; Vincent and Drake, 1976).

A linear-length density contour map showing the aggregate length of lineaments in kilometers per square kilometer was produced by computer using linear features that had been mapped by photointerpretation of a contrast-stretched color composite image of the Saginaw Bay frame. A "trough" of lower linear-length density, trending northeastwards toward Saginaw Bay from the central part of the frame, was located at the approximate position of the hypothesized graben. The northwest-southeast-trending linear length density local high, which appears to close the southwestern end of the "trough," coincides with a known anticline of the same trend which is productive in the area where the anticline and graben intersect (the Porter Field).

Correlation of Seismic Surveys with the Landsat Analysis

In 1977, a 4-mile long segment of a seismic survey line was acquired from Mobil Oil Corporation for the segment where the seismic traverse line crossed one of the small, subtle lineaments that had been mapped in the Pinconning area from Landsat data. It definitely showed evidence of a fault where the lineaments had been mapped.

In 1978, after an agreement had been reached between Geospectra Corporation and the Wiser Oil Company, a comprehensive series of seismic surveys were made in the area of the principal linear feature mapped from the Landsat imagery. Part of the area is shown in Figure 31-137. An MSS band 4/5 ratio image was also used to enhance the principal linear feature and other linear features shown in Figure 31-138.

All the seismic data (50 km of data from the study area) were collected in 1978 and 1979 after leasing in the area had been completed on the basis of the Landsat analysis results. Seismic shot points are shown as small crosses in Figure 31-137, and the dashed line trending northeast is that part of the principal Landsat lineament which lies

Fig. 31-136. Computer-enhanced Landsat 1 image (1320-15525; 8 June 1973) of the Saginaw Bay area and environs in east-central Michigan. The MSS band 7 image has been contrast-stretched to emphasize any subtle features related to the underlying geologic structure. The image covers an area of about 185 km by 185 km. Image courtesy of Geospectra Corporation.

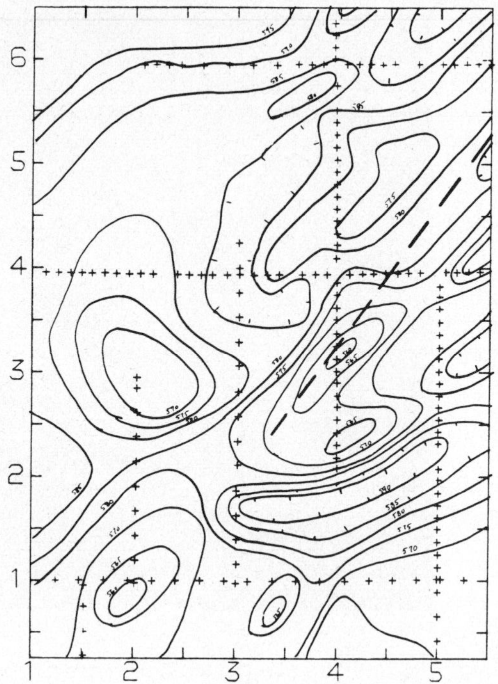

Fig. 31-137. Hand-contoured plot of the top of the Devonian Dundee Formation in the Bay County, Michigan, seismic survey area. The contour interval is 5 milliseconds. Seismic shot points are shown by the small crosses; the principal Landsat lineament, mapped from Figure 31-136, is shown by the dashed line. North is toward the top, and the outside grid numbers are shown in miles. (Contour map courtesy of James Gow, Wiser Oil Company)

inside the seismic study area. The contour lines show seismic return times in milliseconds from the top of the Dundee Formation (Devonian), which is one of the producing horizons in the Kawkawlin oil field about 10 km to the northwest of the study area. The top of the Dundee Formation is estimated to be an average of 1,000 m below surface in this area. This contour map was hand-drafted from the seismic data by Jim Gow, a Wiser Oil Company geophysicist. Note that a Dundee high (560 milliseconds is the shortest return time recorded for the Dundee Formation in the entire study area) occurs immediately adjacent to the principal Landsat linear feature. This contour map confirms that subsurface structure favorable for hydrocarbon accumulation is associated with the principal lineament mapped from Landsat data, an area where the glacial drift is approximately 200 m thick.

The potential for hydrocarbon accumulation is not confined to the Dundee Formation, however. To demonstrate this point, Geospectra Corporation produced three-dimensional plots for the same seismic study area of the top of the Dundee

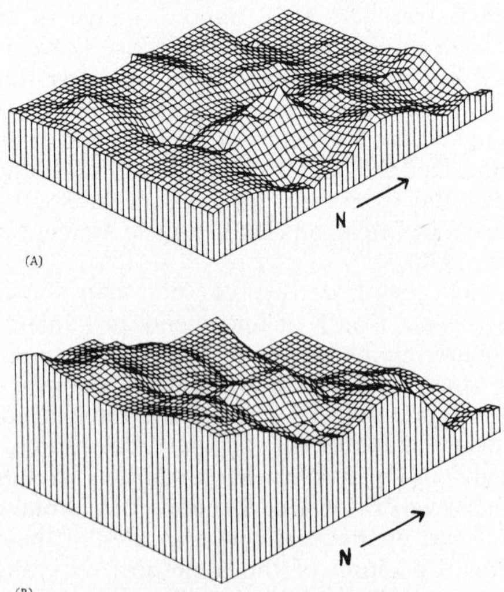

Fig. 31-138. Three-dimensional (isometric) computer plots of the top of the Devonian Dundee Formation (A) and a Cambrian sedimentary unit (B) as calculated from seismic data for the Bay County, Michigan, petroleum exploration area. The tops of the Devonian and Cambrian formations are estimated to be approximately 1,000 m and 3,500 m below the surface, respectively. Computer plots courtesy of Geospectra Corporation

Formation and the top of a Cambrian age unit estimated to be approximately 3,500 m below the surface. These computer-drafted plots, shown in Figure 31-138 as viewed from a southeasterly direction, show highs and lows which are typically skewed north-south and east-west, because those are the directions of the seismic lines. This artificial skew has largely been removed from the hand-drafted contour map of the previous figure. The Cambrian unit also has a structural high (1,450 milliseconds) along the principal lineament, and the Cambrian sediments occur at greater depth toward the northwest, in concurrence with the graben hypothesis. The Detroit River, Niagran, and Trenton Formations, which are prolific oil and gas producers elsewhere in the Michigan Basin, all occur between the Dundee Formation and this Cambrian unit. Therefore, it is likely that favorable structure for hydrocarbon accumulations occurs in all of the most prolific producing horizons below the principal lineaments.

Findings and Conclusions

In 1980, a hole was drilled along the Landsat lineament on the structural high (seismic return time of 560 milliseconds) of the Dundee Formation shown in Figure 31-137 to the underlying Detroit River Formation. Although there was a gas show in the Detroit River Formation (Devonian) sufficiently strong to entice drillers to complete the well, it was later abandoned as non-commercial. The structural high on the Dundee Formation was also confirmed. Two more wells are expected to be drilled soon inside the study area shown in Figure 31-137. One of the new wells is expected to be a deep test down to Ordovician and Cambrian Formations.

A study of Landsat computer-processed images and available published information has led to the discovery of a subsurface structure that has been confirmed by seismic data to be favorable for hydrocarbon accumulation down to a depth of 3500 m, in an area covered by 200 m of Pleistocene glacial drift. The fact that a linear feature related to such deep structure should be observed in Landsat data through such thick glacial deposits is indeed surprising and still remains somewhat of a mystery. The most likely explanation for this phenomenon appears to be the action of groundwater. It is hypothesized that groundwater escaping upward along a fault trace may have caused the glacial till in that area to remain more moist than the surrounding areas, producing a "moisture stripe." This phenomenon may also be responsible for the Quanicassee River following part of the fault.

The principal lineament is expressed in the Landsat image (Figure 31-136) as a dark feature and in the contrast-stretched Landsat MSS color composite image (not shown here) as a bluish feature. Vegetation is an unlikely cause for the appearance of the lineament, because the color composite image displays little or no red in that area. In the MSS band 4/5 ratio image (not shown here) the principal lineament appears light-toned, perhaps indicating less iron oxide in the soil than in the surrounding area. This same effect would be observed if the soil along the lineament were more vegetated or covered by more water than adjacent areas.

Enhanced Landsat images, produced from digital terrain data, and a new automatic *Li*near *R*ecognition and *A*nalysis software package (LIRA) maps linear features directly from Landsat CCT's without image interpretation. This technique has been used by Geospectra geologists to identify several new areas of favorable structures in the Michigan Basin. An additional 200,000 acres (81,000 ha) have been leased as a result of those studies and more studies are underway.

Coal Exploration

Use of Aerial Photography for Geologic and Coal-Bed Mapping in Southwestern Wyoming

Aerial photographs are routinely used in field mapping by personnel of the U.S. Geological Survey (see Ray, 1960). In coal resource studies, the occurrence and thicknesses of coal beds seen in outcrop or in shallow excavations, together with outcrops of other geologic units and data, are generally plotted on topographic maps of a given area, because such maps are the most useful means of depicting the geometry and distribution of the beds at the surface. More complete resource investigations also include the study and correlation of subsurface data obtained from drilling and sample analyses, with the data shown in various ways, such as in tables, diagrams, cross-sections, isopach maps and overburden maps. Nevertheless, the initial phases of investigation are most commonly devoted to the production of geologic and coal bed maps.

The figures included in this article show part of an area in southwestern Wyoming that has been mapped with the aid of aerial photographs and a stereoscope for a coal resource study by the U.S. Geological Survey. The geologic map (Figure 31-139A) includes parts of the Elkol and Warfield Creek 1:24,000-scale quadrangles (M'Gonigle, 1979a and 1979b) and shows lithologic units and coal beds south of the large strip mines near Kemmerer, Wyoming. The map was produced from both 1:20,000-scale black-and-white (Figure 31-139B) and 1:36,000-scale color vertical aerial photographs (Color Figure 31-140 and Figure 31-141). The aerial photographs are mounted here as stereoscopic pairs for comparison with the geologic map and with each other; for ease of presentation in this volume, all three figures are reproduced at about one-half of their original scales. Markings on the photos are those made during mapping in the field and during compilation of the

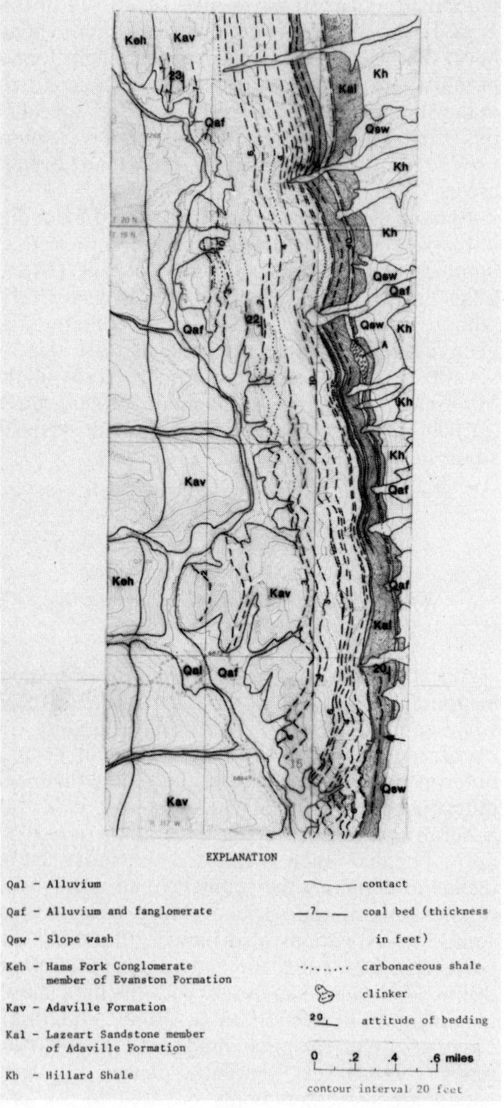

(A)

BBL-19R-10

BBL-19R-9

0 1
Kilometer

(B)

Fig. 31-139. A. Geologic map of parts of the Elkol and Warfield Creek 7½-minute quadrangle maps, southwestern Wyoming. Arrow in bottom right-hand corner shows the direction of view of the oblique aerial photograph shown in Figure 31-142. B. Stereoscopic pair of vertical black-and-white aerial photographs showing part of the area covered by A. The bar scale is 1 km long. Aerial photographs acquired by the U.S. Department of Agriculture on 1 October 1955. Photographs BBL-19R-9 (41°38′08″ north latitude; 110°38′47″ west longitude). Original scale, 1:20,000.

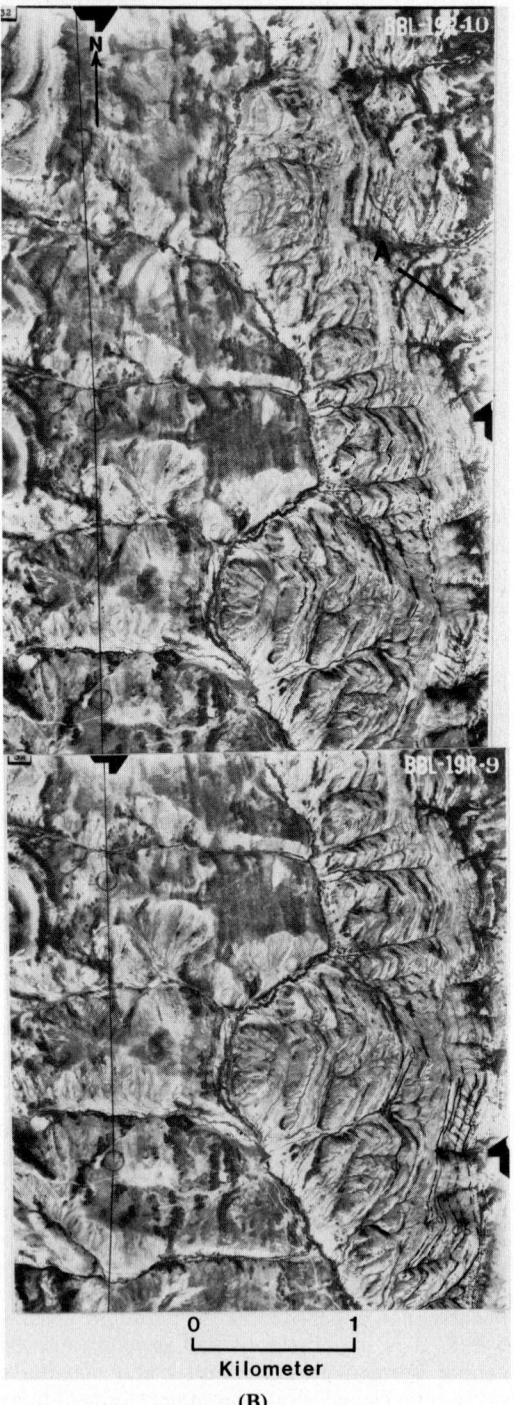

map in the office; contacts and other linear data, such as coal beds, were commonly plotted directly on the aerial photographs, while point data (stations) were marked by pinholes and circled, with data noted on the back of the photograph near each pinhole.

In this particular project, transfer of data from the aerial photographs to topographic maps was generally not done in the field, but later in the office. Most contacts could be accurately located on the topographic maps by inspection of stereopaired aerial photographs viewed with a simple stereoscope. Coal beds, however, were traced with the aid of a Kern PG-2 photogram-

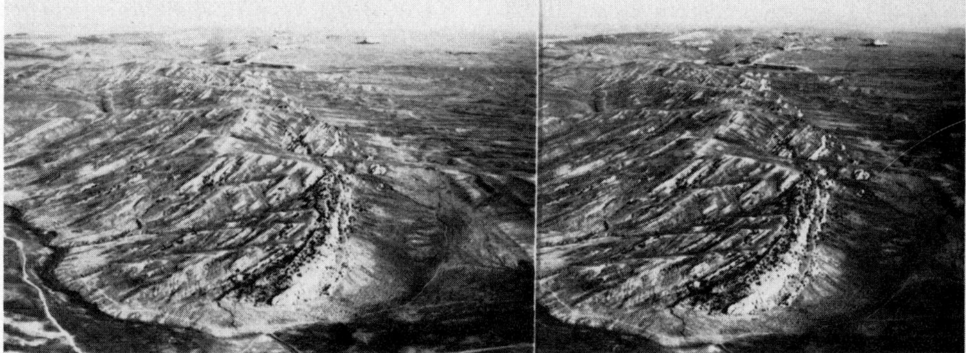

Fig. 31-141. Stereoscopic pair of oblique 35 mm, color-infrared aerial photographs, here reproduced in black-and-white, looking northward at the same area included in Figures 31-139A and B and Color Figure 31-140. The view is approximately along the strike of the bedding in the Cretaceous Adaville Formation. Aerial photographs by John W. M'Gonigle, U.S. Geological Survey.

metric plotter (see the following discussion in this section) in order to obtain the highest possible accuracy in placement of geologic information on the topographic base maps. Some geologic contacts were also located with the PG-2 photogrammetric plotter, particularly in areas of rather featureless (and therefore ambiguous) topography, such as smooth, steep canyon walls or flat plains.

Each type of aerial photograph has its own advantages and disadvantages. Here, as in other mapping projects, color aerial photos were found to be more useful than black-and-white aerial photos in general geologic mapping of the region, because the color greatly aids in discriminating rock types. It is estimated that overall mapping time on this project was reduced by as much as 50 percent when color aerial photographs were used in place of black-and-white, because there were fewer ambiguities in interpretation and correlation. This was particularly noticeable where stratigraphic and structural complications or erosional features interrupted the lateral continuity of rock units.

Where exposed, carbonaceous shales and coal beds are more distinctive on color aerial photographs, and areas of red clinker (where coal beds have burned) are prominent on color aerial photos in contrast to the black-and-white photos (compare point A (clinker) on Figure 31-139A and B and Color Figure 31-140, and see the following discussion in this section).

The 1:20,000-scale black-and-white aerial photographs used in the geologic mapping were found to be more useful than the 1:36,000-scale color aerial photographs in the precise plotting of locations (stations) in the field and in detailed tracing of coal beds. This was primarily because the larger scale and sharper resolution of the former (compare Figures 31-139B and Color Figure 31-140) enhanced the precision and ease of determining and marking position at a specific field location. For example, small bushes (less than 0.5 m in height) and patterns of grass and brush are readily discerned on aerial photographs at the 1:20,000 scale. On the other hand, such a scale is often more cumbersome for general mapping than smaller scale photos, because many more photos are needed for stereoscopic coverage of a given area. In practice, both the black-and-white and color aerial photographs were used together in the coal-bearing parts of the field area, and color aerial photographs were used alone for general geologic mapping in the rest of the region.

Black-and-white aerial photographs, when obtained in print form with a semimatte finish, can be readily marked with pencil or ink, whereas currently available color photos unfortunately generally come with a rather glossy surface, unmarkable except with special wax-based pencils or with ink. The emulsion of most currently available color aerial photographs is more fragile than that of black-and-white aerial photographs; rain drops are ruinous to the color emulsion, whereas black-and-white prints can be blotted dry with no ill effects. In fact, india ink can be removed from black-and-white photos readily with water or alcohol, whereas ink removal from color photographs is extremely difficult. During the course of the mapping, it was noted that even the finest ink lines are a hindrance to accurate photogrammetric plotter work because the prints are opaque and the lines destroy the precision otherwise possible with the machine; therefore, it is recommended that pencil alone be used in marking the aerial photographs in the field (or that the marks be placed on a transparent overlay), if plotter work is contemplated.

Some oblique aerial photographs of this area taken from a small airplane with a 35-mm camera illustrate some other possible uses of aerial photography in geologic mapping. In the stereoscopic pair of 35-mm oblique aerial photographs (Figure 31-141), the stereoscopic view to the north along the general strike of the west-dipping stratigraphic sequence includes the area shown in Figures 31-139A and B in Color Figure 31-140. Such oblique stereoscopic views supplement those provided by vertical aerial photographs, because they show details of cliff faces and perspectives not visible in the vertical view.

In order to obtain a continuous series of aerial photographs of the cliffy outcrops at large scales and without the foreshortening seen in vertical aerial photographs or from the ground below the cliffs, 35-mm oblique aerial photographs of the outcrops in this area were taken from a small airplane flying parallel to the strike of the beds and at such an altitude as to view the beds essentially "down-dip." Figure 31-142 illustrates this viewing technique, where the view is to the west from a location shown on Figure 31-139A. David Lawrence of Yale University has found this photocoverage to be very useful in plotting detailed stratigraphic data for his Ph.D. thesis (David Lawrence, oral communication, 1981).

If aerial photography from small airplanes is contemplated, it is suggested that several types of film and filters be used, because each type has a different spectral response (see Figures 31-44 and 31-47) and hence can potentially provide different information about the rocks and soil. For example, inspection of 35-mm oblique color infrared aerial photographs showed that this type of film, for some reason, generally enhanced the color contrast between sandstone and shale in this region over what can be seen in regular color photos. This is well illustrated by the stereoscopic view of Color Figure 31-143, in which sandstone and shale layers in the foreground (area B) show up with greater contrast than they do on the color aerial photographs (location B on Color Figure 31-140).

Fig. 31-142. Oblique, 35 mm-, black-and-white aerial photograph looking west and down-dip at cliffy exposures of the Cretaceous Adaville Formation. The arrow points to the natural exposure of a thin coal bed. Another coal bed, directly above the thicker upper sandstone (the top of the Lazeart Sandstone Member), is about 3 m thick here, but is covered with a thin layer of slope wash (see Figure 31-139A). Aerial photograph by John W. M'Gonigle, U.S. Geological Survey.

Use of the Kern PG-2 Photogrammetric Plotter in Geologic and Coal-Bed Mapping, Powder River Basin, Northeastern Wyoming

The photogrammetric plotter (Figure 31-144) has proven to be a practical and effective tool for rapidly producing geologic and coal-bed maps directly from vertical aerial photographs. The plotter corrects for tilt and distortion within a stereoscopic model (area of overlap between two aerial photographs); it permits the precise selection of features visible within the stereoscopic model; and it plots their locations automatically onto a basemap (*E*, Figure 31-144). Thus a photogrammetric plotter enables one to compile a precise map concurrently with photointerpretation.

One of the main features of the photogrammetric plotter is the floating mark, a tiny bright spot which can be positioned on any feature within the stereoscopic model being viewed. The elevation of the floating mark, in feet or meters above sea level, is displayed on a digital readout device (*C*, Figure 31-144). Some types of data derived or measured directly from the stereoscopic model, by using the floating mark, are discussed later and in the figure captions. A discussion of photogrammetric plotter operation is included in Figure 31-144.

The plotter has been used in projects designed to investigate and to map, in a relatively short time, large areas of coal-bearing rocks in the western United States. The specific mapping examples discussed here concern recent work in the central part of the Powder River Basin, an area of major coal deposits in northeastern Wyoming.

The Powder River Basin is a large, asymmetric, structural depression between the Bighorn Mountains and the Black Hills. Dips on the long eastern flank of the Basin, where these mapping examples are taken, are typically less than one degree. Locally, and on the steep western flank of the Basin, dips exceed 10 degrees.

About twelve mappable coal beds crop out in the area between Sheridan and Gillette, Wyoming. Coal outcrops range from one to seven meters in thickness. The coal beds are in the Tongue River Member of the Fort Union Formation of Paleocene age and in the overlying Wasatch Formation of Eocene age. A distinctive zone of fresh-water mollusk fossils, between the Roland coal bed and the overlying Arvada coal bed, serves as the contact between the two formations. Other rocks in the study area are mudstone, shale, calcareous siltstone, sandstone, and clinker. The sequence of strata is typically even-banded, with a repetition of light-brown shale and grey mudstone. Color Figure 31-145 shows a typical view of the area.

Red-orange clinker, rock baked and fused by coal beds which have burned in place, is a common feature in the Powder River Basin. This feature stands out vividly on color aerial photographs

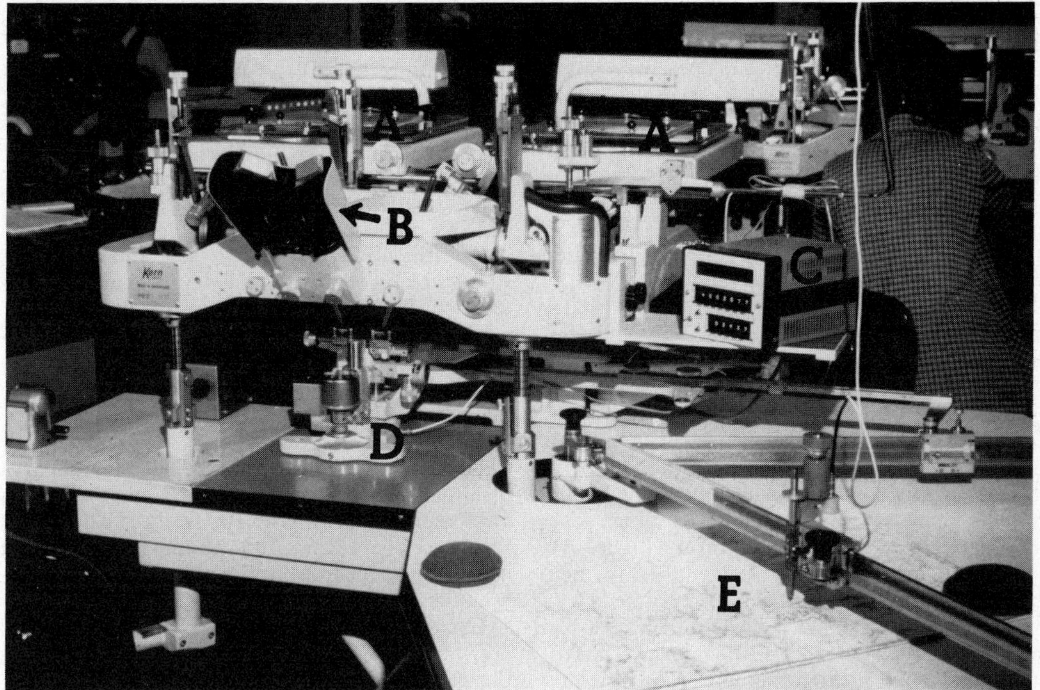

Fig. 31-144. "Kern PG-2" photogrammetric plotting instrument used in the U.S. Geological Survey photogrammetric plotter laboratory:

A) Glass plates to hold aerial photographs
B) Operator's binocular viewer
C) Digital readout device for determining elevation of floating mark
D) Operator's controls
E) Plotting table with basemap and automatically positioned pen

Geologists operate the photogrammetric plotter to produce geologic and coal bed maps directly from vertical aerial photographs. The Kern PG-2 plotter can accommodate either paper prints or film transparencies; these are inserted into the machine at *A*. When the various settings on the machine are adjusted, the operator sees (through *B*) a brightly-lit, three-dimensional image containing a tiny bright spot (floating mark). Three magnifications can be selected with which to view the stereoscopic image.

During the setting of each stereopair the plotter is calibrated (from a topographic map or other elevation control), so that it displays (at *C*) the precise elevation, in feet or meters above sea level, of the floating mark within the stereoscopic model. The location of the floating mark is controlled by the operator, using the controls at *D*. The operator can move the floating mark so that it appears to "climb" up or down a hillside or to remain stationary on a given geologic horizon or feature.

Attached to the body of the plotter is a pantograph arm with pen (*E*); the arm travels over an oriented basemap so that the location of the pen on the map represents the location of the floating dot within the stereomodel. As the geologist looks through the viewer (*B*) and moves the floating mark along a certain feature in the stereomodel, the pen automatically records the path of the floating mark, at the correct elevation, onto the basemap. Thus the geologist compiles an accurate map while he/she views and interprets the stereoscopic image.

Stereoscopic vision is required to operate the photogrammetric plotter. The techniques of calibrating for elevation control and orienting the photographs and the basemap are not difficult to learn and the operator's ability to use these techniques improves with practice. Note: Use of manufacturer's name in this paper is for descriptive purposes only and does not constitute endorsement by the U.S. Geological Survey.

(Color Figure 33-146A). As the coal burned, the roof rocks were baked and fused to resemble porcelain or brick, and collapsed down to the level of the coal ash. Thus the base of the clinker corresponds to the base of the unburned coal.

When color aerial photographs were used in the plotter, the contact between red clinker and light-colored underlying rocks was very easy to follow using the floating mark, and the position of the associated coal bed was drawn automatically onto the basemap. Where the coal beds have not burned, they were mapped by automatically plotting outcrops, diggings, and weathering features seen on the aerial photographs.

The capabilities of the plotter also facilitate the regional correlation of the coal beds. Where coal beds are virtually flat-lying, the floating mark was used to define a horizontal plane, across valleys and through hills, to compare the position of the coal beds and other strata. The horizontal planes were defined with the floating mark at several different elevations or levels; the planes were extended throughout the stereoscopic model to compare outcrops in several locations. In more structurally complex areas, the vertical exaggeration in the stereomodel accentuated the dip of the strata, and the floating mark provided elevation control for calculating dips and strikes and for measuring intervals to marker beds. In these ways it was possible to establish a regional correlation of coals across the Powder River Basin.

It was found very practical to combine plotter work and fieldwork in coal investigations. Examples of how photogrammetric plotting was used to produce maps, plan fieldwork, and resolve field problems are shown in Color Figure 31-146A, Figure 31-146B, and in Figure 31-147A and B. Color Figure 31-146A shows a natural color stereopair (original scale approximately 1:24,000) and the resulting coal bed map (original scale 1:24,000) produced on the plotter (Figure 31-146B). Specific outcrops were identified for field visits to measure and describe the coal beds.

Figure 31-147A shows a portion of a coal bed map initially produced on the plotter; photogrammetric work located problem areas which required field data for resolution. The specific outcrops selected for field visits to resolve the mapping problems are indicated on the map. Figure 31-147B illustrates the map produced after visiting the selected outcrops.

Figures 31-147A and B illustrate how initial plotter work rapidly resolved mapping problems in the field. Through use of the plotter, one prominent coal bed was traced around the northern part of a hill (Figure 31-147A). Further south, a major coal bed was traced at an average of 16 meters higher in elevation. This suggested that there was displacement of a single faulted coal bed, a single downdraped coal bed, or two separate coal beds whose outcrops and clinker coincidentally formed this pattern.

Further plotter study revealed that:

1) There was no apparent faulting evident, because other strata were traced around and through the hill, using the floating mark;
2) the dip of a stratum above the coal, as calculated by floating mark elevations, was about one-half degree to the southwest. This dip supported the idea of two coal beds, if the coal had this same dip.
3) There were outcrops resembling coal at locations *2* and *3* (Figure 31-147A), about 16 meters apart in elevation.

Yet it was still possible that there was only one coal bed, because the dip of the coal itself (as calculated in adjacent stereomodels) was inconsistent, and sandstone channels visible at coal level made the coal position rather unpredictable. Also, a fault nearby had the correct displacement and trend and could conceivably extend into this area.

From correlations in nearby areas, it was known that if there are two coals here, the lower one would be the Roland and the upper one would be the Arvada, and, if so, there would be a zone of mollusk fossils between them. However, the fossiliferous horizon could not be conclusively identified in the stereoscopic model.

Fieldwork in this area began by going directly to the four pre-selected outcrops, indicated by black dots on Figure 31-147A. At *1* a section was measured which contained the lower coal and the fossil zone; the section was continued to outcrop *2,* where, indeed, the upper coal was found. At outcrop *3,* the lower coal was again found, with the mollusk fossils above it. And at outcrop *4* the upper coal was found, with the fossil zone below it, and the lower coal further below.

Thus, because of prior knowledge of this area from plotter work, it was possible to resolve this particular mapping problem in the field in less than one day. Figure 31-147B illustrates the coal bed map produced from Figure 31-147A after visiting the four selected outcrops.

Usually, field time is limited, but even when it is not, fieldwork is made more productive by using the plotter. Before fieldwork was begun precise measurements had been made of strike and dip, fault displacement, locations of coal beds, and elevations of outcrops of interest. The investigators had an accurate preliminary map and a geologic framework in which to fit the information gathered in the field. Thus it was possible to concentrate fieldwork in areas where plotter work had indicated problems in structure or coal correlation. It was found that using the plotter after fieldwork was also very productive. For instance, strata measured in the field could be identified in the stereoscopic model by elevation (if not by appearance) and then mapped into areas not yet visited in the field.

Although it is difficult to quantify the amount of field time saved by starting fieldwork with accurate preliminary maps, experience in the Powder River Basin demonstrated that a day spent mapping with the plotter was equivalent to about five days in the field. Figure 31-148 shows the portion of a 7½-minute quadrangle map that was produced on the plotter during a typical day's work. The area covers about 35 square kilometers and was mapped from two sets of stereopaired aerial photographs.

Certainly there are areas in the western United States where the geology is such that photogrammetric plotting would not proceed as rapidly as depicted in these examples. However, the ability to do precise mapping, at whatever rate, concurrently with photointerpretation is a definite advantage.

Within the U.S. Geological Survey, the photogrammetric plotter is being used also to map a variety of other geologic features, including surficial deposits, complex structural areas, changes in areas of active volcanism (Moore and Albee,

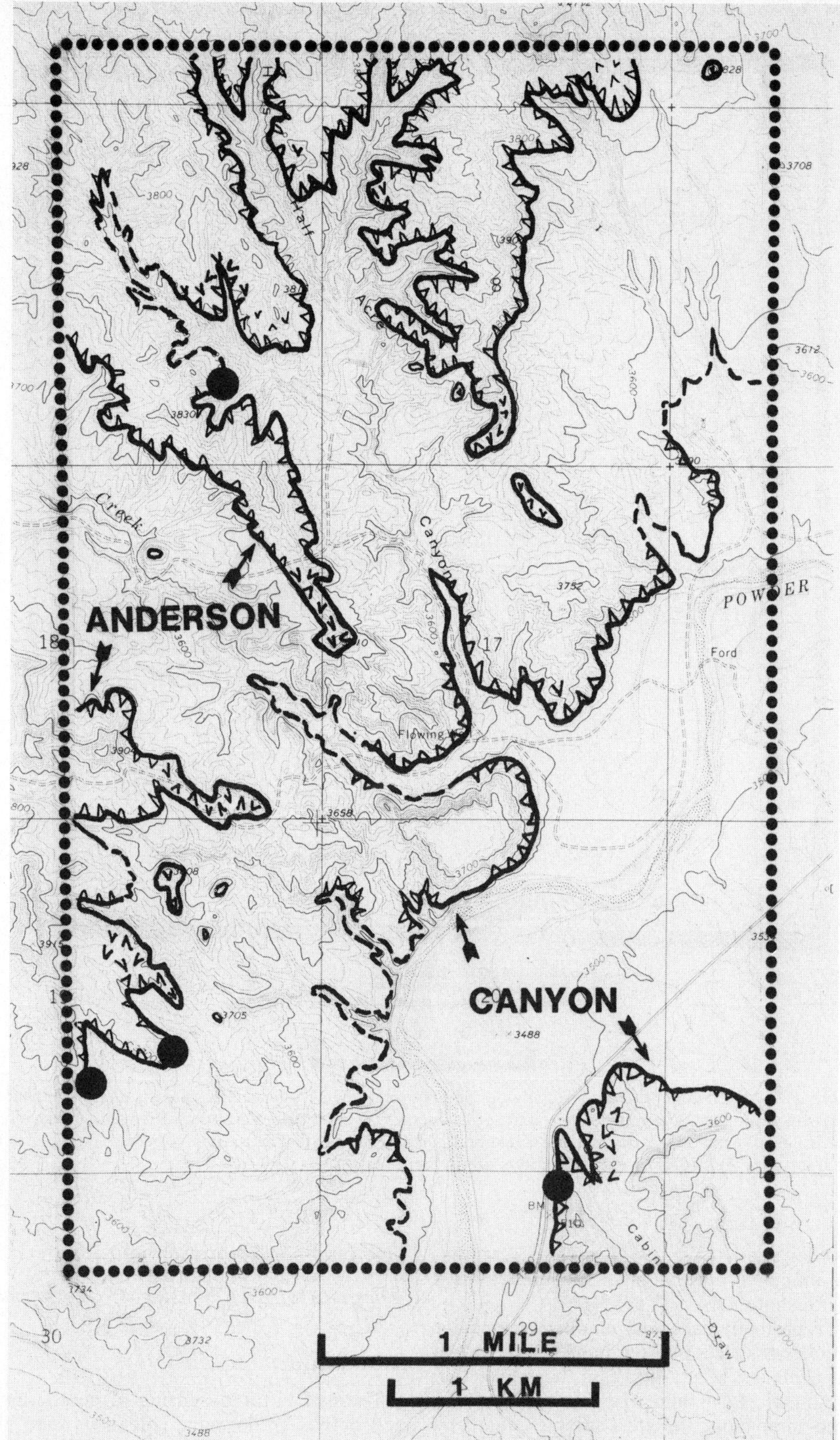

Fig. 31-146B. Coal bed map, produced on the PG-2 photogrammetric plotter, using the stereopair shown in Color Figure 31-146A. Areas of "<> v" marks indicate clinker. Dashed line indicates approximately where coal bed is located. Contour interval is 20 ft (6.1m); map elevations are in feet. The map shows the location of two major coal beds (the Anderson and the Canyon), both of which have burned extensively at the outcrop. It was also possible, with the plotter, to find places where the coal did not burn. These locations are marked by black dots. The geologist may go directly to these pre-selected outcrops in the field to measure and describe the coal beds.

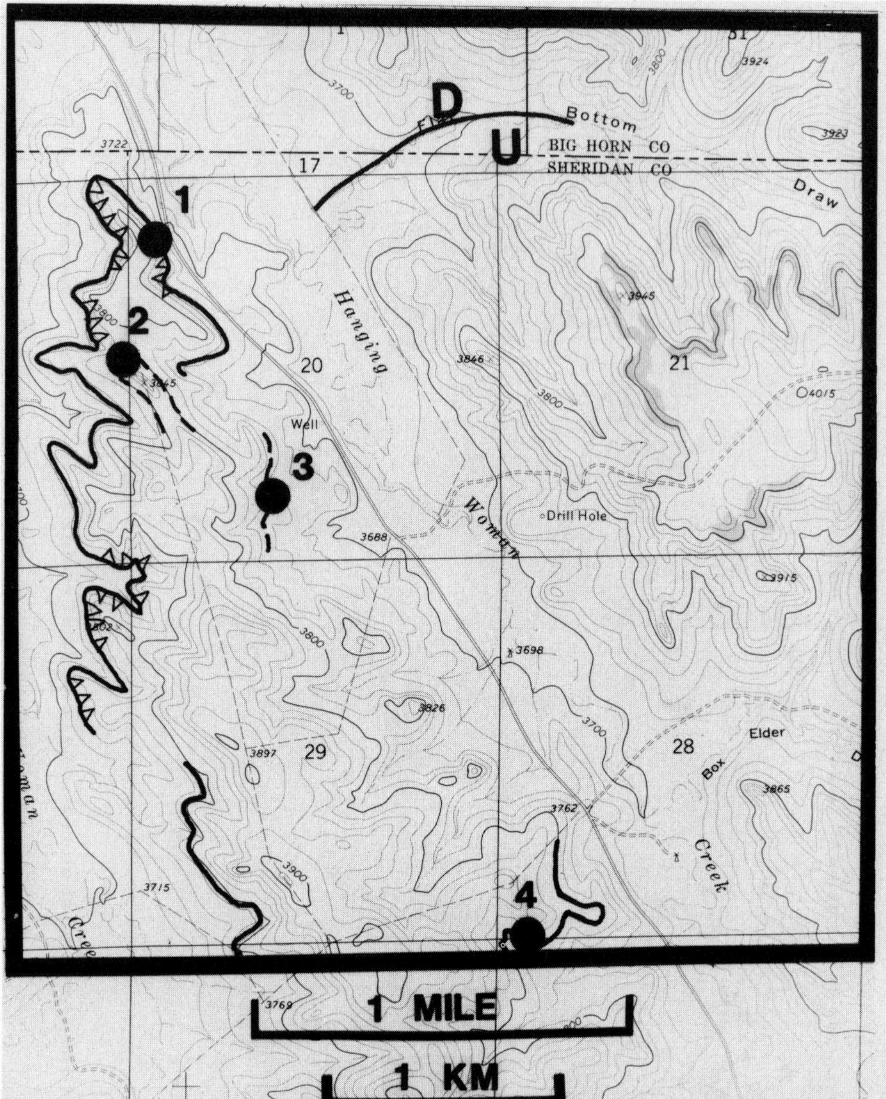

Fig. 31-147A. Coal-bed map produced initially on the photogrammetric plotter. This area is along Hanging Woman Creek, near the Montana-Wyoming border, and approximately 29 kilometers west of the area shown in Color Figure 31-146A. Areas of "< > v" marks indicate clinker. Dashed line indicates approximately where coal bed is located. Contour interval is 20 ft (6.1m); map elevations are in feet.

1982; Jordan and Kieffer, 1982), geologic hazards, surface mining progress, land subsidence, and changes in shorelines.

Work is presently underway on the development of computer-assisted photogrammetric mapping systems (Pillmore et al., 1981). In these systems, non-horizontal planes and irregular surfaces are projected throughout the stereoscopic model by use of a computer-positioned floating mark, constantly maintained on a pre-calculated mathematical surface. The projection, optics, and measuring systems of photogrammetric plotting machines are described in the fourth edition of the *Manual of Photogrammetry* (Slama et al., 1980).

Geothermal Exploration and Related Surveys

Thermal Infrared Sensing of Geothermal and Volcanic Activity

Introduction

Thermal infrared sensing of geothermal areas and active or dormant volcanoes (see also the section on "Remote Sensing of Volcanic Hazards" in this chapter) has effectively used optical-mechanical scanning radiometers from aircraft or satellites (Color Figure 31-149) or fixed-field radiometers on the ground or in an aircraft to record thermal emission from such dynamic phenomena. The motivation behind such ground,

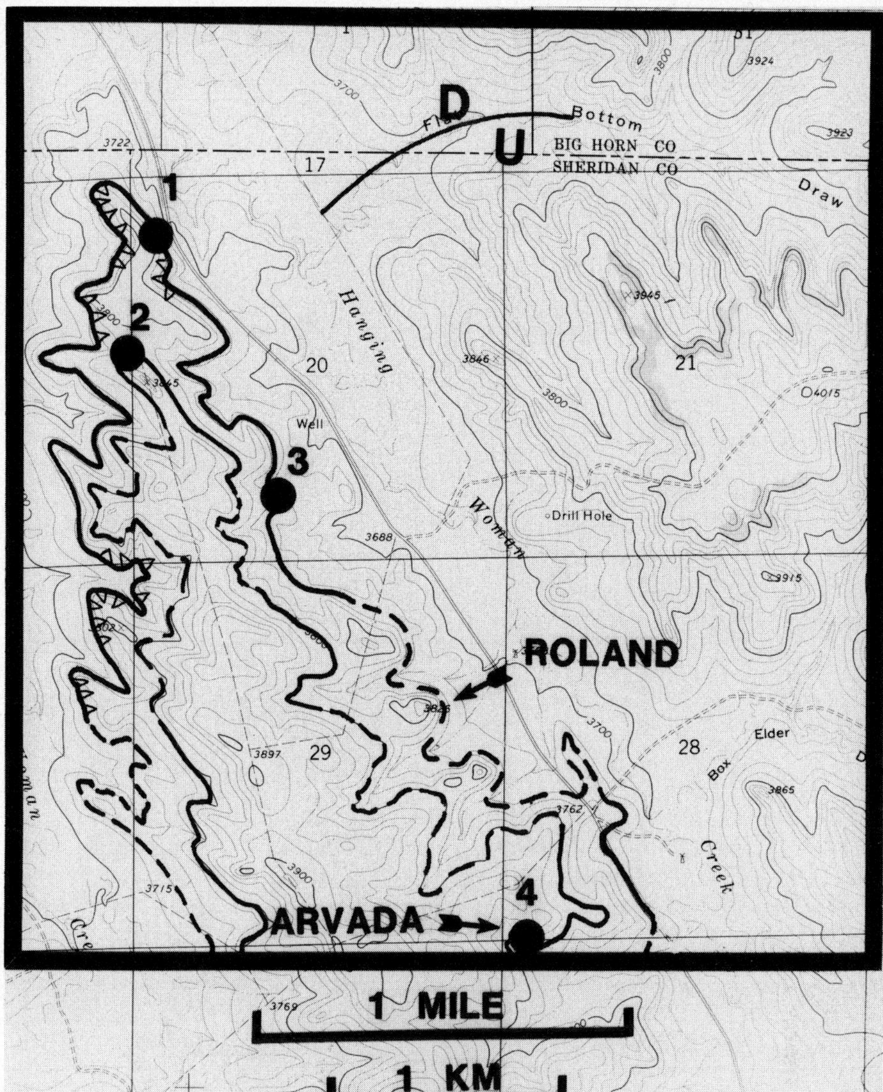

Fig. 31-147B. Coal-bed map produced from Figure 31-147A after field visits to the four outcrops indicated (black dots).

aerial, and satellite surveys has been threefold: (1) *scientific,* to better our understanding of volcanic and geothermal processes; (2) *hazard warning,* to determine if aerial thermographic surveys can lead to a predictive technique (Moxham, 1972), thereby reducing the loss of human life caused by volcanic eruptions; and (3) *geothermal exploration,* to delineate areas of surface geothermal activity as a first step in the development of a geothermal area for power and/or space heating.

The thermal infrared sensors used in these surveys record energy emitted in either the 3 to 5.5 μm or 8 to 14 μm wavelength regions of the electromagnetic spectrum. Images are analyzed in strip form or in mosaics (Williams and Ory, 1967; Williams et al., 1976). A good review of thermal infrared surveys of geothermal areas and volcanoes can be found in "Infrared Sensing of Active Geologic Processes" (Friedman and Williams, 1968), in "Terrestrial Remote Sensing: Applications of Thermal Infrared Scanners to the Geological Sciences" (Williams, 1972a), in "The Use of Broadband Thermal Infrared Images to Monitor and to Study Dynamic Geological Phenomena" (Williams, 1981), and in "Interpretation of Thermal Infrared Images" (Sabins, 1980).

The use of the thermal infrared part of the electromagnetic spectrum by geologists includes more than just the detection of thermal emission for geothermal and volcanic studies (Friedman and Williams, 1968; Williams, 1972a, 1981). Both the middle infrared and the thermal infrared regions have been used for compositional information (Watson, 1971a, 1971b, 1975; Watson et al., 1971; Hunt, 1981b; Kahle, 1980, 1981; Lyon, 1981; and Vincent, 1981a, 1981b), and the infrared (reflected and emitted) regions of the elec-

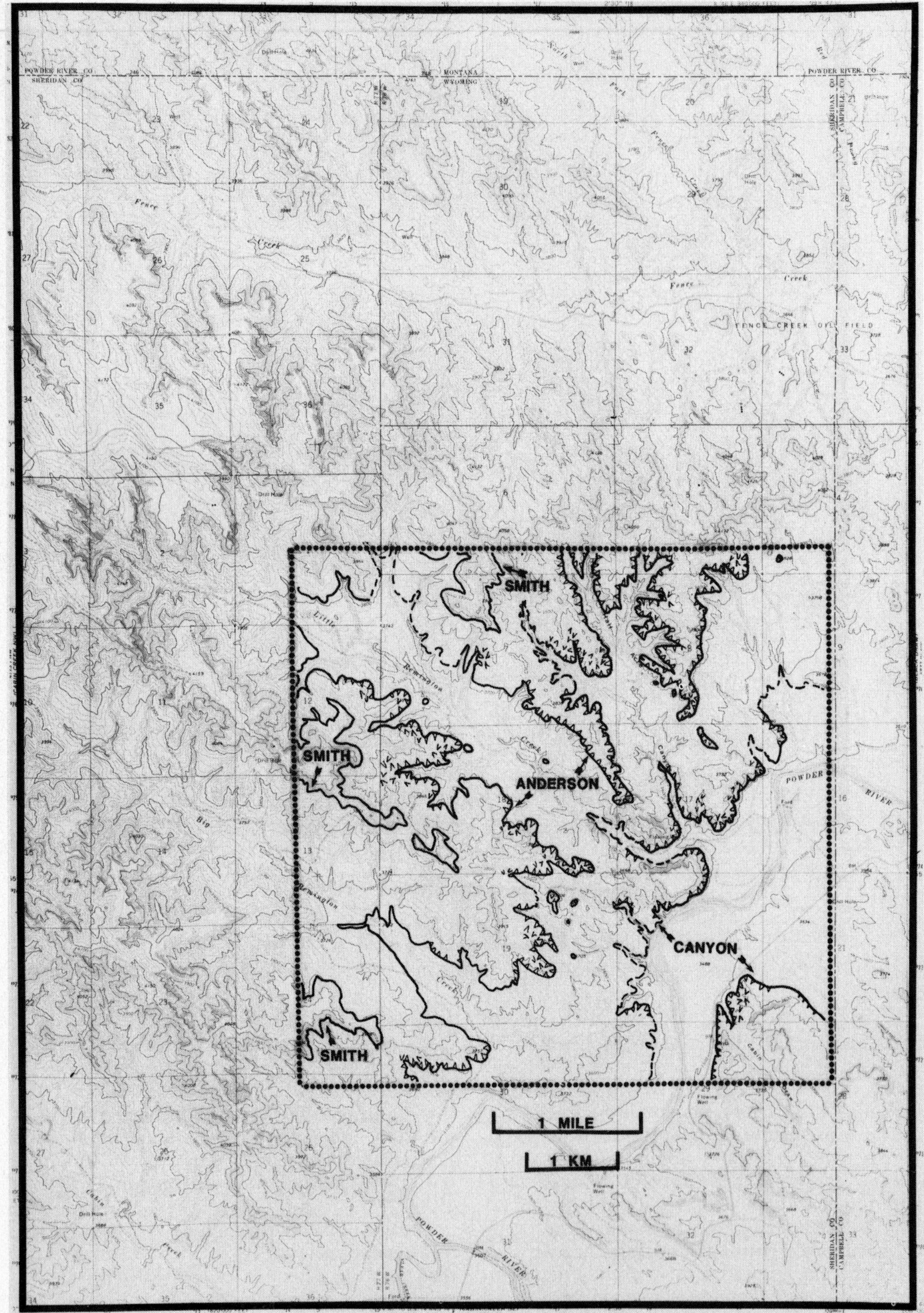

Fig. 31-148. Portion of a 7½-minute quadrangle map that was produced on the photogrammetric plotter during a typical work day. The area outlined covers about 35 square kilometers; it was mapped stereoscopically from two pairs of aerial photographs. The Canyon, Anderson, and Smith coal beds are shown. Areas of "< > v" indicate clinker. Dashed line indicates approximately where coal bed is located. Contour interval is 20 ft (6.1 m); map elevations are in feet.

tromagnetic spectrum will be increasingly exploited by geologists (Goetz and Gillespie, 1981) as new sensors are designed and made available (Brown, 1981; Wright, 1981; and Goetz and Rowan, 1981).

The most up-to-date compilation of information that is of interest to geologists about the use of the thermal infrared region of the electromagnetic spectrum can be found in the collection of papers in the "Workshop on Geological Applications of Thermal Infrared Remote Sensing Techniques" (Settle, 1981). This Lunar and Planetary Institute report also contains a complete bibliography on instrumentation, physical properties, thermal surveys/observations, thermal surveys/theory, thermal inertia surveys, emissivity surveys, atmospheric studies, and planetary studies. Much of the following section is adapted from the paper by Williams (1981) in the Workshop report.

Airborne thermal infrared images (Williams, 1972b) have been used to monitor five principal types of dynamic geological phenomena: geothermal areas; volcanoes; ground water discharge into marine, lacustrine, riverine, and estuarine waters; water currents; and water pollution. In all five of these geologic applications, the phenomenon being studied is generally not manifest and, hence, cannot be mapped on aerial photographs or by satellite imaging systems limited to the visible or reflective infrared wavelengths. Although satellite thermal infrared surveys have been made of erupting volcanoes and marine currents and eddies, the inadequate spatial resolution of thermal infrared sensors on civilian satellites has almost precluded their use in studies of other dynamic geological phenomena.

Infrared Surveys of Geothermal Areas

In contrast to the study of volcanoes, where inaccessibility, large areal dimensions, and the often dynamic nature of the thermal regime make the motivation for study usually scientific, the thermal character of many geothermal areas has been intensely studied under the impetus of economic exploitation.

Airborne thermal infrared surveys have been made of geothermal areas in the United States, including several in Yellowstone National Park (McLerran and Morgan, 1964; Miller, 1966; Williams et al., 1976). Surveys also were made in Italy (Del Bono et al., 1971) and Iceland (Friedman et al., 1969a, 1969b) using infrared systems of U.S. manufacture. Again, the high-temperature geothermal areas of Iceland have been the most comprehensively studied. Airborne thermal infrared imaging surveys were made of many areas in Iceland by the U.S. Air Force Cambridge Research Laboratories (AFCRL) in 1966 and 1968 and by the National Aeronautics and Space Administration (NASA), under an Earth Resources Technology Satellite 1 (ERTS 1) project (Williams et al., 1974, and Williams, 1978a) in 1973. The research in Iceland is now (1982) in its final stages, and a report is in preparation as a cooperative endeavor between the U.S. Geological Survey and the National Energy Authority of Iceland.

The following generalizations governing infrared emission from geothermal areas have been reached:

1. Airborne thermal infrared imaging surveys can be employed to map the surface manifestations (particularly areas of altered ground and anomalously high heat flow) of geothermal activity. It is a most useful technique in remote or poorly known areas—for example, Torfajökull, Iceland (Pálmason et al., 1971) and Kverkfjöll, Iceland (Friedman et al., 1969a, 1969b, 1972; see also Figure 31-152A and B). For those areas where the heat flow is higher than normal but below detectability with conventional aerial thermal infrared surveying techniques, it is necessary to resort to sophisticated modelling techniques. Watson, in his classic study of anomalously high heat flow in the Raft River area, Idaho, reported on the results of this modelling approach in several papers (Watson, 1971a, 1971b, 1973, 1975; Watson et al., 1971).
2. Under certain conditions, thermal infrared techniques may be used to define the areal extent of a geothermal field. Areal extent is an important parameter when one is trying to estimate the power potential of a field before exploitation. These techniques have been used to yield more accurate estimates of surface geothermal anomalies than were previously available (for example, Torfajökull, Iceland; Saemundsson, 1969).
3. The configuration of thermal anomalies and their relation to tectonic features at the surface, such as faults and fissures, may be determined by infrared surveys. This information may be of value in choosing optimum sites for drillholes in areas to be exploited.
4. Infrared techniques have been used to document changes in the thermal regime of two geothermal areas, Reykjanes and Theistareykir, in Iceland's neovolcanic zone. Successive surveys have shown that changes occur in surface thermal activity with time.
5. A marked increase in surface thermal activity is sometimes found to precede volcanic eruptions; for example, the Askja eruption in north-central Iceland in 1961. Periodic surveys to ascertain changes in surface thermal activity may, therefore, be of value in predicting volcanic eruptions.
6. Airborne thermal infrared imaging techniques can probably be used to follow change in surface thermal activity caused by exploitation of geothermal areas for power

production. The natural surface manifestations of thermal activity normally decrease during exploitation.

7. Although it would be most useful to be able to map areas of geothermal emission from orbital altitudes, the spatial resolution of thermal infrared sensors on civilian satellites is too large. These sensors have a large instantaneous field of view (IFOV), which is 1 km for the National Oceanographic and Atmospheric Administration (NOAA) series of weather satellites and 0.6 km for the Heat Capacity Mapping Mission (HCMM) spacecraft (see Color Figure 31-149 and Figure 31-24). Landsat multispectral scanner (MSS) images of Iceland, which have a pixel resolution of about 80 m, were used to identify and to grossly map some areas of geothermally altered ground (for example, the Námafjall area, using the variation in spectral reflectance of the altered ground areas (Williams et al., 1974; Williams, 1978c). The thermal infrared sensor (MSS band 8) on the Landsat 3 spacecraft had a pixel resolution of 240 m, but the sensor malfunctioned and very little, if any, useable scientific data were acquired. A noise equivalent delta temperature (NEΔT) of 1°K and a spatial resolution comparable to the Landsat MSS system (~80 m) would be required for direct measurement of thermal emission from high-temperature geothermal areas.

Infrared Surveys of Volcanic Areas

During the first decade (1960–1970) of the application of airborne thermal infrared imaging techniques to the geological sciences, considerable progress was made in the qualitative study of volcanic thermal processes and geothermal-field convective and conductive heat transfer. Anomalously high infrared emission (in the 3 to 5.5 μm and 8 to 14 μm wavelength regions), indicating active volcanic processes, was detected by airborne and satellite thermal infrared systems in Hawaii (Fischer et al., 1964; Gawarecki et al., 1965), Alaska, the Cascade volcanoes of the western United States (Moxham et al., 1965), Costa Rica, the Philippines (Moxham and Alcarez, 1966), Italy (Del Bono et al., 1971), Antarctica (Mount Erebus volcano), the U.S.S.R., Jan Mayen, and Iceland, to name a few of the principal geographic areas. Of all these areas, Surtsey Volcano, Iceland, and other Icelandic volcanoes (for example, Hekla (Figure 31-13); Askja; and Heimaey (Color Figure 31-174 and Figures 31-177 and 31-178) have probably received the most comprehensive study by airborne or satellite thermal infrared techniques. (See also the section of this chapter on "Remote Sensing of Volcanic Hazards" for a discussion of the excellent aerial thermographic studies at Mount St. Helens volcano, Washington).

The following generalizations were reached regarding thermal infrared emission from volcanic areas and its detection:

1. Thermal emission from both effusive and explosive-type volcanoes has been recorded successfully by aerial and satellite thermal infrared sensors (Williams and Friedman, 1970a; Color Figure 31-150, Figure 31-151, Figure 31-153, and Figure 31-13).
2. Although all active volcanoes surveyed show some indications of abnormally high infrared emission, thermal anomalies have also been reported for several volcanoes generally regarded as dormant (no historical activity) or inactive (known historical activity). It may be necessary and prudent, therefore, to reclassify some inactive volcanoes as dormant if aerial or field surveys locate any abnormally high infrared emission from either the summit crater or flanks of such volcanoes.
3. The stages of volcanic activity studied by infrared techniques range from intereruptive through eruptive to late volcanic. Surface thermal anomalies have been documented by infrared surveys, indicating changes in the thermal regime of several volcanic areas in Iceland's neovolcanic median zone: Kverkfjöll (Figure 31-152), Hekla (Figure 31-13), and possibly Askja. Such changes in thermal regime, while not necessarily indicative of pre-eruptive activity, did indeed precede the Askja eruption of 1961. Active effusive eruptions were imaged by aerial thermographic methods at Surtsey, Iceland (Figure 31-151), and Etna, Italy. Active pyroclastic volcanoes imaged by aerial thermographic methods were Irazú volcano, Costa Rica, and Taal Volcano, the Philippines (Moxham and Alcarez, 1966).
4. The active processes that produce posteruptive and intereruptive thermal anomalies are manifested by radiant emission associated with convective heat transfer from crater or vent areas, convection along fracture and fault systems, porous scoriaceous rocks that emit convective thermal currents, primary and secondary fumaroles and solfateras, and various forms of hydrothermal activity.
5. The most outstanding associated tectonic features are: (a) curvilinear to concentric fractures and fault patterns associated with caldera subsidence or possible resurgence and (b) linear, en echelon fault, and dilation-fissure patterns in planetary rift zones.
6. The main contributions of thermal infrared images (aerial thermographs) have been: (a) determination of the relative degree of activity or intensity of points of thermal emission; (b) determination of the geometric relation of thermal anomalies to tectonic pat-

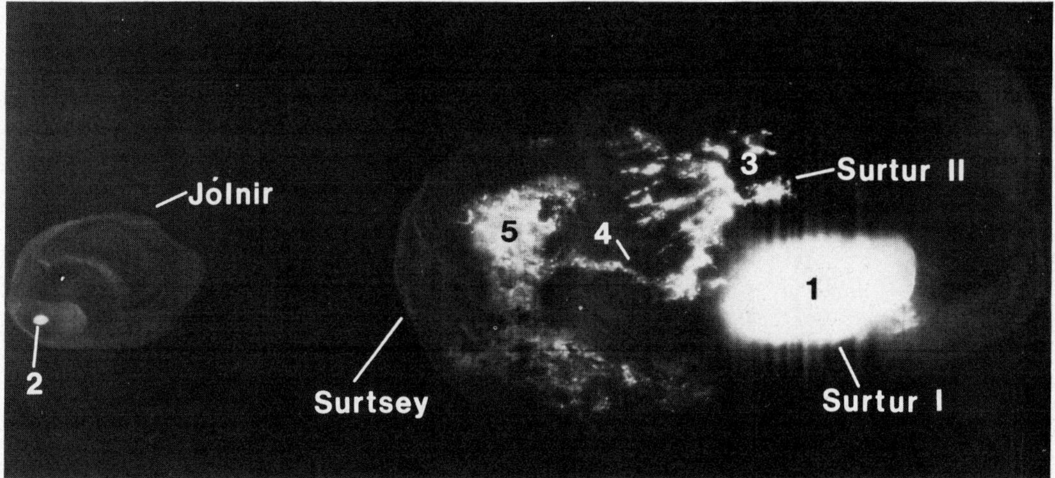

Fig. 31-151. Aerial thermograph of Surtsey and Jólnir, Vestmannaeyjar, Iceland, showing thermal emission from: *1.* The new effusive volcanic eruption which began at about 1100 hr UT on 19 August 1966 in the Surtur I crater (see Color Figure 31-150), *2.* the recently active crater on Jólnir, *3.* the Surtur II crater and, *4.* a lava tube extending from the Surtur II crater to an apparent subsurface lava lake at *5.* The thermograph was acquired by the M1A1 scanner at an altitude of 1,525 m., at 1745 hr UT (Icelandic standard time) with a filtered InSb detector (4.5–5.5 μm). Image courtesy of Air Force Cambridge Research Laboratories. (Modified from Friedman et al., 1969a).

terns; (c) estimation of the proportion of radiant emission to total heat flow in one effusive eruption, Surtsey in 1966 (Williams et al., 1968); (d) determination of the threshold of satellite detectability of volcanic processes by a thermal sensor on an orbiting satellite (Figure 31-153); (e) qualified confirmation of the high energy levels involved in effusive, in contrast to explosive, volcanic activity; and (f) provision of evidence that low spatial-resolution infrared systems with high thermal sensitivity in the 3 to 5.5 μm range are suitable for detection from Earth orbit of effusive basalt lava eruptions (Williams and Friedman, 1970a; Figure 31-153).

7. The predictability of volcanic eruptions by means of airborne infrared images probably depends to a large extent on the detection of changes in the pattern of intereruptive convective heat flow by systematic line-scan imaging techniques, preferably using a calibrated system.

Future Trends

Infrared Surveys of Hydrogeologic Phenomena

The research area of greatest potential application and economic value for airborne thermal infrared surveys using optical-mechanical line-scanner systems is in hydrogeology and related fields. Under certain conditions, airborne scanning radiometers can record ground water discharge into marine (Williams and Fernandopullé, 1972), lacustrine, riverine, and estuarine waters; can depict the flow of water currents (Taylor and Stingelin, 1969); and can often pinpoint the source (outfall) and surface movement of pollutants.

Airborne Applications

Future trends in the application of aerial thermography to geological studies will likely be directed toward hydrogeologic problems and will emphasize the monitoring of water pollution in lakes, rivers, estuaries, harbors, and other coastal areas. Research will continue on the use of aerial thermography to measure surface water currents. If hydrological phenomena could be monitored using thermal infrared sensors on orbiting spacecraft, much more data would be made available. Certainly a first step to reaching the objective of operational satellite surveys would be to establish a solid aerial research program designed to acquire high altitude (>13,000 m) aerial thermography of geothermal, volcanic, and other areas and to assess the utility of such data for regional studies.

The number of scientific studies of the dynamic thermal regime of volcanoes will probably increase; the emphasis will be on monitoring thermal changes in volcanoes that pose a threat to human life such as those in the Cascade Range of the northwestern United States (Lipman and Mullineaux, 1982). Even the dynamic thermal character of those volcanoes which have observatories (for example, Kilauea, Mount Vesuvius, and Mount Etna) is inadequately known. Repetitive surveys of volcanoes may lead to a predictive technique based on the change of surface thermal activity before a volcanic eruption. From a purely scientific basis, aerial thermography represents an

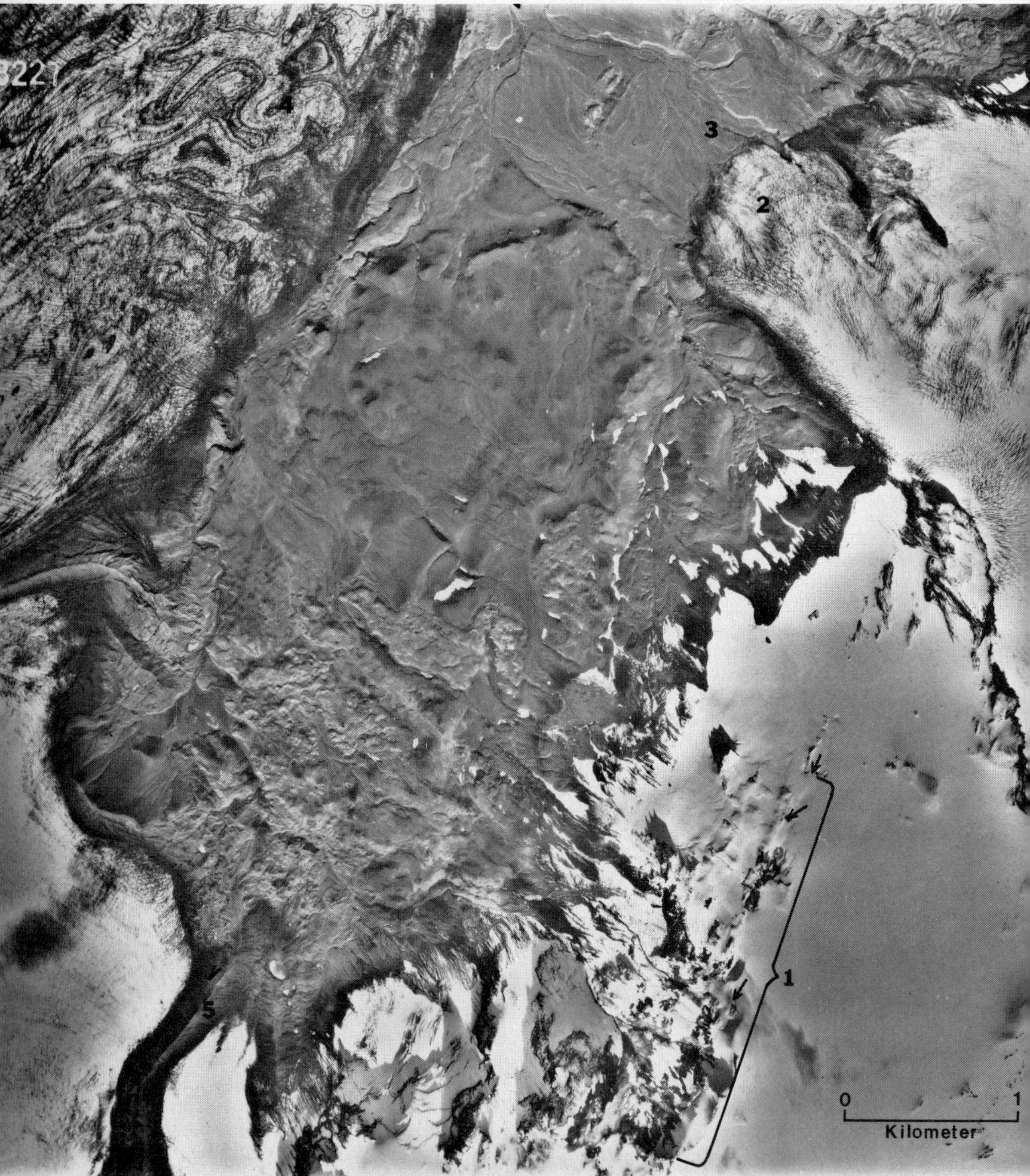

Fig. 31-152. A. Vertical aerial photograph of the northwest margin of the Kverkfjöll subglacial volcanic ridge, a geothermal area on the northern ridge, and a geothermal area on the northern margin of the Vatnajökull ice cap, Iceland (see Color Figure 31-201 and Figures 31-204 and 31-205). *1.* Linear orientation of the subglacial and subaerial Hveradalur thermal area (limits shown by bracket; note also small collapse cauldrons in the Kverkfjöll outlet glacier as shown by arrows); *2.* Kverkfjöll outlet glacier; *3.* stream emerging from subglacial outflow tunnel at the terminus of Kverkföll (note the notch in the terminus); *4.* Eastern margin of the Dyngjujökull lobe; *5.* medial moraine. U.S. Air Force aerial photograph no. 13227, Project 55-AM-3, acquired on 24 August 1960, from an altitude of about 5,500 m. The camera used was a T-11 with a 152.04 mm focal length.

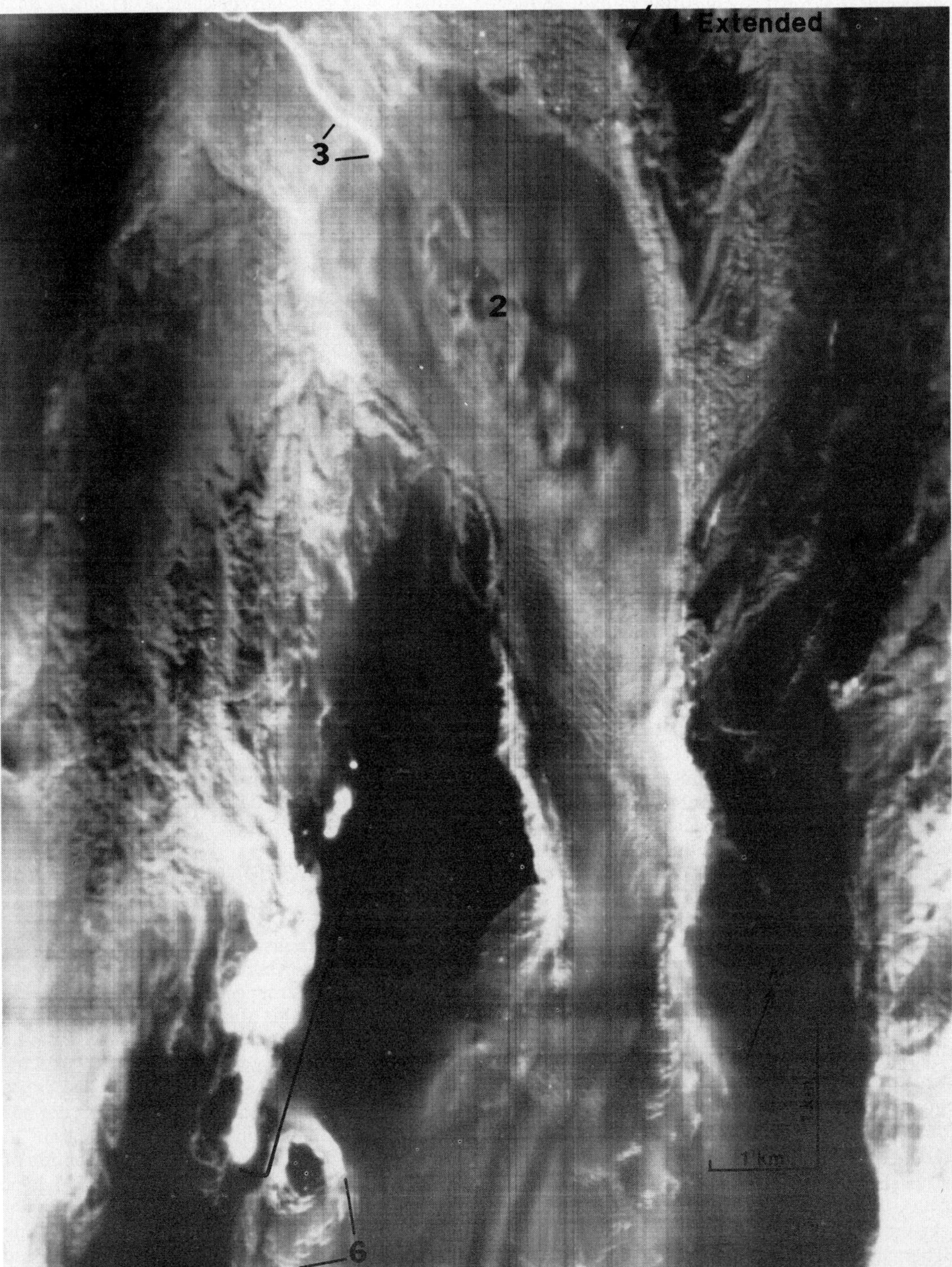

Fig. 31-152B. Aerial thermograph, of the same area as shown in A, acquired on 22 August 1966, at 2357 UT (Icelandic standard time), from an approximate altitude of 6,100 m with the M1A1 infrared scanner, using an unfiltered InSb detector (1–5.5 μm) (Fisher et al., 1965). *1.* Linear orientation of thermal features of the subglacial and subaerial Hveradalur thermal area (limits shown by bracket) and probable northern extension of structure under the Kverkfjöll outlet glacier and to the north (dashed line); *2.* Kverkfjöll outlet glacier; *3.* hot-water stream emerging from subglacial outflow tunnel, whose origin is from a northern subglacial extension of the Hverdalur thermal area. *4.* Eastern margin of Dyngjujökull lobe. *5.* Not within image area. *6.* Thermal emission from concentric fractures around perimeter of large collapse cauldron. Aerial thermograph courtesy of U.S. Air Force Cambridge Research Laboratories. (Figure 31-152A and B modified from Friedman et al., 1969a, 1972).

important tool for the acquisition of data on thermal activity in remote areas such as the Aleutian Islands.

Aerial thermographic surveys of geothermal areas will be increased for scientific reasons and for exploitation of such areas for geothermal power generation and, in colder climes, for hot water distribution for heating and personal and commercial use. The search for geothermal sources of power is being stressed by many countries. Aerial thermography can be used for reconnaissance exploration, assessment of the extent of geothermal areas (particularly at remote sites), and in monitoring changes in the nature of geothermal activity during development and exploitation.

Satellite Applications (Terrestrial)

Geologic applications of orbital thermal infrared scanning radiometers fall into two classes: terrestrial (Merifield et al., 1969a) and extraterrestrial (Merifield et al., 1969b). The terrestrial applications discussed previously are still rather few, because both the spatial resolution and the thermal resolution of the twelve types of orbital scanning radiometers used by the United States are too large (Anonymous, 1981). This is particularly true of the spatial resolution. Future improvements in spatial resolution will likely lead to new geologic applications, particularly in marine geology, oceanography, glaciology, and climatology (Brown, 1981).

Higher resolution satellite thermographic surveys of coastal marine currents will be of particular value in marine geological and oceanographic investigations. Oceanic currents, such as the Gulf Stream (Figure 31-62), and the discharge of major rivers into coastal marine waters have been recorded by both the NOAA series and the HCMM spacecraft. As can be seen from Figure 31-62, however, a much improved spatial resolution of thermal sensors on satellites would produce much more useful information. Figure 31-226, shows large-scale marine eddies off the southwest coast of Iceland imaged on MSS band 4 of Landsat 2 (Williams, 1978d). Whether these eddies can also be delineated thermally is unknown.

Unpublished research by Donald R. Wiesnet (oral communication, 1980) on the Greenland Ice Cap showed that thermal sensors onboard the NOAA series of spacecraft can be used to monitor the thermal regime of glaciers on a seasonal basis. Thermal sensors having a higher spatial resolution would permit such surveys to be conducted seasonally on smaller ice caps, outlet glaciers, and valley glaciers. Of particular interest to glaciologists and climatologists are the varying thermal characteristics of the accumulation and ablation areas of glaciers on a seasonal basis.

As was proven by the studies of the Surtsey Volcano in Iceland and in studies of other volcanoes, it is presently possible to monitor, on a worldwide basis, effusive volcanic activity using orbital thermography (Figure 31-153). Such research is primarily scientific, however, because it is directed at the monitoring of volcanoes remote from inhabited areas.

Satellite Applications (Extraterrestrial)

Instrumentation similar to that used to acquire terrestrial orbital thermography could be modified and placed in orbit around the planet Mars (Kieffer, 1981), the Galilean moon Io, and other planets and moons suspected of being volcanically active. Neither the planet Venus nor Titan, one of Saturn's moons, can be thermally imaged, however, (except, perhaps, by passive microwave sensors) because both their surfaces are shrouded by atmospheres. A system similar to the Nimbus II high resolution infrared radiometer could definitely record an effusive volcanic event from either the Ionian or Martian surface at an altitude of 1,000 km, if the eruption were of a magnitude not less than the August 1966 effusive eruption on Surtsey, Iceland (Merifield, et al., 1969a; Williams, 1972a). An orbiting spacecraft provides the only feasible means of monitoring dynamic thermal phenomena on extraterrestrial surfaces because the areas of most of them are so huge. In the case of Mars, for example, the area to be monitored is equal to the subaerial (dry-land) area of the Earth's surface. Compositional mapping of the Moon's surface has been carried out with thermal sensors from the Apollo 17 spacecraft (Mendell, 1981) and from Earth-based telescopes (Potter and Morgan, 1981).

ENGINEERING GEOLOGY

Introduction

Aerial photographs and other types of remotely sensed data are an important source of information for the engineering geologist (Pettyjohn, 1980, and Chapter 32 of this Manual). Imagery is used as the first source in obtaining new geologic data or to supplement existing data. It is used to plan investigations, cover areas between points of ground observations, detect subtle and not-so-subtle changes in terrain, and to help integrate all types of surficial patterns to permit a deduction of their geologic significance. Remotely sensed data, however, are only supplementary tools and do not stand alone. The final judgment by the engineering geologist rests on careful and comprehensive field observations, subsurface exploration, and laboratory tests. Yet the tools provided from remote sensing technology are very essential, if engineering geology investigations are to be thorough, timely, and economical.

The next chapter in this *Manual,* Chapter 32, *Engineering Applications,* is specifically directed at a comprehensive review of the subject. This

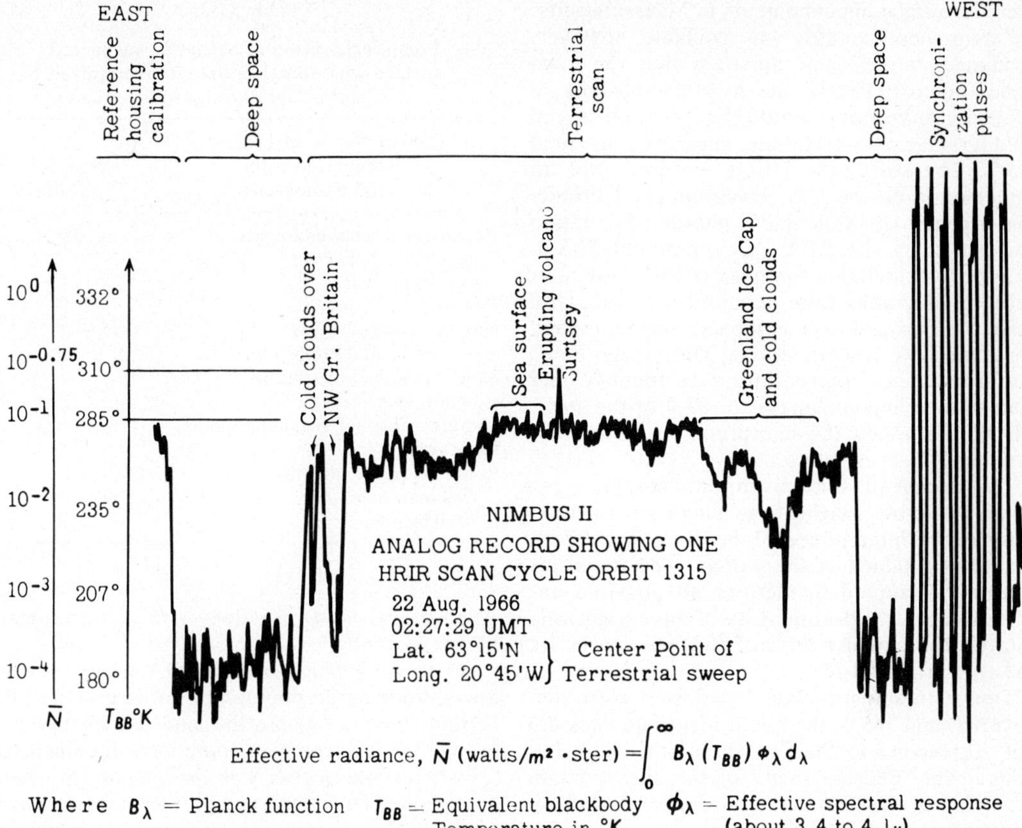

Fig. 31-153. Effective radiance ($\bar{N}$) from the Earth's surface along a profile from Greenland to Great Britain 02:27:29 UMT, 22 August 1966, recorded at an altitude of 1,114 km by the Nimbus II HRIR System. This analog oscillograph record of one scan cycle, orbit 1315, was made from the original interrogation of the Nimbus II spacecraft. The highest positive spike represents infrared radiation in the 3.4 μm to 4.1 μm wavelength band emitted by lava craters and incandescent flows from Surtsey, integrated with infrared radiation from the surrounding 61-km^2 ocean surface.

section of the Geology chapter is limited, therefore, to the presentation of selected case histories, by seven geologists. Collectively, these case histories deal with five different topics of interest to engineering geology: waste disposal, site selection, construction materials, monitoring of surface mining and land reclamation, and coastal engineering.

Waste Disposal

Liquid Waste

Aerial photography has the following three distinct uses in the evaluation and control of hazardous wastes:

1. Recognizing, identifying, and locating presently active hazardous waste disposal areas.
2. Tracing former land use patterns to discover old hazardous waste disposal sites.
3. Aiding in the quantitative evaluation of the problems associated with a particular site.

These uses of photography are best illustrated by the following specific case study.

In the late 1970s, under the safe drinking-water act, a major study was funded by the Environmental Protection Agency (EPA Grant No. F001-173-780) to investigate the environmental problems posed by process waters and ponded wastes deposited in surface lagoons in the United States. Funds were allocated to individual state agencies which then designed their own studies. In Massachusetts, the program became known as the Massachusetts Surface Impoundment Assessment (SIA) and was carried out by the Massachusetts Department of Environmental Quality Engineering, the DEQE (Bowley, 1980). The intent of the study was to determine the number of surface waste impoundments in the State, to make an assessment of the contamination potential of each, and to provide detailed information on a statistically representative sample of the sites.

At the onset of the study, the DEQE had a limited amount of information concerning certain specific surface impoundments in Massachusetts. No state-wide inventory was available, however. It immediately became apparent that the most rapid and cost effective means of establishing an accurate inventory would be through aerial photographic interpretation, supported by field work. Therefore, the DEQE entered into an agreement with the U.S. Environmental Protection Agency (EPA) to have photointerpretation performed by the EPA Environmental Photographic Interpretation Complex (EPIC). All color aerial photographs were acquired by EPIC with their "Enviropod"—a compact, self-contained panoramic two-camera system. Ultimately, EPIC performed photointerpretation to identify and map surface impoundments for 89% of the state. DEQE performed the interpretation for the remaining 11% of the state.

The design of surface impoundments is extremely variable, ranging from a highly engineered complex of interconnected pools to haphazard sludge pits hidden in the woods. The initial inventory attempted to identify all possible impoundments. Subsequent field investigations showed the actual nature of the sites to be as shown in Table 31-35.

Table 31-35

Field verification of actual or suspected surface-waste-impoundments recognized on aerial photographs.

Actual Nature of Feature Recognized from Aerial Photographs	Number of Sites
Actual waste impoundments	316
Concrete lined sewage treatment plant tanks	23
Fire protection ponds	19
Natural features (eutrophicated ponds etc.)	23
Sand and gravel operations (many with wash ponds)	21
Skating rinks and swimming pools	6
Reservoirs	3
Reported impoundments not visible in field	14
Miscellaneous	36

Thus, 316 actual impoundment sites were discovered, and 145 of the photo-identified sites did not correspond to the SIA definition of an impoundment. Because many of the sites contain several impoundments, a total of 1,962 individual impoundments were discovered.

The following characteristics serve to identify surface waste impoundments:

1. Rectangular, obviously engineered ponds adjacent to industrial buildings (Color Figure 31-154).
2. Rectangular and square ponds associated with sewage treatment plants.
3. Irregular ponds associated with other disposal operations. These include sludge pits in municipal dumps and ponds associated with barrel storage areas and junkyards (Color Figures 31-154, 31-155, and 31-156).
4. Ponds associated with unexplained dead vegetation (Color Figure 31-154).
5. Ponds in unnatural locations such as clearings in the woods, adjacent to rivers, and in gravel pits with adjacent roads leading to industrial facilities.

After the identification phase, field studies were undertaken to determine the character of each site. Locating and identifying former impoundments which had been subsequently filled and redeveloped proved exceptionally difficult. Some could be resolved by detailed inspection of the aerial photography of specific areas. Other old impoundments were found by inspection of old aerial photography taken during the 1950s. Use of aerial photography in this way may resemble its use in an archaeological study (see Chapter 26, Applications in Archaeology and Anthropology, of this *Manual*).

In the final phases of the SIA program, new low-altitude aerial photography was procured for certain sites: (1) where the land use history had been extremely complex and poorly documented, (2) where field access was limited, or (3) where unexplained environmental problems were known to exist. This photography was used to make detailed evaluations of these sites. Factors that could be easily evaluated on these photographs include: site drainage patterns; changes in drainage; nature of impoundments, including presence or absence of a liner (Color Figures 31-154 and 31-155); presence and nature of landfill dumps (Color Figure 31-156); abandoned impoundments; and presence of hazard-related materials on the site (such as tank trucks and drums. (See Color Figures 31-155 and 31-156). Several of the sites investigated during the course of this study are or have been under litigation.

Radioactive Waste

The objective of the Paradox basin, Utah and Colorado, remote sensing project of the Geological Survey was to acquire, process, and interpret geologically a variety of satellite and aircraft image products covering the region in order to evaluate factors in the tectonic and geomorphologic environment which might significantly affect the quality of a nuclear waste repository site; particular attention was given to detecting and delineating previously unmapped structures (Friedman and Simpson, 1978, 1980; Friedman and Heller, 1983).

Computerized rose diagrams were prepared uniformly from azimuthal histogram plots of several thousand lineaments of the Moab 1° × 2°

Quadrangle (1:250,000-scale map). These diagrams permitted a detailed statistical comparison of Landsat MSS lineament trends mapped from multispectral scanner scenes with magnetic-field and gravity-field lineaments and fault and fold-axis strike trends.

A good correlation, statistically more than coincidence will allow, was found to exist between major Landsat lineaments, magnetic- and gravity-field lineaments, northwest axes of folds, and northwest fault trends. A similarly good correlation exists between the first-order surface azimuthal trends of all lineaments mapped from Landsat and Precambrian (pre-600 million years before present time) structural discontinuities trending northeast which were mapped from the magnetic and gravity fields. The Precambrian structural discontinuities may be on a strike projection of the pre-Sinyala fault system of northern Arizona.

The good correspondence of azimuthal trends of magnetic and gravity lineaments, presumably of Precambrian basement origin, and surface lineaments and mapped faults suggests that equivalent structural patterns are likely to be present in the intervening Phanerozoic sequence (post-600 million years before present time). The Laramide age (approximately 70 million years before present time) and possibly the Pleistocene age (the period of time from about 1 million years to about 11,000 years before present time) of the above-mentioned faults (dating based on stratigraphic cross-cutting relationships) support this conclusion. The surface lineamentation and faulting are thus interpreted to represent Precambrian basement-controlled structures in the younger Phanerozoic cover (Friedman et al., 1979). The major lineaments mapped (those longer than 20 km) show the best azimuthal correlation with lineaments of the magnetic field and mapped northwest-striking faults and fold axes and probably represent deep and long-lived faults in the Earth's crust.

The laccolithic cluster of the La Sal Mountains and several circular features occur at intersections of some of the major northwest and northeast lineaments of the Paradox basin and are thus of special tectonic significance. Circular features such as the Lockhart basin, below which salt solution and subsidence of the overlying sequence of clastic rocks have occurred, may represent the negative analog of salt diapirism and laccolithic intrusive activity in the Paradox basin. Surface faulting, such as fault clusters of *en echelon* character which lie astride the northeast-trending Precambrian structural discontinuities, may thus have played a significant role in localizing halokinetic processes of salt flowage and dissolution.

A planimetrically rectified multispectral scanner (MSS) band ratioed composite image of the northern Paradox basin prepared of Landsat MSS image 5165–17030 (1 October 1975) (Color Figure 31-157B) shows much of the region under investigation to be a possible site for radioactive waste disposal. Computer compatible tapes of image 5165–17030, (which shows, among other geomorphic features, Salt Valley, Fisher Valley, Castle Valley, Gibson Dome, Harts Point, the La Sal Mountains and the Uncompahgre Plateau), were used in processing pixel data of Landsat MSS spectral band ratios 4/5, 4/6, and 6/7.

Three one-quarter scenes were digitally joined to make planimetrically corrected images of 2100 scan lines, each 1900 pixels long, the maximum-size image capability of U.S. Geological Survey *Optronics* image-processing equipment. Because a 2-percent histogram (of image tonal values) stretch for each ratioed image resulted in scenes of low contrast, a histogram stretch based on a cumulative density function (CDF) was also applied. The CDF stretch increased the contrast among the most frequent brightness values and decreased the contrast among the least frequent values. The CDF-stretched data yielded high-contrast black-and-white images. A color composite was made from the data for these three ratios using the U.S. Geological Survey color-write system and Ektachrome film with filters red (4/5), blue (4/6), and green (6/7).

The colors of the resulting positive transparency (or color print) are related to several mappable features of the Earth's surface in this region, notably, the distribution of semiarid-terrain and forest vegetation and irrigated-land agriculture and outcrop areas or cliff exposures of specific lithologic units (Williams, 1964) of the Phanerozoic sequence in the Paradox basin, as shown in Figure 31-157A and Color Figure 31-157B. The band ratioed color composite image (Color Figure 31-157B) is highly useful for delineating many geologic contact boundaries in this region, but the same image-processing procedures applied to another Landsat MSS image will not necessarily yield similar color tones, particularly for certain lithologic units. Other factors, including differing Sun angle, surface-texture, and reflectance properties, can cause somewhat different color tones in the identified lithologic units. Nevertheless, the suite of green tones strongly represented in this scene is closely related to reflectance characteristics of a variety of iron oxide-bearing red-bed units in a semi-desert geomorphic erosion stage.

Landfills

Recent landfill areas are easily identified on aerial photographs. Typical identifying features include the following:

1. A lack of vegetation.
2. An obvious relief feature unrelated to the geomorphology of the area. Landfills are typically fan shaped with squared boundaries and have an uphill slope toward the

working face at the edge of the fan. Frequently a ridge of cover material is deposited along the crest of the working face. Fan-shaped features of geological origin generally slope down toward the edge of the fan.
3. An association with old sand pits, gravel workings, or quarries.
4. An access road leading to the working part of the landfill and frequently around much of the periphery.
5. Dump-truck loads of cover material systematically placed on the site.
6. A high albedo in vegetation at the edges associated with windblown materials coming from the landfill.
7. A fence line surrounding the parcel or occurring at its juncture with adjacent access points.

These and other characteristics of recent landfills are illustrated in Color Figures 31-154, 31-155, and 31-156 in the previous section on "Liquid Waste."

Aerial photographs may also be used to identify abandoned landfills. Although vegetation and capping operations may have obscured the finer details of the site, the gross morphology remains obvious. Color Figure 31-158 shows the type of features which indicate abandoned landfills.

Site Selection

Reconnaissance is an essential and integral part of virtually all site selection problems; the use of remotely sensed data is one means of conducting the reconnaissance needed for engineering geology investigations. Specific, detailed surveys of several types are also conducted after the reconnaissance phase in many cases.

In 1974, the Tularosa Basin, New Mexico (Figure 31-159), was chosen as a representative study site for similar basins in southwestern United States, where the MX Missile System might be based. Analogous geological terrains were sought in which a variety of nuclear weapon effects tests could be conducted (Neal, 1978). The Landsat 1 image in Figure 31-159 (1333–17102; 21 June 1973) shows a variety of arid-region landforms common to many other basins, in addition to the distinctive sulfate lacustrine facies, the origin of the so-called "white sands." During these studies a relict Pleistocene lake was discovered (visible on Figure 31-159) and named Lake Trinity (Neal, 1976). A transect across this lake basin (Figure 31-160) revealed four discrete and distinctive sedimentary facies which were subsequently verified by boreholes *A, B, C, D*. The four geological settings were chosen for explosive-effects tests, and the results were applied to other basins. Two 630-ton high explosive tests were conducted within and outside the sulfate facies a few miles north of the area shown in Figure 31-160.

Previous studies of arid-basin landforms have relied heavily on remotely sensed data, because of the frequently inhospitable or inaccessible terrain, and because of the inherent advantage of the sparse vegetative cover not masking the terrain. Surface features on and adjacent to playas that are recognizable on remotely sensed data are also reliable indicators of ground-water discharge (Neal, 1972).

The ability to make qualitative judgments regarding ground-water depth using remotely sensed data was especially important in making decisions on sites for the MX Missile System, where predicted weapon effects depended on ground moisture which, in turn, affected the spacing of shelters (Melzer, 1976).

The U.S. Air Force has used playas for auxiliary landing fields from time to time and will probably continue to do so in the future. The initial Space Shuttle mission landed on the 19-km-long playa (Rogers Dry Lake) at Edwards Air Force Base, California. The alternate landing area in the Tularosa Basin, New Mexico, which was used by the second Space Shuttle *Columbia* mission because of wet surface conditions on Rogers Dry Lake, is visible on Figure 31-159, just west and north of the White Sands. Remotely sensed data are especially useful in determining the state of surface moisture on playas used for such landings (Neal, 1972). A similar application of Landsat 1 imagery was described by Krinsley (1976, 1977), wherein a route for an all-weather highway crossing an inhospitable portion of the Iranian Dasht-I-Kavir (salt desert) was proposed.

Remotely sensed data have been an essential ingredient in the siting of nuclear power plants (McEldowney and Pascucci, 1979). The identification of lineaments and the assignment of their origin to faults and fractures are perhaps the principal application, but significant uses are also found in surface mapping, geophysics, and borehole siting. Trenching is often performed across lineaments. A variety of data and imagery is used in power plant siting, including black-and-white, color and color-infrared aerial photography, thermal infrared imagery, side-looking airborne radar imagery, and airborne geophysical surveys (Eichen and Pascucci, 1977).

Construction Materials

Introduction

Construction materials (sand, gravel, and crushed rock aggregate) have become increasingly costly in many areas because of long-haul distances to construction sites, often brought about by consideration of the impact on and alteration of the environment. Locating construction materials near a construction site is obviously desirable, and remotely sensed data are an important part of the exploration for such materials.

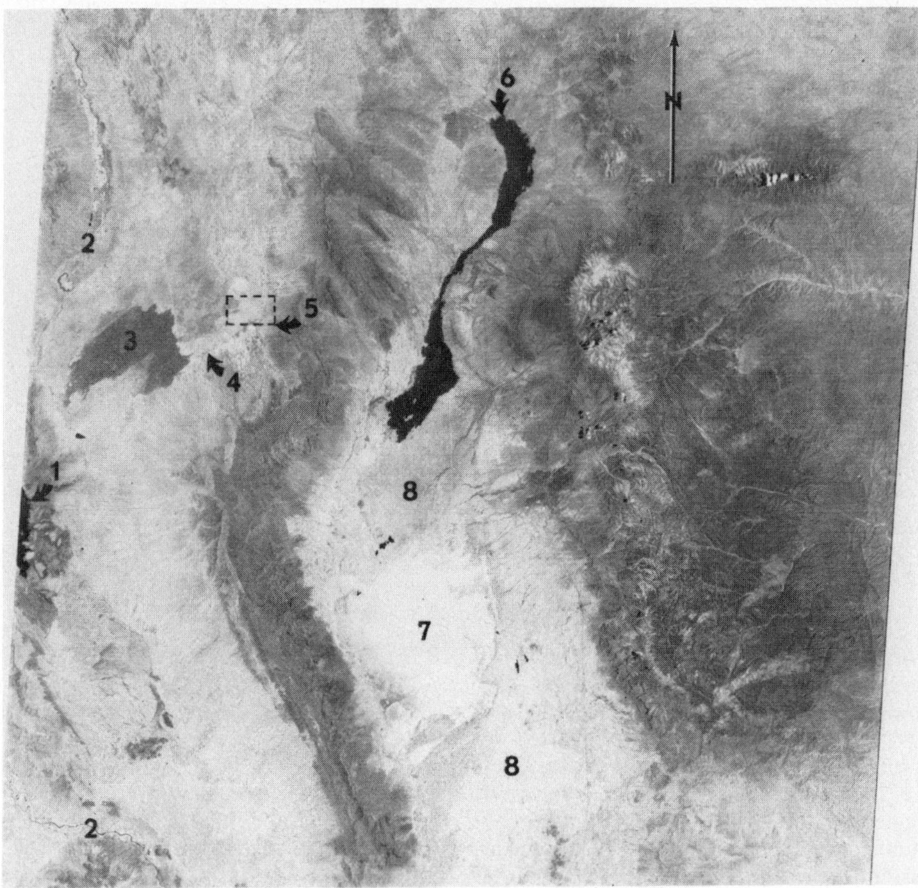

Fig. 31-159. Landsat 1 image (1333-17102; 21 June 1973) of the Lake Trinity basin, New Mexico. Jornada del Muerto valley is on the left, and Tularosa Basin is on right (east): (*1*) Elephant Butte reservoir, (*2*) Río Grande, (*3*) San Marcial lava flow, (*4*) shoreline, Pleistocene Lake Trinity, (*5*) location of Figure 31-160 and boreholes *A*,*B*, *C*, and *D*, (*6*) lava, (*7*) White Sands (windblown gypsum), and (*8*) basin of Pleistocene Lake Otero.

The use of aerial photographs and other types of remotely sensed data for locating construction materials and for other engineering applications has been documented by a number of investigators (Eardley, 1943; Belcher, 1945; Jenkins et al., 1946; Frost, 1946; Hittle, 1949; Lueder, 1951; Schultz and Cleaves, 1955; Mollard and Deshaw, 1956 and 1958; Colwell, 1960; Ray, 1960; Mollard, 1962; Orr and Quick, 1971; Way, 1973; Avci, 1977; Nanda, 1978). Most investigators concluded that photographic interpretation of an area before field-surveys is the most efficient exploration approach because it focuses attention on areas where sand and gravel deposits or bedrock are most likely to occur (Belcher, 1945; Ray, 1960; Schultz and Cleaves, 1955; Colwell, 1960). Thus, field surveys can be conducted in those areas having the highest probability of occurrence of construction materials and thereby exploration costs can be minimized). Nanda (1978) reported a 73-91 percent photographic interpretation accuracy, based on ground verification, in locating hidden aggregate deposits in three areas in India.

Images of the Earth's surface as seen from above, exhibit patterns that can be interpreted in terms of landforms, drainage, and types of surface cover. Landforms are the key element in photographic interpretation of granular materials in some areas because they indicate the results of the geologic processes of erosion, transportation, and deposition (Ray, 1960). The landforms that may contain materials suitable for construction in various physiographic regions depend on the geologic history of the region (Table 31-36).

Gross physical characteristics of surficial materials can be interpreted from the drainage patterns exhibited on aerial photographs. Drainage patterns are indicative of type, distribution, structure, and attitude of near-surface rock materials (Parvis, 1950; Way, 1973), but climatic controls in

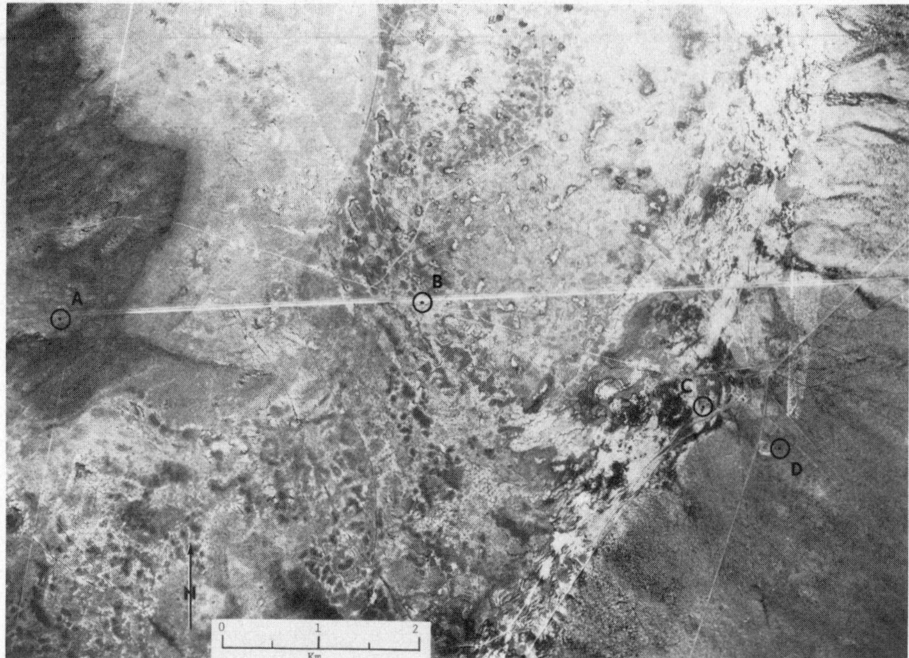

Fig. 31-160. Vertical aerial photograph showing transect across the Trinity lake basin and the four principal facies: windblown sand, sulfate, argillaceous sulfate, and alluvial fan. Boreholes *A, B, C,* and *D* are in each of these four facies, respectively. Photograph acquired by the U.S. Army on 7 April 1963.

the region must be considered when interpreting the significance of these patterns. The density (length and spacing of drainage ways) of the drainage patterns generally is related to the relative permeability of surface materials (Ray, 1960; Way, 1973); coarse-density drainage is diagnostic of granular deposits in some regions (Ray, 1960). Gulley shape is indicative of the granularity and cohesion of surface materials, and V-shaped gullies are characteristic of granular soils (Parvis, 1950; Way, 1973). Eardley (1943) suggests that the occurrence of construction materials in some areas, such as rugged mountains, sandy plains, and swampy basins can best be interpreted by understanding the geologic processes active in the region.

Frost (1946) and Belcher (1948) suggest that light tones on black-and-white aerial photographs are indicative of coarse-grained, well-drained soils. However, photographic tones can also be influenced by topographic position as well as climate and parent rock materials (Jenkins et al., 1946). Light image tones may be related to coarse-grained soils, where geographic and physiographic conditions are considered in the interpretation.

Vegetative cover can be indicative of granular materials. Some trees (such as most species of pine) prefer well-drained soils (Belcher, 1945), and other types of distinctive vegetation can grow on sites that have specific soil and moisture relationships (Tator, 1951). Orr and Quick (1971) found that the vegetation phenology cycle was an important factor to consider in exploring for construction materials in both the Mekong and Mississippi deltas. Time of year must be considered when selecting aerial photographs or planning for new photographic missions.

Methodology

The most efficient approach to locating construction materials consists of several phases of analysis and interpretation (Orr and Quick, 1971). The first phase involves regional analysis of geomorphic and physiographic components to delineate sites that have potential for the occurrence of construction materials. The analysis of landform, drainage, and cover types produces an interpretation of surficial geologic units and processes. Available maps, space images (Color Figure 31-161A), small scale aerial photographs (Color Figure 31-161C), aerial photographic mosaics (Figure 31-161B), and side-looking airborne radar (SLAR) images are useful for these regional interpretations. After delineation of regional units that have potential as sources of construction materials, more detailed analyses of local areas are performed with larger scale images to refine the interpretations and assess the physical characteristics of surficial materials. McKim and Merry (1975) found that 1:33,600-scale

Table 31-36

Examples of Landforms Having Potential as a Source for Construction Materials

[1] Delta and Marginal Delta Plains	[2] Glaciated Areas	Plains Areas	Mountainous Areas
Active Beaches	Alluvial Fans	Active Channels	Active Channels
Stranded Beaches	Eskers	Abandoned Channels	Point Bars
Beach Ridges	Moraines	Point Bars	Bars and Islands
Abandoned Channels	Kames	Bars and Islands	Terraces
Active Channels	Kame Terraces	Terraces	Alluvial Fans
Point Bars	Outwash Areas	Alluvial Fans	Scree Slopes
River Bars and Islands	Outwash Terraces	Bedrock Areas	Bedrock Areas
	Bedrock Areas		

[1] From Orr and Quick (1971)
[2] From McKim and Merry (1975)

black-and-white aerial photographs were adequate to delineate 14 surficial geologic units in a glaciated terrain in Maine. In deltaic environments, color-infrared aerial photographs generally are best for the widest range of geomorphic interpretations because of distinctive hues and contrasts between soil, vegetation, and water cover types (Orr and Quick, 1971) (Color Figure 31-161C). Ordinary color and black-and-white aerial photographs are also useful for detailed analyses.

Hanson and Morganstein (1970) used an eight-channel multispectral scanner system to acquire data in the 0.38- to 1.0-μm-wavelength interval of an area in the Mississippi delta and then developed an automatic procedure for locating construction materials. Feasibility of the technique was adequately demonstrated, but the authors concluded that additional research was needed to document operational capabilities under a wider range of environments and conditions.

Airborne magnetometer and resistivity (E-PHASE) sensors have been used successfully to acquire information on bedrock materials suitable as sources of aggregate. The combined results of the aeromagnetic and E-PHASE survey in Maine (Sellman et al., 1976) provided information on the most suitable rocks by correlating high resistivity with associated mineral content as inferred from aeromagnetic data.

The next phase of an exploration effort is ground-based data acquisition to verify interpretation results. At this point in the approach, a detailed field survey plan can be developed for those areas that have the highest potential for construction materials occurrences. Drill-hole locations, geophysical survey lines, and sites for geological sampling can be selected to optimize the field data collection. McKim and Merry (1975) used seismic data and borehole logs in combination with imagery interpretation results to locate and estimate successfully the volume of available construction material near a dam site in Maine. They concluded that a considerable savings in costs of construction should result from the study because of a reduction in transportation distances from material sources to the construction site.

The targetting of construction materials is an iterative process involving integration of information extracted during all phases of the study. Several scales of analysis of image patterns are possible at a given scale as well as on images of different scales. The convergence of information from the smallest to the largest scale of analysis and from satellite image to high-resolution aerial photograph reinforces the interpretations made at each level (Color Figure 31-161). It is important that the analyst understands the fundamentals of image formation and is able to detect and identify landscape patterns on images. For interpreting the occurrence of granular materials in the shallow subsurface, the analyst must also be able to conceptualize the geologic and geomorphic processes that were responsible for formation of the landscape features (Taranik and Trautwein, 1976).

Conclusions

A number of investigators have concluded that the most efficient approach to exploring for construction materials is the integration of image interpretation results and field studies. A multiple phase approach involving regional, site-specific, and ground-based data analysis provides geologic information in the most timely fashion and at the least cost (Orr and Quick, 1971; McKim and Merry, 1975; Ray, 1960; Schultz and Cleaves, 1955). The use of remote sensing techniques before a field sampling program can enhance success in the exploration effort.

Monitoring of Surface Mining and Land Reclamation

Various Federal and State agencies make use of remote sensing techniques for gathering land-use and environmental data to investigate the impact of large areas of surface disturbance, such as surface coal mining and reclamation on public domain and private lands (Coker et al., 1977; Knuth et al., 1978; Bayne and Lawrence, 1979; Pettyjohn, 1980; and Boldt and Scheibner, 1981). Comparative environmental analysis of surface mining areas must include: (1) spatial—the size of the area that is affected directly or indirectly by surface mining and related activities, (2) dynamic—the rate of change in surface morphology and land use as a result of mining and reclamation, and (3) temporal—yearly and seasonally dependent aspects, such as the density, distribution, and health of native local vegetation, that provide a measure of the rate and success of reclamation efforts. An investigation conducted to test the usefulness of various sources of remotely sensed data (including medium- and high-altitude aerial photography and Landsat imagery) to observe these environmental factors was conducted on Federal land in southeastern Montana (Mamula, 1978). The area, including the Rosebud mine (Rosebud County), was leased by the Federal Government for surface coal mining. The resulting land-use classification of the Rosebud mine area, based entirely on remotely sensed data (and later field verified) included: (1) highwall and bench areas, (2) ungraded spoil piles, (3) graded and recontoured spoil piles, (4) revegetated and recontoured areas, (5) natural and impounded surface water, and (6) miscellaneous areas. This spatial, surface mine, land-use classification scheme, including calculation of respective acreage figures, can be routinely compiled (and figures tabulated) through temporal (qualitative and quantitative) analysis of medium- or high-altitude black-and-white and/or color-infrared aerial photography. However, availability of suitable aerial photography over surface mines on a regular interval is usually not consistent. Enlarged Landsat multispectral scanner (MSS) imagery can be used and is available in a continuous repetitive cycle, although the data are somewhat less effective for studying all but the largest surface mine operations (Rehder, 1976). The 30-meter resolution of the Landsat 3 RBV images and images from the Landsat 4 Thematic Mapper may be of greater assistance for such investigations. Over the long lifespan of many coal surface mines, however, cultural and natural processes and their cumulative environmental impacts can most effectively be monitored by capitalizing on the close correlation between black-and-white, or preferably, color-infrared aerial photography and U.S. Geological Survey 7.5-minute quadrangle map cadastral grid data.

At present, aerial photography acquired on a contract basis by a consortium of several U.S. Federal agencies as part of the National High Altitude Photography (NHAP) program, provides the cyclical repeat coverage needed to monitor and investigate the impact on the physical environment of many eastern and western surface coal mine operations (see also Figure 31-9 and Color Figure 31-10). As initially described (U.S. Geological Survey, 1980), the NHAP Program has been designed so that black-and-white and color-infrared photography is acquired along flight lines centered over USGS 7.5-minute quadrangles. The 152.4-mm (6-in.) focal-length camera is loaded with black-and-white panchromatic aerial film, whereas the 210-mm (8¼-in.) focal-length camera is loaded with color-infrared aerial film. This system provides 23 × 23 cm (9 × 9 in.) black-and-white and color-infrared aerial photography at scales of 1:80,000 and 1:58,000, respectively; coverage is stereoscopic with a base-to-height ratio of approximately 0.57 and a sidelap of about 30 percent. The first and fourth exposures of the color-infrared camera coincide with the first and third exposures of the black-and-white camera to simplify the production of orthophotoquads and facilitate indexing. All coverage is geographically retrievable and Federally maintained and inventoried for any user wanting to purchase reproductions. The EROS Data Center (Department of the Interior) Sioux Falls, South Dakota, and the Agricultural Stabilization and Conservation Service (Department of Agriculture) Salt Lake City, Utah, are the distribution centers for the NHAP photography.

There is no doubt of the utility of medium-altitude aerial photography for a wide variety of mapping and interpretive applications. Sufficient experience has been gained to indicate that the NHAP data base is of immediate and significant value to Federal and State agencies involved in aspects of inspection and monitoring of surface mine leases, especially in the vast expanses of the coal-rich Northern Great Plains. Enlarged color-infrared prints that can be gridded with cadastral and land ownership/lease block data are an extremely valuable asset to field personnel during the mining and reclamation lease inspection process. The established NHAP program may not satisfy all agency needs for aerial photography but it does, for the first time, provide systematic coverage of the conterminous United States at regular intervals. As such, NHAP photography represents an intermediate-scale complement to the Landsat program, providing a comparative data base of high-resolution spectral and textural information for research purposes.

The Minerals Management Service of the Department of the Interior has initiated a research program aimed at capitalizing on the repeat availability of NHAP data over western mineral lease areas. Color-infrared NHAP photographs (1:58,000-scale) are conveniently enlarged and registered to USGS 7.5-minute topographic map grid data (1:24,000); further enlargement to 1:6,000-scale (one inch equals 500 feet) is done for surface mine areas for which a greater amount of detail is required by mining engineers. This type of specialized "orthophotoquad mapping" can be done on approximately a 2- to 3-year repetitive cycle over the same lease area in order to build up a historical file pertaining to the impact of mineral development of that area. There are important geological, engineering, land-use planning, societal, and legal implications for the establishment of a historical aerial photographic map file of prelease through postlease activities. The acquisition time table will probably remain flexible enough so that energy lease lands will receive priority coverage when requested during repetitive coverage cycles of the conterminous United States.

An example of spatial, dynamic, and temporal land use change at the Rosebud mine during the last 6 years is presented in Color Figures 31-162 and 31-163. The comparison is based on a NASA high-altitude aerial photograph acquired on 16 October 1975 and NHAP medium-altitude aerial photography of the same area acquired on 14 September 1980.

Coastal Engineering

Remote Sensing as an Aid to Navigational Improvements on Cape Cod, Massachusetts

Before the availability of aerial photography and other forms of remotely sensed data, the only methods available to measure shoreline and inlet changes were land surveys and hydrographic soundings. These methods are time consuming, costly, and generally cover only limited areas.

The outer shores and inlets on Cape Cod have a long history of instability (Department of the Army, 1968) (See also the "Marine Geology" and the "Mini-Atlas of Image and Related Coverage of Cape Cod, Massachusetts and Environs" sections of this chapter). Erosion and accretion of shorelines, as well as shoaling and migration of coastal inlets are a continuous process (See also Figures 31-40, 31-55, 31-57, and 31-230 and Color Figures 31-42, 31-46, 31-54, 31-227, 31-228, and 31-232). Perhaps nowhere along the shoreline have these natural processes been more of a problem to navigation than at Stage Harbor, on the "elbow" of the Cape, just west of Monomoy Island (See especially Color Figures 31-42, 31-49, 31-50, 31-51, and Figure 31-57).

Stage Harbor is a natural inlet, visited by Bartholomew Gosnold in 1602 and Captain Champlain in 1606. In 1890, a Federal navigation channel 1.8 m (6 ft.) deep and averaging 46 m (150 ft.) wide, was authorized to be dredged from Nantucket Sound, around Harding Beach, and into Stage Harbor. The project was completed in 1901 and last maintained in 1956 (See Figures 31-164A and B and Color Figure 31-165).

The Stage Harbor channel was badly shoaled in 1960, when Monomoy Island was breached by a severe coastal storm, just south of Morris Island. Limited emergency dredging by the town of Chatham was accomplished in 1962 and again in 1963, providing a channel 2.5 m (8 ft.) deep and 31 m (100 ft.) wide through shoaled areas within the limits of the Federal channel, for commercial and recreational boating interests. Shoaling continued at an accelerated rate, however, and limited maintenance dredging was undertaken by the Corps of Engineers in April 1964, to provide a channel only 1.5 m (5 ft.) deep and 31 m (100 ft.) wide in the critical area of the Federal channel around the tip of Harding Beach. It became apparent that periodic maintenance dredging could not be economically justified by the Corps of Engineers, so it was decided to relocate the channel.

Obtaining hydrographic soundings in the continually shoaling area was difficult and costly, and it was decided to utilize vertical aerial photographs as an aid in solving the problem at Stage Harbor. Mosaics were made from U.S. Coast and Geodetic Survey aerial photographs of the breached area: those taken on 17 April 1961 showed beach details; those acquired on 6 October 1963 (Color Figure 31-165) showed submarine details. After the mosaics had been analyzed, it was decided to modify the existing channel to a depth of 3 m (10 ft.) and a width of 46 m (150 ft.) and to relocate it across Harding Beach directly into the deeper water of Nantucket Sound, shown on the photographs. In addition, it was decided to provide a 458-m (1500-ft.) long sand dike between the tip of Harding Beach and Morris Island along shoaled areas shown in the photographs, to prevent future shoaling of the harbor and the relocated channel. The project was completed during July 1965 (see Figure 31-164B).

The use of the detailed aerial photographs of submarine morphology has proved to be an invaluable and economic aid in resolving the navigation problem at Stage Harbor. Present and future, more sophisticated, remote sensing methods will provide qualitative engineering information that will further assist the Corps of Engineers in designing economical protective shoreline and navigation projects in the New England Division and other areas.

Seasat SAR Ocean Observations, Santa Barbara Channel, California

A digitally correlated Seasat SAR image of the Santa Barbara Channel, California (Figure 31-166), documents a variety of valuable ocean surface phenomena (see also Figure 31-30, a SIR-A image of the same area). The image was acquired on 4 August 1978 at 0630 hr. p.s.t. At the time of satellite passage, the eastern half of the channel was obscured by fog, and the western half covered by high cirrus clouds. The reason that features can be delineated on SAR imagery is the result of different radar-reflectance properties of the surfaces being imaged. Some of the ocean-

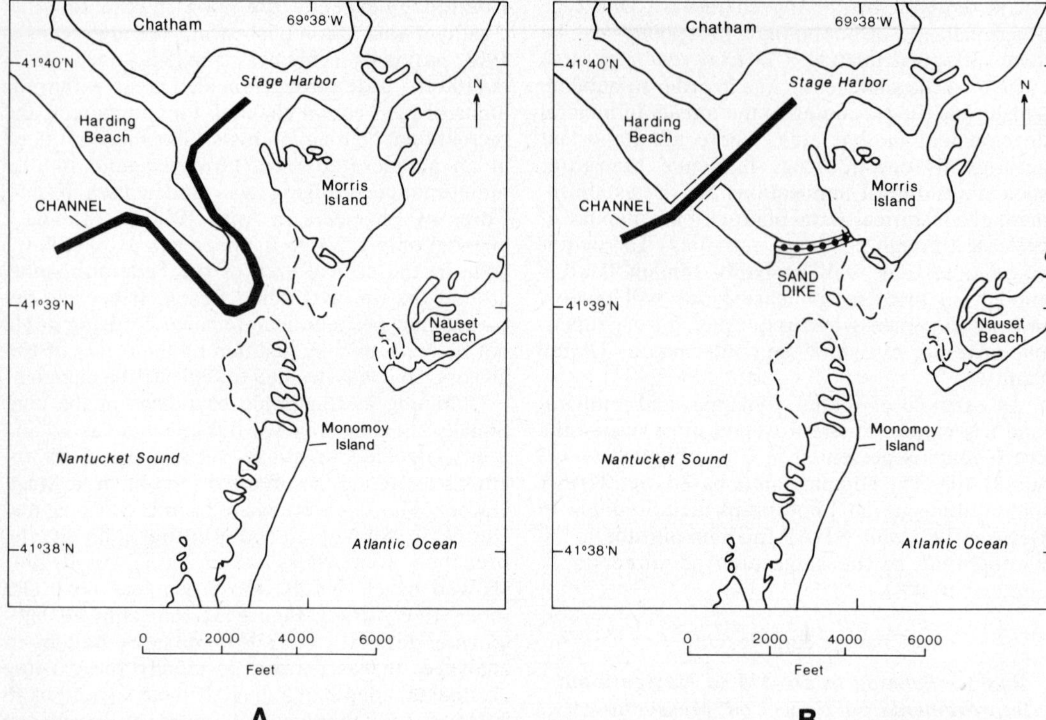

Fig. 31-164. Sketch maps of the Stage Harbor-Harding Beach-Monomoy Island-Nantucket Sound area of southeastern Cape Cod, Massachusetts. A. Map showing the location of channel (original project) from Nantucket Sound to Stage Harbor, between the tips of Harding Beach and Monomoy Island prior to the 1964–65 project to create a new channel across Harding Beach. B. Map showing the location of the new channel (relocated project) from Nantucket Sound to Stage Harbor across Harding Beach and the sand dike between Harding Beach and Morris Island.

surface phenomena recorded include the following: wind slicks, current shears, ships' wakes, internal and surface waves, and wave refraction patterns. Three of the channel islands and several coastal communities have been identified for orientation on Figure 31-166.

During August, the marine currents in the Santa Barbara Channel are in a transitional phase between the upwelling and oceanic current seasons (Pirie and Steller, 1977). The dark area (specular surface on radar) visible off the coast from Gaviota is caused by light offshore winds smoothing out the ocean surface. In the vicinity of Coal Oil Point, Fu and Holt (1982) mapped oil slicks from naturally occurring oil seeps. Upwelling occurs along the coast during these offshore-wind periods. The relatively straight edge of this pattern along its western side is the result of the north-northwest winds being blocked by the land mass in the Point Conception area.

Just below reference point *A* off Gaviota, a ship's wake disturbs the smooth surface pattern. Surface waves resulting from this passing ship have roughened the water surface resulting in detectable radar backscattering. Internal wave packets with wavelengths ranging from 0.8 to 1.2 km are present to the east (right) of the wake. A current-shear boundary trends northwest-southeast adjacent to point *B;* it also parallels the Santa Cruz Submarine Canyon, and this probably affects the surface currents in this area. Adjacent to point *C*, north of Santa Cruz Island, there is a specular surface in the shape of an offset linear streak. The offset is the result of a current shear which is part of the large counter-clockwise gyre which often forms in the channel during August. Surface film from natural oil seeps may also have damped the surface waves in this area (Vesecky and Stewart, 1982). Eight oil drilling platforms located near the coast off Rincon Island are visible at point *D*.

The surface current shears and their resulting patterns can be used to model the surface current dynamics, important in oil-spill trajectory analyses and other applications. This information can be correlated with temperature and water body differences. The wind fields are detectable over relatively large expanses of the ocean surface, and SAR-type information can detect these phenomena under all weather conditions both day and night.

GEOLOGIC HAZARDS

Introduction

Geologic hazards are a common occurrence in most areas of the planet, although some areas, because of their geologic setting, are more prone

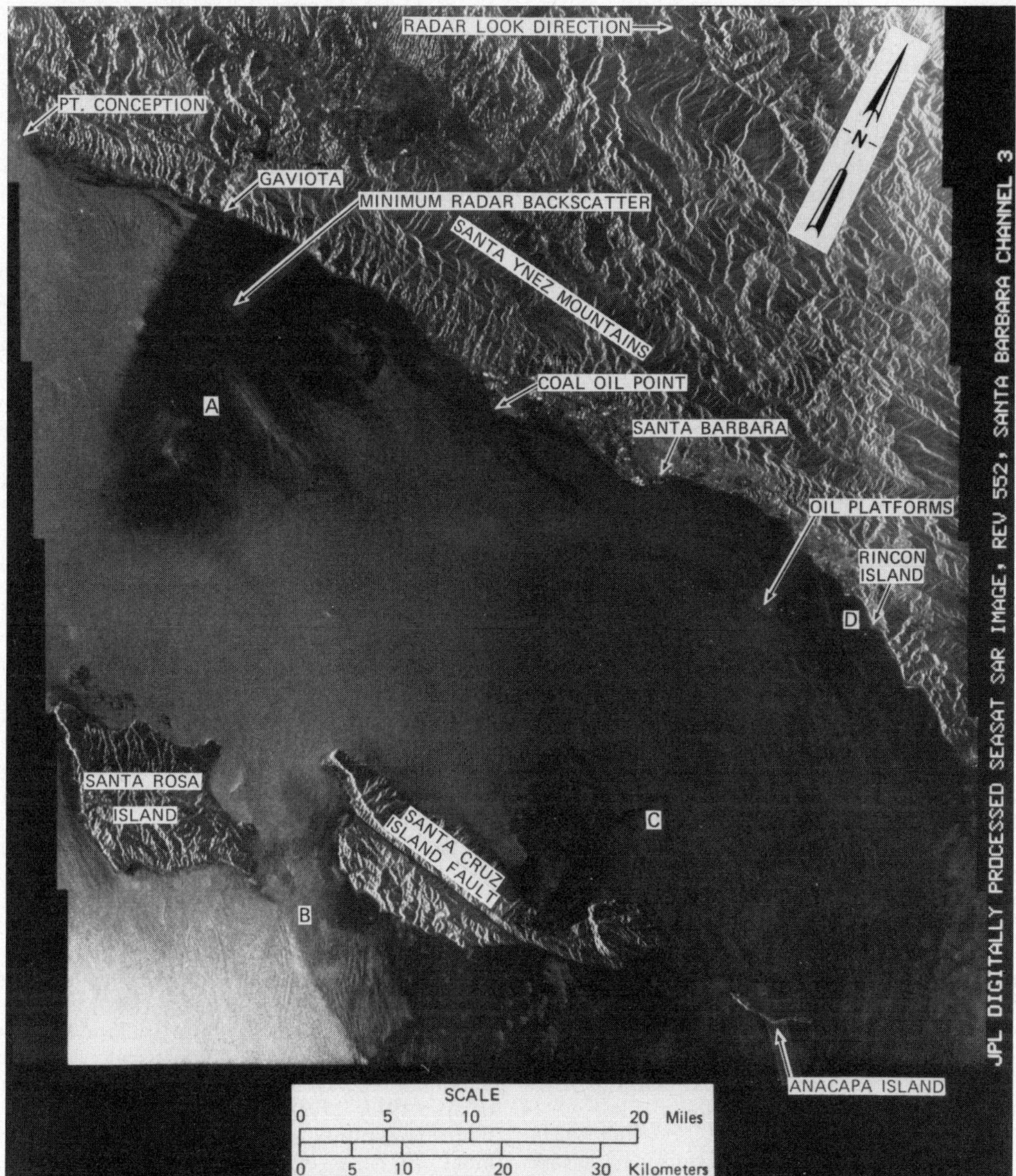

Fig. 31-166. Seasat synthetic-aperture radar (SAR) image of the Santa Barbara Channel, California, acquired on 4 August 1978, showing the following ocean-surface phenomena: wind slick, current shears, ships' wakes, internal and surface waves, and wave-refraction patterns. (Seasat SAR image courtesy of Jet Propulsion Laboratory, California Institute of Technology). (See also Figure 31-30, a SIR-A image of the same area acquired on 13 November 1981. One more drilling platform has been added to the eastern group of platforms giving a new total of 9).

than others to such hazards. The impact of any geologic hazard is magnified, either in terms of economic loss or lives lost, if the hazard occurs in a densely populated area.

In the United States, the major geologic hazards with which we must contend are earthquakes, floods (Pettyjohn, 1980), ground failures, and volcanic eruptions. Hays (1981) discusses the importance of these various hazards in a U.S. Geological Survey Professional Paper, "Facing Geologic and Hydrologic Hazards: Earth-Science Considerations." The U.S. Geological Survey is charged with the national responsibility for "Warning and Preparedness for Geologic-Related Hazards" (Federal Register, 1977). Such hazards, worthy of public notification through established procedures, include the following: earthquakes, volcanic eruptions, landslides, glacier-related phenomena, and subsidence.

In this section, case histories in which remotely sensed data played an important part are presented with respect to five different geologic hazards or variations of these hazards: seismic, geomorphic, volcanic, glaciologic, and man-induced. Man-induced geologic hazards include culm-bank fires, dam failures, and coal-mine collapse.

Nearly all geologic hazards are amenable to study or to monitoring by remotely sensed data, because nearly all geologic phenomena that create a hazardous situation to man or his works leave a telltale sign in the geologic record of a previous occurrence in the area. Stereoscopic aerial photographs are especially valuable in geomorphological studies of an area. Recent faulting, landslides (Nilsen, 1973), floodplains, lava flows and tephra layers, glacial landforms, and solution (karst) features (Newton, 1976) all can be recognized on stereoscopic pairs of aerial photographs by a competent geologist trained in such analysis. Although remotely sensed data are a valuable tool in studies of geologic hazards, each study must be backed up by careful field investigations. Such data can only delineate areas of potential geologic hazards; in other words, they can show where, but not when, such hazards are likely to occur. For additional information on the use of remote sensing techniques in the study of geologic hazards, especially landslide hazards, please refer to Chapter 32, Engineering Applications, of this *Manual.*

Remotely sensed data can also be invaluable to scientists involved in documenting and assessing the post-event impact of a particular geologic hazard in a given area (see Color Figures 31-174 and 31-176). Most such post-event surveys are carried out with aircraft equipped with conventional aerial cameras. As satellite imagery becomes more readily available and of higher spatial resolution, it will be more widely used in post-event disaster assessment (Richards et al., 1983).

Remote Sensing in Assessments of Seismic Hazards

The use of remote sensing in studies of seismic hazards has so far been overwhelmingly dominated by vertical aerial photographs, especially stereopairs of aerial photographs, although side-looking airborne radar (SLAR), thermal infrared images (Wallace and Moxham, 1967), and other remote sensors have been helpful in some tectonic investigations (see Glass and Slemmons, 1978). Aerial photographs are crucial in two broad categories of seismic-hazard studies: assessment of the seismic hazards of a given region (discussed extensively by Glass and Slemmons, 1978), and postearthquake investigations.

Geologic studies of seismic hazard are aimed at the history of late Quaternary seismicity as expressed by active faults, landslides and other ground failures caused by shaking, ground deformation (tectonic uplift, subsidence, and warping), and, earthquake-generated waves along shorelines. Aerial photographs are the most important tool for detecting recently active faults and former ground failures in a given area, and may be helpful in discovering evidence of late Quaternary ground deformation and recent inundation by large waves.

Postearthquake investigations not only use all the techniques of geologic assessment of seismic hazards (to find places of likely renewed activity) but may also use postearthquake vertical aerial photographs to detect, record, and measure surface faulting (for example, Tchalenko and Ambrayses, 1970; Clark, 1972; Brown et al., 1973; Sharp, 1975, 1982), ground failure (for example, Plafker, 1968; McCulloch and Bonilla, 1970; Plafker et al., 1971), and, after some earthquakes, shoreline uplift or subsidence (Plafker, 1967, 1969) and the effects of large waves (Coulter and Migliaccio, 1966, Figures 6, 7; McCulloch, 1966). In addition, aerial photographs have been widely used after earthquakes for damage assessment (for example, Grantz et al., 1964).

Scale is an important consideration in both types of studies. Although assessments of seismic hazards may benefit from the broad view of major faults and other tectonic features as seen from Earth orbit (Allen, 1975; Tapponnier and Molnar, 1977) or high altitudes (Figure 31-167) (Clark, 1969, 1972), most work of hazard investigation uses a much larger working scale that fits the project's objectives. For example, Glass and Slemmons (1978) suggested 1:12,000 as a useful intermediate scale for general studies of active faults. The U.S. Geological Survey used this scale for most of a series of aerial photographs taken in 1966 and 1967 (see Hedel and Villalobos, 1979) for 1:24,000-scale strip maps showing recently active traces of major faults in California (for example, Ross, 1969; Vedder and Wallace, 1969), although some of these maps also successfully used existing (but not optimum) 1:20,000-scale U.S. Department of Agriculture aerial photographs (Clark, 1973, 1982). Detailed analysis of active faults commonly requires larger scales (Slemmons, 1977); for example, the intricate displaced topography of the Garlock fault is better shown at 1:5,000 scale (Color Figure 31-168.).

The particular scale required for postearthquake photographs depends strongly on the project's objectives and the size of the features studied. For surface ruptures, if the photographs merely serve as a base on which field parties map the location and schematic representations of individual cracks, the scale should probably be about 10–50 percent larger than that of the intended map. If, however, the photographs themselves must register the surface rupture, the scale must be large enough to resolve details of interest. For example, investigations of the 1966 Parkfield-Cholame, California, earthquake (M = 5.5; maximum right-lateral slip, about 70 mm) used 1:5,000-scale preearthquake photographs to record fractures and 1:1,200-scale postearthquake

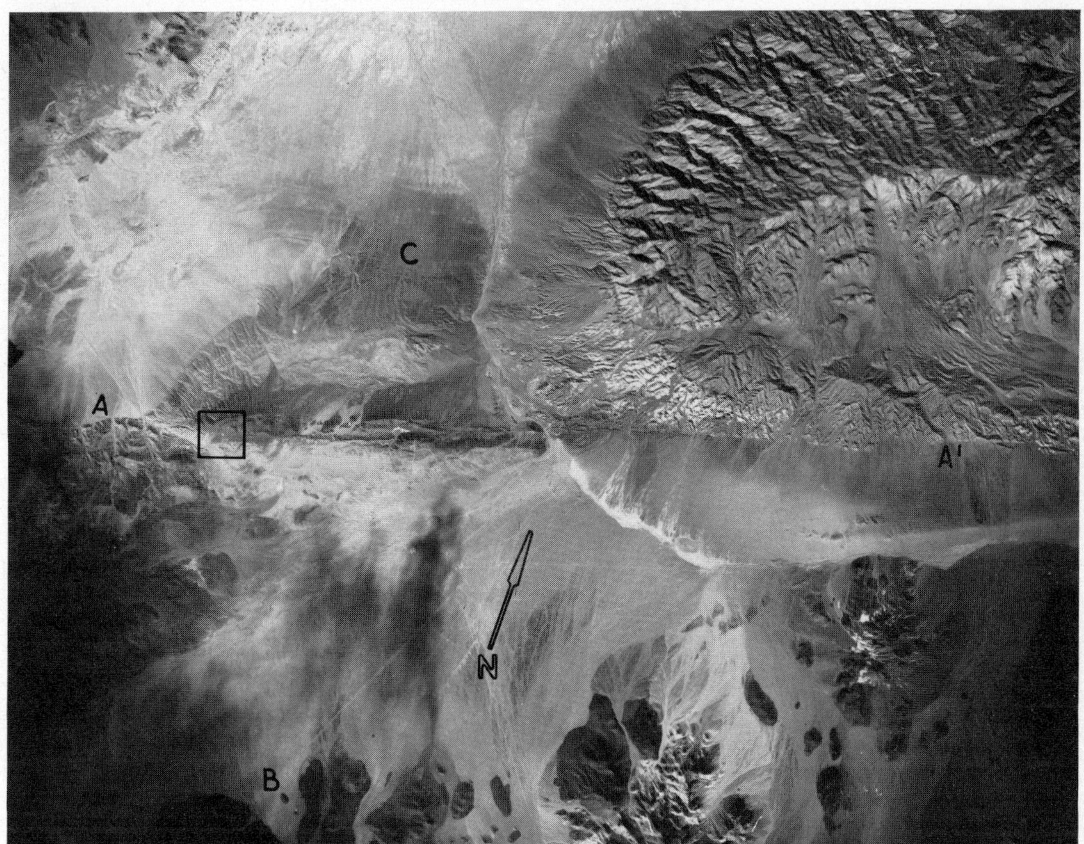

Fig. 31-167. Small-scale (1:120,000 on 9-by 9-in. negative), vertical aerial photograph, of the Garlock fault (*A-A'*) near Searles Valley, California. The linear nature of this strike-slip fault, expressed mainly by scarps, valleys, and ridges, shows clearly here as well as on 1:20,000-scale aerial photographs, which were the primary tool for making a 1:24,000-scale strip map of the most recently active trace (Clark, 1973). In places where the surface trace is missing for several miles, large-scale photographs are less useful than small-scale photographs, which reveal the continuity of the fault throughout its length. Cloud shadows partly obscure volcanic rocks near *B;* shorelines of Pleistocene Lake Searles appear at *C.* (U.S. Geological Survey aerial photograph No. GSAF 744V 067, 29 November, 1967: Sun elevation 34°; Sun azimuth, 186°). Square delineates the area shown in Color Figure 31-168.

photographs to detect cracks about 50 mm wide and a little more than 1 m long (Brown and Vedder, 1967). Field parties mapped ruptures of the 1968 Borrego Mountain, California, earthquake ($M = 6.4$; maximum strike-slip displacement, about 0.4 m across zones less than a few meters wide) on available (1953) 1:20,000-scale U.S. Department of Agriculture aerial photographs that showed most of the preearthquake scarps. Although postearthquake 1:20,000-scale aerial photographs showed few of the ruptures, 1:4,000-scale low-Sun-angle aerial photographs taken 1 year later (Figure 31-169.) showed some of the surviving 50- to 100-mm-wide tectonic cracks and many subsequent 0.5- to 2-m-wide erosional fissures that had formed along the rupture (Clark, 1972; Clark et al., 1972). A series of 1:10,000-scale aerial photographs taken the day of the San Fernando, California, earthquake of February 9, 1971, show almost none of the ground ruptures ($M = 6.4$; maximum left-lateral oblique slip, about 2.8 m, distributed over many cracks in a zone as much as 200 m wide (U.S. Geological Survey staff, 1971) but served as a base for investigators to plot fractures. These photographs also displayed earthquake-triggered landslides very well (Morton, 1971). Because of abundant vegetation and large shadows, even 1:2,400-scale low-Sun-angle aerial photographs taken 3 days after this earthquake failed to record many of the fractures that were less than 50 mm wide. A series of 1:8,000-scale low-Sun-angle aerial photographs of the dominantly strike-slip surface rupture of the Imperial Valley, California, earthquakes of 15 October 1979 ($M = 6.5$; maximum slip, about 0.8 m; Sharp, 1982; Sharp et al., 1982), clearly show many of the positions and details of the narrow rupture zone. For recording details of fracture geometry and displacement in complex rupture zones, 1:8,000 seems to be about the smallest practical scale for postearthquake studies; larger scales are commonly more valuable. Aerial photographs of smaller scale may be useful, however, particularly after earthquakes with large

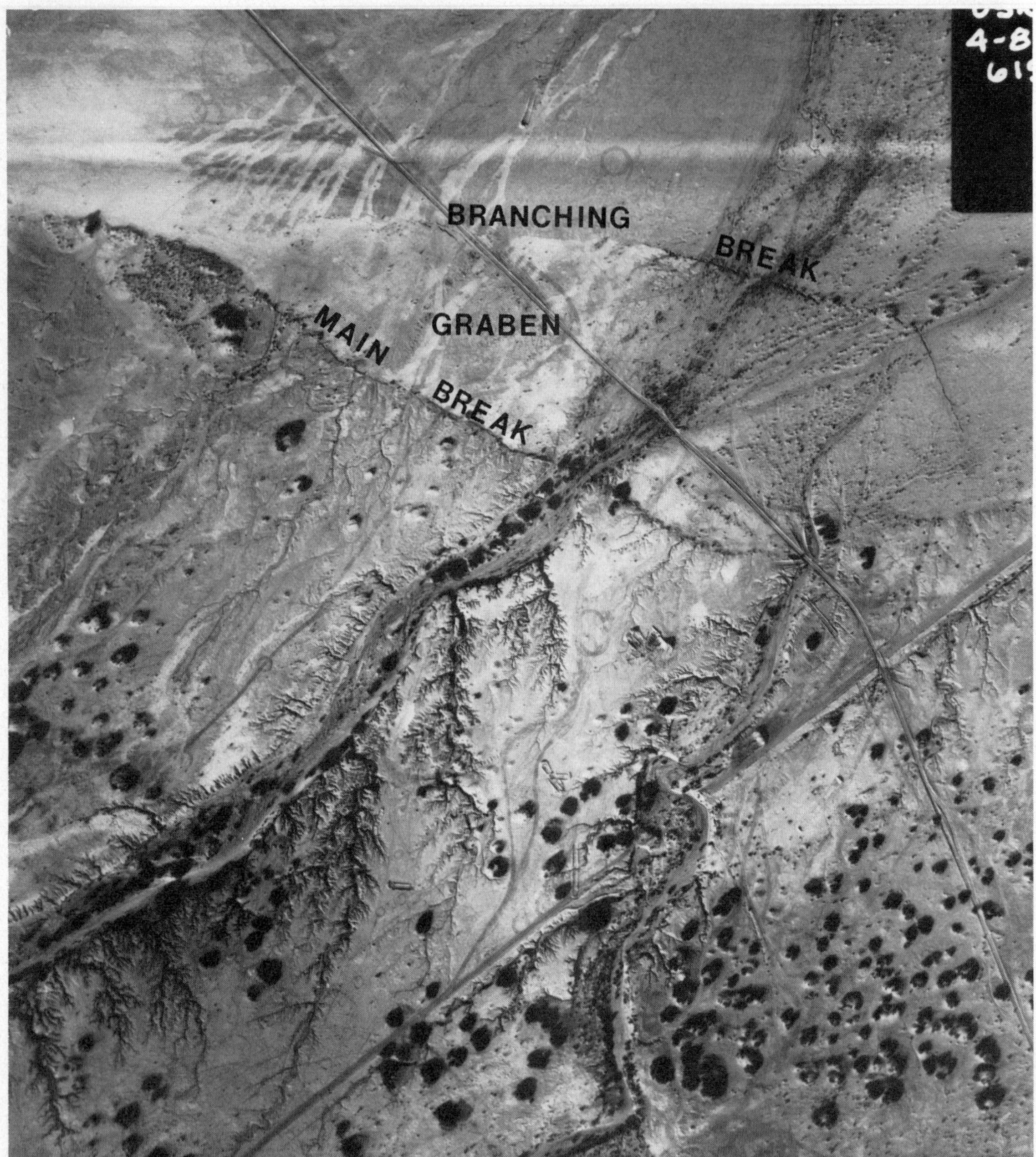

Fig. 31-169. 1:4,000-scale (9-by 9-in. enlargement), low-Sun-angle vertical aerial photograph of the main break and a branching break of the 1968 Borrego Mountain earthquake in southern California, taken 1 year after the earthquake. Maximum displacement in this area in 1968–69 was about 200 mm right lateral and 100 mm vertical (NE. side down) across the main break and about 50-mm vertical (SW. side down) across the branching break. Runoff from heavy rains a few months after the earthquake poured into tectonic fractures and caused spalling and foundering of blocks from walls to form 1-m-wide by 3-m-deep fissures visible here. The late Holocene scarp along the main break is 0.5 to 1.5-m high and is emphasized by shadowing, stream incision of the upthrown block, and local vegetation contrast across the scarp (see Clark et al., 1972). Low Sun angle and large scale permit identification and detailed study of the rupture, although horizontal offset in this earthquake was too small to measure on this aerial photograph. (U.S. Navy aerial photo no. USN 619, 8 April 1969, taken by the U.S. Navy Light Photographic Squadron 63, for the U.S. Geological Survey, on 5-in. roll film; Sun elevation, 15°; Sun azimuth, 268°).

displacement, as shown by a 1:25,000-scale vertical aerial photograph of a faulted glacier in Alaska (Figure 31-170A and B).

Low Sun angle is probably the most important requirement of aerial photographs used for fault and ground-failure studies, both preearthquake and postearthquake. The well-known and important benefit of low Sun angle for detecting faults was pointed out by Richter (1958) and described in more detail and used intensively by Slemmons (1969) and Cluff and Slemmons (1972). Slemmons (1977) and Glass and Slemmons (1978) stressed that the best emphasis of fault scarps comes from near-grazing illumination which just shadows the scarps from a direction normal to them. On some dark surfaces, such as those with desert varnish, illumination from the opposite direction makes the scarp appear as a bright line that is about as conspicuous as a shadowed scarp. Glass and Slemmons (1978) recommended Sun angles between 10° and 25° for most low-relief terrain, and higher angles in more hilly areas to avoid too many shadows. Clark (1971) and Threet (1982) described simple tables, charts, and methods to determine the date and time for specific solar azimuth and altitude at a given place.

Although some investigators have suggested that color film is unsuitable for low-Sun-angle aerial photographs (Sherard et al. 1974; Slemmons, 1977, Glass and Slemmons, 1978), excellent 1:5,000-scale color negative photographs (Color Figure 31-168) were taken of the entire Garlock fault in 1976 and 1977 (shown in black-and-white in Clark, 1978) with Sun angles as low as 10°. These photos successfully combine the benefits of both low Sun angle and color.

Geomorphic Hazards

The use of remotely sensed data for the identification and classification of surficial hazards such as landslides (Briggs et al., 1975; Briggs, 1977; and Pomeroy, 1982), land-subsidence features, and floods is generally confined to large-scale aerial photographs. Landslide identification and classification can be made by recognizing abnormal, subordinate surface forms, such as hummocks, lobes, head scarps, and streams that resurge as large springs along the toe of the slides. Most landslides are less than 150 m in length up slope and in width across the slide. The subordinate features seldom exceed 15 m in size. Because of the small size of the features used to identify and classify landslides, aerial photographs at a scale of less than 1:20,000 are of limited value. The characteristic valley form, however, can generally be identified on aerial photographs at scales as small as 1:80,000. Satellite photographs and other space images and airborne radar images are of little value in identifying landslides except for very large ones, such as the slide on the Gros Ventre River near Kelly, Wyoming. However, radar and

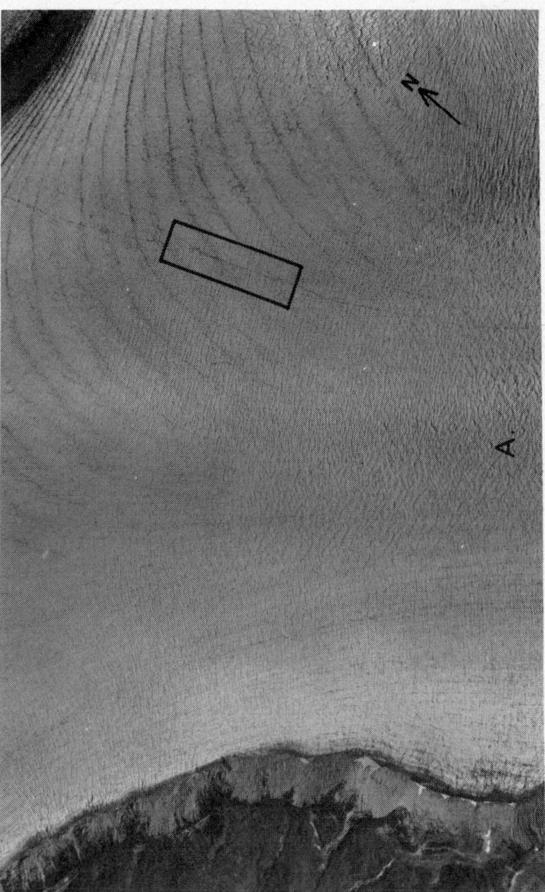

Fig. 31-170. A. Above: Part of a 1:25,000-scale (on 9-by 9-in. negative) aerial photograph taken 13 months after $M = 7.9$; Lituya Bay, Alaska, earthquake of 10 July 1958, showing surface ruptures along Fairweather fault across Grand Plateau Glacier. B. (Next page) shows an 8X enlargement along rupture from which 3.3 m of right offset of crevasses was measured. In addition to recording ruptures and displacement in an area where much of the surface break was obscured by water, ice, snow, and vegetation, this remarkable photograph demonstrates coupling between the glacier and its bed during rapid displacement that accompanied the earthquake and, by its record of subsequent distortion of the rupture, shows how sequential photographs of this rupture would record glacier flow. U.S. Coast and Geodetic Survey aerial photograph no. S-5168, 11 August 1959; Sun elevation, about 45°; Sun azimuth, about 150°. Photographs and interpretation courtesy of George Plafker, U.S. Geological Survey (written communication, 1982).

satellite images commonly show linear features that are guides to geologic structures that may be tied to the development of landslides. Forest cover greatly hampers the identification of old landslides and, therefore, the taking of aerial photographs at the time of minimum leaf-cover is a prime requirement for identification of slides. Successive coverage by aerial photographs made

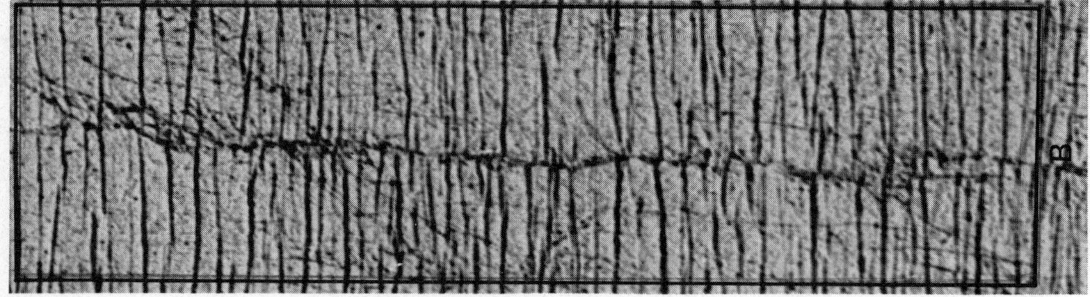

Fig. 31-170 B

over a period of many years may also be used as a basic method of identifying old slides that are later obliterated by revegetation, erosion, and cultivation of the land.

Natural subsidence can occur from solution of carbonate, sulfate, and halite rocks. Terrain undergoing solution is generally referred to as karst, and the subsidence features consist of sinkholes, karst valleys, shafts and open fissures. These features are readily identifiable on aerial photographs from their shapes. Karst areas with sinkholes larger than about 200 m can be identified from Landsat images but with little detail. Because sinkholes generally have bottoms formed of soil or rotten rock spanning large underground openings, they are subject to collapse from fluctuations in groundwater levels. Therefore, a successive series of aerial photographs, covering many years, can give some information on the development of the sinkhole. Subsidence can also result from sediment compaction following ground water and/or hydrocarbon withdrawal (See Figure 31-35), liquefaction of soils, and crustal movements.

Remote sensing has been applied to floods to assess damage, determine duration, and movement of various aspects of floods such as peaks and fronts, and to delineate flood plains and structures on them (Figure 31-171). Satellite images, with repetitive passes, can be used for determining water movement where the flood is of long duration (Deutsch, 1976; Morrison and White, 1976; Berg et al., 1981; Kalensky et al., 1981; Kruus et al., 1981; Lowry et al., 1981; McAdams, 1981), but in most cases floods and flash floods occur over such a short time interval that satellite images are of limited use because the repeat period of the Landsat satellite is greater than the duration of the floods. Satellite data have been successfully used to map maximum flood inundation (Deutsch, 1976).

Remote Sensing of Volcanic Hazards

Introduction

Volcanic eruptions produce a variety of effects in the immediate vicinity of the activity but can also affect areas thousands of kilometers away (tephra falls) (see Figures 31-172, 31-173, 31-175, and 31-177). The major hazards to human life and property include hot avalanches, hot particle and gas clouds (nuée ardentes), mudflows, tephra (ash) falls, lava flows, volcanic gases, mudflows and floods (including tsunamis) (Federal Register, 1977; Sheets and Grayson, 1979; Warrick, 1979).

In this part of the section on geologic hazards, the topic "remote sensing of volcanic hazards" is only used in a restricted sense, with major emphasis given to analysis of data acquired by aircraft and satellite imaging systems or by ground imaging sensors (Cassinis and Lechi, 1974). A complete discussion would necessarily include seismic sensors (Civella et al., 1974), infrasonic arrays, and the use of satellites or other relay systems to transmit thermal data (Friedman and Frank, 1978), seismic signals (Ward and Eaton, 1976), data from tiltmeter arrays or geochemical analyses of volcanic gases from data collection platforms to a central receiving facility for analysis.

Potential volcanic hazards in a given area can usually be readily recognized by an historic (or prehistoric) record of volcanic activity (Crandell and Mullineaux, 1975; Crandell et al., 1979), by the distinctive nature of volcanic landforms (Williams et al., 1981a), or by the presence of geothermal activity in the summit crater (Del Bono et al., 1971; Cassinis and Lechi, 1974; Frank et al., 1977), on the flanks (Friedman et al., 1969a), or in close proximity to volcanoes (Friedman et al., 1972). Before, during, and following volcanic activity within a populated area, oblique aerial photographs (Krimmel and Post, 1982) (Figure 31-172), terrestrial photographs (Voight, 1982), vertical aerial photographs (Sobieralski, 1968; Moore and Albee, 1982; Jordan and Kieffer, 1982), satellite images (Friedman and Williams, 1968; Williams and Friedman, 1970a; Rice, 1981), satellite photographs (Friedman and Heiken, 1977; Simkin and Krueger, 1977), thermal infrared images (aerial thermographs or thermograms; Williams, 1972b), radar imagery (see Figure 31-15), and maps made from image data (including sequential topographic maps) have been effectively used to provide a permanent record or to

Fig. 31-171. Oblique aerial photograph of the Wilkes Barre, Pennsylvania flood caused by the torrential rainfall associated with the passage of Hurricane Agnes in June 1972. U.S. Geological Survey photograph. (See Flippo and Lenfest, 1973).

monitor a specific event or a sequence of events. For example, satellite images can be effectively used to record the areal extent of tephra deposition and lava flows (Lefebvre, 1975) from volcanic eruptions (See Color Figure 31-176). Several of the studies included in U.S. Geological Survey Professional Paper 1250, "The 1980 Eruptions of Mount St. Helens, Washington" (Lipman and Mullineaux, 1982) depended upon remotely sensed data. One such study, "Thermal Infrared Observations of the Eruptions of Mount St. Helens, Washington, 1980 and 1981" is included in this section. The National Aeronautics and Space Administration also acquired high-altitude color-infrared aerial photographs with a U-2 aircraft before (1 May 1980; No. 580002875; frame 5332) and after (19 June 1980, No. 580002891, Frame 0038) the catastrophic 18 May 1980 eruption of Mount St. Helens. These photographs are available from the EROS Data Center for comparative analysis. The other case history presented in this section is "Remote Sensing of Eldfell Volcano, Heimaey, Vestmannaeyjar, Iceland," where sequential aerial photographs played an important role in determining where spraying of sea water should be concentrated in an attempt to stem or stop the flow of lava into the fishing port of Vestmannaeyjar, Iceland (Williams and Moore, 1976a, 1976b).

Beginning in the mid-1960's, satellite and aerial thermography were used extensively to survey active or dormant volcanoes in many parts of the world (Moxham, 1972). Gawarecki et al. (1965) were the first to use satellite thermography in studies of active volcanoes (Hawaii). Friedman et al., 1967; Williams et al., 1968; Merifield et al., 1969b; and Williams and Friedman, 1970a, used the Nimbus II high resolution infrared radiometer (HRIR) to record energy from the renewed effusive volcanic activity on Surtsey, Iceland in 1966. Aerial thermographic surveys of Surtsey were also carried out in late August 1966 (Williams et al., 1967; Williams et al., 1968; Friedman and Williams, 1968) and in August 1968 (Friedman and Williams, 1970a, 1970b). The combination of aerial and satellite thermographic surveys of Surtsey enabled quantitative measurements to be made of the volcanic thermal emission (Williams and

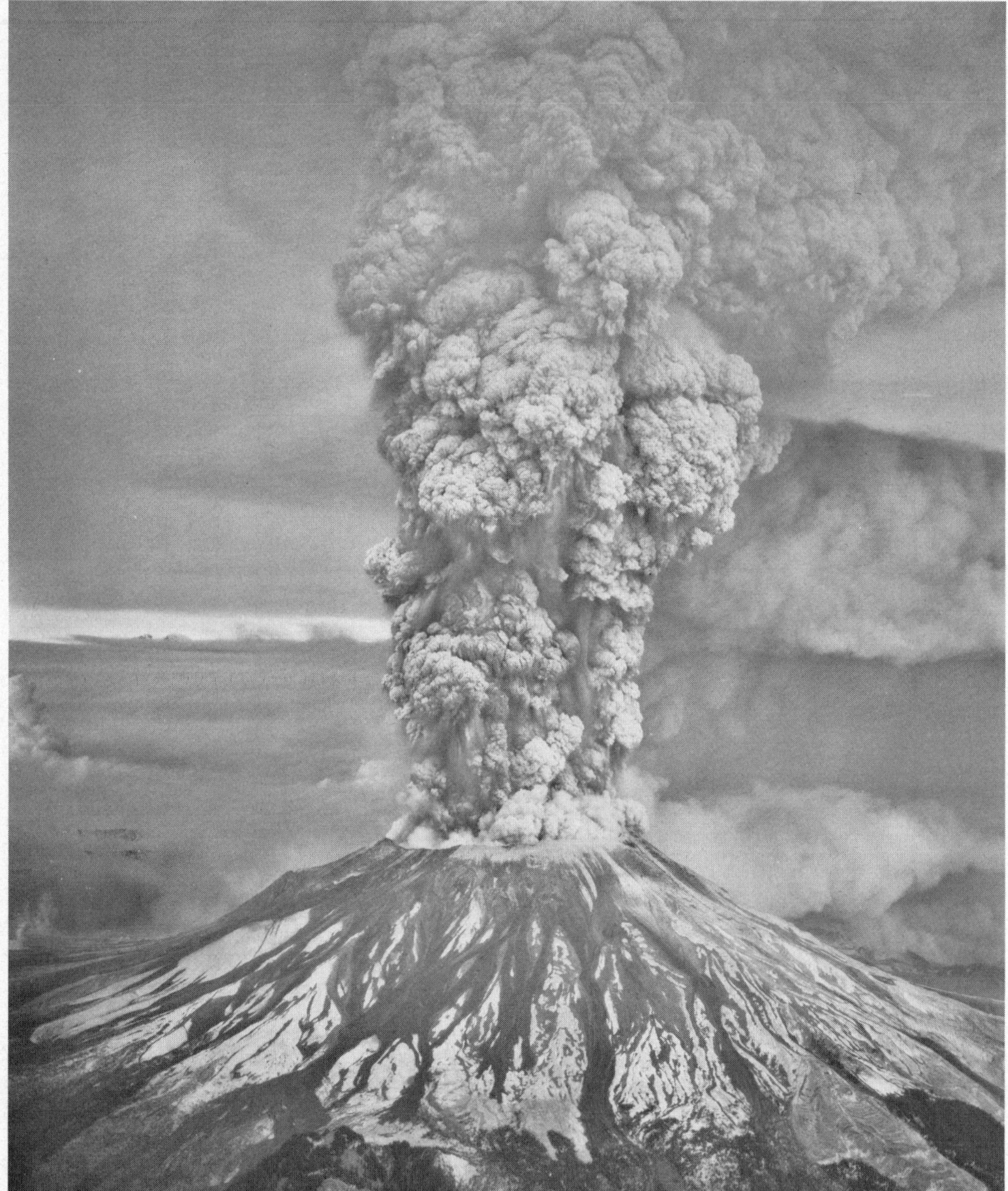

Fig. 31-172. Oblique aerial photograph of Mount St. Helens, Washington, at about noon on 18 May 1980. The billowing eruption plume of volcanic ash (tephra) and other volatiles emanates from the 1.5-km wide summit crater and extends well into the stratosphere. U.S. Geological Survey photograph No. 80-S3-137 by Robert M. Krimmel.

Friedman, 1970a). Additional aerial thermographic surveys were made of Hekla volcano (Friedman et al., 1969c; Friedman et al., 1970) and of other volcanoes in Iceland (Friedman and Williams, 1968; Friedman et al., 1969a).

For a review of airborne and satellite thermal infrared surveys of volcanoes see the proceedings volume of the "Workshop on Geological Applications of Thermal Infrared Remote Sensing Techniques" (Settle, 1981), especially the review paper by Williams (1981), and an earlier review paper (Williams, 1972a).

Landsat images have also been used to document volcanic activity, but only in a time-lapse way because of the 18-day repeat cycle for Landsats 1, 2, and 3. Heiken and Pitts (1975) carried out studies of Sakura-Zima Volcano, Japan, and Stromboli Volcano, Italy. Short et al. (1976) discussed Tya Tya Volcano in the Kurile Islands, U.S.S.R. Williams et al. (1974) discussed the eruption of Eldfell Volcano on Heimaey, Vestmannaeyjar, Iceland. Volcanic emanations can be discerned from the summit of Mount Erebus, Antarctica; an effusive volcanic eruption on Fernandina Island, Galapágos Islands, Ecuador, is indicated on MSS band 7 as a thin bright (incandescent) line leading downslope. Table 31-37 lists the known Landsat images which have documented volcanic activity.

Another hazard related to volcanic hazards is the impact that a large explosive volcanic eruption can have on the Earth's climate. Several volcanic eruptions during the past three decades have ejected volcanic gases and tephra into the stratosphere (Cronin and Williams, 1970) (Figure 31-172). The 1982 series of eruptions from El Chichón Volcano, Mexico, caused death and destruction in the populated area around the volcano (Aldana E. and Garrett, 1982; Weintraub, 1982), but a farther reaching impact may result from the effect on the planet's climate because of the enormous ejection of volcanic material into the stratosphere (Tilling, 1982). Polar-orbiting weather satellites (NOAA series) have already recorded a decrease in the amount of solar energy reaching the Earth's surface; the expansion of the initial eruption cloud was recorded by successive images from a geostationary weather satellite (GOES series).

Remote Sensing of Eldfell Volcano, Heimaey, Vestmannaeyjar, Iceland

Iceland, a small (103,000 km^2) island republic in the North Atlantic, has an abundance of natural hazards for such a small area, and these natural hazards can be grouped into six categories: geologic, hydrologic, glaciologic, meteorologic, oceanographic, and biologic. The most important class is geologic, and it can be further subdivided into geomorphic, geothermic, seismic, and volcanic (Williams, 1980). According to Thorarinsson (1977), disasters during the historical period since Iceland's settlement (about 874 A.D.) have been caused by four different types of volcanic activity: deposition of tephra, damage to plants or animals by noxious gases, flooding caused by jökulhlaups (glacier outburst floods), and inundation by lava flows. Figure 31-173 is modified from Thorarinsson (1977) and shows the relationship between the volcanic hazards of Iceland and the areal extent of damage.

The Lakagígar eruption of 1783 was the largest effusive volcanic eruption recorded anywhere in the world in historic time (Thorarinsson, 1970). Thorarinsson (1968), with the aid of aerial photographs, compiled a geologic map of the Laki lava flows and confirmed Thoroddsen's original estimate of 565 km^2 as the total area covered with lava. The Lakagígar eruption, however, occurred in a lightly populated rural area. Thorarinsson noted that the lava flows covered 2 churches and

Table 31-37

Selected Landsat images upon which volcanic activity (venting, tephra falls, lava flows) is recorded

Volcano	Geographic Location	Landsat Image No.	Date	Activity
Stromboli	Lipari Islands, Italy	1070-09113	10 Oct 1972	venting
Sakura-Zima	Kyushu Island, Japan	1132-01242	2 Dec 1972	venting
Mount Erebus	Ross Island, Antarctica	1174-19433	13 Jan 1973	venting
Eldfell	Heimaey, Iceland	1229-12151	9 Mar 1973	lava flows, venting
Tya Tya	Kurile Islands, U.S.S.R.	1358-00341	16 Jul 1973	venting
Hekla	Southwestern Iceland	1392-12191	19 Aug 1973	tephra fall
Mount Nyiragongo	Zaire	21111-07053	6 Feb 1978	lava flows
Fernandina	Galápagos Islands, Ecuador	30624-15340	19 Nov 1979	lava flows
Mount St. Helens	Washington, U.S.A.	21999-18140	13 Jul 1980	blast impact
Hekla	Southwestern Iceland	22044-12072	27 Aug 1980	tephra fall
Mount St. Helens	Washington, U.S.A.	30942-18031-A	2 Oct 1980	blast impact
Krafla	Northern Iceland	22223-12000	22 Feb 1981	lava flows
Krafla	Northern Iceland	22278-12050	18 Apr 1981	lava flows
Krafla	Northern Iceland	22278-12053	18 Apr 1981	lava flows
Hekla	Southwestern Iceland	22278-12055	18 Apr 1981	lava flows and tephra fall
Krafla	Northern Iceland	31184-11570	1 Jun 1981	lava flows
Krafla	Northern Iceland	31185-12022	2 Jun 1981	lava flows
Krafla	Northern Iceland	31185-12025	2 Jun 1981	lava flows
Hekla	Southwestern Iceland	31186-12090	3 Jun 1981	venting

THE PRINCIPAL VOLCANIC HAZARDS OF ICELAND AND THE AREAL EXTENT OF DAMAGE
(Modified from Thorarinsson, 1977)

AREAL IMPACT	TYPE OF ERUPTION	LAVA FLOWS	TEPHRA FALLS	VOLCANIC GASES	MAAR FORMATION	JÖKULHLAUP	FLOODS
LOCAL < 25 KM	EFFUSIVE	●	●	●	▲	■	■
	MIXED	●	●	●	▲	●	■
	EXPLOSIVE	▲	●	●	■	●	■
REGIONAL > 25 KM	EFFUSIVE	■	■	■	▲	■	■
	MIXED	▲	●	■	▲	●	■
	EXPLOSIVE	▲	●	▲	■	●	■
GLOBAL	EFFUSIVE	▲	▲	■	▲		
	MIXED	▲	▲	▲	▲		
	EXPLOSIVE	▲	■	▲	▲		

● Occurs exclusively or often
■ Rarely occurs
▲ Never occurs

Fig. 31-173. The principal volcanic hazards of Iceland and the areal extent of damage. (Modified from Thorarinsson, 1977.) Two types of floods, jökulhlaup and "other floods," are shown. A jökulhlaup is a glacier outburst flood caused by subglacial geothermal or volcanic activity or by the failure of ice-dammed lakes. The "other flood" category refers to floods caused by volcanic debris (tephra) or laval disrupting or damming of local or regional streams or rivers.

14 farms, and that 30 more farms were badly damaged.

By comparison, in the 1973 Eldfell eruption on Heimaey, 10 times as many buildings were destroyed as were destroyed and damaged by Laki's lava flows, yet the area covered by lava flows was only 0.4% the area of the 1783 lava flows from Lakagígar.

The 23 January–July 1973 eruption of Eldfell on the island of Heimaey, in the Vestmannaeyjar volcanic archipelago, off the southwestern coast of Iceland did considerable damage to the modern fishing port of Vestmannaeyjar (Grove, 1973) (Color Figure 31-174). The initial stage of the eruption, from 23 January to early February 1973, was partly explosive and yielded a large volume of tephra (volcanic ash), some of which was deposited on Vestmannaeyjar (Figure 31-175A) (Thorarinsson et al., 1973). The later stages of the eruption from Eldfell were primarily effusive, and lava flows eventually destroyed over 300 homes on the east side of the town (Figure 31-175B). The last major movement of lava flows to the west ended against the walls of two fish-processing plants (Figure 31-175C; see also Color Figure 31-174).

Pre- and post-eruption Landsat images show the increase in area of the island of Heimaey to the east (about 2.5 km²) (Color Figure 31-176). Nevertheless, low-altitude aerial photographs were the most important type of remotely sensed data used to document the appearance of Heimaey before the eruption, at various times during the eruption, and during recovery operations after the eruption (Figure 31-177).

(A)

(B)

(C)

Fig. 31-175. Three ground photographs, taken on 23 July 1973, showing the impact of the Edfell 1973 eruption on Vestmannaeyjar. *A.* View looking east across homes nearly buried in tephra in the eastern part of the town. Most of the tephra fell from late January to early February 1973. A lava flow, artificially cooled by sea water, looms in the background. *B.*

During the effusive (lava flow) phase of the eruption, Icelandic authorities decided, after consultation with Icelandic scientists, that sea water should be sprayed on the advancing lava flows to stem or to stop the flow of lava into the eastern part of town (Williams and Moore, 1973, 1976a, 1976b; and Williams, 1976c). This lava-cooling operation became a large undertaking (Sigurgeirsson, 1974; Jónsson and Matthíasson, 1974), and over 6 million m^3 of water was eventually pumped onto the lava flows. Scientists and planning authorities benefited from detailed topographic maps rapidly produced from vertical aerial photographs taken of the lava flows at frequent intervals. These maps, supplemented by geodetic surveys, permitted measurement of the rate of lava movement, optimum location of lava diversion barriers (Williams and Moore, 1976b), and determination of the places at which to concentrate the lava-cooling operations (Jónsson and Matthíasson, 1974). Vertical aerial photographs similar to Figure 31-177 were also critical in the planning of administrative and scientific programs.

About a month after effusive volcanic activity had ceased, an aerial thermograph was acquired of Heimaey. Figure 31-178 shows the extent of thermal emission from the still-cooling lava flows on the eastern part of Heimaey. Heat from these hot lava flows is still being extracted and used for space heating as of this writing (December 1982).

Thermal Infrared Observations of the Eruptions of Mount St. Helens, Washington, 1980 and 1981

Thermal infrared observations of Mount St. Helens by the U.S. Geological Survey were carried out between 1966 and 1975 (Friedman and Frank, 1978); the historic thermal areas on the contact margins of the pre-1980 summit lava dome were observed during these aerial surveys, and detailed surface investigations determined their heat flow to the surface. Infrared aerial surveys were reinstituted on 30 March 1980, 3 days after the first 1980 Mount St. Helens eruption. The following case history is taken from Kieffer, Frank and Friedman (1982) and Friedman (1982). The objectives of these observations were geologic (volcanic) hazards prediction, identification of locations of enhanced heat flow that might possibly precede a flank eruption, and quantitative measurements of the thermal changes associated with the sequence of eruptive events.

View looking north at a house engulfed by a lava flow in the eastern part of town. Heimaklettur, a Pleistocene palagonite ridge, appears in the background (See Color Figure 31-174 and Figure 31-177). *C*. View looking southeast from dock area in the northern part of town at edge of fish-processing plants (See Color Figure 31-174). Photographs courtesy of Richard S. Williams, Jr., U.S. Geological Survey.

Aircraft-based instruments used included: uncalibrated film-recording scanners, moderate and high resolution video-recording systems, handheld imaging systems and radiometers, and calibrated digitally recording scanners. The observations of 30 March showed anomalous heat in the summit crater, locally along the southern bounding fault of the newly developed summit graben, in two large fractures in the region of historic thermal emission on the upper north slope that later became the bulge, and at the other historic thermal area high on the southwest slope. In the area of the bulge, infrared anomalies increased in abundance from early April until just prior to the 18 May eruption, when the upper part of the bulge appeared to be perforated by heat leaks of a few to 100 m lateral extent. All these areas of excess thermal emission were obliterated by the 18 May eruption. During periods between eruptions, excess thermal radiation from the summit of Mount St. Helens on 16 May was approximately 3 MW (megawatts). The traces of the first two landslide failure surfaces on 18 May were through clusters of thermal anomalies.

Qualitative infrared images in the 8 to 11.5 μm and 8 to 14 μm spectral regions were obtained over Mount St. Helens by forward-looking and vertical-mount aerial scanning systems on 31 May, 3, 6–8, and 19 June, and on 15 July 1980. Quantitative or calibrated images were obtained on 11–13, 19, and 20 August (Color Figure 31-179 A and B and Figure 31-180 A and B. See also Table 31-38).

Night and predawn infrared images, obtained during times of diurnal surface-temperature minima, depict the spatial patterns of high thermal emission associated with the crater and vent area on 7 and 8 June. Following the eruption of 12 June, the emergence of a dacite dome was confirmed by radar images on 13 June and subsequently on 15 June by visual observation. The infrared images of 19 June show a concentric annular distribution of thermal emission associated with the emergent dome. On 19 June the emergent dome, the rampart, and a southeast-striking fracture, controlling alinement of fumaroles within the crater floor area, were studied in detail, using three different scanning systems: an RS-14A scanner, a Daedalus scanner and a FLIR nutational scanner, all sensitive to emitted radiation in the 8–14 μm band. A large, relatively warm ring appeared in the images of 19 June southeast of the dacite dome; it was about equal in size to the dome and was outlined by a ring of fumaroles. Infrared images of 15 July show the annular and radial fracture pattern of the dome prior to its destruction on 22 July. An *en echelon* set of northwest-striking fractures in the crater and amphitheater region was also clear on the 15 July image. These fractures were related to the location of the first and subsequent lava domes and at least two of three smaller hot spots. The calibrated surveys of 13–17 August gave the temper-

Fig. 31-177. Stereopairs of vertical aerial photographs of Vestmannaeyjar, Heimaey, Iceland, on 31 August 1976 (three years after cessation of volcanic activity) showing position of lava flows on the eastern part of the town and exhumation of buildings previously covered by tephra (see Color Figure 31-174 and Figure 31-175A) but not engulfed by lava flows (Figure 31-175B). The lava flows previously resting against the two fish-processing plant buildings (Color Figure 31-174 and Figure 31-175C) have been removed. Aerial photographs (E-Mynd, Frames 889 and 890) acquired by the Iceland Geodetic Survey from an altitude of about 2,600 m with a Wild-Heerbrugg RC-10 aerial camera (152.10-mm focal length).

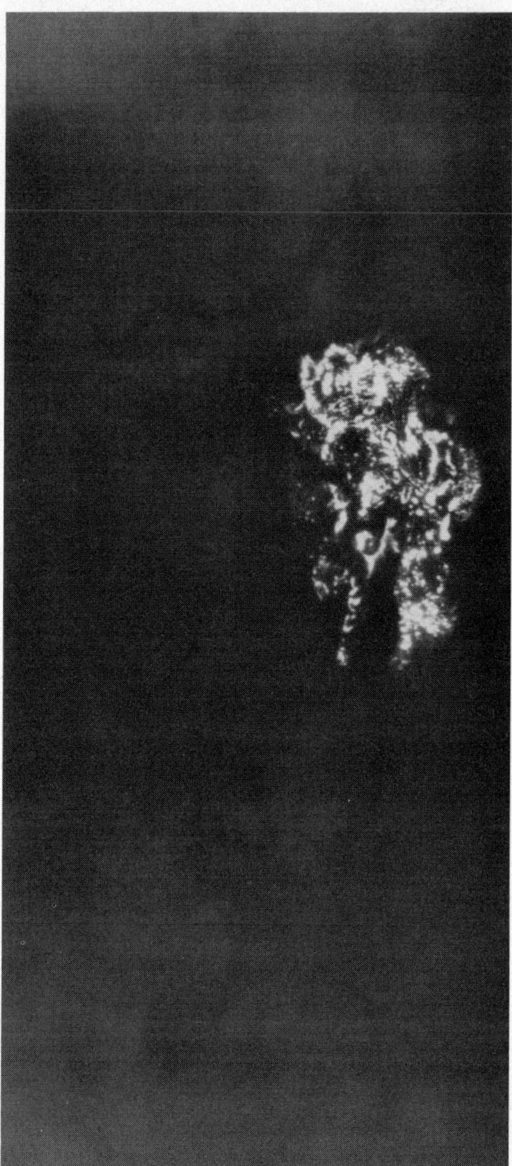

Fig. 31-178. Vertical aerial thermograph of the island of Heimaey, Vestmannaeyjar, Iceland, acquired on 20 August 1973 at 2233 hr universal time (Icelandic standard time) from an altitude of 3,050 m. The thermograph shows the complex pattern of thermal emission from joints and fissures in the cooling lava flows. Compare with Color Figure 31-174, a color-infrared aerial photograph acquired three days earlier. (NASA Mission 253, Flight 20, Line 1, Strip position 38-00-77.)

ature of the partly cooled rind of the August dome and the day-night temperature differences of an array of pyroclastic-flow deposits, as well as the temperature of Spirit Lake, several secondary phreatic fumaroles, and surrounding terrain.

The heat content of the dacite dome of June 1980 was estimated from laboratory measurements of thermophysical properties, supplemented by published data on latent heat of fusion, and on infrared radiance. A crystallization temperature of 970–990° C was inferred from melting experiments and Fe-Ti oxide geothermometry (Melson and Hopson, 1982). The volume of the visible dome was estimated to be 4.6×10^6 m^3 from vertical aerial photographs. The thermal energy of the visible dacite dome was 1.5×10^{16} J (Joule), calculated from data on volume, crystallization temperature, bulk density (2.2 g/cm^3), specific heat (1.059 J/g/K), latent heat of fusion (350 J/g), which yielded a total heat content of 1,385 J/g for solid dacite and 140 J/g for associated volatiles. Including inferred subsurface dome material above the conduit, the total energy yield was approximately 3.4×10^{16} J. Eruption of the dome was comparable to an eruption of intermediate magnitude, intensity III+ on the Tsuya scale, or intensity IV+ on the Hédervári scale. The total thermal energy released at Mount St. Helens because of the May and October 1980 eruptions was estimated conservatively at 1.2×10^{18} J, about 93 percent of the total energy released during this period.

Additions via domal growth to the earlier 1980 volcanic eruption energy brought the cumulative energy yield from 18 May 1980 through May 1982 to about 1.4×10^{25} ergs. Volcanic power was expended from 12 June 1980 through May 1982 at a rate of 2.3×10^3 megawatts, assuming steady-state eruption conditions. This represents a decline from the average power estimate for the period 18 May 1980 (excluding the explosion energy of the Plinian eruption) through 13 August 1980 (which was 4 to 10×10^3 megawatts). The gradual change in eruption style from a Plinian explosive eruption to generally nonexplosive dome growth may thus correspond to the reduction in volcanic power.

Remote Sensing of Glaciologic Hazards

Glaciologic hazards can be placed into four categories: (1) slow (advance or recession) and fast (surge) movement, (2) jökulhlaups (glacier-outburst floods) (3) iceberg discharge, and (4) rapid ice breakoffs (for example, the Mattmark disaster in Switzerland). Generally speaking, the normal advance or retreat of a glacier is relatively slow (a few meters to hundreds of meters per year). Most glaciers in temperate latitudes have been in a general retreat since the late 1880's, although many have reached a stable position and some have begun to readvance during the 1960s and 1970s. Although a gradually advancing glacier may eventually overrun various works of man, the main danger, in both slow advance and retreat, is in the damming of lakes in tributary valleys by advance, or the releasing of the dam because of retreat. Such lakes may rapidly drain, causing widespread flooding downstream (Federal Register, 1977).

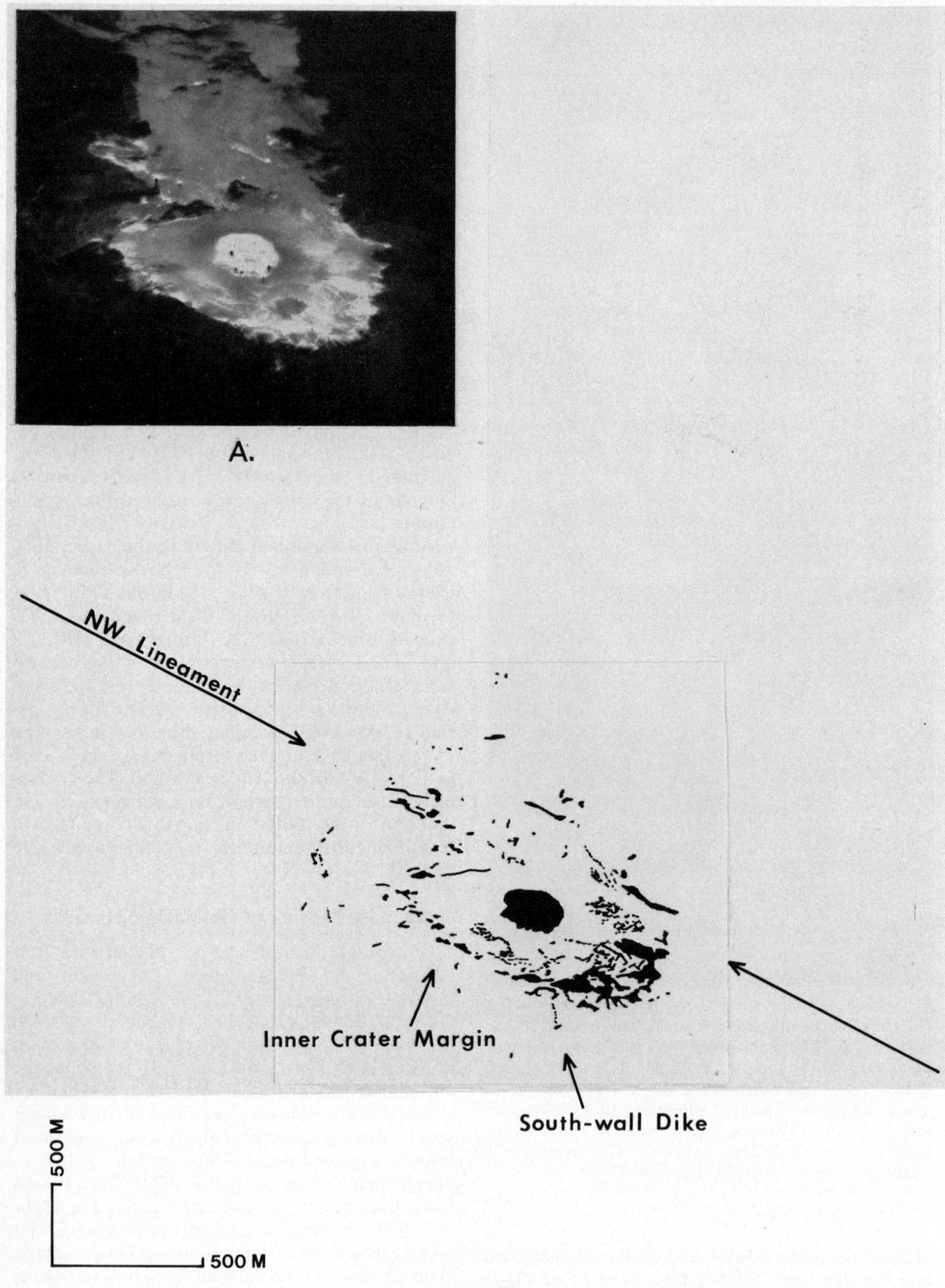

Fig. 31-180. Daedalus multispectral vertical scanner image courtesy of EG&G. 20 August 1980, 0330 PDT; 8–14 μm region. Hottest areas appear white on *A* and black on *B*. (From Figure 174 in Friedman et al., 1982b.)

Table 31-38

Differential radiant exitance in W/m² for three color-coded temperature levels at Mount St. Helens, 13 August, 1948–1951 PDT (From Table 27 in Friedman et al., 1982). (Refers to Figure 31-179B)

Color code for Color Figure 31-179	°C	°K	(W/m²)	$Wd\lambda_1 - Wd\lambda_2$ (279K = 0.437 × 10² W/m²)	W/m²
Yellow	32.7–36.3	306 310	5.116 × 10²	4.643 × 10²	464
Orange	39.9–43.5	313–317	5.600 × 10²	5.127 × 10²	513
Red	55.7–91.4	319–365	8.242 × 10²	7.769 × 10²	777
Incandescent (proper point)	525	798	4.321 × 10³	4.274 × 10³	4,274
Lava	850	1,123	2.967 × 10⁴	2.962 × 10⁴	29,620

The slow retreat of a tidal glacier may suddenly accelerate if it retreats from an end moraine in shallow water, resulting in a rapid disintegration of the lower reach of the glacier. Such a situation is expected to occur at Columbia Glacier, Alaska (Figure 31-181), where the rapid breakup of its terminus will likely discharge a large volume of icebergs into Prince William Sound (Meier et al., 1980). If current, wind, and tidal conditions permit, many of these icebergs will drift into the main shipping channel from Valdez, the southern terminus of the trans-Alaska oil pipeline, and perhaps pose a hazard to the giant oil tankers that frequently transit Prince William Sound (Post, 1977).

The Columbia Glacier has been the subject of an intensive scientific study by the U.S. Geological Survey. The research began in 1974 (Post, 1975), and during the course of the study various types of remotely sensed data have been used, including periodic low-altitude oblique aerial photographs, high- and low-altitude vertical aerial photographs, water-depth determination with hydrographic soundings, Mini-Sparker and Lister Boomer surveys to determine submarine morphology of the terminal-moraine shoal, and airborne and surface radio-echosounding of glacier thickness (Meier et al., 1978). Iceberg plumes have also been mapped from specially enhanced Landsat images.

During the period October 1976 and September 1977, oblique and vertical aerial photographs were acquired of 20 other tidal glaciers in Alaska, in addition to the Columbia Glacier. The aerial photographic data were used to determine rate of flow and to map changes in the termini of these calving glaciers. Because of the heavily crevassed character of the lower reach of the Columbia Glacier (Figure 31-181) and the impossibility of conducting conventional ground surveys, sequential aerial photographs were acquired at intervals ranging from about 6 weeks to 3 months from July 1976 to the present (1982) from an altitude of 5,500 m. Surface ice velocity vectors (Figure 31-182), speed of flow along the centerline of the glacier as a function of time (Figure 31-183), surface deformation tensors, and surface height changes in meters were photogrammetrically determined from those sequential aerial photographs (Meier et al., 1978, 1980; Sikonia and Post, 1979; Rasmussen and Meier, 1982; Sikonia, 1982; Fountain, 1982; and Meier, 1979). These data were then used to develop and calibrate numerical models of glacier flow leading to predictions.

A surging glacier is one in which the movement of the terminus suddenly increases to several times its normal flow, resulting in a sudden advance of several kilometers or more over a period of a few months (Meier and Post, 1969). Such sudden advances may engulf portions of highways, railroads, or other manmade structures and dam adjacent valleys, producing ice-dammed lakes (Krimmel et al., 1976; Meier, 1976; and Post et al., 1976). The failure of ice-dammed lakes can cause serious flooding downstream. Post (1969) reviewed the distribution of surging glaciers in western North America (see also Meier, 1976; Post et al., 1976). Thorarinsson (1964, 1969a) has similarly discussed those glaciers of Iceland which are known to have surged historically.

Glacier advance has been noted on a number of sequential Landsat images of Iceland during the period 1972–1982 (Williams, 1983b). Eyjabakkajökull, an outlet glacier in the northeastern part of Vatnajökull, Iceland (see Figure 31-205 and the section of this chapter on "Glaciology,") began to surge in late August 1972, and had already surged about 1 km by the time of the acquisition of the first Landsat image of the area on 14 October 1972 (1083–12023) (Williams et al., 1974). A 22 September 1973 image (1426–12070) showed additional movement of 1.8 km (Color Figure 31-201). Figure 31-184 is an oblique aerial photograph of Eyjabakkajökull as it appeared on 25 July 1973. Figure 31-185 is a Landsat 3 return beam vidicon (RBV) image (30157–11565-D) of Eyjabakkajökull on 9 August 1978. The Landsat 3 RBV image has nearly three times better resolution than the multispectral scanner (MSS) image and shows considerably more detail (Williams 1979b; Williams and Ferrigno, 1981).

Fig. 31-181. Oblique aerial photo of the terminus of the Columbia Glacier, Prince William Sound, Alaska, in July 1976. The glacier is located 40 km west-southwest of Valdez, Alaska, and is expected to begin a rapid retreat in the 1982–1984 period (Meier et al., 1980). By 1986 the glacier is expected to have receded 8 km, and the resulting discharge of large numbers of icebergs into shipping lanes serving Valdez, the southern terminus of the trans-Alaska pipeline, is likely to pose a continuing hazard to the passage of oil tankers. (Photograph courtesy of Larry Mayo, U.S. Geological Survey.)

Jökulhlaup phenomena (glacier-outburst floods) have been reported from Alaska, Pacific Northwest, South America, and Iceland and other glacierized areas. In Iceland, jökulhlaups occur either as the result of failure of the ice dams of ice-dammed lakes (Thorarinsson, 1940), or as the result of subglacial geothermal and/or volcanic activity (Thorarinsson, 1958a; Björnsson, 1975, 1976). Ground and aerial photographs (Barth, 1950; Rist, 1968; Hannesson, 1975; and Björnsson, 1976) and satellite images (Thorarinsson et al., 1974; Williams and Thorarinsson, 1974; Williams et al., 1974; Williams et al., 1975; Williams, 1976b; Rist, 1974; Björnsson, 1975; and Tómasson, 1975) have been used to document and/or monitor phenomena related to jökulhlaups in Iceland. In Alaska, occasional jökulhlaups occur that are caused by subglacial geothermal and/or volcanic activity (Benson and Motyka, 1978), but the vast majority of glacier-outburst floods in Alaska are caused by the failure of ice-dammed lakes (Post and Mayo, 1971). The Post-Mayo study (1971) was based on a review of maps and the analysis of more than 15,000 aerial photographs. Jökulhlaups have been documented in the northwestern United States (Richardson, 1968); the 18 May 1980 catastrophic eruption of Mount St. Helens included a jökulhlaup component in the debris flow that coursed down the Toutle and Cowlitz River Valleys to the west and south (Brugman and Meier, 1981). All of the glacier-capped volcanoes of the Cascade Mountains have the potential for jökulhlaups. Because of this potential danger, the U.S. Geological Survey has an active program of studying the glaciers on the Cascade volcanoes with radio-echosounding techniques in order to predict the magnitude of the flood hazards.

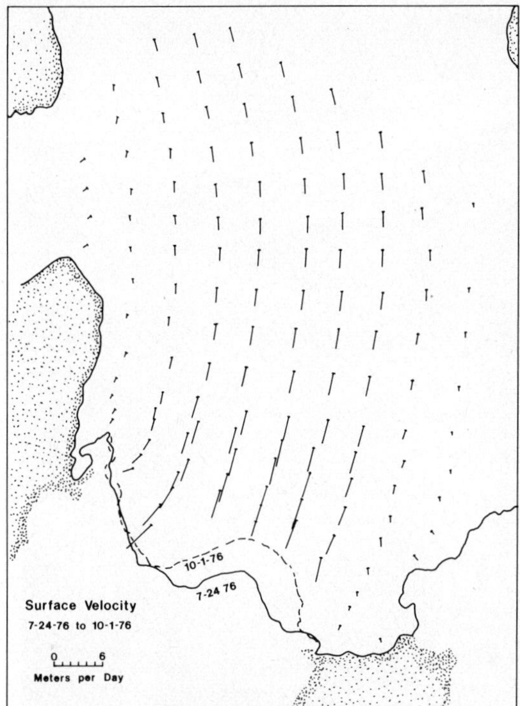

Fig. 31-182. Map of surface velocity vectors on the lower, heavily crevassed part of the Columbia Glacier, Alaska, that were photogrammetrically determined from aerial photographs acquired at about 6-week intervals during the period 24 July 1976 to 1 October 1976. Measurement error is about 0.04 m per day. (From Meier et al., 1978).

Man-Induced Geologic Hazards

Introduction

Most of the geologic hazards created by man are similar to natural ones, except that man-made hazards result from abrupt changes in otherwise stable or quasi-stable natural conditions. Most of the man-induced hazards are small compared to natural ones and are generally best seen on large-scale aerial photographs.

The largest man-made hazards occur because of poorly constructed spoil banks, landfills, and tailing ponds arising from mining and manufacturing. Of less consequence, but still of importance are the hazards caused by poorly designed dredging and hydraulic spoil, cuts and fills, irrigation canals, logging operations, gully erosion, ground failure caused by leaking swimming pools, and installation of groins (jetties) along a coast. Aerial photographs, generally at scales of 1:20,000 or larger and acquired repetitively over the years, are well suited to monitoring stability of slope faces, growth of structures, and changes in and around the structures that can lead to unstable conditions. Large-scale, periodic ground and aerial photographs, as well as multispectral scanner images, thermal infrared images, and other remotely sensed data, can be used to detect illegal dumping of materials and to give clues to improper dumping of toxic wastes (see the Section on "Waste Disposal" in this chapter).

For many types of geologic hazards a sequential series of aerial photographs taken over many years may be needed to establish a base of refer-

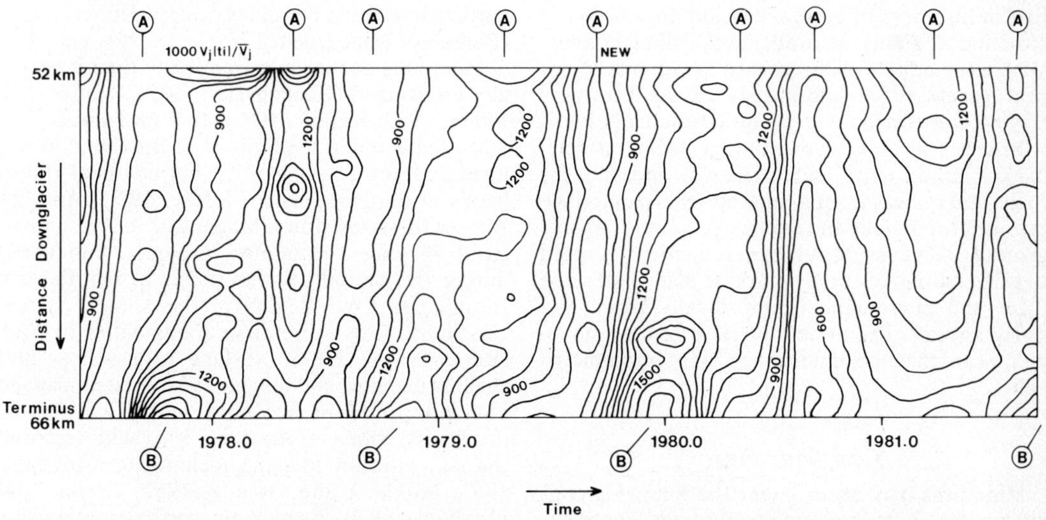

Fig. 31-183. Chart showing variations in surface speed in meters per year (normalized velocity) with time and distance (from the terminus), along the centerline of the lower part of the Columbia Glacier, Alaska. *A.* Denotes annual fluctuations in speed, probably caused by changes in basal water pressure, that occur simultaneously over the entire lower part of the Columbia Glacier. *B.* Denotes peaks in speed originating at the terminus at the end of 1977, 1978, 1979 (absent in 1980), and 1981 which diffuse upglacier and are caused by high summertime calving rates during the period (except during the summer of 1980). (Chart and caption courtesy of Mark F. Meier, U.S. Geological survey, 1982, written communication; from unpublished data.)

Fig. 31-184. Oblique aerial photograph looking south across terminus of the surging glacier, Eyjabakkajökull, northeastern Vatnajökull, Iceland, as it appeared on 25 July 1973, after it had completed a 2.8 km surge (See Color Figure 31-201). Photograph by Richard S. Williams, Jr., U.S. Geological Survey.

ence. Such photographs can be of great significance (scientific and legal) in helping to establish a cause and effect relationship. This is particularly true in instances of coastal erosion after the construction of groins, seawalls, and artificial dunes. Where the effects of the hazard is regional in impact, repetitive satellite images are good for portraying both seasonal and long term changes.

In cases of movement of slope faces on spoil banks, tailing pond embankments, and fills for roads and railways, aerial photographs can reduce the need for tedious and expensive labor-intensive ground surveys. Movement can be detected through motion of survey markers placed on slope faces and photographed periodically from low-flying aircraft. The direction and extent of movement can then be determined photogrammetrically.

Coal-Mine Fires

Mine fires may occur under the following conditions: (1), in underground coal seams where extraction of coal during previous mining operations has been less than 100 percent; and (2), in surface coal-refuse banks (culm banks). Mine fires are caused naturally (generally through spontaneous combustion or from lightning strikes) or by man. The common practice of using open strip-mining cuts and coal-refuse areas for illegal dumping of trash, and the subsequent ignition of this material, most often constitute the source of mine fires, particularly in the populous eastern United States (Bureau of Mines, 1972).

Mine fires occur more frequently than is generally recognized (Rabchevsky, 1979; Johnson and Miller, 1979; Kehrer, 1982). In a nationwide survey, conducted in 1968, the U.S. Bureau of Mines identified and examined 292 burning coal refuse banks in thirteen states (McNay, 1971). Another Bureau of Mines study (Magnuson, 1974) summarized 70 mine fire-control projects, conducted during the period 1949 to 1972, in the Eastern Bituminous Coal Region of the United States. Coal-mine fires were first documented in 1766 (Rabchevsky, 1982). Surface refuse fires and shallow underground mine fires have been mapped and monitored beginning as early as 1963 in the Pennsylvania Anthracite Fields with airborne thermal infrared imaging techniques (Moxham, 1967; Moxham and Greene, 1967; Greene and Moxham, 1968). Systematic and comprehensive thermal imaging of coal-mine fires was initiated by the U.S. Bureau of Mines in 1971 with the publication of Volume I of the *Compendium of Burning Coal Refuse Banks—Pennsylvania Anthracite Region* (Knuth, 1971). Subsequent reports have been directed at surveys of coal-mine fires in the

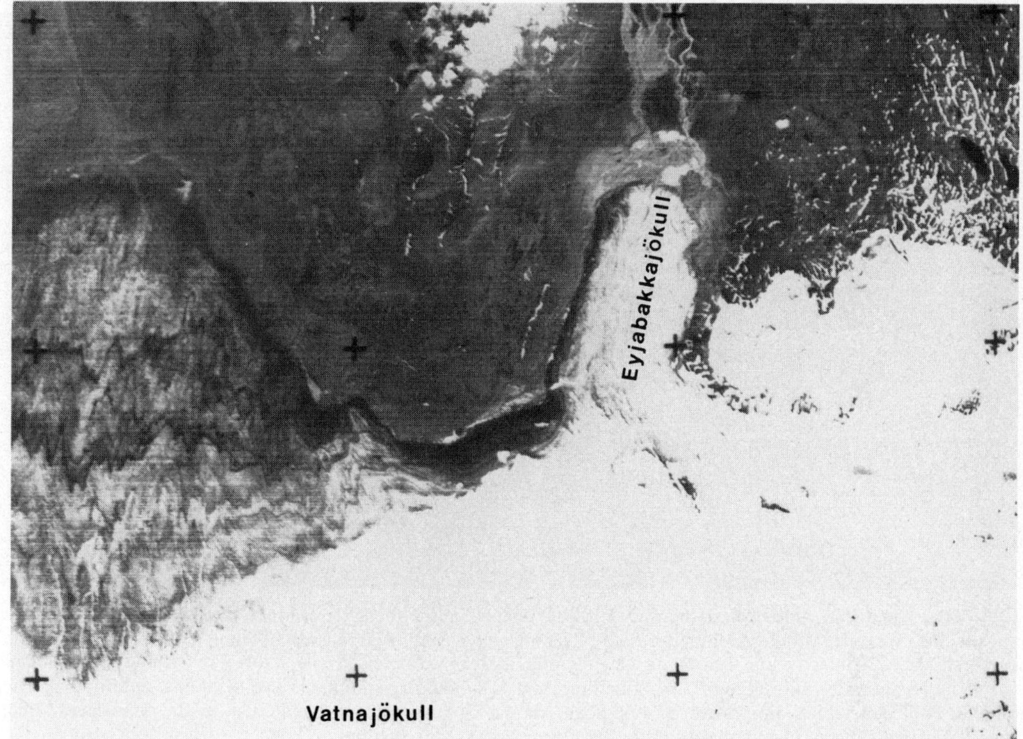

Fig. 31-185. Part of an enlargement of a Landsat 3 return beam vidicon (RBV) image (30157-11565-D) acquired on 9 August 1978. This Landsat 3 RBV image of the surging glacier, Eyjabakkajökull, northeastern Vatnajökull, Iceland, shows the improvement in detail available with such images when compared with Landsat MSS images (See Color Figure 31-201).

western (Rabchevsky, 1972) and eastern (Knuth and Stamm, 1972 and 1975) parts of the United States.

Underground fires are generally a phenomenon of previously mined areas but may also occur in shallow, unmined seams (outcrops), particularly in the western United States, where lightning strikes are often the igniting mechanism. Such mine fires are usually shallow and propagate along the strike of the coal bed, often through abandoned workings. Deep-mine fires in coal seams where overburden exceeds 15 m have not been successfully mapped using airborne thermal imaging techniques. Shallow coal-seam and outcrop fires are, however, readily detectable as shown in Figure 31-186, a thermal infrared image (aerial thermograph) acquired in 1974 of the mine fire near Centralia, Pennsylvania. First imaged in 1971, the Centralia fire has continued to spread, despite efforts to extinguish or contain it. In 1981, the expanding fire even threatened the town of Centralia, itself. More recent aerial thermographs show the spread of the fire along the strike of the coal seam, but burning cannot be traced much beyond the outcrop. Experience has shown that useful thermal infrared imagery of underground mine fires is not generally obtainable at altitudes greater than 3 km above the terrain, or where the spatial resolution exceeds 3 m. Useful imagery is also usually not obtainable when direct sunlight is illuminating the fire area, unless the fire is at the surface. Data collection flights, therefore, are generally scheduled for night, twilight, or for overcast days using a detector sensitive in either the 3- to 5.5-μm or the 8- to 14-μm spectral region of the electromagnetic spectrum (filtered during daytime surveys, unfiltered during nighttime surveys).

Detection and delineation of burning areas in coal-refuse banks is one of the more routine and well documented applications of airborne thermal infrared imaging. Because of the relatively shallow depth of the burning, surface thermal anomalies were detected by the infrared scanning radiometer record, with reasonable accuracy, as was the extent of subsurface burning. Drilling and associated temperature measurements in coal-refuse banks, conducted by the U.S. Bureau of Mines in conjunction with airborne thermal infrared surveys, generally confirm the accuracy of thermal infrared imagery (Stingelin et al., 1971). The thermal infrared image of the Williamstown, Pennsylvania, coal-refuse bank fire (Figure 31-187) illustrates the high quality thermal infrared

Fig. 31-186. Nighttime, unrectified thermal infrared image (aerial thermograph) of the mine fire at Centralia, Pennsylvania (Western Middle Anthracite Field). The image was acquired in early evening during May 1974 from a survey altitude 1,220 m above the terrain, using a filtered (7.8 to 14 μm) MeCdTe detector. The burning zone is located at 40°48′04″N. latitude and 76°20′05″ W. longitude and is advancing along the strike of the outcropping Buck Mountain coal bed which dips north at 40°. Scale shown is measured perpendicular to the scan lines. Image courtesy of HRB-Singer, Inc. Image acquired by a Reconofax XVI system.

imagery that is routinely acquired with airborne thermal infrared scanning radiometers. Advances in scanning system electronics, temperature calibration, signal processing, and signal display permit acquisition of additional quantitative information on mine fires.

Dam Failures

Earthen dams, like other artificial embankments, can be readily monitored for stability through the use of a series of very large scale, repetitive aerial photographs. Movement markers placed on the face of the dam, the abutments, and on the freeboard area of the back side of the dam can be readily surveyed through the use of photographs taken from low-flying aircraft. Changes in the configuration of the face of the dam can also be determined from such photographs. Leakage through or around the dam can be identified through heat and moisture sensors as well as through visual inspection of large scale aerial photos. A permanent record of the construction history of a dam and spillway can also be readily obtained by means of repetitive large scale aerial photographs.

Application of aerial photographs to masonry and concrete dams is similar to that for earthen dams, and additional features, such as cracks and loose masonry blocks, can be detected and monitored. Ground (terrestrial) photographs taken repetitively at key observation points can also be used to monitor those features noted on the aerial photographs.

The pool area behind the dam can be monitored for slope stability and sedimentation by use of aerial photographs. For large reservoirs, satellite images are very useful to monitor pool levels and to determine pool changes in relation to runoff. Aerial photographs play an important part in monitoring development of the flood plain downstream from the dam, especially in reference to large volume spillway discharge and/or flooding by failure of the dam. If failure of a dam occurs, aerial photographs are commonly the primary source for determining pre-failure conditions, history of events leading to the failure, and damage downstream resulting from the failure (Hays, 1981). In exceptional circumstances, ground (terrestrial) photographs have been used to sequentially document dam failure (for example, the Teton Dam failure on 6 June 1976, near Newdale, Idaho).

Coal Mine Collapse

Collapse of coal mines gives rise to several types of surface forms that can be identified on large-scale aerial photographs. These forms include irregular, rolling ground formed of low mounds and swales where the collapsed mine is deep (about 90 m) below the surface. Patterned ground, formed by geometrically arranged depres-

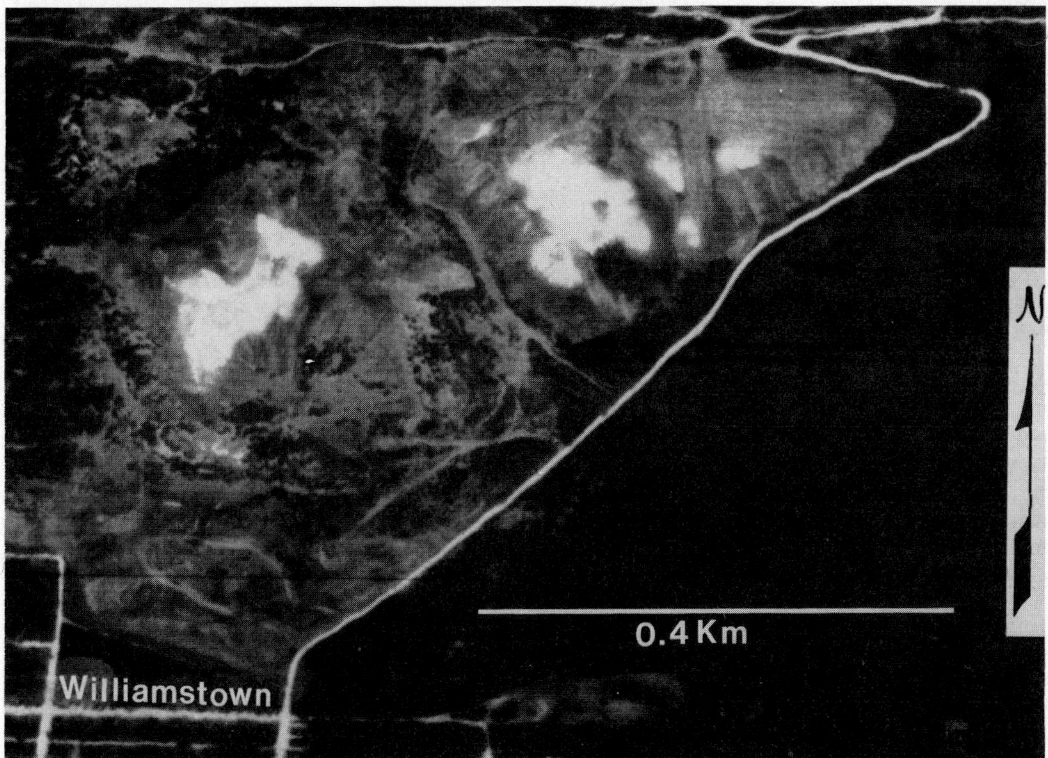

Fig. 31-187. Nighttime (2006 hours e.d.t.), unrectified thermal infrared image (aerial thermograph) of the Williamstown, Pennsylvania coal-refuse bank fire (Southern Anthracite Field). The image was acquired on 28 May 1971 from a survey altitude 610 m above the terrain using an unfiltered Ge:Cu detector. The culm bank is located directly north of Williamstown at 40°35′17″ N. latitude and 76°36′51″ W. longitude. Burning is present in two areas shown as bright (white) patches on the refuse pile. Scale shown is measured perpendicular to the scan lines. Image courtesy of HRB-Singer, Inc. Image acquired by a Reconofax X system.

sions and ridges, reflecting the presence of rooms and pillars, is commonly present over shallow mines (Hays, 1981; see also Figure 31-188). In areas of very shallow mines, the depressed areas are steep-sided and contain intermittent ponds. Distinct zoning of vegetation is also common over collapsed mines with greener and denser grasses developed in the subsided areas over rooms, and yellowish green, sparser grass over the pillars. Black-and-white conventional aerial photographs show some of the vegetation differences, but true color or color-infrared aerial photographs are generally needed for positive identification.

Fig. 31-188. Oblique aerial photograph of surface subsidence caused by collapse of roof supports left from underground mining operations in the Rock Springs area of Wyoming (Photograph by F. W. Osterwald).

GLACIOLOGY

Introduction

Although glaciology usually includes glaciers, sea ice, and snow cover, the latter two topics have already been addressed in Chapter 27, Weather and Climate; Remote Sensing of the Earth's Weather, and in Chapter 29, Water Resources Applications, of this *Manual*. For additional papers on sea ice the reader is referred to Campbell et al. (1975), Zwally and Gloersen (1977), and Zwally et al. (1983). The use of satellites to observe ice and snow is discussed by Gloersen and Salomonson (1975). The use of remote-sensing techniques to study snow cover is discussed in a series of papers by Meier (1973a, 1973b, and 1975).

This section of the chapter will address the use of remote sensing technology, especially aerial photographs and satellite images, for glaciological and related research, including a discussion of specific applications (case histories) to glaciers, glacial geology (Elson, 1980), and one aspect of periglacial phenomena, permafrost. In these topics glaciology is used in a broad sense.

According to the *Glossary of Geology* (Bates and Jackson, 1980), glaciology is "(a) The study of all aspects of snow and ice; the science that treats quantitatively the whole range of processes associated with all forms of solid existing water; and (b) the study of existing glaciers and ice sheets, and of their physical properties". Glacial geology is "the study of the geologic features and effects resulting from the erosion and deposition caused by glaciers and ice sheets. The term periglacial refers to "the processes, conditions, areas, climates, and topographic features at the immediate margins of former and existing glaciers and ice sheets, and influenced by the cold temperature of the ice."

In the United States, glaciology courses are usually taught by geophysicists or atmospheric scientists; courses in glacial geology and periglacial processes are usually taught in geology departments, usually by geomorphologists. In western Europe, the Soviet Union, and Canada the latter two topics are usually taught in geography departments by physical geographers. As can be seen from the preceding definitions there is considerable overlap between the subdisciplines, so much so that it is difficult to study one without at least including some aspects of the others.

In the past decade, geomorphologists-physical geographers in the United Kingdom have published an excellent series of textbooks on the above subjects: "Glacial and Fluvioglacial Landforms" (Price, 1973); "Glacial Geomorphology" (Embleton and King, 1975a); and Periglacial Geomorphology" (Embleton and King, 1975b). In 1976, David Sugden and Brian John did a superb job in synthesizing these topics in their textbook, "Glaciers and Landscape—A Geomorphological Approach" (Sugden and John, 1976). They were also the first to use Landsat images to illustrate different aspects of glaciers in a textbook. In the United States, Richard Foster Flint's work, "Glacial and Pleistocene Geology," has been the basic textbook on the subject (Flint, 1971) from a geologist's viewpoint. Paterson's textbook, "Physics of Glaciers," has been the standard work on glaciology in the United States and Canada (Paterson, 1969 and 1981).

Glaciological research has benefited from the development of sophisticated remote sensing devices carried in aircraft and in satellites (cameras, multispectral scanners, imaging radars, radio-echosounding equipment, radar altimeters, etc.). Before the launch of Landsat-1 both the U.S. (National Academy of Sciences, 1969b) and Canada (Working Group on Ice Reconnaissance and Glaciology, 1971) discussed plans for making effective use of satellite- and/or airborne-sensing of glaciers and/or other hydrological phenomena. Several symposia proceedings, books, and book chapters on the uses of remote sensing for glaciological investigations have been published in recent years.

In 1973, the International Glaciological Society sponsored its first symposium on Remote Sensing in Glaciology, and a number of excellent papers were published in the proceedings volume (Glen et al., 1975), including an excellent paper, "Glacier Applications of ERTS Images" (Krimmel and Meier, 1975).

In 1974, an interdisciplinary symposium, "Advanced Concepts and Techniques in the Study of Snow and Ice Resources," devoted an entire session to remote sensing (Santeford and Smith, 1974). Much of the emphasis was on snow and sea ice, topics well covered in Chapter 29, "Water Resources Applications," of this *Manual,* and in the "Water Resources" chapter (Salomonson and Rango, 1980) of *Remote Sensing in Geology* (Siegal and Gillespie, 1980). The textbook, *Remote Sensing in Meteorology, Oceanography, and Hydrology* (Cracknell, 1981) is another good source book.

Because of the rapid growth in the application of remote sensing technology to glaciology, the International Glaciological Society will hold the Second Symposium on Remote Sensing in Glaciology in Cambridge, England, in 1986.

Remote Sensing of Glaciers

Glaciers cover about 16 million km^2 of the Earth or about 10 percent of the land area of the planet. Most of the glacial ice is in Antarctica (13,586,380 km^2) (Drewry et al., 1982) and Greenland (1,726,400 km^2) (Bauer, 1954), with only 4 percent (700,000 km^2) distributed in highland and high latitude areas throughout the rest of the world (Sugden and John, 1976).

Terrestrial and aerial photographs (Color Figures 31-8 and 31-199; and Figures 31-5, 31-181, 31-184, 31-190, and 31-218) have long been used to document the position of glacier termini, to measure the areal extent of glaciers and ice caps, to illustrate types of glaciers and associated features (Armstrong et al., 1973), and to study landforms produced by glaciers (Mollard, 1974). Terrestrial photographs of glaciers are especially valuable for historical studies of glacier fluctuation, because they are often the only permanent and objective record of the position of a glacier's terminus. In the recent Time-Life book, *Glacier* (Bailey, 1982), terrestrial photographs show the 5-km retreat of the tidal Dawes Glacier, Alaska, between 1941 and 1967. Remote sensing of another Alaskan tidal glacier, the Columbia Glacier, is discussed in the "Geologic Hazards" section of this chapter.

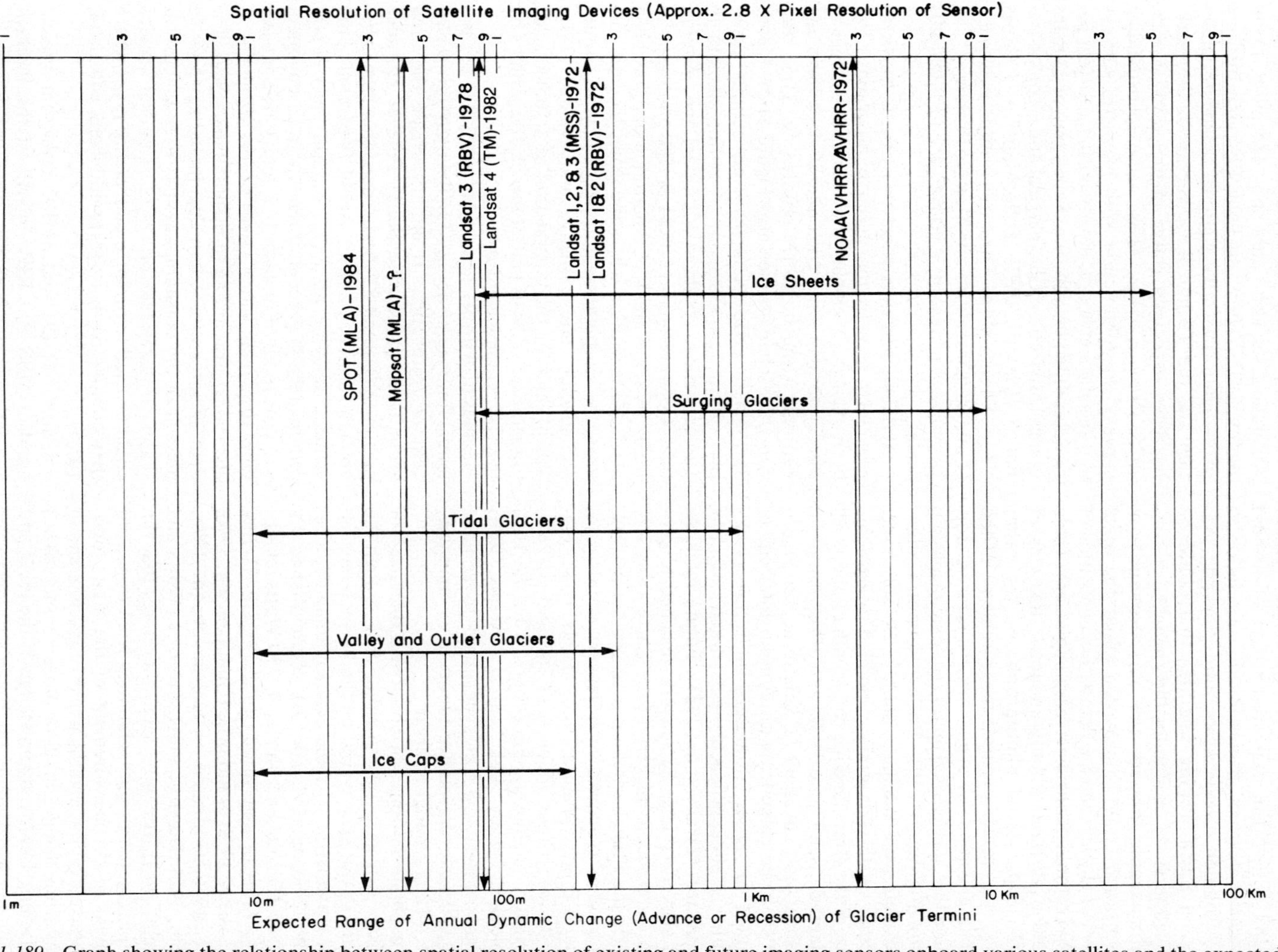

Fig. 31-189. Graph showing the relationship between spatial resolution of existing and future imaging sensors onboard various satellites and the expected range of annual dynamic change of glacier termini of different types of glaciers. Pixel resolution of sensor is defined as the dimension in meters of the area represented on the Earth's surface by each picture element (pixel) contained within a specific satellite image band. The termini of ice sheets includes ice shelves from which large tabular icebergs can calve off, producing large annual changes.

Imaging sensors and radar altimeters mounted in satellites are especially well suited for monitoring large ice caps and ice sheets (Color Figure 31-201, Figures 31-36 and 31-193). Figure 31-189 shows the expected range of annual dynamic change of glacier termini in relation to the resolution of existing or future imaging sensors on Earth-orbiting satellites that could be utilized to study or monitor the changes.

Monoscopic imagery can only provide two-dimensional (x, y) information about the areal extent of glaciers and the fluctuation in position of glacier termini or change in area of glaciers (Table 31-39) over time. Photogrammetric analysis of stereoscopic aerial photographs on the other hand, can provide accurate measurements of the surface elevations (z) of glaciers, the type of information that is presented on topographic maps (see Color Figure 31-219). Radar altimetry measurements from satellites, such as Seasat, have, through successive profiles, enabled topographic maps of the Greenland ice cap (south of about 72° N. latitude; see Figure 31-36) and of the Antarctic ice sheet (north of about 72° S. latitude, see Figure 31-193) to be compiled with contour intervals of about 1 m. Laser altimeters on future polar-orbiting satellites will achieve accuracies of a few centimeters (Zwally et al., 1981). Future satellites will also carry imaging sensors designed to provide stereoscopic imagery of the Earth's surface. SPOT, the French satellite to be orbited in 1984, will be able to acquire such imagery (Centre National d'Études Spatiales, n.d.; Table 31-48; Figure 31-259; Color Figures 31-260 and 31-261). Mapsat, a satellite system proposed by the U.S. Geological Survey (Itek Optical Systems, 1981; Colvocoresses, 1982b), is designed to achieve 1:50,000-scale or smaller topographic mapping, with a 20-m contour interval.

To obtain information about the thickness of glaciers, it is necessary to either drill through the glacier or use geophysical methods, such as seismic, gravimetric, or radio-echosounding devices. Radio-echosounding may be accomplished by surface (O'Neil and Jones, 1975; Watts et al., 1975; Morgan and Budd, 1975; Figure 31-206; Björnsson, 1978) or airborne (Gudmandsen, 1970, 1971, 1976; Gudmandsen and Jakobsen, 1975; Robin, 1975; Drewry, 1981) traverses (Watts and Wright, 1981). It can be seen, therefore, that the use of remote sensing technology is an important element in the measurement of the area, topographic elevation, and thickness of glaciers and in monitoring changes in all of these characteristics of glaciers. Improved knowledge of the areal and volumetric variations in the Earth's glaciers over time is important to a better understanding of changes in the Earth's climate (Campbell, 1979; Allison, 1981; Williams, 1983a).

Remote sensing of Antarctica

Remote sensing technology is being increasingly applied by scientists in their quest to gain a better understanding of the Antarctic ice sheet. Radio-echosounding instrumentation, such as that operated from low-flying aircraft has been and is being used to map the thickness of glacial ice in Antarctica (see also figure on p. 154 and p. 155 in Bailey, 1982). The Scott Polar Research Institute in association with the National Science Foundation and the Technical University of Denmark, has been especially active in this work (Robin, 1975; Drewry, 1981). These radio-echosounding surveys, in conjunction with magnetic surveys and seismic soundings, mostly during a 10-year period, provided data for the preparation of 13 maps in the "Antarctica: Glaciological and Geophysical Folio" (Drewry, 1982). Folio Map No. 4 shows "Ice Sheet Map Thickness and Volume" (1:10,000,000 scale); Folio Map No. 3 shows "The Bedrock Surface of Antarctica" (1:6,000,000 scale).

Aerial photography acquired by the U.S. Navy, especially trimetrogon aerial photography, has been extensively employed by the U.S. Geological Survey to map Antarctica and in glaciologic and geologic (Smith, 1967) studies by U.S. and non-U.S. scientists (see also Figure 31-5). Figure 31-190 is a good example of an oblique aerial photograph (trimetrogon) of outlet glaciers in Taylor Valley, southern Victoria Land, Antarctica. Conventional vertical aerial photography exists of some areas of Antarctica and has been used, in association with supporting geodetic surveys, for morphological investigations of cirques and glacial valleys (Aniya and Welch, 1981a, 1981b), and for the compilation of 1:50,000-scale topographic maps of part of the ice-free valleys by the U.S. Geological Survey.

Landsat, with its near polar orbit, has the potential for imaging about 11 million km^2, or 79 percent of Antarctica. It cannot image the area around the geographic South Pole at latitudes greater than about 81° S., because that area is beyond the orbit of Landsat. About 70 percent of the Landsat imaging area (7.7 million km^2) or approximately 55 percent of the Antarctic continent was covered by Landsats 1, 2, and 3 MSS images with 10 percent cloud cover or less (Williams et al., 1982a). According to Swithinbank (1980), less than 20 percent of Antarctica, including about 50 percent of the coastal areas, has been mapped at scales of 1:250,000 or larger. Landsat images could be used to produce planimetric maps where they do not already exist (Swithinbank and Land, 1977).

Figures 31-191 and 31-192 represent typical Landsat MSS and RBV images of the Rennick Glacier area, northern Victoria Land. Figure 31-192 is one of the few (out of about 1,500 scenes; Ferrigno et al., 1982) usable Landsat 3 RBV images available of Antarctica.

Landsat images could be used to triple the area of Antarctica currently covered by planimetric maps. In addition, Landsat MSS image maps could be prepared to satisfy the need for adequate 1:250,000-scale base maps for geological data

Fig. 31-190. High oblique, trimetrogon aerial photograph (left oblique) looking east towards Ross Island and Mount Erebus Volcano, Antarctica, in the background. Taylor Glacier, Taylor Valley, and various outlet glaciers are visible in the foreground. U.S. Navy aerial photo No. 250 was acquired on 7 November 1959 from an altitude of 6,100 m (20,000 ft) above sea level with a 153.99 mm—focal length camera.

(Wolmarans and Krynauw, 1981) and geophysical data previously gathered from ground traverses, airborne instrumentation, or satellite sensors (Swithinbank and Land, 1977; Williams and Schoonmaker, 1979; Ødegaard and Helle, 1982).

Figure 31-193 is an uncontrolled Landsat MSS image mosaic of the Amery Ice Shelf and the terminus of the Lambert Glacier, East Antarctica, prepared by the U.S. Geological Survey, over which are printed 1 m- and 5 m- contours compiled by GeoScience Research Corporation from Seasat radar altimetry data (Brooks et al., 1982). If sufficient Landsat 3 RBV images exist of an area, 1:100,000-scale image maps can be prepared. Landsat image maps can also serve as the base for aeronautical charts of Antarctica over which the standard aeronautical information can be printed. Figure 31-194 is part of a 1:500,000-scale enlargement of a Landsat MSS image of Mount Takahe, a massive caldera-capped volcano which is partly subglacial (''table-mountain'') and partly subaerial (''shield volcano'') in Marie Byrd Land, West Antarctica. The figure illustrates how well Landsat can portray important geographic features on aeronautical charts. If a sufficient number of image-identifiable ground-control points surveyed by either conventional or Doppler satellite methods (MacDonald, 1976a) are

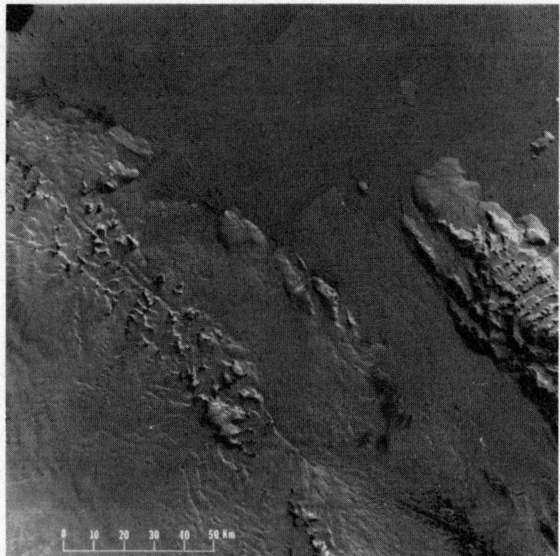

Fig. 31-191. Landsat 1 MSS image (1460-21103; 26 October 1973) of the Rennick Glacier and environs, Oates Coast, northern Victoria Land, Antarctica. The image includes an area of 33,000 km^2 in a 186 × 180 km trapezoidal format and has a pixel resolution of 80 m. Compare with Figure 31-192, a Landsat 3 RBV image of the northeast quadrant.

present on a particular image, a fitted grid can be generated which will convert the Landsat image to a Landsat image map.

Another significant attribute of Landsat images of Antarctica is the precise date (and time) of acquisition. The dynamic nature of the coast of Antarctica, when compared with published maps (MacDonald, 1976b; Swithinbank et al., 1976; Swithinbank and Land, 1977), is readily apparent and can be documented. "Time-lapse" measurements of speed of flow of outlet glaciers can also be accomplished. For example, 12 Landsat MSS images were evaluated in a computation of the speed of flow of the terminus of the Pine Island Glacier, an important outlet glacier on Walgreen Coast, West Antarctica. Only two Landsat images (1185-13530, 24 January 1973; Path 246, Row 114; and 2022-13582, 13 February 1975; Path 249, Row 113) were sufficiently cloud-free and far enough separated in time to determine that the terminus of Pine Island Glacier had moved about 4.5 km during a period of 750 days, or an average speed of flow of 6 m per day. Later measurements on 1982 Landsat images further confirmed this average speed.

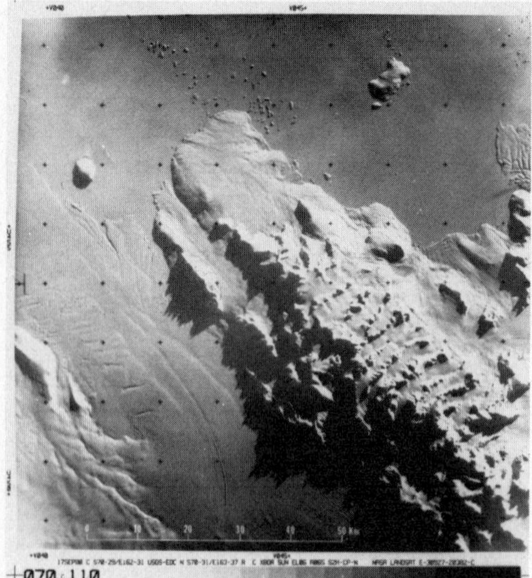

Fig. 31-192. Landsat 3 RBV image (30927-20382-B; 17 September 1980) of the Rennick Glacier and environs, Oates Coast, northern Victoria Land, Antarctica. The image includes an area of 8,100 km^2 in a 90 × 90 km square format and has a pixel resolution of 30 m, suitable for 1:100,000-scale enlargements. Compare with Figure 31-191, a Landsat 1 MSS image of a larger area. Landsat MSS images can normally be enlarged only to a scale of 1:250,000.

The National Oceanic and Atmospheric Administration and the U.S. Geological Survey, with the support of the National Science Foundation, are compiling a 1:5,000,000-scale image mosaic of Antarctica which is based on NOAA 6 meteorological satellite images (Berg et al., 1982). Figure 31-195 is a NOAA 6 satellite image of the McMurdo Sound area and part of the Transantarctic Mountains. NOAA satellite images can be used to monitor some types of dynamic changes in coastal areas, such as calving of parts of ice shelves and outlet glaciers, if the change is more than about 3 km. NOAA satellite images also can be used to image the area around the South Pole not covered by Landsat images.

Except for the work by Rivereau (1978), who mapped lineaments and rock outcrops in the mountain ranges of Antarctica from Landsat images, very little use has been made of Landsat images for geological studies other than as planimetric base maps (Wolmarans and Krynauw, 1981). The Japan National Institute of Polar Research published a 1:200,000-scale "(Working) Map of the Meteorite Ice Field, Yamato Mountains, Antarctica," in 1976, to support geological mapping in the Queen Fabiola (Yamato) Mountains, East Antarctica. Color Figure 31-196 is a Landsat MSS color composite of the Queen Fabiola (Yamato) Mountains showing the large areal extent of exposed glacier ice, the so-called "blue ice." Areas of blue ice, on the upstream side of nunataks and where other glaciological conditions are satisfied, can contain extraordinary accumulations of meteorites. Since 1969, Japanese scientists have discovered 4,813 meteorites in the Queen Fabiola (Yamato) Mountains; U.S. scientists have discovered 1,187 specimens in the Allan Hills area and environs, west of McMurdo Sound, Antarctica, for a total of 6,000 meteorite fragments (Bull and Lipschutz, 1982). The number of meteorites found in Antarctica, in only 13 years, represents about 25 percent of the total number of meteorites in the worldwide collection of meteorites. Landsat and NOAA images could

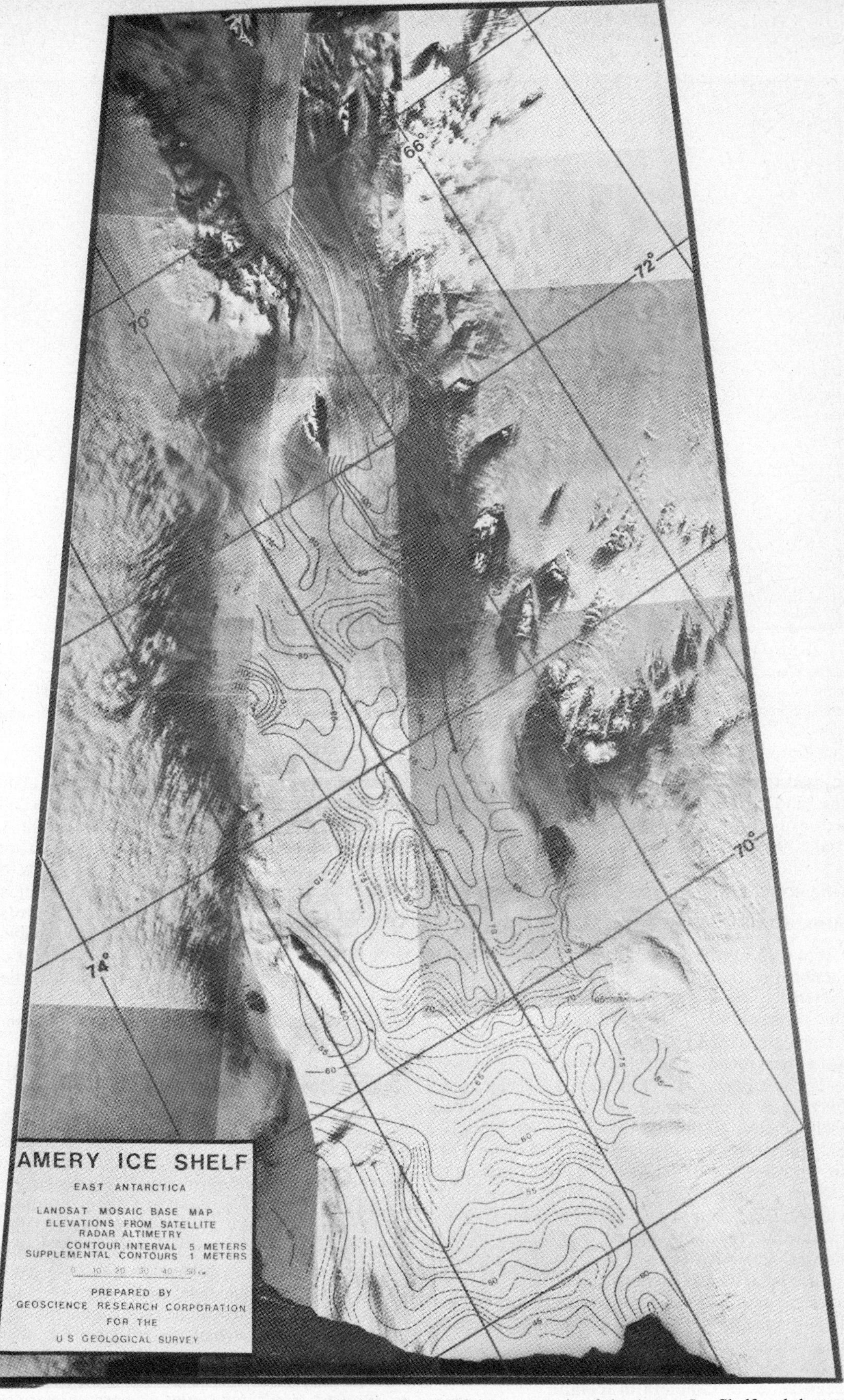

Fig. 31-193. Contours plotted on an uncontrolled Landsat MSS image mosaic of the Amery Ice Shelf and the terminus of the Lambert Glacier terminus, East Antarctica. The 1- and 5-m contours were determined from numerous Seasat radar altimeter traverses across the area. Contours courtesy of GeoScience Research Corporation (Brooks et al., 1982). Landsat images can be used as planimetric base maps, where adequate geodetic control is present, for the plotting of geological (Wolmarans and Krynauw, 1981) and geophysical data.

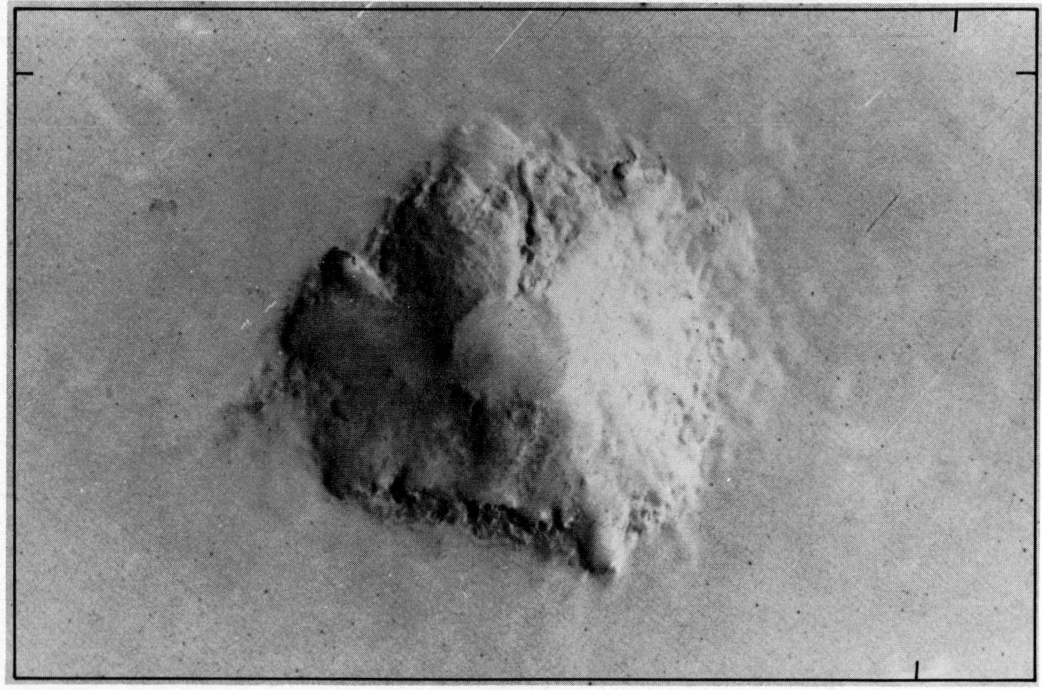

Fig. 31-194. Part of a Landsat 1 MSS image (1119-14280; 19 November 1972) of the Mount Takahe area, Marie Byrd Land, West Antarctica, enlarged to 1:500,000. Landsat images provide important data for the preparation of base maps for geological and geophysical mapping, aeronautical charts, and glaciological studies. Mount Takahe is a partially buried shield volcano, approximately 30 km in diameter, and topped by an 8-km wide, quasicircular caldera.

be used to delineate areas of blue ice in Antarctica, and the most promising areas for further meteorite finds could be inspected (Williams et al., 1982b and 1983b).

Remote Sensing of the Glaciers of Iceland

Introduction

Iceland abounds in dynamic geological phenomena that, for over 200 years, have attracted the attention of geologists. Special scientific emphasis has been directed at: (1) its geothermal areas (See Figures 31-152A and 31-152B), (2) its frequent volcanic activity (See Color Figure 31-174 and Figures 31-175A, B, C) and the great diversity of volcanic landforms, and (3) its glaciers (Color Figure 31-197; Björnsson, 1980a) and glacial landforms (See Figure 31-207 and Color Figure 31-208 A and B). The subject of this section is Iceland's glaciers, and especially the way in which remote sensing technology is being used to provide a permanent record of changes in the physical characteristics of the glaciers such as surface area, ice cap margins and termini of outlet glaciers, and surface features caused by glacier flow or subglacial volcanic and geothermal activity (Williams, 1983b).

Occurrence of Glaciers in Iceland

Glaciers in Iceland occur principally as ice caps or outlet glaciers from ice caps. Figure 31-198 is a sketch map of the 13 principal ice caps of Iceland with areas greater than 20 km^2 (Table 31-39). According to Thorarinsson (1943), there are 33 different geographic place names of ice caps and cirque glaciers in Iceland. A careful review of published maps, books, and journal articles, however, indicates that there are actually a total of 85 separately named outlet glaciers associated with the 13 individual ice caps out of a total of about 330 named and unnamed ice caps, outlet glaciers, and other types of glaciers (mostly cirque-type) that can be identified on maps of Iceland at a scale of 1:100,000 or smaller.

Modern Observations of Iceland's Glaciers

The systematic measurement of the annual variation of the position of outlet glaciers and ice cap margins was begun by the Icelandic meteorologist-glaciologist, Jón Eythórsson, in the 1930s (Eythórsson, 1949 and 1963). With the publication of *Jökull,* beginning in 1951, glacier variation data have been reported nearly every year since. After Eythórsson's death in 1968, Sigurjón Rist, the Icelandic hydrologist-glaciologist, assumed the responsibility for annual reports. In 1951, Eythórsson had reported on the position of 26 outlet glaciers or margins of 8 different ice caps (Eythórsson, 1951). In 1977, Rist reported on 40 of 61 monitored outlet glaciers or margins of 11 different ice caps (Rist, 1977).

Although the annual monitoring of the position

Table 31-39

Areas of the Principal Glaciers of Iceland (km²)

IHD Index[1] Number	Glacier Name	A Bödvarsson[2] (From Gunnlaugsson, 1844) (unpub.)	B Bödvarsson[3] (From Thoroddsen, 1901) (unpub.)	C Thoroddsen[4] (1906)	D Thorarinsson[5]	E Thorarinsson[6] (1958b)	F Williams[7] (From Landsat images, unpub.) and Björnsson (1980b)	G Björnsson,[8] (From Landsat images or aerial photos*) 1980b	H Percentage[9] (E-F or G/E) in Area
14–34 (outlet glaciers)	Vatnajökull	8940	8500	8500	8410	8538	8300	8300	−3
5–7 (outlet glaciers)	Langjökull	1384	1400	1300	1021	1022	953	953	−7
9–11 (outlet glaciers)	Hofsjökull	1570	1400	1350	987	996	925	925	−7
12 & 13 (outlet glaciers)	Mýrdalsjökull	1100	1000	1000	685	701	596	596	−15
	Eyjafjallajökull				101	107	77.5	77.5	−28
2 (outlet glacier)	Drangajökull	708	340	350	204	199	—	160*	−19.6
None	Tungnafellsjökull	115	170	100	50	50	—	48	−4.0
None	Thórisjökull	Included in Langjökull			34.5	33	—	32	−3.0
None	Thrándarjökull	84	112	100	27	27	—	22	−18.5
None	Tindfjallajökull	38	35	25	26	27	—	19	−29.6
None	Eiríksjökull	96	113	100	23.5	23	—	22	−4.3
1 (outlet glacier)	Snaefellsjökull	43	28	20	22	22	—	11*	−50.0
None	Torfajökull	140	112	100	27.5	21	—	15	−28.6

1—In Iceland, IHD (International Hydrological Decade) Index Numbers are assigned only to 34 individual outlet glaciers from 6 ice caps and 3 cirque glaciers. Annual measurements are made of the variation in the position of the termini (or at points along the termini) of these outlet and cirque glaciers, not areal measurements of the entire ice cap. Within the resolution limits of the satellite images used, satellite imagery can permit a frequent areal measurement of each ice cap to be made, thereby providing a measurement of dynamic changes within and at the margins of an entire ice cap, including its outlet glaciers (from Rist, 1967 and 1977; and Williams, 1979c).

2—Unpublished area measurements by Agúst Bödvarsson, former Director, Iceland Geodetic Survey, from Björn Gunnlaugsson's 1844 map of Iceland (1:480,000)

3—Unpublished area measurements by Ágúst Bödvarsson, former Director, Iceland Geodetic Survey, from Thorvaldur Thoroddsen's 1901 Geological Map of Iceland (1:600,000)

4—Thorvaldur Thorodssen's area measurements of Iceland's glaciers were based on Gunnlaugsson's 1844 map and Thoroddsen's 1901 map (based on 1881–1898 field surveys)

5—Based on Danish Geodetic Institute maps (surveyed in 1902–1938)

6—Based on Danish Geodetic Institute maps, including post-World War II editions

7—Area calculations made from 19 August 1973 (1392-12185; 1392-12191) and 22 September 1973 (1426-12070) Landsat images of Iceland (See also Björnsson, 1980b)

8—Area calculations made from 19 August 1973, 22 September 1973, 9 August 1978 (30157-11565-D) Landsat images, and 1960 aerial photographs

9—First five glaciers calculated by Williams (unpub.), the remaining eight by Björnsson (1980b).

Fig. 31-195. NOAA 6 advanced very high resolution radiometer (AVHRR) image (0.8–1.1 μm) of the McMurdo Sound area and environs, Antarctica, acquired on 27 February 1980, at 1812 UT. NOAA 6 AVHRR images have a pixel resolution of about 1.1 km and can be used to monitor dynamic changes along the coast if the changes are sufficiently large (>3 km in dimension). NOAA images also record information in the area not covered by Landsat (poleward of about 81° S. latitude), thus providing coverage of the Transantarctic Mountains. NOAA image courtesy of Donald R. Wiesnet, National Oceanic and Atmospheric Administration (National Environmental Satellite Data and Information Service).

of 40 different glacier termini or ice cap margins represents a significant effort, it includes only about 12 percent of the 330 individual named and unnamed outlet glaciers associated with the various ice caps in Iceland or termini of cirque glaciers which potentially could be monitored annually. It should also be noted that the current position measurements are "spot" measurements and represent only a "sample" of the overall state of Iceland's ice caps, outlet glaciers, and cirque glaciers.

Modern Mapping of Iceland's Glaciers

The modern mapping of Iceland's glaciers were begun by personnel of the Danish Geodetic Institute in 1904, and their plane-table surveys continued until just before World War II, resulting in nearly complete 1:250,000- and 1:100,000-scale map series (See Color Figure 31-219) and some 1:50,000-scale maps. Thorarinsson (1943) compiled a table of the areas of Iceland glaciers based on these pre-World War II maps and on the earlier

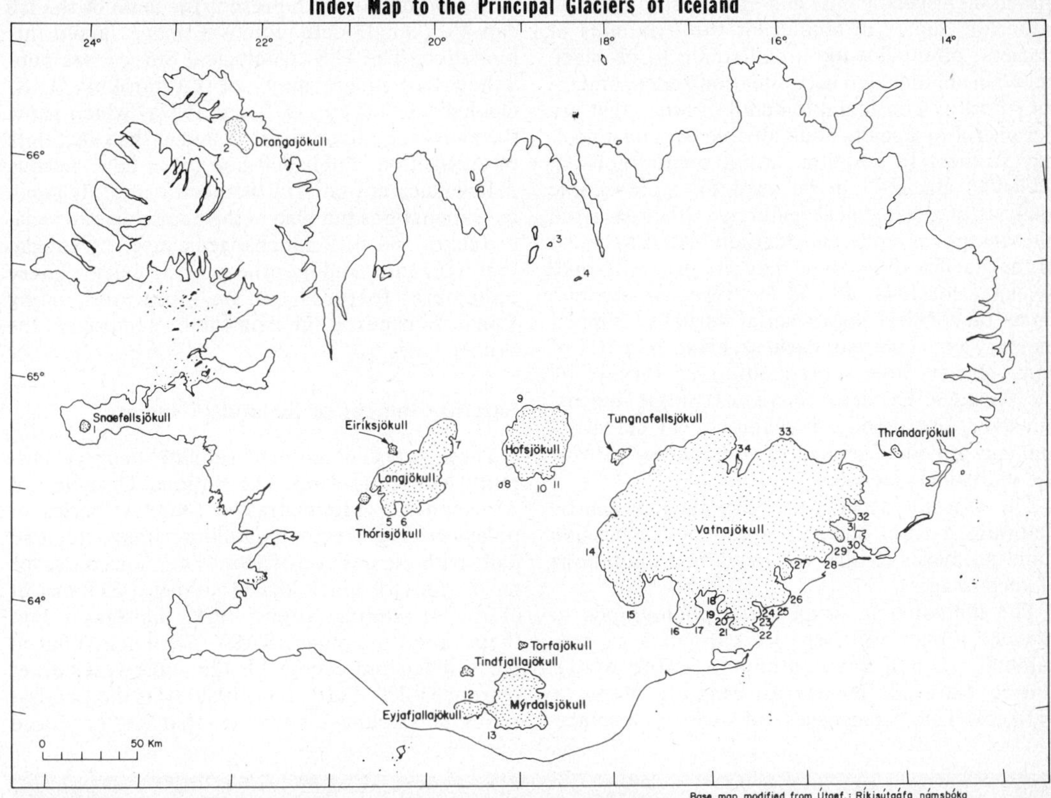

Fig. 31-198. Index map showing the 13 principal ice caps of Iceland and the 34 International Hydrological Decade (IHD) index numbers assigned by Rist (1967 and 1977) to the individual outlet glaciers (of 6 different ice caps) and cirque glaciers (3 different ones) monitored annually (See also Table 31-39).

work of Thorvaldur Thoroddsen (Thoroddsen, 1892 and 1906). The U.S. Army Map Service completed new 1:250,000- (Series C562) and 1:50,000-scale (Series C762) maps of Iceland after World War II, using aerial photogrammetric surveying techniques. The U.S. Defense Mapping Agency and the Icelandic Geodetic Survey are preparing a new series (C761) of 1:50,000-scale maps of Iceland. The Icelandic Geodetic Survey is preparing a 1:10,000-scale orthophotomap series. The Icelandic Geodetic Survey also publishes special-purpose maps at various scales, in addition to periodic revisions of the 1:100,000- and 1:250,000-scale Danish Geodetic Survey maps.

Imaging of Iceland's Glaciers

Although aerial photographs of the glaciers of Iceland had been used by some scientists, such as Iwan (1935), who published oblique aerial photographs of glaciers taken from a Zeppelin, the Danish Geodetic Institute acquired the first aerial photographs of Iceland's glaciers in 1937 specifically for topographic mapping (Norlund, 1938). Six oblique aerial photographs of Vatnajökull, taken by the Danish Geodetic Institute in June and August 1937 from 3,600 m, were published by Ahlmann (1937).

In 1944 and 1945, the U.S. Army Air Force acquired vertical aerial photographs of Iceland to support map revision needs and to support special map projects for other agencies. In 1956 and from 1959 to 1961 (Project 55-AM-3), the U.S. Air Force rephotographed most of Iceland to support a new 1:50,000-scale map series (Series C761).

There have been a number of miscellaneous aerial surveys of Iceland since 1960, mostly in support of special research projects. The U.S. Air Force, the U.S. Navy, and the National Aeronautics and Space Administration (NASA) all conducted limited aerial surveys in Iceland during the 1960s and 1970s (See Figure 31-199).

Beginning in September 1972, the first in the Landsat series of satellites began to acquire Landsat images of Iceland, providing a valuable new source of information about Iceland's glaciers. During 1973 the best series of Landsat images of the ice caps of Iceland were acquired, although a few individual excellent images have been acquired in recent years through the use of Landsat receiving stations in Canada and Sweden.

Aerial photographs and satellite images of glaciers are considerably more useful than conventional maps to glaciologists because: (1) they represent original source material; (2) they are acquired on a specific date and at a specific time, an important factor in studies of the dynamics of glaciers, often allowing a distinction to be made between ablation and accumulation zones; and (3) they portray considerable detail of areas that are peripheral to glaciers and valuable in glacial geology studies. In addition, aerial photographs of Icelandic glaciers can be used for stereoscopic analysis of glaciological features. Although aerial photographs provide considerable detail for most of the glaciers of Iceland, they are generally only available for 1944–45 and for 1959–60, the two times of comprehensive aerial surveys. Supplementary coverage has been archived of parts of some glaciers from subsequent aerial surveys by the Icelandic Geodetic Survey. Landsat images, however, are readily available to all scientists and can provide a sequential (time-lapse) view of the individual glaciers.

The dynamic aspects of these glaciers can be determined from changes noted on successive Landsat images (within the resolution limitations of such images).

The limitation in using aerial photographs to produce a map of a large ice cap, such as Vatnajökull, is in the discontinuous nature of the source material. Nearly all existing maps of Vatnajökull are "composites" of a variety of source material and do not represent the state of the ice cap on a single date. Landsat images avoid this limitation. The U.S. Geological Survey has published two image maps of Vatnajökull (U.S. Geological Survey, 1976 and 1977a) which show the entire ice cap on the date and at the exact time of acquisition. Published line maps have serious deficiencies not only in the portrayal of dynamic ice cap margins but also in the depiction of proglacial lakes and surficial changes caused by subglacial volcanic and geothermal activity. These ephemeral features can be monitored, using Landsat images, if the data can be acquired at the proper time.

Satellite Images of Iceland's Glaciers

Three types of civilian satellite imagery currently exist of Iceland. The National Oceanic and Atmospheric Administration (NOAA) series of polar-orbiting weather satellites image Iceland daily with a resolution of about 1 km, too coarse for most types of glaciological studies (Williams et al., 1974). During August 1978, the Seasat synthetic aperture radar (SAR) instrument imaged most of Iceland except for the southwest corner (Figure 31-200; Ford et al., 1980). It is the Landsat series of satellites, however, that has produced

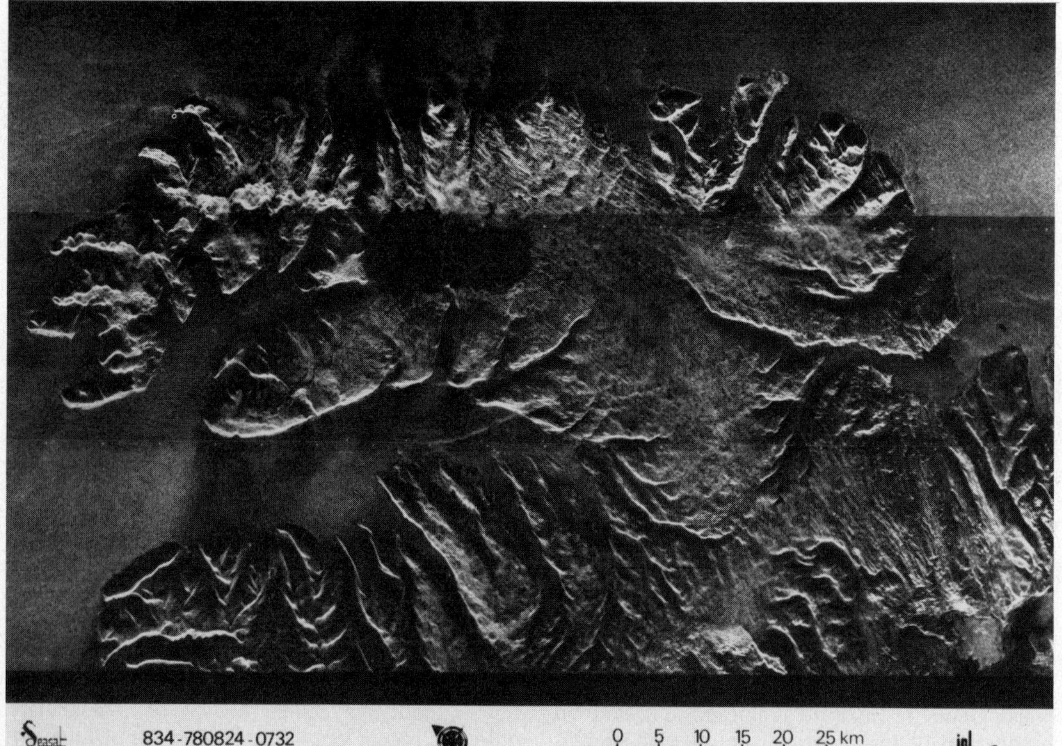

Fig. 31-200. Seasat synthetic aperture radar (SAR) image mosaic of northwestern Iceland (Vestfirdir area) acquired on 24 August 1978 (No. 834-780824-0732) which includes Drangajökull, a 200-km² ice cap shown as a dark area in the left center of the figure. The area is frequently obscured by clouds making it very difficult to acquire cloud-free Landsat images (See Color Figure 31-197 and the index map, Figure 31-198). A Seasat SAR, 1:500,000-scale radar mosaic of most of Iceland is available from Hunting Surveys, Ltd., England (Hunting Geology and Geophysics, Ltd., n.d.). Seasat SAR image courtesy of John P. Ford, Jet Propulsion Laboratory.

the most useful, although discontinuous, coverage of Iceland and its glaciers since 1972 (Williams et al., 1974; Williams and Thorarinsson, 1974; Gudbergsson and Williams, 1983).

From analysis of Landsat images of Iceland, the following types of glaciological phenomena have been observed on individual or successive paired images: (1) glacier advance and recession (including glacier surges), (2) effect on the glacier surface of subglacial volcanic and geothermal activity (Color Figure 31-201 and Figure 31-204), (3) variation in proglacial lakes, (4) effect of jökulhlaups, (5) glacier flow, and (6) ablation phenomena (see Color Figure 31-202 and Figure 31-203).

Glacier advance or recession has been noted on a number of sequential Landsat images of Iceland during the past decade. Eyjabakkajökull, an outlet glacier in the northeastern part of Vatnajökull, began to surge in late August 1972; it has been discussed in the previous section, "Geologic Hazards," of this chapter (See Figures 31-184, 31-185, and 31-205; Williams et al., 1974; Williams, 1979b; Williams and Ferrigno, 1981). Crabtree (1976) compared field observations of Mýrdalsjökull with aerial photographs and three Landsat images (1392-12191, 19 August 1973; 1426-12070, 22 September 1973; and 1446-12180, 12 October 1973) to document advance and recession of the ice cap margin and termini of several outlet glaciers. Crabtree (1976) also used Landsat images to distinguish the soaked zone from the percolation/dry snow zone of the accumulation area (Color Figure 31-202 and Figure 31-203) and noted the difficulties in determining the exact position of the termini of outlet glaciers covered with surface debris.

Subglacial volcanic and geothermal activity is indicated on Landsat images by collapse cauldrons of various diameters and related features. The 31 January 1973 image (1192-12084) (Figure 31-204) and an enhanced 22 September 1973 image (1426-12070) (Color Figure 31-201) of Vatnajökull have been analyzed for a number of investigations. An extension of geothermal activity south into Vatnajökull from Hveradalur in the Kverkfjöll area was discussed by Thorarinsson et al. (1974) (See also Figures 31-152 A and B and Figure 31-205). Jökulhlaups on Skaftá are related to the two collapse cauldrons east of Hamarinn in western Vatnajökull (Figure 31-204) (Williams et al., 1974; Thorarinsson et al., 1974; Williams, 1976b). Thorarinsson et al. (1974) and Björnsson (1975) discussed a line of cauldrons north of Skeidarárjökull which are related to the March 1972 jökulhlaup from Grímsvötn. Tómasson (1975) published a map of the path of this jökulhlaup based on the Landsat image (Figure 31-204). Rist (1974), in his discussion of the August 1973 jökulhlaup from Graenalón, a glacier-dammed lake in southwestern Vatnajökull, used successive Landsat images to calculate a 1.75×10^9 liters reduction in volume of Graenalón after the jökulhlaup.

Landsat images have also been used as substitutes for conventional line maps (U.S. Geological Survey, 1976 and 1977a) and as illustrations for scientific articles: Björnsson (1978) used a Landsat image of Vatnajökull to show the traverse line of his radio-echosounding survey between Tungnaárjökull and the edge of Grímsvötn (Grímsfjall) (Figure 31-206) and for comparison with the cross-section showing ice thickness and subglacier topography.

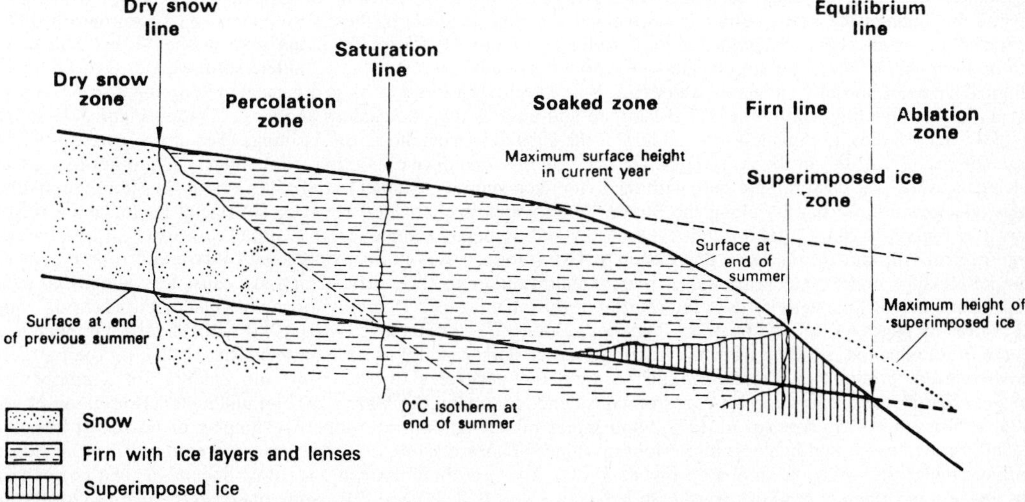

Fig. 31-203. Schematic representation of the zones in the accumulation zone of a glacier, the area up glacier from the equilibrium line. Bare glacier ice is exposed down glacier (to the right of the firn line). The ablation zone extends down glacier from the equilibrium line. See Color Figure 31-202 (from Björnsson, 1971; after Paterson, 1969 and 1981). The concept of zonation in the accumulation area was developed by Benson (1959, 1961, and 1962), from his work in Greenland and by Müller (1962), from his work on Axel Heiberg Island, Northwest Territories, Canada.

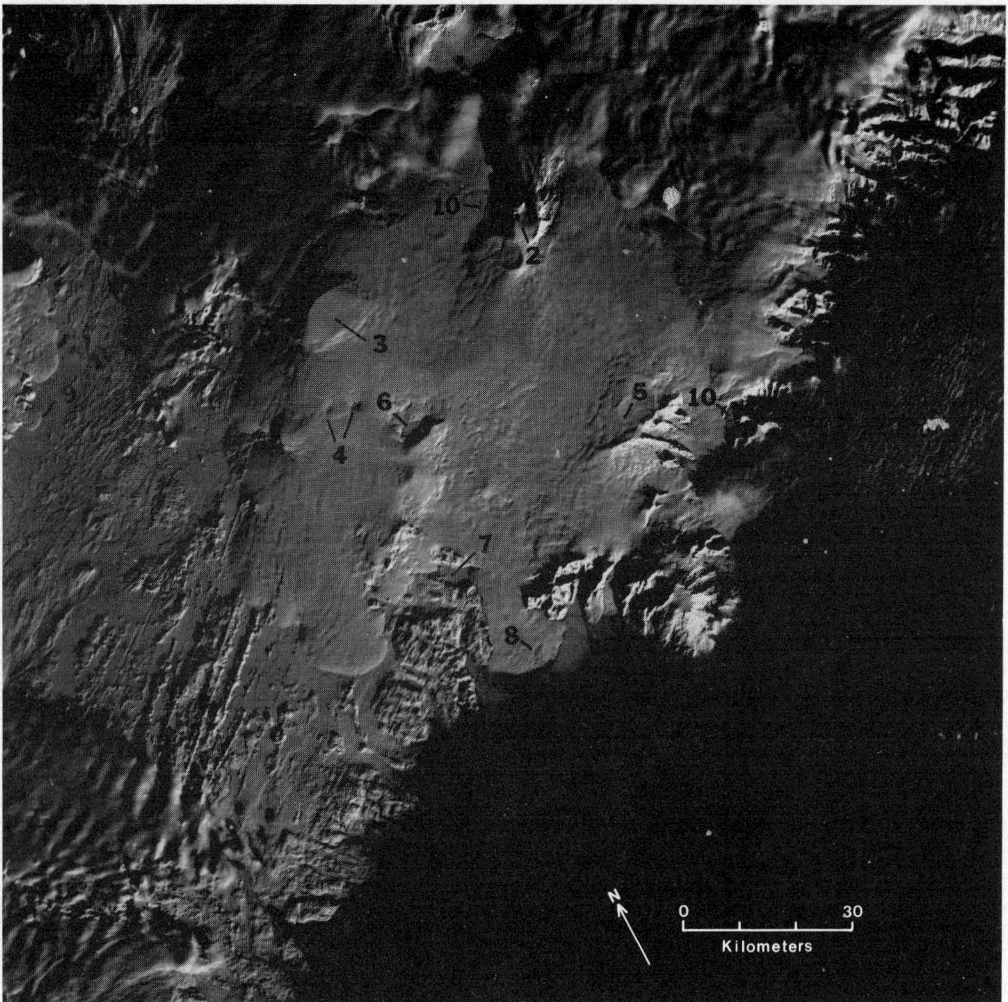

Fig. 31-204. Low solar elevation angle (7°), Landsat 1 image (1192-12084; 31 January 1973) of the snow-covered Vatnajökull area and environs, Iceland, which shows a number of volcanic and glaciologic features within and beyond the limits of the Vatnajökull icecap, Iceland's largest, at about 8,300 km² (Williams and Thorarinsson, 1974; Thorarinsson et al., 1974; Williams, 1976b). Refer to Figure 31-205 for an index map of the Vatnajökull area. Features shown include: (1) elliptical shape of a previously unknown subglacial caldera southwest of Kverkfjöll; (2) elliptical shape of partially subglacial caldera at Kverkfjöll, with trend of its geothermal area on the northwest side (Hveradalur) extending southwest into the icecap and intersecting the caldera shown in (1) (See Figures 31-152A, 31-152B, and 31-205); (3) hyaloclastite caldera of the subglacial volcano at Bárdarbunga (See also Figure 31-194 for large subglacial volcano in the Antarctic); (4) two collapse cauldrons resulting from subglacial geothermal and/or volcanic activity; the easternmost one, with the associated sinuous channel to the southwest is the source of jökulhlaups (glacier outburst floods) along the Shaftá; flood water is channeled between subglacial hyaloclastite ridges shown in Figure 31-206; (5) faint ellipitical features and associated nunataks at Esjufjöll, probably an ice-covered large central volcano; (6) the large depression of Grímsvötn, a well-known subglacial caldera and source of periodic jökulhlaups under Skeidarárjökull across Skeidarásandur to the south, which are caused by subglacial geothermal and/or volcanic activity (See the section of this chapter on "Remote Sensing of Glaciologic Hazards" and the figures on pages 64, 65, and 66 of the Time-Life Book, *Glacier* (Bailey, 1982)); (7) the frozen lake, Graenalón, a source of jökulhlaups resulting from the failure of its ice dam, usually in late summer (Rist, 1974); (8) the partially snow-covered terminus of Skeidarárjökull. Around the periphery of the Grímsvötn caldera are a number of punctate features resulting from ice cauldron subsidence following the March 1972 jökulhlaup (Thorarinsson et al., 1974; Björnsson, 1975; Tómasson, 1975). Southwest of Vatnajökull are superb examples of northeast-trending grabens, crater rows, and hyaloclastite (Móberg) ridges (Thorarinsson, 1974; Thorarinsson et al., 1974); the ridges continue under the icecap as shown on Figure 31-206. Two prominent volcano-tectonic lineaments can be seen on this image. One extends N.45° E. for 80 km between Kverkfjöll ((1) and (2)), along the eastern edge of Grímsvötn (6) to the southwestern edge of Vatnajökull. It approximately follows the concealed contact (Figure 31-205) between late Quaternary and Holocene volcanism and Early Quaternary flood basalts. The second lineament extends N.35° W., just north of Graenalón at (7) and may have important regional tectonic significance (Saemundsson, 1974; Thorarinsson et al., 1974). Concentric recessional moraines in front of Múlajökull, an outlet glacier from Hofsjökull (See Color Figures 31-199 and 31-201) can be seen at (9). Medial moraines are visible on Dyngjujökull (See Figure 31-152A) at (10) and southeast of Esjufjöll at Vedurárrönd. This image was published as a 1:500,000-scale Landsat image map by the U.S. Geological Survey (1977).

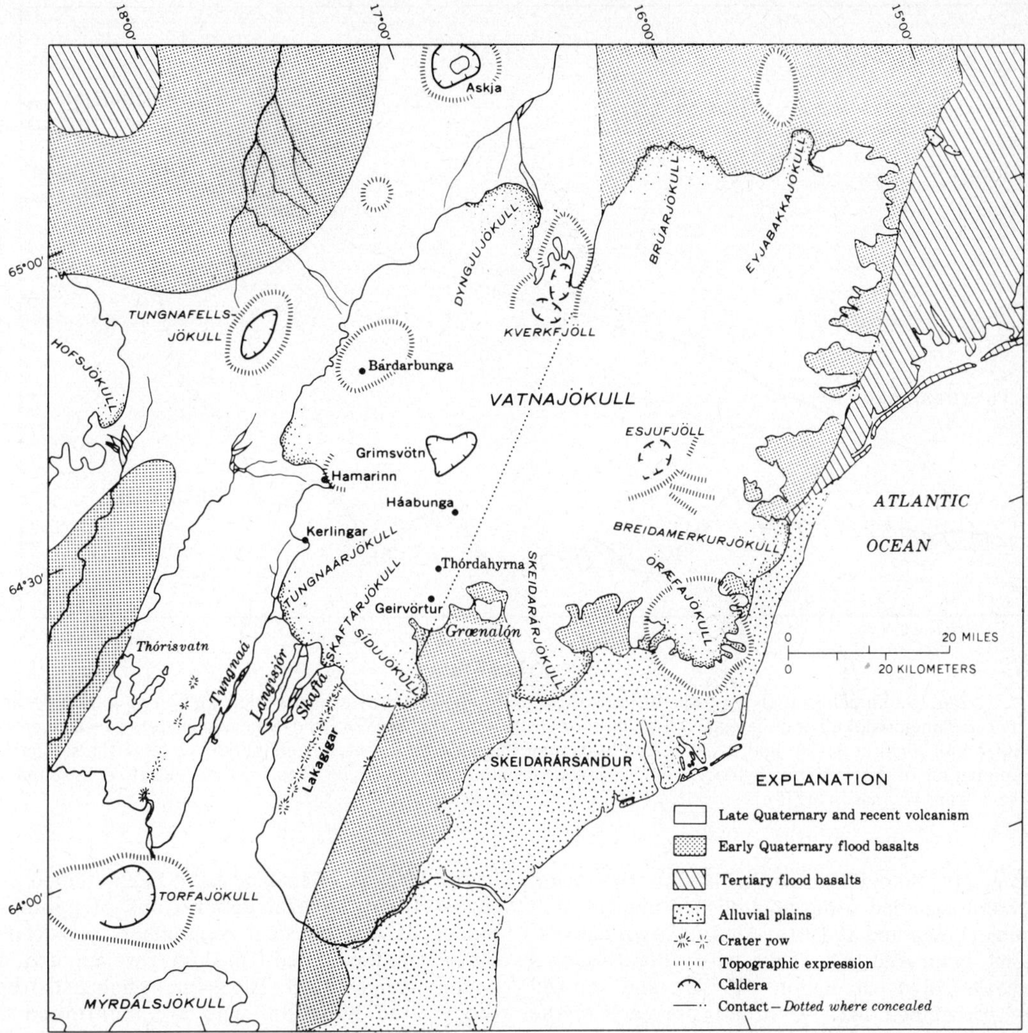

Fig. 31-205. Geologic sketch map of the Vatnajökull area and environs, southeastern Iceland (from Williams, 1976b). See also the index map to Iceland (Figure 31-198) and the three Landsat images of Vatnajökull (Color Figures 31-201 and 31-202 and Figure 31-204).

Unenhanced and specially enhanced Landsat images have been used by several scientists for analysis of geomorphic, structural, and tectonic features concealed by Iceland's glaciers (Williams et al., 1973; Williams and Thorarinsson, 1974; Thorarinsson et al., 1974. Soha et al., (1976) and Williams et al. (1977) discussed how the interpretability of the 22 September 1973 Landsat image of Vatnajökull could be markedly improved by computer-enhancement techniques (Color Figure 31-201). Reflectivity variations were considered by Williams et al. (1979) to portray the accumulation and ablation zones (bare glacial ice, superposed ice, saturated snow or slush, and wet snow) on Vatnajökull (Color Figure 31-202; Figure 31-203). Delineation of the snowlines on Hofsjökull, Langjökull, Mýrdalsjökull, and Eyjafjallajökull was discussed by Williams (1976a). Munzer and Bodechtel (1980) used digital image processing techniques of CCT's to examine the subglacial landforms of, and to map lineaments on, Vatnajökull. Bodechtel et al. (1979) compared Landsat and Seasat synthetic aperture radar (SAR) images of Iceland, including its glaciers, in their analysis of morphologic and tectonic features. Hunting Geology and Geophysics, Ltd. (n.d.) prepared a 1:500,000-scale Seasat SAR mosaic of Iceland.

Contorted medial moraines or tephra layers in Skeidarárjökull, visible on successive Landsat images, have been used to calculate the speed of flow of this outlet glacier east of Graenalón. During an 11-month interval, between 14 October 1972 (1083-12023) and 22 September 1973 (1426-12070), about 600 m of displacement had occurred (Williams et al., 1974 and 1975).

Area calculations from Landsat images for Iceland's glaciers have been carried out by Williams

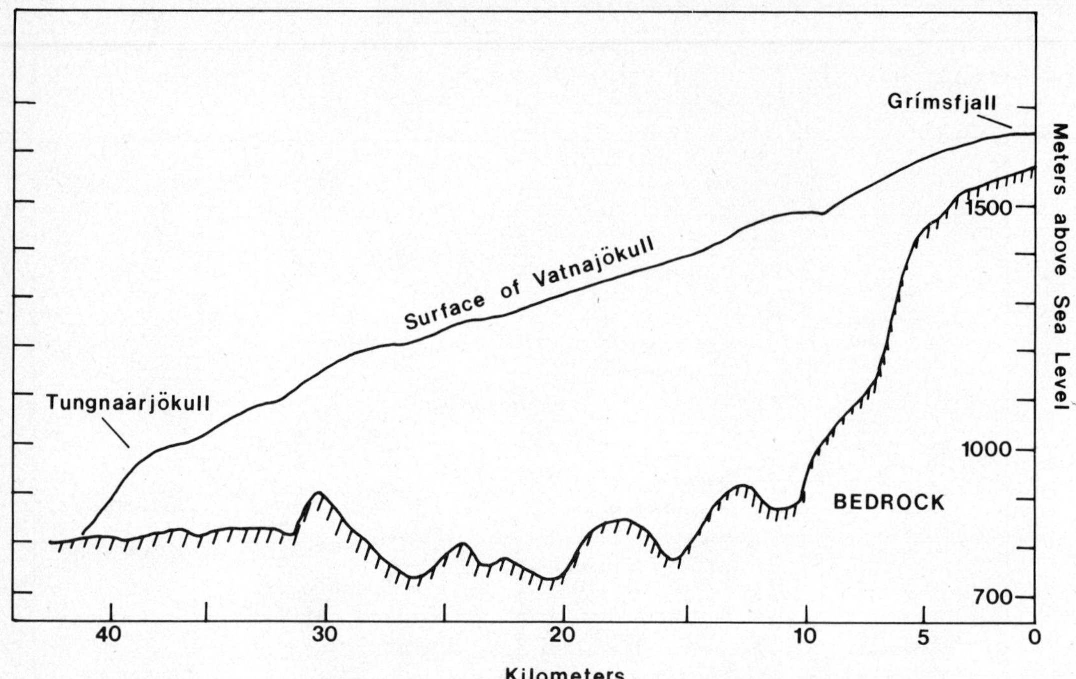

Fig. 31-206. Radio-echosounding profile of the subglacial topography in western Vatnajökull along a traverse line between Tungnaárjökull and the edge (Grímsfjall) of Grímsvötn, the large (45 km^2) volcanic caldera in west-central Vatnajökull (Figures 31-205 and 31-204). The maximum thickness of ice here is about 600 m. Note the subglacial continuation of the Móberg (hyaloclastite) ridges so prominently shown on Figure 31-204. Radio-echosounding profile from Björnsson (1978).

et al. (1975) and Björnsson (1980b). Björnsson (1980b) included some unpublished data of Williams (1983a and 1983b), added his own calculations, compared the results with Thorarinsson's previous area calculations (1958b), and showed the percentage decrease in area for each of the principal glaciers of Iceland. Table 31-39 summarizes excerpts from the work of many scientists who have calculated the area of Iceland's glaciers from published maps, aerial photographs, and Landsat images.

Remote Sensing in Glacial Geology Studies of the Proglacial Area of Breidamerkurjökull, Southeastern Iceland

The subject of glacial geology has produced a diverse scientific literature well documented by Flint (1971). The application of remote sensing technology to glacial geology was discussed by Elson (1980) in *Remote Sensing in Geology* (Siegal and Gillespie, 1980). An excellent set of stereoscopic aerial photographs can be found in Mollard (1974). Rather than summarize the same material covered by Elson (1980) or attempt to distill the enormous amount of glacial geology literature into this small section of the chapter, a review of the important use of aerial photographs in a classic glacial geology study of the proglacial area of Breidamerkurjökull, a large outlet glacier from Vatnajökull in southeastern Iceland (Figures 31-207 and 31-205) is given.

The glaciers of Iceland have been studied extensively by several generations of geomorphologists and physical geographers, especially those from the United Kingdom, Sweden, and, of course, Iceland itself. The geomorphologists from the United Kingdom who are mentioned as textbook authors in the "Introduction" to the "Glaciology" section (Price, 1973; Embleton and King, 1975a, 1975b; Sugden and John, 1976) conducted field work and have published one or more papers on the glaciology and/or glacial geology of Iceland. The late Prof. Hans Ahlmann and Prof. Gunnar Hoppe of Sweden and the late Prof. Sigurdur Thorarinsson of Iceland have studied and published on various aspects of the glaciology and/or glacial geology of Iceland. The outstanding Swedish scientific journal, *Geografiska Annaler,* has been an important publication medium for much of this work.

During the summers of 1965, 1966, and 1967, students and faculty of the Department of Geography, University of Glasgow, Scotland, carried out glacial geology studies of the proglacial area of Breidamerkurjökull (Color Figure 31-201; Figures 31-205 and 31-207). The study serves as an excellent example of the use of sequential aerial photographs for deriving information about glacial processes in an area undergoing rapid change. The scientists used published topographic maps, prepared new topographic and geomorphic maps during field work, and used existing vertical aerial

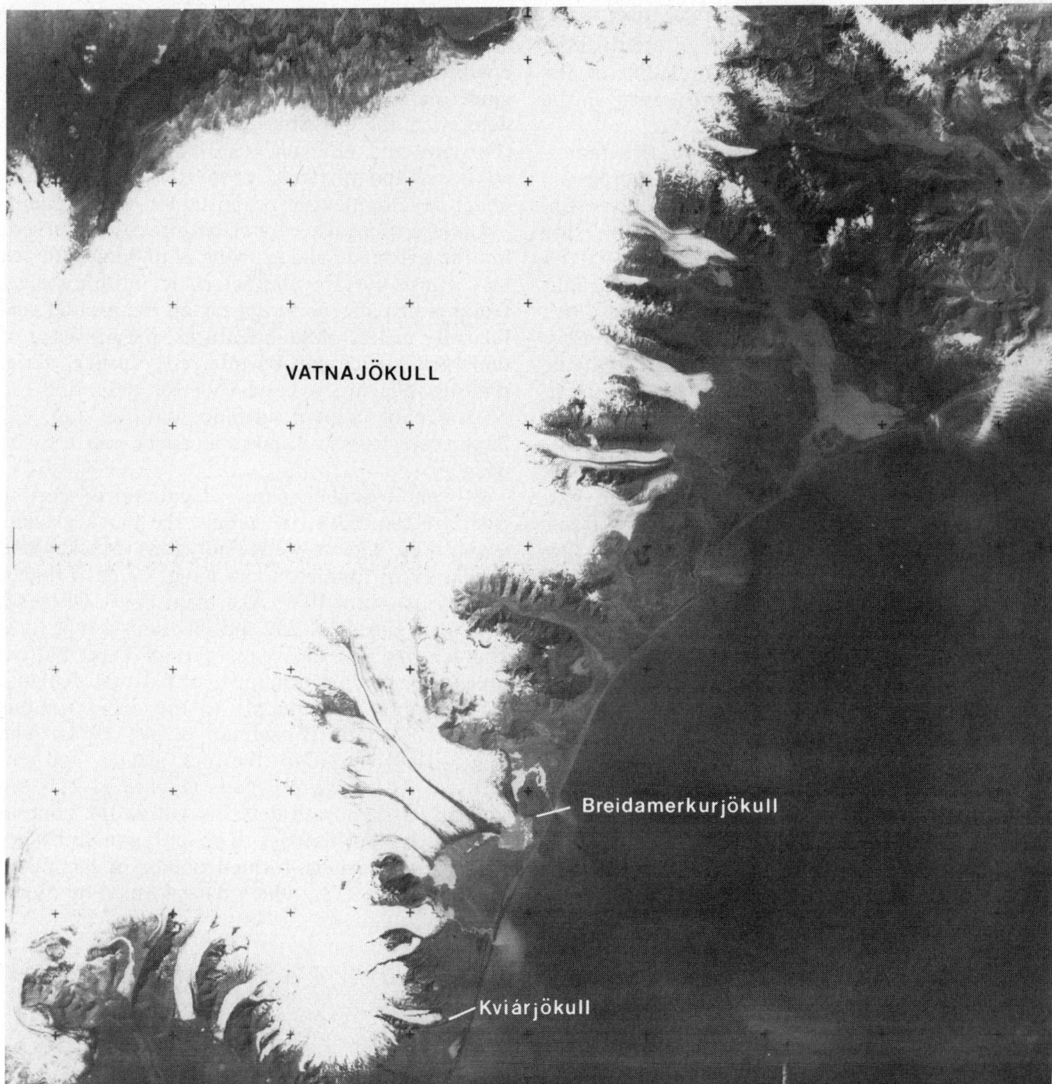

Fig. 31-207. Landsat 3 RBV image (30157-11572-B; 9 August 1978) of the southeast part of Vatnajökull, Iceland, including the outlet glaciers, Breidamerkurjökull, (shown in Color Figure 31-201) and Kvíárjökull (shown in Figure 31-218 and Color Figure 31-219). Compare with Color Figure 31-201, a Landsat MSS color composite of the Vatnajökull ice cap. Landsat MSS images have a pixel resolution of 80 m; Landsat 3 RBV images have a pixel resolution of 30 m (Williams and Ferrigno, 1981).

photographs or acquired new ones especially for the research project (Price, 1969a). Danish Geodetic Institute maps compiled in 1904 from plane-table surveys (See Color Figure 31-219 for example of the type), and a 1951 plane-table map compiled by the Durham University Exploration Society were used. In addition, two new topographic maps were compiled by the University of Glasgow by photogrammetric means from August 1945 U.S. Army Air Force vertical aerial photographs and August 1965 vertical aerial photographs on color, color-infrared, infrared, and black-and-white films of Breidamerkurjökull acquired by the University of Glasgow in cooperation with the Iceland Geodetic Survey and the Icelandic Highways Department (Welch and Howarth, 1968; Price and Howarth, 1971). In addition to the 1945 and 1965 vertical aerial photographs, the research team also used July 1960, July 1961, June 1964, and September 1964 vertical aerial photographs acquired by the Iceland Geodetic Survey. (During August 1965, vertical aerial photographs were also acquired of Surtsey during the Breidamerkurjökull surveys. See the "Geomorphology" section of this chapter for a discussion of Surtsey).

One of the color aerial photographs acquired of Breidamerkurjökull in August 1965 was subsequently published in the *Manual of Color Aerial Photography* with an excellent discussion of the photograph by Roy Welch (Welch, 1968), although the color photograph was inadvertently reversed during printing. Welch (1966, 1967, 1968, and 1969) discussed the use of different aerial films and photogrammetric measurements of glacial landforms (Welch and Howarth, 1968) in the mapping of Breidamerkurjökull.

The geomorphic study illustrated Breidamerkurjökull's changing drainage system across its proglacial area in a superb series of seven sequential maps by Price and Howarth (1971) which are based on maps produced in 1904 and 1951 by ground-survey methods and aerial photographs recorded in 1945, 1960, 1961, 1964, and 1965. In addition to the preparation of various maps, additional studies were undertaken to evaluate the use of photogrammetric and air photo interpretation techniques for studies of glacial landforms (Welch and Howarth, 1968), and to develop a recent historical record of the formation of kames, eskers, moraines, and drainage patterns (Price, 1969b and 1971; Howarth and Price, 1969; Price and Howarth, 1971; Howarth, 1966 and 1971). Many of the above studies are extensions of work described in unpublished Ph.D. dissertations by Welch (1967) and Howarth (1968).

Another result of the research was the publication of two 1:30,000-scale glacial geology maps of Breidamerkurjökull, with 10-m contour intervals, showing the landforms in its proglacial area and the position of the glacier terminus, based on vertical aerial photographs acquired in August 1945 and August 1965 (Howarth and Welch, 1969). In Color Figure 31-208, A and B are sections from the two maps showing the striking changes in the glacier terminus, proglacial lakes, drainage systems, and landform patterns of part of the proglacial area of Breidamerkurjökull, as the glacier retreated approximately 1 km between August 1945 and August 1965. In Color Figure 31-208 A and B, note especially the extensive eskers and kames and kettle topography which developed during the 20-year interval.

Permafrost

In glaciology, the study and monitoring of the accumulation of snow, the distribution and form of ice masses, and the response of snow and ice to changing weather conditions and climatic fluctuation are all topics of study in which aerial and satellite photographs and images are very adequate because of the distinct appearance of glaciers and glaciological features on such data.

There is, however, another environment of ice, not visible on the Earth's surface, but which exists in arctic areas as subsurface, frozen ground ice. In the Russian scientific literature, this world of ice is sometimes called the underground glaciation. During the summer, the surface layer thaws (the active layer), but downwards the ground remains frozen for tens or hundreds of meters (permafrost). Ice-bound material has been recorded at a depth of more than 1,400 m in Siberian drill holes, for example.

Because about 25 percent of the Earth's surface is underlain by permafrost, the environment of underground ice has a great influence on natural conditions and human activity, especially for engineering projects. Thus, an important task is to determine the possible existence of permafrost (Ferrians and Hobson, 1973), including its distribution and physical properties, in any area under development or proposed for development.

Underground ice cannot be directly observed, but the existence and growing of underground ice may cause certain characteristic morphological features or patterns to appear on the ground surface. By means of such features, the presence of underground ice can be inferred. Vertical aerial photographs are a good tool for detecting the presence of critical surface features and, for large-scale forms, Landsat imagery can also be used.

Morphological features diagnostic of permafrost are the palsa, the pingo, the block glacier, thaw lakes, and ice-wedge polygons (Mackay and Black, 1973; Embleton and King, 1975a; French, 1976; Washburn, 1979; Åkerman, 1980). Of these, the palsa (Figure 31-209 and 31-210) is a type form characterized in the outer part of a permafrost area (zone of discontinuous permafrost). Iceland, for example, falls mainly in the discontinuous permafrost zone (Friedman et al., 1971). The pingo (closed system), the rock glacier, and patterns of ice-wedge polygons (Figure 31-211) are indicative of ice-rich deposits within the zone of continuous permafrost. The polygon lines are underlain by wedge-formed bodies of laminated ice (Figure 31-212), which have formed by recur-

Fig. 31-209. Vertical, black-and-white aerial photograph of a typical palsa area in the Varanger district, northeasternmost Norway (Svensson, 1963a). The bog surface is covered by hummocks or ridges of frozen peat (maximally 5 m high), sometimes growing together and developing low, elevated surfaces in the bog. Approximate scale, 1:20,000. Aerial photograph courtesy of Fjellanger-Wideroe A/S, Oslo.

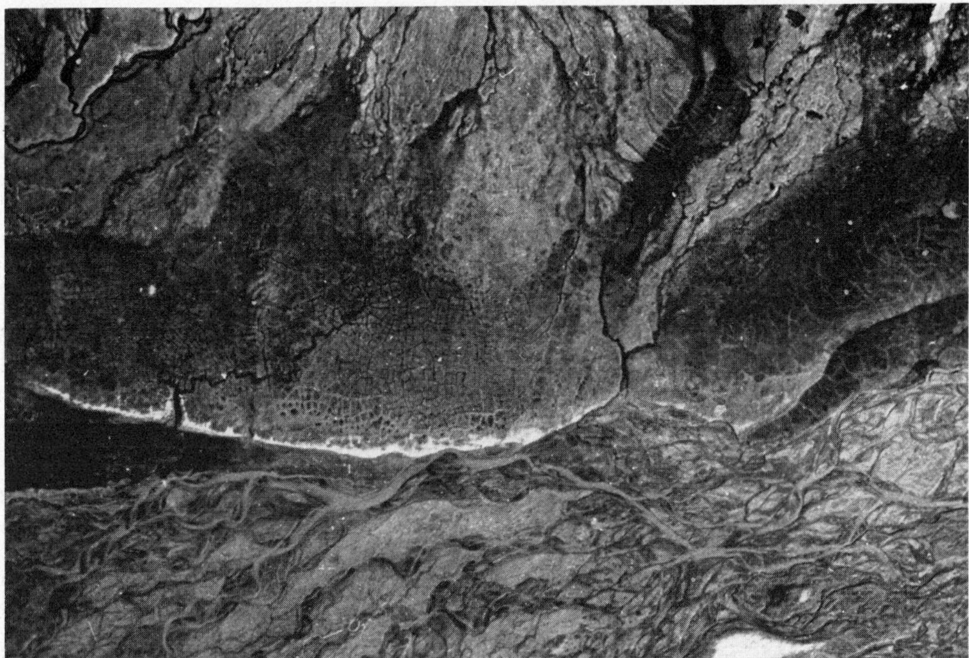

Fig. 31-211. Vertical, black-and-white aerial photograph of a terrace with a network of ice-wedge polygons in the lower course of the river Adventelva, Spitsbergen (Svalbard), Norway. The relief of the polygon surface is, in some places, enhanced by water, which has accumulated in the low polygon furrows (beaded drainage). Approximate scale, 1:56,000. Aerial photograph courtesy of Norsk Polarinstitutt, Oslo.

rent fissuring of the ground caused by thermal contraction.

The melting out of ice because of natural erosion or man-induced disturbance of the thermal equilibrium may cause depressions, the so-called thermokarst basins, in permafrost areas. In general they are easily detectable in aerial photographs as pools or lakes (Figure 31-213).

As with true glaciers, the amount and distribution of underground ice that is associated with permafrost depends on climatic fluctuation. Former areas of permafrost, for example the zone bordering the Wisconsinan (Weichselian) glaciation, are now free of perennial ground ice and are situated far outside the present arctic realm. However, subtle traces of former frozen ground features may still be revealed on aerial photographs of these areas. Usually they are very faint, but at times they are outlined by a low surface relief or by vegetative crop marks in cultivated land areas (See also the section on Geomorphology of this chapter). These relict or fossil features are of great interest for paleoclimatic and paleobotanic reconstruction of the former periglacial environment (Péwé, 1973; Svensson, 1978).

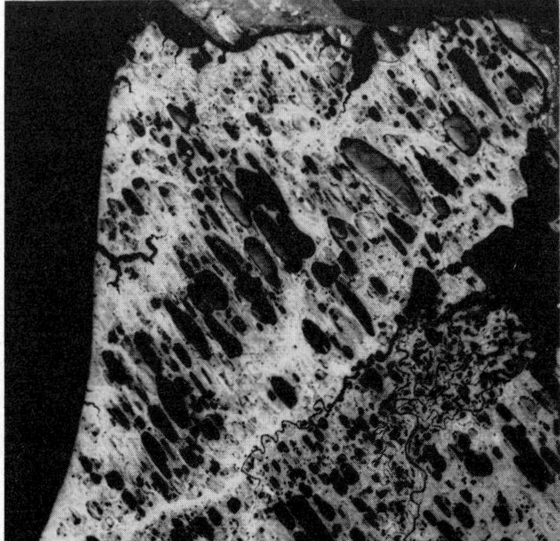

Fig. 31-213. Thaw lakes formed in ice-rich permafrost on the coastal plain south of Point Barrow, northern Alaska. Approximate scale of image is 1:1,000,000 (Landsat MSS band 7 image 2902-21175; 12 July 1977).

GEOMORPHOLOGY

Introduction

Geomorphology is one of the principal subdisciplines of geology in which remotely sensed data play an important role in most scientific investigations, whether they be local, regional, or global. Geomorphology is defined as "The science that treats the general configuration of the Earth's surface; specifically the study of the classification, description, nature, origin, and development of present landforms and their relationships to underlying structures, and of the history of geologic changes as recorded by these surface features." The term is especially applied to the genetic interpretation of landforms, but has also been restricted to features produced only by erosion or deposition. The term was widely applied in Europe before it was used in the U.S., where it has come to replace the term physiography and is usually considered as a branch of geology; in Great Britain (and in Scandinavia), it is usually regarded as a branch of geography" (Bates and Jackson, 1980). Considerable overlap exists between geomorphology and physical geography; both rely heavily on stereopaired aerial photographs and other types of image data in various studies. Verstappen (1977a) published the first specialized textbook entirely devoted to "Remote Sensing in Geomorphology," in which he introduced students to the great variety of types of remotely sensed data available for geomorphological analysis. In addition to traditional studies with stereopaired aerial photographs, he provided examples of side-looking airborne radar (SLAR) images, thermal infrared images (thermographs), sidescan sonar images and mosaics, and satellite images. Verstappen (1977b) also published an atlas of aerial photographs in geomorphological mapping as a chapter in the ITC Textbook of Photo-Interpretation.

The subject of geomorphology (or physical geography) has been addressed in a number of traditional textbooks (Thornbury, 1969; Strahler, 1969), modern textbooks (Scheidegger, 1970; Ritter, 1978; Bloom, 1978), scientific journals (e.g., *Geografiska Annaler, Zeitschrift für Geomorphologie*, etc.), and in both the *Manual of Photographic Interpretation* (Colwell, 1960) and the first edition of the *Manual of Remote Sensing* (Reeves et al., 1975b). Chapter 6, Photo Interpretation in Geology (Tator et al., 1960), of the *Manual of Photographic Interpretation* contains a lengthy section (47 pages) on geomorphology, and the authors state that "The geological interpretation of aerial photographs is essentially applied geomorphology." Several stereopaired aerial photographs are used to illustrate both Chapter 6 and Chapter 16, Terrain and Minerals: Assessment and Evaluation (Reeves et al., 1975a) in the first edition of the *Manual of Remote Sensing*. In Chapter 16 of the Manual, 25 pages are specifically devoted to geomorphology. Lowman and Lattman (1980) give a brief review of the use of remote sensing technology to geomorphology in the book, *Remote Sensing in Geology* (Siegal and Gillespie, 1980).

The use of aerial photographs, single or stereopaired, satellite images, and images which record energy outside the visible part of the electromagnetic spectrum for geomorphological studies is a modern adaptation of H. T. U. Smith's comments about aerial photographs 40 years before the publication of this second edition of the *Manual of Remote Sensing:* "it is in the field of . . . geomorphic interpretation that aerial photographs offer the broadest opportunities" (Smith, 1943; also in Reeves et al., 1975a).

The use of aerial photographs in geomorphic and other types of geologic studies has been discussed previously in the section on "Photogeology and Geological Remote Sensing" of the "Analysis Techniques" part of this chapter. Excellent collections of stereopaired aerial photographs are available in a variety of textbooks and handbooks (for example, Ray, 1960; Miller and Miller, 1961; Mollard, 1974), and the reader is referred to the "Photogeology" section for a discussion of this literature.

The series of Landsat satellites has markedly broadened the availability of image data of the Earth's surface, so geomorphologists now have two-dimensional image information of virtually the entire land area and many shallow-sea areas of the planet. Comparative two-dimensional studies of many of the Earth's varied landforms have become possible for the first time. McKee (1979) has already taken advantage of available Landsat data to complete the first global geomorphic study of sand seas. By the mid-1980s, the U.S. Geological Survey will have published another global study, "Satellite Image Atlas of Glaciers," also dependent on the availability of Landsat images of the entire planet (except for the areas poleward of 81° north and south latitudes). Williams and Carter (1976) and Short et al. (1976) have both published an excellent collection of annotated Landsat images.

In this section of the chapter, geomorphology will be briefly addressed in two aspects, relief and non-relief. Geomorphic studies of relief features are dependent on the availability of stereopaired aerial photographs. Two examples are, respectively, the volcanic island of Surtsey and the terminal moraine of Kvíárjökull, both from Iceland. Geomorphic studies of non-relief features do not depend on stereopaired aerial photographs. They do, however, require images acquired at optimum times (such as in the case of "cropmarks" in archaeologic studies; Hampton, 1975) or multispectral images that enhance subtle patterns on the Earth's surface. The examples presented are all drawn from field studies of periglacial phenomena in Sweden.

Geomorphology of Relief Features

Geomorphic Evolution of Surtsey, Iceland

The volcanic island of Surtsey, one of many such islands in the Vestmannaeyjar archipelago off the southwestern coast of Iceland, first appeared above the ocean's surface on 14 November 1963 (Thorarinsson, 1967). During the next 3½ years Surtsey, or its three satellite volcanoes, Surtla (solely a submarine landform), Syrtlingur, and Jólnir, were volcanically active for varying periods of time (Figure 31-214). The initial volcanic activity in each island was explosive, as sea water came in contact with upwelling magma (Figure 31-216A). Only on Surtsey did this explosive, tephra-producing activity evolve into an effusive eruption, with successive lava flows forming a cap on the initially unconsolidated

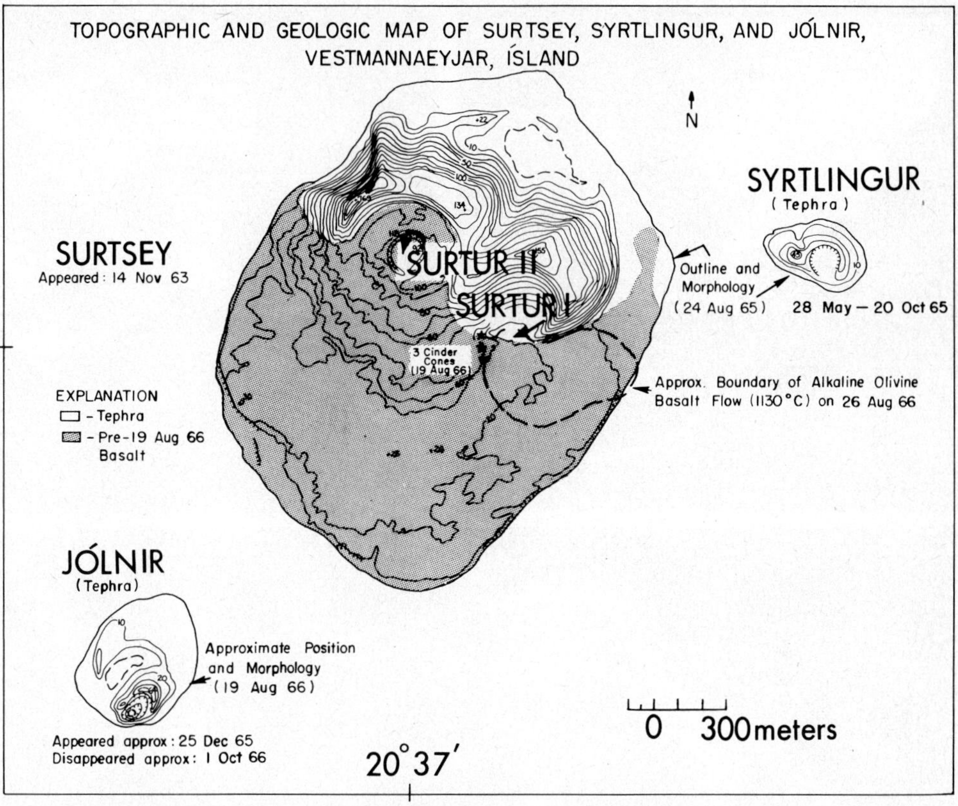

Fig. 31-214. Topographic and geologic map of the volcanic island, Surtsey, and two of its ephemeral satellite volcanic islands, Jólnir (See Figure 31-216D, 31-216E, 31-151, and 31-217), and Syrtlingur. The outline of Surtsey is as it appeared on 24 August 1965. The map was compiled in 1967 from the following map and aerial photographic sources: Iceland Geodetic Survey, Thorarinsson (1965), and field observations by Richard S. Williams, Jr., and Jules D. Friedman of the U.S. Geological Survey.

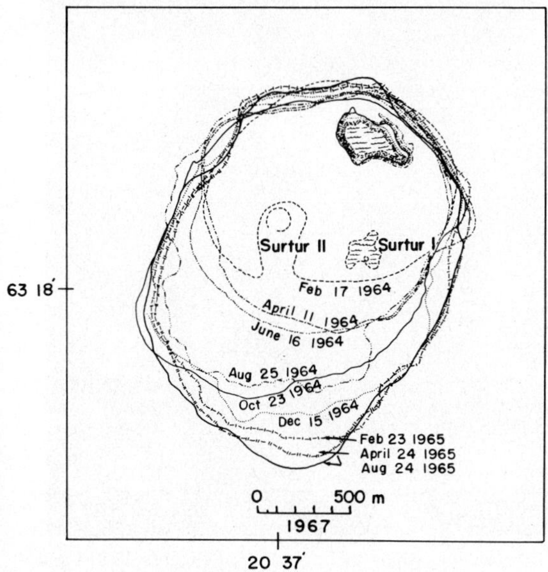

Fig. 31-215. The changing outlines of the volcanic island, Surtsey, Vestmannaeyjar, Iceland, during the period 17 February 1964 (Figure 31-216A) to 24 August 1965 (Figure 31-216C). During this interval of time Surtsey added new land from tephra falls, lava flows, and accretion of its northern coast, and loss of land because of wave erosion on the south (Norrman, 1980). Map modified from Thorarinsson (1965, 1967).

tephra. Between November 1963 and July 1967, Surtsey generally increased in area because of volcanic activity (Figure 31-215). Between July 1967 and the present, the island's area has generally decreased each year (Table 31-40).

Surtsey has been a scientific preserve for geologists and biologists for most of its nearly 20-year history. Biologists, such as the Icelandic botanist, Sturla Fridriksson, have carefully studied the establishment of plant life on the island from very early in its history (Fridriksson, 1975). Geomorphologists have been especially interested in studying the geomorphic evolution of Surtsey. Thorarinsson (1965, 1967, 1969b) discussed the

Table 31-40

Areal coastal changes of Surtsey Island, 1967–75 (hectares) (From Norrman, 1980)

Period	Cliffs	Beaches	Total
1967–68	−13	−2	−15
1968–69	−7	−10	−17
1969–70	−7	+4	−3
1970–72 (2 yrs.)	−9	−5	−14
1972–74 (2 yrs.)	−11	−1	−12
1974–75	−3	+4	+1
1967–75	−50	−10	−60

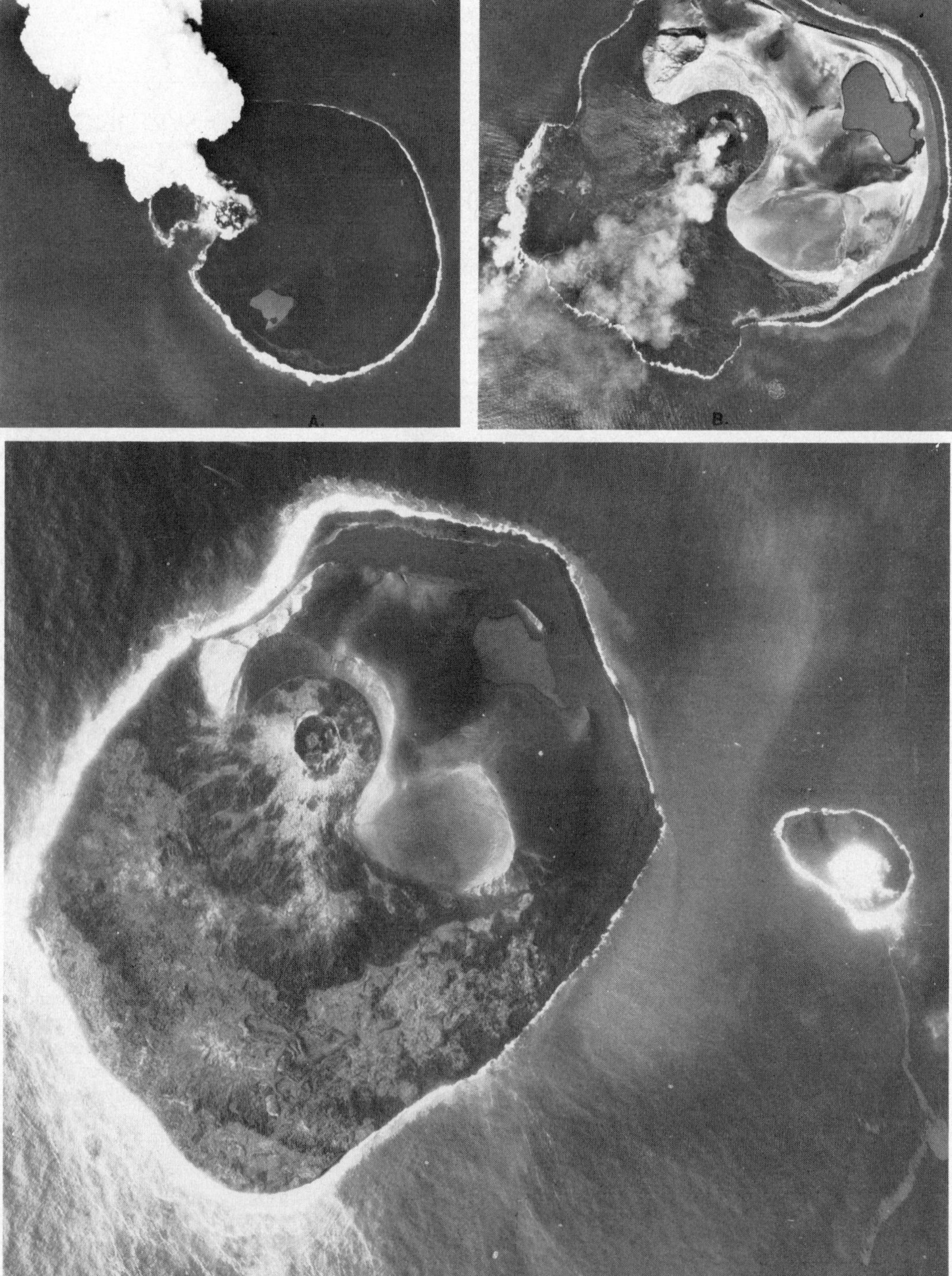

Fig. 31-216. A sequence of vertical aerial photographs of the volcanic island of Surtsey (See Figures 31-214 and 31-215) from 17 February 1964 to 20 July 1979, showing the geomorphic evolution of Surtsey during its active volcanic stage (November 1963–July 1967) and its erosional stage (post-July 1967). All aerial photographs (except Color Figure 31-216H) were acquired by the Iceland Geodetic Survey. A. First vertical aerial photograph of Surtsey, showing an explosive tephra eruption from the Surtur II crater. At this time the island is composed of tephra (17 February 1964, 2,115 m altitude, Frame 6086A). B. Effusive volcanic activity has spread an apron of lava flows in a modified shield volcano from the Surtur II crater (25 August 1964, 2,110 m altitude, Frame 6609A). C. Volcanic

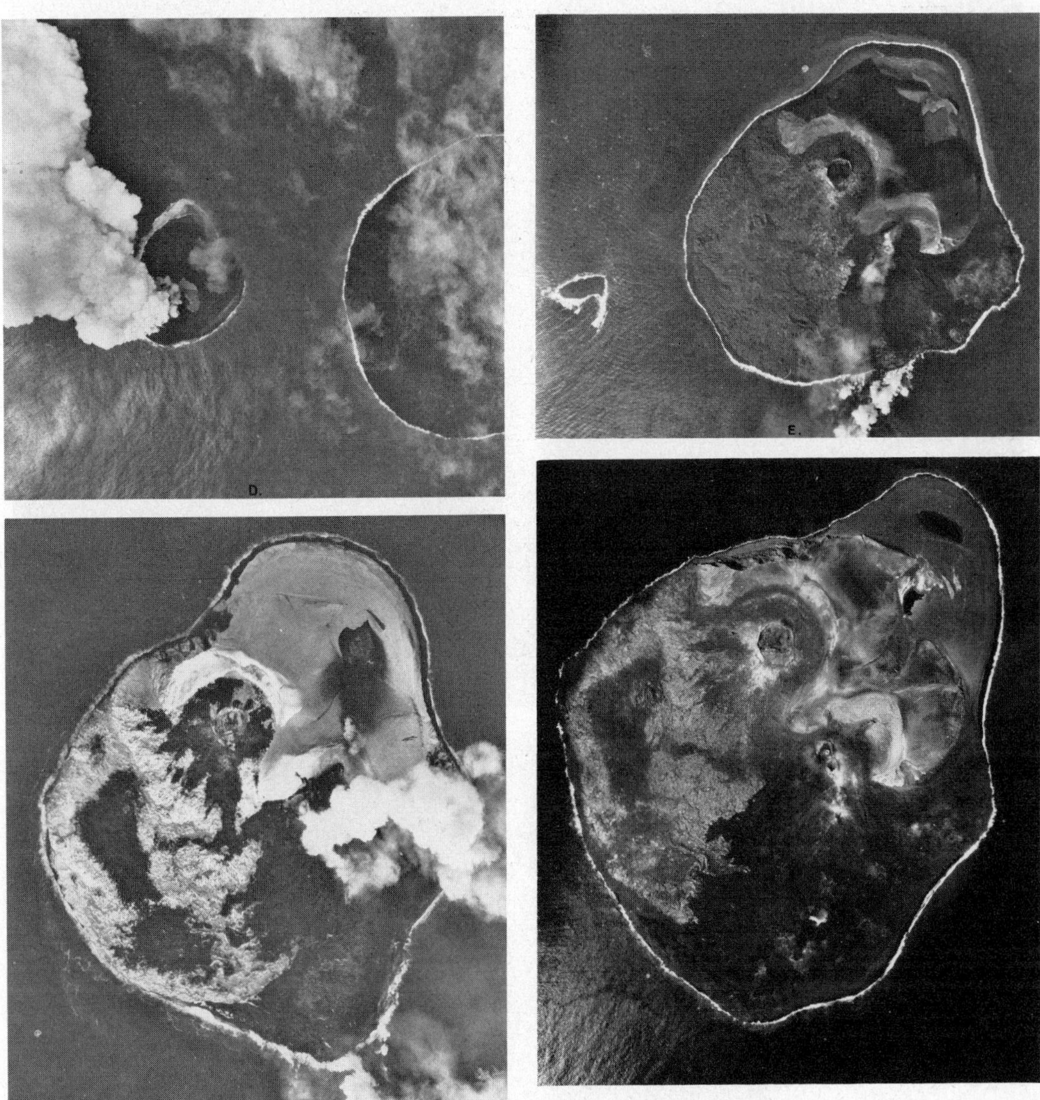

activity on Surtsey had subsided; focus of activity has shifted to its satellite volcano, the ephemeral island of Syrtlingur (see Figure 31-214) (24 August 1965, altitude unknown, Frame 375). D. Volcanic activity on Surtsey is still dormant; focus of activity has shifted to another satellite volcano, Jólnir (See Figure 31-214, 31-216E, and 31-217) (14 June 1966, 2,200 m altitude, Frame 8820A). E. New effusive activity has occurred on Surtsey in the Surtur I crater; Jólnir has nearly been washed away (2 October 1966, 2,310 m altitude, Frame 9122A). F. Effusive volcanic activity is renewed on the south and north slopes of the Surtur I crater. Note how the melt pattern in the light snow cover on the west side of Surtsey reveals the pattern of thermal emission. Compare with the aerial thermograph of 19 August 1966 shown in Figure 31-151. (3 January 1967, 2,200 m altitude, Frame 9508A). G. Volcanic activity on Surtsey has finally ceased early in the month, leaving the island subject only to erosional processes. (17 July 1967, 2,250 m altitude, Frame 243B). H. NASA color aerial photograph, 3 April 1968. See Color Figure 31-216H. I. (next page) Stereopair of vertical aerial photographs of Surtsey showing the extent of erosion on the south coast and deposition of eroded lavas as a cobble beach to the north (23 August 1969, 2,240 m altitude, Frames 3953B and 3954B). J. Surtsey in July 1979 showing the depositional peninsula on its northern coast and the pronounced erosion which has taken place on its southern coast. Eolian erosion has also affected the tephra cones of Surtur I and II, revealing different depositional layers (20 July 1979, 2,230 m altitude, Frame 4411 F). Aerial photographs A, B, D, E, F, G, and I were acquired with a 115.06 mm-focal length camera. Aerial photograph C used a camera with a 152.44 mm focal length.

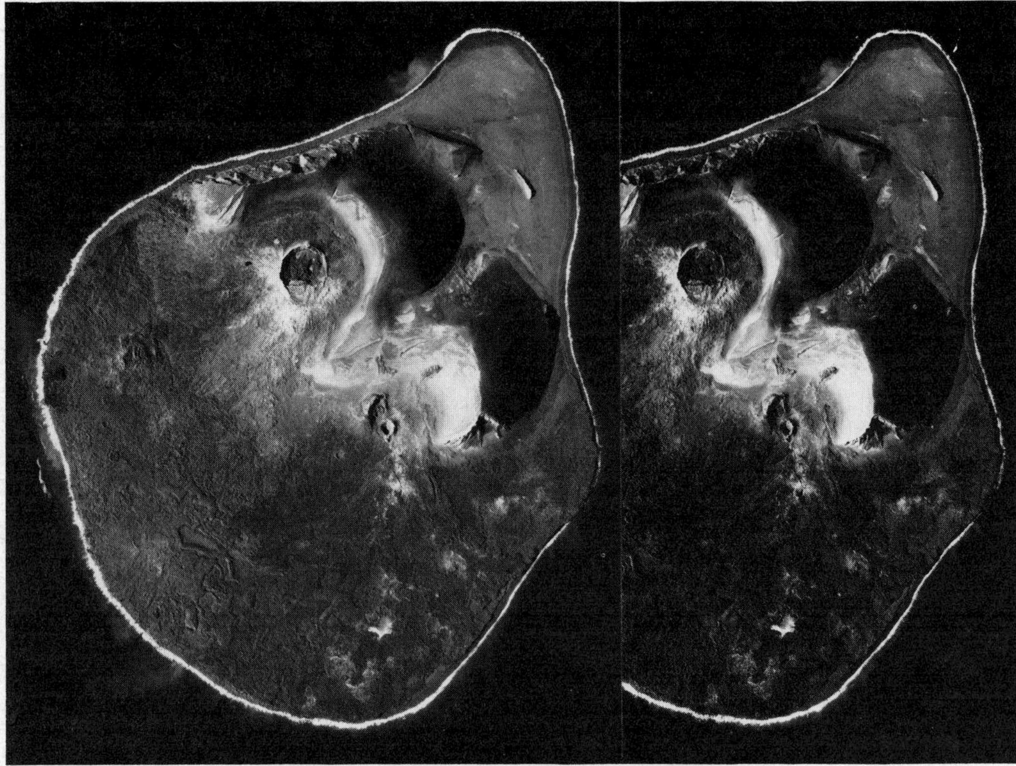

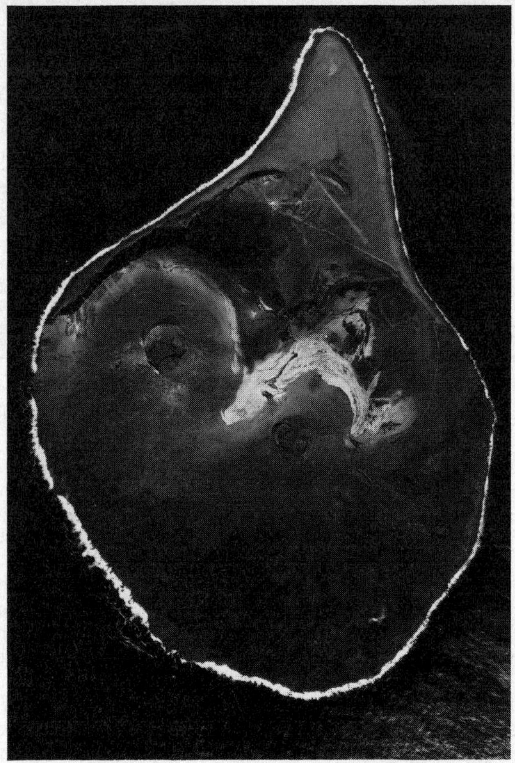

Aerial photographs H and J used a 152 mm and a 152.10 mm (RC-10 camera) focal length, respectively. Ninety-two different sets of vertical and oblique aerial photographs of Surtsey were acquired between 17 February 1964 and 22 July 1981, including one or more each year except 1978. Note: Figures 31-216A–J have all been reduced from their contact print scale for printing in this *Manual*.

various phases of the Surtsey eruption in a series of books and papers. It was the Swedish geomorphologist, John Norrman, however, who has carried out the most intensive and sustained geomorphic studies of Surtsey (Norrman, 1980). Norrman's excellent geomorphic studies are based on field surveys and the preparation of detailed 1:5,000-scale topographic maps of Surtsey by aerial photogrammetric methods. Norrman's map of the terrestrial and submarine topography of Surtsey was published in 1970 from field work carried out in 1968 (Norrman, 1970). Table 31-40 shows the annual areal changes in Surtsey between 1967 and 1975 as measured from aerial photographs.

Before July 1967, the last time in which lava was flowing on the surface, Surtsey was generally increasing in area from successive periods of explosive, then effusive volcanic activity. Figure 31-215 is a map showing the various outlines of Surtsey between 17 February 1964 and 24 August 1965 (Thorarinsson, 1965). Figure 31-216 A–J (H is Color Figure 31-216H) is a sequence of 10 vertical aerial photographs of Surtsey (and its satellite volcanic islands of Jólnir (See Figures 31-214, 31-216D, 31-216E, 31-151, and 31-217) and Syrtlingur (Figure 31-216C) acquired during a 15½-year period (between 17 February 1964 and 20 July 1979), showing the geomorphic evolution as recorded photographically. Between 17 February 1964 and 22 July 1981, 92 different sets of vertical and oblique aerial photographs were acquired of Surtsey, including one or more each year except for 1978. Vertical and oblique aerial photographs (and satellite images) are important sources of data for geomorphologists who, in a time-lapse fashion, study relatively rapid changes in landforms, such as have been occurring on Surtsey during the past 19 years.

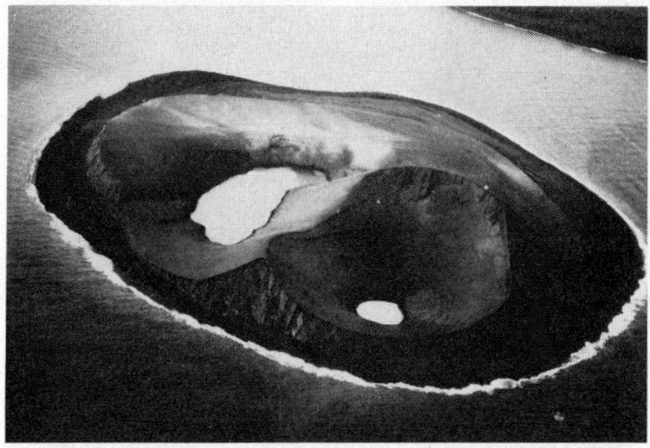

Fig. 31-217. Oblique aerial photograph of the satellite volcanic island of Jólnir as it appeared on 19 August 1966 (Friedman and Williams, 1968). Surtsey is partly visible in the upper righthand corner of the photograph. (See Figure 31-151 for an aerial thermograph of Jólnir on the same date. See also Figure 31-214, 31-216D, and 31-216E). Jólnir was composed of tephra which is easily eroded by wave action. It first appeared on approximately 25 December 1965 and was nearly gone by 2 October 1966 (Figure 31-216E). By July 1973 Jólnir had been further reduced to a 30 m-deep submarine shoal. (Aerial photograph by Richard S. Williams, Jr., U.S. Geological Survey).

Most geomorphic processes produce slow, evolutionary changes in landforms; a few geomorphic processes can cause a catastrophic or rapid change in landforms. The scale of the photograph or image must be sufficient, however, to record any changes. Changes in landforms may be readily apparent on aerial photographs but not evident on Landsat images.

Remote Sensing of Kvíárjökull Outlet Glacier, Iceland

Kvíárjökull is one of about 100 outlet glaciers that emanate from the 8,300 km² Vatnajökull ice cap in southeastern Iceland (Color Figure 31-201; Figure 31-205). Kvíárjökull and its classic end moraine were first mapped by planetable methods in 1904 by the Danish Geodetic Survey. The map was published as Blad 87, Öraefajökull, of the Danish General Staff's 1:100,000-scale map series of Iceland. In 1904, Kvíárjökull was actively forming its terminal moraine, and the map clearly shows the glacier only 1.5 km from the coast. On 30 August 1945, the U.S. Army Air Force, as part of an Army Map Service project to remap Iceland at a scale of 1:50,000, photographed Kvíárjökull. Figure 31-218 is a stereopair of aerial photographs which shows that the glacier has retreated about 0.5 km from its classically shaped terminal moraine. Color Figure 31-219 is a 1972 edition of the 1:100,000-scale map of Kvíárjökull (Blad 87, Öraefajökull) which was revised photogrammetrically by the Iceland Geodetic Survey from vertical aerial photographs acquired during the period 1961–1969. Figure 31-207 is a Landsat 3 RBV image of Kvíárjökull as it appeared on 9 August 1978. Stereopaired vertical aerial photographs, in conjunction with historic maps, permit geomorphologists to accurately study fluctuations of glacier termini and to document and map the often ephemeral features at the margins of glaciers (See Color Figure 31-208A and B).

Geomorphic Studies of Non-Relief Features

For many areas the origin of the landscape, its formation and evolution, is established through many years of field mapping by geologists and landscape analysis by geomorphologists. Scientists, however, always have a desire to know more. In many landscapes, the surface is rich in details that make possible a complete reconstruction of the processes that shaped the surface or subsequently acted upon it. In other landscapes, evidence is limited or appears to be absent entirely, especially in areas of low relief, such as plains. Morphologically, a plain often appears to be of no interest from the vantage point of the field geologist.

By means of aerial photographs, however, this impression may be drastically changed. Contrasts on the soil surface or patterns in the vegetative

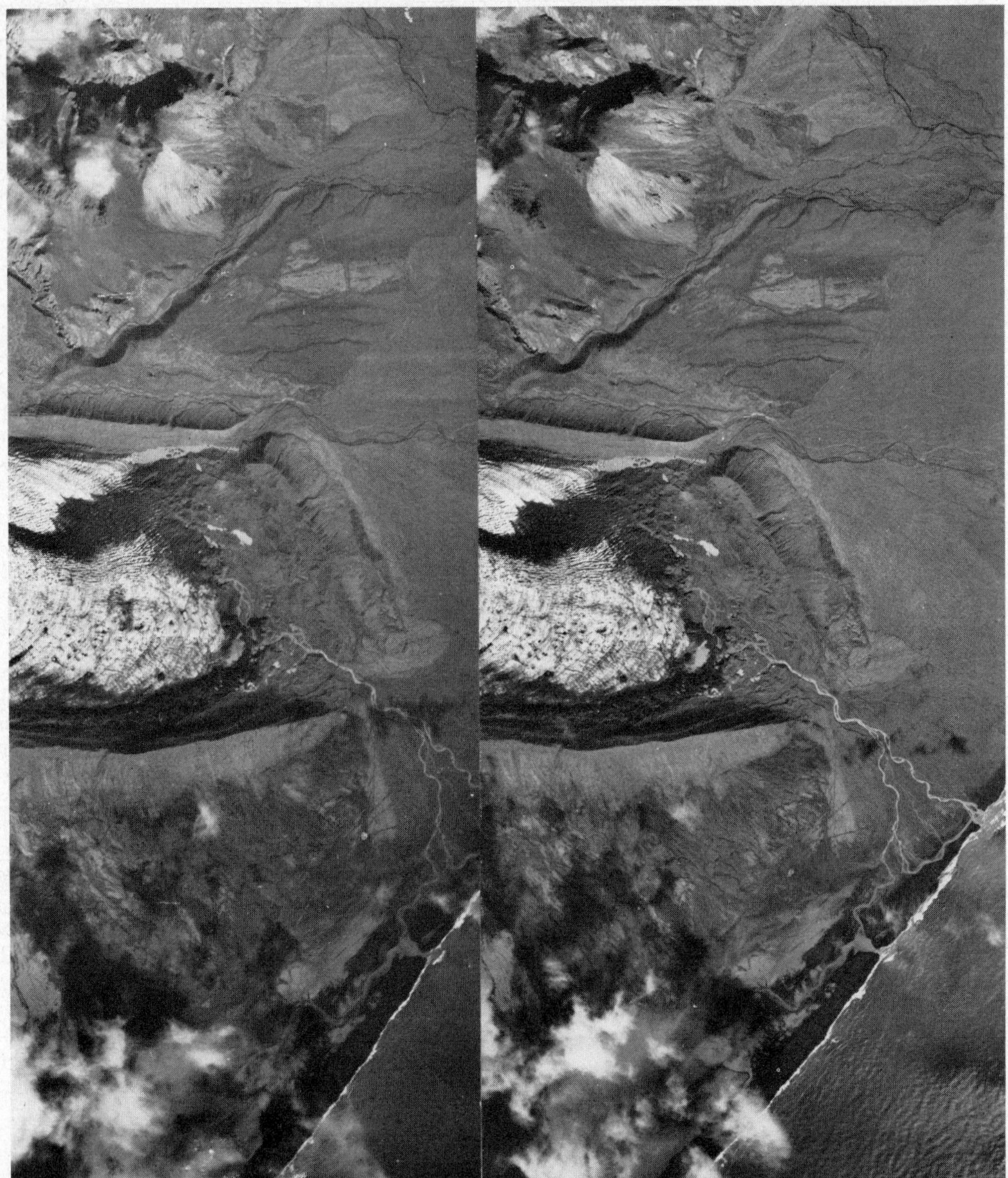

Fig. 31-218. Stereopair of vertical aerial photographs of the terminal moraine of Kvíárjökull, an outlet glacier from the southern margin of Vatnajökull, Iceland (See Color Figures 31-201 and 31-219). In 1904, the 1:100,000-scale Danish Geodetic Institute map (Blad 87, Öraefajökull) showed the glacier tightly against its end moraine. In these aerial photographs (Sortie 1214, Roll 2-1, Frames 23 and 24), acquired by the U.S. Army Air Force on 30 August 1945, from an altitude of 7,000 m, with a 152 mm-focal length camera, the glacier had retreated and breached its terminal moraine in two locations with meltwater streams (See Color Figure 31-219). See also Figure 31-207, a Landsat 3 RBV image of Kvíárjökull and environs acquired on 9 August 1978.

cover often provide new information about the surface and lead the geomorphologist to intuitive conclusions about its origin or stimulate his/her imagination. Many such ground details have no relief and would in their ''non-topographic'' character be impossible to detect from an observation place on the ground or, if detected, would be difficult to explain in terms of a meaningful connection among the details. Especially difficult to detect are subtle features or faint traces of processes that are no longer active, so that their original shape has been diffused by erosion or cultivation. Some examples of such fossil or palimpsest elements are discussed in the next few paragraphs.

In cultivated fields on a coastal plain in southern Sweden scattered patches of sand occur (Figure 31-220). They have no readily apparent re-

Fig. 31-220. Terrestrial photograph of a newly ploughed field on the Vittskövle plain, close to the Baltic Sea in southern Sweden, in which scattered patches of sand stand out clearly because of the distinct contrast to the humus-rich soil of the area. When viewed from the ground perspective, these sand patches appear to have a random distribution. (Photograph by Harald Svensson).

lationship and might be perceived to be the result of eolian transportation from nearby dune fields. The aerial photograph, however, immediately provides the answer (Figure 31-221). The sand patches form part of an anastamosing pattern, typical of a braided stream channel system, in a characteristic triangle form appearing on the surface. Originally, the now fossil delta was formed by fluvial activity in a marshy environment and was later covered by a thin peat layer. The detection of the fluvial pattern by means of aerial photographs makes possible a further and more complete study of the area from both a morphogenetical and a chronological (radio-carbon dating) point of view (Svensson, 1967).

Fig. 31-221. Vertical, black-and-white aerial photograph of part of the Vittskövle plain, southern Sweden. In the cultivated fields, the existence of a fluvial pattern can be clearly seen. The pattern is delineated because of differential growth in the crops; lowest vegetation is in the sandy soil. Approximate scale of the aerial photograph is 1:25,000. Aerial photograph courtesy of Statens Lantmäteriverk, Sweden.

Fig. 31-222. Vertical, black-and-white aerial photograph showing the occurrence of dark, mostly rectilinear pattern (crop marks) in a raised plain, built up by glaciofluvial and marine deposits of Laholm Bay of the west coast of Sweden. Approximate scale, 1:14,000. Aerial photograph courtesy of Statens Lantmäteriverk, Sweden.

In aerial photographs of a plain on the west coast of Sweden, fragments of a line pattern are seen in arable land (Figure 31-222). The photographs were acquired after an extremely dry spring in which the crops became very stressed. During dry seasons, in subsequent years, lines of higher-growing cereals have appeared in the same place where they were recorded on aerial photographs taken in 1947 (See also Figure 31-223). Test pits made in the ground show infillings of material that is more fine-grained than the surrounding matrix, thereby producing higher water capacity and better growth conditions in the lines (Figure 31-224). As a matter of fact, the fragmentary line pattern reveals a network of ice-wedge polygons that had formed under permafrost conditions (See

Fig. 31-223. Terrestrial photograph, taken on 26 June 1964, of part of polygon pattern enhanced by crop marks in a cultivated area of the west coast of Sweden. (Photograph by Harald Svensson).

Fig. 31-224. Terrestrial photograph taken in September 1973, of a vertical section dug through one of the lines recorded by the terrestrial photograph shown in Figure 31-223. The funnel-formed infilling is a cast of a former ice wedge (See also Color Figure 31-212 of this chapter). (Photograph by Harald Svensson).

also Figure 31-211 and Color Figure 31-212 of this chapter) in Late Weichselian time. The faint indications of polygons on the aerial photograph provide conclusive paleomorphological and paleoclimatological evidence regarding the former permafrost environment of the area (Svensson, 1972; 1973).

Studies by Svensson (1963b) and Christensen (1974) have also revealed patterned ground in Denmark through mapping of crop marks. Archaeologists, of course, have used crop marks for many years as a means of locating various types of historic and prehistoric sites of human occupation. For a review of an experiment in multispectral aerial photography the reader is referred to Hampton (1975).

Fig. 31-225. Vertical, black-and-white aerial photograph of the deglaciated area in front of the Löipskartind glacier, Norway, displaying a parallel-line pattern of 25–50-cm high till ridges (fluted moraine). The low relief is enhanced by the shadows due to low angle of solar illumination. Approximate scale, 1:10,000. Aerial photograph courtesy of Fjellanger-Wideröe A/S, Oslo.

Outside the margin of modern glaciers, in nearly uncovered till surfaces, a specific ground pattern, the fluted moraine, may be present. This pattern was first observed in many places by means of aerial photographs (Hoppe and Schytt, 1953; Figure 31-225). The relief of these fluted surfaces is usually quite low and, on the ground, very hard to detect in cultivated areas (Svensson, 1970). The fluted moraine is of great importance for evaluating the thermal properties of the base of the glacier and for calculating the dynamic processes in the ice-rock interface. These conditions are hard to observe directly, but are of great interest in theoretical studies of glacial geomorphology (Svensson, 1962; Boulton 1976; Humlum 1978; and Krüger and Humlum, 1981).

MARINE GEOLOGY

Introduction

Coastal, nearshore, and offshore areas represent some of the most difficult areas in which to conduct accurate geologic and geomorphic mapping. Along many coasts, the coastline is continually changing because of a combination of natural forces and man-made modifications. Under such dynamic conditions topographic base maps are often out of date. Moreover, many of the world's coastal areas and shallow seas have never been properly mapped and, even when dynamic changes are absent, accurate maps or charts simply do not exist.

Aerial photographs or satellite images (Tisdall and El-Baz, 1979) become an especially important tool for the field geologist, because such image data are generally more current than available base maps. The first part of this section discusses an effective use of such data in studies of the evolution of Nauset Spit and in the geomorphic processes which produce such changes. For additional examples of the use of aerial photographs and satellite images in coastal studies, a review of the 24 papers in *Air Photography and Coastal Problems* (El-Ashry, 1977) is recommended.

New technology is being employed to study certain types of dynamic phenomena which cover large areas, to document short-lived phenomena, and to study the morphology of the sea floor (Ewing, 1965; Yentsch et al., 1979). Figure 31-226, a Landsat image of the area just south of the Vestmannaeyjar volcanic archipelago off the southwest coast of Iceland, delineates the complex nature of the area's coastal marine currents (Williams, 1978d). Understanding the dispersal of sediments in the marine environment requires knowledge about coastal marine currents (Hunter, 1976; Carlson, 1976). Color Figure 31-227, a wintertime Landsat false-color composite image of Cape Cod and environs, documents the extent of

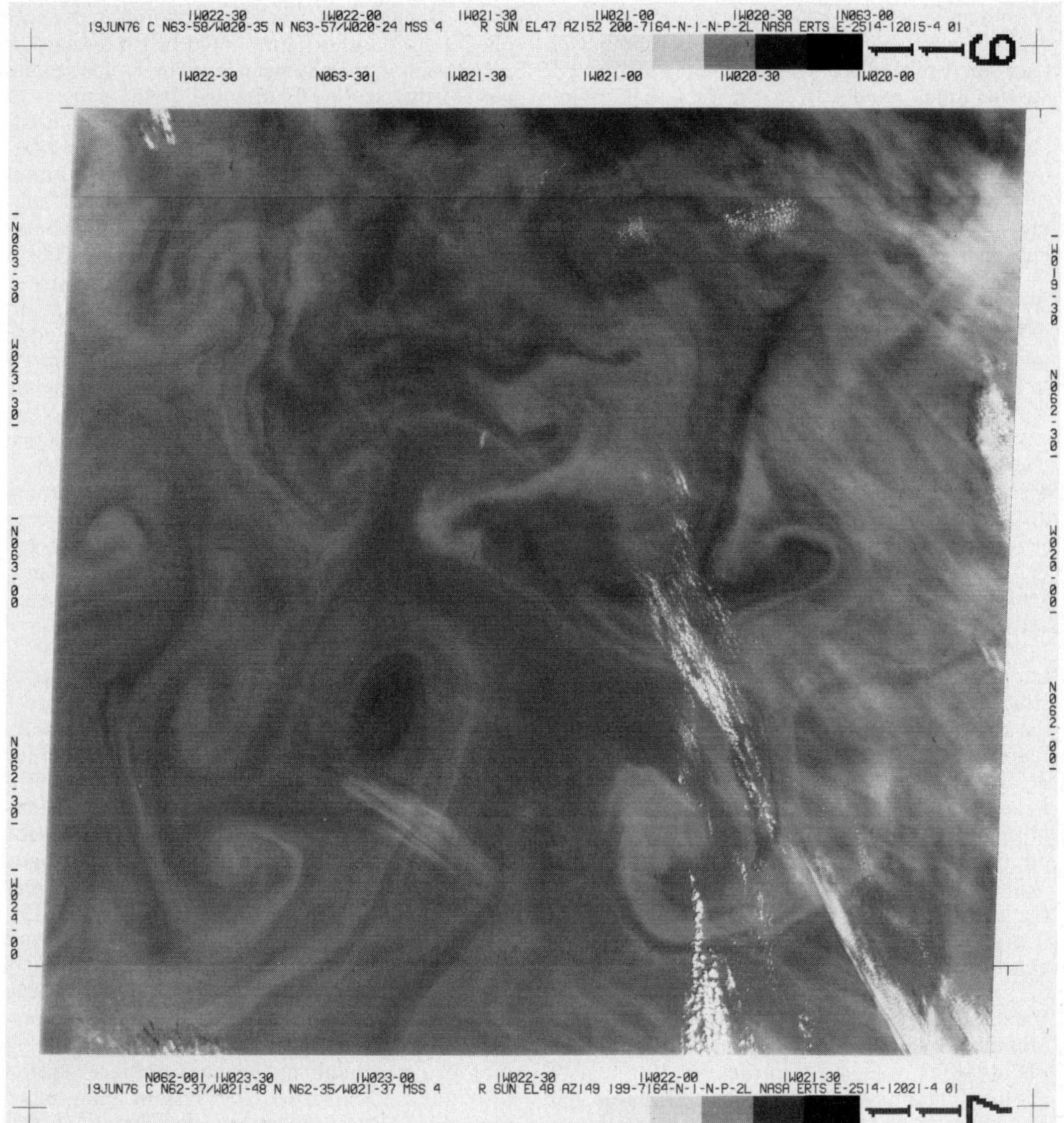

Fig. 31-226. A multispectral scanner (MSS) band 4 (0.5–0.6 μm) image (2514-12021), acquired on 19 June 1976 just south of the Vestmann Islands, Iceland, recorded some near-surface, large-scale marine current patterns. The image encompasses about 33,700 km^2 and is centered at 62° 37′ N. latitude and 21° 48′ W. longitude. Water depths within the image area range from about 100 m to over 2000 m; hence the image straddles the insular slope along the southwest coast of Iceland. Within the area of the image at least 8 well-defined eddies are visible, and at least 3 well-developed double eddies (one turning clockwise, the other counter-clockwise) can be delineated in the near-surface waters. Individual eddies associated with the double eddies have diameters of 20 km to 30 km, and individual stream currents leading to the double eddies extend for 50 km to 70 km. The coastal shelf area to the north of the image is a region of high productivity of phytoplankton and zooplankton which support a large and economically important fish population. Although the light-toned water spectrally resembles sediment-laden water, the distance from the Icelandic coast (125 km to the center of the image) makes it more probable that the light tone of the water is the result of concentrations of phytoplankton. Therefore, the variation in light- and dark-toned water is probably caused by variations in the concentrations of phytoplankton.

sea ice in the harbors and bays of southeastern Massachusetts (Williams et al., 1981). Sea ice can be an important geomorphic agent for modifying the sea floor and for affecting sedimentary processes (Barnes and Reimnitz, 1976; Reimnitz and Barnes, 1976).

In coastal areas it has been traditional for the field geologist involved in geologic quadrangle mapping to stop the mapping at the water-land boundary, even though the geology continues offshore. Through the use of new technology, it is probable that future geologic quadrangle maps of coastal areas will be made more complete, by including the submarine geology and morphology on the map. A related problem in geologic mapping of coastal areas results from the fact that bathymetric charts use a different reference datum (mean low water) than the datum which is used for topographic maps (mean high water) (See Figure 31−235). In coastal areas having a high tidal range, a considerable difference in the "coastline" may result. For geologic mapping, it would be more logical to use a single reference datum (Williams and Friedman, 1970).

In many areas geologists have used self-contained underwater breathing apparatus (SCUBA) to map geologic or geomorphic features on shallow parts of the sea floor. U.S. Geological Survey geologists have studied sea-floor gouges produced by ice on the floor of the Beaufort Sea to assess hazards associated with future offshore drilling for hydrocarbons (Reimnitz and Barnes, 1974). Italian geologists discovered methane seeps in the Tuscan Archipelago during submarine geological studies with SCUBA gear (Del Bono and Giammarino, 1968).

Water penetration by aerial photography is extraordinarily variable, and considerable research still needs to be accomplished to be able to predict the best time of year, time of day, and optimum film-filter combination for a particular coastal area. For example, 30-meter penetration of water is not unusual with Ektachrome and Ektachrome infrared films in tropical marine waters and penetration of over 60 meters has been reported by Willard Vary (oral communication, 1968) in the Pacific Ocean with the GAF 1000 minus-blue film (Current, 1969). At various locations around the coast of Cape Cod, maximum penetration depths of 7 meters were reached with Ektachrome infrared aerial photography. At least in the waters around Cape Cod, Ektachrome infrared aerial photography was most impressive for coastal geologic mapping applications (Williams and Friedman, 1970).

An optimum film-filter combination needs to be used for mapping in coastal and shallow offshore areas. Black-and-white aerial photography is best for the delineation of topographic and cultural features (Figures 31-35 and 31-234A and B). Color and/or color-infrared aerial photography appears best for surficial and bedrock geologic mapping (Color Figure 21-232). Natural color (Color Figures 31-39 and 31-235) or color infrared (Color Figure 31-40) or GAF 1000 minus-blue aerial photography appears to be best for delineation of underwater morphology. Infrared aerial photography is best for delineation of the water-land contact (shoreline) (Figures 31-37, 31-236, and 31-238). Surveys to delineate topographically the shoreline should be conducted at high tide. Submarine morphology is captured better at low tide, when there is less water to penetrate and to scatter light. Obviously, then, while the time of mean high water would be the best for conducting some marine aerial surveys with infrared aerial photography, other films would be preferable for aerial photography at mean low water. The second part of this section is directed at the subject of "Remote Sensing of Submarine Features in Shallow Water," especially the use of various film-filter combinations to optimize the penetration of shallow water for geologic mapping of submarine morphology. Two reef areas, Middle Sambo Reef (Florida Keys) and Cayo Icacos (Puerto Cayo Icacos (Puerto Rico) are used as examples.

In deeper waters marine geologists must resort to the use of research submersibles or side-scan sonar to carry out surveys. Sidescan sonar is being used to image large areas of the Continental Shelf and Slope to delineate sea-floor morphology and to study the results of dynamic processes that are operating in these areas. Submarine morphology is addressed in the third part of this section. Seismic and sonar-array surveys are also conducted to determine submarine morphology and sub-bottom geologic structure in deeper water, especially by petroleum exploration companies.

Coastal Studies

There are four principal approaches to coastal studies depending on time frame and scale: geomorphic, historical, process, and geographic. Remote sensing can provide a unique data set for application to each type of approach, often providing key information for management decisions (El-Ashry, 1977). This is particularly true for dynamic landforms, such as the barrier islands and spits (See the "Mini-Atlas of Image and Related Data of Cape Cod, Massachusetts, and Environs" section of this chapter), which extend along much of the U.S. East and Gulf Coasts (Leatherman, 1979a). While satellites, particularly Landsat, can provide repetitive synoptic information, aircraft-acquired data still remain the best tool for coastal studies, because of higher spatial resolution. The Nauset Spit system along outer Cape Cod, Massachusetts (Color Figure 31-228; See also Figure 31-34) recently has been the subject of comprehensive studies, employing many of the above methods.

Geomorphic studies involve the use of morphological features and their spatial relationships to reconstruct the evolutionary development of a barrier structure. In particular, beach/dune ridges can be utilized to interpret prehistorical geomorphic changes, such as spit extension, barrier accretion, or inlet breaching and downdrift lateral migration (Leatherman, 1979b). For example, the southern end of Monomoy Island is characterized by recurved beach ridges (Figure 31-229). Some of these ridges have formed during recorded times, illustrating the rate and magnitude of spit growth and longshore sand transport to this sedimentary sink (See also Oldale et al., 1971).

Historical photographs (See especially those in the "Mini-Atlas of Image and Related Data of Cape Cod, Massachusetts, and Environs" section

Fig. 31-229. This oblique aerial photograph, looking southwest, of the southern terminus of Monomoy Island, Massachusetts, exhibits distinctive arcuate-shaped beach/dune ridges. These ridges (white in contrast to lower-lying wetlands) represent earlier shoreline positions and thus are literally lines of accretion. Their recurvature indicates the importance of at least two directions of significant wave activity impinging on the spit tip. Photograph by Stephen P. Leatherman.

of this chapter), charts, and maps can provide a general picture of barrier evolution. For Nauset Spit, the earliest accounts extend back nearly 400 years to the French explorer Champlain. While these maps allow a qualitative evaluation of barrier changes through time, the first U.S. charts from which quantitative measurements can be made were produced by the Coast & Geodetic Survey in the mid-1800s. These charts, although quite accurate, have two principal limitations when detailed monitoring of historical barrier changes are desired: (1) the charts are not sufficiently detailed to allow for a thorough analysis of changes in barrier environments, and (2) because the charts are normally only produced at infrequent intervals (40–50 years is common), a detailed understanding and interpretation of short-term physical processes and morphological responses is not possible with this type of data. The long-term rate and nature of barrier migration is governed by the integration of many episodic events (Anders and Leatherman, 1982). More detailed, timely information can be acquired by utilizing vertical aerial photographs.

For most coastal areas, vertical aerial photography is available from the 1930s (Taylor and Spurr, 1973). The amount and quality has greatly improved since the 1950s. For accurate determinations of shoreline changes, aerial photographs should be corrected for distortion (Tanner, 1978), brought to a common scale, and overlaid for a direct comparison of changes.

The northern section of the Nauset Spit system clearly illustrates the rapidity of changes during a 40-year record (Figure 31-230). During this time period, Nauset Inlet migrated northward, in opposition to the long-term prevailing direction of littoral drift. Concurrent with its shortening in length, the northern spit (Nauset Spit-Eastham) also experienced shoreline erosion, averaging about 0.9 m per year between 1938 and 1978, based on comparisons of aerial photographs. This rate of erosion agrees favorably with field determinations by Zeigler et al. (1964) and Leatherman et al. (1981) for the long term rate of erosion along outer Cape Cod.

The distribution and relative percentage of the four primary barrier environments (beach, dune, washover, and marsh) have also changed significantly during this time frame. Historical aerial photographs clearly show the loss of dune and the marsh burial by massive washover deposits resulting from the 6–7 February 1978 "Nor'easter"; this major winter storm essentially engulfed the barrier, flooding all but the highest physiographic features and dissecting the spit into small sections by overwash sluiceways (Figure 31-231).

While remote sensing, particularly aircraft photography, has been of primary application in historical shoreline and environmental analysis, it has also played a role in process measurements (e.g., Dietz, 1947). Overwash and inlet processes in large part control barrier dynamics and serve to move the island/spit landward through time with sea level rise (Leatherman, 1981). Rates of landward transport by storm-generated overwash can be estimated from aerial photographs (Color Figure 31-232). Field surveys are necessary to determine the vertical dimension (thickness) of the washover deposition to facilitate volumetric calculations on a per storm basis. The 1978 New England Blizzard resulted in over 8.4 m^3 of overwash sand deposition per 0.3 m of shoreline along Nauset Spit, with depositional thickness approaching 1.5 m in some areas (Zaremba and Leatherman, 1983). This type of information is critical in determining the role of various transport processes in different environmental settings for tailoring land use and management plans.

The southern spit section (Nauset Spit-Orleans) has elongated since 1940, although there have been several periods of retreat and progression. Aerial photographic comparisons and field surveys showed that annual spit extension averaged 110 m per year between 1969 and 1977. Shoreline erosion was also sporadic along the southern spit section during this period, varying between 1 and 3 m per year (Knutson, 1980).

Aircraft and satellite sensors have also been used effectively to monitor inlet-related erosional and depositional processes and patterns (e.g., El-Ashry, 1979; Wang, 1980). During the 1978 blizzard, Monomoy Island (Figure 31-228) became separated into two distinct barrier sections (Williams, 1979a) when a new inlet was created (See Figures 31-36, -37, and -39). Landsat 3 return beam vidicon (RBV) data are particularly useful for discriminating the land/water interface (See Figures 31-49 and -51) and for monitoring on a regular basis the sequential changes over a short-time period (Williams, 1979b).

Finally, remote sensing can play an expanded

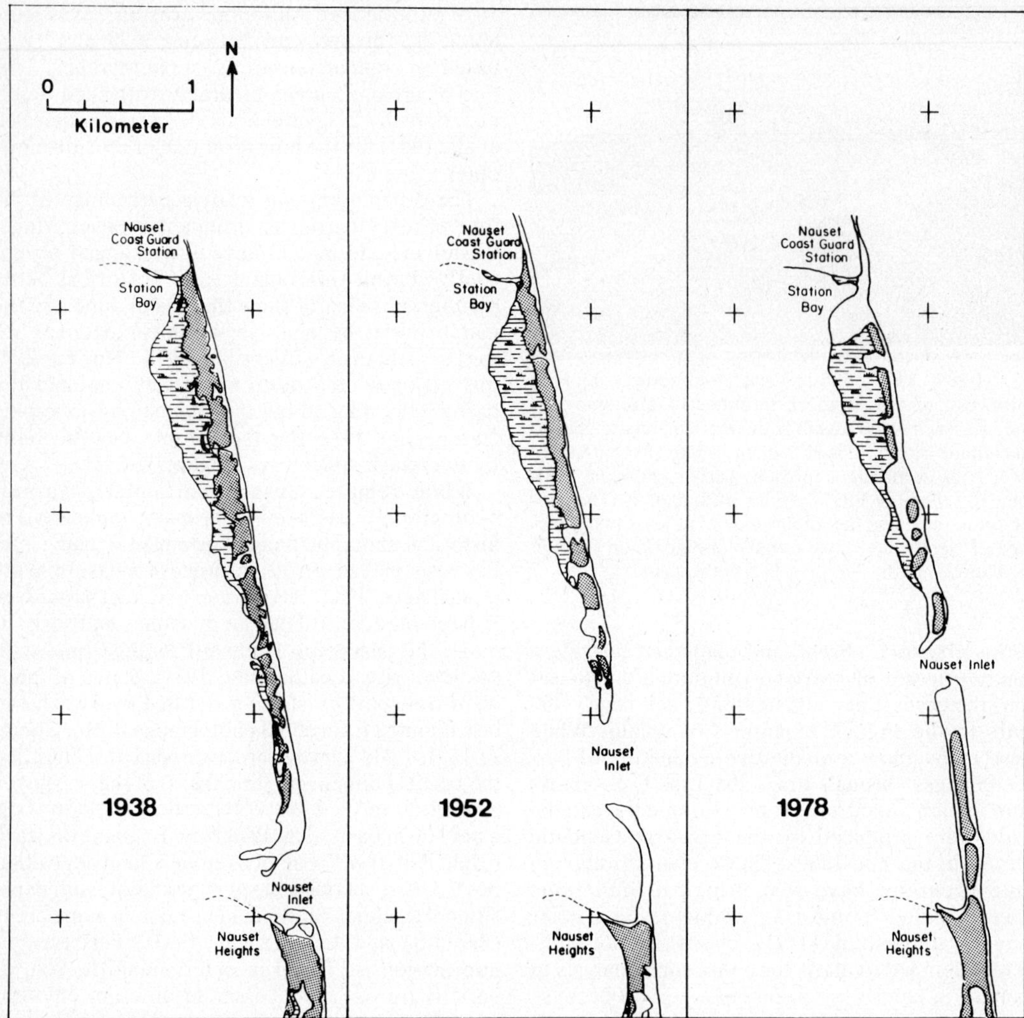

Fig. 31-230. Three sets of U.S. Geological Survey vertical, black-and-white aerial photographs (photo scale 1:20,000) were used to depict the overall changes in barrier morphology as well as to determine the rate of shoreline retreat. This 40 year sequence clearly indicates the trend for the northern spit segment—shortening and perhaps eventual dissolution as presently mapped. The once wide barrier dunes have been eroded and large sections flattened to produce the present situation. The violent 6–7 February 1978 ''Nor'easter'' storm was instrumental in significantly altering barrier physiography and plant community distribution.

role in geographical studies of coastal areas. For example, Landsat MSS or TM images could serve as an excellent data source for a global barrier-island study. At present the worldwide distribution of barrier islands is only partially known. Previous studies (McGill, 1958; Gierloff-Emden, 1961; and Cromwell, 1971) were based solely on available maps and charts that exhibited a wide range in data quality (due to different scales and cartographic accuracy, dates of publication, and source material used). Landsat data can provide an original, uniform data base for classification of barriers; distinct barrier types could reflect the physical factors (geologic, climatic, and oceanic). This classification scheme would have direct application to petroleum exploration (by obtaining knowledge on modern analogs for comparison to now-buried ancient, oil-bearing barriers) as well as providing invaluable information for resource management and land-use planning. Relative barrier stability and vulnerability to storms are primary considerations in the location of human development. Lastly, the classification system would permit a quantitative assessment of the uniqueness of barrier types as an integral element in the selection of coastal biospheres (UNESCO Man and Biosphere Program).

Remote Sensing of Submarine Features in Shallow Water

Accurate mapping of submarine features in shallow water environments has been discussed by a number of authors. Lundahl (1948) was one

Fig. 31-231. Richard C. Kelsey, an aerial photographer from Chatham, Massachusetts, braved the elements and took this remarkable oblique aerial photograph (No. 78-207-17) (and dozens of other extraordinary photographs) on 7 February 1978 during the final stages of this major winter storm, known as the New England Blizzard of 1978. Note that much of the barrier (Nauset Beach) is awash with only the higher dunes above the storm tide. The spit was broken into short sections by the storm-generated overwash surges, which transported large quantities of sea water across the barrier. The associated entrained sand was deposited on the back-barrier environments, resulting in a few hundred meters of bayshore accretion in major washover localities. Aerial photography courtesy of Richard C. Kelsey, Chatham, Massachusetts.

of the first scientists to determine water depth with aerial photographs. Tewinkle (1963) and Harris and Umbach (1972) also discussed the use of aerial photographs for underwater mapping.

Duntley (1963), in his classic research on the transmission of light in sea water, stimulated research by a number of groups on the optimum wavelength for achieving maximum water penetration. Table 31-41 is from theoretical and experimental studies that were carried out by Polcyn and Rollin (1969).

Reef areas represent almost ideal test sites for experimenting with various aerial film and filter combinations to achieve optimum water penetration, because reef-building organisms require a sediment-free environment and relatively warm water for maximum growth. Such conditions also produce a relatively clear water column. U.S. marine geologists and other scientists have frequently used field sites in the Florida Keys, the Bahamas, or Puerto Rico for experiments. Two case histories are presented in this section, one in the Middle Sambo Reef area of the Florida Keys, the other in the Cayo Icacos area, off the northeast coast of Puerto Rico.

Reef accumulations are well represented in the geologic record and often serve as traps for hydrocarbons. Reefs are difficult to map accurately because of varying water depths and the irregular submarine topography. Landsat images and aerial photographs are important tools in the accurate

Table 31-41

Wavelengths of maximum transmittance for ocean water (After Polcyn and Rollin, 1969)

	Wavelength of Maximum Transmittance (μm)
Pure water	0.470
Clearest ocean water	0.470
Average ocean water	0.475
Clearest coastal water	0.500
Average coastal water	0.550
Average inshore ocean water	0.600

mapping of reefs from regional or local viewpoints, respectively. Color aerial photography has special value (See Color Figure 31-235).

In September 1966, the Air Force Cambridge Research Laboratories carried out a series of experimental aerial surveys of the Cayo Icacos area of Puerto Rico. Cayo Icacos is one of a chain of reefoid islands (La Cordillera) that lie off the northeast coast of Puerto Rico (Figure 31-233). Figure 31-234 is conventional black-and-white aerial photography of Cayo Icacos from two survey altitudes, 6,100 m and 2,440 m. Color Figure 31-235 is a color aerial photograph of Cayo Icacos from an altitude of 1,570 m, showing the intricate detail in the fringing reefs. Figure 31-236 is a set of nine uncontrolled mosaics of the Cayo Icacos area acquired by Air Force Cambridge Research Laboratories using the Itek nine-lens camera, in September 1966, one of the earliest experiments to empirically determine the optimum film-filter combination to achieve maximum water penetration for geological mapping in shallow-water areas. Nine discrete bands of the photographic spectrum were recorded between 0.39 and 0.878 μm, from the near-ultraviolet through part of the near-infrared (Table 31-42). Aerial spectrophotography with band 1 (0.39 to 0.445 μm), band 2 (0.445 to 0.495 μm), and band 3 (0.505 to 0.545 μm) of the Cayo Icacos Archipelago had the deepest water penetration. Submarine morphology can be discerned to depths of 20 m or more in all three bands, although band 2 has the best depth penetration. There is a decrease in water penetration in bands 4, 5, and 6 (0.555 to 0.700 μm) and essentially no penetration above 0.700 μm (bands 7, 8, and 9) (Williams, 1971).

Other scientists and groups were also involved in actual tests with various film-filter combinations, including Tuddenham (1968) in Australia, and Cronin (1965), Wenderoth (1969) and Yost and Wenderoth (1970) in the United States. Conrod et al. (1968) carried out a number of important studies in the Bahamas with aerial photography.

In March 1967, the U.S. Naval Oceanographic Office, in conjunction with Air Force Cambridge Research Laboratories and other organizations,

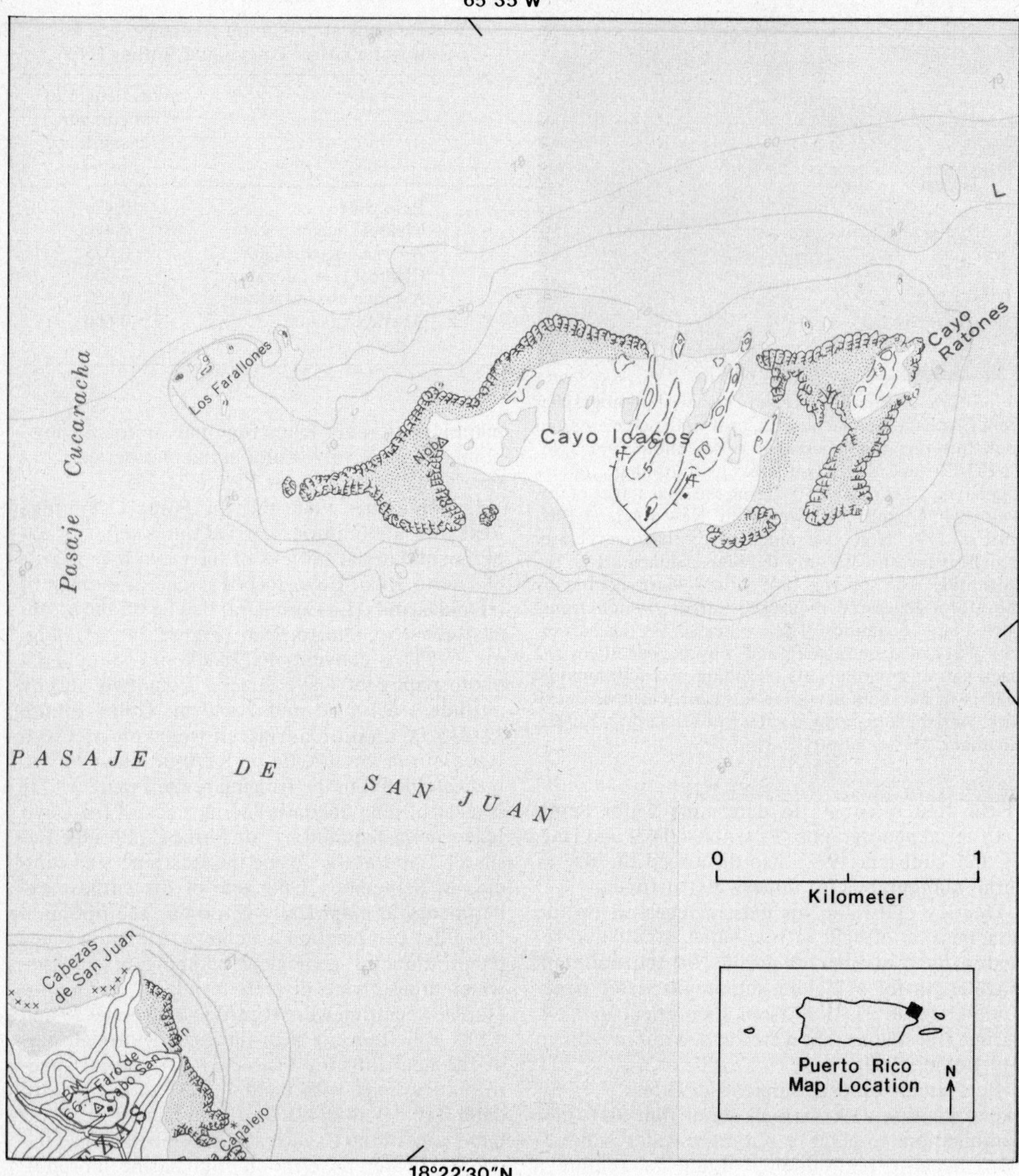

Fig. 31-233. Part of Cayo Icacos 7½-minute series quadrangle map (1:20,000), showing the Cayo Icacos area and environs, Puerto Rico. The topographic contour interval is 10 m with supplemental dashed contours at 5 m using mean sea level as the datum. The depth curves and soundings are in feet (0.305 m per foot) using mean low water as the datum. The mean range of the tide is approximately 0.3 m. The shoreline shown represents the approximate line of mean high water. U.S. Geological Survey map, 1960 edition. The topography was mapped from 1941 aerial photographs by photogrammetric methods and by planetable surveys. The area was field checked in 1953; subsequent revisions were made in 1958. The hydrography was compiled from U.S. Coast and Geodetic Survey charts 917 (1955) and 921 (1952).

carried out a comprehensive series of aerial surveys with different aerial films and filters of several reefs (Figure 31-237) in the Florida Keys (Vary, 1967 and 1969a). Resolution and color targets were emplaced on the shoreward and seaward sides of the reefs, and spectroradiometric measurements were made. One unexpected result of this experiment was that, "false-color infrared film, using the Wratten No. 12 filter over the lens, showed excellent penetration and ocean bottom detail at the 65-foot (20 m) depth (Vary, 1967)." The Itek nine-lens multiband camera was also used in this experiment. A set of nine 70-mm aerial photographs of the Middle Sambo Reef area,

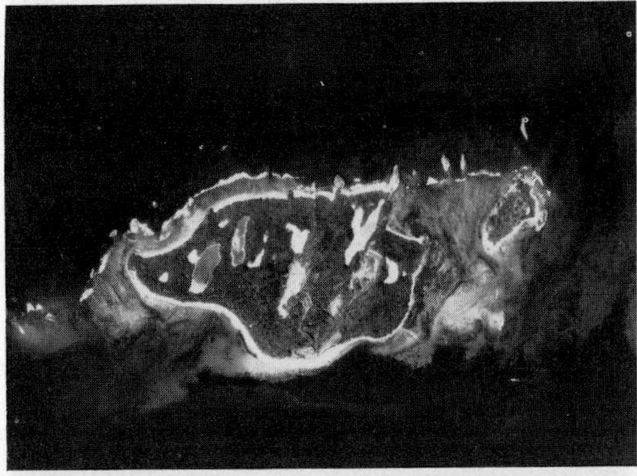

Fig. 31-234. Vertical black-and-white aerial photographs of Cayo Icacos, Puerto Rico, showing the water penetration capability of panchromatic (black-and-white) aerial film used with a Wratten 12 filter at two different survey altitudes. *Top.* U.S. Coast and Geodetic Survey Photograph no. L-1378 acquired on 7 March 1965 from a survey altitude of 6,100 m. Camera focal length, 152.21 mm. Photograph scale, 1:40,000. *Bottom.* U.S. Coast and Geodetic Survey photograph no. W-1596 acquired on 19 October 1961 from a survey altitude of 2,440 m. Camera focal length, 153.02 mm. Photograph scale, 1:8,000. The film-filter combination used in obtaining both photos produces a photographic record of the 0.51 to 0.68 μm spectral band (See Figure 31-35).

Florida Keys, is presented in Figure 31-238. Table 31-43 shows the film-filter combinations used in the aerial surveys of the Middle Sambo Reef area.

Another result of the Florida Keys experiment was the request by Willard E. Vary to the GAF Corporation to produce an experimental blue-insensitive aerial film (Vary, 1969a and b; Current, 1969). This new film, a modified Anscochrome D-500 aerial color film with the blue-sensitive layer omitted, was successfully tested in February 1968 in the Bahamas (Vary, 1969a). Later tests in the Bahamas in July 1968 showed that bottom detail could be identified on aerial photographs to depths of 46 m (150 ft). The 0.56-μm part of the spectrum was also found to achieve the best water penetration and record the most ocean bottom detail (Vary, 1969a). The GAF film was eventually designated GAF 2575 (Anscochrome D-1000 blue-insensitive color film) and was later used by the U.S. Naval Oceanographic Office in April 1973 during aerial and field studies of the southern coast of Iceland. Kodak also developed an aerial color film to be used for mapping shallow submarine environments as reported by Marszalek (1977). The film is called Eastman Kodak SO-224 (Eastman Kodak experimental water penetration film).

With the availability of spacecraft photography and imagery, scientists turned their attention to this new source of data to map shallow water areas. Noble (1970) reported on the use of Apollo and Gemini photographs and photometric data from the R/V *Ben Franklin* submarine experiment to identify "optimum photographic spectral bands for bathymetric mapping by aerospace photogrammetric techniques." Although Noble (1970) determined, from theory and practice, that 0.43 to 0.58 μm would be optimum, atmospheric haze and other considerations led him to recommend that the 0.46- and 0.58-μm spectral band would be the best for oceanographic remote sensing. He also recommended 0.46 to 0.51 μm and 0.51 to 0.56 μm if two spectral bands are available. The Landsat 4 Thematic Mapper system, with its 0.45- to 0.52-μm spectral region for band 1 should prove especially valuable for geological mapping in shallow-water areas. Band 4 of the MSS system on Landsats 1, 2, and 3 had a spectral range of 0.5 to 0.6 μm, although band 1 of the little-used RBV camera on Landsats 1 and 2 had a spectral range of 0.475 to 0.575 μm.

Polcyn and Lyzenga (1973) discuss the updating of coastal and navigational charts with Landsat data. A site in the Bahamas was also used in a joint NASA-Cousteau experiment to establish the water-penetration capabilities of Landsat imagery (Polcyn, 1976). Hammack (1977) described the potential of Landsat to produce better maps of the shallow seas of the world, areas which are notorious for the inadequacy of existing maps. The Hydrographic/Topographic Center of the Defense Mapping Agency, in association with the U.S. Geological Survey and the National Ocean Survey, published a two-image, 1:500,000-scale, experimental image map of the Berry Islands, Bahamas (U.S. Defense Mapping Agency, 1980). A digitally enhanced Landsat MSS false-color composite image and a photographically enhanced Landsat MSS band 4 image depicted shallow water features to depths of about 30 m. The MSS band 4 image was overprinted with standard hydrographic information (including depth curves) from British surveys (between 1836 and 1885) and U.S. Navy surveys to 1963, although the hydrographic data were adjusted to the Landsat image. Colvocoresses (1977a and b) discussed Landsat mapping and charting of shallow seas, and also published the first 1:500,000-scale false-color Landsat image map of the Florida Keys (U.S. Geological Survey, 1977d). The Landsat image used for the Florida Keys image map was en-

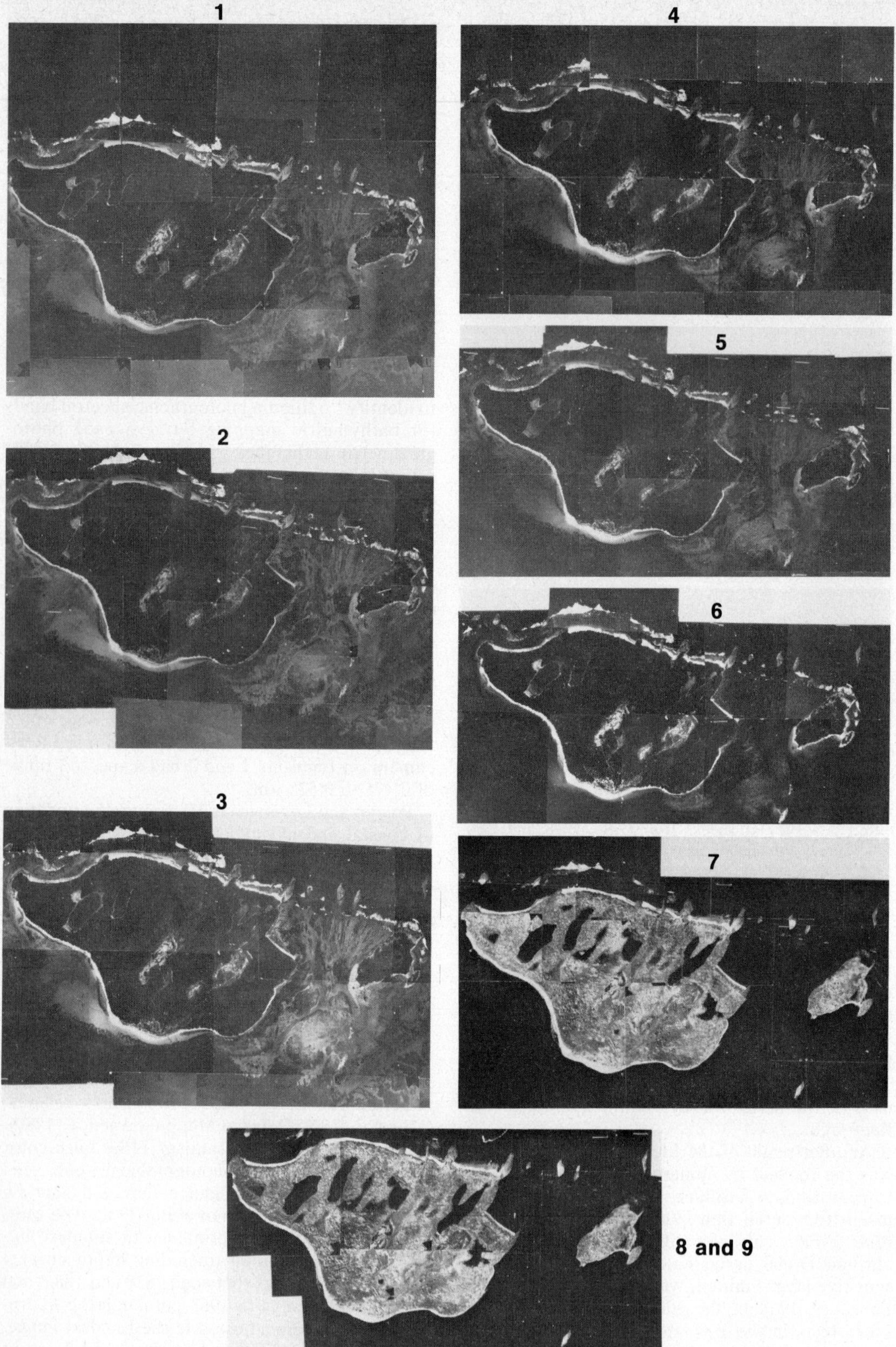

Fig. 31-236. A set of nine uncontrolled mosaics compiled from individual aerial spectrophotographs of the Cayo Icacos area, Puerto Rico, showing variation in water penetration using the Itek nine-lens camera system. The 0.390 to 0.878 μm part of the electromagnetic spectrum has been divided into 9 discrete passbands by appropriate filters (See Table 31-42). The aerial photographs were acquired as part of a U.S. Air Force Cambridge Research Laboratories project on September 1966 to determine the optimum film-filter combination for penetration of tropical marine waters.

Table 31-42

Band matrix, showing transmission wavelengths, filters, and films of aerial spectrophotography of the Cayo Icacos area, Puerto Rico, acquired by the Itek nine-lens multiband camera (152.4-mm focal length) in September 1966.

Band 1	*Band 2*	*Band 3*
0.390–0.445 μm	0.445–0.495 μm	0.505–0.545μm
Wratten 36 & 47B	Wratten 3 & 48	Wratten 12 & 65
Kodak 5401	Kodak 5401	Kodak 5401
Band 4	*Band 5*	*Band 6*
0.555–0.595 μm	0.555–0.595 μm	0.725–0.700 μm
Wratten 22 & 52 & polarized parallel to line of flight	Wratten 22 & 52 & polarized perp. to line of flight	Wratten 92
Kodak 5401	Kodak 5401	Kodak 5401
Band 7	*Band 8*	*Band 9*
0.700–0.878 μm	0.765–0.878 μm	0.820–0.878 μm
Wratten 89B & 1.2 ND	Wratten 87 & 0.9 ND	Wratten 87C
Kodak 5424	Kodak 5424	Kodak 5424

Note: Both the Kodak 5401 and 5424 aerial films have since been replaced by the Kodak 2402 and 2424 aerial films, respectively (Bill Reed, Eastman Kodak Company, oral communication, 1982). ND = neutral density.

hanced to portray features in the areas of water (Colvocoresses, 1977c).

The U.S. Department of the Interior has the responsibility for the administration of the U.S. Trust Territory of the Pacific Islands and other Pacific outlying areas, including the preparation of maps of these areas by the U.S. Geological Survey. Carter and DeNoyer (1978) applied the theoretical and experimental work of Polcyn (1976) with a computer-assisted analysis and classification of shallow marine waters of the Velasco Reef and Palau Islands, Palau District, Western Caroline Islands. Five bathymetric zones could be mapped for the bathymetric intervals 0–4 m, 5–9 m, 10–14 m, 15–19 m, and 20–24 m.

Reed (1981), in a recent study of the coastal waters around Cape Cod, Massachusetts, used seasonal Landsat images, Skylab photography, bathymetric charts, and Seasat SAR imagery to interpret hydrographic features. Figure 31-59 contains a bathymetric chart, a Landsat image, and a Seasat SAR image, of the Nantucket Shoals area which are from Reed's research.

The use of Landsat images by the U.S. Defense Mapping Agency to rechart the Chagos Archipelago in the Indian Ocean is of significance to geologists as well as mariners. The entire reef system was, for the first time, properly positioned and in the process a new reef (Colvocoresses Reef) at about 10 m depth was discovered. The areal distribution of the Chagos reef system as mapped from Landsat images suggests linear structural control along its western boundary which was not obvious from previous charts of the area. Other shallow sea areas are also being recharted by the U.S. Defense Mapping Agency (and others) which are resulting in the discovery of additional submerged reefs and the proper positioning of many others (Hammock, 1977; Colvocoresses, 1982a).

Submarine Morphology

Geologists, archaeologists (Rosencrantz, 1975), and oceanographers interested in features on or morphology of the sea floor have, over the years, acquired many photographs and images of shallow and deeper water areas from a variety of imaging systems that range from handheld cameras used by divers to sophisticated systems in manned and unmanned research submersibles (See Color Figure 31-53) or surface ships (See Figures 31-65 and 31-66). The area of coverage in submarine photographs is, however, severely limited because of the restrictions on natural or artificial illumination. Sidescan sonar is free of such restrictions. Over the past decade, sidescan-sonar systems, originally developed to satisfy classified military requirements, have become available to nonmilitary research scientists and are increasingly being used by geologists to study the submarine morphology of the Continental Shelf and Slope.

Interest in using sidescan sonar to delineate details of submarine morphology derives in part from a need to understand the nature and extent of dynamic processes that shape the sea floor. Of special concern in the work described here are processes that lead to mass slumping and sliding of sediment, movements that could constitute serious potential hazards to the development of oil and gas resources on the Continental Shelf and Slope off the northeastern United States.

Initially, the approach to this problem was to examine the bathymetry and shallow subsurface stratigraphy and structure of the shelf and slope by systematically surveying an area (Figure 31-239) using single-channel, high-resolution, seismic-reflection techniques. Interpretations of these data (Aaron et al., 1980, 1982) indicate that features apparently related to slumping

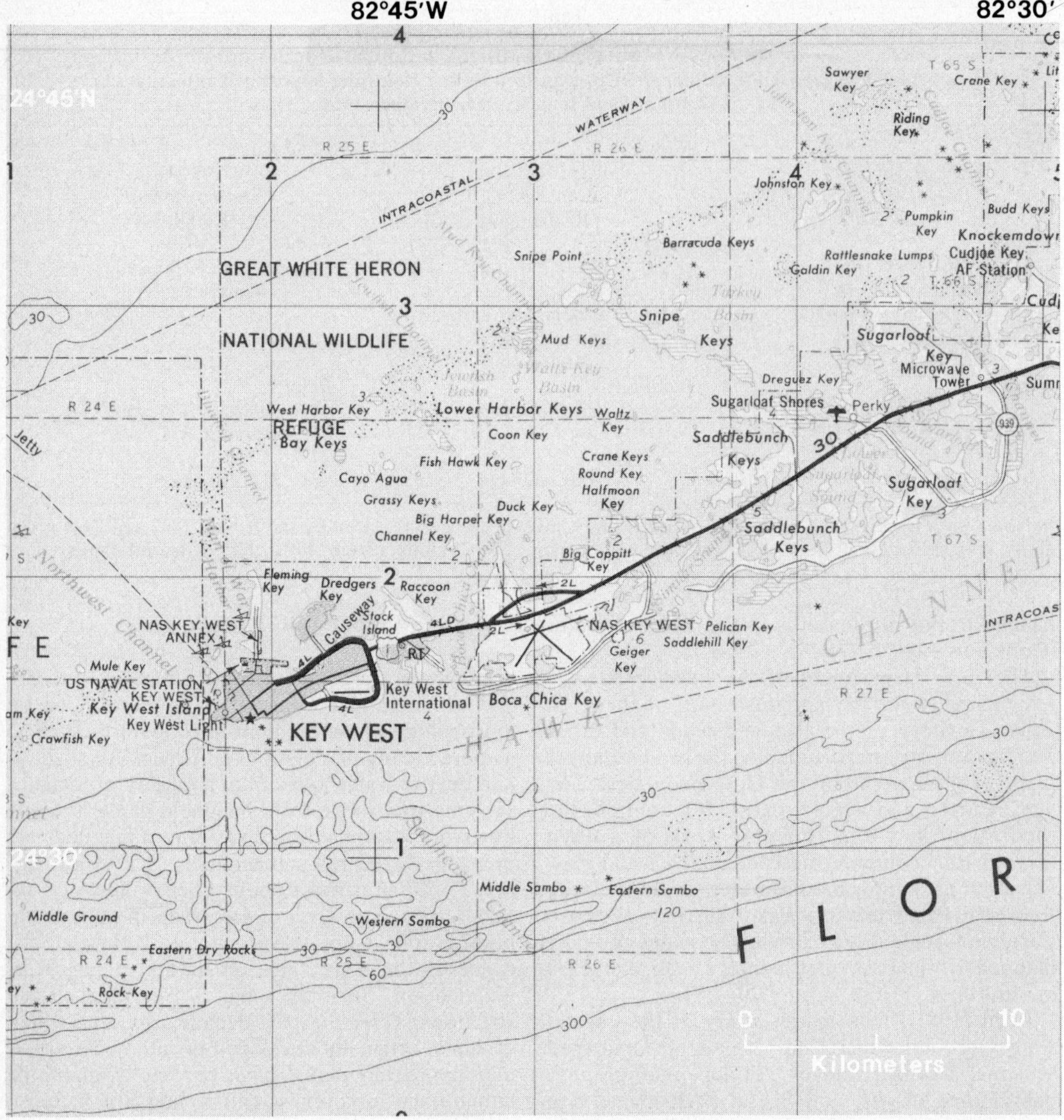

Fig. 31-237. Part of Key West, Florida, 1° × 2° quadrangle map series (NG 17-11, 1:250,000) of the United States, showing the Middle Sambo Reef area, Florida Keys. U.S. Geological Survey map, 1981 edition. Prepared by the U.S. Army Topographic Command (RMGE), Washington, D.C. Compiled in 1955 from United States quadrangles, 1:25,000, 1943–51 and USC&GS charts, 1933–48. Planimetry revised in part from aerial photographs taken 1946–47. Map field checked 1956. Revised in 1972 by the U.S. Geological Survey from aerial photographs taken 1969–72.

and other forms of mass movement may underlie a significant fraction of the Continental Slope. However, the slope is relatively steep (5°–10°) and is strongly dissected along much of its length by large canyons and countless smaller channels and gullies. Seismic-reflection profiles in such rough terrain are characterized by ambiguous diffractions and sideechos that tend to confuse bottom and subbottom reflecting horizons. Accurate interpretations of morphology are further hindered by the essentially linear character of the profile data, making interpolations between profiles extremely questionable.

Sidescan sonar sensing, on the other hand, provides continuous low-angle acoustic-reflection images of the seafloor that permit qualitative characterization of topography and materials to be made over large areas. Digital techniques of data acquisition and manipulation combined with accurate and precise navigation further increase the flexibility and utility of sidescan sonar as an important mapping tool for geologists.

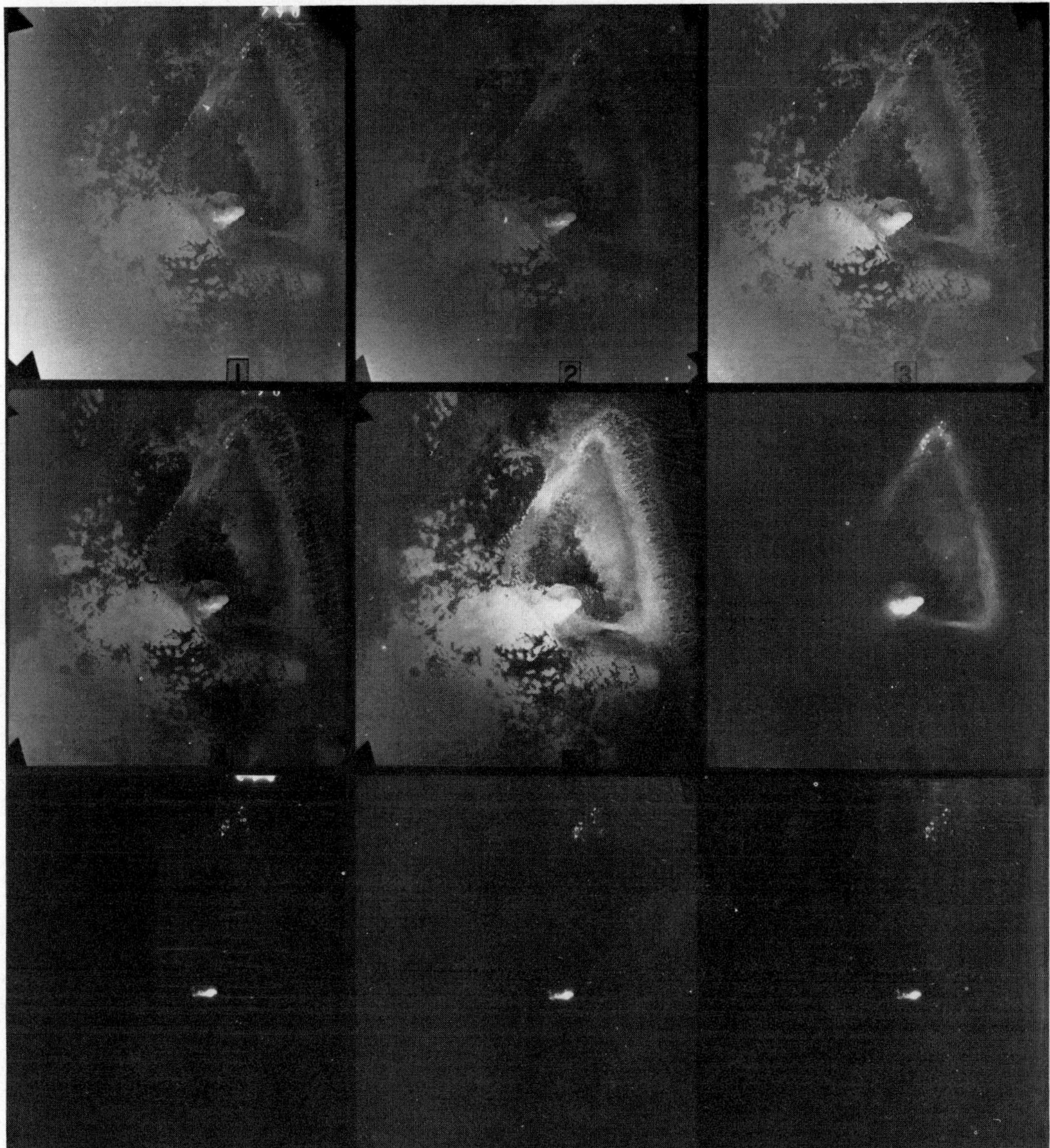

Fig. 31-238. Aerial spectrophotography of Middle Sambo Reef, Florida Keys, Florida, showing variation in water penetration using the Itek nine-lens camera system. The 0.385 to 0.880 μm part of the electromagnetic spectrum has been divided into 9 discrete passbands by appropriate filters (See Table 31-43). The aerial photographs were acquired as part of a joint U.S. Naval Oceanographic Office—U.S. Air Force Cambridge Research Laboratories project on 19 March 1967 between 0800 and 0900 e.s.t. from an altitude of 3,050 m. (Vary, 1967 and 1969a; Stringham and Williams, 1970). Note that the only subaerial feature is the tiny coral island visible on bands 7, 8, and 9, which are the three bands at the bottom of this figure.

Sidescan sonar operates on the same principles and produces essentially the same type image data as side-looking airborne radar (SLAR) (See also Figures 31-58, 31-59, 31-65, and 31-66 of this chapter and Chapter 10, Radar Imaging Systems, in this *Manual*). The major difference in application lies with the marine environment. Modern sidescan-sonar systems incorporate a towed, weight-depressed transducer assemblage (called a "fish"), tethered to the ship by a conducting cable. For deep-tow systems, the hydrodynamic stability of the fish, variations in ship speed and heading, effect of undercurrents, and variations in bottom elevation, must all be controlled or carefully monitored to ensure spatially isometric images along a projected trackline. Navigationai precision of the highest caliber is required if a mosaic is to be made. For long-range systems these variables make synthetic aperture operation a practical impossibility. Acoustic propagation ve-

Table 31-43

Band matrix, showing transmission wavelengths, filters, and films of aerial spectrophotography of the Middle Sambo Reef area, Florida Keys, acquired by the Itek nine-lens multiband camera (152.4-mm focal length) on 19 March 1967 from an altitude of 3,050 m (10,000 feet)

Band 1 0.385–0.450 μm w/vertical polarization Wratten 36 & 47B Kodak 5424	*Band 2* 0.385–0.450 μm w/horizontal polarization Wratten 36 & 47B Kodak 5424	*Band 3* 0.440–0.525 μm Wratten 3 & 48 Kodak 5424
Band 4 0.505–0.570 μm Wratten 12 & 65 Kodak 5401	*Band 5* 0.550–0.625 μm Wratten 22 & 52 Kodak 5401	*Band 6* 0.615–0.690 μm Wratten 92 Kodak 5401
Band 7 0.700–0.880 μm Wratten 89B & 1.2 ND Kodak 5424	*Band 8* 0.755–0.880 μm Wratten 87 & 0.9 ND Kodak 5424	*Band 9* 0.785–0.880 μm Wratten 87C Kodak 5424

Note: Both the Kodak 5401 and 5424 aerial films have since been replaced by the Kodak 2402 and 2424 aerial films, respectively (Bill Reed, Eastman Kodak Company, oral communication, 1982). ND = neutral density.

locities vary with water temperature, pressure, and salinity. Because the deepwater column is not homogeneous, errors of more than 5 percent in slant range can be recorded by long-range sonar (Flemming, 1981). Focussing and cutoff problems arise from thermal stratification in the deepwater column (Stephens, 1970). For short-range sonar imaging (i.e., at ~100 kHz, depths less than 1,000 m, swath widths less than 1 km), the physical problems are negligible or are easily corrected for, and excellent bottom imaging is now routinely accomplished (Prior and Coleman, 1980). Regional survey needs in depths between 500 and 2,500 m require the less tractable long-range and deep-tow mid-range sonars.

In 1979, a long-range sidescan-sonar survey was carried out by the U.S. Geological Survey (USGS) in cooperation with the British Institute of Oceanographic Sciences. The GLORIA (Geologic Long Range Inclined Asdic) II system (Laughton, 1981) was used to survey more than 110,000 km² of the Atlantic Continental Slope

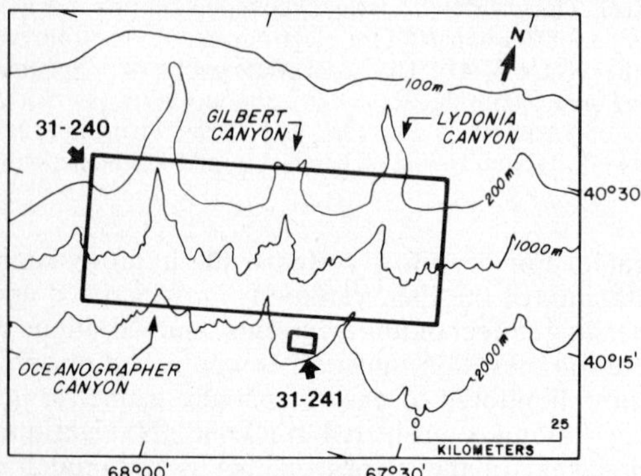

Fig. 31-239. General bathymetry and major canyons of part of the Atlantic Continental Slope south of Georges Bank. Heavy outlines portray areas depicted in Figures 31-240 and 31-241 (Contours from Uchupi, 1965).

between lat. 42° and 27°N. to depths of 5,000 m. Off the coast of New England, traverses oblique to the slope were made to emphasize canyon morphology. In some places, lines were spaced to give 50 percent overlap through the use of two look directions. Also in places, a third transit at right angles was made to resolve foldover problems and to improve control for constructing a mosaic. Along with the images, 10-kHz echo-sounding profiles and single-channel (airgun) subbottom profiles were obtained to provide bathymetric and subbottom structural control.

GLORIA II emits sidebeams of 2.5° in the horizonal direction and 30° in the vertical direction, with acoustic frequencies of 6.2 and 6.8 kHz and a 100-Hz bandwidth (Teleki et al., 1981). A 40-s or a 20-s pulse repetition rate provides optional 60-km and 26-km swath widths, respectively. The 20-s rate produces better spatial resolution. Pixel dimension down range is constant (20 m), but along track it ranges from 80 m (1 km out from the source) to 700 m (15 km out). Dynamic range is about 60 db, but only about 17 db is recoverable on film used to record the data.

The data were stored on analog tape and were subsequently photoreproduced. Spatial distortion along the ship's track was corrected by anamorphic photography (Rusby and Somers, 1977); the images were reduced photographically to construct a mosaic at a scale of 1:250,000 on a British Transverse Mercator grid base. Figure 31-240 is an example of the final sidescan-sonar image obtained for the North Atlantic region (Scanlon, 1982).

The 1979 GLORIA survey provided a nearly synoptic overview of the morphology of the Continental Slope (Scanlon, 1980, 1982). The images showed that the slope is intricately dissected by a variety of incised channels, which generally have a palmate tributary pattern; they also showed that incised features are strongly differentiated according to degree of etching and lateral and lon-

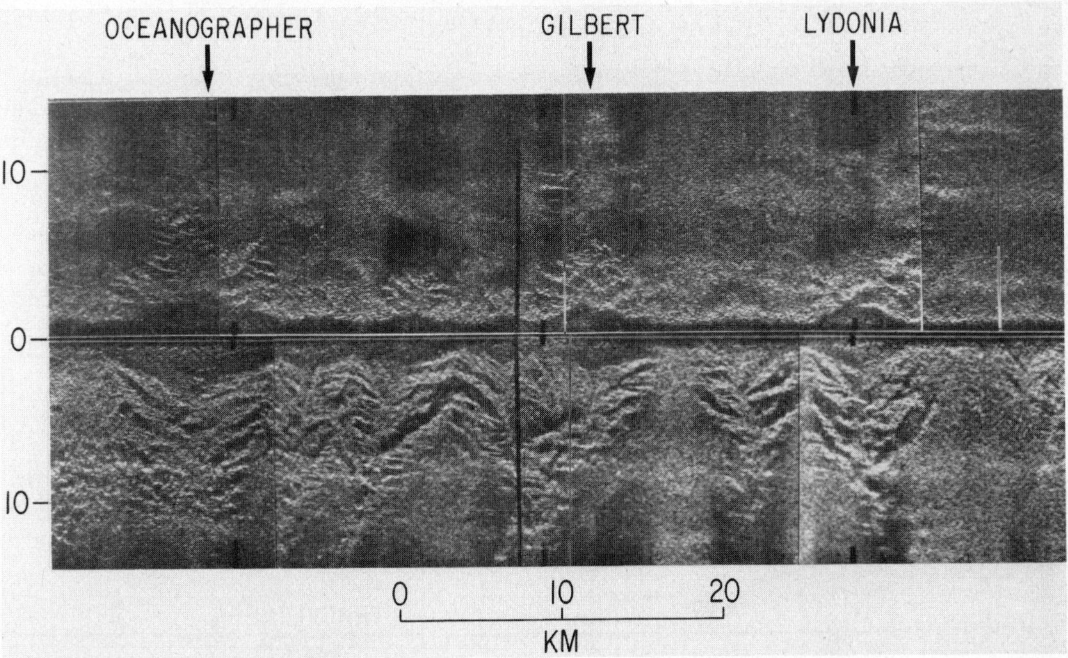

Fig. 31-240. GLORIA II composite sonar image of Atlantic Continental Slope off Georges Banks (See Figure 31-239). Three major canyons (Oceanographer, Gilbert, Lydonia) and many smaller ones are shown. Bright areas are reflecting surfaces facing trackline; dark areas are acoustic shadow. Speckled tone represents combined textural and slope components below resolution. For a group of submarine photographs taken in the vicinity of the head of Lydonia Canyon, see Figure 31-47.

gitudinal extent, suggesting different generations of erosion. Image brightness, especially in the near range, is essentially a function of local slope inclination. Surface texture should influence brightness on a pixel-by-pixel scale, but the images are too "noisy" to determine this. Unfortunately, because of acoustic distortions, the optimum resolving power of GLORIA was not realized. Therefore, the submarine physiographic patterns could not be adequately traced or studied in detail. Images of higher resolution and less acoustic distortion were required to resolve geologic questions concerning the nature of incision and the materials involved.

In 1980–1981, Dennis W. O'Leary and John M. Aaron undertook more detailed surveys with the International Submarine Technology, Ltd. (IST) Sea MARC (Mapping And Remote Characterization) I midrange sonar, in cooperation with W. B. F. Ryan of Lamont-Doherty Geological Observatory. This prototype system was built specifically for use in the search for the S. S. *Titanic* in 1980–1981.

In 1980, an area of about 2,800 km^2 was surveyed on the Continental Slope off Georges Bank between long. 67°30′ and 68°15′W. in water depths of 500 and 2,400 m (O'Leary, 1982). In 1981, an additional 1,8?5 km^2 of slope was surveyed between long. 70°10′ and 70°30′W. The 1980 cruise utilized five parallel tracks along the slope, each swath separated by a narrow gap. The idea was to maximize areal coverage and to emphasize terrain features oblique or at a high angle to canyon axes. Images were closely spaced, however, so that topographic continuity could be established. In 1981, the same pattern was used, but lines were also run along the axes of Alvin and Atlantis Canyons.

The Sea MARC I "fish" is a neutrally bouyant vehicle tethered to a 1,000-kg depressor towed approximately 300 m off the bottom at all times so that a "draped" survey is obtained. A tow speed of approximately 2.5 knots is required. In order to keep track of the fish, which, in deep water can be more than one kilometer behind the ship, and maintain elevation (controlled by taking in or letting out wire), a 4.5-kHz fish-mounted pinger provides continuous bottom profile record and ship-to-fish range. These data, as well as data from a pressure depth gauge and a current meter mounted on the fish, are later processed together with shipboard navigation records to obtain a plot of the actual fish track.

Sea MARC I emits beams of 1.7° in the horizontal direction and 50° in the vertical direction at frequencies of 27 and 30 kHz, with a bandwidth of 5 kHz. A 2.5- or a 5-km swath width, 2,028 pixels across, is optional. This survey used mainly the 5-km swath, with a 4-s pulse repetition rate. The pixel dimension is 2.34 m.

Sea MARC I has a dynamic range of approximately 140 db. Because the electrostatic paper

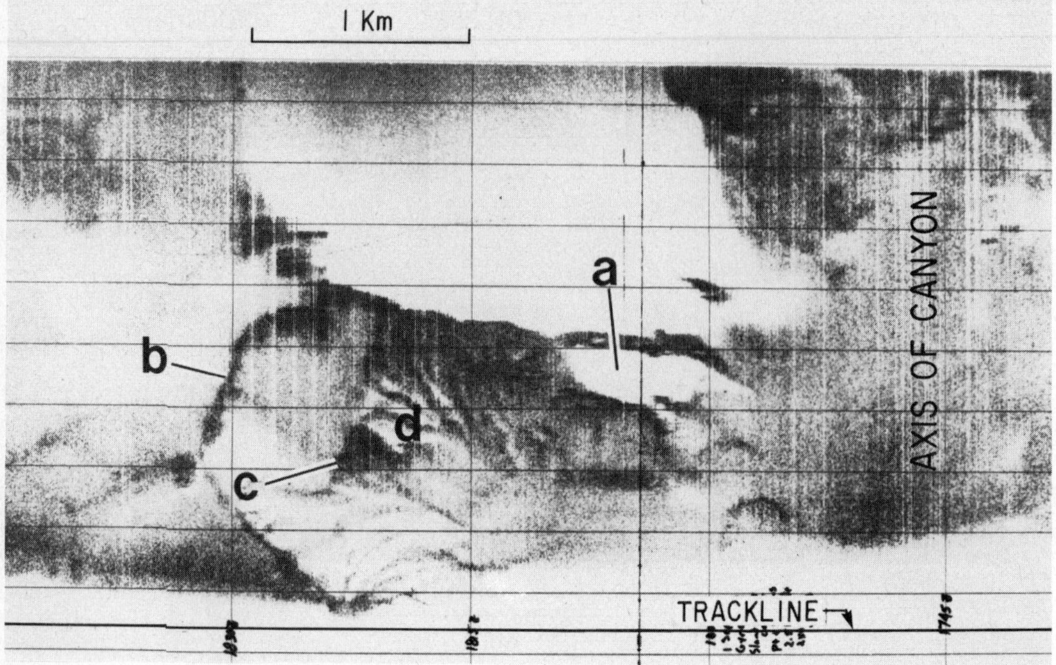

Fig. 31-241. Slant-range-corrected, mid-range sidescan-sonar image (half swath) of etched terrain along west side of Jigger Canyon in the vicinity of long. 67°43′ W. at water depth of approximately 1,800 m (See Figure 31-239). Horizontal grid spacing, 250 m. Darker tones are reflecting surfaces facing trackline; bright areas are acoustic shadow. Gray tones are mainly textural modulation. Image shows a large amphitheater or cirque opening on Jigger Canyon. Detailed features include: a, toreva block; b, steep arcuate headwall; c, younger depression cut in floor of amphitheater; d, rills associated with, or possibly consequent to, the younger depression.

recorders used for analog display only have a 16-db range, transmitted signal returns of 60 db were automatically classified according to 16 equal digital number (DN) interval values. A slant-range correction was also automatically performed. Each array of 1,024 pixels per side was collected, then printed according to proper range calibration. Although an array can be printed as many as four successive times to accommodate spatial distortion along track, this distortion was mainly controlled by manual adjustment of recorder-chart speed with respect to ship speed.

The sonar images were spatially correlated with a bathymetric contour map (constructed from previously acquired closely spaced (1 km) seismic profiles) to provide general control and correction for orientation of landform outlines and to indicate relief. After the superposition of images, bathymetric contours were redrawn to show details of imaged landform outlines and inferred variations in contour-line spacing. This merging of bathymetry and sonar image data results in what is essentially a shaded-relief landform map that can be used for morphometric analyses, correlation with subsurface data, and preparation of coring, bottom photography, or submersible surveys.

The mid-range sonar images show that the Continental Slope is a deeply eroded and sharply etched complex terrain (O'Leary and Twichell, 1981) similar in many ways to subaerial terrain formed under arid climatic conditions. Forms suggestive of mass wasting are especially prominent on the lower slope (Figure 31-241). Subtle variations believed related to surface texture are well displayed in the analog images, although digital tape recording (not available during this survey) would have retrieved more subtle spectral information. The midrange images showed very little spatial distortion; it might be possible, therefore, to make distinctions in the images between brightness related to surface roughness and that related to slope reflection. The midrange data also facilitated interpretation of the GLORIA data by resolving certain geomorphic ambiguities.

These regional sidescan-sonar surveys undertaken by the U.S. Geological Survey demonstrate the utility and possibilities for further application of sonar remote sensing. Quantitative geomorphic studies of the sea floor at regional scale are now possible, and targets for detailed study are clearly identifiable. However, the acquisition of precise and accurate bathymetry by narrow-beam echo-sounding and by selective bottom photography (Color Figure 31-53) is the key to effective utilization of sonar image data. Without good bathymetry and bottom characterization, quantitative studies of sonar images are limited.

GEOBOTANY

Geobotanical Exploration

The surficial expressions of economic deposits of fuels or heavy metals are in some places distinctive enough to have detectable effects upon vegetation shown in imagery. The identification of vegetation anomalies in imagery provides a means for targetting ground-based exploration activity. These geobotanical techniques have exploration potential in any vegetated terrain. They are particularly useful, however, in areas with dense to moderately dense vegetation, where rocks and soils are completely obscured.

Because of the seasonal variation in vegetation, there is in most places a best time of year for acquiring imagery. The best expressions of many botanical features occurs at the peak of the growing season, when the amount of green biomass is at its highest level. On the other hand, some vegetation phenomena that have exploration significance may be apparent only at some other time, such as during senescence. The overall condition of vegetation also varies from year to year because of climatic fluctuations, and thus the imagery acquired in certain years is superior to that acquired in others.

Fortunately, only a few types of vegetation anomalies have been found that are indicative of fuel or mineral deposits: sparse or barren areas, plant community transitions, changes in plant morphology or chlorophyll content, changes in chlorophyll fluorescence (See Chapter 33, Agricultural Applications, and Chapter 34, Forestry Applications, of this *Manual*). Unfortunately, all of these anomalies can also be caused by a variety of environmental factors or by disturbance. The identification and mapping of most of these vegetation anomalies is best accomplished with imagery. The cause of such anomalies commonly cannot be discerned, however, without ground-based field investigations.

Characteristics of Geobotanical Anomalies

Heavy Metals

While minor quantities of B, Mg, P, S, Cl, K, Ca, Mn, Fe, Cu, Zn, and Mo are required for normal plant growth, absorbed excesses of these and other heavy metals have adverse effects on plant metabolism. The more severely toxic elements include Ni, Cu, Cr, Co, and Pb. To become absorbed by the plant roots, an element must be in a soluble form. In general, the availability of metals is determined by the soil Eh-pH conditions, though plants have some ability to modify these factors in the vicinity of their roots. The weathering of metal sulfides produces acidity that in some cases is extreme enough to be directly toxic to vegetation.

There are several ways in which heavy metal toxicity can be expressed. Plant growth may be stunted (Foy et al., 1978) or deformed (Lyon, 1975). Certain species may be excluded from mineralized ground (Bolviken et al., 1977) or found in reduced numbers. In severe cases, barren areas develop (Color Figure 31-242). In contrast, other species (indicator plants) may be found exclusively or in greater numbers in mineralized areas (Cannon, 1957). Heavy metals are also known to cause chlorosis (a deficiency of chlorophyll pigments) characterized by a slight to distinct yellowing of leaves (Foy et al., 1978). Excessive heavy metals in leaves can also disrupt the normal functioning of leaf stomates, resulting in increased leaf temperatures (Horler et al., 1980a).

The exploration target for many metal sulfide deposits consists of argillically altered areas rather than high concentrations of metals. These clay-rich areas commonly have the following characteristics: impeded water infiltration, low pH's, and deficiencies in exchangeable cations and phosphorus (Billings, 1950). The result can be barren areas, sparser vegetation, or transitions to plant communities adapted to more adverse conditions (Milton, 1981).

Hydrocarbons

Outcrops of coal or oil seeps in general act as plant fertilizers. The additional organic matter also improves soil structure so that moisture capacity is increased. The typical result is enhanced vegetative growth or gigantism (Vostokova et al., 1961; Brooks, 1972).

Abnormal concentrations of gases in the root zone are known to have adverse affects on certain plants. Experiments show these effects to be species dependent (Gilman, et al., 1981). Davis (1977) reports on vegetation killed by leaks in natural gas mains. Leone et al. (1977) found stunting and chlorosis in corn due to the lateral migration of gases from a sanitary landfill. It appears that methane is not in itself toxic to plants. Microbial utilization of methane produces carbon dioxide and depletes the soil atmosphere of oxygen (Hoeks, 1972). Flower et al. (1981) conclude that the resulting damage to vegetation is most likely caused by one or more of the following factors: 1) toxicity of CO_2 to the roots, 2) lack of available O_2 for use in root cell respiration, and 3) anaerobic conditions reducing and mobilizing toxic levels of heavy metals such as Fe, Mn, or Zn.

Detection of Geobotanical Anomalies

A variety of sensor types have been used to detect geobotanical anomalies. The important spectral characteristics of vegetation are covered in other chapters of this *Manual,* especially Chapters 33, 34, and 35. In evaluating the utility of a sensor system, consideration should be given to the number, width and placement of the filter bands as well as to the spatial resolution of the resulting imagery. Several vegetation anomalies

TABLE 31-44

Relationships Between Exploration Targets, Possible Vegetation Anomalies, and Corresponding Spectral Anomalies

	Biomass	Chlorophyll	Plant Communities	Morphology	Thermal
Heavy Metals	—	—	yes	yes	+
Argillic Alteration	—	?	yes	?	+/?
Oil Seeps Coal Outcrops	+	?	?	yes	–/?
Methane	—	—	yes	?	+
Filterband Combinations	IR/red and other vegetation indices	"blue shift" green/blue green-blue green-red orange-red	IR/red and ?	IR and ?	Thermal IR

+ = increase, – = decrease, ? = too varied or data insufficient to make generalization.

induced by heavy metal toxicity have been detected with Landsat (Lyon, 1975; Bolviken et al., 1977).

In general, however, the spatial resolution of Landsat (57m × 79m) is too coarse for most geobotanical exploration purposes. Resolutions of 10 to 30 meters are needed for most situations. A summary of the relationships between exploration targets, possible vegetation anomalies, and corresponding spectral anomalies is presented in Table 31-44.

Biomass Anomalies

Several standard vegetation indices can be used to assess biomass (Tucker, 1979). These indices are based on various combinations of red and infrared channels (e.g., IR/red or IR minus red/IR plus red).

Infrared Anomalies

Richardson et al. (1969) found that there are subtle infrared reflectance differences in leaves dependent on the cellular structure. These differences may turn out to be useful in discriminating vegetation types and detecting vegetation stress. Horler et al. (1980b) found that metal stress in field and laboratory plants decreased reflectance in the equivalent Landsat Thematic Mapper (TM) bands at 1.65 and 2.20 μm.

Thermal Anomalies

Because of stunted root growth or obstruction of normal stomatal activity, geochemically stressed vegetation tends to be more susceptible to water stress than normal vegetation. This water stress can result in slight increases in canopy temperatures (Horler et al., 1980a). Such thermal anomalies have been detected in some imagery (Canney et al., 1972; Lefevre, 1980). The effect is ephemeral, however, readily confused with or swamped by normal drought stress, and is dissipated or lost in atmospheric turbulence.

Chlorophyll Anomalies

The chlorophyll content of leaves can be assessed with green/blue, green-blue, green-red or orange-red indices using bands 20- to 50-nm wide (Elvidge and Lyon, 1982). These indices were found to be superior to those available from combinations of Landsat multispectral scanner (MSS) or thematic mapper (TM) bands.

An airborne spectroradiometer with 512 channels from 400 to 1100 nm (Chiu and Collins, 1978) has been used to detect spectral anomalies apparently associated with chlorosis in forest canopies growing in mineralized areas. Birnie and Hutton (1976) detected increased reflected radiance in the 670 nm region. Collins (1978) reports a "blue shift" in the chlorophyll absorption edge in the 700- to 750-nm region. Birnie and Francica (1981) found increased green/blue values.

Remote Detection of Geobotanical Stress

For centuries, observers have recognized that changes in surface vegetation sometimes reflect the subsurface geology and indicate potential mineral deposits. The study of plant species whose growth is uniquely restricted to or absent from definable geologic units is geobotany. Numerous indicator plants have been identified as useful for mineral prospecting (See for example Carlisle and Cleveland, 1958; Cannon, 1960 and 1971; NASA, 1968; Brooks, 1972; and Rose et al., 1979). In addition, the physical state of a given species may change as a result of heavy metal stress. The changes may be manifested as gigantism, dwarfing, or chlorosis (yellow discoloration) (Yost and Wenderoth, 1971).

Greenhouse experiments have been undertaken to measure the reflectance response of various plants under the influence of heavy metal stress. Horler et al. (1980b), summarizing previous work and reporting on their own work, note that heavy metal stress is generally manifested by increased reflectance in the visible part of the spectrum and that this increase is associated with a decrease in the chlorophyll content of the stressed leaves. Schwaller and Tkach (1980) observed increased reflected radiance across the 475–1,650 nm spectral band. Chang and Collins (1980) report a spectral shift of 10–40nm in the chlorophyll absorption band edge resulting from heavy metal stress. Chlorophyll does not actually change its absorption wavelength, but only the strength of that absorption itself, resulting in the apparent "blue shift."

Greenhouse studies have been extended to the study of the reflectance properties of natural vegetation growing in several kinds of mineralized ground. A number of investigators (including Canney et al., 1970; Yost and Wenderoth, 1971; Howard et al., 1971; Press, 1974; and Horler et al., 1980b) have studied a variety of vegetation types in a variety of mineralized settings, and they generally conclude that the result of heavy metal stress is an increase in the visible reflected radiance. Although Horler et al. (1980b) feel that the change in reflectance properties correlates best with total chlorophyll, Howard et al. (1971) suggest that their observed increase in reflected radiance results from a decrease in needle density of stressed vegetation.

To exploit these clear spectral responses of vegetation to heavy metal stress in a remote sensing program, airborne multispectral scanners of the type developed by Chiu and Collins (1978) must be used. This kind of sensor permits rapid coverage of large areas with a high-resolution (spatially and spectrally) scanner. Experiments reported by Birnie and Hutton (1976), Birnie and Dykstra (1978), Collins et al. (1977), Collins (1978), Collins and Chiu (1979), and Birnie and Francica (1981) have used the multispectral scanner and found correlations between spectral data and mineralization. Birnie and his coworkers report increased reflected radiance values in the visible part of the spectrum for lodgepole pine (670 nm) and Douglas fir (560 nm), while Collins and his coworkers report a spectral shift of 10-20 nm in the chlorophyll absorption band edge in a conifer forest canopy.

Other promising methods of remotely detecting geobotanical stress include the measurement of stress induced fluorescence in vegetation (Hemphill et al., 1977). Although work must be done on many different vegetation types under different kinds of heavy metal stress to determine the specific types and causes of the spectral responses, it is clear that remote detection of geobotanical stress is one of the methods that should be used in a mineral exploration program.

Geobotanical Exploration from Landsat Images and Aerial Photographs

A visual interpretation of geobotanical associations using remotely sensed data can benefit geological investigations, in particular those which involve mineral prospecting. Aerial photographs and satellite imagery can record vegetation species composition, plant density, and physiological changes of a vegetation type. More simply, they can record the presence or absence of vegetation within an area. Also, the ability to recognize the association of flora with certain minerals has been the subject of many studies (Cannon, 1957, 1960, and 1971). Vegetation under stress and floral associations are inferred to be caused by anomalous, subsurface geologic phenomena such as mineralization or moisture content.

Stressed vegetation can manifest itself through a change in the color of leaves or needles. Press (1974) noted that heavy metals may interfere with the absorption of Fe, an important element for the synthesis of chlorophyll. This, in turn, interferes with the normal metabolic processes of certain species, and thus may cause chlorosis (yellowing) or dehydration in vegetation. A decreased response in the infrared would be detected because of changes in the internal leaf structure, as well as increased reflectance in the visible due to the decrease in the amount of chlorophyll. Canney et al. (1979) describe small areas of deciduous forests in temperate regions that display autumn colors in advance of the rest of the nearby forest. They speculate that these areas are associated with concealed mineral deposits.

Adverse effects on vegetation density increase the amount of surface (barren rock, soil, and understory vegetation) reflected radiance. Raines and Canney (1980) found that vegetation density differences affect the tone or color of photographs and imagery, even when individual plants are not completely resolvable. Density differences may be the result of disease, insect infestation, competition between plant species for establishment within a given area, mixtures of species, dwarfism, or gigantism and are identified by the size and shape of the canopy of the vegetation.

A successful geobotanical prospecting tool is the knowledge of the preference of certain plant communities or species for particular rock types, soils, or zones of mineralization. Although the identification of individual plants may be beyond the resolvable limits of most remote sensor systems, image or photo tone, texture, and pattern of these plant communities within a mineralized zone may not.

Not all geobotanical relationships are associated with mineralization. Commonly, vegetation may be alined along a fracture or fault serving as a conduit for water flow or acting to juxtapose an aquifuge. Likewise, vegetation can be associated with a particular lithologic unit where the availability of water is conducive to plant growth. These relationships can be recognized by anomalously linear or areal growth patterns on false color infrared imagery. Figure 31-243 and Color Figure 31-244 are a black-and-white index aerial photograph and enlarged parts of three areas on the same photograph compared with enlargements of the same three areas on a Landsat MSS false-color composite, respectively, of an area near Sundance, Wyoming, showing the variation in density of coniferous vegetation with respect to rock type.

Regional Geobotany

In many terrestrial environments, vegetation is sufficiently dense to obscure rocks and soils from direct observation in imagery. When working with such imagery, one must analyze spatial and spectral variation in vegetation for clues about the underlying geology.

The success of geobotany can be attributed to the fact that plants respond to their environment. Plants are sensitive to a wide variety of parame-

Fig. 31-243. Vertical, black-and-white aerial photograph of the Sundance, Wyoming area which provides an index to three study areas. The photograph depicts the relative affinity of coniferous vegetation for certain rock types. In general, igneous rocks support relatively dense stands of conifers while sedimentary rocks support populations that are not as dense. Area 1 is an extrusive igneous rock exhibiting a dense stand of conifers. Areas 2 and 3 are sandstone with some limestone and show less dense stands of conifers with more bare ground visible. (U.S. Geological Survey aerial photograph GS-VCZ, No. 1-41, acquired on 21 September 1954. Approximate scale: 1:40,000). Compare with Color Figure 31-244, which compares three areas on a Landsat MSS false color composite with enlargements of the same three areas on the black-and-white aerial photograph.

ters, only some of which are geologically important.

In most areas of the world, there is a seasonal variation in vegetation, and the expression of geobotanical features varies depending on the seasonal condition of the vegetation. Vegetation also varies from year to year as the climate fluctuates. Many geobotanical features show up best at the peak of green biomass at the height of the growing season, but not always because other features show up best at some other point in the growth cycle, such as during early growth or senescence.

One of the primary environmental factors that vegetation responds to is soil moisture. This is a complex variable that depends on many things: annual precipitation and temperature patterns, slopes, exposures, soil texture, and permeability. Various lithologies develop different soil properties in regard to water availability. This difference in soil properties commonly (though not always) results in vegetation differences between adjacent areas of diverse lithologies. For instance, shales and sandstones can often be distinguished on the basis of vegetation differences. The shale soils tend to be less permeable than the sandstone soils. The sandstone areas are commonly dry at the surface but provide water at depth for deeply rooting species. Limestone soils are usually quite dry and support rich and distinctive assemblages of species.

Structure can also affect vegetation, usually by its influence on topography and moisture availability. Drainages are commonly structurally controlled, and drainage vegetation is usually quite distinctive from adjacent areas. In some places, faulting creates barriers to ground water movement, enhancing vegetative growth on the ponded side.

Besides the effects of moisture availability, lithologies can exert a chemical effect on vegetation. The most extreme example of this sort is the ease with which ultrabasic areas can be distinguished in imagery. Figure 31-245 shows a Landsat MSS band 7 dot-print image centered on a serpentine area in the foothills of the Sierra Nevada in central California near the Melones Reservoir. Color Figure 31-246 is a geologic map of the same area. Serpentine soils have several attributes detrimental to many plants: a low Ca/Mg ratio, high

Fig. 31-245. A Landsat MSS band 7 dot-print image showing a dark serpentine outcrop in the foothills of the Sierra Nevada in central California, near the Melones Reservoir (Landsat image ID No. 1074-18114; 5 October 1972). The ultrabasic area, known as the Red Hills, is composed primarily of serpentine but also has some intermixed gabbro. The vegetation on the Red Hills is a sparse chaparral made up of *Ceanothus cuneatus* with scattered digger pine. The other rock types in the area are typically covered with grassland or oak-woodland. Field surveys indicate that the greatest spectral contrast between the ultrabasics and the other rock types is during the period April to mid-May.

concentrations of Ni, Cr, and Co, and they tend to waterlog during the wet season and become parched during the dry season. As a result, serpentine areas are typically occupied by distinctive plant communities. There are other lithologies that also exert chemical effects on vegetation. Soils developed from granitic rocks are typically low in nutrients and develop depauperate vegetations compared to more basic intrusives. The high pH's that develop in soils on carbonate material limit nutrient availability to certain plant species.

ARCHAEOLOGICAL GEOLOGY

Archaeology seeks to explain and illuminate the lives and activities of past people, groups, and culture through examination and analysis of the places in which they lived and the materials they have left behind. In doing this, archaeologists make use of a wide variety of data, methodology and theories "borrowed" from sciences such as geology, geography, biology, economics, anthropology, and political science, as well as many techniques adapted from the fields of photogrammetric engineering, remote sensing, surveying, photography, and materials sciences. Archaeology is closely allied with the anthropological fields of ethnology and physical anthropology; in recent years the widespread need for the conservation and protection of prehistoric and historic sites and materials has given rise to yet another aspect of archaeology—cultural resources management. Applications of remote sensing to archaeology, anthropology and cultural resources management are discussed and illustrated at length in Chapter 26, Applications in Archaeology and Anthropology, of this *Manual*.

Much of the explanatory emphasis of archaeology centers about man's relationship with the natural world, and much of that is encompassed in the subject of geology. As described in Chapter 26, a process of informed sample stratification, or division into representative zones of a study area to be sampled, is a logical first step in explaining that portion of past peoples' activities that was directly related to the necessities and opportunities presented by their environment. Surface and subsurface geology controls such immediately important natural factors as the availability of surface water, the availability of soil, and the complexities of landforms and can often be measured and mapped by the archaeologist, using remote sensing techniques and data, for input into sample designs.

In other cases, however, geological and geomorphological data are of even greater priority in archaeological analysis and interpretation. The archaeologist, of course, can only use those archaeological data that he/she can see and find on and under the ground, and a large proportion of his/her time is directed toward locating cultural evidence of the past. The sites, sherds, bones, and stone flakes that are available for study are only a small subset of those originally discarded or lost by their users, however. Natural processes are responsible for the disturbance or destruction of archaeological materials in many cases; in others, aggradation of a surface may encapsulate and preserve—but effectively hide—artifacts or features (McCauley et al., 1982). Erosion at a later date can expose these making them available, but there is a fine line between exposure and destruction.

Geomorphological processes are important factors in archaeological preservation, in what the archaeologist sees when looking for past materials, and in the appearance of those materials when they are found. This situation has been illustrated by an experimental project carried out by the Remote Sensing Division of the National Park Service at Chaco Culture National Historical Park in northwestern New Mexico. Located in Chaco Canyon, this relatively isolated area contains a high concentration of archaeological sites, particularly those related to the height of the Pueblo period, between about AD 900 and AD 1200, in the American Southwest. During the project, the geomorphological dynamics of a portion of the broad, alluvial floor of Chaco Canyon were mapped and measured using 1:6,000-scale color transparency aerial photographs, and differences in the occurrence of archaeological materials as noted by archaeologists during ground surveys were compared to the geomorphic zones thus mapped (Figure 31-247). Surface alluvial and erosional processes were found to affect not only the preservation of archaeological materials but also to influence the type of materials found (Ebert and Gutierrez, 1981). For instance, walls standing 0.3 m above the surface after aggradation were far more obvious than potsherds lying at the base of those walls. Such studies also hold promise for the geologist and geomorphologist, of course, for they might permit the dating of relatively recent surfaces and surface changes.

The archaeological analysis of river terraces also promises to be of aid in the relative dating of materials, as well as the investigation of their preservation and visibility. Some of the earliest archaeologically related analyses of river terraces by geologists aided in establishing tentative dates for some of North America's earliest PalaeoIndian materials at the Lindenmeier Site in Colorado (Ray, 1940; Bryan and Ray, 1940). In Ebert and Gutierrez's (1981) study mentioned above, photointerpretation of 1:6,000-scale color aerial photographs (in transparency format), coupled with checking of interpretations using 1:30,000-scale U.S. Geological Survey black-and-white aerial mapping photographs, permitted the correlation of disconnected parts of several fluvial terraces and the association of archaeological remains found on related segments of similar age.

Another important application of geological remote sensing to archaeology, recently explored in

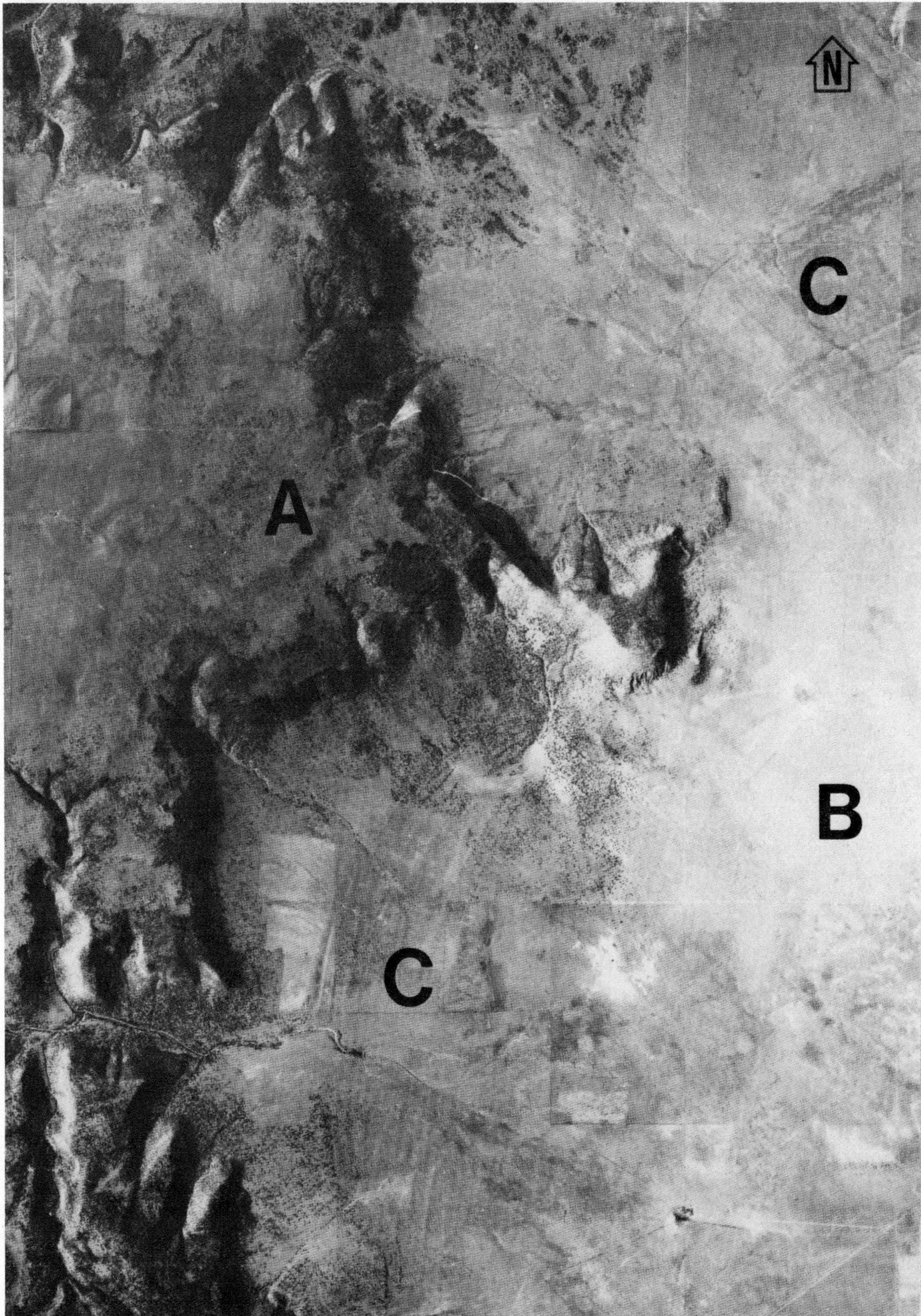

Fig. 31-247. This part of a black-and-white aerial photograph, at an original scale of 1:30,000 and centered at approximately 35°40′ N. and 106°12′W., shows features associated with ancient Lake Estancia. A mesa-top at "A" overlooks the Estancia Basin of New Mexico, bordered by wave-cut cliffs. "B" is the floor of the present Basin, which was also the floor of the ancient lake, and is surrounded by a series of concentric strandlines visible at "C". PalaeoIndian archaeological sites were found above the shorelines seen at "C," while later Archaic materials were found both above and below the shorelines, indicating a later date of deposition and climatic conditions that were much drier than were those during PalaeoIndian times. (Photograph by Defense Mapping Agency, archive designation VV BE M43, 31 January 1954, roll 120, frame 5654.)

both the southwestern United States and in southern Africa, centers about the use of limnological analyses, particularly the mapping of extinct strandlines. During the high stand and recession of the Pleistocene-Holocene ancient Lake Estancia in Central New Mexico, prominent beach terraces were etched along the shoreline and remnants of these and other lacustrine features (bars and spits) still remain (Meinzer, 1911). A study of Early man's occupation of this area began with the mapping of these features using U.S. Geological Survey aerial photographs of the region at a scale of approximately 1:32,000 (Lyons, 1969). It was determined that the Lake was a fresh water body during the Pleistocene with a sill at its southern end, over which the lake emptied into a series of smaller satellite lakes, and that after the retreat of continental glaciers the lake eventually withdrew. PalaeoIndian sites that are dated from approximately 12,000 BP to 8,000 BP are located above the strandlines. Archaic sites occur above and below them because the Archaic tradition extended well into the Holocene. The solution of water balance equations, which are based on lake surface area and estimated evaporation, precipitation, and leakage, also permit palaeoclimatic reconstructions for these two contrasting cultural periods. These climatic differences suggest that the subsistence differences between them were as much a result of prevailing conditions as of "cultural advancement."

A second example of the application of remote

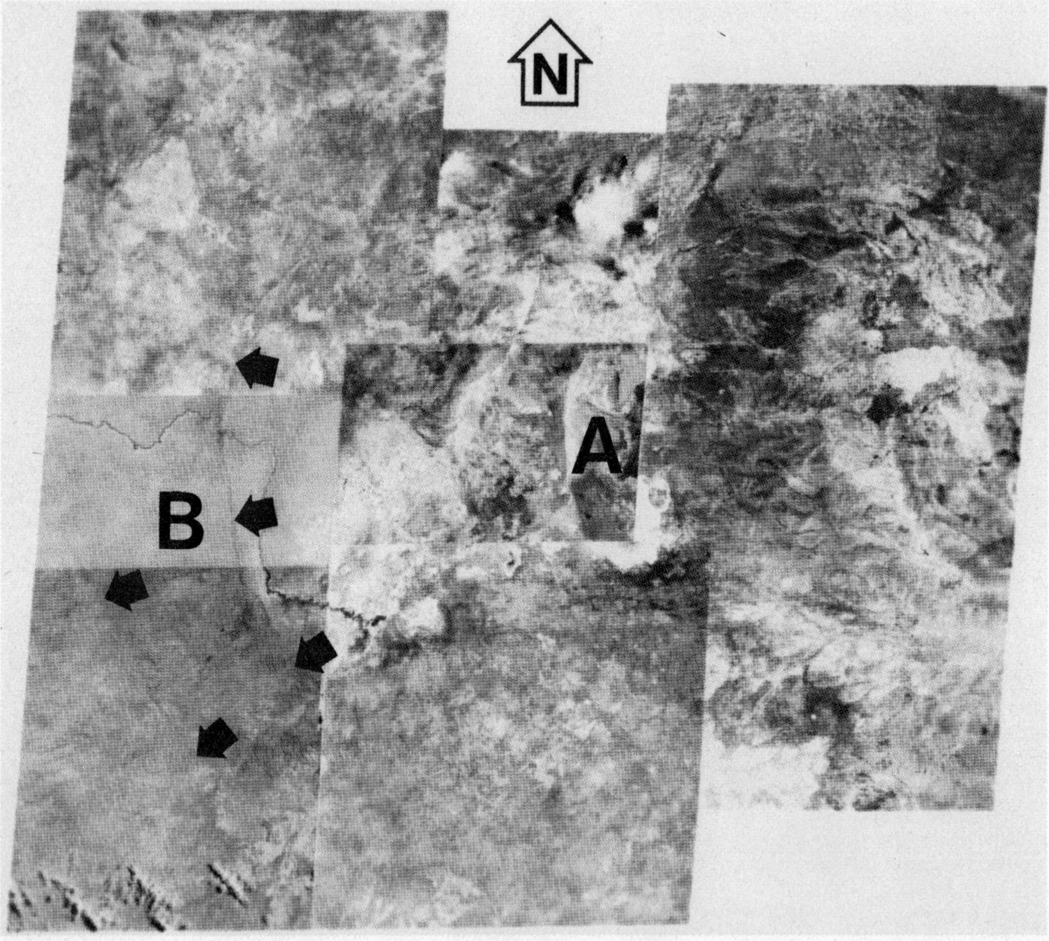

Fig. 31-248. Mosaic of eight Landsat 1 and 2 MSS band 5 images, centered at approximately 21°00′ S. and 25°45′ E., illustrating abandoned shorelines of ancient Lake Makgadikgadi, in the Central District of Botswana (1179-07514, 18 January 1973; 1053-07513, 14 September 1972; 1089-07522, 20 October 1972; 1196-07463, 4 February 1973; 1178-07462, 17 January 1973; 2048-07351, 11 March 1975; 1147-07383, 13 October 1973; 1147-07385, 13 October 1973). The lowest part of the closed basin, sometimes filled with water during extremely wet years, appears at "A" with a prominent curving shoreline. Arrows at "B" indicate the traces of older shorelines or associated offshore features. Little or no topographic expression of these features is discernible in the Kalahari Desert's flat thorn-scrub terrain. The extinct shoreline of Lake Makgadikgadi, mapped using Landsat data like these, were correlated with occurrences of prehistoric artifacts and other materials, thus allowing relative dating and speculation on the climatic conditions under which sites were deposited. (Scale of mosaic is approximately 1:3,425,000; original scale of the mosaic was 1:1,000,000).

sensing to archaeological analysis involved the interpretation and measurement of the abandoned shoreline of a closed-basin lake in Botswana, southern Africa, in 1975–1976. The central basin of southern Africa was occupied at some time or times in the past, most probably before about 12,000 years ago, by a vast closed-basin lake, Lake Makgadikgadi. Shorelines of this lake, mapped with 1:47,000-scale black-and-white aerial photographs and Landsat MSS bands 5 and 7 imagery (Figure 31-248), were associated with Late and Middle Stone age archaeological sites and materials located in Botswana's Central District during an extensive survey there. The approximate ages of the shorelines have been estimated by geologists (Cooke, 1975) and were useful for grouping sites roughly by time period. More important, however, are the palaeoclimatic implications of successive lake levels. Closed-basin lakes are huge, sensitive rain and temperature gauges; the effective available moisture necessary to maintain any specific stillstand of such a lake can be calculated, and the archaeologist thus can relate specific sites to certain environmental conditions, regardless of their age (Ebert and Hitchcock, 1978; Ebert 1978).

A third example is provided from analysis of radar imagery acquired during the OSTA-1 mission of the Space Shuttle. McCauley et al. (1982), in a preliminary analysis of SIR-A imagery of the arid region along the border of the Sudan and Egypt, showed that 1 to 5 m of dry sand was penetrated by the radar energy, revealing buried valleys and "possible Stone Age occupation sites." In their paper, "Subsurface Valleys and Geoarchaeology of the Eastern Sahara Revealed by Shuttle Radar," the authors state that, "The presence of old drainage networks beneath the sand sheet provides a geologic explanation for the locations of many playas and present-day oases which have been centers of episodic human habitation" and that, "continued mapping of the relict Quaternary drainages as additional radar images are obtained from future Shuttle missions should lead to the discovery of numerous additional occupation sites."

There are also theoretical lessons that archaeologists can learn from remotely sensed geological data. The principles of geological stratigraphy have long been applied by archaeologists in excavations of sites of virtually all ages. However, the concept of utilizing time zones that transgress stratigraphic boundaries to refine dates based upon stratigraphic sequences has been paid little heed in cultural studies. Sears et al. (1941), in a classic stratigraphic study of transgressive and regressive marine deposits, demonstrated that, under certain conditions of fluctuating water levels and accompanying migrating depositional zones, time zones do not necessarily correspond with the top of a lithologic unit but instead transgress stratigraphic boundaries. Another example of this phenomenon is the time variations that occur at the base of a migrating dune field—older at point of inception and younger at the leading edge. Such a situation occurs in the southern end of the Estancia Basin, where aerial photographs were used to map a migratory dune field in which Folsom and Archaic artifacts were found (Lyons, 1969 and 1981).

ASTROGEOLOGY

Introduction

Astrogeology applies the methods and principles of the geological sciences to studies of the Moon, the planets, and other celestial objects (Ronca, 1965). Imagery is the most comprehensive source of remote sensing information for geologic interpretation of planetary surfaces (Davies and Murray, 1971). The first stage of geologic exploration of a planet or satellite is to identify, characterize, and outline the distribution of major rock or material units exposed at its surface and to portray these units on geologic maps (Greeley and Carr, 1976; see also Chapter 36, "Extraterrestrial Applications," of this *Manual*). A geologic map incorporates lithologies, age relations, and geometric positions of rock units; it is distinct from a topographic map, which is concerned only with surface form. Spatial resolution of image information determines the detail of interpretation that is possible: high resolution provides more information upon which to base interpretation; low- to moderate-resolution images provide an overview of the present geologic state of a planetary body, and enable mapping of the regional distribution of the main physiographic units. Major superpositions and cross-cutting relations and large tectonic elements can be identified, and crater densities (important for age determination) can be established. Moderate- to high-resolution images reveal smaller geologic features that can be compared with terrestrial analogs, and which furnish information on processes that have modified the planet or satellite.

Because geologic studies of the planets and satellites depend on remotely sensed data, the techniques of stratigraphic and structural analysis differ from those normally used on Earth. On Earth, rock units are classified according to their lithology, determined either in outcrops or by laboratory studies. Their ages are determined by stratigraphic relations, diagnostic fossils, or by absolute geochronologic methods. On the planets and satellites, the units are defined largely by studies of surface landforms: their topography, their albedo, and their superposition and intersection relationship (Wilhelms, 1972). Planetwide chronologies (Figure 31-249) can be established by dating units by cratering histories of their exposed surfaces (Neukum and Wise, 1976; Soderblom, 1977) and with respect to marker horizons; i.e., extensive deposits formed in a short time interval, such as a blanket of ejecta from a large impact basin.

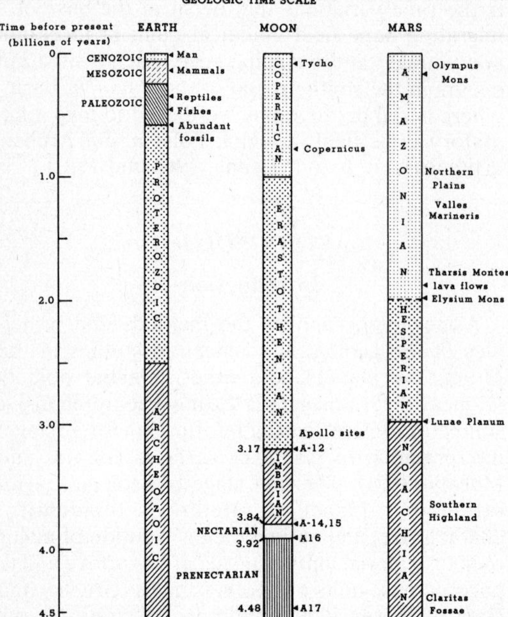

Fig. 31-249. Comparison of geologic events and time scales of Earth, Moon and Mars (Scott, David H., 1982, personal communication).

Rigorous geologic mapping of extraterrestrial surfaces was first demonstrated by Shoemaker and Hackman (1962) for the Moon in early 1960. Their work lead to a systematic geologic mapping program of the Moon, at 1:1,000,000 scale, undertaken by the U.S. Geological Survey under the auspices of NASA in 1961. The first maps were based on telescopic observations from Earth; maps published after 1967 were based on Lunar Orbiter data (Wilhelms, 1970). The entire near side of the Moon has been mapped at this scale. Twenty-seven large-scale (less than 1:250,000) Apollo landing site maps and a composite 1:5,000,000-scale geologic map of the near side (Wilhelms and McCauley, 1971) also have been published. Most of the surface of the Moon has been mapped at 1:5,000,000 scale, based on Apollo and Lunar Orbiter data (Figure 31-250).

A program to map the geology of Mars was begun in 1971 for NASA by geologists of the U.S. Geological Survey, universities, and NASA field centers. Thirty quadrangles have been mapped at 1:5,000,000 scale, using Mariner 9 data (Figure 31-251). Maps of the Viking landing sites were made at larger scales. A 1:25,000,000-scale geologic map of Mars, based on a synthesis of the 1:5,000,000-scale maps, also has been published (Scott and Carr, 1978). Large-scale geologic maps (1:2,000,000-scale) of special interest areas on Mars are planned.

About 45 percent of the surface of the planet Mercury is being mapped at 1:5,000,000 scale, using Mariner 10 pictures (Figure 31-252). This program began in 1975 (Trask and Guest, 1975) and is still in progress.

A preliminary interpretation of the crustal history of Venus has been made from altimetry and radar-image data obtained by the Pioneer Venus spacecraft and by Earth-based radars (Masursky et al., 1980). The imaging resolution of the spacecraft and Earth-based radar are 20 to 30 times lower than the resolution of the early telescopic maps of the Moon. The Soviet Union sampled the venusian surface and made chemical analysis by means of 10 successful landers (Vinogradov et al., 1973).

In 1981, NASA initiated a program, under the management of the U.S. Geological Survey, to map the geology of the satellites of Jupiter (Io, Europa, and Ganymede) utilizing Voyager data.

The strategy of planetary exploration has been to progress from telescopic observation, to fly-by missions, to orbiters, and then to landers. Each spacecraft has had its characteristic capabilities but usually each succeeding spacecraft has had increased resolution of the imaging system (Davies and Murray, 1971). This was the strategy used in exploring Mars. The first mission to Mars, Mariner 4, flew by the planet in July 1965 at a distance of 9,844 km and returned 22 pictures having a maximum resolution of 3 km. This flight was followed in February and March of 1969 by Mariner 6 and 7, which flew by Mars at an approach distance of 3,500 km and obtained a total of 235 pictures having a maximum resolution of several hundred meters. Mariner 9 was placed in orbit around Mars in November 1971, and photographed more than 95 percent of the planet; it returned 7,239 pictures during a year of operation. The Mariner 9 spacecraft carried a low- and a high-resolution camera which achieved maximum resolutions of 1 km and 100 m, respectively. The Viking orbiters, inserted into orbit around Mars in July and August 1976, returned over 50,000 pictures; 1,170 of these taken from an orbital altitude of 300 km had a resolution of 6 to 8 m. Also, over 6,000 pictures with a resolution of about 100 m were taken in systematic sequences for mapping purposes. The Viking landers returned over 4,400 pictures of the martian surface; maximum resolution was better than 1 mm. The landers also carried out a series of complex experiments that analyzed the surface materials for biologic, chemical, and physical properties.

The exploration of the Moon followed a similar strategy except that the first pictures were not taken by a fly-by mission but were a series of approach images [vidicon camera] taken by the Ranger spacecraft in 1964 and 1965, immediately before their destruction by intentional impact on the Moon. The maximum resolution of the last of the series of nested pictures was about 0.5 meter. Ranger was followed between May 1966 and January 1968 by five Lunar Orbiters and by five successful Surveyor unmanned landers. The orbiters returned 1,474 pictures; the landers returned 86,897 pictures as well as chemical analyses of the soil. The culmination of lunar exploration was the six Apollo manned landings between July 1969 and December 1972, all of

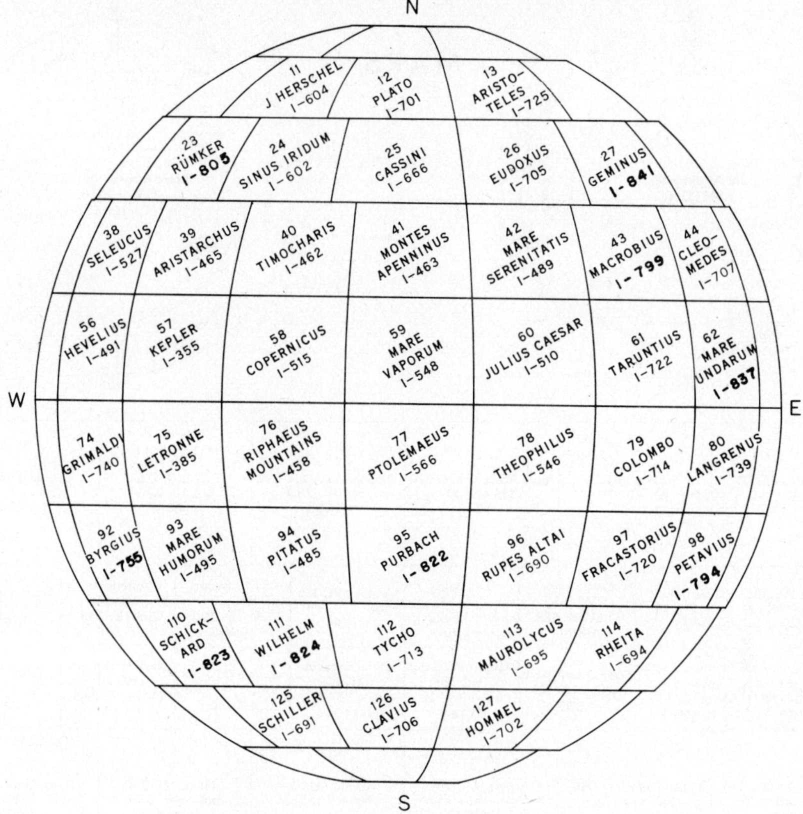

INDEX MAP OF THE EARTHSIDE HEMISPHERE OF THE MOON

Number above quadrangle name refers to lunar base chart (LAC series); number below refers to published 1:1,000,000 geologic map

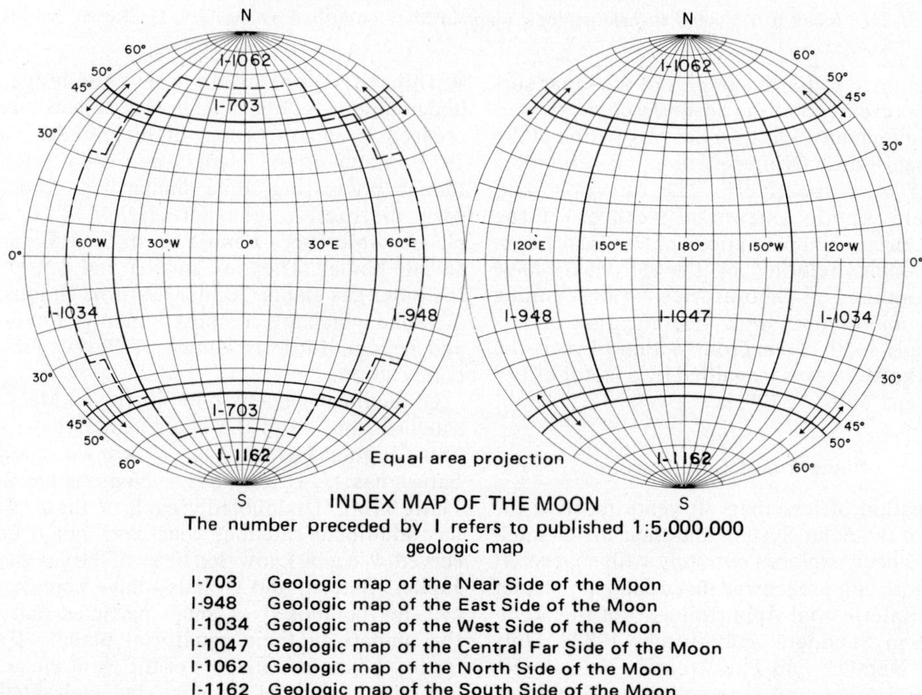

INDEX MAP OF THE MOON

The number preceded by I refers to published 1:5,000,000 geologic map

I-703	Geologic map of the Near Side of the Moon
I-948	Geologic map of the East Side of the Moon
I-1034	Geologic map of the West Side of the Moon
I-1047	Geologic map of the Central Far Side of the Moon
I-1062	Geologic map of the North Side of the Moon
I-1162	Geologic map of the South Side of the Moon

Fig. 31-250. Index to 1:1,000,000-scale and 1:5,000,000-scale geologic maps of the Moon published by the U.S. Geological Survey.

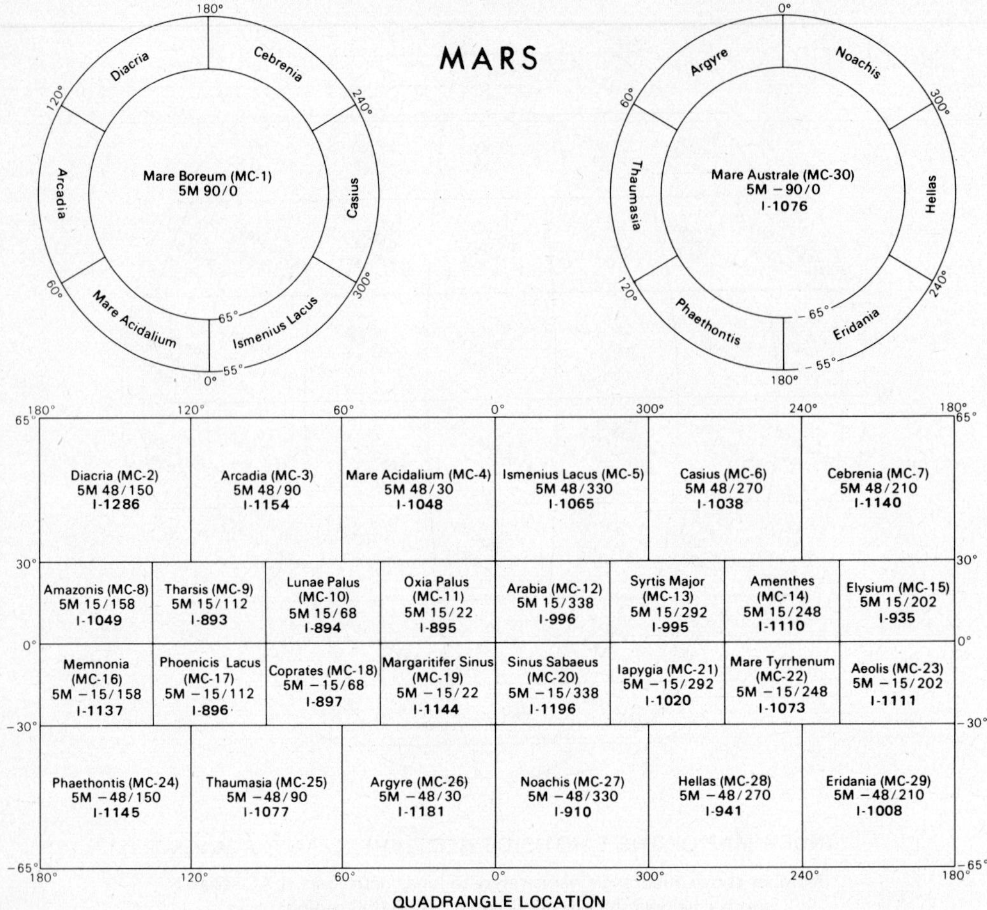

Fig. 31-251. Index to 1:5,000,000-scale geologic maps of Mars published by the U.S. Geological Survey.

which returned to Earth samples of the lunar surface. The resolution of the best telescopic view of the Moon is about 1 km. The best resolution of the Ranger and Lunar Orbiter pictures was about 0.5 m and 1 m, respectively. The Surveyor and hand-held Apollo astronaut pictures of the Moon's surface had resolutions of less than 1 mm. Lunar samples returned by the astronauts have been examined in laboratories with scanning electron microscopes with magnification of less than a thousandth of a micron. Within 10 years the view of the Moon was magnified by a factor of 10^{11} (Davies and Murray, 1971).

Planets and Satellites

In a period of less than 30 years most of the planets of the Solar System and their major satellites have been explored remotely with spacecraft (See a more comprehensive discussion in Chapter 36, "Extraterrestrial Applications," of this *Manual* and in Saunders and Mutch, 1980). Only Uranus, Neptune and Pluto remain unexplored, but even Uranus and Neptune may be imaged by Voyager 2 before the end of this decade. Within this short time our knowledge and understanding of our neighboring planets have increased by many orders of magnitude (Sagan, 1975). Each of the planets and their satellites, though possessing some similarities, have their own distinctive characteristics. The inner planets—Mercury, Venus, Earth, and Mars—are silicate bodies. They are smaller and denser than the outer gas giants, Jupiter, Saturn, Uranus, and Neptune which are composed mainly of hydrogen and helium. Little is known of Pluto's size and composition.

Of the inner planets, only Earth and Mars have satellites, but all of the outer planets have satellites. Jupiter has at least 16 known satellites, Saturn has 17, Uranus has 5, Neptune has 2, and Pluto has 1. Undoubtedly each of these planets has additional satellites that have not been observed. We now know that three of the gas giants—Jupiter, Saturn, and Uranus—have complex ring systems composed of small particles that orbit the planets in their equatorial planes. Except for Io, the major satellites of the giant planets are composed mostly of water ice, but each satellite is different. Tables 31-45 and 31-46 summarize the

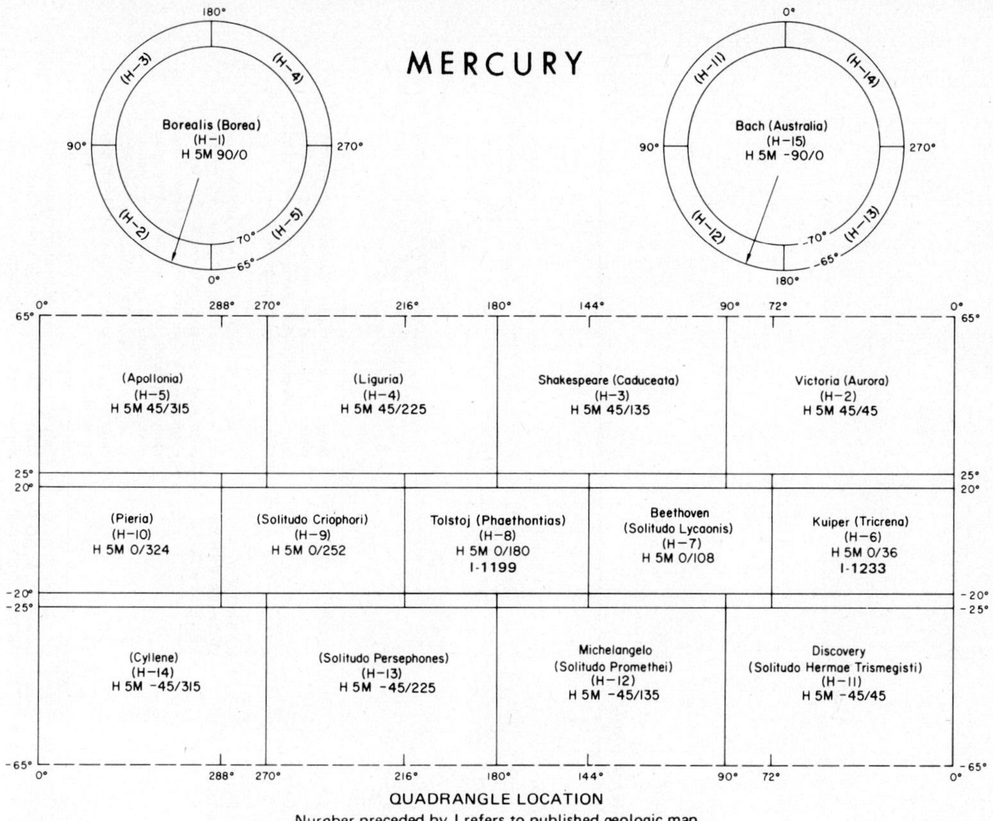

Fig. 31-252. Index to 1:5,000,000-scale geologic maps of Mercury published by the U.S. Geological Survey.

known characteristics of the planets and their satellites.

Terrestrial analogs of extraterrestrial landforms

To recognize and understand processes that have shaped the surfaces of other planets and their satellites we depend on our terrestrial experience; i.e., we observe features on planetary surfaces that resemble those we know on Earth. For example, volcanism has been an important process in forming the surface of Mars. We recognize large features on Mars that have the form of large shield volcanoes on Earth; vast areas of Mars are covered by flows of material whose lobate terminations and textures are similar to lava flows on Earth. From a study of meteorite impact and explosion craters on Earth, craters and the cratering process can be recognized and understood on the Moon and other solid bodies in our Solar System.

Interpretation of landforms on a planetary surface is guided, as well as limited, by the interpreter's experience in observing landforms and geological processes on Earth. The processes that produce landforms on a planet depend on its physical environment—that is, its gravity, temperature and atmosphere. These vary from planet to planet. Similar processes may create different landforms, and different processes similar landforms. The greater the experience of the observer, the greater the probability that the landforms will be identified correctly. It is important for the observer to find diagnostic landforms that uniquely define a process, or multiple landforms that indicate one process (Lucchitta, 1982). Interpretations can be supported by theoretical modelling or by laboratory modelling of processes under simulated conditions (Dwornik, 1976). Analog studies have been particularly useful in furnishing information concerning the processes that are thought to operate on Jupiter's satellites (Figures 31-253 and 31-254). On Mars the recognition of volcanic, eolian (Figure 31-255), glacial (Figure 31-256) and mass wasting (Figure 31-257) activity can be credited mostly to comparison of the martian features to terrestrial analogs of known origin (Lucchitta, 1982).

Landsat images of Earth can be compared to images of other planets, because Landsat sensing systems are similar to the imaging systems of planetary missions. The varying altitude of planetary orbiters or flyby spacecraft produces ground resolution and image scales comparable to the Landsat images. Scene illumination is another important factor for interplanetary image comparisons. Most images of planetary surfaces are

TABLE 31-45

Characteristics of the major planets (1)

	MEAN DISTANCE FROM SUN (MILLIONS OF KILOMETERS)	PERIOD OF REVOLUTION	INCLINATION OF AXIS	EQUATORIAL DIAMETER (KILOMETERS)	MASS (EARTH – 1)	VOLUME (EARTH – 1)	DENSITY (WATER – 1)	ATMOSPHERE (MAIN COMPONENTS)	KNOWN SATELLITES	PLANETARY MISSIONS	SURFACE CHARACTERISTICS
MERCURY	57.9	88 DAYS	<28°	4,880	.056	.06	5.4	NONE	0	MARINER 10	Densely cratered lunar-like surface, smooth plains in intercratered areas. Lobate scarps and scalloped cliffs indicate possible crustal shortening. Has weak magnetic field.
VENUS	108.2	224.7 DAYS	3°	12,104	.815	.88	5.2	CARBON DIOXIDE	0	MARINER 10 PIONEER VENUS VENERA 4 TO 14	Dense CO_2 atmosphere obscures surface. Radar altimetry and image data provide global distribution of topographic relief, and surface roughness. Surface consists of upland rolling plains, highlands, and lowlands. Granitic composition for rolling plains indicated by Venera 8. Highest point is 11.1 km above planetary datum, may be a great volcano. Lowlands may be basaltic lava.
EARTH	149.6	365.26 DAYS	23°27′	12,756	1	1	5.5	NITROGEN, OXYGEN	1	OVER 2400 SPACE CRAFT PLACED IN ORBIT, OVER 1000 STILL IN ORBIT (1981)	Only planet with nitrogen-oxygen atmosphere. Most dynamic planet. Thin mobile crust. Active atmosphere. Temperature, pressure conditions, favor existence of liquid water.

TABLE 31-45 (Continued)

MARS	227.9	687 DAYS	23°59′	6,787	.108	.15	3.9	CARBON DIOXIDE, ARGON (?)	2	MARINER 4, 6, 7, 9 MARS 4, 5, 6 VIKING 1, 2	Thin CO_2 atmosphere. Southern hemisphere mostly ancient cratered lunar-like terrain. Northern hemisphere has younger, more diverse features, giant canyons, large channels possibly cut by water, great volcanoes, polar caps of H_2O and CO_2. Vast sand seas surround the poles.
JUPITER	778.3	11.86 DAYS	3°05′	142,800	317.9	1.316	1.3	HYDROGEN, HELIUM	16	PIONEER 10, 11 VOYAGER 1, 2	No solid surface. Thick dense turbulent atmosphere. Planet composed of hydrogen and helium with small rocky core. Ring of small particles in the equitorial plain extending outward 55,000 km from the top of the atmosphere.
SATURN	1,427	29.45 YEARS	26°44′	120,000	95.2	755	.7	HYDROGEN, HELIUM	17	PIONEER 10, 11 VOYAGER 1, 2	No solid surface. Thick dense turbulent atmosphere. Planet composed of hydrogen and helium with small rocky core. Has an extremely complex ring system over 65,000 km wide, composed of icy particles microns to meters in size.
URANUS	2,869.6	84.01 YEARS	82°5′	51,800	14.6	67	1.2	HYDROGEN, HELIUM, METHANE	5	VOYAGER 2	No solid surface. Composed of hydrogen and helium with rocky core; atmosphere contains more methane than Jupiter or Saturn. Has a complex system of at least 5 dark rings.
NEPTUNE	4,496.6	164.8 YEARS	28°48′	49,500	17.2	57	1.7	HYDROGEN, HELIUM, METHANE	2		No solid surface. Composed of hydrogen and helium with rocky core; atmosphere contains more methane than Jupiter or Saturn.
PLUTO	5,900	247.7 YEARS	7°	6,000(?)	.1(?)	.1(?)	?	NONE DETECTED	1		Unknown

(1) Physical data after Sagan, 1975; Smith et al., 1979a; 1979b.

TABLE 31-46
Characteristics of the major satellites (2)

Planet	Satellite	Distance from planet, (thousands of km)	Sidereal period (days)	Orbit Inclination (deg)	Radius (km)	Density (g cm^{-3})	Surface Characteristics
Earth	Moon	384	27.321	23	1738	3.34	Two major divisions of surface—maria and terra. Terra mostly anorthitic composition. Maria mostly basins filled with basalt. Cratering record indicate early period of intense bombardment prior to 4 billion years. Early intense cratering also recognized on Mercury and Mars. Lithosphere about 1,000 km thick.
Mars	Phobos	6	0.318	1	11	1.92	Densely cratered surface, dark regolith. Grooves 100 to 200 m across, 20 m deep possibly are cracks formed by tidal forces or from an impact that formed large crater. May be carbonaceous chondrodite composition.
	Deimos	21	1.262	2	7	1.67	Smooth light regolith, has no grooves. Loose regolith fills and subdues smaller craters.
Jupiter (16 satellites)	Io	422	1.769	0	1814	3.54 ± 0.3	Most dynamic satellite in Solar System. Eight active volcanoes detected by Voyager. Yellow-orange color due to sulfur compounds. Surface constantly renewed by volcanic activity.
	Europa	671	3.551	1	2569	3.00 ± 0.5	Brightest Jovian satellite, thin ice crust, mantle of water or softer ice and rocky core. Large-scale fracture and ridge system in crust. Scarcity of impact craters on surface implies continual resurfacing process by new ice along fractures.
	Ganymede	1070	7.154	0	2621	1.95 ± 0.5	Largest planetary satellite. Ice crust, rocky core, surface has bright ray craters, light linear stripes, grooved terrain with many faults, regions of dark, heavily cratered terrain. Evidence of tectonic movement of the crust and resurfacing activity.
	Callisto	1883	16.689	0	2400	1.85 ± .05	Thick ice crust; rock core. Surface very ancient and heavily cratered. Large concentric rings indicate remains of several enormous impact basins. No evidence of crustal motion or internal activity.
Saturn (17 satellites)	Mimas	186	0.942	2	196	1.19 ± .05	Densely cratered ice surface. Giant 130 km crater centered on leading hemisphere with central peak. Crater about ⅓ diameter of satellite, maximum size possible without disrupting the body. Grooves up to 90 km long, 10 km wide and 1–2 km deep may be fractures from giant crater or from tidal forces.
	Enceladus	238	1.370	0	250	1.2 ± .04	Ice crust with possible fluid interior. Brightest satellite. Wide diversity of cratered terrains. Bands of smooth terrain with rectilinear pattern of grooves similar to terrain on Ganymede. Late episode of resurfacing.

TABLE 31-46 (Continued)

	Tethys	295	1.887	1	530	1.21 ± .16	Densely cratered ice surface with trough that extends 140°, with average width of 100 km and depth of 3 to 5 km. Outward-facing hemisphere has 400-km diameter crater, about 40 percent of diameter of Tethys. Crater floor has rebounded to circumference of satellite.
	Dione	377	2.736	0	560	1.43 ± .06	Densely cratered ice surface with large regional variations in albedo, complex network of bright wispy linear markings may be controlled by regional system of fractures. Differing crater populations of various terrains imply resurfacing.
	Rhea	527	4.417	0	765	1.33 ± .09	Densely cratered ice surface with regional variation of light and dark terrains. Bright wispy streaks similar to Dione's. Large craters are irregular shaped and polygonal; no flattened craters.
	Titan	1222	15.945	0	2575	1.88 ± .01	Only satellite in Solar System known to have substantial atmosphere. Atmospheric pressure 1.6 bars; nitrogen is dominant gas. Methane (6%), hydrogen (0.2%), and possibly argon (12%) are next abundant gasses. Surface features not visible through dense aerosol atmosphere.
	Hyperion	1481	21.276	1	205 × 130 × 110		Irregularly shaped object with angular features, facets and rounded areas. Probably remnant of larger body destroyed by collision. Large impact craters visible; series of scarps 300 km long may define a spall 200 by 300 km. Darkest object of the Saturnian system.
	Iapetus	3560	79.330	15	730	1.16 ± .09	Leading hemisphere an order of magnitude darker than trailing hemisphere. Surface densely cratered; some craters in light hemisphere have dark floors. Dark material of unknown origin, may be coating derived from other bodies (e.g., Phoebe) or may be extruded from interior.
	Phoebe	12954	550.45	30 R	110		Outermost satellite. Surface albedo approximately the same as dark hemisphere of Iapetus.
Uranus	Aeriel	192	2.520	0	350	7.2	Unknown
	Umbriel	267	4.144	0	250	7.6	Unknown
	Titania	438	8.705	0	500	8.2	Unknown
	Oberon	586	13.463	0	450	6.81	Unknown
	Miranda	128	1.414	(?)	120	1.4	Unknown
Neptune	Triton	353	5.876	20 R	1900	4.9	Unknown
	Nereld	5600	359.88	28	120	4.2	Unknown

(2) Physical data from Allen, 1963; Smith et al., 1979a; 1979b; 1981; Davies and Katayawa, 1971; Smith et al., 1982. R = retrograde.

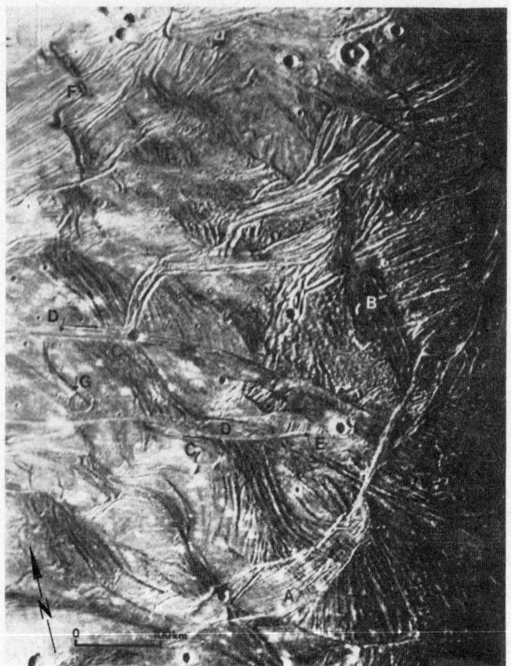

Fig. 31-253. Tectonism on Ganymede. Ridges and grooves are interpreted to be fractures and faults in the icy crust. Note superposition of fault sets at A, truncation at B, and displacement at C. Displacement of groove sets along band D and small cracks at E may indicate left lateral shear. Deep grooves at F may be grabens. Some grooves cut across craters (G). Image center at lat +45° N., long 310° W. Image range is 13,855 km. (Voyager Image 979J1+001) (from Lucchitta, 1980).

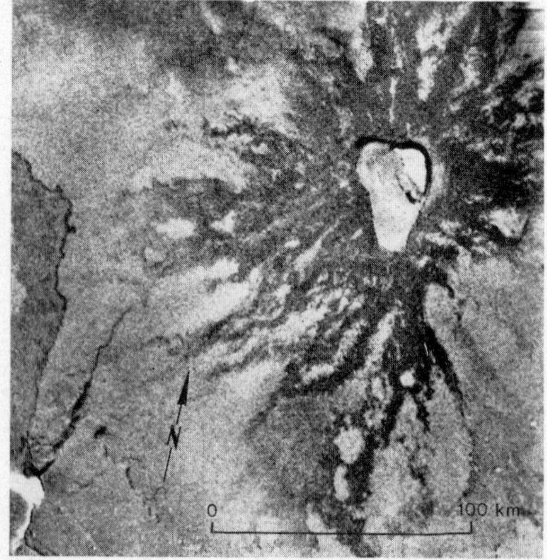

Fig. 31-254. Ma'asaw Patera on Io satellite of Jupiter. Caldera at center is surrounded by low-albedo flows. Flows, of unknown composition, are thought to be either sulfur or sulfur-rich, highly fluid silicate lavas. Image center is at lat −52.6° S., long 330.0° W. Image range is 30,165 km. (Voyager I image 199J1+000) (from Schaber, 1980).

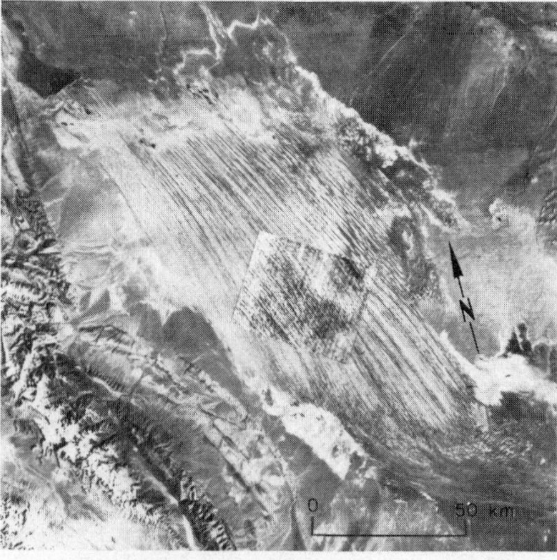

Fig. 31-255. Wind erosion on Mars. Composite of Landsat and Mariner 9 images at the same scale. Mariner frame of the southern Amazonis region of Mars is about 50 km wide (DAS 6823253) and is inset into Landsat image no. 1182-06112 (21 January 1973) of wind-eroded terrain in the Lut Desert, Iran. Image center near lat +30° 30′ N., long 58° 30′ E. (from McCauley, 1976).

taken at low-Sun elevations, in order to enhance relief. Most terrestrial images, by contrast, are taken at high Sun elevations in order to enhance albedo and spectral response and to facilitate stereophotogrammetric mapping. For this reason, Landsat images taken at winter or at high latitudes are preferred for interplanetary comparisons; pictures taken in snow-covered areas on Earth are especially valuable because they simulate the low contrasts, in terms of albedo and color, of many planetary surfaces (Figure 31-258).

FUTURE TRENDS

Geological remote-sensing research usually must rely on the acquisition, by governmental or commercial organizations, of data from aircraft and spacecraft and ready access to such data by geologists. Research may involve studies of static features (such as geologic structure and lithologic mapping) or dynamic geological phenomena (such as shoreline changes and glacier fluctuations). For the study of static features, one-time coverage, at appropriate spatial and spectral resolution, under various illumination conditions, is needed. For the study of dynamic features, repetitive coverage is required (Williams, 1982).

Geological remote sensing research in the future will depend on remotely sensed data already archived and on data to be obtained from newly developed and already developed sensors carried on spacecraft and aircraft. Future U.S. spacecraft will be launched into orbit via the Space Transportation System, the so-called Space Shuttle, rather than by unmanned rockets (Taranik and

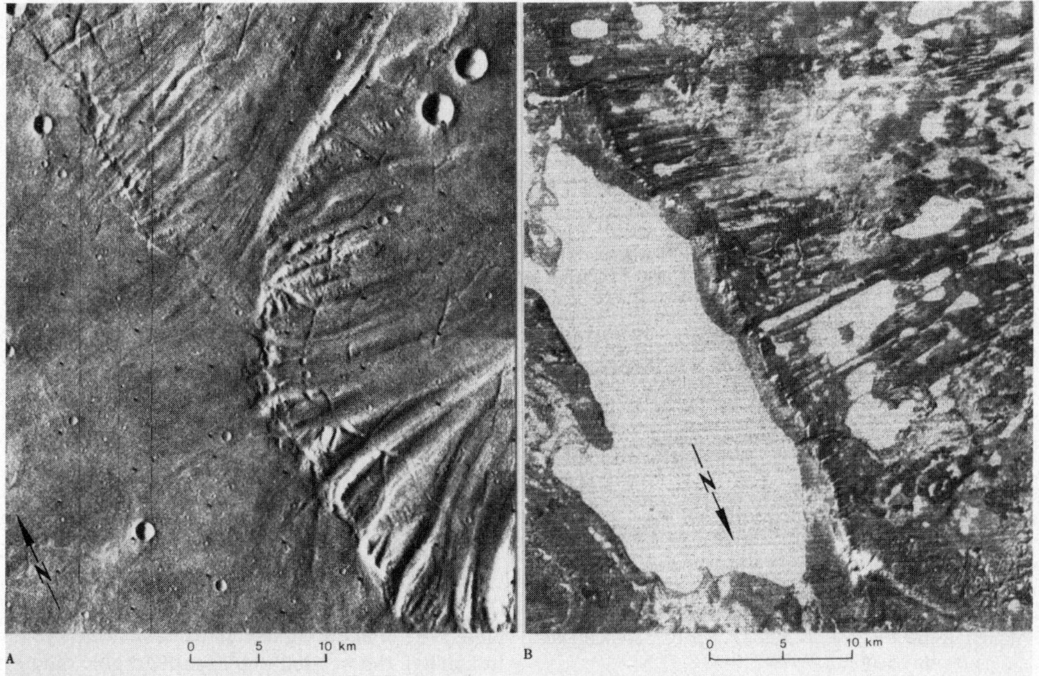

Fig. 31-256. Grooved terrains on Mars and Earth. Diagonal ridges trending across center of images apparently dammed and ponded flowing material, which eventually breached and deeply scoured the terrain at right. Flow was towards the right. (A) Breached wrinkle ridge in Chryse Planitia and channel scour marks. Image center near lat +20° N., long 52° W. Image range is 1,623 km. (Viking Orbiter I image 46A65). (B) Scarp and glacial grooves near Lac Belot, Northwest Territories, Canada. Resemblance to Mars image suggests ice scour for some of the martian channel features. Image Center lat +67° N., long 123° 30′ W. (Landsat image no. 1594-19304-7, 9 March 1974) (from Lucchitta et al., 1981).

Settle, 1981; Settle and Taranik, 1982). Specially designed, unique experimental sensors can be carried on pallets within the Space Shuttle, such as has already occurred with the SMIRR and the SIR-A sensing systems which were carried on the first Office of Space and Terrestrial Applications Space Shuttle Mission (OSTA-1). See the earlier section of this chapter on Types and Availability of Remotely Sensed Data for Geological Investigations. High-performance jet aircraft already routinely carry mapping cameras and other sensing systems (e.g., the thematic mapper simulator), to conduct wide-area surveys at about 13,000 m above mean sea level. An excellent review of future directions in the types of remotely sensed data that will be needed to advance geological remote sensing technology is provided in the paper on "Geologic Remote Sensing" by Goetz and Rowan (1981).

As in the past, geologic remote sensing technology will contribute most in the subdisciplines of geomorphology, structural geology and tectonics, economic geology, and lithologic mapping. Lithologic mapping, however, may not be direct but rather inferential, for example, as in using geobotanical indications of underlying rock type or the presence of abnormally high concentrations of a specific mineral as discussed and illustrated in the "Geobotany" section of this chapter.

The report by the Committee on Earth Sciences of the Space Science Board, "A Strategy for Earth Science from Space in the 1980s; Part I: Solid Earth and Oceans," included an important discussion of geological remote sensing research in Section III, Objectives for Continental Geology (National Academy of Sciences, 1982). In Section III of this document three important scientific objectives are discussed in the following manner:

"The primary scientific objectives for the study of continental geology from space for the next decade, in order of priority, are

1. To determine the global distribution and composition of continental rock units,
2. To determine the morphology and structural fabric of the Earth's land surface,
3. To measure temporal changes in geologic conditions at the Earth's surface."

The Committee on Earth Sciences also made several recommendations in the section relating to geologic remote sensing including:

". . . . the program of research necessary to achieve the scientific objectives for continental geology must strike a balance among space measurements using proven remote-sensing methods, development of space-measurement

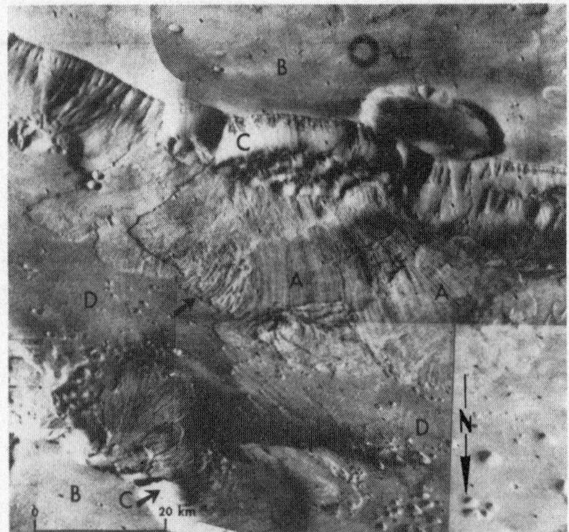

Fig. 31-257. Landslides in Gangis Chasma, Valles Marineris, Mars. Landslides typically have longitudinally grooved debris aprons (A). Plateau surface at B; chasma walls at C; floor at D. Diffuse dark circle near top right is artifact. View to the south. Image center is at lat +8.5° N., long 44.5° W. Image range is 4,991 km. (Viking Orbiter I mosaic 14A29-32, Space Science Data Center) (from Lucchitta, 1979).

systems employing techniques shown to be useful in laboratory and field experiments, and additional laboratory and field work to test the promise for remote sensing of new types of measurement.''

''Considering the state of instrument and technique readiness, the Committee recommends that global imaging in the 1.5 to 1.8 μm and 2.1 to 2.4 μm bands, as well as bands in the shorter-wavelength regions of the visible and near-infrared, be conducted at the earliest opportunity. These passbands will provide new, first-order compositional information not available by other techniques in the near term. The measurements should be made with a digital imaging system to permit optimum information return using image-processing techniques.''

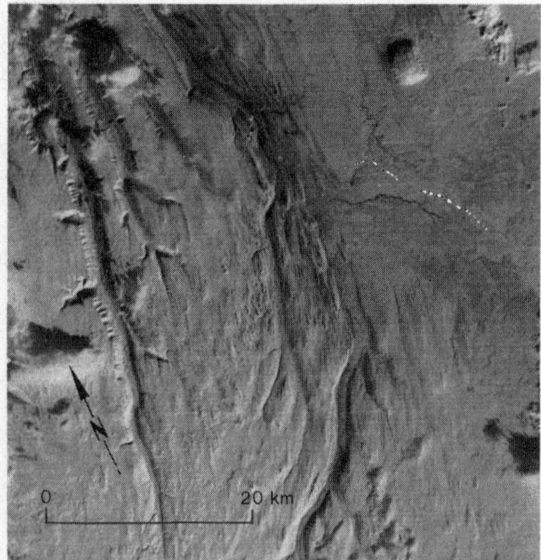

Fig. 31-258. Winter scene in northeastern Iceland, Earth. This landscape resembles landscape on Mars because of low Sun elevation and uniform albedo. Image center is near lat +65° 15′ N., long 17° 30′ W. (Landsat image no. 1229-12142-7, 9 March 1973).

''The space images to date of surface structural and physiographic features of the continents have been two dimensional.—The most useful form for such information in structural and morphologic studies is a three dimensional image. Such a representation can be obtained either by direct stereo imaging or by generating a stereo image from digital topographic data.''

''The Committee recommends that digital topographic data be acquired of all land surfaces as a primary means to determine the morphology and structural fabric of the continental crust. Horizontal spatial resolution should be 30 m or less, and topographic heights should be measured to an accuracy of 10 m or better.''

The U.S. Geological Survey has already completed the conceptual design for a satellite system to achieve the needed digital topographic data for morphological and structural studies of the Earth's surface (Itek Optical Systems, 1981). The section of this chapter which deals with the proposed Mapsat satellite system contains additional information about such a system.

Goetz and Rowan (1981) in their *Science* paper, ''Geologic Remote Sensing,'' also provided five approaches that should be pursued in geological remote sensing technology in the future:

''1) Global stereoscopic image coverage with resolution five or more times greater than Landsat to derive dips of strata, aid in landform interpretation, and reveal evidence for faults.
2) High-spectral-resolution imaging in the region from 2 to 2.5 μm for the identification of layered silicates, clays, and carbonates.
3) Exploitation of the region from 8 to 14 μm through multispectral scanners, active laser spectral reflectance measurements, and orbiting high-spatial-resolution, broadband images.
4) Multifrequency and multipolarization radar systems for increased recognition of lithologic units based on surface roughness characteristics.
5) Active imaging systems in the near- and short-wavelength infrared region, utilizing tunable laser systems to obtain data in very narrow spectral bands not otherwise obtainable with passive systems. Active systems also have potential for obtaining elevation data directly without the aid of stereoscopic images.''

With an increase in spatial and spectral resolution of image data from future sensors there is, of course, the problem of processing huge amounts of data. Each Landsat MSS image has about 7½ million pixels (about 30 million when all four bands are used). The Landsat 4 TM system has about 37½ million pixels in each of bands 1 through 5 and 7 and about 2.3 million pixels in band 6 (about 227 million when all seven bands are used). An increase in spatial resolution and number of spectral bands may not necessarily solve the problems of ambiguities in spectral discrimination of lithologic units, however. Jensen (1982) recently worked with 7.5-m pixels and seven spectral bands of images from a multispectral scanner on a geographic remote sensing project in the Denver, Colorado, area. He reports finding that the increase in spatial resolution only added more spectral units to be discriminated, and that the increased number of spectral bands reduced the number of ambiguities between classes to about the same as was previously achieved with Landsat MSS data of the same area. There is no question, however, that for geological studies the increase in resolution between the Landsat MSS image (79-m pixel size) and the Landsat 3 RBV image (30-m pixel size) has permitted many more structural and morphologic features to be mapped that were just not visible at the Landsat MSS resolution.

Most large petroleum and mineral exploration companies now have special laboratories involved in using geological remote sensing technology in mineral and energy exploration. This trend is expected to continue as geological remote sensing technology expands and improves.

One other future trend that should not be overlooked is international cooperation in the use of remotely sensed data from spacecraft to study (and map and monitor) the Earth's surface. The increase in international cooperation and exchange of information has already begun to take place on several large scale projects. Remotely sensed data in this sense become sort of a scientific *lingua franca* with which scientists from many countries can work together on the same or similar problems using the *same* data source. Landsat images, in particular, have figured prominently in several projects which, to be feasible, needed access to a source of global data. The "Satellite Image Atlas of Glaciers" project of the U.S. Geological Survey (See section of this chapter on Glaciology) and the International Geological Correlation Programme Project 143, "Remote Sensing and Mineral Exploration" (See section of this chapter on Economic Geology) are major cooperative international scientific projects made possible by the ready availability of Landsat images of the Earth's surface. Ed McKee's classic work, "A Study of Global Sand Seas (1979)," was the first geological investigation to effectively use Landsat images in a major comparative geomorphic study of the sand seas of our planet.

A good review of future trends in satellite remote sensing systems is provided in Section 17.10 of Chapter 17, Satellite Photogrammetry, in the fourth edition of the *Manual of Photogrammetry* (Slama et al., 1980) and in the Office of Technology Assessment report (1982). Section 17.10, "Future Satellite Remote Sensing Systems," was authored by Light et al. (1980) and comprehensively discusses current and proposed developments in the United States and other nations. The design of each type of new sensor system and the orbital parameters for each proposed spacecraft are detailed (See also Table 31-50 in this section).

Future sensor systems and spacecraft of particular interest and practical use to geologists are briefly discussed in the following sections, with emphasis on the U.S. Landsat 4, the French SPOT, the U.S. Geological Survey's proposed Mapsat, and the U.S. Large Format Camera (Space Shuttle) programs. Also presented is the U.S. program to photograph the contiguous 48 states from high-altitude aircraft at periodic intervals and a U.S. Geological Survey program to acquire radar imagery of various parts of the United States. More specific information on various sensor systems and spacecraft is given by Light et al. (1980), by Doyle (1982), in Chapters 12 and 13 of this *Manual,* and in Chapter 5, U.S. Civilian Space Program, and Chapter 7, International Efforts in Space, of the report, Civilian Space Policy and Applications, by the Office of Technology Assessment (1982).

LANDSAT 4

Although Landsat 4 is primarily designed as an agricultural applications and crop inventory satellite, on the basis of the choice of the seven spectral bands and orbital altitude, a mid-infrared band (2.08–2.35 μm) and a thermal infrared band (10.4–12.5 μm) are of considerable experimental interest to geologists. Landsat 4 carries a new sensor, the thematic mapper (TM), as well as a multispectral scanner (MSS) that is nearly identical to those on Landsats 1, 2, and 3. Table 31-47 shows the TM bands, wavelengths, and picture elements (pixel) size. Chapter 12, Landsat Satellites, of this *Manual* provides a more complete description of Landsat 4, including both tabular and image illustrations of its characteristics and sensor capabilities.

TABLE 31-47

Thematic Mapper Sensor Characteristics

Band	Wavelength Sensed (in micrometers)	Pixel Size (in meters)
1	0.45–0.52	30
2	0.52–0.60	30
3	0.63–0.69	30
4	0.76–0.90	30
5	1.55–1.75	30
6	10.4–12.5	120
7	2.08–2.35	30

TABLE 31-48

SPOT Sensor Parameters (Modified from CNES, n.d.)

Characteristics of the HRV Sensor	Multispectral mode	Panchromatic mode
Spectral bands	0.50–0.59 μm	0.51–0.73 μm
	0.61–0.68 μm	
	0.79–0.89 μm	
Instrument field of view	4.13°	4.13°
Ground sampling interval (nadir viewing)	20 m × 20 m	10 m × 10 m
Number of pixels per line	3000	6000
Ground swath width (nadir viewing)	60 km	60 km

SPOT

The French "Système Probatoire d'Observation de la Terre (SPOT)" will carry two high [haute] resolution visible (HRV) sensors which can operate in either a multispectral mode or a high-resolution panchromatic mode according to the Centre National d'Études Spatials (CNES, n.d.) of France. Table 31-48 shows the characteristics of the SPOT sensors.

In addition, the field of view can be directed ±27° left or right of the ground track by means of a plane mirror steerable by ground command. This feature makes it possible to view a critical area on several consecutive days and also, by virtue of the different look angles, to achieve stereoscopic coverage.

Figure 31-259 and Color Figure 31-260 are two pairs of SPOT-simulation images prepared by Réjean Simard (1981) of the Canada Centre for Remote Sensing. Color Figure 31-261 shows a pair of simulated SPOT images distributed by CNES. The 10-m panchromatic and 20-m multispectral pixel resolutions, which can be taken in a stereoscopic mode, and the combination of a multispectral scanner image with the higher resolution panchromatic image (see Color Figure 31-261) will provide an important new source of information about the Earth's surface to geologists and will be of particular value to geomorphologists and structural geologists. The French are also planning a military intelligence version of SPOT to be called Satellite Militaire de Reconnaissance Optique (SAMRO).

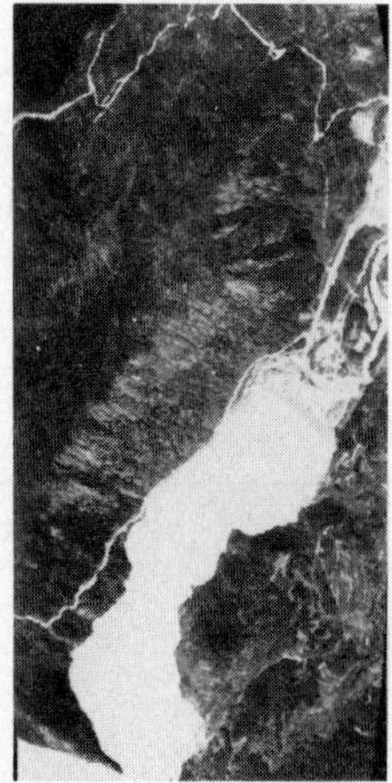

Fig. 31-259. Stereoscopic pair of simulated SPOT images of the Gun Lake area of the Canadian Rockies (in British Columbia). The area covered is 18 km² with a simulated 10-m pixel resolution (Simard, 1981).

MAPSAT

A U.S. Geological Survey-designed integrated satellite system, Mapsat, would offer the capability of acquiring both multispectral and stereoscopic repetitive coverage on a global basis with an accuracy sufficient to produce 1:50,000-scale maps with a 20-m contour interval (Itek Optical Systems, 1981; Colvocoresses, 1982). The Mapsat concept has not yet been adopted by either the U.S. or foreign governments, but it does provide geologists a glimpse of the outer limits of what is feasible to the civilian remote sensing community with the technology available in the early 1980s. The characteristics of the Mapsat sensors are provided in Table 31-49.

Perhaps the most serious limitation of the current Landsat series of spacecraft sensors is the lack of a true stereo capability. To appreciate the significance of the third dimension to the field geologist, please refer to the section of this chapter on Coal Exploration for a discussion on the use of the Kern PG-2 stereoplotter in coal mapping in Wyoming.

The significance of the Mapsat concept to geologists is the provision of the third dimension, at two different base-to-height ratios, one for areas of low-relief (1.0), the other for regions of moderate to high relief (0.5). Availability of stereoscopic images and digital elevation data of

TABLE 31-49

Mapsat Sensor Parameters

Band	Wavelength Sensed (in micrometers)	Pixel Size (in meters)
1	0.47–0.57	10
2	0.57–0.70	10
3	0.76–1.05	10

the land and shallow sea areas of the Earth would revolutionize the geologic subdiscipline of geomorphology (and the counterpart geographic subdiscipline of physical geography). It would also provide an important new source of data for regional studies of geologic structure and for comparative regional geologic structure on a global basis.

SPACE SHUTTLE

Various missions of the Space Shuttle will carry many sensors which will produce data of interest to geologists. These include imaging radar systems, thermal infrared imaging systems, microwave imaging systems (Staelin and Rosenkranz, 1978) and various types of photographic or electro-optical systems operating in the visible and near-infrared part of the electromagnetic spectrum.

Because the Space Shuttle (Orbiter) returns to Earth after each mission, it is particularly well-suited to carry photographic sensors which require recovery of the original exposed film. At least two, and perhaps more, of the Space Shuttle missions will carry the Large Format Camera (LFC), which will acquire high resolution stereo photographs of the Earth. The basic design is directed at achieving cartographic objectives. With the appropriate high-resolution film, the LFC will produce panchromatic and color-infrared photographs with ground spatial resolutions of 10 m and 20 m, respectively. Maps at a scale of 1:50,000, with contour intervals of 20 m, will be possible with photographic data from the LFC.

AIRCRAFT

Unless the image coverage obtained from past military-intelligence reconnaissance satellite systems is made generally available for widespread use by commercial, academic, and governmental scientific users, the principal source of high-resolution stereo photographs in black-and-white panchromatic, black-and-white infrared, color, and color-infrared for geological studies will continue to be those acquired by conventional aerial photographic surveys. The U.S. Geological Survey and other Federal agencies began a long-term program in 1980, as detailed earlier in this chapter, to repetitively cover the contiguous 48 States every 6 years with medium-altitude (about 13,000 m) aerial photographs. Two cameras and film types are used, black-and-white (1:80,000-scale) and color-infrared (1:58,000-scale) as shown in Figure 31-9 and in Color Figure 31-10, respectively. Another interagency program for high-altitude (about 20,000 m) modern aerial photographic coverage of Alaska, by NASA U-2 and RB-57F aircraft, was begun in 1978 with black and white (1:125,000-scale) and color infrared film (1:62,500-scale) as shown in Color Figure 31-8 (Williams and Taranik, 1981). The U.S. Geological survey, since 1980, has also been acquiring, by contract with commercial survey firms, side-looking airborne radar (SLAR) imagery of various areas of the United States, with special emphasis on Alaska (Moore and Sheehan, 1981).

OTHER SPACECRAFT (U.S. AND NON-U.S.)

The review by Light et al. (1980) of "Future Satellite Remote Sensing Systems" in the fourth edition of the *Manual of Photogrammetry* (Slama et al., 1980) and Doyle's more recent review (1982) provide a comprehensive discussion of spacecraft and sensors under consideration for launch by the United States and a number of other countries. Of particular interest to geologists are the Shuttle Imaging Radar (SIR-B) and the multispectral resource sampler (MRS) concept, which uses a multispectral linear array (MLA) as the sensor. It should be recognized, however, that Landsat 4 (and its back-up Landsat D) represent the last sweepscan sensor systems of their type currently planned in the Landsat series. Indeed, the United States has no current plans to launch any Earth resources survey spacecraft after Landsat 4 and D.

The European Space Agency (ESA) plans to put a metric camera and a microwave remote sensor on the first Spacelab mission. Subsequent Spacelab missions may carry an optical mechanical scanner and a multispectral linear array (MLA) sensing system. The first satellite to be launched by the Space Shuttle, the Shuttle Pallet Satellite (SPAS-01) will carry a two-band linear array sensor designed to achieve a 10-m pixel resolution on future U.S. Space Shuttle missions. ESA also has plans for an ESA Resources Satellite (ERS) which will acquire oceanographic, glaciologic, and climatologic data. The advanced version (AERS) will carry a synthetic aperture radar (SAR) and a multispectral sensor for land observation. Brazil, Canada, India (Indian remote sensing satellite, IRS), Japan, and the Soviet Union all have plans to launch Earth resources survey spacecraft. Japan has ambitious plans for launching and operating several varieties of Earth Resources Satellites (ERS): a marine observation satellite (MOS-1) and, of particular interest to geologists, a land observation satellite (LOS-1) designed especially for mineral and energy resources exploration. The USSR MKF-6 multispectral film camera system has been used on manned Soviet spacecraft, particularly on Salyut 6 and 7. As Doyle, a recent President of the International Society of Photogrammetry and Remote Sensing (in Section 17.10 of Light et al., 1980) points out, "The USSR relies primarily upon film return satellites for Earth observations." However, the recent Meteor satellites have carried electro-optical multispectral sensors.

The United States has endeavored to operate two Earth survey programs each with well defined objectives (Office of Technology Assessment, 1982): a civilian program directed at Earth resources surveys and a military program directed

TABLE 31-50

Summary of Earth Resources Satellites—March 1982 (Modified from Doyle, 1982)

Country	Proven Concept	Being Designed	Under Construction	Operating System
USA	MLA	SIR-B	Landsat D′	Landsat 3 & 4
	Mapsat	—	OSTA-3	—
USSR	?	?	Space Station	Meteor
	—	—	—	Salyut 6 and 7
	—	—	—	Cosmos Series
France	SAMRO	SPOT-3 & -4	SPOT-1 & -2	—
ESA	AERS	ERS-1	SPAS-01	—
	Spacelab-D		Spacelab-1	—
Japan	ERS-1	MOS-1	—	—
		LOS-1		
Canada	Radarsat	—	—	—
Brazil	Equatorial	—	—	—
India	IRS-1	—	—	Bhaskara 2
Indonesia	Equatorial	—	—	—
China (P.R.C.)	—	Chinasat 11 & 12	Chinasat 10	—

at strategic reconnaissance. The uncertainty of the continuation of a U.S. Government civilian Earth survey satellite program, after 10 years of near total dominance, looms as a serious hindrance to further development of satellite remote sensing in the United States (Office of Technology Assessment, 1982).

Doyle (1982), in a paper which updates the *Manual of Photogrammetry* Section 17.10 (Light et al., 1980), discusses the present status of satellite remote sensing programs. Table 31-50 is adapted from this recent paper and summarizes which satellites are operating, which are being built, which are being designed, and which are still in the conceptual stage by each nation active in Earth resources-type surveys from space. Finally, in Volume I of this *Manual* more information on future sensor systems and satellite platforms is included in Chapter 1, Development and Principles of Remote Sensing; Chapter 13, Microwave and Infrared Satellite Sensors; and Chapter 14, Meteorological Satellites.

REFERENCES

Aaron, J. M., Butman, Bradford, Bothner, M. H., and Sylwester, R. E., 1980, Maps showing environmental conditions relating to potential geologic hazards on the United States northeastern Atlantic continental margin: U.S. Geological Survey Miscellaneous Field Studies Map MF-1193, 3 sheets.

Aaron, J. M., Butman, Bradford, Bothner, M. H., and Hampson, J. C., Jr., 1982, U.S. Geological Survey environmental studies on the Continental Shelf and Slope off New England: *in Geotechnology in Massachusetts,* (Farquhar, O. C., editor): University of Massachusetts, Amherst, Massachusetts, p. 521–538.

Abdou, I. E., 1978, Quantitative methods of edge detection: Image Processing Institute, University of Southern California, Los Angeles, California, Report 830, 167 p.

Abrams, M. J., 1978, Computer image processing—geologic applications: National Aeronautics and Space Administration, Jet Propulsion Laboratory Publication 78–34, 29p.

Abrams, Michael, and Siegal, B. S., 1980, Lithologic mapping: *in Remote Sensing in Geology* (Siegal, B. S., and Gillespie, A. R., editors), John Wiley and Sons, New York, p. 381–418.

Abrams, M. J., Ashley, R. P., Rowan, L. C., Goetz, A. F. H., and Kahle, A. B., 1977, Mapping of hydrothermal alteration in the Cuprite mining district, Nevada, using aircraft scanner images for the spectral region 0.46 to 2.36 μm: *Geology,* v. 5, no. 12, p. 713–718.

Abrams, Michael, Conel, James, and Lang, Harold, 1983, The joint NASA/Geosat test case project: Jet Propulsion Laboratory, Pasadena, California (in press).

Agterberg, F. P., 1981, Application of image analysis and multivariate analysis to mineral resource appraisal: *Economic Geology,* v. 76, no. 5, p. 1016–1031.

Ahern, F. J., Teillet, P. M., and Goodenough, D. G., 1979, Transformation of atmospheric and solar illumination conditions on the CCRS image analysis system: *in* Proceedings of the 1979 Symposium on Machine Processing of Remotely Sensed Data, Purdue University, West Lafayette, Indiana, p. 34–52.

Ahlmann, H. W., 1937, Vatnajökull in relation to other present day Iceland glaciers; Chapter 4 *in* Vatnajökull. Scientific results of the Swedish-Icelandic investigations 1936–37: *Geografiska Annaler,* v. 19, no. 3–4, p. 212–231.

Ahmed, Waheeduddin, Kahn, S. N., and Schmidt, R. G., 1972, Geology and copper mineralization of the Saindak quadrangle, Chagai District, West Pakistan; U.S. Geological Survey Professional Paper 716-A, p. A1–A21.

Åkerman, Jonas, 1980: Studies on periglacial geomorphology in West Spitsbergen: Medd. från Lunds Universitets Geografiska Institution, Avhandlingar 89, 297 p.

Albert, N. R. D., 1982, Preliminary map showing reconnaissance photogeologic interpretation of SLAR imagery of the Bristol Bay, Ugashik and Karluk quadrangles, Álaska: U.S. Geological Survey

Open-File Report 82-0141, 5 over-size sheets, scale 1:250,000.

Aldana E., Guillermo, and Garrett, Kenneth, 1982, The disaster of El Chichón (Mexican volcano spews death and destruction): *National Geographic,* v. 162, no. 5, p. 654–659.

Allan, J. A., and Bradshaw, M., editors, 1981, Geological and terrain analysis studies by remote sensing: *in* Proceedings of the Eighth Annual Conference of the Remote Sensing Society, Reading, England, 146 p.

Allen, C. R., 1975, Geological criteria for evaluating seismicity: *Geological Society of America Bulletin,* v. 86, no. 8, p. 1041–1057.

Allen, C. W., 1963, Astrophysical quantities: The Althone Press, University of London, 2nd edition, 291 p.

Allison, Ian, editor, 1981, Sea level, ice, and climatic change: International Association of Hydrological Sciences, Publication No. 131, 471 p.

American Geological Institute, 1980, Directory of geoscience departments: United States and Canada: American Geological Institute, Falls Church, Virginia, 19th Edition, 186 p.

American Society of Photogrammetry, 1979, Cover photo: *Photogrammetric Engineering and Remote Sensing,* v. 45, no. 2, p. 115 (caption).

Anders, F. J., and Leatherman, S. P., 1982, Mapping techniques and historical shoreline analysis—Nauset Spit, Massachusetts: *in Geotechnology in Massachusetts* (Farquhar, O. C., editor), University of Massachusetts, Amherst, Massachusetts, p. 501–509.

Anding, D., Kauth, R., and Turner, R., 1971, Atmospheric effects on infrared multispectral sensing of sea-surface temperature from space: Report No. NASA CR-1858, University of Michigan, Ann Arbor, Michigan, 103 p.

Andrews, H. C., 1970, Computer techniques in image processing: Academic Press, Inc., New York, 197 p.

Aniya, Masamu, and Welch, Roy, 1981a, Morphometric analyses of Antarctic cirques from photogrammetric measurements: *Geografiska Annaler,* v. 63, Series A, no. 1–2, p. 41–53.

Aniya, Masamu, and Welch, Roy, 1981b, Morphological analyses of glacial valleys and estimates of sediment thickness on the valley floor: Victoria Valley system, Antarctica: *Antarctic Record,* no. 71, p. 76–95.

Anonymous, 1981, Orbital sensors capable of obtaining thermal infrared measurements of the Earth's surface: *in* Workshop on Geological Applications of Thermal Infrared Remote Sensing Techniques (Settle, Mark, editor), Technical Report No. 81-06, Lunar and Planetary Institute, Houston, Texas, p. 28–29.

Armstrong, Terence, Roberts, Brian, and Swithinbank, Charles, 1973, Illustrated glossary of snow and ice: Second Edition, Scott Polar Research Institute, Cambridge, England, 60 p.

Aronson, J. R., and Emslie, A. G., 1975, Applications of infrared spectroscopy and radiative transfer to earth sciences; *in Infrared and Raman Spectroscopy of Lunar and Terrestrial Minerals* (Karr, C., ed.), Academic Press, New York, p. 143–164.

Ashley, R. P. and Abrams, M. J., 1978, Mapping of limonite, clay mineral and alunite contents of hydrothermally altered rocks in the Cuprite mining district, Nevada, using aircraft scanner imagery for the 0.46–2.36 μm spectral region (abs.): *Economic Geology,* v. 73, no. 2, p. 307.

Ashley, R. P., and Abrams, M. J., 1980, Alteration mapping using multispectral images—Cuprite mining district, Esmeralda County, Nevada: U.S. Geological Survey Open-File Report 80-367, 19 p.

Avci, M., 1977, Airphoto interpretation of granular construction materials for engineering purposes in Tremp Basin, Spain: *in Bulletin of the Mineral Research and Exploration,* Institute of Turkey, Foreign Edition, no. 89, p. 75–80.

Avery, T. E., 1962, Interpretation of aerial photographs; an introductory college textbook and self-instruction manual: Burgess Publishing Company, Minneapolis, 192 p.

Avery, T. E., 1968, Interpretation of aerial photographs: Second Edition, Burgess Publishing Company, Minneapolis, 324 p.

Aviation Week and Space Technology, 1978, Carter confirms recon satellite deployment: *Aviation Week and Space Technology,* v. 109, no. 2, p. 22.

Aviation Week and Space Technology, 1980, Thermal data mapped from Iceland, New York: *Aviation Week and Space Technology,* v. 113, no. 11, p. 52.

Babcock, E. A. and Sheldon, L. G., 1976, Relationships between photo lineaments and geologic structures, Athabaska oil sands area, northeast Alberta: *in* Proceedings of the Second International Conference on Basement Tectonics (Newark, Delaware), p. 177–190.

Bailey, R. H., 1982, Glacier: *in* the "Planet Earth" series, Time-Life Books, Alexandria, Virginia, 176 p.

Baker, Simon, and Dill, H. W., Jr., 1970a, The far West; *in* The Look of Our Land, An Airphoto Atlas of the Rural United States: Agricultural Handbook No. 372, U.S. Department of Agriculture, Economic Research Service, Washington, D.C., January, 48 p.

Baker, Simon, and Dill, H. W., Jr., 1970b, North central; *in* The Look of Our Land, An Airphoto Atlas of the Rural United States: Agricultural Handbook No. 382, U.S. Department of Agriculture, Economic Research Service, Washington, D.C., September, 64 p.

Baker, Simon, and Dill, H. W., Jr., 1971a, The East and South; *in* The Look of Our Land, An Airphoto Atlas of the Rural United States: Agricultural Handbook No. 406, U.S. Department of Agriculture, Economic Research Service, Washington, D.C., April, 99 p.

Baker, Simon, and Dill, H. W., Jr., 1971b, The mountains and deserts; *in* The Look of Our Land, An Airphoto Atlas of the Rural United States: Agricultural Handbook No. 409, U.S. Department of Agriculture, Economic Research Service, Washington, D.C., May, 68 p.

Baker, Simon, and Dill, H. W., Jr., 1971c, The plains and prairies; *in* The Look of Our Land, An Airphoto Atlas of the Rural United States: Agricultural Handbook No. 419, U.S. Department of Agriculture, Economic Research Service, Washington, D.C., October, 85 p.

Ballard, R. D., and Grassle, J. F., 1979, Return to oases of the deep: *National Geographic,* v. 156, no. 5, p. 689–705.

Barnes, P. W., and Reimnitz, Erk, 1976, Flooding of sea ice by the rivers of northern Alaska: *in ERTS-1, A New Window On Our Planet* (Williams, R. S., Jr., and Carter, W. D., editors), U.S. Geological Survey Professional Paper 929, p. 356–359.

Barosh, P. J., 1976, Interpretation of aeromagnetic data in southwest Connecticut, and evidence for faulting along the northern fall line: *in* Proceedings of the Second International Conference on Basement Tectonics (Newark, Delaware), p. 99–110.

Barrett, E. C., and Curtis, L. F., 1976, Introduction to environmental remote sensing: Science Paperbacks, John Wiley and Sons, Inc. (A Halsted Press Book), New York, 336 p.

Barth, T. F. W., 1950, Volcanic geology, hot springs, and geysers of Iceland: Carnegie Institution of Washington, Publication No. 587, 174 p.

Bates, R. L., and Jackson, J. A., editors, 1980, Glossary of geology: Second Edition, American Geological Institute, Falls Church, Virginia, 749p.

Batson, R. M., Edwards, Kathleen, and Eliason, E. M., 1975, Computer-generated shaded-relief images: *Journal of Research,* U.S. Geological Survey, v. 3, no. 4, p. 401–408.

Batson, R. M., Edwards, Kathleen, and Eliason, E. M., 1976, Synthetic stereo and Landsat pictures: *Photogrammetric Engineering and Remote Sensing,* v. 42, no. 10, p. 1279–1284.

Bauer, Albert, 1954, Synthèse glaciologique: *in* Contributions à la Connaissance de l'Inlandis du Groenland (Expéditions Polaires Françaises—Missions Paul-Émíle Victor), Part 2, International Association of Hydrology, Publication No. 39, p. 270–296. (In French)

Bayne, J. N., and Lawrence, H., 1979, Application of satellite data to surface mine monitoring in selected counties of South Carolina: Bureau of Mines OFR 11-80, 441 p. (Available also from NTIS: Accession No. PB 80-144629).

Beall, A. O., Jr., and Squyres, C. H., 1980, Modern frontier exploration strategy, a case history from upper Egypt: *Oil and Gas Journal,* v. 78, no. 14, p. 106–110.

Bechtold, I. C., Reynolds, J. T., and Archer, R. L., 1976, Fracture pattern detection by space imagery analysis of the Kaiparowits Plateau, southern Utah: *in* Proceedings of the Second International Conference on Basement Tectonics (Newark, Delaware), p. 451–462.

Belcher, D. J., 1945, The engineering significance of soils patterns: *Photogrammetric Engineering,* v. 11, no. 2, p. 115–148.

Belcher, D. J., 1948, Determination of soil conditions from aerial photographs: *Photogrammetric Engineering,* v. 14, no. 4, p. 482–488.

Bender, Friedrich, 1967, Geologische Karte von Jordanien: Geological Survey of the Federal Republic of Germany, 1:750,000-scale.

Bender, Friedrich, 1968a, Geologie von Jordanien: Berlin, Gebr. Borntraeger, 230 p.

Bender, Friedrich, 1968b, Geologische Karte von Jordanien, Aqaba-Ma'an: Geological Survey of the Federal Republic of Germany, 1:250,000-scale.

Bender, Friedrich, 1974, Explanatory notes on the geological map of the Wadi Araba, Jordan: *Geologisches Jahrbuch Reihe B,* v, 10, (Hannover: E. Schweizerbart'sche Verlagsbuchhandlung), p. 3–62 (3 sheets; 1:100,000 scale).

Bender, Friedrich, 1975a, Geology of the Arabian Peninsula, Jordan: U.S. Geological Survey Professional Paper 560-1, 36 p.

Bender, Friedrich, 1975b, Geologic map of Jordan: U.S. Geological Survey Professional Paper 560-I, Plate I, 1:500,000.

Benson, C. S., 1959, Physical investigations on the snow and firn of northwest Greenland: 1952, 1953, and 1954: U.S. Army Snow, Ice, and Permafrost Research Establishment (SIPRE), Cold Regions Research and Engineering Laboratory (CRREL), Research Report No. 26, 62 p.

Benson, C. S., 1961, Stratigraphic studies in the snow and firn of the Greenland ice sheet: *Folia Geographica Danica,* v. 9, p. 13–37.

Benson, C. S., 1962, Stratigraphic studies in the snow and firn of the Greenland Ice Sheet: Snow, Ice, and Permafrost Research Establishment (SIPRE), U.S. Army Cold Regions Research and Engineering Laboratory (CRREL) Research Report No. 70, 93 p. (Published version of C. S. Benson's Ph.D. Dissertation (1960), Division of Geological Sciences, California Institute of Technology, Pasadena, California, 213 p.)

Benson, C. S., and Motyka, R. J., 1978, Glacier-volcano interactions on Mt. Wrangell, Alaska: *in* 1977–78 Annual Report of the Geophysical Institute, University of Alaska, Fairbanks, p. 1–25.

Bentz, F. P. and Gutman, S. I., 1977, Landsat contributions to hydrocarbon exploration in foreign regions: U.S. Geological Survey Professional Paper 1015 (Woll, P. W., and Fischer, W. A., editors), p. 83–92.

Berg, C. P., Wiesnet, D. R., and Matson, Michael, 1981, Assessing the Red River of the North 1978 flooding from NOAA satellite data: *in Satellite Hydrology* (Deutsch, Morris, Wiesnet, D. R., and Rango, Albert, editors), American Water Resources Association, Minneapolis, p. 309–315.

Berg, C. P., Wiesnet, D. R., and Legecknis, Richard, 1982, The NOAA-6 satellite mosaic of Antarctica: A progress report: *Annals of Glaciology* (Proceedings of the Third International Symposium on Antarctic Glaciology, Ohio State University, Columbus, Ohio, 7–12 September 1981), v. 3, p. 23–26.

Berlin, G. L., Chavez, P. S., Jr., Grow, T. E., Soderblom, L. A., 1976, Preliminary geologic analysis of southwest Jordan from computer enhanced Landsat 1 image data: *in* Proceedings of the Annual Meeting of the American Society of Photogrammetry, Falls Church, Virginia, p. 545–563.

Bernstein, Ralph, Anuta, P. E., Bajcsy, Ruzena, Hunziker, Raul, and Rosenfeld, Azriel, editors, 1978, Digital image processing for remote sensing: IEEE Press Selected Reprint Series, The Institute of Electrical and Electronic Engineers, Inc.; John Wiley and Sons, Inc., New York, 473 p.

Billings, W. D., 1950, Vegetation and plant growth as affected by chemically altered rocks in the western Great Basin: *Ecology,* v. 31, no. 1, p. 62–74.

Birnie, R. W. and Hutton, M. S., 1976, Reflectance spectra of vegetation growing over the Heddleston copper-molybdenum deposit, Montana (abs.): *in* Abstracts with Programs, Geological Society of America, Boulder, Colorado, v. 8, no. 6, p. 779–780.

Birnie, R. W. and Dykstra, J. D., 1978, Application of remote sensing to reconnaissance geologic mapping and mineral exploration: *in* Proceedings of the Twelth International Symposium on Remote Sensing of Environment, Environmental Research Institute of Michigan, Ann Arbor, Michigan, v. 2, p. 795–804.

Birnie, R. W. and Francica, J. R., 1981, Remote detection of geobotanical anomalies related to porphyry copper mineralization: *Economic Geology,* v. 76, no. 3, p. 637–647.

Björnsson, Axel, Saemundsson, Kristján, Einarsson, Páll Tryggvason, Eysteinn, and Grönvold, Karl, 1977, Current rifting episode in north Iceland: *Nature,* v. 266, no. 5600, p. 318–323.

Björnsson, Helgi, 1971, Hugleiding um jöklarannsóknir á Íslandi: *Jökull,* v. 20 (1970), p. 15–26. (In Icelandic)

Björnsson, Helgi, 1975, Explanation of jökulhlaups from Grímsvötn, Vatnajökull, Iceland: *Jökull,* v. 24 (1974), p. 1–26.

Björnsson, Helgi, 1976, Subglacial reservoirs, jökulhlaups, and volcanic eruptions: *Jökull,* v. 25, p. 1–14.

Björnsson, Helgi, 1978, Könnun á jöklum med rafsegulbylgjum: *Náttúrufraedingurinn,* v. 47 (1977), no. 3–4, p. 184–194. (In Icelandic)

Björnsson, Helgi, 1980a, Glaciers in Iceland: *in Geology of Iceland, Jökull* (Special Issue), v. 29 (1979), p. 74–80.

Björnsson, Helgi, 1980b, The surface area of glaciers in Iceland: *Jökull,* v. 28 (1978), p. 31.

Blanchard, Rollin, 1968, Interpretation of leached outcrops: Nevada Bureau of Mines Bulletin 66, 196 p.

Bliamptis, E. E., editor, 1967, Remote sensing of the geological environment: Special Report, Terrestrial Sciences Laboratory, Air Force Cambridge Research Laboratories, L. G. Hanscom Field, Bedford, Massachusetts, 1 June 1967, 23 p.

Blodget, H. W., 1977, Lithology mapping of crystalline shield test sites in western Saudi Arabia: Ph.D. Dissertation, The George Washington University, Washington, D.C., 190 p.

Blodget, H. W., 1981, Landsat in the search for Appalachian hydrocarbons (abs.): *in* Proceedings of the Forty-Seventh Annual Meeting of the American Society of Photogrammetry, Falls Church, Virginia, p. 259–262.

Blodget, H. W., and Brown, G. F., 1982, Geological mapping by use of computer-enhanced imagery in western Saudi Arabia: U.S. Geological Survey Professional Paper 1153, 10 p.

Blodget, H. W., Brown, G. F., and Moik, J. G., 1975, Geological mapping in northwestern Saudi Arabia using Landsat multispectral techniques: *in* Proceedings of the NASA Earth Resources Survey Symposium, Houston, Texas, v. 1B, p. 971–989.

Blodget, H. W., Gunther, F. J., and Podwysocki, M. H., 1978, Discrimination of rock classes and alteration products in southwestern Saudi Arabia with computer-enhanced Landsat data: U.S. National Aeronautics and Space Administration Technical Paper 1327, 34 p.

Blom, R. G., Abrams, M. J., and Adams, H. G., 1980, Spectral reflectance and discrimination of plutonic rocks in the 0.45–2.45 μm region: *Journal of Geophysical Research,* v. 85, no. B5, p. 2638–2648.

Bloom, A. L., 1978, Geomorphology—a systematic analysis of late Cenozoic landforms: Prentice-Hall, Inc., Englewood Cliffs, New Jersey, 510 p.

Bodechtel, Johann, Hiller, K., and Münzer, Ulrich, 1979, Comparison of Seasat and Landsat data of Iceland for qualitative geologic applications: *in* Proceedings of the Seasat-SAR Processor Workshop, Frascati, Italy (10–14 December 1979) p. 61–67.

Boldt, C. M. K., and Scheibner, B. J., 1981, Application of remote sensing for coal waste embankment monitoring: Bureau of Mines IC No. 8857, p. 40–45.

Bolviken, B., Honey, Frank, Levine, S. R., Lyon, R. J. P., and Prelat, Alfredo, 1977, Detection of naturally heavy-metal-poisoned areas by Landsat-1 digital data: *Journal of Geochemical Exploration,* v. 8, no. 1–2, p. 457–471.

Boulton, G. S., 1976, The origin of glacially fluted surfaces—observations and theory: *Journal of Glaciology,* v. 17, no. 76, p. 287–309.

Bowley, D. R., 1980, Surface waste impoundments in Massachusetts: A survey report: Massachusetts Department of Environmental Quality Engineering, Office of Planning and Program Management, Boston; variously paginated by sections.

Brennan, B., and Bandeen, W. R., 1970, Anisotropic reflectance characteristics of natural earth surface materials: *Applied Optics,* v. 9, no. 2, p. 405–412.

Brew, A. N. and Neyland, H. M., 1980, Aerial photography: *in Manual of Photogrammetry* (Slama, C. C., Theurer, Charles, and Hendriksen, S. W., editors): 4th Edition, American Society of Photogrammetry, Falls Church, Virginia, p. 279–303.

Briggs, R. P., 1977, Environmental geology, Allegheny County and vicinity, Pennsylvania—Description of a program and its results: U.S. Geological Survey Circular 747, 25 p.

Briggs, R. P., Pomeroy, J. S., and Davies, W. E., 1975, Landsliding in Allegheny County, Pennsylvania: U.S. Geological Survey Circular 728, 18 p.

Brock, B. B., 1957, World patterns and lineaments: *Transactions, Geological Society of South Africa,* v. 60, p. 127–160.

Brock, B. B., 1960, A philosophy of mineral exploration: *Optima,* v. 10, no. 3, p. 143–158.

Brockmann, C. E., 1978, Mapa Geológico de Bolivia: Servicio Geológico de Bolivia, La Paz, Bolivia; scale 1:1,000,000.

Brockmann, C. E., 1980, Mapa de lineamientos y cuerpos intrusivos de los Andes Bolivianos; Mapa de las fajas mineralizadas de los Andes Bolivianos; Mapa de metalogenesis de los Andes Bolivianos, Relación con la Placa de Nazca, Servicio Geológico de Bolivia, La Paz, Bolivia; scale 1:1,000,000.

Brooks, R. R., 1972, Geobotany and biogeochemistry in mineral exploration: Harper and Row, Inc., New York, 290 p.

Brooks, R. L., 1981, Terrain profiling from Seasat altimetry: NASA Report CR-156878; 81 N31604, 59 p.

Brooks, R. L., 1982, Satellite altimeter results over East Antarctica: *Annals of Glaciology* (Proceedings of the Third International Symposium on Antarctic Glaciology, Ohio State University, Columbus, Ohio, 7–12 September 1980) v. 3, p. 32–35.

Brooks, R. L., Campbell, W. J., Ramseier, R. O., Stanley, H. R., and Zwally, H. J., 1978, Ice sheet topography by satellite altimetry: *Nature,* v. 274, no. 5671, p. 539–543.

Brooks, R. L., Williams, R. S., Jr., Ferrigno, J. G., and Krabill, W. B., 1982, Amery Ice Shelf topography from satellite radar altimetry (abs.): *in* Abstracts Volume of the Fourth International Symposium on Antarctic Earth Sciences (James, P. R., Jago, J. B., and Olive, R. L., eds.), University of Adelaide, South Australia, p. 23.

Brown, G. D., Jr., Cronin, J. F., Skehan, J. W., Dowling, R. W., and O'Leary, D. W., 1967, Multispectral photographic studies of a red-bed facies, Minas Basin, Nova Scotia: Terrestrial Sciences Laboratory (Project 7628), Air Force Cambridge Research Laboratories, Bedford, Massachusetts, Environmental Research Papers, No. 276 (AFCRL-67-0603), November, 34 p.

Brown, G. F. and Jackson, R. O., 1959, Geologic map of the Asir quadrangle, Kingdom of Saudi Arabia: U.S. Geological Survey Miscellaneous Geologic Investigations Map I-217A, scale, 1:500,000.

Brown, R. D., Jr., and Vedder, J. G., 1967, Surface tectonic fractures along the San Andreas fault, California: *in The Parkfield-Cholame, California, Earthquakes of June-August 1966—Surface Geologic Effects, Water-Resources Aspects, and Preliminary Seismic Data,* U.S. Geological Survey Professional Paper 579, p. 2–23.

Brown, R. D., Jr., Ward, P. L., and Plafker, George, 1973, Geologic and seismologic aspects of the Managua, Nicaragua, earthquakes of December 23, 1972: U.S. Geological Survey Professional Paper 838, 34 p.

Brown, T. J., 1981, Development of an earth resources pushbroom scanner utilizing a 90-element 8-14 micrometer (Hg, Cd) Te array: *in* Workshop on Geological Applications of Thermal Infrared Remote Sensing Techniques (Settle, Mark, editor), Technical Report No. 81-06, Lunar and Planetary Institute, Houston, Texas, p. 33–52.

Brugman, M. M., and Meier, M. F., 1981, Response of glaciers to the eruptions of Mount St. Helens: *in The 1980 Eruptions of Mount St. Helens, Washington* (Lipman, P. W., and Mullineaux, D. R., editors), U.S. Geological Survey Professional Paper 1250, p. 743–756.

Bryan, Kirk and Ray, L. L., 1940, Geologic antiquity of the Lindenmeier Site in Colorado: Smithsonian Miscellaneous Collections, v. 99, no. 2, Publication 3554, 76 p.

Bryan, M. L., 1981, Optically processed Seasat radar mosaic of Florida: *Photogrammetric Engineering and Remote Sensing,* v. 47, no. 9, p. 1335–1337.

Bull, Colin, and Lipschutz, M. E., editors, 1982, Workshop on Antarctic glaciology and meteorites: LPI Technical Report No. 82-03, Lunar and Planetary Institute, Houston, Texas, 57 p.

Bureau of Mines, 1972, Coal fires in abandoned mines and inactive deposits: U.S. Department of Interior, Bureau of Mines Information Leaflet, 20 p.

Burns, R. G., Mineralogical applications of crystal field theory: Cambridge University Press, London, 1970, 224 p.

Campbell, W. J., editor, 1979, ICEX: Ice and climate experiment: Report of Science and Applications Working Group, National Aeronautics and Space Administration. Goddard Space Flight Center, Greenbelt, Maryland, December, variously paginated by sections.

Campbell, W. J., Weeks, W. F., Ramseier, R. O., and Gloersen, Per, 1975, Geophysical studies of floating ice by remote sensing: *Journal of Glaciology,* v. 15, no. 73, p. 305–327.

Canich, M. R., 1976, A study of the Tyrone-Mount Union lineament by remote sensing techniques and field methods: M.S. Dissertation, Department of Geosciences, the Pennsylvania State University, University Park, Pennsylvania, 59 p.

Canney, F. C., Yost, Edward, and Wenderoth, Sondra, 1970, Relationship between vegetation reflectance spectra and soil geochemistry: New data from Catheart Mountain, Maine: *in* Proceedings of the Third Annual Earth Resources Program Review; NASA Johnson Space Flight Center (JSC), Houston, Texas, NASA MSC-03742, v. 1, p. 18-1–18-9.

Canney, F. C., Hessin, T. D., Burge, W. G., 1972, Analysis of thermal patterns of geochemically stressed trees at Catheart Mountain, Maine: *in* Proceedings of the Fourth Annual Earth Resources Program Review, Houston, Texas, NASA, p. 57–61.

Canney, F. C., Cannon, H. L., Cathrall, J. B., and Robinson, K., 1979, Autumn colors, insects, plant disease and prospecting: *Economic Geology,* v. 74, no. 7, p. 1673–1676.

Cannon, H. L., 1957, Description of indicator plants and methods of botanical prospecting for uranium deposits on the Colorado Plateau, U.S. Geological Survey Bulletin, 1030-M, p. 339–516.

Cannon, H. L., 1960, Botanical prospecting for ore deposits: *Science,* v. 132, no. 3427, p. 591–598.

Cannon, H. L., 1971, The use of plant indicators in ground water surveys, geologic mapping, and mineral prospecting: *Taxon,* v. 20, no. 2–3, p. 227–256.

Carey, S. W., 1962, Scale of geotectonic phenomena: *Geological Society of India Journal,* v. 3, p. 97–105.

Carlisle, D., and Cleveland, G. B., 1958, Plants as a guide to mineralization: California Division of Mines Special Report 50, 31 p.

Carlson, P. R., 1976, Mapping surface current flow in turbid nearshore waters of the northeast Pacific: *in ERTS-1, A New Window On Our Planet* (Williams, R. S., Jr., and Carter, W. D., editors), U.S. Geological Survey Professional Paper 929, p. 328–329.

Carter, W. D., 1967, Annotated bibliography of USGS technical letters—NASA on remote sensing investigations: U.S. Geological Survey Open-File Report, U.S. Geological Survey Library Accession No. 1326, December, 46 p.

Carter, W. D., 1981, New maps and journals on space geology from the Academy of Sciences of the USSR: *Geotimes,* v. 26, no. 10, p. 21–23.

Carter, W. D., 1982, Remote sensing for exploration of Precambrian mineral deposits: *in The Development Potential of Precambrian Mineral Deposits,* Natural Resources and Energy Division, U.N. Department of Technical Cooperation for Development, Pergamon Press, New York, p. 365–381.

Carter, W. D., 1983, A strategy for mineral and energy resource independence—The North American Plate Mosaic Project: Remote Sensing and Mineral Exploration—1982: *in* COSPAR Advances in Space Research, Pergamon Press, New York (in press).

Carter, W. D., and DeNoyer, J. M., 1978, Landsat images of part of the U.S. Trust Territory of the Pacific Islands and other Pacific outlying areas: *in* Proceedings of the Twelth International Symposium on Remote Sensing of Environment, Environmental Research Institute of Michigan, Ann Arbor, Michigan, v. 1, p. 755–764.

Carter, W. D., Rowan, L. C., and Huntington, J. F., editors, 1980, Remote sensing and mineral exploration: Committee on Space Research (COSPAR), *Advances in Space Exploration,* Pergamon Press, New York, v. 10, 173 p.

Carter, W. D., Kowalik, W. S., Ballon, Raul, and Brockmann, C. E., 1983, Mapping Andean salar deposits by Landsat radiance values: U.S. Geological Survey Open-File Report (in press).

Cassinis, R., and Lechi, G. M., 1974, The use of infrared radiometry in geothermal areas: *in Physical Volcanology* (Civella, L., Gasparini, P., Luongo, G., and Rapolla, A., editors), Elsevier Scientific Publishing Company, New York, p. 117–131.

Centre National d'Études Spatiales, n.d., SPOT, satellite-based remote sensing system: Information brochure and simulated SPOT images distributed by Centre Spatial de Toulouse, Toulouse, France, 20 p.

Chang, S. H., and Collins, Wm., 1980, Toxic effect of heavy metals on plants (abs.): *in* Official Program of the Sixth Annual William T. Pecora Memorial Symposium, "Integration of Remote Sensing into the Exploration Process," April 1980, Sioux Falls, South Dakota, p. 122–124.

Chapman, W. H., 1979, Surveying from the air using inertial technology: *in* Proceedings of the 39th Annual Meeting of the American Congress on Surveying and Mapping, American Congress on Surveying and Mapping, Falls Church, Virginia, p. 352–365.

Chapman, W. H., 1982, Aerial profiling of terrain system: *in* Proceedings of the Surveying Requirements Meeting, U.S. Army Corps of Engineers, Vicksburg, Mississippi, p. 247–260.

Chavez, P. S., Jr., 1975, Atmospheric, solar, and MTF corrections for ERTS digital imagery: *in* Proceedings of the American Society of Photogrammetry Fall Meeting (Phoenix, Arizona), Falls Church, Virginia, p. 69–69a.

Chavez, P. S., Jr., and Bauer, Brian, 1982, An automatic optimum kernal-size-selection technique for edge enhancement: *Journal of Remote Sensing of Environment,* v. 12, no. 1, p. 23–38.

Chavez, P. S., Jr., Berlin, G. L., Acosta, A. V., 1977a, Computer processing of Landsat MSS digital data for linear enhancement: *in* Mapping with Remote Sensing Data (Proceedings of the Second Annual Wm. T. Pecora Memorial Symposium), Sioux Falls, South Dakota, 25–29 October 1976, p. 235–250.

Chavez, P. S., Jr., Berlin, G. L., Mitchell, W. B., 1977b, Computer enhancement techniques of Landsat MSS digital images for land use/land cover assessments: *in* Proceedings of the Sixth Remote Sensing of Earth Resources Symposium, Tullahoma, Tennessee, p. 259–275.

Chavez, P. S., Jr., O'Connor, J. R., McMaken, D. K., and Eliason, Eric, 1979, Digital image processing techniques of integrated images and non-image data sets: *in* Proceedings of the Thirteenth International Symposium on Remote Sensing of Environment, Environmental Research Institute of Michigan, Ann Arbor, Michigan, v. 1, p. 439–454.

Chavez, P. S., Jr., Berlin, G. L., and Sowers, L. B., 1982, Statistical methods for selecting Landsat MSS ratios: *Journal of Applied Photographic Engineering,* v. 8, no. 1, p. 23–30.

Chiu, H. Y. and Collins, Wm., 1978, A spectroradiometer for airborne remote sensing: *Photogrammetric Engineering and Remote Sensing,* v. 44, no. 4, p. 507–517.

Christensen, Leif, 1974, Crop-marks revealing large-scale patterned ground structures in cultivated areas, southwestern Jutland, Denmark: *Boreas,* v. 3, no. 4, p. 153–180.

Civella, L., Gasparini, P., Luongo, G., and Rapolla, A., editors, 1974, Physical Volcanology: Elsevier Scientific Publishing Company, New York, 333 p.

Clark, M. M., 1969, Geologic utility of small scale airphotos: U.S. Geological Survey Open-File Report 69-40, 30 p.

Clark, M. M., 1971, Solar position diagrams—solar altitude, azimuth, and time at different latitudes; *in* Geological Survey Research 1971: U.S. Geological Survey Professional Paper 750-D, p. D145–D148.

Clark, M. M., 1972, The surface rupture along the Coyote Creek fault: *in The Borrego Mountain, California, Earthquake of April 9, 1968,* U.S. Geological Survey Professional Paper 787, p. 55–86.

Clark, M. M., 1973, Map showing recently active breaks along the Garlock and associated faults, California: U.S. Geological Survey Miscellaneous Geologic Investigations Map I-741, scale 1:24,000.

Clark, M. M., 1978, Finding active faults using aerial photographs: *Earthquake Information Bulletin,* v. 10, no. 5, p. 169–173.

Clark, M. M., 1982, Map showing recently active traces along the Elsinore and associated faults, California, between Lake Henshaw and Mexico: U.S. Geological Survey Miscellaneous Geologic Investigations Map I-1329; scale, 1:24,000.

Clark, M. M., Grantz, Arthur, and Rubin, Meyer, 1972, Holocene activity of the Coyote Creek fault recorded in the sediments of Lake Cahuilla: *in The Borrego Mountain, California, Earthquake of April 9, 1968,* U.S. Geological Survey Professional Paper 787, p. 112–130.

Clarke, P. F., Hodgson, H. E., and North, G. W., 1982, A guide to obtaining information from the USGS 1982: Geological Survey Circular 777, U.S. Geological Survey, Reston, Virginia, 42 p.

Cluff, L. S., and Slemmons, D. B., 1972, Wasatch fault zone—features defined by low-sun-angle photography: *in Environmental Geology of the Wasatch Front,* 1971, Utah Geological Association Publication 1, p. G1–G9.

Coker, A. E., Higer, A. L., and Rogers, R. L., 1977, The application of remote sensing technology to assess the effects of and monitor changes in coal mining in eastern Tennessee (abs.): *in* Proceedings of the First Annual William T. Pecora Memorial Symposium (Woll, P. W., and Fischer, W. A., editors), U.S. Geological Survey Professional Paper 1015, p. 173.

Collins, R. J., McCowan, F. P., Stories, L. P., Petzel, G., and Everett, J. R., 1974, An evaluation of the suitability of ERTS data for the purposes of petroleum exploration: *in* 3rd Earth Resources Technology Satellite—1 Symposium, v. 1, sec. A, U.S. National Aeronautics and Space Administration Special Publication 351, p. 809–821.

Collins, William, 1978, Remote sensing of crop type and maturity: *Photogrammetric Engineering and Remote Sensing,* v. 44, no. 1, p. 43–55.

Collins, William, Chang, S., Kuo, J. T., and Rowan, L. C., 1981, Remote mineralogical analysis using a high-resolution airborne spectroradiometer: Preliminary results of the Mark II system: *in* Digest 1981 International Geoscience and Remote Sensing Symposium (IGARSS '81), 8–10 June 1981, Washington, D.C., v. 1, p. 337–344.

Collins, W. E., Raines, G. L., and Canney, F. C., 1977, Airborne spectroradiometer discrimination of vegetation anomalies over sulfide mineralization—a remote sensing technique (abs.): *in* Abstracts with Programs, Geological Society of America, Boulder, Colorado, v. 9, no. 7, p. 932–933.

Collins, W. E. and Chiu, H. Y., 1979, Signature evaluation of natural targets using high spectral resolution techniques: *in* Proceedings of the Thirteenth International Symposium on Remote Sensing of Environment, Environmental Research Institute of Michigan, Ann Arbor, Michigan, v. 1, p. 567–582.

Collins, W. G., and Van Genderen, J. L., editors, 1978, Remote sensing applications in developing countries: The Remote Sensing Society, Department of Geography, University of Reading, Reading, England, 101 p.

Colvocoresses, A. P., 1974, Remote sensing platforms: Geological Survey Circular 693, U.S. Geological Survey, Reston, Virginia, 75 p.

Colvocoresses, A. P., editor, 1975, Platforms for remote sensors: *in Manual of Remote Sensing* (Reeves, R. G., Anson, Abraham, and Landen, David, editors), American Society of Photogrammetry, Falls Church, Virginia, v. 1 (Janza, F. J., Blue, H. M., and Johnston, J. E., editors), p. 538–588.

Colvocoresses, A. P., 1977a, Landsat mapping of offshore regions: *in* Proceedings of the 1977 Offshore Technology Conference, Houston, Texas, Paper OTC 3043, p. 565–568.

Colvocoresses, A. P., 1977b, Landsat mapping and charting of shallow seas: U.S. Geological Survey, National Mapping Division, Reston, Virginia, Memorandum for the Record, EC-49-Landsat, 1 p. (Includes two reports; "Landsat Goes to Sea" by J. C. Hammack and "Landsat Mapping of Offshore Regions" by A. P. Colvocoresses; and a 1:500,000-scale Landsat Image Map of the Florida Keys)

Colvocoresses, A. P., 1977c, Mapping and charting from Landsat: in Proceedings of the First Annual Wm. T. Pecora Memorial Symposium, U.S. Geological Survey Professional Paper 1015 (Woll, P. W., and Fischer, W. A., editors), p. 43–59.

Colvocoresses, A. P., 1982a, Recent activities reflecting the mapping capabilities of Landsat: Memorandum for the Record, EC-80-Landsat/Mapsat; U.S. Geological Survey Open-File Report 82-227, 15 p.

Colvocoresses, A. P., 1982b, The economic feasibility of operational Earth sensing from space: U.S. Geological Survey Open-File Report No. 82-250, 6 p.

Colwell, R. N., editor, 1960, Manual of photographic interpretation: American Society of Photogrammetry, Washington, D.C., 868 p.

Colwell, R. N., editor, Bousky, Samuel, Casamajor, Paul, Simmons, Herbert, Stone, Kirk, and Welch, Robin, 1960, Procurement of aerial photography: *in Manual of Photographic Interpretation* (Colwell, R. N., editor): American Society of Photogrammetry, Washington, D.C., p. 19–98.

Colwell, R. N., Brewer, William, Landis, Glenn, Langley, Philip, Morgan, Joseph, Rinker, Jack, Robinson, J. M., and Sorem, A. L., 1963, Basic matter and energy relationships involved in remote reconnaissance: Report of Subcommittee I, Photo Interpretation Committee, American Society of Photogrammetry: *Photogrammetric Engineering,* v. 29, no. 5, p. 761–799.

Condit, C. D., and Chavez, P. S., Jr., 1979, Basic concepts of computerized digital image processing for geologists: U.S. Geological Survey Bulletin No. 1462, 16 p.

Conel, J. E., Abrams, M. J., and Baird, K. W., 1980, Uranium: Spectral discrimination of alteration phenomena in sediments: *Modern Geology,* v. 7, no. 2, p. 115–135.

Conrod, Alfred, Kelly, M. G., and Boersma, A., 1968, Aerial photography for shallow water studies on the west edge of the Bahama Banks: Report RE-42, Massachusetts Institute of Technology, Experimental Astronomy Laboratory, Cambridge, Massachusetts, November, 58 p.

Cooke, H. J., 1975, The palaeoclimatic significance of caves and adjacent landforms in western Ngamiland, Botswana: *Geographical Journal,* v. 141, pt. 3, p. 430–444.

Cornillon, Peter, 1982, A guide to environmental satellite data: University of Rhode Island, Graduate School of Oceanography, Ocean Engineering NOAA Sea Grant, Technical Report 79, Kingston, Rhode Island, 469 p.

Coulson, K. L., 1966, Effects of reflection properties of natural surfaces in aerial reconnaissance: *Applied Optics,* v. 5, no. 6, p. 905–917.

Coulter, H. W., and Migliaccio, R. R., 1966, Effects of the earthquake of March 27, 1964, at Valdez, Alaska: U.S. Geological Survey Professional Paper 542-C, p. C1–C36.

Crabtree, R. D., 1976, Changes in the Mýrdalsjökull ice cap, south Iceland: Possible uses of satellite imagery: *Polar Record,* v. 18, no. 112, p. 73–76.

Cracknell, A. P., editor, 1981, Remote sensing in meteorology, oceanography, and hydrology: John Wiley and Sons, New York, 542 p.

Crandell, D. R., and Mullineaux, D. R., 1975, Technique and rationale of volcanic-hazards appraisals in the Cascade Range, northwestern United States: *Environmental Ecology,* v. 1, no. 1, p. 23–32.

Crandell, D. R., Mullineaux, D. R., and Miller, C. D., 1979, Volcanic hazards studies in the Cascade Range of the western United States: *in Volcanic Activity and Human Ecology* (Sheets, P. D., and Grayson, D. K., editors), Academic Press, New York, p. 195–219.

Cromwell, J. E., 1971, Barrier islands distribution: A worldwide survey (abs.): *in* Abstracts of the Second National Coastal and Shallow Water Research Conference (Gorsline, D. S., editor), University Press, University of Southern California, Los Angeles, September, p. 50.

Cronin, J. F., 1965, Unconventional photography and oceanography: *In Oceanography from Space* (Ewing, G. C., editor), Woods Hole Oceanographic Institution Report No. 65-10, Woods Hole, Massachusetts, April, p. 63–72.

Cronin, J. F., 1967, Terrestrial multispectral photography: Terrestrial Sciences Laboratory (Project 7628), Air Force Cambridge Research Laboratories, Bedford, Massachusetts, Special Reports, No. 56 (AFCRL-67-0076), January, 46 p.

Cronin, J. F., and Williams, R. S., Jr., 1970, Volcanism and the upper atmosphere (abs.): *in* Abstracts with Programs, 1970 Annual Meetings of the Geological Society of America, Boulder, Colorado, v. 2, no. 7, p. 529–530.

Cronin, J. F., Rooney, T. P., Williams, R. S., Jr., Molineux, C. E., and Bliamptis, E. E., 1968, Ultraviolet radiation and the terrestrial surface: Terrestrial Sciences Laboratory (Project 7628), Air Force Cambridge Research Laboratories, Bedford, Massachusetts, Special Reports, No. 83 (AFCRL-68-0572), November, 39 p.

Cronin, J. F., Williams, R. S., Jr., and Adams, J. B.,

1971, Geologic sensor studies in the West Indies (abs.): *in* Transactions of the Fifth Caribbean Geological Conference (1968), *Geological Bulletin,* no. 5, Queens College Press, New York, p. 251.

Current, I. B., 1969, A blue-insensitive Anscochrome aerial film: GAF Corp. Report, Binghamton, New York, 12 p.

Daily, Mike, 1983, Hue-saturation-intensity split-spectrum processing of Seasat radar imagery: *Photogrammetric Engineering and Remote Sensing,* v. 49, no. 3, p. 349–355.

Dave, J. V., 1980, Effects of atmospheric conditions on remote sensing of a surface non-homogeneity: *Photogrammetric Engineering and Remote Sensing,* v. 46, no. 9, p. 1173–1180.

Davies, M. E., and Katayama, F. Y., 1981, Coordinates of features on the Galilean satellites: *Journal of Geophysical Research,* v. 86, no. A10, p. 8635–8657.

Davies, M. E., and Murray, C. B., 1971, The view from space: Photographic exploration of the planets: Columbia University Press, New York, 163 p.

Davis, J. C., 1973, Statistics and data analysis in geology: John Wiley and Sons, New York, 550 p.

Davis, S. H., Jr., 1977, The effect of natural gas on trees and other vegetation: *Journal of Arboriculture,* v. 3, no. 8, p. 153–154.

Deirmendjian, D., 1960, Atmospheric extinction of infra-red radiation: *Quarterly Journal of the Royal Meteorological Society,* v. 86, no. 369, p. 371–381.

Del Bono, G. L., and Giammarino, Stani, 1968, Rinvenimento di manifestazioni metanifere nelle praterie a posidonie sui fondi marini prospicenti lo Scoglio d'Africa nell'Arcipelago Toscano: *Atti dell'Instituto di Geologia della Universita di Genova,* v. 6, no. 1, p. 3–11. (In Italian)

Del Bono, G. L., Williams, R. S., Jr., and Cronin, J. F., 1971, Photogeologic and thermal infrared imagery geologic surveys in Italy in 1966: *Bollettino del Servizio Geologico D'Italia,* v. 91 (1970), p. 3–44.

Dellwig, L. F., MacDonald, H. C., and Kirk, J. N., 1968, The potential of radar in geological exploration: *in* Proceedings of the Fifth Symposium on Remote Sensing of Environment, University of Michigan, Ann Arbor, Michigan, p. 747–763.

Denny, C. S., Warren, C. R., Dow, D. H., and Dalo, W. J., 1968, A descriptive catalog of selected aerial photographs of geologic features in the United States: U.S. Geological Survey Professional Paper 590, 79 p.

Department of the Army, 1968, Survey report: Pleasant Bay, Chatham, Orleans, Harwich, Massachusetts: Department of the Army, New England Division, Corps of Engineers, Waltham, Massachusetts, November, 61 p.

Department of the Army, 1979, Remote sensing applications guide: U.S. Army Corps of Engineers, DAEN-RDC, Washington, D.C., 820 p.

Deutsch, Morris, 1976, Optical processing of ERTS data for determining extent of the 1973 Mississippi River flood: *in ERTS-1, A New Window on Our Planet* (Williams, R. S., Jr., and Carter, W. D., editors), U.S. Geological Survey Professional Paper 929, p. 209–213.

Deutsch, Morris and Ruggles, F. H., Jr., 1978, Hydrological applications of Landsat imagery used in the study of the 1973 Indus River flood, Pakistan: *American Water Resources Association Water Resources Bulletin,* v. 14, no. 2, p. 261–274.

Deutsch, Morris, Strong, A. E., and Estes, J. E., 1977, Use of Landsat data for the detection of marine oil slicks: *in* Proceedings of the Offshore Technology Conference, Houston, Texas, Paper OTC 2763, p. 311–318.

Deutsch, Morris, Vollmers, R. R., and Deutsch, J. P., 1980, Landsat tracking of oil slicks from the 1979 Gulf of Mexico oil well blowout: *in* Proceedings of the 14th International Symposium on Remote Sensing of Environment, Environmental Research Institute of Michigan, Ann Arbor, Michigan, v. 2, p. 1197–1211.

Dickson, Paul, 1977, Out of this world: American space photography: Delacorte Press, New York, 158 p.

Dietz, R. S., 1947, Aerial photographs in the geological study of shore features and processes: *Photogrammetric Engineering,* v. 13, no. 4, p. 537–545.

Doyle, F. J., 1982, Status of satellite remote sensing programs: U.S. Geological Survey, National Mapping Division Report (April-Revised), Reston, Virginia, 15 p.

Drake, Benjamin, 1976, Saginaw Bay Graben and its implications for the origin of the Michigan Basin and Pleistocene glaciation (abs.): *in* Abstracts of the Midwestern Regional Meeting of the American Geophysical Union, Ann Arbor, Michigan, p. 760–761.

Drake, Benjamin, and Vincent, R. K., 1975, Geologic interpretation of LANDSAT-I imagery of the greater part of the Michigan Basin: *in* Proceedings of the 10th International Symposium on Remote Sensing of Environment, Environmental Research Institute of Michigan, Ann Arbor, Michigan, v. 2, p. 943–948.

Drewry, D. J., 1981, Radio echo-sounding of ice masses: Principles and applications: *in Remote Sensing in Meteorology, Oceanography, and Hydrology* (Cracknell, A. P., editor), p. 270–284.

Drewry, D. J., 1982, Antarctica: Glaciological and geophysical folio: Series of 11 topical map sheets, ranging in scale from 1:3,000,000 to 1:25,000,000, and 2 explanation sheets; Scott Polar Research Institute, Cambridge, England.

Drewry, D. J., Jordan, S. R., and Jankowski, E., 1982, Measured properties of the Antarctic ice sheet: Surface configuration, ice thickness, volume and bedrock characteristics: *Annals of Glaciology,* v. 3, p. 83–91.

Driscoll, Everly, 1971, The bureaucratic odyssey of a space mapping camera: *Science News,* v. 100, no. 22, p. 362.

Drummond, R. R., 1972, Digest of NASA Earth observation sensors: Report X-733-72-76-464, National Aeronautics and Space Administration, Goddard Space Flight Center, Greenbelt, Maryland, 315 p.

Dueholm, K. S., 1979, Geological and topographic mapping from aerial photographs: Meddelelse No. 10, Instituttet for Landmaling og Fotogrammetri, Danmarks Tekniske Højskole, 204 p.

Duggin, M. J., 1980, The field measurement of reflectance factors: *Photogrammetric Engineering and Remote Sensing,* v. 46, no. 5, p. 643–647.

Duntley, S. Q., 1963, Light in the sea: *Optical Society of America Journal,* v. 53, no. 2, p. 214–233.

Dwornik, S. E., 1976, Earth-based studies: *in A Geological Basis for the Exploration of the Planets,* (Greeley, Ronald and Carr, M. H., editors), U.S. National Aeronautics and Space Administration SP-417, p. 85–88.

Dykstra, J. D., and Birnie, R. W., 1979, Reconnaissance geologic mapping in Chagai Hills, Baluchistan, Pakistan by computer processing of Landsat data: *American Association of Petroleum Geologists Bulletin,* v. 63, no. 9, p. 1490–1503.

Dyring, Eric, 1975, USA-rapport föreslår: Slut på fria satellitbilder-Militären tar over kontrollen (United States proposals on remote sensing policy): *Forskning och Framsteg,* no. 5, p. 26–31. (In Swedish)

Eardley, A. J., 1941a, Interpretation of geologic maps and aerial photographs: Edwards Brothers, Ann Arbor, Michigan, 99 p.

Eardley, A. J., 1941b, Aerial photographs: Their use and interpretation: Harper and Brothers Publishers, Harper's Geoscience Series, New York, 203 p.

Eardley, A. J., 1943, Aerial photographs and the distribution of constructional materials: *in* Proceedings of the National Research Council, Highway Research Board, 23rd Annual Meeting, v. 23, p. 556–568.

Eastern National Monument and Park Association, 1980, Cape Cod and vicinity: Experimental image poster prepared by the U.S. Geological Survey in cooperation with the National Park Service; scale 1:250,000; For sale by Cape Cod National Seashore Agency, P.O. Box 518, Eastham, Massachusetts.

Ebert, J. I., 1978, Remote sensing of playa lakes as a source of climatic data: *in* Proceedings of the American Society of Photogrammetry 1978 Fall Technical Meeting, Falls Church, Virginia, p. 159–175.

Ebert, J. I., and Gutierrez, A. A., 1981, Remote sensing of geomorphological factors affecting the visibility of archaeological materials: *in* Technical Papers of the American Society of Photogrammetry 47th Annual Meeting, Falls Church, Virginia, p. 226–236.

Ebert, J. I., and Hitchcock, R. K., 1978, Ancient Lake Makgadikgadi, Botswana, mapping, measurement and palaeoclimatic significance: *in Palaeoecology of Africa and the Surrounding Islands,* (van Zinderen Bakker, Sr., E. M., and Goetzee, J. A., editors), A. A. Balkema, Rotterdam, v. 10/11 (1975–1977), p. 47–56.

Egyptian Geological Survey, 1981, Geologic maps of Egypt: Egyptian Geological Survey Map 19; scale 1:2,000,000.

Eichen, Leo, and Pascucci, R. F., 1977, Applications of remote-sensing technology for powerplant siting: *in* Proceedings of the First Annual William T. Pecora Memorial Symposium (Woll, P. W., and Fischer, W. A., editors), U.S. Geological Survey Professional Paper 1015, p. 123–135.

Einarsson, Páll, 1976, Relative location of earthquakes in the Tjörnes Fracture Zone: *Vísindafélag Íslendinga,* (Societas Scientiarum Islandica) Greinar, v. 5, p. 45–60.

Elachi, Charles, 1980, Spaceborne imaging radar: Geologic and oceanographic applications: *Science,* v. 209, no. 4461, p. 1073–1082.

Elachi, Charles, 1982, Radar images of the Earth from space: *Scientific American,* v. 247, no. 6, p. 54–61.

Elachi, Charles, Brown, W. E., Cimino, J. B., Dixon, T., Evans, D. L., Ford, J. P., Saunders, R. S., Breed, Carol, Masursky, Harold, McCauley, J. F., Schaber, Gerald, Dellwig, Louis, England, Anthony, MacDonald, Harold, Martin-Kaye, P., and Sabins, Floyd, 1982, Shuttle imaging radar experiment: *Science,* v. 218, no. 4576, p. 996–1003.

El-Ashry, M. T., editor, 1977, Air photography and coastal problems: Benchmark Papers in Geology, v. 38, Dowden, Hutchinson and Ross, Inc., Stroudsburg, Pennsylvania, 425 p.

El-Ashry, M. T., 1979, Use of Apollo-Soyuz photographs in coastal studies: *in* Apollo-Soyuz Test Project; Summary Science Report; Earth Observations and Photography (v. 2; El-Baz, Farouk, and Warner, D. M., editors): NASA SP-412, p. 531–543.

El-Baz, Farouk, and Warner, D. M., editors, 1979, Apollo-Soyuz Test Project; summary science report; Earth observations and photography (v. 2): NASA SP-412, 692 p.

Elson, J. A., 1980, Glacial geology: *in Remote Sensing in Geology* (Siegal, B. S., and Gillespie, A. R., editors), John Wiley and Sons, New York, p. 505–551.

Elvidge, C. D., and Lyon, R. J. P., 1982, Radiometric detection of chlorosis, Stanford Remote Sensing Laboratory, Technical Report 82-1, 42 p.

Embleton, Clifford, and King, C. A. M., 1975a, Glacial geomorphology: John Wiley and Sons, New York, 573 p.

Embleton, Clifford, and King, C. A. M., 1975b, Periglacial geomorphology: Second Edition, John Wiley and Sons, New York, 203 p.

Estes, J. E., and Simonett, D. S., editors; Atwater, S. G., Azevedo, L. H., Branch, M. C., Brooner, W. G., Burnelle, D. N., Davies, H. L., Everett, J. R., Hardoin, T. L., Hooper, J. O., Jensen, J. R., Lewis, A. J., Liang, Ta, MacDonald, H. C., Mel, M. R., Morain, S. A., Moore, R. P., Olsen, C. E., Jr., Pease, R. W., Pettinger, L. R., Pressman, A. E., Rouse, J. W., Jr., Sabins, F. F., Jr., Tinney, L. R., Van Roessel, J. W., Waite, W. P., Welch, R. I., and White, P. G., 1975, Fundamentals of image interpretation: *in Manual of Remote Sensing* (Reeves, R. G., Anson, Abraham, and Landen, David, editors): v. 2 (Bowden, L. W., and Pruitt, E. L., editors), American Society of Photogrammetry, Falls Church, Virginia, p. 869–1076.

Ewing, G. C., editor, 1965, Oceanography from space: Woods Hole Oceanographic Institution Report No. 65-10, Woods Hole, Massachusetts, April, 469 p.

Eythórsson, Jón, 1949, Variations of glaciers of Iceland, 1930–1947: *Journal of Glaciology,* v. 1, no. 5, p. 250–252.

Eythórsson, Jón, 1951, Jöklamaelingar 1950 og 1951; *Jökull,* v. 1, p. 16. (In Icelandic)

Eythórsson, Jón, 1963, Variation of Iceland glaciers 1931–1960: *Jökull,* v. 13, p. 31–33.

Falconer, Allan, Deutsch, Morris, Myers, Lynne., and Anderson, R., 1975, Photo-optical contrast stretching of Landsat data for multidisciplinary analyses of the Lake Ontario Basin: *in* Proceedings of the Third Canadian Symposium on Remote Sensing, Canadian Aeronautics and Space Institute, Ottawa, Canada, p. 173–179.

Falconer, Allan, Deutsch, Morris, Myers, Lynne, and 1981a, Lake Ontario dynamics and water quality observations using thematically enhanced Landsat data: *in Satellite Hydrology* (Proceedings of the 5th William T. Pecora Memorial Symposium on Remote Sensing; Deutsch, Morris, Wiesnet, D. R., and Rango, Albert, editors), American Water Resources Association, Minneapolis, p. 655–661.

Falconer, Allan, Myers, Lynne, and Deutsch, Morris, 1981b, Observations on Lake Ontario Basin hydrogeology from optical enhancements of Landsat imagery: *in Satellite Hydrology* (Proceedings of the Fifth William T. Pecora Memorial Symposium on Remote Sensing; Deutsch, Morris, Wiesnet, D. R., Rango, Albert, editors), American Water Resources Association, Minneapolis, p. 427–436.

Feder, A. M., 1960, Radar geology: M. A. Dissertation, Department of Geology, University of Buffalo, Buffalo, New York, 80 p.

Feder, A. M., 1962, Radar geology can aid in regional oil exploration: *World Oil,* July, p. 130, p. 132, p. 134, p. 136, and p. 138.

Federal Register, 1977, Department of the Interior, Geological Survey: Warning and preparedness for geologic-related hazards: Proposed procedures: *Federal Register* (The National Archives of the United States), v. 42, no. 70 (Tuesday, 12 April 1977), p. 19,292–19,296.

Ferrians, O. J., Jr., and Hobson, G. D., 1973, Mapping and predicting permafrost in North America: A review. North American Contribution. Permafrost: *in* Proceedings of the Second International Conference on Permafrost, Yakutsk, Siberia, U.S.S.R., p. 479–498.

Ferrigno, J. G., Williams, R. S., Jr., and Kent, T. M., 1982, Evaluation of Landsat 3 RBV images for earth science studies in Antarctica (abs.): *in* Volume of Abstracts (James, P. R., Jago, J. B., and Oliver, R. L., editors), Fourth International Symposium on Antarctic Earth Sciences, Adelaide University, South Australia, August, p. 59.

Fischer, W. A., 1960, Spectral reflectance measurements as a basis for film-filter selection for photographic differentiation of rock units: U.S. Geological Survey Professional Paper 400B, p. B136–B138.

Fischer, W. A., 1976, Introduction to Chapter 2, Applications to geology and geophysics: *in ERTS-1, A New Window on Our Planet,* (Williams, R. S., Jr. and Carter, W. D., editors) U.S. Geological Survey Professional Paper 929, p. 48–49.

Fischer, W. A., Moxham, R. M., Polcyn, Fabian, and Landis, G. H., 1964, Infrared surveys of Hawaiian volcanoes: *Science,* v. 146, no. 3645, p. 733–742.

Fischer, W. A., editor, and Anson, Abraham, Badgley, Peter, Cronin, J. F., Hemphill, W. R., Hopkins, J. H., Landen, David, Lowman, P. D., Jr., Orr, D. G., Porcello, L. J., Quackenbush, R. S., Jr., Quick, J. R., Ross, D. S., Tewinkel, G. C., Wenderoth, S. G., Yost, E. F., and Zissis, G. J., 1975, History of remote sensing: *in Manual of Remote Sensing* (Reeves, R. G., Anson, Abraham, and Landen, David, editors): v. 1 (Janza, F. J., Blue, H. M., and Johnston, J. E., editors), American Society of Photogrammetry, Falls Church, Virginia, p. 26–50.

Fischer, W. A., Angsuwathana, Prayong, Carter, W. D., Hoshino, Kazuo, Lathram, E. H., Albert, N. R., and Rich, E. I., 1976, Surveying Earth and its environment from space; *American Association of Petroleum Geologists Memoir 25,* p. 63–72.

Fischer, W. A., Orr, D. G., and Greenlee, D. D., 1978, An example of the merging of Landsat, topographic, and aeromagnetic data in a geologic and hydrologic study of a karst region: Claunch, New Mexico: *in* Proceedings of the 12th International Symposium on Remote Sensing of Environment, Environmental Research Institute of Michigan, Ann Arbor, Michigan, v. 2, p. 805–823.

Fisher, D. F., England, George, and Fisher, D. S., 1965, Airborne infrared scanning system M1A1: University of Michigan Report No. 6517-1-F, for the Terrestrial Sciences Laboratory, Air Force Cambridge Research Laboratories, Bedford, Massachusetts, 91 p.

Fleming, J., and Dixon, R. G., 1981, Basic guide to small-format hand-held oblique aerial photography: Canada Centre for Remote Sensing, Energy, Mines and Resources Canada, Ottawa; User's Manual 81-2, 63 p.

Flemming, B. W., 1981, Some implications of velocity variations and transmission losses of underwater sound on image accuracy and range efficiency of side scan sonar systems: University of Cape Town Marine Geoscience Unit, Technical Report no. 12, p. 153–159.

Flint, R. F., 1971, Glacial and Quaternary geology: John Wiley and Sons, New York, 892 p.

Flippo, H. N., Jr., and Lenfest, L. W., Jr., 1973, Flood of June 1972 in Wilkes-Barre area, Pennsylvania: U.S. Geological Survey Hydrological Investigations Atlas 523, scale 1:24,000.

Flower, F. B., Gilman, E. F., Leone, I. A., 1981, Landfill gas, what it does to trees and how its injurious effects may be prevented: *Journal of Arboriculture,* v. 7, no. 2, p. 43–52.

Fontanel, A., and Voûte, C., editors, 1982, Remote sensing in geology: *Photogrammetria,* v. 37, no. 3–5, 201 p.

Fontanel, A., Blanchet, C., and Lallemand, C., 1975, Enhancement of Landsat imagery by combination of multispectral classification and principal component analysis: *in* U.S. National Aeronautics and Space Administration Earth Resources Survey Symposium Proceedings, Houston, Texas, NASA TM X-58168, p. 991–1012.

Ford, J. P., 1980, Seasat orbital radar imagery for geologic mapping: Tennessee-Kentucky-Virginia; *American Association of Petroleum Geologists Bulletin,* v, 64, no. 12, p. 2064–2094.

Ford, J. P., 1982, Geological mapping from spaceborne imaging radars: Kentucky-Virginia, USA: *in* Proceedings of the International Geoscience and Remote Sensing Symposium (IGAARS '82, Munich, Federal Republic of Germany), v. 2, p. 6.1–6.6.

Ford, J. P., Blom, R. G., Bryan, M. L., Daily, M. I., Dixon, T. H., Elachi, Charles, and Xenos, E. C., 1980, Seasat views North America, the Caribbean, and western Europe with imaging radar: California Institute of Technology, Jet Propulsion Laboratory, Pasadena, California, JPL Publication 80-67 (1 November 1980), 141 p.

Fountain, A. G., 1982, Columbia Glacier photogrammetric altitude and velocity: Data Set (1957–1982): U.S. Geological Survey Open-File Report 82-756, 226 p.

Foy, C. D., Chaney, R. L., White, M. C., 1978, Physiology of metal toxicity in plants: *Annual Review of Plant Physiology,* v. 29, no. 4, p. 511–566.

Francica, J. R., Birnie, R. W., and Johnson, G. D., 1980, Geologic mapping of the Ladakh Himalaya by computer processing of Landsat data: *in* Proceedings of the Fourteenth International Symposium on Remote Sensing of Environment, Environmental Research Institute of Michigan, Ann Arbor, Michigan, v. 2, p. 773–782.

Frank, David, Meier, M. F., and Swanson, D. A., 1977, Assessment of increased thermal activity of Mount Baker, Washington, March 1975–March 1976: U.S. Geological Survey Professional Paper 1022-A, 49 p.

French, H. M., 1976, The periglacial environment: Longman, Inc., New York, 309 p.

Freund, R., 1974, Kinematics of transform and transcurrent faults: *Tectonophysics,* v. 21, no. 1, p. 93–134.

Fridriksson, Sturla, 1975, Surtsey, evolution of life on a volcanic island: Butterworth and Company, London, 198 p.

Friedman, J. D., 1982, Energy of domal evolution at Mount St. Helens, Washington, June 1980 through May 1982 (abs.): *in* Abstracts with Programs, 1982 Annual Meetings, Geological Society of America, Boulder, Colorado, v. 14, no. 7, p. 492.

Friedman, J. D., and Williams, R. S., Jr., 1968, Infrared sensing of active geologic processes: *in* Proceedings of the Fifth Symposium on Remote Sensing of Environment, University of Michigan, Ann Arbor, Michigan, p. 787–820.

Friedman, J. D., and Williams, R. S., Jr., 1970a, Comparison of 1968 and 1966 infrared imagery of Surtsey: *in Surtsey Research Progress Report,* The Surtsey Research Society, Reykjavik, v. 5, p. 88–92.

Friedman, J. D., and Williams, R. S., Jr., 1970b, Changing patterns of thermal emission from Surtsey, Iceland, between 1966 and 1969: *in Geological Survey Research 1970,* U.S. Geological Survey Professional Paper 700D, p. D116–D124.

Friedman, J. D., and Heiken, Grant, 1977, Volcanoes and volcanic landforms; Part A of Chapter 5, Skylab 4 observations of volcanoes; *in Skylab Explores the Earth,* NASA Special Publication, SP-380, National Aeronautics and Space Administration, Washington, D.C., p. 137–171.

Friedman, J. D., and Frank, David, 1978, Thermal surveillance of active volcanoes using the Landsat-1 data collection system: Part 3. Heat discharge from Mount St. Helens, Washington, NTIS Report N78 22435/LL, 24 p.

Friedman, J. D., and Simpson, S. L., 1978, Landsat investigations of the northern Paradox basin, Utah and Colorado: U.S. Geological Survey Open-File Report 78-900, 60 p.

Friedman, J. D., and Simpson, S. L., 1980, Lineaments and geologic structures of the northern Paradox basin, Colorado and Utah: U.S. Geological Survey Miscellaneous Field Studies Map MF-1221; 2 sheets, scales, 1:250,000 and 1:500,000.

Friedman, J. D., and Heller, J. S., 1983, Uncontrolled X-band radar-image mosaic of the northern Paradox basin: U.S. Geological Survey Open-File Report 82-796, (in press).

Friedman, J. D., Williams, R. S., Jr., Miller, C. D., and Pálmason, Gudmundur, 1967, Infrared surveys in Iceland in 1966: *in Surtsey Research Progress Report,* The Surtsey Research Society, Reykjavik, v. 3, p. 99–103.

Friedman, J. D., Williams, R. S., Jr., Pálmason, Gudmundur, and Miller, C. D., 1969a, Infrared surveys in Iceland in 1966: *in Geological Survey Research 1969,* U.S. Geological Survey Professional Paper 650-C, p. C89–C105.

Friedman, J. D., Williams, R. S., Jr., and Pálmason, Gudmundur, 1969b, Infrared emissions from Kverkfjöll subglacial volcano, Iceland (abs.): *in* Volume of Abstracts, Symposium on "Volcanoes and Their Roots", International Association of Volcanology and Chemistry of the Earth's Interior, Oxford, England (addendum, 1 p.).

Friedman, J. D., Williams, R. S., Jr., and Parker, D. C., 1969c, Infrared emission from Hekla volcano (abs.): *in* Abstracts, 50th Annual Meeting, Transactions of the American Geophysical Union, v. 50, no. 4, p. 340.

Friedman, J. D., Williams, R. S., Jr., and Thorarinsson, Sigurdur, 1970, Thermal emission from Hekla Volcano, Iceland, before eruption of 5 May 1970 (abs.): *in* Abstracts with Programs, 1970 Annual Meetings of the Geological Society of America, Boulder, Colorado, v. 2, no. 7, p. 555.

Friedman, J. D., Johansson, C. E., Óskarsson, Níels, Svensson, Harald, Thorarinsson, Sigurdur, and Williams, R. S., Jr., 1971, Observations on Icelandic polygon surfaces and palsa areas. Photo interpretation and field studies: *Geografiska Annaler,* v. 53, (Series A), no. 3–4, p. 115–145.

Friedman, J. D., Williams, R. S., Jr., Thorarinsson, Sigurdur, and Pálmason, Gudmundur, 1972, Infrared emission from Kverkfjöll subglacial volcanic and geothermal area, Iceland: *Jökull,* v, 22, p. 27–43.

Friedman, J. D., Case, J. E., and Simpson, S. L., 1979, Tectonic implications of lineaments of the northern Paradox basin, Utah-Colorado (abs): *EOS,* (Transactions, American Geophysical Union), v. 60, no. 18, p. 396–397.

Friedman, J. D., Olhoeft, G. R., Johnson G. R., and Frank, David, 1982a, Heat content and thermal energy of the June dome in relation to total energy yield of Mount St. Helens, May–October, 1980: *in The 1980 Eruptions of Mount St. Helens* (Lipman, P. W., and Mullineaux, D. R., editors), U.S. Geological Survey Professional Paper 1250, p. 557–568.

Friedman, J. D., Frank, David, Kieffer, H. H., and Sawatzky, D. L., 1982b, Thermal infrared surveys: May 18 crater, subsequent lava domes, and associated volcanic deposits: *in The 1980 Eruptions of Mount St. Helens* (Lipman, P. W., and Mullineaux, D. R., editors), U.S. Geological Survey Professional Paper 1250, p. 179–294.

Froidevaux, C. M., 1978, Tertiary tectonic history of Salawati area, Irian Jaya, Indonesia: *American Association of Petroleum Geologists Bulletin,* v. 62, no. 7, p. 1127–1150.

Froidevaux, C. M., 1980, Radar, an optimum remote-sensing tool for detailed plate tectonic analysis and its application to hydrocarbon exploration (an example in Irian Jaya, Indonesia.): *in Radar Geology: An Assessment,* JPL Publication 80-61, Jet Propulsion Laboratory, California Institute of Technology, Pasadena, California, p. 457–501.

Frost, R. E., 1946, Identification of granular deposits by aerial photographs: *in* Proceedings of the National Research Council, Highway Research Board, 26th Annual Meeting, v. 25, p. 116–129.

Fu, Lee-Lueng, and Holt, Benjamin, 1982, SEASAT views oceans and sea ice with synthetic-aperture radar: Jet Propulsion Laboratory, Pasadena, California, Report 81-120, 200 p.

Gathright, T. M., 1982, Lineament and fracture trace analysis and its application to oil exploration in Lee County, Virginia: Virginia Division of Mineral Resources Publication No. 28, Charlottesville, 40 p.

Gawarecki, S. J., Lyon, R. J. P., and Nordberg, William, 1965, Infrared spectral returns and imagery of the Earth from space and their applications to geologic problems: *Scientific Experiments for Manned Orbital Flight,* v. 4 (Science and Technology Series, American Astronautical Society), p. 13–33.

Gierloff-Emden, H. G., 1961, Nehrungen and Lagunen: *Petermanns Geographische Mitteilungen,* v. 105, no. 2, p. 81–92 and no. 3, p. 161–176. (In German)

Gillespie, A. R., 1976, Directional fabrics introduced by digital filtering of images: *in* Proceedings of the Second International Conference on Basement Tectonics (Newark, Delaware), p. 500–507.

Gillespie, A. R., 1980, Digital techniques of image enhancement: *in Remote Sensing in Geology* (Siegal, B. S., and Gillespie, A. R., editors) John Wiley and Sons, New York, p. 139–226.

Gillespie, A. R., and Kahle, A. B., 1977, The construction and interpretation of a digital thermal inertia image: *Photogrammetric Engineering and Remote Sensing,* v, 43, no. 8, p. 983–1000.

Gilman, E. F., Leone, I. A., and Flower, F. B., 1981, The adaptability of 19 woody species in vegetating a former sanitary landfill: *Forest Science,* v. 27, no. 1, p. 13–18.

Glass, C. E., and Slemmons, D. B., 1978, State-of-the-art for assessing earthquake hazards in the United States, Report II-Imagery in earthquake analysis: U.S. Army Corps of Engineers Miscellaneous Paper S-73-1, p. 32–41.

Glen, J. W., Adie, R. J., and Johnson, D. M., editors, 1975, Symposium on Remote Sensing in Glaciology, Cambridge (England), 16–20 September 1974; *Journal of Glaciology,* v. 15, no. 73, 482 p.

Gloersen, Per, Wilheit, T. T., Chang, T. C., Nordberg, William, and Campbell, W. J., 1973, Microwave maps of the polar ice of the Earth: Report No. X-652-73-269, National Aeronautics and Space Administration, Goddard Space Flight Center, Greenbelt, Maryland, August, 37 p.

Gloersen, Per, and Salomonson, V. V., 1975, Satellites—new global observing techniques for ice and snow: *Journal of Glaciology,* v. 15, no. 73, p. 373–387.

Goetz, A. F. H., and Rowan, L. C., 1981, Geologic remote sensing: *Science,* v. 211, no. 4484, p. 781–791.

Goetz, A. F. H., and Gillespie, A. R., 1981, Some sensitivity considerations for a thermal IR multispectral scanner: *in* Workshop on Geological Applications of Thermal Infrared Remote Sensing Techniques (Settle, Mark, editor), Technical Report No. 81-06, Lunar and Planetary Institute, Houston, Texas, p. 53–62.

Goetz, A. F. H.; Billingsley, F. C.; Gillespie, A. R.; Abrams, M. J.; Squires, R. L.; Shoemaker, E. N.; Lucchitta, Ivo, and Elston, D. P., 1975, Applications of ERTS images and image processing to regional geologic problems and geologic mapping in northern Arizona: California Institute of Technology, Jet Propulsion Laboratory, Technical Report 32-1597, 188 p.

Goetz, A. F. H., Rowan, L. C., and Kingston, M. J., 1982, Mineral identification from orbit: Initial results from the Shuttle multispectral infrared radiometer: *Science,* v. 218, no. 4576, p. 1020–1024.

Gold, D. P., 1980, Structural geology: *in Remote Sensing in Geology* (Siegal, B. S., and Gillespie, A. R., editors), John Wiley and Sons, New York, p. 419–483.

Gold, D. P., and Parizek, R. R., 1976, A study of lineaments, fracture traces and joints in Pennsylvania (abs.): *in* Proceedings of the Second International Conference on Basement Tectonics (Newark, Delaware), p. 142.

Gold, D. P., Parizek, R. R., and Alexander, S. S., 1973, Analysis and application of ERTS-1 data for regional geological mapping: *in* Proceedings of the Symposium on Significant Results Obtained from the Earth Resources Technology Satellite-1, NASA SP-327, p. 231–245.

Gold, D. P., Alexander, S. S., and Parizek, R. R., 1974, Application of remote sensing to natural resource and environmental problems in Pennsylvania: *Earth and Mineral Sciences Bulletin,* The Pennsylvania State University, v. 43, no. 7, p. 49–53.

Grantz, Arthur, Plafker, George, and Kachadoorian, Reuben, 1964, Alaska's Good Friday earthquake, March 27, 1964, a preliminary geologic evaluation: U.S. Geological Survey Circular 491, 35 p.

Greeley, Ronald, and Carr, M. H., editors, 1976, A geological basis for the exploration of the planets: NASA SP-417, 109 p.

Green, A. A., Huntington, J. F., and Roberts, G. P., 1978, Landsat digital enhancement techniques for mineral exploration in Australia: *in* Proceedings of the Twelfth International Symposium on Remote Sensing of Environment, Environmental Research Institute of Michigan, Ann Arbor, Michigan, v. 3, p. 1755–1762.

Greene, G. W., and Moxham, R. M., 1968, Additional infrared surveys of coal mine fires in the anthracite and bituminous fields, Pennsylvania: U.S. Geological Survey, Interagency Report BM-4 (Bureau of Mines), April, 79 p.

Greenwood, W. R., 1980, Reconnaissance geology of the Wadi Matahah quadrangle sheet 181 43D, Kingdom of Saudi Arabia, Ministry of Petroleum and Mineral Resources, Directorate General of Mineral Resources, Map CM 39, Scale 1:100,000.

Greenwood, W. R., 1981, Geology of the Abha quadrangle (sheet 18F), Kingdom of Saudi Arabia: U.S. Geological Survey Saudi Arabia Mission Miscellaneous Document 45 (Interagency Report 385), 49 p.

Grolier, M. J., and Overstreet, W. C., 1978, Geologic map of the Yemen Arab Republic (Şan A'): U.S. Geological Survey Miscellaneous Investigations Series, Map I-1143-B; scale 1:500,000.

Grove, Noel, 1973, A village fights for its life (Volcano overwhelms an Icelandic village): *National Geographic,* v. 144, no. 1, p. 40–67.

Gudbergsson, G. M., and Williams, R. S., Jr., 1983, Landsat images of Iceland: 1972–1982: Department of Geosciences, University of Iceland, Reykjavik (in press).

Gudmandsen, Preben, 1970, Notes on radar sounding of the Greenland ice sheet: *in* Proceedings of the International Meeting on Radioglaciology (Gudmandsen, Preben, editor). Technical University of Denmark, Laboratory of Electromagnetic Theory, Lyngby, p. 124–133.

Gudmandsen, Preben, 1971, Electromagnetic probing of ice: *in Electromagnetic Probing in Geophysics* (Wait, J. R., editor), Golem Press, Boulder, Colorado, p. 321–348.

Gudmandsen, Preben, 1976, Studies of ice by means of radio-echosounding: Technical University of Denmark, Laboratory of Electromagnetic Theory, Lyngby, Report No. 162, 22 p.

Gundmandsen, Preben, and Jakobsen, G. H., 1975, Radar echosounding of the Greenland inland ice: *Europhysics News* (European Physical Society), v. 7, no. 5, p. 2–4.

Gunnlaugsson, Björn, 1844, Uppdráttr Íslands, gjördr ad fyrirsögn Ólafs Nikolas Ólsens eptir landmaelingum Bjarnar Gunnlaugssonar, er stydjast vid Thríhyrningamál og strandamaelingar thaer, sem hid Konúngliga Rentukammer hefir látidgjöra og reiknad hefur Hans Jakob Scheel: gefinn út at enu Íslenzka Bókmentafélagi, Reykjavík og Kaupmannahöfn. Scale: 1:480,000. Four sheets folded in a box. The map has three tintings: physicogeographical, administrative, and hydrographic). (In Icelandic)

Gurney, C. M., 1980, Threshold selection for line detection algorithms, *in* IEEE Transactions on Geoscience and Remote Sensing, April 1980, vol. GE-18, no. 2, p. 204–211.

Hackman, R. J., 1967, Geologic evaluation of radar imagery in southern Utah: *in Geological Survey Research 1967,* U.S. Geological Survey Professional Paper 575-D, p. D135–D142.

Halbouty, M. T., 1976. Applications of Landsat imagery to petroleum and mineral exploration: *American Association of Petroleum Geologists Bulletin,* v. 60, no. 5, p. 745–793.

Halbouty, M. T., 1980, Geologic significance of Landsat data for 15 giant oil and gas fields: *American Association of Petroleum Geologists Bulletin,* v. 64, no. 1, p. 8–36.

Haman, P. J., 1976, Angular and spatial relationships of Landsat lineaments of the United States: *in* Proceedings of the Second International Conference on Basement Tectonics (Newark, Delaware), p. 353–360.

Hamlin, C. L., and Goodenough, W. H., 1978, Archiving remotely sensed data: *Science,* v. 202, no. 4363, p. 9.

Hammack, J. C., 1977, Landsat goes to sea: *Photogrammetric Engineering and Remote Sensing,* v. 43, no. 6, p. 683–691.

Hampton, J. N., 1975, An experiment in multispectral air photography for archaeological research: *in Photography in Archaeological Research* (Harp, Elmer, Jr., editor): University of New Mexico Press, Albuquerque, New Mexico, p. 157–210.

Hannesson, Pálmi, 1975, Oblique aerial photograph of the 27 May 1938 jökulhlaup (Skeidarárhlaup) from Grímsvötn showing estimated peak discharge across Skeidarársandur of 50,000 m³ per sec: *Jökull,* v. 24, cover.

Hanson, D. S., and Morganstein, D. R., 1970, Analysis of multispectral scanner data for location of sand and gravel deposits: U.S. Army Engineer Topographic Laboratories, Ft. Belvoir, Virginia, Final Technical Report, January, 58 p.

Harper, Dorothy, 1976, Eye in the sky: Introduction to remote sensing: Canada Science Series, Multiscience Publications, Ltd., Montréal, 164 p.

Harris, W. D., and Umbach, M. J., 1972, Underwater mapping: *Photogrammetric Engineering,* v. 38, no. 8, p. 765–772.

Hays, W. W., editor, 1981, Facing geologic and hydrologic hazards. Earth-science considerations: U.S. Geological Survey Professional Paper 1240-B, 109 p.

Hedel, C. W., and Villalobos, H. A., 1979, Index maps for large-scale vertical black-and-white aerial photographs along the San Andreas fault, California: U.S. Geological Survey Open-File Report 79-287; scale 1:250,000, 3 sheets.

Heiken, Grant, and Pitts, D. E., 1975, Identification of eruption clouds with the Landsat satellites: Los Alamos Scientific Laboratory of the University of California, Los Alamos, New Mexico, Report LA-UR-75-1848, 14 p.

Hemphill, W. R., 1981, Cooperative role of NASA and the Geological Survey in the development of techniques to measure luminescence: *in* Workshop on Applications of Luminescence Techniques to Earth Resource Studies (Hemphill, W. R. and Settle, Mark, eds.), LPI Technical Report 81-03, Lunar and Planetary Institute, Houston, p. 9–11.

Hemphill, W. R., Watson, R. D., Bigelow, R. C., and Hessin, T. D., 1977, Measurement of luminescence of geochemically stressed trees and other materials: *in* Proceedings of the First Annual William T. Pecora Memorial Symposium (Woll, P. W., and Fischer, W. A., editors), U.S. Geological Survey Professional Paper 1015, p. 93–112.

Henriksen, S. W., editor; Schroeder, S. H., and Brewer, R. K., 1980, Field surveys for photogrammetry: *in Manual of Photogrammetry* (Slama, C. C., Theurer, Charles, and Henriksen, S. W., editors), Fourth Edition, American Society of Photogrammetry, Falls Church, Virginia, p. 413–452.

Hills, E. S., 1961, Morphotectonics and the geomorphological sciences with special reference to Australia: *Quarterly Journal of Geological Society of London,* v. 117, no. 465, p. 77–89.

Hittle, J. E., 1949, Air photo interpretation of engineering sites and materials: *Photogrammetric Engineering,* v. 15, no. 4, p. 589–603.

Hobbs, W. H., 1911, Repeating patterns in the relief and in the structure of the land: *Geological Society of America Bulletin,* v. 22, no. 2, p. 123–176.

Hoeks, J., 1972, Changes in composition of soil air near leaks in natural gas mains: *Soil Science,* v. 113, no. 1, p. 46–54.

Hoeppener, Jon, Kalthoff, E., and Schrader, P., 1969, Zurphysikalischen tectonik: Bruchbildung bei verschiedenen deformationen in experiment: *Geologischen Rundschau,* v. 59, no. 1, p. 179–193. (In German)

Holter, M. R., Nudelman, S., Suits, G. H., Zissis, G. J., and Wolfe, W. L., 1962, Fundamentals of infrared technology: MacMillan, New York, 442 p.

Hood, V. A., n.d., A global satellite observation system for Earth resources: Problems and prospects: National Science Foundation Report No. NSF-RA-X-75-014, 1975(?), 137 p.

Hoppe, Gunnar and Schytt, Valter, 1953, Some observations on fluted moraine surfaces: *Geografiska Annaler,* v. 35, no. 2, p, 105–115.

Horler, D. N. H., Barber, J., and Barringer, A. R., 1980a, Effects of cadmium and copper treatments and water stress on the thermal emissions from peas (*Pisum sativum L.*): Controlled environment experiments: *Remote Sensing of Environment,* v. 10, no. 3, p. 191–199.

Horler, D. N. H., Barber, J., and Barringer, A. R., 1980b, Effects of heavy metals on the absorbence and reflectance spectra of plants: *International Journal of Remote Sensing,* v. 1, no. 2, p. 121–136.

Hovis, W. A., Clark, D. K., Anderson, F., Austin, R. W., Wilson, W. H., Baker, E. T., Ball, D., Gordon, H. R., Mueller, J. L., El-Sayed, S. Z., Sturm, B., Wrigley, R. C., and Yentsch, C. S., 1980, Nimbus-7 coastal zone color scanner: System description and initial imagery: *Science,* v. 210, no. 4465, p. 60–63.

Howard, J. A., Watson, R. D., and Hessin, T. D., 1971, Spectral reflectance properties of *Pinus ponderosa* in relation to copper content of the soil-Malachite mine, Jefferson County, Colorado: *in* Proceedings of the Seventh International Symposium on Remote Sensing of Environment, University of Michigan, Ann Arbor, Michigan, v. 1, p. 285–297.

Howarth, P. J., 1966, An esker, Breidamerkurjökull,

Iceland: British Geomorphological Research Group, Occasional Paper, no. 3, p. 6–9.

Howarth, P. J., 1968, Geomorphological and glaciological studies, eastern Breidamerkurjökull, Iceland: Ph.D. Dissertation, Department of Geography, University of Glasgow, Scotland, 384 p.

Howarth, P. J., 1971, Investigation of two eskers at eastern Breidamerkurjökull, Iceland: *Arctic and Alpine Research,* v. 3, no. 4, p. 305–318.

Howarth, P. J., and Price, R. J., 1969, The proglacial lakes of Breidamerkurjökull and Fjallsjökull, Iceland: *The Geographical Journal,* v. 135, part 4, p. 573–581.

Howarth, P. J., and Welch, Roy, 1969, Breidamerkurjökull, southeast Iceland: Department of Geography, University of Glasgow, Scotland: two map sheets (August 1945 and August 1965), scale 1:30,000.

Humlum, Ole, 1978, Fluted moraine på Omö—isbevaegelsesretning og aflejringsmade: *Årsskrift,* Dansk Geologisk Forening, p. 15–22. (In Danish)

Hunt, G. R., 1977, Spectral signatures of particulate minerals in the visible and near infrared: *Geophysics,* v. 42, no. 3, p. 501–513.

Hunt, G. R., 1978, Assessment of Landsat filters for rock type discrimination, based on intrinsic information in laboratory spectra: *Geophysics,* v. 43, no. 4, p. 738–747.

Hunt, G. R., 1979, Near-infrared (1.3 to 2.4 μm) spectra of alteration minerals: Potential for use in remote sensing: *Geophysics,* v. 44, no. 12, p. 1974–1986.

Hunt, G. R., 1980, Electromagnetic radiation: The communication link in remote sensing: *in Remote Sensing in Geology* (Siegal, B. S. and Gillespie, A. R., editors): John Wiley & Sons, New York, p. 5–45.

Hunt, G. R., 1981a, Identification of kaolins and associated minerals in altered volcanic rocks by infrared spectroscopy: *Clays and Clay Minerals,* v. 29, no. 1, p. 76–78.

Hunt, G. R., 1981b, Emission spectra in the thermal infrared region: *in* Workshop on Geological Applications of Thermal Infrared Remote Sensing Techniques (Settle, Mark, editor), Technical Report No. 81-06, Lunar and Planetary Institute, Houston, Texas, p. 63–71.

Hunt, G. R., and Salisbury, J. W., 1970, Visible and near-infrared spectra of minerals and rocks: I. Silicate minerals: *Modern Geology,* v, 1, no. 4, p. 283–300.

Hunt, G. R., and Salisbury, J. W., 1971, Visible and near-infrared spectra of minerals and rocks: II. Carbonates: *Modern Geology,* v. 2, no. 1, p. 23–30.

Hunt, G. R., and Salisbury, J. W., 1974, Mid-infrared spectral behavior of igneous rocks: U.S. Air Force Cambridge Research Laboratories Technical Report AFCRL-TR-74-0625, Bedford, Massachusetts, 77 p.

Hunt, G. R., and Salisbury, J. W., 1975, Mid-infrared spectral behavior of sedimentary rocks: U.S. Air Force Cambridge Research Laboratories, Technical Report AFCRL-TR-75-0356, Bedford, Massachusetts, 49 p.

Hunt, G. R., and Salisbury, J. W., 1976a, Visible and near-infrared spectra of minerals and rocks: XI. Sedimentary rocks: *Modern Geology,* v. 5, no. 4, p. 211–217.

Hunt, G. R. and Salisbury, J. W., 1976b, Visible and near-infrared spectra of minerals and rocks: XII. Metamorphic rocks: *Modern Geology,* v. 5, no. 4, p. 219–228.

Hunt, G. R., and Salisbury, J. W., 1976c, Mid-infrared spectral behavior of metamorphic rocks: U.S. Air Force Cambridge Research Laboratories Technical Report AFCRL-TR-76-0003, Bedford, Massachusetts, 67 p.

Hunt, G. R., and Wynn, J. C., 1979, Visible and near-infrared spectra of rocks from chromium-rich areas: *Geophysics,* v. 44, no. 4, p. 820–825.

Hunt, G. R. and Ashley, R. P., 1979, Spectra of altered rocks in the visible and near infrared: *Economic Geology,* v. 74, no. 7, p. 1613–1629.

Hunt, G. R. and Evarts, R. C., 1981, The use of near-infrared spectroscopy to determine the degree of serpentinization of ultramafic rocks: *Geophysics,* v. 46, no. 3, p. 316–321.

Hunt, G. R., Salisbury, J. W., and Lenhoff, C. J., 1971a, Visible and near-infrared spectra of minerals and rocks: III. Oxides and hydroxides: *Modern Geology,* v. 2, no. 3, p. 195–205.

Hunt, G. R., Salisbury, J. W., and Lenhoff, C. J., 1971b, Visible and near-infrared spectra of minerals and rocks: IV. Sulphides and sulphates: *Modern Geology,* v. 3, no. 1, p. 1–14.

Hunt, G. R., Salisbury, J. W., and Lenhoff, C. J., 1972, Visible and near-infrared spectra of minerals and rocks: V. Halides, phosphates, arsenates, vanadates and borates: *Modern Geology,* v. 3, no. 3, p. 121–132.

Hunt, G. R., Salisbury, J. W., and Lenhoff, C. J., 1973a, Visible and near-infrared spectra of minerals and rocks: VI. Additional silicates: *Modern Geology,* v. 4, no. 2, p. 85–106.

Hunt, G. R., Salisbury, J. W., and Lenhoff, C. J., 1973b, Visible and near-infrared spectra of minerals and rocks: VII. Acidic igneous rocks: *Modern Geology,* v. 4, no. 3, p. 217–224.

Hunt, G. R., Salisbury, J. W., and Lenhoff, C. J., 1974a, Visible and near-infrared spectra of minerals and rocks: VIII. Intermediate igneous rocks: *Modern Geology,* v. 4, no. 4, p. 237–244.

Hunt, G. R., Salisbury, J. W., and Lenhoff, C. J., 1974b, Visible and near-infrared spectra of minerals and rocks: IX. Basic and ultrabasic igneous rocks: *Modern Geology,* v. 5, no. 1, p. 15–22.

Hunter, R. E., 1976, Movement of turbid-water masses along the Texas coast: *in ERTS-1, A New Window On Our Planet* (Williams, R. S., Jr., and Carter, W. D., editors), U.S. Geological Survey Professional Paper 929, p. 334–336.

Hunting Geology and Geophysics, Ltd., n.d., Seasat-1 radar mosaic (of) Iceland: 1:500,000-scale mosaic constructed by Hunting Surveys, Ltd., Borehamwood, Hertfordshire, England, for the Royal Aircraft Establishment (European Space Agency-Earthnet), from Seasat SAR data acquired by the Royal Aircraft Establishment receiving station, Oakhanger, England. Survey mode optical correlation of the SAR data by the Environmental Research Institute of Michigan.

Isachsen, Y. W., 1973, Spectral geological content of ERTS-1 imagery over a variety of geological terrains in New York State: *in* Symposium Proceedings, Management and Utilization of Remote Sensing Data (Anson, Abraham, editor), American Society of Photogrammetry, Falls Church, Virginia, p. 342–363.

Isachsen, Y. W., 1974, Fracture analysis of New York

State using multi-stage remote sensor data and ground study: Possible application to plate tectonic modeling: *in* Proceedings of the First International Conference on the New Basement Tectonics (Hodgson, R. A., Gay, S. P., Jr., and Benjamins, J. Y., editors), Utah Geological Association Publication no. 5, p. 200–217.

Isachsen, Y. W., and McKendree, W. G. 1977, Preliminary brittle structures map of New York, and generalized map of recorded joint systems in New York: New York State Museum Map and Chart Series 31, Albany 7 maps.

Isachsen, Y. W., Wright, S. F., Revetta, F. A., and Dineen, R. J., 1977, The Panther Mountain circular fracture: A possible buried meteorite crater: *in* Guidebook to Field Excursions (Wilson, P. C., editor), 49th Annual Meeting, of the New York State Geological Association, p. 131–138.

Itek Optical Systems, 1981, Conceptual design of an automated mapping system (Mapsat): Final Technical Report to the U.S. Geological Survey, Reston, Virginia; Itek Report No. 81-8449A-1 (12 January 1981), variously paginated by sections.

Iwan, Walter, 1935, Island: Studien zu einer Landeskunde: Berliner Geographische Arbeiten, Herausgegeben vom Geographischen Institut der Universitat Berlin, No. 7, Kommissionsverlag von J Englehorns Nachf., Stuttgart, 155 p. (In German)

Janza, F. J., editor; Allen, W. A., Foote, R. S., Fung, A. K., Gausman, H. W., Hemphill, W. R., Horton, M. L., Kennedy, J. M., Myers, V. I., Offield, T. W., Rabchevsky, G. A., Rouse, J. W., Jr., Rowan, L. C., Watson, Kenneth, Watson, R. D., and Wiegand, C. L., 1975, Interaction mechanisms: *in Manual of Remote Sensing* (Reeves, R. G., Anson, Abraham, and Landen, David, editors), v. 1 (Janza, F. J., Blue, H. M., and Johnston, J. E., editors): American Society of Photogrammetry, Falls Church, Virginia, p. 75–179.

Jefferis, Robert, and Voight, Barry, 1981, Fracture analysis near the mid-ocean plate boundary, Rekjavík-Hvalfjördur area, Iceland: *Tectonophysics,* v. 76, no. 3–4, p. 171–236.

Jenkins, D. S., Belcher, D. J., Greeg, L. E., and Woods, K. B., 1946, The origin, distribution, and airphoto identification of United States soils: U.S. Civil Aeronautics Administration Technical Development Report, no. 52, 202p.

Jennrich, R., 1977, Stepwise discriminant analysis: *in Statistical Methods for Digital Computers* (Enslein, Kurt, Ralston, Anthony, and Wilf, H. S., editors), v. 3 of Mathematical Methods for Digital Computers, Wiley Interscience, New York, p. 76–95.

Jensen, J. R., 1982, Analysis of improved spectral and spatial resolution remote sensing data for urban applications (abs.): *in* AAG Program Abstracts (Fesenmaier, D. R. and Nostrand, R. L., compilers), Annual Meeting of the Association of American Geographers, San Antonio, Texas, 25–28 April 1982, p. 97.

Johnson, P. L., and Atwood, D. M., 1969, Aerial sensing and photographic study of the El Verde Rain Forest, Puerto Rico: Research Report 250, U.S. Army Corps of Engineers, Cold Regions Research and Engineering Laboratory, Hanover, New Hampshire, December, 19 p.

Johnson, Wilton, and Miller, G. C., 1979, Abandoned coal-mined lands; nature, extent, and cost of reclamation: U.S. Department of the Interior, Bureau of Mines Information Leaflet, 29 p.

Jónsson, V. K., and Matthíasson, Matthías, 1974, Hraunkaelingin á Heimaey—Verklegar framkvaemdir: *Tímarit Verkfraedingafélags Íslands,* v. 59, no. 5, p. 70–81 and p. 83. (In Icelandic)

Jordan, Raymond, and Kieffer, H. H., 1982, Topographic changes at Mount St. Helens: Large-scale photogrammetry and digital terrain models: *in The 1980 Eruptions of Mount St. Helens, Washington* (Lipman, P. W., and Mullineaux, D. R., editors), U.S. Geological Survey Professional Paper 1250, p. 135–141.

Kahle, A. B., 1977, A simple thermal model of the Earth's surface for geologic mapping by remote sensing: *Journal of Geophysical Research,* v. 82, no. 11, p. 1673–1680.

Kahle, A. B., 1980, Surface thermal properties: *in Remote Sensing in Geology* (Siegal, B. S., and Gillespie, A. R., editors), John Wiley and Sons, New York, p. 257–273.

Kahle, A. B., 1981, Remote Sensing of the Earth using multispectral middle infrared scanner data: *in* Workshop on Geological Applications of Thermal Infrared Remote Sensing Techniques (Settle, Mark, editor), Technical Report No. 81-06, Lunar and Planetary Institute, Houston, Texas, p. 72–75.

Kahle, A. B., and Rowan, L. C., 1980, Evaluation of multispectral middle infrared aircraft images for lithologic mapping in the East Tintic Mountains, Utah: *Geology,* v. 8, no. 5, p. 234–239.

Kahle, A. B., Madura, D. P., and Soha, J. M., 1980, Middle infrared multispectral aircraft scanner data: Analysis for geological applications: *Applied Optics,* v. 19, no. 14, p. 2279–2290.

Kahle, A. B., Weill, G., and Carter, W. D., 1981a, Session on remote sensing—1980: COSPAR Advances in Space Research, Pergamon Press, New York, v. 1, no. 10, 314 p.

Kahle, A. B., Schieldge, J. P., Abrams, M. J., Ally, R. E., and Levine, C. J., 1981b, Geologic applications of thermal inertia imaging using HCMM data: National Aeronautics and Space Administration, Jet Propulsion Laboratory Publication 81-55, 199 p.

Kalensky, Z. D., Moore, W. C., and Scherk, L. R., 1981, Flood delineation by Landsat (abs.): *in Satellite Hydrology* (Deutsch, Morris, Wiesnet, D. R., and Rango, Albert, editors), American Water Resources Association, Minneapolis, p. 302.

Kauth, R. J., and Thomas, G. S., 1976, The tasselled cap—A graphic description of the spectral-temporal development of agricultural crops as seen by Landsat: *in* Proceedings of the Symposium on Machine Processing of Remotely Sensed Data, Purdue University Laboratory for Applications of Remote Sensing, West Lafayette, Indiana, p. 4B-41–4B-53.

Kehrer, D. M., 1982, The Earth is burning!: *Science Digest,* v. 90, no. 6, p. 28–29.

Keighley, J. R., Lynn, W. W., and Nelson, K. R., 1980, Use of Landsat images in tin exploration, Brazil: *in* Proceedings of the Fourteenth International Symposium on Remote Sensing of Environment, Environmental Research Institute of Michigan, Ann Arbor, Michigan, v. 1, p. 341–343.

Kelly, T. E., 1961, Photogeology—a quick, economical tool for oil hunters: *The Oil and Gas Journal,* v. 59, no. 47, p. 265–266, p. 268–272, and p. 274.

Kieffer, H. H., 1981, Infrared thermal mapping of Mars:

Design, observation, and analysis (abs.): *in* Workshop on Geological Applications of Thermal Infrared Remote Sensing Techniques (Settle, Mark, editor), Technical Report No. 81-06, Lunar and Planetary Institute, Houston, Texas, p. 77–78.

Kieffer, H. H., Frank, David, and Friedman, J. D., 1982, Thermal infrared surveys: Observations prior to the eruption of May 18: *in The 1980 Eruptions of Mount St. Helens* (Lipman, P. W., and Mullineaux, D. R., editors), U.S. Geological Survey Professional Paper 1250, p. 257–278.

Klass, P. J., 1971, Secret sentries in space: Random House, New York, 236 p.

Kneizys, F. X., Shettle, E. P., Gallery, W. O., Chetwynd, J. H., Abreu, L. W., Selby, J. E., Fenn, R. W., and McClatchey, R. A., 1980, Atmospheric transmittance/radiance: Computer Code LOWTRAN-5: Report No. AFGL-TR-80-0067, U.S. Air Force Geophysical Laboratory, Hanscom AFB, Bedford, Massachusetts, 233 p.

Knuth, W. M., 1971, Compendium of burning coal refuse banks, Pennsylvania: Volume 1, Anthracite Region, U.S. Bureau of Mines, Washington, D.C.: HRB/Singer, Inc.; Bureau of Mines Contract No. S0111882; unpaginated, looseleaf, field handbook.

Knuth, W. M., and Stamm, E. C., 1972, Compendium of burning outcrop and mine fire, Pennsylvania Anthracite region: Supplement A to Volume I, Compendium of Burning Coal Refuse Banks: HRB-Singer, Inc.; Bureau of Mines Contract No. S0122114, 108 p. (plus 124 p. of map supplements).

Knuth, W. M. and Stamm, E. C., 1975, Compendium of burning coal refuse banks, Pennsylvania Anthracite region: HRB-Singer, Inc., Bureau of Mines Contract No. S0144084, 144 p. (plus two map supplements, 77 p. and 281 p., respectively).

Knuth, W. M., Fritz, E. L., and Schad, J. A., 1978, Investigation of color and color infrared aerial photographic techniques for mining and reclamation planning and monitoring: HRB-Singer, Inc., Bureau of Mines Contract No. J0155041, Final Report No. 4936-F. Bureau of Mines OFR 37-79, 215 p. (Available also through NTIS: Accession No. PB-294-707).

Knutson, P. L., 1980, Experimental dune restoration and stabilization, Nauset Beach, Cape Cod, Massachusetts: U.S. Army Corps of Engineers Coastal Engineering Research Center Technical Paper No. 80-5, Fort Belvoir, Virginia, 42 p.

Koch, G. S., Jr., and Link, R. F., 1970, Statistical analysis of geological data: John Wiley and Sons, Inc., New York, 375 p.

Kowalik, W. S., 1981, Atmospheric correction to Landsat data for limonite discrimination: Ph.D. Dissertation, Department of Applied Earth Sciences, Stanford University, Stanford, California, 365 p.

Kowalik, W. S., and Gold, D. P., 1976, The use of Landsat-1 imagery in mapping lineaments in Pennsylvania: *in* Proceedings of the First International Conference on the New Basement Tectonics, Utah Geological Association Publication no. 5, p. 236–249.

Krabill, W. B., and Brooks, R. L., 1979, Land subsidence measured by satellite radar altimetry: *in Satellite Hydrology* (Deutsch, Morris, Wiesnet, D. R., and Rango, Albert, editors), Proceedings of the Fifth Annual Wm. T. Pecora Memorial Symposium on Remote Sensing, Sioux Falls, South Dakota (10–15 June 1979), American Water Resources Association, Minneapolis, p. 683–688.

Krabill, W. B.; Collins, J. G.; Swift, R. N., and Butler, M. L., 1980, Airborne laser topographic mapping results from initial joint NASA/U.S. Army Corps of Engineers Experiment: NASA Technical Memorandum 73287, National Aeronautics and Space Administration, Washington, D.C., 33 p.

Krimmel, R. M., and Meier, M. F., 1975, Glacier applications of ERTS images: *Journal of Glaciology,* v. 15, no. 73, p. 391–401.

Krimmel, R. M., and Post, Austin, 1982, Oblique aerial photography, March–October 1980: *in The 1980 Eruptions of Mount St. Helens, Washington* (Lipman, P. W., and Mullineaux, D. R., editors), U.S. Geological Survey Professional Paper 1250, p. 31–51.

Krimmel, R. M., Post, Austin, and Meier, M. F., 1976, Surging and nonsurging glaciers in the Pamir Mountains, U.S.S.R.: *in ERTS-1, A New Window on Our Planet* (Williams, R. S., Jr., and Carter, W. D., editors), U.S. Geological Survey Professional Paper 929, p. 178–179.

Krinsley, D. B., 1976, Selection of a road alinement through the Great Kavir in Iran: *in ERTS-1, A New Window on Our Planet,* (Williams, R. S., Jr., and Carter, W. D., editors), U.S. Geological Survey Professional Paper 929, p. 296–299.

Krinsley, D. B., 1977, Use of ERTS-1 (Landsat 1) images for engineering geologic applications, North-Central Iran: *in* Proceedings of the First Annual Wm. T. Pecora Memorial Symposium (Woll, P. W., and Fischer, W. A., editors) U.S. Geological Survey Professional Paper 1015, p. 113–121.

Kroeck, Dick, 1976, Everyone's space handbook, a photo/imagery source manual: Pilot Rock, Inc., Arcata, California, 175 p.

Krüger, Johannes, and Humlum, Ole, 1981, The proglacial area of Mýrdalsjökull: *Folia Geográphica Danica,* v. 15, no. 1, 58 p.

Krumpe, P. F., 1976, The world remote sensing bibliographic index: Tensor Industries, Inc., Fairfax, Virginia, 619 p.

Kruus, J., Deutsch, Morris, Hansen, P. L., and Ferguson, H. L., 1981, Flood applications of satellite imagery: *in Satellite Hydrology* (Deutsch, Morris, Wiesnet, D. R., and Rango, Albert, editors), American Water Resources Association, Minneapolis, p. 292–301.

Kunde, V. G., 1965, Theoretical relationship between equivalent blackbody temperatures and surface temperatures measured by the NIMBUS HRIR observations from the NIMBUS I meteorological satellite; NTIS Publ. No. N66-12130-136, p. 23–36.

Lang, H. R., 1982, Finding the Lost River gas field: Lineament density analysis in hydrocarbon exploration: *in* Proceedings of the IEEE International Geoscience and Remote Sensing Symposium (IGARSS '82, Munich, Germany), v. 1, p. 1.1–1.5.

Lattman, L. H., 1958, Technique of mapping geologic fracture traces and lineaments on aerial photographs: *Photogrammetric Engineering,* v. 24, no. 4, p. 568–576.

Lattman, L. H., 1959, Geomorphology: New tool for finding oil: *The Oil and Gas Journal,* v. 59, no. 18, p. 231–236.

Lattman, L. H., 1963, Geologic interpretation of airborne infrared imagery: *Photogrammetric Engineering,* v. 29, no. 1, p. 83–87.

Lattman, L. H., and Parizek, R. R., 1964, Relationship between fracture traces and the occurrence of groundwater in carbonate rocks: *Journal of Hydrology,* v. 2, no. 3, p. 73–91.

Lattman, L. H., and Ray, R. G., 1965, Aerial photographs in field geology: *in* Geologic Field Techniques Series (Jahns, R. H., editor), Holt, Rinehart and Winston, New York, 221 p.

Laughton, A. S., 1981, The first decade of GLORIA: *Journal of Geophysical Research,* v. 86, no. 12, p. 11, 511–11, 534.

Leatherman, S. P. editor, 1979a, Barrier Islands: Academic Press, Inc., New York, 325 p.

Leatherman, S. P., 1979b, Barrier island handbook: National Park Service Cooperative Research Unit, Amherst, Massachusetts, 101 p.

Leatherman, S. P., editor, 1981, Overwash processes: *in* Benchmark Papers in Geology Series, Hutchinson Ross Publication Company, Inc., Stroudsburg, Pennsylvania, 276 p.

Leatherman, S. P., Giese, G., O'Donnell, P., 1981, Historical cliff erosion of outer Cape Cod: National Park Service Cooperative Research Unit Report No. 53, Amherst, Massachusetts, 52 p.

Lee, Keenan, and Weimer, R. J., 1975, Geologic interpretation of Skylab photographs: Remote Sensing Report No. 75-6, NASA Contract NAS 9-13394, Colorado School of Mines, Golden, Colorado, 72 p.

Lefebvre, R. H., 1975, Mapping in the Craters of the Moon volcanic field, Idaho, with Landsat (ERTS) imagery: *in* Proceedings of the 10th International Symposium on Remote Sensing of Environment, Environmental Research Institute of Michigan, Ann Arbor, Michigan, v. 2, p. 951–963.

Lefevre, M. J., 1980, The contribution of remote sensing to geochemical and geologic applications in unexposed regions: Two cases: *in* Official Program of the Sixth Annual William T. Pecora Memorial Symposium, "Integration of Remote Sensing into the Exploration Process," April 1980, Sioux Falls, South Dakota, p. 126–127.

Leitao, C. D., Purdy, C. L., and Brooks, R. L., 1975, GEOS-3 preprocessing report: NASA Report TM X-69357, National Aeronautics and Space Administration, 121 p.

Leone, I. A., Flower, F. B., Arthur, J. J., and Gilman, E. F., 1977, Damage to New Jersey crops by landfill gases: *Plant Disease Reporter,* v. 61, no. 4, p. 295–299.

Lepley, L. K., 1976, Discernment and separate display of regional and nonregional lineament patterns by optical fourier processing: *in* Proceedings of the Second International Conference on Basement Tectonics, (Newark, Delaware) p. 560–570.

Light, D. L., editor; Brown, Duane, Colvocoresses, A. P., Doyle, F. J., Davies, Merton, Ellasal, Atef, Junkins, J. L., Manent, J. R., McKenney, Austin, Undrejka, Ronald, and Wood, George, 1980, Satellite photogrammetry: *in Manual of Photogrammetry* (Slama, C. C., Theurer, Charles, and Henriksen, S. W., editors): Fourth Edition, American Society of Photogrammetry, Falls Church, Virginia, p. 883–977.

Lillesand, T. M., and Kiefer, R. W., 1979, Remote sensing and image interpretation: John Wiley and Sons, New York, 612 p.

Lintz, Joseph, Jr., and Simonett, D. S., 1976, Remote sensing of environment: Addison-Wesley Publishing Co., Boston, 694 p.

Lipman, P. W., and Mullineaux, D. R., editors, 1982, The 1980 eruptions of Mount St. Helens, Washington: U.S. Geological Survey Professional Paper 1250, 844 p.

Livingston, R. G., editor; Berndsen, C. E., Ondrejka, Ron, Spriggs, R. M., Kosofsky, L. J., Van Steenburgh, Dick, Norton, Clarice, and Brown, Duane, 1980, Aerial cameras: *in Manual of Photogrammetry* (Slama, C. C., Theurer, Charles, and Henriksen, S. W., editors): 4th Edition, American Society of Photogrammetry, Falls Church, Virginia, p. 187–277.

Llaverias, R. K., 1970, Remote sensing bibliography for Earth resources, 1966–1967: U.S. Geological Survey Interagency Report No. 189, 127 p.

Llaverias, R. K., and Lowe, D. G., 1970, Remote sensing bibliography for Earth resources, 1968: U.S. Geological Survey Interagency Report No. 203, 246 p.

Longshaw, T. G., 1974, Field spectroscopy for multispectral remote sensing: An analytical approach: *Applied Optics,* v. 13, no. 6, p. 1487–1493.

Lowe, D. S., 1968, Line scan devices and why use them: *in* Proceedings of the Fifth International Symposium on Remote Sensing of Environment, University of Michigan, Ann Arbor, Michigan, v. 1, p. 77–101.

Lowe, D. S., 1980, Acquisition of remotely sensed data: *in Remote Sensing in Geology* (Siegal, B. S., and Gillespie, A. R., editors): John Wiley and Sons, New York, p. 47–90.

Lowe, D. S., editor; Kelly, B. O., McDevitt, H. I., Orr, G. T., and Yates, H. W., 1975, Imaging and nonimaging sensors: *in Manual of Remote Sensing* (Reeves, R. G., Anson, Abraham, and Landen, David, editors): v. 1 (Janza, F. J., Blue, H. M., and Johnston, J. E., editors), American Society of Photogrammetry, Falls Church, Virginia, p. 367–397.

Lowman, P. D., Jr., 1964, A review of photography of the Earth from sounding rockets and satellites: NASA TN No. D-1868, National Aeronautics and Space Administration, Washington, D.C., 25p.

Lowman, P. D., Jr., 1965, Space photography, a review: *Photogrammetric Engineering,* v. 31, no. 1, p. 76–86.

Lowman, P. D., Jr., 1967, Geologic applications of orbital photography: NASA TN No. D-4155, National Aeronautics and Space Administration, Washington, D.C., 37 p.

Lowman, P. D., Jr., 1969a, Apollo 9 multispectral photography: Geologic analysis: Report No. X-644-71-15, National Aeronautics and Space Administration, Goddard Space Flight Center, Greenbelt, Maryland, 53 p.

Lowman, P. D., Jr., 1969b, Geologic orbital photography: Experience from the Gemini program: *Photogrammetria,* v. 24, no. 3/4, p. 77–106.

Lowman, P. D., Jr., 1973, Geologic uses of Earth orbital photography: *in The Surveillant Science* (Holz, R. K., editor) Houghton Mifflin Company, Boston, p. 170–182.

Lowman, P. D., Jr., 1980, The evolution of geological space photography: *in Remote Sensing in Geology* (Siegal, B. S., and Gillespie, A. R., editors): John Wiley and Sons, New York, p. 91–115.

Lowman, P. D., Jr., and Lattman, Laurence, 1980, Geomorphology: *in Remote Sensing in Geology* (Siegal, B. S., and Gillespie, A. R., editors), John Wiley and Sons, New York, p. 485–503.

Lowman, P. D., Jr., McDivitt, J. A., and White, E. H., II, 1967, Terrain photography on the Gemini IV mission: Preliminary report: NASA TN No. D-3982, National Aeronautics and Space Administration, Washington, D.C., 15 p.

Lowman, P. D., Jr., Frey, H. V., Shenk, W. E., and Dunkleman, L., 1973, Manual for 70 mm hand-held photography from Skylab: Report No. X-644-73-

147, National Aeronautics and Space Administration, Goddard Space Flight Center, Greenbelt, Maryland, 81 p.

Lowry, R. T., Langham, E. J., and Mudry, N., 1981, A preliminary analysis of SAR mapping of the Manitoba Flood, May 1979: *in Satellite Hydrology* (Deutsch, Morris, Wiesnet, D. R., and Rango, Albert, editors), American Water Resources Association, Minneapolis, p. 316–323.

Lucchitta, B. K., 1979, Landslides in Valles Marineris: *Journal of Geophysical Research,* v. 84, no. B14, p. 8097–8113.

Lucchitta, B. K., 1980, Grooved terrain on Ganymede: *Icarus,* v. 44, no. 2, p. 481–501.

Lucchitta, B. K., 1982, The use of images and morphologic analogs in space exploration: *in* Proceedings of the Fifteenth International Symposium on Remote Sensing of Environment, Environmental Research Institute of Michigan, Ann Arbor, Michigan, v. 2, p. 73–79.

Lucchitta, B. K., Anderson, D. M., and Shoji, H., 1981, Did ice streams carve martian outflow channels?: *Nature,* v. 290, no. 5809, p. 759–763.

Lueder, D. R., 1951, The preparation of an engineering soil map of New Jersey: *in* Proceedings of the Symposium on Surface and Subsurface Reconnaissance, American Society Testing Materials Special Technical Publication No. 122, p. 73–81.

Lundahl, A. C., 1948, Underwater depth determination by aerial photography: *Photogrammetric Engineering,* v. 14, no. 4, p. 454–462.

Lynn, D., and Allan, J. A., compilers, 1981, Matching remote sensing technologies and their applications: *in* Proceedings of the Ninth Annual Conference of the Remote Sensing Society, Department of Geography, University of Reading, Reading, England, 550 p.

Lyon, R. J. P., 1964, Evaluation of infrared spectrophotometry for compositional analysis of lunar and planetary soils. Part II. Rough and powdered surfaces: Stanford Research Institute (SRI) Final Report, NASA Contractor Report CR-100, Contract NASR 49 (04), 172 p.

Lyon, R. J. P., 1965, Analysis of rocks by spectral infrared emission (8 to 25 microns): *Economic Geology,* v. 60, no. 4, p. 715–736.

Lyon, R. J. P., 1975, Mineral exploration applications of digitally processed Landsat imagery: *in* Proceedings of the First William T. Pecora Memorial Symposium (Woll, P. W. and Fischer, W. A., editors), U.S. Geological Survey Professional Paper 1015, p. 271–292.

Lyon, R. J. P., 1975, Correlation between ground metal analysis, vegetation reflectance, and ERTS brightness over a molybdenum skarn deposit, Pine Nut Mountains, western Nevada: *in* Proceedings of the Tenth International Symposium on Remote Sensing of Environment, Environmental Research Institute of Michigan, Ann Arbor, Michigan, v. 2, p. 1031–1044.

Lyon, R. J. P., 1981, 8-13 micrometer spectra from high altitude (RB57) underflights compared with concurrent S191 data from Skylab SL-3 mission: *in* Workshop on Geological Applications of Thermal Infrared Remote Sensing Techniques (Settle, Mark, editor), Technical Report No. 81-06, Lunar and Planetary Institute, Houston, Texas, p. 79–91.

Lyon, R. J. P., and Patterson, J. W., 1966, Infrared spectral signatures—A field geological tool: *in* Proceedings of the Fourth International Symposium on Remote Sensing of the Environment, University of Michigan, Ann Arbor, Michigan, p. 215–230.

Lyon, R. J. P., Honey, F. R., and Ballew, G. I., 1975, A comparison of observed and model predicted atmospheric perturbations on target radiance measured by ERTS: *Proceedings IEEE,* no. 63, p. 244–249.

Lyons, T. R., 1969, A study of the Palaeo-Indian and desert culture complexes of the Estancia Valley, New Mexico: Ph.D. Dissertation, Department of Geology, University of New Mexico, Albuquerque, New Mexico, 355 p.

Lyons, T. R., 1981, Geological time surfaces and archaeological dating: *in The Artifact: Archaeological Essays in Honor of Mark Wimberly* (Foster, M. S., editor), v. 19, nos. 3&4, p. 222–230.

MacDonald, H. C., 1980, Techniques and applications of imaging radars: *in Remote Sensing in Geology* (Siegal, B. S., and Gillespie, A. R., editors): John Wiley and Sons, New York, p. 297–336.

MacDonald, W. R., 1976a, Geodetic control in polar regions for accurate mapping with ERTS imagery: *in ERTS-1, A New Window on Our Planet* (Williams, R. S., Jr., and Carter, W. D., editors), U.S. Geological Survey Professional Paper 929, p. 34–36.

MacDonald, W. R., 1976b, Glaciology in Antarctica: *in ERTS-1, A New Window on Our Planet* (Williams, R. S., Jr., and Carter, W. D., editors), U.S. Geological Survey Professional Paper 929, p. 194–195.

Mackay, J. R., and Black, R. F., 1973, Origin, composition and structure of perennially frozen ground ice. A review. North American Contribution, Permafrost: *in* Proceedings of the Second International Conference on Permafrost, Yukutsk, Siberia, U.S.S.R., p. 185–192.

Magnuson, M. O., 1974, Control of fires in abandoned mines in the Eastern Bituminous Region of the United States. A supplement to Bulletin 590: U.S. Bureau of Mines, Information Circular 8620, 53p.

Malila, W. A., Gleason, J. M., and Cicone, R. C., 1976, Atmospheric modeling related to thematic mapper scan geometry: Final Report to NASA No. CR 147792, Environmental Research Institute of Michigan No. 119300-5-F, 131 p.

Mamula, Ned, 1978, Remote sensing methods for monitoring surface coal mining in the northern Great Plains: U.S. Geological Survey *Journal of Research,* v. 6, no. 2, p. 149–160.

Mamula, Ned, and Voight, Barry, 1983, Preliminary tectonic analysis of northeast Iceland: A Landsat perspective: *Geological Society of America Bulletin,* (in press).

Marsh, S. E., 1978, Quantitative relationships of surface geology and spectral habit to satellite radiometric data: Ph.D. Dissertation, Department of Applied Earth Sciences, Stanford University, Stanford, California, 225 p.

Marsh, S. E. and Lyon, R. J. P., 1980, Quantitative relationships of near-surface spectra to Landsat radiometric data: *Remote Sensing of Environment,* v. 10, no. 4, p. 241–261.

Marszalek, D. S., 1977, Water penetration aerial color film for geologic and environmental mapping of the coastal zone: *in* Abstracts with Programs, 1977 Annual Meeting, Geological Society of America, Boulder, Colorado, v. 9, no. 7, p. 1085–1086.

Masursky, Harold, Eliason, Eric, Ford, P. G., McGill,

G. E., Pettengill, G. H., Schaber, G. G., and Schubert, Gerald, 1980, Pioneer Venus radar results: Geology from images and altimetry: *Journal of Geophysical Research,* v. 85, no. A13, p. 8232–8260.

Mathisen, D. A., 1978, The unfulfilled promises of remote sensing: *New Engineer,* v. 7, no. 6, p. 14, p. 16–20, p. 23, p. 25, and p. 26–27.

Maul, G. A., and Sidran, M. 1973, Atmospheric effects on ocean surface temperature sensing from the NOAA satellite scanning radiometer: *Journal of Geophysical Research,* v. 78, no. 12, p. 1909–1916.

Maurin, A. F., and Riguidel, M. J., 1978, Eléments de morphologie généralisée: *in* TOTAL, Compagnie Française des Pétroles, Paris, 133 p. (In French)

Mayhew, M. A., 1982, Large-scale crustal magnetization models of the U.S. derived from Magsat data and relations to tectonic provinces (abs.): in Abstracts of the 24th Plenary Meeting of Committee on Space Research, Symposium on Remote Sensing and Mineral Exploration (17–20 May 1982), Ottawa, Canada, p. 88.

McAdams, M. P., 1981, Quantitative analysis of Landsat flood mapping capabilities (abs): *in Satellite Hydrology* (Deutsch, Morris, Wiesnet, D. R., and Rango, Albert, editors), American Water Resources Association, Minneapolis, p. 308.

McCauley, J. F., 1976, The surficial geology of Mars: *in* Proceedings of the Colloquium on Planetary Physics and Geology, Acad. Nazionale Lincei, Rome (1974), p. 141–158.

McCauley, J. F., Schaber, G. G., Breed, C. S., Grolier, M. J., Haynes, C. V., Issawi, Bahay, Elachi, Charles, and Blom, Ronald, 1982, Subsurface valleys and geoarchaeology of the eastern Sahara revealed by shuttle radar: *Science,* v. 218, no. 4576, p. 1004–1020.

McClain, E. P., 1980, Environmental satellites: Reprinted from the McGraw-Hill *Encyclopedia of Environmental Science,* McGraw-Hill Book Company, Inc., New York, 16p.

McCulloch, D. S., 1966, Slide-induced waves, seiching, and ground fracturing caused by the earthquake of March 27, 1964, at Kenai Lake, Alaska: U.S. Geological Survey Professional Paper 543-A, p. A1–A41.

McCulloch, D. S., and Bonilla, M. G., 1970, Effects of the earthquake of March 27, 1964, on the Alaska Railroad: U.S. Geological Survey Professional Paper 545-D, p. D1–D161.

McEldowney, R. C., and Pascucci, R. F., 1979, Application of remote-sensing data to nuclear power plant site investigations: *in Geology in the Siting of Nuclear Power Plants;* Geological Society of America, Reviews in Engineering Geology, v. 4, Boulder, Colorado, p. 121–139.

McGill, J. T., 1958, Map of coastal landforms of the world: *Geographical Review,* v. 48, no. 3, p. 402–405.

McKee, E. D., editor, 1979, A study of global sand seas: U.S. Geological Survey Professional Paper 1052, 429 p.

McKee, E. D., and Breed, C. S., 1976, Sand seas of the world: *in ERTS-1, A New Window on Our Planet* (Williams, R. S., Jr., and Carter, W. D., editors), U.S. Geological Survey Professional Paper 929, p. 81–88.

McKee, E. D., Breed, C. S., and Fryberger, S. G., 1977, Desert sand seas: *in Skylab Explores the Earth,* NASA SP-380, p. 5–47.

McKim, H. L. and Merry, C. J., 1975, Use of remote sensing to quantify construction material and to define geologic lineation: U.S. Army Corps of Engineers, Cold Regions Research and Engineering Laboratory, Special Report 242, Pt. 1, 21 p.

McLerran, J. H., and Morgan, J. O., 1964, Thermal mapping of Yellowstone National Park: *in* Proceedings of the Third Symposium on Remote Sensing of Environment, University of Michigan, Ann Arbor, Michigan, p. 517–530.

McNay, L. M., 1971, Coal refuse fires, an environmental hazard: U.S. Bureau of Mines, Information Circular 8515, 50 p.

Meier, M. F., 1973a, Evaluation of ERTS imagery for mapping and detection of changes of snowcover on land and on glaciers: *in* Symposium on Significant Results Obtained from the Earth Resources Technology Satellite-1, v. I (Technical Presentations), Section A, NASA SP-327, p. 863–875.

Meier, M. F., 1973b, Measurement of snow cover using passive microwave radiation: *in* Proceedings of the Banff Symposia, "The Role of Snow and Ice in Hydrology," International Association of the Hydrological Sciences, v. 1, p. 739–750.

Meier, M. F., 1975, Application of remote-sensing techniques to the study of seasonal snow cover: *Journal of Glaciology,* v. 15, no. 73, p. 251–265.

Meier, M. F., 1976, Monitoring the motion of surging glaciers in the Mount McKinley massif, Alaska: *in ERTS-1, A New Window on Our Planet* (Williams, R. S., Jr., and Carter, W. D., editors), U.S. Geological Survey Professional Paper 929, p. 185–187.

Meier, M. F., 1979, Variations in time and space of the velocity of lower Columbia Glacier, Alaska (abs.): *Journal of Glaciology,* v. 23, no. 89, p. 408.

Meier, M. F., and Post, A. S., 1962, Recent variations in mass net budgets of glaciers in western North America: *in* Proceedings of the Symposium of Obergurgl (Austria), "Variations of the Regime of Existing Glaciers," (Ward, W., editor), International Association of Scientific Hydrology Publication No. 58, p. 63–77.

Meier, M. F., and Post, Austin, 1969, What are glacier surges?: *Canadian Journal of Earth Sciences,* v. 6, part 2, p. 807–817.

Meier, M. F. and Evans, W. E., 1975, Comparison of different methods for estimating snow cover in forested, mountainous basins using Landsat (ERTS) images: *in Operational Applications of Satellite Snow Cover Observations* (Rango, Albert, editor), NASA Special Publication SP-391, p. 215–234.

Meier, M. F., Post, Austin, Brown, C. S., Frank, David, Hodge, S. M., Mayo, L. R., Rasmussen, L. A., Senear, E. A., Sikonia, W. G., Trabant, D. C., and Watts, R. D., 1978, Columbia Glacier progress report—December 1977, U.S. Geological Survey Open-File Report 78-264, 78 p.

Meier, M. F., Rasmussen, L. A., Post, Austin, Brown, C. S., Sikonia, W. G., Bindschadler, R. A., Mayo, L. R., and Trabant, D. C., 1980, Predicted timing of the disintegration of the lower reach of Columbia Glacier, Alaska: U.S. Geological Survey Open-File Report 80-582, 47 p.

Meinzer, O. E., 1911, Geology and water resources of Estancia Valley, New Mexico, with notes on ground-water conditions in adjacent parts of central New Mexico: U.S. Geological Survey Water-Supply Paper 275, 89 p.

Mekel, J. F. M., 1970, The use of aerial photographs in geological mapping: ITC Textbook of Photo-Interpretation, v. 8-1, International Institute for Aerial Survey and Earth Sciences (ITC), Enschede, The Netherlands, 169 p.

Mekel, J. F. M., 1972, The geological interpretation of radar images: ITC Textbook of Photo-Interpretation, v. 8-2, International Institute for Aerial Survey and Earth Sciences (ITC), Enschede, The Netherlands, 64 p.

Melson, W. G., and Hopson, C. A., 1982, Preeruption temperatures and oxygen fugacities in the 1980 eruptive sequence: *in The 1980 Eruptions of Mount St. Helens* (Lipman, P. W., and Mullineaux, D. R., editors), U.S. Geological Survey Professional Paper 1250, p. 641–648.

Melzer, L. S., 1976, Site investigation methodology for multiple aim point missile systems: Air Force Weapons Laboratory, Technical Report No. 76-145, p. 13–18.

Mendell, W. W., 1981, Limb darkening of lunar surface radiance as seen from orbit (abs): *in* Workshop on Geological Applications of Thermal Infrared Remote Sensing Techniques (Settle, Mark, editor), Technical Report No. 81-06, Lunar and Planetary Institute, Houston, Texas, p. 92–93.

Merembeck, B. F., Borden, F. Y., Podwysocki, M. H., and Applegate, D. N., 1977, Application of canonical analysis to multispectral scanner data: *in* Proceedings of the 14th Annual Symposium on Computer Applications in the Mineral Industries, Society of Mining Engineers, American Institute of Mining, Metallurgical and Petroleum Engineers, New York, p. 867–879.

Merifield, P. M., Saari, J. M., Shorthill, R. W. Wildey, R. L., Wilhelms, D. E., and Williams, R. S., Jr., 1969a, Interpretation of extraterrestrial imagery: *Photogrammetric Engineering,* v. 35, no. 5, p. 477–492.

Merifield, P. M., Cronin, J. F., Foshee, L. L., Gawarecki, S. J., Neal, J. T., Stevenson, R. E., Stone, R. O., and Williams, R. S., Jr., 1969b, Satellite imagery of the Earth: *Photogrammetric Engineering,* v. 35, no. 7, p. 654–688.

M'Gonigle, J. W., 1979a, Preliminary geologic map of the Elkol quadrangle, Lincoln County, southwestern Wyoming: U.S. Geological Survey Open-File Report 79-1150, scale 1:24,000.

M'Gonigle, J. W., 1979b, Preliminary geologic map of the Warfield Creek quadrangle, Lincoln County, southwestern Wyoming: U.S. Geological Survey Open-File Report 79-1176, scale 1:24,000.

Miesch, A. T., 1981, Estimation of the geochemical threshold and its statistical significance: *Journal of Geochemical Exploration,* v. 16, no. 1, p. 49–76.

Mikkola, Aimo, 1982, The application of space imagery to mineral exploration in the Baltic shield: *in The Development Potential of Precambrian Mineral Deposits,* Natural Resources and Energy Division, U.N. Department of Technical Cooperation for Development, Pergamon Press, New York, p. 383–395.

Miller, J. B., 1975, Landsat image studies as applied to petroleum exploration in Kenya: *in* Proceedings of the NASA Earth Resources Survey Symposium, Houston, Texas, p. 605–624.

Miller, J. B., 1977, Landsat-1 image studies as applied to petroleum exploration in Kenya: U.S. Geological Survey Professional Paper 1015, p. 137–150.

Miller, L. D., 1966, Location of anomalously hot Earth with infrared imagery in Yellowstone National Park: *in* Proceedings of the Fourth Symposium on Remote Sensing of Environment, University of Michigan, Ann Arbor, Michigan, p. 751–769.

Miller, V. C., and Miller, C. F., 1961, Photogeology: McGraw-Hill Book Company, Inc., New York, 248 p.

Milton, N. M., 1981, Use of reflectance spectra of native plant species for interpreting airborne multispectral scanner data in the East Tintic Mountains, Utah: *in* Proceedings of the International Geoscience and Remote Sensing Symposium (IGARSS '81), Institute of Electronic and Electrical Engineers, Washington, D.C., p. 614–616.

Missallati, A., Prelat, A. E., and Lyon, R. J. P., 1979, Simultaneous use of geological, geophysical, and LANDSAT digital data in uranium exploration: *Remote Sensing of Environment,* v. 8, no. 3, p. 189–210.

Moik, J. G., 1980, Digital processing of remotely sensed images: NASA Special Publication SP-431, National Aeronautics and Space Administration, Washington, D.C., 330 p.

Mollard, J. D., 1962, Photo interpretation in prospecting for granular construction materials: *in* International Society of Photogrammetry, Transactions, Commission VII, Symposium on Photo Interpretation, Delft, Netherlands, p. 514–523.

Mollard, J. D., 1974, Landforms and surface materials of Canada; A stereoscopic airphoto atlas and glossary: A guide to terrain analysis for engineering, geological and environmental studies: J. D. Mollard, Ph.D., Consultant, Airphoto Interpretation, Regina, Saskatchewan, looseleaf notebook, variously paginated by sections.

Mollard, J. D. and Deshaw, H. E., 1956, How airphotos locate granular deposits: *Construction World,* v. 11, no. 12, 20 p.

Mollard, J. D. and Deshaw, H. E., 1958, Ten years of mapping granular deposits from aerial photographs: Highway Research Board Bulletin No. 180, p. 20–32.

Moody, J. D., and Hill, M. J., 1956, "Wrench-fault tectonics": *Geological Society of America Bulletin,* v. 67, no. 9, p. 1207–1246.

Moore, G. K., and Sheehan, C. A., compilers, 1981, Evaluation of radar imagery for geologic and cartographic applications: U.S. Geological Survey Open-File Report 81-1358, 37 p.

Moore, J. G., and Albee, W. C., 1982, Topographic and structural changes, March–July 1980—photogrammetric data: *in The 1980 Eruptions of Mount St. Helens, Washington* (Lipman, P. W., and Mullineaux, D. R., editors), U.S. Geological Survey Professional Paper 1250, p. 123–134.

Moore, R. K., editor; Chastant, L. J., Porcello, L. J., Stevenson, Joseph, and Ulaby, F. T., 1975, Microwave remote sensors: *in Manual of Remote Sensing* (Reeves, R. G., Anson, Abraham, and Landen, David, editors): v. 1 (Janza, F. J., Blue, H. M., and Johnston, J. E., editors), American Society of Photogrammetry, Falls Church, Virginia, p. 399–537.

Morgan, V. I., and Budd, W. E., 1975, Radio-echosounding of the Lambert Glacier basin: *Journal of Glaciology,* v. 15, no. 73, p. 103–111.

Morrison, R. B., and White, P. G., 1976, Monitoring

flood inundation: *in ERTS-1, A New Window on Our Planet* (Williams, R. S., Jr., and Carter, W. D., editors), U.S. Geological Survey Professional Paper 929, p. 196–208.

Morton, D. M., 1971, Seismically triggered landslides in the area above San Fernando Valley: *in The San Fernando, California, Earthquake of February 9, 1971:* U.S. Geological Survey Professional Paper 733, p. 99–104.

Moxham, R. M., 1967, Infrared surveys at Scranton and Laurel Run, Pennsylvania: U.S. Geological Survey, Technical Letter BM-1 (Bureau of Mines-1), 27 March 1967, 5 p.

Moxham, R. M., 1972, Thermal surveillance of volcanoes: *in The Surveillance and Prediction of Volcanic Activity, A Review of Methods and Techniques,* Earth Sciences Series, UNESCO, Paris, v. 8, p. 103–124.

Moxham, R. M., and Alcarez, Arturo, 1966, Infrared surveys at Taal Volcano, Philippines: *in* Proceedings of the Fourth Symposium on Remote Sensing of Environment, University of Michigan, Ann Arbor, Michigan, p. 827–843.

Moxham, R. M., and Greene, G. W., 1967, Infrared surveys of coal mine fires in the anthracite and bituminous fields, Pennsylvania: U.S. Geological Survey, Interagency Report BM-2 (Bureau of Mines), July, 50 p.

Moxham, R. M., Crandell, D. R., and Marlatt, W. E., 1965, Thermal features at Mount Rainier, Washington, as revealed by infrared surveys: *in Geological Survey Research 1965,* U.S. Geological Survey Professional Paper 525-D, p. 93–100.

Muehlberger, W. R., Gucwa, P. R., Ritchie, A. W., and Swanson, E. R., 1977, Global tectonics: Some geologic analyses of observations and photographs from Skylab: *in Skylab Explores the Earth,* NASA SP-380, p. 49–88.

Müller, Fritz, 1962, Zonation in the accumulation area of the glaciers of Axel Heiberg Island, N.W.T., Canada: *Journal of Glaciology,* v. 4, no. 33, p. 302–313.

Münzer, Ulrich, and Bodechtel, Johann, 1980, Digitale Verarbeitung von Landsat-Daten uber Eis- und Schneegebieten des Vatnajökulls (Island): *Bildmessung und Luftbildwesen,* v. 48, p. 21–28. (In German)

Nanda, R. L., 1978, Use of photo interpretation techniques for survey of highway material resources: *Journal of the Indian Society of Photo-interpretation,* v. 6, no. 2, p. 67–74.

National Academy of Sciences, 1969a, Geology: *in* Useful Applications of Earth-Oriented Satellites; Panel 2 of the Summer Study on Space Applications, Division of Engineering, National Research Council, National Academy of Sciences, Washington, D.C., 65 p.

National Academy of Sciences, 1969b, Hydrology: *in* Useful Applications of Earth-Oriented Satellites; Panel 3 of the Study on Space Applications, Division of Engineering, National Research Council, National Academy of Sciences, Washington, D.C., 73 p.

National Academy of Sciences, 1977, Resource sensing from space: Prospects for developing countries: Report of the Ad Hoc Committee on Remote Sensing for Development, Board on Science and Technology for International Development, Commission on International Relations, National Research Council, Washington, D.C., 202 p.

National Academy of Sciences, 1982, A strategy for Earth science from space in the 1980's; Part I: Solid Earth and oceans: Committee on Earth Sciences, Space Science Board, Assembly of Mathematical and Physical Sciences, National Research Council, National Academy Press, Washington, D.C., 99 p.

NASA, 1967, Earth photographs from Gemini III, IV, and V: NASA SP-129, National Aeronautics and Space Administration, Washington, D.C., 266 p.

NASA, 1968a, Earth photographs from Gemini VI through XII: NASA SP-171, National Aeronautics and Space Administration, Washington, D.C., 327 p.

NASA, 1968b, Application of biogeochemistry to mineral prospecting, a survey: National Aeronautics and Space Administration, Washington, D.C., NASA SP-5056, 135 p.

NASA, 1970, This island Earth: NASA SP-250, National Aeronautics and Space Administration, Washington, D.C., 182 p.

NASA, 1971, Data users handbook: NASA Earth Resources Technology Satellite, Goddard Space Flight Center, Greenbelt, Maryland, 15 September 1971, looseleaf notebook with pagination by sections. Various revisions issued on 17 November 1972, 2 September 1976, and 1 April 1977.

NASA, 1973, Advanced scanners and imaging systems for Earth observations: NASA SP-335, National Aeronautics and Space Administration, Washington, D.C., 604 p.

NASA, 1974, Skylab Earth resources data catalog: National Aeronautics and Space Administration, L. B. Johnson Space Center, Houston, Texas; Report No. JSC 09016, 359 p.

NASA, 1977, Skylab explores the Earth: NASA SP-380, National Aeronautics and Space Administration, Washington, D.C., 517 p.

NASA, 1978a, Nimbus 7 CZCS data catalog: National Aeronautics and Space Administration, Goddard Space Flight Center, Greenbelt, Maryland, issued monthly.

NASA, 1978b, The Nimbus 7 Users' Guide: National Aeronautics and Space Administration, Goddard Space Flight Center, Greenbelt, Maryland, August, 263 p.

NASA, 1980, Heat Capacity Mapping Mission (HCMM) data users handbook for Applications Explorer Mission-A (AEM): National Aeronautics and Space Administration, Goddard Space Flight Center, Greenbelt, Maryland, 120 p., (originally issued in December 1978, revised in November 1979, and October 1980).

National Geographic Society, 1976, Portrait USA: National Geographic Supplement, v. 150, no. 1, scale 1:4,560,000, p. 140A.

National Geographic Society, 1979, Strange world without sun: *National Geographic,* v. 156, no. 5, p. 680–688.

Neal, J. T., 1962, Airborne geoscience research: *Photogrammetric Engineering,* v. 28, no. 3, p. 438–441.

Neal, J. T., 1972, Playa surface features as indicators of environment: *in* Playa Lake Symposium (Reeves, C. C., Jr., ed.), ICASALS Publication No. 4, Texas Tech University, Lubbock, Texas, p. 107–132.

Neal, J. T., 1976, Pleistocene Lake Trinity, a sulphate basin in the Northern Jornada del Muerto, New Mexico: *in* Abstracts with Programs, Geological Society of America, Boulder, Colorado, v. 8, no. 6, p. 1026–1027.

Neal, J. T., 1978, Test site selection for nuclear blast

and shock simulation: *in* Proceedings of the Blast and Shock Simulation Symposium, Defense Nuclear Agency Publication No. 4797, Washington, D.C., p. 7–16.

Neukum, G., and Wise, D. U., 1976, Mars: A standard crater curve and possible new time scale: *Science,* v. 194, no. 4272, p. 1381–1387.

Newton, J. G., 1976, Early detection and correction of sink hole problems in Alabama, with a preliminary evaluation of remote sensing applications: Alabama Highway Department, Bureau of Research and Development, Research Report HPR-76, 83 p.

Nilsen, T. H., 1973, Preliminary photointerpretation map of landslide and other surficial deposits of the Concord 15-minute quadrangle and the Oakland West, Richmond, and part of the San Quentin 7½-minute quadrangles, Contra Costa and Alameda Counties, California: U.S. Geological Survey Miscellaneous Field Studies Map MF-493, Scale, 1:62,500.

Noble, V. E., 1970, Potential application of multispectral photography for oceanography: *in* Proceedings of the Geodetic and Research and Development Symposium, Seventh DOD Geodetic-Cartographic-Target Materials Conference, Cameron Station, Virginia, p. 107–130.

Norlund, N. E., 1938, Denmark: *Photogrammetric Engineering,* v. 4, no. 31, p. 119–120.

Norman, J. W., 1976, Photogeological fracture trace analysis as a subsurface exploration technique: Transactions, Institute of Mining and Metallurgy, Section B, v. 85, p. B52–B62, Discussion on above paper, Section B, v. 86, p. B58–B60.

Norrman, J. O., 1970, Trends in postvolcanic development of Surtsey island; Progress report on geomorphological activities in 1968: *in* Surtsey Research Progress Report, no. 5, Reykjavik, p. 95–112.

Norrman, J. O., 1980, Coastal erosion and slope development in Surtsey Island, Iceland: *Zeitschrift für Geomorphologie,* v. 34 (Supplementband), p. 20–38.

Oberlander, T., 1965, The Zagros streams: A new interpretation of transverse drainage in an orogenic zone: *Syracuse Geographer,* Series no. 1, 168 p.

Oberste-Lehn, Deane, 1970a, Passive microwave study of geologic materials in a volcanic province: M.S. Dissertation, Department of Geology, Stanford University, Stanford, California, 107 p.

Oberste-Lehn, Deane, 1970b, Phenomena and properties of geologic materials affecting microwaves—a review: Stanford Remote Sensing Laboratory Technical Report No. 70-10, Department of Geology. Stanford University, Stanford, California, April, 57 p.

Ødegaard, Helge, and Helle, S. G., 1982, Polar mapping using Landsat data. Svalbard and Dronning Maud Land: Joint Final Report, April: International Business Machines A/S and Norsk Polarinstitutt, Oslo, 66 p.

Office of Technology Assessment, 1982, Civilian space policy and applications: Report No. OTA-STI-177 (June), Office of Technology Assessment, Congress of the United States, Washington, D.C., 391 p.

Offield, T. W., 1975, Line-grating diffraction in image analysis: Enhanced detection of linear structures in ERTS images, Colorado Front Range: *Modern Geology,* v. 5, no. 1, p. 101–107.

O'Hara, C. J., and Oldale, R. N., 1980, Maps showing geology and shallow structure of eastern Rhode Island Sound and Vineyard Sound, Massachusetts: U.S. Geological Survey Miscellaneous Field Studies Map 1186; in 5 sheets.

Oldale, R. N., 1976, Geologic history of Cape Cod, Massachusetts: U.S. Geological Survey Scientific Leaflet No. INF-75-6, 24 p.

Oldale, R. N., Friedman, J. D., and Williams, R. S., Jr., 1971, Changes in coastal morphology of Monomoy Island, Cape Cod, Massachusetts: *in Geological Survey Research* 1971, U.S. Geological Survey Professional Paper 750-B, p. B101–B107.

O'Leary, D. W., 1982, Midrange sidescan-sonar data from the Continental Slope off Georges Bank between Lydonia and Oceanographer Canyons: U.S. Geological Survey Open-File Report 82-0600, 4 p.

O'Leary, D. W., and Twichell, D. C., 1981, Potential geologic hazards in the vicinity of Georges Bank basin: *in* Summary Report of the Sediments, Structural Framework, Petroleum Potential and Environmental Conditions of the United States Middle and Northern Continental Margin in Area of Proposed Oil and Gas Lease Sale No. 76 (Grow, J. A., compiler); U.S. Geological Survey Open-File Report 81-765, p. 48–68.

O'Leary, D. W., Friedman, J. D., and Pohn, H. A., 1976a, Lineaments, linear, lineation: Some proposed new standards for old terms: *Geological Society of America Bulletin,* v. 87, no. 10, p. 1463–1469.

O'Leary, D. W., Friedman, J. D., and Pohn, H. A., 1976b, Lineament and linear, a terminological reappraisal: *in* Proceedings of the Second International Conference on Basement Tectonics (Newark, Delaware), p. 571–577.

O'Neil, R. A., and Jones, S. J., 1975, Radio depth sounding on Barnes ice cap (abs.): *Journal of Glaciology,* v. 15, no. 73, p. 458.

Orhaug, Torleiv, and Dyring, Eric, 1973, Stormakternas vakande rymdögon: *Forskning och Framsteg,* no. 6, p. 2–14. (In Swedish)

Orr, D. G., and Quick, J. R., 1971, Construction materials in delta areas: *Photogrammetric Engineering,* v. 37, no. 4, p. 337–351.

Østrem, Gunnar, 1975, ERTS data in glaciology—an effort to monitor glacier mass balance from satellite imagery: *Journal of Glaciology,* v. 15, no. 73, p. 403–414.

Otterman, Joseph, Ohring, G., and Ginsburg, A., 1974, Results of the Israeli multidisciplinary data analysis of ERTS-1 imagery: *Remote Sensing of Environment,* v. 3, no. 2, p. 133–148.

Overstreet, W. C., and Rossman, D. L., 1970, Reconnaissance geology of the Wadi Wassat quadrangle: U.S. Geological Survey, Saudi Arabian Project Report 117, 68 p.

Pálmason, Gudmundur, Friedman, J. D., Williams, R. S., Jr., Jónsson, Jón, and Saemundsson, Kristján, 1971, Aerial infrared surveys of Reykjanes and Torfajökull thermal areas, Iceland, with a section on cost of exploration surveys: *in Geothermics* (1970), Special Issue 2, U.N. Symposium on the Development and Utilization of Geothermal Resources, Pisa 1970, v. 2, pt. 1, p. 399–412.

Parizek, R. R., 1975, On the nature and significance of fracture traces and lineaments in carbonate and other terranes: *in* Proceedings of the U.S.-Yugoslavian Symposium (Karst Hydrology and Water Resources), Dubrovnik, v. 1, p. 3-1–3-62.

Parizek, R. R., 1976a, Lineaments and groundwater: *in* ORSER-SSEL, The Pennsylvania State University, Interdisciplinary Application and Interpretations of

EREP Data Within the Susquehanna River Basin (McMurtry, G. T., and Petersen, G. W., editors), SKYLAB EREP Investigation No. S-475, NASA Contract No. NAS 9-13406, p. 4-59–4-86.

Parizek, R. R., 1976b, Application of fracture traces and lineaments to groundwater prospecting: *in Field Guide to Lineaments and Fractures in Central Pennsylvania* (Gold, D. P., and Parizek, R. R., editors), Second International Conference on Basement Tectonics, (Department of Geosciences, The Pennsylvania State University), p. 38–59.

Parker, D. C., and Wolff, M. F., 1965, Remote sensing: *International Science and Technology,* no. 43 (July), p. 20–31.

Parvis, Merle, 1950, Drainage pattern significance in airphoto identification of soils and bedrock: *Photogrammetric Engineering,* v. 16, no. 3, p. 387–409.

Pascucci, R. F., Smith, A. F., and Pearson, J. E., 1981, Determination of the contribution of side-looking airborne radar to structural geologic mapping: Unpublished contractor-prepared U.S. Geological Survey Report, 61 p.

Paterson, W. S. B., 1969, The physics of glaciers: Pergamon Press, New York, 250 p.

Paterson, W. S. B., 1981, Physics of glaciers: Pergamon Press, New York, 380 p.

Peterson, Rex, 1979, Oil and gas exploration by pattern recognition of lineament assemblages associated with bends in wrench faults: *in* Proceedings of the Thirteenth International Symposium on Remote Sensing of Environment, Environmental Research Institute of Michigan, Ann Arbor, Michigan, v. 2, p. 993–1014.

Pettyjohn, W. A., 1980, Environmental geoscience: *in Remote Sensing in Geology* (Siegal, B. S., and Gillespie, A. R., editors), John Wiley and Sons, New York, p. 635–657.

Péwé, T. L., 1973, Ice-wedge casts and past permafrost distribution in North America: *Geoforum,* v. 15, p. 15–26.

Pillmore, C. L., Dueholm, K. S., Jepsen, H. S., and Schuch, C. H., 1981, Computer-assisted photogrammetric mapping systems for geologic studies—a progress report: *Photogrammetria,* v. 36, no. 5, p. 159–171.

Pirie, D. M., and Steller, D. D., 1977, California coastal processes study—LANDSAT II: NASA Final Report, LANDSAT Investigation No. 22200, 164 p.

Plafker, George, 1967, Surface faults on Montague Island associated with the 1964 Alaska earthquake: U.S. Geological Survey Professional Paper 543-G, p. G1–G42.

Plafker, George, 1968, Source areas of the Shattered Peak and Pyramid Peak landslides at Sherman Glacier: *in The Great Alaska Earthquake of 1964, Hydrology:* U.S. National Academy of Sciences Publication 1603, p. 374–382.

Plafker, George, 1969, Tectonics of the March 27, 1964, Alaska earthquake: U.S. Geological Survey Professional Paper 543-I, p. I1–I74.

Plafker, George, Ericksen, G. E., and Fernandez, C. J., 1971, Geological aspects of the May 31, 1970, Peru earthquake: *Seismological Society of America Bulletin,* v. 61, no. 3, p. 543–578.

Plascyk, J. A., 1975, The MKII Fraunhofer line discriminator (FLD II) for airborne and orbital remote sensing of solar stimulated luminescence: *Optical Engineering,* v. 14, no. 4, p. 339–346.

Plascyk, J. A., and Gabriel, F. C., 1975, The Fraunhofer line discriminator MKII—an airborne instrument for precise and standardized ecological luminescence measurement: *Institute of Electrical and Electronic Engineers* (*IEEE*), v. IM 24, no. 4, p. 306–313.

Podwysocki, M. H., and Gold, D. P., 1979, Some possible surface expression of a regular fracture grid deformed by subsurface structures: *in* Proceedings of the Second International Conference on Basement Tectonics (Newark, Delaware), p. 542–559.

Podwysocki, M. H., and Segal, D. B., 1980, Preliminary digital classification of limonitic rocks, Frisco 15-minute quadrangle, Utah: U.S. Geological Survey Open-File Report 80-19, scale 1:48,000.

Podwysocki, M. H., Segal, D. B., and Abrams, M. J., 1982a, Use of multispectral scanner images for assessment of hydrothermal alteration in the Marysvale, Utah, mining area: U.S. Geological Survey Open-File Report 82-675, 29 p.

Podwysocki, M. H., Pohn, H. A., Phillips, J. D., Krohn, M. D., Purdy, T. L., and Merin, I. S., 1982b, Evaluation of remote-sensing, geological, and geophysical data for south-central New York and northern Pennsylvania: U.S. Geological Survey Open-File Report 82-0319, 179 p.

Podwysocki, M. H., Segal, D. B., and Jones, O. D., 1983a, Mapping of hydrothermally altered rocks using airborne multispectral scanner data, Marysvale, Utah, mining area; Remote Sensing and Mineral Exploration—1982: *in* COSPAR Advances in Space Exploration, Pergamon Press, New York (in press).

Podwysocki, M. H., Segal, D. B., Collins, W. E., and Chang, S. H., 1983b, Mapping the distribution and mineralogy of hydrothermally altered rocks using airborne multispectral scanner and spectroradiometer data, Marysvale, Utah: *in* Proceedings of the Second Thematic Conference: Remote Sensing for Exploration Geology (Dallas, Texas; December 1982), 16th International Symposium on Remote Sensing of Environment, Environmental Research Institute of Michigan, Ann Arbor, Michigan (in press).

Pohn, H. A., 1970, Remote sensor application studies progress report, July 1, 1968, to June 30, 1969—Analysis of images and photographs by a Ronchi grating: U.S. Geological Survey Report, 14 p. (NTIS Report no. PB 197–101).

Pohn, H. A., Offield, T. W., and Watson, Kenneth, 1974, Thermal inertia mapping from satellite—Discrimination of geologic units in Oman: U.S. Geological Survey *Journal of Research,* v. 2, no. 2, p. 147–158.

Polcyn, F. C., 1976, Final report on NASA/Cousteau ocean bathymetric experiment—remote bathymetry using high gain Landsat data: Environmental Research Institute of Michigan Report No. NASA-CR-ERIM-118500-1-F, May, 127 p.

Polcyn, F. C., and Rollin, R. A., 1969, Remote sensing techniques for the location and measurement of shallow water features: University of Michigan, Ann Arbor, Michigan; Willow Run Laboratories SPOC Project, Report No. 8973-10-P, U.S. Naval Oceanographic Office Contract No. N62306-67-C-0243, 72 p.

Polcyn, F. C., and Lyzenga, David, 1973, Updating coastal and navigational charts using ERTS-1 data: NASA SP-351, p. 1333–1346.

Pomeroy, J. S., 1982, Landslides in the greater Pittsburgh region, Pennsylvania: U.S. Geological Survey Professional Paper 1229, 48 p.

Post, Austin, 1969, Distribution of surging glaciers in western North America: *Journal of Glaciology,* v. 8, no. 53, p. 229–240.

Post, Austin, 1975, Preliminary hydrography and historic terminal changes of Columbia Glacier, Alaska: U.S. Geological Survey Hydrologic Investigations Atlas, HA-559, 3 sheets.

Post, Austin, 1977, Reported observations of icebergs from Columbia Glacier in Valdez Arm and Columbia Bay, Alaska, during the summer of 1976: U.S. Geological Survey Open-File Report 77-235, 7 p.

Post, Austin, and Mayo, L. R., 1971, Glacier dammed lakes and outburst floods in Alaska: U.S. Geological Survey Hydrologic Investigations Atlas, HA-455, 3 map sheets, 10 p.

Post, Austin, Meier, M. F., and Mayo, L. R., 1976, Measuring the motion of the Lowell and Tweedsmuir surging glaciers of British Columbia, Canada: *in ERTS-1, A New Window on Our Planet* (Williams, R. S., Jr., and Carter, W. D., editors), U.S. Geological Survey Professional Paper 929, p. 180–184.

Potter, A. E., and Morgan, T., 1981, Remote sensing of lunar rock composition from thermal infrared measurements (abs.): *in* Workshop on Geological Applications of Thermal Infrared Remote Sensing Techniques (Settle, Mark, editor), Technical Report No. 81-06, Lunar and Planetary Institute, Houston, Texas, p. 94.

Premoli, Camillo, 1976, Formation of and prospecting for uraniferous calcretes: *Australian Mining,* v. 68, no. 4, p. 13–16.

Press, N. P., 1974, Remote sensing to detect the toxic effects of metals on vegetation for mineral exploration: *in* Proceedings of the Ninth International Symposium on Remote Sensing of Environment, Environmental Research Institute of Michigan, Ann Arbor, Michigan, v. 3, p. 2027–2038.

Pressman, A. E., 1960, Photogeology speeds up groundwork for oil hunters: *The Oil and Gas Journal,* v. 58, no. 37, p. 162–166, and p. 168.

Price, J. C., 1977, Thermal inertia mapping: A new view of the Earth: *Journal of Geophysical Research,* v. 82, no. 18, p. 2582–2590.

Price, J. C., 1981, The heat capacity mapping mission—system characteristics, data products and interpretation (abs.): *in* Workshop on Geological Applications of Thermal Infrared Remote Sensing Techniques (Settle, Mark, editor), Technical Report No. 81-06, Lunar and Planetary Institute, Houston, Texas, p. 95.

Price, R. J., 1969a, The University of Glasgow Breidamerkurjökull Project (1964–67). A progress report: *Jökull,* v. 18 (1968), p. 389–394.

Price, R. J., 1969b, Moraines, sandar, kames, and eskers near Breidamerkurjökull, Iceland: Transactions, Institute of British Geographers, no. 46, p. 17–43.

Price, R. J., 1971, The development and destruction of a sandur, Breidamerkurjökull, Iceland: *Arctic and Alpine Research,* v. 3, no. 3, p. 225–237.

Price, R. J., 1973, Glacial and fluvioglacial landforms: Hafner Publishing Company, New York, 242 p.

Price, R. J., and Howarth, R. J., 1971, The evolution of the drainage system (1904–1965) in front of Breidamerkurjökull, Iceland: *Jökull,* v. 20 (1970), p. 27–37.

Prior, D. B., and Coleman, J. M., 1980, Sonograph mosaics of submarine slope instabilities, Mississippi River delta: *Marine Geology,* v. 36, no. 1-2, p. 227–239.

Quackenbush, R. S., Jr., editor, Lundahl, A. C., and Monsour, Edward, 1960, The development of photo interpretation: *in Manual of Photographic Interpretation* (Colwell, R. N., editor), American Society of Photogrammetry, Washington, D.C., p. 1–18.

Rabchevsky, G. A., 1979, Coal fires: *Earth Science,* Summer, p. 12–15.

Rabchevsky, G. A., 1972, Determination from available satellite and aircraft imagery of the applicability of remote sensing techniques to the detection of fires burning in abandoned coal mines and unmined coal deposits located in north central Wyoming and southern Montana: Allied Research Associates, Inc., Bureau of Mines Contract No. S0211087, Final Report No. 8G86-F, 99 p.

Rabchevsky, G. A., 1982, Coal fires in mines, refuse banks, and unmined deposits: *Earth Science,* Summer 1982, p. 16–19.

Ragland, T. M., and Chavez, Pat, Jr., 1977, The EROS digital image processing system (EDIPS): A complement to the NASA/GSFC master data processor (MDP): *in* Proceedings of the Second Annual William T. Pecora Memorial Symposium, "Mapping with Remote Sensing Data," American Society of Photogrammetry, Falls Church, Virginia, p. 47–63.

Raines, G. L., and Lee, Keenan, 1974, Spectral reflectance measurements: *Photogrammetric Engineering,* v. 40, no. 5, p. 547–550.

Raines, G. L., and Canney, F. C., 1980, Vegetation and geology: *in Remote Sensing in Geology,* (Siegal, B. S., and Gillespie, A. R., editors), John Wiley and Sons, New York, p. 365–380.

Raines, G. L., Offield, T. W., and Santos, E. S., 1978, Remote-sensing and subsurface definition of facies and structure related to uranium deposits, Powder River Basin, Wyoming: *Economic Geology,* v. 73, no. 8, p. 1706–1733.

Raines, G. L., Frisken, J. G., Kleinkopf, M. D., and de la Fuente-D., Mauricio, 1980, The discovery of the Alcaparroso mineral deposit, Sonora, Mexico: *in* Official Program of the Sixth Annual William T. Pecora Memorial Symposium, "Integration of Remote Sensing into the Exploration Process," April 1980, Sioux Falls, South Dakota, p. 67–69.

Rasmussen, L. A., and Meier, M. F., 1982, Continuity equation model of the drastic retreat of Columbia Glacier, Alaska: U.S. Geological Survey Professional Paper 1285-A, 23 p.

Ray, L. L., 1940, Glacial chronology of the southern Rocky Mountains: *Geological Society of America Bulletin,* v. 51, no. 2, p. 1851–1917.

Ray, R. G., 1960, Aerial photographs in geologic interpretation and mapping: U.S. Geological Survey Professional Paper 373, 230 p.

Reed, C. L., 1981, Interpretation of hydrographic features in the waters off Cape Cod: M.S. Dissertation, School of Natural Resources, University of Michigan, Ann Arbor, Michigan, May, 136 p.

Reeves, R. G., Kover, A. N., Lyon, R. J. P., and Smith, H. T. U., editors; Abdel-Gawad, Monem, Breiner, Sheldon, Dellwig, L. F., Everett, J. R., Friedman, J. D., Gawarecki, S. J., Gazley, Carl, Jr., Gonzalez-S., I. A., Hackman, R. J., Hemphill, W. R., Hunt, G. R., Kirk, J. N., Knepper, D. H., Jr., Lamar, J. V., Lathram, E. H., Lee, Keenan, Le Schack, L. A., Levin, S. B., Lewis, A. J., Luttrell, G. W., MacDonald, H. C., Marrs, R. W., Merifield, P. M., Orr, D. G., Pestrong, Raymond, Rumsey, I. A. P., Sabins, F. F., Jr., Salisbury, J. W., Smedes, H. W., Spencer, J. W., Jr., Stratton, R. H., Wing, R. S., and Webber, F. J., 1975a, Terrain and minerals: Assessment and evaluation: *in Manual of Remote Sensing* (Reeves, R. G., Anson, Abraham, and Landen, David, editors): v. 2 (Bowden, L. W., and Pruitt, E. L., editors), American Society of Photogrammetry, Falls Church, Virginia, p. 1107–1351.

Reeves, R. G., Anson, Abraham, and Landen, David, editors, 1975b, Manual of Remote Sensing; v. 1, Theory, Instruments, and Techniques (Janza, F. J., Blue, H. M., and Johnston, J. E., editors), p. 1–868; v. 2, Interpretation and Applications (Bowden, L. W., and Pruitt, E. L., editors), p. 869–2144: American Society of Photogrammetry, Falls Church, Virginia, 2144 p.

Regan, R. D., 1977, Regional and global geological studies using satellite magnetometer data: U.S. Geological Survey Professional Paper 1015 (Woll, P. W., and Fischer, W. A., editors), p. 229–230.

Regan, R. D., Cain, J. C., and Davis, W. M., 1975, A global magnetic anomaly map: *Journal of Geophysical Research,* v. 80, no. 5, p. 794–802.

Rehder, J. B., 1976, Changes in landscape due to strip mining: *in ERTS-1, A New Window on Our Planet* (Williams, R. S., Jr., and Carter, W. D., editors), U.S. Geological Survey Professional Paper 929, p. 254–257.

Reimnitz, Erk, and Barnes, P. W., 1974, Sea ice as a geologic agent on the Beaufort Sea shelf of Alaska: *in* Proceedings of the Arctic Institute of North America Symposium on Beaufort Sea Coastal and Shelf Research, San Francisco, California, p. 301–353.

Reimnitz, Erk, and Barnes, P. W., 1976, Influence of sea ice on sedimentary processes off northern Alaska: *in ERTS-1, A New Window On Our Planet* (Williams, R. S., Jr., and Carter, W. D., editors), U.S. Geological Survey Professional Paper 929, p. 360–362.

Rice, C. J., 1981, Satellite observations of the Mt. St. Helens eruption of 18 May 1980: Space Sciences Laboratory Report SSL-81 (6640)-1, Aerospace Corporation, Los Angeles, California, 20 p.

Richard, K., and Courtwright, J., 1966, Structure and mineralization at Silver Bell, Arizona: *in Geology of the Porphyry Copper Deposits* (Titley, S. R., and Hicks, C. L., editors), University of Arizona Press, Tucson, Arizona, p. 157–164.

Richard, R. R., Merkel, R. F., and Meeks, G. R., 1978, NS001 MS Landsat-D thematic mapper band aircraft scanner: *in* Proceedings of the 12th International Symposium on Remote Sensing of Environment, Environmental Research Institute of Michigan, Ann Arbor, Michigan, p. 719–728.

Richards, P. B., 1982, The utility of Landsat-D and other satellite image systems in disaster management: Naval Research Laboratory Final Report, NASA DPR S-79677, June, 45 p.

Richards, P. B., Robinove, C. J., Wiesnet, D. R., Salomonson, V. V., and Maxwell, M. S., 1983, Recommended satellite imagery capabilities for disaster management: *in* Proceedings of the Meeting of the International Astronautical Federation, Paris, France, September 1982, Paper IAF-82-103 (in press).

Richardson, A. J., Williams, A. A., and Thomas, J. R., 1969, Discrimination of vegetation by multispectral reflectance measurements: *in* Proceedings of the Sixth International Symposium on Remote Sensing of Environment, University of Michigan, Ann Arbor, Michigan, v. 2, p. 1143–1156.

Richardson, B. F., Jr., editor, 1978, Introduction to remote sensing of the environment: Kendall-Hunt Publishing Co., Dubuque, Iowa, 496 p.

Richason, Don, 1968, Glacier outburst floods in the Pacific Northwest: U.S. Geological Survey Professional Paper 600-D, p. D79–D86.

Richers, D. M., 1981, Geochemical survey of lower Pennsylvanian Corbin sandstone outcrop belt in eastern Kentucky: *in American Association of Petroleum Geologists Bulletin,* v. 65, no. 9, p. 1551–1567.

Richter, C. F., 1958, Elementary seismology: W. H. Freeman, San Francisco, 768 p.

Rist, Sigurjón, 1967, Jöklabreytingar 1964/65, 1965/66 og 1966/67: *Jökull,* v. 17, p. 321–325. (In Icelandic)

Rist, Sigurjón, 1967, Jökulhlaups from the ice cover of Mýrdalsjökull on June 25, 1955 and January 20, 1956: *Jökull,* v. 17, p. 243–248.

Rist, Sigurjón, 1974, Jökulhlaupaannáll 1971, 1972, og 1973: *Jökull,* v. 23 (1973), p. 55–60. (In Icelandic)

Rist, Sigurjón, 1977, Jöklabreytingar 1964/65–1973/74 (10 ár), 1974/75 og 1975/76: *Jökull,* v. 26 (1976), p. 69–74. (In Icelandic)

Ritter, D. F., 1978, Process geomorphology: W. C. Brown Co., Dubuque, Iowa, 603 p.

Rivereau, J. C., 1978, Morphostructural outline of the Antarctic: Inventory of lineaments and outcropping areas from Landsat 1 and 2 imagery: Rueil-Malmaison, Bureau d'Études Industrielles et de Coopération de l'Institut Français du Pétrole (BEICIP).

Robin, G. deQ., 1975, Radio-echosounding: Glaciological interpretations and applications: *Journal of Glaciology,* v. 15, no. 73, p. 49–63.

Robinove, C. J., 1963, Photography and imagery—a clarification of terms: *Photogrammetric Engineering,* v. 29, no. 5, p. 880–881.

Robinove, C. J., 1975, Worldwide disaster warning and assessment with Earth Resources Technology Satellites: U.S. Geological Survey Project Report (IR) NC-47, 65 p.

Robinson, B. F., and Biehl, L. L., 1979, Calibration procedures for measurement of reflectance factors in remote sensing field research: *SPIE* (Society of Photo-Optical Instrumentation Engineers), v. 196 (Measurements of Optical Radiation), p. 16–26.

Ronca, L. B., 1965, Selenology vs. geology of the Moon, etc.: *Geotimes,* v. 9, no. 9, p. 13.

Rose, A. W., Hawkes, H. E., and Webb, J. S., 1979, Geochemistry in mineral exploration: Second Edition, Academic Press, New York, 657 p.

Rosencrantz, D. M., 1975, Underwater photography and photogrammetry: *in Photography in Archaeological Research* (Harp, Elmer, Jr., editor) University of New Mexico Press, Albuquerque, New Mexico, p. 265–309.

Ross, D. C., 1969, Map showing recently active breaks along the San Andreas fault between Tejon Pass and Cajon Pass, Southern California: U.S. Geological Survey Miscellaneous Geologic Investigations Map I-553; scale, 1:24,000.

Ross, S. S., 1978, Remote sensing: How much security?: *New Engineer,* v. 7, no. 6, p. 6.

Rowan, L. C., and Lathram, E. H., 1980, Mineral exploration: *in Remote Sensing in Geology* (Siegal, B. S., and Gillespie, A. R., editors), John Wiley and Sons, New York, p. 553–605.

Rowan, L. C., and Kahle, A. B., 1982, Evaluation of 0.46- to 2.36-μm multispectral scanner images of the East Tintic mining district, Utah, for mapping hydrothermally altered rocks: *Economic Geology,* v. 77, no. 2, p. 441–452.

Rowan, L. C., Offield, T. W., Watson, Kenneth, Cannon, P. J., and Watson, R. D., 1970, Thermal infrared investigations, Arbuckle Mountains, Oklahoma: *Geological Society of America Bulletin,* v. 81, no. 12, p. 3549–3562.

Rowan, L. C., Wetlaufer, P. H., Goetz, A. F. H., Billingsley, F. C., and Stewart, J. H., 1974, Discrimination of rock types and detection of hydrothermally altered areas in south-central Nevada by the use of computer-enhanced ERTS images: U.S. Geological Survey Professional Paper 883, 35 p.

Rowan, L. C., Wetlaufer, P. H., and Goetz, A. F. H., 1976, Discrimination of rock types and detection of hydrothermally altered areas in south-central Nevada: *in ERTS-1, A New Window on Our Planet* (Williams, R. S., Jr., and Carter, W. D., editors), U.S. Geological Survey Professional Paper 929, p. 102–105.

Rowan, L. C., Goetz, A. F. H., and Ashley, R. P., 1977, Discrimination of hydrothermally altered and unaltered rocks in visible and near-infrared multispectral images: *Geophysics,* v. 42, no. 3, p. 522–535.

Rowan, L. C., Goetz, A. F. H., and Kingston, M. J., 1983, Preliminary analysis of shuttle multispectral radiometer data in southern Egypt: Remote Sensing and Mineral Exploration-1982: *in* COSPAR Advances in Space Research, Pergamon Press, New York (in press).

Rudd, R. D., 1974, Remote sensing: A better view: Duxbury Press, Boston, 135 p.

Rusby, J. S. M., and Somers, M. L., 1977, The development of the GLORIA sonar system from 1970 to 1975: *in Voyage of Discovery* (Angel, M. V., editor): Supplement to *Deep-Sea Research,* v. 24, no. 7, p. 611–625.

Rydstom, H. O., 1970, Geologic exploration with high resolution radar: Report No. GIB-9193A, Code 99696, Goodyear Aerospace Corporation, Litchfield Park, Arizona, July, 48 p.

Sabins, F. F., Jr., 1978a, Remote sensing, principles and techniques: W. H. Freeman and Company, San Francisco, 426 p.

Sabins, F. F., Jr., 1978b, Oil exploration: *in Remote Sensing-Principles and Interpretation,* W. H. Freeman and Company, San Francisco, p. 300–313.

Sabins, F. F., Jr., 1980, Interpretation of infrared images: *in Remote Sensing in Geology* (Siegal, B. S., and Gillespie, A. R., editors), John Wiley and Sons, New York, p. 275–295.

Sabins, F. F., Jr., 1981, Remote sensing laboratory manual: One of three sections. The other two sections are: Instructor Guide and a 35 mm slide set; Remote Sensing Enterprises, La Habra, California; looseleaf notebook variously paginated by sections.

Saemundsson, Kristján, 1969, Infrared imagery of Torfajökull thermal area: National Energy Authority, Department of Natural Heat, Report, January 1969, Reykjavik, 22 p.

Saemundsson, Kristján, 1974, Evolution of the axial rifting zone in northern Iceland and the Tjörnes Fracture Zone: *Geological Society of America Bulletin,* v. 85, no. 4, p. 495–504.

Sagan, Carl, 1975, The solar system: *Scientific American,* v. 233, no. 3, p. 22–31.

Salas, G. P., 1975, Relationship of mineral resources to linear features in Mexico as determined from Landsat data: *in* Proceedings of the First Annual William T. Pecora Memorial Symposium, U.S. Geological Survey Professional Paper 1015 (Woll, P. W., and Fischer, W. A., editors), p. 61–82.

Salomonson, V. V., and Rango, Albert, 1980, Water resources: *in Remote Sensing in Geology* (Siegal, B. S., and Gillespie, A. R., editors), John Wiley and Sons, New York, p. 607–633.

Santeford, H. S., and Smith, J. L., compilers, 1974, Advanced concepts and techniques in the study of snow and ice resources: Proceedings of an Interdisciplinary Symposium, Work Group on Snow and Ice, Work Group on Remote Sensing, Work Group on Nuclear Techniques of the U.S. National Committee for the International Hydrological Decade, National Academy of Sciences, Washington, D.C., 789 p.

Saunders, D. F., and Hicks, D. E., 1976, Regional geomorphic lineaments on satellite imagery—their origin and applications: *in* Proceedings of the Second International Conference on Basement Tectonics (Newark, Delaware), p. 326–352.

Saunders, D. F., Thomas, G. E., Kinsman, F. E., and Beatty, D. F., 1973, ERTS-1 imagery use in reconnaissance prospecting—evaluation of the commercial utility of ERTS-1 imagery in structural reconnaissance for minerals and petroleum: Type III Final Report to NASA, U.S. Department of Commerce, National Technical Information Service, Accession No. 174-10345, 4 p.

Saunders, R. S., and Mutch, T. A., 1980, Extraterrestrial geology: *in Remote Sensing in Geology* (Siegal, B. S., and Gillespie, A. R., editors), John Wiley and Sons, New York, p. 659–677.

Scanlon, K. M., 1980, Morphology of the continental slope off New England from GLORIA II long-range side-scan sonographs (abs.): *in* Abstracts with Programs, Geological Society of America, Boulder, Colorado, v. 12, no. 7, p. 515.

Scanlon, K. M., 1982, Geomorphic features of the western North Atlantic Continental Slope between Northeast Channel and Alvin Canyon as interpreted from GLORIA long-range sidescan-sonar data: U.S. Geological Survey Open-File Report 82-728, 8 p. and one map sheet.

Schaber, G. G., 1980, The surface of Io: Geologic units, morphology, and tectonics: *Icarus,* v. 43, no. 3, p. 302–333.

Scheidegger, A. E., 1970, Theoretical geomorphology: Second Edition, Springer-Verlag, Berlin, 435 p.

Schmidt, R. G., 1973, Use of ERTS-1 images in the search for porphyry copper deposits in Pakistani Baluchistan: *in* Proceedings of the NASA Goddard Space Flight Center Symposium on Significant Results Obtained from Earth Resources Technology Satellite-1, New Carrolton, Maryland, March 1973, NASA SP-327, v. 1, sec. A, p. 387–394.

Schmidt, R. G., 1976, Detection of hydrothermal deposits, Saindak area, western Pakistan: *in ERTS-1, A New Window on Our Planet* (Williams, R. S., Jr., and Carter, W. D., editors) U.S. Geological Survey Professional Paper 929, p. 89–91.

Schmugge, Thomas, 1980, Techniques and applications of microwave radiometry: *in Remote Sensing in Geology* (Siegal, B. S., and Gillespie, A. R., editors): John Wiley and Sons, New York, New York, p. 337–361.

Schultz, J. R., and Cleaves, A. B., 1955, Aerial photographic interpretation of soils: *in Geology in Engineering,* John Wiley and Sons, Inc., New York, p. 370–388.

Schwab, W. C., and Frohlich, R. K., 1976, The structural interpretation of aeromagnetic lineaments in northern Rhode Island: *in* Proceedings of the Second International Conference on Basement Tectonics (Newark, Delaware), p. 86–98.

Schwaller, M. R., and Tkach, S. J., 1980, Premature leaf senescence as an indicator for geobotanical prospection with remote sensing techniques: *in* Proceedings of the Fourteenth International Symposium on Remote Sensing of Environment, Environmental Research Institute of Michigan, Ann Arbor, Michigan, v. 1, p. 347–358.

Scott, D. H., and Carr, M. H., 1978, Geologic map of Mars: U.S. Geological Survey Map I-1083 (Scale, 1:25,000,000).

Sears, J. D., Hunt, C. B., and Hendricks, T. A., 1941, Transgressive and regressive Cretaceous deposits in southern San Juan Basin, New Mexico: U.S. Geological Survey Professional Paper 193-F, p. 110–119.

Segal, D. B., 1982, Theoretical basis for differentiation of ferric-iron bearing minerals using Landsat MSS data: *in* Summaries of the Second Thematic Conference: Symposium on Remote Sensing for Exploration Geology (Dallas, Texas; December 1982), 16th International Symposium on Remote Sensing of the Environment, Environmental Research Institute of Michigan, Ann Arbor, Michigan, p. 177–179.

Segal, D. B., and Podwysocki, M. H., 1980, A technique using diazo film for isolating areas of spectral similarity: U.S. Geological Survey Open-File Report 80-1267, 15 p.

Seidman, J. B., 1972, Some practical applications of digital filtering in image processing: *in* Proceedings of the Computer Image Processing and Recognition Symposium (University of Missouri, Columbia, Missouri, 24–26 August 1972), p. 9-1-1–9-1-12.

Sellmann, P. V., Arcone, S. A., and Delaney, A. J., 1976, Airborne resistivity and magnetometer survey of northern Maine for obtaining information on bedrock geology: U.S. Army Corps of Engineers, Cold Regions Research and Engineering Laboratory, CRREL Report 76-37, 19 p.

Settle, Mark, editor, 1981, Workshop on geological applications of thermal infrared remote sensing techniques: Lunar and Planetary Institute Technical Report 81-06, Houston, Texas, 138 p.

Settle, Mark, and Taranik, J. V., 1982a, Application of space technology to global potential field mapping (abs.): *in Abstracts* of the 24th Plenary Meeting of Committee on Space Research, Symposium on Remote Sensing and Mineral Exploration (17–20 May, 1982), Ottawa, Canada, p. 86–87.

Settle, Mark, and Taranik, J. V., 1982b, Use of the Space Shuttle for remote sensing research: Recent results and future prospects: *Science,* v. 218, no. 4576, p. 993–995.

Settle, Mark, and Taranik, J. V., 1983, Mapping the Earth's magnetic and gravity fields from space: Current status and future prospects: *in* COSPAR, Advances in Space Exploration; Remote Sensing and Mineral Exploration-1982, Pergamon Press, Oxford, in press.

Sharp, R. V., 1975, Displacement on tectonic ruptures: *in* San Fernando, California Earthquake of 9 February 1971; *California Division of Mines and Geology Bulletin,* no. 196, p. 187–194.

Sharp, R. V., 1982, Comparison of 1979 surface faulting with earlier displacements in the Imperial Valley: *in The Imperial Valley, California, Earthquake of October 15, 1979:* U.S. Geological Survey Professional Paper 1254, p. 213–221.

Sharp, R. V., Lienkaemper, J. J., Bonilla, M. G., Burke, D. B., Cox, B. F., Herd, D. G., Miller, D. M., Morton, D. M., Ponti, D. J., Rymer, M. J., Tinsley, J. C., Yount, J. C., Kahle, J. E., Hart, E. W., and Sieh, K. E., 1982, Surface faulting in the central Imperial Valley: *in The Imperial Valley, California, Earthquake of October 15, 1979:* U.S. Geological Survey Professional Paper 1254, p. 119–143.

Shcheglov, A. D., editor, 1979, Cosmogeological map of linear and circular structures of the USSR territory: Academy of Sciences, USSR, Moscow, scale 1:5,000,000. In 4 sheets with explanation in Russian and English.

Sheets, P. D., and Grayson, D. K., 1979, Volcanic activity and human ecology: Academic Press, New York, 644 p.

Sherard, J. L., Cluff, L. S., and Allen, C. R., 1974, Potentially active faults in dam foundations: *Geotechnique,* v. 24, no. 3, p. 367–428.

Shih, E. H. H., and Schowengerdt, R. A., 1983, Classification of arid geomorphic surfaces using Landsat spectral and textural features: *Photogrammetric Engineering and Remote Sensing,* v. 49, no. 3, p. 337–347.

Shoemaker, E. M., and Hackman, R. J., 1962, Stratigraphic basis for a lunar time scale: *in* Proceedings of the International Astronomical Union Symposium 14 (Kopal, Zdenek, and Mikhailov, S. K., editors), Academic Press, New York, p. 289–300.

Short, N. M., 1982, The Landsat tutorial workbook; basics of satellite remote sensing: NASA Reference Publication No. 1078, 553 p.

Short, N. M., Lowman, P. D., Jr., Freden, S. C., and Finch, W. A., 1976, Mission to Earth: Landsat views the world: NASA Special Publication, SP-360, National Aeronautics and Space Administration, Washington, D.C., 459 p.

Siegal, B. S., and Gillespie, A. R., editors, 1980, Remote sensing in geology: John Wiley and Sons, New York, 702 p.

Sigurgeirsson, Thorbjörn, 1974, Hraunkaeling: *Timinn,* 19 January 1974, p. 8–9 and p. 13. (In Icelandic)

Sikonia, W. G., 1982, Finite-element glacier dynamics model applied to Columbia Glacier, Alaska: U.S. Geological Survey Professional Paper 1285-B, 74 p.

Sikonia, W. G., and Post, Austin, 1979, Columbia Glacier, Alaska: Recent ice loss and its relationship to seasonal terminal embayments, thinning, and glacier flow: U.S. Geological Survey Hydrologic Investigations Atlas, HA-619, 3 sheets.

Silver, L. T., Anderson, T. H., Conway, C. M., Murray, J. D., and Powell, R. E., 1977, Geological features of southwestern North America: *in Skylab Explores the Earth,* NASA SP-380, p. 89–135.

Simard, Réjean, 1982, Resultats de simulations d'images stereoscopiques HRV SPOT sur le site de Gun Lake, C.B.: *in* Proceedings of the Seventh Canadian Symposium on Remote Sensing, Winnipeg, Manitoba, Canada (8–11 September 1981), p. 541–551. (In French)

Simkin, Tom, and Krueger, A. F., 1977, Summit eruption of Fernandina Caldera, Galápagos Islands,

Ecuador: *in Skylab Explores the Earth,* NASA Special Publication SP-380, National Aeronautics and Space Administration, Washington, D.C., p. 171–173.

Simons, J. H., and Beccasio, A. D., 1964, An evaluation of geoscience applications of side-looking airborne mapping radar: The Raytheon Company, Autometric Facility, Alexandria, Virginia, April, 70 p.

Sinclair, A. J., 1976, Applications of probability graphs in mineral exploration: Association of Exploration Geochemists, Special Volume No. 4, 95 p.

Singer, D. A., and Mosier, D. L., 1981, A review of regional mineral resource assessment methods: *Economic Geology,* v. 76, no. 5, p. 1006–1015.

Skaley, J. E., 1980, Photooptical techniques of image enhancement: *in Remote Sensing in Geology* (Siegal, B. S., and Gillespie, A. R., editors), John Wiley and Sons, New York, p. 119–138.

Skolnik, M. I., 1962, Introduction to radar systems: McGraw-Hill Book Company, New York, 648 p.

Slama, C. C., Theurer, Charles, and Henriksen, S. W., editors, 1980, Manual of photogrammetry: Fourth Edition, American Society of Photogrammetry, Falls Church, Virginia, 1056 p.

Slaney, V. R., 1982, Landsat images of Canada—a geological appraisal: Geological Survey Paper 80-15, Geological Survey of Canada, Ottawa, 102 p.

Slater, P. N., 1975, Photographic systems for remote sensing: *in Manual of Remote Sensing* (Reeves, R. G., Anson, Abraham, and Landen, David, eds.): v. 1 (Janza, F. J., Blue, H. M., and Johnston, J. E., editors), American Society of Photogrammetry, Falls Church, Virginia, p. 235–323.

Slater, P. N., 1980, Remote sensing, optics and optical systems: Addison-Wesley Publishing Company, Reading, Massachusetts, 575 p.

Slemmons, D. B., 1969, New methods of studying seismicity and surface faulting: *EOS* (Transactions, American Geophysical Union), v. 50, no. 4, p. 397–398.

Slemmons, D. B., 1977, Faults and earthquake magnitude: Miscellaneous Paper S-73-1, Report 6, U.S. Army Engineer Waterways Experiment Station, Vicksburg, Mississippi, 129 p.

Smith, B. A., Soderblom, L. A., Johnson, T. V., Ingersoll, A. P., Collins, S. A., Shoemaker, E. M., Hunt, G. E., Masursky, Harold, Carr, M. H., Davies, M. E., Cook, A. F., II, Boyce, J. M., Danielson, G. E., Owen, Tobias, Sagan, Carl, Beebe, R. F., Veverka, Joseph, Strom, R. G., McCauley, J. F., Morrison, D., Briggs, G. A., and Suomi, V. E., 1979a, The Jupiter system through the eyes of Voyager 1: *Science,* v. 204, no. 4396, p. 951–972.

Smith, B. A., Soderblom, L. A., Beebe, Reta, Boyce, J. M., Briggs, G. A., Carr, M. H., Collins, S. A., Cook, A. F., II, Danielson, G. E., Davies, M. E., Hunt, G. E., Ingersoll, A. P., Johnson, T. V., Masursky, Harold, McCauley, J. F., Morrison, David, Owen, Tobias, Sagan, Carl, Shoemaker, E. M., Strom, R. G., Suomi, V. E., and Veverka, Joseph, 1979b, The Galilean satellites and Jupiter: Voyager 2 imaging science results: *Science,* v. 206, no. 4421, p. 927–950.

Smith, B. A., Soderblom, L. A., Beebe, Reta, Boyce, J. M., Briggs, G. A., Bunker, Anne, Collins, S. A., Hansen, C. J., Johnson, T. V., Mitchell, J. L., Terrile, R. J., Carr, M. H., Cook, A. F., II, Cuzzi, J., Pollack, J. B., Danielson, G. E., Ingersoll, A. P., Davies, M. E., Hunt, G. E., Masursky, Harold, Shoemaker, E. M., Morrison, David, Owen, Tobias, Sagan, Carl, Veverka, Joseph, Strom, R. G., and Suomi, V. E., 1981, Encounter with Saturn: Voyager 1 imaging science results: *Science,* v. 212, no. 4491, p. 163–191.

Smith, B. A., Soderblom, L. A., Batson, R. M., Bridges, P. M., Inge, J. L., Masursky, Harold, Shoemaker, E. M., Beebe, Reta, Boyce, J. M., Briggs, G. A., Bunker, Anne, Collins, S. A., Hansen, C. J., Johnson, T. V., Mitchell, J. L., Terrile, R. J., Cook, A. F., II, Cuzzi, Jeffrey, Pollack, J. B., Danielson, G. E., Ingersoll, A. P., Davies, M. E., Hunt, G. E., Morrison, David, Owen, Tobias, Sagan, Carl, Veverka, Joseph, Strom, R. G., and Suomi, V. E., 1982, A new look at the Saturn system: The Voyager 2 images: *Science,* v. 215, no. 4532, p. 504–537.

Smith, H. T. U., 1943, Aerial photographs and their applications: D. Appleton-Century Co., Inc., New York, 372 p.

Smith, H. T. U., 1967, Photogeologic interpretation in Antarctica: *Photogrammetric Engineering,* v. 33, no. 3, p. 297–304.

Smith, J. T., Jr., and Anson, Abraham, 1968, Manual of color aerial photography: American Society of Photogrammetry, Falls Church, Virginia, 550 p.

Smith, N. C., and Renfro, K. M., 1955, Photogeology: Matured exploration tool: *The Petroleum Engineer,* v. 27, no. 10, p. B-94–B-100.

Smith, W. L., editor, 1977, Remote-sensing applications for mineral exploration: Dowden, Hutchinson, and Ross, Stroudsburg, Pennsylvania, 391 p.

Smith, W. L., Rao, P. K., Koffler, Russell, and Curtis, W. R., 1970, The determination of sea-surface temperatures from satellite high resolution infrared window radiation measurements: *Monthly Weather Review,* v. 98, no. 8, p. 604–611.

Sobieralski, V. R., 1968, Volcanic activity (Fig. 10.18): *in Manual of Color Aerial Photography,* American Society of Photogrammetry, Falls Church, Virginia, p. 416–417.

Soderblom, L. A., 1977, Historical variation in the density and distributions of impacting debris in the inner solar system: Evidence from planetary imaging: *in Impact and Explosion Cratering* (Roddy, D. J., Pepin, R. O., and Merrill, R. B., editors), Pergamon Press, New York, p. 629–633.

Soha, J. M., and Schwartz, A. A., 1978, Multispectral histogram normalization contrast enhancement: *in* Proceedings of the 5th Canadian Symposium on Remote Sensing, Victoria, British Columbia, p. 86–93.

Soha, J. M., Gillespie, A. R., Abrams, M. J., and Madura, D. P., 1976, Computer techniques for geological applications: *in* Proceedings of the Caltech/JPL Conference on Image Processing Technology, Data Sources and Software for Commercial and Scientific Applications, California Institute of Technology, Pasadena, California (3–5 November 1976), JPL SP 43-30, p. 4-1–4-21.

Sonder, R. A., 1938, Die lineamenttektonik und ihre problems: *Eclogae Geol. Helvetiae,* v. 31, no. 1, p. 199–238. (In German)

Sonder, R. A., 1947, Discussion of "shear patterns of the Earth's crust": by F. A. Vening Meinesz: *Transactions, American Geophysical Union,* v. 28, no. 6, p. 939–945.

Sonder, R. A., 1956, Mechanik der Erde, Stuttgart, 291 p. (In German).

Staelin, D. H., and Rosenkranz, P. W., 1978, High resolution passive microwave satellites: Applications Review Panel Report, Final Report (14 April 1978),

Research Laboratory of Electronics, Massachusetts Institute of Technology, Cambridge, Massachusetts, variously paginated by sections.

Stafford, Lannon, 1969, Imaging radars for Earth resources: Technical Report LEC/HASD No. 649D-21-006, National Aeronautics and Space Administration, Manned Spacecraft Center, Houston, Texas, August, 31 p.

Stephens, R. W. B., 1970, The sea as an acoustic medium, *in Underwater Acoustics* (Stephens, R. W. B., editor): Wiley Interscience, London, p. 1–21.

Stingelin, R. W., Knuth, W. M., and Stamm, E. C., 1971, An airborne infrared and ground control analysis of ten selected coal refuse banks in the Pennsylvania Anthracite Region: U.S. Bureau of Mines, Washington, D.C., HRB-Singer, Inc. report No. 4480-F, Bureau of Mines Contract No. S0101825, 71 p.

Strahler, A. N., 1963, The earth sciences: Harper, New York, 681 p.

Strahler, A. N., 1969, Physical geography: Third Edition, John Wiley and Sons, New York, 733 p.

Stringham, J. A., and Williams, R. S., Jr., 1970, Applications of reconnaissance concepts to mapping problems: *in Proceedings of the Geodetic and Research and Development Symposium,* Seventh DOD Geodetic-Cartographic-Target Materials Conference, Cameron Station, Virginia, p. 37–105.

Strong, A. E., 1976, Algal blooms in Utah lake: *in ERTS-1, A New Window on Our Planet* (Williams, R. S., Jr., and Carter, W. D., editors), U.S. Geological Survey Professional Paper 929, p. 270–272.

Sugden, D. E., and John, B. S., 1976, Glaciers and landscape, a geomorphological approach: John Wiley and Sons, New York, 376 p.

Svensson, Harald, 1962, Glacier movement as revealed by aerial photographs: *Photogrammetria,* v. 18, no. 4, p. 140–147.

Svensson, Harald, 1963a, Frozen ground morphology of northeastern-most Norway: *in* Proceedings of the Helsinki Symposium on the Ecology of Subarctic Regions 1966, p. 161–168.

Svensson, Harald, 1963b, Some observations in West-Jutland of a polygonal pattern in the ground: *Dansk Geografisk Tidsskrift,* v. 62, no. 2, p. 122–124.

Svensson, Harald, 1964, Aerial photographs for tracing and investigating fossil tundra ground in Scandinavia: *Biuletyn Peryglacjalny,* v. 14, p. 321–325.

Svensson, Harald, 1967, Studies of a ground pattern: *Geografiska Annaler,* v. 49 (Series A), no. 2-4, p. 344–350.

Svensson, Harald, 1970, Spår av "fluted surfaces" från den senaste nedisningen: *Geologiska Föreningen* i Stockholm Förhandlingar, v. 92, p. 79–85. (In Swedish)

Svensson, Harald, 1972, The use of stress situations in vegetation for detecting ground conditions on aerial photographs: *Photogrammetria,* v. 28, no. 3, p. 75–87.

Svensson, Harald, 1973, Distribution and chronology of relict polygon patterns on Laholm plain, the Swedish west coast: *Geografiska Annaler,* v. 55 (Series A), no. 3-4, p. 159–175.

Svensson, Harald, 1978, Ice wedges as a geomorphological indicator of climatic changes: *in* Proceedings of the Nordic Symposium on Climatic Changes and Related Problems, Danish Meteorological Institute, Climatological Papers 4, p. 9–17.

Swain, P. H., and Davis, S. M., editors, 1978, Remote sensing: The quantitative approach: McGraw-Hill, New York, 396 p.

Swanson, Eric, 1978, Remote sensing assisted field studies in the Sierra Madre Occidental, Mexico (abs.): *in* Abstracts with Programs, Annual Meeting of the Cordilleran Section of the Geological Society of America, Boulder, Colorado, v. 10, no. 3, p. 149.

Swithinbank, Charles, 1980, The problem of a glacier inventory of Antarctica: *in* Proceedings of the Riederalp (Switzerland) Workshop on World Glacier Inventory, International Association of Hydrological Sciences Publication No. 126, p. 229–236.

Swithinbank, Charles, and Land, Charles, 1977, Antarctic mapping from satellite imagery: *in* Proceedings of the 28th Symposium of the Colston Research Society, Remote Sensing of the Terrestrial Environment (Peel, R. E., Curtis, L. F., and Barrett, E. C., editors), University of Bristol, Butterworth Publishing Co., London, p. 212–221.

Swithinbank, Charles, Doake, Christopher, Wager, Andrew, and Crabtree, Richard, 1976, Major changes in the map of Antarctica: *in* Notes. *Polar Record,* v. 18, no. 114, p. 295–299.

Switzer, Paul, Kowalik, W. S., and Lyon, R. J. P., 1981, Estimation of atmospheric path-radiance by the covariance matrix method: *Photogrammetric Engineering and Remote Sensing,* v. 47, no. 10, p. 1469–1476.

Tanner, W. F., 1978, Standards for measuring shoreline changes: A study of the precision obtainable and needed in making measurements of changes: Department of Geology, Florida State University, Tallahassee, Florida, 88 p.

Tapponnier, Paul, and Molnar, Peter, 1977, Active faulting and Cenozoic tectonics in China: *Journal of Geophysical Research,* v. 82, no. 20, p. 2905–2930.

Taranik, J. V., 1978a, Characteristics of the Landsat multispectral data system: U.S. Geological Survey Open-File Report No. 78-187, 114 p.

Taranik, J. V., 1978b, Principles of computer processing of Landsat data for geological applications: U.S. Geological Survey Open-File Report No. 78-117, 98 p.

Taranik, J. V., 1981, Heat capacity mapping mission: Interim results and achievements: *in* Workshop on Geological Applications of Thermal Infrared Remote Sensing Techniques (Settle, Mark, editor), Technical Report No. 81-06, Lunar and Planetary Institute, Houston, Texas, p. 23–26.

Taranik, J. V., and Trautwein, C. M., 1976, Integration of geological remote sensing techniques in subsurface analysis: U.S. Geological Survey Open-File Report No. 76-402, 80 p.

Taranik, J. V., and Settle, M., 1981, Space Shuttle: A new era in terrestrial remote sensing: *Science,* v. 214, no. 4521, p. 619–626.

Taranik, J. V., Sheehan, C. A., Carter, W. D., and Reynolds, C. D., 1978, Targeting exploration for nickel laterites in Indonesia with Landsat data: *in* Proceedings of the Twelth International Symposium on Remote Sensing of Environment, Environmental Research Institute of Michigan, Ann Arbor, Michigan, v. 2, p. 1037–1051.

Tator, B. A., 1951, Some applications of aerial photographs to geographic studies in the Gulf Coast region: *Photogrammetric Engineering,* v. 17, no. 5, p. 716–725.

Tator, B. A., editor; Belcher, D. J., Blanchet, P. H., Desjardins, Louis, Fischer, W. A., Gould, D. B.,

Lattman, L. H., McAdams, K. A., Melton, F. A., Olive, W. W., Rich, J. L., Smith, N. C., Van Lopik, J. R., Vann, J. H., and Whitworth, V. L., 1960, Photo interpretation in geology: *in Manual of Photographic Interpretation* (Colwell, R. N., editor): American Society of Photogrammetry, Washington, D.C., p. 169–342.

Taylor, C. E., 1979, Appraisal of photographic and cartographic records maintained by the USGS Library, Denver, Colorado: Special Report to the U.S. Geological Survey, U.S. Geological Survey Library, Reston, Virginia, 15 p.

Taylor, C. E., and Spurr, R. W., 1973, Aerial photographs in the National Archives: Special List No. 25, National Archives and Records Service, General Services Administration, Washington, D.C., 106 p.

Taylor, J. I., and Stingelin, R. W., 1969, Infrared imaging for water resources studies: *Journal of the Hydraulics Division,* American Society of Civil Engineers, v. 95, no. HY1, Proceedings Paper 6331, January, p. 175–189.

Taylor, M. M., 1974, Principal components colour display of ERTS imagery: *in* National Aeronautics and Space Administration Special Publication SP-351 (Third Earth Resources Technology Satellite-1 Symposium; 10–14 December 1973; Washington, D.C.), v. 1 (Technical Presentations), Section B, p. 1877–1897.

Tazieff, Haroun, 1976–77, An exceptional eruption: Mt. Niragongo, January 10th, 1977: *Bulletin Volcanologique,* v. 40, no. 3, p. 189–200.

Tazieff, Haroun, 1979, Nyiragongo (translated from the French): Barron's-Woodbury, New York, 287 p.

Tchalenko, J. S., and Ambraseys, N. N., 1970, Structural analysis of the Dasht-e Bayaz (Iran) earthquake fractures: *Geological Society of America Bulletin,* v. 81, no. 1, p. 41–60.

Teleki, P. G., Roberts, D. G., Chavez, P. S., Somers, M. L., and Twichell, D. C., 1981, Sonar survey of the U.S. Atlantic Continental Slope; acoustic characteristics and image processing techniques: *in* Proceedings—Offshore Technology Conference, no. 13, v. 2, p. 91–102.

Tewinkel, G. C., 1963, Water depths from aerial photographs: *Photogrammetric Engineering,* v. 29, no. 6, p. 1037–1042.

Thompson, A. M., and Baker, R. N., 1982, Integrated geologic and remote sensing mineral exploration in Baja, California: *in Energy Resources of the Pacific Region* (Halbouty, M. T., editor), American Association of Petroleum Geologists, Studies in Geology, no. 12, p. 13–19.

Thorarinsson, Sigurdur, 1940, The icedammed lakes of Iceland with particular reference to their values as indicators of glacier oscillations: *Geografiska Annaler,* v. 21, no. 3-4, p. 216–242.

Thorarinsson, Sigurdur, 1943, Oscillations of the Iceland glaciers in the last 250 years; Chapter 11 *in* Vatnajökull. Scientific results of the Swedish-Icelandic investigations 1936–37–38: *Geografiska Annaler,* v. 25, no. 1-2, p. 1–54.

Thorarinsson, Sigurdur, 1958a, The Öraefajökull eruption of 1362: *Acta Naturalia Islandica,* v. 2, no. 2, 100 p.

Thorarinsson, Sigurdur, 1958b, Flatarmál nokkurra íslenzkra jökla samkvaemt herforingjarádskortunum: *Jökull,* v. 8, p. 25. (In Icelandic)

Thorarinsson, Sigurdur, 1964, Sudden advance of Vatnajökull's outlet glaciers 1930–1964: *Jökull,* v. 14, p. 76–89.

Thorarinsson, Sigurdur, 1965, Sitt af hverju um Surtseyjargosid: *Náttúrufraedingurinn,* v. 35, no. 4, p. 153–181. (In Icelandic)

Thorarinsson, Sigurdur, 1967, Surtsey. The new island in the North Atlantic: The Viking Press, New York, 47 p. (with 54 illustrations).

Thorarinsson, Sigurdur, 1968, Skaftáreldar og Lakagígar: *Náttúrufraedingurinn,* v. 37 (1967), no. 1-2, p. 27–57. (In Icelandic)

Thorarinsson, Sigurdur, 1969a, Glacier surges in Iceland with special reference to the surges of Brúarjökull: *Canadian Journal of Earth Sciences,* v. 6, no. 4, pt. 2, p. 875–882.

Thorarinsson, Sigurdur, 1969b, Sidusta thaettir Eyjaelda: *Náttúrufraedingurinn,* v. 38, no. 3-4, p. 113–212. (In Icelandic)

Thorarinsson, Sigurdur, 1970, The Lakagígar eruption of 1783: *Bulletin Volcanologique,* v. 33, no. 3, p. 910–929.

Thorarinsson, Sigurdur, 1974, On the topography of the volcanic zones in Iceland (abs.): *in Geodynamics of Iceland and the North Atlantic Area* (Kristjánsson, Leó, editor), D. Reidel Publishing Company, Boston, p. 203–205.

Thorarinsson, Sigurdur, 1977, At leve på en Vulkan: *Geografisk Tidsskrift,* v. 76, p. 1–13. (In Danish)

Thorarinsson, Sigurdur, Steinthórsson, Sigurdur, Einarsson, Thorleifur, Kristmannsdóttir, Hrefna, and Óskarsson, Níels, 1973, The eruption on Heimaey, Iceland: *Nature,* v. 241, no. 5389, p. 372–375.

Thorarinsson, Sigurdur, Saemundsson, Kristján, and Williams, R. S., Jr., 1974, ERTS-1 image of Vatnajökull: Analysis of glaciological, structural, and volcanic features: *Jökull,* v. 23 (1973), p. 7–17.

Thornbury, W. D., 1969, Principles of geomorphology: John Wiley and Sons, New York, 594 p.

Thoroddsen, Thorvaldur, 1892, Islands Jøkler i Fortid og Nutid: *Geografisk Tidskrift,* v. 11 (1891–92), no. 5-6, p. 111–146. (In Danish)

Thoroddsen, Thorvaldur, 1901, Geological map of Iceland: Surveyed in the years 1881–1898, Edited by the Carlsberg Fund; Scale, 1:600,000, 2 sheets.

Thoroddsen, Thorvaldur, 1906, Die Gletscher Islands: *in* Island, *Grundriss der Geographie und Geologie, Petermanns Geographische Mitteilungen,* Gotha, Erganzungshefts 152 und 153, pt. 5, p. 163–208. (In German)

Threet, R. L., 1981, Stereographic prediction of grazing solar illumination: *Photogrammetric Engineering and Remote Sensing,* v. 47, no. 3, p. 365–368.

Tilling, R. I., 1982, Volcanic cloud may alter Earth's climate (Mexican volcano spews death and destruction): *National Geographic,* v. 162, no. 5, p. 672–675.

Tisdall, Tracey, and El-Baz, Farouk, 1979, Analysis of water color as seen in orbital and aerial photographs of Cape Cod, Nantucket, and Martha's Vineyard, Massachusetts: *in* Apollo-Soyuz Test Project: Summary Science Report; Earth Observations and Photography (v. 2; El-Baz, Farouk, and Warner, D. M., editors), NASA SP-412, p. 455–480.

Tómasson, Haukur, 1975, Grímsvatnahlaup, 1972 mechanism and sediment discharge: *Jökull,* v. 24 (1974), p. 27–39.

Townsend, W. F., 1980, An initial assessment of the performance achieved by the Seasat-1 radar altimeter: *IEEE Journal of Oceanic Engineering OE-5,* No. 2, p. 80–92.

Trask, N. J., and Guest, J. E., 1975, Preliminary geologic terrain map of Mercury: *Journal of Geophysical Research,* v. 80, no. 17, p. 2461–2477.

Trollinger, W. V., 1968, Surface evidence of deep structure in the Delaware basin: *in Delaware Basin Exploration Guidebook;* West Texas Geological Society Publication 68-55, p. 87–104.

Tucker, C. J., 1979, Red and photographic infrared linear combinations for monitoring vegetation: *Remote Sensing of Environment,* v. 8, no. 2, p. 127–150.

Tuddenham, W. G., 1968, Test photography with MS Ektachrome for water penetration and depiction of the offshore sea floor: University of Sydney, Australia, 38 p.

Turcotte, D. L., 1974, Are transform faults thermal contraction cracks? *Journal of Geophysical Research,* v. 79, no. 17, p. 2573–2577.

Turner, F. J., and Weiss, L. E., 1963, *Structural analysis of metamorphic tectonites:* McGraw-Hill Book Co., New York, 545 p.

Turner, R. E., 1975, Signature variations due to atmospheric effects: *in* Proceedings of the 10th International Symposium on Remote Sensing of Environment, Environmental Research Institute of Michigan, Ann Arbor, Michigan, v. 2, p. 671–682.

Turner, R. E., 1979, Elimination of atmospheric effects from remote sensor data: *in* Proceedings of the 12th International Symposium on Remote Sensing of Environment, Environmental Research Institute of Michigan, Ann Arbor, Michigan, v. 2, p. 783–793.

Turner, R. E., Malila, W. A., and Nalepka, R. F., 1971, Importance of atmospheric scattering in remote sensing: *in* Proceedings of the 7th International Symposium on Remote Sensing of Environment, University of Michigan, Ann Arbor, Michigan, v. 3, p. 1651–1697.

Turner, R. L., Raines, G. L., Kleinkopf, M. D., and Lee-Moreno, J. L., 1982, Regional northeast-trending structural control of mineralization, northern Sonora, Mexico: *Economic Geology,* v. 77, no. 1, p. 25–37.

Twidale, C. R., 1971, *Structural landforms:* M.I.T. Press, Cambridge, Massachusetts, 247 p.

Uchupi, E., 1965, Map showing relation of land and submarine topography, Nova Scotia to Florida: U.S. Geological Survey Miscellaneous Investigations Map I-451, 3 sheets, 1:1,000,000.

U.S. Air Force, 1966, Report of the USAF Scientific Advisory Board Geophysics Panel Ad Hoc Study Group on Remote Sensing and Evaluation of Terrain Data: Department of the Air Force, Washington, D.C., December, 24 p.

U.S. Air Force, 1967, Security classification concerning airborne passive scanning infrared imaging systems: Air Force Regulation, AFR No. 205-39, Department of the Air Force, 10 March 1967, 17 p.

U.S. Defense Mapping Agency, 1980, Berry Islands, The Bahamas: Landsat Image Format Series, N2600W07752, Experimental Printing, 1:500,000-scale, Defense Mapping Agency, Washington, D.C. Also available from U.S. Geological Survey, Reston, Virginia.

U.S. Geological Survey, 1970, Wadi Wassat quadrangle, Kingdom of Saudi Arabia: U.S. Geological Survey Saudi Arabian Project Report 117, 68 p.

U.S. Geological Survey Staff, 1971, Surface faulting: *in The San Fernando, California, Earthquake of February 9, 1971:* U.S. Geological Survey Professional Paper 733, p. 55–76.

U.S. Geological Survey, 1976, Vatnajökull, Iceland (Fall Scene): Landsat Image Format Series, N6359W01723, Experimental Printing, 1:500,000-scale, U.S. Geological Survey, Reston, Virginia.

U.S. Geological Survey, 1977a, Vatnajökull, Iceland (Winter Scene): Landsat Image Format Series, N6359W01723, Experimental Printing, 1:500,000-scale, U.S. Geological Survey, Reston, Virginia.

U.S. Geological Survey, 1977b, Orthophotoquad (map) of the Monomoy Point, Massachusetts quadrangle: 7.5 minute quadrangle series; scale, 1:25,000.

U.S. Geological Survey, 1977c, EROS Digital Image Enhancement System (EDIES) "Fact Sheet": User Services leaflet, EROS Data Center, Sioux Falls, South Dakota, May, 7 p.

U.S. Geological Survey, 1977d, Florida Keys, Florida: Landsat Image Format Series, N2430W08103, Experimental Printing, 1:500,000-scale, U.S. Geological Survey, Reston, Virginia.

U.S. Geological Survey, 1979, Landsat data users handbook: U.S. Geological Survey, EROS Data Center, Sioux Falls, South Dakota; Looseleaf notebook with pagination by sections.

U.S. Geological Survey, 1980, Landsat data users notes: Issue no. 11, U.S. Geological Survey and National Aeronautics and Space Administration, Sioux Falls, South Dakota, 8 p.

U.S. Geological Survey, 1981a, How to order aerial photographs: National Mapping Program leaflet, U.S. Geological Survey, Reston, Virginia, 2 p.

U.S. Geological Survey, 1981b, Understanding color-infrared photographs and false color composites: National Mapping Program leaflet, U.S. Geological Survey, Reston, Virginia, 2 p.

U.S. Geological Survey, 1981c, The aerial photography summary record system: National Mapping Program leaflet, U.S. Geological Survey, Reston, Virginia, 2 p.

U.S. Geological Survey, 1981d, Using APSRS microfiche: National Mapping Program leaflet, U.S. Geological Survey, Reston, Virginia, 2 p.

U.S. Geological Survey, 1981e, Micrographic indexes: National Mapping Program leaflet, U.S. Geological Survey, Reston, Virginia, 2 p.

U.S. Geological Survey, 1981f, Advance materials: National Mapping Program leaflet, U.S. Geological Survey, Reston, Virginia, 2 p.

U.S. Geological Survey, 1981g, Looking for an old aerial photograph: National Mapping Program leaflet, U.S. Geological Survey, Reston, Virginia, 2 p.

U.S. Geological Survey, 1981h, Alaska from space: National Mapping Program leaflet, U.S. Geological Survey, Reston, Virginia, 2 p.

U.S. Geological Survey, 1981i, Intermediate-scale base maps: National Mapping Program leaflet, U.S. Geological Survey, Reston, Virginia, 2 p.

U.S. Geological Survey, 1981j, A selected bibliography on maps, mapping, and remote sensing: National Mapping Program leaflet, U.S. Geological Survey, Reston, Virginia, 2 p.

U.S. Geological Survey, 1981k, Slope mapping: National Mapping Program leaflet, U.S. Geological Survey, Reston, Virginia, 2 p.

U.S. Geological Survey, 1981l, Looking for an old map: National Mapping Program leaflet, U.S. Geological Survey, Reston, Virginia, 2 p.

U.S. Geological Survey, 1981m, How to order maps on microfilm: National Mapping Program leaflet, U.S. Geological Survey, Reston, Virginia, 2 p.

U.S. Geological Survey, 1981n, How to order Landsat images: National Mapping Program leaflet, U.S. Geological Survey, Reston, Virginia, 2 p.

U.S. Geological Survey, 1981o, Manned spacecraft photographs and major metropolitan area photographs and images: National Mapping Program leaflet, U.S. Geological Survey, Reston, Virginia, 2 p.

U.S. Geological Survey, 1981p, Map scales: National Mapping Program leaflet, U.S. Geological Survey, Reston, Virginia, 2 p.

U.S. Geological Survey, 1981q, Map accuracy: National Mapping Program leaflet, U.S. Geological Survey, Reston, Virginia, 2 p.

U.S. Geological Survey, 1981r, National Cartographic Information Center: National Mapping Program leaflet, U.S. Geological Survey, Reston, Virginia, 2 p.

U.S. Geological Survey, 1982, Landsat Data User's Notes: Issue No. 21, U.S. Geological Survey and National Aeronautics and Space Administration, EROS Data Center, Sioux Falls, South Dakota, January, 12 p.

U.S. Geological Survey, n.d., Mini catalog of map data: National Mapping Program poster, U.S. Geological Survey, Reston, Virginia, 2 p.

U.S. Geological Survey, n.d., Map data catalog: National Mapping Program booklet, U.S. Geological Survey, Reston, Virginia, 48 p.

U.S. Geological Survey, n.d., Understanding maps and scale: National Mapping Program poster, U.S. Geological Survey, Reston, Virginia, 1 p.

University of Michigan, 1962, Proceedings of the First Symposium on Remote Sensing of Environment: University of Michigan, Institute of Science and Technology, Infrared Laboratory, Ann Arbor, Michigan (13–15 February 1962), March 1962, (Also available as a Second Revised Printing (April 1964), Report No. 4864-1-x2, 110 p.) 149 p.

VanderBrug, G. J., 1976, Experiments in iterative enhancement of linear features: *in* Proceedings of the Symposium on Machine Processing of Remotely Sensed Data, Purdue University, West Lafayette, Indiana, p. 4A32–4A44.

Vary, W. E., 1967, Preliminary results of tests with aerial color photography for water depth determination: U.S. Naval Oceanographic Office Report, Washington, D.C., 4 October 1967, 14 p.

Vary, W. E., 1969a, Remote sensing by aerial color photography for water depth penetration and ocean bottom detail: *in* Proceedings of the Sixth International Symposium on Remote Sensing of Environment, University of Michigan, Ann Arbor, Michigan, v. 2, p. 1045–1059.

Vary, W. E., 1969b, A new non-blue-sensitive aerial color film: *in New Horizons in Color Aerial Photography,* Seminar Proceedings, American Society of Photogrammetry, Falls Church, Virginia; and Society of Photographic Scientists and Engineers, Washington, D.C., p. 127–130.

Vedder, J. G., and Wallace, R. E., 1970, Map showing recently active breaks along the San Andreas and related faults between Cholame Valley and Tejon Pass, California: U.S. Geological Survey Miscellaneous Geologic Investigations Map I-574; scale, 1:24,000, 2 sheets.

Vedder, J. G., Dibblee, T. W., and Brown, R. D., 1973, Geologic map of the upper Mono Creek-Pine Mountain area, California: U.S. Geological Survey, Miscellaneous Geologic Investigations Map I-752.

Venkataramanan, D., 1979, Some application of Landsat imagery interpretation for petroleum targetting (sic) in India: *in* Proceedings of the Thirteenth International Symposium on Remote Sensing of Environment, Environmental Research Institute of Michigan, Ann Arbor, Michigan, v. 2, p. 911–923.

Verstappen, H. Th., 1977a, Remote sensing in geomorphology: Elsevier Scientific Publishing Company, New York, 214 p.

Verstappen, H. Th., 1977b, An atlas illustrating the use of aerial photographs in geomorphological mapping: ITC Textbook of Photo-Interpretation, v. 7, International Institute for Aerial Survey and Earth Sciences (ITC), Enschede, The Netherlands, 178 p.

Vesecky, J. F., and Stewart, R. H., 1982, The observation of ocean surface phenomena using imagery from the SEASAT synthetic aperture radar: An Assessment: *Journal of Geophysical Research,* v. 87, no. C5, p. 3397–3430.

Viljoen, R. P., Viljoen, M. J., Grootenboer, J., and Longshaw, T. G., 1975, ERTS-1 imagery: An appraisal of applications in geology and mineral exploration: *Mineral Science and Engineering* (South Africa), v. 7, no. 2, p. 132–168.

Vincent, R. K., 1973, Ratio maps of iron ore deposits, Atlantic City District, Wyoming: *in* Proceedings of Symposium on Significant Results Obtained from the Earth Resources Technology Satellite-1, NASA SP-327, v. 1 (Technical Presentations), Section A, p. 379–386.

Vincent, R. K., 1975, The potential role of thermal infrared multispectral scanners in geologic remote sensing: *in* Proceedings of the Institute of Electrical and Electronic Engineers, v. 63, p. 137–147.

Vincent, R. K., 1981a, Historical development of multispectral thermal infrared techniques for geological mapping: *in* Workshop on Geological Applications of Thermal Infrared Remote Sensing Techniques (Settle, Mark, editor), Technical Report No. 81-06, Lunar and Planetary Institute, Houston, Texas, p. 17–21.

Vincent, R. K., 1981b, Multispectral thermal IR experiments in the early 1970's (abs.): *in* Workshop on Geological Applications of Thermal Infrared Remote Sensing Techniques (Settle, Mark, editor), Technical Report No. 81-06, Lunar and Planetary Institute, Houston, Texas, p. 96–97.

Vincent, R. K., and Thomson, F. J., 1972a, Rock-type discrimination from ratioed infrared scanner images of Pisgah Crater, California: *Science,* v. 175, no. 4025, p. 986–988.

Vincent, R. K., and Thomson, Fred, 1972b, Spectral compositional imaging of silicate rocks: *Journal of Geophysical Research,* v. 77, no. 14, p. 2465–2472.

Vincent, R. K., and Drake, Ben, 1976, Potential use of Landsat data for petroleum exploration in the Michigan basin: *in* Proceedings of the Fifteenth Annual Conference of the Ontario Petroleum Institute, Ontario Petroleum Institute, Inc., Chatham, Ontario, Canada, Technical Paper no. 8, p. 1–23.

Vincent, R. K., and Coupland, D. H., 1980, Petroleum exploration with Landsat in Bay County, Michigan—An interim case study: *in* Proceedings of the Fourteenth International Symposium on Remote Sensing of Environment, Environmental Research Institute of Michigan, Ann Arbor, v. 1, p. 379–387.

Vincent, R. K., Thomson, Frederick, and Watson, Kenneth, 1972, Recognition of exposed quartz sand and sandstone by two-channel infrared imagery: *Journal of Geophysical Research,* v. 77, no. 14, p. 2473–2477.

Vincent, R. K., Rowan, L. C., Gillespie, R. E., and Knapp, Charles, 1975, Thermal-infrared spectra and chemical analyses of twenty-six igneous rock samples: *Remote Sensing of Environment,* v. 4, no. 3, p. 199–209.

Vinogradov, A. P., Surkov, Yu. A., and Kirnozov, F. F., 1973, The content of uranium, thorium and potassium in the rocks of Venus as measured by Venera 8: *Icarus,* v. 20, no. 3, p. 253–259.

Voight, Barry, 1982, Time scale for the first moments of the May 18 eruption: *in The 1980 Eruptions of Mount St. Helens, Washington* (Lipman, P. W., and Mullineaux, D. R., editors), U.S. Geological Survey Professional Paper 1250, p. 69–86.

Vostokova, Y. A., Vyshivkin, D. D., Kasynova, M. S., Nesvetaylova, N. G., and Shvyryayeva, A. M., 1961, Geobotanical indicators of bitumen: *International Geological Review,* v. 3, p. 598–608.

Wallace, R. E., and Moxham, R. M., 1967, Use of infrared imagery in study of the San Andreas fault system, California: *in Geological Survey Research 1967,* U.S. Geological Survey Professional Paper 575-D, p. D147–D156.

Walker, A. S., 1982, Deserts of China: *American Scientist,* v. 70, no. 4, p. 366–376.

Wang, Y. H., 1980, Satellite applications on a coastal inlet stability study: *in* Proceedings of the 17th International Coastal Engineering Conference, American Society of Civil Engineers, p. 2581–2594.

Ward, P. L., and Eaton, P. L., 1976, New method for monitoring global volcanic activity: *in ERTS-1, A New Window on Our Planet* (Williams, R. S., Jr., and Carter, W. D., editors), U.S. Geological Survey Professional Paper 929, p. 106–108.

Wark, D. G., Yamamoto, G., and Lienesch, J. H., 1962, Methods of estimating infrared flux and surface temperature from meteorological satellites: *Journal of Atmospheric Sciences,* v. 19, no. 5, p. 369–384.

Warren, C. R., Schmidt, D. L., Denny, C. S., and Dale, W. J., 1969, A descriptive catalog of selected aerial photographs of geologic features in areas outside the United States: U.S. Geological Survey Professional Paper 591, 23 p.

Warrick, R. A., 1979, Volcanoes as hazard: An overview: *in Volcanic Activity and Human Ecology* (Sheets, P. D., and Grayson, D. K., editors), Academic Press, New York, p. 161–194.

Washburn, A. L., 1979, Geocryology, a survey of periglacial processes and environments: Edward Arnold, Publishers, Ltd., London, 406 p.

Watson, Kenneth, 1971a, Geophysical aspects of remote sensing: *in* Proceedings of the International Workshop on Earth Resources Survey Systems, Superintendent of Documents, U.S. Government Printing Office, Washington, D.C., v. 2, p. 409–428.

Watson, Kenneth, 1971b, A thermal model for analysis of infrared images: National Aeronautics and Space Administration, Third Annual Earth Resources Program Review, Houston, Texas, v. 1, Sec. 13, p. 1–16.

Watson, Kenneth, 1973, Periodic heating of a layer over a semi-infinite solid: *Journal of Geophysical Research,* v. 78, no. 26, p. 5904–5910.

Watson, Kenneth, 1975, Geologic applications of thermal infrared images: *in* Proceedings of the IEEE, no. 63, p. 128–137.

Watson, Kenneth, Rowan, L. C., and Offield, T. W., 1971, Application of thermal modeling in the geologic interpretation of infrared images: *in* Proceedings of the 7th International Symposium on Remote Sensing of Environment, University of Michigan, Ann Arbor, Michigan, v. 3, p. 2017–2041.

Watson, Kenneth, Hummer-Miller, Susanne, and Offield, T. W., 1981, Geologic applications of thermal-inertia mapping from satellite: U.S. Geological Survey Open-File Report 81-1352, 73 p.

Watson, Kenneth, Hummer-Miller, Susanne, and Offield, Terry, 1982, Geologic thermal-inertia mapping using HCMM satellite data: *in* Proceedings of the Institute of Electrical and Electronic Engineers, 1982 International Geoscience and Remote Sensing Symposium (IGARSS '82, Munich, Federal Republic of Germany, June 1982), v. 1, p. 2.1–2.6.

Watson, R. D., 1981, Quantification of luminescence in terms of a rhodamine WT standard: *in* Workshop on Applications of Luminescence Techniques to Earth Resource Studies (Hemphill, W. R., and Settle, Mark, editors), LPI Technical Report, 81-03, Lunar and Planetary Institute, Houston, p. 19–22.

Watts, R. D., England, A. W., Vickers, R. S., and Meier, M. F., 1975, Radio-echosounding on South Cascade Glacier, Washington, using a long-wavelength, mono-pulse source (abs.): *Journal of Glaciology,* v. 15, no. 73, p. 459.

Watts, R. D., and Wright, D. L., 1981, Systems for measuring thickness of temperate and polar ice from ice or from air: *Journal of Glaciology,* v. 27, no. 97, p. 459–469.

Way, D. W., 1973, Terrain analysis—a guide to site selection using aerial photographic interpretation: Dowden, Hutchinson and Ross, Inc., Stroudsburg, Pennsylvania, 392 p.

Weeden, H. A., and Bolling, N. B., 1980, Fundamentals of aerial photography interpretation: *in Remote Sensing in Geology* (Siegal, B. S., and Gillespie, A. R., editors): John Wiley and Sons, New York, New York, p. 229–255.

Weintraub, Boris, 1982, Fire and ash, darkness at noon (Mexican volcano spews death and destruction): *National Geographic,* v. 162, no. 5, p. 660–671 and p. 676–684.

Welch, Roy, 1966, A comparison of aerial films in the study of the Breidamerkur glacier area, Iceland: *The Photogrammetric Record,* v. 5, no. 28, p. 289–306.

Welch, Roy, 1967, The application of aerial photography to the study of a glacier area, Breidamerkur, Iceland: Ph.D. Dissertation, Department of Geography, University of Glasgow, Scotland, 252 p.

Welch, Roy, 1968, Color aerial photography applied to the study of a glacial area: *in Manual of Color Aerial Photography* (Smith, J. T., Jr., and Anson, Abraham, editors), American Society of Photogrammetry, Falls Church, Virginia, p. 400–401.

Welch, Roy, 1969, Reflectance characteristics of a glacial landscape and their relation to aerial photography: *in* Proceedings of the Seminar on "New Horizons in Color Aerial Photography," American Society of Photogrammetry (and Society of Photographic Scientists and Engineers), Falls Church, Virginia, p. 17–35.

Welch, Roy, and Howarth, P. J., 1968, Photogrammetric measurements of glacial landforms: *Photogrammetric Record,* v. 6, no. 31, p. 75–96.

Wenderoth, Sondra, 1969, Hydrographic and oceanographic applications of multispectral color aerial photography: *in New Horizons in Color Aerial Photography,* Seminar Proceedings, American Society of Photogrammetry, Falls Church, Virginia; and Society of Photographic Scientists and Engineers, Washington, D.C., p. 115–125.

Wenderoth, Sondra, and Yost, Edward, 1974, Multispectral photography for Earth resources: Science Engineering Research Group, C. W. Post Center, Long Island University, Greenvale, New York; Report to NASA's L. B. Johnson Space Center, Houston, Texas, under Contract No. NAS9-11188, pages variously numbered by sections.

Wheeler, R. L., and Stubbs, J. L., Jr., 1976, Style elements of systematic joints: Statistical analysis of size, spacing and other characteristics: *in* Proceedings of the Second International Conference on Basement Tectonics (Newark, Delaware), p. 491–499.

Wheeler, R. L., Mullennex, R. H., Henderson, C. D., and Wilson, T. H., 1974, Major cross-strike structures of the central sedimentary Appalachians; progress report: *in* Proceedings of the West Virginia Academy of Science, v. 46, no. 2, p. 196–203.

Whittlesey, J. H., 1970, Tethered balloon for archaeological photos: *Photogrammetric Engineering,* v. 36, no. 2, p. 181–186.

Wiesnet, D. R., and Scott, R. B., 1982, The NOAA satellites: A largely neglected tool in the land sciences (abs.): *in* Abstracts of the 24th Plenary Meeting of Committee on Space Research, Symposium on Remote Sensing and Mineral Exploration (17–20 May, 1982) Ottawa, Canada, p. 93.

Wildey, R. L., 1967, Spatial filtering of astronomical photographs. II. Theory: *Astronomical Journal,* v. 72, no. 7, p. 884–886.

Wilhelms, D. E., 1970, Summary of lunar stratigraphy-telescopic observations: U.S. Geological Survey Professional Paper 559-F, p. F1–F37.

Wilhelms, D. E., 1972, Geologic mapping of the second planet: U.S. Geological Survey Interagency Report; Astrogeology 55, 36 p.

Wilhelms, D. E., and McCauley, J. F., 1971, Geologic map of the near side of the Moon: U.S. Geological Survey Miscellaneous Geologic Investigation Series Map I-703 (scale, 1:5,000,000).

Williams, P. L., 1964, Geology, structure, and uranium deposits of the Moab quadrangle, Colorado and Utah: U.S. Geological Survey Miscellaneous Geologic Investigations Map I-360, 2 sheets (Sheet 1, Geology: Sheet 2, Structure and Uranium Deposits; scale, 1:250,000).

Williams, R. S., Jr., 1971, Geological applications of aerial spectrophotography: Preliminary findings from Cayo Icacos, Puerto Rico (abs.): *in* Transactions, Fifth Caribbean Geological Conference (1968), Geological Bulletin, No. 5, Queens College Press, New York, p. 252.

Williams, R. S., Jr., 1972a, Terrestrial remote sensing: Applications of thermal infrared scanners to the geological sciences: *in* Pt. 3, ISA Transducer Compendium, Instrument Society of America, Pittsburgh, Pennsylvania, p. 219–236.

Williams, R. S., Jr., 1972b, Thermography: *Photogrammetric Engineering,* v. 38, no. 9, p. 881–883.

Williams, R. S., Jr., 1975, Scientific rationale for the selection of film-filter combinations in the archaeological remote sensing experiment, Great Britain: Appendix E to Chapter 6, An experiment in multispectral air photography for archaeological research, by J. N. Hampton: *in Photography in Archaeological Research* (Elmer Harp, Jr., ed.), School of American Research Advanced Seminar Series, University of New Mexico Press, Albuquerque, p. 202–210.

Williams, R. S., Jr., 1976a, Dynamic environmental phenomena in southwestern Iceland: *in ERTS-1, A New Window on Our Planet* (Williams, R. S., Jr., and Carter, W. D., editors), U.S. Geological Survey Professional Paper 929, p. 109–112.

Williams, R. S., Jr., 1976b, Vatnajökull icecap, Iceland: *in ERTS-1, A New Window on Our Planet* (Williams, R. S., Jr., and Carter, W. D., editors), U.S. Geological Survey Professional Paper 929, p. 188–193.

Williams, R. S., Jr., 1976c, Diversion of lava by water cooling during the eruption of Eldfell Volcano, Heimaey, Iceland (abs.): *in* Abstracts with Programs, Northeastern Section, 10th Annual Meeting, Geological Society of America, Boulder, Colorado, v. 8, no. 2, p. 300–301.

Williams, R. S., Jr., 1978a, Review of *Manual of Remote Sensing: Economic Geology,* v. 73, no. 2, p. 290–292.

Williams, R. S., Jr., 1978b, Geomorphic processes in Iceland and on Mars: A comparative appraisal from orbital images (abs.): *in* Abstracts with Programs, 1978 Annual Meetings of the Geological Society of America, Boulder, Colorado, v. 10, no. 7, p. 517.

Williams, R. S., Jr., 1978c, Satellite geological and geophysical remote sensing of Iceland: Type III Progress Report to the National Aeronautics and Space Administration, Goddard Space Flight Center, Greenbelt, Md., Earth Resources Technology Satellite 1 Experiment No. SR 9651; 15 January 1973–15 August 1974, 1 January 1978, 66 p.

Williams, R. S., Jr., 1978d, Landsat image of dynamic marine phenomena off the southwest coast of Iceland (abs.): *EOS* (Transactions, American Geophysical Union), v. 59, no. 4, p. 301.

Williams, R. S., Jr., 1979a, Delineation of recent changes in the coastline of Monomoy Island, Cape Cod, Massachusetts, with Landsat 3 images (MSS and RBV) (abs.): *in* Proceedings of the 45th Annual Meeting of the American Society of Photogrammetry, Falls Church, Virginia, v. 1, p. 290–291.

Williams, R. S., Jr., 1979b, Regional geologic mapping using Landsat 3 return beam vidicon images: Examples from Iceland and Cape Cod, Massachusetts (abs.) *in* Abstracts with Programs, 1979 Annual Meetings, Geological Society of America, Boulder, Colorado, v. 11, no. 7, p. 541.

Williams, R. S., Jr., 1979c, Iceland—satellite monitoring of changes of glaciers of Iceland: *in* Glaciological Field Stations (Vivien, Robert, editor), pt. 1; *Glaciological Data,* Rpt. GD-4, World Data Center A for Glaciology (Snow and Ice), Boulder, Colorado, February, p. 72–77.

Williams, R. S., Jr., 1980, Geologic hazards of Iceland: A classification (abs.): *in* Abstracts with Programs, 1980 Annual Meetings, Geological Society of America, Boulder, Colorado, v. 12, no. 7, p. 550.

Williams, R. S., Jr., 1981, The use of broadband thermal infrared images to monitor and to study dynamic geological phenomena: *in* Workshop on Geological Applications of Thermal Infrared Remote Sensing

Techniques (Settle, Mark, editor), Technical Report No. 81-06, Lunar and Planetary Institute, Houston, Texas, p. 98–106.

Williams, Richard, 1982, Remote sensing: *Geotimes,* v. 29, no. 2, p. 53–55.

Williams, R. S., Jr., 1983a, Glaciers—clues to climatic change: U.S. Geological Survey Scientific Leaflet (in press).

Williams, R. S., Jr., 1983b, Satellite glaciology of Iceland: *Jökull* (in press).

Williams, R. S., Jr., and Ory, T. R., 1967, Infrared imagery mosaics for geological investigations: *Photogrammetric Engineering,* v. 33, no. 12, p. 1377–1380.

Williams, R. S., Jr., and Friedman, J. D., 1970a, Satellite observation of effusive volcanism: *British Interplanetary Society Journal,* v. 23, no. 6, p. 441–450.

Williams, R. S., Jr., and Friedman, J. D., 1970b, Geologic mapping applications of coastal aerial photography, Cape Cod, Massachusetts (abs.): *Photogrammetric Engineering,* v. 36, no. 6, p. 597.

Williams, R. S., Jr., and Fernandopullé, Denis, 1972, Geological analysis of aerial thermography of the Canary Islands, Spain: *in* Proceedings of the Eighth International Symposium on Remote Sensing of Environment, University of Michigan, Ann Arbor, Michigan, v. 2, p. 1159–1194.

Williams, R. S., Jr., and Moore, J. G., 1973, Iceland chills a lava flow: *Geotimes,* v. 18, no. 8, p. 14–17.

Williams, R. S., Jr., and Thorarinsson, Sigurdur, 1974, ERTS-1 image of Vatnajökull area: General comments: *Jökull,* v. 23 (1973), p. 1–6.

Williams, R. S., Jr., and Carter, W. D., editors, 1976, ERTS-1, A new window on our planet: U.S. Geological Survey Professional Paper 929, 362 p.

Williams, R. S., Jr., and Moore, J. G., 1976a, Iceland chills a lava flow: *in Focus on Environmental Geology—A collection of Case Histories and Readings from Original Sources* (Ronald Tank, editor), 2nd edition, Oxford University Press, New York, p. 49–58. (Includes addendum by Williams in addition to reprint of *Geotimes* article (Williams and Moore, 1973)).

Williams, R. S., Jr., and Moore, J. G., 1976b, Man against volcano: The eruption on Heimaey, Vestmann Islands, Iceland: U.S. Geological Survey Scientific Leaflet, INF-75-22, 20 p.

Williams, R. S., Jr., and Schoonmaker, J. W., Jr., 1979, Surveying Antarctica: From dogsled to satellite: *Air and Space,* v. 3, no. 1, p. 2–4.

Williams, R. S., Jr., and Ferrigno, J. G., 1981, Satellite image atlas of the Earth's glaciers: *in Satellite Hydrology* (Deutsch, Morris, Wiesnet, D. R., and Rango, Albert, editors), Proceedings of the Fifth Annual Wm. T. Pecora Memorial Symposium on Remote Sensing, Sioux Falls, South Dakota (10–15 June 1979), American Water Resources Association, Minneapolis, p. 173–182.

Williams, R. S., Jr., and Taranik, J. V., 1981, Remote sensing: *Geotimes,* v. 26, no. 2, p. 51–53.

Williams, R. S., Jr., Friedman, J. D., Thorarinsson, Sigurdur, Sigurgeirsson, Thorbjörn, and Pálmason, Gudmundur, 1967, Analysis of 1966 infrared imagery of Surtsey, Iceland (abs.): *in* Program and Abstracts of Papers, v. 7, International Association of Volcanology, XIVth General Assembly of International Union of Geodesy and Geophysics, Zürich, Switzerland, p. 61.

Williams, R. S., Jr., Friedman, J. D., Thorarinsson, Sigurdur, Sigurgeirsson, Thorbjörn, and Pálmason, Gudmundur, 1968, Analysis of 1966 infrared survey of Surtsey, Iceland: *in Surtsey Research Progress Report,* The Surtsey Research Society, Reykjavík, v. 4, p. 177–192.

Williams, R. S., Jr., Thorarinsson, Sigurdur, and Saemundsson, Kristján, 1973, Vatnajokull area, Iceland: New volcanic and structural features on ERTS-1 imagery (abs.): *in* Abstracts with Programs, 1973 Annual Meetings of the Geological Society of America, Boulder, Colorado, v. 5, no. 7, p. 864–865.

Williams, R. S., Jr., Bödvarsson, Ágúst, Fridriksson, Sturla, Pálmason, Gudmundur, Rist, Sigurjón, Sigtryggsson, Hlynur, Saemundsson, Kristján, Thorarinsson, Sigurdur, and Thorsteinsson, Ingvi, 1974, Environmental studies of Iceland with ERTS-1 imagery: *in* Proceedings of the Ninth Symposium on Remote Sensing of Environment, Environmental Research Institute of Michigan, Ann Arbor, Michigan, v. 1, p. 31–81.

Williams, R. S., Jr., Bödvarsson, Ágúst, Rist, Sigurjón Saemundsson, Kristján, and Thorarinsson, Sigurdur, 1975, Glaciological studies in Iceland with ERTS-1 imagery (abs.): *Journal of Glaciology,* v. 15, no. 73, p. 465–466.

Williams, R. S., Jr., Hasell, P. G., Jr., Sellman, A. N., and Smedes, H. W., 1976, Thermographic mosaic of Yellowstone National Park: *Photogrammetric Engineering and Remote Sensing,* v. 42, no. 10, p. 1315–1324.

Williams, R. S., Jr., Mecklenburg, T. N., Abrams, M. J., and Gudmundsson, Bragi, 1977, Conventional vs. computer-enhanced Landsat image maps of Vatnajökull, Iceland (abs.): *in* Abstracts with Programs, 1977 Annual Meetings, Geological Society of America, Boulder, Colorado, v. 9, no. 7, p. 1228–1229.

Williams, R. S., Jr., Thorarinsson, Sigurdur, Björnsson, Helgi, and Gudmundsson, Bragi, 1979, Dynamics of Icelandic ice caps and glaciers: *Journal of Glaciology,* v. 24, no. 90, p. 505–507.

Williams, R. S., Jr., Morris, E. C., and Thorarinsson, Sigurdur, 1981a, Illustrated geomorphic classification of Icelandic volcanoes: *in* Reports of Planetary Geology Program, 1981 (Holt, H. E., compiler), NASA Technical Memorandum, No. 84211 (December 1981), p. 183–185.

Williams, R. S., Jr., Ferrigno, J. G., Kent, T. M., Lind, Elizabeth, Barnes, J. C., and Onysko, Steven, 1981b, Extent of sea ice in the harbors and bays of Cape Cod, Massachusetts, on February 15, 1979 (abs.): *in* Abstracts with Programs, 1981 Annual Meeting, Northeastern Section, Geological Society of America, Boulder, Colorado, v. 13, no. 3, p. 184.

Williams, R. S., Jr., Ferrigno, J. G., Kent, T. M., and Schoonmaker, J. W., Jr., 1982a, Landsat images and mosaics of Antarctica for mapping and glaciological studies: *Annals of Glaciology,* v. 3, p. 321–326.

Williams, R. S., Jr., Meunier, T. K., and Ferrigno, J. G., 1982b, Delineation of blue-ice areas in Antarctica from satellite imagery (abs.): *in* Workshop on Antarctic Glaciology and Meteorites (Bull, Colin, and Lipschutz, M. E., editors), LPI Technical Report No. 82-03, Lunar and Planetary Institute, Houston, Texas, p. 49–50.

Williams, R. S., Jr., Falcone, N. L., Barlow, R. A., and Fitzpatrick-Lins, Katherine, 1983a, New Bedford quadrangle, Massachusetts: A prototype

1:100,000-scale Landsat 3 return beam vidicon (RBV) image map (abs.): *in* Summaries, Seventeenth International Symposium on Remote Sensing of Environment (Ann Arbor, Michigan; 9–13 May 1983), Environmental Research Institute of Michigan, Ann Arbor, Michigan (in press).

Williams, R. S., Jr., Meunier, T. K., and Ferrigno, J. G., 1983b, Blue ice, meteorites, and satellite imagery in Antarctica: *Polar Record,* (in press).

Wolfe, W. L., editor, 1965, Handbook of military infrared technology: U.S. Government Printing Office, Washington, D.C., 906 p.

Wolfe, W. L., and Zissis, G. J., editors, 1978, The infrared handbook: U.S. Government Printing Office, Washington, D.C., variously paginated by sections.

Woll, P. W., and Fischer, W. A., editors, 1977, Proceedings of the First Annual William T. Pecora Memorial Symposium: U.S. Geological Survey Professional Paper 1015, 370 p.

Wolmarans, I. G., and Krynauw, J. R., compilers, 1981, Reconnaissance geological maps of the Ahlmannryggen (sheet 1), Borgmassivet (sheet 2), and Kirwanveggen (sheet 3) areas, Western Dronning Maud Land, Antarctica: South African Committee for Antarctic Research, Pretoria, Series of 3 satellite image map sheets; scale, 1:250,000.

Working Group on Geology, 1971, Geology: Report No. 6 *in* Resource Satellites and Remote Airborne Sensing for Canada, Program Planning Office for Resource Satellites and Remote Airborne Sensing, Interdepartmental Committee on Resource Satellites and Remote Airborne Sensing; The Department of Energy, Mines and Resources, Ottawa, 26 p.

Working Group on Ice Reconnaissance and Glaciology, 1971, Ice reconnaissance and glaciology: Report No. 7 to the Program Planning Office for the Interdepartmental Committee on Resource Satellites and Remote Airborne Sensing for Canada, Information Canada (White, Dennis, editor), Ottawa, 37 p.

Wright, F. L., 1981, Design of an aircraft thermal infrared multispectral scanner (TIMS): *in* Workshop on Geological Applications of Thermal Infrared Remote Sensing Techniques (Settle, Mark, editor), Technical Report No. 81-06, Lunar and Planetary Institute, Houston, Texas, p. 107–116.

Wu, C., Barkan, B., Huneycutt, B., Teang, C., and Pang, S., 1981, An introduction to the interim digital SAR processor and the characteristics of the associated Seasat SAR imagery: JPL Publication 81-26, Jet Propulsion Laboratory, Pasadena, California, 82 p.

Yentsch, C. S., Skea, W., Laird, J. C., and Hopkins, T. S., 1979, Ocean color observations: *in* Apollo-Soyuz Test Project; Summary Science Report; Earth Observations and Photography (v. 2; El-Baz, Farouk, and Warner, D. M., editors), NASA SP-412, p. 441–454.

Yost, Edward, and Wenderoth, Sondra, 1970, Remote sensing of coastal waters using multispectral photographic techniques: Report No. SERG-TR-10 (Contract No. N00014-67-C-0281), U.S. Naval Oceanographic Office, 219 p.

Yost, Edward, and Wenderoth, Sondra, 1971, The reflectance spectra of mineralized trees: *in* Proceedings of the Seventh International Symposium on Remote Sensing of Environment, University of Michigan, Ann Arbor, Michigan, v. 1, p. 269–284.

Zaremba, R., and Leatherman, S. P., 1983, Overwash processes and foredune ecology, Nauset Spit, Cape Cod, Massachusetts: U.S. Army Corps of Engineers Coastal Engineering Research Center (CERC), Technical Report (in press).

Zeigler, J. M., Tuttle, S. D., Giese, G. S., and Tasha, H. J., 1964, Residence time of sand composing the beaches and bars of outer Cape Cod: *in* Proceedings of Ninth Conference in Civil Engineering, American Society of Civil Engineers, p. 403–416.

Zwally, H. J., and Gloersen, Per, 1977, Passive microwave images of the polar regions and research applications: *Polar Record,* v. 18, no. 116, p. 431–450.

Zwally, H. J., Thomas, R. H., and Bindschadler, R. A., 1981, Ice sheet dynamics by laser altimetry: National Aeronautics and Space Administration, Technical Memorandum No. 82128, 11 p.

Zwally, H. J., Comiso, J. C., Parkinson, C. L., Campbell, W. J., Carsey, F. D., and Gloersen, Per, 1983, Antarctic sea ice cover, 1973–1976, from satellite passive microwave observation: NASA Special Publication, National Aeronautics and Space Administration, Washington, D.C., (in press).

Authorship Key for Chapter 31, Geological Applications

Section Title	Author(s)
INTRODUCTION	R. S. Williams, Jr. and S. E. Marsh
DEFINITION OF GEOLOGICAL REMOTE SENSING	
Electromagnetic Spectrum Regions Relevant to Geology	R. S. Williams, Jr.
Theoretical Studies	S. E. Marsh
Laboratory Studies	"
Field Studies and Data Acquisition	S. E. Marsh and R. S. Williams, Jr.
Aircraft Data Acquisition	R. S. Williams, Jr.
Satellite Data Acquisition	"
HISTORICAL CONSIDERATIONS	"
INTERRELATIONSHIP OF GEOLOGIC REMOTE SENSING WITH CLOSELY RELATED DISCIPLINES AND SUBDISCIPLINES	"
TYPES AND AVAILABILITY OF REMOTELY SENSED DATA FOR GEOLOGICAL INVESTIGATIONS	
REMOTELY SENSED IMAGE DATA FROM AIRBORNE PLATFORMS	
Types of Photographic Images	R. S. Williams, Jr. & C. S. Southworth
Availability of Photographic Images	C. S. Southworth
Types of Non-Photographic Images	C. S. Southworth & R. S. Williams, Jr.
Availability of Non-Photographic Images	C. S. Southworth
IMAGING SENSOR SYSTEMS ON SPACECRAFT	"
NON-IMAGING AND/OR HIGH-RESOLUTION SYSTEMS	
Airborne Systems	
Profilometry	R. L. Brooks
Luminescence	W. R. Hemphill, R. D. Watson, & A. F. Theisen
Radiometry	L. C. Rowan
Satellite Systems	
Profilometry	R. L. Brooks
Radiometry	A. F. H. Goetz, L. C. Rowan and M. J. Kingston
MINI-ATLAS OF IMAGE AND RELATED DATA COVERAGE OF CAPE COD, MASSACHUSETTS AND ENVIRONS	
Introduction	R. S. Williams, Jr.
Photographic Data of Cape Cod and Environs	"
Imagery of Cape Cod and Environs	R. S. Williams, Jr. & C. S. Southworth
ANALYSIS TECHNIQUES	
PHOTOGEOLOGY AND GEOLOGICAL REMOTE SENSING	R. S. Williams, Jr.
MULTISPECTRAL DATA	
Optical Image Processing	M. Deutsch
Atmospheric and Radiometric Corrections	S. E. Marsh and W. S. Kowalik
Digital Image Processing and Enhancement for Geology	
Introduction	R. S. Williams, Jr.
Enhancements for lineament identifications	P. S. Chavez, Jr.
Selected methods for digital processing of multispectral scanner data for mapping lithologic units	M. H. Podwysocki and D. B. Segal
COMBINED DATA SETS	C. M. Trautwein
APPLICATIONS OF REMOTE SENSING TECHNOLOGY TO GEOLOGIC SUBDISCIPLINES	
INTRODUCTION	R. S. Williams, Jr. & C. S. Southworth
GEOLOGIC MAPPING	
Introduction	R. S. Williams, Jr.
Lithology	
Lithologic mapping	A. B. Kahle
Computer-enhanced Landsat imagery for lithologic discrimination in southwestern Saudi Arabia	H. W. Blodget
Structure and Tectonics	
Application of remote sensing to structural geology and tectonics	D. P. Gold
Structure and tectonics of northeastern Iceland	B. Voight and N. Mamula, Jr.
Mapping brittle structures	Y. W. Isachsen
ECONOMIC GEOLOGY	
Introduction	W. D. Carter
Mineral Exploration	
Examples of significant mineral resource exploration activities and discoveries based on the use of Landsat data	W. D. Carter
The use of multispectral image band ratios in mineral exploration	"
Airborne remote sensing applications to porphyry copper exploration	M. J. Abrams
Petroleum Exploration	
Introduction to the use of remote sensing technology in the search for oil and gas	W. D. Carter
The use of radar in the search for hydrocarbons	J. P. Ford
Application of Landsat imagery to petroleum exploration in Bay County, Michigan	R. K. Vincent & D. H. Coupland

Chapter 32

Engineering Applications

Author-Editor: OLIN MINTZER

Contributing Editors: SIAMAK KHORRAM and ROBERT RYERSON

Contributing Authors: ANDREW S. BENSON, J. E. COLCORD, ROBERT COLWELL, PHILLIP J. HOWARTH, SIAMAK KHORRAM, GARTH R. LAWRENCE, B. SEN MATHUR, J. D. MOLLARD, HAROLD T. RIB, ROBERT A. RYERSON, RANDALL W. THOMAS, JAMES P. SCHERZ, SHARON L. WALL, JAMES M. WARDLOW, ROBIN I. WELCH.

General Contents: Image analysis, terrain analysis, site investigations and regional planning, water resources engineering, landslide studies, transportation facilities, and additional engineering applications of thermal infrared sensing.

INTRODUCTION

For more than 35 years the interpretation of aerial photographs has been an indispensible technique for use in accomplishing a variety of engineering projects. In recent years, newer multispectral remote-sensing systems, such as infrared and radar, and the availability of large regional views as provided by satellite coverage, have been added to the tools available to the interpreter who wishes to apply remote sensing in various engineering studies.

The areas in which image interpretation techniques are applied to the engineering field and which, therefore, will be treated in this chapter are: (1) image analysis, (2) terrain analysis, (3) site investigations and regional planning, (4) water resources engineering, (5) landslide studies, (6) transportation facilities, and (7) additional applications of thermal infrared imagery.

The information required to operate in this variety of areas in the field of engineering is derived from an analysis of the patterns present on the imagery. These patterns reflect the influence of the type of parent material; the geologic processes undergone; the climatic, biotic and physiographic environment; and man's activity. The variety of patterns developed due to the interplay of these various factors forms the basis for image interpretation. The basic premise for this technique is that materials developed under the same geologic and environmental conditions will have similar patterns on the imagery; dissimilar materials will have dissimilar patterns.

Aerial photographic interpretation techniques applied to engineering studies involve the recognition of basic landforms as indicated by the pattern elements on the photography. The elements comprising the patterns include: topography (size, shape, slope, aspect, elevation); drainage (form, type, texture); erosion (form, type); photo tone-color (shade of gray, or the combination of hue, brightness and saturation as seen on color photography), together with texture, uniformity, sharpness of boundary; vegetation (type, associations); and culture (man's influence on the landscape and his adjustments to variations in landforms and environment).

Similar techniques have been developed for the interpretation of other types of imagery such as thermal infrared, radar and multispectral, and the techniques for interpreting these other media are still in the developmental stage. The integration of various data types, such as conventional map and tabular data with remotely-sensed data, has been made possible recently through the development of geobased information systems. Comprehensive discussions on the fundamentals of image and photographic interpretation are given in Chapters 24 and 25 of this Manual. Details with respect to geobased information systems are given in Chapter 22. Augmenting these chapters, the following a brief summary provides some of the techniques used in engineering practice.

IMAGE ANALYSIS

Success in the overall interpretation of various forms of remote sensing imagery for engineering purposes is dependent on five factors: qualifications of the interpreter; photographic and imagery parameters; natural conditions, the equipment; the analysis techniques used and available supplemental background data. An understanding of the limitations imposed by these factors is important to the proper planning and utilization of interpretation techniques for engineering inventories and analyses. Interpreter qualifications and basic properties of imagery are discussed at length in other chapters of this manual.

The value of imagery interpretation as the means of obtaining information that is needed in engineering programs is recognized, with the result that aerial, or satellite coverage frequently is collected specifically for the purpose of making engineering-related studies. This allows the interpreter more latitude in the selection of a sensor-analyses combination that will be optimum for the making of any such study. The combination can be "tailored" to fit the job, thus increasing the accuracy of the interpreter's product. Some combinations are discussed later in this chapter.

Although there has been a great increase in the variety of sensor systems and analysis techniques available during the past 10 years, aerial photography (including panchromatic, color and color infrared) is still the primary sensor system used. Panchromatic black-and-white film is still used extensively. The sources of this photography are the U.S. Department of Agriculture, the U.S. Geological Survey, private industry, and numerous other organizations.

There is no one scale which will satisfy all of the requirements for the variety of engineering studies to be performed. For example, scales of 1:2,400 to 1:6,000 have been used for pavement-condition surveys; scales of 1:8,000 to 1:12,000 have been used for detailed soils mapping; scales of 1:15,000 to 1:30,000 have been used for terrain analyses; and scales of 1:40,000 or smaller have been used for regional planning studies. In a cost-benefit analysis recently made by Ryerson (1981) it was shown that, as one moves to color or color infrared, smaller scales may be used to collect the same information. The result is that fewer photographs are interpreted and consequently there is a significant savings in time and money.

Another concern is the adequacy of coverage. Sufficient coverage of an area is required in order to determine the extent of local conditions and the expected variations. Numerous instances have occurred where limited area coverage has been obtained for analysis of a selected route location. Some condition was then uncovered which required shifting of the alinement outside the limits of the selected route, and additional coverage was needed to make a study of the alternate alinement. This caused a delay in planning the project and required reflying of the route at a much greater cost than if sufficient area coverage had been obtained originally. Sufficient breadth of coverage should be obtained initially to meet any contingency that might reasonably be expected to arise. For example, a general rule of thumb used for route location is that the width of coverage should be approximately 0.6 of the length. In recent studies, general coverage has been facilitated through an initial study of satellite data, as discussed later in this chapter. Color and color infrared aerial photographs have been reported by several interpreters as providing the best data, and some engineering organizations have gone exclusively to color or color infrared for their interpretation projects. Others have found that the use of drones, or small aircraft with simple systems such as 35-mm cameras also will yield useful site specific data (Mintzer, et al., 1978). Evaluations of newer systems, such as thermal infrared, radar and microwave, have demonstrated that these systems are also practical, but usually only for special situations. For example, infrared is of value for locating major shallow subsurface channels, and distinguishing between massive continuous rock bodies and noncontinuous fractured rock bodies and colluvial slopes.

A major limitation for the use of these newer systems for engineering studies is the cost of obtaining and analyzing the data. For a new system to be advantageous over photography it is necessary to demonstrate that it can provide the same information at a savings in cost, or more information for the same cost. Except in a few special cases, this has not been demonstrated for the newer systems but this picture is changing rapidly as more research results, using these newer systems, are reported. In the descriptions that will be given in this chapter regarding the techniques for performing various kinds of engineering studies, the applicability of the newer systems and analysis techniques will be included where they have been demonstrated to be practical and economical, or where they provide information not obtainable by other methods.

Natural factors must be considered in planning a mission. For example, light planning in a given geographic area may be largely controlled by meteorological conditions; in addition, consideration needs to be given to such environmental conditions as season of the year and time of day.

Landforms and the soils developed on them are relatively stable, but surface features resulting from moisture conditions, vegetation cover, and land use may change rapidly and affect the pattern elements observed. For example, if a project is planned for developing an engineering soils map in a humid-temperature climate, spring is the best time of year in which to obtain coverage, when the water table is usually at its highest level and the tonal contrasts between coarse-textured and fine-textured soils are the greatest. The optimum time for the flight within this period, depending on local considerations, is after the snow has melted, before the leaves on trees have budded, and after the farmer has plowed his fields for spring planting. Likewise, if an effort is being made to locate unstable slopes in the same environment, spring is still an optimum time. The presence of groundwater often highlights the presence of springs and seepage areas and emphasizes the most critical zones. An additional consideration for areas on the slope of a hillslide is the sun angle. The flight should be planned for a time when the hillside is illuminated by the Sun, and not in shadows. If thermal infrared imagery is being considered for locating the seepage zones, a nighttime flight would be optimum; at that time the maximum temperature contrasts usually occur between the warmer surface waters and the cooler seepage waters, and the dryland terrain is cooler than both. If coverage is obtained for these studies after vegetation leafs out, the features that the interpreter wishes to evaluate may be masked by the cover of vegetation and difficult to recognize.

Equipment and techniques available for the making of engineering analyses by remote sensing have been changing rapidly over the past decade. The ones currently being used in engineering and other studies are described in detail elsewhere in this manual and only briefly discussed in the following sections.

ANALYSIS PROCEDURE

Each engineering project is unique in one or more respects, but the chief functions and basic methods of image analysis are similar in all projects. Image analysis for engineering studies contains five basic stages; they are (1) preliminary planning, (2) data collection, (3) data analysis, (4) field verification, and (5) final analysis and presentation of data. The steps required in the performance of each of the stages are outlined below.

Preliminary Planning

A. List the purpose of the engineering survey and the prime factors that need to be evaluated.

B. Review existing imagery and literature coverage that is pertinent in order to determine the extent to which new coverage must be obtained.

C. *Plan for new coverage.* This will include consideration of sensor package and platform, time of year, time of day, scale, flight direction, field support data required, and proposed method of analysis and method of presentation. For large projects a cost/benefit analysis should be made comparing alternate techniques of data acquisition, reduction, analysis and presentation.

Data Collection

A. *Perform preliminary field investigations.* Preliminary investigations are needed to determine reflectance, emittance, and other properties of natural materials and critical terrain features. These data are used for selecting film-filter and other sensor combinations, and the optimum times of day for performing the aerial mission. If topographic maps are to be prepared for the study area and photogrammetric methods are to be used, targets should be placed and field control obtained.

B. *Maintain close coordination with the nearest weather station.* To obtain up-to-date weather forecasts and long-range forecasts a close coordination with nearby ground meteorological stations should be maintained. This step is necessary in order to select the best weather conditions for the aerial mission, and also to keep standby time for the mission to a minimum.

C. *Obtain imagery coverage as planned.* For extensive missions or where a large number of sensors are flown, arrangements should be made to have some immediate playback or developing of a portion of the data collected for inspection before the aircraft leaves the area. This is to insure that the test area was properly covered and that the data appear to be satisfactory.

D. *Collect ground support data simultaneous with the aerial flights.* The ground data should be collected to aid in reducing and interpreting the aerial data. Depending on the sensors flown and the extent of the mission, such data as solar reflectance, sky radiance, infrared radiation, soil temperature, soil moisture, and microclimatic conditions are obtained at a test site along with ground photographs.

Data Analysis

A. *Prepare a mosaic of the entire area.* The mosaic provides an overview of large regional patterns which can be related to physiographic, geologic, climatic, environmental, and cultural factors. The mosaics can be prepared from aerial photography, radar, or satellite coverage where available. They may be uncontrolled, semicontrolled, or fully controlled depending on the specific requirements for the project.

B. *Review existing literature.* This is done primarily to help identify pertinent features and obtain information on the types, sequence, and lateral and vertical extent of the geologic formations present. Examples of the types of literature reviewed are reports pertaining to ground-water and well-log borings; geologic and geophysical reports, including pertinent masters and doctoral theses; agricultural soil surveys; and engineering soil surveys and auger and core boring reports. See a later section on combined techniques.

C. *Develop a regional concept of the study area.* An evaluation of the mosaic, the literature review, and some preliminary stereoscopic examination of the aerial photography and other imagery can be used for developing a comprehensive familiarity with the study area. By this means a three-dimensional concept of the region often can be developed that depicts regional geologic structure; distribution of geologic materials and landforms; vertical thickness and sequence of formations; and interrelationships existing between regional physiography, geology, and cultural and environmental factors.

D. *Perform detailed analysis* of the aerial data

collected. This entails the detailed stereoscopic examination of the pattern elements on the photography (i.e., topography, drainage patterns, erosion patterns, photo tone patterns, native vegetative patterns, and patterns of cultivation. A detailed study of all of the pattern elements together usually is critical in determining the soil and terrain characteristics of landforms.

Analysis of topography gives some indication of the massiveness and hardness of rocks or the texture of unconsolidated materials. Hard massive bedrock is usually indicated by steep slopes and high massive mountains or ridges with little fracturing and dissection by drainage. These materials will form good foundations for large structures but will require blasting if excavation is required. Steep slopes in areas of low hills or fairly level plains may indicate rock or granular materials; where slopes are gentle or have a low gradient they usually indicate fine-grained materials.

The characteristics of the *drainage patterns* provide important clues to types of materials. The presence of a very dense drainage pattern is generally indicative of a very dense drainage pattern is generally indicative of an impervious, fine-grained material, whereas sparse drainage patterns are associated with porous, coarse-grained materials. The type of drainage pattern is also indicative of the types of materials. For example, dendritic drainage patterns indicate the presence of clays or shales and pinnate drainage patterns indicate the presence of loess or silty materials.

The types of *erosion patterns,* and more specifically the types of gullies formed, are especially indicative of the types of soils. The configurations of gullies are associated with certain soil types. Coarse-grained materials are associated with steep, V-shaped gullies, whereas fine grained materials are often represented by shallow, gently sloping, saucer-shaped gullies.

Often, the *photographic tone/or color pattern* suggests changes of vegetation and soil moisture, and in some cases, even the soil type is revealed on both black-and-white and color photographs. Dark-gray photo tones often indicate a high water table and/or depressions containing fine-grained soils. Light-gray tones correspond to low moisture, porous materials, and/or topographically elevated positions containing coarser soils. Frequently, from experience with photo tones and color and the associated vegetation types, the interpreter is able to define the soil type.

Vegetative patterns often can provide information on soil textures and moisture conditions. In the eastern United States, jack pine trees are usually found on coarse-grained soils, whereas black ash, tamarack, and white elm are normally found on wet, fine-grained soils. Deciduous or coniferous forests are recognized from contrasting photo tones or colors.

Cultural patterns (or how man adjusted to the land) often can provide valuable information on soil and terrain conditions. The types of agricultural patterns may be especially valuable indicators. For example, the presence of orchards usually indicates well-drained, coarse soils; the presence of contour farming or an irregular wavy plowing pattern indicates steeper slopes, and (or) erodible soils; and the presence of deep furrows indicates heavy clays and a high water table. If quarries are observed on the photo, the gradient or steepness of slope of the quarry walls often enables the interpreter to determine if rock or gravel is represented. For example, vertical slopes and squared edges often indicate stable bedrock, whereas slopes at a more gradual angle and with scalloped edges often indicate gravels.

Together all of the above elements offer evidence of a given terrain pattern and its associated soil types. The descriptions of the elements are synthesized into a representative pattern for each landform, and the soil and terrain properties are estimated. In deriving information of the various types that are discussed under the above headings, the image analyst usually must make a comparative evaluation of the various kinds of sensor data; he also must develop a correlation of the interpreted data, the field data collected in support of the aerial mission, and the information derived from the literature review. In large projects containing a variety of sensor data, special machine analysis techniques such as density slicing, image enhancement or ditigal image analysis might be employed. These special techniques increase the interpreter's awareness of what items can be uniquely separated on the various data sets and increase his accuracy of interpretation. This increase in the amount and accuracy of information derived through remote sensing decreases the amount and cost of field verification required in subsequent steps.

E. *Develop a classification system.* This system should be relevant to the particular project and define the basic units to be mapped. Based on the detailed analysis of the data and literature review, typical vertical profiles for the basic units are developed and the areal extent of each unit is delineated on the photographs and on the mosaic of the study area. Attempts are also made at this stage to estimate the physical properties of the basic units delineated.

F. *Delineate the basic units classified* on the photographs and mosaic. If necessary, transfer the delineated units to a base map. In classifying and delineating the basic units, one should give consideration to the cost and purpose of the overall study.

Field Verification

A. *Select sampling locations*. These locations should offer the best opportunities for verifying the office interpretation. Points for investigation and routes of travel to the various points are determined from the stereoscopic study of the photography. As the interpreter gains experience in a given region, the amount of field checking will diminish. Some spot checks should be made, however, even in familiar areas.

B. *Perform the field verification*. The amount of field investigation required varies with the experience of the interpreter, the amount of supplementary information available, and the detail and accuracy required for the project. The photographs and/or imagery are brought to the field for correlation of the delineated patterns with the ground conditions.

C. *Collect limited samples for defining basic properties*. Convenient sampling areas such as cut slopes and natural slope exposures are selected. Some limited borings may be required, but hand-held exploration equipment such as augers, probes and peat samplers are usually sufficient at this stage.

D. *Complete laboratory testing of samples collected during field verification*. Determination of engineering properties of the soil and rock units is verified by the laboratory analysis.

Final Analysis and Presentation of Results

A. *Perform the final analysis*. This analysis will be based on combining the results of the field verification and the preliminary analysis which is corrected and completed.

The subsequent steps performed depend on the application to be made of the derived information in relation to the final project. Where information is being developed for preliminary planning purposes, such as in the making of regional inventories or in the early stages of project planning and in the comparison of alternative sites in corridor surveys, the final report can now be developed. Where detailed information is required for design purposes, such as the final location and design of a route, or for specific site-investigation purposes, such as the location of bridge crossing or the design of cut slopes, the information developed to this point forms the basis for planning a detailed subsurface exploration program. The information developed in the subsurface exploration program is then combined with the information developed to this stage in preparing the final report.

B. *Prepare final report*. The report usually includes maps delineating areal extent of the landforms and the soils which are classified, along with typical cross sections and vertical profiles, tables summarizing the test data, and a written text. The format of the final maps depends on the map accuracies required. For small scale regional map coverage, annotated photographs or mosaics, or direct tracings from these media would be satisfactory. Where large scale, accurate maps are required, additional processing is needed in order to transfer the information from the photography to the finished map. This commonly is accomplished at minimum cost by transferring the details from the photography to an existing base map of suitable accuracy through the use of a detail-transfer instrument such as the vertical sketchmaster, transferscope, or reflecting projector. Where suitable base maps are not available, more costly techniques will be necessary such as rectifying the individual prints and preparing controlled mosaics, preparing the base maps by photogrammetric mapping techniques, or by preparing orthophoto maps. Generally, the most useful maps are those which have a photo base. These provide background details that are not usually included on a topographic map.

THE KEY APPROACH

In the use of a "key" approach, a set of keys or photographs depicting the common patterns present in a given region or area must be developed. A detailed description of the pattern elements and materials present is also included. In this approach, the unknown photo patterns are compared to the various keys and a selection is made of the one that is closest in appearance. This matching of patterns then defines the unknown landform and provides a fairly complete description of its expected properties. This approach has not been used extensively because of the great variety of photographic patterns encountered due to variations in climate, geology, culture, and other factors.

Examples of studies where regional keys have been developed are those by Liang (1964) and Feinberg (1964). Liang describes the characteristics and pattern elements for tropical soils. The report by Feinberg includes mosaics and stereograms depicting the pattern elements for landforms found in each of three major physiographic regions of Illinois. The key approach may also be used with satellite or other data. The use of keys designed by a specialist-interpreter allows a less experience individual to make maximum and rigorous use of data and to do so more rapidly than would be possible with an unstructured, unaided analysis. Figure 32-1 is an example of a stereogram key. The following is the description of this stereogram:

1. LOCATION—Scott County (T15N;R13W)

2. PARENT MATERIAL—Wind Blown Silt Loam of Wisconsin Age (loess greater than 2 meters thick). Alluvial soils appear in the valley in the lower portion of the stereogram.

3. LANDFORM—The topography is generally dissected with the upland areas nearly level to gently rolling.

4. REGIONAL DRAINAGE PATTERN—Well developed pinnate drainage system.

5. EROSIONAL FEATURES AND GULLY CHARACTERISTICS—In general, the gullies display a "U" shaped cross-section.

6. COLOR TONE—Nearly uniform light to

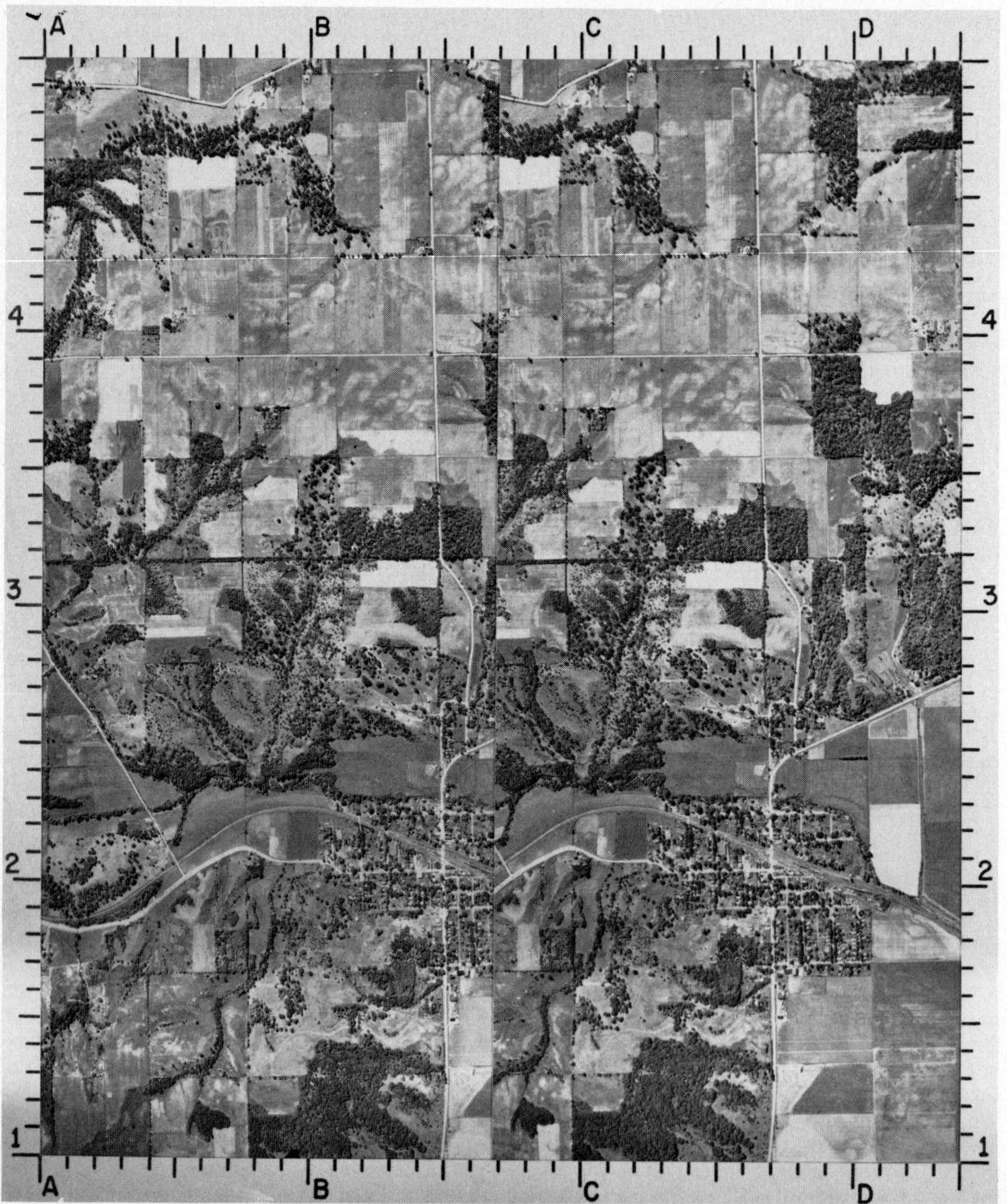

Fig. 32-1. This example of a photo "key" is in the form of a stereogram that illustrates a landform containing loess greater than 2 meters thick. (After Feinberg, 1964, courtesy University of Illinois.)

medium color tones except for the vegetation pattern.

7. SPECIAL FEATURES—The pinnate drainage pattern and "U" shaped gullies.

8. VEGETATION—These soils have developed primarily under a forest vegetation; natural timber stands usually are still evident.

9. PEDOLOGICAL DESCRIPTION—The soils are primarily Gray-Brown Podzolics mapped as Fayette and Sylvan. Bold, a Regosol, appears on some of the steeper slopes where soil profile development has been impeded.

10. SUMMARY—The pinnate drainage pattern and "U" shaped gullies indicate that this area consists of thick loess.

THE COMBINED TECHNIQUE APPROACH

Several studies have been reported where an emphasis was placed on identifying naturally occurring terrain associations in which the same group of landforms recur with the same interrelations. Brink and Partridge (1967) describe a terrain classification system developed in South Africa, in which natural terrain features are grouped into physiographic units, called "land systems", on the basis of the recurring association of individual landforms, called "land facets". These physiographic units then form a framework for economically collecting and storing information on the engineering properties of any recurring landform within the unit and for assessing its suitability for particular purposes. The land-system approach, first developed and employed in Australia in the 1940s, permits selection of basic terrain units by airphoto interpretation, with a minimum of field checking. The land-facet units are delineated on the airphoto. The soil properties and hydrological characteristics are outlined. Intensive field studies and soil profiling are carried out to check key areas of each land facet. Generalized profiles are generated for each land facet along with detailed profiles for the field-test pits.

Odenyo (1979) applied a similar land-system approach using satellite data in the Narok region of Kenya. The nine land units mapped were related primarily to grazing, but were also found to be of use for the Kenya Geological Survey's use for terrain analysis and for updating topographic maps. Single visual analysis of photographic products, color additive viewing and digital image analysis were all found to contribute useful results, depending upon the level of detail required.

In another approach Mintzer and Bates (1975) reported that the terrain analyst combines the following three techniques: (1) using color infrared photography in extracting landform, drainage system, and vegetation data, (2) next using color aerial film in extracting the gully shape and gradient, erosional feature, and natural color tone data, (3) next, using both color and color infrared photography in extracting cultural and special feature data. The photointerpreter then established the engineering soils characteristics from a combination of the above data with other terrain data from appropriate field survey methods, e.g., geophysical and earth auger and/or core-boring surveys at specific locations. This combination of the interpretation of color and color infrared photography and field survey methods results in a cost-effective investigation procedure.

An optimum system that was described by Rib (1966) for performing detailed engineering soils mapping uses data from aerial color transparencies combined with data recorded by the multichannel scanner in the ultraviolet and middle infrared and far infrared regions of the spectrum.

TERRAIN ANALYSIS

ENGINEERING SOILS MAPPING

Almost every major construction project is controlled to a very high degree by the natural terrain conditions. It is a rare situation when construction does not require some adjustments of the terrain, even if this is done merely by leveling off the high areas or filling in the low areas. An optimum location, however, requires a minimum disturbance of the natural landscape.

The cost of moving the natural materials is usually the largest single cost in engineering construction projects. Hence a knowledge of the materials comprising the terrain is important for the proper planning, location, construction, and maintenance of engineering facilities. For example, the cost of excavating hard-rock layers might be five to ten times greater than the cost of excavating unconsolidated materials. Therefore, as a method of minimizing construction costs, some organizations prepare engineering soils maps on a regional basis for planning and location purposes, and perform detailed engineering soil surveys for design purposes.

Remote sensing-aided terrain analysis plays a vital role in acquiring knowledge of terrain features. Its main virtue is that it provides a wealth of detail in the form of a dimensional model of the landscape. Such a model helps to present the relationships existing among climate, geology, soils, vegetation, and the culture of a given environment. No other media of presentation is as useful in portraying these interrelationships.

The same remote sensing-aided approach is employed in locating construction materials and also in defining the regional physiographic setting and delineating the various landforms present. There is one major difference, however. In performing investigations for engineering soils mapping it is necessary to recognize and delineate the soils for all of the landforms, not just for those which might contain construction materials. The next paragraphs develop this application of terrain analysis.

The value of aerial photographic interpretation for delineating engineering soils was demonstrated some four decades ago by Belcher (1943).

Shortly thereafter, through the use of aerial photographs as a basic tool in delineating soils, regional engineering soils-mapping programs were initiated. For example, such programs were initiated in New Jersey and Indiana in 1946 and in Maine in 1948. The basic mapping symbols and techniques developed in these early studies have been expanded and adapted for use in performing numerous engineering soil surveys throughout the world. Regional engineering soils-maps are usually prepared for general planning and site evaluation purposes. Their preparation generally involves information gleaned from the literature, analyzed on aerial photography, and verified by limited field checking, field exploration, and laboratory testing. A second, more detailed level of engineering soil-mapping is also performed; these maps are prepared mainly for final route-location studies and for site investigations where detailed information on soils and their properties are needed for design purposes. Aerial photographic interpretation is used extensively for this level of mapping also.

Several approaches have been used in the performance of engineering soils-mapping, including the development of regional keys depicting typical pattern elements for the various landforms; the mapping of soils classes of significance to engineering construction, the delineation of landform-engineering soil relationships; and the grouping of landforms into physiographic settings or terrain associations having common engineering problems or methods of analysis.

Soil Classes

Figure 32-2 illustrates a portion of an engineering soils-map prepared for a quadrangle area in Maine. This type of map depicts soils classes pertinent to engineering construction. Through the use of such a map, it is relatively easy to group the various soils into the two categories—common and rock excavation—used in estimating construction costs. The report accompanying the map includes discussions of the landforms present and engineering characteristics of the various materials.

Landform Engineering Soil Characteristics

The most extensive form of engineering soil mapping uses the landform-engineering-soils relationship. Surveys of this type have been performed on a statewide, county, area, and strip basis. The delineation of landforms constitutes the foundation of this technique; however, the landforms are subdivided based on their engineering characteristics which include soil texture, drainage conditions and, for some, a slope category.

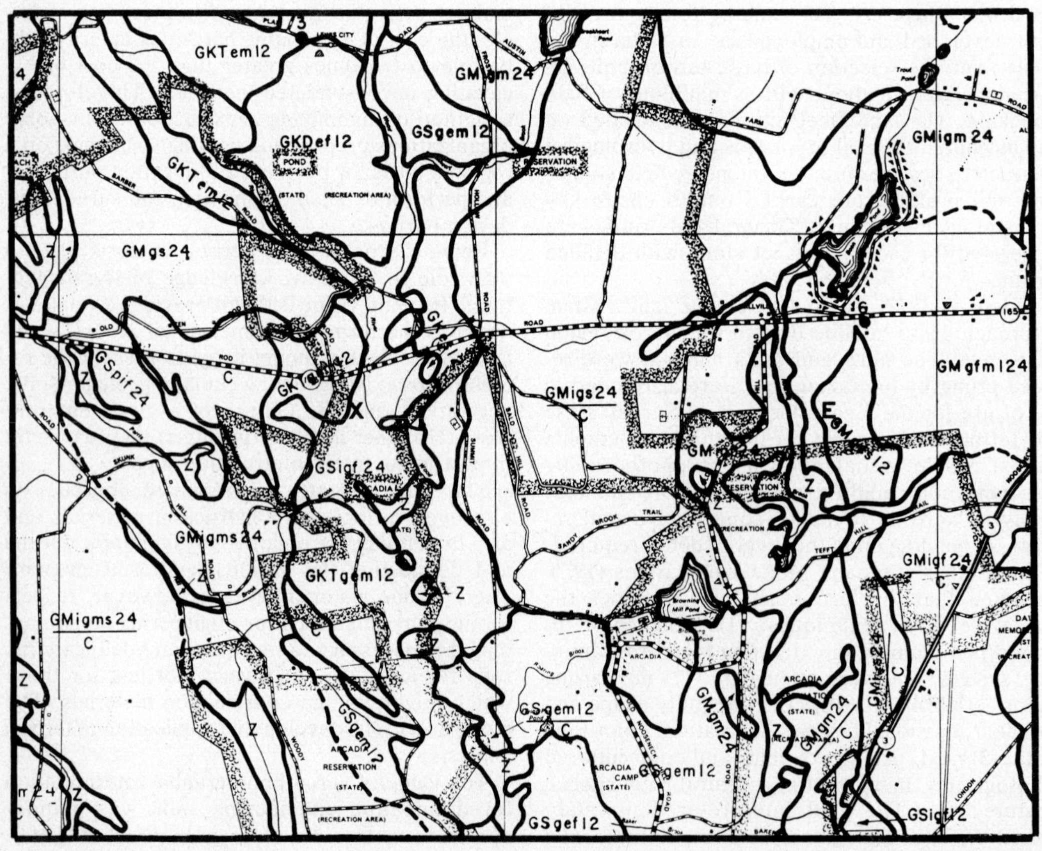

Fig. 32-3. Portion of engineering soils map for Washington County, Rhode Island. (After Moultrop, 1956.)

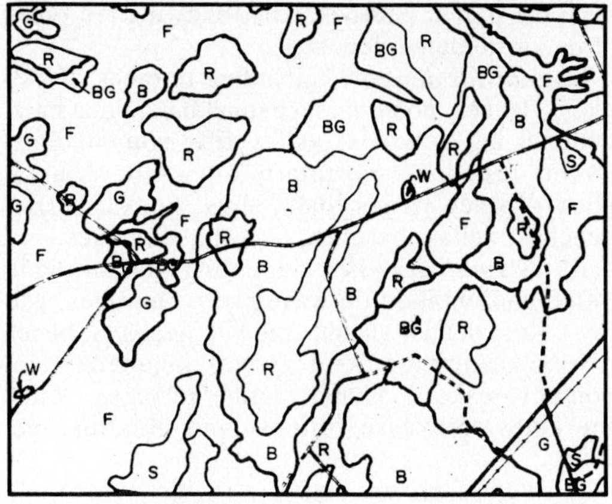

LEGEND

R – ROCK, ledge less than 5' from surface

G – GRANULAR, generally clean gravel or sand

BG – BOULDER GRANULAR, matrix sandy or gravelly

B – BOULDER, matrix silty or clayey

F – FINES, silt or clay

S – SWAMP, saturated sites with surface peat less than 3' thick

P – PEAT, saturated bogs with surface peat more than 3' thick

W – WATER, generally more than 2' deep

Fig. 32-2. Portion of engineering soils map for Gorham Quadrangle, Maine. (Courtesy Maine Department of Transportation.)

The Statewide engineering soil maps that have been completed for New Jersey and Rhode Island are examples. Figure 32-3 shows a portion of an engineering soil map for Rhode Island; as noted in this figure, a two-part symbolic notation is used to describe the mapping unit. For example, for the unit GMgs24, the first part of the symbol, being in the capital letters, GM, indicates the landform—in this case, a glacial ground moraine. The second group of letters and numbers refers to the engineering characteristics. The first letter (or letters) used in this part of the symbol indicates drainage characteristics or depth to water table—e, excellent; g, good, i, imperfect; p, poor. The next characteristic described is the slope—f, flat, 0–3%; m, medium, 3 to 7%; or S, steep, >7%. In the last part of the symbol, the numbers refer to the texture of the soils—the numbers 1 through 7 refer to A-1 through A-7 of the AASHO Classification system—thus, 24 refers to the A-2 and A-4 textural classes. A change of any of these characteristics requires the establishment of a new map unit (e.g., GMigm24). Additionally, if a thin layer of one material overlies another material, both are included as, for example GMigs24/C, which indicates a thin layer of glacial grouund moraine overlying crystalline bedrock. The engineering soil maps for Rhode Island (and similar maps for New Jersey) are incorporated in reports that also include test data for the major soil units delineated, additional details on the properties of the various landforms, and a discussion of the engineering uses or problems associated with the mapped soil units.

This mapping system is extensively used on a county basis and for strip or corridor studies. A number of other states in the United States, and also several other countries, have adopted similar methods with minor variations in presentation. In many corridor studies, for example, photo mosaics are used as map bases. For most corridor studies, the mapping units are displayed on photo mosaics.

CONSTRUCTION MATERIAL INVENTORIES

The availability of suitable construction materials within economic hauling distance is crucial to the planning and execution of most engineering projects. Additionally there is a need to conserve the high-quality materials which are becoming scarce. These conditions have been especially prevalent in highway construction in the United States where an extensive highway program was initiated in 1956 and large quantities of construction materials were needed. As a result, many state highway organizations have instituted regional materials inventories on an area, county, or statewide basis to assist in their planning programs.

Remote sensing imagery, particularly aerial photography, has proved to be of great value in performing inventories for construction materials. The materials successfully mapped from various kinds of imagery range from boulders and quarry rock, through sand and gravel of varying quality, to bituminous-filler and clay-binder materials. Materials of specific characteristics and gradations, such as sand-clay mixtures, well graded or narrow graded mixtures and clean granular materials with low percentages of fines have been located by experienced interpreters using aerial photographic interpretation techniques. This latter level of detail, however, is usually limited to specific site investigations where large-scale photo coverage is obtained and more detailed analysis is performed.

One approach that has been developed for locating construction materials through the use of image interpretation and terrain analysis techniques is worthy of emphasis. The approach includes defining the regional physiographic setting and delineating the various landforms from their photo patterns. Knowledge of the origin of various landforms (the science of geomorphology) is imperative to this approach. An understanding of the processes by which landforms are developed enables one to predict the types of materials to be expected. For example, if sand dunes are recognized, one can expect to find a fairly uniform gradation of sand-size particles because these materials are deposited by wind action which very selectively sorts the materials that it moves. In this approach, a search is made for those landforms which, because of the nature of their origin, are likely to contain the particular materials sought. Once the landforms are delineated, they are analyzed in detail in order to select the most favorable sites for field investigation. Conversely, this approach will also eliminate those areas where the likelihood of obtaining suitable materials is very small, thus allowing the search to be concentrated in the most favorable areas. This technique has been successfully applied for forty years. An example of the degree of success achieved by this technique was reported by Mollard and Dishaw (1958). The results were based on their experience over a 10-year period in locating over 2,000 prospects. For those sites verified by borings, successes ranging from 70 to 100 percent were achieved in locating construction materials.

Granular materials are the most common construction materials sought. These can be found as naturally occurring unconsolidated deposits, or suitable consolidated deposits can be crushed to obtain the required materials. The terrain analysis approach is utilized in locating both types of deposits.

Suitable granular materials can be obtained from numerous landforms, both unconsolidated and consolidated. The landforms, and specific features within the landforms, are listed below. These are subdivided into unconsolidated and consolidated and are further separated by method of deposition or mode of origin. In estimating the potential of a deposit within these landforms one finds it necessary to consider the environment in which it is found, the physiographic setting, and its geologic history.

Now, in addition to the more common panchromatic airborne data, maps for similar purposes are being prepared using color infrared photography, Landsat digital analysis and field work. Singhroy (1980) has prepared sand, gravel and quaternary geology maps for a 25,000 square km area using a combination of methods in the PAS region of northern Manitoba. Although the data sources used do require that minor modifications be made to the methods used, the approach used by Singhroy is still largely the same as those described where only panchromatic data were obtained.

Unconsolidated Deposits

Most granular materials are closely associated with deposition by running water. Running water has the ability to sort materials as a function of the velocity or changes in velocity of the currents. Some sorting is associated with other agents, such as wind and gravity, but such sorting is either too selective (wind deposits usually have a uniform particle size) or very poor (gravity produces a broad range in sizes varying from boulders to clay). In some cases granular materials are found in residual soils which have weathered in place from the parent bedrock. The following are types of unconsolidated deposits:

(1) *Alluvial deposits,* including terraces, floodplains, levees, point bars, channel bars, delta bars, channel lag deposits; valley fill—alluvial fans, basin deposits; mountain outwash; coastal plains—beach ridges, bars, spits, hooks, barrier beaches, deltas, beaches, and alluvial cones.

(2) *Glacial deposits,* including glaciofluvial-pitted and unpitted outwash plains, terraces, valley trains, deltas, glacial-spillway deposits, beach ridges associated with glacial lakebeds; ice-contact—eskers, kames, kame terraces, kame moraines, crevasse fillings, and ablation moraines;

(3) *Wind deposits:* dunes;

(4) Gravity deposits; talus;

(5) residual deposits, including residual granite; lag gravels and other regolith; cinder cones; and

(6) man-made deposits such as mine tailings.

The typical pattern elements for many of the landforms and features listed above are detailed elsewhere in the manual.

Consolidated Deposits

The suitability of consolidated rock formations is more a condition of the physical and chemical properties of the rock than the type of rock. Most types of rock have been used, on occasion and in some form, as a source of aggregate or building stone; however, experience has indicated which rock types are most suitable and which characteristics are desirable. The properties of rocks can only be determined by means of extensive laboratory testing, not by patterns present on imagery. Nevertheless, aerial photographic techniques can be used to locate those rock types which offer the best potential as a source of aggregate. A knowledge of local geologic conditions, and experience with the use of the local rock types are crucial in the performance of this analysis. For example, limestones are used extensively in engineering construction, yet not all limestones make suitable aggregates. The following are types of consolidated rocks used in engineering construction: (1) *sedimentary rocks,* including limestone, dolomite, sandstone, and conglomerate (well indurated); (2) *igneous rocks,* including intrusive-granite, syenite, monzonite, diorite, gabbro, extrusive-rhyolite, trachyte, latite, dacite, andesite, basalt, silis and dikes-diabase, and dolerite; and (3) *metamorphic rocks,* including gneiss, schist, and quartzite having variable properties.

The recognition by image interpretation techniques of the individual rock types based on their mineralogical composition is not possible. What is usually accomplished by interpretation techniques is to classify the rocks into broad classes—e.g., granite (coarse-grained igneous) rocks which include granite, syenite, monzonite, and numerous minor types. Typical pattern elements for some of the major rock types are illustrated elsewhere.

REGIONAL INVENTORIES

Levels of mapping performed in regional materials inventories vary considerably. In some surveys, only inventories of existing pits are prepared. In these surveys, photographs are utilized to correlate the materials in the pits with the landforms in which the pits are located. They also aid in defining the limits of an existing pit and in determining whether that pit can be expanded. In other surveys, the purpose is to inventory existing material pits and to locate other potential material sources. Limited field investigations are performed in conjunction with these surveys to verify the analysis, to determine estimates of quantities of materials available, and to collect samples for laboratory testing to determine the suitability of materials for their intended use.

A typical materials map, as prepared by a highway commission, indicates existing and proposed material sites, various landforms depicted by different colors and stippled patterns, and geologic units shown by standard mapping symbols. Included in an accompanying report are detailed maps showing the locations of pit sites, tabulations of the test results for the pits, and a detailed discussion of the various soil and rock units encountered and the suitability of each for construction uses. Regional inventories have also been prepared by Singhroy and Dixon (1980) from the visual analysis of Landsat data for general features such as organic, alluvial, littoral and glacial deposits, gravel pits, precambrian rock, paleozoic rock and lineaments. More detailed inventories for the same region were done from 1:36,000 scale color infrared photos.

DISCUSSION OF PROCEDURES AND IMAGERY

An abbreviated form of the procedure outlined in the preceding section can be utilized to perform a regional materials inventory. Two reports by the Colorado Department of Highways (1962) and the Wyoming State Highway Department (1963) provided detailed procedures for performing region materials inventories.

For the majority of the materials inventories that have been performed in the past, black-and-white panchromatic photographs, available from the Department of Agriculture or the U.S. Geological Survey at scales of 1:20,000 or smaller, were utilized. In some of the more recent surveys (e.g., in Kansas and Wyoming), panchromatic photographs flown by the State highway organization at scales of 1:15,840 or smaller were used.

In the late 1950s and early 1960s, the Bureau of Public Roads, Denver Regional Office, performed several materials surveys of national parks using color aerial photography (Chaves and Schuster, 1964). These surveys proved very successful and the investigators concluded that color photography was preferable to black-and-white photography for materials surveys. Some of the advantages noted for color were (1) a greater ease in distinguishing between unconsolidated and consolidated deposits and between the various types in each category, (2) an increase in the accuracy of the interpretations, and (3) an increase in the speed at which the analyses were performed. These results have been confirmed and broadened to include some success with Landsat in the work by Ryerson (1981) and Singhroy (1980). In subsequent material surveys, color photography has been used exclusively. Nevertheless, the available panchromatic imagery is the most widely used for localized studies because of cost. Where panchromatic photographs are not available, as in many locations in the underdeveloped parts of the world the use of Landsat data should be considered.

Sensor Comparison for Materials Survey

A search for materials was performed by a team of scientists over a region in the Mississippi Delta (Orr and Quick, 1971). The team evaluated several types of remote sensor data including side-looking radar, thermal infrared imagery, natural and infrared color photography, a nine-channel multispectral scanner, and black-and-white photography. An Apollo-9 photograph which covered the test area also was reviewed. The general consensus of the team was that, for detailed mapping of construction material sources in a delta environment, color infrared photography is the best. (Refer to the comparative analysis shown in Table 32-1). In evaluating the various systems for performing a regional analysis—determining the physiographic setting and identifying landforms—the photoindex sheet prepared from the black-and-white photography is best although radar imagery also can be useful. In this Mississippi Delta study the radar imagery was very helpful in delineating relic shorelines and filled embayments, which focused the attention of the interpreter on certain areas to be evaluated in detail using the larger scale imagery to complete the search for the construction materials.

In a corollary evaluation it was also noted that many of the landforms could be distinguished on the Apollo satellite photography. The recent successful use of Landsat multispectral scanner data by a variety of investigators supports this earlier assessment of the usefulness of Apollo satellite photography.

The results of these studies and many other studies, have demonstrated that, for large regional materials inventories, ultra-high altitude aircraft or sattelite photography and side-looking radar can be useful. The synoptic views provided by these systems enable interpreters to determine regional geologic structure and regional trends not previously noticed on the larger scale photographs or mosaics. This results in a better understanding of the origin of the various landforms and the regional landscape, making it possible to predict with more confidence the types of materials that

TABLE 32-1

Sensor Comparisons For Detailed Area Study of Landforms

Landform	Sensor			
	Pan Photo	Color Photo	Color IR Photo	Thermal IR
Active Beach	Good	Good	Excellent	Good
Chenier	Good	Good	Excellent	Fair
Marsh	Fair	Good	Excellent	Excellent
Terrace	Good	Excellent	Good	Fair
Backswamp	Good	Good	Excellent	No Coverage
Natural Levees	Good	Excellent	Excellent	Fair
Abandoned Channels	Good	Excellent	Excellent	No Coverage
Point Bars	Good	Excellent	Excellent	No Coverage
River Bars & Islands	Good	Excellent	Good	No Coverage
Spoil Banks	Good	Good	Good	Good

[After Orr & Quick (1971), Courtesy U.S. Army Engineer Topo. Labs.]

might be encountered and the possibilities of locating buried deposits. Detailed discussions of the uses of satellite data for mineral-deposit exploration is presented elsewhere in this manual. Recent resource studies reported by Colwell, et al. (1981) and Benson, et al. (1981) in California and Colorado focused attention on image interpretability in the search for resources. Details are given on these studies in the next section.

Image Interpretability for Resources

Two recent examples of interest to engineers and others will serve to illustrate how the relative interpretability of various image types can establish categories of forest resources. The first of two test sites is in California, on the Plumas National Forest, and the second is in Colorado, on the San Juan National Forest. The research was conducted under auspices of the Nationwide Forestry Applications Program of the U.S. Forest Service.

The California Test

The objective of the California test (Colwell, et al. 1981) and (Benson et al. 1981), was to determine the relative interpretability of seven image types in terms of meeting informational requirements of the U.S. Forest Service using the Plumas National Forest in California as a test site. The seven image types consisted of: (1) aeronegative color, (2) aero-negative black-and-white, (3) optical bar color infrared, (4) large format camera (LFC) black-and-white, (5) LFC color infrared, (6) LFC color, and (7) "Geopic" color infrared enhancements made from bands 4, 5 and 7 of the Landsat MSS system. Image types 1 and 2 were obtained from 16,000 feet above the terrain; image types 3 through 6 were obtained from 65,000 feet above the terrain; and image type 7 was obtained at an altitude of 570 miles. The interpretability of these image types was evaluated using three different kinds of photo interpretation tests: (1) nonstereoscopic study by 28 student photo interpreters of 200 points on each of the seven image types,[1] (2) stereoscopic study by nine photo interpreters of 120 points on the first six image types, and (3) delineations of the resource boundaries by three photo interpreters. The latter of these tests compared the boundaries of eight of the resource categories: (conifer, conifer/hardwood, hardwood/conifer, shrub, grassland and bareground), as annotated on the last six image types, with delineations as annotated on the aeronegative color image. The 28 photo interpreters used for the first of these three tests were students who, at the time (1981) were enrolled in a rigorous 4-unit (credit) course in forest photogrammetry and photo interpretation, as presented in the Department of Forestry and Conservation at the University of California, Berkeley. Those interpreters used in the second and third of the above tests were highly experienced foresters employed under the university's Remote Sensing Research Program.

Based on the results of the California Test, the investigators concluded that the large-scale aeronegative color and black-and-white image types are significantly more interpretable than the image types obtained from the high-altitude platforms. The mean percent correct values for the image types ranged from 31 percent (type 7) (Landsat "Geopic" enhancement) to 91 percent (type 1) (stereo aeronegative color). With the exception of the LFC color photography, which in this particular instance, had a rather pronounced bluish cast, the interpretability ratings of the various image types obtained from the U-2 aircraft were not statistically different, one from another, in interpretability. (See Table 32-2).

Although the overall percent correct values were required to rank the image types with respect to relative interpretability, the quality of the

[1] In each instance the 200 points were comprised of 20 examples of each of the following 10 resource categories: conifer, conifer-hardwood, hardwood, hardwood-conifer, shrub, grass, water, rock, bare soil and "other" (mainly agricultural or urban).

TABLE 32-2

Summary of Image Rankings Based on Photo Interpretation Tests Conducted on the Plumas National Forest of California

Rank	Image Type*	Kappa and Significance**	Percent Correct -Range-	Percent Correct -Mean-	Percent Error -Mean-
1	Aeroneg Color	.699	57.5–80.8	73.7	26.3
2	Aeroneg Black & White	.657	60.8–78.3	70.0	30.0
3	Optical Bar CIR	.602	59.2–72.5	65.2	34.8
4	LFC Black-&-White***	.600	64.2–78.3	65.0	35.0
5	LFC CIR	.595	54.2–83.3	64.5	35.5
6	LFC Color	.529	56.7–78.3	58.8	41.2

* All image types were enlarged to a scale of 1:24,000 to match that of the aeronegative photography. Results were based on the stereo interpretation of 120 points per image by nine image analysts for eight resource categories (15 points per category): conifer, conifer/hardwood, hardwood/conifer, shrub, grassland, and bare ground. For further details, see Colwell, et al., 1981.

** The vertical lines to the right of the ranked Kappa values serve to group those image types which are not significantly different at the 0.95 confidence level.

*** LFC refers to the 9 × 18 inch large format mapping camera used in this study.

information present on the respective image types is perhaps best evaluated by the performance of the "best" photo interpreter. The overall conclusion drawn by the investigators is as follows: a skilled forest photo interpreter who is familiar with a particular forest environment can identify the resources present with an accuracy of 80 to 90 percent on any of the six aircraft image types and with an accuracy of 40 to 60 percent on the Landsat-image type. This conclusion is based on the interpretation of a total of 45,680 points in both non-stereo and stereo tests.

The Colorado Test

The Colorado test site (Colwell et al., 1982), where the research was conducted under the direction of the same team of remote sensing scientists as at the California Test site, had the objective of determining the relative interpretability of seven aerospace image types in terms of meeting the informational requirements of engineers and others of the U.S. Forest Service on the San Juan National Forest, Colorado. These image types consisted of: (1) aeronegative color, (2) aeronegative black-and-white, (3) 9 × 9-inch color infrared, (4) AMPS[2] color infrared, (5) AMPS high contrast black-and-white, (6) AMPS low contrast black-and-white, and (7) Landsat "Geopic-enhanced" color infrared images. Of these, image types 1 and 2 were obtained from 16,000 feet above the terrain, image types 3 through 6 from 60,000 feet, and image type 7 from 570 miles. The interpretability of these images was evaluated from stereoscopic photo interpretation tests that were administered as in the California Test, again using 28 students who, at that time (1982) were enrolled in the same basic 4-unit (credit) course as for the California Test site. These interpreters looked at 200 points in each image type and labeled each point as one of the following ten resource categories: aspen, spruce-fir, ponderosa pine, oakbrush, aspen-conifer, oakbrush-conifer, mesic meadow, xeric meadow, willow, and rock/talus. The results from these tests were analyzed and the image types were ranked as to interpretability. Based on the Colorado Test, the investigators concluded that the 9 × 9-inch color infrared photography was statistically the most interpretable with an overall identification accuracy of 75.1 percent.

[2] AMPS is the term applied to a 6-camera multiband system that originally was flown on Skylab but which was made available for this test by NASA in order to obtain matched sets of multiband, high-altitude, aerial photos rather than space photos. The cameras of this system have focal lengths of approximately 6 inches and the frame size for each photograph is approximately 70 × 70 mm.

Analysis of Results

The readers of this section who may be interested in "signature extension" for the image types tested, would ask this question: to what extent can the photographic tones and other image attributes of features in one geographic area be extended to identify similar features of engineering significance in another area? In order to answer this question the investigators compared the results from these California and Colorado tests, attempting to accentuate commonalities as they existed (See Tables 32-2, and 32-3 which show the common conditions). Table 32-2 is a record of results from the California Test providing a summary of the rankings of six image types, based on eight resource categories interpreted. Similarly, Table 32-3 provides, for the Colorado Test, a summary of rankings for seven image types, based on ten resource categories.

In addition to the common factors of the two sites it is noted that there are differences in (1) the environmental conditions, (2) some resource categories and (3) some image types. However,

TABLE 32-3

Summary of Image Rankings Based on Photo Interpretation Tests Conducted on the San Juan National Forest of Colorado

Rank	Image Type*	Kappa and Significance**	Percent Correct -Range-	Percent Correct -Mean-	Percent Error -Mean-
1	9 × 9 CIR	.724	57.5–87.0	75.1	24.9
2	AMPS CIR	.688	46.0–88.5	71.9	28.1
3	Aerong Color	.656	41.0–84.5	69.1	30.9
4	Aeroneg B&W	.615	28.0–79.5	65.4	35.6
5	Landsat Geopic	.548	44.0–73.5	59.4	40.6
6	AMPS B&W "Low"	.542	42.5–74.5	58.8	41.2
7	AMPS B&W "High"	.491	33.0–72.0	54.2	45.8

* All image types were at the scale obtained from contact printing, except for those obtained from the AMPS system which were 2-diameter enlargements.

** The vertical lines to the right of the ranked Kappa values serve to group those image types which are not significantly different at the 0.95 confidence level.

based on their field experience and their familiarity with the image characteristics of the resource categories, the investigators are confident that good inferences of the "signature extension" type can be made. Here we are specifically concerned with utility of the "best" image type observed at the Colorado test site (viz. the 9 × 9-inch color infrared, scale 1/60,000), as compared with that of the "best" image type observed at the California test site, (viz., the 9 × 9-inch aeronegative natural color scale 1/24,000).

In comparing the interpretability of the two "best" image types as they represented two different forest environments, one must account for 1) the relatively moderate climate at the California Site, 2) the resource categories exhibiting a high degree of diversity and vegetation mix, and 3) hypsographic distribution. At the California Site, with the exception of a few stands of true fir found at the higher elevations, there are no extensive stands of any single species of pure conifer. Most conifer stands consist of mixes of Douglas-fir, ponderosa pine, incense cedar and white fir, along with associated hardwoods such as Pacific madrone, tanoak, California blackoak and dogwood. The proportion of each species of conifers is highly variable and depends upon soil type, elevation zone, aspect, and past and present management practices.

Because of the complexity noted above of the vegetation mixes at the California test site, the investigators have concluded, with respect to signature extension, that the 9 × 9-inch color infrared photography image type noted in Table 32-3 would have ranked just below the high spatial resolution aeronegative color and black-white image types listed in Table 32-2. In this mixed conifer type forest, a high resolution image type is needed to separate the various conifer categories (which, collectively, comprise the "conifer" component) and this separation can only be done consistently on the basis of crown shape and percent crown closure for conifer and hardwoods. Within most of the conifer stands at the California site, this would not be observed on the 9 × 9-inch color infrared image type. It is noted that the 9 × 9-inch color infrared photography would be better than the (nominally) higher spatial resolution optical bar color infrared photography for the following reason: for a given 9 × 9-inch color infrared stereo pair, the analyst sees a large area at a greater base-to-height ratio than is possible on a given optical bar stereo pair. In addition, the interpretability at the outer edge of the 9 × 9-inch color infrared image is better than that at the maximum allowable scan angles of the optical bar color infrared. At the greater scan angles, the tree crowns "lay over", to a greater degree than occurs on the 9 × 9-inch conventional color infrared image, to the extent that it becomes very difficult to estimate the relative vegetative mix within a given timber stand.

In contrast with the above, one finds that, on the Colorado test site there tend to be more severe climatic zones and consequently much less diversity in the vegetation species composition in any particular zone. Color infrared film, to a greater extent than natural color film, tends to provide color differences that serve to accentuate the identity of each of the near monotypic blocks of vegetation. Furthermore, as compared with natural color photography, the color infrared film provides superior haze penetration. Hence, these color differences tend to be more discernible, even on a high altitude color infrared image than on a relatively low-altitude, natural color image.

The engineer needs to note that identifying and mapping of the vegetation types is of both direct and indirect importance. On the one hand in respect to logging, the engineer or forester has a direct need to know the location and distribution of the merchantable species of timber. On the other hand, for road design and construction one is more concerned with soils and drainage factors and notices that these factors are difficult to interpret through the forest cover. Usually these factors can be inferred, however, from an analysis of the species composition and vigor of the indicator species of vegetation.

Table 32-4 lists the film-filter combinations that

TABLE 32-4

Photographic Film-Filter Combinations Obtained in the San Juan National Forest Through Use of the Airborne Multispectral Photographic System (AMPS)*

Station	Film Type	Filter
1	EK 2443	EE
2	EK 3412*	BB
3	SO 127*	EE
4	SO 242	FF
5	EK 3414*	BB
6	EK 3414	AA

* Only the three marked with an asterisk were incorporated in the full scale test that involved 28 students. The following relates the listing in Table 32-3 to that contained above:
AMPS CIR = SO127/EE
AMPS B&W "Low Contrast" = EK 3414/BB
AMPS B&W "High Contrast" = EK 3412/BB

were used in the AMPS camera system for the Colorado test site. Also, there were three of these combinations that were incorporated in the full scale test involving the 28 students.

Finally, it is concluded that the best image types are color infrared and aeronegative color for the interpretability of forest resources.

In making these conclusions the investigators noted that the study was subjective and no accounting was made of factors such as 1) costs of photo acquisition, 2) image rectification, and 3) misplaced lines with respect to ground sampling error. Also, it is noted that these circumstances do not invalidate the above conclusions.

It needs to be noted here that in a later section of this chapter and in other chapters in this Manual there are discussions of various sensor and image types, and recommendations are given as to image types to use for given objectives.

COMPARISON OF REMOTE SENSING SYSTEMS

Black-and-white photographic interpretation has been the basic remote sensing method for various regional engineering-soils investigations in the past. Because of the value of the resulting soils maps, extensive investigations have been conducted with a variety of remote sensing systems in an effort to increase the accuracy of interpretation of soil and terrain conditions and decrease the amount of field verification required.

The major conclusion that has been reached by a vast majority of investigators, after evaluating various film types and sensor systems, is as follows: *natural color aerial photography is the best single sensor for interpreting soils*. Color infrared photography is of special value, however, in determining important terrain features such as drainage, land use, and vegetation conditions. Illustrative examples of the value of color aerial photography for soils delineation are demonstrated by Mintzer (1968b). Rib and Miles (1969b), and Parry et al. (1969).

In evaluating other sensor systems, some investigators have indicated that ultra-high-altitude aircraft and satellite photography, and side-looking radar imagery offer unique regional coverage and present the best overview of physiographic settings and terrain associations. On the other hand, pilotless drones containing small cameras or sensors and small tethered balloons have been suggested as constituting useful platforms for common aerial photography where localized conditions are of interest. Other sensor systems, such as those used in acquiring multispectral photography, thermal infrared imagery, and microwave radiometry, often can provide supplemental information that could not have been derived from the analysis of aerial photography or from the interpretation of one system alone. Digital terrain data produced by the Defense Mapping Agency and the U.S. Geological Survey have been used by many investigators for terrain analysis. For example, Khorram and Smith (1980a) used these digital tapes for mapping elevation, slope, and aspect in a wildland area in northern California. In many cases, the information obtained by these other sensors provided confirming evidence that increased the accuracy of the identification of soil and terrain conditions. Some examples of additional or supportive information provided by remote sensing systems, other than traditional aerial photography, are illustrated by the following figures and comments.

Figure 32-4 shows a comparison between black-and-white photography and daytime thermal infrared imagery (Rib, 1966a and b). Although the two types of images were collected a month apart, the tonal relationships noted for most of the selected points remained the same. Comparison revealed several tonal relationships that were of assistance in separating significant engineering soils and rock units which could not be distinguished from the analysis of either one alone.

As noted on the visible-wavelength photography, the glacial till soils on topographic highs (point 5) have light tones because of their low organic content and predominance of well-drained silts. The glacial-till depressional soils (point 6) have dark tones because of high organic content, finer texture, and darker colored soils. These

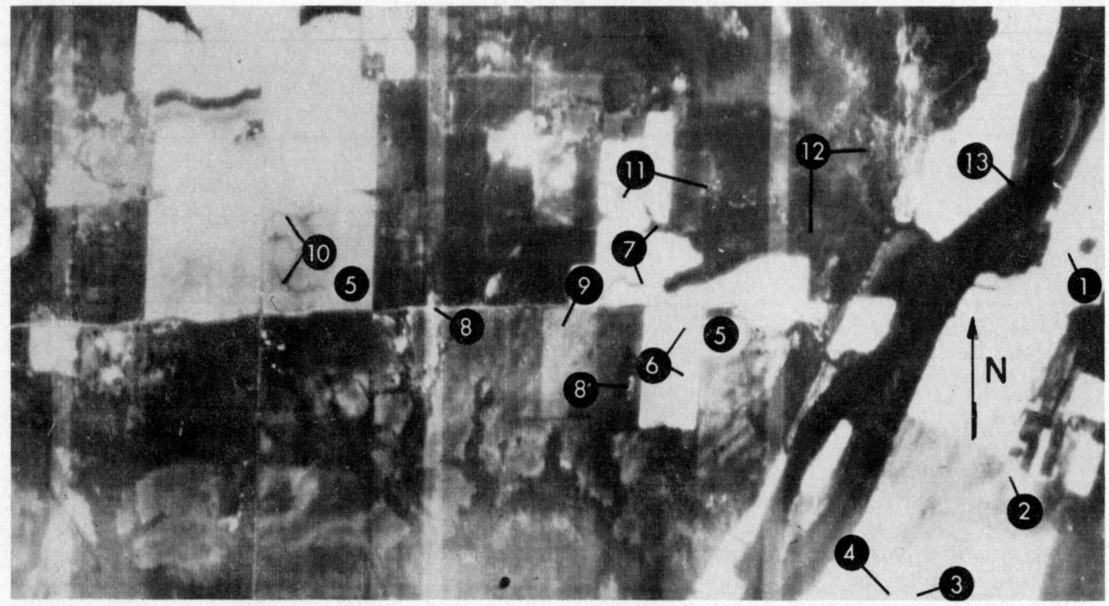

a. Daytime IR (June 2, 1966)

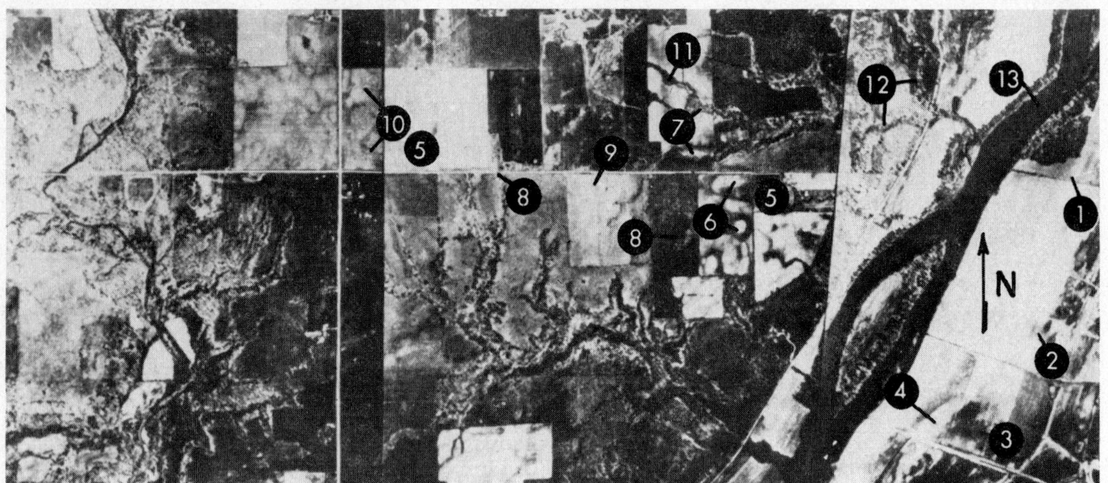

b. B&W photography (May 2, 1966)

Fig. 32-4. Comparison of soils and terrain features on black-and-white photography and daytime infrared imagery. (After Rib, 1966b.)

same areas show a reversal in tones on the infrared imagery: in the infrared region the darker colored soils absorb more heat than the lighter colored soils, emit more energy in the infrared region, and thus have a lighter tone on the imagery. Another example of tonal reversal is shown by point 1, which is a depressional, dark-colored, fine-textured soil in the flood plains. Other factors can change the tonal relationships on the imagery. Examples where tonal reversals do not occur for dark, fine-textured soils are demonstrated by points 3, 7, and 10. Point 3 is a muck deposit and points 7 and 10 are soils located in low drainageways in glacial tills. For these soils, the tones are not light on the infrared imagery, but medium to dark, and for different reasons; the darker tones in areas 3 and 10 are caused by a high moisture content while that of point 7 is caused by the presence of vegetation. Both of these factors cause a cooling effect on the surface soils, resulting in darker tones on the imagery. Similar variations are noted in light-colored soils. Not all light-colored soils on the photography are dark on the infrared imagery. Many are light or medium-light on the infrared (points 2, 4, 8, and parts of the area at 9). Point 2 is sand, 4 is shale bedrock covered by thin alluvium, 8 is sandstone bedrock, and 9 is a field where sandstone is very shallow. These areas remain lighter on the infrared imagery owing to differences in emissivity, heat capacity, thermal conductivity and (or) surface-to-volume ratio. The factor causing the lighter tone cannot be evaluated by comparing the infrared imagery and visible photography alone. The overall effect and not the

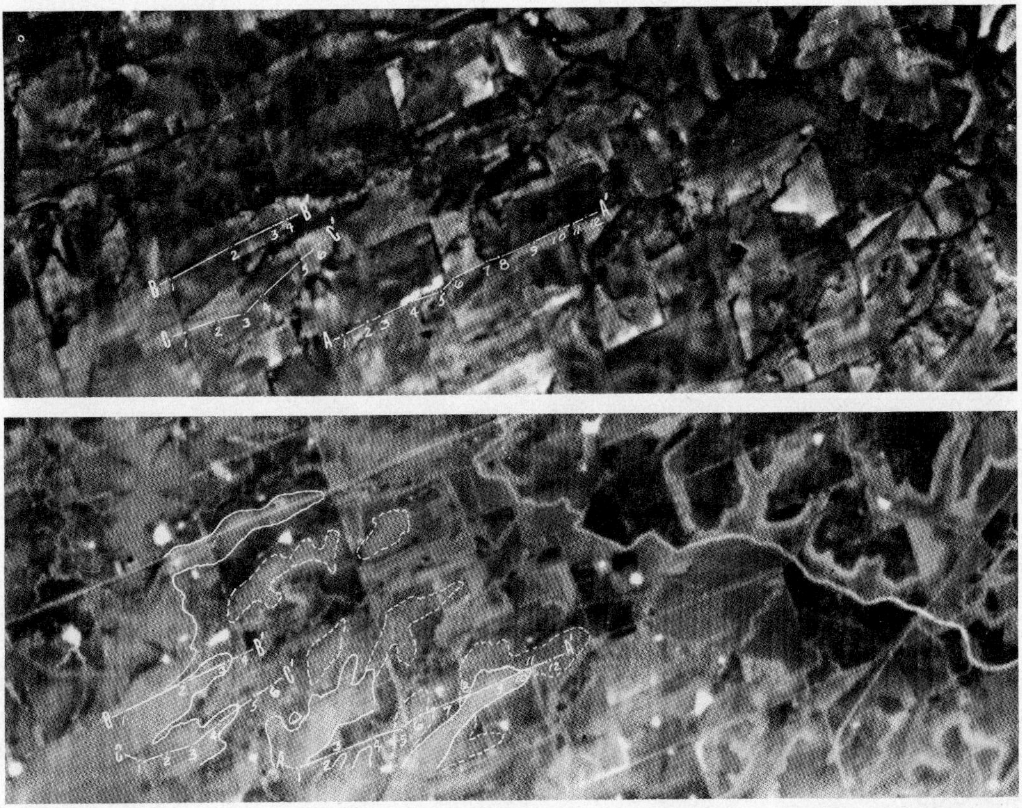

Fig. 32-5. Infrared imagery of the Oskaloosa area, Kansas, illustrating potential for indicating presence of high water table or shallow soil over bedrock. Black indicates cold. *Top* image was acquired in the daytime. *Bottom* image was acquired at night. (After Stallard and Myers, 1972, courtesy State Highway Commission of Kansas.)

causative factor is important, however, and the various soil and rock units indicated can be delineated by comparing the data collected by these two sensors.

Points 11, 12, and 13 were included to show that, in daytime thermal infrared imagery, differentiation is difficult between low vegetation, trees, and water. However, these features can be separated on nighttime thermal infrared imagery.

Figure 32-5 is another example of the complementary information provided by thermal infrared imagery as reported by Stallard and Myers (1972). An evaluation was made of various remote-sensing systems for identifying engineering soils groups within a test site in Jefferson County, Kansas. Marked differences were observed between day and night thermal infrared imagery (8- to 14-μm). The daytime imagery showed contrasts attributable to differences in vegetation and soil conditions. Some very interesting contrasts were observed, however, on the nighttime imagery. The light-toned patterns outlined on the nighttime imagery did not coincide with any delineations or patterns observed on any of the other types of photography or imagery. These light-toned thermal patterns tended to indicate the presence of either a high water table or of a shallow soil cover over bedrock—both of which act as a strong thermal reservoir in areas of intense radiation. This tentative conclusion was based on field drilling and probing along lines A-A′, B-B′, and C-C′. One of the profiles, B-B′, is shown in Figure 32-6.

Myers and Stallard (1975) revealed that certain ground conditions must prevail in order to repeat the significant signatures detected in the earlier imagery. Comparing the two sets of data (Figure 32-7) shows that the thermal patterns of interest on the earlier imagery were not repeated because of differences in surface-moisture conditions. In

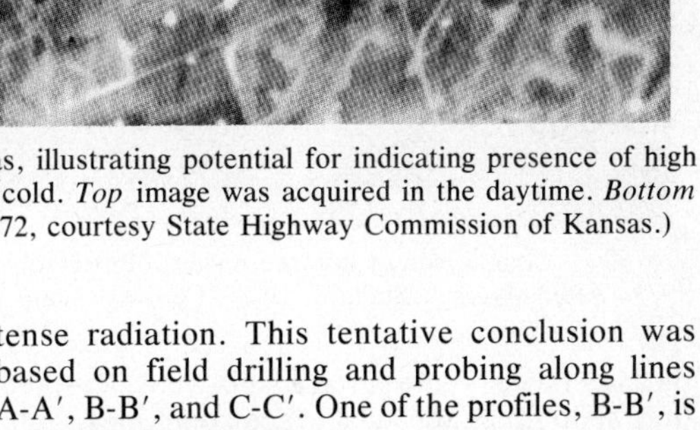

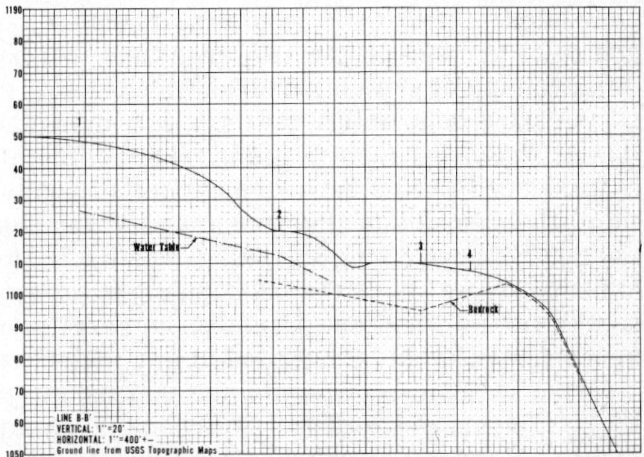

Fig. 32-6. Profile B-B′ for infrared imagery, Oskaloosa area, Kansas (After Stallard & Myers, 1972—Courtesy State Highway Commission of Kansas).

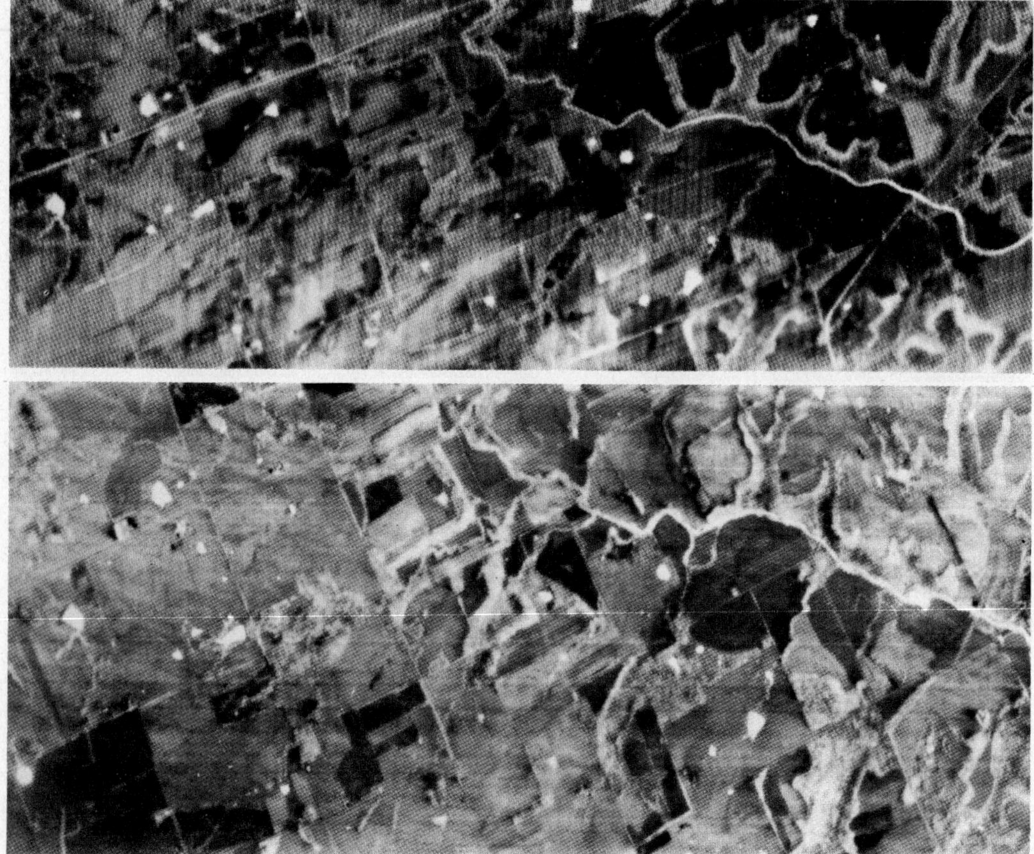

Fig. 32-7. Comparison of infrared images of Oskaloosa area, Kansas taken three years apart. *Top:* 1970; *Bottom:* 1973. (After Myers & Stallard, 1974—Courtesy State Highway Commission of Kansas).

the later imagery, the surface soils were saturated, thus obliterating the subsurface conditions.

Myers and Stallard (1975) reported the use of enhanced thermal infrared imagery. The "Digicolor" process provided a first-generation color enhancement of the imagery from magnetic tape. Both daytime and nighttime enhanced images are shown in Color Figure 32-8 together with a photomap of the area studied. Subtle contrasts and temperature relationships were selected for enhancement and insignificant "hot" and "cold" signatures were eliminated. Comparison of the black-and-white and color versions of the 1973 nighttime thermal infrared imagery shows that areas which appear to be the same temperature on the black-and-white imagery by visual analysis are in reality different, as evidenced by the enhancement. First generation enhancements of this sort are very valuable not only for engineering-soil delineation but in any situation where subtle thermal signatures provide significant information. Usually, however, enhancements should be interpreted in conjunction with the standard black-and-white thermal image of the same area as well as a normal aerial photograph. This approach will maximize the effectiveness of the analysis and minimize the possibility of errors associated with ambiguities found in the enhancements.

The studies by Rib, and also those by Myers and Stallard, suggested that the combined use of color aerial photography and nighttime thermal infrared imagery (8- to 14-μm) rendered the most distinctive evidence for detection, evaluation and mapping of engineering soil groups.

Comments on Accuracy of Mapping

The accuracy assessment of an engineering soils map is very subjective. In referring to accuracy one implies the existence of some standard to gauge against. The actual field conditions are there to check against. The soils are present in such variability, however, and have such gradational or indefinite boundaries, that the true and precise definition of boundaries that correspond to differences in soil properties is a herculean if not an economically unattainable goal. For practical purposes, the accuracy of the mapped units is checked by field observation of the boundaries at selected points and the collection and testing of representative samples of the mapped soil groups. Thus, one is not looking for a 100 percent verification of all boundaries, but a point check of those boundaries which are most critical to the location and construction of an engineering facility (boundaries for areas of soft subsoil, landslides, etc.).

Use of Computer Techniques

Initial efforts in using computer techniques for distinguishing engineering soils were reported by Tanguay (1969). The multisensor data for the analysis were collected by the University of Michigan's multispectral scanner, and the computer facilities and techniques used were those developed by the Laboratory for Agricultural Remote Sensing (LARS) at Purdue University. The method employed digital computer techniques. Results of the study indicated that the LARS system could detect and classify areas showing the distribution of major soil classes, drainage features, wet zones, muck pockets, and bare rock areas. The final interpretation and overall significance, however, had to be assessed by a competent soils interpreter.

More recent evaluations have gone one step farther, but still incorporate competent interpretation into the process of digital analysis. Singhroy et al. (1980a) suggest beginning with a simple photo interpretation analysis of a Landsat image. They then apply the Geologic Package of a system that has been developed by the Canada Centre for Remote Sensing. The package produces contrast-optimized images, linear enhancements and textural enhancements. All three of these outputs are then displayed on a video monitor where they are then further interpreted visually by the engineering specialist before maps can be made. A number of commercial firms in the United States and elsewhere are equipped to produce such digitally enhanced Landsat products at scales of 1:250,000 or larger for terrain analysis.

Digital terrain data have been used by many investigators for making the topographic analyses necessary for some engineering projects. Scientists from the Remote Sensing Research Program (RSRP) of the University of California at Berkeley have used such data in a variety of projects. For instance, Khorram and Smith (1980a) used digital terrain tapes to first map the elevation data for an area in northern California (Shown in Figure 32-9a). Then, to expand the topographic information base, the new elevation map was used as input into algorithms designed to model slope and aspect. The slope-determination routine considers the elevation gradient between data cells that surround the cell for which the slope is being calculated. Based on these elevational differences, a sloping plane is described whose normal vector is calculated and used as a measure of slope. The azimuthal orientation of this normal vector describes aspect. The final maps of elevation, slope, and aspect are shown in Figures 32-9a, b and c. They were produced by using an electro-optical film writer called IGOR (Image Ganged Optical Reproducer) designed and built by personnel of the RSRP.

For verification of the results, 20 sample sites were selected within the study area. The elevation, slope, and aspect values for these sample sites were obtained both from the computer-generated map, based on the described procedure, and from the U.S. Geological Survey topographic map, scale 1:62,500 and compared statistically. The results of this investigation were used in a remote sensing-aided computerized information system for solar and net radiation estimation, hydrologic modeling, large scale planning, and natural resources inventories (Khorram et al., 1977b).

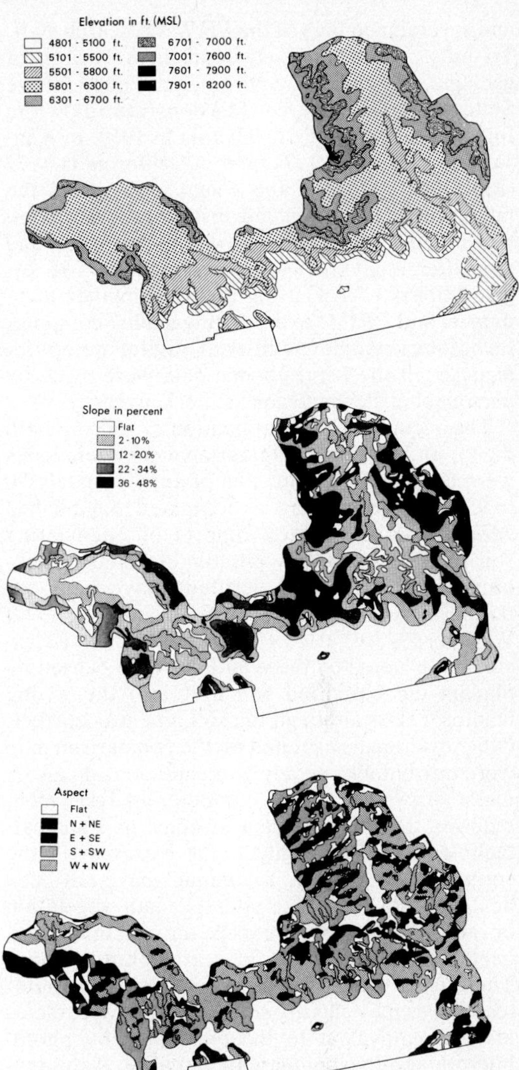

Fig. 32-9. Elevation, slope and aspect maps for major portion of the Middle Fork of the Feather River Watershed in California's Sierra Nevada. All three of these maps were derived from digital terrain data. Color coded maps for this same area, with information pertaining to temperature, solar radiation, and potential evapotranspiration, comprise Color Figure 32-38 and are discussed later in this chapter.

The Federal Highway Administration (FHWA) instituted a research program in 1967 to further evaluate various airborne sensor systems for determining critical soil and terrain conditions, and to develop computer techniques for extraction and analysis of multisensor data. This program was a cooperative effort that included several state highway organizations, several contractors, and several members of the FHWA research staff. To fully evaluate the systems in a variety of geologic and climatic environments, sites in five States were investigated. Missions were flown in Indiana in 1967, in Pennsylvania in 1969, in Kansas and Virginia in 1970, and in California in 1972 (Rib, 1972). Most of the visual analyses of the multisensor data collected in this program were performed by the state highway departments and computer analyses were performed by two organizations: LARS, using digital computer techniques; and ERIM, mainly using analog computer techniques. Some additional digital computer analyses of the Pennsylvania data were made by personnel at Pennsylvania State University.

The accuracies of identification of soils by both digital and analog computer-analysis techniques were fairly high. Accuracies of approximately 70 to 95 percent or higher—as compared to published agricultural soil survey reports or engineering soils maps prepared by photointerpretation techniques and then field verified—have been reported by several investigators (Mathews, 1972; Wagner, 1972; West, 1972). Some of the misclassifications noted by these authors were caused by placing the classified soil unit into the wrong landform class although the soil type was correct. Other discrepancies found on the comparison map were attributable merely to insufficient detail. A major drawback to the computer-analysis technique is that present data systems and analysis techniques identify only items exposed on the surface of the terrain. Techniques have not been developed to recognize soils beneath vegetation or other surface cover, except through inferences used with the traditional aerial photography. Therefore, where soils are exposed, computer techniques may classify soil textures to accuracies at least equivalent to those attained by photointerpretation techniques. In actuality, the potential exists to distinguish soil parameters other than texture, for in most cases several separate spectral signatures are grouped together for mapping each of the soil units. These signatures may be of other soil parameters which cannot be recognized at present. As seen by these cases, the computer-analysis techniques are reported to be valuable supplements for use in preparing regional engineering soils maps.

A study reported by Stallard and Myers (1972) not only evaluated the usefulness of various sensor systems for engineering soils mapping, but also verified the map units and properties of the soils determined by remote sensing techniques. Figure 32-10 shows the engineering soils map that was prepared for one of the three test areas. The approach used was based on the technique previously described for mapping landform-engineering-soil characteristics. Each map unit designation included six items: (1) Landform classification, (2) surface soil classification according to the Unified Soil Classification System, (3) composite soil texture according to the Kansas State Highway Commission classification system, (4) depth to groundwater table, (5) depth to bedrock, and (6) slope. The definitions of the various terms are shown in the legend. All of the items with the exception of slope class were interpreted from the imagery. The slope class was extrapolated from USGS topographic maps. After the soils map had been completed, soil samples were taken in each major soil-map unit on a statistical basis. The field sample data, along with the information obtained from the local office of the Soil Conservation Service, provided a standard against which to evaluate this soil map.

A high degree of accuracy (an average of 91 percent) in the interpretation of landforms was achieved for the three study areas. A varying degree of success was achieved in the use of the interpretative modifiers (items 2 to 5) in the study areas. The low accuracy (an average of 37 percent) for estimating the Unified Classification (item 2) might be attributed to the lack of familiarity with the Unified system. More success was achieved with the Kansas Classification System (item 3) through which average accuracies of 83, 87, and 38 percent were attained. The low accuracy of 38 percent in one of the areas may be attributed to the fact that this area contained highly variable soils. A fair-to-good degree of accuracy was achieved in the interpretation of "depth to groundwater table" and "depth to bedrock" (average accuracies of 84 percent and 77 percent, respectively). In any case, the errors associated with any given map will vary according to the map's detail, source of data, and the experience of the individual who is doing the interpretation. These factors, and the eventual use of the map, should be kept in perspective by those wishing to make and use an engineering soil map with the aid of remotely sensed data.

The use of multisensor missions capable of collecting a great variety of data creates a major data processing problem in that the large volumes of data tend to inundate the interpreter. For example, in the Kansas study previously described, 74 different sets of data were collected over a single test site; furthermore, a total of 140 different sets of data were collected over the 5 test sites investigated. The analysis of this volume of data by visual methods is a long and tedious process. To aid this procedure various methods of data extraction and analysis techniques have been developed (Singhroy, 1980). Some of these systems are now used operationally for engineering soils-mapping programs.

SITE INVESTIGATIONS FOR REGIONAL PLANNING

Remote sensing techniques have been used extensively in site investigations for more than 40 years. Photointerpretation of engineering sites and associated terrain mapping assumed great importance during World War II, and there has been a steady increase in such activities since that time. Today few site investigations are undertaken by

LEGEND:

Item 1.		Item 2.		Item 3.	Item 4.	Item 5.	Item 6.
Landform Classification		Unified Classification		Composite Kansas Soil Textural Classification	Depth to Ground-Water Table	Depth-to-Bedrock Classification	Slope Classification
Fluvial	Fp - floodplain	GW-	clean, well-graded gravels	1. S to SL (sand-sandy loam)	P. poor, 0 to 5 feet	1. Less than 3 feet to bedrock	F- flat, 0 to 3%
	Ft- minor terrace	GP-	clean, poorly graded gravels	2. Si to SiL (silt-silty loam)	I- imperfect, 5 to 10 feet	2. 3 to 10 feet to bedrock	M- moderate, 3 to 10%
	Ftn, Fn- Newman terrace	GM-	gravel-sand-silt mixtures	3. SiCL, SiC, CL, L, SCL, SC	G- good, 10 feet plus	3. More than 10 feet to bedrock	S- steep, over 10%
	Fm- Menoken terrace	GC-	gravel-sand-clay mixtures	(silty clay loam-silty clay-			
	Fo- oxbow and meander scar	SW-	clean, well-graded sands	clay loam-loam-sandy clay loam			
	Fpv(E) floodplain veneer	SP-	clean, poorly graded sands	-sandy clay)			
	(erratic)	SM-	silty sands	4. C (clay)			
		SC-	clayey sands				
Residual	R/Ssh- Severy shale	ML-	inorganic silts and fine sands				
	R/Tls- Topeka limestone	CL-	lean inorganic clays				
	R/Csh- Calhoun shale	OL-	organic silt and silty clay				
	R/Dls- Deer Creek limestone	MH-	inorganic fine sand or silt				
	R/Tsh- Tecumseh shale	CH-	inorganic fat clays				
	R/Lls- Lecompton limestone	OH-	organic clays, medium-to-high				
	R/Ksh- Kanwaka shale		plasticity, organic silts				
Eolian	El- loess						
Glacial	Gd- glacial drift						
Colluvial	C- colluvium						

Fig. 32-10. Engineering soil map for test area in Oskaloosa area, Jefferson County, Kansas (After Stallard and Myers, 1972—Courtesy State Highway Commission of Kansas).

engineers and others without their first interpreting and mapping terrain features from aerial photographs. Within the last decade, site studies initiated from the interpretation of conventional aerial photographs have been integrated with the analysis of Landsat and other types of imagery. Normal color and color infrared photographs are frequently used in making environmental impact studies. In addition, depending on the type of site investigated, and the kinds of factors requiring assessment, the investigator may also choose to interpret thermal or radar imagery.

Among the more common kinds of engineering-related sites studied and referred to here are dams and reservoirs, bridge and pipeline crossings of rivers, airstrips, staging and docking facility sites, and industrial, town, and recreational sites. Two very common kinds of studies often associated with such site investigations and greatly facilitated by remote sensing are the location of groundwater supplies for towns and industries, and the location of construction materials, chiefly sand and gravel. In addition, studies have been made in search of commercial supplies of sand and gravel. Although studies of all of the

types mentioned above are based largely on the analysis of imagery acquired by remote sensing, they usually are augmented by field investigations. A large proportion of these studies have been made in relation to specific site investigations of one kind or another. The vast majority of the case studies used in this section come from approximately fifteen hundred studies done over the last thirty years by one engineering firm.

Regional data base maps for engineering site planning and route corridor-selection are now prepared routinely using remote sensing methods. For example, in studies done since 1977 by a single engineering firm, a total area of some 46,500 mi² (120,000 km²) of terrain has been mapped in southern Saskatchewan and another 166,000 mi² (430,000 km²) in northern Ontario. The mapping was done from the interpretation of low cost, small scale black-and-white panchromatic airphotos, and was correlated with existing map information and a limited amount of data obtained during field reconnaissance. The terrain data base maps were prepared first, and then used to prepare up to eight derivative maps covering such things as site and route selection, foundation quality for light construction, waste disposal, effluent irrigation, groundwater supplies, construction aggregate sources, and hazard-prone lands. Legends for these maps are designed to help regional planners discourage site development in very difficult or hazard-prone terrain and to concentrate site investigations in more favorable environments. An important consideration in these studies has been the development of a suitable terrain-mapping legend based on features that can be interpreted reliably from conventional airphotos at scales ranging from 1:50,000 to 1:80,000. Examples of the legends developed for regional site planning and route corridor selection are discussed below.

In many ways the remote sensing methodology that is used in regional planning is similar to that followed in the making of site investigations and associated terrain evaluations. In site investigations, however, the methodology used varies to a great extent with the individual site and with certain determinants related specifically to it. These determinants can usually be grouped broadly under four headings: (1) economic factors, (2) engineering factors (e.g., topographic, hydrological, geotechnical, and engineering-geology factors), (3) environmental problems and opportunities, and (4) social considerations. Even the most highly experienced remote sensing analyst usually prepares a formal list of the considerations that must be taken into account in carrying out the task at hand. Such a list is often enlarged and modified as interpretive work proceeds on a particular type of project in a specific biophysical environment.

INTEGRATION OF REMOTE SENSORS

A listing should be made of the objectives of a project, its components, and the stages of investigation before very much of the remote sensing activity related to that project is carried out. Reconnaissance, feasibility, and preliminary site investigations are often followed in the office by design and construction planning. One sequence of office and field studies for site selection and development using remote sensing techniques is illustrated in Figure 32-11.

The best types of remote sensing imagery to use, and their appropriate combinations, should also be considered at an early stage. The scales of available or easily acquired imagery must be known because this knowledge determines the alternative approaches that can be taken, probable costs, and level of data that can be reliably interpreted. Ryerson (1981) has demonstrated that costs may be lower with acquisition of new color infrared or color data than with the purchase of existing black-and-white data when interpretation costs are considered. It should therefore be clear how important the early planning stage is. In any case, although many older examples in this section use black-and-white photography, where vegetation parameters are the key factors in interpretation, color infrared imagery at smaller

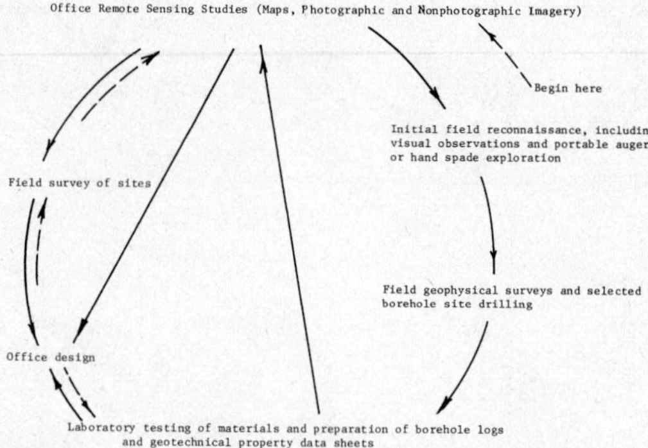

Fig. 32-11. Sequence of office remote sensing and associated field studies for site investigations. Note that the direction of solid black arrow suggests initial or primary direction of flow in studies; the direction of dashed arrow indicates secondary direction of flow in studies.

scales often would have been equally useful, had it been available. Before deciding on the image types to use (such as black-and-white panchromatic airphotos, normal color or color infrared airphotos, thermal IR and radar imagery, and Landsat visual or digital products) the investigator will often make a search of available maps that provide correlative information on soils, geology, topography, land use, and vegetation.

For producing engineering products, many investigators have successfully used visual and digital data acquired by airborne scanners. Examples are the M^2S scanner, the Daedalus scanner, and the Ocean Color Scanner. These scanners commonly are flown at altitudes of 60,000 to 65,000 feet as part of a sensor system that includes conventional cameras as well.

A typical sequence of remote sensing operations used by engineers for a general site investigation in largely inaccessible terrain embodies the following steps:

1. Acquisition and interpretation of 1:1,000,000 scale Landsat images. This study usually entails the examination of color composites made from bands 4, 5 and 7, in the form of either transparencies or paper prints.
2. Acquisition and interpretation of 1:125,000 scale black-and-white airphoto mosaics. (In some cases one could substitute enlarged Landsat images or, more often, high altitude aerial photography.)
3. Correlation of data interpreted from steps 1 and 2 with data shown by, or inferred from, existing maps. Usually these maps provide guides to the kinds of landforms, the types of surface and near surface soil and rock materials, and the topography, drainage conditions, vegetation types, and land use one can expect to find in the study area. Even so, such map data are rarely detailed enough to give the information that is desired. Much of the remote sensing work that is done in the office at this stage entails

the correlation of patterns, greytones, and colors on Landsat imagery with similar attributes as displayed on mosaics assembled from small scale black-and-white panchromatic airphotos. In most site and regional inventory studies, uncontrolled mosaics have been found to be sufficiently accurate.

4. Listing of terrain features and surficial geologic conditions which, based on work performed in the above steps, may be classed as favourable or undesirable in terms of project requirements. (Almost all of the imagery examples that follow were interpreted originally for site investigations.)
5. Stereoscopic interpretation of 1:50,000 to 1:80,000 scale contact black-and-white panchromatic airphotos. (As noted elsewhere in this chapter, color infrared photographs at small scales are often preferable, depending in part upon economics.) Continual cross checks are made with patterns or features having an expression on Landsat imagery.
6. Preparation of a provisional terrain-mapping legend. The legend can be designed to show components of the terrain in small or large areas. Furthermore, it can be designed to show a single terrain component, such as landform, surface material, drainage, or vegetation, or it may be set up to encompass various combinations of these interrelated components of the terrain. An experienced interpreter will often be able to design a legend in the office, hundreds of miles from the area being studied.
7. Selection of sites to be field checked, based on an analysis of the various types of remote sensing data. Sites should be annotated to indicate features and conditions that the interpreter expects to find in the field.
8. Field checking. Reconnaissance checking in the field commonly is carried out by means of fixed-wing aircraft mounted on wheels, floats, or skis. Helicopters, boats, tracked vehicles and foot travel are the other means used.
9. Stereoscopic re-examination of small scale aerial photographs. This step nearly always follows initial field-reconnaissance studies and is directed primarily to the making of more detailed studies of the more promising sites.
10. Selection of types and scales of aerial photography to be flown specifically for the job, and usually covering only the more promising sites.
11. Determination of the usefulness and cost, for purposes of making detailed site analyses, of data acquired by other remote sensing devices, such as thermal infrared scanners, multispectral scanners and side-looking airborne radar (SLAR).
12. Acquisition and analyses of remote sensing data of the types selected in Steps 10 and 11.
13. Concurrent detailed digital analysis of Landsat MSS data, if such data are to be used at this stage.
14. Construction of detailed site maps and tables as necessary to achieve integration of all of the information acquired in the above steps.

Because each site investigation tends to have certain unique aspects, each step has to be tailored to the project at hand and to its specific set of requirements. The sequence given above is only one of many that can be followed. Table 32-5 shows common types and scales of remote sensing imagery used at different stages of site investigations carried out in Canada.

A variety of case histories follow. For the most part they represent operational remote sensing techniques that have been used to advantage in site investigations in Canada. As such, many are not referenced but, with the permission of the investigators, have been taken directly from the files of J. D. Mollard Associates Limited for use here. Most of them have involved the mapping of geological landforms (*e.g.* eskers, dunes, till plains, beach ridges, and bedrock ridges) as well as surface materials. Materials mapped include significant types of bedrock, till, and boulders, as well as gravelly, sandy, silty, clayey, and organic soil materials. Geomorphic and climatic processes affecting their compositional variability, and the physical conditions of earth materials and their geotechnical properties, are normally assessed to the extent considered practical. Besides landform and surface material, information on drainage, erosion, vegetation, and land use is frequently interpreted. Landscapes affected by permafrost or

TABLE 32-5

Types and Scales of Remote Sensors for Different Stages of Site Investigation

Stage of investigation	Type of sensor	Range of scale
Large area evaluation	Airphoto mosaics and LANDSAT imagery (2-D viewing only)	1,000,000 to 1:125,000
Selection and evaluation of alternative sites	High altitude B&W panchromatic aerial photography (stereoscopic viewing)	1:80,000 to 1:40,000
Intermediate stage terrain examination and evaluation of selected site	B&W panchromatic and/or color infrared aerial photography (stereoscopic viewing)	1:40,000 to 1:20,000
Detailed terrain classification and mapping followed by site location	Conventional B&W panchromatic, color, or color infrared aerial photography (stereoscopic viewing)	1:20,000 to 1:10,000

peat accumulation in northern regions are often given special attention.

An additional activity engaged in concurrently with the above, when appropriate, is the development of geobased information systems, thereby bringing about the integration of various kinds of data needed in locating the areas that are suitable for the particular kind of development being considered.

DAM SITES

Some Considerations in Dam Site Investigations

Size and type of dam have an important bearing on site location. For instance, different considerations apply to earth dams than to concrete gravity, arch, or buttress dams. Among the chief physical factors governing the choice of an earth dam over the other types are topography, geology, and hydrology.

Topography affects the surface configuration of the dam site and the reservoir upstream of it; topography also affects accessibility to the site and to sources of construction material needed to build it. In valleys where more than one dam site appears feasible, the influence of topography on the clearing of timber and the relocation of existing routes and sites can be very important for economic, social, or environmental reasons. Valley or canyon shape affects the ratio of dam width to height; and in bedrock terrain this can influence the decision on the best type of dam to build. Valley or canyon shape also has a bearing on abutment irregularities and, therefore, on the kind of site exploration and dam design necessary to ensure safety. Joints and faults, either open or filled with erodible material, when situated at the interface between the embankment and the bedrock foundation, are often critical owing to the flow of water along interfaces, which in turn can affect the safety of a dam. For similar reasons, the geometry of the bedrock at key trench locations likewise can be of great importance.

Geological considerations involve the determination of different soil and rock types and their physical properties. Evidences of jointing, faulting, folding, and the presence of solution features in bedrock terrains are often evaluated first in the office and then verified in the field. In many cases the regional view afforded by Landsat provides valuable clues to unknown faults and related features. Depositional environments, processes and materials—and the loading history of soil materials and their geotechnical qualities—are considered carefully where a mantle of soil material masks the bedrock surface. Usually geological information must be obtained on potential sources of construction materials for building the dam, as well as for constructing highways, railways, and other routes that must be relocated. Geological and topographic types of information are also used to assess hydrologic conditions affecting the location of a dam site.

Rock type and bedrock structure are often more significant in choosing sites for arch and concrete gravity dams than earth dams. On the other hand, depth to solid bedrock and the location, sequence, thickness, and extent of each stratum in the overburden are more important in site investigations for embankment dams. Natural hazards, such as earthquakes and volcanic eruptions, should be considered in all dam types. Indicators of potentially critical slope-stability conditions or groundwater-leakage conditions may be observed using remote sensing techniques; and these studies should be made early in dam site investigations.

Perched water tables, artesian aquifers, and soluble rocks in foundations are related to groundwater hydrology conditions. Other hydrological factors that should be considered include the volume and depth of the water to be stored, the volumes and velocities of controlled and uncontrolled releases, and their influence on river-bed degradation. In many areas hydrological considerations must take into account the water-supply requirements for municipal and industrial usage, irrigation, power, recreation, fish, and wildlife, and also the flood control requirements. Other hydrological considerations include flooding of farmland or built-up areas, and the flooding of commercial timber. They also include the long term effects of reservoir water on shore erosion, the future stability of slopes around the reservoir, and influences on the adjacent groundwater regime. Remote sensing studies have been applied repeatedly to the mapping and appraisal of all of these.

Narrow valleys, good bedrock conditions in foundations, and large stream flows tend to favor the selection of concrete dams. Wide valleys, poor foundations, and small stream flows favor earth dams. The selection of dam type is also influenced by the presence or absence of suitable sites for spillways and outlet structures.

Through the development of computerized geobased information systems (GIS), one can integrate various data types (both remotely-sensed and conventional) for better analysis of all of these conditions. GIS databanks may be used by many engineers for such purposes at much less time and cost than conventional methods.

Of the 104 dam site studies available for inclusion here, over one-half have involved hydroelectric energy. Hydro energy development is relatively clean. Environmentally it does not inflict the adverse side effects that are associated with coal and nuclear power development, such as air pollution caused by burning coal, unsightly strip mining, fear of nuclear accident, or the possibility of passing nuclear wastes along to future generations. Even so, remote sensing techniques are commonly used to study the social and environmental impacts of hydro development.

Gardiner Dam Site, Saskatchewan

There were a number of major considerations in selecting a site for the Gardiner Dam, many of which were based on the results of an analysis of

Fig. 32-12. Small scale airphoto showing Gardiner Dam under construction. Remote sensing studies have been carried out at this dam site and associated reservoir (Lake Diefenbaker) intermittently over a period of 33 years. Location: south central Saskatchewan. Photo A19361-3, taken in 1965. (For features indicated by numbers or letters in this and following figures, the reader is referred to the text. Note also that all photos with A series numbers are courtesy of the National Air Photo Library in Ottawa.)

surficial geological details associated with the site. These included the location and cause of past large scale slumping of valley sides, alternative layouts of major engineering structures (dam, spillway, diversion tunnels, powerhouse site), and the location of potential sources of pervious and impervious fill-borrow and concrete-aggregate supplies. The final consideration included the degree to which the selected site fulfilled the project objectives of hydroelectric power generation, irrigation, flood control, river regulation, and the creation of recreation amenities.

Figure 32-12 is an airphoto showing the Gardiner Dam during construction, in 1965. Figure 32-13 shows the site nearly 20 years prior to construction, in 1944; systematic remote sensing studies for site investigations were begun in 1947. Figure 32-12 shows the dam under construction at 1; the spillway under construction to the left of 2; the powerhouse and diversion tunnel outlet at 3; large areas of slumped valley sides at 4 and 5; major deposits of sand and gravel at 6, 7 and 8; the transition from a rolling till plain to a potentially irrigable, gently sloping glaciolacustrine plain at 9; and scrub-covered sandy land at 10, which was the area recommended for a small provincial park adjoining the reservoir.

Classic examples of gullies in gravel (G) and in clean sand (S) appear in Figure 32-13. Gullies formed in slopes in clean granular deposits are stubby and V-shaped in cross-section, with steep gradients. The natural angle of repose of these materials is about 30°. Granular materials change from cobbly gravel, near the apex of a lower glacial delta at A, to fine gravelly sand at the right side of the photo at C (*see* the faintly expressed abandoned current scars). The deep sand deposit at S accumulated at the distal end of a much higher level glacial delta. Blocks of glacier ice rafted into the glacial lake, along with deltaic

Fig. 32-13. Large scale photo showing slump and granular deposit indicators identified prior to construction of Gardiner Dam. Vicinity of spillway structure (*see* areas 2 to 7 in Fig. 32-12). (Photo A7238-33, taken in 1944.)

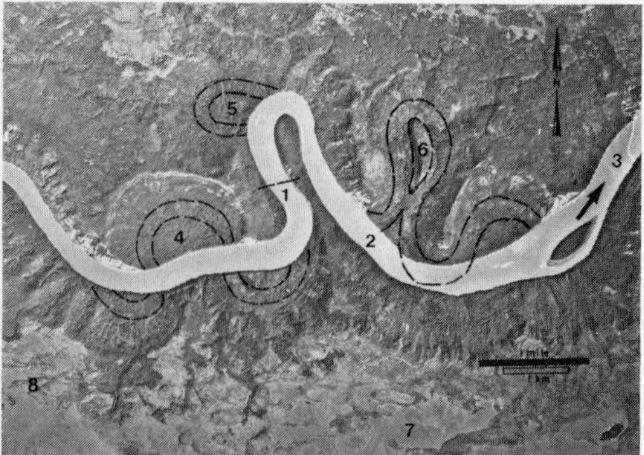

Fig. 32-14. Photo interpretation studies of this deeply incised reach of the Athabasca River were made for hydro power development and for prediction of long term, stable slopes for open pit mining of the tarsands in northern Alberta. (Photo A23884-150, taken in 1974.)

sands, melted to form the kettleholes at K. A short, narrow ridge of sand crosses the bottom of one of them at R. Large glacial boulders to the right and left of B suggest that, during or after formation of the lower glacial delta, glacial meltwaters cut down through the surficial granular materials into a bouldery till, forming a layer of concentrated boulders on the ground surface.

Small white flecks on the sides of the graded sand gullies at S are caused by active wind erosion. Here the exposed subsoil is colonized by creeping ground juniper. Alluvial fans west of F have been built at the bottom end of the gullies. A spring can be seen to the right of F (wooded area). Slump blocks at L are associated with large scale creep movements in bentonitic clay shale underlying thick deposits of surficial sand and gravel.

An earth dam has now been constructed across the valley at D (*see* the area west of 2 in Figure 32-12).

Crooked Rapids Dam Site, Athabasca River, Alberta

Two airphoto interpretation studies were made in the 1960s in the area shown in Figure 32-14. One was a study for a hydro development site, at 1; the other was for a regional statistical study of unfailed, failed, and potentially unstable valley sides. This latter study covered an area of several thousand square miles. Data on slope height and slope inclination, bedrock stratigraphy in valley walls, and slope stability were plotted on graphs to help predict the long term strength of oil-impregnated sandstone strata in the sides of proposed deep open-pit tarsand mines. Rapids caused by limestone in the riverbed appear at 2 and 3 (bands of white water).

Critical features studied along this reach of the Athabasca River included the short seepage path through the east side of the reservoir, just upstream of the proposed dam site at 1; the quality of foundations in the river bed; and the origin of the large bowl-shaped depressions occurring in the 500-foot-high valley walls at 4, 5, 6 and elsewhere.

Initially the indentations in the Athabasca River valley walls were considered to be huge cavities created by old slides. However, closer airphoto inspection revealed that the locations of the depressions match a pattern of old meander cutoffs. In fact, an old meander core can be seen at 6. Muskeg appears on the upland at 7, and sand dunes, surrounded by peat, at 8.

RESERVOIR INVESTIGATIONS

Types of Reservoir Studies

Remote sensing studies concentrated around the shores of proposed and existing reservoirs are often made before and after filling of the reservoir with runoff water. These remote sensing studies include identifying locations of potential reservoir leakage; predicting shore erosion as it affects damage claims, the purchase of shore lands, and recreational development along the shore; and forecasting anticipated changes in the stability of slopes around the reservoir as a result of bank saturation, slow or rapid drawdown of water in the reservoir, and current and wave erosion of shore bluffs. Effects created by the raising of

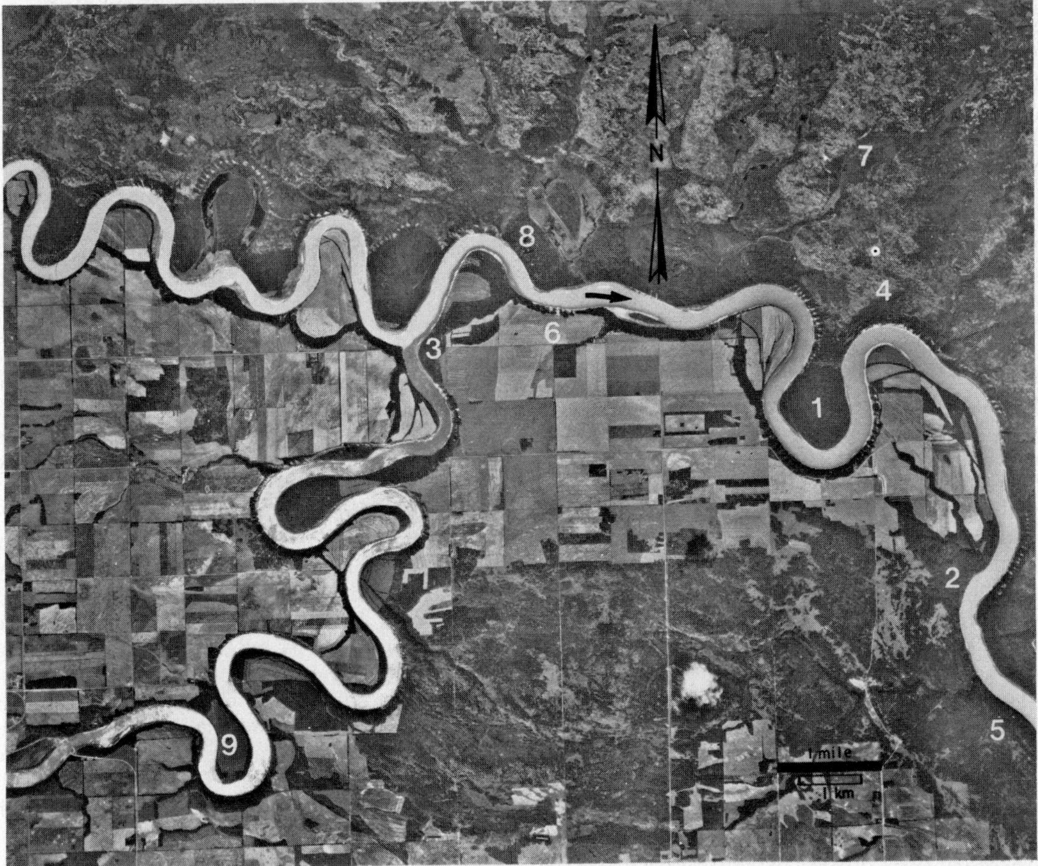

Fig. 32-15. Largely by means of photo interpretation, shore erosion, valley wall stability, and changes in groundwater levels were evaluated for a reservoir located west of a proposed dam site at 1. Remote sensing studies also resulted in the preparation of maps detailing anticipated reservoir effects. (Photo A21664-13, taken in 1970.)

water levels in water wells near the reservoir—such as the recharging of dry sand and gravel strata and the creation of commercially developable aquifers, along with the creation of new springs and seepages—often require assessment from the study of airphotos. Other studies that are amenable to preliminary appraisal using remote sensing methods include reservoir sedimentation problems, estimates of the cost of shoreline protection measures, studies related to the extraction and stockpiling of commercial deposits of sand and gravel that would be inundated by reservoir water, timber clearing and the effects of this on recreation and wildlife, and flooding of existing transportation and communication facilities.

The Forks Reservoir Shoreline Studies, Saskatchewan River

The proposed Forks Reservoir is the farthest upstream of four contiguous reservoirs. The dam associated with one of the four has been built, another dam is being designed and will be constructed shortly, and two more dam and reservoir sites are being investigated in pre-feasibility studies. All reservoir sites are located downstream of the confluence of the North and South Saskatchewan River branches, which can be seen in Figure 32-15. This figure shows the downstream portion of the proposed Forks Reservoir. The proposed dam site associated with this reservoir is located at 1.

Five different sets of airphotos, of varying scales and ages, were examined in an attempt to predict the location, rate, and extent of shore erosion, the evolution of long term stable slopes above the full supply level, the location and extent of the areas affected by anticipated slope failures, and the effect of the new reservoir on groundwater conditions in the area surrounding the reservoir.

The mapping of existing slumped terrain and the identification of recent changes in slope stability were aided by the examination of stereoscopic airphotos spanning a period of 50 years. Estimates of shore erosion were also based on an office interpretation of several sets of airphotos, coupled with field checks at critical locations previously identified in the airphotos.

Empirical and theoretical approaches were used to estimate the ultimate stability of beach and bluff slopes that would be situated above full supply level. In the empirical approach, a large number of stable beach and bluff gradients were measured where the slope-forming materials, stratigraphy, topography, and groundwater conditions were similar to those mapped around the shore of the proposed Forks reservoir. In the theoretical approach, several sets of design charts were used to assess the long term stability of slopes, considering slope geometry (slope height and inclination), strength parameters (the effective cohesion intercept, and the effective angle of internal friction), the unit weight of slope-forming materials, and an assumed range of groundwater conditions in the valley sides.

The proposed Forks dam site axis is located at 1 in Figure 32-15. A power canal, running from the dam site to a powerhouse at 2 is envisaged in order to take advantage of a gain in head associated with rapids and fall in the riverbed between 1 and 2. At the confluence of the North and South Saskatchewan Rivers, at 3, less suspended sediment (darker tone) is indicated in the South Saskatchewan River.

Factors of importance in assessing the evolution of long term stable slopes are the locations of large slides at undercut slopes, as at 4 and 5. Small active slides caused by groundwater piping can be seen in the valley side north of 6. Large swamps at 7 (and elsewhere, as indicated by darker tones east and west of the north point) are associated with a high water table in surficial sands. Here winds have blown lacustrine deltaic fine sand into dunes.

Coarse cobbly gravels were identified in alluvial terraces at 8 and 9, and at several other meander loops on both the North and South Saskatchewan Rivers and at 1 on the Saskatchewan River. All of these deposits must be identified and tested, and the volumes of recoverable material determined, because they contain many millions of tons of granular material in a region where nearly all of the existing sources of sand and gravel on the upland have been depleted.

Field observations revealed dry intertill sand and gravel at undercut river bends that are located at, and just above, the confluence of the two rivers. These presently drained sand and gravel strata will be recharged by reservoir water if the project is built.

RIVER CROSSING SITES FOR BRIDGES AND PIPELINES

Considerations in Selecting River Crossing Sites

Determinants in selecting river-crossing sites for bridges and pipelines are often quite similar to those in dam site selection: engineering, economic, social, and environmental. An important geotechnical consideration is the stability of slopes leading down to and up from the water crossing. Such slopes often include river banks, terrace faces and, above them, valley walls. Preliminary mapping and evaluation of past unstable slopes from airphoto studies have been made on the more than 100 river-crossing sites. Most of these sites were located and assessed from airphotos prior to follow-up field investigations.

The stability of river beds and river banks is of equal importance. Places where significant erosion of the river bed and banks can occur from fast flowing water or river ice should be looked for. General scouring of the river bed over an appreciable length of channel may be caused by river diversion, which adds clear water flow, by upstream damming and entrapment of transported sediment in the reservoir, and by removal of downstream controls, as at bedrock or boulder rapids. Scour can also be caused by the mining of sand and gravel in the river bed, or by a straightening of the channel, thereby creating artificial changes in regime of the river and consequent degradation of the reach of meandering river that was straightened and shortened.

The photointerpreter studying bridge and pipeline crossing sites is also interested in the history of river erosion and sedimentation. Such history will often shed clues needed in locating the sites where scour is most likely to occur. Included are all sharp river bends, all steep undercut banks, channel contractions, reaches below the confluence of rivers, and such man-made obstructions as bridge piers, abutments, and piles. The possibility of scour should be checked downstream of weirs and beneath pipelines installed in river beds that consist primarily of loose silt and fine sand. In assessing these situations in airphotos, one must consider the amount of river channel contraction, the degree of curvature of a river bend, the nature of the bed and bank materials, and flood flows and flow depths.

When selecting bridge sites and pipeline crossings, the remote sensing specialist will often try to predict the behaviour of a river from features that can be identified on topographic maps and aerial photographs. Stream type, suggested largely by channel pattern, should bring to mind a number of interrelated parameters that should be looked for when examining airphotos stereoscopically. An example of common interrelationships among fluvial features and phenomena, which has been used for bridge and pipeline crossings in parts of western and northern Canada, is summarized in Figure 32-16, the patterns of stream types, interpreted from airphotos and used to construct Figure 32-16 are shown in Figure 32-17.

Nelson River Pipeline Crossing

Figures 32-18 and 32-19 show river and ice jam features which the photointerpreter should consider when selecting pipeline river crossings in subarctic Canada. Ice jams frequently develop downstream of rapids sections, easily recognized at 1 and 2 in Figure 32-18 (both upper and lower airphotos). Ice jams may also develop where the river channel narrows considerably, as at 3 in the upper photo in Figure 32-18. This section of channel is shown in greater detail in Figure 32-19. The effects of river ice and associated water erosion at

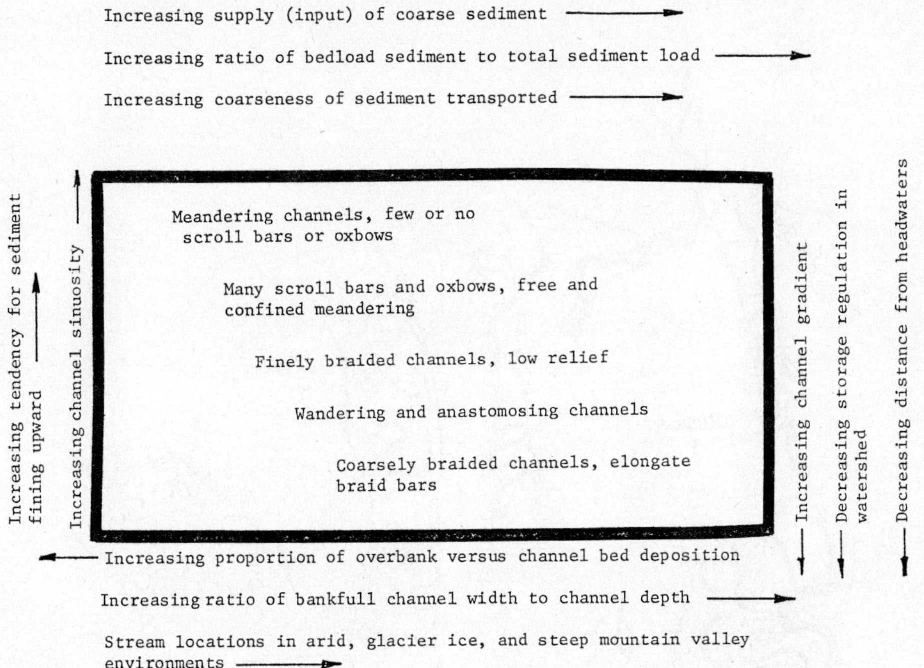

Fig. 32-16. Correlations to look for when interpreting stream type patterns and stream behavior from operational remote sensing studies.

ice jams can also be seen more easily in airphotos of larger scale, as revealed in Figure 32-19. Three deep scour holes appear in till in the floodplain at 1 in Figure 32-19. A high level by-pass channel associated with jamming and raising of the river level appears at 2, and a trail of cobbles can be seen at 3. The cobbles were scraped by river ice from the river bank upstream and from the nearby river bottom. The pothole-like scour holes at 1 often contain what are referred to as "kettle-stones."

Terrain conditions change from densely wooded fluvial sand, unaffected by permafrost, at the terrace west of 4 in Figure 32-19 to peat-covered, ice-rich marine silt and clay on the upland east of 4.

The pipeline crossing investigated at 3 in Figure 32-18 (*see,* also, the larger scaled view in Figure 32-19) is one of five remote sensing studies made in the area shown by Figure 32-18. A new townsite on permafrost-free sand was selected at location 4 on Figure 32-18 (top photo). Cut lines for the partly developed townsite can be observed at 3 in Figure 32-18 (bottom photo). Work on an embankment dam was begun near 1 in Figure 32-18 (bottom photo) and later halted. At this reach a cofferdam as high as the eventual height of the main earth dam is required for dam construction. The region surrounding the area shown in Figure 32-18 was searched at least three times for granular materials and for suitable till borrow for earth dam construction. Airphoto interpretation studies were also carried out, using several ages and scales of black-and-white photography, in a search for an earth-electrode installation site for a DC ground-current return. That search area extended outward from location 4 on Figure 32-18 (bottom photo), and included large areas of terrain dominated by frozen and unfrozen peatland.

Bridge Site, Moisie River, Quebec

In the early 1960s a bridge site was planned to cross the Moisie River a short distance above its confluence with the St. Lawrence estuary. The northeast abutment was to be located southeast of 1 on Figure 32-20. North of here at 5 a whitish, actively-eroding dendritic gully system can be observed. The proposed road right-of-way, leading to the southwest shore of the Moisie River, appears at 2. Forest logging operations (at 3, 4, and elsewhere) no doubt increased the chances of erosion from surface runoff and subsurface piping in thick, loosely deposited fine sand and silt. Following a very heavy rain the tree-like gully system centred around 5 was initiated. Groundwater seepages through relatively pervious fine sand and silt layers in this raised marine delta were fed by water from the rainstorm and also from the large peat bog at 6. Sand that has been eroded from the gully system at 5 forms a crescent-shaped deposit on the floodplain at 7.

A small gully has begun to develop at 8. Sediment from this gully can be seen as a small white patch on the shoreline.

Airphoto interpretation of the topographic setting, of the origin and composition of sediments in the undercut valley wall, and of the effects of extensive logging on the upland can be used to advantage in selecting bridge sites in environmen-

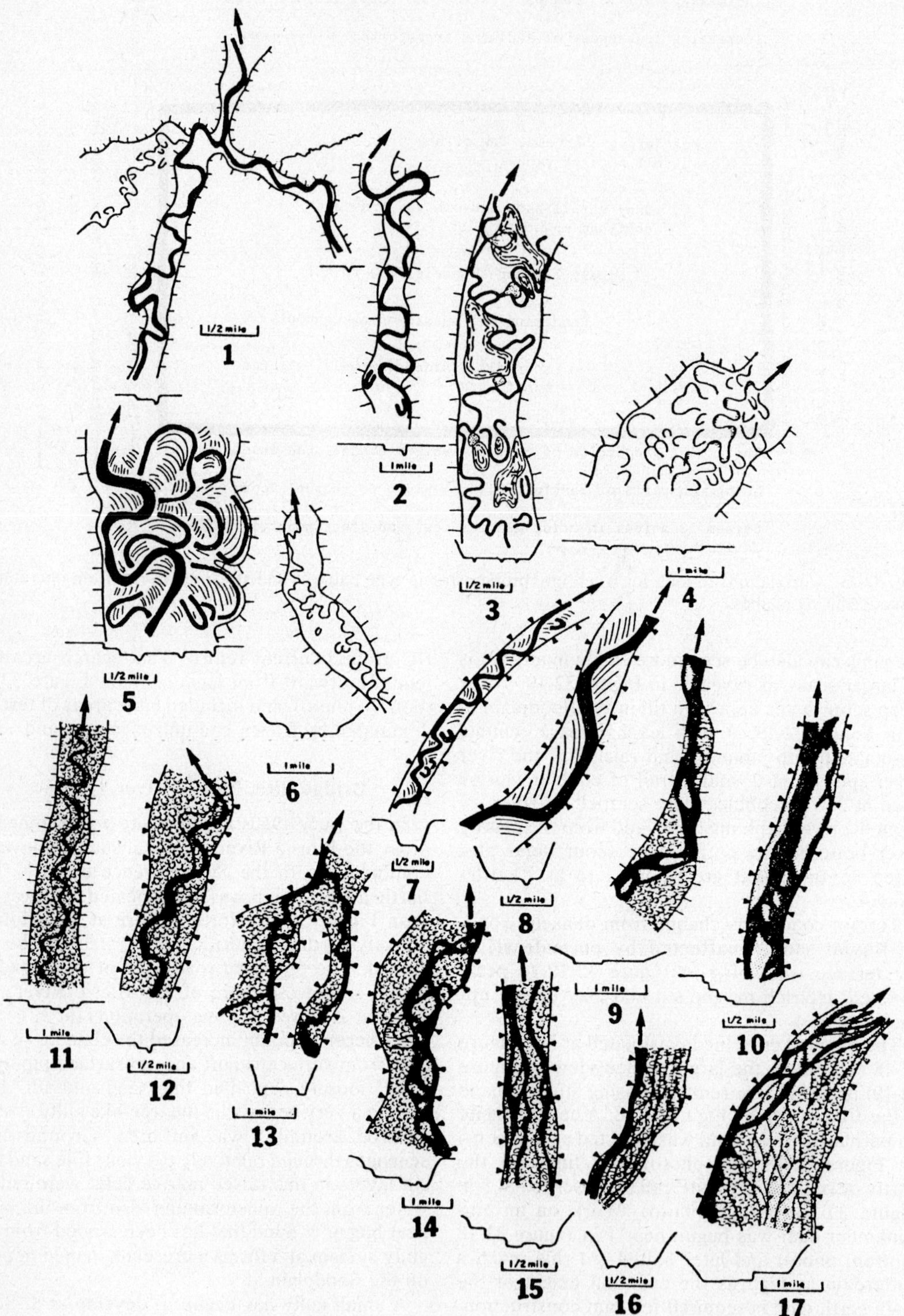

Fig. 32-17. Examples of stream types mapped from airphotos. These examples were used in the construction of Figure 32-16.

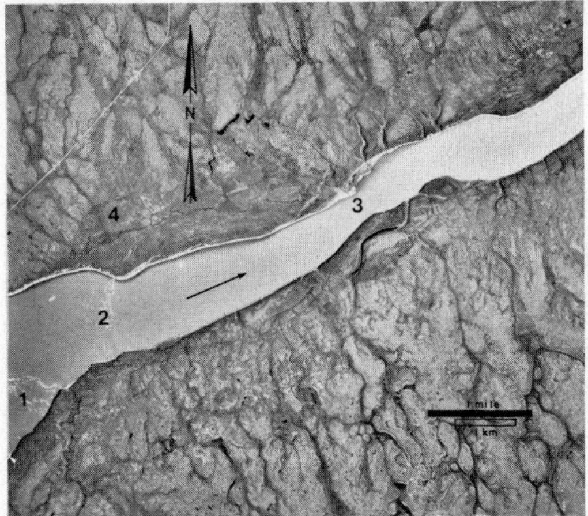

Fig. 32-18. Pipeline-crossing for a town and also a cofferdam and a DC earth electrode site were located in the area shown here through the use of several ages and scales of airphotos. (Top photo: A14126-82, taken in 1954. Bottom photo: A24226-94, taken in 1975.)

Fig. 32-19. Large scale airphoto showing enlarged view of features caused by recurrent ice jams in the Nelson River a short distance downstream of Limestone Rapids. This site was studied for a proposed large diameter pipeline crossing. (Photo A22549-171, taken in 1971).

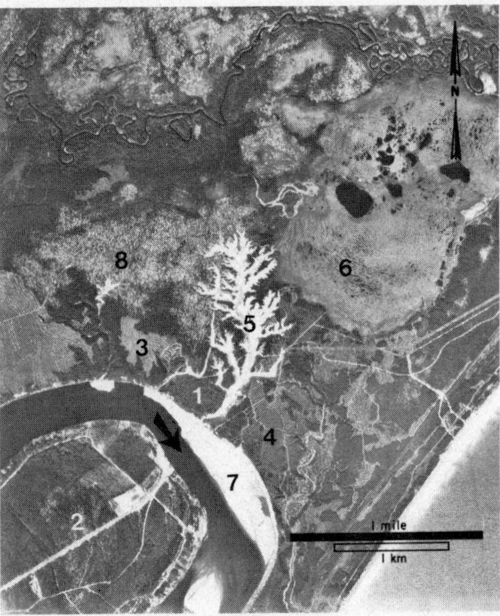

Fig. 32-20. Accelerated gully erosion on a high, terraced, glaciomarine sand delta. Erosion followed extensive logging, cutting of a highway right-of-way, a cloudburst, and high runoff and groundwater discharges. Location: mouth of the Moisie River near its junction with the St. Lawrence estuary (lower right). (Photo A20266-136, taken in 1967).

tally sensitive terrain, such as the one depicted in Figure 32-20.

AIRSTRIP SITES

Considerations in Airstrip Selection and Maintenance

Although remotely sensed data on terrain conditions have been used frequently to select the sites of large commercial airports around major cities, a number of factors besides terrain conditions require careful evaluations. These include the high cost of land, problems associated with air traffic control, travel distance to the central business district of the city served, and traffic congestion between the airport and city. Each factor can have an important economic, social, or environmental bearing on the selection of an airport site. On the other hand, physical factors usually dominate the location of airstrips constructed in remote areas where airstrips are required for new mine and associated townsite development. A good example is the airstrip built by Eldorado Nuclear Ltd, located a few miles southeast of Uranium City in northwestern Saskatchewan.

In the more remote and less accessible regions of Canada, small scale stereoscopic airphotos are often studied to map the distribution of rough Precambrian bedrock terrain, muskeg, permafrost features, and large granular deposits—physical factors that often determine the selection of sites for both commercial and non-commercial airstrips.

Fig. 32-21. Stereogram showing the site of a future airstrip in a remote region, where the terrain consists almost entirely of very rough Precambrian bedrock, permafrost-affected peatland, and scattered patches of fine grained mineral soils. The deposit at 1 is composed of sand and gravel. Sandy beach sediments at 2 have been eroded from these granular deposits. The bedrock ridge at 3 is the site of a future uranium mine. (Photos A8022-12 and 13, taken in 1945).

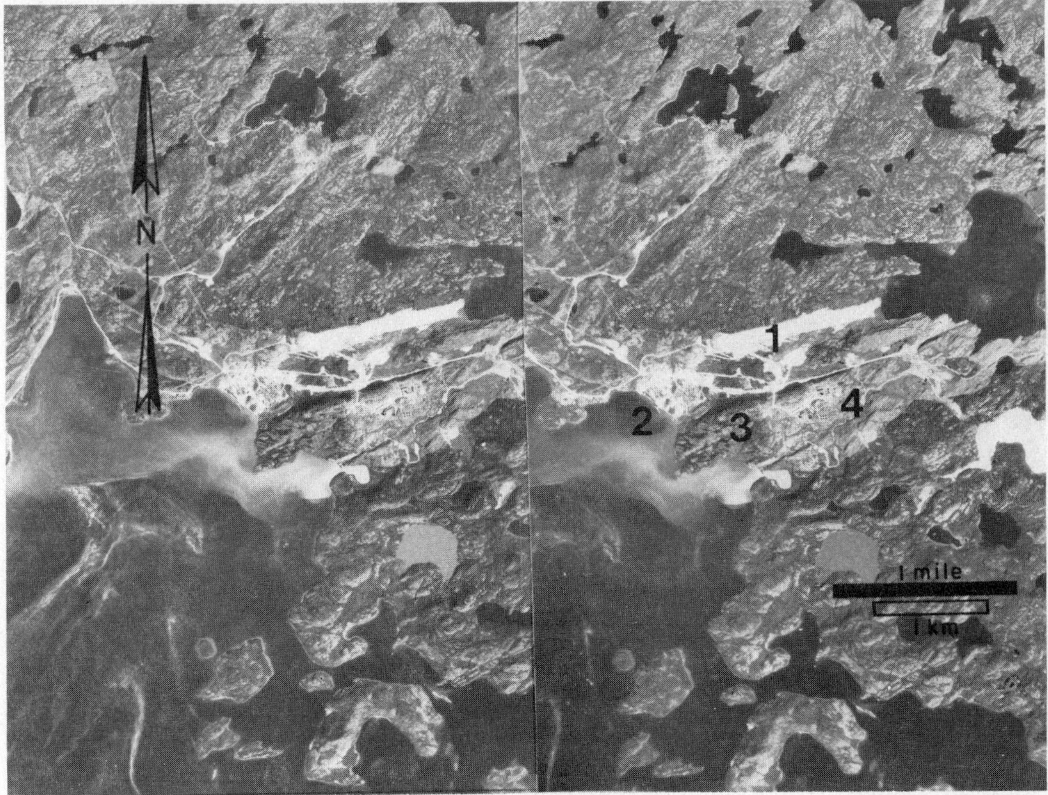

Fig. 32-22. Stereogram assembled from small scale airphotos showing regional terrain conditions around the area shown in Fig. 32-21. Note the lack of good airstrip sites, except for the area at 1. (Photos A14635-143 and 144, taken in 1955.)

Eldorado Nuclear Ltd Airstrip, near Uranium City, Saskatchewan

Figures 32-21 and 32-22 should be viewed alternately. The selected airstrip site can be observed at location 1 in Figure 32-22. Terrain indicators of a thick deposit of sand and gravel are the sandy beach at 2, which is more easily identified on Figure 32-21 than on Figure 32-22. The linear escarpment at 3 in Figure 32-21 is probably associated with ancient faulting and uranium mineralization. Mine buildings can be seen between 3 and 4 on Figure 32-22.

Clues to the existence of a thick deposit of sand and gravel at the present airstrip location at 1 in Figure 32-21 are the whitish nearshore zone at 2 in Figure 32-21, more uniform tree vegetation on sand and gravel than on bedrock or peatland, and smoother and more regular topography. Even for STOL (short takeoff and landing) airstrips, good approach and takeoff sight distances (*i.e.* no obstructions) can be important determinants in site location in areas of rugged terrain. Other factors of importance in airstrip location in many parts of subarctic and arctic areas are low relief, the excavation quality of materials that must be graded (hard rock requiring blasting versus well drained sand and gravel), good internal drainage of foundation soils, the absence of deep peat and difficult permafrost conditions, and proximity of the airstrip to granular materials for contruction.

Office airphoto studies of the region surrounding the area shown in Figures 32-21 and 32-22 have also been made to try to identify potential recreational sites that can be reached by air, canoe, overland hiking, lake ice in winter, and four-wheel drive vehicles. Terrain conditions were also mapped for a study for the expansion of nearby Uranium City itself.

Aggregate Location for Runway Maintenance, Canadian Forces Base Training Establishment, Cold Lake, Alberta

Figure 32-23 shows two of 210 sand and gravel prospects originally identified from the interpretation of 1:80,000 scale black-and-white airphotos. The 210 prospects were outlined on 58 contact-size airphotos. Descriptions were given for each granular material prospect within a 50-mile radius of the runway that required upgrading. High quality aggregate was required for base and wearing-course upgrading because the runway is used in the training of jet pilots. Within one day of obtaining the request for this aggregate search, a second enquiry was received for a similarly large volume of sand and gravel within the same area. In the second request, the granular material was required for the construction of a network of roads and plant buildings needed to extract heavy oil, using stream injection methods, at Cold Lake, Alberta.

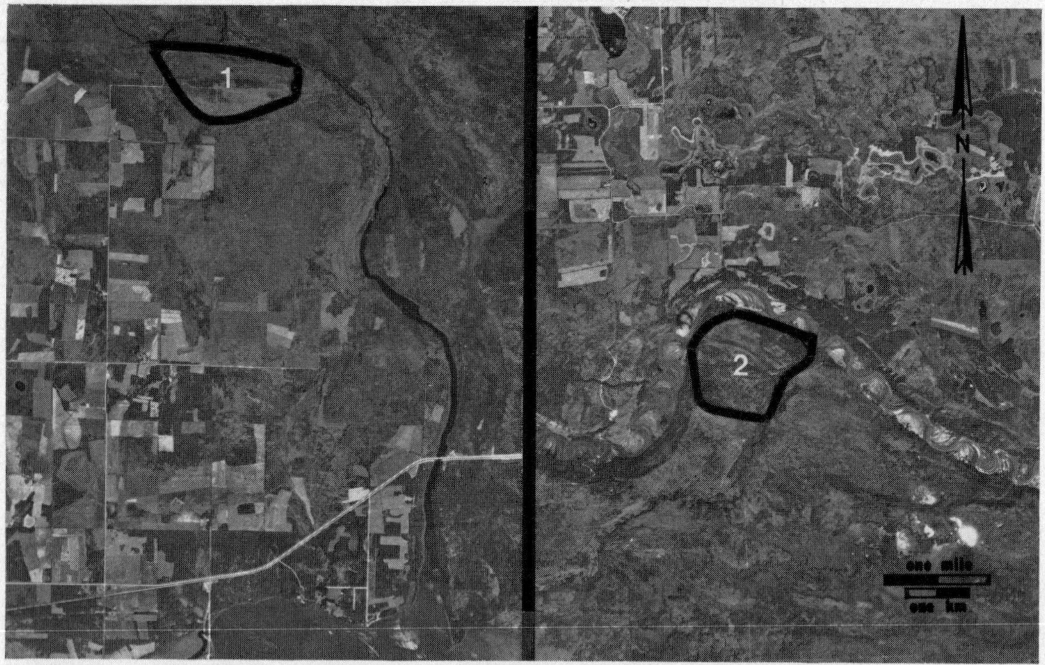

Fig. 32-23. The outlined areas at 1 and 2 are two of 210 sand and gravel prospects mapped within a 50-mile radius of the Canadian Forces Base Airstrip at Cold Lake, Alberta. The original search for sand and gravel deposits was carried out from small scale airphotos. On this project, a one mile saving in haul resulted in a cost saving of about $10 million. (Photos A21799-25 (left), and A21799-33 (right), taken in 1970.)

The estimate of aggregate requirements for both projects totalled approximately 10 million tons. Both clients agreed to a single search, and to share the cost of airphoto studies. From the outset it was apparent that, since hauling costs amounted to 10 cents a ton-mile, a one mile reduction in haul distance would generate a saving of $1 million. The deposits at 1 and 2 on Figure 32-23 are two of the better-looking prospects that were identified during office study of the airphotos and later explored and verified in the field.

The deposit outlined in the left airphoto is about 43 miles from the runway that required upgrading, whereas the deposit outlined in the right photo is some 19 miles distant. Geophysical surveys and testhole exploration confirmed the presence of about 4½ million tons of coarse gravel in the deposit on the left at 1, and some 2 million tons of coarse gravel on the right photo at 2. Based on economics, on aggregate gradation and quality, and on quality requirements for different uses, different granular material prospects will be developed for coarse aggregate and fine aggregate for concrete, for base course and wearing course lifts on the runway, for road building, and for backfill at swampy locations.

As is customary in these kinds of remote sensing studies, all prospects were outlined on airphotos, of various types and/or on Landsat data, both visual and digital. Places were marked for field observation and follow-up exploration. Preliminary descriptions of prospects included remarks on the origin of each granular deposit, expected grain-size composition and variability, anticipated depth and volume, anticipated cleanness of prospect, the chances of finding a high content of deleterious particles in a deposit, groundwater problems, and the identification of water sources for aggregate washing. In addition, all prospects were rated as excellent, good, fair, poor or doubtful in order to guide follow-up field investigations. Many of the sand and gravel prospects mapped were tested because decisions on land purchase and aggregate development must usually take into account a number of social and environmental problems and opportunities.

PIPELINE STAGING AND TANKER DOCKING SITES IN THE ARCTIC

Considerations in Site Selection and Problem Identification

A number of staging sites in the High Arctic have been investigated for the off-loading of large-diameter pipe and compressor-station building supplies. Sites for the construction of tanker docking facilities at possible natural gas liquefaction terminals have also been examined in airphotos, in Landsat imagery, and on the ground. Two examples of the sites examined for staging and docking facilities are located at Basil Bay on Canada's mainland, north of the Eskimo village of Coppermine (pipeline staging site), and Bridport Inlet on the south shore of Melville Island (a natural gas liquefaction terminal and associated docking facility site).

Main considerations are deep water in a bay or inlet that is protected as much as possible from wind-driven ice floes, and from fragments of tabular icebergs called "ice island fragments." Other considerations, besides the hazard of wind-driven ice floes, require the presence of a site for an associated airstrip, a large dry area for stockpiling pipe, a nearby source of construction aggregate (either sand and gravel or suitable bedrock), protection from the ravages of strong winds so common in the High Arctic, deep water near shore, relatively good underwater and shore foundation conditions for the construction of tanker docking facilities, and a large enough bay for the turn-around of large tankers. With respect to ice flows, methods have been developed to measure ice ridges using a laser profilometer. Operational use of this method now provides details, previously difficult to obtain, on ice deformation, including ridge heights, widths and distribution, which can be used in models such as for trafficability through ice by ships (Lowry and Brochu, 1978).

Bridport Inlet, Melville Island, Northwest Territories

Figure 32-24 shows a large inlet opposite the Meacham River delta at 1. A major natural gas liquefaction terminal and associated tanker docking facility have been proposed for this inlet. The bedload sediment that is being transported and deposited by the Meacham River consists mainly of fine sand. When the flows in the Meacham are low, which is most of the year, this river bed is dry and the fine sand on it is blown about by strong onshore and offshore winds.

Bridport Inlet is almost sealed off from the ocean by two long projecting ridges at 2 and 3, and by Dealy Island at 4. The sandstone ridges, marking the crest of an anticline, extend more than three-quarters of the way across the bay, serving to keep ice floes at 5 from moving freely into the bay. These ridges also keep floes that are in the bay, as at 6, from drifting out into the open ocean. A synoptic view of the bay, the two ridges projecting into the bay, and Dealy Island is shown more clearly in a small portion of a Landsat image (*see* inset), which was acquired July 28, 1974. A number of Landsat scenes were studied in an attempt to better assess the break-up and movement of offshore sea ice. Dull brick-red areas on the Landsat image correspond to areas of sparse ground cover—usually areas where the weathered surficial materials have been derived from patches of thin, frost-churned till or from shale of Devonian or Lower Cretaceous age. Because these areas have higher clay contents and are wetter, they support more low-growing vegetation. The curving light bands on the Landsat image or sandstone beds at and near ground surface.

The small, light toned scar at 7 in the airphoto is the horseshoe-shaped cavity of a bimodal flow. Here less than 1 m of fine-grained soil material overlies dirty ground ice.

As the potential site for a docking facility, the Meacham River delta is unattractive because of exposure to high winds and the likelihood that it contains soft silt and clay below the prograding deltaic sand. A better and more protected site occurs at 8, downslope from an actively eroding exposure of Hecla Formation sandstone. Because tankers have deep drafts (18 m), airphotos should be taken when the bay is clear of ice so that shoal water areas, as at 9 and 10, can be identified over the entire area of the inlet. Similarly, studies with advanced sensors, such as laser profiling instruments and airborne scanners, should be carried out during conditions which reflect average to worst cases. This requirement must be met in order to provide information that applies under the circumstances that are of greatest concern to the engineering study.

INDUSTRIAL SITES

Considerations in Mine Site Development

The development of new mines commonly requires large sources of industrial water and small sources of drinking water. Often, too, the quality of water needed for a mining operation is different than that used for domestic purposes. Site development also usually requires a large source of good quality construction aggregate, and the selection of suitable sites for a plant and for employee housing. Sites for man-made reservoirs and waste disposal facilities must be selected as well.

The location of suitable water and aggregate supplies, access roads and railway spurs, is often guided by information generated through remote sensing studies. For example, all of the considerations listed above had to be taken into account at a proposed site for a new sodium sulphate mine at Ingebright Lake, north of Maple Creek, and at a proposed new potash mine east of Rocanville, Saskatchewan. Extensive remote sensing studies and follow-up field investigations were carried out in the area surrounding both of these mine sites in the 1960s and early 1970s.

Groundwater Location for a New Sodium Sulphate Plant at Ingebright Lake, Saskatchewan

Figure 32-25 shows the regional geologic setting at and surrounding Ingebright Lake, which is a salt-covered lake bed. Cone-shaped deposits of sodium sulphate reach depths of 100 feet at 1 in Fig. 32-25. Some 1800 gallons a minute of water were required for the commercial solution-mining of sodium sulphate on the bottom of Ingebright Lake. Preliminary engineering and cost studies suggested that a pipeline to carry water from the South Saskatchewan River, 40 miles to the north, would cost about 1.5 million dollars. Remote sensing studies were then undertaken in an attempt to discover where a water supply might be located that would provide the required 1800 gpm. After thoroughly test drilling an aquifer target

Fig. 32-24. Bridport Inlet on the south shore of Melville Island, Northwest Territories. This area has been investigated from photos such as these to determine its suitability for a deepwater docking facility and a natural gas liquefication terminal. (Photo A17722-6, taken in 1962. Inset is Landsat color image E-1735-1906-2, acquired July 28, 1974.)

identified from the airphotos, the engineers installed two wells at 2 and one well at 3, at a total cost of approximately $80,000.

Figure 32-25 is part of a small scale black-and-white airphoto; it shows the main physical features at the site. The clue to a narrow, winding buried valley containing 250 feet of mostly stratified fine sand and silt is a chain of closed alkali salt depressions. The string of aligned alkali sloughs follows a discontinuous depression that runs from 1 to 2 to 3 to 4. Altogether, 80 testholes were drilled into this depression in order to define the best section of aquifer to develop for use as a solution-mining water supply. Groundwater in the strongly artesian sand aquifer was of very poor quality from the standpoint of human consumption in that it contained approximately 3500 ppm total dissolved solids. Such water, however, is entirely satisfactory for use in the solution-mining of sodium sulphate.

Water pumped over a period of one year is stored in the reservoir between 2 and 3. Water

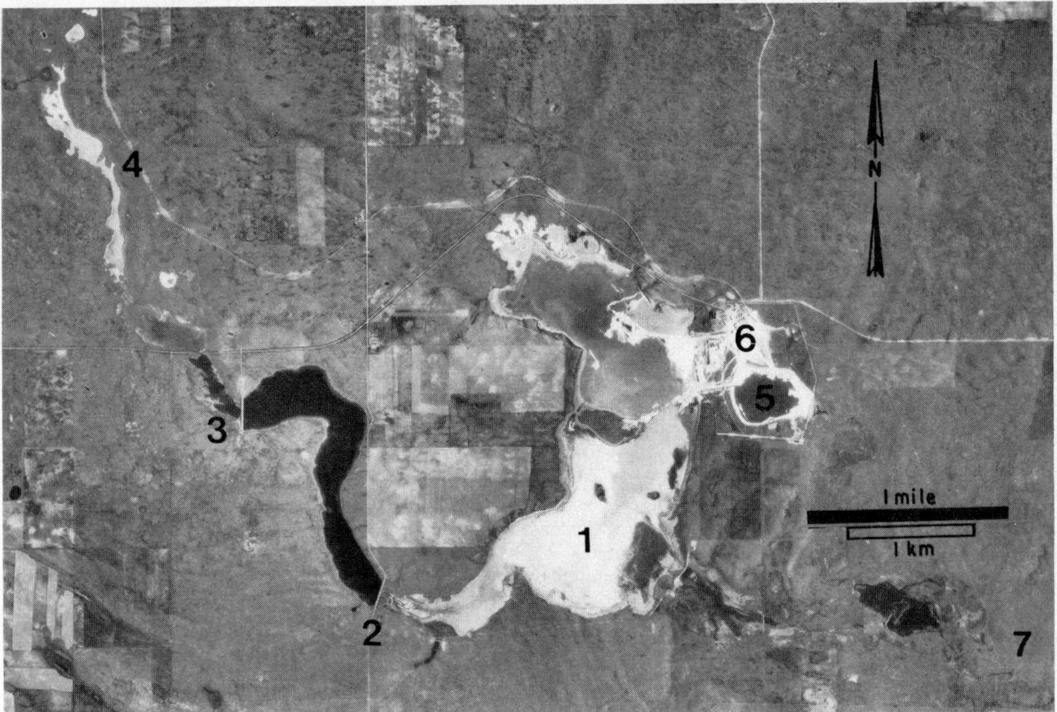

Fig. 32-25. Small scale airphoto showing the regional geologic setting at Ingebright Lake, Saskatchewan, the location of a large sodium sulphate deposit. An 1800 gallon a minute groundwater supply for solution mining was discovered here using remote sensing techniques. (Photo A25139-186, taken in 1979.)

from this storage reservoir is drained onto the salt-covered lake bottom during the hottest period of the summer, and is allowed to dissolve sodium sulphate. After the brine has reached a desired concentration it is drained into the small brine storage reservoir at 5. During the winter the salt precipitates onto the reservoir bottom, and the clear water is drained back into Ingebright Lake to the west. The sodium sulphate is then bulldozed uphill to the plant, located at 6.

Several springs discharging from a near-surface sand aquifer at 7 (southeast of the scale bar) were also identified in the airphotos. Water of good quality is obtained from shallow wells installed here and is piped to the plant and to employee residences near 6 for drinking, toilet, and other domestic uses.

Some common airphoto-identifying clues to prospective sand and gravel aquifers are given in Table 32-6.

Groundwater Location for a New Potash Mine Site, East of Rocanville, Saskatchewan

The landscape shown in Figure 32-26, a color infrared photograph, adjoins the Qu'Appelle Valley just west of the Manitoba-Saskatchewan border. The first of several remote sensing studies was made to try to discover a 1200 gallon a minute groundwater supply for use in the construction and development of a proposed new potash mine. The potash plant is located about one mile south of the Qu'Appelle River.

Streamflow in the Qu'Appelle River is highly variable. In addition, the quality of the river water varies considerably throughout the year owing to high turbidity, algae, and other constituents. As a result, a decision was made to try to locate a suitable groundwater supply.

Color Figure 32-26 shows the potash-plant site, now constructed and in operation at 1, and the Qu'Appelle River at 2. Clues to the locations of three water-table aquifers are evident at 3, 4, and 5. These clues, primarily in the form of major seepages and extensive swampy ground, are more easily observed in large scale black-and-white airphotos. The three aquifers—all located in pervious strata within a buried valley—are separated by layers of either impermeable clay till or lacustrine clay.

Airphotos and photomosaics revealed an extensive deltaic sand plain north of the Qu'Appelle Valley. These surficial sands represent an obvious source of groundwater recharge. A 1200 gallon a minute water well, yielding water of excellent quality, was installed at 6.

Additional remote sensing studies carried out at this potash-mine operation included: (1) a study leading to the location of high quality aggregates for the manufacture of concrete, used to construct the mine shaft; (2) a study of potential leakage from the reservoir at 7; and (3) an environmental assessment study, relating to potential soil and groundwater pollution. There is evidence of unhealthy vegetation at 8, suggested by the appear-

TABLE 32-6

Airphoto Identifying Features of Aquifers in 185 Groundwater Investigations

Surficial features	Number of times helpful in 185 studies
Topography	
a) Appraisal of regional physiographic setting	72
b) Appraisal of local relief setting	81
Vegetation	
a) Phreatophytes and aquatic plants	24
Geological landforms	
a) Modern pervious alluvial terraces and flood plains	29
b) Stratified pervious valley-fill deposits in abandoned spillway channels	11
c) Proglacial glacial outwash and outwash-deltas:	
1) as plains and aprons (both pitted and unpitted)	13
2) along glacial-spillway channels	14
d) Kames and kame moraines	12
e) Coarse-textured dead-ice moraine showing evidence of "submasked" stratified drift in the subsurface	4
f) Esker, esker-kame, and crevasse-filling complexes	8
g) Alluvial fans, both surficial and deeply buried	6
h) Beach ridges	3
i) Partly drift-filled valleys marked by a chain of elongate closed depressions	7
j) Largely masked bedrock valleys cutting across modern valleys, indicated by local non-slumping of weak shale strata in valley sides as well as commonly by springs near the base of the valley side	6
k) Drift-filled valleys in extensively exposed bedrock terrains	4
l) Sand dunes assumed to overlie sandy glaciofluvial sediments	2
Hydrographic features and associated hydrologic environment	
Lakes and streams:	
a) Test sites near large fresh-water lakes that might recharge aquifers	20
b) Test sites near small perennial and intermittent lakes; includes chains or strings of elongate fresh, brackish, or saline lakes and sloughs occurring in inactive drainage systems	12
c) Test sites near perennial rivers and large creeks in valleys, usually having a small active and a larger "fossil" flood plain	26
d) Test sites near small intermittent drainages; includes misfit creeks in abandoned glacial spillways and meltwater channels	19
e) No defined drainage channel in former glacial spillways and meltwater channels	4
Moist depressions and seepages:	
a) Moist depressions, marshy environments, and seepages (significance depends on and interpretation of associated phenomena	9
b) String of alkali-salt flats or brackish and salty lakes along inactive drainage systems	5
c) Salt precipitates (e.g. salt crusts) and localized anomalous-looking "burn-out" patches in the soil and vegetation and associated with salt migration and accumulation	8
Springs (types tentatively inferred from aerial photographs	
a) Depression springs (where the land surface locally cuts the water table or the upper surface of the zone of saturation)	5
b) Contact springs (permeable water-bearing strata—usually along the sides of valleys that cut across the interface between differing geological strata)	12
c) Artesian springs occurring on undulating upland till plains (permeable water-bearing bed between relatively impermeable confining beds, and with enough head to discharge water on the ground surface)	6
d) Artesian springs occurring on or near the base of hillsides, valley slopes, and scarps	6
e) Springs where the type could not be reasonably inferred from aerial photographs	7
f) "Fossil" springs; where the shape of the former spring is evident but the water table has lowered and the former spring no longer discharges	2
Land use	
a) Farming practices, including anomalous waste lands that are too wet or boggy to cultivate	10
b) Roads leading to springs or to existing water wells	2
c) Water-filled dugouts and gravel pits in surficial alluvial and glaciofluvial deposits	3
d) Coal mines	2

ance of the vegetation in this color IR photo. Both normal color and color IR airphotos were taken, and were interpreted for the purpose of pollution surveillance and monitoring.

The color IR photo in Figure 32-26 also shows a main railway line where, at point 9, long-continued slope-stabilization work has been carried out over a period of many years. Abandoned channel scars on a thin gravelly outwash plain are evident at 10. The area to the south, at and surrounding 11, is an undulating till plain.

URBAN SITES

Considerations Affecting the Location and Expansion of Urban Areas

In selecting sites for new towns and villages, especially those in remote regions, the interpreter of airphotos and other remotely sensed data must again consider a variety of social, economic, engineering, and environmental factors. In the case of urban areas, these factors often include aesthetic considerations and recreational amenities.

When a site is being selected for a new mining town, or for the expansion of an existing one, many of the factors in the following list need to be located and evaluated. (The list is compiled from some 60 urban sites, small and large, that have been studied on aerial photographs over the past 35 years.)

(1) Alternative sources of naturally occurring construction material within an economic haul distance of the new town site;

(2) An adequate supply of surface water or groundwater for domestic, commercial and industrial uses, and for the disposal of industrial wastes and sewage;

(3) Areas having desired earthwork and foundation characteristics, such as deep, well-drained sand and gravel or relatively shallow soil materials over competent bedrock (as required for the siting of heavy buildings) and well drained, medium textured soil materials (as preferred in residential areas). Weak or highly compressible foundation materials should be identified and avoided as building sites insofar as possible;

(4) Major transportation/communication routes;

(5) Routes for the pipelines that will be required in order to transport water from reservoirs into town;

(6) Water intake sites on river or lake shores, where potential erosion and sedimentation problems should be avoided or properly assessed if they cannot be avoided;

(7) Sewage lagoon sites;

(8) Specific features of the natural landscape that could adversely affect the feasibility and cost of constructing underground utilities. Such features often include wetland areas and areas in which the soil is either of poor quality from the engineering standpoint or of highly variable thickness over the bedrock;

(9) Potential hazards to life and property—e.g., poorly drained lands that are susceptible to recurrent flooding, high groundwater table areas, landslides and other types of unstable ground, excessive erosion or sedimentation. Aggradation of permafrost can cause excessive differential heaving and permafrost degradation can produce excessive differential settlement. Soil and rock materials that contain fissures or other pervious zones that might affect contamination by pollutants should be considered as they affect domestic, industrial, or radioactive wastes.

One can easily add to the list of features that the remote sensing analyst may have to look for and try to assess. For example, noise and air quality are other factors.

In locating and laying out new town sites from image interpretation, it is common to try and harmonize engineering and development with ecological problems and opportunities. For example, the interpreter will verify whether runoff can be recharged into the ground near the place where rain falls. Generally, vegetation along the sides of existing natural drainage courses should not be cleared; and ponds and reservoirs should be located to minimize runoff. Built-up areas should be laid out to avoid flood-prone land while maintaining the base flow in streams and water levels in ponds and reservoirs. Potentially erodible or unstable hill and valley sides should be mapped. Wetlands and natural woodlands should be identified, and consideration should be given to setting them aside for recreational use or as important areas from the standpoint of wildlife habitat. Wetlands may be used to help reduce stream pollution, the risk of flooding, and depleting water supplies.

Many of the features associated with run-off, soil moisture, vegetation and water levels may be readily interpreted from color infrared aerial photography acquired at a time of year appropriate for the region of interest. Individuals with knowledge of local conditions are the best qualified to identify the best imaging date. Airphoto interpretation studies of the Uranium City area, in the remote northwestern corner of Saskatchewan, addressed a number of these concerns when a search was undertaken for additional land for urban expansion (Figure 32-27). Possible campsite locations were also noted along a 500-mile long highway route study from La Ronge to Uranium City (Figure 32-28); this figure shows the distribution of mineral and organic soil materials as well as bedrock. Study of Figure 32-28 reveals different surface materials, types of topography, surface and groundwater hydrology, and corresponding changes in the natural vegetation. The kettled sand and gravel outwash filling is of course considered to be a most desirable area for urban development.

Another example of the use of modern remote sensing technology as an aid to selecting specific locations for residential development within a vast wildland area is described in a report by Khorram

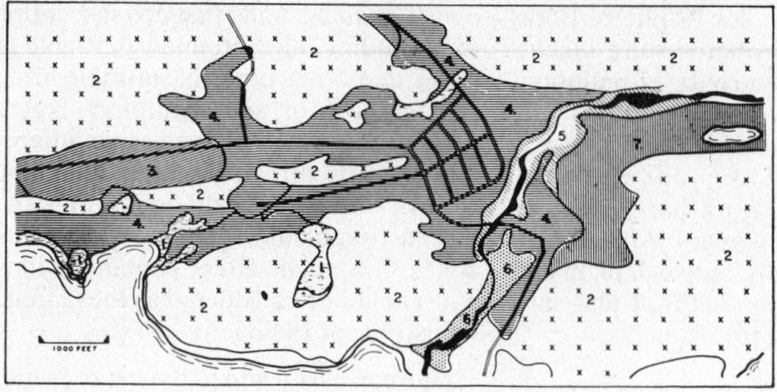

LEGEND

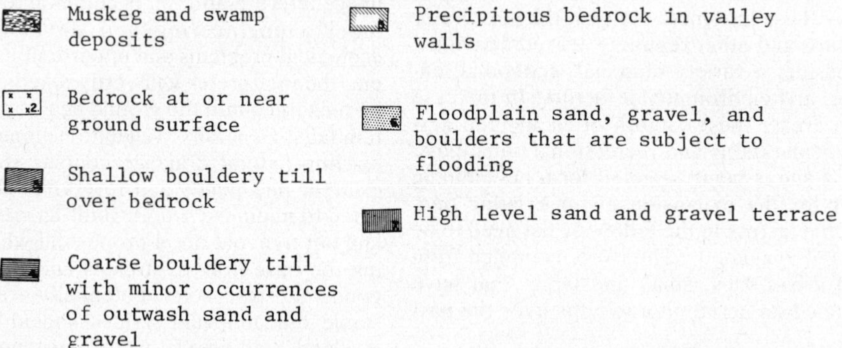

Fig. 32-27. Map of Uranium City area showing important terrain features mapped from airphotos. Terrain types in this area have a marked influence on new housing development and the installation of underground utilities. This map was prepared to help locate areas into which the town might expand in the late 1950s and early 1960s.

(1982). That report is based on work done in California's Sierra Nevada. Because the approach used and the data bank and sensor systems employed differed significantly from the Canadian examples described above, Khorram's work will be presented in the next section.

Residential Suitability Mapping in California

Khorram's (1982b) study describes the manner in which remote sensing facilitated the selection of sites suitable for urban development in a vast area of northeastern California. It provides an excellent example of the successful integration of remote sensors. It also describes the procedures that were used for the development and utilization of a spatially-referenced digital databank to provide the required spatial data for large scale land-use planning and specifically for the selection of the urban development sites by the Plumas County Planning Department. The data included intermediate maps pertaining to slope, ownership, soils type, and potential timber-growth capacity as well as final maps of land developability and potential residential suitability. All of these maps were at a scale of 1:62,500 in color-coded transparency form and of the same registration (coordinate system).

The performance of Khorram's investigation was aided by the direct participation and close cooperation of several other researchers from the Remote Sensing Research Program (RSRP) of the University of California at Berkeley and by various resource managers of the U.S. Forest Service of the Plumas National Forest, and personnel of the Plumas County Planning Department. The results of this study were designed to be used by the County Planning Department primarily for selection of one or more areas for urban development and generally for land use planning. The urban development was aimed at providing a bedroom community for the city of Reno, Nevada.

The study area, selected jointly by the resource managers and the investigators, covers approximately 1500 km² extending from latitude 39° 45′ to 40° 30′ North and longitude 120° 00′ to 120° 35′ West. This area, located in the upper part of the Middle Fork of the Feather River Watershed along the northeastern edge of the Sierra Nevada Mountains in California, is within commuting distance (approximately 40 ± 12 miles) of the city of Reno.

The approach that was used involved the development and utilization of a spatially-referenced digital databank based on both remotely-sensed data and conventional tabulated and map data. The types of remotely-sensed data included: (1) Landsat Multispectral Scanner (MSS) data; (2)

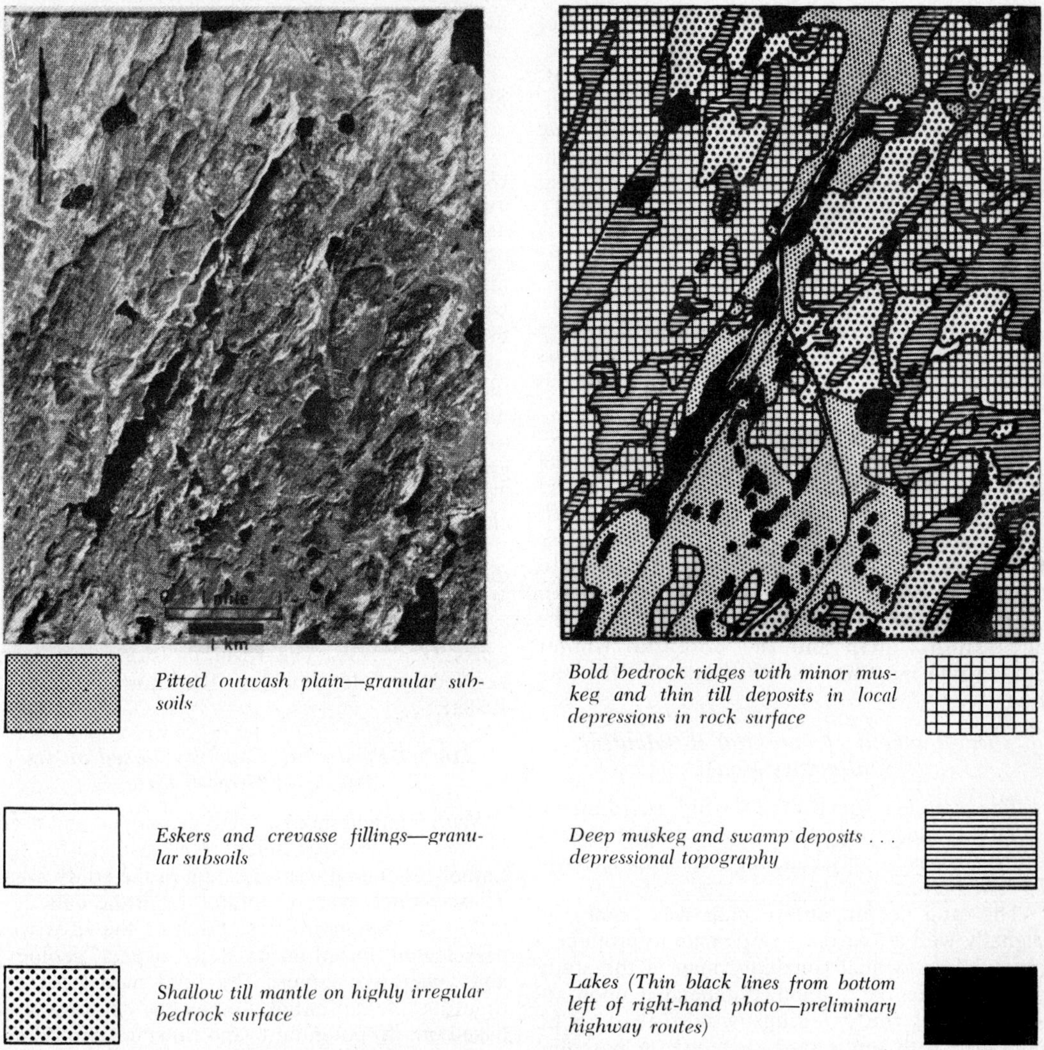

Pitted outwash plain—granular subsoils

Bold bedrock ridges with minor muskeg and thin till deposits in local depressions in rock surface

Eskers and crevasse fillings—granular subsoils

Deep muskeg and swamp deposits . . . depressional topography

Shallow till mantle on highly irregular bedrock surface

Lakes (Thin black lines from bottom left of right-hand photo—preliminary highway routes)

Fig. 32-28. Airphoto and matching terrain map showing a small section of selected highway route and associated campsite development area. This area is along a proposed highway route from LaRonge to Uranium City in northern Saskatchewan. It was studied from airphotos in the late 1950s.

NOAA-5 Satellite Very High Resolution Radiometer (VHRR) data; and (3) U-2 color infrared photography.

The conventional ground-acquired data included the following items: (1) U.S. Geological Survey (USGS) topographic maps; (2) Defense Mapping Agency (DMA)/USGS digital terrain data; (3) "ground truth" data consisting of land cover, vegetation type, vegetation density, and timber-growing capacity; (4) soil plant-available water; and (5) ownership and accessibility maps.

The investigation produced a Land Developability map and a Potential Residential Suitability map of the study area. Both of these maps were produced to the scale of 1:62,500, and were designed to be visually overlayed onto the ownership map and soils permeability map, of the same scale and the same registration, for planning purposes. In order that the reader can adequately appreciate how a multifacted engineering-related study such as this might be performed, the spatially-referenced databank, the models for generation of both land developability and potential residential suitability maps, and the preparation of the required input parameters for the mathematical models will now be discussed.

Development of a Spatially-Referenced Digital Databank

The databank used in this study was a storage and retrieval system, known as MAPIT, developed by personnel of the University's Remote Sensing Research Program. In this system, all remotely-sensed, map-based and ground-acquired data for a given geographic area may be stored as "profiles". A profile represents a relationship between a point and some attribute or condition located at that point. The point is described by an X, Y coordinate pair and the attributes belonging to that point are usually referred to as its Z values.

Structurally, MAPIT consists of a number of distinct programs controlled by the user to accomplish a number of tasks related to manipulating, interpolating and displaying areal data. Mathematical models, such as those used for preparing some of the input parameters as well as the final products in this study, require a variety of areal inputs. These inputs reside in and use MAPIT data storage and organizing capabilities.

Development of Land Developability Model

The criteria for the land-developability models were based on recommendations of the Plumas County Planning Department as will be noted later.

A four-class accessibility map of the study area was combined digitally with a map of its potential timber-producing capacity (based on tree age and height) thereby generating a six-class land developability map. The classes used ranged from Class I (representing the land most developable) to Class VI (representing the land least developable). The procedures used for preparation of both the accessibility map and the potential timber growth-rate map are discussed later in this section.

Development of Potential Residential Suitability Model

The criteria for the potential residential-suitability model were based on recommendations of the Plumas County Planning Department as will be noted later.

The land developability map was combined digitally with a four-class slope map to produce a potential residential suitability map of the study area. (Residential suitability depended upon assuming that the dwellings would need to be equipped with septic tank systems). A potential residential suitability map consisted of eight classes ranging from Class 1 (representing potentially the most suitable land for residential development) to Class 8 (representing potentially the least suitable land for residential development). The procedure used for preparation of the slope map will be discussed presently.

Input Parameters for Land Developability and Potential Residential Suitability Models

The required input data for these models were accessibility, slope, and timber-producing capacity based on potential growth rate.

Accessibility

The four-class accessibility map of the study area, prepared by the Plumas County Planning Department, was digitized in terms of the geographic coordinates and corresponding accessibility, transferred to the Landsat-based coordinate system and stored in MAPIT. The accessibility classes were: Class 1 = accessible during all times of the year; Class 2 = accessible during the summertime with paved roads; Class 3 = accessible during the summer time with dirt roads; and Class 4 = not accessible.

Slope

The slope map was prepared from the elevation data of the study area as generated from the DMA/USGS digital terrain data. See U.S. Department of the Interior (1978) for description of digital terrain tapes. The technique for generation of the elevation databank included: 1) reformatting the digital terrain tapes to be compatible with the RSRP computer system, 2) determining the boundaries of the study area, 3) selecting the control points, and 4) transferring the data into the Landsat-based coordinate system. To determine the slope, the normal vectors were defined to represent the elevational gradients between a given pixel and the four pixels surrounding it. The slope data were then grouped into the four classes and stored in MAPIT.

Slope in percent, one criterion for determining the potential residential suitability, was made up of the following four classes: Class A = 0 to 10; Class B = 10 to 20; Class C = 20 to 30; and Class D = >30. The resultant slope map was verified by field checking (Khorram and Smith, 1980a).

Timber-Producing Capacity Based on its Potential Growth Rate

Mathematical models were developed and used to compute the tree height/tree age ratio for the timber-producing trees existing in the study area. These models were generated from the data collected in 35 sample sites. Each of the sites was preselected, based on its slope, aspect, geology, and vegetative cover. The final model for estimating the potential timber-growth rate was based on: (a) potential evapotranspiration; (2) aspect; and (3) soil plant-available water. For a detailed description of the potential timber-growth rate models see Smith and Khorram (1980a).

A brief discussion of the input parameters to the potential timber-growth rate model is as follows:

(a) Potential Evapotranspiration

The procedure for mapping potential evapotranspiration (ETp) was based on the Jensen and Haise (1963) model, but modified as necessary to permit it to be applied to a wildland area. The inputs to this model were net radiation and average daily surface temperature. The detailed procedure for potential evapotranspiration mapping is discussed by Khorram et al. (1978).

The net radiation was estimated from a series of mathematical models which, in turn, were based on the solar constant together with data on elevation, slope, aspect, latitude, albedo, cloud cover, and surface temperature. The detailed description of the net radiation model is discussed by Khorram (1982a).

The surface temperature data were based on Very High Resolution Radiometer (VHRR) data obtained from the NOAA-5 satellite. This satellite's daytime pass over the study area occurs at approximately 9:30 AM and its nighttime pass occurs at approximately 9:30 PM. The averaged values of these passes, on a pixel-by-pixel basis, was used as the average daily surface temperature. The procedure for surface-temperature mapping is discussed by Khorram and Smith (1980b).

(b) Aspect

Like the slope map, discussed earlier in this section, the aspect map was prepared from the DMA/USGS generated elevation databank of the study area. The aspect data were determined from the normal vectors representing the elevational gradients between a given pixel and the four pixels surrounding it. See Khorram and Smith (1980a) for the detailed procedure.

(c) Soil Plant-Available Water

The soil plant-available water values for each soil type within the study area were determined in the laboratory by analysis of the soil samples collected from the 35 preselected sample sites described earlier.

The soils data for the study area came from combining all available soils data from the soils maps generated by the Soil Conservation Service, the California Department of Water Resources, and the U.S. Forest Service organization in the Plumas National Forest. Information appearing on these soils maps was grouped into classes and digitized; then the data were transferred to the Landsat-based coordinate system. The digitized soil data were converted into the digital soil plant-available water data as determined in the laboratory for the soil classes. See Khorram et al. (1978) for the detailed procedure.

RESULTS

The results included a series of color-coded maps each pertaining to: (1) the primary input parameters such as soil plant-available water, potential timber-growth rate, slope, accessibility, and ownership; (2) secondary input parameters such as elevation, aspect, surface temperature, net radiation, and potential evapotranspiration; and (3) the final outputs-land developability and potential residential suitability. All of the color-coded maps were produced in positive transparency form and to the scale of 1:62,500; thus when overlayed onto one another and onto other maps of the same scale they facilitate the development of resource management plans. The originals of these color-coded maps were produced by using an electro-optical film-writing device called IGOR (Image Ganged Optical Reproducer) designed and built by personnel of the RSRP, University of California, Berkeley. These originals were then photographically enlarged to the desired scale. For the purpose of publication and for ease in highlighting certain attributes of the area, one at a time, on a series of thematic maps, these maps, although initially color-coded, were converted to black-and-white.

Examples of the map products dealing, respectively, with slope accessibility, land developability, potential residential suitability, and ownership are shown in Figures 32-29 a through 32-29e.

The land-developability map was generated (based on criteria shown in Table 32-7) through an integration of the data on the timber-producing capacity and accessibility maps. As previously indicated, this land-developability map contained six classes ranging from Class I (representing the most developable) to Class VI (representing the least developable) areas. In Figure 32-29c two of these classes, viz. Classes V and VI, are combined.

The potential residential suitability map was generated (using criteria shown in Table 32-8) from the land developability map and the slope map. This map contained eight classes ranging from Class 1 (representing the most suitable land) to Class 8 (representing the least suitable land).

The ownership map, originally prepared by the Plumas County Planning Department, was digitized and transferred to the Landsat-based coordinate system. A new version of the ownership map was then produced by the IGOR film-writing device. In this way, the ownership map was prepared at the same scale and used the same coordinate system as all other input and output products. This five-class ownership map included the following categories: Forest Service I, Private, Timberland Preserve, Agricultural Preserve, and Forest Service II. The Forest Service I class included a relatively small area belonging to the Plumas National Forest, one which is surrounded by large privately-owned areas.

The final maps (land developability and potential residential suitability) were designed to be visually overlayed onto political and geographic maps of the same scale and coordinate system (e.g., ownership, zoning, etc.) for selection of urban development sites in this case and, more generally, for regional land use planning purposes.

CONCLUSIONS

Based on the findings in this investigation it was concluded that:

TABLE 32-7

Criteria Used in Producing the Land Developability Map from Data Appearing on (1) the Accessibility (ACC map, and (2) the Potential Growth Rate (PGR) map

PGR \ ACC	Cl. 1	Cl. 2	Cl. 3	Cl. 4
Cl. 1	VI	VI	VI	VI
Cl. 2	VI	VI	VI	VI
Cl. 3	II	IV	V	VI
Cl. 4	I	III	V	VI
Cl. 5	I	III	V	VI
Cl. 6	I	III	V	VI

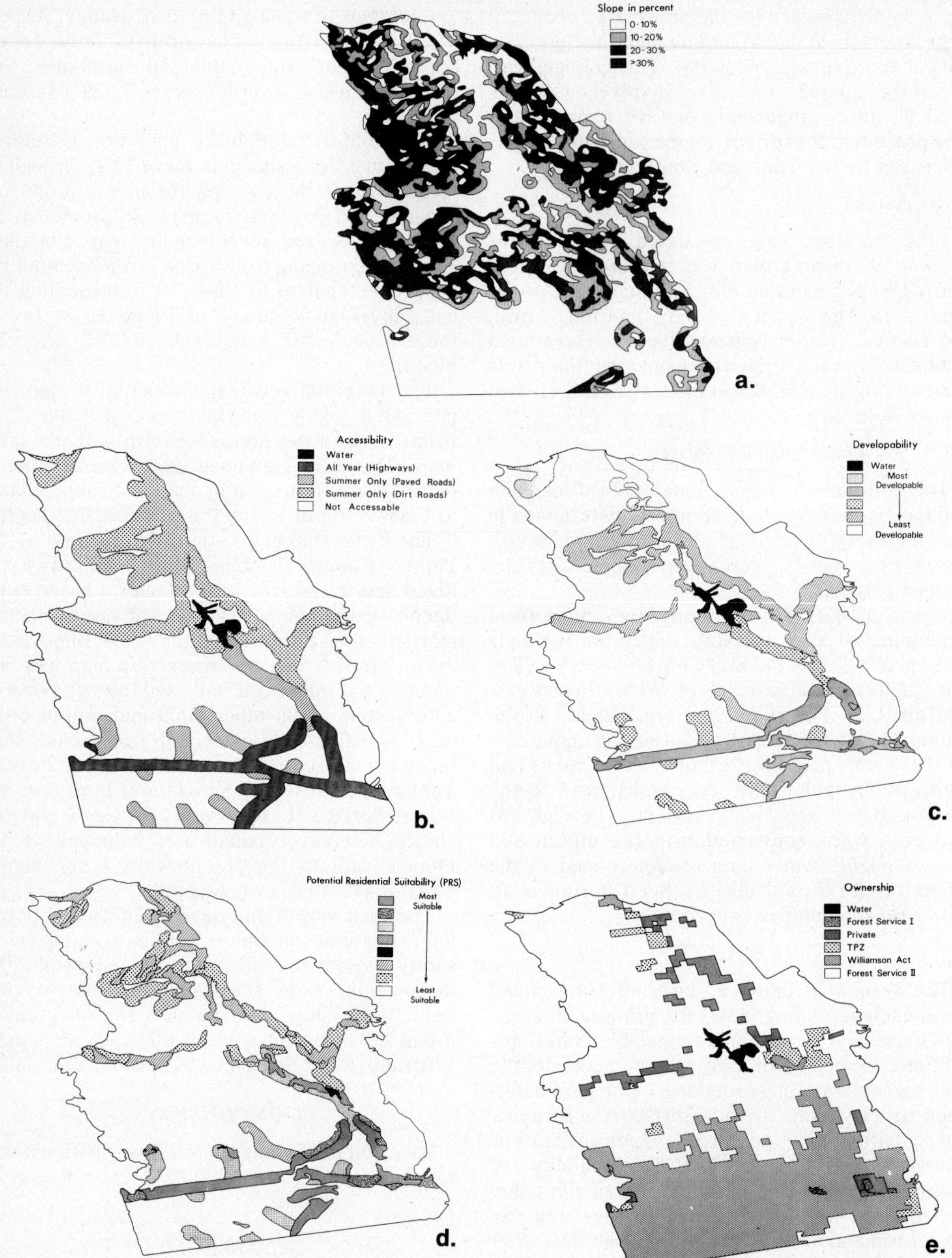

Fig. 32-29. Black-and-white reproductions of color-coded maps for the upper part of the Middle Fork of the Feather River Watershed. As indicated by the legends, these maps deal with the following attributes of the area: slope, accessibility developability, potential residential suitability, and ownership. (After Khorram, 1982b.)

(1) A spatially-referenced databank is useful in combining remote sensing-related-data of various kinds, (viz, satellite, aircraft and conventional imagery; tabulated data; conventional map data; and ground-acquired data) to produce resource-related maps in terms of net radiation, timber-producing capacity, land developability, and potential residential suitability.

(2) Landsat MSS data are useful in mapping the vegetation types and their resultant albedo.

(3) DMA/USGS digital terrain data provide adequate elevation and slope data for this type of analysis.

(4) NOAA-5 satellite VHRR data provide the means for mapping surface temperatures.

(5) Certain intermediate maps (e.g. those per-

TABLE 32-8

Criteria Used in Producing the Map of Potential Residential Suitability (with Individual Septic Tank Systems) from Slope and Developability Data

Slope / DEV	Cl.A	Cl.B	Cl.C	Cl.D
Cl.I	1	1	2	8
Cl.II	2	3	3	8
Cl.III	4	4	5	8
Cl.IV	5	6	6	8
Cl.V	7	7	8	8
Cl.VI	8	8	8	8

taining to elevation, slope, aspect, accessibility, etc.) are useful in the assessment of resource conditions and resource management practices, especially when overlayed or combined with other maps such as those pertaining to the timber-producing capacity of the area.

(6) Remote sensing data, when combined with ground-acquired data, provide a composite data base that is second to none for the selection of urban development sites and for regional land use planning purposes.

Finally, the field checking that was done by Plumas County officials convinced them of the accuracy of the approach detailed above. Consequently the county officials are using the techniques in making the final selection of the suitable sites for urban development of "bedroom communities" for the City of Reno, Nevada.

Planning Engineering Activities in an Urban Area: The Battlefords, Saskatchewan

Several remote sensing site investigations have been carried out during the growth of an urban area known as North Battleford, which is located on the north side of the North Saskatchewan River in Canada. The Town of Battleford is located on the south side of the river (see Figure 32-30). With reference to that figure, the target area for the large capacity water wells that were installed at 1 was identified originally from a study of high altitude airphotos. These wells now supply water to the Town of Battleford.

Terrain types and terrain conditions were mapped for the entire area shown in Figure 32.30. The mapping program was implemented as part of a series of studies for urban planning. Pressure for city building lots in area 2 had already developed because of the commanding view overlooking the scenic North Saskatchewan River, with its wooded islands, sand bars, and winding channels. However, the area at 2 consists of badly slumped terrain, where former slope movements can be easily reactivated.

A small slide developed at the bridge abutment at 3. Movement of this slide created costly maintenance at the south abutment of the bridge. Area 4 is an open pit in a large gravel terrace. Parallel linear features (grooves and low ridges) at 5, west of the north point, are glacial flutings. They indicate the local direction in which glacier flow once took place and, by inference, suggest that material in the subsurface is lodgement till.

Scrub vegetation has colonized the sand dunes at 6. This area is a fragile landscape because it is susceptible to wind erosion when the vegetation is removed, or the ground is much disturbed. Additional landslides and groundwater seepage areas were identified at 7, 8, and 9 in the urban terrain mapping study.

As well as mapping, classifying, and rating the terrain from airphotos, a regional land use and land cover mapping study was undertaken recently (1981) using enhanced Landsat data products. These products included color composite Landsat images (bands 4, 5, 7); a series of image products generated by the Image Analysis System of the Canada Centre for Remote Sensing in Ottawa, Canada; and scaled multiple-theme greytone maps. All Landsat products of the digital analysis were produced at a scale of 1:250,000 to match the existing National Topographic Map Sheets at the same scale.

RECREATIONAL SITES

Considerations in Selecting and Developing Recreational Sites

A number of remote sensing studies have been done on the development of new parks and extensions to existing parks, and for new resort developments and extensions to existing ones. Small-area studies have been carried out on the selection of playground sites, campsites, picnic grounds, boat launching sites, marinas, ski slopes, and the identification of hiking trails. These types of sites may be studied with conventional black-and-white imagery or with small hand-held systems in light aircraft (Fleming and Dixon, 1981) or with pilotless drones. Depth and quality of water in lakes, shoreline type and beach materials, underwater materials in the nearshore zone, and access to a recreational site are usually important first considerations. Topography (slope, scenery), drainage, and natural vegetation are also very important control factors in site location. Locally, however, soils and rocks, and their properties, can have an important bearing on the type and location of the recreational facilities. Soil types and conditions are emphasized in the following paragraphs.

In general wet soils are unsuitable for access roads, campsites, playgrounds, and picnic sites. Soils that pond water and dry out slowly after heavy rains present problems where intensive recreational use is contemplated. Location and identification of such areas at appropriate intervals after rains is done cost-effectively with 35 mm color infrared imagery acquired by using light aircraft to carry the hand-held cameras, or using a recent innovation where small drones carry the cameras loaded with various 35 mm film types. Without watering, one finds it difficult to establish

Fig. 32-30. Black-and-white panchromatic airphoto of The Battlefords area of Saskatchewan, showing the landscape where studies relative to an urban groundwater supply, urban land use classification, and regional planning have been carried out using remote sensing methods. Small scale and large scale black-and-white airphotos, color airphotos, and enhanced Landsat data products were studied at different times during the course of this investigation.

and maintain grass cover for playing fields and golf courses on soils that are coarse in texture, and therefore permeable and droughty. On the other hand, soils that are subject to frequent flooding have severe limitations for campsites and recreational buildings. Such soils should be used for parks, as open space for hiking and nature study, or as wildlife habitat.

The feasibility of many kinds of outdoor activities is determined by such soil properties as texture, stoniness, depth to bedrock, and the ability of a soil to support different kinds of vegetation, which in turn is related to the natural fertility of the soil. Of these factors, soil texture is probably the most significant property because so many other soil characteristics are closely related to it (Table 32-9).

Most playground sites, campsites, and picnic areas require a fairly level surface, good drainage, freedom from flooding, and a well drained soil that will provide a firm surface and will support intensive foot, bicycle, or other traffic. Other desirable features are a good topsoil, a nearby source of sand and gravel, a good water supply, the absence of excessive shore erosion or unstable slopes, the availability of suitable sewage lagoons and septic tank filter fields, and the presence of aesthetically attractive cottage and commercial building sites. Depth to the water table, land surface gradients, flooding susceptibility, depth to bedrock or to hardpan, and stoniness or rockiness, in turn, have a bearing on the selection of sites for sewage lagoons and septic tank filter fields. In this regard, see also the preceding Plumas County example in which the location of areas suitable for septic tanks was of major importance.

The texture of surface and near surface soil materials affects sewage absorption, the stability and bearing strength of road and building foundations, and the rate of internal drainage. In fact, natural soil drainage, which is a feature of both soil and landscape, is a very important consideration when one is selecting parcels of land for residences, roads, underground services, and sewage disposal fields. Percent slope is a feature of the landscape that determines surface drainage, and thus site location.

TABLE 32-9

Qualitative Effect of Soil Texture on Other Soil Characteristics. Used as a Guide in the Selection of Recreational Sites

Some Soil Characteristics	Soil Texture	
	Sandy	Clayey
Water holding capacity	low	high
Permeability or drainability	high	low
Porosity	low	high
Size of dominant pores	large	small
Tendency to shrink and to swell	low	high
Bearing strength when dry	high	high
Bearing strength when wet	high	low
Stability from sliding on slopes	high	low
Plasticity	low	high
Soil fertility	low	high
Erodibility by running water	low	high
Erodibility by wind action	high	low
Workability in wet weather	high	low

Soils on the floodplains of rivers usually still remain subject to flooding and, therefore, should not be used as sites for residences. Depth of soil, which has a direct affect on the amount of effluent that can be disposed of, affects building lot size. Stoniness and rockiness also affect site location. The salt content of soils (corrosion potential) and the depth of organic material (peat and muck) are other characteristics that are identifiable on airphotos. Both can influence the selection and development of attractive recreational sites.

Bonne Bay National Park, Western Newfoundland

Figure 32-31 shows the inland portion of Western Brook Pond in western Newfoundland. The scenery in this part of Newfoundland is highly variable and very beautiful. Before the area shown in Fig. 32-31 was raised some 200 to 300 feet above present sea level, resulting from postglacial isostatic rebound following deglaciation, the coast was cut by a number of deep fiords. These fiords were occupied, enlarged, and deepended by valley glaciers.

An office airphoto study and follow-up field investigation were made to assist in the selection and development of a proposed new national park. That study included the identification of sites for marinas and boat-launching, park access roads, hiking trails, scenic "lookout" points along roads, and nature interpretive centers for park visitors. Water supply and sewage disposal sites were also identified. Terrain and surface geology in the proposed park area were also mapped from the airphotos.

Western Brook Pond from 1 to 2 is spectacular for boating and sailing; towering rugged cliffs at 3, 4, and 5 range from 500 to nearly 1000 feet in height. Alluvial fans, as at 6, have formed along the base of the cliffs. Active talus accumulation is a common occurrence below rocky cliffs, and can be a hazard to hikers. The very rugged topography and rocky surface presented challenges to the photointerpreter in his efforts to locate access road routes.

REGIONAL TERRAIN INVENTORY AND PLANNING STUDIES

Considerations in Regional Planning Studies

As noted above, in this chapter components of the terrain commonly mapped from the photos include landforms, surface materials, topography, and drainage. This section illustrates in detail the application of remote sensing image interpretation to this problem. The legend coding-system developed and applied allows for the mapping of complex as well as simple terrain units. Qualitative geotechnical characteristics can be inferred for each terrain unit. The units can also be rated for

Fig. 32-31. Western Brook Pond in Bonne Bay National Park, western Newfoundland. Preliminary airphoto interpretation studies were designed to select the best area for a park and to identify access road routes, scenic lookout sites, hiking trails, campgrounds, and associated water and sewage disposal facilities. (Photo A12793-144, taken in 1950.)

their suitability for a specific land use and for resource planning and development, based on identifiable terrain types and conditions.

Between 1977 and 1980, the Engineering and Terrain Geology Section of the Ontario Geological Survey contracted out the preparation of 137 engineering geology terrain maps and 110 accompanying reports covering an area of 166,000 mi^2 in northern Ontario. A typical legend coding-system developed for these regional terrain inventory maps in northern Ontario is shown in Tables 32-10 A & B. An example of a general rating of grouped terrain units for certain types of engineering use is illustrated in Table 32-11.

It is possible to derive different types of user information from the terrain data base maps. Two of them are (1) engineering characteristics, and (2) land use suitability. Engineering characteristics are derived from the interpretation of the terrain unit letter codes, and provide a preliminary framework of geotechnical references and information. Land use suitability maps are derived for particular regional engineering and resource planning studies. Digital remotely-sensed data can be very useful in both engineering characteristics determinations and land use suitability mapping. The details are discussed earlier in this chapter.

For making best use of these maps the map user should have an appreciation of the engineering characteristics of the various terrain units and their practical significance. Some geotechnical characteristics that may be inferred in a general but still useful way for each map unit include qualitative and relative descriptions of: (1) shear strength, (2) compressibility, (3) plasticity, (4) permeability, (5) bulk density, (6) natural water content, (7) susceptibility of the soils to frost action, (8) texture of the soil, (9) common slope conditions, (10) approximate depth to bedrock, (11) stratigraphic complexity, (12) approximate position of the groundwater table, and (13) surface drainage conditions (see also Table 32-11).

The engineering significance of a terrain unit can be related to specific uses by means of a second level of derived information, as shown on maps. The preparation of these maps requires an assessment of the geotechnical characteristics for a particular land use. Some background knowledge of geology, engineering, land use planning, and resource development is required for the optimum use of these maps.

The following information can be derived from the terrain data base maps when used in conjunction with qualitative descriptions of geotechnical characteristics of the different terrain units: (1) bearing capacity; (2) earth-borrow suitability; (3) stability of slopes; (4) necessity to dewater excavations; (5) ease of shallow excavation; (6) contaminant-migration hazard; (7) septic-system suitability; (8) construction suitability; (9) relative site-grading rating; (10) aquifer potential (groundwater prospects); (11) susceptibility to flooding; (12) sand and gravel potential; (13) overburden drilling conditions; (14) subgrade suitability for roads; and (15) cut and fill conditions (relative volume of earthwork).

The purpose of these maps is to provide general information so that the map and report user can

TABLE 32-10A

Legend for Regional Planning and Engineering Terrain Maps in Northern Ontario. Prepared for the Engineering and Terrain Geology Section of the Ontario Geological Surveys. (See Accompanying Explanation of Symbol Use and Graphical Symbols on Next Page)

LANDFORM

MORAINAL

- **ME** *End moraine*
- **MG** *Ground moraine*
- **MH** *Hummocky moraine*

GLACIOFLUVIAL

- **GD** *Ice contact delta, esker delta, kame delta, delta moraine*
- **GE** *Esker, esker complex, crevasse filling*
- **GK** *Kame, kame field, kame terrace, kame moraine*
- **GO** *Outwash plain, valley train*

GLACIOLACUSTRINE

- **LB** *Raised (abandoned) beach form*
- **LD** *Glaciolacustrine delta*
- **LP** *Glaciolacustrine plain*

ALLUVIAL

- **AP** *Alluvial plain*

COLLUVIAL

- **CS** *Slope failure*
- **CT** *Talus pile*
- **CW** *Slopewash and debris creep sheet; minor talus*

EOLIAN

- **ED** *Sand dunes*

ORGANIC

- **OT** *Organic terrain*

BEDROCK

- **RL** *Bedrock plateau*
- **RN** *Bedrock knob*
- **RP** *Bedrock plain*
- **RR** *Bedrock ridge*
- **/R** *Bedrock below a drift veneer*

MATERIAL

- **b** *boulders, bouldery*
- **c** *clay, clayey*
- **g** *gravel, gravelly*
- **p** *peat, muck*
- **r** *rubble*
- **s** *sand, sandy*
- **m** *silt, silty*
- **t** *till*

TOPOGRAPHY

LOCAL RELIEF

- **H** *Mainly high local relief*
- **M** *Mainly moderate local relief*
- **L** *Mainly low local relief*

VARIETY

- **c** *channelled*
- **d** *dissected, gullied*
- **j** *jagged, rugged, cliffed*
- **j*** *cliffed volcanic rock signature*
- **k** *kettled, pitted*
- **n** *knobby, hummocky*
- **p** *plain*
- **r** *ridged*
- **s** *sloping*
- **t** *terraced*
- **u** *undulating to rolling*
- **w** *washed, reworked*

DRAINAGE

SURFACE CONDITION

- **W** *Wet*
- **D** *Dry*
- **M** *Mixed wet and dry*
- **h** *Suspected high water table*

judge the general suitability of a particular terrain unit for some use.

In Saskatchewan, 1:250,000 scale data-base maps have been prepared for four regions covering 46,500 mi^2. As in the northern Ontario regional terrain and engineering geology study, maps were prepared from the interpretation of 1:50,000 to 1:80,000 scale black-and-white panchromatic airphotos. In addition to the data-base maps, eight special-use derivative maps were prepared. These maps correspond to National Topographic Series map sheets covering the study area. Each map sheet covers an area of 6,000 mi^2 (15,540 km^2). The following eight special use maps, covering parts of southern Saskatchewan surrounding major urban centers, were prepared:

(1) granular material resource potential;
(2) groundwater potential;
(3) natural hazard prone lands (i.e., lands that are susceptible to flooding, slumping, severe wind or water erosion, etc.);
(4) suitability for light-foundation construction;
(5) suitability for light-foundation structures;
(6) suitability for solid-waste disposal;
(7) suitability for sewage-effluent disposal by infiltration; and
(8) suitability for septic-tank waste disposal.

Table 32-12 to 32-19 show the information contained in the legend of each of the above special use maps in a specific area. These examples also illustrate how the special use maps were derived from the terrain data base maps. Legends for the terrain data base maps for regional planning are similar in many respects to the legends developed for route and site investigations in northern Canada (see Tables 32-20 A and B and 32-21). Tables 32-20 A and B show the legend developed for route-corridor terrain mapping and related site investigations for pipeline-route selection and as-

TABLE 32-10B

Continuation of Table 10A

The letter codes describing the terrain units are made up of four components arranged as follows:

MATERIAL	LANDFORM
TOPOGRAPHY	DRAINAGE

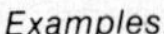

1.

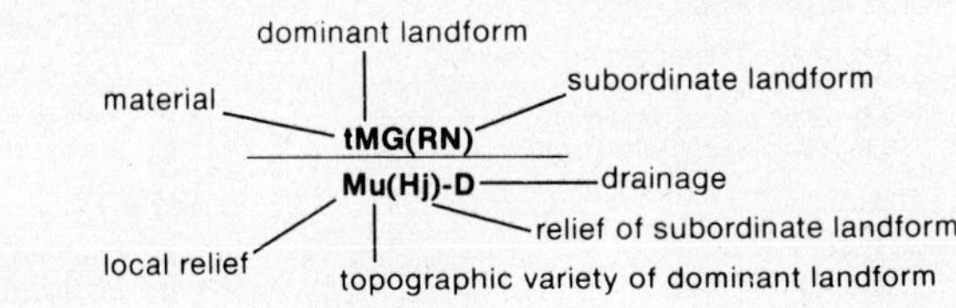

2.

pOT/sGO
Lp-W

slash indicates a veneer of one landform overlying a second landform

SYMBOLS

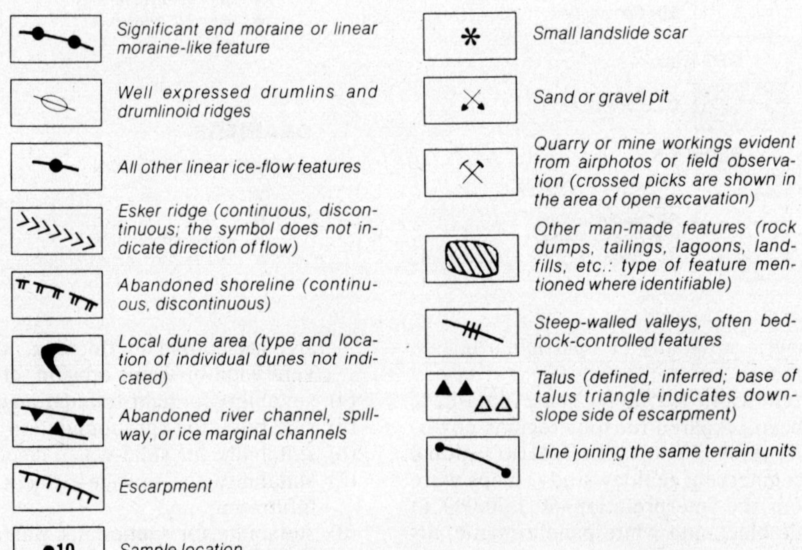

NOTE 1:

This map is a landform inventory, as determined largely by airphoto interpretation, that provides base data for engineering and resource planning. Accuracy of terrain unit boundaries is consistent with map scale. Detailed investigations are required to obtain site specific geotechnical information. Refer to accompanying report for more detailed terrain descriptions and engineering significance.

TABLE 32-11

Rating of Terrain Groups for Special Engineering Uses

PRINCIPAL LANDFORMS AND SURFACE MATERIALS	SIGNIFICANT PHYSICAL PROPERTIES OF TERRAIN GROUPS	Engineering and Construction Use Suitability (Good, Fair, Poor)*																Land Use Planning Hazards or Constraints (Low, Moderate, High)*				
		Vertical and horizontal route alignment	Overland accessibility and trafficability	Suitability for earth dams and dikes	Potential source of high quality aggregate	Potential source of low quality aggregate	Construction workability in wet weather	Foundation quality for heavy structures	Foundation quality for light structures	Excavatability of earth materials	Drainage ditch and canal construction	Underground pipeline and cable construction	Likelihood of small earthwork quantities	Septic system waste disposal	Solid waste disposal (sanitary landfills)	Liquid waste dispoal (unlined lagoons)	Surface water storage (unlined reservoirs)	Contamination of surface and ground water	Flooding hazard from surface runoff	Possibility of high water table position	Slope stability hazard (slides, flows, falls)	Soil erosion potential on fresh man-made cuts
GROUP 1: Organic plain: peat and muck (pOT)	Level to depressional, high water table, low bearing strength, low to very low permeability, high compressibility, subject to flooding	G	P	P	P	P	P	P	P	P	P	P	P	P	P	P	P	H	H	H	L	L
GROUP 2: Colluvial slopes and rubbly beaches (rCT, CW, rLB)	Steep slopes and sloping beaches, active soil creep except on rLB, low frost susceptibility, low shrinkage, low plasticity, moderate to high permeability, variable compressibility	P	P	P	F/G	P	G	P/G	P/G	F/P	P	P/F	P	G	P	P	P	H/M	L	L/H	M/H	L/H
GROUP 3: Alluvial plain: sand, silt, minor clay, gravel, cobbles and boulders; variable texture (sAP, smAP, msAP)	High water table, high permeability, high bearing strength, variable relative density, low compressibility, high to medium frost susceptibility	G	F	P/F	P	F/P	P/F	P	P/F	G	P	P/F	G/F	G/F	P	P	P	H	H	H	L	L
GROUP 4: Sand ridge: fine to medium sand, minor silt in dunes and abandoned beaches (sED, sLB)	Low unconfined strength, low to medium relative density, nonplastic, non-swelling, generally not frost susceptible	G	G	P	P	G	G	G	G	G	P	G/F	G	G	P	P	P	H	L	M	L	L
GROUP 5: Sand plain: sand, minor silt, fine gravel (sGO, sLD, sLP, smLP, smLD)	Well drained, nonplastic, high shear strength and permeability, low compressibility, moderate to high bearing capacity, low frost susceptibility	G	G	P	P	G	G	G/F	G	G	P	G/F	G	G	P	P	P	H	L/M	M/H	L	M
GROUP 6: Silt plain: silt, fine sand, minor clay (mLP, mcLP, msLP)	High frost susceptibility, moderate to low permeability and moderate to high compressibility, low shrink and swell potential, moderate to high shear strength, nonplastic to low plasticity	G	G	F	P	P	P	F	F	G	F	G	G	F/P	F	P/F	P	L/M	M	M	L/M	M/H
GROUP 7: Clay plain: clay, mostly highly plastic; minor silt, cobbles, boulders (cLP, cmLP)	Highly plastic, high shrink and swell potential, poor internal drainage, low bearing strength when wet, low permeability	G	G	G	P	P	P	P	P	G	G	G	G	P	G	G	G	L	L/H	L	L	L/H
GROUP 8: Sand and gravel plain: sand, gravel, minor silt, cobbles, boulders (sgGO, sgLD, sgLP, gsGO, gsLD)	Low compressibility, high permeability, good internal drainage, high shear and bearing strengths, stable in slopes	G	G	P	G	G	G	G	G	F/G	P	F/G	G	G	P	P	P	H	L	L/M	L	L
GROUP 9: Sand and gravel ridges and mounds: sand, gravel, minor silt, surface boulders or till (sg and gs, GD, GE, GK, LB)	Low compressibility, high permeability, good internal drainage, high shear and bearing strengths, stable in slopes	P/F	G/F	P	G/F	G	G	G	G	F/G	P	F/G	P	G	P	P	P	H	L	L/M	L	L
GROUP 10: Till mounds and ridges: very bouldery till; silt, sand, gravel inclusions locally (ts and tm MH, ME)	Moderate permeability, low to moderate internal drainage, low compressibility, high shear and bearing strengths, moderate frost susceptibility, low shrink and swell potential	P/F	F	P	P	F/G	F/G	G	G	P/F	P	P/F	P	F	P	P	P	H	L	L	L	L
GROUP 11: Till plain: bouldery silty to sandy till (tm and tsMG)	Low permeability and compressibility, low plasticity and shrink and swell potential, good internal drainage on rises, high bearing strength	G	G	F/G	P	P	G/F	G	G	G	F	G	G	P/F	G	F	F	L/M	L	L	L	L
GROUP 12: Discontinuous till veneer over bedrock: thin and patchy bouldery till over bedrock; frequent rock outcrops (tsMG)	Variable permeability and water table position, low shrink and swell potential, high shear and bearing capacity	P/F	P/F	F	P	P	P/G	G	G	P/F	P	P	P	P	P	P/F	P/F	L/M	L	L/M	L	L
GROUP 13: Bedrock knobs, ridges, plateaus and plains: (RN, RR, RL, RP)	Variable water table position, high compressive strength, variable permeability dependent on fractures in rock	P	P	P	G/P	P	G	G	G	P	P	P	P/G	P	P	F/P	P	L	L	L	L	L

*Rating combinations such as F/G and L/M read fair to good and low to moderate, respectively.

TABLE 32-12

Granular Construction Material Potential

In this study, terrain and geotechnical factors are used in rating land capability limitations for each map unit. Alternative land use, socio-economic and certain environmental factors are not considered. The system of weighted rating used is illustrated below in a chart; the description of the terrain and geotechnical factors considered is also described below. The rating applies to the classification and mapping of landforms, and to the surface and near surface materials in them. Because of the small scale of the maps, the outlined map units may contain small areas of other map units that are not indicated.

MAP UNITS[1]			TERRAIN AND GEOTECHNICAL FACTORS[2]								WEIGHTED[3] RATING
LETTER SYMBOL	LANDFORM DESCRIPTION	SURFACE MATERIAL AND ASSOCIATED BASIC LANDFORM TYPE AND VARIETY (see legend)	GRADATION	VARIABILITY	EXPLORATION	INDICATORS	QUALITY	OVERBURDEN	WATER TABLE	COARSENESS	
ME	End moraine	tME	2	1	1	4	2	6	6	3	25
MG	Ground moraine	tMG, tMG$_a$, tMG$_c$, tMG$_l$	4	2	1	0	2	3	3	6	21
		tMG$_e$, tMG$_{ec}$	2	1	1	2	1	9	9	9	34
MH	High relief hummocky moraine	tMH, tMH$_s$	2	1	1	2	1	6	9	3	25
MH$_L$	Low relief hummocky moraine	tMH$_L$, tMH$_{Ll}$, tMH$_{Ls}$	4	2	1	0	2	3	3	6	21
LD	Lacustrine delta	sLD, sLD$_k$, sLD$_w$	2	3	1	4	2	6	3	0	21
		smLD, smLD$_w$	2	0	1	2	1	3	3	0	14
LP	Lacustrine plain	mLP, mLP$_c$									NG
		msLP, msLP$_w$									NG
		cmLP, cmLP$_c$, cmLP$_s$									NG
		cLP									NG
LH	High relief hummocky lacustrine terrain	smLH, smLH$_k$, smLH$_w$	4	0	0	2	1	0	6	0	13
		mLH									NG
		msLH									NG
		cmLH, cmLH$_s$									NG
		cLH									NG
LH$_L$	Low relief hummocky lacustrine terrain	smLH$_L$, smLH$_{Ls}$, smLH$_{Lw}$	4	0	1	2	1	3	3	0	14
		mLH$_L$, mLH$_{Ls}$									NG
		msLH$_L$, msLH$_{Ls}$									NG
		cmLH$_L$, cmLH$_{Ls}$									NG
		cLH$_L$									NG
GK	Ice contact knobs and ridges	sGK, sGK$_w$	2	1	1	4	2	3	9	3	25
		sgGK, sgGK$_k$	4	2	1	4	2	6	9	6	34
		gsbGK	4	2	1	4	2	6	9	9	37
GO	Outwash plains and valley trains	sGO, sGO$_k$	2	2	2	4	2	9	6	3	30
		sgGO, sgGO$_c$, sgGO$_s$, sgGO$_{sk}$, sgGO$_l$, sgGO$_k$	6	3	3	6	3	9	6	6	42
		gsGO, gsGO$_c$, gsGO$_k$, gsGO$_{kc}$	6	3	3	6	3	9	6	9	45
		gGO	2	2	2	6	3	9	6	9	39
AP	Alluvial plain	vAP, vAP$_m$	4	0	1	4	1	3	0	3	16
		msAP, msAP$_c$									NG
		sAP	2	1	1	4	1	3	0	3	15
		sgAP	6	1	2	2	3	6	3	6	29
AF	Alluvial fan	vAF	4	0	1	2	1	3	3	3	17
AT	Alluvial terrace	vAT, vAT$_c$	4	0	1	2	1	3	6	6	23
		sAT	2	1	1	2	1	3	6	0	16
		gsAT	6	3	2	4	3	9	6	6	39
ED	Eolian dunes	sED	0	2	3	6	1	6	6	0	24
OT	Organic terrain	pOT									NG
BT	Bedrock terrain	ss, shBT									NG
SL	Slumped valley slopes	vSL									NG
SE	Eroded valley and hill slopes	vSE									NG

CONTRASTING MATERIALS AND LANDFORMS IN THE NEAR SURFACE ZONE

LETTER SYMBOL	SURFACE LANDFORM	BURIED NEAR SURFACE LANDFORM	SURFACE AND BURIED MATERIAL AND LANDFORM	GRADATION	VARIABILITY	EXPLORATION	INDICATORS	QUALITY	OVERBURDEN	WATER TABLE	COARSENESS	WEIGHTED RATING
LD/MG	Lacustrine delta	Ground moraine	sLD/tMG	2	3	1	4	2	6	3	0	21
			smLD/tMG, smLD$_w$/tMG	4	0	1	2	1	3	3	0	14
			sLD$_w$/tMG$_e$	2	2	1	2	1	3	3	9	23
LD/LP	Lacustrine delta	Lacustrine plain	smLD$_w$/cmLP	4	0	1	2	1	3	3	0	14
LD/GO	Lacustrine delta	Outwash plain	smLD/sgGO	6	3	1	2	3	3	6	6	30
LP/MG	Lacustrine plain	Ground moraine	msLP/tMG									NG
			mLP/tMG, mLP/tMG$_l$									NG
			cmLP/tMG, cmLP$_c$/tMG$_c$									NG
			cLP/tMG									NG
			msLP/tMG$_e$	2	1	1	0	1	3	6	9	23
			mLP/tMG$_e$, mLP/tMG$_{ec}$	2	1	1	0	1	3	6	9	23
			cmLP/tMG$_e$	2	1	1	0	1	3	6	9	23
LP/GO	Lacustrine plain	Outwash plain	msLP/sgGO	6	3	1	0	3	3	6	6	28
			mLP/sgGO	6	3	1	0	3	3	6	6	28
			mLP/gsGO	6	3	1	0	3	3	6	6	28
LP/GK	Lacustrine plain	Icecontact knobs&ridges	cLP/sgGK	4	1	0	0	2	0	6	6	19
LH$_L$/MH$_L$	Low relief hummocky lacustrine terrain	Low relief hummocky moraine	smLH$_L$/tMH$_L$	4	0	1	2	1	3	3	0	14
			msLH$_L$/tMH$_L$									NG
			mLH$_L$/tMH$_L$									NG
			cmLH$_L$/tMH$_L$									NG
LH$_L$/LH$_L$	Low relief hummocky lacustrine terrain	Low relief hummocky lacustrine terrain	msLH$_L$/mLH$_L$									NG
												NG
GO/MG	Outwash plain	Ground moraine	sgGO/tMG	6	0	1	4	2	9	9	6	37
AT/MG	Alluvial terrace	Ground moraine	sgAT/tMG$_e$	4	1	2	4	2	9	6	9	37
			smAT/tMG$_e$	2	1	1	2	1	3	3	9	22
			msAT/tMG$_e$	2	1	1	2	1	3	3	9	22
MG/BT	Ground moraine	Bedrock terrain	tMG/BT									NG

GRAPHIC SYMBOL	SURFACE FEATURE	GRADATION	VARIABILITY	EXPLORATION	INDICATORS	QUALITY	OVERBURDEN	WATER TABLE	COARSENESS	WEIGHTED RATING
<<<<	Esker ridge	4	2	1	4	2	6	9	6	34
	Meltwater channel or spillway	4	0	1	4	1	3	0	3	16
	Tunnel valley or stream trench	4	0	1	4	1	3	0	3	16

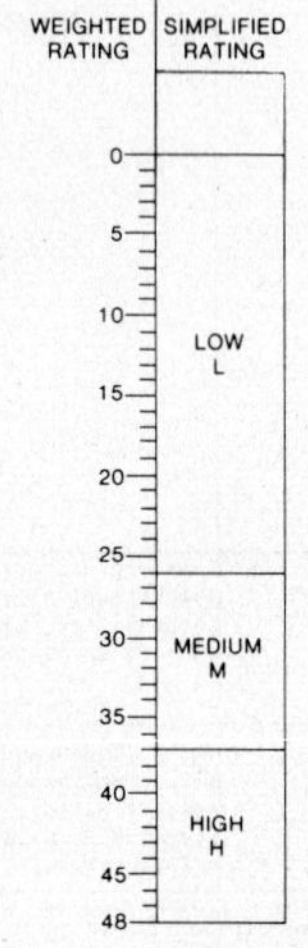

TERRAIN AND GEOTECHNICAL FACTORS	RATING	WEIGHT	WEIGHTED RATING
GRADATION	0 1 2 3	2	0 2 4 6
VARIABILITY	0 1 2 3	1	0 1 2 3
EXPLORATION	0 1 2 3	1	0 1 2 3
INDICATORS	0 1 2 3	2	0 2 4 6
QUALITY	0 1 2 3	1	0 1 2 3
OVERBURDEN	0 1 2 3	3	0 3 6 9
WATER TABLE	0 1 2 3	3	0 3 6 9
COARSENESS	0 1 2 3	3	0 3 6 9

CHART ILLUSTRATING METHOD USED TO DERIVE WEIGHTED RATING

2 1 0 2 4 6 Miles

2 1 0 2 4 6 8 10 Kilometres

MAP UNITS[1]

For description of landform basic types and varieties, and the recurring types of soil materials found in them, which together form the mapping symbols and the map units, see the accompanying report.

TERRAIN AND GEOTECHNICAL FACTORS[2]

Gradation. Refers to the grain size distribution of mineral particles in a natural granular deposit. The rating is based on the interpreted depositional environment of recognizable landforms and the inferred granular materials in them.

Variability. Refers to area-to-area changes in material within a granular deposit. Some landforms are much more variable than others.

Exploration. Refers to the relative amount of time and cost needed to prove up a sand and gravel deposit for some commercial use. Less variable (**i.e.** more homogeneous landforms containing granular deposits require significantly less time and money to explore than do highly variable ones.

Indicators of quality and use. Refers to the chances of seeing surface indicators in aerial photographs or on the ground that can be used to predict the gradation, variability, aggregate quality, and thus the commercial possibilities of a granular deposit.

Quality. Refers to the possible presence and relative amount of such undesirable substances as silt, clay, or shale. High quality aggregate is required for high strength concrete manufacture whereas proper quality granular deposits are satisfactory for low quality concrete aggregate, traffic gravel, and subbase or base course material for blacktopped roads.

Overburden depth. Refers to the anticipated stripping ratio, which is the ratio of overburden depth to thickness of recoverable granular material. Although the overburden depth is of the order of 1 m on most granular deposits, it can be highly variable.

Water table. Refers to the probability of encountering the water table during removal of sand and gravel. The chances of finding a high ground water level is greater near streams and beneath low-lying depressions.

Coarseness. Refers to the maximum size of coarse material that is expected.

WEIGHTED RATING[3]

Denotes a numerical ranking of the map units related to the possibility of finding economically developable sand and gravel deposits for such uses as low quality concrete, subbase and base course aggregate for blacktopped roads, and traffic gravel.

TABLE 32-13

Ground Water Potential

In this study, terrain and geotechnical factors are used in rating land capability limitations for each map unit. Alternative land use, socio-economic and certain environmental factors are not considered. The system of weighted rating used is illustrated below in a chart; the description of the terrain and geotechnical factors considered is also described below. The rating applies to the classification and mapping of landforms, and to the surface and near surface materials in them. Because of the small scale of the maps, the outlined map units may contain small areas of other map units that are not indicated.

MAP UNITS[1]			TERRAIN AND GEOTECHNICAL FACTORS[2]				
LETTER SYMBOL	LANDFORM DESCRIPTION	SURFACE MATERIAL AND ASSOCIATED BASIC LANDFORM TYPE AND VARIETY (see legend)	RECHARGE	EXPLORATION	QUALITY	SURFACE INDICATORS	WEIGHTED[3] RATING
ME	End moraine	tME	3	2	2	0	7
MG	Ground moraine	tMG, tMG_a, tMG_c, tMG_f	3	2	2	0	7
		tMG_e, tMG_{ec}	3	2	2	0	7
MH	High relief hummocky moraine	tMH, tMH_s	6	2	4	0	12
MH_L	Low relief hummocky moraine	tMH_L, tMH_{Lf}, tMH_{Ls}	6	2	2	0	10
LD	Lacustrine delta	sLD, sLD_k, sLD_w	9	6	6	2	23
		smLD, $smLD_w$	9	4	6	2	21
LP	Lacustrine plain	mLP, mLP_c	6	0	0	0	6
		msLP, $msLP_w$	6	2	4	1	13
		cmLP, $cmLP_c$, $cmLP_s$	3	0	0	0	3
		cLP	3	0	0	0	3
LH	High relief hummocky lacustrine terrain	smLH, $smLH_k$, $smLH_w$	9	2	6	1	18
		mLH	6	0	0	0	6
		msLH	9	2	4	1	16
		cmLH, $cmLH_s$	3	0	0	0	3
		cLH	3	0	0	0	3
LH_L	Low relief hummocky lacustrine terrain	$smLH_L$, $smLH_{Ls}$, $smLH_{Lw}$	9	2	6	2	19
		mLH_L, mLH_{Ls}	6	0	0	0	6
		$msLH_L$, $msLH_{Ls}$	6	2	4	1	13
		$cmLH_L$, $cmLH_{Ls}$	3	0	0	0	3
		cLH_L	3	0	0	0	3
GK	Ice contact knobs and ridges	sGK, sGK_w	6	2	6	2	16
		sgGK, $sgGK_k$	6	2	6	2	16
		gsbGK	6	2	6	2	16
GO	Outwash plains and valley trains	sGO, sGO_k	9	6	6	3	24
		sgGO, $sgGO_c$, $sgGO_s$, $sgGO_{sk}$, $sgGO_f$, $sgGO_k$	9	6	6	3	24
		gsGO, $gsGO_c$, $gsGO_k$, $gsGO_{kc}$	9	6	6	3	24
		gGO	9	6	6	3	24
AP	Alluvial plain	vAP, vAP_m	9	4	4	2	19
		msAP, $msAP_c$	9	4	4	3	20
		sAP	9	6	4	3	22
		sgAP	9	6	4	3	22
AF	Alluvial fan	vAF	6	2	4	3	15
AT	Alluvial terrace	vAT, vAT_c	6	2	4	3	15
		sAT	6	6	4	3	19
		gsAT	6	6	4	3	19
ED	Eolian dunes	sED	9	4	6	2	21
OT	Organic terrain	pOT	6	0	0	0	6
BT	Bedrock terrain	ss, shBT	3	2	2	0	7
SL	Slumped valley slopes	vSL	0	0	0	0	0
SE	Eroded valley and hill slopes	vSE	0	0	0	0	0

CONTRASTING MATERIALS AND LANDFORMS IN THE NEAR SURFACE ZONE

LETTER SYMBOL	SURFACE LANDFORM	BURIED NEAR SURFACE LANDFORM	SURFACE AND BURIED MATERIAL AND LANDFORM	RECHARGE	EXPLORATION	QUALITY	SURFACE INDICATORS	WEIGHTED RATING
LD/MG	Lacustrine delta	Ground moraine	sLD/tMG	9	2	2	0	13
			smLD/tMG, $smLD_w$/tMG	9	2	2	0	13
			sLD_w/tMG_e	9	2	2	0	13
LD/LP	Lacustrine delta	Lacustrine plain	$smLD_w$/cmLP	6	0	0	0	6
LD/GO	Lacustrine delta	Outwash plain	smLD/sgGO	9	4	6	2	21
LP/MG	Lacustrine plain	Ground moraine	msLP/tMG	6	2	2	0	10
			mLP/tMG, mLP/tMG_f	3	2	2	0	7
			cmLP/tMG, $cmLP_c/tMG_c$	3	2	2	0	7
			cLP/tMG	3	2	2	0	7
			$msLP/tMG_e$	6	2	2	0	10
			mLP/tMG_e, mLP/tMG_{ec}	3	2	2	0	7
			$cmLP/tMG_e$	3	2	2	0	7
LP/GO	Lacustrine plain	Outwash plain	msLP/sgGO	6	2	6	1	15
			mLP/sgGO	6	2	6	1	15
			mLP/gsGO	6	2	6	1	15
LP/GK	Lacustrine plain	Icecontact knobs & ridges	cLP/sgGK	3	2	6	1	12
LH_L/MH_L	Low relief hummocky lacustrine terrain	Low relief hummocky moraine	$smLH_L/tMH_L$	9	2	2	0	13
			$msLH_L/tMH_L$	6	2	2	0	10
			mLH_L/tMH_L	3	2	2	0	7
			$cmLH_L/tMH_L$	3	2	2	0	7
LH_L/LH_L	Low relief hummocky lacustrine terrain	Low relief hummocky lacustrine terrain	$msLH_L/mLH_L$	6	0	0	0	6
GO/MG	Outwash plain	Ground moraine	sgGO/tMG	9	2	2	0	13
AT/MG	Alluvial terrace	Ground moraine	$sgAT/tMG_e$	9	2	2	0	13
			$smAT/tMG_e$	9	2	2	0	13
			$msAT/tMG_e$	6	2	2	0	10
MG/BT	Ground moraine	Bedrock terrain	tMG/BT	3	2	2	0	7

GRAPHIC SYMBOL	SURFACE FEATURE	RECHARGE	EXPLORATION	QUALITY	SURFACE INDICATORS	WEIGHTED RATING
<<<<	Esker ridge	6	2	6	3	17
	Meltwater channel or spillway	9	2	4	2	17
	Tunnel valley or stream trench	9	2	4	2	17

WEIGHTED RATING	SIMPLIFIED RATING
0–3	MOSTLY UNSUITABLE U
3–11	POOR P
11–18	FAIR F
18–24	GOOD G

TERRAIN AND GEOTECHNICAL FACTORS	RATING	WEIGHT	WEIGHTED RATING
RECHARGE	0	3	0
	1		3
	2		6
	3		9
EXPLORATION	0	2	0
	1		2
	2		4
	3		6
QUALITY	0	2	0
	1		2
	2		4
	3		6
SURFACE INDICATORS	0	1	0
	1		1
	2		2
	3		3

CHART ILLUSTRATING METHOD USED TO DERIVE WEIGHTED RATING

MAP UNITS[1]

For descriptions of landform basic types and varieties, and the recurring types of soil materials found in them, which together form the mapping symbols and the map units, see the accompanying report.

TERRAIN AND GEOTECHNICAL FACTORS[2]

Aquifer recharge. Refers to the possibility and reliability of aquifers receiving significant amounts of recharge from rainfall and snowmelt or seepage from nearby streams, lakes and ponds.

Ground water exploration. Refers to the relative amount of time and cost needed to prove up an economically developable ground water source. In this regard, usually it requires much more time and money to discover and prove up a 5 ℓps (66 Igpm) ground water supply, for example, than a 1 ℓps (13 Igpm) supply.

Ground water quality. Refers to the total dissolved mineral content in ground water. The quality of ground water is usually better in near surface sand and gravel aquifers than it is in deep aquifers, which commonly underlie a thick confining layer of low permeability clayey till or fine grained lacustrine sediments through which the recharge waters move.

Surface indicators. Refers to the likelihood of seeing indicators in aerial photographs or on the ground that may help to locate an economically developable ground water source, e.g. springs, seepages, surface salt patches, hydrophytic plants, and favorable geomorphic settings. The occurrence of these indicators is uncommon in the case of many deep intertill and subtill aquifers.

WEIGHTED RATING[3]

Denotes a numerical ranking of map units related to the chances and costs of discovering relatively small economically developable ground water supplies.

TABLE 32-14

Natural Hazard Potential

In this study, terrain and geotechnical factors are used in rating land capability limitations for each map unit. Alternative land use, socio-economic and certain environmental factors are not considered. The system of weighted rating used is illustrated below in a chart; the description of the terrain and geotechnical factors considered is also described below. The rating applies to the classification and mapping of landforms, and to the surface and near surface materials in them. Because of the small scale of the maps, the outlined map units may contain small areas of other map units that are not indicated.

MAP UNITS[1]			TERRAIN AND GEOTECHNICAL FACTORS[2]								WEIGHTED[3] RATING
LETTER SYMBOL	LANDFORM DESCRIPTION	SURFACE MATERIAL AND ASSOCIATED BASIC LANDFORM TYPE AND VARIETY (see legend)	EXPANSIVE SOILS	FLOODING	SLOPE INSTABILITY	EROSION					
ME	End moraine	tME	6	9	6	1					22
MG	Ground moraine	tMG, tMG_a, tMG_c, tMG_f	6	9	9	3					27
		tMG_e, tMG_{ec}	6	9	9	3					27
MH	High relief hummocky moraine	tMH, tMH_s	6	6	6	2					20
MH_L	Low relief hummocky moraine	tMH_L, tMH_{Lf}, tMH_{Ls}	6	9	9	3					27
LD	Lacustrine delta	sLD, sLD_k, sLD_w	6	9	9	0					24
		smLD, $smLD_w$	6	9	9	0					24
LP	Lacustrine plain	mLP, mLP_c	6	6	9	3					24
		msLP, $msLP_w$	6	9	9	1					25
		cmLP, $cmLP_c$, $cmLP_s$	2	3	9	3					17
		cLP	0	3	9	3					15
LH	High relief hummocky lacustrine terrain	smLH, $smLH_k$, $smLH_w$	6	6	6	1					19
		mLH	6	6	6	2					20
		msLH	6	6	6	1					19
		cmLH, $cmLH_s$	2	6	6	2					16
		cLH	0	6	3	2					11
LH_L	Low relief hummocky lacustrine terrain	$smLH_L$, $smLH_{Ls}$, $smLH_{Lw}$	6	9	9	1					25
		mLH_L, mLH_{Ls}	6	9	9	2					26
		$msLH_L$, $msLH_{Ls}$	6	9	9	1					25
		$cmLH_L$, $cmLH_{Ls}$	2	6	9	3					20
		cLH_L	0	6	9	3					18
GK	Ice contact knobs and ridges	sGK, sGK_w	6	9	9	1					25
		sgGK, $sgGK_k$	6	9	9	2					26
		gsbGK	6	9	9	2					26
GO	Outwash plains and valley trains	sGO, sGO_k	6	9	9	1					25
		sgGO, $sgGO_c$, $sgGO_s$, $sgGO_{sk}$, $sgGO_t$, $sgGO_k$	6	9	9	2					26
		gsGO, $gsGO_c$, $gsGO_k$, $gsGO_{kc}$	6	9	9	3					27
		gGO	6	9	9	3					27
AP	Alluvial plain	vAP, vAP_m	6	3	6	1					16
		msAP, $msAP_c$	6	3	6	1					16
		sAP	6	3	6	1					16
		sgAP	6	3	6	1					16
AF	Alluvial fan	vAF	6	3	9	2					20
AT	Alluvial terrace	vAT, vAT_c	6	6	6	2					20
		sAT	6	6	6	2					20
		gsAT	6	6	6	2					20
ED	Eolian dunes	sED	6	9	6	0					21
OT	Organic terrain	pOT	6	0	9	3					18
BT	Bedrock terrain	ss, shBT	6	9	3	0					18
SL	Slumped valley slopes	vSL	6	9	0	1					16
SE	Eroded valley and hill slopes	vSE	6'	9	3	0					18

CONTRASTING MATERIALS AND LANDFORMS IN THE NEAR SURFACE ZONE

LETTER SYMBOL	SURFACE LANDFORM	BURIED NEAR SURFACE LANDFORM	SURFACE AND BURIED MATERIAL AND LANDFORM	EXPANSIVE SOILS	FLOODING	SLOPE INSTABILITY	EROSION	WEIGHTED RATING
LD/MG	Lacustrine delta	Ground moraine	sLD/tMG	6	9	9	1	25
			smLD/tMG, $smLD_w$/tMG	6	9	9	1	25
			sLD_w/tMG_e	6	9	9	1	25
LD/LP	Lacustrine delta	Lacustrine plain	$smLD_w$/cmLP	4	9	9	1	23
LD/GO	Lacustrine delta	Outwash plain	smLD/sgGO	6	9	9	1	25
LP/MG	Lacustrine plain	Ground moraine	msLP/tMG	6	9	9	1	25
			mLP/tMG, mLP/tMG_f	6	9	9	3	27
			cmLP/tMG, $cmLP_c/tMG_c$	2	9	9	3	23
			cLP/tMG	0	9	9	3	21
			$msLP/tMG_e$	6	9	9	1	25
			mLP/tMG_e, mLP/tMG_{ec}	6	9	9	3	27
			$cmLP/tMG_e$	2	9	9	3	23
LP/GO	Lacustrine plain	Outwash plain	msLP/sgGO	6	9	9	1	25
			mLP/sgGO	6	9	9	3	27
			mLP/gsGO	6	9	9	3	27
LP/GK	Lacustrine plain	Icecontact knobs&ridges	cLP/sgGK	0	9	9	2	20
LH_l/MH_L	Low relief hummocky lacustrine terrain	Low relief hummocky moraine	$smLH_L/tMH_L$	6	9	9	1	25
			$msLH_L/tMH_L$	6	9	9	1	25
			mLH_L/tMH_L	6	9	9	2	26
			$cmLH_L/tMH_L$	2	9	9	3	23
LH_L/LH_L	Low relief hummocky lacustrine terrain	Low relief hummocky lacustrine terrain	$msLH_L/mLH_L$	6	9	9	1	25
GO/MG	Outwash plain	Ground moraine	sgGO/tMG	6	9	9	2	26
AT/MG	Alluvial terrace	Ground moraine	$sgAT/tMG_e$	6	6	6	2	20
			$smAT/tMG_e$	6	6	6	1	19
			$msAT/tMG_e$	6	6	6	1	19
MG/BT	Ground moraine	Bedrock terrain	tMG/BT	6	9	9	3	27

GRAPHIC SYMBOL	SURFACE FEATURE	EXPANSIVE SOILS	FLOODING	SLOPE INSTABILITY	EROSION	WEIGHTED RATING
<<<<	Esker ridge	6	9	9	2	26
	Meltwater channel or spillway	6	9	3	0	18
	Tunnel valley or stream trench	6	9	3	0	18

WEIGHTED RATING	SIMPLIFIED RATING
0–18	HIGH HAZARD POTENTIAL H
18–22	MODERATE HAZARD POTENTIAL M
22–27	LOW HAZARD POTENTIAL L

TERRAIN AND GEOTECHNICAL FACTORS	RATING	WEIGHT	WEIGHTED RATING
EXPANSIVE SOILS	0	2	0
	1		2
	2		4
	3		6
FLOODING	0	3	0
	1		3
	2		6
	3		9
SLOPE INSTABILITY	0	3	0
	1		3
	2		6
	3		9
EROSION	0	1	0
	1		1
	2		2
	3		3

CHART ILLUSTRATING METHOD USED TO DERIVE WEIGHTED RATING

MAP UNITS[1]

For description of landform basic types and varieties, and the recurring types of soil materials found in them, which together form the mapping symbols and the map units, see the accompanying report.

TERRAIN AND GEOTECHNICAL FACTORS[2]

Expansive soils. Refers to excessive shrinkage and swelling characteristics of certain clay materials, which upon volume change produce stresses that can cause serious damage to rigid structures built on them.

Flooding. Refers to landscapes that may be inundated from overbank flooding of perennial or intermittent streams, and from ponding of surface waters in closed depressions during spring runoff or heavy rainstorms. Flooding of basements can also be caused by a seasonally high water table.

Slope instability. Refers mainly to large retrogressive slope failures along valley sides underlain by weak bentonitic shales of Cretaceous age. This type of large-scale slope instability is called slumping, and usually occurs as a slow intermittent creep. Also included are small failures in other types of soil and rock materials and geomorphic settings.

Wind and water erosion. Refers to soil drifting caused by wind action and to sheet, rill, and gully erosion caused by surface runoff of rainfall and snowmelt waters. It also includes lateral stream erosion along undercut river banks.

WEIGHTED RATING[3]

Denotes the numerical ranking of map units based on potential limitations affecting regional land use development.

TABLE 32-15
Suitability for Light Foundation Construction

In this study, terrain and geotechnical factors are used in rating land capability limitations for each map unit. Alternative land use, socio-economic and certain environmental factors are not considered. The system of weighted rating used is illustrated below in a chart; the description of the terrain and geotechnical factors considered is also described below. The rating applies to the classification and mapping of landforms, and to the surface and near surface materials in them. Because of the small scale of the maps, the outlined map units may contain small areas of other map units that are not indicated.

MAP UNITS[1]			TERRAIN AND GEOTECHNICAL FACTORS[2]				WEIGHTED[3] RATING
LETTER SYMBOL	LANDFORM DESCRIPTION	SURFACE MATERIAL AND ASSOCIATED BASIC LANDFORM TYPE AND VARIETY (see legend)	DRAINAGE	TOPOGRAPHY	WORKABILITY	BEARING STRENGTH	
ME	End moraine	tME	2	1	2	6	11
MG	Ground moraine	tMG, tMG_a, tMG_c, tMG_f	3	3	2	9	17
		tMG_e, tMG_{ec}	1	3	0	9	13
MH	High relief hummocky moraine	tMH, tMH_s	1	1	2	6	10
MH_L	Low relief hummocky moraine	tMH_L, tMH_{Lf}, tMH_{Ls}	2	2	2	6	12
LD	Lacustrine delta	sLD, sLD_k, sLD_w	3	3	3	9	18
		smLD, $smLD_w$	3	3	3	9	18
LP	Lacustrine plain	mLP, mLP_c	1	3	1	3	8
		msLP, $msLP_w$	3	3	3	6	15
		cmLP, $cmLP_c$, $cmLP_s$	1	3	1	3	8
		cLP	1	3	1	3	8
LH	High relief hummocky lacustrine terrain	smLH, $smLH_k$, $smLH_w$	2	1	3	9	15
		mLH	1	1	1	3	6
		msLH	2	1	3	6	12
		cmLH, $cmLH_s$	1	1	1	3	6
		cLH	1	1	1	3	6
LH_L	Low relief hummocky lacustrine terrain	$smLH_L$, $smLH_{Ls}$, $smLH_{Lw}$	3	2	3	9	17
		mLH_L, mLH_{Ls}	2	2	1	3	8
		$msLH_L$, $msLH_{Ls}$	3	2	3	6	14
		$cmLH_L$, $cmLH_{Ls}$	1	2	1	3	7
		cLH_L	1	2	1	3	7
GK	Ice contact knobs and ridges	sGK, sGK_w	3	1	3	9	16
		sgGK, $sgGK_k$	3	1	3	9	16
		gsbGK	3	1	3	9	16
GO	Outwash plains and valley trains	sGO, sGO_k	3	3	3	9	18
		sgGO, $sgGO_c$, $sgGO_s$, $sgGO_{sk}$, $sgGO_t$, $sgGO_k$	3	3	3	9	18
		gsGO, $gsGO_c$, $gsGO_k$, $gsGO_{kc}$	3	3	3	9	18
		gGO	3	3	3	9	18
AP	Alluvial plain	vAP, vAP_m	1	3	1	3	8
		msAP, $msAP_c$	1	3	3	6	13
		sAP	1	3	3	9	16
		sgAP	1	3	3	9	16
AF	Alluvial fan	vAF	2	3	1	3	9
AT	Alluvial terrace	vAT, vAT_c	2	3	1	3	9
		sAT	3	3	3	9	18
		gsAT	3	3	3	9	18
ED	Eolian dunes	sED	3	1	3	9	16
OT	Organic terrain	pOT	0	3	0	0	3
BT	Bedrock terrain	ss, shBT	1	1	1	3	6
SL	Slumped valley slopes	vSL	3	1	2	6	12
SE	Eroded valley and hill slopes	vSE	3	1	2	6	12

CONTRASTING MATERIALS AND LANDFORMS IN THE NEAR SURFACE ZONE

LETTER SYMBOL	SURFACE LANDFORM	BURIED NEAR SURFACE LANDFORM	SURFACE AND BURIED MATERIAL AND LANDFORM	DRAINAGE	TOPOGRAPHY	WORKABILITY	BEARING STRENGTH	WEIGHTED RATING
LD/MG	Lacustrine delta	Ground moraine	sLD/tMG	3	3	3	9	18
			smLD/tMG, $smLD_w$/tMG	3	3	3	9	18
			sLD_w/tMG_e	3	3	3	9	18
LD/LP	Lacustrine delta	Lacustrine plain	$smLD_w$/cmLP	3	3	3	6	15
LD/GO	Lacustrine delta	Outwash plain	smLD/sgGO	3	3	3	9	18
LP/MG	Lacustrine plain	Ground moraine	msLP/tMG	3	3	3	6	15
			mLP/tMG, mLP/tMG_f	1	3	1	3	8
			cmLP/tMG, $cmLP_c$/tMG_c	1	3	1	3	8
			cLP/tMG	1	3	1	3	8
			msLP/tMG_e	3	3	3	6	15
			mLP/tMG_e, mLP/tMG_{ec}	1	3	1	3	8
			cmLP/tMG_e	1	3	1	3	8
LP/GO	Lacustrine plain	Outwash plain	msLP/sgGO	3	3	3	6	15
			mLP/sgGO	1	3	1	3	8
			mLP/gsGO	1	3	1	3	8
LP/GK	Lacustrine plain	Icecontact knobs&ridges	cLP/sgGK	1	3	1	3	8
LH_L/MH_L	Low relief hummocky lacustrine terrain	Low relief hummocky moraine	$smLH_L$/tMH_L	3	2	3	9	17
			$msLH_L$/tMH_L	3	2	3	6	14
			mLH_L/tMH_L	2	2	1	3	8
			$cmLH_L$/tMH_L	1	2	1	3	7
LH_L/LH_L	Low relief hummocky lacustrine terrain	Low relief hummocky lacustrine terrain	$msLH_L$/mLH_L	3	2	3	6	14
GO/MG	Outwash plain	Ground moraine	sgGO/tMG	3	3	3	9	18
AT/MG	Alluvial terrace	Ground moraine	sgAT/tMG_e	3	3	3	9	18
			smAT/tMG_e	3	3	3	9	18
			msAT/tMG_e	3	3	3	6	15
MG/BT	Ground moraine	Bedrock terrain	tMG/BT	3	3	2	6	14

GRAPHIC SYMBOL	SURFACE FEATURE	DRAINAGE	TOPOGRAPHY	WORKABILITY	BEARING STRENGTH	WEIGHTED RATING
<<<<	Esker ridge	3	1	3	9	16
	Meltwater channel or spillway	1	3	1	3	8
	Tunnel valley or stream trench	1	3	1	3	8

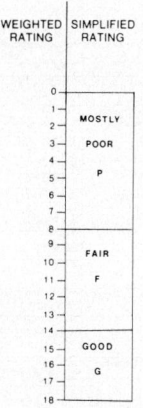

MAP UNITS[1]

For description of landform basic types and varieties, and the recurring types of soil materials found in them, which together form the mapping symbols and the map units, see the accompanying report.

TERRAIN AND GEOTECHNICAL FACTORS[2]

Drainage. Refers to ponding of water in poorly drained closed depressions and along streams subject to overbank flooding. It also includes the likelihood of a high water table, commonly following removal of excess surface waters, and to the need to provide adequate drainage control during construction.

Topography. Refers to costs relating to site levelling and preparation, and thus mainly to cut and fill quantities in earthwork construction.

Workability. Refers to how quickly soils dry out and become workable following wet weather, the tendency of equipment to become bogged down in soils that lose strength and become soft and unworkable under traffic, and near surface soil variability.

Bearing strength. Refers to requirements for providing adequate structural foundations on various terrain types.

WEIGHTED RATING[3]

Denotes the numerical ranking of map units for light foundation construction, such as earthwork excavation for roads and streets, basements, and slab foundation structures.

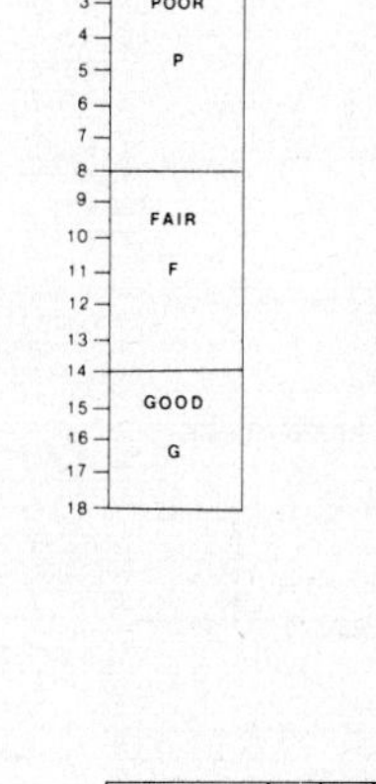

TERRAIN AND GEOTECHNICAL FACTORS	RATING	WEIGHT	WEIGHTED RATING
DRAINAGE	0	1	0
	1		1
	2		2
	3		3
TOPOGRAPHY	0	1	0
	1		1
	2		2
	3		3
WORKABILITY	0	1	0
	1		1
	2		2
	3		3
BEARING STRENGTH	0	3	0
	1		3
	2		6
	3		9

CHART ILLUSTRATING METHOD USED TO DERIVE WEIGHTED RATING

TABLE 32-16

Suitability for Light Foundation Structures

In this study, terrain and geotechnical factors are used in rating land capability limitations for each map unit. Alternative land use, socio-economic and certain environmental factors are not considered. The system of weighted rating used is illustrated below in a chart, the description of the terrain and geotechnical factors considered is also described below. The rating applies to the classification and mapping of landforms, and to the surface and near surface materials in them. Because of the small scale of the maps, the outlined map units may contain small areas of other map units that are not indicated.

MAP UNITS[1]			TERRAIN AND GEOTECHNICAL FACTORS[2]								WEIGHTED[3] RATING
LETTER SYMBOL	LANDFORM DESCRIPTION	SURFACE MATERIAL AND ASSOCIATED BASIC LANDFORM TYPE AND VARIETY (see legend)	DRAINAGE	STRENGTH COMPRESSIBILITY	SHRINK AND SWELL	FROST HEAVE	EROSION				
ME	End moraine	tME	4	9	9	9	2				33
MG	Ground moraine	tMG, tMG$_a$, tMG$_c$, tMG$_f$	6	9	9	9	3				36
		tMG$_e$, tMG$_{ec}$	6	9	9	9	3				36
MH	High relief hummocky moraine	tMH, tMH$_s$	4	6	9	9	1				29
MH$_L$	Low relief hummocky moraine	tMH$_L$, tMH$_{Lf}$, tMH$_{Ls}$	4	9	9	9	3				34
LD	Lacustrine delta	sLD, sLD$_k$, sLD$_w$	6	9	9	9	1				34
		smLD, smLD$_w$	6	9	9	9	1				34
LP	Lacustrine plain	mLP, mLP$_c$	4	6	9	0	3				22
		msLP, msLP$_w$	4	9	9	3	2				27
		cmLP, cmLP$_c$, cmLP$_s$	4	6	3	6	3				22
		cLP	4	6	0	6	3				19
LH	High relief hummocky lacustrine terrain	smLH, smLH$_k$, smLH$_w$	6	9	9	9	0				33
		mLH	4	6	9	0					19
		msLH	4	9	9	3	0				25
		cmLH, cmLH$_s$	4	6	3	6	1				20
		cLH	4	6	0	6	2				18
LH$_L$	Low relief hummocky lacustrine terrain	smLH$_L$, smLH$_{Ls}$, smLH$_{Lw}$	6	9	9	9	1				34
		mLH$_L$, mLH$_{Ls}$	6	6	9	0	3				24
		msLH$_L$, msLH$_{Ls}$	6	9	9	3	2				29
		cmLH$_L$, cmLH$_{Ls}$	6	6	3	6	3				24
		cLH$_L$	6	6	0	6	3				21
GK	Ice contact knobs and ridges	sGK, sGK$_w$	6	9	9	9	1				34
		sgGK, sgGK$_k$	6	9	9	9	2				35
		gsbGK	6	9	9	9	3				36
GO	Outwash plains and valley trains	sGO, sGO$_k$	6	9	9	9	1				34
		sgGO, sgGO$_c$, sgGO$_s$, sgGO$_{sk}$, sgGO$_t$, sgGO$_k$	6	9	9	9	3				36
		gsGO, gsGO$_c$, gsGO$_k$, gsGO$_{kc}$	6	9	9	9	3				36
		gGO	6	9	9	9	3				36
AP	Alluvial plain	vAP, vAP$_m$	0	3	6	3	2				14
		msAP, msAP$_c$	0	6	9	3	2				20
		sAP	0	9	9	9	2				29
		sgAP	0	9	9	9	2				29
AF	Alluvial fan	vAF	4	3	6	3	3				19
AT	Alluvial terrace	vAT, vAT$_c$	6	6	9	6	3				30
		sAT	6	9	9	9	2				35
		gsAT	6	9	9	9	3				36
ED	Eolian dunes	sED	6	9	9	9	1				34
OT	Organic terrain	pOT	0	0	9	3	3				15
BT	Bedrock terrain	ss, shBT	6	6	0	9	0				21
SL	Slumped valley slopes	vSL	6	3	3	9	1				22
SE	Eroded valley and hill slopes	vSE	6	6	9	6	1				28

CONTRASTING MATERIALS AND LANDFORMS IN THE NEAR SURFACE ZONE

LETTER SYMBOL	SURFACE LANDFORM	BURIED NEAR SURFACE LANDFORM	SURFACE AND BURIED MATERIAL AND LANDFORM	DRAINAGE	STRENGTH COMPRESSIBILITY	SHRINK AND SWELL	FROST HEAVE	EROSION				WEIGHTED RATING
LD/MG	Lacustrine delta	Ground moraine	sLD/tMG	6	9	9	9	1				34
			smLD/tMG, smLD$_w$/tMG	6	9	9	9	1				34
			sLD$_w$/tMG$_e$	6	9	9	9	1				34
LD/LP	Lacustrine delta	Lacustrine plain	smLD$_w$/cmLP	6	9	9	9	1				34
LD/GO	Lacustrine delta	Outwash plain	smLD/sgGO	6	9	9	9	1				34
LP/MG	Lacustrine plain	Ground moraine	msLP/tMG	6	9	9	3	2				29
			mLP/tMG, mLP/tMG$_f$	6	6	9	0	3				24
			cmLP/tMG, cmLP$_c$/tMG$_c$	6	6	3	6	3				24
			cLP/tMG	6	6	0	6	3				21
			msLP/tMG$_e$	6	9	9	3	2				29
			mLP/tMG$_e$, mLP/tMG$_{ec}$	6	6	9	0	3				24
			cmLP/tMG$_e$	6	6	3	6	3				24
LP/GO	Lacustrine plain	Outwash plain	msLP/sgGO	6	9	9	6	2				32
			mLP/sgGO	6	6	9	3	3				27
			mLP/gsGO	6	6	9	3	3				27
LP/GK	Lacustrine plain	Icecontact knobs&ridges	cLP/sgGK	6	6	0	6	3				21
LH$_L$/MH$_L$	Low relief hummocky lacustrine terrain	Low relief hummocky moraine	smLH$_L$/tMH$_L$	6	9	9	9	1				34
			msLH$_L$/tMH$_L$	6	9	9	3	2				29
			mLH$_L$/tMH$_L$	6	6	9	0	3				24
			cmLH$_L$/tMH$_L$	6	6	3	6	3				24
LH$_L$/LH$_L$	Low relief hummocky lacustrine terrain	Low relief hummocky lacustrine terrain	msLH$_L$/mLH$_L$	6	9	9	3	2				29
GO/MG	Outwash plain	Ground moraine	sgGO/tMG	6	9	9	9	3				36
AT/MG	Alluvial terrace	Ground moraine	sgAT/tMG$_e$	6	9	9	9	3				36
			smAT/tMG$_e$	6	9	9	9	3				36
			msAT/tMG$_e$	6	9	9	3	2				29
MG/BT	Ground moraine	Bedrock terrain	tMG/BT	6	9	9	9	3				36

GRAPHIC SYMBOL	SURFACE FEATURE	DRAINAGE	STRENGTH COMPRESSIBILITY	SHRINK AND SWELL	FROST HEAVE	EROSION				WEIGHTED RATING
<<<<	Esker ridge	6	9	9	9	2				35
	Meltwater channel or spillway	0	3	6	3	2				14
	Tunnel valley or stream trench	0	3	6	3	2				14

WEIGHTED RATING	SIMPLIFIED RATING
0–25	MOSTLY POOR P
25–30	FAIR F
30–35	GOOD G

TERRAIN AND GEOTECHNICAL FACTORS	RATING	WEIGHT	WEIGHTED RATING
DRAINAGE	0, 1, 2, 3	2	0, 2, 4, 6
STRENGTH AND COMPRESSIBILITY	0, 1, 2, 3	3	0, 3, 6, 9
SHRINK AND SWELL	0, 1, 2, 3	3	0, 3, 6, 9
FROST HEAVE	0, 1, 2, 3	3	0, 3, 6, 9
EROSION	0, 1, 2, 3	1	0, 1, 2, 3

CHART ILLUSTRATING METHOD USED TO DERIVE WEIGHTED RATING

MAP UNITS[1]

For description of landform basic types and varieties, and the recurring types of soil materials found in them, which together form the mapping symbols and the map units, see the accompanying report.

TERRAIN AND GEOTECHNICAL FACTORS[2]

Drainage. Refers to the possibility of having to provide extra flood and high water table protection to remove excess water after construction.

Soil strength and foundation settlement. Refers to the chances of costly repairs to light structures associated with low strength and high compressibility foundation soils, and also with the stability of cut slopes.

Shrink and swell. Refers to the tendency of certain soils to shrink excessively on drying or swell excessively upon wetting, causing costly and often continuing maintenance problems.

Frost heave. Refers to the tendency of some soils to develop ice lenses and swell during frozen ground conditions in the winter. In the spring the ground thaws, softens and settles, forming frost boils and chuck holes in roads, and cracking rigid structures.

Erosion. Refers to both wind and water erosion hazards. Wind erosion is most common on fine sa..dy soils; water erosion occurs widely on loose silty and fine sandy soils, especially those underlying freshly cut, unprotected, sttep slopes.

WEIGHTED RATING[3]

Denotes the numerical ranking of map units from the standpoint of long-term maintenance of light foundation structures.

TABLE 32-17

Suitability for Solid Waste Disposal

In this study, terrain and geotechnical factors are used in rating land capability limitations for each map unit. Alternative land use, socio-economic and certain environmental factors are not considered. The system of weighted rating used is illustrated below in a chart; the description of the terrain and geotechnical factors considered is also described below. The rating applies to the classification and mapping of landforms, and to the surface and near surface materials in them. Because of the small scale of the maps, the outlined map units may contain small areas of other map units that are not indicated.

MAP UNITS[1]			TERRAIN AND GEOTECHNICAL FACTORS[2]									WEIGHTED[3] RATING
LETTER SYMBOL	LANDFORM DESCRIPTION	SURFACE MATERIAL AND ASSOCIATED BASIC LANDFORM TYPE AND VARIETY (see legend)	PERMEABILITY	WATER TABLE	TOPOGRAPHIC SETTING	SLOPE	FLOODING	EXCAVAT ABILITY	COVER MATERIAL	PERMEABLE STRATA	PROXIMITY TO WATER BODIES	
ME	End moraine	tME	6	6	2	4	9	2	2	3	6	40
MG	Ground moraine	tMG, tMG_a, tMG_c, tMG_f	9	6	6	6	9	2	3	6	9	56
		tMG_e, tMG_{ec}	9	6	4	6	6	1	0	6	6	44
MH	High relief hummocky moraine	tMH, tMH_s	6	6	2	4	6	2	1	3	6	36
MH_L	Low relief hummocky moraine	tMH_L, tMH_{Lf}, tMH_{Ls}	9	6	4	6	9	2	3	6	9	54
LD	Lacustrine delta	sLD, sLD_k, sLD_w	0	0	4	2	9	2	1	0	3	21
		smLD, $smLD_w$	0	0	4	2	9	2	1	0	3	21
LP	Lacustrine plain	mLP, mLP_c	9	4	2	4	6	3	3	6	6	43
		msLP, $msLP_w$	6	4	2	4	6	3	3	3	3	34
		cmLP, $cmLP_c$, $cmLP_s$	9	6	2	2	6	2	3	9	9	48
		cLP	9	6	2	2	3	2	2	9	9	44
LH	High relief hummocky lacustrine terrain	smLH, $smLH_k$, $smLH_w$	0	2	2	4	6	2	1	0	3	20
		mLH	9	4	2	4	6	3	3	6	6	43
		msLH	6	4	2	4	6	3	3	3	3	34
		cmLH, $cmLH_s$	9	6	2	4	3	2	3	9	9	47
		cLH	9	6	2	4	3	2	2	9	9	46
LH_L	Low relief hummocky lacustrine terrain	$smLH_L$, $smLH_{Ls}$, $smLH_{Lw}$	0	0	6	4	9	2	1	0	3	25
		mLH_L, mLH_{Ls}	9	4	6	4	9	3	3	6	6	50
		$msLH_L$, $msLH_{Ls}$	6	4	6	4	9	3	3	3	3	41
		$cmLH_L$, $cmLH_{Ls}$	9	6	6	4	9	2	3	6	9	54
		cLH_L	9	6	6	4	9	2	2	6	9	53
GK	Ice contact knobs and ridges	sGK, sGK_w	0	2	2	4	9	2	1	0	3	23
		sgGK, $sgGK_k$	0	2	2	4	9	2	1	0	3	23
		gsbGK	0	2	2	4	9	2	1	0	3	23
GO	Outwash plains and valley trains	sGO, sGO_k	0	0	4	2	9	2	1	0	3	21
		sgGO, $sgGO_c$, $sgGO_s$, $sgGO_{sk}$, $sgGO_t$, $sgGO_k$	0	0	4	2	9	2	1	0	3	21
		gsGO, $gsGO_c$, $gsGO_k$, $gsGO_{kc}$	0	0	4	2	9	1	1	0	3	20
		gGO	0	0	4	2	9	1	1	0	3	20
AP	Alluvial plain	vAP, vAP_m	3	0	4	2	0	2	3	3	0	17
		msAP, $msAP_c$	0	0	4	2	0	2	3	0	0	11
		sAP	0	0	4	2	0	2	2	0	0	10
		sgAP	0	0	4	2	0	2	2	0	0	10
AF	Alluvial fan	vAF	3	2	4	4	6	2	3	0	0	24
AT	Alluvial terrace	vAT, vAT_c	3	2	4	2	6	2	3	0	0	22
		sAT	0	2	4	2	6	2	1	0	0	17
		gsAT	0	2	4	2	6	1	1	0	0	16
ED	Eolian dunes	sED	0	4	2	2	9	2	1	0	6	26
OT	Organic terrain	pOT	0	0	6	0	0	0	0	0	3	9
BT	Bedrock terrain	ss, shBT	6	6	2	0	9	1	0	3	9	36
SL	Slumped valley slopes	vSL	3	6	0	0	9	2	1	0	6	27
SE	Eroded valley and hill slopes	vSE	3	6	0	0	9	2	1	0	6	27

CONTRASTING MATERIALS AND LANDFORMS IN THE NEAR SURFACE ZONE

LETTER SYMBOL	SURFACE LANDFORM	BURIED NEAR SURFACE LANDFORM	SURFACE AND BURIED MATERIAL AND LANDFORM	PERMEABILITY	WATER TABLE	TOPOGRAPHIC SETTING	SLOPE	FLOODING	EXCAVAT ABILITY	COVER MATERIAL	PERMEABLE STRATA	PROXIMITY TO WATER BODIES	WEIGHTED RATING
LD/MG	Lacustrine delta	Ground moraine	sLD/tMG	3	0	6	6	9	2	1	3	6	36
			smLD/tMG, $smLD_w$/tMG	3	0	6	6	9	2	1	3	6	36
			sLD_w/tMG_e	3	0	6	6	9	1	1	3	6	35
LD/LP	Lacustrine delta	Lacustrine plain	$smLD_w$/cmLP	3	0	6	6	9	2	1	6	6	39
LD/GO	Lacustrine delta	Outwash plain	smLD/sgGO	0	0	4	2	9	2	1	0	3	21
LP/MG	Lacustrine plain	Ground moraine	msLP/tMG	6	2	6	6	9	2	3	6	9	49
			mLP/tMG, mLP/tMG_f	9	4	6	6	9	2	3	6	9	54
			cmLP/tMG, $cmLP_c$/tMG_c	9	4	6	6	9	2	3	6	9	54
			cLP/tMG	9	4	6	6	9	2	3	6	9	54
			msLP/tMG_e	6	0	6	6	9	1	3	3	6	40
			mLP/tMG_e, mLP/tMG_{ec}	9	4	6	6	9	1	3	3	6	47
			cmLP/tMG_e	9	4	6	6	9	1	3	3	6	47
LP/GO	Lacustrine plain	Outwash plain	msLP/sgGO	0	0	4	2	9	2	3	0	3	23
			mLP/sgGO	0	0	4	2	9	2	3	0	3	23
			mLP/gsGO	0	0	4	2	9	1	3	0	3	22
LP/GK	Lacustrine plain	Icecontact knobs&ridges	cLP/sgGK	0	2	4	4	9	2	2	0	6	29
LH_L/MH_L	Low relief hummocky lacustrine terrain	Low relief hummocky moraine	$smLH_L$/tMH_L	6	0	6	6	9	2	1	6	6	42
			$msLH_L$/tMH_L	6	2	6	6	9	2	3	6	9	49
			mLH_L/tMH_L	9	4	6	6	9	2	3	6	9	54
			$cmLH_L$/tMH_L	9	4	6	6	9	2	3	6	9	54
LH_L/LH_L	Low relief hummocky lacustrine terrain	Low relief hummocky lacustrine terrain	$msLH_L$/mLH_L	6	4	4	4	6	3	3	6	6	42
GO/MG	Outwash plain	Ground moraine	sgGO/tMG	3	0	6	6	9	1	1	3	6	35
AT/MG	Alluvial terrace	Ground moraine	sgAT/tMG_e	3	0	6	4	6	1	1	3	0	24
			smAT/tMG_e	3	0	6	4	6	2	1	3	0	25
			msAT/tMG_e	6	0	6	4	6	3	1	3	0	29
MG/BT	Ground moraine	Bedrock terrain	tMG/BT	9	6	6	6	9	1	2	3	9	51

GRAPHIC SYMBOL	SURFACE FEATURE	PERMEABILITY	WATER TABLE	TOPOGRAPHIC SETTING	SLOPE	FLOODING	EXCAVAT ABILITY	COVER MATERIAL	PERMEABLE STRATA	PROXIMITY TO WATER BODIES	WEIGHTED RATING
<<<<	Esker ridge	0	2	2	4	9	2	1	0	3	23
	Meltwater channel or spillway	3	0	4	2	0	2	3	3	0	17
	Tunnel valley or stream trench	3	0	4	2	0	2	3	3	0	17

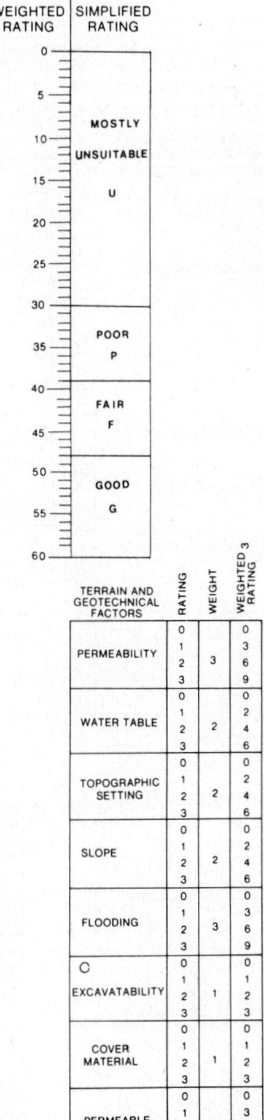

CHART ILLUSTRATING METHOD USED TO DERIVE WEIGHTED RATING

MAP UNITS[1]

For description of landform basic types and varieties, and the recurring types of soil materials found in them, which together form the mapping symbols and the map units, see the accompanying report.

TERRAIN AND GEOTECHNICAL FACTORS[2]

Permeability. The permeability of near surface soils and the slope of the water table govern the rate of leachate movement away from sanitary landfill sites. Homogeneous, low permeability, near surface soil materials are desired.

Water table. The water table should be over 10 m below the bottom of the solid waste deposit to prevent contamination of potable ground water supplies. Sites located on regionally higher land favor this requirement.

Topographic setting. Sites where surface runoff does not pond but drains away from the landfill site are preferred. Runoff from landfill sites should be contained, collected, and allowed to evaporate. Thus shallow closed depressions situated on slightly higher ground are desired.

Slope. Steep slopes foster more rapid migration of leachate from landfills. Ground water containing leachates may migrate downslope for years after the abandonment of a landfill site. Seepage of contaminated ground waters may also discharge on steep hillsides below waste disposal sites.

Flooding. The risk of flooding should be minimized because flooding of landfill sites introduces excess water into the waste material, makes the daily placement of cover material more difficult and costly, and increases the chances of spreading contaminants away from the site.

Excavatability. The ease of excavation of a pit and of cover material affects the time and cost of the operation. Equipment operators may not have heavy enough machinery to excavate hardpan materials or dense concentrations of surface boulders.

Variability of cover material. An ample supply of nearby cover material and topsoil are desired. Coarse clean granular materials, for example, have limited possibilities for rapid topsoil development and do not form a sufficiently impermeable cover over the solid waste.

Permeable strata in the subsurface. Permeable partings, seams, lenses, and layers in the subsurface foster the movement of leachates to nearby areas, and so are undesirable.

Proximity to aquifers or surface water bodies. Landforms that contain economically developable ground water in surficial sand and gravel aquifers or that contain streams and lakes are considered poor solid waste disposal sites.

WEIGHTED RATING[3]

Denotes the numerical ranking of map units related to the prospect of finding suitable solid waste disposal sites.

TABLE 32-18

Suitability for Sewage Effluent Disposal by Infiltration

In this study, terrain and geotechnical factors are used in rating land capability limitations for each map unit. Alternative land use, socio-economic and certain environmental factors are not considered. The system of weighted rating used is illustrated below in a chart; the description of the terrain and geotechnical factors considered is also described below. The rating applies to the classification and mapping of landforms, and to the surface and near surface materials in them. Because of the small scale of the maps, the outlined map units may contain small areas of other map units that are not indicated.

MAP UNITS[1]			TERRAIN AND GEOTECHNICAL FACTORS[2]							WEIGHTED[3] RATING
LETTER SYMBOL	LANDFORM DESCRIPTION	SURFACE MATERIAL AND ASSOCIATED BASIC LANDFORM TYPE AND VARIETY (see legend)	PERMEABILITY	TOPOGRAPHY	SLOPE	FLOODING	GROUND WATER POLLUTION	DEPTH OF SOIL	DEPTH TO WATER TABLE	
ME	End moraine	tME	3	0	0	3	2	6	6	21
MG	Ground moraine	tMG, tMG_s, tMG_c, tMG_f	0	3	3	9	3	6	6	30
		tMG_e, tMG_{ec}	0	3	3	9	2	6	6	29
MH	High relief hummocky moraine	tMH, tMH_s	4	0	0	3	2	6	6	21
MH_L	Low relief hummocky moraine	tMH_L, tMH_{Lf}, tMH_{Ls}	0	1	2	9	3	6	6	27
LD	Lacustrine delta	sLD, sLD_k, sLD_w	12	3	3	9	0	6	4	37
		smLD, $smLD_w$	12	3	3	9	0	6	4	37
LP	Lacustrine plain	mLP, mLP_c	0	3	3	6	3	6	6	27
		msLP, $msLP_w$	8	3	3	9	1	6	6	36
		cmLP, $cmLP_c$, $cmLP_s$	0	3	3	3	3	6	6	24
		cLP	0	3	3	3	3	6	6	24
LH	High relief hummocky lacustrine terrain	smLH, $smLH_k$, $smLH_w$	12	0	0	3	2	6	6	29
		mLH	0	0	0	3	2	6	6	17
		msLH	8	0	0	3	1	6	6	24
		cmLH, $cmLH_s$	0	0	0	3	3	6	6	18
		cLH	0	0	0	3	3	6	6	18
LH_L	Low relief hummocky lacustrine terrain	$smLH_L$, $smLH_{Ls}$, $smLH_{Lw}$	12	2	2	9	0	6	4	35
		mLH_L, mLH_{Ls}	0	2	2	9	3	6	6	28
		$msLH_L$, $msLH_{Ls}$	8	2	2	9	1	6	6	34
		$cmLH_L$, $cmLH_{Ls}$	0	2	2	6	3	6	6	25
		cLH_L	0	2	2	6	3	6	6	25
GK	Ice contact knobs and ridges	sGK, sGK_w	8	0	0	9	2	6	6	31
		sgGK, $sgGK_k$	8	0	0	9	2	6	6	31
		gsbGK	8	0	0	9	2	6	6	31
GO	Outwash plains and valley trains	sGO, sGO_k	12	3	3	9	0	6	4	37
		sgGO, $sgGO_c$, $sgGO_s$, $sgGO_{sk}$, $sgGO_f$, $sgGO_k$	8	3	3	9	0	6	4	33
		gsGO, $gsGO_c$, $gsGO_k$, $gsGO_{kc}$	8	3	3	9	0	6	4	33
		gGO	8	3	3	9	0	6	4	33
AP	Alluvial plain	vAP, vAP_m	4	3	3	0	0	6	2	18
		msAP, $msAP_c$	8	3	3	0	0	6	2	22
		sAP	12	3	3	0	0	6	2	26
		sgAP	12	3	3	0	0	6	2	26
AF	Alluvial fan	vAF	4	3	3	3	1	6	4	24
AT	Alluvial terrace	vAT, vAT_c	4	3	3	3	1	4	4	22
		sAT	12	3	3	3	1	4	4	30
		gsAT	12	3	3	3	1	4	4	30
ED	Eolian dunes	sED	12	0	0	9	0	6	6	33
OT	Organic terrain	pOT	12	3	3	3	2	4	0	27
BT	Bedrock terrain	ss, shBT	0	0	1	6	3	6	6	22
SL	Slumped valley slopes	vSL	0	0	0	9	3	6	6	24
SE	Eroded valley and hill slopes	vSE	0	0	0	9	3	6	6	24

CONTRASTING MATERIALS AND LANDFORMS IN THE NEAR SURFACE ZONE

LETTER SYMBOL	SURFACE LANDFORM	BURIED NEAR SURFACE LANDFORM	SURFACE AND BURIED MATERIAL AND LANDFORM	PERMEABILITY	TOPOGRAPHY	SLOPE	FLOODING	GROUND WATER POLLUTION	DEPTH OF SOIL	DEPTH TO WATER TABLE	WEIGHTED RATING
LD/MG	Lacustrine delta	Ground moraine	sLD/tMG	12	3	3	9	2	0	0	29
			smLD/tMG, $smLD_w$/tMG	12	3	3	9	2	0	0	29
			sLD_w/tMG_e	12	3	3	9	2	0	0	29
LD/LP	Lacustrine delta	Lacustrine plain	$smLD_w$/cmLP	12	3	3	6	2	0	0	26
LD/GO	Lacustrine delta	Outwash plain	smLD/sgGO	12	3	3	9	0	6	4	37
LP/MG	Lacustrine plain	Ground moraine	msLP/tMG	8	3	3	9	2	2	2	29
			mLP/tMG, mLP/tMG_f	0	3	3	9	3	6	4	28
			cmLP/tMG, $cmLP_c$/tMG_c	0	3	3	9	3	6	4	28
			cLP/tMG	0	3	3	9	3	6	4	28
			msLP/tMG_e	8	3	3	9	2	0	0	25
			mLP/tMG_e, mLP/tMG_{ec}	0	3	3	9	3	6	4	28
			cmLP/tMG_e	0	3	3	9	3	6	4	28
LP/GO	Lacustrine plain	Outwash plain	msLP/sgGO	8	3	3	9	0	6	4	33
			mLP/sgGO	8	3	3	9	0	4	4	31
			mLP/gsGO	8	3	3	9	0	4	4	31
LP/GK	Lacustrine plain	Icecontactknobs&ridges	cLP/sgGK	0	2	2	3	1	6	4	18
LH_l/MH_L	Low relief hummocky lacustrine terrain	Low relief hummocky moraine	$smLH_L$/tMH_L	12	1	2	9	2	0	2	28
			$msLH_L$/tMH_L	8	1	2	9	2	2	2	26
			mLH_L/tMH_L	0	1	2	9	3	6	4	25
			$cmLH_L$/tMH_L	0	1	2	9	3	6	4	25
LH_L/LH_L	Low relief hummocky lacustrine terrain	Low relief hummocky lacustrine terrain	$msLH_L$/mLH_L	8	1	2	9	2	2	2	26
GO/MG	Outwash plain	Ground moraine	sgGO/tMG	12	3	3	9	2	0	0	29
AT/MG	Alluvial terrace	Ground moraine	sgAT/tMG_e	12	3	3	3	1	0	0	22
			smAT/tMG_e	12	3	3	3	1	0	0	22
			msAT/tMG_e	8	3	3	3	2	0	0	19
MG/BT	Ground moraine	Bedrock terrain	tMG/BT	0	3	3	9	3	6	6	30

GRAPHIC SYMBOL	SURFACE FEATURE	PERMEABILITY	TOPOGRAPHY	SLOPE	FLOODING	GROUND WATER POLLUTION	DEPTH OF SOIL	DEPTH TO WATER TABLE	WEIGHTED RATING
<<<<	Esker ridge	8	0	0	9	2	6	6	31
	Meltwater channel or spillway	4	3	3	0	0	6	2	18
	Tunnel valley or stream trench	4	3	3	0	0	6	2	18

WEIGHTED RATING	SIMPLIFIED RATING
0–30	MOSTLY UNSUITABLE TO POOR U
30–34	FAIR F
34–40	GOOD G

TERRAIN AND GEOTECHNICAL FACTORS	RATING	WEIGHT	WEIGHTED RATING
PERMEABILITY	0 1 2 3	4	0 4 8 12
TOPOGRAPHY	0 1 2 3	1	0 1 2 3
SLOPE	0 1 2 3	1	0 1 2 3
FLOODING	0 1 2 3	3	0 3 6 9
GROUNDWATER POLLUTION	0 1 2 3	1	0 1 2 3
DEPTH OF SOIL	0 1 2 3	2	0 2 4 6
DEPTH TO WATER TABLE	0 1 2 3	2	0 2 4 6

CHART ILLUSTRATING METHOD USED TO DERIVE WEIGHTED RATING

MAP UNITS[1]

For description of landform basic types and varieties, and the recurring types of soil materials found in them, which together form the mapping symbols and the map units, see the accompanying report.

TERRAIN AND GEOTECHNICAL FACTORS[2]

Permeability. Medium permeability silty and fine sandy materials are the best materials for effluent disposal. If the permeability is too high, natural filtration of the effluent will be limited. On the other hand if the permeability is too low, effluent application rates must also be low, otherwise surface ponding will occur.

Topographic setting. Gently undulating topography is desirable. If the land surface is flat or roughly undulating, ponding is likely to occur in poorly drained depressions.

Slope. The slope desired depends on the near surface soil permeability. On steep slopes underlain by low permeability soils, the effluent may run off over the land surface before it can percolate into the soil. Long, uniform, gentle slopes are favored.

Flooding. Lands subject to overbank flooding of streams and to ponding of rainfall and snowmelt in closed depressions are not favored for effluent disposal by infiltration. Ponded waters may dissolve contaminants from polluted soils and carry them into nearby lakes and streams.

Ground water pollution. Lands suitable for sewage effluent disposal by infiltration may be underlain by aquifers containing potable drinking water. In these situations the possible pollution of ground water poses a concern. Thus a decision must be made whether the land in question should be used to provide a potable source of water or to dispose of sewage effluent by infiltration.

Depth of permeable near surface soil. The depth of permeable near surface soil over a low permeability layer affects the amount of sewage effluent that can be stored in the subsurface.

Depth to water table. The application of sewage effluent over a high water table area may raise the ground water level to or above ground surface, thereby concentrating salts at and near ground level or ponding effluent on the surface.

WEIGHTED RATING[3]

Denotes the numerical ranking of map units for sewage effluent disposal by infiltration.

TABLE 32-19

Suitability for Septic Tank Waste Disposal

In this study, terrain and geotechnical factors are used in rating land capability limitations for each map unit. Alternative land use, socio-economic and certain environmental factors are not considered. The system of weighted rating used is illustrated below in a chart, the description of the terrain and geotechnical factors considered is also described below. The rating applies to the classification and mapping of landforms, and to the surface and near surface materials in them. Because of the small scale of the maps, the outlined map units may contain small areas of other map units that are not indicated.

MAP UNITS[1]			TERRAIN AND GEOTECHNICAL FACTORS[2]								WEIGHTED[3] RATING
LETTER SYMBOL	LANDFORM DESCRIPTION	SURFACE MATERIAL AND ASSOCIATED BASIC LANDFORM TYPE AND VARIETY (see legend)	PERMEABILITY	SURFACE DRAINAGE	DEPTH TO WATER TABLE	FLOODING HAZARD	DEPTH OF PERMEABLE MATERIALS	PROXIMITY TO WATER BODIES			
ME	End moraine	tME	0	6	6	6	0	3			21
MG	Ground moraine	tMG, tMG$_{a}$, tMG$_{c}$, tMG$_{l}$	0	4	6	6	0	3			19
		tMG$_{e}$, tMG$_{ec}$	0	4	4	6	0	3			17
MH	High relief hummocky moraine	tMH, tMH$_{s}$	0	4	6	4	0	2			16
MH$_{L}$	Low relief hummocky moraine	tMH$_{L}$, tMH$_{Ll}$, tMH$_{Ls}$	0	4	6	4	0	3			17
LD	Lacustrine delta	sLD, sLD$_{k}$, sLD$_{w}$	12	6	4	6	6	3			37
		smLD, smLD$_{w}$	12	6	4	6	6	3			37
LP	Lacustrine plain	mLP, mLP$_{c}$	4	2	6	4	2	3			21
		msLP, msLP$_{w}$	8	4	6	6	4	3			31
		cmLP, cmLP$_{c}$, cmLP$_{s}$	0	2	6	4	0	3			15
		cLP	0	2	6	2	0	3			13
LH	High relief hummocky lacustrine terrain	smLH, smLH$_{k}$, smLH$_{w}$	12	6	6	6	6	2			38
		mLH	4	4	6	6	2	2			24
		msLH	8	6	6	6	4	2			32
		cmLH, cmLH$_{s}$	0	2	6	4	0	2			14
		cLH	0	2	6	4	0	2			14
LH$_{L}$	Low relief hummocky lacustrine terrain	smLH$_{L}$, smLH$_{Ls}$, smLH$_{Lw}$	12	6	4	6	6	3			37
		mLH$_{L}$, mLH$_{Ls}$	8	4	6	4	2	2			26
		msLH$_{L}$, msLH$_{Ls}$	8	4	6	4	4	2			28
		cmLH$_{L}$, cmLH$_{Ls}$	0	2	6	4	0	2			14
		cLH$_{L}$	0	2	6	4	0	2			14
GK	Ice contact knobs and ridges	sGK, sGK$_{w}$	12	6	6	6	6	2			38
		sgGK, sgGK$_{k}$	8	6	6	6	6	2			34
		gsbGK	8	6	6	6	6	2			34
GO	Outwash plains and valley trains	sGO, sGO$_{k}$	12	6	4	6	6	3			37
		sgGO, sgGO$_{c}$, sgGO$_{s}$, sgGO$_{sk}$, sgGO$_{t}$, sgGO$_{k}$	8	6	4	6	6	3			33
		gsGO, gsGO$_{c}$, gsGO$_{k}$, gsGO$_{kc}$	8	6	4	6	6	3			33
		gGO	8	6	4	6	6	3			33
AP	Alluvial plain	vAP, vAP$_{m}$	8	2	2	2	4	0			18
		msAP, msAP$_{c}$	12	2	2	2	4	0			22
		sAP	12	2	2	2	4	0			22
		sgAP	12	2	2	2	4	0			22
AF	Alluvial fan	vAF	8	2	2	4	4	1			21
AT	Alluvial terrace	vAT, vAT$_{c}$	8	4	4	4	4	1			25
		sAT	12	6	4	4	4	1			31
		gsAT	12	6	4	4	4	1			31
ED	Eolian dunes	sED	12	6	4	6	6	1			35
OT	Organic terrain	pOT	0	0	0	0	0	0			0
BT	Bedrock terrain	ss, shBT	0	2	6	6	0	3			17
SL	Slumped valley slopes	vSL	0	4	6	6	0	3			19
SE	Eroded valley and hill slopes	vSE	0	4	6	6	0	3			19

CONTRASTING MATERIALS AND LANDFORMS IN THE NEAR SURFACE ZONE

LETTER SYMBOL	SURFACE LANDFORM	BURIED NEAR SURFACE LANDFORM	SURFACE AND BURIED MATERIAL AND LANDFORM	PERMEABILITY	SURFACE DRAINAGE	DEPTH TO WATER TABLE	FLOODING HAZARD	DEPTH OF PERMEABLE MATERIALS	PROXIMITY TO WATER BODIES			WEIGHTED RATING
LD/MG	Lacustrine delta	Ground moraine	sLD/tMG	12	4	0	6	2	3			27
			smLD/tMG, smLD$_{w}$/tMG	12	4	0	6	2	3			27
			sLD$_{w}$/tMG$_{e}$	12	4	0	6	2	3			27
LD/LP	Lacustrine delta	Lacustrine plain	smLD$_{w}$/cmLP	12	4	0	6	2	3			27
LD/GO	Lacustrine delta	Outwash plain	smLD/sgGO	8	6	4	6	6	3			33
LP/MG	Lacustrine plain	Ground moraine	msLP/tMG	8	4	0	6	2	3			23
			mLP/tMG, mLP/tMG$_{l}$	0	4	6	6	0	3			19
			cmLP/tMG, cmLP$_{c}$/tMG$_{c}$	0	4	6	6	0	3			19
			cLP/tMG	0	4	6	6	0	3			19
			msLP/tMG$_{e}$	8	4	0	6	2	3			23
			mLP/tMG$_{e}$, mLP/tMG$_{ec}$	0	4	6	6	0	3			19
			cmLP/tMG$_{e}$	9	4	6	6	0	3			19
LP/GO	Lacustrine plain	Outwash plain	msLP/sgGO	8	6	4	6	6	3			33
			mLP/sgGO	8	6	4	6	6	3			33
			mLP/gsGO	8	6	4	6	6	3			33
LP/GK	Lacustrine plain	Icecontact knobs&ridges	cLP/sgGK	4	6	4	6	6	3			29
LH$_{L}$/MH$_{L}$	Low relief hummocky lacustrine terrain	Low relief hummocky moraine	smLH$_{L}$/tMH$_{L}$	12	4	0	6	2	3			27
			msLH$_{L}$/tMH$_{L}$	8	4	0	4	2	2			20
			mLH$_{L}$/tMH$_{L}$	0	4	6	4	0	2			16
			cmLH$_{L}$/tMH$_{L}$	0	4	6	4	0	2			16
LH$_{L}$/LH$_{L}$	Low relief hummocky lacustrine terrain	Low relief hummocky lacustrine terrain	msLH$_{L}$/mLH$_{L}$	8	4	2	6	2	3			25
GO/MG	Outwash plain	Ground moraine	sgGO/tMG	12	4	0	6	2	3			27
AT/MG	Alluvial terrace	Ground moraine	sgAT/tMG$_{e}$	12	6	0	4	2	1			25
			smAT/tMG$_{e}$	12	6	0	4	2	1			25
			msAT/tMG$_{e}$	8	6	0	4	2	1			21
MG/BT	Ground moraine	Bedrock terrain	tMG/BT	0	2	6	6	0	3			17

GRAPHIC SYMBOL	SURFACE FEATURE	PERMEABILITY	SURFACE DRAINAGE	DEPTH TO WATER TABLE	FLOODING HAZARD	DEPTH OF PERMEABLE MATERIALS	PROXIMITY TO WATER BODIES			WEIGHTED RATING
<<<<	Esker ridge	8	6	6	6	6	1			33
	Meltwater channel or spillway	8	2	2	2	4	0			18
	Tunnel valley or stream trench	8	2	2	2	4	0			18

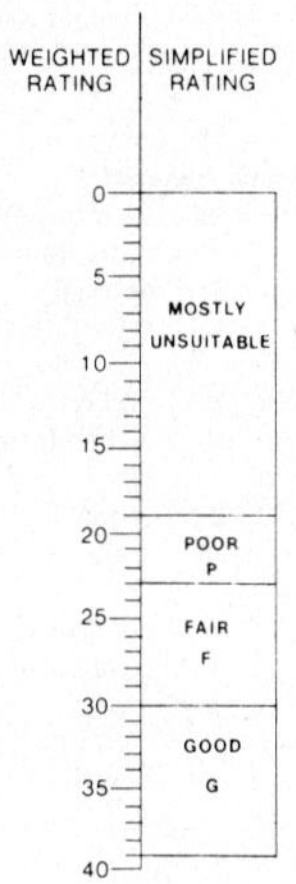

TERRAIN AND GEOTECHNICAL FACTORS	RATING	WEIGHT	WEIGHTED RATING
PERMEABILITY	0	4	0
	1		4
	2		8
	3		12
SURFACE DRAINAGE	0	2	0
	1		2
	2		4
	3		6
DEPTH TO WATER TABLE	0	2	0
	1		2
	2		4
	3		6
FLOODING HAZARD	0	2	0
	1		2
	2		4
	3		6
DEPTH OF PERMEABLE MATERIALS	0	2	0
	1		2
	2		4
	3		6
PROXIMITY TO WATER BODIES	0	1	0
	1		1
	2		2
	3		3

CHART ILLUSTRATING METHOD USED TO DERIVE WEIGHTED RATING

MAP UNITS[1]

For description of landform basic types and varieties, and the recurring types of soil materials found in them, which together form the mapping symbols and the map units, see the accompanying report.

TERRAIN AND GEOTECHNICAL FACTORS[2]

Permeability. Soil materials having a permeability greater than about 1×10^{-3} cm/sec (clean and silty very fine and fine sand) should exist below the depth of frost penetration for the successful operation of underground seepage disposal fields.

Drainage. Proper surface drainage control is required to prevent ponding of runoff waters on the disposal field. Good subsurface drainage is required to prevent progressive buildup of the water table and waterlogging of the disposal field. Low-lying marshy and swampy depressions should be avoided.

Depth to water table. The water table should be more than 10 m below the disposal field for proper natural filtration of septic wastes.

Flooding. Flooding from streams and ponding of surface waters in depressions build up the water table and produce an unfavorable disposal setting. Floodwaters can also carry septic wastes to nearby lakes and streams and pollute them.

Depth of permeable materials. In order to prevent water table buildup above an impermeable substrate, up to 5 m of moderately permeable soil material is required above the impermeable substrate for the necessary dissipation of septic wastes.

Proximity to surface water bodies. Septic wastes moving through soils near streams, lakes and sloughs may pollute water in them. Large numbers of septic tank disposal fields near streams and lakes are therefore undesirable. However, single disposal fields may be tolerated provided that the fields are more than 50 m away from a water body and are not too close together. Public Health regulations in effect in Saskatchewan govern this aspect of the pollution hazard.

WEIGHTED RATING[3]

Denotes the numerical ranking of map units for purposes of septic tank waste disposal.

TABLE 32-20A

Terrain-Type Legend

THE LEGEND COMPRISES FOUR MAIN COMPONENTS ARRANGED AS FOLLOWS:

1 MATERIAL	2 LANDFORM
3 TOPOGRAPHY	4 DRAINAGE

1 MATERIAL

Bedrock material

ca	carbonate rocks (limestones/dolomites)	**ss**	sandstones/conglomerates
x	crystalline rocks (igneous/metamorphic rocks)	**sh**	shales/siltstones

Overburden material

b	boulders	(dense concentration of boulders at ground surface and/or deposits containing many boulders)
c	clay	(laminated; admixed with some silt and fine sand)
d	diamicton	(mixed, unsorted to crudely sorted, fine and coarse debris of non-glacial origin, such as slopewash or solifluctеd material)
g	gravel	(layered; admixed with sand and cobbles)
m	silt	(laminated; admixed with fine sand and some clay, mostly between 0.002 and 0.2 mm in diameter)
o	organic material	(as a substantial component, i.e. 20 to 50 per cent, in mineral soils)
p	peat	(forest peat type; Sphagnum moss, or bog, type; and sedge-reed, or fen, type)
r	rubble	(loose, angular frost-shattered bedrock; largest size in Shield areas)
s	sand	(layered; poorly and well graded sand; interbedded fine to coarse sand, mostly between 0.2 and 4.8 mm in diameter)
t	till	(commonly very bouldery in Shield areas)

2 LANDFORM

A ALLUVIAL
- c channel
- d delta
- f fan
- p floodplain
- t terrace

M MORAINE
- d drumlinized moraine
- g ground moraine
- h hummocky moraine
- r ridged moraine/end moraine
- w washboard/ribbed moraine

C COLLUVIAL
- f flow
- m slopewash/solifluction sheet
- s slide/slump
- t talus/fall

O ORGANIC
- b bog, patterned (peat plateau/palsa/collapse scar)
- f fen, patterned/ribbed
- u unpatterned and undifferentiated peatland

R BEDROCK
- d dipping beds
- f foliated rocks
- h horizontal beds
- j jointed/fractured rocks
- m massive rocks

G GLACIOFLUVIAL
- c crevasse filling
- d delta
- e esker
- k kame
- o outwash plain
- v valley train

E EOLIAN
- l loess blanket
- d dunes

L LACUSTRINE
- b beach ridge, abandoned
- d deltaic plain
- n near shore/offshore plain

W MARINE
- b beach ridge, abandoned
- d deltaic plain
- n near shore/offshore plain

MODIFICATIONS TO LANDFORMS
- a terraced
- b braided/channelled
- d dissected
- e wind-eroded
- g glacier-streamlined
- k kettled/pitted
- p patterned ground
- r stream-eroded
- s smoothed/striped by mass wasting
- t modified by thermokarst activity
- w wave-eroded

3 TOPOGRAPHY

Forms	Varieties
H HILLY	d depressed
K KNOBBY	e elevated
L LEVEL	g gently
M ROLLING	h hummocky microrelief
R RIDGED	j jagged/irregular
S SLOPING	m moderately
U UNDULATING	r rounded/smoothed
	s steeply/strongly
	v very little/slightly

4 DRAINAGE

Forms	Varieties
D DRYLAND (seasonal)	b beaded drainage
W WETLAND (seasonal)	g gullies and ravines
M MIXED DRY AND WET LAND (seasonal)	n nival (snowpatch) forms
	r rills
	t thermokarst forms

TABLE 32-20B

Terrain Type Legend Used for Regional Corridor Studies for Large Diameter Pipeline Routes, Compressor Stations, and Airstrip Sites in Northern Canada. Prepared Using Integrated Remote Sensing Methods for Polargas, Ontario

SYMBOLS

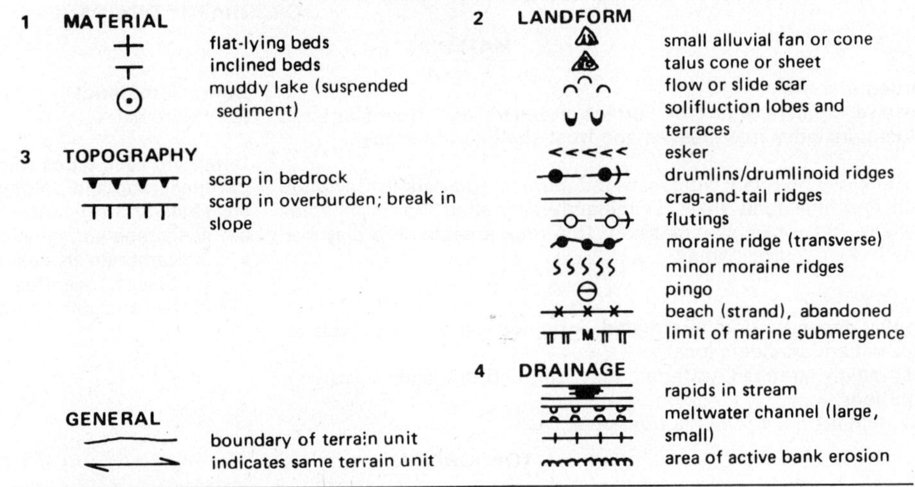

EXAMPLES OF COMBINED SYMBOLS

1 MATERIAL

m,c	Silt dominant, clay subordinate
s, g	Sand dominant, gravel subordinate
d, r	Diamicton dominant, rubble subordinate
b,t	bouldery surface layer over till or bouldery till

3 TOPOGRAPHY

Mg,Rg	Gently rolling dominant, gently sloping ridges subordinate
Rj	Ridged terrain, jagged/irregular surface

2 LANDFORM

Go-d	Glaciofluvial outwash-delta (hybrid form)
Ge-k	Glaciofluvial esker-kame (complex association)
Mg,s	Ground moraine modified by slope wash and/or solifluction processes
Ac,b	Alluvial channel, braided

4 DRAINAGE

Wt	Wetland (seasonal) with thermokarst forms
D,Wr	Dryland (seasonal) dominant, wetland (seasonal) with numerous wet rills during runoff subordinate

sociated regional environment studies in northern Canada. Table 32-21 is a simplified version used by pipeline engineers for estimating costs at the feasibility level.

Landsat Image, Mackenzie Delta, District of Mackenzie, Northwest Territories

Since 1972 Landsat images have been used for planning offshore exploration for oil and gas in the Beaufort Sea, and for site and route investigations in the Mackenzie Delta and adjoining upland area. Figure 32-32 is one of the many Landsat scenes studied during the course of these investigations. An early study entailed preliminary assessment of the significance of the dense concentration of deep thermokarst basins around 1 (upper right quadrant of scene). A large block of terrain, centered around 2 (west of 1), was later examined in Landsat imagery for indications of postglacial fault activity.

Scattered ice flows can be seen northwest of the large plume of suspended sediment, the concentration of which diminishes notably beyond 3. Exposed bluffs of Tertiary sand and gravel at 4 were mapped originally from airphotos and were later studied in the field and in Landsat imagery. Campsite, granular fill borrow, and river crossing investigations were centered around 5. Investigations for a large diameter pipeline crossing were made at 6. In these and other studies Landsat images were interpreted along with small scale photomosaics and small scale (1:60,000) B&W stereoscopic airphotos.

Digital manipulations of Landsat data such as contrast stretches and simple digital displays on a color television monitor produces images for site investigations. In the early stages of these investigations the use of geobased information systems combines Landsat digital data with tabulated background map data for use in making a more comprehensive assessment of the existing conditions.

TABLE 32-21

Simplified Version of Table 32-20a for Use in Preliminary Pipeline Cost Estimating in Northern Canada. (Prepared for Polargas, Ontario)

TERRAIN LEGEND

Symbols in mapped terrain units are arranged as follows: $\frac{\text{MATERIAL}}{\text{TOPOGRAPHY DRAINAGE}}$

MATERIAL

Overburden material

B Excessive concentrations of surface boulders and frost-shattered boulders; includes frost-heaved and frost-shattered bedrock

T Till. Generally bouldery with scattered surface boulders in Canadian Shield and limestone areas. Dominantly silty sand matrix in areas of Rx and Rss rocks, silty matrix in Rca rock areas, and a clay-rich matrix in Rsh bedrock areas

G Granular deposits. Undifferentiated, stratified sand and gravel; includes silt and boulders locally

F Fines; mainly stratified, waterlaid clay, silt, and fine sand in varying proportions

P Peat; includes fen and Sphagnum peat

Bedrock material

R Bedrock

Dominant rock types (modifier)

ca limestone and dolomite
sh shale and siltstone
ss sandstone and conglomerate
x Precambrian crystalline rocks (gneisses, granites, greenstones and allied hard rocks)

TOPOGRAPHY

Relief appearance

H Hilly
K Knobby
L Level
M Rolling
R Ridged
S Sloping
U Undulating

Modifier

c choppy microrelief
d depressed
e elevated
g gently
j jagged or irregular
k kettled
m moderately
p patterned ground
r smoothly rounded
s steeply or strongly

DRAINAGE

Surface drainage condition

D Topographically higher and better drained sloping areas, commonly with a wet active layer in permafrost areas; mostly smooth ridges and upper slopes in till areas

W Topographically lower areas; includes poorly drained flats and depressions in till areas, most low relief F areas, and all P areas

Modifier

g gullies
r rills
t thermokarst basins

Thermal Infrared Image, Northern Okanagan Lake, British Columbia

The scene in Fig. 32-33 shows part of northern Okanagan Lake in British Columbia. This 8-level-mode thermal IR image was acquired by Intertech, Calgary; it shows 6 grey tones, corresponding to surface temperature differences of 1°C or less. These thermal IR images are taken several times a year and are correlated with ambient air temperature for purposes of predicting the amount of evaporation off the lake. Thermal IR images may also be used to discover points of groundwater inflow to the lake. Good estimates of evaporation losses and surface and groundwater flow into the lake are needed because the depth of usable storage on the lake, for such uses as irrigation and channel water maintenance, is only about 4 feet.

Light tones, as at 1, are warmer than darker tones, as at 2. This map of changing surface water temperatures of part of Okanagan Lake is useful for locating the best spots in which to troll for sport fish and also for locating warmer waters (e.g. on the west side of the lake) indicative of the sites that provide better swimming conditions and better microclimate for resort development.

The small tributary at 3 is Shorts Creek, which has built a delta into the west side of Okanagan Lake some 13 miles southwest of Vernon, British Columbia. The dark, narrow, jagged line at 4 is caused by the wake of a boat. The propeller has churned the lake water, bringing deeper-lying colder water to the lake surface.

Similar imagery has been used in the Niagara Peninsula of southern Ontario to aid in producing a 1:50,000 scale map of grape climatic zones for site selection of vineyards (Wiebe and Anderson,

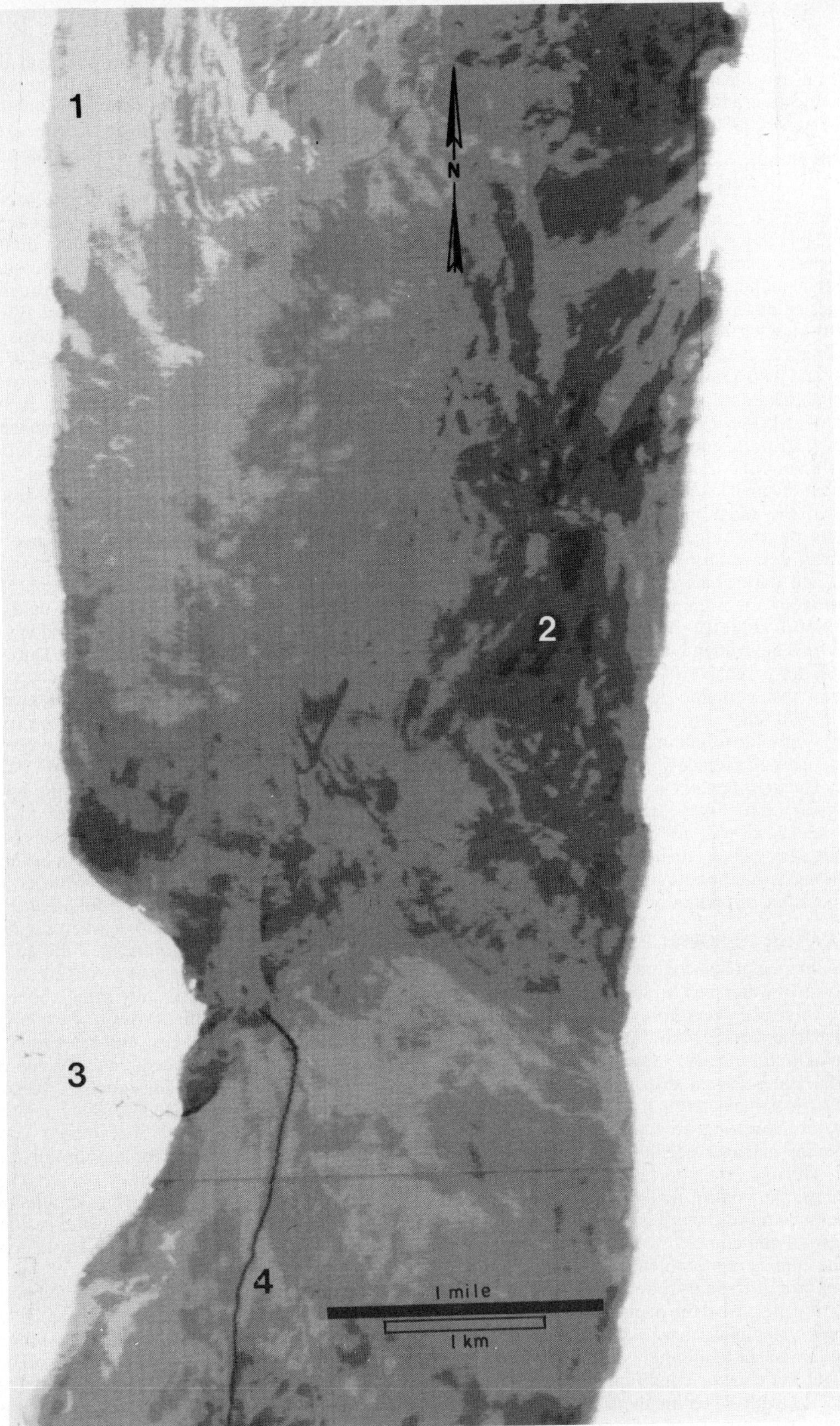

Fig. 32-33. Eight-level mode thermal IR image showing part of northern Okanagan Lake, British Columbia. Used to predict evaporation losses. (Courtesy of Intertech, Calgary.)

1976). Similar methods have also been used for other microclimatic sensitivity studies related to site selection and forest regeneration (Lawrence and Banner, 1980).

SLAR Image, Stony Mountain Upland, South of Fort McMurray, Alberta

The west-looking SLAR mosaic shown in Figure 32-34 has a scale of 1:250,000. The radar imagery was acquired in November 1979 as part of a regional mapping of landscape features. Radar images emphasize relief features, which is evident in this view of the northeastern part of the Stony Mountain Upland. The relief on large glaciated hills, and also a number of linear terrain features, are highlighted by radar shadows. The short linear feature at 1 (upper center of scene) is, however, a landing strip. It should not be confused with natural lineaments in the relief and drainage. A number of small lakes and large ponds on the west side of the radar mosaic appear black. Aspen stands on the ridges at 2 and 3 indicate well drained, dominantly sandy soils. The mixture of light and dark areas east of these ridges corresponds to an area of mixed hardwood and softwood—chiefly intermingled spruce and poplar. The wetland area around west of 4 is largely an extensive black spruce and tamarack swamp that contains pockets of sphagnum bog and ribbed fen.

It is difficult to distinguish the small lake at 5 from the radar shadow just to the north of the lake. The very fine linear feature east of 2 and 3 is a railway right-of-way. Because radar imagery visualizes the overall architecture of large areas of terrain very well, it can often be used to advantage with small scale photomosaics and high altitude stereoscopic airphotos of the same area.

WATER RESOURCES ENGINEERING

Water-resources engineering begins with the analysis of watershed or drainage basins and ends with water use. Therefore, a water-resources engineer is concerned with water-related parameters such as water quality, water quantity, water loss, spatial and seasonal distribution, and water demand. In moving water from a watershed to the user, or from one area to another, the water-resources manager needs to evaluate the collection, storage, transportation, and distribution systems, the control facilities, and the impact of various water-engineering measures on both the terrestrial and aquatic environments.

The engineering problems pertaining to surface water are different than those pertaining to ground-water. Also the management strategies for surface water and ground water vary considerably from one basin to another, depending on physiographic and climatic conditions. For example, the problems with water in the western part of the United States tend to be different than those in the eastern part. While western managers may be primarily concerned about water quantity, eastern managers may be primarily concerned about water quality and water pollution. Therefore, water-resources engineers are concerned with the assessment of existing hydrologic conditions such as watershed layout, hydrologic landcover topography, and precipitation.

Remotely-sensed data collected from satellites and aircraft are directly applicable to watershed analysis, surface-water mapping, snow quantification, near real-time flood-damage assessment, and even ground water assessment. Through the development of geobased information systems (GIS) one may bring together, store, analyze, and combine data of various kinds. The types of data that are incorporated into a GIS are those that have a geographic or spatial component. A list of all of the potential water-resources engineering applications of a GIS would be very large and the users would include a great many Federal, state, and local government agencies and many kinds of private industry. For making engineering analyses of water-resources projects one may use efficiently those GIS systems that are based on remotely-sensed data such as Landsat MSS, NOAA satellite VHRR, digital terrain data, and high-to-low altitude aerial photography along with ancillary data in tabular or map form. These ancillary forms of data commonly pertain to watershed boundaries, road networks, and climate. Repetitive remotely-sensed data can also provide a temporal dimension for use in monitoring dynamic water-related phenomena.

Historic Landsat data are useful in the assessment of changes in the watershed area that are likely to have an impact on water-resources engineering projects. With Landsat-4 data becoming operational, the Thematic Mapper with its additional channels and with nearly 3-fold higher resolution (30m), as compared to earlier Landsats, shows promise of greatly increasing the applicability of remote sensing in water-related studies. Such improved capabilities may bring about new applications, including the analysis of water quality in smaller streams, lakes, and reservoirs, the analysis of smaller watersheds, and the monitoring of thermal pollution. Consequently, these improved capabilities may lead to the operational use of Landsat data in water-resources control and monitoring (e.g. water pollution and water-quality monitoring).

This section provides several case studies of applications of remote sensing to water resources in engineering-related studies in the North American context. In addition, the results of some of these investigations already have been found to be directly applicable in many other geographic areas as well. For example, the suspended sediment models developed in one continent are being applied to other continents also (T. Alföldi, personal communication, 1974). For the technical and physical bases of water-related remote sensing work, and for a more comprehensive statement of applications, the reader is referred to the

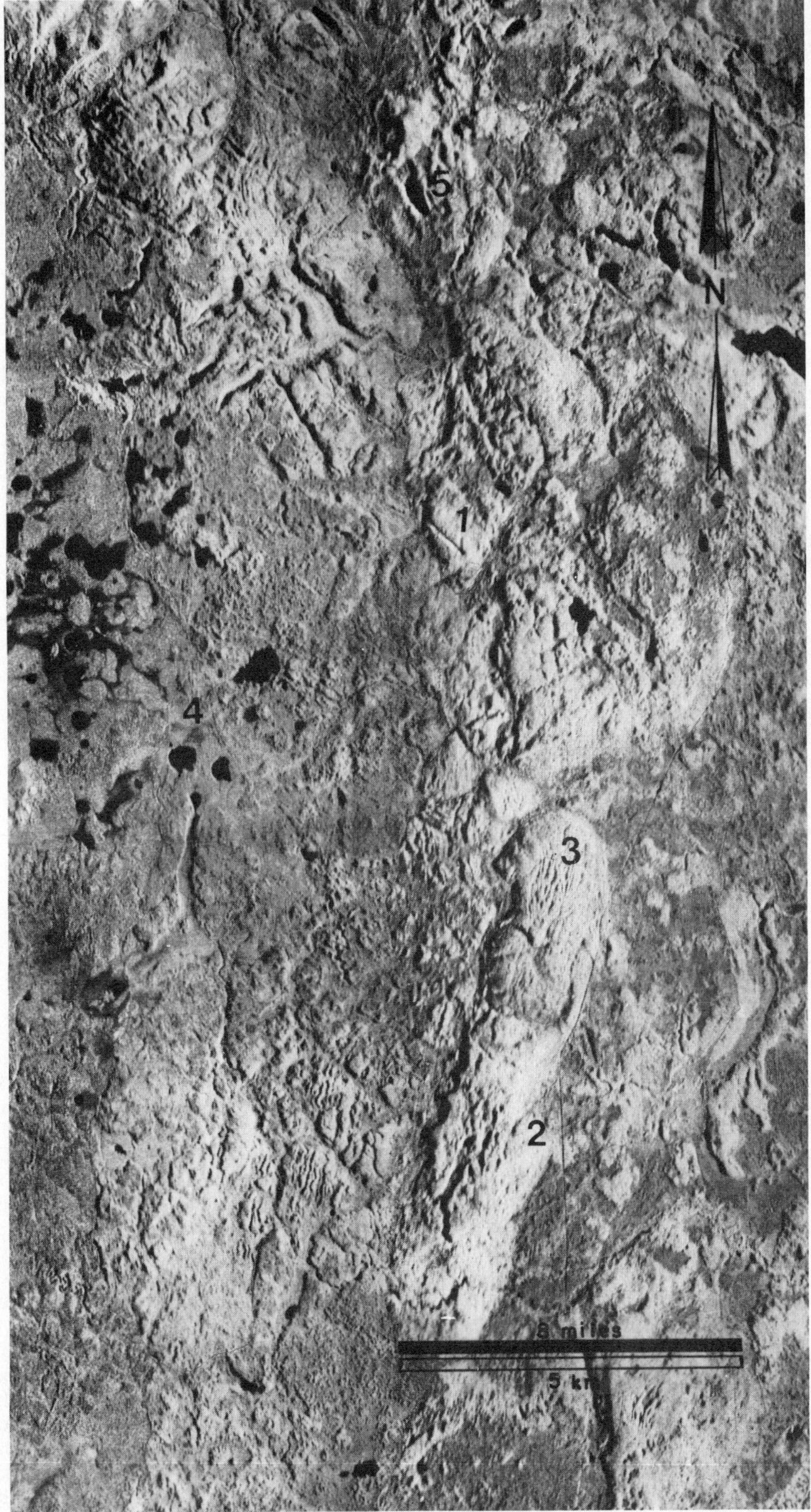

Fig. 32-34. SLAR mosaic showing hilly glacial terrain south of Fort McMurray, Alberta. Used for regional exploration and planning in inaccessible terrain. (SLAR image courtesy of Mars Ltd. Aerial Remote Sensing, Calgary, Alberta.)

chapters in this manual dealing with these subjects—especially Chapter 29.

The examples provided here deal with (1) remote sensing for water management, planning and operations in California; (2) monitoring the effects of dam construction; (3) remote sensing-aided forecasts of water yield; (4) methods of water quality mapping and monitoring; (5) Landsat imagery used in a hydrologic study, and (6) surface and near-surface thermal infrared applications.

Remote Sensing for Water Management, Planning and Operations in California

The State of California Department of Water Resources (DWR) embarked on detailed planning for the State Water Project immediately following World War II in order to formulate plans for the development of water supplies needed to meet the demands of expanding agriculture. The population growth that was expected to take place during the 50s and 60s, prompted the critical examination of relationships that exist in California between water use and water supplies. With respect to water use, the basic information needed included that pertaining to the location and areal extent of agricultural lands being farmed (both irrigated and nonirrigated) and the type of crops being grown. These data, along with estimates of the rate of water use by each crop, provided the basis for estimating the amount of water currently used. With respect to the estimation of California's water supply, even in those earlier days, some use was made of aerial photographs, as detailed in the hydrology chapter of the Manual of Photographic Interpretation (American Society of Photogrammetry, 1960). Thus the combination of these water use and water supply estimates provided a basis for estimating future supplemental needs for water and the amount of shortages to be expected. Other sources of crop data, such as the County Agricultural Commissioner's reports, the Crop and Livestock Reporting Service reports, and the Census of Agriculture data, were not by themselves sufficient for this purpose because these sources provided statistics on a total county basis only and did not give the specific location of crops, which is necessary in order to relate water use to specific water supplies. In the late 40s, DWR initiated a land-use survey designed to provide the specific crop-location information needed for this and other DWR studies and activities.

The early surveys by DWR were mapped on 1:20,0000 scale black-and-white photography purchased from the U.S. Commodity Stabilization Service. Typically, the photographs were several years old, requiring field inspection of each parcel. Other engineering data were also collected at the time of survey, such as well locations, sources of water, and the type of water diversion that was being employed. The data from completed photographs were transferred to 1:24,000 scale United States Geological Survey (USGS) quadrangles to correct for scale differences between photographs so that acreages could be determined.

In the 1970s a major change was made in forecasting the water yield for the California Water Project. This change involved the use of remote sensing. The detailed procedure, as applied on the Plumas National Forest, is discussed in a preceding section of this chapter.

Paralleling this development, a major change was made beginning in 1967 in the survey system for estimating water demand. This system continues to be the principal survey procedure used by DWR in estimating water demand. In late June of early July of each year 35mm Ektachrome vertical aerial photographs of the area to be surveyed are taken from a light aircraft (Cessna 182). These photos are obtained at an elevation of approximately 5,000 feet above ground level using a Nikon camera with a wide-angle lens having a focal length of 28mm. The camera has a motor drive and an intervalometer so that the interval between successive exposures is synchronized with the speed of the aircraft to give about a 20 percent end overlap (forward lap) of photographs. Each of the resulting transparencies covers about one square mile at a scale of approximately 1:62,500. The resulting 35mm color slides are numbered by flights and are indexed on USGS 7½-minute quadrangle map sheets, scale 1:24,000, for ease of location and use. The analyst, in the process of making the surveys, projects the transparencies onto a screen and interprets the field boundaries. To the extent possible, the analyst also identifies the crop in each field. The parcel boundaries are delineated on fade-out blue copies of the USGS quadrangles. The quadrangle map sheets are then taken to the field and each parcel is checked by trained agriculturalists for positive identification of the type of crop growing in each field. To facilitate processing of the data, ozalid copies are made of the completed field sheets. The background lines disappear when the fade-out blue map is reproduced on an ozalid machine, thereby leaving only the delineated field boundaries and crop identifications. For each field, the boundaries are then digitized, along with the crop identification code. Acreages are computed from the digitized data. Listings by crop type are then made by irrigation districts, water districts, hydrographic areas, census districts, counties, cities, and by any other areas that may be needed for engineering studies.

The current land use survey program covers the State over a seven-year period (approximately one-seventh of the State is surveyed each year).

The survey is of primary value in determining the location, nature, and amount of current water use and for monitoring water use. The latter information is needed in order to identify developing water supply shortages and to determine the availability of supplies for local use and/or export to other areas of need. In addition, DWR finds many other engineering-related uses for these data, including the following:

(1) estimates of the location and amount of future irrigation water use and the ability of users to pay for water;
(2) estimates of the location of future urban development and the impacts of such growth on regional agricultural production and related water use;
(3) quantity and nature of water use (including the timing of applications and the routing of water) in order to optimize State and federal reservoir operations. This is a continuing study required by the ever changing amount and nature of water use;
(4) estimates of available local surface and ground water supplies through the derivation and use of hydrologic models;
(5) the nature and characteristics of water use, including prevailing irrigation practices, in order to determine potentials for water savings through increased irrigation efficiency. This is required for development and implementation of programs to promote water conservation;
(6) location, nature, and amount of water-quality degradation due to agricultural chemicals (fertilizers, pesticides, and herbicides);
(7) evaluation of water rights;
(8) assessment of flood damage and facilitating of flood-plain zoning;
(9) evaluation of the impact, on specific areas, of proposed water facilities and associated water-management actions;
(10) determination of areas of opportunity for use of reclaimed waste water;
(11) identification of current and potential future soil-water drainage problems as required for planning major drainage-waste disposal systems; and
(12) assessment of the potential impact of water-quality changes on agriculture in specific areas.

In its efforts with Landsat technology to develop the identity of irrigated lands and, eventually, the identity of all major crop types, DWR has been working with (1) the National Aeronautics and Space Administration's Ames Research Center; (2) the Remote Sensing Research Program of the University of California Space Sciences Laboratory at Berkeley and, more recently, with (3) the University's Geography Remote Sensing Unit at Santa Barbara; and (4) the United States Department of Agriculture. Systems using both manual and digital analysis of Landsat multi-date data have been developed for operational use. The figures on page iv in the front of this volume are illustrative of these efforts.

Another DWR program is concerned with remote sensing techniques to monitor water-quality, wherein aerial photographs are used to determine the extent and the source of water quality problems.

Mention should be made of the following additional uses of remote sensing that are made by DWR and its associates in California:

(1) Flood-plain management and zoning activities use aerial photographs to document the areal extent of flooding along with the damage caused by flooding. DWR finds that the general public relates better to photographs than to line maps. Hence, remote sensing is a valuable tool for demonstrating the damage caused and the extent of flooding.
(2) The Snow Surveys program uses remote sensing to document the extent of snow cover. This information is used in estimating water runoff and in documenting where gages have been or should be placed in certain inaccessible places in the high Sierras.
(3) In the process of photogrammetric mapping, aerial photographs in the form of stereo pairs are placed in plotters to make contour maps of all project sites. Completed reservoirs are mapped by this method and the areas and volumes between contours are determined in order to arrive at area-capacity curves for operation of the various State water facilities.
(4) The Riverbank Protection Program uses remote sensing to measure and record the changes in stream courses over the years.
(5) Wildlife-habitat protection has been facilitated through the use of remote sensing to document changes in riparian vegetation that are likely to affect wildlife.
(6) Various geologic programs in California continue to use remote sensing to locate lineaments and other geological features that might affect the construction of dams and aqueducts. These features often are difficult or impossible to locate on the ground but usually are readily identifiable from remote sensing.
(7) Litigations involving DWR and other State agencies use aerial photography to demonstrate and document conditions at certain specific times. In these instances, as in many others, remote sensing is very important in recording facts and documenting happenings.
(8) The forecasting of weather and flood conditions makes use of data acquired by the SMS/GOES satellite system. In the process, a Laserfax Photographic Recorder is used to receive high quality weather images every half-hour. This information, because of its great usefulness is forecasting weather and floods, might be regarded as constituting the very heart of the State Flood Forecasting Center.
(9) The Instream Water Use Studies constitute a program that employs remote sensing to locate areas suitable for river access and to define areas for various recreational uses, such as fishing, rafting, and hiking.

This brief summary has shown that remote sensing has been and continues to be used in the planning and development of California's State Water Project and also in the many other engineering and management activities of the California Department of Water Resources.

MONITORING THE EFFECTS OF DAM CONSTRUCTION

The Peace Athabasca Delta

This case study is taken from the work of Howarth et al. (1982). The Peace Athabasca Delta is situated at the western end of Lake Athabasca in northeastern Alberta (lat. 58° 45′ N and long. 111° 30′ W). It is a large wetland complex covering an area of approximately 4000 km^2. Since glacial times, a delta has been produced by fluvial deposition from the Peace and the Athabasca rivers. The delta contains two large, interconnected, shallow lakes, as well as small, perched lake basins, numerous small creeks and several larger, meandering rivers with levees. There is little relief in the area, and there are important ecological relationships between the relief, the hydrologic regime and the vegetation communities.

In the Peace-Athabasca Delta, the changes that occurred were produced unintentionally. The establishment of the Bennett Dam disrupted the flow regime of the Peace River and, in particular, prevented high spring flows in the Peace River downstream from the dam. In the past, these high flows had effected a hydrologic dam, causing outflow from the Delta to cease. This resulted in a back-up of water in Lake Athabasca and a temporary flooding of the Delta. Such flooding was important for the renewal of nutrients in the Delta's ecosystem.

Low flows in the Peace River in 1969, 1970, and 1971 led to low water levels in the Delta and subsequent shifts in the vegetational composition over much of the area. In 1971, a weir was constructed near the outlet of the Delta, designed to hold what water and runoff there was available in the Delta and thus to reduce the effects of increasingly lower water levels. By 1974, through a combination of the dam and a series of natural events, much of the Delta was experiencing wide-spread flooding (Figure 32-35). Finally, in 1976, under the influence of a new weir, the water level fluctuations were restricted to more normal ranges. The vegetation once again readjusted to accommodate these hydrologic conditions. However, the previous hydrologic regime has not been fully re-established and monitoring of the delta on a long-term basis is necessary so that the need for further remedial measures can be evaluated, as required.

Although the water level changes appear to be insignificant, the area is so flat and the vegetation so intimately tied to the hydrologic regime that even minor fluctuations can have a considerable effect on the vegetation. As the area is part of the Wood Buffalo National Park and an important habitat for the staging and migrating of waterfowl, as well as for the survival of several thousand free-roaming bison, there is obviously concern that the environment should not be disrupted. The extent of the changes that occurred in the period 1973 to 1976 is clearly illustrated in Figure 32-35. The aim of the work reported in this section was to establish a method to use Landsat digital data to not only detect but also determine the nature of the hydrologic and vegetation changes in the delta.

In their studies of the Peace-Athabasca Delta, Howarth and Wickware used a digital approach involving both enhancement and classification of Landsat data. Early work involved standard Landsat Computer Compatible Tapes (CCTs), but in the more recent studies, use was made of products produced by the Canada Centre for Remote Sensing's (CCRS) through use of its Digital Image Correction System (DICS) (Guertin et al., 1979). From their studies, Howarth and Wickware have recommended a five-step procedure for determining hydrologic and vegetation change in a wetland environment such as the Peace-Athabasca Delta. The amount of detail required influences the number of stages through which the analyst should work.

The first stage is preprocessing. This involves the production of DICS data to ensure accurate registration of images from two or more dates. Preprocessing also involves the application of solar and atmospheric corrections to attempt to standardize these factors for the images being studied. In the work undertaken at CCRS in Ottawa, Howarth and Wickware used the solar and atmospheric correction programs developed by Ahern et al. (1977a and 1977b).

The second stage is image enhancement. Various combinations and ratios of Landsat bands from two dates are displayed on the cathode ray tube (CRT) monitor of the image analysis system using different color guns. In general, it is found that combinations of Bands 5 and 7 data display the most information, but the combinations will vary depending upon ground conditions. From a comparison of 1973 and 1976 Landsat data, it was found that, of several combinations tested, the Band 5 ratio (1973/1976) being displayed with the red gun and the Band 7 ratio (1973/1976) being displayed with both the blue and green guns gave the best indication of change. When working with 1975 data, however, these investigators found that the above ratio was limited in its display. The best single combination was found to result when Band 5 (1975) was assigned to the blue gun, a ratio of Band 5 (1976/1975) was assigned to the green gun and Band 7 (1975) was assigned to the red gun. In both cases, the different colors generated on the display system showed areas of different types and degrees of change. Even these simple enhancements are able to provide the resource man-

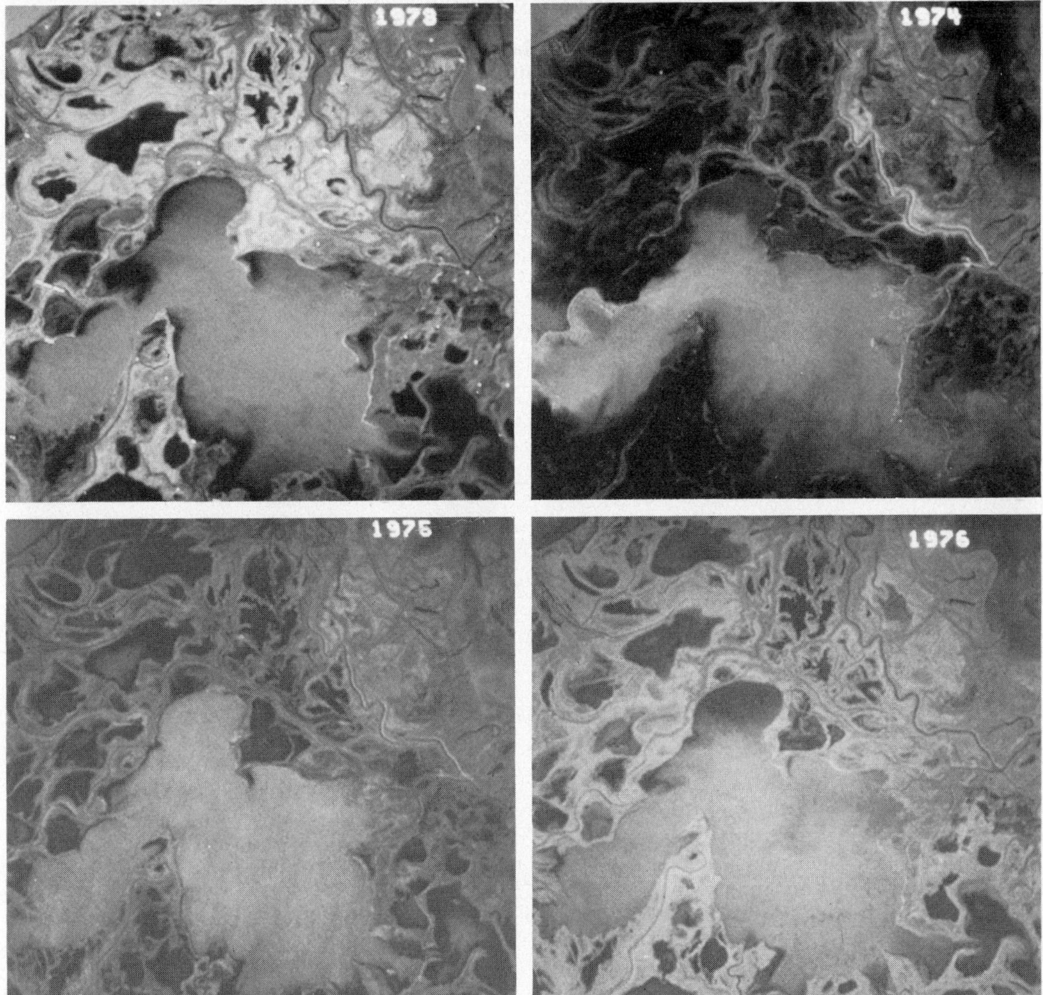

Fig. 32-35. This Figure shows the usefulness of multi-year Landsat imagery for mapping changes in water level. The area shown is the Peace Athabasca Delta in Alberta, Canada. (From Howarth et al., 1982.)

ager with information showing where field checking of change should be carried out. Such enhancements also aid in the location of transects for detailed field observations and for low-level aerial photography.

If information is required from Landsat data as to the nature of water-related environmental changes, the analyst must move to the third stage involving classification procedures. Supervised classification is definitely preferred over unsupervised and only a small area should be displayed on the CRT monitor so that training sites for the classification can be precisely identified. In early work, areas measuring 512 × 512 pixels were analyzed. With the standard Landsat pixels of 58 m × 79 m, the areas of study were approximately 30 km × 40 km on the ground. For later studies with DICS data and a 50 m^2 pixel, an area of 170 × 170 pixels (8.5 km × 8.5 km) was selected for analysis. Training-site selection could be done much more effectively in this latter case when each pixel on the CRT monitor was displayed approximately nine times.

Details of the classes identified are shown in Table 32-22. As each vegetation type in the classifications is well defined and spectrally distinct, there is no spectral overlap between the classes. Even allowing for the fact that a large portion of each image consists of water, a high percentage of the pixels are classified. From a comparison of the classifications with 1:7,000 scale 70 mm color photographs acquired along selected transects in the study area, accuracy of classification has been determined as high.

Following classification, it is possible to undertake the fourth stage, known as post-classification change detection. As shown in Table 32-22, the percentage change in area may be determined for each class. Although this procedure indicates that change has occurred, it does not identify the nature of the change. This can be done using a change matrix in which the changes

TABLE 32-22

Percentage Areas of the Water and Vegetation Classes Identified for the 170 × 170 Pixel Test Area in the Peace-Athabasca Delta. (Source: Wickware and Howarth, 1981)

Class Name	Percent Area	
	1975	1976
Turbid Open Water	53.91	60.07
Less Turbid Open Water	13.41	5.77
Scolochloa—Carex[1]	7.23	2.92
Scolochloa—Carex[2]	7.02	12.87
Calamagrostis	3.06	4.55
Immature Fen/*Scirpus*	—	3.60
Fen/*Salix*	1.95	—
Carex	—	7.24
Sparsely Vegetated Mudflats	6.55	0.28
Unclassified	6.87	2.70
Totals	100.00	100.00

on a pixel-by-pixel basis are calculated. Results are usually displayed as percentage change between each combination of classes from the two dates (e.g., Table 32-22).

Although the results show the nature of the change, they do not show the spatial pattern. This can be achieved using binary theme prints on which addition and subtraction of appropriate classes from two dates can be displayed by an electrostatic printer. Such data can be generated at a scale suitable to overlay on existing map bases. Alternatively, line printer output can be generated to give a conflict character-assignment map which shows in alphanumeric form what the change, if any, has been for each individual pixel between the two dates being compared.

The fifth and final stage is signature-file extension. If only a small area (e.g., 170 × 170 pixels) has been analyzed, it is obviously important to try and extend it to other parts of the overall area being studied. This can be done by signature-file extension in which the spectral ranges established for the training sites are applied outside the initial area of study. As long as the environment remains similar, errors will not be introduced. Away from the initial study site, some areas were not classified, but the majority of these were not represented in the initial site.

In conclusion, the results noted in Table 32-23 from Wickware and Howarth (1981) suggest that the five stages outlined here form an excellent sequence for analyzing hydrologic and vegetation changes over a large area. The investigators state that the amount of information required will depend on the complexity of the problem and the level and type of resource management decision to be made (Howarth and Wickware, 1981).

The LG2 Reservoir

The LG2 Reservoir is located approximately at latitude 51° 30′ N and longitude 67° 00′ W (Figure 32-36), on LaGrande Riviere, which drains westward from part of northern Quebec and flows into the eastern side of James Bay. In this area of typical "Shield" country with an undulating topography of glaciated igneous and metamorphic rocks, la Société de'énergie de la Baie James selected sites for a series of reservoirs to provide water for the generation of hydro-electric power. The first reservoir in the complex to be filled with water was LG2. Filling commenced on 27 November 1978, and the reservoir reached its maximum depth of approximately 175 m in the fall of 1979. At full capacity, the surface area of the water is approximately 2800 km^2, an extremely large reservoir by any standards.

In 1976, La Direction Environment de la Société de'énergie de la Baie James established an ecological monitoring network to follow the growth of the LG2 Reservoir and the biotic, physical and chemical quality of the water. This network consisted of monitoring stations set up throughout the complex. At the same time, a complementary program to assess the capabilities of Landsat imagery for monitoring changes was established. La Société conducted this study. The aims of the demonstration project were to record the growth of the reservoir, to observe the distribution of suspended sediment, organic matter, floating peat bogs and debris such as trees, and to monitor the ice cover on the reservoir during the winter months.

Of 81 possible Landsat 2 and 3 images, only 11 had no clouds or very little cloud cover; in addition, 10 were only partially useable. Weather observations from Environment Base Camp at Lake Hélene (near LG2 Reservoir) aided the selection of suitable dates for imagery and rush orders were placed to ensure rapid delivery. The data were usually received within 3–4 days. Visual analysis was generally used with the Landsat images in the form of paper prints. For more important dates, analysis involved the use of color composites on a multispectral viewer.

Laframboise and Bachand (1980) showed a sequence of four images covering the area of the

TABLE 32-23

Change Matrix Indicating Percentage of Habitat Overlay Between 1976 and 1973 (from Wickware and Howarth, 1981)

	1973 Habitat class						
1976 Habitat Class	Turbid open water	Less turbid open water	*Scolochloa/ Scolochloa-Carex*	Wet *Carex* fen	*Carex/ Calamagrostis* fen	Shrub fen/fen *Salix-Alnus-Populus*	*Picea-Populus*
Turbid open water	77.9	12.1	—	—	—	—	—
Less turbid open water	8.1	65.1	1.2	3.0	2.5	11.7	—
Scolochloa/Scolochloa-Carex	—	13.9	9.8	3.2	6.6	51.6	—
Wet *Carex* fen	—	1.8	2.7	15.1	33.2	28.0	—
Carex/Calamagrostis fen	—	1.7	2.4	18.3	33.0	24.0	—
Shrub fen	—	2.4	3.5	2.9	9.0	70.4	—
Salix-Alnus-Populus	—	—	—	3.5	14.1	48.9	2.0
Picea-Populus	—	8.6	—	—	—	31.0	51.6

[1] Percentage overlaps are expressed relative to 1976 habitats.

LG2 Reservoir from prior to, and during, the fillings. Figure 32-36 here shows exactly the same portion of the region on Landsat digital data before and after filling, as displayed on the image analysis system of the Canada Centre for Remote Sensing. Of particular note is the use of the geometrically corrected Landsat images, which allow perfect date-to-date registration. With the availability of such data, temporal comparisons are facilitated. Further details and illustrations are available elsewhere (Thompson et al., 1982).

With respect to turbidity, no variations in the water were observed for the summer image on either the visual or digital data. Further analysis of turbidity was thus was not warranted.

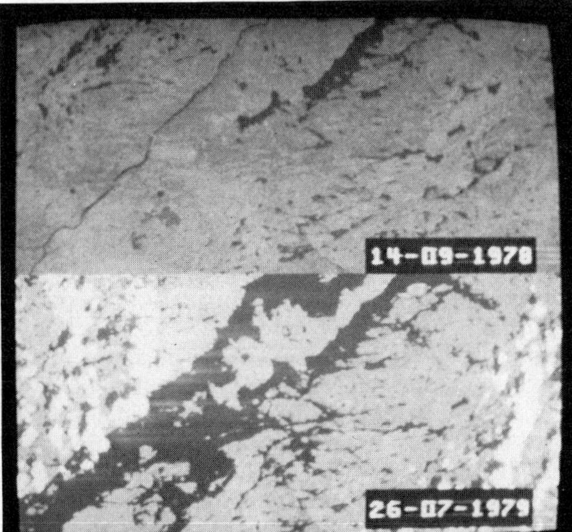

Fig. 32-36. LG-2 Reservoir before and after filling. The split screen shows exactly the same area of about 12 × 25 km on the two dates indicated. (Howarth et al., 1982: Courtesy of the Canada Centre for Remote Sensing.)

Insofar as environmental factors are of concern, several features of the vegetation were of particular interest. First, through the use of visual analysis, it was possible to observe partially inundated vegetation at the borders of the lake, due to the change in radiance produced by the combination of vegetation and water. As the water level rose during the winter, the forest cover was lifted vertically by the ice and stayed in its original position of growth. This made it virtually impossible to determine the exact boundary of the lake at this stage. The winter image was thus used, along with a topographical map and Landsat images from preceding years, to evaluate the area affected by uplift of vegetation, and to determine the type of vegetation and degree of uplift. This was done using visual analysis, but it could have been done in more detail using digital processing techniques. A final feature of the vegetation was floating peat bogs and other debris, particularly tree trunks. As pointed out by Laframboise and Bachand (1980), the identification of patches of floating debris depends on their dimensions, their form and their contrast with their surroundings. For the LG2 Reservoir, the generally small size of floating peat bogs and debris made them impossible to detect, although larger ones had been identified in an experiment carried out on an older reservoir (Howarth et al., 1982).

REMOTE SENSING-AIDED FORECASTS OF WATER YIELD

This section is devoted to a case study involving the application of remotely-sensed data to the preparation of inputs to a water yield forecast model. The obvious engineering application for such a capability is that it will permit dams, aqueducts and other water-related structures to be more intelligently located, designed and constructed.

The development of an area's water resources is an excellent example of the application of human knowledge and experience to the management of an environment. Biswas (1970) traces the history of hydrology to about 3200 B.C. It has been established that the early inhabitants of the American Southwest built well-developed systems for the utilization of water resources (Garstka, 1972).

As the population increases there is an inevitable increase in the value of water and in the price people will be willing to pay to secure water. In the state of California, where agriculture is the primary industry and where, therefore, water is by far the most valuable resource of wildlands, there is, in addition, a massive engineering effort to ensure that humans will continue to enjoy an adequate supply of high quality water for their own domestic needs. The concentrations of population that are found in the Los Angeles Metropolitan area and San Francisco Bay area, for example, are assured of domestic and municipal water because of the existence of very large carry-over storages in various reservoirs that are hundreds of miles away from the points of water demand.

As indicated by many investigators, our knowledge of the environment, especially with respect to the climatology and hydrology of the water yielding areas, is inadequate. Relatively little hydrologic research has been conducted in the water-producing drainage basins in the wildland areas. Because watershed analysis, in particular, plays an important role in water resources management, research of the type described in this section is of great potential importance to engineers and others who are charged with developing and managing water resources.

The fundamental physical data upon which water resource developments are based consist essentially of hydrologic, meteorologic, climatologic, geologic, ecologic, biologic, and pedologic observations. The collection of such data requires a great expenditure of time and effort in the water-yielding drainage basins and in project-development areas.

The development of water-yield forecast models has been furthered recently because advances in computers have made it possible to analyze more complex water-information systems in a much shorter time. River-forecast models are used as a tool in reservoir-operations management. The outputs of these models are used for flood control, optimization of water supply and water demand, generation of hydroelectric power, and regulation of reservoirs for recreational uses. Major rivers in the United States are operated by the U.S. Army Corps of Engineers, the Department of the Interior, various State agencies and by combinations of these. All water resource managers in the United States use the output of water yield forecast models as the basis for their reservoir operations and decision making (Billingsley, et al., 1976).

In most developed countries, the river-forecast models that are used rely heavily on historical records of the stream gage data. Lack of adequate stream-flow data can lead to the making of major errors in water-yield forecasting. The amount of water storage, infiltration, and deep percolation to the ground-water table can vary greatly from one area within a watershed to another. Experience has shown that the accuracy of river-forecast models can be greatly improved if the modeling is done separately for each sub-basin and the results then pooled together. When historical stream gauge data are not available for sub-basins, remote sensing can provide spatial data for each sub-basin and greatly reduce the error in runoff forecasts that are made for the entire basin.

Empirical equations are often used for predicting the peak runoff for flood control purposes in simpler models. More complex models, such as the River Forecast Center Model, the Standford Model, and the Storm Model, require quantification of parameters at many points in the hydrologic cycle. The inputs to such models generally consist of large amounts of data regarding precipitation and also about watershed physiographic characteristics. As a rule of thumb, the more complex the model, the larger the number of coefficients, and the more difficult it becomes to quantify the inputs to the model. Remote sensing data provide hydrologists with the tools to quantify some of these inputs more accurately and on a sub-basin basis.

The following is an example of the use of remotely-sensed data, combined with conventional data, in preparing the inputs to a complex model developed in the early 1970's by Burnash, Ferrali and McGuire and operated jointly by the National Weather Service and the California Department of Water Resources. This model was analyzed by Algazi et al. (1977) for determining the utility of the use of remote sensing data as an input. They pursued the objectives of the model-behavior analysis through simulation and the making of sensitivity analyses that were designed to study the effects on predicted runoff that resulted from varying the dynamic inputs and internal parameters of the model.

In the sensitivity analyses, a study was made of the effect on monthly volume-runoff resulting from variations in the parameters representing precipitation (rainfall and snowmelt), evapotranspiration, lower zone and upper zone tension-water capacity, the percent imperviousness of the watershed, and the percent of the watershed in riparian vegetation, streams and lakes. This study was based on historic data on precipitation and runoff from 1962 to 1969. The most sensitive and critical parameters were found to be precipitation during the entire year and evapotranspiration, principally for the spring regime of the model. From these results, they concluded that precipitation and evapotranspiration were the two most important parameters to acquire by remote sensing techniques.

In the following section, we describe the procedures that were developed by the Remote Sensing Research Program (RSRP) of the University of California at Berkeley (Khorram et al., 1978) for more accurate estimation of the several major inputs to the River Forecast Center model, including snow areal extent, snow water content, and potential evapotranspiration procedures. The procedures were applied to the water source basin for the California Water Project. This area is known as the Feather River Watershed, FRW, (780,000 hectares). The snow water-content procedure described here was applied to the Spanish Creek Watershed (48,000 hectares), a sub-basin of the Feather River Watershed. The methodology for estimating evapotranspiration (ET) is being applied to the Middle Fork of the Feather River Watershed (300,000 hectares). All of these watersheds are shown in Figure 32-37.

Snow Quantification

The use of remotely-sensed data for snow mapping is discussed in detail in the Water Resources chapter of this manual. In the present section, only the work that was applied to the above-mentioned specific case study will be presented.

Snow Areal Extent Estimation Procedure

Imagery obtained from the Landsat MSS has provided the raw data for use in estimating the areal extent of snow as reported in Chapter 29. It also was used in the engineering-oriented study that is described in this section.

The method used at the RSRP to estimate areal extent of snow is based upon the analysis of imagery defined by artificial units (grids) using environmental considerations (Katibah, 1975). This procedure allows the image analyst to make decisions in discrete units of the imagery as to the areal extent of snow, based upon factors effecting the snowpack.

Three cloud-free dates in the spring of 1973, April 4th, May 10th, and May 28th, covering the Feather River Watershed, were used for this snow-cover inventory. Landsat imagery in the form of simulated color infrared enhancements of bands 4, 5, and 7 was utilized for the interpretation procedures (Katibah, 1973). On these three dates random transects were flown across the watershed using a 35 mm camera to acquire large scale photography required as an aid in determining the actual snow conditions on the ground.

To estimate the areal extent of snow, the Landsat images were gridded with image sample units (ISU's), each equalling approximately 400 hectares. These image sample units were then transferred to the large scale photography where applicable. The image sample units on the areal photography were divided into five classes containing from 0 to 100 percent snowcover.

The gridded Landsat color-enhanced images were then interpreted, sample unit-by sample unit, and coded using the following method to account for vegetative cover and density and, to some degree, aspect and elevation. Scale-matched simulated color infrared enhancements of Landsat MSS imagery were produced for April 4, 1973; May 10, 1973; May 28, 1972 and also for August 31, 1972 in reflection print form. The April and May dates represented the snowpack and were gridded, while the August, 1972 data, representing a cloud-free summer image, was not gridded. The purpose of the August imagery was to provide a clear, snow-free, vertical view of actual ground relationships among vegetation/terrain features. The August imagery was superimposed with each of the snow-pack dates, in turn, using a mirror stereoscope. By using this technique the image analyst could observe what conditions actually existed on the ground in the image sample-unit he was interpreting for snowpack.

The large scale photography was used to calibrate the Landsat data where applicable. The sample-unit-by-sample-unit interpretation of the Landsat imagery was then used to find the estimate for the areal extent of snowcover in the watershed. Summation of each of the individual snow cover classes was used to estimate the areal snowcover of each image sample unit on the ground. By addition of these totals the areal extent of snowcover for the entire area was estimated.

This estimation of the areal extent of snow was based only on the Landsat image-interpretation results. To correct this estimate, the image sample-units where snow areal-extent "ground truth" was obtained (from large scale aerial photography) were compared with the same image sample-units on the Landsat imagery. The relationship between the snow areal extent values on the corresponding Landsat and "ground truth" sample units served as the basis for applying the ratio-estimator statistical technique (Cochran, 1963). This technique not only provides a correction for the original interpretation estimate, but also allows for an estimate of the precision of this estimate through the application of confidence intervals. The 95 percent confidence intervals around the areal extent of snow estimates were then calculated. For a description of the detailed procedure, see Khorram et al. (1978).

Snow Water Content Estimation Procedure

The rate at which the snow cover becomes depleted in the spring and summer months provides an index that is inversely related to the snow water-equivalent and snowmelt runoff (Rango and Salomonson, 1975; Khorram, 1977a). The procedure used in this study was designed to generate an estimate of watershed-wide snow water-content and an associated statement of precision. This system employed a stratified double sampling technique based on Cochran (1963) and Raj (1968) and used both ground snow-course and Landsat data. Its objective was to combine snow water-content information for the whole watershed, as obtained inexpensively from Landsat data, with information gained from a much smaller and more expensive sample of ground-based measurements at snow courses. This method is described below:

Black-and-white Landsat MSS imagery for April 4, May 10, and May 28, 1973 covering the Feather River Watershed (7,800 Km^2) was obtained and transformed into a simulated infrared color composite. Bands 4, 5, and 7 were used for this purpose. In the color-combining process, an ISU grid was randomly placed over each image so as to cover the watershed of interest. Each ISU in this study represented an area of about 400 hectares.

Estimates of snow areal-extent by Landsat ISU for previous year(s) and current snow build-up dates were made. Each ISU was interpreted manually as to its average snow areal-extent cover class according to a snow environment-specific technique (Khorram, et al., 1976). Estimates of snow areal-extent by ISU for Landsat snow season date were then calculated.

The snow areal-extent data were transformed to snow water-content data. The snow water-content index was estimated from the following first order, time-specific model:

$$X_i = \sum_{j=1}^{J} (M_{ij})(G_j)\, K_i$$

where

X_i = estimated snow water-content for image sample unit i,

M_{ij} = snow cover midclass-point based on photo interpretation; expresed on a scale of 0.00 to 1.00 for image sample unit i on the jth Landsat snow-season date,

G_j = weight assigned (0.00–1.00) to a past M_{ij} according to the date of a current estimate,

K_i = the number of times out of j that sample unit i has greater than zero percent snow cover, and

J = total number of snow season dates considered.

To insure reasonably high correlation between X_i and corresponding ground acquired snow water-content values, there usually should be at least three snow season dates considered (j = 3). Normally, one or two dates of Landsat imagery would be required during the early snow accumulation season. Occasionally, j may be only two, such as when the first date consists of an April 1 snow water-content map based on the past year's Landsat data. In all cases the sample unit grids on all dates must be in common register with respect to a base date grid-location.

All of the image sample units were stratified into Landsat snow water-content index classes. The number of ground sample units (GSU's), by stratum or snow courses, required to satisfy the allowable error criterion for the basin snow water-content estimate was then calculated. The number of required ground samples may be determined (Thomas and Sharp, 1975) for individual strata according to the snow survey direct cost budget for the watershed of interest and according to the following stratum-specific statistics: relative stratum size, Landsat-derived snow water-content variability, Landsat-to-ground correlation, and Landsat-to-ground sample-unit cost ratio. This study employed six snow water-index strata, with water-content index values ranging from less than 0.1 to over 8. Such stratification was used to control the coefficient of variation of the overall basin snow water-content estimate.

The GSU's were allocated among snow water-content strata with equal probability within strata in accordance with stratified random sampling requirements.

The final product was the watershed-wide estimate of snow water-content according to a summation of strata-wide snow water-content estimates generated from regression equations relating the Landsat snow water-content index data in each stratum to the corresponding sample of ground snow water-content measurements.

Potential Evapotranspiration Estimation Procedure

The following procedure is based on the work of Khorram et al. (1978). Evaporation may be defined as the transfer of water vapor from a non-vegetative surface on the earth into the atmosphere. Evapotranspiration is the combined evaporation from all surfaces and the transpiration of plants. Except for the omission of a negligible amount of water used in various metabolic activities, evapotranspiration is the same as the "consumptive use" of the plants. The fact that the rate of evapotranspiration from a partially wet surface is greatly affected by the nature of the ground leads to the concept of potential evapotranspiration. Penman (1948) defined potential evapotranspiration as "the amount of water transpired in unit time by a short green crop, completely shading the ground, of uniform height and never short in water."

The approach involved a geobased information system that employed both remotely-sensed and ground-acquired information. The remotely-sensed data included Landsat Multispectral Scanner (MSS) data, NOAA-5 satellite Very High Resolution Radiometer (VHRR) data, and aircraft multiband photography. The ground-acquired data included USGS topographic maps, DMA/USGS digital terrain data, ground truth data, and climatic data obtained from ground stations. This databank provided the required data (input data) to mathematical models for estimating potential daily evapotranspiration (output product). The information system used in this study was the RSRP's MAPIT System (discussed earlier in this chapter).

The required input data for potential evapotranspiration (ET) models were composed of the locational, physiographic, and climatic characteristics of the watershed. The physiographic data included elevation, slope, aspect, and hydrologic land use. The climatic data included cloud cover, temperature, and solar radiation.

The methodology used for generating elevation, slope, and aspect data employed USGS digital terrain tapes. The technique for elevation mapping included the study-area boundary determination, control-point selection, elevation interpolation (if necessary), and elevation classification into desired zones. The determination of slope required establishment of the normal vector to a plane representing the elevational gradient between a given pixel and the four pixels surrounding it. Aspect was derived from the azimuthal orientation of the normal vector. This detailed procedure is discussed by Khorram and Smith (1980a).

The hydrologic land-use determination was derived from a computer-assisted classification of multispectral data. Basically, this classification involved two steps: (1) an unsupervised classification of selected sample sites using a clustering algorithm to develop class training-statistics; and (2) a supervised classification of the entire study area (using a maximum-likelihood classification algorithm) based on the training statistics developed by the unsupervised classification. The final product was a land-cover map of the watershed. The detailed procedure is discussed by Khorram et al. (1978).

Cloud-cover data were obtained from two ground meterological stations (Mt. Shasta and Blue Canyon) located north and south of the study area, respectively. The temperature map was based on Very High Resolution Radiometer (VHRR) data obtained from the NOAA-5 satellite. NOAA-5 is an environmental satellite operating in a sun-synchronous near-polar orbit at an altitude of about 1450 km. Its daytime pass over the study area occurs at approximately 9:30 AM and its nighttime pass occurs at approximately 9:30 PM. The averaged data of these passes

$$\frac{(\text{daytime} + \text{nighttime})}{2}$$

were used in this study. The nominal resolution of the data is approximately 0.9 km at the satellite subpoint. The detailed procedure for mapping surface temperature based on NOAA-5 satellite data is discussed by Khorram and Smith (1980b).

The basis for the evapotranspiration model used in this study was that of energy conservation. The model used was a semiempirical one, originally developed by Jensen and Haise (1963). This model is represented by the equation:

$$ET_p = (0.014\ \mathrm{T} - 0.37)Q_n$$

where

ET_p = is potential daily evapotranspiration in inches of water,

T is the mean daily temperature expressed in °F, and

Q_n is net radiation in inches of evaporation equivalent.

The net radiation, Q_n, was calculated from the equation

$$Q_n = (1-\alpha)Q_s + Q_{nL}$$

where α is albedo, Q_s is the total flux of shortwave radiation from the sun and the sky, and Q_{nL} is the net flux of longwave radiation. The albedo values were calculated from the land use/land-cover map of the study area as prepared from Landsat MSS data. The total flux of shortwave radiation was estimated from a modified form of the Angstrom equation (Linacre, 1967). The net longwave radiation was calculated from surface temperature and cloud-cover data. For detailed procedures, see Khorram (1982a).

RESULTS

The results obtained, in terms of snow areal extent, snow water content, and evapotranspiration, will now be discussed separately.

The estimates of the areal extent of snow and their confidence intervals for April 4, May 10, and May 28 are shown in Table 32-24. The appropriate statistical parameters, such as standard deviations and population ratio-estimator values, are also presented in Table 32-24. Additional Chi-square tests indicated that, on all dates, the experimental set-up was adequate at the 2.5 percent significance level.

The results of Landsat-aided snow water-content estimates, based on Spanish Creek Watershed data, are shown in Table 32-25. This set of data is used to represent Feather River Basin values, due to the similarity in snow water-content class distribution between the two basins for the snow-season dates investigated. The correlation coefficient between the average ground-based and Landsat-based values of snow water-content indices are 0.85 and 0.77 for three dates and two dates of analysis, respectively. Since more than two dates probably will be available in most operational snow water-content estimation situations, a conservative value of 0.80 was selected as the correlation coefficient to be used in the sample-size analysis.

A side-by-side comparison of an operational system and this Landsat-aided snow water-content estimation system was facilitated by a blending of statistical and economic theory. This comparison, using 1974 data, was based on the analysis of 2200 image sample-units at 15¢ each and 26 ground sample units at $150 each. The analysis indicated a decided advantage for the Landsat-aided method (Sharp and Thomas, 1975).

The areal distribution of certain input parameters to the ET model for the study area are shown in Color Figures 32-38a, b, and c,viz. those pertaining to temperature, net radiation, and potential

TABLE 32-24

Summary of Results from Areal Extent of Snow Estimation (in Hectares) Along with the Confidence Intervals

	April 4, 1973	May 10, 1973	May 28, 1973
Landsat-1 estimate of the areal extent of snow	511,378	205,768	60,516
Estimate of the true areal extent of snow	501,355	195,644	57,847
Standard deviation of the areal extent of snow estimate	12,776	14,526	17,126
Population ratio estimator	.9804	.9509	.9559
Total number of hectares inventoried	879,642	813,014	798,340
Total number of image sample units inventoried	2,218	2,050	2,013
Total number of image sample units sampled	80	52	49
Confidence intervals (95%)	$485,940 \leq Y_R \leq 536,816$	$176,601 \leq Y_R \leq 234,935$	$26,075 \leq Y_R \leq 94,958$

evapotranspiration, respectively. With respect to the areal distribution of potential evapotranspiration value for May 16, 1975, (Color Figure 32-38c), the results ranged from 0.05 to 0.41 inches of water. These results were in agreement with the expected values in this watershed. Based on the results of this case study it was concluded that:

1. Landsat MSS imagery was adequate for snow areal-extent estimation.
2. The use of Landsat MSS imagery was a cost-effective method in estimating snow water-content.
3. The combination of Landsat MSS data and aerial photography was useful in land use/land cover and subsequent albedo mapping for hydrologic modelling and watershed analysis.
4. ET is a major input component to most hydrologic models and yet, to the best of our knowledge, this is the first remote sensing-aided attempt to prepare watershed-wide ET estimates.
5. NOAA-5 satellite data can provide a good alternative for the mapping of surface temperature and radiation components required in hydrologic modeling.
6. DMA/USGS digital terrain data can provide an adequate database for topographic analysis.
7. Landsat and aircraft data can be very useful in hydrologic land-use classification and consequently in albedo mapping.
8. Because remote sensing can be very useful for both snow quantification and ET estimation it is of great potential value in water yield forecasting; such a capability is, therefore, of importance to those engineers

TABLE 32-25

Landsat-Based Snow Water Content Statistics Based on Spanish Creek Watershed Data for April 4, May 10, and May 28, 1973

Stratum Index	Landsat-1 Snow Water Content Estimate Range	Ave. Snow Water Content Index Per Image Sample Unit	Standard Deviation for X_{hi}	Coefficient of Variation	Total Snow Water Content Index	Stratum Weight Based on Snow Water Content	Number of Image Sample Units for the Spanish Creek Watershed	Number of Image Sample Units for the Feather River Watershed
1	0.00–0.10	0.0000	0.0000	0.00	0.00	0.0000	32	503
2	0.10–0.35	0.1833	0.1194	65.14	7.15	0.0304	39	614
3	0.35–1.00	0.7808	0.1883	24.12	10.15	0.0432	13	205
4	1.00–3.00	2.0480	0.4404	21.50	51.20	0.2178	25	393
5	3.00–5.00	3.9557	0.4525	11.44	55.38	0.2356	14	220
6	5.00	6.1750	0.9672	15.66	111.15	0.4729	18	283
							N = 141	N = 2,218

who are responsible for designing, constructing and operating such structures as dams and aqueducts.

METHODS OF WATER QUALITY MAPPING AND MONITORING

In this next section there are several remote sensing methods described for analyses, monitoring and classifying water quality in streams and lakes, but additional details on water quality methods are in other chapters, such as those dealing with The Marine Environment and with Water Resources. Here, the methods include analysis of river turbidity, analysis of lake turbidity, monitoring and classifying lake quality, analysis of water quality in an estuarine environment, water quality mapping, oil-spill detection and monitoring, analysis of sediment concentrations, and classification of water and land cover.

Analysis of River Turbidity Plumes using Aerial Photographs

In July 1977, the U.S. Army Corps of Engineers, in conjunction with the U.S. Environmental Protection Agency (EPA), State agencies, and university personnel from Minnesota and Wisconsin, made an extensive study of turbid plumes on the Mississippi River south of St. Paul, Minnesota. Part of the plumes were caused by dredging operations in the river while others were caused by normal river-barge traffic.

Air photos, taken simultaneously with water samples collected by a crew in a boat, were analyzed on a color microdensitometer. Water samples, once they had been subjected to laboratory measurements of turbidity, were used to calibrate each photo. With the aid of these calibrations, turbidity maps were made of the plumes.

The location of the turbidity study site was about 7 miles SE of St. Paul, Minnesota near Lower Grey Cloud Island. There the bottom muds were dredged up in the river channel (at the dredge site), loaded onto barges, and deposited upstream on an island called the disposal site. Figure 32-39 locates the dredge and disposal sites and also the sites where surveying measurements were made in order to provide photographic control, as needed for making the turbidity maps.

A trial run was made on July 21 to ascertain the precise direction and speed of the river current and to optimize the aerial photography of the river water. On July 25 and 26, water aerial photography of the river water. On July 25 and 26, water and aerial photos were taken at the disposal and dredge site, respectively.

Settling Tests and Light Penetration Analyses

The samples collected from near the barge plume were shaken up and placed in a column having a depth of two feet. Settling tests were then run on each sample by checking the turbidity of

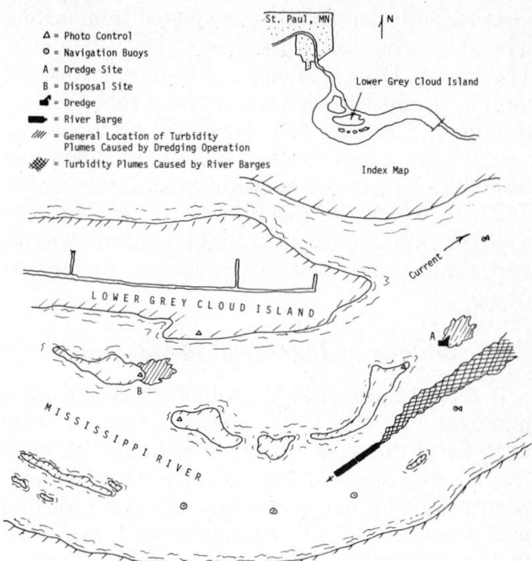

Fig. 32-39. Map showing dredge and disposal sites near Grey Cloud Island.

the water at mid-point of the column at different times. The initial turbidity of the sample was 40 FTU (approximately equal to JTUs). After 30 minutes, a turbidity of 12 was recorded. A residual turbidity of 7 JTUs existed after 2 hours and about 6 JTUs after 4 hours. With barges passing on an average of about every 2 to 4 hours, one would expect background turbidity values in the river to be in the range of about 6 to 7, as indeed they were recorded (see Figure 32-40).

Sunlight penetrates the water (and returns to the camera) only to a certain depth. In any given

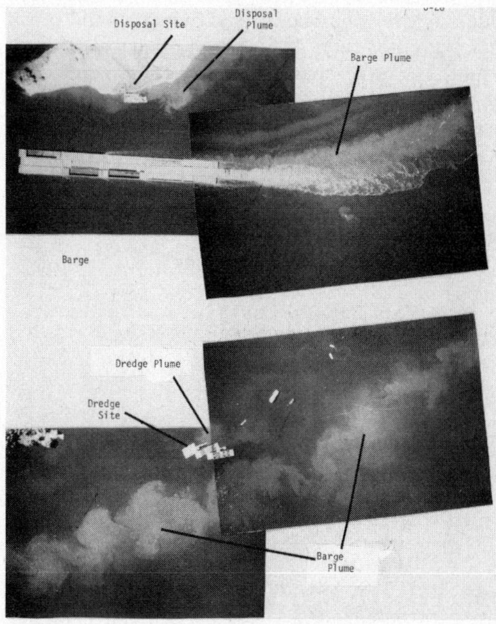

Fig. 32-40. Turbidity plumes near dredge and disposal sites in vicinity of Grey Cloud Island.

instance, this depth can be estimated from turbidity values obtained from Secchi Disc Readings. The Secchi Disc is a white disc that is lowered until it disappears. In each instance this "extinction depth" is recorded. Figure 32-41 shows the relationship between Secchi Disc readings and turbidity. This curve is useful in ascertaining the depth to which the penetration of light into the river for various surface turbidities is sufficient for underwater features to be recorded on the air-photo.

Analysis of the Aerial Photography

In this experiment, each roll of aerial film had a photographic step wedge exposed onto it. From analysis of the image of the step wedge (through the use of a spectral microdensitometer) a curve portraying film density versus relative exposure was produced. It is to be emphasized, however, that such analyses cannot be meaningfully undertaken except with carefully calibrated aerial photographs. An alternate procedure would be one of relying on digital multispectral scanning. Also in this experiment, density readings were taken in a grid fashion throughout the image of the water. Density readings were also taken on the images of the white panels appearing on the photos. These density readings were then corrected for lens falloff and for possible sun angle effects on the panel whenever conditions warranted. In judging when the conditions do, indeed, warrant the making of such corrections, one must consider the fact that skylight surface-reflectance also increases significantly for angles of greater than 40° from the vertical. However, a normal or long focal length camera (as was used in this case) does not exceed the 40° from the vertical. If a wide-angle lens had been used (which would sense angles greater than 40° from the vertical) then it would also have been necessary to consider corrections for skylight radiance from the water surface.

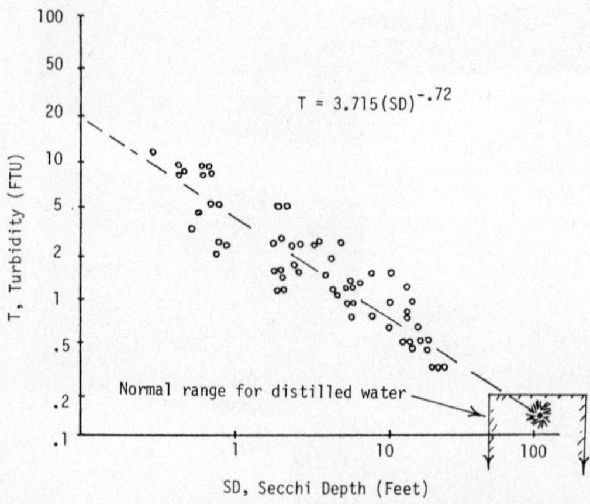

Fig. 32-41. Relationships between Turbidity (T) and Secchi Disc readings (SD) for Wisconsin Lakes sampled during 1974/1975. (Scherz, et al., 1977.)

The planimetric base map that was needed in the step described above was made through the use of conventional photogrammetric techniques. An aerial image showing three control points was projected onto a grid where these control points were plotted to the desired scale. Then a rectified enlargement was made showing correct planimetric locations on the shoreline boats and other features. A tracing of this enlargement constituted the base map.

The X and Y positions of the control points, panels, shorelines and boats on the photo were also recorded during microdensitometer analysis. These locations, as well as the corrected densities, were then plotted on a transparent grid in their correct locations. This density grid was then projected onto a standard base map and the data manually transferred to this map. This density grid was then ready for tracining of turbidity "contours". First it was necessary, however, to ascertain which film densities corresponded to the turbidity contours desired.

The spectral responses, as recorded for the selected points on calibrated color and color infrared photos, were analyzed with a color microdensitometer. This analysis revealed that the algae bloom on the river nearby could be easily differentiated from the stirred-up lake bottom muds. This analysis also showed that, in the visible red part of the spectrum, there was good differentiation between very turbid and slightly turbid water and that the values obtained there were significantly above any noise levels existing in the remote sensing system.

Obtaining Film Density Values Corresponding to Desired Turbidity Values

To obtain the turbidity "contour" lines for any given scene from the density grid, several curves were used.

First the density of the reflectance panel for that scene was superimposed on the D-Log E curve for that film. This gave a working value for exposure E_p corresponding to the radiance from the panel (P″) which exposed that part of the film. A value of 0.5 E_p gave a corresponding exposure of A:″ = 0.5, etc. Such an analysis of various points resulted in a workable Density-AP″ curve for that scene, as indicated in Figure 32-42a.

The next step was to determine the location of the AP″-Turbidity curve for that scene. This was done by comparing the water turbidity-values with the AP″ values of the water from the site where the sample was collected, and AP″ values having been determined previously from analysis of the film densities. Figure 32-42b shows an actual AP″-Turbidity curve for one of the scenes. The upper limit of the curve (a) obviously could not be greater than 100 percent. The minimum level of the curve (b) was the minimum theoretical value of AP″ expected on a particular day. For a calm clear day this minimum is about 0.5 percent.

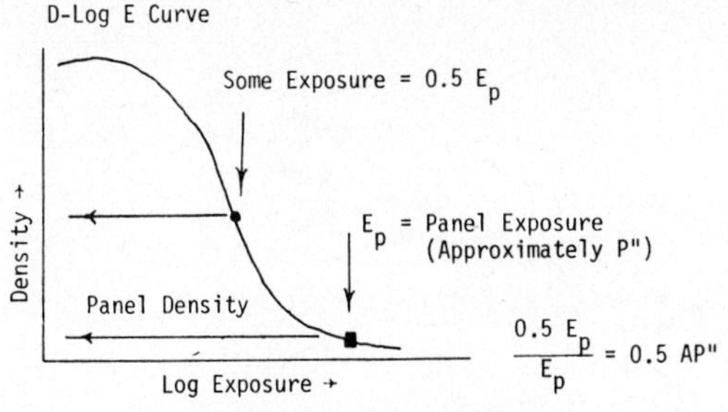

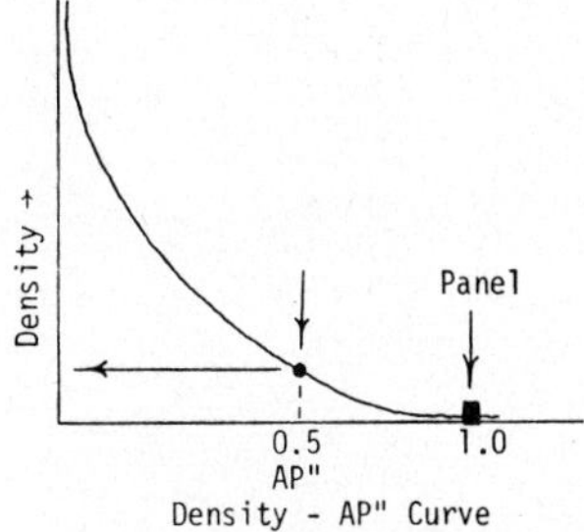

Fig. 32-42a. Obtaining Density-AP″ Curve from analysis of D-LogE Curve and Panel Density.

For a calm, completely overcast day the minimum theoretical value is about 5.0 percent.

Once the correlation curve had been positioned for that scene, the desired turbidities to be contoured (such as 10, 15, and 20 JTUs) were related to AP″ values through the AP″-Turbidity curve.

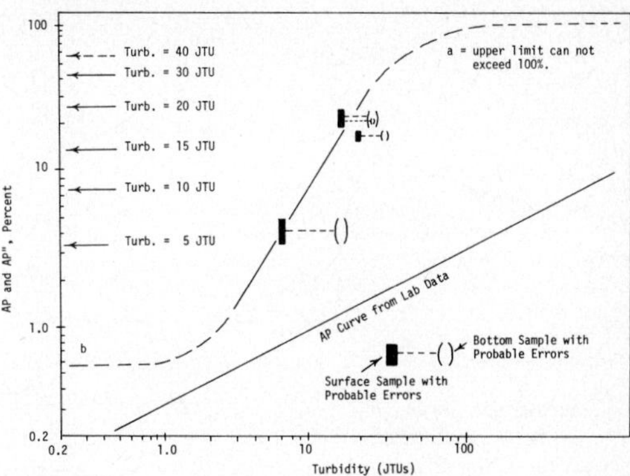

Fig. 32-42b. AP″-turbidity curve for photo of dredge site at 3:54 pm, 26 July. b = minimum value of skylight reflectance.

The resulting values of AP″ were, in turn, related to film densities through the Density-AP″ curve, as illustrated in Figure 32-42A. Once the density values had been determined for the desired turbidity values, the density grid was contoured.

Figure 32-43 shows a turbidity map of the dredge site and an approximate turbidity map of a plume caused by an approaching barge. Also, Figure 32-44 shows the turbidity map of the dredge site as of 0.5 hour after the barge had passed.

Turbidity Caused by Barges

From observations of Figures 32-43 and 32-44 it is obvious that the barge traffic caused high values

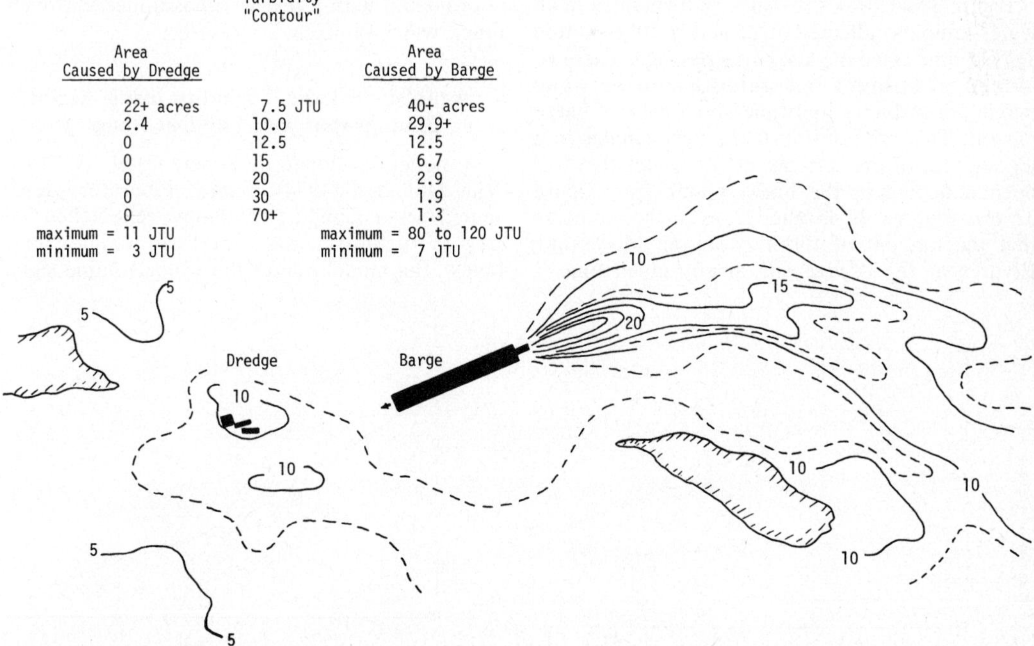

Area Caused by Dredge	Turbidity "Contour"	Area Caused by Barge
22+ acres	7.5 JTU	40+ acres
2.4	10.0	29.9+
0	12.5	12.5
0	15	6.7
0	20	2.9
0	30	1.9
0	70+	1.3
maximum = 11 JTU		maximum = 80 to 120 JTU
minimum = 3 JTU		minimum = 7 JTU

Fig. 32-43. Approximate surface turbidity map of barge approaching the dredge site at 3:15 pm, 26 July 1977. No simultaneous water samples on barge plume. (Data for barge plume obtained by "bridging" from photo with samples.)

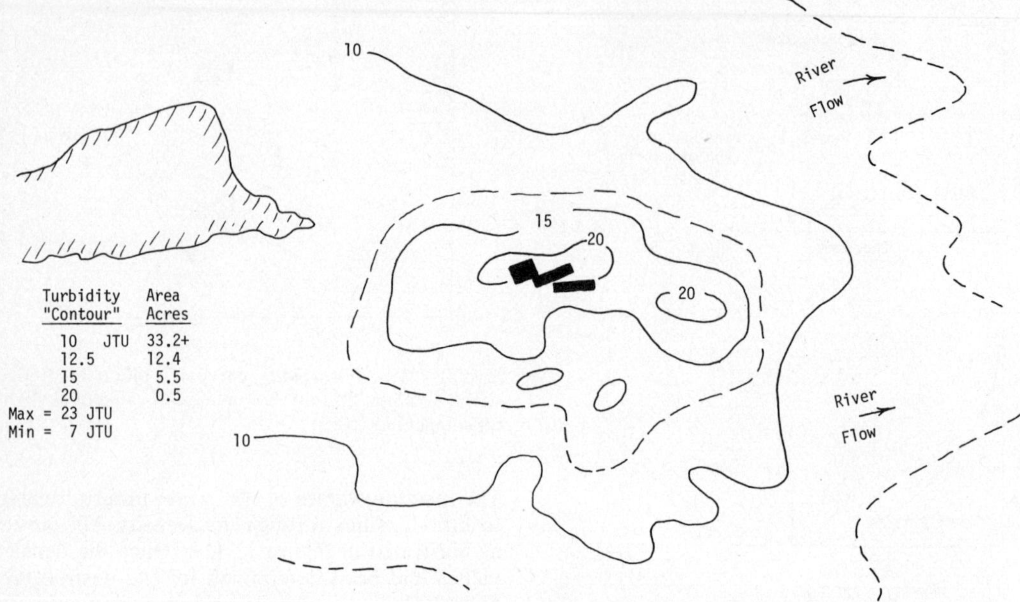

Fig. 32-44. Surface turbidity map for dredge site at 3:54 pm, 26 July (0.5 hrs. after barge passage.)

of turbidity over large areas of the river. The turbidity that had been stirred up by the barges lingered in the area until the bottom materials either settled or drifted away. The current was very low at both the dredge and disposal sites (less than 30 ft/min or 0.35 miles/hour), so settling was the prime factor for clearing of the water after a barge had passed.

Figure 32-45 shows the surface turbidities of all water samples collected on 25 and 26 July, plotted against time since the last barge passage. There is, indeed, a straight line relationship between maximum turbidity and time since the last barge passed. This relationship looks very similar to a log-log plot of the settling test for water that had been collected in the barge plume (see Figure 32-46). Figures 32-45 and 32-46 further indicate that the amount of turbidity in the Mississippi River near the dredge site at any given time is primarily determined by the length of time that has elapsed since a barge passed.

The above observations demonstrate that the effect of barge traffic on both the magnitude and areal extent of turbidity in the river is many times greater than the effects of either the dredging or disposal operations. These conclusions were reached by careful analysis of aerial photo data coordinated with a few well-positioned and well-timed water samples.

Analysis of Lake Turbidity using Aerial Photographs and Landsat Imagery

A model developed by Scherz (et al. 1977) and Van Domelen (1974), relates laboratory-determined water albedo and volume reflectance values to turbidity. Also, at boat and airborne levels, the model relates the water-volume signal

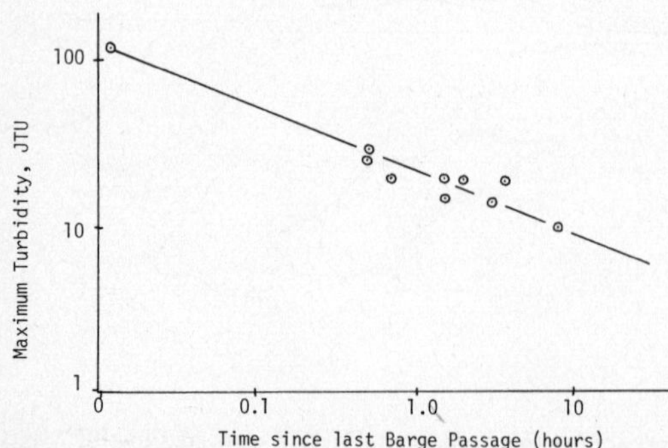

Fig. 32-45. Surface turbidities in river versus time since last barge passage.

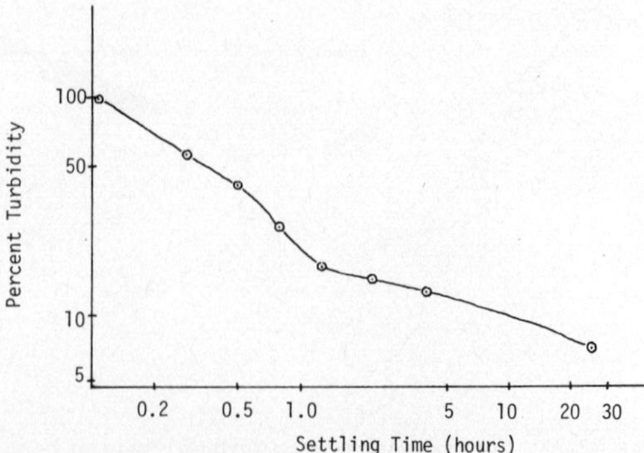

Fig. 32-46. Settling test results plotted on log-log scale (Water sample collected in barge plume).

to that from the water surface and from atmospheric backscatter.

Aerial images of water bodies readily show the presence of turbidity that is caused by algae, silt and pollution materials suspended in the water. In principle, the brightness of the energy reflected from such materials in the water can be related to turbidity. The relationship is complicated, however, by the variable effects that are due to different light conditions from scene to scene and by the effects of atmospheric backscatter and the reflection of skylight and sunlight from the water surface. These variables, must, somehow, be cancelled or subtracted out from the total energy received by the airborne camera or sensor. Some of these variables can be cancelled out by ratioing radiance (light energy per unit solid angle) from the water to that from a standard reflectance panel. This ratio is called Apparent Reflectance and is designated as AP. Varying light conditions can also be cancelled out by the ratio of albedo under one light condition to albedo under another, perhaps standard, light condition.

In the field, at boat level, a water-surface signal is added to the water volume signal. The surface signal is caused by skylight and sunlight reflecting from the water surface. With an airborne sensor the upwelling signals are attenuated by the atmosphere and an atmospheric backscatter signal is also added. All of these modifications create "noise" added to the desired signal from the water volume, which alone relates to water quality.

On aerial photos of water bodies, if a standard reflectance panel appears somewhere in the scene, the turbidity of the water volume can still be mapped with a fair degree of accuracy if at least one turbidity reading is taken in the water at about the same time as the aerial photos are taken. On satellite imagery, standard reflectance panels are not normally visible. The turbidity of lakes and their spectral reflectance signatures (similar to those obtained in the lab) can still be determined, however, provided that a clear-water lake is used as a reflectance standard on the satellite imagery in a manner parallel to the way a reflectance panel is used on an aerial photo.

The turbidity of water, as measured in the laboratory, can be related to energy backscattered from a laboratory sample tube. Figure 32-47 shows the relationship between turbidity and the amount of red energy backscattered from a 2-ft long water-sample tube which has a black bottom. One ratio is apparent reflectance (AP), which is calculated by comparing the radiance from the water to that of a white barium sulfate standard reflectance panel. The reflectance of the panel dealt with in Figure 32-47 was 0.39. The other ratio shown in Figure 32-47 is the volume albedo (A).

Figure 32-48 contains some curves which give total surface albedo as a function of sun elevation and sky conditions. The solid curves are from real observations made of a large clear lake in Russia under normal conditions with surface waves of various sizes (large waves modify the values somewhat). For the extremely unusual conditions of a glassy calm surface on a large lake, the surface albedos would be lower than for normal con-

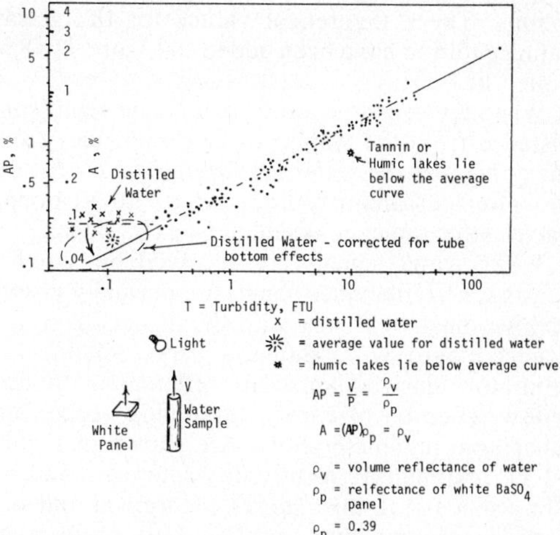

Fig. 32-47. Laboratory apparent reflectance, AP, and laboratory albedo, A, plotted against turbidity for 127 different lake samples collected over three years. This curve is for red light (0.65 micrometers). The average value of 50 laboratory distilled water samples is also shown (modified from Scherz, et al., 1977, p. 68). For very clear distilled water the bottom effects of the short laboratory sample tube begin to override the signal from the water volume and must be corrected for as shown.

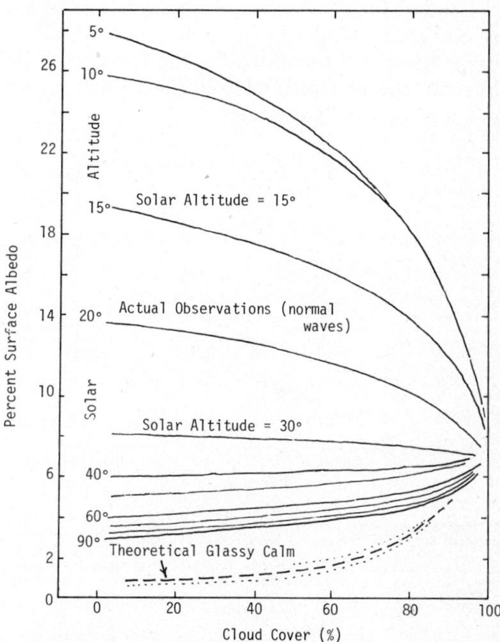

Fig. 32-48. Total surface albedo as measured from a water body as a function of cloud cover and solar altitude as reported by Ter-Markaryants (from Van Domelen, 1974). Theoretical values for a "glassy" calm day are shown in dotted lines.

ditions. These theoretical values for the glassy calm condition have been added to Figure 32-48 in dotted lines.

When the sensor is airborne at some significant distance from the water, surface radiances from the panel and the water are attenuated by the atmospheric transmittance. Also, an airborne backscatter radiance is added to each signal.

When water signals are received by satellite sensors, no reflectance panels are normally visible and we must work just with the signals from the water. Figure 32-47 shows a curve relating the laboratory-derived apparent reflectance to turbidity. Theoretical curves (assuming certain atmospheric parameters) also are shown in Figure 32-49, relating AP″ to turbidity (curves a and b). The same figure also shows theoretical and observed satellite data modified to precent of maximum signal (curve c). Although the satellite data in Figure 32-49 match the shape of the theoretical curve, c, the actual AP″ curves, made from analyzing aerial photos, deviate from the theoretical curves. This deviation, which occurs when white field panels are used, results from our presuming that the photo optimally exposed for both "a" and "b", which pertain to theoretical AP″ curves for a calm, clear day and a calm, overcast day, respectively, with low aircraft altitude. Curve "c" is a theoretical curve for satellite data, expressed in percent of maximum satellite signal, which is 2.00 mw/cm^{2}sr^{-1}.

Estimating Turbidity from Satellite Imagery

Landsat images of clear-water lakes can be used as radiance standards for obtaining turbidity of other lakes and, theoretically, the Landsat images of such lakes can also be used for estimating various atmospheric parameters.

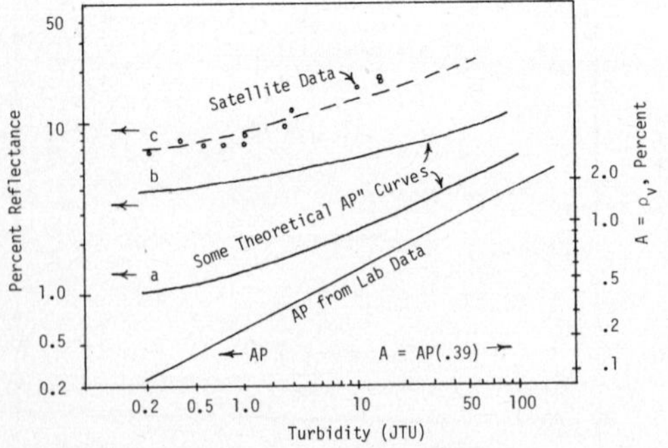

"a" and "b" are theoretical AP" curves for a calm clear day and a calm overcast day, respectively with low aircraft altitude.

"c" is a theoretical curve for W" (for satellite data) where W" is expressed in percent of maximum satellite signal which is 2.00 mw/cm^{2}sr^{-1}.

The theoretical curves were computed from:

$AP'' = \rho_v C_5 + C_6$ and $W'' = \rho_v C_2 + C_3$ where C_2, C_3, C_5 and C_6 are assumed to be atmospheric constants for the aerial image being investigated.

Fig. 32-49. Theoretical curves relating turbidity to airborne reflectance signals. Red energy (0.65 micrometers). The dots associated with curve "c" pertain to actual satellite data acquired of Lake Superior on 30 September 1973.

The satellite radiance from a clear-water lake can be used to determine the position of the satellite correlation-curve for that day, as shown by Figure 32-50, (which can be approximated by straight lines). Given the satellite-based radiance of any other lake one can, by the use of this correlation curve, determine the turbidity of the lake in question.

Figure 32-51 shows a turbidity map of Lake Superior made from an analysis of brightness values of various parts of the lake as imaged by the Landsat MSS. For a particular kind of material in water, there will be various other water-quality parameters which may correlate to turbidity. (For example, for steel mill waste in water, the amount of Fe may correlate to turbidity; similarly, for algae in water, chlorophyll may correlate to turbidity, etc.). Usually, however, such correlations to turbidity will be different for different materials in the water.

Analyzing Aerial Film

The straight-line correlation curves shown in Fig. 32-52 closely approximate the theoretical curves obtained from Landsat MSS imagery because of the linear response of the satellite sensors. For aerial films, however, the data do not usually closely match the theoretical curves. This is because of the non-linear responses of aerial film and the fact that a white panel often has reflectance values that fall within the saturated portion of that curve.

If one or more turbidity samples (or Secchi Disc readings) are obtained at about the time when the aerial photos are taken, the position of the correlation curve (relating turbidity to AP″ for that photo) can be determined (see Figure 32-52). With the AP″ turbidity curve determined, turbidity values can be mapped from the photo. This analysis, of course, assumes that no signals come from the bottom of the water body to cause noise to the airborne signal. If the depth to bottom is greater than the depths at which the Secchi Disc readings are made, bottom signals are no problem.

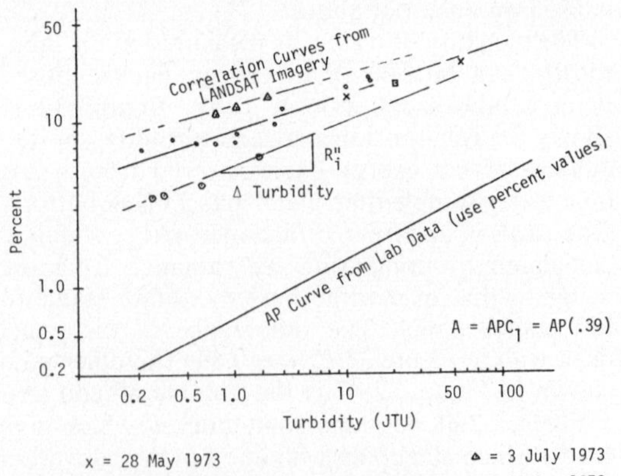

Fig. 32-50. Comparison of correlation curves from Landsat imagery with AP curve, as obtained from laboratory data.

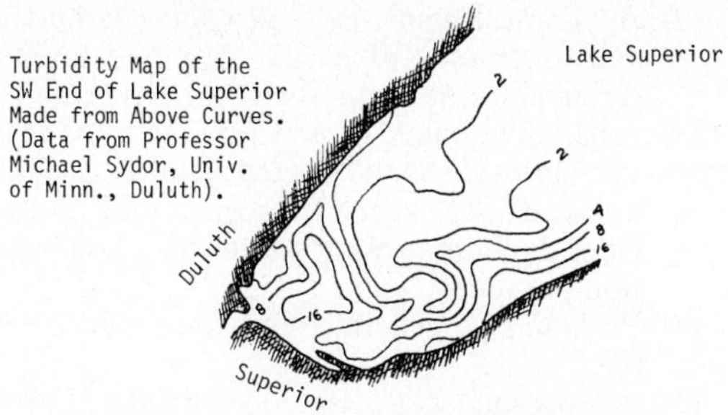

Turbidity Map of the
SW End of Lake Superior
Made from Above Curves.
(Data from Professor
Michael Sydor, Univ.
of Minn., Duluth).
Lake Superior
Duluth
Superior

Fig. 32-51. Laboratory apparent reflectance (AP) and W″ reflectance data from Landsat imagery for water in Lake Superior near Duluth. The satellite correlation curves are W″, expressed in percent of maximum signal, which is 2.00 mw/cm²sr⁻¹.

Monitoring and Classifying Lake Quality using Landsat MSS Data

Federal legislation (PL92-500 of the Federal Water Pollution Control Act Amendments of 1972, Section 314) and state legislation (Wisconsin's Lake Protection and Rehabilitation Law, Chapter 301 of Assembly Bill 766, 1973) require the Wisconsin Department of Natural Resources (DNR) to monitor and/or classify all significant lakes in the state. This requirement has resulted in a cooperative program being developed between the DNR and the University of Wisconsin-Madison (UW-MSN) and an assessment being made of the trophic status of all of the significant inland lakes in the state of Wisconsin, using Landsat data. The analysis technique employs a semi-automatic data acquisition and data handling system which, in conjunction with an analytical categorization scheme, is used to classify each inland lake into one of seven categories of eutrophication, as well as into four problem types.

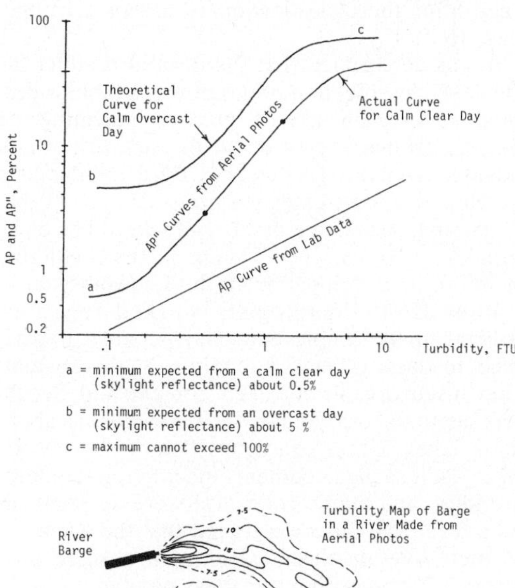

Fig. 32-52. Correlation curves relating aerial data to turbidity.

The classification procedure involves three steps: (1) "navigation" of the Landsat tapes, (2) extraction of the relevant Landsat data, and (3) categorization of the Landsat data. The navigation procedure is the only labor-intensive portion of this activity.

The categorization algorithm involves using data from three different dates of Landsat data which have been normalized for atmospheric effects. The output from the categorization technique is listed by county, lake's name, type of water quality problem, and trophic class. The state of Wisconsin first completed a state-wide lake assessment for the year 1976 and then repeated the process using Landsat data from 1979/1980. Previously, in the spring of 1974, a joint project funded by the DNR was begun between the University of Wisconsin-Madison and the DNR Bureau of Water Quality to investigate the feasibility of using Landsat imagery for lake classification. The water quality data collected included information on such parameters as Secchi disk depth, color, algae concentration, and chlorophyll content.

Landsat Imagery Analysis

The designation of lake quality, as derived from Landsat MSS data, was the goal for the project. More specifically, the objective was to obtain by such means an overall designation related to the trophic state of the lake. Because determining the trophic state of a lake is more complex than just determining turbidity, other studies were initiated in the summer of 1975. These included a time-series analysis of satellite-data variation during the open water season. The time-series analysis was performed to evaluate the variability of lake reflectance as the growing season for algae progressed from the early Spring through the Summer to the Fall. Twenty southeastern Wisconsin lakes were identified on four Landsat scenes. These lakes ranged from oligothrophic to fairly eutrophic. Band 5 brightness values from each of the lakes for each of the dates were found. As might be expected, the scene brightness values of the lakes in Band 5 increased in almost all cases as the algal turbidity levels increased during the summer months and then decreased in the fall.

Through time-series investigations made during the growing season, both the ground calibration and satellite data indicated that no one-time remote sensing sampling technique would be adequate to monitor something as dynamic as a water body. The primary source of turbidity in these lakes was, indeed, algae and the level of the algae population was found to be highly indicative of the trophic state for these lakes. Dissimilarities existed in the turbidity-versus-time plots, however, both from one lake to another and from one year to another for the same lake. These dissimilarities showed the necessity for monitoring the lakes at least two or three times during the open water season.

An abbreviated atmospheric model was developed to characterize how the atmosphere affected the sensed energy and a scheme was devised to correct for its effects. The signal recorded by Landsat is a function of a number of parameters. The atmospheric correction is discussed in detail elsewhere in this manual. Two different approaches were employed for atmospheric correction. One approach has been to use very clear water within the scene as the closest approximation to this black surface. The clear lake data are subtracted band-for-band from every other lake in the scene, resulting in the subtraction of approximately the atmospheric reflectance and the reflectance of the clear water.

The second approach entails finding the atmospheric scattering component involved using the histogram of the entire scene for each band. Since the scattering term is added to all signals in the scene, in our studies of these lakes the minimum signal value found in the histogram was assumed to be the value of the atmospheric scattering. Substituting the minimum point in the histogram for the clear lake value in the above equation gave the corrected Landsat signal for a lake in this scheme. This latter technique was found to be more reliable and easier to implement on an operational basis.

A number of other algorithms were investigated to help correct for atmospheric effects. The two most successful entailed the use of a normalized ratio of bands and a chromaticity ratio, respectively (Alfoldi, et al., 1978). Neither of these was completely successful; hence the normalization techniques described above were employed in the state-wide assessment.

Semi-Automatic Technique

The requirement that lake classification be derived from the analysis of up to three dates of Landsat imagery mandated the development of an automated data-extraction procedure. The scale of the project required enough computation to warrant batch-mode data processing with as little human intervention as possible. Since the procedure was operational and was to be used by people whose background was not in computer operation, programs were needed that were easy and simple to use. The Madison Academic Computing Center's Univac 1100/82 was available to both the University and the DNR, so programs were designed for use of this existing facility.

An integrated system of programs and files was developed to meet the goals of the project. The major components included:

(1) A master "lakes file", ACCESS, which stores geographic information and other data for all the lakes of interest, with a "linked list" structure which allows its use in a variety of modes.
(2) A "control-point file", providing latitudes and longitudes of easily identified points throughout the state, for scene navigation.
(3) A data file, linked to ACCESS, to provide a storage place for extracted data. It is built up gradually, scene-by-scene, until sufficient data are present to run the classification program.
(4) Programs to generate, test, and edit these files.
(5) A program, CONTROL, to estimate control point locations and produce microfische "pictures" to allow manual determination of Landsat coordinates for scene navigation.
(6) A navigation program, SATNAV, to produce coefficients to convert Landsat coordinates to latitude-longitude, and vice versa.
(7) A data-extraction program, EXTRACT, to locate and file data for all lakes covered by a single Landsat tape file.
(8) A classification program which uses the Landsat data filed from EXTRACT for producing a trophic designation for each lake.

Categorization Scheme

An important part of any lake-classification program is that of determining the type of product desired. Early in the investigation it was determined that thematic representations of the results of the classification were not desired. If a pixel-by-pixel classification of the water quality is made, an individual lake is not likely to have a uniform class assigned. If this type of scheme were used for state-wide lake classification, each of the lakes would have to be separately categorized after the classification to assign a trophic class to it.

It was determined that the desired product for this lake-classification program was a tabular designation for each lake of interest, indicating the lake-water characteristic and the severity of any existing problem. To this end, DNR limnologists developed a lake-categorization scheme (Table 32-26) and classified a number of lakes. The class numbers were assigned for use in the predictive model to be developed from the Landsat data.

Four DNR limnologists worked together, utilizing their sample data and personal experience, to classify 39 lakes according to this system. They used organic nitrogen content and Secchi disk depth to help validate their conclusions about these lakes. Other characteristics of the various lakes such as algae content, macrophyte content, turbidity, and humic color also were determined. As a result the limnologists felt that the accuracy of their overall classification, on a scale of 1 through 7, was within one class number.

Experimental Results

Landsat data from 1975 and 1976 were extracted from data tapes and corrected for atmo-

TABLE 32-26

Representative Lake Classification: Waukesha County, Wisconsin

Lake No.	Lake Name	Class	Type	Dates
1	Ashippun	4	Algae	3
2	Beaver	3	Algae	3
3	Big Muskego	7	Algae	3
4	Cornell	6	Algae	3
5	Crooked	5	Macrophyte	3
6	Denoon	5	Algae	3
7	Dutchman	4	Algae	3
8	Eagle Spring	5	Algae	3
9	Forest	4	Macrophyte	3
10	Fowler	4	Algae	3
11	Golden	5	Algae	3
12	Hunters	5	Algae	3
13	Keesus	4	Algae	3
14	Lac LaBelle	4	Algae	3
15	Little Muskego	5	Algae	3
16	Lower Genesee	4	Algae	3
17	Lower Nashotah	3	Algae	3
18	Lower Nemahbin	4	Algae	3
19	Lower Phantom	5	Algae	3
20	Merton Millpond	5	Algae	3
21	Mid Genesee	3	Algae	3
22	Montery Millpond	4	Algae	3
23	Moose	4	Algae	3
24	Nagawicka	4	Algae	3
25	North	3	Algae	3
26	Oconomowoc	2	Algae	3
28	Ottawa	6	Algae	3
29	Pewaukee	4	Algae	3
30	Pine	4	Algae	3
31	Pretty	3	Algae	3
32	Rainbow Springs	5	Algae	3
33	School Section	6	Algae	3
34	Silver	3	Algae	3
35	Spring	7	Algae	3
36	UN S23, 14T8R17	5	Algae	3
37	Upper Genesee	5	Algae	3
38	Upper Nashotah	3	Algae	3
39	Upper Nemahbin	3	Algae	3
40	Upper Oconomowoc	6	Algae	3
41	Upper Phantom	5	Algae	3
42	Waterville Millpond	5	Algae	3
43	Wood	5	Algae	3

spheric effects. A nine-parameter regression was used to correlate the Landsat data with the DNR categorization. Regression parameters were: (1) the average of the atmospherically corrected data over the three days for each band (these parameters comprise the average spectral signature for the lake); (2) the average variance of the corrected signal values within the lake over the three days for each band; and (3) the variance in the spectral signature from the average signature over time.

These nine values were calculated for each lake from the Landsat data and regressed with the trophic class numbers supplied by the DNR limnologists. A number of different regressions (Scarpace, et al., 1979) were investigated. The use of Landsat data from 1975 and 1976 and the data for other lakes provided very good correlations between the Landsat data and the trophic classes. Figure 32-53 is typical of the results found in the study.

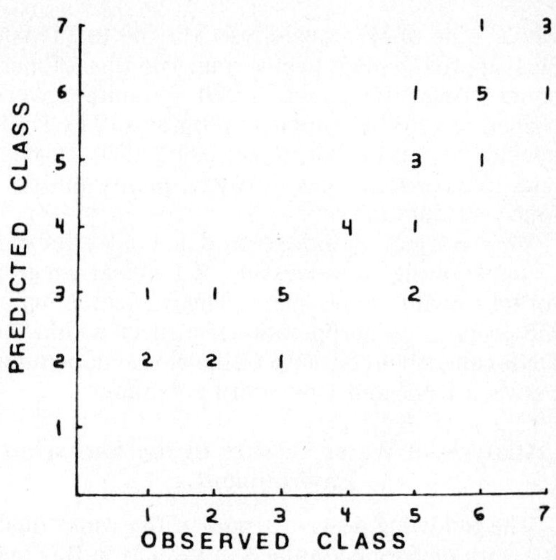

Fig. 32-53. The relationship between observed trophic class and predicted trophic class for the regression using 34 lakes from the 1976 Landsat imagery.

Lake Classification Technique

Both "parallelepiped" and "elliptical table look-up" classifiers were investigated. The parallelepiped classifier met with only limited success; hence the present configuration of the lake-classification programs uses the elliptical classifier for the lake-type determination. Satisfactory training sets for the classifier were found for clear lakes, macrophyte lakes, tannin lakes, and algae lakes. Most of the problem lakes in the state are algal.

These techniques differ little from land-cover classification algorithms in use by other Landsat investigators. (Boland, 1976) (Rogers et al., 1977). In present study the lake type was determined based on statistics from all the lake pixels, not on a pixel-by-pixel classification (Scarpace et al., 1979).

Summary and Conclusions

During 1978 and 1979, Landsat data from 1976 were extracted for at least three dates for every significant lake in the state of Wisconsin. This resulted in the classification of approximately 5000 lakes. Table 32-26 shows a representative lake classification for one country. Included in the tabulation by county are (1) lake number, (2) lake name, (3) predicted trophic class, (4) lake characteristic, and (5) the number of dates used to predict the trophic class and lake type.

Correlations between Landsat-derived class numbers and DNR-supplied trophic class numbers are very good. Virtually all of the lakes with good "ground calibration" were classified correctly within the DNR criteria. Results of the lake-classification program received favorable review.

The computer-number and personnel-costs for executing the lake-classification programs were modest. The cost of classifying the lakes in the whole state of Wisconsin was $13,000 to $14,000 plus approximately twelve man-months of personnel time. Wisconsin DNR personnel were trained to run the computer programs. The DNR decided to repeat the project using 1980 Landsat data to assess changes in water quality during a four-year duration.

This project demonstrated a cost-effective example of digital processing of Landsat imagery for lake quality assessment. The implementing of the program as an operational project within the DNR came about because of the close cooperation between DNR and University personnel.

Analysis of Water Quality in the Estuarine Environment

The following describes part of the water quality work performed in the San Francisco Bay and Delta by personnel of the Remote Sensing Research Program (RSRP) of the University of California at Berkeley (Khorram, 1981a and 1981b).

The San Francisco Bay and the associated delta that is formed by the confluence of the Sacramento and San Joaquin Rivers constitute one of California's most important aquatic resources. Hence engineers are among those who are most vitally interested in obtaining remote-sensing-derived information relative to water quality in the San Francisco Bay and Delta region.

Within this area there is a region of high biological activity because of certain physical forces. Specifically, there is an up-welling of the heavier salt water that enters the Delta from the Pacific Ocean via San Francisco Bay. This, in turn, stimulates a downwelling of fresh water that enters the Delta from the Sierra Nevada Mountains via the Sacramento and San Joaquin Rivers. Consequently, there is a strong circulation of water currents and their associated nutrients.

Programs for monitoring the quality of water in this Delta and elsewhere have demonstrated that the concentrations of phytoplankton, chlorophyll, particulate organic nitrogen and phosphate, turbidity particles, and inorganic suspended solids tend to be significantly higher in the region of high biological activity than in adjacent upstream or downstream areas (Arthur, 1975; Ball, 1975). To some scientists this region has also become known as the "entrapment zone" (Arthur and Ball, 1979 and Siegfried et al., 1979). Location of this region may thus have important effects upon the Delta fisheries, whose contribution to the economy of the area is officially estimated to exceed $10 million annually (Siegfried et al., 1978). This region of high biological activity can be located by comparison of suspended solids, chlorophyll, turbidity, and conductivity determinations along the longitudinal axis of the estuary (Arthur, 1975, 1977; Khorram, 1981a, 1981b).

Many investigators have been limited by the inability to view the entire or even a large portion of the San Francisco Bay-Delta Estuary at one time. Its large size and spatial variability create both theoretical and logistical problems in any entrapment zone study. Relying upon a series of sampling stations distributed throughout the estuary, investigators have had to interpolate parameter values between stations, and have been forced to extrapolate in the large shallow bays where research boats cannot penetrate. The patchiness, evidenced by the use of a continuous-flow fluorometer, illustrates the inaccuracy of such interpolation and extrapolation.

Remote sensing allows scientists to obtain a synoptic "snap-shot" of the entire San Francisco Bay and Delta region—a view that is otherwise impossible to obtain. By determining ground level parameters simultaneously at several sampling locations and calibrating multispectral data obtained at the time of sampling, one finds it possible to obtain a reasonably accurate representation of the variations in water quality that occur both between and beyond these sampling stations. Additionally, because of the instantaneous synoptic view that remote sensing provides of the entire area, investigators have the ability to assess chlorophyll, salinity, suspended solids, and turbidity parameters without the need of correcting for the time required to travel between stations.

The objective of the effort reported upon in this section was to investigate the use of remotely sensed data, combined with in situ data, for assessing water quality parameters and for locating the region of high biological activity within the San Francisco Bay Delta. The water quality parameters of interest included suspended solids, chlorophyll concentration, turbidity, and electrical conductivity (salinity). The remote sensing data included that obtained with the Ocean Color Scanner (OCS) and through the use of conventional cameras containing color infrared film. Both kinds of data were obtained from a NASA U-2 aircraft. In addition, Landsat multispectral scanner (MSS) data were acquired at almost the same time. Specifically, all of the remotely sensed data and boat-acquired data obtained between the hours of 9:00 a.m. and 10:30 a.m., local times, on 14 September, 1978. The techniques utilized in collecting and analysing the data are described in the following sections.

Collection and Laboratory Analysis of Water Quality Samples

Water samples for use in determining various water quality parameters were collected at 29 predetermined sites in the San Francisco Bay-Delta Region. These water quality parameters, as measured on the 29 sample sites, included salinity, turbidity, and suspended solids. In addition, chlorophyll concentration was measured at nine of the 29 sites. These sample sites as shown in Color Figure 32-54, were located along the channel through the Delta. The chlorophyll data were only collected on the first nine sites as the sample boats progressed from San Pablo Bay toward the Delta region.

Water samples used for determining suspended solids were collected through a weighted hose. Samples were packed in ice in the field and taken to the laboratory for analysis. In the laboratory, suspended materials were collected by vacuum filtration on pre-weighed 0.45 screen filters, dried at 105°C, and reweighed on an analytical balance. Chlorophyll samples were collected by vacuum filtration on glass-fiber filters, pre-treated with a $MgCO_3$ suspension, and analyzed by the fluorometric method. Turbidity and salinity were determined in situ by a turbidity meter and a refractometer, respectively.

Acquisition and Analysis of the Ocean Color Scanner (OCS) Data

The ocean color scanner is a ten channel line-scanning radiometer having a 90° total field of view and a 3.5 milliradian spatial resolution. This scanner was flown on a NASA U-2 aircraft at an altitude of 65,000 feet over the study area on September 14, 1978. The calibrated computer-compatible tapes containing the OCS data were obtained from NASA-Goddard Space Flight Center (GSFC). The following procedure was used in processing and analyzing the OCS data.

Reformatting and Replacement of Bad Data Lines

The OCS data were reformatted in order to provide a format compatible with the RSRP image processing system, located at the Berkeley campus of the University of California. As the OCS data contained a number of bad data lines (referred to here as dropouts), algorithms for replacement of these dropouts with new data were developed by personnel of the RSRP (Khorram, 1981a). As an example, the data in channel 10 of the OCS, before and after replacement of the dropouts, are shown in Figure 32-55.

OCS Channel Selection

The OCS data comprised of six usable channels; Channels 1, 2, 8, and 9 were not used. Because of the excessive amount of light scattering caused by atmospheric haze particles at short wavelengths, Channels 1 and 2 contained inappropriate data for water quality studies; Channel 8 was not acquired over the study area; and Channel 9 sensitivity was designed for ground terrain. A correlation matrix was examined to determine the dependency of the six remaining channels on each other. Based on this correlation matrix, Channels 4 and 6 were eliminated and only four channels were chosen for analysis. For reasons indicated above, these remaining (usable) channels were 3, 5, 7 and 10. The peak wavelengths and wavelength ranges (based on Blaine et al., 1977) for these channels are shown in Table 32-27.

Sample Site Location

The 29 water quality sampling sites were located in the OCS data coordinate system by applying a coordinate transformation equation between the Universal Transverse Mercator (UTM) coordinate system and the OCS scanner coordinate system. The UTM coordinates of sample sites were then used to compute the OCS coordinates for three sites. To insure the correct correspondence (or registration) of the spectral data to the sample sites, the radiance values for a nine-pixel square surrounding each predicted sample

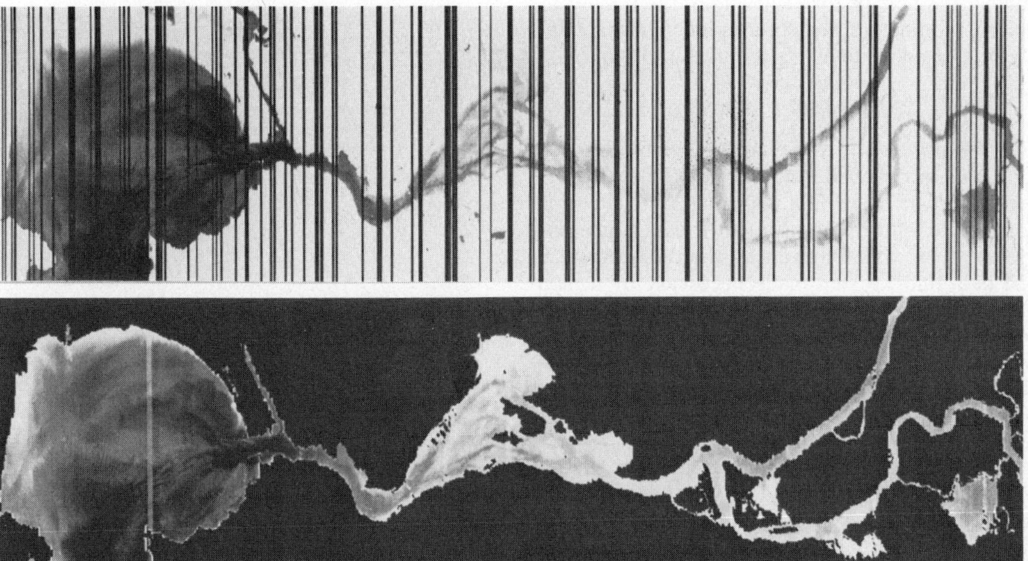

Fig. 32-55. Display of digital Ocean Color Scanner (OCS) data in Channel 10 prior to rectification of the dropout regions (top) and after rectification of the dropout regions and masking (bottom). (After Khorram, 1981a.)

TABLE 32-27

Peak Wavelengths and Ranges for the Four OCS Channels Used in the San Francisco Bay/Delta Water Quality Studies

	Wavelength, μm	
	Peak	Range
OCS Channel 3	.506	.475–.530
OCS Channel 5	.586	.550–.610
OCS Channel 7	.667	.645–.690
OCS Channel 10	.778	.745–.800
Landsat Band 4	—	0.5–0.6
Landsat Band 5	—	0.6–0.7
Landsat Band 6	—	0.7–0.8
Landsat Band 7	—	0.8–1.10

site were extracted from OCS Channels 3, 5, 7, and 10.

Cell Mean Radiance Value Calculation

The average (mean) radiance value of the nine-pixel cell encompassing each sample site was computed. These average cell values were used as independent variables in the regression models.

Development of Regression Models for Estimating Water Quality Parameters from OCS Data

A number of statistical models were examined for determining the best relationships between each water quality parameter and the mean radiance values generated for Channels 3, 5, 7 and 10. In addition, several numerical functions of these channels were examined. Based on the evaluation of correlation coefficients (R), "F" values, and the significance levels of these "F" values, the best fit for each one of the four water quality parameters was determined. These models are discussed in detail by Khorram, 1981a.

Data Acquired by the MSS and OCS

In the processing and analyzing of the Landsat CCT data as obtained from the EROS Data Center, the CCT data were first reformatted to a format compatible with that of the RSRP data processing system. This was followed by the replacement of bad data lines, a step made necessary by a moderate amount of malfunctioning of the Landsat MSS at the time of data acquisition. These lines were replaced, pixel-by-pixel in each instance, by the use of averaged values of the data from immediately adjacent pixels in the lines above and below the bad data line.

In the concluding step, the regression models (based on the OCS or Landsat Data) were applied to the entire study area. In this step, the regression models that had been developed between (1) the water quality measurements from boats from the 29 sample sites, and (2) the mean radiance values as obtained from either the OCS data or the Landsat data, were extended to the entire study area for mapping the desired water-quality parameters. The extension of these models to the entire study area was accomplished through the use of a simple linear discriminant function. By applying this function to each pixel in the study area and then grouping the values thus obtained for each continuous water-quality variable into discrete classes, the investigators performed the classification. These discriminant functions were applied to the OCS and Landsat data, respectively, to product color-coded water-quality maps for the parameters of interest.

Results

The results of laboratory analysis for the 29 water-quality sample sites are shown in Table 32-28. The best regression model for each water-quality parameter was selected, based on calculated R^2 values, "F" values, partial R values, and the residuals between the observed and predicted values. The correlation coefficients and "F" values for these models are shown in Table 32-29.

Based on the results of the statistical analysis, the following models were selected to represent the relationship between the water quality measurements obtained from the sample sites and the mean radiance values of the OCS data (and, correspondingly, the Landsat data) for those sites. Four out of eight best models were used for mapping water quality parameters, as discussed below.

TABLE 32-28

Results of Laboratory Analysis for Water Quality of Samples Collected on September 14, 1978 in the San Francisco Bay/Delta Region

Sample Site #	Suspended Solids mg/l	Turbidity, mg/l SiO_2	Salinity, %	Chlorophyll, mg/l
1	29.8	13.0	22.0	.0038
2	37.4	23.0	23.0	.0044
3	39.4	20.0	18.0	.0040
4	36.2	18.0	19.0	.0051
5	44.4	14.0	15.0	.0062
6	28.9	21.0	16.0	.0138
7	47.1	43.0	12.0	.0201
8	—	44.0	10.0	.0451
9	65.4	41.0	10.5	.0520
10	69.8	58.0	6.0	
11	67.9	41.0	3.0	
12	73.0	53.0	4.5	
13	64.9	50.0	3.0	
14	69.3	50.0	3.0	
15	75.6	47.0	1.0	
16	74.5	50.0	1.5	
17	43.1	41.0	1.5	
18	69.2	47.0	0.0	
19	56.4	41.0	0.0	
20	53.5	37.0	0.0	
21	24.5	41.0	1.0	
22	80.0	53.0	1.5	
23	46.2	52.0	1.0	
24	46.0	45.0	0.0	
25	45.0	37.0	0.0	
26	21.8	31.0	0.0	
27	29.9	23.0	0.0	
28	19.7	18.0	0.0	
29	8.1	16.0	0.0	

TABLE 32-29

Correlation Coefficients (R) and "F" Values Relating Remotely Sensed Data to "Ground Truth" for the Four Water Quality Parameters Estimated Using OCS and Landsat Data

Water Quality Parameter	R		F*	
	OCS	Landsat	OCS	Landsat
Salinity	0.948	.863	73.5	24.3
Chlorophyll	0.934	.905	11.4	59.2
Suspended Solids	0.609	.674	4.9	6.9
Turbidity	0.791	.907	10.0	60.2

* All "F" values are significant at the 0.01 level.

Chlorophyll model

$$y_{CH} = a + b/X_7 + cX_{10} + dX_{10}^2$$

where

y_{CH} = chlorophyll concentration expressed in mg/l,

X_7 = the mean radiance value in Channel 7 of OCS data

X_{10} = the mean radiance value in Channel 10 of OCS data and

$a = -0.2919$, $b = 0.8065$, $c = 0.0951$, and $d = -0.0093$.

Similarly, the following models were selected to represent the best relationship between the water quality measurements for sample sites obtained from boats and the corresponding mean radiance values of Landsat data.

Suspended solids model

$y_{ss} = A + b(Z_6)^{1/2} + c/(Z_7)^2 + d(Z_5)^{1/3}$, where

$Z_5 = X_5/2\cdot 8132$,

$Z_6 = X_6/2\cdot 7002$,

$Z_7 = (X_7 - 0\cdot 5524)/0\cdot 4265$,

y_{ss} = suspended solids expressed in mg/l,

X_5 = the mean radiance value in band 5 of Landsat data,

X_6 = the mean radiance value in band 6 of Landsat data,

X_7 = the mean radiance value in band 7 of Landsat data,

$a = 399\cdot 850$, $b = 136\cdot 787$, $c = -0\cdot 0115$ and $d = -321\cdot 630$.

Turbidity model

$$\ln|y_T - 12\cdot 9| = a + b\ln|X_4 - 17\cdot 43| + c\ln|X_5 - 14\cdot 32| + d\,|\ln X_6 - 4\cdot 88|.$$

Therefore,

$$y_T = 12\cdot 9 + e\exp(a + b\ln|X_4 - 17\cdot 43| + c\ln|X_5 - 14\cdot 32| + d\ln|X_6 - 4\cdot 88|),$$

where

y_T = turbidity expressed in mg/l SiO_2,

X_4 = the mean radiance in band 4 of Landsat data,

X_5 = the mean radiance value in band 5 of Landsat data,

X_6 = the mean radiance value in band 6 of Landsat data,

$a = 1.053$, $b = 0\cdot 613$, $c = 0\cdot 529$, $d = 0\cdot 860$.

Salinity model

$$y_{EC} = a + bx_4 + cx_6 + dx_7$$

where

y_{EC} = salinity expressed in parts per thousand

x_4 = the mean radiance value of Landsat MSSR

x_6 = the mean radiance value of Landsat MSS6

x_7 = the mean radiance value of Landsat MSS7

$a = 59.96$, $b = -1.228$, $c = 3.004$, and $d = 8.981$.

To prepare the color-coded maps of estimated water quality parameters, the regression models established between the 29 water quality measurements from boats and remote sensing data (i.e. the OCS data and Landsat data) were extended to the entire study area, using a simple linear discriminant function. Applications of these functions resulted in the production of a series of maps which were then color-coded to represent the four water quality parameters. The color-coded maps of the entire study area showing salinity, chlorophyll concentration, suspended solids and turbidity (based on the OCS data) appear in Color Figures 32-56a through 32-56d. Similarly, the color-coded maps for these same four parameters, but based on Landsat MSS data, are shown in Color Figures 32-57a through 32-57d.

Evaluation of the various correlation coefficients showed that all of the "F" values were significant at the 0.01 level of significance. This indicates that variations in spectral response account for a significant portion of the variations in water quality parameters.

In each of the color-coded maps, the values of the given water quality parameter increase in accordance with the following sequence of colors: dark blue, blue, green, yellow, orange, red, brown

(low concentration) → (high concentration)

Analysis of the chlorophyll, suspended solids, and turbidity maps indicates the existence of a region with high chlorophyll content, high suspended solids, and high turbidity, referred to as one of high biological activity. This region is shown in red and brown colors in Color Figures 32-56b, 32-56c, 32-56d, respectively, and also in Color Figures 32-57b, 32-57c, and 32-57d, respectively. As previously indicated, to some scientists this region has also become known as the "Entrapment Zone."

With respect to salinity, the quantitative values obtained in these tests conformed logically with the qualitative inferences that one might make; that is to say, the salinity values in the study area increased in the downstream direction. They increased constantly with the flow of fresh water of the Sacramento and San Joaquin Rivers through the San Pablo Bay portion of San Francisco Bay toward the saline waters of the Pacific Ocean. The lowest salinity values were observed in the west side of San Pablo Bay due to the fresh water inflow from Petaluma and Sonoma Creeks to the Bay.

Conclusions

Based on the results and the associated analyses that were made of the various kinds of imagery, the following conclusions are indicated:

(1) Areas having relatively high chlorophyll concentrations were clearly discernible on digitally enhanced imagery made from OCS and Landsat data. Such areas also were characterized by high turbidity and a high concentration of suspended solids.

(2) The high quality aerial photography of both conventional color or infrared-sensitive color films did not provide information on water quality in this study area.

(3) Comparing the results of the OCS and Landsat data there are indications that:

a. Chlorophyll concentrations can be estimated from either the OCS or Landsat MSS data with approximately equal accuracy.
b. Salinity can be better estimated from the OCS data.
c. Turbidity can be better estimated from the Landsat MSS data; and
d. Neither the OCS data nor the Landsat data provided a highly reliable basis for estimating suspended solids.

(4) Remote sensing data derived from the Ocean Color Scanner and/or the Landsat Multispectral Scanner provided useful mapping water quality parameters associated with determining both the size and the migration of the "region of high biological activity" (otherwise known as the "entrapment zone") in the San Francisco Bay and its associated Delta.

Water Quality Mapping of San Francisco Bay and Delta

The work on water quality that has just been described was extended to the entire San Francisco Bay and Delta, largely for the purpose of acquiring information that might affect water resource management and engineering decisions throughout a far greater geographic entity than had been studied in the initial investigation. An additional objective of this follow-on study was to investigate the usefulness of Landsat Multispectral Scanner (MSS) digital data for mapping selected water-quality parameters throughout a vast area. There also was interest in determining where the area of high biological activity was located within the Delta region at the time when this new set of remote sensing data was acquired. As in the previous study, the water-quality parameters of interest included salinity, turbidity, suspended solids, and chlorophyll concentrations.

The project was performed jointly by a team comprised of visiting scientists from North Carolina State University and their counterparts from the University of California at Berkeley and Davis. Cooperating government agencies in this greatly enhanced water quality project included the USGS-Water Resources Division, the Bureau of Reclamation of USDI, the National Marine Fisheries Service of NOAA, the California Departments of Water Resources and Fish and Game, and the National Aeronautics and Space Administration.

The approach that was involved: (1) Simultaneous acquisition of water quality samples and Landsat MSS data for 73 predetermined sample sites; (2) laboratory analyses of water quality samples; (3) location of sample sites on Landsat data; (4) computation of mean radiance values for blocks of 3 pixels-by-3 pixels surrounding each sample site; (5) development of regression models between water quality measurements from boats and Landsat data for sample sites; (6) verification of models and accuracy assessment; and (7) application of models to the entire study area. A total of 73 valid samples were collected from 11 boats between the hours of 9:30 and 10:30 a.m., local time, on October 27, 1980. The Landsat overpass was at approximately 10:00 a.m. on the same day. Through use of essentially the same procedures as were described previously for the smaller scale study the four best mathematical models were developed, one for each water quality parameter (See Table 32-30). The models were based on 50 sample sites for these four parameters and were selected through an analysis of their R^2 values, "F" values, residuals between observed and predicted values, and their simplicity.

The four selected models were applied to the 23 remaining sites for verification. The simulated values, based on the selected models and the water quality measurements for 23 sample sites, (along with their correlation coefficients) were used to evaluate the performance of the selected models. There was an unusually great diversity in the physical and environmental conditions within this study area. Hence, the form of the models for this area is different from that of the models for the delta, described earlier in this chapter. For a description of the detailed procedure that was employed and of the models that were used, see Khorram et al. (1982).

The coefficients of determination and the correlation coefficients for water quality models and their verifications are shown in Table 32-31. The final results included a series of color-coded maps, each pertaining to a water-quality parameter accompanied by statistical summaries and accuracy-assessment tables. For example, two of

TABLE 32-30

Mathematical Models Derived from Landsat MSS Data and Corresponding Surface Data for the Entire San Francisco Bay and Delta Region

Salinity Model
$Y_{EC} = 92.0 - 19.7(\ln X_5) - 11.8(\ln X_6)$, where
Y_{EC} = salinity expressed in parts per thousands
X_5 = the mean radiance value in band 5 of Landsat data
X_6 = the mean radiance value in band 6 of Landsat data

Turbidity Model
$Y_T = 3.70 - 0.04(X_4)^2 + 0.08(X_5)^2 + 0.09(X_6)^2 - 0.57(X_7)^2$, where
Y_T = turbidity expressed as Nephelometric Turbidity Units
X_4 = the mean radiance value in band 4 of Landsat data
X_5 = the mean radiance value in band 5 of Landsat data
X_6 = the mean radiance value in band 6 of Landsat data
X_7 = the mean radiance value in band 7 of Landsat data

Suspended Solids Model
$Y_{SS} = -79.20 + 34.20(\ln X_5) + 0.35(\ln X_7)$, where
Y_{SS} = total suspended solids expressed in mg/l
X_5 = the mean radiance value in band 5 of Landsat data
X_7 = the mean radiance value in band 7 of Landsat data

Chlorophyll Model
$Y_{CH} = -40.90 + 2.70(X_4) - 2.60(X_5) + 1.90(X_6) + 2.10(X_7) - 0.07(X_4)^2 + 0.30(X_5)^2 - 0.25(X_6)^2 - 0.19(X_7)^2$, where
Y_{CH} = chlorophyll concentration expressed in mg/l
X_4 = the mean radiance value in band 4 of Landsat data
X_5 = the mean radiance value in band 5 of Landsat data
X_6 = the mean radiance value in band 6 of Landsat data
X_7 = the mean radiance value in band 7 of Landsat data

these maps pertaining to salinity and turbidity for the study area, are shown in Color Figures 32-58 and 32-59 respectively.

The distribution of salinity, turbidity, suspended solids, and chlorophyll values throughout the Bay and the Delta regions were in general agreement with the expected values of these parameters and the reported values in the literature (Conomos and Peterson, 1974; Arthur and Ball, 1979; Cloern, 1979; Orsi and Knutson, 1979; and Khorram, 1981a, b).

TABLE 32-31

Coefficients of Determination and Correlation Coefficients for Water Quality Models and Their Verifications

Water Quality Parameter	R^2 for best fit models based on 50 samples	"R" for verifications of models based on 23 remaining samples
Salinity	0.910	0.64
Turbidity	0.903	0.88
Susp. Solids	0.810	0.62
Chlorophyll *a*	0.760	—

All of the "F" values were significant at the 0.01 level, indicating that the variations in Landsat MSS data account for a significant portion of the variations in water quality parameters.

Oil Spill Detection and Monitoring

The pollution of marine environments can be very harmful, because it affects the ecological balance governing life. The level of danger depends on the level of pollution reached, the nature and diversity of polluted discharges, and the difficulty of enforcing regulations. Remote sensing facilitates action designed to reduce or prevent pollution because it provides an effective means of monitoring urban and industrial discharges.

Early investigations in the infrared, ultraviolet, visible, and microwave areas of the electromagnetic spectrum demonstrated that the detection, identification, and quantification of oil spills, as well as the monitoring of water quality, can be accomplished through the proper use of modern remote sensing techniques. Based on this research, it was recognized that a multi-sensor system should be developed to achieve performance in all conditions of weather and lighting. These investigations led to development by the U.S. Coast Guard of an airborne, real-time, all-weather, day/night remote sensing system, having the capability to detect oil pollutants and identify the vessels from which the pollutants were being discharged.

The system, known as AIREYE, was designed for installation in medium-range surveillance jet aircraft. AIREYE includes Side Looking Airborne Radar, Infrared and Ultraviolet Line Scanners, an aerial reconnaissance camera, an active-gated television sensor system, an airborne data-annotation device, and a surveillance-system operator console.

Recently, the U.S. Coast Guard has also developed an airborne Forward Looking Infrared (FLIR) System to be installed on its short-range recovery helicopters. This FLIR was developed for use in helicopter search and rescue operations under conditions of darkness and diverse weather, but also for other missions such as marine environment control, law enforcement, and disaster relief operations.

In a similar recent research effort, low cost sensors were used by the "Istituto di Macchine", Department of Engineering, at the University of Catania in Italy to determine levels of pollution in Sicily's coastal waters. Professor A. L. Geraci, from that University, conducted an extended investigation using remotely sensed data for the evaluation of the degree of pollution in the eastern Sicily's coastal zone, and for the assessment of oil pollution levels within Augusta Bay and the associated coastal area, (Geraci, 1982).

The approach consisted of simultaneous acquisition of data from: (1) reflectance infrared photography and (2) thermography. Most of the aerial explorations were conducted from helicopters, that could reach locations not accessible to fixed-wing aircraft, and could maintain low altitudes more safely. The helicopter employed was an Augusta Bell 204/B ASW, supplied by an Italian Naval Air Station located in Catania.

The reflectance infrared photos recorded tonal values or false color renditions of near-visible infrared radiation up to 1.2 μm. Conventional 35-mm cameras were used, because of their compactness, and ease of handling. They were equipped with 55-mm lenses. Also, a Kodak Wratten filter no. 12 (deep yellow) was used over the camera lens to penetrate the haze and absorb the blue light. An Airborne AGA Thermovision System (AATS) was used to detect and record infrared radiation from 2 to 5.6 μm. The AATS was operated through an open door of the helicopter. The thermograms, as displayed on the color monitor, were copied with a 35-mm camera. Figure 32-60 shows these remotely-sensed imagery examples that were acquired during the investigation.

The results included a series of color-coded maps, each pertaining to the two kinds of remote sensing techniques used (reflective and thermal infrared). Based on these results and the associated analyses that were made of the various kinds of imagery, the following conclusions were indicated: (1) highly polluted areas and illegally discharging vessels can be clearly discernible on suitably enhanced imagery made from remote sensing data of the type acquired in this study; (2) thermal infrared imagery must be used in conjunction with aerial photography in order to provide accurate interpretation of the data; and (3) the possibility of finding a meaningful correlation between remotely sensed data in coastal zones, and the urban and industrial agglomerations bordering them, merits further exploration.

Analysis of Sediment Concentrations in a Marine Environment

The Bay of Fundy is located on the east coast of Canada. At its head, the Bay of Fundy divides into two arms; the southernmost, with an area of 1100 km^2, is the Minas Basin. The Bay of Fundy is famous for the largest tides in the world with an extreme range of 16.30 m at Burntcoat Head in the Minas Basin (McWhirter, 1982).

Within the Minas Basin, much of the bedrock consists of unresistant sandstones, often overlain by sandy till. The rapid tidal movement of water into and out of the Minas Basin, actively eroding the sandstone, means that a wide range of suspended sediment concentrations may be encountered in different parts of the basin. Thus, it is an ideal site for studying methods of determining suspended sediment concentrations using remote sensing.

The aim of the Bay of Fundy study was first to develop and test a method for measuring suspended sediment concentrations using Landsat digital data. Once proven, the aim was then to apply this procedure to acquire suspended sediment concentrations for parts of the Bay of Fundy (in particular the Minas Basin). The data were then to be input to a numerical model for assessing the impacts of a proposed tidal barrage to generate electric power.

According to Munday et al. (1979) "appropriate data reduction techniques for Landsat measurement of suspended sediment have been sought by many investigators since the launch of Landsat-1 in 1972." The work reported herein covers a series of experiments that began in 1974 and concerned the development of a methodology for detecting suspended sediment concentrations using Landsat digital data. Alfoldi and Munday (1978) reported that, although classification techniques are often suitable for discrete spectral classifications of vegetation, similar techniques are not readily available for water. "Because water is a dynamic fluid, contaminants are continuously distributed and many contaminants are colored . . ."

The technique, known as chromaticity analysis, originates from methods of color measurement initially applied to Landsat data by Munday (1974a, 1974b). Further developments led to the procedure described in detail by Alfoldi and Munday (1977, 1978) and by Munday et al. (1979).

Chromaticity analysis makes use of the normalized radiance values for Landsat MSS Bands 4, 5, and 6. Chromaticity "y" is calculated from N5/(N4 + N5 + N6), while chromaticity "x" is obtained from N4/(N4 + N5 + N6). Data can then be plotted on a chromaticity diagram (Figure 32-61). With changing suspended sediment concentrations, the data points in chromaticity space describe a sediment locus, as shown in Figure

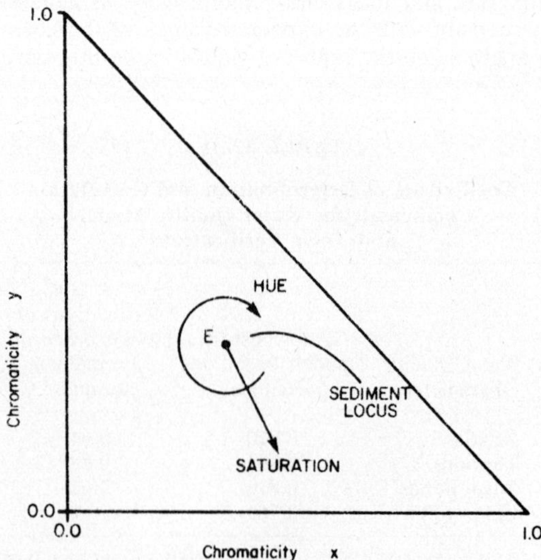

Fig. 32-61. Plot showing x and y chromaticity values and also values pertaining to hue and saturation that are centered around the sediment locus, as explained in text.

32-61. As atmospheric conditions change, the position of the locus is also changed, radially from E, but it is possible to adjust data to a standard locus for a particular area. By this means, the data from more than one Landsat pass can be analyzed, without reference to surface data on each occasion.

In the early stages of the work, correlation of the X-chromaticity value with suspended sediment concentrations measured at the exact time of the Landsat pass was an important element in verifying the procedure. Surface water samples were acquired by Amos (1976) within 10 minutes of an overpass using Knudsen or Nalgene bottles lowered from a helicopter at surveyed sample stations. Samples were analyzed within hours of collection. Over the period of the experiment, samples were acquired from 108 points in the Bay of Fundy. "Correlation between satellite and surface data for the combined data sets (after relative atmospheric adjustments) is 96 percent and the absolute error of the calibrated satellite measurements is approximately 44 percent over a range of 1 mg per liter to 1,000 mg per liter (Munday et al., 1979).

In summary, the procedure involves several steps for the user. First, sets of Landsat digital values in Bands 4, 5, and 6 are recorded and averaged for a series of 4-by-6 pixel arrays, selected to provide good coverage for areas of interest. The MSS digital values are converted to radiances using Landsat sensor calibration factors. Second, the chromaticity transformation involves the plotting of the sediment locus for each date being analyzed. In the third step, a standard locus is selected and other loci are adjusted to the standard locus. The fourth step is the calculation of a regression between the reference-loci chromaticities and the data pertaining to suspended-sediment concentrations, as acquired in the field. Location of the field points is obviously important so that the correct pixels are used in the regression. Finally, the regression is applied so that values of suspended sediment concentration may be obtained for selected scene locations. Given this information it is possible to map the distributions, should such an information-display be required. A number of maps have been produced which show suspended sediment concentrations for the Minas Basin.

As previously indicated, the Bay of Fundy and its extension, the Minas Basin, were chosen for the studies partly because a tidal barrage has been proposed for the generation of electric power. Given the high suspended-sediment concentrations in the Minas Basin, there was concern that disruption of normal flow could lead to siltation and rapidly make the system inoperative. Through chromaticity analysis, detailed data have been generated for use in a predictive numerical model for the Fundy system. The model shows that "no significant sedimentation is to be expected from the proposed tidal barrage during the design lifetime of the project, confirming the viability of the tidal power plan" (Munday et al., 1979).

Classification of Water and Land Cover Using Landsat Data

Often the first step is that of producing a map that will accurately portray the area's water and land-cover attributes. The percent reflectance values at different wavelengths from each of the different water and land cover-types at ground level tend to be unique. However, by the time the signals reach the satellite, the unique spectral reflectance signatures for these types are greatly altered by atmospheric effects that change from day to day. A detailed discussion of this problem and its solution can be found in other chapters of this manual.

If a clear lake appears in the Landsat scene, the reflectance signature from it is predictable and usually is close to zero. If we assume that reflectance from the water, itself, is zero then the apparent reflectance from this lake is essentially the total atmospheric effect recorded by a Landsat scene. Calibration values can be subtracted from the recorded spectral reflectance values of all other categories of features also, thereby leaving the unique ground-level reflectance signature of the feature in each instance. Once the corrected spectral signatures have been correlated with the corresponding water and land-cover categories, color-coded maps can readily be produced to portray these categories. We also can assume in most instances that the water quality of clear-water lakes (unlike eutrophic, mesotrophic or algal lakes) does not change during the seasons of the year nor from year to year. Consequently, no ground truth collection, made simultaneously with a Landsat overflight, is necessary.

For this case study, a Bendix MDAS Computer was used to create the color-coded maps. Since the eye can reliably recognize and distinguish only about 10 or 12 different colors on a color-coded map, only 12 color categories were used (See Figure 32-62a). It will be noted from the following list that more than half of the color categories used differentiated water bodies on the basis of water quality, and the color code was printed for categories as shown in parenthesis:

(1) clear-water lake (dark blue)
(2) marl or clay in lake (turquoise)
(3) medium algal lake (light green)
(4) heavy algal lake (dark green)
(5) tannin water and bottom effects in lakes (brown)
(6) red clay or iron-mill waste in water (red)
(7) bare rock and concrete (pink)
(8) non-forest open (white)
(9) disturbed open, (pink)
(10) hardwood forest (yellow)
(11) conifer forest (grey green)
(12) mixed deciduous-conifer forest (light grey)

In 1977 the Minnesota Pollution Control Agency funded a test project to classify lakes and land cover in Northern Minnesota by use of the multispectral analysis of Landsat data. Three Landsat MSS scenes were involved: one-half of a Landsat image (Color Figure 32-62a) acquired on 12 August, 1972 and two images acquired on 6 and 7 September, 1976, respectively. A comparison of the scenes from 1972 with those from 1976 showed significant changes in the lake water and land cover in the area.

The maximum seasonal growth of algae in a typical Minnesota lake occurs in a 3-week period from mid-August to early September. Hence, all of the Landsat data used in this study were recorded during this period.

Aerial reconnaissance was done over the area to locate deep, clear water lakes (with no bottom showing) which could be used as a standard. Others were located for use as training lakes for various categories that were to be recognized (such as tannin lakes, lakes with bottoms showing—overriding noise—and lakes with different types of algae present). Also, for training purposes, examples of the various land areas were chosen, (e.g. natural open, disturbed open, bare rock, conifer, and mixed forests). Once the corresponding spectral signatures had been determined from the analysis of Landsat MSS digital data, the training areas were no longer needed in all scenes because the Landsat signatures from one scene could be "bridged" to other scenes. Deep, clear-water lakes were selected as standards in all scenes, however. The Landsat residuals from the training lakes were later analyzed to see if their spectral signatures were, indeed, those of the categories desired. The precise spectral signatures were already known for these lakes from laboratory analysis of different water types (i.e., marl lakes, algal lakes, tannin lakes, etc.).

Signatures for Water Types

Figure 32-62b shows the signature for different types of lakes; the clear-water lakes were assigned the dark blue color on the map. The spectral signatures appearing in Figures 32-62 are those obtained through use of the Bendix Modification to the Landsat signals. The Bendix Modification multiplies the signals in MSS bands 4, 5, and 6 by 4 and the signals in band 7 by 2. Band 4 corresponds to green (0.55 micrometers), band 5 to red (0.65 micrometers) and bands 6 and 7 to near-infrared wavelengths centered around 0.75 and 0.95 micrometers, respectively.

Figure 32-62b also shows the signature of marl lakes (blue-water lakes), which have small particles of marl in the water. These lakes reflect an unusual amount of blue-green energy and give the otherwise clear water a distinct blue or turquoise color. Glacial lakes (such as Lake Louise in Canada) have tiny suspended particles of rock flour. Such lakes give spectral signatures that are

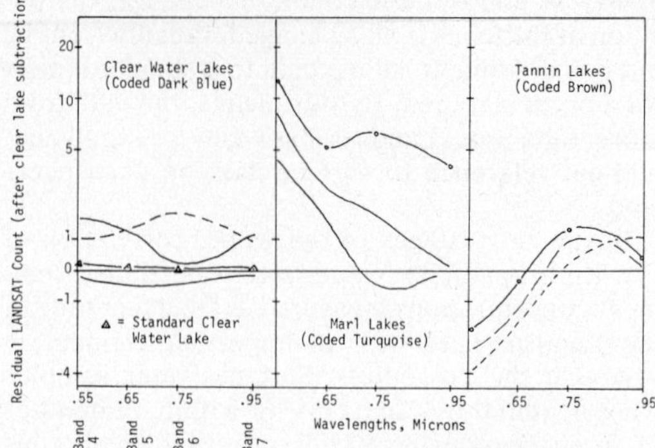

Fig. 32-62b. Residual satellite spectral signatures for various types of lakes, as indicated (Bendix Modification).

similar to those of marl lakes. Marl lakes are identified only with difficulty by the testing of water samples in the lab, but they are readily detected and categorized by means of Landsat mapping, as Figure 32-62b shows. These lakes were printed out as turquoise.

Figure 32-62b also shows the signature of tannin or humic lakes. In such lakes the tannic acid and other humic matter (called yellow stuff, gelbstuff, etc. in the literature) causes an absorption of blue light giving the water a yellow or brown cast. The nature and concentration of this material are measured in the lab for any given lake by the test called "color". Such lakes were printed out as brown on the color-coded map produced by the computer as shown in Figure 32-62a.

Figure 32-63 shows the spectral signatures for lakes containing various amounts of algae. Lakes with medium to heavy amounts of algae were printed as green, and those with light amounts of algae were printed as light green.

Some lakes have bottom noise present (e.g. from sand or mud bottoms). These noise signals can override the signals from the water itself. Figure 32-63 shows signatures from several such

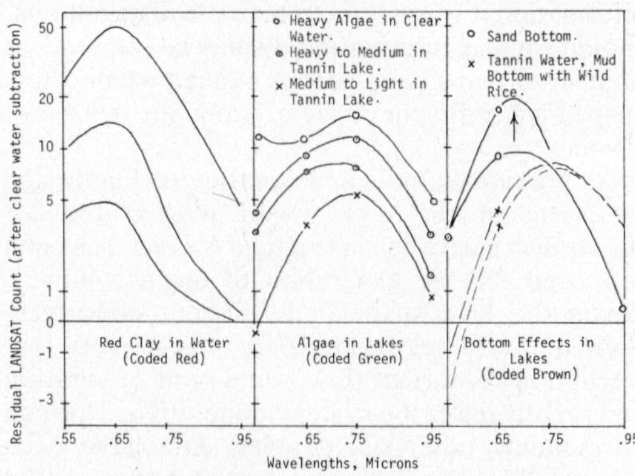

Fig. 32-63. Residual satellite spectral signatures for various kinds of lakes, as indicated (Bendix Modification).

lakes. Although a separate color code could have been used on the map to portray bottom lakes, because of the limit to the number of useable colors, bottom lakes were assigned a brown color, the same as the tannin lakes. Later, when water parameters were related to tannin lakes, these bottom lakes (here given the same color code as tannin lakes) caused some confusion in the subsequent analysis. The use of another color, such as yellow for bottom effects of lakes, would have alleviated this problem.

The erosional runoff from red clay, and from similar red materials that are exposed during iron mining operations, had similar signatures as shown in Figure 32-63. Lakes having this type of water were color-coded as red on the map.

Signatures for Land Cover Types

Figure 32-64 shows the spectral signatures for bare rock, such as the rock that characterizes most open-pit mines. The signature for bare concrete also is given. The spectral signature for areas disturbed by cultural activities (called "disturbed open") also will be found in Figure 32-64.

The category designated as "non-forest open" includes open swamps and prairies or fields with natural grass cover.

Forests of small conifers and large white-pine trees could be differentiated from other forest types and from each other. Both of these categories were color-coded grey-green. Spectral signatures for these categories are given in Figure 32-63. Figure 32-64 also gives spectral signatures for hardwood and mixed hardwood/conifer forests. These categories were color-coded light grey on the final output map. Figures 32-63 and 32-64 show examples of the above mentioned forest land-cover categories.

The final products resulting from this study were delivered to the Minnesota Pollution Control Agency. Three scenes comprised the final product. These were called Scene 12 (from 12 August, 1972), Scene 6 (from 6 September, 1976), and Scene 7 (from 7 September, 1976). Two versions of Scene 12 were produced. One version printed hardwoods, conifers, and mixed forests as three categories. The other version combined hardwood and hardwood/conifer-mixed forests into one category. The latter was preferable because, as previously indicated, there is a rather stringent limit to the number of colors that the human eye can reliably detect on a color-coded map (See Figure 32-62a).

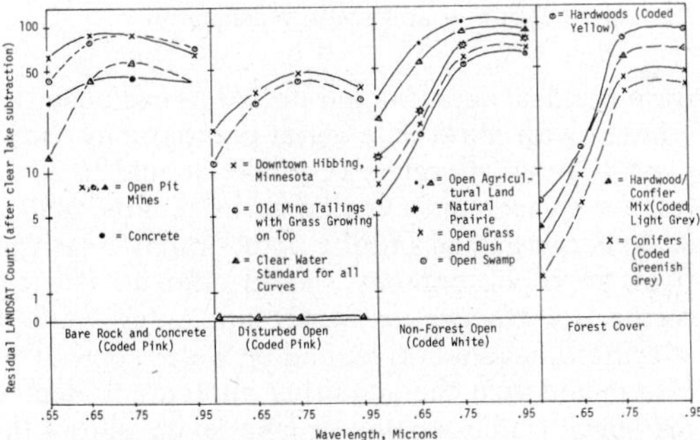

Fig. 32-64. Residual Landsat MSS spectral signatures for various kinds of land cover, as indicated (Bendix Modification).

LANDSAT IMAGERY USED IN A HYDROLOGIC STUDY

A hydrologic study, under a USGS contract, was conducted on the Nisqually River Basin, Washington, through the use of Landsat imagery (Colcord et al., 1974). This area, shown in figure 32-65, is situated generally south and east of Puget Sound. The Nisqually River is essentially glacial- and runoff-fed and flows generally westerly from Mt. Rainier, elevation 4400 m, to the Nisqually Delta region at sea level. The total study was of greater scope than is to be described in this section, which will pertain primarily to uses made of Landsat's synoptic view/and multidate capabilities in the above-named basin.

Hydrologic Data

The present trend is to utilize digital simulation models for river basin studies (Layton, 1973). Currently, the most important model inputs are those pertaining to precipitation, evapotranspiration and area/topography, based on river sections. The major remote sensing-aided hydrologic parameters that were considered in this case study are shown in Table 32-32.

Available stream flow data in the area were obtained from the USGS (1972) and showed ranges from 5 to 200 m^3/sec, with a rapid fluctuation of over 100 m^3/sec in 8 days. From the U.S. Environmental Data Service it was established that the average daily rainfall was about 1 cm per day in the rainy season. The effects of snow melt were also high during the "dry" months of May-June producing stream flows in excess of 50 m^3/sec. Other normal input data sources were USGS maps (scale 1:250,000), a geologic map (scale 1:500,000), and a U.S. Department of the Army map (scale 1:750,000) showing hydrologic elements of the State of Washington.

For this case study, use was made of Landsat imagery (scale 1:1,000,000) in MSS bands 4, 5, 6, and 7. Among the specific objectives or questions that were identified as items of concern to the survey engineer and possibly compatible with the Landsat imagery were those shown in Table 32-33. As discussed below, these items can be logically grouped under two main headings, viz. those pertaining to watershed area and those pertaining to land use and runoff characteristics.

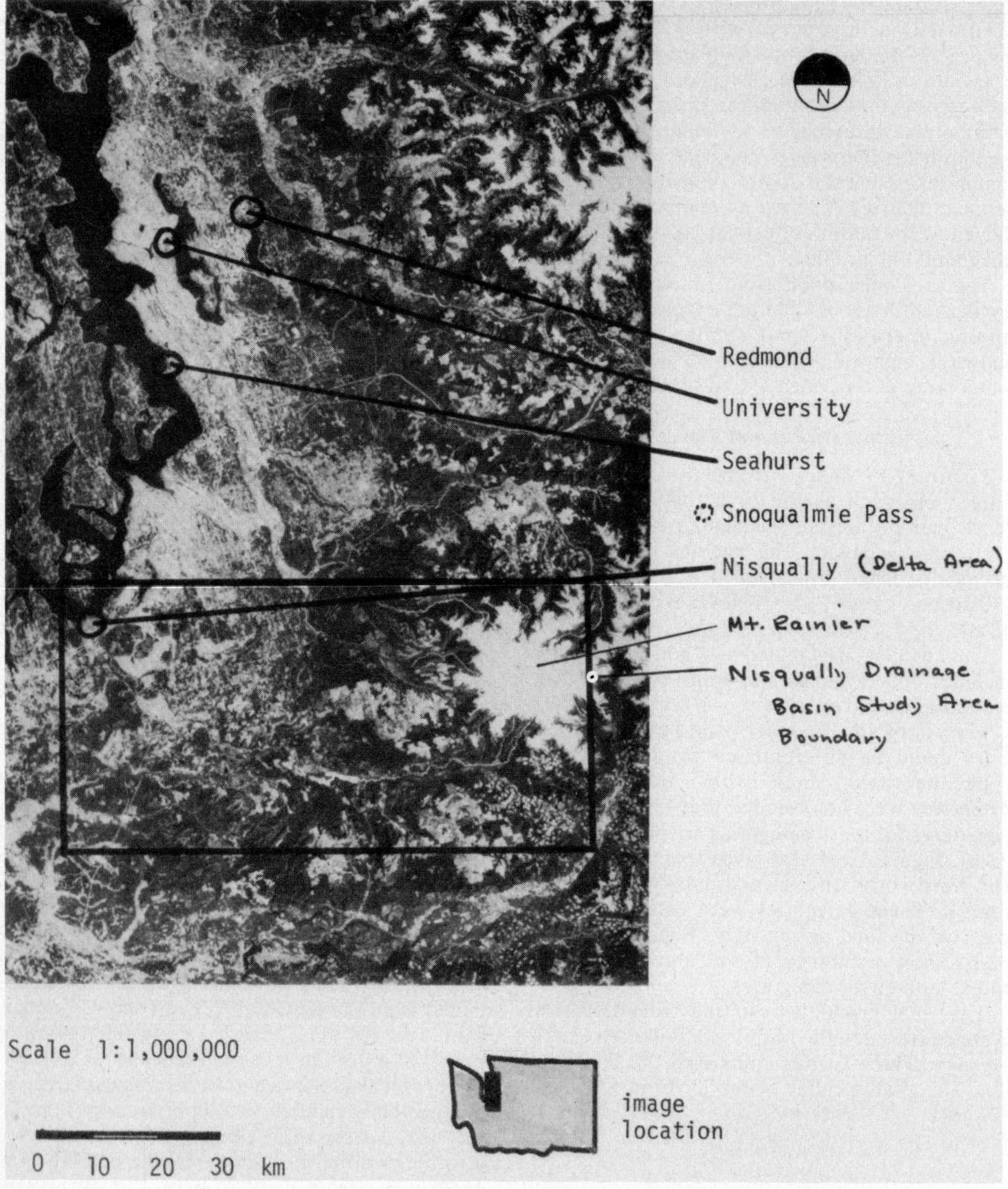

Fig. 32-65. Landsat MSS band 5 image of the test site location in northwest Washington.

Determination of Watershed Area

One of the considerations governing the accuracy of runoff predictions obviously is that of areal determination. Since Landsat MSS bands 6 and 7 are in the near-infrared region, where the dark tone of water bodies contrasts sharply with the light tone of vegetated lands, the water-land interface usually is sharp in those bands. Hence simple tests were made utilizing bulk multi-date imagery from those bands. The expected distortions of images occurred and it was demonstrated that, unless care is taken in controlling distortion, shoreline changes may be inferred incorrectly from Landsat data. Similar problems can occur, of course, with multi-date aerial photography; however, a given shoreline of a lake should have no relief displacement even on aerial photos, and, if scale is consistent and the photography near vertical, such photography should offer an efficient method of tracing the water-land interface at a specific time and corresponding water stage. This also means that the acquiring photograph depicting specific tide heights or lake levels allows the land surveyor to delineate the local tidal/lake datum. This study concluded that, with Landsat MSS bulk imagery being generally available and showing sharp water/land boundaries, water sur-

TABLE 32-32

Four Categories of Hydrologic Parameters, Including Designation (•) of Possible Remote Sensing Input

1. *Topography*
   (•) Drainage basin (watershed) area
   River cross section
   River slope
   (•) Reservoir and lake retention area
2. *River Characteristics*
   Water velocity
   Stream discharge
   (•) Water quality, water pollution
   (•) Springs, groundwater sources
   (•) Bloom, algae
   (•) Sedimentation
   Floods
   Tidal effect
3. *Run-off Characteristics*
   (•) Land use
   (•) Geology
   (•) Vegetation, crops
   (•) Snow pack and snow line
   Environmental factors
4. *Atmospheric Data*
   Precipitation
   Snow fall, snow melt
   Evaporation
   Evapotranspiration

face areas can be determined through the use of simplistic mapping devices such as the Bausch and Lomb Zoom Transferscope, with accuracies of greater than 97 percent. A further evaluation of bulk images at a scale of 1:1,000,000 also was made in which the image was compared to a 1:250,000 scale USGS map. Again, an accuracy of greater than 97 percent was indicated. From these results the investigator concluded that, for discrete areas, the bulk imagery from Landsat MSS provided areal accuracies for bodies of water and land that are roughly equivalent to those obtained through the use of 1:250,000 scale cartographic products.

This study pointed out certain problems encountered in wooded, mountainous areas at moderately high latitudes (e.g. 45 degrees). Landsat imagery at the scale of 1:1,000,000. The major problems found with Landsat under these conditions were:

a. Only major rivers can be seen (i.e. those having an average flow rate of at least 25 m^3/sec.) due to the small scale, extensive vegetation cover and shadow;
b. Passes and watershed ridge locations are difficult to discern without stereo images—a problem that can be partly solved through geomorphological inference, but made additionally difficult by shadow and snow.
c. In forested flat land, the problem of locating watershed boundaries is especially pronounced.

As mentioned by several researchers (Colcord et al., 1974), the best results from Landsat MSS imagery usually are obtained using band 7. Band 5 appeared to give better results along tributary streams in shadow areas due to effects of skylight. Comparing the above results with those of the Hydrologic Elements Map prepared by the Department of the Army it is noted that the true accuracy of Landsat-based determinations of watershed areas was better than 93 percent.

A more detailed study of the Nisqually River Watershed area (actually three basins combined into one due to construction of dams) was carried out using Landsat MSS imagery and a zoom Transferscope at a scale of 1:250,000. Values obtained were compared to those based on the USGS map 1:250,000 scale and also to those tabulated in a paper dealing with Water Resources Data for Washington (USGS, 1972). The results for three basins are shown in Table 32-34a.

TABLE 32-33

Objectives of the Nisqually, Washington Study

A. *Watersheds*
   Determination of watershed from Landsat imagery.
   Area and accuracy of watersheds.
   Possibility of monitoring changes in river or reservoir area.
B. *River Characteristics*
   Check for sources of change of input to river (springs or pollution).
   Check for sedimentation by changes in image density.
   Mapping of floods on the basis of satellite imagery.
C. *Land-Use Characteristics*
   Determination of number of classes identifiable on Landsat imagery.
   Monitor change in snow pack and snow line and correlate to discharge.

Determination of Land Use and Runoff Characteristics

On the Nisqually project, since the forest land was intensively logged, Landsat-1 MSS imagery provided a more up-to-date interpretation of the land use status than could be obtained from

TABLE 32-34a

Drainage Basin Areas (Km^2) of the Nisqually

Basin	Areas using Landsat-1 and ZTS	Areas using USGS Map 1:250,000	Area from Water Resource's Data (USGS, 1972)
Alder Dam	790	753	740
La Grande Dam	10	15	8
Nisqually Delta	850	787	—
Total	1,650	1,555	748

existing military topographic maps. Under the water category, major streams and waterways could be classified on band 7 to a certain extent. At this latitude (45 degrees) there is approximately 40 percent sidelap between adjacent Landsat MSS swaths. When conjugate images, as shown on these two swaths, are viewed stereoscopically, the "vertical exaggeration" is no exaggeration at all. To the contrary, it is only about 0.6 according to LaPrade, 1972. Nevertheless it is of some help in this context. August data should be used as part of the multi-date imagery because imagery acquired in August defines most clearly the permanent snow. The land use maps of the study area based on 1:250,000 scale USGS maps and Landsat imagery are shown in Figures 32-66 and 32-67.

With the data obtained, the comparative runoff values (see Table 32-34b) from the Alder Dam watershed were analyzed using the Rational Formula ($Q = Ci\ A$). For this watershed the overall accuracy was greater than 95 percent. Significantly lower accuracy values were obtained, however, in estimating runoff values for some of the individual watersheds. The areas in which very substantial changes in land use had occurred, such as recent clearings, may also have contributed significantly to the changing siltation flow and other environmental factors. Consequently, for the analysis of stream flow, up-to-date imagery should be used. It is also possible to "calibrate" streams so that the projected impact of extensive clearing on stream flow in a given watershed can be accurately estimated. This calibration is best accomplished from an analysis of sequential (before-and-after) Landsat images of other watersheds that appear on the remote sensing imagery to be compared and for which stream-flow values are available both before and after those watersheds were cleared.

SURFACE AND NEAR-SURFACE THERMAL INFRARED APPLICATIONS

A primary concern of siting studies for power plants, nuclear or other, is the potential effect of cooling water on the local river, lake or marine environment. The thermal environment that exists at the outset can be readily mapped with a thermal infrared line-scanner. Once the decision to build has been made, an application of the infrared technology is to monitor the thermal effluent's extent, movement and nature as shown in Figure

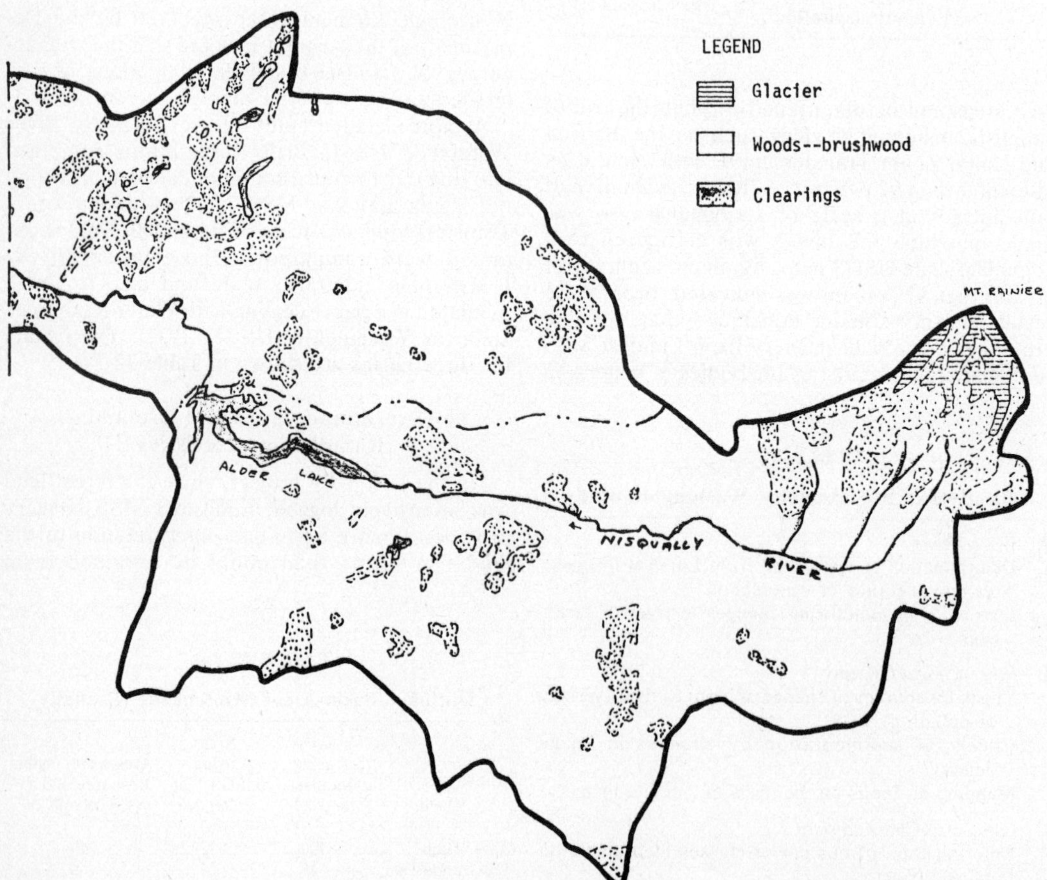

Fig. 32-66. Nisqually River land use classification, scale 1:250,000. (Source USGS maps of Hoquiam and Yakima.)

32-68a. Although in situ methods can be used, these tend to be labor-intensive and therefore costly. In addition they tend to yield only localized point data. Furthermore, on-the-ground-monitoring equipment is expensive to transport from site-to-site.

A comparison of in situ versus remote sensing-based monitoring of thermal effluents has been made in Ontario, Canada. Such work has demonstrated that airborne thermal infrared scanning has many advantages over in situ techniques in measuring the horizontal extent of waste-heat plumes. The most important of these advantages are the cost and the ability to define temperature distribution instantaneously within the entire plume (Ryerson, 1981).

With thermal data one can locate in situ measurements stations with more confidence and with more nearly optimum station density than before, Time saved in acquiring in-situ measurements is important because the data pertain to a dynamic feature and must therefore be collected in as short a time as possible. In addition, the alternative of data collection by ship can be very expensive.

Thermal infrared scanning can also readily provide information on the behavior of thermal plumes under a variety of discharge and environmental conditions and therefore be of assistance in the investigating of waste-heat impact on the biota and in determining whether the operator is complying with objectives and guidelines (Ross et al., 1976).

Thermal infrared imagery used to monitor groundwater seepage has proved a valid means to provide information of the water movement. Figure 32-68b shows a reservoir dam and the effluent of groundwater moving under the dam.

Although remote sensing for thermal plume analysis is not fully operational (in part because it does not yield a profile by depth), it has contributed to the study of plumes and has yielded useful data. For example, a prohibited waste-heat discharge was found at one site during construction. Current studies focus on using digital thermal infrared line scanner data for waste-heat dispersion models, including the study of plume behavior in winter. Figure 32-68b shows the effectiveness of the thermal infrared image to locate the effluent along with its areal extent, movement, and nature.

SUMMARY

Although only a limited number of water resource studies have been presented, they do show

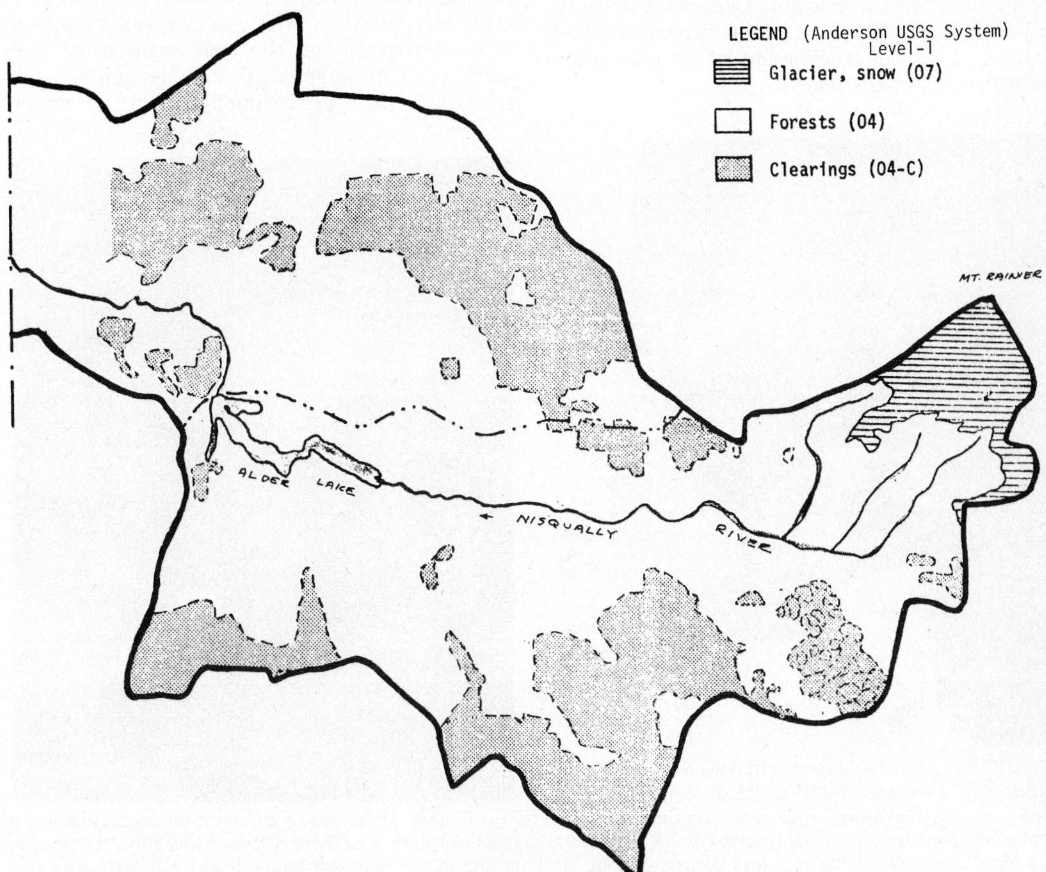

Fig. 32-67. Nisqually River land use classification based on Landsat MSS imagery.

Table 32-34b

Land Use/Runoff (Q) Prediction in Alder Dam Watershed

Land Use	C-Factor* Range	USGS Map Area km²	Landsat-1 Area km²	C	Comparing Q Contribution Map	Landsat	% Diff. Run-off from Map
(09) Permanent snow/ice	.6−.9	62	48.5	.75	46	36	−22
(04-C) Clearing, meadow	.05−.2	65	160	.15	10	24	+140
(04) Forested land	.05−.2	622	543	.10	62	54	−13
(05) Lakes, reservoirs	1.0	12.5	11.5	1.0	12	11	−8
			Total		130	125	−4

* (See References Wilson, 1969 and Wolf, 1966).

some engineering applications of satellite data, aerial photography, and thermal infrared remote sensing.

A major aspect of remote sensing for water quality studies is in its synoptic view: such a view contributes spatial data to complement in situ measurements for engineering applications. A secondary contribution is that remote sensing makes available temporal observations of changing phenomena. These well-planned studies have demonstrated how integrated in-situ, airborne and/or satellite data (such as described above) apply to current conditions. Depending upon the problem, the material on remote sensing methods cited in this chapter show important applications to water resource studies.

LANDSLIDE STUDIES

Landslides are slope movements. The presence or potential occurrence of landslides can constitute a major concern in the location of engineering projects. Remote sensing can play an important role in landslide investigations, both in the evaluation of landslide susceptibility and in the analysis of specific landslide events. In applications of the former type, site and route location efforts will benefit by the avoidance of areas that appear to be particularly susceptible to landslides. In addition, remote sensing often can minimize the cost of on-site soil testing. Also, measures aimed at erosion-control and the improvement of soil-stability over large areas can be guided cost-effectively through the intelligent use of remote

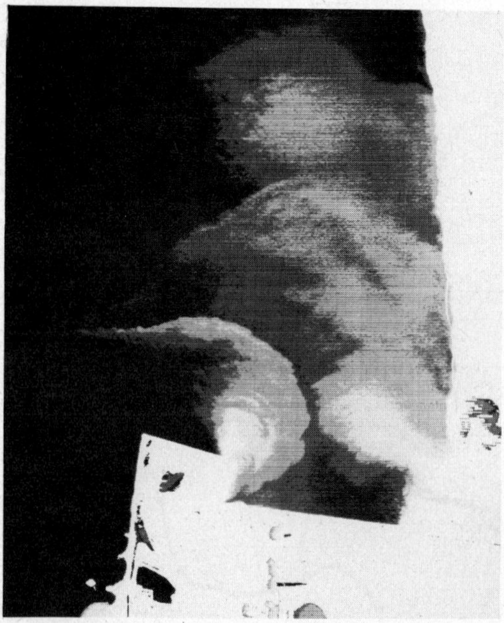

Fig. 32-68a. Dual discharge of effluents from a pulp and paper mill. This level sliced image shows the effluent plume in discrete tones. Each tone represents a 2°C change in temperature. Aerial thermography provides a complete overview of the size and distribution of the plume. The cooling to ambient temperature is clearly visible. From the thermography, center line decay can be calculated.

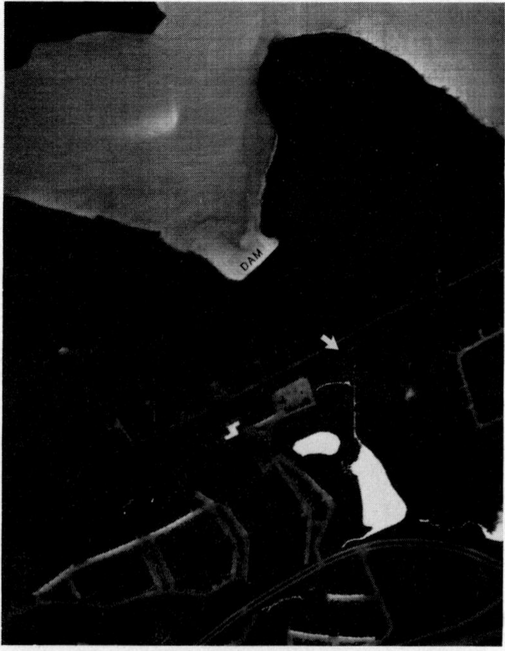

Fig. 32-68b. Detection of groundwater seepage below a reservoir dam. The thermal image was collected in early spring in the predawn hours. The air temperature was 37° F. Under clear skies the ground temperature was at or near freezing. The flying altitude was 1200 feet above ground level.

sensing methods. For example, the comparative analysis of Landsat imagery that has been acquired over an extended period of time can provide a method for noting changes in the landscape and its surroundings. If such imagery is multiseasonal it can facilitate the recognition of certain important linear features and landforms that are indicative of potential hazards, such as landslides and sinkholes.

The analysis of an individual landslide event can be greatly facilitated if remote sensing coverage taken both prior to and after that event is available. If the pre-slide coverage was acquired only a short time before the slide occurred, the probability is increased of being able to make a "before-and-after" study, the better to identify the causes of the landslide. Such a study can provide information of great potential importance in legal proceedings and/or insurance considerations. Similarly, if remote sensing data are collected very soon after a landslide event has occurred, they may be especially useful in documenting likely causes of the slope movement. In contrast to personal on-site observations, which can be faulty, narrow in scope, or subject to personal biases, remotely sensed data are quantitative, objective and open to re-interpretation by anyone. They may be used for updating maps after the occurrence of a landslide and for making dimensional estimates of the volume of earth moved or surface area of land affected. In addition, such data can identify other environmental changes, such as the realignment of drainage patterns, the destruction of standing forests and other vegetation, and the interference of landslide debris with roads, utilities, and buildings.

No other technique can readily provide a three-dimensional overview of the terrain from which the interrelations that exist among slope, drainage, surface cover, rock type and sequence, and human activities on the landscape can be effectively viewed and evaluated.

REGIONAL APPROACH

Analyzing the regional geology of the terrain is the appropriate way in which to initiate preliminary landslide investigations. Geomorphologists have divided the formations of the earth's surface into physiographic regions, that is, into regional areas based upon geological history, especially with respect to the method of deposition of earth materials and soils and thus with respect to the landforms which, in consequence, have been produced.

Early efforts to rate the landslide severity of the various physiographic regions of the United States were reported by Baker and Chieruzzi (1959). Their ratings of landslide severity were based on information gathered for the various physiographic regions with regard to frequency of occurrence, size of moving mass, and construction dollars expended per year. Baker and Chieruzzi further noted in their development of the correlation of landslide severity to physiographic regions that specific geologic formations were usually associated with the landslides in particular regions. A listing of some of the more common landslide-susceptible formations is included in their report.

A map by Radbruch-Hall et al. (1976) is another example of rating the severity of landslides in the United States. This map shows areas of relative incidence of landslides and areas susceptible to landslides. These investigators discuss the slope-stability characteristics of the physiographic regions of the United States, the geologic formations, and the geologic conditions that favor landsliding in the various physiographic provinces.

The information provided by Radbruch-Hall, et al. and Baker and Chieruzzi, should be used as a guide during the preliminary evaluation of the landslide potential as related to the regional concept. The regional concept is not intended to depict precise boundaries and conditions. The delineation of areas of low incidence means not that extensive landslides do not occur in those areas but that they are negligible in comparison with occurrences and magnitudes of those in the other areas. Even though, as noted in these reports, certain geologic formations are normally associated with landslides, no formation has developed slides throughout the entire extent of its outcrop area. A more detailed investigation is required to pinpoint the actual slides or vulnerable locations.

LANDFORMS SUSCEPTIBLE TO LANDSLIDES

Following the regional analysis, the individual landforms which comprise the physiographic regions are evaluated. The designation of a specific landform connotes both a genetic classification and a type of landscape. For example, a sand dune landform denotes deposits by wind movement and sorting, which form unconsolidated, smooth, flowing hills and ridges. An appreciation of the genetic aspects of a given landform enables one to estimate its potential susceptibility for movement. The type of landscape of each landform provides a basis for separating the various landforms and thus recognizing those most prone to sliding. In this section, landscape characteristics of landforms are used as the basis for recognizing landslides and landslide-prone areas.

Landslides can occur in almost any landform if the conditions are right (e.g., steep slopes, high moisture level, sparsely vegetated, or having no vegetative cover). Conversely, landslides may not occur on the most landslide-susceptible terrain if certain conditions are not present (e.g., clay shales on flat slopes with low moisture levels). Experience in observing and working with various landforms, however, indicates that landslides are common in some landforms and rare in others. Table 32-35 provides a key to landforms and their susceptibility to landslides. The subdivisions are

TABLE 32-35

Key to Landforms and Their Susceptibility to Landslides (Rib, H. T. and T. Liang, Ch. 3)

Topography	Landform or Geologic Materials	Landslide Potential*
I. Level terrain		
A. Not elevated	Floodplain	3
B. Elevated		
1. Uniform tones	Terrace, lake bed	2
2. Surface irregularities, sharp cliff	Basaltic plateau	1
3. Interbedded-porous over impervious layers	Lake bed, coastal plain, sedimentary plateau	1
II. Hilly terrain		
A. Surface drainage not well integrated		
1. Disconnected drainage	Limestone	3
2. Deranged drainage, overlapping hills, associated with lakes and swamps (glaciated areas only)	Moraine	2
B. Surface drainage well integrated		
1. Parallel ridges		
a. Parallel drainage, dark tones	Basaltic hills	1
b. Trellis drainage, ridge-and-valley topography, banded hills	Tilted sedimentary rocks	1
2. Branching ridges, hilltops at common elevation		
a. Pinnate drainage, vertical-sided gullies	Loess	2
b. Dendritic drainage		
(1) Banding on slope	Flat-lying sedimentary rocks	2
(2) No banding on slope		
(a) Moderately to highly dissected ridges, uniform slopes	Clay shale	1
(b) Low ridges, associated with coastal features	Dissected coastal plain	1
(c) Winding ridges connecting conical hills, sparse vegetation	Serpentine	1
3. Random ridges or hills		
a. Dendritic drainage		
(1) Low, rounded hills, meandering streams	Clay shale	1
(2) Winding ridges connecting conical hills, sparse vegetation	Serpentine	1
(3) Massive, uniform, rounded to A-shaped hills	Granite	2
(4) Bumpy topography (glaciated areas only)	Moraine	2
III. Level to hilly, transitional terrain		
A. Steep slopes	Talus, colluvium	1
B. Moderate to flat slopes	Fan, delta	3
C. Hummocky slopes with scarp at head	Old slide	1

* (1) susceptible to landslides; (2) susceptible to landslides under certain conditions; (3) not susceptible to landslides except in dangerous locations.

based on topographic expression and, in the case of hilly terrains, also on drainage patterns. This table gives only those landforms in which landslides are most common and is not meant to be all-inclusive. Almost all landforms rated as highly susceptible to landslides are composed of alternate layers of pervious and impervious materials (rock or soil), a fact that needs to be specifically recognized. Descriptions of landslide-susceptible landforms and stereoscopic illustrations of some of these landforms subject to containing landsliding are shown by Schuster and Krizek (1978). Descriptions of some of the pertinent landforms listed in Table 32-35 are included elsewhere in this chapter.

Vulnerable Locations

Following a regional evaluation of landslide-susceptible terrain and a study of the individual landforms where slides are most common, a detailed analysis is made of certain vulnerable locations that are conducive to sliding. Typical vulnerable locations include areas of steep slopes, cliffs or banks being undercut by stream or wave action, areas of drainage concentration and seepage zones, areas of hummocky ground, and areas of fracture and fault concentrations. Special attention should be directed to such locations when maps or aerial photographs are examined and field studies are performed. In addition, areas that have recently slid usually require immediate and close scrutiny because additional movement may occur.

STEEP SLOPES

If slopes are steep enough, movement can occur on any landform. However, on landforms highly susceptible to landslides, other factors being equal, the steepest slopes obviously are the most vulnerable locations. Only slopes of similar materials should be compared. For example, a slope that has been cut in earth or talus should not be compared with a rock cliff in adjacent landform, and slopes in bedrock generally are more stable, even though steeper, than slopes in adjacent soil areas.

The most common cause of the large number of slides that occur on steep slopes is the instability of residual or colluvial soils on a bedrock surface. The loose, unconsolidated soils cannot maintain as steep a slope as the underlying rock surface. Consequently, these two components tend to be in equilibrium. Any of several factors, such as a sudden heavy rainfall or an excavation at the toe of the slope (Figure 32-69) may result in sliding of the overlying soil mass. A study of cut-slope failures in North Carolina (Leith et al., 1964) reported that about two-thirds of the slides occurred in weathered soil materials and one-third occurred in rock slopes.

CLIFFS AND BANKS UNDERCUT BY STREAMS OR WAVES

Landslides are common in cliffs or banks that are subject to attack by streams or waves. If the banks are made up of soil or other unconsolidated materials, the weakest (and hence the most favorable) slide position is often located at the point of maximum curvature of the stream. In areas of rock outcrops, on the other hand, the exposure at and near the point of maximum curvature is often hard rock, and the weak spots are to be found upstream and downstream of this point. These conditions are shown in Figure 32-70.

Many landslides occur along the edges of oceans and lakes because of under-cutting by waves. Locating the point of maximum water impact is more complex and difficult along lake and ocean shores than along stream banks. Factors to be considered include shape and slope of the shoreline, direction of wave action, and frequency and magnitude of storms producing large waves. Data obtained at different periods of time are often of value in the analysis of these factors.

Areas of Drainage Concentration and Seepage

A survey conducted by the Federal Highway Administration of major landslides on the Federal-aid highway system in the United States

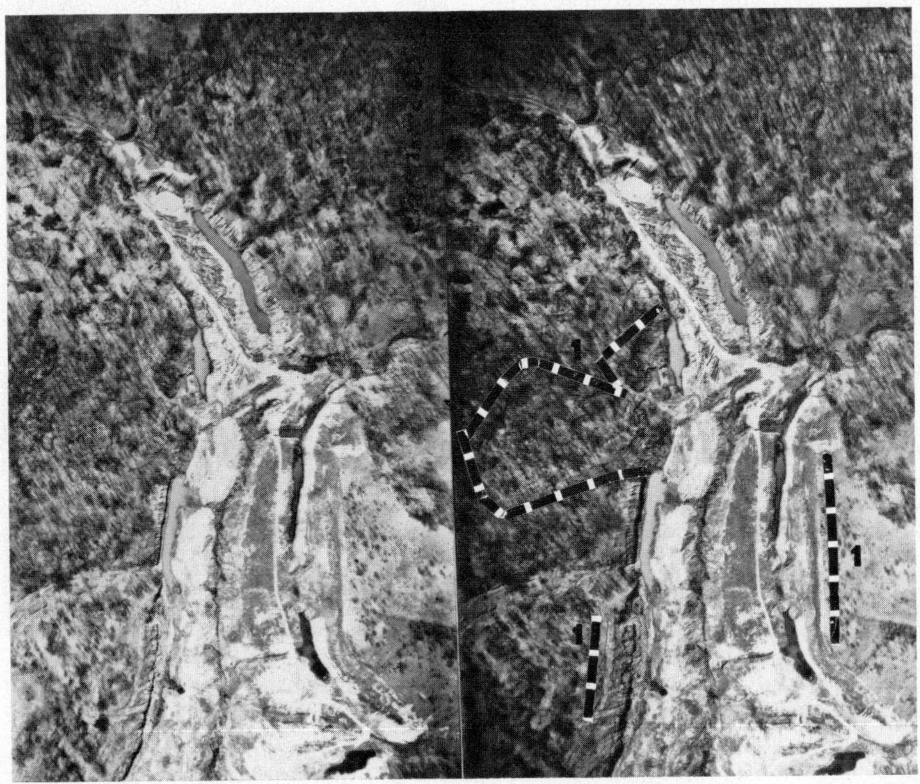

Fig. 32-69. The landslides at (1) were caused by undercutting of slopes by coal stripping operations in Washington County, Ohio. (After Norell, 1965, courtesy Ohio Department of Transportation.)

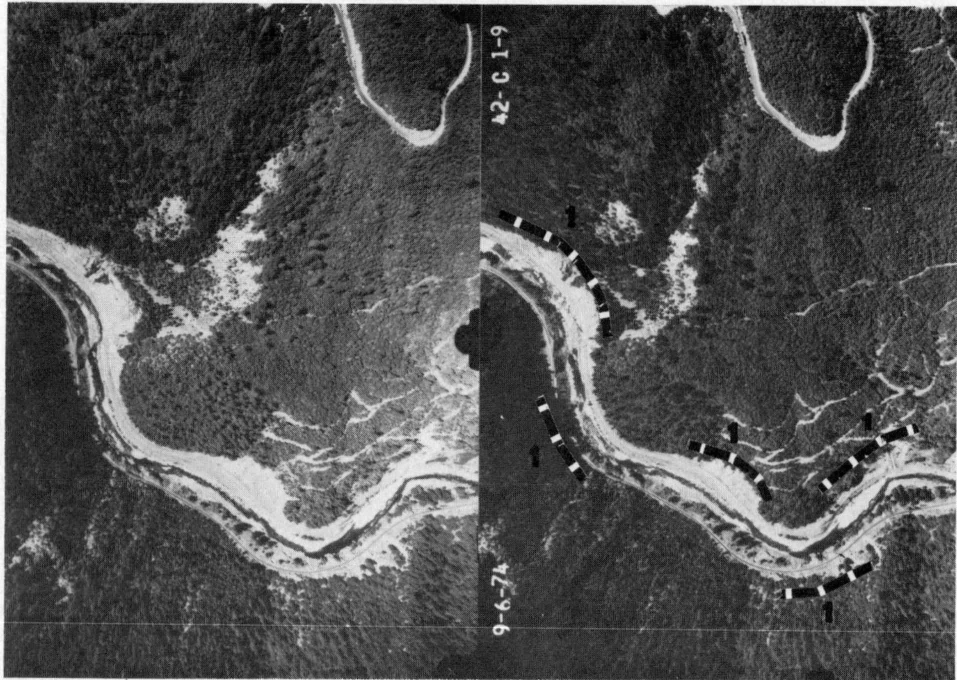

Fig. 32-70. These landslides in Douglas County, Oregon, were caused by undercutting of slopes by stream. Slides largely composed of weathered colluvial materials are evident at almost every bend of the stream where it undercuts naturally steep slopes (1). (Courtesy Federal Highway Administration.)

revealed that water is either the controlling factor, or a major contributing factor, in about 95 percent of all landslides (Chassie and Goughnour, 1976). Thus, careful study of the drainage network and areas of concentration of outfall or water based on aerial photographs of appropriate scale is extremely important. Close scrutiny of existing slide scars often indicates that a line connecting the scars points to drainage channels on higher grounds. Such drainage may appear as seepage water, which is responsible for the damage. An example of this conditions is shown in Figure 32-71.

Seepage with subsequent sliding is likely to occur in areas below ponded depressions, reservoirs, irrigation canals, and diverted surface channels (Figure 32-71). Such circumstances are sometimes overlooked on the ground because the water sources may be far above the landslide itself, but they become obvious in aerial photographs. For example, the associated vegetation growth, at early stage, is very apparent on color infrared photos. The importance of recognizing the potential danger in areas that are situated below diverted surface drainage, especially in fractured and porous rocks, needs particular emphasis. Extensive field experience has proved repeatedly that, within an unstable area, one of the most dangerous sections is the lower part of an interstream divide through which surface water seeps from the higher to the lower stream bed. The recognition of seepage is sometimes aided by the identification of near-surface channels, wet areas, tall vegetation on the slope, and displaced or broken roads adjacent to the slope. Delineating the drainage network, and especially the presence of seeps or springs, is extremely important for

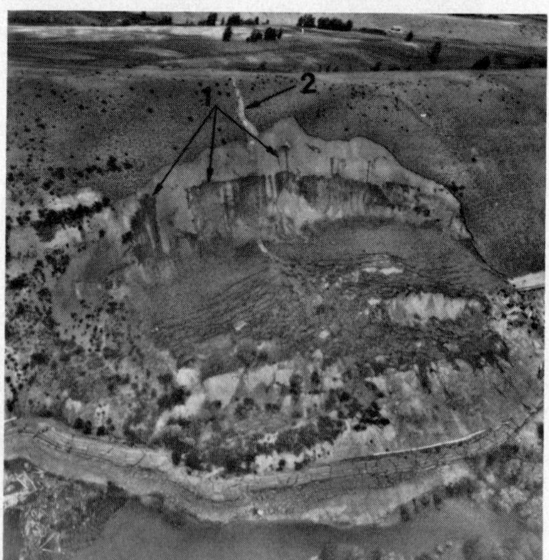

Fig. 32-71. Oblique aerial view of extensive seepage zones (1) on the scarp face of a massive landslide in Kittitas County, Washington. Seepage was the major cause of slides in these unconsolidated sediments. A remnant of the dry sand flow (2) is seen above the scarp face and is the result of seepage at higher level. Note the landslide-disrupted highway at base of slope and a water canal carried through the slope. (Courtesy Washington State Department of Transportation.)

planning new construction. Many highway fills have failed because the natural drainage was blocked by the fill and allowance was not made for drainage. Figure 32-72 shows such a situation.

Areas of Hummocky Ground

The presence of hummocky ground, wherein characteristics are inconsistent with those of the general regional slopes, and also the presence of a scarp surface (sometimes not very distinct) at a higher elevation are often indications of an existing landslide. The older the landslide is, the more established the drainage and vegetation become on the slide mass. The drainage and vegetation thus help in determining the relative age and stability of the slide.

Once an old landslide is found, it serves as a warning that the general area has been unstable in the past and that new disturbances may start new slides. However, such a warning should not discourage construction unconditionally, because

Fig. 32-72. Landslide in highway fill section of Ohio—22 in Jefferson County. A report by the staff of the Ohio Department of Transportation attributed failure to saturation of colluvial materials on side slopes beneath fill by springs outcropping on the hillsides. The few small drainageways commencing part way down the slope (1) and the small slide (2) on the natural hillside indicate the presence of seepage zones and instability of the natural slopes. (Courtesy Ohio Department of Transportation.)

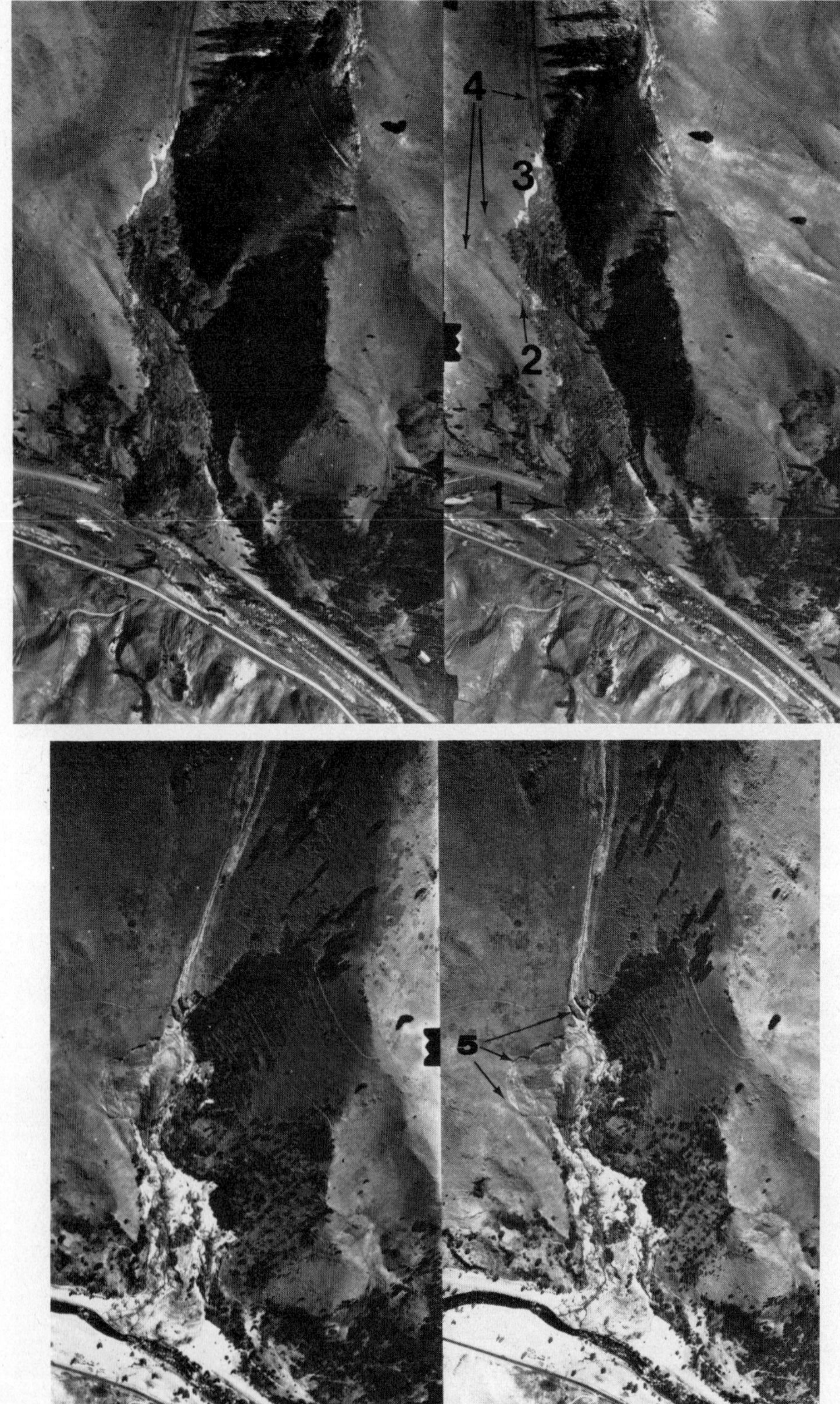
4
3
2
1
5

the unstable condition of the past may not necessarily exist today. In some parts of the western United States, for example, railroads built in extensive old landslide areas have been stable for a long time. Nevertheless, special care should be taken in construction on old slides.

Areas of Concentration of Fractures and Bedding Planes

Movements of slopes may be structurally controlled by surfaces or planes of weakness, such as faults, joints, bedding planes, and areas of foliation. These structural features can divide a rock mass into a number of individual units which may act independently of one another. The result can be an incorrect slope design because the designer considered the rock to be one continuous mass rather than a series of individual blocks. These planes of weakness also provide egress for water and vegetation, which further weaken the individual units by wedging action, frost heave, and the reduction of sliding friction. A careful search should be made to locate areas with close spacing of faults and joints, especially where they cross and divide the rock mass into smaller blocks.

Recent Landslides

The occurrence of a landslide does not mean that final adjustments to the unstable conditions have occurred and no further movement will occur. In many cases, especially in unconsolidated deposits, the materials present in the scarp face remain in an unstable condition because they are on a very steep slope. The scarp face rapidly retrogrades uphill by continued slumping until a more stable condition occurs. Thus, a new landslide should be investigated as soon as possible not only to determine corrective measures but also to look for evidence of possible continued movement. The most significant sign of possible further instability is the presence of cracks on the crown of the slide. Figure 32-73 (top) shows some telltale signs at a slide in Idaho. Figure 32-73 (bottom) shows the same slide area 5 months later. Additional movement occurred in the area where the telltale signs were evident in the earlier photograph.

USES OF AERIAL PHOTOGRAPHY

The interpretation of aerial photography has proved invaluable in the study of landslides. Most of the types of slides found in nature, whether they are old, stabilized landslides, active landslides, or even potentially unstable areas, can be recognized on photographs by experienced interpreters. Factors that govern the ability to recognize landslides on aerial photographs include scale of photography, time of year, and film type used.

Scale of photography can be very critical in the interpreter's ability to identify landslides because it influences his ability to denote topographic features that are indicative of landslides. In some cases, the landslides are so massive sometimes involving over four cubic kilometers (1 mi^3) of material—that they are virtually impossible to detect on the ground or on large scale photographs (Dishaw, 1967). Small scale photographs (1:40,000 or smaller) are required in order to stereoscopically view the full extent of massive slides. However, most landslides encountered in engineering construction are small.

Cleaves (1961) stated: "Within the scope of the author's experience, more than 90 percent of all landslides requiring control or correction have been small, averaging less than 500 feet in width from flank to flank, and 100 to 200 feet in length from crown to toe." Hence, on normal scale (1:20,000) photography, most landslides would not be detectable. To identify these small slides, large scale photography (1:4,800 or larger) usually is required. For regional inventories two scales of photography might be useful—small scale (1:60,000 or smaller) and medium scale (1:15,000 to 1:30,000). For investigating specific sites for engineering construction, a large photographic scale (1:8,000 to 1:12,000) would be necessary for evaluating alternate sites, and a very large scale (1:4,800 or larger) would be required for the final evaluation of landslide susceptibility.

The evaluation of the drainage network and prevailing moisture conditions is important for delineating existing landslides as well as for estimating the potential of sliding in a given area. The time of year and type of photography obtained are important in evaluating these elements. The optimum time of year for delineating drainage conditions is when the water table is highest and vegetative cover is at a minimum. In this situation, the areas of highest moisture or the wettest soils, and also the presence of seepage zones on the slopes, are indicated by dark photo tones and/or luxuriant growth of vegetation, the latter being observed as dark toned areas on conventional black-and-white aerial photos. In the humid, temperate climate of the Northern Hemisphere, the optimum time would be in the spring of the year. The timing of photo flights would vary in other climatic zones.

Fig. 32-73. Sequential photography for use in evaluation of recent slide development along US-95 in Idaho County, Idaho, and investigation of possible further movements. Aerial photography (*top*), taken in February 1974 shortly after occurrence of slide (1), shows features present at crown of slide that forewarn of further movement: fissures above scarp face (2), loose colluvial materials on steep slopes at crown (3), and evidence of water feeding into these loose materials (4). Photography obtained 5 months later in July 1974 (*bottom*) shows that additional slide movement (5) occurred in areas indicated as potentially unstable in photograph above. (Courtesy Federal Highway Administration.)

For many years panchromatic photography has been used successfully for delineating landslides and is still the most economical type of photography for locating landslides of large areal extent. Color photography, however, is the optimum film type for locating small slides and potential slides. Ektachrome color is favored by many interpreters because the natural color tones and major changes in color and texture make it easier to delineate slip zones and differentiate rock types. Such photography provides good detail in shadow areas and in water-covered areas. Color infrared films are most helpful for delineating the presence of water on the surface and for giving clues to subsurface water conditions by showing the vigor of the surface vegetative cover. The contrast of the luxuriant vegetative cover (indicated in bright red on the color infrared film) in areas of seeps or high moisture is striking in comparison to drier vegetated and bare ground areas. This has made color infrared film especially valuable for locating the presence of seepage zones at or near the surface.

Diagnostic Features of Landslides on Aerial Photographs

The following features, discernible on conventional panchromatic black-and-white aerial photography, are typical of landslides or landslide-susceptible terrain, but not all features are evident for each landslide. Most but not all of these features are shown in Figure 32-74. The numbers in Figure 32-74 correspond to those in the list below.

1. Land masses undercut by streams (see Figure 32-70 also);
2. Steep slopes having large masses of loose soil and rock;
3. Sharp line of break at the scarp (head end) or presence of tension cracks or both;
4. Hummocky surface of the sliding mass below the scarp;
5. Unnatural topography, such as spoon-shaped trough in the terrain;
6. Seepage zones;

Fig. 32-74. Typical characteristic features of a landslide as shown in aerial photographs. Slide occurred in Yakima County, Washington. Numbered items are discussed in text. (After Cleaves, 1961).

7. Elongated undrained depressions in the area;
8. Closely spaced drainage channels;
9. Accumulation of debris in drainage channels or valleys;
10. Appearance of light tones where vegetation and drainage have not been reestablished (see Figure 32-70 also).
11. Distinctive change in photograph tones from lighter to darker, the darker tones indicating higher moisture content;
12. Distinctive changes in vegetation indicative of changes in moisture; and
13. Inclined trees and displaced fences or walls due to creep.

In some cases the slide itself is not discernible, but indirect evidence of its existence is noted. For example, where a highway is built on unstable soil, the irregular outlines and nonuniform tonal patterns on the photo represent broken or patched pavements.

Identification of Vulnerable Locations

Many slides are too small to be detected readily in photography at the scales normally available (i.e., 1:15,000 to 1:40,000). Consequently, the photographs should be closely examined for signs that indirectly indicate the presence of slides or, if signs are not visible, for the vulnerable locations where slides usually occur. Typical vulnerable locations have been described previously.

Aerial photographs are valuable aids in identifying the vulnerable locations. The shape and slope of the terrain are readily discernible from stereoscopic examination of the photographs. In fact, the vertical appearance of the terrain is exaggerated when viewed through a stereoscope. Moderate slopes appear steep, and steep slopes appear almost vertical, making them easier to delineate. In addition, the slopes can be measured on the photographs by using simple measuring devices, such as an engineer's scale or a parallax bar, and by applying photogrammetric principles.

Drainageways, seepage zones, fractures, and fault zones are also readily evident on aerial photographs. By means of stereoscopic examination, the complete drainage network can be mapped, including the intermittent streams and small gullies. The presence of wet zones or seepage areas is evidenced by darker tones caused by the higher moisture content in the soils or by a more luxuriant vegetative growth over the wet areas. Areas of drainage or water concentration above a slope should be closely examined because they are vulnerable locations. Subsurface seepage from these areas can lead to slope failures; Figure 32-71 shows such a case. Fracture and fault zones are indicated on the photographs by dark linear or curvilinear lines. The darker tones are usually due to the better growth of vegetation along the fracture zones where it is easier for the roots of plants to grow and where moisture levels are usually higher. In the delineating of fracture zones, care must be taken not to interpret man-made features, such as fence lines or field boundaries, as fracture zones. Generally, features having almost perfectly straight lines, right-angle intersections, and standard geometric patterns are man-made.

Procedure for Landslide Investigations

The following procedure is recommended for photographic studies of landslides (Schuster and Krizek, 1978).

1. Lay out sites of planned facility on photographs.
2. Delineate areas that show consistent characteristics of topography, drainage, and other natural elements and classify into landform types. Large and obvious slides are identified at this stage.
3. Evaluate the general landslide potential of the landform types, using Table 32-35 as a guide.
4. Make a detailed study of cliffs or banks adjacent to river bends and all steep slopes. In doing so, compare slopes within the same landform type: For instance, slopes in bedrock landforms are more stable, even though steeper, than slopes in adjacent soil areas. Because slides are usually small, as imaged on photographs, look carefully and inspect slopes in minute detail. Give particular attention to the following features:

Slide	*Feature*
Existing	Hillside scarps and hummocky topography
	Parallel spoon-shaped dark patches on hillsides
	Irregular outlines of highways and random cracks or patches on existing pavements
Potential	Ponded depressions and diverted drainageways
	Seepage areas suggested by faint dark lines, which may mean near-surface channels, and fan-shaped dark patches, probably reflecting wet vegetation

Relatively new slides appear in light tones because vegetation and drainage are not well established. The parallel spoon-shaped dark patches on hillsides are likely to reflect vegetation in minor depressions. Lines drawn through the axes of the scarps in the slides often point to drainageways on higher ground that contribute to landslide movement.

5. Ground check all suspected slides.

USES OF OTHER REMOTE SENSING SYSTEMS

Multisensor investigations for terrain analysis and landslide investigations have been performed by several investigators, including Alföldi (1974),

Gagnon (1975), Rib (1966a and b), and Tanguay and Chagnon (1972). The general consensus of such studies is that large-scale aerial photography (most preferred color) provides the most information on terrain conditions, specifically for landslide detection. Several of the reports did indicate, however, that satellite imagery and infrared imagery, both reflective and thermal, offered some unique information that would prove useful for landslide investigations.

Satellite Imagery

At the small scale of satellite imagery, only extremely large landslides can be identified directly. Since most landslides are small, they are not directly identifiable on satellite imagery. However, the value of satellite imagery, as noted by Alföldi (1974) and Gagnon (1975), is that the landslide susceptibility of an area can be determined indirectly from some of the features that are identifiable at those scales. Regional physiography, geologic structure, and most landforms as well as land-use practices and the distribution of vegetation are evident on the satellite imagery. These features in conjunction with the tonal patterns present on the imagery provide clues to the types of surface materials present, the surface moisture conditions, and the possible presence of buried valleys. Correlating these factors to geology and topography and using local experience in a region make it possible to rate the landslide susceptibility. Recently, through the development of computerized databanks, satellite data may be integrated with topographic and land cover data for determining which areas are susceptible to massive landslides. For example, Alföldi noted in his study of landsliding in eastern Ontario that on the satellite image the clay plains were easy to spot because they are almost 100 percent cultivated; the till plains were recognizable because they form a poorer agricultural area with the field and forest sections intermixed; and the elevated sand plains of the old Ottawa River delta (which overlay the clay plains and are most susceptible to sliding) are mostly in forest.

An additional advantage noted by Alföldi for satellite imagery was the repetitive coverage available. Seasonal changes in vegetative cover and moisture levels—as indicated by tonal changes—can be evaluated to increase the accuracy of interpretation of terrain conditions. Also, any changes noted during the year in the landslide-susceptible zones, such as urban expansion, clear-cutting of forest, forest fires, and drainage of swamps, might presage renewed or new landslide activities. This could alert the interpreter to the necessity for a more detailed investigation in these areas. Experience has indicated that both the band 5 (0.6 to 0.7 μm) and the band 7 (0.8 to 1.1 μm) Landsat Multispectral Scanner data and also the infrared color composite are most beneficial

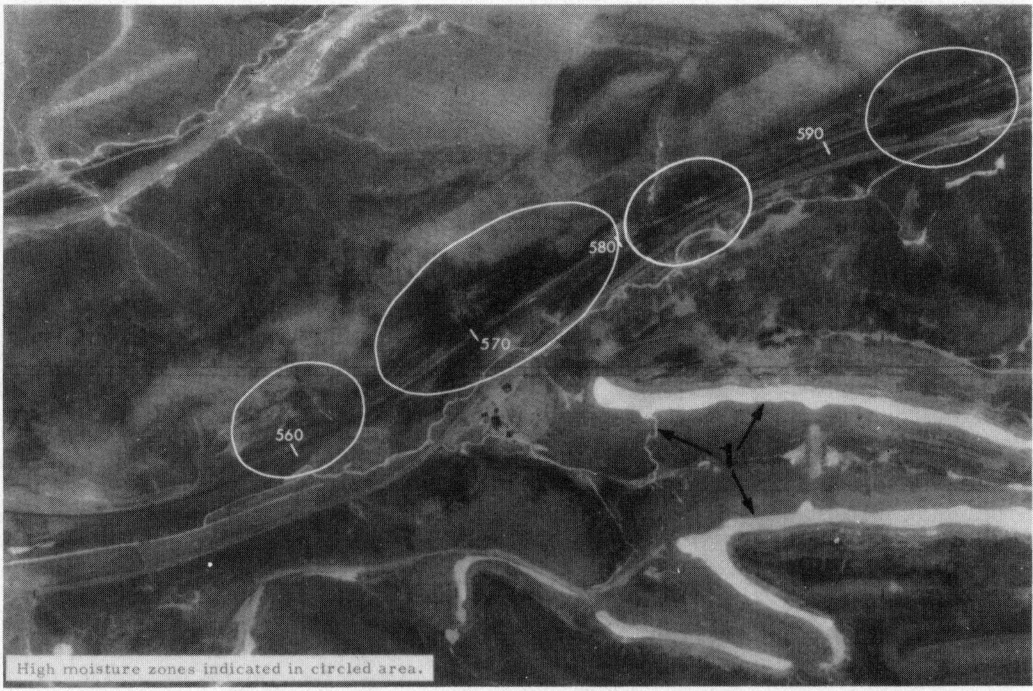

Fig. 32-75. Nighttime thermal infrared imagery, acquired at 0645 h local time, showing stations 560 to 580, and environs of the Muskingum Mine Railroad in Ohio. Dark-toned regions within circled areas are interpreted as zones of seepage and high moisture levels and have been associated with slides along the railroad right-of-way. Light-toned areas (1) represent standing water on surface. (Courtesy of D. Rubin, American Power Service Corporation, and F. J. Buckmeier, Texas Instruments, Inc.)

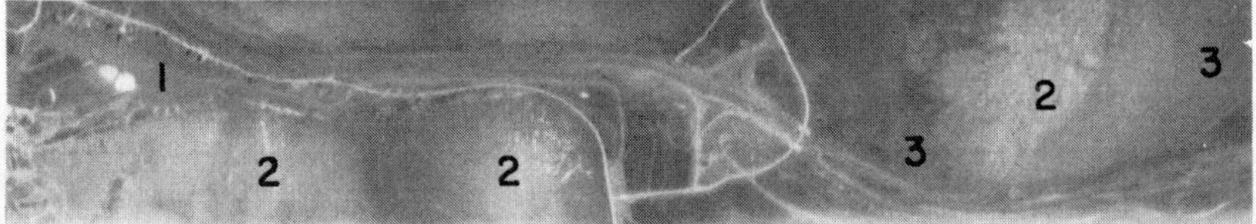

Fig. 32-76. Nighttime thermal infrared imagery of tiled sedimentary rocks in Schuylkill County, Pennsylvania. Lightest tones (1) are water or roads; next lightest (2) are shallow soils over continuous bedrock; medium gray tones (3) in bedrock area generally indicate more fractured rock and colluvial slopes.

for landslide investigations. Now that the Landsat-4 Thematic Mapper data have become available, it is expected that additional channels and higher resolution data will be much more useful in landslide studies. The detailed description of such data is discussed elsewhere in this manual.

Thermal Infrared Imagery

Thermal infrared imagery offers some unique information that cannot be obtained directly from the analysis of aerial photography. The combination of aerial photography and thermal infrared imagery provides a more accurate and complete portrayal of terrain conditions than can be obtained from either system alone. Thermal infrared imagery provides the following highly useful types of supplemental information for evaluating existing landslide and landslide-susceptible terrain:

1. Surface and near-surface moisture and drainage conditions;
2. Indication of the presence of massive bedrock or bedrock at shallow depths;
3. Distinction between loose colluvial materials of the type that are present on steep slopes and are susceptible to landslides, and the

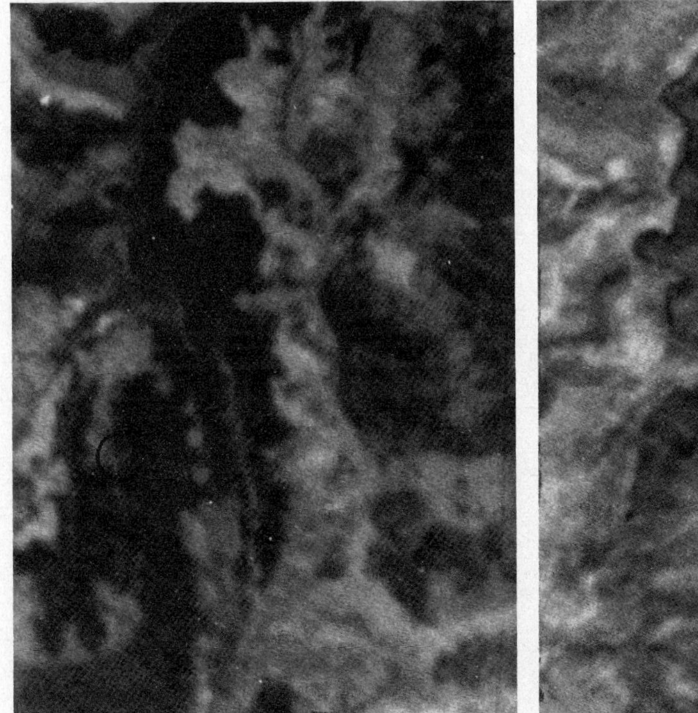

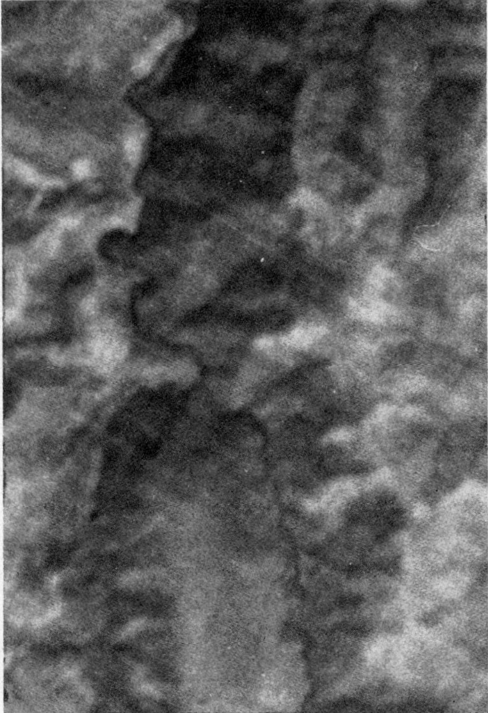

Fig. 32-78. The above images, reproduced here at a scale of approximately 1/125,000, were acquired by the Landsat MSS system on April 6, 1975 in band 5 (left image) and band 7. Band 5 appears to be very useful in monitoring changes in land use pattern, location and extent of recent harvesting activity, snow pack location, and sediment plume activity. However, these parameters are generally more readily detectable and identifiable on Landsat MSS color composites, though at two to three times the single band cost. For instance, note the difficulty in differentiating the areas harvested six to eleven years ago from other land categories in the above figure. Band 5 also appears to be the best of the 4 MSS bands to use in detecting large slide areas in forest vegetation types. This situation can be illustrated by reference to the large slide within circled area above. On the other hand, information on landform characteristics and the extent of bays, lagoons, and large stream channels is more readily extractable from Band 7 imagery. However Band 7 is generally not useful, in the region examined, as an information source for land use, vegetation type, period since forest harvest, or sediment plume detection.

1
J
2
3
4
E
H
A
6
B
D
C
5
7
8
F
9
G

massive bedrock that is more stable on steep slopes; and

4. Diurnal temperature changes that occur in soil masses (these provide clues to the soil-water mass conditions).

Tanguay and Chagnon (1972) demonstrated the usefulness of thermal infrared imagery and aerial photography for evaluating the moisture and drainage regime associated with a landslide. A flow slide had occurred within the crater of a former slide in clay-lake beds in the vicinity of Saint-Jean-Vianney, Quebec. To plan a drilling program, as necessary to evaluate the potential of further movements, required that the areas of seepage, water runoff, and wet soils be identified. By means of photography alone, these features could not be uniquely separated from areas of standing water, topographic shadows, and dense vegetation (brush and forested zones) because they all produced similar dark tonal patterns. However, the combination of photography and daytime and nighttime thermal imagery made it possible to separate these various features and identify the critical items for planning the drilling program.

Figure 32-75 shows a nighttime (predawn) thermal infrared image of an area along a railroad line being investigated for locating potential areas for landslides. The railroad had been plagued for years with landslide problems. The circled darker areas were interpreted as zones of seepage and high moisture levels—potential areas for landslides. Based on this analysis, the circled areas were drained, and no further slides have occurred in those areas.

Another example of the value of thermal infrared imagery is shown in Figure 32-76 which illustrates nighttime thermal infrared imagery of an area of tilted sedimentary rocks. The massive bedrock areas (point 2) are indicated by light tones. The fractured rock zones and colluvial slopes, which are more susceptible to landslides, are indicated by medium dark tones (point 3). These types of data from infrared imagery are useful in conjunction with aerial photography for rating the landslide susceptibility of the terrain.

A REMOTE SENSING-BASED STUDY OF REDWOOD NATIONAL PARK'S LANDSLIDE AND EROSION CONDITIONS

An area subjected to effects of severe landslide occupies certain wildlands of northern California that are quite valuable, viz. those in and around Redwood National Park. The nature and extent of the landslides and erosion in that area can be appreciated from a study of the photographic illus-

Fig. 32-81. This aerial photo mosaic shows a portion of the Redwood Creek Unit of Redwood National Park as photographed on March 9, 1972 from an altitude of 30,000 feet by using a camera with a negative size of 9 × 9″ and a focal length of 12″. The photographs were obtained immediately following a major storm of infrequent recurrence interval. North is at the top and the approximate photo scale is 1/30,000. The white dashed line is the approximate boundary for this southernmost portion of the Park and delimits the so-called "Emerald Mile" through which a section of Redwood Creek, known as "The Worm", flows from south to north. The dotted line is the approximate perimeter of the proposed 800 foot wide "buffer strip" in which the type of logging that would be permitted in the future would be carefully controlled by Park authorities. However, a large part of this potential buffer has been at least partly cutover. Letters and numbers appearing on this mosaic are keyed to illustrations in the body of the report. Areas of particular interest include the following:

"2" A timber stand known as "Big Tree Grove" which reportedly contains the world's tallest tree is at "2".

"3" An area in which sedimentation of the stream is readily apparent even on photography of this small scale (the mouth of Tom McDonald Creek) is at "3".

"5" The mouth of Bridge Creek is at "5".

"C" A portion of the Bridge Creek drainage unit, in which clear-cut logging has recently taken place, is at "C".

"6" A portion of the Bridge Creek Drainage unit in which logging has not yet occurred is immediately north of "6".

"7" The northernmost of four conspicuous "patch cut" areas is at "7".

"8" A very extensive "slip-out" area which has occurred in virgin timber, despite the fact that man has never disturbed the area, is at "8".

"D" A grass-covered area known as "Dolason Hill Prairie" is at "D".

"E" One of the largest remaining areas of virgin redwood is at "E". A smaller landslide has occurred in uncut timber adjacent to the south fork of Harry Weir Creek, southwest of "E".

"F" A grass-covered area known as "Count Hill Prairie" is at "F".

"9" A large "slip-out" area which is contributing much silt and large aggregates to Redwood Creek is at "9".

Areas which have recently been clear-cut, or nearly so, include those at "H", "C" and those immediately east of "2" and "9". In addition, the Arcata cut at and beyond the northeastern corner of this photograph at "J" has proceeded up to the edge of the proposed 800-foot-wide buffer and the Georgia Pacific cut near the mouth of Bridge Creek has gone into the proposed buffer and approaches the boundary of the established Park.

Fig. 32-82. This photo mosaic has been reduced to a scale of 1/40,000 in order to include much of the Redwood National Park area that is to the north of that shown in Figure 32-81. It is made from the March 1972 1/30,000 scale aerial photo coverage of the Park. Note that Big Tree Grove, also seen in Fig. 32-81, is at "1". Also note the extent of the cut at "2" which reaches the proposed 800 foot wide buffer strip indicated by the dotted line. The dashed line indicates the park boundary.

Fig. 32-83. Not all erosion in Redwood National Park and its environs is directly attributable to man's activity. The stereograms shown in these figures constitute a series designed to illustrate the progression of a naturally occurring land slide from 1948 to 1972. The above stereograms include Point 8 in Figure 32-81.

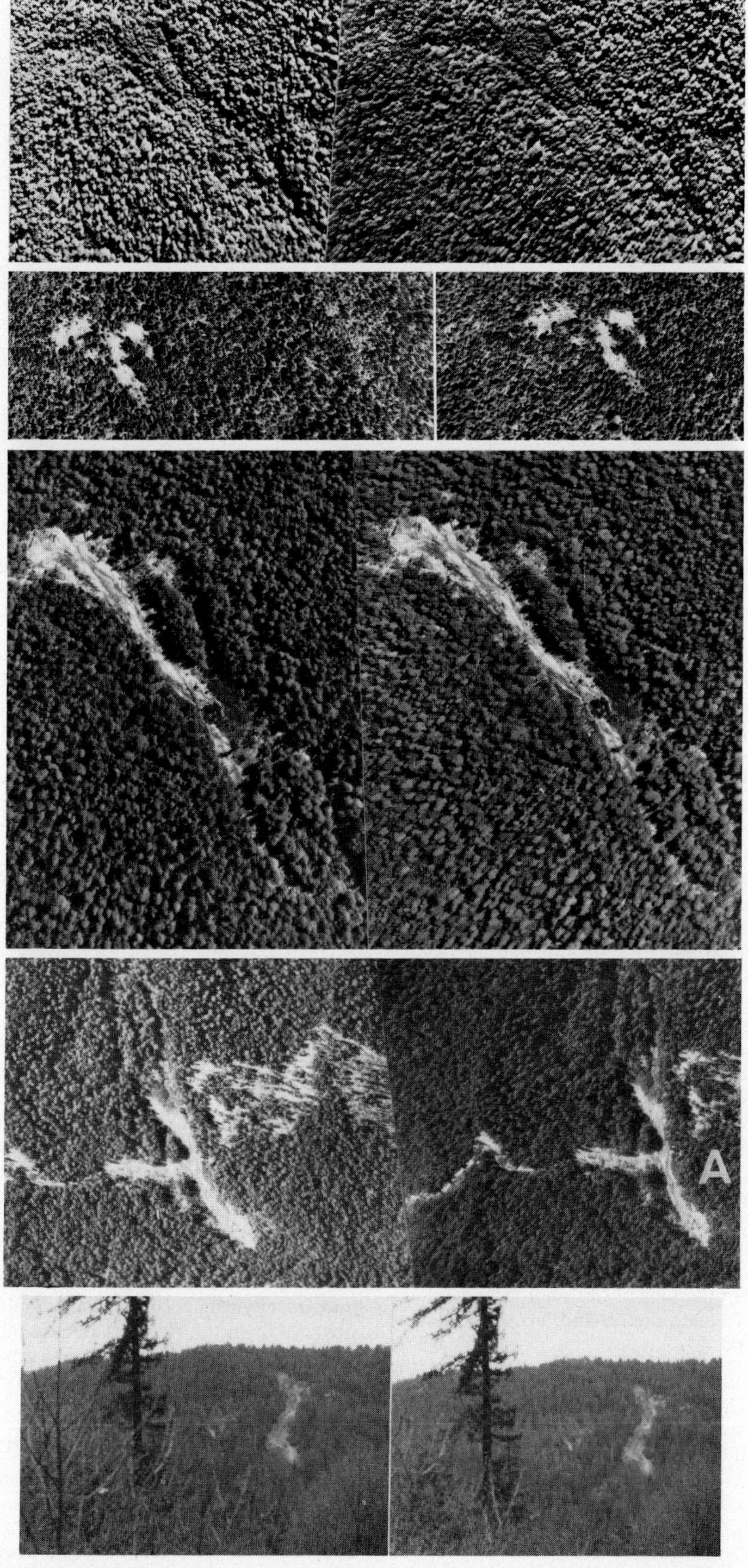
A

Fig. 32-84. Active cable logging north of the Redwood Creek unit of the Park. Note the spar pole and related machinery at "a" used to transport logs to the landing via a cable system. Note the small amount of soil disturbance along the slope at "b" as compared with that found in certain tractor-logged areas. Logging engineers have found that, by this method of yarding, soil disturbances caused by dragging logs along the surface of the ground are minimized.

trations found in Figures 32-77 through 32-88 in this section. These Figures are arranged in a series; Figures 32-77, 32-79, and 32-85 through 32-88 are in color. The series progresses from satellite imagery (acquired by both Skylab and Landsat) through high- and medium-altitude photography, to low-altitude oblique and ground photography. Hence, the reader may find it instructive to make a comparative study of this series in its entirety.

From a study of figures 32-77 through 32-88 it will become apparent that the use of remote sensing-derived information will be an advantage in the planning and implementation of various measures that will help control landsliding and erosion in and around Redwood National Park. In addition, the study illustrates the usefulness of remote sensing for the integration and documentation of many different kinds of information pertaining to erosion and landslide conditions in a given area.

The material appearing in this case study is based primarily upon work performed under two projects as reported in detail by Colwell, et al. (1973a, b). There were two considerations that prompted the obtaining of this photographic documentation of conditions in and around Redwood National Park:

(1) Unusually severe and serious landslide problems exist throughout the area. As is clear from the accompanying photos, these landslide-related problems pose a threat to what is perhaps the most impressive stand of virgin timber still present on the face of the earth—a stand that reportedly includes the world's tallest tree.

(2) A charge has been given to the U.S. National Park Service, mandating that such problems be solved. Specifically, Public Law 90-545, under which the Redwood National Park was established, states:

> In order to afford as full protection as is reasonably possible to the timber, soil and streams within the boundaries of the park, the Secretary of Interior is authorized. . . . to acquire land from, and to enter into contracts and cooperative agreements with, the owners of land on the periphery of the park, and on watersheds tributary to streams within the park, designed to assure that the consequences of forest management, timbering, land use and soil conservation practices, will not adversely affect the timber, soil and streams within the park.

The scenes, Figures 32-77 through 32-88, provide a large number and variety of high quality image types that over the years have been acquired showing conditions in the Redwood National Park and its environs. The reader should be able to observe from the figures and caption information the usefulness of each of these image types in documenting landslide and erosion conditions generally.

TRANSPORTATION FACILITIES

The selection of a route—whether for a highway, railroad, pipeline, transmission line, or drainage canal—requires several stages of analysis and the evaluation of numerous factors.

Although only long, narrow, linear strips need be located, large regions must be investigated in order to select possible corridors for evaluation and to arrive at the best and most economical route.

For more than 35 years, aerial photographic interpretation and, more recently, the interpretation of other sensor systems, have proved to be of value for assessing many important factors that influence the location of transportation systems. The most extensive application of image interpretation has been in the highway-construction field, although similar route problems have been discussed elsewhere in this chapter for pipelines, transmission lines, etc. Large highway-construction programs initiated in the United States—for example, the planning, design, and construction of 66,000 km (41,000 mi) of interstate highways—and elsewhere necessitated the use of time- and money-saving techniques, such as image interpretation, to accomplish the work. Many of the same techniques used in highway programs are obviously equally applicable to other transportation systems. Highways will be used as the primary example in this section.

Some of the major areas of highway engineering for which image interpretation techniques can be applied are: (1) highway planning, including condition and inventory surveys, and traffic surveys, (2) highway location surveys including corridor evaluation and environmental analysis, (3) construction surveys, (4) maintenance surveys, and (5) special applications such as claims and litigations, and research. Information gathered and analyzed during the highway planning phase are necessary for determining the adequacy of the present road system, the number of kilometers and types of new highways needed, and control areas through which the highways must pass. Based on these surveys, the locations of new highways are scheduled. The actual selection of the route is usually accomplished in four steps: (1) reconnaissance of area to determine feasible routes; (2) reconnaissance of route alternatives to select a route; (3) preliminary survey of selected route; and (4) location survey and staking of the route on the ground. This is then followed by the construction of the highway. Maintenance surveys are performed on the completed highway systems for ascertaining damages caused by sudden catastrophes such as floods, earthquakes, and landslides, as well as for ascertaining actions required to maintain the highway as nearly as possible in its original condition. In addition to these more standard applications, image interpretation techniques are applied to numerous special studies, a few of which are described in this section.

HIGHWAY PLANNING

Highway planning includes the orderly and continuing collection of information including history, condition, use effects, and needs, and analyses of these data for efficient and economic development of highway systems. Proper planning must be comprehensive and coordinated. It should be part of an overall master plan for area or regional development and the development of transportation facilities must be coordinated with the planning activities of other agencies and organizations. Remote sensing systems offer unique advantages for regional planning and coordination. Many examples of planning activities for regional and urban analyses are discussed in detail in other chapters of this book. Some discussion of pertinent applications to highway planning activities will be included in this section.

Condition and inventory surveys are performed in order to compile statistics on the following: (1) length of the various types of highways; (2) kind and number of structures; (3) road-surface types, widths, and condition; (4) information on land uses bordering the highway; and (5) other data. Periodic evaluation and updating of highway conditions are required for the proper planning of a highway system. Some of the detailed data needed in the inventory, such as sight distances, riding quality, superelevations and clearances can be determined best by field methods. Large amounts of the necessary data, however, can be obtained by aerial remote sensing techniques, field-photologging techniques, or combinations of these systems.

Traffic is an important factor in the planning of new highways and in evaluating the capacities and needs of existing highways. Some of the parameters required in traffic surveys include volume counts, classification (i.e., size, weight, vehicle type), speed, spacing, and origin and destination surveys, as well as the evaluation of the regional land use for possible generation of traffic.

Remote Sensing Techniques

Medium and large scale photographs play an important role in acquiring data and analyzing conditions pertinent to highway planning and to the subsequent making of highway location surveys. The availability of satellite imagery and more extensive coverage of high altitude photography has provided additional tools in accomplishing some highway-planning functions.

Aerial Photography

Medium and large scale aerial color photographs have proved very useful for identifying many of the items to be inventoried, such as road types, road widths, road conditions, land use bordering the highways, drainage conditions, and kinds and number of structures. Advantages of color over black-and-white photography for this application include better discrimination of smaller objects at a given scale and the fact that identification of the objects is more positive and more rapid on the color photography. Some investigators prefer natural color photography because the natural appearance of the objects makes them easier to identify. Others prefer infrared photography because road boundaries, road conditions (e.g., patching, break-up) and drainage features (e.g., drainage channels, seepage areas, sediment, pollution) are more discernible on this film type.

Limited availability of high altitude aerial photography (taken at a scale of 1.50,000 or smaller), until recently, has minimized the use of this potentially valuable data source in highway planning and in other areas of highway engineering. The large regional coverage per photograph and excellent resolution characteristics of the film—especially color infrared film—make it possible to evaluate both regional factors and local detailed features. Interpreters can readily identify regional factors such as geology, drainage, land use and material sources on high altitude aerial photography while analzying a minimum number of photographs. For example, to perform a stereoscopic evaluation of an area that is 40 km (25 miles) long by 6.4 km (4 miles) wide requires approximately 7 photographs at a scale of 1:80,000 versus approximately 70 photographs at a scale of 1:20,000. Detailed information such as road lengths, type and number of structures, and drainage conditions can also be obtained from these small scale photographs through use of high magnification viewers or by enlarging the photographs. Woodman and Farrell (1975), in evaluating various kinds of satellite imagery and aerial photography for highway uses, reported obtaining the same basic regional hydrologic information from 1:125,000 scale high altitude U-2 photography as they usually obtained from the 1:20,000 scale conventional aerial photography.

To overcome problems created by a lack of aerial photographic coverage in the United States, the U.S. Geological Survey through its National Cartographic Information Center (NCIC), and supported by many other Federal and some State government organizations, has recently developed, and is now implementing, a National High Altitude Program (NHAP). The goal of this program is to obtain high altitude photographic coverage of the conterminous United States on a 6-year cycle. Simultaneous coverage is obtained with 9″ × 9″ black-and-white photography at a scale of 1:80,000, and 9″ × 9″ color-infrared photography at a scale of 1:58,000. During the first 3 years since initiation of the NHAP, contracts have been awarded covering approximately 55% of the conterminous United States.

Satellite Imagery

Landsat satellite coverage of the United States and most other parts of the world has been obtained at frequent intervals since July 1972. Repetitive coverage of an area of interest is possible every 18 days from a single satellite, or approximately every 9 days when two satellites are operational. The common visual products used in a remote sensing analysis are the 23 cm (9 in) format prints at a nominal scale of 1:1,000,000 produced from the multispectral scanner (MSS) data. A black-and-white print can be obtained for each of the four bands collected—i.e., green, red, and two reflective infrared bands—or a false color composite usually can be obtained, as produced from three of the bands. Alternate products available are enlargements up to a scale of 1:200,000, or computer-compatible magnetic tapes from which computer analysis can be performed. The typical Landsat scene covers an area that is 185 Km (115 miles) on a side. It has an effective resolution of about 80 m (approximately 250 ft.), although long linear features, such as roads as narrow as 10 m (33 ft), are sometimes visible where they are in sharp contrast to their surroundings (Short et al. 1976). The Thematic Mapper data aboard Landsat-4 affords more effective resolution.

The small scale of the satellite images and their limited ground resolution have generally limited the use of satellite data to the planning and early highway location stages. Landsat images have been used in a variety of regional surveys such as (1) development of broad regional classifications of natural resources and land use; (2) evaluation of regional geology and structure; (3) development of regional physiographic settings as an initial step in developing engineering-soils units; (4) the performance of regional materials inventories; and (5) the making of inventories of drainage networks and catchment areas where the terrain features, such as topography, drainage, vegetation and land use, stand out clearly. They are also very useful in areas where topographic mapping is inadequate (Gaydos and Newland, 1978; Beaumont and Beaven, 1977).

The use of satellite date in the initial stages of highway engineering has been greatest in those parts of the world where existing map coverage and support data are inadequate, and where field evaluation is extremely difficult. Much less use has been made of satellite data by highway organizations in the more developed countries because topographic, geologic, and agricultural maps and aerial photography are readily available. The satellite data still provide very useful information regarding changes and trends in cultural items because of the repetitive and up-to-date coverage available.

Enhancement of satellite multispectral data has been especially useful for analyzing regional features such as land use, geology, landscape units and potential construction materials, and has been applied in several highway engineering projects (Beaumont, 1978). Computer-assisted analysis of satellite data has been performed using the computer-compatible tapes. This technique has resulted in a greater distinction of tonal patterns present in the satellite scenes, but the details interpreted are still limited by the coarse resolution of the scanner system (80 m). Thus this technique has not been applied to any extent in highway studies.

Condition and Inventory Surveys

One of the problems in performing condition and inventory surveys by remote sensing techniques is the need to cover large areas while identifying small features. Many of the objects can be identified on photography at scales of 1:15,000 to 1:40,000, using stereoscopes with large powers of magnification. Some items, such as the determination of pavement condition, require very large scale photography—1:6,000 to 1:2,400 or larger.

Land Use Surveys

Critical to any planning function is the need to know the existing land uses bordering the highways so that the impact of various proposals can be evaluated. The use of satellite imagery and aerial photography for determining land use and land cover conditions is adequately documented in other parts of this manual (e.g., chapter 30). The land use data can be integrated with other data types such as digitized ownership and topographic data, for road network determinations.

The sophistication and level of detail for land-use classification systems, as analyzed by remote sensing techniques, are governed by requirements of the project and the scale of the available remote sensing data. Table 32-36 illustrates the various land use classes developed for a project in Ontario, Canada. This level of detail was achieved utilizing 1:15,840 scale photography. Figure 32-89 includes a portion of the land use map developed for this project. The various applicable land use classes are presented on a mosaic which is used as the base map.

Photologging

Some features that must be inventoried are not discernible even on large-scale aerial photographs; consequently many States perform ground photographic surveys. The ground technique most widely used by highway organizations is the photologging system. The photologging system consists of a 35- or 16-mm camera mounted in a car or panel truck. The camera is situated approximately at the driver's eye-level and is aimed slightly to the right to obtain good coverage of the highway and roadside. Sequential color photographs are taken from the moving vehicle at equal increments of distance. An odometer impulse device that pulses at any desired increment permits the vehicle to travel at any speed while recording. Normally, a camera with a dual lens system is used. In such instances a primary lens provides a view of the highway and a secondary lens enables ancillary information, such as date, route description, direction and milepoint, to be recorded across the top or bottom of each frame. Figure 32-90 shows a view of a second-generation camera and control box mounted in a vehicle, and Figure 32-91 depicts a typical second generation photolog frame. The second-generation photolog system, in addition to providing pictures of the roadway, its environment, and data about the highway such as milepoint, route number and direction, facilitates the functions of taking measurements and recording information on magnetic tape. Measurements made include bearing, degree of curvature and gradient.

Approximately 41 States, the Forest Service, the Federal Highway Administration, and several municipalities in the United States and at least nine other countries are involved in photologging programs (Baker, 1980). Photolog surveys have been utilized in several stages of highway engineering, including condition and inventory surveys, and surveys of traffic, construction, and maintenance conditions. Examples of some uses of photologging are summarized in Table 32-37. This list is not all inclusive as new uses are continuously being developed.

Traffic Surveys

Aerial surveillance techniques for studying traffic were used in 1927 by Johnson (1928), who stated: "The aerial photographic survey of traffic affords a means of making far better observations than is possible otherwise." This statement is still true over 50 years later. Traffic data collected from fixed positions on the road by standard methods, such as visual counting or use of traffic counters, do not provide continuous traffic-movement information related to changing conditions of speed, volume, density, acceleration, headways in space and time, lane interactions, accordian actions, or propagation of disturbances. Aerial survey methods and special analysis techniques have been developed to provide information on these and other pertinent traffic parameters.

Aerial surveillance techniques have been used for a variety of traffic studies. These include (1) the evaluation of regional traffic patterns; (2) analysis of traffic patterns for major arteries between urban centers and within urban areas; (3) analysis of traffic patterns at tunnels, bridge crossings, on bridges, and at major airports; (4) parking studies; and (5) origin and destination surveys.

Systems used for traffic studies range from periodic visual observation of traffic conditions from the air at key locations in a highway network to the use of various camera systems, including 35-, 70-, 8-, and 16-mm movies used in a time-lapse mode; mapping cameras with 5 × 5 in. and 9 × 9 in. format; Sonne continuous strip cameras; and TV cameras with videotape recording of the data. Generally, the platforms for the cameras have been fixed-wing light aircraft and helicopters. More recently, flight-height restrictions and greater cost of aircraft have resulted in wider use of ground survey methods, including mounting

TABLE 32-36

Example of Land Use Mapping Units in the W. Flamborough-Ancaster Area, Ontario, Canada[1]

1. RESIDENTIAL—existing and under construction.
   - RH High Density Residential—areas where individual lots occupy less than ⅛ acre.
   - RM Medium Density Residential—areas where individual lots occupy from ⅛ acre to ½ acre.
   - RL Low Density Residential—Lot size greater than ½ acre but where individual houses are close enough to be grouped conveniently.
   - r Individual Residence—where houses are too widely separated to be grouped under RL—includes strip and estate residences.
2. COMMERCIAL
   - Cc Shopping Centres, Malls or Plazas.
   - C Strip Development—areas where businesses are spread along one or both sides of a roadway; individual businesses—isolated and not easily grouped.
3. INSTITUTIONAL
   - Ich Churches
   - Ic Cemeteries
   - Is Schools
   - Ih Hospitals
   - Ig Government—All levels, includes Jails, Arenas, Power Plants, Sanitary Land Fill, etc.
4. TRANSPORTATION
   - Th Highways, Roads
   - Tr Railways
   - Ta Airports, Public, Private and Float Plane Facilities.
   - Tl Power Transmission Lines
   - Tp Pipelines—Oil, Gas
5. INDUSTRIAL
   - Ie Extractive—Sand, Gravel, Clay, Stone or Mineral
   - Ip Processing—Reworking of raw material only by mechanical, heat or chemical processes to produce materials from which goods can be made.
   - Ihf Heavy Fabrication—Plants utilizing products from processing industries that require heavy lifting.
   - Ilf Light Fabrication—Plants utilizing products from processing industries that do not require heavy lifting equipment and whose products are neither very bulky nor extremely heavy.
6. AGRICULTURAL
   - Ao Vineyards and Orchards
   - As Specialty Farms—Market Gardening, Sod, Tobacco, Poultry, Fur, etc.
   - Am Mixed—Beef or Dairy operation, includes cropland, improved pasture and fallow.
7. RECREATIONAL
   - R Public—Parks, Beaches, Conservation Areas, Private Golf Courses, Ski Clubs, plus money making ventures, Miniature Golf Resorts, etc.
   - Rc Cottages—Private shoreline development.
8. OPEN SPACE
   - O Woodland, Reforested Land, Permanent Pasture, Abandoned Land or Land impossible to include in the crop rotation cycle due to physical and topographic problems, land with rock outcrops.
   - Os Swamps or marshes.

[1] Level of detail achieved utilizing 1:15,840 scale aerial photography, 1972 Conditions.

cameras at fixed stations or in moving vehicles. Not included in this discussion is the use of TV cameras at fixed stations on a highway for simultaneous monitoring of traffic conditions at several points. This real-time observation permits rapid response to emergency conditions but does not provide the information for detailed analysis of the traffic situation (unless it is recorded and analyzed at a later time).

Methods of data collection depend on the traffic parameters evaluated and the analysis technique utilized. Time-lapse photography can be controlled by repeatedly flying over a test area, by hovering over or observing from fixed points and recording traffic passing underneath, or by following a platoon of cars over the highway system. Some systems sample only a small portion (less than 10 percent) of the traffic while others sample 100 percent of the traffic over a given time period. The accuracies attained by these methods, as compared to ground collection methods, have been very high. Variances of 1 percent or less have been reported for determining traffic volume, and at the 5 percent level of significance, no significant difference was noted between the aerial and the ground methods. Similar results have been reported for vehicle speed comparisons—i.e., there is no significant difference between aerial and ground techniques at the 5 percent level of significance. Additionally, many parameters can be evaluated only by aerial surveillance techniques.

A few case studies best demonstrate how remote sensing systems can be applied.

Regional Traffic Surveys

A study was performed by the Port of New York Authority in their Sky Count Program to measure the current traffic demands imposed on a critical street network during peak traffic periods (Jordon, 1965). The approximately 2.6 sq. km. area was at the southern end of Manhattan Island where the World Trade Center complex with space for 50,000 persons was to be constructed. To fulfill a need for intensive studies of existing and future traffic characteristics, data were collected with an aerial camera with a 6-inch focal-length lens and a 9″ × 9″ format at an altitude of 2,100 m. Five observations of the study area were obtained at 15-minute intervals between 8 a.m.

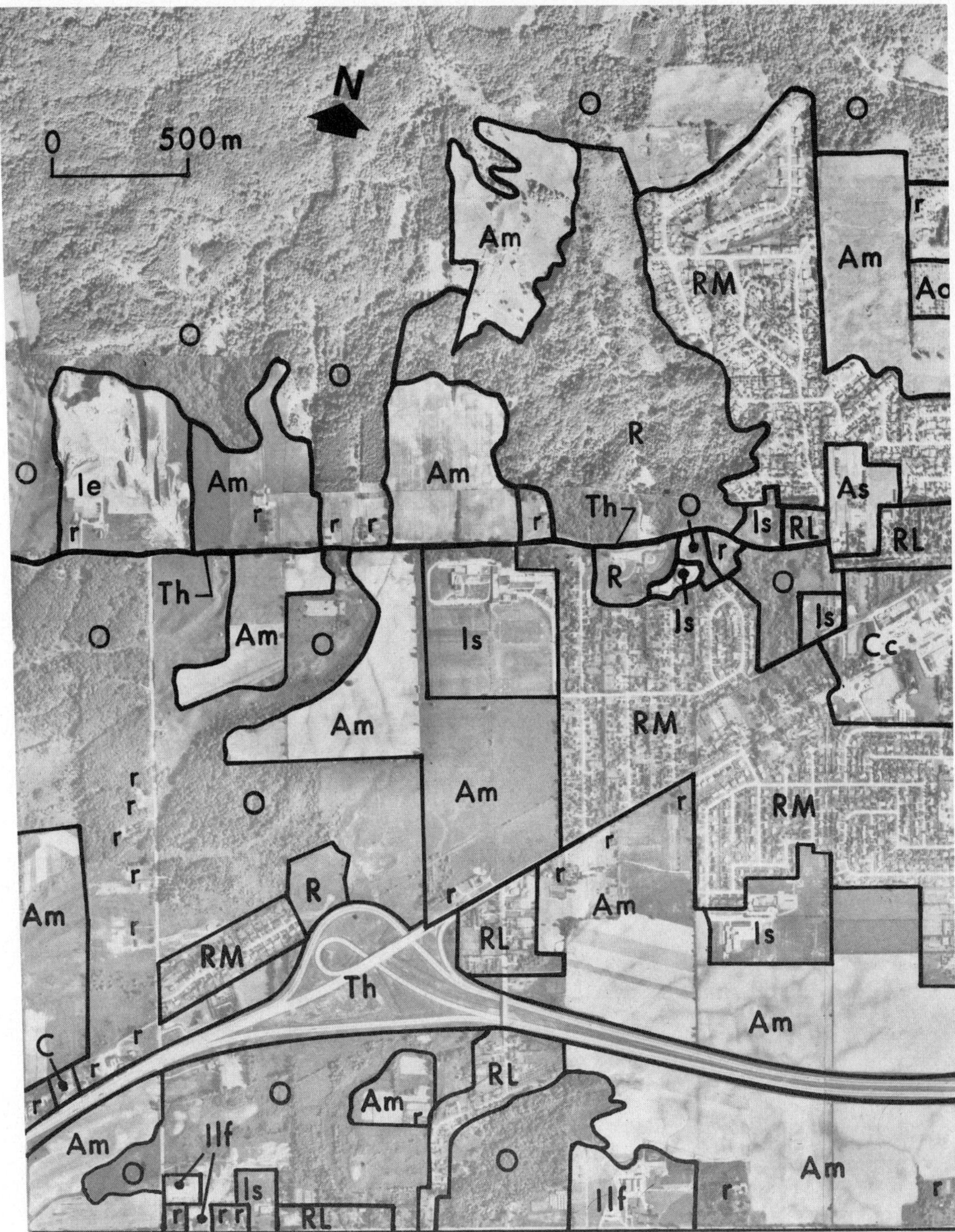

Fig. 32-89. Land use map on mosaic base for portion of W. Flamborough-Ancaster Area along Highway 403, Ontario, Canada. (Courtesy Ontario Ministry of Transportation and Communication.)

and 9 a.m. A grid segment (approximately 1 square mile)[3] containing 180 cells (approximately 400 × 400 feet in size) on a 12 × 15 array was superimposed on the negatives. Two items were recorded for each cell: the number of active unparked vehicles observed (referred to as "traffic demand level"), and the number of active vehicles found in queues (referred to as "congestion level"). These data were recorded on punch cards and an average of five observations for each cell

[3] Although the metric SI system ordinarily is used in this Manual, United States highway engineering characteristically uses the English system. All of the data in the following sections were taken in the English system, and conversion would be neither practicable nor desirable.

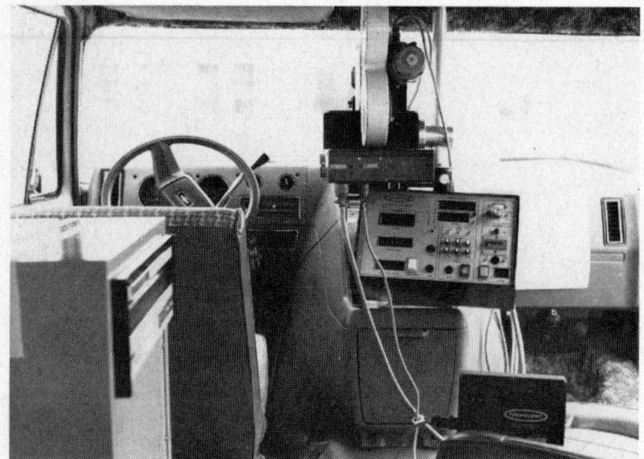

Fig. 32-90. View of second generation photologging camera and control box mounted in vehicle. (Courtesy William T. Baker, Federal Highway Administration, 1980.)

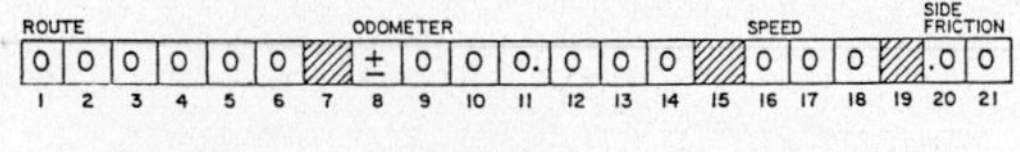

Fig. 32-91. Second generation photolog frame with data display. (Courtesy Techwest.) TOP: Graphic representation of data display on photolog frame. BOTTOM: Second generation photolog frame.

was displayed on a computer printout. Based on these data, a contour map depicting average traffic demand was prepared. This is illustrated in Figure 32-92. The heaviest demand levels (20 or greater) are indicated by the darkest shaded area in the figure. Evaluation of these two factors and a ratio of the factors—congestion level/traffic demand level (referred to as "delay ratio")—indicated that, at present traffic levels, the study area could accommodate considerably more traffic without general saturation. This technique could be applied to note the effect of a change in signalization, street directional patterns or other aspects of traffic management on the traffic flow. The value of this technique is that it does not involve consideration of traffic speed, travel times, and flow rates. It requires minimal data reduction, gives a single measure of traffic conditions over a wide area, and results can be displayed readily by computer techniques in the form of contour intervals drawn on a conventional street map.

Analysis of Major Arteries

Two notable analysis techniques have been developed for evaluating traffic parameters based on

TABLE 32-37

Examples of Photologging

Stage	Use
1. Condition and inventory	(a) reviewed in combination with aerial photos to review roadway sections relative to needs and maintenance programs. (b) compilation of historical record of changes over a period of time. (c) inventory of critical features of roadway geometrics, roadway appurtenances and land utilization adjacent to roadway.
2. Traffic	(a) accident location identification. (b) inventory and evaluation of effectiveness of traffic signals, sign locations, road lighting, channelizations, and speed zone studies. (c) analysis for indication of traffic generators and access points. (d) establishing origin and destination interview locations.
3. Construction	(a) before and after construction highway conditions. (b) use in estimating cost of repair or reconstruction in deteriorated areas.
4. Maintenance	(a) study effectiveness of various landscape designs. (b) use before and after photolog and photographic records as basis for estimating damages due to natural causes. (c) evaluation of effectiveness of various programs such as roadside spraying, snow plowing, patching, etc.

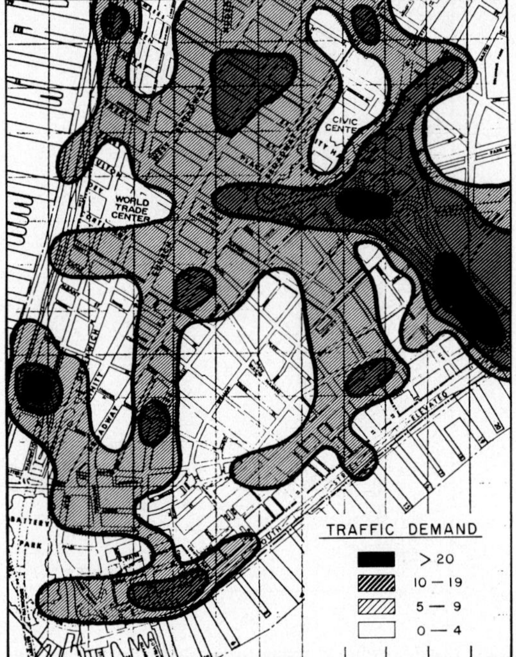

Fig. 32-92. Contour analysis of average traffic demand for an area in downtown Manhattan, New York, during 0800 to 0900 weekday peak traffic period. (After Jordan, 1965, courtesy Port of New York Authority.)

aerial data collection; these techniques are density-contour maps and vehicle-trajectory maps. The density-contour map depicts the variation of traffic density (vehicles per lane-mile) with time and distance for a given highway segment. Figure 32-93 is an example of a density-contour map illustrating a period of heavy traffic congestion (Wagner and May, 1963). The traffic-density values were determined from time-lapse photography by establishing subsections ranging from ¼ to ½ mile in length and counting the number of cars in a segment at each time of coverage. The result was converted to traffic density by the relationship.

$$\text{Traffic} = \frac{\text{vehicle count}}{\text{subsection length} \times \text{number of lanes}}$$

Road overpasses or other distinct highway features were used as subsection boundaries; they could be quickly identified on the film negative and exact distances to these features were available from plans. These data were plotted on a graph depicting the traffic density occurring in each subsection at each interval of time measured. Contour lines joining points of equal traffic density were then constructed. On Figure 32-93 densities of less than 40 vehicles per lane-mile indicate steady flow conditions and are shown in white. Densities between 40 and 60 vehicles per lane-mile depict areas of impending congestion. Densities exceeding 60 vehicles per lane-mile indicate congestion. Traffic information which can be determined by reviewing traffic density contour maps of this type include: (1) the approximate locations which act as sources of congestion; (2) the duration of congestion at any point along the highway; and (3) the length of highway under congestion at any time during the period analyzed. For example, Figure 32-93 indicates that congestion had its source near the 4¾ mile point starting at approximately 4:20 p.m., congestion persists at this point for more than 2 hours, and congestion extends backward from this point nearly 3½ miles and forward movement of the traffic is affected for another ¾ mile.

Vehicle-trajectory maps depict the movement of a platoon of vehicles as they progress along a highway. The aerial data are collected by a fast-cycling camera mounted in a helicopter which traverses the road at a fixed height and in synchronization with a moving platoon of cars. In order to obtain vehicle velocities with desired accuracy, the positions of ground-control points along the highway are obtained. In reducing the data, an analytical plotter is used to obtain the X and Y coordinates of each ground-control point, the front center of each vehicle, and the center of the photograph on each of the photographs. The data are printed out and then processed by an electronic computer to obtain map coordinates for a traffic analysis. Time-distance diagrams, such as that shown in Figure 32-94 are constructed from the computer output for each lane (Treiterer and Taylor, 1966). For each photograph, the accumulated distance to each vehicle is plotted in a vertical line at the corresponding time. The vehicle trajectory is drawn by connecting the consecutive accumulative distance points for each vehicle. An alternate method of preparing a vehicle trajectory map, reducing the time spent in data processing and replotting, includes printing only the pertinent photographic data, stripping it from the photograph, and fitting the strips on a suitable coordinate plot at the correct time axis. The fixed points are fitted to position coordinates on the time axis to form bar charts. By photographic interpretation, unique displaced objects are connected by

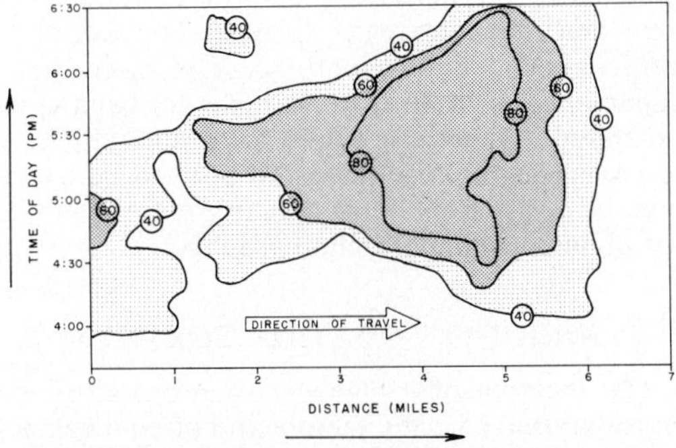

Fig. 32-93. Traffic-density contour map showing heavy congestion. (After Wagner and May, 1963, courtesy Transportation Research Board.)

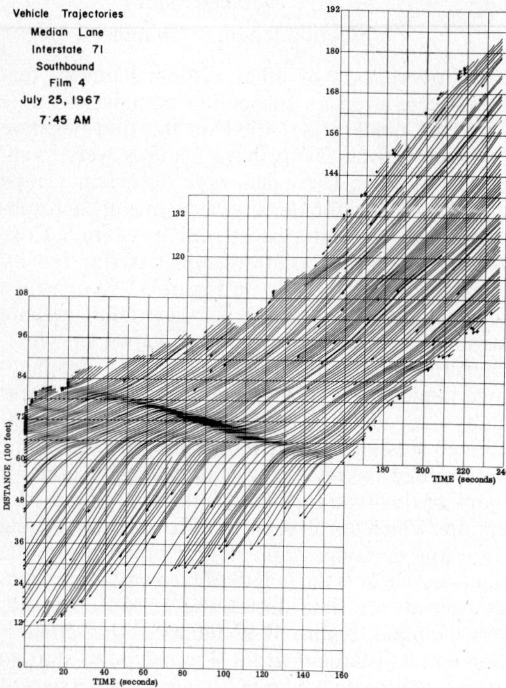

Fig. 32-94. Vehicle trajectory map. (Courtesy J. Treiterer, Ohio State University.)

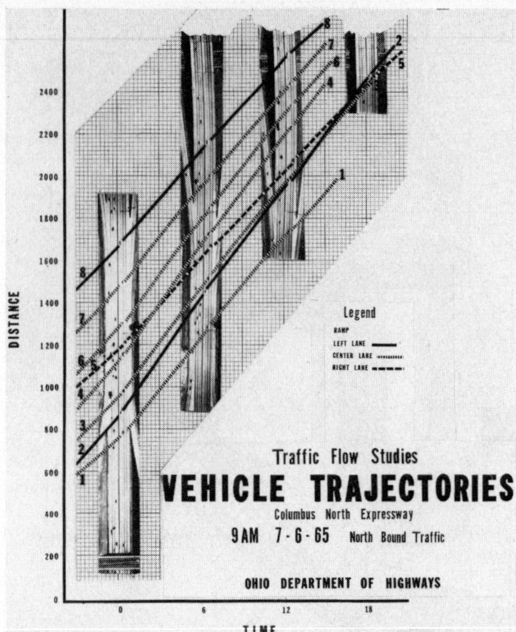

Fig. 32-95. Vehicle trajectories using photographic strips (After Herd, 1968, courtesy Ohio Department of Transportation.)

identifying trajectory curves. Less easily identified points are then joined and each trajectory is numbered for identification. Contours or grade and alignment data are plotted on one or more straight-line position bar charts. Figure 32-95 illustrates this method (Herd, 1968). By selection of different colored lines or symbols, the trajectory of a vehicle can be indicated even while the vehicle weaves between lanes. Analysis of vehicle-trajectory curves provides significant information on traffic parameters (See Table 32-38).

Origin and Destination Study

A study was performed by the Port of New York Authority to determine the origin of traffic entering specific ramps of the George Washington Bridge and the destination of traffic leaving specific ramps. Data were collected by a combination of aerial techniques and post card questionnaires handed out to motorists using the bridge. Aerial data were collected from a hovering camera-equipped helicopter at three different locations in the study area. Continuous photographic sequences were obtained during 5-minute sampling periods spaced uniformly throughout the study period. This provided approximately 50 percent time-coverage of all traffic activities at each of the three locations. A composite of the traffic flow from New Jersey to New York during the morning peak-hour traffic and a breakdown of the origin and destination flows for an individual route feeding the bridge was prepared (see Figure 32-96). The combination of these two survey techniques provided traffic engineers and planners with actual data for analyzing and improving traffic flow on the George Washington Bridge ramp system (Caggiano, 1967).

Comments on Remote Sensing Data Used

Black-and-white panchromatic film is the most widely used film type for performing traffic studies. Natural-color photography has proved valuable in studies where it was necessary to identify particular vehicles and follow their paths through the test area. Some investigations have been performed to evaluate other sensor systems—e.g., thermal imagery for nighttime surveys and radar for regional traffic inventories or vehicle-speed determinations. The operational capabilities of these systems have not been demonstrated to date. The scales have generally been too small or the resolution too poor to be useful and the cost for obtaining data is greater than that for obtaining standard photography.

HIGHWAY LOCATION SURVEYS

The location of a highway route requires the consideration of many factors and often a balancing of conflicting features; the goal is to obtain a route at minimum cost, with minimum disturbance to the natural terrain and cultural features,

TABLE 32-38

Significant Information Provided by Analysis of Vehicle Trajectory Curves

1. Vehicle speeds	The slope of the trajectory indicates vehicle velocity. Changes in slope thus indicate changes in velocity, i.e., acceleration indicated by steepening of the slope, deceleration by a flattening of the slope.
2. Density counts and headway	Analyzing a vertical section at a specific time provides the number of vehicles in a selected mile and in each lane, and the spacing between vehicles at that time. Using the method of incorporating photo slices of the traffic, information on the type of vehicle and lane changing can also be noted.
3. Volume counts	Analyzing a horizontal section at a specific point on the highway provides the number of vehicles passing the point within a given time period and the headway or gap between vehicles.
4. Weaving and disturbances	Visual analysis of photographic strips and curves provides information on weaving maneuvers, and the generation, magnitude, propagation and causes of disturbances.

and with minimum requirements for maintenance. The process of selecting a route is usually accomplished in four steps: (1) reconnaissance survey of an area to determine feasible routes; (2) reconnaissance survey of all possible or practicable routes, to select the best route; (3) preliminary survey of the selected route; and (4) location survey and staking of the designed route on the ground.

The number of factors to be considered for any route location is voluminous; a list of all possible factors is beyond the scope of this chapter. Some of the factors considered are given in Table 32-39. The factors are grouped under four basic subject areas and several major features. These areas are considered in determining the impact of a proposed route on the environment. It is evident that several specific factors are important in evaluating more than one feature (e.g., topography). These factors can be further subdivided because several specific items must be evaluated for each factor. For example, the following breakdown illustrates some of the items considered for several important factors: (1) Geology—rock types, strike and dip, massiveness, landslides, faults, fractures,

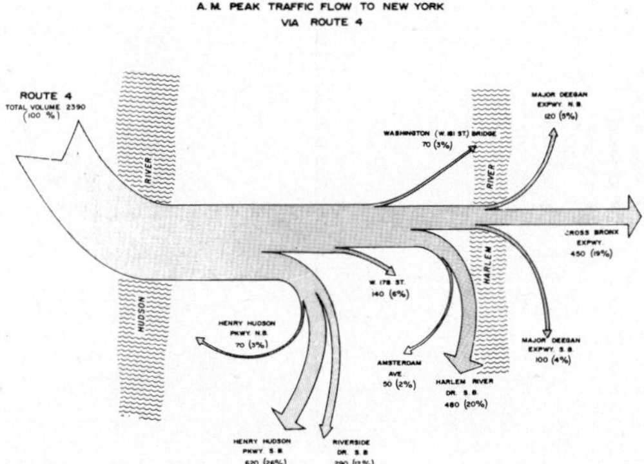

Fig. 32-96. Morning peak flow for Route 4 traffic portion of origin/destination survey for George Washington Bridge, April 1965 (After Caggiano, 1967, courtesy Port of New York Authority.)

subsidence, earthquake risk; (2) soils—presence of weak soils, swamps, erodible soils, wet soils; (3) drainage-drainage patterns, structure needs, watershed areas, flooding, high-water table, seepage, infiltration and runoff; and (4) land use—type of use, presence of historical and archeological sites, cost, severances, access. Not all factors or features are given equal consideration; they vary from job to job. Most factors, however, are considered.

Role of Image Interpretation and Corollary Techniques

Image interpretation techniques have been more extensively applied to highway location surveys than to any other stage of highway engineering. They have been proven to be of value in each of the steps as the survey progresses from reconnaissance of an area to the final staking on the ground.

Panchromatic photography is the major type of remote sensing data utilized for a route location. Color photography has been used to a limited extent, mostly for detailed analysis of selected routes, but its use has been increasing. The use of nonphotographic sensor systems has been attempted in only a few selected cases. Aerial photography also is used for preparing the topographic maps by photogrammetric methods, and for the preparation of mosaics on which the results of the various interpretative analyses are presented. The availability of satellite imagery and high altitude photography provides an excellent base for performing the regional evaluation of possible routes. Such imagery can also be used to develop the base maps for presentation of the possible alternatives.

One technique for displaying possible route alternatives for further analysis is illustrated in Figure 32-97. This presentation was developed as a hypothetical example demonstrating how various route alternatives can be illustrated in stereoscopic correspondence on the photography for further evaluation.

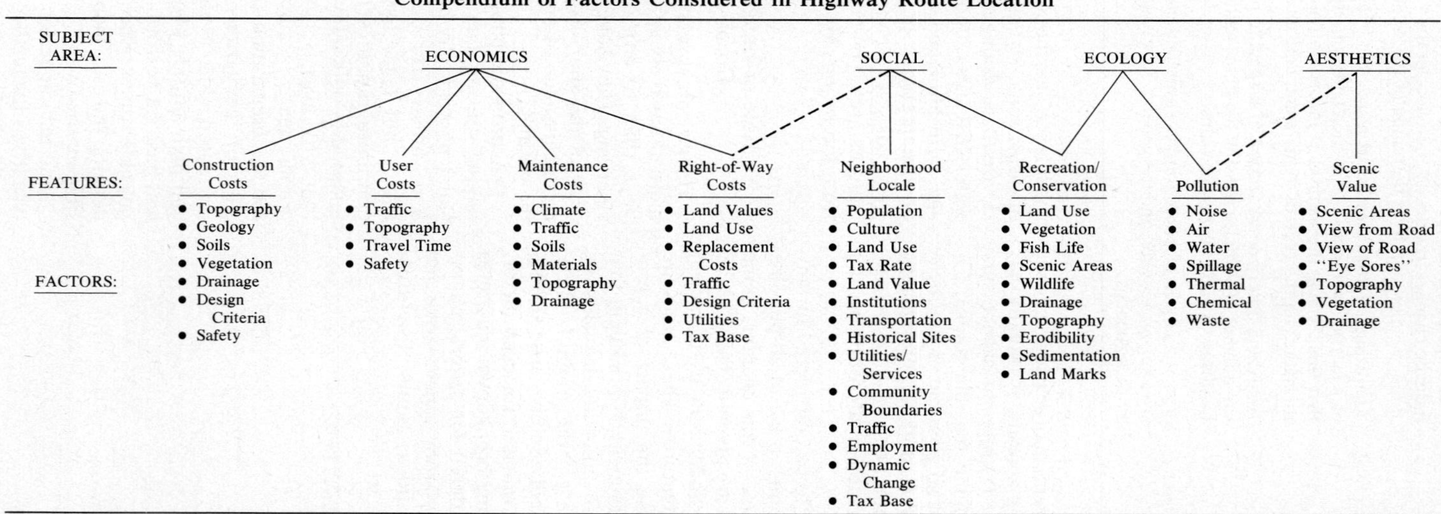

TABLE 32-39

Compendium of Factors Considered in Highway Route Location

SUBJECT AREA:	ECONOMICS				SOCIAL	ECOLOGY		AESTHETICS
FEATURES:	Construction Costs	User Costs	Maintenance Costs	Right-of-Way Costs	Neighborhood Locale	Recreation/ Conservation	Pollution	Scenic Value
FACTORS:	• Topography • Geology • Soils • Vegetation • Drainage • Design Criteria • Safety	• Traffic • Topography • Travel Time • Safety	• Climate • Traffic • Soils • Materials • Topography • Drainage	• Land Values • Land Use • Replacement Costs • Traffic • Design Criteria • Utilities • Tax Base	• Population • Culture • Land Use • Tax Rate • Land Value • Institutions • Transportation • Historical Sites • Utilities/ Services • Community Boundaries • Traffic • Employment • Dynamic Change • Tax Base	• Land Use • Vegetation • Fish Life • Scenic Areas • Wildlife • Drainage • Topography • Erodibility • Sedimentation • Land Marks	• Noise • Air • Water • Spillage • Thermal • Chemical • Waste	• Scenic Areas • View from Road • View of Road • "Eye Sores" • Topography • Vegetation • Drainage

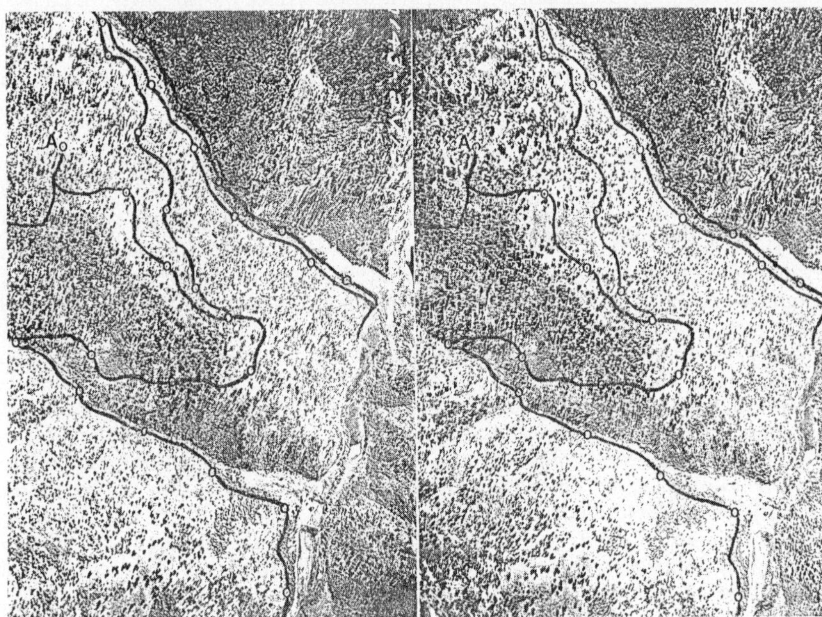

Fig. 32-97. As shown here, possible alternate corridors to be evaluated can be placed on the aerial photography in stereo correspondence for further discussion and analysis. As a hypothetical example, several possible routes at 10 percent grade were developed from Point A, near the crest, to the road in the lowlands. The technique includes the basic photogrammetric principles.

Computer techniques have been applied in some recent surveys to quantify and evaluate the relative influence of various factors on route selection, and to generate perspective views of the proposed highway in order to evaluate the aesthetic and safety qualities of the road alignment as viewed by the driver. Computer techniques have also been used to determine earth quantities by photogrammetric methods for designing a road, comparing costs of alternative alignments and for determining final pay quantities during the construction of a road. In these applications, photography or information interpreted from photography has been the basic data source.

Digitized data also have been used for accessibility mapping of undeveloped areas. Such a map can be a useful product in large scale road network and land-development planning. Khorram (1982b), studied the potential residential suitability of a wildland area in Northern California by integrating various kinds of information including the accessibility map. This map was produced by digitizing the road network and registering the data within a geobased information system compatible with other data types used in this study (see Site Investigations and Regional Planning sections of this chapter).

The following sections include discussions and some case studies demonstrating the application of image interpretation, in some instances coupled with other analysis techniques, for route location and for evaluating environmental impacts of highways.

Corridor Studies

Various State highway organizations have prepared strip maps for corridor studies using the same format and techniques as those used for performing regional inventories. For example, New Mexico prepared aggregate-resources maps and soils maps for interstate corridors; Wyoming has prepared materials maps for interstate corridors; Indiana and Louisiana have prepared engineering soils maps for interstate corridors; Maine has prepared drainage maps and engineering soils maps for route location studies; and Florida has prepared maps depicting land use, drainage, soils, vegetation, and key features for recent corridor studies. Examples of region-wide studies are illustrated in some of the earlier sections of this chapter.

Environmental Analysis—Case Studies

The need to prepare environmental impact studies and to obtain a greater degree of public involvement in the highway route selection process provided a fertile field for the application of remote sensing techniques. The use by highway organizations varied widely. Those organizations previously making extensive use of remote sensing relied heavily on remote sensing techniques in evaluating a variety of environmental factors. Some developed special methods using remote sensing techniques for depicting sensitive areas such as wetlands and wildlife habitats; for evaluating noise-impact and soil-erosion characteristics; and for detailing terrain features (soils, geology, topography, drainage) and cultural features (archeology, utilities, land use). In organizations where remote sensing was not used extensively, aerial photography was used mainly to provide a view of the areas under investigation and to indicate possible route alternatives (Rib, 1977).

Corridor Analysis, Butler County, Kansas

Factors evaluated in this study (Kansas State Highway Commission, 1974) are grouped under two major features—physiography (soil and geologic conditions, topography, drainage areas) and cultural features (archeology, utilities, land utilization, cultural improvements). Black-and-white photography at a scale of 1:24,000 was interpreted to delineate major soil types and estimate soil-mantle thickness; to determine the boundary of a critical geologic member; to indicate the drainage divides; to locate major utility lines (aided by use of utility maps and some field reconnaissance); and to delineate land use and cultural features.

Wiscassett By-pass, U.S. Route 1, Maine

Three possible corridors within the study areas were considered (Maine Department of Transportation, 1972). The factors evaluated included technical data (traffic, safety, route requirements, design and physical features, construction costs), man-made environment (land use, socioeconomic, recreation, air quality, noise quality, aesthetics) and natural environment (topography, geology, soils, surface water, vegetation and wildlife). The impact of each of the corridors investigated on these factors was evaluated. Many of these factors were interpreted directly from aerial photography or indirectly through related features which were identifiable on aerial photography.

Figure 32-98 is an example of a photo-identifiable feature that is used as a secondary indicator for evaluating environmental impact. In this case it is used to determine the impact of the corridors on wildlife which cannot be directly detected on the photography. This figure depicts areas of tidal wetlands. Details of vegetative growth found in this area are based on spot field observations.

In evaluating the tidal wetlands, it is noted that waterfowl and other wildlife depend on the luxuriant vegetation found in the saline tidal mudflats as a food source. These are the only areas open for food production during the critical spring and winter seasons. Construction along this route would change these from saline to freshwater marshes which would remain frozen during the critical spring and winter seasons. Therefore, of the three possible routes, Corridor B, which would involve filling of approximately 7 to 8 acres of tidal marshland, would have the most serious adverse effect on the tidal wetlands. Corridor C also presents problems; it crosses a mudflat favored by marine-worm diggers. The loss of this mudflat area could have an adverse economic impact on this industry.

Application of Aerial Mapping in the Development of Tollways and Turnpikes

Two factors evaluated in this study (Hawkes and Brown, 1973) which are of special interest and for which aerial photographic techniques are used are noise-impact and soil-erosion characteristics. Noise contours are developed by computing acoustic noise levels from basic alignment geometrics, traffic volumes, and vehicular mix. The noise contours are superimposed on terrain contours developed by photogrammetric mapping and adjustments are made for topographic relief (e.g., whether a cut section or fill section). Noise-sensitive objectives are located on the map and determinations are made as to where special acoustic barriers are required. Seismic vibration was also evaluated in this study.

The degree of erosion hazard is assessed by evaluating soil textures, existing slope class and amount of vegetation cover. After a determination has been made of the areas of high erosion potential, locations of temporary and permanent erosion control devices are depicted on the photogrammetric maps. The quantities of materials required to construct the various devices are then estimated for use in developing construction costs.

Surveillance of Environmental Impact, Ontario, Canada

Construction of any new transportation facility brings about alterations in the physical and biological environment during and after its development. Least understood is the amount of disturbance or impact occurring at a site as a result of any highway engineering activity. A long-term study being conducted by the Ontario Ministry of Transportation and Communications deals with the role of airborne remote sensing systems in monitoring and predicting the interaction of highway location, environmental effects and engineering actions before, during and after construction in a given area. (Mathur, 1979).

The parameters studied by remote sensing include:

(a) landforms
(b) soils: (1) organic material, (2) inorganic—textural type
(c) topography: (1) areal, (2) site
(d) surface drainage
(e) subsurface drainage: field tiles
(f) hydrology: flood plain boundaries
(g) natural vegetation: (1) species, (2) stress condition
(h) land use: (1) farm types, (2) urban, (3) recreation and culture

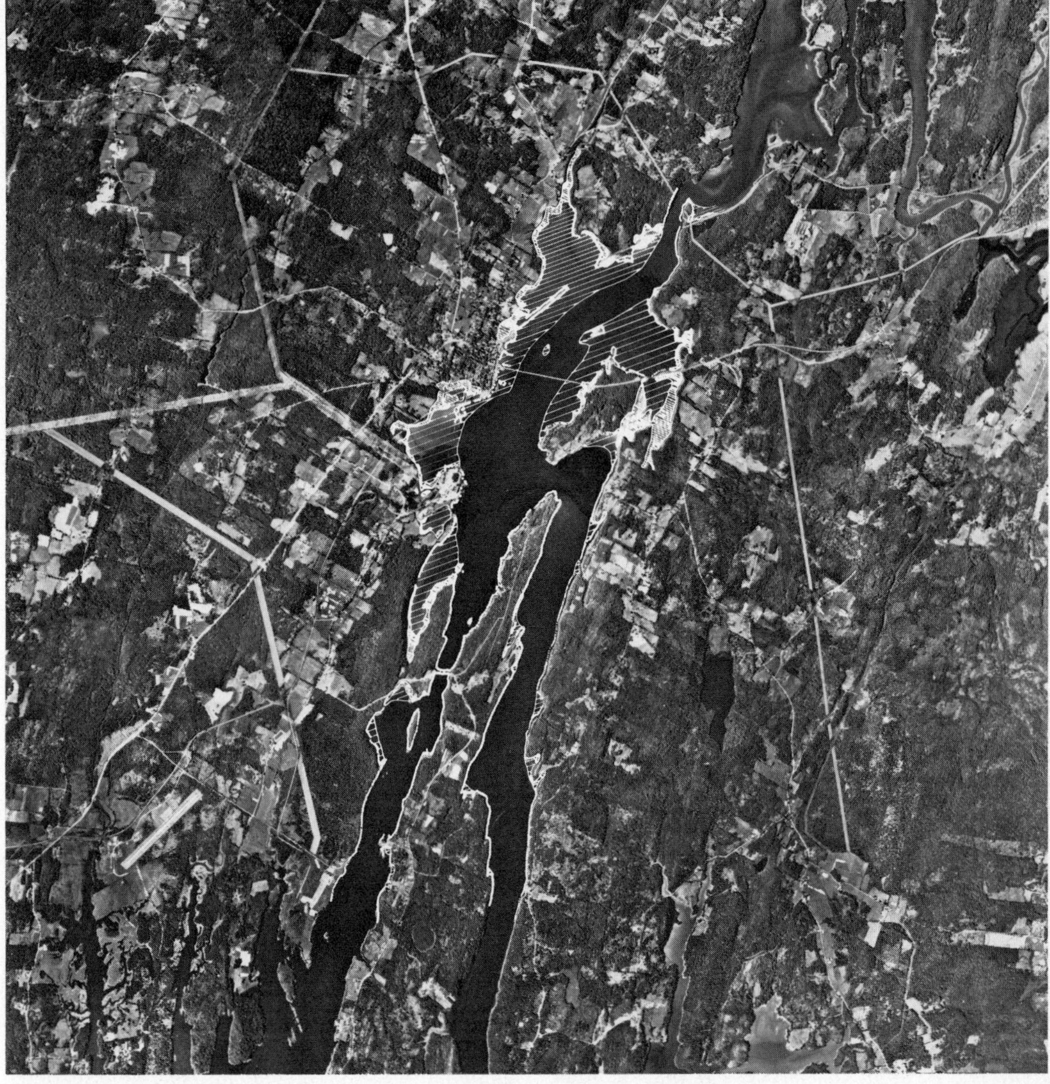

Fig. 32-98. Portion of environmental impact study photo map depicting tidal wetlands. (Courtesy Maine Department of Transportation, 1972.)

(i) crop types
(j) erosion

The sensors selected for data collection were guided by the types of environmental parameters to be studied. Experience has shown that parameters on physical terrain, ground cover and land use can be adequately studied with conventional aerial photography. Others, such as soil quality and water quality parameters, require more specialized techniques. Therefore, a sensor package consisting of a metric camera, two multispectral camera systems, an infrared line scanner and a radiometer was assembled.

Conventional aerial photography is the major source of data and provides for a rapid, economic analysis of the two test sites. Color photography at a scale of 1:5,000 shows a vast amount of detail. Color infrared photography at a scale of 1:5,000 enables vegetative stress to be studied in detail. Each batch of color infrared film used should be assessed beforehand as to its infrared balance according to the standards developed by Flemming (1979). Early work in this context showed varying degrees of balance of the various batches of film used for test purposes. Color photography at a scale of 1:10,000 and black-and-white photography at a scale of 1:25,000 are useful for detecting and analyzing such parameters as landforms and flood plains, which require extensive coverage. For a project of this magnitude, a combination of these photographs is required to meet the requirements of all the selected parameters.

To meet the requirements in data collection and analysis, photography and imagery are acquired periodically for the before, during and after construction stage of the highway. For each stage, coverage is required seasonally i.e. spring, summer, fall and winter. For spring, summer and fall 2 flights each are acquired: one during the day, to operate all the primary and supplementary sensors, and one at night for the infrared line scanner. For winter, only color infrared photography is acquired. A limited amount of field work, in the form of sampling and ground observations, is carried out during and subsequent to the acquisition of the remote sensing coverage.

Analysis of the data collected to date indicates that the remote sensing techniques are highly reliable, economical, and hence essential, in the assessment of many environmental parameters pertaining to highway construction and use. Reflected and emitted portions of the electromagnetic spectrum, when used in suitable combinations, can provide the required information at various levels of details. Figure 32-99 and Color Figure 32-100 illustrate some of the information and the level of detail being cataloged for this study.

Public Hearings

The use of remote sensing images has proved to be very beneficial in assisting highway organizations in their interaction with the public. Enlarged natural color photographs and color mosaics have been well received at public hearings. It is easier for the lay public to relate to a photograph showing the actual existing conditions, than to graphical representations of the topography, land use and proposed alignments.

One technique that has worked exceptionally well at public hearings and other conferences dealing with the selection of route alternatives is the use of photomontages. This technique involves the overlaying of a computer perspective plot of a proposed route over an oblique photograph of the area. The basis of this process is that of determining the camera position and line-of-sight for the oblique photograph from which the perspective plot can then be properly merged at the proper scale and positioned so that it appears to be part of the original landscape scene. Once the perspective plot is overlaid on the photograph an artist can enhance the plot to make it more realistic. Color Figure 32-101 is an example of a photomontage prepared to illustrate how the completed facility would merge into the landscape and land use. This process is almost as good as, and a lot less expensive than, building a model of the area to depict what the final route would look like after construction. An alternate to this method is that developed by the Arizona Department of Transportation which shows the perspective plot in stereo correspondence on a stereo pair of photographs (Figure 32-102).

Evaluation of Alternate Corridors

The selection of an optimum corridor from among the various corridors evaluated can be a difficult step. Several of the factors considered often may be in opposition. For example, the route with the minimum construction costs may have the most adverse effect on recreation or wildlife factors. Several methods have been developed in attempts to determine the influence of various factors and to note the overall effect if the relative influences of these factors were varied. Generally these methods consist of (1) reducing the various factors to a common basis (or base map), (2) establishing a relative ranking of the various items delineated for each factor, and (3) developing some method of quantifying and displaying the data so that various factors can be compared. In the various methods developed, remote sensing techniques provide a means of obtaining and displaying many of the factors under consideration. Three case studies are included to provide a sampling of methods that have been developed.

Estimation of Construction Costs, Ontario, Canada

A new alignment was to be constructed to replace an existing section of a highway in the Precambrian Shield area of Southern Ontario, Canada. Several sets of terminals (and therefore several new alignments) were to be considered. The schedule was such that information on each alignment and its estimated cost was required in a very short time. More importantly, final selection of the terminals and the corresponding alignment depended on the estimated costs. Locations of alignments were performed routinely, considering only such engineering parameters as topography, soil, drainage, etc. Because of time restrictions, only those photos already in the file could be used. In this case, they were black-and-white photos at a scale of 1:15,840, taken in summer. To estimate the construction cost, each proposed alignment was first divided into one-mile sections. Within each section, the clearing, grading and major drainage requirements were estimated and then compared to reference values developed from actual construction costs of similar types of highway in a similar geographical area to arrive at an estimated cost. Summation of the costs for each section gave the total for the alignments. Table 32-40 lists the information developed for one of the alignments. Costs obtained in this manner may differ from the actual "as built" costs but when several alignments are compared, the relative values should be correct. The entire study (from receipt of the request to submitting the report) was completed in about one week. A new alignment was selected and has since been constructed, all within schedule.

Comparison of General Routes by Terrain Appraisal Methods, New York

The factors evaluated in this study (Hofmann and Fleckenstein, 1960) were earthwork costs and terrain classes. Since approximately one-third of

Fig. 32-99. Surface drainage parameter, an indication of the level of drainage details required for the Ontario Environmental Surveillance Project. (Courtesy Ontario Ministry of Transportation and Communication.)

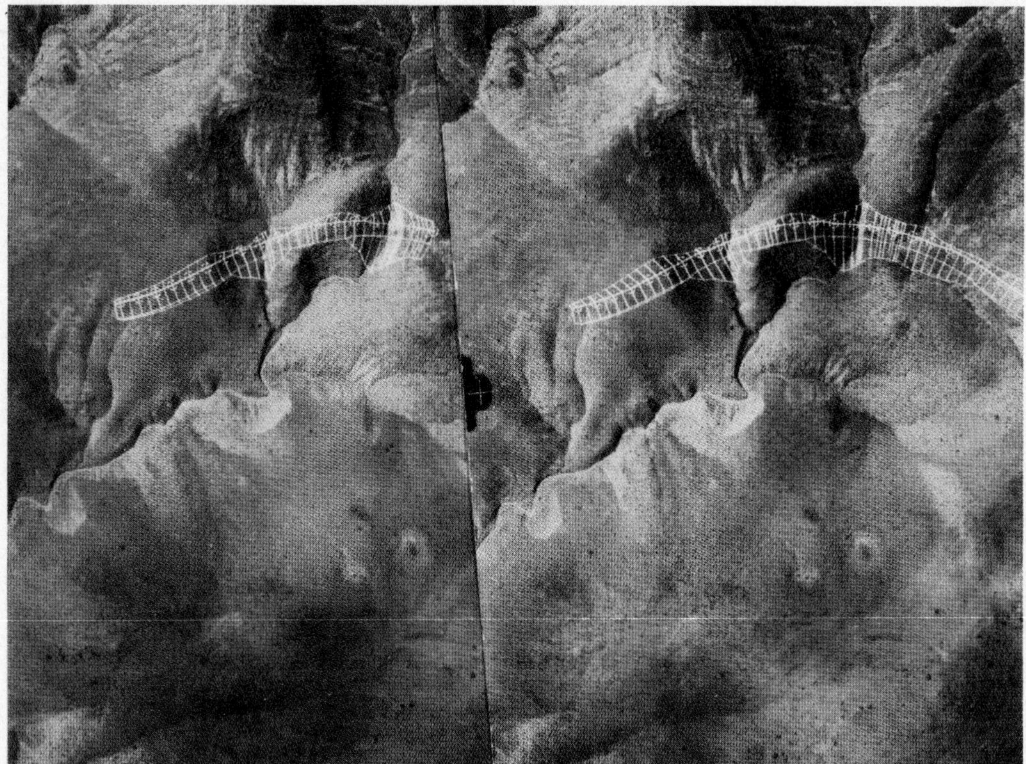

Fig. 32-102. This figure illustrates another of the various techniques which have been developed to portray the appearance of a proposed engineering structure in the natural landscape setting. The location of the highway segment was determined from topographic maps compiled by photogrammetric methods. The coordinates of the highway cross sections for this segment were measured on the topographic map; the photographic coordinates for the highway sections were then calculated by use of special computer programs. This permitted the superposition of the highway segment cross-sections directly onto the photographs for viewing in three dimensions. Areas requiring the construction of cuts and fills and drainage structures are readily apparent. The aesthetic appearance of the road and location of scenic views from the road can also be determined. (Courtesy Arizona Department of Transportation.)

the cost of modern highways is spent on earthwork, it was felt that, other factors being equal, the route-traveling terrain requiring the least earthwork should be selected. Figure 32-103 illustrates the terrain-class map developed to evaluate the two routes under consideration. This map was developed from a review of literature, analysis of aerial photography, and field inspection. The relative earthwork costs for each of these terrain classes were derived by comparing the actual earthwork costs for similar terrain classes on a road recently constructed to the same standards. From these data, earthwork terrain-factors were developed which indicated the relative cost per mile for each terrain class. Once these factors were established, it was necessary only to multiply each terrain factor by the number of miles in the corresponding terrain class and add up the total for each route. The route with the lowest total was the least costly route. Using this method, it was determined that costs for the western route would be 7 percent less than for the other route.

Computer-Aided Analysis of Alternate Route Locations, Southern Tier Expressway, New York–Pennsylvania

A comprehensive study was instituted for the Southern Tier Expressway, which extends approximately 100 miles from Hinsdale, New York, to I-90 northeast of Erie, Pennsylvania. The objective of the study was to evaluate all potential highway corridors (Turner, 1978). It was important that potential corridors selected for detailed cost evaluation and environmental impact assessment be identified on the basis of social, economic and ecological considerations as well as engineering feasibility. To evaluate the large number of factors, a computer-aided system containing two companion programs was used: Generalized Mapping Analysis and Planning Systems (GMAPS) and Generalized Computer-Aided Route Selection (GCARS). To develop as comprehensive and complete a group of highway alternatives as possible the GMAPS-GCARS

TABLE 32-40

Estimation of Construction Costs for One Route Alternative[1]

Mile	Clearing	Grading	Major Water Crossing	Remarks	Estimated Costs (1975)
0–1	0.25 mi. light 0.25 mi. heavy 0.50 mi. medium	0.25 mi. light 0.75 mi. medium	1–50 ft.±	—intersects hydro lines at mile 0.2 —0.3 miles over glacial till —traverse 500 ft. ± of organic deposits	$ 320,000
1–2	heavy	light–medium		—some rock outcrops —traverses about 200 ft.± organic deposits	$ 330,000
2–3	0.2 mi. light	0.5 mi. light		—intersects one access road	$ 315,000
	0.8 mi. heavy	0.5 mi. light		—traverses 100 ft. organic deposit	
3–4	light	0.5 mi. medium 0.5 mi. heavy		—traverses 300 ft.± organic deposits —alignment over bedrock outcrops	$ 355,000
4–5	heavy	medium–heavy		—alignment close to one beaver pond —traverses some shallow organic deposits	$ 345,000
5–5.8	heavy	medium	1–10 ft.±	—portions of alignment on shallow till over bedrock —some rock outcrops	$ 230,000
				Total	$1,895,000

[1] Ontario Ministry of Transportation and Communication.

analyses were checked by an independently conducted conventional transportation analysis.

The GMAPS programs utilize a composite computer-mapping technique. This technique is basically equivalent to the manual procedure involving the construction of successive transparent overlays on which values are represented by graduated tones or colored shadings that indicate a relative value for a particular factor. In composite computer-mapping, the overlaying of tonal transparencies is replaced by the algebraic combination of two or more matrices whose elements have numerical values corresponding to the grey toned densities.

Data for 22 baseline maps (Table 32-41), describing a variety of engineering, cultural, economic and environmental factors, were plotted on a 1:62,500 scale base. These data were converted to a cellular matrix representation—cell size 650 feet by 520 feet—and entered into computer storage via the GMAPS program. This program also allows for the manipulation of databases to create new models by overlaying or "compositing" techniques.

From the baseline data listed in Table 32-41 new derivative, determinant and composite maps were produced by composite computer-mapping. Each derivative map provides a more refined or specific description of some aspect of the regional en-

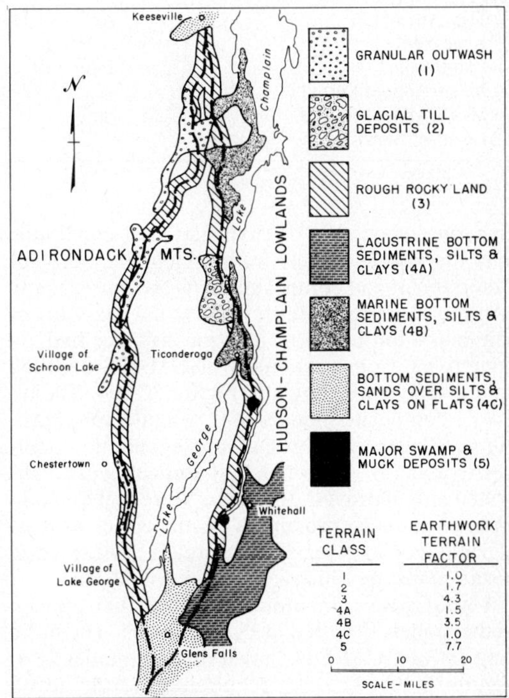

Fig. 32-103. Terrain class map. (After Hofmann and Fleckenstein, 1960, courtesy New York State Department of Transportation.)

TABLE 32-41

Baseline, Derivative, Determinant, and Composite Data Displays (After Turner, 1978)

Baseline	Derivative	Determinant	Composite
Accessibility	Land Values		
Land Use			
Pipelines and Transmission Lines	Erosion Potential	Land Acquisition Cost	
Existing Transportation			
Landforms	Existing Maintenance	Construction Cost	Highway Site
Soil Types	Conditions		Feasibility
Slope			
Mean Annual Rainfall	Geotechnical Factors	Maintenance Cost	
Mean Annual Snowfall			
Water Bodies	Drainage Potential		
Ecologically Sensitive Areas	Water Quality Sensitivity		
Groundwater Yield			
Water Bodies	Erosion Potential		
Recreation		Ecological Impact	
Land Use	Vegetation Types		
Landforms		Land Use Impact	
Slope	Scenic Sensitivity		
Agricultural Districts		Energy Utilization	Environmental
Areas of Highway Needs	Agricultural Produc-	Impact	Impact
Soil Types	tivity		
Historical, Archeological, and	Noise Sensitivity	Cultural and Social	
Cultural Sites		Impact	
Population Density	Air Quality Sensitivity		
Level of Service			
Outmigration	Areas of Economic		
Unemployment	Need		
Land Use	Institutions, Recreation		
Average Family Income	and Commercial	Stimulate Regional	
Recreation	Areas	Economy	Social and
Population Density	Growth Centers		Economic
Trip Attraction	Areas of Highway	Improve Accessibility	Benefits
Accessibility	Need		
Landforms	User Costs	User Benefits	
Mean Annual Snowfall			
Mean Annual Rainfall	Safety		
Level of Service			

gineering, economic or environmental conditions than the baseline maps from which it is derived. Determinant and composite maps are increasingly complex models depicting some aspect of suitability for highway location as specified by their titles. Some examples of GMAPS baseline data displays are shown in Figure 32-104. The ultimate composite models of (1) engineering feasibility, (2) improving social and economic conditions, and (3) environmental impact were calibrated and approved; then they were combined in various ratios to produce a sequence of (4) total highway-corridor feasibility models. These composite maps are illustrated in Figure 32-105.

The GCARS computer system was then applied to the models depicted in Figure 32-105. The basic concept of GCARS is shown diagrammatically in Figure 32-106. The first requirement is the development of base maps for each factor evaluated. The maps are then digitized, thus converting a graphical measure to a numerical measure. A numerical rating system is developed for each of the items delineated for the factors mapped. This step is essentially completed by the GMAPS procedure. The numerical measures for each factor are then converted to values by applying the rating scale; thus numerical cost models are developed. These factors are then depicted as solid three-dimensional surfaces (See Figure 32-106) where the "highs" indicate high costs and the "lows" indicate low costs. In actual practice, models are stored as matrices within a computer. Then, through the use of established programs, minimum paths can be selected across each model. Desirable routes will follow the "valleys" across such "cost models." The most desirable route combines "directness" and low "elevations" so that the lowest "total cost" may be obtained. Less desirable routes follow other valleys and passes over the intervening high-cost areas. Sometimes a shorter route that has a higher cost-per unit length, may be the most desirable. Routes are compared in terms of overall length and total cost. Models for several factors can be superimposed and summed to produce cost models for any desired combination of factors. Before sum-

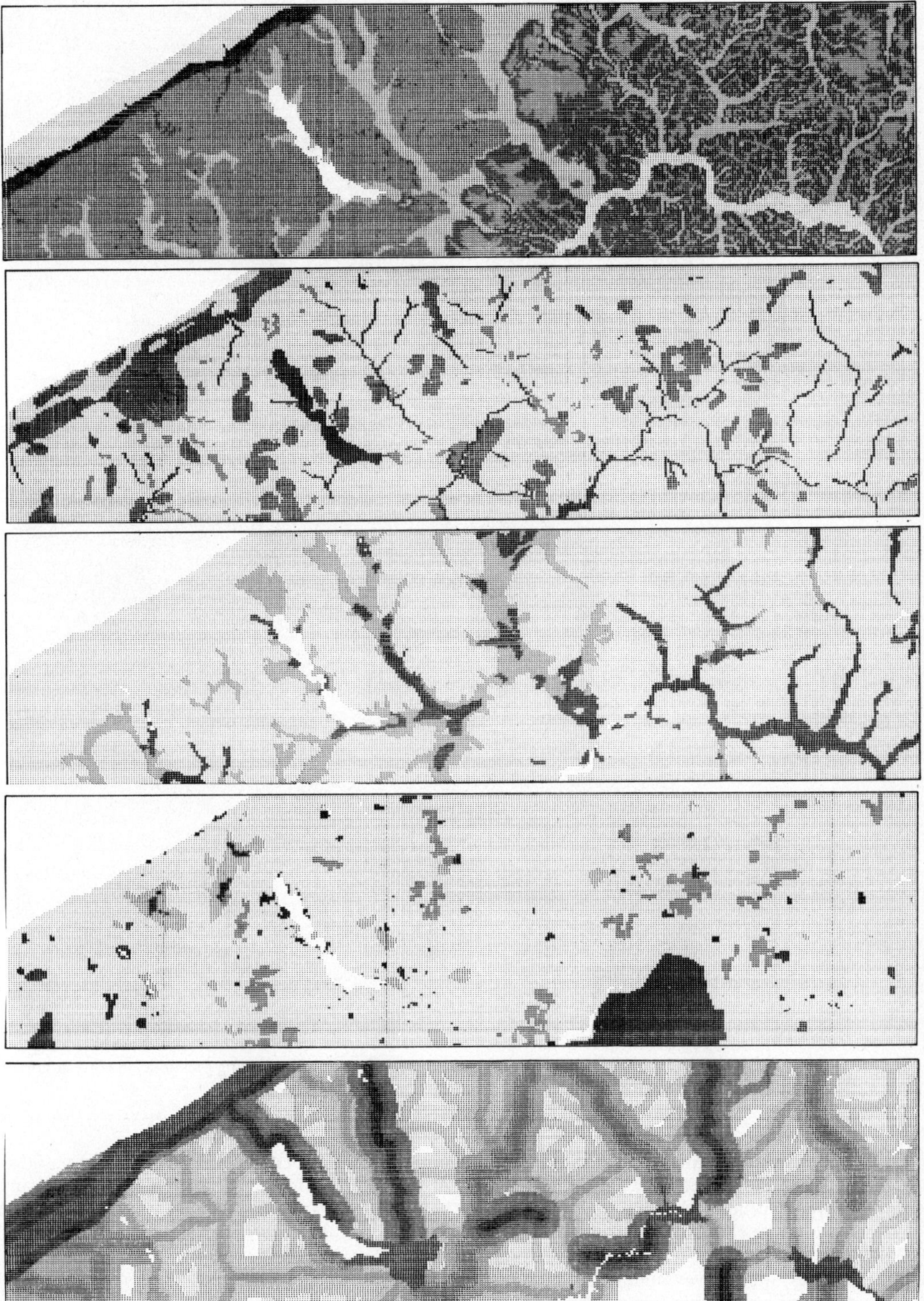

Fig. 32-104. Five GMAPS maps displays. (After Turner, 1978.) a. Landforms, b. Ecologically sensitive areas, c. Groundwater yield, d. Recreational sites, e. Road accessibility

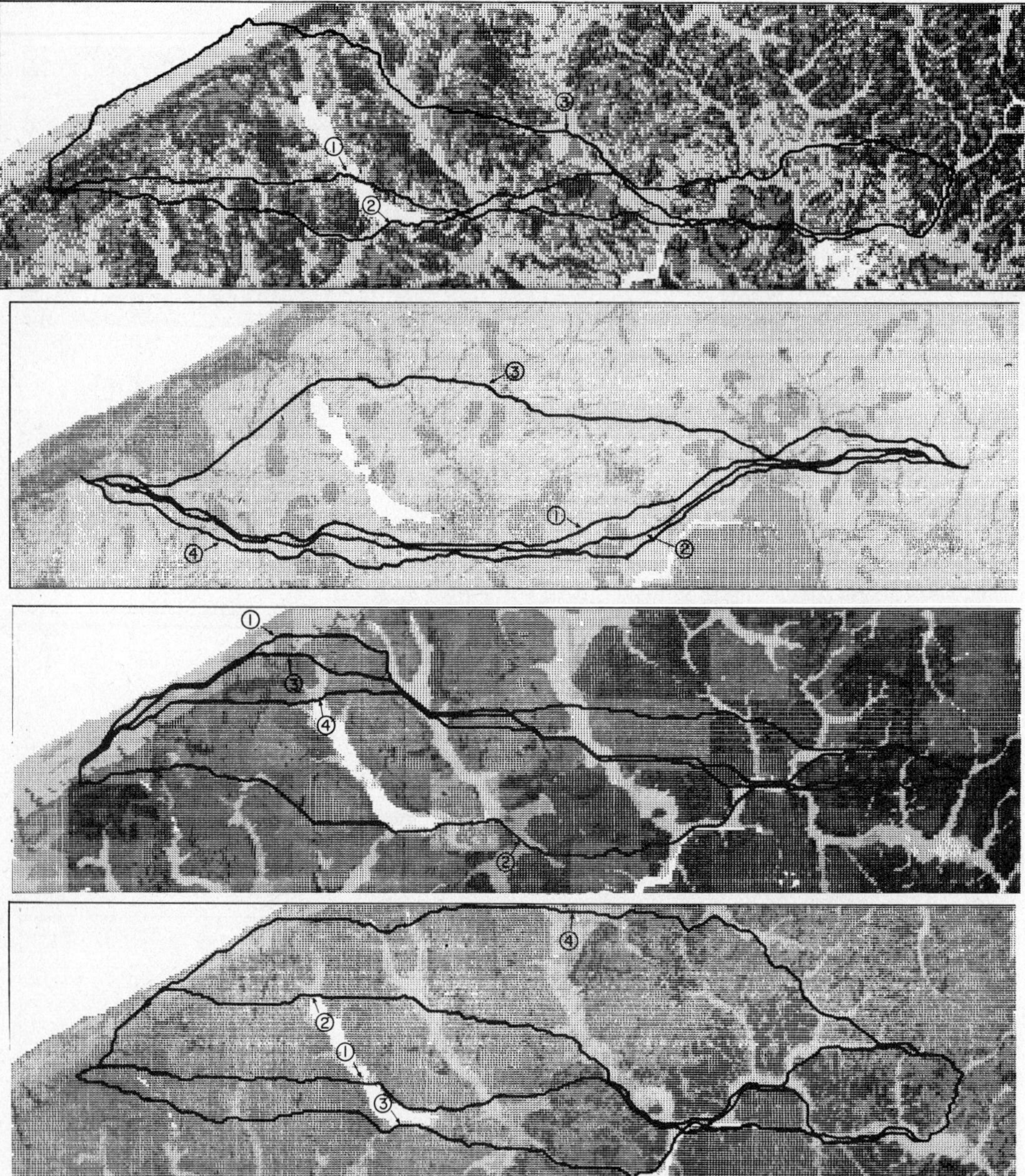

Fig. 32-105. Four GCARS alternate route analyses. (After Turner, 1978.) a. Engineering feasibility b. Environmental impact c. Socio-economic benefit d. Total highway feasibility

mation, each model can be multiplied by a weighting factor allowing it to be enhanced to any desired degree (Turner and Miles, 1971).

After many runs had been made using the GCARS procedure, a general pattern emerged with five major alternatives dominating. Four alternatives were illustrated in Figure 32-105. These results were integrated with the routes obtained by conventional techniques, and 12 detailed alternate corridor locations were selected, using various combinations of 31 sections (refer to Figure 32-107). These corridor locations were re-entered into the GMAPS-GCARS programs. Computation of the levels of impact of each alternative of any baseline map or derived model was possible, and allowed for the rapid comparison of all alternatives.

Use of the computer-aided technique reduced the time required for the study and allowed a more rapid generation of a large number of alternatives. This resulted in the preparation of the draft and final Environmental Impact Statement within 1½ years of project initiation rather than requiring the normal 2–3 year time period.

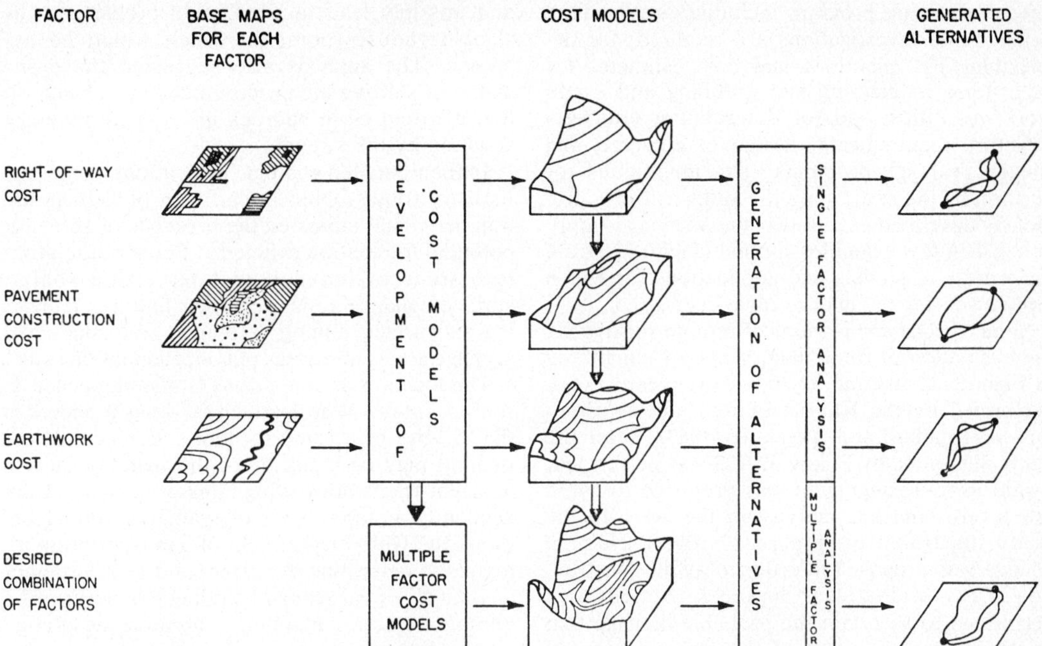

Fig. 32-106. Basic concepts of the Generalized Computer-Aided Route Selection System (After Turner and Miles, 1971, courtesy Transportation Research Board.)

Application in Preliminary and Final Route Location

Remote sensing techniques have been used extensively in the last steps of route location. Detailed analyses of large scale photographs are performed to provide information on topography, soils and geology, drainage conditions, land use, construction materials, vegetation, cultural and archeological features, and many other environmental factors. This information is useful in establishing final alignment and grade; in planning a

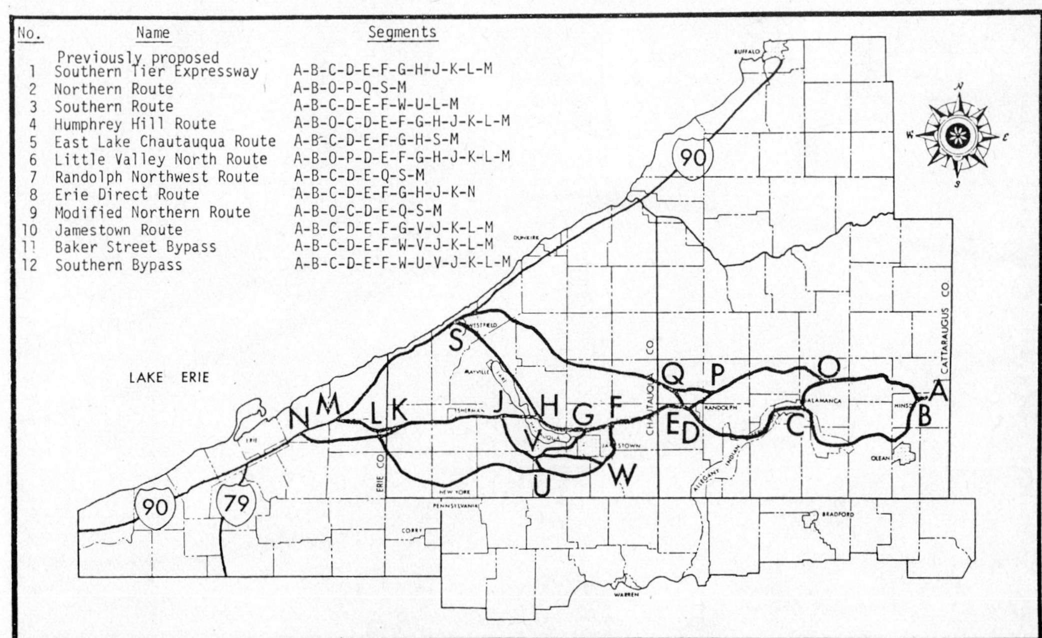

Fig. 32-107. Twelve alternate routes under consideration for completion of the southern tier expressway. (Courtesy New York Department of Transportation.)

field exploration program including areas where detailed site investigations are required; for determining bid quantities and cost estimates for such items as clearing and grubbing and earthwork quantities; and for determining drainage-structure requirements, design of culverts, and special drainage problems. The techniques for performing the studies are the same as those previously described except that the analysis is more detailed. A few examples are included to illustrate the variety of studies and applications for which remote sensing techniques have been used.

Some special applications of remote sensing for the evaluation of route alignments are illustrated in Figures 32-108 and 32-109—examples of work performed by the Kansas State Highway Commission (Stallard and Anschutz, 1963). After the study illustrated by Figure 32-108 was performed, a shift in route alignment was proposed to avoid the severe landslide activity in the area. In the study illustrated by Figure 32-109, a channel change was proposed in order to avoid construction of two bridges over the creek. A study was performed to ascertain the probable damage that might occur as a result of the channel change. Analysis of the stream regime indicated that shortening the stream would increase its gradient to such an extent as to result in a very significant increase in its erosive action. This change would endanger the bridge at point B, increase bank sloughing in areas where it was presently active, and possibly lead to a pollution problem at the slaughterhouse (point E) which would be bypassed. The analysis also suggested the desirability of shifting the proposed channel change so that it would be in bedrock in order to minimize scouring by the stream.

In their detailed geologic evaluations of the preliminary routes, Ohio Department of Transportation personnel indicated the presence of actual or potential foundation problems. Some of the problem areas included landslides; mine shafts; surface-collapse zones, or other features indicating subsurface mining activities; wet zones and seepage areas; and areas of poor foundation soils.

The location of coal seams is also important in route location. Where a mineral deposit occurs in the vicinity of a route location, the value of the mineral may be many times the value of land in adjacent tracts not having mineral deposits. Consequently, in the process of evaluating route locations, the Ohio Department of Transportation attempts to determine the extent and overburden of any minable coal seams. Norell (1966) described a photogrammetric mapping technique developed by the State for defining the original coal base. Strip mines, drift-mine entries, tipples, and test pits are located by photographic interpretation. These features and available borings establish the control for photogrammetric mapping of the seams.

Because major foundation problems which

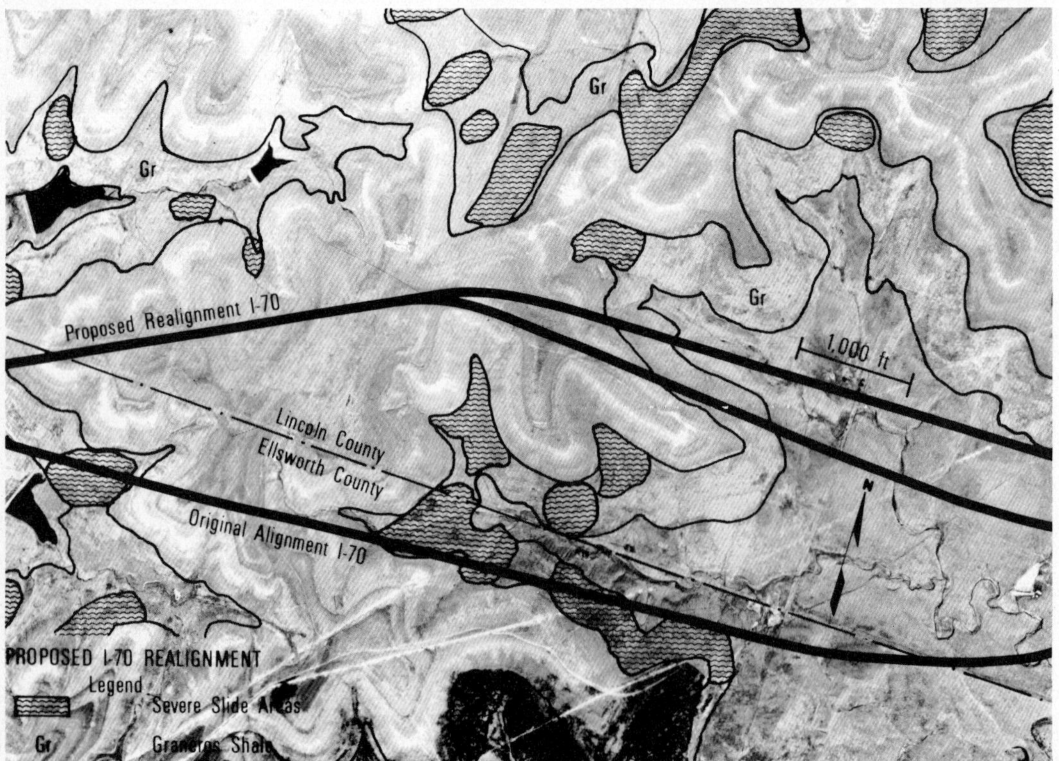

Fig. 32-108. Example of the use of remote sensing techniques as an aid in minimizing the effects of landslide terrain on a preliminary route location. (Stallard and Anschutz, 1963, courtesy State Highway Commission of Kansas.)

Fig. 32-109. Example of the use of remote sensing techniques to evaluate a proposed channel change during a preliminary route-location investigation. (Stallard and Anschutz, 1963, courtesy State Highway Commission of Kansas.)

greatly influence route location occur in much of the State, the Ohio Department of Transportation has developed a series of manuals for training State engineers and stereo-plotter operators. The manuals cover landslides, subsurface mining and organic deposits (Norell, 1965, 1968, 1973a and b); they contain stereo photographic sets illustrating airphoto patterns for diagnostic features that are indicative of these major foundation problems.

Another special technique that has been developed to aid in the final highway location and design is that of producing highway perspectives. It is largely a computer technique for developing views of the highway and surrounding landscape as seen from a "driver's eye" viewpoint. Animated movie sequences can be developed which show a view of the road as seen by the driver going at various speeds (Feeser, et al., 1971). These aid the designers in evaluating the aesthetic impact of a road and help develop a profile that would blend with the landscape. This technique is similar to that of preparing photomontages, as discussed earlier.

CONSTRUCTION SURVEYS

Remote sensing techniques have not been applied as extensively in construction surveys as they have in the previous stages, but there are several areas where they are of value. Because a photograph provides an excellent record of conditions existing at a given time, it may be a valuable asset in a construction survey. Photographs taken prior to construction can be used to locate material sources, to determine major construction problems that might be encountered, and to estimate bid quantities and prepare contract bids. Photographs can also be used for planning construction sequences, planning traffic flow patterns during construction, determining access routes to construction sites for equipment and material, and even for planning equipment requirements. Several photographs secured before, during, and after construction provide an invaluable construction-progress record and show whether construction practices have caused pollution, property damage, changes in natural drainage, or other detrimental conditions. "Before and after" photo coverage has also proved valuable in documenting cases for court actions.

Large-scale panchromatic photography, both vertical and oblique, has been the major form of data obtained. There has been an increase, however, in the use of color photography in this stage because of the ease of interpretation and the greater amount of information that can be extracted.

MAINTENANCE SURVEYS

Maintenance surveys are performed to evaluate the condition of a highway system and facilities, and to determine the corrective measures needed to bring the system back to its original (or an acceptable) condition. The types of surveys performed during this stage include: (1) condition and inventory surveys to determine the condition of the roadway and its appurtenances and to prioritize the system for performing maintenance; (2) surveys on damage caused to the highway system or abutting property by water, salt, wind, landslides, earthquakes, or subsidence; and (3) surveys to locate suitable construction materials for maintenance use.

To perform damage surveys, data are needed for specific sites as well as for large (regional) areas. Information is required immediately after a catastrophe so that plans can be made and actions taken to repair the facilities. The fastest response to this need is to obtain aerial photography as soon as the weather conditions permit. Usually two scales are obtained—a medium to small scale for regional coverage and a large scale for detailed evaluation of specific sites. Color photography provides the best detail for these types of surveys. If weather conditions do not clear immediately after a catastrophe, radar coverage can provide suitable regional detail, but not the detail required for evaluation of specific sites. Satellite imagery can be very valuable for evaluating damage occurring over large areas as well as for providing before and after coverage. This was well demonstrated in the study of the Mississippi River basin flooding in the vicinity of St. Louis in the spring of 1973. Guss (1974) reported on the use of Landsat MSS data to investigate a 1,200-mile reach of the Mississippi River from St. Louis to the Gulf of Mexico. Before- and after-coverage was evaluated to determine the extent of flooding.

For State maintenance programs, there must be large quantities of materials at reasonable hauling distances from material maintenance sheds. Materials survey information can be readily documented from an existing statewide materials survey (e.g., Maine, 1964), or from a more limited materials inventory of the area within a reasonable distance of each maintenance shed.

Several highway organizations have developed special techniques to inventory their highway system and determine existing conditions of the highway as part of their overall program to prioritize highway segments for maintenance. Photologging is one technique used by some organizations for this purpose. This technique has been discussed in an earlier portion of this section. Examples of other techniques are described in the following case studies.

Pavement Inventories

In Kansas and Maine special techniques have been developed for using color aerial photography to inventory pavement conditions. Natural color photography prints made from color negative film at a scale of 1:1,200 is used to determine the condition of staining of concrete pavements in Kansas (Stallard 1965). The presence of stains of a particular color provides advance indication of pavement disintegration before any visible cracks or distress are indicated on the surface. Early stages of staining are more difficult to identify and some insignificant stains may be erroneously recorded, but at advanced stages of staining, detection accuracies of about 95 percent are attained. A stain-classification system was developed using this technique for rating the condition of the concrete pavement based on the type and number of stains appearing in each panel. The pavement-condition problem identified by this technique is unique to certain midwestern States and hence the method is applicable only to this region.

In the technique developed in Maine, color infrared transparencies are used for rating pavement conditions (Stoeckeler, 1970). To perform a detailed pavement survey, a scale of 1:2,400 or larger is required. The features mapped in the detailed pavement survey include the following: (1) transverse cracks—sealed and unsealed, (2) longitudinal cracks—sealed and unsealed, and (3) vertical deformations—rutting (longitudinal), heaves (transverse) and ripples. Figure 32-110, a black-and-white enlargement of a color-infrared stereo photo, illustrates some of these features.

For performing reconnaissance-level surveys, which do not require information on unsealed cracks or pavement relief, a scale of 1:6,000 is adequate and panchromatic photography can be used. The reconnaissance-level survey was used to develop a technique for relating crack patterns to statewide highway-sufficiency-rating classes. This technique is used as a guide for rating the

Fig. 32-110. Stereopair of an unsealed transverse reflection crack and associated frost heave developed over a culvert in a recently placed pavement overlay. The crack was about 2.5 cm (1″) wide and the heave was a little over 1.3 cm (1/2″) high and 46 cm (18″) wide on March 11, 1968, when the photo was taken. The crack extends from the man at A to the passing lane shoulder at B. The ripple or washboard pattern C is distinct in the passing lane but not in the travel lane. The stereogram is constructed with negative black-and-white enlargements made from Ektachrome Infrared Aero positive transparencies having an original scale of about 2.5 cm = 60 m or 1:2400. (After Stoeckeler, 1968, courtesy Maine Department of Transportation.)

needs of pavement sections for replacement or overlay.

Skid Resistance

A photo-interpretation method was developed by Schonfeld (1974) of the Ontario Ministry of Transportation and Communications to determine the skid resistance properties of a pavement surface from stereo photographs. For routine testing operations, the photo-interpretation test method has significant advantages over the traditional skid trailer. Sites with difficult alignments, such as curves and intersections, can be investigated and testing can proceed during the winter months, providing the pavement surface is clean and dry. In addition, the various surface parameters provide a greater understanding of the reasons for a deficiency in skid resistance at a particular site. This technique was adopted by the American Society for Testing and Materials (1981) (ASTM) as the "Standard Test Method for Classifying Pavement Surface Textures," ASTM standard, Designation: E770-80. A manual describing the procedures and listing tentative minimum skid-resistance values for actual pavements was prepared by the Ontario Ministry (Holt and Musgrove, 1977).

This method consists of analyzing the pavement surface-texture by developing measurements and classifying the macroprojections and the microprojections as illustrated in Figure 32-111. The six texture parameters determined include:

1. 'A' - height of macroprojections
2. 'B' - width of macroprojections
3. 'C' - angularity of macroprojections
4. 'D' - density of distribution of macroprojections
5. 'E' - harshness of macroprojection surfaces
6. 'F' - harshness of microprojections or matrix surfaces

When examining the actual pavement surface or the stereo photographs, the photo-interpreter gives the surface a six-digit code referring to each of the above parameters. Once a section of pavement has been coded, (a) the surface texture can be fully described by the parameter numbers, and (b) the skid number of that surface can be calculated.

The stereo photographs are taken with a 35 mm single-lens reflex camera with a 50 mm macro (close-up) lens mounted in a special box equipped with an electronic flash (illustrated in the ASTM Standards, Part 15, 1981). Reference scales for estimating texture dimensions are mounted on a metal or plastic slider plate placed at the bottom of the camera box. Figure 32-112 is a typical example of a stereo photograph showing the plate and reference scales and pavement surface.

Two types of equipment may be used to analyze the pavement stereo pairs: A microstereoscope for colored slides and a mirror stereoscope for enlarged panchromatic prints. Color slides have been established as the Ontario Ministry's standard test method, due to the superior optical qualities of the transparency and the greater ease with which the photo-interpreter can interpret the fine-texture parameters.

The 35 mm color slides are analyzed by coding 10 random centimeter squares within the stereo

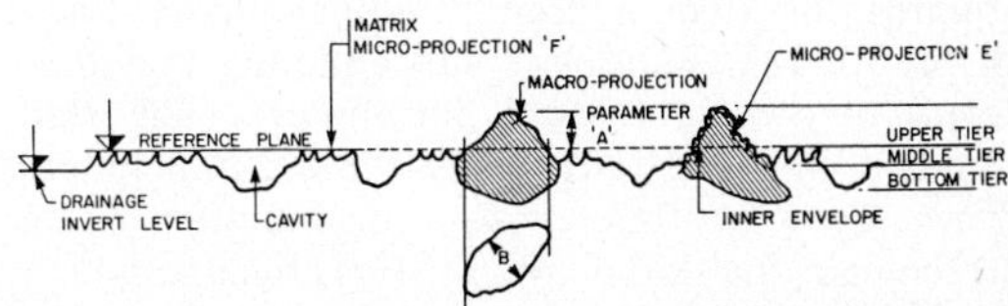

Fig. 32-111. Pavement surface profile indicating the six texture parameters evaluated. (After Holt and Musgrove, 1977; courtesy Ontario Ministry of Transportation and Communication.)

Fig. 32-112. Typical stereogram of pavement surface showing plate and reference scales. (After Holt and Musgrove, 1977; courtesy Ontario Ministry of Transportation and Communication.)

overlap area of approximately 210 cm^2 (30 sq. in.). A transparent grid is placed on top of the left-hand picture. The grid, marked off in one-centimeter squares, has 10 numbered, random squares. Each of the random squares is subsequently coded in terms of the six texture parameters. The skid number (SN) is a function of the six parameters for a given centimeter square. For any given stereo pair, the skid number (SN) is the average of the skid numbers for the 10 random squares that have been coded on that stereo pair.

During the interpretation, reference is made to the vertical reference wedge, visible in the stereo pairs (Figure 32-112), at approximately pavement-surface level, in order to classify the heights of the macroprojections. The wedge is graduated from 0.25 mm at the lower end to 6 mm at the upper end, with the sides of the wedge having a slope of 45°.

LITIGATION AND CLAIMS

One of the special areas of application where aerial photography and interpretation techniques have been applied extensively is that involving litigation. Areas of use include condemnation cases to aid in determining the value of the land and appurtenances; damage suits as a result of some highway related event; and environmental

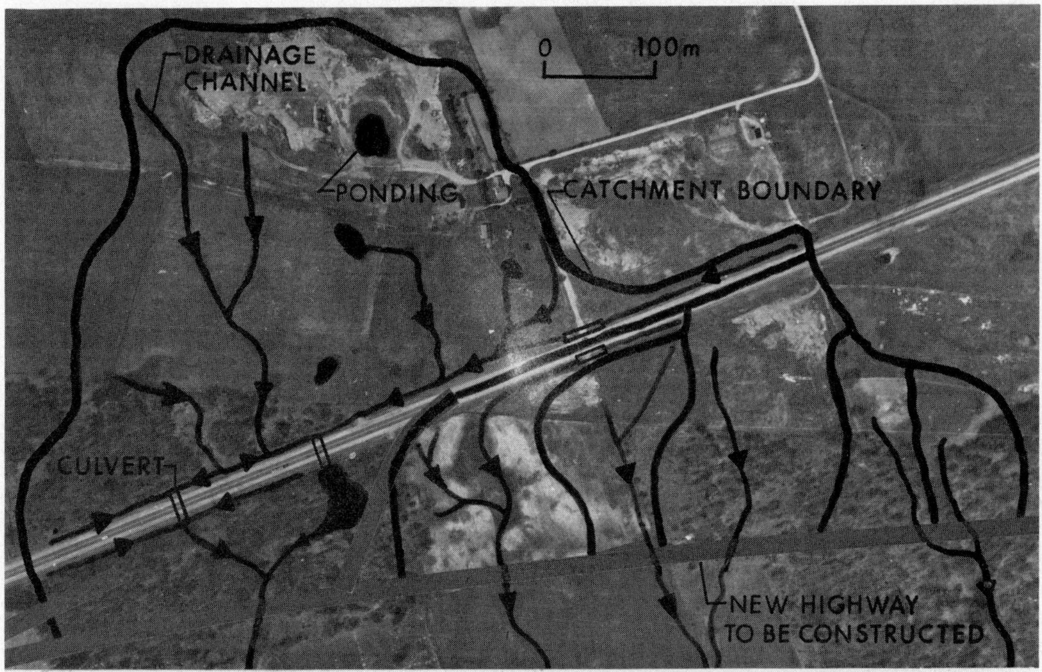

BEFORE CONSTRUCTION : 1962 CONDITION

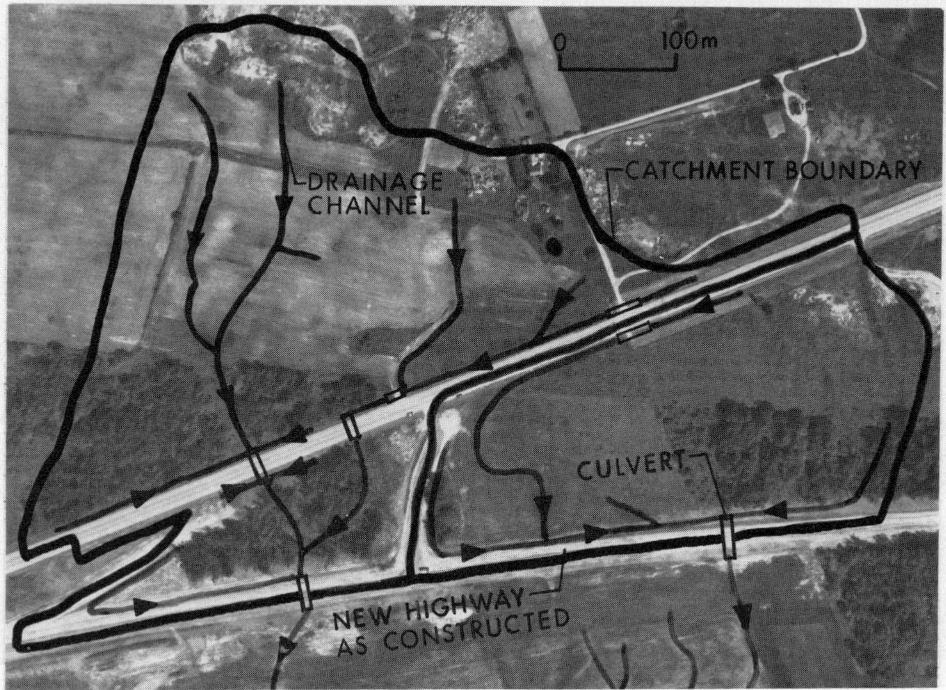

Fig. 32-113. Delineation of drainage conditions existing before and after construction accomplished by stereoscopic analysis of the aerial photography obtained at these two times. *Top:* Before construction, 1962 condition. *Bottom:* After construction, 1972 condition.

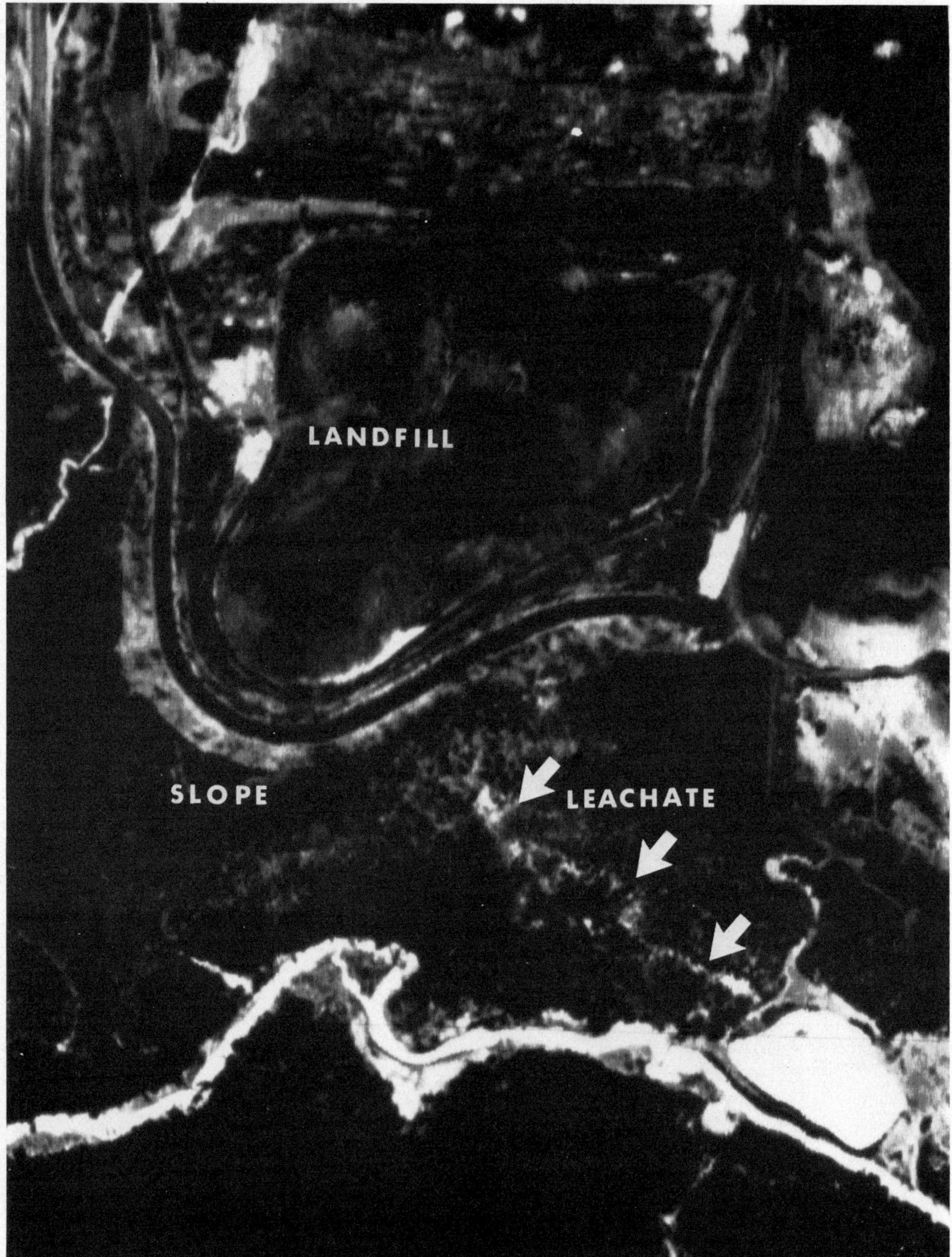

Fig. 32-114. Thermography of landfill sites is used for detection of leachate springs and leachate-contaminated water movement from the site. In this image the actual site is elevated approximately 50′ above the stream. Leachate was detected in the pond but the transport mechanism was not known. This early spring thermography shows a clear connection between leachate discharge at the base of the valley slope and the pond. The leachate is seen to move parallel to the river and into the pond. This movement was not visible on simultaneously acquired color and color infrared photography or from ground observations.

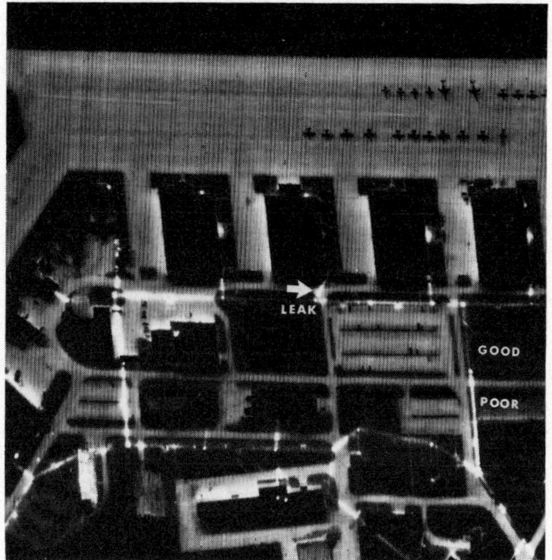

Fig. 32-115. Thermography is an ideal method for surveying central heating distribution systems. Aerial thermography provides a complete overview of the streamlines. The imagery shows line leaks; damaged lines; confirms problems; shows unsuspected problems and confirms repairs. On this image, both good lines, showing low heat loss and poor lines, showing high heat loss, are visible. Also, several line leaks are present. Note the change in line brightness as it travels beneath different surface materials. Interpretation of steamline imagery should only be done in conjunction with the physical plant people in charge of the system under study. This image was acquired in the predawn hours from 1200 feet above ground. The air temperature was 25°F.

impact hearings. Figure 32-113 illustrates before and after conditions.

Aerial photographs are permanent and unbiased records of all natural and cultural elements which constitute the environment. Photographs previously taken constitute historic records while up-to-date photography portrays existing conditions. Together, they provide a chronological record.

A photo interpreter, through systematic and logical analysis of aerial photographs, provides the necessary pertinent information on physical factors and cultural activities and, at the same time, accounts for those sensitive interactions which existed at the time of photography between the physical presence of the highway and the "environment." *Provided that suitable photography is available,* the photo interpreter can provide complete and accurate information on natural and cultural conditions *before, during and after* a particular event. It has been shown that only through photo interpretation can all the facts on progressive changes in an area be collected, identified, analyzed and correlated to arrive at an inexpensive yet sound conclusion.

In the early stages of construction of the Interstate highway network in the United States, the use of aerial photography for condemnation cases was more common. Caruthers (1963) listed several examples wherein aerial photography was used in court cases to assist the Kansas State Highway Commission in condemnation cases—largely as illustrations to indicate the existing conditions and value of the land in dispute.

In later periods, the emphasis shifted to damage claims due to highway construction, and more recently, to environmental impacts. The Remote Sensing Section of the Ontario Ministry of Transportation and Communications reported assisting in 88 claims against the Ministry between 1971 and 1974. These investigations were concerned with either (1) determining conditions existing prior to any changes, e.g. land use and cover, topography, drainage; (2) changes temporarily introduced during construction, e.g., changes in drainage patterns, vegetation damage, erosion and siltation, detours, stockpiling, damage by construction machinery, and access problems; or (3) permanent changes after construction, e.g. drainage diversion, flooding, damage to surface and subsurface outlets, erosion and siltation of agricultural and recreation land, and damage to crops by pollution, flooding and erosion.

Figure 32-113 illustrates the use of photography and photo interpretation techniques in one of the Ontario Ministry's cases. A new highway was constructed in Southern Ontario and, subsequently, it was alleged that the construction had resulted in drainage problems in a property. The "before" construction photography was used to evaluate the drainage conditions existing prior to construction. Such aspects as drainage channel location, height-of-land, and catchment area were determined. Similarly, the "after" construction photography was studied. Comparison of the two indicated that, apart from some ponds drying up and the shifting of some channels and heights-of-land, one culvert on the new highway was used to drain what previously were 3 catchment areas. The effect of such a change was not evident on the aerial photographs but this information was provided for further analysis.

In this case, although the photographs did not specifically reveal any deterioration of drainage conditions, they provided sufficient indications of the changes that had occurred for the design engineers and claim adjusters to concentrate their investigation to arrive at a realistic appraisal of the original allegation.

In most of the cases where photography is used, photographic enlargements or mosaics are prepared so that the overall perspective of the area under contention and it surroundings is shown. The photography should be taken as close to the date of the event of highway involvement as possible. A minimum number of markings should be added. The expert witness who is called upon to verify the photographs, enlargements or mosaics should collect all of the background information possible on flight height, who took the photography, its purpose, the equipment used and so forth.

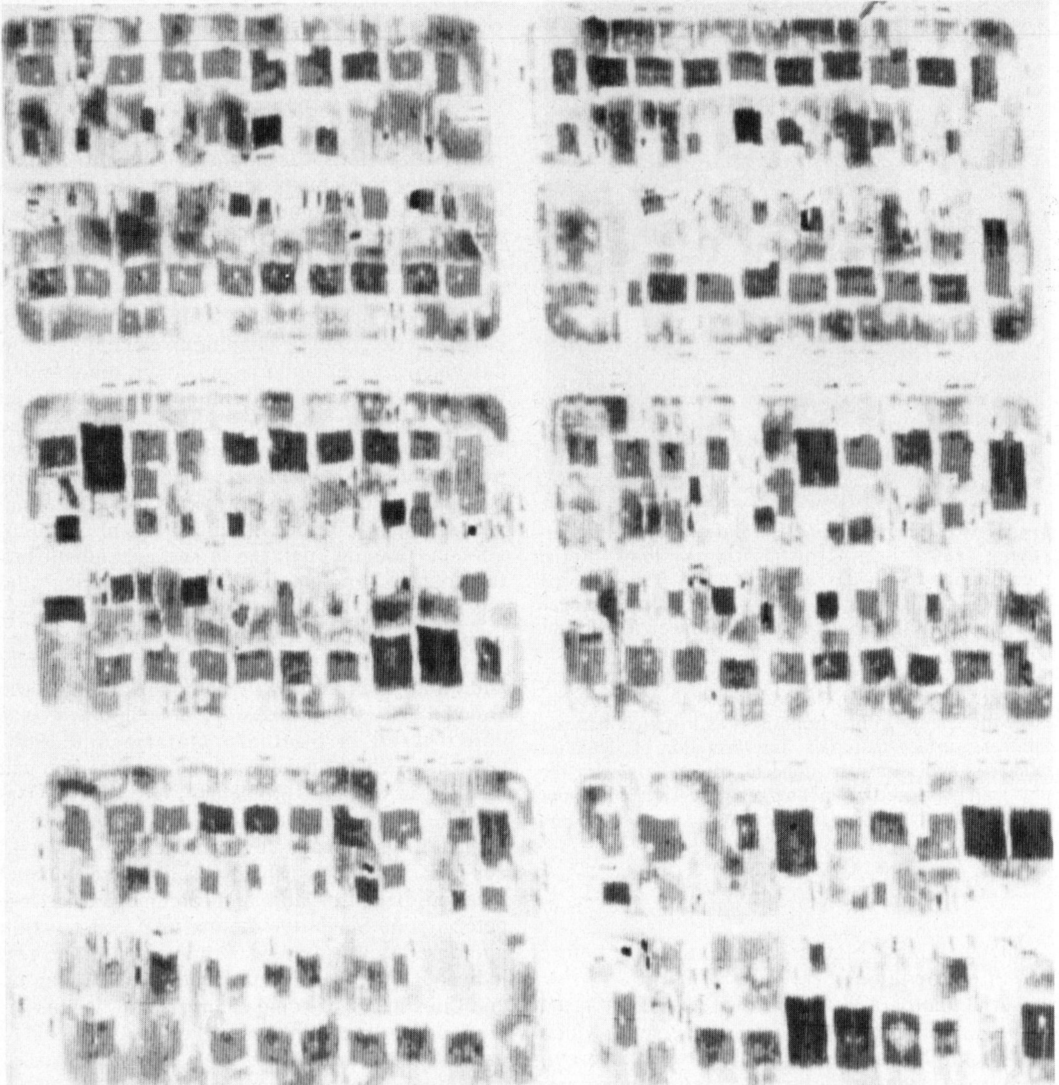

Fig. 32-116. Aerial thermography for the detection of residential heat loss. This thermograph of house rooftops shows a wide distribution of rooftop temperatures. This level-sliced image "categorizes" rooftops into six distinct grey tones. The lighter the tone the warmer the rooftop. Interpretation of the imagery is based on house type, number of floors, age and roof construction as well as the tone on the thermography.

The advantage of photography for use in court is best summarized by the statement by Irving Goldstein in his book *Trial Technique,* "A good exhibit will continue to argue the merit of the attorney's cause long after his voice has been stilled. A jury may forget some of the oral testimony, but members of the jury cannot very well overlook or forget the exhibit which serves as an ever present reminder of the truth of testimony contentions."

ADDITIONAL APPLICATIONS OF THERMAL INFRARED IMAGERY

These additional applications appearing in Figures 32-114 through 32-116 do not have a place within anyone of the major headings for this chapter. As with many of the preceding illustrations the caption accompanying each of these figures provides an explanation for conveying the engineering significance.

REFERENCES

Adsit, J. H. and E. B. Olson, 1965, Construction materials survey, Interstate 80, Granger Junction to Flaming Gorge Junction section: Wyoming State Highway Dept. in cooperation with Federal Highway Admin.

Ahern, F. J., D. G. Goodenough, S. C. Jain, V. R. Rao and G. Rochon, 1977a, Use of clear lakes as standard reflectors for atmospheric measurements. Proc. 11th International Symposium on Remote Sensing of Environment, Ann Arbor, MI, pp. 731–755.

Ahern, F. J., D. G. Goodenough, S. C. Jain, V. R. Rao and G. Rochon, 1977b, Landsat atmospheric corrections at CCRS. Proc. 4th Canadian Symposium on Remote Sensing, Quebec City, pp. 583–594.

Aird, W. J., 1980, Utilization of aerial photography for monitoring environmental impacts from pipeline construction. Sixth Canadian Symposium on Remote Sensing, Halifax, Canada, pp. 431–432.

Alfoldi, Thomas T., 1974, Regional study of landsliding in eastern Ontario by remote sensing, Master of Science Thesis, Department of Civil Engineering, University of Toronto, 79 pp.

Alfoldi, T. T. and J. C. Munday, Jr., 1977, Progress toward a Landsat water quality monitoring system, Proc. 4th Canadian Symposium on Remote Sensing, Quebec City, pp. 325–340.

Alfoldi, T. T. and J. C. Munday, Jr., 1978, Water quality analysis by digital chromaticity mapping of Landsat data, Canadian Journal of Remote Sensing, 4(2): 108–126.

Algazi, R. et al., 1977, Water supply studies by the Davis Campus Group. In R. N. Colwell (Prin. Inv.), Integrated Study of Earth Resources in the State of California Using Remote Sensing Techniques, Space Sciences Laboratory, Series 18, Issue 44, University of California, Berkeley, 43 pp.

American Society for Testing and Materials, 1981, Standard Test Method for Classifying Pavement Surface Textures, ASTM Designation E 770-80, Annual Book of ASTM Standards, Part 15, pp. 1219–1232.

American Society of Photogrammetry, 1960, Manual of Photographic Interpretation. George Banta Co. 868 pp. Illus.

Amos, C. G., 1976, Suspended sediment analysis of seawater using Landsat imagery, Minas Basin, Nova Scotia, in Report of Activities, Part C, Geological Survey of Canada, Paper 76-1C, pp. 55–60.

Arizona Highway Department, 1965, Arizona materials inventory, Yuma County: Ariz. Highway Dept., in cooperation with Federal Highway Admin., Feb.

Arthur, J. F., 1975, Preliminary studies on the entrapment of suspended materials in Suisun Bay, San Francisco Bay—Delta Estuary. In Proceedings, Workshop on Algae Nutrient Relationships in the San Francisco Bay and Delta, San Francisco Bay and Estuarine Assoc., pp. 17–36.

Arthur, J. F., 1977, The null zone: 1976 measurements and possible ecological significance. Presented at the winter meeting of the San Francisco Bay and Estuarine Association.

Arthur, J. F., and M. D. Ball, 1979, Factors influencing the entrapment of suspended material in the San Francisco Bay estuarine system. Proceedings, San Francisco Bay: The Urbanized Estuary, edited by T. J. Conomos (American Association for the Advancement of Science), pp. 143–175.

Auschutz, G. and A. H. Stallard, 1967, An overview of site evaluation: Photogramm, Eng., vol. 33, pp. 1381–1396.

Baker, Robert T. and R. Chieruzzi, 1959, Regional Concept of Landslide Occurrence. Highway Research Board Bulletin 216, pp. 1–16.

Baker, W. T. and A. M. Wlute, editors, 1973, New developments in optical instrumentation—a problem solving tool in highway and traffic engineering. Photo-Optical Instrumentation Eng. Proc., vol. 37, Washington, D.C., April 11–12.

Baker, W. T., 1980, Photologging: A Close Range Photogrammetry Tool for Highway Engineers, International Archives of Photogrammetry, Vol. B-10, pp. 21–33, XIVth Congress of the International Society of Photogrammetry, Hamburg, Germany.

Ball, M. D., 1975, Chlorophyll levels in the Sacramento-San Joaquin Delta to San Pablo Bay. Proceedings, Workshop on Algae Nutrient Relationships in the San Francisco Bay and Delta, San Francisco Bay and Estuarine Assoc., pp. 54–102.

Barr, D. J. and R. D. Miles, 1970, SLAR imagery and site selection. Photogramm. Eng., vol. 36, pp. 1115–1171.

Barr, D. J. and M. D. Hensey, 1974, Industrial site study with remote sensing. Photogramm. Eng., vol. 40, pp. 79–86.

Barringer, A. R. and J. P. Shock, 1966, Progress in the remote sensing of vapors for air pollution: Geologic and oceanographic applications. Fourth Symp. on Remote Sensing of Environment, April 12–14, Univ. Michigan, Proc., pp. 779–791.

Barringer, A. R., B. C. Newbury and A. J. Moffatt, 1968, Surveillance of pollution from airborne and space platforms. Fifth Symp. on Remote Sensing of Environment, 16–18 April, Univ. Michigan, Proc., pp. 123–156.

Beaumont, T. E. and P. J. Beaven, 1977, The Use of Satellite Imagery for Highway Engineering in Overseas Countries, Transport and Road Research Laboratory Supplementary Rpt. 279, Berkshire, England.

Beaumont, T. E., 1978, Remote Sensing for Transport Planning and Highway Engineering in Developing Countries, Transport and Road Research Laboratory Supplementary Rpt. 433, Berkshire, England.

Belcher, D. J., 1943, The engineering significance of soil patterns. Ann. Meeting of Highway Research Board 23rd Proc., pp. 569–598.

Belcher, D. J., 1955, Selection of the five sites most favorable for the location of the new capitol of the United States of Brasil. Donald J. Belcher and Assoc., Ithaca, New York.

Belcher, D. J., J. D. Mollard and W. T. Pryor, 1960, Photo interpretation in engineering. Chapter 6, in Manual of photo interpretation, American Society Photogramm.

Belcher, D. J., R. D. Miles, A. H. Stallard and J. McLaurin, 1972, Evaluation of remote sensing by engineering/cartography panel. Seminar on Operational Remote Sensing, Feb. 1–4, Am. Soc. Photogramm., pp. 287–299.

Benson, A. S., K. J. Dummer, and J. T. Hardin, 1981, Accuracy Assessment of Remote Sensing Derived Information in Wildland Environments. *In* Proceedings, ASP-ACSM Fall Technical Meeting, pp. 584–590.

Berrill, J. B. and L. J. Ferrer, 1973, Photo-computer plot montages for highway design. Highway Research Board, Highway Research Record No. 437, pp. 1–8.

Billingsley, F. C., M. R. Helton, and V. M. O'Brien, 1976, Landsat follow-up: A report by the applications survey groups. Jet Propulsion Laboratory, Technical Memorandum, 33–803, Vol. II.

Bird, S. J. G., 1973, Environmental criteria for recreationally oriented highway planning. Highway Research Board, Highway Research Record No. 452, pp. 19–28.

Biswas, A. K., 1970, History of hydrology. American Elsevier Pub. Co. Inc., 336 pp.

Blaine, L. R., C. J. Murphy III and W. L. Barnes, 1977, The ocean color scanner experiment. NASA Goddard Spaceflight Ctr., Paper X-941-77-153, 51 pp.

Boland, D. H., 1976, Trophic classification of lakes

using Landsat-1 multispectral scanner data. Report to U.S. Environmental Protection Agency, Corvallis, Oregon, E. P. A. - 600/3-76-03, 245 pp.

Brink, A. B. A. and T. C. Partridge, 1967, Kyslami land system: an example of physlographic classification for the storage of terrain data. Proc. Fourth Regional Converence for Africa on Soil Mechanics and Foundation Engineering, Cape Town, South Africa, pp. 9–14.

Brown, B. P., 1973, Remote Sensing Activities by California's Department of Water Resources. Unpublished paper presented at International Symposium on Remote Sensing and Water Resource Management at Centre for Inland Waters, Burlington, Ontario, Canada.

Bruce, R. L. and I. Scully, 1966, Manual of landslide recognition in Pierre Shale, South Dakota. Final Report, Research Project 615(64), South Dakota Dept. of Highways and South Dakota Geol. Survey in cooperation with Federal Highway Admin.

Caggiano, F. N., 1967, Project Sky Count: An Airborne Reconnaissance System for Transportation Analysis. Airborne Photo-optical Instrumentation, Feb. 20–21, Soc. of Photo-optical Instrumentation Eng., Proc., pp. X-1 to X-13.

California Department of Water Resources, 1974, California State Water Project. Bulletin No. 200. Vol. I, 1973 pp. Illus.

Caruthers, O. E., Jr., 1963, Use of Aerial Photography in the Kansas Courts, Highway Research Board, Highway Research Record No. 65, pp. 86–94.

Chassie, Ronald G. and R. D. Goughnour, 1976, National Highway landslide experience. Highway Focus, vol. 8, No. 1, pp. 1–9, Federal Highway Administration, January.

Chaves, J. R. and R. L. Schuster, 1964, Use of color photography in materials survey. Highway Research Board, Highway Research Record No. 63, pp. 1–9.

Chaves, J. R., 1968, Color photos for highway engineering. Photogramm. Eng., vol. 34, pp. 375–379.

Chickering, H. G., 1948, A proposed method of reconnaissance for aerial and field surveys. Photogramm. Eng., vol. 14, pp. 551–555.

Clasna, V. J., Jr., 1963, Photogrammetry and comprehensive city planning for small communities. Photogramm. Eng., vol. 29, pp. 681–684.

Cleaves, A. G., 1961, Landslide investigations: A field handbook for use in highway location and design. Bureau of Public Roads, 67 pp.

Cloern, J. E., 1979, Phytoplankton ecology of the San Francisco Bay System: The status of our present understanding. In T. J. Conomos (Ed.), Proc. San Francisco Bay: The Urbanized Estuary, American Assoc. Adv. Sci., Washington, D.C., pp. 247–263.

Cochran, W. G., 1963, Sampling techniques. Second Edition, John Wiley & Sons, Inc., New York, 413 pp.

Colcord, J. and Bernath, H., 1974, Engineering Uses of EROS (Final Report), University of Washington, Seattle, Sept. (USGS Contract No. 14-08-0001-12865).

Colcord, J. E., 1975. Landsat I—imagery in hydrologic studies. Proc. Amer. Soc. Photogramm., Phoenix.

Colorado Department of Highways, 1962, Procedures for location of granular materials. Prepared in cooperation with Federal Highway Administration.

Colwell, Robert N. (Editor), 1960, Manual of photographic interpretation. American Society of Photogrammetry, Falls Church, Virginia, 868 pp.

Colwell, Robert N., 1982, Irrigated Lands Assessment for Water Management, Annual Report. Space Sciences Laboratory, Series 22, Issue 23; University of California, Berkeley, 330 pages.

Colwell, R. N., D. M. Carneggie and R. W. Thomas, 1974, ERTS-1 data as an aid to wildland resource management in California. University of California, Berkeley, Space Sciences Laboratory Final Report, Series 16, Issue 22, 256 pp., illus.

Colwell, Robert N., Andrew S. Benson, Kevin J. Dummer and James T. Hardin, 1981, Evaluating large format aerospace systems for coordinated resource planning. Final Report, Forest Service Contract No. 53-3187-0-70. Remote Sensing Research Program, Department of Forestry and Resource Management, University of California, Berkeley, California.

Colwell, R. N., A. S. Benson, K. J. Dummer and J. T. Hardin, 1982, The analysis of wildland data acquired by advanced high altitude systems on the San Juan National Forest, Colorado. Final report on work done under Forest Service Contract No. 53-8137-1-40, 95 pages, illus.

Congalton, Russell G., Roy A. Mead, Richard G. Odewald and Joel Heinen, 1981, Analysis of forest classification accuracy. Cooperative Research Report between U.S. Forest Service and Virginia Polytechnic Institute and State University, Blacksburg, Virginia, 24061, Cooperative Agreement No. 13-1134.

Conomos, T. J. and D. H. Peterson, 1974, Biological and chemical aspects of the San Francisco Bay Turbidity maximum. Proceedings, Symposium International Relations Sedimentaries entre Estuaries et Plateux Continentaux, Institut de Geologie du Basin d'Aquitaine, 15 pp.

Cyra, D. J., 1971, Traffic data collection through aerial photography. Highway Research Board, Highway Research Record No. 375, pp. 28–39.

DeLoach, W. C., 1973, Remote-sensing applications of environmental analysis. Highway Research Board, Highway Research Record No. 452, pp. 29–39.

Dill, H. W., Jr., 1963, Airphoto analysis in outdoor recreation-site inventory and planning. Photogramm. Eng., vol. 29, pp. 67–70.

Dishaw, H. E., 1967, Massive landslides. Photogramm. Eng., vol. 32, pp. 603–609.

Earth Satellite Corporation, 1973, Remote sensing of mined area reclamation application inventory. Earth Satellite Corp., Washington, D.C., for NASA Task 160-75-73-04-10, Report No. NASA CR-124608.

Eckel, Edwin B., 1958, Landslides and engineering practice. Highway Research Board, Special Report No. 29, Washington, D.C., 232 pp.

Eyre, L. A., B. Adolphus and M. Amiel, 1970, Census analysis and population studies. Photogramm. Eng., vol. 36, pp. 460–467.

Feeser, L. J., J. D. Meyer and J. D. Cutrell, 1971, Simulating the Driver's View of New Highway Before Construction, Public Roads, Vol. 36, No. 7, pp. 141–147.

Feinberg, A. S., 1964, Airphoto interpretation of Illinois soils. Masters Thesis, Eng. Exp. Sta., Univ. Illinois, Project IHR-12.

Fenneman, Nevin M., 1949, Physical division of the United States. U.S. Geological Survey map.

Fisher, L. T. and F. L. Scarpace, 1975, The use of an interactive graphics terminal for analysis of ERTS imagery. Proc. ASP Annual Meeting.

Flemming, J., 1979, Standardization techniques for aerial color infrared film. Interdepartmental Committee on Aerial Surveys, EMR, Ottawa, Canada.

Flemming, J. and R. G. Dixon, 1981, Basic guide to small-format hand-held oblique aerial photography. CCRS Users Manual 81-2, Canada Centre for Remote Sensing, EMR, Ottawa, Canada, 66 pp.

Frost, R. E., 1945, Identification of granular deposits by aerial photographs. Highway Research Board, Proc. vol. 25, pp. 116–129.

Frost, R. E. and K. B. Woods, 1948, Airphoto patterns of soils of the western United States—as applicable to airport engineering. U.S. Dept. of Commerce, CAA Tech. Development Rept. No. 85.

Frost, R. E., 1953, Factors limiting the use of aerial photographs for analysis of soil and terrain. Photogramm. Eng., vol. 19, pp. 427–436.

Gagnon, Hughes, 1975, Remote sensing of landslide hazards on quick clays of eastern Canada. Proceedings of the 10th International Symposium of Remote Sensing of Environment, 6–10 October, pp. 803–810. Environmental Research Institute of Michigan, Ann Arbor.

Garofalo, D. and F. Wobber, 1974, Solid waste and remote sensing. Photogramm. Eng., vol. 40, pp. 45–60.

Garstka, U. Walter, 1972, Water resources in the west. In Proceedings of A Symposium on Watersheds in Transition, American Water Resources Association and the Colorado State University, Fort Collins, Colorado, pp. 8–15.

Gaydos, L. and W. L. Newland, 1978, Inventory of Land Use and Land Cover of the Puget Sound Region Using Landsat Digital Data, Journal Research, USGS, Vol. 6, No. 6, pp. 807–814, November–December.

Geraci, A. L., 1978, Jonian sea water exploration and evaluation interim report. Machinery Institute, Department of Engineering, University of Catania, Italy.

Geraci, A. L., R. N. Colwell and S. Khorram, 1980, Remote sensing of water quality for estuarine environments. Space Sciences Laboratory, University of California, Berkeley, Special Report, Series 21, Issue 23.

Geraci, A. L., 1981, Remote sensing techniques aid in water pollution evaluation. Sea Technology, Vol. 10, pp. 20–21.

Geraci, A. L., 1982, Determining levels of pollution in Sicily's coastal waters. (Personal communication.)

Giever, P. M., 1966, Needs for remote sensing data in the field of air and water pollution control. Fourth Symp. on Remote Sensing of Environment, April 12–14, Univ. Michigan, Proc., pp. 21–23.

Guertin, F., T. J. Butlin and R. G. Jones, 1979, La correction geometrique des images Landsat au Centre Canadien de Teledetection. Canadian Journal of Remote Sensing, 5(2), pp. 118–127.

Guss, P. F., 1974, ERTS-1 data analysis of the 1973 Mississippi River flood. Fortieth Annual Meeting of Am. Soc. of Photogramm., St. Louis, Missouri, March 10–15, Proc., pp. 402–403.

Hausmanis, I. and A. K. Turner, 1971, Computer-aided analysis of alternative route locations within the Guelph-Dundas Test Area. Project T-28, Dept. of Soil Engr., Univ. of Toronto for Ontario Ministry of Transportation and Communication.

Hawkes, T. W. III and D. A. Brown, 1973, Application of Aerial Mapping to Development of Highways, Highway Research Board, Highway Research Record No. 452, pp. 10–18.

Herd, L. O., 1968, An aerial photographic technique for presenting displacement data. Highway Research Board, Highway Research Record No. 232, pp. 64–67.

Hittle, J. E., 1946, The application of aerial strip photography to highway and airport engineering. Highway Research Board, Proc., vol. 26, pp. 226–235.

Hofmann, W. P. and J. B. Fleckenstein, 1960, Comparison of General Routes by Terrain Appraisal Methods in New York State, Highway Research Board, Proc. Vol. 39, pp. 640–649.

Holman, W. W., R. K. McCormack, J. P. Minard and A. R. Jumikis, 1957, Practical applications of engineering soil maps. Engineering Soil Survey of New Jersey, Rept. No. 22, Rutgers Univ. Press, New Brunswick, New Jersey.

Holmquist, K., 1977, The Landsat lake autrophication study. Unpublished UW-MSN Masters Degree Report.

Holt, F. and G. Musgrove, 1977, Skid Resistence: Photo Interpreters' Manual, Ontario Ministry of Transportation and Communications, First Ed., November.

Howarth, P. J. and G. M. Wickware, 1981, Procedures for change detection using Landsat digital data. International Journal of Remote Sensing, Vol. 2, no. 3, pp. 277–291.

Howarth, P. J. et al., 1982, Landsat for monitoring hydrologic and coastal change in Canada. In M. D. Thompson (Editor) Landsat for Monitoring the Changing Geography of Canada, Canada Centre for Remote Sensing, Energy Mines and Resources, Ottawa, Canada.

Howarth, P. J. and G. M. Wickware, In Preparation, Landsat analysis of hydrologic and vegetation change in the Peace-Athabasca Delta. I. Preprocessing and enhancement.

Howarth, P. J. and G. M. Wickware, In Preparation, Landsat analysis of hydrologic and vegetation change in the Peace-Athabasca Delta. III. Change detection and signature file extension.

Howe, R. H. L., H. R. Wilke and D. E. Bloodgood, 1956, Application of airphoto interpretation in the location of groundwater. Jour. American Water Works Assoc., vol. 48, no. 11.

Hsu, S., 1971, Population estimation. Photogramm. Eng., vol. 37, pp. 449–454.

Hunter, G. T. and S. J. G. Bird, 1970, Critical terrain analysis. Photogramm. Eng., vol. 36, pp. 939–954.

Indiana Engineering Experiment Station Joint Highway Research Project, 1959, Atlas of county drainage maps of Indiana. Purdue Univ., Engineering Extension Dept., Extension Series No. 97.

Jensen, M. E. and H. R. Haise, 1963, Estimating evapotranspiration from solar radiation. Paper 3737, Proc. ASCE. J. Irrig. and Drain. Div. 89 (IR-4).

Johnson, A. N., 1928, Maryland aerial survey of highway traffic between Baltimore and Washington, Eighth Ann. Meeting of the Highway Research Board, Proc., pp. 106–115.

Johnson, G. O., 1973, Compiling preliminary foundation data from existing information on soils and geology. Highway Research Board, Highway Research Record no. 426, pp. 1–6.

Jordon, T. D., 1963, Development of the sky count technique for highway traffic analysis. Highway Research Board, Highway Research Record No. 19, pp. 35–46.

Jordon, T. D., 1965, Sky Count of Traffic Congestion and Demand, Traffic Engineering and Control, Vol. 7, No. 5, pp. 312–315.

Kansas State Highway Commission, 1974, Corridor Analysis for Butler County, Remote Sensing Section, Location and Design Concepts Department.

Karns, D., 1980, Photogrammetric Cadastral surveys and GLO corner restorations. Submitted to ACSM.

Katibah, G. P., 1964, Photogrammetric survey of section corners. California Division of Highways, Sacramento, California.

Katibah, E. F., 1973, A simple photographic technique for producing color composites from black-and-white multiband imagery with special reference to ERTS-1. Forestry Remote Sensing Laboratory, University of California, Berkeley, 9 pp.

Katibah, E. F., 1975, Areal extent of snow estimation using Landsat-1 satellite imagery. In An Integrated Study of Earth Resources in the State of California Using Remote Sensing Techniques, Space Sciences Laboratory, Series 16, Illue 34, University of California, Berkeley, 15 pp.

Khorram, S., E. F. Katibah and R. W. Thomas, 1976, Remote sensing-aided procedures for water yield estimation. In R. N. Colwell (Prin. Inv.), Integrated Study of Earth Resources in the State of California Using Remote Sensing Techniques, Space Sciences Laboratory, Series 18, Issue 5, University of California, Berkeley, 64 pp.

Khorram, S., 1977a, Remote sensing-aided systems for snow quantification, evapotranspiration estimation, and their applications in hydrologic models. Proceedings, Eleventh International Symposium on Remote Sensing of Environment, Environmental Research Institute of Michigan, Ann Arbor, Michigan, pp. 795–807.

Khorram, S., 1977b, Water supply studies. In R. N. Colwell (Prin. Inv.), An Integrated Study of Earth Resources in the State of California Using Remote Sensing Techniques, Space Sciences Laboratory, Series 18, Issue 44, University of California, Berkeley, 97 pp.

Khorram, S., H. G. Smith, E. F. Katibah, R. W. Thomas, J. M. Sharp and A. Kaugars, 1978, Procedural manuals. Remote sensing as an aid in watershed-wide estimation of solar radiation, water loss to the atmosphere, areal extent of snow, and water content of snow. In R. N. Colwell (Prin. Inv.), An Integrated Study of Earth Resources in the State of California Using Remote Sensing Techniques, Space Sciences Laboratory, Series 17, Issue 53, University of California, Berkeley, 164 pp.

Khorram, S. and H. G. Smith, 1980a, Use of digital terrain data in topographic analysis. In Case Studies of Applied Advanced Data Collection in Management, American Society of Civil Engineers, New York, New York, pp. 368–383.

Khorram, S. and H. G. Smith, 1980b, Surface temperature mapping by NOAA-5 satellite. In Case Studies of Applied Advanced Data Collection in Management, American Society of Civil Engineers, New York, New York, pp. 286–300.

Khorram, S., 1981a, Use of ocean color scanner data in water quality mapping. Journal of Photogrammetric Engineering and Remote Sensing, Vol. 47, no. 5, pp. 667–676.

Khorram, S., 1981b, Water quality mapping from Landsat digital data. International Journal of Remote Sensing, Taylor & Francis Ltd., London, England, Vol. 2, No. 2, pp. 145–153.

Khorram, S., 1982a, A remote sensing-aided procedure for site-specific estimation of net radiation over large areas. Journal of Applied Photographic Engineering, Society of Photographic Scientists and Engineers, Vol. 8, No. 1, pp. 31–35.

Khorram, S., 1982b, Development of a remote sensing-aided digital databank for large scale land use planning. Proceedings, Sixteenth International Symposium on Remote Sensing of Environment, 13 pp.

Khorram, S., A. W. Knight and S. D. DeGloria, 1982, Water quality mapping of the entire San Francisco Bay and Delta from Landsat multispectral scanner data. In R. N. Colwell (Prin. Inv.), Coastwatch Report, Space Sciences Laboratory Series 23, Issue 6, University of California, Berkeley, 33 pp.

Kiefer, R. W., 1972, Sequential aerial photography and imagery for soil studies. Highway Research Board, Highway Research Record No. 421, pp. 85–92.

Laframboise, P. and A. Bachand, 1980, Les images Landsat dans la surveillance et l'etude du'reservoir LG2 durant la periode de remplissage. Sixth Canadian Symposium on Remote Sensing, Halifax, Canada, pp. 453–457.

LaPrade, G., 1972, Stereoscopy . . . A more general theory. Photogrammetric Eng., vol. 38, no. 12.

Lawrence, G. and A. Banner, 1980, The application of thermography for locating potential frost pockets in forest cutovers. Sixth Canadian Symposium on Remote Sensing, Halifax, Canada, 369–376.

Layton, J. 1973, Personal communication. Consulting Engineer with CH_2M/Hill, Bellevue, Wn., Consultant to RIBCO, State of Washington.

Leith, C. J., C. P. Fisher, C. S. Deal, C. P. Gupton and C. A. Yorke, 1964, An investigation of the stability of highway cut slopes in North Carolina. North Carolina State University, Raleigh, Project ERO-110-U, 129 pp.

Liang, Ta, 1952, Landslides—An aerial photographic study. Cornell University Ph.D. Thesis, 274 pp.

Liang, T. and D. J. Belcher, 1958, Landslides and engineering practice: Airphoto interpretation. Highway Research Board Special Rept. 29, Washington, D.C., Ch. 5.

Liang, T., 1964, Tropical soils: Characteristics and airphoto interpretation. Final Rept. Project No. 8623, Cornell Univ. for Air Force Cambridge Research Lab., Contract No. AF 19(628)-291.

Lillesand, T. M. and R. W. Kiefer, 1979, Remote Sensing and Image Interpretation, John Wiley and Sons, N.Y.

Linacre, E. T., 1967, Climate and the evaporation from crops. ASCE, IR-4:61–79.

Lobeck, A. K., 1939, Geomorphology. McGraw-Hill Book Co., Inc., New York, 731 pp.

Lovelace, A. D., et al., 1962, Aggregate resources and soils study New Mexico Interstate Route 10. New Mexico Highway Department in cooperation with Federal Highway Administration.

Lowry, R. T. and C. J. Brochu, 1978, An interactive correction and analysis system for airborne laser profiles of sea ice. Canadian Journal of Remote Sensing, vol. 4, no. 2, pp. 149–160.

Lueder, D. R., 1959, Aerial photographic interpretation: principles and applications. McGraw-Hill, New York, Toronto, London, 462 pp.

Lyon, J. 1978, An Analysis of Vegetation Communities in the Lower Columbia River Basin. Proceedings of the PECORA IV Symposium on Wildlife Habitat Assessment, Sioux Falls, SD, p. 321–327.

Madson, Carlisle, 1975, The public land survey—lost, obliterated and rediscovered. ACSM Journal, June, pp. 155–158.

Maine Department of Transportation, 1972, Wiscasset

By-pass Corridor Location Analysis and Environmental Review, U.S. Route 1, Bureau of Highways, Location and Survey Division, Project No. 26-1(511).

Maine State Highway Commission, 1964, Materials Survey Maintenance Sand Study, Division 3, Soil Mechanics Series Tech. Paper 64-2(B).

Maruyasu, T., M. Nishio and M. Kawahara, 1964, Application of aerial photographic interpretation for civil engineering in Japan. Journal Japan Soc. of Photogramm. Special vol. no. 1, pp. 49–69.

Mathews, H. L., 1972, Application of multispectral remote sensing and spectral reflectance patterns to soil survey research. Ph.D. thesis, Pennsylvania State Univ., 110 pp.

Mathews, H. L., et al., 1973, Application of multispectral remote sensing to soil survey research in southeastern Pennsylvania. Soil Science Soc. of Am., Proc., vol. 37, no. 1, pp. 88–93.

Mathur, B. S. and J. F. Gartner, 1968, Principles of photo interpretation in highway engineering practice. Materials and Testing Div., Ontario Dept. of Highways, 236 pp.

Mathur, B. S., 1979, Remote Sensing Sensors for Environmental Studies, American Society of Civil Engineers, Transportation Engineering Journal, Vol. 105, No. TE 4, pp. 439–455, July.

Meyers, L. D. and A. H. Stallard, 1975, Soil identification by remote sensing techniques in Kansas. Part II, State Highway Commission of Kansas in cooperation with Federal Highway Admin.

McWhirter, N. (Editor), 1982, Guinness Book of records. Edition 28, Guinness Superlatives Ltd., London, 350 pp.

Miles, R. D., 1970, Remote sensing and development of annotated aerial photographs as master soils plans for proposed highways. Joint Highway Research Project, Purdue Univ., Final Rept., no. 15.

Millsaps, J. L., 1981, Forward Looking Infrared (FLIR) System for U.S. Coast Guard Search and Surveillance. Office of Research and Development, U.S. Coast Guard Headquarters, Washington, D.C.

Mintzer, O. W. and R. A. Struble, 1965, Terrain investigation techniques for highway engineers. Ohio State Univ., Eng. Exp. Sta. Rept. No. 196-2 and Appendix 1.

Mintzer, Olin W., 1966, Application of photo interpretation to highway engineering design. Final Report No. EES 196, Engineering Experiment Station, The Ohio State University, Columbus, Ohio, October, 213 pp.

Mintzer, O. W., 1968a, A comparative study of photography for soils and terrain data. U.S. Army Engineer Topographic Laboratories, Fort Belvoir, Virginia, Tech. Report 38-TR.

Mintzer, O. W., 1968b, Photographic interpretation for color aerial photographs. In Manual of Color Aerial Photography, Am. Soc. of Photogramm. pp. 425–430.

Mintzer, Olin and Bates, David, 1975, Use of Aerial Color Photography for Highway Applications, Final Report, 3849-1, DOT-FH-11-8234, Federal Highway Administration, U.S. Department of Transportation, Washington, D.C. 20590.

Mintzer, O. W. and Spragg, D., 1978, Mini-Format Remote Sensing for Civil Engineering, Transportation Engineering Journal, American Society of Civil Engineering, New York, N.Y.

Mollard, J. D., 1952, Aerial photographic studies on the Central Saskatchewan Irrigation Project. Cornell Univ., Ph.D. Thesis, Ithaca, New York, 303 pp.

Mollard, J. D., 1958, Damsite studies from aerial photographs. Engineering Geology Case Histories No. 2, Division of Engineering Geology, Geological Society of America, March, p. 21.

Mollard, J. D. and H. E. Dishaw, 1958, Ten years of mapping granular deposits from aerial photographs. Highway Research Board Bulletin no. 180, Highway Research Board, Washington, D.C., 11 pp., reprinted.

Mollard, J. D. and H. E. Dishaw, 1960, Airphoto interpretation in municipal engineering. The Engineering Institute of Canada, Western Zone Technical Converence, Banff, Alberta, October 2 and 3, 1959, Engineering Journal, September issue, 6 pp., reprinted.

Mollard, J. D., 1962, Photo analysis and interpretation in engineering-geology investigations: A review. Geol. Soc. Am., Rev. in Eng. Geology, vol. 1, pp. 105–127.

Mollard, J. D., 1967, Role of Photo interpretation in finding ground-water sources in Western Canada. Paper delivered at the Second Symposium on Air Photo Interpretation, Ottawa, March, 9 pp., reprinted.

Mollard, J. D., 1968, Landform analysis. Section 10.13 in Manual of Color Aerial Photography, Am. Soc. of Photogramm., pp. 406–407.

Mollard, J. D., 1969, Photo-interpretation studies in the location of prairie groundwater supplies. Paper presented to 22nd Annual Canadian Soil Mechanics Conference, Kingston, Ontario, December 8 and 9, reprinted in Canadian Geotechnical Journal, vol. 7, no. 2, 1970, 9 pp.

Mollard, J. D., 1970, Photointerpretation studies in the location of prairie groundwater supplies. Canadian Geotechnical Jour., vol. 7, no. 2, pp. 127–135.

Mollard, J. D., 1975, The integration of different aerial remote sensors and map data in making engineering and resource studies. Presented in Third Canadian Symposium on Remote Sensing, Edmonton, Sept. 22–24.

Mollard, J. D., 1976, Some regional landslide types in Canada. Chapter contribution prepared for GSA Volume on Landslides, Geological Society of America, Boulder, Colorado (in press).

Mollard, J. D., 1978, Landforms and surface materials of Canada: A stereoscopic airphoto atlas and glossary, Sixth edition. 677 stereograms, 1986 contact airphotos, map of Canada, 13 chapters, over 2500 terms in glossary, 336 pp., J. D. Mollard Associates, Regina, Saskatchewan, Canada, S4P 0R7.

Moredock, K., 1962, Preparation of right-of-way plans from aerial mosaics. Highway Research Board Bull. 354, pp. 60–67.

Moultrop, K., 1956, Engineering soil survey of Rhode Island. Eng. Expt. Sta., Univ. Rhode Island, Bull. no. 4.

Moultrop, K., 1964, Rhode Island aggregate survey. Univ. Rhode Island, Eng. Bull. no. 6, sponsored by Rhode Island Dept. of Public Works in cooperation with Federal Highway Admin., June.

Munday, J. C., et al., 1971, Oil slick studies using photographic and multispectral scanner data. Proc., Seventh Internat. Symp. on Remote Sensing of Environment, May 17–21, Univ. Michigan, 3 vol., pp. 1027–1044.

Munday, J. C., Jr., 1974a, Lake Ontario water mass determination from ERTS 1. Proc. 9th International

Symposium on Remote Sensing of Environment, Ann Arbor, Mich., pp. 1355–1368.

Munday, J. C., Jr., 1974b, Water quality of lakes of southern Ontario from ERTS 1. Proc. 2nd Canadian Symposium on Remote Sensing, Guelph, Ontario, pp. 77–85.

Munday, J. D., Jr., T. T. Alfoldi and C. L. Amos, 1979, Bay of Fundy verification of a system for multidate Landsat measurement of suspended sediment. In Deutsch, M., D. R. Wiesnet and R. Rango (Eds.), Satellite hydrology. Proc. 5th Annual William T. Pecora Memorial Symposium on Remote Sensing, pp. 622–640.

Neumaier, G. and F. Silvestro, 1969, Measurement of pollution using multiband and color photography. Proc. New Horizons in Color Aerial Photography, Am. Soc. Photogramm., June 9–11, pp. 47–58.

Noble, D. F., 1972, Virginia's use of remote sensing in the preliminary aerial survey-highway planning stage. Final Rept., VHRC 71-R19, Virginia Highway Research Council in cooperation with the Federal Highway Admin.

Norell, W. F., 1965, Air Photo Patterns of Landslides in Southeastern Ohio, Ohio Dept. of Transportation in Cooperation with Federal Highway Admin., 2 vols., 44 plates.

———, 1966, Coal Outcrop and Overburden Mapping with Kelsh Plotter, Highway Research Board, Highway Research Record No. 109, pp. 39–48.

———, 1968, Air Photo Patterns of Subsurface Mining in Ohio, Ohio Dept. of Transportation in Cooperation with Federal Highway Admin., 2 vols., 50 plates.

———, 1973a, Air Photo Patterns of Organic Accumulation Sites, Ohio Dept. of Transp. in Cooperation with Federal Highway Admin., 2 vols., 72 plates.

———, 1973b, Measuring and depicting trouble areas in stereo-models. Highway Research Record No. 452, Highway Research Board, pp. 45–54.

Odenyo, V. A. O., 1979, Application of Landsat data to the study of land and range resources in the Napok Area, Kenya. Canada Centre for Remote Sensing Research Report 79-4, EMR, Ottawa, Canada.

Olson, C. E., et al., 1969, Inventory of recreation sites. Photogramm. Eng., vol. 35, pp. 561–568.

Ontario Ministry of Transportation and Communication, 1971, Preliminary area study for proposed East-Metro Freeway. Photogrammetry Office, Dept. of Highways.

Orr, D. G. and J. R. Quick, 1971, Construction materials in delta areas. Photogramm. Eng., vol. 37, pp. 337–351.

Orsi, J. J. and A. C. Knutson, Jr., 1979, The role of mysid shrimp in the Sacramento-San Joaquin estuary and factors affecting their abundance and distribution. Proceedings, San Francisco Bay Symposium: The Urbanized Estuary. T. J. Conomos (Ed.), Amer. Asso. Adv. Sci., pp. 401–409.

Parry, J. T., W. R. Cowan and J. A. Heginbottom, 1969, Soils studies using color photos. Photogramm. Eng., vol. 35, pp. 44–57.

Parvis, M., 1948, Development of drainage maps from aerial photographs. Highway Research Board, Proc., vol. 26, pp. 150–163.

Penman, H. L., 1948, Natural evaporation from open water, bare soil, and grass. Proc. Roy. Soc., (London) A. 193, pp. 120–145.

Petersen, G. E., J. Riordan and M. O. Cummings, 1974, Construction materials inventory of Shawnee Co., Kansas. State Highway Commission of Kansas, Construction Materials Inventory Rept., No. 26.

Piech, K. R. and J. F. Walker, 1972, Outfall inventory using airphoto interpretation. Photogramm. Eng., vol. 38, pp. 907–914.

Port of New York Authority, 1968, Cameras aloft! Project sky count: Analysis of transportation patterns and land usage from aerial photography. Operations Service Dept., Operations Standards Div.

Radbruch-Hall, D. H., R. B. Colton, W. E. Davies, B. A. Skipp, I. Lucchitta and D. J. Varnes, 1976, Preliminary landslide overview map of the conterminous United States. U.S. Geological Survey, Miscellaneous Field Studies Map MF-771.

Raj, Des, 1968, Sampling theory. McGraw-Hill Book Company, San Francisco, 302 pp.

Rango, A. and V. V. Salomonson, 1975, Employment of satellite snowcover observations for improving seasonal runoff estimates. In Proceedings of a workshop on Operational Applications of Satellite Snowcover Observations, South Lake Tahoe, California, pp. 157–174.

Ray, Richard G., Aerial photographs in geologic interpretation and mapping. Geologic Survey Professional Paper 373, 230 pp.

Reeves, Robert G. (Editor-in-Chief), 1975, Manual of remote sensing. American Society of Photogrammetry, Falls Church, Virg., 2 vols., 2144 pp.

Rib, H. T., 1966a, An optimum multisensor approach for detailed engineering soils mapping. Joint Highway Research Project No. 22, 2 vols., Purdue Univ., 406 pp.

Rib, H. T., 1966b, Utilization of photo interpretation in the highway field. Highway Research Board Bull. No. 109, pp. 18–25.

Rib, H. T., 1968, Remote sensing applications to highway engineering. Public Roads, vol. 35, no. 2, U.S. Dept. of Transportation, Federal Highway Admin., pp. 29–36, June.

Rib, H. T. and R. D. Miles, 1969a, Automatic interpretation of terrain features. Photogramm. Eng., vol. 25, pp. 153–164.

Rib, H. T. and R. D. Miles, 1969b, Multisensor analysis for soils mapping. Highway Research Board, Special Rept. 102, pp. 22–37.

Rib, H. T., 1972, Partnership in research: A cooperative remote sensing research program. Highway Research Record No. 421, Highway Research Board, pp. 33–40.

Rib, H. T., 1977, The Role of Remote Sensing in the Building of the Interstate Highway System, Environmental Research Institute of Michigan, Proceedings of the Eleventh International Symposium on Remote Sensing of Environment, 25–29, April, pp. 379–383.

Rib, H. T., J. M. Spencer, Jr., J. F. Koca, C. P. Falls, 1977, Evaluation of Aerial remote sensing systems for detecting subsurface cavities in Kansas. Report No. FHWA RD-75-119, Federal Highway Administration.

Rib, H. T. and T. Liang, 1978, Landslides: analysis and control, Transportation Research Board, Special Report 176, Chapter 3.

Robinson, Charles S., Fitzhugh T. Lee, et al., 1972, Geological, geophysical, and engineeeing investigations of the Loveland Basin Landslide, Clear Creek County, Colorado, 1963–1965. Geological Survey Professional Paper 673, 43 pp.

Rogers, H. H., K. Peacock, and N. Shah, 1973, A technique for correcting ERTS Data for solar and atmospheric effects. Proc., 3rd ERTS Symposium, Washington, D.C.

Rogers, H., N. J. Shah, J. B. McKeon and V. E. Smith,

1977, Computer mapping of water quality in Saginaw Bay with Landsat digital data. Proc., Spring ASP Convention.

Ross, D. I., R. M. Chaterjee and N. D. Herzog, 1976, Evaluation of airborne thermal infrared mapping of cooling water discharges from generating stations. Water Resources Branch, Ontario Ministry of the Environment, Toronto.

Ross, D. I. and J. D. Kinkead, 1976, Selected aspects of operational thermal IR remote sensing. 19th Conference on Great Lakes Research, Univ. of Guelph, Guelph, Ontario, Canada.

Royster, D. L., and F. W. Davis, 1965, Summer County engineering soil survey and highway materials inventory. Tennessee Dept. of Highways, Research Project No. HPS-HPR 1(24).

Rula, A. A., W. E. Grabau and R. D. Miles, 1963, Forecasting trafficability of soils: Airphoto approach. U.S. Army Engineer Waterways Expt. Sta., Tech. Memo. No. 3-331, Report 6, Corps of Engineers, 2 vols., 218 pp.

Ruth, B. E., 1972, Interpretation of site conditions in the Appalachian Plateau. Proc., 38th Ann. Meeting of the Am. Soc. Photogramm., Washington, D.C., March 12–17, pp. 73–81.

Ryerson, R. A., 1981, Results of a benefit cost analysis of the CCRS airborne program. Canada Centre for Remote Sensing Research Report 81-1, EMR, Ottawa, Canada.

Scherz, J. P., M. Adams, F. I. Scarpace and W. J. Woelkerling, 1977, Assessment of aquatic environment by remote sensing. IES report 84, Univ. Wisconsin—Madison.

Schonfeld, R., 1974, Photo-Interpretation of Pavement Skid Resistance, Ontario Ministry of Transportation and Communication, Research Report 188.

Schoonmaker, J. W., 1974, Geometric evaluation of MSS imagery from ERTS-1. Proc., ASP 40th Ann. Meeting, St. Louis.

Schuster, R. L., and R. J. Krizek, (Eds.), 1978, Landslides: Analysis and control. Transportation Research Board, Special Report 176, 234 pp.

Scully, John, 1973, Landslides in Pierre Shale in central South Dakota. Final Report, State Study No. 635(67), Prepared for South Dakota Dept. of Transportation and Federal Highway Administration, Pierre, South Dakota, Dec., 707 pp.

Scarpace, F. L., and K. Wade, 1974, Monitoring the trophic status of inland lakes. Proc., Fall ASP Convention, Sept.

Scarpace, F. L., K. Holmquist and L. T. Fisher, 1979, Landsat analysis of lake quality. Photogrammetric Eng., and Remote Sensing, vol. 45, no. 5., pp. 623–633.

Sharp, J. M., and R. W. Thomas, 1975, A cost-effectiveness comparison of existing and Landsat-aided snow water content estimation systems. In Proceedings of the Tenth International Symposium on Remote Sensing of Environment, Environmental Research Institute of Michigan, University of Michigan, Ann Arbor, Michigan, 7 pp.

Sharpe, C. F. Stewart, 1938, Landslides and related phenomena. Columbia University Press, New York, 137 pp.

Shelton, Robert J., Personal communication. Robert J. Shelton & Associates, 228 West Main, Missoula, Montana, 59801.

Short, N. M., et al., 1976, *Mission to Earth: Landsat Views the World,* NASA SP-360, National Aeronautics and Space Administration.

Siegfried, C. A., A. W. Knight, and M. E. Kopache, 1978, Ecological studies in the western Sacramento-San Joaquin Delta during a dry year, Water Science and Engineering Paper, No. 4506, University of California, Davis, Calif., 121 pp.

Siegfried, C. A., M. E. Kopache, and A. W. Knight, 1979, The distribution and abundance of Neomysis Mercedis in relation to the entrapment zone in the Western Sacramento-San Joaquin Delta. Transactions of the American Fisheries Society, Vol. 108, No. 3, pp. 262–270.

Singhroy, V., 1980, Sand and gravel resources and quaternary geology of the Pas Region. Geological Report GR-80-2, Manitoba Dept. of Energy and Mines, Winnipeg, Manitoba.

Singhroy, V., W. D. Bruce and G. R. Stevens, 1980a, Geologic and terrain analysis of the Annapolis County Region of Nova Scotia: An application of digital Landsat, radar and color infrared data. Sixth Canadian Symposium on Remote Sensing, Halifax, Canada, pp. 501–512.

Singhroy, V. and R. Dixon, 1980b, Resource inventory and terrain analysis in Northwestern Manitoba from remote sensing data. Sixth Canadian Symp. on Remote Sensing, Halifax, Canada, pp. 513–523.

Skousen, Stanley J., 1971, Section subdivision by photogrammetry. U.S. Forest Service, Region I, Missoula, Montana.

Society of American Foresters, 1980, Forest cover types of the United States and Canada. F. H. Eyre (Ed.), Washington, D.C., 20014.

Smith, H. G. and S. Khorram, 1980, Remote sensing aided procedure for conifer growth modeling in Northeast Sierra Nevada. Proceedings, Fall Technical Meeting, American Society of Photogrammetry. RS-2F-1 to RS-2-F-14.

Smith, John T., Jr., (Editor-in-Chief), 1968, Manual of color aerial photography, American Society of Photogrammetry, Falls Church, Virginia, 550 pp.

Smith, P. C., 1951, Appraisal of soil and terrain conditions for part of the Natchez Trace Parkway. Public Roads, vol. 26, no. 10.

Stafford, D. B., and J. Langfelder, 1971, Air photo survey of coastal erosion. Photogramm. Eng., vol. 37, pp. 565–575.

Stallard, A. H., and G. Anschutz, 1963, Use of the Kelsh Plotter in geoengineering and allied investigations in Kansas, Highway Research Board, Highway Research Record No. 19, pp. 53–107.

Stallard, A. H., 1965, An evaluation of color aerial photography for engineering purposes, Kansas State Highway Commission, Special Rept. 1, prepared in cooperation with Federal Highway Admin.

Stallard, A. H. and T. A. Witty, Jr., 1966, The use of photo interpretation in archeological salvage progress in Kansas. State Highway Commission of Kansas, Special Report No. 2, Project 64-6.

Stallard, A. H. and R. R. Biege, Jr., 1966, Evaluation of color aerial photography in some aspects of highway engineering. Highway Research Record No. 109, Highway Research Board, pp. 18–26.

Stallard, A. H., 1972, Use of remote sensors in highway engineering in Kansas. Highway Research Board, Highway Research Record No. 421, pp. 50–57.

Stallard, A. H., and L. D. Myers, 1972, Soil identification by remote sensing techniques in Kansas. Part I, State Highway Commission of Kansas in cooperation with Federal Highway Admin.

Standberg, C. H., 1964, An aerial water quality reconnaissance system. Photogramm. Eng., vol. 30, pp. 46–54.

Stoeckeler, E. G. and W. R. Gorrill, 1959, Airphoto

analysis of terrain for highway location studies in Maine, Highway Research Board Bull. No. 213, pp. 29–43.

Stoeckeler, E. G., 1968, Use of color aerial photography for pavement evaluation studies in Maine. Maine Dept. of Transportation, Materials and Research Tech. Paper 68-6R, prepared in cooperation with Federal Highway Admin.

Stoeckeler, E. G., 1970, Use of color aerial photography for pavement evaluation studies, Highway Research Board, Highway Research Record No. 319, pp. 40–57.

Strandberg, C. H., 1966, Water quality analysis. Photogramm. Eng., vol. 32, pp. 234–248.

Strandberg, C. H., 1967, Aerial discovery manual. John Wiley & Sons, Inc., New York.

Sydor, Michael, 1973, Satellite Data from Lake Superior, Duluth, University of Minnesota, Duluth.

Tanguay, M. G., 1969, Aerial photography and multispectral remote sensing for engineering soils mapping. Purdue Univ., Joint Highway Research Project, No. 13, 308 pp.

Tanguay, M. G. and R. D. Miles, 1970, Multispectral data interpretation for engineering soils mapping. Highway Research Board, Highway Research Record No. 319, pp. 58–77.

Tanguay, Marc G., J. Y. Chagnon, J. Y., 1972, Thermal infrared imagery at the St. Jean-Vianney Landslide. First Canadian Symposium on Remote Sensing, pp. 387–402.

Terzaghi, K., 1950, Mechanism of landslides. Geol. Soc. Am., Application of Geology to Engineering Practice, Berkeley, vol. 1, pp. 83–123.

Thomas, Morris M. (Editor-in-Chief), Manual of Photogrammetry. 3rd Ed., 2 Vols., Society of Photogrammetry, Falls Church, Virginia, 1199 pp.

Thomas, R. W. and J. M. Sharp, 1975, A comparative cost effectiveness analysis of existing and Landsat-aided systems for estimating snow water content. In An Integrated Study of Earth Resources in the State of California Using Remote Sensing Techniques, Space Sciences Laboratory Series 16, Issue 34, University of California, Berkeley.

Thompson, M. D. (Editor), P. Howarth, R. Ryerson, and F. Bonn (authors-editors), 1982, Landsat for monitoring the changing geography of Canada. Canada Centre for Remote Sensing Contribution to COSPAR, Energy, Mines and Resources, Ottawa, Canada, 84 pp.

Thornbury, William D., 1954, Principles of Geomorphology. John Wiley & Sons, New York, 618 pp.

Treiterer, J. and J. I. Taylor, 1966, Traffic Flow Investigations by Photogrammetric Techniques, Highway Research Board, Highway Research Record No. 142, pp. 1–12.

Turner, A. K., 1968, Computer-assisted procedures to generate and evaluate regional highway alternatives. Purdue Univ., Joint Highway Research Project No. 32, 281 pp.

Turner, A. K. and R. D. Miles, 1971, the GCARS system: A computer-assisted method of regional route location. Highway Research Board, Highway Research Record No. 348, pp. 1–15.

Turner, A. K. and I. Hausmanis, 1973, Computer-aided transportation corridor selection in the Guelph-Dundas Area of Ontario. Highway Research Board Special Report No. 138, pp. 55–70.

Turner, A. K., 1978, A Decade of Experience in Computer Route Selection, Photogrammetric Engineering and Remote Sensing, Vol. 44, No. 12, pp. 1561–1576, December.

U.S. Department of Interior, 1978, Digital Terrain Tapes User's Guide, Geol. Survey, Reston, Va. 12 pp.

U.S. Forest Service, 1963, Application of aerophotogrammetry to dependent corner restoration surveys. North Central Region, Milwaukee, Wisconsin.

U.S. Geological Survey, Department of the Interior, 1972, Water Resources Data for Washington, Surface Water Records.

Van Domelen, J. F., 1974, Photographic Remote Sensing—A Water Quality Management Tool, Ph.D. Thesis, University of Wisconsin, Madison.

Van Lopik, J. R., A. E. Pressman and R. L. Ludim, 1968, Mapping pollution with infrared. Photogramm. Eng., vol. 34, pp. 561–564.

Voute, C., 1964, Contributions of photo-interpretation to engineering projects in various stages of execution. Photogrammetria 19, 1962–1964, 5 pp. 179–191.

Wagner, F. A., Jr., and A. D. May, Jr., 1963, Use of Aerial Photographs in Freeway Traffic Operations Studies, Highway Research Board, Highway Research Record No. 19, pp. 24–34.

Wagner, R. R., 1963, Using airphotos to measure changes in land use around highway interchanges. Photogramm. Eng., vol. 29, pp. 645–649.

Wagner, T. W., 1972, Multispectral remote sensing of soil areas: A Kansas study. Highway Research Board, Highway Research Record No. 421, pp. 71–77.

Wall, S. L., R. W. Thomas, C. E. Brown and E. H. Bauer, 1982 A Landsat-based inventory procedure for agriculture in California, *In* 1982 Symposium on Machine Processing of Remotely Sensed Data, Purdue University.

Wall, S. L., R. W. Thomas, C. E. Brown, M. Eriksson, and E. H. Bauer, 1982. A Landsat-based inventory for the estimation of irrigated land on arid areas. Proc. International Symposium on Remote Sensing of Environment—First Thematic Conference: "Remote Sensing of Arid and Semi-Arid Lands, Cairo, Egypt. pp. 523–532.

Walters, W. C., L. M. Lasserre, and H. D. Salassi, Jr., 1963, Louisiana engineering soils map report. No. 2, Lafayette-Port Allen Strip: Louisiana Dept. of Highways in cooperation with Federal Highway Admin.

Wardlow, James M., 1982, The use of remote sensing to answer engineering questions for water management planning and operations. Unpublished special report. 10 pages.

Way, D. S., 1973, Terrain analysis: A guide to site selection using aerial photographic interpretation. Dowden, Hutchinson & Ross, Inc., Stroudsburg, Penn., 392 pp.

Way, D. S., 1978, Terrain Analysis, McGraw-Hill Co., New York. 438 pp.

Welch, Robin I. and R. N. Colwell, 1972, An aerial photographic documentation of conditions in Redwood National Park and adjoining areas. Special Report prepared by Earth Satellite Corporation in cooperation with an *ad hoc* Redwood National Park Study Team, 126 pp., illus.

West, T. R., 1972, Engineering soil mapping from multispectral imagery using automatic classification techniques. Highway Research Board, Highway Research Record No. 421, pp. 58–65.

White, J. R., R. E. Schmidt, and W. E. Plage, 1979, The AIREYE remote sensing system for oil spill surveillance. Proceedings of the 1979 Oil Spill Conference, pp. 301–304.

Wickware, G. M., 1978, Wetland mapping and environmental monitoring using digital Landsat data. Proc. 5th Canadian Symp. on Remote Sensing, Victoria, B.C. pp. 150–157.

Wickware, G. M. and P. J. Howarth, 1981, Change detection in the Peace-Athabasca Delta using digital Landsat data. Remote Sensing of Environment, Vol. 11, No. 1, pp. 9–25.

Wickware, G. M. and P. J. Howarth, In preparation, 1982, Landsat analysis of hydrologic and vegetation change in Peace-Athabasca Delta. II. Classification.

Wiebe, J. and E. T. Andersen, 1976, Site selection for grapes in the Niagara Peninsula (includes map Grape climatic zones in Niagara). Horticultural Research Institute of Ontario, Vineland Station, Ontario, Canada.

Williams, G. P. and H. P. Guy, 1973, Erosional and depositional aspects of Hurricane Camille in Virginia, 1969. Geological Survey Professional Paper 804, 80 pp.

Wilson, E., 1969, Engineering Hydrology, MacMillan, London,

Wolf, P., 1966, Comparison of methods of flood estimation. River Flood Hydrology, The Institution of Civil Engineers, London.

Woodman, R. G. and R. S. Farell, 1975, Multidisciplinary analysis of Skylab photography for highway engineering purposes, Maine State Highway Commission, Final Report August 1973—February 1975, prepared for National Aeronautics and Space Administration.

Woods, K. B., (Editor-in-chief), 1960, Highway engineering handbook. McGraw-Hill, New York.

Wyoming State Highway Department, 1963, Procedure manual for statewide materials location survey. Prepared in cooperation with Federal Highway Admin.

Zaruba, Q., and V. Mencl, 1969, Landslides and their control. American Elsevier Publishing Co., Inc., New York, 202 pp.

CHAPTER 33

Remote Sensing Applications in Agriculture

Author-Editor: VICTOR I. MYERS

Contributing Authors: MARVIN E. BAUER, HAROLD W. GAUSMAN, WILLIAM G. HART, JAMES L. HEILMAN, ROBERT B. MACDONALD, ARCHIBALD B. PARK, ROBERT A. RYERSON, THOMAS J. SCHMUGGE, FREDERICK C. WESTIN.

GENERAL CONTENTS: Agriculture resource surveys: soils identification; factors influencing interpretation of soils; surveys; Vegetation information: physiological factors influencing electromagnetic radiation, crop canopies, crop identification, agricultural resource surveys and information systems: Agricultural management and production: water management, identifying irrigated lands, high water tables, salinity, insects and disease, arid and semi-arid land monitoring, wind and water erosion; Agriculture management and planning; Change assessment and yield estimates: Corn Blight Watch Experiment, CITARS, LACIE: Agricultural Identification mechanisms: spectral, spatial, and temporal resolutions; outlook for the future; references.

INTRODUCTION

Resource managers require rapid and accurate methods for acquiring and interpreting data for the development and management of our natural resources. Agricultural remote sensing involving crops and soils is extremely complex because of the dynamic nature and inherent complexity of biological materials and soils, yet remote sensing technology offers numerous advantages over traditional methods of conducting agricultural and other resource surveys. Advantages include the potential for accelerated surveys; capability to achieve a synoptic view under relatively uniform lighting conditions; availability of multispectral data providing increased information; capability of repetitive coverage to depict seasonal and long-term changes; the relatively inexpensive cost of monitoring from space; the opportunity of integrating existing surveys into an updated monitoring system; the change detection capabilities needed by regulatory programs for updating information on vegetation/terrain conditions; availability of imagery with minimum distortion, thereby permitting direct measurement of important agrophysical parameters; and the fact that remotely sensed data provide a permanent record.

Remote sensing of earth resources utilizes electromagnetic energy which ranges from short wavelength ultraviolet through visible, near infrared, and thermal infrared, to the longer wavelength active radar and passive microwave systems. All of these wavelengths are applicable to and useful for agricultural remote sensing. Data from the several wavelength regions should be used together for many applications. One should bear in mind that studies in the agricultural disciplines can be far more comprehensive if a wide array of data is considered in both the analysis and the applications phases. Useful ancillary data include weather and temperature data provided by NOAA weather satellites.

Both photointerpretation and computer classification have broad application in the interpretive process. A great advancement in the application of computers to this science is the development of the capability of storing vast and varied information, ranging from historical information and aerial photography to spacecraft data, ground reference, and other forms of ancillary data—all in the form of a highly useful data base/information system.

Specific resources and conditions which can be monitored include soils, vegetation, land use, erosion, soil moisture, high water tables, salinity, and desertification. Some of the principal resource-monitoring applications that pertain specifically to agriculture are described in this chapter.

AGRICULTURAL RESOURCE SURVEYS

SOILS IDENTIFICATION

Introduction

Man's increasing pressure on the land masses of the world dictates the need for research that will lead to better and more rapid methods of delineating soil boundaries, and for quantifying different chemical-physical parameters of the world's soil resources, as well as for characterizing and mapping soils.

Aerial photographs have been traditionally used in soil surveys for soil-boundary detection, landform analyses and visual perception of tonal qualities associated with the spatial patterns of soils. Landsat data have recently been shown to be most useful for generalized soil and land-system surveys. The application of Landsat data to soil investigations has progressed beyond the research stage to its use in extensive soil surveys, generally conducted at the reconnaissance level.

Factors Influencing Interpretation of Soils

Much knowledge has been accumulated about the chemical and physical characteristics of soils relative to the soil-forming factors of parent material, relief and climate. However, there is only limited knowledge of how these factors relate to reflected and emitted radiation, although that body of knowledge is growing. A number of studies have shown that information about the spectral properties of soils is useful in their identification and characterization (Westin, 1973a; Obukhov and Orlov, 1964; Bauer, et al., 1978a).

Physical, chemical and engineering determinations of most soil properties follow well-established laboratory-based analysis procedures. Certain of these soil properties have been related by empirical and experimental correlation to spectral response.

Soil Color

Soils often can be distinguished from one another by their photographic tone and/or color characteristics—factors which derive from properties of the soils materials, themselves. In addition electromagnetic radiation can be sensed by detectors that respond in spectral regions beyond those discernible by the human eye—specifically in the ultraviolet, infrared and microwave regions.

Color is often diagnostic of major soils classifications (Soil Survey Staff, 1951). Soil color is included in the description of a soil profile. The Munsell color system is generally used, which utilizes a descriptive system of hue, value and chroma. Hue is the dominant spectral color and corresponds to the wavelength region used in remote sensing measurements. Value refers to the relative brightness of the color and is a function of the total amount of light reflected. Chroma is the relative purity or strength of the spectral color.

Surface color that differs from that of the parent material is usually an indication of the processes involved in soil formation and may also be indicative of other factors such as excessive soluble salts (Myers et al., 1963) or erosion.

Obukhov and Orlov (1964) found that all of the soils that they investigated had spectral reflectance characteristics related to soil color. Minimum reflectance occurred in the blue-violet portion of the spectrum and ranged from 13 percent for the A horizon of thick chernozem to 18 percent for the same horizon of sod-podzolic soils. Maximum reflectance was in the red region of the spectrum where the reflection coefficient of the same samples increased from 15 to 44 percent. Obukhov and Orlov concluded that the visible red region and the near infrared region are the most favorable for a qualitative and quantitative description of soils.

In order to use remote sensing techniques to map soil it is necessary to understand the relationship between soil properties and soil color. Johannsen and Barney (1976) cited that the most important factors influencing soil color are mineralogy and chemical constituents, soil moisture, soil structure, particle size and organic matter content. da Costa (1979) quantified several important relationships for a number of soils from a broad climatic area covering 10 states when he published regression equations relating clay content, organic carbon, water retained at 15 bars and cation exchange capacity with soil color.

da Costa (1979) gives an excellent summary of qualitative factors influencing soil color and describes studies conducted to quantify relationships between soil properties and soil color. As he points out, the accuracy of soil properties estimated using soil color is not extremely high but spectroradiometers can be used to improve the accuracy of these estimates. Such instruments, of course, are not readily available to soil scientists in the field. However, the use of imagery gathered in discrete wavebands can be useful in that regard.

Mineral Content

Coblentz (1962) published an enormous number of infrared reflectance spectra for various minerals. Reflectance from materials depends upon the intermolecular vibration of the molecules present near the surface of the material. The optimum spectral range for use in detecting and identifying various minerals is from 8 μm to 14 μm because the fundamental SI-0 vibration occurs in this range.

The reflectance properties of a mineral are determined largely by the integrated contribution from each surface grain. Surface roughness, particle size, water film, contaminants, dust, dew, and physical discontinuities occurring on a surface have a very substantial effect on the reflectance. Hovis (1966), using techniques similar to those of Colbentz established that infrared reflectance spectra of minerals of the carbonate, sulfate, nitrate, and silicate families exhibit spectral absorption band patterns that can be detected by remote sensor. Lyon (1965) studied the emittance of minerals and developed infrared techniques for analyzing the composition of soils. Samples were analyzed by traditional mineralogical means and then spectral signatures were established for rocks and soils in the 8 μm to 13 μm wavelength region. By matching the incoming spectrum with standard curves in the memory of a computer, one can establish the bulk composition of the rock surfaces.

Limited research has been accomplished concerning the ultraviolet reflectance spectra of soils. Substantial difficulties are imposed on such imaging by atmospheric and systems limitations.

Yet, laboratory measurements by Hemphill and Carnahan (1965) indicate that some minerals would exhibit more image contrast when photographed or imaged in the ultraviolet region rather than at longer wavelengths. One of the primary investigators in this area, da Costa (1979) summarized results from several authors who had dealt with the spectral features of some common minerals in the visible and near-infrared portions of the electromagnetic spectrum. His summary revealed that several minerals, including quartz and feldspar, are spectrally featureless. Others such as kaolinite, gibbsite, and muscovite, exhibit spectral response due to hydroxyl effects. Oxides and hydroxides of iron, aluminum, and titanium are important for soils in general. Water bands are frequently spectral features resulting from water inclusions. Montmorillonite clays, which absorb water molecules, exhibit absorption of electromagnetic energy in those bands. Still other minerals are opaque such as ilmenite and magnetite and are spectrally featureless.

Certain minerals, including those that are predominant in ferrous and ferric ions, exhibit high response in the red region of the spectrum. These include limonite, rutile, brotite, hematite and goethite.

In a summary literature survey of spectral features of some common minerals, da Costa (1979) states that the features are grouped in two processes—electronic and vibrational. The electronic process is due to iron, regardless of form. The ferric iron response bands are at approximately 0.40, 0.70, and 0.87 μm, and the ferrous ion response bands are at approximately 0.43, 0.45, 0.51, 0.55 and 1.0 μm. Divalent manganese in octahedral coordination is found at bands 0.34–0.37, 0.41–0.45, and 0.45–0.55 μm. Most rocks and minerals contain silicates but the silicate cannot be observed in the visible and near-infrared spectral regions. If water is present in the spectra of minerals, two characteristic response bands appear which are at 1.4 and 1.9 μm, respectively. Five response wavelengths are characteristic of carbonate: 1.90, 2.00, 2.16, 2.35, and 2.55 μm.

Obukhov and Orlov (1964) found that soils and horizons with an elevated content of iron are easily distinguished by the reflection characteristics for pure Fe_2O_3 using spectrophotometry. The intensity of the reflection in the region from 0.50 to 0.64 μm is inversely proportional to the iron content. This can be expressed by a linear equation for artificial mixtures of SiO_2 and Fe_2O_3 having iron oxide contents of 2–9 percent where R (percent) is reflection at 0.64 μm and C is the percent of Fe_2O_3.

$$R(\%) = 84 - 4.9 \times C \qquad (33\text{-}1)$$

This demonstrates an excellent relationship for developing a quantitative determination of iron in soils.

Bauer et al. (1979) point out that soil mineralogy influences soil reflectance in various manners. Soils with gypsic mineralogy reflect highly because of the inherent reflectance properties of gypsum. However, montmorillonitic soils, often associated with higher organic matter levels, show low reflectance attributable to this high organic matter content.

Organic Matter

Organic matter in soils has a profound influence on soil color. However, under different climatic and management conditions the organic matter constituents will vary significantly. Kristof and Zachary (1971) and Baumgardner et al. (1970) used multispectral data and computer-assisted analyses to delineate and map spectral classes generated into five different ranges of organic matter content for mineral soils containing from 1.5 to 7 percent organic matter (Figure 33-1). Prairie soils, forest soils and transitional soils were included in this study. Since the organic matter content affects the color, heat capacity, water holding capacity, cation exchange, structure, and erodability of soils, a rapid and easy assessment of this parameter may be of great utility to the soil surveyor and the land use planner.

The dark color in the surface horizons is generally associated with organic matter content. In a summary of other investigations da Costa (1979)

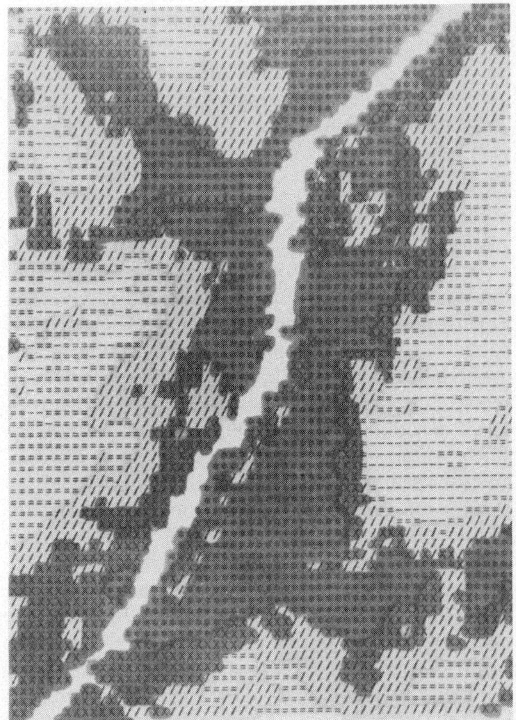

Fig. 33-1. Computer printout displaying five levels of soil organic matter.

described results obtained by other investigators in which they expressed soil color as a function of organic matter and iron, pH and type of clay. He also cited work by McKeague et al. (1971) who found significant relationships between many pairs of variables such as color value and organic matter, and organic matter and cation exchange capacity at pH 7. The correlations were obtained for 115 soils studied. Furthermore, they showed that color value can be used to estimate organic matter content of soil, but that this correlation is more pronounced in some soils than in others.

Studies conducted by da Costa (1979) showed organic carbon to be negatively correlated with Landsat MSS bands 4, 5 and 6; however, correlation coefficients higher than 0.50 were concentrated within the 0.50 to 0.74 μm range and the correlation decreased from band 4 to band 6. Thus, the interval of best correlation corresponded to the visible and a small part of the reflected infrared region of the electromagnetic spectrum. This demonstrates the importance of organic matter in visual surface-soil color-determinations. In the studies described, the number of soil properties that correlated with band 4 were very low compared to other bands, except in the case of organic carbon.

Although increased organic matter content has been seen to decrease soil reflectance in mineral soils, knowledge as to the decomposition stage of organic material is more important in understanding reflectance properties of organic soils (Bauer et al., 1979). Less decomposed organic materials have higher reflectance in the near-infrared region because of the enhanced reflectance that is attributable to the remnant cell structure of well preserved fibers. In contrast, very highly decomposed organic materials show very low reflectance throughout the 0.5 to 2.3 μm wavelength region.

Particle Size

Soil particle size influences reflectance and thermal diffusivity of soils and, indirectly, soil moisture and other measurements. The effect of particle size on reflectance of energy, as measured in the laboratory by spectrophotometry, has been reported by several authors (Bowers and Hanks, 1965; Shockley et al., 1962; and Orlov, 1966). These studies conclude that increasing particle diameter results in a decrease of reflectivity. This conclusion is correct only for the laboratory case of dispersed soils. The artificial breakdown of aggregates usually leads to an increase in the reflection coefficients caused by the character of the mutual position of aggregates (Orlov, 1966). Fine particles fill the volume more completely and give a more even surface. Coarse aggregates, having an irregular shape as a rule, form a very complex surface with a large number of interaggregate spaces (i.e. pores, cracks).

Decreasing particle size has been seen to increase soil reflectance among sandy-textured soils, possibly by forming a smoother surface with fewer voids to trap incoming light. The inverse appears to be true with medium to fine-textured soils, however, possibly because increased moisture content and organic-matter content associated with higher clay contents lead to lower reflectance.

Values of reflection in terms of aggregate diameter were determined by Orlov (1966). Computations showed that the dependence of the reflection coefficient (R) on aggregate diameter (d) can be expressed by an exponential equation of the form:

$$R = k \cdot 10^{-nd} + R\infty \qquad (33\text{-}2)$$

where k, n and $R\infty$ are the constants of equation (1) that are independent of the aggregate diameter and characteristic for each soil group or horizon; k and n are constants describing the shape of the curve: and $R\infty$ is the reflection coefficient of aggregates of maximum diameter. In general the parameters n and k characterize the shape of the aggregates and their packing. The higher the value of k, the more marked is the difference in this respect between the fine and coarse fractions. The higher the value of n, the more gradually do the shape and packing of the aggregates change as they become coarser. Computations were made by Orlov for a combination of important soils and horizons.

It is significant that the reflectivity of all soils and horizons obeys the same trend, and only the parameters of the equation—k, n, and $R\infty$—change. This emphasizes that the determining factors are the diameter of the aggregates and the form of their surface, and not their chemical composition. The latter finding agrees well with the data of Bowers and Hanks (1965) who studied the kaolinite fraction with a constant chemical and mineralogical composition.

Mineralogy and soil particle-size appear to have an effect on the microwave diffuse-transmission density (Estes et al., 1977; Schmugge et al., 1977). Janza (1976) also states that brightness temperatures of microwave radiometers are functions of the electrical, compositional and textural properties of the surface and of the radiometer used for the measurements.

Size of soil particles has a considerable influence on heat transfer. Experiments by Chudnovskii (1962) show a variation of 400% in thermal conductivity of soils ranging in texture from sand to loam. The greater the content of clay particles relative to sand, the smaller the thermal conductivity will be for the solid phase. As soil moisture increases, the air content decreases and soil conductivity increases, since the calorimetric conductivity of air is many times lower than that for soil and water. The volumetric heat capacity of clay relative to other soils is least in the dry state and highest in the moist state because of the greater porosity of clay.

Soil Texture

Soil texture, as governed by the size of different soil particles, appears to play some role, along with other characteristics, on spectral response. Johannsen and Baumgardner (1968) illustrated this point with the data shown in Figure 33-2. The Princeton silt, which is a soil that contains in excess of 90 percent silt, has almost no organic matter and, therefore, was found to have the highest reflectance at all wavelengths. The Pembroke clay (60 percent clay) had the most nearly unique curve with a high response in the visible reflective region due to its reddish color (limestone parent material) and its distinct water-absorption bands at 1.45 and 1.95 μm. This curve closely resembles the curve for a relatively pure kaolinite which, interestingly, is the dominant clay mineral in this soil.

The reflectances of varying mixtures of clay and sand were measured by Gerberman and Neher (1979) at five wavelengths from 0.40 to 0.86 μm. Soil samples with low sand levels (10–30%) had the lowest reflectance while pure sand had the highest reflectance. As percent sand was increased from 0 to 100 percent, the amount of sand needed to give a statistically significant higher reflectance decreased. The percent reflectance for a particular level of sand increased as wavelength increased. Any one of the five wavelengths could be used for discriminating among sand levels in a clay soil.

Cation-exchange capacity is frequently seen to have a high negative correlation with reflectance, especially in the 2.08–2.32 μm middle-infrared region. Although there is no direct physical basis for this relationship, it seems that cation-exchange capacity is acting as a natural integrating factor for clay type and content as well as for organic matter content,—soil parameters which exhibit inherent spectral behavior.

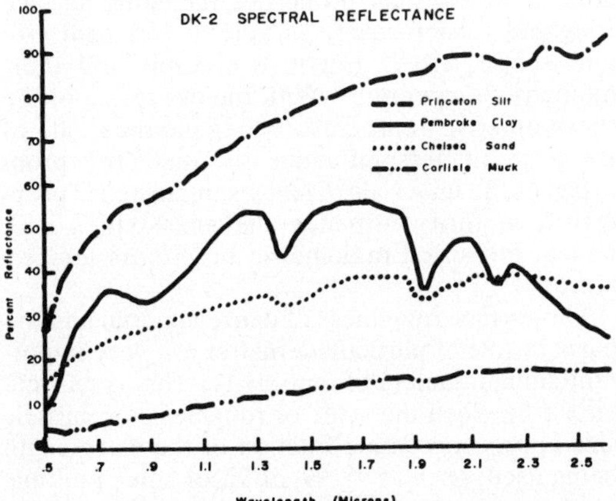

Fig. 33-2. Reflectance measurements from four Indiana soils at similar moisture levels. (Johannson and Baumgardner, 1968).

Al Abbas et al. (1972) have shown a relationship between multispectral response and clay content of surface soils (Figure 33-3). Research indicates that mulltispectral analysis and pattern recognition techniques may be used to delineate and map gross textural differences in surface soil.

Lundien (1966) and Simonett et al. (1967) indicate that radar is less sensitive to soil texture variations within a field than to crop and soil moisture and other differences between adjacent fields. Bare dry sand is a good absorber of energy in the radar wavelengths; hence it yields very little backscattered return.

Microwave radiometers measure the thermal emission from soils in the frequency range of 1–30 GHz. The magnitude of this emission depends on the temperature of the soil and on the dielectric or emissive properties of the soil. Due to the differing amounts of water that can be tightly bound to soil particles there is a dependence of a soil's dielectric properties on its texture. This dependence has been observed in laboratory mea-

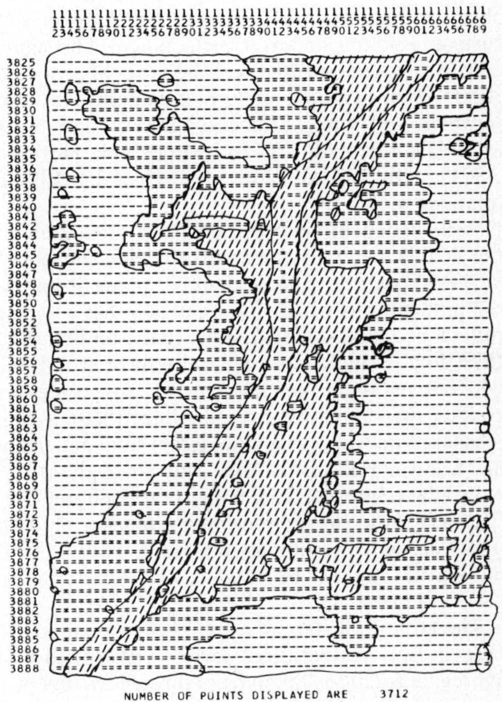

Fig. 33-3. Computer display of three levels of clay content produced with output from a linear prediction equation. Analysis of the information content of this figure and of the information depicted in Figure 33-1 is suggestive of a degree of correlation between the effect of organic content and clay content on the multispectral response characteristics of soil.

surements of the dielectric constant of soils and in both active and passive microwave observations of soil moisture directly. To obtain an absolute measurement of soil moisture content with a microwave remote sensor some knowledge of the moisture characteristics for the soil will be required. Alternatively it has been shown that the state of the moisture in the surface layer of the soil, expressed as a percent of field capacity (FC), can be measured directly. This latter information may, in some applications, be more important than the absolute content (Schmugge, 1980). The relations between soil moisture and microwave response are discussed in later sections of this chapter.

Structure and Surface Roughness

Surface roughness (soil aggregation), as governed by tillage treatments, has a substantial influence on the reflectance of soils which tend to aggregate. Heilman and Wiegand[1] demonstrated that soil surfaces of a Hidalgo sandy clay loam, disturbed by several tillage treatments, had a substantial influence on reflectance from the soil as observed on aerial photographs. The rougher soil surface resulted in lower reflectance. The same general conclusion was reached by Myers and Allen (1968) in a field study in a medium-textured silt-loam soil with surface roughness variations artificially created, and by Stoner and Horvath (1971) with soil surfaces disturbed by tillage treatments.

Reflectance of undisturbed soils measured in the field is generally the inverse of that measured in the laboratory. This is readily apparent on aerial photographs which show sands to have higher reflectance than silts and clays. In the undisturbed case, fine-textured soils generally have structure, which gives them the characteristic of aggregates coarser than sand (Myers and Allen, 1968). When measured on a spectrophotometer, sand generally has a lower reflectance than clay throughout the wavelength interval from 0.45 μm to 2.50 μm. Reflectance from the identical soils, measured in an undisturbed condition in the field, shows sand having the highest reflectance and fine-textured soil the lowest.

The structural state of a soil plays an important role in soil reflectance. Orlov (1966) studied aggregate fractions taken from three horizons of a Sod-Podzolic soil. He found that the influence of structure was dominant over that of texture. Belonogova and Tolchel'nikov (1959) have demonstrated that the reflectance of soil minerals depends on their dispersion in the soil. Structureless soils reflect 15 to 20 percent more light than soil with well-defined structure. Reflectance varies with particle diameter but the shape of the spectral curve remains the same. Reflectance of soils with their natural structure is the same as the composite curves of their aggregate for fractions from 0.25 mm to 7 mm in diameter. The reflectance of aggregates 2 mm and greater in diameter is practically constant.

Variations in tillage treatments also influence thermal characteristics of soils. Such treatments resulted in small but significant differences in thermal response in studies of seven different tillage practices by Wiegand et al. (1968). Attempts were made to relate both relative image tone and exitance against parameters of roughness, average clod size, and deviations from a plane surface. Their conclusion was that tillage treatments caused variable bulk density changes which would cause differential infiltration and availability of soil moisture. They concluded also that tillage effects on surface temperatures could be accounted for by considering thermal conductivity, aerodynamic roughness, soil water evaporation, and emission characteristics of tilled soil.

Surface roughness is prominently included among properties of targets governing the differential scattering coefficient from active microwave systems (Moore, 1976). In making a comparison between passive and active microwave systems, Janza (1976) states that both sensors are responsive to roughness. It is encouraging to note that Batlivala and Ulaby (1977) minimized the effects of surface roughness of bare fields by the proper choice of radar parameters. They concluded that the optimum parameters were 4 GHz, HH or VV polarization, and 7–17° angle-of-incidence range.

In general, radar return responds to both surface and subsurface roughness and to moisture content. For the application of radar as a remote sensing tool in mapping soil moisture content, it is imperative that the choice of sensor parameters be made such that the effects of soil type and roughness on the scattering coefficient are minimized with little or no loss of sensitivity to soil moisture.

Microwave brightness is also dependent upon a number of physical properties including surface roughness (microrelief, particle or fragment size and shape, etc.), moisture content and thermometric temperature. With the exception of the thermometric temperature dependence, all of these parameters influence the dielectric properties of the materials. Changes in dielectric properties, in turn, influence the emissivities and, hence, measured radiometric brightness temperatures.

The surface roughness of dense material and the particle size of particulate matter are very important subparameters of emissivity. This is particularly true when the scale of roughness or particle size approaches some fraction of the wavelength being used, such as ½, ⅓, or ¼, or some multiple of the wavelength up to about 10. Thus, for a 1.0-mm wavelength system, the concern is with roughness ranging in size from 0.25 and 10.0 mm, and for a 13-cm system the concern is with rough-

[1] Unpublished data from the U.S. Department of Agriculture, Weslaco, Texas (1975).

ness varying from about 4 cm up to 130 cm (Lintz et al., 1976).

Cultural practices can affect the spectral response characteristics of soils. Many variations of spectral reflectance and emittance are not due to difference in soils, but can frequently be related to differences in field cultivation or other patterns of human activity. It is impossible to control these variations so as to have greater uniformity of the area to be mapped. Ground observations should be made during the day of the satellite or aircraft overpass to provide an understanding of the variations in pattern present on the imagery.

Soil Emissivity

Thermal emissivity is defined as the ratio of energy radiated at the surface of a material to that radiated by a blackbody at the same absolute temperature (see Chapter 3 Volume I.) The Stefan-Boltzman equation (also discussed in Volume I Chapter 3) shows that no instruments can yield a correct estimate of the surface temperature if the emissivity of the surface is not considered. Taylor (1979) studied the thermal emissivities of exposed soil surfaces for some major soil subgroups. He noted the important point of knowing the emissivity for the specific spectral interval sensed by the remote sensing radiometer. He found that brightness temperature sensed by narrow-band infrared radiometers (10.4–12.6 μm) as used on SMS-GOES spacecraft would not deviate from thermodynamic temperatures by more than −2°C as a result of surface emissivity. Emissivity effects on the brightness temperature determined with broad-band radiometers (5–15 μm) however, could be as great as −6°C. Obviously ground-truth temperature measurements should be made with a narrow-band radiometer operating in a spectral band nearly matching the spectral band to which the aircraft or spacecraft scanner responds.

Fuchs and Tanner (1968) reported a decreasing emissivity of a Plainfield sand as the surface moisture was also decreasing. Their measurements of moisture content in the top 2.5 cm of soil varied from 8.4 to 0.7 percent while emissivity changed from 0.94 to 0.88. In additional findings they reported that rapid surface drying to a depth of 0.2 cm would change the moisture content very rapidly with a corresponding change in emissivity. Thus, it would appear that soil temperature interpretations should consider how the measurements were obtained.

Although accurate measurements of temperatures is made more complex by the dependence of emissivity on changing properties of materials, accurate determination of these differences in emissivity may provide a means for the detection of at least one transient condition, that of soil moisture. The emissivity of a sandy soil at wavelengths of 8 to 14 μm may be influenced only very little. The emissivity of soils may vary from 0.88 or less for sands to 0.98 for clay or loam (Idso et al., 1975c). Since the emissivity of water is about 0.97 in the infrared region, the emissivity of sandy soil could be expected to be a function of water content. Variations in emissivity of that magnitude could be expected to introduce a measuring differential of nearly 8°C.

Emissivity in the microwave region is more complicated. It is composed of two principal components, namely the dielectric constant and surface roughness. The dielectric constant is dependent upon soil moisture and layering. At the microwave wavelength of 1.55 cm the emissivity of water is 0.40, that of a dry soil 0.94 and that of wet soil, 0.60. Schmugge et al. (1977) show that the emissivity contrast between dry and wet soils can be used to measure variations in the relative amount of moisture in the soil.

Soil Exitance

Soil Temperature. The temperature of the soil is a variable of great importance in agriculture. It is related to moisture losses through evaporation, to rate of weathering and chemical dissolution, to microbiological activity and decomposition of organic matter, to germination of seeds, and to rate of plant growth. Thermal infrared sensing holds considerable promise for identifying some subsurface soil conditions.

Temperature sensing of surface soils with an airborne thermal line scanner was used by Myers and Heilman (1969) for detection of certain soil characteristics. An area of a typical alluvial floodplain is shown in Color Figure 33-4, which is an Ektachrome infrared photograph. The soil surface and subsurface vary in texture from loam to sandy loam. The light-colored soils of the figure are sandy loams and the darker soils are loams.

Diurnally-flown thermal imagery of the area appearing in Figure 33-4 appears in Figure 33-5. The temperatures are significant in relation to soil properties. Thermal imagery of soils has the unique characteristic of being influenced by subsurface characteristics. Surface soil equivalent blackbody temperatures in field E (Figure 33-4) were shown to be related to certain profile characteristics. Soil-moisture contrasts in bare soil in fields D and E produced a 3°C difference between surface-soil temperatures as observed on imagery flown at 0600 hours. Diurnal surface-soil temperatures in a dry field varied an average of 48°C on 1 June, 1966. At the same time surface temperatures varied an average of 41°C in a field with a higher soil moisture content.

Factors Influencing Interpretation of Thermal Imagery. Soil moisture and soil air humidity are the most important factors influencing thermal characteristics of the soil. As a soil becomes more moist, its conductivity increases. However the increase becomes gradually less marked as moisture content continues to rise. This is due to the circumstance that changes of temperature conductivity depend on simultaneous changes in heat ca-

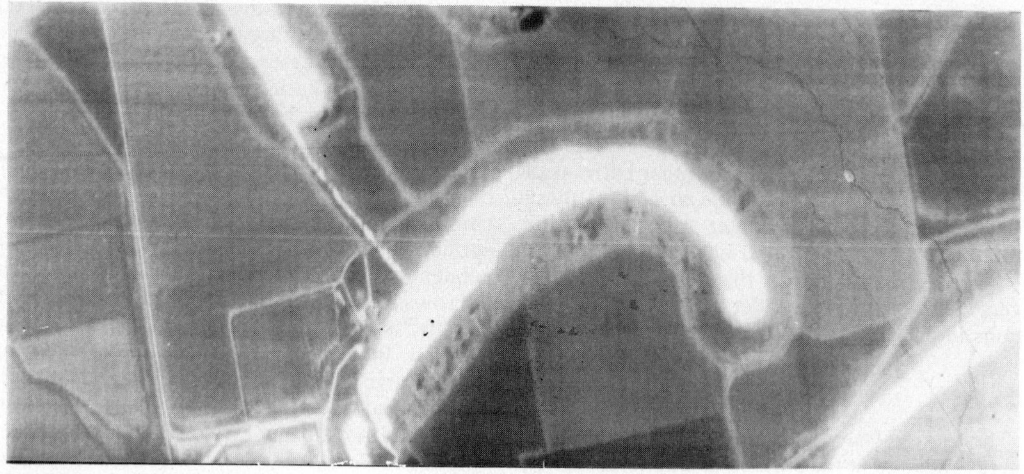

0 6 0 0 HRS

1 4 0 0 HRS

1 9 0 0 HRS

Fig. 33-5. Diurnally flown thermal imagery (8–14 μm) of the area in Figure 33-4, June 1, 1966, Weslaco, Texas.

pacity. The transfer of heat in the soil profile, as well as addition and loss of heat, is complex because a soil's thermal conductivity depends on its moisture content which changes with depth and time. Furthermore, during the warm season the distribution of temperatures in the soil is determined by the thermal properties of the upper soil layer which may, depending on cultivation practices, be a poor conductor of heat. In dry weather a high temperature gradient within a superficial surface layer may be due to decreased temperature conductivity of this layer, caused by lesser compactness and lower moisture content. In spring, clay soils are cooler than sandy ones; in autumn, they are warmer.

The presence of a crop affects the soil temperature. A grass cover does not greatly alter the mean daily and annual soil temperature, but it does reduce the daily and annual range. The size of soil particles has a considerable influence on heat transfer. The greater the content of clay particles relative to sand, the smaller the thermal conductivity will be for the solid phase.

Factors Influencing Time of Sensing. The minimum soil depth at which there is constant daily temperature, and also that of constant annual temperature, are in the relation of the square roots of oscillation periods. Since these are 24 hours and 1 year, respectively, the depth of damping of annual oscillation exceeds the depth of damping of diurnal oscillation by a factor of 19 (Chudnovsky, 1962).

During daylight the variations of heat exchange follow those of solar radiation. The time when heat exchange becomes negligible coincides more or less with sunrise and sunset. Thermal sensing of soils at night, in contrast to daytime sensing, shows greatest surface contrasts indicative of profile conditions. After sundown, heat flow is upward and varies with soils, depending on their respective heat-storage capacities and thermal conductivities.

The times of the maxima and minima daily temperature lag with depth. The retardation is proportional to depth. The amplitudes decrease with depth. As depth increases in arithmetic progression, the amplitude decreases in geometric progression and is damped at a certain depth.

Diurnal oscillations of temperature in a moist soil are less than in a dry soil. Thermal imagery produced over an area of relatively uniform soil may show temperature differentials at a time after rainfall when there is a moisture differential between well-drained and poorly drained topsoils.

The annual course of heat exchange in soil shows certain peculiarities. The greatest positive heat exchange occurs in spring and in early summer; the greatest negative heat exchange, in early winter. Imagery produced by thermal sensing in the late fall or early winter, depending on latitude and likely weather conditions, can be most productive for detecting soil conditions by taking advantage of uniform crop- and soil-moisture conditions that exist at that time of year over wide areas.

Polarizing Properties

Measurements by Coulson (1966) of the reflecting and polarizing properties of various soils, sands, and vegetation in the visible and near-infrared spectral regions showed that dark surfaces polarize reflected radiation strongly while highly reflecting surfaces have relatively weak polarizing properties. He found that the reflectance of mineral surfaces increases with increasing angle of incidence and with increasing wavelengths to about 2.2 μm. There was no specular reflection from the surfaces measured. Over highly reflecting surfaces such as deserts, upward radiation from the surfaces measured will be dominated by surface-reflected radiation in the red and near-infrared regions. Atmospheric scattering affects principally the reflectance from the short wavelengths and dark surfaces. Coulson (1966) published directional reflectance data for the following types of mineral surfaces: gypsum sand, New Mexico; bean sand, Florida; red clay, Pennsylvania; limonite, Alabama; limestone, Pennsylvania; yolo loam, California; and black loam, Iowa. All these materials were found to exhibit distinct directional reflectance properties.

Soil reflection is primarily diffuse, whereas reflection from water is strongly specular. The value of visible light to detect these extremes has been demonstrated by Curran (1978) and others. They show that high polarization values produced at large phase angles, in the order of 50° to 60° (Brewsters angle for water is 53°), are highly dependent on soil moisture, with only slight dependence on atmospheric haze, surface roughness, and, within certain limits, surface slope, soil type, and clouds.

Soil moisture change was monitored by Curran (1979) using polarized visible light recorded on panchromatic film, and visible and infrared reflectance recorded on color infrared film. Polarized reflectance successfully recorded a fairly wide, but not unlimited, range of changing soil-moisture conditions, thus suggesting its limited suitability as a monitoring technique. Figure 33-6 shows the relationship between the polarization of visible light and soil moisture for a sandy loam soil.

Surveys

Accurate classification of soils depends on many indirect as well as direct observations. Indirect factors may often be observed more easily from space or aircraft imagery than from ground observations because of the synoptic view it provides.

Soils with their variable profile characteristics and variable productivity relate directly to crop-

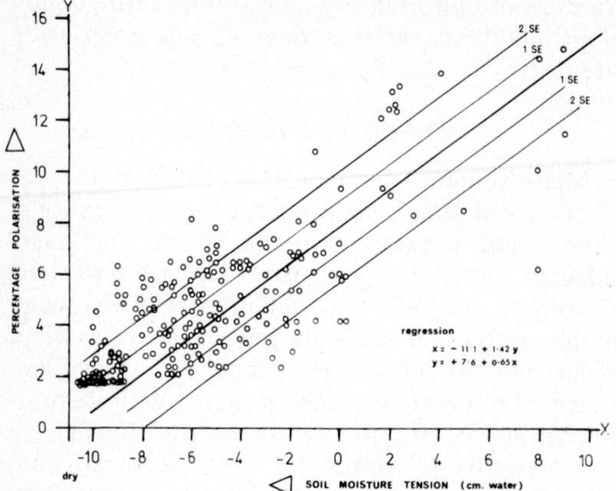

Fig. 33-6. The use of polarized panchromatic and false color infrared film in the monitoring of soil surface moisture.

land production. Therefore, mapping of soils on an appropriate level of detail is vital for agricultural planning and for estimating production. However, variability of soil characteristics observed by remote sensors also results in confusion when observing and classifying range vegetation and crops. Orderly study of soils and vegetation together in the context of remote sensing interpretations is rapidly solving many of the earlier difficulties that were experienced.

To satisfactorily serve all demands for evaluating land, three intensities of classification related to specific purposes have been established; reconnaissance, semi-detailed and detailed. The reconnaissance and semi-detailed categories are designed to provide generalized information which may be all that is required for low intensity agricultural or rangeland uses, or which may be preliminary for identifying potential areas of high productivity. The detailed classification provides information for detailed land use planning and development of projects.

Identification of Soils

Although a large body of knowledge has been accumulated about the physical and chemical characteristics of soils as they are influenced by the soil forming factors of climate, parent material, relief, biological activity, and time, there is less knowledge of how these factors relate to the reflected radiation from surface soils. Modern soil classification systems emphasize the importance of information about the quantitative composition of soils. In order to differentiate among soil groups it is necessary to rely on laboratory and/or field measurements of selected soil properties.

Shockley et al. (1962) demonstrated the value of a soil moisture signature in identifying soils. Some soils that were otherwise difficult to differentiate could be distinguished from each other when using values of reflectance measured with variable moisture contents. If reflectance at six wavelengths—1.40, 1.75, 1.94, 2.25, 4.00, and 4.50 μm—were known, any soil that they tested could be identified.

Obukhov and Orlov (1964) stress that wetting and pulverization of the soil surface bring the reflectivity of soils closer to each other. Because of this, the most contrasting photographs should be obtained at a low moisture content. A low contrast can also be expected with the sun at a high angle above the horizon.

Classification of soils with respect to their curve shape has been done successfully by Condit (1970). Spectral reflectance values extending from 0.32 to 1.00 μm were obtained for 160 soil samples collected from 36 states. Measurements were made of both wet and dry samples, which varied widely in color and reflectance. Procedures were established for classifying the soils into three general types with respect to their curve shape. As a result of a characteristic vector analysis of the spectral reflectance data it was concluded that reflectance data for all wavelengths should be predictable from measurements at five wavelengths—0.45, 0.54, 0.64, 0.74, and 0.86 μm.

Reflectance of disturbed soils measured in the laboratory results in a typical increasing reflectance pattern throughout the wavelength interval from 0.5 to 2.5 μm similar to several of the curves shown in Figure 33-2. Spectrometer studies involving reflected radiation from the solar source are of interest because the environmental and radiation factors experienced in the field have very substantial influence on reflection spectra. Therefore, field spectra may differ substantially from spectra measured on a spectrophotometer under laboratory conditions. Also, the capacity to measure both visible and infrared adds a valuable dimension to the use of soil spectra to explain many soil characteristics and to predict soil response to different treatments, management, and variations in climate. Reflectance measurements in the near and middle infrared often reveal textural, structural, mineralogical and/or other significant differences which may not be detectable by standard color observations (Figure 33-7) (Stoner, 1979) (Stoner et al., 1980).

It is difficult to measure the reflectance from soils under natural field conditions because

1) O_2 and CO_2 and water vapor absorption reduce incoming solar radiation in certain wavelength bands;
2) illumination from the sun varies in intensity with atmospheric conditions and solar radiation;
3) radiance from soils is affected by soil structure and other factors; and
4) the intensity of the sun peaks at about 0.5 μm falling off rapidly at shorter and longer wavelengths.

Relative spectral radiance spectra for agricultural scenes, measured under natural lighting con-

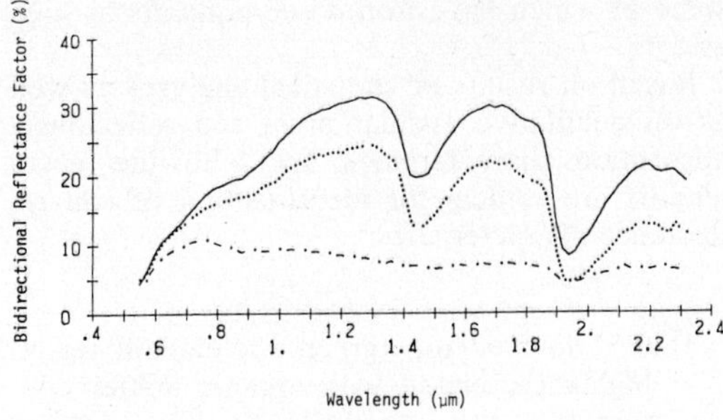

Key to Soils Data

Soil	Curve	% Organic Matter	% Fe_2O_3
Dill (Oklahoma, USA)	——	0.6	0.87
Arroyo (Spain)	••••	1.28	2.00
Londrina (Brazil)	-.-.-	2.28	25.6

Fig. 33-7. Reflectance curve for three dark red surface soils having moist Munsell color notations 2.5 YR 3/6. (Stoner, 1979).

ditions are shown in Figure 33-8 (Holmes, 1970). Contrast these spectra with the curves of Figure 33-2. It is evident that the soil spectra from the field peak in approximately the red part of the spectrum which is frequently the case for soils. Greatest soil contrasts are most often found in the wavelength interval from 0.6 to 0.7 μm. It is for that reason that band 5 of Landsat imagery is frequently superior to a color composite or to other

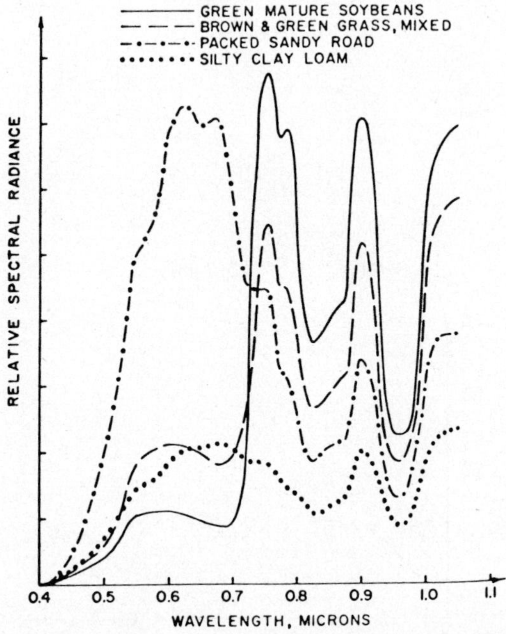

Fig. 33-8. Relative spectral radiance spectra for agricultural scenes in the range of 0.4–1.05 μm, August 30, 1966, 11:50 to 11:56 a.m. Curves should be compared in functional form only: gain settings were changed between spectra.

wavelength bands for identifying soils and soil boundaries.

Several studies have shown that information concerning spectral properties of soils may be useful in their identification and characterization. Specific soil properties can be estimated using spectral responses of different bands using a spectroradiometer. da Costa (1979) calculated correlation coefficients for a large number of soil properties and spectral responses. For example, clay content and water retained in soil at 15 bars may be estimated with a large degree of confidence in the wavelength region from 1.90 to 2.32 μm, and organic carbon could be estimated in the wavelength region from 0.50 to 0.74 μm. Clay and sand content correlations were negative with spectral reflectance while silt was positive. Calcium, magnesium and potassium are related to spectral response negatively. Sodium content is postively related to spectral response. The author simulated the four Landsat bands and found a good correlation with soil properties, although correlation coefficients for band 4 were lower than for other bands for the same soil property except for organic carbon. Total nitrogen is not correlated with spectral reflectance, however C/N ratio relationships are negatively correlated up to 1.89 μm. Iron content of soils is positively related to spectral response in the spectral region 1.15–1.39 μm.

If present satellite sensors and the improved sensor systems planned for future satellites are to be used most effectively in the preparation of land use capability maps and soil productivity ratings, as these relate to crop production, it is crucial to define quantitatively the soil variables related to productivity which can be measured by or correlated with multispectral radiation from the surface soil. Bauer et al. (1978a) made soil spectral measurements on a number of disturbed soils from many locations throughout the U.S. An artificial light source was used for the measurements, therefore, the results can only be used for comparison among soils and would not necessarily resemble field spectra as pointed out elsewhere in this chapter. The diversity of soil spectral response is evident from the soil curves presented in Figure 33-9. All ten soil orders of the U.S. Soil Taxonomy were represented in the measurements, including four Oxisols from Brazil which were included to contrast with what has conventionally been thought of as the "typical" soil reflectance curve. Each pair of curves for a given soil series represents the duplicate field samples which may vary in certain soil characteristics within an allowable range permitted for that soil series. For example, small differences in clay content or organic matter content may in themselves affect soil reflectance while at the same time directly influence the moisture weight percentage in soil samples equilibrated at an equipotential moisture tension. It is generally recognized that increased moisture weight percentage results in decreased soil reflectance throughout the 0.4 to 2.5 μm wavelength range. This soil moisture influence can be seen in some of the duplicate samples in Figure 33-9, although moisture alone is not responsible for this overall "darkening" effect.

Aside from the obvious soil water absorption bands at 1.44 and 1.94 μm, a small yet distinct absorption band can be seen at 2.2 μm in some soils. This band has been attributed to the hydroxyl group and is especially prominent in kaolinite and montmorillonite clays. Another broad but well defined absorption band is that at 0.9 μm, attributed to iron oxides in the soil. This band is obvious in the Oxisols from Parana State, Brazil, some of which have iron oxide contents as high as 25%.

Based on results of statistical analyses as well as on qualitative evaluation of soil reflectance/absorption characteristics, the following wavelengths are critical for identification of soil reflectance characteristics:

1) 0.52 to 0.62 μm (green wavelength region highly correlated with organic matter content;
2) 0.7 μm and 0.9 μm (ferric iron absorption wavelengths);

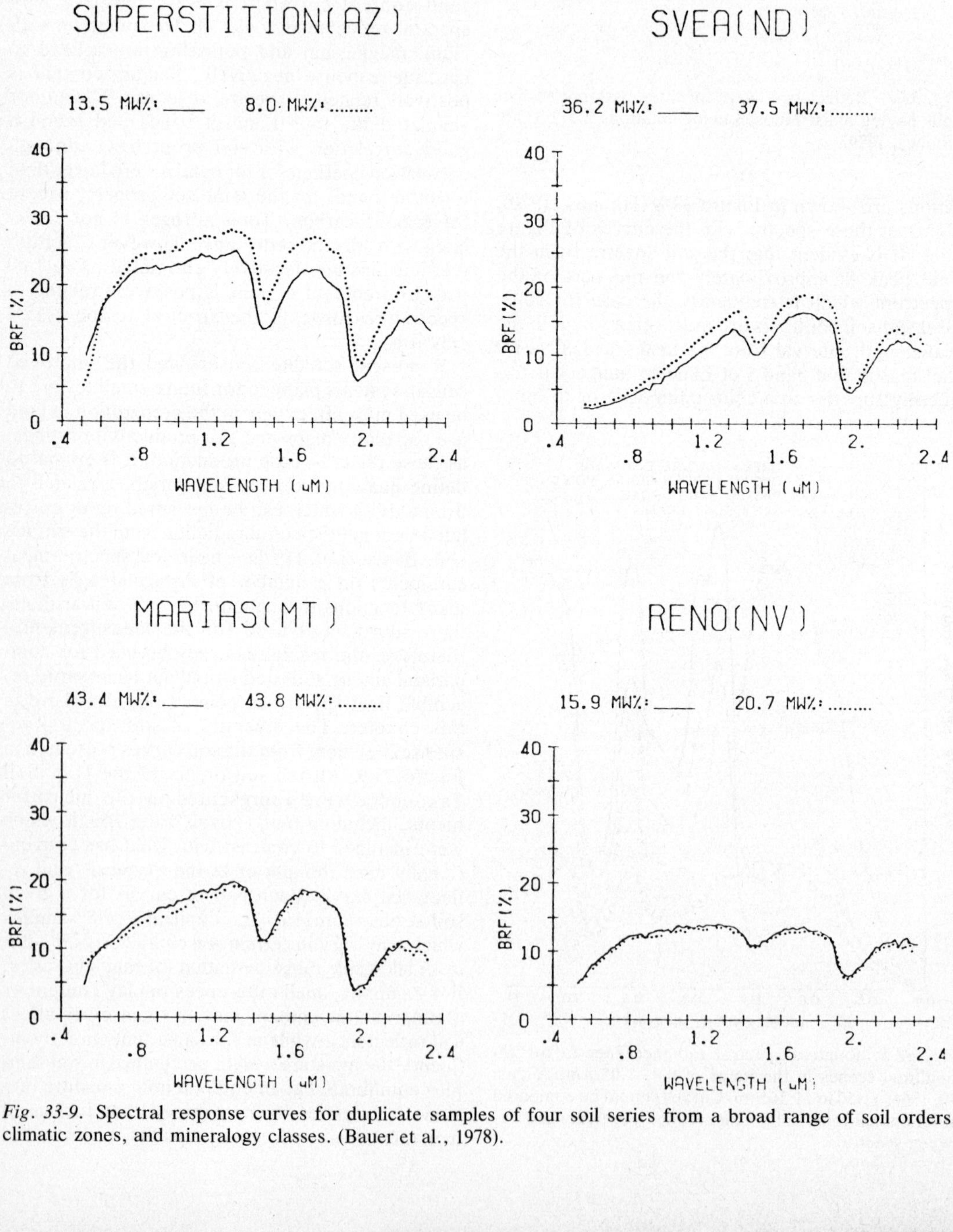

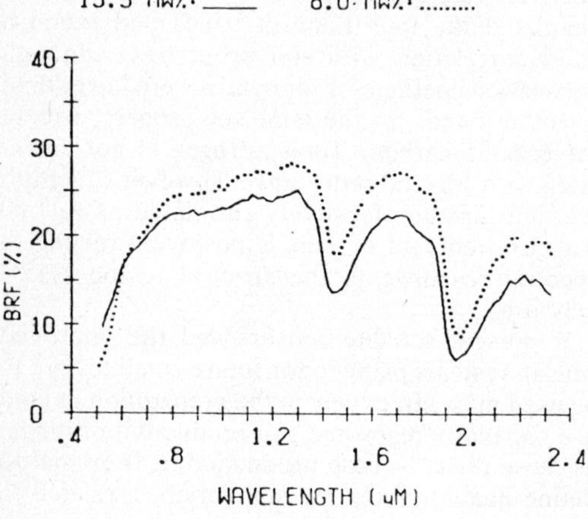

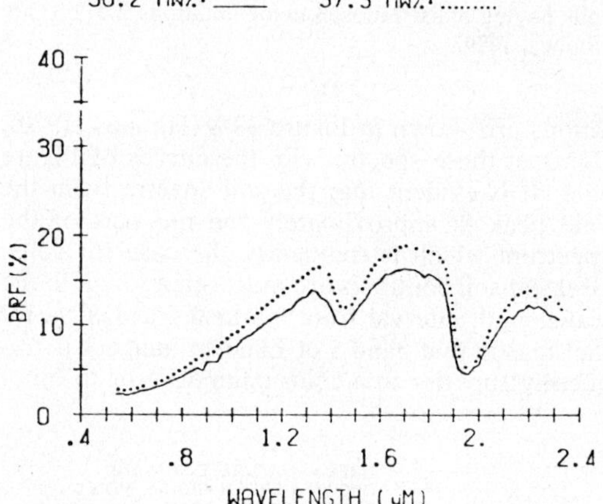

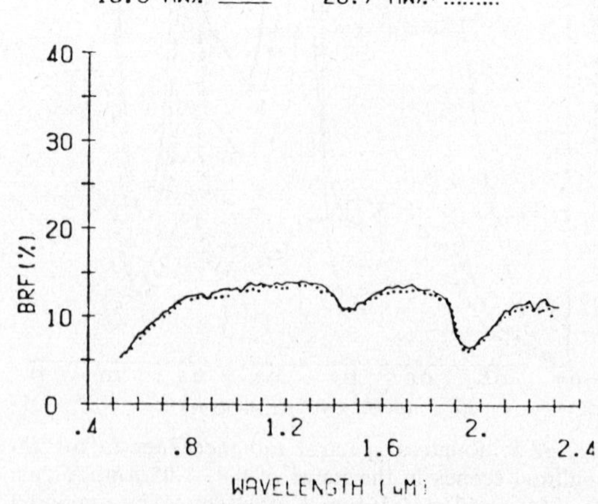

Fig. 33-9. Spectral response curves for duplicate samples of four soil series from a broad range of soil orders, climatic zones, and mineralogy classes. (Bauer et al., 1978).

3) 1.0 μm (ferrous iron and hydroxyl gibbsite absorption wavelength);
4) 1.22 to 1.32 μm and 1.55 to 1.75 μm (regions of highest reflectance for many soils, correlated with many soil properties); and
5) 2.08 to 2.32 μm (region of highest correlations with soil moisture).

Although spectral bands for the Thematic Mapper sensor include 0.52 to 0.60 μm, 1.55 to 1.75 μm and 2.08 to 2.35 μm, the 0.76 to 0.90 μm near infrared wavelength band is too broad for specific iron oxide studies in soils, a fact that could limit its usefulness in erosion studies as well as soil productivity surveys (Bauer et al., 1979).

It can be seen that soils with widely differing physiochemical and site characteristics are no more similar in their reflectance properties than are different species of plants throughout their growth cycles. To treat soil reflectance as a constant, unchanging characteristic from location to location and from date to date is to ignore the well-ordered physical and chemical relationships that impart diverse spectral reflective character to soils.

Prediction models for certain soil parameters using only soil reflectance data as inputs account for a large proportion of the variance when soils are grouped by specific climatic zone (Bauer et al., 1979). The importance of visible, near infrared, and middle infrared reflectance data is seen repeatedly. Iron oxide has high predictive values only in the humid frigid and some arid regions. Results are highly climate specific. As in previous studies of this type, cation exchange capacity often accounts for more variance than other soil parameters which are known to exhibit inherent spectral behavior. Cation exchange capacity seems to be acting as a natural integrating factor for several other soil parameters such as organic matter and particle size.

Engineering properties of four soils were spectrally compared by Bauer et al. (1978a). The Abbott soil is a fine, montmorillonitic clay from an arid zone of Millard County, Utah. Dia, a fine loam from the arid zone of Churchill County, Nevada, and Brackett, a loam from the subhumid climate of Bell County, Texas exhibit moderate liquid limits and volumetric shrinkage readings, and are well graded. The Fort Wingate soil is a friable loam from the semiarid zone of McKinley County, New Mexico. The latter soil exhibited a very low volumetric shrinking reading and has a low cation exchange capacity. Bauer et al. (1978a) list 15 engineering properties of each of the soils in the cited publication, some of which are also of interest to agriculturalists. Figure 33-10 shows spectral reflectance measurements for the four soils.

Extending these results to the level of airborne remote sensors, it is likely that reflectance data from carefully selected wavelength bands could be used to extract information from bare soil areas that could be related to levels of organic matter, soil moisture, iron oxide content, particle size content, or used as an indicator of potential productivity such as cation-exchange capacity for certain specified climatic areas. Where prior information is available about soil drainage and parent-material classes, even better correlations can be expected within more homogeneous areas of soil inference.

Soil Normalization

Classification of physiographic areas, using aircraft and satellite multispectral scanner data in crop yield estimation, has been hampered by many site factors. Levels of accuracy are influenced by soil background-reflectance, agricultural diversity, physiographic region, climate, and soil type. Spectral data have been shown to relate to vegetation-density indicators such as plant density, leaf area index and biomass. Temporal and spatial extension of signatures continues to be a problem that may be minimized by application of procedures that account for some of these variables. This will advance the time when operational use of Landsat data may become routine for agricultural monitoring. Several researchers have noted the effect of soils on Landsat-derived vegetative signatures (Richardson et al., 1975; Tucker and Miller, 1977) and of vegetation on Landsat-derived soil signatures (Siegal and Goetz, 1977; Westin and Lemme, 1978).

Recent studies have evaluated the benefits of better crop analysis methods. There are numerous remote sensing efforts in the U.S. to improve crop inventories; some of these include attempts to take into account soil background confusion factors. One of the most notable recent studies is the Large Area Crop Inventory Experiment (LACIE) which is described elsewhere in this chapter.

Dalsted et al. (1978) conducted a study in South Dakota to determine the value of stratifying Landsat digital data to contend with and buffer background reflectance due to soils, topography, etc. Stratifications considered reflectance zone criteria (surface soil anomalies, land position, geographic area, etc.) for major areas. Training for classification was then implemented within each zone. This stratification approach compensated for soil, topographic and other anomalies which confound reflectance.

Richardson and Wiegand (1977) developed a potentially significant procedure for "normalizing" the effect of soils. They found that MSS digital data for bare soil, cloud tops, and cloud shadows followed a highly predictable linear relation (soil background line) for MSS bands 5 and 7, and for bands 5 and 6. Increasing vegetation development as indicated by leaf area index (LAI), for grain sorghum crops was associated with displacement of sorghum MSS digital counts perpendicularly away from the soil background line. This procedure resulted in an index of plant vegetative

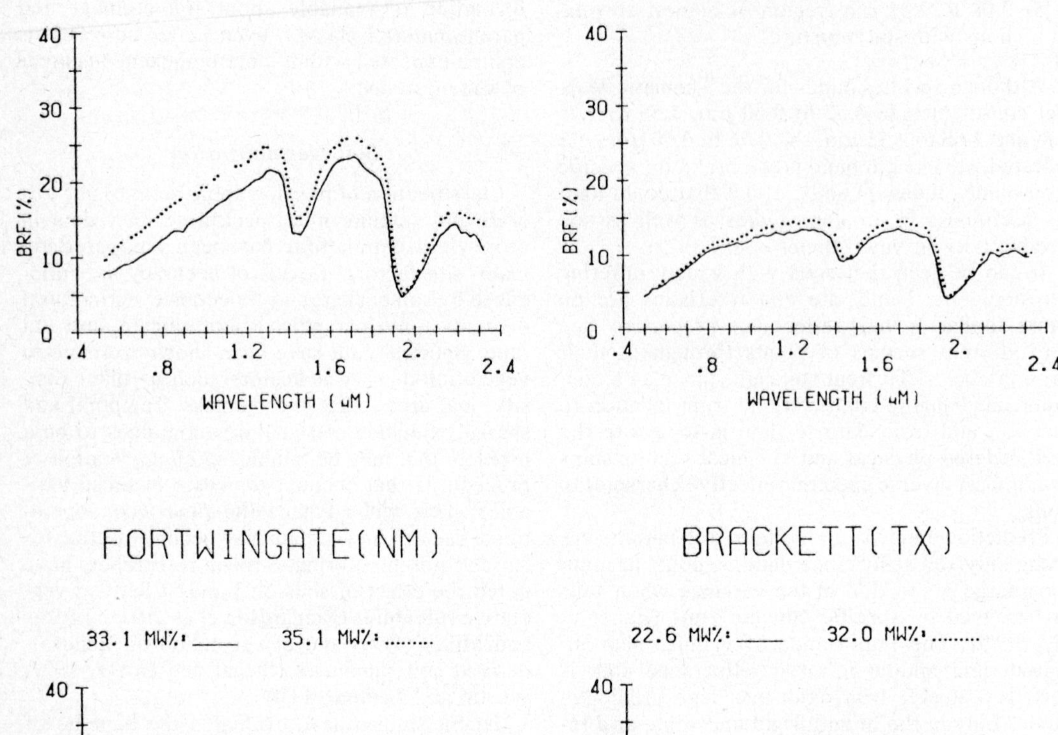

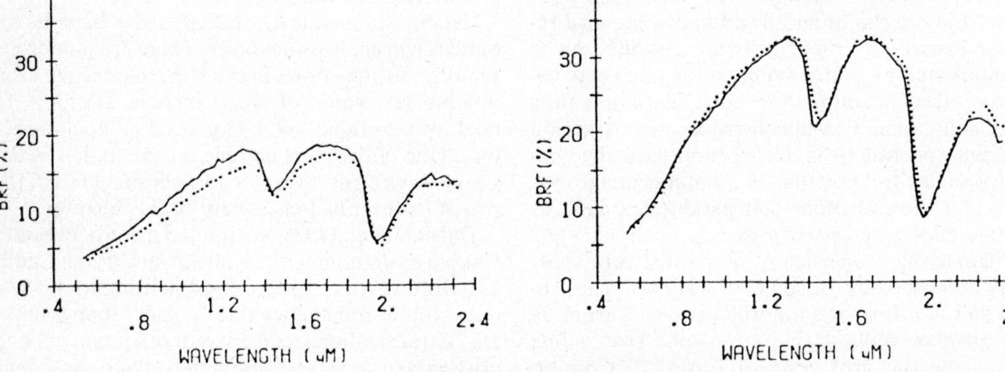

Fig. 33-10. Comparison of four soils with differing engineering properties. (Bauer et al., 1978).

development. It was demonstrated that, by using a table lookup-procedure and printer symbols for each decision region, Landsat study-areas or scenes could be gray mapped to meaningfully display vegetation density and soil condition categories without prior knowledge of local crop and soil conditions.

The effects of soils and vegetation on Landsat spectral characteristics were investigated by Westin and Lemme (1978). Six soil associations and several crops were analyzed. Imagery from two dates was studied to assess the influence of soil association on the spectral signature of vegetation and bare ground. The investigators found that soils did influence all vegetative spectral differences to some degree. Data from the six soil associations were used to separate four categories of land use with an accuracy of 94 percent.

Results of this study indicate that soil associations could not consistently be recognized with high accuracy within the data of a single vegetative type. Some of this was attributed to farming practices—different crop varieties, dates of planting, rates of fertilization, etc. In identifying crops, therefore, it would seem that training sets would be more representative of a study area if they contained data points from all soil associations present. The best procedure for normalizing soils in relation to crop identification is to delineate the soil associations on Landsat imagery in the spring under bare soil conditions and then use the delineated areas for later stratification of areas for purposes of improving crop classification.

In a study to determine wheat acreages from Landsat imagery in Australia, Parson and Fitzpatrick (1978) experienced limited success in identifying planted areas. As a result of their studies, the following three guidelines were established for an operational system using Landsat data to determine wheat acreage.

1) the area studied must first be subdivided into areas in which land forms, soil type and agricultural practices are relatively uniform;
2) for each area, acreage estimates must be made separately, using multi-temporal Landsat data at appropriate stages of the phenologic cycles of the area's crops, to obtain data that are most likely to provide accurate results; and
3) adequate ground data must be collected for verification.

The influence of underlying soil spectra on the vegetated composite-canopy spectral reflectance has been the subject of recent investigations by Tucker and Miller (1977). The objective was to determine the contribution of the underlying soil spectra to permit accurate analysis of remotely sensed data for vegetated surfaces. They found early in their study that soil or background spectra dominated low-biomass grass-canopy spectral reflectance. They observed that the incoming spectral irradiance will interact with the grass canopy and, depending upon the vegetational density or biomass, can also interact with the soil background. The interactions with the soil background decrease as the vegetational density or biomass increases until the asymptotic spectral radiance or reflectance is nearly reached. Thereafter, increases in the vegetational density or biomass effect no change in the canopy spectra. This can be explained because the canopy is of sufficient density and thickness to prevent the penetration of the incident spectral irradiance to the ground. Hence, the incident spectral irradiance is attenuated before it interacts with the bottommost biomass or the soil background. As the vegetational density increases to the point where the spectral reflectance closely approaches the asymptote at a given wavelength, the soil-spectra contribution to the canopy spectra is minimal at that wavelength. When the canopy is of sufficient density or biomass to result in the asymptotic spectral reflectance, there is no soil spectral-reflectance contribution to the composite-canopy spectral-reflectance. Thus the relative contribution of the soil spectra to the composite-canopy spectra is inversely related to the biomass or vegetation density.

The asymptotic spectra for green grass canopies were quite different from those for the soil surface at the study site (Figure 33-11). As plant growth and development result in increasing amounts of green plant material above the dry soil surface, the canopy spectra change. In regions of the spectrum where absorption occurs, the composite-canopy spectra decrease and approach the asymptotic green reflectance spectra. In the near infrared region of 0.71 to 0.74 μm, the composite canopy spectra do not change appreciably. In spectral regions where minimal or no absorption occurs, such as the 0.74 to 1.20 μm region, the composite canopy spectra increase and approach the asymptotic green reflectance spectra. Discrimination of vegetation biomass, for example, is strongly dependent upon the soil surface-vegetation spectral reflectance or radiance contrast. For this reason, some wavelengths are far superior to others for discrimination of green vegetation biomass (Tucker and Maxwell, 1976). At 1.65 and 2.20 μm dry soils are much more reflective than living plants (Leamer et al., 1978) making bands centered around these wavelengths promising for improving the discrimination between cropped and fallowed fields and for assessing vegetative cover or density.

The effectiveness of some wavelengths decreases while that of others increases when the soil surface is wet (Figure 33-11). The soil-vegetation reflectance contrast decreases in the red and blue and water absorption band (1.4 to 2.5 μm) regions while it increases in the near infrared region. This has also been reported by Colwell (1974).

Computer Classification of Soils

With the urgency throughout the world to complete soil surveys as expeditiously as possible, new tecnhiques and tools are being devised to carry out soils mapping programs. One such new mapping tool is Landsat data showing the spectral characteristics of soils.

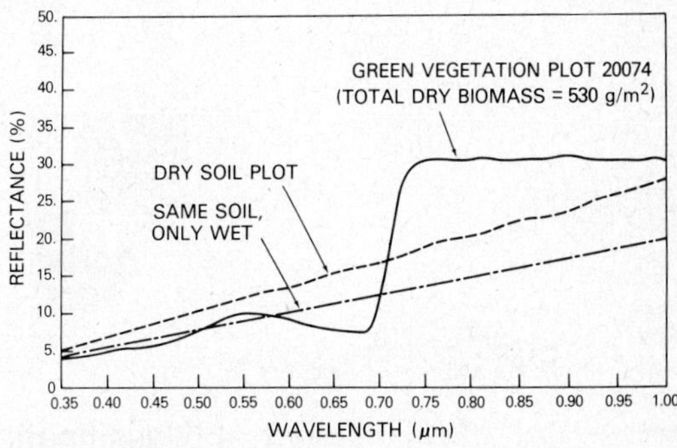

Fig. 33-11. Spectral reflectances for dry soil, wet soil, and the asymptotic green reflectance. The dry soil and wet soil are for five bare soil plots measured when dry and wet, respectively. The asymptotic green reflectance curve is from a plot of blue grama grass having a total dry biomass of 530 g/m². (Tucker and Miller, 1977).

Spectral maps were prepared by Weismiller et al. (1979) which depict the pattern and boundaries of the spectral characteristics of soils occurring in Jasper County, Indiana (Figure 33-12). The spectral information was produced using computer-aided analysis of Landsat-1 MSS data collected on June 9, 1973. Mapping for Jasper County was carried out on halftone film positive mylar sheets. The sheets show the aerial photographic image of the mapping area and can be used to overlay the spectral maps. Thus the soil scientist has the benefit of both conventional aerial photography and soil spectral characteristics to guide him in delineating map unit boundaries. At a scale of 1:15,840 each Landsat pixel represents 0.20 hectare and is depicted by a distinctive symbol on the spectral maps.

In the Jasper County soil survey, the spectral classes represented were correlated with soil drainage classes. Although it was sometimes possible to correlate soil properties such as color, surface texture and organic matter content with the spectral classes, these correlations did not prove as consistent as those with drainage characteristics. Soils underlying the vegetation class were inferred without the benefit of field data and thus the interpretations made may exhibit discrepancies.

Field observations for studies similar to the above should be taken at locations with similar spectral response, and extrapolation done only after checking surface and subsurface horizons. Spectral maps can speed up the soil-mapping procedure if the soil scientist has a knowledge of spectral variability. The satellite data available are useful to help in soil-mapping tasks, but the data can also be used to make some predictions about physico-chemical properties of the soil.

Although preliminary soil maps produced from multispectral scanner data can speed up the soil-survey process to a considerable extent, Kristof and Zachary (1974) and Evans et al. (1976) point out that inasmuch as conventional soil classes are based on both surface and subsurface soil characteristics, soil classification using multispectral scanner data can be expected only to augment and not replace traditional soil mapping.

Piech and Walker (1974) illustrated that additional information on soil characteristics can be extracted from conventional aerial imagery through careful analysis of image photometric properties. The authors used an empirical reflectance ratio obtained from the red and blue spectral bands of conventional color aerial photographic images to delineate relative soil moisture and texture patterns. If the darker soil element has a greater red-to-blue reflectance ratio than the lighter soil element, the tonal variation between the soil elements is caused principally by moisture. If the darker soil element has a smaller red-to-blue reflectance ratio, the soil elements differ principally due to texture. To perform the ratio analyses, the interpreter must relate soil image densities to reflectances by calibrating the color imagery through density variations in shadow images. The authors emphasize that the procedure cannot be quantified at this point.

Principles used in the multispectral analysis ap-

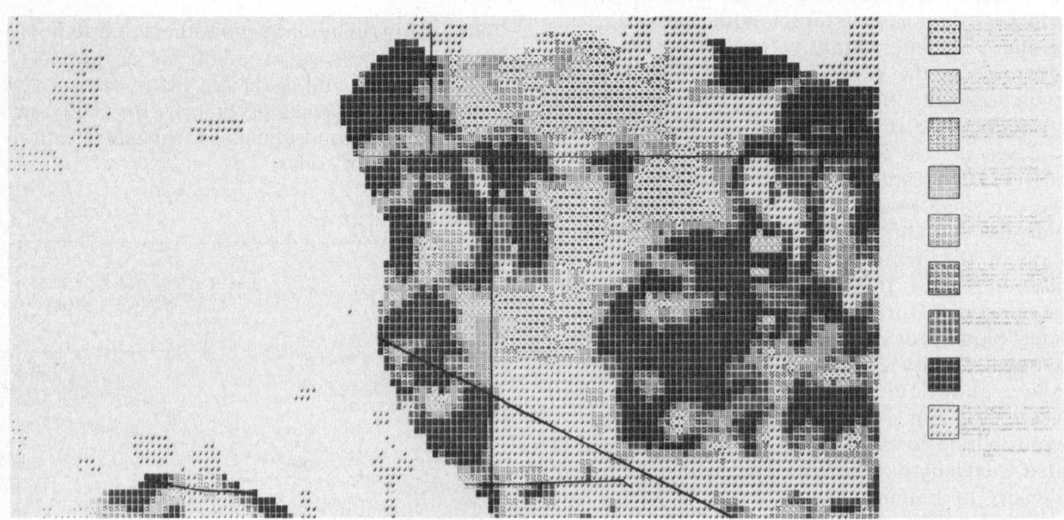

Fig. 33-12. Spectral classification of soil characteristics, Atlas sheet no. 61, Outwash area, Jasper County, Indiana. (Weismiller et al., 1979).

proach differ significantly from those of traditional photo interpretation. The photo interpreter attempts to identify from air photos the recognizable surface characteristics and soil forming factors of particular soils and deduces certain characteristics based on his experience. The multispectral approach attempts to identify directly the surface characteristics of particular soil types.

Wong et al. (1977) point out that it is generally recognized that a reliable soil-identification technique cannot be based entirely on multispectral analysis. The accuracy of soil identification using either spectral analysis or microwave-radiation techniques has been considerably degraded by data noise caused by vegetative cover, atmospheric conditions, and other factors. These authors demonstrated that small differences in soil types and soil characteristics can be distinguished by the use of quantitative terrain data. Their aim was to develop a reliable method of automatic soil identification based on the combined application of quantitative terrain factors and remote sensing data.

Digital analyses of Landsat multispectral scanner data were made by Kirschner et al. (1977) to show that the data can provide detail and definition of soil features not readily discernible through visual interpretation of Landsat imagery. The authors pointed out that digital analysis of their data provided quantification of soil mapping. The percentage of soil-mapping-unit inclusions were readily ascertained according to soil-moisture regimes, and soil complexes were easily quantified. The studies reported here were on a limited area of 430 hectares but with variable soils and topography.

Detailed and Semi-Detailed Surveys

Detailed surveys which necessitate intensive ground investigations require information on the soil composition (kind and amount of each soil) of soil-mapping units delineated during a soil survey. This estimate of soil composition for each soil-mapping unit is described in the soil-survey report.

Detailed soil data are used for determination of irrigable areas, water requirements, irrigation and drainage systems, land development, land appraisal, payment capacity, project benefits and costs, establishment of assessments, and proper land use. The demands for soil surveys increase every year as more policy decision-makers gain an appreciation for soil-survey data and their use. Unfortunately, the time that a soil surveyor has for mapping soils becomes less and less every year because of other soil-survey-related duties and activities, such as special requests for soils assistance by cities, counties, and individuals. Therefore any remote sensing techniques which would allow more efficient use of the soil surveyor's time would have wide application for any organizations or agencies which survey soils.

Landsat data and imagery have been used effectively to provide a base for the preparation of semi-detailed soils maps at scales ranging from 1:50,000 to 1:250,000 (National Academy of Sciences, 1977). Based on the demonstrated capability of the present Landsats, it should be possible to provide many of the developing nations with semi-detailed soils maps that would be impossible to obtain in the foreseeable future by conventional methods. The areas where this cannot be done are those where dense tropical vegetation prevents any observation of the soil, although some information can be gained by mapping vegetation types which have an affinity for particular kinds of soil.

The capability to overlay Landsat multispectral data onto topographic maps and/or other two-dimensional data opens new possibilities for producing land-use capability maps and soil-productivity ratings. A good land-use capability map can help a developing country select areas for possible priority agricultural development.

Significant use of Landsat imagery is foreseen for land and soil surveys in the developing world. In a project now underway in Syria, digitally enhanced Landsat imagery, and small-area soil and plant-community ground observations of research personnel in the field are being used to provide a basic land-system evaluation of the country at scales of 1:100,000, and smaller. Similar work has also been done by Odenyo (1979) for the Narok area of Kenya using both digital and visual analysis of Landsat data.

Reconnaissance Resource Surveys

Soils, although existing at the surface of the earth, represent a zone of physical, biological and chemical alteration of the underlying geologic material. Many of the processes which have been active in the formation of the soil have also been significant in the formation and alteration of the landscape. At some places on the landscape there is a continuous tearing down by the processes of erosion and at other places a subsequent building up by the processes of sedimentation. It is the balance between erosion and sedimentation that forms and alters the landscape in concert with climate and living organisms, as conditioned by time, and in turn influences to a large degree where certain soils will occur and what their properties will be (Worcester and Moore, 1978).

The interrelationships among geologic material, landscape, land use and soil help form the basis for large-area soil delineation. The distribution of soils and characteristic soil properties on the landscape exhibit a certain unique universality which provides the areal predictability necessary for reliable delineation.

The large-area small-scale resource inventories needed for general surveys can best be accomplished using Landsat and other space imagery. Remote sensing technology provides a means for the rapid collection of land-use and other resource data for a variety of planning purposes. Even in the U.S., where detailed surveys have been underway for many years, there are still extensive areas without adequate resource surveys. Also, where adequate surveys already exist, temporal considerations demand updating from time to time. In developing countries where the resource data base is meager, surveys can provide planning information never before available.

Reconnaissance surveys are valuable in two primary ways: (1) they provide a rapidly acquired information base for planning purposes, and (2) they provide spatial information that can be used to plan and execute detailed surveys and to minimize return from the often limited financial resources available.

For areas where little or no soils information is available, remote sensing procedures have been established by Frazee (1975), which detail methods and interpretation techniques for conducting soil surveys on the soil-association level. These can be conducted rapidly.

Conducting soil surveys using Landsat data poses a variety of problems arising from: (1) heterogeneity of soils and landscapes, (2) lack of basic data concerning soils, (3) inaccuracy of interpretation, and others. These difficulties can be minimized by inspecting various soil characteristics as they relate to capabilities in various parts of the electromagnetic spectrum.

A knowledge of soils in an area, even on a reconnaissance or soil association level, can aid in the interpretation of remotely sensed data for determining development potentials. Soil associations are broad geographical associations of one or several soils. Soil-association maps show the spatial relationships of land units developed in unique climatic, geologic, and topographic environments, and having characteristic slopes, soil depths, textures, water-holding capacities, and the like (Westin, 1973a).

The application of Landsat data to soil investigations has progressed beyond the research stage to its use in extensive operational soil surveys, generally conducted on the reconnaissance level. Especially relevant is the size of area covered in each scene, over 3 million hectares. The advantage this has in soil-association identification is that it provides a regional view. Since soil associations occur in repeating patterns, the delineations in general can be observed to see if a change in land use, slope, drainage pattern, hydrology or other feature has occurred. Landsat scenes can be enlarged up to a scale of about 1:60,000 for use in identifying soil associations and determining their use.

In a resource survey Landsat color composites together with band 5 black-and-white imagery can be utilized for wet and dry seasons. Preliminary soil-association maps can be prepared from UN Food and Agriculture Organization (FAO) and other general soil studies using Landsat imagery as a base. Final corrected soil maps may then be prepared by synthesizing background material with observations and data obtained in the field. Final descriptions of each mapping unit and complete soil properties and soil interpretation tables are prepared as well as final interpretive maps showing rangeland potential, agricultural potential, erosion hazard, etc.

Soil landscapes which are intrepreted from Landsat imagery are classified and shown on soil-association maps. Soil landscapes exhibit a characteristic surface geometry, kind and density of vegetation, and hydrology (Westin and Frazee, 1976). Landsat imagery shows patterns from which inferences must be drawn by a qualified soil scientist based on at least limited field studies and sampling.

Small-scale soils maps (1:5,000,000) are now available for all continents and 1:1,000,000 scale mapping is quite common. However, for national and local agricultural purposes, soil-association maps of a much larger scale (at least 1:200,000 and even 1:100,000) are essential. For specialized studies of irrigable land, 1:10,000 or 1:25,000 scale maps may be needed. Relatively few soil maps at scales greater than 1:1,000,000 are available for the developing world.

Countries need to know the distribution, composition, characteristics, and genesis of the various soil types they possess. These data profiles enter into the determination of land capability or land-use potential, a vital assessment for the rational and efficient use of land for agricultural or other purposes.

Use of Landsat to Recognize Soil Survey Boundaries. Soil boundaries caused by climatic and vegetative differences usually cannot be observed on conventional aerial photographs. No one photograph covers enough area to show soil differences due to changes in climate and native vegetation, since these boundaries are diffuse (unless there is an abrupt change in elevation). However, the Landsat scenes cover an area large enough to observe climatic boundaries.

Figure 33-13 is an area in eastern South Dakota and western Minnesota, approximately along latitude 45°15′. Shown are two positive prints (scale 1:500,000) of part of one scene. One scene is of June 17, 1973, the other is of August 28, 1973. Both are of the same area and are MSS 7. The tonal differences of the east and west parts are due to reflective differences of soils and vegetation. In this area the climate gradually becomes more humid toward the east. The western part of the scene receives about 53 centimeters of annual precipitation, while the eastern part receives about 63 centimeters. More cereal crops and alfalfa are grown in the drier western part of this area, while more corn and soybeans are grown in the more humid eastern part.

On June 17, the cereal crops and alfalfa are near their peak of green growth, thus showing light tones on the June 17 print of MSS 7. Corn and soybeans do not yet cover the ground, and the

Fig. 33-13. Diffuse soil boundaries on Landsat due to climatic change. Upper scene is of 17 June 73 1329-16440-7. Lower of 28 Aug. 73 1401-16430-6. Corn and soybeans are the dominant crops on the Udoll soils on the right while cereals dominate on the left on the drier Ustolls. In June the cereals are returning strong radiances in the IR (light tones). While mostly bare soil reflectance (dark tones) is being returned where corn and soybeans do not yet cover the soil. In August the situation is reversed. Scene along S.D.-Minn border along 45 degree latitude.

dark tones are those of nearly bare soil. In this approximate area the line is drawn between Udoll soils (Prairie soils) on the east and Ustolls soils (Chernozem soils) on the west. The August scene has dark tones on the west where much of the crop has been harvested and light tones in the east where corn and soybeans are growing vigorously.

While soil boundaries caused by climatic and native vegetation differences usually are diffuse, boundaries due to a change in soil parent-material generally are sharp. Figure 33-14 is a positive print at a scale of 1:500,000 of MSS 7 data from southwestern South Dakota. Loess, sand, and limestone soil parent-materials all are apparent on the scene.

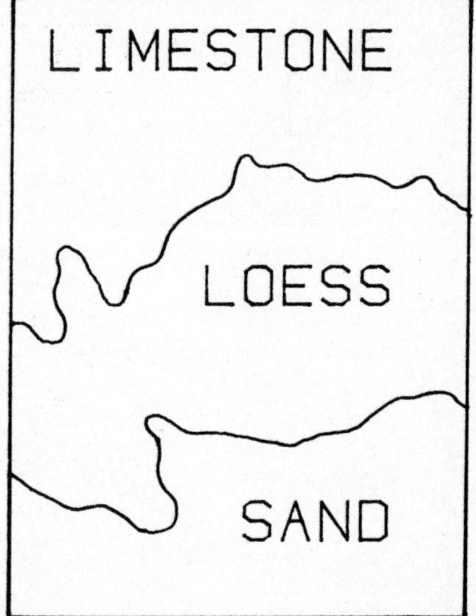

Fig. 33-14. Sharp boundaries due to soil parent material differences. Southwest South Dakota. Landsat scene 3 June 1973. 1315-17073-5.

Stream patterns, imagery tone, land-use characteristics, and landform types all give clues for these separations.

Soil boundaries on the Great Plains caused by relief differences are especially apparent on imagery when sun angle is low over fine-textured soils. Most of the precipitation on soils of this texture runs off because of low infiltration rates. The result is a network of closely spaced drains. These relief features are accentuated in the winter at low sun angle. Figure 33-15 includes positive prints of MSS 7 data of scale 1:250,000 from the Cretaceous Pierre Shale area in central South Dakota. Both summer (upper part of figure) and winter (low sun elevation) scenes are shown.

Temporal Character. Figure 33-16 includes MSS 7 prints on two dates of an area in southeast South Dakota. Both scenes are positive prints of scale 1:250,000. Two principal soil associations are shown: A, a nearly level, immaturely dissected, glacial plain of silty soils formed in a thin loess cap over glacial till; and B, a dissected plateau having rolling, deep loessial soils. The association of rolling soils (B) is easily distinguishable on the May 30 scene on the left but is less easy to see on the August 28 scene.

Vegetation, or the seasonal lack of it, plays a large part in distinguishing these associations. The rolling Area B is characterized by two features—considerable erosion and more close-growing crops (alfalfa, grass and cereals). This causes a mottled tone. The mostly light areas in the positive print made from MSS 7 data are crops, and the splotches of white are the light colored, eroded soils. The flat soils of A in late May are bare or only partially covered with recently planted corn and soybeans. Thus, they appear as light-toned, geometric shapes on this positive print. In the late August scene the two soil association areas appear almost uniformly light toned since practically all the soils are mantled with some kind of growing vegetation on this date.

This May scene thus is superior to the August scene to show these soil-association boundaries.

Multispectral Capabilities. Dark, shale-derived soils occur in central South Dakota on steep slopes leading down to the Missouri Reservoir. In places these soils are actively eroding. The black eroding soil in small gullies adjacent to grass-covered interfluves on gentler slopes provides a reflectance contrast recorded on MSS 7 (on the right in Figure 33-17) but not on MSS 5.

Figure 33-17 shows MSS 5 and 7 prints of scale 1:250,000 of the same area. The eroding soils (see arrows on figure), appear as a dark fringe on the positive print of MSS 7. Thus in this area eroding areas can be detected on MSS 7 but not on MSS 5.

Soil surveyors are concerned about separating open water from marshland. The panchromatic film used for most soil-survey base maps records energy in the visible spectrum comparable to MSS 4 and 5 of Landsat. Panchromatic images of emergent vegetation in marshes and open water reflect about the same on MSS 5 on the right of Figure 33-18, but open water (L for lake) is easily identified in MSS 7 on the left and separable from marshland

Fig. 33-15. Soil association boundaries caused by topography differences can be seen more clearly at low sun elevation. The top scene is of 19 July 74. 1726-16414-7 Sun E1.56. The lower scene is of 10 Dec 74. 1870-16364-7, Sun E1.19. South Central South Dakota.

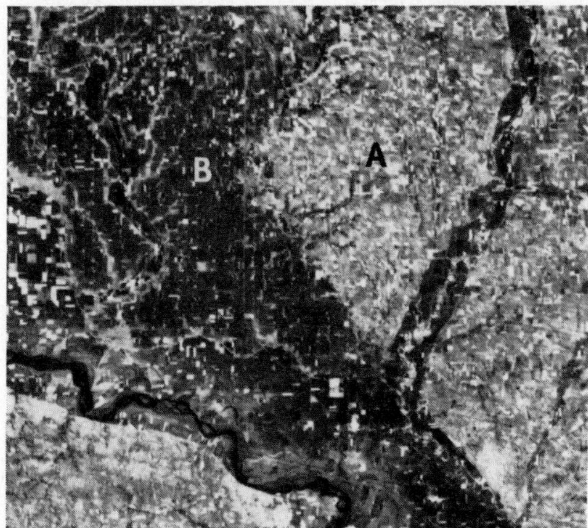

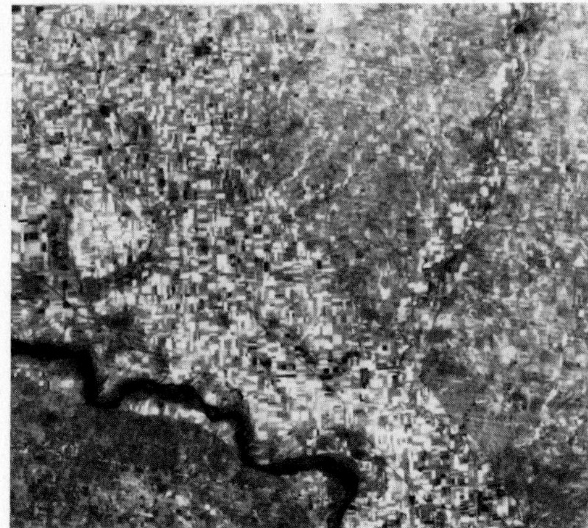

Fig. 33-16. Temporal change useful in identifying soil associations. The left scene is of 30 May. 1311-16444. On this date soil association A and B are easily distinguished. The right scene is of 28 Aug. 1401-16433-7. On this date soil areas A and B are difficult to distinguish. The area is in SE South Dakota.

(M). Figure 33-18 is an example from glaciated eastern South Dakota. These are positive prints of scale 1:500,000. What appears to be four lakes on MSS 5 are shown to be two lakes and two marshes on MSS 7.

Near-Orthographic Character of Landsat Images. The geometric quality of Landsat MSS images permits joining with very little distortion. Moreover, overlays of controlled base-maps fit Landsat images. Overlays can be prepared for geology, soils, cultural features, drainage, and the like to assist users in orienting themselves on Landsat images for planning purposes.

Computer Compatible Tapes (CCT). Most of the work with Landsat data has been with images converted from the electronic signals received by the multi-spectral scanner. The digital data themselves have more dynamic range than can be accommodated by a photograph.

The effect of soils and vegetation on Landsat spectral properties was investigated using a CCT. This study was described earlier under "Soil Normalization" (Westin and Lemme, 1978). Landsat scenes were examined for categories of agricultural land use and for determining the influence of soil association on the spectral signatures of vegetation and bare ground. Computer-classification programs used were the K-class Classifier (Serreyn and Nelson, 1973) and a stepwise discriminant-analysis program from a package of biomedical (BMD) computer programs (Dickson, 1974).

The April 19 data were useful to separate cropland from grassland and to locate areas of open

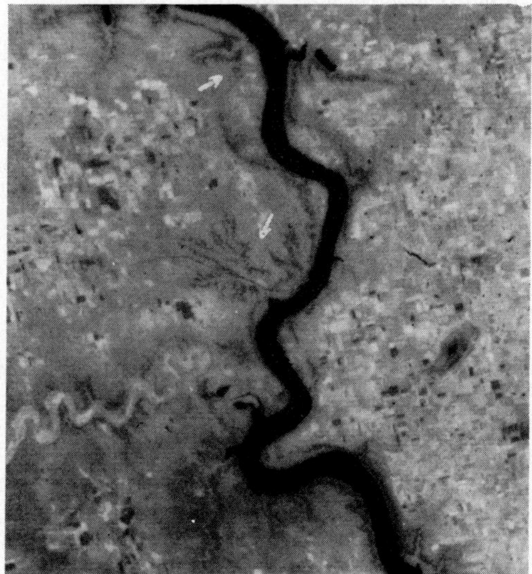

Fig. 33-17. Multispectral differences in the appearance of erosion above the Missouri River reservoir in South Dakota. The dark, fringe-like area identified by an arrow on the right scene. MSS-7. is bare eroding soil. The eroding soil is not apparent on the left scene. MSS-5. 17 Aug. 72 1025-16551.

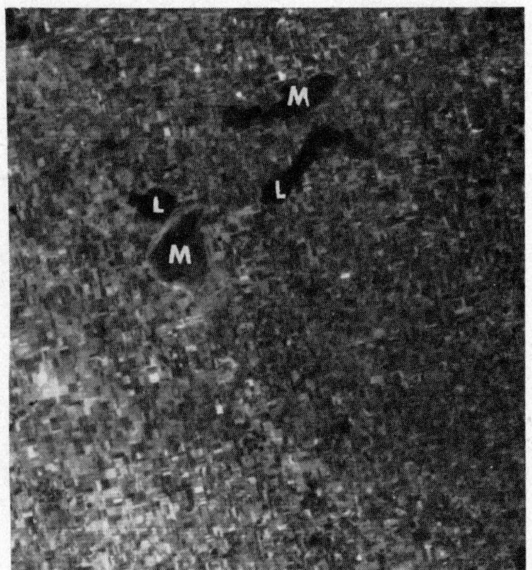

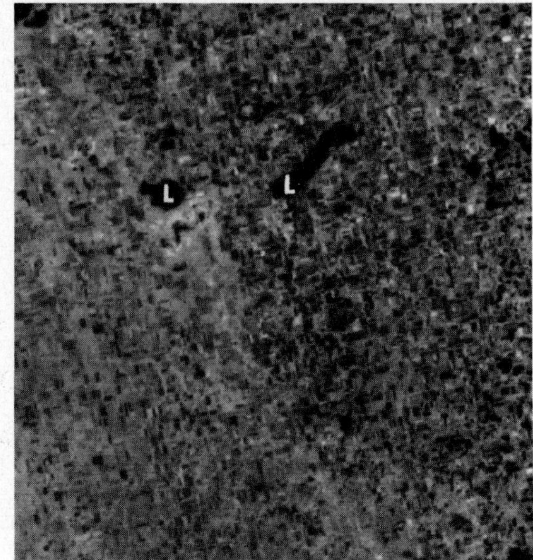

Fig. 33-18. Multispectral differences in the appearance of lakes (L), and marshes (M), on Landsat. What appears to be four lakes on MSS-5, on the left, are shown to be two lakes and two marshes MSS-7, shown on the right. Glacial Till Plain, East Central S.D. 21 Sept. 1060-16491.

water. Soil differences had a more pronounced influence on the spectral properties of grassland than on cropland. The June 20 data showed that, although soil associations could not consistently be separated within the data of a single vegetative type, soils did influence all vegetative spectral reflectances to some degree.

A generalized training set containing data points from each of the six soil associations was used to separate four categories of agricultural land use in the 12,950-hectare test area at an accuracy of about 94 percent.

Use of Landsat for a Low Intensity Soil Survey

In many localities, agricultural land is assessed for taxation according to its ability to produce agricultural crops or native grass. A method based on soil inventory data and land sales was developed by Westin et al. (1974) for use by assessors in South Dakota. Soil inventory data necessary to use this method are available for only about half of the counties in South Dakota. General soils information, such as low intensity soil surveys which are inexpensive to make, can provide the data needed for the remaining counties until detailed soil surveys are completed.

Through the use of many of the characteristics of the Landsat imagery, a low intensity soil survey of the Great Plains part of Pennington County, South Dakota has been completed. Landsat color-composite transparencies, single-band transparencies, and enlargement prints were interpreted to produce a soilscape[2] map for 400,000 hectares. Areas of similar photographic characteristics were delineated on mylar over a color composite using a light table and a three-power magnifying glass. The field checking was done by a resource team of soil, geology, and range science specialists.

The time necessary to map and field check the general soils for 400,000 hectares using Landsat imagery was six weeks. The soilscape map and land sales data were used to prepare a land value map for the portion of Pennington County east of the Black Hills.

The characteristic features observable on the Landsat color composite transparency which were used for interpreting the soilscape boundaries were tone, color, land use patterns, and drainage patterns. Color-composite transparencies were most useful for interpretation of boundaries between soilscape areas. The interaction among the individual bands provided useful clues for interpretations. Most soilscape boundaries could be delineated on the Landsat imagery. Areas such as flood plains that were too small to delineate using the color-composite transparency at a scale of 1:1,000,000 were mapped using the 1:250,000 scale prints.

The map with major soil boundaries was interpreted in the office by photo interpretation of the color-composite transparencies, single-band positive transparencies, and the enlarged prints. This map was transferred to a USGS 1:250,000 scale topographic map for field checking. Each area delineated was visited; the soil, vegetation and geologic materials were described as well as the surface features responsible for the reflectance patterns apparent on the Landsat imagery. Examination of the Landsat imagery in the field indicated several additional boundaries. These were delineated and a final map prepared.

[2] The term "soilscape" is a contraction of "soil landscape" described in Buol, Hole and McCracken in Soil Genesis, p. 300 as the assemblage of soil bodies on a land surface in a particular landscape.

The cost for this low-intensity soil survey was roughly 2¢ per hectare. The Landsat imagery for the area of the county covered cost $100, and travel was $100. The office interpretation, the field check, and the final drafting took about 26 man-days.

Soil Suitability Ratings and "Intensity of Use as Cropland" Ratings for Crop Inventories

Agrophysical units (APUs) are defined by Murphy et al. (1978) as "geographic areas having definable/comparable agronomic and physical parameters which reflect a range in agricultural use and management." Murphy et al. (1978) report that, during the Large Area Crop Inventory Experiment (LACIE) 1975–76 crop year, APUs were developed and tested over the Great Plains area of the United States in an effort to reduce variability in large-area crop inventories. The APUs reduced the number of sample segments needed to meet a coefficient of variation of 5 percent. Wheat yields were found to be more homogenous within APUs than yields within Crop Reporting Districts. Other benefits included a high correlation between historical acreage and APUs and improved methods of monitoring episodal events.

In the 1975–76 period APUs were derived from interpreting and synthesizing Landsat imagery, soil maps, and meteorological data. Landsat imagery was used to determine if the soil essentially was cropland or not (agriculture/nonagriculture overlay).

Methods for soil-suitability ratings were developed by Westin (1976) and Downs et al. (1976). The following soil characteristics were used in the soil suitability ratings: texture, depth, water-holding capacity, drainage, salinity, and slope. Climatic data were used to set precipitation and temperature variation with an APU.

During this period APUs were developed by LACIE technicians for parts of the spring wheat regions of Russia and Canada and the spring and winter wheat regions of an area of the U.S. Great Plains.

In addition to soil-suitability ratings, an interpretation of Landsat images called "Intensity of Use as Cropland" (IUC) has been provided for the LACIE project and also for another project called the Crop Condition Assessment Division (CCAD) of the Foreign Agricultural Service (FAS) of the United States Department of Agriculture (USDA).

Area Covered. Soil-suitability ratings and IUC ratings have been developed for areas in Brazil, Argentina, Uruguay, Mexico, the states of Montana and North and South Dakota in the U.S., and areas in Russia, China, India and South Africa (Figure 33-19). The total area covered exceeds 26 million square kilometers.

Soil-Suitability Rating. The sources of soils information for developing the soil-suitability ratings are the World Soil Maps (WSM) and legends of the USDA Soil Conservation Service. These materials consist of maps drawn on Operational Navigation Charts (ONCs) at a scale of 1:1,000,000. The soil-map units are in terms of the 1938 USDA Yearbook of Agriculture Soil Classification scheme. A descriptive legend is available in which the profile and landscape characteristics and occurrence of the soils are described.

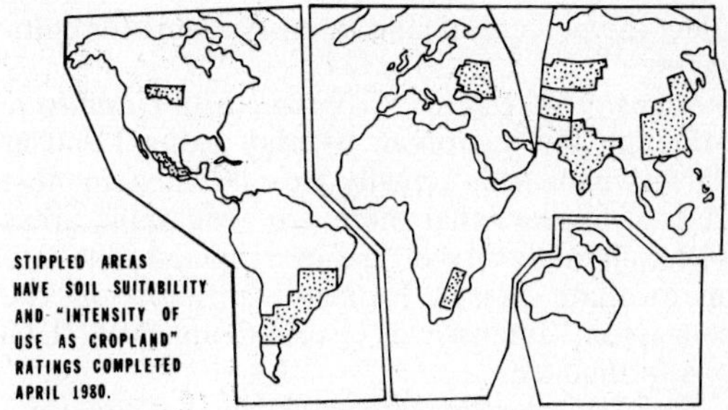

STIPPLED AREAS
HAVE SOIL SUITABILITY
AND "INTENSITY OF
USE AS CROPLAND"
RATINGS COMPLETED
APRIL 1980.

Fig. 33-19. Areas in the world for which soil suitability ratings and IUC ratings have been developed.

A major step in using the WSM materials is the conversion of the soil-map units to the U.S. Department of Agriculture, Soil Conservation Service Comprehensive System of Soil Taxonomy (1975). This can be done only in general terms since the soil-map units are soil associations or landscapes of soils. The steps in the conversion process included 1) determination of moisture and temperature regimes from climatic data, and 2) determination of taxonomic class where possible. The taxonomic class of the dominant soil present is determined down to an analogous soil series in the U.S. system. If there is no analogous soil series the family classification is determined.

The reason for conversion to the comprehensive system is that more precise soils interpretations are possible if moisture and temperature-regime information is known. For example, a common soil-mapping-unit term used in the WSMs is "Chernozem". Chernozems are defined as dark-colored soils of subhumid or semiarid grasslands. They extend over a moisture range from subhumid to semiarid and a temperature range that includes both winter and spring cereals. In the comprehensive system two temperature and two moisture regimes subdivide the former chernozem-soil region into smaller soil areas about which more precise interpretations can be made about crop adaptability and yields.

After converting the WSM soil map units to the nomenclature of the comprehensive system the soils are given a soil-suitability rating of 1, 2, 3, and 4, with 1 being suited and 4 poorly suited for crop growing.

Even though the subdivision of soil-map units accomplished by use of the comprehensive system improved the homogeneity of the units, it still is not certain that soils apparently suited for cultivation are, in fact, being cultivated. Slope and other differences exist within delineations that affect the percent of land that is suitable for cultivation.

Intensity of Use as Cropland From Landsat. Although the ag/non-ag overlay using Landsat shows which lands actually are cultivated, in most cases it appears that there are geographic areas where the intensity of use as cropland falls into intermediate classes. From this observation, the concept of "Intensity of Use as Cropland" (IUC) was formulated.

Five IUC classes were established. They are:

1) 70–100 percent cultivated;
2) 50–70 percent cultivated;
3) 30–50 percent cultivated;
4) 5–30 percent cultivated; and
5) less than 5 percent cultivated.

Land with an IUC class 1 is essentially without limiting factors. In class 2 clumps of fields or scattered fields dominate over grassland or forest land. In class 3 grassland or other non-crop uses are more common than cropland, although much cropland still exists. Class 4 land is dominantly grazing or forest land although there is some farming. Class 5 land is almost totally non-cultivated.

The ability of image interpreters to separate lands into five IUC classes on Landsat imagery depends in part on the scale of the imagery and the time available. The procedure described calls for manual interpretation of film color-composites of a scale of 1:1 million and a time frame of 4 to 6 hours per scene. These constraints were imposed because mapping of the study area of over 26 million square kilometers was scheduled for completion within a 3-year period.

Establishing Standards for Intensity of Use as Cropland Classes. Standards for establishing the percentage ratings for the five IUC classes were set using imagery from South and North Dakota and Montana. Together these three states have an area of about 760,000 square kilometers or 76 million hectares. Fifty-nine Landsat scenes cover this area and its borders, but 127 scenes were examined in temporal coverage. The total time involved to make this interpretation and the transfer to the ONC chart was about 8 weeks for one man.

The Landsat interpretation into the five IUC classes for this 76 million hectare area was digitized to obtain the hectares in each class. These percentage figures, representing the percent cultivated land in each class, are as follows (in terms of the bottom limit for each class): class 1, 80 percent; class 2, 60 percent; class 3, 40 percent; class 4, 10 percent; class 5, 1 percent. The Landsat interpretation indicated 22,025,135 hectares of cropland, compared to 22,896,576 hectares estimated by the Conservation Needs Inventory (CNI) of the Soil Conservation Service of the USDA[3]. Because the Landsat estimate was close to the CNI figure (about 4 percent below), and considering all factors, these results were acceptable for this project.

[3] The CNI inventory data for North and South Dakota and Montana were supplied by the Soil Conservation Service state conservations of each state.

Procedure for Determining Intensity of Use as Cropland. The specific techniques used to recognize IUC classes include the use of color, tone, and pattern along with evidences of restrictive topography, and restrictive wetness, aridity, and saline conditions. Keys were established for the recognition of features that restrict cropping (Table 33-1).

The symbols used in the interpretation are numbers 1 through 5, indicating the IUC class, followed by one or more letters for land classes 2 through 5 to indicate the soil factors thought to be responsible for the land class. The letter symbol gives a rational reason for the rating.

Soils rated 1 appear to have no serious limitation for use as cropland; however, the symbol FP (floodplain) separates class 1 alluvial land from other class 1 lands, setting apart those soils which may have drainage problems. Column 1 in Table 33-1 lists soil limitations visible on Landsat. With unfavorable topography an effort is made to characterize the topography as m (mountainous), h (hilly) or r (rolling). Column 2 lists the clues visible on Landsat for the featured limitation. Column 3 gives the soil inferences that can be drawn. Column 4 tells why the soil inferred is not adapted for cropping, while column 5 lists the estimated soil taxonomic unit. Column 5 indicates that a correlation exists between soils and land use and that soils data can aid in determining the IUC classes. For example, soils classed as aridisols rarely are cropped unless irrigated. Likewise, Salorthids (saline soils), Psamments (sands), and Entisols (thin soils) usually are not cropped. Other soils such as Aridic Ustolls (dry grassland soils) are mostly in grassland, while Orthods (cool forest soils) are mostly in trees. Still other groups of soils, such as the Udic Ustolls (warm humid grassland soils), generally are cropped intensively for cereal grains, specifically winter wheat.

Large fields of cropland can usually be recognized by regular boundaries and high reflectance if the scene was recorded when crops were growing. If fields are small and irregular, as in China, and pixels yield mixed signatures, cropped fields yield an unbroken mottled appearance that contrasts with non-cropped areas.

The interpretation was done using a block-mapping approach. The scenes from several paths are laid out utilizing temporal coverage as available. This permits establishing standards to use with the recognition keys. The interpreter then is prepared for change as he moves to new areas and environmental factors change and affect the use of the land.

The interpretation results are delineated, by means of a felt-tip pen, on transparent plastic overlays to Landsat film color-composites, viewed over a light table. After a scene has been interpreted the overlay is placed under an Opera-

TABLE 33-1

Recognition Keys–Landsat Intensity-of-Use-as-Cropland (IUC) Classes.

Feature and Symbol	Recognition Key	Soil Inferences	Soil Deficiency	U.S. Taxonomic Class
Unfavorable upland topography (m,h,r)	Closely spaced drains, visible relief, shadows	Thin, possible rocky soils	Low water availability, erodable, difficult to farm	Entisols, lithic sub-groups
Aridity (d)	Low IR reflectance, light tones, few streams	Thin soils, low in organic matter	Low water availability, low organic matter	Aridisols
Wetness (w)	Dark mottled tones, irregular patterns, banding	Water logged soils, organic soils	Lack of oxygen for roots and micro-organisms	Fluvaquents, "aqu" subgroups of Mollisols, Alfisols,
Salinity (s)	Light tones, mottled patterns, white rims around water bodies	Saline soils	Low ability to supply water	"Salic" great groups
Sand (sa)	Light tones, dunes, blurred boundaries	Sandy soils	Low water holding capacity, erodable	Psamments, Spodosols
Fine textured bedrock (b)	Dark uniform tones, close drain spacing	Clayey soils, churning soils	Poor physical condition, low oxygen supply	Fine and very fine families of Ustolls, Orthids, Vertisols
Erosion (e)	Light tones, gullies, mining activity	Most productive part of soil lost	Low organic matter, poor physical condition	Eroded phases

Note: The presence of grasslands and forests often indicate soils unsuitable for cropland—too steep, rocky, wet, dry, cold or stony. The vegetation in these areas masks the soil feature responsible for this non-cropland use. Grasslands and forests under these circumstances are collectively identified by the symbol "ve". Low temperatures as a deterrent to cropping are identified with the letter "c" if identifiable.

tional Navigation Chart (ONC) of 1:1,000,000 scale and the boundaries transferred to the ONC chart.

As an aid in recognizing IUC and soil boundaries the following characteristics of Landsat have been found useful: (1) the synoptic view of 3.5 million hectares, on which the condition of soils and stage of vegetation growth can be compared because the data are recorded at nearly the same moment; (2) the temporal feature, permitting study of soils and vegetation as they change with time, weather, and climate; (3) the multi-spectral capabilities, which increase the possibilities of unique signatures for vegetation and soils; and (4) the near-orthographic character of the scenes, allowing for the construction of mosaics when using block-mapping procedures.

Soil information System. Apart from assistance in developing APUs, the soil map and IUC maps developed from Landsat and other sources have a utility for general users interested in soils and land use.

Often a method of showing the interaction of several kinds of data is required. One such method is the use of an information system such as the Area REsource Analysis Systems (AREAS) program being developed by Wehde (1980) at the South Dakota State University Remote Sensing Institute. In this system digital images and/or outline maps are georeferenced and entered into the system. The system has the capability for multiple map overlay and thematic generalization. Outputs are in the form of area tabulations and data maps for line printer, drum plotter or film recorder.

Color Figure 33-20 is a film-recorder output map of IUC classes for South Dakota. The IUC classes are depicted in color. Such maps show at a glance the spatial relationships of cropland areas. Color Figure 33-21 is a film recorder output map of IUC classes for a region in China and Color Figure 33-22 is for a region in Russia.

The AREAS program permits study of the relationships of soils and IUC. Acreage of each of the 5 IUC classes can be determined. Relationships can be studied between soils and IUC and the environmental factors of climate, soil parent material, gemorphology of occurrence, and moisture and temperature regimes.

VEGETATION IDENTIFICATION

Physiological Factors Influencing Electromagnetic Radiation

Light reflectance from plant canopies is an integrated response to various plant structures, soil background, and surface adherents to plants' vegetative and reproductive appendages. Nature does a thorough job of providing heterogeneous landscapes. Even though these differences sometimes cause difficulty in interpretation, there are responses which persist and therefore aid interpretation. Many of the vegetation factors affecting energy responses will be described in this section. For more detail on general energy interactions the reader should look to Chapters 2 through 5.

General Review of Literature

Many reviews of the interaction of light with plant leaves and their appendages are available, including remote sensing applications, instrumentation and physics of light interaction. Among such reviews are those by Shull, 1929; Seybold, 1932; Mestre, 1935; Billings and Morris, 1951; Moss and Loomis, 1952; Gates and Tantraporn, 1952; Gates et al., 1965; Aboukhaled, 1966; Knipling, 1967, 1970; Pearman, 1966; Weber and Olson, 1967; Thomas et al., 1967; Allen and Richardson, 1968; Scott et al., 1968; Sinclair, 1968; Myers and Allen, 1968; Allen et al., 1969; Gausman and Cardenas, 1969a; Wiegand et al., 1969, 1972, 1979; Carlson et al., 1971; Gausman et al., 1971a, 1971b, 1974; Woolley, 1971; Kumar, 1972; Tucker, 1977a, 1977b, 1979; and Bunnik, 1978.

The reflectance of leaves has been related to yield (Thomas et al., 1967). Leaf anatomy has been related to photosynthetic activity (Hesketh and Baker, 1969), and to optical constants (Gausman et al., 1970a, 1970b, 1971b; Gausman and Allen, 1973). Leaf optical properties have been related to their energy-balance and water-use efficiency (Aboukhaled, 1966) and soil moisture (Dadykin and Bedenko, 1961). In addition, leaf spectra have been compared with the spectra of leaf extracts, potato tuber tissue, glass beads in water, and frozen leaves (Woolley, 1971). Also, light relations in plant canopies have received considerable attention (Allen et al., 1970c; Idso and DeWit, 1970; Suits, 1972a; Suits and Safir, 1972b; Bunnik, 1978; and Chance and LeMaster, 1978).

Mechanism of Light Reflectance

Willstätter and Stoll (1918) explained leaf light-reflectances and transmittances on the basis of critical reflection of visible light at the cell wall-air interface of spongy mesophyll tissue. Sinclair (1968) hypothesized that leaf reflectance derives from the diffuse characteristics of cell walls in the leaf. Gausman et al. (1969b, 1969c, 1970a) quantitatively related near-infrared reflectance to number of intercellular air spaces. Reflectance increased with an increase in numbers of air spaces, because diffused light passed more often from a high (hydrated cell walls, about 1.4) to a low (intercellular air, 1.0) refractive index.

Diffuse reflectance and transmittance of a compact leaf such as corn (*Zea mays*), a leaf impregnated with water, and an immature cotton (*Gossypium hirsutum*) leaf immediately after it unfolds (Gausman et al., 1969b) can all three be predicted from a plate theory (Allen et al., 1969). Maturation of a cotton leaf is characterized by development of intercellular air spaces in the mesophyll that increase light reflectance and reduce light transmittance of the leaf. Generalization of the plate theory (flat-plate model) to include the effect of intercellular air spaces leads to the concept of void area index (VAI) of a leaf. When a leaf is regarded as a pile of N compact layers separated by infinitesimal air spaces the VAI is given by N-1. The VAI of a compact leaf is zero. The VAI is roughly related to the average number of air cavities penetrated by a ray passing through a leaf.

Allen and Richardson (1968) used the Kubelka-Munk (K-M) theory (Kubelka and Munk, 1931) to describe near-infrared reflectance and transmittance of leaves as measured when the leaves were stacked in a spectrophotometer. Basic entities in application of the K-M theory to leaves are the reflectance and transmittance of a single leaf.

Parameters that emerge from the flat-plate theory include a measure of the water and air in the leaf and the effective index of refraction n and absorption coefficient k. The effective index of refraction of a typical leaf is not inconsistent with the refractive index of epicuticular wax. The effective absorption coefficient of a typical leaf is a superposition of the absorption coefficients of chlorophyll and pure liquid water. The plate model of a leaf is used to determine moisture content from reflectance and transmittance measurements. The absorption of a compact leaf can be simulated closely over the 1.35- to 2.5 μm wavelength interval (WLI) by absorption of an equivalent water thickness. The flat-plate model has been generalized to include the effect of intercellular air spaces in the leaf.

Leaf Structure

Air spaces in the leaf mesophyll usually develop schizogenously (separation of neighboring cell walls) (Fahn, 1967), although in banana (*Musa*) leaves, development may be lysigeonous (disintegration of cells in the place where the space develops). Intercellular spaces usually develop markedly after a leaf is one-fourth to one-third final size. In tobacco (*Nicotiana tabacum*) leaves, epidermal cells continue enlarging after cells of the middle and lower mesophyll have stopped growing (Avery, 1933). Stresses result that pull mesophyll cells apart, giving rise to spongy mesophyll tissue with intercellular spaces. Palisade cells (upper mesophyll) are also pulled apart by enlarging epidermal cells. Although palisade parenchyma tissue appears to be compact in transections (cross sections) of leaves, paradermal sections (parallel to the leaf) reveal that surface areas of palisade cells are exposed to air (Slatyer, 1967). The internal surface area of a leaf is usually greater than the external surface area; the palisade region usually has a larger internal exposed surface than the spongy parenchyma (Turrell, 1936 and Esau, 1965). The volume of air spaces is smaller in xerophytic than in mesophytic leaves. However, Fahn (1967) found that the ratio between the internal free surface of the leaf and its external surface is small in a mesophytic and large in a xerophytic leaf. The ratio of the volume of intercellular spaces to the total volume of leaves is between 77:1000 and 713:1000 (Sifton, 1945).

Compact, dorsiventral, and isolateral leaves will be considered here. A compact leaf, such as corn (*Zea mays*), has a mesophyll comprised of relatively compact chlorenchyma (chloroplast-containing cells) with few but large intercellular spaces (air voids); a dorsiventral leaf, such as cotton (*Gossypium hirsutum*), has palisade parenchyma (cells of elongated form perpendicular to the leaf surface) on one side of the blade (lamina), and spongy parenchyma on the other; and an isolateral leaf, such as the river red-gum eucalyptus (*Eucalyptus camaldulensis*), has palisade parenchyma cells on both sides of the blade.

The structure of a dorsiventral leaf (typical of an orange leaf, *Citrus sinensis*) is shown in Figure 33-23. The top layer of cells is the upper epidermis. The epidermal cells have a cuticular layer on their upper surfaces that diffuses but reflects very little light. The long narrow cells below the upper epidermis are palisade cells. They house many chloroplasts with chlorophyll pigments that absorb visible light. The cells below the palisade cells are spongy-mesophyll cells. They have many air spaces among them (intercellular air spaces). It is here that oxygen and carbon dioxide exchange takes place for photosynthesis and respiration. The lower epidermis is like the upper epidermis, except a stoma or port is present where gases enter and leave a leaf.

The spongy mesophyll is important in remote sensing because it scatters near-infrared light. For a mature citrus leaf, approximately 55 percent of the incoming near-infrared is reflected from the leaf, 40 percent is transmitted through the leaf, and 5 percent is absorbed within the leaf. The effects of the structural components of plant leaves

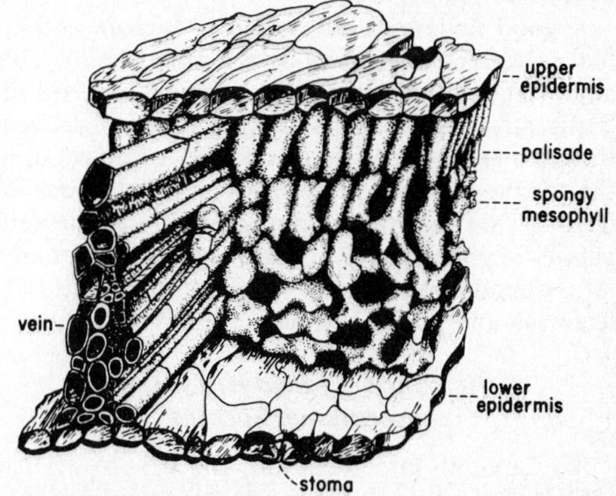

Fig. 33-23. Three-dimensional drawing of a leaf structure that is similar to the structure of a citrus leaf (redrawn from *Van Nostrand's Scientific Encyclopedia,* 1947).

on reflectance are varied. Sinclair (1968) showed that the pathway of light transmitted through leaves was affected by the cellular structure. Gates and Tantraporn (1952) studied the infrared-reflectance spectrum from 1.5 to 25 μm of many plants. In this study the wavelengths beyond 2 μm were reflected more from the upper surface of leaves than from the lower surface, older leaves reflected more than younger leaves, and shaded leaves more than sunlit leaves. Previous work by Billings and Morris (1951) showed that the inverse was true for visible light apparently because of the absence of palisade cells on the lower side of the leaf.

In an early work Shull (1929) reported, measuring the reflectance of monochromatic light of different wavelengths from the upper and lower surfaces of leaves of a number of plant species. He found that the under surfaces reflect more light than the upper surfaces of the same leaves. This same result was obtained by Moss and Loomis (1952) with poplar (*Populus*) leaves. Shull also determined the reflectance of a young leaf as compared with a more mature leaf. His results showed a decreased reflectance with maturity which he attributed to a greater development of chlorophyll. Spectral effects of structural differences between monocotyledonous and dicotyledonous leaves were studied by Hoffer and Johannsen (1969). The dorsiventral structure of the dicotyledonous leaves usually gave a higher reflectance.

Monteith (1959) reported that the reflection coefficients given in the literature for the same crop varied greatly. He showed that reflectance from the soil and stage of crop maturity were the primary causes of these variations. Pearman (1966) measured the reflectance in the visible region from the leaves of 32 Western Australian species and concluded that crop maturity greatly affected his results. Plants reaching the late stages of maturity with less chlorophyll had higher responses in the visible region than younger plants with more chlorophyll.

A good review of physiological factors and effect on reflection is given by Knipling (1970). This author stated that when physiological stresses directly affect the reflectance properties of leaves, the most pronounced initial changes often occur in the visible spectral region rather than in the infrared. This change was due primarily to the sensitivity of chlorophyll. Sometimes recognition of a stress condition is due to a reduction in the total leaf area and not due to pigmentation changes.

Reflectance, Transmittance, and Absorptance Spectra

To facilitate interpretation, the 0.5- to 2.5 μm wavelength interval (Figure 33-24) is divided into three categories:

1) the 0.5- to 0.75-μm visible light absorptance region, dominated by pigments—primarily chlorophylls a and b (chlorophyll c occurs in brown algae), carotenes, and xanthophylls;
2) the 0.75- to 1.35-μm near-infrared region, a region of high reflectance and low absorptance affected considerably by internal leaf structure; and
3) 1.35- to 2.5-μm, a region influenced some by leaf structure, but affected greatly by water concentration in the tissue with strong water absorption bands occurring at both 1.45- and 1.95-μm.

The peaks following these bands at 1.65- and 2.2 μm, respectively, decrease as leaf-tissue water-content increases.

Figure 33-24 shows spectrophotometrically measured (Beckman Model DK-2A spectroreflectometer) reflectance, transmittance, and absorption of mature citrus leaves (orange, *Citrus senensis*). Reflectance, transmittance, and absorptance were 10, 40, and 50 percent respectively, at the 0.55-μm green peak within the 0.5- to 0.75-μm band of the EM spectrum. Absorption was primarily caused by pigments. Within the 0.75- to 1.35-μm near-infrared wavelength band, there was about 55 percent reflectance, 40 percent transmittance, and 5 percent absorptance. Above 1.35 μm, absorptance increased greatly because of water absorption of light energy. *In general, the spectral transmittance curves for all mature and healthy leaves are similar to their special reflectance curves over the 0.5- to 2.5-μm wavelength/ band, but slightly lower in magnitude.*

A typical plant leaf can be specified by four optical parameters: (1) the equivalent water thickness (EWT), a number that specifies the amount of water in a leaf; (2) the void area index (VAI), a measure of the intercellular air space in a leaf; (3) the effective index of refraction, n; and, (4) the effective absorption coefficient, k (Allen et al.,

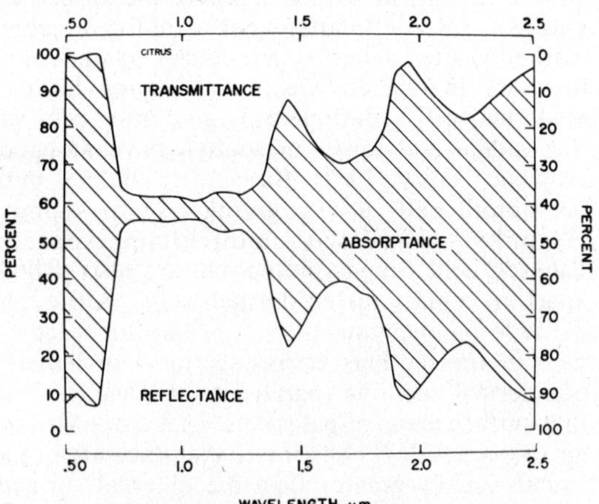

Fig. 33-24. Diffuse reflectance (data corrected for MgO standard decay), transmittance, and absorptance [100 − (percent transmittance + percent reflectance)] or the upper (adaxial) surface of a mature orange leaf. Each spectrum is an average of 10 leaves.

1969, 1971; Gausman et al., 1970a, 1971a). Between 1.35- and 2.5-μm, the absorption spectra of leaves are not appreciably different from pure liquid water. Leaf reflectance differences among plant leaves between 0.5- and 1.35-μm are caused principally by Fresnel reflections at external and internal leaf surfaces and by plant pigment absorption. Reflectance in the 1.35- to 2.5-μm wavelength interval is influenced largely by Fresnel reflections and absorption by water. The effective index of refraction of a leaf can be approximated by a cubic equation.

Optical parameters can be subjected to statistical pattern-recognition techniques to discriminate among kinds of leaves. The electromagnetic signatures predicted from measured optical constants can be tested against quantitative reflectance measurements from ground- or airborne optical-mechanical scanners.

Leaf Maturation

Theoretically, a leaf has one continuous air space (maze or labyrinth effect); but here the air space will be considered compartmentalized among mesophyll cells to give intercellular air spaces. Maturity of leaves is related to their phyllotaxis—arrangement on the axis or shoot (Hammond, 1941). Young leaves at the plant apex have compact mesophylls, primarily with small protoplasmic cells, while full-grown leaves further down the stem are lacunose (loosely arranged mesophyll structure) with large vacuolated cells.

The influence of age on orange (*Citrus sinensis*) leaf structure is shown in Figure 33-25. The top transection represents a very young citrus leaf (fifth leaf from apex of new growth flush), and the bottom transection represents a mature citrus leaf (eighth leaf from apex of previous growth flush). The young leaf was compact with few air spaces in its mesophyll, while the older leaf was "spongy" and had many air spaces. The spongy mature leaf, as compared with the compact younger leaf, had about 5 percent less reflectance in the visible (0.5 to 0.75 μm) and 15 percent more in the near-infrared (0.75 to 1.35 μm) (Figure 33-26). The "spongy" effect in the mature leaf increased reflectance because there were more intercellular air spaces. Scattering of light within leaves occurs at cell wall (hydrated cellulose)-air cavity interfaces that have refractive indices of about 1.4 and 1.0, respectively.

Leaf Maturation in Relation to Optical Parameters. As previously indicated, a leaf with intercellular air spaces can be regarded as a pile of N compact layers separated by air spaces (Allen et al., 1969; Gausman et al., 1970a). The void area index (VAI) of a leaf with intercellular spaces (a noncompact leaf) is given by N-1 where N is not necessarily an integer. An equivalent water thickness (EWT) is the thickness of a water layer of thickness D that would yield the same reflectance and transmittance of the leaf.

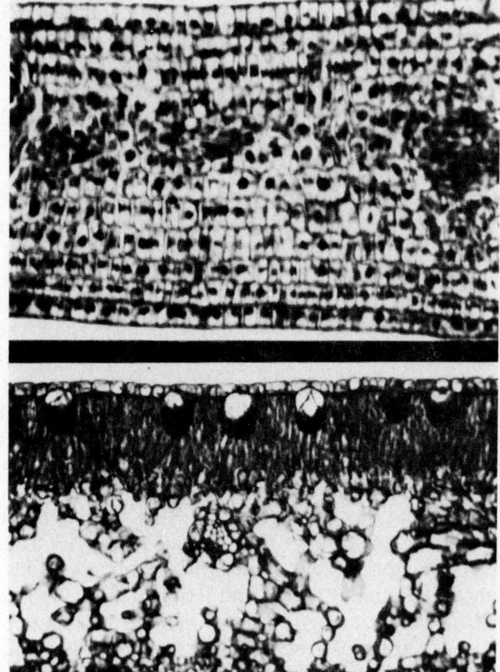

Fig. 33-25. Photomicrographs of leaf transections of young (top, 64× and mature (bottom, 33×) orange leaves.

A young nonexpanded cotton (*Gossypium hirsutum*) leaf, for example, has a D/N value of about 180 μm which is essentially the leaf thickness. As intercellular air spaces develop during leaf expansion, D/N decreases to about 130 μm. Finally the leaf cells increase in size with essentially no increase in intercellular air spaces; this phase is characterized by a D/N of about 140 μm. During the leaf expansion period, reflectance increases about 5 percent for laboratory-grown plants and

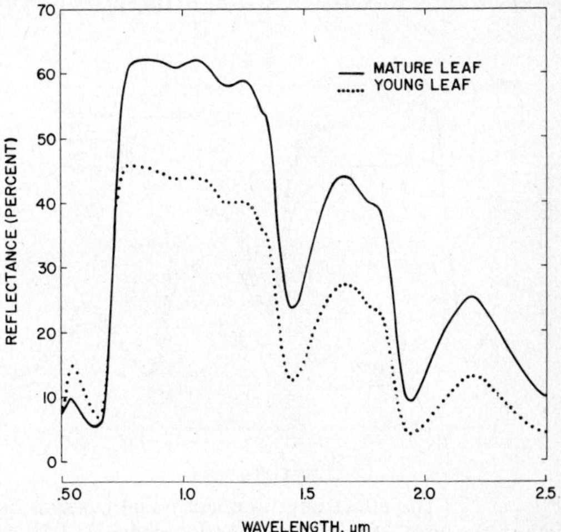

Fig. 33-26. Diffuse reflectance of the upper (adaxial) surfaces of young (bottom dotted line spectrum) and mature (upper solid line spectrum) orange leaves.

about 15 percent for field-grown plants in the 0.75- to 1.35-μm wavelength band of the EM spectrum. Maximum reflectance corresponds to a minimum value of D/N.

Influence of Phyllotaxis (leaf nodal position) on Reflectance and Infinite Reflectance. Figure 33-27 represents spectra obtained from spectrophotometric measurements on single cotton (*Gossypium hirsutum*) leaves for nodes 2 to 6 over the 0.5- to 2.5-μm wavelength interval.

Youngest leaves, second node down from plant apices, had approximately 10 percent less leaf reflectance between 0.75- and 1.35 μm than the older leaves. However, reflectance of the older leaves increased 2 to 3 percent as the leaves aged from upper node three down to six. There was little change in reflectance among all leaves in the visible range (0.5 to 0.75 μm) and among leaves of nodes three to six above 1.35 μm.

Stacking leaves against a soil background within a plant canopy (Myers and Allen, 1968) or over the port of the spectrophotometer's reflectance attachment (Allen and Richardson, 1968) increases reflectance as the number of leaf layers increase (because of multiple transmission and reflection from leaves) until a relatively constant value is reached called infinite reflectance, R_∞. Infinite reflectance is analogous to infinite thickness.

Allen et al. (1970a, 1979b) found that the absorption spectra of leaves do not differ appreciably from the spectrum of pure water between 1.35- and 2.5 μm. Absorptance increases between 1.35- and 2.5-μm, as the water content of a leaf increases. Hence, R_∞ may be considered a quantity that is influenced by the internal leaf structure in subdividing water.

The R_∞ from stacking four leaves of different maturity from nodes 2 through 6 is shown in Figure 33-28 for the 0.5- to 2.5-μm wavelength band. Between 0.5 and 0.75 μm and between 1.35- and 2.5-μm, R_∞ was reached for a full-grown leaf when one or two leaves were stacked on a spectrophotometer for measurement or by the time plants growing in the field reached a leaf area index (LAI) of 2. LAI is the cumulative one-sided leaf area per unit ground area projected from the canopy top to a plane at a given distance above ground level (Wiegand et al., 1969). Between 0.75- and 1.35-μm, stacking of six to eight leaves was necessary to attain R_∞.

Figure 33-28 shows that the youngest leaves (second node) had the lowest R_∞, particularly between .75- and 1.35-μm; R_∞ increased approximately 5 percent for older leaves of the third node, and then progressively increased at a slower rate for leaves of the fourth, fifth, and sixth nodes. These results indicated that the upper planes of leaves of a plant canopy have a lower R_∞ than successively lower planes of the same dimension.

Infinite reflectance, R_∞, is a promising term for use in crop discrimination procedures, yield prediction, and modeling in remote sensing (Wiegand et al., 1979). Crop reflectance increases as numbers of leaf layers increase within a plant canopy until a stable reflectance value R_∞ is reached. In the 0.5- to 0.75 μm, 0.75- to 1.35 μm, and 1.35- to 2.5-μm wavelength intervals, LAI's of 2, 8, and 2, respectively, are required to reach R_∞. For example, a typical mature cotton leaf reflects about 48 percent of the incident light between 0.75- and 1.35-μm, and transmits about the same amount of light to leaves below it on the plant, that, in turn reflect about half and transmit about half of the light. Multiple transmission and reflection from leaves in a plant canopy results in an approximate 75 percent reflectance of incident light between 0.75- and 1.35-μm.

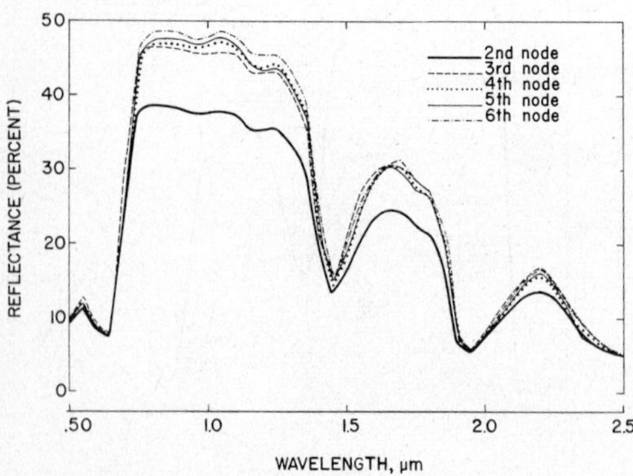

Fig. 33-27. The effect of cotton leaf nodal position on light reflectance. Leaves representing nodes 2, 3, 4, 5, and 6 basipetally had after emergence ages of 20–25, 29, 33, and 35 days, respectively. Each spectrum is an average of seven leaves.

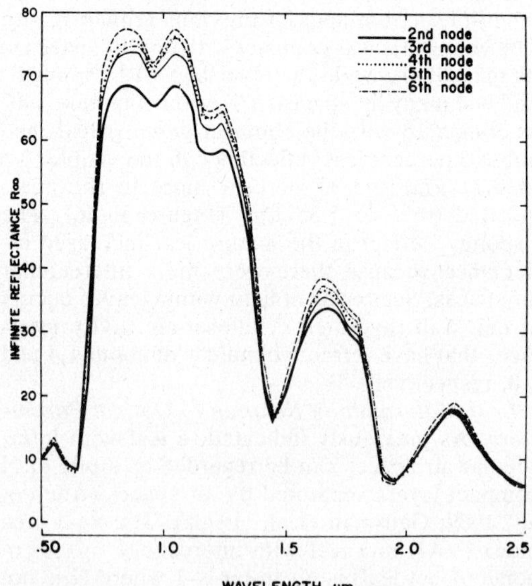

Fig. 33-28. Effect of cotton leaf nodal position on infinite reflectance, R_∞. Leaves representing nodes 2, 3, 4, 5, and 6 basipetally had after emergence ages of 20, 25, 29, 33, and 35 days, respectively.

Pigments

The chlorophyll of green leaves usually absorbs 70 to 90 percent of the light in the blue (about 0.45 μm) or red part (about 0.68 μm) of the spectrum (Kleshnin and Shul'gin, 1959). Absorptance is smallest in the wavelength region around 0.55 μm, where a reflection peak of usually less than 20 percent occurs from upper leaf surfaces. Low pigment content results often in higher reflectance (Rabideau et al., 1946; Carter and Myers, 1963; Myers et al., 1963; Benedict and Swidler, 1961). For example, Rabideau et al. (1946) found that light green leaves of cabbage (*Brassica oleracea*) and lettuce (*Lactuca sativa*) had 8 to 28 percent higher reflectance than the average of six darker green species.

Leaf pigments also include carotenoids (carotenes and xanthophylls) and anthocyanins, and they markedly affected the light absorptance and hence reflectance of cotton (*Gossypium hirsutum*) plant leaves (Figure 33-29).

Early work on leaf pigmentation is recorded by Coblentz (1913) and Shull (1929). Shull used a prism spectrophotometer to measure the visible spectra in the 0.42- to 0.7-μm wavelength band. He showed a depression in the reflective curve at 0.68 μm which is attributed to absorption of light energy by chlorophyll.

Hoffer and Johannsen (1969) illustrated marked differences in spectral response due to different leaf pigments in *Coleus* leaves (Figure 33-30). The solid line curve in this figure is a green leaf with dominant chlorophyll pigmentation that showed a peak of 0.55 μm (due to the green), a low reflectance in the red (due to chlorophyll absorption) and then a sharp increase at about 0.7 μm to the reflectance infrared. The white *Coleus* leaf without any apparent pigmentation had a high level of reflectance throughout the 0.5- to 0.9-μm region.

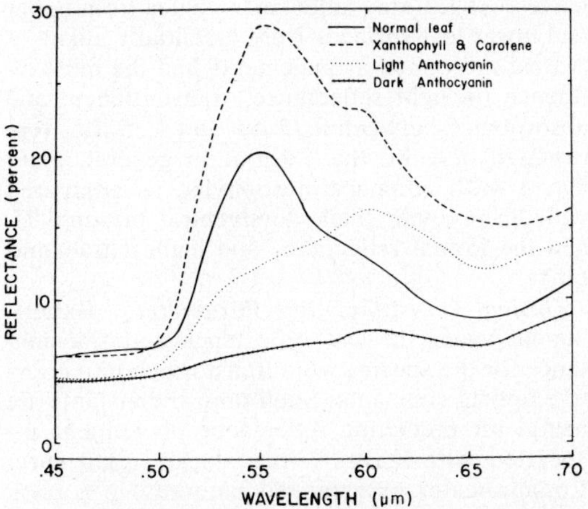

Fig. 33-29. Reflectance produced by leaves of four genetic strains of cotton high in chlorophyll (normal green leaf), xanthophyll and carotene (yellow), light anthocyanin (light red), and dark anthocyanin (dark red), respectively. Each spectrum is an average of five leaves.

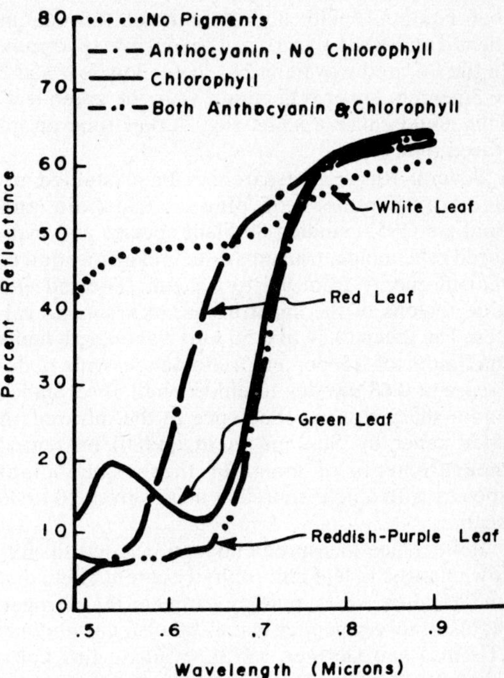

Fig. 33-30. Spectral response (.5 to .9 μm) from four pigmentation conditions of *Coleus* leaves.

All of the leaves had similar responses beyond 2.6-μm.

Spectra for leaves of tuliptree (*Liriodendron tulipifera*) are shown in Figure 33-31 (Hoffer and Johannsen, 1969). Both leaves were quite succulent and at normal high moisture content. However, one leaf was a normal, deep green color and the other was a bright yellow. This yellow coloration was caused by the normal autumn breakdown of the chlorophylls, which were not reformed, thus allowing the presence of the carotenes and xanthophylls to become evident. These carotenoid pigments were present before the chlorophyll breakdown, but were masked by the chlorophylls. These spectra showed the usual curve for a green leaf, but the yellow leaf had a very sharp increase in reflectance starting at 0.5 μm, and

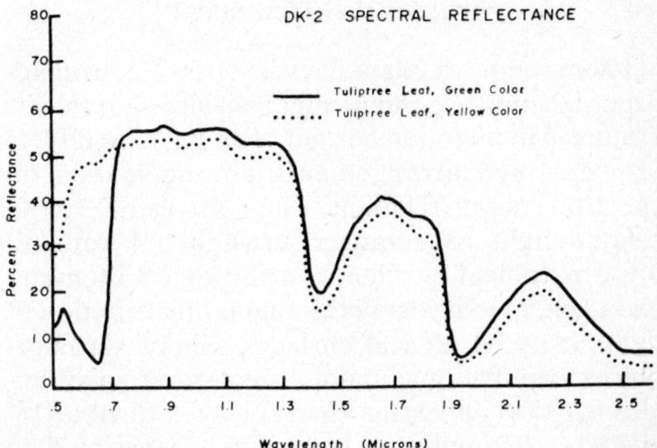

Fig. 33-31. Spectra from green and yellow Tuliptree leaves.

then continued with high reflectance throughout the red and green portions of the visible spectrum. In the infrared wavelengths, the yellow leaf had 2 to 3 percent lower reflectance than the green leaf. This could cause a somewhat darker tone on infrared film.

Several Russian researchers have studied reflective responses of plants. Kleshnin and Shul'gin (1959) studied 80 plant species and measured reflectance, transmission, and absorption of radiant energy from 0.5 to 0.8 μm. The red and blue regions of the spectra had very similar values. The green (0.54 to 0.56 μm) wavelength had a maximum of 15 percent reflectance with a decrease at 0.68 μm due to chlorophyll absorptance and a sharp rise in reflectance in the infrared. A later paper by Shul'gin et al. (1960) presented measurements of many of these same plant species with a light source at angles from 10 to 80 degrees.

Reflectance measurements can be used to: follow changes in leaf chlorophyll content (Benedict and Swidler, 1961), quickly estimate the nitrogen status of sweet pepper leaves (*Capsicum annum*) (Thomas and Oerther, 1972), evaluate turf color (Birth and McVey, 1968), and measure amounts of green and dry biomass (Tucker et al., 1975). Tucker reported that carotenoids (carotenes and xanthophylls) were useful in measuring dry biomass. He also found through *in situ* measurements that both the 0.45- and 0.68-μm wavelengths were more sensitive than 0.55-μm to the chlorophyll concentration of a short-grass prairie. Laboratory-measured hemispherical reflectance of leaves of six crops was inversely related to each crop's leaf chlorophyll and carotenoid concentrations (Thomas and Gausman, 1977a). However, of the 0.45-μm, 0.55-μm, and 0.67-μm wavelengths tested, 0.55-μm seemed superior for individually relating the two pigments to leaf reflectance. Later in the growth period, carotenoid pigments may become more effective in determining reflectance, however. Sanger (1972) found that during senescence the chlorophyll concentration decreased at a faster rate than carotenoid.

Mesophyll Arrangements (Internal Structural Differences)

Interactions of plant leaves with electromagnetic radiation are empirically considered in the literature, but attention has not been given to effects of mesophyll arrangements among leaves on spectral energy relations. Shul'gin et al. (1960) related light reflectance between 0.4-μm and 0.6-μm to leaf-surface morphologies of many plant genera to measure the angular distribution of light scattered off leaf surfaces. Shiny xeromorphic leaves had maximum reflection at small angles (to 15°), pubescent (hairy) leaves to 10 to 15° and 60 to 70°, and wrinkled and dull leaves at 70 to 80°. Rao et al. (1979) reported that crops have a different leaf orientation when viewed at a low-oblique angle (<45 degrees) than when viewed vertically. Howard (1966) studied spectral relations of isobilateral eucalyptus leaves, but other types of leaves were not included.

Leaf mesophylls among 30 plant species have been compared with:

1) spectrophotometrically measured percent reflectances and transmittances, and calculated absorptances [100 − percent reflectance + percent transmittances] of the leaves between 0.5 μm and 2.5 μm;
2) percent leaf water contents;
3) leaf thickness measurements; and
4) optical and geometrical leaf parameters.

For each species, data are given as the averages of 10 leaves (Gausman et al., 1970b, 1971b).

Spectral Measurements. Spectral measurements were made on both lower (abaxial) and upper (adaxial) leaf surfaces of banana, begonia, corn, crinum, eucalyptus, ficus, hyacinth, ligustrum, oleander, rose, and sedum leaves.

Lower leaf surfaces of dorsiventral leaves had higher reflectance values than upper leaf surfaces, indicating that the spongy parenchyma contributed more to light scattering than did the palisade parenchyma of the leaf mesophyll. This was substantiated by equal reflectance values of upper and lower surfaces of compact leaves.

Spectrophotometrically-measured transmittance values were lowest when light was passed from the top through the leaves compared with passing light from the bottom through the leaves. The difference in transmittance was caused by greater light diffusion by upper leaf surfaces, since the spectrophotometer used irradiated the specimen with direct light. At 0.55-μm, reflectance was greater from the bottom than from the top of dorsiventral leaves, indicating that the chloroplasts in the palisade cells absorbed light. Bottom and top reflectance values were the same for the compact leaves. At 1.0-μm, reflectance values from upper and lower leaf surfaces were essentially alike.

Leaf mesophyll arrangements had the most influence on light reflectance, transmittance, and absorptance between 0.75 μm and 1.35 μm, represented here by the 1.0-μm. In general, plant leaves with compact mesophylls, as compared with leaves with thick dorsiventral mesophylls, had the lowest reflectance and highest transmittance.

Optical Constants and Parameters. Experimental values of leaf reflectance and transmittance for the species were transformed into effective optical constants. Such optical constants are useful for predicting reflectance phenomena associated with leaves either stacked in a spectrophotometer or arranged naturally in a plant canopy. The index of refraction n was plotted against wavelength to obtain dispersion curves. The values for the absorption coefficient k were tabulated for the various crops.

The dispersion curves of most of the plant

leaves were remarkably similar. The dispersion curves were characterized by similar shapes and relatively close confidence bands. Most of the plants were analyzed to obtain the equivalent water thickness (EWT). There was no highly statistically significant difference between water obtained experimentally and water determined theoretically. The limiting value of reflectance from leaves piled sufficiently deep is characterized by infinite reflectance, R∞. Inifinite reflectance was tabulated at 1.65 μm for the crop species.

Leaf Damage

Colwell (1956) suggested using infrared film to record any disease that interfered with the internal reflection of light within leaves. Keegan et al. (1956) studied effects of stem rust (*Puccinia graminis tritici*) and leaf rust (*Puccinia triticina* or *Puccinia rubigovera tritici*) of wheat on light reflectance. Data showed that severe as compared with low rust infestation caused a rounding of the shoulder of the reflectance plateau between 1.0- and 0.75-μm. The same response in reflectance was noted by Gausman and Cardenas (1969a) after hair removal on upper leaf surfaces of the velvet plant (*Gynura aurantiaca*).

Cellular discoloration within leaves may be useful in detecting nonvisual symptoms of plant maladies. However, previsual detection has had variable success. Manzer and Cooper (1967) found that late blight of potatoes (*Solanum tuberosum*) could be detected by aerial photography three to five days before visual symptoms became apparent; tobacco (*Nicotiana tabacum*) ringspot virus could be detected about one day before visual symptoms were evident (Burns et al., 1969); and ozone-damaged leaf areas were detected photographically 16 hours before the damage was visible (Gausman et al., 1978b). In contrast, Heller (1968) found that beetle damage could not be predetected, and Meyer (1967) reported that biological variability interfered with previsual detection of tree disease. Caution should be exercised when suggesting the use of remote sensing for previsual detection of plant stress.

To study effects of internal leaf damage on energy spectra, cotton (*Gossypium hirsutum*) leaves were treated with anhydrous ammonia, (Cardenas et al., 1969–1970). Spectrophotometric measurements on treated leaves showed increased absorptance and reduced reflectance and transmittance between 0.5- and 1.35 μm. Apparently the brownish discoloration increased leaf opaqueness.

Some conditions that caused decreased reflectances (rounding of the plateau) have been studied (Figure 33-32). Reflectance was reduced by severe rust infection on Westar wheat (*Triticum aestivum*) leaves (Keegan et al., 1956), benzene vapor on cotton leaves, natural freezing of *Coccolobis uvifera* (sea grape) leaves, ammonia treatment of cotton leaves, and hair removal by shaving velvet-plant leaves (Gausman and Cardenas, 1969a).

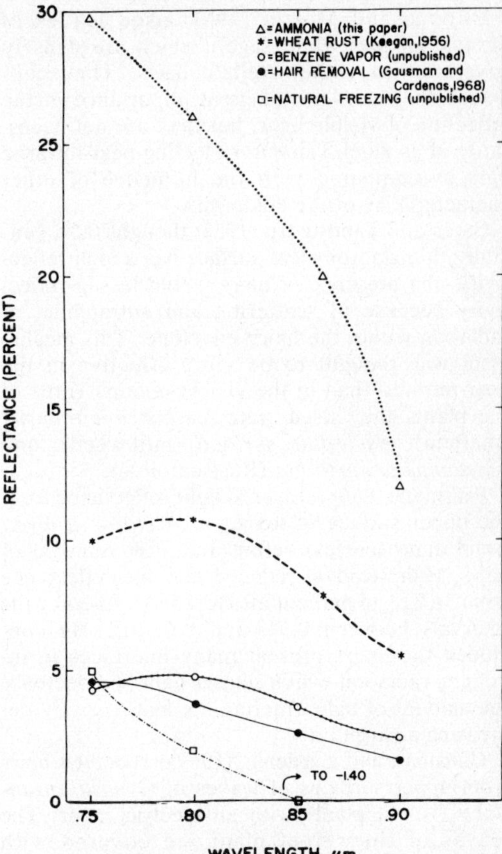

Fig. 33-32. Influence of treatment of cotton leaves with ammonia (Δ) and benzene gases (o), rust infection of Westar wheat leaves (*), hair removal from the velvet plant by shaving (●), and natural freezing of sea grape leaves (□) on light reflectance from their upper surfaces over the wavelength internal .75 to .90 μm. Values plotted in the figure are decreasing in reflectance compared with corresponding experimental controls.

Leaf Pubescence (Hairiness)

Some studies have shown inconsistencies in effects of pubescence (hairiness) on light reflectance. Shull (1929) concurred with Coblentz (1913) that hairs on *Verbascum thapsus* (mullein) did not increase reflection of light to any marked extent in comparison with leaves on nonhairy plants. Shull also found this with *Abutilon theophrasti* (velvet leaf). Tomentose hairiness, however, on the under surface of *Populus alba* (silver-leaved poplar) and *Magnolia acuminata* (cucumber tree) increased reflection. Moss (1951) found that the hairy lower surface of *P. alba* leaves, as compared with their upper surface, reflected about 15 percent more incident light between 0.4- and 0.7-μm.

Billings and Morris (1951) used leaves of *Eurotia lanata* (white sage), which are densely covered with white, stellate hairs. They concluded that hairs are of great importance in the reflection of visible light, but they are not necessarily of as much value in reflecting near-infrared light as compared with the influence of other characteristics of the epidermis.

Gates and Tantraporn (1952) thought that, generally, if a glabrous leaf surface has a high reflectivity, the presence of hairs would lessen reflectivity because of scattering and entrapment of radiation within the hairy envelope. This mechanism was thought to be more effective in the near-infrared than in the visible region. Three of the plants they used were *Verbascum thapsus* (mullein), *Asclepias syriaca* (milkweed), and *Elaeagnus angustifolia* (Russian olive).

Pearman (1966) measured light reflectance from the upper surface of leaves of *Arctotheca nivea* (sand dune species) before and after removal of hairs. Hair removal reduced average reflectance from 31.7 to 15 percent measured at 0.02-μm wide intervals between 0.34- and 0.62-μm. He concluded that hairs present many interfaces to incoming radiation which scatter light and decrease the amount of light entering the leaf, thereby decreasing absorption.

Gausman and Cardenas (1969a) removed hairs from upper surfaces of leaves of *Gynura aurantiaca* (velvet plant) with an electric razor. The leaves of the velvet plant are covered with velvet-like, purple hairs containing anthocyanin. The hairs are multicellular and unbranched.

Hairiness increased total and diffuse reflectance between 0.75- and 1.0-μm but decreased total and diffuse reflectance between 1.0 and 2.5-μm. Removal of hairs had little, if any, influence on total or diffuse reflectance in the visible range, 0.5- to 0.7 μm. This is not in agreement with Billings and Morris (1951) or Pearman (1966) whose studies indicated that the pubescence of leaves enhanced the reflection of visible light. However, hairs of *Gynura aurantiaca* as compared, for example, with white, stellate hairs of *Eurotia lanata* (Billings and Morris, 1951), are purple because they contain anthocyanins. According to Gilliam et al. (1962), the absorption band of the anthocyanins, although somewhat pH-dependent, usually occurs within the range 0.50- to 0.55-μm. Thus, absorption of light by the anthocyanins may modify the reflectance of light which normally reaches its maximum peak in the visible portion of the spectrum at 0.55 μm, when measured spectrophotometrically.

Removal of hairs by shaving increased absorptance of near-infrared light between 0.75- and 1.0-μm. The maximum increases were 4.4 and 4.2 percent at 0.75-μm and 0.80-μm respectively. Theoretically, unshaven leaves were highly transparent to near-infrared light, 0.75- to 1.0-μm. After removal of hairs, leaves apparently became more opaque and light absorptance was increased. The increase in opacity was apparently caused by discoloration of the exudate on the stumps after their removal. The discoloration of browning of the exudate was probably caused by phenol oxidase (Bonner and Galston, 1952).

Pubescence may be important in remote sensing for the detection of silverleaf sunflower (*Helianthus argophyllus*) (Gausman et al., 1977c). This plant is a weed in the sandy soils of south and southeast Texas. The young plant parts are densely white-tomentose. This pubescence greatly increased reflectance between 0.5- and 2.5-μm as compared with the sparsely-hairy leaves of common sunflower (*Helianthus annus*). Color Figure 33-33 shows that this increased reflectance caused silverleaf sunflower's image on Eastman Kodak Aerochrome infrared-color (type 2443) film to be "pinkish" as compared with darker magenta images for other plant species. Therefore, aerial photography may be useful to distinguish silverleaf sunflower plants from other plant species to locate its endemic areas, monitor its spread, and delineate areas needing weed control.

Sun and Shade Leaves

Sharma and Sen (1971) reviewed research on differences among sun and shade leaves. Shade leaves are thinner than sun leaves with a greater volume of air space, thinner palisade cells, and fewer stomata. Leaves differentiating in the shade (shade leaves) have weaker development of palisade tissue (Esau, 1965) than leaves exposed to light during differentiation (sun leaves). Thus, differences in mesophyll structure occur in leaves at different levels of the same plant because of variable light conditions that occurred during leaf development.

The influence of sun and shade leaves from an avocado (*Persea americana*) tree on near-infrared light (0.75 to 1.35 μm) reflectance is shown in Figure 33-34. At 1.0-μm, old sun leaves with a well

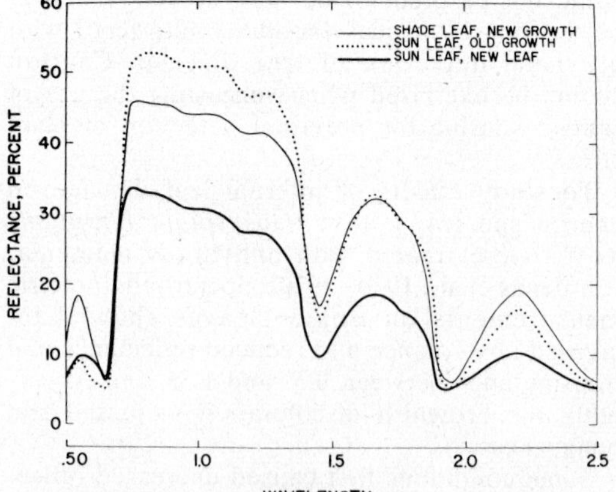

Fig. 33-34. Influence of sun and shade leaves from new growth and sun leaves from old growth of an avocado tree on light reflectance for the 0.5- to 2.5 μm wavelength interval. Each spectrum is an average of measurements of five leaves.

differentiated mesophyll had the highest reflectance (50.7 percent), a new shade leaves had intermediate reflectance (43.9 percent), and new sun leaves with a compact mesophyll had the lowest reflectance (30.1 percent). Differences are also present between 0.5- and 0.75-μm in the visible part of the spectrum. At the 0.55-μm green reflectance peak, new shade leaves with a low chlorophyll concentration had the highest reflectance as compared with the old and new sun leaves. Reflectances of the old and new sun leaves were approximately the same at 0.55-μm. The broadened peak for the new sun leaf was caused by an apparent greater concentration or proportion of anthocyanin pigments than was present in the old sun leaf.

Leaf Water Content

Young leaves contain less water than mature leaves because immature cells in young leaves are primarily protoplasmic with little vacuolate water storage (Lundegardh, 1966). During cell growth, cell water-filled vacuoles develop that may later coalesce to a central sap cavity, and the protoplasm covers only the cell wall in a thin layer.

Figure 33-35 shows the influence of progressive drying of excised cotton (*Gossypium hirsutum*) leaves on their spectrophotometrically-measured light reflectance in the laboratory between 0.5- and 2.5-μm. It is evident that dehydration increased reflectance greatly over the entire 0.5- to 2.5-μm wavelength interval.

Thomas et al. (1966) found that reflectance increased as relative turgidity (Namken, 1965) decreased below values of 80 percent at selected wavelengths of 0.54-, 0.85-, 1.65-, and 1.45-μm. Relative turgidity is used to measure water stress in plants. It is the actual leaf water-content expressed as a percentage of the turgid or saturation water content. Regression equations were also calculated (Thomas et al., 1971) to express the reflectance of incident light from the upper (adaxial) surface of single leaves as a function of their relative turgidity and water content. Reflectances at the 1.45- and 1.95-μm water absorption bands were related to the leaf's relative turgidity or water content. However, because of variations in internal leaf structure associated with the availability of water during leaf development, the ability to predict the leaf water-status from reflectance measurements was poor. With cotton, the greatest change in reflectance occurred when the relative turgidity was below 70 percent and the leaves were visibly wilted. Within the relative turgidity range, 70 to 80 percent, reflectance changes were small, and they may not be definable for predictive purposes because of variation among leaves of field-grown cotton caused by age differences and osmotic stresses. Carlson et al. (1971) found that the leaf reflectances of corn (*Zea mays*), sorghum (*Sorghum bicolor*), and soybean (*Glycine max*), were highly correlated with relative water content at two strong water-absorbing bands, 1.45 μm and 1.95 μm, and two bands of lower absorptivity, 1.1 μm and 2.2 μm.

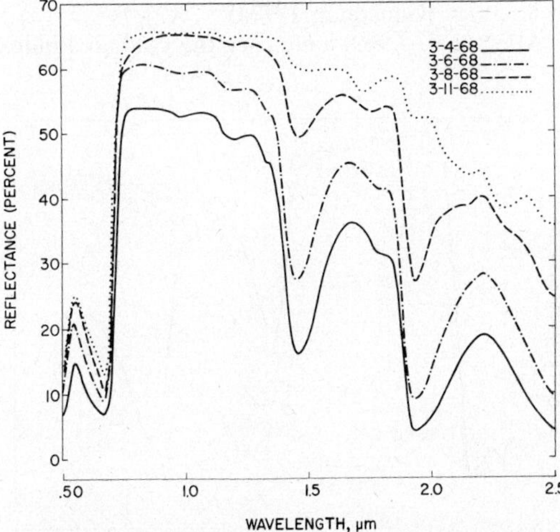

Fig. 33-35. Effect of progressive leaf drying on spectrophotometrically measured reflectance of upper surfaces of cotton leaves at four dates over the .5- to 2.5-μm wavelength interval. Each spectrum is an average of five leaves.

In general, the correlation of leaf water content with reflectance is strongest in the near-infrared region of the spectrum. However, Dadykin and Bedenko (1961) have related reflectance of oak leaves to different moisture regimes between 0.4 and 0.8-μm. Knipling (1969) has stated that any physiological disturbance to a leaf usually results in an increase in visible light reflectance.

Effects of pigment changes and leaf moisture are closely related and sometimes cannot be separated. Johannsen (1969) measured the reflectance of corn (*Zea mays*) and soybean (*Glycine max*) leaves at decreasing soil moisture contents. His results showed that the water absorption bands (1.45 and 1.95 μm) are inversely related to leaf moisture. The green color response and chlorophyll absorption (0.53 μm and 0.64 μm respectively) also showed a high negative correlation with leaf moisture. This indicated that changes in leaf moisture were apparently affecting the pigments in the leaf within a very short period of time.

Robinowitch (1951) reported that the scattering of light in leaves was decreased by the injection of water into the air channels. This condition is important since leaves at full turgor are shown to have a lower reflectance than nonturgid leaves.

Moss and Loomis (1952) attempted to determine the effect of the water content in leaves by infiltrating selected leaves with water. They received a decrease in absorption throughout most of the visible region and a sharpening of the band at 0.68 μm. Pearman (1966) reported a decrease in reflectance from intercellular infiltration. He demonstrated that a decrease in water content was found to increase reflectances, as evidenced by his observing 32 Western Australian plant species. In general, he pointed out that the normal drying of leaves causes an increase in reflectance in the visible and that special care must be taken to minimize water loss from leaves after collection and during experiments.

Hoffer and Johannsen (1969) showed the close relationship between water absorption and reflectance for a healthy, turgid green leaf (Figure 33-36). At wavelengths where water absorption was high, leaf reflectance was low. This was most apparent in the primary water-absorption bands centered at 1.45 and 1.95 μm. There were also slight increases in water absorption at approximately 0.96 and 1.2 μm. These minor water-absorption bands caused slight decreases in leaf reflectance. As the next section shows, the 1.65- and 2.2-μm peaks, following the 1.45- and 1.95-μm water absorption bands, respectively, had decreased leaf reflectance with increased leaf water content, and they offer promise for discrimination in remote sensing.

Succulent Plants. Succulent plants have water-storage tissue developed in their leaf mesophyll (Fahn, 1967). Therefore, they have a higher water content and absorb more radiation in the near-infrared water-absorption region (1.35 to 2.5 μm) than nonsucculent plants. Peperomia (*Peperomia obtusifolia*) is an example of a succulent plant that has a water-storage tissue called the hypodermis. Figure 33-37 shows the absence of the typical near-infrared light reflectance peak at about 2.2-μm from the upper leaf surface of peperomia. This absence was caused by light absorptance by water stored in the cells of Peperomia's hypodermis (Gausman et al., 1977b).

Practically, this phenomenon may be useful to distinguish succulent plants from crop- and woody-plants (Gausman et al., 1978a) by using sensor bands to encompass either the 1.6- or 2.2-μm wavelengths. These wavelengths were predicted by Richardson et al. (1969) to be useful for plant species discrimination by remote sensing. Also, Leamer et al. (1978) have shown them to be useful to distinguish between soil and vegetation—soil was more reflective than vegetation.

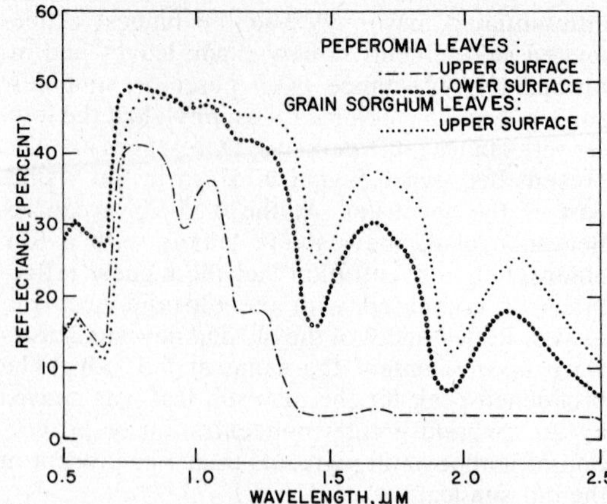

Fig. 33-37. Total diffuse light reflectance spectra between 0.5- to 2.5-μm of grain sorghum's upper and Peperomia's upper and lower leaf surfaces.

Leaf Air Spaces

Much of the light reflectance from leaves occurs internally since it is appreciably reduced by infiltrating leaves with water (Pearman, 1966; Moss, 1951) or with oil mixtures (Woolley, 1971). Figure 33-38 shows that vacuum infiltration of citrus (*Citrus sinensis*) leaves markedly reduced the reflectance over the 1.35- to 2.5-μm wavelength interval by eliminating the role of hydrated cell wall-air interfaces. Woolley (1971), however, predicted that internal discontinuities other than air-cell interfaces must be responsible for a part of the light reflectance by a leaf. It has been shown that refractive discontinuities among cell membranes, crystals, cell walls, and surrounding protoplasm contribute to the reflectance of near-infrared light at 0.85-μm (Gausman, 1977a).

Allen et al. (1969) have used the void area index

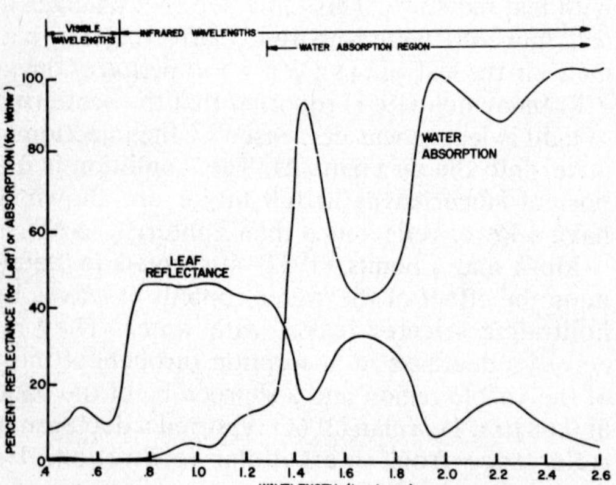

Fig. 33-36. Relationship between leaf reflectance and water absorption in the 0.4- to 2.6-μm wavelength region.

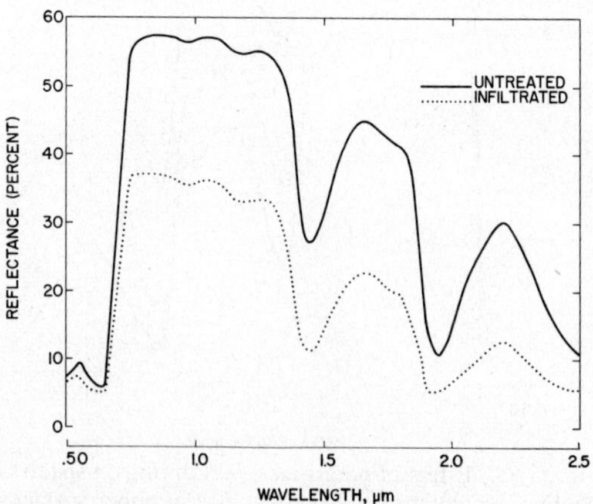

Fig. 33-38. Reflectance of untreated and vacuum-infiltrated (water) citrus leaves. Each spectrum is an average of five leaves.

(VAI) to measure the intercellular air spaces in both compact and noncompact leaves. The VAI is strongly correlated with laboratory measurements of intercellular air in mesophylls of leaf transections (Gausman et al., 1970a). Since air cavities develop with leaf growth, the VAI can be expected to vary from zero, when the leaf first unfolds or is compact, to a maximum value when the leaf reaches full growth.

Leaf Senescence

Senescence is deterioration in plant leaves, flowers, fruits, stems, and roots as they near the end of their functional life (Salisbury and Ross, 1969). In many perennial plants, the above-ground vegetation dies yearly, but the crown and roots remain alive. Leaves die on deciduous woody perennials, but stems and roots remain viable. Most herbaceous annual plants have a progressive senescence of their leaves from the older to the younger, followed by death of both stem and roots.

During leaf senescence, starch, chlorophyll, protein and ribonucleic acid (RNA) components are degraded, and catabolic products may be translocated to anabolically-active areas of plants. The fall coloration of leaves is caused partly by the unmasking of yellow and orange carotene and red anthocyanin pigments when the green chlorophyll pigments are lost. Leaf senescence is augmented by elevated temperatures, droughts, short photoperiods, and nutrient deficiencies.

As broad leaves senesce, their light reflectance usually increases markedly in the green visible light wavelength region peaking at 0.55 μm, because of chlorophyll degradation (Knipling, 1967). If an abundance of anthocyanin or carotene pigments are present after the chlorophyll has been lost, there may be relatively high reflectances in the red and near the blue region of the visible spectrum, respectively.

Leaf senescence decreased near-infrared light reflectance over the 0.75- to 1.35-μm wavelength interval in experiments conducted by Knipling (1967). The decrease in infrared light reflectance however, is not nearly as great as the increase in the reflectance of visible light. With some plant species, particularly leaves of forest trees and cereal crops, the infrared plateau at about 0.75 μm is reduced and rounded off considerably. This is characteristic of leaves with damaged cells (Cardenas et al., 1969–1970).

Salinity and Nutrient Levels

Salinity. The effect of salinity on internal structure of leaves is well documented. Morphological studies indicated that plants from natural or induced saline environments had thicker leaves, more developed palisade parenchyma, fewer chloroplasts, less chlorophyll, smaller intercellular spaces, and fewer stomata per unit area (Lesage, 1890; Harter, 1908; Chermezon, 1910; Wuhrmann, 1935; Hayward and Long, 1941; Uphof, 1941; Hayward and Bernstein, 1958). Salinization also depressed cell division in leaves of cotton (*Gossypium hirsutum*) (Strogonov, 1962), and reduced the rate of cell enlargement and of protein and RNA syntheses in bean leaves (*Phaseolus vulgaris*) (Nieman, 1965). Spectrophotometric studies have shown that individual cotton plant leaves affected by salinity have reduced reflectance and increased transmittance compared with unaffected leaves of the same chronological age (Gausman et al., 1969b). The salinity-stressed leaves are stunted with a more compact cell arrangement than non-osmotically stressed leaves.

Leaf maturity is very important in evaluating the influence of salinity on light reflectance of leaves. For example, in a field study, leaves were sampled from third and fourth nodes down from apexes of cotton plants to simulate what an overhead remote sensor would see. Third and fourth node leaves from osmotically-stressed plants were older, because such plants were not growing as rapidly as the nonstressed plants. In this case, the salinity-stressed leaves had higher reflectance and lower transmittance than nonstressed leaves.

Cotton plants affected by salinity appeared darker red on Kodak Ektachrome Infrared Aero 8443 (CIR) transparencies and prints compared with normal appearing plants (Thomas, 1970). This was caused by a higher chlorophyll concentration in high-salt than in low-salt leaves. Leaves with high chlorophyll concentration induced a darker red tone than leaves with low chlorophyll because high chlorophyll increased red light absorptance, decreased its reflectance (less radiation impinging on the film), and caused a more saturated image in the magenta dye layer which allowed less green light transmittance, and thus produced a darker appearance. Less chlorophyll caused higher red light reflectance, less magenta dye, and a lighter appearance. To understand this, the basics of CIR film must be reviewed. The CIR film has three image layers individually sensitized to green, 0.5- to 0.6-μm; red, 0.6- to 0.7-μm, and near-infrared radiation, 0.7- to 0.9-μm, instead of to blue, green, and red radiation as in Ektachrome film (Fritz, 1967). A yellow filter is used on the camera to absorb the blue radiation, to which all three of these layers are also sensitive. Upon processing, yellow, magenta, and cyan positive images are formed in the green-, red-, and infrared-sensitive layers, respectively. The overall impression to an observer viewing the finished print or transparency will depend upon which of the positive images in the dye layers predominate with respect to visual appearance.

Because the eye-sensitivity peaks in the green, the magenta layer generally contributes most to the subjective impression of lightness or darkness in a color print or transparency. For example, healthy leaves, with high near-infrared as compared with low near-infrared reflectance for unhealthy leaves, record red because a light-toned cyan image (less dense or less saturated) results, which allows the transmittance of more red radiation in the viewing.

Salinity-affected areas are usually easy to detect in cotton fields of the Lower Rio Grande Valley of Texas, although it is sometimes difficult to distinguish saline areas from cotton root-rot (*Phymatotrichum omnivorum*) areas (Nixon et al., 1975). Color Figure 33-39 shows a cotton field that has both saline and cotton root-rot areas. The healthy cotton plants appear magenta, saline areas are whitish, and root-rot areas are darker blotches with a sharper demarcation around their perimeters. This information has been useful to inform growers about the extent of their problems with salinity and root rot.

Nutrients. Thomas (1970) and Thomas and Oerther (1972) conducted studies to determine the feasibility of using spectrophotometrically measured diffuse reflectance from upper (adaxial) leaf surfaces of sweet pepper (*Capsicum annum*), cabbage (*Brassica oleracea*), and spinach (*Spinacia oleracea*) leaves to quickly estimate their N (nitrogen) status. Leaf absorptance in the visible region of the electromagnetic spectrum is primarily dependent on the concentrations of chlorophylls a and b and carotenoids (carotene and xanthophyll) in components (grana) of the chloroplasts. Ordinary green leaves absorb 75 to 90 percent of the light in the blue (about 0.45 μm) and red part (about 0.68 μm) of the spectrum. Absorptance is smallest in the wavelength region around 0.55 μm, where a reflection peak of usually less than 20 percent occurs. It was surmised that since N nutrition of plants markedly affects pigment concentrations and subsequent leaf color, limiting N would reduce pigment concentrations and therefore increase reflectivity because of decreased radiation absorptance. This tenet was substantiated: reflectance was inversely correlated with the leaf N content of three plant species. Regression equations were developed expressing reflectance as a function of the leaf N content of greenhouse-grown plants. These functions can be used to estimate the N content of field crops. With field-grown sweet peppers, for example, the difference between Kjeldahl-determined and reflectance-estimated N contents was less than 0.7 percent. Results indicated that leaf reflectance can be used to quickly estimate the N status of crop plants.

Thomas and Oerther (1977b) used aerial photography to evaluate the effects of N fertilizer on growth and yield of sugarcane (*Saccharum officinarum*) (Color Figure 33-40). Strong linear relationships between the optical density of CIR film, plant density, percent vegetation cover and sugarcane yields suggested that yields should decrease as does the ratio of transmission of light through film in stressed and nonstressed areas. Estimated yields were calculated as the product of the maximum or potential yield and the transmission ratio.

Remote sensing may be useful to determine some nutrient deficiencies. It was determined, for example, that multispectral data from Landsat-1 could be used to detect differences in chlorophyll concentration between chlorotic (iron deficient) and green (normal) grain sorghum (*Sorghum bicolor*) plants (Gausman et al., 1975). Band 5 data were used, representing the chlorophyll absorption band at 0.65 μm. Color Figure 33-41 shows that chlorotic sorghum areas that were 2.8 acres (1.1 hectare) or larger were identified on a computer printout of band 5 data. This resolution is sufficient for practical applications in detecting chlorotic areas in otherwise homogeneous grain sorghum fields.

Surveys

Crop Canopies

The comprehensive physiological studies of individual leaves contribute to an understanding of the processes involved. The interaction of electromagnetic energy with individual leaves, though complex, becomes increasingly so when one considers an assemblage of leaves in a crop canopy.

Parameters in addition to hemispherical leaf reflectance which may be very important in determining the image tone (reflectance) of a vegetation canopy include:

1) transmittance of leaves;
2) amount and arrangement of leaves;
3) characteristics of other components of the vegetation canopy (stalks, trunks, limbs);
4) characteristics of the background (soil reflectance, amount of leaf litter);
5) solar zenith angle;
6) look angle; and
7) azimuth angle.

Other plants such as weeds and volunteer plants from earlier year crops may also contribute to the reflectance.

Most vegetation targets are mixtures of different components, including leaves, other plant structures, background and shadow. These components are oriented at many different angles with respect to the source of incident radiation, so the irradiance on them varies. In addition, the projected area of each component illuminated and viewed depends on the solar zenith angle and the look angle and azimuth. The assemblage of vegetation components is the vegetation canopy. It is the bidirectional reflectance of the vegetation canopy (also called canopy reflectance) that determines the relative tone on remote sensing imagery collected in the short-wave (0.3–3.0-μm) part of the spectrum.

Once light strikes a crop canopy, reflectance, transmittance, scattering, and absorptance all influence disposition of the incident energy. Model and field studies by Myers et al. (1966b) and Colwell (1974) have shown that near-infrared spectrophotometer studies of single leaves can be very misleading for predicting reflectance from crops.

This is due to near-infrared light being transmitted through the top of the crop canopy, changes in light quality within a canopy, multiple internal reflections within the canopy, and reinforcement of reflectance from the top of the canopy. Clearly there are parameters other than hemispherical reflectance of individual leaves that are important in determining canopy reflectance and tone.

Reflectance from a crop canopy is a difficult measurement because:

1) O_2 and CO_2 and water vapor absorption reduce incoming solar radiation in certain wavelength bands;
2) illumination from the sun varies in intensity with numerous conditions;
3) radiance from field crops is affected by crop geometry, background soil reflectance, and other factors; and
4) the intensity of the sun peaks at about 0.5 μm, falling off rapidly at shorter and longer wavelengths.

Differences among crop species and dynamic changes due to growth, development, stress, and varying cultural practices also cause differences in the reflectance spectra of crops (Bauer et al., 1978a).

Interaction of Light with Plant Canopies. Spectrometer studies involving reflected radiation are of special interest to agriculturalists because the dominant environmental factors experienced in the field have very substantial influence on reflection spectra. Field spectrometer spectra have important contrasts with reflectance spectra from individual leaves measured on a spectrophotometer in the laboratory.

A comparison between the spectra from a cotton field, measured with a field spectroradiometer, and that from a cotton leaf using a laboratory spectrophotometer can be made, giving results as diagrammed in Figure 33-42. Field reflectance was obtained from a height of 300 cm over the crop canopy. Laboratory measurements of leaf-reflectance were obtained with a Beckman DK-2A spectrophotometer.

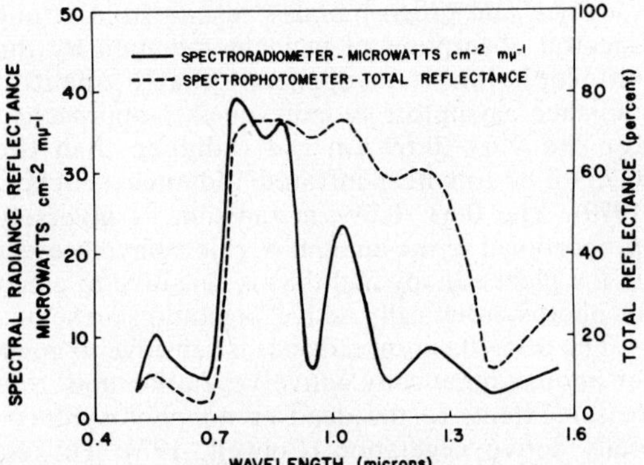

Fig. 33-42. Comparison of reflected energy from cotton, measured in the field with a spectroradiometer, and total reflectance from a cotton leaf, measured with a Beckman DK-2A laboratory spectrophotometer.

Bonner and Galston (1952) conducted research which shows that, in a plant or in an array of plants, light transmitted by the top layer of leaves is incident upon lower leaves, and so on. In general, as light passes through an assemblage of leaves, it is absorbed according to Lambert's law, and light intensity falls off exponentially with path length through the absorbing assemblage. A crop ordinarily produces enough layers of leaves so that the final light intensity that emerges at the soil level is very low.

Some radiation is scattered among leaves of a crop canopy by multiple reflection so that the reflectance (albedo) for the canopy as a whole is less than for single leaves and seldom exceeds 25 percent. The amount of scattering increases with the irregularity of the leaf surface and with solar elevation, because sunlight penetrates further into the canopy as the sun approaches the zenith. Monteith (1959) summarized measurements of reflectance obtained with a small solarimeter designed for field work. In Israel his measurements were made from a helicopter hovering about 10 m above the surface of the vegetation, but elsewhere the instrument was held or mounted at a height of about 1 m. Values of reflectance at solar elevations of 20° to 40° and 40° to 60° for several crops were, respectively, grass 27 percent and 24 percent, kale 30 percent and 26 percent, barley 26 percent and 23 percent, and beans 28 percent and 24 percent.

Davis (1957) has shown that the reflectance of grass varies with the altitude of the sun. His values varied from 22 percent at noon to about 43 percent at sunrise and 48 percent at sunset. Halstead (1957) found that the reflectance of plants also varies with color; for example, the values for green and dry grasses are reported as 15 percent and 25 percent, respectively.

A measurement of light interception by a cotton-crop canopy was made by Baker and Meyer (1966). Great change during the day was observed in the relative percent interception in all stands when the crop was young. Percent interception began to level off at a leaf area index of about 3.

Robertson (1964) described studies in which measurements were made of sky radiance under various conditions and of light under the vegetative canopies of several different crops. The spectral composition of light was determined for five different types of vegetative canopies.

De Wit (1959) developed a method of calculating growth as a function of the incident light energy. One of the assumptions in the method is that the reflection and absorption of light by crop leaves is independent of angle of incidence, and that the leaf orientation is random. Solar radiation has a changed spectral composition after transmission through vegetation and, therefore, the fraction of light reaching the ground beneath the crop will differ from the fraction of solar radiation transmitted. Stanhill (1962) found that the extent of the difference in the quality of transmitted solar radiation will depend on the proportion of radiation transmitted through the leaves and the proportion that reaches the ground as unaltered sun flecks pass through gaps in the crop canopy.

The ratio of reflected to incident radiation is known as the albedo or reflectivity of the surface. Landsberg and Blanc (1958) measured representative values of the albedo of various surfaces as follows: high dry grass, 0.32; green grass, 0.20; planted fields, 0.15; wet sand 0.09; forest (dense), 0.07; water surface (average), 0.06; desert sand, 0.25.

Reifsnyder and Lull (1965) have summarized albedo of some natural surfaces from Russian studies as follows: fresh dry snow, 0.80 to 0.95; dry light sandy soils, 0.25 to 0.45; dry clay or gray soils, 0.20 to 0.35; moist gray soils, 0.10 to 0.20; dark soils, 0.05 to 0.15; meadows, 0.15 to 0.25; rye and wheat fields, 0.10 to 0.25; deciduous forest, 0.15 to 0.20; coniferous forest, 0.10 to 0.15.

Plants affected by drought, disease, and other factors may also influence reflectance from crop canopies. In studies of plants affected by drought, Molga (1962) states that the upper leaves of the plant, which are the ones detected by aerial sensors, stay in good shape the longest by drawing water from leaves positioned lower, the result of which is that the latter are first to wilt or dry up during a drought. Thomas et al. (1966) conducted field and greenhouse experiments to determine the effects of plant height, percentage of ground cover, and soil salinity on the spectral characteristics of cotton. Reflectance of individual leaves was affected by leaf age, moisture content, nitrogen fertilization, and salinity.

Even though the outward appearance of plant leaves may remain much the same, many influences of man and nature cause physiological and ecological changes to take place. Many changes occur that are not readily apparent. The remarkable adaptability of plants to their environment, as well as the ability of plants to make compensation for many influences, frequently brings about alteration of the plant mechanisms. This, in turn, influences the relationship among the various amounts of reflected, transmitted or absorbed light.

In agriculture and rangeland management there is a continual effort to achieve efficient management of our renewable resources. Vegetation conditions are dynamic and correct appraisal of conditions is essential for forecasting trends. The unique characteristics of space imagery provide the means for measuring vegetation characteristics on a spatial, spectral and temporal basis. The change in the field condition becomes the major clue of identification when use is made of multidate imagery in conjunction with phenological changes in a region.

Methods for utilizing remote sensing for evaluating the presence and condition of vegetation have been developed. Landsat imagery as color composites and MSS band 5 alone, at a convenient scale of 1:125,000 or smaller, shows the general distribution and condition of vegetation. Imagery from different seasons can sometimes provide additional information. (See later sections for vegetation applications involving Landsat imagery).

Measuring and understanding the effects of less than all of the important parameters may prove unsatisfactory to explain canopy reflectance and its variability under changing conditions. Individual parameters may be quantitatively much more important in one situation than in another. In addition, certain parameters may be positively correlated with each other (mutually amplifying) in one situation and negatively correlated (dampening) in a different situation.

Soil-Plant Interactions and Vegetation Indices. Classification of physiographic areas using aircraft and satellite multispectral scanner data in crop and yield estimation has been hampered by many site factors. Levels of accuracy are influenced by soil background reflectance, agricultural diversity, physiographic region, climate and soil type. Spectral data have been shown to relate to vegetation density indicators such as plant density, leaf area index and biomass. Temporal and spatial extension of signatures continues to be a problem that may be minimized by application of procedures that account for some of these variables. This will advance the time when operational use of Landsat data may become routine for agricultural monitoring.

Since the advent of the space applications program, substantial research and development has been directed toward establishment of various indices for identification and quantification of resource phenomena such as crop development, crop identification, soil moisture changes, effect of background soil radiance on vegetation, reflectance, etc. Rouse et al. (1974), Wiegand et al. (1974), Kauth and Thomas (1976), Richardson and Wiegand (1977), Lautenchlager and Perry (1981), and Wiegand and Richardson (1982). Hielkema (1978) and Tucker (1979) summarize indices developed recently which have the greatest theoretical and practical value.

The relationship between the 0.63–0.69-μm radiance and green biomass results from strong spectral absorption of incident radiation by the chlorophylls. It is apparent that a spectral radiance asymptote is more quickly approached for the 0.63–0.69-μm red radiance than the 0.75–0.80-μm near-infrared radiance (Tucker, 1979). The 0.63–0.69-μm radiance is inversely proportional to the amount of chlorophyll present in the plant canopy and thus is sensitive to green or photosynthetically active vegetation present.

The 0.75–0.80-μm radiance is sensitive to green or photosynthetically active vegetation and, to a lesser extent, to the dead or nonphotosynthetically active vegetation (Colwell, 1974; Tucker, 1977c, 1978).

The relationship between the 0.75–0.80 μm near-infrared radiance and biomass results from the lack of appreciable spectral absorption in the 0.74–1.20 μm region and the high degree of intra- and interleaf scattering in the plant canopy. In the absence of spectral absorption, proportionally more incident spectral radiance escapes from the canopy than is absorbed. Thus the spectral radiance in the 0.74–1.20 μm region is said to be enhanced or increased over the level of radiance of the background material.

The earliest experiments involving transformation used the IR/red ratio (Jordan, 1969; Pearson and Miller, 1972; Colwell, 1973, 1974; Carneggie et al., 1974; Rouse et al., 1973, 1974. Rouse et al. (1973, 1974), then developed a transformation called the vegetation index (VI) where:

$$VI = \frac{MSS\ band\ 7 - MSS\ band\ 5}{MSS\ band\ 7 + MSS\ band\ 5} \quad (33\text{-}2)$$

Later a transformed vegetation index (TVI) was used to avoid working with negative ratio values:

$$\mathrm{TVI} = \sqrt{\mathrm{VI} + 0.5} \quad (33\text{-}3)$$

More recently, the TVI was modified by using MSS band 6 instead of MSS band 7. Tucker (1977c) explains this as due to an apparently greater soil-green vegetation contrast in MSS band 6 than in MSS band 7.

Richardson and Wiegand (1977) developed a perpendicular vegetation index (PVI) that accounts for soil background variations:

$$\mathrm{PVI} = \sqrt{(\mathrm{Rgg5} - \mathrm{Rp5})^2 + (\mathrm{Rgg7} - \mathrm{Rp7})^2} \quad (33\text{-}4)$$

where:

- PVI is the perpendicular distance between the candidate vegetation point and the soil background line,
- Rp is the reflectance of a candidate vegetation point for Landsat bands MSS 5 and MSS 7, and
- Rgg is the reflectance of soil background corresponding to a candidate vegetation point.

Kauth and Thomas (1976) developed a technique for transforming Landsat MSS information in four-dimension data space using the four MSS bands. From this, a soil-brightness index (SBI) and green-vegetation index (GVI) were calculated as follows:

$$\mathrm{SBI} = 0.43{*}\mathrm{MSS\ 4} + 0.63{*}\mathrm{MSS\ 5} + 0.59{*}\mathrm{MSS\ 6} + 0.26{*}\mathrm{MSS\ 7} \quad (33\text{-}5)$$

and

$$\mathrm{GVI} = -0.29\ \mathrm{MSS\ 4} - 0.56\ \mathrm{MSS\ 5} + 0.60\ \mathrm{MSS\ 6} + 0.49\ \mathrm{MSS\ 7}. \quad (33\text{-}6)$$

Note that all the SBI independent variable coefficients are positive, while the GVI independent variables, MSS 4 and MSS 5, are negative. The SBI establishes the data space of soils and the GVI departs from it, in a negative or absorptive fashion with MSS 4 and MSS 5, approximately the same coefficients for MSS 6 for both models, and a positive departure for MSS 7.

The foregoing indices were developed using Landsat data. The reduced atmospheric scattering from space in bands 5 and 7 accounts for the use of the longer wavelength MSS bands. Results obtained by Kanemasu (1974) using a ground handheld radiometer showed that the 0.545 and 0.655 μm bands gave good results regardless of crop type. This indicated that monitoring vegetation from aircraft altitudes might be done successfully with bands in shorter wavelengths.

The use of vegetation indices for crop monitoring, range forage estimates, drought detection and other purposes is proving to be a powerful tool in providing rapid quantitative information on crop and range vegetation anomalies, with a minimum of human interpretation. However, variable results obtained by investigators indicate that it may be difficult to predict which, if any, vegetation index is superior to others in any particular circumstance.

Hielkema (1978) points out that one of the main problems for the effective use of machine-generated vegetation indices for repetitive large area monitoring is the lack, at present, of adequate correction procedures for temporal and spatial variations in target spectral signatures caused by external factors. The sun angle and haze corrections developed by the Environmental Research Institute of Michigan (ERIM) are techniques which address this problem.

Vegetation indices involving temporal studies, either diurnal or seasonal, must take into account the dependence of spectral reflectance on solar elevation. Jackson et al. (1979) illustrated this by planting wheat in three plots, one plot a solid canopy, a second with north-south rows and a third east-west, the latter two with rows planted 0.3 meter apart. Figure 33-43 shows the ratio MSS 7/MSS 5 as related to solar elevation for the three plot conditions. The decrease of IR and the increase of visible bands will change markedly with solar angle as illustrated in the figure. The E-W plot showed a relatively small decrease with increasing sun angle; however, the N-S plot showed a considerable decrease. The crossover point occurred at about 49° solar elevation. The corresponding azimuth angle was about 130°. The crossover was near the Landsat overpass time, which leads to the conclusion that band ratios of Landsat data may be similar for E-W and N-S rows when the solar azimuth is near 130°. Under those conditions the ratios would be near a minimum for rows oriented northwest to southeast and near a maximum for rows oriented northeast to southwest.

Phenology. Comparison of data interpreted on air and space imagery taken at different times, or at intervals of several years, serves as a criterion for observation of the dynamics of vegetation and the environment (Vinogradov, 1977). On succes-

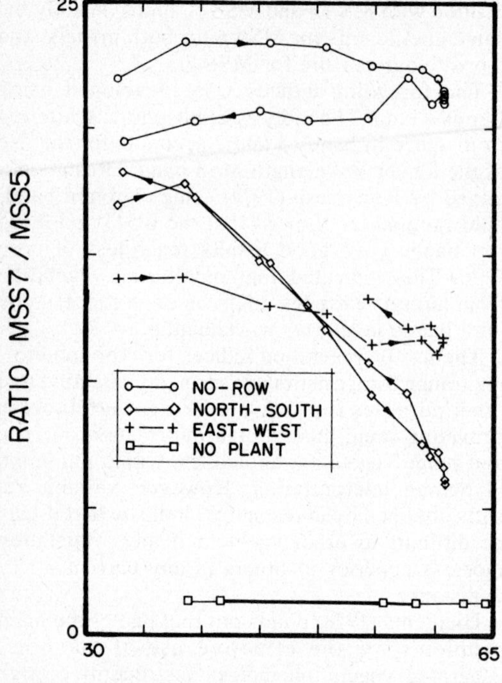

Fig. 33-43. The ratio MSS 7/MSS 5 as related to solar elevation for a dense no-row plot of wheat, for a north-south and an east-west plot at 0.3 m row spacing, and for a no-plant plot. The three wheat plots were at the same growth stage (Jackson et al., 1979).

sive space imagery, various dynamic changes in vegetation are revealed, caused by overgrazing, changes in land use, climate-induced arid conditions, atmospheric pollution, irrigation-systems construction and others. Thus, dynamic models of changes in the vegetation and environment, prepared from repeated ground, air and space imagery and measurements can be extrapolated for predictions over a number of years.

A major trend in remote sensing methods in ecological botany is the determination of rhythm and dynamics of vegetation. Phenological observations were conducted by Dethier et al. (1973) which included repeated ground, aircraft and satellite phenological measurements along the principal meridianal corridors of the USA and Canada. Two phenological rhythms were observed: "the green wave"—spring or early summer emergence of foliage sprouting in agricultural communities and growth of pasture vegetation; and "the brown wave"—late summer or autumnal ripening and harvesting in agricultural communities, disappearance of vegetation, and drying of pasture lands.

Park (1979) reported on a study entitled LAMP (Landsat Agricultural Monitoring Program) which was a research effort that focused on the problems of designing a capability to react quickly to a reported agricultural event which was expected to seriously impact production. The preferred approach was that of acquiring coverage of a reported event after the fact by comparing the acquired Landsat scene of the area against a previously acquired Landsat data base. It was somewhat difficult to implement because it was difficult to collect a "normal" data base.

Park points out that in a recent research report dealing with a simulated Thematic Mapper (TM), data taken over winter- and spring-wheat, oats, barley, alfalfa, and pasture show that the following stages of growth can be determined for cereals:

1) bare soil (pre-planting);
2) emergence;
3) booting or pre-bud (pre-heading);
4) heading;
5) ripening; and
6) harvest (pre-harvest).

Correlation of spectral signals from successive photographs of identical areas of the vegetative ground cover reveal optical shifts in the imagery of the latter (Vinogradov, 1977). In order to separate out the optical shifts caused by phenological changes in the vegetation and to make these changes independent of sun angle, use is made of the coefficient K. Vinogradov calls this the zonal brightness of the vegetation, given by

$$K = \frac{R7 - R5}{R7 + R5} \qquad (33\text{-}7)$$

where:

$R5$ = reflectance in the red region of the spectrum;

$R7$ = reflectance in the near-infrared region.

This coefficient is used elsewhere as a vegetation transformation for detection of growing vegetation and for estimating biomass (see equation 33-2). The seasonal trend of the regional brightness coefficient illustrates well the phenological rhythms of the vegetation; K decreases in autumn and winter, and increases in spring and summer. Improved identification can be obtained by stratifying agricultural areas by selected parameters, by using temporal imagery, by using natural phenological indicators, through greater knowledge of local agronomic practices and by other methods. The stratification technique was found to be useful in an Australian study by Parson and Fitzpatrick (1978) in which they mapped areas of wheat. They found that success was dependent on stratification of the total survey area into smaller areas in which crop phenology, terrain type and soil type and agricultural practice are relatively uniform.

Biomass. Spectral methods for estimating biomass involve the measurement of reflected spectral radiance which results from the interaction between the incident solar spectral irradiance and the vegetation canopy (Hielkema, 1978). Fundamentally canopy reflectance is due to:

1) optical properties (reflection and transmittance) of leaves and other canopy components;
2) reflection of soil background;
3) canopy geometry—leaf angle distribution and leaf area index;
4) illumination angle, atmosphere, etc.

Wavelength regions generally used for spectral biomass estimations are the red (0.63–0.70 μ) and near infrared (0.75–1.00 μm) wavelengths. The red spectral interval corresponds to the region of maximum chlorophyl absorption. The near-IR spectral interval corresponds to maximum reflectance of incident light by living vegetation. Hielkema states that spectral biomass-estimation techniques have been found to be accurate for low to medium biomass quantities, but are of little value for biomass values over 5000 kg/ha. The biomass- and primary-productivity-related parameter obtained by the spectral method is the photographic IR/red ratio or some similar transformation.

The IR and red radiances exhibit important relationships with respect to biomass (Tucker, 1979). The red radiance exhibits a non-linear inverse relationship between integrated spectral radiance and green biomass, while the near-infrared component exhibits a non-linear direct relationship.

Discrimination of vegetation biomass is strongly dependent upon the soil surface-vegetation spectral reflectance or radiance contrast. For this reason, some wavelengths are far superior to others for discrimination of green vegetation biomass. The green region has a low soil-vegetation reflectance contrast. This results from the fact that chlorophyll is slightly absorptive in the green region (absorption coefficient $\simeq$ 10), while much more absorptive in the red region (absorption coefficients of $\simeq$ 40–90). Figure 33-44 shows red and IR radiance plotted against total wet biomass (Tucker, 1979). It can be concluded that red and IR bands are superior for estimating biomass.

Crop Identification

Reflectance Detection. The estimation of crop acreage, recognition of crop stress, and timely and accurate prediction of crop yield are matters of critical interest everywhere even though such data are particularly difficult to obtain in developing countries (National Academy of Science, 1977). Perhaps no information is more basic for agricultural planning, export-import negotiations of agricultural commodities, and the making of yield predictions, than data on crops being grown in a region or a country. The major problem encountered by the photo interpreter in studies of crops is the development of techniques which provide both speed and accuracy in field mapping.

Studies aimed at perfecting techniques for identifying crops and estimating acreage and yield of crops have intensified with improvements in the technology and the increased availability of space imagery. The procedures for identifying crops and for estimating yields, utilizing remote sensing procedures, are frequently complex. Yet, accuracy of crop identification with present Landsat data has been reported as being 90 percent or higher in studies of areas where there are large, homogeneous, rectilinear fields with few competing crops, such as irrigated rice in California, potatoes in New Brunswick, Canada, oilseeds in western Canada and bare ground being readied for wheat in autumn in Kansas, Oklahoma, and Texas. The accuracy of Landsat data for crop identification in certain U.S. wheat-growing areas

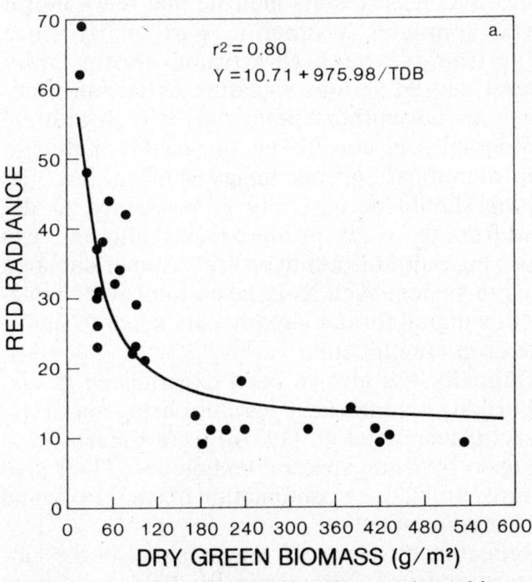

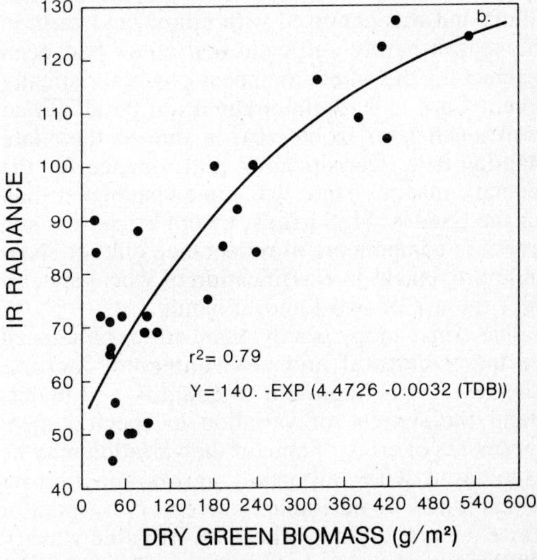

Fig. 33-44. Radiance plotted against total wet biomass for the (a) 0.63–0.69 and (b) 0.75–0.80 μm intervals for the June data. Similar results were obtained for total dry biomass, leaf water content, total wet biomass, and total chlorophyll content for this sampling time. The total wet biomass was predominantly green and contained little dead vegetation (Table 1).

has been studied in various LACIE and AgRISTARS experiments. Relatively few areas in other parts of the world, however, are structured as simply as wheat fields in the United States. In the developing countries cropland frequently is interspersed with noncropland, fields are small and irregularly shaped, and numerous crops have similar spectral responses. In such complex environments a single Landsat image may not provide enough data to be useful for purposes of crop identification.

The application of remote sensing to crop inventories relies upon the ability to detect and identify the crops of interest from spectral data. Unless correct crop identification can be consistently made, the goal of accurately predicting crop acreage and production cannot be achieved. Bauer et al. (1978) and Hixson et al. (1978) studied some of the factors which contribute to inaccuracies in crop identification in intensive studies in Kansas and North Dakota. Spectral plottings of the data were made in two ways: (1) several crops on the same measurement date, and (2) a single crop over several dates. The plots gave qualitative indications of what crops may be separable at any given time and also show changes in a crop over the growing season (Hixson et al., 1978). Analyses for crop identification were conducted using simulated Landsat MSS and thematic mapper wavelength bands. The analyses included data from the Williams County, North Dakota, and Finney County, Kansas, intensive test sites. Spring wheat in North Dakota was confused with pasture by the Landsat MSS bands when only single date information was used. The two cover types were separable when multitemporal information was used. The training method used was found to have a greater effect on classification results than did the amount of training. Winter wheat in Kansas was occasionally confused with alfalfa and was confused with fallow land early in the season before sufficient soil cover had been reached for the wheat to appear characteristically green. Corn and sorghum could not be identified until much later in the season due to their late planting date. Classification performance for the thematic mapper bands was somewhat higher than for the Landsat MSS bands. Use of brightness and greeness components of reflectance did not show an improvement in classification of wheat spectra over the use of two Landsat bands.

The crop canopy is a dynamic entity influenced by many cultural and environmental factors. Therefore, it is important to quantify and understand the sources of variation in spectral measurements of crops. Some of the variation may be associated with important agronomic factors which it may be desirable to inventory or monitor (for example, dryland wheat vs. irrigated wheat). On the other hand it is also important to know the magnitude of variation associated with a factor such as cultivar which we would most likely not want to identify or monitor.

Examples of wheat spectra acquired in 1976 at the Williston, North Dakota, Agricultural Experiment Station are shown in Figure 33-45 to illustrate some of the effects of agronomic treatments on the spectral response of spring wheat (Bauer et al., 1978a). Further experiments using an improved experimental design were conducted in 1977 to determine the effects of the various agronomic treatments (soil moisture availability, planting date, nitrogen fertilization, and cultivar) on the reflectance of spring wheat canopies. It was found that, early in the growing season, planting date is the primary agronomic factor influencing the reflectance of spring wheat.

The spectral differences are primarily due to differences in the amount of vegetation present. Later in the season, at the heading to ripening stages, the level of soil moisture becomes the most important factor. Wheat grown on land with higher levels of available soil moisture had a greater percent soil cover, leaf area index, and biomass causing increased near-infrared reflectance and reduced visible reflectance. In these experiments, cultivar and nitrogen fertilization had relatively little effect on the spectral response of spring wheat. The primary difference in the two cultivars was in plant height, rather than in leaf area or biomass. The addition of nitrogen fertilizer had only minor effects on the growth of wheat because the soils of this area of North Dakota are relatively high in nitrogen supplying capacity.

The uniformity of Landsat digital spectral data may avoid distortion of a type that is frequently characteristic of photography. Jensen et al. (1978) compared the utility of multidate crop classification using microdensitometer-scanned color-infrared high-altitude photography (original scale 1:120,000) and Landsat digital data for a 140-square-kilometer study area in Kern County, California. Their results indicate that the Landsat digital approach is superior, particularly since vignetting in the high-altitude-photography dataset caused serious signature extension problems. As the authors point out, if high altitude photography is considered for digital multidate crop identification, any image with serious vignetting should be carefully preprocessed or deleted from the study. A more logical alternative is to use high-altitude multispectral scanner data or a satellite system such as Landsat to provide imagery in a digital format already conducive to multidate crop identification.

Difficulty has always been experienced in distinguishing among citrus varieties using aerial imagery. Gausman et al. (1977d) were successful in doing so by using spectral techniques. Their procedures utilized the combination of laboratory and field measurements.

Reflectance spectra for single leaves in the laboratory and tree canopies in the field were measured with a spectroradiometer and aerial infrared-color photos were taken for three citrus varieties [Valencia and Marrs oranges (*Citrus*

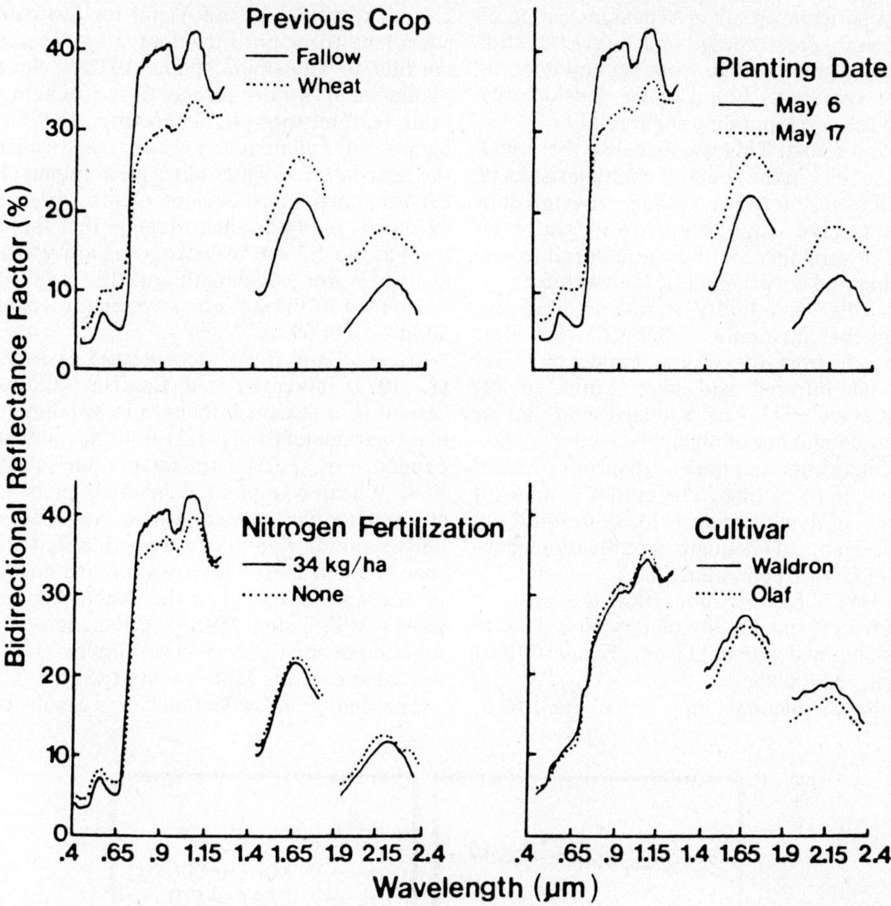

Fig. 33-45. Effects of agronomic treatments on the spectral reflectance of spring wheat. Spectra were measured on June 18, 1976, during the stem extension stage of development, except for the spectra of cultivars which were measured on July 16 after heading. (Bauer et al., 1978).

sinensis (L.) Osbeck) and Redblush grapefruit (*Citrus paradisi* Macf.)] to determine whether they were distinguishable based on their reflectance and photographic characteristics.

Redblush had the highest and Marrs had the lowest field- and laboratory-measured visible (0.5 to 0.75 μm) reflectance; Redblush and Valencia had the highest and lowest near-infrared (0.75 to 2.5 μm) reflectances, respectively. Differences in reflectance among the varieties were larger for field measurements of trees than for laboratory measurements of single leaves over the entire 0.5 to 2.5 μm waveband.

Crop identification of citrus varieties has been particularly difficult using Landsat imagery. Gausman et al. (1977c) used Landsat-1 imagery to distinguish between Redblush grapefruit (*Citrus paradisi* Macf.) and orange (*Citrus sinensis* (L.) Osbeck) citrus varieties and to estimate their hectarages satisfactorily. Accordingly, Landsat-1 MSS data for a December 11, 1973, overpass were used in conjunction with imagery analysis of Productive Properties' 600 ha citrus farm in Hidalgo County, Texas. Computer-aided variety classification accuracies for the farm, using MSS data, were 83 percent, 91 percent, and 86 percent for Redblush grapefruit, orange, and total hectarages, respectively. The percentage comparisons of computer and farm manager's farm inventory estimates for Redblush grapefruit, orange, and total hectarages were 16.9 percent underestimate, 13.9 percent overestimate, and 2.4 percent underestimate, respectively. These classification and hectarage comparison accuracies indicate that there is a good potential for computer-aided inventories of grapefruit and orange citrus orchards with satellite MSS data. This projected use will become more realistic with further refinements in MSS ground resolution, and data acquisition and processing.

Many investigators of separability of agricultural crops have concentrated on Landsat MSS bands because no other means for band separation were available. More recently Ahern et al. (1979) have evaluated the Landsat-D TM bands using simulated data and have found a significant improvement in crop discrimination over the Landsat MSS bands. Kumar (1977) evaluated twelve spectral channels in the visible, near infrared, middle infrared and thermal infrared for discriminating agricultural cover types consisting of corn, soybeans, green forage and forest. Overall separability of green forage from several agricultural cover types was found to be considerably lower than the corresponding separability of corn, soybeans and forest. This was because there was natural variability in the spectral characteristics of hay as well as pasture. The author, studying data from the twelve channels, found that the maximum separability of the agricultural cover types is obtained by using all of the twelve channels. Yet, this separability is not significantly better than the separability achieved when four spectral bands from the visible, reflectance, infrared, middle infrared, and thermal infrared, respectively, were employed. Kumar found that an increase in the number of channels used in a classification algorithm requires a disproportionate increase in computer time. The author concluded that a subset of five channels is likely to fulfill the dual requirements of adequate classification accuracy and moderate computer time.

Kumar (1977) further states that the greatest overall statistical separability of agricultural cover types was obtained with data from channel 7 (0.61 to 0.70 μm, red channel).

Sensor bands encompassing either the 1.6- or 2.2-μm wavelengths are useful for distinguishing succulent from nonsucculent plant species according to Gausman et al. (1978a). Succulent plants have water-storage tissue developed in their leaf mesophyll. Therefore, they have a higher water content and absorb more radiation in the near-infrared water absorption region (1.35 to 2.5 μm) than nonsucculent plants (Allen et al., 1970b). It is important to realize that the energy level at the 2.2 μm wavelength is lower than that at the 1.6 μm wavelength and that a sensor that responded to the 2.2 μm wavelength would need additional cooling.

Effect of Sun Angle. Recent work by Jackson et al. (1979) indicated that Landsat studies concerned with seasonal changes in remotely-sensed plant parameters must take into account the dependence of spectral reflectance on solar elevation. Wheat was planted in three plots for this study. One plot termed no-row was planted to a dense stand; two plots were planted in rows spaced 0.3 m apart with rows running north-south in one and east-west in the other. Figure 33-46 shows MSS 7 and MSS 6 reflectance data as a function of solar elevation and figure 33-47 shows the same data for MSS 5 and MSS 4.

The change in reflectance was small for bare

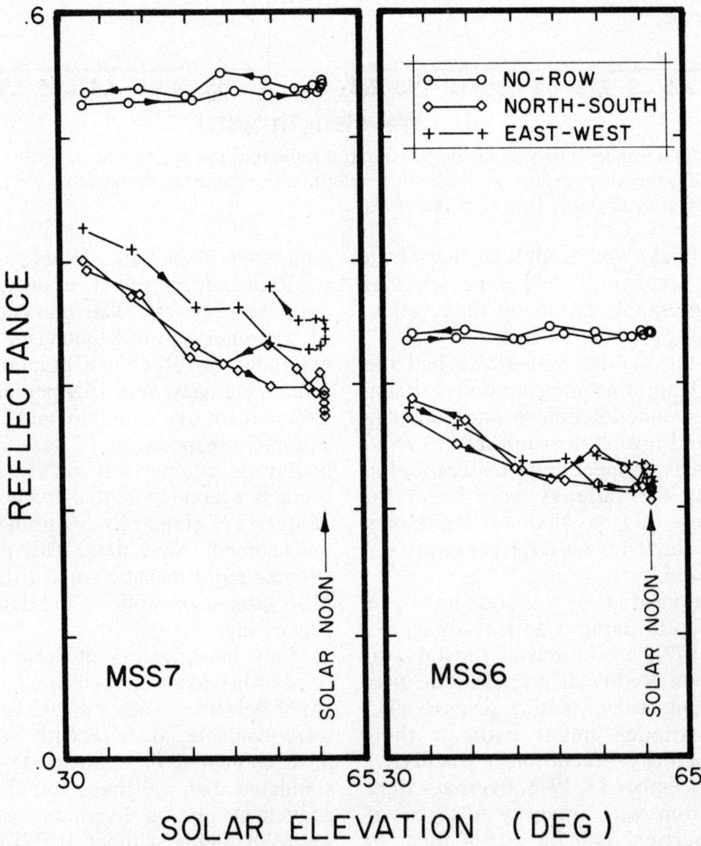

Fig. 33-46. Reflectance in bands MSS 7 and MSS 6 of wheat at the same growth stage but with different row spacing and row direction for various solar elevations.

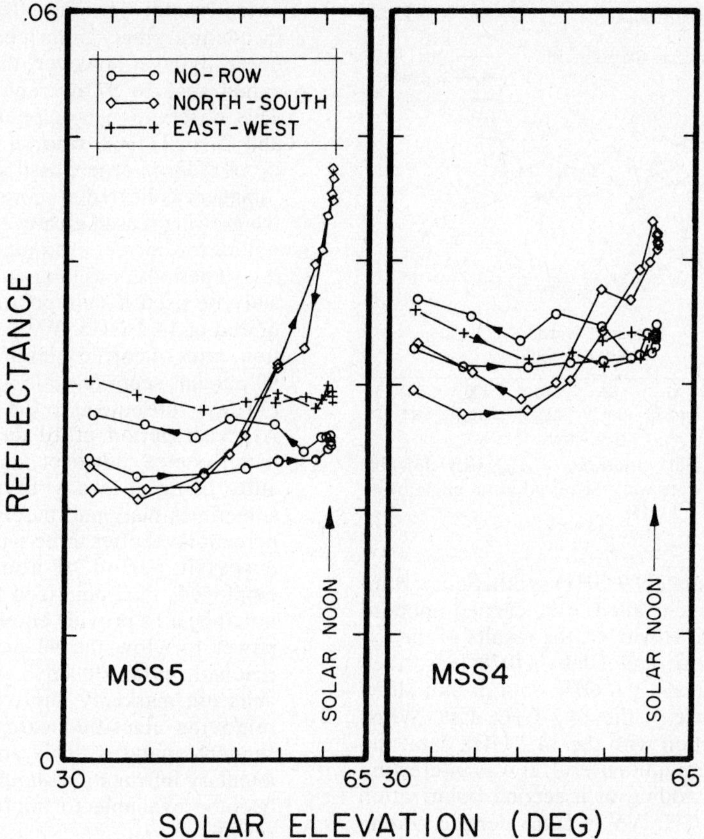

Fig. 33-47. Same as Figure 33-46 but for bands MSS 5 and MSS 4.

soils and for dense plant canopies with change in solar angle. However, for incomplete canopies with row structure typical of most wheat fields, the reflectance can be highly dependent on solar elevation and the difference between azimuth and row direction. Considering the rather large changes in reflectance factors and ratios that were observed for a single variety at the same stage of growth, the authors concluded that crop configuration is a major determinant of spectral reflectance of wheat, possibly sufficient to obscure varietal differences, and that it should not be neglected in interpreting multidate imagery that spans a significant time period.

Radar Detection. A number of researchers have recently focused attention on the development of sensors and data-interpretation techniques for aiding in providing more accurate and timely crop inventories than possible using traditional inventory methods. The problem of avoiding the effects of cloud cover has plagued many of these efforts which use optical sensors for gathering data.

Radar, operating as it does in the microwave frequencies, offers a solution to the cloud-cover problem in that radar-image acquisition is relatively independent of cloud cover. Although several studies have been conducted to evaluate the value of radar as a crop classifier, Bush and Ulaby (1978) were among the first to quantitatively determine the information content of the temporal aspect and the significance of the choice of sensor parameters, particularly frequency. Their studies included fields sown in corn, milo, soybeans, wheat and alfalfa. The data base developed from the studies contained angular, polarization, spectral and temporal dimensions.

As an example, Figure 33-48 presents the temporal response of $\sigma_{VV}°$ (dB) of the five crop species as measured at 14.2 GHz at an off-nadir angle of 50°.

To aid in determining which of twelve (six frequencies × two polarizations) measurement dimensions provided the most discriminating power, F-ratios (between-class to within-class variance) were formed on 24 days spaced by five days over the 120-day test period. The most striking feature of their results was that, regardless of the frequency chosen, data acquired with a system having a vertical transmit, vertical receive (VV) antenna configuration had larger F-ratios on the average than did the corresponding HH data. The 14.2 GHz, VV dimension was deemed best for crop classification purposes.

Currently the Space Shuttle Orbiter is available for space-related research activity. In addition to the L-band system, which has already flown, an

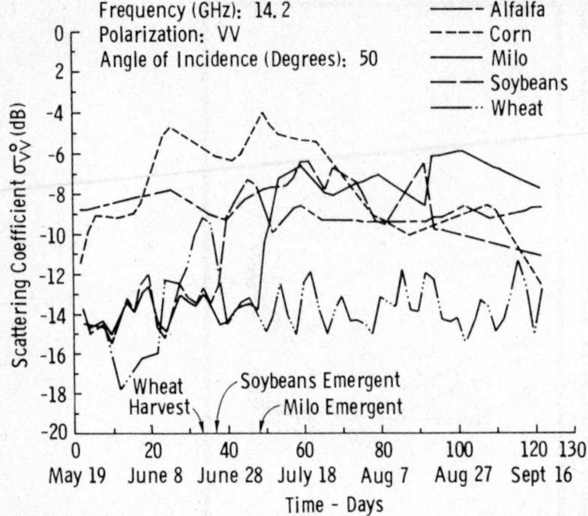

Fig. 33-48. Temporal variations of $\sigma_{vv}{}^{\circ}$ (dB) for five crop types. These data were acquired at an angle of incidence of 50° at 14.2 GHz.

X-band (approximately 9 GHz) synthetic-aperture imaging radar is scheduled to be carried onboard on several flights. However, the results of the experiments by Bush and Ulaby (1978) described here indicate that the 9.0 GHz data do not show nearly the promise of the 14.2 GHz data. When used in combination with the 14.2 GHz data, the 9.0 GHz data seem quite useful. It was interesting to find that the adding of a second polarization channel to 14.2 GHz VV is equivalent to adding another frequency channel at 9.0 GHz so far as crop classification is concerned. From a practical standpoint, the dual polarized, single frequency system must be considered preferable to the dual-frequency system. The following conclusions were reached by the authors:

1) data in the 13–16 GHz band seem to contain the greatest discriminating power. Moreover, VV data are better suited for a classification task than are HH data, regardless of the frequency (between 8 and 18 GHz) chosen. This study found 14.2 GHz, VV to be optimum for crop classification purposes. If an additional frequency/polarization configuration is available, the 9.0 GHZ, horizontal transit, horizontal receive (HH) configuration seems most suitable (of those configurations tested) to use in combination with the 14.2 GHz, VV configuration. Addition of an HH polarization channel to the 14.2 GHz, VV channel, however, provides the same improvement as the addition of the 9.0 GHz HH channel;
2) multi-date data must be employed in the classification process if rates of correct classification in excess of 90 percent are to be obtained. Through the testing of data containing variance due only to scintillation, for example, it was found that the average daily rate of correct classification was only 82.7 percent using dual-frequency, dual-polarized data. Through the use of single-frequency singly-polarized data acquired every 10 days, however, the rate of classification rose to 97 percent within 30 days. This observation is supported by Schwarz and Caspall (1968) who, in studying the utility of radar in crop classification, noted that "imagery collected at more than one time in the growing season greatly enhances the probability for correct crop identification"; and
3) revisit periods of as long as 15 days may possibly be used if dual-polarized data are acquired at 14.2 GHz. With such a configuration, rates of correct classification exceeding 90 percent seem feasible after three revisits (30 days subsequent to the first sample day). A revisit period of 10 days is probably a better choice, however, as classification results using a 15-day revisit period were sometimes marginal, barely reaching the 90 percent level after three target visits. Unless a revisit period of about five days is employed, dual-polarized 9.0 GHz data do not appear to provide enough discrimination power to allow the 90 percent level to be reached within 30 days. Classification results are markedly improved, however, by removing classified categories from subsequent analyses. This and other types of ancillary information should be used as they become available to improve classification results.

Ahern et al. (1979) demonstrated that seven forage classes could be separated to an average accuracy of 79.8 ± 9.6 percent, using a 13.3 GHz polarization scatterometer. They also showed that accuracies increased when data from an optical sensor were included in the analyses.

AGRICULTURAL RESOURCE SURVEYS AND INFORMATION SYSTEMS

Resource Surveys and Interpretations

Resource managers and decision makers are in need of logical methods for collecting, recording, and interpreting data for the development and management of our natural resources. Remote sensing offers appreciable advantages over current methods of conducting resource surveys.

In undertaking extensive studies involving natural resources, it is usually desirable to accomplish a task of great magnitude in a relatively short time. Progress is being made in many countries toward completing natural resource and related surveys using traditional methods. These surveys pertain to soils, vegetation, geology, demography, and other resource-related attributes. However, completion of these surveys is scheduled for many years in the future. The same situation exists even in the United States where activity on similar surveys has been underway for many years.

Landsat imagery has proven useful for accelerating reconnaissance resource surveys. For reconnaissance-level studies Landsat imagery provides the advantages of synoptic, near orthographic and repetitive coverage, and is useful for complementing more detailed photographic studies from aircraft. The high resolution required for urban planning is generally not required for reconnaissance-level planning with respect to agricultural resources. In developed countries, planners are interested in frequent updating of broad-scale changes in the nature of land cover. In developing countries, large area resource inventories are vitally needed in planning for utilization of resources. In either case, savings in time and cost by having accurate reconnaissance maps are realized by locating regions of highest potential for development and information needs. The reconnaissance-level information is used to plan and carry out detailed surveys, thus maximizing returns from frequently limited financial resources.

Soil associations, including gross physical characteristics, landforms, and land-use potential and limitations, can be fairly rapidly delineated using Landsat imagery. Vegetation information that can be monitored includes extent of vegetative communities, productivity, biomass, vulnerability to and extent of damage, evidence of mismanagement, and gross changes in land cover. Soil erosion and erosion hazard can be monitored by satellite. The presence of shallow groundwater, high water tables and salinity can be detected using direct and indirect indicators of these conditions. The occurrence of soil moisture, which is important to the development and utilization of resources, can be detected under favorable conditions. Future satellite systems with improved and/or additional sensors will improve the opportunity for sensing resources and their limitations.

Sudan Resource Studies

A reconnaissance resource inventory of the 160,000 km^2 Sudd region of the upper Nile River in southern Sudan was completed in 1977 using Landsat imagery (Abdel Hady et al., 1978; Worcester and Moore, 1978).

Portions of 14 Landsat scenes were photographically contrast-enhanced and tone-matched to prepare 1:500,000 scene-matched mosaics of wet- and dry-season black-and-white (band 7) and false-color infrared imagery (Best and Smith, 1978). The mosaic was used as a base map to summarize historical resource data and for mapping photolandscape units (Color Figure 33-49).

Photo interpretation keys were prepared for use in synthesizing the basic resource data and in interpreting the differences observed on the imagery. The task was to relate color tones, textures, and geometric patterns to specific and consistent differences on the ground. Landform, vegetation, drainage patterns, geology, and soil color were all clues used to relate the differences observed on the imagery. These observations were recorded for later use in making interpretations from imagery.

Aircraft flights at 1000-m altitude above ground level were conducted to further describe the landscape, especially in regions that were inaccessible on the ground. Ground surveys were conducted where communication paths permitted.

Finally, a new potential land-use map that could be employed in aerial reconnaissance and field studies was developed utilizing all of the materials prepared and information obtained in the preparatory steps. This required careful study and evaluation of all the resource data, relating it to differences observed on the Landsat imagery. Seven months were required to conduct surveys and prepare a preliminary soil map with a legend based upon soil landscape units. Guides for planning and conducting large-area surveys were prepared.

Resource Interpretations

For resource studies in areas where experimental and production data are available, a tabulation is made of important soil properties and basic soil interpretations, such as yields of important crops, livestock carrying capacity, etc. for each of the principal kinds of soil in the area. This information is used in making the first draft of a potential land-use map and for making predictions about soil behavior under a moderately high level of technology for each of the soil units identified as existing in the area.

Interpretive maps were produced from the final potential land-use map in the Sudan studies. Predictions were made about soil limitations and soil potentials for rainfed agriculture, and for the irrigation of cultivated crops common in the area (Figures 33-50 to 33-53). Also, predictions were made with respect to suitable areas for growing paddy rice, for engineering uses, and for soil limitations of wetness. Also soil slopes were defined.

The resource study of the Sudd region involved complex evaluations of natural resources and environmental impact relationships in light of the broad interpretations that were made from Landsat imagery. Development of techniques was necessary to enable planners to rapidly manipulate large amounts of geographic data, to perform desired tabulations and calculations, and to display the results in a timely and effective manner.

The multitemporal and multispectral Landsat data are normally available for any region of the world and are of suitable spatial and spectral resolution for reconnaissance inventories. Landsat data are now being used operationally for large-area inventories involving many thousands of square kilometers. A minimum of training of resource personnel and planning is required for agencies to utilize the technology by their own personnel in their respective agencies.

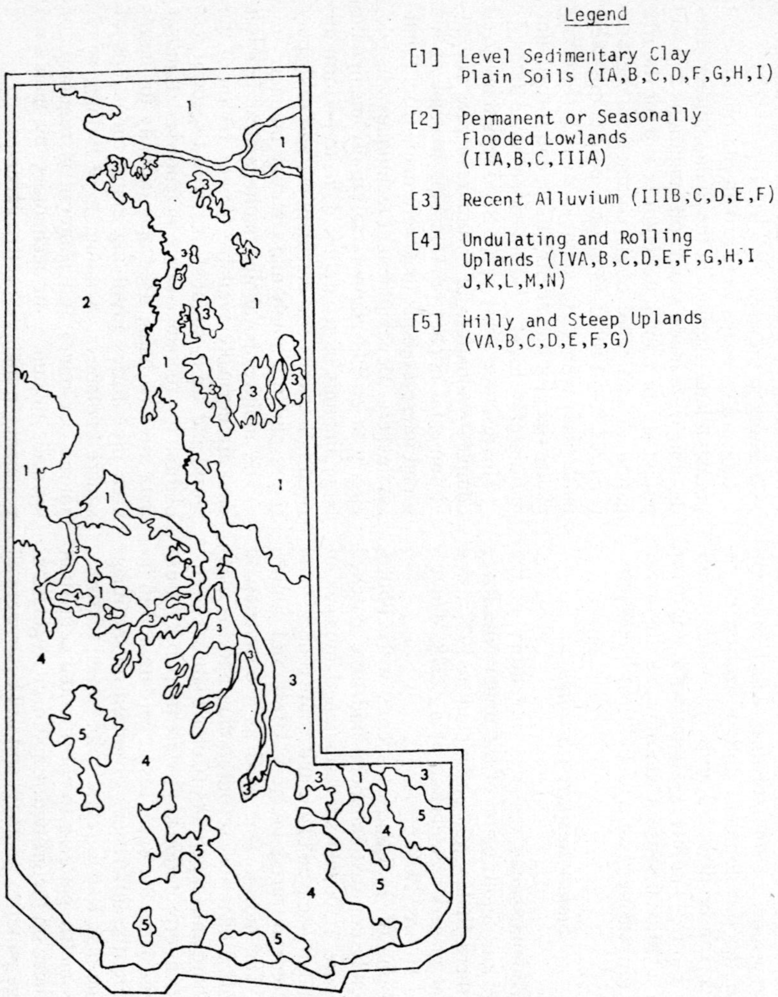

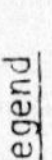

Legend

[1] Level Sedimentary Clay Plain Soils (IA,B,C,D,F,G,H,I)

[2] Permanent or Seasonally Flooded Lowlands (IIA,B,C,IIIA)

[3] Recent Alluvium (IIIB,C,D,E,F)

[4] Undulating and Rolling Uplands (IVA,B,C,D,E,F,G,H,I J,K,L,M,N)

[5] Hilly and Steep Uplands (VA,B,C,D,E,F,G)

Fig. 33-50. Generalized interpretive soils map.

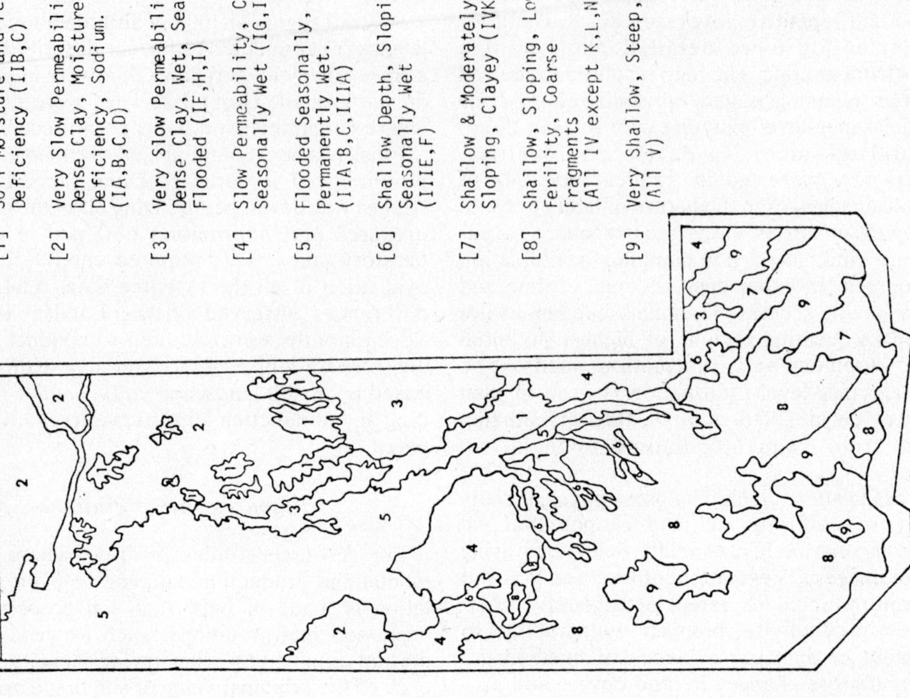

Legend

[1] Soil Moisture and Fertility Deficiency (IIB,C)

[2] Very Slow Permeability Dense Clay, Moisture Deficiency, Sodium. (IA,B,C,D)

[3] Very Slow Permeability Dense Clay, Wet, Seasonally Flooded (IF,H,I)

[4] Slow Permeability, Clayey, Seasonally Wet (IG,IIID,IVN)

[5] Flooded Seasonally, Permanently Wet. (IIA,B,C,IIIA)

[6] Shallow Depth, Sloping, Seasonally Wet (IIIE,F)

[7] Shallow & Moderately Deep, Sloping, Clayey (IVK,L)

[8] Shallow, Sloping, Low Ferility, Coarse Fragments (All IV except K,L,N)

[9] Very Shallow, Steep, Stony (All V)

Fig. 33-51. Map of interpreted soil limitations for rainfed agriculture.

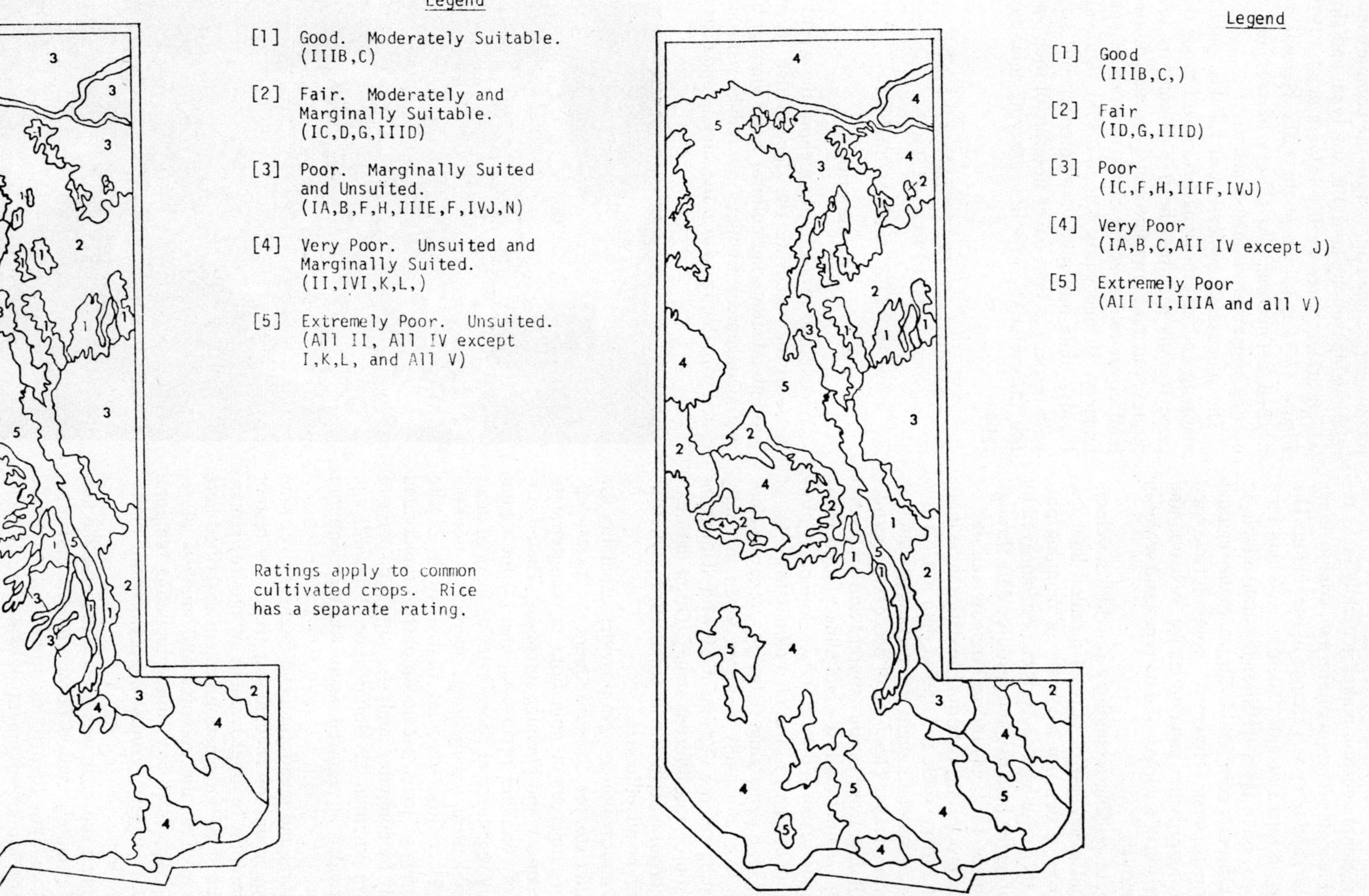

Fig. 33-52. Interpretive map of soil potential for rainfed agriculture.

Fig. 33-53. Interpretive map of soil potential for irrigation of common cultivated crops.

Agricultural Information Systems

The regional decision maker generally lacks relatable basic information on the use, composition, character, and temporal change of the region. The most basic forms of data, such as soils and land use maps, have been traditionally unavailable in formats directly comparable for use in the regional planning process. However, computer-assisted processing provides capabilities for inclusion and comparison of a variety of data types and scales in land analysis.

Remote sensing technology provides a vehicle for rapid collection of current detailed land use and resource data for a variety of planning purposes. Its role in land analysis can be improved by integrating the land use and resource data through assignment to a ground-referenced cellular or polygon processing system. The data can be input in digital form, such as Landsat data in CCT format, or as manually digitized map images such as soil survey maps. This type of processing permits input of data from various sources and scales and provides spatial analyses including plotted maps at various scales and quantitative summary information in tabular form. It also eliminates many of the steps (i.e., enlarging, reducing, preparing clear overlays, etc.) usually required when manually integrating remote sensing data with data from other sources at different scales. Often data can be incorporated in the original form, whether numeric or mixed codes.

The information system provides capability for computer storage, for subsequent rapid retrieval, and for numerous combinations of resource data for inventory and planning purposes. The advantages of computer processing are: (1) the digital data are registered to coordinates and stored in an accessible form for updating, overlay or monitoring; (2) tedious manual tasks, such as the measuring of areas, are accomplished rapidly and reliably by computer analysis; and (3) displays of maps of potential development or of hazards can be provided at required scales via the computer plotter.

Natural and man-influenced changes over time are important for recognizing trends of further deterioration or improvement of our resources. The data-base information system described here lends itself to the updating and analysis of resource data from any source including temporal Landsat imagery, ground surveys, aerial photographs, etc.

Such an approach was used to increase the utility of remotely sensed interpretations and other resource data to provide quantitative spatial analysis for improvement of water quality and land management. Remotely sensed data, soils, slope and drainage data, were integrated through a computerized georeferenced information system to assist in identification of point- and nonpoint-source pollution problems within the Lake Herman watershed in southeastern South Dakota. The information system provided an efficient method for detailed analysis of the complex interrelationships among land use, soil type, and slope which affect water quality (Myers et al., 1979). Figure 33-54 is a map of the Lake Herman watershed delineating two classes of cropland requiring conservation management.

The successful application of Landsat data for mapping and inventorying land cover types over large regions requires the use of analysis techniques that take into account large variations in land cover and spectral characteristics of space imagery. In one study Rhode (1978) subdivided Landsat data into relatively homogeneous strata with respect to land cover in terms of their spectral characteristics for inventorying land-cover types.

AGRICULTURAL MANAGEMENT AND PRODUCTION

WATER MANAGEMENT

Introduction

Widespread droughts throughout the world repeatedly demonstrate the dependence of man upon an adequate agricultural water supply. Certain management practices can reduce risks associated with fluctuations in the amount and distribu-

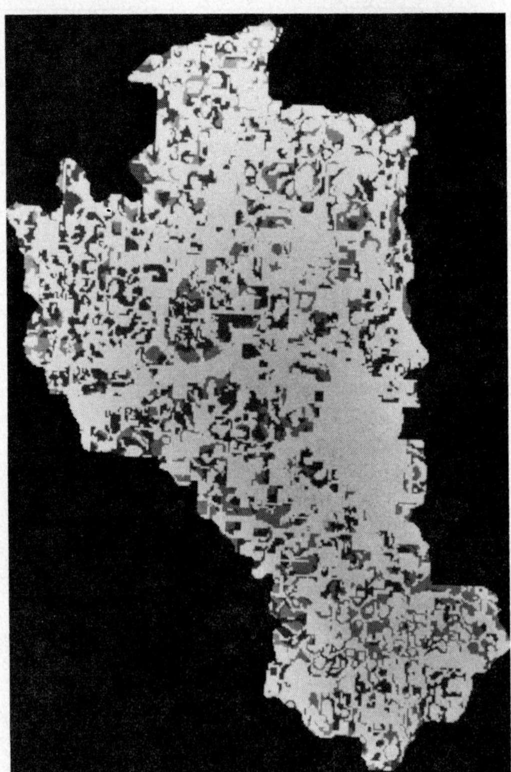

Fig. 33-54. Cropland in Lake Herman watershed requiring conservation management. Scale ½″ = 1 mile, dark grey = severe, medium grey = moderate, light grey = other.

tion of precipitation. However, many management practices carry risks of their own. Adoption and implementation of appropriate techniques for water conservation and management require both qualitative and quantitative information on plant- and soil-water status. Remote sensing can be a useful tool for providing such information.

Soil Moisture

The amount of water stored in the soil profile is extremely important in agricultural management. Soil moisture is essential for seed germination and for the growth and development of agricultural crops. It is also important for partitioning rainfall and irrigation water into runoff, infiltration and redistribution, drainage, and storage. Soil moisture influences the development of insects that spend part of their life in the soil; soil moisture also influences soil-borne plant pathogens. In addition, soil moisture affects soil erosion and evapotranspiration, and it is a critical input in models used in irrigation management and yield prediction.

Classical methods of measuring soil moisture, such as gravimetric sampling, neutron moisture-probes, etc., are useful where point measurements are sufficient to approximate the water content of small surrounding areas. However, it is difficult to extrapolate point measurements to larger areas because of variations in soil texture and rainfall. Remote sensing methods which cover large areas within a short period of time offer an alternative to classical techniques (Schmugge, 1978).

Remote sensing of soil moisture depends upon the measurement of electromagnetic energy that has been reflected or emitted from the surface. Variation in the intensity of this radiation depends on either its dielectric properties (e.g. the index of refraction), or its temperature, or a combination of both. The property that is important depends on the wavelength region that is being considered as indicated in Table 33-2.

Reflectance Methods

Reflectance in the visible and near-infrared regions of the spectrum is sensitive to differences in surface characteristics at the soil-plant-air interface. Numerous laboratory measurements have shown that soil spectral reflectance increases between 0.25 and 1.00 μm (Condit, 1970; Von Minnus, 1967; Blanchard et al., 1974). When water is added, spectral reflectance decreases (Von Minnus, 1967; Condit, 1970; Kanemasu, 1974). The absolute magnitude of reflectance varies considerably because of differences associated with texture, structure, roughness, organic matter, mineral content, and illumination geometry.

Idso et al. (1975b) measured bare soil albedo (0.3 to 2.5 μm) and water content for a smooth Avondale loam at various depth intervals from 0 to 10 cm. After correcting for effects of sun angle, they found a linear relationship with soil moisture over the range of 0 to 0.18 cm^3/cm^3 in the 0 to 0.2 cm layer. They also found similar relationships for deeper layers apparently due to the close correlation between soil moisture at the surface and at deeper layers. Reginato et al. (1977) evaluated aircraft data collected over smooth and rough Avondale-loam plots at different moisture contents and confirmed the albedo-water content relationship established by Idso et al. (1975a). Spectral band ratios were not related to soil water-content in the 0.45 to 1.03 μm region.

Widespread application of the albedo technique is restricted by two major drawbacks. First, correlation with deeper soil water-content breaks down for light applications of water. Second, it is unlikely that a universal relationship can be established because of the great difference in albedo that exists among different soils.

Evans (1979) evaluated aerial photos collected over bare soils and found that color changes related to soil moisture varied with different soils. He concluded that it was unlikely that changes in soil water-content could be estimated using photo density.

Application of bare-soil reflectance techniques to conditions of vegetation cover is more difficult because of problems in extracting soil reflectance from composite-canopy reflectance. However, vegetation may be used as an indicator of soil water-content. Werner et al. (1971) investigated the relationship between film density and available soil moisture in the root zone for sorghum. The highest correlation was obtained in the 0.59 to 0.70 μm wavelength band.

TABLE 33-2

Electromagnetic Properties for Soil Moisture Sensing.

Wavelength Region	Property Observed
Reflected visible and infrared 0.3–3 μm	Reflectance/index of refraction
Thermal infrared 10–12 μm	Temperature
Active microwave 1–50 cm	Backscatter coefficient/ dielectric properties
Passive microwave 1–50 cm	Microwave emission/dielectric properties and temperature

Moore et al. (1975) correlated Skylab S-192 multispectral scanner measurements (0.56 to 0.61, 0.68 to 0.76, 0.78 to 0.88, 1.55 to 1.75, and 2.10 to 2.35 μm) with soil moisture content of three different layers (0 to 2, 2 to 14, and 10 to 30 cm) for 13 fields (vegetated and fallow). The most highly significant correlation was obtained for the 2.10 to 2.35 μm band and for a depth of 0 to 2 cm. Wavelengths greater than 2.1 μm were required to reliably separate wet and dry bare surfaces.

One of the most useful applications of reflectance techniques may be to provide estimates of plant cover (leaf area index, biomass, percent cover, etc.) as inputs into soil moisture budget/ water management models. Heilman et al. (1976) evaluated soil water content in the top 150 cm of soil for winter wheat fields in a five-state region of the Great Plains using an evapotranspiration model with solar radiation, air temperature, precipitation and leaf area index estimated from Landsat as inputs. They found that soil moisture estimates compared favorably with the traditional Crop Moisture Index and could be interpreted in terms of yield.

Thermal Methods

While soil reflectance-measurements are sensitive to soil moisture at the very surface of soil, surface soil-temperatures are influenced by deeper soil conditions. This is illustrated in Figure 33-55, which shows computer-enhanced visible and thermal images of three plots differing in moisture regime (Reginato et al., 1976). Plot 1 was continually dry, plot 2 was wet initially and allowed to dry, and plot 3 was continually wet. Plots 1 and 2 appear light in the visible region while plot 3 is dark. In the thermal infrared region, there is a temperature difference among all three plots with plot 1 the warmest, 2 intermediate, and 3 coolest.

Surface soil temperatures are influenced by both internal and external factors. The internal factors are thermal conductivity (K), and heat capacity (C), where $P = (KC)^{1/2}$ defines what is known as "thermal inertia." The external factors are primarily meteorological—solar radiation, air temperature, relative humidity, cloudiness, and wind. The combined effect of these external factors represents the driving function for diurnal variations of surface temperature. Thermal inertia is an indication of the soil's resistance to this driving force. Since both the heat capacity and thermal conductivity of a soil increase with an increase in soil moisture, the resulting thermal inertia will increase.

A complicating factor is the effect of surface evaporation in reducing the net energy input to the soil from the sun. Evaporation complements the other effects of water in soil by reducing the amplitude of the diurnal surface-temperature cycle. As a result, the day-night temperature difference is an indicator of some combination of soil moisture and surface evaporation.

Cihlar et al. (1979) conducted an experiment on diurnal surface-temperature variations and near-surface soil water-content using four missions in early May, early June, early July and mid September. The study yielded the following results:

1) under clear sky condition, an inverse linear relationship existed between the day-minus-night surface-temperature differential (ΔT), and soil water-content expressed in percent of field capacity (*PFC*) in the top 2 to 4 cm of soil;
2) with the exception of the post-harvest Mission 4, near-surface water-content was the most important ground variable, even under a considerable amount of straw-mulch cover;

Fig. 33-55. Visible and infrared imagery of three fields in different stages of drying.

3) the variability of the "ΔT vs soil-water" relationship was reduced appreciably when soil texture was taken into account;
4) the temporal stability of the "PFC vs ΔT" relationship did not improve when diurnal air-temperature variations were taken into account; and
5) the relationship between PFC and ΔT improved to various degrees when apparent radiances from the visible spectrum were also included; correlation coefficients approached 0.8 (0.7) for the top 2 cm (4 cm) of the soil, respectively.

Results of this study indicate that the diurnal surface temperature variations correspond fairly closely to near-surface content in fallow fields with various straw mulch, roughness, soil texture and seasonal conditions. Effects of other variables (e.g. surface slope, variable cloud cover) have not yet been examined. Myers and Heilman (1969) reported that wet soils generally were cooler than dry soils during the day but were warmer than dry soils at night. This suggests that a thermal-inertia approach could be used to evaluate soil moisture.

Idso et al. (1975b) conducted ground-based experiments on a bare Avondale loam and found that volumetric water contents of 2 to 4-cm thick surface layers of soil were a function of the amplitude of the diurnal surface-temperature wave for clear day-night periods (Figure 33-56). They also found that water content was a function of the daily maximum value of the surface soil-air temperature differential. Tests on three additional soils ranging

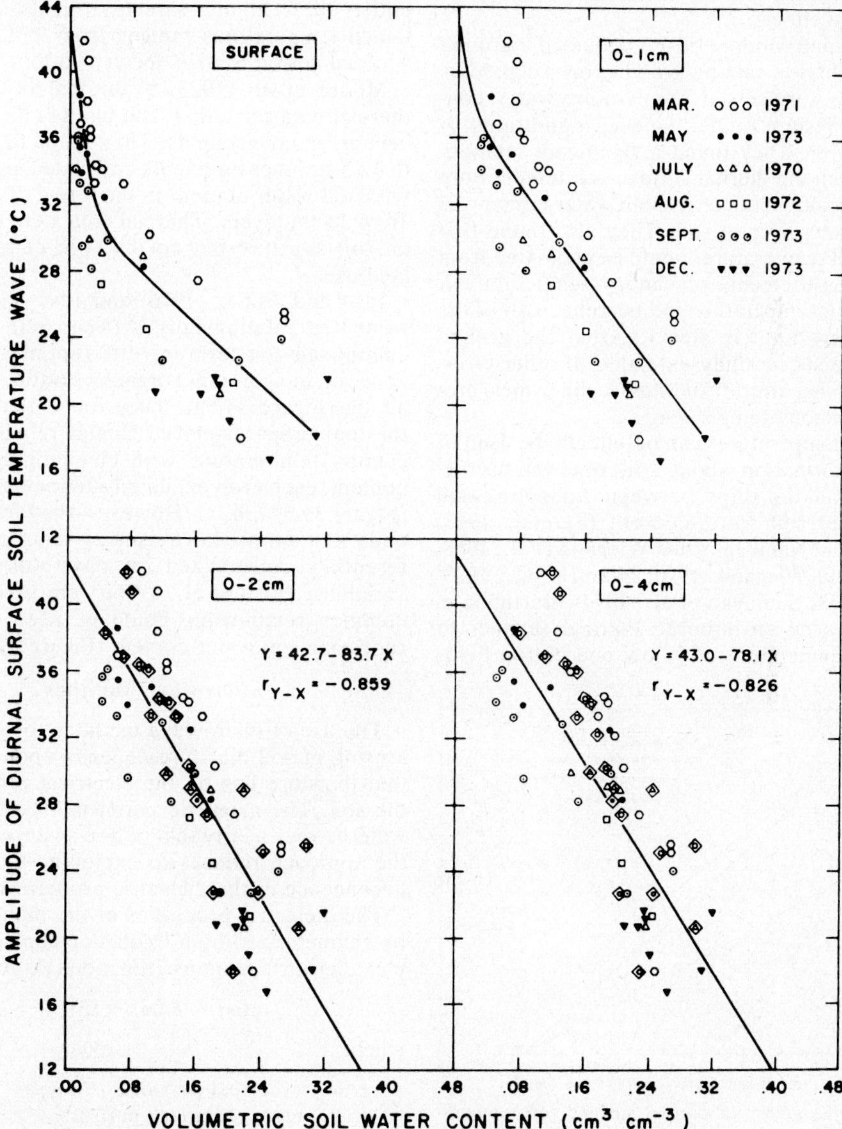

Fig. 33-56. Summary of results for the diurnal temperature variation versus mean daylight soil moisture (Idso et al., 1975).

from sandy loam to clay indicated that relationships could not be extended to the other soil types (Figure 33-57). They found, however, that expressing soil water-content in terms of a pressure potential reduced dependence on soil type (Figure 33-58). In the absence of pressure-potential data, textural dependence can be reduced by expressing soil water-content as a percent of field capacity (Schmugge et al., 1978). Using aircraft observations of surface-soil temperature. Reginato et al. (1976) and Schmugge et al. (1978) obtained results similar to those of Idso et al. (1975b).

Temperature and water-content relationships are complicated by evaporation and by associated environmental factors such as wind, relative humidity, and air temperature. A normalization procedure proposed by Idso et al. (1976) that uses air temperature will compensate for some, but not all, of these effects.

Heilman and Moore (1980) conducted a ground study in a barley canopy growing on Volga loam to evaluate surface-soil temperature-water content relationships under varying conditions of canopy cover. They found a significant relationship between the diurnal surface-soil temperature difference and volumetric soil water-content in the 0 to 4-cm layer of soil. They also found that surface-soil temperatures could be estimated from remote measurements of canopy temperature if minimum air temperature and percent cover of the canopy were known. Results from the ground study were successfully extended to other crops and soils using aircraft thermal-scanner measurements of canopy temperature.

Canopy temperatures can potentially be used to extract information about soil water-status because of relationships between transpirational cooling and soil water-content (Tanner, 1963; Wiegand and Namken, 1966; Wiegand et al., 1968; Thomas and Wiegand, 1970; Nixon et al., 1973; Ehrler, 1973; Sumayao et al., 1977). Bartholic et al. (1972) used an airborne thermal scanner to measure temperature of fallow and cotton fields under different moisture treatments. They obtained temperatures ranging from 29°C for well-watered cotton to 37°C for dry cotton.

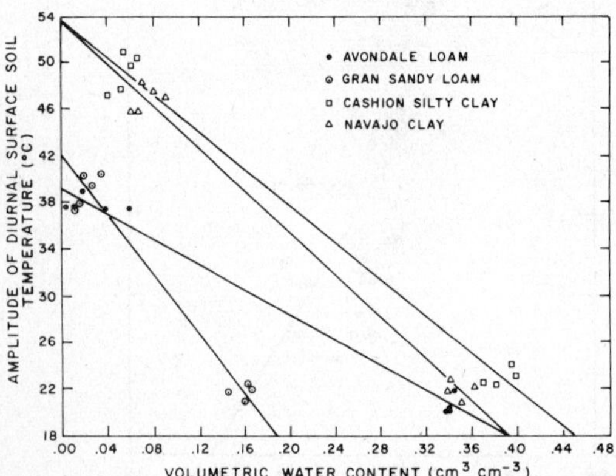

Fig. 33-57. Summary of diurnal surface soil temperature variation versus mean daylight volumetric soil water content of the 0- to 2-cm depth increment for four different soils (Idso et al., 1975).

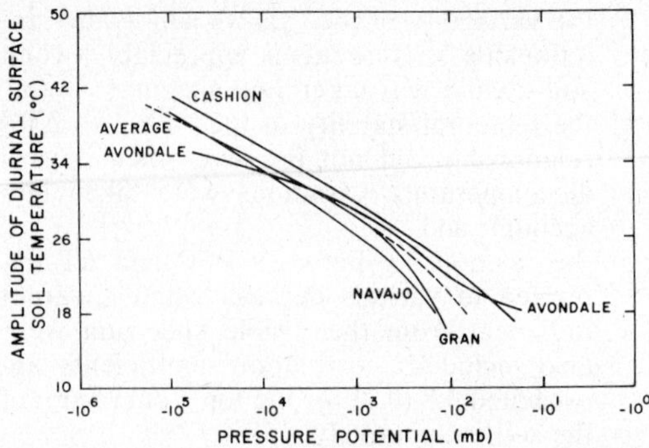

Fig. 33-58. Diurnal surface soil temperature variations versus the mean daylight soil water pressure potential of the 0- to 2-cm depth increment for four different soils (Idso et al., 1975).

Moore et al. (1975) evaluated Skylab S-192 thermal data for fallow and alfalfa (10 to 89 percent green cover) fields. They found that the 10.2 to 12.5 μm measurements correlated significantly with soil water-content in the 0 to 2, 2 to 10, and 10 to 30-cm layers. Thermal data were dependent on soil moisture but not on type of agricultural land use.

Idso and Ehrler (1976) and Idso et al. (1978) evaluated relationships between mid-afternoon canopy air-temperature differentials and root-zone soil moisture for sorghum, cotton, wheat and alfalfa (Figure 33-59). They found that although the four crops displayed similar relationships of canopy temperature with changing soil water-content, each crop exhibited a unique relationship (Figure 33-59a, b, c). However, they found that if midafternoon-predawn canopy-temperature differentials, normalized for day-to-day weather variability (Idso et al., 1976) were used, a single common relationship could be used to specify root-zone soil water-content (Figure 33-59d).

Microwave Methods

The use of microwave methods for the remote sensing of soil moisture depends upon the effect that moisture has on the electrical properties of the soil. The magnetic permeability of soils and water is very nearly that of free space and, hence, the approach reduces to exploiting the moisture dependence of the dielectric properties of soil.

The dielectric properties of the moist soil may be characterized by a frequency-dependent complex dielectric response-function (Bottcher 1952):

$$\xi(\omega) = \xi_r(\omega) + i\xi_i(\omega) \qquad (33\text{-}8)$$

where

$\xi_r(\omega)$ = the real part of ξ
$\xi_i(\omega)$ = the imaginary part of ξ
i = square root of -1

and ω is the (angular) frequency.

The function $\xi_r(\omega)$ is nearly constant from $\omega = 0$ to the neighborhood of the relaxation frequency ω_R of dipoles in the medium. The time ω_R^{-1} is the time constant for the decay of polarization, when the electric field is removed. Beyond ω_R, the function ξ_r decreases until it reaches the visible region of the spectrum, and is equal to the index of refraction squared. The real part of the dielectric response-function is a measure of the energy stored by the dipoles aligned in an applied electromagnetic field. When the frequency is greater than ω_R, the dipoles can no longer follow the field and the ability of the medium to store electric-field energy decreases.

The function $\xi_i(\omega)$ is a measure of the energy-dissipation rate in the medium. Viewed as a function of frequency, and starting from low ω, it rises to a peak at ω_R and, thereafter, decreases. The behavior described is due to the permanent dipoles in the soil medium. In complicated heterogeneous media, there may be more than one relaxation mechanism and more than one absorption peak. Furthermore frequencies above the medium may show further dispersion and absorption regions due to direct molecular excitations. The frequency ω_R will generally lie in the microwave range (18 GHz in H_2O), whereas the latter molecular excitations will be in the submillimeter or infrared regions of the spectrum (Bottcher, 1952; Hasted, 1974). In soils ξ_r is reduced to around 1 GHz due to the binding of the water molecules to the soil particles (Hoekstra and Delaney, 1974).

The preceding description generally applies to all dispersive media. In a soil, the values of ξ_r are typically between 3 and 5, whereas the value of ξ_r for water is about 80. Hence, relatively small amounts of free water in a soil will greatly affect its electromagnetic properties. This dependence is shown in Figure 33-60, which presents the results of laboratory measurements at the wavelength of

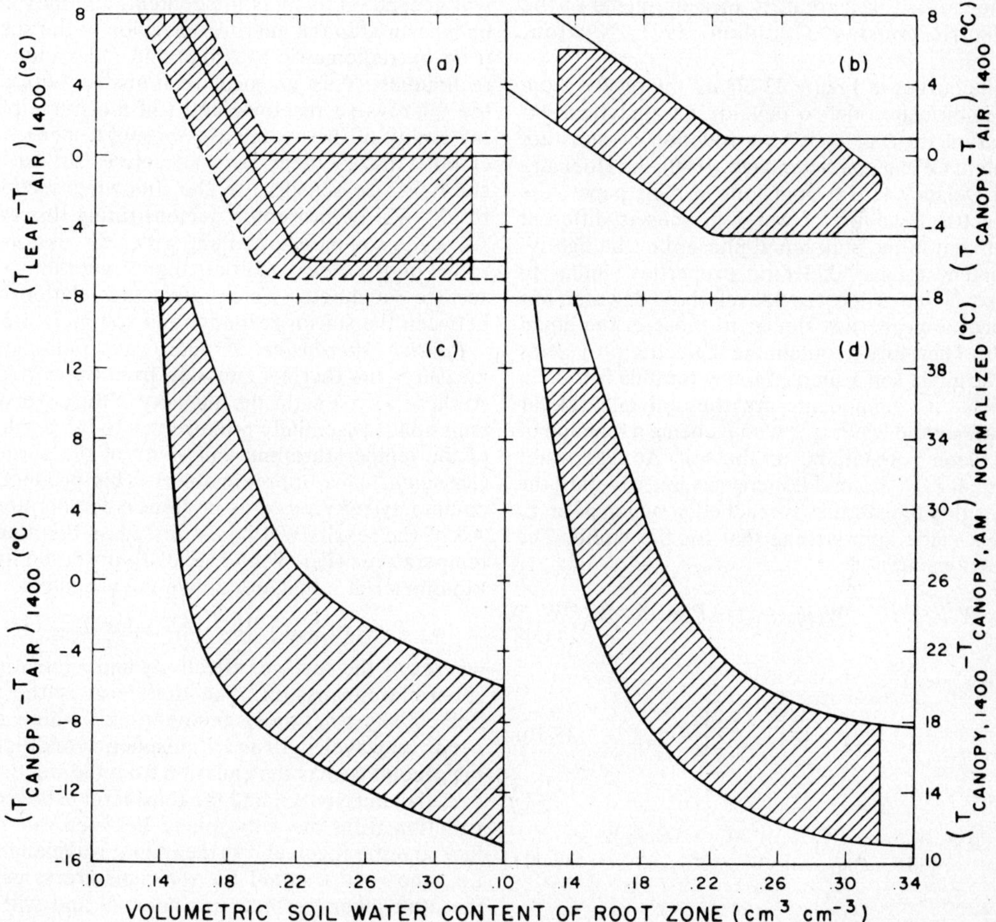

Fig. 33-59. (a) The midafternoon (14.00 h) leaf-air temperature differential of individual sunlit cotton and sorghum leaves vs. the volumetric soil water content of the crop's root zone. The solid lines (Idso and Ehrler, 1976) are for leaf temperatures obtained by thermocouple, and the dashed lines for leaf temperatures obtained by infrared thermometer. (b) A similar relation for wheat utilizing infrared thermometer-derived canopy temperatures. (c) A similar relation for alfalfa utilizing infrared thermometer-deriver canopy temperatures. (d) The midafternoon-presunrise canopy temperature differential of wheat and alfalfa normalized to remove effects of climatic variability as described in the text vs. the volumetric water content of the crops' root zones (Idso et al., 1978).

21 cm (frequency of 1.4 GHz) for three soils ranging in texture from a light sand to a heavy clay.

At low moisture levels there is a slow increase in ξ_r with soil moisture. However, above a certain point, or transition moisture, the slope of the curve increases sharply due to a change in the behavior of the water in the soil. When water is first added to a soil it is tightly bound to the soil particles. In this state the water molecules are not free to become aligned and the dielectric properties of this water resemble those of ice. As the layer of water around the soil particle becomes larger, the binding to the particle decreases and the water molecules behave as they do in the liquid state; hence, the greater slope at the higher soil moisture values. The transition moisture depends on the soil texture, i.e., particle-size distribution. It is less for a sand and larger for a clay and has been found to be linearly dependent on the wilting point for the soil. This effect has been demonstrated in laboratory measurements of the dielectric constant (Lundien, 1971; Newton, 1977b).

The curves in Figure 33-60 are the results from an empirical model to develop an analytical expression for ξ of soils as a function of moisture content (Wang and Schmugge, 1980). As Hoekstra and Delaney (1974) point out in their paper, the dielectric behavior of water in soils is different from that in the bulk-liquid phase, i.e. the tightly-bound water has dielectric properties similar to those of ice while the loosely-bound water has dielectric properties similar to those of the liquid state. Therefore to obtain the dielectric properties of the moist soil a simple mixing formula is used in which the components are the soil mineral (or rock), air and water (ξ_x) with ξ_x being a function of the water content, W_c, in the soil. At zero water content $\xi_x = \xi_{ice}$ and it increases linearly until the transition moisture w_t is reached at which point ξ_x has a value approaching that for the liquid. The equations are:

$$\xi = W_c\xi_x + (P - W_c)\xi_a + (1 - P)\xi_r, \text{ for } W_c \lesssim W_t \quad (33\text{-}9)$$

with

$$\xi_x = \xi_1 + (\xi_w - \xi_i)\frac{W_c}{W_t} \vdots \cdot \gamma \quad (33\text{-}10)$$

and

$$\xi = W_t\xi_x + (W_c - W_t)\xi_w + (P - W_c)\xi_a + (1 - P)\xi_r, \text{ for } W_c > W_t \quad (33\text{-}11)$$

with

$$\xi_x = \xi_i + (\xi_w - \xi_i)\gamma \quad (33\text{-}12)$$

where P is the porosity of the dry soil; ξ_a, ξ_w, ξ_r and ξ_i, are the dielectric constants of air, water, rock and ice respectively, and ξ_x stands for the dielectric constant of the initially-absorbed water. In Wang and Schmugge (1980) the values of W_t and γ, as functions of the soil wilting point (WP), were determined for 18 soils by a least squares fit to the data. The results are:

$$W_t = 0.49\ WP + 0.16$$
$$\gamma = -0.57\ WP + 0.48.$$

Based on the above research, it appears that reasonable estimates of the dielectric constant for soils can be made both as a function of moisture content and microwave frequency if the knowledge of the soil texture or moisture characteristic is available. The frequency-dependence is contained in the dielectric constant for water which is well understood. It is assumed that there is no frequency dependence of W_t within the microwave spectral region.

The dielectric properties of a medium determine the propagation characteristics for electromagnetic waves in the medium, and as a result they affect the emissive and reflective properties at the surface. Thus these latter two quantities for a soil will depend on its moisture content, and they can be measured in the microwave region of the spectrum by radiometric (passive) and radar (active) techniques. This physical relationship between the microwave response and soil moisture, plus the ability of the microwave sensors to penetrate clouds, makes them very attractive for use as soil-moisture sensors. In the following sections results will be presented demonstrating this sensitivity to soil moisture, along with a discussion of some of the noise factors, e.g. vegetation and surface roughness, which affect the relationship between the sensor response and soil moisture.

Passive Microwave. A microwave radiometer measures the thermal emission from the surface. At these wavelengths the intensity of the observed emission is essentially proportional to the product of the temperature and emissivity of the surface (Rayleigh-Jeans approximation). This product is commonly referred to as brightness temperature. All of the results will be expressed as brightness temperatures (T_B). The value of T_B observed by a radiometer at a height h above the ground is

$$T_B = \tau(rT_{sky} + [l - r]\ T_{soil}) + T_{atm} \quad (33\text{-}13)$$

where r is the surface reflectivity and τ the atmospheric transmission. The first term is the reflected sky-brightness temperature, which depends on wavelength and atmospheric conditions; the second term is the emission from the soil ($l - r = e$, the emissivity); and the third term is the contribution from the atmosphere between the surface and the receiver. At the longer wavelengths, i.e., those best suited for soil moisture sensing, the atmospheric effects are minimal and will be neglected in this discussion.

Thermal microwave emission from soils is generated within the soil volume. The amount of energy generated at any point within the volume depends on the soil dielectric properties (or soil moisture) and the soil temperature at that point. As energy propagates upward through the soil volume from its point of origin, it is affected by the dielectric (soil moisture) gradients along the path of propagation. In addition, as the energy crosses the surface boundary it is reduced by the effective transmission coefficient (emissivity), which is determined by the dielectric characteristics of the soil near the surface.

The emission from the soil surface can be expressed as:

$$T_B = e \int_{-\infty}^{0} T(z) \propto (z)\, exp(-\int_{z}^{0} \propto (z')dz')dz \qquad (33\text{-}14)$$

where $T(z)$ is the temperature profile and $\propto (z)$ is the absorptivity as a function of depth, which depends on moisture content. Results from numerical solutions to this equation have been presented by Njoku and Kong (1977), Wilheit (1978) and Burke et al. (1979). These papers have included results which indicate that the models do a good job of predicting T_B for a smooth surface. One of the most significant results from these models is that the effective sampling depth is on the order of only a few tenths of a wavelength (Wilheit, 1978). For a 21-cm-wavelength radiometer this is about 2 to 5 cm.

The range of dielectric constant presented in Figure 33-60 produces a change in emissivity from greater than 0.9 for a dry soil to less than 0.6 for a wet soil, assuming an isotropic soil with a smooth surface. This change in emissivity for a soil has been observed by truck-mounted radiometers in field experiments (Poe et al., 1971), (Blinn and Quade, 1972), (Schanda et al., 1978), (Newton, 1977b) and by radiometers in aircraft (Schmuge et al., 1974), (Burke et al., 1979), (Choudhury et al., 1979) and satellites (Eagleman and Lin, 1976), (Schmugge et al., 1977). In no case were emissivities as low as 0.6 observed for real surfaces. This is primarily due to the effects of surface roughness which generally increase the surface emissivity.

As shown in Figure 33-60, there is a greater range of dielectric constant for soils at the 21-cm wavelengths. This fact, combined with a larger soil moisture sampling depth and better ability to

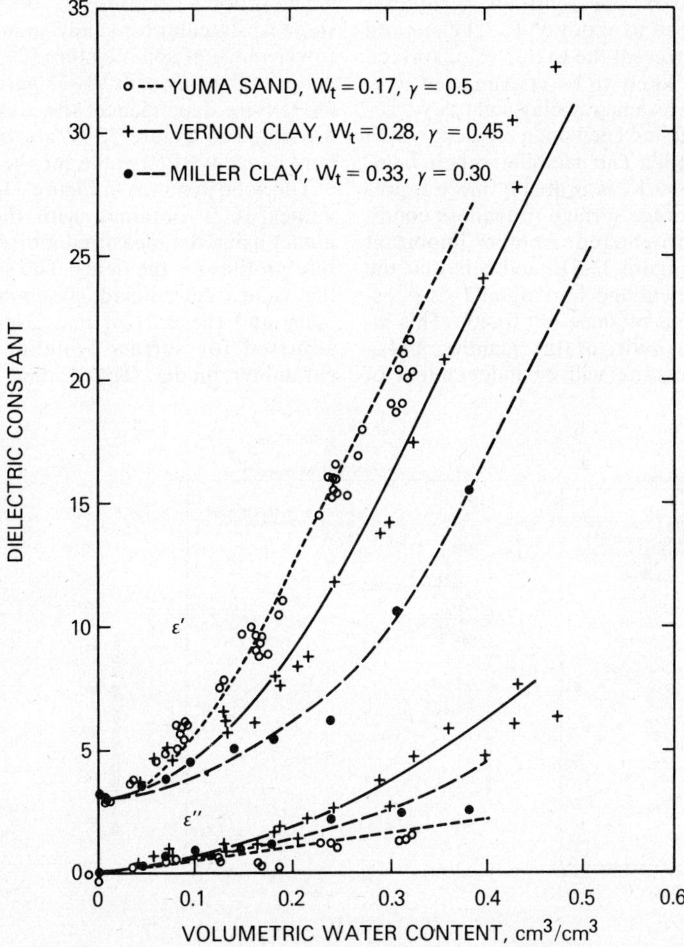

Fig. 33-60. Laboratory measurements of the real and imaginary parts of the dielectric constant for three soils as a function of moisture content at a wavelength of 21 cm. The data for Yuma Sand and Vernon Clay Loam are from Lundien (1971) and those for Miller Clay are from Newton (1977). (From Wang and Schmugge, 1980).

penetrate a vegetative canopy, makes the longer wavelength sensors better suited for radiometric soil-moisture sensing.

In Figures 33-61a and b, the field measurements of Newton (1977a) are plotted versus angle of observation for various moisture contents and for three levels of surface roughness. The horizontal polarization is that for which the electric field of the water is parallel to the surface and the vertical polarization is perpendicular to it. These results indicate the effect of moisture content on the observed values of T_B and the effect of surface roughness which is to increase the effective emissivity at all angles and to decrease the difference in T_B for the two polarizations at the larger angles.

For the smooth field there is a 100 K change in T_B from wet to dry soils and clearly this range is reduced by surface roughness. The effect of the roughness is to decrease the reflectivity of the surface and to increase its emissivity. For a dry field the reflectivity is already small (<0.1) so that the resulting increase in emissivity is small. As seen in Figure 33-61b, surface roughness has a significant effect for wet fields where the reflectivity is larger (>0.4). The range of T_B for the rough field is reduced to about 60 K. The smooth and rough fields represent the extremes of surface conditions that are likely to be encountered, e.g. the rough surface was a heavy clay soil (clay fraction 60 percent) that had been deep plowed, which produced large clods. The medium rough field, with a T_B range of 80 K, is probably more representative of the average surface roughness condition that will be encountered. Another important observation from Figure 33-61a and b is that the average of the vertical and horizontal T_B's is essentially independent of angle out to 40°. This indicates that the sensitivity of this quantity, $1/(T_{BV} + T_{BH})$, to soil moisture will be independent of angle. This result will be useful if the radiometer is to be scanned to provide an image.

When the brightness temperatures for the medium-rough field are plotted vs soil moisture in the 0 to 2-cm layer, there is an approximate linear decrease of T_B (Figure 33-61c). As the thickness of the layer increases, both the slope and intercept of the linear regression also increase. This is because the moisture values for the high T_B cases increase, whereas they remain essentially the same in the low T_B or wet cases. This type of behavior was also seen in the results obtained from aircraft platforms and leads to the conclusion that the soil-moisture sampling depth is within the 2 to 5-cm range for the 21-cm wavelength. This agrees with the predictions of theoretical models of radiative transfer in soils (Wilheit, 1978; and Burke et al., 1979).

Results from a 1975 aircraft experiment over irrigated agricultural fields are presented in Figure 33-62 (Choudhury et al., 1979). These results were obtained over fields with a range of soil textures from sandy loam to heavy clays. In the analysis it was observed that there was a textural dependence of the T_B reponse to soil moisture, i.e. the slope was greater for sandy soils which had a narrower range of soil moisture (0–20 percent) compared to the clay soils (0–35 percent). To account for texture dependence, the soil moisture values presented in Figure 33-62 are normalized to the field capacity (FC) value for the particular soil.

The solid symbols in Figure 33-62 are calculated values of T_B obtained with the Wilheit (1978) model using the measured moisture and temperature profiles for the fields. The solid line connects the values determined, assuming a smooth surface, and the dashed line connects the values adjusted for surface roughness using a one-parameter model. The dashed line fits the ob-

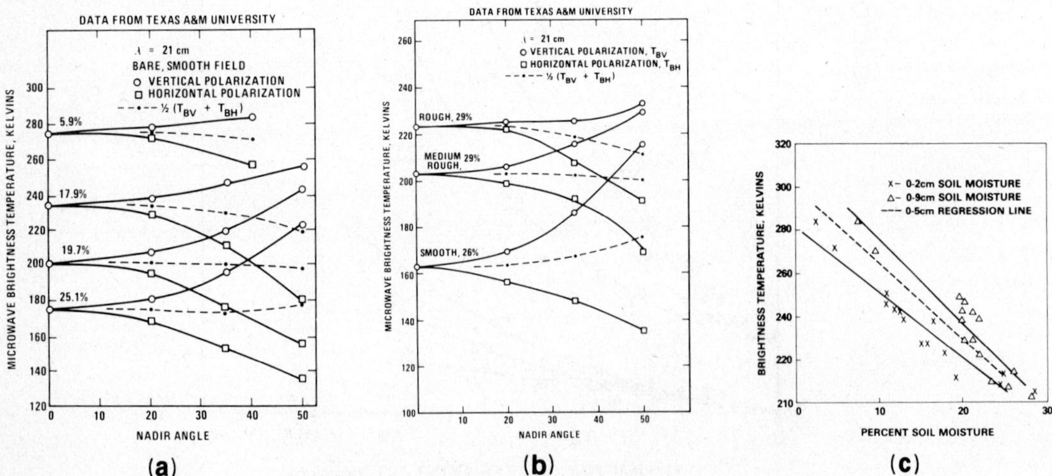

Fig. 33-61. Results from field measurements performed at Texas A&M University: (a) T_B versus angle for different moisture levels; (b) T_B versus angle for different surface roughness at about the same moisture level; (c) T_B versus soil moisture in different layers for the medium rough field (Newton 1977).

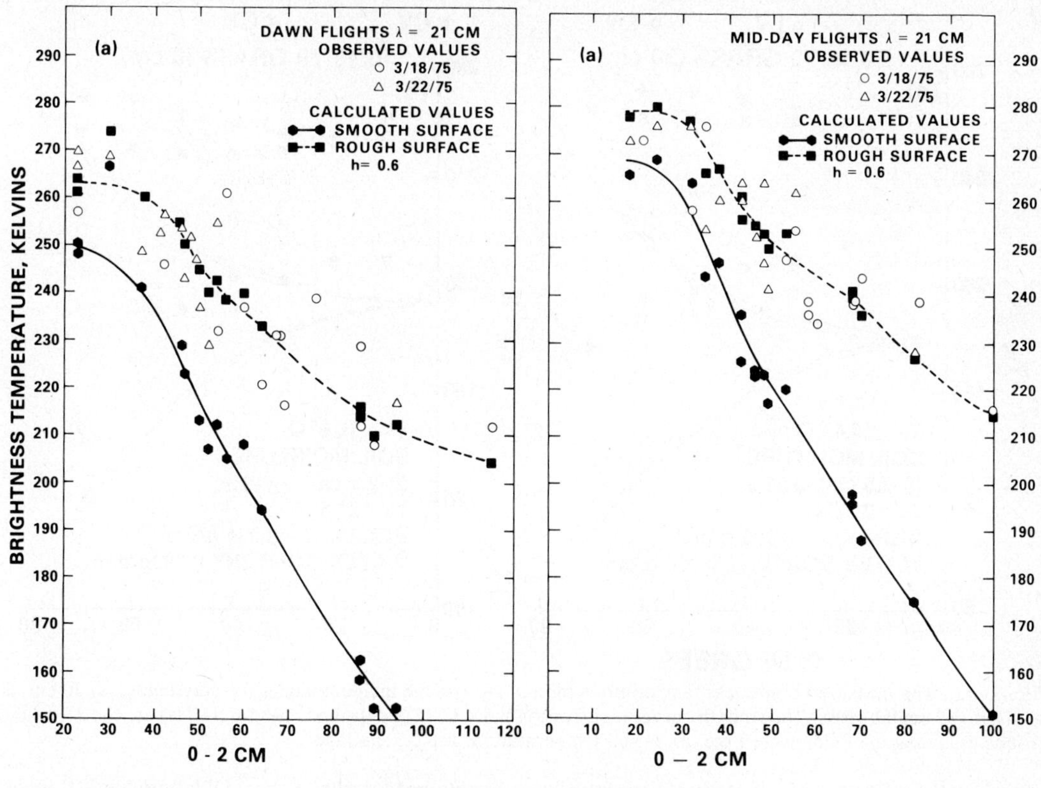

Fig. 33-62. Aircraft observations of T_B over agricultural fields around Phoenix, Arizona from March 1975 flights for both early morning and midday flights.

served values quite well. The values of the parameter were selected empirically to give a best fit to the data and it is clear that the same value works well for both the dawn and midday flights. The effects of soil temperature are seen in the T_B differences between the dawn and midday flights.

There is little change in T_B for soil moisture values for the 0 to 2-cm layer out to about 30% of field capacity. When the data are replotted vs the 0 to 5-cm layer, the flat region extends out to about 50% of Field Capacity (FC) with a steeper slope beyond that value. In Figure 61 the dielectric curves break at moisture levels which are about 40 to 50% of FC. These aircraft results support the conclusion that the sampling depth is about 2 to 5 cm.

A vegetative canopy acts as an absorbing layer which absorbs some of the upwelling radiation from the soil and also emits radiation at its own temperature. The magnitude of the effect depends on the amount of vegetation and the wavelength of observation. A thick canopy would approximate a Lambertian black body, i.e., it would have an emissivity close to one and show no angular or polarization effects. Basharinov and Shutko (1978) and Kirdiashev et al. (1979) have reported on observations made in the USSR over the 3 to 30 cm wavelength range for a variety of crops. Their results indicate that for small grains and grasses the sensitivity to soil moisture is 80 percent to 90 percent of that expected for bare ground at wavelengths greater than 10 cm. Broadleaf cultures, like mature corn or cotton, transmit only 20 percent to 30 percent of the radiation from the soil at wavelengths shorter than 10 cm and about 60 percent at the 30 cm wavelength. The observed 30 percent to 40 percent sensitivity for a forest at the 30 cm wavelength, although they did not mention the type or height of trees.

In Figure 33-63 results from field experiments for grass-covered fields are presented at the 6 (C-band) and 21 (L-band) cm wavelengths (Wang et al., 1980). Data for two grass heights are presented: 30 cm in Figure 33-63a and 12.5 cm in Figure 33-63b. There is little or no change of T_B with angle observed at the 6 cm wavelength for the 30 cm tall grass and T_B is about that which was observed for a dry field. Also, there is little difference between the values obtained at different polarizations, as would be expected for a thick canopy. The 21 cm results display angular and polarization effects similar to those seen in Figure 33-61 for the bare fields. However, T_B has increased: a bare field with the same moisture content had a T_B = 190 K compared to 220 K for the 30 cm grass field. In comparison a dry field would be expected to have T_B = 270 to 280 K. Thus, the dynamic range between wet and dry fields is re-

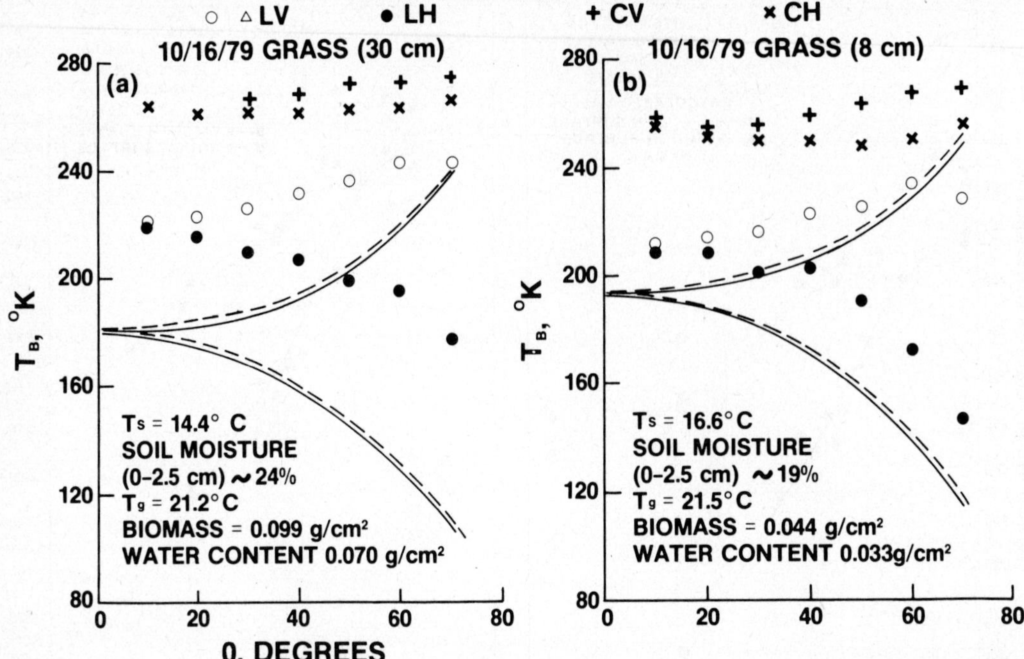

Fig. 33-63. The measured brightness temperatures plotted against the incident angles for grasslands; (a) 30 cm tall grass; (b) 10 cm tall grass. The smooth curves (solid ones for 1.4 GHz and dashed ones for 5 GHz) are the calculated results assuming the fields were bare. Ts is soil temperature at top 2.5 cm layer.

duced by the presence of vegetation from 80 K to about 50 K. Similarly, the polarization difference at $\theta = 40°$ is reduced from 38 K to 21 K. Both of these factors indicate that for a field with a dense 30 cm grass cover, sensitivity to soil moisture was reduced to about 60 percent of the bare soil case, which is a little less than the transmissivity reported by the Russians. The quantification of vegetation in terms of wavelength and biomass parameters is a near term objective which should be achieved shortly.

The Earth Resources Experiment Package (EREP) onboard Skylab contained a 20-cm radiometer. This sensor was non-scanning with a 115 km field of view between half power points. With this coarse spatial resolution, it would be difficult to directly compare sensor response and soil moisture measurements. However, there have been two reports of indirect comparisons. McFarland (1976) showed a strong relationship between the Skylab 21-cm brightness temperatures and Antecedent Precipitation Index (API) for data obtained during a pass starting over the Texas and Oklahoma panhandles and proceeding southeast toward the Gulf of Mexico.

Eagleman and Lin (1976) carried the analysis of the Skylab data a step farther and compared the brightness temperature with estimates of the soil moisture over the radiometer footprint. The soil-moisture estimates were based on a combination of actual ground measurements and calculations of the soil moisture using a climatic water-balance model. They obtained a correlation of 0.96 with data obtained during five different Skylab passes over Texas, Oklahoma and Kansas. This result is quite acceptable considering the difficulty of obtaining soil-moisture information over a footprint of such a size and considering the fact that the brightness temperature was averaged over the broad range of cultural conditions that occurred over the area.

These results from space, supported by the more detailed aircraft and ground measurements presented earlier, strongly support the possibility of using microwave radiometers for soil-moisture sensing. A difficulty with this approach is that the spatial resolution is limited by the size of the antenna which can be flown. For example, at a wavelength of 21 cm, a 10 m × 10 m antenna is required to yield 20-km resolution from a satellite altitude of 300 km. It is possible to use the coherent nature of the signal in active microwave systems (Synthetic Aperture Radar, SAR) to obtain better spatial resolutions (Moore, 1976) and it is this approach which will be discussed next.

Active Microwave. The backscattering from an extended target, such as a soil medium, is characterized in terms of the target's scattering coefficient σ^0. Thus, σ^0 represents the link between the target properties and the scatterometer responses. For a given set of sensor parameters (wavelength, polarization and incidence angle relative to 0°), σ^0 of bare soil is a function of the soil surface-roughness and also of dielectric properties, which depend on the moisture content. The variations of σ^0 with soil moisture, surface roughness, incidence angle, and observation frequency have been studied extensively in ground-based experiments conducted by scientists at the University of Kansas (Ulaby, 1974; Ulaby et al., 1974; Batlivala and Ulaby, 1977) using a truck-mounted 1 to 18 GHz (30 to 1.6 cm wavelengths) active microwave system.

To understand the effects of incidence angle and surface roughness consider the plots of σ^0 versus angle presented in Figure 33-64 for five fields with essentially the same moisture content but with considerably different surface roughness. At the longest wavelength (1.1 GHz, Figure 33-64), σ^0 for the smoother fields is very sensitive to incidence angle near nadir, while for the rough field σ^0 is almost independent of angle. At an angle of about 5°, the effects of roughness are minimized. As the wavelength decreases, Figure 33-64b and 33-64c, all the fields appear rougher, especially the smooth field, and as a result the five curves intersect at larger angles. At 4.25 GHz, they intersect at 10°, and it was this combination of angle and frequency that yielded the best sensitivity to soil moisture independent of roughness (Ulaby and Batlivala, 1976; and Ulaby et al., 1978).

These experiments were performed in both 1974 and 1975. The first experiment was performed on a field with high clay content (62%), whereas for the second, the clay content was lower. Although both experiments provided the same specifications of the radar parameters for soil moisture sensing, i.e., frequency around 4.75 GHz and a 7–17° nadir angle, the observed sensitivity of σ to soil moisture in the 0 to 1 cm layer was different for the two experiments (See Figure 28-7 in the Water Resources Chapter). When the soil-moisture content was expressed as a percent of field capacity to account for textural differences, the sensitivities became almost identical (Figure 28-18) with a correlation of 0.84. This dependence on the percent of field capacity resembles that observed with the thermal-inertia and passive-microwave techniques. Similarly the sampling depth for active microwave sensors also seems to be limited to the surface few centimeters of the soil for the wavelengths considered in the Kansas study (Ulaby et al., 1978).

Although no detailed airborne investigations have yet been reported on the active microwave response to the soil-moisture content beneath a vegetation canopy, differences between dry fields and fields undergoing irrigation have been de-

WET SOILS: 0.34 – 0.4 g/cc IN TOP cm

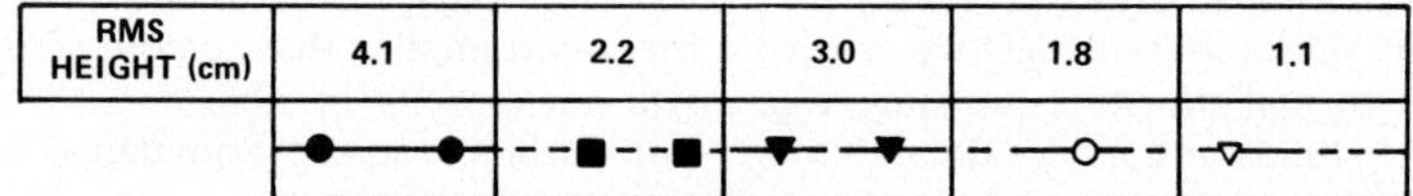

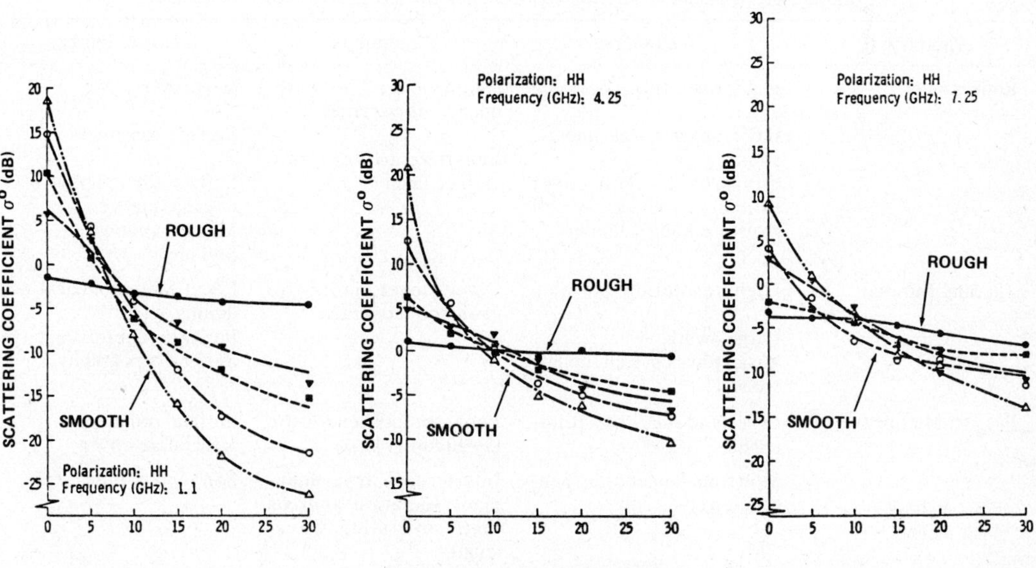

Fig. 33-64. Angular response of scattering coefficient for the five fields for high levels of moisture content at: (a) L-band (1.1 GHz-27 cm); (b) C-band (4.25 GHz-7 cm); (c) X-band (7.25 GHz-4.1 cm). 1975 soil moisture experiment (Batlivala and Ulaby, 1977).

tected using radar observations. During a flight by the NASA/JSC/SPP3A aircraft over a test site near Garden City, Kansas, using a 13.3 GHz scatterometer, Dickey et al. (1974) measured several fields, each of which contained sections into which irrigation water was flowing and sections ready for irrigation but not yet wetted. In one corn field the effect of the irrigation on the radar return seemed to produce a difference of about 7 dB at angles within 40° from nadir between the irrigated and non-irrigated sections. The differences in σ can only be attributed to the effect of moisture since all ground conditions, except soil water content, were similar over the entire field.

The presence of a vegetation canopy over the soil surface reduces the sensitivity of the radar backscatter to soil moisture by: 1) attenuating the signal as it travels through the canopy down to the soil and back, and 2) contributing a backscatter component of its own. Moreover, both factors are, in general, a function of several canopy parameters, including plant shape, height and moisture content, and vegetation density. The effect of the vegetation cover on the radar response to soil moisture is to reduce the sensitivity by about 40% when the bare soil and vegetation-covered responses are compared as a function of percent of FC in the top 5 cm. The vegetation-covered response represents data for several crops—wheat, corn, soybeans, and milo—covering the wide range of growth conditions (Ulaby et al., 1977).

There are many similarities in the two microwave approaches to soil moisture sensing, e.g. ability to penetrate clouds and moderate amounts of vegetation and the limitation to sampling only the surface 2–5 cm of the soil. The major difference is that of spatial resolution. For passive systems the resolution is limited by the size of the antenna. On the other hand, through the use of synthetic aperture techniques, the obtaining of a spatial resolution of 100 m or better is possible from space, e.g. 25-m resolution was obtained through use of the 18 cm synthetic-aperture radar on the Seasat satellite. The problems with the latter approach are the difficulty in getting an absolute calibration for the SAR, the strong sensitivity to surface roughness and look angle, and the large amount of data that would have to be handled in any operational context.

The sensitivity to soil moisture of the various remote sensing approaches discussed here has been demonstrated in field or aircraft experiments and to a certain extent from spacecraft platforms. These experiments have also indicated some of the problems associated with each approach (Table 33-3). Some of the limitations listed are of a fundamental nature, such as cloud-cover effects using thermal infrared, whereas others could be reduced or eliminated by more advanced technology, such as larger antennas to achieve improved radiometer resolution or the development of SAR calibration techniques. There is a fundamental limitation which applies to all of the approaches, i.e. they seem to be sensing the moisture content in a layer only 5–10 cm thick at the surface. This limitation implies that remote sensing approaches will not directly be able to satisfy those applications which require knowledge of the moisture

TABLE 33-3

Comparison of Remote Sensing Approaches for Estimating Soil Moisture

Approach	Advantages	Limitations	Noise Sources
Reflectance	High resolution possible	Cloud cover limits frequency of coverage	Vegetative cover
	Basic physics well understood		Surface roughness
		Sensitive to very thin surface layer	
	Can provide plant cover estimates for use in soil moisture budget models		Surface topography Organic matter Mineral content Sun angle
Thermal Infrared	High resolution possible	Cloud cover limits frequency of coverage	Local Meteorological conditions
	Large swath Basic physics well understood		Partial vegetative cover Surface topography
Passive Microwave	Independence of atmosphere	Poor spatial resolution (5–10 km at best)	Surface roughness Vegetative cover
	Moderate vegetation penetration	Interference from man-made radiation sources limits operating wavelengths	Soil temperature
Active Microwave	Independence of the atmosphere	Limited swath width	Surface roughness Surface slope
	High resolution possible	Calibration of SAR	Vegetative cover

conditions in the root zone of the soil. However, remote measurements in combination with appropriate soil-moisture budget models can be used to characterize root-zone soil moisture.

Evapotranspiration

Evapotranspiration (ET) is an important consideration in many of the management practices in both dryland and irrigated agriculture. It influences time of seeding, pesticide application, and many tillage practices. Information on ET is particularly important in irrigation scheduling where it is used to estimate soil water depletion. ET is also important in watershed hydrology for predicting runoff and groundwater recharge.

Because of the dynamic nature of evapotranspiration, many of the traditional models are difficult to use on a regional basis as a result of difficulties in obtaining the input parameters. To overcome these difficulties, it is desirable to use a model that can incorporate remote sensing inputs to utilize the synoptic and repetitive view available from remote sensing. Several approaches have been developed to use remote sensing, primarily reflectance and thermal technqiues, for estimating ET. These approaches range from simple empirical methods to complex energy balance/aerodynamic models. Idso et al. (1975a) discussed an empirical method for estimating potential evaporation using remote sensing. They developed the equation for bare soils

$$E = 1.72 \times 10^{-2} \left[S_M + 1.56 (R_A - R_G) + 156\right] \quad (33\text{-}15)$$

where E (mm dav^{-1}) is the 24-hr evaporation rate, S_N is daily net solar radiation, and R_A and R_G are 24-hr totals of incoming and outgoing thermal radiation, respectively. R_A was calculated using screen level air temperature (Idso and Jackson, 1969). R_G was calculated using the Stefan-Boltzmann equation and the average of the daily maximum and minimum surface temperature. Idso et al. (1977) reported that equation (15) may work for crops as well as for bare soil (Figure 33-65).

Jackson et al. (1976) presented a method for using albedo measurements of bare soil to partition the fraction of soil exhibiting energy-limiting (potential) evaporation and the fraction exhibiting soil-limiting evaporation during the transition phase (energy limiting to soil limiting). During the transition phase, a fraction of a field has a dry surface that evaporates at the soil-limiting rate, while the remainder evaporates at the potential rate. They defined a partitioning factor β; as

$$\beta = (\alpha_d - \alpha)/(\alpha_d - \alpha_w) \quad (33\text{-}16)$$

where α_d is dry soil albedo, α_w is wet soil albedo, and α is the soil albedo for a particular day. β varies from 1 for a wet surface to 0 for a dry surface.

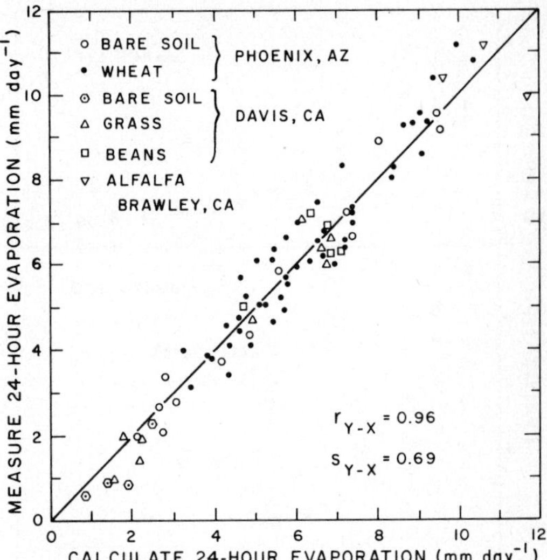

Fig. 33-65. Measured vs. calculated daily evaporation for different crop and soil surfaces in Arizona and California at various times of the year (From Idso et al., 1977).

The combined actual evaporation rate (E_c) was calculated using the equation

$$E_c = \beta E_p + (1 - \beta) E_s \quad (33\text{-}17)$$

where E_p and E_s are potential and soil-limiting evaporation, respectively. Jackson et al. (1976) used the Priestly-Taylor (1972) relationship

$$E_p = a (Rn - G) 1c/(1c + \gamma) \quad (33\text{-}18)$$

to calculate potential evaporation where "a" is a constant (1.32), Rn is net radiation, G is soil heat flux, s is the slope of the saturation vapor pressure/air temperature curve, and γ is the psychrometric constant. A regression equation developed from measurements on bare, wet soil was used to estimate Rn from measurements of incoming solar radiation. Incoming solar radiation can be estimated from geostationary satellite data (Tarpley, 1979; Gautier et al., 1980). Soil limiting evaporation E_s was estimated using the equation of the form (Ritchie, 1972; Black et al., 1969)

$$E_s = Ct^{-1/2} \quad (33\text{-}19)$$

where C is a constant which varies with soil type and t is time in days from the start of the soil-limiting phase. Figure 33-66 compares calculated and lysimetrically determined evaporation using the albedo technique.

Kanemasu et al. (1977) estimated ET of winter wheat using a model which incorporated Landsat estimates of leaf area index (LAI). In addition to LAI, model inputs were daily estimates of solar radiation, maximum and minimum air temperature, and precipitation. Model outputs were daily estimates of potential ET, transpiration, soil evaporation, and root-zone soil moisture.

Potential ET was calculated using the Priestly-Taylor equation discussed earlier where "a" is

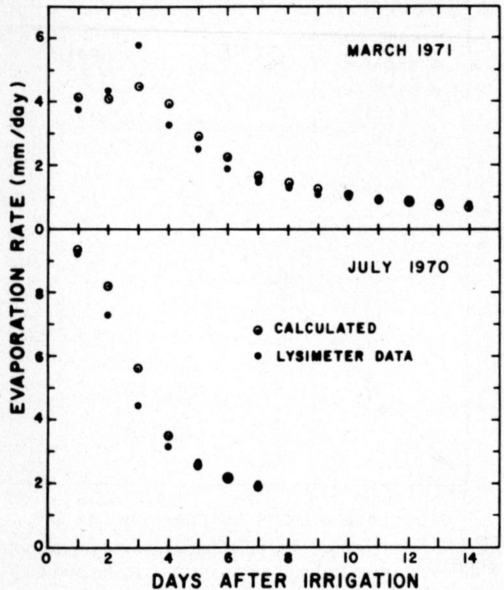

Fig. 33-66. The calculated and lysimetrically determined evaporation rates for March 1971 and July 1970 (From Jackson et al., 1976).

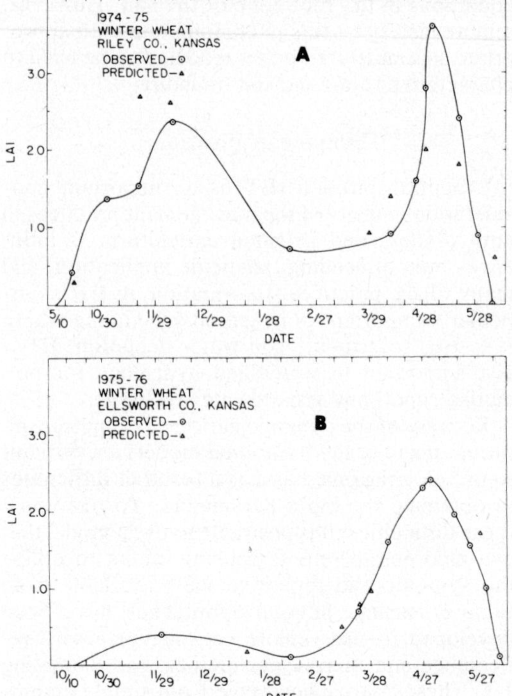

Fig. 33-67. Comparison of observed leaf area index with Landsat-predicted values (From Kanemasu et al., 1977).

1.35 for winter wheat. Rn was estimated from solar radiation using regression equations.

During the energy-limiting phase, soil evaporation was calculated using the equation

$$E_o = (\tau/a)\, ET_p \qquad (33\text{-}20)$$

where E_o is daily soil evaporation, ET_p is the Priestly-Taylor estimate of potential *ET*, τ is an energy transmittance term equal to exp (-0.737 *LAI*). During the soil-limiting phase, evaporation was calculated using the equation

$$E_s = ct^{1/2} - c(t - l)^{1/2} \qquad (33\text{-}21)$$

where c depends on the hydraulic properties of the soil, and t is days from the start of the soil-limiting phase.

Kanemasu et al. (1977) calculated transpiration using the equation

$$T = a_v\, (l - \tau)\, ET_p \qquad (33\text{-}22)$$

for crop cover less than 50% and

$$T = (a - \tau)\, ET_p \qquad (33\text{-}23)$$

for crop cover greater than 50% were $a_v = 1.56$. When the available soil water content was less than 30% of field capacity, they linearly decreased T to zero at zero available moisture.

Leaf area indices were estimated from Landsat data using a regression equation (Figure 33-67). Kanemasu et al. (1977) found that monthly *ET* rates estimated with Landsat predicted LAI agreed favorably with estimates using observed *LAI*. Wiegand et al. (1979) discussed the implication of using Landsat estimates of *LAI* for *ET* modeling. They concluded that *LAI* estimates from Landsat can extend models to as many fields as are of interest in an area or, when not used as a direct model input, can serve as a way to reinitialize models periodically during the growing season.

Khorram and Smith (1979) investigated the use of remote sensing data from weather satellite, Landsat and aircraft data for estimating potential *ET* in a 700 km² study area in the Middle Fork of the Feather River Watershed along the northeastern edge of the Sierra Nevada Mountains in California. They used the Jensen-Haise (1963) model

$$ET_p = (0.014T - 0.37)R_s \qquad (33\text{-}24)$$

where ET_p is daily potential *ET* in day^{-1}, T is mean daily temperature in °F, and R_s is daily incoming solar radiation in evaporation equivalent of day^{-1}. Temperatures were obtained from NOAA-5 satellite Very High Resolution Radiometer data. Solar radiation was derived from a technique utilizing temperature, day length, total incoming radiation, cloud cover, albedo, slope, and aspect (Khorram, 1977a; Khorram, 1978b). Land use, as estimated from Landsat and aircraft data, was used to estimate albedo.

Another approach for estimating ET with remote sensing is to use a form of the energy balance equation which incorporates the canopy-air temperature differential in an aerodynamic expression of sensible heat flux. A general form of this equation is

$$ET = R_n - G - f(u)\, C\, (T_c - T_a) \qquad (33\text{-}25)$$

where $f(u)$ is a function of windspeed, C is the volumetric heat capacity of air, T_c is canopy temperature, and T_a is air temperature. The reliability of equation (25) as a predictor of ET has been demonstrated by Brown and Rosenberg (1973), Stone and Horton (1974), Blad and Rosenberg (1976), and Heilman and Kanemasu (1976).

Heilman et al. (1976) used thermal scanner measurements of canopy temperature for corn, soybean and millet in a resistance form of equation (25) to estimate ET. The resistance equation they used was

$$ET = R_n - G - C\,(T_c - T_a)/r_H \qquad (33\text{-}26)$$

where r_H, the thermal diffusion resistance, is a complex function of wind velocity, temperature, and surface roughness. Measurements of wind-velocity and air-temperature profiles were used to determine r_H (Heilman and Kanemasu, 1976). Model estimates of the instantaneous ET flux using remote measurements of canopy temperature differed from lysimetric measurement by −0.40 to 0.17 ly/min. The large relative errors were caused primarily by effects of atmospheric attenuation on the scanner measurements.

Soer (1980) used equation (26) and aircraft scanner measurements of T_c to estimate ET for a grassland area in the Netherlands. The resistance was estimated using measurements of windspeed, air temperature, and the assumption that roughness lengths for momentum and heat were identical and equal to 0.13 h where h is crop height (Monteith, 1973).

Instantaneous fluxes of ET were converted to 24-hr estimates using the Tergra model (Soer, 1977) which simulated the daily course of crop surface temperature, ET, R_n, G, and dew formation. Boundary conditions were temperature and soil water pressure at a reference level in the soil, surface energy balance, and the temperature and vapor pressure at a reference level in the atmosphere.

Soer (1977) analyzed the accuracy of the ET calculations by assuming that the equations were correct and the only errors introduced in the calculations originated from measurement errors of the input parameters. He concluded that the main attention must be given to obtaining accurate estimates of crop temperature and roughness length for heat transfer.

Determination of the wind function $f(u)$ in equation (25) is tedious and is dependent on crop type and location. However, Jackson et al. (1977) found that for their experimental conditions in Phoenix, Arizona, wind was not of major importance in calculating ET. They speculated that effects of wind may be somewhat compensated for by changes in canopy temperature. Therefore, they combined the wind factor into a constant that multiplies the canopy-air temperature difference in an equation of the form:

$$ET = R_n - B\,(T_c - T_a) \qquad (33\text{-}27)$$

where ET and Rn represent daily values, $T_c - T_a$ is the canopy-air temperature difference measured between 1300 and 1400 hours, and B is the composite constant that must be determined. For 24-hr periods, soil heat-flux G is negligible. For their environmental conditions, Jackson et al. (1977) obtained a value of $B = 0.064$. Their results indicate that if daily values of Rn are available, ET can be estimated reasonably well using equation (27). Use of equation (27) at other location requires local calibration, a constraint shared by many ET models.

Estimating ET using remote sensing data requires that an appropriate model be chosen so that a minimum of ground measurements are needed. Many of the models assume a uniform plane of sources and sinks and thus are not applicable to conditions of partial vegetative cover. Also, a number of models require local calibration. Since only a selected number of models were discussed, reviews by Tanner (1967), Jensen (1974), and Kanemasu et al. (1979) are recommended for a more complete discussion of ET equations.

Irrigation Scheduling

Efficient use of irrigation water requires application of proper amounts of water at the right time. Many of the current techniques for scheduling irrigation use meteorological and soil measurements with no direct measurements of crop canopies. Remote sensing of a crop's spectral response to changes in plant water-content offers the potential of a rapid, noncontact method or evaluating crop water-status which could lead to improved methods of irrigation scheduling.

Stomatal closure as a result of water stress results in elevated plant temperatures (Ehrler, 1978a, b). Possible uses of canopy temperature to evaluate crop water-status have been discussed by Tanner (1963), Wiegand and Namken (1966), Erhler and van Bavel (1967), Aston and van Bavel (1972), Erhler (1973), Nixon et al. (1973), Moore et al. (1975), and Blad and Rosenberg (1976). Hiler and Clark (1971) and Hiler et al. (1974) described a "stress day index" for use in irrigation scheduling, and suggested that the leaf-air temperature difference may be an indicator of crop water stress.

Idso et al. (1977) and Jackson et al. (1977) introduced the stress degree day (SDD), which is the daily value of the canopy-air temperature difference ($T_c - T_a$) measured at the time of maximum surface temperature. Generally, $T_c - T_a$ will be near zero or negative for canopies with adequate water, and positive for water-stressed canopies. Thus, the sum of positive values of $T_c - T_a$ may be an index of when to irrigate. Jackson et al. (1977) defined the quantity

$$SDD_{pos} = \Sigma_{n=i}^{N}\,(T_c - T_a)n \qquad (33\text{-}28)$$

where the index i is the first day after irrigation, and N is the number of days required for SDD_{pos} to reach a prescribed value. Negative values of $T_c - T_a$ are set equal to zero. In a study of six differentially irrigated aurum wheat plots at Phoenix, Arizona using hand-held radiometer measurements of T_c, Jackson et al. (1977) proposed a

SDD_{pos} of 10 as an index of when to irrigate for their experimental conditions.

Millard et al. (1978) collected thermal scanner data over the Phoenix wheat plots at 5:53 a.m. and 2:06 p.m. local time on April 1, 1976. Temperature differences among the plots are shown in Color Figures 33-68 and 33-69. Air temperature at 5:53 a.m. was 7.2°C. Color Figure 33-68 shows that water-stressed plots 1 and 3 were at ambient air temperature, while temperatures of the remaining plots were 3 to 4°C lower than ambient. At 2:06 p.m., air temperature was 28.3°C. Color Figure 33-69 shows that plots 1 and 3 were as much as 8° and 6°C, respectively, warmer than ambient. The other non-stressed plots were cooler than ambient.

Color Figure 33-70 shows a "difference" image obtained by subtracting the morning data from the afternoon data. The stressed plots had diurnal temperature variations of about 10°C greater than the well-irrigated plots. Color Figure 33-71 is an image of a thermal inertia form of the *SDD* (Idso et al., 1977), defined as

$$SDD = \frac{\text{(p.m. − a.m.) crop temperature}}{\text{(p.m. − a.m.) air temperature}} \times 18\text{-}18 \qquad (33\text{-}29)$$

The air temperature difference was used to normalize the *SDD* for environmental variability. Color Figure 33-71 shows that plots 1 and 3 had positive values of *SDD*.

Spectral reflectances which are related to crop cover can potentially be used to estimate crop coefficients that are used in meteorologically-based computer-scheduling techniques. The crop coefficient (K_c), which is the ratio of actual to potential evapotranspiration, was defined by Wright and Jensen (1978):

$$K_c = K_{co}K_a + K_s \qquad (33\text{-}30)$$

where K_{co} is a mean crop coefficient derived from experimental data under non-limiting soil moisture conditions, K_a is a coefficient related to available soil moisture, and K_s is a coefficient to adjust for increased soil evaporation after irrigation or rain. K_c and estimates of potential *ET* are used to calculate actual *ET*.

The crop coefficient is closely related to crop growth since actual *ET* is low at emergence, reaches a maximum at full cover, and then decreases as senescence begins. Jackson et al. (1980) found a similarity between K_c and the remotely sensed perpendicular vegetation index (PVI) of Richardson and Wiegand (1977) (Figure 33-72). The K_c data were obtained from spring small grains in Idaho, and the PVI was obtained from spectral measurements on full-planted wheat in Arizona.

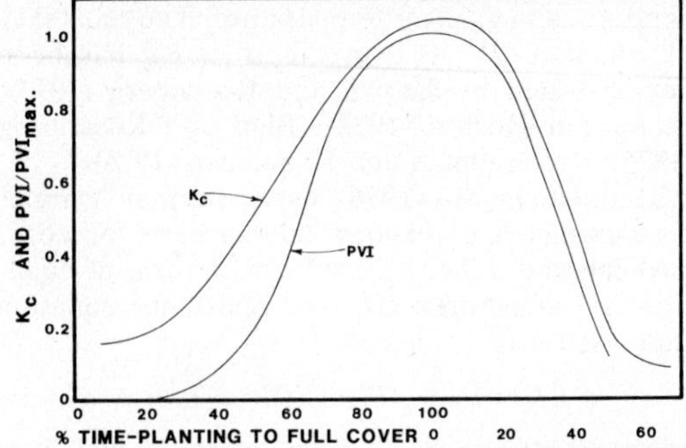

Fig. 33-72. Crop coefficient for small grains and the ratio of the perpendicular vegetation index (PVI) of wheat to the PVI at full cover as related to the percent of time from planting to full cover and the days after full cover (Jackson et al., 1980).

IDENTIFYING IRRIGATED LANDS

A number of studies have been conducted in recent years in various river basins to estimate current and projected water use and availability. Results of various studies for the same areas have differed substantially. The principal reason for the differences are that broad-based agreement cannot be reached on water withdrawal and consumptive-use rates for irrigation and determination of irrigated land acreages.

Several investigators have been successful in determining irrigated versus nonirrigated acreage using Landsat and aerial imagery (Calabrese, 1979). Manual techniques and photointerpretation, along with digital techniques, have been used in conjunction with multidate imagery and conventional data to provide statistical estimates of irrigated acreage by county and by basin in addition to irrigated land maps.

Landsat imagery was used by Hoffman (1979) to economically identify and locate land irrigated by center-pivot irrigation systems. Through the use of high altitude IR photographs a count of 2,057 center-pivots was obtained. Using Landsat data for the same sections and time periods, 2,154 areas Hoffman classified as being irrigated by center-pivot systems.

In preparation for more extensive studies in the Columbia River Basin Johnson and Loveland (1979) conducted studies in several representative areas. Project objectives included determination of irrigated acres, annual rate of irrigation expansion, growth rate of center-pivot irrigation, type of crops being irrigated and crop water-requirements for irrigated lands. The analysis effectively demonstrated the usefulness of a geobased information system used in combination with remotely-sensed data for predicting the location of irrigation development.

Visual interpretation techniques provided a quick economical method of estimating total irrigated lands in the Klamath Basin of California and Oregon (Draeger, 1979). Williams (1979) points out on the other hand that digital image-analysis techniques have several advantages compared with visual techniques. They are generally considered faster, provide objective decisions, and can readily employ multiband and/or multidate image data, thus allowing crop identification in addition to irrigation identification. The digital nature of the source image-data is also directly compatible with geographic information databases. Digital techniques, however, are more expensive and may be cost-prohibitive for small efforts.

Results of investigations for determining extent of irrigated lands by Hoffman (1979), and Poracsky (1979), indicate that Landsat band 5 is best for this use because of the characteristic of actively growing vegetation to absorb visible red light. Irrigated areas are typically located where there is sparase rainfall during the growing season, and the only plants growing rapidly in midsummer under dry conditions are those being irrigated. Band 5 shows the difference between absorbed and reflected red light, thus separating rapidly growing plants from relatively dormant plants. It appears that use of one of the vegetation transformations for detecting live vegetation, described elsewhere in this chapter might be useful.

HIGH WATER TABLES

Introduction

Irrigation provides the lifeblood for many arid lands throughout the world. The mineral richness of desert soils virtually assures good crops when irrigation is utilized. However, if the water is allowed to accumulate and evaporate, it leaves behind the soluble salts carried in the water and concentrates the residual salts already in the soil. A lack of drainage has resulted in millions of acres of irrigated lands going out of production or becoming dangerously saline. Design of new drainage systems, maintenance of old systems, and management of cropping systems depends upon a knowledge of the depth to the continually fluctuating water table. Excessive water in the soil profile occurs in several ways, the most common is that of high water-tables resulting from an accumulation of excessive water over a restricting layer in the profile. In many cases benefits resulting from removal of excess water have been very conspicuous and have led to a hundred-fold increase in land values. The term "high water table" as used here refers to an accumulation of gravitational water that at some time during the year interferes with or influences agricultural plant growth. Frequent monitoring, such as can be provided by remote sensing, can provide necessary information for managing water tables for optimum crop production. Management, in turn, can provide optimum water table depths and prevent fluctuations to avoid damage to root systems.

Research Investigations

Thermal-infrared aircraft missions for detecting shallow groundwater have been successful in South Dakota (Myers and Moore, 1972; Moore and Myers, 1972). These studies showed that a surface-emittance anomaly could be used to locate near-surface groundwater in glacial drift. However, the principles of thermal inertia involved also apply to shallow unconfined water tables under conditions experienced in many agricultural areas. The authors concluded that predawn August thermography conducted under certain prescribed meteorological conditions was optimum for interpretation of near-surface groundwater occurrence.

In a study of high water tables on the Kansas Bostwick Irrigation project, the Remote Sensing Institute of South Dakota State University cooperating with the U.S. Bureau of Reclamation (Ryland et al., 1975), used remote sensing technology to determine depth to water tables. The study, using corn as an indicator crop, showed that both Landsat-1 and aircraft remote sensing data correlated significantly with water-table depth measured in observation wells throughout the District. Landsat-1 MSS band 6 for May and MSS band 7 for August correlated significantly (0.01 level) with water-table depth for 144 observation wells.

Abdel Hady and Korbs (1971) evaluated diurnal thermal measurements of soil surfaces for soils having water-table depths from 0 to 1.2 m. They found that water-table depths of 0.0 m and 1.2 m produced a maximum blackbody temperature-difference of $\simeq$ 20 C. They also found that the maximum apparent surface-temperature differences associated with water-table depth-differences occurred between 1400–1600 Central Standard Time (CST) with an inversion of the relationship after midnight.

Research studies in Kern County California (Estes et al., 1975) were directed toward detection and delineation of water tables that are perched near the surface. Nearly 3000 hectares in the county have perched water tables within 3 meters of the surface. Investigations included both high altitude color infrared photography and Landsat multispectral imagery to determine their utility for this task. The use of thermal infrared imagery was also being investigated for this problem.

Thermal infrared imagery used on a Jefferson County Kansas test site revealed that anomalous areas of more intense thermal radiation coincided with relatively shallow depths to groundwater or bedrock (Myers and Stallard, 1974).

Spectral and Spatial Relations

The only intensive research investigation that has explored the reflective portion of the electromagnetic spectrum for detecting high water tables is the one described earlier by Ryland et al. (1975). In that study, data from Landsat band 6 in May and band 7 in August correlated well with water-table depths using corn as an indicator crop. The specific effects of water table on physiological characteristics of crops have not been investigated; therefore it can only be surmised that the influences are related to effects of moisture availability, aeration, and perhaps the indirect effects of salinity. Temperature measurements of crops in relation to water-table depths were significant, indicating that the thermal band of about 8.7 to 11.5 μm is suitable for this purpose.

On the Kansas Bostwick-project studies described by Ryland, thermal IR daytime and nighttime imagery was obtained over corn fields affected by water tables at various depths. Correlations of temperature and water-table depth data were highly significant indicating that evapotranspiration of crops during the peak transpiration period of 1400–1600 CST is ideal for measuring plant temperatures in relation to high water tables. The time of 1100 CST for the overpass projected for the Thematic Mapper, though not ideal, is acceptable. Any earlier time of day, such as the 0930 CST time of overpass of Landsat-1 and 2, is unacceptable because the solar insolation at that sun angle does not normally stress the crop. Also, nighttime thermal imagery is needed for detecting surface-temperature anomalies on bare soils that may be related to high water tables.

Studies by Myers and Moore (1972), Moore and Myers (1972), and Abdel Hady et al. (1971), indicate that thermodynamic conditions creating surface-temperature anomalies for bare soils are dissimilar under conditions of water tables near the surface on the one hand and deeper than about 1.5 meters on the other hand. This results in the apparent respective optimum times for thermal sensing of predawn for deeper water tables and midafternoon for very shallow water tables.

Various crops used as indirect indicators of water-table depths respond differently because of variable physiologic response of different crops to environmental conditions. Cotton has been found to be extremely responsive to changes in environment as exhibited in spectral and thermal characteristics and is an excellent indicator crop for high water table, salinity, and evapotranspiration measurements. Sorghum is intermediate in response and corn is one of the poorer indicators of stress conditions. Estimates of depth to water table made from data collected on remote sensing missions should be made based on measurements in fields of the same crop. This will also minimize differences in emissivity.

Spatial resolution of the thematic mapper (30 m for the reflective bands and 120 m for the thermal IR band) should be adequate for mapping water tables on a general reconnaissance level.

The season in which the missions are scheduled for the mapping of water tables is an important consideration. If crops are used as indicators of subsurface conditions the best response can be obtained when the canopy nearly or entirely covers the bare soil surface, and when the root system is developed to a substantial depth. This will generally be in July or August in most northern latitudes. Further research may show that thermal sensing of bare soils may yield additional information when done in late spring when water tables are usually lowest.

The main requirement for ground truth for Landsat interpretation are observation wells placed in and around the area being evaluated. Depth-to-water-table measurements would serve as calibration data for initial and continuing checks. A crop inventory of the area would be required, at least to the extent that crops in target fields are identified.

Meteorological data that should be available include (1) ambient temperature, since an important parameter for correlation and comparison purposes is leaf temperature minus ambient temperatures; and (2) rainfall.

Remote sensing may provide much of the information necessary for implementing corrective measures in areas needing drainage and for managing fluctuating water tables in areas where drainage is not needed or may not be desirable. On irrigation projects in the Western States having substantial acreages of alfalfa, it may be possible to relate emitted radiances from alfalfa fields to water-table depths. It is anticipated that evaluations will be made of dominant problems from spacecraft using broad survey methods and from aircraft using more detailed methods.

SALINITY

Saline soil conditions reduce the value and productivity of considerable areas of land throughout the world. Salinity commonly occurs in irrigated soils because of accumulations of soluble salts introduced with the irrigation water in soil profiles over extended periods.

Irrigation water always contains salts that may be deposited in the soil when the water is evaporated or used by plants. Irrigation intervals and methods must be manipulated to compensate for the reduced ability of salt-affected plants to utilize moisture. Management of the salt balance is required to maintain high crop-production rates. Management includes application of excessive irrigation water for leaching excess salts, providing soil drainage, and using proper agronomic practices such as growing salt-tolerant crops. These requirements are seldom met; therefore the worldwide soil-salinity problem can be expected to worsen. It must be realized that management of salts cannot be separated from management of water tables which is discussed in an earlier section of this chapter. Abandonment of formerly productive soils in many areas around the world has resulted from continued use of irrigation waters containing high or moderate quantities of dissolved salts. Thus the ability to determine and monitor the effects of salts on soils and plants is of great importance to agriculture.

Many soils on irrigated projects may acquire accumulations of salts within the rooting profile, with no evidence of their occurrence appearing on the soil surface. Where salts occur, the wide range of salinity effects among crops can be an aid to salinity detection through use of indicator plants. Salinity occurrence, as well as its severity, may be apparent from various crops. However, not all crops exhibit the same degree of expression of salinity damage. Barley, cotton and sugar beets are very salt-tolerant. Most other cereal and forage crops are moderately salt-tolerant. Fruit crops are salt-sensitive.

The Kern County Water Association in California, a typical productive irrigated area, has indicated that the existence of saline-alkali soils in a study area affects crop yields (Estes et al., 1975). The Geography Remote Sensing Unit, University of California, Santa Barbara (GRSU) first identified the dimension of the saline-alkali agricultural damage based on photo interpretation and estensive ground truth. Later a quantitative procedure for assessing salinity damage, based on field tonal (density) values, was developed that is highly correlated with a field average of electrical conductivity values.

Ektachrome infrared film has been used in applications of photo interpretation in estimating the severity and extent of known salt-affected areas in cotton fields on nonirrigated farms in the lower Rio Grande Valley of Texas (Myers et al., 1963). This same film has been used by Crown (1978) to evaluate and demonstrate the effectiveness of deep plowing as a solonetzic soil-amelioration practice. This film, one layer of which is sensitive in the near-infrared wavelengths, emphasizes the infrared reflectance of healthy green vegetation, which appears bright red or pink on the photographs. Cotton plants affected by salinity appear as darker shades of red and, when seriously affected, very dark. White areas on the photo are accumulations of salt on bare soil-surface areas.

It is significant that spectrophotometer reflectance curves from cotton plants affected by various degrees of moisture stress are different at wavelengths less than 1.35 μm, but are nearly identical at longer wavelengths. It has been shown by Myers et al. (1966b) that leaves from different cotton plants affected by various amounts of salinity have reflectance curves with substantial differences throughout the entire range of the spectrum from 0.5 to 2.5 μm. Such contrasting reflectance characteristics may provide the means for remote detection and identification of these phenomena.

In establishing a correlation between photographic color contrast and the average salinity in the soil profile, using color-infrared film, one must select a depth increment that is fairly representative of the rooting depth. Also, timing of the measurement must be correlated with the maximum contrasts of plant color and height. In the case of cotton in south Texas, the ideal timing is usually just before the cotton bolls begin to open. In a reasonably uniform deep soil, cotton plants at the boll stage are utilizing moisture and nutrients to a depth of about 1.5 m. Photographs of an area affected by salinity can be studied without special equipment, and a great deal can be deduced from the observations. It is desirable, however, to take advantage of instrumentation and techniques that have been developed to automate the procedures.

The Skylab investigations by Richardson et al. (1976) in Texas, report that 1.09 to 1.19 μm digital data correlated significantly with the salinity of fallow cropland. Unfortunately, the December date for the Skylab overpass in their investigation did not permit the collecting of reflectance and emittance data for salt-affected crops since very few crops are growing in the area in December.

A salinity problem not directly associated with irrigation faces a vast agricultural area in the Northern Great Plains and an adjacent area in southern Canada. The permeable soils of the area are underlain by an impermeable layer of salt-laden shales. During fallow seasons, and during periods of high rainfall, excess water moves to the salty impervious substrata picking up salt and gradually moving downslope. This transport of saline water leads to an accumulation of salt in low areas. The salt accumulations are called saline seeps and have the potential for affecting an area of about 590,000 square km in the U.S. Early identification of these saline seep source areas can lead to corrective measures such as planting salt-tolerant alfalfa to use excess water before it causes the saline-seep problem.

Saline-seep investigations reported by Dalsted and Worcester (1979) and Horton et al. (1977) identified methods for detecting various stages of saline seeps. Crops growing in the test sites were used as indicator plants for detecting saline sites of various degrees of seriousness. Also bare soils, in places, were indicative of serious salinity associated with saline seeps. Seeps in more advanced stages were best identified using color infrared photography, and emergent or incipient seeps were best identified using thermal data. Microwave sensing of saline seeps has produced positive preliminary results in a partially completed study in South Dakota by Carver and Bush (1979). The microwave analyses conclude that the maximum microwave sensitivity to soil salinity is to be found at L-band or lower frequencies and for the lowest soil moisture.

The remote sensing of plant canopy temperatures for detection of salinity also is feasible (Myers et al., 1966a). If dissolved salts are present in the soil solution, the osmotic potential of the soil water is increased. Consequently, the soil water is rendered less available to plants. A reduction in the rate of water uptake by roots and a decrease in the water content of the stalks are among the first detectable plant responses to salinity. Thus the transpiration rate is reduced and the accompanying canopy temperature increases.

The relationship of cotton leaf-temperature to the salinity of the soil is shown in Figure 33-73. The data indicate that the range in leaf temperature associated with variations in soil salinity from 0.5 to 15 millimhos cm^{-1} was 2.7°C on one date in early June and 5.4°C on a date in late June.

Experience has indicated that remote sensing of

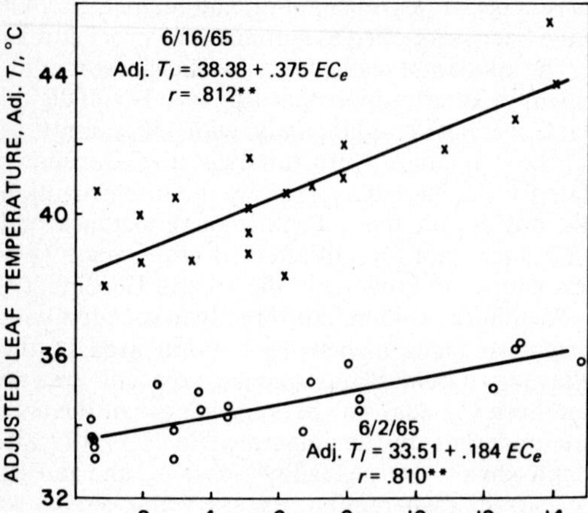

Fig. 33-73. Effect of salinity of the soil (expressed as electrical conductivity of saturated soil extracts, EC_e) on cotton leaf temperature adjusted to mean solar radiation (Adj. T_l) during the measurement period of two sampling days. Simple correlation coefficient (r) and regression equations as given on the figure. 0–4 ft profile. **Indicates significance at the 1 percent level.

plant temperatures is most successful when some physiological moisture stress has developed, especially under high evaporative-demand conditions. Hence, differences between salt-affected and salt-unaffected crops will be more evident if one delays measurements for a week or more following a soaking rain or irrigation and if the measurements are made around noon or in the early afternoon when incident solar radiation and air temperatures are near their maximal daily values.

In making ground measurements, one should always make enough observations to verify that the observed thermal or photographic image-pattern is due to salinity and not to other factors. Among these other factors are topographical undulations (which would cause high areas to suffer moisture stress first and might result in crops drowning out in low areas), soil-texture or soil-series patterns, variations in the management of different fields, and other causes not apparent from the spectral signatures alone.

The spatial resolution of thermal imagery is less than that of photographic imagery. Thus the fine detail in photographic coverage can be used to verify stand conditions resulting from uneven plant emergence; to verify disease, insect damage, or nonuniformity of fertility that would affect stand and plant size; to verify percent ground cover differences due to differences in planting date and species; and to verify other features that could complicate the interpretation of the thermal imagery. For these reasons, simultaneous coverage with reflective bands, of the area that is to be thermally scanned, is recommended.

The times for acquiring both reflectance and thermal data are important. This is true to a greater degree for thermal data, however, because phenomena being measured generally influence surface temperatures on a diurnal as well as a seasonal and annual cycle. Different physical manifestations at the same site may be helpful in measuring a single phenomenon, such as thermal response, from vegetation during the afternoon peak of evapotranspiration and soil-temperature measurements at predawn. Temperature contrasts are frequently subtle and may be influenced by meteorological factors which are frequently changing, yet these very dynamic characteristics create the changing thermal environment necessary for detection in the thermal infrared band. The upcoming Thematic Mapper will provide an opportunity to obtain thermal data that are not feasible to obtain at present.

Landsat-1 and 2 when they were operational passed over the same area every 18th day, and with their passes staggered by 9 days, gave 9-day coverage in areas where they were both operating. The obtaining of approximately that same frequency of coverage by the thematic mapper will be sufficient to monitor changes in vegetative cover as the growing season progresses. Passage over the same position at the same hour of the day each time minimizes illumination variations and other sun-angle effects.

INSECTS AND DISEASE

As previously indicated in this chapter, remote sensing offers great promise for the detection of insects, diseases, and nematodes which damage crop plants since it affords an opportunity to survey large areas rapidly, with considerable savings in manpower and expenses. Because of the great economic importance of these agents, further discussion is needed here. Any pest which supplies sufficient plant stress to significantly distort the reflectance signal is a candidate for detection by means of remote sensing. Most of the work conducted to date indicates that aerial color-infrared photography offers the greatest promise for the operational use of remote sensing in recognition and survey of insects, mites, diseases, and nematodes. Identification of host plants of insect pests with aerial color-infrared photography also offers an excellent means of planning large area control or eradication programs. The density and distribution of the hosts can be used to plan trapping or population monitoring systems.

Insects

Distortions of the reflectance characteristics of crop foliage by insects may result from feeding injury, as for example in the form of foliage deposits from the end products of insect metabolism and, secondarily, from fungus growth on these end products. Feeding injury can cause discoloration of the foliage, geometric distortion of leaves and of the general shape of the plant (e.g. tree crown), and defoliation, depending on what portion of the plant is being subject to stress. Damage can vary in intensity, so that quantitative measurement of the degree of damage can be made and can possibly be correlated with the density of insect infectations. Most insect surveys are now conducted by laborious, time-consuming, and expensive methods. Where extensive areas are planted to a single crop, the surveys sometimes require so much time that the areas surveyed initially may be significantly changed by the time the survey is completed. This factor is particularly important when insects and mites with short life cycles are involved that exhibit rapid and extreme fluctuations in numbers. Ground surveys for pests of this type usually involve the sampling of a very small portion of the total area and the application of statistical methods to estimate total or representative populations. Since insect pests are subject to rapid increases and decreases in density, it is frequently necessary to make repeated or continuous ground surveys in order to properly evaluate the population dynamics of the pest under study.

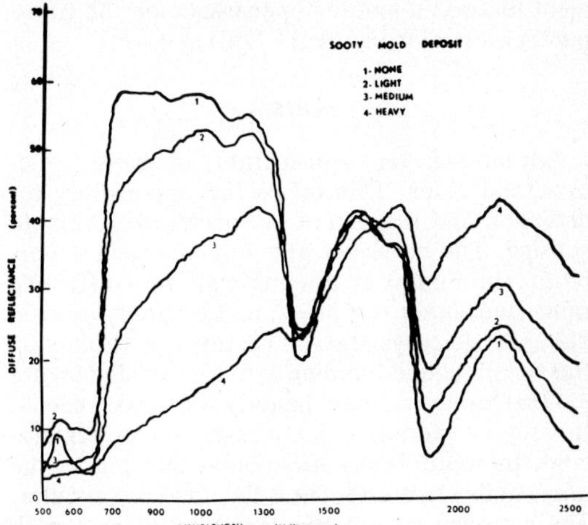

Fig. 33-75. Effect of sooty-mold on reflectance (percent) of leaves coated with varying amounts of the fungus resulting from infestations of brown soft scale insects.

Sooty-Mold Producers

The production of sooty-mold by insects comes about as a result of the excretion of "honeydew", a highly proteinaceous material. The "honeydew" deposits give a glistening appearance to the affected foliage. These deposits serve as a host medium for the sooty-mold fungus, *Capnodium citri* (Berk. and Desm.). The black hyphae spread across the surfaces of the leaves soon after the honeydew becomes established and, if the insects continue to fluorish, the foliage becomes covered with a thick black fungal mat. Reflectance measurements showed that citrus blackfly, brown soft scale, and citrus mealybug were indirectly associated, respectively, with thinnest, intermediate, and thickest sooty-mold deposits on citrus leaves. Since reflectance decreased as the sooty-mold deposit increased, this verifies that remote sensing with photography is useful to distinguish among early infestations by these three insects (Gausman and Hart 1974).

Hart and Myers (1968) demonstrated with brown soft scale, *Coccus hesperidum* L., on citrus that this darkening of the foliage can be detected with aerial color-infrared photography (Color Figure 33-74). They also showed the effect that varying levels of the fungus have on the reflectance characteristics of citrus leaves (Figure 33-75). Hart and Ingle (1969) later found that population levels of the brown soft scale as low as 1 percent, in terms of the number of leaves infested, could be detected using this technique (Color Figure 33-76). It is estimated that the use of this remote sensing method in place of ground surveys for control of brown soft scale would reduce costs by at least 50 percent.

A second sooty-mold producer, the citrus blackfly, *Aleurocanthus woglumi,* (Ashby), (Hart et al., 1973), was also found to be detectable (Color Figure 33-77a), but medium to heavy infestations are required before this pest is identifiable, since it does not produce honeydew as abundantly as the brown soft scale. The injury to citrus leaves caused by the feeding of the citrus blackfly, resulted in damage to cells of the lower epidermis, such as thickened cell walls, loss of cellular contents, and the appearance of chlorotic areas (Hart et al., 1976). The spectral properties of the leaves, reflectance, transmittance, and absorptance, were all significantly different in the visible region (500 to 700 μm) of the spectrum from those of healthy leaves. This pest is a constant threat to the citrus industry in the United States since is it found over widespread nearby areas of Mexico. Ground surveys for detection have been replaced by aerial infrared-color photography, a significant operational use of remote sensing. This has resulted in an annual savings of about $44,000 per year in surveying for this pest in the 56,400 hectares of citrus in northeast Mexico (1973 cost figures).

Other sooty-mold producers, such as citrus mealybugs, *Planococcus citri* (Risso), (Hart et al., 1971) and yellow pecan aphids, *Monellia* spp., (Payne et al., 1971) have also been found to be detectable.

Since there are a large number of insects that produce honeydew, and subsequently sooty-mold, precise identification may be difficult where infestations of these species overlap. It has been found, however, that some species may be separated because their location on the trees causes differences in the patterns of sooty-mold development (Hart et al., 1971). For example, the brown soft scale attached itself on upper leaf surfaces and stems out to the end of the terminals. This results in a solid coating of sooty-mold over the infested area. The citrus mealybug, however, is found within the crown of the tree on fruit clusters in crotches and where limbs cross. This results in a discontinuous pattern of sooty-mold development that has a mottled appearance on the transparencies (Color Figure 33-77b).

Ants

Ant mounds vary considerably in shape, structure, and color. This offers the opportunity for detection and separation of species with remote sensing. The harvester ant, *Pogonomyrmex barbatus* (F. Smith) is a significant pest affecting rangeland, other turf areas, and a variety of crop plants. It destroys grass in pastures by turning up the soil in mound-building activities (Color Figure 33-78a), and it feeds heavily on grass seeds, thereby interfering with the reseeding of pastureland. In south Texas the mound that marks the home of this harvester ant is flat and very circular. The entrance hole at the center provides a dark spot. Aerial color-infrared photographs clearly show the distinct circular pattern with the dark spot at the center (Color Figure 33-78b) (Hart et al., 1971).

The Texas leaf-cutting ant, *Atta texana* (Buckley) is also damaging to pastures. In addition, it can be a severe pest in citrus groves, where it has been known to defoliate a citrus tree overnight. This ant builds mounds that are irregular in outline with multiple entrances. These irregular clusters of mounds show up readily on color-infrared film (Color Figure 33-79) (Hart et al., 1971) and can be easily separated from the pattern produced by harvester ants and many other ant species.

Studies conducted in the Texas coastal plains from July 1971 through August 1972 demonstrated that mounds of the red imported fire ant, *Solenopsis invicta* (Buren), can be detected on color, color-infrared (CIR), and black-and-white infrared (BWIR) aerial photographs of ant-infested land (Green et al., 1977).

Analysis of aerial photographs taken with a modified Fairchild K-37 camera, and also with a Hasselblad camera, showed that up to 90 percent of the total imported fire ant mounds present in research plots could be visually detected on photographs having scales of 1:2000. CIR and BWIR film types appear to be superior to regular color film in terms of the ability to visually detect fire ant mounts on resulting photographs. December proved to be the optimum month for detecting red imported fire ant mounds in Texas with the aerial camera systems used in the study.

The spread of the imported fire ant (*Solenopsis invicta*) throughout the Southwestern United States and along the Gulf Coast from Florida to Texas has become a major concern to agricultural interests and the general public. Currently, imported fire ant infestations are spread over 126 millions of acres. Present ground survey procedures cover only a small portion of the infested area (2 percent) and are very costly. Aerial infrared-color photography has been proven useful as a survey tool for establishing the location, number and size of fire ant mounds within an area at a much lower cost, with 80 percent accuracy, at an altitude of 3000 feet. An improved photographic technique (Ingle et al., 1975) has been developed by exposing the film when the sun was at a low angle to the mounds. This technique provides detection of fire ant mounds with greater accuracy (Color Figure 33-80). Such photography emphasizes the shadows of the mounds, thereby providing a more positive identification and an extension in the range of size of mounds that can be detected; it also eliminates flat objects. Low sun-angle photography for imported fire ant infestations is superior to the existing ground survey because a larger area can be surveyed at a reduced cost. Also permanent records concerning each mound are obtained. It is also superior to maximum sun-angle photography since it improves the accuracy of photointerpretation due to the increase in the number of items making up the signature of a fire ant mound. The mound is represented by a larger area on the film because its shadow makes possible better detection of smaller mounds and because there is an area of high contrast on the film for each mound, which makes it easier to see the mounds. The photographs can now be taken from a higher altitude, which results in a reduction of cost and a saving in time.

A comparison of films and filters for detection of ant mounds revealed that Kodak Aerochrome Infrared film, 2443, with a Kodak Wratten Filter No. 15 plus 20B, provided the optimum contrasts and the most effective definitions.

Crown Distortion

The larvae of root borers, *Prionus* sp., feed on the roots of pecan trees: roots are girdled and sometimes severed by their activities (Payne et al., 1971). Heavy feeding results in a gradual limb-by-limb death of the pecan tree. Foliage on the declining trees is thinner, more irregular and lighter green than foliage on healthy trees. Aerial surveys using Kodak Ektachrome Infrared 8443 and Kodak Aerochrome Infrared 2443 films produced photographs showing a fingerlike appearance (Color Figure 33-81) where shadows of the infested trees occurred. This latter characteristic can be emphasized with oblique photographs or those taken with decreased sun angle.

Color Changes of Foliage

The black pecan aphid, *Tinocallis caryaefoliae* (Davis) feeds on pecan leaves causing an initial yellowing of the foliage which later turns dark brown (Payne et al., 1971). When infestations become heavy, premature leaf drop may result. The optimum time for survey is during August and September, since injury levels are usually highest at that time. Observations of transparencies prepared from Kodak Ektachrome infrared 8443, with an orange filter revealed that trees affected by the black pecan aphid appeared lighter in color and softer in texture. The differences observed between healthy and infested trees were quite subtle on the transparencies. However, the reflectance characteristics clearly showed the effect of the feeding injury and differences in susceptibility of pecan varieties to black pecan aphids.

The European red mite, *Panonychus ulmi* (Koch), damages foliage of peach trees by sucking juices from the leaves. Reduction of chlorophyll causes a loss of green color. When injury becomes severe the upper leaf surfaces develop a bronze or silver appearance (Payne et al., 1971). Injury to citrus leaves caused by the Texas citrus mite, *Eutertranychus banski* (McGregh), is very similar and Hart and Ingle (1969) found that injury by this mite resulted in an increase in the diffuse reflectance (Figure 33-82). Aerial color-infrared photograhs of a peach orchard with European red mite showed that infested trees had a ragged appearance and were lighter red than noninfested trees.

Defoliation

Harris et al. (1976) in work with the walnut caterpillar demonstrated that they could: 1) identify all pecans in the area photographed (Color Figure 33-83), 2) assess the amount of pecan defoliation, 3) determine the cause of pecan defoliation, 4) correlate causal factors associated with pecan defoliation with the defoliated as well as foliated trees, and 5) assess the long term effects of defoliation on the affected trees.

The investigation of walnut caterpillar defoliation has resulted in the development of techniques to identify pecans in mixed-tree native stands in South Central Texas through the use of color-infrared photography. This capability should be of value to the pecan industry in identifying total acreage for potential pecan production, and in locating pecans that may be endangered by exogenous insects. It may also be of value to the ecologist in describing the tree associations under natural conditions, and to the entomologist in future investigations, as well as others.

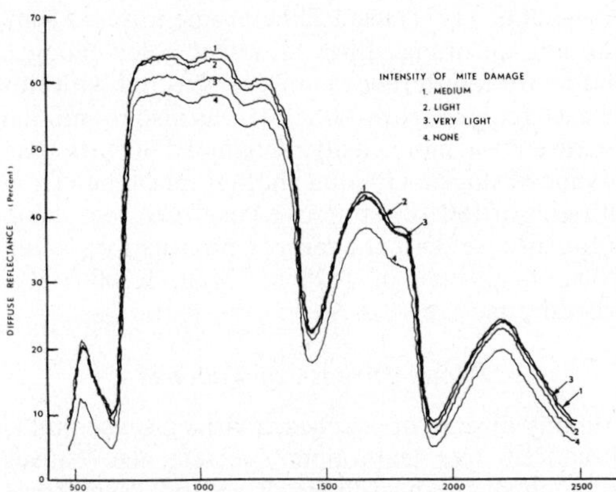

Fig. 33-82. Effects of varying amounts of damage by the Texas citrus mite on reflectance (percent) of citrus leaves.

Cultural Control

Aerial infrared photography was demonstrated to be effective in early determination of the effectiveness of plant-growth regulators applied to remove late-season fruiting forms of cotton, *Gossypium* spp., and associated diapause larvae of the pink bollworm, *Pectinophora gossypiella* (Saunders) (Henneberry et al., 1979). In addition, response to fertilizer, differences in cultivars, and disease were readily identified.

Color-infrared aerial photography is a unique tool for monitoring the application of growth regulators to cotton. Effectiveness can be determined before it can be detected on the ground and as early as two days after treatment. Also, the coverage obtained when a growth regulator is applied can be determined. If total coverage is not obtained, the overall effectiveness of the treatment is diminished. With the aerial photographic method described, large areas of cotton could be monitored, and coverage could be determined at a cost of 1 cent/ha. Also, color-infrared photography can be used to identify many of the cultural and varietal characteristics important in cotton culture.

Hart et al. (1978) found that the density and distribution of various host plants of tropical fruit flies can be detected with color-infrared aerial photography. Studies conducted in Hawaii, El Salvador and Mexico demonstrated that hosts of melon flies, oriental fruit flies and Mediterranean fruit flies can be identified at a scale of 1:10,000 (Color Figure 33-84). This work offers considerable promise for improved survey, control and eradication programs for these pests. The technique is particularly useful in rugged, inaccessible terrain, which fruit flies frequently inhabit. The procedure can be applied for approximately 1.0 cent/ha.

Radar Location of Insects

Glover et al. (1966) conducted experiments designed to extend basic knowledge of the radar backscattering properties of insects in free flight and to gain knowledge of the characteristics of these flights, in order to distinguish echoes from insects from those due to clear air phenomena. Four different species of insects were studied at altitudes from 1.6 to 3.0 kilometers and temperatures from 7° to 13°C. The species were hawkmoths, tobacco budworm moths, honeybees, and dragonflies. Three high powered, high-sensitivity radars of differing wavelengths (10.7 cm., 3.2 cm., and 71.5 cm.) were used. The observations confirmed the view that insects are highly localized point targets which can be tracked for long periods.

In the area of entomology, the utility of these experiments lies in the demonstration that radar can be successfully used to measure entomologically significant parameters which have heretofore been considered largely unmeasureable. It appears from these studies that radar is of value to the entomologist in tracking insect flight and insect behavior patterns and migration; in determining free-flight azimuth, elevation, range and velocity of single insects; and in making studies to determine the uniqueness of cross-section fluctuation spectra for a given species.

Plant Disease and Nematodes

Plant stress caused by plant diseases and nematodes provides signatures similar to those resulting from insect damage. Crown distortion, color changes in foliage, defoliation and textural differences are also associated with these organisms.

Maine Potato Late Blight

Control of potato late blight, can be both expensive and ineffective unless applications of the fungicides are properly timed (Manzer and Cooper, 1967). Timing of these applications is dependent upon accurate knowledge of both weather conditions and inoculum potential. Potato dump-piles are the most important source of primary inoculum in Maine; but once late blight becomes established in fields, its progress there largely determines the inoculum potential. Conventional ground evaluation of disease development in the field has not been satisfactory because it is too slow to be practical as a source of useful current information. Aerial disease survey, utilizing infrared film, offers a practical solution to the need for speed.

Fig. 33-85. A black and white infrared photograph with a long rectangular fungicide test plot running longitudinally through the center of the frame. Late blight of potatoes is seen as dark areas. The large, black square to the right of the test plot was used as the source of inoculum. (Photo furnished by Prof. G. Cooper and F. Manzer, Depart. of Bot. and Pl. Path., Univ. of Maine).

The illustrations show the appearance of late blight in test plots at the Maine State Agricultural Experiment Station Farm in Presque Isle, Maine. Figure 33-85 is representative of the appearance and detail one could expect on a black-and-white infrared photograph of the field while Color Figure 33-86 is the same field as shown in the now discontinued Kodak Ektachrome Infrared 8443. While these photographs show the disease in a state that was fairly well advanced, it was possible to pick up primary infection approximately 2–3 days before the symptoms were visible on the ground surveys. This, of course, allows careful timing of spray applications from the basic knowledge of the inoculum potential observed in the photographic survey of the fields.

Clitocybe Root Rot of Pecans

Payne et al. (1971), using Ektachrome 8443 color-infrared film showed that pecan trees infected with the fungus known as clitocybe root rot, *Clitocybe tabescens* (Bres) were stunted and appeared lighter red than healthy trees. The stunting caused irregular canopy images (Color Figure 33-87) giving the trees a starlike appearance. Stunting is the only above-ground symptom of this disease and infected trees may appear healthy although much of the root system has been destroyed by the disease. For this reason the problem is probably widespread in the southeast but has gone undetected because there is a lack of characteristic symptoms on visual observation.

Bunch Disease of Pecans

Bunch disease, an ailment of pecans of undetermined cause and spread, causes color changes in the foliage. The color of the diseased leaves varies from dark to chlorotic and production of abnormally large numbers of lateral buds results in a bushy growth of slender, willowy shoots. Payne et al. (1971) used Ektachrome infrared 8443 film and an orange filter to detect color changes and textural differences in trees affected with this disease (color Figure 33-88). It was also found that the disease is more easily recognized in early than advanced stages. Oblique photographs taken from altitudes of 180–275 m proved useful for detection of the disease. Direct overhead photographs taken from an altitude of 1525 m (scale 1:5000) also proved effective.

Phony Disease of Peaches

Phony disease of peaches, a virus disease that is spread by five leafhopper species, also causes color changes and foliage distortion. Symptoms are shortened internodes, profuse lateral branches, and flattened leaves. This disease is difficult to detect visually by ground observation. Aerial photographs with Kodak Ektachrome infrared 8443 film, with an orange filter (Payne et al., 1971) greatly facilitated early detection. The images of the diseased peach trees were pink, while the healthy trees were red on the color-infrared film. Since early identification and removal of the diseased trees reduces the possibility of spread by insect vectors, this remote sensing technique may provide the opportunity for early roguing and a slowing of the spread of disease.

Southern Corn Leaf Blight

The feasibility of detecting various degrees of corn leaf blight, *Helminthosporium maydis,* infection with color infrared photography was undertaken by Bauer et al. (1971). Symptoms of this disease are small, in the form of brown lesions, usually on the lower leaves. The lesions range in size from small specks up to 2.5 cm in diameter. The larger lesions may coalesce to form large areas of dead tissue. As the disease progresses the entire plant may be killed within two weeks of the initial infection. A Zeiss RMK camera with 30.5 cm focal length was used with Kodak Ektachrome Aero 8442, and Kodak Aerochrome Infrared 2443 films. Aerial photography demonstrated that the disease was detectable in late August and early September from altitudes of 912 to 18,240 m. Three degrees of severity (none or slight, moderate, and severe) were identified and accurately classified on 1:60,000 scale transparencies with 10× magnification. Color infrared photography was found to be more effective than regular color photography.

Miscellaneous Diseases

Bacterial blights and root rot in field beans were identified by Philpotts and Wallen (1969) using color-infrared photography. In addition, the importance of using blight-free Idaho seed produced plots of a deep pink (disease free) uniform tone in contrast to the clearly defined infection foci located in bean fields which had been produced from seed grown in southwestern Ontario for a number of years. It was found that all the interpretations made at the contact scale of 1:3600 can also be carried out at the smaller scale of 1:8400.

Meyer and Calpuzos (1968) used Kodak Ektachrome Infrared 8443 film for detection of crop diseases. Plots of sugar beets which had been inoculated with the *Cercospora* leaf spot pathogen were photographed from various flight altitudes in July, August and September of 1966. The progress of the disease was clearly discernible—four levels of infection intensity were discriminated on the photographs. In addition, a developing epidemic of late blight disease in a potato field was detected with the film during the same period.

Pratt et al. (1973) surveyed cotton and alfalfa fields from the air in two agricultural areas in California, for the root rot caused by *Phymatotrichum omnivorum* (Shear) Duggar. *Phymatotrichum*-infested areas in alfalfa could be recognized by the roughly circular ring pattern with sinuous margins. The ability to detect, on aerial photographs, areas affected by *Rhizoctonia* root canker was confirmed by laboratory examinations of infected roots. Spots related to salt or water conditions were oval and confined by the irrigation borders.

Dutch Elm Disease. Dutch elm disease has been responsible for the deaths of many thousands of elms in the nation (Waltz 1970). Bark beetles carry a fungus under the elm bark causing the tree to produce gums which plug the xylem vessels that carry water to the tree leaves. Once infected, trees cannot be saved. The answer lies in sanitation, the disposition of dead wood, where bark beetles can be harbored. Spraying with DDT was previously effective in controlling the beetle, but that is no longer permitted.

Early detection of Dutch elm disease permits prompt removal of diseased trees and limits further contamination and spread of the disease-carrying beetles. Meantime, a rapid detection and control program will help keep elms alive.

Spreading Decline of Citrus. Spreading decline of citrus trees is caused by the burrowing nematode, *Radopholus similis* (Cobb). It is a major problem on Florida citrus that causes extensive damage to the root structure of infested trees. Norman and Fritz (1965) demonstrated that damage caused by this nematode was detectable with aerial photography using Kodak Ektachrome color-infrared 8443 film. Foliage stress caused by damage to the root system was visible at early stages (Color Figure 33-89). The authors showed that the technique offered promise for an inexpensive and effective means of surveying for this important citrus pest.

The Reniform Nematode on Cotton. Heald et al. (1972) demonstrated that damage to cotton plants from the reniform nematode, *Rotylenchulus reniformis* (Linford and Oliveira) could be detected with aerial photographs using Ektachrome Infrared Film 8443. At scales of 1:4000 and 1:8000, areas treated with the nematocide 1.3 dichloropropene and untreated areas were clearly distinguishable (Color Figure 33-90). The infested areas appeared much lighter in color. In this study the authors also used infrared-color photography to show that crop rotation from cotton to grain sorghum can significantly reduce the infestation level of the reniform nematode in cotton fields. This pest is known to infest at least 65 different plants, including many of the major vegetable crops.

ARID AND SEMI ARID LAND MONITORING

Arid and semi arid regions cover more than one third of the earth's land surface. Most of these areas are between the extremes one normally thinks of as desert, ranging from sand dunes to lush irrigated areas. They range from thinly vegetated desert fringes that sustain only nomadic peoples and their grazing animals to larger areas comprising much of central Africa, Latin America and Asia that support a settled population, crops and livestock.

Less than half the world's drier zones are unproductive deserts for climatic reasons, but land usable for grazing and agriculture is increasingly experiencing a temporary or permanent decline in productivity because of a combination of pressures from man and from radical changes in weather.

The causes of deterioration of resources in critical arid areas includes not only possible climatic changes but also overgrazing, erosion, salinity and alkalinity, high water tables, cutting of woodlands on mountain slopes, and others. Desertification studies must consider human populations along with desert dynamics since humans are largely accountable for the problem and humans must find an answer to the problem.

The effect of mishandling natural resources, such as occurred in the severe Sahel drought of a few years ago, has a devastating effect on human populations. According to U.N. estimates (A Conf. 74/1, 1977) an area of 6,850,000 km^2 was severely affected in the drought of 1968–1973 in sub-Saharan Africa. This affected a total population of 16,165,000 people of which 19 percent were urban based, 37 percent cropping based, and 44 percent animal based.

An assessment of the resources of the critical areas, and monitoring of changes, whether they be improvements or further deterioration, can best be made using remote sensing procedures. Landsat imagery and supplemental aircraft imagery from limited selected areas have been shown to be effective for the assessment of grazing lands such as those involved in critical areas of desertification.

Vegetation can be used as an indicator for interpreting the remote sensing imagery since many components of a geographical landscape (soils, rocks, salinity, erosion, groundwater, etc.) may often leave an indirect mark on aerial photographs by modifying plant cover. Vinogradov (1961) points out that vegetation is best used as an indicator in interpreting factors of habitat such as soil conditions, hydrogeology, and geology.

The occurrence and condition of grasslands is a good indicator of the status of many areas involved in the process of desertification. This is especially true in the shortgrass prairies which occupy the drier lands that are often transitional to deserts or dry woodlands.

Soil association maps used in conjunction with vegetation surveys on a reconnaissance level are valuable for assessing current and potential agricultural and grazing land production in areas affected by desertification. Soil associations are broad geographical associations of one or several soils and can best be delineated using small scale Landsat imagery which covers large areas and has sufficient detail for the purpose (Westin, 1973a).

The temporary and more permanent resource decline, the latter referred to as desertification, can be halted and reclamation procedures initiated only if there is a physical assessment of the processes involved, an estimate of the magnitude of the problem, and the specific location of problem areas along with areas of potential for reclamation. Myers et al. (1978c) outlined a number of justifications for a monitoring program in which the advantages of remote sensing technology are emphasized.

There is a need for identifying critical indicators of desertification and indicators of disaster resulting from shorter term droughts and misuse of land by human populations. Remote sensing can provide most of the information needed for these situations. The purposes of critical indicators (Reining et al., 1978) are to:

1) assess vulnerability to desertification;
2) provide an early warning to action and emergency relief agencies concerning impending disaster;
3) predict the onset of desertification before it starts; and
4) assess the effects of desertification processes and the effectiveness of rehabilitation efforts.

Some of these early warning indicators are:

1) rainfall trends;
2) higher than normal temperatures;
3) extension of grazing patterns in the vicinity of water holes;
4) numbers of dust storms;
5) receding water levels in wells;
6) changes in sediment load in streams, lakes and reservoirs;
7) extension of cultivated areas into unsuitable drylands;
8) destruction of vegetation for fuel and construction;
9) occurrence of surface soil crusts;
10) changes in soil organic matter; and
11) buildup of livestock numbers during wet years.

These physical indicators would normally be considered along with social and economic indicators.

Remote sensing technology can provide many of the early diagnostic indicators of human disaster related to misuse of lands and climate changes in arid and semi arid lands. Also, it can provide an assessment of land resources and potential.

Research and development studies have illustrated the great advantage of remote sensing applied to resource evaluations and interpretations coupled with desertification-indicator studies (Myers et al., 1978c). A limited study was made of short, medium and long term indicators of desertification in the Sahel (Worcester et al., 1978a).

Figures 33-91 and 33-92 are Landsat scenes of a

Fig. 33-91. Landsat scene of an area along the Mauritania-Senegal border, 15 February, 1976. This scene along with that of Figure 92 exhibits a climatic-dependent, seasonal grazing region. The influence of short- and long-term climatic effects are portrayed. Temporal changes can be noted by comparing the same region in different years. A) interdune area where ephemeral vegetation can support some livestock grazing in normal years. Dark tones indicate vegetation while lighter tones are non-vegetated or barren dune areas, B) seasonality of flooding exemplified by lake expansion or contraction, C) long-term dune encroachment on cropping practices, D) example of variable amount of water in main Senegal River channel.

Fig. 33-92. Same area as Figure 91 but a later date, 6 November 1977. (Worcester et al., 1978).

climatic-dependent, seasonal grazing region in Mauritania. Temporal changes can be noted by comparing the same region in different years and seasons. Landsat data are shown to be an effective tool for assessment of desertification. In particular, the study concluded that correlation of Landsat data with ground observations could lead to the development of techniques for predicting the real time and near future reaction of arid and semi arid regions in relation to desertification-related stresses. The authors made a particularly useful contribution by associating short and long term indicators in relative degree of expression to soil order.

WIND AND WATER EROSION

Soil erosion is a serious problem in many parts of the world. The loss of soil, plant nutrients and soil water-storage capacity from fields reduces yields and increases the cost of crop production. Water erosion also results in accumulation of sediment in fields, in waterways and in reservoirs.

Wind erosion is a costly damage that affects millions of km^2 of land throughout the world every year. It is the most serious element of desertification. Wind erosion is part of a series of cyclic processes. This cycle is part of a much broader cycle of weathering which includes disintegration, decomposition, movement, and sorting of rock and soil materials.

Soil-destabilization processes are the basic cause of wind erosion. Devegetation, i.e., depletion or destruction of vegetation cover, is the most important basic cause of wind erosion. Drought at times has reduced or stopped vegetative growth but drought alone is seldom the cause of severe erosion. Plant cover, tillage practices, disposition of residues, crop sequence and general management efficiency all influence the extent of erosion.

Methods of erosion detection and assessment are based on tonal, textural and physiographic recognition features. On Landsat imagery erosion is temporally apparent as a change in soil color, in appearance of sand and gravel and bare rock, in accentuated dendritic drainage patterns, and others. Multispectral differences in the appearance of erosion areas of black, shale-derived soils above a Missouri River Reservoir in South Dakota are shown in Figure 33-17.

According to Tueller and Booth (1975) erosion features identifiable from space imagery are:

1) erosion potential associated with changes in vegetation and litter;
2) changes in soil type and soil color;
3) occurrence of dendritic soil patterns;
4) occurrence of sand dunes;
5) definition between bare soil or rock; and
6) vegetative cover.

Resource surveys conducted according to procedures described by Klingebiel and Myers (1974) contain basic soil and resource information related to erosion prevention. Soil properties important to land use and land management are estimated and described for the different soils. Using these basic data, predictions about soil behavior can be made, such as erosiveness, and recommendations can be made as to recommended agronomic and management practices for reducing erosion. Also, areas can be identified where erosion-control measures would yield greatest benefits.

AGRICULTURAL MANAGEMENT AND PLANNING

Agriculture contrasts with many resource disciplines because it varies with climate, economics, and many other external forces which determine what a farmer will plant and how the crop will grow. A high level of management is obviously an important factor in successful farming. Furthermore, a single assessment of agriculture is not feasible to provide management information because disasters, or average or over-abundant production have a certain probability of occurring every year.

A fortunate aspect of U.S. agriculture is that rapid transfer of information is available through the Cooperative Extension Service (Moore and Myers, 1977). The actual farming population is small and generally well educated in accepting the information with the capability to react where action could provide a benefit. Therefore, both the public and private sectors of agriculture have a suitable information system to accept and use information provided by remote sensing. In contrast, in regions of subsistence farming, 50–90 percent of the population may be agrarian living in rural areas with poor communication with population centers. These regions need information at higher levels of government. For many countries or states, agriculture is the largest economic commodity in their budget. Therefore, the impact of technological advances in agriculture can have a large effect on the growth and development of countries along with the stability of their peoples and governments.

Moore and Myers (1977) outlined examples of remote sensing users on various levels ranging from global users to the individual farmer. Each of these levels has unique requirements that affect dissemination of information.

Global

As outlined by MacDonald and Hall (1978), worldwide food production estimates are needed on a timely and accurate basis. The Large Area Crop Inventory Experiment (LACIE) was established as a cooperative project of the United States Department of Agriculture (USDA); National Aeronautics and Space Administration (NASA); and, National Oceanic and Atmospheric Administrations (NOAA). The project established the capability of estimating area, yield, and production of wheat in the U.S. and other major wheat-producing regions of the world. The LACIE project is described in detail elsewhere in this chapter. The effort required to develop procedures operational for global efforts is considerable.

Desertification is a global problem prevalent on nearly all continents. Presently, feasibility studies are being sponsored by the United Nations concerning the use of Landsat for monitoring desertification. Overgrazing, a cause of desertification, can potentially be assessed using proven procedures (United Nations, 1977).

International

Remote sensing can be important where information is required in adjacent countries or in a region. Extensive areas of Mexico are infested with the citrus blackfly. Hart et al. (1973) found that medium to high concentrations of citrus-blackfly fungus in tree tops were detectable using aerial color-infrared photography. Surveys must be conducted periodically to determine the potential of infestation in the United States. A procedure using aerial photography rather than ground surveys saved about $44,000 (U.S.) per year to inventory 56,400 hectares of citrus in northeast Mexico (1973 cost figures).

National

Efforts to inventory regions on a national scope appear two sided. In developed countries, reconnaissance-level data are available. Therefore, the developed country will benefit from remote sensing by monitoring those dynamic features such as land use. Systematic approaches as proposed by Anderson et al. (1971) can be applied to maintain uniformity in the types and levels of information extracted. Many states have completed landuse inventories using remote sensing approaches.

However, in developing countries, reconnaissance data generally are unavailable. For example, the 280-km Jonglei Canal is to be built in southern Sudan to channelize approximately one-third of the White Nile water which is normally lost in the Sudd region, a large marsh, due to evapotranspiration and groundwater seepage. Adequate resource data of the surrounding 160,000 km^2 do not exist to facilitate the development of the canal. Even though detailed ground investigations are required to locate and design the canal, location of the closest sources of suitable aggregate, location of soils which could be irrigated, assessment of vegetation before project development, and measurements of drainage intensities and directions are not known at this time.

State

Many states have applied Landsat data to landuse inventories. In most instances for a level I landuse inventory Landsat provides the only economical source of data available on a state-wide basis.

Imagery can be used to provide a spatial awareness for disseminating information contained in line maps. For example, Westin (1973a, 1974) used Landsat data to provide information on conceptual dollar values of various regions in South Dakota and summaries of soil texture, soil slope, and soil-test information. Many users do not understand soil variation mapped by a professional soil scientist but they can relate to the mapping units on the basis of pH or phosphorous level which may alter certain of their management procedures in the application of chemical soil additives. At the county level Westin (1974) has provided a map of soil texture, organic matter content, and soil pH which is used to partially determine application rates of herbicides for weed control. The line map on a Landsat background at 1:250,000 scale can be used by applicators to locate actual fields within the region of interest.

Region

In vast areas of the western U.S., saline seeps are emerging through natural processes but the rate of formation is being accelerated by man through certain of his cultural practices such as fallowing land to build up soil-moisture reserves. Miller and Bahls (1976) estimated that 57,000 hectares of land in Montana are not productive due to saline seeps and that their growth is approximately 10 percent per year. Worcester and Brun (1977) estimated that the growth of seeps in western North Dakota has been 700 to 1400 percent during the past 25 years. An earlier section in this chapter describes saline-seep research studies conducted through regional cooperation.

In agriculture, the important and economical uses of remote sensing are those which can provide base data to alter management decisions. In regional problems, even though the application may be used at the individual enterprise or farm level, the regional economic impact is large.

Individuals

The society of agriculture consists of a large group of private enterprises, independently owned and operated. These private enterprises often combine to form cooperatives for purchasing, selling, or planning purposes. The enterprises are also represented by many county, state, or federal governmental bodies. Billingsley et al. (1976) illustrated that the bulk of users who need information frequently can not accept highly sophisticated user products (Figure 33-93). Unfortunately, present design characteristics of

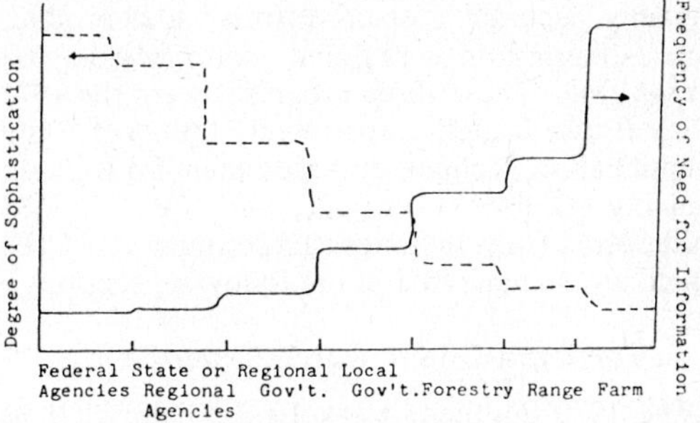

Fig. 33-93. The need and level of information required for various agricultural users of remote sensing technology. (After Billingsley et al., 1976).

Landsat satellites require detailed digital processing to fully utilize data at "large" scales required by individual producers. Wiegand (1974) demonstrated that reflectance anomalies of chlorotic areas of a sorghum field which were visually apparent in larger-scale aircraft data were observed in digitally processed Landsat data. Until a mechanism is established to provide these detailed analyses on a routine and cost-effective basis or until the scale of interpretation product is suitable for visual inspection of individual fields, use by the individual farmers will be limited.

CHANGE ASSESSMENT AND YIELD ESTIMATES

The following narrative is concerned with multidisciplinary efforts directed toward assessment of agricultural conditions on a global basis. Additional important elements of crop-yield forecasting are included in the first edition of this *Manual of Remote Sensing* (Reeves, 1975).

Advances in Agricultural Remote Sensing in the 1970's

In the late 1960's, NASA recognized the need for usable and transferable aerospace technology to address earth resources problems caused by environmental impacts such as catastrophic weather events and biological effects. One of the most obvious remote sensing applications was found to be the assessment of agricultural conditions on a global basis. The NASA approach to this technology development and transfer to the user community began rather modestly and progressed significantly. There were three distinct yet interrelated experiments that were conducted to achieve the present global agricultural-assessment program; an experiment to map and assess corn-blight effects in 1971; a program to assess and develop reliable and repeatable remote sensing data classification techniques and procedures; and lastly, a demonstration of the operational feasibility of using satellite sensor data and automated data classification techniques to conduct acceptably accurate crop assessment and production estimates on a regional, countrywide and global basis. These three programs were the 1971 Corn Blight Watch Experiment, the 1973 Crop Identification Technology Assessment for Remote Sensing (CITARS) Project, and the 1974–78 Large Area Crop Inventory Experiment (LACIE) which are summarized in the following sections.

The Corn Blight Watch Experiment

The 1970 Southern Corn Leaf Blight (SCLB) caused widespread crop damage, starting in the southern states and spreading throughout the Corn Belt (Figure 33-94). The predicted yield for corn in 1970 was 4.82 billion bushels, but the devastating SCLB fungus (*Helminthosporium maydis*) ruined 15 percent of the crop. Indiana alone suffered a loss of 95 million bushels.

Agricultural specialists anticipated that SCLB spores would survive the winter to attack susceptible varieties of corn again in 1971. Late in the summer of 1970, various scientific groups gathered to discuss the problem and to design a remote-sensing experiment to monitor the spread of SCLB.

Based upon previous experiments, the investigators flew aircraft over blighted corn fields in central Indiana to test the usefulness of remote sensors for detecting losses in crop vigor. The cooperating agencies then began to make arrangements for the 1971 Corn Blight Watch Experiment, the first regional application of remote sensing by rapid aerial reconnaissance ever undertaken for agriculture.

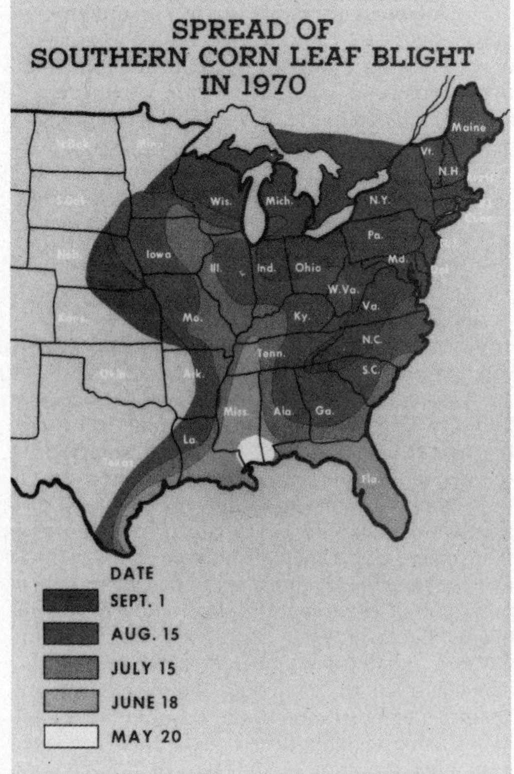

Fig. 33-94. Spread of southern corn leaf blight in 1970.

Plant pathologists have found that SCLB is caused by a fungus propagated by windblown spores. Under warm, moist conditions, the fungus rapidly attacks the leaves of susceptible varieties of corn. Significantly, only strains of corn with Texas male-sterile (T) cytoplasm (a genetic characteristic that eliminates the need for detasseling in hybrid seed production), are highly susceptible to SCLB; corn with normal cytoplasm is resistant to the fungus. Approximately 85 percent of the corn planted in 1970 was of susceptible varieties.

The first symptom of the disease is the appearance of small, spindle-shaped lesions on the lower leaves. The lesions spread to cover and kill the leaves, moving progressively up the leaf canopy until the entire plant is affected. Low levels of blight infection do not severely influence production of grain; however, in the more severe levels, the grain, if produced, is chaffy and light. Yields are also reduced by secondary infections associated with SCLB, such as ear and stalk rot.

Until the relationship between plant maturity, blight severity, and yield was more fully understood, it was not possible to establish the effects of different levels of blight at different maturity stages. Remote sensing investigators, therefore, designed their experiments to detect six levels of infection that they were confident included all significant levels.

The Experiment

The Corn Blight Watch Experiment was formally initiated in April 1971 by USDA. The experiment was supported by NASA with cooperation from Purdue University, University of Michigan, U.S. Air Force, and agricultural agencies of the seven states in the experimental region. The objective of the experiment was to evaluate the use of advanced remote sensing techniques and concepts to:

1) Detect the development and spread of corn blight during the growing season across the Corn Belt region;
2) assess different levels of infection present in the Corn Belt;
3) amplify information acquired by ground visits to better assess current blight status and the probable impact on crop production; and
4) estimate through extrapolation the applicability of these techniques to similar situations occurring in the future.

Sampling Plan

The Statistical Reporting Service (SRS) of the USDA developed a sampling plan to allow inferences to be made from remote-sensor data and supporting field observations. Each 1- by 8-mile sample segment over the seven states covered by the remote sensing experiment was selected to be representative of similar segments in the entire region. Large portions of non-agricultural land were excluded before selection of the test segments.

The SRS statisticians selected six to 10 representative corn fields in each test segment to be visited by ground observers in conjunction with each remote-sensor overflight. With this ground verification, the remote-sensor data could be interpreted for all of the test segments and could be extrapolated to reflect crop conditions over the entire seven-state region.

Test Area

The 210 representative segments within the seven-state Corn Belt region included approximately 1806 corn fields along 30 flight lines in the states of Ohio, Illinois, Indiana, Missouri, Iowa, Nebraska, and Minnesota (Figure 33-95). In addition, eight overlapping flight lines were established over an intensive study area in eastern Illinois and western Indiana.

Data Acquisition

The NASA Johnson Space Center (JSC) made its Earth Survey-3 (ES-3) aircraft available to obtain high-altitude photographs. In addition, a C-47 aircraft operated by the University of Michigan was deployed for multispectral scanner coverage from lower altitudes. The range and altitude of the C-47 were adequate for the intensive study area. For larger areas, the RB-57F was the ideal platform because of its 60,000-foot operating altitude. The principal sensors used on the RB-57F were RC-8 cameras using a variety of film/filter combinations.

For the initial land-use mapping and photo-index preparation the cameras used panchromatic film. To detect and map SCLB, the cameras used color-infrared film which is sensitive to plant vigor. It records vigorous crops as bright red, and unhealthy or dead plants as light pink to brown to gray. A skillful photo interpreter can recognize several levels of corn blight infection recorded on this film (Figure 33-96). Interpretation of the photographs was performed by persons with photo-analysis skills and agricultural backgrounds assigned to the experiment by USDA, state agricultural experiment stations, and several universities.

The experiment was divided into three sequential phases covering the spring and summer of 1971. Phase I, April 15 to April 30, was devoted to land-use mapping using the black-and-white baseline photography. Phase II, May 10 to May 30, provided information about soil conditions. Phase III, June 14 to October 1, involved actual detection of corn blight by remote sensors, including the cameras containing color-infrared film and also including the University of Michigan 12-channel multispectral scanner.

Field Observations

In coordination with the remote-sensor overflights, more than 500 enumerators from the ASCS, Agricultural Extension Service (AES), Cooperative Extension Service (CES), and personnel representing the states involved reported field observations from designated sample points. When blight was suspected, leaf samples were collected and analyzed for the occurrence of symptomatic lesions. These data were compiled by the Statistical Reporting Service in Washington, D.C., and sent to the Laboratory for Applications of Remote Sensing (LARS) Interpretation Center at Purdue University for incorporation into the blight-detection program. These field observations gave the interpreters confirmation regarding blight occurrence as interpreted from remote-sensor data.

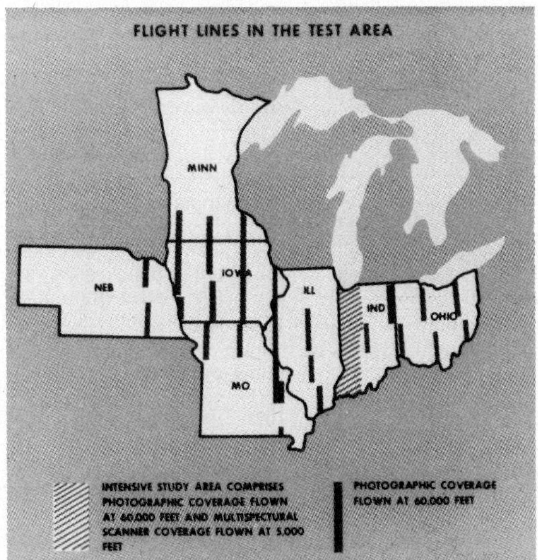

Fig. 33-95. The SCLB test area showing flight lines and the intensive study area.

Fig. 33-96. Color IR image of corn fields showing an area affected by corn blight.

Data Reduction

The photographic data acquired by the RB-57F were processed at JSC in Houston and delivered to the LARS Interpretation Center at Purdue University. A team of photo-analysts at LARS reduced the data on a bi-weekly basis, classifying the corn of each segment into blight or non-blight categories. Blighted fields were classified by levels of severity and by the uniformity of occurrence of blight within the field.

Multispectral scanner data acquired over the intensive study area were analyzed by pattern-recognition programs at LARS and at the University of Michigan Willow Run Laboratories (WRL). The LARS used a fully digital computer system and WRL employed an analog/digital hybrid computer system to identify blighted fields and classify them by levels of blight severity (Figure 33-97).

Experimental Results

Overall effects of SCLB on corn production were greatly reduced in 1971 because of increased acreage planted in corn, increased use of blight-resistant varieties, early planting, and a timely cool period in mid-July. However, blight occurred throughout the region in the early summer. Although the levels of severity reported were low, SCLB was detected in 600 counties by mid-July. Then, at about the time of tasseling, came several weeks of cool, dry weather. This unexpected respite from the anticipated warm, humid conditions tended to check the spread of SCLB and to inhibit the severity in plants that had been infected.

USDA used information from the Corn Blight Watch Experiment to evaluate reports of widespread SCLB. For example, field observations from the Watch showed that very few fields in the sample areas had suffered more than mild infections as of late July. Thus, Corn Blight Watch information was used in 1971 as a tool in evaluating all possible indications. It could not be used as an estimator of production because no parameters relating early blight infections to final production had been developed. Consideration of moderate to severe levels were later found to be important in estimating. Several weeks before harvest, preliminary results of the Corn Blight Watch Experiment indicated the absence of conditions that would seriously reduce total corn production. As anticipated, a bumper crop of corn was harvested in 1971.

Fig. 33-97. Computer printout map classifying corn fields by blight level reveals severely blighted area. Same area as Figure 33-96.

In several regions where susceptible T-cytoplasm varieties of corn were heavily planted, SCLB significantly reduced yield. Preliminary studies relating yield to level of blight severity and maturity indicated that only two or three levels of blight were significant in predicting crop yield. Early in the season, detection of mild degrees of blight was used to trace its extent and spread but, from the middle to the end of the growing season, only moderate to severe levels of blight were used in estimating corn production for the year.

At the end of the 1971 season, corn yields were obtained from sample units in fields that had been observed throughout the summer. These samples were used to establish relationships between levels of infection at different crop maturity stages and yield reductions. The results of these studies are shown in Table 33-4.

Summary

The 1971 Corn Blight Watch Experiment was highly successful in many respects. It was an excellent demonstration of the operational capabilities of remote sensing techniques. Extremely large quantities of data were collected over the seven-state experimental area on a bi-weekly basis. These data were processed, reduced, eval-

TABLE 33-4

Average Yield (Bushels/Acre[a]) as Influenced by Time and Severity of Blight Infection.

Date	Blight severity level					
	0	1	2	3	4	5
	All cytoplasm types[b]					
July 26 to 30	114	114	112	84	69	—
Aug. 9 to 13	115	114	114	104	86	92
Aug. 23 to 27	115	112	117	114	100	77
Sept. 6 to 10	108	113	113	122	115	104
	T-cytoplasm[c]					
July 26 to 30	100	102	90	76	57	—
Aug. 9 to 13	98	98	106	98	83	81
Aug. 23 to 27	99	94	102	105	92	78
Sept. 6 to 10	100	86	104	104	101	92

[a] Preliminary calculations show coefficients of variation of 5 to 15 percent.
[b] Average yield, 113 bushels/acre
[c] Average yield, 98 bushels/acre

uated, and reported within these short 2-week cycles.

Some of the benefits of the experiment were unexpected. For example, the fact that SCLB was being carefully monitored appeared to dampen corn price fluctuations on the commodity market.

Several conclusions can be drawn from these data:

1) yield reduction is related to early season blight infection;
2) slight to moderate blight infection occurring in late August and early September had little (if any) effect on yield; and
3) the blight levels that significantly affected yields are the same levels that can be detected by remote sensing.

The Corn Blight Watch Experiment was the first time that a sound statistical design was used for a large-scale, remote sensing program. Accuracy of corn identification by remote sensing exceeded 90 percent throughout the experiment, and the correlation between field observations of blight levels and classification of blight levels using remote-sensor data was quite good. Graphs showing these correlations illustrate the potential of remote sensing as a technique for monitoring crops for disease and other stresses (Figure 33-98).

The following statistics reflect the scale of the undertaking:

1) 38,500 flight-line miles were flown by data-acquisition aircraft;
2) 400 hours of RB-57F and C-47 aircraft flight time were expended;
3) 18,000 frames of photographs were taken, processed, and reduced;
4) 53,000 square miles of Corn Belt farmland were photographed biweekly (1,800 square miles of which were evaluated biweekly);
5) 800 people were involved in the experiment;
6) 14 different state and federal organizations were involved in one aspect or another;
7) 2,500 man-hours were expended at LARS in photographic interpretation during the experiment; and
8) 1,806 fields were visited biweekly by ground observation personnel.

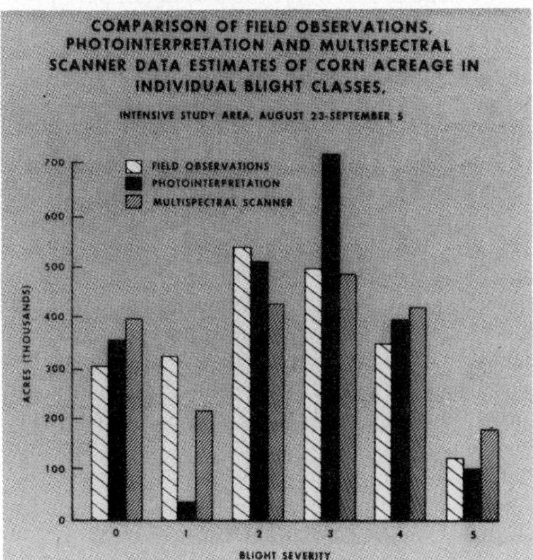

Fig. 33-98. Comparison of field observations, photointerpretation and multispectral scanner data estimates of corn acreage in individual blight classes.

From the standpoint of quantity and quality of data collected, the 1971 experiment was the largest and most successful remote sensing project attempted to that time in agriculture.

Crop Identification Technology Assessment for Remote Sensing (CITARS)

Introduction and Objectives

CITARS (Crop Identification Technology Assessment for Remote Sensing) was an experiment to quantify the crop identification performance (CIP) achievable with several automatic data processing (ADP) classification techniques which used satellite remote sensor data. It was conducted from April 1973 to April 1975. Participants were the Earth Observations Division (EOD) of NASA's Johnson Space Center, the Environmental Research Institute of Michigan (ERIM), the Laboratory for Applications of Remote Sensing (LARS) of Purdue University, and the Agricultural Stabilization and Conservation Service (ASCS) of the U.S. Department of Agriculture. Crop identification performances for corn and soybeans were assessed in six sites in Illinois and Indiana. Digital remote sensor data were collected by the ERTS-1 (now called Landsat-1) multispectral scanner (MSS) periodically throughout the 1973 growing season. The classification procedures were predefined at EOD, ERIM, and LARS to minimize subjective analyst judgment and interaction in the crop identification process.

The five specific objectives of CITARS were:

1) To assess the effects of Landsat data acquisition during the corn and soybean growing season on crop-identification performance;
2) To assess the effects of differing geographic locations having differing soils, weather, management practices, crop distributions, and field sizes on crop-identification performance;
3) To quantify the variation in crop-identification performance using the differing automatic-data-processing (ADP) classification procedures;
4) To test the ability to extend training signatures, selected within a test area, to train the classifiers in other areas (known as signature extension); and
5) To assess crop identification benefits to be derived from classifying with multiple Land-

sat data acquired multi-temporally during the crop growing season.

Technical Approach

To accomplish the objectives, six major tasks were completed. These were:

1) design of a sampling scheme in Illinois and Indiana for corn and soybeans using Landsat MSS data;
2) acquisition and preparation of a Landsat-1 data set with ancillary data sufficient to support the experimental objectives and design;
3) computer-aided processing of this data set with the selected classification algorithms and procedures;
4) quantification of the crop-identification performances to evaluate the ability of these procedures to satisfy agricultural-applications requirements;
5) a statistical analysis to quantitatively evaluate the impact of major factors known to affect crop-identification performance; and
6) an interpretation of the results to ascertain the underlying factors responsible for the results and to draw inferences as to the status of the technology as it relates to agricultural applications.

For sampling to be representative of corn and soybeans in the two states, Illinois and Indiana, six primarily agricultural counties were randomly selected to span the geographical distribution of the region's agriculture. Within each county a 5 × 20-mile (8 × 32 km) segment was selected (see Figure 33-99). Additionally, each of the six segments was located in an overlap zone of Landsat Imagery so that coverage was available on two successive days on each 18-day Landsat cycle.

Quantification of crop-identification performance is somewhat complicated by the size of Landsat pixels. Boundary pixels containing more than one signature are common in areas with small fields. Therefore, classifier-performance quantification was divided into two categories; pure pixel performance and performance based on determinations of crop-area proportion (the percentage of a crop in the region being classified). Also, to remove the analyst's crop-identification decision-errors from classification performance, crop signatures from known fields using aerial photography were used to train the classifiers.

Periodic crop observations of the fields used to train the classifiers were made by USDA/ASCS throughout the growing season. Photointerpretation of multidate aerial photography was successfully used to increase the size of the data base. Photointerpretation data evaluated pure pixel classification accuracy and crop area proportion estimates. Determinations as to accuracy of photointerpretation results, as summarized in Table 33-5, were judged as indicating that Landsat-data classification by such means is warranted.

To evaluate crop-identification performance, three categories or classes of data were defined. The first two were the major crops, corn and soybeans, while the third, called "other", included all other ground covers. Analysis of wheat-recognition possibilities through the use of Landsat data acquired early in the year was attempted, but the amount of wheat in the segments was too small and the reliability of its photointerpretation too low to support meaningful conclusions.

Multiple passes of Landsat data were registered with an average root mean square (rms) error of less than one-half pixel, enabling multitemporal classifications of the data and eliminating the need to manually locate field- and section-coordinates in each Landsat scene.

The need to maximize the number of pure pixels selected from the relatively small-sized fields present in several of the segments made selection of field coordinates more difficult than expected. Manual selection methods were found to be inadequate for the job, but a computer-aided method of transforming digitized photomap coordinates to Landsat line and column coordinates was developed and successfully used (NASA, JSC, 1974).

A key task prior to the start of ADP classifications was to define and document data-analysis procedures which were repeatable, easily followed, and capable of incorporating the judgment and skill of experienced analysts. Although there was concern that crop-identification performance might be reduced by restricting analyst decisions, it was necessary that variability due to analysts be minimized for meaningful comparisons of results.

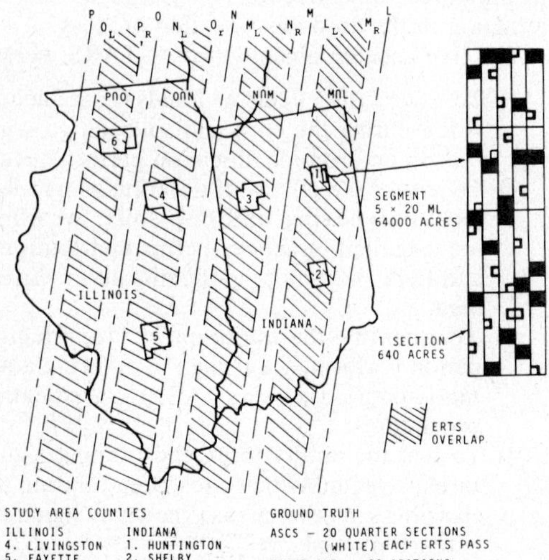

Fig. 33-99. CITARS Landsat data set design.

Results and Discussion

Statistical analyses provided a key to the quantitative assessment of remote sensing technology for crop identification, for both pure pixels and

TABLE 33-5

Comparison of crop type identifications made by ASCS and by photointerpretation.

Cover Type	Statistic	ASCS Total	Photointerpreted Total	Photointerpreted Correct	Commission Error
Corn	Number of Fields	50	51	46	5
	Percentage	100.0	102.0	92.0	10.0
	Number of Acres	1,181	1,197	1,165	32
	Percentage	100	101.3	98.6	2.7
Soybeans	Number of Fields	65	66	61	5
	Percentage	100.0	101.5	93.8	7.7
	Number of Acres	1,550	1,540	1,523	16
	Percentage	100.0	99.3	98.3	1.1
Other	Number of Fields	108	106	99	7
	Percentage	100.0	98.1	91.7	6.5
	Number of Acres	879	874	838	36
	Percentage	100.0	99.4	95.3	4.1

crop-area-proportion estimation. Previous results were confirmed in some instances, while in others unanticipated results led to reconsiderations and new insights into certain aspects of the technology. This section summarizes the major results from the CITARS experiments, as related to the five specific objectives.

Effects of Landsat Data Acquisition. The time of Landsat data acquisition during the growing season was found to be an important factor influencing crop identification performance, because of the phenological development cycles (or crop calendars) of the major ground covers. Peak accuracy for pure pixel classification was 75 to 80 percent in mid-August, as shown in Figure 33-100. At this time, the variability within the major crops (corn and soybeans) was low and the amount of ground cover was high.

The solid line in Figure 33-100 represents the expected performance for the average of all single-date procedures, assuming no interaction between the factors: location (test site) and time. The use of a noninteractive model for computing only one site had Landsat data for more than two of the six time periods. The individual points marked on the graph represent actual performances by the various procedures. The variability present is an indication that factors other than time also influenced pure pixel-classification performance.

A similar expected-performance time profile also was calculated for crop-area-proportion estimates over the aggregation of whole test sections. This profile showed roughly the same rms proportion error for all of the time periods, except mid-July when the error was substantially greater. In mid-July, the variability among corn fields and among soybean fields was high, and the amount of ground cover was low. It is noted, however, that variability in performance among procedures at any given time was much greater for area-proportion estimation than in the case of pure pixel classification.

Effects of Test Site Characteristics. Missing Landsat data hampered the analysis when comparisons were made between sites. Nevertheless, area-proportion estimation-accuracy was found to be much more test site dependent than was pure pixel classification accuracy, when expected responses were computed. The only major site characteristics which were found to be correlated with area-proportion estimation-accuracy were average field size and proportion of corn and soybeans in the segment. As shown in Figure 33-101, proportion-estimation errors were smallest for the site with the largest average field size. Similarly, the proportion-estimation error was found to be smallest for the size with the greatest percentage of corn and soybeans among its ground-cover types.

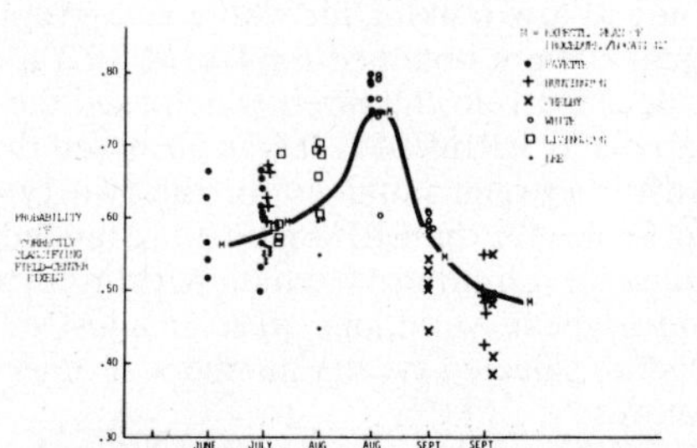

Fig. 33-100. Variability of crop identification performance during growing season.

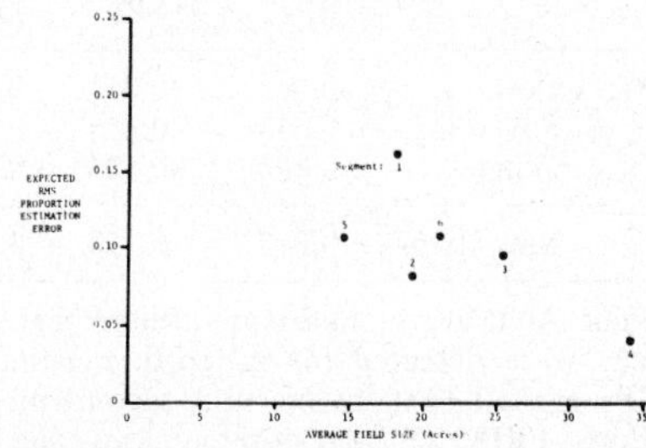

Fig. 33-101. Effects of field size (no. boundry pixels) on proportion error.

The correlation of field size and the accuracy of crop area-proportion estimates is attributed primarily to the decrease in the percent of pixels containing mixtures of crops or other ground cover as field size increases. In addition, it has been observed that large fields tend to be more uniform, and areas having larger fields have relatively fewer fields of other covers. The influence of these factors on crop-identification performance will be discussed further below.

Effects of ADP Procedure on Local Recognition With Single Time Data. EOD, ERIM, and LARS each defined a principal ADP classification procedure which was used for the major comparisons of crop-identification performance using data from single Landsat passes; alternate procedures also were defined and tested by ERIM and LARS. In this section these procedures are compared for local recognition, that is, when the test data were located in the same site and Landsat pass as the training data. Non-local recognition is considered in a following paragraph. Identical training field, test field, and test section coordinates were used with all procedures.

Major differences were found in results for the three principal ADP procedures. Local performance results with these procedures are summarized in Table 33-6, where overall average-performance figures are given, as well as the number of specific analysis of variance (ANOVA)

TABLE 33-6

Comparison of principal procedures for local recognition.

A. Mean Classification Accuracy for Pure Pixels (15 cases)

Class	LARS/SP1	ERIM/SP1	EOD/SP1
Corn	0.66	0.70	0.62
Soybeans	0.59	0.68	0.61
Other	0.50	0.53	0.46
Overall	0.58	0.64	0.57

For seven ANOVA comparisons where significant differences were detected for pure pixel performance, ERIM/SP1 ranked first in six.

B. Mean Proportion Estimation Biases and Rms Error for Whole Areas (15 cases)

		LARS/ SP1	ERIM/ SP1	EOD/ SP1
Bias For:	Corn	0.063	0.064	0.025
	Soybeans	0.033	0.059	0.081
	Other	−0.096	−0.124	−0.106
	Rms Error	0.095	0.150	0.108

For eight ANOVA comparisons where significant differences were detected for proportion estimation, LARS/SP1 ranked first in one and second in seven comparisons, ERIM/SP1 first in three cases, and EOD/SP1 first in four comparisons.

comparisons for which significant differences were detected (out of a total of ten comparisons). The ERIM/SP1 procedure was best for pure pixel classification, while LARS/SP1 was the most consistent for whole area proportion estimation and had the lowest average rms proportion error. These results indicate that pure pixel classification, which has commonly been used to evaluate crop identification performance, is not a reliable indicator of the accuracy of proportion estimates for whole areas.

Primary differences between the ADP procedures lie in the training procedures and decision rules used. Yet, there are some characteristics which they share that can contribute to the observed results.

First, there is inherent bias in proportion estimates based on aggregated counts of maximum likelihood pixel-by-pixel classifications. Bias exists because the expected performance of a classifier depends on the true crop proportions present, as well as on its performance matrix for individual pixels. Table 33-6, illustrates that all three principal procedures employed in this analysis consistently underestimated the proportion of "other" in the test data. Furthermore, the expected rms error in proportion estimation was found to be correlated with the percent of "other" in a test site.

Second, the whole areas included pixels which contain mixtures of two or more ground covers. Mixture pixels were determined to be a major source of biased proportion estimates by a special analysis, as well as by the fact that expected proportion errors tended to be largest in segments with the smallest average field size (Figure 33-101) and, therefore, with the greatest number of mixture pixels.

The three principal procedures tested differed in two ways. Both LARS/SP1 and EOD/SP1 used a clustering procedure to define training statistics (usually several classes for each major crop) and employed a quadratic decision rule. ERIM/SP1, on the other hand, formed a single signature for each major crop, used a variable number of signatures for "other", and used a linear decision rule. The differences in performance among the three procedures were determined to be due to the method of training rather than the decision rule used since similar results (high ranking for pure pixel classification and low ranking for whole area proportion estimates) were obtained for ERIM/SP2, a quadratic decision-rule classifier, which used the same signatures as ERIM/SP1. It was observed that the disparity between rankings for the two types of performance for the ERIM procedure tended to be reduced or eliminated when within-corn and within-soybean variations were smallest and the procedure selected greater numbers of other-class signatures.

An attempt was made to correct for classifier bias in proportion estimates by using the performance matrix for pure pixels, but the attempt was unsuccessful both on a section-by-section and an aggregated segment basis.

A linear decision rule optimized on a class-pairwise basis was used in ERIM's principal procedure and a quadratic decision rule, similar to those of LARS/SP1 and EOD/SP1, in its alternate procedure. It was found that the accuracy of results with the linear rule were equal to or better than those with the quadratic rule using the same signatures. Resource constraints of CITARS did not permit similar comparisons with signature sets obtained by a different procedure, but such or similar comparisons were recommended.

Another CITARS result which, on the surface, seems surprising is the lack of improvement of LARS/SP2 (non-equal major class prior probabilities) over LARS/SP1 (equal prior probabilities). Theoretically, apart from boundary pixels, the Bayesian classifier should produce its minimum error rate when "correct" values for the frequency of occurrence of each spectral class are utilized as parameters in the classification rule. LARS/SP2 included a procedure for estimating the prior probability of each spectral class based on existing agricultural statistics. For CITARS classifications the LARS/SP2 procedure utilizing unequal prior probabilities did not produce an improvement over assuming them to be equal. This is attributed in part to the fact that the agricultural statistics used were at a county level only, differing by as much as 20 percent from the true proportions in the test sections which were subsets of the county and not randomly located within it. Boundary pixels are another possible cause. Use of prior probability information in the form of class weights should not be discouraged, based solely on the CITARS analysis, since it does not constitute a definitive test. Instead, it was recommended that further tests be made to determine the sensitivity of the maximum-likelihood classifiers to class weights.

In other experiments, LARS showed that significant differences in classification performance can be obtained with different training sets and that training-set size alone does not determine the adequacy of a training set. These results and results discussed earlier point out the dependence of crop-identification performance on the development of training statistics.

Effects of Non-local Recognition and Preprocessing. In non-local recognition, the training statistics are used to classify data from a different location and/or a different Landsat pass (signature extension). Such procedures are desirable and/or necessary in order to reduce the cost of obtaining ground-identification information for training classifiers in operational applications. The segments from which training data are selected are called training segments whereas, the non-local segments to which these data are applied are called recognition segments. The effect of non-local recognition on performance with the single-time procedures was evaluated for 20 pairings of the 15 data sets.

Comparisons of classification performance indicated that average pure-pixel performance for the three principal procedures in nonlocal recognition was reduced by 22 percent with respect to that obtained locally. For crop area-proportion estimates, the average rms error of non-local estimates was 23 percent greater than that for local estimates. The degradation associated with non-local classification performance was shown to be correlated ($r = -0.77$) with differences in atmospheric optical depth (a measure of haze level) between the training and recognition segments. Other differences present in the data sets were those of soil type, agricultural practice, crop maturity, scene composition, training-data selection, and MSS scan-angle, all of which can affect the representativeness of signatures. The results clearly indicate problems in successfully applying training statistics to different locations and/or Landsat passes.

One way of extending the realm of applicability of signatures is to transform them radiometrically so they better represent observation conditions at recognition segments. Preprocessing with a mean-level adjustment algorithm (ERIM/SP1), which is a relatively simple preprocessing algorithm, was found to be of some help in improving nonlocal recognition performance. Overall, the preprocessing procedure ranked above the three principal procedures for both crop area-proportion estimation and pure pixel performance and substantially reduced the correlation between optical depth differences and pure pixel performance (from $r = -0.77$ to $r = -0.28$), but was not consistent in its performance, especially for whole areas.

The mixed results obtained in specific analyses of variance indicate that differences in composition of training and test areas are important factors affecting nonlocal recognition. Additional research was determined to be required to improve upon the signature-adjustment algorithm tested and to better account for spectral variability due to scene composition. A limited test of a more complex signature-extension algorithm at ERIM, in an effort supplementary to CITARS, indicated that improved results are possible.

Effects of Multitemporal Data. One CITARS segment (Fayette) had several clear Landsat overpasses which were spatially registered and then analyzed and processed multitemporally with the EOD/SP1 procedure. Significant increases in crop-identification performance were obtained, compared to the best single-date performance. Use of multitemporal data increased pure pixel-classification accuracy from 81 percent to 89 percent correct and halved the rms error in proportion estimation. These substantial improvements in performance were obtained for this one segment by using basically the same data-analysis procedures as for single-date data; nevertheless, new analysis procedures taking into account the increased complexity of multitemporal scenes needed to be researched and developed. Although use of multi-temporal data requires a more complex data-processing system (registration, increased data-base size, and more complex data-analysis procedures), the increased complexity may well be justified by increased performance.

Additional Results on Relation of Crop and Sensor Characteristics. Two key factors influencing crop identification with remote sensor data are:

1) the nature of the spectral variation among and within the classes to be identified; and
2) the capability of the sensor to measure the spectral variation.

An understanding of the relationship of these factors may help explain the levels of crop-identification performance obtained in CITARS. In several instances it was found that accurate identification of corn, soybeans, and "other" was not possible even when all the fields analyzed were used to train the classifier. This may have been due to a lack of differences in the spectral characteristics of the three classes or to an inability of the Landsat MSS to resolve and precisely measure the differences present. The latter is suspected to account for at least a part of the problem since crop classifications made during the 1971 Corn Blight Watch Experiment, using MSS data with more spectral bands, narrower bands, and greater sensitivity and dynamic range, showed that these same cover types could be more accurately identified. Additional comparisons of Landsat and aircraft-acquired MSS or other high spectral-resolution data such as would be available from the LACIE (Large Area Crop Inventory Experiment) field measurements project (MacDonald et al., 1975) would be needed to verify this point.

Conclusions

CITARS has provided a quantitative assessment of 1973-era technology for remote identification of major agricultural crops. The use of quantitative measures of classification performance and statistical evaluations of the results have been important parts of the technology assessment. The major conclusions from the CITARS experiments were:

1) crop-identification performance for corn and soybeans varied throughout the growing season, with pure pixel-classification accuracy being maximum in late August;
2) the probability of correct classification of pure pixels was not well correlated with, and thus was not a reliable indicator of, proportion-estimation performance;
3) proportion-estimation accuracy was strongly correlated with both average field size and proportion of major crops in the segment, but pure pixel-classification accuracy was not. Mixture pixels containing two or more cover types were determined to be major contributors to the bias in proportion estimates;
4) the manner in which ground cover classes were selected and used to train the classifier strongly influenced the amount of bias in proportion estimates;
5) probability of correct classification and proportion estimation accuracy both were decreased when training statistics developed for a different location or date were used;
6) a mean level adjustment algorithm for first order adjustments to training statistics used for nonlocal classifications increased the probability of correct classification of pure pixels, but did not improve proportion estimates for whole areas;
7) the use of multidate data improved both proportion estimation accuracy and probability of correct classification.

In addition it was shown that relatively automatic data analysis procedures can be defined which produce repeatable results, are suited for processing relatively large data volumes, and incorporate, to a large degree, the judgment and expertise of experienced analysts.

CITARS provided valuable direction for future research and development of remote sensing technology and guidelines for the design of operational crop production survey systems utilizing remote sensing technology thereby contributing significantly to the LACIE (discussed next).

Large Area Crop Inventory Experiment (LACIE)

Introduction

International trade decisions based on inadequate information about global food supplies have had severe economic and social effects. In 1972 and again in 1977, advance knowledge of the shortfall in the Soviet grain crop could have had a positive effect on the U.S. economy, rather than the negative effect that resulted from the lack of accurate timely information concerning the status of the Soviet crop.

While the United States publicizes accurate forecasts of its wheat crop, many other nations either do not make reliable estimates of their crop or do not release their figures until annual purchases are completed. Such organizations as the U.N. Food and Agriculture Organization (FAO) and the U.S. Department of Agriculture (USDA) are chartered to provide information on global food production, but their reports have been heavily reliant on information generated by the countries themselves.

In 1974, the Large Area Crop Inventory Experiment (LACIE), a joint effort of NASA, the USDA, and the National Oceanic and Atmospheric Administration (NOAA), began to apply satellite remote sensing technology on an experimental basis to forecasting harvests in important wheat production areas. Completed in 1978, the LACIE program was designed to demonstrate that remote sensing from earth orbiting satellites can provide accurate, timely information on foreign commodity production, and this information is significantly more accurate than data generated by existing data-collection methods.

Three years of intensive evaluation of LACIE estimates for the U.S. crop and 2 years of experience in estimating the Soviet crop indicated that accuracy commensurate with USDA performance goals for foreign wheat-production forecasting was achievable in regions where fields are sufficiently large to be resolved by Landsat. In a 1977 quasi-operational test, the LACIE in-season forecast of a 30 percent shortfall in the 1977 Soviet Spring wheat crop came within 10 percent of official Soviet figures released several months after harvest. LACIE midseason winter-wheat forecasts also predicted, within 7 percent, a 23 percent above-normal Soviet winter-wheat crop several months before harvest. Although an operational error caused winter wheat estimates later in the season to be inflated some 10 to 14 percent, LACIE total wheat-estimates were within 6 percent of the final Soviet figures 6 months before their release. The coefficient of variation of the LACIE total wheat-estimate for the U.S.S.R. was 3.8 percent, well within accuracy goals established for the project.

These experimental wheat-forecasting results have spurred the USDA, NASA, and NOAA to expand their efforts over the next several years to develop and evaluate space remote sensing technology for other major commodities and global crop regions.

The material which follows discusses the need for improved crop forecasts, describes the remote sensing approach to global crop forecasting, and provides a summary of key LACIE results (NASA, 1978a; NASA, 1978b; and NASA 1978c).

Current Forecast Systems

Agricultural information should have the qualities of objectivity, reliability, timeliness, adequacy of coverage, efficiency, and effectiveness. Production statistics in many countries do not meet any of these standards. Fewer than ten nations have systems that provides adequate crop production estimates. A larger number have systems providing only annual production data for major crops. The United States, which recently started issuing measures of precision of its domestic crop-production forecasts, is the only country that publishes information on survey methodology and reliability of estimates. Reasons for the absence of quality agricultural-production statistics include: lack of funds for collecting and tabulating data; an inadequate technical capability to formulate sound sampling and data-collection procedures; absence of a suitable sampling frame, and difficulty in quantifying the benefits of improved crop information. Table 33-7 shows the current accuracy of USDA forecasts of foreign commodity-production and the accuracy goals for 1985. For example, the accuracy of at-harvest estimates for the U.S.S.R. is 65/90. This means that only 65 percent of the time will the USDA at-harvest estimate be within 10 percent of the final U.S.S.R. estimate. Note that the most accurate system is that for the United States.

The frequency and magnitude of the differences between the early-season, at-harvest, and final estimates can be explained in part by the fact that the in-season estimates generally incorporate the assumption that historical trends in weather and planting patterns will prevail. Usually, these estimates are based on the foreign government's own reports of planted hectarage and the historical average for yield. Because weather patterns differ widely from year to year, the probability is low that in any one year actual hectarage, actual yield, or actual production will be within 10 percent of its average value.

Elements of Crop Production Forecasts

Accurate crop-production forecasts require accurate forecasts of the hectarage for harvest, its geographic distribution, and the associated crop yield as determined by local growing conditions. Both crop hectarage and yield are sufficiently variable from year to year and within a year to require periodic monitoring. These variations are created by slowly changing factors, such as irrigation, fertilization, and climate, and by rapidly changing factors, such as weather, market price, and government policy.

To quantify the complex effects of these factors on crop production, both hectarage and yield must be assessed at subregional levels (strata) where the limited ranges and simple interactions of the factors permit successful modeling and estimation. For example, a yield forecast stratum should be sufficiently homogeneous in soil type, crop variety, land use, and climate to preclude the necessity for hopelessly complex yield-forecast models. Hectarage and yield of significantly different crop subclasses should be individually considered. For example, there are two major growth habits of wheat (Figure 33-102). Winter wheat, planted in the fall, can have twice the yield of spring wheat, but it is subject to freeze damage during its dormancy period. Thus, despite its lower yield, spring wheat is often planted in severely cold regions. A single yield model cannot adequately describe the response of both types of wheat to such a wide range of weather conditions.

It should be emphasized that existing opera-

TABLE 33-7

USDA Current Forecast Accuracies and 1985 Goals for Wheat Production Estimates in Six Countries.

Country	Forecast: Early Season*	Mid-Season†	Pre-Harvest‡	At Harvest
	Current accuracy			
Argentina	46/90		61/90	64/90
Brazil	8/90		31/90	31/90
Canada	26/90		45/90	94/90
India	57/90	64/90	88/90	
U.S.S.R.	23/90	31/90	34/90	65/90
United States	90/90§	100/90	100/90	100/90
	1985 goal			
Argentina	60/90		75/90	80/90
Brazil	30/90		50/90	60/90
Canada	50/90		60/90	95/90
India	70/90	75/90	90/90	90/90
U.S.S.R.	50/90	60/90	65/90	85/90
United States	90/92	95/95	99/95	99/95

* From 90 to 120 days before harvest
† From 45 to 60 days before harvest
‡ From 15 to 30 days before harvest
§ Winter wheat only, 1 June

tional crop-forecast systems do not really predict the future. Weather forecasts and the vagaries of policy and human factors being what they are, crop forecasters must be content to assess as accurately as possible the impacts of preceding and current conditions on future harvest production. In this context, the ideal forecast system is one that can accurately assess current crop status and

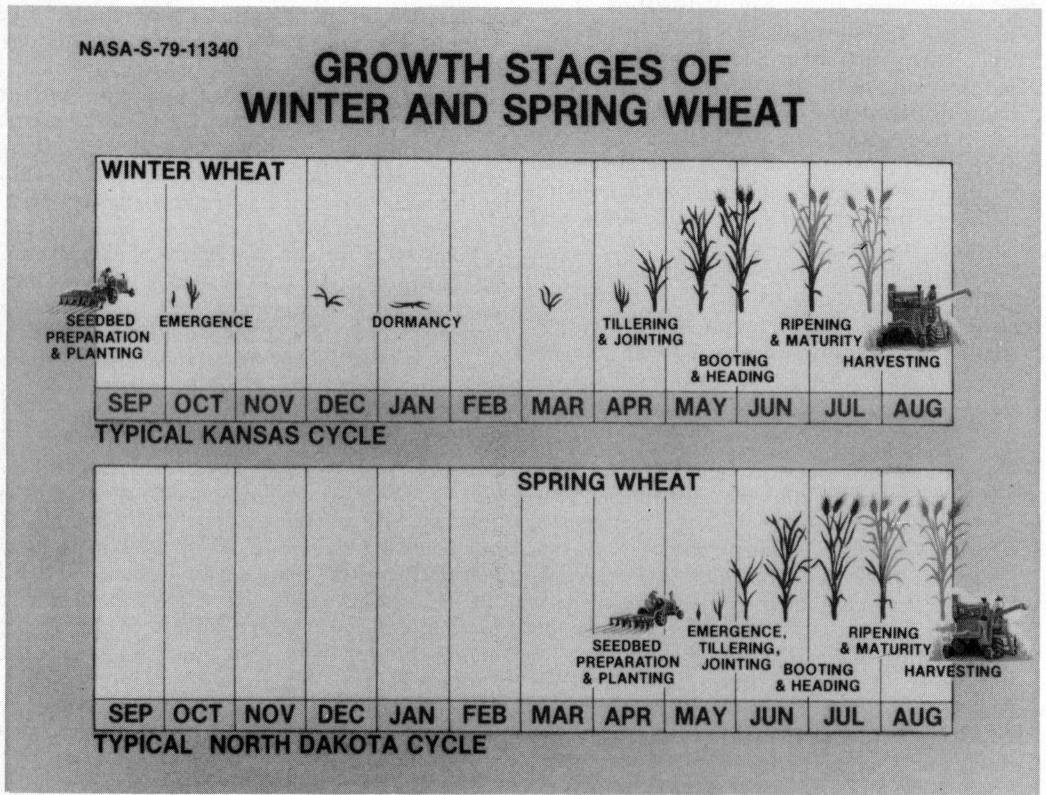

Fig. 33-102. Growth states of Winter and Spring wheat.

can detect and rapidly respond to changes in relevant conditions.

During the first half of the crop year, hectarage information has traditionally played a dominant role in market price because early in the season crop hectarage is more predictable than is crop yield. The forecast accuracy of even the perfect yield model is limited by uncertainty about future events. As harvest time approaches, yield information increases in value because plant processes are closer to completion and the chances of a major perturbation by an unforeseen event are reduced. Thus, a crop forecast system should aim to produce reliable crop hectarage information early in the season and then concentrate on increasingly accurate yield and production forecasts as the crop nears harvest.

The Large Area Crop Inventory Experiment

Although the need for improved crop forecasts had been recognized for some time, prior to Landsat, the means did not exist for USDA personnel to make such forecasts in a number of inaccessible regions of the globe. Major technological developments included:

1) multispectral scanners;
2) pattern recognition techniques;
3) high-speed digital computers;
4) the 1972 launch by NASA of the first of the Landsat series of polar-orbiting satellites;
5) the development of a global weather-reporting network by the World Meteorological Organization (WMO); and
6) the development of models capable of relating weather to crop yields.

A proof-of-concept experiment, LACIE was designed:

1) to assimilate this remote sensing technology;
2) to apply the resultant experimental system to monitoring wheat production world-wide;
3) to isolate and rank key technical problems;
4) to modify the approach as necessary and conceivable; and
5) to demonstrate the technical and cost feasibility of a global agricultural monitoring system.

In the LACIE design, heavy emphasis was placed on an objective, quantitative evaluation of the technology under as many representative global agricultural conditions as possible. Timeliness and accuracy goals were established. The LACIE experimental inventory system was designed to achieve monthly at-harvest production estimates that would converge to within 10 percent of the true estimate at the national level with a confidence of 90 percent. To evaluate the LACIE system, an extensive accuracy assessment effort was incorporated into the design (Houston et al., 1978). In addition, peer groups consisting of recognized experts in industry, the academic community, and government were periodically invited to review in depth the technical approach and results.

LACIE was conducted in three phases, each covering a global crop year. In phase I, beginning in November 1974, existing remote sensing technology was tested over the nine-state U.S. Central Plains region. Test results were sufficiently encouraging to expand testing in phase II to include wheat regions in the U.S.S.R. and Canada. In this same period, technology problems uncovered in phase I were addressed by the supporting research effort. In phase III, a second-generation technology, developed in phases I and II, was used to forecast the 1977 Soviet wheat crop at the country level. Evaluations were continued in the U.S. nine-state region, where detailed ground observations and USDA crop estimates were available for comparison. A limited amount of ground-observed data were collected in Canada as well. The project also conducted exploratory studies in India, China, Australia, Argentina, and Brazil.

Crop Forecast Technology

The remote sensing crop-forecast system developed and evaluated by LACIE used Landsat multispectral scanner data to identify crops and estimate their hectarage for harvest and used global weather data from the WMO ground network to forecast yield for harvested hectares. Instead of complete coverage by Landsat, a stratified random sample was employed. This 2 percent statistical sample of the data incurs a sampling error of less than 2 percent.

Locations of the sample unit sites and temporal windows for Landsat data acquisitions were specified by LACIE personnel at the Johnson Space Center (JSC) in Houston and transmitted to Goddard Space Flight Center in Greenbelt, Maryland. Acquisitions were spatially registered to previous acquisitions, and transmitted to Houston. During LACIE, nearly 40,000 acquisitions were transmitted. These data were also examined for episodic event detection and crop-area estimation.

Global meterological data were acquired from the Global Telecommunications System of the World Meteorological Organization (WMO), the U.S. Air Force Environmental Technical Applications Center, and the NOAA Environmental Satellite Service. These data were stored in a computer data-base at the National Meteorological Center in Suitland, Maryland. These data were processed for yield forecasting and crop maturity stage-estimation by NOAA's Center for Climatic and Environmental Assessment at Columbia, Missouri. Yield forecasts were transmitted to JSC for input to the production forecasts. Regional meteorological and qualitative crop-condition summaries, used in crop identification, were provided by NOAA personnel at JSC.

Although ground-acquired data on crop identification and crop condition were not used directly to estimate crop area in the LACIE system, such data were used to develop techniques and to assess the accuracy of LACIE crop forecasts. To support the development of techniques, spectrometer and other field measurements were acquired at several intensive study sites.

Sampling and Aggregation

Within the LACIE program a stratified random sample of Landsat data was used to estimate area. In the United States the strata were counties, in the U.S.S.R. they were oblasts, and in Canada census districts. The sample unit was a segment measuring 5 by 6 nautical miles. The Landsat data acquired from a sample segment were used to estimate the areal proportion of wheat growing in that segment. An average of all segments in each stratum was used to compute the stratum wheat hectarage. Stratum hectarages were then aggregated to "zones" selected to be relatively homogeneous with respect to wheat distribution, climate and soils. Meteorological data from the primary weather stations were used as input to zone-specific yield models to compute zone yields. In the U.S. Great Plains there were 12 such zones. Production was computed as the product of the zone wheat-hectarage and the zone yield. The variances of these estimates were also computed. Estimates of area, yield, and production, as well as variance estimates, could then be summed to obtain estimates and their precisions for any aggregate of zones. Figure 33-103 summarizes the LACIE production-estimation scheme. Aggregations were made monthly throughout the growing season. Reports for the U.S. Great Plains, the U.S.S.R., and Canada were mailed to the USDA before the USDA released independent estimates for the same regions. The variance estimates, as well as comparisons with the USDA estimates, were used to assess the accuracy of LACIE estimates in terms of the 90/90 criterion.

Landsat Data Analysis Procedures

An important segment of the LACIE program involved the development of machine-assisted image-analysis procedures. To identify the crops in a data segment took an analyst 2 weeks or more during the Corn Blight Watch Experiment discussed previously (NASA, 1974). JSC personnel felt that LACIE, with anticipated data loads in

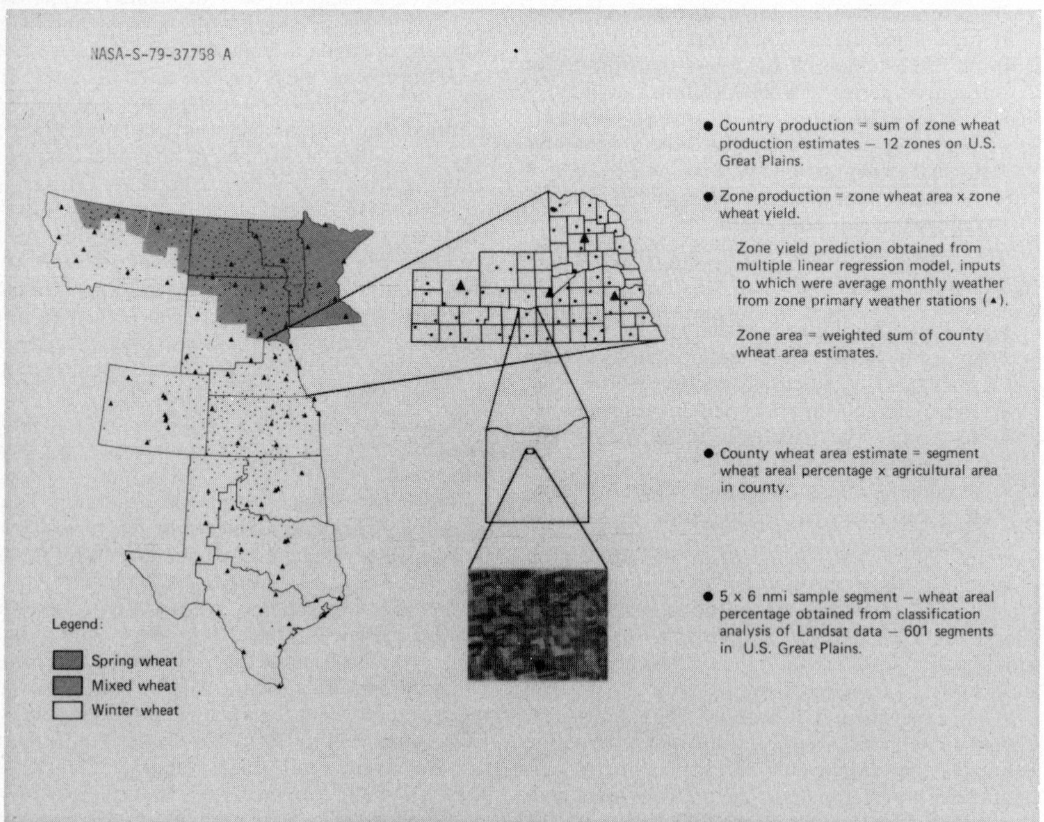

Fig. 33-103. LACIE production estimation from sampling. Zone wheat area (estimated from Landsat sample segments) times zone yield (modeled from meteorological data) equals zone production. Production for any aggregate of zones, such as for the U.S. Great Plains (12 zones), equals the sum of zone production estimates. Shown at the lower right is a sample segment (5 by 6 nautical miles) used for classification analysis of Landsat data; there are 601 segments in the Great Plains. Triangles denote zone primary weather stations.

excess of 50 segments a day, needed to reduce manual analysis requirements. An analysis procedure was needed which would allow one person not highly trained in machine processing theory to analyze a segment in no more than 1 day. Much effort was expended before and during LACIE to achieve these ends. These efforts culminated in a machine-processing procedure named Procedure 1. Procedure 1 was a four-step process.

First, the Landsat computer-generated film imagery and ancillary data were prepared and assembled into packets to be used by analysts to identify crops. Landsat data included available full-frame (100 by 100 nautical miles) color-infrared (CIR) film and segment-level CIR film products generated at JSC from digital data, as well as graphic and numerical representations of multispectral scanner data. Ancillary data included such historical agronomic information as crop-maturity calendars, cropping practices, and field size, as well as modeled adjustments to the normal wheat-crop calendar in response to the current year's weather (Robertson, 1968) and summaries of the meteorological and crop conditions for the current year.

Next, analysts used image interpretation procedures to label as "small grain" or "other" about 100 Landsat pixels (picture elements, each representing about 1 acre) preselected at random from a grid covering the segment. With spectral data from a proper sequence of Landsat acquisitions and estimates of the stages of development of small grains and confusion crops at those dates, the analyst could, in most instances, identify small grains and distinguish them from other crops. Analysts depended primarily on temporal differences in the visible and infrared reflectances of small grains and other crops. These temporal differences result from differences in the densities and colors of the various crops growing and maturing at different rates. The analyst used ancillary agricultural data to determine what other crops had to be separated from small grains. Meteorological summaries were employed so that the analyst could note unusual conditions such as droughts or floods that could affect crop appearance.

Third, about 40 of the 100 analyst-labeled pixels were used to train clustering and maximum-likelihood classification algorithms to classify as small grain or other each of the segment's 23,000 pixels. Analyst and computer identifications for the remaining 50 analyst-labeled pixels were then compared to estimate the frequency of agreement between analyst and computer. This frequency of agreement was used to "correct" the percentage of pixels computer-classified as small grain to estimate the proportion of the segment area where small grains were growing. A projection of the hectarage ratio of wheat to small grain, based on econometric models (Umberger, 1978) was used to convert the derived estimate to wheat.

Finally, analysts evaluated the acceptability of the result before applying production aggregation procedures. Thus, the Landsat data analysis can be characterized as a manually-assisted machine processing approach. Analysts typically spent 2 to 3 hours on a segment to label 100 pixels. The machine labeled all 23,000 pixels in the segment in 2 to 3 mintues, based on the training data supplied by the analyst interpreters.

Yield Forecast Procedures

The yield that would be obtained from harvested hectares was forecast in LACIE through the use of regression models that incorporated weather variables obtained from stations in the WMO network. These agrometeorological models (Sakamoto and LeDuc, 1977), were based on multiple linear regression of historical yields and monthly averages of temperature and precipitation effects (Figure 33-104). In the U.S. Great Plains yardstick region, nine winter wheat and five spring wheat models covered 12 zones. The data series used to develop the U.S. models was approximately 45 years long. In the U.S.S.R., 15 winter wheat and 16 spring wheat models covered 33 zones. The data series available to develop these models was only 10 years long. In both the United States and the U.S.S.R., the historical yield data for each zone consisted of a hectarage-weighted sum of the data for the smallest reporting areas within the zone—for example, the average monthly temperatures for the counties weighted by each county's proportion of the hectarage.

Accuracy Assessment and Evaluation

An extensive accuracy-assessment program was conducted to evaluate the performance of the LACIE technology; error contributions from the various technology components were estimated, as were the effects on performance of key agricultural and climatological factors. The ac-

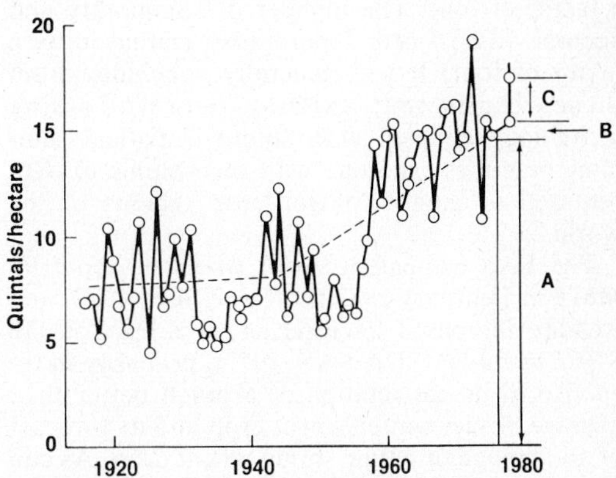

Fig. 33-104. LACIE yield determination by mathematical models. Yield—A (preceding year yield for average weather) + B (yearly adjustment for technology trend) + C (effects of current weather).

curacies of area, yield, and production estimates were established by statistical comparisons to ground observations and independent USDA figures. Statistical approaches were used to test the hypothesis that relative differences between LACIE at-harvest production estimates and actual production would be no greater than 10 percent in 9 out of 10 years. Testing such a hypothesis, without actually operating for many years, required that several assumptions be made.

First, this test was based on the assumption that, over a period of many years, the production estimates would be normally distributed. Second, the variance of this normal distribution was assumed to be related to estimates of the area and yield variances for a single year.

Since yield was modeled at the zone level, it could not be evaluated at the segment level as could area. A statistical test, utilizing historical data at the zone level, was developed to test these models. Each model was tested for its ability to predict zone yields with sufficient accuracy to support the 90/90 criterion for ten independent test years within a historical series of yield data. These tests are reported in detail in Houston et al. (1978) which also contains information on all LACIE accuracy-assessment methods.

System Performance

The efficiency and capacity of the LACIE system improved markedly during the 3 years of operation. The scope increased fourfold in the number of active segments (700 in phase I to 3,000 in phase III), ninefold in the number of Landsat acquisitions (2,000 in phase I to 18,000 in phase III), and fivefold in the number of segments that were machine processed (1,100 in phase I to 5,000 in phase III). The number of analysts declined slightly. This was possible because more efficient analysis procedures reduced the analyst contact-time to two hours per segment, one-fourth the phase I level. Yield estimates, weather summaries, and crop-calendar estimates increased by a factor of four. The number of commodity and accuracy-assessment reports also increased by a factor of four. It was generally concluded from these results that existing data-processing technology coupled with sound statistical sampling practices can cope with the volume of data required to monitor major crop regions of the world.

The FAS estimated Soviet wheat at about 97 MMT in February 1977 (USDA ESCS, 1977) and steadily increased its forecast to a high of 110 MMT in July (USDA FAS, 1977), primarily in response to its expectation of a much better than average Soviet winter-wheat crop and its forecast of an average or better spring wheat crop. As can be seen in Figure 33-105, FAS began to decrease its U.S.S.R. forecast in August. The final FAS release, on October 20, 1977 carried a wheat estimate of 90 MMT.

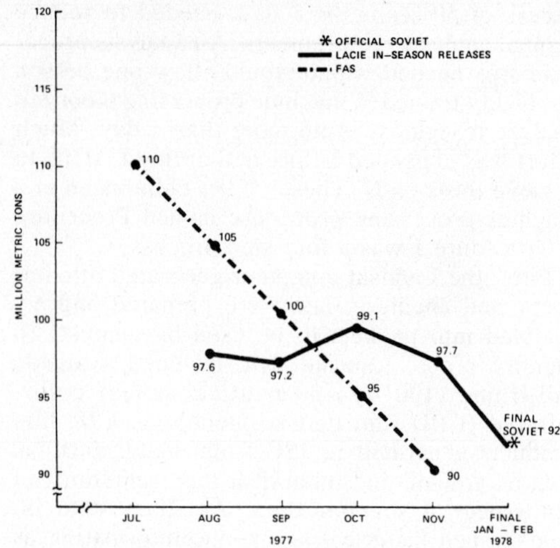

Fig. 33-105. LACIE phase III total U.S.S.R. wheat production results for 1977 compared to FAS and official Soviet estimates.

In late January 1978, the U.S.S.R. announced its 1977 wheat production as 92 MMT−51.9 MMT of winter wheat (9.8 MMT above average) and 40.1 MMT of spring wheat (8.1 MMT below average). The LACIE final winter and spring wheat area, yield, and production estimates did not differ significantly from the Soviet figures. These accuracies are also consistent with the 90/90 criterion.

As Figure 33-106 shows, the LACIE winter-wheat forecast increased from the May to the August report. Previous experience indicated that the increase from May to June was the result of steadily increasing visibility to Landsat of the emerging winter-wheat canopy in March and April. However, since winter wheat had completely emerged by June, the continued increase in the winter-wheat hectarage estimate through July and August had no known physical basis. The

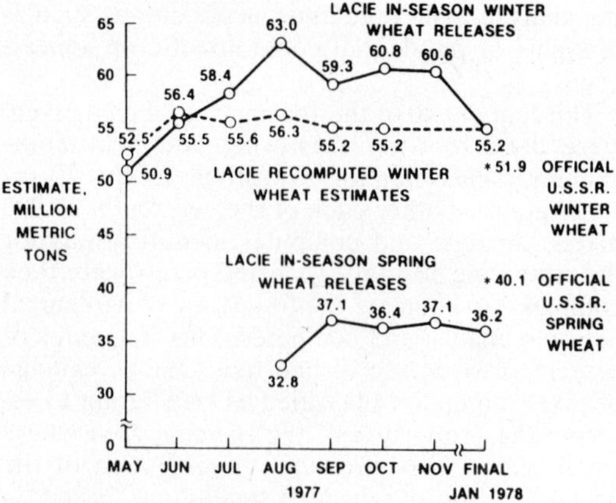

Fig. 33-106. Contributions of LACIE estimates of U.S.S.R. winter and spring wheat production for 1977 to total production results.

spring wheat estimate stabilized as expected. An analysis of the winter-wheat problem indicated that, for about 20 percent of the winter-wheat segments, Landsat data were mistakenly not acquired during March and April. This period is critical for distinguishing between the two major grain types because winter grains are greening after their dormancy and spring grains have not yet emerged. Estimates were being made for segments with May Landsat passes only, and analysts could not separate spring and winter grains. To determine the effect of this error on the estimates of hectarage, estimates in the affected segments were treated as total-grain hectarages. They were reduced to winter-wheat hectarages by multiplying them by historical ratios of winter wheat to total-grain hectarage for the local region.

The dashed line of Figure 33-106 shows the recomputed winter-wheat production estimates obtained in this way. The recomputed estimates represent the seasonal forecasts that would be expected operationally. As winter wheat completed its emergence, the production estimates stabilized. Fluctuations thereafter were dominated by changing estimates of yield. A similar behavior can be noted in the real-time estimates of Soviet spring wheat, which were unaffected by the Landsat data-order error.

From the meteorological inputs (monthly average temperature and precipitation) to the LACIE yield models, a clear pattern of drought emerged, the apparent result of a shortage in available soil moisture. Available soil moisture, a predominant term in the yield models, was estimated for a Soviet crop region as the difference between the monthly precipitation and the potential evapotranspiration (Thornthwaite 1948) for the region. Figure 33-107, a display by crop region of the percent deviation from normal in estimated available soil moisture, indicates a clear pattern of potential drought in the heart of the Soviet spring wheat region. Consequently, LACIE yield estimates for these regions decreased, as shown in Figure 33-108.

This pattern of drought and consequent yield reduction was corroborated by the Landsat data. In each LACIE sample segment, the level of drought stress was estimated from Landsat digital data by the method of Thompson and Wehmanen 1977. Figure 33-109 indicates the geographic area for which the Landsat data were indicating severe drought stress. This area overlaps every region where the LACIE technology was forecasting a below-normal yield.

For the northern crop regions, however, LACIE was forecasting above-average yields. The total effect on production of these counterbalancing tendencies could be assessed only if the wheat area in each of these crop regions was adequately known. Thus, the LACIE hectarage estimates for each crop region were multiplied by the yield forecasts to obtain production estimates. When these individual production figures were summed, the overall estimate of spring wheat production was 36.3 MMT, about 21 percent below normal, which, as discussed earlier, compared quite favorably with the Soviet estimate.

Results and Discussion

Phase III Soviet Results. In 1977, LACIE monitored U.S.S.R. wheat production from early season through harvest. Monthly commodity production forecasts were sent to the USDA in Washington the day before the corresponding public release by its Foreign Agricultural Service (FAS). LACIE made its first forecast of Soviet winter wheat production on April 1, 1977, and its first forecast of spring and total wheat on August 8, 1977. Figure 105 shows the LACIE in-season forecasts for U.S.S.R. total wheat, the FAS forecasts, and the official Soviet estimate.

The initial LACIE in-season forecast of total U.S.S.R. wheat production was 97.6 million metric tons (MMT), about 11 percent below the FAS projection and 6 percent above the final U.S.S.R. figure of 92.0 MMT. The final LACIE estimate of 91.4 MMT differed from the final Soviet figure by

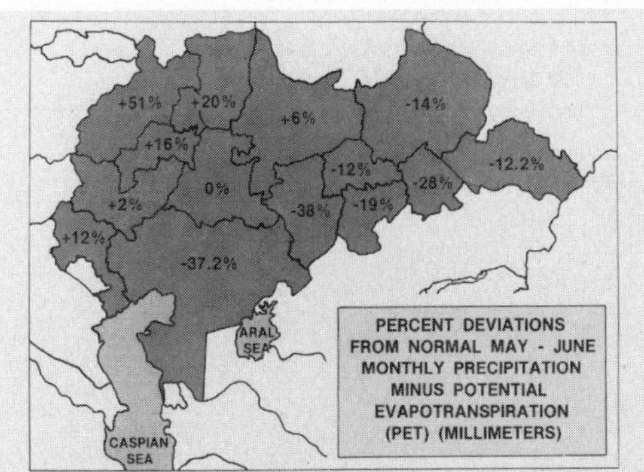

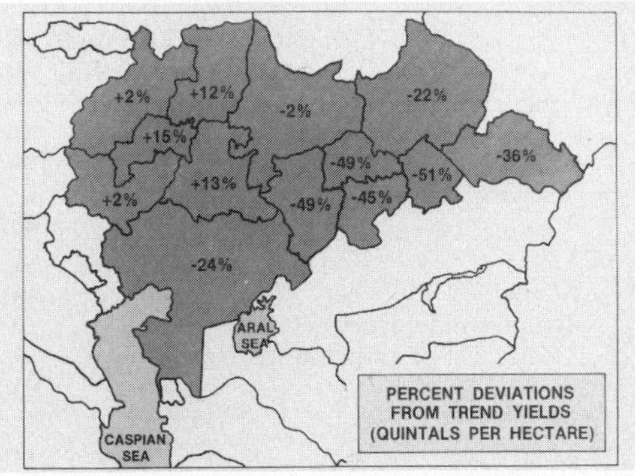

Fig. 33-107. Percent deviation from normal May–June monthly precipitation minus potential evapotranspiration (computed by the Thornthwaite method) for U.S.S.R. spring wheat regions during 1977.

Fig. 33-108. Percent deviation from trend yields in 1977, as forecast by the LACIE yield models for U.S.S.R. spring wheat.

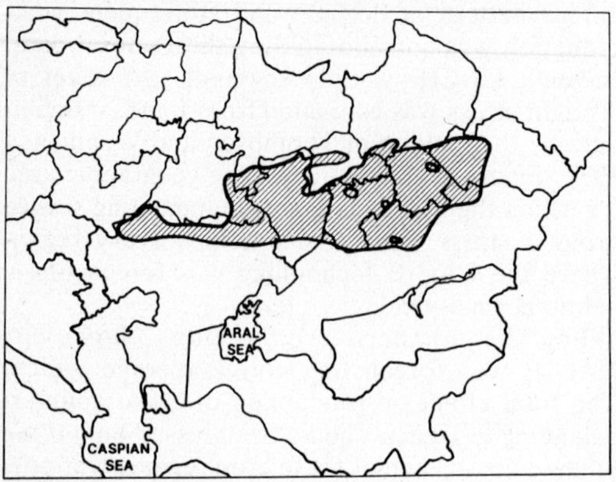

Fig. 33-109. Stressed vegetation areas (shaded) mapped from Landsat radiometric measurements of U.S.S.R. spring wheat areas in July 1977.

about 1 percent. The estimated coefficient of variation in the final LACIE estimate was 3.8 percent. This difference and CV are consistent with those required for the 90/90 accuracy criterion.

The early-season LACIE forecasts for U.S.S.R. winter wheat ranged from 51 to 55 MMT, indicating a near-record winter wheat crop (Figure 33-106). The LACIE April and May winter-wheat area estimate of about 21 million hectares indicated that U.S.S.R. planting was 15 percent above average and 22 percent above the 1976 figure. Moreover, LACIE yield forecasts stood at 25.5 quintals per hectare, 11 percent above the Soviet average. By July, spring wheat fields had grown to the point of detectability by Landsat. The August area estimate of 39 million hectares indicated that the U.S.S.R. spring-wheat planting was almost 9 percent below average. This, combined with the LACIE yield-model forecast of a 20.5 percent decline in yield from average, indicated that the Soviet spring-wheat production would suffer a major reduction, falling 30 percent below average (USDA, FAS 1977). These trends held and LACIE correctly forecast that the U.S.S.R. would achieve only an average total wheat crop.

Phase II Soviet Results. In the initial test of the LACIE technology for the Soviet Union, a portion of their wheat crop was monitored over "indicator regions" during the second year of LACIE. The analysis procedures, however, were not those used in phase III but those used in phase I. The comparison of LACIE estimates with Soviet estimates was also complicated by the lack of complete Soviet figures at the indicator region level; however, as can be seen from Table 33-8, the LACIE final estimates of area, yield, and production were in reasonable agreement with FAS estimates for these regions. Although differences between these estimates cannot be considered statistically significant, a tendency to overestimate Soviet winter wheat area was observed. In general, however, because of the much larger fields on the Soviet state farms, analysts judged the Soviet Union an easier analysis task than the United States and Canada, with their much smaller farms and fields. The phase II results were sufficiently encouraging that the LACIE Soviet test was expanded to the full-country level in order to obtain a more reliable set of comparison statistics and encounter the full range of variability in the Soviet wheat crop.

U.S. and Canadian Experiments. The U.S. test area was comprised of winter wheat and spring wheat regions in the U.S. Great Plains. The spring-wheat region included an area where no winter grains were grown and an area where spring and winter grains were mixed. LACIE technology was also evaluated throughout Canada, a country that grows only spring wheat with practices similar to those used in the U.S. northern Great Plains. In these two countries, independent and very reliable data were available for comparison at the regional, state, sample segment, and field levels. The LACIE estimates of winter- and spring-wheat hectarage for test sites in

TABLE 33-8

Comparison of FAS and LACIE Phase II Estimates for U.S.S.R. Winter and Spring Wheat Indicator Regions, 1976.

Estimate	Area (10^6 ha)	Yield (quintal/ha)	Production (MMT)
	Winter wheat		
FAS	11.3	27.6	31.2
LACIE	14.2	24.6	34.9
Relative difference, percent*	20.4	−12.2	10.6
Coefficient of variation, percent	6	5	7
	Spring wheat		
FAS	17.1	11.3	19.3
LACIE	19.1	10.5	20.1
Relative difference, percent*	10.5	− 7.6	4
Coefficient of variation, percent	4	8	9

* Calculated as [(LACIE − FAS) ÷ LACIE] × 100.

the reference region were compared to ground observations. The results for the 1976, 1977, and 1978 crop years are shown in Figure 33-110. Substantial improvement was realized with increased experience and better procedures. At the Great Plains level, the LACIE estimates of winter-wheat production, yield, and hectarage ranged from satisfactory to excellent in replications of the experiment in the reference region over three crop years (Table 33-9).

Although significant improvements were also realized in the U.S. spring-wheat estimates over the three crop years, the accuracies achieved were much lower than those for the U.S. winter-wheat region. The poorer area estimation accuracies were due to the predominance of narrow fields and to confusion crops. LACIE significantly underestimated the hectarage of spring wheat in both the United States and Canada in 1975 and again in 1976. The ground-observed sample sites clearly showed that a primary source of the error was the limited resolution of the Landsat multispectral sensor. The sensor was not able to resolve the narrow strip-fallow wheat fields (Figure 33-111). As a result, these fields were misclassified and the area of wheat underestimated.

While this strip-fallow practice is widely followed in the northern United States and Canada, it is not prevalent in the U.S.S.R. Therefore, the Landsat resolution was found adequate for the U.S. hard red winter-wheat region, with its relatively large fields, and for the still larger fields in both the winter- and spring-wheat regions of the U.S.S.R. Thus, while the U.S. and Canadian investigations substantiated the performance of the LACIE approach used in the U.S.S.R., they also revealed that, where field width is on the order of the current Landsat sensor resolution (80 meters), better sensor resolution, such as that to be provided on Landsat D, is required.

Another factor contributing to wheat area-estimation errors in some regions was the presence of certain confusion crops such as spring barley, which looks very much like spring wheat. In the spring wheat region, a two-step process was used to estimate wheat hectarage. The area of spring grains was first estimated from Landsat data and then econometric models were used to infer wheat hectarage. These models used a time series of crop hectarages to predict the hectarage ratio of wheat to small grains for the current year.

The performance of U.S. ratio models was reasonably good. However, such econometric models are inapplicable to many foreign situations. Therefore, there continued to be a need to estimate wheat directly from the Landsat data. Work done after the completion of LACIE has shown that barley can be discriminated from wheat, given Landsat coverage after barley begins to ripen and before wheat begins to ripen, about 2 weeks later. During this period, the preharvest gold of ripening barley can be readily distinguished from other, still green vegetation such as wheat.

Evaluations of the U.S. yield models using 10 years of historical data indicated a performance consistent with the 90/90 criterion except for years with extreme agricultural or meteorological conditions. The models were developed with data for the 45 years preceding each of the test years. A nonparametric statistical test employed to analyze these data did not reject the 90/90 hypothesis;

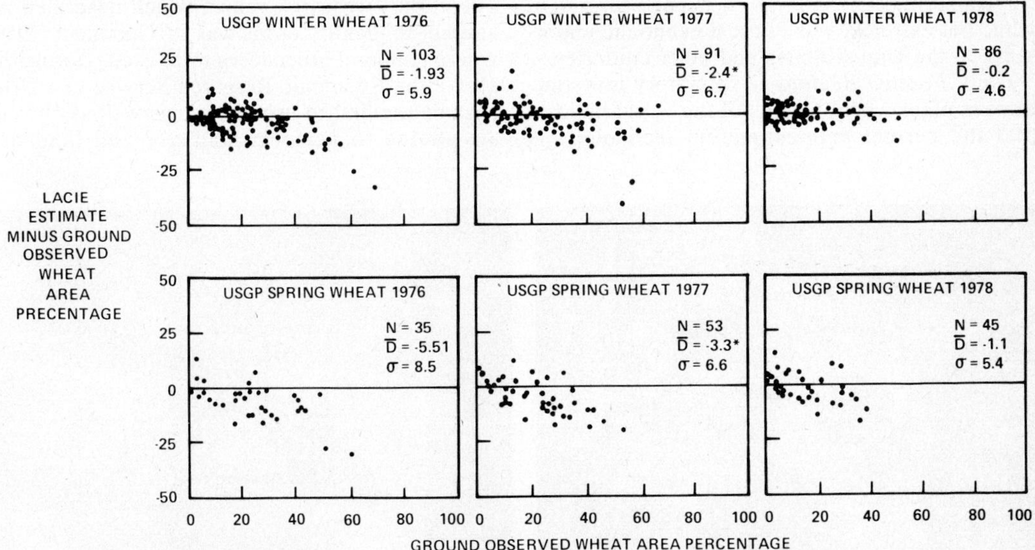

Fig. 33-110. LACIE estimates compared with ground observations for test sites in the U.S. Great Plains (1978 data are from LACIE follow-on). *N* is number of blind sites for which Landsat acquisitions were adequate to make estimates; D is the average of the sequential differences between LACIE estimates; ●, ground observations; σ, standard deviation of the difference; *, significant bias at a 10 percent level of confidence.

TABLE 33-9

Relative Differences Between LACIE and USDA Final Figures For the U.S. Great Plains With Coefficients of Variation.

Period		Area	Yield	Production
		Winter wheat reference region		
Phase I	(1975)	−0.1 ± 7.0	4.2 ± 2.6	5.0 ± 7.0
Phase II	(1976)	−7.3 ± 5.0	0 ± 5.0	−7.2 ± 7.0
Phase III	(1977)	4.6 ± 3.2	−8.2 ± 5.1	−3.4 ± 5.8
		Spring wheat reference region		
Phase I	(1975)	−30 ± 9.8	4.7 ± 4.1	−25 ± 11
Phase II	(1976)	−26 ± 6.0	3.4 ± 7.0	−22 ± 10
Phase III	(1977)	−8.5 ± 3.5	−16 ± 8.4	−26 ± 9.0
		Total wheat−spring and winter		
Phase I	(1975)	−11 ± 5.7	4.3 ± 4.0	−5.6 ± 5.9
Phase II	(1976)	−14 ± 4.0	1.1 ± 4.0	−12 ± 5.0
Phase III	(1977)	0.4 ± 2.4	−10 ± 4.3	−10 ± 4.8

however, had the models exceeded the tolerance bounds in at least one more year, as they appear to have done in 1977, the 90/90 hypothesis would have been rejected. In addition, the root mean squared error (RMSE) of 1.9 bushels per acre is larger than desirable for a 90/90 estimator. It should be noted that 1974 was a very dry year in the U.S. Great Plains and wheat yields were very poor. The LACIE yield models failed to respond to this deviation and overestimated the yield by 4.6 bushels per acre. Without 1974, the RSME would drop from 1.9 to 1.3 bushels per acre, which is not significantly different from that required for a 90/90 estimator. It thus appears that the yield models may satisfy the 90/90 criterion in years without extreme departures from normal yield. As reported earlier, the LACIE yield models were responsive to the departure of the 1977 Soviet spring wheat crop, a departure which, while not extreme, was of great economic importance to the United States and other countries.

Other Foreign Regions. Exploratory investigations in other wheat regions of the world indicate that the current area-estimation technology is generally applicable to regions in the Southern Hemisphere such as Australia and Argentina. The improved sensor resolution of Landsat D will be required for regions such as China and India, where fields are small. Yield model tests in these countries indicate that models less dependent on historical data will be needed for Australia, China, Argentina, and Brazil.

AGRICULTURAL IDENTIFICATION MECHANISMS

INTRODUCTION

The history of remote sensing in agriculture contains a comprehensive record of exploitation of panchromatic photos, largely by USDA, and use was virtually exclusively for mensuration. The earliest research report which described an attempt to identify crops was by Goodman (1959) where temporal procedures were used. During the 1950's the Economic Research Service of USDA became the first to use historic as well as current air photos to perform land use and land use

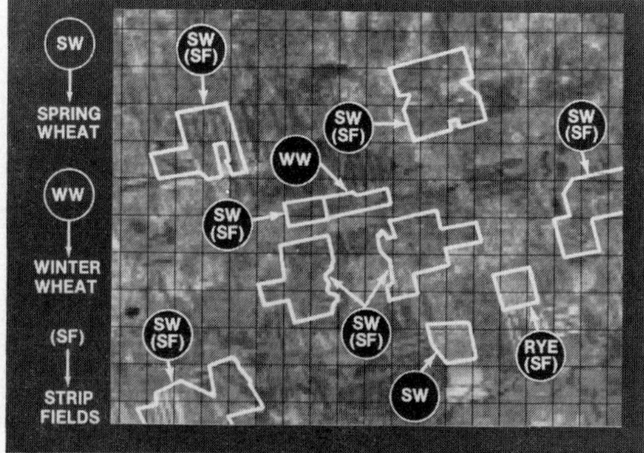

Fig. 33-111. Comparisons of (a) low-altitude aerial photograph of strip fields with (b) Landsat image of the same area. Abbreviations: SW, spring wheat; WW, winter wheat; SF, strip fields.

change-analysis in an operational context. Their technique exploited all the information available in panchromatic photography; tone, texture and spatial patterns of natural and cultural features.

In the last twenty-five years the technology of identification has advanced from utilization of broad band black-and-white to black-and-white infrared, then to narrower band color and then to color-infrared film. The value of color and color infrared has been adequately covered in the literature. The major uses of these films for agriculture monitoring are documented in this section.

In the late 1960's agriculturalists evaluated narrow-band multispectral data. In a separate development the exploitation of thermal-infrared technology occurred. The two technologies were combined in the 18-channel University of Michigan scanner.

Starting with the Mercury and Gemini space programs of NASA, photography of the earth demonstrated the value of a parameter heretofore unavailable; that of context. In space photos one could see the field in the context of the entire watershed, or a county in the context of the entire Crop Reporting District. Soils and vegetation are prominent features of terrain that can be easily identified on a reconnaissance level using Landsat imagery. Identification of more detailed features such as soil limitations and specific plant species, and spectral recognition of other anomalies is more difficult.

Research has demonstrated that there are certain fundamental wavelengths for operational earth sensing that are best suited to agricultural needs. However, it is obvious from the discussions of Tucker (1978), and Colvocoresses (1977), for example, that scientists from various disciplines cannot reach absolute agreement on specific wavelength bands that will suffice for all. Economic and political, along with technical considerations, will eventually dictate the specific configuration of operational systems. When one considers how identification is done with our current technology it is necessary to consider three descriptions of the system; its spectral resolution, its spatial resolution and finally its temporal resolution.

SPECTRAL RESOLUTION

There are two components to the spectral dimension of the instrument, viz. (1) the spectral band pass and (2) the radiometric accuracy. On a theoretical basis one can compare a 4 band 6 bit Multispectral Scanner (MSS) with a 7 band 8 bit Thematic Mapper (TM). G. F. Hughes, as reported in correspondence with the IEEE Transactions on Information Theory in January 1968, studied the relationship between classifier performance (probability of correct classification) and a quantity defined as *measurement complexity*. For MSS and TM this quantity can be expressed as the product of the number of spectral bands times the signal-quantization precision, expressed in terms of the number of discrete bits used. In a well designed sensor system this is closely related to the signal-to-noise ratio (SNR) of the data. Hughes found theoretically that, for the case of two equally likely classes, with unlimited amounts of training data, the plot of classifier performance versus measurement complexity increased monotonically to the right. There is a definite knee in the curve however and for the practical case, i.e. with finite training data, the curve is not monotonic but has a maximum occurring at a certain measurement complexity. Studies by Landgrebe et al. (1978) have shown that six bands in the visible, near- and middle-IR appear to provide the optimum number of bands. These results were obtained in two independent studies; one in which data-compression schemes showed the dimensionality of the visible and near-IR to be four and the area beyond one micrometer to add another two. The second study dealt with a retrospective analysis of classification results using an 18 channel scanner. The F statistic ranking showed that for any target four bands were adequate and that for all targets six bands were adequate. This latter work excluded the thermal infrared band. Landgrebe in a recent personal communication[4] stated that many workers have reported results confirming the Hughes postulate. MSS is considered by most agricultural authors to be a two-dimensional instrument. Hixson, Bauer and Scholz, found that MSS bands 5 and 7 produced results not significantly different from the use of all bands. It is of interest that these findings are consistent whether one picks a single date or uses all four dates, i.e. 4 bands versus 16 bands. These results compare well with General Electric (1978) studies in land-cover classification. In processing the 81 counties which make up the U.S. drainage of the Great Lakes, MSS bands 5 and 7 were found to be as accurate as all 4 bands. Preliminary analysis by the Agristars research staff suggest that TM can be expected to double the MSS capability to a dimensionality of 4[5]. It is anticipated that this will be expressed in two ways; first, through higher classification accuracies and second by improved techniques for discrimination between crop species that were difficult or impossible with MSS, e.g. cereal grains.

The TM bands provide a good framework within which to review research concerning waveband selection. The justification for this is that waveband boundaries generally occur at transition points on the typical reflectance-

[4] Personal communication from D. E. Landgrebe to A. B. Park.

[5] Technical requirements for first generation operational Landsat scanner.

wavelength curve. Very discreet physical phenomena in soil-plant-radiation interactions cause the peaks and valleys characteristic of reflectance curves. The transition wavelengths seldom vary. The magnitude of the peaks and valleys does vary and provides the basis for soil, crop and crop-condition discrimination.

TM-1 (Thematic Mapper Channel 1-0.45 to 0.52 μm)

The chlorophyll of green leaves usually absorbs 70 to 90 percent of the light in the blue part of the spectrum. Absorptance is smallest and reflectance greatest around the 0.55 μm wavelength. Tucker et al. (1975) found that the 0.45 μm wavelength was more sensitive than the 0.55 μm wavelength to the chlorophyll concentrations of a short grass prairie using *in situ* measurements.

Soils also have characteristic reflectance patterns at specific wavelengths. Condit (1970) concluded from reflectance data for 160 soil samples collected from 36 states that reflectance at all wavelengths between 0.32 and 1.00 μm should be predictable from measurements at five wavelengths—0.45, 0.54, 0.64, 0.74 and 0.86 μm. Procedures were established for classifying the soils into three general types with respect to their curve shape.

TM-2 (0.52 to 0.60 μm)

Low pigment content, which includes chlorophyll, carotenoids and anthocyanins, often results in higher reflectance. The chlorophyll absorptance of green leaves is smallest in the wavelength region around 0.55 μm where a reflection peak of usually less than 20 percent occurs from upper leaf surfaces. Thomas and Gausman (1977a) found in laboratory measurements that hemispherical reflectance of leaves of six crops was inversely related to each crop's leaf chlorophyll and carotenoid concentrations. However, of the 0.45-, 0.55-, and 0.67-μm wavelengths tested, the 0.55-μm wavelength seemed superior for individually relating the two pigments to leaf reflectance.

Knipling (1969) has stated that any physiological disturbance to a leaf usually results in an increase in visible light reflectance. Senescence of broad leaves results in light reflectance increases principally in the green wavelength region peaking at 0.55 μm, because of chlorophyll degradation. This contrasts to a decrease as a result of senescence in the 0.75- to 1.35-μm region.

Nitrogen status of several vegetable crops can be quickly measured in this wavelength region using spectrophotometry (Thomas 1970; Thomas and Oerther 1972). Limiting nitrogen reduces pigment concentrations and therefore increases reflectivity because of decreased radiation absorptance.

TM-3 (0.63 to 0.69 μm)

Relative radiance spectra for agricultural scenes, including soils and agricultural cover types, show the greatest contrast in the 0.6 to 0.7 μm wavelength interval. The chlorophyll of green leaves usually absorbs about 70 to 90 percent of the light in the red part (about 0.68 μm) of the spectrum.

Soil spectra from the field peak in approximately the red part of the spectrum, which is frequently the case for soils. Greatest soil contrasts are most often found in the wavelength interval from 0.6 to 0.7 μm.

TM-4 (0.76 to 0.90 μm)

The near-infrared wavelength region is the best spectral band in which to distinguish plants and plant conditions. Reflectance in the 0.70 to 1.35 μm wavelength interval is caused by the lack of pigment absorption and by the lack of absorption by liquid water. Reflectance changes are associated primarily with changes in the size and shape of cells and intercellular spaces and, perhaps, with other physiological changes in leaf structure. Healthy leaves have high infrared reflectance as compared with low infrared reflectance for unhealthy leaves.

In general, the correlation of leaf water-content with reflectance is strongest in the longer wavelengths of the near-infrared region of the spectrum. For example, Thomas et al. (1966) found that reflectance increased as relative turgidity decreased below values of 80 percent as selected wavelengths of 0.54, 0.85, 1.45 and 1.65 μm.

Keegan et al. (1956) studied effects of stem rust and leaf rust of wheat on light reflectance. Severity of rust infestation caused a rounding of the shoulder of the reflectance plateau between 0.75 and 1.0 μm. Cardenas et al. (1969–1970) also detected the rounding-off of the shoulder of the infrared plateau which the authors attribute to damaged cells. The study showed that in the wavelength intervals 0.5 to 0.75 μm and 0.75 to 1.35 μm, internal leaf damage in cotton leaves resulted in decreased reflectance. Apparently the brownish discoloration increased opaqueness.

Leamer et al. (1978) produced an interesting method for differentiating wheat- from background soil-reflectance by showing that soil is much less reflective than green vegetation at 0.9 μm and much more reflective at 1.65 and 2.2 μm, making each of these latter wavelengths valuable for distinguishing vegetation from bare soil and for assessing cover or density.

TM-5 (1.55 to 1.75 μm)

Plant stresses have been detected in the 0.75 to 2.5 μm band in the near-infrared region through changes in leaf structure and water content. Research in USDA at Weslaco, Texas , (Allen et al.,

1970; Gausman et al., 1972) has indicated that 1.65- and 2.2 μm are candidate wavelengths for plant-species discrimination.

Gausman et al. (1978a) and Hoffer and Johannsen, (1969) showed that infrared reflectance in the 1.55- to 1.75 μm wavelength region decreased with increased water content.

Succulent plants have water-storage tissue developed in their leaf mesophyll (Fahn, 1967). Therefore they have a higher water content and absorb more radiation in the near-infrared water absorption region (1.35 to 2.5 μm) than nonsucculent plants.

It is significant that spectrophotometer reflectance curves from cotton plants affected by various degrees of moisture stress are different at wavelengths less than 1.35 μm but are nearly identical at longer wavelengths. It has been shown by Thomas et al. (1966) that leaves from different cotton plants affected by various amounts of salinity have reflectance curves with substantial differences, between 0.5 and 2.5 μm, throughout the entire spectrum. Such contrasting reflectance characteristics may provide the means for remote detection and identification of these phenomena.

Salinity, which causes plant structural changes, also results in an increase in solute concentration within the cell cytoplasm which may be associated with reflectance contrasts at wavelengths longer than 1.35 μm Myers et al. (1970).

TM-7 (2.08 to 2.35 μm)

Reflectance from 2.08 to 2.35 μm decreases as leaf-tissue water content increases. Accordingly, Gausman et al. (1978) were able to readily distinguish succulent from woody species at the 1.3-, 1.6- and 2.2-μm wavelengths. The significance of these findings lies in two areas. Some of the succulent species are important feed crops in desert environments and in any case are confusion crops in a variety of agricultural scenarios for which the Rio Grande River Valley of Texas is a homologue.

Skylab investigations by Moore et al. (1975) showed that wavelengths greater than 2.1 μm in the reflective infrared were required to reliably distinguish between wet and dry bare soils.

TM-6 (10.4 to 12.5 μm)

Skylab investigations by Moore et al. (1975) showed that thermal data provide a better estimate of soil moisture than data from reflective bands. Also, many environmental factors are correlated with crop-canopy temperatures; these include plant height, percentage of ground cover, relative leaf turgidity, soil salinity and total plant moisture stress (Myers et al., 1966a). The many effects of salinity on plants (pigmentation; leaf thickness, size, and structure; hydration; and transpiration) can affect the conversion of electromagnetic energy into thermal energy and the dissipation of this energy.

Grain yield-predictions have been made with adequate success by Idso et al. (1979), using a combination of the stress-degree-day concept of plant water stress-assessment with the growing degree-day concept of plant phenological development.

Use of the thermal band has been demonstrated by a variety of authors. Reference is made to earlier sections of this chapter. Some results are summarized here in Table 33-10. Note that Landsat D, with a thermal resolution of 120 meters, is cited as having high utility in that context. As for time of day and frequency of observation the system is far from optimum. The Landsat D thermal band can only be considered as part of an overall data-acquisition strategy involving either the exploitation of a dedicated satellite like HCMM or one or both of the Metsats. Landsat D can provide a high-resolution image as a temporal sample of the low resolution, more continuous coverage of other satellites. In addition to time in terms of days, Landsat also provides a temporal sample at a unique time of day. This capability provides an additional measurement for model-fitting purposes and is useful in the context that some models, particularly soil-moisture models which run daily, use multiple meteorological observations made during that day. This is an area for substantial additional research.

In addition to data peculiar to individual bands, there is the value of new spectral information that is intrinsic to the individual bands. Tucker (1978) has provided excellent information concerning vegetation parameters. Others have treated these parameters in more analytic terms. Richardson et al. (1969) noted that in TM bands 5 and 6 infinite reflectance was reached at an earlier stage of growth than in band 4. Indeed the plant was transparent in band 4. The significance of this was not

TABLE 33-10

Area/Farm

Corn	–	28 hectares	Rye	–	14 hectares
Sorghum	–	44 hectares	Cotton	–	56 hectares
Soybeans	–	36 hectares	Tobacco	–	2 hectares
Peanuts	–	20 hectares	Potatoes	–	12 hectares
Wheat	–	48 hectares	Hay	–	20 hectares
Oats	–	12 hectares	Vegetables	–	16 hectares
Barley	–	32 hectares	Orchards	–	16 hectares

reported by the authors but in the context of plant-phenology modeling a most important unknown is planting date. If one can, by using spectral information, establish planting date retrospectively during the period from planting to heading, the accuracy of the model can be substantially improved. An algorithm to use this information has already been developed by Crist and Malila (1980).

SPATIAL RESOLUTION

There are three attributes of spatial resolution important in agriculture. There is the obvious requirement to measure the area of the field. Second, one needs a certain population of pixels in order to determine the multispectral statistics of the field, and third, one strives to detect small anomalies within the field.

From the perspective of mensuration, pixel size is a measure of uncertainly. If one has an 80-meter pixel then there is an 80-meter uncertainty as to the actual location of each end of a line. Mensuration errors are therefore reduced as resolution gets better. In order to achieve accuracies of even 10% of the field, size must be very large for an 80-meter system (≃40 hectares). Using modern comparators it is possible for an analyst to center the cross hairs to within 1/5 of a resolution element. Thus for the 400,000 m^2 the residual 16 meter error can cause him to measure the field as 632 ± 32 meters in the worst case. This yields an error of ±40,000 m^2 or 4 hectares which equals 10 percent.

Second, a smaller pixel size yields larger pixel populations per field. Moreover, fewer of the pixels are border pixels. For their study of the Thematic Mapper the National Academy of Science (1977) stated that 60 pixels were required to make a proper radiometer measurement of the target. If one thinks of this as approximately an 8 × 8 array these field sizes per resolution are straightforward

80 meters	40.96 hectares
30 meters	5.76 hectares
10 meters	0.64 hectares

To place this in the proper perspective one asks the question: given the crops of interest what are the field sizes that the operational system must observe? The obvious global crop of interest

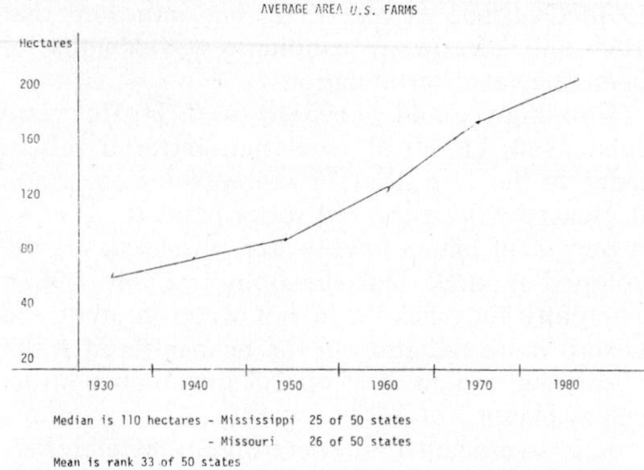

Fig. 33-112. Growth in average acreage on U.S. farms since 1930.

known to be grown in small fields in most countries is rice. However domestic data, which are much more reliable, illustrate the problem equally well. Figure 33-112 shows the growth in average acreage on U.S. farms since 1930. However, when one examines the data it is noted that the mean (≃180 hectares) is ranked 33 of 50 and that the median is ≃ 110 hectares or approximately 60 percent of the mean. Table 33-11 shows the average hectarage of crops on farms, indicating that for mensuration purposes 80-meter systems are marginal if that. If one applies the same logic to the crop averages, i.e., taking 60 percent of the average, (see Table 33-12) one sees that even the 30-meter system does not do the entire job. Nonetheless the 30-meter system is recommended as the global system with a 10-meter option for selected coverage. Table 33-13 shows the Agristars crops and their distribution. It is clear from just these data that we will be dealing with many states where the median field size is a much better criterion, since the actual distribution of fields is better realized.

Finally there is the issue of detection. It is straightforward to consider the case where one has an anomally (one corner of a field under moisture stress or insect attack) and is using a sensor with sufficient spatial resolution so that there are many resolution elements across the target. Small pixels detect small areas. Consider, however, the case where a small anomalous area occurs within a single pixel. In order to detect this

TABLE 33-11

Median Area/Farm

Corn	–	18 hectares	Rye	–	9 hectares
Sorghum	–	29 hectares	Cotton	–	37 hectares
Soybeans	–	24 hectares	Tobacco	–	1 hectare
Peanuts	–	13 hectares	Potatoes	–	8 hectares
Wheat	–	32 hectares	Hay	–	13 hectares
Oats	–	8 hectares	Vegetables	–	10 hectares
Barley	–	21 hectares	Orchards	–	10 hectares

TABLE 33-12

Agristars Crops–Corn, Wheat, Soybeans, Rice

Corn ranks in top 4 crops in 24 states
Wheat ranks in top 4 crops in 19 states
Soybeans rank in top 4 crops in 20 states
* Rice ranks in top 4 crops in 2 states
Soybeans and rice rank in top 4 crops in 2 states
Pairs of corn, wheat or soybeans rank in top 4 crops in 14 states
Corn, wheat and soybeans rank in top 4 crops in 5 states

* In California and Texas, cotton outranks rice

pixel as being different from the surrounding ones, precise radiometry is required. If the pixel is 80 meters on a side and the radiometer has a sensitivity (NEP) of 3 percent we can calculate the detection scenario(s). Suppose 10 percent of the pixel, i.e. 64 m^2, is affected to the extent that its reflectance in MSS band 7 is 30 percent different than that of the surrounding area. Theoretically that pixel will be 3 percent different than its neighbors and detectable.

TEMPORAL RESOLUTION

When the advantages of various remote sensing platforms are compared it is clear that repetitive coverage is unique to the spacecraft since it is comparatively much more expensive to acquire the necessary coverage for agriculture with aircraft. Studies done for USDA showed that if one wants to map 2,500,000 Km^2 once or monitor an area repetitively so that the sum exceeds 2,500,000 Km^2 then satellites are cheaper than aircraft. It is in fact difficult for the satellite system. The "real" scenario starts with the premise that the *requirement* for coverage is 4 times during the growing season. This is complicated further because the "growing season" is unique to *both* geographic location (latitude and altitude) and to the crop. Additionally there is the problem of clouds which is often treated as a problem from 30 percent to 50 percent of the time. This is acceptable only for the temperate zones of the agricultural world. Tropical and semitropical crop environments can have cloud problems at least 75 percent of the time. Rice, which is second only to wheat in economic terms, has the bulk of its growing area in such environments.

Finally, there is the issue of monitoring the crops to assess vigor and predict yield. While it is true that most of the research supports the thesis that one only needs to measure the crop response at the reproductive stage of development for yield prediction purposes, this does not relieve the observation-frequency problem. One must still monitor for crop-condition assessment. USDA in the documentation of their own requirements has provided the spectral, spatial and temporal specifications for their own prioritized applications. Crop-condition assessment, which has the highest priority, requires observations weekly under optimum conditions. Ten days is acceptable; 14 days is a maximum. The importance of the transient phenomenon, which can exert its influence on yield and then disappear, is best exemplified by moisture stress during the vegetative phase. The ability of the reflective-infrared spectrum to detect wilt (moisture stress) is one of the oldest observables in remote sensing. On the other hand the plant can and does recover from a wilted condition, usually overnight. Crop stress models discount yield when moisture stress occurs and, even though remote sensing can detect such stress, there is no present system short of geosynchronous development that can observe it except by chance. Nonetheless repetitive-coverage frequen-

TABLE 33-13

Planting and Harvesting Dates, Missouri.

Planted Hectares	Crop	Planting	Harvest	Crop Reporting District (CRD)
9	Barley	Sept. 10–Oct. 1	June 5–June 15	4, 5, 6, 7, 9
1041	Corn	April 20–June 1	Oct. 10–Nov. 15	State
117	Cotton	April 20–June 1	Oct. 1–Nov. 1	9
894	Hay		June 5–Sept. 10	State
68	Oats	March 1–April 25	June 25–July 10	State
2	Rice	May 1–May 20	Oct. 5–Oct. 25	8, 9
6	Rye	Aug. 15–Oct. 20	June 15–June 25	State
86	Sorghum	May 15–June 20	Oct. 15–Nov. 15	State
1260	Soybeans	May 1–June 20	Oct. 10–Nov. 5	State
414	Winter Wheat	Sept. 20–Nov. 1	June 15–July 1	State

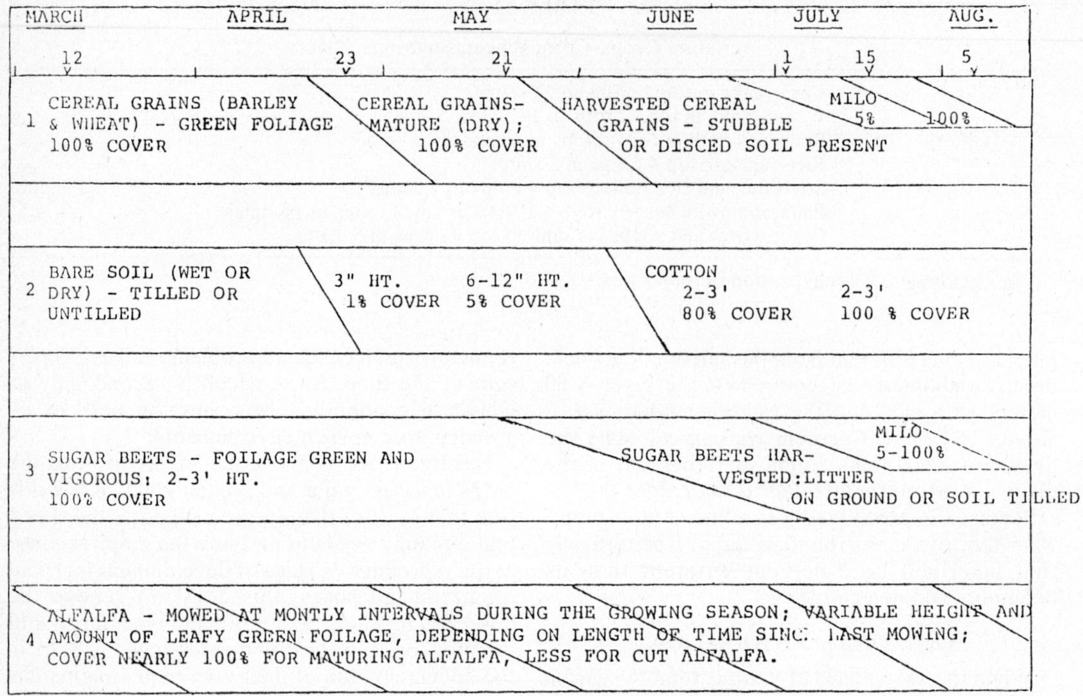

Fig. 33-113. Crop calendar—1969 sequential color. Mesa test site.

cy-studies by the Agristars staff have led to a formal recommendation for a 4 day cycle which creates an average observation frequency of 8 days for the temperate zones. This is marginally acceptable for rice. Just as it is not necessary to look outside of the U.S. for small fields, one can study the temporal problem at home where the data are more accurate. Within the U.S., Missouri is the median State in terms of field size and it presents an interesting problem for data-acquisition-cycle analysis. Table 33-14 shows the planting and harvesting dates taken from USDA publication #283. If one takes the heading event as a critical observation for yield determination and assumes that this event occurs in much the same way that the well known *green wave* occurs, then for a single species that is grown statewide the heading event occurs 50 percent to 65 percent of the way through the growth period of the plant, depending on the species, and *moves from south to north at a rate of approximately 15 miles a day*. Thus in Missouri the heading event can begin and end in a period of approximately 20 days. If there is a system which looks once in 8 days and if one of those looks is obscured by clouds there is one chance to make the critical observation. If one then examines the dates for rice, which grows only in Crop Reporting Districts (CRD's) 8 and 9, we find a maximum window length of 3 days; and even with a 4 day repeat cycle the event may be missed. Fortunately research shows that although the actual reproductive event for any individual plant occurs in a few hours, the within-field variance spreads that to a few days and the between-field variance further spreads that to approximately 1 week for an ecozone. Thus, a 4-day acquisition cycle is acceptable.

Finally, there is a most important consideration not previously given much weight. That is the exploitation of time as a discriminant. In the past there was a considerable effort to identify the most nearly optimum time for identification of species A versus the surrounding "confusion" crops. These parameters are shared by virtually all plant species and are in fact related to plant growth and conditions. Thus there can be as much spectral difference between two different stages of growth for some species as there is between species.

Studies in the use of time as a discriminant were a part of the very early research. Figure 33-113 is a crop calendar that was developed in 1969 to exploit the simplest of classification capabilities—green vegetation versus bare soil. With a knowledge of the temporal distribution of the species in an area one can achieve very high classification accuracies—>95% with four observations during the growing season. Virtually all of the errors were attributable to border-pixel allocation. A more sophisticated technique has been suggested for Agristars. A temporal/spectral classification approach—one which compensates for different planting dates, has been developed by Badhwar[6]

[6] A semi-automatic technique for multitemporal classification of a given crop. July 31, 1979.

of the Agristars research staff and uses 4 observations during the growing period. Since this approach is essentially a quantitative change-detection method, spectral accuracy is considered an enabling technology.

FUTURE AGRICULTURAL REMOTE SENSING SURVEYS

The instruments of the TIROS and NIMBUS series in the 60's which offered a ground resolution of the order of 1000 m were the first to reveal the potential of space as a vantage point for agriculture and other resources. Second generation remote sensing instruments, with a resolution of the order of 100 m were developed during the 70's for the Landsat series of satellites. The Landsat-D thematic mapper of the 80's will provide spatial resolution of approximately 30 m and spectral SPOT system, conceived by the French, will introduce satellite-borne instruments with resolution of the order of 10 m. The fine resolution provided by the newer systems was a basic design decision dictated by the small-scale subdivision of much of the agricultural land in many parts of the world.

Significant advantage can be derived from the simultaneous use of complementary capabilities of space systems of the 80's. While Landsat-D will procure a wealth of improved spectral information, SPOT will provide higher resolution with a capability for stereo coverage and frequent access to user-specified sites. For many applications data from the resource satellites, Landsat-D and SPOT, should be used together with weather and temperature data from NOAA weather satellites and from ground data.

It is essential that Federal and State agencies, universities and private enterprise intensify research programs directed to solving the more complex relations of resources-atmosphere and electromagnetic energy. At the same time extensive research and development efforts will continue refinement of the technology and transfer to operational test systems. Finally will come adaptation of the technology to forecasting production of numerous food and fiber crops.

A technology development program has been initiated to support the possible implementation of an operational global monitoring system based on the Secretary of Agriculture's Initiative to develop improved uses of aerospace technology for agricultural purposes. The initiative sets the following priorities:

1) early warning of changes affecting production and quality of renewable resources;
2) commodity production forecasts;
3) land-use classification and measurement;
4) renewable resources inventory and assessment;
5) land productivity estimates;
6) conservation practices assessment; and
7) pollution detection and impact evaluation.

While all seven requirements are of major importance to the USDA, the first two requirements essentially capture the department's most urgent need for better, more timely information on world crop conditions and expected production. In response to the Secretary's Initiative, federal agencies have established the Agriculture and Resource Inventory Surveys Through Aerospace Remote Sensing (AgRISTARS) Program, a six-year program of research, development, evaluation, and application of aerospace remote sensing which has been initiated in Fiscal Year (FY) 1980.

The goal of AgRISTARS is to determine the usefulness, cost, and extent to which aerospace remote sensing data can be integrated into existing or future USDA systems to improve the objectivity, reliability, timeliness, and adequacy of information required to carry out USDA missions. The overall approach is comprised of a balanced program of remote sensing research, development, and testing which addresses domestic resource management, as well as commodity production information needs.

It is in the developing Third World that remote sensing seems likely to make the biggest difference—that is the significance of synoptic images may be more important in some fields and as planning land use and assessing agricultural problems, than in nations that are better mapped or more thoroughly explored.

In developing countries, 50–90% of the population may be agrarian. These regions need information at higher levels of government. For many countries, agriculture is the largest economic commodity in their budget. Therefore, the impact of technological advances in agriculture can have a large effect on the growth and development of countries along with the stability of their people and government.

As the advanced technology of the 1980's becomes a reality, resource monitoring and crop forecasting to meet global as well as particular country needs will gain significantly from advances in meteorological satellites and in microwave sensors, plus improvements which can be expected in data processing and satellite transmission and delivery.

The extension of capabilities that it is possible to foresee in agriculture in the early and mid 1980's period are described below (National Academy of Sciences, 1977).

1) improved recognition of soil/crop contrast and sharper discrimination of soil boundaries;
2) substantial improvement in statistical sample designs for estimating acreage of major food grains generally, including those in small-field, multicrop areas;
3) significant improvements in yield estimation for various crops through merging the following data in an appropriate computer modeling system: (a) crop yield/stress mod-

els for crops; (b) meteorological satellite data; (c) ground meteorological data; (d) ground crop type and condition data in a proper sample design; (e) ground observations on yield estimates; (f) imagery with improved resolution such as the Thematic Mapper and SPOT data; and (g) imaging radar data if available;

4) development of useful capability to detect and regularly monitor regional crop disease;
5) refinement of capability to estimate soil moisture. If successful, regional assessment of water yield and drought stress should be possible on much finer spatial and temporal scales than is now available. Future resource satellites, with the addition of microwave sensors, and meteorological satellites will play various roles in the estimation process;
6) determination of evapotranspiration, which at present is inadequately measured, for irrigation scheduling, for crop yield estimates and for many other purposes.

REFERENCES

A/CONF 74/1 UN Conference on desertification, 1977, Desertification an overview.

Abdel Hady, M. and H. H. Korbs, 1971, Depth to groundwater table; J. of Irrig. and Drain. Div., Proc. of Am. Soc. of Civil Engineers, 97:355–367.

Abdel Hady, M., Younis, H. A., M. E. El Shazly, M. A. Hamid, K. Khalil, Y. Bishara, B. K. Worcester, A. A. Klingebiel, M. DeVries, 1978, Soil resources and potential for agriculture development in Bahr El Jebel area, southern Sudan; Remote Sens. Center, Acad. of Sci. Res. and Tech. Cairo, Egypt.

Aboukhaled, A., 1966, Optical properties of leaves in relation to their energy-balance, photosynthesis, and water use efficiency; Ph.D. Thesis, University of California Library, Davis, 139 p.

Ahern, F. J., D. G. Goodenough, A. L. Grey, R. A. Ryerson, R. J. Vibikaitis and M. Goldberg, 1979, Simultaneous microwave and optical wavelength observations of agricultural targets; Canadian J. of Remote Sensing, 4(2), 127–142.

Al Abbas, H. H., P. H. Swain and M. F. Baumgardner, 1972, Relating organic matter and clay content to the multispectral radiance of soils; Soil Science.

Allen, W. A. and A. J. Richardson, 1968, Interaction of light with a plant canopy; J. Opt. Soc. Am., v. 58, p. 1023–1028.

Allen, W. A., H. W. Gausman, A. J. Richardson and J. R. Thomas, 1969, Interaction of isotropic light with a compact leaf; J. Opt. Soc. Am., v. 59, p. 1376–1379.

Allen, W. A., H. W. Gausman, A. J. Richardson, 1970a, Mean effective optical constants of cotton leaves; J. Opt. Soc. Am., v. 60, p. 542–547.

Allen, W. A., H. W. Gausman, A. J. Richardson and C. L. Wiegand, 1970b, Mean effective optical constants of thirteen kinds of plant leaves; Appl. Opt., v. 9, p. 2573–2577.

Allen, W. A., T. V. Gayle and A. J. Richardson, 1970c, Plant canopy irradiance specified by Duntley equations; J. Opt. Soc. Am., v. 60, p. 327–376.

Allen, W. A., H. W. Gausman, A. J. Richardson and R. Cardenas, 1971, Water and air changes in grapefruit, corn, and cotton leaves with maturation; Agron. J., v. 63, p. 392–394.

Anderson, J. R., E. E. Hardy and J. T. Roach, 1971, Land use classification system for use with remote sensor data; U.S. Geol. Survey Circular 671, Superintendent of Documents, Washington, D.C.

Aston, A. R. and C. H. M. van Bavel, 1972, Soil surface water depletion and leaf temperatures; Agron. J. 64:368–373.

Avery, G. S., 1933, Structure and development of the tobacco leaf; Am. J. Bot., v. 20, p. 565–592.

Baker, D. N. and R. E. Meyer, 1966, Influence of stand geometry on light interception and net photosynthesis in cotton; Crop Science 6, p. 15.

Bartholic, J. F., L. N. Namken and C. L. Wiegand, 1972, Aerial thermal scanner to determine temperatures of soil and of crop canopies differing in water stress; Agron. J. 64:603–608.

Basharinov, A. Ye and H. M. Shutko, 1978, Determination of the moisture content of the earth's cover by superhigh frequency (microwave) radiometric methods: a review; Radiotckhmika i Electronika, v. 23, p. 1778–1791, NASA Tech. Trans. TM 75411.

Batlivala, P. P. and F. T. Ulaby, 1977, Estimation of soil moisture with radar remote sensing; Proc. Eleventh Int. Symp. of Remote Sens. Environ., v. II, p. 1557–1566.

Bauer, M. E., R. P. Mroczynski, R. B. MacDonald and R. M. Hoffer, 1971, Detection of southern corn leaf blight using color infrared aerial photography; Proc. Third Biennial Workshop on Color Aerial Photography in the Plant Sciences, March 2–4, 1971, Gainesville, Florida, p. 114–126.

Bauer, M. E., M. M. Hixon, L. L. Biehl, B. F. Robinson and E. R. Stoner, 1978a, Final report Vol. 1, agricultural scene understanding; prepared by LARS, Purdue Univ., for NASA, JSC, Contract Report No. 112578.

Bauer, M. E., M. M. Hixson, B. J. Davis and J. B. Etheridge, 1978b, Area estimation of crops by digital analysis of Landsat data; Photogram. Eng. and Rem. Sens., v. 44, no. 8, p. 1033–1043.

Bauer, M. E., L. L. Biehl, C. S. J. Daughtry, B. F. Robinson and E. R. Stoner, 1979, Final report agricultural scene understanding and supporting field research; Prepared by LARS, Purdue Univ. for NASA, Contract Report No. 112879.

Baumgardner, M. F., S. J. Kristof, C. J. Johannsen and A. L. Zachary, 1970, Effects of organic matter on multispectral properties of soils; Proc. Ind. Acad. Sci. 79:413–422.

Belonogova, I. N. and J. S. Tolchel'nikov, 1959, On the dependence of the spectral brightness of minerals on the degree of dispersion; Izv. Akad. Nauk., SSSR, Ser. Geo. 11, p. 98–101.

Benedict, H. M. and R. Swidler, 1961, Nondestructive method for estimating chlorophyll content of leaves; Science, v. 133, p. 2015–2016.

Best, R. G. and J. R. Smith, 1978, Photographic contrast enhancement of Landsat imagery; Photogram. Eng. and Rem. Sens., v. 44, no. 8, p. 1023–1026.

Billings, W. D. and R. J. Morris, 1951, Reflection of visible and infrared radiation from leaves of different ecological groups; Am. J. Bot., v. 38, p. 327–331.

Billingsley, F. C., M. R. Helton and V. M. O'Brien, 1976, Landsat follow-on: a report by the applications survey groups; National Aeronautics and

Space Administration Tech. Mem. 33-803, v. II, Jet Propulsion Lab, Pasadena, Calif.

Birth, G. S. and G. R. McVey, 1968, Measuring the color of turf with a reflectance spectrophotometer; Agron. J., v. 60, p. 640–643.

Black, T. A., W. R. Gardner and G. W. Thurtell, 1969, The prediction of evaporation, drainage, and soil water storage for a bare soil; Soil Sci. Soc. Am. Proc. 33:655–660.

Blad, B. L. and N. J. Rosenberg, 1976, Measurement of crop temperature by leaf thermocouple, infrared thermometry, and remotely sensed thermal imagery; Agron. J. 68:635–641.

Blachard, M. B., R. Greeley and R. Toettleman, 1974, Use of visible, near infrared and thermal infrared remote sensing to study soil moisture; Proc. 9th Int. Symp. on Remote Sens. of Environ., Ann Arbor, Michigan.

Blinn, J. C. and J. C. Quade, 1972, Microwave properties of geological materials: studies of penetration depth and moisture effects; 4th Annual Earth Resource Program Review, NASA Johnson Space Center, Houston, Texas, January 17–21, 1972.

Bonner, J. and A. W. Galston, 1952. Principles of plant physiology; Freeman, San Francisco, Calif., 499 p.

Bottcher, C. F., 1952, Electric polarization; Elsevier, New York.

Bowers, S. S. and R. J. Hanks, 1965, Reflection of radiant energy from soils; Soil Sci. 100, p. 130–138.

Brown, K. W. and N. J. Rosenberg, 1973, A resistance model to predict evapotranspiration and its application to a sugar beet field; Argon. J. 65:341–347.

Brown, R. J., K. P. B. Thomson, F. J. Ahern, K. Staenz, J. Cihlar, B. Klumph and C. Pearce, 1981, Landsat MSS applied to rangeland management in western Canada; 15th Int. Symp. on Remote Sens., Ann Arbor, Michigan, 1981.

Bunnik, N. J. J., 1978, The multispectral reflectance of shortwave radiation by agricultural crops in relation with their morphological and optical properties; H. Veenman & Zonen B. V., Wageningen, 176 p.

Buol, S. W., F. D. Hole and R. J. McCracker, 1959, Soil Genesis and Classification, Iowa State University Press, Ames.

Burke, W. J., T. Schmugge and J. F. Paris, 1979, Comparison of 2.8 and 21 cm microwave radiometer observations over soils with emission model calculations; J. of Geophysical Research, v. 84, p. 287–294.

Burns, E. E., M. J. Starzyk and D. L. Lynch, 1969, Detection of plant virus symptom with infrared photography; Trans. Illinois State Acad. Sci., v. 62.

Bush, T. F. and F. T. Ulaby, 1978, An evaluation of radar as a crop classifier; v. 8, Remote Sens. of Environ., p. 15-36.

Calabrese, M. A., 1979, A review of future remote sensing capabilities; Symp. Proc., Identifying Irrigated Lands Using Remote Sensing Techniques, Missouri Riv. Basin Comm. and USGS.

Cardenas, R., H. W. Gausman, W. A. Allen and M. Schupp, 1969–1970, The influence of ammonia-induced cellular discoloration within cotton leaves (*Gossypium hirsutum* L.) on light reflectance, transmittance and absorptance; Remote Sens. of Environ., v. 1, p. 199–202.

Carlson, R. E., D. N. Yarger and R. H. Shaw, 1971, Factors affecting the spectral properties of leaves with special emphasis on leaf water status; Agron. J., v. 63, p. 486–489.

Carneggie, D. M., S. D. deGloria and R. N. Colwell, 1974, Usefulness of ERTS-1 and supporting aircraft data for monitoring plant development and range conditions in California's annual grassland; BLM Final Report 53500-CT3-266(N).

Carter, D. L. and V. I. Myers, 1963, Light reflectance and chlorophyll and carotene contents of grapefruit leaves as affected by Na_2SO_4, NaCl and $CaCl_2$; Proc. Am. Soc. Hort. Sci., v. 82, p. 217–221.

Carver, K. R. and T. F. Bush, 1979, Airborne multispectral remote sensing of saline seeps: the 1978 Harding County experiment; Final Report on Contract NAS9-15421, Physical Science Lab., New Mexico State Univ., 118 p.

Chance, J. E. and E. W. LeMaster, 1978, Plant canopy light absorption model with application to wheat; Appl. Opt., v. 17, p. 2629–2636.

Chermezon, H., 1910, Recherches anatomiques sur les plantes littorales; Ann Sci. Nat. Bot., v. 12, p. 117–129.

Choudhury, B., T. Schmugge, R. W. Newton and A. Chang, 1979, Effect of surface roughness on the microwave emission from soils; NASA Tech. Mem. 79606, J. of Geophysical Research, v. 84:565.

Chudnovskii, A. F., 1962, Heat transfer in soil; Published for the National Science Foundation, and the Dept. of Ag. by the Israel Program for Scientific Translation.

Cihlar, J., T. Sommerfeldt and B. Paterson, 1979, Soil water content estimation in fallow fields from airborne thermal scanner measurements; Canadian J. of Remote Sens. 5(1), 18–32.

Coblentz, W. W., 1913, The diffuse reflecting power of various substances; Bur. Standards Bul., v. 9, p. 283–325.

Coblentz, W. W., 1962, Investigations of the infrared spectra; Republication under the joint sponsorship of the Coblentz Society and the Perkin-Elmer Corporation, Reprinted by permission of the Carnegie Institution of Washington, p. 145.

Colvocoresses, A. P., 1977, Proposed parameters for an operational Landsat; Photogram. Eng. and Remote Sens., v. 43, no. 9, p. 1139–1145.

Colwell, J. E., 1973, Bidirectional spectral reflectance of grass canopies for determination of above ground standing biomass; Ph.D. Thesis, Univ. of Michigan, University microfilm 75-15, 693, 174 p.

Colwell, J. E., 1974, Vegetation canopy reflectance; v. 3, Remote Sens. of Environ., p. 175–183.

Colwell, R. N., 1956, Determining the prevalence of certain cereal crop diseases by means of aerial photography; Hilgardia, v. 26, p. 223–286.

Condit, H. R., 1970, The spectral reflectance of American soils; Photogram. Eng., v. 36, p. 955–966.

Coulson, K. L., 1966, Effect of reflection properties of natural surfaces in aerial reconnaissance; Appl. Opt., v. 5, p. 905–917.

Crist, H. and W. Malila, 1980, An algorithm for estimating crop calendar shifts of spring small grains using Landsat spectral data; Agristars SR-EO-00459.

Crown, P. H., 1978, Remote sensing: soil problems in crop production; Alberta Remote Sensing Center, Pub. 79-2, Edmonton, Alberta, 19 p., 6 color.

Curran, P. J., 1978, A photographic method for the recording of polarized visible light for soil surface moisture indications; Remote Sens. of Environ., v. 7, p. 305–322.

Curran, P. J., 1979, The use of polarized panchromatic and false color infrared film in the monitoring of soil surface moisture; Remote Sens. of Environ., v. 8, no. 3, p. 249–266.

da Costa, L. M., 1979, Surface soil color and reflectance as related to physico-chemical and mineralogical soil properties; Ph.D. dissertation, Univ. of Mo., Columbia.

Dadykin, V. P. and V. P. Bedenko, 1961, The connection of the optical properties of plant leaves with soil moisture; Dokl. Acad. Sci., USSR, Bot. Sect., v. 134, p. 2–214.

Dalsted, K. J., B. K. Worcester and M. E. DeVries, 1978, Influence of soils on Landsat spectral signatures of corn; J. Article no. SDSU-RSI-J-78-14.

Dalsted, K. J. and B. K. Worcester, 1979, Detection of saline seeps by remote sensing techniques, Photog. Engr'g, Vol. 45, No. 3, pp. 285–291.

Davis, P. A., 1957, Exploring the atmosphere's first mile; Pergamon Press, Ltd., London, p. 377–383.

Dethier, B. E., M. D. Ashley, B. Blair and R. J. Hopp, 1973, Phenology satellite experiment; Symp. Significant Results Obtained from ERTS-1, v. 1, Wash. D.C., p. 157–165.

DeWit, C. T., 1959, Potential photosynthesis of crop surfaces; Netherlands J. Agr. Sci. 7, 141.

Dickey, F. M., C. King, J. C. Holtman and R. K. Moore, 1974, Moisture dependency of radar backscatter from irrigated and non-irrigated fields at 400 MHz and 13.3 GHz; IEEE Trans. on Geosci. Elect., v. GE-12, no. 1, p. 19–22.

Dickson, W. J., 1974, BMD, biomedical computer programs; Univ. of Calif. Press, Berkeley, Calif.

Downs, J. M., J. D. Murphy and D. R. Thompson, 1976, Use of agrophysical maps in large area crop inventories; 68th Meeting, Am. Soc. of Agron., Houston, TX, Nov. 28–Dec. 2.

Draeger, W. C., 1979, Monitoring irrigated land acreage in the Klamath River Basin of Oregon using Landsat imagery; Symp. Proc., Remote Sens. of Irrigated Lands, Missouri River Basin Commission, EROS Data Center.

Eagleman, J. and W. Lin, 1976, Remote sensing of soil moisture by a 21 cm passive radiometer; J. Geophysical Research, v. 81, p. 3660.

Ehrler, W. L. and C. H. M. van Bavel, 1967, Sorghum foliar responses to changes in soil water content; Agron. J. 59:234–246.

Ehrler, W. L., 1973, Cotton leaf temperatures as related to soil water depletion and meteorological factors; Agron. J. 65:404–409.

Ehrler, W. L., S. B. Idso, R. D. Jackson and R. J. Reginato, 1978a, Wheat canopy temperatures: relation to plant water potential; Agron. J. 70:251–256.

Ehrler, W. L., S. B. Idso, R. D. Jackson and R. J. Reginato, 1978b, Diurnal changes in plant water potential and canopy temperature of wheat as affected by drought; Agron. J. 70:999–1004.

Esau, K., 1965, Plant anatomy; Wiley New York, 2nd edition, p. 767.

Estes, J. E., J. R. Jensen, L. R. Tinney, M. Rector, 1975, Remote sensing inputs to water demand modeling; Research Report under NASA NGL.

Estes, J. E., M. R. Mel, and J. O. Hooper, 1977, Measuring soil moisture with an airborne imaging passive microwave radiometer, Photogram. Eng. and Rem. Sens., Vol. 43, No. 10, p. 1273–1281.

Evans, R., J. Head and M. Dirkzwager, 1976, Air photo-tones and soil properties: implications for interpreting satellite imagery; Remote Sens. of Environ. 4:265–280.

Evans, R., 1979, Air photos for soil survey in Lowland England: factors affecting the photographic images of bare soils and their relevance to assessing soil moisture content and discrimination of soils by remote sensing; Remote Sens. Environ. 8:39–63.

Fahn, A., 1967, Plant anatomy; Pergamon Press, New York, p. 584.

Frazee, C. J., 1975, Soilscapes interpreted from Landsat-1 imagery, Pennington County, South Dakota; Remote Sensing Institute, SDSU, Public. No. SDSU-RSI-75-04.

Fritz, N. L., 1967, Optimum methods for using infrared-sensitive color film; Photogram. Eng. v. 33, p. 1128–1138.

Fuchs and Tanner, 1968, Surface temperature measurements of bare soils; J. of Appl. Meteor. 7(2) p. 303–305.

Gates, D. M. and W. Tantraporn, 1952, The reflectivity of deciduous trees and herbaceous plants in the infrared to 25 microns; Science, v. 115, p. 613–616.

Gates, D. M. and H. J. Keegan, J. C. Schleter and V. R. Weidner, 1965, Spectral properties of plants; Appl. Opt., v. 4, p. 11–20.

Gausman, H. W. and R. Cardenas, 1969a, Effect of leaf pubescence of *Gynura aurantiaca* on light reflectance; Bot. Gaz., v. 130, p. 158–162.

Gausman, H. W., W. A. Allen and R. Cardenas, 1969b, Reflectance of cotton leaves and their structure; Remote Sens. Environ., v. 1, p. 19–22.

Gausman, H. W., W. A. Allen, V. I. Myers and R. Cardenas, 1969c, Reflectance and internal structure of cotton leaves, *Gossypium hirsutun* L.; Agron. J., v. 61, 374–376.

Gausman, H. W., W. A. Allen, R. Cardenas and A. J. Richardson, 1970a, Relation of light reflectance to histological and physical evaluations of cotton leaf maturity; Appl. Opt., v. 9, 545–552.

Gausman, H. W., W. A. Allen, M. L. Schupp, C. L. Wiegand, D. E. Escobar and R. R. Rodriguez, 1970b, Reflectance, transmittance, and absorptance of electromagnetic radiation of leaves of eleven plant genera with different mesophyll arrangements; Texas A&M, Univ. Tech. Monograph No. 7, 38 p.

Gausman, H. W., W. A. Allen, R. Cardenas and A. J. Richardson, 1971a, Effect of leaf nodal position on absorption and scattering coefficients and infinite reflectance of cotton leaves, *Gossypium hirsutum* L.; Agron. J., v. 63, p. 87–91.

Gausman, H. W., W. A. Allen, C. L. Wiegand, D. E. Escobar, R. R. Rodriguez and A. J. Richardson, 1971b, The leaf mesophyll of twenty crops, their light spectra, and optical and geometrical parameters; U.S. Dept. of Agric. Tech. Bul. 1465, 59 p.

Gausman, H. W. and W. A. Allen, 1973, Optical parameters of leaves of 30 plant species; Plant Physiol., v. 52, p. 57–62.

Gausman, H. W., 1974a, Leaf reflectance of near-infrared; Photogram. Eng., v. 40, p. 183–192.

Gausman, H. W. and W. G. Hart, 1974, Reflectance of four levels of sooty-mold deposits produced from the honeydew of three insect species; J. Rio Grande Valley Hort. Soc., 28:131–136.

Gausman, H. W., A. H. Gerbermann and C. L. Wiegand, 1975, Use of ERTS-1 data to detect chlorotic grain sorghum; Photogram. Eng. and Remote Sens., v. 41, no. 2, p. 177–179.

Gausman, H. W., 1977a, Reflectance of leaf components; Remote Sens. Environ., v. 6, p. 1–9.

Gausman, H. W., D. E. Escobar and E. B. Knipling, 1977b, Relation of *Peperomia obtusifolia's* anomalous leaf reflectance to its leaf anatomy; Photogram. Eng. and Rem. Sens., v. 43, no. 9, p. 1183–1185.

Gausman, H. W., D. E. Escobar, A. J. Richardson, R. L. Bowen and C. L. Wiegand, 1977c, Use of Landsat-1 data to distinguish grapefruit from orange trees and estimate their hectarages; J. Rio Grande Valley Hort. Soc. 31:139–144.

Gausman, H. W., D. E. Escobar and C. L. Wiegand, 1977d, Reflectance and photographic characteristics of three citrus varieties for discrimination purposes; F. Shahrokhi (ed.), Remote Sens. of Earth Resources, Univ. of Tennessee, Tullahoma 6:341–355.

Gausman, H. W., R. M. Menges, D. E. Escobar, J. H. Everitt and R. L. Bowen, 1977e, Pubescence affects spectra and imagery of silverleaf sunflower (*Helianthus argophyllus*); Weed Sci., v. 25, p. 437–440.

Gausman, H. W., D. E. Escobar, J. H. Everitt, A. J. Richardson and R. R. Rodriguez, 1978a, Distinguishing succulent plants from crop and woody plants; Photogram. Eng. and Remote Sens., v. 44, no. 4, p. 487–491.

Gausman, H. W., D. E. Escobar, R. R. Rodriguez, C. E. Thomas and R. L. Bowen, 1978b, Ozone damage detection in cantaloupe plants; Photogram. Eng. and Remote Sens., v. XLIV, no. 4, p. 481–486.

Gautier, C., G. Diak and S. Masse, 1980, A simple physical model to estimate incident solar radiation at the surface from GOES satellite data; J. Appl. Met. 19:1005–1012.

General Electric Company, 1978, Land cover analysis of the U.S. drainage of the Great Lakes; Great Lakes Basin Commission Report.

Gerbermann, A. H. and D. D. Neher, 1979, Reflectance of varying mixtures of a clay soil and sand; Photogram. Eng. and Remote Sens., v. 45, no. 8, p. 1145–1151.

Gilliam, A. E., E. S. Stern and E. R. H. Jones, 1962, An introduction to electronic absorption spectroscopy in organic chemistry; Arnold, London, 326 p.

Glover, K. M., K. R. Hardy, T. G. Konrad, W. N. Sullivan and A. S. Michaels, 1966, Radar observations of insects in free flight; Science, v. 154, no. 3752.

Goodman, M. S., 1959, A technique for the identification of farm crops on aerial photographs; Photogram. Eng., v. 28, p. 984–990.

Green, L. R., J. K. Olson, W. G. Hart and M. R. Davis, 1977, Aerial photographic detection of imported fire ant mounds; Photogram. Eng. and Remote Sens., v. 43, no. 8, p. 1051–1057.

Halstead, M. H., 1957, Forecasting micrometeorological variables; Final Report, Contract AF 19(604)–1117.

Hammond, D., 1941, The expression of genes for leaf shape in *Gossypium hirsutum* L. and *Gossypium arboreum* L. Am. J. Bot., v. 28, p. 124–138.

Harris, M. K., W. G. Hart, M. R. Davis, S. J. Ingle and H. W. Van Cleave, 1976, Aerial photographs show caterpillar infestation; The Pecan Quarterly, 10(2), 12–18.

Hart, W. G. and V. I. Myers, 1968, Infrared aerial color photography for the detection of population of brown soft scale on citrus groves; J. Econ. Entomol. 61(3):617–624.

Hart, W. G. and S. J. Ingle, 1969, Detection of anthropod activity on citrus foliage with aerial infrared photography; Proc. of the Workshop on Infrared Color Photography in the Plant Sciences, Gainesville, Florida, March 5–7, 1969.

Hart, W. G., S. J. Ingle, M. R. Davis, C. Mangum, A. Higgins and J. C. Boling, 1971, Some uses of infrared aerial color photography in entomology; Proc. of the Workshop on Color Aerial Photography in the Plant Sciences, March 2–4, 1971, Gainesville, Florida, p. 99–113.

Hart, W. G., S. J. Ingle, M. R. Davis, C. Mangum, 1973, Aerial photography with infrared color film as a method of surveying for citrus blackfly; J. Econ. Entomol., 66(1):190–194.

Hart, W. G., H. W. Gausman and R. R. Rodriguez, 1976, Citrus blackfly feeding injury and its influence on the spectral properties of citrus foliage; J. Rio Grande Valley Hort. Soc., 30:37–43.

Hart, W. G., S. J. Ingle, M. R. Davis, and J. A. Diez-Perez, 1978, The use of color infrared aerial photography to detect plants attacked by tropical fruit flies; Proc. of the 12th Int. Symp. on Remote Sens. of the Environ., v. 2, p. 1409–1411.

Harter, L. L., 1908, The influence of a mixture of soluble salts principally sodium chloride, upon the leaf structure and transpiration of wheat, oats, and barley; U.S. Dept. of Agr. Bur. Plant Ind. Bul. 134.

Hasted, T., 1974, Aqueous dielectrics; Chapman Hall, London.

Hayward, H. E. and E. M. Long, 1941, Anatomical and physiological responses of the tomato to varying concentrations of sodium chloride, sodium sulphate, and nutrient solutions; Botan. Gaz., v. 102, p. 437–462.

Hayward, H. E. and L. Bernstein, 1958, Plant-growth relationships on salt-affected soils; Botan. Rev., v. 24, p. 584–635.

Heald, C. H., W. H. Thames, C. L. Wiegand, 1972, Detection of *Rotylenchulus reniformis* infestation by aerial infrared photography; J. of Nematology, v. IV(4), Gainesville, Florida.

Heilman, J. L. and E. T. Kanemasu, 1976, An evaluation of a resistance form of the energy balance to estimate evapotranspiration; Agron. J. 68: 607–611.

Heilman, J. L., E. T. Kanemasu, N. J. Rosenberg and B. L. Blad, 1976, Thermal scanner measurement of canopy temperatures to estimate evapotranspiration; Remote Sens. Environ. 5:137–145.

Heilman, J. L. and D. G. Moore, 1980, Thermography for estimating near-surface soil moisture under developing crop canopies; J. Appl. Meteor.

Heilman, M., and C. L. Wiegand, 1975, Unpublished data, U.S. Department of Agriculture, Weslaco, Texas.

Heller, R. C., 1968, Previsual detection of Ponderosa pine trees dying from bark beetle attack; Proc. 5th Symp. Remote Sens. Environ., Univ. of Michigan, Ann Arbor, p. 387–434.

Hemphill, W. R. and S. V. Carnahan, 1965, Ultraviolet absorption and luminescence investigations progress report; NASA Earth Resources Survey Program Tech. Letter NASA 6.

Henneberry, T. J., W. G. Hart, L. A. Bariola, D. L. Kittock, H. F. Arle, M. R. Davis and S. J. Ingle, 1979, Parameters of cotton cultivation from infrared aerial photography; Photogram. Eng. and Remote Sens., v. 45, no. 8, p. 1129–1133.

Hielkema, J. V., 1978, Advanced training and research

on satellite remote sensing techniques and applications in the United Kingdom and the United States; AGLT/RSU Series 2/79, Foot and Agr. Org. of the U.N., Rome, Italy.

Hiler, E. A. and R. N. Clark, 1971, Stress day index to characterize effects of water stress on crop yields; Trans. Am. Soc. Agr. Eng. 14:757–761.

Hiler, E. A., T. A. Howell, R. B. Lewis, and R. P. Boos, 1974, Irrigation timing by the stress index method, Amer. Soc. Agr. Eng. 17:393–398.

Hixson, M. M., M. E. Bauer and L. L. Biehl, 1978, Crop spectra from LACIE field measurements; LACIE-00469, JSC-13734, and LARS Contract Report 011578.

Hixson, M. M., M. E. Bauer and W. Scholz, 1980, An assessment of Landsat data acquisition history on identification and area estimation of corn and soybeans; Proc. Symp. on Machine Processing of Remotely Sensed Data, Purdue Univ.

Hoekstra, P. and A. Delaney, 1974, Dielectric properties of soils at UHF and microwave frequencies; J. of Geophysical Research 79:1699–1708.

Hoffer, R. M. and C. J. Johannsen, 1969, Ecological potential in spectral signature analysis; IN: Remote sensing in ecology; Univ. of Georgia Press, Athens, Georgia, p. 1–16.

Hoffman, Richard O., 1979, Identifying and locating land irrigated by center-pivot irrigation systems using satellite imagery; Symp. Proc., Remote Sens. of Irrigated Lands, Missouri River Basin Commission, EROS Data Center.

Holmes, R., 1970, Edit., chapters on remote sensing handbook; Nat. Acad. of Sciences, Washington, D.C.

Horton, M. L., et al., 1977, Remote sensing applications for detection of saline seeps; Final Report for Old West Regional Commission, Grant No. 10570035.

Houston, A. G., A. H. Feiveson, R. J. Chikara and E. M. Hsu, 1978, Independent peer evaluation of the large area crop inventory experiment, (LACIE); Proc. of Tech. Sessions, NASA JSC-16015, NTIS, Springfield, VA.

Hovis, W. W., Jr., 1966, Infrared spectral reflectance of some common minerals; Appl. Optics, v. 5. p. 245–248.

Howard, J. A., 1966, Spectral energy relations of isobilateral leaves; Aust. J. Biol. Sci., v. 19, p. 757–766.

Idso, S. B. and R. D. Jackson, 1969, Thermal radiation from the atmosphere; J. Geophysical Res., v. 74, pp. 5397–5403.

Idso, S. B. and C. T. DeWit, 1970, Light relations in plant canopies; Appl. Optics, v. 9, p. 177–184.

Idso, S. B., R. D. Jackson and R. J. Reginato, 1975a, Estimating evaporation: a technique adaptable to remote sensing; Science 189:991–992.

Idso, S. B., R. D. Jackson and R. J. Reginato, 1975b, Detection of soil moisture by remote surveillance; Am. Scientist, v. 63, Sept.–Oct.

Idso, S. B., T. J. Schmugge, R. D. Jackson and R. J. Reginato, 1975c, The utility of surface temperature measurements for the remote sensing of soil water status; J. Geophys. Res. 80:3044–3049.

Idso, S. B. and W. L. Ehrler, 1976, Estimating soil moisture in the root zone of crops: a technique adaptable to remote sensing; Geophys. Res. Lett. 3:23–25.

Idso, S. B., R. D. Jackson and R. J. Reginato, 1976b, Compensating for environmental variability in the thermal inertia approach to remote sensing of soil moisture; J. Appl. Meteor. 15:811–817.

Idso, S. B., R. J. Reginato and R. D. Jackson, 1977, An equation for potential evaporation from soil, water and crop surfaces adaptable to use by remote sensing; Geophys. Res. Lett. 4:187–188.

Idso, S. B., R. D. Jackson and R. J. Reginato, 1978, Remote sensing for agricultural water management and crop yield prediction; Agric. Water Manage. 1:299–310.

Idso, S. B., J. L. Hatfield, R. D. Jackson and R. J. Reginato, 1979, Grain yield prediction: extending the stress-degree-day approach to accomodate climatic variability; Remote Sens. of Environ., v. 8, no. 3, p. 267–272.

Ingle, S. J., M. R. Davis and W. G. Hart, 1975, An improved method for detection and survey of the imported fire ant; Proc. of the Workshop on Infrared Color Photography in the Plant Sciences, 67–77, August 19–21, 1975, Sioux Falls, South Dakota.

Jackson, R. D., S. B. Idso and R. J. Reginato, 1976, Calculation of evaporation rates during the transition from energy-limiting to soil-limiting phases using albedo data; Water Resour. Res. 12:23–26.

Jackson, R. D., R. J. Reginato and S. B. Idso, 1977, Wheat canopy temperature: a practical tool for evaluating water requirements; Water Resour. Res. 13:651–656.

Jackson, R. D., P. J. Pinter, Jr., S. B. Idso and R. J. Reginato, 1979, Wheat spectral reflectance: interactions between crop configuration, sun elevation, and azimuth angle; Appl. Optics, v. 18, no. 22, p. 3730–32.

Jackson, R. D., S. B. Idso, R. J. Reginato and P. J. Pinter, Jr., 1980, Remotely sensed crop temperatures and reflectances as inputs to irrigation scheduling; Proc. ASCE Specialty Conference, Boise, Idaho.

Janza, F. J., 1976, Passive microwave systems; Chapter in Remote Sens. of Environ., Addison-Wesley Pub. Co., Reading, Mass.

Jensen, J. R., J. E. Estes and L. R. Tinney, 1978, High-altitude versus Landsat imagery for digital crop identification; Photogram. Eng. and Remote Sens., v. 44, no. 6, p. 723–733.

Jensen, M. E. and H. R. Haise, 1963, Estimating evapotranspiration from solar radiation, Journal of Irrigation and Drainage Division, American Society of Civil Engineers, Vol. 89, pp. 15–41.

Jensen, M. E., 1974, Consumptive use of water and irrigation water requirements; Am. Soc. of Civil Engineers.

Johannsen, C. J., 1969, The detection of available soil moisture by remote sensing techniques; Ph.D. Thesis, Purdue Univ., 266 p.

Johannsen, C. J. and T. W. Barney, 1976, A review of soil color; Unpublished report for Lockheed Electronics, Inc., 11 p.

Johannsen, C. J. and M. F. Baumgardner, 1968, Remote sensing for planning resource conservation; Proc. of 1968 Ann. Meet. Soil Cons. Soc. of Am., p. 149–155.

Johnson, G. E. and T. R. Loveland, 1979, The Columbia River and tributaries irrigation withdrawals analysis project; Symp. Proc., Remote Sens. of Irrigated Lands, Missouri River Basin Commission, EROS Data Center.

Jordan, C. F., 1969, Derivation of leaf area index from quality of light on the forest floor; Ecology 50:663–666.

Kanemasu, E. T., 1974, Seasonal canopy reflectance

patterns of wheat, sorghum and soybean; Remote Sens. of Environ. 3:43–57.

Kanemasu, E. T., J. L. Heilman, J. O. Bagley and W. L. Powers, 1977, Using Landsat data to estimate evapotranspiration of winter wheat; Environ. Manage. 6:515–520.

Kanemasu, E. T., M. L. Wesely, B. B. Hicks and J. L. Heilman, 1979, Techniques for calculating energy and mass fluxes; IN: B. J. Barfield and J. F. Gerber (eds.), Modification of the aerial environment of plants, ASAE monograph No. 2, Am. Soc. of Ag. Engineers, p. 156–182.

Kauth, R. J. and G. S. Thomas, 1976, The tasselled cap—a graphic description of the spectral temporal development of agricultural crops as seen by Landsat; Proc. of the Symp. Machine Processing of Remote Sens. Data, LARS, Purdue.

Keegan, H. J., J. C. Schleter, W. A. Hall, Jrs. and G. M. Haas, 1956, Spectrophotometric and colorimetric study of diseased and rust resisting cereal crops; Nat. Bur. Stds. Rpt. 4591.

Khorram, S., 1977a, A solar energy estimation procedure using sensing techniques; Proc. 6th Ann. Remote Sens. of Earth Resources Conf., Univ. of Tennessee Space Inst., Tullahoma.

Khorram, S., 1977b, Remote sensing-aided systems for snow quantification, evapotranspiration estimation, and their application to hydrologic models; Proc. 11th Int. Symp. of Remote Sens. Environ., v. I, p. 795–806.

Khorram, S., 1978, Use of Landsat-1 multispectral and NOAA-4 infrared data in estimating solar radiation components over the Middle Fork of Feather River Watershed; Proc. Conf. on Climate and Energy: Climatological Aspects and Industrial Operation, Am. Meteor. Soc., Ashville, N.C., p. 51–56.

Khorram, S. and H. G. Smith, 1979, Use of Landsat and environmental satellite data in evapotranspiration estimation from a wildland area; Proc. 13th Int. Symp. on Remote Sens. of Environ., Ann Arbor, Michigan, p. 1445–1453.

Kirdiashev, K. P., A. A. Chukhlanisev and A. M. Shutko, 1979, Microwave radiation of the earth's surface in the presence of vegetation cover; Radiotekknika i Elektronika, v. 24, p. 256–264, NASA Tech. Trans. TH-754.

Kirschner, F. R., S. A. Kaminsky, E. J. Hinzel, H. R. Sinclair and R. A. Weismiller, 1977, Quantification of soil mapping by digital analysis of Landsat data; Proc. 11th Int. Symp. of Remote Sens. Environ., v. II, p. 1567–1574.

Kleshnin, A. F. and I. A. Shul'gin, 1959, The optical properties of plant leaves; Dokl. Akademcii Nauk SSSR, v. 125, p. 1158, Translation: A.I.B.S. Doklady, v. 125, p. 108–110.

Klingebiel, A. A. and V. I. Myers, 1974, Soil study of Tampico and Isthmus areas of Mexico using ERTS imagery to determine potential land use; Washington, D.C., World Bank (Unpublished).

Knipling, E. B., 1967, Physical and physiological basis for difference in reflectance of healthy and diseased plants; Clearinghouse for Fed. Sci. and Tech. Inf., U.S. Dept. of Comm., Ad 652 679, 24 p.

Knipling, E. B., 1969, Leaf reflectance and image formation on color infrared film; Remote Sens. in Ecology, P. O. Johnson, ed., Univ. of Georgia Press, Athens, Georgia, p. 17–29.

Knipling, E. B., 1970, Physical and physiological basis for the reflectance of visible and near-infrared radiation from vegetation; Remote Sens. of Environ., v. 1, p. 155–159.

Kristof, S. J. and A. L. Zachary, 1971, Mapping soil types from multispectral scanner data; Proc. 7th Int. Symp. on Remote Sens. of Environ., v. III, p. 2095–2108.

Kristof, S. J. and A. L. Zachary, 1974, Mapping soil features from multispectral scanner data; Photogram. Eng., v. 40, no. 12, p. 1427–1434.

Kubelka, V. P. and F. Munk, 1931, Ein Beitrag zur Optik der Farbanstriche; Z. Tech. Physik., v. 11, p. 593–601.

Kumar, R., 1972, Radiation from plants—reflection and emission, a review; AA&ES 72-2-2, Purdue Univ., Lafayette, Indiana, 88 p.

Kumar, R., 1977, Evaluation of spectral channels and wavelength regions for separability of agricultural cover types; Proc. 11th Int. Symp. of Remote Sens. Environ., v. II, p. 1081–1090.

Landsberg, H. G. and M. L. Blanc, 1958, Interaction of soil and weather; Soil Sci. Soc. Am., Proc. 22, 491.

Leamer, R. W., J. R. Noriega and C. L. Wiegand, 1978, Seasonal changes in reflectance of two wheat cultivars; Agron. J., v. 70, p. 113–118.

Le Duc, S. K., 1976, Wheat yield model, LACIE; NASA Tech. Memo. TM 74833.

Lesage, P., 1890, Les modifications des feuilles chez les plantes maritimes, Chapitre III, Conclusions de la partie experimentale; Rev. Gen. Botan., p. 168–170.

Lintz, J., P. A. Brennan and P. E. Chapman, 1976a, Contributing chapter in *Remote Sensing of Environment,* p. 420.

Lundegardh, H., 1966, Plant physiology; Am. Elsevier Pub. Co., New York, 549 p.

Lundien, J. R., 1966, Terrain analysis by electromagnetic means: radar responses to laboratory prepared soil samples; U.S. Army WES Tech. Report 3-3639, Report 2, 55 p.

Lundien, J. R., 1971, Terrain analysis by electromagnetic means; Tech. Report 3-693, Report 5, U.S. Army Waterways Exp. Sta., Vicksburg, Mississippi, 85 p.

Lyon, R. J. P., 1965, Analysis of rocks of spectral infrared emission (8 to 25 microns); Economic Geology 60, 715–736.

MacDonald, R. B., F. G. Hall and R. B. Erb, 1975, The large area crop inventory experiment (LACIE)—an assessment after one year of operation; Proc. of 10th Int. Symp. on Remote Sens. of Environ., Ann Arbor, Michigan.

MacDonald, R. B. and F. G. Hall, 1978, LACIE: An experiment in global crop forecasting; Proc. of LACIE Symp. (JSC-14551), Oct. 13–26, 1978, NASA, JSC, Houston, Texas.

Manzer, F. E. and G. R. Cooper, 1967, Aerial photographic methods of potato disease detection; Maine Agric. Exp. Sta. Bul. 646, p. 1–14.

McFarland, M. H., 1976, The correlation of Skylab L-band brightness temperatures with antecedent precipitation; Proc. of the Conf. on Hydrometeorology, Fort Worth, Texas, Meteor. Soc., p. 60–65.

McKeague, J. A., J. H. Day and J. A. Shields, 1971, Evaluating relationships among soil properties by computer analysis; Can. J. Soil Sci. 51:105–111.

Mestre, H., 1935, The absorption of radiation by leaves and algae; Cold Spring Harbor Symp. Quant. v. 3, p. 191–209.

Meyer, M. P., 1967, No. title; Proc. Workshop Infrared Color Photography in the Plant Sciences, Winter Haven, Florida, Part V, p. 5–7.

Meyer, Merle P. and L Calpouzos, 1968, Detection of crop disease identifications; Photogram. Eng., v. 36, p. 1116–1125.

Millard, J. P., R .C. Goettelman, R. D. Jackson, R. J. Reginato and S. B. Idso, 1978, Crop water-stress assessment using an airborne thermal scanner; Photogram. Eng. and Remote Sens., v. 44, no. 1, p. 77–85.

Miller, M. R. and L. L. Bahls, 1976, An overview of saline seep programs in the states and provinces of the Great Plains; Regional Saline Seep Control Symp. Proc., p. 13–17, Billings, Montana.

Monteith, J. L., 1959, The reflectance of short-wave radiation by vegetation; Quart. J. Roy Meteorol. Soc., 85:386–392.

Monteith, J. L., 1973, Principles of environmental physics; Am. Elsevier, New York.

Moore, D. G. and V. I. Myers, 1972, Environmental factors affecting ground water mapping; Report SDSU-RSI-72-06 to U.S. Geological Survey.

Moore, D. G., M. L. Horton, J. J. Russell and V. I. Myers, 1975, Evaluation of thermal X/5 detector Skylab S-192 data for estimating evapotranspiration and thermal properties of soils for irrigation management; Proc. of the NASA Earth Resources Survey Symp., Houston, TX, NASA TM X-58168, p. 2561–2583.

Moore, D. G. and V. I. Myers, 1977, Progress and needs in agricultural research, development, and applications programs; Proc. 11th Int. Symp. of Remote Sens. Environ., v. I, p. 257–266.

Moore, R. K., 1976, Active microwave systems; Chapter from *Remote Sensing of Environment,* Addison-Wesley Publishing Co., Reading, Mass.

Moss, R. A., 1951, Absorption spectra of leaves; Unpublished Ph.D. Thesis, Iowa State University Library, Ames, Iowa, 68 p.

Moss, R. A. and W. E. Loomis, 1952, Absorption spectra of leaves. I. The visible spectrum; Plant Physiol., v. 27, p. 370–391.

Murphy, J. D., G. A. May and J. M. Downs, 1978, An agricultural data base to support diverse remote sensing applications; Proc. Int. Symp. on Remote Sens. for Obs. and Inventory of Earth Resources. v. III, Freiburg FRG.

Myers, L. D. and A. H. Stallard, 1974, Soil identification by remote sensing techniques in Kansas; Part I, State Highway Commission of Kansas, Report to U.S. Department of Transportation, Federal Highway Administration.

Myers, V. I., L. R. Ussery and W. J. Rippert, 1963, Photogrammetry for detailed detection of drainage and salinity problems; Trans. ASAE, v. 6, p. 322–334.

Myers, V. I., D. L. Carter and W. J. Rippert, 1966a, Remote sensing for estimating soil salinity; J. Irrig. Drain. Div., Am. Soc. Civ. Eng. 92(IR 4):59–68.

Myers, V. I., C. L. Wiegand, M. D. Heilman and J. R. Thomas, 1966b, Remote sensing in soil and water conservation research; Proc. 4th Symp. on Remote Sens. of Environ., Univ. of Michigan, Ann Arbor, Michigan.

Myers, V. I. and W. A. Allen, 1968, Electrooptical remote sensing methods as nondestructive testing and measuring techniques in agriculture; Appl. Opt., v. 7, p. 1818–1838.

Myers, V. I. and M. D. Heilman, 1969, Thermal infrared for soil temperature studies; Photogram. Eng., v. 35, p. 1024–1032.

Myers, V. I., M. D. Heilman, R. J. P. Lyon, L. N. Namken, D. Simonett, J. R. Thomas, C. L. Wiegand, J. T. Woolley, 1970, Chapter on Soil water, and plant relations; IN: *Remote Sensing,* Nat. Acad. of Sci.

Myers, V. I. and D. G. Moore, 1972, Remote sensing for defining aquifers in glacial drift; Proc. 8th Int. Symp. on Remote Sens. of Environ., Willow Run Labs, Univ. of Michigan, p. 715–728.

Myers, V. I., et al., 1978a, Remote sensing applications to resource problems in South Dakota; SDSU-RSI-78-14, Remote Sensing Institute, South Dakota State Univ., Brookings.

Myers, V. I., J. L. Heilman and D. G. Moore, 1978b, Remote soil moisture measurement: need, present methods and obstacles; Microwave Remote Sens. Symp., Houston, TX, NASA and Texas A&M University, p. 23–29.

Myers, V. I., D. G. Moore, M. DeVries and B. Worcester, 1978c, Remote sensing for monitoring resources for development and conservation of desert and semi-desert areas; South Dakota State University, SDSU-RSI-78-08.

Myers, V. I., et al., 1979, Remote sensing applications to resource problems in South Dakota; Ann. Prog. Report to NASA, Grant No. NGL 42-003-007, SDSU-RSI-79-14.

Namken, L. H., 1965, Relative turgidity technique for scheduling cotton (*Gossypium hirsutum*) irrigation; Agron. J., v. 57, p. 38–41.

NASA, Johnson Space Center, 1974, Corn blight watch experiment; Experiment results, v. III.

NASA, Johnson Space Center, 1974, Corn blight watch experiment; Summary report, v. 3, NASA SP-353, NTIS, Springfield, Va.

NASA, Johnson Space Center, 1977, LACIE: wheat yield models for the United States; NASA JSC-00431, rev. A, NTIS, Springfield, Va.

NASA, Johnson Space Center, 1978a, Independent peer evaluation of the large area crop inventory experiment (LACIE); NASA JSC-14550, NTIS, Springfield, Va.

NASA, Johnson Space Center, 1978b, Independent peer evaluation of the large area crop inventory experiment (LACIE); Proc. of Plenary Session, NASA JSC-14551, NTIS, Springfield, Va.

NASA, Johnson Space Center, 1978c, Independent peer evaluation of the large area crop inventory experiment (LACIE); Proc. of Tech. Sessions, NASA JSC-16015, NTIS, Springfield, Va.

National Academy of Sciences, 1977, Resource sensing from space: prospects for developing countries; Prepared for Agency for Int. Dev., Wash., D.C., PB-264 171.

Newton, R. W., 1977a, Advances in passive microwave techniques of remotely estimating soil water content; Microwave Remote Sensing Symp., Houston, TX, NASA and Texas A&M Univ., p. 30–49.

Newton, R. W., 1977b, Microwave remote sensing and its application to soil moisture detection; Tech. Report RSC-81, Remote Sensing Center, Texas A&M Univ., College Station, TX, 500 p. (Univ. microfilms no. 77-20,398).

Nieman, R. H., 1965, Expansion of bean leaves and its suppression by salinity; Plant Physiol., v. 40, p. 156–161.

Nixon, P. R., L. N. Namken and C. L. Wiegand, 1973, Spatial and temporal variations of crop canopy temperatures and implications for irrigation scheduling; IN: Shahroki, F. (Ed.), Remote Sensing of Earth Resources, Univ., of Tennessee, Tullahoma, TN, p. 643–657.

Nixon, P. R., S. D. Lyda, M. D. Heilman and R. L. Bowen, 1975, Incidence and control of cotton root rot observed with color infrared photography; Texas Agric. Exp. Sta. Misc. Pub. 1241, 4 p.

Njoku, E. G. and J. A. Kong, 1977, Theory for passive microwave remote sensing of near-surface soil moisture; J. of Geophysical Research, 82:3108–3118.

Norman, G. G. and N. L. Fritz, 1965, Infrared photography as an indicator of disease and decline in citrus; Proc. Fla. State Hort. Soc., 78:59–63.

Obukhov, A. I. and Orlov, D. C., 1964, Spectral reflectivity of the major soil groups and possibility of using diffuse reflection in soil investigations; Pochvovedeniye, No. 2.

Oden I. Influence of particle (aggregate) size on reflectivity; Diklady Soil Sci., No. 13, Supplement.

Odenyo, V. A. O., 1979, Application of Landsat data to the study of land and range resources in the Narok area, Kenya, Canada Centre for Remote Sensing, Research Report 79-4, CCRS, EMR, Canada.

Orlov, D. S., 1966, Quantitative patterns of light reflection on soils: I. Influence of particle (aggregate) size on reflectivity; Diklady Soil Sci., No. 13. Supplement.

Park, A. B., 1979, Agricultural monitoring and assessment; Env. Research Inst. of Mich., Seminar on Remote Sens. Appl.

Parsons, A. J. and E. A. Fitzpatrick, 1978, Determination of wheat acreages from Landsat imagery in the Narrabri District of New South Wales, Australia; Proc. 12th Int. Symp. on Remote Sens. of Environ., v. III, p. 1849–1857.

Payne, J. A., W. G. Hart, M. R. Davis, L. S. Jones, D. J. Weaver and B. D. Horton, 1971, Detection of peach and pecan pests and diseases with color infrared aerial photography; Proc. 3rd Biennial Workshop on Color Aerial Photography in the Plant Sciences, March 2–4, 1971, Gainesville, Fla., p. 216–230.

Pearman, G. I., 1966, The reflection of visible radiation from leaves of some western Australian species; Australian J. Biol. Sci., v. 19, p. 97–103.

Pearson, R. L., and L. D. Miller, 1972, Remote mapping of standing crop biomass for estimation of the productivity of the shortgrass prairie, Eighth International Symposium on Remote Sensing of Environment, University of Michigan, Ann Arbor, Mich., 1357–1381.

Philpotts, L. E. and V. R. Wallen, 1969, IR color for crop disease identifications; Photogram. Eng., v. 35, p. 1116–1125.

Piech, K. R. and J. E. Walker, 1974, Interpretation of soils; Photogram. Eng., v. 40, no. 1, p. 87–94.

Poe, G. A., A. Stogryn and A. T. Edgerton, 1971, Determination of soil moisture content using microwave radiometry; Final Rep. No. 1684 FR-1, DOC Contract 0-35239, Aerojet-General Corp., El Monte, CA, 169 p.

Poracsky, J., 1979, Irrigation mapping in western Kansas using Landsat; Symp. Proc., Remote Sens. of Irrigated Lands, Missouri River Basin Comm., EROS Data Center.

Pratt, R. M., G. F. Snow and T. R. Carpenter, 1973, Phymatotrichum root rot of cotton and alfalfa in an aerial survey; Phytopathology, v. 53, no. 10.

Priestley, C. H. B. and R. J. Taylor, 1972, On the assessment of soil heat flux and evaporation using large parameters; Mon. Weather Rev. 100:81–92.

Rabideau, G. S., C. S. French and A. S. Holt, 1946, The absorption and reflection spectra of leaves, chloroplast suspensions and chloroplast fragments as measured in an Ulbricht-sphere; Am. J. Bot., v. 33, p. 769-777.

Rao, V. R., E. J. Brach and A. R. Mach, 1979, Bidirectional reflectance of crops and the soil contribution; Remote Sens. of Environ., v. 8, p. 115–125.

Reeves, R. G., Editor-in-Chief, 1975, ASP Manual of remote sensing, chapter on crops and soils; p. 1739–1744.

Reginato, R. J., S. B. Idso, J. F. Vedeer, R. D. Jackson, M. B. Blanchard and R. Goettelman, 1976, Soil water content and evaporation determined by thermal parameters obtained from ground-based and remote measurements; J. Geophys. Res. 81:1616–1620.

Reginato, R. J., J. F. Vedder, S. B. Idso, R. D. Jackson, M. B. Blanchard and R. Goettelman, 1977, An evaluation of total solar reflectance and spectral band ratioing techniques for estimating soil water content; J. Geophys. Res. 82:2101–2104.

Reifsnyder, W. E. and H. W. Lull, 1965, Radiant energy in relation to forests; Tech. Bul. No. 1344, U.S. Dept. of Agric., Forest Serv.

Reining, Priscilla, 1978, Handbook on desertification indicators; Am. Assoc. for the Advancement of Sci., Wash., D.C.

Rhode, W. G., 1978, Improving land cover classification by image stratification of Landsat data; Proc. 12th Int. Symp. on Remote Sens. of Environ., v. I, p. 729–741.

Richardson, A. J., W. A. Allen and J. R. Thomas, 1969, Discrimination of vegetation by multispectral measurements; Proc. 6th Int. Symp. on Remote Sens. of Environ., p. 1143–1156.

Richardson, A. J., C. L. Wiegand, H. W. Gausman, J. A. Cuellar and A. H. Gerbermann, 1975, Plant, soil and shadow reflectance components of row crops; Photogram. Eng. and Remote Sens., v. 41, no. 11, p. 1401–1407.

Richardson, A. J., A. H. Gerbermann, H. W. Gausman and J. A. Cuellar, 1976, Detection of saline soils with Skylab multispectral scanner data; Photogram. Eng. and Remote Sens., v. 42, no. 5, p. 679–684.

Richardson, A. J. and C. L. Wiegand, 1977, Distinguishing vegetation from soil background information; Photogram. Eng. and Remote Sens., v. 43, no. 12, p. 1541–1552.

Richardson, A. J., C. L. Wiegand, R. J. Torline and M. R. Gautreaux, 1977b, Landsat agricultural land use survey; Photogram. Eng. and Remote Sens., v. 43, no. 2, p. 207–216.

Ritchie, J. T., 1972, Model for predicting evaporation from a row crop with complete cover; Water Resour. Res. 8:1204–1213.

Robertson, G. W., 1964, Agricultural Meteorological Tech. Bul. 4.

Robertson, G. W., 1968, A biometeorological time scale of a cereal crop involving day and night temperatures and photoperiod; Int. J. Biometeorol., 12, p. 191.

Robinowitch, E. I., 1951, Photosynthesis and related processes, v. II, Parts I and II; Interscience Publishers, Inc., New York.

Rouse, J. W., R. H. Haas, J. A. Schell and D. W. Deering, 1973, Monitoring vegetation systems in the great plains with ERTS; 3rd ERTS Symp., NASA SP-351, I:309–317.

Rouse, J. W., R. H. Haas, J. A. Schell, D. W. Deering and J. C. Harlan, 1974, Monitoring the vernal advancement and retroradiation (greenwave effect) of natural vegetation. Report No. RSC 1978-4, Remote Sensing Center, Texas A&M Univ., College Station, TX.

Ryland, D. W., F. A. Schmer and D. G. Moore, 1975, Investigation of remote sensing to detect near-surface groundwater on irrigated land; Report No. SDSU-RSI-75-10, Remote Sens. Inst., South Dakota State Univ.

Sakamoto, C. M. and S. K. Le Duc, 1977, Wheat yield model, LACIE: NASA Tech. Memo. TM-74834.

Salisbury, F. B. and C. Ross, 1969, Plant physiology; Wadsworth Pub. Co., Belmont, CA, 747 p.

Sanger, J. E., 1972, Quantitative investigations of leaf pigments from their inception in buds through autumn coloration to decomposition in falling leaves; Ecology, v. 52, p. 1075–1089.

Schanda, E., R. Hofer, D. Wyssen, A. Musy, D. Meylan, C. Morzier and W. Good, 1978, Soil moisture determination and snow classification with microwave radiometry; Proc. 12th Int. Symp. on Remote Sens. of Environ., Manila, P.I.

Schmugge, T. J., P. Gloersen, T. Wilheit and F. Geiger, 1974, Remote sensing of soil moisture with microwave radiometers; J. of Geophysical Research, 79:317–323.

Schmugge, T. J., J. M. Meneeley, A. Rango and R. Neff, 1977, Satellite microwave observations of soil moisture variations; Water Res. Bul., Am. Water Res. Assoc., v. 13, no. 2, p. 265–281.

Schmugge, T. J., 1978a, Remote sensing of surface soil moisture; J. of Appl. Meteor., v. 17, p. 1549.

Schmugge, T. J., B. Blanchard, A. Anderson and J. Wang, 1978b, Soil moisture sensing with aircraft observations of the diurnal range of surface temperature; Water Res. Bul. 14:169–178.

Schmugge, T. J., 1980, Effect of soil texture on the microwave emission from soils; NASA Tech. Mem. 80632, Goddard Space Flight Center.

Schwarz, D. E. and F. R. Caspall, 1968, The use of radar in the discrimination of agricultural land use; Proc. 5th Symp. on Remote Sens. of Environ., Univ. of Michigan, Ann Arbor, p. 233–247.

Scott, D., P. H. Menalda and R. W. Brougham, 1968, Spectral analysis of radiation transmitted and reflected by different vegetations; New Zealand J. Bot., v. 6, p. 427–449.

Serreyn, D. V. and G. D. Nelson, 1973, The K-Class classifier; Remote Sensing Inst., South Dakota State University, RSI 73-08.

Seybold, A., 1932, Uber die optischen Eigenschaften der Laublätter; Planta, v. 16, p. 195–226, v. 18, p. 479–508.

Sharma, K. D. and D. N. Sen, 1971, Sun and shade tolerance in the ecophysiology of *Solanum surattense* Burm.; Z. Pflanzenphysiol., v. 64, p. 263–266.

Shockley, W. G., S. J. Knight and E. B. Lipscomb, 1962, Identifying soil parameters with an infrared spectrometer; Proc. 2nd Symp. on Remote Sens. of Environ., Univ. of Michigan Press, Ann Arbor.

Shul'gin, I. A., V. S. Khazanov and A. F. Kleshnin, 1960, On the reflection of light as related to leaf structure; Translated from Dokl. Akademii Nauk SSSR, v. 134, p. 471–474.

Shull, C. A., 1929, A spectrophotometric study of reflection of light from leaf surfaces; Bot. Gaz., v. 87, p. 583–607.

Siegal, B. S. and A. F. H. Goetz, 1977, Effect of vegetation on rock and soil type discrimination; Photogram. Eng. and Remote Sens., v. 43, no. 2, p. 191–196.

Sifton, H. B., 1945, Air-space tissue in plants; Bot. Rev., v. 11, p. 108–143.

Simonett, D. S., J. E. Eagleman, A. B. Erhart, D. C. Rhodes and D. E. Schwarz, 1967, The potential of radar as a remote sensor in agriculture: I. A study with K-band imagery in western Kansas; CRES Report 61-21, 13 p.

Sinclair, T. R., 1968, Pathway of solar radiation through leaves; M. S. Thesis, Purdue Univ., Library, Lafayette, Indiana.

Slatyer, R. O., 1967, Plant-water relationships; Academic Press, Inc., New York, 366 p.

Soer, G. J. R., 1977, The Tergra model, a mathematical model for the simulation of the daily behavior of crop surface temperature and actual evapotranspiration; NIWARS Publ. 46, Delft.

Soer, G. J. R., 1980, Estimation of regional evapotranspiration and soil moisture conditions using remotely sensed crop surface temperatures; Remote Sens. Environ. 9:27–45.

Soil Survey Staff, 1951, Soil survey manual; Bur. of Plant Ind. Soils, and Agric. Eng., USDA Handbook No. 18.

Stanhill, G., 1962, The effect of environmental factors on the growth of alfalfa in the field; Netherlands J. Agri. Sci. 10:247.

Stone, L. R. and M. L. Horton, 1974, Estimating evapotranspiration using canopy temperatures; Field evaluation, Agron. J. 66:450–454.

Stoner, E. R. and E. H. Horvath, 1971, The effect of cultural practices on multispectral response from surface soil; Proc. 7th Int. Symp. of Remote Sens. of Environ., v. III, p. 2109–2113.

Stoner, E. R., 1979, Physiochemical, site, and bidirectional reflectance factor characteristics of uniformly-moist soils; Ph.D. Thesis, Purdue Univ.

Stoner, E. R., M. F. Baumgardner, L. L. Biehl and B. F. Robinson, 1980, Atlas of soil reflectance properties; Research Bul. 962, Agric. Exp. Sta., Purdue Univ.

Strogonov, B. P., 1962, Physiological basis of salt tolerance of plants; Daniel Davey and Co., Inc., New York, Translation by Israel Program for Scientific Translations, 1964.

Suits, G. H., 1972a, Calculation of the directional reflectance of a vegetative canopy; Remote Sensing of Environment, v. 2, no. 2, Am. Elsevier, New York, p. 117–125.

Sumayao, C. R., E. T. Kanemasu and T. Hodges, 1977, Soil moisture effects on transpiration and net carbon dioxide exchange of sorghum; Ag. Meteor. 18:401–408.

Tanner, C. B., 1963, Plant temperatures; Agron. J. 55:210–211.

Tanner, C. B., 1967, Measurement of evapotranspiration; IN: R. M. Hagan, H. R. Haise and T. W. Edminister (eds.). Irrigation of agricultural lands; Am. Soc. Agron., Madison, WI, p. 534–574.

Tarpley, J. D., 1979, Estimating incident solar radiation at the surface from geostationary satellite data; J. Appl. Meteor. 18:1172–1181.

Taylor, S. E., 1979, Measured emissivity of soils in the southeast United States; Remote Sens. of Environ., 8:359–364.

Thomas, J. R., V. I. Myers, M. D. Heilman and C. L. Wiegand, 1966, Factors affecting light reflectance of cotton; Proc. 4th Symp. Remote Sens. of Environ., Inst. of Sci. and Tech., Univ. of Michigan, Ann Arbor, Michigan, p. 305–312.

Thomas, J. R., C. L. Wiegand and V. I. Myers, 1967, Reflectance of cotton leaves and its relation to yield; Agron. J., v. 59, p. 551–554.

Thomas, J. R., 1970, Contributing author, Soil, water and plant relations; IN: Remote Sens., Nat. Acad. of Sci., Wash., D.C., p. 264–267.

Thomas, J. R. and C. L. Wiegand, 1970, Osmotic and matric suction effects on relative turgidity, temperature and growth of cotton leaves; Soil Sci. 109:85–91.

Thomas, J. R., L. N. Namken, G. F. Oerther and R. G. Brown, 1971, Estimating leaf water content by reflectance measurements; Agron. J., v. 63, p. 845–847.

Thomas, J. R. and G. F. Oerther, 1972, Estimating nitrogen content of sweet pepper leaves by reflectance measurements; Agron. J. p. 11–13.

Thomas, J. R., and H. W. Gausman, 1977a, Leaf reflectance vs. leaf chlorophyll and carotenoid concentrations for eight crops; Agron. J., v. 69, p. 799–802.

Thomas, J. R. and G. F. Oerther, Jrs., 1977b, Estimation of crop conditions and sugar cane yields using aerial photography; Proc. Am. Soc. Sugar Cane Technology, v. 6, p. 93–99.

Thompson, D. R. and O. A. Wehmanen, 1977, The use of Landsat digital data to detect and monitor vegetation water deficiencies; Proc. 11th Int. Symp. on Remote Sens. of Environ., 45:201.

Thornthwaite, C. W., 1948, An approach toward a rational classification of climate; Geogr. Rev. 28:55.

Tucker, C. J., L. D. Miller and R. L. Pearson, 1975, Shortgrass prairie spectral measurement; Photogram. Eng. and Remote Sens., v. 41, no. 9, p. 1157–1162.

Tucker, C. J. and E. L. Maxwell, 1976, Sensor design for monitoring vegetation canopies; Photogram. Eng. and Remote Sens., v. 42, no. 11, p. 1399–1410.

Tucker, C. J., 1977a, Asymptotic nature of grass canopy spectral reflectance; Appl. Opt., v. 16, p. 1151–1157.

Tucker, C. J., 1977b, Spectral estimation of grass canopy variables; Remote Sens. of Environ., v. 6, p. 11–26.

Tucker, C. J., 1977c, Use of near infrared/red radiance ratios for estimating vegetation biomass and physiological status; Proc. 11th Int. Symp. of Remote Sens. of Environ., v. I, p. 493–494.

Tucker, C. J. and L. D. Miller, 1977, Soil spectra contributions to grass canopy spectral reflectance; Photogram. Eng. and Remote Sens., v. 43, no. 6, p. 721–726.

Tucker, C. J., 1978, A comparison of satellite sensor bands for vegetation monitoring; Photogram. Eng. and Remote Sens., v. 44, no. 11, p. 1369–1380.

Tucker, C. J., 1979, Red and photographic infrared linear combinations for monitoring vegetation; Remote Sens. of Environ., v. 8, p. 127–150.

Tueller, P. T. and D. T. Booth, 1975, Large scale color photography for erosion evaluations on rangeland watersheds in the great basin; Proc. of Am. Soc. of Photogram., Phoenix, Ariz., p. 708–753.

Turrell, F. M., 1936, the area of the internal exposed surface of dicotyledon leaves; Am. J. Bot., v. 23, p. 255–264.

Ulaby, F. T., 1974, Radar measurement of soil moisture content; IEEE Trans. on Antennas and Propagation, v. AP-22, p. 257–265.

Ulaby, F. T., J. Cihlar and R. K. Moore, 1974, Active microwave measurements of soil water content; Remote Sens. of Environ., v. 3, p. 185–203.

Ulaby, F. T. and P. P. Batlivala, 1976, Optimum radar parameters for mapping soil moisture; IEEE Trans. on Geosci. Elect., v. GS-14, no. 2, p. 81–93.

Ulaby, F. T., G. A. Bradley, M. C. Dobson and J. E. Barc, 1977a, Analysis of the active microwave response to soil moisture: Part II: vegetation-covered ground; RSL Tech. Rept. 264-18, Univ. of Kansas Center for Research, Inc., Lawrence, KS.

Ulaby, F. T., P. P. Batlivala, M. C. Dobson, 1978, Microwave backscatter dependence on surface roughness, soil moisture and soil texture; Part I—Bare Soil, IEEE Trans. on Geosci. Elect., v. GS-16, p. 286–295.

Umberger, D. E., M. H. Proctor, J. E. Clark, L. M. Eisgruber and C. B. Braschler, 1978, Independent peer evaluation of the large area crop inventory experiment (LACIE); Proc. of Tech. Sessions, NASA JSC-16015, NTIS, Springfield, Va.

United Nations, 1977, Transnational project to monitor desertification processes and related natural resources in arid and semi-arid areas of southwest Asia feasibility report; United National Environ. Prog., Nairobi, Kenya.

Uphof, J. C. Th., 1941, Halophytes; *Botan. Rev.*, v. 7, p. 1–58.

U.S. Department of Agriculture Yearbook, 1938, Soils and men; House Document No. 398, U.S. Gov. Print. Off., Wash., D.C.

U.S. Department of Agriculture, Soil Conservation Service, 1975, *Soil Taxonomy;* Agr. Handbook 436 U.S. Gov't Printing Off., Wash. D.C.

U.S. Department of Agriculture, 1977a, Econ. Res. Serv. Foreign Agric. Econ. Rep., Second forecast of 1977 Soviet grain crop; Rep. FG 10–77.

U.S. Department of Agriculture, 1977b, Econ. Res. Serv. Foreign Agric. Econ. Rep., Wheat situation; WS-239.

U.S. Department of Agriculture, 1977c, USDA Econ. Res. Serv. Foreign Agric. Econ. Rep. 132.

Vinogradov, B. V., 1961, Vegetation as an indicator in the interpretation of aerial photographs of desert landscapes in western Turkmania, Izvestiya Vsesoyuznogo Geograficheskogo Obshchestra, Reference in Soviet Geog. II(5), 7 p.

Vinogradov, B. V., 1977, Remote sensing in ecological botany; Remote Sens. in Environ., v. 6, p. 83–94.

Von Minnus, E., 1967, Spektrale remission unbewachsener bode als laktor bei der luftobeldinterpretation; Selbstverlag der Bundesanstalt fur Landeskunds and Raumforschung, Bad Godesberg, 41 p.

Waltz, F. A., 1970, Multidiscipline remote sensing research for hydrology and agriculture; J. of Remote Sens., v. 1, no. 4.

Wang, J. R. and T. J. Schmugge, 1980, An empirical model for the complex dielectric permittivity of soils as a function of water content; IEEE Trans. Geosci. and Remote Sens., G.E.-18, p. 288–295.

Weber, F. P. and C. E. Olson, Jr., 1967, Remote sensing implications of changes in physiologic structure and functions of tree seedlings under moisture stress; Ann. Prog. Rep. for Remote Sens. Lab for Natural Resources Prog., NASA, by the Pacific Southwest Forest and Range Exp. Sta., 61 p.

Wehde, M. W., 1980, Natural resource applications of computerized data processing; the AREAS Example, J. Soil and Water Cons. Jan-Feb issue, Soil Cons. Soc. Am., Ankeny, IA.

Weismiller, R. A., F. R. Kirschner, S. A. Kaminsky, E. J. Hinzel, 1979. Spectral classification of soil characteristics to aid the soil survey of Jasper County, Indiana; LARS Tech. Rep. 040179, Purdue Univ.

Werner, H. D., F. A. Schmer, M. L. Horton and F. A. Waltz, 1971, Application of remote sensing techniques to monitoring soil moisture; Proc. 7th Int. Symp. on Remote Sens. of Environ., Ann Arbor, Michigan, p. 1245–1258.

Westin, F. C., 1973a, ERTS-1 MSS imagery: A tool for identifying soil association; SDSU-RSI-J-73-03.

Westin, F. C., 1973b, ERTS mosaic of South Dakota; AES Info Series No. 5 and Remote Sens. Inst. SDSU-RSI-73-17, South Dakota State Univ., Brookings, South Dakota.

Westin, F. C., 1974a, Soil textures and landforms on ERTS-1 imagery; AES Info Series No. 8, South Dakota State Univ., Brookings, South Dakota.

Westin, F. C., M. Stout, Jrs., D. L. Bannister and C. J. Frazee, 1974, Soil surveys for land evaluation; Assessors J., Oct. 1974.

Westin, F. C., 1976, Agrophysical mapping; Final Report Contract NAS 9-15178, NASA/JSC, Houston, TX.

Westin, F. C. and C. J. Frazee, 1976, Landsat data, its use in a soil survey program; Soil Sci. Soc. Am. J. 40:81–89.

Westin, F. C. and G. D. Lemme, 1978, Landsat spectral signatures: studies with soil associations and vegetation; Photogram. Eng. and Remote Sens., v. 44, no. 3, p. 315–325.

Wiegand, C. L. and L. N. Namken, 1966, Influences of plant moisture stress, solar radiation, and air temperature on cotton leaf temperature; Agron. J. 58:582–586.

Wiegand, C. L., M. D. Heilman and A. H. Gerberman, 1968, Detailed plant and soil thermal regime in agronomy; Proc. 5th Symp. on Remote Sens. of Environ., Willow Run Laboratories, the Univ. of Mich., Ann Arbor, Michigan, April 16–18, 1968, p. 325–334.

Wiegand, C. L., H. W. Gausman, W. A. Allen and R. W. Leamer, 1969, Interaction of electromagnetic energy with agricultural terrain features; Proc. Earth Resources Prog. Status Review, Earth Resources Div., NASA, Manned Spacecraft Center, Houston, TX.

Wiegand, C. L., H. W. Gausman and W. A. Allen, 1972, Physiological factors and optical parameters as bases of vegetation discrimination and stress analysis; Proc. Seminar on Operational Remote Sens., Am. Soc. Photogram., Falls Church, Virginia, p. 82–102.

Wiegand, C. L., 1974, Reflectance of vegetation, soil, and water; Type III Final Report to NASA, No. 5-70251-AG. Agricultural Research Serv., USDA, Weslaco, TX.

Wiegand, C. L., A. J. Richardson and E. T. Kanemasu, 1979, Leaf area index estimates for wheat from Landsat and their implications or evapotranspiration and crop modeling; Agron. J. 71:336–342.

Wilheit, T. T., 1978, Radiative transfer in a plane stratified dielectric; IEEE Trans. on Geosci. Elect., G. E.-16, 138–143.

Williams, T. H. L., 1979, Irrigation mapping in western Kansas using Landsat, Part II Practices and problems; Symp. Proc., Identifying Irrigated Lands Using Remote Sens. Tech., Missouri River Basin Comm. and EROS.

Wilstatter, R. and A. Stoll, 1918, Untersuchungen uber die assimilation der Kohlensaure; Springer, Berlin, p. 122–127.

Wong, K. W., T. H. Thornburn and M. A. Khoury, 1977, Automatic soil identification from remote sensing data; Photogram. Eng. and Remote Sens., v. 43, no. 1, p. 73–80.

Woolley, J. T., 1971, Reflectance and transmittance of light by leaves; Plant Physiol., v. 47, p. 656–662.

Worcester, B. K. and L. J. Brun, 1977, Growth and development of saline seeps; North Dakota Farm Research, Fargo, North Dakota.

Worcester, B. K., K. J. Dalsted and D. G. Moore, 1978a, Sahel-Sudano desertification study; Prepared for Dept. of State, South Dakota State Univ., Remote Sensing Inst., SDSU-RSI-78-07.

Worcester, B. K. and D. G. Moore, 1978b, Delineation of soil-landscapes in the Sudd region of Sudan on Landsat imagery; Proc. 12th Int. on Remote Sens. of Environ., v. II, p. 1155–1166.

Wright, J. L. and M. E. Jensen, 1978, Development and evaluation of evapotranspiration models for irrigation scheduling; Trans. Am. Soc. Agr. Eng. 21:88–96.

Wuhrmann, K., 1935, Die ionenaufnahme der pflanzen and ihre morphogenetis che wirkung; *Eidg. Tech. Hochschule,* Zurich, Cited by Hayword, H. E., and E. M. Long (1941).

CHAPTER 34

Forest Resource Assessments

Author-Editors: ROBERT C. HELLER and JOSEPH J. ULLIMAN

Contributing Authors: ROBERT C. ALDRICH, ROBERT C. HELLER, GERD HILDEBRANDT, ROGER M. HOFFER, DONALD T. LAUER, LEO SAYN-WITTGENSTEIN, and DONALD A. STELLINGWERF

GENERAL CONTENTS: Inventory; classification; forest volume; products; pulpwood; sawlogs; temperate zone (United States, Canada, Europe, Southern Hemisphere); tropical zones; damage assessment; fire; insects; disease; air pollution; storm damage; forest management; forest recreation; computer-aided processing of digital data; references.

INTRODUCTION[1]

A large portion of the Earth's renewable natural resources is made up of forests. All projections indicate that wood use will continue to increase because of the shrinking land base, increasing world population and greater use of wood for energy and fuel. The forest manager must know where the natural resources occur and what their condition is in order to decide how best to balance the many demands on the resource for best utilization of future generations.

In this scheme of best use, remote sensing plays a key role to assist in making accurate, timely and cost effective resource evaluations. Foresters were among the first to use aerial photographs and remote sensing data for applications to inventory, damage assessment, management and recreation. In this chapter not only are North American forestry applications described but also applications in Europe, the Southern Hemisphere and the tropical zones.

Foresters are using more color films, high altitude photography, small format photography, panoramic photography, multi-stage and multiphase sampling, satellite images and digital analysis of satellite data than they were when the first edition of the *Manual of Remote Sensing* was published in 1975. Examples of such applications are described in this chapter.

[1] In portions of this chapter, much of what was published in Chapter 17 on Forest Lands: Inventory and Assessment in the Manual of Remote Sensing, 1975 (Editor-in-Chief, Robert G. Reeves; Associate Editors, Abraham Anson and David Landen) has been wholly or partly incorporated in the present chapter. We wish to acknowledge authors who contributed to the 1975 edition of the Manual of Remote Sensing and who had made no changes in this edition. They include: G. A. Thorley, Chapter author editor; David A. Bernstein, G. Ross Cochran, Robert F. Kruckberg, Bernard Hostrop, Hartmut Kenneweg, Merle P. Meyer, and Charles E. Olson.

INVENTORY OF FOREST LANDS

TEMPERATE ZONE: UNITED STATES AND CANADA

Classification of Forest Lands

Professional land managers have learned that much of the information desired about forest lands can be readily seen on various types of aerial and space imagery. And for those features and conditions not discernible on the imagery, information frequently can be acquired about them on the ground more efficiently with the aid of imagery. Thus, systems have evolved for rapidly and accurately classifying forest lands in terms of variables which can be estimated or measured directly on the images themselves. This concept of classifying the forest according to what can be seen directly on the imagery, which gives the best estimate of true ground conditions, is basic to making the fullest possible use of remotely sensed imagery.

The simplest classification method is one that merely discriminates vegetated from nonvegetated lands. Other methods may involve detailed descriptions of cover types, individual species, density classes, age classes, size classes or complex plant associations. Classification of forest lands on images may involve visually identifying homogeneous categories, drawing boundaries between categories, point sampling with a list of points by categories, or automatically categorizing a matrix of picture elements using computer assisted techniques. Regardless of which method is used to classify forest land, or in what part of the world the work is being done, it is important to emphasize that images are used to complement, improve, or reduce field work, rather than take its place. For example, when Fang (1980) reported that the forest inventory of all of the People's Republic of China had been completed one or more times, he noted "... to carry out a large area forest inventory with aerial photography is the correct way. Advantages of this method are to hasten and reduce field work. . . ." Thus, a desirable classification procedure is one which maximizes the use of aerial or space imagery and minimizes the amount of field work in a manner that meets forest management or planning objectives.

Analysis Procedures

Successful classification of existing vegetation types by means of visual interpretation techniques normally requires the services of a highly skilled imagery analyst. (Procedures for classifying forest lands using machine-assisted techniques are presented later in this Chapter.) The analyst must be fully aware of the many factors that govern the interpretability of imagery and, in addition, be a trained professional land manager familiar with the forest environment to be classified. The experienced analyst can quickly become familiar with the image characteristics (i.e., size, shape, color or tone, shadow, pattern, location and association) correlated with each set of features or conditions to be classified. Training aids or reference materials, such as an image-interpretation key, are often useful for familiarizing the analyst with these identifiable characteristics. Colwell (1965) noted that an interpretation key to forest types should contain at least four elements:

(1) oblique view photo illustrations, which reveal the ecological site preferred by each type;
(2) vertical view stereograms of each type similar to what the interpreter would ordinarily be called upon to interpret;
(3) a word description which sets forth in some systematic fashion the image recognition features for each type; and
(4) a statement describing the significance of each type.

An example of a dichotomous or "two-branched" key for several tree types found in the southeastern United States is presented in Table 34-1 (Krumpe, 1971). A summary of the advantages and limitations of interpretation keys, including the dichotomous type, was published by Roscoe, et al. (1955).

During the phase of familiarization and training, in which interpretation keys may be utilized, the analyst should be aware of the many factors that affect the quality of remotely sensed imagery,

TABLE 34-1

A Dichotomous Photointerpretation Key to Several Forest Tree Species in Eastern Tennessee Using Color Transparency Film.

1. Branching is layered, radially triangular; crown margin is serrate, crown foliage is light green to moderate green	WHITE PINE
1. Branching not radially triangular; crown margin is not serrate	Go to 2.
2. Leaves mostly inconspicuous, tree branches virtually bare	Go to 3.
2. Leaves present in crown	Go to 5.
3. No foliage present; dark colored bole and branches completely bare	WHITE BASSWOOD
3. Very little foliage remaining (<5 percent)	Go to 4.
4. Branching gives crown a fine textured appearance	WHITE ASH or BLACK WALNUT
4. Branching appears medium textured	YELLOW BUCKEYE
5. Crown foliage thinning; trees losing a significant portion (40 percent) of leaves in early fall	Go to 6.
5. Crown foliage is dense or full; leaves abundant on branches	Go to 8.
6. Branching appears finely divided or dissected; crown margin shape is circular or oval and usually large. Branches are a silver grey color. Crown foliage is finely textured, crown color is a moderate orange yellow to dark orange yellow	AMERICAN BEECH
6. Branching appears more massive and is moderately divided; crown shape and size is variable	Go to 7.
7. Crown apex domed or tufted, crown margin moderately sinuate; crown foliage colors are a moderate red and, or moderate reddish orange	BLACKGUM
7. Crown apex rounded, crown size small, crown color dark pink to greyish red	SWEETGUM
8. Crown margin shape circular or oval and generally entire	Go to 9.
8. Crown margin shape is generally irregular with medium to large sinuations; crown apex is domed, tufted, or billowy	Go to 10.
9. Crown texture fine and feathery; crown small with random lineation. Predominant crown colors are moderate olive green to yellow green	SHORTLEAF PINE or VIRGINIA PINE
9. Crown texture very fine, crown apex rounded to broadly oval, small sized crowns with tufted or parted appearance, crown color is light yellowish green	BLACK LOCUST
10. Large masses of foliage divided and part crown. Crown foliage is moderate yellow green to moderate yellowish green	WHITE OAK
10. Predominant crown colors are light greyish red, greyish red or dark yellowish pink	SCARLET OAK

(Abridged from Krumpe, 1971).

such as film-filter combination (or detector sensitivity), image scale, season of year, time of day, and image processing, because the appearance of forest lands is greatly influenced by these factors (see Figure 34-1).

Once the analyst is familiar with the area to be classified, the classification scheme to be employed, and the image characteristics of each feature or condition to be discriminated, one prepares the imagery to the extent necessary (plotting effective areas, property boundaries, etc.) and classifies by delineating directly on the imagery a small but representative portion of the forest. The analyst attempts to stratify the area into homogeneous strata, homogeneity being dictated by

(a) the complexity of the forest environment,
(b) the proper number of classes that best combine differences between classes with accuracy of interpretation,
(c) the quality of imagery, and
(d) the ability of the analyst to be accurate and consistent.

The analyst must then field check these preliminary interpretations. This can be done either on the ground or from the air in a low flying aircraft or helicopter (Johnson, 1952; Choate, 1953). Once the analyst is convinced that the level of accuracy of his preliminary interpretation is acceptable, he can return to the office to delineate and classify the remainder of the forest land—continuing to field check all questionable areas.

Upon completion of the interpretation task, an extensive amount of field work is generally required. The amount depends on the objectives of the classification project. If the objective is timber inventory, important parameters not seen on the imagery must be measured in the field, such as stem diameters, tree heights, species, defects, causes of defects, and percent cull. If the objective is to make a soil-vegetation survey, data are collected in the field on such items as soil depth, soil color, parent material, pH, texture, and granularity.

A final step in the process of forest-land classification is to determine the area of each delineated and identified stratum. Ordinarily standard area-measurement techniques are used to accomplish this task, either directly on the imagery when scale is accurately known and relief displacement is not severe, or on planimetric or orthophoto maps which contain the image interpretation results. These methods include the use of dot grid, line transect, cut-and-weigh, or automatic coordinate-count equipment.

The analysis procedure generally described above has been applied in all forest regions of the world using many different types of imagery. Also, numerous practical classification systems for use with remotely sensed imagery (e.g., Anderson, et al., 1976) have been developed and tested for mapping existing forest vegetation. A portion of one such system, used by Roberts and Dana (1981) is shown in Table 34-2. This classification system was developed for use in the Rocky Mountain region with 1:24,000 scale color infrared photographs and was applied as follows:

1. The minimum mapping unit area is 2 hectares. A condition or feature must be 40 meters wide to be mapped. Exceptions to the minimum area rule are water bodies, where 0.5 hectares will be the minimum.

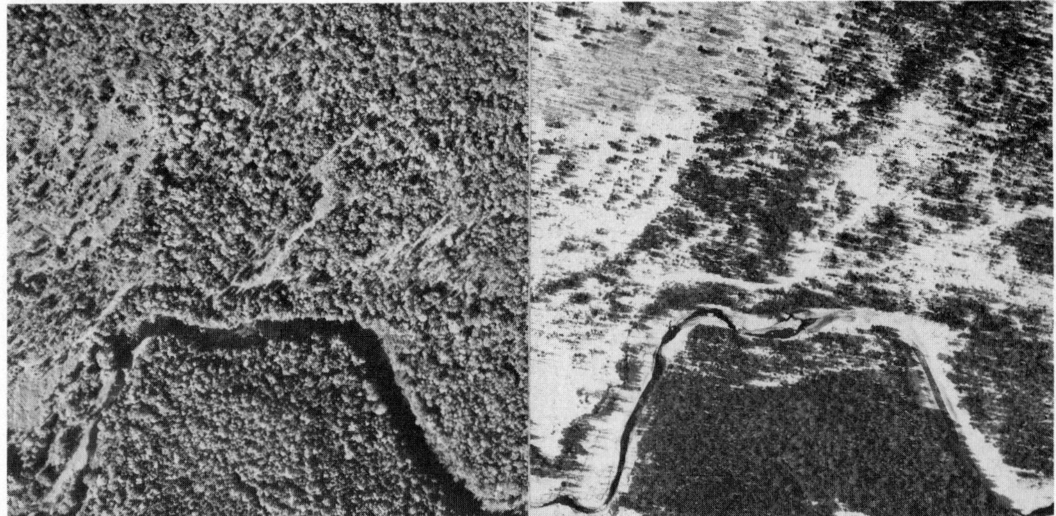

Fig. 34-1. Seasonal panchromatic photographs, original scale 1:5,000, of northern hardwoods and spruce-fir in Eastern Maine—summer on the left and winter on the right. Note that the crowns of leafless hardwoods, abundant in upper portion of view, are difficult to distinguish on the winter photo, but are easily seen on the summer photo. The dark-toned conifers contrast well against all other features in winter. Also note details of the stream meandering through lower part of view and evidence of logging roads in right, middle portion of the winter photo. (Photos courtesy of James W. Sewall Co. (right) and Forest Service, USDA (left)).

TABLE 34-2

Codes and Mapping Symbols for the Classification of Forest Lands in Grand County, Colorado

Classification	Type	Code
11 Land		
21 Vegetated		
31 Wildland		
41 Tree Covered		
51 Evergreen		
61 Needleleaf		.001
	Lodgepole pine	.002
	Spruce	.003
	Subalpine fir	.004
	Juniper	.005
	Mixed conifer	.006
62 Broadleaf	None	
52 Deciduous		
61 Needleleaf	None	
62 Broadleaf		.010
	Aspen	.011
	Cottonwood	.012
	Willow	.013
	Cottonwood-Willow	.014
	Mixed Deciduous	.015
53 Evergreen-Deciduous		
61 Needleleaf-Broadleaf		.020
	Conifer/Aspen	.021
	Lodgepole/Aspen	.022
	Spruce/Aspen	.023
	Subalpine fir/Aspen	.024
	Mixed Conifer/Aspen	.025
62 Broadleaf-Needleleaf	None	
54 Deciduous-Evergreen		
61 Needleleaf-Broadleaf	None	
62 Broadleaf-Needleleaf		.030
	Aspen/Conifer	.031
	Aspen/Lodgepole pine	.032
	Aspen/Spruce	.033
	Aspen/Subalpine fir	.034
	Aspen/Mixed conifer	.035
42 Nontree cover		
51 Shrubland		
61 Sagebrush complex		.040
	Sagebrush	.041
	Sagebrush/grassland	.042
	Grassland/sagebrush	.043
62 Other shrub complex		.050
	Other shrub	.051
	Other shrub/grassland	.052
	Grassland/other shrub	.053
63 Mixed shrub complex		.060
	Mixed shrub	.061
	Mixed shrub	.062
	Grassland/Mixed shrub	.063
64 Grassland complex		.070
	Meadow	.071
	Grassland	.072
32 Developed land		
22 Nonvegetated		

(Abridged from Classification System developed by the U.S. Forest Service, Rocky Mountain Forest and Range Experiment Station, Fort Collins, Colorado)

2. Vegetated land is land on which more than 2 percent of the area is occupied by vegetation of some kind.

3. Tree covered land is land on which the tree component covers more than 25 percent of the area *based on tree crown closure*.

4. Nontree covered land is land on which the tree component covers less than 25 percent, based on tree crown closure. When tree cover comprises 10 to 25 percent of the land, and there is a dominant nontree class, the tree class will be added to the nontree cover class (low density tree cover).

5. Tree covered land has two categories: "woodland" (25 to 60 percent crown closure) and "closed forest" (60 to 100 percent crown closure).

6. Pure tree cover types are those in which 70 percent or more of the crown closure is occupied by either: (1) a single tree species or (2) two tree species when a tree cover class is described as including two species in combination (e.g., spruce-fir).

7. A mixed tree cover type is one in which two or more species make up the stand composition but no single species comprises over 70 percent of the stand. This is not to be confused with 6. (2) above.

8. Tree covered classes (25 to 100 percent crown closure) will carry nontree cover classes when there is evidence of a nontree class existing (e.g., Grassland).

9. North slopes in dark shadows will be mapped as separate land units but with the appropriate cover class annotated.

A selective image-interpretation key, which generally describes each of the forest types, accompanies the above set of procedures. An example color infrared photograph and classification map derived from the photograph are shown in Color Figure 34-2.

An example of a classification system used by the Forest Service of the U.S. Department of Agriculture in heavy timber-producing forests in the western United States is given in Figure 34-3

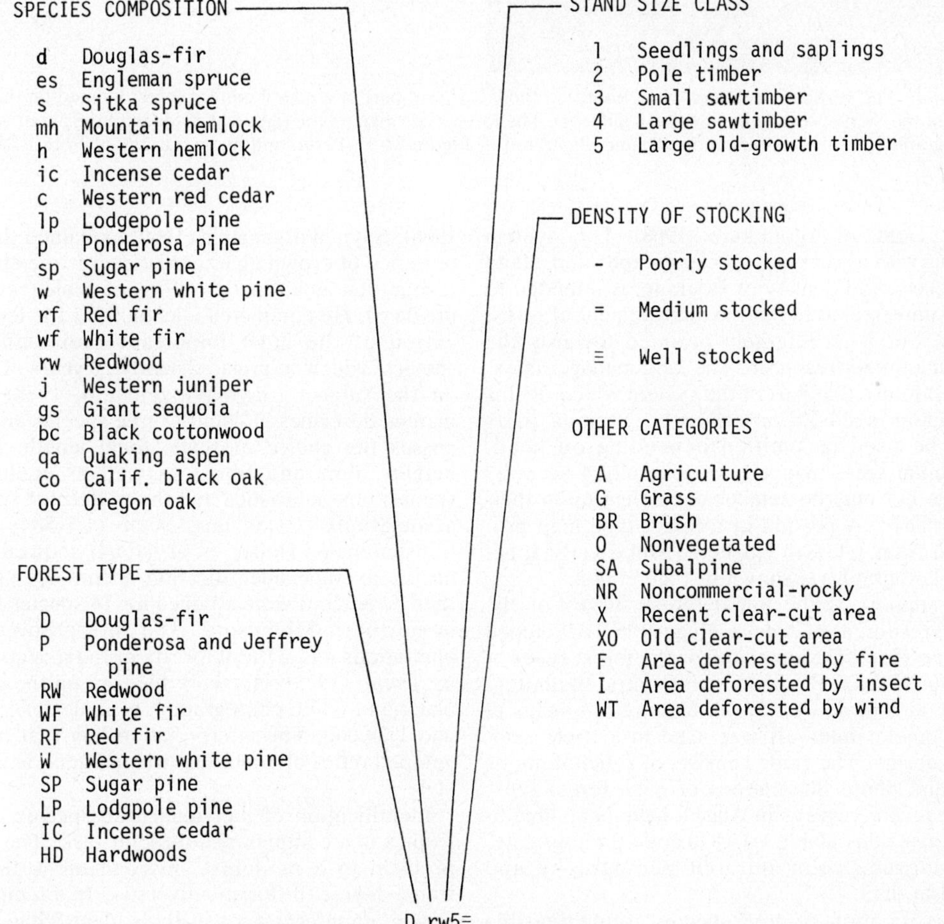

Fig. 34-3. This classification scheme is currently in use in heavily timbered areas of western United States. The example symbol, D, rw5≡, indicates: Forest type—Douglas-fir (D); Species composition—Plurality of basal area in Douglas-fir with 20% or more of basal area in redwood (rw); Stand size class—Plurality of basal area in large over growth sawtimber (5); and Density—well stocked (≡). (From U.S. Dept. of Agriculture, 1968.)

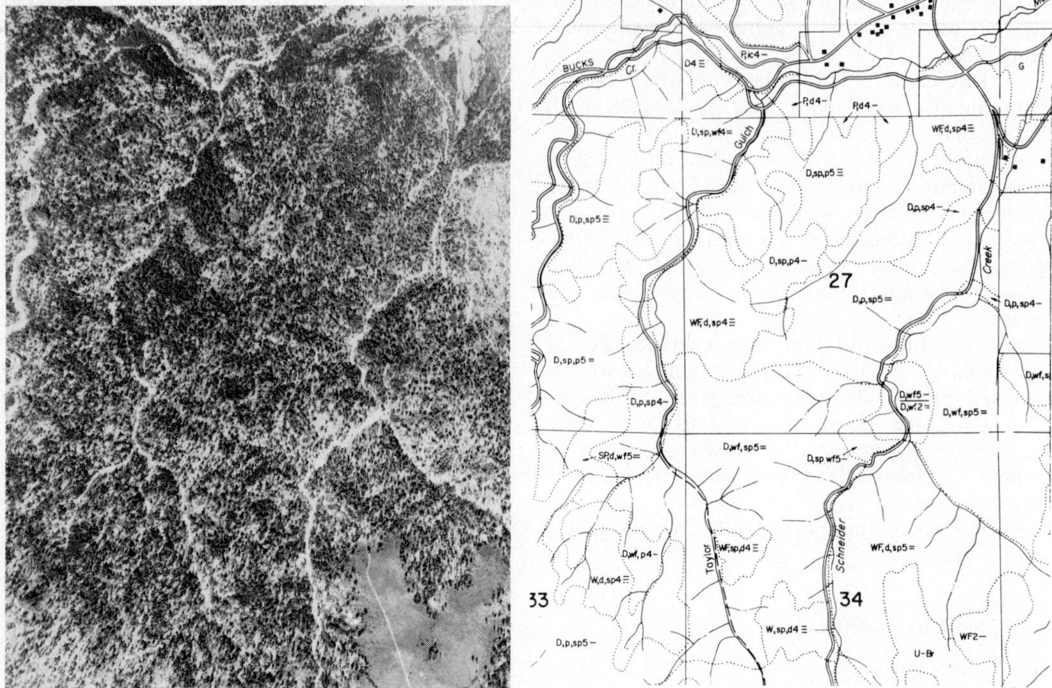

Fig. 34-4. The vertical black-and-white photo on the left shows part of a mixed conifer forest located on the west slope of the Sierra Nevada in northern California. The forest type map on the right was made with the aid of vertical photography and the classification scheme illustrated in Figure 34-3. (Photo and map courtesy of Forest Service, USDA.)

(U.S. Dept. of Agriculture, 1968). This system dealing with forest type, species composition, stand size class, and density of stocking, is intended to be of universal utility in the management of forest lands, but it is definitely oriented towards the existing timber resource. The land manager is expected to use that part of the system which fits his particular needs. Consequently, various parts may be used or omitted depending on need. Minimum areas mapped are generally 4 hectares in size but may be smaller depending upon their importance. A portion of a forest-type map produced from 1:15,840 black-and-white vertical aerial photography is shown in Figure 34-4.

In support of the initial forest survey of the forest resources of Alaska, Hegg (1967) developed a photo-identification guide for the forest types of interior Alaska. Black-and-white infrared photography at a scale of 1:5,000, acquired in strips at 50-kilometer intervals was used in a triple sampling design. The guide consists of a dichotomous key and photo illustrations of each forest type. More recent surveys in Alaska have been able to make use of available 1:15,840 color photography, high-altitude color infrared photography and Landsat data.

Classification of tree species, often feasible only on large-scale imagery, normally involves a highly subjective interpretation process and is dependent on image quality, interpreter skill, and interpretation equipment (Zsilinsky, 1966; Lauer 1966). Sayn-Wittgenstein (1960) explained the importance of crown characteristics for tree-species recognition as a function of image scale (see Figure 34-5). He completed a manual for the identification of the most important Canadian tree species, which culminated nearly 20 years of work on the subject (Sayn-Wittgenstein, 1978). The manual describes in detail 43 tree species and discusses the choice of scale, focal length, flying height, films and filters as well as ecological species characteristics and the pattern of species associations. Using large-scale (1:1,584) color transparencies Heller, et al. (1964) showed that, on the average, identification accuracies greater than 80 percent were attained for 14 species found in northern Minnesota. The interpretation of plantations and natural forests in India is covered by Tiwari (1975). His work focused on the use of black-and-white photography, at scales of 1:5,000 and 1:20,000. The interpretation key that he developed relies on tone, texture and general crown style.

Identification of individual tree species in the tropics is a complex and difficult task. The basic problem in a rainforest environment is the extreme degree of floristic diversity. In a temperate region, an interpreter usually is identifying only a dozen or more individual species; however, in a rainforest there may be several hundred different species present. Consequently, aerial photographs acquired at medium to small scales in tropical areas

Fig. 34-5. Identification of individual tree species often requires large scale photography flown to optimum specifications. To identify the three tree types illustrated here, which occur in eastern Canada, an interpreter must evaluate characteristics of tree form, such as crown shape and branching habit (Sayn-Wittgenstein, 1960). On the right are vertical and horizontal drawings of crowns for white pine, balsam fir and aspen. The matching stereograms show these trees. These photographs were made with panchromatic film, scale 1:600; the top two stereograms were taken in the summer while the bottom one was taken in the spring. (Photos and drawings courtesy of Forest Management Institute, Canadian Department of Environment.)

have been used primarily for making generalized forest cover type maps. Some modest success at tree species identification in the tropics has been demonstrated using specialized large-scale photography. DeMilde and Sayn-Wittgenstein (1973) used 1:1,000 to 1:5,000 scale photos in Surinam and found color and color infrared to be much superior to black-and-white. Photo interpreters, working with color-transparency film taken at a scale of 1:2,000 over a tropical rainforest in northern Australia, were able to correctly identify 55 out of 111 different rainforest species at least once (Myers and Benson, 1981). Moreover, twenty-four species were identified with more than 75 percent accuracy and 11 of these were correctly identified with 100 percent accuracy.

It is necessary to note that many forest-classification projects do not require information on location and distribution of each important forest type but are concerned only with the total area occupied by each type. Under these circumstances, a time-consuming and costly forest-type map, similar to that shown in Figure 34-4, is not prepared; instead, the total area occupied by

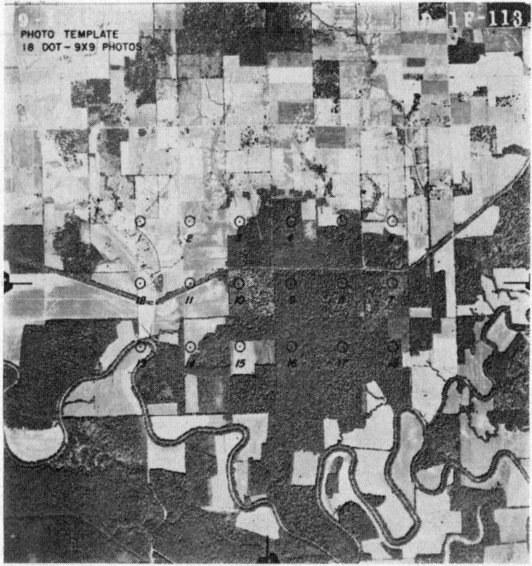

Fig. 34-6. A templet with 18 dots per photo oriented over one photograph of the stereo pair. In this case 9 of the 18 dots fall on forest detail. Under stereo the number of dots in sawtimber and in smaller size classes could be determined. One-acre (0.4 ha) circles around dots are used as minimum areas in classification.

each type is determined by classifying a large number of sample points located systematically or at random on the imagery (see Figure 34-6). The area of each type is calculated as follows:

$$\frac{\text{Area of}}{\text{Type}} = \frac{\text{Total}}{\text{Area}} \times \frac{\text{Number of points in type}}{\text{Total number of points}} \qquad (34\text{-}1)$$

Procedures for selecting the number, size, shape and location of points were outlined by Bickford (1952) and Langley (1962).

The advantage of a sampling over a mapping procedure is that it eliminates the necessity of defining boundaries and, therefore, is relatively fast. In 1967, under the direction of the U.S. Forest Service's Nationwide Forest Survey, 19 million hectares of forest land in the United States were inventoried by a point-classification sytsem (Aldrich, 1968). The goal for the Forest Survey is to resurvey each state every 5 to 10 years. Aldrich pointed out that the accuracy of forest-land classification varies from region to region. Ages of the available photography, major land-use patterns, and the ability of the interpreter all have an effect. He noted examples, however, where in northern Washington 2 percent of the land classifications were wrong and in southwestern Georgia the error associated with estimated forested area was only ±0.7 percent. For an area of 0.69 million hectares in Maine, Young, et al. (1963) showed that when nine strata were used, photo classifications were correct 99 percent of the time.

Langley (1962) noted that, for The Forest Survey in California, the point-classification system is preferred to the complete-delineation system for several reasons:

(1) the network of photo points provides a population from which to select a random sample for field examination;
(2) it is easier and more accurate to classify the small area around a point into land use class, stand size class, volume class, and cover type than it is to estimate an average for these classes within a larger delineated area;
(3) point information is easily transferred into a machine-readable format which greatly facilitates computational procedures compared to the tedious counting necessary with area delineation; and
(4) point sampling is faster and less expensive than total area delineation involving transfer from photo to base map and area determination from the map.

The methods discussed above for extracting classification information from imagery have dealt solely with manual techniques. At present there is no substitute for the trained land manager who can scan an image and pick out significant classification information through his complex, parallel multiprocessor computer, the brain—but in many respects the brain is slow. The analyst can perform many of the simpler classification tasks; however, machine-assisted image classification is much faster. Moreover, computer processing enables man to extend his capability and perform tasks not otherwise possible. For example, automatic image-classification techniques may not only make decisions regarding forest classification, but can also

(1) perform first approximation interpretations, calling attention to areas needing further analysis by the human,
(2) combine and integrate remotely sensed data gathered in portions of the spectrum beyond the visible,
(3) extract additional information from imagery by amplifying small differences in radiance which are below the human eye's threshold, and
(4) assist in the process of stratifying forest cover types for allocating sample units on which detailed information is collected.

A separate section in this Chapter is devoted to an in-depth treatment of computer-assisted data analysis for forest resource assessment, and only one practical application will be discussed here. The U.S. Bureau of Land Management (BLM) requires a low-cost, quick, and accurate means of inventorying and mapping vegetation on millions of hectares of public land in the western U.S., including Alaska. A procedure was developed and successfully tested in Mojave County, Arizona in which vegetation classes were mapped from Landsat digital data using a BLM vegetation-classification system (Rohde and Miller, 1981).

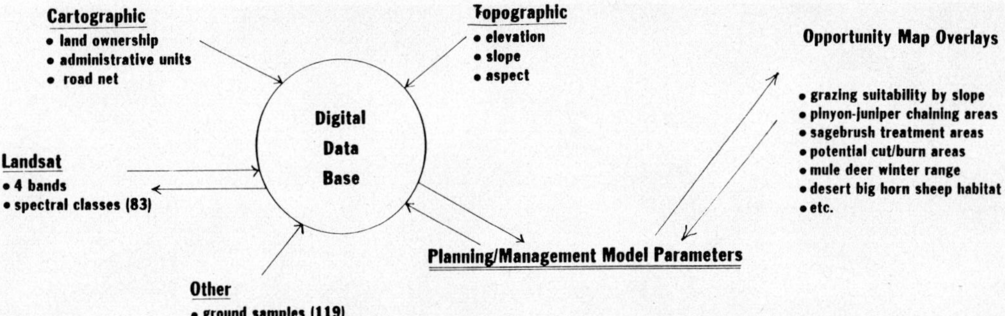

Fig. 34-7. U.S. Bureau of Land Management personnel in Arizona used digital image analysis and spatial data processing techniques to produce management-opportunity map-overlays. All data sources were registered to a common geographic reference, slope and aspect were derived from elevation, Landsat brightness values were classified into spectral classes and assigned to vegetation cover classes, vegetation-terrain relationships were identified and vegetation cover classes were adjusted accordingly. Example map products are illustrated in Fig. 34-8.

Furthermore, overlays for resource management and planning were made by the BLM by combining into a digital data base the vegetation data with terrain, road network, ownership and other forms of disparate data. Figure 34-7 illustrates how advanced concepts of digital image analysis and information processing were used in Arizona by the BLM to determine arid lands resource planning and management opportunities. Management-opportunity maps were prepared at a scale of 1:126,720 for an entire BLM planning unit. These maps included the following: (a) grazing suitability by slope, (b) pīnon-juniper chaining areas, (c) sagebrush treatment areas, (d) potential cut/burn areas, (e) mule-deer winter range, and (f) desert big-horn-sheep habitat (see Color Figure 34-8).

Advancements in machine-assisted analysis have given the forest land manager the opportunity to make, over large areas, meaningful and spatially accurate vegetation cover maps, as well as digital data bases for use in geobased information systems. Miller, et al. (1981) recently reported that spatially referenced digital-data bases can be manipulated using geoprocessing techniques to provide resource managers with planning-level information by:

(1) producing maps with varying scales and resolution,
(2) performing search and proximity analysis,
(3) generating simple statistics and tabular summaries,
(4) overlaying data sets for composite mapping,
(5) predicting occurrence of events through time with simulation and predictive modeling, and
(6) updating with additional layers of information.

Types of Imagery

Black-and-White Photography

The degree to which forested areas can be classified on black-and-white photography depends heavily on the quality, scale, and season of photography, film-filter combination used, and the analyst's background and training. Photographs most commonly available throughout the world are 22.8- × 22.8-cm black-and-white photo prints (see Figure 34-9). An excellent guide to using black-and-white photography is the USDA Forest Service's publication, Forester's Guide to Aerial Photo Interpretation (Avery, 1978). Also, Johnson and Sellman (1974, 1977 and 1979a,b) developed a series of dichotomous interpretation keys for mapping forest cover types in the southeastern U.S. on 1:20,000 black-and-white photography. This was the first approach that developed systematic keys for a large region.

Research has been performed on determining the best film/filter combinations for evaluating forested environments, but the debate between two types, panchromatic minus-blue and infrared minus-blue, has not been concluded. For example, in the case of tree-species identification in areas in the western United States, panchromatic photography has been deemed superior (Jensen and Colwell, 1949). In the east, however, it is necessary to distinguish between hardwoods and conifers which can occur in equal quantities in many forest regions. For these areas, Spurr and Brown (1946), and Seeley (1949) concluded that infrared minus-blue is more effective in providing a tonal separation between hardwoods and conifers but less valuable for detecting individual species (see Figure 34-10). Late spring and early fall are preferable seasons to make these distinctions when using an infrared minus-blue film-filter combination in the southern United States. In the Great Lakes States, Chase and Korotov (1947) and Steigerwaldt (1948) preferred infrared minus-blue. Ulliman and Meyer (1971) reported that small scale black-and-white infrared photographs at scales as small as 1:90,000 could be used to classify forest stands in Minnesota as accurately as photographs at scales of 1:15,840. However interpreters preferred photographs at a 1:31,680 scale. In Alaska, Stone (1950) used panchromatic photos successfully, but Hegg (1967) preferred infrared minus-blue. Obviously, transferring the results from one region to another, as is sometimes attempted when selecting the proper film/filter combination for determining species

Fig. 34-9. Intensively managed forest in the Forest District of Königsbrunn (Fed. Rep. Germany). Beech and spruce stands of different ages, structure, and form of regeneration and treatment are illustrated. Aerial photo, original scale 1:10,000 taken with ZEISS RMK A30/23, October 12, 1965, 2 P.M., panchromatic film. (Courtesy of the Public Ministry of Interior, Baden-Württemberg)

composition, is not satisfactory. Nevertheless, optimum photo specifications can be determined, based upon the purpose of the survey, and are affected by the many factors discussed earlier in this section.

Color and Color Infrared Photography

The quality of color and color infrared films has improved substantially in the last fifteen years, and prints made from color negative film cost little more than conventional black-and-white prints. Consequently, the use of color photography for purposes of forest land classification is rapidly expanding. Becking (1959) determined that aerial color photography offered greater possibilities for mapping forest vegetation/soil types and discriminating between tree species than conventional black-and-white photography, since features and conditions were represented in a wide range of colors distinct in hue, saturation and brightness. Likewise, Anson (1970) showed that color infrared photography provided considerably more information than black-and-white photography when mapping vegetation types in South Carolina. Heller, Doverspike and Aldrich (1964)

Fig. 34-10. Vertical aerial stereograms, scale 1:30,000, showing stands of short leaf pine, Virginia pine and upland hardwoods in northeastern Georgia. The top examples were made with infrared-sensitive film while the bottom examples were made with panchromatic film and a minus-blue filter. Note that in this case the black-and-white infrared film is superior to the panchromatic film for discriminating between stands of pure hardwoods, at A, pure pine, at B, and mixed pine-hardwood at C. (Photos courtesy of School of Forestry, Northern Arizona University.)

showed that tree-species identifications in Minnesota made using color transparencies were significantly more accurate than those made using black-and-white photo prints.

For vegetation types occurring in the California Coast Range, Lauer (1968a,b) showed that color infrared photography provided the image analyst with significantly more information on forest types than conventional black-and-white panchromatic photography, especially when attempting to locate and identify those forest types having similar image characteristics. Interpretations of the panchromatic and color infrared photos rendered overall type-identification accuracies of 77 percent and 89 percent, respectively. In addition to providing more accurate results, the analysis of color infrared photography required 25 percent less interpretation time than the black-and-white photography. Hudson et al (1976) also indicated that forest species groups could be classified on scales of 1:36,000 to 1:120,000 using color infrared photographs. But apparently these smaller scales are not acceptable to the forest land manager on the ground who prefers the 1:15,840 scale on a standard paper print for operational programs (Marshall and Meyer, 1978).

A comparison was made in New Zealand between color, color infrared and black-and-white photography for distinguishing vegetation types and other features (Stephens, 1976). Stephens found that the color infrared was superior to the other film types because it offered a greater range in color signatures for cover types. Aldrich (1979) noted that the photointerpreter's job today versus 20 years ago is easier, more satisfying and done with greater confidence because many color combinations as well as infrared sensitivity are available.

The U.S. Forest Service routinely acquires color photography at scales of 1:15,840, 1:20,000, and 1:24,000 over national forests. Color-reversal negative film generally is used, from which either black-and-white or color prints can be made for laboratory and field use. Similarily, the U.S. Bureau of Land Management has embarked on a program to acquire color infrared photography of all public lands at an average scale of 1:31,250. The U.S. Fish and Wildlife Service is a prime user of the BLM color infrared photography whereby

a) 46 vegetative cover classes are interpreted and delineated with a 2-hectare minimum mapping unit,
b) vegetation cover classes are transferred to a 1:24,000 base map,

c) map units are digitized and entered into a geobased information system,
d) map units are digitally merged with other types of disparate map data, and
e) digital geoprocessing and analysis are performed to support the U.S. Fish and Wildlife Service's Rapid Assessment Methodology as it is applied to coal development areas in the western U.S. (VanDerwalker, 1979).

Large-scale Photography

The most common scale of imagery used by forest-land managers today is medium-scale photography (1:15,840 to 1:24,000). Larger scales of imagery often are procured for special purposes where greater image detail is needed. It has been noted earlier that accurate tree-species identification often requires scales as large as 1:2,000 or greater, since morphological as well as phenological characteristics of trees must be resolvable on the photographs. Nearly 20 years ago, Canadian foresters were developing methods to acquire low-cost, large scale photo coverage of forest lands (Zsilinszky, 1966). Lyons (1967) demonstrated a method for acquiring sample photography at large scale for assessing forest cover types in British Columbia, Canada. He used 70mm fixed air-base photos from a helicopter. Species composition can be estimated as reliably from the resulting stereo photos as from the ground (Waelti, 1978).

Large-scale color or color infrared photography taken with a 35mm camera is a very reliable, low cost and simple-to-use tool. Meyer (1973) and Meyer and Grumstrup (1978) developed specifications for, and applied, a 35mm system to support field-level vegetation resource assessments in Montana. It should be noted, however, that 35mm, large-scale photography is not a substitute for, but rather a supplement to, conventional forms of resource photography taken over large areas with larger format camera systems. The large format photography provides a controlled frame of reference for the 35mm.

Small-scale Photography

In cases where less image detail can be accepted to provide greater area coverage by each image, small-scale imagery may be most efficient. For example, in California, Lauer and Benson (1973) compared conventional black-and-white imagery, at a scale of 1:15,840, with color infrared imagery, at a scale of 1:120,000 taken by an RB-57F aircraft flying at 20,000 meters above the terrain. For each type of imagery, forest-land classification results for a 20,000 hectare area were compared in terms of accuracy of boundary placement, type identification, and interpretation time. It was found that in nearly every instance the small-scale color infrared imagery provided results comparable with those derived from the much larger scale black-and-white imagery. The most important finding, however, was that interpretation time decreased by more than 50 percent when using the small-scale imagery.

Conversely, when Marshall and Meyer (1978) tested medium and small-scale black-and-white and color infrared photography for purposes of vegetation-type identification in Minnesota, they found scales smaller than 1:24,000 were unacceptable. They noted that the small-scale color infrared transparencies were overly cumbersome for day-to-day use. Despite this fact, very high-quality photography can be obtained from high altitudes (Gut and Hohle, 1977); however, this type of photography will not replace certain types of conventional aerial photography. Instead, it will be useful primarily for making generalized forest classification maps of large areas quickly and inexpensively.

Panoramic Photography

High-resolution, medium-scale panoramic aerial photography has been increasingly available to resources managers through the U.S. Forest Service's National Forestry Application's Program, and the National Aeronautics and Space Administration's Airborne-Instrumentation Research Project. An Optical Bar Panoramic Camera mounted in a U-2 aircraft flying at approximately 20,000 meters above terrain can cover a swath about 70 kilometers wide and 3½ kilometers long in a single frame of photography. Each frame of panoramic photography varies in photo scale from approximately 1:30,000 to 1:60,000, but resolution is good enough on the central portion of each frame to resolve medium contrast objects as small as 2/3 meter wide. Thus, panoramic photography can be described as very high-quality, medium-scale, color or color infrared transparency imagery, which can be obtained quickly over large areas. Panoramic distortions in the geometry of each frame limit the usefulness of the imagery for making precise measurements of forest land features; however, forest cover types can be easily, accurately and rapidly detected and identified over vast areas. In addition, users of panoramic photography have found that it lends itself well to the tracing of roads and trails that do not appear on other resource photographs or maps (Bowlin, 1979, and Befort, et al., 1980).

Side-looking Airborne Radar (SLAR) Imagery

Among the early studies designed to determine the utility of Side-looking Airborne Radar (SLAR) imagery for purposes of evaluating forested environments was the work by Morain and Simonett (1966) in Utah, Oregon and Alaska. (For a detailed discussion of radar fundamentals and SLAR please refer to Chapter 9.) They found that a skilled analyst working with SLAR imagery could, by means of tonal and texture comparisons combined with basic geographic knowledge of the area,

a) prepare regional vegetation-type maps,
b) delimit vegetation zones as they vary with elevation,
c) delimit old burn patterns,
d) delimit timberline, and
e) identify species in areas characterized by near-monospecific stands.

Other studies (Viksne, et al., 1970; Hardy, et al., 1971; Daus and Lauer, 1971) have suggested that at best only broad classification of forest lands can be accomplished with radar imagery, and the major limiting factor to this type of imagery is that the tones and textures exhibited by various vegetation/terrain features are greatly influenced by topography. Most often the presence or absence of forest is the only vegetation parameter consistently identifiable on SLAR imagery (Henderson, 1979).

SLAR imagery is most useful in tropical regions where the acquisition of aerial and space imagery is difficult or impossible due to adverse weather conditions (Sicco Smit, 1971 and Francis, 1976). Banyard (1979) reviewed the difficulties and limitations associated with the use of SLAR imagery for forest type mapping in a tropical region in Indonesia. He showed that with a "photo-truth" key, it was possible to delineate six forest types—hill forest, dry lowland forest, wet lowland forest, swamp forest, mangrove forest and shifting cultivation/secondary forest.

Orthophoto Maps

In North America and Europe orthophoto maps are becoming routinely available for use by forest land managers. According to Stellingwerf (1979), orthophoto maps at scales of 1:10,000 and 1:20,000, are used as forest-management tools in Sweden, Austria and West Germany. Cover-type mapping is done on the aerial orthophoto stereopairs, cover-type information is easily transferred to the orthophoto map, and sample units can be identified and allocated from the orthophoto map. An orthophoto map provides an excellent base map upon which many forest types can be discerned due to tone and texture differences, but many cover-type categories cannot be identified. However, when the orthophoto map is supplemented with sample strips of larger scale 35 mm color infrared photography, vegetation mapping can be done efficiently, accurately and at very low cost (Mead and Gammon, 1981).

Landsat Imagery

An evaluation of the utility of Landsat data for assessing forest resources, with emphasis on machine-assisted analysis, is presented later in this chapter. The utility of Landsat images using visual analysis is briefly described here. Please refer to Chapter 12 for a more detailed discussion. A comprehensive study of Landsat Multispectral Scanner (MSS) data, at approximately 80-meter resolution, for wildland cover classification was conducted by the U.S. Forest Service, in cooperation with the National Aeronautics and Space Administration. It was concluded from the study that Landsat MSS images were primarily valuable as a source of first level information for multistage inventories of forest and range resources (Heller, 1975). Krumpe, et al. (1973) demonstrated that broad wildland cover-types, such as conifer forest, hardwood forest, chaparral, grassland, cultivated and pasture land, and desert shrub, could be mapped on Landsat MSS images with an overall mapping accuracy of approximately 80 percent, with the greatest amount of confusion occurring between the chaparral and hardwood types. Benson and Lauer (1973) reported that identification of forest density classes (non-stocked, low-stocked, medium-stocked, and high-stocked) was not possible on Landsat MSS images. Sayn-Wittgenstein and Kalensky (1975) noted that Landsat MSS images did not yield accuracies of land classification that were useful beyond the reconnaissance level, and that interpretations based mainly on spectral information have some severe limitations. They felt that spatial information should be given more attention. These studies demonstrate both the advantages and limitations of conventional Landsat MSS images for mapping wildland resources.

The interpretability of current Landsat data can be improved considerably when presented to the image analyst as color composite images (i.e., three spectral channels), at approximately 40-meter ground resolution, and in stereo. Lauer and Todd (1981) used computer-processing techniques to merge Return Beam Vidicon (RBV) and MSS digital data to form a 40-meter resolution, three channel, color composite image (see Color figure 34-11). Further processing was done using digital terrain data to introduce parallax into a "left conjugate" image, and thus create a stereo pair. These RBV/MSS stereo images were analyzed to obtain information about a wildland area in the Coastal Range of California. Interpretations of vegetation types, densities, canopy characteristics and boundaries were improved when using the RBV/MSS stereo image compared to the MSS or RBV image alone. However, greater than 40-meter resolution images are needed to consistently detect and identify canopy characteristics related to forest stand structure. The location of, and association between, vegetative features as influenced by topography were readily seen on the RBV/MSS stereo pair, factors which aided in mapping vegetation types. Analysis of the RBV image alone rendered results on vegetation type and boundary placement which were almost as good as those derived from the merged RBV/MSS image. However, when interpreting the merged image, which combined all available spectral and spatial information, the interpreter more quickly recognized image characteristics and rapidly identified vegetation types and subtle boundaries between types. Consequently, less time was spent

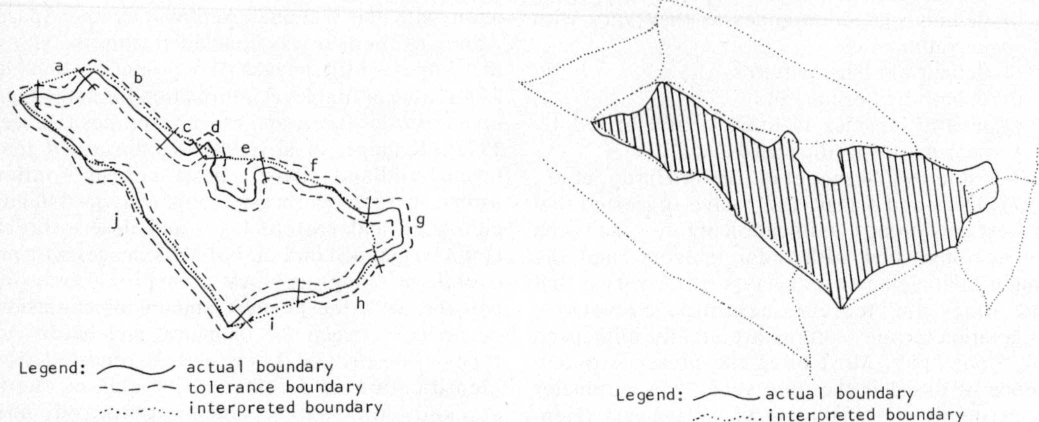

Fig. 34-12. These maps indicate how boundary coincidence (left) and area coincidence (right) can be determined. On the left, segments b, d, f, h, and j (as measured along the actual boundary) are considered to be coincident with the actual boundary, because within these segments the interpreted boundary falls within a prescribed tolerance interval. Likewise, the shaded area, on the right, is the coinciding area between actual and interpreted areas. Both procedures allow for a quantitative evaluation of interpretation results. (Courtesy of Forestry Remote Sensing Laboratory, University of California.)

by the interpreter pondering interpretation decisions when using the merged RBV/MSS image as compared to the RBV or MSS image alone. The results of the work reported by Lauer and Todd suggest that data acquired from the proposed Mapsat sensors, the planned Thematic Mapper on Landsat D, High Resolution Visible imaging instrument on the French SPOT Satellite and the Large Format Camera on the Space Shuttle, will allow for improved classifications of forest lands, using visual interpretation techniques, as compared to classifications performed using data currently available from Landsats 1, 2, and 3.

Evaluation of Results

Forest-land classification accuracy is not easily defined; therefore, interpretation results derived from remote sensor imagery are often difficult to evaluate. The most common method of evaluation is to select point samples throughout the areas that have been classified. Ground data collected at these point locations are then compared, usually in quantitative fashion, with corresponding classification information for the unit within which the points fall. However, forest lands are generally made up of heterogeneous units and delineated classes are identified only by averaging what is seen within the class. The selected points may not be representative of the generalized class in which they fall—unless a large number of ground points are visited.

Jaakkola and Draeger (1971) discovered, while interviewing various users and producers of forest maps, that few guidelines exist at the present time for an objective evaluation of classification accuracy or its importance. Consequently, an acceptable map showing classification information is one which is "sufficiently reliable and usable." The Soil Survey Staff (1960) described land classification as ". . . contrivances made by men to suit their purposes . . . a perfect one would have no draw backs when used for the purposes intended."

Studies which have attempted to assess quantitatively the accuracy of image-interpretation results include Pomerening and Cline (1953); Webster and Beckett, (1964), and Vermeer, (1968), Webster and Wong, (1969); Kuhl, (1970); Rudd, (1971). After reviewing the techniques used by these investigators, Jaakkola and Draeger suggested that a *combination* of area comparison used by Pomerening and Cline, and Kuhl, and boundary comparisons used by Vermeer would be most helpful for evaluating forest-classification accuracy. Comparisons of boundaries are made by superimposing two maps (one based on ground data and one on image interpretation) and measuring with a map reader, proportions of boundaries falling inside or outside a prescribed "tolerance" zone (Fig. 34-12). These data would then give an indication of the quality of the delineation but not the identification inside the boundaries. Thus, it is also necessary to measure relative area of matching identification (Fig. 34-12). For further discussion on determination of mapping accuracy see chapter 30 and recent article by Rosenfield et al. (1982).

Forest Volume Inventory

The gross volume of wood in standing timber can be estimated almost as closely on aerial photographs as it can be on the ground; the accuracy depends on the quality and scale of the photographs, the experience of the interpreter, the quality of measurement devices, and the accuracy of photographic-scale determinations.

Aerial photographs have been used in volume estimation since the late 1920's with some of the earliest work being done by Canadians. Seeley (1929) used oblique photographs to measure tree heights from shadows and related tree heights to timber volume. Heights, tree counts, and crown density measurements made on aerial photographs were described by Andrews and Trorey (1933), Andrews (1936), and Nash (1948).

In the United States, one of the earliest references to the successful use of aerial photographs in forest inventory was made by Foster (1934) but it was not until the late 1940's that forest volume estimation became a widespread practice. Spurr (1945, 1946, 1948) spearheaded early developments and brought aerial photography to the attention of forest-land managers at a time when improved survey techniques were badly needed to speed-up the inventory of the Nation's forest resources. The work of Rogers (1946, 1947, 1949) and Moessner (1948, 1949) contributed much to the promotion of photointerpretation and photo measurements in the nationwide Forest Survey.

Developments in photo volume estimation since the early 1950's have been built upon the work of these pioneers. Many studies tested the accuracy of photo measurements; others related tree and stand measurements on photographs to timber volume on the ground and resulted in both tree and stand volume equations for many species and forest types in North America. Stratifying forest land by volume classes on aerial photographs also increases the efficiency of forest inventory (MacLean, 1963, 1972; Moessner 1963b; Bickford, 1963). The popularity of these predictors of volume is the result of rising costs of traditional forest inventory techniques and the need for more relevant and more frequent information. As new techniques and more efficient sampling designs become available, estimating tree and stand volume on photographs will become even more widely practiced.

Components of Volume

Cubic-foot or board-foot volume of wood contained in individual trees or stands of trees can be predicted by using one or more measurements made on aerial photographs. The components of photo volume are tree height, crown diameter, crown closure, crown area, and number of trees, which can all be measured quite accurately on stereoscopic photographic coverage; crown diameter can be measured on nonstereoscopic photographic coverage. These measurements are related to volume measured on the ground by regression analysis techniques, and the results are summarized in photo volume tables. Tables can be developed for all tree species, for combinations of species, and for forest types.

The component that can best be related to volume is height. Tree or stand height can be measured on stereoscopic photographic coverage in two ways: (1) by shadow lengths, or (2) by parallax measurements.

Shadow length as a measure of tree height was explored by Seeley (1929) and Rogers (1947, 1949); aside from these initial studies the technique has found little practical use. Perhaps the principal objection to this method is the time required to prepare graphs for the conversion of photo shadow lengths into feet. A different set of conversions is required for every degree of latitude, for every month of the year, and for every hour of the photographic day. Furthermore, it is almost impossible to find a full tree shadow in dense stands. When the sun is at its peak altitude, shadows are usually much too short to measure.

Parallax measurements are used more often than other methods to measure height. The simplest parallax measuring device is the parallax wedge (Spurr, 1945; Rogers, 1946; Moessner, 1962; Wert and Myhre, 1967; Lund, 1971). Converging lines of the wedge are examined under a stereoscope. At the points where the two converging lines intersect at the base of the tree, the parallax measurement is observed and recorded; a similar parallax measurement is made at the top of the tree. The difference between the two parallax readings (differential parallax) is multiplied by a conversion factor to give the total tree height (see Fig. 34-13). Parallax is also measured with instruments employing the "floating" dot. These instruments range from the complex and expensive stereocomparators to the simple family of parallax bars sometimes referred to as stereometers or height finders.

Parallax bars are most frequently used for tree and stand height measurements. They are simple to use, and many interpreters prefer them to the parallax wedge. All parallax bars have three things in common—a fixed dot, a movable dot, and a micrometer. The fixed dot is placed over an image on the left-hand photo of the stereo pair, and the movable dot is placed over the same point on the right-hand photo. With both dots on the exact same image point on the two photographs, in which case they appear fused in stereo and at the appropriate elevation, the micrometer reading is recorded. The procedure is repeated for a second point, and the micrometer read again. The difference between the two readings is equal to the differential parallax and is converted to difference in height between the two points using equation 34-2.

Moessner (1961) tested three parallax-measuring devices—a parallax wedge, an Abrams height finder (floating dot), and an Austin Photo interpretometer (floating dot). He found no significant difference in accuracy between the parallax wedge and the height finder, but both were faster and more accurate than the interpretometer. All three devices are used with a pocket stereoscope.

Until recently, a parallax bar was the only parallax measuring device that could be used with a mirror stereoscope. This was changed when

Fig. 34-13. For measurement of tree heights, the parallax wedge is oriented over the stereo pair. In this example the intersection of the sloping, graduated line of the wedge with the top of the tree was read at point 1. Then the wedge was shifted and the intersection of the sloping line with the ground level equal to that at the tree base was read at point 2. The parallax difference, in this case .076 inches, when multiplied by the factor of 2.2 feet per .001 inch, indicates a stand height of 167 feet (50.1 m), Approximate scale: 1:3000.

Lund (1971) modified the parallax wedge. In his modification, Lund separated the converging lines of the wedge to accommodate 21-cm average separation of photo-image points. This separation is required by most mirror stereoscopes. The parallax bar is still preferred to the wedge by many interpreters because they find it easier to follow floating dots than the intersections of converging lines of the parallax wedge. By using either of these instruments on medium-scale (1:12,000 to 1:24,000) photographs, it is possible to classify stands within 3-meter height classes (Spurr and Brown, 1946b). On large-scale (1:1,000 to 1:4,000) 70-mm photographs, tree heights are measured consistently to within 1 meter of ground measurements by using a parallax bar or parallax wedge (Weber, 1965; Kippen and Sayn-Wittgenstein, 1964; Aldred, 1976) (Fig. 34-14).

Parallax measurements are converted to tree height by using one of several equations. One simplified and more generally used equation is:

$$h = \frac{H_b \cdot dp}{P_t} \quad (34\text{-}2)$$

where:

h = height of tree
H_b = height of the photo exposure station above the base of the tree
dp = differential parallax
P_t = parallax measured to the tree

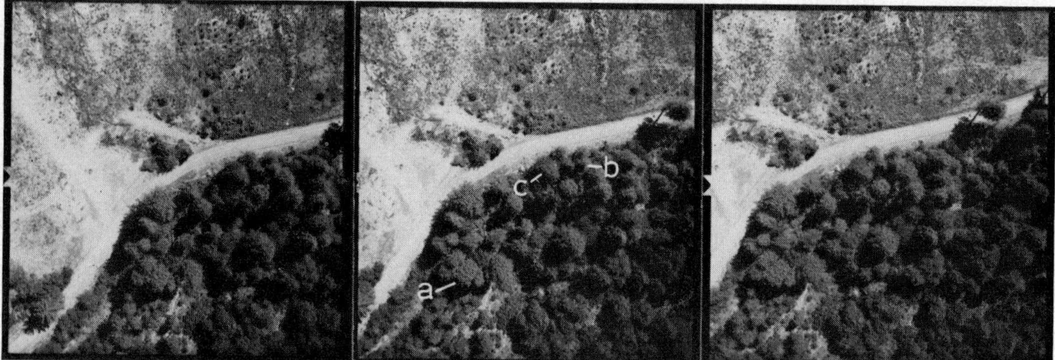

Fig. 34-14. Large-scale 70-mm photographs such as these are becoming popular in volume inventories. Notice the tree detail that can be interpreted and how easy it would be to make tree measurements by species: (a) ponderosa pine (Pinus ponderosa Laws.), (b) sugar pine (Pinus lambertiana Dougl.), and (c) Douglas-fir (Pseudotsuga menziesii (Mirb.) Franco var. menziesii). (Photos courtesy of Forest Service, USDA.)

According to Schut and Van Wijk (1964), the accuracy of computed tree heights depends to a large extent upon the choice of camera, flying height, and focal length. But from their tests they indicate that only in exceptional cases is it possible to obtain a value of less than 0.3 meters as a standard error in determining tree heights from parallax measurements.

Burns (1979) showed that most students can learn to calculate timber volume per acre from aerial photographs within 15% of the ground estimate, provided they are familiar with the ground conditions. They do this by estimating tree height ocularly rather than computing the height with a parallax measuring instrument.

Crown diameters are not closely correlated with stand volume. However, they have been found to be correlated with the trunk diameter at breast height (*dbh*) and consequently with tree volume (Bonner, 1964, 1964a; Feree, 1953). Nevertheless, many photointerpreters still find that the average crown diameter contributes enough to volume to be included as an independent variable in photo stand volume tables.

Crown diameters are usually measured in at least two directions and then averaged for the tree. In this way irregular crowns are taken into account. As photographic scales increase, more side branches become visible, and it becomes more difficult to determine the periphery of the crown. The usual procedure is to draw an imaginary crown line touching the branch tips. Measurements between opposite crown lines in the longest and the shortest dimension can then be averaged. An exception to this rule is where the radial displacement of trees causes distortion of the crowns, particularly on the edges of the photograph. This distortion is greatest on photographs taken with short focal-length camera lenses. The result is that the crown dimension in the direction of the radial displacement is an oblique or sideview of the crown. Measurements should be made perpendicular to this radial distortion.

Instruments for measuring crown diameters range from ordinary opaque rulers to transparent micrometer scales to parallax bars. Opaque rulers are ineffective for photomeasurements because they are cumbersome and the markings so crude that they cover much of the tree image to be measured. Micrometer wedges are printed in fine lines on positive film; they are satisfactory for making accurate crown-width measurements (Losee, 1956). One type of micrometer wedge has a series of graduated dots that are fitted over the crown until the correct size is found (Moessner, 1960b). Another type of wedge is used to read crown width according to graduations along one of two converging lines. A similar device, called a pole scale, was developed by Losee (1956); this consists of a graduated series of lines increasing in length by 1/200 of an inch. Less frequently used is a technique utilizing a parallax bar where the movable mark is set on one side of the image of the

Fig. 34-15. Measurement of crown diameters by dot type and micrometer wedges. Tree No. 1 measures .035- to .037-inch by dot type wedge. Tree No. 2 measures .036- to .037-inch by the micrometer wedge. Conversion to feet on the ground at the scale of these photographs would place both these pines in the 12-m to 15-m (40- to 50-foot) crown-diameter class.

tree crown and the micrometer scale is read in this position. The movable mark is then moved to the other side of the crown; the micrometer scale is read again. The difference between micrometer readings is directly related to the crown width. Regardless of the technique used, photo measurements are converted to diameters in meters by using the photographic scale ratio expressed in millimeters on the photographs to meters on the ground (see Fig. 34-15).

The accuracy of crown-diameter measurements varies according to the scale of the photograph. On 1:20,000-scale photographs, Moessner (1950) could place crown diameters in 1.5-m classes. Worley and Meyer (1955) found that they could measure crowns on 1:12,000-scale photographs within 1 m of the ground measurements. Using 1:1,200-scale photographs, Losee (1953) measured the average crown diameters for stands within 30 cm of the average measured on the ground.

Crown closure is a measurement of the percentage of an area covered by tree crowns when the crown perimeters are projected vertically to the ground. Losee (1956) described two methods for measuring crown closure. The first method uses a count of all trees including doubtful trees and then another count of all trees minus the doubtful trees. An average of the two counts is used as a measure of crown density. However, tree counts are difficult and time-consuming to make, and are economical and accurate only on medium- or large-scale photographs of good quality for which all factors affecting tree counts have been evaluated. Tree counts can be considered reliable only when photographs clearly resolve crowns of large trees in open stands.

A second method for measuring crown closure described by Losee (1956) eliminates the need for counting individual tree crowns. Crown closure is estimated using one of four techniques: (1) by visual estimate, (2) by crown density scale, (3) by direct comparison with stereograms, and (4) by measurement using dot grids.

Visual estimates of crown closure are accurate if the interpreter follows some well-defined guidelines. Two methods have been offered by Pope (1960). The first is called "tree cramming." The interpreter mentally divides the forest plot into sections. Trees are ocularly moved from sections of scattered trees to fill holes in denser sections. When all trees are visually crammed together, the interpreter estimates the proportion of the plot occupied by crowns. The second method suggested by Pope is the "tree counting method." All trees of the average crown diameter are counted. Then the number of trees of average-size crowns needed to mentally fill in the gaps are counted. The ratio of the total tree count to 100-percent count is equal to crown-closure percent.

Crown-density scales are probably used in some form or another by most interpreters. Some scales are made up with black dots on light backgrounds and others with light dots on dark backgrounds. These scales can be made up to represent different ranges of crown density (Moessner, 1949, Aldrich and Norick, 1969). Crown closure is estimated by comparing the standards with the photo plot or stand in question (Figure 34-16). Another approach is a scale with 10-percent crown-closure classes represented in 1.5-m crown-diameter classes. Aldrich (1967) found this device to be particularly useful for classifying forest plots into volume classes based upon crown closure and mean crown diameter, using non-stereo photo coverage.

A method developed in Canada for checking photomeasured crown closure on the ground uses a "moosehorn" and a procedure described by Robinson (1947). Using the "moosehorn," Losee

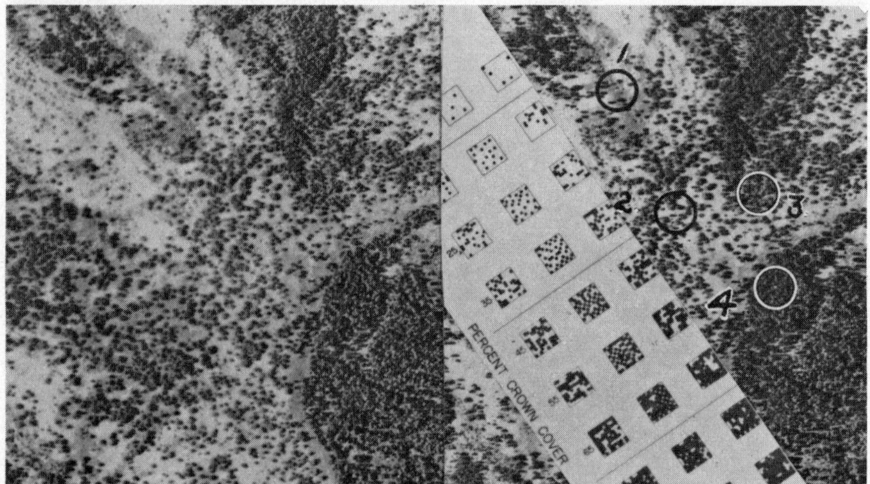

Fig. 34-16. The Crown Density Scale is viewed by the interpreter as he examines the photos in stereo, and readings are made by comparison. Plot No. 1 has approximately 10 percent; Plot No. 2, 35 percent; Plot No. 3, 65 percent; and Plot No. 4, 85 percent crown coverage.

(1953) measured crown closure on the ground to compare with photographic measurements. He found that on 1:7,200-scale photos the average error for 12 stands was −1.3 percent (±9.9 percent) and on 1:1,200 scale the average error was −0.3 percent (±5.5 percent). Bonner (1968) compared estimates of crown closure for 18 softwood and 18 hardwood stands. Ground measurements were made with the "moosehorn"; photo measurements were made on 1:15,840-scale aerial photographs. The results for both softwood and hardwood indicated that photo estimates are about 10 percent less than ground estimates. Crown closure can also be checked by plane-table mapping. Both methods are slow, and for this reason the recent trend has been to accept estimated crown closures made on the aerial photographs.

Crown area is another component of volume; this component was used as an independent variable for predicting volume of individual trees on large-scale (1:1,200 to 1:2,000) photographs in Canada (Sayn-Wittgenstein and Aldred, 1967). The combined variable height (H) times the log of crown area (CA) is the most significant variable used to estimate tree volumes on large-scale aerial photographs. Crown area is usually measured with a dot grid; the number of dots is converted to area in square meters through the use of a photo-scale conversion factor.

Another variable used along with individual tree volume estimates is the tree count. Unfortunately tree counts become increasingly inaccurate as the photographic scales become too small to resolve individual crowns. On 1:1,200-scale photographs, however, it might be possible to count 80 percent or more of the dominant, codominant, and intermediate trees. As photographic scales increase from 1:15,840 to 1:3,500, Young (1953) found that tree count accuracy improved only 20 percent for any given set of conditions. The highest accuracy was about 85 percent on 1:3,500-scale photos in open stands; the lowest was about 20 percent in extremely dense stands. Thus, the accuracy of tree counts depends on the photographic scale, resolution, stand density, and the homogeneity of the stand.

Volume Tables

Two types of tables useful in volume inventories are the single-tree volume table and the stand volume table. Single-tree volume tables are important in large-scale photo sampling because individual trees are easily and accurately counted (Lyons, 1966; Aldred and Kippen, 1967). Stand volume tables are used most often with medium-scale photography (1:12,000 to 1:24,000) in extensive forest inventories. Many volume tables have appeared in the literature during the past 20 years.

Single-tree volume tables usually relate the independent variable crown diameter to diameter at breast height (*dbh*) and then relate *dbh* and total height to volume. Computation is done for each species by measuring crown diameter on the aerial photograph and then relating this measurement to *dbh* using a linear regression equation constructed for this purpose (Nyyssonen, 1955; Minor, 1951, 1960; Feree, 1953; Dilworth, 1959). Bonner (1964a) found that crown diameter and height were the strongest variables for regression equations to predict *dbh*. With estimates of *dbh* and height, regular local volume tables can be entered for rough estimates of tree volume. Stand volumes can be found by adding individual tree volumes together (Bonner, 1964b).

Another effective method of estimating volume is to prepare special tree aerial volume tables. Usually crown diameters and tree heights are measured on the photographs and related to tree volumes measured on the ground, using multiple regression techniques.

A different approach was reported by Sayn-Wittgenstein and Aldred (1967). They related the combined variable height times the logarithm of crown area to volume and found that it was the most powerful volume predictor for trees on large-scale 70-mm photographs.

Regardless of individual techniques used, single-tree volume equations have been constructed for several important tree species and can easily be constructed for any species in a similar way (Bonner, 1964a; Feree, 1953; Lyons, 1966).

The number of photo stand volume tables available for timber types in the United States and Canada has increased remarkably during the past 20 years (Allison and Breadon, 1958, interior British Columbia; Morris, 1957, black spruce type for the northeastern coniferous zone; Pope, 1961, Douglas-fir in Pacific Northwest; Weber, 1965, white spruce and balsam fir in Minnesota; Duffy and Meyer, 1962, lodgepole pine in west-central Alberta; Gingrich and Meyer, 1955, upland oak; Hanks and Thomson, 1964, Iowa hardwoods; Moessner et al. 1951, hardwood stands in Central States; Bonner, 1966, several forest types in Canada; Moessner, 1960a, ponderosa pine type in the Rocky Mountains; Chapman, 1965, California timber types; Avery, 1958, southern pine and hardwoods; Moessner, 1963a, conifer stands in Mountain States; Avery, 1959, northern Minnesota). There are many volume tables that have not been published because they were constructed to satisfy a special requirement.

Stand photo volume tables vary to some extent by construction technique and according to tree species. For instance, some investigators have found that volume is correlated best with height and crown closure (Moessner, 1963a; Bonner, 1966; Pope, 1962). Although they agree on the components, they disagree on variables. For instance, Moessner found that both height and crown diameter and their products raised to the second power increased the value of his stand volume table. Bonner, on the other hand, found that only height, crown diameter, and the product of height and crown diameter contributed significantly to volume. Pope found that height squared times crown closure was the best single variable for estimating volume—accounting for over 80 percent of the variation. In another example, Moessner et al. (1949) found that the three components—height, crown diameter, and crown closure—worked best in his Central States hardwood volume tables. These differences and others indicate that for each situation there may be a separate set of variables. For example, Sayn-Wittgenstein and Aldred (1967) found that every tree species they tested required a different regression equation to give the lowest standard error of the estimate. Some of the conditions that change the outcome include photographic scale, whether stands are pure or mixed, and whether stands are even-aged or uneven-aged.

One variation in volume estimation introduced in recent years is the use of volume classes based on measurements of crown closure and crown diameter on single, nonstereo aerial photographs (Aldrich and Norick, 1969). If the objective is volume stratification, this technique can be effective. Aldrich and Norick, 1969, tested two nonstereo, five nonstereo, and eight stereo volume classes for effectiveness in reducing the variation in volume in sampling extensive forested areas. Nonstereo volume classes based on crown diameter and crown closure proved to be more efficient than the stereo volume classes.

Other Volume Estimators

Volume can be estimated visually on aerial photographs provided the photointerpreters have sufficient experience and know the local conditions (Loetsch and Haller, 1964). Stereograms constructed for known volume per unit area and stand conditions are often used as guides in ocular estimation of volume and other volume-related classes (Moessner, 1956; Avery, 1967a; Aldrich, 1967; Busch, et al., 1979). If the individual stands are small and their boundaries can be easily identified, volumes can be estimated directly without use of photo plots. All components of volume are considered as whole, and a volume class is determined by comparison with stereograms. Large, extensive stands with indistinct boundaries require the use of sample plots to guide the interpreter.

Photographs taken from space have opened new possibilities for forest-volume prediction. These photographs, with ground resolutions of less than 100 meters, require either monocular interpretation techniques or automated film density measuring techniques for volume prediction. In monocular interpretation, volume is related to area of forest with the aid of a stereomicroscope. The premise here is that the greater the forest area within prescribed sample units, the greater the timber volume (Langley, Aldrich, and Heller, 1969; Aldrich, 1971). This theory has a weakness: heavy cutting in predominantly timber-producing areas and differences in growth rates caused by local physiographic conditions create variations in volume not associated with area. If these conditions can be stratified, then volume predictions based upon ocular forest-area estimates may be useful for first-level information in broad forest-volume inventories.

Film density measured by monocular comparison with reference levels or by microimage density-measuring scanners may in the future permit predictions of forest volumes from space photography. In Finland, Kuusela and Poso (1970) used a picture taken by the meteorological satellite, ESSA 8, on March 21, 1969. They found that forested areas could be distinguished by their greater darkness against a snow background at that time of year. Then, using mean volume of growing stock per areal unit for 46 subregions in southern Finland, they related these data to grades of darkness on the satellite pictures. Independent variable used were: (1) distance between the center of the subregion and the center of the satellite picture from east to west, (2) distance between the center of the subregion and the center of the satellite picture from north to south, and (3) the mean darkness of the region multiplied by 100. As a result, they found a highly significant dependence between mean volumes obtained by regression equations and those measured in the forest.

Film density measured with an automatic scanning and recording microdensitometer may someday permit delimitation of forest areas by volume. However, it will probably be many years before this technique is developed to a stage of operational proficiency (Aldrich, 1971).

Another unique approach to photo volume estimation utilizes a high-order photogrammetric mapping instrument to plot stand profiles. Smith (1969) used a Kelsh plotter to make a profile of the ground along a selected scan line and then made a second profile of the tree crowns of the forest along the same line. These profiles provide a cross section of the forest. Using sample cross sections of known width throughout the forest, he determined the volume of the space occupied by the forest and related it to volume of wood. Smith pointed out conditions that should be investigated, including stand structural arrangements, relative timber stand density, and automation of the method.

Timber Volume Estimation

Timber volume might be estimated for any number of reasons, such as determining the value of a property for sale, settling estates, taxes, or information required to help decision-making by resource managers. Regardless of the purpose, the techniques used can be grouped into three categories: ground sampling, photo sampling, or combinations of ground and photo sampling. Although ground sampling is still most economical for small tracts, it may be advantageous to use aerial photo estimates directly without ground sampling, or indirectly with ground sampling, for tracts exceeding 260 hectares (1 sq. mi).

Direct Volume Estimation

Mean volume per areal unit can be estimated directly from aerial photographs for each timber type without field sampling. This method requires aerial volume tables for major species or groups of species which can be identified on the photographs. Photointerpretation keys may also be useful but are not advisable except where tables cannot be obtained or constructed; estimates made from keys are subjective and often inaccurate. The direct method is not often used in the United States, but in parts of Canada that are still accessible only by air, canoe, pack train, or foot, this method is used for rapid reconnaissance of large areas (Seeley, 1955).

Only a few reports have been published that compare direct photo estimates of volume and estimates based upon ground methods alone. Rogers (1956) showed that in only 6 of 35 comparisons were the differences greater than 10 percent. The remaining 29 cases were almost evenly balanced between positive and negative differences. However, the 6 cases with more than 10-percent difference would require adjustment after ground checking. Certainly one disadvantage to direct estimation is the inability to estimate the magnitude of the bias from using stand aerial-volume tables.

Moessner (1964) compared the ability of short-course trainees with university students and forester photointerpreters. He found that repeated estimates of total volume are nearly as precise as repeated ground estimates when both are made by inexperienced or only partially trained personnel. According to Moessner, foresters who have acquired controlled precision in stereo measurement techniques can improve the accuracy of their estimates with comparatively little increase in experience. On the average, foresters with on-the-job training were within 15 percent of the ground estimates 90 percent of the time. Short-course trainees and university students in mensuration scored 46.5 and 50.5 percent, respectively, for the same 15-percent acceptable limit of accuracy.

The direct method of volume estimation was used by the Forest Branch of the Canadian Department of Northern Affairs and Natural Resources for two extensive surveys. According to the *Manual of Photo Interpretation* of the American Society of Photogrammetry (1960), direct estimation in these two surveys were considered adequate for economic development of the coniferous and hardwood forests of northern Canada, where there are few species and the stands tend to be uniform in density and height. The direct method depends on aerial volume tables and, according to Seeley (1955), these tables will provide valuable estimates for mixed stands of very low density on photographs taken during winter. On winter photographs, the interpreter can count and measure all trees that contribute to volume.

Large Scale Aerial Photography

An early example of direct estimation using large-scale 70-mm photographic samples was reported by Westby, Aldred, and Sayn-Wittgenstein (1968). The Canadian Forestry Branch required a preliminary estimate of the timber volume and tree-size distribution for the spruce forests on a 310,800-hectare (1,200 mi^2) area in the MacKenzie River Delta north of the Arctic Circle. The area was difficult to reach except by navigable rivers and streams. The MacKenzie River Delta survey relied almost entirely on large-scale aerial photographs. First, the area was divided into seven equal blocks, and each block was sampled with two strips of photographs. The photographs were taken at a scale of 1:1,200 and the scale was checked by radial line plot. Nine rectangular sample plots (10 $\times$ 70 meters on the ground) were established on the photographs at random locations on each flight strip. The *dbh* and volume of each tree in the photo plots were estimated using functions of height, crown area, and number of trees growing in a circular area surrounding the subject tree, and the number of trees among the six nearest neighbors that were taller than the tree under consideration. These functions were derived from only 100 trees measured on the ground. The precision of the survey was calculated by using the method outlined by Schumacher and Chapman (1954, p. 94).

Large-scale aerial photography has been used on major operational forest inventories completed by the provincial forestry agencies of Ontario and Alberta. The streamlined procedures employed are described by Aldred and Lowe (1978) and Nielsen, Aldred and MacLeod (1979). The last-named report gives the results of a trial in the Yukon Territories.

The British Columbia Ministry of Forests has made a major commitment to large-scale aerial photography, based on helicopter photography acquired with a twin-camera boom and interpretation and analysis using systems which combine a Zeiss Stereocord and a Hewlett Packard desktop computer.

Tropical applications of large-scale aerial photography have been the subject of further investigation. Promising applications are discussed by Nielson and Aldred (1978) and results of photomeasurement and species identification trials in Surinam are reported by Aldred (1976) and Sayn-Wittgenstein et al. (1978).

Indirect Volume Estimation

Indirect estimation is used most often for management planning inventories and on extensive forest inventories. It is more accurate than direct photo volume estimation and leads to information of greater precision.

The first step in indirect estimation is usually division of the area into homogeneous volume classes. There should be less volume variation within a class than for the whole area. Bickford (1961) listed three requirements for the most efficient sampling: (1) classes must be defined independently of the samples to obtain class averages; (2) class averages must show real differences; and (3) observations must be properly distributed by classes.

Forest type, site, stand size, and other condition classes are associated with stand volume, but volume classes are best for volume estimation. This was borne out in a study by Moessner (1963b), who found that cubic volume classes obtained by aerial photo measurements and photo volume tables were the best of 18 schemes tested. He also found that use of volume classes could reduce field survey time by about 60 percent.

Basically, two methods for dividing forest land by volume classes are available. The classes may be delineated directly on aerial photographs and the area within each class measured with a planimeter, a dot grid, or other method. This procedure is usually used for management planning inventories in which maps are a required by product needed in local land management decisions. Another method of dividing forest land utilizes many systematically distributed photo sample plots. Volume is estimated for each photo plot within each type or condition class. This is the most efficient method for extensive forest volume inventories where accurate information is not required for precise locations. Information within extensive geographical or political boundaries is accurate, and useful for broad land management decisions and policy making.

The principal difference between direct photo volume estimation and indirect photo volume estimation is the use of ground samples in the indirect procedure. From the samples, accurate measurements are obtained of tree volume, tree-size distribution, species composition, growth, stocking, mortality, site conditions, and other characteristics not measurable on aerial photographs. These data are related to the photo sample plot data and then expanded to the total survey area. Procedures for selecting ground samples are covered in many forest inventory and statistical sampling textbooks (e.g., Schumacher and Chapman, 1948; Spurr, 1952; Avery, 1967b; Loetsch and Haller, 1964; Cochran, 1953).

No single standard sampling procedure is applicable to all forest volume inventories. According to Langley (1969), a design should take into account the population parameters being estimated, the distribution of the population variables used to estimate the parameters, existing information related to the variables, and the optimal allocation of funds available for the survey. Shiue and John (1962) stated that " . . . a good sampling procedure is dependent upon the ability and ingenuity of the designer in incorporating various basic sampling systems with measurement techniques to provide an efficient design which gives valid estimates (with known precision) for a particular inventory problem."

Forest inventory sampling designs that include the use of aerial photographs are varied. Several designs have appeared in the literature during the past 10 years (Shiue and John 1962; Rogers, 1960, 1961; Bickford, 1963; Bickerstaff and Hirvonen, 1969). Probably the most unique sampling design was presented by Langley (1969). It is unique in that it offers a way to use photographs taken from orbital or suborbital altitudes as a first level of information in a multistage design with variable probability sampling theory.

In the first phase of Langley's sampling design, photographs taken from space are subdivided into subunits by using a grid template. For example, an area of 2,000,000 hectares (5-million acres) in the states of Louisiana, Mississippi, and Arkansas covered by an Apollo-9 photography was subdivided into 6.4 kilometer square blocks (Figure 34-17). These units were small enough to be easily identified from an aircraft during the subsampling phase, and yet they were large enough that the variation between units was small and controllable by image interpretation. Each subunit was examined under 7× magnification and the area of forest in each unit was related to volume—the more forest area in the unit, the greater the timber volume. Subunits in predominantly upland pine/hardwood could also be separated from the subunits in bottomlands. Subunits were then selected from each class, using variable probability theory, i.e., the greater the predicted volume in each subunit, the greater the probability of its selection.

The second phase consists of four subsampling stages—two photographic stages and two ground stages. The first sampling stage requires photographing a sample of subunits in each class identified on the space photograph. In the Apollo-9 study, five subunits were selected—two in the upland pine/hardwood category and three in the bottomland hardwood category. The subunits were then photographed at a 1:60,000 scale; Polaroid film was used so that imagery could be obtained and interpreted and the next sample stage selected while the surveyors were still airborne. A strip grid was superimposed on the Polaroid prints; each strip represented an area 0.5 by 6.4 km on the ground. The relative timber volume was predicted for each strip by photointerpretation. Then two sample strips were selected from each subunit, with probability proportional to predicted volume.

In the second photographic sampling stage, each sample strip selected in the subunits was photographed in color at a much larger scale. In the example, two 70-mm cameras operating simultaneously photographed each strip; one photographed the entire strip at a 1:12,000-scale, while the other camera photographed clusters of 1:2,000-scale photo triplets at 10-second intervals

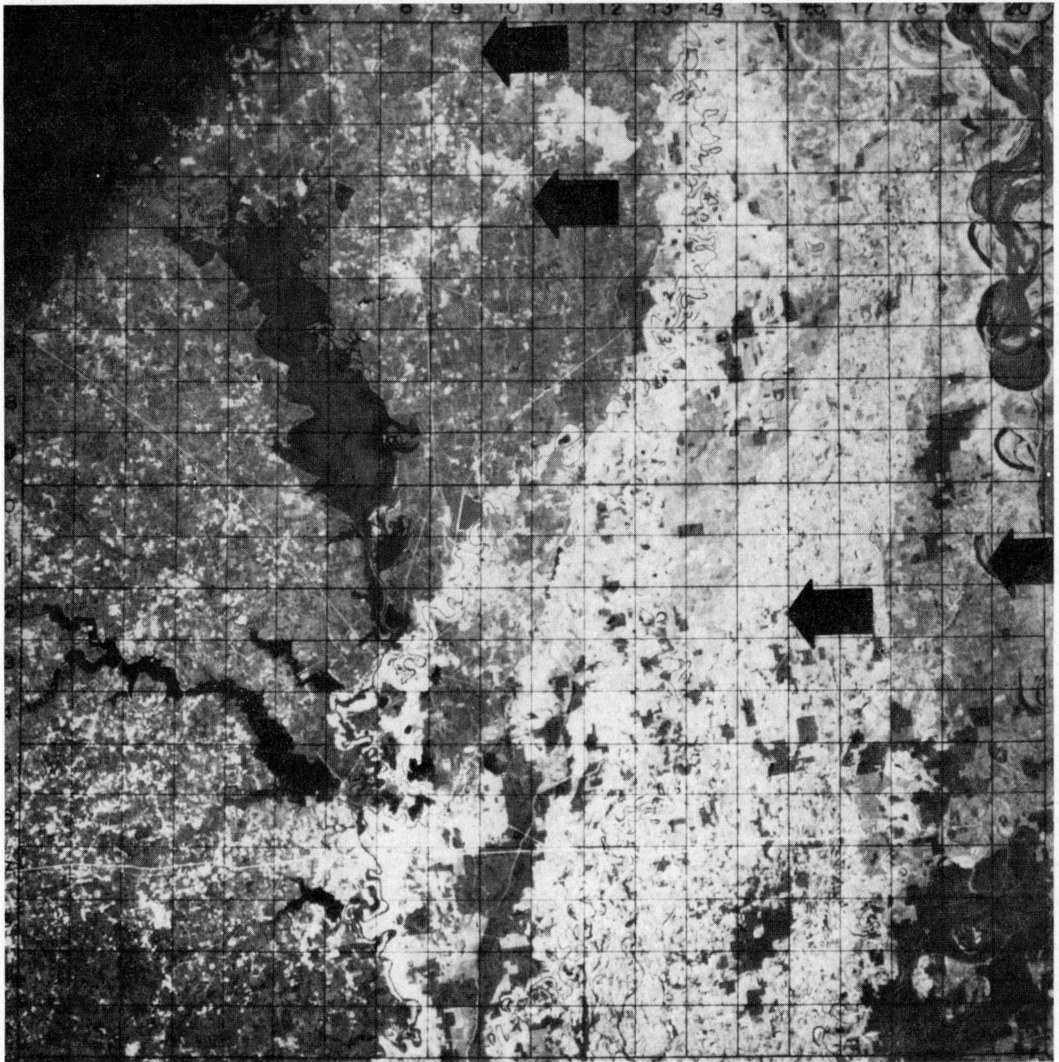

Fig. 34-17. The interpreted portion of an Apollo-9 color-infrared frame (printed in black-and-white) for the Mississippi area is shown with grid template attached. Arrows point to first-stage sample blocks. (Photo courtesy of NASA.)

along the strips. The center photo of these triplets was subdivided into four 0.24 hectare (0.6-acre) photo plots. Each photo plot was examined under a stereoscope and the cubic-foot volumes predicted, using conventional volume measurement techniques. The 1:2,000-scale photography provided a measure of the strip area; the ratio of this area to the photo cluster area provided an expansion factor for the 1:2,000-scale photo volumes. One photo plot was selected in each sample strip as a ground sample, using probability proportional to the predicted volumes.

Using enlarged photographs, a field crew located the precise corners of each photo plot selected for ground cruising. In this third sampling stage, diameters of all trees were measured and the volume in each tree predicted from a volume table. The predictions were used to select four to six trees on which boles were precisely measured. A precision optical dendrometer was used in this last sampling stage to measure the bole characteristics for the sample trees. These measurements were then used to compute the solid wood volume in each tree.

According to Langley, to obtain timber volume estimates applicable to the entire survey area, the measured tree volumes are expanded back through the sampling formula by using the probabilities and area expansion factors computed at each sampling stage. In the Apollo-9 study the estimate for each stratum was

$$v = \frac{1}{m} \sum^{m} \frac{1}{p_i\, n_i} \sum^{m} \frac{1}{p_j} \cdot \frac{A_j}{a_c} \cdot \frac{1}{p_p\, t_p} \cdot \sum^{tp} \frac{v_k}{p_k} \tag{34-3}$$

where:

- v_k is the measured volume of the kth sample tree on a selected ground plot,
- p_k is the probability of selecting the kth sample tree,
- p_p is the probability of selecting the pth plot from the cluster of plots delineated on the 1:2,000-scale 70-mm photos in a strip,
- p_j is probability of selecting the jth sample strip in a selected 6.4- × 6.4-km (4 × 4 mile) square area,
- p_i is the probability of selecting the ith sample square,
- a_c is the area covered by the cluster of 1:2,000-scale 70-mm photographs within a strip,
- A_j is the total area of the jth sample strip,
- t_p is the number of samples trees measured on the pth plot,
- n_i is the number of sample strips in the ith 6.4 × 6.4 km square,
- m is the number of 6.4 × 6.4 km squares.

While this sampling design may not have been optimal for the Apollo-9 forest volume inventory, it was considered successful. Langley reported that the space photos served to reduce the sampling error among the first-stage sampling units by a factor of 2.5. With only 10 ground samples totaling 2.4 hectares (6 acres) out of the 2 million hectares (5 million acres), the total cubic volume of timber was estimated to be 63-million gross cubic meters (2.225-billion gross cubic feet) with an estimated sampling error of 13 percent.

Titus et al (1975) reported a similar timber inventory based upon manual and automated analysis of data from Landsat-1 (ERTS-1) and from supporting aircraft to provide stratified multistage sampling. The objective was to estimate several parameters, including the standing volume of merchantable timber within the 465,000 ha (1,148,550 acres) of the Plumas National Forest in California. A four-stage sampling design was used; the first stage involved automatic classification of Landsat-1 data tapes and selection of primary sampling units, the second stage involved acquisition of large-scale aerial photos over selected primary sampling units and selection of photo plots based on manual interpretation; the third stage involved visiting selected photo plots on the ground and selecting sample trees by dbh class; and the fourth stage was a subsample of these trees for precise volume measurement. Local volume equations were developed from these measurements for species and dbh classes and applied to the sample tree tally. When the tree volumes were expanded through the various stages of the sample design, the average timber volume per hectare for the Plumas National Forest was estimated to be 41.5 MBF with a sampling error of 7.8 percent. The cost for the multistage system was 65 percent less than an inventory of comparable precision based on ground plots alone.

Two similar inventories in Oregon and Washington (Oregon State Department of Forestry, 1978; Harding and Scott, 1978) met with less success.

From the foregoing examples it is concluded that Landsat can supply accurate data and that there are adequate computer systems to handle the data, but software and procedures must be refined. Considerably more work is required to make the Landsat-based inventory process both operational and competitive with existing systems.

Special Remote Sensing Considerations

No longer is it possible to use an old photograph to solve today's and tomorrow's environmental problems. For instance, changes are occurring so rapidly that even those factors affecting timber volume cannot be measured within acceptable limits of error on photographs that are more than three years old.

Depending on the survey-accuracy goal and the resource information required, photographic scale is critical in volume inventories. Probably 1:12,000- or 1:15,840-scale color or black-and-white photographs are large enough for most indirect volume estimation or volume stratification. However, if tree volume is required, larger scale (from 1:1,600 to 1:4,000) photographs are necessary. If the large-scale photographs are used in combination with the medium-scale resource photography, a great deal more information can be extracted and less ground work will be necessary. Figure 34-18 shows a comparison of four photographic scales useful in volume inventory.

When extremely small-scale or microscale photographs are enlarged, the interpretation of volume classifications and area-related volume characteristics is improved. However, because of very limited parallax and relatively poor ground resolution of microscale photographs, enlargements cannot always improve tree and stand measurements.

Some new remote sensing tools may benefit volume estimation in the next decade. One of these tools is a radar or laser altimeter. In Canada a foliage penetrating radar altimeter has been built especially for use with large-scale 70-mm forest sampling photography (Fig. 34-19). Because height above ground is printed directly on each 70-mm frame, the interpreter can compute tree and stand measurements more accurately than even before. Tree-height measurements within 5-percent accuracy of the true height appears to be within range of the instrument at a flying height of 300m above the ground. Laser profilers should also be useful for photographic surveys as well as for making height profiles of the stands. As height is related to volume, volume stratification should be possible.

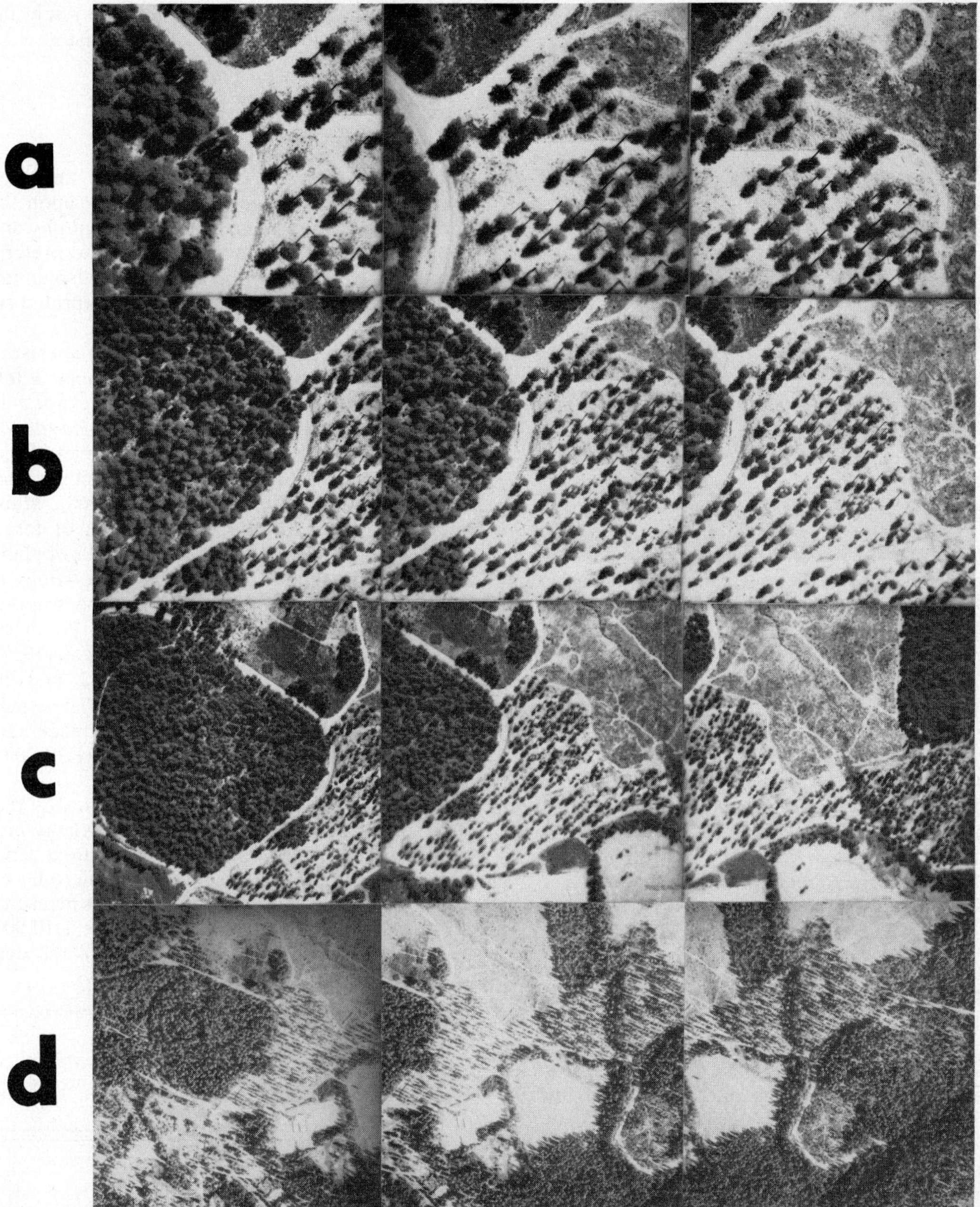

Fig. 34-18. This example shows photographs of the same forested area at four different scales—(a) 1:2,000, (b) 1:4,000, (c) 1:8,000, and (d) 1:16,000. Notice the increasing crown detail and the improved stand-condition information as the scale increases. (Photos courtesy of Forest Service, USDA.)

During summer, 1981, an inertial guidance system was used to direct a Forest Service, USDA photographic plane to pre-selected latitude and longitude sample locations in Alaska. Stereo triplets (23-by-23-cm in size) were obtained at a scale of 1:3,000 over the sample areas by flying the aircraft to latitude/longitude readings provided by the guidance instrument on the pilot's instrument panel. Revisits by air may be made in future years by using similar guidance systems, thus insuring remeasurement of the identical sample plots (Personal communication from R. S. Driscoll, Forest Service, USDA, Ft. Collins, Colorado).

Volume of Forest Products

Aerial and ground photographs have been used to estimate, by photogrammetric methods, the volume of pulpwood storage piles, the volume of

Fig. 34-19. The foliage-penetrating radar altimeter used in large-scale photography has a narrow beam (left) and measures aircraft altitude above ground. Conventional radar altimeters have a broad beam (right) and give results affected by the intervening crown canopy. (Illustration courtesy of the Canadian Forestry Service.)

pulpwood and sawlogs in booms and water storage areas, and the volume of pulpwood in trucks, tiers and conveyors. Since little new information has been published since 1975, the reader is referred to pages 1373–1376 of the first edition of the *Manual of Remote Sensing* (1975).

TEMPERATE ZONE: EUROPE

With few exceptions, forest-land management in Europe has long been identified with intensive management based on sustained-yield and multiple-use concepts. This has led to well cared for, productive forests and to extremely small units of management, particularly in middle Europe (see Figure 34-9). During more than 200 years of European forestry history, an intensive system of management for individual forest enterprises has been developed. Detailed periodic inventories and continuous mapping are the foundations for the planning of all the forestry activities and management within the forest enterprises.

The first example of forest-land photointerpretation was published in the *Berliner Tageblatt* on September 10, 1887. A German forester, asked to map and describe a forest, took photos from a captive balloon. Practical application began directly after World War I when aerial photos were utilized in several European countries for the mapping of forest areas and the detailed listing of forest conditions.

Since 1950, aerial photos have been regularly used in most European countries for the creation and updating of forestry maps, as well as to serve as an information source and storage system for forest inventories within forest enterprises.

Classification of Forest Lands

Analysis Procedures

The identification of stand types and tree species on aerial imagery is dependent upon the photo scale, film type, season of photography and the general quality of the image. The interpreter's familiarity with the local conditions is also an important factor in the quality of the interpreted results.

For the interpretation of stand characteristics and the classification of forest lands only, a few interpretation keys have been developed in Europe, such as those appearing in the *Handbook for Forest Surveyors* for use in the Soviet Union (Tretjakov, et al., 1953) and those published for use in Czechoslovakia by Cěrmak (1960). Many interesting hints for the interpretation of forest land, however, can be found in textbooks of photo interpretation such as the Swedish *"Tolkning av flygbilder"* (1955), the textbook of the International Training Center (ITC) for Aerial Photography in Enschede (1964), those of Baumann (1957), Kurth, et al., (1962), Samojlovič (1952) and the booklets from Stellingwerf (1966, 1968a). These publications discuss primarily the use of black-and-white photography of medium scales (1:10,000–1:30,000).

Research by Akça (1969–1970) indicates that the use of scanning microdensitometers may provide another means for classifying forest land. Figures 34-20 and 34-21 show some microdensograms of beech stands and other features measured on black-and-white aerial photos, scale 1:10,000. Using the average spatial frequency of the den-

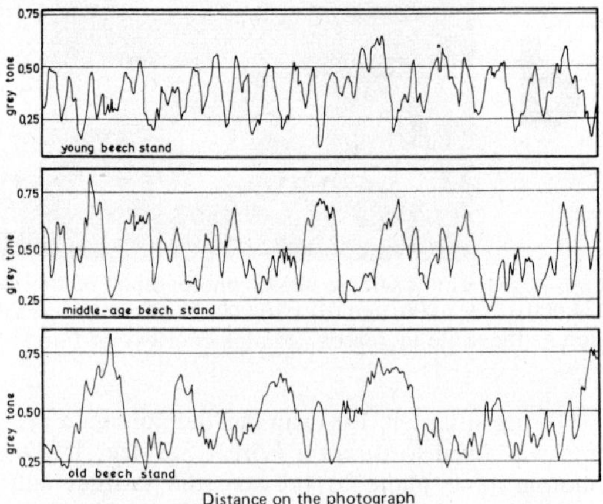

Fig. 34-20. Densograms of young, middle-aged and old beech stands. The shading between crowns, especially marked by old stands, can be seen in the densograms as peaks. With increasing age the frequency of the grey density variations declines. This is attributable, at least in part, to the increasing diameter of individual tree crowns. (Akca, 1969/1970.)

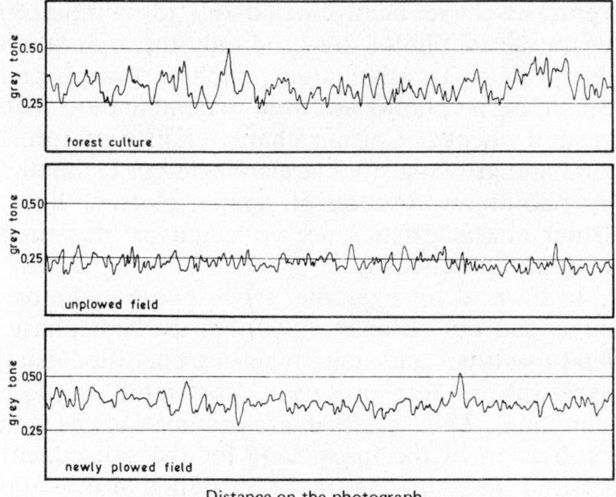

Fig. 34-21. Densograms of forest culture, unplowed and newly plowed fields. (Akca 1969/1970.)

sogram curves, their mean quadratic deviation, and arithmetic mean, it is possible by discriminant analysis to separate many land-use and forest types. Accuracy of separation ranges from 87 to 100 percent.

Usually it is necessary to provide information for the area of each class delineated, such as the sizes of areas covered by certain species, age classes, and forest or land-use types. In inventories of forest enterprises such quantities can be measured after first defining boundaries on the aerial photos and then photogrammetrically mapping these delineations in order to eliminate image displacements. In national surveys, however, quantitative area information can only be obtained using photo sampling. When this is done, however, the following errors can result: (1) Random error (representative error) which is influenced by the size or number of sample plots; (2) classification error, which can develop either from interpretation error or through land-use changes between the time the photos were taken and the time the inventory is made; (3) photogrammetric error caused by scale differences and displacements of features as imaged in the individual photos; and (4) photogrammetric error caused by scale differences between sets of photos, each set covering a large part of the total inventory area.

Photogrammetric errors can lead to different "weights" of individual sample plots; hence they are more important with large photo scales and when large height differences occur within the survey area. These errors can be avoided if the aerial photo sample plots are selected from orthophotos or if a stereo plotter is used whereby the sample plot units are transferred from a map to the stereo model. The use of a stereo plotter is quite expensive, however. A simpler method of eliminating photogrammetrical errors has been suggested by Tomašegović (1963). Also, the radial line method has proved to be particularly effective (Loetsch and Haller, 1964; Hildebrandt and Schindler, 1966; Schindler, 1967).

Types of Imagery

A large number of investigations have been carried out in Europe to determine the optimum remote-sensing system for the identification of tree species and stand types (Belov and Berezin, 1958; Hildebrandt, 1963; Lackner, 1966; Nyyssönen, Poso and Keil 1968; Pohorley, 1958; Rabenau, 1969). Unfortunately, the results of these studies do not always agree in all details. On the following items, however, there appears to be agreement (assuming equal image quality, and equal skill and local knowledge of the interpreter): (1) The reliability of tree species identification increases with an increase of the photo scale. (2) Using large and medium scales of photography, color and color infrared aerial photos give better results for species identification than do black-and-white photos. (3) Stand type classifications, including the delineation of the different types is more reliable when using medium-scale (approximately 1:15,000) aerial photos, than with large- or small-scale photos. In forest regions with only a few species (e.g., northern and northeastern Europe) stand-type classification is possible on 1:30,000 photos. (4) Independent of the scale, for stand-type classification, color infrared and black-and-white infrared photos are preferred, if aerial photos are taken in summer.

Black-and-white aerial photos (scales 1:10,000 and 1:15,000) presently are the primary tools for forest-land classification and mapping. These photos are usually of very high quality, thereby allowing detailed interpretations to be made (Figure 34-9). At times, large-scale aerial photography obtained with 70-mm or 35-mm cameras is used. An important new development is the introduction of two-scale/two-film aerial photography for forest inventory in the Soviet Union (Demidov, 1970).

Color and color infrared photographs have opened new possibilities for classifying forest lands. Since the mid-50's foresters in the Soviet Union and more recently in East Germany have used the two-layer Spectrozonal film SN 2(IR-color) (Koxlovsij, 1957; Jordanskij, 1957; Wolff, 1967a,b and 1970; Demidov, 1970). In Western European countries, three-layer Kodak Color Infrared Film is being investigated for possible use in the detection and inventory of diseases and air pollution, as well as in plant-cover research (Hildebrandt and Kenneweg, 1968a,b; Pollanschütz, 1968; Stellingwerf, 1968b; Rossetti, et al., 1966; Kenneweg, 1970, 1971).

If only black-and-white aerial photos with subtle grey-tone differences between species are available, it may be possible by means of the equidensity film AGFA-CONTOUR to obtain a distinct identification of certain species groups or vegetation types. The AGFA-CONTOUR film allows the production of both black-and-white and colored equidensities. The enhancement of desired grey-tone differences becomes possible through the transformation of such tone differences into remarkable color differences. Color Figure 34-22 shows a summer forest with beech and spruce in various mixtures. The original black-and-white aerial photo from which this enhancement was made does not allow the beeches and spruces to be reliably separated. The color equidensity image produced from the same negative shows—for the forest land—the mature spruces in blue, the beeches in red, young spruces in whitish red, and areas of predominantly bare soil in yellow.

Forest Volume Inventory

In extensively managed forests, for example in parts of northern Russia, the various timber stands are described by means of photointerpretation. The interpreted results are controlled: by (1) visual inspections from aircraft flying at heights of 200 to 400 m (640 to 1,280 ft) above the ground, and (2) sample plots visited in the field.

In forests with average management intensity, various methods of ground sampling in combination with photointerpretation are in use. Ground cruise lines are established and the forest areas between the cruise lines are incorporated into the survey by using only aerial photointerpretation of such characteristics as stand type, species, age-class, average tree height, site index, density, and volume.

In the intensively managed forests of central Europe, every stand is included in an inventory. Each inventory is carried out within permanent compartments (15 to 20 ha) and subcompartments created by combining aerial-photo and site maps of those areas that have the same site and stocking conditions. For these areas, using photointerpretation and other available information, the species, stand structure, density and other conditions within each stand are described. The results are confirmed by a field survey and are modified as necessary (for example, by adding information on timber quality and damage, natural regeneration, and ground cover). The standing volume is normally measured within field survey plots. Based on information gathered from time studies, it is known that a 30- to 60-percent saving of time can be realized when the forester uses photointerpretation along with his ground work (Hildebrandt, 1957). The actual amount of time saved depends on the quality and scale of the aerial photos, stocking and site conditions, and the capability of the interpreter. Detailed mapping in connection with the forest enterprise inventory is done with various photogrammetric methods, depending on conditions of relief, required accuracy, or quality of existing maps. Sketchmasters and stereo plotters, as well as high-quality orthophoto instruments are used for the production of orthophoto maps.

Aerial photographs have also been used for national forest inventories in European countries. In Sweden, Finland, and Norway, where such inventories have been carried out for almost 50 years, aerial photos are used only for orientation in the field. In other areas, such as France and Spain, aerial photos are used in combination with ground checks. Certain characteristics of forest land and growing stock are collected or estimated by photointerpretation of sample plots or lines. Other characteristics are collected or measured with ground checks.

In France, for example, a two-phase inventory is carried out (Brenac, 1962). In the first phase, land-use types and stand types are classified using aerial-photo interpretation (restricted random sampling). This results in area estimations as well as division of the forest land for the subsequent ground measurements of growing stock and growth increment. A predetermined number of photo sample plots within each category is selected and ground checked. Growing stock and growth increment are measured in the forest.

Species classes, height classes, and density classes have been successfully recognized by photointerpretation (Brenac, 1962; Sandberg, 1963; Wolff, 1960), and treatment classes have been discussed (Nyyssönen et al, 1968). Black-and-white aerial photos at scales between 1:10,000 and 1:20,000 have, as a general rule, provided acceptable results. Various possibilities of photointerpretation and ground-checking combinations in forest-inventory work have been presented in *A System for Forest Inventories* by Hildebrandt (1964). A complete presentation of aerial-photo usage for forest inventories including associated statistical problems, was made by Loetsch and Haller (1964).

Components of Volume

Components of photo volume as reviewed in the section on United States and Canada, include estimates of stand density, crown diameter, and tree height. For the determination of stand density, methods used include visual estimates, counting by means of a dot grid, or estimates with the aid of a set of model stereograms or comparison photographs. Klier (1969) suggested two other estimation methods:

(1) Using the following equation:

$$\text{Stand Density} = \frac{\text{average crown area} \times \text{crown count per hectare}}{10} \quad (34\text{-}4)$$

where the average crown area can be calculated from the measurement of n crown diameters, d_k, within the aerial photo

$$\frac{\pi \Sigma d_k}{4n}$$

(2) Using the Bitterlich principle: A wedge with a certain angle opening (Table 34-3 and Fig. 34-23) is printed on transparent film. The film is placed

TABLE 34-3

Angle Opening	k
36°52′	10
16°16′	50
11°26′	100

over the aerial photo and the point of the wedge is held in place on the photo with a pin. By turning the model through all four quadrants, one can count each tree whose crown touches or is cut by both legs of the wedge.

The stand density is equal to $Z \cdot 1/k$, where Z is the number of counted crowns, and k is stand density factor for a given angle. The best results have been obtained with photos at a scale of 1:5,000, when the wedge has an angular opening of 16° 16′. Ground controls have shown that the estimated stand density is low by about 10 percent. A good correlation exists between the results of the photointerpretation and the ground check.

As a result of many investigations in several European countries, it is known that trees of the same species with equal stem diameters at breast height can have many different crown-diameter measurements. Nevertheless, it is possible to establish a relationship between crown diameter and stem diameter enabling, at least, an estimation to be made of the *dbh*. Investigations by Wodera (1948) in Switzerland and in Saxony established the following relationships:

Species	*dbh*
Spruce	$-33.6 + 26.6\ D - 2.5\ D^2$
Fir	$-5.2 + 6.8\ D$
Pine	$-3.5 + 8.1\ D + 0.31\ D^2$
Beech	$-6.4 + 8.7\ D + -\ 0.4\ D^2$

(D = crown diameter measured from the aerial photo)

By grouping pine trees in 2-meter height classes, Ilvessalo (1950), found in Finland that a correlation between 0.85 ± 0.013 and 0.77 ± 0.047 existed (depending on the height class), between the maximum crown diameter and the *dbh*. In birch, this correlation is lower (0.79 to 0.64); in spruce, it is weak and uncertain (0.55 to 0.39).

There also exists a correlation between the crown diameter and the stem volume, provided the trees are classified according to heights. Again the correlation is fairly high in pine, lower in birch, and both weak and uncertain in spruce.

Wolff (1966) found good correlations between the mean *dbh* of stands (pine, spruce, oak) and the mean crown diameter, as well as the number of trees. Further research about such relationships was done in Germany by Eule (1959) in beech stands and by Klier (1970) in spruce stands, by Tomašegovič (1963) in spruce and fir stands in Yugoslavia, and by Samojlovič (1940) and Berezin and Trunov, (1957, 1963) for several tree species in the Soviet Union.

For surveys of stand volume in regions of intensive forest management, it is often necessry to know the distribution of stems and standing volume by diameter classes. In spite of the fact that trees with the same *dbh* may have different crown diameters, Hildebrandt and Kenneweg (1969) showed with beech stands that it is possible to estimate the distribution of stems by *dbh*-classes, using crown-diameter distribution as measured from aerial photos (Fig. 34-24).

Knowing the percentage standing volume distribution as a function of the *dbh* classes, the volume distribution in beech stands can be also estimated from crown measurements made on aerial photos (Fig. 34-25). For crown measurements and for the establishment of crown-size distribution, the Zeiss "Teilchengrössen-Analysator" (Fig. 34-26) has proven itself very worthwhile. Haenel et al (1972) developed an automatic method for the determination of the number of crowns and average crown diameter of stands. Local radial gradients of photographic density (grey tones) are used by a computer for identification delineation,

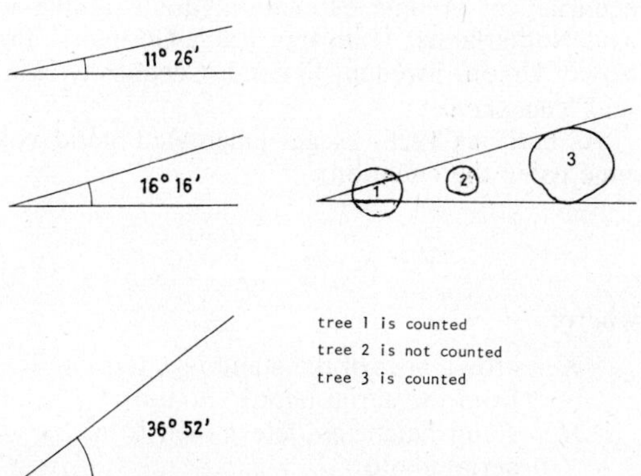

Fig. 34-23. Examples of wedges with varying angles used in calculating stand density by the Bitterlich principle.

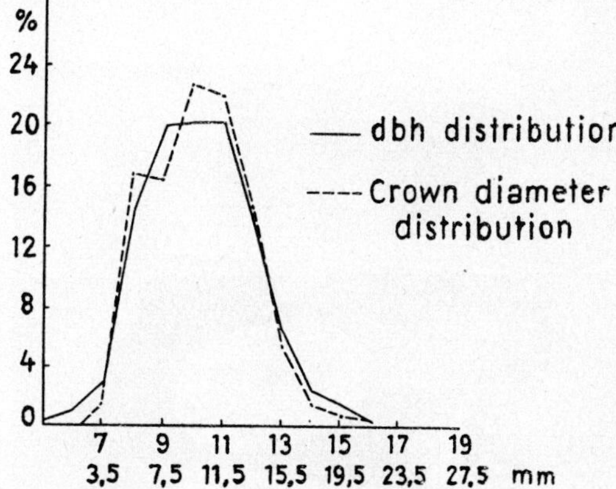

Fig. 34-24. Diameter breast height (dbh) and crown-diameter distribution in a beech stand (Hildebrandt and Kenneweg, 1969).

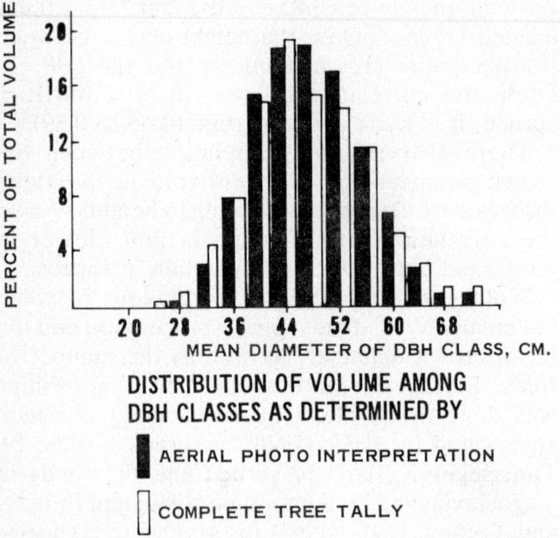

Fig. 34-25. volume distribution of dbh steps (Hildebrandt and Kenneweg, 1969).

and measurements of crowns. Table 34-4 shows the results of the technique when applied to a 55-year-old pine stand.

The exactness of stereo-photogrammetric tree-height measurements has been investigated many times. The results in Table 34-5, selected from tests of various European investigators, have been established by the use of stereomicrometers and mirror stereoscopes.

Based on the research results of Akça et al (1971), it can be inferred that color and color infrared photos give the same accuracy of tree height measurements. Theoretical investigations about errors of observation and measurement on aerial photos (Schultz, 1970; Akça et al., 1971) have given valuable insight as to the reliability of photo measurements of tree height. Schultz has shown that photo measurements of tree height can be made in the best cases to within ± 0.4 m, and in the poorest to within ±3.2 m, of the actual tree heights.

Along with mirror stereoscopes and stereomi-

Fig. 34-26. Zeiss Teilchengrössen-Analysator.

TABLE 34-4

Feature	Ground Data	Image Analysis: Conventional by man	Automatic by computer
Number of tree crowns	66	63	62
Average crown diameter	2.0 m	1.9 m	1.9 m

crometers, pocket stereoscopes with micrometers have been developed which allow photogrammetric tree-height measurements to be made in the field (Ackerl, 1966). The accuracy of measurement is approximately equal to that obtained with mirror stereoscopes in the laboratory.

A method of tree-height measurement, similar to that which employs the well known parallax wedge, is based on the parallax-disc that Perlwitz (1963) developed in combination with a tree-height calculator.

Timber Volume Estimation

Many problems of statistics, forest mensuration, growth, and yield are touched upon when volume estimation by aerial-photo interpretation and measurement are investigated. Along with discussions of special problems of interest in this field some publications are especially useful as basic reference tests. In one particularly thorough text Bočarov and Samojlovič (1964) deal with certain mathematical and mensurational fundamentals of forest photointerpretation based on Russian data.

Estimates of growing stock, using aerial-photo interpretation, have had little practical use in Europe because of the intensive forest management practiced and the necessary accuracy that is required. These requirements have created doubt or skepticism regarding the idea of making volume estimations from aerial photos (Nyyssönen, 1967). Nevertheless, from the beginning of aerial photography usage in European forestry, many investigations have been carried out to determine the accuracy of volume estimation—for example in The Netherlands, Hungary, Italy, Germany, the Soviet Union, Sweden, Finland, Czechoslovakia, and Yugoslavia.

As early as 1928, Zieger calculated stand volume using the equation:

$$V = K \cdot H \cdot \frac{F}{Z_{m^2}} \qquad (34\text{-}5)$$

where:

K = crown area of the stand is determined from the aerial photo

H = stand height as determined from the aerial photo

F = stand form factor (from tables)

$$Z_m = \frac{\text{crown diameter of the mean tree.}}{dbh \text{ of the mean tree}}$$

TABLE 34-5

Accuracy of Height Measurements on Aerial Photography–A Summary of Recent European Studies

Author	Film	Scale	Average Height Error (Meters)	Species
Stellingwerf 1962	panchromatic	1:10,000	1.20	pine
"	panchromatic	1:20,000	1.70	pine
Akça et al, 1971	panchromatic	1:5,000	1.10	conifer
"	panchromatic	1:5,000	1.27	hardwood
"	color infrared	1:5,000	1.54	conifer
"	color infrared	1:5,000	0.66	hardwood
"	panchromatic	1:10,000	1.86	conifer
"	panchromatic	1:10,000	1.32	hardwood
"	color infrared	1:10,000	1.23	conifer
"	color infrared	1:10,000	1.57	hardwood
"	color	1:10,000	1.52	conifer
"	color	1:10,000	1.64	hardwood
Bátkai et al, 1971	panchromatic	1:8,500	1.44	oak/conifer
"	panchromatic (Winter)	1:5,200	3.09	hardwood
"	panchromatic (Winter)	1:5,200	2.09	poplar

Example:

size of stand = 1 hectare, H = 20 meters, K = 9,000 square meters, F = 0.5; Z_m = 2.8 meters/0.2 meter

$$V = 459 \text{ m}^3 \text{ solid volume}$$

The error resulting from such a volume calculation is estimated to be ± 10 percent.

A more sophisticated method is the measuring of "Wuchsraumprofilen" (growing space profiles) from aerial photos using a stereo plotter. The "growing space", R, of the stand is calculated as follows:

$$R = a \cdot q_i \cdot S \tag{34-6}$$

where:

a = distance between profiles
q_i = area of the profile
S = density of the crown canopy.

The relationship between the "growing space" and the volume is calculated for each stand type by means of ground checks so that stereo photogrammetrically measured "growing space" can be correlated to the estimated stand volume.

Stellingwerf (1962) used the relationship between volume and stand height, and developed a volume equation for certain species using the height of the dominant trees and the number of trees as variables.

Example: Scots pine in The Netherlands

$$\text{stand volume} = 61.88 + 0.027\,X_1^3 + 0.12\,X_2 \tag{34-7}$$

where X_1 = average height of dominant trees within the stand and X_2 = number of trees per hectare.

Tandon (unpublished investigations) used the relationship between actual volume and tree counts on aerial photos. He carried out this study for the middle- and old-aged forest stands of both hardwood and conifer species in the Black Forest adjoining Freiburg, Germany. A good correlation (0.73–0.84) was found to exist between ground observations of total volume and tree counts made from aerial photos. This correlation was stronger than that between volume and total tree counts made on the ground. A possible explanation is that some of the smaller trees in the understory, which contribute very little towards the total volume, were difficult to identify on the aerial photos and therefore were eliminated, resulting in lesser deviations between the two variables. In subsequent testing, three age-classes (young, middle, and old) were also interpreted on aerial photographs and were combined with tree counts on photos to develop a new independent variable. A much stronger correlation (0.90–0.95) was found to exist between this new variable and the volume. It appears that even more encouraging results are possible if the forest stands are divided according to their age classes.

Bogyay (1970) found the following relation for 10- to 60-year-old pine stands in Hungary:

$$\log V = 1.37 - 0.02 \log K + 1.026 \log H \tag{34-8}$$

where:

V = total actual volume per hectare
K = age of the stand
H = average stand height, provided by stereo-photogrammetric measurements.

The correlation coefficient was found to be as high as 0.96. However, it was necessary to adjust the result if the density was 80 percent or less.

An intriguing method for volume estimation in exploitation areas of the Soviet Union was proposed by Gordeev (1954). Using oblique aerial photos (scales between 1:5,000 and 1:1,000) taken during the winter, the tree heights, the diameter at breast height, and the number of trees are measured or counted. The volume is then calculated using volume tables. The result of this volume calculation is adjusted by means of a coefficient which is necessary because one tree can cover another. The error resulting from such a volume calculation is estimated to be ±10 percent.

Some interesting research has been performed on the use of aerial photography as an aid to planning ground timber cruises (for example, Wolff, 1960; Sandberg, 1963; Nyyssönen, Poso and Keil, 1968). An attempt was also made to systematize the forest-inventory methods with regard to the manner of combination of photointerpretation, ground work, and sampling procedure (Hildebrandt, 1964).

The combination of "list sampling" or "Three-Pee sampling" (Grosenbough, 1965) with aerial-photo interpretation for forest inventories has been suggested by Loetsch (1970). He illustrated that, for growing stock inventories, stand volumes estimated by photointerpretation are suitable for compiling the "list."

Use of Modern Remote-Sensing Techniques

In the countries of western Europe investigations are still in their early stages to determine the usefulness of line scanners for thermal IR and multispectral images, and of microwave and radar imagery for forestry purposes. Some test flights have been carried out, but the preliminary results for the most part are poor, their greatest value being ideas on how to plan further research (Bodechtel and Kritikos 1971; Svensson, 1971; Vermeer, 1971).

Most of the equipment for obtaining line scan imagery which is used in western European institutes is bought in North America. Some institutes have designed their own equipment and procedures for special purposes such as for image enhancement and for automated recognition of features and patterns (Bodechtel and Kritikos, 1971; Haenel et al, 1972; Wieczorek, 1972; Tzschupke, 1973), and also radiometers and airborne thermal line scanners (Zenker, 1971; Künzi, et al, 1971). European commercial companies may soon offer more remote-sensing instruments.

Progress in the use of multiband techniques in remote-sensing research is much slower in most European countries than in North America. This is because of the lack of large unmanaged areas as a necessary challenge for research and to the existence of relatively small research institutes which cannot afford the expensive instrumentation necessary for advanced remote sensing. For the same reasons, conventional aerial photography continues to be the most applied type of imagery for research and operations in the field of forest inventory and assessment.

In eastern Europe the possibilities for application of new remote sensors, such as line scanners, are even more limited. Only in the USSR is the development of inventory techniques based on modern remote-sensing systems strongly promoted. It is difficult to get extensive information about remote-sensing activities in that country, because publications are not readily available. However, it is known that an important part of Russian investigations is in basic research; an immense amount of ground data has been collected very systematically and comprehensively during the last 15 years (Steiner and Gutermann, 1966; Hodarev et al, 1971). Today the main interest in the USSR has moved toward such modern techniques as thermal IR and multiband line scanning and to the use of electronic digital computers for image evaluation (Kondratyev et al, 1971; Tscherkasov, 1971).

TEMPERATE ZONE: SOUTHERN HEMISPHERE

Classification of Forest Land

North American methods of image analysis have been adapted to local conditions for forest surveys in countries of the temperate zone of the Southern Hemisphere (e.g., Chile, Australia, and New Zealand).

In Australia and New Zealand emphasis in the use of remote sensing has been directed primarily towards broadly classifying timber resources. Black-and-white aerial photos at a range of scales (often 1:31,000 and smaller) have been used to classify Australian forests into three classes: productive and semiproductive forests, and low-grade woodlands. For large areas of indigenous forests in New Zealand, small-scale (1:40,000) vertical aerial photography remains the basic scale for forest classification. During the last 20 years, increasing use has been made of medium-scale (1:10,000, 1:15,840, 1:20,000) photography for forest classifications and volume studies; various larger scales (1:4,000 to 1:10,000) and some 1:15,840 have been employed for detailed ecological studies.

The New Zealand Forest Service has prepared a series of forest-type maps of central North Island from black-and-white aerial photos and ground sampling. These show forest type, species composition, stand size, and density of stocking. This classification is similar to that used by the Forest Service, U.S. Department of Agriculture, for dense forests in the western United States (Forest Service Handbook, 1968).

Analysis Procedures

There are important structural and floristic differences between Northern Hemisphere and

Southern Hemisphere temperate forests. Consequently, image characteristics differ considerably. Modifications of many accepted Northern Hemisphere forest-survey techniques are required for successful application in southern temperate forests. Certain special properties of southern forests demand new approaches and new techniques in classifications and analyses.

Temperate forests in Australia, Chile, and New Zealand are evergreen and predominantly broadleaf. An important exception is an area dominated by a deciduous southern beech (*Nothofagus*) in Chile. All except one (*N. gunnil,* Tasmania) of the other numerous species of *Nothofagus* in the three countries are evergreen. Five distinctive temperate forest formations occur in these countries.

Forest Formations	*Country*
1. Indigenous: Evergreen	
Broadleaf sclerophyll	Australia
Broadleaf-podocarp (sub-tropical)	New Zealand, Chile
Broadleaf (sub-antarctic)	
Indigenous: Deciduous	
Broadleaf	Chile
2. Exotic: Evergreen	
Needleleaf (conifers)	Australia, New Zealand, Chile

Temperate forests of these three countries have a varied infusion of tropical (Indo-Malayan and Pan-Tropic) and antarctic elements in their indigenous floras.

Sub-antarctic Forests

Temperate subantarctic forests are of simple structure dominated by various species of the *Nothofagus* genus. They occur in montane situations at higher altitudes in all three countries and at successively lower elevations as latitudes increase in southern New Zealand, and southern Chile. For image analysis and forest land mensuration, these formations most closely approximate Northern Hemisphere conifer forests and some single-stand broadleaf deciduous communities.

Sclerophyll Forests

Most temperate forests in Australia are dominated by some 400 different species of the single, myrtaceous genus, *Eucalyptus*. "Wet" and "dry" sclerophyll forests (Schimper's Summer Evergreen Formation) are the most important economic forests of Australia.

Classification maps prepared from panchromatic photos, at scales of 1:31,000 upwards, for a national inventory of Australian forest lands divided productive forests into five stand-class sizes: over-mature, mature, poles, saplings and regeneration. Height classes were: over 52 m, 40–52 m, 27–40 m, 15–27 m, 3–15 m, and under 3 m. Density classes, based on crown cover (over 80%, 50–70%, 20–40%, and under 10%) are considered important for volume measurements in Australian forest classification. Principles for making volume measurements directly on aerial photos do not vary greatly from those already discussed. Tree height, stand height, crown diameter, crown closure, tree number, and *dbh* all have application. Several examples of volume inventories in different types of forests in Australia were given by Howard (1970).

Subtropical Forests

Forest physiognomy of New Zealand and Chilean broadleaf-podocarp temperate rainforest has no counterpart in Northern Hemisphere's temperate forests. In terms of structural complexity, floristic richness, diversity and range of growth forms (life-forms), and profusion and luxuriance of growth, these complex multilayered forests are closely akin to tropical rainforests (Richards, 1952). *Subtropical* is the term normally used in New Zealand in contrast to subantarctic for the predominantly single stand *Nothofagus* forests.

Many of the various techniques and problems discussed in a following section on the tropical zone apply to these Southern Hemisphere broadleaf-podocarp (subtropical) forests. Both the broadleaved endemic conifer, kauri, *Agathis australis* (Auricariaceae), and several species of narrower and smaller-leaved podocarps (*Dacrydium* and *Podocarpus* spp.) are New Zealand's economically most important indigenous timber trees. In indigenous forestry, attention is focused primarily upon podocarps.

When young "ricker" (poles) kauri grow in stands, their slender pyramidal form and uniform height provide a distinctive texture that can be readily identified on medium- to small-scale black-and-white aerial photos. Mature kauri is readily recognized even on small-scale black-and-white photos because of its large crown diameter and distinctive photographic tone. Photo-interpreters have experienced no difficulty in identifying another *Agathis* species by its distinctive crown on photos of temperate-type forests in highland New Guinea (Boon, 1956).

Crowns are smaller on dakua (*Agathis vitiensis*) in the highlands of western Viti Levu in Fiji. This may explain the difficulty experienced by Rees (1972) in identifying these on black-and-white infrared and on color photographs. Also, they are less emergent than the New Zealand species. Podocarps can be identified only with difficulty and low accuracy on medium-scale panchromatic photography. They are more readily mapped on large-scale photos. Experimental use of color infrared photography demonstrates that the podocarp, rimu (*Dacrydium cupresinoides*), is readily identified and easily differentiated from broadleaf trees at medium to large scales.

Exotic Conifer Forests

Plantations of exotic conifers, notably the Monterey pine, *Pinus radiata,* have been planted for more than 50 years over hundreds of thousands of hectares in New Zealand. Much smaller areas are planted in parts of southern Australia. The conifer forests in New Zealand differ in several ways from Northern Hemisphere conifer forests. The New Zealand forests are very intensively managed, single stand, even age, 15- to 25-year rotation forests. They are essentially intensive monoculture tree crops.

Air-photo mensuration techniques are well developed. Many local modifications, however, are required to monitor specific management practices. For instance, one current silvicultural practice is to plant trees at 2-meter intervals, both along and between rows. In addition to periodic trimming, these trees are subsequently thinned at various stages for chipboard, pulpwood, fence posts, and round log and sawn timber. Both the close spacing and vigorous growth of pines in New Zealand result in crown closure well before 10 years. Photo mensuration studies use stand height rather than tree height until forests are thinned.

Types of Imagery

The general conclusions about the usefulness of remote-sensor imagery discussed previously for temperate forests of North America and Europe apply broadly to Southern Hemisphere sclerophyll and subantarctic forests. Forest characteristics, problems of identification and measurement of broadleaf-podocarp forests more closely approximate those of tropical forests. The most commonly used remote sensor imagery are panchromatic photo prints of 17.8- × 17.8 cm (7 × 7 in.) or 22.8- × 22.8 cm (9 × 9 in.) format at scales of 1:15,840 and smaller.

Very little research on best film-filter combinations for forest-land management has been carried out until recently. For 20 years the only available aerial photos for large areas of both Australia and New Zealand were those flown as an emergency war measure during the early 1940's. Photograph scales and quality varied greatly. Higher quality panchromatic photos at larger scales flown more recently and at frequent intervals are used for the most intensively managed forests. Panchromatic photos (1:15,840 to 1:20,000) have proved suitable for identification and measurement of kauri and for *Pinus* forests in New Zealand. They are much less suitable for broadleaf-podocarp forests. Species identification and volume studies of eucalypt forests are possible on panchromatic photos at scales of 1:15,840, but accuracy is not high. Accuracy of both identification and volume analyses improve greatly at scales of 1:10,000 and larger.

Black-and-white infrared aerial photography does not provide a good tonal contrast between New Zealand's economically important podocarps (gymnosperms) and the less important broadleaf trees (angiosperms). The film has not been exhaustively tested because of an increasing economic emphasis on exotic softwood forests where use of conventional panchromatic photography is quite satisfactory. The high economic return from these forests justifies use of medium- to large-scale panchromatic photography.

Many species of the genus *Eucalyptus* exhibit very strong ecological site preferences, especially those associated with altitude, aspect, and drainage (Cochrane, 1968a). In addition, many of the economically important timber trees, e.g., karri (*E. diversicolor*) and jarrah (*E. marginata*) in western Australia, stringybark (*E. obliqua*) and alpine ash (*E. delegatensis*) in New South Wales and Victoria, and mountain ash (*E. regnans*) in Tasmania and Victoria have distinctive crown characteristics (texture, shape, tone) and growth-density patterns resulting from a tendency to grow in near-pure stands. This assists photo identification (Cochrane, 1969). All of these species and many others, such as marri (*E. calophylla*), blackbutt (*E. patens*) wandoo (*E. redunca*), peppermint (*E. radiata*), *E. goniocalyx, E. baxteri,* and blue gum (*E. globulus*) are readily recognized and mapped on black-and-white photographs at scales upwards from 1:15,840.

Color aerial photography assists forest typing in Australian sclerophyll forests because many eucalypts have distinctive crown hues, particularly at times of new growth (Cochrane, 1968b, 1970). Atmospheric haze on photos taken above 3,000 m limited the wide use of color for Australian forestry to low-altitude, large-scale surveys until the advent of the negative-positive process (Sims and Benson, 1966, 1967).

TROPICAL ZONE

To understand the application of aerial remote sensing in studying tropical forests some knowledge of these forest communities is essential. A brief description is therefore given of the most important forest community; the tropical rainforest. Of course the description cannot be complete but reference is made to textbooks written by Richards (1952), Haig et al (1958), and others. These textbooks also describe other tropical-forest communities.

The tropical rainforest is a climax vegetation which develops under very humid conditions; rainfall of 1500 to 4000 mm is evenly distributed throughout the year; a less-rainy season, with as little as 50 to 75 mm of rainfall per month, does not last longer than 3 months; temperature is quite constant, varying only from about 20° to 26° C; and light intensity under the trees is approximately 1/30 to 1/240 of that in the open.

Three main tropical rainforest regions are generally recognized: (1) The Amazon region is dominated by the Amazon basin with an area of

3,000,000 km^2. Also included are other parts of the east coast of Brazil, and coastal areas from Equador to Mexico. (2) The Southeast Asia region, includes Indonesia, Papua, Territory of New Guinea, Malaysia, Burma, Thailand, North and South Vietnam, and the Philippines. (3) The African region covers an area from Gambia to Liberia, Southern Nigeria, Cameroon, Gabon, Congo to the Great Lakes, and the eastern slopes of Madagascar.

The tropical rainforest contains a richness of species and families. In vertical profile it consists of three layers. The top canopy, with its emergent trees, is open. Except under special climatic and edaphic conditions, no single species is of outstanding importance, either ecologically or economically. These forests do not contain an inexhaustible source of timber. Their botanical richness can be considered inversely proportional to their commercial value. It has been suggested that the tropical rainforest is the botanist's dream and the forester's nightmare. The enormous number of species and genera hampers resources assessment, initial utilization, and management. Volume assessment, without identification, creates difficulties. It follows that means of identification, such as appearance of woody tissue, vegetative organs, and even bark, as well as influorescences, are important. Regeneration after exploitation is difficult because all representatives of a chosen species in the logging area generally are felled.

Le Ray (1958) wrote about the meteorological conditions which are considered to be necessary for obtaining useful aerial photographs in Cameroon, Gabon, and Madagascar. The data may also be true, for most other tropical rainforest areas. Favorable "photographic-weather" conditions specify no haze and less than 1/8 cloud cover. Le Ray, and others, have stated that, to date, no rules exist which permit reliable forecasting of favorable periods of good photographic weather. Their only definite conclusion is that in the rainy season aerial photography is impossible. In addition to clouds and haze, smoke and dust give unfavorable atmospheric conditions for good photography.

The most likely period for good "photographic weather" follows closely after the rainy season, when the ground vegetation is too damp to burn, the air is free of dust, and cloud cover is reduced.

Conventional Photography and Visual Observations From the Air

In addition to aerial photography, ordinary visual observations from a plane or preferably from a helicopter are still considered to be of great importance for many forest-land reconnaissance purposes. Another advantage of light aircraft is that they offer a means of transport essential for field-inventory crew and equipment, especially for areas difficult of access. Hindley (1969) and others mentioned that views from helicopters give a good impression of forest-land conditions of altitudes of 60 to 100 m over the tree tops or slightly higher in turbulent air conditions. Some examples of costs involved in visual air reconnaissance were given for Guatemala and for Malaysia (Hindley, 1969). Compared to ground-transport costs (including onboard fuel, boat depreciation, etc.), the expenses for using a helicopter were about 60 percent less. The time saved, as compared to terrestrial access, amounted to about 30 percent. Flight programs should be planned in detail on photographs, maps, and mosaics, considering the high charter costs.

Low-flying planes are also occasionally used for visual type reconnaissance. Twin-engine types are preferred for safety reasons in remote and inaccessible areas. Several authors, including Le Ray (1958, 1962), Francis (1957, 1960), and Stellingwerf (1969) reported on the use of such planes for reconnaissance inventories. Le Ray (1962) mentioned the combination of aerial photographs, scale 1:50,000 covering 1,500,000 ha, and aerial visual observations, in addition to a 1-percent field enumeration in equatorial Africa. Percentage costs for each of the components of the survey were: (1) aerial photographs 20 to 25 percent; (2) aerial observations 3 to 4 percent; and (3) fieldwork 70 to 75 percent. For an area of 30,000 ha (equal to one exploitation unit) with 1:10,000-scale photographs, the percentage costs were: (1) aerial photographs 45 to 60 percent; (2) aerial observations 1 to 2 percent; and (3) fieldwork (5-percent enumeration) 40 to 50 percent.

Planes or helicopters are sometimes used in the tropics in order to register visually the relative percent of individual forest types by means of regularly spaced point observations from the air. The system works in a similar way to the dot-grid method used on aerial photographs or maps.

Because cost figures for surveys using aerial photography are important, but absolute figures have only limited value, an attempt is made in Table 34-6 to analyze the relative costs for separate activities from flying to fieldwork (the cost for photographs, scale 1:70,000 per 100 km^2 equals 1). It is emphasized that the figures must be considered relative; they are an average of many data which diverge widely and should, therefore, be used only as a rough guide. Prices for areas smaller than 500 km^2 are relatively higher; therefore only areas greater than 500 km^2 are considered in Table 34-6.

In addition, the costs of the two field surveys are indicated only for small- and large-scale photographs respectively, as reconnaissance field surveys are usually made using small-scale photographs, and exploitation surveys are made using large-scale photographs. Positioning charges depend on flying distances to operation areas. In the tropics these costs vary greatly (Nyyssönen, 1961).

The cost figures in Table 34-6 clearly indicate the necessity of employing small-scale photo-

TABLE 34-6

Relative Costs of Tropical Surveys Using Various Scales of Photography

Reciprocal of Scale:	Photography	70,000		40,000		20,000		10,000	
	Map	150,000		100,000		50,000		25,000	
Aerial Survey relative cost		a[1]	b[2]	a	b	a	b	a	b
(1) Flying		16	16	44	44	108	108	242	242
(2) Photography[3]		1	2	3	6	12	24	48	96
(3) Interpretation[4]		1.4	1.4	4	4	17	17	68.4	68.4
(4) Radial triangulation[5]		1.2	—	4	—	16	—	62.4	—
(5) Aerial triangulation with adjustment		—	12	—	36	—	148	—	600
(6) Orientation and plotting[6]		1.2	36	4	44	13.2	128	108	413.2
(7) Cartography[7]		18	28	18	28	40	72	100	192
Total		38.8	95.4	77	162	206.2	497	628.8	1,618.6
Field cruising cost									
Reconnaissance (0.005-0.01%)		80		80					
Exploitation (2%)						2,000		2,000	

[1] Planimetric map.
[2] Topographic map with contours.
[3] For topographic mapping, diapositives are needed for use in plotting instruments; in addition paperprints are needed for interpretation.
[4] At a rate of seven models per man day.
[5] At a rate of 1.5 man hours per photograph for slotted template method.
[6] Based on 3 man hours per model for photogrammetric plotters and 0.25 man hours per photograph for transfer instruments such as epidiascope.
[7] No contours are drawn on planimetric maps.

graphs for item 5-aerial triangulation and item 6-orientation and plotting. The additional cost of large-scale aerial photographs becomes important. Their use for several disciplines seems advisable where feasible, in order to share the costs of flying plus processing. The figures given also indicate that flying with two different focal length cameras in one plane should be tried when useful results are expected with respect to scale and coverage. The small scale photographs, giving complete coverage, are used for mapping and interpretation, and the large scale photographs are only used for interpretation.

In tropical areas is is possible to obtain aerial photographs from 4,000-m altitude on a greater number of days than from 8,000-m altitude; hence the use of super-wide-angle cameras, e.g., 7.6-cm (3-inch) focal length and 22.86- × 22.86-cm (9- × 9-inch) negative size, may be preferable for certain conditions. In this way 1:50,000-scale photographs can be obtained at an altitude of 4,400-m. Increasing the forward and side overlap may be necessary, however, in order to minimize the areas in which excessive relief displacement occurs (Zarzycki, 1968).

The requirements for maps for special inventories of limited areas, and therefore the expenditure for these maps, differ from those for large regional-development projects of which forestry forms only a part. Ground control will usually make up the greatest part of the cost. In unexplored areas, a Doppler navigation system, which does not require ground stations, can ensure photographs without gaps in the side lap areas. Also a statoscope, horizon camera, and airborne profile recorder may be used in the aircraft in order to reduce ground control. Geodetic ground-control equipment together with the above-mentioned system may be limited to six aerodist remote stations spaced at distances of 100 to 200-km. In this manner: (1) first-order geodetic control can be obtained, which may have an accuracy of the order of 1:150,000, when the height of the aircraft, and position and elevation of ground stations are given, and (2) coordinates of the nadir point of each aerial photograph can be computed with an accuracy of ±6m (Zarzycki, 1968). Mapping can be accomplished by these means in a fraction of the time and at lower cost than by more conventional methods. *Ad hoc* reconnaissance and exploitation surveys may not justify such requirements and may therefore be carried out with less accurate maps.

As reported by Asmoro, et al (1976), the National Resource Survey and Mapping Project of Indonesia, together with its remote sensing program, is providing Indonesia with photo maps at 1:25,000 scale for densely populated islands such as Java, and at 1:50,000 scale for the rest of Indonesia. In addition, selected areas are being covered by photo maps at 1:250,000 scale, mainly based on SLAR imagery, and to a small extent on

Landsat imagery. The latter will only be obtainable for limited areas from the Alice Springs receiving station in Australia. The project covers 1,959,500 sq.m., where forests are the major resource. The integrated resource map at 1:50,000 scale consists of the compilation of the base maps, the resources data and the resources maps.

Base maps will be constructed in flat terrain from panchromatic rectified photographs (scale 1:100,000 except for Java 1:50,000). For hilly and mountainous terrain the orthophoto mapping procedure has been selected. The doppler-satellite positioning technique will give geodetic coordinates of ground control points. Further densification of vertical ground control will be obtained from supplementary flights with an Airborne Profile Recorder. Coordinates of the photo control points, in the coordinate system of the ground control points, will be determined by aerial triangulation applying block adjustment. Forestry resource data will be extracted from 1:60,000 scale color infrared photographs, exposed simultaneously with the 1:100,000 panchromatic photographs using a dual camera system, i.e. one with a superwide angle (1:100,000 scale) and one with a wide angle lens (1:60,000 scale).

Classification of Forest Lands

The greatest value of remote sensing in tropical forests lies in the possibility which it offers to differentiate and sometimes to recognize various plant communities. Presently, this task is best performed on aerial photographs. Photo communities are the smallest distinct assemblages of plant species discernible in stereopairs of aerial photographs at a specific scale. The differences between them are based on photophysiognomy, meaning the external structure of the vegetation as recorded on the photographs. Photocommunities may conform to plant formations or subformations at scales of 1:20,000 to 1:80,000; to subformation associations at scales of 1:10,000 to 1:15,000; and to smaller assemblages of vegetation at scales of 1:5,000 and larger (Howard, 1970). The recognition of single trees belonging to different species, even on large-scale photography, is more an exception than a rule in tropical forests, particularly in rainforests.

The most common photography in use for tropical-forest inventory is panchromatic, although this is sometimes used in combination with black-and-white infrared (Le Ray, 1962). Taylor and Stewart (1958) described the classification of vegetation in Papua and the Territory of New Guinea based on field surveys and interpretation of 1:40,000-scale photographs. Paymans (1970) published the results of a field survey and photo-interpretation for vegetation classification using 1:36,000- and 1:86,000-scale photographs of central and west Papua. He obtained similar results with both scales. The following types were distinguished: (1) saline environment: mangrove (2 species); (2) brackish environment: *Metroxylon sagu* (sagopalm), *Nipa fructicans* (nipa palm), mangrove, fresh water tree genera and pandanus palm mixtures; (3) beach ridges and swales; (4) coastal back plain with tall forests having open canopy and scattered large-crowned emergents on poorly drained soils, and denser well developed crowns on better drained sites, *Campnosperma* swamp forest and *Melaleuca* sp. on temporarily inundated soils; and (5) hill and mountain forest with altitudinal zones in which crown size and density classes are related to broad volume classes. In addition to differences in flora, it is of primary importance to distinguish broad units such as coastal areas, flood plains, and hills, as they usually indicate different forest types due to different growing conditions.

The delineation of dense and open *Araucaria* stands, together with tropical and subtropical forest, on 1,763 super-wide-angle 1:70,000-scale photographs (covering 7-million ha), with subsequent field sampling, was undertaken during a reconnaissance survey in Brazil. Results obtained proved the usefulness of small-scale photography for estimating the remaining exploitable dense *Araucaria* stands. Another example is the use of 1:60,000 scale photographs, together with mosaics, in the Dominican Republic for determining forest types, such as mixed broadleaved forest, low broadleaved forest, shrub forest and thornbush, pine forest (2 density classes), and swamp and mangrove forests (Randall, 1969).

In Africa, IGN (Institut Géographique National, Paris) used 1:50,000-scale photographs taken with a 125-mm focal length camera (adapted to French plotting instruments) for topographic mapping purposes. Photographic scales of 1:25,000 and 1:10,000 have been used in a forest test area in Gabon. Since water courses may be detected with more ease on infrared film and as roads and tracks and inner detail of forest structure can be detected better on panchromatic film, IGN from 1960 onward used both film types simultaneously for work in Africa. The use of 1:50,000-scale photographs was determined by meteorological conditions and mapping specifications (Lanly, 1970).

Clement (1974) describes the successful detection of an important tree species, okoumee (Aucoumea klaineana) in Gabon, by using color, crown structure and texture characteristics available on color photographs (scale 1:5,000).

In Surinam, 1:40,000-scale photographs were used (camera f = 115 mm) for mapping and also for forest, geology, and soil reconnaissance surveys. Broad vegetation units could be distinguished. These photographs also clearly showed the position of rapids, beyond which exploitation is not considered economically feasible. Heinsdijk (1953) distinguished on the panchromatic photographs with the subsequent field sampling of the following types: (1) mangrove forest (2 species); (2) swamp forest: 2.1) *Virola surinamensis,*

Symphonia globulifera forest, 2.2) *Triplaris surinamensis* forest, 2.3) *Dimorphandra conjugata* forest, 2.4) *Hura crepitans* forest, 2.5) *Erythrina glauca* forest; (3) marshland forest: 3.1) *Mora excelsa* forest, 3.2) *Carapa guinensis* forest, 3.3) *Eperua falcata;* (4) dryland forest: 4.1) *Eperua falcata* forest, 4.2) *Groupia glabra* forest, 4.3) *Mora gongrijpii* forest. (See also Dillewijn [1957], Boon [1955], and Stellingwerf [1966].)

The Canadian Forestry Service, the Surinam Forestry Service, the Surinam Central Bureau for Aerial Survey and the FAO Forestry Development Project conducted a test in Surinam, S.A. to determine the feasibility of identifying tropical tree species (Sayn-Wittgenstein, et al, 1978). It was concluded that the identification of several important species is feasible, provided that good color photographs at scales larger than 1:4,500 are available. However, for most species, additional work is required to confirm observed key features and to estimate the consistency with which they can be recognized. The more unusual crown forms are more consistently recognized. Though crown shape, size and dominance are important features for recognition, color is very important in identification. Color infrared is less desirable than normal color becuse of dark shadows and the difficulty in relating the many red and pink hues to the natural colors observed in the field.

In Guyana, Swellengrebel (1959) used 1:30,000-scale panchromatic photographs for classifying five different forest types of which two contained greenheart (*Ocotea rodiaei*). In Sarawak, Francis and Wood (1955) distinguished 16 vegetation types of which 12 were forest types. In West Irian (Indonesia), Swellengrebel applied 1:20,000-scale panchromatic photographs for a forest reconnaissance and exploitation survey, covering 4,000 sq km. This scale proved to be useful for the selection of stands in which *Agathis labillardieri* and *Araucaria cunninghamii,* when predominant, could be recognized. Furthermore, the photographs, together with the compiled map, gave information on areas which were valuable with respect to timber and accessibility. Studying photographs of areas where inventories had previously been made of round wood volumes, allowed identification and demarcation of four classes which were defined by their round wood volume per hectare. Also delineated were usually wet, marshy flat land; dry, moderately hilly land; and very steep hills where access is difficult.

Successful recognition of trees on aerial photographs in tropical forests depends partly on gregarious growing habit and partly on recognizable crown form and texture, as well as on tonal appearance. It is only under exceptional conditions that individual tree species can be recognized on photographs with scales of 1:10,000 to 1:15,000. *Araucaria cunninghamii* is recognized at high elevations in West Irian because of its pencil-shaped crown and its white appearance, the latter feature probably due to lichens (Boon, 1955). The recognition of baboen (*Virola surinamensis*) on 1:10,000-scale panchromatic aerial photographs of Surinam coastal swamp areas is another example. The trees are recognized by their light tone and roundish crowns (Stellingwerf, 1966). Maripa palms (*Maximiliana maripa*) in swamps in Surinam and Casuarina trees (*Casuarina equisetifolia*) on sandy beaches in the Far East are both recognizable because of their characteristic crown shape. Zahir-ud-Din (1954) reported that Gurjan, a principal dipterocarpus species in Pakistan, is recognizable on 1:20,000-scale photographs by its characteristic crown. Yang (*Dipterocarpus alatus*) in Thailand has been reported by Loetsch (1957) to be recognizable on photographs by its large emerging bright-toned crown. Another dipterocarp, *Shorea curtisii,* is reported to be depicted on photographs by its white crown and its growing habitat on ridges. The same is true for *Dipterocarpus intricatus* in high open dipterocarp forests in Cambodia and Thailand. Fast-growing *Lagerstroemia,* when flowering, has been identified on aerial photographs of Vietnam (Rollet, 1960). In India, teak (*Tectona grandis*) is visible by its cottony texture on 1:15,000-scale photographs. *Isoberlinea doka* woodland can be delineated easily on 1:40,000-scale photographs due to its habit of growing on lateritic hill caps and appearing to form fungi-like patterns on the photographs (Francis, 1960).

For further descriptions of interpretations of vegetation types and individual trees, refer to Nyyssönen (1961), Zonneveld, et al., (1952), Howard (1970), Boon (1955), and Miller (1957).

In general, 1:30,000 to 1:70,000 scales are used for inventories of the reconnaissance type, for general land-use classification, and for broad vegetation mapping. Scales ranging from 1:15,000 to 1:20,000 are applied for exploitation inventories of limited areas. The large 1:10,000 scale is occasionally used for very limited areas in which single tree recognition is essential and, of course, possible. Large-scale photographs are unsuitable for differentiation into communities and types because the great amount of registered detail and the limited area covered by one photograph hamper making a distinction between communities.

Timber Volume Estimation

Although aerial photographs are mostly used for type delineation and enumeration is done in the field within the types, measurement on photographs has occasionally been made to predict timber volume. Area measurement is the same as has been already described for the temperate zones. Problems are encountered because the nature of tropical forests makes correct plot location, identical to photo location, very difficult. Moreover, relationships usually have proved to be weak, and photographic scales are often too small.

Counting easily recognizable tree crowns in

more homogeneous stands on large-scale photographs is relatively simple. This has been done with *Virola surinamensis* trees in swamp forest in Surinam on 1:10,000-scale photographs. Regions having an average density of 1 to 7 trees per hectare were found to be suitable for exploitation. Location of trees is another important factor for a decision prior to exploitation. Approximately 70 percent of the actual number of *Virola* trees were correctly identified. Photointerpretation thus yielded a conservative estimate of actual stand density. Interpretation precision increased with tree diameter; all trees with a *dbh* > 40 cm (merchantable) were visible on the panchromatic photographs.

Paelinck (1958) in the Congo (Brazaville) could not, with acceptable precision, evaluate timber volume of limba (*Terminalia superba*) by counting crowns because the exploitable timber (≥ 60-cm good quality timber) could not be separated from others on 1:50,000-scale photographs. Rollet (1960) made measurements of crown diameter on 1:40,000-scale photographs of five species belonging to the Dipterocarpaceae in Cambodia and Vietnam, but found a weak correlation of the relationships between crown-diameter and *dbh,* and crown diameter and volume. In Malili-Celebes (Indonesia), before estimating timber volume from aerial photographs, the relationship between *dbh* and crown diameter of upper canopy trees was first investigated. As species identification was impossible on 1:10,000-scale photographs, all species were included in the test. The regression equation was found to be:

$$d = 3.5\,C + 12.3 \qquad (34\text{-}9)$$

in which d = *dbh* (cm), and C = crown diameter (m).

The standard deviation of *dbh* estimates as derived from crown-diameter measurements proved to be 30 percent (r = +0.61). This means a standard deviation of 60 percent in volume per tree. Table 34-7 gives a rough estimate of total timber volume, through aerial photographic measurements of crown diameter and number of crowns (using the *dbh*-volume relation established in the field; Paymans, 1951).

In Guyana, Swellengrebel (1959) calculated the relationship between *dbh,* as measured in the field (with 10-cm interval), and crown diameter, as measured on 1:10,000-scale photographs (1.5 m interval). He points out that in strongly mixed tropical forest, when species identification on photographs is impossible or uncertain, only trees with a *dbh* ≥ 50 cm for *Eperua falcata* and *dbh* ≥ 70 cm for *Mora excelsa* can be identified on the 1:10,000-scale photographs. Heinsdijk (1957) and Heinsdijk and Glerum, (1967) showed with experiments in the Amazon region that gross volume of the forest could be estimated from photographs (1:40,000 scale) measuring crown diameters and number of crowns of upper-story trees with *dbh* ≥ 35 cm, taking into account the ratio between the

TABLE 34-7

Timber Volume Estimates

	Number of trees per hectare in the diameter class of highest frequency		
	20	40	60
Forest quality	Timber volume in m³ per hectare		
Poor	20	40	60
Moderate	70	140	210
Good	100	400	600

volume of upper-story trees and total volume per hectare. Use of the table for the Amazon region in Linhares forest near the Atlantic coast gave under- and over-estimations in low-volume and high-volume patches, respectively.

Tree heights in tropical rainforest are extremely difficult to determine from aerial photographs as a closed canopy of the combined stories prevents the ground from being seen. Height, in addition to crown closure, has been used to classify woodland from sclerophyll formation (Howard, 1959). Crown closure has been used for defining types (de Rosayro, 1959; Paelinck, 1958; Howard, 1970).

Swellengrebel (1961) followed a different approach in determining the volume of green heart (*Ocotea rodiaei*) in Guyana. He proved the correctness of the assumption that within each forest type the volume of *Ocotea* per unit area is constant. He did this by a linear regression between timber volume and the areas of green heart-bearing forest types (from photographs at a 1:30,000 scale). In order to obtain a 1-year production for a mill, 10 photo-blocks of 810 ha each had to be utilized. The error then equals about 1.5 percent. Swellengrebel pointed out that the accuracy of photo-volume estimates is low for smaller areas. This creates difficulties in the decision about the standard, and therefore the cost, of road construction.

Swellengrebel's method will save time-consuming counting and measuring of individual upper-story crowns. Only an area determination must be made after interpretation is completed. All methods of volume determination on photographs, however, involve field enumerations of total volume and commercial volume and are hampered by difficulties in finding the exact plot or area location on the photo. It must be concluded that, despite high quality photographs, only rough estimates can be made of relative height classes and broad crown closure classes in tropical forests.

Other Remote Sensing Techniques

Viksne, Liston, and Sapp (1970) and Sicco Smit (1971) reported on the use of side looking airborne radar (SLAR) for forestry purposes (for a more detailed discussion of radar systems please refer to Chapter 9). A great advantage of radar is that operations can be started and finished on schedule regardless of the weather. The first authors describe briefly the mapping of vegetation over an area of 17,000 km² in Panama. Coverage with overlap was obtained in approximately 4 hours with a YEA-3A aircraft with a ground speed of 350 knots. The AN/APQ 96 SLAR, which operates in the K-band, was chosen for this area, as near-perennial cloud cover limits the application of aerial photography. Because the K-band does not penetrate vegetative cover it enables the evaluation of various vegetation types based on their radar-return characteristics, according to Viksne, Liston, and Sapp (1970). Smit stated that only very broad regions can be delineated in Colombia from radar imagery (scale approximately 1:220,000) of the area between Turnaco, Barbacoas, and Guapi, covering approximately 1,250,000 ha. The following regions, based on physiographic differences, were interpretable on the radar imagery: (1) coastal zone influenced by the sea; (2) alluvial plains and low terraces subject to inundation; (3) terraces intersected by low hills; (4) high hills and high plains. It is emphasized that only subsequent interpretation of small-scale aerial photographs (in Colombia, approximately 1:40,000), visual aerial reconnaissance, and knowledge of the local vegetation, permitted further subdivision of these regions into 2 to 5 vegetation types.

For Nicaragua, information on 1:250,000-scale radar imagery, flown in 1971 by Hunting, was given by Francis (1972). The Westinghouse AN/APQ 97 radar, operating in the Ka-band, reveals similar units to those described above, but also showed *Pinus caribeae* stands to be darker than the other units. It was possible to distinguish three density classes in the pine stands. For areas larger than 15,000 km², radar was less expensive than black-and-white photography. A disadvantage is that a large aircraft is required.

Perhaps the most ambitious use of radar in tropical forest-land inventory is in Brazil. When completed, project RADAM (Radar Amazon) will have acquired radar imagery of nearly 5-million km² of the Amazon Basin. Radar mosaics at a scale of 1:200,000 will have been produced from the Goodyear, synthetic-aperature, SLAR images. Brazilian scientists interpret the imagery and conduct ground investigations in order to produce maps of the geology, geomorphology, hydrology, vegetation cover, soil types, and land-use potential of this vast area. These maps will be used to select priority areas for more detailed remote sensing and ground survey studies (Azevedo, 1971).

Sicco Smit (1978) conducted a case study of the Mahogany forest region, state of Goias, Brasil in 1971–72 using SLAR imagery. He found that forest and non-forest areas could be accurately delineated but the Mahogany forest types could not be differentiated from non-Mahogany, either by tone or physiographic position.

The Centre Technique Forestier Tropical (CTFT) of France has, in cooperation with another French organization (Geotechnip), carried out experiments with 1:5,000- and 1:10,000-scale full-color, and color-infrared aerial photographs to test whether or not individual trees in the upper canopy of tropical rainforest in Gabon and Cameroon could be identified. The identification of commercial species, such as *Cynometra hankei, Erythophleum invorense, Lophira alata* and *Pycnanthus angolensis* and others, proved to be disappointing because of the polymorphic structures of their crowns. The phenological variations of the species, which take place over a long period–the first young red leaves of Azobé, for example, appear in January and the last ones in April–are important factors in the failure to identify individual trees of these species (Lanly, 1970).

The International Institute for Aerial Survey and Earth Sciences (ITC) in the Netherlands in cooperation with KLM Aerocarto carried out similar test flights in October 1970 in Surinam, operating four films simultaneously: panchromatic Aviophot Pan 30 Agfa, Kodak black-and-white infrared, full color CN17 Agfa, and Kodak Ektachrome Infrared. Four Hasselblad 70 mm cameras (15 cm focal length) mounted in one frame were used, providing 1:5,000 and 1:10,000 scales (see color Figure 34-27). Like the French experiment results, however, phenological variations between individuals of a species hampered recognition in this experiment. To date, only two species have been identified, both having no commercial value, on full color and color infrared film only. In both cases color and crown texture were the important features for recognition.

Unsuccessful attempts to distinguish tree species in Fiji, particularly *Agathis vitiensis,* on color and infrared photographs are mentioned by Rees (1972), even though the compact arrangement of cells in the *Agathis* leaf should produce a different reflectance pattern. Black-and-white negatives and film positives of each of the layers of the color films, as well as prints from various combinations of these film negatives and positives, were evaluated.

Photographic color prints of 1:50,000 scale of color infrared film covering an area in Sumatra (Indonesia) made by KLM Aerocarto show that types such as mangrove, swamp forest, and dry-land forest are easily recognizable.

The application of Landsat imagery for forestry in the humid and arid tropics is reported by several authors. Landsat imagery, just as conventional aerial photography, is strongly influenced by the unfavorable weather conditions in the tropics. The frequent coverage of the same area by Landsat, however, permits the study of at least some areas in several countries such as Thailand, Cameroon, the Philippines and others which may not have good coverage. Temporal coverage, even with Landsat, may not be possible in these areas.

For the following data, extensive use has been made of an unpublished report of Baltaxe (1979): In Thailand, Landsat MSS imagery was used in the form of 1:1,000,000 diazochrome additive-color composites and 1:500,000 black-and-white images of bands 4, 5 and 7 and of bands 5 and 7. Together with additional information from the field and from aerial photographs, maps made from the Landsat images were used for determination of the total forest cover. Comparing this information with forest-cover data either from different dated aerial photographs or Landsat imagery, permitted a rough calculation of the reduction of the forest cover over large areas, at relatively low cost (Morain and Klankamsoon, 1978). Also, computer print-outs of Landsat imagery covering the years 1972 through 1977 were used for determining the expansion of shifting cultivation area in northeastern Thailand. Information from additional 1:20,000 to 1:60,000 scale aerial photographs on shifting cultivation, irrigated rice, hill evergreen forest and other forest types grouped together helped to compile a geometrically rectified map. Mapping of the difference values for MSS band 7, displayed by assigning different grey levels to various levels of difference in tone (scene brightness), permitted detection of shifting cultivation at one year intervals. The difference maps of MSS band 5 showed where permanent agriculture was encroaching on the forest (Miller, et al, 1978). Computer-processed Landsat-1 data, using principal components analysis were successfully used for mangrove-forest recognition.

In the Philippines, 1:1,000,000 scale black-and-white prints of Landsat band 5 served as an index and a guide for terrain and survey locations. This information supplied spectral signatures for the computer classification. Thirty thematic color-copies, scale 1:500,000, and area tabulations by forest classes were obtained in this way for closed and open dipterocarp forest, pine forest, mossy forest and mangrove forest (Lachowski et al, 1978 and 1979).

In Bangladesh, computer-processed Landsat data, with additional data from 1:30,000 scale aerial photographs, permitted two mangrove species to be distinguished at a 71 percent accuracy-level.

From the above studies, it can be stated that Landsat imagery for forestry in the tropics is useful as a tool only in reconnaissance surveys, a limitation which is partially due to the nature of the tropical forest itself. Area information on mapped Landsat data concerning forest types cannot be considered to be reliable. Its use should mainly be seen as an indicator of the presence, absence or change of total forest area. This information is of high value and is obtained at relatively low cost. This is a significant contribution in addition to that of aerial photography. The latter tool facilitates general vegetation mapping, the location of timber, and layout of roads, the determination of distances to mills and outlets and the provision of other information essential to tropical forest land-management.

DAMAGE ASSESSMENT

The use of remote sensing in the detection of the effects of damaging agents on a forest precedes most other remote sensing forestry uses. For example, in 1921 F. C. Craighead found slow-moving seaplanes useful for visually assessing the extent of damage by a defoliator, spruce budworm (*Choristoneura fumiferana* Clem.), to spruce and fir stands in the northeastern United States and Canada (Eaton, 1942). In 1925, F. P. Keen flew in Army biplanes over remote parts of California to photograph widespread infestations by the western pine beetle (*Dendroctonus brevicomis* Lec.). Both Craighead and Keen were prominent entomologists who foresaw the value of viewing vast areas in a short period of time.

In the United States, insects and diseases account for a timber loss equal to our annual growth, and this loss exceeds that from fire by seven times (USDA, 1965). Elsewhere, the effects of these two agents are almost as devastating. From 1963 to 1965, for example, 40 percent of the pine resource in Honduras was lost to an epidemic of southern pine beetle (*Dendroctonus frontalis* Zimm.) (Beal et al, 1964). Because these damaging agents are dynamic forces, entomologists and pathologists find that remote sensing techniques are most valuable when they are used at critical periods of stress, and when the imagery or data output can be made available to them quickly.

The most commonly used sensors are: (1) eyes for direct aerial observation; (2) aerial photography (at various scales and primarily with color or color infrared films); and (3) multispectral scanners. The scanners have proven useful for fire mapping but have not yet proven successful for the detection of early stages of stress.

An account of the early experimental work in damage assessment is found in the *Manual of Photographic Interpretation* (1960). In the last decade, many improvements in quality of films, interpretation techniques, equipment, and operational procedures have been made.

A comprehensive guide to the recognition of forest-damage syndromes on aerial photographs in Canada has been completed by Murtha (1972a). Forest damage is defined as any type and intensity of an effect, on one or more trees, produced by an external agent, that temporarily or permanently reduces the financial value, or impairs or removes the biological ability of growth and reproduction. Four basic damage types are recognized on the basis of their appearance on aerial photographs: (1) trees that appear completely defoliated; (2) trees that show some defoliation through the presence of bare branches; (3) trees that show foliage as a color inconsistent with the normal foliage color of the species; and (4) trees that show no visible sign of damage, but have a deviation from their normal reflectance pattern in the non-visible light range.

One damage-causing agent may produce a number of damage syndromes and, conversely, a damage syndrome may have been caused by any one of a number of agents. Hence there is always a need for ground checking and for good background references (e.g., Davidson and Prentice, 1967) before photointerpreters can make reliable estimates of the nature of the damage-causing agent. Manuals, such as that by Murtha (1972a) and photointerpretation keys (Figure 34-28) are aids to successful photointerpretation.

For economic reasons the operational application of remote sensors for forest protection purposes in Europe is more limited than in North America (only in the USSR is the situation comparable). Intensive forest land management and easy access to forest land areas eliminate the need for more costly remote sensing methods and techniques. In Scandinavia, where forests are managed less intensively, forest protection generally is not a very serious problem. In most of the other European countries forest lands are too small or too widely scattered for economic aerial control systems.

FIRE CONTROL

Although aerial photographs have been used in fire control since the late 1920's, there is very little information in the literature regarding their use. In 1941 the use of aerial photographic techniques for fuel-type mapping was described, and many of the U.S. Forest Service's regional fire plans and planning handbooks have called for techniques using aerial photography intelligence gathering, planning, and damage assessment during and following fires. Quick access to information on prints from the Polaroid Land Cameras was recognized early: "Aerial pictures can be taken with this camera and dropped to the fire camp in a matter of minutes," (Johnson and Thomas, 1951). Yet, Arnold (1951) felt that few fire-control investigators recognized the potential of aerial-photo applications.

Late in 1961, the Northern Forest Fire Laboratory of the U.S. Forest Service, at Missoula, Montana, began a fire-detection research program; the primary objective was to develop a system capable of detecting fires, both man-caused and lightning-caused, day or night, through all normally encountered atmospheric conditions. At the time this research was begun, a number of remote sensing techiques were explored, primarily the use of electronic devices. Hirsch discussed the use of infrared, radar, and television. Results of the project indicated that the best approach was to use airborne thermal IR line-scanning devices. Research since that time has been confined to the same techniques and progress has been reported periodically (Hirsch 1963, 1965, 1968, 1971; Wilson et al, 1971).

Shortly after Project Fire Scan research began, experiments with IR scanners were started in Canada, primarily using equipment developed by the Computing Devices of Canada. The results of the development of the AFDS-2 scanner have been reported by Williams and Kourtz (1965), Williams (1966), and Felton (1967).

Pre-fire Planning

A successful fire management program is dependent on information gathered and action taken before a fire starts. Pre-fire planning requires: (1) an inventory of existing facilities and features, such as firebreaks, heliports, transportation systems, and timber and fuel types, (2) knowledge of special hazards, land and resource values, special problem areas such as wilderness areas, and weather and storm patterns; and (3) a record of major ridges, canyons, and other topographic features which have a tactical value for line construction purposes, plus natural fuelbreaks and barriers.

The most commonly used operational remote sensing tool for fire management is the 1:15,840 vertical, panchromatic minus blue aerial photograph. In recent years, developments in photographic techniques, and improvements in camera equipment and film, have led to more and more use of low level and oblique photographs, both panchromatic and color (Cochrane, 1970). Timeliness and availability are sometimes a problem and, although not specifically referring to fire use, Klein (1970) suggested the use of 35-mm aerial photography as a substitute for larger photographs.

Fire Detection and Mapping

The function of a fire detection system is to find fires before they get large enough to cause serious suppression problems. Hirsch (1964) gave the attributes of an ideal detection system: detection of a fire the instant it starts; effective operation day and night, and despite low visibility caused by smoke, fog, or dense timber cover; and ability to distinguish potentially dangerous fires from those of no concern to fire control forces.

The most important output of a fire mapping system is a picture of the fire edge in relation to such ground features as ridgetops, valley bottoms, streams, and prominent landmarks. Such pictures must show sufficient detail to enable the interpreter to determine the precise location of the fire edge, hotspots, and spot fires in relation to fuel-type changes and fuel breaks.

Properly interpreted, recent photographs can satisfy most requirements for fire management. However, remote sensing systems that operate in the visual portion of the spectrum are only efficient during daylight hours when the atmosphere is clear. As a result of tests in 1964 on a prototype

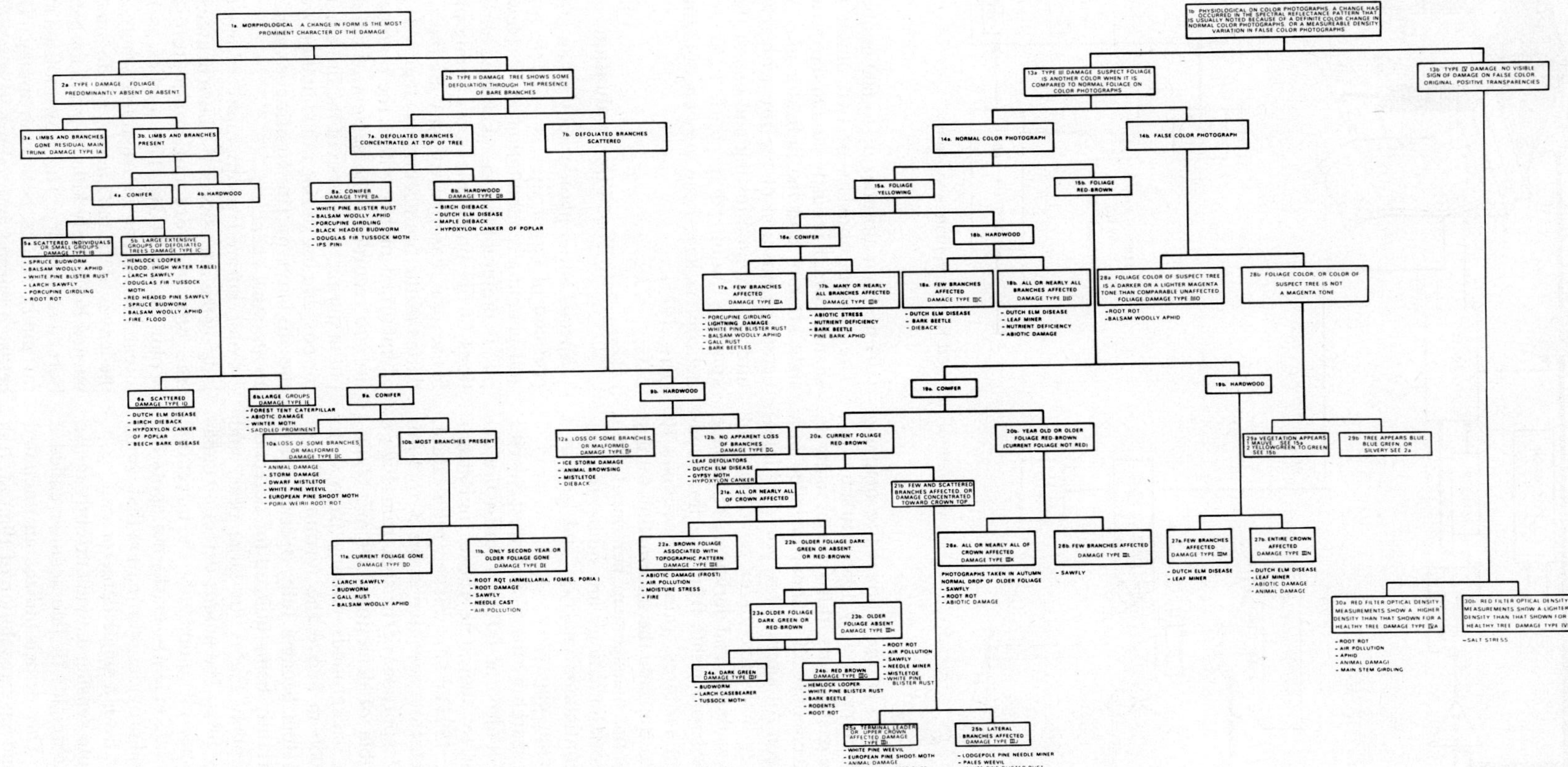

Fig. 34-28. Key to air-photo interpretation of forest damage (Murtha, 1972).

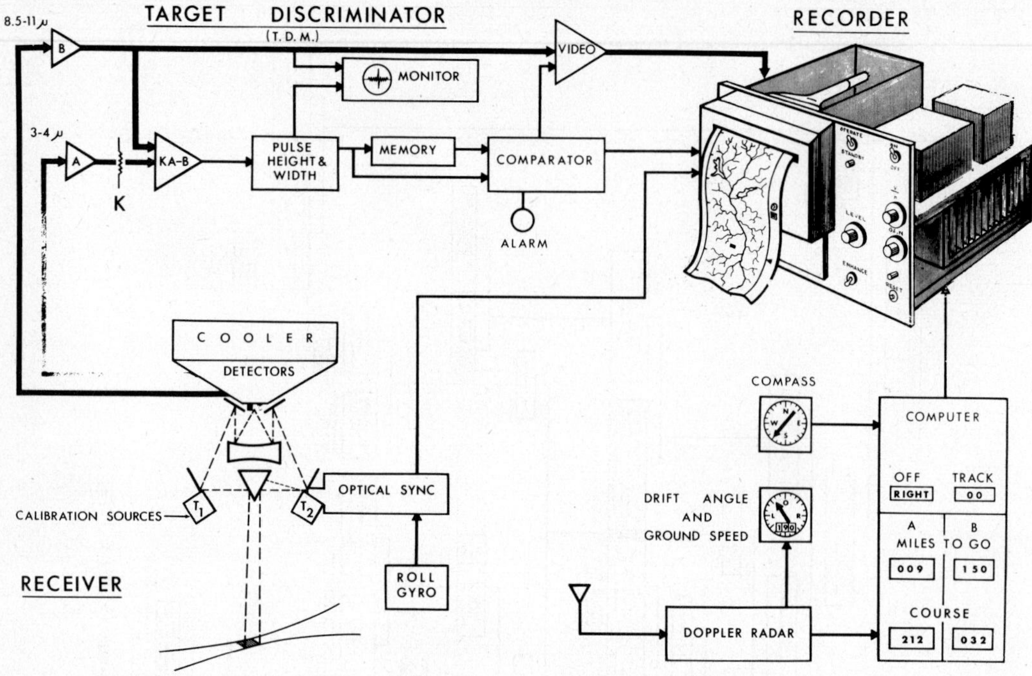

Fig. 34-29. Block diagram of infrared fire-surveillance system.

thermal IR system, an operational fire mapping system was developed in July 1966 (Hirsch et al, 1968). Over the past 5 years, this system, using an HRB-Singer Reconofax XI scanner installed in a Beechcraft Queen Air aircraft, has provided intelligence on over 300 fires under conditions of smoke, limited visibility, and darkness. The image produced on Polaroid prints from the line scanning equipment requires trained specialists to interpret, and to translate the information to map form usable by fire management personnel. This system's greatest limitations are nonrectilinearized and noncontinuous imagery, and lack of a small-fire detection capability.

Continued research by the U.S. Forest Service has produced an airborne infrared forest fire detection system to overcome these limitations (Hirsch et al, 1971). This bispectral system, utilizing two detectors sensitive in the 3- to 4-μm and 8.5- to 11-μm portions of the spectrum (Figure 34-29), is capable of detecting 0.09 m^2, 600°C. targets against backgrounds ranging from 0° to 50°C. from 5,000 m above the terrain (Figure 34-30). This system became fully operational in 1971 and is currently being utilized for detection patrols, fire-mapping flights for large fires, and mop-up on large fires. The unique detection capability of this bispectral equipment—the target discrimination module (TDM)—marks hot spots too small to print and provides information that would not be available from a single detector unit.

Another remote-sensing device currently in use is a small, light-weight, IR line scanner called a "fire spotter." This device utilizes an uncooled InSb detector and has a resolution of 10mrad. It is capable of detecting fires with areas as small as 0.09 m^2 from altitudes up to 600 m above terrain. The system consists of two units: the optical scanner which is easily mounted externally on any light aircraft, and the indicator unit, which indicates the position of the fire with respect to the aircraft. Limited testing and use of the small fire spotters mounted on both light aircraft and helicopters have been very encouraging (Kruckeberg, 1971).

Post-Fire Evaluation

Special aerial photography taken after the suppression of larger fires is valuable for mapping, and planning rehabilitation work. New photography is preferable to the sketching aerial observations of fire boundaries on existing photography for several reasons. Not only are the actual boundaries of the fire accurately depicted, but islands of unburned timber within the boundaries of the fire are recorded. The system of bulldozer trails put in during the course of the fire can be mapped for possible use during rehabilitation work and for the division of fire-fighting expenses among cooperating agencies.

Color infrared photography is most useful in judging the intensity of the burn in areas where the tree crowns were not completely consumed (Color Figure 34-31). This applies particularly to forest fired areas in Australia where commonly the leaves of the tall *Eucalyptus* dominants are not burned but merely killed by radiant heat from the burned understory. The crown foliage may remain for many weeks before dropping. Differences between burned and unburned Australian forests are often difficult to detect on black-and-white ae-

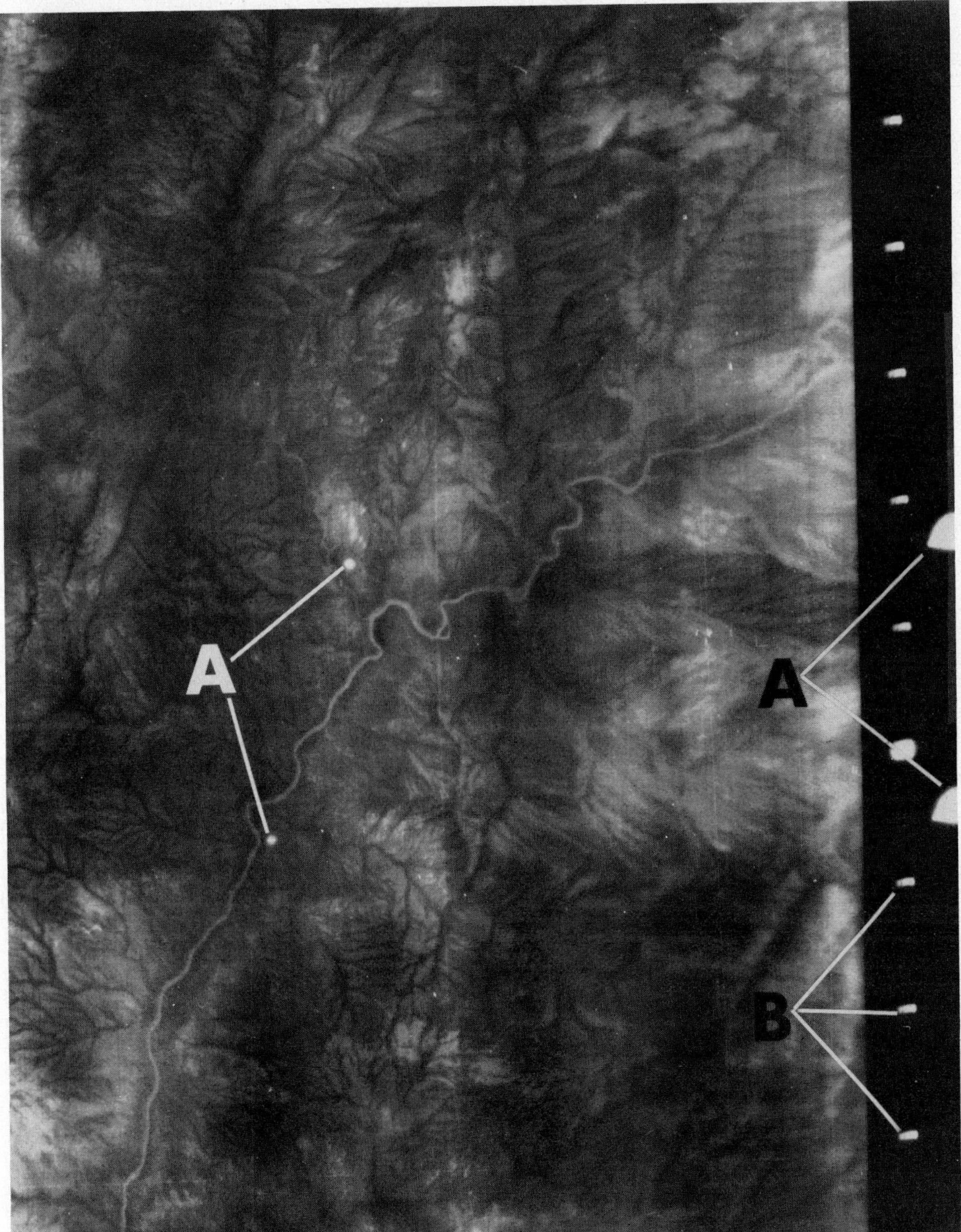

Fig. 34-30. Image produced during a high-altitude patrol flight for detecting small fires. (A) Marks placed electronically on the image to indicate targets too small for print. Fires are to the left of the marks. (B) Doppler radar-signal marks, 1 nautical mile apart.

rial photographs. Assessment of intensity of burn or severity of damage is even less satisfactory. Contrasts are much more obvious with color photography and strikingly dramatic with color infrared photos (Benson and Sims, 1967).

In a project evaluating the impact of fire suppression on the vegetation in the San Bernardino Mountains of Southern California, wildfire damage to chaparral, woodlands, and conifer forest during the Big Bear Fire (November, 1970) was interpreted from color infrared aerial photography at a scale of 1:20,000 (Minnich, 1974). The long-term change or persistence of certain vegetation types could be analyzed because the area had been floristically mapped from imagery flown prior to the burn (Minnich, et al, 1970).

Photographs from Gemini and Apollo earth-orbiting spacecraft have shown forest fires. In remote areas, such as Cape York Peninsula in Australia and in Northern Hemisphere boreal forests, these sometimes gave the first evidence of forest fires. LANDSAT-1 imagery provides similar information very rapidly, and has proved an accurate source for mapping the extent of forest fires and for evaluating the degree of damage.

An analysis by Lauer and Krumpe (1973), comparing a computer-generated map of a burned area from a LANDSAT-1 image with a ground survey operational map, showed a 10 to 1 cost advantage in mapping in favor of the LANDSAT-1 imagery. Equally important, much higher accuracy of actual damaged area was obtained from the LANDSAT-1 imagery. There was a 12 km^2 difference in the two estimates amounting to a 22 percent error in burned area measured by conventional ground methods. The computer map also recorded considerable areas of unburned forest within the fire perimeter that were overlooked by ground-crew mappers.

INSECTS

Forest insects cause symptoms of tree and forest injury which are more easily recognized than those caused by forest diseases or air pollution. For example, defoliators of coniferous or hardwood trees frequently cause the foliage to change color from a normal green-yellow to yellow or dark yellow-red. These changes are readily visible, occur over large areas, and can be mapped by direct observation or with color aerial photography.

Bark Beetles

Bark beetles of the genus *Dendroctonus* kill more mature coniferous trees each year than any other group of forest insects in the United States. These beetles are predominantly in pines (*Pinus* spp.), but also occur in spruce (*Picea* spp.) and larches (*Larix* spp.). In most cases, the beetles attack the largest and most valuable trees and kill them by sheer force of numbers—several hundred to several thousand beetles may attack one tree. Usually a symbiotic fungus carried by the beetles is deposited in the beetle galleries. This fungus grows and causes occlusion within the water-conducting tissues of the tree. The combination of girdling by the beetles and occlusion of the wood by fungal hyphae causes a rapid decline in tree vigor and eventual death.

Because many trees are attacked at one time and begin showing signs of stress by changes in foliage color, they can be differentiated from healthy trees by remote sensing methods. Often a unique pattern of tree killing permits an interpreter to follow successive attacks by the beetles through the changes in foliage color. The pattern of newly dying pines (yellow) is often adjacent to older infestations (yellow-red), which may abut still older infestations (gray trees with no foliage); all infestations may be surrounded by green and green-yellow healthy trees. Color and color infrared photographs are particularly useful in locating such infestation centers.

Heller et al (1959) reported that color film was a more accurate sensor than panchromatic or black-and-white infrared films for detecting southern pine-beetle infestations in the southeastern United States. Ciesla et al (1967) used color infrared film for evaluating southern pine-beetle infestations at a scale of 1:3,960. They found color infrared to be superior to color because it separated hardwoods from conifers and also penetrated the atmospheric haze better than color film. Since 1967, the Southeastern Area of State and Private Forestry (Forest Service, U.S. Department of Agriculture) has been photographing up to 20 epidemics per year by using stereo pairs of 22.8 cm format transparencies in a sampling scheme which covers from 7 to 10 percent of epidemic areas.

Ciesla (1977) has aptly summarized the optimum uses of color and color infrared films for forest insect damage in the United States. He identifies which film to use for different insects, forest types and atmospheric conditions.

Since the mid-1960's, the mountain pine beetle (*Dendroctonus ponderosae* Hopkins) has been in epidemic status in the northern Black Hills of South Dakota and killed thousands of ponderosa pine trees (*Pinus ponderosa* Dougl. ex Laud). In film-filter studies conducted by the Pacific Southwest Forest and Range Experiment Station (Forest Service, U.S. Department of Agriculture) under the sponsorship of the Earth Resources Survey Program, NASA, Heller et al (1969) reported the following: (1) Color or color infrared (CIR) films taken at large scale (1:1,584) discriminated healthy from stressed trees only when discoloration had begun. Neither film at any scale could be used as a previsual sensor of stress of ponderosa pine. (2) For areas similar to the Black Hills, where one generation of beetles emerges per year, correct interpretation of newly killed trees on color or CIR transparencies most often occurred during periods of maximum discoloration—August to October. Although photointerpreters using May imagery discovered only 40 percent of the known dying trees, they were able to detect three times as many dying pines as the ground observers. Apparently, the slight yellowing tinge to the foliage was registered on the color films, but could not be discerned visually. (3) Small-scale (as small as 1:174,000) color and CIR aerial photography was useful for detecting large infestations during periods of maximum foliage discoloration.

At how small a scale can one detect infestations of various sizes? By relating known infestation sizes with resolution capabilities of sensors used on high-flying aircraft, it is possible to arrive at expected accuracy levels. On a study area measuring 1.6 × 5 km in the Black Hills National Forest, the infestations and trees were distributed as shown below:

Size Class (Number of discolored trees per infestation)	Average Largest Dimension m	(ft)	Number of Infestations	Percent of Total
1–3	5	(16)	109	52
4–10	13	(43)	49	23
11–20	24	(80)	27	13
21–50	45	(149)	16	8
51–100	53	(175)	3	1
100+	122	(403)	7	3
			211	100

Color and CIR aerial photographs were obtianed at six scales (Color Figure 34-32), ranging from 1:7,920 to 1:174,000. Careful photointerpretations were made by three specialists, beginning with the smallest scale and progressing to the largest (Figure 34-33). Student's "t" test showed no significant difference in the two films, and, as might be expected, detection success fell off as the scales and images became smaller.

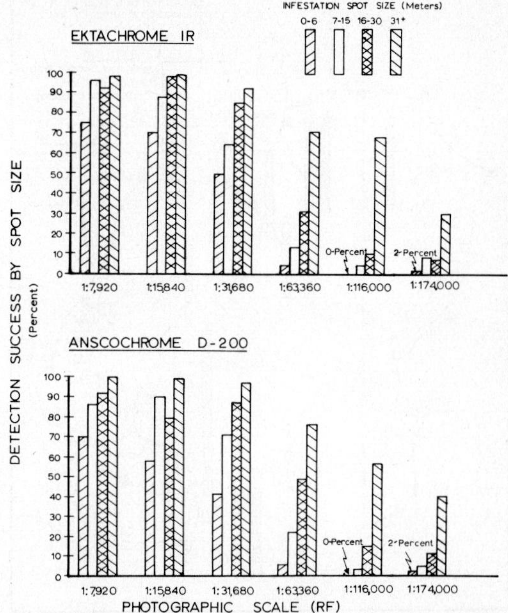

Fig. 34-33. Detection successes (mean of results from three photointerpreters) expressed as a percent of four infestation size classes, six photographic scales, and two films. Detection success was high on 1:31,680-scale transparencies—even for infestations 7 to 16 meters in diameter (average of 4 to 10 trees). It dropped considerably at scales smaller than 1:31,680.

These results suggest that small infestations, 1 to 3 trees (5 meters), are not likely to be detected with acceptable accuracies when scales are smaller than 1:31,680. However, infestations of 4 to 10 trees (13 meters) can be detected at scales around 1:100,000 with an accuracy of about 20 percent. For determining stress in the early stages of an epidemic, good resolution is needed (approximately 5 meters). For detecting these small, low-contrast targets (one or two pines that are just changing color from a green-yellow to yellow), scales of 1:8,000 are needed.

Landsat 1 was launched near the end of the above mentioned Black Hills study which permitted us to determine the likelihood of detecting large bark beetle infestations on satellite data (Heller, 1975). Both photo interpretation of false color images and computer-aided processing methods failed to detect several yellow-red beetle infestations that were over 300 meters in their longest dimensions. Only when infestation areas covered large contiguous areas—such as entire watersheds (e.g. Glacier National Park in Montana, 1979–80), could Landsat data be used as an aid to forest-pest management. Early detection of bark-beetle infestations by present Landsat sensors is unlikely because wavebands are too broad to help vegetation-damage analysts discriminate between healthy and dying vegetation (Heller, 1978). For example, most vegetation under stress begins to appear chlorotic, i.e. yellow to orange (yellow-red). These yellow-red wavebands (0.58 to 0.62 μm) are integrated into the top of the green Landsat waveband (0.50 to 0.60 μm) and the bottom of the red waveband (0.60 to 0.70 μm). This conclusion was also reached by Weber and Polcyn (1972) when using aircraft multispectral scanners. The Landsat D (thematic mapper) has a yellow-orange filter (0.58 to 0.62 μm) as an alternative waveband in the multispectral scanner system. Vegetation-damage analysts believe data from this waveband will enhance damage detection.

A multistage probability sampling scheme has been developed for evaluating stand damage. It incorporates visual observation, sample strip color photography (scale 1:8,000 on 22.8 cm format), and ground visits of timber losses. A study using this system was performed in the Black Hills National Forest and provided estimates with a sampling error of only 12 percent (Heller and Wear, 1969). A sketch map survey conducted in the fall of 1968 indicated the extremities of the outbreak and an increasing number of dying trees. This survey was used to select flight lines for photography. The outbreak area of 5,400 km^2 was gridded into 211 flight strips, each 1.6 × 16 km (1 × 10 miles) and numbered, and the number of infestations was totaled by flight strip. Strip lengths of 16 km were chosen instead of stereo pairs or random triplet locations because the strips simplified aerial navigation, reduced flight time, and permitted the film to be interpreted stereoscopically in the roll.

Because ground sampling of almost inaccessible infestations is costly, 30 samples were expected to be adequate. To estimate variances, a minimum of two ground visits per flight strip was needed; three per strip were decided upon because it is usually more economical to make several visits in one lo-

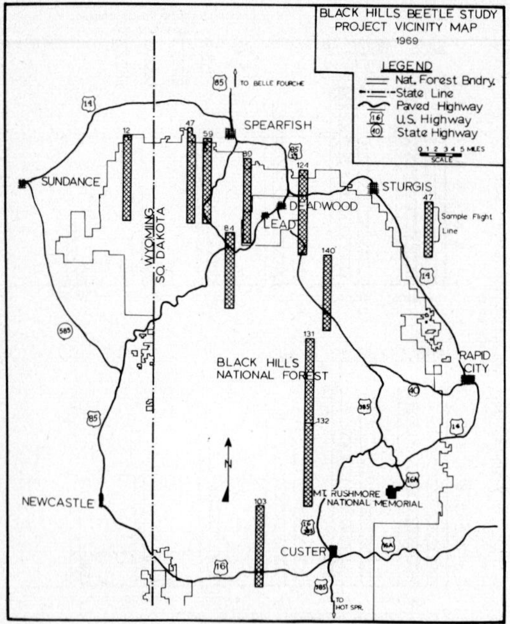

Fig. 34-34. Portion of Black Hills National Forest with locations of 10 flight lines, each 1.6 × 16 km. The lines were selected according to probabilities based on predictions made from a previous sketch-mapping survey.

cality than to move to different localities. Therefore, a total of 10 flight strips (Figure 34-34) were selected (three ground visits/strip × 10 strips) based on the probability that more strips with many infestations would be chosen than those with few infestations. Color transparencies (Anscochrome D-200) were obtained of the strips with a Zeiss RMK 21/23 at a scale of 1:8,000.

Infestations were plotted directly on the transparencies. Each infestation was numbered, the trees within it were counted, and these data were listed. The three infestations to be ground checked were selected with the probability of their being chosen based on the number of trees counted within each infestation. Thus, it was more likely that ground inspections would be made at large infestations than at small ones.

The number of killed trees and their timber volume were estimated from photointerpretation and ground visits. The estimates from the sample were expanded to the entire (whole) epidemic area. The sampling error was ±12 percent at one standard deviation.

Judging from results reported from this and other multistage probability surveys (Langley, 1969; Wert and Roettgering, 1968), this method is accurate and reasonable in cost. It provides estimates of timber killed and trend of epidemics, and is applicable for other bark-beetle surveys.

Another interesting technique, involving the use of 35mm aerial color photographs, has been used to evaluate the impact of the mountain pine beetle in lodgepole pine, (*Pinus contorta* Dougl.). Mortality, as discerned on photographs even by inexperienced interpreters, was highly correlated with actual ground counts (Klein, 1973).

The western pine beetle attacks primarily ponderosa and Coulter pine (*Pinus coulteri* D. Don.) in Oregon, Washington, Idaho, and California. This insect pest produces from two to three generations per year. In the four States, ponderosa pine is a valuable commercial tree species, contributing to the economy of many rural communities. Miller (1960) estimated that the average annual loss from the western pine beetle is one billion board feet, the equivalent of 100,000 five-room houses. Becaue of the high value of timber and the sporadic nature of epidemic occurrences, several investigators have explored the use of aerial color photography. Caylor and Thorley (1970) found that 85 to 90 percent of recent stand mortality could be detected on sequential color photographs (scale 1:5,000); early fading (crown discoloration) was detectable with greater accuracy than by ground observation. They recommended the use of two flights per year to produce accurate overlay maps of beetle-caused tree mortality and to indicate seasonal changes in insect activity. They emphasized that because green hues of living tree crowns vary, and may overlap initial stages of mortality discoloration, isolated single photographic flights must be considered inadequate for early detection.

Sequential color photography (scale 1:8,000) is being used to follow the effectiveness of a chemically synthesized beetle pheromone (sex attractant) at two locations in California (Bass Lake, 5,600 ha, and McCloud Flats, 13,000 ha). The pheromone is being used as a manipulative agent to attempt control of bark beetle populations. On such large areas it is impossible to count the hundreds of dying host pine trees by ground methods. Careful photointerpretations are being made after photography in June, August, and October to follow the impact of the sex attractant on the host tree population.

Sequential color photographs (scale 1:5,000) also were used effectively to record stand mortality in the Sequoia-Kings Canyon National Park (Thorley et al, 1965). About 2,500 hectares of the park were subdivided into 1,376 2-hectare units for interpreting tree mortality. Most of the tree volume lost was from a fir engraver (*Scolytus ventralis*). Sequential analyses of stand mortality trends were performed of this high-use recreational area for three successive years at a very modest cost.

Damage to Douglas-fir (*Pseudotsuga menziesii* (Mirb.) Franco) by the Douglas fir beetle (*Dendroctonus pseudotsugae* Hopkins) has been successfully detected and appraised by several investigators (Wear et al, 1966; Wert and Roettgering, 1968; Ciesla et al, 1971; McGregor, et al, 1972). Coastal Douglas fir from California, Oregon, and Washington provides one-fourth of the sawtimber cut for wood products in the United States.

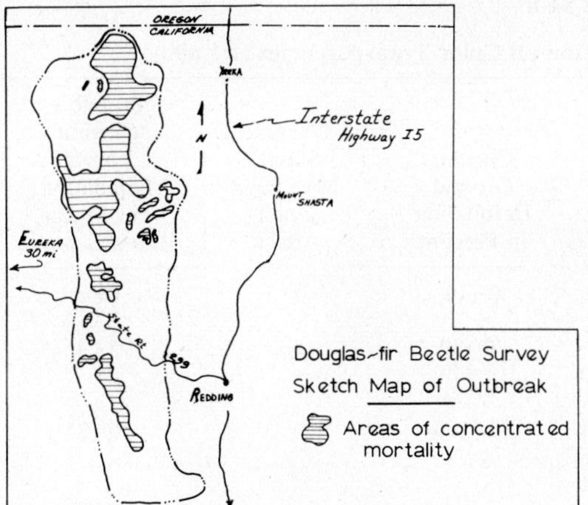

Fig. 34-35. Sketch map prepared from aerial reconnaissance survey of Douglas-fir beetle infestations. Outlined area represents approximately 12,200 sq. km. (4,700 acres).

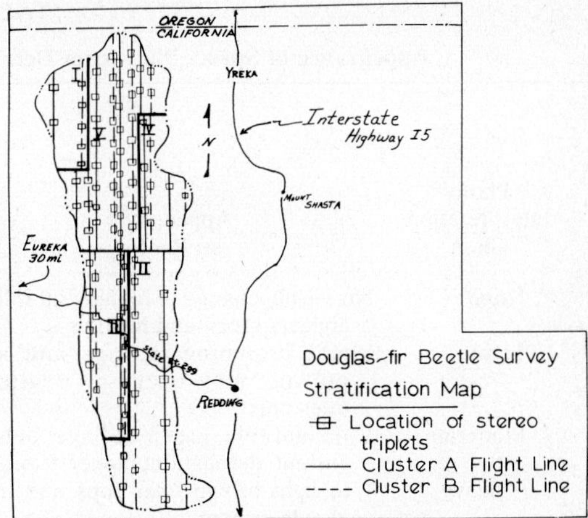

Fig. 34-36. The infested region in northwestern California was divided into five areas of Douglas-fir mortality concentrations (I & II-light, III & IV-medium, V-heavy) which were sampled at different frequencies. The heavy areas were sampled four times as much as the light, and the medium areas 2.66 times as much as the light. Stereo triplets and flight lines in each cluster were premarked on flight maps.

A survey conducted in the spring of 1967 in northern California deserves some elaboration because it was the first in which ground surveys and aerial photography were combined with a new probability sampling technique which has many other applications for aircraft and space imagery (Aldrich, 1971; Langley, 1969). After washouts from storms in 1964, Douglas-fir beetle populations built up rapidly and began attacking live trees in 1965 and 1966. A visual observation flight was made in November 1966, and the extent and seriousness of the mortality was roughly estimated. Foresters and statisticians examined the sketch map (Figure 34-35) prepared in a fall 1966 reconnaissance flight and established the sampling procedure for use in this stratified, two-stage, cluster sampling design. The first-stage units consisted of a systematic sample of aerial photographs with multiple random starts. The second-stage units were drawn at random, with probability proportional to number of dead trees as estimated from interpretation of the medium scale (1:8,000) stereo triplet color transparencies (Figure 34-36). The sampling error at one standard deviation was ±13 percent for count of dead trees, ±12 percent for gross volume mortality, and ±11 percent for net volume mortality. The benefit-cost ratio was 100:1 in favor of using this kind of sampling photographic survey over other methods.

A useful handbook was prepared for estimating damage by insects in western forests (Wear et al, 1966). Subjects covered include: (1) when to use the photographic method, (2) estimating sample size, (3) surveys to measure mortality trends, and (4) modifications for specific insect problems. Discussion of the use of double sampling with regression for stand mortality surveys is presented.

Optical-mechanical scanners have been used successfully for fire detection and mapping (Hirsch et al, 1968) and are described in more detail in Chapters 8 and 12. However, the scanning systems optimum for fire detection are not sensitive enough to detect stress in trees whose temperature seldom exceeds 5°C above normally healthy pines. In connection with the Black Hills photographic tests mentioned above (Heller et al, 1969), biophysical studies were carried out to see if it is possible to detect tree stress previsually and to relate ground observations to airborne sensors. To date, Weber and Polcyn (1972) have found that by using multispectral scanners flown at moderate altitudes (500 to 2,500 meters) and data processed on combinations of digital and analog computers they can separate healthy from discolored dying pines and old-killed pines, and pine from hardwood and rock outcrops. However, detection of dying, but still green, pines at acceptable levels was not possible. Part of the difficulty was the inability to combine the visible channels with the reflective and thermal infrared channels.

A new generation of multispectral scanners, which incorporate a common field stop (or common instantaneous field of view), has been developed and promises to improve signature recognition. Certainly the advantages of automatic interpretation by multispectral analysis offer an inviting prospect for detecting and mapping trees under stress which are invisible to the eye or which cannot be registered on aerial color films.

Defoliators

Defoliation of needles or leaves by caterpillars (family *Lepidoptera*) causes a thinning of foliage, usually accompanied by a discoloration from green to yellow to yellow-red (or brown). The spruce budworm, one of the most serious coniferous defoliators, feeds on the terminal shoots of

TABLE 34-8

Appearance of Spruce Budworm Defoliation on Color Transparencies (1:1,600).

Photo Interpretation Class	Appearance on Photos	Current Ground Defoliation in Percent	Years Stands have been under Attack	Probable Current Larval Population Per 38 cm Branch sample
None	No visible change in foliage; all foliage appears green and healthy	0–10	0–1	0–2
Light	Slight browning in tops and side crowns; may occur on scattered trees only	20–40	1–3	3–4
Moderate to heavy	Considerable green foliage behind current defoliation; appears orange to light brown over tops and most of side crowns	50–100	1–3	10–25
Severe	Entire crown appears gray behind dark brown color of current heavy defoliation; occasional to many dead fir trees in stand	50–100 (plus heavy previous defoliation)	3–5	5–15

balsam fir *Abies balsamea* (L.) Mill. and Douglas-fir branches to an extent which can be detected on large-scale aerial photographs (1:1,584) taken during early summer.

Interpreters viewing large scale color photographs usually can detect foliage damage more accurately than can ground observers. This is because the damage occurs in the tops of the fir trees and along the inverted cone-shaped crowns, and the vertical view shows more of the discoloration than the ground view (Color Figure 34-37). Studies conducted near Ely, Minnesota, by the Pacific Southwest Forest and Range Experiment Station show that defoliations as low as 15 percent of the total crown can be detected. Broad defoliation levels became more evident in terms of tree damage and budworm populations when separated into the classes shown in Table 34-8.

By using the criteria in Table 34-8 and combining measurements of tree height and crown closure percent of dead fir and spruce, Weber (1964) constructed a volume table for large-scale, 70-mm color photography. He used this volume table, and 49 photo and 29 ground plots to estimate that some 5.5 million cubic meters of spruce-fir timber were killed over 220,000 hectares in northern Minnesota in 1962. This technique of sample photography for arriving at volume estimates of mortality was time saving and much less expensive than ground methods.

The Canadian Forestry Service has also successfully employed aerial photography at a scale of 1:10,000 to estimate stand mortality due to attacks by spruce budworm. Four levels of damage are identified: (1) mortality over 50 percent; (2) 25 to 50 percent mortality; (3) 10 to 25 percent mortality; and (4) 1 to 10 percent mortality (Color Figure 34-38).

Aldrich and Heller (1969) were able to show the changes in a forest community resulting from high spruce budworm populations over a 10-year period by using large-scale (1:1,600) color transparencies. These provided a good assessment of changes in balsam-fir density and permitted identification of boreal tree species with high accuracy. Phytographs based on photointerpretation showed differences in species frequency composition, and crown cover. One phytograph (Color Figure 34-39) shows change in species occurring as a result of high budworm population.

Gypsy moth *Porthetria dispar* (L.) larvae feeding on hardwood foliage in the northeastern United States causes foliage thinning. When the tree species favored by the gypsy moth grows in pure stands, defoliations can be so severe that the trees appear to be in a midwinter condition. They have no leaves and can be identified quite accurately from the air in late spring. In mixed hardwood stands however, many nonfavored species grow, and several levels of defoliation may be present. Therefore, remote sensing evaluations of such stands must be made carefully and with more frequent ground sampling.

Although epidemic outbreaks of insects in Europe may be detected on the ground cheaply and early enough for control and salvage, there are some reports about operational application of CIR photointerpretation for stand damage inventories. Wolff (1967b) first studied damages caused by *Semania diniana* on Norway spruce *Picea abies* (L.) Karst. Another East German investigation of damages on Scots pine *Pinus silvestris* L. by *Diprion pini* (Wolff, 1970) showed that, compared to ground survey, better information and a reduction of work time of 28 percent was recorded. By use of the interpretation results, it was possible to correct a decision about chemical application and thus to save considerable costs. In addition, the mortality rate was accurately predicted. Wolff recommends the spectro-

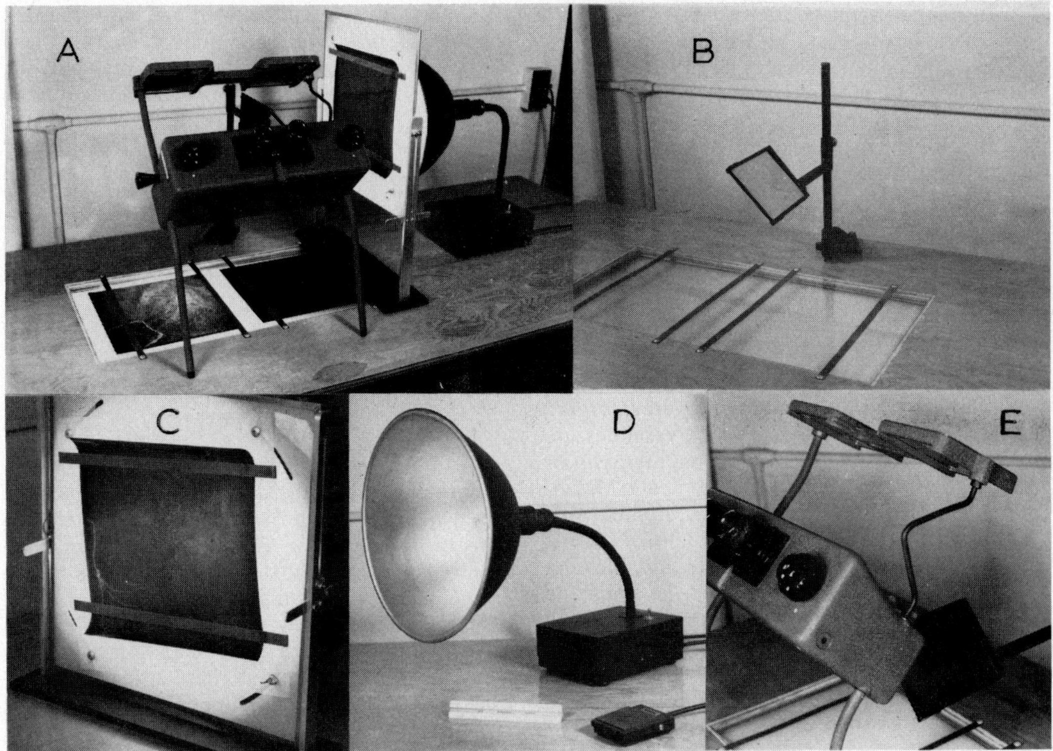

Fig. 34-40. A split-beam plot transfer device for use with an Old Delft Scanning Stereoscope; (A) the complete setup, (B) the split-beam mirror, (C) the photo easel, (D) the photo easel illuminator, and (E) the modified mirror arm. This device is useful for comparing successive years of aerial photography for detection of changes. By adjustment of the distance between (B) and (C), the two photos are easily superimposed and plot data can be quickly transferred from one photo to the other. (Photos courtesy of Forest Service, USDA.)

zonal film SN-6, scale of 1:5,000 to 1:7,000 for this purpose. It was advantageous to interpret sample units each consisting of 20 trees.

Sucking Insects

The balsam woolly aphid (*Chermis picea* Ratz), a European import, attacks the true firs (*Abies* spp.) on both the east and west coasts of Canada and the United States. Bajzak (1967), Pope (1957), and Heller et al (1966) have reported success in using color films to assess stem killing and gouting, which causes deformation of the top, twigs, and branches. Stem killing can be diagnosed on color transparencies at scales of 1:4,000 to 1:8,000. To detect gouting and top deformation accurately, large scales (1:1,200 to 1:1,600) are needed. Color film was preferred for studies in Maine and Vermont, and the timing of photography was determined to be best either just before or just after coloration of hardwood foliage in the fall. CIR film seemed best to Bajzak (1967) for diagnosing stem killing in Newfoundland.

Aldrich and Drooz (1967) followed the trend of balsam woolly aphid mortality over a four year period (1960 through 1963) on Fraser fir *Abies fraseri* (Pursh) Poir. in the vicinity of Mt. Mitchell, North Carolina. They sampled an 8,800-hectare area by using 215 0.4 hectare photo plots and 30 ground plots stratified on 1:7,920-scale aerial color photographs by mortality classes. Each year the 215 photo plots were transferred to new color photos by using a transfer device developed by Aldrich (Figure 39-40). From the photo and ground counts of mortality and live trees, regression lines were developed to adjust photo estimates made from the larger photo sample. In 1960, the first year of the survey, the adjusted estimate of dead Fraser fir per acre was 13.9 ± 0.8 trees (1 standard deviation). By 1963, the upward trend in mortality reached 21.3 ± 1.0 dead fir per acre (1 standard deviation). Since balsam woolly aphid damage is cumulative, equations were developed to compute the index of accumulated fir mortality on a rating scale of 0 to 4. This survey design was three times more efficient than a simple random sample of equal intensity.

Risk Rating Models

For effective forest or pest management, we need the ability to predict spatial and temporal changes in pest populations and resulting damage to the forest resource. The manager must be able to forecast where, when and to what extent a particular pest will cause damage. The development of risk-rating systems based on forest site and stand characteristics is a first step toward attaining these capabilities (Mika et al., 1980).

Many risk-rating systems have been developed and used by entomologists working solely on the ground. Risk-rating models, based on forest stand and site conditions derived by the interpretation of aerial photographs, have been developed by Sader and Miller (1976), Miller and Heller (1978) and Anderson (1981). In most cases, aerial photographs already available to the forest manager are used to determine the site and stand variables. For example, the variables usually associated with Douglas-fir tussock moth (*Orgyia pseudotsugata* McD) defoliation are: elevation, aspect, slope, topographic position, crown diameter, crown density and radiation index. These variables are all measurable on resource aerial photography or maps. A logistic probability-function called RISK (Hamilton, 1974) whose dependent variable was the presence or absence (1 or 0) of defoliation was identified with the independent site variables described above. When one enters the site and stand values for a forest stand into the RISK regression equation, the calculations produce the likelihood of a stand being defoliated, expressed as a percentage. This is similar to a probability statement made by weather forecasters on the likelihood of precipitation.

When the forest manager has this kind of information, he can decide whether or not silvicultural treatments are needed. Heller and Sader (1980) produced a handbook on the use of the probability model and the measurement of the independent variables on aerial photography.

High Altitude (Small Scale) Photography

Small scale (1:32,000 to 1:120,000 and smaller) color and color IR aerial photography became available to remote sensing investigators from the first NASA aircraft in 1965 and continues to the present time. Forest-insect damage can be mapped on these photos with certain restrictions. For example, Ciesla (1974) had access to U-2 color IR photos (scale 1:120,000) over western Montana where two defoliators (western spruce budworm, *Choristoneura occidentalis* Freeman and pine butterfly, *Neophasia menapia* (F. and F.) and also the mountain pine beetle were present. No spruce budworm infestations were detected and only severe pine-butterfly infestations were evident when pure stands of ponderosa pine had a grass understory. This small scale CIR photography was only partially effective in identifying forest stands suffering heavy tree mortality due to bark-beetle infestations. As noted earlier, Heller et al (1969) found either color or CIR photography at a scale of 1:32,000 sufficient only if the manager could overlook small mountain-pine-beetle infestations of one or two trees.

In 1979 and 1980, the U-2 aircraft of NASA's Airborne Instrumentation Research Project flew missions in Colorado and Wyoming with the Itek KA-80A Optical Bar Panoramic Camera. These missions were flown for the purpose of evaluating large mountain-pine-beetle infestations on the 1:32,000 Color IR photos. The panoramic transparencies were used to define sample locations for larger scale color photography (1:6,000–1:8,000) and ground inspections (Klein et al., 1980).

Befort et al (1980) describe how the forest manager may more effectively use the unconventional format (12.7 by 101 cm) Color IR photography. Attendant with use of small-scale photos are the efficiencies in cost of acquisition and need for fewer photos during the photo interpretation process.

DISEASE

Most visible symptoms of forest disease are evident only when the disease is far advanced in the host tree. As with insect damage, manifestations of disease show as discolorations and thinness of foliage. In most cases, symptoms of disease damage are less uniform than those of insect damage and frequently occur only in single trees. Several investigators have had some success using aerial photography to detect disease symptoms which occur in the upper portions of tree crowns.

Ash dieback is a disease of white ash *Fraxinus americana* L. in the northeastern United States which has plagued forest pathologists since the early 1940's. Although no causal agent has been identified, most pathologists agree that prolonged drought conditions appear to induce this and other hardwood declines. In a study near Poughkeepsie, New York, Croxton (1966) found that interpretations from color photographs (1:1,600 scale) could be used to determine whether ash trees were healthy, moderately affected, or dead. Accuracy was about 80 percent once ash had been separated from other similar-appearing hardwood species. The principal symptoms which aided diagnosis were changes in foliage color, absence of leaves, and reddening of dead branches (Color Figure 34-41). Photography scales smaller than 1:1,600 were found to be unfit for identifying ash.

Dutch elm disease *Ceratocystis ulmi* (Buisman) C. Moreau on American elm *Ulmus americana* L. causes dying and discoloration in the tops and terminal branches throughout the growing season. Meyer and French (1966) used CIR transparencies taken at a scale of 1:9,600 over the cities of St. Paul and Minneapolis, Minnesota, to plot diseased American elms along the city streets. They interpreted transparencies taken in early summer when the dying tops of affected trees are most apparent. They reported 75 percent success in correctly identifying infected trees and identified many trees not previously suspected of being diseased from ground inspection. In a photographic study to detect Dutch elm infected trees in Denver, Colorado, LaPerriere and Howard (1971) reported that CIR transparencies (scale 1:7,000) represented an important management tool for detecting and monitoring Dutch elm disease. They stressed that an investigator should be familiar with tree morphology and pathology, factors of climate and soil, landscaping practices, and patterns of urban development to do effective photointerpretation.

Oak wilt *Ceratocystis fagacearum* (Breitz) Hunt is caused by a fungus which occludes the water-conducting tissues of oak (*Quercus* spp.), primarily red oaks. The symptoms show up as dying-back of the top and discoloration and wilting of foliage. Aerial photography (color and CIR films) was used in 1960 over areas of wilting oak trees in eastern Tennessee by Roth et al (1963) at scales of 1:7,920, and 1:3,960. Color transparencies were superior to CIR for detecting oak wilt and a scale of 1:3,960 was more accurate than a scale of 1:7,920.

Other studies have confirmed the utility of detecting oak wilt using aerial photography (Meyer and French 1967, Latham et al. 1969, Ulliman and French 1977). Ulliman and French found that smaller scales (1:15,840 to 1:31,680) were as good for detecting infection centers and that color infrared was better at the higher altitudes required for those scales. True color was better, however, for altitudes below about 4000 feet. Commission errors (i.e. calling other features oak wilt that were not) were a serious problem, especially at the larger scales.

The effects of *Fomes annosus* Fr. (CKe) on conifers in North America resemble those of small bark-beetle infestations (*Ips* spp. and *Dendroctonus* spp.) and must be verified on the ground. Hanson and Lautz (1969) found they could identify trees in plantations killed by *F. annosus* on 1:4,000-scale CIR transparencies. The affected short-leaf pines *Pinus echinata* Mill. were 35 years old and 15 meters in height in thinned plantations in southern Illinois. At that scale, about one-third of the photo counts were commission errors. Smaller scales were even less accurate. Black locust *Robinia pseudoacacia* L. infested with locust leaf miner, other hardwoods, and other dead pines were confused with the dead shortleaf pines. To date, it has not been possible to detect *F. annosus* on either color or CIR transparencies before foliage discoloration occurs.

Weber and Wear (1970) made detailed biophysical studies of *Poria weirii* (Murr.) Murr.-affected Douglas-fir trees in 1969–1970 near Wind River, Washington, to learn whether spectral and temperature differences exist between healthy and affected trees. Multispectral line scanners (including thermal bands) were flown over the test site, and the data reprocessed into photo-like images. Temperature and spectral differences were too small for airborne scanners to be depended upon as a means of previsual detection. However, Wear (1971) has found that small scale CIR transparencies (1:32,000) were an effective means of detecting the unusually shaped openings in the fir timber occurring along the Cascade Mountains of Oregon. The ringworm-like circular patterns (Figure 34-42) were positively associated with the occurrence and spread of the root rot disease. Because some of the infection centers are so large and the pattern is so unusual, very small-scale imagery from high altitude aircraft or satellites may detect their presence.

Houston (1969) has used large-scale (1:1,000) color and CIR films over a 3-year period to evaluate beech *Fagus grandifolia* Ehrh. bark disease in the northeastern United States. Both types of color films detected foliar symptoms of the disease equally well. Houston suggested using both films simultaneously because more information was gained thereby. Color photography proved to be superior to panchromatic and color infrared in forest pathology studies monitoring *Dothistroma pinii* on New Zealand pines. Color photography is also superior to black-and-white for tracing outbreaks of the wasp, *Sirex noctilio* F., in pines in Australia and New Zealand. Recognition of the browning of the needles is easier on the color photos and can be detected sooner allowing more rapid treatment and control.

Fungus diseases in conifer stands in Central Europe usually cause greater damages and losses than do insect outbreaks. The detection of a very damaging disease to European conifers, the "red root rot" (*Fomes annosus*) has not been successful. In Europe, this fungus does not directly kill the tree or cause discoloration of the foliage. Fir tree rot or Hallimasch (*Armillaria mellea*) causes great damage to trees of all species and age groups. The appearance of damage does not show any characteristic crown symptom, though ailing or dead trees can easily be recognized on the aerial photo; the particular fungus itself, however, cannot be determined from the aerial photo. Wolff (1970) reported that in 10- to 12-year-old stands of Scots pine on SN-2 aerial photos (scale 1:4,000) 60 percent of the trees with foliage, but already dead, could be recognized.

Damages caused by *Cronartium ribicola* Fischer on *Pinus strobus* L. are easy to detect on medium scale color and color infrared aerial photos. In West Germany, besides healthy and dead crowns, two levels of damage to single crowns could be differentiated. On the basis of single crown interpretation, a spot infestation percent could be ascertained for the stand. The susceptibility of this tree species to infestation by fungi can even serve as an indicator of the species (Hildebrandt and Kenneweg, 1969). Another fungus, which can inflict great damage in young stands of the Scots pine is *Lophodermium pinastri* (Schrad.). Stands in moist sites protected from wind are especially vulnerable. In northern Germany the use of color infrared film showed that there is a strong correlation between damages and moisture (Kenneweg, 1971).

Beech stands *Fagus silvatica* L. are adversely

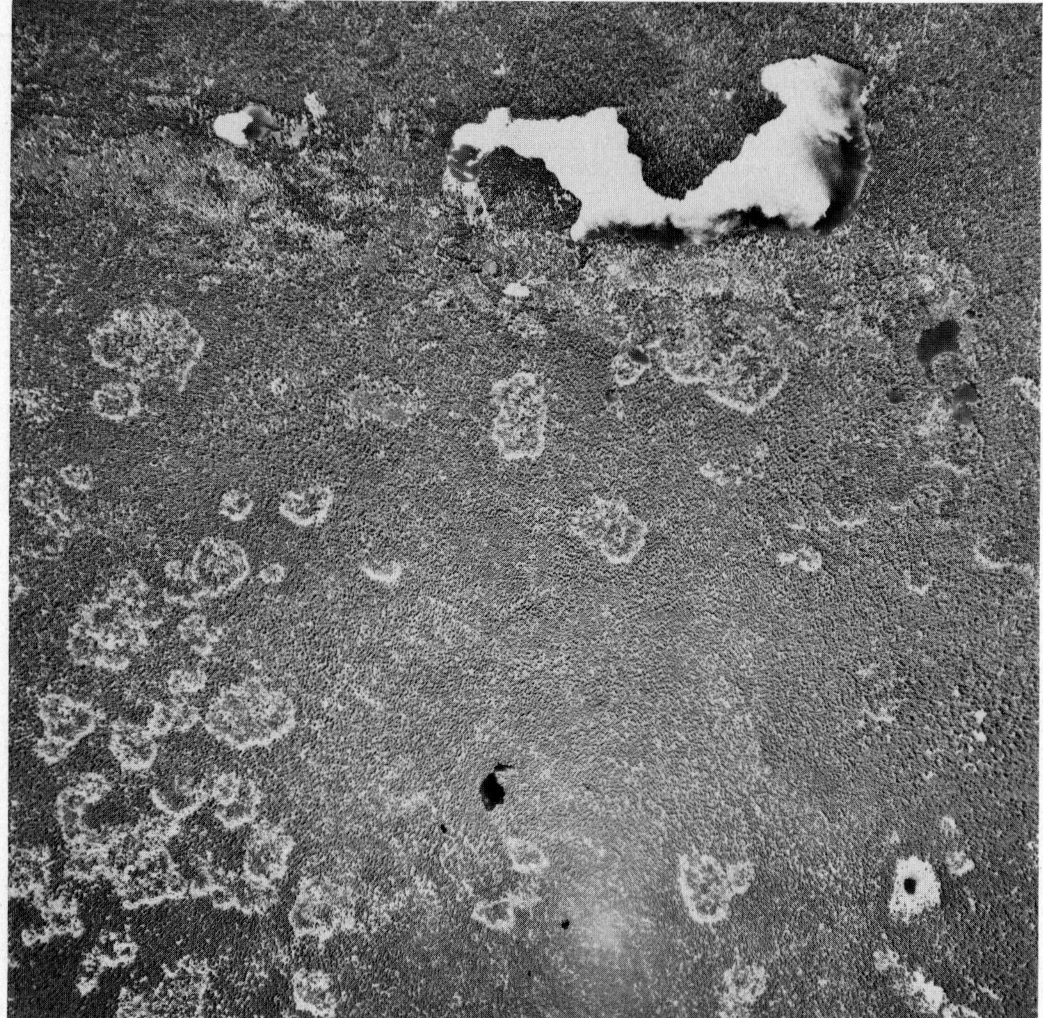

Fig. 34-42. Centers of large *Poria weirii* root-rot infections have been identified on 1:15,840-scale panchromatic aerial photographs. The area is near Waldo Lake in Oregon's Cascade Range. The "ringworm" appearance of the spreading infection was verified on the ground to be caused by P. weirri. (Photo courtesy of Forestry Service, USDA.)

affected following dry years by various fungi and bacteria species; the disease is designated as the "beech-bark-dieback." In the early infestation stage, a yellowish leaf discoloration takes place. In midsummer this phenological feature of the illness can be detected in mature stands on color infrared film on photo scales as small as 1:12,000 (Hildebrandt and Kenneweg, 1968a; Wolff, 1970).

Forest damage is not always sudden, initially catastrophic, or in need of early detection—some tree disease infection centers, for example, move extremely slowly. In such cases, monitoring can best be accomplished through long term sequential aerial photographic coverage, often of the kind normally provided for forest management purposes. An example of how conventional panchromatic 1:20,000-scale aerial photography of black spruce stands can be used to monitor the spread of existing dwarf mistletoe centers, and provide acceptable estimates of timber losses, is shown in Figure 34-43 (Meyer and French, 1966). The sequential mortality maps, together with local aerial volume table estimates of timber losses, proved to be the first index to the seriousness of this disease in the forests of Minnesota.

AIR POLLUTION

One of the most successful uses of color film has been for detection of air pollution injury to forest trees. Air oxidants, including ozone, fluorides, nitrous oxide, peroxacetyl nitrate (PAN), copper oxides, and sulphur dioxide cause foliage discoloration, dropping of leaves, and eventual death. Some of these symptoms are subtle and affect the same species of trees selectively; that is, some trees are more resistant to air-oxidant injury than others. An apparently healthy tree can be

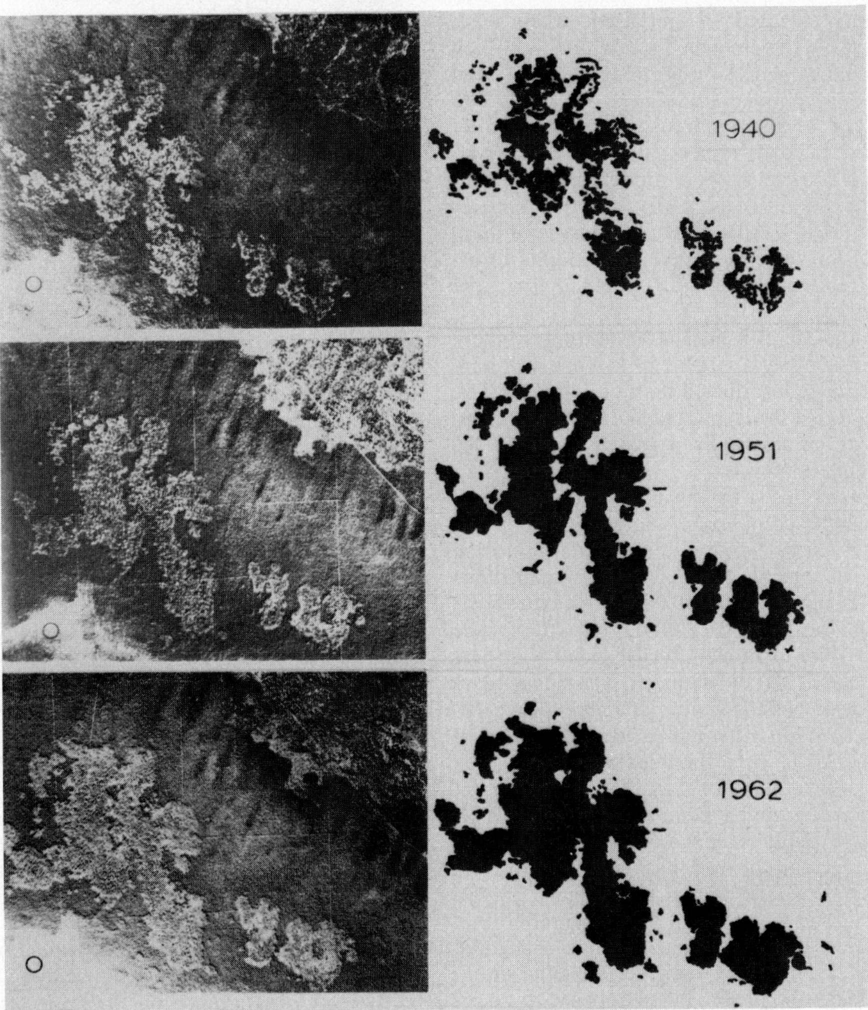

Fig. 34-43. Portions of sequential aerial photographs of a dwarf mistletoe-infected spruce stand in Minnesota. Progressive losses were determined on the basis of area and volume measurements made directly on the photographs. (Photos and drawings courtesy of School of Forestry, University of Minnesota.)

growing adjacent to a neighbor of the same species in its last stages of decline. Air-oxidant injury is frequently slight at first and can be easily overlooked on black-and-white films, but is detectable on color films.

Miller et al (1969), Heller (1969), and Wert (1969b) have studied the effects of oxidant air pollution on ponderosa pine foliage in the Los Angeles, California area and reported on how the foliar damage can be evaluated on color films. The symptoms most useful in identifying affected pines were color, low density and shortness of needles, and high frequency of bare branches. Munsell color notations made both on the ground and on color films identified healthy foliage as a green-yellow hue (2.5 GY) and affected foliage as yellow (10 Y to 2.5 Y) to yellow-red (7.5 YR) hues. Healthy pines usually have up to 5 years of needle retention, whereas smog-affected foliage may have only the current growth of needles. These needles may be short and discolored. Finally, because of the absence of lush foliage, many bare branches are visible to the photointerpreter and permit him to identify positively an affected tree.

In a feasibility study of four films and five scales (1:1,600 to 1:32,000), the best combination for detecting air pollution damage was Anscochrome D/200 color film with a didymium rare-earth filter, taken at large scale (1:1,600). This film-filter-scale was used in a two-stage probability sampling survey over 40,000 hectares of the San Bernardino National Forest northeast of Los Angeles in May, 1969. Results show that half of all ponderosa pine trees were affected by air pollution to some degree. Since ponderosa pine constitutes 80 percent of the forest species in this area, the concern over the fate of these trees is considerable.

Randomly selected 70 mm flight strips were flown to derive the estimate that about 1.3 million ± 25 percent ponderosa pine trees were affected to some degree by air oxidants. Wert (1969b) found that of the affected trees, about 3 percent of the pines were dead, 15 percent were severely affected, and 82 percent were moderately affected.

A similar survey was made over the Angeles National Forest in April, 1970; the degree of damage was found to be less severe. Reflights to assess the trend of damage are planned for three year intervals.

The United States is not the only country where air oxidants pose a problem to forest lands. Murtha (1972b) describes a method whereby SO_2 damage to forest lands in Canada can be evaluated on scales of photography as small as 1:160,000 (Color Figure 34-44).

In Germany most air oxidants originate from large industrial complexes, such as those along the Ruhr Valley. Sulphur dioxide and fluorine gases are responsible for most of the coniferous injury. CIR films (scale 1:5,000) are favored for identifying various levels of damage to spruce (*Picea* spp.) For example, on CIR, healthy spruce appears purple, lightly damaged spruce appears purple-pink or pink, heavily damaged spruce appears gray or green-gray, and dead spruce appears blue-green. Also, when the needles are red, they appear strongly yellow on CIR. Because many of the older needles have fallen off from earlier fumigations, dark shadows appear on affected spruce; this effect is emphasized by the high contrast characteristic of CIR film (Color Figure 34-45). Hildebrandt and Kenneweg (1969) also noted that interpretation on CIR transparencies along the edge of the Ruhr Valley indicated that the intensity of damage to spruce stands appears to be related to altitude, even at long distances from the source of air oxidants. Thus, vitality of Norway spruce stands could be used as an indirect indicator of the distribution of oxidants. The CIR film depicted all levels of damage from old dead snags to less severely damaged trees. Darkening of foliage of moderately affected spruce can be seen on black-and-white IR prints, but are hardly detectable on panchromatic prints. Because of the silvicultural intensity and multiple use of most European forests, information about air-oxidant injuries is requested for almost every small planning unit, in addition to estimated damage for large areas. Thus, large-scale coverage and interpretation of numerous test circles or test lines, based on tree counts, is preferred to the much less expensive multi-stage sampling.

In East Germany (German Democratic Republic), where more than 5 percent of the entire forest area suffers severely from air-oxidant damages, detailed smoke-damage inventories are an essential part to the ten year forest management plan in some forest regions. In most cases these air pollution surveys are carried out with two-stage probability sampling (Wolff 1970, Lux 1965). The first stage involves interpretation of SN-6 spectrozonal aerial photography at a scale of 1:5,000 to 1:8,000; the second stage is a ground check of randomly selected plots.

STORM DAMAGE

Storm damage (windthrow, in particular) is a vexing problem to the forest manager because of the relatively short period during which economic salvage can be effected, coupled with the fact that older, more valuable stands are often the most severely affected. Evaluation of storm damage is difficult to accomplish on the ground, but can usually be efficiently detected and assessed by means of aerial photography (Rhody, 1962; Tokmanoglu, 1969; Neustein, 1971). Location of the damage and its areal extent are necessary for ground examination and possible salvage operations. Where good aerial photographic coverage already exists, relatively crude hand-held, low oblique 35 mm photographs can provide data on new storm damage and the new information can be suitably annotated on the existing photo coverage.

When damage is extensive and involves high-value forest, standard 22.8- × 22.8 cm (9- × 9 in.) metric camera aerial photo coverage may be warranted. However, a system of vertical 35 mm aerial photography, described in detail by Neustein and Waddel (1972), proved extremely effective in the detection and assessment of serious forest windthrow in portions of Scotland in 1968 (Neustein, 1971). This technique provided the desired information at relatively low cost and with little delay because simple, easily available equipment was used.

In many cases, an adequate evaluation of the scope and importance of damage can be made directly from photographs on the basis of personal judgment by an experienced forester. Where more specific loss data are required, however, ocular judgments made from photographs must be verified and adjusted by means of more quantitative data obtained by photo and (or) field measurements.

Ground enumeration of forest damage, particularly in the case of mature forest windthrow, is difficult and expensive. A few days spent probing jackstraw piles of down, broken, and twisted timber (Figure 34-46) are usually sufficient to send even the hardiest forest land manager searching for appraisal alternatives. Fortunately, there are a number of possibilities involving remote sensing applications which will often suffice.

One of the simplest approaches is the determination of the forest cover types, type boundaries, and type areas in the damage zones by means of the damage detection photography—usually in conjunction with previous photography and cover type maps of the area. Once the type boundaries, respective areas, and degrees of damage within the cover types are established, volumes per unit area can be assigned from ground samples in the damaged area, from previous ground inventories in the adjoining (similar) cover type, or from

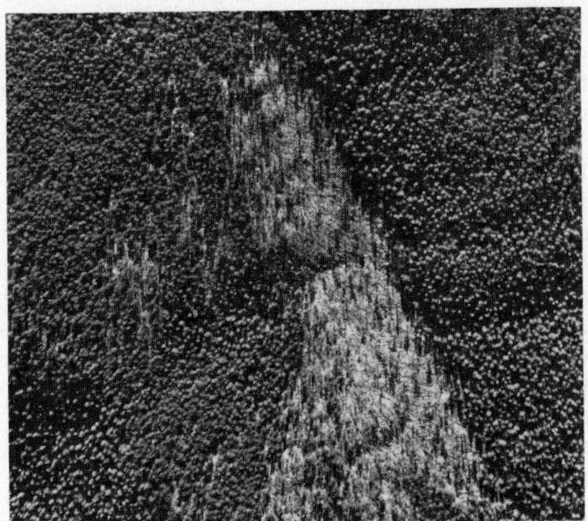

Fig. 34-46. Windthrown Douglas-fir type. Gifford Pinchot National Forest.

ground samples taken in the undamaged timber adjacent to the damaged portion of the cover type.

A number of investigators have attempted to reduce ground enumeration even further by employing aerial volume tables for damage loss estimates. Their manner of employment will depend upon the nature of the damage. Weber (1965) and Wert (1969a), in dealing with standing insect killed timber, devised aerial volume tables for the purpose of evaluating losses of this type through direct photomeasurements of the dead trees. Meyer and French (1966) also investigated dead timber; some was standing and not measurable on the aerial photography and the remainder was lying on the ground. They used aerial volume estimates taken in the adjoining live timber, and applied them to the damaged area. The estimate of total loss was found to compare quite favorably with local ground inventories in the same forest cover type and condition class.

FOREST MANAGEMENT

The use of remote sensing in forest land management is one of the most important applications in the field of wildland resource management and yet it is one of the least documented. Unlike extensive timber volume inventories and damage surveys where techniques and results are precisely defined, its chief value lies in routine day-to-day applications in on-the-ground management of forest land. It is here that remote sensing techniques save the wildland manager time and give him a more complete picture for making his management decisions. This kind of application was well summarized recently by a forester in the Southern Pine Region.

> "They [aerial photographs] are probably the greatest time-saving device a forester on a large property can possess. Without photographs, one can spend a great deal of time showing personnel where to go and what to do. With photographs, timber markers, surveyors, and other personnel can be dispatched in a short while, with a clear conception of the day's job. Little or no assistance is needed in learning the property. Land corners, roads, timber and other features can be located without difficulty. Working daily with photographs gives the user a mental picture of the timber and terrain that could not possibly be acquired without long experience on the property. Errors can be readily discovered in land descriptions. Logging operations, planting, road construction, fire suppression and investigation, and timber stand improvement work can be planned with a great deal more efficiency and less time. Aerial photographs serve as a permanent record of timber conditions at the time of flight and can be compared with subsequent flights."

TIMBER HARVEST PLANNING

The use of remote sensing in timber harvest planning on large privately owned forested areas can start several years in advance of the actual harvest. First, the overall plan for development is reviewed and a decision is made as to the general area where cutting will take place in the next 5 to 10 years. A number of possible routes for roads through the selected area can be sketched on aerial imagery. Details to be considered in planning roads include bluffs, rock outcrops, slides and critical slopes, saddles, draws, switchback flats, drainage and vegetation (see Chapter 32 for additional details). An approximation of grade between check points can be obtained by measuring elevation differences directly on the photography with a parallax bar. The final selection of the routes is made after ground examination of the most promising of those indicated on the photography.

After route selection, aerial photographs are examined for possible locations of cutting areas. If it is known that the selected area has recently been subject to severe wind, insect, or disease damage, then up-to-date photography should be obtained. This new photography would pinpoint the areas where salvage operations are appropriate. Otherwise, existing photography, along with available type maps, help locate areas of merchantable timber and show their extent.

Once the general area of the proposed timber harvest is selected, topographic features that influence cutting boundaries, logging methods, soil stability, and road layout can be studied more intensely on the aerial photography.

For example, much of the planning of the clear-cut type of timber harvest can be done on aerial photography. Sketching several tentative cutting units on the photos would be performed first. Consideration should be given at this time to the future development of the area (for example, where subsequent cutting units will be). With the use of remote sensing, the development of the area as a whole can be visualized, minimizing the possibility of ending up with a group of illogically shaped units on the final harvest cut.

After the units are located, aerial photographs are helpful in deciding what kind of logging will be practiced. Areas for landings, spar trees, spur roads, and swing locations can be first located on photos and then checked in the field. Aerial photography has been put to effective use in designing Skyline logging systems (Carson et al, 1970). Skyline logging is an overhead cable system which has the capability of lifting a log free of the ground and bringing it into a central yarding point. A forester may wish to explore several alternate locations before installing the system. By use of a stereoplotter and a computer program this task can be quickly performed. The Skyline designer prepares equipment information and specific road descriptions on the aerial photography. The stereoplotter operator, through the use of a digitizer, takes terrain point data directly on the stereophotograph model and prepares data cards describing each Skyline road. The output will tell the forester whether the proposed route is a feasible one for Skyline logging.

Should a proposed harvest area be put up for open bid, aerial photography is an excellent way of showing a sale area to a prospective purchaser. Aerial photography can also be shown to operators when explaining points in the sale contract. Enlargements are helpful in many cases of discussions between operators and owners. The enlarged picture is more easily understood by people not familiar with the stereo viewing of photos.

Aerial photography can also be helpful to logging companies that do not own forest property. It is difficult for the timber buyers from logging companies to visit the wood lot of each farmer or small landowner who wishes to sell timber. To avoid unnecessary travel, offers can be screened by reviewing aerial photography. In the United States, the Agricultural Stabilization and Conservation Service, the Forest Service, and the Soil Conservation Service of the Department of Agriculture, as well as other Federal agencies sell aerial photographic coverage at a nominal cost. County assessors' offices in heavily timbered counties in the Western States often maintain a file of photos for assessment purposes.

MONITORING LOGGING AND REFORESTATION

The professional forest land manager has obligations to tomorrow as well as today and no matter how well-engineered and tailored the forest harvest operation may be for current logging operations *per se*, it should also relate to: (1) logging impact upon the environment on a day-to-day basis (e.g., erosion problems, effects upon water quality, damage to residual stands and reproduction); (2) anticipated and desired condition of the land, wildlife, water, soil, and vegetation at the termination of logging operations; and (3) reforestation plans, both artificial and natural.

Even when recent air-photo coverage is available at the inception of a major timber harvest, along with forest stand type maps prepared during the inventory preceding it, both become quickly outdated as cutting and roadbuilding progress. When logging operations cover extensive areas, the system of sequential metric camera aerial photography and mapping described by Catto (1965) is a desirable one. His study involved extensive boreal forest cutover operations in which ground monitoring was prohibitively expensive. By periodically obtaining 1:15,840 scale, panchromatic, 22.8- × 22.8 cm format, aerial photo coverage of the operating area, it was possible to delineate progress in cutting and road development, and transfer this to base maps at an acceptable level of accuracy and cost.

When more extensive areas are involved (e.g., multi-country, region, or province), and delineation of species groups and stand classes to the degree described by Catto is not required, sequential high-altitude, small-scale aerial photography, or near-orbital or orbital earth satellite scales of imagery may suffice. Evidence for this possibility lies in the fact that relatively small changes in cutting boundaries of black-spruce stands in northern Minnesota are clearly discernible on color infrared, imagery, 1:120,000-scale, metric camera coverage flown in 1971 by the NASA RB57F photo aircraft from an altitude of 20,000 meters. Panchromatic and black-and-white infrared, 70 mm, 1:500,000 scale photography, exposed simultaneously on the same overflights also clearly portrayed boundaries of relatively small (7–10 hectares), recent timber cuts (Douglas, 1973).

Because of problems of cost and (or) photo aircraft availability at the proper time and place, a number of simple remote sensing techniques have been successfully developed in lieu of metric camera coverage and application methods. Of particular note is the system developed by Zsilinszky (1969), utilizing a conventional motor driven 35 mm camera. Locally available aircraft are employed which are capable of performing photographic missions at an altitude range of 200 to 4,000 meters. Sixteen of these camera units are now in operation using a variety of lens/film combinations to monitor cutovers, road development (see Figure 34-47), and other day-to-day operations in Ontario Province, Canada.

A similar application for monitoring logging progress and road clearing on a smaller operational area has been made by Cook (1969) in the form of a unit involving a conventional hand operated 35 mm camera side-mounted on a light aircraft. A battery operated tape recorder permits the photographer/observer to record in-flight observations about specific areas of photo coverage. Use of color infrared or color film in 35 mm camera units of this type also may provide an inexpensive, expedient tool for monitoring erosion problems or changes in sediment load in streams caused by logging operations.

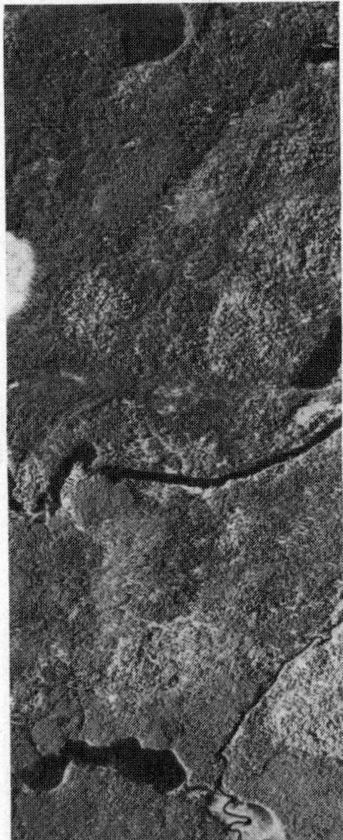
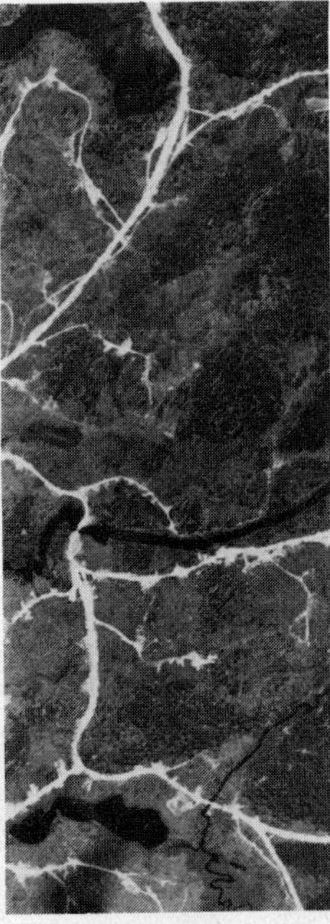

Fig. 34-47. Left: portion of a conventional panchromatic (original scale 1:15,840) vertical photograph exposed prior to forest cutting and road construction. Right: view of the same area enlarged from a 1:100,000-scale 35-mm negative exposed after logging and road establishment. (Courtesy of Ontario Dept. of Lands and Forests.)

Another feature of logging operations, often a subject of concern to the forest manager, is the amount, character, and distribution of logging slash which may have importance in terms of fire hazard (i.e., fuel buildup) or relationship to future reforestation of the cutover. Morris (1970) has developed techniques for the application of 70 mm large-scale aerial photography to the inventory and assessment of logging slash. In both cases, two helicopter mounted simultaneously activated 70 mm cameras are used. Meyer, et al (1971), using a vertically positioned 70 mm camera in a light aircraft, showed both color and color infrared medium-scale photography to be useful in planning and analyzing slash distribution patterns. These patterns relate to prescribed future burning designed to control dwarf-mistletoe disease infection of black spruce and prepare the site for natural regeneration (see Color Figure 34-48). In future operations, a 35 mm unit similar to that described by Cook will be employed in place of the 70 mm camera.

The three general operational categories of remote-sensing applications to reforestation are: (1) location and evaluation of potential sites, (2) planning and execution of the actual reforestation, and (3) regeneration and plantation survival surveys. Because of time and financial constraints, conventional forms of panchromatic or black-and-white infrared aerial photography are the primary remote sensing tools in use. However, the formats and scales of photography in use vary greatly; they range from 22.8 × 22.8 cm metric camera coverage at scales of 1:4,000 to 1:15,000, to 70 mm and 35 mm from large scale to medium scale coverage obtained with helicopters or light fixed-wing aircraft from relatively low altitudes.

Potential sites will usually be clearcut areas without significant advance reproduction, or areas which are differentially understocked due to such factors as silvicultural treatment, rodent problems, insect epidemics, disease, and fire. The screening of recent aerial photographs for potential reforestation sites is not usually a particularly demanding task, but their evaluation with respect to whether or not reforestation is necessary and feasible, and the extent to which it is required, demands a high degree of professional photo interpretation skill. Depending upon the situation, the photographs must usually be capable of providing a reasonable amount of such information as the location and intensity of current stocking, location of natural seed sources and their anticipated boundaries of effectiveness, tree species best adapted to the site, possible rodent problems, and the presence of terrain and/or vegetation features which might seriously affect seeding, planting or seedling establishment (e.g., dense brush, down timber and/or logging residues, rocks, steep slopes, and topographic hazards to aerial seeding).

Regardless of the type and scale of aerial photography used in determining location and intensity of current stocking, the particular seedling size-class detection threshold of the photography in use must be ascertained, Swantje (1957), found his minimum seedling size-class detection limit on exposed Douglas-fir seedlings to be a 0.6- to 1-meter height-class when using growing season 1:15,840 scale aerial photography. However, by taking the photographs in the fall, he found that he could also detect seedlings of the same minimum size under dense willow and fern cover. Reinhold (1967) tested 1:4,000 scale panchromatic and infrared photography on 2- to 6 year old pine plantations and found that, for average conditions, he achieved his best detection level with panchromatic photographs.

What is obvious from the work of Swantje, Reinhold, and a number of other investigators is the need not only to obtain regeneration survey photography of a suitable type, quality, and scale, but also to secure it at a time when vegetation conditions optimize the chances for seedling recognition. Color infrared photography posseses unique capabilities in this respect, particularly in discriminating coniferous seedlings from surrounding herbaceous and deciduous plant cover. Color Figure 34-49 illustrates the successful use of both color and color infrared 70 mm medium scale photography to assess pine-plantation survival and condition (Meyer et al, 1971). Some advantages of color and color infrared over black-and-white photography in this case was the marked tree background color differential due to the timing of the overflight in autumn when the herbaceous plant materials were largely dormant, making plantation survival patterns easily visible. A further dividend of this photography was the ability to detect current plantation losses due to *Armillaria* root rot.

Once the boundaries of the sites to be seeded or planted are established on the aerial photographs, their areas can be determined, the amount of seed or number of required seedlings can be calculated, and working plans and budgets can be developed. An interesting approach to this operational phase was developed by Page (1969a) who used a 70 mm camera positioned vertically from a light aircraft to photograph clearcut areas. Initially, he enlarged the photographs to 1:4,800 scale and prepared mosaics which he used to calculate cutover surface area, amount of seed required, rate of seeding, and flight-line locations. He now uses the same camera from higher altitudes to obtain smaller scales of photography which work equally as well as large scale photographs (Page, 1969b) and, in addition, eliminates the cost of mosaics (Figure 34-50).

Under natural conditions in the Western Cascades, detection of Douglas-fir and other conifer seedlings has had mixed success. Haapala and Neumann (1972) used 1:1200 and 1:2400 scale spring color infrared photography at a two-times

Fig. 34-50. Use of an enlarged 70-mm aerial photograph to control an aerial seeding operation in New Zealand (Photo courtesy New Zealand Forestry Research Institute).

enlargement to detect mixed conifer reproduction (0–6″ dbh). They could detect the trees greater than four feet tall and having a crown diameter of at least three feet. Bernstein (1974) on the other hand found 1:1000 and 1:500 scale color-negative prints inadequate for detecting Douglas-fir seedlings. He also noted that the aerial photo survey costs more.

Nelson (1977) reported on a forest-plantation assessment-system being instituted on Weyerhaeuser lands in North Carolina. For the assessment and measurement of sites prior to plantation establishment, 70 mm color infrared film is exposed at scales of 1:12,000 to 1:24,000. For periodic detailed evaluations of established plantations, starting after the third growing season, larger scales (1:2,400) are needed to determine survival, height growth and area needing fertilizer treatment.

FOREST RECREATION

RECREATION RESOURCE INVENTORY

There is an increasing demand for development of new recreation areas and more intensive management of existing areas. Usually the selection of new areas requires evaluation of several alternative sites, and often requires a search for alternative sites for potential acquisition. Inventory techniques utilizing aerial photographs can provide the needed information more rapidly, and less expensively, than ground techniques alone (MacConnell and Stoll, 1968; Olson, Tombaugh, and Davis, 1969). More intensive use of existing areas requires careful consideration of the impact of one use on other potential uses of the same area, and of the possible adverse effect of new uses of one area on existing uses of adjacent areas. The need to plan for complementary uses and to minimize conflicting uses has resulted in the reevaluation of large, existing recreation areas in the light of changing recreational demands with an eye to design and implementation of comprehensive zoning plans. Aerial photographs and other remotely sensed imagery provide ideal tools for terrain analysis and cover type mapping, both of which are vital to intensive recreation planning. Such techniques are not peculiar to outdoor recreation but are simply applications of resource inventory techniques described elsewhere in this volume.

Intensive recreation planning is usually performed by a team of planners and the team approach should be utilized in design of recreation inventory specifications. In designing recreation inventories it should be remembered that regional inventories cannot be completed by a single person in a realistic time period, nor are all areas equally useful for recreation development. Site standards which can be uniformly applied to a diversity of landforms by a number of interpreters are essential. In actual practice, standards vary from one managing agency to another, and tend to be design oriented rather than inventory oriented. Design-oriented standards usually require quality judgements and management decisions from interpreters. Experience has shown that it is better to clearly separate the inventory and management decision making functions (Olson, Tombaugh, and Davis, 1969).

RECREATION RESOURCE MONITORING

Management of large outdoor recreation areas presents several resource monitoring problems that seem at least partially amenable to solution through remote sensing. Some problems require policing actions, such as detection of illegal roads "pushed" into wilderness and other restricted areas by trespassers. Other problems require assessment of the rate of change due to legal use. Soil compaction from heavy use of local areas, such as camp sites, picnic areas, water access points, and back country trails, often results in serious damage to tree roots. Remote sensing research on previsual detection of declining tree vigor may provide a means of monitoring such damage and determining when rotation of heavy use among alternative sites is needed to preserve the tree cover. Opportunities also exist to monitor shoreline erosion, ski-slope stability, and a host of other dynamic terrain variables.

Recent increases in the use, and abuse, of snowmobiles, dune buggies, jeeps, and all-terrain vehicles have revealed a need to monitor the location and concentration of these vehicles. The presence of hot engines, and the need to monitor their use during hours of darkness, have raised interest in the possible application of thermal sensors for this task. Except in areas with dense coniferous overstories, thermal sensors are capable of detecting snowmobiles and dune buggies, and other vehicles are also detectable when they are not overtopped by a green-vegetation canopy or overhanging-rock outcrop.

The importance of body contact water sports in outdoor recreation is so great that all methods of monitoring water quality or of mapping the spread of pollutants is of potential recreational use.

Recreation management practices need to be responsive behavior patterns, particularly the density of users per unit area. Due to the extensive areas involved, such as wilderness areas, it is nearly impossible to monitor user density patterns without intruding, and possibly altering, user behavior and satisfaction. Many managers have expressed interest in utilizing remote sensors to monitor use patterns of individual recreationists, but no reports of such efforts are known.

COMPUTER-AIDED PROCESSING OF DIGITAL DATA

BACKGROUND

The advent of multispectral scanner (MSS) systems has been a particularly significant step in the development of remote sensing technology, since data from MSS systems are obtained in digital format and therefore are ideally suited to computer-aided analysis techniques. Multispectral scanner data were first obtained for nonmilitary research in 1964 when a University of Michigan system was flown over test sites at Purdue University for the purpose of evaluating this type of instrument system for agriculture and forestry applications (Lowe et al., 1965; Hoffer, 1967). These early experiments indicated the potential for identifying various vegetative cover types, based solely upon reflectance differences in the various wavelength bands of the scanner data, but it was also apparent that some type of computer-aided analysis technique would be required to effectively analyze the subtle differences in spectral reflectance being measured. Pattern-recognition theory was tested and successfully applied to multispectral scanner data in 1966 (Holter et al., 1970; Landgrebe and staff, 1967). This first use of pattern recognition techniques for identifying vegetative cover types involved agricultural crops. Field patterns were easily defined and each field represented a single cover type, usually of fairly uniform density, and without significant spectral variation due to topographic effects—a much simpler situation than is usually encountered when working with wildland vegetation. During the period from 1966 to 1971, most efforts were directed at developing and refining methodologies to use in applying pattern-recognition techniques to multispectral scanner data. Much of this work was funded by NASA, and was conducted at the Laboratory for Applications of Remote Sensing (LARS), Purdue University, and the Willow Run Laboratories, University of Michigan (later incorporated as the Environmental Research Institute of Michigan, or ERIM).

Early Studies Using Aircraft MSS Data

Although much of the early work was devoted to agricultural crops, there were some notable studies involving attempts to classify forest cover using multispectral scanner data and computer-aided analysis techniques. These studies included one by Smedes et al. (1970) for a rugged 31-km square forested area in Yellowstone National Park. Using the Purdue University LARSYSAA programs, computer-classification techniques were applied to aircraft data and computer maps were generated which showed the distribution of eight vegetation/terrain classes, including forest, bog, glacial-till meadows, glacial-kame meadows, tallus, vegetated bedrock rubble, exposed bedrock, and water. Overall mapping accuracy was determined to better than 80%.

Computer identification of tree species was attempted for a 32-hectare area of forest land in Michigan by Rohde and Olson (1972). This study site contained cottonwood, aspen, willow, elm, sugar maple, white cedar, black locust, white oak, red oak, black walnut, spruce, and pine. The University of Michigan Spectral Processing and Recognition Computer (SPARC) was used to classify the test site. Using six wavelength bands of data in the $0.4-1.0$ μm portion of the spectrum, they were able to separate coniferous and broad-leaved species and, at a more detailed level, red oak, white oak, black locust, black walnut, and sugar maple were also successfully separated. Although discrimination among individual species of conifers was not as successful as for broad-leaved species, they did get consistent separation between pine and spruce. The authors stated that "approximately 85 percent of all trees within the Saginaw Forest property were correctly identified."

Multispectral scanner data obtained over a test site in California were classified by Lent et al. (1969) who attempted to discriminate between nine cover-type groups, including open; mixed-species brush fields; dense snow brush; dense manzanita; mixed conifer stands with some hardwoods; cleared brush fields; roadways and some bare areas; lake shore-line; water; and snow. Through the use of a maximum-likelihood classification algorithm, the final classifications were rated as being successful and in only one case were two classes consistently found to be inseparable, these being roadways and lake shoreline. Most of the misclassified areas in the classification were found to involve shadow areas and varying spectral response due to the undulating terrain.

A study by Coggeshall and Hoffer (1973) for an area of forest and agricultural land in southern Indiana showed that multispectral scanner data obtained in mid-summer could be used to classify and map major forest and agricultural cover types with a high degree of accuracy. The University of Michigan multispectral scanner that had been used in the three studies cited above was again the source of the MSS data. The scanner obtained data in seven wavelength bands in the visible portion of the spectrum, two bands in the near infrared ($0.7-1.3$ μm), two bands in the middle infrared ($1.3-3.0$ μm), and one band in the thermal infrared ($9.3-11.7$ μm). The cover types classified included deciduous (a mixed forest containing mostly tulip poplar, oak, hickory, maple, and ash), conifer (white pine), corn, soybeans, forage, and water. When all twelve channels of scanner data were utilized, quantitative evaluation of the classification results based on test areas indicated that deciduous and coniferous forest cover could be classified with an accuracy of over 95%, and that the overall classification performance was approximately 95%.

A detailed waveband evaluation was included in the same study, and indicated that all four spectral regions (visible, near infrared, middle infrared, and thermal infrared) are valuable for mapping forest and other cover types. The visible portion of the spectrum alone could do a good job of separating forest from other cover classes but resulted in poor differentiation between deciduous and coniferous forest. The near- and middle-infrared portions of the spectrum were effective for separating deciduous and coniferous forest classes, but resulted in confusion between deciduous forest and other classes of vegetation. Thus, a combination of visible and either the near- or middle-infrared wavelength region was required to obtain accurate mapping of forest cover versus other vegetative cover types, as well as an accurate differentiation between deciduous and coniferous cover types.

This study also indicated that, as the number of wavelength bands utilized in the classification is increased, the classification accuracy increases rapidly at first but then the incremental improvement becomes relatively small (see Figure 34-51). As more and more wavelength bands were used, however, the amount of computer time required increased very rapidly. The authors concluded that five wavelength bands provided the best compromise between classification accuracy and computer time required. A feature-selection processor indicated that when the classifier was restricted to five channels, the best over-all classification performance would be provided by two channels in the visible wavelength region, one in the near infrared, one in the middle infrared, and the thermal infrared channel (Coggeshall and Hoffer, 1973).

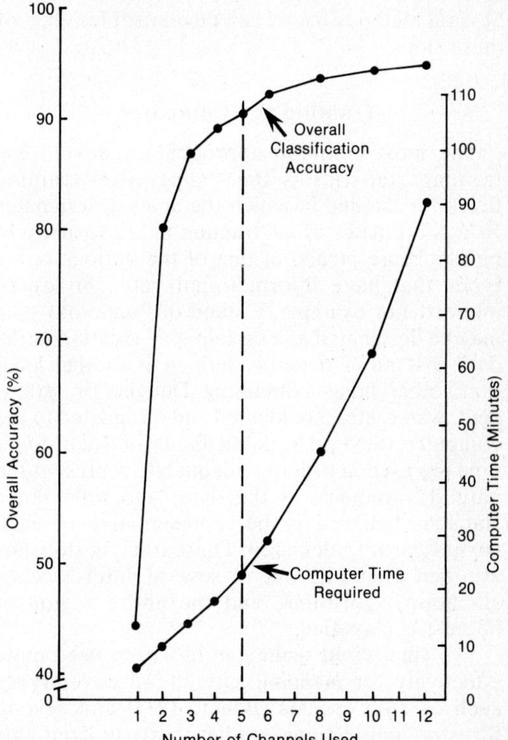

Fig. 34-51. Overall classification accuracy and computer time required as a function of the number of wavelength bands or channels of MSS data used (from Coggeshall and Hoffer, 1973).

TECHNIQUES FOR ANALYZING LANDSAT MSS DATA

Introduction

Following the launch of Landsat-1 in 1972, many studies were conducted to evaluate the capabilities as well as the limitations of this type of data for mapping forest cover. Many of these efforts involved manual interpretation, but there were also many in which the same pattern recognition techniques that had been developed for airborne MSS data were tested.

Most of these early studies utilized the "supervised" analysis technique, but in some cases the unsupervised or clustering technique was applied because the vegetative cover types were spectrally or spatially complex, or there were insufficient reference data. As the advantages and limitations of these different techniques became more apparent, modifications were tested, new classification algorithms were developed, and different analysis techniques were defined. Before further discussion of the results of computer-aided analysis of Landsat MSS data, it seems appropriate to review the fundamentals of the most common analysis procedures utilized.

The concept behind most CAAT (computer-aided analysis techniques) involves a man/machine interaction, in which the man "trains" the computer to recognize spectral patterns, i.e., specific combinations of numbers that represent reflectance measurements in each of several wavelength bands, for the cover types of interest.[2] This training process usually involves rather limited areas for which accurate information exists concerning the type and condition of the ground cover. After a representative set of training statistics has been developed, the computer is programmed to classify the reflectance values for each resolution element or pixel in the entire MSS data set of interest. In this way, the speed of the computer is used to advantage, and a large geographic area can be mapped and acreage tabulations obtained much more rapidly than would be possible using standard image-interpretation techniques. Additional detail regarding computer-classification procedures may be found in Chapter 18 of this Manual.

There are three major steps in computer-aided analysis procedures: (1) training the computer, (2) classifying the data, and (3) evaluating the results.

[2] The acronym CAAT is used to clearly indicate that these techniques involve significant analyst interaction with the computer and the data, as opposed to the use of the acronym ADP (automatic data processing) which is more suitably applied to activities such as banking, bookkeeping, etc. in which the analyst has very little freedom to influence the data processing sequence.

Several methods have been developed for each of these steps.

Training the Computer

The most common approach for developing training statistics is the "supervised training fields" technique in which the analyst designates X-Y coordinates of a "training field" (usually of nine or more pixels) of one of the various cover types that have informational value or are of interest. For example, a stand of Ponderosa pine may be designated at a certain X-Y location in the data; a stand of Aspen is defined at another location; other areas containing Douglas-fir, grassland, water, etc., are located and designated to the computer. Several training fields of each cover type are used in order to adequately represent the natural variability in the data, and reflectance statistics believed to be representative of each cover type are calculated. These training statistics are then used with one of several different classification algorithms, and the entire region of interest is classified.

The supervised technique has been used quite effectively for mapping agricultural cover types such as crop species (Bauer, 1975) and several forestry applications studies have utilized this technique with varying degrees of success (Kan and Dillman, 1975; Dodge and Bryant, 1976; Williams, 1976; Fleming and Hoffer, 1977; Lee, et al., 1977; Shimabukuro, et al., 1980; Mead and Meyer, 1977, and others). In wildland areas, where the cover types of interest are often not spectrally homogeneous, use of this supervised technique sometimes does not yield acceptable accuracy or reliability. The primary reason for this is that the analyst often has insufficient knowledge concerning the spectral characteristics of the different cover types to effectively define locations in the data that represent all significant variations in spectral response for every cover type of interest. This difficulty can be overcome, in part, through effective use of a different technique for training the computer referred to as "clustering."

Clustering (sometimes referred to as the "nonsupervised" technique) is frequently used to develop training statistics for forestry applications. There are several different clustering algorithms available, such as ISOCLS, LARSYS CLUSTER, and CLASSY, each of which has different analyst-defined input parameters. In essence, the analyst simply designates the areas to be clustered and the appropriate input parameters required, such as the number of spectrally distinct groups into which the data should be divided. The computer is programmed to iteratively group or cluster the data into "similar" spectral classes, and will print out or display a map indicating which resolution elements in the data belong to which spectral class. The analyst must then relate this spectral map to aerial photos, type maps, or ground reference data, and determine which cover types are represented by which spectral group or groups (e.g., group 1 is Aspen, groups 3, 7, and 12 are Ponderosa pine, groups 2 and 16 are exposed rock, etc.).

The clustering approach has been shown to be effective in overcoming some of the difficulties analysts have experienced in using the supervised approach. However, when working with large geographic areas the amount of computer time involved in the iterative clustering sequence can make this technique prohibitively expensive if every pixel of data in the entire area of interest is utilized. In addition, the number of spectral groups defined is often very large, since a single cover type of interest is usually represented by several spectral groups. In areas where the vegetative cover is complex (e.g., small stands, species mixtures, variations in stand density, etc.), the spatial resolution of Landsat data often makes it difficult to reliably relate each of the spectral groups defined by the computer to a particular vegetative cover type of interest.

Due to the difficulties experienced in using either the supervised or unsupervised technique for developing training statistics on forest and rangeland areas, several hybrid approaches have been developed which incorporate various aspects of both the supervised and the unsupervised techniques. These hybrid techniques have sometimes been referred to as "guided clustering". One of these hybrid techniques, called "multicluster blocks" by Fleming and Hoffer (1977), has been shown to be particularly effective in minimizing the amount of computer time as well as the analyst time required, while at the same time enabling a higher classification accuracy to be obtained than could be achieved using any other technique for developing training statistics. In using this technique, the analyst designates several relatively small but very heterogeneous blocks of data, each of which contains several different cover types and many spectral classes (see Color Figure 34-52). Each block is clustered individually, and aerial photos are used to relate the spectral groups defined by the computer to the informational classes of interest to the user. The spectral groups for all of the different cluster blocks are then combined into a single set of spectral training statistics using specially designed computer software. This final set of spectral statistics is then used to classify the entire geographic area of interest, using one of the different classification algorithms available. Another hybrid technique, called "mono-cluster blocks" involves clustering all of the designated cluster blocks together, thereby simplifying the process of combining the spectral groups into a single set of training statistics.

In essence, these hybrid techniques are simply methods for using the computer to aid the analyst in discovering the natural spectral groupings present in the scanner data, so that the analyst can more effectively correlate the spectral classes with the desired informational classes, such as forest cover types. The key point is that knowledgeable use of the most appropriate technique for developing training statistics appears to be a very critical step in the effective utilization of satellite data and computer-aided analysis techniques. In defining which training technique is most appropriate, the analyst must consider the spectral and spatial complexity of the area to be classified, the level of detail desired, the size of the area involved, and the software available.

Classifying the Data

After the training statistics have been developed, the next step involves the actual computer classification of the data. There are a number of different algorithms which have been successfully used in a wide variety of studies. The Gaussian maximum-likelihood algorithm, implemented in LARSYS and many other software sysems, is one of the most common and effective algorithms available. However, there are several other algorithms available that are much more efficient in terms of the amount of computer time required to classify a data set. The parallelepiped algorithm, implemented on the G.E. Image 100 system, has been widely used. A minimum distance algorithm has also been used with success by some investigators. One classification technique that has been tested and found to be especially fast involves a "table look-up" procedure (Shlien and Goodenough, 1974). Several of these algorithms are discussed in detail in Chapter 12 (Remote Sensor Data Systems, Processing, and Management) of this manual, and also by Swain and Davis (1978), and elsewhere in the literature. Descriptions of the various algorithms are beyond the scope of this section. However, some comparisons among results obtained using different algorithms will be described later in this chapter. It is important to note that, from an applications standpoint, the analyst often has relatively little choice as to which algorithm is used, but must simply apply whichever algorithm has been incorporated into the software of the analysis system available to him.

Evaluating the Results

The third step in a computer classification of MSS data involves the evaluation of the classification results. Generally, one of three procedures is followed in such an evaluation. The first procedure involves a qualitative evaluation of the classification results, which can be obtained by visually comparing the classification-output map or image display to aerial photos or to an existing cover-type map of the classified area. Although this procedure is subjective, it does provide a quick and often reasonably good estimate of the general accuracy of the classification obtained. However, quantitative evaluation procedures are often desired since they allow a more definitive evaluation of the computer-classification results.

The quantitative evaluation procedure most commonly used involves a sample of specific, well defined areas of known cover types which are designated as "test areas". To avoid any possible bias on the part of the analyst, such test areas should be located by means of an appropriate statistical sampling design, and the locations should be defined prior to the classification. The computer-classification results for the various test areas are tabulated and compared to the actual cover types present. The actual cover type is determined through interpretation of large scale aerial photos or from existing cover-type maps (assuming they are accurate), or by on-site examination in the field. The result is an accuracy-assessment table, often referred to as a classification-error matrix, confusion table, or contingency table.

In addition to simply tabulating the classes into which each pixel was classified, the percentage of correctly classified pixels is usually calculated for the individual cover types and for all test data. A basic assumption in this procedure is that the test data-set provides a reliable, representative sample of the various cover types in the entire area that was classified. The accuracy assessment for individual classes is calculated as:

$$C_i\ (\%) = \frac{N_i}{T_i}\ (100)$$

where

C = the percent correct for class i
N_i = the number of pixels correctly classified into class i
T_i = the total number of pixels in the test data-set actually belonging to class i

(34-10)

The overall classification performance can be calculated using the equation:

$$P\ (\%) = \frac{\sum_{i=1}^{n} N_i}{\sum_{i=1}^{n} T_i}\ (100) \qquad (34\text{-}11)$$

where

P = the overall classification-performance, or accuracy, expressed in percent

$\sum_{i=1}^{n} N_i$ = the total number of correctly classified pixels in all classes

$\sum_{i=1}^{n} T_i$ = the total number of test pixels in all classes

Table 34-9 is an example of an accuracy-assessment table, including the accuracy per-

TABLE 34-9

An Example of a "Performance Evaluation Table"[1]

Cover Type	No. of Samples[2]	No. of Samples Classified as: Coniferous	Deciduous	Grassland	Barren	Water	Shadow[3]	Percent Correct
Coniferous	9,634	9,110	22	53	21	96	332	94.6
Deciduous	1,475	113	1,286	76	0	0	0	87.2
Grassland	3,677	49	129	2,988	510	0	1	81.2
Barren (rock or soil)	35	0	0	1	34	0	0	97.1
Water	1,349	6	0	0	0	1,334	9	98.9
Totals	16,170	9,278	1,437	3,118	565	1,430	342	

Overall Performance = (9,110 + 1,286 + 2,988 + 34 + 1,334)/16,170 = 91.2%

[1] (from Hoffer et al., 1975a).

[2] Each "sample" is a Landsat resolution element. The column labelled "No. of Samples" indicates the total number of resolution elements present in the test areas. The cover type in each test area was determined by photo interpretation and field visits to the test site locations.

[3] One of the 14 spectral classes that had been defined and used for the classification involved areas of topographic shadows, but since this was not an actual cover type, any resolution elements belonging to this spectral class were considered as errors in the classification.

centages for the individual classes and the overall classification-accuracy. In this example, errors of omission (pixels which should have been classified into a particular class but were not) are shown in the rows, while the columns indicate the errors of commission for each class (pixels which were included in that class that should not have been). Assuming that the test data set is a statistically valid representation sample of the cover types present, such accuracy-assessment tables should provide a reasonable indication of classification accuracy, and are very useful in assessing the causes of misclassification.

In some studies, accuracy-assessment tables have been developed using the training data from which the statistics used to train the classification algorithm were derived, rather than a statistically defined independent set of test data. A tabulation of the classification results for the training data does not constitute a valid evaluation of the classification results over the entire area, but simply provides an indication of the potential for spectrally differentiating the cover types involved. High classification accuracies are often obtained for the training data, but if the training data do not adequately represent the spectral characteristics of the various cover types present in the entire data set, the actual classification may be considerably less accurate than is indicated by the training-data results. A tabulation of the training data therefore may serve as a preliminary indication of the potential for obtaining a reasonably accurate classification, but should not be used to indicate the accuracy achieved in the classification of the entire area.

In addition to simply calculating accuracy percentages, Hord and Brooner (1976) and Todd et al. (1980) describe procedures for determining confidence intervals, and Congalton et al. (1981) described a statistical method for making comparisons among various error matrices. Kalensky and Scherk (1975) have proposed a "mapping accuracy" measure to express the error in class location on a map, since the "overall classification-accuracy" only takes errors of omission into account. Further development of these and other methods for assessing classification performance and comparing among different classification results will provide better standards for evaluating a variety of computer-classification techniques.

In addition to accuracy-assessment tables (i.e., error matrices or confusion tables), another commonly used quantitative method to evaluate classification results involves a comparison of acreage estimates obtained from the computer classification of the satellite data to acreage estimates obtained from an existing source of information or obtained by some conventional method, such as manual interpretation of aerial photos and use of dot grids or planimeters. Several investigations have shown that acreage estimates from computer classification of satellite data compare favorably with acreage estimates derived from other data sources (Heller [Tech. Coord.] 1975; Hoffer and Staff 1975a; Dodge and Bryant, 1976; Bryant et al. 1978; Hoffer, Noyer and Mroczynski, 1978; Rohde et al. 1979; and others). Some of these results will be described later.

The spatial resolution of Landsat data is such that each pixel represents an area on the ground of approximately 0.45 hectares. Therefore, one of the major advantages of computer classification of Landsat MSS data lies in the potential to obtain acreage measurements by simply having the computer tabulate the number of pixels classified into the various informational classes, and multiplying by 0.45 hectares to obtain the total acreage of that particular cover type. Often, however, errors of omission and errors of commission in the classification will cancel each other in such acreage tabulations (i.e., a pixel that should have been classified as conifer was actually classified as deciduous, an error of omission—will be offset in the acreage tabulation by a pixel that was erroneously included in the conifer class that was actually deciduous forest, an error of commission). This results in acreage tabulations that are frequently more accurate than would be anticipated, based upon the actual classification performance. Such situations occur most often when acreages have been tabulated over large geographic areas. However, if the classification performance is not relatively high, as the area involved in the acreage tabulation becomes smaller the accuracy of the acreage estimate decreases significantly. For example, Bryant et al. (1978) classified a forestland area in Maine into six cover-type groups, including classes of softwood, hardwood, and mixed-wood. For the entire Ashland District (194,124 hectares), the Landsat-based, computer-derived estimate of total forest acreage was only 1.6 percent higher than the estimate obtained by the Seven Islands Land Co., using conventional techniques. Acreage differences for the hardwood, softwood, and mixed-wood classes are within 5 percent for the entire district, but when the district was divided into 29 townships, there was an average of 10 to 16 percent difference in acreage estimates for the different cover types. The authors concluded that a quantitative evaluation of classification performance indicated only about a 60 percent accuracy, which would explain much of the decreased acreage estimates of the forest-cover types in the individual townships as compared to the entire district. However, since their classification of forest and non-forest categories was fairly high, total forest-acreage comparisons were still within 1.8 percent agreement, even at the township level.

LANDSAT ANALYSIS RESULTS

As previously indicated, after the launch of Landsat-1 a great deal of interest was generated in evaluating the potential for using Landsat satellite data to map forest cover types and condition. In a study using the first digital tape of Landsat-1 MSS data ever obtained, the entire frame was classified into eleven spectral classes using a clustering algorithm to define the spectral training classes, and the Gaussian maximum likelihood algorithm was used to classify the data (Landgrebe et al., 1972). Because of the immediate delivery of the data tape from NASA to the research team, the data reformatting and classification was completed in less than 3 days after the Landsat scanner system first started collecting data on July 25, 1972. Subsequent field checking of the area indicated that forest cover had been successfully delineated with a high degree of reliability. An area that had been sprayed with 2,4,5-T as part of a rangeland improvement project could be clearly defined in the infrared channels of the Landsat data, even though the area could not be detected on the visible wavelength bands nor was it evident to the field team who flew over the area in a small airplane (see Figure 34-53 and Color FIgure 34-54). The analysis of these Landsat data also revealed that variations in slope and aspect of the terrain caused distinct differences in the reflectance values measured by the satellite, even though there were only minimal differences in the type and condition of the forest cover. After the field observations of the area, a more detailed classification of these data resulted in the definition of twenty spectral classes representing eleven informational categories. Figure 34-53 shows the individual wavelength bands of a portion of this first frame of Landsat-1 MSS data and Color Figure 34-54 shows the computer classificaiton of the same area.

Two of the most detailed early studies directed at the use of Landsat data and computer-aided analysis techniques for classifying forest lands were recorded by Heller [Tech. Coord.] (1975) and Hoffer and Staff (1975a). The study reported by Heller was conducted by the U.S. Forest Service on test sites located in Georgia, Colorado, and South Dakota, and included an evaluation of both manual and computer-aided analysis techniques. In the Georgia test site, three different blocks were classified into both Level-I and Level-II (Anderson et al., 1976) groups. The Level-I included forest, non-forest and water, and the Level-II cover types included pine, hardwood, grassland, bare soil, wild vegetation, and water. Computer classifications were carried out using both a Minimum Distance to the Means and a Gaussian Maximum Likelihood algorithm. The classification results were evaluated using all three techniques described previously (i.e., a qualitative comparison between classification maps and reference data, quantitative point-by-point accuracy tabulations, and acreage comparisons). The qualitative evaluation indicated that the classifications appeared to be reasonably accurate. This was confirmed by the acreage comparison which showed that computer classification of forest land was within 6 percent of the acreage estimated by conventional techniques. Point-by-point comparison indicated that forest land was classified with a 90 percent accuracy with the Maximum Likelihood algorithm and 80 percent with the Nearest Neighbor algorithm. Computer-aided analysis of two blocks of data in the Black Hills, S.D. indicated a high degree of accuracy for distinguishing coniferous forest from other cover types using either classification algorithm. However, the classification performances for the other cover types present in the test site were much lower at either a Level-II or Level-III degree of detail. The authors concluded that a more detailed breakdown of cover type could not be accomplished with an acceptable degree of accuracy using either computer-aided analysis tech-

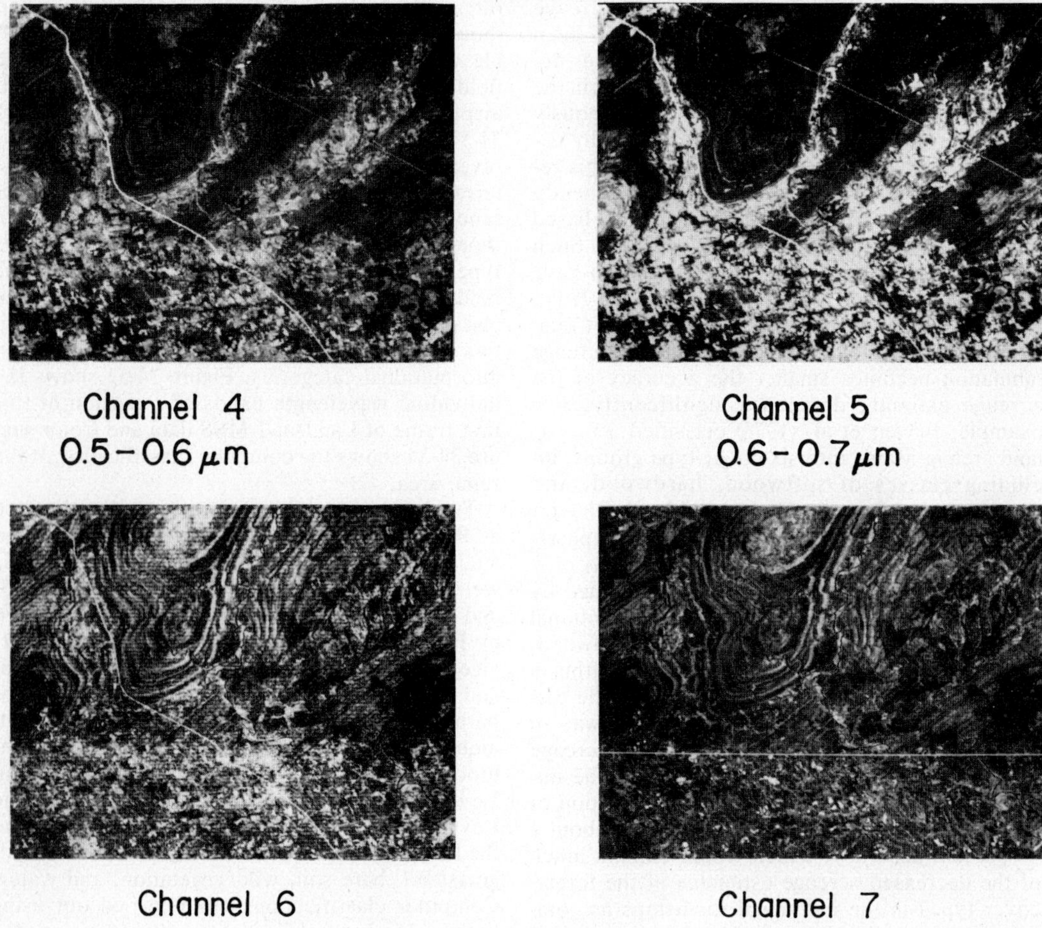

Fig. 34-53. Individual bands of a portion of the first frame of CCT Landsat-1 data recorded. The area shown is of a portion of the Ouachita Mountains in Oklahoma.

niques or with manual interpretation procedures. They stated that, in general, Landsat data provide a "good Level-I forest classifier and can be useful for detection of U.S. Forest Survey inventory plot changes or disturbances where there is much human activity." They also stated that "Forest acreages estimated by computer classification for relatively small areas are better than the accuracies now required by the Forest Survey for counties," but raised several questions regarding the operational use of Landsat data. Another conclusion was that "Classification can be done most effectively by computer, but photointerpretation produces equally accurate results. Choice would depend on availability of trained people and equipment". This study also indicated that stress conditions in ponderosa pine due to a mountain pine-beetle infestation could not be detected on the Landsat data using either manual or computer-assisted analysis techniques, although large stands of eucalyptus trees dying from stress due to extremely low temperatures could be discriminated quite accurately.

The study by Hoffer and Staff (1975a) involved four test sites in the San Juan Mountains and the Front Range in Colorado. The test sites varied in size from 14,378 to 1,011,740 hectares. In each site, training statistics were developed using a multi-cluster blocks procedure and the data were classified using the Gaussian Maximum-Likelihood algorithm. The classification results were compared to type maps developed by researchers from the University of Colorado who had not been involved in the computer-aided analysis. Quantitative evaluations were also conducted using both test fields and acreage comparisons. At least 2,000 pixels were included in each of the test-data sets. Level-II classifications (coniferous, deciduous, grassland, barren, tundra, and water) of each of the four test sites resulted in 91 percent to 94 percent overall classification accuracies. On one site, the data were classified at a Level-III degree of detail (ponderosa pine, Douglas-fir/Englemann spruce/sub-alpine fir, Gambel oak, aspen, grassland, barren, cropland, and water), which resulted in an overall classification

TABLE 34-10

Color Codes and Category Grouping Related to Figure 34-54.

Category	Spectral Class	Color Code	Cover Type
1	13, 14	Dark Green	Forested areas
2	12, 15	Light Green	Forested areas having high response due to topographic position.
3	16	Black	Forested areas having low infrared response due to stress conditions and topographic position.
4	17, 19	Dark Red	Forest area that has been sprayed with 2, 4, 5-T for a range control project. Vegetation has been killed.
5	18, 20, 21	Magenta	Forested areas with low infrared response due to 2, 4, 5-T application last year. (Area could not be distinguished from surrounding vegetation when flying over in a light aircraft).
6	1	White	River Area—water and surrounding vegetation.
7	4	Very Pale Yellow	Highly reflecting objects including I-75, limestone outcrops, and scattered points of highly reflective material (particularly soils).
8	3	Yellow	Agricultural areas having large amounts of highly reflective soil exposed.
9	5, 6, 7, 8	Orange	Agricultural areas with moderately reflective soils.
10	9, 10	Brown	Agricultural areas with relatively low reflecting soils.
11	2, 11	Blue	Range land

accuracy of 76 percent. Some of the individual species were classified with only a 62 percent accuracy, which was not considered to be satisfactory for meeting user needs. Comparisons between Level-I or Level-II acreage estimates obtained by computer analysis of satellite data and those obtained by the planimetering of type maps prepared from aerial photos indicated very good agreement (correlation coefficients were over 0.97 in each of the four test sites). On the large San Juan Mountain test site, acreages were compared for each of seven 7½′ U.S.G.S. quadrangles for Level-I cover types (forest, grassland, barren, and water). A maximum of 10 percent difference was found for any cover type within any of the quadrangles, and for all seven quadrangles together, less than 6 percent difference was found between the acreage estimates for any of the cover types.

In addition to evaluating classification techniques and performances, this study also found that elevation and aspect could be related to significant variations in spectral response for several forest cover types, and that stand density caused a significant variation in the spectral response of ponderosa pine. A cost comparison indicated that for a relatively small test site (23,000 hectares), computer classification was twice as expensive as photo-interpretation (2.2¢/hectare vs. 1.1¢/hectare, respectively), but that for a relatively large area (e.g., 1,011,740 hectares) the computer classification was approximately one-fourth as expensive as photo interpretation (i.e., 0.25¢/hectare vs. 1.1¢/hectare, respectively). Thus, the size of the area involved, the level of detail of the classification, and the spatial and spectral complexity of the area can all have a significant impact on the relative costs of computer classification versus photo interpretation. The authors concluded that, for areas of over 40,000 hectares, computer-aided analysis may be more cost-effective than photo interpretation for obtaining maps and acreage estimates at a Level-I degree of detail.

Overall, the results of this study indicated coniferous and deciduous forest-cover types as well as forest and non-forest categories could be effectively identified and mapped using satellite data and computer-aided analysis techniques, even in areas of mountainous terrain where the vegetative cover and the terrain cause the data to be spectrally and spatially complex (Hoffer and Staff, 1975a).

A portion of the Sam Houston National Forest was used as the test site in a study reported by Heath (1974). Three different computer-analysis techniques were attempted, and the results obtained using a clustering technique were reported. The researchers attempted to identify hardwood, pine, cut-over hardwood, cut-over pine, regenerated pine, and site-prepared cover classes, as well as eight non-forest land-use classes. An overall classification performance of 74 percent was obtained for the fourteen classes included in the analysis. Heath concluded that computer-aided analysis techniques were better than conventional image interpretation in classifying these data, largely because the interpreters could not separate all of the fine gradations of tone that were required for a detailed classification. This study did detect one instance where approximately two hectares of pine had been killed by pine bark-beetles. They also reported that an area subjected to a prescribed burn could be clearly delineated on Landsat data obtained approximately five weeks after the burn had taken place. The burned area showed up as a distinct black color on a color infrared composite image as compared to the surrounding healthy vegetation. Burned over areas were also defined and mapped using computer-aided analysis techniques in Landsat investigations reported by Hitchcock and Hoffer (1974), Lawrence and Herzog (1975), and Lee et al. (1977).

One of the most significant advantages for spacecraft data involves the synoptic view. Satellite data can be utilized very effectively to stratify major cover type groups, and several papers have described various approaches to multi-stage sampling, using Landsat or other satellite data in the first stage of the analysis (Langley, 1969; Gialdini et al., 1975; Barker and Fethe, 1975; Nichols et al., 1973 and 1976; Johnson et al., 1979; Rohde et al., 1979; Strahler et al., 1979; and Todd et al., 1980). Although the specific techniques used and the results obtained can vary considerably, there is definite agreement that multistage sampling procedures offer a cost-effective method for obtaining reasonably accurate and reliable resource information of various types.

A paper by Johnson et al. (1979) pointed out that the Washington State Department of Natural Resources (DNR) is responsible for managing two million acres of forest land. Approximately 12¢ per acre is spent each year by the DNR on the collection, storage, and retrieval of forest-inventory data. After some of the early research work indicated the potential for utilizing Landsat data and computer-aided analysis techniques, the Washington DNR developed a project to evaluate satellite technology and multistage sampling for forest inventory (Nichols et al., 1976). Three test sites in S.W. Washington were involved. A clustering algorithm was used to aid in defining training statistics and a Gaussian Maximum-Likelihood algorithm was used to classify the data into cover types which included timbered land, open areas, agricultural areas, water, and swampland. These results were then used for an initial stratification of the data into forest and non-forest groups in order to reduce the number of ground samples required. Statistical analysis indicated that, without the stratification, 595 ground plots would have been required, but with stratification (utilizing Landsat data and computer analysis), only 96 plots were needed. After allowing for the cost of the stratification procedure, it was shown that stratification would be over 3.6 times more cost effective.

This study also evaluated the relative usefulness of the Landsat computer-aided analysis results and results obtained by photo interpretation of Landsat data. The correlation between photo interpretation of U-2 aerial photography and both Landsat manual interpretation and computer-processing results were compared. They found a correlation coefficient of 0.68 between the U-2 and the Landsat photo interpretation results, and 0.78 between the U-2 and the Landsat computer-aided analysis results, indicating the higher relative value of the computer-aided analysis (Nichols et al., 1976).

As a follow-on to this Washington DNR multistage forest inventory project, Johnson et al. (1979) described a study in which the DNR computer-based Gridded Resource Inventory Data System (GRIDS) was used to aid in developing training statistics for computer classifications of Landsat data. This project also examined the potential for using Landsat data and computer-aided analysis techniques to separate forest stands into discrete forest-resource classes. Landsat data were geometrically adjusted to a State Plane Coordinate system and registered to the GRIDS plot locations on two test sites. The GRIDS data were then used to characterize spectral clusters developed from Landsat data. The results showed that separability among the more detailed resource classes was not as reliable as among the more general resource classes. Overall agreement between detailed classes was only 56 percent, but most of the classification error was between similar cover-type classes such as recent clearcuts and planted clearcuts. If the age of reproduction had reached five years, however, reproduction could be reliably distinguished from clearcuts in this western hemlock zone, as indicated by classification accuracies of 85 percent and 87 percent for clearcuts and reproduction, respectively. The classification accuracy of established conifer forest was 91 percent, indicating that conifer forest could be reliably separated from reproduction and clearcuts in that area. Second-growth and old growth sawtimber stands were classified with 64 percent and 70 percent accuracy, respectively. However, in the second test site area—a well-established second-growth conifer forest—they found that reproduction, poletimber, and sawtimber size classes could not be reliably separated using Landsat data. They determined that the spectral response was heavily influenced by the topography and tree-crown canopy-closure, and concluded that future efforts would need to utilize more sophisticated data-processing techniques.

Rohde et al. (1979) described stratification and multistage sampling procedures that were used in conjunction with computer-aided analysis of Landsat data to determine the area of six forest and rangeland cover types. The multistage sample included 1:6000 scale aerial photos to obtain the acreage estimates of the various cover types. The results indicated that the combination of Landsat data and large-scale aerial photos can be effectively used to inventory wildland vegetation.

The "Ten-Ecosystem Study" by the U.S. Forest Service was a major effort in the late 1970's to evaluate the utility of Landsat data and computer-aided analysis techniques for mapping forest and rangeland cover in ten broadly-defined ecosystems throughout the U.S. (Kan and Weber, 1978; Mazade et al., 1981). The ecosystems studied included: (1) Coastal Range and Rocky Mountain Conifers, (2) Northern Hardwoods, (3) Northern Conifer, (4) Chaparral-Piñyon-Juniper, (5) Oak-Pines, (6) Southeastern Pines, (7) Boreal, (8) Rangeland, (9) Pacific Coast and (10) Central Hardwoods. The first phase involved evaluating classifications that were based on training data in order to determine the maximum information content in the Landsat data. The results showed that the potential to discriminate between softwood, hardwood, grassland, and water cover types is very good (i.e., accuracies for the training data were in the high 90 percentile range), but discrimination between species or forest types varied considerably, with accuracies ranging from 9 to 98% for the training data. Next, Level II (hardwood, softwood, grassland and water) classifications were performed using training data developed with the aid of aerial photos from a representative, contiguous 10 percent portion of the test site. Evaluation of these classifications, using test data, showed accuracies ranging from 70–94 percent. These investigators reported that, for sites where the vegetation was homogeneous and occurred in large contiguous patterns, accuracies were 85 percent or higher and area estimates were reasonably accurate. However, where classification accuracies fell below 80 percent, the area of major cover classes tended to be underestimated.

Differences in crown-canopy closure have been shown to cause significant differences in spectral response on both Landsat and Skylab MSS data. Stand density was found to be related to spectral response in stands of ponderosa pine by Hoffer and Staff (1973, 1975a, and 1975b), Heller [Tech. Coord.], 1975; Lawrence and Herzog, 1975; Strahler et al., 1978, and Mead et al. (1979), and for Southern pine forests by Williams (1976). Differences between closed and open canopies also caused distinct differences in spectral response in Douglas-fir, and red and white fir stands in California (Strahler et al., 1978). In all eight studies, as the canopy became more open, there was an increase in reflectance measured by the Landsat or Skylab scanner. Williams (1976) indicated that the increased reflectance was most apparent in the infrared channels due to the high reflectance of the lush ground vegetation prevalent in the Southeast on the August data being analyzed. The ponderosa pine stands generally had a light yellowish-brown grass understory that became more predominant in the Landsat or Skylab pixel as the canopy closure decreased, again causing increased reflectance (Hoffer and Staff, 1975b).

Costs involved in computer processing of Landsat data have been reported by several investigators, but varied considerably depending on the hardware and software used, the method of evaluating costs, the level of detail and type of results obtained, and the size of the area involved. Bryant et al. (1978) indicated a total cost of 3.5¢/hectare for classifying approximately 0.5 million areas into 6 cover-type categories using the GISS software system. Using a clustering algorithm to develop training statistics and a maximum-likelihood algorithm for classifying the data, Rohde (1978) indicated that Landsat data can be used to accurately map the extent of strip-mine disturbance in deciduous forest cover-type at a cost of 3¢/acre (7¢/hectare). Rohde (1978) also described the classification of 293,000-acre area in Alaska. The Landsat data were classified into one of nine land-cover types at a cost of 4.0¢/hectare, with a resultant overall classification-accuracy of 84.5 percent ± 4.2 percent at the 0.95 probability level, as estimated by a stratified random-sampling procedure. Kan and Weber (1978) determined that their analysis costs ranged from 12¢ to 25¢/hectare for some of the Ten-Ecosystems study sites, but it should be noted that the study involved several phases in the analysis sequence, so these costs included more than just a single classification of the data. Mazade et al. (1981) stated that the final results of the Ten-Ecosystems Study indicated that, using their approach, an operational capability to map softwood, hardwood, grassland, and water classes with an accuracy of 70 percent or better would cost 12¢/hectare. Hoffer et al. (1975a) reported costs of 2.5¢/hectare for classifying a 22,800-hectare test site, based on a Gaussian Maximum-Likelihood classifier, and only 0.25¢/hectare when the 982,400-hectare San Juan Mountain test site was classified to a Level II degree of detail. Using the same techniques, Krebs and Hoffer (1976) cited costs of 4.0¢/hectare, including topographic overlay, for an area of over one million acres, and pointed out that if the size of the study area had been significantly increased, the unit cost would have been lower, and vice-versa.

In addition to the studies directed at classification of various forest cover-types, there has been considerable interest in the potential for using Landsat data to monitor changes in the forest-resource base throughout the world. Williams and Miller (1979) summarized the results of several studies in which the authors had been involved for areas in the Republic of China, Nigeria, Haiti, the Dominican Republic, and Thailand. Each of these studies involved digital analysis of Landsat data, although they varied in complexity and the type of analysis procedure used. In the Republic of China, the results showed the encroachment of agricultural land into forest land areas on steep slopes. Classification of Landsat data of Nigeria indicated that significant portions of virgin forest reserves established by the Nigerian government have been illegally logged and converted to brushland and permanent agriculture.

A frame of Landsat data along the border between Haiti and the Dominican Republic was analyzed using a Euclidean Distance algorithm. A total of 17 spectral classes were used to identify five informational categories—forest, brushland, agricultural areas, wetlands, and barren areas. Identification of the barren areas was of particular concern. The authors state that "The economy of the area [in the Dominican Republic] is depressed and an ever-increasing amount of the remaining forest land within the province is being cleared to plant agricultural crops and to provide charcoal for cooking and heating. This phenomenon has occurred in an even more accelerated fashion just across the border in neighboring Haiti, with the result that the Haitian mountain slopes are practically devoid of forest vegetation and are severely eroded" (Williams and Miller, 1979). The classification results depicted the high level of forest exploitation that has occurred, particularly on the Haitian side of the border. These investigators also determined that, at the province level, a sampling of as little as one or two percent of the Landsat cells that had been actually classified provided an adequate representation of land cover. They indicated that image sampling prior to classification may be feasible for reducing processing costs in the monitoring of large-scale forest alteration on a national or worldwide basis.

In describing the situation in Thailand, the authors state "All forest types are being cut and burned on cycles of something like six or seven years, replacing the past process which selectively used more resilient forest types on cycles of greater than 20 years. Suddenly all of the forests are disappearing and the abundant tropical rainfall and unavoidable erosion of these large non-forested areas is making the process irreversible. Large scale examples of the results of this disastrous process already exist in the extensive desert areas which have been created in the northeastern portion of the Amazon Basin" (Williams and Miller, 1979). The analysis involved Landsat data of Thailand for five consecutive anniversary dates. The data were preprocessed and "difference images" were generated for consecutive pairs of dates, using the same spectral band each time. The resulting image products provided an effective mechanism for reconnaissance-level detection of areas of change that had occurred between consecutive years.

In Brazil, Landsat data are being utilized on an operational basis to map, evaluate and analyze the rate of deforestation of the forest cover of Brazil, with particular emphasis on the Amazon region. Landsat data obtained between 1973 and 1975 were used to do a preliminary assessment of deforestation of the Amazon region until that time. It was found that 2.86×10^6 hectares had been deforested through 1975 (approximately 1.02 percent of the forested portion of the entire Brazilian Amazon region). Three years later, Landsat data were again analyzed to determine the rate of deforestation. It was found that, during this three year interim, approximately 4.86×10^6 hectares of additional forest land had been deforested. Thus, as of 1978, over 7.7×10^6 hectares of the Brazilian Amazon region had been deforested (Carneiro, 1981). Although this analysis of the deforested area has involved manual interpretation for the most part, computer-aided analysis techniques have been involved in mapping areas of reforestation on an annual basis in order to identify, evaluate and analyze the effectiveness of a tax incentive program for forest plantations. As of 1980, almost 4 million hectares of forest, primarily of *Eucalyptus spp.* and *Pinus spp.* has been established, with over 400,000 hectares of new plantations being established annually (Carneiro, 1981). This monitoring program, which is primarily based upon the use of Landsat data, may be the world's largest operational project for forest-cover mapping.

COMPARISON STUDIES AND REFINEMENTS IN ANALYSIS PROCEDURES

As a result of the early studies which showed many of the limitations as well as the capabilities for computer-aided analysis of Landsat data, much additional work has been conducted to evaluate how to best utilize single or multidate Landsat data and to develop alternative or more sophisticated methods for processing Landsat data. In some cases ancillary data have been used in conjunction with the Landsat data to develop more effective analysis techniques.

Time of Year, Multi-temporal Data, and Change Detection

One of the earliest recognitions of a major source of variability in Landsat data involved the changes in spectral response due to season. Aldrich (1975) stated that the season of the year can be a much more critical factor in working with the relatively low-resolution Landsat data than with high-altitude aerial photographs. He found that spring data were best, and late-fall to late-winter was his second choice for detecting disturbances in a forest environment in Georgia. Late June is also the best time of year for obtaining data to monitor gypsy-moth defoliation in Pennsylvania, according to Williams (1975), who showed that Landsat data could be used successfully for this purpose. However, to detect stress conditions caused by southern pine beetle in ponderosa pine in South Dakota, June is the worst possible time of year, whereas late summer would provide much better data, according to Heller [Tech. Coord.] (1975). Even on late summer data, however, stress conditions could not be detected.

Barker (1981) indicated that, for purposes of classifying coniferous forest cover, winter data were much better than any other season, since they were working in the southwestern U.S., and were not interested in general cover-type mapping. Williams (1976) and Kourtz (1977) also indicated that winter data were best for obtaining an accurate differentiation between pine and hardwood. However, Williams (1976) indicated that summer data provided some differentiation among three levels of pine-canopy closure, whereas Kourtz (1977) stated that the winter data provided a great deal of coniferous density and age class information. Kan and Dillman (1975) indicated that early- or late-spring phenological conditions could be better than winter season for separating softwoods, hardwoods, and regeneration in the Southeastern U.S. In evaluating the best season to obtain Landsat data for the purpose of mapping general forest-cover types, Kan and Weber (1978) indicated that late spring and/or early summer were frequently chosen as best, but this was not consistent among the ten different ecosystems. They pointed out that the selection of best dates reflected the season when vegetative growth was vigorous, when shadow effects would not adversely affect the analysis, and when snow cover was minimal.

Landsat data were used to study phenologic changes taking place in forest canopies in a study by Blair and Baumgardner (1977). The results, based upon analysis of data from oak stands in fourteen sites indicated that the progression of the brown- and green-wave effects in the fall and spring could be clearly detected and quantified spectrally, due to differences in foliage cover, leaf senescence, and regrowth. They pointed out that, because leaf growth in the spring could be completed within 11 to 14 days, such rapid leaf development could be completed between the 18-day cycles of Landsat data. Ashley and Rea (1975) used data from the Vermont test site and found that ratios (R) of the Band 5 and Band 7 data, using the formula R = Density Band 5 − Density Band 7/Density Band 5 + Density Band 7, could be used to indicate foliage condition. Typical ratios for forest vegetation were given as 0.25−0.50 for full leaf development or the all-green condition; 0.10−0.25 for some intermediate stage of leaf development or coloration; and below 0.10 for complete leaf-off condition. Williams et al. (1979) tested eight vegetative indexes in relation to gypsy-moth defoliation, and found that all were highly correlated and that, in all cases, the index values tended to increase as the severity of defoliation decreased.

Seasonal effects on data obtained in the San Juan Mountains of Colorado were found to be significant in a Skylab study by Hoffer and Staff (1975b). They found that, for vegetative-mapping purposes, satellite color-infrared photography obtained on August 8 was superior to photos obtained on June 5 over the same test site. However, at a Level III degree of detail, computer-aided classification of the Skylab S-192 MSS data obtained simultaneously with the photography gave much better results for the June data (71 percent) than the August data (48 percent). This difference was due to the poor quality of the August data set, even though the "best" (in terms of signal/noise ratio) four wavelength-bands were used in both classifications. These results point to the importance of MSS data-quality on computer-aided analysis for forestry applications.

Such differences in seasonal effects on spectral reflectance clearly indicate the need to define the purpose for which the data are to be used, to evaluate the data quality, and to be knowledgeable about the spectral characteristics of the cover types and conditions of interest in relation to the spectral characteristics of other features present in the scene.

Because seasonal effects can have such a significant impact on the spectral characteristics of various cover types, one of the potential applications for computer-aided analysis techniques involves the analysis of digitally-combined data sets from different dates. Landsat data from three dates (September 5, 1972; June 3, 1973; and October 6, 1973) were used by Kalensky and Scherk (1975) to classify coniferous forest, deciduous forest and agricultural land for a test site in southeastern Ontario. A multidate classification using all three dates resulted in a classification performance of 83 percent, whereas the classification accuracy for single-date analysis varied significantly as a function of the date of the image acquisition. The September 5 data resulted in the highest single-date classification performance (81 percent), while the October 6 date resulted in lowest classification accuracy (67 percent), largely because the deciduous forest was at the peak of fall coloration. Several combinations of wavelength bands and dates were tested, and these investigators found that the multidate classifications were consistently above 80 percent in accuracy.

Kan and Dillman (1975) also combined three dates (winter, early spring, and late spring) of Landsat data for the Sam Houston National Forest, and found that a temporal analysis using the early- and late-spring data could improve classification accuracy by as much as 11 percent over single-date analysis. The best results in this study were 79 percent classification accuracy of test fields for softwood, hardwood, and regeneration. However, in reporting some of the preliminary results of the Ten-Ecosystems Study, Kan and Weber (1978) reported that these classifications did not indicate any definite advantage for multitemporal versus single-date classifications at the Level II degree of detail. Using a test site in Minnesota, Mead and Meyer (1977) found that registered May and July imagery did not provide more

TABLE 34-11

A Summary of the Categories Discriminated Using Winter, Summer and Multi-Temporal Data Sets (from Williams, 1976).

Category	Category Discrimination		
	Feb 74	Aug 73	Feb 74 + Aug 73
Hardwood	YES	NO	YES
Closed Canopy Pine	NO*	YES	YES
Partial Crown Closure	NO*	YES	YES
Open Canopy Pine	NO*	YES	YES
Regeneration	YES	YES	YES
Clearcut	YES	YES	YES
Change (veg. to clearcut)	NO	NO	YES

* An overall pine category was extracted using 26 Feb. '74 data, but a breakdown into meaningful subcategories was not possible.

reliable classification results than the use of May imagery alone. A multiple-date data set was also used by Strahler et al. (1978), but they did not indicate the relative value of using more than one date in the classification.

In addition to using multi-temporal data in an attempt to increase classification accuracies of the various cover types, there has been considerable interest in registering multiple sets of Landsat data and assessing changes in cover type (i.e., clearcutting, burned areas, regeneration, etc.) or condition of the forest cover (i.e., insect or disease situations).

Lee et al. (1977) indicated that summer data obtained one year apart (i.e., "anniversary data") were effective in defining clearcut areas and also a burned-over area. In both cases, only Channels 5 and 6 (0.6–0.7 μm and 0.7–0.8 μm, respectively) were used in conjunction with a supervised procedure for developing training statistics and a parallelepiped classification-algorithm. The burned-over area determined from the Landsat data was considerably less than had been reported by the British Columbia Forest Service because unburned areas had been included within the outline of the entire burned-over area of the Forest Service report, and also because a portion of the area had involved a ground fire that did not kill most of the trees, so had not been defined on the Landsat data.

Williams (1976) combined winter and summer Landsat data, and showed that the combination allowed all of the categories of interest to be defined. Hardwood could be separated from pine most effectively on winter data; stand-density classes could be defined better on summer data; regeneration and clearcut could be defined on either date; and the combination of dates allowed for the possibility of change detection. His conclusions are summarized in Table 34-11, where "YES" indicates ability for category separation and "NO" the inability to separate categories.

The potential for using Landsat data and computer-processing techniques to monitor gypsy-moth defoliation in Pennsylvania was evaluated by Williams (1975) and Williams et al. (1979). The severity and extent of gypsy-moth defoliation reaches a peak in late June through early July. If defoliation has been only light to moderate in previous years, the deciduous trees will produce new foliage by late July. Therefore, the timing of data collection from both the satellite and on the ground is extremely important. In this study, Landsat data obtained on 19 July 1976 were digitally registered to Landsat data obtained nearly a year later, on 27 June 1977. Heavy gypsy-moth defoliation was apparent in the June 1977 imagery for several thousand acres of land which had been a healthy, full-foliated forest in July 1976. These researchers found that accurate definition of areas of defoliation could be achieved by a simple enhancement which involved subtracting the corresponding wavebands of the "before" and "after" data sets to yield "difference" images. However, because leaf-biomass changes in agricultural crops showed very similar differences in spectral response for the two dates, the areas of forest defoliation and areas of agricultural crop-cover tended to be confused. To overcome this problem, the 1976 imagery was used to classify the data into forest and non-forest categories, and a "binary mask" was defined to eliminate all areas of cover types other than forest. Then, by comparing the 1977 to the 1976 imagery and creating the difference image, they found that areas in which spectral differences in leaf biomass were indicated could be related to defoliation levels. Williams pointed out, however, that the low resolution of the Landsat satellite-data makes such data appropriate only for detecting defoliation or mortality over large areas.

Improved Classification Techniques

As previously indicated, variations in spectral response in the Landsat data due to topographic effects were obvious on the very first frame of Landsat data collected. Some of the early Landsat studies indicated that coniferous and deciduous cover types could be differentiated in spite of the spectral variability due to the topography, but individual species could not be differentiated reliably. To evaluate the potential for improving the classification of forest species or cover types through the use of topographic data in conjunction with Landsat data, Hoffer and Staff (1975b) registered 13 wavelength bands of Skylab S-192 MSS data to a DMA (Defense Mapping Agency) digital-data tape containing elevation data digitized from 1:250,000 U.S.G.S. topographic maps. Aspect and slope data were interpolated from the elevation data. Classification results, using a combination of the best three bands of Skylab data plus elevation data, showed an improvement in classification accuracies of over 10 percent for several forest-cover types. Based on this initial positive result, a detailed study was conducted using Landsat MSS and DMA topographic data (Hoffer et al., 1979; Fleming and Hoffer, 1979). A topographic distribution model was developed to statistically characterize the topographic distribution of the various species present in the test site. Three methods of combining the Landsat and topographic data were evaluated. A layered classification technique (discussed below) was determined to provide the best results, with the overall classification-performance increasing by 15% for a Level III classification. However, even with use of the topographic data, overall classification-accuracy was only 66 percent, although much of the confusion occurred between "pure" cover-type classes (e.g., Douglas-fir, or Englemann spruce/subalpine fir) and intermediate-elevation "mixed" classes (e.g., Douglas-fir *and* Englemann spruce/subalpine fir).

In northern California, Strahler et al. (1978) followed a similar procedure in registering Landsat data to a DMA digital-elevation data-tape and in generating slope and aspect data through an interpolation process. Two techniques were used for classifying the combined Landsat/topographic data. The first, defined as the "logical channel" approach, treated elevation and aspect as additional channels of Landsat data. The second approach, termed the "probabilistic approach" used the topographic information to incorporate prior probabilities into the maximum-likelihood decision rule. The cover types included open and closed canopies of Ponderosa pine, Douglas-fir, white fir, and red fir, as well as classes of mixed small trees, meadow, barren, hardwoods, grass and shrubs, unclassified, and roads. The results indicated a 58 percent overall accuracy when no terrain information was included, but the prior-probabilities approach, with Landsat data plus elevation data, gave a 71 percent accuracy; elevation-plus-aspect increased the accuracy to 77 percent. Use of the "logical channel" resulted in classification accuracies of 82 to 85 percent. They concluded that the different methods did not seem to make a significant difference in the classification results, but the incorporation of terrain information into a forest-classification system could increase the accuracy of classification of species-specific cover types by as much as 27 percent.

The discussion thus far has suggested that several different classification algorithms are available to the analyst. Among the more common algorithms that have been used for forestry applications are the Gaussian Maximum-Likelihood, Parallelepiped, and Minimum-Distance-to-the-Means algorithms. There have been several other classification algorithms developed over the past few years, some of which have been successfully evaluated for forestry applications. One of these, called the "Layered Classifier," is designed to allow the analyst to interface with the data at several critical points in the analysis sequence (Swain et al., 1975). This algorithm is also designed to be a relatively inexpensive classifier to use, and to be effective in situations where many wavelength bands are available for use in the classification. In using the Layered Classifier, one must first classify the data into a relatively few, spectrally distinct and easily separated classes based upon a user-defined subset of the wavelength bands available. In the second and subsequent steps or layers in the analysis sequence, only the pixels classified into one of the classes defined by the previous layer are involved, and at each layer the analyst specifies the particular combination of wavelength bands that will most effectively identify the particular cover types of features of concern at that level in the classification sequence (see Figure 34-55).

As mentioned previously, the Layered Classifier has been used in conjunction with a set of Landsat data that had been digitally registered and combined with a set of topographic data (elevation, slope, and aspect) (Hoffer et al., 1979; Fleming and Hoffer, 1979). The first layer used Landsat data to classify major cover types (coniferous forest, deciduous forest, grassland, and water), and then the second-layer topographic data were used to classify individual forest-cover types (see Figure 34-56). Employment of the combined Landsat and topographic data to classify cover types using this technique increased the overall classification-accuracy by 15 percent over that obtained using Landsat data alone.

There have also been techniques developed to provide alternative approaches to the classification of individual pixels of data, one at a time. Most classification algorithms classify the data on an individual pixel-by-pixel basis. These are the so-called "per-point" classifiers. Such individual-pixel classifiers often result in a "salt-and-pepper" effect in the classification map obtained. Foresters sometimes object to the resulting classification map because it contains more detail than is actually required or desired. To overcome this objection, computer programs have been developed to post-process the data after the classification has been completed, allowing the data to be "smoothed," thereby eliminating much of the salt-and-pepper effect. Kan (1976) describes

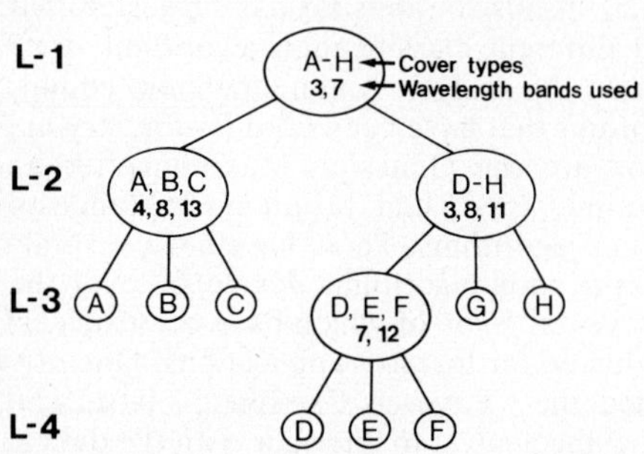

Fig. 34-55. The Layered classification technique concept. Letters indicate the cover types, and numbers indicate wavelength bands used in the classification.

two such programs, called "CLEAN" and "GETMIX", which were designed to map mixed features and remove the spotty appearance in single-pixel classification maps. Through a series of iterations, isolated pixels and small groups of pixels that were classified differently from the surrounding pixels were merged into the surrounding cover type. The user can specify the size of the group of pixels that would be subject to such merging, e.g., 10 acres, in order to produce a map looking more like a forest type-map based on a 10-acre-minimum mapping unit. Todd et al. (1980) described a similar "smoothing" procedure.

Other approaches to obtain more generalized maps without the salt-and-pepper effect involve classification techniques that utilize the spatial variability in the spectral data within the classification algorithm. Sadowski et al. (1978) indicated that multi-element classification rules, such as the "nine-point" rules developed at ERIM, have considerable potential for improving classification performance. These multi-element rules determine the classification of individual resolution elements on the basis of the spectral data from that resolution element and its eight immediate neighbors. Such an approach attempts to improve classification performance by incorporating the likelihood that a specific resolution element belongs to the same cover type as its neighbors. The influence of neighboring pixels can be varied, depending on the particular nine-point rule being applied. Four rules were tested, using aircraft MSS data having 32 m resolution, and resulted in classification performances ranging from 13 to 25 percent better than when the pixels had been classified individually. Sadowski's paper also describes a "proportion space" technique that had been tested on a set of aircraft data with good results, indicating another approach that may prove effective in working with the higher-resolution data that will be available from future satellite systems.

The relationship between spatial resolution of the MSS data and classification accuracy has been examined by several researchers, and in each study, the results demonstrated that classification accuracies of per-point classifiers will decrease as the spatial resolution of the data increases (Kan et al., 1975; Sadowski et al., 1978; and Latty and Hoffer, 1981). This is particularly true for forest or other wildland cover types since they usually have more spectral variability from one pixel to the next, whereas agricultural crops are more uniform. This is shown in Figure 34-57 which compares classification performance by cover types using MSS data at four resolutions ranging from 15 m to 80 m. Except for the relatively smooth-crowned stands of tupelo, the forest-cover classes all have distinct decreases in classification performance as the spatial resolution increases.

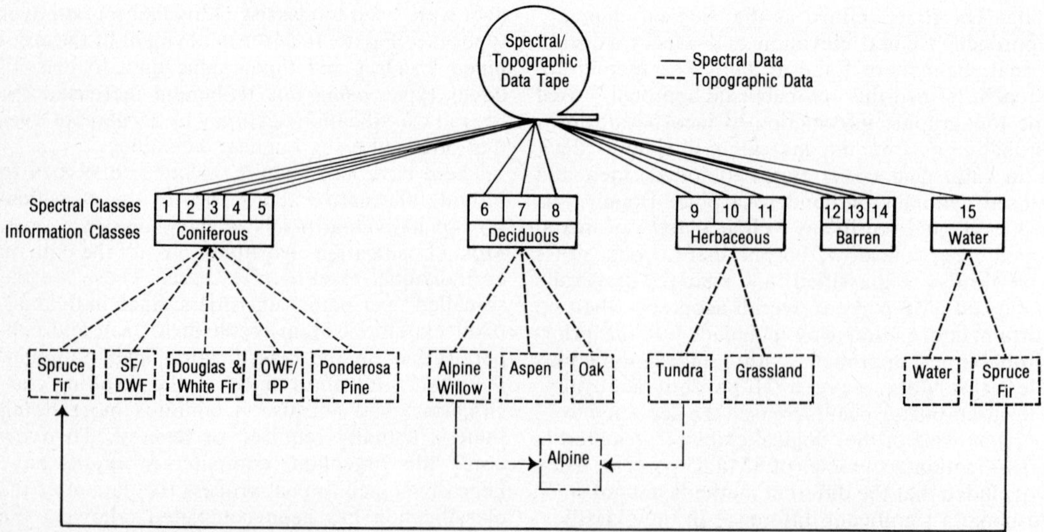

Fig. 34-56. Example of decision tree used with the Layered Classifier to combine spectral and topographic data (from Fleming and Hoffer, 1979).

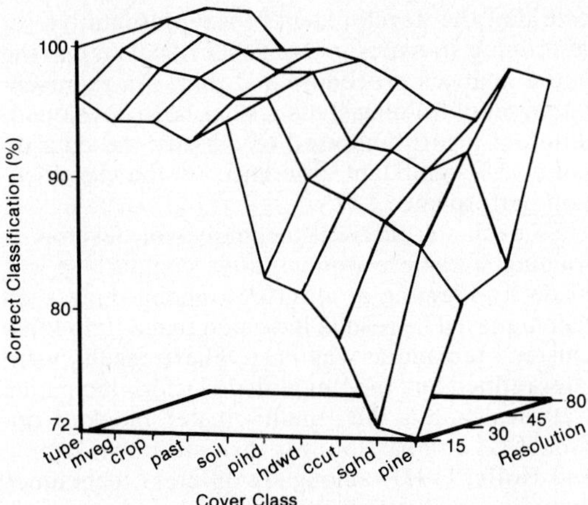

Fig. 34-57. Classification accuracy obtained for various cover types in relation to spatial resolution of MSS data (from Latty and Hoffer, 1981).

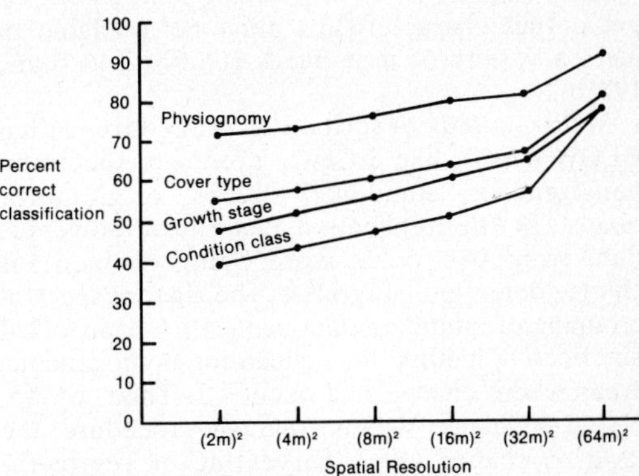

Fig. 34-58. Classification accuracy in relation to spatial resolution and the physiognomic hierarcy of the forest cover classes (from Sadowski, et al., 1978).

Whereas many of these differences were statistically significant for the forest-cover classes, none of the agricultural classes had significant differences in classification accuracy as a function of spatial resolution. The impact of spatial resolution on classification performance is also shown in Figure 34-58. This figure not only shows the distinct increases in percent correct-classification as spatial resolution decreases, but also illustrates that, as the level of detail being derived from the data increases, the overall classification-performance will decrease. In this case, the physiognomic classes involved were pine sawtimber and pine regeneration; the cover type classes were shortleaf pine, loblolly pine, and pine regeneration; the growth stage divided mature and immature pine sawtimber (regardless of species); and the condition class separated the mature and immature pine sawtimber according to species (Sadowski et al., 1978).

The studies just described each utilized aircraft data to study the impact of scanner resolution on classification performance and demonstrated that, as resolution increases, variance in spectral-response values increases and classification accuracies decrease, especially for forest-cover types. Such results clearly indicate that techniques must be developed to effectively utilize the increased spatial resolution that will be available from future satellite systems. However, even the 80-meter spatial resolution of the Landsat-1, 2, and 3 MSS data has been shown to contain distinct pixel-to-pixel variations in spectral response in areas of forest cover. Strahler et al. (1979) described an interesting approach to characterize the spatial information content in Landsat data by calculating the standard deviation in spectral response of Landsat Band 5 data for a three-by-three moving window. Once calculated, these values were scaled, associated with the center pixel of the three-by-three window, and output in image format representing the texture of the data. These texture values were calculated for every pixel in the data set, thereby creating a data tape having four channels of spectral data plus one channel of textural data. They found that use of the texture channel in an unsupervised classification on the study site in northern California produced many site-specific volume-homogeneous classes.

A different approach to the use of textural data has been incorporated into the "ECHO" (Extraction and Classification of Homogeneous Objects) classifier which was used by Hoffer and Staff (1975b) to classify forest and other cover type classes in an area in S.W. Colorado. The computer is programmed to define the boundary around an area that has generally similar spectral characteristics, and then the entire area within that boundary is classified as a unit into one of the cover-type classes defined by the training statistics. The key aspect of this algorithm is that the boundaries of the forest stands in the area to be classified are defined by the computer. In addition, the ECHO algorithm utilizes both the average spectral response and the variance in spectral response—what a photointerpreter would call "texture"—in making the classification decision. The result is an output map that has an appearance somewhat like a standard forest-type map (see Color Figure 34-59). In comparing the classification accuracy of the ECHO classifier and a Gaussian Maximum-Likelihood per-point classifier, the researchers found that each of the Level III cover types was classified more accurately by the ECHO classifier—some, such as ponderosa pine and pasture, by as much as 10 and 15 percent, respectively. In qualitative evaluations of this type of classification map, some users expressed a preference for the ECHO map output, whereas others who were involved in different applications, preferred the Per-Point classification output-maps. The authors concluded that there is probably no single classification algorithm that would be completely satisfactory for all user requirements, so different algorithms having different output characteristics must be available to meet a variety of user needs (Hoffer and Staff, 1975b).

Malila (1980) described a procedure called BLOB which, like ECHO, produces spectrally homogeneous "stand-like" "blobs" or groups of pixels. In the application described, however, data from two dates were being utilized for change-detection purposes. The spatial/spectral grouping of multidate data using BLOB provided an effective method for delineating homogeneous areas where change had occurred. Then, by applying a Change Vector Analysis procedure, the type of change (e.g., harvesting or regrowth) could be determined. Of particular significance was the use of BLOB to reduce the effects of spatial misregistration in the multi-temporal data, thereby preventing the appearance of many erroneous single-pixel "changes" in the resulting map.

In considering different users and their particular information needs, one should not assume that it is always necessary to apply classification algorithms to MSS data in order to produce the most useful output. Interesting studies by Kourtz (1977) and Kourtz and Scott (1978) involved the development of forest-fire fuel maps using Landsat data and computer-analysis techniques. They found that neither supervised nor unsupervised classification results were satisfactory because these techniques group the data into a relatively few, well-defined categories, and important transition areas were not evident on the classification output maps. However, through the use of some rather sophisticated computer-enhancement procedures, output imagery could be produced that showed key features of interest such as vegetated areas, road networks and water bodies, and which also showed the transition zones among vegetation types. These enhanced output products were much more satisfactory to the field personnel, and therefore have been adopted as the standard product for operational use in the Canadian forest-fire management program. Lee et al. (1977) also indicated that image enhancement using a principal-components technique was very effective in producing imagery on which logged-over areas, coniferous forest, urban areas, roads, power lines, water, and cultivated lands could be defined using manual interpretation.

Comparison of Analysis Techniques

A variety of techniques have been developed for defining the training statistics, and there are many algorithms available for classifying the MSS data. Questions are often raised concerning the comparative effectiveness and efficiency of the different techniques and algorithms. Relatively few studies have been conducted, however, in which only the classification algorithm or training technique was varied in order to compare the various methods. Results of the few studies that have been conducted to compare methods indicate that the development of a representative set of training statistics is the most critical step in the entire analysis procedure and that, if a representative set of training statistics has been developed, different algorithms used to classify the data do not cause significant differences in the classification performance.

To evaluate different techniques of developing training statistics, a quantitative comparison was made by Fleming et al. (1975) among three such techniques. The results indicated that a "modified cluster" technique was best. These results were later refined and the "modified cluster" technique was designated the "multi-cluster blocks" approach. A comparison was then made by Fleming and Hoffer (1977) among six different techniques including the supervised, clustering, and four-hybrid techniques, referred to as multi-cluster blocks, mono-cluster blocks, multi-cluster fields, and mono-cluster fields. The multi-cluster blocks technique involved clustering several heterogeneous blocks separately and then combining the data into a single set of training statistics, whereas the mono-cluster blocks started with the same set of heterogeneous blocks but clustered them together in a single cluster-group. Multi-cluster fields used homogeneous, supervised training fields and clustered them separately to determine their spectral characteristics and then combined the data into a single set of training statistics, whereas the mono-cluster fields started with the same set of supervised training fields and clustered them together. After the training statistics were developed, the same set of data was classified using a Gaussian Maximum-Likelihood classifier, and the same test areas were used for the comparison among each of the six techniques. The results indicated that the method of developing training statistics can have a statistically significant impact on (a) the accuracy of classification achieved, (b) the analyst time required for developing the training statistics, (c) the computer time required, and (d) the ease and effectiveness of the analyst/data interface. Of particular significance was the finding that the supervised technique resulted in the lowest classification accuracy, (64.7 percent), which was 14 percent lower than that obtained by the Multi-cluster Blocks technique. The Unsupervised Clustering technique used approximately twice as much computer time as the supervised or any of the hybrid techniques. The Mono-cluster Blocks and Multi-cluster Blocks techniques required the least amount of analyst time (about 1/4 of that required by the Supervised technique, and 1/2 the amount required by the Unsupervised Cluster technique). The authors concluded that, for areas of complex vegetative cover, the Multi-cluster Blocks technique is the most effective procedure for developing training statistics.

Lee et al. (1977) evaluated both the Supervised and Unsupervised Clustering techniques, and concluded that the Unsupervised Clustering approach did not produce satisfactory results, whereas Beaubien (1979) stated that "For areas as large and complex as in this study, an unsupervised method (classification or enhancement) must be used, especially if the degree to which forest types can be distinguished or the factors which affect classification are to be determined." Thus, there is considerable disagreement concerning the most appropriate method for developing training statistics, although a majority of authors working in complex areas of forestland tend to use some form of unsupervised or guided clustering (e.g., multi-cluster or mono-cluster blocks) technique. However, there are several different clustering algorithms in use, and some of these have very distinct differences in the algorithm as well as their input and output characteristics.

The "ISOCLS" (Interactive Self-Organizing Clustering System) and "CLASSY" clustering procedures were compared by Werth (1981) for a test site on the Clearwater National Forest in Idaho. In this study, the clustering algorithms were used to actually classify an entire 7½′ U.S.G.S. quadrangle (as opposed to clustering relatively small areas to develop training statistics and then classifying the entire area of interest with the Gaussian Maximum-Likelihood, or some other, classification algorithm). An attempt was made to define individual forest-cover types but it was determined that none of the forest types in the study area could be defined by a unique spectral class or classes. Therefore, the results were based on coniferous forest, cut-over, grass, and water classes. The results were reported on the basis of both overall classification-accuracy (LARS Staff, 1970) and mapping accuracy as defined by Kalansky and Scherk (1975). Overall classification-accuracies were 81 percent for the ISOCLS algorithm and 77 percent for the CLASSY algorithm, while overall mapping-accuracies were 69 percent for ISOCLS and 67 percent for CLASSY. The author indicated that the overall results were not significantly different, although there were greater differences among some of the individual classes. Because ISOCLS required many trial-and-error runs to find the proper parameters, he concluded that "CLASSY appears to show more promise for forest stratification than ISOCLS and shows more promise for consistency" (Werth, 1981).

ISOCLS was also involved in a study that compared ISOCLS to the Gaussian Maximum-Likelihood algorithm (Nelson and Hoffer, 1980). A Procedure-1 approach, developed at NASA's Johnson Space Center for agricultural applications, was used in conjunction with ISOCLS to develop one set of training statistics, whereas the Multi-cluster Blocks technique, described earlier, was used to develop a second set of training statistics. Both sets of training statistics were used to classify the same set of data, using the Gaussian Maximum-Likelihood classification algorithm. The results, compared at a Level II degree of detail, showed an 88.3 percent overall classification-accuracy for the Procedure-1 ISOCLS technique and 87.4 percent for the Multi-cluster Blocks technique—a difference that was not statistically different at the 0.10 α-level. The authors indicated, however, that the ISOCLS clustering-procedure was rather difficult to use, requiring considerable effort to develop appropriate clustering parameters for a particular site, and that a particular set of ISOCLS parameters seems to be rather site- and date-specific.

Four different classification algorithms were compared in a study reported by Hoffer (1979) using Landsat data obtained over a mountainous, mostly forested test site in southwestern Colorado. A single set of training statistics had been developed, and the results were compared using the same set of test areas, which included 3,704 acres—a 10 percent sample of the area classified. Classifications were conducted using the Gaussian Maximum-Likelihood algorithm, the Minimum Distance to the Means algorithm, the ECHO (Extraction and Classification of Homogeneous Objects) classifier (discussed previously) and a Layered classifier. Classifications were conducted at both Level II (coniferous forest, deciduous forest, grassland, barren, and water) and Level III (ponderosa pine, spruce/fir with <80 percent crown closure, spruce/fir with >80 percent crown closure, mixed coniferous/deciduous, aspen, oak, rangeland, agricultural land, barren, and water) degrees of detail. The results showed that each of the four classifiers produced classification accuracies that were approximately the same—between 92.8 percent and 93.8 percent at Level II, and between 73.0 percent and 75.9 percent at Level III. However, the cost of computer time varied significantly, with the Gaussian Maximum-Likelihood classifier requiring approximately 2½ times more computer time than the Minimum-Distance-to-the-Means classifier.

The author concluded that if a good, representative set of training statistics has been defined, the particular algorithm used to classify the data may not be a critical factor in obtaining satisfactory results. However, the spectral complexity of cover types involved and the degree of detail into which the data are being classified also must be taken into account. If the cover types have distinct, easy-to-separate spectral characteristics, perhaps one of the simpler classification algorithms can be utilized effectively rather than the more costly but more robust Gaussian Maximum-Likelihood algorithm. A key summary statement is that "no matter which classification algorithm is utilized, the training statistics must effectively represent the spectral characteristics of the various cover types present in that data set! If the training statistics are not representative, the classification results will not be satisfactory" (Hoffer, 1979).

Increasingly, mathematical models using remote sensing and ground data are being applied to forest management problems. Three such models were developed for the Feather River area (500 km^2) in northeastern California—a well documented test area used by investigators from the Space Sciences Laboratory in Berkeley, California. These models incorporated various kinds of remote sensing and ground data as independent variables and were used to predict thematic dependent variables such as:

1) timber production capability maps
2) daily potential evapotranspiration and
3) average daily radiation.

From these data, thematic maps were produced illustrating each of the dependent variables. For example, Smith and Khorram (1980) investigated the generation of timber production capability maps on the Feather River test area. They used a geo-based information system (developed earlier by Khorram et al. 1978) using albedo, topography, surface temperature net radiation, potential evaporation and soil/plant available water as input parameters for this predictive model. From the model they developed a potential growth rate map (Figure 34-60) that could be useful to forest managers.

Daily potential evapotranspiration (ET) was also calculated by Khorram (1979) on a larger test area in northeastern California by developing a model using several independent variables which included data from: 1) Landsat MSS, 2) the NOAA-5 very high resolution radiometer, 3) aircraft multiband photography, 4) USGS topographic maps, 5) DMA/USGS digital terrain tapes and 6) climatic ground instruments. A daily potential evapotranspiration map was prepared from the evapotranspiration model (Color Figure 34-61) showing potential ET in inches of water. The method shows promise for water resource managers and foresters concerned with making silvicultural decisions.

Over the same watershed described above, Khorram (1982) used many of the same independent variables but also included: 1) solar radiation, 2) the solar constant, 3) time of year, 4) solar declination, 5) slope and aspect to derive a net radiation map (Color Figure 34-62). Remote sensing showed a potential for mapping the site-and-time specific radiation components over large areas. Khorram and Katibah (1981) also produced a vegetation/land cover type map (Color Figure 34-63) that used classification schemes outlined earlier by Hoffer. This map illustrates the complexity of the Feather River Watershed and was used as input data for the other studies described above.

SUMMARY ON THE STATE-OF-THE-ART OF COMPUTER-AIDED ANALYSIS TECHNIQUES AS APPLIED TO FOREST COVER MAPPING

In addition to a few studies prior to 1972, the launch of Landsat-I has led to numerous studies involving the use of computer-aided analysis

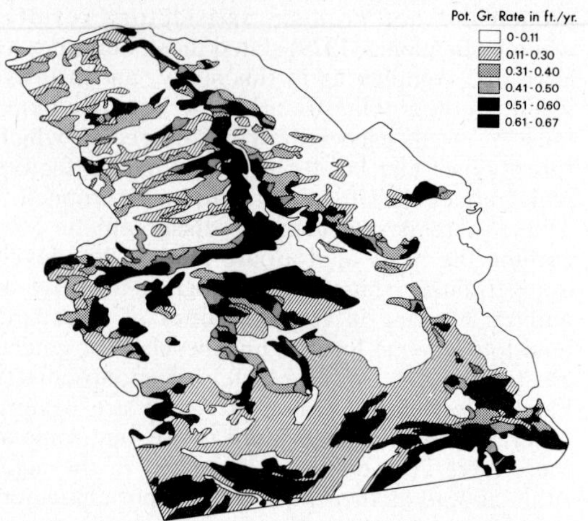

Fig. 34-60. Potential Growth Rate map of Conifer over the study area.

techniques for mapping forest and other cover types. The results of these studies tend to be quite varied, however. In examining the diversity of results obtained, it becomes clear that (a) the methodologies used are far from "standardized," and (b) the spectral, spatial, and temporal characteristics of the forest scene are complex. Many of the studies to date have involved relatively limited test sites, so it is difficult to extrapolate the results reliably to larger geographic areas. In some studies the level of detail involved in the classification was quite simple, such as forested *vs.* non-forested areas, whereas in other studies the classification involved individual forest cover-types. Different methods for developing the training statistics and the use of different classification algorithms may explain some of the variation in results obtained. Perhaps one of the major reasons for variability in the assessment of the capabilities and limitations of computer-aided analysis techniques is in the diversity of methods used to evaluate the classification results. Some studies reported only a single qualitative evaluation, others reported quantitative evaluations based on acreage estimates, others erroneously used training data as a basis for evaluating the classification results, and others used various methods to define a set of test data. Different levels of experience among the analysts, as well as differences in the capabilities of the hardware and software utilized, were also probable causes of some of the diversity in the results obtained. One should also remember that the definitions of various cover-type classes, in addition to the characteristics and quality of the reference data can influence the evaluation of the classification results. For example, the definition of what constitutes "forest land" or "brushland" can influence the decision as to whether a particular location in the data was classified correctly or incorrectly. In addition, the computer-classification results are often compared to data obtained through manual interpretation of aerial photos. Since photo interpretation (or most any other method of interpreting "ground truth") is not always 100 percent accurate, it is possible that various degrees of uncertainty may have been involved in the evaluation of classification results due to identification errors in the reference data.

In spite of the differences in individual results obtained by computer-aided analysis of Landsat data, a number of studies did indicate general agreement concerning many of the capabilities as well as the limitations of this technology. Forest land could be distinguished from non-forest with over 90 percent accuracy in studies by Heller [Tech. Coord.] (1975), Hoffer and Staff (1975a and 1975b), Kan and Weber (1978), Heath (1974), and others. Level II (Anderson et al., 1976) classifications of coniferous forest, deciduous forest, and other cover types such as cropland, rangeland, water, exposed rock, etc., generally resulted in accuracies of 80 percent or better, according to Kalensky and Scherk (1975), Hoffer et al. (1973, 1975a, 1975b, 1978), Kan and Dillman (1975), Kan and Weber (1978), Kalensky and Wightman (1976), Williams (1976), Krebs and Hoffer (1976), and Strahler et al. (1978).

A number of researchers attempted to classify individual forest cover-types with varying degrees of success. In general, most of them concluded that the results were quite variable and that the classification accuracies of many classes using single-date Landsat data were not accurate enough to provide operationally useful data. Use of multiple-date data or the combining of Landsat with topographic or other reference data sometimes improved overall classification accuracies for individual forest cover-types considerably, although there were often significant differences in performance among the cover types present (Kan and Dillman, 1975; Kan and Weber, 1978; Hoffer et al., 1975a and 1975b; Krebs and Hoffer, 1976; Lee, 1976; Strahler et al., 1978; and Mead and Meyer, 1977).

Acreage measurements of forest/non-forest categories or of coniferous forest/deciduous forest/other cover types can be obtained with a relatively high degree of accuracy (Bryant et al., 1978; Dodge and Bryant, 1976; Kan and Weber, 1978; Hoffer and Staff, 1975a and 1975b; Hoffer, Noyer, and Mroczynski, 1978; Rohde et al., 1979; and Lachowski et al., 1979).

Burn areas have a distinct, low spectral response on Landsat digital data (which shows up as black on false-color composite Landsat imagery), and there have been numerous studies indicating that burn areas could be delineated and mapped very effectively by manual interpretation of Landsat imagery. There have also been several studies which showed that computer-aided analysis techniques could be used to map burn areas (Heath, 1974; Hitchcock and Hoffer, 1974; Lawrence and Herzog, 1975; and Lee et al., 1977). Some of these studies indicated that the computer-generated maps obtained using Landsat data might be more accurate than the maps prepared by conventional techniques because pockets within the perimeter of the burn area were classified as forest rather than burn, whereas on the conventional maps such areas were not delineated.

Both manual techniques and computer-aided analysis of Landsat data are effective in delineating and mapping recent clearcuts. Recent clearcut areas within the forest can be separated from reproduction and established conifer forest with about 80–85 percent accuracy according to Johnson et al. (1979). Bryant et al. (1979) stated that computer classification allowed clearcuts as small as three hectares to be identified. They also found that clearcut areas were quite variable in spectral response—much of this variability being related to the number of years (up to 5) since the area had been clearcut. Williams (1976) found only a 54 percent classification accuracy for clearcut, but Lawrence and Herzog (1975), Kourtz (1977), and Lee et al. (1977) indicated that clearcuts could be mapped with reasonable accuracy using Landsat MSS data and CAAT.

Reproduction could be reliably distinguished from clearcuts only after the reproduction was at least five years of age, according to Johnson et al. (1979). Hawley (1979) stated that the potential to detect regeneration using Landsat data is related to tree height (3–5 meter minimum), crown-width (2–3 meter crown-diameter minimum), percentage ground cover (25 percent minimum), and appropriate spatial distribution over the area of interest. Kan and Dillman (1975) concluded that Landsat data could be used effectively to discriminate between hardwood, softwood, and regeneration.

Several studies have shown that satellite MSS data can be combined with topographic data to classify and map individual forest cover types more accurately (e.g., 15–23 percent) than by using the satellite MSS data alone (Hoffer and Staff, 1975b; Strahler et al., 1978; Hoffer et al., 1979; and Fleming and Hoffer, 1979).

Classification techniques to define and utilize spatial as well as the spectral information content in MSS data demonstrate considerable potential (Hoffer and Staff, 1975b; Hoffer et al., 1978; Sadowski et al., 1978; Malila, 1981). Such "spectral/spatial" classifiers will be of particular value in working with higher resolution MSS data obtained from satellite systems of the future, according to the studies by Sadowski et al. (1978) and Latty and Hoffer (1981). Post-processing techniques that "smooth" the data to minimize the "salt-and-pepper" characteristics of per-pixel classifiers are also available (Kan, 1976). Such techniques should also be useful in producing output maps that are more acceptable to many users.

One of the areas of greatest value for remote sensing technology as applied to forestry involves multi-stage sampling. Several investigators have utilized Landsat data in conjunction with (1) small-, medium-, and/or large-scale aerial photography and (2) ground sample data to derive various types of forest inventory and other resource information (Gialdini et al., 1975; Nichols et al.,

1976; Rohde, 1978; Rohde et al., 1979; Johnson et al., 1979; and Barker, 1981). Nichols et al. (1976) gave a fairly detailed description of the sampling procedures involved using computer-analyzed Landsat data, aerial photos, and ground samples to estimate stand volume in several condition classes of forest cover in Washington. They showed that computer analysis of the Landsat data would enable them to stratify the data much more efficiently—only 1/6th the number of ground plots were required if computer stratification was used (as compared to no stratification), whereas a photo interpretation stratification only enabled the number of ground plots to be reduced by 1/2.

One of the greatest potentials for operational utilization of Landsat satellite data and computer-aided analysis techniques involves the capability to monitor changes in the forest resource-base. Heller (1975) showed that Landst data "provide a good Level I forest classifier and can be useful for detection of U.S. Forest Survey inventory-plot changes or disturbances where there is much human activity."

The potential for utilizing Landsat data as input to an annual update of their Forest Resource Information System (FRIS) led St. Regis Paper Company, Southern Timberlands Division, to develop an operational capability to analyze Landsat data by computer-aided analysis techniques. St. Regis personnel state that effective use of the Landsat data provides them with rapid and more accurate updates of their Administrative Working Units than can be obtained by other methods. They have also found that the Landsat data enable them to develop more efficient ground-sampling strategies, thereby saving a considerable amount of field-crew time (Barker, 1981).

Several studies involving alteration of the forest cover, particularly in tropical regions have shown that Landsat data have considerable potential for monitoring large-scale changes. This is particularly important in view of the significant amounts of deforestation occurring in the tropics in various countries throughout the world. Talbot (1981) stated that "without a doubt, the hottest issue confronting the global community is deforestation in the tropics. The implications of this activity have been cited on numerous occasions to have serious repercussions for the global carbon-dioxide balance, for local rainfall patterns, and for the welfare of human populations in those areas, especially with regard to the loss of options for more sustainable uses of forested lands." According to Williams and Miller (1979) "the greatest potential for the sustained application of space systems in the forest management process is the monitoring of areas where forest cover is subjected to alteration by either natural or man-made activities."

REFERENCES

Ackerl, F., 1966, Ein Taschenstereoskop mit Mikrometer für die Ausbildung und den praktischen Gebrauch; Allg. Forstzeitung, pp. 229–230.

Akça, A., 1969/1970, Eine Untersuchung zur Unterscheidung und Identifizierung einiger Objekte auf Schwarz-weiss-Luftbildern durch quantitative Beschreibung der photographischen Tuxtur; Diss. Freiburg i. Br.

Akça, A., 1971, Identification of land use classes and forest types by means of microdensitometer and discriminant analysis; Application of Remote Sensors in Forestry, Joint Rept., Freiburg i. Br., pp. 147–164.

Akça, A., Hildebrandt, G., and P. Reichert, 1971, Baumhohenbestimmung aus Luftbildern durch einfache Parallaxenmessung; Forstwiss, Centralblatt, pp. 201–215.

Aldred, A. H. 1976, Measurement of tropical trees on large-scale aerial photographs; Can. For. Serv., For. Man. Inst. Inf. Report FMR-X-86.

Aldred, A. H., and F. W. Kippen, 1967, Plot volumes from large-scale 70 mm air photographs; Forest Sci., vol. 13, no. 4, pp. 419–426.

Aldred, A. H., and J. J. Lowe, 1978, Application of large-scale photos to a forest inventory in Alberta; For. Man. Inst. Inf. Report FMR-X-107, Canadian Forestry Service, Ottawa, Canada.

Aldrich, R. C., 1967, Stratifying photo plots into volume classes by crown closure comparator; U.S. Forest Service Research Note PSW-151.

Aldrich, R. C., 1968, Remote sensing and the forest survey—present applications, research and a look at the future; Symposium on Remote Sensing of Environment, 5th, Ann Arbor, Michigan, University of Michigan, pp. 357–372.

Aldrich, R. C., 1971, Space photos for land use and forestry: Photogramm. Eng., vol. 37, no. 4, pp. 389–401.

Aldrich, R. C., 1975, Detecting disturbances in a forest environment; Photogrammetric Engineering and Remote Sensing, vol. 41, no. 1, pp. 38–48.

Aldrich, R. C., 1979, Remote sensing of wildland resources: a state-of-the-art review; USDA Forest Service, Rocky Mountain Forest and Range Exp. Sta., Fort Collins, Colorado, Gen. Tech. Rep. RM-71.

Aldrich, R. C., and A. T. Drooz, 1967, Estimating Fraser fir mortality and balsam woolly aphid infestation trend using aerial color photography, Forest Sci., vol. 13, no. 3, pp. 300–313.

Aldrich, R. C., and R. C Heller, 1969, Large-scale color photography reflects changes in a forest community during a spruce budworm epidemic, *in* Remote sensing in ecology; Athens, Univ. of Georgia, pp. 30–45.

Aldrich, R. C., and N. X. Norick, 1969, Stratifying stand volume on nonstereo aerial photos—reduces errors in Forest Survey estimates: U.S. Forest Service Research Paper PSW—51, 14 p.

Allison, G. W., and R. C. Breadon, 1958, Provisional aerial photo stand volume tables for interior British Columbia: Forest Chron., v. 34, pp. 77–83.

American Society of Photogrammetry, 1960, Manual of photographic interpretation: Washington, D.C., Amer. Soc. Photogramm., 868 p.

Anderson, H. N., 1981, Photointerpretation techniques establish hazard rating criteria for western spruce budworm-susceptible forest stands; Master's Thesis, College of Forestry, Wildlife and Range Sciences, University of Idaho, Moscow, Idaho.

Anderson, J. R., Hardy, E. E., Reach, J. T., and Witmer, R. E., 1976, A land use and land cover classification system for use with remote sensor data; U.S. Geological Survey Professional Paper 964.

Andrews, G. S., 1936, Tree heights from air photographs by simple parallax measurements; Forestry Chron., vol. 12, no. 2, pp. 152–197.

Andrews, G. S., and L. G. Trorey, 1933, The use of aerial photographs in forest surveying; Forestry Chron., vol. 9, no. 4, pp. 33–59.

Anon., 1887, Verwendung der Ballonphotographie zu forstwirtschaft-lichen Zwecken; Berliner Tageblatt, 10.

Anson, A., 1970, Color aerial photos in the reconnaissance of soils and rocks; Photogramm. Eng., vol. 36, pp. 343–354.

Arnold, K, 1951, Uses of aerial photographs in control of forest fires; Jour. Forestry, vol. 49, pp. 631.

Ashley, M. D., and J. Rea, 1975, Seasonal vegetation differences from ERTS imagery; Photogrammetric Engineering and Remote Sensing, vol. 16, no. 6, pp. 713–719.

Asmoro, Pranoto, Kardono Darmoyuwono, Jacub Rais, and Z. D. Kalensky, 1976, Integrated resource mapping by multistage and multisensor remote sensing in Indonesia; National Coordination Agency for Surveys and Mapping (Bakosurtanal) Jakarta, Indonesia.

Avery, T. E., 1958, Composite aerial volume table for southern pines and hardwoods; Jour. Forestry, v. 56, pp. 741–745.

Avery, T. E., 1959, Volume tables for aerial timber estimating in northern Minnesota; U.S. Forest Service, Lake States Forest Expt. Sta. Paper 78.

Avery, T. E., 1966, Forester's guide to aerial photo interpretation; Agriculture Handbook 308, U.S. Department of Agriculture, Forest Service.

Avery, T. E., 1967a, All sorts of stereograms; Photogramm. Eng., vol. 33, pp. 1397–1401.

Avery, T. E., 1967b, Forest measurements; New York, McGraw-Hill, pp. 198–205.

Azevedo, L. H. A. de, 1971, Radar in the Amazon; Internat. Symp. on Remote Sensing Environment, 7th, Univ. of Mich., Proc., vol. 3

Bajzak, Denes, 1967, Detection and appraisal of damage by balsam wooly aphid on *Abies balsamea* (L.) Mill, by means of aerial photography; Ph.D. thesis, New York State Coll. Forestry, Syracuse.

Baltaxe B., 1979, Unpublished Report on the Use of Landsat Imagery for Tropical Forestry, FAO.

Banyard, S. G., 1979, Radar interpretation based on photo-truth keys; ITC Journal, no. 2, 1979, pp. 267–276.

Barker, G. R., and T. P. Fethe, 1975, Operational considerations for the applications of remotely sensed forest data from Landsat or other airborne platform; Proceedings of the NASA Earth Resources Survey Symposium, Houston, Texas, NASA TM X-58168, vol. 1-A, pp. 115–133.

Barker, G. R., 1981, Operational FRIS (Forest Resource Information System); Conference on Space Technology and Industrial Forest Management, St. Regis Paper Co., Jacksonville, Fla.

Bátkai, S., M. Csapo, and G. Jancso, 1971, Genauigkeitsuntersuchung der mittels Stereotop durchgeführten Höhenbestimmung; Berichte d.III. Int. Symp. für Photointerpretation, vol. 1, pp. 215–228.

Bauer, M. E., 1975, The role of remote sensing in determining the distribution and yield of crops; Advances in Agronomy, vol. 27, pp. 271–304.

Baumann, H., 1957, Forstiliche Luftbildinterpretation: Schriftenreihe Landesforstverwaltung Baden-Württemberg, Bd. 2.

Beal, J. A., W. H. Bennett, and D. E. Ketcham, 1964, Beetle explosion in Honduras; Am. Forests, vol. 70, no. 11, pp. 31–33.

Beaubien, J., 1979, Forest type mapping from Landsat digital data; Photogrammetric Engineering and Remote Sensing, vol. 45, no. 8, pp. 1135–1144.

Becking, R. W., 1959, Forestry applications of aerial color photography; Photogrammetric Engineering, vol. 25, pp. 559–565.

Befort, W. A., Heller, R. C, and J. J. Ulliman, 1980, Updating forest road maps with panoramic aerial photography; Forest, Wildlife and Range Exp. Sta., University of Idaho, Moscow, Idaho, Tech. Report 12.

Befort, W. A., R. C. Heller, and J. J. Ulliman, 1980, Viewing and handling panoramic aerial photographs; Tech. Report 7. Forest, Wildlife, and Range Exper. Sta. University of Idaho, Moscow, Idaho.

Belov, S. V., and A. M. Berezin, 1958, Značenie uslovij aerofotografirovanija i razlicnych tipov aeroplenok dlja izucenija lesov; Trud. Lab. Aeromet.; Izd. Akad. Nauk SSR, Moskau-Leningrad Nr. 6, pp. 146–175.

Benson, A.S., and Lauer, D. T., 1973, Testing multiband ERTS-1 imagery for classification of commercial conifer forests; Fourth Biennial Workshop on Color Aerial Photography in Plant Sciences and Related Fields, July 1973, Proceedings: University of Maine, Orono, Maine, pp. 57–76.

Benson, M. L., and W. G. Sims, 1967, False-colour film fails in practice: J. For., v. 65, p. 904.

Berezin, A. M., and I. A. Trunov, 1957, Korreljacionnaja svjaz diametrov derev'ev na vysote gradi s diametri kron dlja drevostoev različnych rajonov Sovetskogo Sojuza; Uc. Zap. Gruppy Lab. Aeromet. Akad. Nauk SSSR.

Berezin, A. M., 1963, Issledovanie tocnosti izmeritel' nogo desifrirovanij dešifrirovanija lesov na aerosnimkach: Sb. ''Metody desifrovanija lesov po aerosnimkam'', Moskau Izd. Nauk SSSR.

Bernstein, David A., 1974, Are reforestation surveys with aerial photographs practical? Photogrammetric Engineering, vol. 40, no. 1, pp. 69–73.

Bickerstaff, A., and R. P. Hirvonen, 1969, Forest inventory practices of Canadian provincial and federal agencies; Dept. Fish and Forest., Can. For. Serv., Forest Manage. Inst. Inform. Rept. FMR-X-19.

Bickford, C. A., 1952, The sampling design used in the forest survey of the northeast; Jour. Forestry, vol. 50, pp. 290–293.

Bickford, C. Allen, 1953, Increasing the efficiency of air photo forest surveys by better definition of classes; Northeastern Forest Experiment Station, Station Paper No. 58.

Bickford, C. A., 1961, Stratification for timber cruising: Jour. Forestry, vol. 59, pp. 761–763.

Bickford, C. A., 1963, An efficient sampling design for forest inventory: The northeastern forest resurvey; Jour. Forestry, vol. 61, pp. 826–833.

Blair, B. O., and M. F. Baumgardner, 1977, Detection of the green and brown wave in hardwood canopy covers using multi-date, multispectral data from Landsat-I. Agronomy Journal, vol. 69, pp. 808–811.

Bočarov, M. K., and G. C. Samojlovič, 1964, Matematicheskie osnovy dešifrirovanija aerosnimkov lesa: Izd. Lesnaja Promyslennost, Moskau.

Bodechtel, J., and G. Kritikos, 1971, Quantitative image

enhancement of photographic data for earth resources: Internat. Symp. on Remote Sensing of Environment 7th, Ann Arbor, Proc., v. 1, p. 469–486.

Bogyay, J., 1970, Möglichkeiten für die Verwendung von Luftbildern bei der Vorratsaufnahme von ungarischen Kiefernwäldern: Berichte III, Int. Symp. für Photointerpretation, Dresden, v. 1, p. 229–241.

Bonner, G. M., 1964a, A tree volume table for red pine by crown width and height; Forestry Chron., vol. 40, pp.

Bonner, G. M., 1964b, The influence of stand density on the correlation of stem diameter and crown width and height for lodgepole pine, Forestry Chronicle, vol. 40, pp. 347–349.

Bonner, G. M., 1966, Provisional aerial stand volume tables for selected forest types in Canada; Canada Dept. Forestry and Rural Devel., Forestry Br. Dept. Pub. no. 1175.

Bonner, G. M., 1968, A comparison of photo and ground measurements of canopy density: Forestry Chron., vol. 44, no. 3, pp. 12–16.

Bonner, G. M., 1977, Forest inventories with large-scale aerial photographs: an operational trial in Nova Scotia; Forest Management Institute Information Report FMR-X-96, Canadian Forestry Service, Ottawa, Ontario.

Boon, D. A., 1955, Report on W. G. 4: Interpretation of vegetation, Com VII ISP; Photogramm. Eng., vol. 26, pp. 283–302.

Boon, D. A., 1956, Recent developments in photo interpretation of tropical forests; Photogrammetria, vol. 12, pp. 382–286.

Bowlin, H. L., 1979, Optical bar camera imagery and technology applications evaluation; timber salvage program, Timber Mgt. Plans Section, Pacific Southwest Region, USDA Forest Service, Mimeo.

Brenac, L., 1962, L'utilisation des photographies aériennes pour l'inventaire des torêts francaises; Bull. Soc. Francaise Photogrammetric 8, pp. 2–31.

Bryant, E. S., A. G. Dodge, Jr., and S. D. Warren, 1978, Satellites for practical natural resource mapping? A forestry test case; Proceedings of the National Workshop on Integrated Inventories of Renewable Natural Resources, Rocky Mountain Forest and Range Experiment Station General Technical Report RM-55, U.S.D.A. Forest Service, Fort Collins, Colorado, pp. 219–226.

Bryant, E., A. G. Dodge, and M. G. E. Eger, 1979, Small forest cuttings mapped with Landsat digital data; Proceedings of the Thirteenth International Symposium on Remote Sensing of Environment, Ann Arbor, Michigan, pp. 971–981.

Burns, Paul Y., 1979, A simple method of determining timber volume on aerial photographs of evenaged stands; Proceedings, Forest Resource Inventories Workshop, Colorado State University, Ft. Collins, CO, pp. 432–435.

Busch, William L., Claude McLean, and John Bell, 1979, Estimate and compare cruise procedure; Proceedings, Forêst Resource Inventories Workshop, Colorado State University, Ft. Collins, CO., pp. 423–431.

Carneiro, C. M. R., 1981, The national forest cover monitoring programme of Brazil; Proceedings of the XVII IUFRO (International Union of Forest Research Organizations) World Congress, Kyoto, Japan, (in press).

Carson, W. W., D. D. Studies, and W. M. Thomas, 1970, Digitizing topographic data for skyline design programs; PNW Research Note—132, Pacific Northwest Forest and Range Expt. Sta., U.S. Dept. of Agriculture, Forest Service.

Catto, A. T., 1965, Aerial photography for mapping cut-over areas; Pulp and Paper Mag. Canada, vol. 66, no. 3, pp. 120–124.

Caylor, J. A., and G. A. Thorley, 1970, Studies on the population dynamics of the western pine beetle, *Dendroctonus brevicomis* Le Conte (*Coleoptera:* Scolytidae); Univ. California Div. Agr. Sci., part 2, sec 2, pp. 10–32.

Cérmak, V., 1960, Kluč na citanie leteckej snimky; Vyskum. Ust. Lesn, Hosp. Zvolen.

Chapman, R. C., 1965, Preliminary aerial photo stand-volume tables for some California timber types; U.S. Forest Service Research Note PSW-93.

Chase, C. S., and J. Korotov, 1947, Key to forest types in Marinette County, Wisconsin, on infrared minus-blue filter at 1:12,000, autumn photography; Processed Copy in the Langlois Library, Catholic University, Washington, D.C.

Choate, G. A., 1953, More on the use of aircraft in checking forest photo interpretation; Jour. Forestry, vol. 51, no. 4, pp. 291–293.

Ciesla, W. M., 1974, Forest insect damage from high altitude color-IR photos; Photogramm. Eng., vol. 40, no. 6, pp. 683–689.

Ciesla, W. M., 1977, Color versus color IR photos for forest insect surveys; Proceedings of 6th Biennial Workshop Aerial color photography in the plant sciences and related fields, Colorado State University, Ft. Collins, CO. August 9–11, 1977, pp. 31–42.

Ciesla, W. M., J. C. Bell, Jr., and J. W. Curlin, 1967, Color photos and the southern pine beetle; Photogramm. Eng., vol. 33, pp. 883–888.

Ciesla, W. M., M. M. Furniss, M. D. McGregor, and W. F. Bousfield, 1971, Evaluation of Douglas-Fir beetle infestations in the North Fork Clearwater River Drainage, Idaho—1971; Rept. no. 71-46, Northern Region, Forest Service, USDA.

Clement, J., and J. Guellec, 1974, Utilisation des photographies acriemes au 1/5000 en couleur pour la détection de l'Okoumé dans le forêt dense du Gabon; Renve Bois et Forêts des Tropiques.

Cochran, W. G., 1953, Sampling techniques, New York, Wiley.

Cochrane, G. R., 1968a, Biogeography with Australian applications; Dury, G. H., and Logan, M. L., eds., Studies in Australian geography; Heinemann, London, pp. 37–69.

Cochrane, G. R., 1968b, Fire ecology in southeastern Australian sclerophyll forests; Proc. Ann. Tall Timbers Fire Ecol. Conf., 7th, vol. 8, pp. 15–40.

Cochrane, G. R., 1969, Ecological valence of mountain ash (*Eucalyptus regnans* F. Muell.) as a key to its distribution; Victorian Nat., vol. 86, pp. 6–26.

Cochrane, G. R., 1970, Colour and false-colour aerial photography for mapping bushfires and forest vegetation; Proc. New Zealand Ecol. Soc., vol. 17, pp. 96–105.

Coggeshall, M. E., and R. M. Hoffer, 1973, Basic forest cover mapping using digitized remote sensor data and ADP techniques; LARS Technical Report 030573, Purdue University, West Lafayette, Indiana.

Colwell, R. N., 1965, Aids for the selection and training of photo interpreters; Photogramm. Eng., vol. 31, no. 2, pp. 326–339.

Congalton, R. G., R. A. Mead, R. G. Oderwald, and J. Heinen, 1981, Analysis of forest classification accuracy; Remote Sensing Research Report 81-1, School of Forestry and Wildlife Resources, Virginia Polytechnic Institute and State University, Blacksburg, Virginia.

Cook, C. F., 1969, The use of light aircraft in forest inventory and mapping; Pulp and Paper Mag. Canada, vol. 70, no. 13, pp. 69–74.

Croxton, R. J., 1966, Detection and classification of ash dieback on large-scale color aerial photographs; U.S. Forest Service Research Paper PSW-35.

Davidson, A. G., and R. M. Prentice, 1967, Important forest insects and diseases of mutual concern to Canada, the United States and Mexico; Canada Dept. Forest. and Rural Develop., Publ. no. 1180.

Daus, S. J., and D. T. Lauer, 1971, Testing the usefulness of side-looking airborne radar imagery for evaluating forest vegetation resources; Rept. of research performed on Contract no. CRINC 1775-9 for Center for Research, Inc., Univ. of Kansas, Forestry Remote Sensing Laboratory, Univ. of California, Berkeley, California.

Demidov, E., 1970, Die Anwendung der farbigen spektrozonalen Luftbildmessung für die forsttaxatorische Interpretation in den USSR; Berichte Ill. Int. Symp. für Photointerpretation, Dresden 1970, vol. 1, 1971, pp. 243–255.

DeMilde, R., and L. Sayn-Wittgenstein, 1973, An experiment in the identification of tropical tree species on aerial photographs; Symposium on Remote Sensing, Freiberg, Germany, 1973, Proceedings; IUFRO Subject Group S 6.05, pp. 21–37.

Dillewijn, F. J. van, 1957, Sleutel voor de interpretatie van begroei-ingsvormen met luchtfoto's 1:40,000 van het noordeljk deel van Suriname, Dienst 's Landsbosbeheer, Paramaribo, Suriname.

Dillworth, J. R., 1959, Aerial photo mensuration tables; Oregon State Coll. Agr. Expt. Sta., Forest Research Div., Research Note. no. 2

Dodge, A. G., Jr., and E. S. Bryant, 1976, Forest type mapping with satellite data; Jour. Forestry, vol. 74, no. 8, pp. 526–531.

Douglas, R. W., 1973, Use of high altitude photography for forest disease detection and vegetation classification within the sub-boreal forest region; PhD thesis, University of Minnesota, St. Paul, Minnesota.

Duffy, P. J. B., and M. P. Meyer, 1962, A preliminary study of aerial volume table construction for lodgepole pine in west-central Alberta; Forestry Chron., vol. 38, no. 2, pp. 212–218.

Eaton, C. B., 1942, The adaptation of aerial methods to the forest loss survey; U.S. Forest Service, Pacific Southwest Forest and Range Expt. Sta., Berkeley, California.

Eule, H. W., 1959, Verfahren zur Baumkronenmessung und Beziehungen zwischen Kronengrösse, Stammstarke and Zuwachs bei Rotbuche; Allgemeine Forstund Jagdzeitung, pp. 185–201.

Fang, You-Ching, 1980, Aerial photo and Landsat image use in forest inventory in China; Photogrammetric Engineering and Remote Sensing, vol. 46, no. 11, pp. 1421–1424.

FAO, 1969, Survey of agricultural and forest resources, Nicaragua; FAO/SF: 49/NIC-2, vol. 1, general.

Felton, G. C., 1967, Use of the AFDS-2 airborne infrared forest fire detection and mapping system; (a report on). Ontario Dept. of Lands and Forests.

Feree, M. J., 1953, A method of estimating timber volumes from aerial photographs; New York Univ., Coll. Forestry at Syracuse.

Fleming, M. D., J. Berkebile, and R. M. Hoffer, 1975, Computer-aided analysis of LANDSAT-1 MSS data: a comparison of three approaches including the modified clustering approach; Proceedings of the Symposium on Machine Processing of Remotely Sensed Data, Purdue University, W. Lafayette, Indiana, pp. 1B:54–61.

Fleming, M. D., and R. M. Hoffer, 1977, Computer-aided analysis techniques for an operational system to map forest lands utilizing Landsat MSS data; LARS Technical Report No. 112277, Purdue University, W. Lafayette, Indiana.

Fleming, M. D., and R. M. Hoffer, 1979, Machine processing of Landsat MSS data and DMA topographic data for forest cover type mapping; Proceedings of the 1979 Symposium on Machine Processing of Remotely Sensed Data, Purdue University, West Lafayette, Indiana, pp. 377–390.

Foster, Ellery, 1934, The use of aerial photographs in mapping ground conditions and cruising timber in the Mississippi River bottomlands; U.S. Forest Service, Southern Forest Expt. Sta., Occas. Paper 37.

Francis, D. A., 1957, The use of aerial photographs in tropical forests; Unasylva 11, pp. 103–109.

Francis, D. A., 1960, Interim report to the Government of the Sudan on forest inventory; FAO Interim Report nr. 59/10/7837, Rome.

Francis, D. A., 1972, Personal correspondence.

Francis, D. A., 1976, Possibilities and problems of radar image interpretation for vegetation and forest types; proceedings IUFRO Congress, Remote Sensing in Forestry, Oslo, Norway, pp. 79–86.

Francis, E. C., and G. H. S. Wood, 1955, Classification of vegetation in North Borneo from aerial photographs; Malayan Forester, vol. 18, no. 1, pp. 38–44.

Gialdini, M., S. Titus, J. Nichols, and R. Thomas, 1975, The integration of manual and automatic image analysis techniques with supporting ground data in a multistage sampling framework for timber resource inventories: three examples; Proceedings of the NASA Earth Resources Survey Symposium, Houston, Texas, NASA TM X-58168, vol. 1-B, pp. 1377–1387.

Gingrich, G. F., and H. A., Meyer, 1955, Construction of an aerial stand volume table for upland oak; Forest Sci., vol. 1, no. 2, 140–147.

Gordeev, P. K., 1954, Izmčenie Lesosyrjevych baz pri promošci aksonometričes koj aerofotos-emki; Lesn. Prom., vol. 14, pp. 4–9.

Grosenbough, L. R., 1965, Three-Pee sampling theory and program "ThPP" for computer generation of selection criteria; U.S. Forest Service Research Paper PSW-21.

Gut, D. and J. Hohle, 1977, High altitude aerial photography: aspects and results; Photogrammetric Engineering and Remote Sensing, vol. 43, n. 10, pp. 1245–1255.

Haapala, Fred H. and Paul O. Neumann, 1972, Determining Stocking of Conifer Reproduction with Large Scale Color Infrared Aerial Photography; DNR Report No. 28, State of Wash.

Haenel, S., W. Perlwitz, and P. Trepte, 1972, Die Bestimmung forstlicher Bestandesdaten durch automatische Luftbildauswertung mittels Digitaltechnik; XII, ISP-Congress, Ottawa.

Haig, L. T., M. A. Huberman, and U. Aung Din., 1958, Tropical silviculture, vol. 1, FAO, Forestry and Forest Products Studies, no. 13.

Hamilton, D. A. Jr., 1974, Event probabilities estimated by regression; USDA Forest Service Res. Paper

INT. 152, Intermtn. Forest and Range Exper. Sta., Ogden, Utah.

Hanks, L. F., and G. W. Thomson, 1964, Aerial stand volume tables for Iowa hardwoods; Iowa State Jour. Sci., vol. 38, no. 4.

Hanson, J. B., and William Lautz, 1969, Infrared photography for estimating tree mortality caused by *Annosus* root rot and notes on color infrared photography to assess insect damage; Workshop on Aerial Color Photography in the Plant Sciences, Gainesville, Proc., pp. 89–92.

Harding, R. A. and R. B. Scott, 1978, Forest inventory with Landsat: Phase II of the Washington Forest Productivity Study; Division of Technical Services, State of Washington, Department of Natural Resources, Olympia, Washington.

Hardy, N. E., J. E. Coiner, and W. O. Lockman, 1971, Vegetation mapping with side-looking airborne radar; Yellowstone National Park, presented at Conference on Propagation Limitations in Remote Sensing, Advisory Group for Aerospace Research and Development, NATO, Neuilly-Sur-Seine, France, Pre-print no. 90, pp. 11-1 to 11-9.

Hawley, D. L., 1979, Forest inventory of clearcuts utilizing remote sensing techniques; Proceedings of the 13th International Symposium on Remote Sensing of Environment, Ann Arbor, Michigan, pp. 1385–1407.

Heath, G. R., 1974, ERTS data tested for Forestry applications; Photogrammetric Engineering, vol. 40, pp. 1087–1091.

Hegg, K. M., 1967, A photo identification guide for the land and forest types of interior Alaska; Northern Forest Expt. Sta., Forest Service, USDA, Research Paper NOR-3.

Heinsdijk, D., 1953, Bosbouwkundige Foto-Interpretatie; Publ. CBL, Paramaribo, no. 13.

Heinsdijk, D., 1957, Report to the government of Brazil on a forest inventory in the Amazon Valley region Rio Tapajos and Rio Xingu; FAO, Rept. no. 601, Rome.

Heinsdijk, D., and B. B. Glerum, 1967, Inventories and commercial possibilities of Brazilian forests; Turrialba, vol. 17, no. 3, pp. 337–347.

Heller, R. C., 1969, Large-scale color photo assessment of smog-damaged pines; Am. Soc. Photogramm. and Soc. Photog. Sci. and Eng., New York, Proc., pp. 85–98.

Heller, R. C., 1973, Analysis of ERTS imagery—problems and promises for foresters; Proceedings, Symposium IUFRO S 6.05, International Union of Forest Research Organizations, Freiburg, Germany, pp. 373–393.

Heller, R. C., (Tech. Coord.), 1975, Evaluation of ERTS-1 data for forest and range-land survey; USDA Forest Service Research Paper PSW-112, Pacific S.W. Forest and Range Experiment Station, Berkeley, Calif.

Heller, R. C., 1978, Case applictions of remote sensing for vegetation damage assessments; Photogramm. Eng., vol. 44, no. 9, pp. 1159–1166.

Heller, R. C., R. C. Aldrich, and W. F. Bailey, 1959, An evaluation of aerial photography for detecting southern pine beetle damage; Photogramm. Eng., vol. 15, pp. 595–606.

Heller, R. C., R. C. Aldrich, W. F. McCambridge, F. P. Weber, and S. L. Wert, 1969, The use of multispectral sensing techniques to detect ponderosa pine trees under stress from insect of pathogenic organisms; Ann. Prog. Rept., Forestry Remote Sensing Lab. for Earth Resources Survey Program, NASA, by Pacific Southwest Forest and Range Expt. Sta., Berkeley, California.

Heller, R. C., R. C. Aldrich, F. P. Weber, R. W. Dana, and N. X. Norick, 1972, Monitoring forestland from high altitude and space; Final Report, NASA Supporting Research and Technology Program.

Heller, R. C., G. E. Doverspike, and R. C. Aldrich, 1964, Identification of tree species on large scale panchromatic and color aerial photographs; Forest Service, USDA, Agriculture Handbook no. 261.

Heller, R. C., J. H. Lowe, Jr., R. C. Aldrich, and F. P. Weber, 1966, A test with large-scale aerial photographs to sample balsam woolly aphid damage in the Northeast: Jour. Forestry, vol. 65, no. 1, pp. 10–18.

Heller, R. C., and S. A. Sader, 1980, Rating the risk of tussock moth defoliation using aerial photographs; USDA Agriculture Handbook No. 569, Washington, D.C.

Heller, R. C., and J. F. Wear, 1969, Sampling forest insect epidemics with color films; Internat. Symp. on Remote Sensing of the Environment, 6th, Ann Arbor, 1970, Proc., pp. 1157–1167.

Henderson, F. M., 1979, Land-use analysis of radar imagery; Photogrammetric Engineering and Remote Sensing, vol. 45, no. 3, March 1979, pp. 295–307.

Hildebrandt, G., 1957, Forsteinrichtungsarbeiten mit Hilfe von Luftbildern; Forst und Jagd, pp. 58–64.

Hildebrandt, G., 1963, Ein Vergleich der forstlichen Luftbild-interpretation panchromatischer und infraroter Bilder; Arch. Internat. Photogramm., vol. 14, pp. 239–244.

Hildebrandt, 1964, Systematik der Waldinventurmethoden unter dem Gesichtspunkt der Luftbildverwendung; Arch. Internat. Photogramm., vol. 14.

Hildebrandt, G., and H. Kenneweg, 1968a, Einige Anwendungsmoglich keiten der falschfarbenphotographie in forstlichen Luftbildwesen; Allg. Forst-U, Jagdzeitung, pp. 205–213.

Hildebrandt, G., and H. Kenneweg, 1968b, Beispiele forstlicher Interpretationsmöglichkeiten falschfarbiger Luftbilder; Internat. Kongr. Photogrammetrie, vol. 11.

Hildebrandt, G., and H. Kenneweg, 1969, Information über die waldvegetation aus farbigen Luftbilderen; Erfahrungen und Erwartungen, Bildmessung und Luftbildwesen, pp. 165–170.

Hildebrandt, G., and Chr., Schindler, 1966, Radiallinien als Strichprobeeinheiten bei Flachenermittlungen verschiedener Landnutzungs—arten aus Luftbildern; Internat. Arch. Photogramm., vol. 16, no. 3, pp. 29–36.

Hindley, E. H., 1969; Forest inventory and photointerpretation in Guatemala; FAO report.

Hindley, E. H., 1971, A progress report from the forest industries development project on inventory design, sampling costs and quality assessment in the mixed dipterocarp forest of Sarawak, Malaysia; Paper presented at the 25th IUFRO World Congress, Florida, U.S.A.

Hirsch, Stanley N., 1963, Applications of remote sensing to forest fire detection and suppression; Symp. Remote Sensing of Environment, 2nd Proc., 1962, pp. 295–308.

Hirsch, S. N., 1964, Forest fire detection systems; Western Forest Fire Research Council, Proc., 1964, pp. 3–5.

Hirsch, S. N., 1965, Preliminary experimental results with infrared line scanners for forest fire surveillance; Symp. Remote Sensing of Environment, 3rd, 1964, Proc., pp. 623–648.

Hirsch, S. N., 1968, Project fire scan-summary of 5 years' progress in airborne infrared fire detection; Symp. Remote Sensing of Environment, 5th, 1968, Proc., pp. 447–457.

Hirsch, S. N., 1971, Application of infrared scanners to forest fire detection; Internat. Workshop Earth Resources Survey Systems Proc., vol. 2, pp. 153–169.

Hirsch, S. N., R. L. Bjornses, F. H. Madden, and R. A. Wilson, 1968, Project Fire Scan fire mapping final report, April 1962 to Dec. 1966; USDA Forest Service Research Paper INT-49, Intermountain Forest and Range Exp. Sta., Ogden, Utah.

Hirsch, S. N., R. F. Kruckeberg, and F. H. Madden, 1971, The bispectral forest fire detection system; Symp. Remote Sensing of Environment, 7th, Proc. 1971, pp. 2253–2272.

Hitchcock, H., and R. M. Hoffer, 1974, Mapping a recent forest fire with ERTS-1 MSS data; Proceedings of the 3rd Annual Remote Sensing of Earth Resources Conference, UTSI, Tullahoma, Tennessee.

Hodarev, Ju. K., et al, 1971, Some possible uses of optical and radio-physical remote measurements for earth investigations; Internat. Symp. on Remote Sensing of Environment, 7th, Ann Arbor, Proc., vol. 1, pp. 99–118.

Hoffer, R. M., 1967, Interpretation of remote multispectral imagery of agricultural crops; Laboratory for Agricultural Remote Sensing, Volume No. 1 (Annual Report), Research Bulletin No. 831, Agricultural Experiment Station, Purdue University, Lafayette, Indiana.

Hoffer, R. M., and LARS Staff, 1973, Techniques for computer-aided analysis of ERTS-1, data, useful in geologic, forest and water resource surveys; Proceedings of the Third Earth Resources Technology Satellite-1 Symposium, vol. 1, Section A, NASA, Goddard Space Flight Center, Washington, D.C., pp. 1687–1708.

Hoffer, R. M., and Staff, 1975a, Natural resource mapping in mountainous terrain by computer analysis of ERTS-1 satellite data; Agricultural Experiment Station Research Bulletin 919, Purdue University, W. Lafayette, Indiana.

Hoffer, R. M., and Staff, 1975b, Computer-aided analysis of SKYLAB multi-spectral scanner data in mountainous terrain for land use, forestry, water resource and geologic applications; (Final Report on Contract No. NAS9-13380, SKYLAB EREP Project 398.), LARS Contact Report 121275, Purdue University, W. Lafayette, Indiana.

Hoffer, R. M., S. C. Noyer, and R. P. Mroczynski, 1978, A comparison of Landsat and forest survey estimates of forest cover; Proceedings of the Fall Technical Meeting of the American Society of Photogrammetry, Albuquerque, N.M., pp. 221–231.

Hoffer, R. M., R. E. Joosten, R. G. Davis, and F. R. Brumbaugh, 1978, Land use and cartography; Skylab EREP Investigations Summary, NASA SP-399, National Aeronautics and Space Administration, Washington, D.C., pp. 7–77.

Hoffer, R. M. 1979, Computer-aided analysis of remote sensor data—magic, mystery, or myth?; Proceedings of the Symposium on Remote Sensing for Natural Resources: An International View of Problems, Promises, and Accomplishments, University of Idaho, Moscow, Idaho, pp. 156–179.

Hoffer, R. M., M. D. Fleming, L. A. Bartolucci, S. M. Davis, and R. F. Nelson, 1979, Digital processing of Landsat MSS and topographic data to improve capabilities for computerized mapping of forest cover types; LARS Technical Report 011579, Laboratory for Applications of Remote Sensing, Purdue University, W. Lafayette, Indiana.

Holter, M. R., H. W. Courtney, and T. Limperis, 1970, Research needs: the influence of discrimination, data processing, and system design; Remote Sensing, with special reference to agricultural and forestry, National Academy of Sciences, Washington, D.C., pp. 385–387.

Hord, R. M., and W. Brooner, 1976, Land use map accuracy criteria; Photogrammetric Engineering and Remote Sensing, vol. 42, pp. 671–677.

Houston, D. R., 1969, Comparison of infrared color and true-color aerial photography for studying beech bark disease; Workshop on Aerial Color Photography in the Plant Sciences, Gainesville, Proc., pp. 76–77.

Howard, J. A., 1959, The classification of woodland in Western Tanzania for the mapping from aerial photographs; Empire Forest Rev. 38, pp. 348–64.

Howard, J. A., 1970, Aerial photo ecology; Faber and Faber Ltd., London.

Hudson, William D., R. J. Amsterburg, Jr. and W. L. Meyers, 1976, Identifying and Mapping Forest Resources From Small Scale Color-Infrared Airphotos, Michigan State Univ. Ag. Exp. Sta. Research Report 304.

Ilvessalo, Y., 1950, On the correlation between the crown diameter and the stem of trees; Comm. Inst. Forest Fenniae, vol. 38, pp. 5–320.

Jaakkola, S. P. and W. C. Draeger, 1971, Techniques for evaluating forest stand delineation; Analysis of remote sensing data for evaluating vegetation resources; Ann. Prog. Rept., Remote Sensing Applications in Forestry, OSSA/NASA, Earth Resources Survey Program, by the Forestry Remote Sensing Laboratory, Univ. California, Berkeley, California.

Jensen, H. A. and R. N. Colwell, 1949, Panchromatic versus infrared minus-blue aerial photography for forestry purposes in California: Photogramm. Eng., vol. 15, pp. 201–223.

Johnson, E. W., 1952, Using aircraft in checking forest photointerpretation; Jour. Forestry, vol. 50, pp. 853–855.

Johnson, C. E. and L. R. Thomas, 1951, The Polaroid camera in fire control; Forest Service Fire Control Notes, vol. 12, no. 2, pp. 24–25.

Johnson, G. R., E. W. Barthmaier, T. W. D. Gregg, and R. E. Aulds, 1979, Forest stand classification in western Washington using Landsat and computer-based resource data; Proceedings of the 13th International Symposium on Remote Sensing of Environment, Ann Arbor, MI. pp. 1681–1696.

Johnson, Evert W. and Larry Sellmann, 1974, Forest cover photointerpretation key for the Piedmont forest habitat region in Alabama. Auburn University, Ag. Exp. Sta. Forestry Dept. Series No. 6.

Johnson, Evert W., and Larry Sellmann 1975, Forest cover photointerpretation key for the mountain forest habitat region in Alabama; Agri. Exp. Stn., Auburn University, Auburn, Alabama, For Dep. Ser. 7.

Johnson, Evert W. and Larry Sellmann, 1977, Forest cover photointerpretation key for the ridge and valley forest habitat region in Alabama, Auburn University. Ag. Exp. Sta. Forestry Dept. Series No. 9.

Johnson, Evert W. and Larry Sellman, 1979a, Forest cover photointerpretation key for the cumberland plateau forest habitat region in Alabama, Auburn University, Ag. Exp. Sta. Forestry Dept. Series No. 10.

Johnson, Evert W. and Larry Sellmann, 1979b, Forest cover photointerpretation key for the warrior basin forest habitat region in Alabama, Auburn University, Ag. Exp. Sta. Forestry Dept. Series No. 11.

Jordanskij, A. N., 1957, Spektrozonal'naja fotografija; Žur. Nauč, i Prikl. Fotogr, i Kinem, vol. 2.

Kalensky, Z., and L. R. Scherk, 1975, Accuracy of forest mapping from Landsat computer compatable tapes; Proceedings of the 10th International Symposium on Remote Sensing of Environment, Ann Arbor, Michigan, pp. 1159–1167.

Kalensky, Z., and J. M. Wightman, 1976, Automatic forest mapping using remotely sensed data; Proceedings of the Symposium on Remote Sensing in Forestry held during the XVI IUFRO (International Union of Forest Research Organizations) World Congress, Oslo, Norway, pp. 115–135.

Kan, E. P., D. B. Ball, J. P. Basu, and R. L. Smelser, 1975, Data resolution versus forestry classification and modeling; Proceedings of the Second Symposium on Machine Processing of Remotely Sensed Data, Purdue University, West Lafayette, Ind., pp. 1B:24–44.

Kan, E. P., and R. D. Dillman, 1975, Timber type separability in Southeastern United States on Landsat-1 MSS data; Proceedings of the NASA Earth Resources Survey Symposium, Houston, Texas, NASA TM X-58168, vol. 1-A, pp. 135–157.

Kan, E. P., 1976, A new computer approach to map mixed forest features and post process multispectral data, Proceedings of the Fall Convention of the American Society of Photogrammetry, Seattle, Washington, pp. 386–401.

Kan, E. P., and F. P. Weber, 1978, The ten-ecosystem study: Landsat ADP mapping of forest and rangeland in the United States; Proceedings of the 12th International Symposium on Remote Sensing of Environment, Ann Arbor, MI., pp. 1809–1825.

Kenneweg, H., 1970, Auswertung von Farbluftbildern für die Abgrenzung von Schädigungen an Waldbeständen, Bildmessung u. Luftbildwesen, pp. 283–290.

Kenneweg, H., 1971, Color and false color photography: its growing use in forestry—a European view, *in* Application of remote sensors in forestry; Internat. Union of Forest Research Organizations, Joint Rept., sec. 25, pp. 57–73.

Kippen, F. W. and Leo Sayn-Wittgenstein, 1964, Tree measurements on large-scale, vertical, 70-mm air photographs: Forest Research Branch Dept. of Forestry, Pub. no. 1053, Ottawa, Canada, p. 16.

Klein, W. H., 1970, Mini-aerial photography; Jour. Forestry, vol. 68, pp. 475–478.

Klein, W. H., 1973, Beetle-killed pine estimates; Photogramm. Eng. vol. 39, pp. 385–388.

Klein, W. H., D. D. Bennett, and R. W. Young, 1980, Evaluation of panoramic reconnaissance aerial photography for measuring annual mortality of lodgepole pine caused by the mountain pine beetle; USDA For. Serv., FI&DM-/MAG, Rep. 80-2, Davis, CA.

Klier, G., 1969, Zur Bestimmung des Kronenschlussgrades im Luftbild Archiv für Forstwesen, pp. 871–876.

Klier, G., 1970, Aerophotogrammetrische Messung an Einzelbaumen bei der Holzart Fichte, Archiv für Forstwesen, pp. 543–553.

Kondratyev, K.Ja., O. B. Vasilyev, and Z. F. Mironova, 1971, On a procedure of coding the optical spectral reflectance of natural formations; Proc. Internat. Symp. on Remote Sensing of Environment, 7th, Ann Arbor. vol. I, pp. 647–661.

Khorram, S., and E. F. Katibah, 1981, Use of Landsat multispectral scanner data in vegetation mapping of a forested area, Proceedings, 1981 Annual Convention, American Society of Photogrammetry, Washington, D.C., 10 pp.

Khorram, S., and H. G. Smith, 1979, use of Landsat and environmental satellite data in evapotranspiration estimation from a wildland area, Proceedings, Thirteenth International Symposium on Remote Sensing of Environment, Ann Arbor, Michigan, pp. 1445–1554.

Khorram, S., 1982, A remote sensing-aided procedure for site-specific estimation of net radiation over large areas, Journal of Applied Photographic Engineering, vol. 8, no. 1, pp. 31–35.

Khorram, S., H. G. Smith, E. F. Katibah, R. W. Thomas, J. M. Sharp, and A. Kaugers, 1978, An integrated study of earth resources in the state of California using remote sensing techniques, Space Sciences Laboratory, Series 19, Issue 53, University of California, Berkeley, 163 pp.

Kourtz, P. H., 1977, An application of Landsat digital technology to forest fire fuel type mapping; Proceedings of the 11th International Symposium on Remote Sensing of Environment, Ann Arbor, Michigan, pp. 1111–1115.

Kourtz, P. H., and A. J. Scott, 1978, An improved image enhancement technique and its application to forest fire management, Proceedings of the Fifth Canadian Symposium for Remote Sensing, Victoria, B.C.

Koxlovskij, B. A., 1957, Sire primenjat′ cvetnuju aerofotos-emku pri lesoustrojstve; Lesn. Choz., Nr. 1 and 10.

Kruckeberg, Robert F., 1971, No smoke needed; USDA Forest Service Fire Control Notes, vol. 32, no. 2, pp. 9–11.

Krumpe, P. F., 1971, The delineation and prediction of forest cover and site parameters by multiband remote sensing on Wilson Mountain, Morgan County, Tennessee: Dept. of Botany, M.S. Thesis, Univ. Tennessee, Knoxville, Tennessee, 120 p.

Krumpe, P. F., 1973, A regional approach to wildland resource distributional analysis utilizing high altitude and earth orbital imagery: Proc. Ann. Meeting, Soc. of Photogramm., 39th, Washington, D.C., March 11–16.

Krumpe, P. F., J. D. Nichols, and D. T. Lauer, 1973, ERTS-1 analysis of wildland resources using manual and automated techniques, *in* Symposium on Management and Utilization of Remote Sensing Data, October 1973, Proceedings; U.S. Geological Survey, EROS Data Center, Sioux Falls, South Dakota, pp. 50–66.

Krebs, P. V., and R. M. Hoffer, 1976, Multiple resource evaluation of Region 2 U.S. Forest Service lands utilizing Landsat MSS data; Type III Final Report, NASA Goddard Space Flight Center, Greenbelt, Md.

Kuhl, A. D., 1970, Color and IR photos for soils; Photogramm. Eng. vol. 36, pp. 475–482.

Künzi, K., M. Wuthrich, and E. Schanda, 1971, A MM-wave scanning radiometer for terrain mapping;

Proc. Internat. Symp. on Remote Sensing of Environment, 7th, Ann Arbor, vol. 2, pp. 865–867.

Kurth, A., et al, 1962, Die Anwendung des Luftbildes im schweizerischen Forstwesen; Mitteilungen der schweizerischen Anst.f.d.Forstl. Versuchswesen 38, H.1.

Kuusela, Kullervo and Simo, Poso, 1970, Satellite pictures in the estimation of the growing stock over extensive areas; Photogramm. Jour., Finland, vol. 4, no. 1.

Lachowski, H. M., D. L. Dietrich, R. M. Umali, E. A. Aquino, and V. A. Basa, 1978, Landsat-assisted forest inventory of the Philippine Islands; Proc. 12th Intern. Symp. Rem. Sens. Environment, Ann Arbor, Mich., pp. 1401–1408.

Lachowski, H. M., D. L. Dietrich, R. Umali, E. Aquino, V. and V. Basa, 1979, Landsat assisted forest land-cover assessment of the Philippine Islands. Photogrammetric Engineering and Remote Sensing vol. 45, no. 10, pp. 1387–1391.

Lackner, H., 1966 Vergleich von 9 Film-Mabstabskombinationen für die Holzarten-Interpretation; Mitteilungen d.forstl.Bundesversuchsanstalt Wien, 72. Heft.

v.Laer, W., 1962. Aerophotogrammetrische Hohenzuwachsmessungen von Waldbestanden; Allg. Forstzeitschrift, pp. 33–34.

Landgrebe, D. A., and Staff, 1967, Automatic identification and classification of wheat by remote sensing; Research Progress Report 279, Agr. Expt. Sta., Purdue University, W. Lafayette, In.

Landgrebe, D. A., R. M. Hoffer, F. E. Goodrick, and Staff, 1972, An early analysis of ERTS-1 data; Proceedings of the NASA Symposium on Preliminary Results of ERTS Data Analysis, Goddard Space Flight Center, Greenbelt, Maryland, September 29, 1972, pp. 21–38.

Langley, P. G., 1962, Aerial photo interpretation manual for the integrated forest survey and timber management inventory in California; Pacific Southwest Forest and Range Expt. Sta., Forest Service, USDA, Berkeley, California.

Langley, P. G., 1969, New multi-stage sampling techniques using space and aircraft imagery for forest inventory; Internat. Symp. on Remote Sensing of Environment, 6th, Ann Arbor, 1969 Proc., pp. 1179–1192.

Langley, P. G., R. C. Aldrich, and R. C. Heller, 1969, Multi-stage sampling of forest resources by using space photography; Ann. Earth Resources Aircraft Program Status Review, 2nd, Houston, 1969 Proc.

Lanly, J. P., 1970, L'utilisation des photographies aériennes à l'Afrique noire francophone; Personal correspondence.

LaPerriere, L. R., and W. A. Howard, 1971, Discriminating previsual symptoms of stress associated with Dutch elm disease through color infrared photography; Univ. Denver, Dept. Geog. Tech. Paper no. 71-1, p. 135.

LARS Staff, 1970, Remote multispectral sensing in agriculture; Laboratory for Applications of Remote Sensing, Annual Report, Research Bulletin 873, Agricultural Experiment Station, Purdue University, West Lafayette, Ind, vol. 4.

Latham, R. P., D. W. French, and M. P. Meyer, 1969, Detecting oak wilt by false color infrared aerial photography; Journal of the Minnesota Academy of Sciences, vol. 36, no. 1, pp. 14–15.

Latty, R. S., and R. M. Hoffer, 1981, Computer-based classification accuracy due to the spatial resolution using per-point and per-field classification techniques; Proceedings of the 7th International Symposium on Machine Processing of Remotely Sensed Data, Purdue University, W. Lafayette, Indiana, pp. 384–392.

Lauer, D. T., 1966, The feasibility of identifying forest species and delineating major timber types in California by means of high altitude small scale aerial photography; Ann. Prog. Rept., Remote Sensing Applications in Forestry, OSSA/NASA, Earth Resources Survey Program, by the Forestry Remote Sensing Lab., Univ. California, Berkeley, California.

Lauer, D. T., and A. S. Benson, 1973, Classification of forest lands with ultra-high altitude, small scale false-color infrared photography; Presented at the Internat. Union of Forestry Research Organizations (IUFRO) Symp. on Remote Sensing in Forestry, Freiberg, West Germany, September 17–21.

Lauer, D. T. and P. E. Krumpe, 1973, Testing the usefulness of ERTS-1 imagery for inventorying wildland resources in northern California; obtained from ERTS-1 1A, NASA/GSFC March 5–9, 1973, pp. 97–104.

Lauer, D. T. and Todd, W. J., 1981, Land cover mapping with merged Landsat RBV and MSS stereoscopic images, *in* ASP Fall Technical Meeting, September 1981, San Francisco, California; American Society of Photogrammetry.

Lawrence, R. D., and J. H. Herzog, 1975, Geology and forestry classification from ERTS-1 digital data; Photogrammetric Engineering and Remote Sensing vol. 41, no. 10, pp. 1241–1251.

Lee, Y. J., 1976, Computer-assisted forest land classification in British Columbia and the Yukon Territory: a case study; Proceedings of the 1976 Fall Convention of the American Society of Photogrammetry, Seattle, Wash., pp. 240–250.

Lee, Y. J., F. Towler, H. Bradatsch, and S. Finding, 1977, Computer assisted forest land classification by means of several classification methods on the CCRS Image 100; Proceedings of the 4th Canadian Symposium on Remote Sensing, Loews Lee Concorde, Quebec, pp. 37–46.

Lent, J. D., 1969, Automatic image classification and data processing, in Analysis of Remote Sensing Data for Evaluating Forest and Range Resources, by R. N. Colwell et al. Ann. Prog. Rept., Earth Resources Survey Program, Forestry Remote Sensing Lab., Univ. California, Berkeley, California.

Le Ray, J., 1958, Aspects et possibilités de la photographie aérienne en zone de forêt dense Africaine; Revue Bois et Forêts des Tropiques, no. 61, pp. 27–33.

——— 1962, L'intrepretation des photographies aeriennes verticales et les problemes de L'exploitation forestiere tropicale; Bull. no. 8, Soc. Francaise de Photogrammetrie, pp. 23–33.

Le Schack, L. A., 1971, ADP of forest imagery, spatial distribution information of the reflecting trees was extracted from aerial imagery and serves to delineate unambiguously five forests from each other: Photogramm. Eng. vol. 37, pp. 885–896.

Le Schack, L. A. and Long, J.B., 1971, Transportation studies show best way to breach jungle mining areas; Eng. and Mining Jour., vol. 172, no. 2.

Loetsch, F., 1957. A forest inventory in Thailand; Unasylva 11, pp. 174–180. 1962, Die Bedeutung des Luftbildes bei Waldinventuren in den Tropen:

Allgemeine Forstzeitschrift, 17 Jg. no. 1/2, pp. 9–17.

———1970, Ein neuer Weg der bestandesweisen Holzmassenabschatzung aus Luftbildern fur Grossrauminventuren; Allg. Forstzeitschrift, pp. 730–732.

Loetsch, F., and Haller, E., 1963, Der Mabstabsadjustierungsfehler bei der Flachenbestimmung durch Punktstichprobe aus dem Luftbild; International Arch. Photogramm., vol. 24, pp. 217–222.

———1964, Forest Inventory Volume 1; Bayr, Landw. Verlag, Munchen.

Losee, S. T. B., 1953, Timber estimates from large scale photographs; Photogramm. Eng., vol. 19, pp. 752–762.

———1956, Measurement of stand density in forest photogrammetry; Canadian Inst. Surveying and Photogramm., (paper presented).

Lowe, D. S., F. C. Polcyn, and R. Shay, 1965, Multispectral data collection program; Proceedings of the Third International Symposium on Remote Sensing of the Environment, University of Michigan, Ann Arbor, Michigan. pp. 667–680.

Lund, H. G., 1971, Mirror stereoscope parallax wedge; U.S. Forest Service Research Note PNW-140.

Lux, H., 1965, Die grossräumige Abgrenzung von Rauchschadenszonen im Einflussbereich des Industriegebietes um Bitterfeld: Wissenschaftl. Zeitschrift d.T.U. Dresden, vol. 14, no. 2, pp. 433–442.

Lyons, E. H., 1966, Fixed air-base 70 mm photography, a new tool for forest sampling; Forestry Chron., vol. 42, no. 4, pp. 420–429.

Lyons, E. H., 1967, Forest sampling with 70 mm fixed air-base photography from helicopters; Photogrammetric Engineering, vol. 22, pp. 213–231.

MacConnell, W. P. and G. P. Stoll, 1968, Use of aerial photographs to evaluate the recreational resources of the Connecticut River in Massachusetts; Bull. no. 573, Expt. Sta., Coll. of Agriculture, Univ. Massachusetts, Amherst, Mass..

MacLean, Colin D., 1963, Improving forest inventory area statistics through supplementary photo interpretation, Journal Forestry vol. 61, pp. 512–516.

MacLean, Colin D., 1972, Improving inventory volume estimates by double sampling on aerial photographs, Journal Forestry vol. 70, pp. 739–740.

Manual of Photo Interpretation, 1960, Photo interpretation in forestry, chap. 7; Washington, D.C., Am. Soc. of Photogramm., pp. 474–482.

Manual of Remote Sensing, 1975, Forest Lands: Inventory and Assessment, chap. 17; Falls Church, Virginia, Am. Soc. of Photogramm. pp. 1353–1426.

Malilia, W. A., 1980, Change vector analysis: an approach for detecting forest changes with Landsat; Proceedings of the Sixth International Symposium on Machine Processing of Remotely Sensed Data, Purdue University, W. Lafayette, Indiana, pp. 326–336.

Marshall, J. R. and M. P. Meyer, 1978, Field evaluation of small scale forest resource aerial photography; Photogrammetric Engineering and Remote Sensing vol. 44, no. 1, pp. 37–42.

Mazade, A. V., and Others, 1981, The ten-ecosystem study: final report; Report LEMSCO-13491, Lockheed Engineering and Management Services Co., Inc., Houston, Texas.

McCormack, R. J., 1967, Land capability classification for forestry; The Canada Land Inventory Rept. no. 4, Dept. of Forestry and Rural Development, 26 p.

McGregor, M. D., W. E. Bousfield, and D. Almos, 1972, Evaluation of the Douglas-Fir beetle infestation in the North Fork Clearwater River Drainage, Idaho, Rept. no. 1-72-10; Northern Region, Forest Service, USDA.

Mead, R. A., R. S. Driscoll, and J. A. Smith, 1979, Effects of tree distribution and canopy cover on classification of ponderosa pine forest from Landsat-1 data; Research Note RM-375, Rocky Mountain Forest and Range Experiment Station, U.S. Forest Service, Ft. Collins, Colo.

Mead, R. A. and Gammon, P. T., 1981, Mapping wetlands using orthophotoquads and 35 mm aerial photographs; Photogrammetric Engineering and Remote Sensing, vol. 47, no. 5, May 1981, pp. 649–652.

Mead, R., and M. Meyer, 1977, Landsat digital data application to forest vegetation and land-use classification in Minnesota; IAFHE RSL Research Report 77-6, University of Minnesota, St. Paul, Minnesota.

Meyer, M. P., 1973, Operating manual—Montana 35 mm aerial photography system; Institute of Agriculture Remote Sensing Laboratory, University of Minnesota, St. Paul, Minnesota, December 1973, IARSL Research Report 73-3.

Meyer, M. P. and D. W. French, 1966, Forest disease spread; Photogramm. Eng., vol. 32, no. 5, pp. 812–814.

Meyer, M. P. and D. W. French, 1967, Detection of diseased trees; Photogrammetric Engineering vol. 32, no. 9, pp. 1035–1040.

Meyer, M. P., D. W. French, R. Latham, C. Nelson, and R. Douglass, 1971, Vigor loss in conifers due to dwarf mistletoe; Ann. Prog. Rep. for Earth Resources Survey Program, NASA, by Univ. Minn. Coll. For.

Meyer, M. P. and Grumstrup, P. D., 1978, Operating manual for the Montana 35 mm aerial photography system (2nd ed.); College of Forestry, University of Minnesota, St. Paul, Minnesota, IAFHE RSL Research Report 78-1.

Mika, P. G., R. C. Heller, and K. J. Stoszek, 1980, Application of models developed to risk rate forest sites and stands to Douglas-fir tussock moth defoliations; Proceedings Hazard Rating Systems in Forest Insect Pest Management, University of Georgia, Athens, Georgia, July 31–August 1, 1980.

Miller, J. M., and F. P. Keen, 1960, Biology and control of the western pine beetle; U.S. Dept. Agr. Misc. Pub. 800.

Miller, L. D., K. Nualchawee, and C. Tom, 1978, Analysis of the dynamics of shifting cultivation in the tropical forests of northern Thailand using landscape modelling and classification of Landsat imagery; NASA, Tech. memo 79545, Greenbelt, MD.

Miller, P. R., J. R. Parmeter, Jr., B. H. Flick, and C. W. Martinez, 1969, Ozone dosage response of ponderosa pine seedlings; Air Pollution Control Assoc. Jour., vol. 19, pp. 435–438.

Miller, R. G., 1957, The use of aerial photographs in forestry in British Colonies; 7th Brit. Commonw. For. Congr.

———1960, The interpretation of tropical vegetation and cops on aerial photos; Photogrammetria, vol. 16, no. 3, pp. 232–240.

Miller, W. A., and R. C. Heller, 1978, Remote sensing approach to identifying preferred Douglas-fir tussock moth (*Orgyia pseudotsugata* McD.) sites;

Proceedings of Remote Sensing Damage Assessment, ASP. Seattle, Washington.

Miller, W. A., M. B. Shasby, W. G. Rohde, and G. R. Johnson, 1981, Developing in-place data bases by incorporating digital terrain data into the Landsat classification process; Proceedings for In-Place Resources Inventories National Workshop, Orono, Maine, August 9–14, 1981.

Minnich, R. A., L. W. Bowden, and R. W. Pease, 1970, Mapping montane vegetation in Southern California from infrared imagery; USDA Contract 14-08-0001-10674. Status Rept. 3. Tech. Rept. 3.

———1974, The Impact of fire suppression on Southern California conifer forests: A case study of the Big Bear fire, November 13–16, 1970; Proc. of Symp. Living with the Chaparral, March 30–31, 1973.

Minor, C. O., 1951, Stem-crown diameter relations in southern pine; Jour. Forestry, vol. 49, no. 7, pp. 490–493.

Minor, C. O., 1960, Estimating tree diameter of Arizona ponderosa pine from aerial photography; Forest Service, U.S. Department of Agriculture, Rocky Mountain For. and Range Exp. Station, Research Note No. 46.

Moessner, K. E., 1948, Photo classification of forest sites; Soc. Am. Foresters, 1948 Proc., pp. 278–291.

———1949, A Crown density scale for photo interpreters; Jour. Forestry, vol. 47, no. 7, pp. 569.

———1950, Principal uses of air photos by the Forest Service; Photogramm. Eng., vol. 16, pp. 301–304.

———1956, Combined vertical and horizontal stereograms; U.S. Forest Service, Intermtn. Forest and Range Expt. Sta. Research Note no. 36.

———1960a, Aerial volume tables for ponderosa pine type in the Rocky Mountains; U.S. Forest Service, Rocky Mtn. Forest and Range Expt. Sta., Research Note no. 76.

———1960b, Estimating timber volume by direct photogrammetric methods; Proc. Soc. of Am. Forestry, San Francisco Meeting 1959, pp. 148–151.

———1960c, Training handbook; U.S. Forest Service, Intermtn. Forest and Range Expt. Sta.

———1961, Comparative usefulness of three parallax measuring instruments in the measurement and interpretation of forest stands; Photogramm. Eng., vol. 27, pp. 705–709.

———1962, Parallax wedge improved; U.S. Forest Service, Intermtn. Forest and Range Expt. Sta., Research Note no. 94.

———1963a, Composite aerial volume tables for conifer stands in the mountain states; U.S. Forest Service Research Note INT-6.

———1963b, A Test of aerial photo classifications in forest management—volume inventories; U.S. Forest Service Research Paper INT-3.

———1964, Learning to estimate stand volume from aerial photos; U.S. Forest Service Research Note INT-25.

Moessner, K. E., D. F. Brunson, and C. E. Jensen, 1951, Aerial volume tables for hardwood stands in the Central States: U.S. Forest Service, Central States Forest Expt. Sta., Tech. Paper no. 122.

Morain, S. A. and D. S. Simonett, 1966, Vegetation analysis with radar imagery. Center for Research Inc., Univ. Kansas, CRES Report 61-9.

Morain, S. A. and B. Klankamsoon, 1978, Forest mapping and inventory techniques through visual analysis of Landsat imagery: examples from Thailand; Proc. 12th Intern. Symp. Rem. Sens. Environ., Ann Arbor, Michigan, pp. 417–426.

Morris, A. W., 1957, Aerial volume table for black spruce type for the northeastern coniferous zone; Canadian Pulp and Paper Assoc. Woodlands Section, Index no. 1650.

Morris, W. G., 1970, Photo inventory of fine logging slash; Photogramm. Eng., vol. 36, pp. 1252–1256.

Murtha, P. A., 1972a, A Guide to aerial photographic interpretation of forest damage in Canada; Canada Forest. Serv., Dept. of the Environment, Publication no. 1292.

———1972b, Sulfur dioxide damage delineation on high-altitude photographs; Can. Symp. Remote Sensing, First Ottawa, Canada, Centre Remote Sensing, Feb. 7, 8, 9.

Myers, B. J. and Benson, M. L., 1981, Rainforest species on large-scale color photos: Photogrammetric Engineering and Remote Sensing, vol. 47, no. 4, pp. 505–513.

Nash, A. J., 1948, Some volume tables for use in air survey; Forestry Chron., vol. 24, no. 1, pp. 4–14.

Nelson, H. A., 1977, Small format color photography for forest plantation planning in the Southeast; Proceedings of the 6th Biennial Workshop "Aerial Color Photography in the Plant Sciences and Related Fields", Colorado State University, Ft. Collins, CO. Aug 9–11, 1977, pp. 141–145.

Nelson, R. F., and R. M. Hoffer, 1980, Procedure 1 and forestland classification using Landsat data; Proceedings of the Sixth International Symposium on Machine Processing of Remotely Sensed Data, Purdue University, W. Lafayette, Indiana, pp. 319–325.

Neustein, S. A., 1971, Damage to forests in relation to topography, soil and crops—Windblow of Scottish forests in January, 1968; Sec. 11, Bull. For. Comm., Edin.

Neustein, S. A. and J. Waddell, 1972, Some investigations in the use of 35 mm aerial photography; July, 1972.

Nichols, J. D., M. Gialdini, and S. Jaakkola, 1973, A timber inventory based upon manual and automated analysis of ERTS-1 and supporting aircraft data using multistage probability sampling; Proceedings of the Third Earth Resources Technology Satellite-1 Symposium, NASA Goddard Space Flight Center, Washington, D.C., pp. 145–157.

Nichols, J. D., R. A. Harding, R. B. Scott, and J. R. Edwards, 1976, Forest inventory of western Washington by satellite multistage sampling; Proceedings of the Fall Convention of the American Society of Photogrammetry, Seattle, Washington, pp. 180–216.

Nielsen, U. and A. H. Aldred, 1978, New developments for tropical surveys prove successful; Proceedings of International Symposium on Remote Sensing for Observation and Inventory of Earth Resources and the Endangered Environment, (Freiburg, W. Germany, July 2–8, 1978) Sponsored by Commission VII ISP and Subject Group 6.05, IUFRO.

Nielsen, U., A. H. Aldred, and D. A. MacLeod, 1979, A forest inventory in the Yukon using large scale photo sampling techniques; For. Man. Inst. Inf. Report FMR-X-121. Canadian Forestry Service, Ottawa, Canada.

Nielsen, Udo and Sayn-Wittgenstein, 1970, The forestry radar altimeter tested over steep topography;

Canadian Forestry Service, Dept. Fisheries and Forestry, Internat. Rept. FMR-18.

Nyyssönen, A., 1955, On the estimation of the growing stock from aerial photographs; Comm. Inst. For. Fem., vol. 46.

———1961, Survey methods of tropical forests; FAO (Food and Agriculture Organization of the United Nations), Rome.

———1967, Photogrammetric volume estimation in forest inventory. Papers of the XIV IUFRO Congress Munchen, v. 6, p. 1–11.

Nyyssönen, A., S. Poso, and Chr. Keil, 1968, The use of aerial photographs in the estimation of some forest characteristics; Acta Forestalia Fennica 82.

Olson, C. E., Jr., L. W. Tombaugh, and H. C. Davis, 1969, Inventory of recreation sites; Photogramm. Eng., vol. 35, pp. 561–568.

Oregon State Department of Forestry, 1978, Douglas County forest condition mapping and forest volume inventory project, Final project report submitted to Pacific Northwest Regional Commission, Salem, Oregon.

Paelinck, P., 1958, Note sur L'estimation du volume des peuplements à limba (Terminalia superba) au Mayumbe, à l'aides des photos aériennes; Bull. agric. Congo Belge, 49, pp. 1045–54.

Page, A. I., 1969a, Use of large-scale mosaics for planning and control of aerial seeding; New Zealand Jour. Forestry, vol. 14, no. 1, pp. 96–97.

———1969b, High-altitude photography for control of aerial seeding operations; New Zealand Jour. Forestry, vol. 14, no. 2, pp. 239–241.

Paymans, K., 1970, Land evaluation by air photo interpretation and field sampling in Australian New Guinea; Photogrammetria, vol. 26, pp. 77–100.

Perlwitz, W., 1963, Hilfsmittel fur die Baumhöhenbestimmung aus Luftbilern-Parallaxenmeßscheibe und Baumhöhenrechner; Sozial. Forstwirtschaft, pp. 150–152.

Pohorly, M., 1958, Posonzeni moznosti vylisovani lesniho detailu pomoci leteckych snimka; Vysckeho Uceni Techn. vol. Prace, pp. 475–488.

Pollanschütz, J., 1968. Erste Ergebnisse uber die Verwendung eines Infrarot-Farbfilms in Osterreich fur die Zwecke einer Rauchschadensfestellung: Centralbl.f.d.ges. Forstwesen, pp. 65–79.

Pomerening, J. A., and Cline, M. G., 1953, The accuracy of soil maps prepared by various methods that use aerial photograph interpretation; Photogramm. Eng., vol. 19, pp. 809–817.

Pope, R. B., 1957, The role of aerial photography in the current balsam woolly aphid outbreak; Forestry Chron., vol. 33, no. 3, pp. 263–264.

———1960, Ocular estimation of crown density on aerial photos: Forestry Chron., vol. 36, no. 1, pp. 89–90.

———1961, Aerial photo volume tables for Douglas-fir in the Pacific Northwest; U.S. Forest Service, Pacific Northwest Forest and Range Expt. Sta. Research Note no. 214.

———1962, Constructing aerial photo volume tables; U.S. Forest Service, Pacific Northwest Forest and Range Expt. Sta. Research Paper no. 49.

Rabenau, 1969, Holzarteninterpretation aus Luftbildern mit statistischer Prüfung des Einflusses von Film and Maßstab. Hochschule für Bodenkultur Wien.

Randall, A. C., 1969, Forest surveys for economic development. Physical resource investigations for economic development; A case book of OAS field experience in Latin America. Gen. Secr. Org. of Am. States, pp. 183–227.

Rees, T. I., 1972, Directorate of overseas surveys, U.K.; personal correspondence.

Reinhold, A., 1967, Large-scale aerial photos as an aid in assessing the silvicultural condition of pine plantations and thickets, Archiv. Forestwesen, vol. 16, pp. 905–910.

Rhody, B., 1962, Methods of estimating storm damage with the aid of aerial photographs, exemplified by the Werdenberger forest. St. Gallen Rheintal; Schweiz. Z. Forstwesen, vol. 114, pp. 314–332.

Richards, P. W., 1952, The tropical rainforest, Cambridge Univ. Press. pp. 1–25, 372–4, 399.

Roberts, E. H., and Dana, R. W., 1981, Comparison of three types of remote sensing data in mapping Grand County, Colorado; Proceedings for In-Place Resources Inventories National Workshop, Orono, Maine.

Robinson, M. W., 1947, An instrument to measure forest crown cover; Forestry Chron., vol. 13, no. 3, pp. 222–225.

Rogers, E. J., 1946, Use of parallax wedge in measuring tree heights on vertical aerial photographs; U.S. Forest Service, Northeastern Forest Expt. Sta., Forest Survey Note No. 1.

———1947, Estimating tree heights from shadows on vertical aerial photographs; U.S. Forest Service, Northeastern Forest Expt. Sta., Sta. Paper No. 12.

———1949, Estimating tree heights from shadows on vertical aerial photographs; Jour. Forestry, vol. 47, no. 3, pp. 182–191.

———1956, Photogrammetry research in forest surveys; Internat. Soc. Photogramm., working group IV, Commun. VII.

———1960, Forest survey design applying aerial photographs and regression techniques for the Caspian Forest of Iran; Photogramm. Eng., vol. 26, no. 3, pp. 441–443.

———1961, Application of aerial photographs and regression technique for surveying Caspian Forests of Iran; Photogramm. Eng., vol. 27, pp. 811–816.

Rohde, W. G., and C. E. Olson, 1972, Multispectral sensing of forest tree species, Photogrammetric Engineering vol. 38, no. 12, pp. 1209–1215.

Rohde, W. G., 1978, Potential applications of satellite imagery in some types of natural resource inventories; Proceedings of the National Workshop on Integrated Inventories of Renewable Natural Resources, Rocky Mountain Forest and Range Experiment Station General Technical Report RM-55, USDA Forest Service, Fort Collins, Colorado, pp. 209–218.

Rohde, W. G., W. A. Miller, K. G. Bonner, E. Hertz, and M. F. Engel, 1979, A stratified-cluster sampling procedure applied to a wildland vegetation inventory using remote sensing; Proceedings of the Thirteenth International Symposium on Remote Sensing of Environment, Ann Arbor, Michigan, pp. 167–179.

Rohde, W. G., and Miller, W. A., 1981, Arizona vegetation resource inventory (AVRI) project final report; U.S. Geological Survey, EROS Data Center, Sioux Falls, South Dakota, unpublished report.

Rollet, 1960, Emploi des photographies aériennes au 1:40,000 pour l'interprétation de la végétation et les inventaires forestiers au Cambodge et au Vietnam. Bois et Forêts Tropiques nr. 74.

Rosayro, R. A. de 1959, The application of aerial photography to stockmapping and inventories on an ecological basis in rainforest in Ceylon. Emp.; For. Rev. 38, pp. 141–147.

Roscoe, J. H., L. D. Black, H. Weiner, H. D. Young, P. Maynard, F. C. Whitmore, and R. N. Colwell, 1955, Photo interpretation keys; Photogrammetric Engineering, vol. 21, pp. 703–724.

Rosenfield, G. H., K. Fitzpatrick-Lins, and H. S. Ling, 1982, Sampling for thematic map accuracy testing, Photogramm Eng. and Remote Sensing, vol. 48, no. 1, pp. 131–137.

Rossetti, C., P. Kowalski, and N. Havé, 1966, Relations entre les charactéristiques de reflexion spectrale de quelques espèces végétales et leurs images sur des photographies en couleur, terrestres et aériennes; Int. Arch. Photogramm vol. 16, no. 2, pp. 27–50.

Roth, E. R., R. C. Heller, and W. A. Stegall, 1963, Color photography for oak wilt detection; Jour. Forestry, vol. 61, no. 10, pp. 774–778.

Rudd, R. D., 1971, Macro land use mapping with simulated space photos; Photogramm. Eng., vol. 37, pp. 365–372.

Sadar, S. A., and W. F. Miller, 1976, Development of a risk rating system for southern pine beetle infestations in Capiah County, Mississippi; Proc. of Remote Sensing of Earth Resources, S.I.F. Shahrok, ed.; University of Tennessee, Tulahoma, Tn., vol. 5.

Sadowski, F. G., W. A. Malila, and R. F. Nalepka, 1978, Applications of MSS systems to natural resource inventories; Proceedings of the National Workshop on Integrated Inventories of Renewable Natural Resources, Rocky Mountain Forest and Range Experiment Station General Technical Report RM-55, USDA Forest Service, Fort Collins, Colorado, pp. 248–256.

Samojlovič, G. G., 1940, Opyt izučenija vzaimosvjazi meždu diametrami kron i nekotorymi taksacionnymi priznakami derev'ev dlja operedelenija ich po aerosnimkam; Sb. Statej Porolžkogo Lesotechn. Inst. No. 2, Joškar-Ola, Margosizdat.

———1952, Lesnoe desifrirovanie aerosnimko. Izd., Lesprojekt, Leningrad.

Sandberg, B., 1963, Provytetaxering i flygbilder med jämförande fältkontroll Inform. No. 21, Nämnden för Skoglig Fotogrammetri, Stockholm.

Sayn-Wittgenstein, L., 1978, Recognition of tree species on aerial photographs; For. Man. Inst. Inf. Report FMR-X-118, Canadian Forestry Service. Ottawa, Canada.

Sayn-Wittgenstein, L., and A. H. Aldred, 1967, Tree volumes form large-scale photos; Photogramm. Eng., vol. 33, pp. 69–73.

Sayn-Wittgenstein, L., R. deMilde, and C. J. Inglis, 1978, Identification of tropical trees on aerial photographs; Canadian For. Serv., Environm. of Canada.

Sayn-Wittgenstein, Leo, and Z. D. Kalensky, 1975, Interpretation of forest patterns on computer compatible tapes; Canadian Symposium on Remote Sensing, Proceedings of Remote Sensing Society, Ottawa, Ontario.

Schindler, Chr., 1967, Möglichkeiten der Strichprobenahme aus Luftbildern unter besonderer Anwendung der Radiallinie zum Zwecke von Flächeniventuren. Diss. Freiburg.

Schultz, G., 1970, Die Baumhöhe als photogrammetrische Messgrösse. Allg. Forestzeitschrift, pp. 754–756.

Schumacher, F. X., and R. A. Chapman, 1948, Sampling methods in forestry and range management; Duke Univ. Press Durham, N.C. Bull. 7, 2nd ed.

———1954, Revised: Duke Univ. Sch. of Forestry Bull., No. 7.

Schut, G. H. and M. C. Van Wijk, 1964, The determination of tree heights from parallax measurements; Canadian Surveyor, vol. 19, pp. 415–427.

Seeley H. E., 1929, Computing tree heights from shadows on aerial photographs; Forestry Chron., vol. 5, no. 4, pp. 24–27.

———1949, Air photography and its application to forestry; Forestry Air Survey Publication No. 6, Canada Dept. of Mines and Resources, Ottawa, Canada.

———1955, A forest survey method; Canada Dept. Northern Affairs and Natl. Resources, Forest Research Div. Tech. Note No. 8.

Shimabukuro, Y. E., Pilho, N. F., Coffler, and S. C. Chen, 1980, Automatic classification of reforested pine and eucalyptus using Landsat data; Photogrammetric Engineering and Remote Sensing vol. 46, no. 2, pp. 209–216.

Shlien, S., and D. Goodenough, 1974, Quantitative methods of processing the information content of ERTS imagery for terrain classification; Proceedings of the Second Canadian Symposium on Remote Sensing, Guelph, Ontario, pp. 245–273.

Shiue, Cherng-Jiann, and H. H. John, 1962, A proposed sampling design for extensive forest inventory; Double systematic sampling for regression with multiple random starts; Jour. Forestry, vol. 60, no. 6, pp. 607–610.

Sicco Smit, G, 1971, Aplicación de las imagenes de radar en la foto-interpretación de bosques humedos tropicales; Public. of CIAF (Centro Interamericano de Foto-interpretación), Colombia.

Sicco Smit, G., 1978, SLAR for forest typeclassification in a semi-deciduous tropical region; ITC Journal vol. 3, pp. 385–400.

Sims, W. G., and M. L. Benson, 1966, Colour aerial photography in forest photo interpretation; Australian Forest Research vol. 2, no. 2, pp. 43–48.

———1967, Atmospheric haze penetration in colour aerial photography, Forest Resources Newsletter vol. 3, pp. 54–55.

Smedes, H. W., K. L. Pierce, M. C. Tanguey, and R. M. Hoffer, 1970, Digital computer terrain mapping from multispectral data; Journ. Spacecraft and Rockets vol. 7, no. 9, pp. 1025–1031.

Smith, D. U., 1969, Timber volume with a Kelsh plotter; Photogramm. Eng., vol. 35, no. 4, pp. 363–365.

Smith, H. G. and S. Khorram, 1980, Remote sensing aided procedure for conifer growth modeling in northeast Sierra Nevada, Proceedings, Fall Technical Meeting, American Society of Photogrammetry, 1980, RS-2F-1 to RS-2-F-14.

Soil Survey Staff, 1960, Soil classification, a comprehensive system, 7th approximation, Soil Conservation Service, USDA.

Spurr, S. H., 1945, Parallax wedge measuring devices; Photogramm. Eng., vol. 11, pp. 85–87.

———1946, Volume tables for use with aerial photographs; Petersham, Massachusetts, Harvard Forest Offset.

———1948, Aerial Photographs in Forestry; New York, Ronald.

———1949, Films and filters for forest aerial photography; Photogramm. Eng. vol. 15, pp. 473–481.

———1952, Forest Inventory; New York, Ronald, pp. 372–442.

Spurr, S. H., and C. T. Brown, Jr., 1946a, Specifications for photographs used in forest management; Photogramm. Eng. vol. 12, pp. 131–141.

———1946b, Tree height measurements from aerial Photographs; Jour. Forestry, vol. 44, no. 10, pp. 716–721.

Steigerwaldt, E. F., 1948, Outline of the state of Wisconsin forest inventory; Unpublished manuscript, Forest Protection Headquarters, Tomahawk, Wisconsin.

Steiner, D., and Gutermann, T., 1966, Russian data on spectral reflectance of vegetation, soil and rock types; Dept. Geography Univ. Zurich.1

Stellingwerf, D. A., 1962, Holzmassenbestimmung von Pinus silvestris auf Luftbildern in den Niederlanden; Allg. Forstzeitung, pp. 29–30.

———1966, Practical applications of aerial photographs in forestry and other vegetations studies; ITC Delft Publications B 36–38.

———1968a, Practical applications of aerial photographs in forestry and other vegetations studies; ITC Delft Publications B 46–48.

———1968b, The usefulness of Kodak Ektachrome Infrared Aerofilm for forestry purposes; XI Int. Congr. Photogramm., Lausanne, Switz.

———1969, Vegetation mapping from aerial photographs; East African Agric. and For. Jour., vol. 34, Sp. issue pp. 80–86.

Stellingwerf, D. A., 1979, Orthophoto maps and/or Landsat print-outs for forestry: aspects of a quantification problem; ITC Journal, no. 4, pp. 499–518.

Stephens, P. R., 1976, Comparison of color, color infrared, and panchromatic aerial photography; Photogrammetric Engineering and Remote Sensing, vol. 42, no. 10, pp. 1273–1277.

Stone, K. H., 1950, Aerial photographic interpretation of natural vegetation in the Anchorage, Alaska area: Survey and Mapping, vol. 10, no. 3, pp. 261.

Strahler, A. H., T. L. Logan, and N. A. Bryant, 1978, Improving forest cover classification accuracy from Landsat by incorporating topographic information; Proceedings of the 12th International Symposium on Remote Sensing of Environment, Ann Arbor, Michigan, pp. 927–942.

Strahler, A. H., T. L. Logan, and C. E. Woodcock, 1979, Forest classification and inventory system using Landsat, digital terrain, and ground sample data; Proceedings of the 13th International Symposium on Remote Sensing of Environment, Ann Arbor, Michigan, pp. 1541–1557.

Svensson, H., 1971, Wind action displayed by thermal imagery; Berichte III. Internat. Symp. Photointerpretation, Dresden 1970, V.I, pp. 697–700.

Swain, P. H., C. L. Wu, D. A. Landgrebe, and H. Hauska, 1975, Layered classification techniques for remote sensing applications; Proceedings of the NASA Earth Resources Survey Symposium, Houston, Texas, NASA TM X-58168, vol. 1-B, pp. 1087–1097.

Swain, P. H., and S. M. Davis (eds.), 1978, Remote Sensing: The Quantitative Approach; McGraw-Hill, Inc.

Swantje, H., 1957, Photogrammetric methods in reforestation surveys; Photogramm. Eng., vol. 23, pp. 789–790.

Swellengrebel, E. J. G., 1959, On the value of large scale aerial photographs in British Guiana forestry; Empire Forestry Ref., vol. 38, no. 1, pp. 54–64.

———1961, Estimation of greenheart volume from small scale aerial photographs; Empire Forestry Rev., vol. 40, no. 2, pp. 104, 162–171.

Taylor, B. W., and G. A. Stewart, 1958. Vegetation mapping in the territories of Papua and New Guinea conducted by CSIRO; Proc. of the Kandy Symp. UNESCO, pp. 127–136.

Talbot, J. J., 1981, What's happening to the world's forest resources; Proceedings of the Seventh International Symposium on Machine Processing of Remotely Sensed Data, Purdue University, W. Lafayette, Indiana, pp. 587–592.

Thorley, G. A., J. D. Lent, and R. N. Colwell, 1965, An evaluation of insect induced mortality by aerial color photography in the Giant Forest Lodgepole area of the Sequoia-Kings Canyon National Park; Univ. California, Final Rept. on project WR-34-64-633, National Park Service.

Titus, S., M. Gialdini, and J. Nichols, 1975, A total timber resource inventory based upon manual and automated analysis of Landsat-1 and supporting aircraft data using stratified multistage sampling techniques; Proceedings of the 10th International Symposium on Remote Sensing of Environment, Ann Arbor, Mich., vol. 2, p. 1157.

Tiwari, K. P., 1975, Tree species identification on large scale aerial photographs at New Forest; Indian Forester, vol. 101, no. 2, pp. 791–807.

Todd, W. J., D. G. Gerhing, and J. F. Haman, 1980, Landsat wildland mapping accuracy; Photogrammetric Engineering and Remote Sensing vol. 46, no. 4, pp. 509–520.

Tokmanoglu, I., 1969, Estimating storm damage in forests; Istanbul Univ. Orm. Fak. Derg., vol. 19A, no. 1, pp. 105–130.

Tomašegovič, Z., 1963, Statistische Messungen in der Forst- und Landwirtschaft mittels stereophotogrammetrischer Methoden; Bildmessung und Luftbildwesen, pp. 48–54.

Tretjakov, N. V., P. V. Gorskij, and G. G. Samojlovič, 1953, Spravocnik taksatora (part VIII aerial photointerpretation); Goslesbumisd. Moskau-Leningrad.

Tscherkasov, I. A., 1971, Up-to-date conditions and automation prospects of aerial photo interpretation processes by the method of optical image filtering; Berichte III. Internat. Symp. Photointerpretation, Dresden, 1970, vol. 1, pp. 821–834.

Tzschupke, Wolfgang, 1973, Farbmessungen an Luftbildern fur vegetationskundliche Zwecke; Bildmessung und Luftbildwesen, pp. 12–20.

Ulliman, J. J. and D. W. French, 1977, Detection of oak wilt with color IR aerial photography; Photogrammetric Engineering and Remote Sensing, vol. 43, no. 10, pp. 1267–1272.

Ulliman, J. J. and C. R. Hatch, 1978, Test of an electronic planimeter; Journal of Forestry. vol. 76, no. 6, pp. 346–347.

Ulliman, J. J. and M. P. Meyer, 1971, The Feasibility of Forest Cover Type Interpretation Using Small Scale Aerial Photographs; Proc. 7th Internat'l Symp. on Remote Sensing of Environment, Ann Arbor, Michigan, pp. 1219–1230.

U.S. Dept. of Agriculture 1968, Timber management plan inventory handbook; Forest Service, USDA, 2441.1 R5 Supplement no. 65.

U.S. Dept. Agriculture, 1965, Timber Trends in the United States; Forest Resources Report No. 17.

VanDerwalker, J., 1979, Proceedings of the rapid assessment methodology: a demonstration conference; Western Energy and Land Use Team, U.S. Fish and Wildlife Service, Fort Collins, Colorado, December 1979, W/CRAM-79/W36.

Vermeer, J., 1968, Results of an objective comparison of film-filter combinations applied to an example of photo interpretation for soil survey; Proceedings of the Eleventh Congress of the International Society for Photogrammetry, For Commission VII. Lausanne, Switzerland.

———1971, Interpretation of radar and thermal-infrared images; Berichte III. Internat. Symp. Photointerpretation, Dresden 1970, vol. 1, pp. 701–709.

Viksne, A., T. C. Liston, and C. D. Sapp, 1970, SLAR reconnaissance of Panama; Photogramm. Eng., vol. 36, no. 3, pp. 253–259.

Waelti, H, 1978, Low-level, fixed base aerial photography for resource management, proceedings of Symposium IUFRO S 6.05, Freiburg, Germany, pp. 163–178.

Wear, J. F., 1971, Monitoring forest land from high altitude and space; Ann. Prog. Rept. Forestry Remote Sensing Lab. for Earth Resources Program, NASA, by Pacific Southwest Forest and Range Expt. Sta., Berkeley, California.

Wear, J. F., R. B. Pope, and P. W. Orr, 1966, Aerial photograph techniques for estimating damage by insects in western forests; U.S. Dept. Agr., Forest Service, Pacific Northwest Forest and Range Expt. Sta.

Weber, F. P., 1964, An aerial survey of spruce and fir volume killed by the spruce budworm in northern Minnesota; U.S. Forest Service Research Note WO-2.

———1965, Aerial volume table for estimating cubic foot losses of white spruce and balsam fir in Minnesota; Jour. Forestry, vol. 63, no. 1, pp. 25–29.

Weber, F. P., and F. C. Polcyn, 1972, Remote sensing to detect stress in forests; Photogramm. Eng., vol. 38, no. 2, pp. 163–175.

Weber, F. P., and J. F. Wear, 1970, The development of spectro-signature indicators of root disease impacts on forest stands; Ann. Prog. Rept., Forestry Remote Sensing Lab. for Nat. Resources Program, NASA, by Pacific Southwest Forest and Range Expt. Sta., Berkeley, California.

Webster, R., and P. H. T. Beckett, 1964, A study of the agronomic value of soil maps interpreted from air photographs; Trans. of the Internat. Cong. of Soil Science, 8th, vol. 5, pp. 795–803.

Webster, R., and I. F. T. Wong, 1969, A numerical procedure for testing soil boundaries interpreted from air photographs; Photogrammetria, no. 24.

Wert, S. L., 1969a, Revised aerial volume table for estimating spruce and fir mortality in Minnesota; Jour. Forestry, vol. 67, no. 5, pp. 334–335.

———1969b, A system for using remote sensing techniques to detect and evaluate air pollution effects on forest stands; Internat. Symposium on Remote Sensing of Environment, 6th. Ann Arbor 1969, Proc., pp. 1169–1178.

Wert, S. L., and R. J. Myhre, 1967, Wedge measures parallax separations—on large-scale 70 mm aerial photographs; U.S. Forest Service Research Note PSW-142.

Wert, S. L., and B. Roettgering, 1968, Douglas-fir beetle survey with color photos; Photogramm. Eng., vol. 34, pp. 1243–1248.

Werth, L. F., 1981, An evaluation of ISOCLS and CLASSY clustering algorithms for forest classification in northern Idaho; Proceedings of the Seventh International Symposium on Machine Processing of Remotely Sensed Data, Purdue University, W. Lafayette, Indiana, pp. 11–17.

Westby, R. L., A. H. Aldred, and L. Sayn-Wittgenstein, 1968, The potential of large-scale air photographs and radar altimetry in land evaluation; Land evaluation: Australia, MacMillan.

Wieczorek, Ulrich, 1972, Der Einsatz von Aquidensiten in der Luftbildinterpretation und bei der quantitativen Analyse von Texturen; Münchener Geographische Abhandlungen, Bd. 7, München.

Williams, D. E., 1966, A report on the 1965 field tests of the airborne fire detection sytem; AFDS-2 developed by computing devices of Canada Limited. Canada. Dept. of Forestry, Forest Fire Research Inst. Misc. Paper.

Williams, D. E., and P. H. Kourtz, 1965, An assessment of the airborne infrared fire detection system; Canada, Dept. of Forestry, Forest Fire Research Inst. Internal rept. No. FF-2.

Williams, D. L., 1975, Computer analysis and mapping of gypsy moth defoliation levels in Pennsylvania using Landsat-I digital data; Proceedings of the NASA Earth Resources Survey Symposium, Houston, Texas, NASA TM X-58168, vol. 1-A, pp. 167–181.

Williams, D. L., 1976, A canopy-related stratification of a southern pine forest using Landsat digital data, Proceedings of the 1976 Fall Convention of the American Society of Photogrammetry, Seattle, Wash. pp. 231–239.

Williams, D. L., and L. D. Miller, 1979, Monitoring forest canopy alteration around the world with digital analysis of Landsat imagery, National Aeronautics and Space Administration, Goddard Space Flight Center, Greenbelt, Maryland.

Williams, D. L., M. L. Stauffer, and K. C. Leung, 1979, A forester's look at the application of image manipulation techniques to multitemporal Landsat data, Proceedings of the Symposium on Machine Processing of Remotely Sensed Data, Purdue University, West Lafayette, Indiana, pp. 368–375.

Wilson, R. A., S. N. Hirsch, B. J. Losensky, and F. H. Madden, 1971, Airborne infrared forest fire detection system; Final rept. USDA Forest Service Research Paper INT-93, Intermountain Forest and Range Exp. Sta., Ogden, Utah.

Wodera, H., 1948, Die Holzmassenermittlung nach Luftbildern Allg. Forst- und Holzwirtschaftszeitung; Heft 13, 14, 15, 16.

Wolff, G., 1960, Zur Verbesserung der Methodik von Holzvorratsinventuren mit Hilfe des Luftbildes; Archiv Forstwesen, pp. 365–379.

———1966, Moglichkeiten und Grenzen der Messinterpretation forstlicher Luftbilder; Archiv Forstwesen, pp. 169–181.

———1967a, Schwarz-weisse und falschfarbige Luftbilder als diagnostische Hilfsmittel für operative Arbeiten beim Forst-schutz (Rauchschaden) und bei der Waldbestandsdüngung; Internat. Arch. Photogramm. vol. 16, no. 2, pp. 85–95.

———1967b, Kronenschäden an Fichte im StFB Marienburg und ihre Diagnose im Falschfarben-Luftbild; Die Sozialistische Forstwirtschaft, pp. 148–151.

———1970, Die Erkennung biotischer Schäden im Falschfarbenluftbild und ihre Bedeutung fur die Forstschutzpraxis; Beiträge fur die Forstwirschaft, 4 Jg., III. pp. 30–38.

Worley, D. P., and Meyer, H. A., 1955, Measurement of crown diameter and crown cover and their accuracy on 1:12,000 scale photographs; Photogramm. Eng., vol. 21, pp. 372–375.

Young, H. E., 1953, Tree counts on air photos in Maine: Photogramm. Eng., vol. 19, pp. 111–116.

Young, H. E., F. M. Call, and T. C. Tryon, 1963, Multimillion acre forest inventories based on air photos: Photogramm. Eng., vol. 29, pp. 641–644.

Zahir-ud-Din, A. S. M., 1954, Aerial survey of Chittagong hill tracts forests; Pakistan Jour. Forestry vol. 4, pp. 237–240.

Zarzycki, J. H., 1968, Remarks on planning and execution of mapping projects in tropical areas; Paper, 11th Cong. ISP Comm IV, Lausanne.

Zenker, S., 1971, Remote sensing activities in Sweden; Internat. Symp. on Remote Sensing of Environment, 7th, Ann Arbor, Proc. vol. 1, pp. 61–70.

Zieger, E, 1928, Ermittlung von Bestandesmassen aus flugbildern mit Hilfe des Hugershoff-Heydeschen Autokartographen, (Determination of stand volume from aerial photographs with the help of the Hugershoff-Heyde Autocartograph.); Mitteilungen aus der sächsischen forstlichen Versuchsanstalt Zu Tharandt vol. 3, pp. 97–127.

Zonneveld, Beltman, Heinsdijk, Cohen, and Van der Eijck, 1952, The use of aerial photographs in a tropical country (Surinam); Photogramm. Eng., vol. 18, pp. 144–157.

Zsilinszky, V. G., 1963, Photographic interpretation of tree species in Ontario; Ontario Dept. of Lands and Forests.

———1969, Supplementary aerial photography with miniature cameras; Photogrammetria, vol. 25, pp. 27–38.

CHAPTER 35

Rangeland Applications

Authors: DAVID M. CARNEGGIE, BARRY J. SCHRUMPF and DAVID A. MOUAT

GENERAL CONTENTS: Basic remote sensing considerations for rangeland resource assessment: define resource problem; determine remote sensing data required; ensure staff qualification and training; correlate ground conditions with remotely sensed data; develop classification scheme and legend; define interpretation procedure and equipment; accomplish field verification and data collection; rectify and produce resource map; perform accuracy assessment. Rangeland resource inventory: objectives, costs and design; mapping and sampling; rangeland inventory using panchromatic aerial photographs; land cover and soil surveys using color and color infrared photographs; inventory applications using Landsat imagery; rangeland mapping using Landsat MSS digital data; sampling applications using remote sensing data. Rangeland resource monitoring: monitoring ecological range conditions; monitoring seasonal range conditions. Wildlife habitat assessment: habitat assessment with aerial photography; habitat assessment with satellite imagery. Animal census: aerial visual observations; aerial photographs for animal census; aerial thermal infrared scanning for animal census; radio telemetery for studying animal populations. Desertification of rangelands. Future outlook and conclusion. References.

INTRODUCTION

Rangelands are naturally occurring areas of grasses, forbs, shrubs, and open stands of trees. The plants and soils of rangeland ecosystems comprise greater than forty percent of the earth's landscape and are recognized as natural grasslands, meadows, tundra, deserts, shrublands, steppes, savannas, and woodlands (Williams et al, 1968). On these lands, herbage and browse provide forage for domestic and wild animals. Rangelands also represent areas for the development of more intensive agriculture, opportunities for recreation, location of housing sites, and sources for minerals, oil, and water.

The vast expanses of rangelands, the numerous benefits that may be derived from them and the acceleration of changes imposed on them combine to create unique challenges for those who are responsible for the management of these lands. In the development of management programs, range resource managers rely heavily upon basic ecological principles and an understanding of the biological and physical resources comprising rangelands. Such an approach requires interdisciplinary coordination in the development of resource inventories and monitoring programs—one which focuses upon the major components of the ecosystem; namely the plants, soils, and animals. The acquisition of sufficient and timely information about these resource components can be a formidable task. Fortunately, the task can often be effectively accomplished with the use of remote sensing technology.

Central to the problem of rangeland resources assessment is the question of resource value versus assessment costs. The value of vegetation and animal products produced on rangelands dictates an extensive approach to their management and imposes similar restrictions on the level of effort that can be expended in gathering data for policy, planning, and management purposes. These restrictions are being alleviated by introducing remote sensing techniques into inventory, monitoring, and analysis procedures. By combining the use of remotely-sensed data with accepted methods of collecting field data, the range manager can: 1) enlarge the area over which data are acquired (an increase the area assessed/person); 2) increase the level of detail collected for specific sites (increase the data/unit area); 3) improve the timing and frequency of conducting and/or updating inventories in relation to changes in the resource and 4) decrease the elapsed time between making of measurements and availability of the derived information with respect to significant changes in the resource base.

In this chapter the authors are concerned with the basic resource information requirements of range and wildlife managers and the remote sensing techniques that may be used to meet these requirements. The organization of chapter subsections therefore follows traditional applications; namely: resource inventory, resource monitoring, habitat assessment, animal census, and assessment of desertification. The emphasis, however, is on the remote sensing techniques available for gathering and analyzing the resource data. These techniques represent only one of many approaches for resource inventory, monitoring, and analysis. Generally, the integration of two or more techniques, e.g. remote sensing, statistics, and field measurements, will produce the most advantageous results. The authors will attempt to provide a useful distinction between approaches that are practical and usable today and those which, based primarily upon research, hold promise for the future. The reader is referred to the Range Resources Chapter (18) of the first edition of the Remote Sensing Manual for a historical perspective of the use of remotely-sensed data for rangeland resource assessment.

BASIC REMOTE SENSING CONSIDERATIONS FOR RANGELAND *RESOURCE ASSESSMENT*

The increased attention that has been given to developments in remote sensing technology during the past ten years should not be construed to mean that remote sensing is the only approach for gathering resource information. In fact, a remote sensing approach is incapable of collecting all the information needed for resource management decision making. Rather, the remote sensing technology does offer decided advantages for the collection of many well defined types of information varying from the general to the specific, especially when: a) there are high costs associated with conventional ground surveys; b) information is required in a short amount of time; c) little or no current resource information exists; and d) accessibility to a resource area is limited. Remote sensing procedures for a variety of rangeland resource assessment applications will be discussed in subsequent sections. The generalized approach that will be presented identifies most of the important questions, considerations, and procedures necessary to capitalize on the advantage of using a remote sensing approach for resource assessment. (A variation of these procedures was presented by Poulton, et al, 1975, in the First Edition of the Manual of Remote Sensing).

Define Resource Assessment Problem

The most important question to be answered before gathering range resource information is "What is the most effective methodology for collecting the resource information?" To answer this question and determine if a remote sensing approach is appropriate, one must first carefully define the kind of information needed, the level of detail to which it should be collected, the desired accuracy of the results, and the nature of the final output product. Remotely-sensed data are particularly useful for most general mapping and detection objectives when applied with a systematic field-data collection strategy, but may be less useful or inappropriate for some applications where detailed identification of small features (e.g., plant species information and ground sampling) is required. If remote sensing techniques can be applied to the problem of interest, a sampling design that is specific to the problem should first be formulated. For example, if the objective is cover mapping, it is important to question and then specify the level of detail to be mapped, the categories or classes of cover to be mapped within a classification scheme, the minimum mapping unit, and the map scale needed to display the results. Answers to these questions will affect the procedures for using a remote sensing approach, particularly of remotely-sensed data to be interpreted, methods for analyzing the data, and the kind of ground sampling needed to validate the interpretive maps. The amount of money available for the assessment will also affect the analysis approach used, the amount of ground sampling required, and the format of an output product.

Determine Remote Sensing Data Required

With the assumption that a remote sensing approach with adequate ground sampling provides an acceptable and efficient method for collecting resource information, one must next consider the following questions: What remotely-sensed data are already available? What is the data scale or resolution, and is this adequate for extracting the defined information? Are contact image scales adequate or will enlargements be needed? What is the appropriate image or data format? (e.g., paper photo print, photo transparencies, negatives?) Are data available at the proper season of year to maximize interpretability of features?

If adequate remotely-sensed data are not readily available one must define the specifications and quality of remotely-sensed data to be acquired by an aerial survey contractor. Considerations with respect to sensor type, image type (film and filter combination), image scale, season of year, stereoscopic coverage, image format, and image quality need to be specified to insure that the remotely-sensed data, once acquired, will enable the resource-assessment objectives to be met.

A large array of sensors and film-filter types are available for data acquisition. Details and characteristics of many of these are presented in Chapter 6. The sensor systems most commonly used for rangeland resource assessment are aerial cameras and the spaceborne multispectral scanner (MSS) aboard Landsat. Aerial camera systems with 35mm, 70mm, and 9 × 9 inch formats can acquire both very large-scale detailed photographs of sample plots and transects as well as medium to small-scale photographs suitable for the mapping and inventory of rangelands. During the past decade, there has been a noticeable trend towards the use of color and color infrared photographs acquired for resource mapping and inventories. For example, the National Aeronautics and Space Administration (NASA) has acquired color and color infrared photos at photo scales of 1:60,000 and 1:120,000 over large areas of the United States. The U.S. Department of Interior, Bureau of Land Management, has acquired color and color infrared aerial photographs for several states in the Western United States at 1:32,000, and the U.S. Forest Service is acquiring most of its resource photography in color at a scale of 1:24,000. A consortium of federal and state agencies in the State of Alaska is acquiring color infrared and panchromatic aerial photographs of the entire state for general resource inventories at a scale of 1:60,000 and 1:120,000 respectively and, in the conterminous United States, a consortium of federal agencies has begun a national program to acquire color infrared and panchromatic, quad-centered photographs at scales of 1:60,000 and 1:80,000, respectively. The use of small-scale Landsat data for range inventory is on the rise, especially in countries where detailed inventories have not been conducted. Table 35-1 presents an overview of the common image scales and the uses of the data for various inventory needs.

Photo quality control also is an important factor if one expects to maximize the information extracted from aerial photographs. Poor print quality is a major cause for poor interpretation of vegetation from black-and-white photographs. The photo-lab technician seldom observes the area photographed and thus cannot be expected to prepare the best rendition in either color or black-and-white. For this reason, it is desirable for the photo interpreter to consult with the photo-lab technician on the proper color balance or tone contrast when the working photo prints are being prepared. Frequently, it is desirable to enhance contrasts thereby improving the differentiation of many vegetation and soil types. This can be accomplished only through direct collaboration between the photo interpreter and the photo-lab technician. For a detailed discussion on image-enhancement techniques, please refer to Chapter 17.

Ensure Staff Qualification and Training

The assessment of rangeland resources requires specialized training in field ecology and understanding of the interrelationships among vegetation, soils, water, landforms, and grazing animals, as well as a knowledge of the impacts of animal and human use upon vegetation, soil stability and watershed quality. One should also have a thorough knowledge of the flora and plant-community ecosystems of the areas to be inventoried. Anyone embarking upon a rangeland assessment with the intention of using remotely-sensed data, also should be skilled in the procedures of image interpretation and possess an understanding of the characteristics of the remotely-sensed data being interpreted. The latter type of training can be acquired through a college-level course in photo interpretation and remote sensing, through government sponsored photo interpretation and remote sensing courses, military photo interpretation training or through operational experience gained by working with remotely-sensed data.

The amount of information extracted through image interpretation may be affected by numerous system and human factors, and is directly related to the interpreter's experience with and understanding of the area. For example, an image in-

TABLE 35-1

Scales of Remotely-Sensed Data Used by Rangeland Resource Managers with Suggested Rangeland Uses.*

Very Large-Scale	1:100 to 1:500	Species identification including grasses and seedlings, erosion estimates, rodent activities, assessing surface soil factors including litter.
Large-Scale	1:600 to 1:2000 (1:5000)	Species measurements, erosion estimates, productivity estimates, condition and trend assessment.
Medium or Conventional-Scale	1:6000 to 1:32,000	Detailed vegetation mapping, rodent activities, erosion features. Condition and trend assessment. Vegetation mapping of plant community, larger erosion features, some planning within allotments.
Small-Scale	1:40,000 to 1:125,000	Planning for range management, vegetation and soil unit mapping on a pasture and/or allotment basis, multiple use planning.
Very Small-Scale (taken from orbital altitudes)	1:200,000 to 1:1,000,000	The synoptic view for planning rangeland use within the multiple use framework, mapping vegetation by zones covering large areas such as mountain ranges.

* Adapted from Tueller, 1980a, and Heller, 1970.

terpreter's ecological understanding of landscapes enables him to detect physical and biological components from the remotely-sensed data; he then can integrate the relationships of these components using associated and convergent evidence to make inferences or conclusions (to be verified on the ground) regarding such phenomena as successional status, range condition and trend, impact of resource use, degree of utilization, productivity and resource stability. A skilled image analyst with detailed knowledge of rangeland ecosystems also can correctly identify and delineate vegetation and soil types by careful analysis of subtle image characteristics, (tone, shape, shadow, pattern, size, and texture), and landscape features. Proper attention to training is an essential step to a well planned and executed resource assessment.

Correlate Ground Conditions with Remotely-Sensed Data

Successful image interpretation cannot proceed without establishing a correlation between the remotely-sensed data and the corresponding ground features. Procedures here can take many forms including: a) preliminary ground or aerial reconnaissance of an area, with imagery in hand, to learn the characteristics of ground features and their corresponding image characteristics; b) systematic field surveys to identify all possible resource types and conditions, and to determine the variability associated with each; c) development of training aids (stereograms and associated image descriptors) and image interpretation keys for all features to be detected, delineated and/or identified (Driscoll, 1970; Seher and Tueller, 1973; Poulton, 1970; and Poulton et al, 1971); and d) collection of all available resource-related information and maps that exist in the literature.

Develop Classification Scheme and Map Legend

When remotely-sensed data are to be used for resource mapping, a classification scheme and a map legend should be developed before proceeding with image interpretation. A discussion of the various techniques for preparing hierarchical classification schemes goes beyond the scope of this chapter. However, the UNESCO classification of vegetation, and the soil taxonomy classifications developed by the U.S. Department of Agriculture's Soil Conservation Service, are well known examples of existing hierarchical classification schemes for vegetation and soil mapping, respectively. Generally, one is advised to examine and possibly to modify existing classification schemes before developing a new scheme.

When developing a map legend, it is important for one to recognize three components, viz. symbolic, descriptive and interpretive components. The latter two can be in the form of narrative and/or tabular-statistics and are commonly combined to produce the information that truly makes the map or inventory usable. The symbolic legend component (Figure 35-1) serves as a shorthand method for cartographic identification of delineated map units (resource classes) by using alpha-numeric symbols as an annotated means of representing the map unit. The map may also be expressed entirely in numeric symbols that require references to interpretive legends to understand their map significance. Such legends are becoming more widespread with the increasing use of computers and geobased information system for storage and retrieval of resource map information (Poulton et al, 1971 and Poulton, 1972).

The descriptive legend is a narrative that explains the map unit, its identifying parameters, and its variability. These legends are usually finalized near the end of an inventory project, and ideally are effectively illustrated. The interpretive legend explains the significance of each legend class in terms of range site; probable climax or ecological condition; expected topographic and soil relationships inherent to the ecosystem unit represented by the legend class; biomass, herbage or forage production-potential; responses to animal use; watershed qualities and influences; land- and resource-use limitations and potentials; suitability for unusual or unique uses; and the most suitable management practices for the sites represented by the legend class. Outstanding examples of these kinds of legends are found in the range-site guides of Anderson (1959) and of Hall (1967), range scientists with the Soil Conservation Service and Forest Service, respectively, of the U.S. Department of Agriculture.

Define Interpretation Procedure and Equipment

The interpretation procedure and equipment to use will be based upon the assessment objective that has been defined and the remotely-sensed data being analyzed. Five interpretation/analysis approaches are listed to illustrate the diversity of approaches from which one can choose.

1) Search procedures for detecting and count-

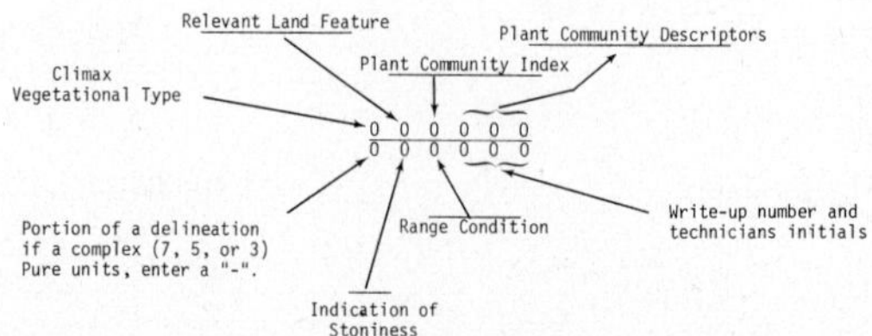

Fig. 35-1. A symbolic legend format used to record delineation characteristics mapped in the Oregon State Land Board range survey exemplified in Figure 35-3. The appropriate symbols and other photo identifiable information were recorded for each delineation number by the interpreter on a special tabulation form. These policy and improvement/management-related interpretations were confirmed by an 18 percent ground check.

ing specific features such as plant indicator species, animals, springs, critical habitat, and cultural features.

2) Manual mapping procedures for producing a resource-type map.

3) Digital mapping procedures for the computer-assisted mapping of resource types.

4) Change detection procedures that will combine search and mapping techniques in order to detect areas or features that have changed and to update an existing inventory or map.

5) Measurement procedures to quantify such parameters as height and area. The use of these analysis procedures is documented in subsequent sections for specific rangeland-assessment examples; however, manual vegetation mapping techniques will be discussed in further detail here because of their importance to rangeland resource inventory.

As one approaches operational vegetation/land-cover mapping, it is important to differentiate between a taxonomic unit and a map unit. A taxonomic unit is a basic unit of the landscape, such as a specific site or a plant community throughout which the soil and environmental characteristics are similar. The map unit is the landscape area delineated according to the legend unit which most nearly describes the taxonomic unit. Ideally, a map unit contains one taxonomic unit, but often more than one taxonomic unit is included within a map unit boundary. However, the map-unit boundary is often arbitrarily located within a continuum zone between different plant communities and thus includes area and plant species that are not necessarily characterized by the legend unit.

There may be two kinds of mapping units (Figure 35-2). A *simple* mapping unit is only one taxonomic unit, with or without allowable inclusions. A *complex* mapping unit consists of two or more taxonomic units within the same boundary, so treated because their intricate pattern prevents practical separation into simple mapping units at the mapping or final compilation scale. A properly made inventory does not average the conditions in the complex mapping units; instead it specifies the individual components of each complex, and estimates the percentages of the unit that each comprises.

In a land-cover mapping project it is important to determine the minimum map-unit area. Minimum map-unit area will depend upon the scale and resolution of imagery interpreted and also upon the final map compilation scale. For broad vegetation maps at scales of 1:125,000 and 1:250,000, the minimum map-unit area may be 160

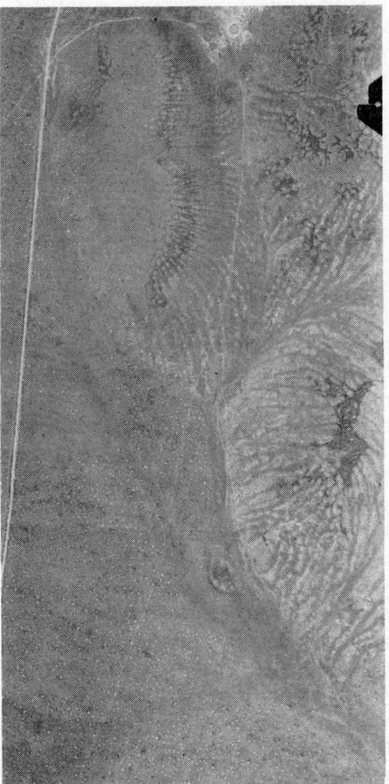

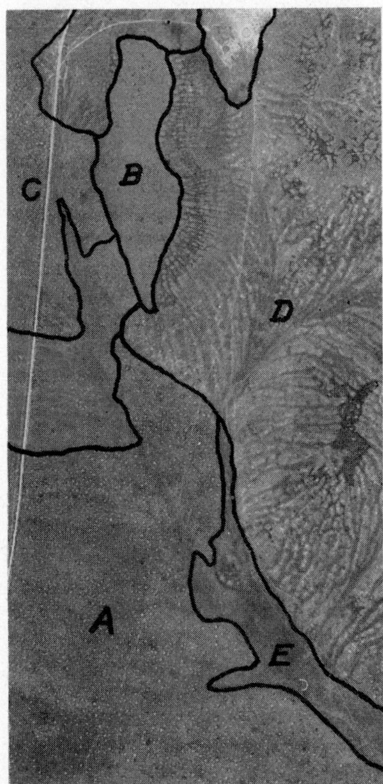

Fig. 35-2. This stereogram illustrates the mapping units that may be allowed in vegetation inventories. Areas A, B, and C, are Simple Mapping Units, or pure types. Area D is a Complex Mapping Unit of two contrasting vegetation types in an extremely intricate pattern. An inclusion of rockland (dark areas) may be observed in Area D, and Area E contains a small, contrasting vegetational inclusion. (Original photoscale 1:20,000)

to 640 acres. For more detailed maps at scales of 1:63,360 and 1:31,680, the minimum map-unit area may be 10 to 40 acres. The minimum map-unit area also may vary for a given map scale depending upon the importance or value of a particular unit to the inventory; high value units may be delineated to a smaller minimum map-unit area.

The image interpreter, hopefully being thoroughly familiar with the vegetation of the area to be mapped, delineates the vegetation units either directly on the image or on an overlay thereof, and identifies each unit according to the map legend. Through careful stereoscopic study, the experienced interpreter can discern relevant detail, ignore misleading image clues, and focus on those that signal important ecological relationships.

Various kinds of stereoscopes are available, each offering advantages or disadvantages for particular interpretation tasks. Their characteristics, advantages and limitations are discussed in Chapter 24. For most mapping tasks involving aerial photographs both scanning and mirror stereoscopes are very satisfactory.

The interpretation of Landsat imagery generally does not require a stereoscope as there is no relief displacement for features as imaged along the satellite track. It is possible to view Landsat imagery in stereo if the area to be interpreted is imaged from adjacent satellite tracks or if, through special image-processing techniques, a "synthetic stereo" image is created. This is accomplished by employing terrain elevation information to calculate topographic displacement. However, use of Landsat data, either false color composite (FCC) prints or digital image representations at larger scales, such as 1:250,000, can facilitate interpretation.

As the delineation and identification are performed, (using either a systematic approach of mapping outward from a single point, or mapping first the known and conspicuous vegetation type) the interpreter assigns, where possible, a symbolic legend to each map unit thereby providing information as to its classification and canopy cover. The interpreter also takes note of the doubtful or questionable interpretation decisions so that these may receive special attention as the ground checking and acquisition of ground sample data are planned.

Accomplish Field Verification and Data Collection

Field verification of interpretive information and the collection of auxilliary data are integral parts of the remote sensing approach for any rangeland resource assessment. The field survey enables the analyst to: a) verify the accuracy of the interpretations; b) observe, identify and interpret phenomena that were difficult to interpret or were not interpretable; and c) correct erroneous interpretations from work he has done in his office. For some mapping projects, several field checks may be required before the final map can be completed. With each check, the ability to make subtle interpretations on the part of the photointerpreter improves, validations of the accuracy of interpretations can be made, and corrections may be made to update the resource map.

The field survey also provides the opportunity to collect auxilliary information that could not be extracted through the interpretation process; information such as percentage composition of plant species, quantitative data on foliar cover, information on soil characteristics and soil stability, and measures of herbage production and utilization. In the course of a rangeland inventory, these data are generally systematically or randomly collected on the ground at predetermined locations, as determined by initial stratification and mapping of the remotely-sensed data. Thus, the remote sensing approach to rangeland assessment is designed to combine the interpretation of remotely-sensed data with field survey methodologies to achieve the final output product.

In the planning of the field survey, some of the statistical questions that might be asked are: Should sample points/plots be randomly or systematically located? How many sample points/plots are required to achieve a specified level of accuracy? What are the appropriate criteria for use in allocating sample points? Considerations with respect to the cost of collecting the field data, the accessibility and hence time required to collect field data, and the specified level of accuracy desired will all govern the design of the field survey needed to achieve inventory or mapping objectives.

Rectify and Produce Resource Map

Interpretative resource maps made directly from aerial photographs or satellite imagery do not achieve their full information potential, especially when mapping objectives require accurate area and distance estimates, until they are: a) compiled in map format at a specified map scale; or b) entered into a geobased information system for further manipulation and retrieval for decision making. Transformation for map generation is generally required to: a) correct for inherent distortions of the photographic images (especially where terrain elevation varies), b) correct for distortions inherent in the system and orbital characteristics; and c) reduce or enlarge the interpretative maps, made at the scale of the photographs, in order to match the scale of the final map. Methods used to accomplish this transformation include: a) the manual transfer of delineations from the interpretive map to the final map using conspicuous cultural and topographic features for reference points; b) the manual transfer of delineations using a transfer scope and/or projection equipment to change scales from that of the photo to the map; c) the transfer of interpretive information to orthophoto maps (photo images digitally rectified to show features in true planimetric position); d) the manual transfer of information using sophisticated stereoplotting equipment; and e) digitizing of the interpretive maps to allow their digital transformation to the final map scale. For most purposes, the interpretive maps made from Landsat imagery at a scale of 1:250,000 (both MSS and RBV) can be transferred to 1:250,000 USGS quadrangles with a high degree of accuracy. Landsat data in computer compatible format (CCT) can be digitally rectified and enlarged to any map scale with acceptable accuracy. The amount of time required to effect the photo-to-map transformation is also an important consideration in planning the procedures to accomplish a resource-assessment objective. The costs, for example, of manually interpreting images and manually transforming the data to a map base may equal or exceed the costs of using a digital approach to image analysis and map-data transformation.

More and more rangeland inventories are incorporating data into digital geobased information systems. Hence, the time and cost of transforming resource information to a format compatible with the digital information system must be considered in the initial project planning and budget, and may ultimately influence the analysis approach followed.

Perform Accuracy Assessment

Today's resource manager demands to have a quantitative or statistical statement about the accuracy of information relied upon for management decisions. The accuracy of resource maps that were produced before the availability of remotely-sensed data has been questioned because of the limitations imposed on the map maker by his inability to observe the true ground relationship of resource types. The current use of remotely-sensed data, combined with modern cartographic techniques, has greatly improved the locational accuracy of map features, but resource maps still are not 100% accurate, partly because of inherent interpretation errors.

Within budgetary limits the resource manager must be able to specify the minimal accuracy requirements that are acceptable for his resource maps and develop procedures for producing such maps that will balance the use of remotely-sensed data and ground sampling.

The value of an accuracy assessment must be weighed against the intended use of the map. Mead and Szajgin (1982), report that "one should not lose track of the differences between the usefulness of a specific product and its estimated accuracy. A quantitative accuracy assessment results in a numerical summary which may or may not represent the usefulness of the product or how well it compares with map products which were previously available." Furthermore, an accuracy assessment may serve little purpose unless it has been carried to the point of an analysis of errors. Many remote sensing studies present error matrices or confusion tables that compare the interpretative class with the ground description for that class. This display of errors allows the user to assess the impact of errors on the usefulness of the product.

If a quantitative assessment of accuracy is to be conducted, consideration should be given to the proper statistical design when one is devising techniques for collecting the data for the assessment. It is important that data collected for the accuracy assessment be acquired simultaneously with that used for map preparation to minimize the cost of data collection and, most importantly, to insure that the same descriptions and definitions of resource classes are used for both.

Topics pertaining to accuracy assessments are discussed at several locations in this Manual. (See, for example, Chapters 23 and 30).

RANGELAND RESOURCE INVENTORY

OBJECTIVES, COSTS AND DESIGN

A rangeland resource inventory provides descriptive and quantitative data designed to contribute towards a basis for decision making. The required inputs for decision making will dictate the items to be inventoried, such as *kind, amount, condition,* and *location* of vegetation, as well as various soil and animal resource features. These inputs will also be the basis for specifying other data attributes such as accuracy, precision, timeliness and scale. Consideration of these items indicates the need for an appropriate inventory technique that can include remotely-sensed data in coordination with sampling and ground survey procedures.

The making of a resource-management decision imparts value to the inventory and consequently impacts the level of effort that may be expended in the acquisition of data. An inadequate inventory could lead to incorrect decisions and, in rangeland resource management, to serious, long term repercussions. High monetary cost may also be unacceptable and detrimental. As pointed out by Avery (1975), "an inventory should be designed to obtain the desired information—no more and no less—for the lowest possible expenditure."

Various strategies exist for determining an appropriate level of expenditure for an inventory. These strategies range from arbitrary to objective and can include procedures that combine cost and loss functions (Husch, 1981). Frequently the budgets of other similar inventories may be used as a guide for estimating the cost of an new inventory. Quantitative data may also be adapted from the few published accounts of costs for inventories that have incorporated the use of remotely-sensed data (e.g. Bonner, 1981; Enslin and Hill-Rowley, 1977; Mouat et al, 1981; and Rohde and Bale, 1975).

The process of determining the value of information, and consequently the amount to be spent for an inventory, is difficult; so also is the process for obtaining accurate and specific statements of information need. Optimization of an inventory design, however, requires as a minimum the following information:

1) Itemization of the resource features to be inventoried, and their attributes, including detailed descriptive and quantitative specifications for their identification and the units of measurement by which they will be expressed in the inventory.
2) Intended use(s) for the inventory and the data formats, with specifications, required for each use.
3) The geographic area over which the inventory will be compiled, and subareas for which data will be summarized, tabulated and/or displayed.
4) Required compatibility of the inventory with other data bases.
5) Subsequent analyses to be performed on the data and, in consequence, the data formats and content required for these analyses.

Given the above, consideration may then be made as to the costs and time required to implement various alternative procedures that will combine remotely-sensed data with field measurement.

MAPPING AND SAMPLING

Often the objectives of an inventory can be satisfied by the preparation of a resource map. Maps have long been a primary tool for locating those areas of the landscape having similar environments, as indicated by the distribution of plant communities. Wherever the same combination of plant species can be found growing together, the same narrow range of plant-growth conditions is likely to occur. (Daubenmire, 1980). Areas thus identified have similar potentials for and responses to resource management, and are therefore basic to decision making in resource allocation and management planning (Poulton, 1970).

Range management information needs can often be satisfied, at least in part, through the cartographic presentation of data. Hardy (1982), reported that 90% of the requests for land cover and resource information specify a preference for maps. Maps display the identity and distribution of features. Maps can be used for qualitative and quantitative assessments of spatial relationships: size, shape, area, frequency, distribution, (see Lewis, 1977; and Cliff et al, 1975). Successive mapping at appropriate time intervals provides a basis for studying the temporal and spatial characteristics of change (Johnson, 1982).

Determinations of area, size, shape, frequency, and the analysis of change can also be approached with properly designed sampling techniques, frequently with greater accuracy and less cost. Therefore, if knowledge of specific location is not a requirement for decision making then sampling alternatives should be considered.

Cochran (1977), discusses the frequent advantages associated with a well devised sampling method: a) reduced cost, due to measuring only a small fraction of the population; b) greater speed; c) greater scope; and d) greater accuracy, due to a reduction in the volume of work, and an increase in quality control. Detailed discussions regarding the advantages of sampling are provided by Freese (1962), and Cunia (1982).

Mapping and sampling techniques are often integrated in an inventory design. The map provides a stratification for the allocation of sample units, while a sampling frame provides for the selection of sample units and the basis for expansion of sample data. This concept will be developed more fully later in this section. (See also the discussion of sampling as applied to forest resources, Chapter 34).

RANGELAND INVENTORY USING PANCHROMATIC AERIAL PHOTOGRAPHS

For years aerial photographs have been used as a "map" to:

1) visually assess the heterogeneity of the landscape
2) select areas to be sampled
3) ascertain the best routes to follow
4) determine the extent and boundaries of a "type"
5) record the site visit (by pin pricking the location of a field, verifying it as a ground truth site and writing its identifying number on the back of the photo print).

These uses are still valid today. In a study of 66 Australian soil surveys in a variety of terrain conditions, Bie and Beckett (1971), showed a 45% reduction in survey effort (man-days/unit area) by aerial photo interpretation. Soil boundaries mapped by photo interpreters were commonly more accurate than those located by interpolation of ground observations.

The extraction of considerably more information about rangelands from aerial photos is also possible. This was demonstrated by an inventory and analysis of state owned rangelands in southeastern Oregon conducted for the Oregon State Land Board (Poulton and Isley, 1970). Nine hundred parcels, ranging in size from 0.18 to 14,459 acres and totaling 606,697 acres, were scattered (generally sections 16 and 36 in each township) over 26,011,520 acres of sparsely populated and mostly roadless shrub steppe. Most of the parcels did not have vehicle access. Information about vegetation types, their condition, acreage and management potentials, and also information about springs, improvements, and access were needed for each parcel in detail suitable for planning improvements, working out details of management with lessees and/or initiating land appraisal leading to land exchanges or sales.

A feasibility study made by Poulton and Isley (1970) estimated that it would cost $1/acre (1966 costs) to conduct the inventory using conventional ground survey techniques. As an alternative, an aerial photo-based inventory was designed that called for 1) an initial field survey to establish a vegetation classification, mapping legend and photo interpretation aids (annotated stereo pairs); 2) photo interpretation of available 1:20,000 black-and-white aerial photos of the remaining state owned land, and compilation of data, and 3) an accuracy check of the photo interpreted acreage. The actual cost incurred using the photo-based inventory was approximately 12¢/acre. In the conduct of this inventory, boundaries of the state-owned rangeland parcels were transferred from planimetric ownership maps to the aerial photographs. The entire rangeland area of Oregon was subdivided into ecological provinces (Anderson, 1956), and these were further stratified into similar physiographic/vegetational provinces on a photo-index mosaic. An eight percent sample of parcels was selected for purposes of correlating ground information with the panchromatic images and developing an ecological mapping legend. Manual photo interpretation procedures, including stereoscopic analysis, were used to delineate plant communities on the remaining 92 percent of the state-owned rangeland. In the process parcels were categorized into three groups: those requiring field verification of delineated plant community boundaries, those not requiring field verification, and those comprising the sample to determine map accuracy. By the end of the project, 18 percent of the parcels had been examined or sampled on the ground.

Figure 35-3 shows an interpretive vegetation map from this project for a four square mile parcel (2,560 acres or 1,037 hectares). The delineations were made by an experienced photo interpreter who also assigned a legend symbol (illustrated in Figure 35-1) to each map unit based upon analysis of the image characteristics for each unit. The climax vegetation types designated include: grassland, shrubland, and open coniferous forest. Relevant land features include: bottomland, level to rolling uplands, fans, and terrace. A plant community index was assigned to indicate relative productivity of the environment (Anderson, 1956). The plant community descriptor used in the legend is a numerical symbol, the broadest vegetation classes being:

100 Salt desert	600 Juniper
200 Silver sagebrush	700 Grasslands
300 Low sagebrush	800 Coniferous forests
400 Big sagebrush	900 Deciduous forests

By adding digits to the right, each of these classes was broken into subsets such that the third digit identified a highly specific kind of plant community. For example, 311 indicates a plant community of *Artemisia rigida/Poa secunda/Sitanion hystrix* on very shallow, stony soil—a highly photo-identifiable community. The designator 320, on the other hand indicated low sagebrush communities dominated by either *Artemisia arbuscula, Artemisia nova,* or *Artemisia longiloba* —three ecologically distinct community dominants that were not photo-separable.

The symbolic map legend (refer to figure 35-1) provides a location in the denominator to indicate the presence of rock outcrop or surface stone seen on the photographs. Range condition was judged from photo interpretation based upon tones indicating the relative amount of ground cover. Alpha symbols P, F, G and E, used in the denominator of the symbolic legend, indicated poor, fair, good and excellent range condition, respectively. These condition judgments were reasonably reliable only for a few climax vegetational types (indicating the difficulty of making these interpretations solely from aerial photographs).

Where landscapes could not be delineated into pure types, complex map units were delineated. Each plant community within a complex map unit was described with its own symbolic legend and the proportion of it within the unit indicated, for example 30%, 50%, or 70%. The interpretative vegetation maps were transferred to topographic maps and the acreage by type estimated using a planimeter. From the vegetation maps prepared, area-by-type information was compiled for each parcel and for parcels in each county. Tables 35-2, 35-3, and 35-4 illustrate how the acreage information was compiled, based upon vegetation-type by county, land uses by county, and potential for improvements by county.

At the conclusion of the mapping an additional random five percent sample was drawn from among those parcels that had not been field examined. Based on field checking of this sample it was determined that average accuracy of all statements made from photo interpretation was 69 percent. Accuracy levels for some types ranged as high as 92 percent and as low as 42 to 50 percent where interpretation was intentionally pushed beyond reasonable limits to acquire a useful estimate of range-management related conditions and parameters (Poulton and Isley, 1970).

Among the early uses of aerial photographs by the USDI Bureau of Land Management, was the employment of panchromatic aerial photograph mosaics for determining broad vegetation boundaries, judging necessary sampling intensities and portraying resource and cultural information needed for management (Henriques, 1949). The USDA Soil Conservation Service used uncontrolled mosaics at scales as large as 1:7,290 and 1:15,840 for cartographic presentation of range site and condition surveys, and the details of ranch management programs (Figure 35-4).

Photo enlargements and/or mosaics serve a functional purpose in coordinated resource plan-

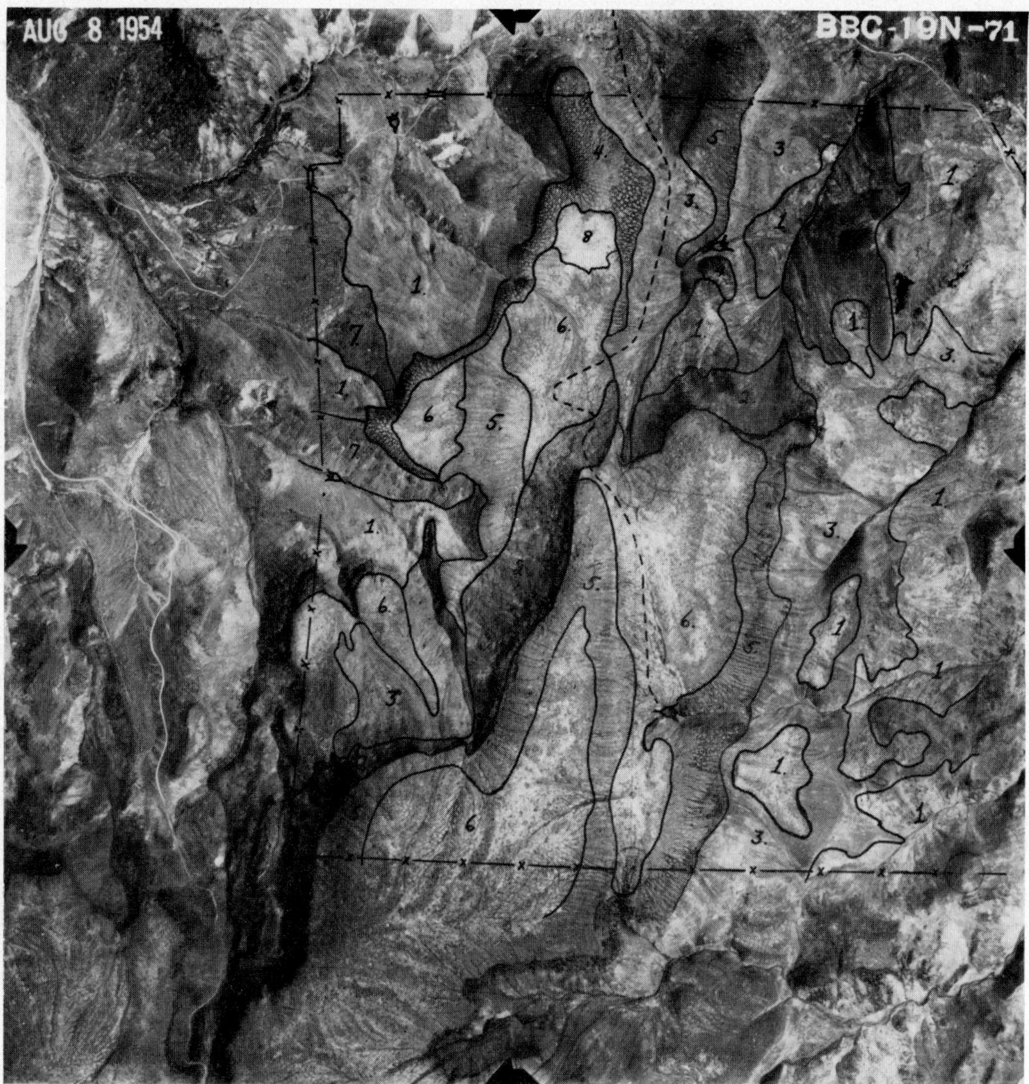

Fig. 35-3. Vegetation units have been delineated on a panchromatic photograph (scale 1:20,000) of state-owned rangeland in Oregon. Ecologically equivalent units have the same numerical label. Descriptive information for each unit was recorded in the field and would accompany the legend for the vegetation maps produced from these interpretative overlays. The area delineated contains four sections or approximately four square miles. North is toward the top of the photo. (Photo courtesy of Department of Range Management, Oregon State University, Corvallis)

ning. In this activity the aerial photograph provides an excellent means for: (1) focusing attention on areas and issues of critical concern, (2) getting all participants to see and talk about the same features, (3) showing spatial relationships among resources, (4) locating roads, springs, corrals, fences, salt blocks and other developments, (5) locating hazards and sensitive areas, and (6) developing and displaying management alternatives (Figure 35-5).

Among soil scientists there seems to be a consensus regarding the utility of very small-scale black-and-white aerial photographs in field survey. Photographs at scales of 1:63,000 and 1:80,000 have been enlarged to serve as satisfactory field sheets, and to correspond with the format and scale of topographic maps. High altitude photographs are also used for reconnaissance surveys—Order 3 or Third Order surveys (Cooperative Soil Survey, Committee 1, 1974; Cooperative Soil Survey, Committee 2, 1976). Small-scale panchromatic aerial photographs (1:40,000) are used routinely in Australia for preparing rangeland inventories, (Wilcox & McKinnon, Dept. of Agriculture, Perth, Western Australia).

LAND COVER AND SOIL SURVEYS USING COLOR AND COLOR INFRARED AERIAL PHOTOGRAPHS

Aldrich (1981), lists eight inventory data elements that are common to most resource systems

TABLE 35-2

Acreage Estimates for Each Vegetation Type by County for the State-Owned Rangeland Inventory Conducted in Southeastern Oregon

County	Shrub	Juniper	Grassland	Conif. Tree	Decid. Tree	Mixed Conif.
Crook	16,411	8,052	319	265	0	0
Deschutes	17,021	5,600	85	728	0	0
Harney	196,423	8,962	3,003	2,989	209	0
Klamath	1,778	1,514	205	3,781	0	25
Lake	56,492	2,904	2,086	1,453	15	0
Malheur	246,582	2,920	7,257	88	180	0
Total	534,707	29,952	12,955	9,304	404	25
Approx. %	91.0	5.1	2.2	1.6	.07	.004

TABLE 35-3

Acreage Estimates for Four Land Use Classes and One Barren Class by County Within the State-Owned Rangeland Inventory Conducted in Southeastern Oregon

County	Range	Range Seeding	Cropland & Hayland: Dry	Cropland & Hayland: Irrig.	Cropland & Hayland: Aband.	Free Water Surface	Essent. Barren
Crook	24,757	0	137	33	279	0	29
Deschutes	22,916	73	2	8	590	0	594
Harney	205,469	0	0	21	32	72	10,133
Klamath	7,583	0	0	5,560	0	420	431
Lake	61,196	0	0	17	0	126	2,927
Malheur	254,039	271	55	876	3	959	6,310
Total	575,960	344	194	6,515	904	1,577	20,424
Approx. %	95.05	.06	.03	1.07	.15	.27	3.37

TABLE 35-4

Acreage Estimates of Acres for Potential Improvement and Development for Six Land Use Activities[1]

County	Suit.[2]	Range Seeding No.	Range Seeding Acres	Brush Control No.	Brush Control Acres	Waterspread No.	Waterspread Acres	Wildlife No.	Wildlife Acres	Water Develop. No.	Recreation No.
Crook	H	38	3,807	36	5,421	4	155	16	2,786	18	5
	M	28	2,339	38	7,889	3	612	53	12,175	22	25
Deschutes	H	42	7,537	50	10,864	1	8	18	2,881	8	1
	M	37	4,483	33	4,579	5	272	50	10,208	13	27
Harney	H	143	26,253	195	48,058	8	887	89	18,840	98	19
	M	172	30,984	238	61,988	22	4,363	255	78,367	156	115
Klamath	H	9	691	6	476	0	—	36	5,957	9	10
	M	8	292	5	311	2	98	29	7,146	3	43
Lake	H	36	9,541	45	15,368	1	2	25	5,375	27	6
	M	56	13,394	53	12,896	5	1,130	51	13,873	48	35
Malheur	H	80	13,309	122	30,928	9	671	85	13,769	79	17
	M	210	47,164	317	94,940	22	2,881	329	112,596	252	99
Totals	H	348	61,138	454	111,115	23	1,723	269	49,608	239	58
	M	511	98,656	684	182,603	59	9,356	767	234,365	494	344
	H + M		159,794		293,718		11,079		283,973	733	402
	%		26.3		48.4		1.8		46.8	—	—

[1] The estimates are given for two categories (high and medium) of suitability for development and for each of the counties in the state-owned rangeland inventory conducted in Southeastern Oregon.

[2] Suitability (High or Medium) for further investigation.

Fig. 35-4. This uncontrolled photomosaic is an ideal base for presentation of range site and condition survey results and for range and ranch management planning. This mosaic was supplied by the U.S.D.A., Soil Conservation Service. Landsat MSS imagery enlarged to a scale of 1:250,000 is serving a similar function.

Fig. 35-5. Coordinated resource planning in Oregon often involves several land management agencies interested in mapping several types of resources. In this figure, range, forest and soil scientists from the Soil Conservation Service, and the U.S. Forest Service, and ranchers from the Marr Flat Grazing Association are using enlarged aerial photographs for identifying land cover, orientation and planning of management alternatives.

and that can be extracted through the interpretation of aerial photographs: horizontal distribution of cover, biomass, landform, mortality, vertical distribution of cover, volume (timber), forage, and changes (disturbances to cover class over a period of time). Aldrich (1979), also provides a comprehensive review of aerial photographic systems and current recommendations for film types and scales to obtain specified data requirements.

In arid and many semi-arid rangeland regions the natural sparcity of vegetation cover makes vegetation mapping from photographs difficult. Frequently, one does not see a photo-signature solely for vegetation; rather one sees an integrated vegetation-soil signature that is strongly influenced by the geologic parent material and associated soils development. Under these conditions, color photographs facilitate the concurrent mapping of vegetation and soils. Color Figures 35-6 and 35-7 are examples of vegetation and soils maps made from conventional 1:20,000 scale color aerial photographs. The map information is a result of intensive photo interpretation followed by validation in the field and subsequent map revisions.

Color photographs are considered extremely useful for soil survey because color contrast generally enables the delineation of different parent materials and plant communities. At a scale of 1:15,840, color aerial photos are well suited for use in Order 2 and Order 1 surveys (Cooperative Soil Survey, Committee 1, 1974).

Color and color infrared slides have also been found helpful in soil survey work. Color slides studied in conjunction with black-and-white field photo-sheets at the premapping stages of a survey aided in distinguishing some soil types in areas of exposed surface. The color infrared slides provided indirect inferences of soil texture, depth and drainage through indications of stage and vigor of plant growth. The slides were useful for identifying soil unit boundaries and estimating the percentage of inclusions within soil complexes (Cooperative Soil Survey, Committee 1, 1974).

The use of color slides in low intensity (Order 3) surveys was tested in Idaho and reported by Harrison (1982). The determination of map unit composition from photo measurements on 1:1,200 and 1:600 scale slides compared well with ground transect results. It was emphasized that the photo interpreter must be able to recognize known soil features and vegetation patterns before he can predict soil taxa from the large-scale photographs. Harrison cautioned that field verification is still necessary.

Krupin (1980), encouraged the use of color and color infrared photographs at scales of 1:24,000 to 1:12,000 to detect areas that can be developed for range-water supplies. Deciduous and herbaceous riparian vegetation types are the key indicators and are readily identified with color infrared photographs in arid regions. The Bureau of Land Management (BLM, USDI) is using color infrared photographs at a scale of 1:32,000 to map vegetation and assess range resources for grazing environmental impact statements in Arizona, Idaho, Wyoming and Montana. The BLM has also mapped extensive areas in Montana using 1:80,000 scale color infrared photos (Batson and Elliot, 1977).

In recent years small-scale (1:60,000 to 1:120,000) color infrared aerial photographs have been used to search for unique and typical natural ecosystems that have been little disturbed by human activities. The photographs are suited for searching large regions, identifying vegetation types (given prior knowledge of the candidate types in the region), assessing the condition (level of disturbance vs. naturalness) of an area and, on the basis of that assessment, either retaining the area for reserve candidacy or rejecting it from further consideration (Mairs, 1976). This application demonstrates the processes of systematic searching and sorting. Such processes permit the complete enumeration of a large area, yet limit costly ground surveys to those relatively few locations where they were needed.

The final application discussed within this section combines the use of small-scale color infrared photographs with images generated from the Landsat multispectral scanner data. The application is to land use mapping for very large regions (statewide) to show the location of rangeland, forest, and irrigated and nonirrigated cropland, including annual fallow (Hall, 1981). The landuse categories were mapped to a ten-acre minimum mapping unit and delineations were transferred to 1:24,000 scale orthophotoquads. As proved to be the case in this application, small scale aerial photographs (1:62,500 and 1:125,000) and Landsat images are quite compatible. The former, with 8–12 foot resolution, provided the spatial detail for identifying major structural features, the arrangement of irrigation systems and the shape of irrigated fields, thus enabling their discrimination from wetlands that are often closely associated. The aerial photographs also provided sufficient spatial detail for recognizing irrigation patterns on bare soil, thus enabling the discrimination of the irrigated from nonirrigated bare soil fields in the same locality.

The Landsat images provided necessary temporal information. Often the aerial photographs had not been acquired on a date suitable for discriminating the irrigated from nonirrigated crops. Landsat images were acquired at key times during the year to facilitate that distinction. Additionally, the aerial photographs were acquired over several years. The Landsat images were used to update the data base to a selected base year.

The land-use map compiled on a 1:24,000 planimetric base (orthophoto or topographic map) was produced at a cost of \$1.61/square mile (0.25¢/acre; 0.62¢/ha). Costs were distributed as follows: salaries and payroll expenses, 64%; supplies and services, 33%; and travel, 3%. The number of aerial photos interpreted was in excess of 1,500. The project continued over three years. Table 35-5 identifies in outline form the activity elements that constitute such a project and the issues or concerns that accompany each activity (adapted from Hall, 1981).

INVENTORY APPLICATIONS USING LANDSAT IMAGERY

Landsat images have been used to map rangeland vegetation and soil since Landsat 1 was launched in 1972. The characteristics of this satellite system, its sensors and imagery characteristics are discussed in detail in Chapter 12.

Landsat MSS images have been used primarily for general land-cover mapping. The techniques used to visually interpret Landsat images are similar to those used with aerial photographs. Stereoscopic viewing is generally not used, although Poulton (1973), noted that such viewing, within the sidelap area common to two adjacent Landsat swaths, facilitated recognition of the association of vegetative cover (darker tones in MSS Band 5) with landform and relief. The smallest observable height difference perceptible through such stereo viewing of unenhanced Landsat images has been calculated as approximately 100m (Hilwig, 1980).

Landsat MSS images are available as black-and-white prints, corresponding to each of the four spectral bands (green, red, and two near-infrared bands), and as false color composites (FCC) (Color Figure 35-8). Landsat color composites (FCC) are superior and used in preference to the black-and-white images for most rangeland mapping projects (Johnson et al, 1974; Tueller et al, 1975; Graetz et al, 1976; Hironaka et al, 1976; Watson, 1977). Landsat MSS images are available at standard scales of 1:1,000,000, 1:500,000, and 1:250,000. Enlargements, made to scales of 1:100,000 and 1:62,500, are possible, but the quality of the image deteriorates markedly as one enlarges beyond a scale of about 1:100,000. Many manual interpretations of Landsat imagery are performed using FCC's enlarged to a scale of 1:250,000.

Selection of the image type, scale and date can have a direct bearing on interpretation results. Hilwig (1980), noted seasonal differences in the interpretability of both static (e.g., drainage patterns, lineaments and landforms) and dynamic features of the landscape. Dynamic features included drainage condition, vegetation and land use. Hilwig listed six factors to consider when selecting imagery: image quality, cloud cover, climate (soil moisture), crop calendar/vegetation phenology, snow cover, and side lap between images (if stereo viewing is to be used).

The MSS minimum field of view is approximately 1 acre. Consequently, rangeland cover types generally need to be 20–40 acres in size to be detected. Moreover, interpreters of Landsat imagery generally use a minimum map unit of 160 to 640 acres in the study of rangeland features. Finer spatial details can be identified and recorded by transferring interpretations to a larger scaled photographic or map base. The level of map detail which can be interpreted from Landsat imagery is dependent upon several factors: (1) the subject being mapped (e.g., land systems, land forms, soil, vegetation), (2) the spatial complexity of the landscape, (3) the spectral contrast among the subjects being mapped, and (4) the date of the imagery. A thorough knowledge of the area to be mapped (Driscoll and Francis, 1975), and as much supporting information as can be gathered, is required; the latter is commonly in the form of existing cover maps and aerial photographs. Existing small scale vegetation maps, however, often don't show the spatial detail that can be evident in a satellite image (Poulton, 1968).

For most rangeland areas, vegetation can be mapped to Level II as described in USGS Professional Paper 964, by Anderson et al, (1976). At Level II the map-unit legend may include three categories of forest (conifer, deciduous, mixed), three categories of woodland (evergreen, deciduous, mixed), and three categories of shrub and herbaceous. Mapping to Level II and to even greater map-unit detail has been documented by using time-sequential Landsat images to view the terrain when certain vegetation types are phenologically distinctive. The mapping technique was revised by Bonner (1979) and the following has been adapted from that source.

Tueller et al, (1975) found that time-sequential Landsat images (Landsat images of the same areas but acquired on different dates), acquired in different seasons, improved the mapping of deciduous and coniferous forest, grassland, marshland and mountain brush in Nevada. In the same study, winter Landsat scenes, with snow visible on the ground, aided the mapping of pinyon-juniper and chained pinyon-juniper types. Hironaka et al, (1976) found that time-sequential Landsat imagery made it possible to map areas recently burned and to distinguish those successfully reseeded to crested wheatgrass (*Agropyron desertorum*) from those dominated by annual grasses such as cheatgrass (*Bromus tectorum*). In addition, Hironaka could distinguish between a big sagebrush/Idaho fescue type (*Artemisia tridentata/Festuca idahoensis*) from a big sagebrush/bluebunch wheatgrass type (*A. tridentata/Agropyron spicatum*) by comparatively interpreting Landsat images obtained in the spring and summer. Colwell et al, (1975) stated that repetitive Landsat images improved the accuracy of mapping major cover types.

The ease of mapping selected generalized range vegetation classification-units from Landsat images at a scale of 1:1,000,000 was reported by Tueller (1980b). "Ease" was ranked on a scale from one to five ranging from "with ease" to "with extreme difficulty or impossible." Open pine forest and pinyon-juniper woodland were

TABLE 35-5

Steps Entailed in the Development of a Photo Interpretation Project.

A. Planning
  1. Define the overall task
    a) review the classification and legend
    b) if standard, review a previous application of it
    c) if experimental, confer about definitions; allow extra time to resolve problems that will occur
  2. What is the best way to approach the task?
    a) what tasks are simultaneous? sequential?
  3. What is the duration of the project?
    a) are there deadlines for specific phases?
    b) what rate of production is necessary to complete work in time allowed?
  4. What is the work to be done?
    a) how much work space and storage space is necessary?
    b) what equipment is necessary?
  5. Who will do the work?
    a) who is available; what are their qualifications?
    b) what skills are necessary?
    c) how much training time will be available?
    d) how much training time will be required?
  6. Resolution
    a) what is the minimum mapping unit required?
    b) can it be accommodated with available materials and equipment?
    c) will secondary phases of project (e.g. transferring) tend to dilute accuracy of minimum mapping unit?

B. Choosing the right materials, staff and schedule
  1. Scale of imagery
    a) what is the final product? does it need to fit in a publication? is it reproducible?
    b) if it is a map, what scale? working materials should not be smaller scale than final product
    c) what imagery is available?
  2. Type of imagery
    a) CIR, B&W, orthophotos, prints or transparencies?
    b) weigh qualities of each in context of task to be done
  3. Date of imagery
    a) what is the most advantageous season/date of imagery for problem? (e.g. during growth season for crops)
    b) how many dates of imagery are necessary? does change need to be monitored?
    c) does imagery need to be updated? Landsat provides timely coverage
  4. Compatibility
    a) are materials to be used compatible with equipment available for project use? (e.g. film positives—light tables)
    b) are scales of materials compatible?
  5. Transferring data
    a) will original delineations need to be transferred to a different data base for production of final product?
    b) is necessary equipment available?
  6. Staff
    a) are sufficient staff available to handle required tasks and meet production schedule?
    b) does staff have familiarity with geographic region to be studied?
    c) how can photointerpreter applicants be pretested?
    d) what is an effective way to train staff for the specific jobs?
  7. Scheduling
    a) will it be possible to ground truth delineations while photointerpretation is being accomplished?
    b) what tasks can be done concurrently?
    c) what materials need to be ordered far in advance?

C. Technical methodology
  1. Initial organization
    a) set up space and equipment
      1) allot work space and storage space
      2) reserve equipment time
    b) implement detailed project procedures for:
      1) monitoring availability of Landsat data if necessary for updating
      2) ordering, receiving, labeling, storing and using map and image materials
      3) gathering and organizing collateral data
      4) coordinating major concurrent production tasks (photointerpretation, transferring, Landsat updating, planimetering, tabulation of data, field work/quality control)
      5) monitoring project status
    c) prepare photointerpretation aids to go along with classification and legend and organize the training program
    d) solicit applicants, screen, and hire if staff is not already available

TABLE 35-5—*Continued*

2. Photointerpretation preparation
   a) delineate project unit boundaries on base maps, label areas
   b) prepare film
      1) organize by project unit (i.e. cut rolls apart if necessary), label
      2) select work areas and stereo pairs; plot effective areas of photos
      3) prepare overlays, apply to photos, register overlays to photos, label overlays
      4) delineate map boundaries on overlays
   c) purchase Landsat imagery, if necessary; prepare color composites
   d) organize collateral data for ready reference
      1) study collateral data
      2) organize it by project unit
      3) select essential background reading for staff members to read
   e) prepare project status sheets
3. Training
   a) make sure workers are able to handle all major tasks involved in project (photointerpretation, transferring, updating, etc.)
   b) discuss mechanical consistency with staff members; all labels should be the same format, all letters the same type face, etc.
   c) make up area scales (for acres, hectares, etc.) that comply with minimum mapping units at photo and map scales
   d) review classification and legend
      1) discuss imagery signatures (variation and consistency)
      2) discuss variation in dates of imagery and how it affects interpretation
   e) collateral data use
      1) read selected materials
      2) discuss use and application of special materials, i.e. tide tables, crop calendars, soil maps, etc. with staff
   f) documentation procedures and quality control
      1) record production time by completed unit (by photo, map, etc.)
      2) maintain continuous project log; not only time by task but order in which work accomplished (valuable for project review, i.e. aiding in building greater efficiency into the system)
   g) training site review
      1) take slides of selected field sites in study area
      2) plot these sites on imagery and maps
      3) describe signatures of sites that appear in slides
      4) contrast and compare ground and image signatures
   h) photointerpretation practice
      1) set up dual stereoscope viewing (possible with Old Delft Scanning Stereoscopes)
      2) have each worker delineate overlay for same area and compare work
4. Production procedures (these will vary depending on the materials used and the methodology applied)
   a) proceed with work in an organized manner unit by unit
   b) delineate classes on images maintaining minimum mapping unit
   c) when beginning a new work area start by "tying-in" all borders with delineations from adjacent completed work areas (this way the final product will be a continuous surface of delineations)
   d) confer with other workers as to questionable areas, confusing photo signatures
   e) review collateral data as necessary
   f) update as necessary with Landsat or other data source
   g) edit delineations on photos
   h) transfer updated delineation to mapping base (use zoom transfer scope, overhead projector, pantograph, lantern slides or other means)
   i) tie-in edges of maps or mapping base after transfers
   j) edit delineations on maps
   k) tabulate data (use planimeter or sampling technique)
   l) field check questionable sites as budget allows
      1) map sites to be checked and plot route
      2) take photos and maps into the field
      3) compare photo signature with ground data; take notes
      4) photograph areas for future reference

D. Quality control
   1. Training
      a) assure uniform training for all workers
      b) frequently check accuracy and consistency of work of all team members
   2. Confer regularly about project problems and status
   3. Editing delineations on aerial photographs
      a) check all edge tie-ins for accuracy and completeness
      b) check all polygons
         1) no polygons with more than one label, and different from neighbors
         2) all polygons closed

TABLE 35-5—*Continued*

4. Editing-technical
   a) each worker should check his own interpretation when complete
   b) each should check another's interpretations when complete
5. Check final products for completeness
   Use field data to improve accuracy of interpretations

E. Monitoring and reporting
1. Monitor progress of project
   a) compose and complete status sheets for work units
   b) record time spent per task per person per day (this makes it possible to monitor production of each worker and estimate production rates by task)
   c) keep track of production rates by task so future budgets will be easier to estimate
2. Report to funding/contracting agency
   a) keep up on contracted reports (monthly, quarterly, annual, summary/final); collect data as project proceeds
   b) keep aware of project status; know if schedules are being met
3. Final products
   a) retain copies of final products
   b) when shipping materials include list of contents on packing slip; insure and/or register
   c) include a copy of the classification legend with the final products and make a list of any assumptions made that might conflict with or further define the categories
   d) critically review contract; have all obligations been met?

mapped with slight difficulty, meadow-riparian was mapped with slight to moderate difficulty, and grasslands and alpine herblands were mapped with moderate difficulty. All the northern and southern desert-shrub-dominated vegetation units of the Great Basin region of western United States (sagebrush, saltbrush and creosote bush types) were mapped with moderate to extreme difficulty.

In areas where the natural vegetation has been disturbed either completely by land conversion or partially by over utilization by grazing animals, vegetation mapping is made more difficult because the boundaries of disturbed areas are frequently more distinct than those of the natural vegetation. Differences in range condition, for example, in the midwest plains grassland and in portions of Australia, obscure natural vegetation boundaries and make it difficult to map vegetation from Landsat imagery (Graetz et al, 1976; Myers et al, 1974; and Gibbens and Lewis, 1975).

The ease of mapping disturbance areas was utilized by Martinko (1982), to quantify loss of potential pronghorn-antelope habitat. Black-and-white prints of Landsat MSS band 5 images acquired over a period of five years were visually interpreted to map the expansion of cropland in rangelands in Kansas.

The mapping of vegetation types through the use of Landsat images, is difficult in arid environments where the vegetation is sparse and does not correlate well with underlying soil and geology. The interpreter will be inclined to delineate the more conspicuous landform and soil boundaries which can be an asset when they correlate well with vegetation types or a hindrance where they do not (Colwell et al, 1975; Graetz et al, 1976; Hironaka et al, 1976; and Johnson et al, 1974).

Hilwig (1980), utilized interpretations of landform, terrain relief, drainage pattern and structural geology to prepare a physiographic analysis and map of the Dehra Dun region of northern India. He found strong relationships between vegetation and elevation in mountainous areas and between vegetation and soil moisture on alluvial plain areas. But the low resolution and small-scale of Landsat imagery made direct identification and mapping of vegetation very difficult over most arid regions. Mouat and Hutchinson (1981), and Mouat and Johnson (1981), used similar techniques to map vegetation and terrain. Interpretive maps of terrain variables, such as elevation, slope, and aspect, as well as parent materials and drainage density that can influence microclimate and particularly soil moisture, were used to facilitate mapping vegetation types (Mouat, 1974).

Johnson et al, (1974), mapped landscape classes in Arizona using Landsat images and supplemented the classification of the units with interpretation of aerial photographs and field surveys. Graetz et al, (1976), compared Landsat interpretations with existing landsystem maps in Australia and deduced that "landsystems possessing broad scale characteristics could be easily mapped. These included hills, mulga (*Acacia aneura*) and spinifex (*Triodia sp.*) sand plains, flood plains and drainage systems with eroded or bare surface . . . however, landsystems having less characteristic vegetation/soil/topography were impossible to map."

Documentation of the accuracy and cost of producing vegetation maps based upon manual interpretation of Landsat MSS imagery is not plentiful, although a few reports are available. Tueller et al, (1975), reported that interpreters correctly identified Nevada vegetation types 75 to 85 percent of the time. With appropriate dates of Landsat FCC images, Driscoll and Francis (1975), achieved 92 percent and higher accuracies when visually identifying the vegetation categories of conifer forest, deciduous forest, and grassland in a mountainous area of central Colorado. Bonner (in Linden, Rohde and Bonner, 1981) interpreted evergreen woodland, Mohave desert shrub, Great Basin desert shrub and other minor vegetation components in northwestern Arizona from 1:250,000 scale enlargements of a Landsat MSS image prepared as a standard FCC and an enhanced FCC (Color Figure 35-8). The enhanced image reduced radiometric variability and increased contrasts for some areas in comparison to the standard FCC. Cover types were interpreted with 83 percent accuracy from the standard FCC (Color Figure 35-9) and with 88 percent accuracy from the enhanced image (but with no statistically significant difference) (Color Figure 35-10). He found, however, a substantial bias in area estimates derived from interpretation of the standard FCC; the bias was nearly eliminated, however, through interpretation of the enhanced FCC.

It is evident from the literature that the usefulness and the ease of producing rangeland maps from the manual interpretation of Landsat MSS data varies depending on intended use, characteristics of the environment to be mapped, and prior knowledge of that environment by the image analyst. The greatest use of Landsat MSS images for producing resource maps is in areas where: (1) no other imagery is available; (2) resource maps do not exist, and (3) the Landsat data are used for initial stratification for sample allocation in a more intensive rangeland-resource inventory. The last application listed will be developed later in this Rangeland Inventory section. Landsat has been used widely in Africa to produce a variety of vegetation and land-use maps used for economic development and improved resource management (Mouat et al, 1978; Cooley and Turner, 1975; MacLeod, 1973; Myers et al, 1978; Pacheco, 1974; Poulton, 1977, and Worcester and Moore, 1978).

RANGELAND MAPPING USING LANDSAT MSS DIGITAL DATA

Landsat multispectral scanner (MSS) digital data have been used to map rangeland resource characteristics, including vegetation cover, in a diverse set of environments. Mapping activities have been conducted for research projects, technique development and testing, and to meet operational requirements for information. Examples of these activities include, but are not limited to: mapping in southern Texas, Everitt et al, (1981); mapping in western Alaska, George and Scorup (1981); mapping in central Colorado, Hoffer et al, (1975); mapping in eastern Colorado, Maxwell (1976); mapping in southern California, McLeod and Johnson (1981), mapping in eastern Idaho, Pettinger (1980); mapping in northern Arizona, Rohde and Miller (1981); and, mapping in Australia at various locations, Tueller et al, (1978).

Analytical Procedures

During the past decade procedures for computer-assisted analysis of Landsat multispectral scanner (MSS) data have evolved rapidly and continuously in the quest for reliable, timely and cost-effective techniques for mapping and tabulating earth-surface features. The reduction of multispectral data into useful information classes involves classification decision-rules that are part of the data-analysis methodology called pattern recognition (Swain and Davis, 1978).

The procedures for computer assisted analysis of Landsat MSS data involve the following five primary areas of activity (Hoffer, 1980):

1) Data reformatting and preprocessing,
2) Definition of training[1] statistics for the classifier,
3) Computer classification of data,
4) Information display and tabulation, and
5) Evaluation of results.

Details regarding the activities that constitute the above five are provided by Hoffer and Fleming (1978), and Johnson and Rohde (1981). Such details, with specific reference to forestry, also appear in Chapter 34 of this Manual. Computer assisted analysis procedures are also discussed in Chapter 24 of this Manual, but in a much broader context.

Hoffer (1980), reviewed the results of tests conducted by Fleming and Hoffer (1977), and Nelson and Hoffer (1979), that compared several procedures for defining training statistics. Hoffer concluded that "a) the technique used to develop the training statistics can have a significant impact on the results obtained, and b) the availability and type of reference data may dictate which technique for developing training statistics would be most efficient and effective in a given situation." Hoffer emphasized that it is most crucial for the training statistics to effectively represent the spectral characteristics of the various cover types that comprise the data set.

Comparisons among classification algorithms have also been conducted (Bauer et al, 1977; Nelson and Hoffer, 1979; and Scholz et al, 1979), and reviewed by Hoffer (1980). Those comparisons revealed little difference among the accuracies achieved with each classifier; however, the costs involved varied considerably depending on the classifier used. The cost per unit area classified with a minimum distance classifier was only 40 percent of the cost to run a maximum-likelihood classifier.

These results have particular significance for rangeland-resource inventory. Inventory costs have to be minimized due to the relatively low monetary return from traditional rangeland resource products. Also, the definition of training statistics for rangeland mapping is particularly difficult given the usually large spectral variability

[1] "Training" is a term derived from "training the computer" to "look" for the best match between an unassigned measurement vector (the four band brightness values for each Landsat MSS pixel) and a set of prototype spectral classes.

TABLE 35-6

Co-occurrence Matrix that Relates Photo Interpretation Class to Spectral Class as Applied to Wildland Vegetation.*

Photo Interpretation classes	Spectral Class																													
	A	E	I	J	L	M	O	P	Q	R	S	X	Y	0	1	6	-	/	)	$	,	.	≡	[	≠	<	<	>	≥	–
B2AD																			1						1				1	
B3AD						1		1								1			1										1	
B3AN						1		1				1													1				1	
B3BD						1																								
B3BN			1								1	1			1				1	1					1	1				
b3AN												1																		
C1AC		1																												
C2BD						1																			1					
C3AN												1																		
C3BD						1							1		1					1					1					
C3BN			1	1			1		1		1	1	1				1	1					1	1		1	1	1		1
C3BP																													1	
L2AC						1																								
P3AN	1	1			1					1				1							1	1								

* This co-occurrence matrix shows the combinations (indicated by a "1") of spectral and photointerpretation classes for which there was a significant association (a rate of pairing that substantially exceeded what would be expected by chance using a Chi-square test) when a Landsat cluster image and a resource cover map produced from photointerpretation were registered. The data set shown in this table contains the significant associations for half of the 60 spectral classes derived from the cluster image and all 14 resource classes (six lodge-pole pine categories, B---; six mixed conifer categories, C---; one brush category, L2AC; and one dry meadow category, P3AN). Each resource (photointerpretation) class has been defined in terms of its capacity to provide forage, thermal protection and hiding cover for elk; e.g. B3BN represents lodge-pole pine ⩾40 feet in height, with crown cover ⩾60%, with both overstory and understory canopy layers and no disturbances. B3BN provides thermal protection. The significant association data were used to aggregate spectral classes into valid habitat classes that expressed the forage, thermal protection and hiding potential of the site (Isaacson et al. 1982).

and spatial heterogeneity of vegetation and soil in semi-arid and arid lands. Tueller (1980a), pointed out that supervised training procedures were very time consuming and left substantial portions of spectral variations unrepresented; both points were corroborated by Hoffer (1980). Clustering techniques can provide representation of all the spectral diversity, but they also can increase, sometimes undesirably, the number of spectral classes; interpretation of these classes is an essential, albeit time-consuming task (Tueller, 1980). Most of the digital-mapping activities previously referenced used clustering techniques to define training statistics.

In the California Desert Conservation Area project (McLeod and Johnson, 1981), clustering techniques were used on portions of ten Landsat scenes to define 1,993 spectral clusters from a 0.5 percent sample of a 25 million acre area. It was expected that the sets of statistics would embrace the entire range of spectral reflectance intensities and patterns found in the area, and that the patterns were related in a rational way to resource characteristics of interest. The spectral clusters were divided into four groups; very high vegetation component, high vegetation component, low to no vegetation component on dark surface, and low to no vegetation component on light surface. The effort to reduce this large number of clusters to a manageable amount and to define the resource characteristics associated with each cluster was extremely time consuming, but enabled these investigators to retain the integrity of the spectral clusters for high vegetation responses. The effort was also necessary in order to classify such a large area.

Other investigators have sought to develop more efficient and effective methods to generate sets of clustering statistics for the classifier (Goldberg and Shlien, 1976; and Shlien and Smith, 1975).

Some studies have shown it possible to utilize the clustering statistics from a previous project as a beginning point for the development of statistics for a different digital data set of the same area (Murray, 1982). This approach applies concepts from earlier work in pattern recognition (Hall, 1965; Sebestyen and Edie, 1966). It requires a suitable, existing training set of statistics and the exception that the plant phenological status and moisture content of plants and soils were similar for both data sets. Once the new data set is completely classified, the final clusters are double checked for their resource-class identities and the identities of new clusters are established (Murray, 1982).

Another procedure that uses prior information is the "Procedure-1" (P-1) technique developed at NASA's Johnson Space Center for the Large Area Crop Inventory (LACIE) program (P-1 was reviewed by Hoffer, 1980). This method employs

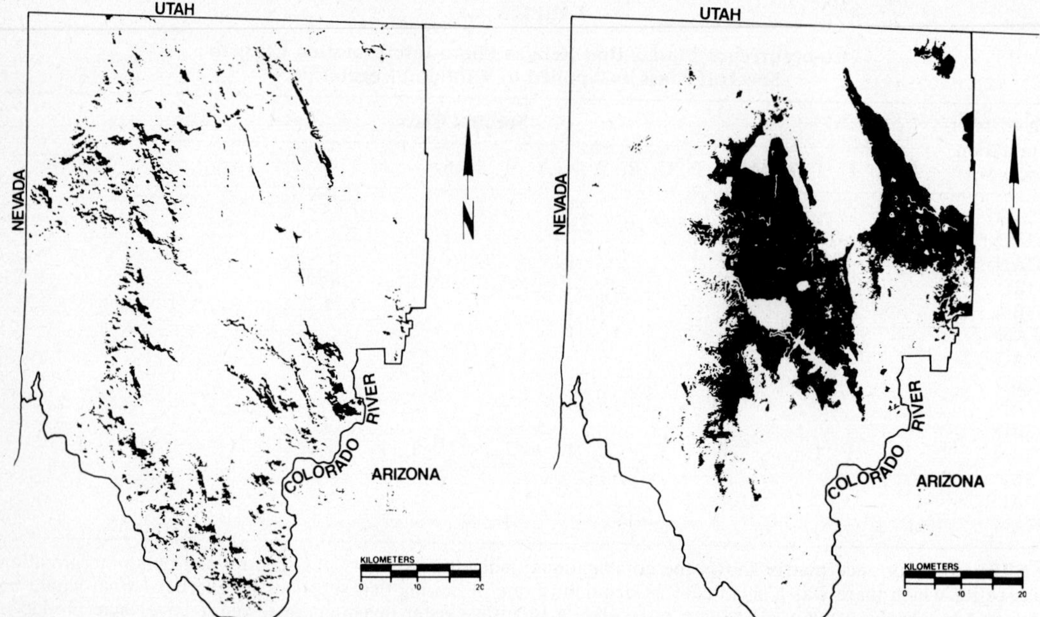

Fig. 35-11. Resource maps derived from selected combinations of Landsat land cover classifications (Figure 35-13) and digital terrain data which show areas that meet the criteria for potential habitat or management opportunity. The left map shows potential mule deer winter range. The mapping criteria were: vegetation-evergreen woodland, Great Basin Desert shrub; terrain—3500 to 5500 feet elevation, 10–50% slope, SE, S, SW aspect. The map on the right shows potential sagebrush treatment areas (areas which could be successfully converted to grassland for increased forage production). The criteria were: vegetation—Great Basin Desert shrub; terrain—5000 to 7000 feet elevation, less than 10% slope.

an array of individual resolution elements of known cover type to "seed" the clustering processor.

When an area is being mapped for the first time with Landsat MSS digital data, a systematic methodology for examining the interrelationships between the spectral classes and the resource classes is needed. Isaacson et al, (1980, 1982), presented methodologies that statistically correlated cover types from aerial photographs (photointerpretation classes) with spectral classes from the Landsat data (Table 35-6). Generally, the associations between spectral groups and resource classes are performed by a resource specialist who is familiar with the character and location of resource classes in the area to be mapped and who had access to aerial photographs and ground data that provided positive verification of the resource classes in question.

Incorporating Ancillary Data

Various types of ancillary data can be incorporated with the Landsat data during the analysis procedures to improve the information content of the final rangeland-resource map. An instructive review of the use of ancillary data for pre-classification scene stratification, post-classification class sorting, and modification of the classifier has been provided by Hutchinson (1982).

Stratification and class sorting have been used to facilitate or improve several rangeland-resource inventories. The 25 million acre California Desert Conservation Area was divided into smaller subareas to create smaller data sets. Bryant et al, (1979) and Pettinger (1980) used environmental stratification (mountainous uplands, lowlands, and agricultural land) in a post-classification context to separate different resource classes that were classified as belonging to the same spectral class. Miller et al (1981) used elevation data to assign pixels within a spectral class (class sorting) to two or more resource classes. In each study where environmental stratification and class sorting was used the ambiguities within resource classes were greatly reduced and mapping accuracies increased.

Kalensky (1974) merged three dates of Landsat MSS data to form a single data set in an effort to improve classification accuracy. Kalensky produced a land-cover classification for the combined data sets and for each of the three data sets that had been merged. He found that classification accuracy was only marginally better on the merged data set relative to the accuracy of the best single-date data set.

By far the most common ancillary data merged with Landsat data for land cover classification are digital terrain data (Hoffer et al, 1979; Shasby et al, 1981; Strahler et al, 1978; and Miller et al, 1981). Not only are the terrain variables of eleva-

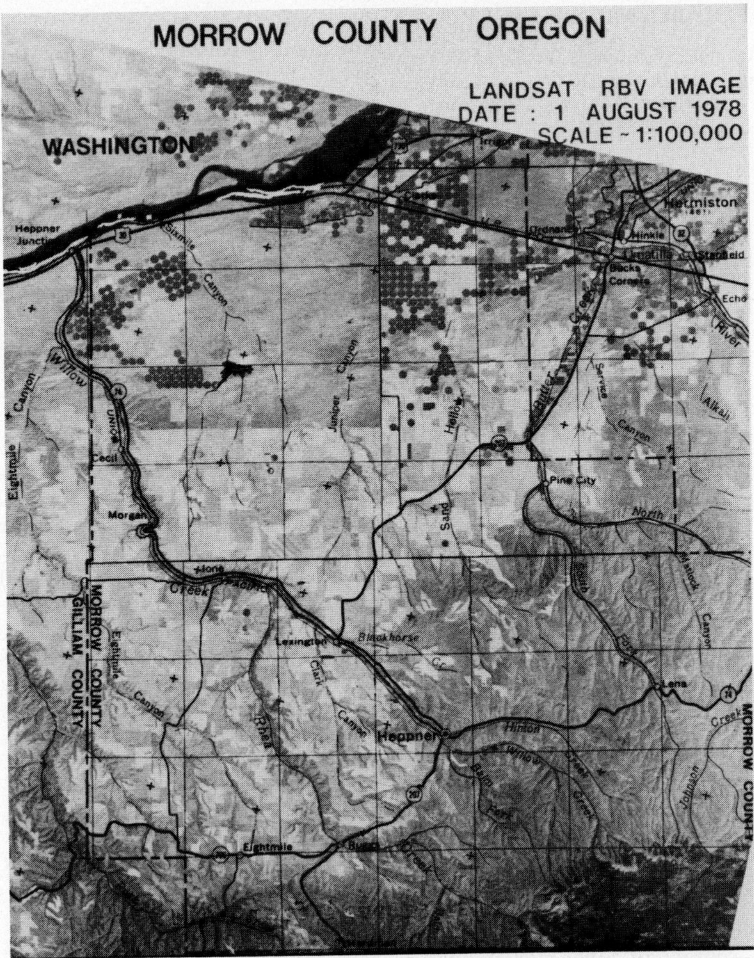

Fig. 35-12. A composite map of a Landsat 3 RBV image and the USGS Oregon Base Map. A transparency was made of the Base Map, overlaid on the RBV and rephotographed. The map shows land net (township and range grid), county boundaries, roads and geographic names in conjunction with a picture of physiography, vegetation cover, land use and urban centers. The product is useful for several activities in planning and executing a rangeland resource inventory. (Photo courtesy ERSAL, Oregon State University, Corvallis)

tion, slope and aspect important for post classification sorting to improve the accuracy of the land cover classification; they also are used in combination with selected land-cover classes to form new interpretative maps. Rohde and Miller (1981), produced several interpretative resource maps, such as mule-deer winter habitat, Bighorn Desert sheep habitat, sagebrush treatment areas and pinyon-juniper sites that can be converted to grassland, for the Bureau of Land Management in Arizona (Figure 35-11).

Murray et al (1982) discuss in detail the use of various types of satellite imagery, including Landsat RBV data (Figure 35-12) Landsat MSS black-and-white imagery, aerial photographs and orthophoto quads to support computer assisted analyses of Landsat MSS digital data. The most important uses of these supporting data sets are for stratification and for identifying the resource class associated with the spectral clusters.

Land Cover Classification Accuracies

Conclusions about the accuracy and usefulness of rangeland cover maps produced by digital analysis of Landsat data are difficult to make because of the variability of results in the literature. The results vary, based upon environment, training, classification procedures, level of detail mapped, and technique employed for determining accuracy. Most of the earlier publications provided only a qualitative assessment of accuracy.

One standard with which to compare level of detail for digital classification results is the U.S. Geological Survey Classification System by Anderson et al, (1976). Land cover is described by levels from the general (Level I) to the specific (Level II, III, etc.). Most of the Landsat digital-classification papers dealing with rangeland mapping contend that Level I categories can be classified with an overall accuracy that ranges between

TABLE 35-7

Classification Accuracy Estimates for Eight Cover Types Mapped in Northwestern Arizona Using Landsat data, and Digital Analysis Techniques (see Figure 35-13).*

	Accuracy			
	Without Terrain Data		With Terrain Data	
Cover Type	Correct	Standard Error	Correct	Standard Error
Cropland/ Pasture	15%	17%	19%	23%
Coniferous Forest	72%	7%	81%	12%
Evergreen Woodland	77%	6%	81%	5%
Deciduous Woodland	4%	4%	70%	23%
Mohave Desert Shrub	74%	6%	96%	2%
Great Basin Desert Shrub	57%	4%	68%	4%
Mountain Shrub	38%	14%	58%	15%
Grassland	1%	1%	2%	1%
OVERALL	54%	5%	73%	5%

* The accuracy figures and accompanying sampling error for each type are given before and after digital terrain data was used to separate confusion within and between classes.

75 and 90 percent. Efforts to classify Level III categories generally result in overall accuracy figures of less than 70 percent, a figure often considered unacceptable.

Classification accuracy of cover maps tends to be higher, of course, in environments where there are only a few cover types, each tending to have distinct spectral characteristics. Such environments are generally mesic, characterized by relatively dense vegetation. Classification accuracy of cover maps produced for arid environments tend to be low especially if there is marginal correlation between vegetation and associated soil and geology. Miller et al (1981) however, have demonstrated that classification accuracy in arid environments can be improved by post-classification class sorting with decision rules based on elevation classes. By this process classification accuracy was increased from 54 percent to 73 percent (Table 35-7). George and Scorup (1981) reported overall accuracies of 77 percent for their mapping of 1.6 million hectares of reindeer range on the Seward Peninsula of western Alaska.

SAMPLING APPLICATIONS USING REMOTE SENSING DATA

Remotely-sensed data and statistical sampling procedures have been incorporated for rangeland resource inventories for many years. Harris (1951) developed a procedure for the efficient and accurate sampling of vegetation on large range areas. The methods he devised utilized conventional aerial photos delineated into major range-vegetation types. Sampling points or units were selected by randomly selecting individual photos from within a major range-vegetation type and then randomly selecting a point from a dot grid overlay. Through use of this process the selected sample points or units, were randomly chosen and stratified by type. Each sample unit was then located on the ground, where measurements were made from a cluster of two or more plots. The plots also were drawn independently and at random. The number of plots per cluster was determined so as to minimize the sampling error of a given cost of securing information. Green weight of herbage by species was estimated at each plot. Harris advised that cluster sampling was advantageous when the cost to travel between sites was at least twice as much as the cost of acquiring data at a plot. Optimum allocation of clusters among range types was also possible.

Waller et al (1978) and Heintz et al (1979) tested the use of a motor-driven 35mm camera for acquiring aerial photos that were used to stratify a very complex grassland in western South Dakota. Enlarged color infrared photographs (1:300 scale) were used to delineate three range sites: a clayey site with a dense stand of western wheatgrass; a thin claypan site with little or no western wheatgrass and a high number of forbs; and a transition zone having a sparse stand of western wheatgrass. Plots were randomly located within each site and the vegetation clipped and sorted. From their analyses they concluded that, in order to sample a standing crop of western wheatgrass to within ten percent of the mean and with 95% probability, would have required 430 plots if the effect of site had been ignored; only thirty plots were required, however, when the stratification was employed.

Tueller (1977), reviewed techniques and costs involved in: a) acquiring 70mm aerial photographs at scales ranging from 1:800 to 1:10,000 and b) the information detail that can be extracted from those scales of photographs. Tueller and Booth (1975), described methods for systematically sampling range ecosystems to evaluate erosion features from large scale aerial photographs.

During the past decade the availability of Landsat data has further encouraged the utilization of remote sensing data in inventory design. Sampling techniques using Landsat data have been applied for rangeland vegetation mapping (Miller et al, 1981), preparing area estimates (Rohde et al, 1979 and Linden et al, 1981), estimating rangeland productivity (Thomas et al, 1975), and for estimating noxious weed infestation (Isaacson and Schrumpf, 1979). Several recent articles provide a good understanding of techniques for integrating remotely-sensed data and statistical sampling procedures.

Peterson et al (1982) discussed the effective use of thematic information derived from a Landsat MSS classification for stratification, extension, developing selection probabilities, expansion

TABLE 35-8

Results of Acreage Estimate and Sampling Error for Each Cover Type in the Land Cover Map (Color Figure 35-13) Produced for Northwestern Arizona Using Landsat Digital Data and Digital Analysis Techniques.*

Cover Types	Acreage Estimates	Sampling Error (acres)
Coniferous Forest	17,104	4,186 (24%)
Evergreen Woodland	678,653	70,864 (10%)
Deciduous Woodland	5,336	1,114 (21%)
Mohave Desert Shrub	504,842	38,940 (8%)
Great Basin Desert Shrub	1,135,091	71,420 (6%)
Mountain Shrub	110,925	33,975 (31%)
Grassland	37,190	22,440 (60%)

* Estimates were based on a stratified cluster sample involving the Landsat land cover classification which were corrected with digital terrain data (Table 35-6), and the interpretation of cover type on 1:6,000 color photographs (Color Figure 35-14, Rohde and Miller, 1981).

factors and projections modeling. Wensel and Eriksson (1981) discussed and demonstrated the use of a survey planning model for achieving an optimal allocation of sampling effort (forage production estimation was their example). Bonner and Morgart (1981) demonstrated the use of classified Landsat MSS data as a sampling frame. Thomas and DeGloria (1979) discuss the use of large-scale aerial photographs as a sample stage in wildland resource inventories with specific reference to design considerations, sample designs, sample size and aerial photograph acquisition and measurement.

Miller et al (1981) used decision rules in a post classification class sorting procedure that was developed from vegetation/terrain data acquired by sampling. In this manner the sampling contributed to the final reclassification and mapping of the Landsat MSS data. Rohde et al (1979) and Rohde and Miller, (1981) have demonstrated that a digital land-cover classification produced from Landsat data is an excellent starting point for designing a sampling strategy that uses aerial photographs and ground sampling to statistically and quantitatively describe each resource class. Moreover, the additional sampling provides a statistical estimate of the acreage of each resource class. This approach is advantageous in that it permits calculation of the classification error associated with each resource class and acreage estimate with selected confidence intervals.

Color Figure 35-13 shows a digital cover map produced by Landsat data for northwestern Arizona. Classification accuracies and acreage estimates for each resource class are presented in Table 35-7 and Table 35-8. Color Figure 35-14 shows one of 129 sample plots used to determine classification accuracy and make acreage estimates. Each sample plot is comprised of 64 Landsat picture elements (pixels). Each pixel was interpreted on the photo to determine resource class, and a subsample of pixels was sampled on the ground to cross-check the photo interpretation results. The costs for producing the cover map and the accompanying acreage estimate was $0.15 per hectare.

Isaacson and Schrumpf (1979), utilized the land cover classifications from Landsat MSS data of ten million acres in western Oregon as the sampling frame for a two-stage sample for estimating the infestation of Tansy Ragwort, a noxious weed. Sample units were selected within sixteen counties so that independent estimates for most of these counties could be calculated. There were clear differences among the infestations for various land cover/use types and among counties, thereby providing a basis for evaluating control programs and focusing additional control efforts.

Multistage Sampling for Forage Production Estimation: Case Example

Thomas et al (1975), utilized a multistage sampling design (Figure 35-15), to estimate rangeland productivity on two Bureau of Land Management planning units in northeastern California and northwestern Nevada. The planning units were Willow Creek, containing 191,000 ha (472,000 ac) and Cal-Neva, containing 330,000 ha (816,000 ac). Based on this example, the basic approach used in applying a multistage sampling framework within a rangeland environment will now be described in general terms to show the reader how the different kinds of remotely-sensed data and ground information are interrelated. The procedures used are designed to provide at a low cost information of sufficient quality for planning and management purposes.

Range productivity estimates having a minimum standard error for a given budget may be obtained within a multistage sampling framework that combines area-wide Landsat data with a subsample from medium-scale (1:30,000) color infrared aerial photographs, large-scale (1:400) color aerial photographs, and a minimal amount of ground data (Color Figure 35-16). The range-productivity estimates from Landsat are calibrated with those from the medium-scale aerial photographs that are, in turn, calibrated with data acquired from a smaller and more expensive sample of large-scale photographs; similarly, in

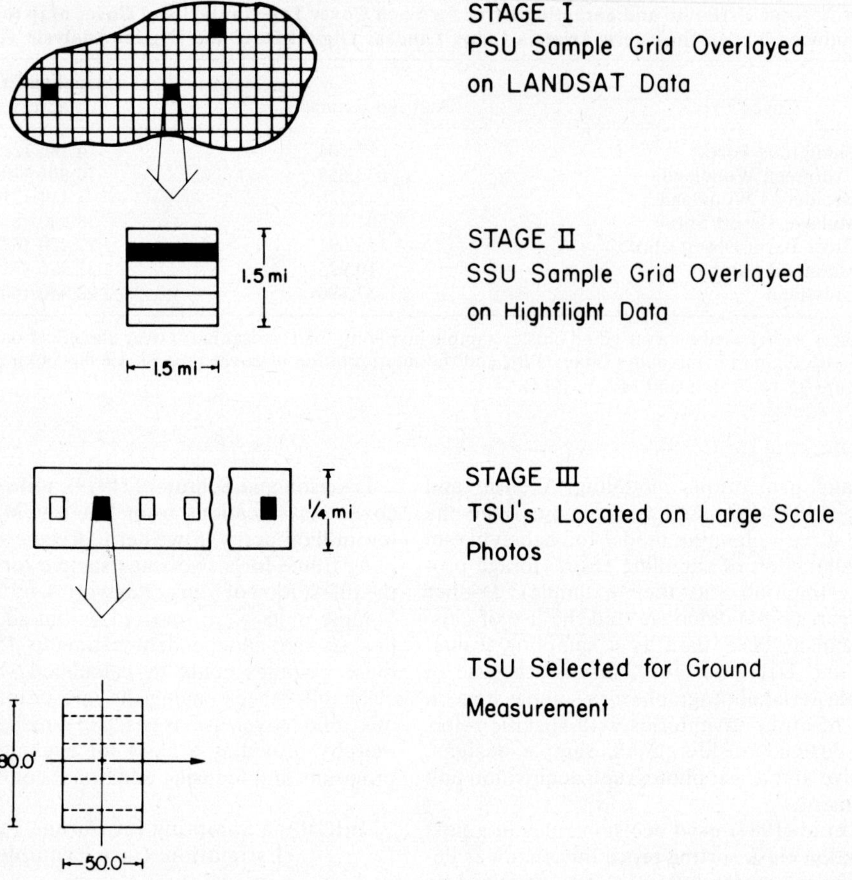

Fig. 35-15. The multistage sampling frame used to estimate forage production within two BLM planning units in northeastern California (Thomas & DeGloria, 1979).

the last step, the range-productivity estimates from the large-scale photographs are, in turn, calibrated with a smaller and more expensive sample of corresponding ground plots.

Implementation of the multistage-sampling framework requires specification of the area to be sampled, the number of stages required to prepare an estimate to within a prescribed allowable error, and also the size, shape and number of sample units at each sample stage, which coresponds to the number of data types sampled. In this example, the Planning Units were the areas for which a range-productivity estimate was made. Each Unit was divided into a matrix of square sampling blocks or primary sample units (PSU) 1.5 miles on a side. The size and shape of the PSU's should be designed to minimize the variance of the estimate and be practical in terms of cost and efficiency of subsampling using aerial photographs and groun data. The first stage estimate from the PSU's was based upon probability of BLM ownership, but other criteria such as probability based on rangeland area or probability based on range productivity may be more effective in reducing the sampling error.

The PSU's were subdivided into six rectangular sample units measuring 0.25 miles wide and 1.5 miles long. These units constituted the second stage sample units (SSU), and estimates of range productivity within these units were derived from color infrared aerial photographs at a photoscale of 1:30,000.

The size and shape of the SSU's were determined by the area required to obtain ten third-stage sampling units (TSU's) along the long axis of the SSU.

The area of the SSU had been cost efficient for subsampling in a prior forest inventory study (Thomas et al, 1975). A light aircraft and a 35mm camera acquired ten color stereo photo-triplets at a scale of 1:400 within the selected SSU's. The location of the TSU's was the middle photograph of the stereo-triplet and measured 50 by 80 feet. Color photographs at a scale of 1:3,700 were also acquired simultaneously to aid in locating the TSU's.

Ground data were obtained for a sample of the TSU's. The ground data, consisting of species composition and a measure of plant canopy cover, were acquired from 250 sample points located at

TABLE 35-9

Number of Samples Selected and Measured at Each Stage of a Multistage Sampling Framework Applied in Two Bureau of Land Management Planning Units in Northeastern California and Northwestern Nevada to Estimate Forage Production.*

Sampling Stage	Sample Size (Willow Creek Planning Unit)	(Cal-Neva Planning Unit)
Stage I (Landsat)	8 of 400 PSU's	19 of 600 PSU's
Stage II (small-scale photographs)	8 of 48 SSU's	19 of 114 SSU's
Stage III A (large-scale photographs)	71 of 71 TSU's	161 of 161 TSU's
Stage III B (ground)	13 of 71 TSU's	30 of 161 TSU's
Productivity Estimates	18,177 AUM's	33,716 AUM's
Standard Error (68% level of confidence	3,719 AUM's	5,175 AUM's
Sampling Error	20.5%	15.4%

* The source of the sample data (Landsat image, aircraft photographs or ground) is listed for each stage. The right columns indicate the number of samples actually measured or interpreted out of the total number allocated at each stage. For example, only 13 of 71 large-scale photo plots (Tertiary sampling units, TSU) were randomly selected and measured on the ground in the Willow Creek Unit. The size of the primary, secondary and tertiary sample units (PSU, SSU and TSU respectively) are given in Figure 35-15. The productivity estimates for each planning unit are expressed in Animal Unit Months, (AUM's) and apply only to the BLM land within each unit. (Adapted from Thomas et al, 1975)

0.2 foot intervals along each of four 50 foot transect lines located within the TSU.

The number of sample units selected at each sampling stage is presented in Table 35-9. In their report (Thomas, et al, 1975) these investigators state: "Ideally, sample size at any stage of the multistage sample design should be large enough to minimize estimate variance for fixed survey cost or to minimize cost for a specified estimate precision. The upper limit on variance specified by the land manager may refer to more than one resource type, but more often in practice refers to the resource type of primary importance. Optimal sample sizes generally can be calculated after a planning unit has been sampled once, and an estimate has been obtained of resource-parameter variance and data-collection costs".

The PSU's were selected randomly from a list of PSU's based on cumulative acreage of BLM land. One out of every six SSU's from each PSU was selected at random for analysis and all TSU's for which large-scale photographs were acquired were interpreted. Approximately two TSU's from each SSU were selected for obtaining the ground measurements to estimate range production.

Estimates of forage productivity for cattle, expressed in terms of carrying capacity on an animal-unit-month (AUM) basis, were obtained for each sample stage. Forage production within the ground sample area was expressed in acres required to support an animal (cow plus calf) for a month. This estimate was obtained by dividing the animal-month forage requirement by the forage/acre available. The forage requirement supplied by each plant species was predetermined by the BLM and forage/acre estimates were based upon the known relationship between foliar-cover data by species and available forage. Forage estimates at the third stage were made similarly to the ground data estimates by relying upon photo analysis of foliar cover from 25 photo points. At each photo point, the interpreter determined whether the cover type was: a) live or dead shrub; b) live or dead trees; c) forbs; d) annual grass; e) perennial grass; f) persistent litter; g) large rocks; or h) bare soil. At the second stage, forage-production estimates were arrived at by first delineating and computing the areas of vegetation types on the 1:30,000 scale color infrared photographs and then computing the average number of acres per animal-unit-month for each cover type, using the following equation:

Total AUM's per SSU =

$$\sum_{i=1}^{N} (\%\ \text{area}_i)\ \frac{\text{Acreage of Sample Unit}}{\text{Ac/AUM}_i} \qquad (35\text{-}1)$$

Average Ac/AUM values for each of the N cover types were obtained from the ground data and ground-calibrated photo TSU's falling in each cover type. Productivity estimates from the PSU's were obtained similarly to those from the SSU; however, the cover types and acreage summaries were derived from digital classification of the Landsat data.

The productivity estimators that served in this example to link the estimates of forage production from each sample level are as described by Thomas et al (1975). The productivity estimates that were obtained for each Planning Unit are shown in Table 35-9. Although the sampling error ranged from 15 to 20 percent, the investigators concluded that they could reduce the error to approximately 10 percent by doubling the number of primary sample units. The projected average operational cost of this inventory was \$.045/acre (1975 costs).

RANGELAND RESOURCE MONITORING

Monitoring rangeland resources involves the observation and measurement of changes between and among components of the resource base, and the determination of their significance. Most of the monitoring has been directed towards determining seasonal and ecological conditions of the range, and measuring changes in these conditions.

The monitoring of seasonal conditions generally involves periodic assessment of plant, animal and environmental phenomena that affect stocking rates and the seasonal health of the range during a growing season. Plant development, green biomass, animal numbers, and hydrologic conditions are among the more important components that are monitored.

The monitoring of ecological conditions generally involves assessment of plants and soils at longer intervals, such as between seasons, to determine whether management practices are contributing either to sustained productivity and long-term health of the range, or to its deterioration. Changes in plant-species composition, foliar cover and soil characteristics within and between vegetation units are indicators of changes in ecological range condition and are important to monitor for an assessment of range trend.

Common to all resource monitoring is the establishment of a base line of information about current conditions. Subsequent observations and measurements are made at appropriate intervals and compared with the base-line data to determine the type, magnitude, and significance of the change. Baseline data can be in the form of statistical data, detailed maps, or annotated photographs.

Range managers have traditionally relied upon visual observations to assess seasonal range conditions, and upon ground-sampling methodologies, including ground photographs, to assess ecological range condition and range trends. More recently, range managers have utilized ground photographs, aerial photographs, and satellite images as a base to record rangeland conditions at one date, and repetitive remotely-sensed data to monitor rangeland changes occurring between two or more dates. This section describes some of the techniques wherein remotely-sensed data are being used to monitor seasonal and ecological range conditions.

MONITORING ECOLOGICAL RANGE CONDITIONS

The term *ecological range condition* refers to the current health or condition of a range-plant community relative to the potential or climax condition of that community that could develop or that had previously existed in an undisturbed state. Dyksterhaus (1949), developed a method for determining range condition that is in standard use by the U.S. Department of Agriculture, Soil Conservation Service. The methodology requires that each plant species in a plant community be identified and labelled an "increaser", a "decreaser" or an "invader" depending upon its response to site disturbance. The determination of excellent, good, fair, or poor range condition is based upon the percentages of increasers, decreasers and invader plant species in the plant community relative to the theoretical or climax community that once occupied the site. Other factors such as soil condition and amount of litter are also considered in the determination.

Although visual observations can provide an acceptable estimate of species composition and foliar (canopy) cover, the determination of most range conditions requires detailed, quantitative measurements of plant composition acquired through labor-intensive ground sampling using line transects, point frames, and clipped sample plots. The proposed use of remotely-sensed data acquired at low altitudes is directed at a determination of dominant plant species and the detection of changes within a plant community but, at higher altitudes, to the detection of major changes to vegetation units resulting, for example, from fire, or man-made conversion.

Ground Photographs for Monitoring Range Conditions

Since the turn of the century ecologists have periodically photographed established ground plots to obtain a permanent, visual record of vegetation changes (Hastings and Turner, 1965). Since the 1950's, observations and measurements of plants and soil along permanent line transects have been used to evaluate changes in ecological rangeland conditions (range trend) in the United States. Various line-transect methodologies are reviewed by Parker (1951), Tueller et al, (1972), and Reppert and Francis (1973). Figure 35-17 contains two oblique, panchromatic photographs showing plant changes during a ten year period. They provide only a qualitative measure of change, unless accompanied by accurate annotations of species composition and quantitative ground data.

Within the last two decades, vertical ground photographs have been used to make measurements of foliar cover and to identify plants species. (Claveran, 1966; Wimbush et al; 1967; Pierce and Eddleman, 1970; and Wells, 1971). The photographs have been taken either with one or two cameras (35 mm or larger frame camera), from a tripod or four-legged stand (Figure 35-18 and Figure 35-19). Wells (1971), devised an interpretation system for estimating foliar cover from a vertical ground photograph. A photo

Fig. 35-17. Panchromatic photographs taken in 1952 and 1962 of the same rangeland area are a permanent record of range conditions and show changes in plant composition, plant structure and soil surface conditions. These photographs are typical of those taken along permanent transects as part of the Parker 3-step method for collecting data to assess changes in range condition. (Courtesy of U.S. Forest Service, Repport & Francis, 1973.)

transparency was placed on a viewing platform and moved at specified increments beneath a cross-hairs imbedded in a viewing lens (magnified). Each time the photograph was moved, a record was made of the plant, soil, or litter material seen beneath the cross-hairs. Other techniques for estimating foliar cover from ground and large-scale aerial photographs include a) use of a dot-grid; and b) measurement of line intercepts (Carneggie et al, 1971).

In some plant communities where plant numbers and density are great, plant canopies often overlap; furthermore, where species of substantial stature (exceeding the height of the cameras) are present, it is difficult to identify and count plant species on large-scale vertical ground photographs. However, such photographs may be useful as a base for constructing a plant species map. Plant communities having an open canopy structure and a small number of low-growing plant species lend themself to direct assessment of species composition and foliar cover using ground photographs. For best results, the ground photographs should be enlarged and taken back into the field where the location and identity of plants and ground conditions can be verified and recorded. Photographs taken at established ground plots at a later date can then be compared with the initial photographs and plant maps to show changes in composition, numbers, location and foliar cover.

Fig. 35-18. Two 35mm cameras mounted for vertical ground stereophotography. Note dual shutter release mechanism. The frame is made of drilled ¼ inch aluminum plate for lightness with strength (from Wells, 1971).

Large-Scale Aerial Photographs for Monitoring Range Conditions

For several decades, foresters, agriculturists, and engineers have used aerial photographs taken from low flying aircraft, helicopters, and balloons. Foresters in particular, have used large-scale photographs for identifying tree species, estimating timber volumes and appraising insect and dis-

Fig. 35-19. Dual mounted 35mm cameras in stand for obtaining stereo ground photographs of a quadrat in northwestern New South Wales, Australia (from Wells, 1971).

ease damage (Heller et al, 1964, 1967; Avery, 1958; Aldrich, 1966; Lyons, 1967; refer to Chapter 34).

Since the late 1960s range managers have experimented with large-scale aerial photographs in studies of range condition, rangeland monitoring and forage-production estimation (Carneggie and Reppert, 1969; Driscoll and Reppert, 1968; Lorain, 1970; Carneggie et al, 1971; Tueller et al, 1972; Carneggie, 1972; Colwell et al, 1975; Driscoll et al 1970; Thomas & DeGloria, 1979; and Miller et al, 1981). In these studies the aerial photographs were used to: 1) detect and identify plant species; 2) measure plant height, foliar cover, plant density and numbers; 3) quantify changes in plant parameters at intervals of one or more years; and/or 4) detect and quantify other surface soil characteristics.

The extent to which precise interpretations and measurements can be made with sufficient accuracy to be practical for monitoring changes in plant communities depends upon: 1) the adequacy of high quality, large-scale aerial photographs taken at two or more dates over the same ground site; 2) the techniques used for making interpretations and measurements on the photographs; 3) the kind and amount of supporting observations and measurements made on the ground; and 4) the characteristics of the plant community being monitored.

Acquiring high quality aerial photographs over a permanently established site at intervals of one or more years presents some unique challenges and requires good mission planning. Among the steps and criteria to consider are the following:

1) *Establish permanent plots or transects.* Determine dimensions of plots or transects so that they can be seen within the photo coverage area of a single photograph. Lay out ground markers so that the pilot and the photographers can see the plot from the air, and so that there are ground reference points at known distances to determine photo scale (Francis, 1970).
2) *Acquire the sequential photographs at approximately the same scale, resolution, exposure, time of day, and season.* For purposes of change detection, photos from different dates should be enlarged to the same scale.
3) *Select a camera and lens system that meet project objectives* for image resolution, photo scale and area photographed in a single frame. 35mm, 70mm and 9 × 9 inch camera systems are commonly used; 35mm and 70mm cameras usually are preferred because they are small, compact, lightweight and easy to handle. Focal lengths for common lenses used in such cameras range from 35 to 200mm. Costs for system operation and film are relatively low compared with those incurred when using 9 × 9 inch cameras. The advantage of a 9 × 9 inch camera is the large ground area photographed per frame. Common lens focal lengths for such a camera range from 100 to 300mm. Techniques and equipment for using 35mm aerial camera systems are documented by Meyer and Grumstrup (1978) and Miller (1974).
4) *Specify type of aircraft and flying altitude.* Single-engine aircraft have the advantage of lower operating costs, lower flying speeds and greater maneuverability. Twin-engine aircraft are more stable, have a longer operating range and fly rapidly between base and study area. Most large-scale aerial photographs acquired of rangeland areas for plant identification and measurement have been taken at altitudes between 300 and 2000 feet. The accuracy of plant identification is strongly affected by the scale of the photographs, the size of the plants, plant structure, and the appearance of the plant at the time the photographs are acquired. Driscoll and Reppert (1968) found that mature shrubs were readily identified on photographs taken from 475 to 1200 feet, but immature shrubs were difficult to identify even on photographs acquired at lower altitudes. Small herbaceous plants often could not be seen or identified unless the photographs were taken at altitudes below 475 feet, or the plants pos-

TABLE 35-10

Percent Correct Identification of Shrubby and Herbaceous Species on Large-Scale, 70mm Color Infrared (CIR) and Color Aerial Film (Data from Driscoll et al, 1970).

Species	Film Type: CIR (Percent)	Film Type: Color (Percent)
Shrubs[1]		
Rothrock sagebrush	100	98
Big sagebrush	90	93
True mountain mahogany	100	92
Parry rabbitbrush	60	56
Green rabbitbrush	56	50
Snakeweed	73	88
Bitterbrush	78	50
Cinquefoil	83	79
Snowberry	60	55
Herbaceous[2]		
Pussytoes	80	10
Blue grama	10	10
Mountain muhly	10	10
Arizona fescue	90	80
Fringed sage	85	25
Trailing fleabane	60	85

[1] Average of four interpreters with varying degrees of photointerpreter experience. Based on July photography.

[2] Average of two interpreters. Based on June photography.

TABLE 35-11

Photo Interpretation Results from Three Dates of Color and Color Infrared Large-Scale Photographs Showing the Percent Correct Identification of Five Dominant Shrubby and Semi Woody Plants in a Plant Community within the Lassen National Forest in northeastern California.*

Color					Color-Infrared			
Plant Species	% Correct	% Error Type I	% Error Type II		Plant Species	% Correct	% Error Type I	% Error Type II
A	73	27	37	JUNE 10, 1967	A	68	32	51
B	50	50	63		B	57	43	39
C	50	50	56		C	38	62	74
E	37	63	39		E	45	55	41
L	13	87	82		L	20	80	64
All 5 Species	45	55	55		All 5 Species	46	54	54
Plant Species	% Correct	% Error Type I	% Error Type II		Plant Species	% Correct	% Error Type I	% Error Type II
A	80	20	33	JULY 25, 1967	A	70	30	28
B	90	10	23		B	100	0	4
C	53	47	24		C	60	40	36
E	97	3	0		E	100	0	6
L	73	27	24		L	87	13	0
All 5 Species	79	21	21		All 5 Species	83	17	17
Plant Species	% Correct	% Error Type I	% Error Type II		Plant Species	% Correct	% Error Type I	% Error Type II
A	83	17	29	OCTOBER 25, 1967	A	80	20	31
B	80	20	15		B	80	20	25
C	93	7	15		C	83	17	24
E	77	23	12		E	73	27	12
L	73	27	21		L	73	27	12
All 5 Species	81	19	19		All 5 Species	78	23	23

* The results are the average of three photointerpreters. Note that the best results were from color infrared photos in July, but that certain species were identified more accurately on color photos in October. Type I errors are omission errors; Type II errors occur when the interpreter incorrectly assigns a plant to the wrong species. (Data from Carneggie, 1972). A = Big sagebrush, (*Artemisia tridentata*); B = Bitterbrush, (*Purshia tridentata*); C = Rabbitbrush (*Chrysothamnus sp*); E = Buckwheat (*Eriogonum umbellatum*) and L = Wherry (*Leptodactylon pungens*).

sessed more distinguishable characteristics when photographed. Carneggie and Reppert (1969) reported obtaining reasonable identification results for five woody species from photographs acquired between 300 and 500 feet.

5) *Select film type and acquisition date*. There is general agreement that color infrared film provides a slight edge over color film for general plant identification and vigor assessment. This has been demonstrated in a number of plant communities in California (Carneggie and Reppert, 1969; Carneggie, 1972), in Colorado (Driscoll and Reppert, 1968; Driscoll et al, 1970), in Nevada (Lorain, 1970; Tueller et al, 1972; Tueller, 1977) and in Western Australia (Carneggie et al, 1971). Some plants, however, could be identified with the same or better accuracy on color photographs (Color Figure 35-20). Most plants could be identified with greater accuracy from photographs taken during the middle to the latter part of the growing season. Tables 35-10 and 35-11 show typical results from plant identification tests in Colorado and California, respectively.

6) *Select appropriate vegetation units to study*. Rangeland plant communities with less than 50% foliar cover and containing relatively large plants, such as shrubs, are more amenable to an analysis of change in ecological condition using large-scale aerial photographs. Since it is virtually impossible to account for all plants, it is important to select a vegetation unit that has a few dominant species that can be readily identified and measured from photo-analysis.

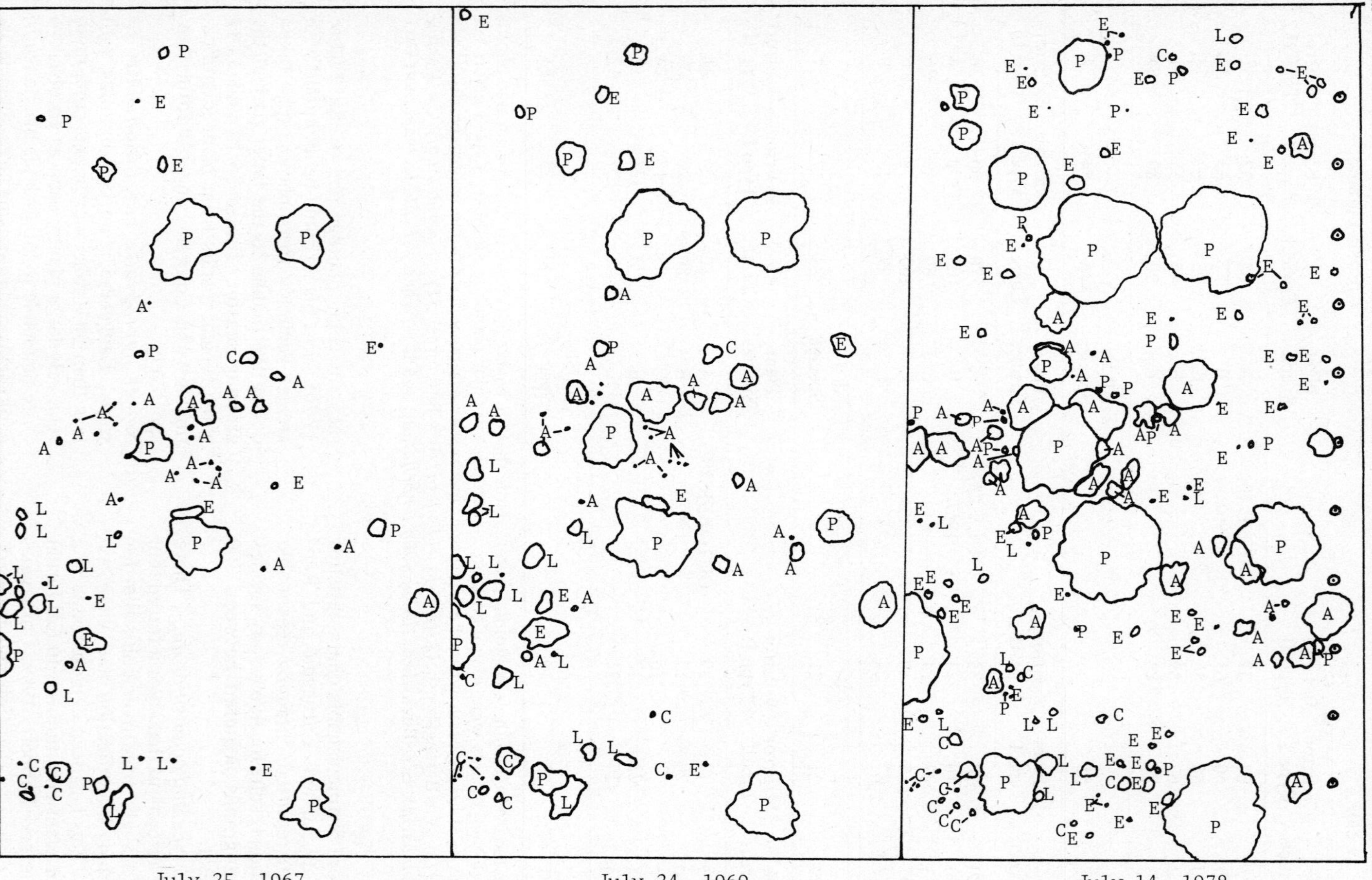

Fig. 35-22. Plant maps of the dominant woody vegetation photographed in July 1967, 1969, and 1978, in Color Figure 35-21. The location and size of plants for these maps were taken from the photographs, and verification of a plants presence and its identity were obtained on the ground at the time the photographs were acquired. Note that these maps provide information on changes in the number of plants as well as the size of the plants over an eleven year period. The plant species include: A-Big sagebrush, (*Artemisia tridentata*); B-Bitterbrush, (*Purshia tridentata*); C-Rabbitbrush, (*Chrysothamnus sp.*); E-Buckwheat, (*Eriogonum umbellatum*); and L-Wherry, (*Leptodactylon pungens*). These maps show only a portion of a larger sample site measuring approximately 52 × 75 feet.

TABLE 35-12

Number and Canopy Area of Bitterbrush and Big Sagebrush Plants in 52′ × 75′ study Plot in northeastern California in July of 1967, 1969, and 1978.*

Number of Plants	Dates of Photographs 1967	1969	1978
Bitterbrush			
Actual number ≤6″ diameter	3	0	44
Actual number > 6″ diameter	25	28	52
Average correctly detected on aerial photographs	25	28	41
Big Sagebrush			
Actual number ≤6″ diameter	18	22	7
Actual number >6″ diameter	37	51	100
Average correctly detected on aerial photographs	36	48	84
Plant Canopy Area Estimates	**1967**	**1969**	**1978**
Bitterbrush			
Photo estimate (ft²)**	203	359	817
Photo estimate (% cover in plot)	4.2%	7.5%	17.2%
Ground estimate***	—	—	18.2%
Big Sagebrush			
Photo estimate (ft²)**	175	229	346
Photo estimate (% cover in plot)	3.6%	4.8%	7.3%
Ground estimates***	—	—	7.7%

* The results of photo detection tests indicate that interpreters can readily detect the dominant shrubby species when they are greater than 6″ in diameter. Photo estimates of canopy cover show a substantial increase in the two dominant species during the eleven year period. A comparison of photo and ground estimates of canopy cover in 1978 indicates the photo estimates are good approximations of canopy cover and are sensitive indicators of change in canopy area. These data were derived from the color infrared aerial photograph in Color Figure 35-21 which show only a portion of the study plot sampled.

** Photo estimates made using a 100 dot/sq. in. dot grid on color infrared aerial photos enlarged to a scale of 1:120.

*** Ground estimates made using eleven-75 foot line transects within the 52′ × 75′ plot.

7) *Specify ground measurements to be made in the field.* In order to develop a monitoring capability using large-scale photographs one must initially collect base-line information about the permanent ground plot. Carneggie (1972), mapped the location and identity of all plants belonging to five woody species for each date on which large-scale aerial photographs were taken in California (Color Figure 35-21). Maps derived from these photographs appear in Figure 35-22. The detailed plant maps in themselves are an excellent record of change in plant species, numbers and size. However, their primary value is for determining the accuracy of plant identification, plant detection, and canopy-cover estimates made from the photographs. Ground estimates of cover could also be made to compare with photo measurements of canopy cover. Table 35-12 shows the photo estimates of canopy cover for two major shrub species at three dates. Note that the ground and photo estimates of canopy cover compare favorably for measurements taken in 1978. Van Lee and Bonner (1981), working in the same study area, determined that there was a 92 percent correlation between aerial photographic and ground measurements of cover when all cover classes were considered simultaneously in multiple regression analysis.

8) *Specify photo interpretation and photo measurement techniques.* The identification of plants on large-scale aerial photographs should be performed while viewing the original transparencies stereoscopically with magnifying lenses, or while using a stereoscope to view enlarged photographs, in opaque print form, made from the original transparencies (Color Figure 35-23). Photo interpretation keys that describe the photo image characteristics of plant species are generally helpful for identifying plants. Stereoscopic viewing is also recommended for the interpretation of plant numbers. Photo measurement of canopy cover could be performed using a dot grid (Carneggie, 1972) or a line transect method (Carneggie et al, 1971), directly on the enlarged photographs. Photographs from different photo missions over a plot should be enlarged to

the same scale for analysis (see Color Figure 35-21). A quick comparison should reveal any changes of plant numbers or canopy area. If large changes are detected, a more careful interpretation will be warranted. The same interpretation or measurement techniques should be applied to photographs from different missions.

9) *Repetitive photographic coverage should be obtained at two to five year intervals.* The subsequent photo missions should duplicate the initial photo mission. If near-vertical photos are taken, it is not necessary to rectify them (correct for geometric distortions) before proceeding with photo analysis.
10) *The photographic record should be filed so that it will not be damaged or deteriorate with age.* The photographs provide a visual and permanent record of plant species, location, and size (as viewed from a vertical position) and can be referred to at any time, in either the field or office, and in comparison with photos taken at other dates to determine the nature and magnitude (both relative and absolute) of changes in plant parameters and surface conditions.

Observations and results extracted from the literature dealing with large-scale aerial photographs demonstrate the potential as well as the limiting factors encountered when using large-scale aerial photographs to monitor vegetation changes. Among the most important are:

1) Some plant species are more readily detected and identified than others and hence lend themself to monitoring with time-lapse photographs. These species may provide clues to changes within the vegetation unit if ecological studies can show them to be valid indicator species, i.e. show that they belong to each group of decreasers, increasers and invaders. The plants most easily detected and identified exhibit conspicuous and recognizable characteristics such as size, shape, color, form, shadow or flowering parts. The plants not readily detectable or identifiable are either small, appear similar to other associated plants, or do not possess recognizable attributes such as flowers or unique foliage color at the time the photographs are taken.
2) An absolute plant count cannot be made, even for the most readily recognized species, because small and juvenile plants cannot be seen. Furthermore, plants may be hidden by adjacent plants or their shadows.
3) Plant count errors occur for the following reasons:
   a. Small plants in close proximity to mature plants will become ingrown and not recognized as separate individuals
   b. Small plants often establish themselves beneath the canopy of large plants and are not seen;
   c. Very small plants cannot be detected (see Table 35-12);
   d. Plants with similar characteristics cannot be identified accurately; and
   e. Small plants may be obscured by the shadows of adjacent larger plants.
4) Photo interpreters can detect the relative direction of changes in plant numbers, although the magnitude may not always be correctly determined. Careful photo comparison of a plant community should reveal a decline or increase in the number of plants even though the species may not be identifiable.
5) Photo-estimates of foliar cover are valid for those species, such as shrubs, that can be detected and identified with reasonable accuracy. Photo estimates for shrub canopy cover compare favorably with ground estimates because small and juvenile plants that cannot be detected do not greatly affect the total estimate of cover (see Table 35-12). Large-scale aerial photographs do not yield good estimates of foliar cover for graminoid plants and their use for this purpose is not recommended.
6) Changes in canopy cover of larger plants may be indicative of changes within the community, whereas changes in numbers of smaller plants may also indicate a change within the plant community (see Table 35-12).
7) Procedures for using aerial photographs for monitoring rangeland changes would be improperly applied if they did not provide for field mapping, verification and sampling when the aerial photographs are first acquired. Intensive data collection would be reduced for subsequent photo missions because most of the change analysis could be performed by photo interpretation and photo measurement. Field surveys would be required when unusual or inexplicable changes were detected by photo interpretation.

Although the monitoring techniques just described have emphasized the detection and measurement of changes in plant composition, plant numbers, and canopy cover, they apply to other surface conditions such as factors contributing to surface roughness, presence, size and distribution of stones, and disturbances by livestock trampling, frost heaving, and rodents. A qualitative evaluation of soil erosion can be made by observing the presence of rills and gullies, change in texture or roughness of soil crust, change in surface color, and exposure of surface rocks (Tueller and Booth, 1975). Certain types of rodent populations can be monitored by the detection of burrows, casts, and diggings (piles) of subsurface soil material.

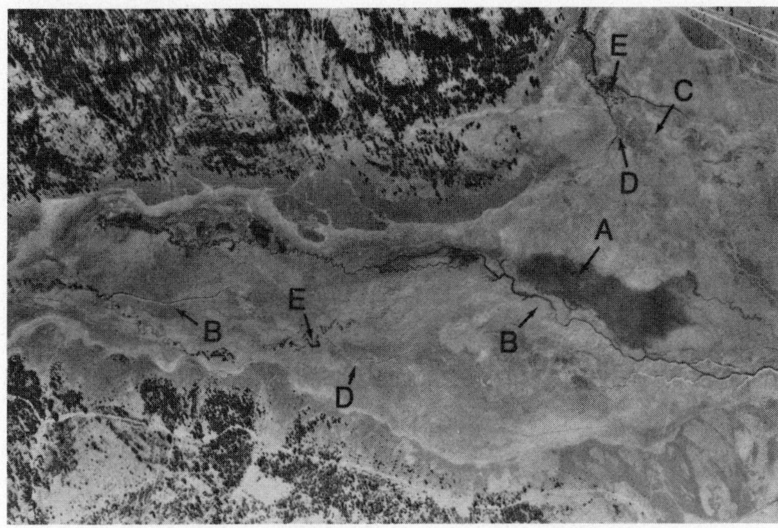

September, 1944

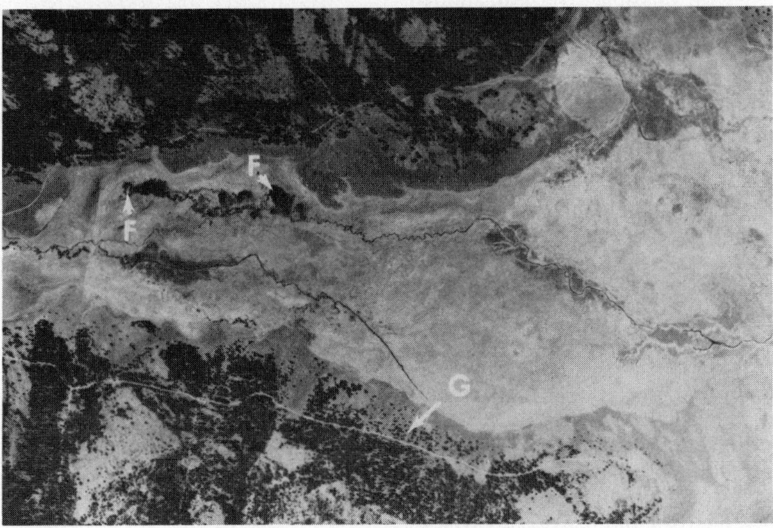

September, 1968

Fig. 35-24. Conventional panchromatic aerial photographs taken 24 years apart show many important changes have taken place in this rangeland and forestland environment in the Sierra Nevada, California. The photo scale of the original photos was 1:15,840. For a description of areas designated by letters, see text. North is towards top of photos. (Photos from Forestry Remote Sensing Program, Univ. of California, Berkeley.)

Medium to Small-Scale Aerial Photographs for Monitoring Range Conditions

Medium to small-scale panchromatic aerial photographs of the type that were acquired several decades ago for making topographic maps and early resource inventories provide a record of past rangeland conditions. Although photo quality and photo scale may not be optimal for analysis of plant-community composition and structure, the photographs show major changes in plant communities and physical landscape attributes that are indicators of long term changes in range condition and past management practices. The value of older photographs is not in the direct determination of ecological range condition, but rather the identification of plant communities in which there has been substantial change over time.

A comparative study of conventional panchromatic photographs taken in 1944 and 1968 in the Sierra Nevadas, California, demonstrates the role of aerial photographs in identifying rangeland areas where there has been a significant modification of range conditions (Carneggie, 1968).

Figure 35-24 shows a portion of the 1944 and 1968 panchromatic aerial photographs with annotations depicting areas of conspicuous change. The photo interpretations that were verified by ground data collected in 1968 include: a) a wet meadow has been converted to a more xeric grassland type at Area A; b) deep gully erosion and a change in stream channels can be observed at Area D; c) big sagebrush communities have become established along stream courses at Area B. (These were not present in 1944; big sagebrush communities have expanded in area and density at Area C); d) riparian vegetation present in 1944 is absent in 1968 at Area E; e) timber species have encroached on the rangeland at Area G and have increased in number and size; and f) the number and location of springs at Area F seems unchanged.

Because of the lack of historical data the factors operating within the valley during the 24 years are difficult to reconstruct; however, it appears that the regrowth of timber species has resulted in reduced surface and subsurface runoff of available water to the valley. The reduction in the water table, combined with possible overutilization of the forage, has resulted in conditions favorable to the encroachment of sagebrush, an undesirable species for livestock. The overall effect may be a reduction in forage production for cattle and wildlife.

Other types of interpretations that can be made from medium to small-scale aerial photographs, but not illustrated by this example include: detecting changes in land use; detecting new man-made developments such as recreational facilities, roads, fences, houses, barns, reservoirs, surface mines, energy development facilities, pipelines, and transmission lines; and detecting the impact and area of fires, insect damage, and flooding.

Although medium-scale aerial photographs show only the more conspicuous changes, they contain clues to subtle changes in range condition. Whereas current photographs of similar type, scale and season as the older photographs are desirable for the comparison of change, this is not a prerequisite. Other film types and scales may be more desirable not only to observe where changes have occurred, but for purposes of evaluating the change in greater detail. Moreover, long term changes can be detected by comparing old photographs with current conditions in the field. Finally, photo interpretation does not always provide conclusive explanations for change, but areas on the ground that warrant closer examination can be identified. Through this process, it is possible to more effectively utilize field crews for determining the causes of the changes.

Landsat Data for Monitoring Range Conditions

Hacker (1980), reports that "of all the potential applications of Landsat data in Australia rangelands, long term monitoring of ecological range conditions is considered the most important." Such monitoring implies a need to discriminate between changes in soil and vegetation that are induced by management and those that are attributable to changes in seasonal conditions.

Unfortunately, there has been too little research to determine the utility of Landsat data for this application. It is clear, though, that Landsat data cannot be used to identify species composition, despite the usefulness of such data for land-cover classification.

Graetz et al (1976) compared Landsat imagery from two dates to locate pastures and areas around watering holes where vegetation density was substantially reduced due to heavy grazing. Deteriorated rangeland could be detected by the higher reflectance from soils where vegetation cover had been reduced around watering holes and along paddock boundaries.

Tupper (1981) performed digital manipulations of Landsat data in an effort to determine the utility of Landsat to assess range conditions in semi-arid woodlands in New South Wales, Australia. Tupper computed two vegetation indices including the MSS 7/MSS 5 ratio and found relatively high correlations (0.80 to 0.84) with several shrub-tree biomass classes. Tupper concluded, however, that it would be difficult to use Landsat data to monitor changes in range condition in semi-arid woodlands because the surface soil has a major influence on reflectance and hence on the correlation with biomass. Tupper did feel that Landsat data could be used to monitor shrub infestations.

Several authors have experimented with surface albedo measurements derived from Landsat MSS data to quantify surface changes due to overgrazing (Otterman and Fraser, 1976; Robinove, 1981; Robinove et al, 1981). Robinove et al. (1981) calculated the albedo value for each Landsat picture element to create an albedo image. By the registering of albedo images from different Landsat scenes, areas of increased or decreased albedo were identified. These areas were found to be correlated with changes in soil moisture, erosional deposition following flooding, and changes in the density of perennial and annual plants. Robinove et al. (1981) recommended the experimental use of this technique for monitoring terrain changes in arid and semi-arid regions.

MONITORING SEASONAL RANGE CONDITIONS

The monitoring of seasonal range conditions includes the determination of plant development, forage quality and quantity, location of available forage, degree of forage utilization, and other ephemeral conditions. The range manager uses this information for determining the time and numbers of livestock and wildlife that can graze without damaging plants, determining the nutritive value of the plants, and assessing the relationship of plant development on forage production, animal movement, and animal behavior.

Traditionally, range and ranch managers monitor their range from horseback or terrain vehicles, making qualitative observations or measurements from randomly located step-transects. Periodic observations about plant development, forage abundance and utilization relative to the present climatic conditions would be compared with observations about seasonal conditions experienced from previous years. Management decisions are generally based upon the manager's long standing experience with the range and its capacity to produce forage under a variety of climatic conditions.

Within the last decade, several remote sensing studies have been conducted to evaluate seasonal range conditions. Central to these studies is the requirement to detect stages of plant development and the presence of green biomass, and to estimate forage production from remotely-sensed data. This requirement has been partially successful for grasslands, where cover and biomass are relatively high and where there is a high correlation between plant canopy reflectance and green biomass. The use of these techniques is limited to appropriate rangelands.

Remote sensing studies to estimate biomass can be grouped into four categories: those conducted on the ground to establish correlations between spectral reflectance and biomass; those made by aerial observers; those using aerial photographs or multispectral data; and those using satellite (Landsat) data to establish correlations between brightness values from the remotely-sensed data and ground measurements of biomass.

Ground Measurements of Biomass

Investigators in Colorado used a hand-held radiometer to measure canopy radiance or reflectance of undisturbed short grass prairie. They found these measurements to be highly correlated with biomass from clipped plots (Pearson et al, 1976). The wavelengths recorded, 0.68 and 0.80 micrometers, were judged optimal based upon previous studies by Tucker et al. (1975), and Colwell (1974). The physiological explanation for the strong, statistically significant and inverse relationship between radiance at 0.68 micrometers and biomass results from chlorophyll absorption. However, the statistically significant direct relationship between radiance at 0.80 micrometers and the amount of biomass results from the absence of chlorophyll absorption.

Aerial Observations of Seasonal Range Conditions

Aerial observation techniques designed for animal census (Norton-Griffiths, 1975; Stelfox and Kufwafwa, 1978; Mbugua and Stevens, 1977; and Gwynne and Croze, 1975) were used to collect data about seasonal variations in range conditions by personnel of the Kenya Rangeland Ecological Monitoring Unit (KREMU, Kenya Ministry of Tourism and Wildlife). Data about plant phenology (greenness), plant physiognomy (cover, height), surface water, fire, soil, cultivation and human settlement were recorded by an aerial observer from an altitude of 300 feet above ground. Environmental data were collected along transects through pre-gridded 5 × 5 km sub-units. Observations were obtained from within a 400 × 400 meter area that was representative of the larger sub-unit. Herbaceous plants were judged on foliar cover (classes included less than 2%, 2–20%, 21–40%, 41–80%, and greater than 80%), height (classes included less than 0.5m, 0.5–1.0m, and greater than 1.0m), and greenness (classes included dry, dry except for green patches, continuous green, and lush green). Low and tall shrubs were judged on cover (same classes as for herbaceous plants) and phenology (classes included no leaves, 1–25% in leaf, 26–50% in leaf, 51–75% in leaf, and 76–100% in leaf).

Some preliminary conclusions by Stelfox and Kufwafwa, (1978) based upon a comparison of observations from two seasons, include: a) plant cover and height were noticeably greater during the lush growing period than in the non-growing dry period; b) observers could detect the removal of herbaceous forage between seasons; c) there was considerable variation in values recorded for permanent attributes, which should not have changed. This was attributed to the heterogeneity of the sub-unit and the difficulty of making sequential observations from the same location within a sub-unit; and d) where observations for permanent features were similar between seasons, they reflected the homogeneity of the conditions within the sub-unit.

Whereas the data collected may not be as quantitative as required for some monitoring purposes, these data were judged adequate for understanding the distribution and movement of wild herbivores over large grazing districts within Kenya. Such methods, together with oblique or vertical aerial photographs, may be useful for assessing seasonal conditions for those rangelands in which there is little existing information about short term vegetation changes and where there are too few scientists and limited budgets to collect these data over large areas.

Aerial Remote Sensing Data for Biomass Estimation

Benson et al, (1973) found that herbage yield on mid- and short-grass prairie in southwestern South Dakota was significantly and highly correlated with brightness values measured from color infrared and panchromatic aerial photographs. However, sun angle and vignetting were factors that limited the use of aerial photographs for estimating herbage yield on a broad regional scale. Driscoll and Francis (1971) measured the amount of light transmitted through color infrared film for short grass pastures in Colorado that had been given different fertilizer treatment. They found that there was a high correlation between film density and forage production estimates made on the ground for the various pastures. In another study of short grass prairie vegetation in Colorado, Pearson et al. (1976) used aerial multispectral scanner data to produce a biomass map showing yield classes of 400, 250, 100, and 50 grams per meter square, and bare soil. There was high correlation (0.98) between the scanner data and ground biomass samples. Each of these studies required substantial ground sampling to calibrate the aerial data with ground data. Further research is needed to determine how far these results can be extrapolated to other rangeland environments.

Landsat Data for Seasonal Monitoring

Manual and digital analysis techniques have been used with Landsat data to estimate biomass and assess plant development. Manual analysis of false color composite Landsat images has resulted in qualitative determinations of the presence or absence and location of green biomass, and the relative amount of standing green biomass.

Several authors have used Landsat imagery acquired at different dates during the growth season to monitor phenological development of forage and assess the seasonal condition of the range, (Carneggie and DeGloria, 1972; Dethier et al, 1973; Rouse et al, 1974; Carneggie, 1975; and Woodzick and Maxwell, 1977). One example of such work appears in Color Figure 35-25.

Carneggie (1975) proposed that an interpreter could predict the relative seasonal conditions of a rangeland during any given year by comparing a Landsat image taken that year during the growing season with a Landsat scene of corresponding date showing the appearance of the rangeland in a year having average or normal conditions. The basis for making such predictions involves selecting Landsat scenes that show, for a normal season, the typical appearance of the range at the outset of plant growth, at peak green foliage development, and at the termination of the growth period. Deviations in appearance, as interpreted from Landsat imagery, would indicate whether seasonal conditions were normal, below normal or above normal.

Gwynne (1977) measured the optical density of Landsat images of Kenya rangelands to show the relative differences in green biomass throughout the region at a given date, and used these measurements to show relative differences in standing green biomass at different times during the year and between years.

Building upon the relationships between spectral reflectance and biomass reported by Pearson et al (1976), Tucker et al (1975), and Colwell (1974), several authors have worked with Landsat digital brightness values from the red and near-infrared bands to form ratios and mathematical indices that are highly correlated with biomass.

Carneggie and DeGloria (1974) and Carneggie (1975) plotted changes in the band ratio MSS 7/MSS 5 for two sites in the California annual grassland, during the 1972-73 growing period. Measurements of dry green biomass were obtained simultaneously with the Landsat passes. The authors showed that the band ratios from uncorrected Landsat MSS bands 7 and 5 changed in response to increases and decreases in the amount of green plant material throughout the growth season. Differences between the band ratios of the two sites were explained by the relative difference in biomass production at each site. The band ratios also were indicators of changes associated with the initiation of green foliage production, the peak period of biomass greenness and the period of foliage senescence. Other authors have reported high correlations between biomass and MSS 7/MSS 5 (Maxwell and Johnson, 1974; Tappan, 1978) and MSS 6/MSS 5 (Tappan, 1978; McDaniel and Haas, 1982; and Jaques, 1982).

Deering et al (1975), Rouse et al (1974), Haas et al (1975), and Tappan (1978) examined the relationship between Landsat data, corrected for haze and sun angle, and biomass and vegetation cover in the Great Plains prairie grasslands. Deering et al (1975) and Rouse et al (1974) developed mathematical indices using Landsat MSS bands 7 and 5, and 6 and 5, which they referred to as Transformed Vegetation Index (TVI) and Transformed Vegetation Index 6 (TVI6) respectively.

$$\mathrm{TVI} = \sqrt{\frac{(\mathrm{MSS}\ 7 - \mathrm{MSS}\ 5)}{(\mathrm{MSS}\ 7 + \mathrm{MSS}\ 5)} + .05} \qquad (35\text{-}2)$$

$$\mathrm{TVI6} = \sqrt{\frac{(\mathrm{MSS}\ 6 - \mathrm{MSS}\ 5)}{(\mathrm{MSS}\ 6 + \mathrm{MSS}\ 5)} + .05} \qquad (35\text{-}3)$$

Both indices were highly correlated to green biomass, but the TVI6 index was the more sensitive to biomass changes. McDaniel and Haas (1982), also found TVI6 to be a more sensitive indicator of green vegetation cover, green biomass, and vegetation moisture relative to the ratio MSS 6/MSS 5.

Tappan (1978) compared optical density measurements from Landsat MSS bands 5, 6, and 7, acquired throughout the 1980 growing season, with biomass and vegetation cover. Tappan compared several indices, including MSS 7/MSS 5, MSS 6/MSS 5, TVI, and TVI6, using the Landsat optical density measurements. Tappan concluded that the ratio MSS 6/MSS 5 produced the highest correlations with biomass and vegetation cover.

Hacker (1980) reviewed the literature regarding the use of band ratios, mathematical indices such as the TVI, regression relationships to predict biomass levels, and multivariate analysis of Landsat digital data for predicting biomass and enhancing the detection of green biomass. He concluded that these techniques have operational potential within appropriate rangeland environments.

The appropriate environments for applying these biomass-estimating techniques still need to be defined by further research. Most of the studies reported were conducted in grasslands with relatively high biomass and vegetation cover. The literature indicates that Landsat MSS ratios are most useful in grasslands where biomass ranges between 500 and 4000 kg. per ha. (Pearson et al, 1975; Jaques, 1982). Lower values are not useful, primarily due to overwhelming reflectance from soil. For values greater than 4000 kg/ha, reflectance response is no longer linear. The extent to which these indices are valid in shrubland or wooded rangeland environments is questionable. Further examination and research are required.

WILDLIFE HABITAT ASSESSMENT

The assessment of wildlife habitat includes the inventory of vegetation and animal populations, the monitoring of habitat changes, and the measurement of habitat parameters including the diveristy, interspersion, and juxtaposition of habitat types. These data are needed to determine if the food, water, cover and breeding requirements of a wildlife species can be met (Giles, 1978). Special challenges are presented for collecting these data because the area requirements for many species overlap, and because wildlife species differ greatly in their mobility, the size or range of their territory, and in the mixture of environments comprising their habitat.

Remote sensing techniques, combined with ground sampling and habitat modeling, are gaining wide acceptance for collecting and analyzing data to assess wildlife habitat. The remote sensing techniques available for assessing wildlife habitat are similar to those discussed for inventory and monitoring of rangelands, and for animal censusing (discussed in the Animal Census section of this chapter).

An extensive collection of references on remote sensing applications for wildlife-habitat assessment and animal census appears in the Proceedings of the Pecora IV Symposium: Application of remote sensing data to Wildlife Management (1978). Anderson et al (1980) present a general discussion of the use of remote sensing in wildlife-habitat assessment in the Wildlife Management Techniques Manual (Schemnitz, 1980). The reader is referred to Carneggie et al (1980) for a bibliography of remote sensing applications in wildlife management.

Wildlife-habitat assessment is generally considered at two levels of analysis: site specific and regional. Site-specific studies are usually conducted in a small area for which the analysis of aerial photographs is appropriate. Regional assessments, covering large areas, are more appropriately made from a combination of satellite images and aerial photographs.

HABITAT ASSESSMENT WITH AERIAL PHOTOGRAPHY

Panchromatic aerial photographs of medium scale (1:15,000 to 1:40,000) have long been available for land-cover mapping. More recently, there has been a trend toward increased use of color and color infrared photographs at large, medium, and small scales for assessing wildlife habitat and monitoring habitat change.

Techniques and procedures to map vegetation for habitat analysis using color and color infrared photographs are presented by Merchant and Waddell (1974); Scheierl and Meyer (1976); Myer (1977); and Treadwell and Mouat (1977). The use of color and color infrared photographs for mapping and assessing wetland habitats are presented by Anderson and Wobber (1973), Austin and Adams (1980), Best (1978), Best and Moore (1977), Cowardin and Myers (1973), Seher and Tueller (1973), and Bonner (1981). Montanari and Wilen (1978) discuss the photo interpretation and mapping of wetlands and riparian habitats by the U.S. Fish and Wildlife Service for the National Wetland Inventory Project. It is apparent from the literature that there are no uniform classification schemes for mapping wildlife habitats or wetlands. Some classification systems characterize the wetland condition and only indirectly refer to the associated vegetation, while other systems directly describe the vegetation occupying the site.

Large scale, small format (35mm and 70mm) aerial photographs are becoming common for assessment of critical habitats. Meyer's publications (Meyer, 1975; Meyer and Grumstrup, 1978) present both the applications and procedures for using 35mm camera systems for habitat assessment. Ulliman et al (1979) used 70mm 1:2,000 scale natural color photographs to predict habitat type based upon recognition of overstory and some understory characteristics. The accuracy afforded by photo analysis compared favorably with ground plots, and at a lower cost. Greentree and Aldrich (1978) used color and color infrared, 70mm aerial photographs at three scales (1:600; 1:1,584; and 1:6,000) to evaluate trout habitat. They found that 1:1,584 scale color photos were optimum for classifying trout-stream habitat conditions based upon the form, shape, and color of instream and streambank features. Hertz (1978) enlarged 35mm color infrared photographs (1:2,000 and 1:3,958) to assess mongoose and blue-face booby habitat on Leduc in the U.S. Virgin Islands. Strong (1980) made microdensitometer measurements from large scale (1:1,200) color infrared photographs obtained in late summer to estimate phytomass production of four sagebrush (*Artemisia tridentata*) steppe habitat types for an assessment of mule-deer winter range.

Medium to small scale (approximately 1:12,000 to 1:120,000) aerial photographs have been used more frequently than large scale (greater than 1:12,000) photographs for habitat assessment. The Pecora IV Symposium Proceedings contain over a dozen articles that report on the use of medium to small scale photographs for this purpose. Most of these articles report using color and/or color infrared photographs.

Rekas (1978) used 1:20,000 scale black-and-white aerial photographs, and integrated slope and soil information to map vegetation types and determine potential black-tailed prairie-dog habitat in Colorado. Potential habitat was derived from a composite map showing vegetation, soils and slopes favored by prairie dogs. Subsequent analysis of actual prairie dog towns (also on 1:20,000 black-and-white photos) showed that 93 percent of the existing prairie dog towns were located within the area identified as potential habitat.

Sugarbaker (1978) used black-and-white aerial photographs (1:20,000 scale, from 1971) and color infrared aerial photographs (1:15,840 and 1:68,000 scale, from 1977 and 1976, respectively) to map vegetation before and after a large fire that occurred in 1976 on the Seney National Wildlife Refuge, Michigan, and to determine its effect on wildlife habitat. Sugarbaker was able to accurately map the fire perimeter, including unburned areas within the fire boundary, and to show vegetation cover-type changes. The post-fire vegetation types were more difficult to identify, thus requiring field verification, but for the most part were an improvement of habitat for most wildlife species due to the creation of more edge and the desirable setback of successional development.

Treadwell (1978), in mapping wildlife habitat on the Three-Bar Wildlife area in central Arizona, interpreted aerial photographs acquired at three scales, namely 1:120,000 and 1:24,000 scale, color infrared, and 1:2,000 scale, 35mm color photographs. The small scale photographs were used to identify general land-resource systems. The 1:24,000 scale photographs were used to identify and delineate plant associations within desert shrub, chaparral, woodland, and forest vegetation types. The large scale, 35mm photographs were used to determine the distribution of dominant plant species used by javelina (*Pecari tajacu*), black bear (*Euarctos americanus*), and desert mule deer (*Odocoileus hemionus*). Treadwell developed a vegetation-classification system oriented to wildlife to map vegetation types and vegetation characteristics that are significant for wildlife habitat assessment. These included plant density, canopy cover, utilization, and disturbance. The final maps included information related to habitat suitability and management recommendations.

Roller (1978) discussed a technique involving the use of aerial photographs from five dates to determine changes in habitat quality over large areas. Roller digitized the vegetation cover type maps produced from each date to automate the location and display of areas that had changed. He then found that the size of the over-winter deer herd decreased as the area and numbers of seedlings and tree species increased.

Pettinger et al (1978) used small scale (1:120,000), color infrared, aerial photographs to measure habitat parameters for the Habitat Evaluation Procedures (HEP) being developed by the U.S. Fish and Wildlife Service (FWS). The purpose of the study was to determine whether or not the measurements from remotely-sensed data alone, or those from remotely-sensed data combined with field measurements, were sensitive enough to provide a measure of habitat value for elk (*Cervus elaphus*) and sage grouse (*Centrocercus urophasianus*). The study area was a proposed mine development in southeastern Idaho where habitat would be lost due to strip mining. Measurements of sixteen habitat parameters were required for the models developed to compute habitat suitability for elk and sage grouse (Table 35-13). Aerial photographs were useful for measuring eight of eleven elk-habitat parameters, but only one of six sage-grouse habitat parameters. Figure 35-26 illustrates the quantitative relationships between four habitat parameters and a Habitat Suitability Index (HSI) for elk in an evergreen-shrub habitat type. Figure 35-27 illustrates one of the vegetation cover maps used to derive habitat characteristics in the study.

Pettinger et al (1978) concluded that small-scale aerial photographs can be used to measure many of the parameters needed to determine habitat suitability for some wildlife species. The use of medium to large scale photographs would extend the list of habitat parameters that are measurable from remotely-sensed data. Many of the parameters (distance to water, distance to cover, diversity, distance to openings, and edge characteristics) are more readily measured on aerial photographs than in the field. The study concluded that an optimal combination of aerial photographic analysis and ground measurement can provide a cost-effective approach to the assessment of habitat suitability for many wildlife species.

HABITAT ASSESSMENT WITH SATELLITE IMAGERY

Landsat imagery has become widely used in wildlife habitat analyses. Landsat imagery was used in over 40 percent of the papers in the Pecora IV Symposium Proceedings on remote sensing applications for wildlife management. Satellite applications for wildlife habitat assessment use techniques developed for other resource applications (especially forest and range management) and make use of additional information and techniques.

A large number of satellite-oriented studies involve the use of aerial photographs, especially when the ground features studied are too small to be seen on Landsat data. Fraser (1980), for example, divided Landsat applications into two broad categories: simple interpretations of phenomena within large areas, and more detailed mapping within smaller areas. An example of the first category is the use of Landsat imagery to determine when snow no longer covers the nesting areas of several species of geese. An interpreter could detect unusually early snow melt or late snow cover,

TABLE 35-13

Habitat Parameters Which are Required in Models to Determine a Habitat Suitability Index for Elk and Sage Grouse.

Parameter	Method of Measurement[1]	Species: sage grouse	elk
Ground cover of forbs (%)	G	X	
Diversity of forbs (#)	G	X	
Sagebrush canopy cover (%)	G	X	
Av. sagebrush height (in.)	G	X	
Distance to strutting grounds (mi.)	G	X	
Distance to water (mi.)	A	X	X
Composition of understory	G		X
Distance to opening (ft.)	A		X
Perceived sight distance (ft.)	G		X
Av. Stand width (ft.)	A		X
Av. tree height (ft.)	A-G		X
Tree canopy closure (%)	A-G		X
Size of stand (acres)	A		X
Distance to road (mi.)	A		X
Distance to forest cover (ft.)	A		X
Herbaceous ground cover (%)	G		X

[1] A = aerial photographs; G = ground measurements.

which is related to high or low populations of geese (Reeves, 1978). In the second category, Fraser concluded that Landsat data have been less useful in an evaluation of waterfowl habitat in which ponds were too small to be detected. Kerbes (1978), used Landsat data to detect potential Ross goose habitat, viz. shallow lakes with islands. He could not always detect small islands but could differentiate shallow and deep lakes on MSS band 6. In these examples, a combination of aerial photographs and Landsat data was needed to assess geese habitat.

A number of papers have discussed techniques that use only Landsat data for wildlife-habitat assessment. Thompson et al (1980) used Landsat imagery to delineate broad vegetation patterns in a 90,000 square km area of the southern District of Keewatin, Northwest Territories, Canada, to determine their relative importance as caribou habitat. For ground truth, the relative importance of each vegetation type was determined in the field from winter and summer pellet-group counts.

Löffler and Margules (1980) used digitally enhanced Landsat imagery to search for areas denuded of vegetation. Such areas could be clues to the distribution and spread of wombats in South

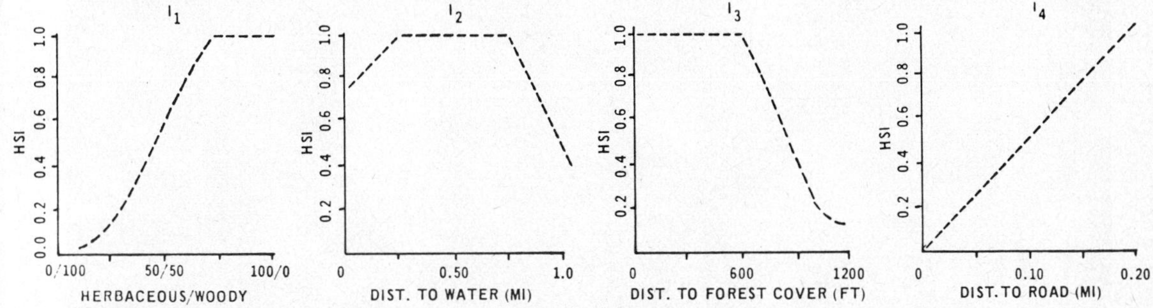

LIFE REQUISITE EQUATIONS

FOOD VALUE (X_1) = I_1

WATER VALUE (X_2) = I_2

COVER VALUE (X_3) = I_3

SPECIAL CONSIDERATIONS VALUE (X_4) = I_4

Fig. 35-26. Quantitative relationships of habitat characteristics to Habitat Suitability Index (HSI) for elk in evergreen scrub (Pettinger et al, 1978). These relationships were developed by the U.S. Fish and Wildlife Service, Ft. Collins, Colorado, for their Habitat Evaluation Procedures (HEP).

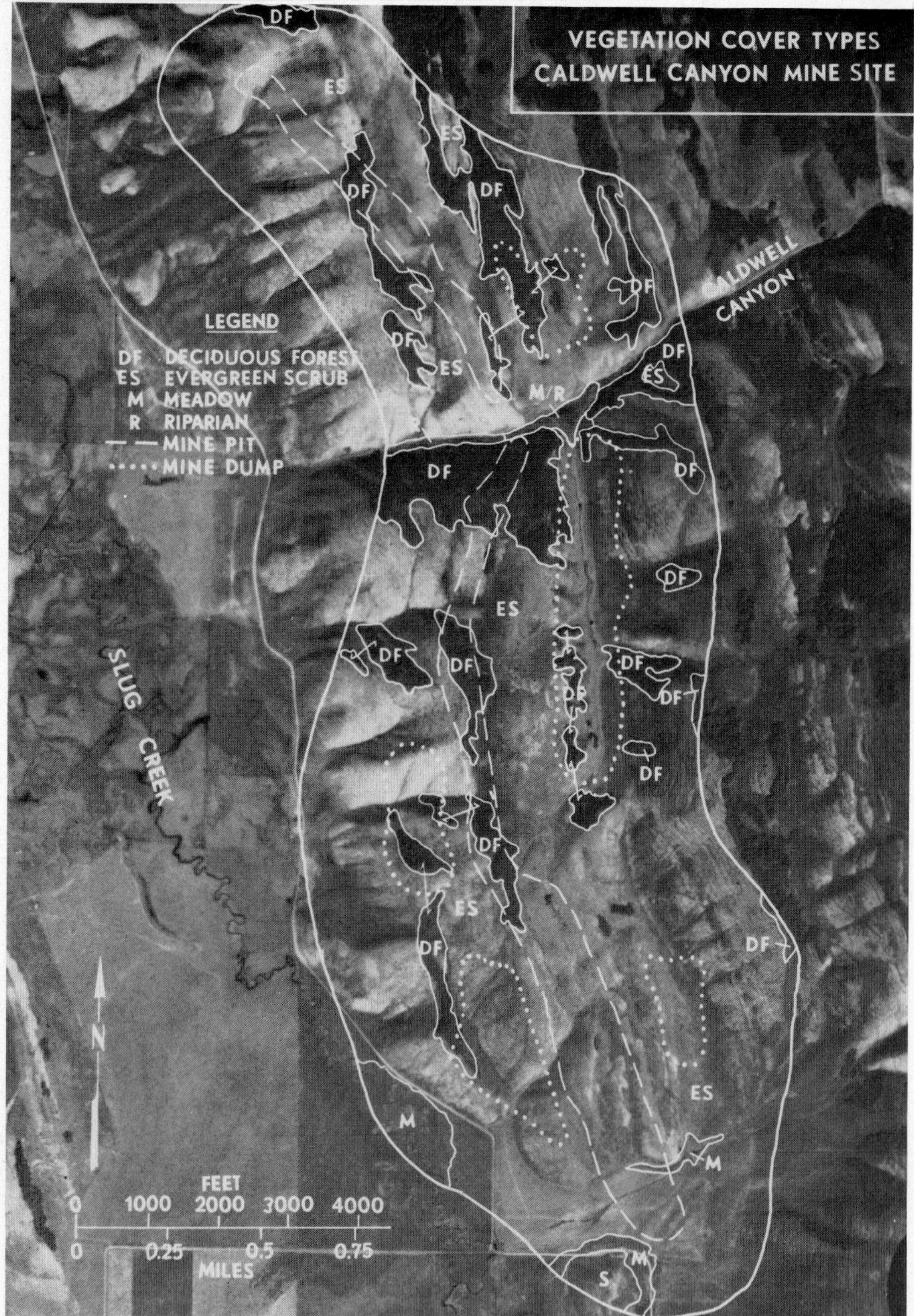

Fig. 35-27. Black-and-white copy of enlarged high altitude color infrared aerial photograph showing vegetation cover map of a proposed phosphate strip mine. This map and these photographs were used in the process of deriving habitat characteristics in the study described by Pettinger et al, 1978.

Australia. They found that wombats were largely responsible for denuding large areas of the Nullarbor Plain.

Best and Sather-Blair (1978) used Landsat imagery to monitor the availability and spatial distribution of winter wildlife habitat in eastern South Dakota. They mapped six classes of winter habitat with an overall classification accuracy of 76.6 percent. Only those classes that had sufficient vegetation density to be apparent on the Landsat imagery and were not covered with snow could be considered to provide winter habitat.

Gilmer et al (1978) used Landsat data to evaluate waterfowl habitat in the prairie pothole region of southcentral Canada and northcentral United States. Landsat data were used to assess pond size, crop type, land cover, and water characteristics. Multidate Landsat imagery greatly increased the capability to discriminate the habitat parameters.

Hill and Falconer (1978), used Landsat data to map six broad habitat zones for the grey kangaroo in southern Queensland, Australia. Landsat data also were useful for monitoring the seasonal status of forage used by kangaroo on cleared rangeland. Together these data were useful for developing a kangaroo-management plan.

Mayer and Fox (1980) used digital Landsat data to map timber types on the McCloud Ranger District, U.S. Forest Service, in northern California. They subsequently determined the wildlife habitat values of each timber type and were able to regroup the original timber type classes to form new habitat maps for the District.

Another remote sensing technique for wildlife-habitat assessment involves the use of Landsat data in digital and image format together with high resolution aerial photographs. This allows for large area analysis in an efficient and cost-effective manner and provides sufficient detail for site-specific management.

Olson (1978) used Landsat data, three scales of color and color infrared aerial photographs, and ground survey to assess estuarine conditions and fish habitat in Queensland, Australia. He found that the Landsat imagery was most useful for providing a regional overview of the 3,250 miles of coastline studied in the project. It also allowed for the delineation of algal and mangrove growth. The aerial photographs were used to analyze vegetation structure and classification, terrain classification, foliage cover, and tidal scour.

Merchant and Waddell (1974) used small scale (1:127,000) color infrared aerial photographs and Landsat imagery to develop a data base for a wildlife habitat inventory in Jefferson County, Kansas. They mapped cover types, measured perimeters and interspersions, and estimated acreages for each cover type for general wildlife management using the aerial photographs. They used Landsat data for mapping broad classes of habitat on a regional scale and for monitoring change.

Isaacson and Leckenby (1981) and Isaacson et al (1982), in a study of Rocky Mountain elk habitat in the Blue Mountains of Oregon (Color Figure 35-28), digitally integrated information from aerial photographs and Landsat digital data to produce cover maps for elk research and management. Isaacson et al (1982) applied digital analysis procedures to Landsat MSS data to produce approximately 60 spectral classes, which were subsequently grouped into general resource cover classes. The authors, using an existing classification scheme, also mapped vegetation types for corresponding areas on color infrared and color photographs (scale of 1:6,000). This cover map was transformed into grid-cell data and registered to the spectral class data derived from Landsat MSS data at a common map scale of 1:24,000. Co-occurrence matrices were developed to show the correspondence between the photo data and the MSS data (Table 35-6). This point-to-point registration provides a means for accurately describing spectral classes and grouping them into appropriate resource classes. The final products from this analysis procedure include a file containing the ungrouped spectral classes with descriptors provided from the cover map produced from aerial photographs, and a generalized habitat map (showing forage, thermal protection and hiding cover potential) based upon logical groupings of spectral classes. The data in digital format can be retrieved and manipulated to produce additional habitat maps based upon different groupings of spectral classes that satisfy management requirements. Figure 35-29 diagrams the analysis procedures used in this habitat inventory.

Roller (1978) developed a model that generates numerical ratings of wildlife habitat quality based upon the analysis of digital remote sensing data. The model integrates attributes of the habitat whose presence and condition are known to affect the survival and welfare of wildlife. The attributes include vegetation and terrain types that affect the food- and cover-requirements of wildlife, and that provide measures of interspersion (spatial distribution of the cover types) and juxtaposition (accessibility of the cover types). Land-cover classifications produced from digital Landsat data and digitized cover-type maps from small scale aerial photographs are primary sources of input data. The general form of the model follows:

$$\mathrm{WHQ} = k_1 \cdot \mathrm{FC}\,[(k_2 \cdot \mathrm{INT}) + (k_3 \cdot \mathrm{JUX})] \tag{35-4}$$

where,

WHQ = Wildlife Habitat Quality
FC = Food and Cover Rating
INT = Interspersion of Cover Types
JUX = Juxtaposition of Cover Types
k_1, k_2, k_3 = Specific calibration coefficients that eliminate any redundancy between variables

The Food and Cover Rating (FC) is derived from the following equation:

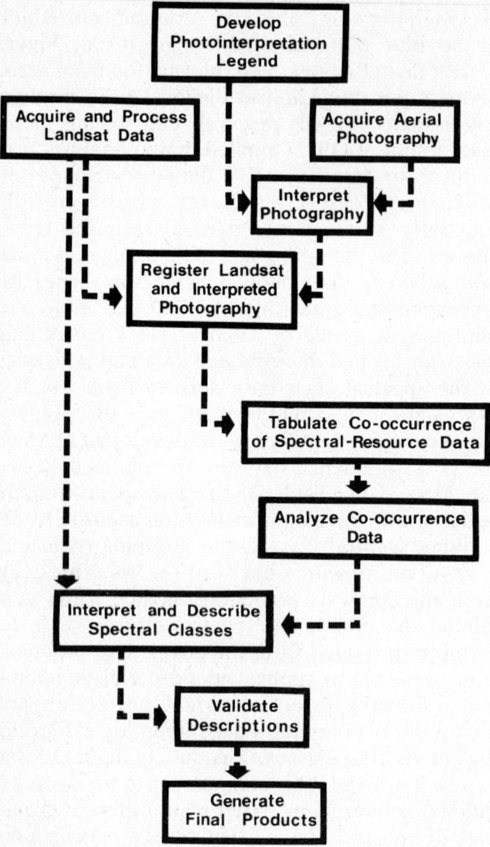

Fig. 35-29. A generalized diagram of inventory procedures that include a systematic and quantitative evaluation of the interrelationships among spectral classes and resource classes (Isaacson et al, 1982). These procedures were used in a study to map cover and habitat for elk in the Blue Mountains of Oregon.

$$FC = \sum_{i=1}^{N} \left[\frac{(T_i/A)}{O_i} \cdot \text{Max score for } T_i \right] \tag{35-5}$$

where,

T_i = area (in m^2) of a given cover type
A = area of the habitat unit under consideration (in m^2)
O_i = optimal relative abundance of a given cover type
N = number of cover types occurring in the habitat unit

Land-cover mapping using panchromatic and color infrared aerial photographs is still the most common application involving remotely-sensed data for wildlife-habitat assessment (Carneggie and Marmelstein, 1978). Indeed, aerial photographs will, in all likelihood, be used more frequently in the future to measure parameters required to determine wildlife habitat suitability. Yet, not all parameters relevant to habitat suitability can be measured or interpreted from aerial photographs. For those that can, however, the use of aerial photographs typically provides a more accurate and efficient method than ground measurement. The use of small format aerial photographs for habitat-suitability studies continues to increase as does the use of Landsat data. The use of Landsat data, in combination with aerial photographs, is also increasing as managers discover that Landsat's synoptic and repetitive coverage is extremely important for analyzing and monitoring habitats over large areas. Finally, resource managers are also finding that integration of aerial photographs with Landsat digital data allows for more precise definition of Landsat-derived resource classes and the production of cover maps whose detail and spatial relationships are ideally suited for assessing wildlife-habitat values in large management areas (such procedures are being used operationally in Alaska to collect data for comprehensive planning on all National Wildlife Refuges).

ANIMAL CENSUS

The inventory of animal populations (kind, number, location, season of use, condition, and sex) and determination of population trends are as important to range-resource management as are data on vegetation type and condition. One of the basic objectives for range and wildlife management is to balance an animal's needs for food and cover with the physiological and ecological requirements of the plants so that plant health and productivity can be maintained at a high level.

Good estimates of wild and domestic animal populations are required to assure that stocking rates are compatible with the habitat's capacity to produce forage. Most of the current techniques used to study wild animal populations have been reviewed by Schemnitz (1980), in the Wildlife Management Techniques Manual. Most involve direct counts of animals, animal signs and related objects, marked animals or, in the case of domestic animals, the use of enumerative procedures.

Many procedures for animal inventory are ground based as well as time consuming and tedious. Often estimates of animal numbers are in error due to the difficulty of finding animals, the difficulty of counting large groups of animals, and the difficulty of determining the proportion of the population that is actually observed (Watson and Scott, 1956; Siniff and Skoog, 1964; Thomas, 1967; and Pennycuick, 1969).

During the last two decades there has been considerable interest and research to develop aerial surveillance techniques for counting wildlife populations. Some of the more common aerial approaches for counting and studying animal populations include: aerial visual observations, use of relatively large-scale aerial photographs, use of thermal infrared scanning systems, and the use of radio telemetry. Details of these techniques follow in subsequent sections.

AERIAL VISUAL OBSERVATIONS

The most common aerial approach for animal census is the visual counting of various game and waterfowl species from aircraft. The most reliable counts occur when: (a) the aircraft is following prescribed flight lines at an appropriate altitude allowing observers to spot the animals; (b) the animals are not obscured by vegetation; (c) there is good contrast between the animal and its background; and (d) the observers are experienced in counting animals from the air.

The U.S. Fish and Wildlife Service uses aerial visual counts to estimate waterfowl numbers along North America's Central Flyway, (Henny et al, 1972), The Kenya Rangeland Ecological Monitoring Unit has developed an elaborate sampling design using straight-line belt transects to count both wild herbivores and four types of domestic stock (Stelfox and Kufwafwa, 1978). (See Color Figure 35-30).

Aerial visual census is not without its problems. Evans et al (1966) reported that forest cover caused a serious visibility bias in an aerial census of moose. The background against which animals are observed is also critical. Gilbert and Grieb (1957) and Evans et al (1966) concluded that a snow cover increased the accuracy of aerial big game counts and in some cases was essential. Bear (1970) also cited the importance of background on the accuracy of antelope counts. Other problems associated with aerial visual census include light conditions, the animal behavior-pattern as a response to time of day, the determination of a sample area, and observer eye-fatigue.

Gill (1969) developed a sampling technique for estimating live animal numbers by visual observation from a helicopter; however, mortality estimates were made on the ground. His procedure for estimating mortality was to stratify the winter range (deer) into three areas: high, medium, and low over-winter mortality, based on historical record. Within each stratum, transects were located (1.6km by 0.4km), and the number of deer carcasses in each transect was determined by ground search.

AERIAL PHOTOGRAPHS FOR ANIMAL CENSUS

Despite the superiority of a photo-census, aerial photographs have been used less than visual observation because of costs and equipment requirements. Siniff and Skoog (1964) cited the difficulty of locating animals on photographs where there is heavy vegetative cover or a low contrast background. Grzimek and Grzimek (1960) rejected photographs because of the very large number of exposures that would have been required to cover the area they studied in Tanzania. Thomas (1967) mentioned the importance of a snow background for air photo counts of caribou. Colwell (1964) discussed the importance of time-of-day for photographing animals out in the open where they can be seen.

The main value of aerial photography is the instantaneous recording of groups of animals that can be examined more closely and counted at a later time. The photographs effectively freeze animal movement to facilitate the counting of animals in groups. Heyland (1978), lists several recent studies whereby vertical and oblique aerial photographs have been used for wildlife census:

Flamingoes (*Phenoenicopterus ruben*)—(Grzimek and Grzimek, 1960; Bartholemew and Pennycuick, 1973)
Pelicans (*Pelicanus onocrotalus* and *P. rufescens*)—(Bartholomew and Pennycuick, 1973)
Penguins (*Aptenodytes patagonica* and *Eudyotes chrysolophus*)—(Bauer, 1963)
Sandhill Cranes (*Grus canadensis canadensis*)—(Leonard and Fish, 1974)
Waterfowl—(Heyland, 1972 and 1973)
Seals—(Sergeant, 1965; Mansfield, 1970; Vaughan, 1971; Lavigne, 1976; and Lavigne and Ronald, 1975)
Caribou (*Rangifer tarandas*)—(Brassard and Potvin, 1973)
Elephants (*Loxodonta africana*)—(Croze, 1972)
Wildebeest (*Connochaetes taurinus albojubatus*)—(Norton-Griffiths, 1973)
Pocket gophers (*Thomomys talpoides*)—(Driscoll and Watson, 1974)
Beluga Whales (*Delphinapterus leucas*)—(Heyland, 1974)
Humpback Whales (*Megaptera novaeongliae*)—(Scott and Winn, 1978)
Red Salmon (Eicher, 1953)
Deer Carcasses (Driscoll, 1971)

Techniques for using aerial photographs to inventory livestock have been developed by Huddleston and Roberts (1968), Perkins (1971), Dudzinski and Arnold (1967), Colwell (1964), and Crofton (1958). Specifications for using aerial photographs to inventory livestock can also be used to census large wild animals; hence, those used by the U.S. Department of Agriculture's Statistical Reporting Service are presented here to demonstrate aerial photographic techniques for animal census.

Either panchromatic black-and-white film or color transparency film are recommended for animal census. While panchromatic film is less expensive, color films have an advantage for identifying animal type and breed. Panchromatic film, when exposed with a red filter, may increase contrast between animal and background and emphasize the animal shadow, but dark shadows cast by trees or other features can obscure animals standing within the shaded areas.

The photo-scale is very important to the success or failure of the animal census. A photo-scale of 1:8,000 is usually satisfactory for detecting larger animals such as cattle, moose and elk, but a minimum scale of 1:5,000 is needed for making consistently accurate census of these larger animals. Larger scale photographs are required to achieve acceptable counting accuracy for smaller animals, such as sheep, deer, and antelope. A photoscale of 1:3000 is minimum for this purpose. A disadvantage of such large scales is the number of photographs required for a large aerial census.

Tests indicate that animals are more accurately counted and identified on stereo photographs, provided that animal movement between exposures is minimal. Animals are often identified by their appearance on one photo when they could not be identified on the conjugate stereo image. Animal movement is sometimes an advantage for verifying the presence of an animal. Where stereo photographs are not required, the number of photographs and photographic handling requirements and costs are lessened.

The season during which to conduct animal census using photographs should be dictated by animal behavior, migratory patterns and the phenology of the vegetation comprising the habitats. For aerial photo census of livestock, there are two seasonal states: dry vegetation and green vegetation. Green vegetation appears dark and uniform on panchromatic film and provides a good background for counting light-colored (or toned) animals such as sheep and an adequate contrast for the identification and counting of cattle. Dry vegetation appears light in tone on panchromatic film and provides a poor contrast for identifying light-colored animals. The optimum time of year to secure photographs for livestock census is during the growing season of the range or pasture vegetation.

Photo-census of many large wild ungulate species is not desirable during the plant growth season. Deer and elk, for example, are widely dispersed, or occupy mountainous areas where forest cover blocks them from view, thus requiring more photographs. Photo-census during the winter season has some advantages: the ranges have less tree cover, deciduous trees have lost their leaves, and the animals tend to concentrate in smaller areas. A snow background also provides high photo contrast and increased counting accuracy.

Many animal species are most active just after sunrise and just before sunset, seeking the shade and protection of a forest cover during the day. Others may cluster into large groups and bask in the sun during the middle of the day. Thus the optimum time of day for censusing must be based upon knowledge of animal behavior. Livestock generally seek shade during the heat of the day in summer. Therefore, aerial photographs should be taken just after sunrise and just before sunset, when the animals are most likely to be visible to the camera.

Photographs taken during the early morning and late-afternoon hours have some disadvantages. The low sun angle provides poor lighting, which requires longer exposures, often resulting in blurred or "fuzzy" images. Shadows from natural and structural features are longer and obscure a larger area in which animals may be found. Despite these disadvantages, the photographs must be taken when the animals are in the open. Moreover, animal shadows are often an aid to animal identification. The time of day is not as critical during winter months since large wild animals are normally active throughout the daylight hours.

A dual camera photo-acquisition system and census procedures were developed for livestock inventory in California (Huddleston and Roberts, 1968). A 9-inch camera with a 6-inch focal length lens and a 70mm camera with a 12-inch focal length lens simultaneously acquired panchromatic photographs and color transparencies respectively (Color Figure 35-31). The photo-scales were 1:5000 and 1:2500 respectively. The field-of-view of both cameras was centered on the same area, and the shutters were triggered simultaneously from the same intervalometer impulse. The panchromatic photographs had a forward overlap of 60 percent for good stereo-viewing, whereas the color transparencies overlapped by only 10 percent and showed a larger scale view of a sample strip paralleling the flight line. The large-scale color transparencies were used to replace ground subsampling since animals are seen clearly on these photographs. The animal counts made from the color transparencies were then compared with the counts made in the corresponding area from the panchromatic photos. This ratio was used to correct animal counts from the entire panchromatic photograph.

Counting animals on aerial photographs should follow carefully specified search techniques. One method involves superimposing a line grid on the photographs, thereby compartmentalizing the photo area into smaller blocks. The interpreter then systematically searches for animals in each of the small block areas.

When more than one animal species or breed is being counted, it may be necessary to develop interpretive keys based upon: (a) the morphological characteristics of the animal as seen on the photographs; (b) the characteristics of the habitat in which animals are most likely found; and (c) other characteristics associated with a particular population, e.g., tendency to be grouped or dispersed, and pattern assumed when startled by low flying aircraft. If it is impractical to obtain ground photographs simultaneous with the aerial photographs for making interpretive keys, a skilled interpreter should be able to identify representative examples of different animals and different habitats on the panchromatic and color aerial photographs.

Aerial census of small animals such as gophers, squirrels, and prairie dogs is difficult because of their size. However, their presence and level of activity can be inferred from aerial photographs taken of their mounds or soil disturbance. Reid et al (1966) determined that the number of late-summer earth mounds, made by northern pocket gophers in some sections of the western United States, was directly related to population density and that periodic changes in numbers of mounds reflected population fluctuations. Counts of northern pocket-gopher mounds made from large-scale (1:600) color or color infrared aerial photographs are 97 percent accurate for estimating rodent populations when adjustments are made based upon the relationship between photo counts and absolute ground counts (Watson, 1973). Since not all mounds can be identified from the photographs, photointerpreted mound counts must be normalized to actual ground counts from a few sample areas to determine the relationships between the two counts. This is done by ratioing the ground counts to photo counts. This information is then applied to the equation:

$$\hat{Y} = 0.6582 \sqrt{RM} \log (RM + 1) \quad (35\text{-}6)$$

where:

$\hat{Y}$ = estimated population density
R = normalized mound count
M = photo identified mound counts per unit area

to provide an estimate of animal density per unit area.

Interpreters favored color infrared photos for discriminating between live vegetation and non-vegetated areas. Earth mounds were not identifiable with acceptable accuracy when photo-scales smaller than 1:8000 were used.

Tietjen et al (1978) have demonstrated the use of conventional panchromatic aerial photographs (1:15,840 scale) for locating prairie-dog colonies and estimating acreage disturbed by their mounds. The intent for future study is to use aerial photographs to assess annual changes in prairie-dog colony dynamics and develop population models that are the key to the development of management plans.

AERIAL THERMAL INFRARED SCANNING FOR ANIMAL CENSUS

Whenever there is a detectable temperature difference between an animal and its background there is a potential for using thermal infrared scanners for animal census (Figure 35-32). Airborne, thermal infrared scanners, have been tested with limited success to detect white-tailed deer, *Odocoileus virginianus,* (Croon et al 1968; McCullough et al, 1968, and Graves et al, 1972); Rocky Mountain mule deer, *O. hemionus,* (Parker, 1972); polar bear, *Ursus maritimus,* (Brooks, 1970); harp seals, *Pagophibes groenlandicus,* (Lavigne and Ronald, 1975); bison, *Bison bison,* and elk, *Cervus canadensis* (Wride and Baker, 1977); moose, *Alces alces,* (Garvin et al, 1964; Wride and Baker, 1977); and boar, *Sus scrofa* (Lenco, 1976).

There are several reasons why thermal infrared scanning has met with limited success for animal census: (a) there must be an effective temperature difference between the animal and its background during the acquisition of the imagery. Parker (1972) reports that the effective radiant temperature of deer did not always exceed the effective radiant temperature of vegetation, soil or rock. A complete snow cover, a rarity in many areas of the country, substantially increases the opportunity for censusing deer from an altitude of 500 to 1000 feet (Marble, 1967; Croon et al 1968). Even so, most animals have evolved effective methods for insulating themselves and thereby minimize the temperature differential between their surface and the environment (Ray and Wartzok, 1975). Heyland (1978), however, reports that some species have a sufficiently high radiative temperature, relative to the ambient air, to permit thermal detection of the animals; for example, this appears to be the case with polar bear (Øritsland et al., 1974), and Pacific walrus (Ray and Wartzok, 1975); (b) Animal census is generally restricted to altitudes below 1000–2000 feet. Animals can be difficult to count when they are bunched in closely spaced groups (Ray and Warzok, 1975); and (c) Animals cannot be detected if obscured by vegetation. They must be in the open or visible to the sensor through the branches of deciduous shrubs or trees.

Fig. 35-32. Example of a thermal infrared image taken with a multispectral scanner during the day of two bison against a snow background. The animals most likely would be obscured if they were beneath the adjacent hardwood cover. (Imagery courtesy of M. Wride, Canada).

In summary, thermal infrared scanning is a promising census technique, but it is not a simple, practical tool to be used for all animal census work. The technique requires specialized knowledge of the sophisticated sensing equipment and of environmental factors influencing its success, including diurnal temperature variations of the animal and its background material, animal behavior and location, weather conditions, navigation and altitude of aircraft. As is frequently the case with new resource measurement methods, the intuitive simplicity of thermal scanning is not realized in practice, but research results indicate thermal scanning may prove to be a valuable technique to future wildlife management.

RADIO TELEMETRY FOR STUDYING ANIMAL POPULATIONS

Most radio telemetry systems have been used primarily for determining the behavior, movement and location of animals within their habitat; not for animal census. Many have been used to monitor the physiological responses of animals within their habitat. Most radio telemetry systems contain a small radio transmitter, attached to an animal by implantation or a collar, and a receiving antenna, which may be operated from the ground (fixed or mobile tracking station) or air (from aircraft or satellite tracking stations).

A list of animal, bird and fish species that have been studied with radio telemetry techniques appears in Table 35-14, accompanied by the author-references. This list is not exhaustive, but illustrates the growing use of radio telemetry to study animal behavior.

Radio-telemetry techniques offer many advantages over direct visual observations for monitor-

TABLE 35-14

References to Studies of Birds and Mammals Equipped with Telemetry Transmitters to Locate Them and Record Behavior (List is not Exhaustive).*

Species	Reference
Birds	
Common crow (*Corvus brachyrhynchos*)	1
Mourning dove (*Zenaida macroura*)	1
Golden eagle (*Aquila chrysaetos*)	1, 2
Sage grouse (*Centrocercus urophasianus*)	2
Sharp-tailed grouse (*Pedioecetes phasianellus*)	2
Red-tailed hawk (*Buteo jamaicensis*)	1, 13
Barn owl (*Tyto alba*)	1
Great horned owl (*Bubo virginianus*)	1, 13
Screech owl (*Otus asio*)	1
Pheasant (*Phasianus colchicus*)	7
California quail (*Lophortyx californicus*)	1
Common raven (*Corvus corax*)	1
Wild turkey (*Meleogris gallopavo*)	6
Turkey vulture (*Cathartes aura*)	1
Mammals	
Pronghorn antelope (*Antilocarpa americana*)	2
Badger (*Taxidea taxus*)	1
Black bear (*Ursus sp.*)	12
Bobcat (Lynx *rufus*)	1
Coyote (*Canis latrans*)	1
Mule deer (*Odocoileus hemionus*)	2
White-tailed deer (*Odocoileus virginianus*)	9
Black-tailed prairie dog (*Cynomys ludovicianus*)	11
Elk (*Cervus canadensis*)	4
Gray fox (*Urocyon cinereoargenteus*)	1
Monkey	5
Opossum (*Didelphis marsupialis*)	1
Raccoon (*Procyon lotor*)	1
Striped skunk (*Mephitis mephitis*)	1, 3
Fin whale (*Balaneoptera physalus*)	10

References

1—Hegdal and Gatz, 1978
2—Biggins and Pitcher, 1978
3—Verts, 1963
4—Denton, 1973
5—Anderson and deMoor, 1971
6—Porter et al, (1978)
7—Dumke and Pils, 1973
8—Heezen and Tester, 1967
9—Jacobsen and Stuart, 1978
10—Ray et al, (1978)
11—Lund, 1974
12—Craighead et al, (1971)
13—Dunstan, 1973

* Cochran (1980), presents a similar table listing references to more than 50 avian species which have been equipped with transmitters.

ing wildlife populations. A few of the major advantages, together with referenced examples, are as follows:

(a) Animals can be located within their habitat despite darkness and vegetation cover. The accuracy of locating an animal is a function of the type of tracking station, the distance of the animal from the tracking station, the number of tracking systems, and the topography. A fixed tracking station was found to be more accurate than mobile or aerial tracking antennae (Biggins and Pitcher, 1978). Methods and accuracies for establishing animal location using ground tracking systems are discussed by Heezen and Tester (1967), Anderson and deMoor (1971), Denton (1973), Biggins and Pitcher (1978), Hegdal and Gatz (1978), and Porter et al (1978). The accuracy of locating animals from aerial and satellite tracking systems is discussed by Denton (1973), and Sebesta and Lund (1978), respectively. Werber (1970), has assembled a bibliography of wildlife tracking systems;
(b) Animals can be located without relying upon a visual sighting; and
(c) A specific animal can be observed throughout an extended time interval.

Some of the disadvantages are given below:

(a) The weight and bulk of the transmitter may interfere with normal behavior. Biggins and Pitcher (1978) reported that nine months was the longest known period of survival among 55 instrumented grouse, with most mortality due to predation. For small or sensitive wildlife populations, such destructive sampling is unacceptable, forcing use of visual techniques;
(b) Battery life limits the time period of monitoring; and
(c) The cost of telemetric observations, which includes collecting the animals, attaching transmitters to them, and tracking them, may exceed the cost of conducting visual observations for some species (Biggins and Pitcher, 1978).

Radio telemetry also has been used to monitor physiological responses of animals. Craighead et al (1971) implanted transmitters in black bears to monitor temperature during hibernation. Lund (1974) and Jacobson and Stuart (1978) used telemetry to measure heart rates of black-tailed prairie dogs (*Cynomys ludovicianus*) and white tailed deer (*Odocoileus virginianus*).

The growing use of telemetry for monitoring animal movement, location, distribution, physiology and behavior can be attributed to the miniaturization of circuitry for transmitters and the increased longevity of transmitter power cells. Utilization of this technology also has grown rapidly due to the increased need to understand behavior of animal populations. Specific location and the length of time that an animal spends in specific areas of a habitat are important considerations when assessing the value of specific components of the habitat to meet the animal's food, water, and cover requirements.

DESERTIFICATION OF RANGELANDS

Desertification is the extension of typical desert landscapes and conditions to areas where they did not occur in the recent past. It is considered a form of land degradation, a condition occurring mostly in arid and semiarid regions (Le Houérou, 1975; Paylore, 1976). Desertification is a significant process for investigation owing to the large amount of land and numbers of people affected, the rate at which it occurs, and its implications for the future. It arises when fragile dryland ecosystems are subjected to excessive use by humans and animals resulting in a loss of productivity and the inability to recover. While desertification may develop from natural causes (such as climatic fluctuations), the degradation associated with human use systems is of principal interest (Reining, 1978). It is the excessive use of land during periods of drought that exacerbates the process of desertification.

Desertification manifests itself in several ways. Its major symptoms include: desolation of native vegetation, declining water tables, salinization of topsoil and water, reduction of surface waters, and high soil erosion. An area suffering from desertification can have all five symptoms, but any one of them can indicate that it is undergoing the process. Frequently, the symptoms are interconnected. For example, the destruction of native plants is quickly followed by soil erosion (Sheridan, 1981).

Desertification can be intensified through such activities of man as irrigation and dryland farming, mining, firewood gathering, settlement, and livestock grazing (Reining, 1978). However, the two most important causes of desertification in the Sahel Region of North Africa are overgrazing and wood gathering (Sheridan, 1981). It is within the rangelands having semiarid and arid climates, and not within the extremely arid deserts, that the risk of desertification is greatest (Owen, 1979).

A plan of action to combat desertification should be comprised of two basic elements: 1) monitoring and inventory; and 2) land use policy and planning (Tolba, 1979). Such a plan should also include measures for reducing grazing, especially surrounding waterholes, stabilizing sand dunes by seeding them with selected species, water harvesting, and regulating the movements of nomadic pastoralists (Owen, 1979).

Remote sensing data acquired from satellite, aircraft and ground sources can play a key role in: 1) the identification of areas subject to desertification; 2) the assessment of rangeland conditions subject to desertification; and 3) monitoring changes of land use and condition over time. Satellite imagery is an especially appropriate source for use in identifying factors that indicate the regional extent of desertification. The repetitive aspect of Landsat data is readily suited to monitoring the dynamic process of desertification (Hutchinson, 1982). The multispectral digital nature of Landsat data facilitates the detection and assessment of various subtle (but highly significant) spectral changes that commonly are associated with degrading rangelands. Landsat data from different dates can be registered for quantification of change (Robinove et al, 1981) and can be integrated with other environmental factors for analysis and assessment in a geographic information system.

Aerial photographs are used to provide more detailed data for verifying observations made on the satellite imagery. Systematic reconnaissance flights (Croze et al, 1978) often at very low altitudes, can be used to provide detailed point data to augment a satellite-based survey.

Three significant conditions that can be detected by remotely-sensed data for assessing the factors or indicators of desertification are desolation of vegetation, soil erosion and sedimentation, and salinization.

Most current work involves the inventory, assessment, and monitoring of vegetational degradation. A significant effect of range-vegetation deterioration and subsequent soil erosion is a concomitant increase in surface brightness or albedo. The resultant increase in albedo means an increase in reflectance of solar radiation and, therefore, a cooling of the lower atmosphere. This in turn suppresses convection that might lead to rain showers. The feedback mechanism is clear (Idso, 1976). Albedo appears to be the single most important indicator of the process of desertification over large areas (Berry and Ford, 1977).

Landsat data can be used to derive albedo and monitor gross changes in desert terrain-conditions. Atmospheric and sun angle corrections are applied to the digital numbers of each pixel in order to estimate brightness levels or albedo. Changes in albedo over time can be attributed to changes in vegetation type and density, changes in bare ground, and changes in the distribution of aeolian material (Robinove and Chavez, 1979).

Use of relative albedo as a measure of range condition (e.g., Robinove and Chavez, 1979; Vinogradov, 1969; Wagner and Colwell, 1969) suggests that procedures that have worked fairly well in monitoring grassland vegetation in temperate environments (e.g., Landsat "green measures") may not be equally effective in arid environments, where vegetation reflectance properties are different from those in temperate environments. Explicit verification of this situation has been reported. For example, Graetz, et al (1982) noted that a simple Landsat "green measure" (e.g., MSS7/MSS5) "will not in any way enhance or separate out cover classes" in arid regions of Australia. McCoy (1981) noted that "many [conventionally-produced] Landsat vegetation maps in arid areas were useless". Colwell (1981) noted that "variation in density of vegetation [in arid lands] is difficult to distinguish spectrally from variation in soil characteristics."

Graetz, et al (1982) now use a procedure for analyzing rangeland cover based on the apparent darkening of bare soil that occurs with an increase in shadowing and obscuration of the soil that occurs with an increase in vegetation cover. This procedure may work reasonably well in areas where large areas with uniform soil can be found, since variation in soil albedo could be confused with increased shadowing from vegetation. Where soil albedo is less uniform, however, other procedures will have to be developed.

Colwell (1981) has developed a procedure that should be helpful in areas with spatial variability in soil albedo. The procedure is based on the different amount of shadowing that occurs at different solar elevation angles for different amounts of vegetation cover. The procedure results in separate measures of vegetation cover (Normalized Vegetation Shadow Measure) and soil albedo (Corrected Soil Brightness Measure). The procedure specifically separates the effects of spatial variation in soil and vegetation condition so that they can be examined independently.

The development and spread of desert locust (*Schistocerca gregaria*) swarms can cause serious damage to range vegetation over large areas, thereby contributing to desertification. Hielkema (1980), working with FAO's Plant Protection Service of the Plant Production and Protection Division, Rome, has demonstrated that meteorological and earth-resource satellite data can offer the possibility for detecting and monitoring the environmental conditions that are favorable for the rapid development of locust populations over large areas. TIROS-N/NOAA 6 imagery, combined with meteorological ground station observations, permits a continuous assessment of the occurrence and distribution of rainfall in a given project area. From this information, areas can be selected for detailed monitoring of vegetation biomass-development with multitemporal Landsat data, enabling the precise location of potential locust-breeding habitats (Hielkema, 1980). Hielkema experimented with a number of analytical techniques, such as principal components analysis and band ratioing, to enhance the detection and classification of areas having high vegetation biomass. The resultant information is subsequently used for calculating a Potential Breeding Activity Factor (PBAF) that can be used by the responsible organization in planning field control (spraying) operations.

Mainguet et al (1979) used Landsat imagery and black-and-white aerial photographs to assess land degradation in the Zinder Region of Niger. Aerial photographs from two dates were used to map zones of land degradation surrounding settlements and villages caused by improper land use practices.

For a further discussion of the potential applications of remote sensing to problems of desertification see the Agriculture and Soils Chapter of this Manual. Also, a bibliography on desertification has been prepared by Paylore and Mabbutt (1980).

FUTURE OUTLOOK AND CONCLUSIONS

Since the publication of the first edition of the Manual of Remote Sensing in 1975, there has been a notable increase in the utilization and acceptance of remotely-sensed data for assessing rangeland resources. This fact is reflected in the present Chapter by the emphasis given to existing techniques and data that have grown in use, and by the addition of material that documents the uses being made of new data and new techniques. Among the topics receiving greater emphasis is the use of color and color infrared medium to small-scale photographs for range inventory and wildlife-habitat assessment. This is due in part to endorsements by government agencies for acquiring these types of photography. It also is due to increased awareness of the improved information content of these data, and of their value for resource assessment. The Wildlife Habitat Section, in particular, draws attention to numerous applications for color and color infrared photographs, ranging from small format, large-scale to large format, small-scale, and not only for traditional mapping of vegetation types, but for assessment and measurement of vegetation parameters that contribute to habitat assessment for wildlife species.

This substantial revision of the previous Range Resources chapter also gives considerably greater attention to the uses of Landsat MSS imagery, interpreted manually, for land cover and land-use mapping on a general level, not only in the United States, but in other countries as well. For most range-related purposes, the Landsat false-color composite is the most interpretable of the Landsat data and is gaining wide acceptance as a map base for various kinds of resource information when enlarged to a scale of 1:250,000. In this format, it is replacing the old panchromatic photo-mosaics as a medium for displaying and orienting geographical information about a management area, and for general management planning.

The increased use of low cost aircraft and satellite data in photographic format illustrates that one does not necessarily need to use sophisticated analysis techniques and advanced computer systems to obtain valuable resource information from remotely-sensed data.

New sections have been added to emphasize the growing use of Landsat data in computer-compatible tape format for producing land cover maps. In the past seven years we have seen digital analysis techniques emerge from research centers and become operational. Examples include the growing number of digital analysis projects conducted by the Bureau of Land Management and the Bureau of Reclamation, in Denver, and by the Department of Natural Resources, the U.S. Fish and Wildlife Service and the Bureau of Land Management, in Alaska. These agencies are producing land-cover maps to which they often merge digital terrain data to produce other resource maps that are proving to be valuable for wildlife-habitat assessment and for other management decisions that derive information from the combination of vegetation and terrain variables. These land-cover maps also have been used in multilevel sampling schemes aimed at estimating forage production and acreages of vegetation types. This latter application offers much greater opportunity for range inventory than has heretofore been possible. Another application of digitally produced land-cover classifications that has yet to be fully developed is utilizing the spatial information about vegetation cover to compute such habitat characteristics as interpersion, diversity and juxtaposition. There is great potential for applying these techniques in vast rangelands and wildlife habitat areas such as Alaska, Australia and Africa.

The animal census section of this chapter has been expanded to reflect the increased use of radio telemetry for studying the behavior and physiology of an animal within its habitat. These techniques are contributing to our understanding of what constitutes important habitat characteristics. The popularity and accuracy of ground-based telemetry has resulted in far greater use of these systems relative to airborne and spaceborne telemetry systems.

The Range Monitoring Section has been expanded to include evolving research techniques that use Landsat data to compute vegetation indices, and ratioed and albedo images for the assessment of seasonal conditions, such as monitoring greenness and biomass, and ecological conditions occurring over a longer period of time. Some of the latter techniques may play an important role in the future monitoring of desertification.

Finally, a section has been added on desertification of rangelands to highlight both this serious and important arid land problem, and the opportunity to use newly developed remote sensing techniques to study various aspects of the problem.

The increased use and acceptance of remotely-sensed data is due in part to the strong research and development programs of the late sixties and seventies. The authors have drawn heavily from the literature dealing with this research, and from their own experience in updating the Range Resources chapter. The literature, unfortunately, does not provide a good measure of the day-to-day operational uses of remotely-sensed data, nor of the innovative approaches developed by the users, which go largely unreported but contribute substantially to the awareness and use of the technology. It remains to be seen how many of the promising applications documented by current research will become operational.

In the near future we can expect to see increased use of new Landsat data products. Among those are enhanced color composites from MSS data, geometrically corrected, scaled to 1:250,000 and formatted to correspond with USGS quadrangles, also scaled to 1:250,000; and high resolution RBV data merged with map information from USGS quadrangles. Also anticipated is great use for data acquired by the Landsat D Thematic Mapper. The increased spatial and spectral resolution of such data, including data from a thermal infrared channel, should spur greater utilization of satellite data for rangeland-resource assessment. Many new applications will no doubt arise from the added information afforded by the thermal remote sensing as it relates to the assessment of vegetation development, soil moisture and surface-water conditions.

We can also look forward to data from the French and Japanese earth resource satellites.

If the momentum for using remotely-sensed data for rangeland-resource assessment is to continue there must be a supporting research and development effort such as was seen in the seventies. Among the areas that are especially in need of further research are:

1) Determine the lower limits for the detectability of vegetation biomass in shrub, open grassland, and semiarid and arid rangelands using aircraft and satellite data.
2) Investigate the effects of system and seasonal variables on the determination of changes in the ecological condition of rangelands, using remotely-sensed data.
3) Continue research that uses surface-reflectance indices to monitor changes in surface conditions.
4) Develop improved techniques for collecting field data to calibrate observations from satellite data.
5) Determine optimum dates for acquiring remotely-sensed data in all types of rangeland.
6) Develop techniques for predicting forage production and seasonal condition.
7) Develop software to convert spatial information of Landsat data into useful measures of habitat characteristics such as interspersion, diversity, edge, and juxtaposition.
8) Assess the effect of improving the spatial and spectral resolution of satellite data on land-cover mapping accuracy and costs.
9) Conduct further operational testing of promising research techniques to determine actual costs in an operational mode.
10) Combine ancillary data (terrain, soils, vegetation) with remotely sensed data into a geographical information system to improve rangeland resource assessment.

The authors encourage user agencies and professional societies to accept a greater responsibility for training and documentation on the operational uses of remotely-sensed data for inventory and monitoring, with emphasis on the specific details of procedures, costs, accuracies, and standards for output products.

REFERENCES

Aldrich, R. C., 1966. Forestry applications of 70mm color: Photogrammetric Engineering, Vol. 32, pp. 802–810.

Aldrich, R. C., 1979. Remote sensing of wildland resources: a state-of-the-art review. U.S. Forest Service General Technical Report RM-71, pp. 56.

Aldrich, R. C., 1981. Limits of aerial photography for multiresource inventories. Proceedings of Arid Land Resource Inventories: Developing Cost-Efficient Methods, pp. 221–229. U.S. Forest Service General Technical Report WO-28.

Anderson, E. W., 1956. Some soil-plant relationships in Eastern Oregon. Journal of Range Management, Vol 9, No. 4, pp. 171–175.

Anderson, E. W., 1959. Range site handbook for the Columbia Basin land resource area of Oregon: U.S. Department of Agriculture, Soil Conservation Service, Portland, Oregon. (mimeo) (exemplary of a set of nine such range site handbooks covering the ecological provinces of Eastern Oregon).

Anderson, F., and P. P. de Moor, 1971. A system for radio-tracking monkeys in dense bush and forest. Journal of Wildlife Management. 35(4):636–643.

Anderson, J. A., E. E. Hardy, J. T. Roach, and R. E. Witmer, 1976. A land use and land cover classification system for use with remote sensor data. U.S. Geological Survey Professional Paper 964, pp. 28.

Anderson, R. R. and F. J. Wobber, 1973. Wetlands Mapping in New Jersey. Photogrammetric Engineering 39(4):353–358.

Anderson, William H., W. Alan Wentz, and B. Dean Treadwell, 1980 A guide to remote sensing information for wildlife biologists. *In* Schemnitz, Sanford D. *ed.* 1980. Wildlife Management Techniques Manual. Washington, D.C. The Wildlife Society. 4th ed. 686 p. (Ch. 18 pp. 291–303).

Austin, A. P. and R. W. Adams. 1980. Aerial color and color infrared survey of marine plant resources. Photogrammetric Engineering and Remote Sensing 44(4):469–480.

Avery, T. E., 1958. Helicopter stereo-photography: Photogrammetric Engineering, Vol. 24, pp. 617–624.

Avery, T. E., 1975. Natural Resource Measurements, Second Edition, McGraw Hill Book Company.

Bartholomew, G. A. and C. J. Pennycuick, 1973, The flamingo and pelican populations of the rift valley lakes in 1968–69, E. African Wildlife Journal, 11:189–198.

Batson, F. T., and J. C. Elliott, 1977. Surface resource inventory of eastern Montana rangelands utilizing high altitude color infrared photographs, *in* Biennial Workshop on Color Aerial Photography in the Plant

Sciences, 6th, Sioux Falls, SD, August 19–21, 1977, Proceedings: Falls Church, VA, American Society of Photogrammetry, pp. 105–112.

Bauer, A., 1963, Utilization de la photographie verticale a l'etude ornithologique des iles Australe: Denombrement des Manchotieres de l'ile aux Cochons (Archipel de Crozet) et de l'ile de Kerguelen, TAFF., No. 25.

Bauer, M. E., L. F. Silva, R. M. Hoffer, and M. F. Baumgardner, 1977. Agricultural scene understanding. LARS Contract Report 112677. Purdue University, West Lafayette, IN, pp. 61–65.

Bear, G. D., 1970. Evaluation of Aerial Antelope Census Technique, Colorado Division of Game, Fish and Parks, Game Information Leaflet, 69.3p.

Benson, L. A., C. J. Frazee, F. A. Waltz, C. Reed, K. L. Carey and J. L. Gropper, 1973. Remote sensing techniques for mapping range sites and estimating range yield: Remote Sensing Institute, South Dakota State University, Brookings, SD SDSU-RSI-73-19, pp. 40.

Berry, L. and R. B. Ford. 1977. Recommendations for a system to monitor critical indicators in areas prone to desertification. Worcester, Mass., Clark University.

Best, R. G. and D. G. Moore. 1977. Inventory of wetlands using remote sensing for the proposed Oahe Irrigation Unit in Eastern South Dakota. Brookings, South Dakota State University. Remote Sensing Inst., SDSU RSI 77-03 30p.

Best, Robert G. 1978. Utilization of color infrared aerial photography to characterize prairie potholes. *In* Proceedings of the 4th Annual Pecora Symposium. pp. 180–187.

Best, Robert G. and Signe Sather-Blair. 1978. The interpretation of winter wildlife habitat in Eastern South Dakota on Landsat imagery. *In* Proceedings of the 4th Annual Pecora Symposium. pp. 50–56.

Bie, S. W. and P. H. T. Beckett, 1971. Quality control in soil survey, II. The costs of soil survey. Journal of Soil Science, Vol. 22, No. 4, pp. 453–465.

Biggins, D. E. and Pitcher, E. J., 1978. Comparative efficiencies of telemetry and visual techniques for studying ungulates, grouse, and raptors on energy development lands in Southeastern Montana: In Proceedings, Pecora IV Symposium: Applications of Remote Sensing to Wildlife Management, October 10–12, 1978. Sioux Falls, SD, National Wildlife Federation, Washington, D.C. Scientific and Technical Series 3. pp 188–193.

Bonner, K. G., 1979. Mapping rangeland vegetation in northwestern Arizona from Landsat imagery. U.S. Geological Survey, EROS Data Center, Sioux Falls, S. Dakota. Unpublished paper p. 36.

Bonner, K. G., 1981. Riparian vegetation mapping in Northeastern California using high altitude color infrared aerial photography. Eighth Biennial Workshop on Color Aerial Photography in the Plant Sciences and Related Fields, pp. 29–37, American Society of Photogrammetry, Falls Church, VA.

Bonner, W. J., Jr. and J. Morgart, 1981. Landsat: a sampling frame for arid land inventories. Proceedings, Arid Land Resource Inventories: Developing Cost-Efficient Methods, pp. 230–239, November 30-December 6, 1980. U.S. Forest Service General Technical Report WO-28.

Brassard, J. M. et F. Potvin, 1973. Etude biometrique d'un troupeau de caribous a partir de photographies aeriennes verticales. MS. Biological Research Service. Quebec Department of Tourism, Fish and Game.

Brooks, J. W., 1970. Infrared scanning of polar bear: In Bears and Their Management. IUCN Pub. No. 23, Calgary, November 6–9.

Bryant, N. A., R. G. McLeod, A. L. Zobrist and H. B. Johnson, 1979. California desert resource inventory using multispectral classification of digitally mosaicked Landsat frames. Proceedings, 1979 Symposium on Machine Processing of Remotely-Sensed Data, pp. 69–79. Purdue University, West Lafayette, IN.

Carneggie, D. M., 1968. Analysis of remote sensing data for range resource management: Annual Progress Report, Remote Sensing Applications in Forestry. For Earth Resources Survey Program, OSSA/NASA. By the Forestry Remote Sensing Laboratory, University of California, Berkeley.

Carneggie, D. M., 1972. Large-scale 70mm aerial photographs for evaluating ecological conditions, vegetational changes, and range site potential, Ph.D. Thesis, Berkeley, University of California, p. 174.

Carneggie, D. M., 1975. Usefulness of Landsat for monitoring plant development and range conditions in California annual grassland, *In* NASA Johnson Space Flight Center, Earth Resources Survey Symposium, Houston, TX, 1975, Proceedings: Vol. 1, Sec. A, pp. 19–42.

Carneggie, D. M. and A. Marmelstein. 1978. A perspective on remote sensing for wildlife management. *In* Proceedings of the 4th Annual Pecora Symposium. pp. 392–397.

Carneggie, D. M., D. G. Wilcox, and R. B. Hacker, 1971. The use of large-scale aerial photographs in the evaluation of Western Australian rangelands: Department of Agriculture, Western Australia, South Perth, Technical Bulletin 10, p. 37.

Carneggie, D. M., D. O. Ohlen, and L. R. Pettinger, 1980. A selected bibliography: remote sensing applications in wildlife management. U.S. Geological Survey, EROS Data Center, Sioux Falls, SD, pp. 30.

Carneggie, D. M. and J. N. Reppert, 1969. Large-scale 70mm aerial color photography: Photogrammetric Engineering, Vol. 35, pp. 249–257.

Carneggie, D. M. and S. D. DeGloria, 1972. The usefulness of ERTS-1 and supporting aircraft data for monitoring plant development in rangeland environments, *In* International Symposium on Remote Sensing of Environment, 8th, Michigan, 1972, Proceedings: Ann Arbor, Environmental Research Institute of Michigan, pp. 1471–1476.

Carneggie, D. M. and S. D. DeGloria, 1974. Determining range condition and forage production potential in California from ERTS-1 imagery, *In* International Symposium on Remote Sensing of Environment, 10th, Michigan, 1974, Proceedings: Ann Arbor, Michigan, Environmental Research Institute of Michigan, Vol. 2, pp. 1051–1061.

Carneggie, D. M., S. D. DeGloria, and R. N. Colwell, 1974. Report of Research Performed under Contract No. 53500-CT3-266-N, Bureau of Land Management.

Claveran, A. R., 1966. Two modifications to the vegetation photographic charting methods: Journal of Range Management, Vol. 19, No. 6, pp. 371–373.

Cliff, A. D., P. Haggett, J. K. Ord, K. A. Tassett, and R. B. Davis, 1975. Elements of Spatial Structure, Cambridge University Press, pp. 258.

Cochran, W. G., 1977. Sampling Techniques, Third Edition. John Wiley and Sons, pp. 428.

Cochran, W. W. 1980. Wildlife telemetry. Chapter 29 In Schemnitz, S.D. ed. Wildlife Management Techniques Manual. Washington D.C. The Wildlife Society 4th ed 686p. (Chapter 29 pp. 507–520).

Colwell, J. E., 1974. Vegetation canopy reflectance: Remote Sensing of Environment, Vol. 3, pp. 175–185.

Colwell, J. E., 1974. Grass canopy bidirectional spectral reflectance, *In* International Symposium on Remote Sensing of Environment, 9th Michigan, 1974, Proceedings: Ann Arbor, Environmental Research Institute of Michigan, Vol. 2, pp. 1061–1087.

Colwell, J. E. 1981. "Landsat Feature Enhancement." Proceedings of the Fifteenth International Symposium on Remote Sensing of Environment, pp. 599–621, Ann Arbor, Michigan.

Colwell, R. N., 1964, Uses of Aerial Photography for Livestock Inventories. *In* Proceedings. Annual Meeting, Agriculture Research Institute Research Council.

Colwell, R. N., S. D. DeGloria, S. J. Daus, R. W. Thomas, and D. M. Carneggie, 1975. Spacecraft and aircraft remote sensing for integrated unit resource inventory and analysis in Northeastern California and Northwestern Nevada: Remote Sensing Research Program, Berkeley, University of California, p. 352.

Cooley, M. E. and R. M. Turner, 1975. Applications of ERTS products in range and water management problems, Sahelian Zone; Mali, Upper Volta and Niger: U.S. Geological Survey, Project Report (IR)WA-4.

Cooperative Soil Survey, Committee 1, 1974. Improving soil survey techniques. Report to the Western Regional Technical Work Planning Conference. San Diego, CA, January 21–25.

Cooperative Soil Survey, Committee 2, 1976. Improving soil survey techniques. Report to the Western Regional Technical Work Planning Conference. Phoenix, AZ, February 9–13.

Cowardin, Lewis and V. I. Myers. 1973. Remote sensing for identification and classification of wetland vegetation. Jour. of Wildlife Management, 38(2):308–314.

Craighead, J. J., F. C. Craighead, Jr., J. R. Varney, and C. E. Cote, 1971. Satellite monitoring of black bear: Bioscience, 21(24):1206–1211.

Crofton, H. D., 1958. Nematode parasite populations in sheep on lowland farms. VI. Sheep behaviour and nematode infections. Parasitology, 48(3 & 4): 251–260.

Croon, G. W., D. R. McCullough, C. D. Olson, Jr., and L. M. Queal, 1968. Infrared scanning techniques for big game censusing: Journal of Wildlife Management, Vol. 32, No. 4, pp. 751–759.

Croze, H., M. Norton-Griffiths and M.D. Gwynne. 1978. Ecological monitoring in East Africa. New Scientist 77(1088):283–285.

Croze, J., 1972. A modified photogrammetric technique for assessing age-structures of elephant populations and its uses in Kidepo National Park. East African Wildlife Journal, 10(2):91–115.

Cunia, T. 1982. The needs and basis of sampling. Proceedings, In-Place Resource Inventories: Principles and Practices, pp. 315–325, August 9–14, 1981. SAF 82-02, Society of American Foresters, Bethesda, MD.

Daubenmire, R. 1980. The scientific basis for a classification system in land-use allocation. The Scientific and Technical Basis for Land Classification, A Joint Technical Session of Working Groups of the Society of American Foresters, Spokane, WA. October 7.

Deering, D. W., J. W. Rouse, Jr., R. H. Haas, and J. S. Schell, 1975. Measuring forage production of grazing units from Landsat MSS data, *In* International Symposium on Remote Sensing of Environment, 10th, Michigan, 1975, Proceedings: Ann Arbor, Environmental Research Institute of Michigan, Vol. II, pp. 1169–1178.

DeGloria, D. S., S. J. Daus, N. Tosta, and K. G. Bonner, 1975. Utilization of high altitude photography and Landsat-1 data for change detection and sensitive area analysis, *In* International Symposium on Remote Sensing of Environment, Environmental Research Institute of Michigan, Ann Arbor, Vol. 1, pp. 359–368.

Denton, J. W., 1973. A radio telemetry system for elk: its use and efficiency. M.S. Thesis. University of Montana, Missoula. p. 77.

Dethier, B. E., M. D. Ashley, B. Blair, and R. J. Hopp, 1973. Symposium on significant results obtained from the ERTS-1; NASA, Goddard Space Flight Center, Vol. 1, Section A, p. 157–167.

Driscoll, R. S., 1970. Identification and measurement of shrub type vegetation on large-scale aerial photographs: Proceedings Third Annual Earth Resources Aircraft Program Review, NASA/MSC, Houston, TX, Vol. 2, pp. 32-1 to 32-15.

Driscoll, R. S., 1971. Color aerial photography—a new view for range management: USDA Forest Service Research Paper RM-67, Rocky Mountain Forest and Range Experiment Station, Ft. Collins, CO, p. 11.

Driscoll, R. S., and J. N. Reppert, 1968. The identification and quantification of plant species, communities and other resource features in herbland and shrubland environments from large-scale aerial photography: Annual Progress Report, Earth Resources Survey Program, OSSA/NASA. By Rocky Mountain Forest and Range Experiment Station, Forest Service, Ft. Collins, CO.

Driscoll, R. S., J. N. Reppert, R. C. Heller, and D. M. Carneggie, 1970. Identification and measurement of herbland and shrubland vegetation from large-scale aerial color photographs. In XI International Grassland Congress Proceedings, University of Queensland, Australia, pp. 95–98.

Driscoll, R. S., and R. E. Francis, 1971. Multistage, multiband, and sequential imagery to identify and quantify non-forest vegetation resources: Ft. Collins, CO, Rocky Mountain Forest and Range Experiment Station, Forest Service, U.S. Department of Agriculture, p. 75.

Driscoll, R.S. and R.E. Francis, 1975. Range inventory. Classification of plant communities. Evaluation of ERTS-1 Data for Forest and Rangeland Surveys, R. C. Heller, Technical Coordinator. U.S. Forest Service Research Paper PSW-112.

Driscoll, R.S. and T.C. Watson, 1974. Aerial photography for pocket gopher populations. Proceedings of Symposium on Remote Sensing and Photo Interpretation, International Society of Photogrammetry Commission VII.

Dudzinski, M. L. and G. L. Arnold, 1967. Aerial photography and statistical analysis for studying behaviour patterns of grazing animals. Journal of Range Management, v. 20, No. 2, p. 77–83.

Dumke, R. T. and C. M. Pils, 1973. Mortality of radio-tagged pheasants on the waterloo wildlife Area. Wisconsin department of Natural Resources Technical Bulletin 72, p. 52.

Dunstan, T. C. 1973. A tail feather package for radio-tagging raptorial birds. Inl. Bird-Banding News 45(1):6–10.

Dyksterhaus, E. J., 1949. Condition and management of rangelands based on quantitative ecology. Journal of Range Management 2:104–115.

Eicher, G. J., 1953. Aerial methods of assessing red salmon population in Eastern Alaska. Journal of Wildlife Management, v. 17, No. 4, P521-527.

Enslin, W. R. and R. Hill-Rowley, 1977. Michigan resource inventories: characteristics and costs of selected projects using high altitude color infrared imagery. Proceedings, First Conference on the Economics of Remote Sensing Information Systems, pp. 194–212. Department of Economics, San Jose State University, San Jose, CA.

Evans, C. D., W. A. Troyer, and C. J. Lensink, 1966. Aerial census of moose by quadrat sampling units: Journal of Wildlife Management, Vol. 30, No. 4, pp. 767–776.

Everitt, J. H., A. J. Richardson and C. L. Wiegand, 1981. Inventory of semiarid rangelands in South Texas with Landsat data. Proceedings, Machine Processing of Remotely-Sensed Data, pp. 404–415. Purdue University. IEEE Cat. No. 81 CH 1637-8 MPRSD.

Fleming, M. D. and R. M. Hoffer, 1977. Computer-aided analysis techniques for an operational system to map forest lands utilizing Landsat data. LARS Technical Report 112277. Purdue University, West Lafayette, IN, pp. 236.

Francis, R. E., 1970. Ground markers aid in procurement and interpretation of large-scale 70mm aerial photography: Journal of Range Management, Vol. 23, No. 1, pp. 66–68.

Fraser, D. 1980. Remote sensing in wildlife management and research. *In* Remote Sensing Symposium. Canada-Ontario Joint Forestry Research Committee Symposium Proceedings O.P.-8. Coord. C.R. Mattice. Great Lakes Forest Research Center, Sault Ste. Marie, Ontario. pp. 11–22.

Freese, F., 1962. Elementary forest sampling. USDA, Forest Service, Agricultural Handbook No. 232, pp. 91.

Graetz, R. D., M. R. Gentle, R. P. Pech, and J. F. O'Callaghan. 1982. "The Development of a Land Image-Based Resource Information System (Libris) and Its Application to the Assessment and Monitoring of Australian Arid Rangelands." Presented at "Remote Sensing of Arid and Semi-Arid Lands." Cairo, Egypt. January 19–25, 1982.

Garvin, L. E., F. D. Beatty, and A. J. Zanon, 1964. Infrared detection of moose, Interpretation and Analysis Section, Rome Air Development Center, p. 40.

George, T. H. and P. C. Scorup, 1981. Reindeer range inventory: use of winter Landsat imagery for stratification of digital classification. Proceedings, Machine Processing of Remotely-Sensed Data, pp. 416–427. Purdue University. IEEE Cat. No. 81 CH 1637-8 MPRSD.

Gibbens, R. P. and J. K. Lewis, 1975. Use of ERTS multisensor data for range resource inventory. South Dakota State University Department of Animal Science, A.S. Series 75-8, pp. 104–19.

Gilbert, P. F. and J. R. Grieb, 1957. Comparison of air and ground deer counts in Colorado: Journal of Wildlife Management, v. 21, No. 1, p. 33–37.

Giles, Robert H., Jr. 1978. Wildlife Management. San Francisco. W. H. Freeman & Co.

Gill, R. B., 1969. Middle Park deer study-population productivity and mortality: Colorado Division of Game Reserve, Federal Aid Project, W-38-R-23, Game Reserve Report Julyu 1969, Pt. 1, p. 105–122.

Gilmer, David S., John E. Colwell and Edgar A. Work. 1978. Use of Landsat for Evaluation of Waterfowl Habitat in the Prairie Pothole Region. *In* Proceedings of the 4th Annual Pecora Symposium. Sioux Falls, National Wildlife Federation. Scientific and Technical Series 3. pp. 197–203.

Goldberg, M. and S. Shlien, 1976. Computer implementation of four-dimensional clustering algorithm. Research Project 76-2. Canada Centre for Remote Sensing, Department of Energy, Mines and Resources, Ottawa, pp. 31.

Graetz, R. D., D. M. Carneggie, R. Hacker, C. Lendon, and D. G. Wilcox, 1976. A qualitative evaluation of Landsat imagery of Australian rangelands: Australian Rangeland Journal, Vol. 1, pp. 53–59.

Graves, H. B., E. D. Bellis, and W. M. Knuth, 1972. Censusing white-tailed deer by airborne thermal infrared imagery. Journal of Wildlife Management. 36:875–884.

Greentree, Wallace J. and Robert C. Aldrich. 1978. Measuring trout habitat as an indication of population on large scale aerial color photographs. *In* Proceedings of the 4th Annual Pecora Symposium. pp. 65–71.

Grzimek, M. and G. Grzimek, 1960, Census of plains animals in the Serengeti National Park, Tanganyika: Journal of Wildlife Management, v. 1, No. 24, p. 27–37.

Gwynne, M. D., 1977. Landsat image studies of green flush events in Kajiado District, Kenya. UNDP/FAO KEN/71/526 Project Working Document No. 15.

Gwynne, M. D. and H. Croze, 1975. East African habitat monitoring practice: a review of methods and application. Proceedings of the International Livestock Centre for Africa Seminar on the Evaluation and Mapping of Tropical Rangelands, Bamako, Mali, 1975, pp. 95–135.

Haas, R. J., D. W. Deering, J. W. Rouse, Jr., and J. A. Schell, 1975. Monitoring vegetation conditions from Landsat for use in range management, *In* NASA earth resources survey symposium—technical session presentations, agriculture and environment, Houston, TX, 1975, Proceedings, Vol. 1, Section A, pp. 43–53.

Hacker, R. B., 1980. Prospects for satellite applications in Australian rangelands: Tropical Grasslands, Vol. 14(3):288–295.

Hall, F. C., 1967. Allotment analysis mapping criteria and range condition standards, Blue Mountains, Region 6: U.S. Department of Agriculture, Forest Service, Portland, Oregon (written communication).

Hall, G. H., 1965. Data analysis in the social sciences: what about the details? Proceedings, Fall Joint Computer Conference, pp. 533–560. Spartan Books, Washington, D.C.

Hall, M. J., 1981. Oregon statewide landuse inventory. Final project report to Oregon Department of Water Resources. Environmental Remote Sensing Applications Laboratory, Oregon State University, Corvallis, Vols I & II, pp. 186.

Hardy, E. E. and R. S. Senykoff, 1982. Design concepts for natural resource inventories. Proceedings, In-Place Resource Inventories: Principles and Practices, pp. 98–107, August 9–14, 1982. SAF 82-02, Society of American Foresters, Bethesda, MD.

Harris, R. W., 1951. Use of aerial photographs and sub-sampling in range inventories. Journal of Range Management 4(4):270–278.

Harrison, W. D., 1982. Remote sensing applications in soil survey-map unit design and quality control. Western Regional Technical Work Planning Conference, National Cooperative Soil Survey. San Diego, CA, February 8–12.

Hastings, J. R. and R. M. Turner, 1965. The Changing Mile. The University of Arizona Press, Tucson, Arizona, pp. 317.

Heezen, K. L., and J. R. Tester, 1967. Evaluation of radio-tracking by triangulation with special reference to deer movements. Journal of Wildlife Management. 31(1):124–141.

Hegdal, P. L. and T. H. Gatz, 1978. Technology of radio-tracking for various birds and mammals: In Proceedings, Pecora IV Symposium: Applications of Remote Sensing to Wildlife Management, October 10–12, 1978. Sioux Falls, SD, National Wildlife Federation, Washington, D.C. Scientific and Technical Series 3. pp 204–206.

Heintz, T. W., J. K. Lewis and S. S. Waller, 1979. Low level aerial photography as a management research tool for range inventory. Journal of Range Management 32(4):247–249.

Heller, R. C., 1970. Imaging with photographic sensors: Chapter 2 in Remote sensing with special reference to Agriculture and Forestry, National Academy of Sciences. Washington D.C. p. 424 (Chapter 2 p. 35–72).

Heller, et al, 1967. A test with large-scale aerial photographs to sample balsam woolly aphid damage in the Northeast: Journal of Forestry 65(1):10–18.

Heller, R. C., G. D. Doverspike, and R. C. Aldrich, 1964. Identification of tree species on large-scale panchromatic and color aerial photographs: U.S. Department of Agriculture, Forest Service, Agricultural Handbook No. 261.

Henny, C. D., D. R. Anderson, and R. S. Pospahala, 1972. Aerial surveys of water fowl production in North America 1955–1971: U.S. Fish and Wildlife Service Special Science Report—Wildlife No. 160, p. 48.

Henriques, D. E., 1949. Practical application of photogrammetry in land classification as used by the Bureau of Land Management. Photogrammetric Engineering, Vol. 15, No. 4, pp. 540–545.

Hertz, Elizabeth. 1978. Application of small format aerial photography for wildlife habitat mapping. *In* Proceedings of the 4th Annual Pecora Symposium. pp. 72–77.

Heyland, J. D., 1972. Vertical aerial photography as an aid in wildlife population studies. Proceedings of First Canadian Symposium on Remote Sensing.

Heyland, J. D., 1973, Increase the accuracy of your airborne censuses by means of vertical aerial photographs. Trans. 27th Northeast Fish and Wildlife Conference.

Heyland, J. D., 1974. Aspects of the biology of beluga (*Delphinapterus leucas* Pallas) interpreted from vertical aerial photographs. Proceedings of Second Canadian Symposium on Remote Sensing.

Heyland, J. D., 1978. Imaging remote sensing systems for animal census: in Proceedings, Pecora IV Symposium: Applications of Remote Sensing to Wildlife Management, October 10–12, 1978. Sioux Falls, SD, National Wildlife Federation, Washington, D.C. Scientific and Technical Series 3. pp 162–170.

Hielkema, J. U. 1980. Remote sensing techniques and methodologies for monitoring ecological conditions for desert locust population development. Rome, United Nations Food and Agricultural Organization, Final Technical Report.

Hill, G. J. E. and A. Falconer. 1978. Relevance of Landsat to kangaroo management in Queensland, Australia. *In* Proceedings of the 4th Annual Pecora Symposium. pp. 287–293.

Hilwig, F. W., 1980. Visual interpretation of multitemporal Landsat MSS data for inventories of natural resources by integrating static and dynamic image elements. Proceedings of Remote Sensing for Natural Resources, September 10–14, 1979, pp. 126–155. University of Idaho, Moscow, ID.

Hironaka, M., E. W. Tisdale, and M. A. Fosberg, 1976. Use of satellite imagery for classifying and monitoring rangelands in southern Idaho. University of Idaho, College of Forestry, Wildlife and Range Sciences, Bulletin No. 9, pp. 7. Moscow, ID.

Hoffer, R. M. and staff, 1975. Natural resource mapping in mountainous terrain by computer analysis of ERTS-1 satellite data. Research Bulletin 919. Purdue University, West Lafayette, IN, pp. 124.

Hoffer, R. M. 1980. Computer-aided analysis of remote sensor data—magic, mystery, or myth? Proceedings, Remote Sensing for Natural Resources Conference, September 10–14, 1979, pp. 156–179. College of Forestry, Wildlife and Range Sciences, University of Idaho, Moscow.

Hoffer, R. M. and M. D. Fleming, 1978. Mapping vegetative cover by computer-aided analysis of satellite data. Proceedings, Integrated Inventories of Renewable Natural Resources Workshop, January 8–12, Tucson, AZ, pp. 227–237. U.S. Forest Service, General Technical Report RM-55.

Hoffer, R. M., M. D. Fleming, L. A. Bartolucci, S. M. Davis, and R. F. Nelson, 1979. Digital processing of Landsat MSS and topographic data to improve capabilities for computerized mapping of forest cover types. Laboratory for Applications of Remote Sensing, Purdue University, West Lafayette, IN. LARS Technical Report 011579, pp. 159.

Huddleston, H. F. and E. H. Roberts 1968. Use of remote sensing for livestock inventories. Proceedings of Fifth Symposium on Remote Sensing of Environment. University of Michigan, Ann Arbor, p. 307–323.

Husch, B., 1981. How to determine what you can afford to spend for inventories. Proceedings, Arid Land Resource Inventories: Developing Cost-Effective Methods, pp. 98–102. U.S. Forest Service, General Technical Report, WO-28.

Hutchison, C. J., 1982. Techniques for combining Landsat and ancillary data for digital classification improvement. Photogrammetric Engineering and Remote Sensing, 48(1):123–130.

Idso, S. B. 1976. Atmospheric dust and surface albedo: effect on desertification. *In* Paylore, P. and R. A. Haney, Jr., eds., Desertification: Process, Problems, Perspectives. University of Arizona, Tucson.

Isaacson, D. L. and B. J. Schrumpf, 1979. Distribution of tansy ragwort in western Oregon. Proceedings, Symposium on Pyrrolizidine (senecio) Alkaloids: Toxicity, Metabolism and Poisonous Plant Control Measures, pp. 163–169. Department of Animal Science, Oregon State University, Corvallis.

Isaacson, D. L., C. J. Alexander, B. J. Schrumpf and R. Murray, 1980. Analysis of association of Landsat spectral classes with ground cover classes in wildland inventories. Proceedings, Remote Sensing for Natural Resources, September 10–14, 1979, pp. 180–191. College of Forestry, Wildlife and Range Sciences, University of Idaho, Moscow.

Isaacson, D. L. and D. A. Leckenby. 1981. Remote sensing inventory of Rocky Mountain elk habitat in the Blue Mountains. Fall Technical Convention, San Francisco. Falls Church, Virginia. Amer. Soc. of Photogrammetry. pp. 282–291.

Isaacson, D. L., D. A. Leckenby and C. J. Alexander, 1982. The use of large-scale aerial photography for interpreting Landsat digital data in an elk habitat analysis project. Journal of Applied Photographic Engineering Vol 8(1) 51–57.

Jacobsen, N. K. and J. L. Stuart, 1978. Telemetered heart rates as indices of physiological and behavioral status of deer: In Proceedings, Pecora IV Symposium: Applications of Remote Sensing to Wildlife Management, October 10–12, 1978. Sioux Falls, SD, National Wildlife Federation, Washington, D.C. Scientific and Technical Series 3.

Jaques, D. R., 1982. Landsat imagery to estimate biomass and range condition in Southwestern Alberta, Canada: In Abstracts for 35th Annual Meeting, Society for Range Management, Calgary, Alberta, February 7–12, 1982.

Johnson, G. R. and W. G. Rohde, 1981. Landsat digital analysis techniques required for wildland resource classification. Proceedings, Arid Land Resource Inventories: Developing Cost-Efficient Methods, November 30–December 6, 1980, La Paz, Mexico, pp. 204–213. U.S. Forest Service General Technical Report WO-28.

Johnson, J. R., B. J. Schrumpf, D. A. Mouat, and W. T. Pyott, 1974. Inventory and monitoring of natural vegetation and related resources in an arid environment: Type III report for National Aeronautics and Space Administration, Goddard Space Flight Center, Greenbelt, Maryland, by Corvallis, Oregon State University, Rangeland Resources Program, p. 328.

Johnson, R. R., 1982. Determining what needs to be sampled and what needs to be mapped. Proceedings, In-Place Resource Inventories: Principles and Practices, pp. 66–68, August 9–14, 1981. SAF 82-02, Society of American Foresters, Bethesda, MD.

Kalensky, Z., 1974. ERTS thematic map from multidate digital images. Proceedings, Symposium on Remote Sensing and Photo Interpretation, pp. 767–785, International Society for Photogrammetry, Commission VII. Canadian Institute of Surveying, Ottawa, Ontario, Canada.

Kerbes, R. H., 1978. Identification of Ross' goose colonies from Landsat imagery. In Proceedings, Pecora IV, Application of Remote Sensing to Wildlife Management, October 10–12, 1978. Sioux Falls, SD, National Wildlife Federation, Washington, D.C. Scientific and Technical Series 3.

Krupin, P. J., 1980. Aerial photo exploration for open range water supplies. Rangelands, Vol 2, No. 5, p. 192.

Lavigne, D. M., 1976. Counting harp seals with ultraviolet photography, Polar Record, Vol. 18(114), pp. 269–277.

Lavigne, D. M. and K. Ronald, 1975. Improved remote sensing techniques for evaluating seal populations. ICES C.M. 1975/N:12.

LeHouérou, H. N. 1975. The nature and causes of desertization. Proceedings of the International Geographical Union on Desertification, Cambridge University, England, 23 Sept. 1975.

Lenco, M. 1976. Essai de dénombrement du gros gibier en forêt de plaine par télédétection aérienne a basse altitude en infra-rouge thermique. Min. de la Culture et de l'environnement. Unpublished MS 8 pp. 37 + photos.

Leonard, R. M. and E. Fish, 1974. An aerial photographic technique for censusing lesser sandhill cranes. Wildlife Society Bulletin, 2(4):191–195.

Lewis, P., 1977. Maps and Statistics. John Wiley and Sons. pp. 318.

Linden, D. S., W. G. Rohde and K. G. Bonner, 1981. Estimating area of vegetation types with Landsat and ancillary data. Proceedings, Arid Land Resource Inventories: Developing Cost-Efficient Methods, pp. 279–286. November 30–December 6, 1980, LaPaz, Mexico. U.S. Forest Service General Technical Report WO-28.

Löffler, E. and C. Margules. 1980. Wombats Detected from Space. Remote Sensing of Environment. 9:47–56.

Lorain, G. E., 1970. 70mm Large-scale aerial photography for range resources evaluation: University of Nevada, Reno, M.Sc. Thesis.

Lund, Gordon F., 1974. Time and energy budgets by telemetry of heart rate from free ranging black-tailed prairie dogs in natural and in model environments. Ph.D. Thesis, Department of Zoology, University of Iowa, Iowa City.

Lyons, E. H., 1967. Forest sampling with 70mm fixed airbase photography from helicopters: Photogrammetria, Vol. 22, pp. 213–231.

MacLeod, N. H., 1973. Applications of remote sensing (ERTS) to resource management and development in Sahelien Africa (Republic of Mali): *In* Symposium on Significant Results obtained from the Earth Resources Technology Satellite-1, Vol. I, Section B, New Carrollton, Maryland, 1973, Proceedings: NASA Goddard Space Flight Center, NASA SP-327, pp. 1475–1481.

Mainguet, M., L. Canon-Cossus and M. C. Chemin. 1979. Dégradation dans les Région Centrales de la Républic du Niger. Travaux de l'Institute de Géographie de Reims, No. 39–40, pp. 61–73.

Mairs, J. W., 1976. The use of remote sensing techniques to identify potential natural areas in Oregon. Biological Conservation, Vol. 9, pp. 259–266.

Mansfield, A. W., 1970. Population dynamics and exploitation of some Arctic seals. *In* Antarctic Ecology, v. 1, *Ed.* M. W. Holage, Academic Press, New York.

Marble, H. P., 1967. Radiation from big game and background: a control study for infrared scanner census: Univ. of Montana, M.Sc. Thesis, p. 86.

Martinko, E. A., 1982. Monitoring agricultural growth in pronghorn antelope habitat. Proceedings, Pecora VII Symposium. Remote Sensing: An Input to Geographic Information Systems in the 1980's, pp.

210–216. Sioux Falls, SD, October 18–21, 1981. American Society of Photogrammetry, Falls Church, VA.

Maxwell, E. L., 1976. A remote rangeland analysis system. Journal of Range Management, 26(1):66–73.

Maxwell, E. L., and G. R. Johnson, 1974. A remote rangeland analysis system: Ft. Collins, Colorado State University: Report No. 1885-F, p. 240.

Mayer, Kenneth E. and Lawrence Fox, III. 1980. Mapping wildlife habitat on the McCloud Ranger District with Landsat digital data. Pacific Southwest Region. U.S. Forest Service USDA Contract No. 53-98A28-0-3045.

Mbugua, S. W., and W. E. Stevens, 1977. Distribution and densities of cattle, shoats (sheep and goats), camels, donkeys, and wildlife in the rangelands of Kenya: Ministry of Tourism and Wildlife, Kenya Rangeland Ecological Monitoring Unit (KREMU), Nairobi, p. 24.

McCoy, R. M. 1981. "Models in Remote Sensing: An Approach to Mapping Vegetation in Arid Lands." Pecora VII Symposium, Sioux Falls, South Dakota.

McCullough, D. R., C. E. Olson, Jr., and L. M. Queal, 1969. Progress in large animal census by thermal mapping in Johnson, P. L., ed., Remote Sensing of Ecology: Athens, University Georgia Press, pp. 138–147.

McCullough, G. W., C. E. Olson, and L. M. Queal, 1968. Progress in large animal census by thermal mapping. *In* Remote Sensing in Ecology. University of Georgia Press. Ed. P. O. Johnson.

McDaniel, K. C. and R. H. Haas, 1982. Assessing mesquite-grass vegetation condition from Landsat: Photogrammetric Engineering and Remote Sensing, Vol 48(3):pp. 441–450.

McLeod, R. G. and H. B. Johnson, 1981. Resources inventory techniques used in the California Desert Conservation Area. Proceedings, Arid Land Resource Inventories: Developing Cost-Efficient Methods, pp. 260–271, November 30–December 6, 1980, La Paz, Mexico. U.S. Forest Service General Technical Report, WO-28.

Mead, R. A., and J. Szajgin, 1982. Landsat classification accuracy assessment procedures: Photogrammetric Engineering and Remote Sensing, Vol 48(1) pp. 139–141.

Merchant, J. W., Jr., and B. H. Waddell. 1974. The use of high altitude photography and ERTS-1 imagery for wildlife habitat inventory in Kansas. *In* Proceedings of the 1974 Fall Convention, Washington, D.C., Sept. 10–13, 1974: Falls Church, Virginia, American Soc. of Photogrammetry. pp. 220–231.

Meyer, M. P. 1975. 35mm aerial photography applications to wildlife population and habitat analysis. *In* Workshop on Remote Sensing of Wildlife, Quebec, Canada. Quebec Service of Biological Research. pp. 31–45.

Meyer, M., and P. Grumstrup, 1978. Operating manual for the Montana 35mm aerial photography system—2nd revision. University of Minnesota Coll. For. IAFHE RSL Res. Rep. 78-1, St. Paul p. 62.

Miller, W. A. 1974. Application of the 35mm aerial photography system to resource lands: Boulder, Colorado, Western International Commission for Higher Education, p. 50.

Miller, W. A., W. J. Bonner, W. G. Rohde, and L. P. Schwartz, 1981. Digital Landsat and terrain data applied to an arid land resource inventory. Proceedings, Arid Land Inventories: Developing Cost-Efficient Methods, pp. 589–592, November 30–December 6, 1980, LaPaz, Mexico. U.S. Forest Service General Technical Report WO-28.

Montanari, J. H. and B. O. Wilen, 1978. Techniques developed and presently being used to conduct the National Wetlands Inventory Project: In Lund, H. G. et al, editors Integrated Inventories of Renewable Natural Resources; Proceeding of the January 8–12, 1978, Workshop, Tucson, Arizona. General Technical Report RM-55. Rocky Mountain Forest and Range Experiment Station, U.S. Forest Service, USDA, pp. 192–198.

Mouat, D. A., 1974. Relationships between vegetation and terrain variables in Southeastern Arizona. PhD Dissertation, Oregon State University, Corvallis, OR pp. 242.

Mouat, D. A. and C. F. Hutchinson, 1981. Sensores remotos y mapas de vegetación en regiones semiáridas. Desierto y Ciencia II(3):4–10.

Mouat, D. A., J. B. Bale, K. E. Foster and B. D. Treadwell, 1981. The use of remote sensing for an integrated inventory of a semi-arid area. Journal of Arid Environments, Vol. 4, pp. 169–179.

Mouat, D. A., J. S. Conn, R. Dodoo, and K. E. Foster, 1978. The application and evaluation of remote sensing in the Tamne River Basin, Ghana: a report to the Council of Scientific and Industrial Research, by Office of Arid Lands Studies, University of Arizona, Tucson, Arizona, p. 34.

Mouat, D. A. and R. R. Johnson, 1981. Vegetation inventories and interpretation of environmental variables for resource management: Grand Canyon National Park, a case study. Proceedings, Arid Land Resource Inventories: Developing Cost-Efficient Methods, pp. 300–308. November 30–December 6, 1980, LaPaz, Mexico. U.S. Forest Service General Technical Report WO-28.

Murray, R., 1982. Personal communication. Environmental Remote Sensing Applications Laboratory (ERSAL), Oregon State University, Corvallis.

Murray, R., C. J. Alexander and M. J. Hall, 1982. Using photography to support computer-assisted analysis of Landsat MSS data. Unpublished paper, Environmental Remote Sensing Applications Laboratory, Oregon State University, Corvallis.

Myers, V. I. 1977. Applications of remote sensing technology in South Dakota to access wildlife habitat change, describe meandering lakes, improve agricultural censusing, map aspen, and quantify. Brookings, South Dakota State University. Remote Sensing Inst. SDSU-RSI 77-17, 80p.

Myers, V. I., D. G. Moore, M. DeVries and B. Worcester, 1978. Remote sensing for monitoring resources for development and conservation of desert and semi-desert areas: *In* American Society of Photogrammetry, Fall Technical Meeting, Albuquerque, New Mexico, Proceedings: pp. 396–422.

Myers, V. I., F. C. Westin, M. L. Horton, and J. K. Lewis, 1974. Effective use of ERTS multisensor data in the Northern Great Plains, Type III Final Report for the period June 12, 1972, to July 26, 1974, prepared for the National Aeronautics and Space Administration. South Dakota State University. Remote Sensing Institute Publication 74-09, p. 121.

Nelson, R. F. and R. M. Hoffer, 1979. Computer-aided processing of Landsat MSS data for classification of forestlands. LARS Technical Report 102679. Purdue University, West Lafayette, IN, pp. 95.

Norton-Griffiths, M., 1973. Counting the Serengeti migratory wildebeest using two-stage sampling. East African Wildlife Journal, 11:135-149.

Norton-Griffiths, M., 1975. Counting animals: African Wildlife Leadership Foundation, Publication No. 1, December 1975, pp. 110.

Olsen, H. F. 1978. Remote sensing of fish habitat areas. *In* Proceedings of the 4th Annual Pecora Symposium. pp. 328–334.

Øritsland, N. A., J. W. Lentfer, and K. Ronald, 1974. Radiative surface temperatures of the polar bear, J. Mammal. 55(2).

Otterman, J. and R. J. Fraser, 1976. Earth atmosphere system and surface reflectivities in arid regions from Landsat MSS data: Remote Sensing of Environment, 5:247–266.

Owen, D. F. 1979. Drought and desertification in Africa: Lessons from the Nairobi Conference. Oikos 33(2):139–151.

Pacheco, R. A., 1974. Terrain interpretation using satellite data; El Fula Area-Sudan: sponsored jointly by Soil Survey Administration, Soil Conservation, Land Use and Water Programming Administration, Forest Department, and Water Corporation, 1974, p. 23.

Parker, H. D., Jr., 1972. Airborne infrared detection of deer: Colorado State University, Ph.D. Thesis, p. 186.

Parker, K. W., 1951. A method for measuring trend in range condition on National Forest ranges: U.S. Department of Agriculture. U.S. Forest Service, Division of Range Research Report.

Paylore, P., ed. 1976. Desertification: A World Bibliography. University of Arizona, Tucson, Office of Arid Lands Studies. 644 p.

Paylore, P. and J. A. Mabbutt, eds. 1980. Desertification: World Bibliography Update, 1976–1980; University of Arizona, Tucson, Office of Arid Lands Studies. 196 p.

Pearson, R. L., C. J. Tucker, and L. D. Miller, 1976. Spectral mapping of shortgrass prairie biomass: Photogrammetric Engineering, Vol. 42, No. 3, pp. 317–323.

Pennycuick, C. J., 1969. Methods of using light aircraft in wildlife biology: Proceedings, workshop on the use of light aircraft in wildlife management in East Africa, December 1968, Special Issue: East Africa Agriculture and Forestry Journal.

Perkins, D. F., 1971. Counting sheep, cattle and ponies on Dartmoor by aerial photography. The Application of Aerial Photography to the Work of the Nature Conservancy. *Ed.* R. Goodier.

Peterson, D. L., J. A. Brass and D. H. Card, 1982. Incorporating remote sensing in inventory design. Proceedings, In-Place Resource Inventories: Principles and Practices, pp. 428–433, August 9–14, 1981. SAF 82-02, Society of American Foresters, Bethesda, MD.

Pettinger, L. R., 1980. Environmental stratification—a method to improve Landsat digital analysis accuracy and land cover map utility. Proceedings, Fourteenth International Symposium on Remote Sensing of Environment, pp. 1587–1599. ERIM, Ann Arbor, MI.

Pettinger, Lawrence R., Adrian Farmer, and Mel Schamberger. 1978. Quantitative wildlife habitat evaluation using high-altitude color infrared aerial photographs. *In* Proceedings of the 4th Annual Pecora Symposium. pp. 335–345.

Pierce, W. R., and L. E. Eddleman, 1970. A field stereophotographic technique for range vegetation analysis: Journal of Range Management, Vol. 23, pp. 218–220.

Porter, W. F., D. B. Siniff, and D. A. Hamilton, 1978. Radio-telemetry techniques for the investigation of the behavior and demography of wild turkeys: In Proceedings, Pecora IV Symposium: Applications of Remote Sensing to Wildlife Management, October 10–12, 1978. Sioux Falls, SD, National Wildlife Federation, Washington, D.C. Scientific and Technical Series 3. pp 214–218.

Poulton, C. E., 1968. The feasibility of inventorying native vegetation and related resources from space photography. Earth Resources Aircraft Program Status Review, Vol II, pp. 40–4 to 40–24. NASA, JSC, Houston, TX.

Poulton, C. E., 1970. Practical applications of remote sensing in range resources development and management. Range and Wildlife Habitat Evaluation, A Research Symposium, USDA, Forest Service, Misc. Publication Number 1147.

Poulton, C. E., 1972. A comprehensive remote sensing legend system for the ecological characterization and annotation of natural and altered landscapes: Proceedings, Eighth International Symposium on Remote Sensing of Environment, Willow Run Labs, Environmental Research Institute, Michigan, Ann Arbor, pp. 393–498.

Poulton, C. E., 1973. The advantages of side-lap stereo interpretation of ERTS-1 imagery in northern latitudes. Proceedings of the Earth Resources Technology Satellite-1 Symposium, pp. 157–161, September 29, 1972, NASA, Goddard Space Flight Center, Greenbelt, MD. X-650-73-10.

Poulton, C. E., 1977. Remote sensing and the ecological management of rangelands in developing nations: Food and Agriculture Organization of the United Nations, Rome, Italy: presented at the Annual Meeting, Society for Range Management, Portland, Oregon, 1977.

Poulton, C. E. and A. G. Isley, 1970. An analysis of state-owned rangeland resources for multiple-use management in Southeastern Oregon. Range Management Program, Oregon State University, Corvallis, pp. 125.

Poulton, C. E., B. J. Schrumpf, and J. R. Johnson, 1971. Ecological resource analysis from high-flight photography for land use planning in applied remote sensing of earth resources in Arizona: Proceedings, Second ARETS Symposium, Arizona, University of Tucson, November 2–4, 1971, pp. 35–43.

Poulton, C. E., D. P. Faulkner, and N. L. Martin, 1971. A procedural manual for resource analysis: Application of ecology and remote sensing in the analysis of range watersheds: Final Project Report, Range Management Program, Agricultural Experiment Station, Oregon State University, Corvallis, in cooperation with USDI, Bureau of Land Management, p. 97.

Poulton, C. E., et al, 1975. Range Resources: Inventory, Evaluation and Monitoring. Chapter 18 in Manual of Remote Sensing, R. G. Reeves, Editor Amer. Soc. of Photogrammetry, Falls Church, VA pp. 1427–1478.

Ray, G. C. and D. Wartzok, 1975. Synergistic remote sensing of walrus and walrus habitat. Workshop on Remote Sensing of Wildlife, Quebec. Ed. J. D. Heyland. pp 145–150.

Ray, G. C., D. Mitchel, D. Wartzok, V. Kozicki, and R. Maiefski, 1978. Radio tracking of a fin whale

(Balaenoptera physalus). Science vol 202:521–24.

Reeves, Henry M. 1978. Applications of satellite imagery to management of Arctic nesting geese. *In* Proceedings of the 4th Annual Pecora Symposium. pp. 219–226.

Reid, E. H., R. M. Hansen, and A. L. Ward, 1966. Counting mounds and earth plugs to census mountain pocket gophers; Journal of Wildlife Management, Vol. 30, pp. 327–334.

Reining, P. 1978. Handbook of Desertification Indicators. Based on the Science Association's Nairobi Seminar on Desertification. AAAS, Wash. D.C. 141 p.

Rekas, Anthony M. B. 1978. Inventory of wildlife habitat: an approach and case study. *In* Proceedings of the 4th Annual Pecora Symposium. pp. 346–352.

Reppert, J. N., and R. E. Francis, 1973. Interpretation of trend in range condition from three-step data: USDA Forest Service Research Paper RM-103, Rocky Mountain Forest and Range Experiment Station, Fort. Collins, CO, p. 15.

Roberts, E. H. and R. N. Colwell, 1968. The application of remote sensing to the inventory of livestock and the identification of crops. Report to Statistical Reporting Service, USDA, by forestry Remote Sensing Laboratory, Univ. of Calif. Berkeley. Sept 30, 1968.

Robinove, Charles J., 1981. The logic of multispectral classification and mapping of land. Remote Sensing of Environment, Vol II, PP. 231–244.

Robinove, C. J., 1981. Efficient arid land monitoring using Landsat images. Proceedings, Arid Land Resources Inventory: Developing Cost-Efficient Methods, U.S. Forest Service General Technical Report WO-28, LaPaz, Mexico, November 30–December 6, 1980 pp. 256–259.

Robinove, C. J. and P. S. Chavez. 1979. Landsat albedo monitoring method for an arid region. Proc. of the AAAS International Arid Lands Conference on Plant Resources, Oct. 1978, Lubbock, Texas.

Robinove, C. J., P. S. Chavez, Jr., D. Gehring and R. Holmgren, 1981. Arid land monitoring using landsat albedo difference images: Remote Sensing of Environment, Vol II:pp. 153–156.

Rohde, W. G. and J. B. Bale, 1975. Future applications of satellite data for comprehensive land use planning. Proceedings, 14th Annual Meeting, American Society of Photogrammetry.

Rohde, W. G., and W. A. Miller, 1981. Arizona Vegetation Resource Inventory (AVRI) project final report: U.S. Geological Survey, EROS Data Center, Sioux Falls, SD, unpublished report, pp. 200.

Rohde, W. G., W. A. Miller, K. G. Bonner, E. Hertz and M. F. Engel, 1979. A stratified-cluster sampling procedure applied to a wildland vegetation inventory using remote sensing. Proceedings, Thirteenth International Symposium on Remote Sensing of Environment, pp. 167–179. ERIM, Ann Arbor, MI.

Roller, Norman E. G. 1978. Quantitative evaluation of deer habitat. *In* Proceedings of the 4th Annual Pecora Symposium. pp. 137–146.

Rouse, J. W., Jr., R. H. Haas, J. A. Schell, and D. W. Deering, 1973. Monitoring vegetation systems in the Great Plains with ERTS, *In* NASA Goddard Space Flight Center, Earth Resources Technology Satellite-1 Symposium, Washington, D.C. 3rd, December 1973, Proceedings, Vol. 1, Section A, pp. 309–319.

Rouse, J. W., R. H. Haas, J. A. Schell, D. W. Deering, and J. C. Harlan, 1974. Type III Final Report, NASA Goddard Space Flight Center, Greenbelt, Maryland.

Scheierl, R. and M. Meyer. 1976. Evaluation and inventory of waterfowl habitats of the Copper River Delta, Alaska, by remote sensing. St. Paul, University of Minn. Inst. of Ag, Forestry, and Home Economics. Remote Sensing Laboratory Research Report 76-3. 46p.

Schemnitz, S. D. ed., 1980. Wildlife Management Techniques Manual; published by the Wildlife Society, Washington, D.C., Fourth Edition, pp. 686.

Scholz, D., N. Fuhs and M. Hixson, 1979. An evaluation of several different classification schemes: their parameters and performance. Proceedings, Thirteenth International Symposium on Remote Sensing of Environment, pp. 1143–1147. ERIM, Ann Arbor, MI.

Scott, G. P. and H. E. Winn, 1978. Assessment of humpback whale (*Megaptera novaeangliae*) stocks using vertical photographs. In Proceedings, Pecora IV pp. 235–243.

Sebesta, P. D. and G. F. Lund, 1978. Overview of NASA wildlife sensing projects: In Proceedings, Pecora IV Symposium: Applications of Remote Sensing to Wildlife Management, October 10–12, 1978. Sioux Falls, SD, National Wildlife Federation, Washington, D.C. Scientific and Technical Series 3. pp 261–268.

Sebestyen, G. S. and J. Edie, 1966. An algorithm for non-parametric pattern recognition. IEEE Transaction in Electronic Computers, Vol. EC-15, pp. 908–915.

Seher, J. S., and P. T. Tueller. 1973. Color aerial photos for Marshland. Photogrammetric Engineering. 39(5):489–499.

Sergeant, D. E., 1965. Exploitation and conservation of harp and hood seals. The Polar Recordings, 12(80):541–551.

Shasby, M. B., R. R. Burgan and G. R. Johnson, 1981. Broad area forest fuels and topography mapping using digital landsat and terrain data. Proceedings, 1981 Machine Processing of Remotely-Sensed Data Symposium, pp. 529–538.

Sheridan, D. 1981. Desertification of the United States. Council on Environmental Quality. U.S. Government Printing Office.

Shlien, S. and A. Smith, 1975. A rapid method to generate spectral theme classification of Landsat imagery. Remote Sensing of Environment 4(1):67–77.

Siniff, D. B., and R. O. Skoog, 1964. Aerial censusing of caribou using stratified random sampling: Journal of Wildlife Management, Vol. 28, No. 2, pp. 391–401.

Stelfox, J. G., and J. W. Kufwafwa, 1978. Objectives and achievements of KREMU's aerial survey section, 1976 to September 1978. Ministry of Tourism and Wildlife, Kenya Rangelands Ecological Monitoring Unit, Nairobi, Aerial Survey Technical Report Series No. 5, p. 61.

Strahler, A. H., T. L. Logan and N. A. Bryant, 1978. Improving forest cover classification accuracy from Landsat by incorporating topographic information. Proceedings, Twelfth International Symposium on Remote Sensing of the Environment, pp. 927–942.

Strong, Laurence L. 1980. Estimating phytomass production of habitat types on sagebrush steppe. Ft.

Collins, Colorado State Univ. Master of Science Thesis. 133p.

Sugarbaker, Larry J. 1978. Analysis of vegetation change following wildfire, and its effects upon wildlife habitat. *In* Proceedings of the 4th Annual Pecora Symposium. pp. 153–158.

Swain, R. H. and S. M. Davis (eds.), 1978. Remote Sensing: The Quantitative Approach. McGraw-Hill International Book Company, pp. 396.

Tappan, G., 1978. The monitoring of rangeland vegetation cover in the Kansas Flint Hills from Landsat data. Masters Thesis. University of Kansas, Lawrence, Kansas.

Thomas, D. C. 1967. Population estimates of barren-ground caribou, March to May, 1967: Canadian Wildlife Service Report 9, p. 44.

Thomas, R. W., K. G. Bonner, S. D. DeGloria, and N. Tosta, 1977. Application of multistage sampling procedures for range resource quantification: (unpublished manuscript).

Thomas, R. W., N. Tosta, D. D. Noren, K. G. Bonner and S. D. DeGloria, 1975. Utilization of multistage sampling procedures for rangeland productivity estimation. Chapter 4, In: Spacecraft and Aircraft Remote Sensing for Integrated Unit Resource Inventory and Analysis in Northeastern California and Northwestern Nevada, R. N. Colwell, Principal Investigator, Remote Sensing Research Program, College of Natural Resources, University of California, Berkeley.

Thomas, R. W. and S. D. DeGloria, 1979. Large-scale aerial photography as a sample stage in wildland resource inventories. Proceedings, Seventh Biennial Workshop on Color Aerial Photography in the Plant Sciences and Related Fields, American Society of Photogrammetry, Falls Church, VA pp. 15–33.

Thompson, Donald C., Gary H. Klassen and Josef Cihlar. 1980. Caribou habitat mapping in the Southern District of Keewatin, N.W.T.: an application of digital Landsat data. Jour. of Applied Ecology. Vol. 17. pp. 125–138.

Tietjen, H. P., J. F. Glahn, and K. A. Fagerstone, 1978. Aerial photogrammetry: A method for defining black-tailed prairie dog colony dynamics: In Proceedings, Pecora IV Symposium: Applications of Remote Sensing to Wildlife Management, October 10–12, 1978. Sioux Falls, SD, National Wildlife Federation, Washington, D.C. Scientific and Technical Series 3. pp 244–247.

Tolba, M. K. 1979. What could be done to combat desertification? *In* Bishay, A. and W. G. McGinnies, eds., Advances in Desert and Arid Land Technology and Development, Vol. 1., papers presented at the Int'l. Conf. on the Applications of Sci. and Tech. for Desert Development. The American Univ. in Cairo, Egypt. Sept. 9–15, 1978. New York, Harwood Academic Publishers.

Treadwell, B. Dean. 1978. Wildlife habitat mapping from color aerial photography in Central Arizona. *In* Proceedings of the 4th Annual Pecora Symposium. pp. 377–384.

Treadwell, B. Dean and David A. Mouat. 1977. Applications of high and medium altitude color aerial photography for vegetation mapping in Arizona. Proceedings of the 6th Biennial Workshop on Aerial Color Photography in the Plant Sciences. Fort Collins. Colorado State University.

Tucker, C. J., L. D. Miller, and R. L. Pearson, 1975. Shortgrass prairie spectral measurements: Photogrammetric Engineering, Vol. 41, No. 9, pp. 1157–1162.

Tueller, P. T., 1977. Large-scale 70mm photography for range resource analysis in the Western United States. Proceedings, Eleventh International Symposium on Remote Sensing of Environmental, pp. 1507–1514. ERIM, Ann Arbor, MI.

Tueller, P. T., 1980. a) Rangeland remote sensing interpretation problems. Proceedings, Remote Sensing for Natural Resources Conference, September 10–14, 1979, pp. 450–465. College of Forestry, Wildlife and Range Science, University of Idaho, Moscow.

Tueller, P. T., 1980. b) Remote Sensing for range management. Presented, Remote Sensing for Resource Management Conference, October 23–30, Kansas City, MO. U.S. Soil Conservation Service and the National Aeronautics and Space Administration.

Tueller, P. T., and D. T. Booth, 1975. Large-scale color photography for erosion evaluations on rangeland watersheds in the Great Basin. Proceedings, ASP/ASCM Fall Convention, pp. 708–753. American Society of Photogrammetry, Falls Church, VA.

Tueller, P. T., F. R. Honey and I. J. Tapley, 1978. Landsat and photographic remote sensing for arid land applications in Australia. Proceedings, Twelfth International Symposium on Remote Sensing of Environment, pp. 2177–2191, ERIM, Ann Arbor, MI.

Tueller, P. T., G. Lorain, K. Kipping, and C. Wilkie, 1972. Methods for measuring vegetation changes on Nevada rangelands: University of Nevada, Reno, Agricultural Experiment Station Technical Bulletin 16, p. 55.

Tueller, P. T., G. Lorain, R. M. Halvorson, and J. M. Ratliff, 1975. Mapping vegetation in the Great Basin from ERTS-1 imagery. Proceedings, American Society of Photogrammetry 41st Annual Meeting, Washington, D.C., pp. 338–370. American Society of Photogrammetry, Falls Church, VA.

Tupper, G. J., 1981. A preliminary evaluation of the applicability of Landsat for the assessment of range conditions in the semi-arid woodland in the Cobar District of New South Wales: CSIRO Division of Land Resources Management, Rangeland Research Centre, Deniliquin, New South Wales, Australia, Technical Memorandum 81/2.

Ulliman, Joseph J., Edward O. Garton and Jeffrey A. Keay. 1979. Wildlife habitat classification using large-scale aerial photography. *In* Forest Resource Inventories. Ft. Collins, Colorado State University.

Van Zee, C. and K. G. Bonner, 1981. Estimating rangeland cover proportions with large scale, color infrared aerial photographs. *In* Proceedings of Eighth Biennial Workshop on Color Aerial Photography in the Plant Sciences and Related Fields. Luray, Virginia, April 21–23, pp. 73–82.

Vaughan, R. W., 1971. Aerial photography in seal research. The application of aerial photography to the work of the Nature Conservancy. Ed. R. Goodier.

Verts, B. J., 1963. Equipment and techniques for radio-tracking striped skunks. Journal of Wildlife Management Vol. 27(3):325–339.

Vinogradov, B. V., 1969, "Remote Sensing of the Arid Zone Vegetation in the Visible Spectrum for Studying the Productivity," Proceedings of Sixth International Symposium on Remote Sensing of

Environment, University of Michigan, Ann Arbor, Michigan.

Wagner, T. W. and J. E. Colwell, 1969, "An Investigation of Grassland Resources Using Multispectral Processing and Analysis Techniques," Willow Run Laboratories Report No. 34795-1-F, University of Michigan, Ann Arbor, Michigan.

Walker, A. S. and C. J. Robinove. 1981. Annotated Bibliography of Remote Sensing Methods for Monitoring Desertification. U.S.G.S. Circular 851, 25 p.

Waller, S. S., J. K. Lewis, M. A. Brown, T. W. Heintz, R. I. Butterfield and F. R. Gartner, 1978. Use of 35mm aerial photography in vegetation sampling. Proceedings, First International Rangeland Congress, pp. 517–520, August 14–18. Society for Range Management, Denver, CO.

Watson, E. K., 1977. A remote sensing based multilevel rangeland classification for the Lac-Du-Bois rangelands, Kamloops, British Columbia: Masters thesis (soils), Vancouver, University of British Columbia, p. 85.

Watson, G. W. and R. F. Scott, 1956. Aerial censusing of the Nelchina caribou herd. Trans. North American Wildlife Conference, v. 21, p. 499–510.

Watson, T. C., 1973. Aerial photos and pocket gopher populations: Colorado State University. M.Sc. Thesis, p. 56 (also on file at the Rocky Mountain Forest and Range Experiment Station, Fort Collins, CO).

Wells, K. F., 1971. Measuring vegetation changes on fixed quadrats by vertical ground stereophotography: Journal of Range Management, Vol. 24, No. 3. pp. 233–236.

Wensel, L. C. and M. Eriksson, 1981. Survey planning model: Application for arid land resource inventories. Proceedings, Arid Land Resource Inventories: Developing Cost-Efficient Methods, pp. 136–141, November 30–December 6, 1980. U.S. Forest Service General Technical Report WO-28.

Werber, Morton, 1970. A bibliography of wildlife movements and tracking systems. The George Washington University Medical Center, Department of Medical and Public Affairs, Washington, D.C. 20009 (NASA Contract NSR 09 010 027).

Wilcox, D. G. and E. A. McKinnon, no date. A report on the condition of the Gascoyne Catchment: Department of Agriculture, South Perth, Western Australia.

Williams, R. E., B. W. Allred, R. N. Denis, and H. A. Paulsen, Jr, 1968. Conservation development and use of the world rangelands, Journal of Range Management, Vol 21(6), pp. 355–360.

Wimbush, D. J., M. D. Barrow, and A. S. Costin, 1967. Color stereophotography for the measurement of vegetation: Ecology, Vol. 48, pp. 150–152.

Woodzick, T. L., and E. L. Maxwell, 1977. Interim report under Experiment Station Project Nos. 15-1482 0076 and 15-1482-0150. Earth Resources Department, Colorado State University.

Worcester, B. K. and D. G. Moore, 1978. Delineation of soil-landscapes in the Sudd Region of Sudan on Landsat imagery; *In* International Symposium on Remote Sensing of Environment, Manila, Philippines, 1978, Proceedings, Environmental Research Institute of Michigan, Ann Arbor, Michigan, Vol. II, pp. 1155–1166.

Wride, M. C., and K. Baker, 1977. Thermal imagery for census of ungulates, *In* International Symposium on Remote Sensing of Environment, 11th, Michigan, Proceedings: Ann Arbor, University of Michigan, pp. 1091–1100.

CHAPTER 36

Terrestrial Moons and Planets

Author-Editor: RAYMOND E. ARVIDSON

Contributing Authors: ELLIOTT C. LEVINTHAL, R. STEPHEN SAUNDERS, PETER H. SCHULTZ

GENERAL CONTENTS: Study of planetary surfaces; application to lunar stratigraphy and surface processes; remote sensing of Mars; radar observations of Venus; observations of the larger satellites of Jupiter and Saturn; the future of planetary remote sensing; sources of lunar and planetary data; reference.

INTRODUCTION

Several trillion bits of remote sensing data covering the moons and planets of our solar system have been returned over the past two decades during various deep space missions. In addition, an equivalent amount of data for the moons and planets have been acquired over the past several decades during the course of Earth-based observations. Most of the Earth-based observations have dealt with the composition, structure, and dynamics of planetary atmospheres, or with the broad-scale characterization of the topography or surface materials of various planetary bodies. In this chapter, we review how analyses of a diverse range of remote sensing data acquired both from spaceborne and Earth-based systems have inceased our understanding of the surfaces of the terrestrial moons and planets. A terrestrial body is a moon or planet that is composed of rocky, metallic, or icy materials, and therefore one that preserves a geologic record. Major terrestrial bodies are Mercury, Venus, Earth and its Moon, Mars (Table 36-1a), and the larger satellites of Jupiter and Saturn (Table 36-1b). Remote sensing of planetary atmospheres or of the giant, gaseous planets is not directly addressed, since a recent review of these subjects can be found in Ingersoll et al. (1979).

The chapter begins with a discussion of the scientific objectives and with the methods of acquisition of remote sensing data for the moons and planets. Next, a number of examples are discussed where analyses of remote sensing data have significantly increased our understanding of the character and evolution of several distinctly different terrestrial bodies: the Moon, Mars, Venus, and the larger satellites of Jupiter and Saturn. The reader is referred to Strom (1979) for an extensive review of remote sensing data and results for Mercury. A discussion of the major scientific objectives behind the probable planetary missions for the near future is then included. The chapter ends with a section on how to acquire NASA-sponsored lunar and planetary data, a subject that is of crucial interest to researchers, and a subject that is often overlooked.

STUDY OF PLANETARY SURFACES

SCIENTIFIC OBJECTIVES

The basic objective behind the acquisition of remote sensing data for the terrestrial moons and planets is to increase our basic knowledge of the geologic histories of these bodies (Toksoz, 1977). Acquisition of such data has been designed in part to better understand the characteristic morphologies of planetary surfaces—the presence or absence of various landforms, such as craters, orogenic zones, and fluvial systems, that may provide clues as to the processes that have shaped the surfaces and interiors of given bodies. Another important objective is the determination of the composition and mineralogy of surface materials and how these characteristics vary spatially on a given body.

A major thrust in planetary sciences has been generating a data set that is useful for comparative studies of all terrestrial bodies. For instance, it can be shown to first order that the heights of lava flows on Earth can be used to infer the lava yield strength, which in turn is related to lava composition (Hulme, 1974). Measurements of flow heights on the Moon and Mars and comparison to terrestrial flows show lavas to be rheologically similar to basaltic flows on Earth (Moore et al. 1978) (Figure 36-1). Such a result places a first-order constraint on the composition of the surfaces of the Moon and Mars. In fact, the interpretation of a basaltic composition is confirmed or strengthened by other sources of remote sensing data for lava flows on these two bodies. As another example, comparison of the major terrain types on Mercury, the Earth, the Moon, and Mars demonstrates that, with the notable exception of the Earth, large percentages (up to 80%) of the surfaces of these bodies are covered with a heavily cratered crust that is 4.0 to 4.5 billion years in age (Figure 36-2). Such a crust has been largely destroyed by erosion and by tectonism on Earth. The fact that such primitive crusts remain on Mercury, the Moon, and Mars points to fundamental differences in the dynamic histories of the Earth opposed to these bodies. In particular,

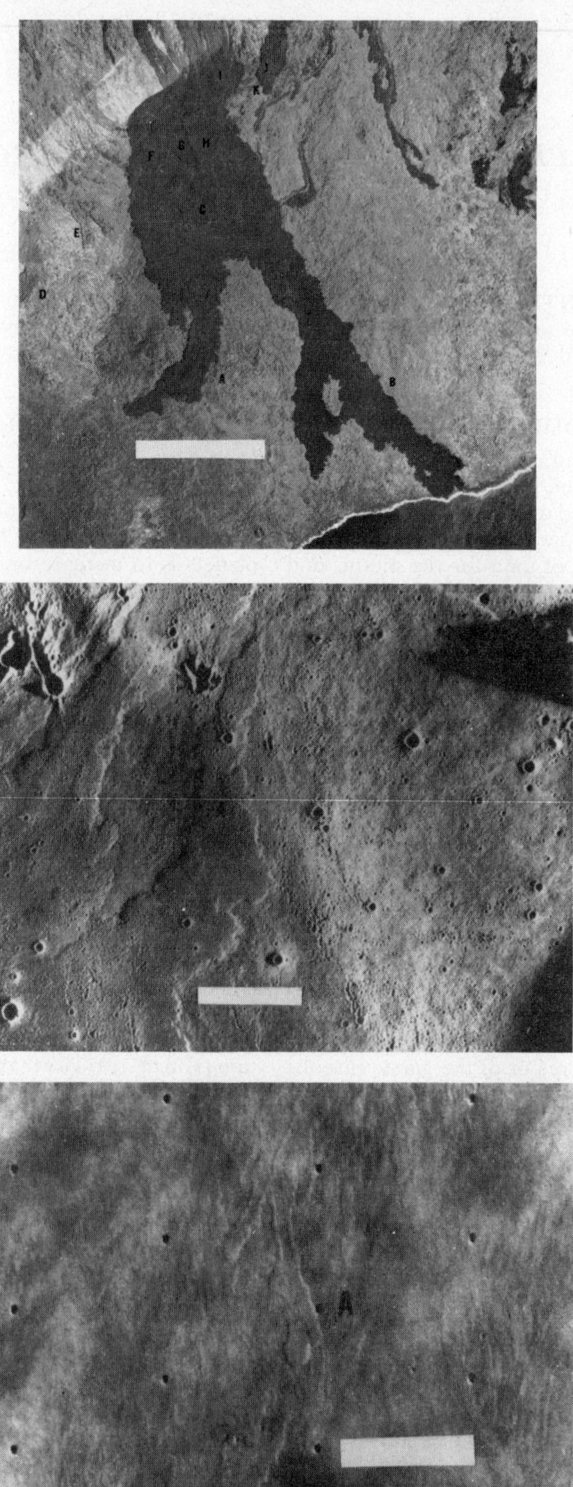

Fig. 36-1. Lava flows on the Earth, Moon, and Mars, are shown at the top, middle, and bottom frames, respectively. The terrestrial flow is an aa basalt flow south of Mauna Ulu, Hawaii. The lunar flow is located in the Imbrium region of the Moon, while the Martian flow is located on the flanks of the volcano called Olympus Mons. Lunar photo is Apollo metric frame 15-M1556; Martian photo is Viking Orbiter frame 047B21. The white bars are 1 km, 10 km, and 10 km long, respectively. Letters delineate individual flows. After Moore et al. (1978).

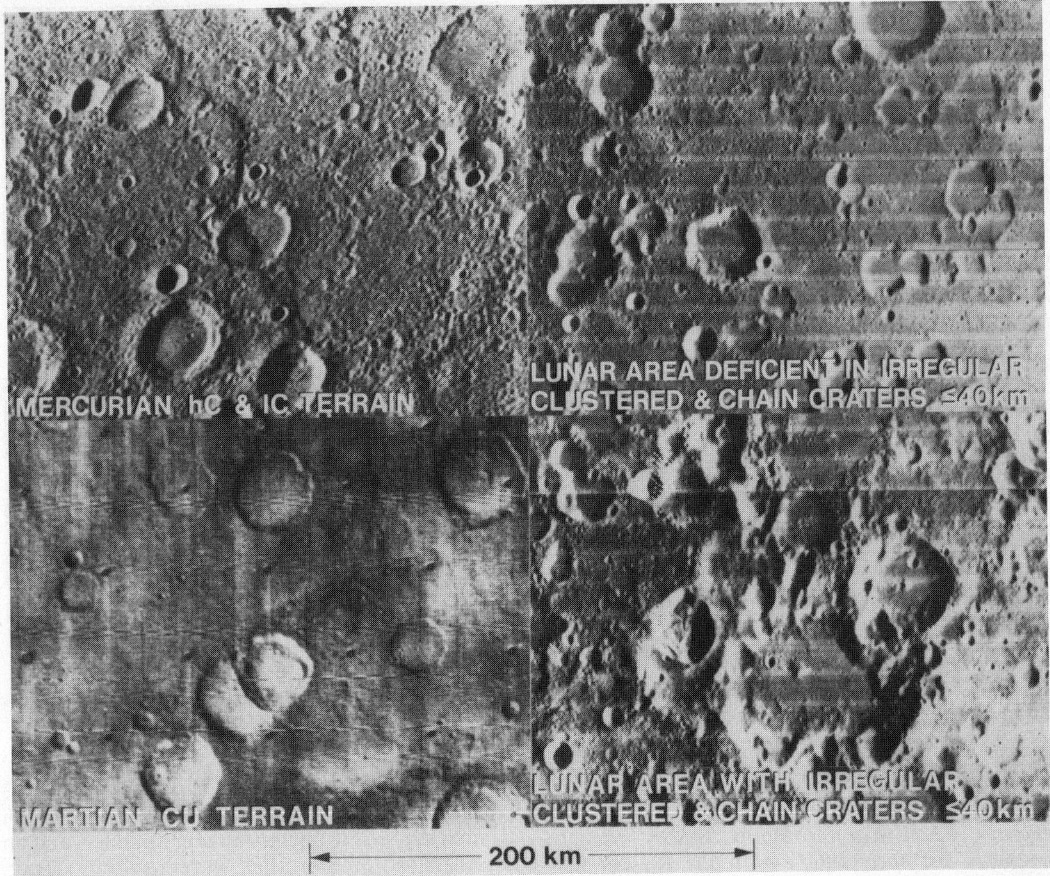

Fig. 36-2. Views of the heavily cratered terrains of Mercury, the Moon, and Mars, shown at the same scale. The lunar terra views, shown on the two right-hand frames, correspond to an area (top view) far from major impact basins, and to an area (bottom view) that probably contains a number of craters that were produced by impact of ejecta from basin-forming events. These secondary craters are characteristically aligned along chains and clusters. The Moon preferentially exhibits an excess of large secondaries because the relatively low gravitational acceleration allows longer flight times for ejecta. As a result, secondaries are widely dispersed on the Moon, relative to Mercury or Mars. However, those areas of heavily cratered lunar terra that are located far from the large impact basins are remarkably similar to the Mercurian and Martian heavily cratered terrains. These surfaces have survived from an early period of geologic time, providing a record that has been erased on the Earth. After Oberbeck et al. (1977).

Mercury, the Moon, and Mars seem to have been unable to support the vigorous convection systems needed to drive lithospheric break-up, creation, and consumption (Solomon, 1978). In general, the thesis of comparative planetology is that the study of the surfaces of all terrestrial bodies enhances manyfold the interpretability of any given body, when compared to examination of any one of the bodies, including the Earth, in isolation.

METHODS

Earth-based observations have played, presently play, and will continue to play an important role in remote sensing of moons and planets. Such observations include tracking changes in the appearance of planetary atmospheres and surfaces over time (Baum, 1974) (Figure 36-3) and spectrophotometric observations that are designed to characterize diagnostic electron transition and charge transfer absorption features in the reflectance spectra of planetary surfaces (e.g. McCord and Clark, 1979). As another example, sophisticated Earth-based radar observations designed to obtain planetary radii and the characteristic surface reflectivity, topography, and roughness have added appreciably to our understanding of Venus (Campbell and Burns, 1980; Jurgens et al. 1980). Unfortunately, except to some extent for the Moon, observations of planetary bodies from Earth or Earth-orbital perspectives cannot produce the spatial resolution and global coverage needed to resolve and understand geologic features, such as faults, craters, and valleys. Even the Space Telescope will only be able to resolve features larger than approximately a couple tens of kilometers in breadth on Mars and Mercury.

In general, spacecraft observations are needed

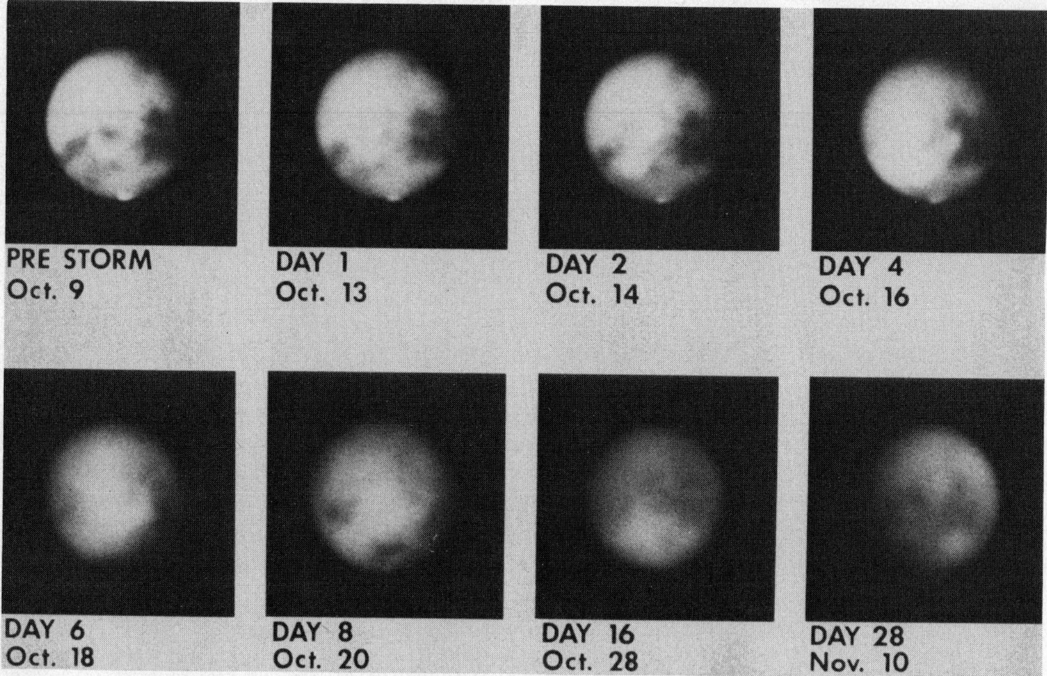

Fig. 36-3. Series of Earth-based telescopic photographs of Mars, showing the progressive obscuration of the surface during the 1971 dust storm. After Baum (1974).

to acquire remote sensing data for the moons and planets at resolutions high enough to distinguish geologic detail. For the Moon, NASA missions began in the early 1960's with the Ranger missions. Between 1964 and 1965, Rangers 7, 8 and 9 telemetered back TV images as the craft approached the lunar surface for eventual hard impact (Heacock et al. 1966). These missions were followed by five soft-landed surveyor Missions in 1965–67, where first-order chemical and physical analyses of the lunar surface were conducted, in addition to acquisition of a large number of TV images (Batson et al. 1974). At approximately the same time, five Lunar Orbiter spacecraft acquired nearly complete coverage of the moon with a best resolution of a couple meters per line pair (Kosofsky and El-Baz, 1970). The culmination of U.S. lunar exploration was the manned Apollo program, where 842 pounds of rocks were returned to Earth (Taylor, 1975).

Photographic remote sensing devices used during the Apollo missions included hand-held cameras that were used for photographing the lunar surface both from orbit and during surface excursions, and high resolution panoramic and metric mapping cameras that were mounted on the orbiting command modules (Apollos 15–17). The geometric fidelity of the panoramic and metric photography, together with the ancillary data acquired by a stellar camera and laser altimeter, provided the base needed to construct detailed orthophotographs and topographic maps for regions under the ground tracks of the Apollo 15–17 command modules. In addition, the Apollo 15 and 16 command modules contained an X-ray spectrometer, an alpha particle spectrometer, a gamma ray spectrometer, a radar sounder, and an infrared radiometer that were used to obtain geophysical and geochemical data for the lunar surface (see: Head et al. 1978). Lunar exploration has proceeded from essentially a reconnaissance mode, to a mode of intensive study of both locally and globally distributed surface materials. Much of the pre-Apollo exploration was designed to assure successful landing of a man on the lunar surface. Nevertheless, such exploration provided fundamental information about the geologic history of the Moon.

Exploration of Mars has proceeded from the flyby reconnaissance mode, with Mariner 4 imaging about 2% of the surface in 1964, and Mariner 6 and 7 acquiring about 25% coverage in 1969, to an orbital mapping mode with Mariner 9 in 1971–72 covering nearly the entire surface with several kilometers/pixel resolution (Figure 36-4). The two Viking Orbiters, beginning in 1976, continued orbital mapping at a somewhat higher resolution than available from Mariner 9 (Snyder, 1979). The Mariner 9 and Viking Orbiters also carried out detailed observations of the moons of Mars, Phobos and Deimos. In addition, the two Viking Landers that touched down on the surface during the summer of 1976 provided detailed information for two separate locations on the planet's surface (Soffen, 1976). In essence, Martian exploration has also proceeded from the reconnaissance to a more science-intensive mode of operation.

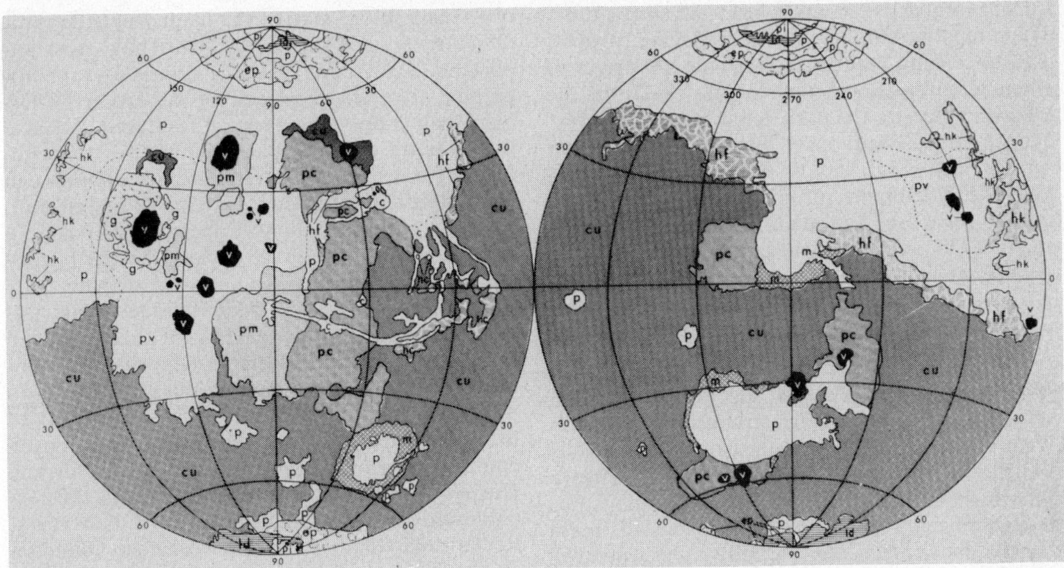

Fig. 36-4. Geologic map of Mars plotted on a Lambert equal area base. The units are defined on the basis of Mariner 9 images. Polar units include pi (permanent ice), ld (layered deposits), and ep (etched plains). Volcanic units include v (volcanic constructs), pv (volcanic plains), pm (moderately cratered plains), and pc (cratered plains). Modified units include hc (hummocky terrains, chaotic), hf (hummocky terrain, fretted), hk (hummocky terrain, knobby), c (channel deposits), p (plains, undivided), and g (grooved terrain). Ancient units include cu (cratered terrain, undivided) and m (mountainous terrain). Note that Mariner 4, 6, and 7 data largely covered the ancient units. After Mutch et al. (1976).

Exploration of Venus has also proceeded along roughly similar lines, beginning with the Mariner 2 flyby in 1974, and most recently culminating with the Pioneer-Venus Orbiter and Probes. The next planned step in the exploration of Venus is the Venus Radar Mapper, which would include an imaging radar instrument capable of mapping the entire surface with a radar cell size of approximately 1 km in width.

Finally, exploration of Jupiter is presently in the reconnaissance mode, with the Pioneer 10 and 11, and the Voyager 1, 2 flybys having occurred in 1973–74 and in 1980, respectively. After the Jupiter encounter, Pioneer 11 encountered Saturn in September 1979, while Voyager 1 encountered Saturn in November 1980 and Voyager 2 encountered Saturn during August of 1981. Voyager 2 has been targeted for an encounter with Uranus in 1986 and Neptune in 1989. The next planned step in the exploration of Jupiter is the Galileo mission of the mid-1980's, which includes both an orbiter to acquire remote sensing data for Jupiter and its satellites and a probe that will return data on the characteristics of the Jovian atmosphere.

Thus, spaceborne acquisition of remote sensing data for moons and planets has proceeded in steps, with the first objective usually being reconnaissance—understanding the fundamental terrain types on the body, together with determining first-order characteristics of any atmospheric constituents. In the case of the Moon, Mars, and Venus, orbital missions were then carried out, with a general aim of more detailed characterization of the surface. Finally, in-situ observations with probes or landers were conducted for these three bodies and, in the lunar case, samples were returned to Earth. Such an exploration philosophy has the unfortunate distinction of forcing the exploration of a given body to become increasingly more complex and costly with each step. This problem will be addressed again, after discussions of how analyses of remote sensing data have significantly increased the understanding of the geologic histories of the Moon, Mars, Venus, and the larger satellites of Jupiter and Saturn.

APPLICATION OF REMOTE SENSING TO LUNAR STRATIGRAPHY AND SURFACE PROCESSES

INTRODUCTION

A vast data base exists for studying the lunar surface. As discussed by Head et al. (1978), that base includes:

- Photography acquired in visible light from Earth, from lunar orbit, and from the lunar surface
- Earth-based spectrophotometric observations in the visible through the thermal infrared
- Orbital observations in the thermal IR
- Earth-based and orbital observations with radar
- Orbital observations of net surface radioactivity and selected elemental abundances

Lunar research is presently in an advanced state, where most researchers utilize a diverse range of remote sensing data to understand the origin of given features (Soderblom et al. 1977). In the following section, the basic lunar stratigraphic sequence is discussed, based on analyses of remote sensing and lunar sample data. This section is followed by examples of how knowledge of lunar surface processes and materials has been greatly enhanced by analysis of the large range of existing remote sensing data.

LUNAR STRATIGRAPHY

The classic work of Shoemaker and Hackman (1962), using Earth-based photographs, established the basis for a lunar stratigraphic column. Their work centered about the crater Copernicus and the Imbrium basin (Figure 36-5). They were able to show that the ejecta from Copernicus lies stratigraphically above the dark, smooth maria and that these ejecta also partially cover a number of other post-mare craters. They also mapped the extent of ejecta deposits that formed as a result of the impact event that produced the multi-ringed Imbrium basin and showed that these deposits overlay most of the surrounding terra materials. The lunar stratigraphy has been considerably revised since Shoemaker and Hackman's (1962) initial work. Wilhelms and McCauley (1971) have synthesized a number of lunar geologic maps and other data into a map of the lunar frontside, produced at a scale of 1:5,000,000. Their most important data source was the continuous photographic coverage produced by the Lunar IV Orbiter, which provided a spatial resolution increase that was an order of magnitude better than obtained from Earth. The stratigraphic column can be simplified into four systems (Pre-Imbrian, Imbrian, Eratosthenian, Copernican), where a system represents a given period of lunar history during which given rock units were emplaced.

The Pre-Imbrian System includes materials emplaced before the impact event that produced the Imbrian basin and ejecta deposits. The Imbrian event occurred approximately 3.9 billion years ago, based on ^{39}Ar-^{40}Ar ages of Apollo 14 breccia samples that are thought to have been emplaced during the impact event (Schaefer and Husain, 1974). The Pre-Imbrian system thus includes rocks older than about 3.9 billion years. Rocks older than 3.9 billion years are abundant on the lunar surface and comprise most of the lunar terra. The terra, which occupy 80% of the lunar surface and almost the whole of the lunar backside (Figure 36-6), consist of heavily cratered surfaces composed of anorthosites, norites, and troctolites that have been intensely brecciated and mixed by numerous impact events (Taylor, 1975). Radiometric ages for such rocks, collected during the Apollo missions, range from about 3.9 to 4.5 billion years, with a tendency for a clustering of ages at 3.9 to 4.2 billion years (Tera et al. 1974). It is thought that the terra crust was formed during the last stages of lunar accretion by progressive crystallization of a magma ocean which was produced by the intense heating of the outer parts of the moon during accretion (Wood et al. 1975).

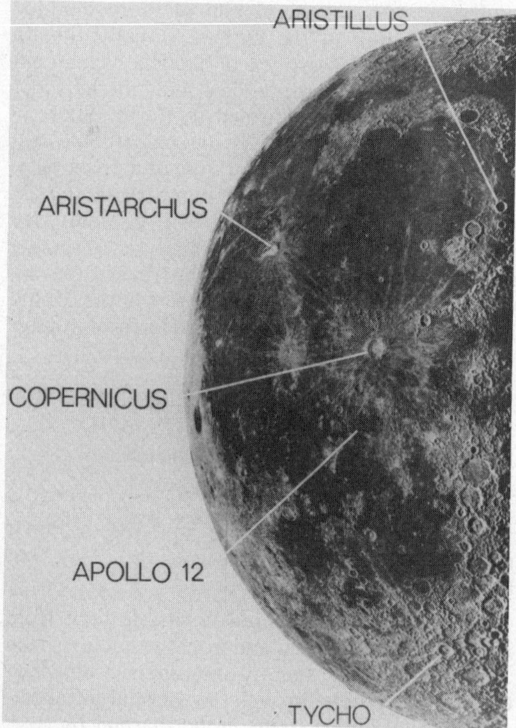

Fig. 36-5. View of the frontside of the lunar surface, showing Copernicus, several other craters, and the Apollo 12 landing site. Mare Imbrium is the dark circular area to the north of Copernicus.

Fig. 36-6. This view of the heavily cratered lunar terra was acquired by the Apollo 16 metric camera during transearth coast to the Moon. Mare Crisium is near the horizon on the upper left of the frame. Apollo 16 metric frame 3023.

The abundance of craters records either: (a) the sweeping-up of the last of the debris during accretion or (b) a late heavy bombardment of the lunar surface by projectiles perturbed into the inner solar system from orbits beyond Jupiter (Wetherill, 1975). In either case, the ages demonstrate that such intense bombardment began to decay to lower rates by about 4 billion years ago. Combining the crater abundances at the landing sites with sample ages shows that the torrential bombardment exponentially decayed with time and that the cratering flux has been constant for about the past 3.0 billion years (Figure 36-7).

The Imbrian System begins with rocks that were emplaced during the impact event that produced the 1300 km diameter Imbrium basin on the lunar nearside. Ejecta deposits extend for several hundred kilometers from the margins of the Imbrium basin. Oberbeck et al. (1975) showed that the kinetic energy contained in the ejecta deposits from such events could allow more mass to be excavated by the secondary craters produced by impact of the ejecta than the mass of ejecta originally excavated by the impact of the asteroid or comet that formed the basin. The material excavated from the secondaries would form part of a ballistic debris surge moving radially outward from the primary crater. For basin-sized events, they postulate that regions close to the basin rim would be scoured and striated by such a process, while regions farther away would be mantled as the debris was deposited. In fact, the ejecta deposits associated with the Imbrian event form a linearly sculpted terrain (the Imbrium sculpture) close to the basin, whereas numerous smooth areas (terra light plains) exist farther away (Figure 36-8). Another consequence of the debris surge is that a great deal of material excavated during impact of the ejecta would be included in the resultant deposits. Considering the vast numbers of terra craters and basins, such processes must have thoroughly disrupted and mixed the terra crust.

The Imbrian System also includes most of the volcanic materials that comprise the lunar maria. These deposits are composed of basaltic lavas that preferentially filled the larger basins of the lunar nearside (Mutch, 1972; Taylor, 1975). Photogeologic analyses demonstrate that the maria exhibit the following volcanic landforms: lobate scarps, sinous rilles, domes, cones, collapse craters, kipukas, lava terraces, mare ridges, and volcanic complexes (Head, 1976) (Figure 36-9). Based on crystallization ages of samples returned from the maria, extensive volcanism began about 3.8 billion years ago and extended over about 1.8 billion years in duration (Taylor, 1975). However, there is some evidence that basaltic volcanism began before the end of heavy bombardment. In particular, some terra regions have impact craters with dark ejecta deposits. The low albedo material was probably excavated from volcanic deposits that were covered by terra debris during basin-sized impact events (Schultz and Spudis, 1979).

The first well exposed Imbrian-aged maria consist of relatively dark, bluish materials presently outcropping near the edges of a number of maria

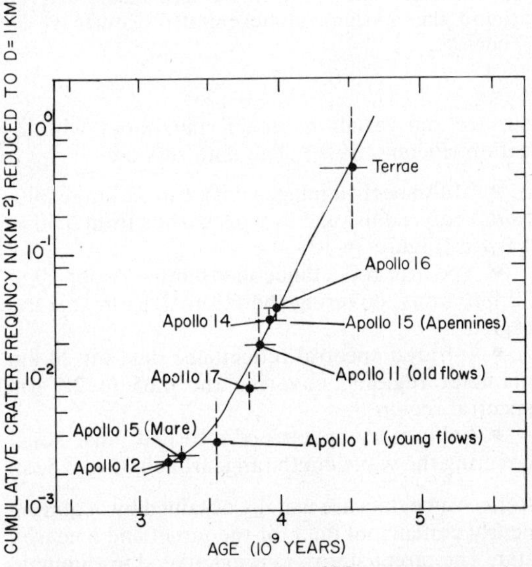

Fig. 36-7. The abundances/area of Moon craters greater than 1 kilometer in diameter are plotted against the ages of the Apollo landing sites in this figure. The derivative of the fit through the data points gives the rate of change of the cratering rate with respect to time. Results demonstrate that the Apollo sites record the ending period of early, torrential bombardment of the Moon. These data can also be used to date the ages of other lunar surfaces, based on crater abundances/area. After Neukum et al. (1975).

Fig. 36-8. Apollo 16 metric camera view of the moon looking south. The large, shallow crater in the middle of the frame is Ptolemaeus, which is about 120 km in diameter. Ptolemaeus is filled with light plains material, probably of an impact origin. The lineated terrain running from upper left to lower right is part of the Imbrium sculpture, with an origin related to impact of ejecta from the Imbrian event. Apollo 16 metric frame 2478.

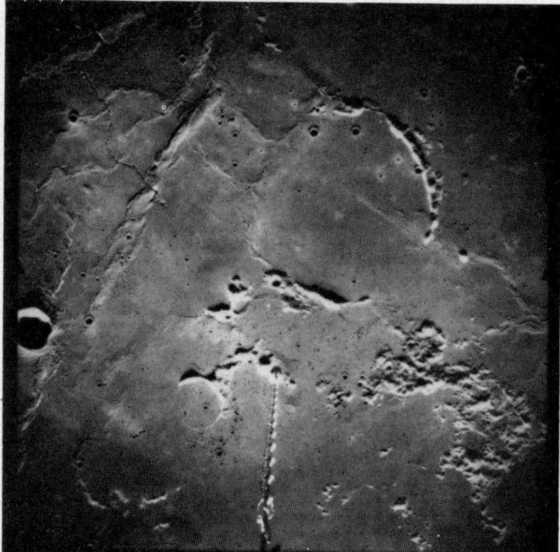

Fig. 36-9. Low sun angle view of the lunar maria with a number of ridges, arches, and wrinkle ridges located in Oceanus Procellarum. Arrows point to two craters filled to their brims with mare materials. Width of view is about 200 km. Apollo 16 metric frame 2836.

COLOR - DIFFERENCE PHOTOGRAPH
OF THE MOON
6100 Å minus 3700 Å

Fig. 36-10. Composite UV (.38 μm)—IR (.78 μm) Earth-based photograph of part of the lunar nearside, showing Mare Serenitatis and Mare Tranquillitatis. Serenitatis is the circular maria. The photograph was made by combining a positive transparency of the IR photograph with an ultraviolet negative. Thus, redder areas appear brighter in the resulting photograph. Note that the bluer maria, which are also older, outcrop around the margins of Serenitatis. Courtesy, E. Whitaker.

regions (Figure 36-10). Based on crystallization ages of Apollo samples, together with color and geologic mapping, this early volcanic phase lasted from 3.8 to 3.5 billion years ago and covered most of the eastern maria. Age estimates have been obtained for mare regions beyond the landing sites by comparing the extent of crater rim and wall degradation relative to the extent of degradation of craters at the landing sites (Soderblom and Lebofsky, 1972; Boyce et al. 1974).

Head (1976) recognized a second period of volcanism, ranging from 3.5 to 3.0 billion years ago (middle to late Imbrian), and covering much of the remaining maria. These deposits are characteristically redder and brighter than the earlier deposits. Finally, a third period of volcanism producing relatively dark, bluish deposits can also be recognized. These relatively young units cover parts of Mare Imbrium and the western maria. The probable age range for the youngest units is 3.0 to 2.5 billion years. Stratigraphically these deposits form part of the Eratosthenian System. The Eratosthenian System also includes materials associated with craters superimposed on the maria that are still fresh-appearing, but lacking extensive ray systems. The youngest system on the Moon is the Copernican, which includes craters with extensive ray systems, such as Copernicus and Tycho.

SPECTROPHOTOMETRIC OBSERVATIONS

Numerous Earth-based observations of the Moon have been obtained covering the visible and near-infrared parts of the spectrum (Adams et al. 1981). Four sets of data have been used extensively to understand the mineralogy of the lunar surface and variations in soil mineralogy with location (Pieters, 1978). The data sets are:

- Multispectral images with 1 to 50 km resolution, acquired using 2 to 5 passbands from 0.30 to 1.0 μm (Figure 36-10)
- Spectral reflectance measurements for 10 to 20 km areas, covering the 0.3 to 1.1 μm spectral region
- Infrared spectral reflectance data for 20 km diameter regions, covering the 0.65 to 2.5 μm spectral region
- Laboratory spectra of returned lunar soils, covering the wavelength range from 0.35 to 2.5 μm

Reflectance data are usually obtained by acquiring nearly concurrent data for the target and a nearby star. The target/star ratio is calculated to eliminate instrumental and atmospheric effects and the flux received from the target is then obtained by normalizing to a standard star and dividing by the solar flux (Adams et al. 1981).

The normal albedo is defined as the reflectance of a surface at normal incidence relative to that of a Lambertian surface, for some wavelength interval. A more detailed discussion of albedo can be found in Chapter 3. The two major geochemical units on the Moon, the terra and maria, can be readily distinguished on the basis of their normal albedoes in the visible part of the spectrum (Pohn and Wildey, 1970). In fact, the albedo of the lunar surface has been shown to correlate directly with the Al/Mg ratio, with the more feldspathic terra being brighter than the basaltic maria.

A primary way of detecting the older, blue maria deposits, the intermediate-aged red deposits, and the younger blue deposits is through the value of the ratio of reflectances in the ultraviolet (0.40 μm) as opposed to the visible (0.56 μm) part of the spectrum. The dark, relatively blue deposits have a high UV/visible ratio, while the brighter, redder deposits have a lower ratio. As shown in Figure 36-11, the ratio has been shown in laboratory analysis of lunar samples to be dependent on the TiO_2 content of the lunar soils. The change in the UV/visible ratio seems to be largely due to the effects of charge-transfer absorptions in Ti- and Fe-bearing glasses and opaque minerals (Charette et al. 1974). Since FeO in lunar soils varies by only about 25%, while TiO_2 varies by a factor of 10, changes in the UV/visible ratio are largely due to variations in TiO_2 contact. The early deposits are TiO_2-rich, the intermediate deposits are relatively depleted in TiO_2, while the youngest deposits are again enriched in TiO_2.

Another important method for interpreting lunar spectrophotometric data concerns the depth, symmetry, and center wavelength of absorption features seen near 1 μm and the depth and center wavelength of the 2 μm absorptions seen in the reflectance spectra of some fresher lunar soils (Figure 36-12). A summary of the locations of these diagnostic features is given in Figure 36-13. The 1 μm feature is due to Fe^{+2} in sixfold coordination within the crystal lattice of pyroxenes, olivine, and plagioclase feldspars (Adams, 1974). Homogeneous glasses also exhibit a broad, symmetric band with a center wavelength located near 1.05 μm (Adams, 1974). Pyroxenes have a relatively narrow absorption feature that ranges between 0.91 and 0.98 μm, depending on the composition. The 2 μm band is also due to pyroxene. Olivines exhibit a band that ranges from 1.03 to 1.06 micrometers as the Fe^{+2} content increases. The olivine band is quite asymmetric. Finally, plagioclase, which can contain FeO as a trace element, often exhibits a broad absorption centered between 1.25 to 1.30 μm. Techniques

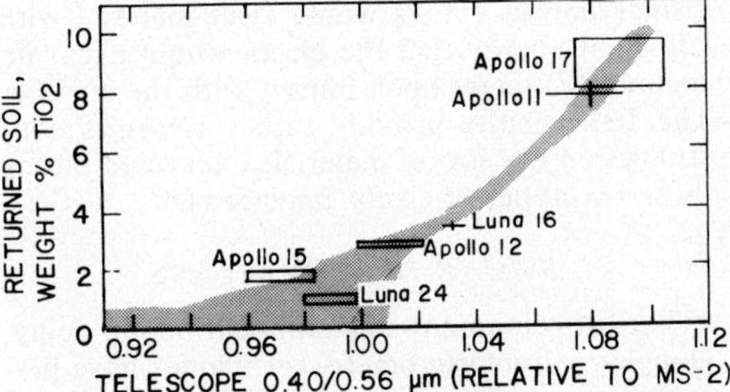

Fig. 36-11. Relation between the TiO_2 content of lunar soils returned during the Apollo missions and the telescopically-derived UV/Visible ratio of the reflectance spectra for the sites. Note that the ratio increases as the TiO_2 content increases. After Charette et al. (1974).

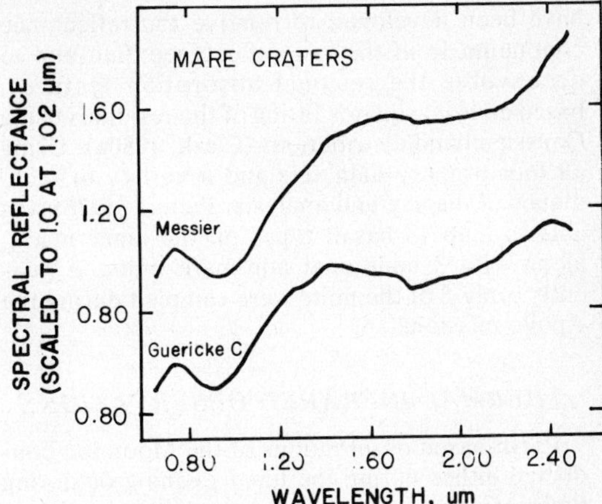

Fig. 36-12. Telescopic reflectance spectra covering two fresh mare craters, Messier and Guericke C. The absorptions centered near 1 and 2 microns can be used to infer the major mafic mineral component. Pyroxene dominates both spectra, with the material in Messier being enriched in calcium relative to material exposed in Guericke C. After Adams et al. (1981).

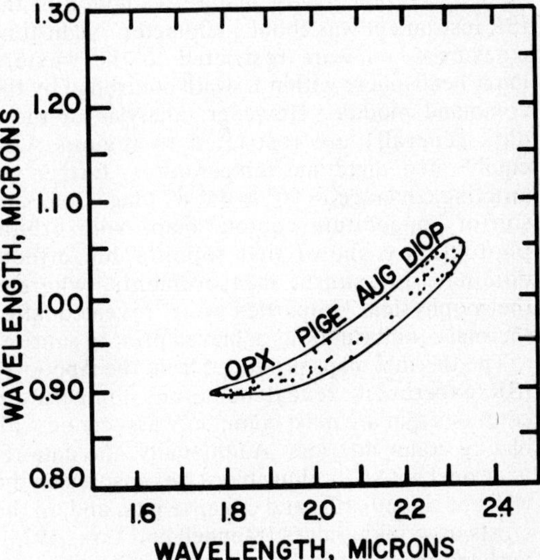

Fig. 36-13. Plot of the center wavelength for the 1 and 2 micrometer absorptions in pyroxenes. For orthopyroxenes, the absorptions shift to longer wavelengths with increasing Fe^{+2} content, while for clinopyroxenes the absorptions shift to longer wavelengths as the Ca^{+2}/Mg^{+2} content increases. OPX = orthpyroxene, AUG = augite, PIGE = pigeonite, DIOP-SAL = diopside-salite. After Adams (1974).

have been developed to remove the reflectance continuum from the raw reflectance data and to deconvolve the residual absorption features, based on least squares fitting of the residuals using Gaussian band distributions (Clark, 1980a). Using all four primary data sets and a variety of techniques of display and analysis, Pieters (1978) was able to map 13 basalt types on the lunar maria, along with 3 additional non-mare units. Apparently, only 5 of the units were sampled during the Apollo missions.

THERMAL INFRARED OBSERVATIONS

Earth-based observations of the Moon are conducted either during the lunar evening or during times of solar eclipse. Eclipse measurements are much easier to acquire than lunar nighttime measurements, mainly because of much higher radiant fluxes emitted from the relatively warm surface during eclipse periods (Saari and Shorthill, 1972). The generally rapid cooling of the lunar surface during eclipse periods demonstrated the existence of a dust layer of low thermal inertia long before the Surveyor or Apollo landings, along with blocky regions in and around many of the younger lunar craters.

The Apollo 17 Infrared Scanning Radiometer (ISR) experiment provided temperature scans of the lunar surface during orbit (Mendell and Low, 1975). Because the time spent at the Moon was only about 6 Earth days, complete cooling histories were not possible to obtain. Rather, the mission objective was to obtain nighttime temperatures. The practical or feature resolution of the ISR instrument was about 1 kilometer. Night-time measurements were restricted to the western lunar hemisphere within a swath controlled by the command module. However, analysis of these data generally are restricted to regions with equilibrated nighttime temperatures, further restricting coverage to 90° to 45° w. long. Comparison of temperature contour maps with orbital photography shows that topographic effects dominate pre-sunset measurements, whereas thermophysical properties were revealed after thermal equilibrium was achieved prior to sunrise.

The thermal maps produced from the Apollo 17 ISR experiment reveal numerous temperature contrasts that are most commonly associated with blocky crater interiors. Additionally, the data reveal local hot spots along blocky exposures on the walls of sinuous rills and collapse pits, and on the crests of wrinkle ridges (Mendell and Low, 1975). Perhaps more intriguing was the revelation that crater-related hot spots were typically associated with very small craters (≲3 km) and with the interiors and inner ejecta deposits of larger fresh craters (Figure 36-14). The broad hummocky ejecta, secondary craters, and rays associated with larger craters usually do not appear to be much hotter than surrounding regions (Schultz and Mendell, 1978). This observation suggests that the continuous and discontinuous ejecta deposits of large craters (≳3 km) are uniformly deficient in blocks (relative to the average mare) larger than 30 cm, whereas the ejecta of small craters (≲3 km) are dominated by large blocks. Such data provide important clues for processes associated with impact cratering. For example, Oberbeck et al. (1975) argue that craters smaller than some critical size would eject material that would be deposited as a blanket of soil and blocks. Larger impact events would eject material with sufficient energy that the ejecta would excavate secondary craters upon impact with the surface. The ISR results provide rather rigorous constraints on the size of materials excavated during these two types of ejecta emplacement.

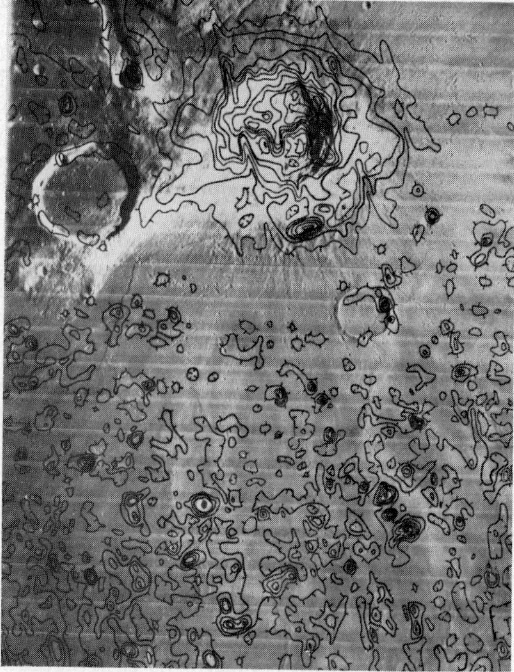

Fig. 36-14. Apollo infrared Scanning Radiometer night-time temperature data are plotted over a section of a Lunar Orbiter frame of the Aristarchus region of the moon. Aristarchus is the bright crater at the top. Aristarchus is 36 km in diameter. The base temperature around Aristarchus is 96 K and the contour interval is 3 K. The floor and walls of Aristarchus are warmer than surroundings. After Schultz and Mendell (1978).

RADAR OBSERVATIONS

The development and application of delay-Doppler and interferometry techniques have permitted relatively detailed Earth-based observations of the Moon at radar wavelengths (Pettengill et al. 1974a). Scattering of radar waves by the lunar surface provides important information about the physical nature of the near surface at scales generally unresolved by the best orbital photography. Moreover, radar penetrates below the surface to a depth about 10 to 50 times the wavelength used, thereby probing hidden characteristics (Thompson, 1979). During the Apollo 17 mission, a long wavelength (2, 20, and 60 meters) radar experiment called the Lunar Sounder extended radar observations of the Moon to lunar orbit, revealing discontinuities believed to represent buried strata and faults (Peeples et al. 1978).

The delay-Doppler technique takes into account both the time delay in radar returns associated with different positions on the lunar globe and the change in frequency (Doppler shifts) associated with lunar libration and Earth-related diurnal librations (Pettengill et al. 1974a). The combined librations cause the sub-Earth point to move through the radar beam during the course of radar observations, thereby generating Doppler shifts. A wide radar beam covering the whole Moon results in two areas exhibiting the same Doppler shift and the same radar time delay. This ambiguity is resolved by combining the delay-Doppler technique with interferometry. Another antenna, physically separated from the first, observes the same areas. Phase differences can then be used to isolate the returns from the two regions (Thompson, 1979). Alternatively, the delay-Doppler technique can be used to image small areas of the Moon so that only one of the two areas with the same delay and Doppler shift are in view at any given time. The reader is referred to Chapter 9—Radar Fundamentals, for a more extensive discussion of these techniques. Topographic mapping with a 10–100 meter vertical accuracy and a 1 to 3 km lateral resolution can also be accomplished by using delay-Doppler and interferometry (Zisk, 1972).

Two surface characteristics generally control the backscattered radar echo from the Moon: diffuse scattering by wavelength-size roughness and quasi-specular scattering by slopes. Three radar wavelengths are currently employed in Earth-based observations of the Moon. The wavelengths are 3.8 cm, 70 cm, and 7.5 m, which effectively span the radio window limited at short wavelengths by atmospheric absorption and at long wavelengths by the ionosphere (Thompson, 1979). The 7.5 m data show that the maria backscatter only ¼ to ½ as much energy as the terra. The 70 cm data show enhanced echoes from a number of lunar craters, including Copernicus and Tycho, as do the 3.8 cm data (Figure 36-15). Thompson et al. (1974) considered thermal IR, radar, and photogeologic data for 51 lunar craters. They suggested that the higher eclipse temperatures and enhanced radar backscatter of Copernican craters are consistent with a blocky, pristine nature of the crater rims and interiors. Older craters only had radar enhancements at longer wavelengths and terra craters did not exhibit any enhancements. Such a sequence is consistent with progressive destruction of larger and larger features by impacts that are small compared to the size of the feature being degraded. Such a sequence seems to dominate the destruction of lunar surface features (Soderblom and Lebofsky, 1972).

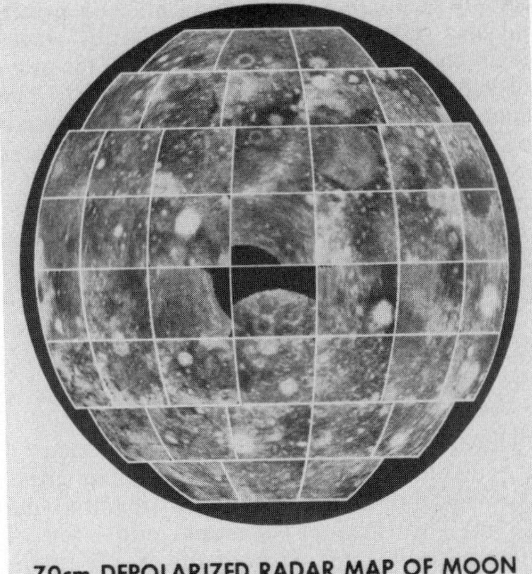

Fig. 36-15. Mosaic of de polarized lunar radar maps acquired with 70-cm wavelength radar. The data are normalized to show departures from the scene average. The resolution cell size varies from 5 to 10 km. Bright returns are due to blocky surfaces in younger craters. After Thompson (1979).

Schaber et al. (1974) pointed out that changes in the dielectric constant of the lunar soil can also locally influence the returned radar power. He suggested that the reduction of returned power may be directly related to electric losses due to increased FeO and TiO_2 contents in the bluer maria units. Finally, Pieters et al. (1973) were able to associate a local 3.8 cm radar low with dark mantle deposits, which are FeO and TiO_2-rich, and that may be fine-grained and therefore absorbing.

Radar sensing of the lunar surface and subsurface was also performed from orbit during the Apollo missions. Bistatic radar experiments on Apollo 14, 15, and 16 spacecraft permitted measurements of surface roughness at two wavelengths (13 cm and 116 cm). The bistatic radar experiment involved transmitting radio waves in relatively close proximity to the surface during lunar orbit and receiving the backscattered signal on the Earth (see Tyler and Howard, 1973). Analysis of depolarized radar echoes showed that topography and slopes at all scale lengths down to 0.5 cm substantially controlled the recorded amplitudes. Differentiation of these effects from the effects due to scattering by near-surface blocks, and to variable electromagnetic absorption is not straightforward.

As discussed, the Lunar Sounder experiment (Apollo 17) employed a coherent synthetic aperture radar system with transmitting and receiving systems on board the Apollo command module. Two long wavelength channels (20 meters and 60 meters) recorded both surface and subsurface specular reflections, whereas a shorter wavelength channel (2 meters) was designed for near-surface reflections (Peeples et al. 1978). The Sounder experiment revealed hidden subsurface structure in several lunar regions. Two interfaces were found in Mare Serenitatis at 0.9 km and 1.6 km depths and a single interface was recorded in Mare Crisium at 1.4 km depth. These interfaces are interpreted as relatively continuous layers of regolith buried by later volcanic materials.

APOLLO ALPHA PARTICLE, X-RAY, FLUORESCENCE AND GAMMA-RAY OBSERVATIONS

Three experiments on the Apollo 15 and 16 command modules permitted mapping gross geochemical differences along the orbital ground tracks. The three experiments consisted of alpha-particle, gamma-ray, and X-ray spectrometers. The combined area covered for the Apollo 15 and 16 groundtracks included about 20% of the lunar surface. The alpha particle experiment was designed to detect the concentration of ^{238}U and ^{232}Th in the lunar soil (Gorenstein and Bjorkholm, 1972). Alpha particles are emitted during the decay of ^{222}Rn and ^{220}Rn. These gases, in turn, are daughter products of ^{238}U and ^{232}Th, respectively. Alpha particles formed less than 10 μm below the surface would have been detected by the alpha-particle spectrometer. Results from this experiment include low resolution (several hundred km) maps of U and Th contents for regions beneath the command module ground tracks (Gorenstein and Bjorkholm, 1972). The gamma-ray spectrometer recorded the decay products of naturally radioactive isotopes (K, Th, U) and the products produced by cosmic-ray induced nuclear reactions with lunar surface materials. Cosmic ray interactions with surface materials produce gamma rays of high energies, where the energy level is diagnostic of such elements as O, Si, Ti, Fe, and Mg. Results from this experiment included maps of net radioactivity and maps of iron and titanium concentration (Arnold et al. 1977; Davis, 1980; Metzger and Parker, 1980). Such data correspond to the concentrations of radioactive elements, iron, and titanium, within about the first ten centimeters below the surface.

Results from the alpha particle and gamma ray experiments demonstrate that the maria have a higher concentration of radioactive elements than do the terra (color Figure 36-16). The western maria exhibit higher radioactivity values and greater variations in radioactivity than do the eastern maria. Moreover, certain regions exhibit particularly high radioactivity that may be indicative of local concentrations of KREEP rocks. KREEP rocks may be volcanic in origin and are enriched in potassium (K), rare earth elements (REE), and in phosphorous (P) (Taylor, 1975). KREEP rocks may be indicative of pre-mare volcanic deposits that were largely destroyed and reworked during the early torrential bombardment of the lunar surface. The Apennine Bench region, for example, is a radioactive high in the radioactivity data (Spudis, 1978). Other such concentrations were detected in (a) the ejecta deposits of the crater Aristarchus and the low-albedo plateau on which it occurs, and (b) the Fra Mauro plains region (ejecta deposits of Imbrium event) and (c) the plains areas within the old basin, Balmer. The distribution maps of Fe concentrations reveal, as expected from other techniques, that Fe concentrations are greatest in the maria and lowest in the terra. Similarly, maps of Ti reveal variations in the maria that are roughly consistent with spectral reflectance data.

The X-ray spectrometer recorded fluorescent X-rays emitted from elements (atomic number less than 14) in response to interaction of solar X-rays with lunar surface materials. The energy distribution and intensity of the solar X-rays in the energy range measured allowed Al, Si and Mg concentrations to be determined over 10% of the lunar surface (Andre et al. 1977). Owing to surface roughness and to solar X-ray flux variations, the data typically are presented as ratios, e.g., Al/Si and Mg/Si (Color Figure 36-17). These ratios clearly distinguish between mafic mare units (low Al/Si) and non-mafic terra units (high Al/Si) at resolutions approaching 15–30 km in certain areas. In addition, variations in Mg/Si ratios reveal contrasting basalt types with, for instance, the older, peripheral maria basalts exhibiting slightly higher Mg/Si ratios than other maria regions (Andre et al. 1979).

LUNAR CONSORTIUM

The diversity of formats, resolutions, and coverage of the various remote-sensing products (and potential field data—Bouguer gravity, magnetic anomalies) for the Moon has created serious difficulties in attempting global comparisons of the different data sets. As a result, a consortium of lunar scientists was formed to develop a data-processing scheme whereby all data sets could be reformatted and presented on a common base. Initial results are illustrated as frontispieces in Volume I of the Proceedings of the Eighth Lunar Science Conference. The color-coded, computer-generated maps permit easy comparison of the various data sets. The data include gamma-ray data, X-ray data, geological maps, gravity maps, magnetometer surveys, electron reflectance measurements, visible and near-infrared spectral reflectance maps, mare relative age maps, albedo maps, and laser altimetry profiles. Smoothing techniques were also developed that allowed direct comparison of data acquired with differing resolution (Eliason and Soderblom, 1977). The Lunar Consortium results are a milestone in the use of data processing oriented toward production of directly comparable data of widely varying types.

REMOTE SENSING OF MARS

PHOTOGEOLOGIC ANALYSES

Photogeologic interpretations of the Martian surface began in 1965 with imaging data acquired during the Mariner 4 flyby. Mariner 4 acquired 22 images, covering about 2% of the surface, with a best spatial resolution of about 5 km/pixel. The image data showed an ancient, highly cratered surface that appeared to have been modified to a greater extent than the lunar terra (Leighton et al. 1965). In 1969, Mariner 6 and 7, also flyby spacecraft, acquired near-encounter data over about 20% of the Martian surface. Most of these data also pointed to an ancient, heavily cratered crust, although some of the images pointed to a collapsed, chaotic terrain that hinted at a variegated and complex surface (Leighton et al. 1969). In 1971–72, Mariner 9, an orbiter, acquired about 7000 images of the atmosphere and surface, with nearly 100% coverage at a resolution of about 1 to 3 km/pixel and 1% coverage at 0.1 to 0.3 km/pixel (Masursky, 1973). The increase in coverage and in resolution profoundly altered views of the geologic evolution of the Martian surface. Mariners 4, 6, and 7, unfortunately, were targeted to acquire imaging data largely in the southern hemisphere, a region dominated by an ancient, heavily cratered crust (Figure 36-4). The northern hemisphere, as shown by Mariner 9, consists of a wide variety of terrain types, mostly sparsely cratered and thus younger than the ancient crustal units exposed in the south. The Viking Mission (2 orbiting and 2 landed spacecraft) appreciably increased our understanding of the origin of various terrain units, in addition to allowing the characterization of a complex of surface phenomena that were not evident in the Mariner data. Part of the reason is that the Viking Orbiter cameras, returning about 50,000 images over their 4 year lifetime, had a somewhat better resolving power than the Mariner cameras (Snyder, 1979). In addition, the periapsis was lowered to approximately 300 km for the Viking Orbiters and scan platform motion was used as an image smear compensation system. Thus, resolutions of as small as 8 meters/pixel were obtained in limited areas (Figure 36-18).

A fairly coherent picture of the global geologic evolution of Mars can now be constructed, largely by combining new data from the Viking mission, with the global distribution of geologic units inferred from the Mariner 9 data (Carr, 1980; Arvidson et al. 1980a). Martian geologic evolution has involved extensive differentiation of the interior, extrusion of major quantities of volcanic materials, both extensional and compressional tectonics, the formation of geomorphic features by periglacial processes, fluvial erosion, and aeolian activity. Approximately half of the Martian surface is occupied by ancient cratered terrain, a surface that has a crater abundance per unit area roughly similar to the highly cratered surfaces of the Earth's Moon and Mercury (Oberbeck et al. 1977). As discussed, the lunar terra, based on the radiometric ages of samples returned during the Apollo missions, seems to have formed between 4.0 and 4.5 billion years ago. The broader scale topography on the Martian cratered terrain is

Fig. 36-18. (a) Viking Orbiter 1 view of relatively young volcanic plains located at 21° S. latitude, 139° W. longitude on the southwest flank of the Tharsis plateau, Mars. The flows have covered most but not all, of the rugged cratered terrain in this region. (b) High-resolution Viking Orbiter 1 view of the flow front shown in the white box in Figure 36-18a. Analysis of the heights of these flows, assuming that they behaved as Bingham fluids, suggests that the lavas are basaltic in composition. Figure 36-18a is VO 1 056A02; Figure 36-18b is VO 1 806A60. The high resolution view covers about 15 km across. After Arvidson et al. (1980a).

probably of similar age. This ancient cratered terrain is separated from the extensive, sparsely cratered plains to the north by a great circle inclined to the equator by about 35 degrees (Mutch et al. 1976). The boundary between the cratered terrain and the plains, where not directly covered by volcanic materials, consists of a complex of faulted escarpments (Figure 36-19). One of the major problems of Martian geologic evolution is that of establishing the reason for, or the processes by which, this major hemispheric asymmetry came about (Wise et al. 1979). The northern hemisphere is covered with plains, largely of volcanic origin, and exhibiting a variety of crater abundances and thus a variety of ages (Neukum and Hiller, 1981). The surfaces probably range in age from several billions to several hundreds of millions of years or younger. A reasonable hypothesis for formation of the asymmetry is that, before or during core formation, convective overturn within the interior led to extension in the north, thereby fracturing and thinning the northern cratered terrain crust. The crustal thinning and extension eased the rise of magmas, leading to extensive volcanic activity. Most of the southern cratered terrain proper was also mantled but only by a relatively thin cover of volcanic materials during the time period corresponding to the massive volcanic resurfacing in the northern plains (Figure 36-20).

The Tharsis plateau of Mars, occupying about 4000 km in breadth, 4 to 10 km in height, and straddling the equator, is covered with a vast expanse of relatively young volcanic materials (Figure 36-21). The plateau also dominates the lower

Fig. 36-20. Viking Orbiter 1 frame of Martian cratered terrain, located at 14° S. latitude, 9° W. longitude. The frame covers about 250 km across. The spectrum of crater morphologies evident in this frame indicates that a significant amount of crater degradation has taken place. The process responsible for the degradation has probably been volcanism. This frame contains one of the few possible flow fronts (demarcated by arrows) within the intercrater plains. Viking Orbiter colorimetric data suggest that the plains in this area are much bluer than exposures of primitive cratered terrain crust. The population of small bowl-shaped craters postdates any crater obliteration event. The channel system located on the right edge of the frame merges to a single channel which then cuts through the cratered terrain for several hundred kilometers. VO 1 frame 84A47. After Arvidson et al. (1980a).

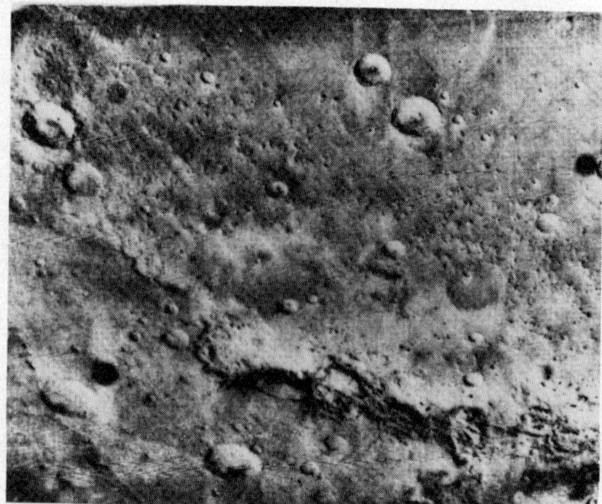

Fig. 36-19. Mariner 9 perspective view of the Martian fretted terrain located at 5° N. latitude, 236° W. longitude. Cratered terrain can be seen in the bottom quarter of the frame. Fretted terrain extends from the escarpment, which marks the edge of cratered terrain, to the top of the image. Fretted terrain may represent the disrupted cratered terrain crust in the northern hemisphere that was not covered by younger volcanic materials. The bottom of the frame is about 600 km across. Mariner 9 A-frame, REV/DAS: 165/07507468. After Arvidson et al. (1980a).

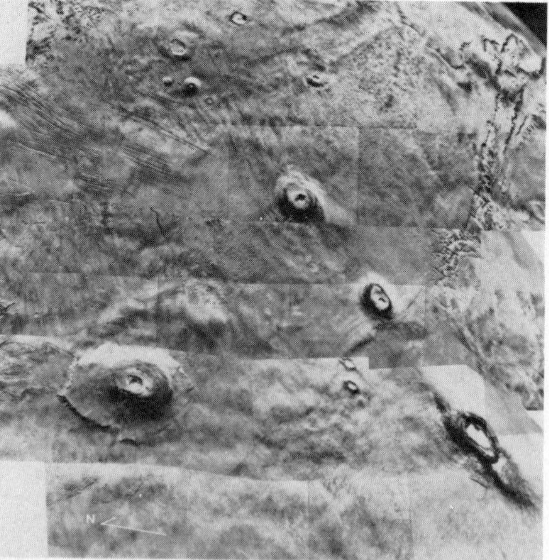

Fig. 36-21. Oblique view of Tharsis Plateau, Mars, showing the volcanoes Olympus Mons (foreground), the Ceraunius Tholus-Arsia Mons line of volcanoes, and the Labyrinthus-Noctis and Valles Marineris canyons (extending to horizon). Olympus Mons is about 600 km wide. Viking Orbiter mosaic composed of frames 753A84 to 90 and 759A71 to 90.

order terms of the Martian gravity field (Phillips and Lambeck, 1980). The topography and the gravitational signature make the plateau the second major asymmetry on Mars. Photogeologic mapping shows that volcanic activity in this region extended over a large portion of geologic time (Neukum and Hiller, 1981). Furthermore, exposures of ancient crust within the plateau, together with mapping of lava thickness based on the largest craters that are nearly filled to their brims, show that the plateau height may be largely caused by relief within the lithosphere and asthenosphere, rather than by thick sequences of lava flows (Plescia and Saunders, 1979). The plateau may be underlain by a thin crust and by a mantle that is slightly less dense than typical (Sleep and Phillips, 1979).

A great surprise from photogeologic studies was the discovery of a number of valleys or channels cut into the Martian surface. These features are clustered in the older plains and the cratered terrain regions (Figure 36-22). Most investigators, based on comparison to terrestrial systems, together with consideration of the physics involved in channel formation, feel that the major channels were cut by torrential floods of water or by water-mud slurries (Masursky et al. 1977; Baker, 1979). The most likely source of the water would have been catastrophic break-out from underground reservoirs. Morphological comparisons between the distinctive and much smaller integrated Martian valley networks and terrestrial fluvial systems demonstrate that the Martian systems need not have formed by rainfall and runoff (Pieri, 1980). In contrast to terrestrial systems, the Martian networks do not exhibit junction angle (angle between branch and trunk) patterns that are indicative of downhill flow of water. Pieri (1980) suggests that the Martian networks are due to groundwater sapping. An alternative hypothesis is that the integrated channels systems formed early in Martian history, when an ammonia- and methane-rich atmosphere supported a greenhouse heated atmosphere that was warm enough to allow rainfall and runoff (Pollack, 1979). Other geologic features that suggest the presence of large reservoirs of water, more than predicted from scaling the abundances of noble gases in the atmosphere (Arvidson et al. 1980a), include:

- The fretted and chaotic terrains, which are reminiscent of collapse following geothermal melting of ice bodies
- The peculiar multi-lobed ejecta deposits surrounding many craters that may indicate the combined effects of impact-melted crustal ices and interaction of the ballistic ejecta with the atmosphere (Figure 36-23)
- Patterned or polygonal ground covering much of the northern high latitudes, which may be dessication features or areas underlain with freeze-thaw polygons (Figure 36-23), and
- Numerous lines of evidence, both theoretical and experimental, that suggest that much of the Martian soil consists of weathered materials consisting of chemically bound hydroxides and water, as well as carbonates, together with adsorbed and absorbed water and carbon dioxide.

The presence of a thin atmosphere on Mars provides for another set of geologic processes that involve transfer of condensates to and from the surface, together with mechanical and chemical

Fig. 36-23. This Viking Orbiter frame of Mars is located at 43° N. latitude, 15° W. longitude within an area of debris flow ejecta that must have come from the crater located to the left side of the frame. Also, polygonal ground can be seen on the right side of the image. The crater may have vaporized and excavated ice upon impact, providing a medium to transport the ejecta as a ground-hugging flow. The polygonal ground can be interpreted as freeze-thaw pologons or as desiccation polygons, both of which involve water. VO frame 9A42. The frame covers about 40 km across.

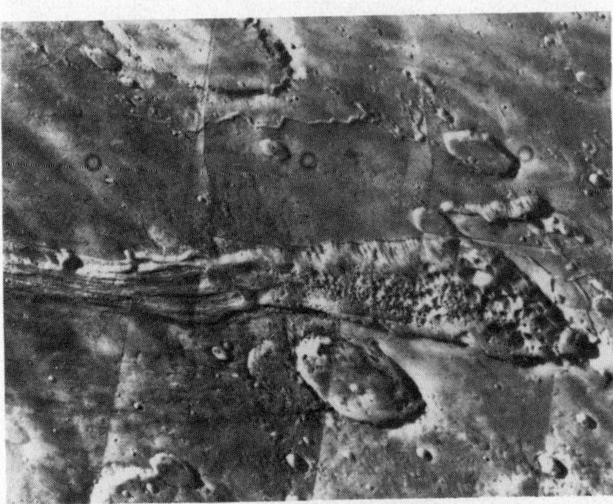

Fig. 36-22. This Viking Orbiter mosaic of Mars is centered at 1° S. latitude, 44° W. longitude and shows a large collapse feature, with a complex channel leaving one end of the valley floor. The most probable explanation for this feature is the rapid release of large quantities of groundwater. The canyon floor is about 10 km wide at the junction with the channel. VO 1 frames 14A65-14A70.

coupling (i.e. aeolian processes) of the surface and the atmosphere. The Martian atmosphere consists primarily of about 5 to 10 mb of CO_2, together with smaller amounts of H_2O, argon, and nitrogen (Owen et al. 1977). Water is thermodynamically stable only as a solid or gas, although recent radar observations of the surface suggest the presence of minute amounts of liquid water at certain times of the day and year in some locations (Zisk and Mouginis-Mark, 1980).

During the winter season in a given hemisphere a mixture of carbon dioxide and water ices condense onto the surface near the poles, forming a thin (meters) seasonal polar cap. The temporary deposits evaporate in the northern polar cap area during the ensuing spring, exposing a 1 to 3 km thick sequence of layered dust and ice deposits (Cutts et al. 1976) (Figure 36-24). The Viking Orbiter Thermal Mapper recorded brightness temperatures as high as 210 K over these deposits during the northern summer (Kieffer et al. 1976). In addition, the Viking Orbiter Water Vapor Mapper showed an atmosphere saturated with water vapor over the northern summer cap (Farmer et al. 1977). Both results are consistent with water-ice as the dominant volatile in these deposits, since carbon dioxide or clathrates would have evaporated at 148 K and 152 K, respectively.

Seasonal CO_2 and H_2O condensates were seen on the surface from the Viking 2 Lander, which appears to have landed in a location (48 degrees N. Lat.) that is just on the inside of the southernmost extent of the northern winter polar cap (Jones et al., 1979) (Figure 36-25). The condensates formed during the two winter seasons of Viking observations. The deposits remained on the surface for about 200 Mars days, finally evaporating as the spring seasons progressed. After the condensates had evaporated, the surface was found to be considerably brighter and redder, probably because of the accumulation of a thin (several micrometers) layer of red dust (Guinness et al. 1979). It seems likely that formation of the condensates and the dust accumulation are linked—the condensates probably formed in the atmosphere by accumulation of water and carbon dioxide ice around tiny dust particles. The increasing particle size as more ice condensed led to

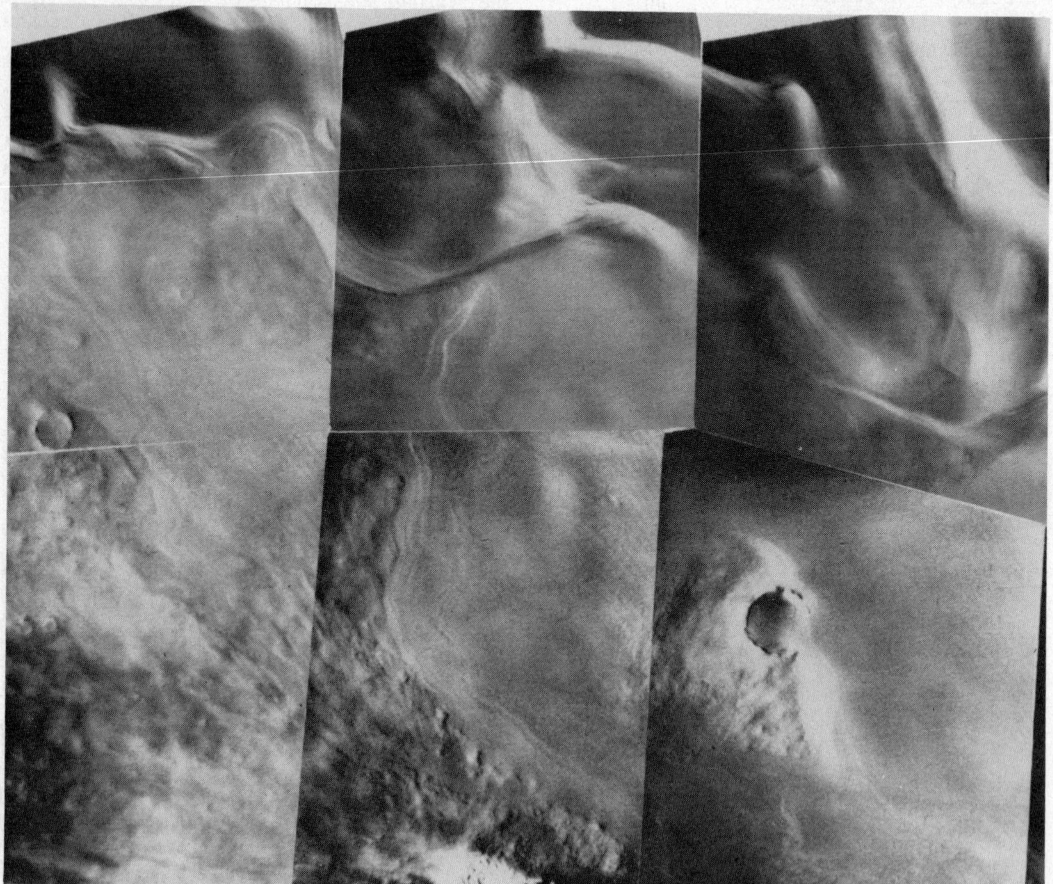

Fig. 36-24. Mosaic of Viking Orbiter 2 frames of Mars located at 70° N. latitude, 270° W. longitude at the edge of the north polar permanent cap. The layered cap deposits, which are a mixture of ice and dust, can be seen covering part of a crater and associated ejecta deposits. The mosaic covers about 300 km in width.

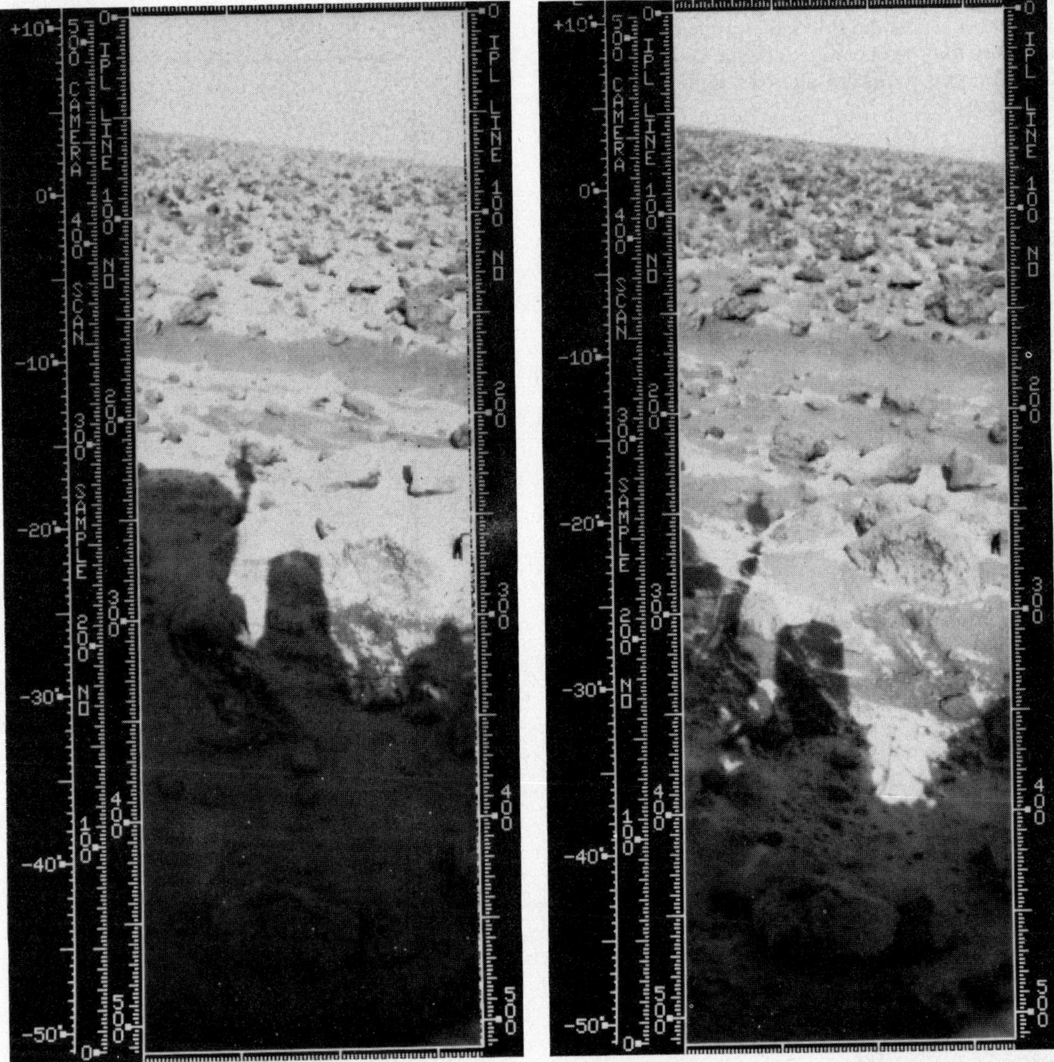

Fig. 36-25. Two Viking Lander 2 frames, both acquired at about 1 pm Mars time, on the 341 and 365 Mars days after landing. Both images were acquired in blue light. Shadows in the foreground are due to the Lander. These frames document the formation and subsequent evaporation of a thin layer of condensates that formed during the winter and evaporated during the ensuing spring. VL 2 frames 21E113 and 21E153.

higher settling velocities and thus a greater chance of being deposited onto the surface. Such a process would be even more enhanced over the colder polar regions, providing a plausible mechanism for accumulation of the layered deposits of dust and ice.

A thick sequence of dust and ice deposits also exists in the south pole, although the composition of the ice is uncertain because global dust storms obscured the deposits and modulated atmospheric temperatures during those southern summer seasons when the Viking Orbiters conducted observations (James et al. 1979). The normal albedo of the south polar deposits has been estimated to be about 0.7, which may make it difficult for all the CO_2 and H_2O deposited during the winter to evaporate during the following spring and summer seasons (James et al. 1979). Thus, the south polar cap deposits may be composed of a mixture of carbon dioxide, water, clathrates, and dust.

Global dust storms generally occur during the summer season in the southern hemisphere. The reason seems to be that Mars is at the perihelion position in its orbit during the southern summer and, as a consequence, receives about 50% more solar energy than it does when it is in its aphelion position (Mutch et al. 1976). However, periodic changes in the orbital obliquity and eccentricity, together with precession of the pericenter and spin axis, should lead to, among other effects, cyclic alternations of which hemisphere is in its summer season during perihelion (Toon et al. 1981). The net result of such orbital variations may be that the asymmetry in polar deposits between the water-ice dominated northern deposits and the carbon dioxide-ice dominated southern deposits may cyclically exchange over periods of one hundred thousand and one million years, respectively.

OBSERVATIONS IN THE VISIBLE TO REFLECTED IR

Earth-based observations, using narrow-band filters, have been acquired of Mars for the wavelength range from 0.3 to 2.5 μm (Adams et al. 1981). The regions resolved with these observations are approximately 200 km across, under the best of observing conditions. These data are detailed enough in spectral resolution, however, to provide important and, at this point, unique information on the mineralogical make-up of surface materials. The Viking Orbiter imaging-system acquired data in the 0.45–0.59 micrometers parts of the spectrum (Soderblom et al. 1978). These data provide several tens of km/pixel resolution of surface features, providing information on the geographic variability of surface materials that is two orders of magnitude better than that obtained from Earth-based studies. Finally, the Viking Lander facsimile cameras acquired colorimetric data in six channels ranging from 0.4 to 1.0 micrometers in wavelength (Huck et al. 1977). These data provide information on the characteristics of Martian soils at the scale of centimeters. The Earth-based data are first discussed to provide a mineralogical framework for Martian surface materials. Following that is a discussion of the distribution of surface materials, based on Orbiter data. The section ends with a review of how the multispectral observations from the Landers were used to infer how the soils exposed at the sites fit into the overall spectral variety seen for the planet.

Mars exhibits a bimodal albedo distribution (Kieffer et al. 1977). The two albedo modes correspond to the classical bright and dark areas that can be seen from Earth. After removal of CO_2 atmospheric effects, bright area spectral reflectances are characterized by strong Fe^{+3} charge-transfer absorptions from the UV to 0.75 microns due to ferric oxides (Figure 36-26) (Singer et al. 1979). A weaker Fe^{+3} absorption, also due to charge transfer absorptions in ferric oxides occurs near 0.87 microns. Between 1.4 and 1.7 micrometers there is a broad absorption band which can be interpreted as being due to H_2O either in ice form or in a hydrated form. Dark area spectra are substantially different (Figure 36-26). The slope from the UV to the red is reduced and the 0.87 μm feature is subdued. Both effects in the dark area spectra can be attributed to a lower ferric iron content. Dark regions also show Fe^{+2} absorptions near 1.0 μm. The depth and band center of this feature shifts from location to location on the planet, probably reflecting differences in mafic mineralogy, mostly in the contents of pyroxene

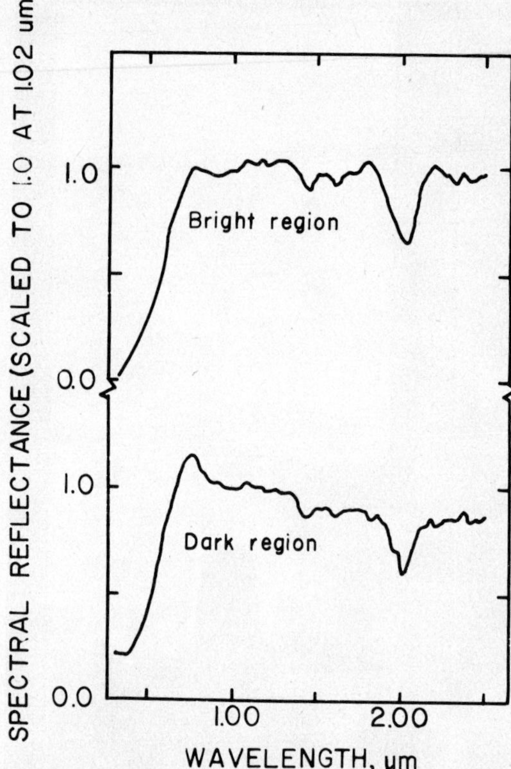

Fig. 36-26. Representative Earth-based specta of Martian bright and dark areas, scaled to unity at 1.02 microns. After Singer et al. (1979).

and basaltic glass (Singer et al. 1980). In contrast to bright areas, dark spectra have a distinct peak near 0.75 μm and slope fairly uniformly downwards from 1.1 to 2.5 μm. Bright areas, which tend to be spectrally uniform from place to place, are probably dominated by soils rich in ferric oxides and hydroxides and clay minerals. Dark areas display a much greater spectral variety, with spectral characteristics consistent with a combination of unweathered dark rock or rock fragments and fine-grained Fe^{+3}-rich weathering products.

The Viking Orbiter imaging system acquired data in three passbands (0.45 μm; 0.53 μm; and 0.59 μm), with a best resolution of several km/pixel. Because the three passbands fall on intense Fe^{+3} absorptions, multispectral images derived from these data primarily reflect differences in Fe^{+3} mineralogy and content (Soderblom et al. 1978). Even so, significant spectral variety has been detected, mainly in darker areas. For instance, on a set of images acquired during the approach phase of the mission, a classical dark area located within the ancient cratered terrain between the equator and about 40 degrees S. latitude was found to consist of two units (Color Figure 36-27): (1) A highly cratered region, which is among the reddest on the planet, and (2) volcanic plains filling in the older terrain, with a very low red component. Volcanic constructs, including those within the Tharsis Plateau, are relatively dark and red.

Lander color data are acquired with large emission angles and usually, with large phase angles relative to Earth-based or orbital perspectives, since the Lander cameras are located only 1½ meters above the surface (Huck et al. 1977). As a consequence, a photometric function, where the bidirectional reflectance of the surface is treated as a function of incidence, emission, and phase angles, must be used to correct the Lander data to viewing and lighting conditions similar to those seen from Earth and from orbit before intercomparisons of the data can be done. Guinness (1981) has used the Hapke function (Hapke, 1981), which explicitly includes the backscatter peak at small phase angles, to model the photometry of soils exposed near the Landers. the soil data show that the ratio of brightness in red as opposed to blue wavelengths changes by 33% as the phase angle changes from 5 to 75 degrees. Guinness was able to explicitly remove the effects of atmospheric skylight color by considering the brightness of shadows cast by rocks. The color variations seen at the landing sites, after correcting for lighting and viewing geometries, is smaller than seen for whole Mars. It appears as if the soils at the landing sites are composed of bright area materials that have been homogenized by winds on a global scale. The variations seen are probably primarily related to particle size, with the darker, less red soils being coarser-grained than the brighter, redder materials. Such an interpretation is consistent with the Viking Lander X-ray fluorescence results for the soils, which show very similar chemistries both between and among the sites (Baird and Clark, 1981). The secular and seasonal variations in the color and albedo of Mars, variations that have been tracked from Earth for nearly a century, thus appear to be related to condensates and to wind erosion and deposition of a thin deposit of bright, red dust.

OBSERVATIONS IN THE THERMAL IR

Spacecraft observations of Mars have also been acquired in the thermal IR part of the spectrum. The Mariner 9 spacecraft carried an IR interferometric spectrometer that measured the emission spectra of the dust cloud that covered Mars for the first three months of the mission in 1971–72 (Hanel et al., 1972). The instrument covered the spectral range from 5 to 50 μm. Quantitative analysis of the spectra, most recently by Toon et al., (1977), shows that a good match to the cloud dust particles is a mixture of materials dominated by igneous silicate minerals rich in SiO_2 or by weathering products such as clays. They also concluded that at most only a few percent of carbonate minerals could exist in the cloud.

The Infrared Thermal Mapper instrument that was carried on board the Viking Orbiters had six spectral bands that covered the range from 0.3 to 24 μm (Kieffer et al. 1977). One of the bands, from 0.3 to 3.0 μm, was designed to obtain the data needed to calculate the normal albedo of the surface, data that would also be necessary to obtain the surface thermal inertia. The albedo data reinforce the bimodality of Martian surface units, with the albedo of bright areas clustering near .26 and the albedo of dark areas clustering near .15 (Kieffer et al. 1977). When these data are plotted against surface thermal inertia it can be shown that as the albedo increases, the thermal inertia decreases (Figure 36-28). The thermal inertia of the Martian surface appears to be most dependent on thermal conductivity of surface materials, which is controlled by particle size and packing. In the simplest interpretation, bright areas consist of materials of finer-grain size or smaller degree of packing (lower thermal inertia) than material found in dark areas. The thermal inertia varies by a factor of eight, which is much less than the factor of 40 variation predicted between bare bedrock and soil exposures on Mars. Clearly there are very few areas covered solely by bedrock. Such results are consistent with the analyses of spectral reflectance data for bright and dark regions as described in the last section.

EARTH-BASED RADAR STUDIES

Radar observations of the Martian surface have been directed toward obtaining topographic profiles and surface roughness characteristics. The geometry of the Martian orbit and the obliquity of the spin axis relative to the ecliptic plane permit any point on the planet's surface which is located within the belt bounded by approximately 25 degrees in latitude about the equator to be scanned by radar from Earth. In the course of a typical observing run the subradar point on Mars traces an arc of nearly constant latitude on the planet. The characteristics of the returned echo can be used to construct an altitude profile along the arc. The first radar observation of Mars during the 1969 opposition encircled the planet at about 22° N. Latitude. Analysis of the data showed three broad, high regions that could later be identified with Tharsis, Elysium, and Syrtis Major regions (Pettengill et al. 1969). The maximum observed topographic variation was about 12 km.

The more favorable oppositions of 1971 and 1973, coupled with the improved instrumentation, offered an opportunity to observe the areas south of the equator with unprecedented sensitivity. The ranging experiments resulted in the acquisition of Mars radar topographic data with an areal resolution of 1.3 degrees in latitude by 0.8 degrees in longitude at MIT's Haystack Observatory (Pettengill et al. 1974b), and an areal resolution of 1.3 degrees by 0.16 degrees at the Goldstone receiving station (Downs et al. 1975), with average altitude uncertainties of 75 m, and less than 100 m,

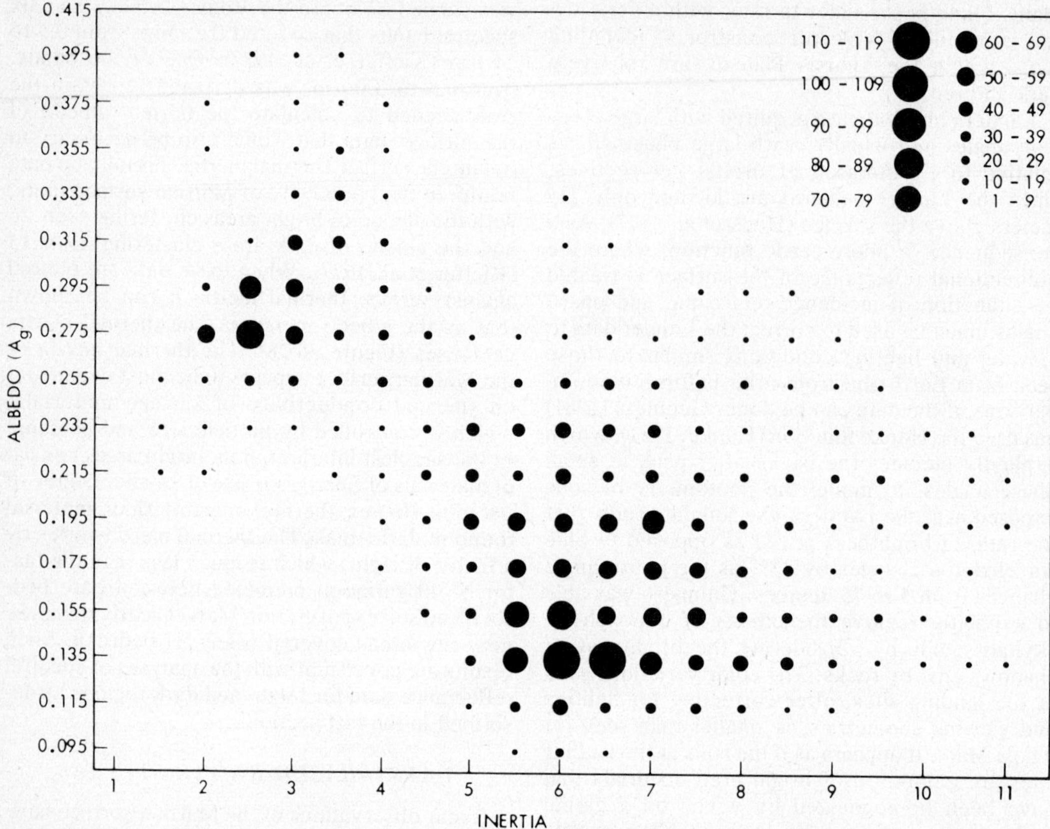

Fig. 36-28. Plot of albedo, corrected for variations in phase angle, verses thermal inertia (in cgs units) for the Martian surface, based on 3345 data points. Data acquired from the Viking Orbiter Infrared Thermal Mapper. Darker areas have a higher thermal inertia. After Kieffer et al. (1977).

respectively. The 1971 and 1973 Goldstone data sets are so far the most complete and have the best areal resolution to date. The wealth of topographic information contained in these data began to be appreciated when viewed in association with the Mariner 9 and Viking Orbiter Imaging data. For example, Figure 36-29 shows topographic profiles across the crater Boeddicker. This crater is significantly shallower than its lunar and Mercurian counterparts. Topographic measurements based on the radar data have also been used to characterize a number of other craters, conclusively showing that Martian craters are significantly shallower than Mercurian and Lunar craters (Roth et al. 1978).

Bistatic spacecraft-to-Earth measurements of surface roughness have been conducted both from the Viking Landers and from the Orbiters (Simpson et al. 1979). The technique, as in the lunar case, involved transmitting a radio pulse from the spacecraft to the surface and then measuring the character of the returned signal from Earth. The Lander bistatic data imply a surface with a dielectric constant close to that of pumice, while the Orbiter bistatic data indicate a quite variable surface in terms of roughness. The smoothest location seen from Orbiter bistatic observations is over the south polar cap deposits.

THE MARS CONSORTIUM

The Mars Consortium consists of a group of researchers with an interest in comparison of the diverse remote sensing data sets that now exist for Mars. As reported in Kieffer et al. (1981), the data consist of:

- Viking Orbiter color images,
- the topography of the planet,
- pre-dawn brightness temperatures,
- the gravity field,
- catalogs of Earth-based spectral reflectances for various regions,
- gelogic maps,
- normal albedoes,
- thermal inertia estimates,
- crater abundance data, and
- data on the character and abundance of aeolian deposits.

The goal of the Consortium is to utilize the data as an integrated package when conducting analyses of, for instance, the distribution of surficial soil

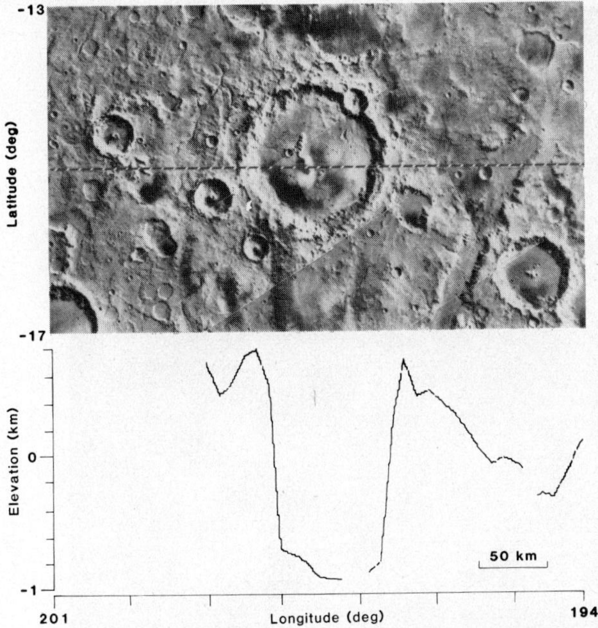

Fig. 36-29. Radar altitude profile across the Martian crater Boeddicher. The radar scan was centered at 15.04°S. Lat. Viking Orbiter frames 596A27, 29, 31, 46, 48 and 50.

units. To facilitate such comparisons, the data sets have all been reformatted to a common base, using the Lunar Consortium data sets as a guide. Viking Lander multispectral data are also included in the Consortia package as ground truth data.

RADAR OBSERVATIONS OF VENUS

EARTH-BASED OBSERVATIONS

Venus is a prime target for radar observations since its surface is obscured from optical and infrared observations by a thick atmosphere and dense clouds. Earth-based imaging of the surface of Venus makes use of the delay-Doppler interferometry mapping technique discussed under the section dealing with radar observations of the Moon.

Earth-based observations are only acquired in the equatorial to midlatitude portion of the planet, centered near 320 degrees east longitude. The reason is that this area of venus is always toward Earth at inferior conjunction. Earth-based, near-equatorial radar reflectivity images of Venus have been produced by both the Arecibo (Campbell and Burns, 1980) and Goldstone Facilities (Jurgens et al. 1980). These data are somewhat complementary since Arecibo maps extend to middle latitudes but miss the equatorial regions, which Goldstone maps in great detail. Typical spatial resolutions for the Earth-based data are several km/resolution element.

Both the Arecibo and Goldstone radars now operate at 12.5 cm wavelengths. Older, coarser radar maps of Venus have been produced at 70 cm by Goldstone and 3.8 cm wavelengths by the Haystack Observatory.

An Earth-based radar image of Venus from Arecibo data is shown in Color Figure 36-30 as an overlay to the Pioneer-Venus altimetry data. The color of an element on the image is proportional to the surface roughness at the centimeter scale, except at lower latitudes, where meter scale roughness influences the returned signal. Perhaps the most dramatic features are the blue Lakshmi Plateau and the red Maxwell area. The center of Maxwell contains a circular depression which may be caldera, which in turn would make this feature a rather large shield volcano. Beta is another prominent broad rise, about 1000 km across. Several areas of valley and ridge topography that are similar to the surface expression of terrestrial basin and range faulting can be seen. Alpha is the most obvious example. A probable fracture zone, some 1000 km long, and 90 km wide, can be seen just to the north of the equator. Finally, a number of circular features that may be craters can be detected. In fact, the abundance of circular features larger than 80 km in diameter, if they are impact craters, suggests that the Venusian surface is approximately a billion years old (Campbell and Burns, 1980).

PIONEER VENUS RADAR OBSERVATIONS

The Pioneer Venus mission carried a radar altimeter (Pettengill et al. 1980). This instrument measured the distance to the surface along the ground track of the spacecraft's orbit, in addition to measuring the scattering characteristics of the surface out to a distance of about 64 km on either side of the beam. The Pioneer-Venus spacecraft is in elliptical orbit that is inclined at an angle of 74 degrees to the equator. The altimeter experiment provided the data to construct the first global (93% of surface) topographic map of Venus. This map has a vertical accuracy of approximately 200 meters and a horizontal "footprint" of about 100 kilometers, about one degree on the surface of Venus (Pettengill et al. 1980). This map, in shaded relief form, provided the base for Figure 36-30.

The topographic map of Venus shows a planet that has a topography which is quite different from the Earth's (Masursky et al. 1980). Perhaps the most fundamental difference consists of the difference in the equatorial bulge for Venus as opposed to the Earth. Due to rotation, the equatorial radius of the Earth is 42 km larger than the polar radius. Venus, in contrast, is a perfect sphere, albeit somewhat lumpy, to within the accuracy of the Pioneer altimeter. Eighty percent of the Venusian topography falls within a topographic range of two kilometers, while only about 5% of the surface is occupied by high plateaus. These relationships can be clearly seen on (Figure 36-31).

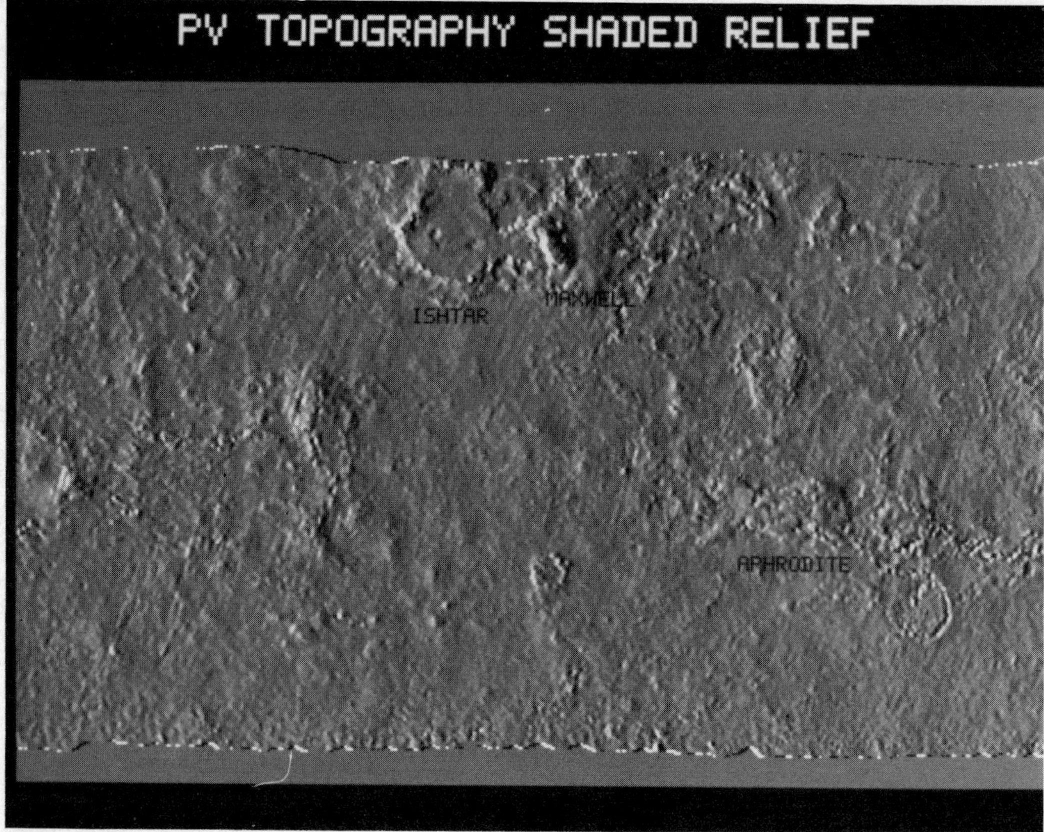

Fig. 36-31. Shaded relief map of Venus based on topographic data acquired by the Pioneer Venus Orbiting altimeter. The northern highlands (the Ishtar Terra) are as large as Australia. The high plateau located on the left-hand portion of Ishtar is the Lakshmi Plateau and the mountains on the right correspond to the radar-rough Maxwell feature seen from Earth. The central highlands area (Aphrodite Terra) is half as large as Africa. After Pettengill et al. (1980).

Venus lacks the strong bimodality between continental and oceanic crust found on Earth, suggesting fundamental differences between the geologic evolution of the two planets. Gravity maps, obtained from tracking spacecraft Doppler accelerations, combined with results for the topography, suggest a correlation between gravity and topography. Such a correlation does not exist on Earth, again suggesting differences in the interiors and geologic evolutions of these two bodies (Phillips et al. 1981).

OBSERVATIONS OF THE LARGER SATELLITES OF JUPITER AND SATURN

EARTH-BASED OBSERVATIONS

Four of the satellites of Jupiter, known as the Galilean Satellites, are roughly the size of the Earth's moon (Table 36-1). Proceeding outward from Jupiter, they are Io, Europa, Ganymede, and Callisto. Except for Io, the bulk density of these bodies, known from mass inferences based on orbits together with volume estimates, indicate that substantial quantities of water exist within their interiors. In fact, both Callisto and Ganymede should have about 50% by mass of water and Europa should contain about 5% water by mass. The remaining materials are thought to consist of alumino-silicates. Pre-Voyager thermal evolution models suggested that Europa, Ganymede, and Callisto should have differentiated in such a manner so as to produce rocky cores surrounded by water or water ice mantles and icy crusts (Lupo and Lewis, 1979). Earth-based spectral reflectance observations supported this interpretation in that all three bodies showed strong absorption features in the IR due to water (Clark, 1980b). Io, which is known from bulk density considerations to be a rocky body, shows an anhydrous surface with a reflectance spectrum characterized by an extremely sharp drop-off in the UV, leading to a yellow-reddish coloration of the surface (Fanale et al. 1979). In addition, Earth-based imaging of Io and Jupiter in the sodium ion emission part of the spectrum shows a cloud of singly and doubly ionized sodium atoms that form a torus in Io's orbital wake (Matson et al. 1974). It was thought that ions are sputtered from the surface by the influx of charged particles from Jupiter's magnetosphere. The sharp UV drop-off in the reflec-

TABLE 36-1a

Bulk Properties of Terrestrial Moons and Planets

Planet	Semimajor AXIS Distance (A.U.)	Equatorial Diameter, Km	Mass (Earth = 1)	Density (gm/cm³)	Uncompressed Density (gm/cm³)	Magnetic Fieldstrength (Gamma)
Mercury	.387	4880	.055	5.44	5.4	220
Venus	.723	12100	.815	5.27	3.9–4.7	0–30
Earth	1.000	12756	1.000	5.52	4.0–4.5	31,000
Mars	1.524	6794	.107	3.93	3.7–3.8	0–60

Note: A.U. stands for astronomical units. 1 A.U. is, by definition, the semimajor axis radius of Earth's orbit, which is equal ot 1.40×10^8 km. Data from Newburn and Matson (1978)

TABLE 36-1b

Bulk Properties of Larger Satellites

Satellite	Equatorial Diameter, Km	Mass (Earth's moon = 1)	Density (gm/cm³)	Mass Fraction Rock	Orbital Radius (Multiples of Planetary Radius)
Earth's Moon	3476	1.00	3.34	1.00	60.27
JUPITER'S GALILEAN SATELLITES					
Io	3640	1.21	3.53	1.00	5.90
Europa	3130	0.66	3.03	.95	9.40
Ganymede	5280	2.03	1.93	.51	14.99
Callisto	4840	1.45	1.79	.54	26.33
SATELLITES OF SATURN					
Mimas	390	?	1.2 ± 0.1	.31	3.12
Enceladus	500	?	1.1 ± .6	?	3.98
Tethys	1050	?	1.0 ± 0.1	.13	4.92
Dione	1120	?	1.4 ± 0.1	.47	6.28
Rhea	1530	.037	1.3 ± 0.1	.37	8.75
Titan	5140	1.91	1.9 ± 0.06	.52	20.3
Hyperion	290	?	?	?	24.9
Iapetus	1440	.035	1.2 ± 0.5	?	59.9

Note: Earth's Moon = .0123 of Earth's Mass. Equatorial Radius of Jupiter = 71,398 km; Equatorial radius of Saturn = 60,330 km. Mass fraction of rock after Tyler et al., 1981, where a core of anhydrous chondritic rock was assumed to be surrounded by water ice. Data from Smith et al. 1979; 1981; Stone and Miner, 1981.

tance spectrum, together with the sodium cloud data, led some researchers to suggest that radiation-damaged salts, left over from Io's degassing, may dominate the surface materials (Fanale et al. 1979).

The surface of the Saturian satellite, Titan, cannot be seen because of a thick, dense atmosphere. Surfaces of the other Saturnian satellites, based on spectral reflectance studies, are composed of water ice (Fink et al. 1976). In all cases, the satellites have low densities, indicating a high mass fraction of water (Table 36-1).

CALLISTO AND GANYMEDE

The Voyager 1 and 2 encounters with Jupiter and its satellites in 1979 provide the data necessary to begin to understand the geologic histories of the Galilean satellites (Color Figure 36-32). The histories of these four bodies are of special interest since their internal composition and structure are quite distinct from the other terrestrial bodies. The Voyager imaging system acquired data in 4 broad bands from the UV to the visible parts of

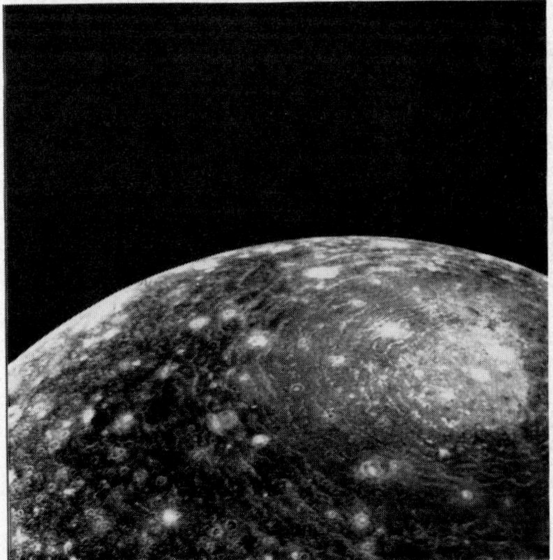

Fig. 36-33. Ring structure on Callisto. This feature is about 1200 km in width and is probably the annealed remnant of an impact basin. Voyager frame 1759 J1 + 000.

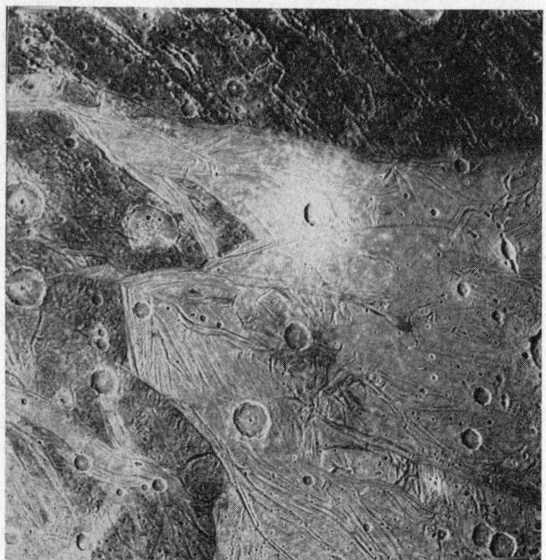

Fig. 36-34. View of the contact between the ancient, heavily cratered, dark terrain (top and left), and the younger striped terrain on Ganymede. Ring structures in the older crust can be seen trending from upper left to lower right on the top of the frame. The frame covers about 1000 km in width. Voyager frame 0550 J2-001.

Fig. 36-36. A complex caldera on Io, about 225 km in width, appears in this view to be the source of a number of branch-like lava flows that are dark relative to the surrounding materials. Voyager 1 frame 0199 J1 + 000.

the spectrum (Smith et al. 1979). In addition, a radiometer and an interferometer were utilized to acquire visible albedo and thermal IR spectra (Hanel et al. 1979). Flybys past all four Galilean Satellites yielded high resolution (1 to 4 km/pixel) views in color and black-and-white of about 1/3 of each satellite's surface. Callisto is the outermost satellite and has the lowest density of the four bodies. The surface is also the darkest of the satellites, although the absolute albedo is still a factor of 2 higher than the Earth's moon. Voyager observations show this body to be the most heavily cratered surface in the Galilean system (Figure 36-33). The abundance of craters is similar to that found on the lunar terra, indicating that the icy crust has remained fairly intact for about 4 billion years. As shown in Figure 36-33 one of Callisto's hemispheres seems to be dominated by a large multi-ringed structure, although that feature is distinctively different in detail from impact basins on the Earth's moon. In addition, a number of other large impact structures can be seen. The shallow appearance of the larger of these features is consistent with viscous relaxation of crater structures, with larger features relaxing to a greater extent than smaller features. The infrared spectrometer indicated that the average temperature of Callisto's surface is about 140 K (Hanel et al. 1979), which makes the icy crust behave rheologically very much like rock. That may explain why the crust has been able to remain intact for such a seemingly long time.

Ganymede is Jupiter's largest Galilean satellite and Voyager observations show it to have rather a unique surface. Darker regions on Ganymede are occupied by a highly cratered crust, much like the icy crust on Callisto. One section of the crust even exhibits evidence for the presence of a large ring structure (Figure 36-34). Brighter striped regions were found to be cutting across the darker crustal regions. The stripes are characterized by a complex of closely spaced, shallow grooves that form networks running roughly parallel to the boundaries of the stripes. The grooved terrain divides the heavily cratered crust into polygons that range in size from several hundred to approximately one thousand kilometers across. The data strongly suggest that the older crust has been disrupted and consumed, gradually being replaced by the grooved terrain. The grooved terrain may have formed by injection of water magma along dikes. The method of consumption of the older crust, either by subduction, assimilation, or burial, remains a mystery. The IR data suggest that the average temperature of Ganymede is about 150 K. Again, the low temperature may explain why a dominantly icy crust has remained intact for so long, although the marked absence of topographic relief (nowhere greater than about 1 km) does imply relaxation by slow flow of ice.

EUROPA AND IO

Europa displays no obvious features related to either ring basins or to impact craters (Figure 36-32). Relief is very subdued.The dominant surface features are a complex set of fractures. Increasing resolution reveals fractures at an increasingly finer scale. The lack of craters implies a youthful surface, i.e., a surface that reflects an active interior capable of destroying the ancient crusts that are so well preserved on Callisto and partially preserved on Ganymede.

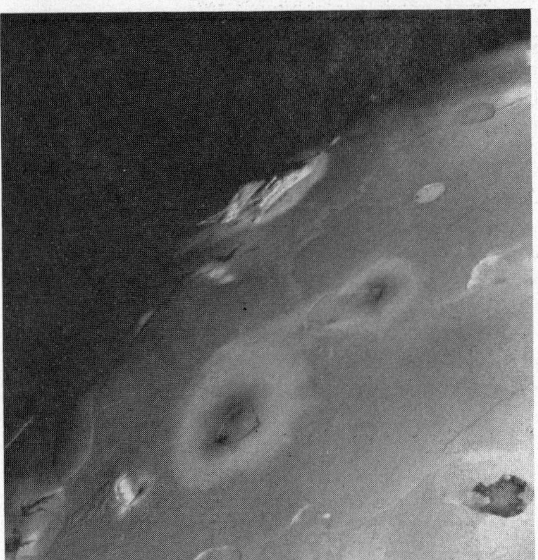

Fig. 36-37. Voyager 1 frame showing a series of escarpments and erosional terrains on Io. Rapid release of SO_2 gas may provide an energy source for such a process (McCauley et al. 1979). Voyager frame 0079 J1 + 000.

Io is perhaps the most interesting of the satellites in that Voyager discovered seven active volcanic vents, spewing gas and ash up to 270 km above the surface (Morabito et al. 1979; Strom et al. 1979) (Color Figure 36-35). The reflectance spectrum, with the sharp UV drop-off and a deep absorption at 4.0 microns, is now thought to be dominated by sulfur dioxide, together with allotropes of sulfur (Fanale et al. 1979; Hapke, 1979). The surface exhibits more than 100 calderas, some more than 200 km in width (Figure 36-36). A number of calderas have narrow (10's km wide), long (100's km) flows extending from them. The absence of relief along the flows implies lavas of rather low viscosity. Much, if not all, of the surface of Io is covered by volcanic materials. About 10% of the surface was beneath the seven volcanic plumes seen by the Voyager spacecraft. In addition, IR observations of an area with a particularly dark caldera are best explained by having the dark area be about 150 K hotter than the surrounding regions (Hanel et al. 1979). This interpretation suggests that warm volcanic material were sensed. The resurfacing rate calculated from the observed volcanic activity corresponds to about 10^{-2} to 10^{-4} cm/year, making the surface activity rate comparable to the Earth's (Johnson et al. 1979). The reason that both Io and Europa are so active may be because of a forced tidal resonance betwen Io, Europa, and Ganymede, causing the orbits of Io and Europa to be rather eccentric. Because of the eccentricity, a considerable amount of tidal energy is transferred from Jupiter to the interiors of these two satellites by tidal flexing (Peale et al. 1979). In fact, it is calculated that tidal heating may dominate the thermal histories of these two bodies, making them unique among solar system objects observed thus far.

A number of escarpments can be seen extending over the surface of Io (Figure 36-37). Voyager showed a number of discrete white clouds, thought to be composed of sulfur dioxide crystals, extending from the escarpments (McCauley et al. 1979). These clouds are probably due to crystallization of SO_2 gas emanating from the interior and escaping along zones of weakness, such as fractures. This observation also suggests that the energetics of this process may provide a unique mechanism for mass movements on Io.

Fig. 36-38. Voyager 1 view of Saturn and its rings. Voyager frame 1682S1-003.

LARGER SATELLITES OF SATURN

Voyager 1 and 2 imaged Saturn, the rings, and a number of Saturn's satellites during the encounters (Smith et al. 1981) (Figure 36-38). Color Figure 36-39 is a mosaic of six of the larger satellites, showing the global appearance of these icy bodies. Titan was found to have a 1 bar atmosphere with a thick haze that obscured the surface. Enceladus is the most reflective of the Saturnian moons, probably because of the presence of relatively clean water ice. Like Io and Europa, Enceladus is in a forced eccentric orbit about Saturn, with the resultant tidal flexing perhaps providing the requisite internal energy needed for resurfacing of any ancient, heavily cratered crust (Figure 36-40). Both Mimas and Tethys are heavily cratered, with a number of deep, bowl-shaped craters. In fact, one crater on Mimas spans a diameter of approximately one-third of the satellite's width. A 750 km long fracture system dominates one hemisphere of Tethys. Dione exhibits a series of bright, "wispy" regions that occupy its trailing hemisphere. The background material for this hemisphere is slightly lower in albedo than the deposits located on the leading hemisphere (Figure 36-39). Clearly, not all regions are equally cratered on Dione, indicating times in the past of local resurfacing. Rhea and Iaepetus

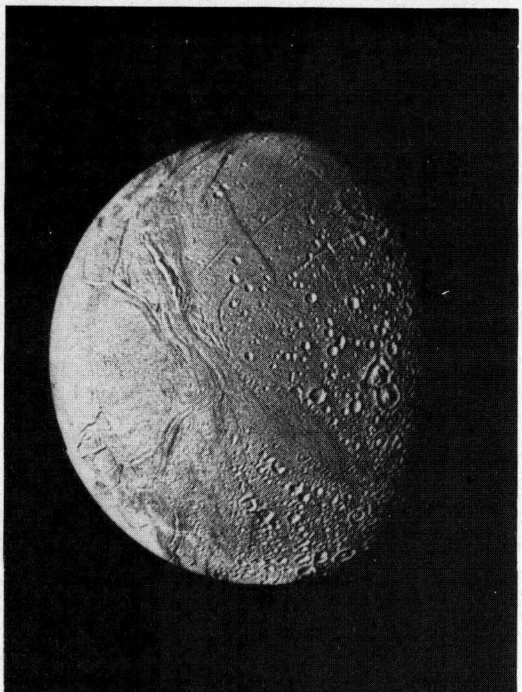

Fig. 36-40. Voyager 2 frame covering Enceladus. After Smith et al. (1981).

are the two largest icy satellites seen by Voyager. Rhea shows distinct areal variations in the density of impact craters and Iaepetus presents two rather different hemispheres in terms of albedo. The trailing hemisphere of Iaepetus is five times as reflective as the leading hemisphere, a value consistent with Earth-based observations.

THE FUTURE OF PLANETARY REMOTE SENSING

INTRODUCTION

Earth-based observations will continue to play an important role in remote sensing of the Moon and other planetary bodies. A definite advantage which Earth-based observers possess is their ability to observe changes in planetary conditions that occur over long timescales. Spacecraft can usually observe only for a limited number of months, while Earth-based observers can detect changes that occur on the timescales of years. Earth-based observers can also wait for infrequent events to occur, while a spacecraft cannot. Finally, Earth-based observations are not constrained by the severe payload weight considerations imposed upon deep space missions. In addition to continuing Earth-based observations, there are several spacecraft missions in the 1980's that are either funded (Galileo) or under serious consideration for funding. In addition, Voyager 2 is targeted to encounter Uranus in 1986 and Neptune in 1989, before accompanying Voyager 1 out of the solar system. We discuss probable future missions to Jupiter, Venus, and Mars to illustrate the expected pattern of future solar system exploration.

JUPITER

The Galileo mission consists of an orbiter to acquire data about Jupiter and its satellites, together with an instrumented probe that will be sent into the giant planet's atmosphere. The scientific objectives are to study the chemical composition and physical state of Jupiter's atmosphere and selected Jovian satellites, and to study the structure and dynamics of the Jovian magnetosphere. The launch was originally scheduled for early 1982 as the scientific successor to the Voyager mission to Jupiter. However, due to delays in Space Shuttle development, Galileo has been rescheduled for a launch in the mid 1980's.

The orbiter, carrying 11 scientific instruments and weighing about 2,660 kilograms at launch, including 1,420 kilograms of propulsion fuel, will transmit scientific and engineering data at rates up to 115,000 bits per second. The probe would penetrate the atmosphere after it is released by the carrier near Jupiter. Once the probe's work is done, the mission operations emphasis will revert back to the orbiter. The primary mission is scheduled to end about 20 months after probe arrival at Jupiter. Scientific teams of 114 investigators have been assembled for 10 nations to interpret the data from the varied instruments to be carried by the orbiter and the probe. The orbiter probably will carry the following instruments:

- A new camera system with a 1.5-meter focal-length lens that makes use of solid-state charge-coupled devices (CCD) instead of the vidicon tubes flown on previous imaging missions to Mars, Mercury, and Jupiter. The CCD system can provide broader spectral response and higher resolution than earlier cameras. Encounters with the larger satellites should yield pictures with a best resolution on the order of 30 to 50 meters.
- A near-infrared mapping spectrometer to provide images (similar to those from Landsat) and reflectance spectra of Jupiter's satellites. The instrument will also measure reflected sunlight and heat emission from Jupiter's atmosphere to study composition, cloud structure and temperature profiles of the Jovian atmosphere.
- An ultraviolet spectrometer to study the composition and structure of the upper atmospheres of Jupiter and its satellites.
- A photopolarimeter-radiometer to measure temperature profiles and the energy balance of the Jovian atmosphere, together with the characteristics of Jovian clouds.

VENUS

The next probable mission to Venus is a Venus radar mapper. The mission might consist of a single spacecraft, weighing 4,000 to 4,500 kilograms, and containing a synthetic aperture radar (SAR) imaging system, together with an altimetry capability. The SAR instrument will probably be similar to the instrument flown on the Seasat mission. The mapper mission would obtain near-global coverage of Venus, providing radar images with a radar cell size of about 1 km. The radar instrument could also take spot images of selected regions—about 2% of the surface—at higher resolution, providing a radar cell size of about 100m.

The radar mapper mission would provide data needed to understand the geologic history of Venus. Such data include radar images of the surface that would be considerably more detailed than seen from Earth or from the Pioneer-Venus altimeter. Such data would resolve the geologic patterns (orogenic zones, channels, volcanoes, impact craters) that are only vaguely suggested by presently available data. Venus is a particularly important target since its size and mass are very close to the size and mass of the Earth. Differences in the geologic histories of these two bodies would therefore strongly reflect differences in the evolutionary paths of the respective interiors. Such differences could tell us much about the interiors of both planets.

OTHER POSSIBILITIES

The exploration of Mars has proceeded from flybys (Mariners 4, 6, 7) to Orbiters (Mariner 9) to combined Orbiter-Lander missions (Viking). Missions under consideration to Mars in the future include rovers that could traverse several hundred kilometers across the surface, penetrators or hard landers to be used for seismic and meteorological networks, and an orbiter that would carry geochemically-oriented sensors. A returned sample mission is another, albeit costly, mission. The future exploration of Mars may be taken as one model of the exploration process, whereby site intensive, detailed studies of the planet become increasingly important. However, this approach entails increasing complexity and cost as the exploration process continues. An alternative strategy to follow is to consider that different planetary bodies offer unique possibilities for exploration. The asteroids, long observed from Earth, are an example of a unique exploration situation. Earth-crossing asteroids often have orbits that are easily accessible from Earth. Some are easier to get to than the Earth's moon. Thus, they do not generally require a flyby, and a rendezvous can be done inexpensively with or without an orbital phase. A relatively inexpensive sample return is also possible. Comets also provide unique opportunities for exploration. The European Space Agency, for example, has given approval to a mission (Giotto) that consists of a single spacecraft that would flyby the comet Halley in the mid 1980's. The real prognosis for the future of spaceborne remote sensing of planets and moons is, in the final analyses, in the hands of the governmental agencies. The real future, in fact, depends on the longterm commitment to exploration of the solar system.

SOURCES OF LUNAR AND PLANETARY DATA

The large number of images already acquired during lunar and planetary missions, together with the large number expected from future missions, such as Galileo and the Venus Radar mapper missions, makes it prohibitively expensive for each Principal Investigator involved in lunar and planetary research to have an extensive data set. Some 500,000 separate images presently exist, along with extensive sets of shaded relief, topographic, and geologic maps. The National Space Science Data Center, Goddard Space Flight Center, Greenbelt, Md., is the prime depository for these products. However, it is sometimes difficult to know what to order from this depository, even with the extensive catalogues available from NSSDC. NASA has recently helped to alleviate this problem by helping to set up eight Regional Planetary Image Facilities at the Jet Propulsion Laboratory, Pasadena, CA.; University of Arizona, Tucson, AZ.; Astrogeology Branch, U.S.G.S., Flagstaff, AZ.; Lunar and Planetary Institute, Houston, TX.; Washington University, St. Louis, MO.; Cornell University, Ithaca, N.Y.; Brown University, Providence, R.I.; University of London, London, Britain; and the Center for Planetology, Rome, Italy. The Facilities are designed to provide browse files for examining image data and ancillary products for interested users. A complete listing of addresses for the NSSDC and the Regional Facilities is given in Table 36-2a. A list of addresses for the European Scientific Community is given in Table 36-2b.

Each Planetary Image Facility has a full set of images of the moons and planets, together with all associated maps and other products, such as catalogs. A computerized search capability has been initiated, consisting of an interactive link to the institution's computer, together with the software necessary to conduct interactive interrogation of image engineering data such as picture center latitude, longitude, slant range, etc. (Arvidson et al. 1980b). Videodisk players are used as "quick-look" displays of image data. Each videodisk stores approximately 104,000 frames, with each frame containing the equivalent of about 525 × 525 picture elements. The videodisk players are linked to the interactive terminals so that searches

TABLE 36-2a

Addresses of NASA Regional Planetary Image Facilities and the National Space Science Data Center. The Image Facilities are designed to provide regional access to complete data sets, while the Space Science Data Center is the primary source of lunar and planetary data.

National Space Science Data Center
Goddard Space Flight Center
Greenbelt, Maryland 20771

Spacecraft Planetary Imaging Facility
Center for Radiophysics and Space Research
Cornell University
Ithaca, New York 14855

Planetary Data Center
Department of Geological Sciences
Brown University
Providence, Rhode Island 02912

Planetary Image Facility
Department of Earth and Planetary Sciences
Washington University
St. Louis, Missouri 63130

Planetary Image Facility
Lunar and Planetary Institute
3303 NASA Road 1
Houston, Texas 77058

Planetary Data Facility
U.S. Geological Survey—Branch of Astrogeologic Studies
2255 North Gemini Drive
Flagstaff, Arizona 86001

Space Imagery Center
Lunar and Planetary Laboratory
University of Arizona
Tucson, Arizona 85721

Planetary Image Facility
Mail Stop 264-11
Jet Propulsion Laboratory
4800 Oak Grove Drive
Pasadena, California 91103

TABLE 36-2b

The European Scientific Community Can Work Through the Following Offices:

World Data Center
Goddard Space Flight Center
Greenbelt, Maryland 20771

Planetary Image Facility
Observatory Annex, 33/35 Daws Lane
University of London Observatory
Mill Hill Park
London NW 7 4 SD, England

Planetary Image Facility
Reparto Planetologia
Viale Universita
11-00185, Rome, Italy

could be conducted on the engineering data base, and those pictures of interest could then be displayed interactively. Such a capability considerably reduces the time that a researcher needs to spend in finding selected data sets. The facilities, which were set up at sites where there is a long-term science interest in planetary sciences, are meant to serve as a regional data base for use by the scientific community. Researchers visit a given Facility, quickly and efficiently search through the entire data base, and then order a subset of the required data needed from NSSDC.

REFERENCES

Adams, J. B., 1974, Visible and near-infrared diffuse reflectance spectra of pyroxenes as applied to remote sensing of solid objects in the Solar System, J. Geophys. Res., V. 79, p. 4829–4836.

Adams, J. B., C. M. Pieters, A. E. Metzger, I. Adler, T. B. McCord, C. R. Chapman, T. V. Johnson, M. J. Bielefeld, 1981. Remote sensing of basalts in the Solar System, in Basaltic Volcanism on the Terrestrial Planets, Lunar and Planet. Inst., p. 439–490.

Andre, C. G., M. J. Bielefeld, E. Eliason, L. A. Soderblom, I. Adler, J. A. Philpotts, 1977, Lunar surface chemistry: a new imaging technique, Science, V. 197, p. 986–989.

Andre, C. G., R. W. Wolfe, I. Adler, 1979, Are early magnesium-rich basalts widespread on the Moon; Proc. Lunar Planet. Sci. Conf. 10th, p. 1739–1751.

Arnold, J. R., A. E. Metzger, R. C. Reedy, 1977, Computer-generated maps of Lunar composition from gamma-ray data; Proc. Lunar Sci. Conf. 8th, p. 945–948.

Arvidson, R. E., K. A. Goettel, C. M. Hohenberg, 1980a, A post-Viking view of Martian geologic evolution; Rev. Geophys. Space Physics, V. 18, p. 565–603.

Arvidson, R. E., L. Bolef, E. Guinness, P. Norberg, 1980b, BIRP-Software for Interactive Search and Retrieval of Image Engineering Data; NASA CR-3299, 71 p.

Baird, A. K. and D. C. Clark, 1981, On the original igneous source of Martian fines; Icarus, V. 45, p. 113–123.

Baker, V. R., 1979, Erosional processes in channelized water flows on Mars; J. Geophys. Res., V. 84, p. 7985–7993.

Batson, R. M., R. Jordan, K. B. Larson, 1974, Atlas of Surveyor 5 Television Data; NASA SP-341, 597 p.

Baum, W. A., 1974, Earth-based observations of Martian albedo changes; Icarus, V. 22, p. 363–370.

Boyce, J. B., A. L. Dial, L. A. Soderblom, 1974, Ages of Lunar nearside light plains and maria; Proc. Lunar Sci. Conf. 5th, p. 11–23.

Campbell, D. B. and B. A. Burns, 1980, Earth-based radar imagery of Venus; J. Geophys. Res., V. 85, p. 8271–8281.

Carr, M. H., 1980, The morphology of the Martian surface; Space Sci. Rev., V. 25, p. 231–284.

Charette, M. P., T. B. McCord, C. Pieters, J. B. Adams, 1974, Application of remote spectral reflectance measurements to Lunar geology—classification and determination of titanium content of Lunar soils; J. Geophys. Res., V. 79, p. 1605–1613.

Clark, R. N., 1980a, A large-scale interactive one dimensional array processing system; Pub. Astron. Soc. Pacific, in press.

Clark, R. N., 1980b, Ganymede, Europa, Callisto, and Saturn's rings: compositional analysis from reflectance spectroscopy; Icarus, V. 44, p. 388–409.

Cutts, J. A., K. R. Blasius, G. A. Briggs, M. H. Corr, R. Greeley, H. Masunsky, 1976, North polar region of Mars, imaging results from Viking 2; Science, V. 194, p. 1329–1337.

Davis, P. A., 1980, Iron and titanium distribution on the Moon from orbital gamma-ray spectrometry with implications for crustal evolutionary models; J. Geophys. Res., V. 85, p. 3209–3224.

Downs, G. S., P. E. Reichley, R. R. Green, 1975, Radar measurements of Martian topography and surface properties: The 1971 and 1973 oppositions; Icarus, V. 26, p. 273–312.

Eliason, E. and L. A. Soderblom, 1977, An array processing system for Lunar geochemical and geophysical data; Proc. Lunar Sci. Conf. 8th, p. 1163–1170.

Fanale, F. P., R. H. Brown, D. P. Cruikshank, R. N. Clarke, 1979, significance of absorption features in Io's IR reflectance spectrum, nature, V. 280, p. 763–766.

Farmer, C. B., D. W. Davies, A. L. Holland, D. P. LaPorte, P. E. Doms, 1977, Mars: water vapor observations from the Viking orbiters; J. Geophys. Res. V. 82, p. 425–4268.

Fink, U., H. P. Larson, T. N. Gautier, R. R. Treffers, 1976, Infrared spectra of the satellites of Saturn: identification of water ice on Iapetus, Rhea, Dione, and Thethys; Astrophys. J., V. 207, p. L63–L67.

Gornstein, P. and P. Bjorkholm, 1972, Alpha-particle spectrometer experiment; Apollo 15 Preliminary Sci. Rept.; NASA SP-289, Section 18, p. 18-1 to 18-7.

Guinness, E. A., R. E. Arvidson, D. C. Gehret, L. K. Bolef, 1979, Color changes at the Viking landing sites over the course of a Mars year; J. Geophys. Res., V. 84, p. 8355–8364.

Guinness, E. A., 1981, spectral properties (.4 to .75 microns) of soils exposed at the Viking 1 landing site; J. Geophys. Res., V. 86, p. 7983–7992.

Hanel, R., B. Conrath, W. Hovis, V. Kunde, P. Lowman, W. Maguire, J. Pearl, J. Pirraglia, C. Prabhakara, B. Schlachman, G. Levin, P. Straat, T. Burke, 1972, Investigation of the Martian environment by infrared spectroscopy on Mariner 9; Icarus, V. 17, p. 423–442.

Hanel, R., B. Conrath, M. Flasar, V. Kunde, P. Lowman, W. Maguire, J. Pearl, J. Pirraglia, R. Samuelson, D. Gautier, P. Gierasch, S. Kuman, C. Ponnamperuma, 1979, Infrared observations of the jovian system from Voyager 1; Science, V. 204, p. 972–976.

Hapke, B. A., 1979, Io's surface and environs: a magmatic-volatile model; Geophys. Res. Lett., V. 6, p. 799–802.

Hapke, B. A., 1981, Bi-directional reflectance spectroscopy: 1. Theory; J. Geophys. Res., V. 86, p. 3039–3054.

Heacock, R. L., G. P. Kuiper, E. M. Shoemaker, H. C. Urey, E. A. Whitaker, 1966, Ranger VIII and IX, Part II, experimenter's analyses and interpretations; JPL Tech. Rept. No. 32-800, 382 p.

Head, J. W., 1976, Lunar volcanism in space and time; Rev. Geophysics and Space Physics, V. 14, p. 265–300.

Head, J. W., C. Pieters, T. B. McCord, J. B. Adams, S. Zisk, 1978, Definition and detailed characterization of Lunar surface units using remote observations; Icarus, V. 33, p. 145–172.

Huck, F. O., D. J. Jobson, S. K. Park, S. D. Wall, R. E. Arvidson, W. R. Patterson, W. D. Benton, 1977, Spectrophotometric and color estimates of the Viking landing sites; J. Geophys. Res., V. 82, p. 4401–4411.

Hulme, G., 1974, The interpretation of lava flow morphology; Geophys. J. Roy. Astron. Soc., V. 39, p. 361–383.

Ingersoll, A. P., A. R. Dobrovolskis, B. M. Jakosky, 1979, Planetary atmospheres; Rev. Geophys. Space Physics, V. 17, p. 1722–1735.

James, P. B., G. Briggs, J. Barnes, A. Spruck, 1979, Seasonal recession of Mars' polar cap as seen by Viking; J. Geophys. Res., V. 84, p. 2889–2922.

Johnson, T. V., A. F. Cook, C. Sagan, L. A. Soderblom, 1979, Volcanic resurfacing rates and implications for volatiles on Io; Nature, V. 280, p. 746–750.

Jones, K. L., R. E. Arvidson, E. A. Guinness, S. L. Bragg, S. D. Wall, C. D. Carlston, D. G. Pidek, 1979, One Mars year: Viking lander imaging observations of sediment transport and H_2O condensates; Science, V. 204, p. 799–806.

Jurgens, R. F., R. M. Goldstein, H. R. Rumsey, R. R. Green, 1980, Images of Venus by three station interferometry-1977 results; J. Geophys. Res. V. 85, p. 8282–8294.

Kieffer, H. H., S. Chase, T. Martin, E. Miner, F. Palluconi, 1976, Martian north pole summer temperatures: Dirty water ice; Science, V. 194, p. 1341–1344.

Kieffer, H. H., T. Z. Martin, A. R. Peterfrend, B. M. Jakosky, E. D. Miller, F. D. Palluconi, 1977, Thermal and albedo mapping of Mars during the Viking primary mission; J. Geophys. Res. V. 82, p. 4249–4291.

Kieffer, H. H., P. A. Davis, L. A. Soderblom, 1981, Mars global properties: Maps and applications; Proc. Lunar Planet. Sci. Conf., p. 1395–1417.

Kosofsky, L. J. and F. El Baz, 1970, The Moon as Viewed by Lunar Orbiter; NASA SP-200, 152 p.

Leighton, R. B., B. Murray, R. Sharp, J. Allen, R. Sloan, 1965, Mariner IV photography of Mars: Initial results; Science, V. 149, p. 627–630.

Leighton, R. B., N. Horowitz, B. Murray, R. Sharp, A. Herriman, A. Young, B. Smith, M. Davies, G. Leovy, 1969, Mariner 6 and 7 television pictures: Preliminary analysis; Science, V. 166, p. 49–67.

Lupo, M. J. and J. S. Lewis, 1979, Mass-radius relationships in icy satellites; Icarus, V. 40, p. 157–170.

McCauley, J. F., B. A. Smith, L. A. Soderblom, 1979, Erosional scarps on Io; Nature, V. 280, p. 736–737.

McCord, T. B. and R. N. Clark, 1979, The Mercury soil-presence Fe^{+2}; J. Geophys. Res., V. 84, p. 7664–7668.

Masursky, H., 1973, An overview of results from Mariner 9; J. Geophys. Res., V. 78, p. 4009–4030.

Masursky, H., J. Boyce, A. Dial, G. Schaber, M. Strobell, 1977, Classification and time of formation of Martian channels based on Viking Data; J. Geophys. Res., V. 82, p. 4016–4038.

Masursky, H., E. Eliason, P. Ford, G. McGill, G. Pettengill, G. Schaber, G. Schubert, 1980, Pioneer-Venus radar results: geology from images and altimetry; J. Geophys. Res., V. 85, p. 8261–8270.

Matson, D. L., T. V. Johnson, F. Fanale, 1974, Sodium

D-Line emission from Io: Sputtering and resonant scattering hypotheses; Astrophys. J., V. 192, p. 43–46.

Mendell, W. W. and F. J. Low, 1975, Infrared orbital mapping of Lunar features; Proc. Lunar Sci. Conf. 6th, p. 2711–2719.

Metzger, A. E. and R. E. Parker, 1980, The distribution of titanium on the Lunar surface; Earth Planet. Sci. Lett., V. 45, p. 155–171.

Moore, H. J., D. W. G. Arthur, G. G. Schaber, 1978, Yield strength of flows on Earth, Mars, and Moon; Proc. Lunar Planet Sci. Conf. 9th, p. 3351–3378.

Morabito, L. A., S. P. Synnott, P. N. Collins, 1979, Discovery of active extra-terrestrial volcanism; Science, V. 204, p. 321.

Mutch, T. A., 1972, Geology of the Moon: A Stratigraphic View; Princeton Univ. Press, 400 p.

Mutch, T. A., R. E. Arvidson, J. W. Head, K. L. Jones, R. S. Saunders, 1976. The Geology of Mars; Princeton Univ. Press, 400 p.

Neukum, G., B. Konig, H. Fechtig, 1975, Cratering in the Earth-Moon system: Consequences for age determination by crater counting; Proc. Lunar Sci. Conf. 6th, p. 2597–2620.

Neukum, G. and K. Hiller, 1981, Martian ages; J. Geophys. Res., V. 86, p. 3097–3121.

Newburn, R. and D. Matson, 1978, Planetary and satellite data; Vols. 2, 3, Proc. Lunar Planet. Sci. Conf. 8th, Prontispiece.

Oberbeck, V. R., F. Horz, R. H. Morrison, W. L. Quaide, D. E. Gault, 1975, On the origin of the Lunar smooth plains; The Moon, V. 12, p. 19–54.

Oberbeck, V. R., W. L. Quaide, R. E. Arvidson, H. R. Aggarwal, 1977, Comparative studies of Lunar, Martian, and Mercurian craters and plains; J. Geophys. Res., V. 82, p. 1681–1689.

Owen, T. B., K. Biemann, D. Rushneck, J. Biller, D. Howarth, A. Lafleur, 1977, The composition of the atmosphere at the surface of Mars; J. Geophys. Res., V. 82, p. 4635–4639.

Peale, S., P. Cassen, R. Reynolds, 1979, Melting of Io by tidal dissipation; Science, V. 20, p. 892–894.

Peeples, W. J., W. Sill, T. May, S. Ward, R. Phillips, R. Jordan, E. Abbott, T. Killpack, 1978, Orbital radar evidence for Lunar subsurface layering in Maria Serenitatis and Crisium; J. Geophys. Res., V. 83, p. 3459–3468.

Pettengill, G. H., C. C. Counselman, L. P. Rainville, I. I. Shapiro, 1969, Radar measurements of Martian topography; Astron. J., V. 74, p. 461–482.

Pettengill, G. H., J. F. Chandler, D. B. Campbell, D. M. Wallace, 1974, Martian surface properties from recent radar observations; Bull. Amer. Astron. Soc., V. 6, p. 372.

Pettengill, G. H., S. H. Zisk, T. W. Thompson, 1974, The mapping of Lunar radar scattering characteristics; The Moon, V. 10, p. 1–16.

Pettengill, G. H., E. Eliason, P. G. Ford, G. B. Loriot, H. Masursky, G. E. McGill, 1980, Pioneer-Venus radar results—altimetry and surface properties; J. Geophys. Res., V. 85, p. 8261–8270.

Phillips, R. J. and K. Lambeck, 1980, Gravity fields of the terrestrial planets—long wavelength anomalies and tectonics; Rev. Geophys. Space Phys., V. 18, p. 27–76.

Phillips, R. J., W. Kaula, G. McGill, M. Malin, 1981, Tectonics and evolution of Venus; Science, V. 212, p. 879–887.

Pieri, D. C., 1980, Martian valleys: morphology, distribution, age, and origin; Science, V. 210, p. 895–897.

Pieters, C., T. B. McCord, S. Zisk, J. B. Adams, 1973, Lunar dark spots and the nature of the Apollo 17 landing area; J. Geophys. Res., V. 78, p. 5867–5875.

Pieters, C., 1978, Mare Basalt Types on the Frontside of the Moon: A summary of spectral reflectance data; Proc. Lunar Planet. Sci. Conf. 8th, p. 2825–2849.

Plescia, J. B. and R. S. Saunders, 1979, Geologic evolution of the Tharsis volcanoes (abstract); Proc. Lunar Planet. Sci. Conf. 10th, p. 989–991.

Pohn, H. and R. L. Wildey, 1970, A photogeologic-photographic study of the normal albedo of the Moon, U.S.G.S. Professional Paper (map) 559-E.

Pollack, J. B., 1979, Climatic change on the terrestrial planets; Icarus, V. 37, p. 479–553.

Roth, L. E., C. Elachi, R. S. Saunders, 1978, Radar depths of large Martian craters, (abs), Lunar Planet. Sci. Conf. 9th, p. 976–978.

Saari, J. M. and R. W. Shorthill, 1972, The sunlit Lunar surface, The Moon, V. 5, p. 161–178.

Schaber, G. G., T. W. Thompson, S. H. Zisk, 1974, Lava flows in Mare Imbrium, Part II: Evaluation of anomalously low earth-based radar reflectivity: U.S. Geol. Survey Interagency Rept: Astrology 64, 31 p

Schaefer O. A., and L. Husain, 1974, Chronology of lunar basin formation, Proc. Lunar Sci. Conf. 5th,, p. 1541–1555.

Schultz, P. H. and P. D. Spudis, 1979, Evidence for ancient Mare volcanism; Proc. Lunar Planet. Sci. Conf. 10th, p. 2899–2918.

Schultz, P. H. and W. W. Mendell, 1978, Orbital infrared observations of Lunar craters and possible implications for impact ejecta emplacement; Proceed. Lunar Planet. Sci. Conf. 9th, p. 2857–2883.

Shoemaker, E. M. and R. J. Hackman, 1962, Stratigraphic basis for a Lunar timescale, in The Moon, IAU Symposium, No. 14 (Z. Kopal and Z. K. Mikhailove, eds.), Academic Press, p. 290–300.

Simpson, R. A., G. L. Tyler, J. P. Brenkle, M. Sue, 1979, Viking bistatic radar observations of the Hellas basin on Mars: preliminary results; Science, V. 203, p. 45–46.

Singer, R. B., T. B. McCord, R. N. Clark, J. B. Adams, R. L. Huguenin, 1979, Mars: Surface composition from reflectance spectroscopy: a summary; J. Geophys. Res., V. 84, p. 8415–8426.

Singer, R. B., 1980, The composition of the Martian dark regions: I. visible and near-IR spectral reflectance of analog materials and interpretation of telescopically observed spectral shape; J. Geophys. Res., in press.

Sleep, N. H. and R. Phillips, 1979, An isostatic model for the Tharsis province, Mars; Geophys. Res. Lett., V. 6, p. 803–806.

Smith, B., L. Soderblom, R. Beebe, J. Boyce, G. Briggs, M. Carr, S. Collins, A. Cook, G. Danielson, M. Davies, G. Hunt, A. Ingersoll, T. Johnson, H. Masursky, J. McCauley, D. Morrison, T. Owen, C. Sagan, E. Shoemaker, R. Strom, V. Suomi, J. Veverka, 1979, The Galilean satellites of Jupiter-Voyager 2 imaging science results; Science, V. 206, p. 927–950.

Smith, B., L. Soderblom, R. Beebe, J. Boyce, G. Briggs, A. Bunker, S. Collins, C. Hansen, T. Johnson, J. Mitchell, R. Terrile, M. Carr, A. Cook, J. Cuzzi, J. Pollack, G. Danielson, A. Ingersoll, M.

Davies, G. Hunt, H. Masursky, E. Shoemaker, D. Morrison, T. Owen, C. Sagan, J. Veverka, R. Strom, V. Suomi, 1981, Encounter with Saturn; Voyager I imaging science results; Science, V. 212, p. 163–191.

Snyder, C. W., 1979, The Extended mission of Viking; J. Geophys. Res., V. 84, p. 7917–7933.

Soderblom, L. A. and L. A. Lebofsky, 1972, Technique for rapid determination of relative ages of Lunar areas from orbital photography; Proc. Lunar Sci. Conf. 3rd, p. 1191–1199.

Soderblom, L. A., J. R. Arnold, J. M. Boyce, R. P. Lin, 1977, Regional variations in the Lunar Maria: age, remanent magnetism, and chemistry; Proc. Lunar Sci. Conf. 8th, p. 1191–1199.

Soderblom, L. A., K. Edwards, E. M. Eliason, E. M. Sanchez, M. P. Charette, 1978, Global color variations on the Martian surface; Icarus, V. 34, p. 446–464.

Soffen, G. A., 1976, Scientific results of the Viking missions; Science, V. 194, p. 1274–1276.

Solomon, S. C., 1978, On volcanism and thermal tectonics on one-plate planets; Geophysical Res. Lett., V. 5, p. 461–464.

Spudis, P. D., 1978, Composition and origin of the appennine bench formation, Proceed. Lunar Planet. Sci. Conf. 9th, p. 3379–3394.

Stone, E. C. and E. D. Miner, 1981, Voyager 1 encounter With the Saturnian system; Science, V. 212, p. 159–163.

Strom, R. B., R. J. Terrile, H. Masursky, 1979, Volcanic eruption plumes on Io; Nature, V. 280, p. 733–736.

Strom, R. B., 1979, Mercury: A Post-Mariner 10 assessment; Space Sci. Rev., V. 24, p. 3–70.

Taylor, S. R., 1975, Lunar Science-A post Apollo View; Pergamon Press, 372 p.

Tera, F. D., A. Papanastassiou, G. J. Wasserburg, 1974, Isotopic evidence for a terminal Lunar cataclysm; Earth Planet. Sci. Lett., V. 22, p. 1–21.

Thompson, T. W., H. Masursky, R. W. Shorthill, G. L. Tyler, S. H. Zisk, 1974, A comparison of infrared, radar and geologic mapping of Lunar craters; The Moon, V. 10, p. 87–117.

Thompson, T. W., 1979, A review of Earth-Based radar mapping of the Moon; Moon and Planets, V. 20, p. 179–198.

Toksoz, N. M., 1977, Report of the Terrestrial Bodies Science Working Group; V. 1, Executive Summary, JPL Publication 77-51, 23 p.

Toon, O. B., J. B. Pollack, and C. Sagan, 1977, Physical properties of the particles composing the Martian dust storm of 1971–72; Icarus, V. 30, p. 663–696.

Toon, O. B., J. B. Pollack, W. Ward, K. Bilski, 1981, The astronomical theory of climate change on Mars; Icarus, in press.

Tyler, G. L., and H. T. Howard, 1973, Dual-Frequency bistatic radar investigations of the Moon with Apollo 14 and 15; J. Geophys. Res., V. 78, p. 4852–4874.

Tyler, G. L., V. R. Eshleman, J. D. Anderson, G. S. Levy, G. F. Lindal, G. E. Wood, T. A. Croft, 1981, Radio science investigations of the saturn system with Voyager 1: preliminary results; Science, V. 212, p. 201–206.

Wetherill, G. W., 1975, Late heavy bombardment of the Moon and terrestrial planets; Proc. Lunar Sci. Conf. 6th, p. 1539–1561.

Wilhelms, D. E. and J. F. McCauley, 1971, Geologic Map of the Nearside of the Moon, U.S. Geological Survey Map I-703.

Wise, D. U. and M. P. Golombek, G. E. McGill, 1979, Tectonic evolution of Mars; J. Geophys. Res. V. 84, p. 7434–7439.

Wood, J. A., 1975, Lunar petrogenesis in a well-stirred magma ocean; Proc. Lunar Sci. Conf. 6th; p. 1087–1102.

Zisk, S. H., 1972, Lunar topography: first radar interferometer measurements of the Alphonsus-Ptolemaeus-Arzachel region; Science, V. 178, p. 977–980.

Zisk, S. H. and P. J. Mouginis-Mark, 1980, Anomalous region of Mars: implications for near surface liquid water; Nature, V. 288, p. 735–738.

Combined Index to Volumes I and II

Compiler: G. C. TEWINKEL

A

B

C

D

E

F

G

H

I

J

K

L

M

N

O

P

Q

R

S

T

U

V

W

X, Y, Z

Other Publications of The American Society of Photogrammetry

MANUAL OF PHOTOGRAMMETRY

Some 100 leading members of the photogrammetric community discuss the mensuration aspects of photogrammetry, leaving the photointerpretation and remote sensing aspects to the other A.S.P. manuals. The 19 chapters cover photogrammetric subjects relating to mathematics, optics, photography, project planning, automation, space applications and non-topographic applications. Fourth Edition, 1980, 1072 pages

HANDBOOK OF NON/TOPOGRAPHIC PHOTOGRAMMETRY

Applications of photogrammetry outside the realm of topographic mapping are thoroughly discussed and illustrated in this volume. Specific applications are described in the fields of architecture; surveys of historical monuments and sites; archaeological surveys; biomedical and bioengineering science; construction of machines, buildings and ships; mining engineering, traffic engineering; accident surveys; crime detection; underwater surveys; x-ray technology; holographic technology, and moire technology. 1979 210 pages

PHOTOGRAMMETRIC ENGINEERING AND REMOTE SENSING

The Society's monthly journal contains technical articles and reviews; announcements of scientific meetings and other events of interest to members; reports of society committees and officers; rosters of members, sustaining members, and regional chapters; news concerning photogrammetrists and the profession; and other appropriate items. The substantial advertising sections contain a wealth of useful technical information on photogrammetric equipment and services.

TECHNICAL PAPERS

The society publishes bound volumes containing the papers presented at major meetings, seminars and symposia. The technical papers of the Spring and Fall Conventions are generally published as preprint volumes which are available to registrants at the time of the convention and which may also be purchased subsequently.

Detailed information on prices and ordering is available from:

American Society of Photogrammetry
210 Little Falls Street
Falls Church, Virginia 22046